KU-384-658

# Grasslands, systems analysis and man

# THE INTERNATIONAL BIOLOGICAL PROGRAMME

The International Biological Programme was established by the International Council of Scientific Unions in 1964 as a counterpart of the International Geophysical Year. The subject of the IBP was defined as 'The Biological Basis of Productivity and Human Welfare', and the reason for its establishment was recognition that the rapidly increasing human population called for a better understanding of the environment as a basis for the rational management of natural resources. This could be achieved only on the basis of scientific knowledge, which in many fields of biology and in many parts of the world was felt to be inadequate. At the same time it was recognized that human activities were creating rapid and comprehensive changes in the environment. Thus, in terms of human welfare, the reason for the IBP lay in its promotion of basic knowledge relevant to the needs of man.

The IBP provided the first occasion on which biologists throughout the world were challenged to work together for a common cause. It involved an integrated and concerted examination of a wide range of problems. The Programme was co-ordinated through a series of seven sections representing the major subject areas of research. Four of these sections were concerned with the study of biological productivity on land, in freshwater, and in the seas, together with the processes of photosynthesis and nitrogen fixation. Three sections were concerned with adaptability of human populations, conservation of ecosystems and the use of biological resources.

After a decade of work, the Programme terminated in June 1974 and this series of volumes brings together, in the form of syntheses, the results of national and international activities.

INTERNATIONAL BIOLOGICAL PROGRAMME 19

# Grasslands, systems analysis and man

EDITED BY

A. I. BREYMEYER

*Associate Professor of Ecology,*
*Institute of Geography,*
*Polish Academy of Sciences*

AND

G. M. VAN DYNE

*Professor of Biology,*
*College of Forestry and Natural Resources,*
*Colorado State University*

CAMBRIDGE UNIVERSITY PRESS

CAMBRIDGE
LONDON NEW YORK NEW ROCHELLE
MELBOURNE SYDNEY

Published by the Press Syndicate of the University of Cambridge
The Pitt Building, Trumpington Street, Cambridge CB2 1RP
32 East 57th Street, New York, NY 10022, USA
296 Beaconsfield Parade, Middle Park, Melbourne 3206, Australia



First published 1980

Printed in Great Britain at the
University Press, Cambridge

*Library of Congress Cataloguing in Publication Data*

Main entry under title:

Grasslands, systems analysis and man.

(International Biological Programme; 19)
Includes bibliographies and index.
1. Grassland ecology. 2. Grassland ecology – Mathematical models.
3. International Biological Programme.
I. Breymeyer, A. I., 1932– . II. Van Dyne, George M., 1932– . III. Series.

QH541.5.P7G74 574.5.'264 77–28249

ISBN 0 521 21872 1

# Contents

## Part II. Systems synthesis



## Part III. System utilization

# Table des matières

# Содержание

# Sumario

# Contributors

Andrzejewska, L. Institute of Ecology, Polish Academy of Sciences, PO Lomianki 05 150, Dziekanów Leśny, Poland.

Arnold, G. W. Division of Land Resources Management, CSIRO, Private Bag, PO Wembley, WA 6014, Australia.

Bazilevich, N. I. Dokuchaev Soil Institute, Pyzhevski 7, Moscow-17, USSR.

Bowman, G. B. School of Forest Resources and Conservation, Institute of Food and Agricultural Sciences, University of Florida, Gainsville, FL 32611, USA.

Bowman, R. A. USDA-ARS Soil Fertility and Management Research, W106 Plant Science, Colorado State University, Fort Collins, CO 80523, USA.

Breymeyer, A. I. Institute of Geography and Spatial Organization, Polish Academy of Sciences, Krakowskie Przedmiescie 30, 00 927 Warsaw, Poland.

Brockington, N. R. The Grasslands Research Institute, Hurley, Maidenhead, Berkshire S16 5LR, UK.

Clark, F. E. Agricultural Research Service, US Department of Agriculture, Fort Collins, CO 80521, USA.

Cole, V. C. Agricultural Research Service, US Department of Agriculture, Fort Collins, CO 80521, USA.

Coleman, D. C. Natural Resource Ecology Laboratory *and* Zoology/Entomology Dept, Colorado State University, Fort Collins, CO 80523, USA.

Dash, M. C. Postgraduate Department of Biological Science, Sambalpur University, Burla, Sambalpur, Orissa, India.

Dommergues, Y. ORSTOM, PO Box 1386, Dakar, Senegal.

Duek, J. J. Institute of Applied Statistics and Computing, University of the Andes, Mérida, Venezuela.

Godron, M. P. — CNRS Louis Emberger Centre for Phytosociological and Ecological Studies, PO Box 5051, Route de Mende, 34033 Montpellier, France.

Gyllenberg, G. — Zoological Institute, Helsinki University, N. Jarnvagagatan 13, Helsinki, Finland.

Harris, L. D. — School of Forest Resources and Conservation, Institute of Food and Agricultural Sciences, University of Florida, Gainesville, FL 32611, USA.

Hinds, W. T. — Battelle, Pacific Northwest Laboratories, PO Box 999, Richland, Washington 99352, USA.

Hunt, W. H. — Biology Department, West Virginia University, Morgantown, West Virginia 26506, USA.

Hutchinson, K. J. — CSIRO Pastoral Research Laboratory, Armidale, NSW 2350, Australia.

Innis, G. S. — Department of Wildlife Sciences, Utah State University, Logan, Utah 84321, USA.

Kajak, A. — Institute of Ecology, Polish Academy of Sciences, PO Lomianki 05-150, Dziekanów Leśny, Poland.

King, K. L. — CSIRO Pastoral Research Laboratory, Armidale, NSW 2350, Australia.

Marshall, J. K. — CSIRO, Western Australian Labs, Division of Land Resource Management, Private Bag, PO Wembley 6014, Western Australia.

Noy-Meir, I. — Botany Department, Hebrew University of Jerusalem, Jerusalem, Israel.

Paul, E. A. — Saskatchewan Institute of Pedology, University of Saskatchewan, Saskatoon, Saskatchewan, Canada S7N 0W0.

Redmann, R. E. — Department of Plant Ecology, University of Saskatchewan, Saskatoon, Saskatchewan, Canada S7N 0W0.

Ribic, C. A. — Department of Ecology and Behavioral Biology, University of Minnesota, St Paul, MN 55108, USA.

Risser, P. G. — Botany Department, University of Oklahoma, Norman, Oklahoma 73069, USA.

Sasson, A. — Division of Biology, UNESCO, Place de Fontenoy, Paris 75007, France *and* Dept of Biology, University of Mohammed V, Rabat, Morocco.

| | |
|---|---|
| Schaefer, R. | Plant Ecology Laboratory, Faculty of Sciences, University of Paris-XI, 91045 Orsay, France. |
| Seligman, N. G. | Agricultural Research Organization, The Volcani Center, PO Box 6, Bet Dagan 50200, Israel. |
| Singh, J. S. | Department of Botany, Kumaun University, Nainital-263002, India. |
| Szocs, Z. | Research Institute for Botany of the Hungarian Academy of Sciences, H-2163 Vacratot, Hungary. |
| Titlyanova, A. A. | Institute of Soil Science and Agrochemistry, USSR Academy of Sciences (Siberian Branch), Sovetskaya 18, Novosibirsk 99, USSR. |
| Trlica, M. J. | Department of Range Science, College of Forestry and Natural Resources, Colorado State University, Fort Collins, CO 80523, USA. |
| Ulehlova, B. | Institute of Botany, CASU, Stara 18, Brno, Czechoslovakia. |
| Van Dyne, G. M. | Department of Range Science, College of Forestry and Natural Resources, Colorado State University, Fort Collins, CO 80523, USA. |
| Zlotin, R. I. | Institute of Geography, USSR Academy of Sciences, Moscow W-17, Staromonetny 29, USSR. |

# Foreword

The first official mention of the Grasslands Biome studies in IBP is in a report of a meeting of the Productivity of Terrestrial Ecosystems section – IBP(PT) – held in Warsaw 30 August to 6 September 1966. The main purpose of the technical sessions at Warsaw was to consider '...the principles and methodology of secondary productivity of terrestrial ecosystems'. Among the working groups which met during that meeting was one devoted to the study of grassland ecosystems. As IBP News No. 7 (1966) records: 'The working team concerned with grassland ecosystems discussed the project of a fully coordinated research programme. Five specialists from different countries were concerned with this during the meeting, and it was decided to elaborate the project subsequently by mail.'

The grasslands studies became Theme 1 of IBP(PT). Later, when the distinction was sharpened between major studies on ecosystems and investigations on special topics such as certain groups of consumers, the grasslands projects became collectively known as the Grasslands Biome Studies. By 1972, when the majority of IBP field studies were terminated, some thirty nations had shared in the work of the grasslands group, and several hundreds of scientists had been involved along with a much greater number of support staff.

The general aims of the grasslands theme were summarized in IBP News No. 13 (1969) as follows: 'Grasslands are one of the most important of terrestrial ecosystem types. Large areas occupy the interior of the principal continents, and provide, when managed for crop or meat production, a major source of man's food. The object of the IBP programme is to learn more about the organic production and overall energy flow of the world's grassland areas. Wherever feasible, concurrent studies with those on natural grasslands will be made on managed areas (pastures and croplands) and successional grasslands in the same region. This will make possible the provision of ecologically sound recommendations for the future utilization of the world's grassland areas.'

The considerable task of drawing up detailed plans and later of ensuring a high degree of coordination among the various projects was performed by a small Working Group. It met for the first time in 1967 under the chairmanship of Dr R. G. Weigert and was attended by scientists from six countries. It was held, appropriately, at Saskatoon, Canada, where, under the direction of Dr R. T. Coupland, a Canadian IBP grasslands team was already at work.

Over the period 1967–74 the Biome Coordinating Committee, with various changes in chairmanship, met on fifteen occasions.

In 1972 the Scientific Committee of IBP (SCIBP) appointed Chief Editors who, in association with coordinators or associate editors, were to be responsible for the preparation and completion of the synthesis volumes.

IBP brought together scientists from countries with different types of government; with very different attitudes towards the place of science in their culture; and with very different levels of scientific development. In addition, the acceptance of English as the language of communication within IBP placed an added burden on many scientists throughout the world. These strains were compounded because those of us with ready access to tape recorders, typewriters, secretaries, copying machines, a superabundance of paper and large, if not unlimited, postage accounts, produced a mass of voluminous reports and correspondence that must have added many hours of labour to scientists beyond the English-speaking world. Now, having studied once again many of the reports received during the course of the grasslands studies, I want to place on record the appreciation of the English-speaking contingent to those many workers who had to struggle with a language which lacks the logic of a formalized grammar. Not least among their problems was that of appreciating those subtleties of meaning which, although not clearly defined in a mother tongue, are, through custom and practice, understood by those born to it.

Whilst the basic approaches to scientific discovery may appear universal, the degree of universality tends to be exaggerated and this is certainly the case in a diverse and very often diffuse subject such as ecology. Ecology was, and to a considerable extent remains, a highly personalized science. In *The Evolution of IBP* (IBP Synthesis Series 1, 1975), I quoted a statement of Margalef: 'Ecosystems reflect the physical environment in which they have developed, and ecologists reflect the properties of the ecosystems in which they have grown and matured. All schools of ecology are strongly influenced by a *genius loci* that goes back to the local landscape.'

Nowhere else in IBP(PT) was the 'ecologist–locale' link more in evidence than in the Grasslands Biome Studies. On many occasions the energy expended in arguments on aims, methods and significance of results appeared to be approaching an explosion level but compensating control systems, usually in the form of good humour, succeeded in keeping the 'reactors' below the critical point of disintegration. It is not the first time in the history of biology that heated discussions have paved the way for scientific clarification. IBP productivity studies in particular have been fraught with arguments arising from differences in philosophy, definitions, and the complexity of the organic systems being studied. It is worth recalling a comment by Professors K. Petrusewicz and A. Macfadyen in their preface to *Productivity of terrestrial animals* (IBP Handbook 13, 1970): 'It has not always been easy to agree on what to say or how to say it. Our discussions have been many and prolonged and the final outcome has usually been based, we believe, on understanding rather than on compromise.' It can certainly be said that discussions within the grasslands theme were many and prolonged.

The USA and Australian IBP grasslands teams were composed largely of scientists whose background and interests centred on domestic animal production. The European, Japanese and Canadian teams on the other hand were less concerned with 'large' consumers and devoted considerable attention to the invertebrate and small vertebrate components of grassland ecosystems. In the early stages of IBP many European ecologists viewed with some scepticism the wholehearted devotion of their North American confrères to the 'systems approach'. They pointed to many unresolved taxonomic problems; to the paucity of knowledge on the feeding habits of key organisms; to the fragmentary information available on the seasonal changes in the contribution of known species to different trophic levels; and, above all, to the need for a thoroughgoing analysis of production within species or groups of organisms as distinct from obtaining knowledge on yields or standing crops.

I believe that it is important to recall that the operations phase of IBP(PT) reached its maximum in 1971, the year when ecologists should have been celebrating the centenary of the birth of A. G. Tansley who in 1935 added the term *ecosystem* to the ecologist's vocabulary. Until the advent of the computer and the creative imagination of a new generation of ecologists with the ability to use mathematical techniques to the full, the ecosystem as conceived by Tansley remained a mental construct, valuable and certainly genuine, unlike the Philosopher's Stone, but seemingly beyond total analysis.

Thirty years after Tansley's publication, IBP provided the opportunity for a thorough examination of the interaction of organisms and environment which together, to use Tansley's phrase, formed one physical system. First, the Canadian IBP(PT) Committee, and later the USA National Committee, decided on a systems approach in their programmes, using large multidisciplinary teams. In both cases grasslands were selected as the first ecosystems to be studied intensively, using the still-embryonic ideas of systems ecology.

At the time that IBP(PT) embarked upon a systems approach, it should be remembered that the systems which were to be studied were largely unknown. In engineering and economics, where systems dynamics had become a way of life, the systems could be simulated with a high degree of precision and the relationship between individual components and particular outputs could be explored. The systems ecologist, at the beginning of IBP, was faced with a very different situation. Knowledge of the component parts and processes was fragmentary even in the most thoroughly explored ecosystems. Thus, the systems ecologist was forced to look at his systems in a manner very different from that followed by his colleagues in aeronautics or in business management.

Natural and cultivated grasslands provide man with the major part of his food supply. They cover at least 23% of the land surface of the globe. Whilst some parts are becoming desert, elsewhere forests are being felled and, on balance, the grasslands biome is on the increase. Because of its extensive nature and diverse vegetational form, the grasslands biome has straddled three biomes in IBP – grasslands, tundra and arid lands. Scientists from all three biomes met at various workshops to discuss common problems, especially those arising from the utilization of the biomes for grazing. In 1969, largely as a result of the successful application of the systems approach in the IBP Grasslands Biome Studies, it was proposed that grasslands specialists from IBP and FAO should meet to explore how far the knowledge gained from IBP studies could be applied and developed further for use on a world scale. The discussions took place and other international organizations showed interest, particularly UNESCO. A proposal for an International Grazing Lands Programme was approved by SCIBP on behalf of the IBP(PT) biome investigators and it received the support of FAO and UNESCO, but it failed to obtain adequate financial support. Nevertheless the exploratory studies made by the tundra, arid lands and grassland biomes were not completely discarded. The MAB Project No. 3 on the *Impact of human activities and land use practices on grazing lands: savanna, grassland (from temperate to arid areas), tundra* owes much to these earlier discussions initiated by IBP.

The two volumes produced on the IBP grassland investigations can be regarded as independent approaches to the problems of studying grassland ecosystems. One volume, titled *Grassland ecosystems of the world* (ed. R. T. Coupland), concentrates on the structure, development and utilization of the world's grasslands, largely by producing extensive word models of the types of grassland which exist. The volume ends with a summary of the major components of the world's grassland ecosystems. The other volume, *Grasslands, systems analysis and man* (ed. A. Breymeyer & G. M. Van Dyne), is a synthesis of the massive amount of data collected during IBP grasslands studies. Its main purpose is to emphasize the dynamic aspects of the grassland ecosystem by giving emphasis to those processes related to productivity. It provides models and sub-models which permit the assessment of the effects of changes or stresses within a grassland ecosystem.

All who have contributed to these volumes are aware of gaps in their attempt to produce a synthesis. The end result is a distillation of numerous points of view and the volumes reflect differences of emphasis, of approach and, above all, disagreements – may it be remembered that disagreements in science are a way of exploring weaknesses and, when accepted in a constructive manner, are the lifeblood of scientific development.

It is always easy to say 'If only it had been possible to . . .'. All of the major biome studies were inadequately funded during the synthesis phase. This limited the opportunities to bring editors, chapter authors and contributors together for discussion. Furthermore, some contributors were unable to give enough time to the synthesis operation because they had to become involved full-time in other occupations. Nevertheless, in spite of these and other difficulties the grassland biome studies have advanced our knowledge of grassland ecosystems and they are a major contribution towards the formulation of a theory of ecosystems. These are major achievements. Taken together, they should ensure that the management of the world's grasslands can be placed on a more rational basis than has been possible in the past. The results are especially timely because the world may be facing changes in weather patterns which could well reduce food yields below today's levels. Thus, IBP Theme 1 has made a contribution towards the fulfilment of the overall aim of IBP to investigate *The biological basis of productivity and human welfare*.

J. B. Cragg
Killam Memorial Professor,
Faculty of Environmental Design
University of Calgary, Canada

# Contents of IBP 18*



* *Grassland ecosystems of the world*, ed. R. T. Coupland. Cambridge University Press, 1979.

**Part III. Semi-natural temperate meadows and pastures**
Subeditor: *M. Rychnovská*



**Part IV. Tropical grasslands**
Subeditor: *R. Misra*

# Introduction

A. I. BREYMEYER

Our book has a long and difficult history, and it is now four years since we decided to prepare it. Originally it was planned to prepare one volume with five editors; there are now two volumes and three editors. This volume is an attempt to present a synthesis of the knowledge on grassland ecosystems. We will analyse the structure and functioning of grasslands, which are well defined ecological systems existing in different geographical regions but which, we believe, always function in a similar and characteristic way. We shall try to follow regularities in their structure and, if possible, construct mathematical models of ecosystem processes or a model of the functioning of the whole ecosystem. Is it possible to develop such a model which is both specific and general at the same time, involving the most important features of the grassland ecosystem? We believed that it was.

Now that I have in my hands the completed manuscript it is possible to see how our preliminary ideas have been realized. The general pattern of ecosystem structure and function proposed in our early deliberations was very simple and commonly known (Fig. 1). It assumed that:

(1) the main stream of energy and matter cycling within the ecosystem follows the pattern described;

(2) the ecosystem is made up of a known and rather limited number of subsystems functioning in parallel;

(3) the organisms which comprise these subsystems utilize the same food resources and transform consumed food in a similar way.

This pattern of ecosystem structure and function determined the outline of this volume, which presents an analysis of the functioning of particular subsystems and of the grassland ecosystem as a whole.

About 40 authors were invited to co-operate in preparing the book. Some were unable to take part and had to be replaced by others and some resigned during the work. All these changes were preceded by long discussions, frequently hot and vigorous. This occurred despite the fact that all authors and editors enthusiastically accepted the proposal to cooperate in writing the book. Everyone was interested in working within a large international team, in which the exchange of ideas between scientists from different ecological schools and cultural circles would be possible. Realization of the idea of close co-operation among ecologists from various parts of the world seemed to be

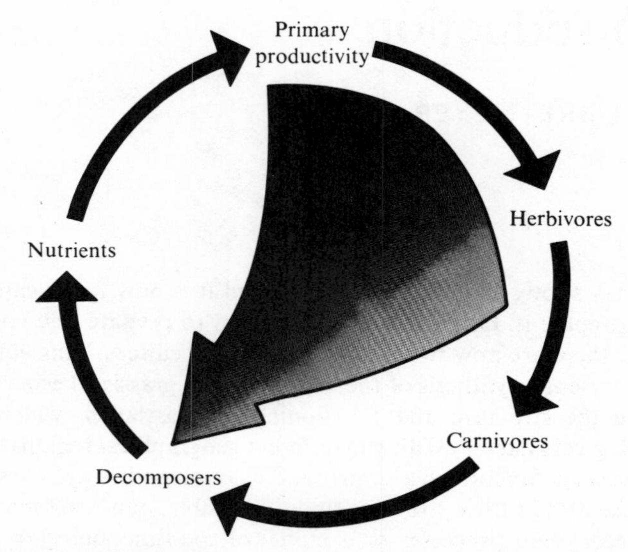


**Fig. 1. Simplified model of energy and matter cycling in an ecosystem.**

attained readily, and, indeed, the first and only meeting of almost all the authors, in Warsaw, took place in an atmosphere of full understanding. Differences of opinion and difficulties in co-ordinating the various viewpoints appeared later and resulted in the withdrawal of some participants. Why did this happen? I think there were several reasons. First, our task was difficult. We were trying to achieve a synthesis of very new and fresh knowledge. How should it be done? Which approaches were reliable enough while also holding promise for further development? Opinions were divided and frequently dependent on the state of knowledge in particular fields. Secondly, direct co-operation was impossible for a variety of reasons. Our book would be more mature and unified if we could have discussed it more often at common meetings. Unfortunately, such discussions were very rare because of financial limitations, and the exchange of views by letter was rather difficult, especially as native languages could not be used. This lack of discussion handicapped our work considerably and, although a common spirit can be felt throughout the text, it represents to some extent a review of different approaches to the problems involved. Thirdly, it was only when successive drafts of our book were being prepared that we realized how great such differences were. We come from different schools and from different cultural backgrounds, so not only were our scientific views different but also our understanding of each others' aims. For some authors 'synthesis' meant the review of all information available on a particular subject and the very preliminary arrangement of it. Others went further and tried to formulate general statements on the basis of the data collected. Some authors preferred the form of an essay and their chapters are not overcrowded with facts. Methods of data processing are also different. We made efforts to use systems analysis and modelling techniques wherever possible and it seems that we do use this modern method of description rather frequently. It does not imply, however, that we consider systems analysis as the only way of processing the ecological material – all good ideas are acceptable.

Work was divided between the two editors so that each was responsible for particular chapters. The chapters allocated to each editor were selected according to the proximity of the topics as well as the proximity to the authors – a factor of great importance as we co-operated with people from several continents. Dr Van Dyne was thus responsible for chapters dealing with abiotic conditions (Chapter 1), primary production (Chapter 2), large herbivores (Chapter 4) and modelling (Chapters 10 and 13). The present author (AIB) was responsible for the chapters on small and large predators (Chapters 5 and 6), small herbivores (Chapter 3), decomposers (Chapter 7) and also for the chapter on 'Comparative studies of ecosystem function' (Chapter 9). Some chapters were prepared by both editors, as for example the chapters on 'Nutrient cycling' and 'Management impacts on structure and function of sown grasslands'. Authors' first versions were discussed with me, but the final versions were prepared in co-operation with Dr Van Dyne. The 'Introduction', 'Trophic structure and relationships' and 'Reflections and projections' were prepared by each of the editors independently, as it was thought that the editors should also have the opportunity to express their views without limitations and without the need to subordinate their possibly contradictory ideas.

The authors performed their task in different ways. Some chapters have coordinators, this being probably the best solution when a number of authors were involved. Here the different authors only submitted contributions and the complete text was written by the coordinator. The chapter on decomposition is an example of such a procedure. It is one of the 'most international' chapters and has nine authors.

The topics of the book are arranged according to the following outline. Part I describes structures and processes observed in subsystems of the grassland ecosystem; Part II contains the general characteristics of the whole ecosystem; and Part III deals with the utilization of ecosystem. During the course of the work some 'supplements' have been introduced into this logical scheme. Thus, in Part I, herbivore and predator subsystems are subdivided into small and large herbivores and vertebrate and invertebrate predators, respectively, which may appear rather naive. This is not, of course, the result of a conviction that uniform trophic groups of animals should be divided according to their body size. In fact it is clear that both the herbivore and predator subsystems are single units from the point of view of ecosystem functioning and, moreover, that their division on the basis of body size is harmful to the purity of the concept of the ecosystem as comprising a limited number of subsystems. This division resulted from necessity, as not only did the methods differ but the authors of these particular sections were widely separated geographically. It is almost certain that if it had been possible to bring these authors together for a month or so then this division of the two subsystems would not have been necessary. There was a similar situation with Chapters 12 and 13, which partly overlap.

The book has been written by 38 authors. Their names are listed below with some additional information to introduce them and, also, to give some idea of the scope and quality of the field from which the facts and concepts presented in the book have been taken.

These 38 authors contribute to the book the spirit of the 15 universities and the 15 institutions in which they work at present. This is a rather impressive number of minds directed to a common aim.

I think that our first success lies in the fact that we have completed our work. Is the book itself a success? The readers will judge that. Personally, I now feel that what we have tried to bring together on the following pages is only the beginning, the first step towards the synthesis of information available on grassland ecosystems. We should do our best to continue the co-operation which has been built up between ecologists all over the world so that it yields a more comprehensive and elegant synthesis.

Preparation of this book has been expensive. Besides the costs covered by particular authors, the final preparation of the manuscript was possible largely through the support of two institutions: US National Academy of Sciences and Polish Academy of Sciences, Division of Biological Sciences. We are grateful to both of them. IBP Central Office, later IBP Publication Committee, with our patient guardian angel Mrs Gina Douglas supported us (both with money and in spirit) probably more than was reasonably possible. We also appreciate the enormous support given by our Convenor, Professor J. B. Cragg, who expended much energy to settle excessively lively disputes among the editors and who also spent much time on the revision of the manuscripts.

We have become friends during these several years of co-operative work and I wish to express my warm thanks to the whole team. Let us look forward to meeting again.

## List of authors

| Name | Degree from | Present position at | Specialization |
|---|---|---|---|
| Andrzejewska, Lucyna | Warsaw University | Institute of Ecology, Dziekanów Leśny, Poland | Invertebrate herbivores |
| Arnold, Graham W. | University of London | CSIRO, Wembley, Western Australia | Agricultural systems |
| Bazilevich, Natalia I. | Moscow University | Dokuchaev Soil Institute, Moscow, USSR | Soil biogeochemistry |
| Bowman, G. Bruce | University of Florida (candidate) | University of Florida, USA | Predator–prey relations |
| Bowman, Rudolf A. | University of California | Colorado State University, USA | Nutrient budgets |
| Breymeyer, Alicja I. | Warsaw University | Institute of Geography, Warsaw, Poland | Trophic structures and relations |
| Brockington, Dick (N.R.) | University of London | Grasslands Research Institute, Hurley, UK | Grassland management |
| Clark, Francis E. | Colorado State University | Agricultural Research Service, Fort Collins | Microbiology |
| Cole, C. Vernon | University of Wisconsin | Agricultural Research Service, Fort Collins | Plant nutrition |
| Coleman, David C. | University of Oregon | Colorado State University, USA | Decomposition processes |
| Dash, Madhab C. | University of Calgary | Sambalpur University, India | Soil ecology, energetics |
| Dommergues, Yvon | University of Paris | ORSTOM, Dakar, Senegal | Nitrogen fixation |

[continued

| *Name* | *Degree from* | *Present position at* | *Specialization* |
|---|---|---|---|
| Duek, Jacobo J. | Humboldt University | University of the Andes, Mérida, Venezuela | Plant ecology |
| Godron, Michel | University of Montpellier | University of Montpellier, France | Plant ecology, modelling |
| Gyllenberg, Goran G. | University of Helsinki | University of Helsinki, Finland | Animal ecology, modelling |
| Harris, Larry D. | Michigan State University | University of Florida, USA | Wildlife ecology |
| Hinds, Warren T. | University of Washington | Pacific Northwest Laboratory, Richland, USA | Physical ecology |
| Hunt, William H. | University of Texas | West Virginia University, USA | Systems ecology |
| Hutchinson, Keith J. | University of New England | Pastoral Research Laboratory, Armidale, Australia | Sheep-pasture relations |
| Innis, George S. | University of Texas | Utah State University, USA | Application of mathematics to ecology |
| Kajak, Anna | Warsaw University | Institute of Ecology Dziekanów Leśny, Poland | Invertebrate predators |
| King, Kathleen L. | University of Queensland | Pastoral Research Laboratory, Armidale, Australia | Pasture microarthropods |
| Marshall, John K. | University of Cambridge | CSIRO, Wembley, Western Australia | Plant autecology, modelling |
| Noy-Meir, Immanuel | Australian National University | Hebrew University of Jerusalem, Israel | Grazing ecosystems, modelling |
| Paul, Elder A. | University of Alberta | University of Saskatchewan, Canada | Soil microbiology |

| | | | |
|---|---|---|---|
| Redmann, Robert D. | University of Illinois | University of Saskatchewan, Canada | Primary production processes |
| Ribic, Christine A. | University of Minnesota | University of Minnesota, USA | Wildlife ecology |
| Risser, Paul G. | University of Wisconsin | University of Oklahoma, USA | Plant ecology and physiology |
| Sasson, Albert | University of Paris | UNESCO, Paris, France | Microbiology |
| Schaefer, R. | University of Strasbourg | University of Paris-XI, Orsay, France | Microbial ecology |
| Seligman, No'am | Hebrew University | Agricultural Research Organization, Bet Dagan, Israel | Range management |
| Singh, Jai S. | Banaras Hindu University | Kumaun University, India | Plant ecophysiology |
| Szocs, Zoltan | University of Budapest | Research Institute for Botany, Budapest, Hungary | Biomathematics |
| Titlyanova, Argenta A. | Leningrad University | Institute of Soil and Agrochemistry, Novosibirsk, USSR | Biogeochemistry |
| Trlica, Milton J. | Utah State University | Colorado State University, USA | Range science, ecophysiology |
| Ulehlova, Blanka | Charles University, Prague | Institute of Botany, Brno, Czechoslovakia | Decomposition processes |
| Van Dyne, George M. | University of California | Colorado State University, USA | Range management, systems analysis |
| Zlotin, Roman I. | Moscow University | Institute of Geography, Moscow, USSR | Ecology and geography of grasslands |

# Part I. Processes and productivity

# 1. Abiotic subsystem

W. T. HINDS & G. M. VAN DYNE

## Contents



## Introduction

*It's not true that life is one damn thing after another . . .*
*it's one damn thing over and over . . .*
(*Edna St Vincent Millay*)

This poignant insight into the human condition rings true in ecology as well, for ecosystems require energy and water everywhere and forever. Energy clearly is involved in every process, whether as chemical free energy gradients in cellular processes or as environmental energy fluxes determining climates. Water is the ubiquitous arbiter of growth, sustenance and structure, especially in terrestrial communities and most particularly in semi-arid or arid climates. These processes are familiar, and have been described in detail in many monographs and texts over the past decade or so (e.g. Slavík, 1974; de Vries & Afgan, 1975; Monteith, 1975; Gates & Schmerl, 1975). Not a great deal of work was done on abiotic processes as such in the IBP grassland studies. Much of what is put forward here is a uniform approach to energy and water budgets that is consistent with IBP grassland modelling tasks and philosophy.

Water and energy can be conveniently thought of as occupying contrasting pathways in ecosystems: water cycling around the system, while energy transits the system. Our interest in specific operations of the ecosystem determines the degree of subtlety required in any analysis (or model). Therefore, to display some ramifications following from differing depths of analysis, we have contrasted two scales of complexity of conceptual models: simple (or low-resolution) and not-so-simple (or moderate-resolution) models. It should become clear that these differences in resolution allow glimpses into altogether different worlds. It still remains to be seen how much these worlds overlap our own.

Another objective of this chapter is to provide a conceptual approach, compatible with and in part developed in, the IBP grassland modelling efforts which encompass driving variables, system state variables, rate processes and site constants. Prototype conceptual models of water and heat flow in the abiotic part of the ecosystem will be illustrated here along with indications of couplings to the biotic subsystems.

Some nutrient flows are in effect abiotic flows but are discussed herein in Chapter 8.

## Water flows

Generally, in preparing a conceptual model of an ecological system one distinguishes between driving variables and system state variables. However, whether a given variable is considered to be a driving variable or to be a system state variable depends on the scope and resolution of the model. For example, in modelling a grassland ecosystem the boundaries of the system often may be only a few metres above the plant canopy and a few metres deep in the soil. Precipitation would be a driving or external variable in that instance. If the upper boundary were many kilometres above the plant canopy, the components of the atmosphere of that system volume would be system state variables and precipitation would be a process. This more general view is taken in the two conceptual models of water flow which follow.

### *A low-resolution view*

Low-resolution models possess a desirable characteristic: 'economy of infrastructure'. This is not to say that subsequent analysis is either simple or unreal, just conceptually economical. Consider a low-resolution model for the hydrological cycle (Fig. 1.1). Four processes are involved: (i) precipitation, (ii) evapotranspiration, (iii) infiltration and (iv) unsaturated flow. This description of hydrology separates water movement into liquid flow down the gravitational vector and both capillarity and a change in density for flow against it. This clearly limits the variety of soil water geometries that can be

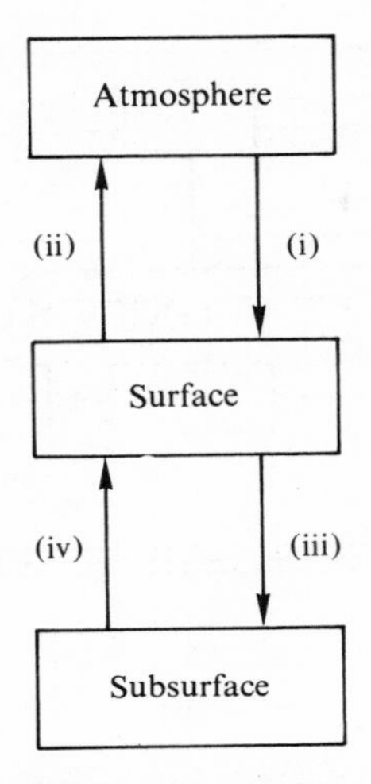


Fig. 1.1. A low-resolution model diagram for water flow in a grassland ecosystem. See text for discussion of compartments and processes.

considered, since saturated conditions (rice paddies, sub-irrigated pastures, and so forth) are excluded.

This model (Fig. 1.1) is not a strict compartment model, since the surface compartment has no capacity. Its role is as an exchange boundary, somewhat analogous to the role surfaces play in radiative energy exchange. That is, the net flux across the surface is a function of remote conditions on both sides of the surface, as well as of the surface itself. However, any simpler representation would be physically unreal since precipitation (clearly independent of soil physics) must be separated from infiltration (which is strongly dependent upon soil physics; Philip, 1975).

### A high-resolution view

A higher resolution model is required to differentiate between phases of water in either the atmosphere or the soil. Water potential profiles in the soil or a plant, or water that has been physically intercepted and diverted into transient reservoirs, likewise require additional compartments. This is particularly true if proper water potential gradients are to be matched with water available for evaporation or flow. Fig. 1.2 presents such a model, with water capacities distributed throughout a botanical skeleton and adjacent soil and air. This model still is a considerable simplification of real life.

The interactions indicated here for atmospheric water are phase changes, represented rather naively as minimally interacting, well-mixed reservoirs. Two phase changes are plainly ignored (vapour–ice exchanges). Well-mixed

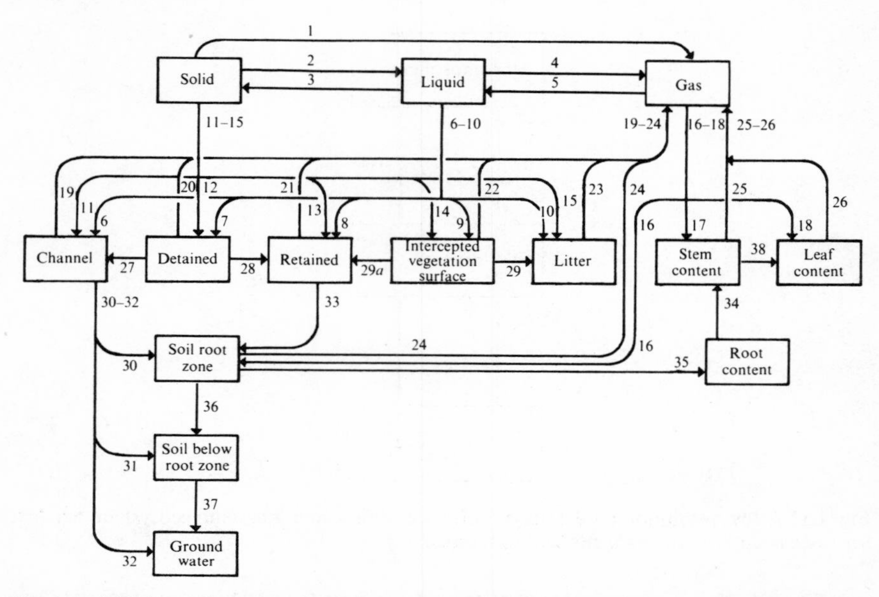


**Fig. 1.2. A high-resolution model diagram for water flow in a grassland ecosystem. Numbers refer to processes listed in Table 1.1.**

reservoirs clearly prevent considerations of profiles, lapse rates, or other gradients with height. It may not be quite so clear that advection is likewise excluded, so that only one-dimensional phenomena can occur. Consequently, the basic character of this model is identical with the simpler one – essentially beads of water strung along the gravitational vector.

Entry into the biosphere requires obvious solid, liquid and vapour pathways, but they each have conceptual problems. For example, condensation of atmospheric vapour onto leaves as dew is common and obvious, but condensation onto litter is complex. In dense grasslands, distillation from the soil surface to leaves or litter is more likely than condensation from the free air. Now, distillation is not a turbulent phenomenon, while dew deposition is (Monteith, 1957). Logically, then, water in the layer of air involved in distillation should be distinguished from that in the layer of air involved in dew formation. Dynamically realistic diurnal cycles could require such a distinction.

Separation of surface waters into several compartments raises potentials for hydrological hierarchies of flow. In the low-resolution model (Fig. 1.1), infiltration had to be indefinitely rapid, as necessary. In Fig. 1.2 hydrometeors can separate into several temporary storages, five including intercepted water, and six if a bare soil compartment were to be added. This proliferation of transient reservoirs should allow realistic infiltration rates, as well as a variety of opportunities for evaporation as the stored amounts increase.

Table 1.1. *A hierarchy of water flows in grassland ecosystems*

| Low resolution: general flow | Fig. 1.2 number | Higher resolution: specific flow |
|---|---|---|
| Within-atmosphere | 1 | Sublimation |
| | 2 | Melt |
| | 3 | Freeze |
| | 4 | Evaporation (virga) |
| | 5 | Condensation |
| Precipitation | 6 | Liquid precipitation to channel |
| | 7 | Liquid precipitation to detention |
| | 8 | Liquid precipitation to retention |
| | 9 | Liquid precipitation to interception |
| | 10 | Liquid precipitation to litter |
| | 11 | Solid precipitation to channel |
| | 12 | Solid precipitation to detention |
| | 13 | Solid precipitation to retention |
| | 14 | Solid precipitation to interception |
| | 15 | Solid precipitation to litter |
| | 17 | Gaseous water to stems |
| | 18 | Gaseous water to leaves |
| Evaporation | 19 | Evaporation from channel |
| | 20 | Evaporation from detention |
| | 21 | Evaporation from retention |
| | 22 | Evaporation from interception |
| | 23 | Evaporation from litter |
| | 25 | Gaseous loss from stems |
| | 26 | Gaseous loss from leaves |
| At surface | 27 | Overland flow, detained to channel |
| | 28 | Overland flow, detained to retained |
| | 29 | Stemflow to litter |
| | 29*a* | Stemflow to retained |
| Infiltration | 30 | Interflow, channel to soil root zone |
| | 31 | Interflow, channel to below-root zone |
| | 32 | Interflow, channel to ground water |
| | 33 | Infiltration |
| Within plant | 34 | Conduction, roots to stems |
| | 38 | Conduction, stems to leaf |
| Unsaturated flow | 35 | Movement to roots |
| Within subsurface | 36 | Percolation, to below-root zone |
| | 37 | Percolation, to ground water |
| Atmosphere–subsurface | 16 | Vapour diffusion, atmosphere to soil |
| | 24 | Vapour diffusion, soil to atmosphere |

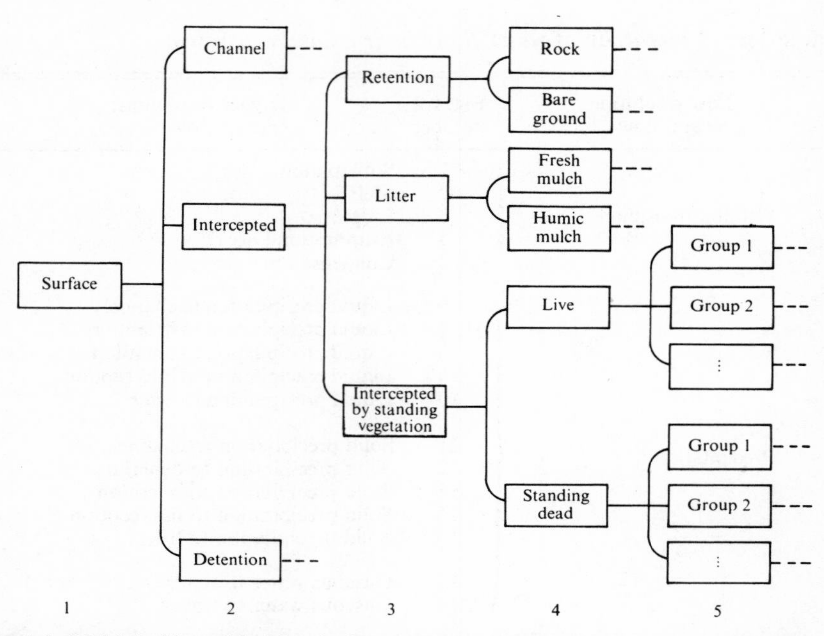


**Fig. 1.3. Multiple hierarchical levels for abiotic compartments. Level 1 corresponds to Fig. 1.1 and level 3 to Fig. 1.2.**

The water flow processes of Fig. 1.1 and 1.2 are compared individually in Table 1.1. The comparison is made easier if one considers the three main 'macrocompartments' of Fig. 1.1 to be subdivided in Fig. 1.2 into three, seven and four 'subcompartments' respectively. This requires definition of new processes, both within and between the three macrocompartments of atmosphere, surface and subsurface.

Only two hierarchical levels have been demonstrated in Figs. 1.1 and 1.2 and Table 1.1 whereas it is possible to conceptualize more levels. Some of these levels have been indicated above in the discussion of Fig. 1.2 and would be introduced by the disaggregation of some of the compartments in that figure. An idea of the hierarchy of subcompartments, and thus of subprocesses, is given in Fig. 1.3. Similar subdivisions could be made at other points in the model diagrams of Figs. 1.1 and 1.2.

A certain amount of precipitation may be caught and held on the standing live and dead vegetation surfaces. We assume, however, this water is adsorbed, not absorbed. And this, too, is a simplification. Rain drops will spread over and be held on the vegetation surface, particularly on broad-leafed plants. As saturation occurs, any excess drips from the leaves or flows along the stems to the soil surface (retention) or to litter. Some of this water, of course, is lost through evaporation during and following rainfall. In some instances this water may be held on the standing vegetation in the solid form as in snow, sleet and ice. Eventually it is converted to the liquid form through melting, though we consider the solid and liquid form separately in the model diagram of Fig. 1.2 (but still neglecting sublimation).

Some transfers between these important temporary storages are easy to visualize, but difficult to measure. For example, transfers between litter water and soil water clearly imply a distinct boundary between litter and mineral soil, an entity we could be hard-pressed to locate in swards or meadows. Furthermore, to describe water movement through litter as 'infiltration' is clearly an unsupportable improvisation (Couturier & Ripley, 1973). Another example could be transfers from detained water to soil water (via infiltration) or to retained water and channel water (via runoff). An elegant, topograpically-mediated dynamic equilibrium (between precipitation rate and detained water) is implied, but it is not often measurable in a muddy field (Grace & Eagleson, 1966).

In some types of grasslands, and in some seasons, the amount of snowpack will be large. Thus, the water content of the five surface storage compartments will be large. Such instances require more elaborate considerations, such as the transfer from the solid to the liquid phase (snowmelt), removal by wind (drifting, an advective process not considered in Fig. 1.2), and solid-to-gas transfer (sublimation).

The soil profile implicit in this moderate-resolution model is only indistinctly resolved. Clearly, the model is based on plant morphology, rather than soil morphology. A simultaneous recognition of soil horizon and soil physical properties is probably unnecessary since water movement is directly dependent upon the physical conditions (Gardner & Ehlig, 1962). The descriptive nomenclature of soil horizons seem to be dispensable (as assumed in this model).

The movement of water into and through the soil profile is due primarily to soil water potential gradients. Processes involved include desorption of the initially wetted zone and sorption of the dryer soil zones below. Redistribution in the profile depends on the depth of the initially wetted zone, the soil water potential gradient between the layers, the conductivity of the soil, and the amount and nature of plant roots. Numerous approaches are available for simulating these hydrological processes (Slavík, 1974; Flemming, 1975).

Hydrologically active zones of the soil profile are divided here into three regions: root zone, below-root zone, and ground water zone. The model in Fig. 1.2 considers only one soil layer in the root zone. This is already a simplification because it is necessary in many models (e.g. see Innis, 1978) to account for evaporation from the top few centimetres of the soil profile as well as water uptake by roots. Multiple transfer routes exist between adjacent zones – vapour diffusion, capillarity and infiltration – except for routes from the surface aimed downward; that is, channel, detained and retained water can infiltrate only. The subsequent fate of soil water is limited to evaporation, transpiration or runoff. The possible existence of phreatophytes in this model is unsettled.

Simple plant physiological processes underlie the simple plant of this model: a root, a stem and a leaf. Each of these organs must act as a reservoir as well as a conduit, so further differentiation into a larger variety of tissues is clearly credible.

Whatever advantages accrue from this moderate level of complexity, one is struck with its depauperate biological content. This may be an inevitable result of the state of the art of 'ecosystem dissection'. Aggregated compartments for the abiotic 'habitat' are more universally accepted at the moment than for the biotic 'community'. It is not yet clear what level of model complexity might allow plant water to be differentiated into xylem water, vacuole water and other functional reservoirs of water storage. For example, regarding guard cells as compartments rather than valves might be profitable (Waggoner & Zelitch, 1965). Should the moderate level of complexity inherent in Fig. 1.1 recognize a 'stoma' between the 'leaf' and the 'atmosphere'? Without stomata, these model plants are simply passive 'blotters', a far cry from reality.

## Energy flows

Energy, in the context of this discussion, refers to heat transfer (latent or sensible) and radiant energy transfer which is not coupled with biological processes of photosynthetic carbon fixation.

### *Abiotic energy flows*

A simple low-resolution model is exhibited in Fig. 1.4. One is immediately struck by a fundamental difference between Figs. 1.1 and 1.4; for energy, the source does not coincide with the sink. For water flows, of course, both source and sink reside in the atmosphere.

In low-resolution energy models, relations between the solar source and the surface can exclude atmospheric interactions. This approach has a long history and it is simultaneously meteorologically naive and ecologically viable (Lemon, Stewart & Shawcroft, 1971). By emphasizing the response of the receptor (the total surface compartment), the source of input energy is almost trivial.

It is not quite so clear that the spectral quality of the input can be so cavalierly dismissed. The band of wavelengths between 400 and 700 nm has long been understood to bracket photosynthetically important wavelengths, but gross energy measurements have tended to dominate light relations.

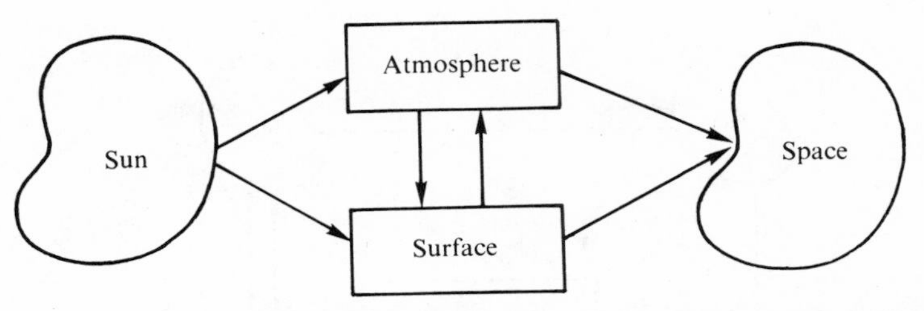


Fig. 1.4. A low-resolution conceptual model for energy flow in grassland ecosystems.

Recently discussions of light relations have reflected a wide acceptance of the concept of photosynthetically active radiation (McCree, 1972). This change is laudable, since light quality is so easily altered by atmospheric circumstance. Unfortunately, to represent the total solar spectrum in piece-wise fashion, more than one 'sun' would be required in Fig. 1.4. Perhaps very low-resolution energy models ought not to exist any longer.

Still, low levels of resolution can be made to yield a wide variety of results ranging from simple analyses of gross energetics (Hinds, 1975) to complex geometrical relations (cf. Verhagen, Wilson & Britten, 1963), depending upon the analyst's skill and intentions. But even with biological surfaces in the model, this low level of complexity can refer only to radiant processes, leaving out entirely all latent and sensible heat transfers. Clearly, then, low-resolution models cannot describe a multidimensional climate; they are restricted to light relations, albeit in a broad sense.

It is useful at this point to clearly distinguish between convection, conduction and radiation (Smith, J. A., 1973). Convection is the transfer of heat by the actual transfer of hot material from one place to another by motion of the material. Conduction is the transfer of heat from warm areas to cooler areas by the action of the molecular motion of warm areas in exciting neighbouring molecules by contact. Heat flow by radiation is accomplished by the continuous emission of energy from any surface in the form of electromagnetic waves. Conduction exchanges often are small in ecosystems, relative to radiative or convective energy exchanges. Conduction exchanges can be included with convective transfer by modifying the convective heat transfer coefficients.

Radiant energy exchanges occur more or less independently from fluid dynamic exchanges (sensible or latent heat transfers). Consequently, adding the fluid flow exchanges must increase model complexity. Fig. 1.5 presents a moderately complex model with them added. Atmospheric processes of energy transfer (convection, advection and evaporation) can parallel radiant energy pathways only to a limited extent. They are constrained by their corporeal embodiment in fluids. For example, no convective analogue exists for long-wave energy transfer directly between leaf and litter. For

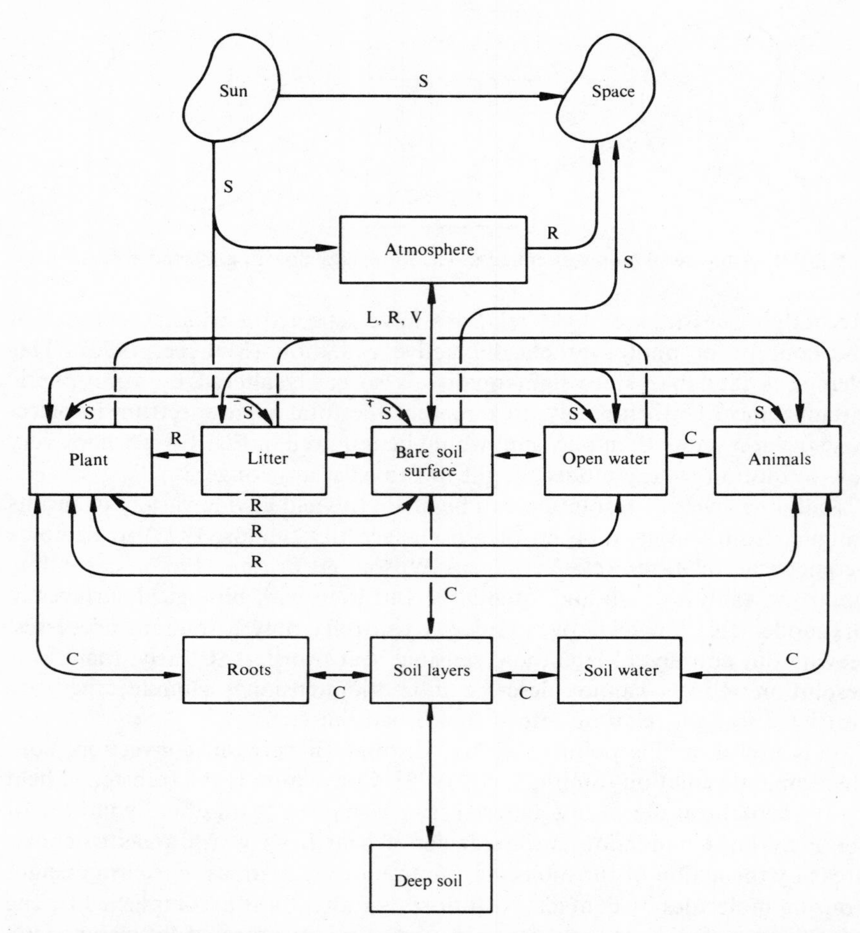


Fig. 1.5. A higher-resolution model diagram for energy flow in a grassland ecosystem. C, conduction; R, long-wave radiation; S, short-wave radiation; L, latent heat; V, convection.

convection, the atmosphere is the essential medium. Adding latent and sensible energy exchanges thus increases model complexity, but less so than adding more compartments to the ecosystem. That move spawns myriad permutations of long-wave radiation exchanges.

The ecosystem in Fig. 1.5 lies between two sources (sun and soil) and between two sinks (space and soil). Perhaps the strongest effect of antipodal sources (and sinks) follows from differences in what we can loosely call their response times. The space sink has an infinite response time, implying that its influence is constant. On the other hand, the deep soil sink has a response time of several days to weeks, producing a considerable lag in annual soil temperature cycles (de Vries, 1975). Since water potentials are affected by temperature, particularly low temperatures, the presence of the deep soil sink in the model adds (indirectly) another level of resolution to the soil profile. Similar considerations hold for the energy sources, but in shorter time intervals. The solar energy response time is a few minutes (diurnally) or a few days (seasonally), much faster than deep soil. This is the fundamental cause of hysteresis in soil temperatures. In a sense, the deep soil acts as a damper during energy transition from the intense and rapidly varying solar input to the infinitely slow chill of space.

Transfers from the sun to the surface and from the surface to space are both radiative. However, they differ in their spectral characters due to the different temperatures of the emitters. The high-temperature sun emits 'short-wave' radiation, while the low-temperature earth–atmosphere system emits 'long-wave' radiation. The difference is important because the atmosphere is relatively transparent to short-wave radiation and relatively opaque to long-wave radiation (part of the so-called 'greenhouse effect').

Short-wave radiation faces several potential diversions on its journey to a terrestrial surface and back into space if reflected. How many of these fates need to be specified in a moderately complicated model? For example, scattering by sub-micrometre atmospheric particles yields a different light quality (Fenn, 1976) than refraction from water droplets (List, 1951) or reflection between leaf layers (Anderson, 1964). Pursuit of atmospheric elegance in these matters will probably leave ecologists hoist on a petard of compounded structural detail. It is far easier to measure many radiation fluxes than to construct elaborate calculations based on either geometry of leaf whorls or profiles of atmospheric particulates.

Reflectivity, on the other hand, can usually be treated benignly. Most biologically important surfaces exhibit a strong tendency toward specular reflection at high angles of incidence (low elevation angles) (Idso, Baker & Blad, 1969). In some circumstances, such as dawn or dusk, or on steep slopes, this can be crucial. In most cases, though, reflectivity can be generalized simply into a fraction of incident energy, as implied in this model.

Long-wave energy exchanges are often numerous and sometimes obscure. They occur constantly, so their relative impact fluctuates diurnally, from important during daylight hours to unique at night. Usually they cannot be ignored, but in some circumstances they may be easily approximated. For example, profiles of all-wave flux resemble profiles of short-wave flux, due to differential absorption and emission of energy at various wavelengths (Cowan, 1968). This implies that long-wave flux profiles might be similar as well, a welcome first-order approximation, if true.

Long-wave radiation enters ecosystems only as a thermal effect, without triggering a variety of activities such as various short-waves do in plant physiology (Galston, 1974). Consequently, long-wave radiation can be represented in relatively simplistic or aggregated fashion, even in complex models.

Sensible heat transfers can likewise be more simply treated in ecological modelling than in meteorological modelling. Gross sensible flux into the earth has attained a respectable degree of understanding and realistic analysis (de Vries, 1975), and analyses of conduction and convection from simple shapes, such as leaves and lizards, are likewise advanced (Gates & Schmerl, 1975). However, sensible flux from a complex conglomeration of many individual organs – a tree, for instance – is not so well understood. Using the same principle described for radiation fluxes we might expect to find it easier to measure physical-ecological fluxes rather than apply atmospheric turbulence equations to dozens of leaf microenvironments, 'thereby standing the conventional micrometeorological approach firmly on its head' (Legg & Monteith, 1975).

Latent heat transfer is dependent on water yet is a major heat transfer process above and below the earth's surface. The relation between latent and sensible heat fluxes tends to be complementary. Evaporation proceeds at the maximum rate possible through the existing hydraulic resistances; any remaining input energy is converted to sensible heat (Monteith, 1965). It also functions as the link between water models and energy models. This link may allow latent heat to appear more important in ecosystem functions than is actually the case. Latent heat flux is water's entry into only the atmospheric vapour phase of a multipartite ecological reservoir. The liquid phases (atmospheric water, soil water and plant water) need no latent heat considerations until a phase change occurs.

Atmospheric influences on energy transfer would be better represented by splitting the atmosphere into at least a three-phase fluid: water vapour, air and condensed water. This approach to reality could weld together water and energy models to produce a moderately complex environmental model. Unfortunately, it still would refer only to a habitat rather than a community. Representing a polyspecific community in a multiclimate habitat is clearly a challenging task.

### *Interrelations of water, heat, autotrophs and heterotrophs*

The above sections show clearly one must simultaneously consider water and energy flow when modelling abiotic features in the grassland ecosystem. Most will agree also that it is necessary to consider simultaneously water and energy flows when discussing autotrophs and heterotrophs. Some relationships among abiotic and biotic components can be discussed briefly here as a prelude to later chapters. This discussion will be aided by reference to Fig. 1.6, where each of the compartments represents a model such as we have

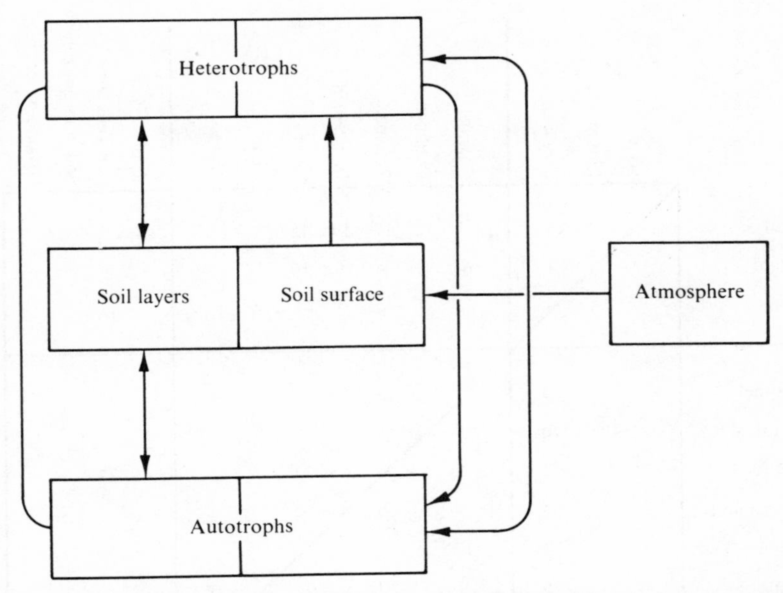


Fig. 1.6. A simplified view of transfer processes between biotic and abiotic components in the grassland ecosystem.

discussed here. A brief discussion of the interrelations follows (after Smith, J. A., 1973).

These other models provide some controls on energy transfers for some compartments and, in return, energy transfers partly govern the rates of flow for these models. For example, leaf area index modifies the incoming radiative flux to the soil, and evaporation of water from plants to the atmosphere is partly determined by soil heat flux to the atmosphere. The atmosphere acts here as an intermediary, both source and sink. Deep soil layers may also act in this dual capacity.

The total heat transfer between the atmosphere and an animal involves several physical processes. The net exchange is dependent on the total heat load of the animal, the total heat production of the animal, and a number of physical and environmental parameters which vary in space and time. Various models for defining the heat transfer process between animals and the environment exist and show promise, with proper modification, for application to the grasslands.

| | Solid | Liquid | Gas | Channel | Detained | Retained | Intercepted | Litter | Stems | Leaves | Roots | Soil–root zone | Soil–nonroot zone | Ground water |
|---|---|---|---|---|---|---|---|---|---|---|---|---|---|---|
| Solid | | 2 | 1 | 11 | 12 | 13 | 14 | 15 | | | | | | |
| Liquid | 3 | | 4 | 6 | 7 | 8 | 9 | 10 | | | | | | |
| Gas | | 5 | | | | | | | 17 | 18 | | 16 | | |
| Channel | | | 19 | | | | | | | | | 30 | 31 | 32 |
| Detained | | | 20 | 27 | | 28 | | | | | | | | |
| Retained | | | 21 | | | | | | | | | 33 | | |
| Intercepted | | | 22 | | | 29*a* | | 29 | | | | | | |
| Litter | | | 23 | | | | | | | | | | | |
| Stems | | | 25 | | | | | | | 38 | | | | |
| Leaves | | | 26 | | | | | | | | | | | |
| Roots | | | | | | | | | 34 | | 35 | | | |
| Soil–root | | | 24 | | | | | | | | | | 36 | |
| Soil–nonroot | | | | | | | | | | | | | | 37 |
| Ground water | | | | | | | | | | | | | | |

**Fig. 1.7. A 'coupling matrix' for the model diagram of Fig. 1.1 for water flow in a grassland ecosystem. Numbers refer to processes listed in Table 1.1.**

## Modelling abiotic processes

### *Matrices of variables, flows, processes and controls*

Numerous flows couple the variables in the conceptual models in Figs. 1.1 to 1.6. The flows are also listed individually in the right-hand column of Table 1.1. Some 39 flows are shown among the 14 variables of the simplified water model in Fig. 1.2. As the number of flows and variables increases it is a useful book-keeping procedure in modelling ecosystems to construct a 'coupling matrix' as in Fig. 1.7. The rows and columns in the coupling matrix are the system state variables of that particular subsystem. An entry into a cell of the matrix, a number, shows the existence of a flow which in effect is a rate process. The coupling matrix in Fig. 1.7 is partitioned to show

| CONTROLLING FACTORS | | Driving variables | System state variables : water flow | Other subsystem state variables | Site constants |
|---|---|---|---|---|---|
| | | Temperature, 1, 2, ..., Wind speed, 1, 2, ... | Solid, Liquid, Gas, Channel, Detained, Retained, Intercepted, Litter, Stems, Leaves, Roots, Soil–root zone, Soil–below root, Ground water | Live shoot biomass, Dead shoot biomass, Litter biomass, Crown biomass, Root biomass | Soil textures, Soil depths |
| Within atmosphere | 1 | | | | |
| | 2 | | | | |
| | 3 | | | | |
| | 4 | | | | |
| | 5 | | | | |
| Precipitation | 6 | | | | |
| | 7 | | | | |
| | 8 | | | | |
| | 9 | | | | |
| | 10 | | | | |
| | 11 | | | | |
| | 12 | | | | |
| | 13 | | | | |
| | 14 | | | | |
| | 15 | | | | |
| | 17 | | | | |
| | 18 | | | | |
| ESS | | | | | |
| Evaporation | 19 | | | | |
| | 20 | | | | |
| | 21 | | | | |
| | 22 | | | | |
| | 23 | | | | |
| | 25 | | | | |
| | 26 | | | | |
| Surface processes | 27 | | | | |
| | 28 | | | | |
| | 29 | | | | |
| | 29*a* | | | | |
| | 38 | | | | |
| Infiltration | 30 | | | | |
| | 31 | | | | |
| | 32 | | | | |
| | 33 | | | | |
| Flow | 34 | | | | |
| Within subsurface | 35 | | | | |
| | 36 | | | | |
| | 37 | | | | |
| Atmosphere–subsurface | 16 | | | | |
| | 24 | | | | |

**Fig. 1.8. A 'flow-effects' matrix structure example showing the general relationships among flows and the factors influencing them including driving variables, state variables of the subsystem of concern, other subsystem state variables in the ecosystem model and site constants.**

flows in and between the group components of the abiotic, water flow submodel – atmosphere, surface and subsurface.

A next step in modelling, after definition of the system variables and their couplings, is the determination of factors affecting the flows. This determination is simplified by means of a 'flow-effects' matrix whose format is illustrated in Fig. 1.8. Here the rows of the matrix represent the flows as given previously in Fig. 1.7. Usually, several factors influence a given flow and are used in the construction of an operational mathematical model derived from the conceptual and diagrammatic models noted thus far. Some of the driving variables noted in Fig. 1.8 are grouped for convenience. Clearly, a different temperature would be influencing processes at the atmospheric level, and at various heights within the atmosphere, as compared to the surface level. Generally, the temperatures at the surface level and at the subsurface levels would be considered state variables of a heat flow submodel of the overall ecosystem rather than driving variables.

### *Formulations for abiotic processes*

A large number of mathematical expressions have been used in characterizing abiotic flows in grassland ecosystems. It is beyond the scope of the present effort to discuss these in detail. We have summarized some of the abiotic process descriptions for grassland and grazingland systems in Table 1.2. General groups of processes are identified in Table 1.2. Both water flow and energy flow processes are included and in some instances they are inseparable. To condense the information on abiotic processes, we have given either general or specific information on the mathematical form of the model or model segment. We have not standardized all the symbols, but instead define the usage of various authors. The original papers should be referred to for a determination of the generality or limitation of the specific process descriptions.

Several abiotic models, particularly in hydrology, are well known worldwide and some are available commercially. An interested reader should see Flemming (1975) regarding overall discussion of hydrological simulation models.

## Some generalizations regarding key variables and processes

The key variables affecting the water flow process in an ecological system include the driving variable precipitation and the state variable soil water. The key process is probably the evapotranspiration process, which is an energy flow as well as a water flow, so another driving variable, i.e. solar radiation, should be added. Thus, one can visualize a compartment called soil water with an input of precipitation and an output of evapotranspiration.

Table 1.2. *A summary of abiotic processes used in simulating grazinglands and related systems*

| Reference | Factors affecting | Model structure | Comments |
|---|---|---|---|
| *Precipitation* | | | |
| Armstrong (1971) | Rainfall | | Model used to compare management strategies of set stocking and rotational grazing |
| Arnold & Galbraith (1974) | Rainfall | | Model used to predict liveweight changes and wool production responses with sheep and cattle |
| Arnold & Campbell (1972) | Daily rainfall | | Model used to predict liveweight changes and wool production while grazing on an annual clover pasture |
| Arnold *et al.* (1974) | Daily rainfall | | Simulation model to assess the effects of a grazing management on pasture and animal production |
| Baker & Horrocks (1974) | Rainfall | | Mathematical model simulating energy and gas exchange at plant–air interface and management considerations |
| Baumgartner (1965) | Rainfall | | Heat, water and $CO_2$ budget of plant cover |
| Brockington (1971) | Rainfall | | Model used to calculate water available to the sward by means of a soil–water budget relating soil moisture to grass production |
| Byrne & Tognetti (1969) | Rainfall | | Simulation model of a pasture environment |
| Caskey (1963) | Daily rainfall | $P_n = (1-P_{n-1})$ | Markov chain – precipitation prediction model |
| Chebotarev (1966) | Rainfall | | Theoretical and empirical approaches to mathematical models in theory of stream runoff |
| Chudleigh & Filan (1972) | Average monthly rainfall | $x = -(Um)(\log_e r)$ | Model of a pastoral property in Wales for sheep; where $x$ = monthly rainfall, $Um$ = average monthly rainfall and $r$ = rainfall |
| Cloud *et al.* (1968) | Daily rainfall | | Model assessing the effect of the forage harvest system |
| Curry & Chen (1971) | Daily rainfall | | Dynamic simulation of plant growth, utilizing actual daily water data |

Table 1.2 (*cont.*)

| Reference | Factors affecting | Model structure | Comments |
|---|---|---|---|
| Eagleman (1971) | Rainfall | | Statistical model derived from experimental data for actual evapotranspiration |
| Fitzpatrick & Nix (1969) | Weekly rainfall | $R = (S_i - P_i)/S_{max}$ | Model for simulating soil water regime in alternating fallow/crop systems; where $R$ = relative available water, $S_i$ = prior water, $P_i$ = precipitation, $S_{max}$ = soil water storage |
| Freer *et al.* (1970) | Daily rainfall | | Simulation model of summer grazing; Green material is a function of rainfall |
| Gabriel & Newman (1962) | Daily rainfall | | Markov chain model for daily rainfall occurrence at Tel Aviv |
| Gardner (1965) | Rainfall | $Q = (P + l_a)^2/(P_a - l_a + S)$ | Discussion of the main components of the hydrological cycle; where $Q$ = runoff, $P$ = precipitation, $l_a$ = amount of precipitation before $T_0$, $S$ = soil parameters |
| Goodall (1970) | Daily rainfall | Markov chain | Model calculates a moisture budget by random sample of daily precipitation; results used to compare different management practices |
| Grace & Eagleson (1966) | Rainfall | | Modelling of overland flow |
| Hopkins & Robillard (1964) | Daily rainfall | Markov chain | Precipitation prediction model for Canadian prairie provinces |
| Jensen *et al.* (1971) | Rainfall | $D = \sum_{i=1}^{n} (E_t - R_e - I + W_d)$ | Model estimating soil moisture depletion from climate, crop and soil data; where $D$ = soil moisture depletion, $E_t$ = energy, $R_e$ = rainfall, $I$ = irrigation, $W_d$ = percolation |
| Jones & Verma (1971) | Rainfall | | Digital simulation of soil moisture stress |
| Jones *et al.* (1972) | Daily rainfall | | Simulated environmental model of weather |
| Lowry & Guthrie (1968) | Daily rainfall | Markov chain | Probability models (of order greater than 1) for precipitation prediction |

| | | | |
|---|---|---|---|
| Mein & Larson (1973) | Rainfall | | Modelling infiltration during a steady rain |
| Parton (1978) | Daily rainfall | | Ecosystem-level (ELM) two-part abiotic grassland model |
| Rasmusson (1968) | Rainfall | | Atmospheric water vapour transport model computes flux divergence as an estimate of evapotranspiration |
| Ritchie (1972) | Daily rainfall | | Model for prediction of evaporation from a row crop with incomplete cover |
| Rose *et al.* (1972) | Rainfall | | Use of potential evapotranspiration to predict net primary productivity of shoot systems |
| Weiss (1964) | Daily rainfall | Markov chain | Model used to predict sequences of wet or dry days |
| Williams (1952) | Daily rainfall | Logarithmic series | Sequences of wet and of dry days considered in relation to series |
| Winkworth (1970) | Rainfall | $S_i + ppt = S_f + RO + E$ | Soil water regime of an arid grassland; where $S_i$ = initial water storage, $ppt$ = precipitation, $S_f$ = final water storage, $RO$ = runoff, $E$ = evaporation |
| Wiser (1966) | Daily rainfall | Monte Carlo | Precipitation – frequency analysis using Monte Carlo methods |
| Zahner & Stage (1966) | Daily rainfall | Regression model | $Y = b_1 y_{i-1} + B_2 S_i + B_3 T_i + B_4 P_i + B_5 S_{i-1} + B_6 T_{i-1} + B_7 P_{i-1} + B_8 D$; a procedure for calculating daily moisture stress on tree growth |
| *Interception* | | | |
| Baumgartner (1965) | Temperature of the air, relative humidity, wind, radiation | | Interception related to evaporation from plant surfaces; part of heat budget over plant canopy |
| Corbet & Crouse (1968) | Precipitation, vegetation cover area | $PISC = \mathrm{PA} + 0.025 PRIP + 0.37$<br>$PILT = (0.015 PRIP + 0.025)\ D$ | Interception model for standing crop and litter used in ELM; where $PISC$ = precipitation on crop, $PRIP$ = precipitation, $PILT$ = surface litter precipitation |
| Freeze & Harlan (1969) | Precipitation, energy balance | | Blueprint for a physically based, digitally simulated hydrological model |

Table 1.2 (*cont.*)

| Reference | Factors affecting | Model structure | Comments |
|---|---|---|---|
| Horton (1919) | Precipitation, projected plant cover, evaporation | Loss $= d - S_i + K_1 E_r t$ | Rainfall interception model; where $d$ = total interception loss, $S_i$ = storage, $K_1$ = evaporation to area ratio, $E_r$ = evaporation, $t$ = time |
| *Sublimation* | | | |
| Branton *et al.* (1972) | Synoptic weather changes | | Agricultural implications of sublimation gains and losses during winter |
| *Infiltration* | | | |
| Asseed & Kirkham (1968) | Soil properties | $y = Kt$, $y = Et^a$, $y = At^{\frac{1}{2}} + Bt$ | Equations describing the horizontal and vertical advance of irrigation water; where $A$ and $B$ are coefficients and different equations represent various soil textures, $y$ = cumulative infiltration, $K$ = Darcy coefficient, $t$ = time, $E$ = evaporation, $a$ = coefficient |
| Biswas *et al.* (1966) | Initial wetting depth, soil properties | $P_d - P_i + X_i = [K_w \partial p/\partial x - K_w]$ $X = X_1 \int_0^{X_1} \mathrm{d}x/kd(x)$ | Redistribution of soil moisture after infiltration, lab analysis; where $P_d$ = pressure dry sfc, $P_i$ = pressure layer $i$, $X_i$ = soil depth, $K$ = capillarity conductivity, $w$ = wetting |
| Black *et al.* (1969) | $\theta$, $k$, height, diffusivity | $\partial F/\partial t = D(\partial\theta/\partial t) + k$ | Prediction of evaporation, drainage and soil moisture storage, infiltration rate discussed in relation to soil moisture storage; where $\theta$ = volume water content, $k$ = conductivity, $F$ = depth, $D$ = diffusivity |
| Childs (1967) | Height, diffusivity | $-v = q = K + D(\mathrm{d}c/\mathrm{d}z)$ | Soil moisture models water movement in soils from precipitation using Darcy's equation as basis; where $K$ = soil characteristic constant, $z$ = depth, $D$ = diffusivity, $v$ = velocity, $c$ = concentration |
| Colwick & Bowen (1974) | Precipitation | | Cotton plant model giving dynamic description of soil moisture flux in top 6 inches of soil |

| | | | |
|---|---|---|---|
| Freeze & Harlan (1969) | Precipitation, energy balance | | Model of entire hydrologic cycle; infiltration is considered; blueprint discussion |
| Gardner (1965) | Initial soil moisture storage, soil properties | $I = At^{\frac{1}{2}} + Bt$ | Model of main parts of hydrologic cycle; where $I$ = infiltration rate, $A, B$ = coefficients, $t$ = time |
| Holtan (1965) | Initial and final soil moisture storage | $f = aS_r^n + f_c$ and $S_r = k(p - ASM) - F$ | Model concept of infiltration capacity where $f$ = infiltration capacity, $a, n$ = constants, $S_r$ = storage potential, $f_c$ = infiltration constant rate, $k$ = basal area of vegetation, $ASM$ = antecedent soil moisture |
| Horton (1940) | Soil properties, initial and final soil moisture storage, plant root systems, vegetation cover | $f = f_c + (f_0 - f_c e^{-K_f t}$ | Model of infiltration capacity as maximum rate soil can absorb water as rain falls; where $f$ = infiltration capacity, $f_c$ = infiltration capacity constant rate, $f_0$ = initial infiltration rate, $K_f$ = time constant, $t$ = time |
| Jones and Verma (1971) | Precipitation intensity | | Simulation model of soil moisture stress; rainfall intensity determines infiltration rate with saturation completed |
| Linsley *et al.* (1958) | Precipitation, soil moisture stress, soil properties | $f_p = f_c + (f_0 - f_c)\, e^{-Kt}$ | Valid when $i_s \geqslant f_p$ throughtout storm: general text describing models of hydrologic cycle where notation follows that of Horton (1940) above |
| Mein & Larson (1973) | Soil properties, soil moisture stress | $f_p' F_s = K_b[1 + (M_d(S(F))]$<br>$f_s = S_{av} M_d / [(I/K_s) - 1]$ | Two-stage model of infiltration during and after rainfall and $1 \geqslant K$; where $f_p$ = infiltration cap, $K_s$ = saturation conductivity, $M_d$ = moisture deficit, $S$ = capillary suction, $F$ = cumulative infiltration capacity, $F_s$ = amount until surface saturation, $S_{av}$ = capillary suction average, $I$ = rainfall intensity |
| Parton (1978) | Precipitation, evapotranspiration, interception, soil properties, terrain | | ELM grassland model water flow has infiltration rate as function of runoff |
| Philip (1964) | Soil properties | $f = S_0 t^{\frac{1}{2}} + AT$ | After Holtan (1965); where $f$ = infiltration rate, $S_0$ = capillary suction, $t$ = time |

Table 1.2 (*cont.*)

| Reference | Factors affecting | Model structure | Comments |
|---|---|---|---|
| Smith (1971) | Soil moisture stress, saturated, unsaturated | $f = ak_n S_0 + f_c$ | Water flow in hydrologic cycle; two-parameter infiltration, e.g. for uniform soil conceptual infiltration model based upon three storage elements, generates excess precipitation, infiltration recovery; where $f$ = infiltration, $a$ = constant, $k_n$ = constant, $S_0$ = storage capacity, $f_c$ = final infiltration rate constant |
| *Percolation* | | | |
| Childs (1967) | Soil properties | $v = -K - D(\mathrm{d}c/\mathrm{d}z)$ | Soil moisture theory models with Darcy flow equation $v = -K - D(\mathrm{d}c/\mathrm{d}z)$, where $v$ relates to rainfall intensity, $K$, $D$ and $z$ are as defined in Childs (1967) on p. 30, and $v$ = velocity |
| Gardner (1965) | Precipitation, soil properties | $W_d = 0.256zt^{-0.128}$ | Models main components of hydrologic cycle; where $W_d$ = percolation, $z$ = depth changing with time ($t$) |
| *Baseflow* | | | |
| Freeze & Harlan (1969) | Precipitation, energy balance | | Blueprint for a physically based digitally simulated hydrologic model |
| Freeze & Witherspoon (1966) | Soil properties | | Math model to represent steady state regional ground water flow |
| Toth (1963) | Topography | $z_t = z_0 + x \tan \alpha + a[\sin (bx/\cos \alpha)]/\cos \alpha$ | Theoretical model of ground water table, only small part participates in hydrocycle; where $z_t$ = depth of water table, $z_0$ = depth impermeable layer, $x$ = horizontal distance, $a$, $b$ = coefficients |
| Yen & Hsie (1972) | Soil properties | | Lab theoretical analysis of laminar flow simulating ground water movement using Darcy's equation |

| | | | |
|---|---|---|---|
| *Interflow* | | | |
| Biswas *et al.* (1966) | Soil moisture | $\partial\theta/\partial t = \partial/\partial x[(K(\partial\phi/\partial x)]$ and $-K_d \mathrm{d}\phi/\mathrm{d}x = K_w(\mathrm{d}\phi/\mathrm{d}x)$ | Redistribution of water after infiltration lab analysis, flow rate given by Darcy's equation; where $x$ = soil depth, $K_d$ = capillary conductivity drying, $K_w$ = capillary conductivity wetting, $\theta$ = volume water content, $\phi$ = precipitation – soil depth |
| Childs (1967) | Soil properties | $v = (\mathrm{d}Q/\mathrm{d}t)A$ | Darcy's law related to velocity of water movement in soils; where $v$ = velocity, $Q$ = flow rate |
| Freeze & Harlan (1969) | Precipitation, energy balance | | Models entire hydrologic cycle, blueprint discussion |
| Gardner (1965) | Soil moisture stress, soil properties | $q = K(\mathrm{d}\Psi/\mathrm{d}x)$ | Models main components of hydrologic cycle; where $x$ = a space variable, $q$ = soil flow, $K$ = hydraulic conductivity, $\Psi$ = water potential |
| Swartzendruber (1966) | Hydraulic conductivity | | Soil water behaviour as described by transport coefficient and functions |
| *Capillary action* | | | |
| Childs (1967) | Soil properties | $e = -K - D(\mathrm{d}c/\mathrm{d}z)$ | Soil moisture theory models with Darcy flow equation; notation as in Childs (1967) on p. 30 |
| Colwick & Bowen (1974) | Soil properties, temperature and relative humidity | | Cotton plant models water in top 6 inches of profile response to gradient |
| *Conduction* | | | |
| Woo *et al.* (1966) | Soil moisture storage, temperature, temperature of air, relative humidity, wind | | Conduction of water through the plant simulated by model of Ohm's law. Differential equations and transfer functions in terms of water suction variables |

Table 1.2 (*cont.*)

| Reference | Factors affecting | Model structure | Comments |
|---|---|---|---|
| *Evaporation, transpiration, evapotranspiration* | | | |
| Aase *et al.* (1973) | Radiation, soil flux, saturation vapour pressure, wind | $Et_0 = [(\Delta CR_n - G)/\Delta + \gamma] + [\gamma/(\Delta + \gamma)]\, a(b + cW)(e_s - e_d)$ | Model for estimating soil water constant based on Penman combination for potential evapotranspiration; where $\Delta$ = slope of saturated vapour curve, $\gamma$ = psychrometric constant, $a$, $b$, $c$ = coefficients, $e$ = saturated vapour pressure for saturated and dry soil ($e_s$ and $e_d$), $R_n$ = net radiation, $G$ = soil flux, $W$ = wind |
| Almeyda & Rasmussen (1972) | Precipitation, temperature of air, radiation, soil flux, sensible heat long-wave radiation | $R_n = G + H + LE = (Q + q)(1 - \alpha) - I$ | Energy budget approach to evapotranspiration; where $R_n$ = net radiation, $G$ = soil heat flux, $H$ = sensible heat, $LE$ = latent heat, $Q$ = convective rate, $q$ = convective rate, $\alpha$ = albedo $I$ = long-wave radiation |
| Armstrong (1971) | Precipitation, soil moisture | $EVR = A(1.0 - e^{a \cdot SM})$ | Model used to compare management strategies of set stocking and rotational grazing; where $EVR$ = evaporation, $A$ = leaf area, $a$ = saturated vapour pressure, $SM$ = soil moisture |
| Arnold & Galbraith (1974) | Temperature of air, radiation, precipitation, soil moisture | | Model used to predict pasture growth and animal production responses |
| Arnold *et al.* (1974) | Temperature of air, radiation, precipiation, soil moisture | | Use of a simulation model to assess effects of grazing management on pasture and animal production |
| Baier & Robertson (1966) | Temperature maximum and minimum, wind, duration of sunshine, saturation vapour pressure, day length | | Correlation of evaporation with daily record of weather data |
| Baier & Robertson (1966) | Standard meteorodata | $AET = \sum_{j=1}^{n} K_j(S_{i(i-1)}/S_j)\, Z_j$ $PET \cdot e^{-w(PET_j - \overline{PET})}$ | Estimates daily soil moisture on zone by zone basis from standard meteorological data; where $AET$ = actual evapotranspiration, $PET$ = potential evapotranspiration, $S$ = soil moisture by layer, $K_j$ = constant and $Z_j$ = depth |

| | | | |
|---|---|---|---|
| Baumgartner (1965) | Heat budget, soil moisture | | Methods and measurement of heat, water and carbon dioxide budget of plant cover |
| Black *et al.* (1969) | Diffusivity, $\theta$ | $E = 2(\theta_i - \theta_0)\,(Dt/\pi)^{\frac{1}{2}}$ | Evaporation ($E$) prediction from unsaturated flow equation for bare soil; where $D$ = diffusivity, $\theta$ = volume water content |
| Blaney & Morin (1942) | Temperature, relative humidity | $E = (0.0167T(114-H)$ | Empirical formula from lake water data; where $E$ = evaporation, $T$ = temperature, $H$ = sensible heat |
| Bogardi (1966) | | | Discusses most evapotranspiration formulas, lists them by function |
| Brockington (1971) | Radiation, saturation vapour pressure, sensible heat precipitation | $ET = [(\Delta/y)\,Rn + Ea]/[(\Delta/y)+1]$ | Soil water budget of sward, management strategies, Penman formulation; where $Rn$ = net radiation, $Ea$ = evaporation, $\Delta$ = slope saturated vapour curve, $y$ = constant |
| Brutsaert (1965) | Friction velocity, molecular diffusivity | | Evaporation as a molecular diffusion process into a turbulent atmosphere |
| Brutsaert (1965) | | | Comparison of methods of Penman, Blaney–Morin, Thornthwaite and Blaney–Criddle; best agreement Penman with open pan |
| Byrne & Tognetti (1969) | Precipitation | | Simulation model of a pasture environment |
| Colwick & Bowen (1974) | Radiation, maximum and minimum air temperature, soil moisture | | Cotton model production systems |
| Crawford & Linsley (1966) | Precipitation, cover type, soil moisture, storage, daily pan evaporation | | Sanford–Kentucky watershed storage model predicts evapotranspiration from interception, soil moisture and channel storages |
| Curry (1971) | Wind, radiation, air temperature, soil moisture | | Dynamic simulation of plant growth using corn |
| Curry & Chen (1971) | Radiation, wind, air temperature, soil | $\frac{dET}{dt} = \frac{\{\Delta(In-C)+pc[es(t)-e]/ra\}}{\{\lambda[\Delta+\gamma(1+rs/ra)]\}}$ | Expand Curry (1971) management strategies for corn; where $In$ = net radiation flux, $G$ = soil heat flux, $pc$ = pressure, $es$ = saturated vapour pressure, $t$ = time, $e$ = saturated vapour pressure, $ra$ = resistance, $\lambda$ = latent heat, $\Delta$ = slope saturated vapour curve, $\gamma$ = psychrometric constant, $rs$ = resistance |

Table 1.2 (*cont.*)

| Reference | Factors affecting | Model structure | Comments |
|---|---|---|---|
| Denmead & Shaw (1962) | Radiation, air temperature, relative humidity, wind | | Availability of soil water to plants |
| Eagleman (1965) | Weather data, soil moisture | | Relates evapotranspiration to soil moisture conditions |
| Eagleman (1971) | Precipitation, soil moisture, weather data, plant | $AET/PET = A+B(MR) + C(MR)^2 + D(MR)^3$ | Experimental data from several climates gives regression equation for actual to potential evapotranspiration (ET); where $AET$ = actual ET, $PET$ = potential ET, $A, B, C, D$ = constants, $MR$ = moisture ratio |
| Ferguson (1952) | Radiation, air temperature, absolute humidity, wind | $E = [Q+(p_e-p_a)h/\Delta]/(1+0.5/\Delta)$ | Factors determining evaporation ($E$) from open water shallow pond; where $Q$ = net radiation, $p_e$ = saturated vapour pressure, $p_a$ = saturated vapour pressure, $h$ = height, $\Delta$ = slope of saturated vapour curves |
| Fitzpatrick & Nix (1969) | Crop, soil moisture, precipitation | | Model for simulating soil water regime in alternating fallow-crop systems |
| Feer *et al.* (1970) | Precipitation | | Simulation of summer grazing |
| Fritschen (1965) | | | Evapotranspiration rates with Bowen ratio versus lysimeter data |
| Gardner (1965) | Volume water content, diffusivity, soil moisture depth | $E = D(\theta)W\pi^2/4L^2$ | Evaporation ($E$) rates related to hydraulic conductivity; where $\theta$ = volume water content, $W$ = critical water content, $L$ = soil depth wetted, $D$ = diffusivity |
| Gates (1962) | Radiation, conduction, sensible heat, short-wave radiation | $S_1+R_1+LE+G+C+S_z = 0$ | Computing latent heat of evaporation by energy exchange equation; where $S_1$ = solar flux, $R_1$ = radiation heat, $LE$ = latent heat, $G$ = soil storage, $C$ = convection, $S_z$ = storage |

| | | | |
|---|---|---|---|
| Gates (1965) | Air temperature, leaf temperature, relative humidity, diffuse sky light, resistance | $E = [sp_l T_l - RHsp_a T_a]/R$ | Flow of moisture from stomatal cavity to free air or atmosphere; where $E$ = flow of moisture, $R$ = total resistance to diffusion, saturated water vapour density ($spT$), $T_a$ = free air, and $T_l$ = leaf temperatures, $RH$ = relative humidity |
| | | $H+E+P = a_s S + r_s a_s(S+s) + a_t(R_n - R_g) - E\sigma T^4$ | Energy balance of a leaf; where $S$ = direct solar radiation, $s$ = diffuse sky light, $r_s$ = resistance, $a_t$ = leaf absorbance, $R_n$ = net radiation, $R_g$ = radiation to ground, $E\sigma$ = Boltzman constant |
| Gindel (1967) | Evaporation, air temperature, wind, relative humidity, soil | | Relationship of transpiration to factors in Israel |
| Goltz *et al.* (1970) | Evaporation, air temperature, relative humidity, wind | $F_r = \lambda[qw + q'w']$ | Evaporation measurements by an eddy correlation technique; where $F_r$ = latent heat flux, $\lambda$ = latent heat of vapour, $q$ = vapour concentration, $w$ = wind |
| Goodall (1970) | Daily precipitation | $E = \max[0, V(1-e^{bs+c})(1-e^{dA+f})]$ | Uses first-order Markov chain-model for comparison of different management practices; where $V$ = evaporation rate of open water surface, $E$ = evaporation, $e$ = saturated vapour pressure, $b$, $c$, $d$, $f$ = constants, $s$ = soil moisture, $A$ = per cent ground cover |
| Goudrian & Waggoner (1972) | Air temperature, relative humidity, temperature, soil temperature | $S = h + v$ | Simulation model describing daily course of microclimatic characteristic of solar canopy+soil; where $h$ = sensible flux, $v$ = latent loss, $S$ = total soil water absorbed |
| Hellmuth (1971) | Photosynthesis | $TR = TF(\mathrm{DW})/Pn(\mathrm{DW})$ | Water use efficiency model called transpiration ratio relates transpiration to photosynthesis or $CO_2$ assimilated; where $TR$ = transpiration, $TF$ = foliar transpiration, $Pn$ = photosynthesis, DW = dry wt |
| Jensen *et al.* (1971) | Precipitation, irrigation, drainage | $ET = Rn - G - H$ | Energy balance approach to determine soil moisture depletion |

Table 1.2 (*cont.*)

| Reference | Factors affecting | Model structure | Comments |
|---|---|---|---|
| Johns (1974) | Soil moisture | $E_t - E_a = A(SWD)^B(E_t)^C$ | Simulation model to predict water use and herbage production on swards; where $E_t$ = evaporation, $E_a$ = actual water use, $A^t$ = area, $SWD$ = soil water deficit, $B$ = constant, $C$ = constant |
| Jones & Verma (1971) | Precipitation, evaporation, soil moisture, soil temperature, soil properties | $E = \Delta Z\langle\{[M_1(0) - M(0) + M_1(L) - M(L)]/2\} + \sum_{z=1}^{L-1} [M_1(Z) - M(Z)]\rangle$ | Simulation model for soil moisture stress; depth, content; $L$ refers to the layers and $M$ to moisture |
| Jones *et al.* (1972) | Air temperature, precipitation | | Simulation of environmental weather model system of temperature, evaporation, precipitation and soil moisture |
| Knoerr & Gay (1965) | Leaf temperature, air temperature, radiation, sensible heat, soil flux, photosynthesis | $\alpha_s S + \alpha_L - R \pm H \pm E \pm Q \pm P_n = 0$ | Energy budget model for a tree leaf; where $\alpha_s$ = absorptivity of soil, $\alpha_L$ = absorptivity of leaf, $S$ = solar radiation, $L$ = long-wave radiation, $R$ = reradiation, $H$ = convection, $E$ = latent heat, $Q$ = storage, $P_n$ = metabolic heat |
| Kohler & Richards (1962) | Air temperature, dew point temperature, wind, radiation | $E = \exp\{(T_a - k_1)(k_2 - k_3 \ln R_n) - k_4 + k_5 \Delta_p k_6(k_7 + k_8 w)/k_9 + (T_a k_{10})^{-2}(k_{11} \cdot 10^{10} \times \exp[k_{12}/(T_n + k_{10})]\}$ | Multicapacity basic accounting for predicting runoff from storm precipitation; where $k_i$ = constants, $R$ = radiation, $w$ = wind, $T_i$ = temperature, $E$ = evaporation, $\Delta_p = f(T_a$ and $T_d)$ |
| Lang (1973) | Height, air temperature, relative humidity, wind, radiation, soil flux, sensible heat | $R_n = C + H_0 + LE_0$ | Measurement of evapotranspiration in presence of advection by means of a modified energy balance procedure; standard notation |
| Lemon *et al.* (1971) | Radiation, air temperature, wind, soil moisture, relative humidity | $Rn + LE + H + G + \lambda P = 0$ | SPAM photosynthesis model regulated by light interception, energy balance solved by leaf layers; standard notation |
| Lettau (1969) | Precipitation, radiation, soil moisture storage | | Water balance model, numerical prediction for evapotranspiration, radiation, soil moisture |

| | | | |
|---|---|---|---|
| Liakopoulos (1966) | Soil moisture storage, pressure in soil | | Model for theoretical prediction of evaporation and water loss |
| Loomis (1965) | Temperature of leaf, radiation | $H = Tr + 2E$ | Model for leaf energy budget, where $H$ = sensible heat, $Tr$ = transpiration, $E$ = evaporation loss by conduction and radiation |
| Miller & Tieszen (1972) | Radiation, wind | $aS + EIR = IR_l \pm H + LE$ | Energy budget model approach to primary productivity of arctic tundra; where $a$ = constant leaf area, $S$ = solar incident radiation, $E$ = emissivity, $I$ = long-wave radiation, $R$ = radiation, $R_l$ = radiation from leaf |
| Nkemdirim & Yamashita (1972) | Radiation, soil heat flux, sensible heat flux | $Rn = E + H + G + \phi$ | Model expressing energy balance over a grassland; standard notation |
| Nkemdirim & Haley (1973) | Radiation, air temperature, height, relative humidity, sensible heat, soil heat flux | | Grassland model to predict evapotranspiration loss; standard notation |
| Rasmussen (1971) | Radiation, air temperature, wind precipitation | $E = 1.6\,(10T/I)^a$ | Model for abiotic processes of Pawnee National Grasslands; where $E$ = evaporation, $T$ = temperature, $I$ = rainfall intensity |
| Ritchie (1972) | Radiation, precipitation, LAI | $E_t = E_s + E_b$ | Model predicting evapotranspiration from a row crop with incomplete cover; $E_t$ = total evaporation, $E_s$ = evaporation from soil, $E_b$ = evaporation from plants |
| Rosenzweig (1968) | Air temperature, precipitation, relative humidity, wind, soil moisture storage | | Model predicting net primary productivity of terrestrial ecosystems, shoot production |
| Seginer (1969) | Radiation, albedo, sensible heat, temperature of soil | $-LE = H + G(1-\alpha)R + E_s(R_{ts} + T_s^4)$ | Model of albedo effect upon evapotranspiration; where $R_{ts}$ = LW sky, $T_s$ = temperature of soil, $E_s$ = evaporation |
| Slatyer (1966) | Water vapour concentration, leaf resistance, wind | $E = (C_l - C_a)/(r_a + r_l)$ | Model of internal leaf transpiration rates; where $E$ = evaporation, $C_l$ = water vapour concentration at evaporating surface, $C_a$ = water vapour concentration in air, $r_a$, $r_l$ = resistances of air and leaf |

Table 1.2 (*cont.*)

| Reference | Factors affecting | Model structure | Comments |
|---|---|---|---|
| Van Bavel (1966) | Radiation, air temperature, $\Delta/\gamma$, $B_v$, wind | $LE = (\Delta/\gamma H + LB_v d_a)/((\Delta/\gamma)+1)$ | Potential evaporation rate model using the Penman combination approach; where $B_v$ = turbulent transpiration coefficient, $LE$ = potential evaporation rate, $\Delta/\gamma$ = wet bulb temperature, $d_a$ = bulb depression |
| Van Bavel *et al.* (1962) | Temperature of air, wind | | Transpiration model for sudan grass system |
| Waggoner & Reifsnyder (1968) | Temperature of air, relative humidity, LAI, leaf resistance, radiation, wind, sensible heat, height | | Transpiration model for sudan grass system; simulation model to predict evaporation profile in leaf canopy |
| Waggoner *et al.* (1969) | Temperature of leaf and air, sensible heat, radiation, relative humidity, leaf resistance wind, height | $E_k - E_{k-1} = (\theta_k - \theta_{k-1})(c_p p)$ | Simulation of microclimate in forest canopy; $c_p p$ = volume heat capacity of air, $\theta_k$ = increase in leaf temperature in $k$th stratum, $E_k$ = sensible heat |
| Woo *et al.* (1966) | Temperature of air, relative humidity, wind | | Simulation model of transpiration investigating various environmental conditions |
| Woo *et al.* (1966) | Light, temperature of air, relative humidity, wind | | Model of the stomatal control mechanism in the transpiration process |
| Wright & Brown (1967) | Temperature of air, relative humidity, wind, radiation | $Rn = H + LE + G$ | Energy balance approach model to determine vertical heat transfer in a crop canopy; standard notation |
| *Overland flow* | | | |
| Freeze & Harlan (1969) | Precipitation, energy balance | | Blueprint for a physically based, digitally simulated hydrologic model |
| Grace & Eagleson (1966) | Precipitation, topography | | Modelling overland flow using partial differential momentum and continuity equations |
| Holtan (1965) | Precipitation, potential storage of land | $Q = (P - k_1 s_p)^2/(P + k_2 s_p)$ | Model computes watershed retention from soil parameter; where $P$ = precipitation, $k_i$ = constants, and $s_p$ = potential storage |

| | | | |
|---|---|---|---|
| *Stemflow-drip* | | | |
| Clark (1940) | Leaf area index | | Interception of precipitation by prairie grasses, weeds and crop plants |
| *Energy exchange* | | | |
| Almeyda & Rasmussen (1972) | Radiation, conduction convection, latent | $Rn = G+H+LE = (Q+q)(1-\alpha)-I$ | Grassland energy budget approach to evaluate evapotranspiration; standard notation; where $Q, q$ = convective rates, $I$ = long-wave radiation, $\alpha$ = albedo |
| Anderson (1966) | Radiation | $I = I_o e^{-KF}$ | Stand structure and light penetration model related to leaf area index; where $I$ = light intensity stand, $I_o$ = light intensity open, $K$ = extinction coefficient, $F$ = leaf area index |
| Anderson & Denmead (1969) | Radiation, leaf area index, foliage inclination | | Direct and diffuse radiation math formulations to compute radiation on plant surfaces |
| Brown & Rosenberg (1972) | Radiation, soil heat flux, temperature, wind, vapour pressure, crop resistance | $LE = \{f[(R_n-S-LE)(\rho Cp)^{-1}r_a + T_a]-e_a\}\{\rho L_p M_w(r_a+r_l)]\rho M_a\}^{-1}$ | Resistance model to predict evapotranspiration of sugar beets; where $R_n$ = net radiation, $S$ = soil heat flux, $LE$ = latent heat, $C_p$ = specific heat of air, $\rho$ = density of air, $r_a$ = air diffusion resistance, $T_a$ = air temperature, $e_a$ = saturated vapour pressure, $r_l$ = crop resistance, $M_w$ = mol. wt water, $M_a$ = mol. wt air, $L_p$ = leaf density |
| Brown & Covey (1966) | Radiation, temperature, leaf | $K_z = (R_n-S)\{-P_a(C_p\partial T/\partial z + (M_w/M_a)/\rho(L_p)\,\partial e/\partial z\}$ | Energy budget approach to transfer processes in a corn field |
| de Vries (1958) | Conduction, temperature gradient | | Model of simultaneous transfer of heat and moisture in a porous medium |
| de Wit (1965) | Radiation, sun position | | Photosynthesis in leaf canopies |
| Dingman (1972) | Temperature of air, water, radiation, wind, cloud cover | | Equilibrium temperature $T_e = [(Q_r-Q_0)/q]+T_a$ of water surfaces related to air temperature and radiation |

Table 1.2 (*cont.*)

| Reference | Factors affecting | Model structure | Comments |
|---|---|---|---|
| Garnier & Ohmura (1968) | Sun position, air | $I_d = I_0 \int_{H_1}^{H_2} p^m \cos(XAS)\, dH$ | Model for calculating direct short-wave radiation incoming on slopes, where $I_d$ = direct radiation, $I_0$ = solar constant, $X$ = slope angle, $S$ = declination, $H$ = sensible heat, $P$ = mean zenith path transmittivity of atmosphere, $m$ = optical air mass |
| Gates (1962) | Radiation, conduction, sensible heat | $S_1 = R+LE+G+H+S_2 = 0$ | Model of energy flow in biosphere (to earth and organisms); standard notation; where $S_1$ = solar flux radiation, $R$ = radiant heat, $S_2$ = storage flux |
| Gates (1962) | Radiation, temperature of leaf and air | $a_s S + a_s s + r_s a_s(S+s) + a_t(R_a+R_g) - ZET_e^4 = H+E+P_n$ | Energy balance model of single leaf; see Gates (1962) on p. 36 |
| Idso *et al.* (1969) | Radiation, temperature filter factor | $E = [R_o - f(T_s)B_s]/[(1/\pi) f(T)\sigma T^4 - f(T_s)B_s]$ | Model for determining infrared emittances of leaves; where $E$ = infrared leaf emittance, $R_0$ = energy by $IR$ thermometer, $T_s$ = temperature of soil, $B_s$ = total hemisphere radiation, $T$ = temperature filter function, $f(T_s)$ = function of soil temperature |
| Idso *et al.* (1966) | Temperature gradient | $E = s\rho_L(T_L) - \rho_a(T_A)/R$ | Energy exchange model using temperature gradient and total plant resistance to compute evaporation (see above) where $s$ = soil, $\rho_L$ = density of leaf, $T_L$ = temperature of leaf, $\rho_a$ = density of air, $T_A$ = temperature of air |
| Kasanaga & Monsi (1954) | Radiation | $I = I_0 e^{-KF}$ | On the light transmission of leaves; see also Anderson (1966) notation |
| Knoerr & Gay (1965) | Temperature of leaf (upper and lower) | $h_c = (H_u+H_l)/2(T_L-T_a)$ | Mean convection heat transfer coefficient computation for energy balance of a leaf; where $h_c$ = convection coefficient, $H_u$ = sensible heat upper leaf, $H_L$ = sensible heat lower leaf, $T_L$ = temperature of leaf |

| | | | |
|---|---|---|---|
| Linacre (1968) | Radiation, temperature, cloud cover | | Radiation intensity major controlling influence on water loss from irrigated crops |
| Monteith (1972) | Radiation (sensible), evaporation, conduction, storage heat flux | $R+M = H+\lambda E+J+G$ | Energy balance model for organisms; where $R$ = radiation, $M$ = metabolic flux, $H$ = sensible heat, $\lambda E$ = evaporation, $J$ = stored heat, $G$ = soil heat flux |
| Murphy *et al.* (1972) | Radiation (sensible, latent, boundary layer), temperature, specific humidity, height | | Dynamic energy balance model to investigate exchange processes |
| Parkhurst *et al.* (1968) | Temperature, area of leaf | $Q_c = h_c A(t_1-t_e)$ | Model to calculate convection coefficient for broad leaves, wind tunnel model; where $h_c$ = sensible flux coefficient, $Q_c$ = sensible flux, $A$ = leaf area, $T_1$ = initial temperature, $t_e$ = temperature of air |
| Parton (1978) | Radiation, temperature of canopy, soil and air, precipitation slope, soil type | | ELM grassland model of water flow and temperature flow |
| Philip (1964) | Density of air, specific heat of air, precipitation, slope, soil type, latent heat | $H = H(0)+\int_0^z \alpha\rho C_p K(\mathrm{d}A/\mathrm{d}z) \times (T_f-T_a)\,\mathrm{d}z$ | One-dimensional model for sensible heat transfer in foliage layers; where $H(0)$ is small, $C_p$ = specific heat of air, $T_a$, $T_f$ = temperature of air or foliage, $A$ = leaf area index, $\rho$ = air density, $z$ = depth, $\alpha$ = absolute humidity, $K$ = constant |
| Portman *et al.* (1961) | Wind, temperature, relative humidity, sensible heat, radiation, water storage flux | $N = (I+G)-(R+N)$ | Model of energy exchange between the atmosphere and water surfaces; where $N$ = total energy exchange, $I$ = incoming solar total, $G$ = incoming atmospheric radiation, $R$ = reflected radiation, $W$ = emitted radiation from water surface |
| Raschke (1956) | (Complex equation) | | Model of turbulent diffusivities, boundary layer resistance; energy balance main function of wind in canopy |

Table 1.2 (*cont.*)

| Reference | Factors affecting | Model structure | Comments |
|---|---|---|---|
| Rose (1966) | Wind | $(u^*)^2 = \tau_0/\rho_a$ | Model of downward transfer of momentum as a function of eddy shearing stress; where $u^*$ = friction velocity, $\tau_0$ = shear stress, $\rho_a$ = density of air |
| Sellers (1965) | Radiation, temperature of air, height | | Solar radiation formulation for ELM |
| Swinbank (1968) | Wind, gravity, height, temperature | $(\partial T_p/\partial z) = k_1(u)^{*2})(qz^2)^{-1}(z/L)^{k_2}$ | Model to estimate vertical flux and horizontal momentum; $u^*$ = friction velocity, $T_p$ = potential temperature, $z$ = depth, $k$, and $k_2$ = coefficients |
| Tanner (1960) | Radiation (sensible) | $E = (R_n - G)/(1+\beta)$ | Energy balance model approach to evapotranspiration from crops; standard notation |
| Van Wijk & Derksen (1966) | Heat conductivity, heat capacity, thermal diffusivity | $a = \lambda/c$ | Model of heat propagation in the soil; where $a$ = thermal diffusivity, $\lambda$ = conduction, $c$ = volume heat capacity |
| Viskanta & Toor (1972) | Radiation | | Model predicting radiant flux per unit volume for a layer of water |
| *Decay-shatter* | | | |
| Uresk (1971) | Precipitation, wind, insect, trampling, faecal excretion | | Litter increase dynamics of blue grama in a shortgrass ecosystem |
| Witkamp (1966) | Temperature, time, relative humidity, precipitation, $CO_2$ | | Model of decomposition of leaf litter in relation to environment by regression analysis |

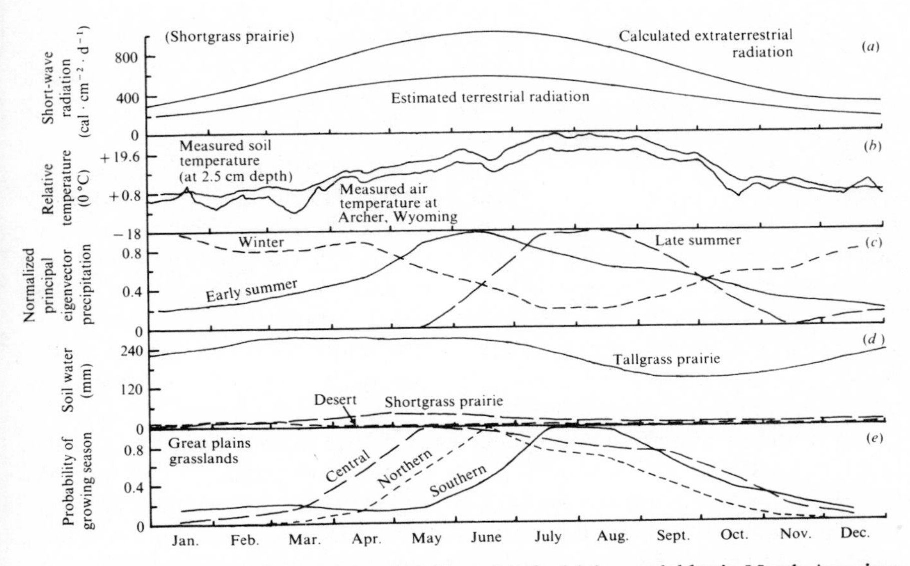


**Fig. 1.9.** Examples of the annual cycle of key abiotic driving variables in North American grasslands based on data from US/IBP grassland biome study sites and nearby locations.

This simplification meets the characteristics of most arid and semi-arid grasslands and some subhumid grasslands because there is little transfer of water through the profile in the grassland soil into deep drainage. Thus, the water input, over an annual cycle, must equal water output. Many other compartments, such as water in plants, are important, of course, but are of small magnitude compared to the soil water store. A simplified view of the magnitudes and dynamics of these variables and processes is shown in the following figures for North American grasslands.

The key driving variables of temperature, solar radiation input and precipitation are noted in Fig. 1.9(*a*)–(*c*). A generalized view of calculated extraterrestrial radiation and estimated terrestrial radiation on a shortgrass prairie is noted in Fig. 1.9(*a*). Solar radiation input in general can be estimated by a sine wave function. It can also be estimated fairly well from position on the earth's surface. The values here match approximately 40° north latitude. Corrections, of course, must be made for attenuation due to cloud cover and atmospheric particles and other factors.

Two key variables in determining water flow relationships are the air temperature and the soil temperature. A plot is shown in Fig. 1.9(*b*) for measured air and soil temperature at 2.5 cm at a site approximately 25 km from the location for which the radiation data are plotted in Fig. 1.9(*a*). Both air and soil temperature in a grassland undergo a considerable seasonal fluctuation. Note here, however, that we are concerned with soil temperature at 2.5 cm. There is a 'damping' effect on soil temperature fluctuations as one goes deeper in the soil profile (see p. 21). Notice in Fig. 1.9(*b*) that there are intervals when air temperature fluctuates considerably but there is no major change in soil temperature. This is especially so in the winter.

Precipitation is a key driving variable in grassland ecosystems. In Fig. 1.9(*c*) we have presented some analyses of precipitation patterns to show general types of rainfall patterns in grasslands in the North American continent. Statistical clustering techniques were used to analyse data from a series of grassland sites in Canada, the United States and Mexico. Three precipitation groups were derived – 'winter, early summer and late summer' (Smith, 1972). A subsequent comparison was made using the principal eigenvectors to reveal the underlying infrastructure of the annual rainfall regime with respect to periods of major rainfall. The normalized principal eigenvectors are plotted in Fig. 1.9(*c*).

The winter precipitation group includes grasslands from the Mediterranean type, shrub-steppe and northern Rocky Mountain regions of the United States. In this type there are peaks of rainfall events in fall, winter and early spring. In some of the mountain sites, of course, the major contribution of the annual precipitation will be as snow. The early-summer precipitation group includes all the Great Plains sites in Canada and the United States. In this group most rainfall is received during May and June and snow does not contribute greatly to the pattern of annual precipitation. The late-summer precipitation group includes the desert grassland sites in southwestern United States and northern Mexico. These sites receive their rainfall in the summer with the peak occurring in July and August.

The dynamics of soil water in grassland ecosystems is illustrated in Fig. 1.9(*d*) with examples for tallgrass prairie, shortgrass prairie and desert grassland in North America. These data are calculations based on long-term average information from weather stations adjacent to grassland research sites. The soil water storage term, $W_2$, in a Thornthwaite–Mather model, is linked closely to water balance and to energy balance, as well as to the ecosystem producers through the transpiration process. Peak soil water storage is dictated primarily by the depth of soil (Rasmussen, 1971). The slope of the water discharge in Fig. 1.9(*d*) illustrates two factors: (i) evaporation and transpiration are removing water, dependent upon the temperature and the amount of plant cover; (ii) precipitation is adding water during the growing season in some sites. Note the comparison in Fig. 1.9(*c*) and (*d*) of the interrelationship between soil water storage and time of precipitation. Both the tallgrass prairie and shortgrass prairie follow the early-summer precipitation pattern of Fig. 1.9(*c*). The desert grassland follows the late-summer precipitation pattern. The time of precipitation and the soil water storage match more for the shortgrass prairie than for the other two types of grassland.

Since both precipitation, via soil water, and temperature influence the growing

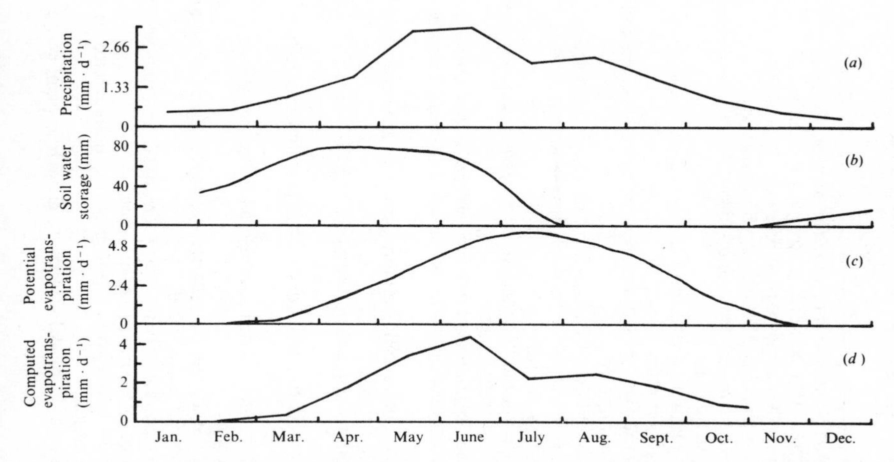


**Fig. 1.10.** Example dynamics of key abiotic features for a single mixed prairie grassland site based on long-term climatic data.

season for grassland, an analysis has been made of the probability of growing seasons for different sites in the North American Great Plains (Smith, 1972). In this analysis, a potential thermal growing season was calculated as the intervals between dates at which the 15-day running mean air temperature was $\geqslant$ 4.4 °C. For combining temperature and precipitation, a threshold temperature of 0 °C and a threshold of 0.25 mm rainfall were used. The joint probability of having a sufficiently high temperature and a sufficiently high rainfall was then used to determine the probability of a growing season.

Desert grasslands comprise the 'southern' group of Fig. 1.9(*e*) whereas the remainder of the Great Plains sites fall into the 'central' group (e.g. the US/IBP Pawnee Site) and the 'northern' group (e.g. the Canadian Matador Site). Annual grasslands and shrub-steppe sites do not have patterns of joint probability similar to these groups. The plot in Fig. 1.9(*e*) shows the values of the principal eigenvectors based on the probability of rain times the probability of temperature for each month in each of these site locations. It is expected that when sufficient data are available for analysis, primary production patterns will follow these probabilities of growing seasons for various locations (see Chapter 2).

The above are generalized characteristics for key variables and processes for different grasslands. In Fig. 1.10 are the results of an analysis of some of the key processes and variables related to the water balance and energy balance for a single grassland site. The location here fits the mixed prairie in the central Great Plains of North America. The analyses are based on 30-year average data (Rasmussen, 1971). Fig. 1.10(*a*)–(*d*) shows the variation in precipitation, soil water storage, potential evapotranspiration and computed

Table 1.3. *Photosynthesis models with respect to abiotic variables*

| Reference | Factors affecting | Comments |
|---|---|---|
| Baker & Horrocks (1974) | $CO_2$ concentration, water vapour, temperature, light | Cornmod: dynamic simulator for corn production |
| Beeman (1966) | Temperature, light, LAI, relative humidity | Tobacco model expresses plant growth in terms of photosynthesis and respiration rates |
| Botkin (1969) | Temperature, light | Oak tree model predicting net photosynthesis |
| Brown (1974) | Soil moisture stress, phenology, radiation, temperature | Mechanistic model determined from $CO_2$ exchange data for two prairie grasses |
| Brown (1969) | Light, $CO_2$ concentration | Model of photosynthesizing leaf for several crop species, predicts Pn rates |
| Chan *et al.* (1969) | Relative humidity, light, temperature | Model for plant growth description based upon Pn and respiration rates |
| Connor (1973) | Radiation | Gromax: model of Pn rate for a total grassland ecosystem |
| Connor *et al.* (1974) | Soil moisture stress, temperature, light, LAI | Model of Pn rates for productivity prediction on the shortgrass prairie |
| Curry (1971) | Light, $CO_2$ concentration, wind, temperature, soil moisture stress | Plant growth model tested on corn predicts Pn, respiration rates and transpiration |
| Duncan *et al.* (1971) | Sun position, light $CO_2$ concentration | Community photosynthesis model for stratified crop canopies |
| Field & Hunt (1974) | Temperature | Alfalfa production model based upon seasonal temperature change |
| Idso & Baker (1968) | Temperature, relative humidity, wind, light | Model of maize and sorghum growth as a function of net photosynthesis |
| Innis (1978) | Temperature, relative humidity, wind, light, precipitation | Community photosynthesis model (ELM) |
| Ledig (1969) | Light, temperature, soil moisture stress | Math model to predict dry matter growth of loblolly pine seedlings |
| Lemon *et al.* (1971) | Sun position, light temperature, wind, $CO_2$ concentration, relative humidity | SPAM: model of corn photosynthesis, energy balance approach |
| Miller & Tieszen (1972) | Light, wind, temperature | Primary arctic tundra productivity model based upon energy balance for a single leaf |
| Rosenzweig (1968) | Evapotranspiration | Primary productivity model of terrestrial communities based upon potential evapotranspiration |
| Stephens & Waggoner (1970) | Light | Model of tropical rain forest photosynthesis related to illumination |
| Waggoner (1969*b*) | Light, temperature, relative humidity, wind, crop extinction coefficient, canopy architecture, leaf angle, resistances | Photosynthesis model simulating growth in plant stands |

evapotranspiration calculated by the Thornthwaite–Mather model using long-term average data. Typical of many grassland stations, these curves show evapotranspiration to be coincident with the precipitation curve from the middle to the end of the year. Only in the spring does the precipitation exceed the evapotranspiration. Much of the soil water which has been stored early in the year is depleted through evapotranspiration in the interval of May to July. This is because of the high energy available for evaporation and because the actively growing plants provide a mechanism for removal of the water from the soil via transpiration.

Further examples of process rates in grasslands are summarized by Ripley & Redmann (1975) for the IBP Matador Site, a mixed prairie in Saskatchewan, Canada.

Photosynthesis is discussed in detail in Chapter 2, but it is important to note briefly here how abiotic factors influence photosynthetic rate. A cross-section of examples of use of abiotic factors in photosynthesis (or plant production) models is given in Table 1.3. The models shown here are for a variety of plant community types; those for domestic or wild grasses and grazinglands include Baker & Horrocks (1974), Connor, Brown & Trlica (1974), Idso & Baker (1968), Innis (1978), Lemon *et al.* (1971) and Miller & Tieszen (1972).

The abiotic variables used in calculating photosynthesis or plant production, in order of relative frequency of use in the models listed in Table 1.3, include: light or radiation or both, temperature, wind, relative humidity, soil water content, carbon dioxide concentration, sun position, water vapour, evapotranspiration, stomatal resistance, and soil solution nitrogen. Abiotic variables occur in such calculations with eight to nine times the frequency of biotic variables. In fact, in many models the photosynthesis or plant production rates are calculated only from abiotic factors.

## Concluding remarks

Compartment models may lend themselves to excessive pathway identification: extra arrows in a conceptual diagram draw no penalty. But many energy transfers that seem clearcut on paper are actually irrelevant or nonexistent in the field, such as convection transfer from terrestrial plants to open water, or conduction transfer between roots and animals. The best guarantee of critical structuring of a new model is a sound physical analysis and a sceptical audience.

A sound physical and biological analysis is important for low-resolution models too, but for a different reason. Compartment models put state variables (i.e. quantities) in the compartments with fluxes between compartments. Noncapacitative compartments indicate trouble, since they allow transitions between state variables to occur without control on the rates.

Therefore, 'incapacitated' compartments in a model hint at defective descriptions. These subtle fugitives from analysis deserve obliteration.

The authors thank Dave Nittmann for his assistance in the review of literature used in preparing this chapter and Freeman M. Smith for his review of the manuscript. Van Dyne's work on the chapter was supported in part by National Science Foundation Grant DEB73-11139.

## References

Aase, J. K., Wight, J. R. & Siddoway, E. H. (1973). Estimating soil water content on native rangeland. *Agricultural Meteorology*, **12**, 185–91.

Almeyda, G. F. & Rasmussen, J. L. (1972). *Surface energy and water budget over the grassland, 1970. US/IBP Grassland Biome Technical Report Number 202.* Colorado State University, Fort Collins.

Anderson, M. C. (1964). Light relations of terrestrial plant communities and their measurement. *Biological Review*, **39**, 425–86.

Anderson, M. C. (1966). Stand structure and light penetration. II. A theoretical analysis. *Journal of Applied Ecology*, **3**, 41.

Anderson, M. C. & Denmead, P. T. (1969). Shortwave radiation on inclined surfaces in model plant communities. *Agronomy Journal*, **61**, 867–72.

Armstrong, J. S. (1971). Modelling in a grazing system. *Proceedings of the Ecological Society of Australia*, **6**, 194–202.

Arnold, G. W. & Campbell, N. A. (1972). A model of a lay farming system, with particular reference to a sub-model for animal production. *Proceedings of the Australian Society of Animal Production*, **9**, 23–30.

Arnold, G. W. & Galbraith, K. A. (1974). Predicting the values of lupines for sheep and cattle in cropping and pastoral grazing systems. *Proceedings of the Australian Society of Animal Production*, **10**, 383–6.

Arnold, G. W., Carbon, B. A., Galbraith, K. A. & Biddiscombe, E. G. (1974). Use of a simulation model to assess the effects of a grazing management on pasture and animal production. *Proceedings of the XII International Grassland Congress*, pp. 47–52.

Asseed, M. & Kirkham, D. (1968). Advance of irrigation water on the soil surface in relation to soil infiltration rate: a mathematical and lab model study. *Iowa Agricultural and Home Economics Experiment Station Research Bulletin*, **565**, 293–316.

Baier, W. & Robertson, G. W. (1966). A new versatile soil moisture budget. *Canadian Journal of Plant Science* **46**, 299–315.

Baker, C. H. & Horrocks, R. D. (1974). An overview of CORNMOD: a dynamic simulation of corn production. *Simulation*, **23**, 105–8.

Baumgartner, A. (1965). The heat, water, and carbon dioxide budget of plant cover, methods and measurements. In *Methodology of plant ecophysiology*, Proceedings of the Montpellier Symposium, pp. 495–512. UNESCO, Paris.

Beeman, J. F. (1966). Growth dynamics of small tobacco plants. PhD Thesis, North Carolina State University, Raleigh, NC.

Biswas, T. D., Nielsen, D. R. & Biggar, J. W. (1966). Redistribution of soil water after infiltration. *Water Resources Research*, **2**, 513–24.

Black, T. A., Gardner, W. R. & Thurtell, G. W. (1969). The prediction of evaporation, drainage, and soil water storage for a bare soil. *Proceedings of the Soil Science Society of America*, **33**, 655–60.

Blaney, H. F. & Morin, K. V. (1942). Evaporation and consumptive use of water empirical formulas. *Transactions of the American Geophysical Union*, **23**, 76–83.

Bogardi, M. E. (1966). Estimation of evaporation and evapotranspiration from climatic data. MS Thesis, Colorado State University, Fort Collins.

Botkin, D. B. (1969). Prediction of net photosynthesis of trees from light intensity and temperature. *Ecology*, **50**, 854–8.

Branton, C. I., Allen, L. D. & Newman, J. E. (1972). Some agricultural implications of winter sublimation gains and losses at Palmer, Alaska. *Agricultural Meteorology*, **10**, 301–10.

Brockington, N. R. (1971). Using models in agricultural research. *Span*, **14**, 1–4.

Brown, L. F. (1974). Photosynthesis of two important grasses of shortgrass prairie as affected by several ecological variables. PhD Thesis, Colorado State University, Fort Collins.

Brown, K. W. (1969). A model of the photosynthesizing leaf. *Physiologia Plantarum*, **22**, 620–37.

Brown, K. W. & Covey, W. (1966). The energy budget evaluation of the micro-meteorological transfer processes within a corn field. *Agricultural Meteorology*, **3**, 73–96.

Brown, K. W. & Rosenberg, N. J. (1972). A resistance model to predict evapotranspiration and its application to a sugar beet field. In *Research in ET 1969–1972 OWRR*. USDI Nebraska Water Resources Research Institute Section 5. (Also *Agronomy Journal*, **65**, 341–7.)

Brutsaert, W. (1965). A model for evaporation as a molecular diffusion process into a turbulent atmosphere. *Journal of Geophysical Research*, **70**, 5017–5024.

Brutsaert, W. (1965). Evaluation of some practical methods of estimating ET in arid climates at low latitudes. *Water Resources Research*, **1**, 187–191.

Byrne, G. F. & Tognetti, K. (1969). Simulation of a pasture–environment interaction. *Agricultural Meteorology*, **6**, 151–63.

Caskey, J. E. (1963). A marker chain model for the predictability of ppt occurrence in intervals of various lengths. *Monthly Weather Review*, **91**, 298–301.

Chan, L. H., Huang, B. K. & Splinter, W. E. (1969). Developing a physical–chemical model for a plant growth system. *Transactions of the American Society of Agricultural Engineers*, **12**, 698–702.

Chebotarev, N. P. (1966). *Theory of stream runoff*. US Department of Communications Clearinghouse for Federal Scientific and Technical Information, Springfield, Virginia. Israel Program for Scientific Translations, Jerusalem.

Childs, E. C. (1967). Soil moisture theory. *Advanced Hydroscience*, **4**, 73–117.

Chudleigh, P. D. & Filan, S. J. (1972). A simulation model of an arid zone sheep property. *Australian Journal of Agricultural Economics*, **16**, 183–94.

Clark, O. R. (1940). Interception of rainfall by prairie grasses, weeds, and certain crop plants. *Ecological Monographs*, **10**, 243–77.

Cloud, C. C., Frick, G. E. & Andrews, R. A. (1968). *An economic analysis of hay harvesting and utilization using a simulation model. USDA Agricultural Experiment Station UNH Bulletin 495.*

Colwick, R. F. & Bowen, H. D. (1974). Modelling cotton production systems from seedbed to market. In *Beltwide Cotton Production Research Conference Proceedings*.

Connor, D. J. (1973). *Gromax: a potential productivity routine for a total grassland ecosystem model. US/IBP Grassland Biome Technical Report No. 208*. Colorado State University, Fort Collins.

Connor, D. J., Brown, L. F. & Trlica, M. J. (1974). Plant cover, light interception and photosynthesis of shortgrass prairie: a functional model. *Photosynthetica*, **8**, 18–27.

Corbett, E. S. & Crouse, R. P. (1968). *Rainfall interception by annual grass and chaparral. USDA Forest Service Research Paper PSW-58.*

Couturier, D. E. & Ripley, E. A. (1973). Rainfall interception in a mixed grass prairie. *Canadian Journal of Plant Science*, **53**, 659–63.

Cowan, I. (1968). The interception and absorption of radiation in plant stands. *Journal of Applied Ecology*, **5**, 367–79.

Crawford, N. H. & Linsley, R. K. (1966). *Digital simulation in hydrology: Stanford watershed model IV. Technical Report No. 39.* Department of Civil Engineering, Stanford University, Palo Alto, California.

Curry, R. B. (1971). Dynamic simulation of plant growth. Development of a model. *Transactions of the American Society of Agricultural Engineers*, **14**, 946–9, 959.

Curry, R. B. & Chen, L. H. (1971). Dynamic simulation of plant growth. Incorporation of actual daily weather data and partitioning of net photosynthate. *Transactions of the American Society of Agricultural Engineers*, **14**, 1170–4.

Denmead, P. T. & Shaw, R. H. (1962). Availability of soil water to plants as affected by soil moisture content and meteorological conditions. *Agronomy Journal*, **54**, 385–90.

de Vries, D. A. (1958). Simultaneous transfer of heat and moisture in porous media. *Transactions of the American Geophysical Union*, **39**, 909–16.

de Vries, D. A. (1975). Heat transfer in soils. In *Heat and mass transfer in the biosphere*, part I, *Transfer processes in the plant environment* ed. D. A. de Vries & N. H. Afgan, pp. 5–28. Wiley, New York.

de Vries, D. A. & Afgan, N. H. (eds.) (1975). *Heat and mass transfer in the biosphere*, part I, *Transfer processes in the plant environment*. Wiley, New York.

de Wit, C. T. (1965). *$P_n$ of leaf canopies. Agricultural Research Report 663.* Centre for Agricultural Public Documents, Wageningen, the Netherlands.

Dingman, S. L. (1972). Equilibrium temperature of water surfaces as related to air temperature and solar radiation. *Water Resources Research*, **8**, 42–9.

Duncan, W. G. & Barfield, B. J. (1971). Description of photosynthesis within plant canopies. *Transactions of the American Society of Agricultural Engineers*, **14**, 960–3.

Eagleman, J. R. (1971). An experimentally derived model for actual evapotranspiration. *Agricultural Meterology*, **8**, 385–94.

Eagleman, J. R. & Decker, W. L. (1965). The role of soil moisture in evapotranspiration. *Agronomy Journal*, **57**, 626–9.

Fenn, R. W. (1976). Optical properties of aerosols. In *Handbook on aerosols*, ed. R. Dennis, pp. 66–92. Technical Information Center, Office of Public Affairs, US Energy Research and Development Administration, Washington, DC.

Ferguson, J. (1952). The rate of natural E from shallow ponds. *Australian Journal of Scientific Research*, **5**, 315–30.

Fitzpatrick, E. R. & Nix, H. A. (1969). A model for simulating soil water regime in alternating fallow crop systems. *Agricultural Meteorology*, **5**, 303–19.

Flemming, G. (1975). *Computer simulation techniques in hydrology*. American Elsevier Publishing Company, New York.

Freer, M., Davidson, J. L., Armstrong, J. S. & Donnelly, J. R. (1970). Simulation of summer grazing. In *Proceedings of the XI International Grassland Congress*, p. 913–17. University of Queensland Press, St Lucia.

Freeze, R. A. & Harlan, R. L. (1969). Blueprint for a physically based, digitally simulated hydrologic response model. *Journal of Hydrology*, **9**, 237–58.

Freeze, R. A. & Witherspoon, P. A. (1966). Theoretical analysis of regional ground-water flows. I. Analytical and numerical solutions to the mathematical model. *Water Resources Research*, **2**, 641–56.

Fritschen, L. J. (1965). Accuracy of ET determinations by the Bowen ratio method. *International Association of Scientific Hydrology Bulletin*, **10**, 38–48.

Gabriel, K. R. & Newman, J. (1962). A Markov chain model for daily rainfall occurrence at Tel Aviv. *Quarterly Journal of the Royal Meteorological Society*, **88**, 90–5.

Galston, A. W. (1974). Plant photobiology in the last half-century. *Plant Physiology*, **54**, 427–36.

Gardner, W. R. (1960). Dynamic aspects of water availability to plants. *Soil Science*, **89**, 63–73.

Gardner, W. R. (1965). Rainfall, runoff and return. *Meteorological Monographs*, **6**, 138–48.

Gardner, W. R. & Ehlig, C. F. (1962). Impedance to water movement in soil and plant. *Science*, **138**, 522–3.

Garnier, B. J. & Ohmura, A. (1968). A method of calculating the direct shortwave radiation income of slopes. *Journal of Applied Meteorology*, **7**, 706–800.

Gates, D. M. (1962). Water balance in terrestrial ecosystems. In *Transport phenomena in atmospheric and ecological systems*. American Society of Mechanical Engineers, New York.

Gates, D. M. (1965). Energy, plants, and ecology. *Ecology*, **46**, 1–13.

Gates, D. M. (1965). Heat, radiant and sensible, chapter 1, Radiant energy, its receipt and disposal. *Meteorological Monographs*, **6**, 1–26.

Gates, D. M. & Schmerl, R. B. (eds.) (1975). *Perspectives of biophysical ecology*. Springer-Verlag, New York.

Gindel, I. (1967). The relationship between transpiration and six ecological factors. *Oecologia plantarum*, **2**, 227–35.

Gindel, I. (1970). *The ecodynamics of forest–water relationships in arid climates. Range Science Series No. 1*. Range Science Department, Colorado State University, Fort Collins.

Goltz, S. M., Tanner, C. B. & Rhurtell, G. W. (1970). Evaporation measurements by an eddy correlation method. *Water Resources Research*, **6**, 440–6.

Goodall, D. W. (1970). Use of computers in the grazing management of semiarid lands. In *Proceedings of the XI International Grassland Congress*, pp. 917–18. University of Queensland Press, St Lucia.

Goudriaan, J. & Waggoner, P. E. (1972). Simulating both aerial microclimate and soil temperature from observations above the foliar canopy. *Netherlands Journal of Agricultural Science*, **20**, 104–24.

Grace, R. A. & Eagleson, P. S. (1966). The modelling of overland flow. *Water Resources Research*, **2**, 399–403.

Hellmuth, E. O. (1971). Ecophysiological studies on plants in arid and semiarid regions in western Australia. III. Comparative studies on photosynthesis, respiration and water relations of ten arid zone and two semiarid zone plants under winter and late summer climatic conditions. *Journal of Ecology*, **59**, 225–58.

Hinds, W. T. (1975). Energy and carbon balances in cheatgrass: an essay in autecology. *Ecological Monographs*, **45**, 367–88.

Holtan, H. N. (1965). A model for computing watershed retention from soil parameters. *Journal of Soil Water Conservation*, **20**, 91–4.

Hopkins, J. W. & Robillard, P. (1964). Some statistics of daily rainfall occurrence for the Canadian prairie provinces. *Journal of Applied Meteorology*, **3**, 600–2.

Horton, R. E. (1919). Rainfall interception. *Monthly Weather Review*, **47**, 603–23.

Horton, R. E. (1940). Approach toward a physical interpretation of infiltration-capacity. *Proceedings of the Soil Science Society of America*, **4**, 399–417.

Idso, S. B. & Baker, D. G. (1968). The naturally varying energy environment and its effects upon net photosynthesis. *Ecology*, **49**, 311–16.

Idso, S. B., Baker, D. G. & Blad, B. L. (1969). Relations of radiation fluxes over natural surfaces. *Quarterly Journal of the Royal Meteorological Society*, **95**, 244–57.

Idso, S. B., Jackson, R. D., Ehrler, W. L. & Mitchell, S. T. (1969). A method for determination of infrared emittance of leaves. *Ecology*, **50**, 899–902.

Innis, G. S. (ed.) (1978). *Grassland simulation model. Springer-Verlag Ecological Studies* 26. Springer-Verlag, New York.

Jensen, M. E., Wright, J. L. & Pratt, B. J. (1971). Estimating soil moisture depletion from climatic, crop, and soil data. *Transactions of the American Society of Agricultural Engineers*, **14**, 954–63.

Johns, G. G. (1974). A soil water : water use relationship for incorporation in models simulating dryland herbage production. In *Proceedings of the XII International Grassland Congress*, pp. 61–8. Moscow.

Jones, J. W., Colwick, R. F. & Threadgill, E. D. (1972). A simulated environmental model of temperature, evaporation, rainfall, and soil moisture. *Transactions of the American Society of Agricultural Engineers*, **15**, 366–92.

Jones, J. W. & Verma, B. P. (1971). A digital simulation of the dynamic soil moisture stress. *Transactions of the American Society of Agricultural Engineers*, **14**, 664–6.

Kasanaga, H. & Monsi, M. (1954). On the light-transmission of leaves and its meaning for the production of matter in plant communities. *Japanese Journal of Botany*, **14**, 304–24.

Knoerr, K. R. & Gay, L. W. (1965). Tree leaf energy budget. *Ecology*, **46**, 14–24.

Kohler, M. A. & Richards, M. M. (1962). Multi-capacity basic accounting for predicting runoff from storm ppt. *Journal of Geophysical Research*, **67**, 5187–97.

Lang, A. R. G. (1973). Measurement of evapotranspiration in the presence of advection by means of a modified energy balance procedure. *Agricultural Meteorology*, **12**, 75–81.

Ledig, F. T. (1969). A growth model for three seedlings based on the rate of photosynthesis and the distribution of photosynthate. *Photosynthetica*, **3**, 263–75.

Legg, B. & Monteith, J. L. (1975). Heat and mass transfer within plant canopies. In *Heat and mass transfer in the biosphere*, Part I, *Transfer processes in the plant environment*, ed. D. A. de Vries & N. H. Afgan, pp. 167–86. Wiley, New York.

Lettau, H. (1969). Evapotranspiration climatonomy. I. A new approach to numerical prediction of monthly ET, runoff, and soil moisture storage. *Monthly Weather Review*, **97**, 691–9.

Lemon, E., Stewart, D. W. & Shawcroft, R. W. (1971). The sun's work in a cornfield. *Science*, **174**, 371–8.

Liakopoulos, A. C. (1966). Theoretical prediction of evaporation losses from groundwater. *Water Resources Research*, **2**, 227–40.

Linacre, E. T. (1968). Estimating the net-radiation flux. *Agricultural Meteorology*, **5**, 49–63.

Linsley, R. K., Jr, Kohler, M. A. & Paulhus, J. L. H. (1958). *Hydrology for engineers*. McGraw-Hill, New York.
List, R. J. (1951). *Smithsonian meteorological tables*, 6th revised edn. Smithsonian Institution, Washington, DC.
Loomis, W. E. (1965). Absorption of radiant energy by leaves. *Ecology*, **46**, 14–17.
Lowry, W. P. & Guthrie, D. (1968). Markov chains of order greater than one. *Monthly Weather Review*, **96**, 798–801.
McCree, K. J. (1972). Test of current definitions of photosynthetically active radiation against leaf photosynthesis data. *Agricultural Meteorology*, **10**, 443–53.
Mein, R. G. & Larson, C. L. (1973). Modelling infiltration during a steady rain. *Water Resources Research*, **9**, 384–94.
Miller, P. C. & Tieszen, L. (1972). A preliminary model of processes affecting primary production in the arctic tundra. *Arctic Alpine Research*, **4**, 1–18.
Monteith, J. L. (1957). Dew. *Quarterly Journal of the Royal Meteorological Society*, **83**, 322–41.
Monteith, J. L. (1965). Evaporation and environment. *Symposia of the Society for Experimental Biology*, **19**, 205–34.
Monteith, J. L. (1972). *Principles of environmental physics*. Edward Arnold, London.
Monteith, J. L. (ed.) (1975). *Vegetation and the atmosphere*, Vol. 1. Academic Press, New York & London.
Murphy, C. E., Jr, Mankin, J. B. & Knoerr, K. R. (1972). A dynamic energy balance and microclimate model of a forest ecosystem. In *Proceedings of the 1972 Summer Computer Simulation Conference*, pp. 1169–75. Simulation Councils Inc., La Jolla, California.
Nkemdirim, L. C. & Haley, P. F. (1973). *An evaluation of grassland evapotranspiration*. Department of Geography, University of Calgary, Alberta, Canada.
Nkemdirim, L. C. & Yamashita, S. (1972). Energy balance over prairie grass. *Canadian Journal of Plant Science*, **52**, 215–25.
Parkhurst, D. F., Duncan, P. R., Gates, D. M. & Kreith, F. (1968). Wind tunnel modelling of convection of heat between air and broad leaves of plants. *Agricultural Meteorology*, **5**, 33–47.
Parton, W. J. (1978). *Abiotic section of ELM*. In *Grassland simulation model*, ed. G. Innis, pp. 31–53. Springer-Verlag, New York.
Philip, J. R. (1957). Evaporation and moisture and heat fields in the soil. *Journal of Meteorology*, **14**, 354–66.
Philip, J. R. (1964). Sources and transfer processes in the air layers occupied by vegetation. *Journal of Applied Meteorology*, **3**, 390–5.
Philip, J. R. (1975). Water movement in soil. In *Heat and mass transfer in the biosphere*, Part I, *Transfer processes in the plant environment* ed. D. A. de Vries & N. H. Afgan, pp. 29–48. Wiley, New York.
Portman, D. J., Elder, F. C. & Rygnar, E. (1961). *Research on energy exchange processes*, pp. 96–109. *University of Michigan Great Lakes Research Publ. 7*.
Raschke, K. (1956). Über die physikalischen Beziehungen zwischen Wärmeübergangszahl, Strahlungsautausch, Temperatur und Transpiration eines Blattes. *Planta*, **48**, 200–38.
Rasmussen, J. L. (1971). Abiotic factors in grassland ecosystem analysis and function. In *Preliminary analysis of structure and function in grasslands*, ed. N. R. French, pp. 11–34. *Range Science Department Science Series No. 10*. Colorado State University, Fort Collins.
Rasmussen, E. M. (1968). Atmospheric water vapor transport and water balance of North America. II. Large-scale water balance investigations. *Monthly Weather Review*, **96**, 720–34.

Ripley, E. A. & Redmann, R. E. (1975). Grassland. In *Vegetation and the atmosphere*, ed. J. L. Monteith, vol. 2, pp. 349–98. Academic Press, New York & London.
Ritchie, J. T. (1972). A model for predicting evaporation from a row crop with incomplete cover. *Water Resources Research*, **8**, 1204–13.
Rose, C. W. (1966). *Agricultural physics*. Pergamon Press, Oxford.
Rose, C. W., Begg, J. E., Byrne, G. F., Torssell, B. W. R. & Goncz, H. (1972). A simulation model of growth-field development relationships for Townsville stylo pasture. *Agricultural Meteorology*, **10**, 161–83.
Rosenzweig, M. L. (1968). Net primary productivity of terrestrial communities: prediction from climatological data. *American Naturalist*, **102**, 67–74.
Seginer, J. (1969). The effect of albedo on the evapotranspiration rate. *Agricultural Meteorology*, **6**, 5–31.
Sellers, W. D. (1965). *Physical climatology*. University of Chicago Press.
Slavík, B. (ed.) (1974). *Methods of studying plant water relations*. Springer-Verlag, Berlin.
Slayter, R. O. (1966). Some physical aspects of internal control of leaf transpiration. *Agricultural Meteorology*, **3**, 281–92.
Smith, F. M. (1971). Volumetric threshold infiltration model. PhD Thesis, Colorado State University, Fort Collins.
Smith, F. M. (1972). Comparative abiotic investigation in the US/IBP Grassland Biome. Paper presented at the Annual Meeting of the AIBS, 27 August–1 September, Minneapolis, Minnesota.
Smith, F. M. (1973). Abiotic functions in grassland ecosystems. In *Analysis of structure, function, and utilization of grassland ecosystems*, principal investigator G. M. Van Dyne, vol. 2. A Progress Report. Colorado State University, Fort Collins.
Smith, J. A. (1973). Abiotic subsystems: heat and water models. In *Process studies workshop report*, ed. D. A. Jameson & M. I. Dyer, pp. 12–121. *US/IBP Grassland Biome Technical Report No.* 220. Colorado State University, Fort Collins.
Stephens, G. R. & Waggoner, P. E. (1970). Carbon dioxide exchange of a tropical rain forest. *Bioscience*, **20**, 1050–3.
Swartzendruber, D. (1966). Soil water behavior as described by transport coefficients and functions. *Advances in Agronomy*, **18**, 327–70.
Swinbank, W. C. (1968). A comparison between predictions of dimensional analysis for the constant flux layer and observations in unstable conditions. *Quarterly Journal of the Royal Meteorological Society*, **94**, 460–7.
Tanner, C. B. (1960). Energy balance approach to evapotranspiration from crops. *Proceedings of the Soil Science Society of America*, **24**, 1–9.
Toth, J. (1963). A theoretical analysis of ground water flow in small drainage basins. *Journal of Geophysical Research*, **68**, 4795–812.
Uresk, D. W. (1971). Dynamics of blue grama within a shortgrass ecosystem. PhD Thesis, Colorado State University, Fort Collins.
Van Bavel, C. H. M. (1966). Potential evaporation: the combination concept and its experimental verification. *Water Resources Research*, **2**, 455.
Van Bavel, C. H. M., Fritschen, L. J. & Reeves, W. E. (1962). Transpiration by sudan grass as an externally controlled process. *Science*, **141**, 269–70.
Van Wijk, W. R. & Derksen, W. J. (1966). Thermal properties of a soil near the surface. *Agricultural Meteorology*, **3**, 333–42.
Verhagen, A. M. W., Wilson, J. H. & Britten, E. J. (1963). Plant production in relation to foliage illumination. *Annals of Botany, New Series*, **27**, 627–40.
Viskanta, R. & Toor, J. S. (1972). Radiant energy transfer in waters. *Water Resources Research*, **8**, 595–608.

Waggoner, P. E., Furnival, G. V. & Reifsnyder, W. E. (1969). Simulation of the microclimate in a forest. *Forest Science*, **15**, 37–45.

Waggoner, P. E. & Reifsnyder, W. E. (1968). Simulation of the temperature, RH and evaporation profiles in a leaf canopy. *Journal of Applied Meteorology*, **7**, 400–9.

Waggoner, P. E. & Zelitch, I. (1965). Transpiration and the stomata of leaves. *Science*, **150**, 1413–20.

Warren Wilson, J. (1967). Stand structure and light penetration. III. Sunlit foliage area. *Journal of Applied Ecology*, **4**, 159–65.

Wassink, E. C. & Stolwijk, J. A. J. (1952). Effects of light on narrow spectral regions on growth and development of plants. I, II. *Proceedings, Kongress Nederlands Akademia Wetensch. Ser. C55*, pp. 471–88.

Weiss, L. L. (1964). Sequences of wet or dry days described by a Markov chain probability model. *Monthly Weather Review*, **92**, 169–76.

Williams, C. B. (1952). Sequences of wet and of dry days considered in relation to the logarithmic series. *Quarterly Journal of the Royal Meteorological Society*, **78**, 91–6.

Winkworth, R. E. (1970). The soil water regime of an arid grassland (*Eragrostis eriopoda* Benth.) community in central Australia. *Agricultural Meteorology*, **7**, 387–99.

Wiser, E. H. (1966). Monte Carlo methods applied to precipitation frequency analysis. *Transactions of the American Society of Agricultural Engineers*, **9**, 538–40, 542.

Witkamp, M. (1966). Decomposition of leaf litter in relation to environment. Microflora and microbial respiration. *Ecology*, **47**, 194–201.

Woo, K. B., Boersma, L. & Stone, L. N. (1966). Dynamic simulation model of the transpiration process. *Water Resources Research*, **2**, 85–97.

Woo, K. B., Stone, L. N. & Boersma, L. (1966). A conceptual model of the stomatal control mechanism. *Water Resources Research*, **2**, 71–84.

Wright, J. L. & Brown, K. W. (1967). *The energy at the earth's surface: comparison of momentum and energy balance methods of computing vertical transfer within a crop. Technical Report ECOM2-671-1.* Fort Huachuca, Arizona.

Yen, B. C. & Hsie, C. H. (1972). Viscous flow model for ground water movement. *Water Resources Research*, **8**, 1299–306.

Zahner, R. & Stage, A. R. (1966). A procedure for calculating daily moisture stress and its utility in regression of tree growth on weather. *Ecology*, **47**, 64–73.

# 2. Autotrophic subsystem

J. S. SINGH, M. J. TRLICA, P. G. RISSER,
R. E. REDMANN & J. K. MARSHALL

## Contents

## Carbon uptake, distribution and utilization

Autotrophic higher plants have in common the physiological processes of photosynthesis, translocation and respiration. The photosynthetic process allows the plant to combine radiant energy with carbon dioxide and, through a series of complex biochemical reactions, to produce an organic molecule which has a relatively higher energy content. This molecule, frequently a sugar, can then be utilized in the plant organ where photosynthesis occurs or it can be translocated to other parts of the plant where it is utilized. Utilization implies that the molecule is converted to other compounds, usually compounds with a lower energy value. The overall process of respiration involves the breakdown of these relatively high-energy compounds with the release of energy, some of which is used by the plant and some of which is lost as heat. This chapter discusses the basic processes of photosynthesis, translocation, respiration and mortality from several points of view. The processes are discussed from a physiological basis and then they are related to the entire producer function in the ecosystem concept. Although the biochemical and physiological attributes are reasonably well known, the application of these processes at the ecosystem level is less definite and still requires extrapolation from a relatively small number of laboratory studies.

### *The process of photosynthesis and its interrelationships with other ecosystem functions*

The process of photosynthesis is a unique process in the ecosystem and is performed by those organisms called the primary producers. Essentially all fixed carbon in the ecosystem originates from the process of photosynthesis or more specifically biochemical steps within this process. As will be seen in the following sections, the actual process may occur through one or more biochemical pathways, depending upon the particular species of plant. Further, different plant species possess photosynthetic steps which respond differently to environmental factors or abiotic influences. All the plant species in an ecosystem, even though they may respond differently to external stimuli, fix carbon which then becomes the energy source for not only the producers, but all the other trophic levels. Thus the photosynthetic process is the basis for all the energy transfer which is a part of the function of any ecosystem.

### *Biochemical description*

In the process of photosynthesis green plants utilize incident solar energy and carbon dioxide to produce oxygen and complex organic materials. Photosynthesis involves a combination of biophysical and biochemical processes driven by radiation of wavelengths approximately 400 to 700 nm which are absorbed by the chloroplasts. Electromagnetic energy is converted to chemical free energy which is then used for biosynthesis of organic molecules (Kamen, 1963).

Although the term photosynthesis implies the necessity of light, in fact several steps of the process can occur in the dark. The series of steps which constitutes the photosynthetic process is usually separated into the light and dark reactions, although the reactions are interdependent. The light reactions involve two photosystems which are composed of photosynthetic pigments and electron carriers. Chlorophyll *a*, which occurs in several forms each with a different photoactivity and absorption maximum in the red region of the spectrum, is the most common pigment and occurs in both photosystems I and II. The chlorophyll *a* form which absorbs longer wavelengths predominates in photosystem I, while the short wavelength absorbing forms together with most of the chlorophyll *b*, *c* or *d* and other accessory pigments (carotenoids and phycobilins) are components of photosystems II (Sěsták, Čatský & Jarvis, 1971). The energy trap for photosystem I is a porphyrin pigment known as P700 and that for photosystem II is the still-hypothetical P690 (or P680 or chl-$a_2$). The trapping molecule is excited by the absorbed energy which results in the transfer of an electron against a gradient of redox potential. This change of energy status results in the production of highly active oxidants and reductants. When photosystem II is sensitized by light, photoreaction 2 forms a primary oxidant $X^+$ and a primary reductant $Q^-$. The primary oxidant causes splitting of water and the evolution of oxygen with the consequent restoration of P690 in the unexcited form. The electron from the primary reductant is transferred through a chain of electron carriers that joins the two photosystems and consists of plastoquinone, cytochromes and plastocyanin (Sěsták *et al.*, 1971). During these steps part of the energy from the electron is utilized in the formation of ATP.

A similar process occurs in photosystem I where P700 accepts light energy and donates an electron to an acceptor Z, a ferrodoxin-reducing substance. The oxidized P700 returns to its ground state by accepting the low-energy electron from photosystem II. The reduced acceptor ($Z^-$) transfers the electron through ferrodoxin and ferrodoxin-NADP-reductase to $NADP^+$ where $NADPH_2$ is formed (Sěsták *et al.*, 1971). Thus utilizable energy in the form of ATP and $NADPH_2$ results from the light reactions of photosynthesis and is used in the light-independent reactions.

The actual fixation of carbon dioxide may be by one of the two major pathways, the $C_4$ (Hatch and Slack) or $C_3$ (Calvin–Benson). In the $C_4$ pathway, which is found predominantly in species of tropical origin, carbon dioxide is bound to phosphoenolpyruvate (PEP), while in the $C_3$ pathway it is attached to ribulose-1,5-diphosphate (RuDP). The two pathways differ in terms of specific compounds but after phosphorylation 1,3-diphosphoglyceric acid is produced and is reduced in both pathways by $NADPH_2$, which results in the primary carbohydrate.

It should be remembered that the formation of two molecules of 3-carbophosphoglyceric acid is the basis of all photosynthetic pathways. The $C_4$ cycle, termed the $C_4$-dicarboxylic acid cycle (Hatch & Slack, 1966, 1968, 1970; Hatch, 1971), enhances the $C_3$ cycle by acting as a mechanism for concentrating carbon dioxide for further processing by the $C_3$ pathway. In the initial $C_4$ reaction, occurring in the mesophyll cells, phosphoenolpyruvic acid accepts carbon dioxide and forms an intermediate carbon compound, oxaloacetate, which is immediately converted to either malate or aspartate. Both malate and aspartate form carbon dioxide in subsequent reactions in the thick-walled bundle sheath cells of $C_4$ plants. This carbon dioxide is not lost because of the thick walls that act as physical barriers to its diffusion. Therefore, carbon dioxide is concentrated for the initial reactions of the $C_3$ cycle. This results in high rates of carbon dioxide fixation at high light intensities and temperatures and at low ambient carbon dioxide concentrations. In addition, reductions in the detrimental effects of high plant water stress are often observed in $C_4$ plants (Downton, 1971).

The $C_3$ cycle requires three ATP molecules and two NADPH molecules to reduce one molecule of carbon dioxide. In contrast, the $C_4$ cycle requires five ATPs and three NADPHs for reduction of each molecule of carbon dioxide. Superficially, the additional energy requirement for the $C_4$ cycle suggests that the $C_3$ cycle is more efficient than the $C_4$ cycle. This is not the case, however, because the $C_4$ plants are capable of utilizing higher light intensities than the $C_3$ plants and in addition show little or no photorespiration.

Grass leaves have been divided into two major anatomical groups, panicoid and festucoid (Brown, 1958). The chlorenchyma cells of panicoides are radially arranged around the vascular bundle. This arrangement is directly associated with the $C_4$ pathway of carbon dioxide fixation. In addition, the bundle sheath cells of panicoides have numerous chloroplasts. The chlorenchyma cells of the festucoides are irregularly arranged between adjacent vascular bundles and this irregular arrangement is associated with the $C_3$ pathway.

There has been a persistent indication in the literature that the bundle sheath cell chloroplasts of $C_4$ plants do not have grana. There is extreme structural dimorphism in the chloroplasts of $C_4$ dicotyledons as well as monocotyledons and there is no causal relationship between chloroplast structural dimorphism and $C_4$ photosynthesis. Further, there is no evidence of any relationship between 'agranal' chloroplasts with or without photosystem II and $C_4$ photosynthesis. The degree of development of dimorphic chloroplasts in the bundle sheath is probably involved with their function in synthesizing and storing starch (Laetsch, 1974).

There are several lines of evidence which indicate that the $C_4$ plants evolved from the $C_3$ plants. It has been speculated by Downton (1971) that the cell walls of the mesophyll could have thickened causing a concomitant reduction in mesophyll air space and a consequent radial arrangement of the mesophyll cells. Another indication suggested by Downton, Berry & Tregunna (1969) comes from the fact that certain members of the *Dichanthium* subgenus of *Panicum* behave as $C_3$ plants even though they belong to the predominantly $C_4$ group. Evans (1971), in a thorough assessment of the taxonomic distribution of plants, also concluded that the $C_3$ cycle was the more primitive photosynthetic pathway. Certainly one of the more important considerations involved in determining the evolution of $C_4$ plants must be that all plants today rely on the $C_3$ mechanism for the ultimate steps in carbon dioxide fixation. The greater net carbon dioxide fixation rates of $C_3$ plants are probably important in terms of survival and adaptation, but, more importantly, the performance of these plants under extreme conditions of stress is definitely of selective advantage in many parts of the world (Bjorkman, 1971). The $C_4$ species survive and may be better adapted than many $C_3$ species under conditions of high water stress, high temperature, high oxygen concentrations, low carbon dioxide concentrations and high irradiances (Bjorkman, 1971). These conditions were probably not typical during the evolution of higher plant life on the planet. The $C_3$ plants must have been the first higher plants to evolve in the low oxygen, high carbon dioxide atmosphere which characterized the earth at that time. The $C_4$ pathway probably evolved as an adaptive mechanism of plants originally native to moist, tropical climates and presumably these plants then immigrated to temperate regions and more temperate climates within the tropical regions. Because of the evolutionary trends, grasses possessing the $C_4$ cycle are commonly referred to as warm-season or tropical grasses, whereas those exhibiting the $C_3$ cycles are referred to as cool-season or temperate grasses (Downton, 1971). The $C_4$ pathway is known to exist in hundreds of monocotyledon and dicotyledon species comprising nearly 100 genera and at least 10 plant families (Bjorkman & Berry, 1973).

There may be some plants which possess features intermediate between the syndrome of features characteristic of either the $C_3$ or $C_4$ pathway. Thus, Kennedy & Laetsch (1974) reported that common carpetweed (*Mollugo verticillata*) has characteristics of both $C_3$ and $C_4$ plants. This species is intermediate between $C_3$ and $C_4$ plants in at least four features generally used to separate those two plant groups: leaf anatomy, cell structure, photorespiration and primary photosynthetic products.

Respiration rates in both $C_3$ and $C_4$ species are usually greater in light than in darkness, but photorespiration cannot be measured directly by conventional carbon dioxide exchange techniques under normal conditions. The biochemical source of photorespiratory carbon dioxide is glycolate which is probably synthesized from ribulose-1,5-diphosphate (RuDP). Light is necessary for the regeneration of RuDP in all plants; thus light controls the production of photorespiratory carbon dioxide. According to Samish & Koller (1968), the lack of measurable photorespiration for $C_4$ plants is also caused by low mesophyll resistance to carbon dioxide diffusion. This low mesophyll resistance of $C_4$ plants is associated with the greater amount of energy available to them for carbon dioxide fixation. Also the $C_4$ plants readily reassimilate any photorespiratory carbon dioxide. The lack of apparent photorespiration in $C_4$ plants might account for the greater net photosynthetic rates observed in these species.

Downton & Tregunna (1968) pointed out that plants with high photosynthetic rates, that is between 40 and 60 mg $CO_2 \cdot dm^{-2}$ leaf area $\cdot h^{-1}$, have carbon dioxide compensation points very near 0 ppm. Plants with low photosynthetic rates, about 20 to 30 mg $CO_2 \cdot dm^{-2} \cdot h^{-1}$, have carbon dioxide compensation points of 39 ppm or greater. The characteristics which distinguish the two photosynthetic pathways are shown in Table 2.1.

In spite of the foregoing discussion comparing photosynthetic pathways, the arbitrary division into either a $C_3$ type or $C_4$ type should be regarded with some caution. For example, Williams & Markley (1973) found that fringed sagewort (*Artemisia frigida*) was a difficult species to place in either class on the basis of either carbon dioxide compensation point or primary photosynthetic products. Other similar exceptions described in the literature are that certain $C_3$ species do not exhibit saturation of photosynthesis at low light intensity (common sunflower (*Helianthus annuus*), Hesketh & Moss, 1963; *Aegilops* species; and some diploid *Triticum*, Evans & Dunstone, 1970), and that old and shade-grown leaves of a $C_4$ species (*Amaranthus edulis*) exhibit light saturation of photosynthesis at levels similar to $C_3$ species (El-Sharkawy, Loomis & Williams, 1968). Common cattail (*Typha latifolia*), a $C_3$ species, was shown to have photosynthetic rates comparable to $C_4$ species (McNaughton & Fullem, 1970). Zelitch (1966) has shown that the carbon dioxide compensation point is greater at high temperatures in plants exhibiting the $C_3$ pathway.

In conclusion, the 'process' of photosynthesis is really a connected system of complex reactions which are directly related to one another and are affected by both abiotic and biotic factors. These reactions include the dehydrogenation of water and the evolution of oxygen, the transfer of electrons of hydrogen atoms from water to $NADP^+$, the production of ATP by cyclic and noncyclic photophosphorylation, and finally the reduction of carbon dioxide to carbohydrate by $NADPH_2$. All of these reactions proceed

Table 2.1. *Characteristics which distinguish the $C_3$ and $C_4$ plants*

| Characteristics | $C_3$ plants | $C_4$ plants |
|---|---|---|
| Leaf anatomy | No significant differentiation in mesophyll and bundle sheath cells | Bundle sheath cells containing large number of chloroplasts and other organelles |
| Net photosynthesis versus light intensity | Saturation at irradiances $< \sim 300$ W·m$^{-2}$ | No saturation |
| Maximum rate of net photosynthesis (mg $CO_2$·dm$^{-2}$ leaf area·h$^{-1}$) | 15–35 | 40–80 |
| Optimum temperature for net photosynthesis (°C) | 15–25 | 30–45 |
| $CO_2$ compensation concentration in photosynthesis (ppm $CO_2$) | 30–70 | 0–10 |
| Inhibitory effect of $O_2$ ($< 1\%$) on photosynthesis and growth | Yes | No |
| Photorespiration (glycolate oxidation) | High | Low |
| Transpiration rate (g $H_2O$·g$^{-1}$ dry wt) | 450–950 | 250–350 |
| $^{13}C/^{12}C$-discrimination ($\delta^{13}C/^{12}C$, ‰ standard) | $-22$ to $-33$ | $-10$ to $-20$ |
| Major pathway of $CO_2$ fixation in light | Reduced pentose phosphate cycle | $C_4$-dicarboxylic acid pathway plus reduced pentose phosphate cycle |
| Growth rate (g dry wt·dm$^{-2}$ leaf area·d$^{-1}$) | 0.5–2 | 3–5 |

After Kanai & Black (1972).

in a stoichiometric ratio, with the exception that ATP formation in cyclic photophosphorylation may result from the activity of photosystem I when the stomata are closed and there is no gas exchange (Sěsták *et al.*, 1971).

## Net photosynthesis and primary production

In essence, primary production is the fixation of carbon or energy and the consequent increase in plant biomass or energy. Biomass units and energy units are interconvertible, and biomass can be expressed as the amount of carbon. An estimate can be made of the amount of carbon or energy fixed by the plant. Therefore, primary production can be measured as carbon or energy taken up by the plant or as the amount of increase in plant biomass, carbon or energy. Gross production minus that energy or carbon lost simultaneously by respiration is known as net production. However, net production results calculated from data of harvested dry matter also include small amounts of nutrients absorbed from soil unless the values are corrected for ash content.

The actual measurement of the rate of photosynthesis can be accomplished by a number of approaches (Šesták *et al.*, 1971):

(*a*) change in energy;
(*b*) consumption of water;
(*c*) oxygen efflux;
(*d*) carbon dioxide influx;
(*e*) dry matter accumulation;
(*f*) accumulation of photosynthetic products;
(*g*) accumulation of energy;
(*h*) rate of formation of energy-rich intermediates;
(*i*) changes in the properties of the photochemical apparatus.

The only methods suitable for routine ecosystem analysis are those from (*c*) through (*g*) in the above list. Methods (*c*) and (*d*) are gasometric and constitute relatively easy methods for obtaining both gross and net photosynthesis. Models (*e*), (*f*) and (*g*) are commonly used to measure net photosynthesis and are frequently repeated at intervals over the season or experimental period. The gasometric methods permit the characterization of both gross and net photosynthesis, but since in large plants the whole organisms cannot be measured, there is difficulty in extrapolating from portions of the plant to the production rate of a whole canopy or even a single plant.

The rates of photosynthesis are not of particular interest except to establish the capacity of a plant to fix carbon. More often, it is important to establish how these rates respond to existing or changing environmental variables such as light, temperature, carbon dioxide levels, etc.

The measurement of gas exchange, either carbon dioxide uptake or oxygen evolution, has been used for years as a measure of the rate of photosynthesis. Not only is this a fairly simple laboratory procedure, but equipment now exists which can be taken to the field where measurements can be made under reasonably normal conditions. Primary production data have largely accumulated as the result of destructive sampling or growth measurements. If the rate of photosynthesis determines the rate of growth or biomass production, theoretically it should be possible to relate gas exchange measurements to primary production. In practice, this relationship is not always easy to establish, probably because (1) it is difficult to measure gas exchange rates for a whole plant over the same time period when destructive production measurements are made, and (2) production measurements inherently include the results of translocation, whereas gas exchange measurements are largely short-term indications of carbon fixation rates.

Field carbon dioxide exchange determinations were made on a shortgrass prairie sod using a transparent dome which enclosed approximately 0.3 $m^2$ of vegetation (Brown & Trlica, 1974). An open system was utilized so ambient air was pumped into the system from a source atmosphere 6 m above the soil surface. A sample of this air was routed through an infrared gas analyser for the determination of a baseline. The remainder of the air was routed to the dome assimilation chamber and a smaller sample of this air was passed through the sample chamber of the analyser. The difference in carbon dioxide concentration between ambient and dome air samples, along with the flow rate of the ambient air entering the dome, was used to calculate the carbon dioxide exchange rate of the sod under the dome. Dark respiration rates were measured by covering the dome and an attempt was made to separate soil respiration by using a correction factor based on laboratory studies.

Carbon dioxide exchange rates in the Canadian mixed-grass prairie were measured using a dome and a closed system (Redmann, 1973). Air was continuously recirculated through the chamber and changes in carbon dioxide concentration were measured with an infrared gas analyser. No attempt was made to separate physically the aboveground and belowground portions. Instead, soil-litter respiration was measured simultaneously on an area from which green shoots had been removed by clipping, and on an adjacent area of intact sward.

In some early experiments where the carbon dioxide absorption, the dry weight increments and the carbon content in the dry matter produced were measured, the calculations based on both carbon dioxide absorption and on carbon content gave the same results; that is, dry matter is made up of 41.4% carbon. This value corresponds with the carbon content of disaccharides. There were, however, other studies which indicated that the dry weight increment was greater or less than the amount theoretically likely from the carbon dioxide absorbed (McCree & Troughton, 1966; data of L. Natr & J. Gloser).

The Natr & Gloser (1967) values of the transformation coefficient used in comparisons of gas exchange estimates of photosynthesis with gravimetric ones have confirmed that 0.64 is a suitable value. On the other hand, the established dependence of the coefficient on the duration of the experiment indicates that the transformation of the dry weight increments into carbon dioxide absorption values will yield quantitatively differing results, according to circumstances. Studies on vegetation types other than grasslands have shown that gas exchange methods overestimated productivity as measured by weight increases, but comparisons were within 25% (Bate & Canvin, 1971; Botkin, Woodwell & Temple, 1970).

### *Variation in net production between and within species*

Like almost all other physiological processes in plants, different species demonstrate different photosynthetic rates. As has been noted previously, the $C_3$ plants with high rates of photorespiration usually exhibit leaf net

Table 2.2. *Range of net photosynthesis rates in single leaves*

| Species | Net photosynthesis ($mg\ CO_2 \cdot dm^{-2} \cdot h^{-1}$) | Reference |
|---|---|---|
| *Zea mays* | 46–63 | Hesketh (1963)<br>Hesketh & Moss (1963)<br>Waggoner, Moss & Hesketh (1963)<br>El-Sharkawy & Hesketh (1965) |
| *Saccharum officinarum* | 42–49 | Hesketh (1963)<br>Hesketh & Moss (1963) |
| *Syricum vulgare* | 55 | El-Sharkawy, Loomis & Williams (1967) |
| *Cynodon dactylon* | 35–43 | Murata & lyama (1963) |
| *Amaranthus edulis* | 58 | El-Sharkawy *et al.* (1967) |
| *Agropyron dasystachyum* | 24 | Redmann (1974*b*) |
| *Typha latifolia* | 44–69 | McNaughton & Fullem (1970) |
| *Dactylis glomerata* | 13–24 | Hesketh (1963)<br>Murata & Iyama (1963) |
| *Triticum aestivum* | 17–31 | Murata & Iyama (1963)<br>Hesketh (1967) |
| *Bouteloua gracilis* | 24–65 | Brown & Trlica (1977*a*) |
| *Festuca arundinacea* | 17–33 | Asay, Nelson & Horst (1974) |

photosynthesis in the range of about 15 to 35 mg $CO_2 \cdot dm^{-2} \cdot h^{-1}$ while $C_4$ plants exhibit higher rates. Those plants with high rates include crop plants such as maize (*Zea mays*) and sugarcane (*Saccharum officinarum*) (Table 2.2).

Various races or varieties within a species may demonstrate different photosynthetic rates (Asay, Nelson & Horst, 1974). Thus, Heichel & Musgrave (1969) recorded a range of 28 to 85 mg $CO_2 \cdot dm^{-2} \cdot h^{-1}$ net photosynthesis for 15 inbreds and hybrids of maize and Chandler (1969) reported a range of 34 to 62 mg $CO_2 \cdot dm^{-2} \cdot h^{-1}$ in 50 varieties of cultivated rice (*Oryza sativa*).

## Environmental control of net photosynthesis and primary production

Plant productivity is affected by (Zelitch, 1971):

(1) the soil including soil water, nutrient availability and carbon dioxide release during soil respiration;

(2) the properties of the leaves such as age, size, shape and angle, their stomatal numbers and behaviour, response of the mesophyll to irradiance, reflectance and transmission properties, effect of temperature on dark respiration and photorespiration, and their physical resistances and carboxylation characteristics;

(3) the architecture of the stand including the total leaf area that covers a unit area of soil, leaf distribution along the stem, and the angle of the leaf elevation from the horizontal;

(4) ambient climatic factors such as air temperature, wind speed, carbon dioxide concentration, relative humidity, angle of the sun, and whether irradiation is diffuse or direct; and

(5) the duration of photosynthesis, changes in photosynthesis with leaf size, efficiency of carbon dioxide assimilation, rapidity with which the leaf area enlarges to absorb the available irradiation, plant height (especially in stands of mixed species), photosynthesis by organs other than leaves, and the efficiency of transport of photosynthate to sink tissues.

This last category involves such diverse organs as roots, rhizomes, stolons, tubers, stems, grain, fibres and fruits. The productivity of species also differs when measured under optimal conditions of growth, and once again the photosynthetically efficient plants with a low rate of photorespiration have the highest net productivity rates, about 50 g dry wt$\cdot$m$^{-2}$ leaf area$\cdot$d$^{-1}$. This is equivalent to a net carbon dioxide uptake of about 735 mg $CO_2\cdot$dm$^{-2}$ ground area$\cdot$d$^{-1}$. Most crop species, however, have maximum growth rates of 20–30 g$\cdot$m$^{-2}$ leaf area$\cdot$d$^{-1}$ (Zelitch, 1971).

The growth rate of plants has been expressed as the rate of increase in dry weight per unit leaf area or the net assimilation rate (NAR). Because of the dependence of the net assimilation rate on the leaf area index (LAI), the rate of dry matter production per unit area of land has been examined with a parameter called crop growth rate, $C$, which is equal to the NAR $\times$ LAI and is usually expressed as g dry wt$\cdot$m$^{-2}$ ground area$\cdot$d$^{-1}$ (Zelitch, 1971). Major factors affecting photosynthesis and productivity will be discussed forthwith.

### *Irradiance*

The visible region of the solar spectrum includes the wavelengths which provide the energy utilized in photosynthesis. However, only the radiation in the 400 to 700 nm waveband should be considered as photosynthetically active irradiance (McCree, 1972). The solar constant is the flux density of total solar radiation at the boundary of the atmosphere at the earth's mean distance from the sun, and it equals 8.1 J$\cdot$cm$^{-2}\cdot$min$^{-1}$. In passing through the atmosphere, much of the radiation is absorbed and scattered by atmospheric water vapour, carbon dioxide, ozone and dust, so that an average of about 5.4 J$\cdot$cm$^{-2}\cdot$min$^{-1}$ of radiation reaches the earth's surface. About 52% of this radiation is in the infrared region at wavelengths greater than 700 nm and 4% is in the ultraviolet region. Since the atmosphere does not greatly affect the solar spectrum between 400 and 700 nm, a surface perpendicular to the sun's rays at the earth's surface will receive about 44% of the total solar radiation as photosynthetically active radiation on a clear day. Therefore, about 2.5 J$\cdot$cm$^{-2}\cdot$min$^{-1}$ of incident radiation are available for photosynthesis.

In single maize leaves, less than 10% of the solar radiation in the blue and red regions is transmitted through the leaves and 15% in the green. While 10 to 20% of the incident radiation is reflected, about 50% of the near infrared radiation is also transmitted (Zelitch, 1971).

At low irradiance, photosynthesis increases linearly with irradiance; hence the photochemical processes are limiting. Photosynthesis is at its most efficient at these low irradiance levels and the efficiency of light energy conversion, or joules of carbohydrate formed per calorie of incident visible irradiance, is about 3%. At higher light intensities the rate of increase in photosynthesis declines until, in $C_3$ plants, the saturation point is reached. Light saturation varies greatly with species, as does the rate of net carbon dioxide absorption, and the higher the light intensities for saturation the greater the net carbon dioxide uptake (Zelitch, 1971).

Chen, Brown & Black (1969) have shown that $C_4$ plants require a higher light intensity for the saturation of photosynthesis than do $C_3$ plants. In fact, light saturation is not achieved at an intensity twice that of sunlight in confirmed $C_4$ species like maize, sugarcane and *Syricum vulgare*. The higher amount of light required to saturate photosynthesis in the $C_4$ plant may be caused by the presence of smaller photosynthetic units. Black & Mayne (1970) have shown that the ratio of $P_{700}$ to total chlorophyll is higher in $C_4$ plants. The presence of a higher density of photosynthetic units may account for the greater amount of carbon dioxide fixed in the light per unit leaf area in $C_4$ plants compared with $C_3$ plants.

The average total solar radiation in the United States for the 100-day summer growing season beginning 1 June ranges from 1600 to 2800 $J \cdot cm^{-1} \cdot d^{-1}$ depending upon the location. Using a modest value of 2100 $J \cdot cm^{-2} \cdot d^{-1}$ for total solar radiation on a bright day, 928 $J \cdot cm^{-2}$ photosynthetically useful radiation between 400 and 700 nm would reach a canopy of leaves. If this radiation were entirely absorbed, as in a dense stand, about 12% of this radiation would be available to provide chemical energy (111 $J \cdot cm^{-2} \cdot d^{-1}$ or 1112 $kJ \cdot cm^{-2}$ ground area $\cdot d^{-1}$, since 15.5 kJ needed per gram carbohydrate). The photochemical energy available might produce $1112/15.5 = 72$ g carbohydrate $\cdot m^{-2}$ ground area $\cdot d^{-1}$. This is about 40% greater than the maximum growth rates which have been shown for crops even though no corrections or losses by respiration have been included. Assuming respiration loss to be 33% of gross photosynthesis, maximum growth rates have been estimated to be 51 g dry wt $\cdot m^{-2}$ ground area $\cdot d^{-1}$ for species like maize (Monteith, 1969).

Since the efficiency of light energy conversion, that is the energy captured by the leaf as organic matter per joule of visible radiation received, is greatest at low irradiance, photochemical efficiency is not directly related to productivity. Clearly the total yield increases greatly with increasing radiation and decreasing photochemical efficiency.

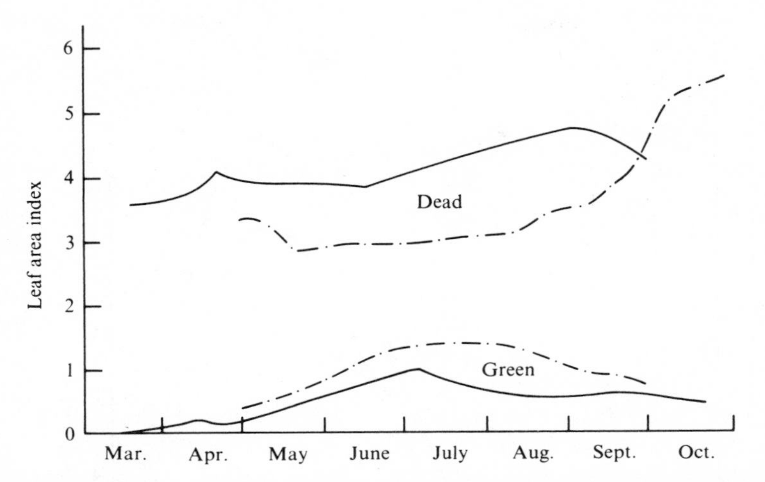


Fig. 2.1. Variation in true green and dead leaf area index (LAI) through 1970 and 1971 in the mixed prairie of Canada. Note that in this figure the top solid curve is for 'dead, 1971', followed by a broken curve for 'dead, 1970'. The bottom broken curve is for 'green, 1970', followed by a solid curve for 'green 1971'. (From Ripley & Redmann, 1976.)

Daily changes in the solar radiation determine, for the most part, the diurnal changes in the rates of photosynthesis and transpiration. In addition to the diurnal changes in solar radiation, the vertical gradient of radiant flux in a canopy is a measure of the absorption of energy by foliage at various heights above the ground surface (Monteith, 1969).

The leaf area index (LAI) is defined as the ratio of leaf area (one side of the leaf) to ground surface area. The quantity and quality of light absorbed by a single leaf is very important in determining the photosynthetic rate of the leaf. However, when productivity of a community is considered, the amount of light available for photosynthesis is also dependent on the LAI. Using measurements of the radiant flux at different canopy heights in conjunction with leaf area measurements, it is possible to calculate an extinction coefficient which relates the light attenuation to leaf area. Leaf angle also becomes an important factor for production because the quantity of light intercepted is dependent upon the angle of the leaf in relation to the angle of incident radiation. Leaf angle interacts with leaf area and its relation to productivity. Pearce, Brown & Blaser (1967) found that leaf angle had little effect on net photosynthetic rate of six-rowed barley (*Hordeum vulgare*) up to an LAI of 2.5. But as the LAI increased above 2.5, the net photosynthetic rate was higher for more vertically oriented leaves.

Monsi & Saeki (1953) obtained the vertical distributions of the LAI and the associated illumination profiles for a large number of forest and herbaceous communities. They were able to fit their observations to $L$ (leaf area

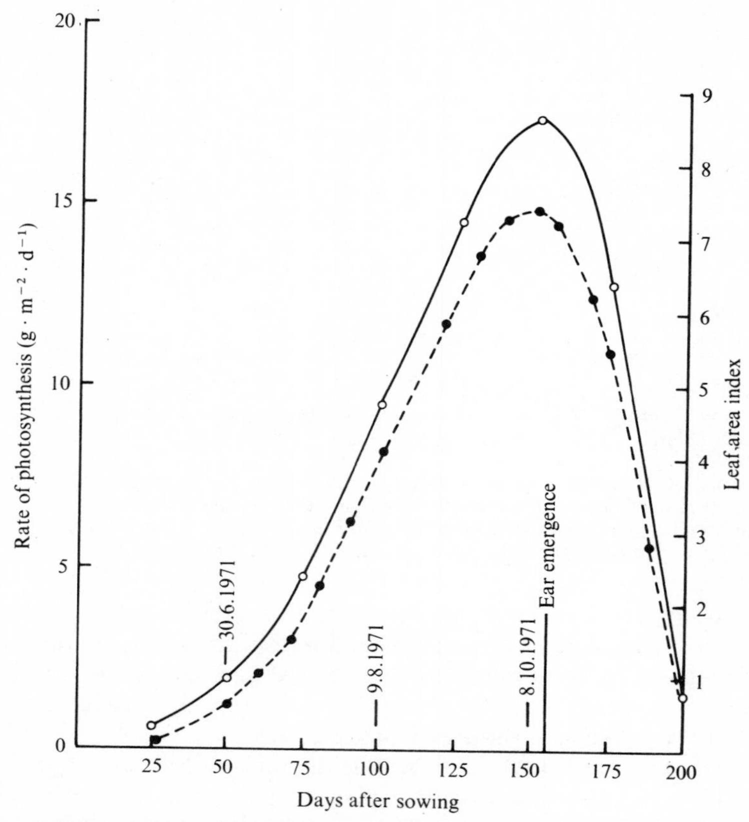


**Fig. 2.2.** Rate of photosynthesis and leaf area index in a Sherpa wheat crop at Rutherglen, Australia. (After Paltridge *et al.*, 1972.)

index) in the Bouguer–Lambert law: $I = I_0 e^{-KL}$, where $I$ and $I_0$ refer to the ilumination on a horizontal surface within and above the canopy, respectively, and $K$ is the extinction coefficient. The fresh weight of leaves was used to estimate the value of $L$ (e.g. Fig. 2.1). For grass-type communities, $K$ was generally in the range of 0.3 to 0.5, while for a canopy with more horizontal leaves $K$ approached 1.0 (Loomis, Williams & Duncan, 1967). The highest crop growth rates have generally been observed with monocotyledonous communities with a high LAI and with tendencies toward erect leaves and low extinction coefficients. Most measurements in crops have shown the maximum value for the LAI to reach about 4.0, although there are some reports as high as 12.0 (Pearce *et al.*, 1967).

The optimum LAI for net photosynthesis is determined by the stand architecture including the average leaf angle, light interception, mutual shading

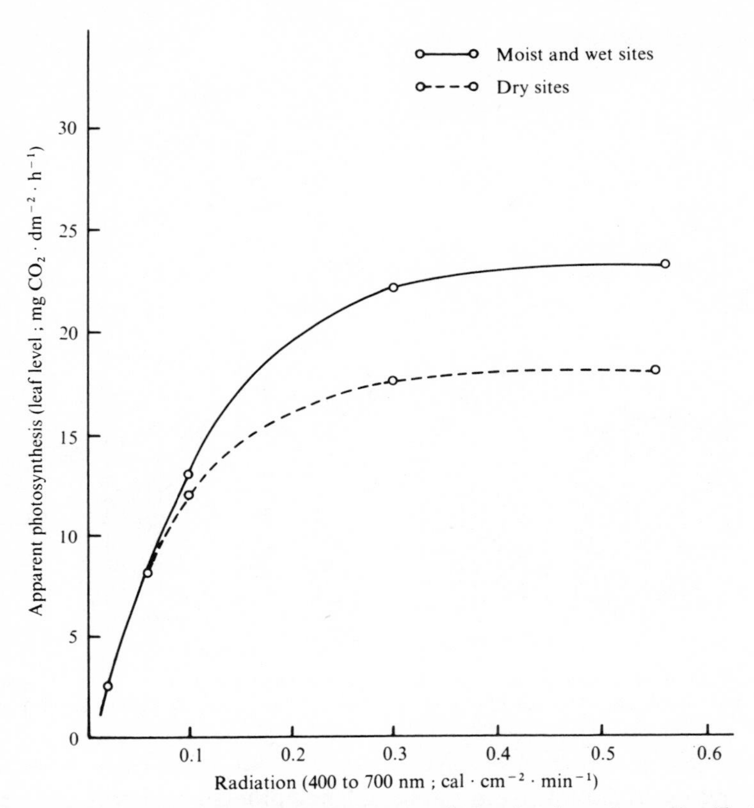


Fig. 2.3. Relationship of photosynthesis and radiation for three sites at Lanžhot, Czechoslovakia. (M. Rychnovská, unpublished IBP data.)

Table 2.3. *Parameters of light curves for net photosynthesis (leaf temperature $21 \pm 1$ °C, $CO_2$ conc. $290 \pm 8$ ppm) at Lanžhot, Czechoslovakia*

| Species | | Parameters | | | | |
|---|---|---|---|---|---|---|
| | | $a$ | $I_0$ | $I_{(0\cdot5)}$ | $P_{N(\max)}$ | $P_{N(315)}$ |
| *Glyceria maxima* | $\overline{X}$ | 0.25 | 5.0 | 74.0 | 26.4 | 22.6 |
| | $s_{\bar{x}}$ | 0.02 | 2.4 | 4.9 | 3.5 | 3.1 |
| *Phalaris arundinacea* | $\overline{X}$ | 0.25 | 5.0 | 53.0 | 20.1 | 18.3 |
| | $s_{\bar{x}}$ | 0.02 | 1.1 | 5.3 | 2.1 | 1.5 |
| *Alopecurus pratensis* | $\overline{X}$ | 0.32 | 7.0 | 49.0 | 20.8 | 19.4 |
| | $s_{\bar{x}}$ | 0.09 | 1.4 | 3.9 | 2.5 | 1.8 |
| *Festuca sulcata* | $\overline{X}$ | 0.10 | 3.0 | 116.0 | 14.7 | 11.4 |
| | $s_{\bar{x}}$ | 0.01 | 1.0 | 34.3 | 1.9 | 1.7 |

$a$, initial slope of the light curve (mg $CO_2 \cdot dm^{-2} \cdot h^{-1} \cdot W^{-1} \cdot m^{-2}$).
$I$, density of radiation flux ($W \cdot m^{-2}$, 400–700 nm).
$I_0$, $I$ at which $P_N = 0$ (light compensation point) ($W \cdot m^{-2}$).
$I_{(0\cdot5)}$, $I$ at which $P_N = 0.5 P_{N(\max)}$ ($W \cdot m^{-2}$). $P_N$, net photosynthesis (mg $CO_2 \cdot dm^{-2} \cdot h^{-1}$).
$P_{N(\max)}$, extrapolated value of $P_N$ for $I \to \infty$. $P_{N(315)}$, $P_N$ at $I = 315\ W \cdot m^{-2}$.

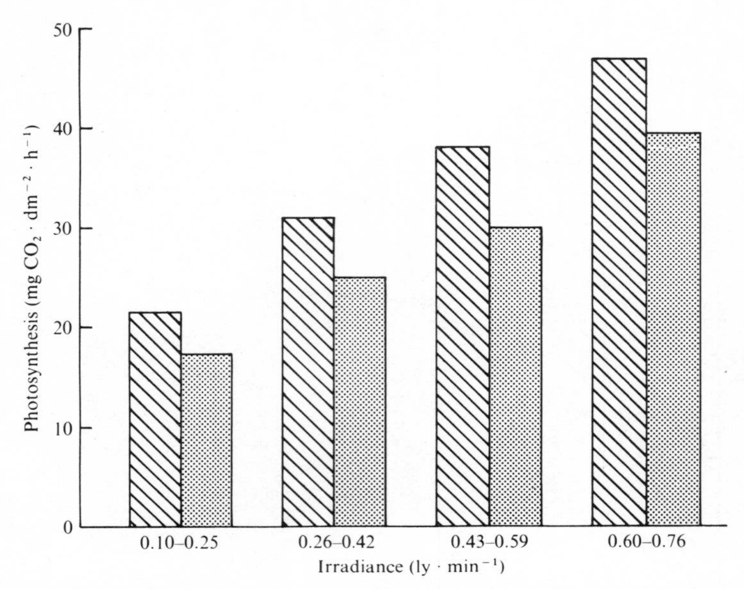


Fig. 2.4. Effect of visible irradiance on the gross (hatched columns) and net (stippled columns) photosynthetic rates for blue grama *in situ*. (After Brown & Trlica, 1977*b*.)

that may occur so as to affect dark respiration and photorespiration, and internal and external diffusive resistances. The optimum LAI will thus vary because of differences in these properties between different plant species and varieties. The total productivity would vary with the time required to initiate optimal conditions for the net assimilation rate and LAI (Zelitch, 1971).

At low light intensities a single exposed leaf shows a greater rate of photosynthesis than does a canopy covering 3.4 times the ground surface area (Zelitch, 1971). The shaded leaves in a stand fix carbon dioxide more slowly than brightly illuminated leaves and may not show positive net photosynthesis if the shade is deep. The canopy requires more light than a single horizontal leaf in order to exceed the light compensation point.

Knight (1973) evaluated the sunlight interception and LAI of shortgrass prairie in Colorado and found that the peak LAI (0.55) occurred in mid-June. Aboveground productivity at the time was about 4.5 $g \cdot m^{-2} \cdot d^{-1}$. Although different grazing treatments had little effect on the LAI, irrigation alone increased the LAI values by about two-thirds and irrigation plus nitrogen fertilization increased the values approximately tenfold.

In an Oklahoma tallgrass prairie, Conant & Risser (1974) found that live material had a maximum LAI of 6.8 on an ungrazed treatment and 5.4 on a grazed treatment with the maxima occurring in late summer. The LAI for dead plant leaves peaked in October with values of 8.4 and 5.8 for the ungrazed

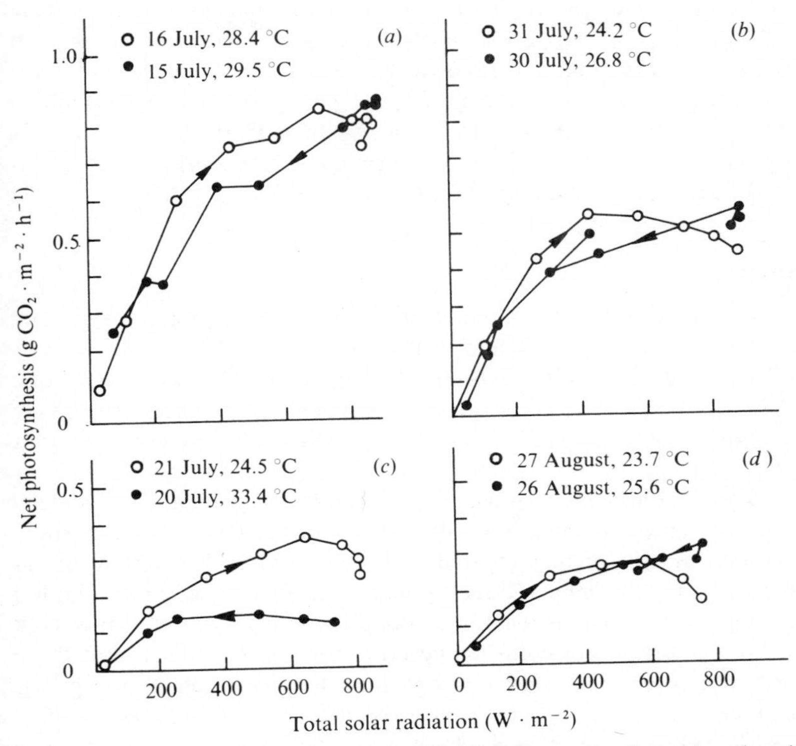


Fig. 2.5. Relationship between net photosynthesis and global radiation based on 1970 field data from a Canadian mixed-grass prairie. (Redmann, 1974*a*.)

and grazed treatments, respectively. Similarly in a mixed-grass prairie in Canada the LAI for dead leaves was higher than that for green leaves (Fig. 2.1; Ripley & Redmann, 1976). The peak LAI for live biomass in tropical grasslands in India ranges from 3.4 (angleton grass (*Dichanthium*) community) to 7.9 (rattail grass (*Sehima*) community) (Mall, Misra & Billore, 1973). Paltridge *et al.* (1972) have reported a similar pattern of changes in the rate of net photosynthesis and LAI for a Sherpa wheat crop grown at Rutherglen, Australia (Fig. 2.2).

Studies on moist meadow (dominated by *Alopecurus pratensis*), wet meadow (dominated by *Glyceria maxima*) and dry meadow (dominated by *Festuca sulcata*) at Lanžhot, Czechoslovakia, indicate a curvilinear relationship between the rate of net photosynthesis and incident solar radiation (Fig. 2.3) (M. Rychnovská, IBP data). The rate of net photosynthesis was considerably higher for the moist and wet meadows at higher light intensities. Parameters for the light intensity/net photosynthesis curves for four major meadow species at Lanžhot, Czechoslovakia, are given in Table 2.3 (J. Gloser, IBP data).

Field observation on the rates of gross and net photosynthesis of blue grama (*Bouteloua gracilis*), a $C_4$ species, indicates increasing rates with increases in irradiance (Fig. 2.4) (Brown & Trlica, 1977*a*). Further, the percentage of gross photosynthesis accounted for by net photosynthesis remained fairly constant with increasing light intensity. In a Canadian mixed-grass prairie the net photosynthesis showed a curvilinear relationship with irradiance (Fig. 2.5; Ripley & Redmann, 1976).

#### *Temperature*

Different plant species demonstrate a wide range of optimum photosynthetic temperatures. General optimal temperature ranges for $C_3$ and $C_4$ species are 10 to 25 °C and 30 to 40 °C, respectively. Most $C_3$ plants become chlorotic around 35 °C (Downton, 1971). Conversely some $C_3$ species are quite active at 5 °C, while most $C_4$ species are photosynthetically inactive at that temperature.

As Bjorkman (1968) has shown with woundwort goldenrod (*Solidago virgaurea*), the optimum light intensity can change and a plant can show limited adaptation to a change in light regime. Likewise, Woledge & Jewiss (1969) found that tall fescue (*Festuca arundinacea* (Screb.) Celak.) adapted to the temperature regime in which it was grown. Plants grown at a high temperature had a high optimum temperature for photosynthesis and those grown at low temperatures had low optimum temperatures. When plants grown at high temperatures were transferred to cooler growing conditions, the optimum temperature decreased as the plants adapted to the cooler environment. Taylor & Rowley (1971) measured the effects of chilling stress under various light and time treatments of assorted $C_3$ and $C_4$ species. The photosynthetic rates of all species decreased when subjected to the chilling stress of 10 °C, but the photosynthetic rates of $C_4$ species declined to negligible levels after 2 or 3 days whereas those of $C_3$ species decreased more slowly and maintained a positive net photosynthesis at 10 °C.

In the Canadian mixed-grass prairie, Redmann (1974*a*) found that net photosynthesis increased with temperature to a maximum around 30 °C (Fig. 2.6). However, at any temperature, the photosynthetic rates were higher when the relative humidity was greater than 50% and the maximum photosynthetic rates were lower under drier air conditions. The percentage of net photosynthesis in the meadow vegetation of Czechoslovakia (Fig. 2.7) also increases with increase in temperature to a peak value and declines with further increases in temperature (M. Rychnovská, IBP data).

Both the gross photosynthesis and net photosynthesis rates increase with increasing temperature to a peak and thereafter the rates decline with further increases in temperature. This response is exemplified by the experiments on blue grama swards (Brown & Trlica, 1977*b*; Fig. 2.8) as well as in individual

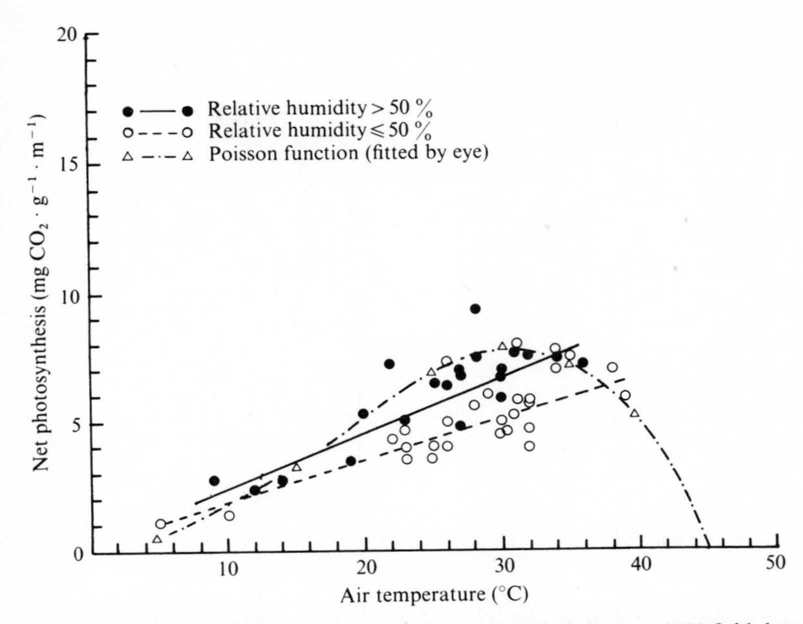


Fig. 2.6. Net photosynthesis in relation to air temperature based on 1970 field data from a Canadian mixed-grass prairie. Ranges of humidity are specified. Global radiation, > 30 $ly \cdot h^{-1}$; soil water, 22–29 %. (Redmann, 1974*a*.)

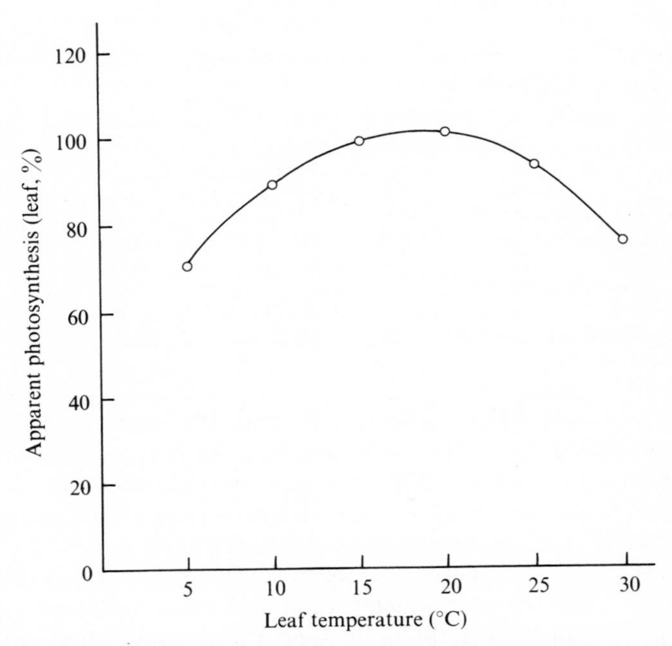


Fig. 2.7. Relationship of photosynthesis and temperature for the meadow site at Lanžhot, Czechoslovakia. (M. Rychnovská, unpublished IBP data.)

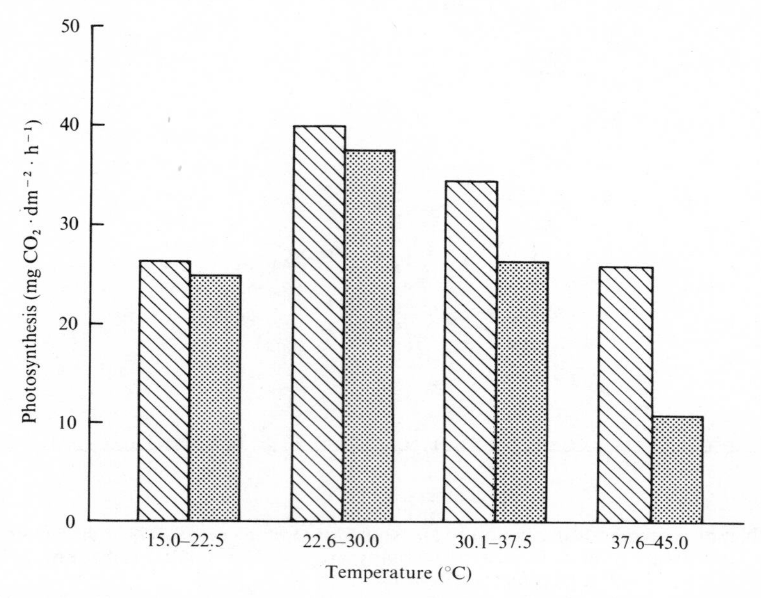


Fig. 2.8. Effect of temperature on the gross (hatched columns) and net (stippled columns) photosynthetic rates for blue grama *in situ*. (Brown & Trlica, 1977*b*.)

Table 2.4. *Temperature dependence of apparent photosynthesis (relative values) in meadow species in Czechoslovakia*

| Species | Leaf temperature (°C) | | | | |
|---|---|---|---|---|---|
| | 10 | 15 | 20 | 25 | 30 |
| *Glyceria maxima* | 0.86 | 0.96 | 1.0 | 0.93 | 0.86 |
| *Phalaris arundinacea* | 0.85 | 0.96 | 1.0 | 0.98 | 0.91 |
| *Alopecurus pratensis* | 0.88 | 0.97 | 1.0 | 0.85 | 0.78 |
| *Festuca sulcata* | 0.88 | 0.98 | 1.0 | 0.92 | 0.75 |

J. Gloser, IBP data.

meadow species (Table 2.4) (J. Gloser, IBP data). Brown & Trlica (1977*b*) noted a significant divergence of the net photosynthesis rates in relation to the gross photosynthesis rates with increasing temperature (Fig. 2.8).

Williams (1974) subjected field-collected sods of western wheatgrass (*Agropyron smithii*) and blue grama to pretreatment temperature regimes of 20/15 °C, 30/15 °C and 40/15 °C day/night temperatures for 50 days and then evaluated their net photosynthesis at analysis temperatures of 20 to 40 °C and 5 deg C intervals (Fig. 2.9). Increasing the pretreatment day temperatures from 20 to 40 °C resulted in decreased net photosynthesis in western wheatgrass

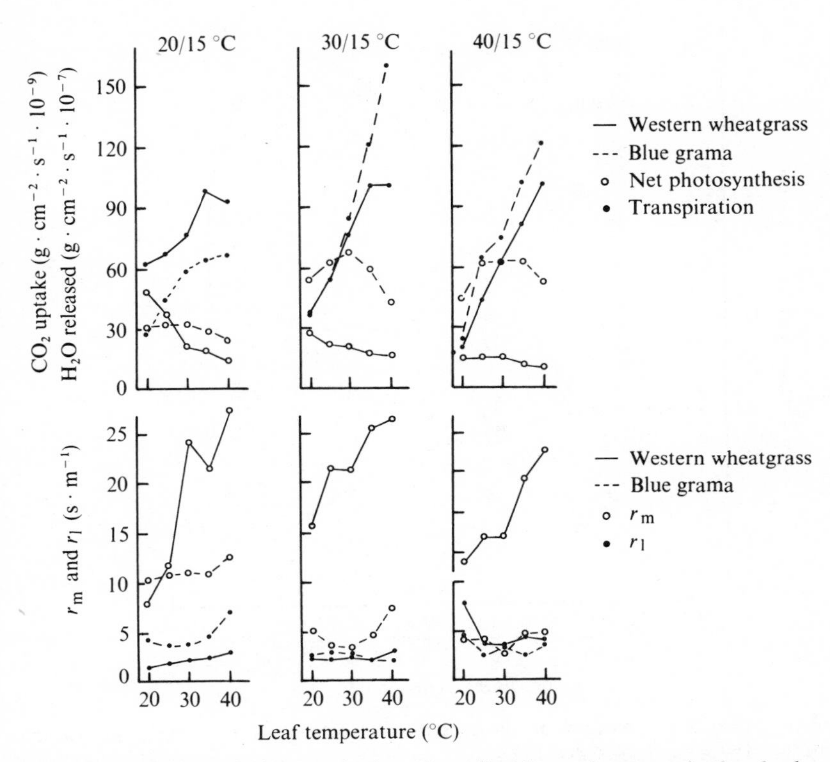


Fig. 2.9. Dependence of photosynthetic carbon dioxide uptake, transpirational release of water, resistance to water vapour transfer by stomata and leaf boundary layer ($r_l$), and remaining physical and metabolic resistances to carbon dioxide uptake ($r_m$) on analysis temperature for blue grama and western wheatgrass. (See text for experimental details.) Species curves by pretreatment growth conditions (20/15 °C, 30/15 °C, and 40/15 °C) were obtained by gas analysis of three plants of each species from each growth condition over the analysis range of temperature, and the standard error of the mean (SE) was calculated for each. Average SE and greatest SE for western wheatgrass 20/15 °C pretreatment were 3.5 and 6.8 for net photosynthesis; 6.1 and 11.6 for transpiration; 0.3 and 0.3 for $r_l$; and 2.8 and 7.7 for $r_m$. Average SE and greatest SE for western wheatgrass 30/15 °C pretreatment were 1.3 and 2.9 for net photosynthesis; 54 and 12.6 for transpiration; 0.3 and 0.6 for $r_1$; and 1.3 and 3.4 for $r_m$. The values for western wheatgrass 40/15 °C pretreatment were 1.1 and 1.4 for net photosynthesis; 15.6 and 36.9 for transpiration; 1.1 and 2.5 for $r_l$; and 4.3 and 7.0 for $r_m$. The SE values for blue grama 20/15 °C pretreatment were 1.6 and 2.6 for net photosynthesis; 3.8 and 7.9 for transpiration; 0.6 and 0.9 for $r_l$; and 1.3 and 1.7 for $r_m$. The values for blue grama 30/15 °C pretreatment were 2.9 and 5.9 for net photosynthesis; 3.2 and 9.0 for transpiration; 0.1 and 0.1 for $r_l$; and 0.4 and 0.6 for $r_m$. The values for blue grama 40/15 °C pretreatment were 4.1 and 6.1 for net photosynthesis; 12.2 and 16.9 for transpiration; 0.5 and 0.7 for $r_l$; and 1.2 and 1.7 for $r_m$. (Williams, 1974.)

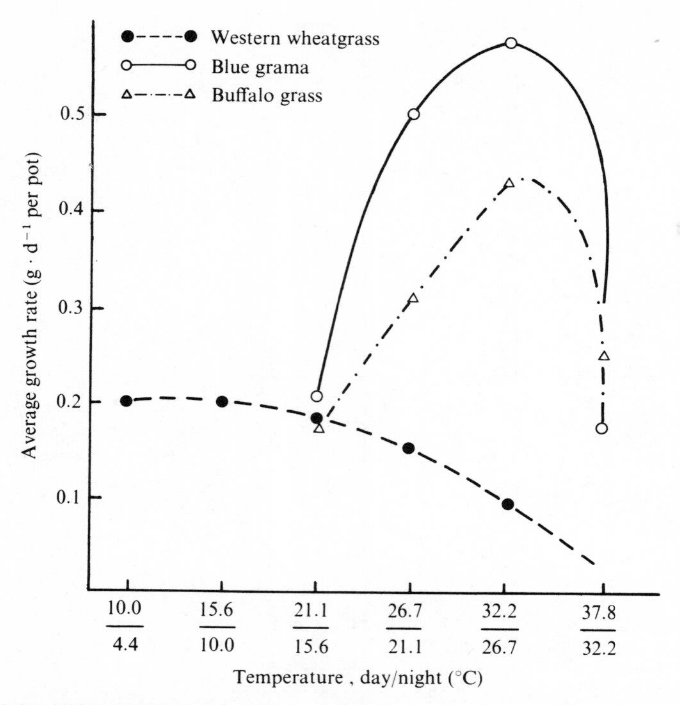


Fig. 2.10. Effect of temperature on the growth of buffalo grass, blue grama and western wheatgrass. (Knievel & Schmer, 1971.)

($C_3$) while in blue grama ($C_4$) net photosynthesis was increased. The effect on photosynthesis of increasing analysis temperatures was the same as observed by increasing pretreatment temperatures. Phenology of the species in the shortgrass prairie is such that western wheatgrass has its greatest growth activity during the cool portion of the growing season whereas blue grama is most active in the warm portion. Thus, photosynthetic adaptation to temperature is strongly suggested as a strategy for ecosystem utilization by reduction of interspecific competition.

Knievel & Schmer (1971) in some greenhouse experiments noted that average growth rate of western wheatgrass was greater at lower temperatures (10 to 15 °C daytime) while the warm-season grasses, blue grama and buffalo grass (*Buchloe dactyloides*), demonstrated higher net production near 30 °C (Fig. 2.10).

Spittlehouse & Ripley (1974) measured diurnal temperature profiles in nine mixed prairies (Fig. 2.11) and Old (1969) measured diurnal temperature profiles in a mixed prairie in both the summer and winter (Fig. 2.12). As would be expected, there was a smaller day/night temperature range in the

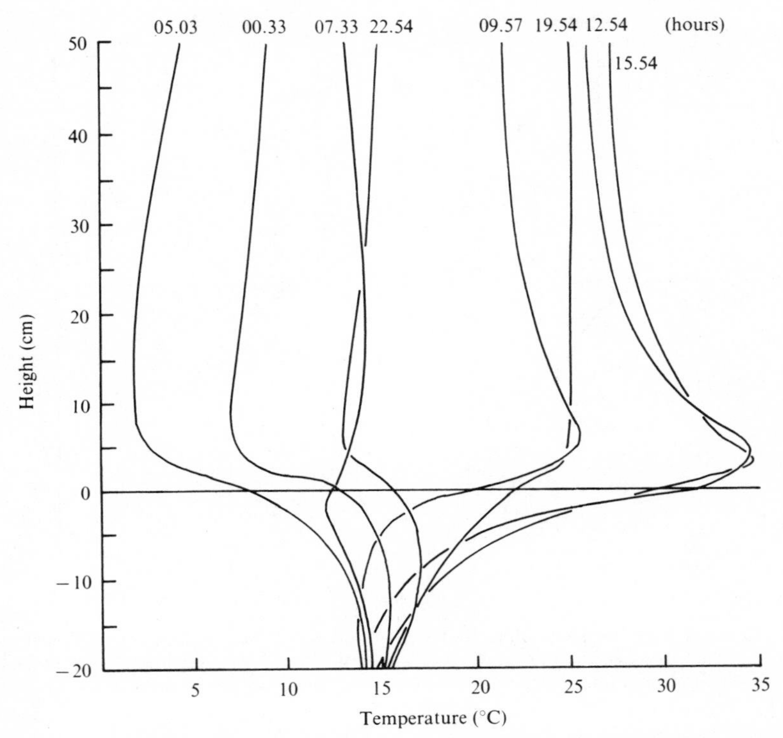


Fig. 2.11. Diurnal temperature profiles on 28 June 1972 in a Canadian mixed-grass prairie. (Spittlehouse & Ripley, 1974.)

canopy during the winter. In July, the greatest temperature differential of about 4 deg C within the canopy occurred at 19.00 hours. While seasonal temperatures affect rate of net production, it appears that temperature differences within the canopy profile are not particularly significant when compared with leaf area index, leaf hydration or irradiance.

### *Carbon dioxide*

Since the light saturation point can be moved upward by increasing the carbon dioxide concentration, this suggests that diffusive resistance limits photosynthesis under high light intensities. At saturating light, net photosynthesis increases in a nearly linear fashion up to about 300 ppm of carbon dioxide. In most plants carbon dioxide saturation occurs around 1000 to 1500 ppm and net photosynthesis doubles or trebles between 300 ppm and these high carbon dioxide levels (Zelitch, 1971). Even in plants like maize, which presumably have high internal diffusive resistances, photosynthesis was 50%

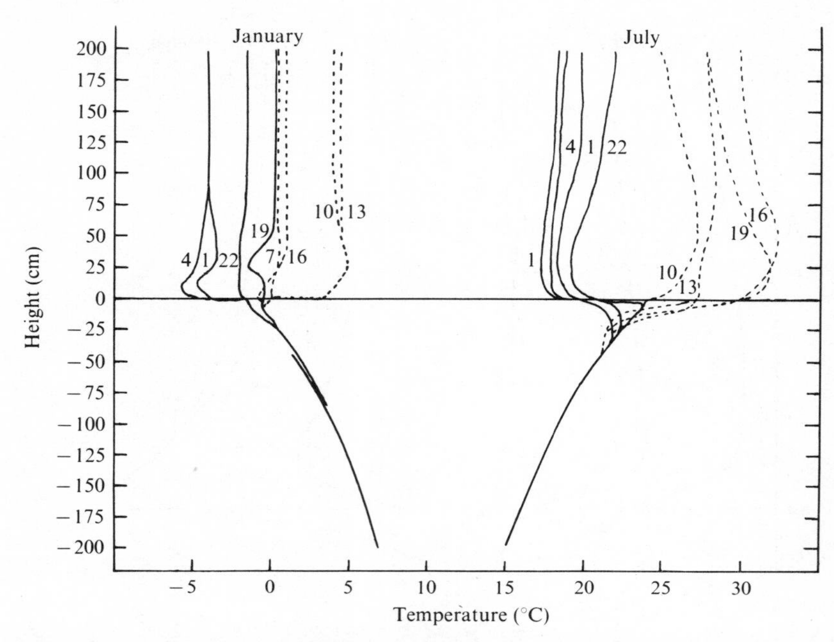


Fig. 2.12. Diurnal temperature profiles in January and July 1967. Values for each height represent a weekly average for the hour specified. Solid lines represent night profiles and dotted lines daylight profiles. Numbers on the lines are hours on the 24-hour clock. (From Old, 1969.)

greater at 500 ppm $CO_2$ than at 300 ppm (Hesketh & Moss, 1963). High concentrations tend to close the stomata, but if the ambient carbon dioxide level is high, sufficient carbon dioxide is still available and will not be limiting (Zelitch, 1971).

As carbon dioxide concentration increases, there is a diminution in the difference between net photosynthetic rates of $C_3$ and $C_4$ plants. Although the $C_4$ plants are more efficient at carbon dioxide uptake and utilization, as the carbon dioxide concentration of the air increases this advantage decreases. The lower the carbon dioxide concentration in the air, the greater the advantage for the $C_4$ plant.

The boundary layer resistance is primarily regulated by wind speed and it does not greatly limit plant growth when plants have an adequate supply of water. This resistance is generally only a small fraction of the total diffusive resistance to photosynthesis in small leaves. The turbulence of the bulk air within the canopy of leaves will decrease exponentially with decreasing light intensity; probably poor ventilation does not ordinarily restrict carbon dioxide assimilation because the light intensity is simultaneously low at positions in the canopy where boundary layer resistance might become large.

Since air within the canopy is moving when wind speeds above the canopy are slow, the diffusive resistance of the bulk air between the layers of leaves in a stand is likewise not very great. Thus the carbon dioxide transfer in the bulk air surrounding the leaves is probably very rapid in comparison with the rate at which it can be removed from the ambient atmosphere (Zelitch, 1971).

Carbon dioxide concentrations within the canopy are a complex function of wind speed, photosynthetic activity, crop roughness, soil and plant respiration, and other factors. The soil's contribution of carbon dioxide photosynthesis, derived from respiration of soil micro-organisms or roots, may be much greater than that of the shoots, but it is negligible compared with that of the atmosphere (Zelitch, 1971). Preliminary data from a Canadian mixed-grass prairie site indicate that soil respiration represents about 85% of the total (soil plus shoots) respiration (Ripley, 1974). Mean fluxes from the soil averaged 0.41 g $(CO_2) \cdot m^{-2} \cdot h^{-1}$ at night and 0.47 g $(CO_2) \cdot m^{-2} \cdot h^{-1}$ during the daylight period. These values may be compared with a mean daytime net atmospheric flux of 0.25 g $(CO_2) \cdot m^{-2} \cdot h^{-1}$.

During photosynthesis a concentration gradient exists between the carbon dioxide in the ambient air and that in the chloroplasts. The flux of carbon dioxide or net photosynthesis is determined by the size of this gradient and by a series of resistances to the diffusion of carbon dioxide. The rate of exchange of matter between leaves and the atmosphere surrounding them can be expressed by the equation $F = \Delta C / \Sigma R$, where flux ($F$) is proportional to the concentration gradient ($\Delta C$) between the interior of the leaf and the ambient air (in the case of carbon dioxide and water) and to the total resistance ($\Sigma R$) to transport from the free air into the leaf. The resistances to diffusion are both within and outside the leaf. The diffusion of carbon dioxide has two resistances in common with the diffusion of water vapour during transpiration, namely the air boundary layer and the stomatal resistance. Photosynthesis has additional resistances beyond the substomatal cavities and the intercellular spaces of the leaf; hence, the pathway for diffusion of carbon dioxide must have a greater total resistance than that for water vapour (Zelitch, 1971). Minimal stomatal resistance in meadow species in Czechoslovakia ranged from 1.5 to 5.9 $s \cdot cm^{-1}$, while the minimal mesophyll resistance varied from 4.2 to 10.9 $s \cdot cm^{-1}$ (Table 2.5). The rates of net photosynthesis were lowest in the species exhibiting maximum resistances.

Williams (1974) studied the effect of pretreatment temperatures of 20/15 °C, 30/15 °C and 40/15 °C on the resistance to water vapour transfer by stomata and leaf boundary layer ($r_l$) and the physical and metabolic resistance to carbon dioxide uptake ($r_m$) for blue grama and western wheatgrass at analysis temperatures ranging from 20 to 40 °C at 5 deg C intervals (Fig. 2.9). The increased rates of net photosynthesis of blue grama plants from warmer pretreatments were generally accompanied by reduced resistances. The resistances, $r_l$ and $r_m$, observed for western wheatgrass were greatest at higher

Table 2.5. *Leaf resistances to carbon dioxide diffusion in Czechoslovakian meadow species*

| Species | $P_N$ | $C_C$ | $r_{sm}$ | $r_{Mm}$ |
|---|---|---|---|---|
| *Glyceria maxima* | 22.3 | 57 | 1.5 | 4.2 |
| *Phalaris arundinacea* | 18.1 | 45 | 1.9 | 6.1 |
| *Alopecurus pratensis* | 18.4 | 50 | 1.5 | 6.0 |
| *Festuca sulcata* | 8.3 | 87 | 5.9 | 10.9 |

J. Gloser, IBP data.

$P_N$, net photosynthesis rate (mg $CO_2 \cdot dm^{-2} \cdot h^{-1}$).
$C_C$, $CO_2$ compensation point (ppm).
$r_{sm}$, minimum stomatal resistance to $CO_2$ ($s \cdot cm^{-1}$).
$r_{Mm}$, minimum mesophyll (= total substomatal) resistance to $CO_2$ ($s \cdot cm^{-1}$).

analysis temperatures. The data indicate that the $r_m$ plays a significant role in controlling photosynthesis rates and has a greater response to pretreatment temperature than does the resistance of the stomata and $r_l$.

### *Oxygen*

The well-known 'Warburg effect' describes the effect of oxygen concentration on photosynthesis. This inhibition of photosynthesis by oxygen has several characteristics summarized by Zelitch (1971): (1) high levels of carbon dioxide diminish the inhibitory effect of oxygen, (2) the inhibition is rapidly reversed by reducing the partial pressure of oxygen or restored by increasing it, and (3) the oxygen effect is associated with the synthesis of glycolate and high rates of photorespiration and, therefore, the effect is not observed in all plants to the same degree. Specifically, $C_4$ plants do not show an increase in net photosynthesis with a decrease in oxygen concentration.

### *Water*

The effect of water stress on photosynthesis is very complex. Perhaps the most important single effect of low soil water potential is low cell turgor which leads to stomatal closure and eventually to reduced leaf area. Water stress therefore reduces the capacity of the plant to carry on photosynthesis. Translocation is also affected by water stress, as this is related to transpiration which is highly correlated with photosynthetic rate (Zelitch & Waggoner, 1962).

Bielorai & Mendel (1969) found that the rates of both photosynthesis and transpiration in citrus seedlings gradually decreased as the soil water stress increased from −0.2 bars to −3.0 bars but rapidly decreased as soil water decreased from −3.0 to −15.0 bars. Very low soil water potentials have a greater effect on photosynthesis than on transpiration (Zelitch, 1971). Maize growing in soil was allowed to deplete the soil water to various contents from 24 to 10% several days before experiments were carried out and the carbon dioxide uptake and transpiration were then compared. The leaves had not yet reached the wilting point even at the lowest soil water; photosynthesis was inhibited 80% and transpiration only 50% when compared with plants maintained at the highest soil water. Therefore, as the soil dried the internal resistances to carbon dioxide assimilation increased more rapidly than the stomatal diffusive resistance (Zelitch, 1971). Carbon dioxide uptake by wild sorghs (*Sorghum bicolor*) decreased to near the compensation point at water potentials near −25 bars (Shearman, Eastin, Sullivan & Kinbacher, 1972).

Majerus (1975) studied the shoot and root growth response of blue grama, western wheatgrass and little bluestem (*Andropogon scoparius*) to soil water potential. There was a considerable difference between the species, but in all cases both shoots and roots stopped growing with less water in the upper soil layers than in the lower layers. Blue grama shoot growth stopped when the soil water potential was −80 bars at a 5 cm depth and −8 bars at 35 cm. Similar values for little bluestem were −24 and −3 bars, while western wheatgrass was intermediate at −30 and −15 bars.

Saturated soil conditions usually produced reduced growth in most plants, though there are notable exceptions such as cultivated rice. Susceptibility to flooding together with low soil oxygen and sometimes the accumulation of toxic materials may be quite different between taxonomically closely related species. For example, in an experimental comparison between spreading panic (*Panicum dichotomiflorum*) which normally grows in wet sites and Texas panic (*Panicum texanum*) which grows on well-drained soils, under wet conditions the growth of the latter was inhibited by 50% as compared to spreading panic (Hoveland & Buchanan, 1972).

Since carbon dioxide must diffuse through open stomata which depend upon guard cell turgor for their opening, a sufficient decrease in the relative leaf water content will close stomata and result in a lowered net photosynthesis. Moisture stress in leaves results when the water required by the plant as determined by the atmosphere and diffusive resistance to transpiration exceeds the supply furnished by the soil water. Leaves of many species wilt when their water content is about 80 to 85% of their fully turgid condition and stomata usually begin to close as the water content decreases even before wilting is observed.

Net photosynthesis is reduced progressively by water stress until negative photosynthetic rates may develop with severe dehydration (Slatyer, 1967). Besides the general effect, there are also more specific effects of water stress that curtail production when the inadequate water supply occurs at critical periods of the growing season. Thus, water stress applied during limited

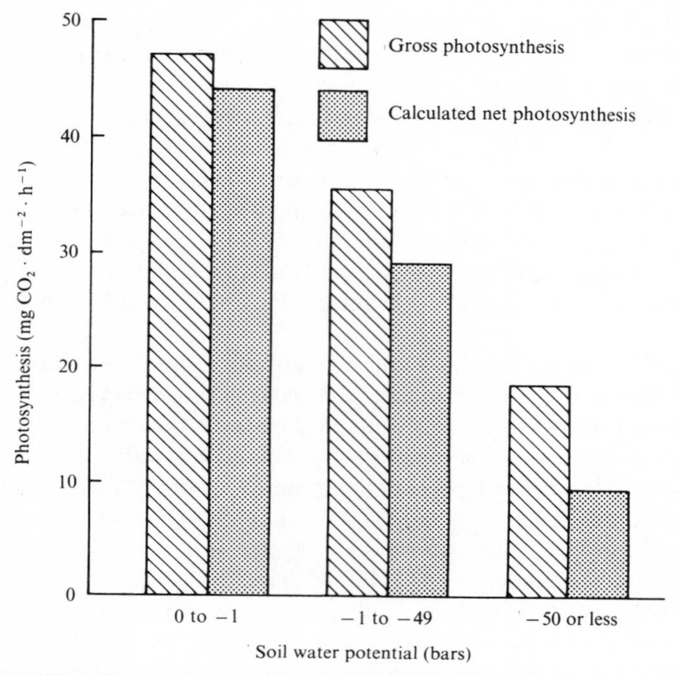


Fig. 2.13. The effect of soil water potential at 10 cm depth on the gross and net photosynthetic rates for blue grama *in situ*. (Brown & Trlica, 1977*b*.)

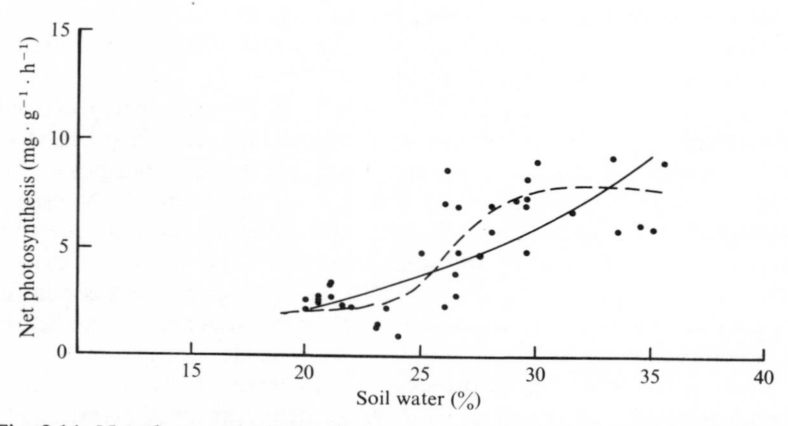


Fig. 2.14. Net photosynthesis in relation to soil water based on 1970 field data from a Canadian mixed-grass prairie. Hyperbolic and arctangent functions fitted using least-squares technique. Ranges of global radiation and air temperature are specified. (Redmann, 1974*a*.)

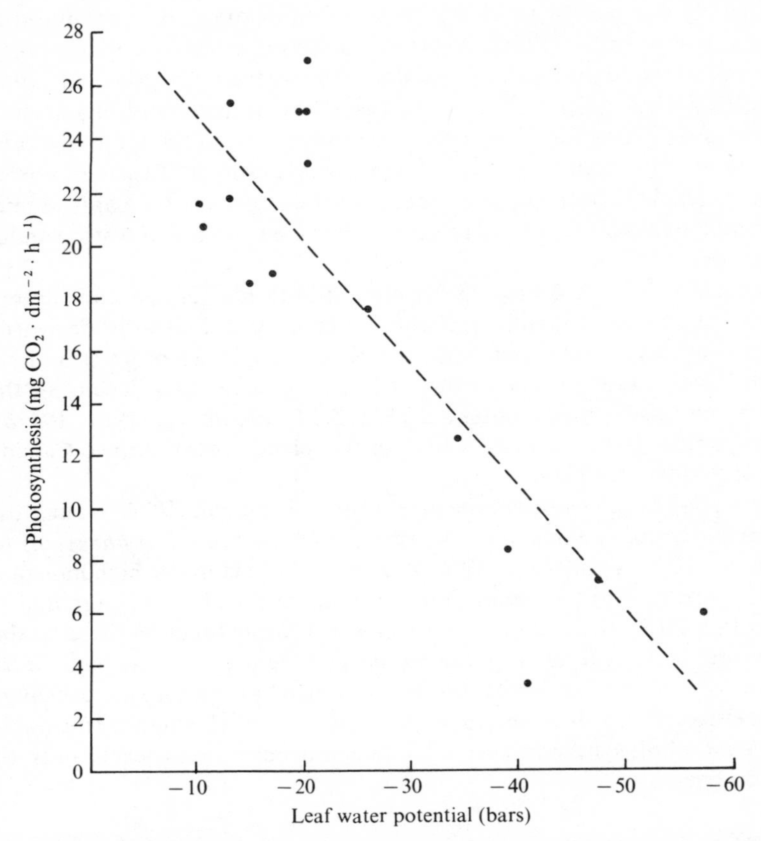


Fig. 2.15. Net photosynthesis of thick spike wheatgrass related to leaf water potential Dashed line fitted using a least-squares technique. (Redmann, 1974*b*.)

periods diminished grain yield in drought-susceptible maize as follows: tassel emerging period, 6%; pollination period, 67%; and ear initiation, 48%. Water stress, therefore, gave the greatest reduction in yield when it occurred during the 5-day period of pollination (Shaw & Laing, 1966).

Soil water potential also affects plant water potential. Boyer (1970*a*) found that leaf enlargement was inhibited earlier and more severely than was photosynthesis or respiration as the leaf water potential decreased in maize, *Glycine hispida*, and common sunflower. Boyer (1970*b*) also found that photosynthesis in *Glycine hispida* was not reduced until leaf water potential dropped below −11 bars, while photosynthesis of corn was affected anywhere below −3.5 bars; therefore, corn, which has a $C_4$ pathway, was more sensitive to the desiccation. Usually, however, the $C_4$ pathway plants have a greater drought resistance than $C_3$ plants. Generally, $C_4$ plants require approximately one-half as much water per gram of dry matter produced as do $C_3$ plants (Downton, 1971). The low mesophyll resistance of $C_4$ plants permits relatively high stomatal resistance to carbon dioxide and water vapour diffusion. A combination of the two-stage stomatal and biochemical apparatus of $C_4$ plants for carbon dioxide reduction, and the mesophyll and bundle sheath cells, maintains a very large partial gradient of carbon dioxide from the atmosphere to the bundle sheath. Therefore, many $C_4$ plants require less stomatal area than $C_3$ plants to permit the same volume of water vapour diffusion out of the leaf.

Brown & Trlica (1977*a*) have shown that in both blue grama and western wheatgrass the carbon dioxide exchange rates were significantly decreased with decreases in soil water potential. Measurements *in situ* of gross and net photosynthesis in blue grama swards indicated a significant decline in the rates with decreasing water potential (Fig. 2.13; Brown & Trlica, 1977*b*). Somewhat similar trends were indicated for the mixed-grass prairie in Canada (Fig. 2.14; Redmann, 1974*a*).

Under high light intensities in the greenhouse, Redmann (1974*b*) found the leaf water potential in thick spike wheatgrass (*Agropyron dasystachyum*) to range from $-10$ bars under low water stress to $-57$ bars under high moisture stress. The relationship between photosynthesis and leaf water potential is shown in Fig. 2.15. Although the figure shows a linear function fitted to the data, a more likely type of response would involve a plateau at high water potentials, with a threshold potential below which carbon dioxide exchange declines (Hsiao, 1973). It is apparent that thick spike wheatgrass is capable of significant photosynthesis even at tissue moisture stress levels near or above $-50$ bars.

### *Nutrients*

The photosynthetic process responds to deficiencies in a number of nutrients, though these effects may be indirect. For example, stomatal openings of potassium-deficient maize leaves were only about one-fifth the width of those in normal leaves (Peaslee & Moss, 1968). Plants deficient in magnesium and many other minerals are usually chlorotic which reduces the rate of photosynthesis. Iron is apparently necessary for the production of chlorophyll, so when it is present in deficient quantities the plants will have a sub-optimal photosynthetic rate.

There is a demonstrable relationship between nitrogen and photosynthesis. Murata (1969) has reviewed the literature, which shows that under certain circumstances nitrogen shows a positive correlation with photosynthetic capacity. Cultivated rice leaves containing 2% nitrogen fixed carbon dioxide at about 8.5 $mg \cdot dm^{-2} \cdot h^{-1}$ while the rate for leaves with 5% nitrogen was approximately 15 $mg \cdot dm^{-2} \cdot h^{-1}$. However, higher nitrogen supply does not always cause higher photosynthetic activity, because with high nitrogen, growth rates are stimulated. This means that the actual nitrogen content per gram of leaf may not be increased greatly and there may also be a resulting relative deficiency of other nutrients. Presumably, nitrogen deficiency adversely affects the chloroplasts since this is the location of a large portion of the leaf protein. The fundamental responses of primary production to nutrients have been adequately reviewed by a number of authors including Fried & Broeshart (1967), Rorison (1969) and Whitehead (1970).

Rogler & Lorenz (1957) studied the response of the North Dakota grazed mixed-grass prairie to nitrogen applications over a period of 6 years. In the heavily grazed pasture, where the range had changed from a typical mid-grass type to almost pure blue grama because of heavy grazing, yields were low the first year but were doubled and trebled by the application of 3.4 and 10 g of nitrogen per square metre. Remnant western wheatgrass stands made a rapid recovery and accounted for most of the yield increase. Yields from the moderately grazed plots were increased by the additional nitrogen, but this response was not so marked because the vegetation was in excellent condition at the start of the experiment. In subsequent years of the experiment, yield response varied with available moisture, but forage increased with nitrogen application. Depending on the time of application and climatic conditions of the particular season, forage yield response is frequently associated with a change in species composition (Reardon & Huss, 1965).

Not all species demonstrate the same response of primary production to nutrients. Wuenscher & Gerloff (1971) examined the growth of little bluestem and Kentucky bluegrass (*Poa pratensis*) to varying levels of phosphorus. These species were growing in phosphorus-deficient soil and the yield of little bluestem increased as added phosphorus was increased from 0 to 5 ppm, but there was no significant response to further increases of 10 and 20 ppm. The yield of Kentucky bluegrass increased significantly with additional phosphorus and the maximum yield had not been reached at 20 ppm. The phosphorus concentration in the shoot tissue of little bluestem did not increase significantly with additional phosphorus until the 20 ppm level. In contrast, phosphorus absorption by Kentucky bluegrass continued to increase as the level of added phosphorus increased. From the same soil and under the same treatment, the amount of phosphorus absorbed by Kentucky bluegrass was only 5 to 10% that absorbed by little bluestem, although the percentage of phosphorus in the tissue was similar in both species.

### *Factor interactions*

Numerous interactions among abiotic and biotic factors may affect photosynthesis and respiration of native plants in grassland ecosystems. Interactions among these factors may affect productivity in ways that studies of

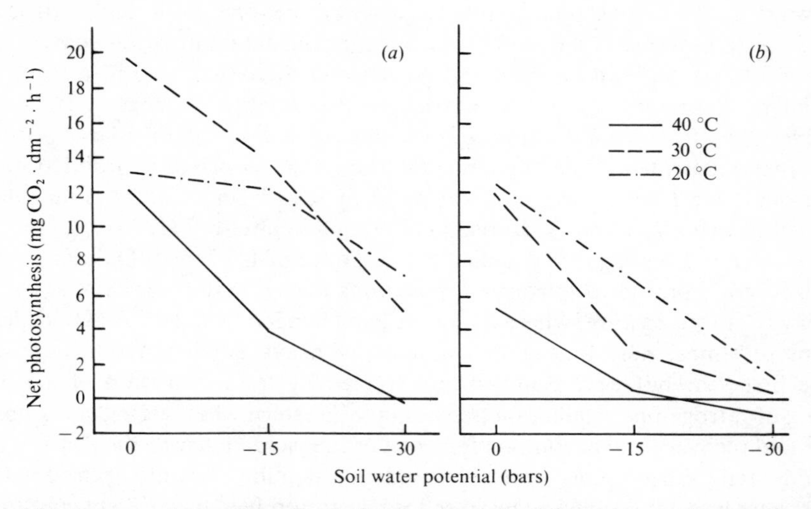


Fig. 2.16. Two-way interaction effects of soil water potential and temperature on the net photosynthetic rates of (*a*) blue grama and (*b*) western wheatgrass. (Brown & Trlica, 1977*a*.)

single-factor effects might not explain. Unfortunately, only a few studies have been undertaken to examine some of these more important interactions affecting primary productivity.

Brown & Trlica (1977*a*) studied the interaction effects of soil water potential, irradiance and temperature on photosynthesis of blue grama and western wheatgrass. They found that light saturation of blue grama, a $C_4$ species, was evident at very high irradiances accompanied by high temperatures and soil water stress (Fig. 2.16). The $C_3$ species, western wheatgrass, was lightly saturated at relatively low irradiances and was more sensitive to water stress. The optimum photosynthetic temperature for blue grama was influenced by water stress and ranged from about 26 to 33 °C. Aboveground dark respiration for both species increased with temperature and decreased with increasing soil water stress. Somewhat similar interactions among soil water potential, irradiance and temperature were noted for photosynthesis of blue grama in a field study (Brown & Trlica, 1977*b*). The reproductive phenological stage was associated with reduced rates of photosynthesis.

In a similar fashion, Redmann (1974*a*) constructed 'response surfaces' for the Canadian mixed-grass prairie. These are three-dimensional graphs which combine two independent variables and show the effect of these variables on net photosynthesis. The combined effects of global radiation and temperature (Fig. 2.17) reveal optimum conditions at maximum global radiation levels and air temperature around 30 °C. The combined effects of global radiation and soil water (Fig. 2.18) show plateaus at both low and high

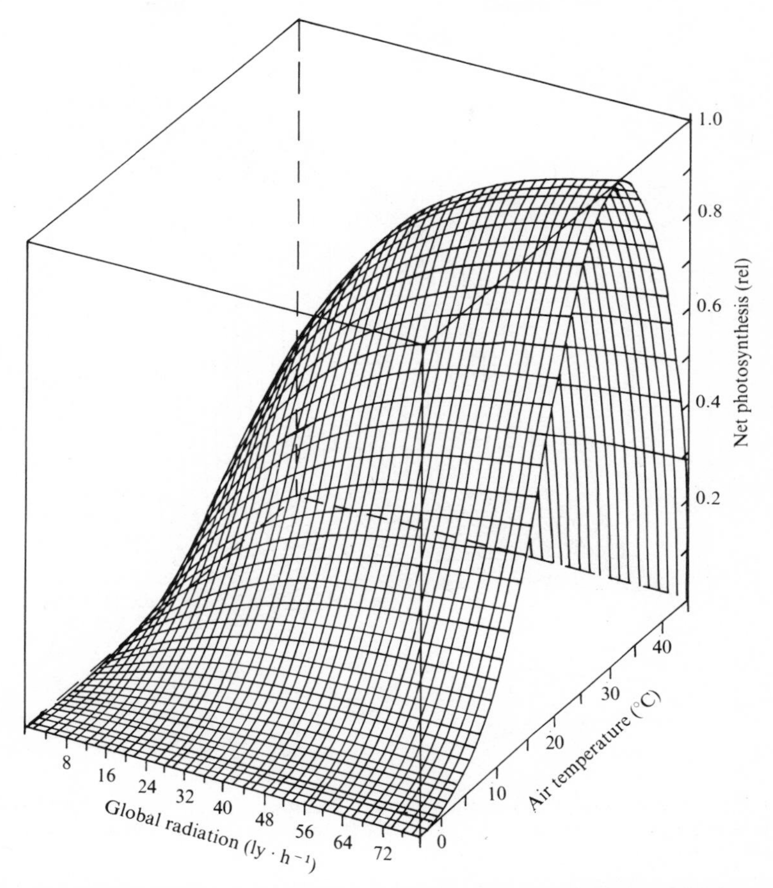


Fig 2.17. Three-dimensional graph showing the relationship of net photosynthesis to air temperature and global radiation based on 1970 data from a Canadian mixed-grass prairie. (Redmann, 1974*a*.)

soil water content and a rapid decline in photosynthesis which appears to occur over a narrow range of intermediate soil water conditions.

### *Biotic controls of primary production*

The previous section has described the effect of various environmental controls on the rate of production. There are also biotic controls on the rate of production which are functions of the plants themselves as opposed to external environmental conditions. It should be clear that this distinction is far from complete. For example, stand morphometry is a biotic control

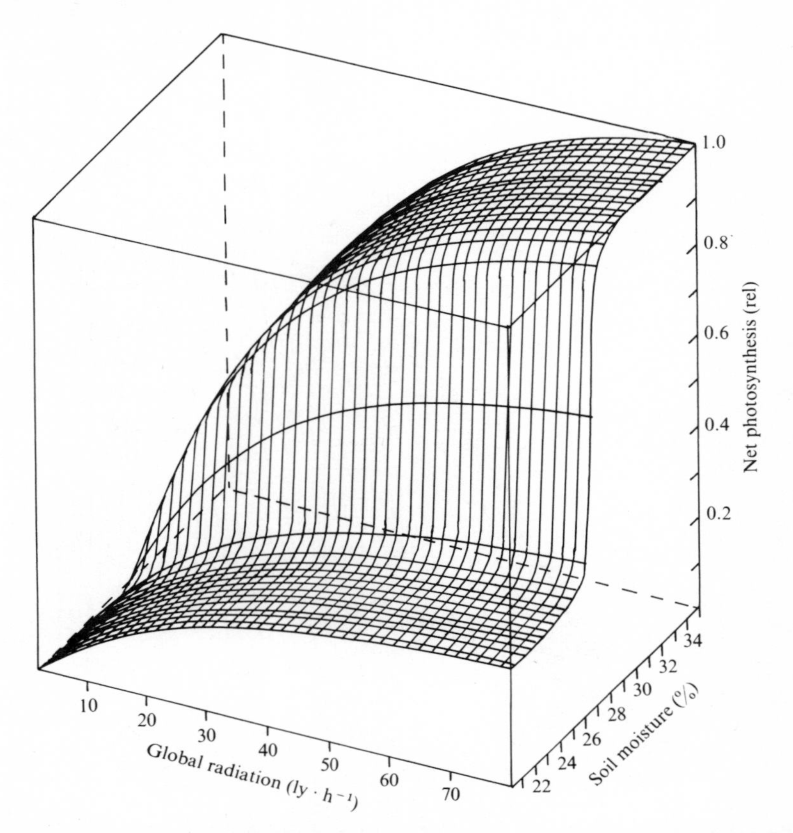


Fig. 2.18. Three-dimensional graph showing relationship of net photosynthesis to global radiation and soil water based on 1970 data from a Canadian mixed-grass prairie. (Redmann, 1974*a*.)

because it is a characteristic of the plants. Nevertheless, the morphometry may be affected by the environmental variables and certainly affects the amount of light received, etc.

Grazing not only reduces photosynthetically active leaf area, but may also cause increased respiration. Increased respiration rates (wound respiration) may then reduce net photosynthesis per unit leaf area remaining. In addition, herbivores usually select the younger, more productive leaves in preference to older mature leaves that are less productive.

Disease and other physiological stresses usually cause reductions in net photosynthesis. This may be caused by increased respiration or damage to the photosynthetic machinery.

Although disease levels of the infestation order of magnitude have been recognized as having a significant effect on primary production, low levels have generally been ignored. In a study of species from a Canadian mixed-grass prairie, Morrall & Howard (1974) produced some preliminary data on the effects of leaf spot disease which is caused by a fungus, *Pyrenophora tritici-repentis*. Experiments in the growth chamber with artificially inoculated plants gave inconsistent results, but there was some suggestion that the disease could cause biomass losses of 20 to 30% in wheatgrass species, even when the leaf area infected was small, i.e. 1 to 5%. The growth chamber studies also suggested that disease accelerated the rate of leaf death, but these results were not supported by a field experiment in which healthy and diseased leaves were tagged and observed through time.

These authors also noted that in the biomass loss experiments, there were significant differences between clones of thick spike wheatgrass and western wheatgrass. In one trial, clones of western wheatgrass showed more resistance than the thick spike wheatgrass. In another trial, differences in biomass loss were revealed even between different clones of western wheatgrass.

### *Phenology*

In general the young vegetative shoots have markedly higher rates of photosynthesis as compared to mature or reproductive shoots. Brown & Trlica (1977*b*) studied the gross and net photosynthesis of blue grama at vegetative and reproductive stages in the shortgrass prairie in Colorado. Their results indicated that the gross and net photosynthesis rates were 46 and 40 mg $CO_2 \cdot dm^{-2} \cdot h^{-1}$, respectively, at a vegetative stage of phenology and 22.5 and 16.5 mg $CO_2 \cdot dm^{-2} \cdot h^{-1}$, respectively, at a reproductive stage of phenology. Thus, the percentage of gross photosynthesis accounted for by net photosynthesis was much less for the reproductive phenological stage illustrating the high metabolic cost of reproduction. Singh & Coleman (1977) have also noted that young shoots of blue grama assimilated more $^{14}C$ as compared to mature shoots.

### *Assimilate concentration*

If assimilate is not transported from the leaf to the sinks, the rate of photosynthesis is decreased, and if new sinks are provided, the photosynthetic rates increase. Beevers (1969) summarizes the effects on photosynthesis caused by the level of assimilate:

(*a*) Photosynthetic rate decreases when leaves are detached, thereby removing the sink.

(*b*) There is a midday depression of photosynthesis when assimilate concentration is high.

(*c*) Photosynthetic rate increases in the remaining leaves following partial defoliation.

(*d*) There is a decreased rate of photosynthesis following interference with translocation or the removal of a sink.

(*e*) There is an influence on the photosynthetic rate dependent on the grafting of different sinks.

It is not clear just how the increased concentration of assimilate inhibits photosynthesis, but present evidence suggests that the photosynthetic rate is decreased when assimilate accumulates to a high concentration in the leaf.

### *Dark and photo-respiration*

As discussed previously, primary production is the result of net photosynthesis where the latter represents gross photosynthesis minus respiration. It would appear that net production could be increased or decreased if the amount of dark respiration and/or photorespiration could be decreased or increased. As most crops mature through the growing season, proportionately more energy may be lost through respiration. However, respiration rates may also be proportional to photosynthetic rates which would indicate decreased respiration as the canopy matures (McCree & Troughton, 1966).

The same trend can be observed in maturing individual plants, as reported by Ishizuka (1969) for cultivated rice plants. When growth of a new leaf had been completed, it was assumed that the amount of respiration carried on by this leaf represented the basal metabolism or the amount of respiration required to maintain the leaf. This value was about 40% of the respiration during the period when the leaf was undergoing rapid growth. Therefore, it was concluded that respiration for growth was 60% of the total respiration of a rapidly growing leaf. If the growth efficiency at this stage is 60%, then it can be shown that the growth efficiency, or the percentage of respiration directly related to growth, decreases as the plant matures.

Brown & Trlica (1977*a*) have compared the aboveground dark respiration of a $C_4$ grass, blue grama, and a $C_3$ grass, western wheatgrass, under various environmental conditions. The overall means of the respiration rates for blue grama and western wheatgrass were 4.5 and 2.0 mg $CO_2 \cdot dm^{-2} \cdot h^{-1}$, respectively. The respiration in both the species decreased with increasing soil water stress and increased with increasing ambient temperature (Fig. 2.19). In blue grama the decrease in the rate was most rapid with the increase of soil water stress from 0 to –15 bars. At 20 °C and at –15 and –30 bars soil water potential, the respiration rates of western wheatgrass were greater than those for blue grama.

In examining the gas exchange rates of single leaves of thick spike wheatgrass, Redmann (1974*b*) found that dark respiration rates at 20 °C and low moisture stress averaged $0.8 \pm 0.4$ mg $CO_2 \cdot dm^{-2} \cdot h^{-1}$. Lowering the temperature to 10 °C resulted in no significant change in the dark respiration rate, whereas raising it to 30 °C doubled the rate to 1.6 mg $CO_2 \cdot dm^{-2} \cdot h^{-1}$. Dark

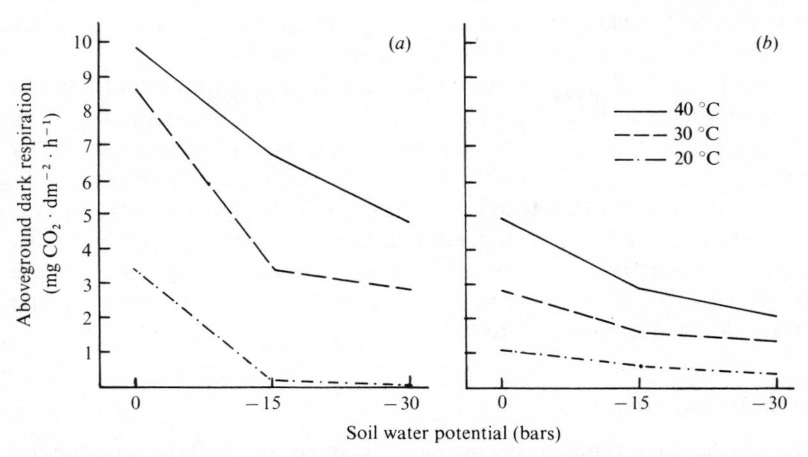


Fig. 2.19. Two-way interaction effects of soil water potential and temperature on the aboveground dark respiration rates of (*a*) blue grama and (*b*) western wheatgrass. (Brown & Trlica, 1977*a*.)

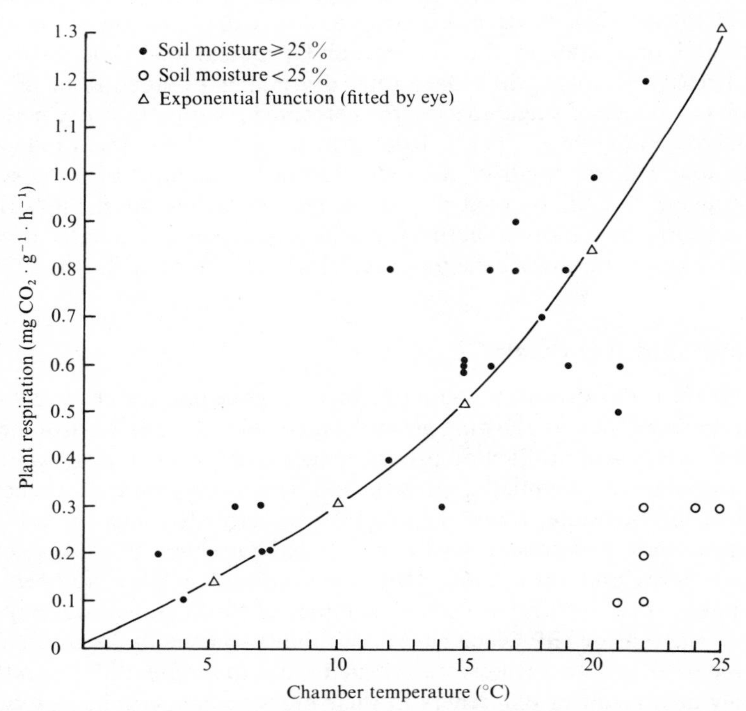


Fig. 2.20. Dark respiration in relation to air temperature based on 1970 field data from a Canadian mixed-grass sward. Ranges of soil water are specified. Exponential function is fitted by eye. (Redmann, 1974*a*.)

respiration rates for swards were about 1.2 mg $CO_2 \cdot g^{-1} \cdot h^{-1}$ for a temperature of about 25 °C as shown in Fig. 2.20 (Redmann, 1974*a*).

Warembourg & Paul (1977) determined the pattern of evolution of labelled carbon from root and microbial respiration in a mixed-grass prairie using $^{14}CO_2$. They found that when the soil was moist, use of labelled carbon by the roots, and consequently $^{14}CO_2$ evolution, was extended over a 3- to 4-week period. Under drier conditions nearly all the labelled carbon was respired during 1 week following assimilation. Respiration of roots plus associated micro-organisms accounted for approximately 9 to 15% of the assimilated carbon. Root-respired $^{14}CO_2$ represented 20 to 29% of the carbon translocated from shoots to roots.

### *Impact of herbage removal*

The effect of herbage removal on primary production has been investigated over the years by a number of authors (Crider, 1955; Neiland & Curtis, 1956; Troughton, 1957; Jameson & Huss, 1959; May, 1960; Frame & Hunt, 1971; Younger, 1972). The general concept has been that too frequent removal of foliage leads to declining yields and that the effect occurs first in the roots and only later in the aboveground production. While close or frequent herbage removal will reduce total dry matter production of individual plants or stands of single species, the percentage of digestible dry matter may be increased (Younger, 1972). Root growth of most species is reduced by defoliation, a direct result of the reduction in the amount of photosynthetically active tissues. In contrast, some recent studies on the US/IBP Grassland Biome have shown that root biomass is as great or greater under the grazed treatment as under the ungrazed treatment (Sims & Singh, 1971).

## Modelling photosynthesis

In recent years a considerable amount of effort has gone into the construction of various types of models allowing generalizations about plant–environment interactions. Models of production processes such as photosynthesis, respiration, partitioning of assimilates, growth and senescence, and death have appeared in the literature. Some models have attempted to incorporate all of these processes, and others have concentrated on single processes such as photosynthesis and respiration. IBP has produced a large number of models; however, we will review some examples of those models developed both inside and outside IBP which have applicability to grassland ecosystems.

The models of photosynthesis described in the literature differ greatly, presumably as a result of differences in their purpose or the points of view and expertise of their designers. Some models are highly mechanistic, in the sense that attempts are made to simulate quantitatively actual processes such

Table 2.6. *A comparison of controlling or driving variables for various canopy or stand simulation photosynthesis models*

| Simulation models and species considered | Control or driving variables | Reference |
|---|---|---|
| Various communities | Irradiance, extinction coefficient and cumulative leaf area index | Monsi & Saeki (1953)<br>Budyko & Gandin (1964)<br>Chartier (1967) |
| *Zea mays* | All of above (except that irradiance is subdivided into direct and diffuse), latitude, time of year | de Wit (1959) |
| *Z. mays* | Light extinction through canopy, leaf area index for each layer, leaf angle for each layer, leaf reflectivity, leaf transmissivity, number of leaf layers, light curve for upper leaves, solar elevation, solar intensity, sky-light brightness, respiration constant 40% of net photosynthesis | Duncan *et al.* (1967) |
| SIMPLE | | |
| *Chloris gayana* | Direct and diffuse radiation penetration into canopy, distribution pattern of foliage, leaf area index | Connor & Cartledge (1970) |
| *Dupontia fischeri* and other arctic species | Energy budget for single leaves, radiation and wind profiles in canopy, leaf area index, air temperature, solar radiation | Miller & Tieszen (1972) |
| ELM | | |
| Primary producer submodel, mixed grassland species | Producer category, soil water potential, average photoperiod temperature, shoot concentration of nitrogen and phosphorus, intercepted insolation, quantum efficiency, light saturation level and phenology | Sauer (1978) |
| Matador | | |
| *Agropyron* spp. | Absorbed radiation, leaf temperature, leaf resistances, leaf area indices of green and dead shoots, leaf water status | Saugier *et al.* (1974) |
| ELCROS | | |
| *Z. mays* | Irradiance, irradiance distribution, photosynthetic curve, leaf area, temperature, age of tissue, availability of water and temperature | de Wit *et al.* (1970) |

as the diffusion of carbon dioxide into the leaf. Other models are semimechanistic or nonmechanistic; their approach may be to fit equations to real data by regression analyses. Precise predictions may be made with the latter type of model, but at the sacrifice of generality and understanding of fundamental mechanisms. Various controlling or driving variables which have been used in modelling photosynthesis are summarized in Table 2.6 for simulation models and in Table 2.7 for semimechanistic and statistical models.

Table 2.7. *A comparison of controlling or driving variables for various canopy or stand semimechanistic and statistical photosynthesis models*

| Semimechanistic and statistical models | Control or driving variables | Reference |
|---|---|---|
| Regression type *Festuca rubra* | Irradiance, temperature, water vapour concentration difference, chlorophyll content and time | Ruetz (1973) |
| Regression type *Bouteloua gracilis* and *Agropyron smithii* | Irradiance, soil water potential and temperature | Brown & Trlica (1974) |
| Integrating $P_N$ over time *Lolium perenne* | Irradiance and temperature | Leafe (1972) |
| Mechanistic and statistical *Panicum obtusum* and *Hilaria mutica* | Temperature, $CO_2$ diffusion resistance, soil water potential, photosynthesis coefficients | Cunningham *et al.* (1974) |
| Mathematical functions and data summations *B. gracilis* | Soil water potential, temperature, visible irradiance, phenology and translocation | Brown & Trlica (1977*c*) |
| Mathematical *B. gracilis* | Leaf area index, leaf angle, irradiance, photosynthesis/ radiation conversion ratio, temperature, and soil water potential | Connor *et al.* (1974) |

Monteith (1972) categorizes studies of crop–weather relationships on the basis of whether 'financial' or 'electrical' analogues are used. He believes that production of dry matter can be accurately estimated using models which are 'financial' statements of the energy, water and carbon balance of plants, but these models can be used only when input data on community structure are known. Changes in plant community structure cannot be predicted until 'we can unravel the interdependent effects of temperature, soil water availability and nutrient supply on plant growth in general, and on the behavior of meristematic tissue in particular' (Monteith, 1972). Electrical analogues are mechanistic in the sense that they describe real fluxes (currents) of carbon dioxide, heat and water vapour which flow in response to concentration (potential) gradients. Resistances to diffusion along these gradients can be calculated.

In the following sections we describe several models at various levels of organization including the molecular and organ levels (leaf models) and the community level (plant canopy models). In addition, some representative statistical models are described.

Table 2.8. *Symbols used in photosynthesis models. Modified from original papers in order to conform with recommended usage*

| Symbol | Definition |
|---|---|
| $P_G$ | Gross photosynthesis |
| $P_N$ | Net photosynthesis, apparent photosynthesis |
| $F$ | $CO_2$ flux into or out of leaf |
| $R$ | Total plant respiration |
| $R_P$ | Peroxisomal (light) respiration |
| $R_D$ | Mitochondrial (dark) respiration |
| $R_L$ | Total respiration in the light ($R_P+R_D$) |
| $C_a$ | $CO_2$ concentration in free air |
| $C_s$ | $CO_2$ concentration in mid stoma |
| $C_i$ | Intracellular $CO_2$ concentration |
| $C_y$ | $CO_2$ concentration 'at point where respiratory $CO_2$ is evolved' (Chartier, 1972) |
| $C_R$ | $CO_2$ concentration at the sites of respiration (Lommen *et al.*, 1971) |
| $C_c$ | $CO_2$ concentration at the chloroplast surface |
| $C_x$ | $CO_2$ concentration at the site of carboxylation |
| $r_a$ | Boundary layer resistance of leaf |
| $r_s$ | Stomatal resistance |
| $r_l$ | Sum of boundary layer and stomatal resistance |
| $r_{sm}$ | Minimum stomatal resistance |
| $r_i$ | Resistance inside junction of $R_L$ and $F$ |
| $r_o$ | Resistance between stomatal cavity and junction of $R_L$ and $F$ |
| $r_m$ | Mesophyll resistance (liquid phase transfer) |
| $r_{sn}$ | Maximum increment attained by an infinite change in $C_s$ |
| $r_x$ | Carboxylation 'resistance' |
| $r_e$ | Excitation 'resistance' |
| $r_{CO_2}$ | Total $CO_2$ transfer resistance |
| $r_{subscript}$ | Other specific resistances defined in the text |
| $I$ | Irradiance |
| $I_0$ | Irradiance at the top of the canopy |
| $T$ | Temperature |
| $T_a$ | Air temperature |
| $T_g$ | Green leaf temperature |
| $W$ | Dry weight of living plant material |
| LAI | Leaf area index |
| $L$ | Downward cumulative leaf area index |
| $Q_{subscript}$ | The rate change of process with 10 deg C increase in temperature |
| $\alpha$ | Maximum efficiency of light energy conversion (initial slope of light response curve) |
| $\beta$ | Mean leaf angle in stand |
| PhAR | Photosynthetically active irradiance (400–700 nm) |
| $K_{subscript}$ | Specific coefficients and parameters defined in text |
| C | Carbon |
| $CH_2O$ | Carbohydrates |
| $\psi_{subscript}$ | Water potential defined in text |
| $t$ | Time |

From Sesták *et al.* (1971).

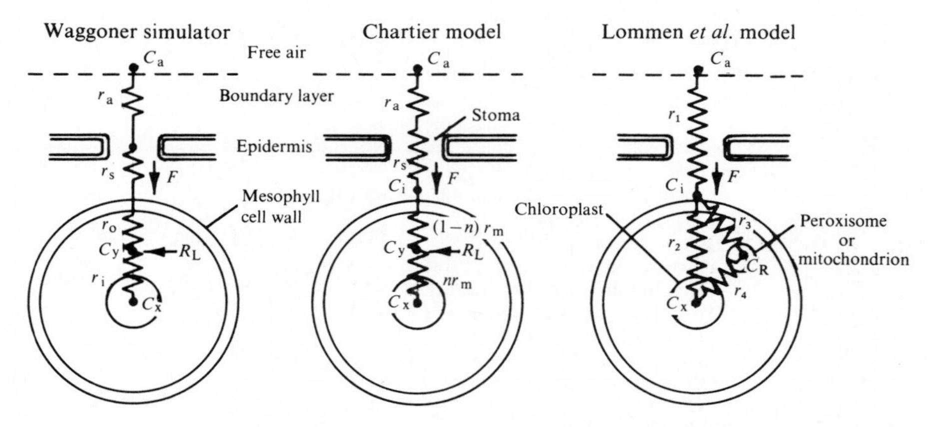


Fig. 2.21. Comparisons of three models depicting carbon dioxide transfers in leaves. Symbols are defined in Table 2.8 and in the text. (Modified from Waggoner, 1969; Chartier, 1972; and Lommen *et al.*, 1971.)

## *Leaf models*

One of the earliest models of carbon dioxide flux and transpiration was that of Penman & Schofield (1951). Later, Gaastra (1959) published a widely cited paper describing the resistances of individual leaves to carbon dioxide and water vapour flux in terms of an electrical analogue. Nearly all subsequent models were based on networks of resistance to gas flow in individual leaves (Lake, 1967; Brown, 1969; Waggoner, 1969; Chartier, 1970, 1972; Lommen, Schwintzer, Yocum & Gates, 1971; Penning de Vries, 1972; Van Bavel, DeMichele & Ahmed, 1973) or in whole plant communities (Monteith, 1963; Duncan, Loomis, Williams & Hanau, 1967; Lemon *et al.*, 1971). The pathways for transfer of carbon dioxide and water and techniques for the measurement of resistances are reviewed by Jarvis (1971). Terminology used in describing various primary productivity models is summarized in Table 2.8. Three leaf models are illustrated in Fig. 2.21.

### *The Waggoner simulator*

In the 'simulator' described by Waggoner (1969) net carbon dioxide flux is calculated from seven other variables using the following equation:

$$F = \frac{C_a - R_L(r_i + r_e)}{r_a + r_s + r_o + r_i + r_x},$$

where $C_a$ is the carbon dioxide concentration in the free air; $r_a$ is boundary layer resistance; $r_s$ is stomatal resistance; $r_e$ is excitation resistance; $r_x$ is carboxylation resistance; $r_o$ is the resistance between stomatal cavities and the junction of $R_L$ and $F$; and $r_i$ is the resistance inside that junction.

This model incorporates a respiration current which was not included in the model of Gaastra (1959). The resistance pattern is modified from Lake (1967). The respiration current ($R_L$) is the sum of dark (mitochondrial) respiration ($R_D$), which is considered to be the same rate in the light and dark, and light (peroxisomal) respiration ($R_P$). The effect of irradiance ($I$) on $R_P$ is described in terms of a Michaelis–Menten equation:

$$R_P = (R_{PX} I)/(K_{RP} + I),$$

where $R_{PX}$ is the maximum light respiration at 20 °C and $K_{RP}$ is the irradiance at half of $R_{PX}$. The effect of temperature ($T$) on both $R_D$ and $R_P$ is expressed through an Arrhenius equation, which for dark respiration is:

$$R_D = R_{DX} \exp\,[(9000)\,(1/293 - 1/T)\,\ln\,(Q_{RD})],$$

where $R_{DX}$ is the maximum dark respiration and $Q_{RD}$ is the $Q_{10}$ for dark respiration (about 2; the $Q_{10}$ for light respiration is considered the same).

Resistances are calculated using data from the literature. Boundary layer resistance is taken as 30 $s \cdot m^{-1}$. Stomatal resistance is made dependent on light and carbon dioxide but independent of temperature:

$$r_S = \left[ r_{sm} + r_{sn} \frac{C_S}{K_{sc} + C_S} \right] \left[ \frac{K_{sI} + I}{I} \right],$$

where $r_{sm}$ is minimum stomatal resistance and $r_{sn}$ is the maximum increment attained by an infinite change in carbon dioxide concentration in mid stoma ($C_S$); $K_{sc}$ is the $C_S$ at half $r_{sn}$; and $K_{sI}$ is the irradiance that increases stomatal conductance ($1/r_S$) to half of maximum.

The mesophyll resistance is considered to be made up of $r_o$, $r_i$ and $r_X$. The first two resistances are assumed independent of environment and made constant at 200 $s \cdot m^{-1}$. The final resistance ($r_X$) is chemical rather than physical in nature. Waggoner (1969) uses the Michaelis–Menten equation to express the response of $r_X$ to light and carbon dioxide:

$$r_X = (K_{cX} + C_X)\,[(K_{cI} + I)/I]\,\mathrm{Arrh}(P_G, Q_P, T)^{-1},$$

where $C_X$ is the carbon dioxide concentration at the site of carboxylation. Other terms not defined previously are $K_{cX}$, the concentration of carbon dioxide for half of the maximum gross photosynthesis at 20 °C ($P_G$); $K_{cI}$, the irradiance for half $P_G$; and $Q_P$, the $Q_{10}$ for photosynthesis (2). The last term is an Arrhenius expression used to incorporate a temperature effect. Measurement of these parameters is extremely difficult because the plant must have no respiration and no resistance but $r_X$. Waggoner (1969) used data for maize to obtain the best estimate for $r_X$.

The next steps involve the calculation of $r_s$, $r_X$ and $F$ based on assumed values of $C_s$ and $C_x$. In turn, $C_s$ and $C_x$ can be calculated using the estimated $F$:

$$C_s = C_a - F(r_a + r_s/2),$$

and

$$C_x = (F + R_L)\, r_x.$$

The estimates of $r_s$, $r_x$ and $F$ are improved by iteration until the last revision of $F$ changes no more than a small amount.

After testing the model to ensure that it performed realistically, it was used to predict the outcome of changes in physiological parameters. The model predicted changes to net photosynthesis of more than 20% as a result of reasonable changes in light respiration. When dark respiration was varied over the ranges observed in nature, net photosynthesis was significantly affected. Both a decrease in the resistance between the respiratory current and the photosynthetic site ($r_i$), and a decrease in the minimum stomatal resistance ($r_{sm}$) were effective in increasing net photosynthesis.

Among the factors affecting chemical aspects of the model, the maximum gross photosynthesis ($P_G$) was surprisingly ineffective. On the other hand, changes in the parameters affecting the curvature of the carbon dioxide and light response curve ($K_{cx}$ and $K_{cI}$) were surprisingly effective.

The model had the advantage of incorporating many of the interactions encountered within a leaf which complicate predictions of net photosynthesis based on changes in individual factors in conventional laboratory and field experiments. Waggoner (1969) concluded that the simulator behaved like a real leaf, demonstrating the fairly satisfactory state of knowledge of carbon dioxide assimilation.

### *The Lommen* et al. *model*

Lommen *et al.* (1971) designed a model for predicting net photosynthesis of leaves for a variety of environmental conditions. The model is based on an electrical analogue describing carbon dioxide diffusion and a Michaelis–Menten equation describing carbon dioxide fixation. A simplified version without respiration and a more complex form including respiration are presented.

Both versions are based on two fundamental relationships common to all mechanistic models described in the literature. The first of these is Fick's law of physical diffusion:

$$F = \frac{C_a - C_c}{r_{CO_2}},$$

where the carbon dioxide flux is related to its concentration gradient between the free air outside the leaf ($C_a$) and at the chloroplasts ($C_c$) divided by the total resistance to its diffusion ($r_{CO_2}$) along the pathway from the atmosphere to the chloroplasts. The second fundamental relationship describes the chemical process of carbon dioxide fixation in terms of the Michaelis–Menten equation for the rate of an enzymatic reaction:

$$F = F_c/(1+K/C_c),$$

where $F_c$ is the carbon dioxide flux at saturating $C_c$, and $K$ is a constant equal to the carbon dioxide concentration in the chloroplast at which $F = F_c/2$. The two equations are combined after solving the second for $C_c$, and the resulting quadratic equation is solved for $F$. The maximum rate of photosynthesis ($F_c$) is made a function of irradiance and temperature:

$$F_c(I, T) = F_{CIT}G(T)/(1+K_I/I),$$

where $F_{CIT}$ is carbon dioxide flux at saturating $C_c$, saturating $I$ and optimum $T$; $G(T)$ is a term for relative values of the maximum rate of photosynthesis as a function of temperature derived from data in the literature; and $K_I$ is the irradiance at which carbon dioxide flux is half the value at saturating $I$.

The resistance network in the first version of the model of Lommen *et al.* (1971) is relatively simple, consisting only of $r_1$ (made up of boundary layer and stomatal resistances) and $r_m$ (all liquid phase resistances). Boundary layer resistance changes with wind speed, leaf size and leaf orientation; stomatal resistance is a function of light, following the inverted Michaelis–Menten equation described in Waggoner's (1969) simulator. No dependencies of $r_s$ on temperature, carbon dioxide, or leaf water status are included because according to Lommen *et al.* (1971) these relationships are complicated and relatively unclear. The exclusion of water status effects is perhaps the greatest problem in directly applying most published models to the grassland situation, where water stress is usually a dominant factor. An attempt by Saugier, Ripley & Lueke (1974) to include water status effects will be described in a subsequent section.

A more complex version of the model by Lommen *et al.* (1971) incorporates respiration (Fig. 2.21). With the addition of respiration, the resistance network (Lake, 1967) becomes more complicated. Resistance $r_1$ remains the same as in the simple version. Resistances $r_2$ and $r_3$ are the direct resistances between the intercellular air spaces and the sites of photosynthesis and respiration, respectively. The direct resistance between the sites of respiration and the chloroplasts is $r_4$. The indirect diffusion paths from the intercellular spaces to the chloroplasts are represented by $r_3+r_4$. Thus, following the electrical analogue, the net diffusive resistance between the intercellular air spaces and the chloroplasts is $r_2$ in parallel with $(r_3+r_4)$.

A criticism of this resistance network was offered by Samish & Koller (1968), who objected to introducing three resistances where it is not even possible to measure one well. Lommen *et al.* (1971) conceded that this criticism was valid. In their model $r_2$ is assumed to equal $r_3$, since the sites of both respiration and photosynthesis are located in the relatively thin layer of cytoplasm near the cell wall and the diffusion paths are of similar length. Because $r_4$ does not contain a cell wall component, and the chloroplasts and peroxisomes are often in close proximity, it is considered less than $r_2$. Assumed values for $r_2$ and $r_3$ were 500 s · m$^{-1}$ and for $r_4$ 100 s · m$^{-1}$.

Lommen *et al.* (1971) cite evidence from the literature that $R_D$ is almost completely suppressed in the light except at low light intensities; a negative exponential relationship between $R_D$ and light is used in the model. $R_D$ is also made a function of temperature, but is considered independent of carbon dioxide and oxygen concentrations in the air. Light-induced respiration is made a function of light, temperature and oxygen concentrations. Light dependence of $R_P$ is given the Michaelis–Menten form, whereas temperature dependence is exponential. $R_P$ increases linearly with increasing oxygen concentration in the air.

Output of the model was illustrated as curves based on ranges of values of environmental and physiological parameters. The curves were similar to experimental results appearing in the literature. As examples: the light compensation point decreased with increasing atmospheric carbon dioxide concentrations; light respiration lowered maximum rates of net photosynthesis and resulted in negative net photosynthesis at low carbon dioxide concentrations in the air. The shapes of the curves relating net photosynthesis to carbon dioxide concentration were dependent on whether carbon dioxide diffusion or carbon dioxide fixation was limiting. Curves made up of two linear portions connected by a sharp elbow ('Blackman curves') resulted when resistance to carbon dioxide uptake controlled the rate of photosynthesis, e.g. when $K$ (the constant in the Michaelis–Menten equation) was very small or $r_{CO_2}$ was large. 'Michaelis–Menten curves' resulted when biochemical reactions controlled the rate of photosynthesis (large $K$ or very small $r_{CO_2}$).

Behaviour of certain aspects of the model such as light-induced respiration was difficult to verify because of the difficulty of experimentally determining the magnitude of total respiration. Recycling of carbon dioxide within the leaf appears to be a major problem in this and other models.

### *The Chartier model*

The model of Chartier (1972) also describes the biochemical and physical aspects of the photosynthetic mechanism in leaves. The carbon dioxide which is used in chemical reactions is considered to be equal to the amount which arrives by diffusion from the air and the respiration sites.

The model (Fig. 2.21) includes the concentration of carbon dioxide at the point 'where the respiratory $CO_2$ is evolved' ($C_y$) and its concentrations at the carboxylation sites ($C_x$).

Net carbon dioxide flux ($F$) is written following Gaastra's (1959) method, as:

$$F = \frac{C_a - C_y}{r_a + r_s + (1-n) r_m},$$

and

$$F + R_L = \frac{C_y - C_x}{n r_m},$$

where $n$ is that fraction of the mesophyll resistance ($r_m$) between the sites where respiratory carbon dioxide is evolved and the sites of carboxylation (equivalent to the $r_i$ of Waggoner).

The concentration $C_x$ is not fixed at zero, or considered equal to the carbon dioxide compensation point, but varies as follows:

$$C_x = C_a - F(r_a + r_s + r_m) - nR_L r_m.$$

An equation for photosynthesis at the chemical reaction level is given as:

$$F + R_L = (\alpha I C_x / r_x)/(\alpha I + C_x/r_x),$$

where $\alpha$ is the maximum efficiency of light energy conversion. The equation is essentially identical to the Michaelis–Menten equation used by Lommen *et al.* (1971). The expression for $C_x$ is substituted in the last equation, yielding a nonrectangular hyperbolic relationship between $F$ and $I$:

$$F + R_L = \frac{\alpha I [C_a - F(r_a + r_s + r_m) - nR_L r_m / r_x]}{\alpha I + [C_a - F(r_a + r_s + r_m) - nR_L r_m / r_x]}.$$

Under conditions of light saturation, this equation reduces to:

$$F = \frac{C_a - R_L(n r_m + r_x)}{r_a + r_s + r_m + r_x}.$$

In the absence of photorespiration the equation is:

$$F = \frac{C_a}{r_a + r_s + r_m + r_x}.$$

The experimental procedure whereby transfer ($r_m$) and chemical ($r_x$) components of intracellular resistance are separated was explained by Chartier (1970).

In the case of $C_3$ plants, where the intensity of photorespiration is great, Chartier (1970) concluded that $(r_x + n r_m)$ was relatively high compared with $[r_a + r_s + (1-n) r_m]$. In $C_4$ plants, which exhibit higher net photosynthesis, the $(r_x + n r_m)$ term was not low enough to give the fast recycling which could completely account for the greater net photosynthesis. Rather, it was the absence of photorespiration which appeared to produce the higher apparent photosynthesis. Waggoner (1969) indicated that substantial increases in photosynthesis could be obtained by reducing the $[r_i(n r_m)]$ term.

Intracellular transfer resistance is lower in $C_4$ than in $C_3$ plants. For example, average values of $r_m$ and $r_x$ for three $C_4$ grasses (elephant-grass (*Pennisetum purpureum*), guinea grass panic (*Panicum maximum*) and buffed sandbur (*Cenchrus ciliaris*) were 34 and 27 $s \cdot m^{-1}$, respectively; whereas values for a $C_3$ legume (*Calopogonium mucumoides*) were 221 and 24 $s \cdot m^{-1}$, respectively (data from Ludlow, 1971). Chartier (1972) believed that the differences in intracellular transfer resistance between $C_4$ and $C_3$ plants could be related to the location of carbon dioxide acceptors, and corresponding enzymes, and their different affinities for carbon dioxide molecules. The structure of the chloroplasts would be important in this regard.

This approach, involving the calculations of $r_m$ and $r_x$ as in the Chartier (1972) model and that of Koller (1970), is appealing because it incorporates much information about carbon dioxide assimilation in a form that helps give a clearer picture of a complex process, particularly the partitioning of resistances. However, Van Bavel (1975) points out that these models include certain unverifiable assumptions and concludes that attempts to model detailed cellular processes might be premature. Laetsch (1974) has pointed out that much more information is needed before a clear picture of structure–function relationships in $C_3$ and $C_4$ plants can be obtained.

### *Leaf models – other examples*

Brown (1969) described a model of the relationship between net carbon dioxide flux and irradiance at a given concentration of carbon dioxide in the air. The model predicts the sum of diffusion resistances, the capacity of the leaf to fix carbon dioxide ($K$), the carbon dioxide concentration at the site of carboxylation, and the respiration rate ($R_L$). Brown derived a general hyperbolic equation relating carbon dioxide flux to irradiance and carbon dioxide concentration:

$$F = (KIC_a - R_L)/(KI/r_{CO_2} + 1).$$

Resistance to the transfer of carbon dioxide from the site of respiration to the site of photosynthesis is assumed to be negligible, and respiration in the light is assumed independent of irradiance. In Brown's analysis a constant $r_{CO_2}$ was assumed, because all data were obtained at intensities above the minimum required to induce stomatal opening.

The object of Brown's (1969) analysis was to evaluate $C_x$, $K$, $r_{CO_2}$ and $R_L$, using $F$ against $I$ data from the literature. The unknowns were solved by using an iterative procedure. $C_x$ at high light intensities ranged from 32 to 144 $\mu l \cdot l^{-1}$, depending on the species. Resulting levels of mesophyll resistance ($r_m$) were lower than those calculated using the assumption that $C_x$ equals zero. Plants which had higher net photosynthetic rates at a given light intensity and $C_a$, had greater values of $C_x$ than those with lower net photosynthetic rates. The $r_{CO_2}$ was inversely related to the rate of net photosynthesis. Values of $r_m$ calculated for cotton (*Gossypium hirsutum*) were inversely related to $C_a$, suggesting that the transfer of carbon dioxide in the cell may involve a concentration-dependent chemical reaction in addition to, or rather than, a physical diffusion process. Calculated respiration rates of wheat (*Triticum aestivum*) in the light were twice those measured in the dark.

Brown's (1969) model was basically an attempt to put data from the literature into a theoretical framework. Although variability in data because of differences in nutrition, water stress, age, and light and temperature preconditioning complicated the approach, predicted and actual results agreed well. Van Bavel (1975) has used the same basic approach as Brown (1969), describing and verifying two equations for calculating carbon dioxide flux density of a leaf based on irradiance, ambient carbon dioxide level and leaf diffusion resistance. The equations are 'primarily intended for use in comprehensive models of plant and canopy behavior, and of productivity and water use of plant communities' (Van Bavel, 1975).

A steady-state simulation model of a leaf based on the physical and physiological processes of water, carbon dioxide and energy exchange between the leaf and the environment was developed by Van Bavel *et al.* (1973). The effects of leaf water potential, leaf temperature and leaf internal carbon dioxide level on the stomatal aperture and diffusion resistance were incorporated in this model. The model also accounts for the carbon dioxide compensation point and the responses of the photosynthetic process to light, temperature and carbon dioxide. Model output suggested that air humidity and ambient carbon dioxide level were the principal environmental parameters that significantly affected water use efficiency. Light and total radiation load were also important factors. A major limitation of the model is that it is based on steady-state conditions, whereas the plant usually experiences ever-changing conditions. Therefore, tests of the model can only be made over short periods of time and in a steady, controlled environment.

Charles-Edwards & Ludwig (1974) describe a model for leaf photosynthesis and respiration by $C_3$ plant species based on relationships between light flux density and concentrations of carbon dioxide and oxygen in the mesophyll tissue. Coefficients in the model were selected so as to produce response curves similar to those observed experimentally. The model is empirical; however, the authors point out that the approach does have a basis in enzyme kinetics.

### *Modelling plant respiration*

In earlier models of plant productivity, respiration was neglected or taken as a constant (Davidson & Philip, 1958; Ross, 1964; Budyko & Gandin, 1964). However, in more recent models, the effect of temperature on respiration (de Wit, 1966; Ross & Bichele, 1969; Kuroiwa, 1970), the different rates of respiration in sunlit and shaded leaves (Duncan *et al.*, 1967) and the influence of leaf age and leaf adaptation to light (Ross & Bichele, 1969) have been considered.

McCree & Troughton (1966) studied photosynthesis and respiration of white clover (*Trifolium repens*) and found that the rate of respiration became proportional to the rate of photosynthesis when light intensity was varied. They were, therefore, critical of models where respiration was considered to be proportional to leaf area and independent of growth rate or light level. They found that output from the model of de Wit (1959) in which respiration was considered to be a constant proportion of photosynthesis agreed with their study results. In a later paper, McCree (1970) quantified the dependence of respiration rate ($R_L$) on photosynthesis rate, dry weight and time as:

$$R_L = 0.25P_N + 0.015W,$$

where $P_N$ is net photosynthesis. That is, at any time, respiration rate was 25% of the photosynthetic rate plus 1.5% of the dry weight of living plant material ($W$). The first term represents growth respiration, whereas the second term represents maintenance respiration. Changes in $R_L$ lagged several hours behind changes in $P_N$. Changes in $P_N$, $R_L$ and $W$ were highly dependent on the rates of death and regrowth of the various plant organs. It was felt that these factors were important in computer simulations of production.

De Wit, Brouwer & Penning de Vries (1970), using a similar approach to respiration in their ELCROS model, found that the coefficients for growth respiration and maintenance respiration derived from McCree's (1970) data gave good results for the simulated growth of corn. Loomis (1970), in discussing the respiration component of the ELCROS model, pointed out that (1) it would be a mistake to assume that the McCree (1970) equation holds for other situations, and (2) in nature respiration may be generally much lower, relative to photosynthesis, than in laboratory situations.

Penning de Vries (1974) considered the problem of respiration in relation to growth and maintenance in plants which arises during efforts to model carbon balance in plant stands. The major processes involving oxygen consumption or carbon dioxide release were judged to be:

(1) Biochemical conversion of photosynthetically produced substrate into biomass (growth respiration). These biochemical processes are well known; the amount of biomass produced can be calculated using the reaction equations. Penning de Vries (1974) cited experimental evidence to confirm that this approach was valid.

(2) Maintenance of already existing cells and their structures (maintenance respiration). The intensity, especially in rapidly growing tissue, appears to be low compared to animals and micro-organisms. The cost of maintenance

processes probably depends on external conditions such as temperature, salinity and water stress. However, more research in this area is needed.

(3) Active transport of organic compounds across cell membranes and in phloem vessels, especially into growing cells which do not produce their own substrate material. The transport cost may be high when large distances are involved.

(4) Photorespiration, which is deducted from gross photosynthesis.

There is abundant evidence that, in plants, respiration is regulated to meet changing metabolic demands through an intimate coupling of respiration to phosphorylation (Beevers, 1961). Future efforts in modelling must recognize the complexity of the respiration process, but still try to get overall relationships which can be expressed in quantitative terms. The work of Penning de Vries (1974, 1975) appears to be an important step in that direction.

### *Canopy and stand models*

Canopy and stand models differ significantly from leaf models in that additional levels of complexity are involved (Table 2.9). No longer are we concerned with photosynthesis of an individual leaf, but of the dynamic changes in photosynthesis of an entire plant community. Therefore, such factors as solar elevation, light penetration into the canopy, spatial distribution of leaves, leaf angles, leaf age and physiological conditions, carbon dioxide concentrations and temperature gradients within the canopy, species differences and contributions to the total canopy, etc., assume paramount roles in community productivity. In dealing with these additional levels of complexity, most modellers have chosen to simplify the more basic mechanisms of carbon dioxide diffusion, photochemical processes and biochemical reactions within leaves that were evident in leaf models. These simplifications are probably justified in dealing with the more complex interactions between the plant canopy and environment.

Monsi & Saeki (1953) and Davidson & Philip (1958) were some of the first researchers to develop models of photosynthesis in plant communities that predicted, for a given light condition, that there was an optimum leaf area for community photosynthesis. Increasing leaf area to the optimum leaf area index caused increased rates of production of organic matter. However, increasing leaf area beyond the optimum caused a decline in productivity, presumably resulting from mutual shading of leaves and the increased respiratory burden.

In several models of plant and canopy photosynthesis (Monsi & Saeki, 1953; Budyko & Gandin, 1964; Chartier, 1967), the photosynthetically active radiation (PhAR) flux inside the canopy is given by the exponential formula:

$$I(z) = I_0 \exp[-KL(z)],$$

Table 2.9. *Hierarchical ordering of common autotrophic processes for comparisons of photosynthesis models*

| Process | Controlling factors for leaf models | Controlling factors for community or stand models |
|---|---|---|
| $CO_2$ diffusion | Various leaf and boundary layer resistances | Leaf, boundary layer and canopy resistances |
| $CO_2$ fixation ($P_G$ or $P_N$) | Irradiance, temperature, $CO_2$, $O_2$, leaf water potential, resistances and nutrients | Irradiance and temperature within various canopy layers, $CO_2$, $O_2$, leaf or soil water potentials, leaf and canopy resistances, nutrient availability, leaf angles and distributions, proportion of living and dead photosynthetic material, phenology or leaf ages, wind speed, humidity, previous environmental conditions, physiological status and species composition |
| Respiration (light and dark) | Irradiance, temperature, resistances, leaf water potential and concentration of substrates | Irradiance and temperature within various canopy layers, leaf or soil water potential, proportion of living and dead material, phenology, concentration of substrates, physiological status and species composition |

where $I$ is the irradiance at height $z$; $I_0$ is the irradiance at the top of the canopy; $L$ denotes the downward cumulative leaf area index from the top of the canopy to the level $z$; and $K$ is the extinction coefficient. This formula was derived utilizing the assumption that the radiation field is isotropic and homogeneous in the horizontal plane and that the distribution of photo-elements in the stand is expressed by the Poisson law.

One of the simplest models of the photosynthetic plant community is that of Kasanaga & Monsi (1954). The foliage of the plant consisted of a number of leaf-planes. In a leaf-plane, there were $u$ leaves (mean leaf area $f$) per unit land area, and these were situated horizontally, without shading each other, and regularly spaced. In this model, the practical profile of the photosynthetic system of a plant community, based on results gathered by a stratified clip technique, could be obtained by changing the area and number of leaves in a leaf-plane, and the number and distance of leaf-planes.

Net photosynthesis ($P_N$) per unit leaf area in the above model was expressed by Monsi (1968) by the following equation:

$$P_N = bI_{1+aI} - R_L,$$

where $R_L$ is respiration in the light per unit leaf area, $I$ is irradiance and $a$ and $b$ are coefficients determining the shape of the curve. Total net photosynthesis would then be a summation of all the net photosynthesis for the various leaf-planes. This simplified model was effective in elucidating the general feature of foliage photosynthesis in relation to leaf area index. It also gave some idea of the influence of leaf arrangement on foliage photosynthesis, but no idea of the impact of leaf inclination on dry matter production of the plant community. Iwaki, Nakajima & Monsi (1964) stated that total photosynthesis calculated by taking into account the mean photosynthetic activity of mature leaves only was accurate enough compared with that obtained after correcting for leaf aging.

De Wit (1959) was probably the first to use a model in which the absorption of diffused skylight was treated separately from direct incoming radiation. In his model, the foliage was treated as randomly oriented leaves of a single layer sufficiently thick to intercept all the light. Given the solar elevation, totals for direct and diffuse radiation, and a Blackman-style light response curve approximating the behaviour of many crop species, he developed tables of potential productivity for various latitudes and times of the year. Stanhill (1962) found a satisfactory fit with experimental data using the model developed by de Wit (1959). However, Williams, Loomis & Lepley (1965) observed that the model substantially underestimated their experimental values for photosynthesis in maize under high levels of incoming radiation.

A simulation model for photosynthesis of plant communities was developed by Duncan *et al.* (1967). Their model was used for estimation of photosynthesis for each hour of the day with different values for one or more variables of the plant community or the environment. Plant community variables included leaf area index, leaf angle for each layer, leaf reflectivity, leaf transmissivity, number of layers, and a light curve for upper leaves. Meteorological variables included solar elevation, solar intensity and skylight brightness. The model assumed that 40 % of net photosynthesis would be lost to respiration. They compared model outputs with observed rates of dry matter production for maize. Correlation analysis of data for all observed and computed points from the actual field data and model-predicted data indicated a highly significant $r$ value. The computer simulations were remarkably similar to the actual data, both in absolute values and in qualitative behaviour.

Connor & Cartledge (1970) developed a model to relate to photosynthetic contribution of the leaves of Rhodes grass (*Chloris gayana*) to direct and diffuse radiation and to the distribution pattern of the foliage. Their mathematical models of photosynthetic rates were developed using photosynthetic response functions from actual data. Predicted rates of photosynthesis, utilizing the model, were shown to agree with photosynthetic rates measured in the field. They found that communities with a leaf area index (LAI) greater than 2.5 could be adequately represented by a single, continuous, horizontal foliage layer. Calculation of the photosynthesis of plant communities from photosynthetic light response functions of individual leaves required a detailed treatment of penetration of radiation into the canopy because the form of the response function was markedly nonlinear over the range of light intensity usually incident upon the leaves. However, in the case of tropical grasses, in which the photosynthetic response is approximately linear over a wide range of light intensities, light saturation of leaves in the canopy was less likely. With the exception of two communities of low LAI ($<2.0$), the strategy of their SIMPLE program provided an adequate representation of the photosynthetic surface of the community because of the linearity of the photosynthetic response of individual leaves of Rhodes grass at these radiation flux densities. The canopy structure of these tropical grass communities could, therefore, be simply represented as a continuous horizontal photosynthetic surface for which a representative photosynthetic response function was available. However, the strategy of their SIMPLE program might have to be modified in order to handle communities which do not have an LAI of at least 2.5.

Miller & Tieszen (1972) developed a preliminary model of physical processes affecting primary production in the Arctic tundra. Their model utilized the energy budget equation for single leaves, equations expressing the radiation and wind profiles in the canopy, and a simplified concept of the allocation of photosynthate to stems and leaves. The model was validated with data from an Arctic ecosystem at Point Barrow, Alaska, consisting primarily of grass species. Their model indicated that calculated net production increased as leaf area increased in spite of decreasing solar radiation through the growing season. Production also increased with decreased standing dead material, increased LAI, increased solar radiation and increased air temperature up to 10 °C. The model could, therefore, be used to generate a surface of net production as a function of solar radiation and air temperature. However, seemingly subtle changes in the input information often caused large changes in the trend in the relations between variables. In addition, their model indicated that the distinction between sunlit and shaded leaves was not critical for the Arctic tundra because the temperatures of these leaves did not vary greatly.

### *ELM submodel*

Photosynthesis of grassland species was simulated in the producer section of the ELM model by Sauer (1978). The ELM simulation model was developed by personnel of the US/IBP Grassland Biome and is a comprehensive grassland ecosystem model that is highly nonlinear and that can be adapted to most grasslands through changes in parameters and initial values. The objectives of the producer section of the ELM model constructed by Sauer were to provide biologically realistic phenology and biomass (carbon) responses to abiotic and biotic variables of grassland ecosystems. The phenology submodel simulated the developmental state of each producer category and thus allowed for seasonal differences in the activity of several producers. The carbon submodel simulated the herbage dynamics of several producer categories as affected by both abiotic and biotic variables.

Sauer's (1978) objective for the carbon submodel was to simulate the interseasonal carbon dynamics of up to 10 producer categories. Producer categories were designated in this particular submodel by Sauer as warm-season grasses, cool-season grasses, forbs, shrubs and cacti. The response of the submodel was to be biologically reasonable for weather conditions that varied from the worst drought that might be expected in a 10-year period to conditions where soil water, nitrogen and phosphorus were nonlimiting. In addition, responses to grazing were supposed to show the expected trends from data and experience. Multi-year simulations to evaluate the cumulative effects of various treatments were desired; therefore, the submodel was required to have an overwintering capacity. Atmospheric carbon moved, by way of the process of photosynthesis, to the shoot system where it was distributed to the live shoots, standing dead, storage structures (crowns), seeds and roots. In the root system, root respiration returned carbon to the atmosphere directly and the process of root death moved live carbon to a belowground litter compartment. The shoot system included leaves and stems of various producer categories. A flow diagram of Sauer's producer submodel of ELM is shown in Fig. 2.22.

The daily $P_G$($g\ C \cdot m^{-2} \cdot d^{-1}$) of each producer category was a function of the effect of soil water potential, average photoperiod temperature, shoot concentration of nitrogen (N), shoot concentration of phosphorus (P), intercepted insolation, quantum efficiency, light saturation level and phenology. Gross photosynthesis was calculated by taking the minimum value of from 0.0 to 1.0 for the effect of the following variables: soil water potential, average photoperiod temperature, N concentration, P concentration and phenology. Therefore, any one of the five preceding variables could be limiting gross photosynthesis. This minimum value of one of the above scaling factors was then multiplied by a minimum insolation value for the day which was either a minimum value of insolation ($cal \cdot cm^{-2} \cdot d^{-1}$) for a maximum gross photosynthetic rate, or the actual insolation available for interception by the canopy on that particular day, whichever was lowest. These values were then multiplied by the fraction of insolation actually intercepted by the vegetation times the quantum efficiency for the producer category ($mol\ C \cdot Einstein^{-1}$) for each producer. Gross photosynthesis was then converted from $cal \cdot cm^{-2}$ to $g\ C \cdot m^{-2}$ by using a constant of 1.032. The concept of gross photosynthetic rates being limited by available photosynthetically active incident radiation used in this submodel was based largely on the

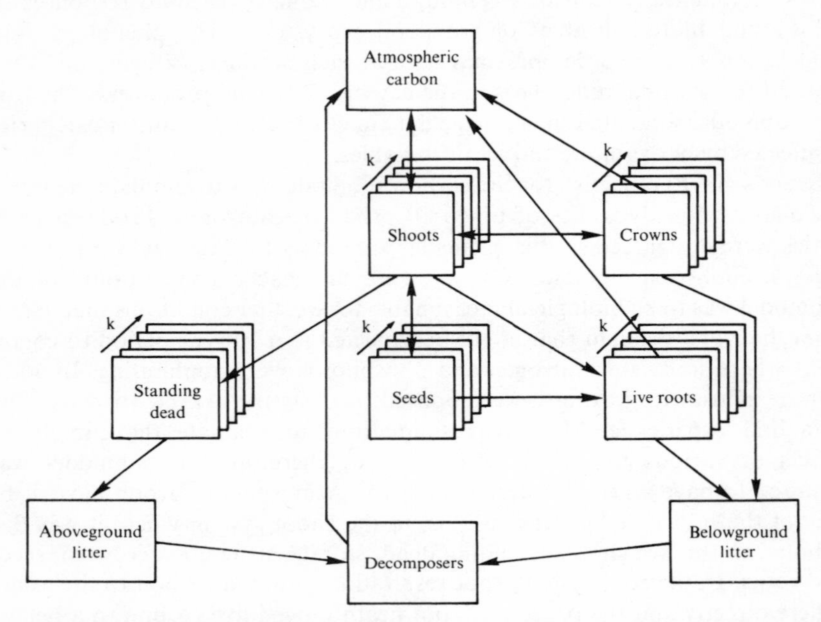


Fig. 2.22. Flow diagram for carbon through the compartments of the primary producer submodel for ELM. k, producers 1 to 10. All compartments are species-specific except the dead root and litter compartments which are common to all producers. (Sauer, 1978.)

work of Connor (1973). The actual gross photosynthesis equation utilized by Sauer is shown below:

$$PGROS_k = \text{minimum of} \begin{cases} PESM_k \\ PEAT_k \\ PNS_k \cdot \text{minimum of} \begin{cases} PSAT_k \\ PAIE \end{cases} \cdot PILS \cdot PQEFF_k \cdot 1.032, \\ PPS_k \\ PEP_k \end{cases}$$

where $PGROS_k$ is the daily photosynthetic rate ($g\ C \cdot m^{-2} \cdot d^{-1}$); $PESM_k$ is the effect of soil water potential; $PEAT_k$ is the effect of average photoperiod temperature; $PNS_k$ and $PPS_k$ are the effects of N and P shoot concentrations, respectively; $PEP_k$ is the effect of phenology; $PSAT_k$ is the minimum value of insolation ($cal \cdot cm^{-2} \cdot d^{-1}$) for a maximum gross photosynthetic rate; PAIE is the insolation ($cal \cdot cm^{-2} \cdot d^{-1}$) available for interception by the canopy; $PILS_k$ is the fraction of insolation intercepted (dimensionless); $PQEFF_k$ is the quantum efficiency (mol C $\cdot$ Einstein$^{-1}$); and k is the producer. The constant 1.032 converts $cal \cdot cm^{-2}$ to $g\ C \cdot m^{-2}$ (Loomis & Williams, 1963).

Sauer (1978) is probably the first person who has simultaneously combined the effects of these various factors described above to regulate photosynthesis on the level of species for producer groups. Choosing a minimum value from a list of variables appeared to work well when the time step was in days and one was interested in seasonal rather than daily changes.

The temperature value used in regulating daily gross photosynthetic rates was the average photoperiod canopy temperature. Average canopy temperature was determined utilizing photoperiod length and maximum and minimum air temperatures. For each producer category, minimum, optimum and maximum photoperiod temperatures for net photosynthetic rates were assigned. A linear interpolation between the minimum and optimum values, using the average photoperiod temperature as the argument, gave a value for the effect of average photoperiod temperature which was scaled between 0.0 and 1.0.

The effects of shoot concentration of N and P on primary producer activity of the primary producer submodel (Sauer, 1978) depended upon the dynamics of N and P as simulated in the nutrient section of ELM (Reuss & Innis, 1978; Cole, Innis & Stewart, 1978). Again, both the N and the P stress terms for each producer category varied between 0.0 and 1.0. Sauer assumed the Michaelis–Menten form for equations which were representative of the relationship between N and P shoot concentrations and photosynthetic rates. A minimum shoot concentration of N for shoot response (N starvation level) was assigned as $0.007\ g\ N \cdot g^{-1}$ dry wt shoot and for phosphorus $0.001\ g\ P \cdot g^{-1}$ dry wt shoot. If either N or P level decreased below these limits, no photosynthesis occurred.

The effect of phenology on gross photosynthesis for each producer also varied from 1.0 to 0.0 as the mean phenophase varied from 1.0 to 7.0. Early stages of phenological development allowed the plant to photosynthesize at maximum capability; however, as flowering began, the phenology effects caused a reduction in photosynthetic rates. As the senesecence phase of phenology was approached, photosynthetic rates declined to zero.

To determine the insolation effect on gross photosynthesis, the abiotic section of ELM was utilized because this section produced simulated values of both diffuse and direct insolation. Sauer (1978) assumed that all diffuse radiation was available for interception. Direct radiation was divided into scattered and directly available fractions, similar to those of de Wit (1965), and the scattered fraction was determined. The scattered fraction was equally divided into upward and downward components. Sauer assumed that nearly all of the downward component (90%) was absorbed by the soil surface and therefore unavailable for photosynthesis. Of the upward-scattered components, he assumed that 10% escaped from the canopy because of the relatively low leaf area index (LAI) which occurred on the shortgrass prairie. The potentially available insolation for interception was then calculated by adding the diffused plus the nonscattered direct radiation to the downward-scattered radiation and multiplying this value by 0.1, plus upward-scattered radiation times 0.9. The constant coefficients for forward scattering and reflection from downward and upward directions were simplifications and could depend on canopy characteristics such as LAI and orientation. Variable coefficients may then be required when diverse canopies are simulated with the ELM model.

A distinction between $C_3$ and $C_4$ species was made by utilizing a minimum value of insolation required for a maximum rate of photosynthesis. In the Pawnee version of ELM, Sauer (1978) utilized blue grama as the $C_4$ species, representing warm-season grasses, and allowed a saturation level of 450 cal $\cdot$ $cm^{-2} \cdot d^{-1}$ for this species. The other noncactus species were $C_3$ species and were given saturation values of 400 $cal \cdot cm^{-2} \cdot d^{-1}$. The saturation level for cacti was assumed to be 500 $cal \cdot cm^{-2} \cdot d^{-1}$ because succulents, in general, are not light-limited.

Insolation interception was then calculated from cover and LAI to estimate the fraction of insolation available for photosynthesis and thereby estimate the potential maximum rate of gross photosynthesis. Light, of course, was intercepted by both live and dead shoots, but only the fraction of light intercepted by the live portions was available for photosynthesis. Sauer (1978) assumed that the cover of each species was not overlapping when LAI was low and bare soil was present. Both basal area and leaf area for each producer category was calculated by utilizing shoot weight information.

In Sauer's (1978) primary producer submodel, light interception for canopies with LAI greater than 1 were estimated by utilizing the Anderson & Denmead (1969) relationship. This fraction was then distributed to each producer according to the ratio of that producer's leaf area to the total leaf area of the canopy. He assumed that distribution of the photosynthetic surface of each species was uniform throughout the canopy, which may be an oversimplification.

Quantum efficiencies were another means of distinguishing between $C_3$ and $C_4$ species. Mean maximum values for quantum efficiency in the photosynthesis portion of Sauer's (1978) submodel ranged from 0.1 for $C_4$ species to 0.06 for $C_3$ species (Ludlow & Wilson, 1971). The quantum efficiency of cactus species was assumed to be approximately 8% that of the $C_4$ species, or 0.008 (Sěsták *et al.*, 1971).

Shoot respiration rates in Sauer's (1978) submodel were temperature-dependent functions of gross photosynthetic rates. He assumed that shoot respiration increased linearly from 0 to 30% of the gross photosynthetic rate as average photoperiod temperature rose from the minimum to the optimum temperature for net photosynthesis. As the average photoperiod temperature increased from the optimum to the maximum temperature for net photosynthesis, respiration increased linearly from 30 to 100% of the gross photosynthetic rate (Dye, Brown & Trlica, 1972). Respiration rate, therefore, was indirectly a function of the variables regulating gross photosynthesis and of the effect of temperature on respiration rate.

The net photosynthetic rate of each producer was, therefore, the difference between the gross photosynthetic rate and the shoot respiration rate (Sauer, 1978). Gross photosynthesis and shoot respiration were simulated separately to permit the study of different functional relationships with temperature, soil water, phenology, etc. However, values for gross photosynthesis, net photosynthesis and shoot respiration were not illustrated as functions of time in the publication by Sauer. Only biomass dynamics of the primary producers were illustrated in the simulation. However, Sauer did integrate gross photosynthetic values to come up with an integrated yearly value of 820.5 $g\,C \cdot m^{-2}$ for 1970 and 792.5 $g\,C \cdot m^{-2}$ for 1971 for gross photosynthesis of shortgrass species. Integrated net photosynthesis values for 1970 and 1971 were 557.5 and 553.0 $g\,C \cdot m^{-2}$, respectively.

### *Matador model*

Saugier *et al.* (1974) described a model of grassland plant growth and water use developed in the Matador Project of the Canadian IBP. The model was designed to simulate canopy and soil microclimate, and plant and soil evaporation (including changes in soil water content) as well as net assimilation and dynamics of the biomass of green and dead shoots. The model uses seven driving variables describing the weather on an hourly basis. The model is mechanistic and has generality in the sense that it can be run for any year for which input data are available and for any mixed grassland plant species for which the physical structure and physiological responses to the environment are known. It was felt that the model might help in understanding the relationships between microclimate, growth and water use in the grassland and aid in determining fundamental questions which still need to be answered.

The model of Saugier *et al.* (1974) was constructed in two interdependent parts: the heat and water flow section, and the carbon dioxide flux and production section. The model is dynamic in that it includes several storage compartments that are updated each hour, e.g. heat and water storage in soil layers, and storage of biomass in green and dead shoots. These variables are fed back as inputs to the model.

A block and arrow diagram of the carbon dioxide flux and production (Fig. 2.23) shows that radiation absorbed, green leaf temperature and stomatal resistance are fed into the photosynthesis/respiration submodel from the heat and water balance subsection (Saugier *et al.*, 1974). In turn, there is feedback of the LAIs of green and dead shoots to the radiation and energy balance submodels.

The photosynthesis/respiration submodel of Saugier *et al.* (1974) is based

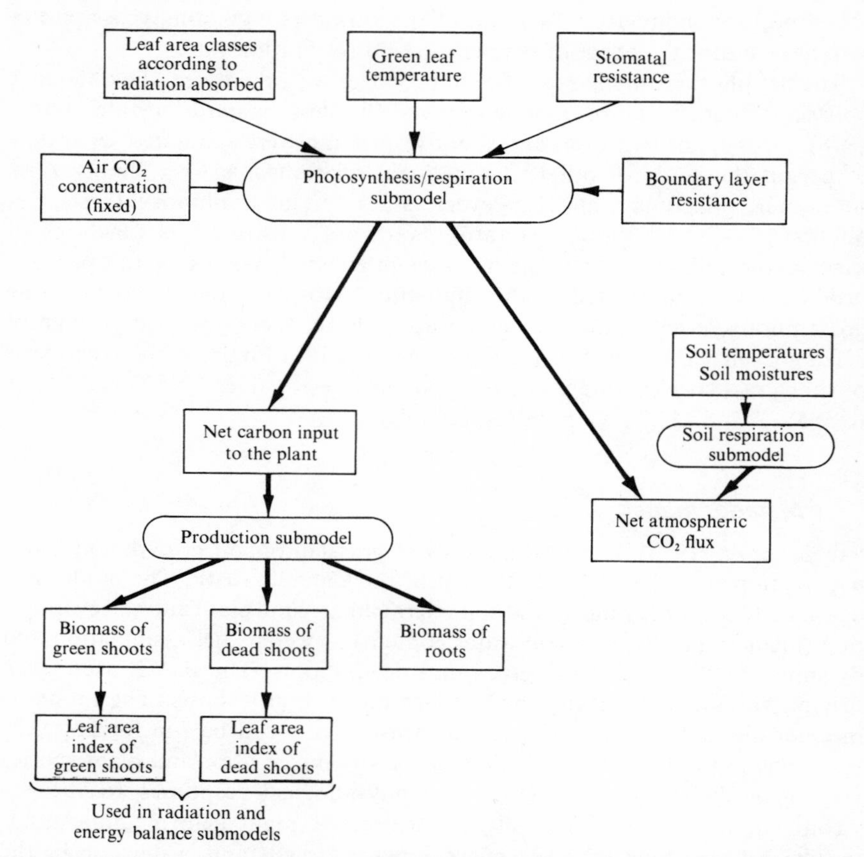


Fig. 2.23. Block diagram of the production section of the model by Saugier *et al.* (1974). The soil respiration submodel is not included in the present version. No provision is made for root death so root biomass is simply allowed to accumulate.

on that of Chartier (1972) discussed earlier, with the addition of water stress and leaf temperature effects. The operative equations are:

$$C_x = C_a - F(r_a + r_s + r_m) - R_L(1-\gamma)\, r_m,$$

and

$$F + R_L = \alpha I/(1 + \alpha I\, r_x/C_c),$$

where all terms are the same as in the Chartier model, except $n = 1-\gamma$. Using the two equations, $F$ can be calculated using irradiance, atmospheric carbon dioxide concentration ($C_a$), resistances ($r_a$, $r_s$, $r_m$, $r_x$), respiration ($R_L$) and the parameter $\gamma$. The term $\gamma$ was calculated to be very small, and was taken as 0, i.e. the respiratory carbon dioxide was considered to enter the photosynthetic resistance pathway in the intercellular spaces.

Stomatal resistance for carbon dioxide is calculated from $r_s$ for water vapour transport. The latter is determined in the water balance section of the model where it is a function of leaf water status and light. The boundary layer resistance is also determined from its value for water vapour transport. Mesophyll resistance is assumed constant and equal to 300 $s \cdot m^{-1}$. Resistance $r_X$ is mainly a physiological resistance related to the activity of carboxylation enzymes. Its minimum value (140 $s \cdot m^{-1}$) corresponds to the temperature optimum for the enzymes; it increases at higher or lower temperatures in the model.

Respiration rate ($R_L$) is the sum of dark respiration and light respiration. Dark respiration is handled in the same way as in Lommen *et al.* (1971). Photorespiration is closely coupled with photosynthesis and is therefore given the same functional relationship to light and temperature as photosynthesis:

$$R_P = R_P\text{MAX} \cdot f(I)\, g(T_g),$$

where $R_P$MAX is the maximum rate of light respiration at atmospheric oxygen level and $f$ and $g$ are functions expressing dependence of photosynthesis on irradiance ($I$) and green leaf temperature ($T_g$), respectively, with maxima equal to 1.0.

Net photosynthesis, or net carbon dioxide assimilated, is fed into the production submodel which is designed to simulate green and dead shoot dynamics and total dry matter production. The expansion of new leaves is more sensitive to water stress than photosynthesis (Hsiao, 1973). As a result the partitioning of photosynthates is dynamic; root : shoot production ratios increase with water stress as has been reported in the literature. Leaf death was made a function of leaf water status; more leaves die during periods of water stress. The transfer from standing dead to litter was considered a function of rainfall intensity. The expressions for the transfer from green to dead and from dead to litter were considered empirical and rather inexact ways of expressing growth and death processes.

The model output was compared with real measurements taken on individual days with opposite extremes of heat and drought, and with seasonal changes in several measured parameters. Simulation of net photosynthesis was considered good, with the possible exception of periods with temperatures above 35 °C, which were infrequent in the study area. The simulation of growth processes showed agreement during portions of the growing season, but failure of agreement resulted from uncertainties of the relationships between leaf dynamics and environment.

Precise modelling of plant growth would require a much more mechanistic model of growth processes. In its present form, Saugier *et al.* (1974) felt that the model was useful for understanding the influence of weather on grassland plant production, mainly through the interaction between precipitation, soil water and plant water status, and through the effects of plant water stress on photosynthesis and production.

### *Other examples*

De Wit *et al.* (1970) developed a model (ELCROS) to simulate the growth of a maize crop. The photosynthetic rate of the crop surface was calculated from the amount of light, the light distribution, photosynthetic function, leaf area and temperature. The leaves grew at the expense of photosynthetic products. Growth rate was dependent on the age of tissue, the amount of reserve products and the availability of water and temperature. Respiration was the sum of growth respiration (based on growth rate of the plant) and maintenance respiration. Results of the simulation agreed reasonably well with field trials.

In the model of de Wit *et al.* (1970) calculation of instantaneous photosynthetic rate was dependent upon the light distribution over the individual leaves of the crop, which had to be calculated. Then, the photosynthetic rate of the crop was obtained by addition of the photosynthetic rates of all leaves calculated using their photosynthesis function. Evaluation of the light distribution was purely a geometrical problem. The influence of the photosynthesis function of individual leaves was very important in calculating growth because of the large variability encountered in this parameter.

The leaf photosynthetic rate in the ELCROS model was:

$$\frac{I}{0.356+I} \cdot 84.5 \text{ kg carbohydrate} \cdot \text{ha}^{-1} \cdot \text{h}^{-1},$$

in which visible irradiance was in $\text{cal} \cdot \text{cm}^{-2} \cdot \text{min}^{-1}$. It was assumed that the instantaneous effect of temperature could be accounted for by a multiplication factor which amounted to 0.0, 0.0, 0.28, 0.54, 0.79, 1.03, 1.08, 1.08, 0.99 and 0.45 for temperatures from 0 to 45 °C at 5 deg C intervals. It was presumed in the model that the temperature of the leaves equalled the temperature of air at standard height. The leaf distribution function was assumed to be slightly plagiophile or of broad-leaf type (de Wit, 1965). The scattering coefficient was assumed to be 0.2.

In the ELCROS model, a time-step of an hour was used (de Wit *et al.*, 1970). Hence, to advance one day in time, all calculations were performed 24 times. The main purpose of ELCROS was to simulate the growth of the crop under optimum conditions.

A mathematical model of productivity was derived by Tooming (1970) by solving a variation problem based on the assumption that in a plant community, adaptation processes tend to provide maximum photosynthesis possible under any given environmental conditions. He concluded that the maximum productivity of a plant community or crop was attained when all leaves of all the species present carried out photosynthesis at a PhAR density which was close to the daily mean value of IDA (irradiation density of adaptation). Tooming utilized the well-known formula (Tamiya, 1951;

Monsi & Saeki, 1953) which contained only two parameters characterizing the photosynthetic apparatus:

$$P_G = P_{GM}I_a/(P_{GM}/a+I_a),$$

where $I_a$ is the photosynthetically active irradiance absorbed by the leaf layers of the community, $P_{GM}$ is the maximum rate of gross photosynthesis ($P_G$), and $a$ is the initial slope of the light response curve. He approximated the respiration ($R$) of the unit layer in the community as follows:

$$R = K_1P_{GM}+K_2P_G,$$

where $K_1$ is the coefficient combining all losses of energy for the maintenance of photosynthetic and nonphotosynthetic organs of a plant and $K_2$ is the coefficient combining energy used for various forms of synthesis and for growth of the organs.

A simple mathematical model describing the effects of nitrogen and light on the vegetative growth of a grass–legume sward was presented by Ross, Henzell & Ross (1972). Their model consisted of a set of first-order differential equations based on the mechanisms involved and provided a conceptual framework for further study of grass–legume swards. The model was capable of reproducing experimental results found in the literature, although proper validation with additional measurements is still needed. Two areas of uncertainty were the processes of transferring legume nitrogen to an associated grass and the mechanisms by which lack of nitrogen limits grass growth.

### *Semimechanistic and statistical models*

The literature contains a large number of studies relating carbon dioxide exchange to environmental conditions. A common approach is to do statistical tests to determine which environmental factors significantly affect gas exchange, or to fit equations to data relating photosynthesis to light, temperature or other factors. In this section we will review a few of the studies of grassland models which have taken this approach.

Ruetz (1973), working in the Solling Project of the German IBP, used climatized cuvettes to follow gas exchange of red fescue (*Festuca rubra*). Multiple regression analysis was used to relate photosynthetic rates to irradiance, temperature, water vapour concentration difference, chlorophyll content and time. A cubic regression equation based on daily irradiance alone explained 85 and 87 % of the variation for the fertilized and the unfertilized plants, respectively. Maximum photosynthetic rates were found between 14 and 22 °C; rates declined at temperatures above 24 °C. The conclusion was that the photosynthetic response of red fescue was limited primarily by irradiance with moisture and temperature playing minor roles.

Increased nutritional status resulted in increases in photosynthesis as a result of increased leaf area and high photosynthetic rates of the individual leaves.

A multiple linear regression model was utilized by Brown & Trlica (1974) for predicting photosynthetic rates of blue grama and western wheatgrass. They found that from 81 to 91% of the variation in observed photosynthetic rates of these two grasses could be accounted for by variations in irradiance, soil water potential and temperature. In addition, interactions among these three important abiotic driving variables significantly affected photosynthetic rates.

Leafe (1972) used a field enclosure technique to measure the carbon dioxide exchange in a sward of perennial ryegrass (*Lolium perenne* L.) in Britain. The goal of the study was a better understanding of the reasons for the seasonal pattern of productivity. The data were used to construct response surfaces relating net photosynthesis to radiation ($I$) and time ($t$). The equations had the form:

$$P_N = (a_0 + a_1 t + a_2 t^2 + a_3 t^3 + a_4 t^4) + (b + b_1 t + b_2 t^2 + b_3 t^3 + b_4 t^4) \log_{10}(I).$$

Dark respiration was also measured and related to temperature ($Q_{10} = 2$).

The approach provided a means for integrating $P_N$ over time using the response surfaces and hourly integrals of radiation. Dark respiration was computed by assuming a linear change in the intervals between measurements and integrating over the dark period using hourly mean temperatures.

The results for different growth periods were examined, and reasons for the three phases of growth (increasing, linear and decreasing) during each period were discussed. $P_N$ was highest during the linear phase; afterward it declined, corresponding to the attainment of a ceiling yield. There was no evidence that declining growth rate was caused by increased respiration; rather it appeared to be related to the lowered photosynthetic efficiency of the crop canopy. The lower efficiency apparently was a result of unfavourable spatial distribution of leaves and age composition.

Cunningham, Balding & Syvertsen (1974) describe a mathematical model developed in the US/IBP Desert Biome study which allowed prediction of net photosynthesis of grass with the $C_4$-dicarboxylic acid pathway of carbon dioxide reduction. The model is mechanistic in the sense that net photosynthesis is considered a temperature-limited, enzyme-catalysed reaction with denaturation of the enzymes at high temperatures. The carbon dioxide supply to the chloroplasts was made proportional to a concentration gradient (carbon dioxide concentration in the chloroplast was assumed to be 0), divided by a total resistance. The latter was the sum of an irradiance-dependent resistance component and a resistance to carbon dioxide flux dependent on soil water potential.

This partitioning of total resistance in the model by Cunningham *et al.*

(1974) is artificial in the sense that irradiance and soil water actually influence a single resistance: stomatal resistance. Another weakness is that the resistance should be a function of leaf water status rather than soil water status.

The operative equation in the model of Cunningham *et al.* (1974) is:

$$P_N = \exp K_1 T_g (K_2 - \exp K_3 T_g) \frac{C_a}{K\{K_4/1 + [K_5 - \exp K_6(K_7 + \psi)]\}},$$

where $\psi$ is soil water potential. All $K$ terms are species-specific coefficients, the values of which were found by using the equation as a multiple nonlinear regression model and determining values for the coefficients which gave the best fit to the measured net carbon dioxide exchange rate data. The model accounted for 83% of the variability in carbon dioxide exchange data for vine mesquite panic (*Panicum obtusum*) and 85 % of the variability in tobosa hilaria (*Hilaria mutica*) data.

This model of Cunningham *et al.* (1974) is basically statistical and without generality, in the sense that coefficients must be calculated from measured field data for the species to be studied. It does, however, allow the prediction of rates of net carbon dioxide exchange as a function of environmental variables and the comparison of effects of each variable.

A computer program was constructed by Brown & Trlica (1977*c*) to simulate production dynamics for blue grama during a single growing season. Previously determined steady-state gross photosynthetic rates, as affected by soil water potential, temperature, visible irradiance and phenological stage, were utilized to control directly the primary flow of carbon in the model. Rates of photosynthesis and respiration had been determined *in situ* on native blue grama sods on the shortgrass prairie. Other directly applicable data and estimates were used to control translocation, shoot and root litter accumulation, and root respiration flows in the model. Total net primary production for blue grama of 714 g carbohydrate $\cdot$ $m^{-2}$ area $\cdot$ $yr^{-1}$ was predicted from the model output. Model predictions compared favourably with 744 g carbohydrate $\cdot$ $m^{-2}$ area $\cdot$ $yr^{-1}$ determined through harvesting techniques for the same growing season, but which also included biomass of other species on the shortgrass prairie site (Lauenroth, 1973).

The model described by Brown & Trlica (1977*c*) was nonmechanistic in that vast amounts of actual field data were required. However, the model was useful in quantifying results which might be obtained by perturbations of abiotic driving variables. For example, the addition of 5 deg C to all observed temperatures through the growing season reduced net primary production from 714 to 34 g carbohydrate $\cdot$ $m^{-2}$ area $\cdot$ $yr^{-1}$; whereas the reduction of all temperatures by 5 deg C increased net primary production to 1107 g carbohydrate $\cdot$ $m^{-2}$ area $\cdot$ $yr^{-1}$. The ecophysiological implications of these results are that net primary production of the shortgrass prairie could possibly be greater during a cooler growing season, even though it is dominated by a

$C_4$ species with a fairly high optimum photosynthetic temperature. Sensitivity analyses indicated a need for additional research on factors affecting translocation and root respiration.

A functional model to describe the relationship between community photosynthesis, leaf area index, irradiance, ambient temperature and soil water potential was proposed and tested with field data collected on blue grama (the dominant species of the shortgrass prairie) by Connor, Brown & Trlica (1974). The model involved an assessment of radiation interception, photosynthetic response under optimum conditions, and the introduction of proportionality factors to account for the effects of nonoptimum conditions of ambient temperature and soil water stress. In their approach, parts of the complex geometrical models of community photosynthesis that have been used with success by other authors (de Wit, 1965; Duncan *et al.*, 1967; Connor & Cartledge, 1970; Paltridge, 1970; Cartledge & Connor, 1971) were utilized. The general equation for photosynthesis was:

$$P_N = I(1-\exp[-\mathrm{LAI}\cos\beta])\,\alpha\cdot\mathrm{ETP}\cdot\mathrm{EWP},$$

where LAI is the leaf area index, $\beta$ is the mean leaf angle, $\alpha$ is the slope of the linear portion of the leaf photosynthesis/irradiance relationship, ETP is the effect of temperature on photosynthesis and EWP is the effect of soil water potential on photosynthesis. The model was compared with several statistical models derived from the same experimental data. Comparisons indicated the functional model was superior to the statistical models beyond the limited data set. By comparison, the functional model provided what can be identified as biologically reasonable predictions. The use of a functional approach to develop predictive equations for ecosystem modelling purposes, therefore, had significant advantages over a purely statistical one.

## Translocation

The carbon fixed through the process of photosynthesis is utilized for a variety of necessary plant functions, such as respiration, elaboration of new vegetative and reproductive structures, and storage. Since survival and growth of a plant and of different organs within the plant's body depend upon these functions, the photosynthate must be translocated to various organs. This transport from the site of synthesis to the sites of utilization is called translocation.

### *Tissues concerned with translocation*

As early as 1679, Malpighi (Richardson, 1968) demonstrated that removal of a complete ring of bark from stems of a number of trees blocked the movement of sap down the trunks. Mason & Maskell (1928*a*) observed an accumulation of sugars above the girdle when they ringed phloem in cotton stems. In a more elaborate study, Rabideau & Burr (1945) fed $^{13}CO_2$ to single leaves of bean plants (*Phaseolus vulgaris*). Before the actual feeding occurred, the phloem in stems of different sets of plants was killed above, below, and both above and below the fed leaf. The blockage of phloem below the leaf resulted in only upward transport of $^{13}C$, blockage above the leaf resulted in only downward transport; in plants where phloem was killed both above and below the fed leaf, $^{13}C$ was not transported out of the blocked zone. Later experiments, involving excised stylet tips of aphids parasitic on the translocation stream, also established phloem as the principal tissue involved in translocation of organic compounds (Kursanov, 1963; Crafts & Crisp, 1971).

Nelson, Perkins & Gorham (1959) presented evidence of different kinds of concurrent translocation of photosynthetically assimilated $^{14}C$ in soybeans (*Glycine max*). These authors found that there were two kinds of translocation. One involved slow translocation of large quantities while the other involved rapid translocation of small quantities of assimilates. Girdling experiments indicated that the slow translocation occurred within the phloem, but rapid translocation of small amounts occurred through xylem as well as some living tissues other than phloem.

### *Compounds involved in translocation*

Chemical analysis of phloem exudate, secreted when sieve cells were punctured, and studies involving isotopes such as $^{14}C$ have indicated that a large variety of compounds is translocated through phloem, such as sugars, amino acids, organic acids, viruses, minerals, hormones, enzymes and steroids (Crafts & Crisp, 1971). However, sucrose is the principal compound involved in bulk transport in the majority of plants and often makes up 90% of the entire mass of assimilates undergoing translocation (Kursanov, 1963). In some cases, sucrose condenses with one or more molecules of galactose to form raffinose, stachyose or verbascose. Other derivatives of sugars like sugar alcohols (Trip, Krotkov & Nelson, 1963; Trip, Nelson & Krotkov, 1965) and sugar phosphates (Kursanov, 1963) may also be important translocation products in certain plants.

According to Arnold (1968), sucrose is specially suited for translocation because it is less reactive and thus is able to move over long distances. Further, sucrose is relatively easily hydrolysable which is desirable for a translocatory compound. Thus, the properties of sucrose and the characteristics of the translocation machinery are well compatible (Crafts & Crisp, 1971).

## *Mechanism of translocation*

There are three phases involved in the process of assimilate translocation: (1) transfer of assimilates from mesophyll cells into phloem, (2) transport through phloem, and (3) transfer of translocates from phloem into receiving cells of the sink. We will briefly discuss the mechanism involved in these three phases, because this may have some bearing on the mathematical modelling of the process.

### *Transfer of assimilates from the mesophyll cells into phloem*

There is considerable evidence that the transfer of assimilates from the mesophyll cells (site of synthesis) into phloem cells is an active process requiring expenditure of energy. The evidence can be briefly summarized as follows:

(1) There may be hundreds of living cells between the mesophyll cell and the sieve cells. Thus, there exists a formidable system of semipermeable membranes, and further the actively metabolizing, intervening cells may interact with the substances being transported.

(2) This transport often occurs against a concentration gradient (Mason & Maskell, 1928*a*, *b*; Kursanov & Turkina, 1954). In fact, in the sieve cells of phloem the concentration of sucrose is often greater than 0.3 M, much higher than in mesophyll cells. Canny (1961) also demonstrated that the rate of influx of assimilates into the conducting system was independent of their concentration in the leaf blade.

(3) This transfer also seems to be faster than the speed of free diffusion. Within 2–3 minutes after feeding an adjacent section of the mesophyll with a radioisotope, radioactivity could be detected in the conducting system.

(4) The transfer from mesophyll cells into the conducting tissue seems to be a selective process. Thus, more recently formed sugars are the first to be transferred as compared with sugars stored in vacuoles (Nelson, 1962; Hofstra & Nelson, 1969). Only a few organic acids enter the conducting system, while a majority of them accumulate in the surrounding cells (Kursanov, Brovchenko & Pariiskaya, 1959). The transfer is especially selective with regard to amino acids. For example, Kursanov *et al.* (1959) found that threonine was the most mobile amino acid in rhubarb (*Rheum rhaponticum*).

(5) Kursanov & Brovchenko (1961) demonstrated that the movement of assimilates from the mesophyll cells into the sieve cells was accelerated when the leaves were enriched with ATP. However, AMP and ADP had no effect on this movement, indicating that it was the third phosphate group of ATP which activated sugars by way of a hexokinase reaction.

On the basis of the above evidence, Kursanov (1963) proposed a scheme for the transfer of assimilates from the mesophyll cell into the adjacent conducting sieve tubes through the involvement of carriers. According to this scheme, the phosphorylated hexose sugars form sucrose phosphate with the help of uridine triphosphate in the mesophyll cells. The phosphorylated sucrose is then carried, after forming a complex with some carrier, across the cytoplasm and the cell membrane into the sieve tubes. Alternatively, the phosphorylated hexose sugars may themselves cross the membrane with carriers and once inside the sieve tube may condense to form sucrose. According to Crafts & Crisp (1971), protoplasmic streaming may also be involved in symplastic movement of assimilates from mesophyll to phloem and, in as much as protoplasmic streaming is a manifestation of cell metabolism, it may accelerate the diffusive movement of solutes from mesophyll to phloem cells.

### *Transport of assimilates in the phloem*

A number of hypotheses have been proposed to explain the movement of assimilates in the sieve tubes of phloem. The most popular of these is the mass-flow hypothesis proposed by Münch (1930). According to this hypothesis, the assimilates are transported by way of the sieve tubes of the phloem from the site of synthesis to regions of utilization along a turgor pressure gradient which is developed osmotically. The sieve tubes in the leaves are constantly supplied with assimilates and this causes a flow of water from xylem into sieve tubes creating a high turgor pressure. At the opposite end, where assimilates undergoing translocation enter the surrounding tissue for storage or utilization, the osmotic pressure and the turgor pressure in the sieve tubes are low. The assimilates thus follow a mass flow from a region of high turgor pressure to a region of low turgor pressure through a system of communicating vacuoles in a long series of sieve cells.

The mass-flow hypothesis has been supported from time to time through collateral experimental results. For example, Mittler (1953) reported that, after a sieve cell was pierced by an aphid stylet, sucrose solution was excreted from the damaged cell for a long time. The chemical composition of the excretion was similar to the composition of phloem sap. Mothes & Engelbrecht (1957) found that the outflow from punctured sieve cells may continue for several weeks, indicating that the damaged cells perhaps continued to receive assimilates from other sieve cells through mass flow. Canny (1961) demonstrated that, if the leaf above an aphid stylet insertion was labelled with $^{14}CO_2$, the liquid released from the punctured sieve cells contained labelled compounds.

Measurements of flow rate of sap from stylet puncture ($\sim 100\ cm \cdot h^{-1}$; Weatherly, Peel & Hill, 1959) and rate of transfer of heat through sieve tubes ($35$–$70\ cm \cdot h^{-1}$; Ziegler & Vieweg, 1961) are in the range of recorded rates of translocation. Recent results indicating open sieve plates and living, continuous, stationary, filamentous reticulum in functioning sieve tubes (Esau & Cheadle, 1961) are compatible with the mass-flow hypothesis (Crafts & Crisp, 1971).

Various other hypotheses, such as protoplasmic streaming (Curtis, 1935; Canny, 1962*a*,*b*; Thaine, 1962), activated diffusion (Mason & Maskell, 1928*a*,*b*), surface migration (Mangham, 1917; Honert, 1932) and metabolic movement (Spanner, 1958; Fensom, 1959; Kursanov, 1963) have also been proposed to account for assimilate translocation.

The hypothesis that transport of assimilates in plants occurs through protoplasmic streaming was suggested by De Vries (1885) and was partially supported by Curtis (1935), Biddulph (1941), Swanson & Böhning (1951) and Thaine (1961, 1962). Thaine (1961, 1962) reported the occurrence of transcellular cytoplasmic tubular strands which transverse the sieve tube lamina and extend through the sieve pores into the next cell. These transcellular strands were reported to be composed of a boundary membrane, longitudinally oriented threadlike structures, microscopic and submicroscopic particles, and a matrix of fluid endoplasm (Thaine, 1964). According to the protoplasmic streaming theory, the food particles are transported through the transcellular strands of protoplasm; the streaming itself is driven by metabolic energy while the translocation is regulated by the pressure or concentration gradients developed outside of the actual pathway of long-distance transport. However, the observed velocity of movement of particles through cyclosis (3–5 $cm \cdot h^{-1}$) is much lower than the measured translocation velocities (50–300 $cm \cdot h^{-1}$) (Crafts & Crips, 1971). Further studies by Esau, Engleman & Bisalputna (1963) revealed that protoplasmic streaming was limited to phloem parenchyma and did not occur in mature sieve elements. They failed to find transcellular strands in the sieve tube cells as observed by Thaine (1961, 1962, 1964).

Recognizing the presence of a concentration gradient in the direction of assimilate movement, Mason & Maskell (1928*b*) believed that the translocation of solutes in phloem was brought about by a process of activated diffusion or accelerated diffusion (Mason & Phillis, 1937). While certain substances like chlorine exhibit 'free diffusion', others such as sugars and phosphorus exhibit 'oriented diffusion' because of the attraction by sinks (Penot, 1965). Canny (1962*a*,*b*), on the basis of the strand structure reported by Thaine (1961), proposed a mechanism which involved the exchange of sugar between strands and a stationary sap through which they passed; the system thus had all the properties of a diffusion-analogue mechanism.

Canny & Askham (1967) analysed the distribution pattern of $^{14}C$ when introduced in a source–sink system and found that the tracer could diffuse in all directions and might not follow the source–sink polarity. In fact, they demonstrated that $^{14}C$ could move against the mainstream of sugar and could accumulate in aphids feeding on the phloem of mature unlabelled leaves. Thus they also supported the view that the translocation system was a diffusion-analogue system. Canny (1971, 1973) has further supported the activated diffusion theory primarily on the basis of translocation kinetics. The studies by Mason & Maskell (1928*a*,*b*) had shown that the maximum dry weight transfer through translocation and the direction of transfer were controlled by the gradient of sucrose concentration and followed the equation given below:

$$M = k\,\mathrm{d}s/\mathrm{d}x,$$

where $M$ is the dry weight transfer, $s$ is the sucrose concentration, $x$ is the distance and $k$ is a translocation coefficient. This equation was identical with the steady-state form of Fick's law for the transfer of mass by molecular diffusion (Canny, 1971). If all the quantities in this equation were represented in cgs units, then the value of $k$ was 0.07 $\mathrm{cm^2 \cdot s^{-1}}$. Canny (1971) calculated the value of $k$ using data on translocation profiles from a large number of independent studies and noted a general correspondence with the above value. Further, Canny (1971, 1973) argued that the accelerated diffusion hypothesis could also reasonably explain the simultaneous movement of two substances in opposite directions if their concentration gradients were reversed. However, the discrepancy between normal diffusion of sugar in water (diffusion coefficient $= 5 \times 10^{-6}\ \mathrm{cm^2 \cdot s^{-1}}$) and activated diffusion of sugar in phloem ($k \sim 7 \times 10^{-2}\ \mathrm{cm^2 \cdot s^{-1}}$), together with the long distance over which the translocate is moved at sustained high speeds, are yet to be satisfactorily explained. In accelerated diffusion the rapid transfer of mass is possible only over short distances.

Visualizing the cytoplasm continuum in sieve tubes brought about by the protoplasmic tubules which pierce the sieve plates, Mangham (1917) proposed that transfer of sugars absorbed at the cytoplasmic surfaces occurred through the waves of disturbance and readjustment of equilibrium propagated along this cytoplasmic continuum. According to Honert (1932), the assimilates move along a continuous interfacial surface between water and cytoplasm in the phloem and hence translocation of assimilates is essentially a surface migration phenomenon. Evert & Mjrmanis (1965) expressed the view that the energy-releasing processes occurring in the sieve tube protoplasm and the cytoplasmic strands were the driving forces for the movement of assimilates along the strands throughout the phloem. However, there is not much experimental evidence to support this hypothesis. Also, Crafts & Crisp (1971) have pointed out the inadequacy of surface migration to account for the known translocation rates.

Kursanov (1961, 1963) strongly supported the view that the translocation of assimilates in phloem is an active process requiring expenditure of energy. He also demonstrated the high respiration intensity of fibrovascular bundles as compared with mesophyll or cortex tissues, as well as the presence of enzymes of glycolysis and oxidation in addition to the acid-soluble nucleotides such as AMP, ADP, ATP, UMP and UDPG in these fibrovascular tissues (Kursanov, 1963). The adverse effect of anoxia (anaerobic conditions) on the movement of assimilates (Geiger & Christy, 1971; Qureshi & Spanner, 1973) seems to support the contention that translocation is an active process. The probable use of energy in assimilate movement in phloem was discussed by Kursanov (1963) who visualized the sieve plate region acting as a glandular cell, absorbing substances from the vacuolar solution, fixing them with appropriate carriers, transporting these complexes through the connecting strands of the protoplasm through sieve pores, and finally releasing them again into the vacuolar solution of the next cell. When this hypothesis was put forward the sieve plates were thought to be plugged, and hence it was necessary to explain the crossing of these zones by the translocates. The fact that some of the known metabolic inhibitors are also readily transported if introduced into the translocation stream of phloem, is not compatible with this proposition.

It is apparent from the above discussion that, out of all the hypotheses, the mass-flow hypothesis has the most experimental support. Evidently, however, much work is still needed to understand fully the translocation mechanism, which may well call for a combination of some of the above hypotheses.

### *Transfer of assimilates from phloem into receiving tissue*

As in the case of transfer of assimilates from mesophyll cells to sieve cells, the transfer from phloem into receiving cells seems to be an active process involving phosphorylation and an expenditure of energy. Bieleski (1960) demonstrated that the sugar concentration in mature internodes of sugarcane was 20% as compared with 4 to 10% in young internodes and 2 to 3% in leaves. This finding illustrated that the unloading of sieve tubes occurs against a concentration gradient. Further, Bieleski (1960) reported that both the loading and unloading of phloem was inhibited by metabolic inhibitors such as potassium cyanide, sodium and monoiodoacetate.

Kursanov (1963), following the scheme of Glasziou (1960), illustrated the unloading processes noted below:

Outer space

$$\text{Sugar} + (\text{R–X}) \rightarrow [\text{sugar–X}] + \text{R}$$
$$[\text{sugar–X}] + \text{carrier} \rightarrow [\text{sugar–carrier}] + \text{X}$$
$$[\text{sugar–carrier}] \rightarrow \text{sugar} + \text{carrier}$$

Inner space

According to Kursanov (1963), the R–X may be ATP and sugar–X may be a phosphate ester. The activated sugar molecule is possibly transported across the membrane into the inner space (receiving cell) with the help of a carrier which could be the endoplasmic reticulum (Crafts, 1961).

### *Bidirectional movement in phloem*

Several authors have reported bidirectional movement in phloem (Curtis, 1935; Biddulph & Cory, 1960). Biddulph & Cory (1965) reported that this bidirectional movement occurred in separate phloem bundles. On the other hand, Palmquist (1938) had earlier demonstrated the movement of fluorescein and carbohydrate in opposite directions in the same phloem bundle. Later, Trip & Gorham (1968) reported that two streams of sugars ($^{14}C$ assimilate and tritiated glucose) could move in opposite directions within the same sieve tube. While bidirectional movement through separate bundles or separate sieve tubes can be explained by the mass-flow hypothesis, the two-way movement within the same sieve tube is hard to explain on the basis of source–sink relations unless it is assumed that one of the two substances moving in opposite directions has tangentially moved from a second sieve tube into the first sieve tube (Peterson & Currier, 1969). However, the radial or tangential movement between sieve tubes is very slow and if a rapid interchange between sieve tubes occurs, the net movement of any substance could be zero (Canny, 1971). Accelerated diffusion, as discussed earlier, offers the best explanation for the bidirectional movement of solutes in the same sieve tube on the basis of reversed concentration gradients.

### *Rates of translocation*

Measured translocation rates are generally quite rapid. The rates measured as specific mass transfer range from 0.14 to 4.8 g dry wt $\cdot$ $cm^{-2}$ phloem $\cdot$ $h^{-1}$ and the translocation velocities range from 1.5 to 7200 cm $\cdot$ $h^{-1}$. The translocation velocity of 39 to 109 cm $\cdot$ $h^{-1}$ as reported by Wardlaw (1965) for wheat may be taken, according to Canny (1971), as the centre of a distribution of velocities measured by many people in many ways. Different compounds also seem to move with different rates. For example, Biddulph & Cory (1957) reported that in the bean plant, sucrose moved at a rate of 107 cm $\cdot$ $h^{-1}$, while tritiated water moved at the rate of 87 cm $\cdot$ $h^{-1}$. Similarly, Nelson & Gorham (1959) reported a range of translocation rates for amino acids from 370 cm $\cdot$ $h^{-1}$ (asparagine, glutamine) to 1370 cm $\cdot$ $h^{-1}$ (alanine, serine).

### *Factors affecting translocation*

#### *Water*

Weatherly *et al.* (1959) and Plaut & Reinhold (1965) reported that water deficiency reduced assimilate movement, although the conducting system remained stable as demonstrated by continued movement of assimilates from the green stem to the roots of bulb canary grass (*Phalaris tuberosa*) after defoliation resulting from severe water deficit (McWilliam, 1968). Roberts (1964) and Wiebe & Wihrheim (1962) reported that high water stress resulted in greater acropetal movement as compared with plants under low water stress. Wardlaw (1967) also observed an increased movement of assimilates from the lower leaves to the spikelet in water-stressed wheat plants. Sosebee & Wiebe (1971) found that translocation in crested wheatgrass (*Agropyron cristatum*) and barley grown without adequate soil water was chiefly to the underground portions of the plants. This relative downward movement of assimilates seemed to be especially influenced by defoliation. In plants with adequate soil water, clipping appeared to stimulate upward movement of assimilates. They postulated that water stress reduced shoot meristematic and cell enlargement activity, thereby reducing the demand for translocation to shoots. Wardlaw (1968*a*) found that the velocity of assimilate movement through a water-stressed leaf in poison ryegrass (*Lolium temulentum*) could be maintained at normal rates if adjacent leaves were removed to increase the demand of assimilates from the stressed leaf. It appears, therefore, that under most conditions the effect of water stress on photosynthesis and growth outweighs that on translocation.

### *Temperature*

Burr *et al.* (1958) reported that cooling the roots of sugarcane reduced assimilate transport by 50 and 82% in 24 h and 80 h, respectively. Warmer soil temperatures are usually more congenial to translocation to roots (Fujiwara & Suzuki, 1961). Habeshaw (1973) also demonstrated reduced translocation to cooled roots. Whitehead, Sansing & Loomis (1959) found reduced or no assimilate transfer to roots of tomatoes (*Lycopersicum esculentum*) when the stem was cooled to 5 °C. Ford & Peel (1967) reported that increased viscosity of sap upon cooling may result in decreased assimilate movement. Geiger (1966) and Swanson & Geiger (1967) also found that when a 2-cm portion of petiole of *Beta vulgaris* was cooled, the assimilate transport was reduced by 65 to 95%.

### *Light*

Rohrbaugh & Rice (1949) found that translocation was greater during the light period as compared to the dark period. Fujiwara & Suzuki (1961) reported a direct correlation between the length of the photo-period and the translocation rate. Under low light intensities, stem growth may compete with grain production for assimilates in wheat (Carr & Wardlaw, 1965). Butcher (1965) reported decreased translocation of $^{14}C$ assimilates in *Beta vulgaris* during the dark period as compared with the light period. However, Nelson & Gorham (1957) fed $^{14}C$-labelled sugars to primary leaves of soybean seedlings and reported that only 4% of the total uptake of $^{14}C$ by leaves was translocated to roots in the light as compared to 16% in the dark. Thus, different plants seem to react differently with regard to the effect of light on translocation.

#### *Oxygen*

Mason & Phillis (1936) have indicated the necessity of oxygen for assimilate transport in phloem. Geiger & Christy (1971) found a rapid decline in the import rate of labelled assimilates when the sink region was subjected to anoxia. More recently, Qureshi & Spanner (1973) have supported the findings of Mason & Phillis (1936). They found that anoxia imposed on the stolon connecting the sink with the source greatly reduced the assimilate transfer from source to sink. Willenbrink (1957) and Ullrich (1961), however, could find no evidence for any effect of anoxia on the transport of fluorescein through the petiole.

### *Partitioning of assimilates and source–sink relations*

Cells that have no chloroplasts and lead an essentially heterotrophic existence can be termed sinks. Examples of belowground sinks are roots, rhizomes, tubers and other storage organs. Aboveground sinks are stems, buds, flowers, fruits and young developing leaves. The activity and demand of sinks are controlled by phenology and environmental factors.

The young shoots are relatively more efficient in carbon fixation (Dahlman & Kucera, 1968, 1969; Singh & Coleman 1977). During this time there is also greater retention of the assimilates in the shoots. Thus, Singh & Coleman (1977) found that within 3 days after labelling there was 31% retention of photoassimilated $^{14}C$ in shoots of blue grama (a semi-arid shortgrass species) during early growth, 20% retention during maturity, and only 15% retention during later stages of growth. Balasko & Smith (1973) reported that in timothy (*Phleum pratense*) the loss of radioactivity from leaf blades as a result of translocation and respiration within the first week after labelling was 70% at initiation of stem elongation, 86% at ear emergence and 87% at anthesis.

A majority of reports on the distribution of assimilates into various plant organs concern laboratory or culture conditions. In perennial ryegrass and poison ryegrass, Ryle (1970*a*) found that 44 to 50% of $^{14}C$ moved to roots. In Italian ryegrass (*Lolium multiflorum*) (Marshall, 1966) and in wheat (Doodson, Manners & Myers, 1964) up to one-half of the labelled carbon which left the source leaf was found in roots. Annual plants, in general, transfer less assimilate to roots (Ryle, 1970*b*). In a semi-arid mixed-grass prairie under natural conditions, Warembourg & Paul (1973) reported that 46 to 54% of the photofixed $^{14}C$ moved to belowground parts. Results from a semi-arid shortgrass prairie indicated that from 38 to 71% of the total

Table 2.10. *Partitioning of photoassimilated $^{14}C$ in blue-grama-dominated shortgrass prairie 3 days after labelling (% of total recovered in plants)*

| Plant part | May | July | September |
|---|---|---|---|
| Shoots | 31.2 | 20.5 | 14.6 |
| Crown | 31.1 | 21.0 | 14.4 |
| Roots | 37.7 | 58.5 | 71.0 |

From Singh & Coleman (1977).

plant $^{14}C$ was located in roots 3 days after labelling (Table 2.10). Bamberg, Wallace, Kleinkopf & Vollmer (1972) estimated that only 10 to 20% of the annual production of photosynthate was translocated to root systems of eight desert species. They found only 4 to 22% of the originally fixed $^{14}C$ in the roots of these desert plants.

Singh & Coleman (1977) found that of the total $^{14}C$ recovered from blue grama plants 3 days after labelling, 38, 58 and 71% was located in roots, while 31, 21 and 14% was in crowns during May, July and September, respectively (Table 2.10). Soil respiration during the 3-day period between labelling and sampling accounted for approximately 9 to 19% of total assimilated labelled carbon in July and September.

Warembourg & Paul (1973) studied carbon distribution in an *Agropyron–Koeleria* grassland association in Canada. Labelling of virgin prairie was done during June and August. They found that in these native grassland species, approximately 50% of the labelled carbon remained above ground and 50% was transported to the root system. This indicated a rapid turnover in the aerial portion of the plants compared with that in the roots. The distribution pattern of the labelled carbon throughout the root system showed that there were differing growth rates during the two labelling periods. In June, the labelled belowground carbon was concentrated in the roots and rhizomes 0–10 cm below the surface and decreased rapidly with depth. In August large amounts of carbon were deposited in the 0–10 cm zone even though soil water was less than −15 bars. The relative root growth per unit of biomass in August was as high at a depth of 55 to 85 cm as it was near the soil surface. In June, it decreased with depth. This indicated an increase of root biomass at depth during the dry period in August which could be attributed to the activities of roots in an area with greater soil water availability.

The time required for photosynthate to reach the roots at various depths ranged from 1 to 5 days and the amount of material deposited in the roots changed with time and soil water availability (Warembourg & Paul, 1973). Soil $^{14}CO_2$ flux indicated that root respiration during the 8 days following a 2-day labelling period accounted for 24% of the labelled carbon translocated to the roots.

Measurement of production and turnover of roots has been one of the most difficult problems in terrestrial ecosystem studies. This has been particularly true in perennial grasslands where root : shoot ratios may exceed 8 : 1 (Sims & Singh, 1971).

Singh & Coleman (1977) evaluated functional root biomass and translocation of photoassimilated $^{14}C$ in a shortgrass prairie ecosystem. Their study was designed to determine the proportion of roots that were functional, the proportion of photosynthate that was channelled belowground and incorporated into roots or lost in respiration, and seasonality of translocation and root activity. Labelling of blue grama plots with $^{14}CO_2$ was done during three stages of vegetation development: early, middle and late growing season. Three days after tagging, samples of aboveground live biomass, crowns and roots were taken and analysed for radioactivity. They found that functional root biomass varied from about 29 to 75% of all roots in different depth zones. The top 20 cm had more functional roots than lower zones. Total root biomass to a depth of 60 cm averaged 62% functional and this proportion was relatively constant from late May through September. With the advance in the growing season, there was more storage of assimilates in roots at shallow depths. Thus, in late May 75% of the total photoassimilated $^{14}C$ was channelled to roots at 0 to 20 cm, in July 84% was recovered from this upper rooting zone, and in mid-September 92% of assimilates went to this shallow root layer. Singh & Coleman believed that new growth, at the advent of the growing season before commencement of photosynthesis, occurred at the expense of reserves in the upper parts of the old roots. With the progress of the growing season, and with increased supply of photosynthate, more roots grow in the upper zones and command a large share of the assimilates. They felt that it was also possible that reserves in the old upper roots, which were depleted earlier in the season, were then replenished. The sum total of these events was that more translocate was retained in upper zones (0–20 cm) in the latter part of the growing season.

Ares (1976) used root observation windows to study the dynamics of the root system of blue grama, and observed that root differentiation and growth started very early in the growing season, presumably at the expense of reserves in the crowns and in older roots of the plant. He also noted that the functional root biomass became concentrated in the top layers of the soil as the growing season proceeded and that 30 to 60% of the newly formed roots died and decomposed within the growing season.

Warembourg & Paul (1977) labelled native grassland with $^{14}C$ at the Matador Site of the Canadian IBP in southwestern Saskatchewan. The vegetation of this study area was an *Agropyron–Koeleria* association. At varying times following labelling, plant material was sampled by clipping the shoots and taking soil–root cores. Plant carbon was measured by dry combustion and radioassays were done by scintillation counting of an aliquot of sodium hydroxide solution used to trap the carbon dioxide during the combustion (Warembourg & Paul, 1973). Six separate areas of the site were labelled during the 1971 growing season from May through September. The soil atmosphere was sampled by taking small aliquots of soil air and determining $^{14}C$ and total carbon dioxide. The production of $^{14}CO_2$ in the soil was attributed to root and associated microbial respiration. When the soil atmosphere no longer contained measurable levels of radioactivity, it was assumed that movement of $^{14}C$ from the foliage to roots had ceased and the root carbon had been deposited in growth or in storage. At this stage, the $^{14}C$ content of above- and belowground plant material was assumed to represent the net biomass increase attributable to $^{14}C$ assimilation by the plants.

Slightly more radiocarbon was retained by foliage than was stored in the roots (Warembourg & Paul, 1977). The maximum aboveground concentration of $^{14}C$ occurred in the growing season when 65% of the plant $^{14}C$ was retained above ground. This corresponded to the period of most active vegetative growth when the green biomass increased by 50 $g \cdot m^{-2}$. In the drier period of the later summer and fall, 53 and 57% of the labelled carbon remained in the shoots. In the 0–10 cm soil depth 73 to 77% of the labelled roots were recovered.

As much as 50% of the labelled carbon in the roots measured in May was not accounted for in September (Warembourg & Paul, 1977). Losses of $^{14}C$ were reduced when labelling was conducted during later parts of the growing season, but as much as 20% of the labelled carbon in roots which had been labelled in August had disappeared by September. Root $^{14}C$ disappeared at an exponential rate with time. Warembourg & Paul (1977) calculated that the half-life for roots during this growing season was 107 days. Belowground production averaged 43% of the net total production, or about 100 $gC \cdot m^{-2}$.

Warembourg & Paul (1977) determined $^{14}C$ in new shoots which had grown on an area previously clipped in May and then sampled in September. They found that approximately 6% of the $^{14}C$ in roots that was present at the end of May had been utilized in growth of these new shoots. These data, therefore, indicated that root-to-shoot transfers were not negligible. They hypothesized that this transfer would certainly be important in early spring when plants were developing new shoots.

Sinks also compete with each other for their supply of assimilates. Wardlaw (1965) found that when two-thirds of the developing grains in wheat were removed, the translocation to crowns and roots increased markedly. In general, the upper leaves form the source of assimilates for flowers, fruits or developing stems while lower leaves supply the crowns and roots. However, if demand is too intense, all leaves may supply to the same sink (Kursanov, 1963). Clipping or grazing generally results in an initial reduction of assimilate transfer to roots because the regrowing shoots become intense sinks (White, 1973).

Numerous interactions among abiotic and biotic factors can cause changes in the source and sink relationships. For example, any factor which favours high rates of photosynthesis and production of assimilates will increase products for export to sinks. Any factor which inhibits assimilate production will reduce the concentration of carbohydrates at the source.

As growth and phenological development progress, tissue which was at one time a sink may become a source of carbohydrate production (Wardlaw, 1968*b*). Reserve carbohydrates in plant storage organs may be utilized for new growth of aboveground plant parts (Cook, 1966*a*). Hence these storage organs represent a source of available energy for plant growth. However, once photosynthate is produced in excess of demand, these same storage organs may be considered as a sink for receiving the excess carbohydrates. Therefore, the relationships among sources and sinks are quite complex and are never static.

### *Transformation of translocates within the sinks*

Transformation of translocates within sinks often occurs. Sugars translocated to underground parts may be converted into starch or fructosans (Weinmann, 1952). Many factors can influence the conversion of one carbohydrate to another. McCarty (1938) found that cool autumn temperatures resulted in the transformation of reserve carbohydrates to more soluble forms in mountain brome (*Bromus carinatus*). Sucrose concentrations increased at the expense of starch reserves in the autumn. Dodd & Hopkins (1958) found similar transformations in blue grama. Smith & Leinweber (1971) found that changes in morphological development of little bluestem tillers were associated with changes in the concentration of different carbohydrate fractions. Suzuki (1971) found that potassium played an important role in the accumulation of fructosan in timothy. He also found that defoliation stimulated the decomposition of long-chain fructosans to sugars without noticeable accumulation of shorter chain fructosans. Several enzymes play an important role in the transformation of carbohydrates within plants and their activity can be affected by both external environmental factors such as light, water, temperature and nutritional status, and internal conditions such as concentration of certain sugars, cellular hydration and pH. Just as enzymes are formed in the plant, they may likewise be inactivated by formation of internal inhibitors (Kramer & Kozlowski, 1960).

The form of carbohydrates stored varies among plant species. Perennial grasses native to temperate latitudes appear to accumulate fructosans and sucrose in their stem bases and rhizomes (Weinmann & Reinhold, 1946; Ojima & Isawa, 1967). Perennial grasses native to semi-tropical or tropical areas appear to store starch and sucrose (DeCugnak, 1931; Hunter, McIntyre & McIlroy, 1970). Okajima & Smith (1964) found that fructosan was the principal storage carbohydrate in several North American grass species studied, except for smooth bromegrass (*Bromus inermis*) where sucrose was the main storage carbohydrate. In all of the other species except bromegrass, starch was present to the extent of 3 to 6% of the dry matter. The principal reserve substances of grasses are, therefore, sugars, fructosans, dextrins and starch (Weinmann, 1952; Smith & Grotelueschen, 1966).

## Creation and distribution of reserves

The organic compounds of plant tissues that are usually considered to be reserve substances are normally classified as carbohydrates, fats, oils or proteins. However, all organic compounds of plants contain energy and most could probably be metabolized for maintenance or growth under certain circumstances. Cook (1966*a*) stated that reserve substances consist largely of carbohydrates. McIlvanie (1942) found that total nitrogen in roots of bluebunch wheatgrass (*Agropyron spicatum*) was depleted 53% during early vegetative growth, whereas carbohydrates were depleted 70%. He considered carbohydrates to be the most important reserve substances. In some plants, however, water-soluble protein nitrogen may be a significant factor in the initiation of early spring growth (Sheard, 1968).

A broad review of nitrogen distribution and utilization in plants is given by Taylor (1967). There is ample evidence that nitrogen is stored in various tissues, mobilized, and translocated to growing points during periods of high demand. It also appears that during periods of active protein synthesis, nitrogen may be translocated as amino acids to some metabolic sink, built into protein, and then be hydrolysed and retransported back to storage regions prior to the loss of the tissue.

Carbon assimilated by green plants in excess of demands for maintenance, growth and reproduction may be considered as surplus carbon or carbohydrate reserves. These reserve compounds may be stored and later utilized during periods when there are high construction demands (Mooney, 1972). In most climates these demands arise when conditions are favourable for growth or regrowth. However, demands may also come at unpredictable times when shoots are partially or entirely destroyed through defoliation by grazers or fire. Thus carbohydrate reserves are required any time that photosynthesis cannot meet the demands of the plant (Menke, 1973). In some cases, the construction of reproductive parts may also occur from reserves rather than from current photosynthate.

Jameson (1964) believed that carbohydrate reserves were only needed for the production of new photosynthetic tissue. Once a plant became a net producer of carbohydrate, he found no evidence to indicate that there would be an increase in growth rate from additional stored carbohydrates. May (1960) and Alcock (1964) have also questioned the value of plant reserves for plant growth, as most studies showed only indirect evidence for use of reserves in growth and little evidence existed for controls and mechanisms for translocation and utilization of carbohydrates in regrowth. However, apical growth of perennial plants is believed by many to take place at the expense of stored materials, whereas cambial growth may rely on current photosynthate (Kozlowski & Keller, 1966). The pattern of carbohydrate utilization often varies with the species and environmental conditions. The environment can affect the ratio of currently produced carbohydrates to stored carbohydrates that are used in primary growth (Merrill & Kilby, 1952; Mochizuki & Hanada, 1958). For some species, internode elongation is so rapid that it is completed before the leaves are fully expanded. When this occurs, utilization of stored materials is a necessity (Kozlowski & Keller, 1966). Similarly, initiation of new root growth in grasses during early growth before the beginning of leaf expansion must come from stored reserves (Singh & Coleman, 1974; Ares, 1976).

Nonstructural reserve carbohydrates may be stored temporarily in living parenchyma cells in most perennating plant parts. Storage organs for carbohydrate reserves occur both above and below ground. These organs include roots, rhizomes, tubers, stolons, seeds, crowns, leaf bases and stem bases of herbaceous species, and also stems and twigs of woody species (Cook, 1966*a*). Balasko & Smith (1973) located organs for accumulation of nonstructural carbohydrates in timothy by feeding $^{14}C$ to plants at three phenological stages. Roots and leaves were the primary recipients of labelled carbohydrates after 1 week following exposure to labelled carbon dioxide ($^{14}CO_2$) at initiation of stem elongation. Stems and inflorescences accumulated most of the labelled carbohydrates when labelling was done at the time of inflorescence emergence. One week following feeding at anthesis, stems and roots contained 73% of the total $^{14}C$.

Sprague & Sullivan (1950) found that although roots had lower concentrations of carbohydrate reserves than other storage organs in some grasses, the total quantity of carbohydrates stored in roots was greater because the mass of roots was greater. Priestly (1962) stated that it was desirable to determine the total amount of carbohydrate reserves in storage organs rather than only the concentration of carbohydrate reserves in these organs. Total available carbohydrates (TAC) or total nonstructural carbohydrates (TNC) can be utilized as a readily available source of energy (Smith, Paulsen & Raguse, 1964). They include reducing and nonreducing sugars, starch, dextrins and fructosans. They do not include the structural carbohydrates, hemicellulose, cellulose or pentosans.

### *Diurnal variations in nonstructural carbohydrates*

Diurnal variations of total nonstructural carbohydrates (TNC) in several plant parts of switchgrass (*Panicum virgatum*) were studied by Greenfield & Smith (1973). They found that there was no diurnal change in reducing sugar percentages in any shoot component. Total sugar percentages, however, increased from 6.00 a.m. to 6.00 p.m. and then decreased for all shoot parts, except for the lowest three internodes which had similar percentages at 12.00 noon, 6.00 p.m. and 12.00 midnight. Starch percentages were also highest at 6.00 p.m. in the inflorescence, leaf blades, and top leaf sheath and internode. Second and third leaf sheaths were similar in starch at 6.00 p.m. and 12.00 midnight, but starch was highest in the bottom two sheaths and all internodes, except the top one, at 12.00 midnight. Highest percentages of nonstructural carbohydrates occurred at 6.00 p.m. in the inflorescence, leaf blades, leaf sheaths and top two internodes. The bottom three internodes were highest in TNC at 12.00 midnight.

Lechtenberg, Holt & Youngberg (1971) studied the diurnal variations in nonstructural carbohydrates in alfalfa and reported that all nonstructural carbohydrates began to accumulate after 6.00 a.m. Hexoses increased first in leaves and then in stems, followed by sucrose and then starch. These reserves were depleted in the same order in the afternoon and at night. Diurnal variations in the starch content of stems were minor. However, starch appeared to act as a temporary storage product in leaves.

Diurnal cycling of carbohydrate reserves were also studied by Weinmann & Reinhold (1946), Holt & Hilst (1969) and Lechtenberg, Holt & Youngberg (1972). Cycling of individual carbohydrate fractions and total nonstructural carbohydrates were related. The cycles usually began with a low carbohydrate level in the morning hours, a peak in the afternoon and a decline during the evening and night. In general, there was a conversion of simple sugars to disaccharides and then conversion from disaccharides to starch or inulin. Holt & Hilst (1969) observed that almost one-third of the daily production of total nonstructural carbohydrates in smooth bromegrass was utilized during the night.

### *Redistribution and utilization of reserves*

The level of carbohydrates in storage organs of plants is constantly changing. Diurnal and seasonal variations have been observed by numerous investigators and have been partially explained by or associated with the natural phenological development of species and the various abiotic factors which affect growth such as temperature, soil water and soil nutrients.

### *Temperature*

Near-optimum daytime temperatures that promote growth or regrowth tend to reduce carbohydrate reserve levels (Alberda, 1957; Brown & Blaser, 1965, 1970; Auda, Blaser & Brown, 1966; Blaser, Brown & Bryant, 1966; Colby, Drake, Oohara & Yoshida, 1966; Smith, 1968; Davidson, 1969). High temperatures increase respiration rates and consequently cause reductions in available carbohydrates.

During active spring growth, temperature was found to be the most important factor affecting the carbohydrate reserve levels of orchard grass (*Dactylis glomerata*) (Colby *et al.*, 1966). Low temperatures promoted the accumulation of carbohydrates whereas high temperatures caused declines in carbohydrate reserve levels. Feltner & Massengale (1965) accounted for more of the variability in TNC levels with maximum temperatures than with any other single measure of temperature. The deleterious effects of frequent defoliations on TNC reserves were greatest during periods of warm temperatures.

### *Water*

Periods of water stress, in general, tend to increase the level of carbohydrate reserves in plants (Eaton & Ergle, 1948; Brown & Blaser, 1965, 1970; Blaser *et al.*, 1966; Maranville & Paulsen, 1970; Sosebee & Wiebe, 1971; Trlica, 1971; Dina & Kilkoff, 1973).

Carbohydrates increased in Bermuda grass (*Cynodon dactylon*) herbage and orchard grass stubble and herbage when growth was retarded by water stress (Blaser *et al.*, 1966). Fall regrowth of crested wheatgrass and Russian wild rye (*Elymus junceus*), stimulated either by natural rainfall or irrigation, resulted in reductions in TNC stores (Trlica & Cook, 1972). Favourable soil water that promotes plant growth or regrowth tends to lower carbohydrate reserves.

Brown & Blaser (1970) indicated that an increase in percentage of carbohydrate under water stress might be caused by a decrease in the protein and amino acid content of a storage organ. Their study of several species showed that water stress reduced protein formation or caused degradation of protein already formed. Therefore, with decreases in other constituents, the percentage of carbohydrates would be increased.

Dina & Kilkoff (1973) found that starch content of big sagebrush (*Artemisia tridentata*) did not change significantly in water-stressed plants, although sugar content significantly increased in leaves, stems and roots of these plants. Leaf nitrogen content significantly decreased in water-stressed plants, while the stem nitrogen content increased. They believed that sugar increases in the leaves, stems and roots, and nitrogen accumulation in the stems of water-stressed plants might be of significance in protecting the RNA–DNA complex, as well as enzymes, during water stress. Also, if sugars did not increase with drought conditions, cellular injury might occur. They also hypothesized that during water stress conservation of nitrogen occurs through storage in the stem and as water stress became less severe, the nitrogen pool within the stems might be mobilized and utilized for renewed cellular growth.

If accumulation of sugar is an important factor in the resistance of plant organs to water stress, continued translocation would appear to be important in survival of storage organs such as rhizomes and roots which are distant from the site of assimilation. However, need for the protection of the photosynthetic apparatus would necessitate the slowing down of the transfer of assimilates away from the chloroplasts (Santarius, 1967; Santarius & Heber, 1967). It could be that the balance between retention of sugars by the chloroplasts and the transfer of sugars to the conducting system will in part dictate the response of the plants to water stress.

### *Nutrients*

Nutrient enrichment of soils that promotes growth or regrowth often lowers carbohydrate reserves (Benedict & Brown, 1944; Burton, Jackson & Knox, 1959; Brown & Blaser, 1965; Adegbola & McKell, 1966*a*; Blaser *et al.*, 1966; Colby *et al.*, 1966; Paulsen & Smith, 1968; Reynolds, 1969; Zanoni, Michelson, Colby & Drake, 1969; Lechtenberg *et al.*, 1972). Adegbola & McKell (1966*a*) found that the concentration of sucrose and fructosan in stems, stolons, roots and rhizomes of Bermuda grass decreased with increasing rates of nitrogen fertilization. Adegbola & McKell (1966*b*) also found that the regrowth of Bermuda grass fertilized the previous season was closely related to the content of reserve carbohydrates in the stubble and rhizomes.

When heavy application of nitrogen to orchard grass was followed by a period of hot, dry weather, carbohydrate reserves often were reduced to critical levels (Colby *et al.*, 1966), resulting in poor recovery of plant growth and serious injury. Turner (1969) found that nitrogen fertilization of couch grass (*Agropyron repens*) resulted in increased rates of utilization of carbohydrates in rhizomes.

Contrary to the above, several investigators have reported that nutrient enrichment of soils has little or no effect on the carbohydrate reserves of plants, or that it causes increased carbohydrate reserves of plants over long time periods (Barnes, 1960; Matches, 1960; Bommer, 1966; Paulsen & Smith, 1968; Zanoni *et al.*, 1969; White, Brown & Cooper, 1972). Percentage of TNC in roots of alfalfa (*Medicago sativa*) increased with increased rates of potassium fertilization (Matches, 1960). Nitrogen fertilization of three grasses resulted in increases in the carbohydrate reserves of stubble and belowground organs (Bommer, 1966).

In some recent studies, nitrogen fertilization has resulted in decreased carbohydrate reserves during early growth of some species, but in increased reserve levels after more photosynthetic tissue has been produced. Pettit &

Fagan (1974) found lowered TNC reserves in buffalo grass during vegetative and reproductive growth after fertilization with nitrogen. However, they found that TNC accumulated more rapidly during later phenological stages in plants that had received heavy nitrogen applications (90 to 120 kg $N \cdot ha^{-1}$). Nitrogen fertilization resulted in greater depletion of TNC after defoliation of smooth bromegrass, but had no effect on the maximum level of carbohydrates accumulated (Paulsen & Smith, 1968). This probably resulted from the greater photosynthetic area produced by fertilized plants. Application of nitrogen caused a decrease in total nonstructural carbohydrates in stem bases of green needlegrass (*Stipa viridula*) only from the time of growth initiation until the second leaf was formed (White *et al*., 1972).

Plants endowed with a large capacity to store carbohydrates may be better adapted for survival during unfavourable environmental conditions (Wilson, 1944). McCarty & Price (1942) found that high concentrations of sugars in the storage organs of plants were associated with cold resistance and survival. Smith (1964) found a positive relationship between plant carbohydrate reserves and winter survival. He indicated that plants low in reserves could not develop a high level of frost hardiness. Hanson & Stoddart (1940) showed that plants with root systems low in carbohydrate reserves were more susceptible to drought injury. Survival of alpine species near their lower altitudinal limits was shown by Mooney & Billings (1965) to be enhanced by the large amounts of stored nonstructural carbohydrates.

### *Biotic factors affecting redistribution and utilization of reserves*

Numerous comprehensive reviews concerning carbohydrate reserves of plants in relation to phenological development, vigour, production and defoliation have been written (Graber, 1931; Weinmann, 1948, 1961; Troughton, 1957; May, 1960; Priestly, 1962; Jameson, 1963, 1964; Cook, 1966*a*,*b*; Coyne, 1969; White, 1973). Depletion of reserves by excess defoliation results in reduced vigour and herbage growth, and in extreme cases can result in death of the plants (Weinmann, 1948).

#### *Growth and phenological development*

The level of carbohydrate reserves in storage organs of plants varies during the annual growth cycle. The establishment of relationships among the carbohydrate reserve cycles and phenological stages of development of plants allows one to relate the physiological changes occurring in plants to the phenological stages of development.

Carbohydrate reserves usually undergo significant drawdown during initiation of spring growth. McCarty & Price (1942) observed that approximately 75% of the total carbohydrates stored in autumn were consumed in winter respiration and for producing about 10% of the herbage growth in mountain forage plants. Mooney & Billings (1960) found that 50% of the reserves in rhizomes of the alpine plant American bistort knotweed (*Polygonum bistortoides*) were used in a 1-week period during early growth. McCarty (1938) and McIlvanie (1942) found that from 10 to 45% of the annual growth was produced before root reserves ceased to decline. Donart (1969) related carbohydrate reserve cycles of six mountain species to growth and development. Minimum root reserves were reached during early spring growth after approximately 15% of the total annual growth had been produced. Therefore, the duration of the decline and the extent of reserve depletion during early growth differs significantly among individual species.

The storage of carbohydrates in forage plants is often inversely related to herbage growth (Aldous, 1930; Sampson & McCarty, 1930; McCarty, 1935, 1938; Weinmann, 1940; McIlvanie, 1942; Kinsinger, 1953; Mooney & Billings, 1960; McConnell & Garrison, 1966; Coyne & Cook, 1970; Menke, 1973). After the low point in reserve stores is reached, some species immediately start replenishing their carbohydrate reserves and continue throughout the remainder of the growing season until they reach maturity or become dormant. Carbohydrates begin to accumulate when the photosynthetic rate overtakes the carbohydrate utilization rate. McDonough (1969) found that the carbohydrate content of the caudices of tall bluebell (*Mertensia arizonica*) continually increased until shoot senescence. An upward trend in root reserves of crownvetch (*Coronilla varia*) was maintained from late spring through summer (Langille & McKee, 1968).

Total available carbohydrates in the storage regions of crested wheatgrass attained a maximum at the time of flowering and decreased steadily during the remainder of the year (Trlica & Cook, 1972; Sosebee & Wiebe, 1973). Although the downward translocation of assimilates was greatest during the fall, the amount being accumulated was apparently less than the portion utilized in root and tiller growth (Sosebee & Wiebe, 1973). In contrast, McConnell & Garrison (1966) and Menke (1973) found that TNC in roots of bitterbrush (*Purshia tridentata*) declined until seed formation and then accumulated until leaf fall. Menke (1973) also found that the lowest level of reserves in roots of prickly pear cactus (*Opuntia* sp.) was at the flowering stage of development.

Many species of plants exhibit additional drawndown in reserves prior to flowering or during fall regrowth. Crested wheatgrass showed a decrease in carbohydrate reserves just prior to flowering (Hyder & Sneva, 1959). Seven of the eight salt-desert species studied by Coyne & Cook (1970) exhibited carbohydrate reserve drawdown during fall regrowth. This same trend was also evident in crested wheatgrass and Russian wild rye (Trlica & Cook, 1972). However, the general trend of reserve accumulation usually continued upward until plants became quiescent or dormant. Coyne (1969) studied seasonal trends in total available carbohydrates with respect to phenological stage of development for eight salt-desert range species and believed that maximum plant vigour in relation to carbohydrate reserves depended upon storage at the end of the growing season.

Translocation and storage of reserves generally is more active in autumn when plants are completing their annual cycle. A decline usually occurs in stored reserves during the quiescent or dormant period because of respiration and slight growth (Cook, 1966*a*). A considerable proportion of stored carbohydrates may be lost to respiration both during periods of quiescence or dormancy and during formation of new tissue (Weinmann, 1961). Generally, however, carbohydrate reserves in storage organs of perennial plants follow what might be termed a V- or U-shaped annual cycle. Theoretical carbohydrate reserve cycles are illustrated in Figs. 2.24 and 2.25. Plants that exhibit somewhat of a V-shaped seasonal cycle in carbohydrate reserve levels usually have rapid drawdown of reserves for initiation of spring growth followed by a rapid accumulation of reserves after the low point in reserve levels has been reached (Fig. 2.24). Plants that have a U-shaped carbohydrate reserve cycle normally maintain low levels of reserves during active growth, with reserve stores being replenished only after growth rates decline as plants approach maturity (Fig. 2.25). However, the seasonal variation of carbohydrate reserves differs with climatic conditions and among species. For many species the lowest reserve levels occur during early growth, but in other species the reserve level is lowest after seed ripening (Jameson, 1963). The changes through a year in carbohydrate reserve levels are similar in most perennial grasses but may be influenced by the growth behaviour of the species and by weather (Weinmann, 1952).

### *Grazing and defoliation*

A number of studies on the effects of defoliation on plant vigour, production and carbohydrate reserves have established that frequency, intensity and season of defoliation can explain most plant responses to defoliation (Aldous, 1930; Sampson & McCarty, 1930; Graber, 1931; Biswell & Weaver, 1933; Bukey & Weaver, 1939; Hanson & Stoddart, 1940; McCarty & Price, 1942; McIlvanie, 1942; Sullivan & Sprague, 1943, 1953; Weinmann, 1943, 1949; Holscher, 1945; Smith & Graber, 1948; Blaisdell & Pechanec, 1949; Sprague & Sullivan, 1950; Waite & Boyd, 1953; Thaine, 1954; Brougham, 1956; Neiland & Curtis, 1956; Welch, 1968; Trlica, Buwai & Menke, 1977). In general, too heavy, too early, or too frequent grazing or defoliation resulted in declining vigour of vegetation (Hedrick, 1958).

Defoliation at any time during the year usually affects plant carbohydrate reserve levels. Plants are, however, affected more by defoliation at certain phenological stages of development than at others. Donart & Cook (1970)

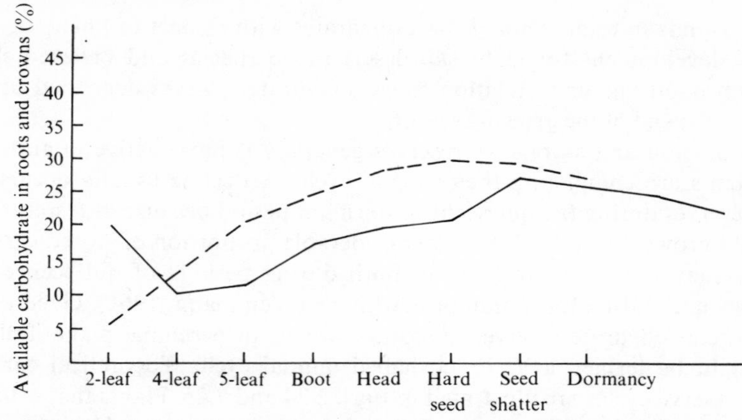


**Fig. 2.24. Carbohydrate balance (solid line) and herbage yield (dashed line) of a typical grass species throughout the annual cycle. (Cook, 1966*a*.)**

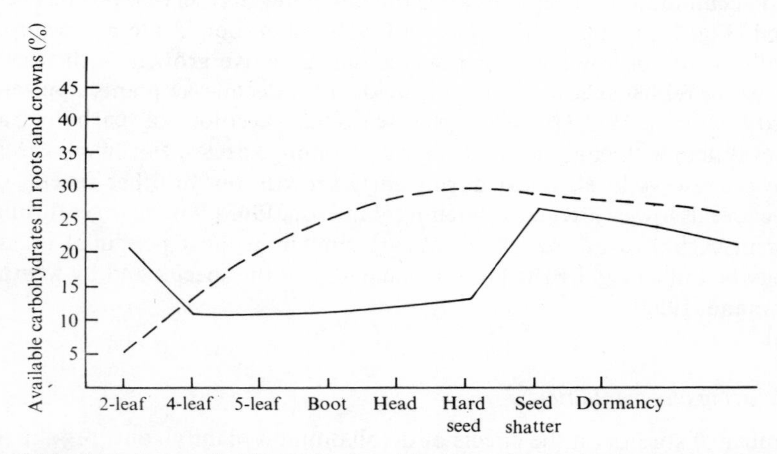


Fig. 2.25. Carbohydrate level (solid line) and herbage yield (dashed line) of a forage species that does not replenish its reserve appreciably until seed formation. (Cook, 1966*b*.)

found that 90% defoliation of two mountain grasses and two forbs early in the growing season, when carbohydrate reserves were low, was more detrimental than defoliation late in the growing season, when carbohydrate reserves were high.

Crested wheatgrass and Siberian wheatgrass (*Agropyron sibericum*) were found to be tolerant of spring grazing but beardless wheatgrass (*Agropyron inerme*) and its close relative bluebunch wheatgrass were sensitive to spring grazing (Hyder & Sneva, 1963*b*). Crested and Siberian wheatgrass accumulated total water-soluble carbohydrates several weeks earlier than did beardless wheatgrass.

Cooper & Watson (1968) found that frequent cutting of alfalfa and sainfoin (*Onobrychis viciaefolia*) was not detrimental if the last cutting was made early enough to provide time for carbohydrate storage in the autumn before killing frosts. Late clipping of big bluestem (*Andropogon gerardii*) depressed reserves available for winter dormancy and resulted in subsequent depression of spring regrowth (Owensby, Paulsen & McKenrick, 1970). Trlica & Cook (1971*a*,*b*, 1972) and Menke (1973) found that severe defoliation of several semi-arid range species was especially detrimental if defoliation occurred when plants were rapidly replenishing reserve stores. This normally coincided with a near-maturity stage of phenological development. Plants that were defoliated late in the growing season usually produced little regrowth, hence little photosynthetic tissue was present during the normal carbohydrate storage period.

Blue grama and western wheatgrass were found to be tolerant of defoliation to 5.0 and 7.5 cm stubble heights, respectively, under several clipping treatments (Fisher, 1966). Everson (1966) found that the concentration and total yield of soluble sugars and starch in storage organs of western wheatgrass were lower under intense clipping treatments. Cook, Stoddart & Kinsinger (1958) found that roots of crested wheatgrass plants clipped at a low stubble height contained a lower percentage of reducing sugars. Roots from plants clipped frequently were low in reducing sugars, sucrose and fructosans. However, Kinsinger & Hopkins (1961) reported that moderately clipped western wheatgrass plants had higher carbohydrate reserves than did unclipped plants. Ogden & Loomis (1972) reported that growth during the fall enabled intermediate wheatgrass (*Agropyron intermedium*) that had been clipped during the summer to recover total water-soluble carbohydrate and root weight to very nearly the same level as control plants.

Willard & McKell (1973) found that clipping little rabbitbrush (*Chrysothamus vicidiflorus*) and snowberry (*Symphoricarpos vaccinoides*) caused reduction in carbohydrate reserves. All three intensities (30, 60 and 90% foliage removal) of defoliation reduced carbohydrate levels. Reserves decreased as the intensity of defoliation increased. Greatest reductions in reserves were noted when plants were clipped during the summer growth period. However, one complete year of rest following clipping in the summers was sufficient to allow full recovery.

One year of rest after severe defoliation may not be sufficient for recovery of many semi-arid and arid range species (Trlica *et al.*, 1977). Menke (1973) found that defoliated plants of fourwing saltbush (*Atriplex canescens*), bitterbrush and fringed sagewort required more than 1 year of rest for recovery of vigour and reserve stores. Cook & Child (1971) indicated that desert plants, when defoliated to the extent that vigour was moderately reduced, required more than 7 years of nonuse for recovery of vigour.

Regrowth following a dormant period or defoliation has been correlated with the carbohydrate reserve status of the plant (Baker & Garwood, 1961; Ward & Blaser, 1961; Reynolds, 1962; Smith, 1962; Hyder & Sneva, 1963*a*; Alcock, 1964; Alberta, 1966; Wolf, 1967; Albuquerque, 1968; Steinke & Booysen, 1968; Laycock & Conrad, 1969; Ueno & Smith, 1970; Greub & Wedin, 1971; Trlica & Cook, 1971*a,b*, 1972). In general, the rate of regrowth was slower and the amount of regrowth was less for plants with lower carbohydrate reserve levels.

Poor regrowth of smooth bromegrass and timothy after the first harvest appeared to be associated with low carbohydrate reserves and inactive basal buds (Reynolds, 1962). Regrowth after defoliation of reed canarygrass (*Phalaris arundinacea*) was correlated with the carbohydrate reserve levels of the plant (Wolf, 1967). Tiller growth rates for tall fescue were associated with amounts of carbohydrate reserves when all laminae were removed (Albuquerque, 1968). Orchard grass tillers with high carbohydrate reserves produced more total dry matter for the first 25 days after defoliation than did tillers with low carbohydrate reserves (Ward & Blaser, 1961).

Davidson & Milthorpe (1965) stated that labile carbohydrate reserves contributed significantly to growth only during the first few days following defoliation. They believed that the extent of the contribution of carbohydrate reserves to growth depended upon the severity of defoliation and on the level of the environmental factors influencing growth and photosynthesis. Conditions optimal for regrowth of perennial ryegrass after repeated defoliation often gave rise to a depletion of the carbohydrate reserves (Alberda, 1957). Factors such as optimal temperature or high nitrogen fertilization which promoted leaf growth tended to lower the soluble carbohydrates after repeated defoliation. Factors which tended to reduce rapid leaf growth or which increased apparent photosynthesis caused increases in carbohydrate reserves.

The size of the reserve pool could influence subsequent regrowth for a significant period of time by influencing the rate of formation of new photosynthetic tissue. Davidson (1969) found a significant positive correlation between root weight and percentage total soluble carbohydrates and hypothesized that root growth rates might be proportional to concentrations of total soluble carbohydrates in roots.

Regrowth potential of couch grass was related to carbohydrate reserves and the number of dormant buds on rhizomes (Schirman & Buckholtz, 1966). Treatments that reduced apical dominance stimulated dormant buds and resulted in enhanced shoot growth. Activation of buds and vigorous regrowth caused a rapid depletion of carbohydrate reserves.

Several researchers have reported a positive correlation between carbohydrate reserve levels of roots of woody plants and the degree of sprouting (Baker, 1918; Aldous, 1935; Jones & Laude, 1960; Tew, 1970). The degree of bud growth may depend on food reserves. However, Wright & Stinson (1970) showed that regrowth following cutting was not in direct proportion to the total carbohydrate reserves in the roots of mesquite (*Prosopis glandulosa*).

Steinke (1969) found that defoliated plants of weeping lovegrass (*Eragrostis curvula*) translocated labelled materials from storage organs to all new leaves where these materials were utilized as the respiratory substrate and for the production of structural material. On the basis of his results, he speculated that immediately after defoliation the reserves were utilized to produce photosynthetic material. Once these first few leaves had been produced, the plant drew on its reserves no longer. However, until downward translocation from new photosynthetic material had begun, the roots were deprived of their regular supply of carbohydrates (Steinke & Booysen, 1968). It was only when the plant was not using carbohydrates in excess of current requirements for growth and respiration that reserves could be restored. Steinke (1969) suggested, therefore, that carbohydrate reserves were important only in determining initial regrowth, but that it was the general carbohydrate status of the plant which determined the overall response to defoliation.

Other researchers have questioned the role of root carbohydrate reserves for regrowth of perennial grasses after defoliation (Mitchell, 1954; Baker, 1963; Moore & Biddiscombe, 1964; Marshall & Sagar, 1965). They believed that reserve carbohydrates were of little importance for regrowth of shoots following defoliation.

## Modelling of translocation and partitioning of assimilates

Several attempts have been made to produce mathematical models to test hypotheses about the mechanisms of translocation. Many of these models have been oversimplified and have not dealt realistically with phloem anatomy and sieve tube structure (Crafts & Crisp, 1971). Dixon (1923), Mason & Maskell (1928*a*), Mason & Phillis (1937), Crafts (1931, 1932, 1933) and Curtis (1935) were some of the earlier workers who utilized the mathematical approach. More recently, Spanner (1958, 1962, 1963) and Spanner & Prebble (1962) utilized more detailed analyses. Canny (1960*a,b*, 1961, 1962*a,b*, 1973), Canny & Phillips (1963) and Canny & Ashkam (1967) utilized the diffusion-analogue concept in translocation calculations. However, Horwitz (1958) took an unbiased approach and used a number of proposed mechanisms. Webb & Gorham (1965) concluded that at present insufficient and conflicting quantitative information on translocation exists. It was, therefore, doubtful whether the process could at present be expressed in mathematical terms. They believed that a mathematical analysis of such a complicated physiological system should extend beyond the relatively simple treatments of Horwitz (1958), Canny (1962*a*) and Evans, Epert & Moorby (1963).

Although many models of photosynthesis do indeed accurately predict photosynthetic rates under varying environmental conditions, what happens to the photoassimilates once they have been produced is more obscure. Some of the photosynthates are naturally utilized in respiration processes. However, the fate of remaining photosynthates has usually not been specified unless they are simply added to biomass of aboveground plant organs. The partitioning of new photosynthates among various plant activities is a key aspect for the development of photosynthetic systems (Loomis, 1970). Monsi & Murata (1970) made this clear with a simple model with constant distribution ratios where large differences in total weight and constituents of the productive system could result from varying the distribution ratios. In their experiments, they showed that the distribution ratios are not constant, but that change occurred continuously during a season.

A vital part of whole plant or plant community models is concerned with partitioning of available assimilates among various plant organs. Assimilate distribution and utilization is frequently dealt with in an empirical or semi-empirical manner and any explicit role that the processes of translocation and utilization may have in determining the pattern of growth has usually been ignored (Thornley, 1972*a*). Such an approach has the disadvantage that when the performance of the simulated plant or community is compared with the actual data, the agreement or lack of agreement between the two may have been caused by the structural relationships that were built into the model or to empirical adjustments. This greatly weakens the use of the model for predictive purposes or as a tool for exploring the importance of underlying mechanisms (Thornley, 1972*a*).

Monsi (1968) described a model whereby gross production was distributed for construction of new photosynthetic and nonphotosynthetic systems ($\Delta F$ and $\Delta C$) including constructive respiration and maintenance respiration of each system. His calculations indicated the importance of $\Delta C/\Delta F$ ratios in affecting growth rate. He believed that investigations of the internal physiology of plants as related to the distribution ratio and translocation of assimilates were needed. Ničiporovič (1968) pointed out that differences in behaviour from one plant to another gave rise not only to quantitative differences in production, but also to differences with respect to distribution and utilization of photosynthetic products.

Distribution of assimilates was governed by allotment factors in simple models described by Monsi & Murata (1970). Changes in production and dry matter distribution were then evaluated by varying the values of these allotment and distribution ratios. Their models and experimental data demonstrated that the distribution of assimilates, particularly the portion allotted to the photosynthetic system, could determine the growth and ultimately the yield of plants and plant communities. However, the factors which regulated the distributions remained as subjects for future research.

The distribution of carbon within the plant is determined by internal control, environment and the stage of plant development. Some of the photosynthate translocated from leaves to stems and roots must be respired for maintenance. Therefore, the net photosynthate available for growth ($P_{ng}$) may be determined by:

$$P_{ng} = P_{gg} - (R_l L + R_s S + R_r R),$$

where $P_{gg}$ is gross photosynthate, and $R_l$, $R_s$, and $R_r$ are maintenance respiration per unit of leaves ($L$), stems ($S$) and roots ($R$), respectively (Curry & Chen, 1971).

Distribution factors of biomass were programmed in a model of plant growth by Curry & Chen (1971) as function generators. These function generators were triggered by an integrated environmental factor, heat units. The heat unit was believed to control the developmental stage and was calculated by:

$$\text{Heat unit} = \left(\frac{T_{max} + T_{min}}{2}\right) - 10 \times \text{day}.$$

In the ELCROS model of de Wit *et al.* (1970), leaves grow at the expense of photosynthetic products. The growth rate was made dependent upon age of the tissue, the amount of reserve products and the availability of water and temperature. In this model, separation of structural tissue and available reserves was made in the distribution of dry matter accumulation. Unfortunately, few models at present have attempted to make this separation.

A model developed by Thornley (1972*a*) was used to describe the partitioning of assimilates in plants and how partitioning patterns would respond to a parameter that was partially determined by the environment, namely gross photosynthetic rate. The model was based on two assumptions: (1) utilization of substrate for growth was dependent upon local substrate concentration according to the Michaelis–Menten equation, and (2) the transport of substrate between two organs in the plant was determined by substrate concentration difference divided by a resistance. Qualitatively, it was established that the response of the model was of the right sort. The main disadvantage of the present model was that growth was dependent upon a single substrate supplied by leaves, which could lead to an imbalance between the plant parts. However, this disadvantage was corrected in a later model (Thornley, 1972*b*) in which plant parts were made interdependent. In the later model, growth of different plant parts was made explicitly dependent on the activity of other plant parts and exhibited a more balanced response to environmental changes.

Total net production of leaves was allocated between belowground roots and rhizomes for storage and growth, and the canopy for producing new leaves and stems, in a preliminary model described by Miller & Tieszen (1972). Net production was allocated so that the aboveground : belowground ratio approached a previously defined figure. The rationale for this constraint was that as leaf area increased, total photosynthesis would increase either until shading within the canopy resulted in low or negative photosynthesis at the lower canopy levels or until the transpiration rate exceeded the rate of water uptake and leaves wilted. Material was allocated to each canopy level in proportion to the net production of that level. Any surplus material was used to create a new level at the top of the canopy. The model calculated a steady increase of aboveground material through the growing season and a steady decrease in belowground material. Miller & Tieszen believed that the steady decrease in belowground material was unrealistic and that possibly a switch function could be used to divert material belowground during certain periods of the growing season.

Transfer of carbon from shoots to roots in the primary producer submodel of ELM was regulated by seven factors: shoot weight, live root weight, and the effects of soil water potential, nitrogen, phosphorus, phenology and grazing (Sauer, 1978). Most of these variables were scaled between 0.2 and 0.9. These scaling factors were then adjusted, depending upon environmental conditions, so that a seasonal average of approximately 85% of the net photosynthate under ungrazed situations was translocated to roots and crowns. If plant stress was being caused by one of the factors, shoot-to-root transfer was set at the maximum value. As plants matured, the fraction of net photosynthate flowing to the roots and crowns linearly increased to the maximum level. Lower translocation rates during vegetative growth resulted in faster shoot growth rates and a more rapid approach to peak leaf area, while the increased transfer with maturity replenished reserves.

The actual transfer of net photosynthate to the roots and crowns was determined by choosing the largest of the independently determined fractions (Sauer, 1978). The largest, rather than the smallest, of the fractions was used because the flow of carbon to the roots was based on conditions which increased, not decreased, root : shoot ratios. The partitioning of the translocates to the crowns and roots was a constant, adjusted to maintain crowns and roots in relative weights comparable with initial conditions.

Crowns in the primary producer submodel by Sauer (1978) represented storage capacity. At the beginning of the growing season before shoot weight exceeded 2 to 5% of a producer's total live carbon weight, carbon in crowns was translocated to the shoot system as a function of soil water potential and soil temperature. Increasing soil temperature increased crown metabolism, which increased translocation to shoots. The ratio of live shoot to total live weight was used to stop crown-to-shoot transfer of carbon. This ratio was also utilized to initiate translocation from shoots to crowns.

A mathematical model of the root biomass dynamics of a shortgrass prairie was developed by Ares & Singh (1974). In this model a pool of carbon located in the aboveground plant organs was available for translocation to the root system. This compartment was assigned a high initial value in the simulation, and the flow of carbon to roots and reserves in the crowns was determined by four coefficients and by the magnitude of the recipient compartment. Sensitivity analysis of the model indicated that the rate of translocation of carbon from the aboveground carbon pool to the nonsuberized roots, their rate of respiration and death, and the rate of their aging were crucial in changing the pattern of carbon distribution in the whole shortgrass ecosystem. Additional research in this area was, therefore, indicated.

Miller & Tieszen (1972) found that most models of photosynthesis or primary production usually had not included: (1) the effect of temperature on respiration and photosynthesis; (2) the relationship of leaf absorbancy to solar radiation; (3) the effects of transpiration on leaf water stress, of leaf water stress on resistance to carbon dioxide exchange, and of leaf resistance to transpiration and photosynthesis; (4) the effect on photosynthetic parts of shading by nonphotosynthetic parts; and (5) multi-species canopies. They pointed out that these omissions may not be important for calculating primary production of agricultural crops when water and temperature are usually not limiting and leaf area of the species of interest comprises the majority of the canopy mass. However, primary productivity of species occurring naturally is often limited by the various driving variables and interactions among biological components of the ecosystem. They believed that these factors should be considered in any primary production model for natural ecosystems. Fortunately, many of the models developed in the IBP programs have utilized additional driving variables and have considered interactions among components of the ecosystem. These models have, therefore, advanced our understanding of various ecosystem processes.

In summary, it appears that certain processes have been sufficiently studied to allow construction of realistic mechanistic models. Electrical analogues for diffusion processes appear to be firmly established in modelling theory. For example, the concepts of resistance to gas diffusion now appear in most mechanistic models. Even here, however, the concepts of chemical resistance and intracellular liquid phase resistances are not entirely clear. Biochemical processes are generally less well known and, therefore, are more difficult to model.

Moving from individual process modelling to whole plant and stand simulation introduces many new problems. The theoretical approach to assimilate distribution is less advanced. Monteith (1972) proposes an extension of the electrical analogy to the movement of assimilates through the plant. Better models of plant development, leaf growth and death in relation to environment are needed. More knowledge of root processes is also greatly needed in future models.

The argument may be made that concern over details of processes is not necessary when modelling at higher levels of organization, where a simple 'black-box' approach to carbon dioxide assimilation and other processes may be adequate. However, if a model is to be explanatory, detail at lower levels of organization is essential. Models combining several levels of organization probably should be the major goal of modelling in the future. In this way, the results can have application both in practical terms of management of ecosystems and in terms of our basic understanding of those systems.

### *Shoot mortality*

The amount of plant material transferred to the standing dead compartment depends upon the amount of initial biomass present as well as the amount elaborated during the growing season and the longevity of the shoots. Warembourg & Paul (1977) labelled mixed-grass prairie plots in Canada with $^{14}CO_2$ and determined the distribution of the aboveground labelled carbon between green, standing yellow and grey material, and litter as a function of time following labelling. As expected, the green plant material contained 100% of the aboveground label immediately after assimilation of $^{14}CO_2$. Minimum transfer of carbon between the above categories occurred in June and July when only 8% of the $^{14}C$ moved from green to yellow material in 28 days. The rate of transfer from green to other categories was maximum when the soil water was low and during later summer months.

Kelly *et al.* (1969) have determined the transfer from green shoot compartment to standing dead compartment by summing the positive increases in standing dead biomass through the year, and by summing the losses from the live compartment. They found that values obtained through the above two methods were considerably different. Thus in an *Andropogon* community summing the losses of live material yielded a value of 417 $g \cdot m^{-2}$, while the sum of the positive increases in actual standing dead was only 288 $g \cdot m^{-2}$. In this case it would appear that not all the material lost from the live compartment is necessarily going to the standing dead, but some is instead being translocated to the roots, directly to the litter, or remaining in the standing dead compartment for such a short period that it is not detected in the sampling scheme used.

Singh & Yadava (1974) and Sims & Singh (1978) have calculated shoot mortality by adding the aboveground net production value for the sampling period to the initial live biomass and subtracting from this the value of live biomass at the end of the season. Transfer rates for 10 North American grasslands calculated as above are given in Table 2.11 (Sims & Singh, 1978). The mean daily rates of transfer of material into the standing dead compartment average 0.95 and 0.86 $g \cdot m^{-2} \cdot d^{-1}$ for the ungrazed and grazed treatments, respectively, in these grasslands. The minimum amount of accumulation in standing dead occurred in the desert grassland (Jornada: 35 to 118 $g \cdot m^{-2} \cdot yr^{-1}$) while the maximum amount occurred in the tallgrass prairie (Osage: over 429 $g \cdot m^{-2} \cdot yr^{-1}$).

In the tropical grasslands in India, Singh (1968, 1973) and Singh & Yadava

Table 2.11. *Net seasonlong accumulation (SA) and accumulation rates (AR) of aboveground standing dead (SD) and litter (L) compartments of some US grasslands*

| | | Ungrazed | | | | Grazed | | | |
|---|---|---|---|---|---|---|---|---|---|
| | | SD | | L | | SD | | L | |
| Site | Year | SA[a] | AR[b] | SA | AR | SA | AR | SA | AR |
| ALE, | 1971 | 69 | 0.31 | 88 | 0.40 | 15 | 0.07 | 4 | 0.02 |
| Washington | 1972 | 103 | 0.48 | 114 | 0.54 | 84 | 0.39 | 131 | 0.62 |
| Bison, Montana | 1970 | 246 | 1.23 | 274 | 1.37 | 153 | 0.77 | 149 | 0.75 |
| Bridger, | 1970 | 73 | 1.16 | 89 | 1.41 | 88 | 1.40 | 75 | 1.19 |
| Montana | 1972 | 97 | 1.35 | 127 | 1.76 | 88 | 1.22 | 125 | 1.74 |
| Cottonwood, | 1970 | 188 | 0.87 | 153 | 0.71 | 159 | 0.74 | 176 | 0.82 |
| South Dakota | 1971 | 209 | 0.97 | 114 | 0.53 | 201 | 0.93 | 90 | 0.42 |
| | 1972 | 261 | 1.16 | 196 | 0.87 | 146 | 0.65 | 105 | 0.47 |
| Dickinson, North Dakota | 1970 | 263 | 1.78 | 166 | 1.12 | 252 | 1.70 | 95 | 0.64 |
| Hays, Kansas | 1970 | 363 | 1.19 | 394 | 1.30 | 371 | 1.22 | 376 | 1.24 |
| Jornada, | 1970 | 90 | 0.50 | 52 | 0.29 | 92 | 0.51 | 79 | 0.44 |
| New Mexico | 1971 | 118 | 0.34 | 109 | 0.32 | 35 | 0.10 | 21 | 0.06 |
| | 1972 | 100 | 0.36 | 131 | 0.47 | 81 | 0.29 | 73 | 0.26 |
| Osage, | 1970 | 331 | 1.42 | 0 | 0.00 | 421 | 1.81 | 202 | 0.87 |
| Oklahoma | 1971 | 396 | 1.79 | 170 | 0.26 | 466 | 2.11 | 250 | 1.13 |
| | 1972 | 290 | 1.40 | 707 | 3.42 | 367 | 1.77 | 395 | 1.91 |
| Pantex, | 1970 | 111 | 0.80 | 103 | 0.74 | 103 | 0.74 | 82 | 0.59 |
| Texas | 1971 | 248 | 1.40 | 151 | 0.85 | 172 | 0.97 | 116 | 0.66 |
| | 1972 | 174 | 0.71 | 194 | 0.80 | 174 | 0.71 | 197 | 0.81 |
| Pawnee, | 1970 | 101 | 0.66 | 146 | 0.96 | 70 | 0.46 | 78 | 0.51 |
| Colorado | 1971 | 128 | 0.66 | 89 | 0.46 | 2 | 0.01 | 32 | 0.16 |
| | 1972 | 48 | 0.40 | 72 | 0.61 | 28 | 0.24 | 101 | 0.85 |

From Sims & Singh (1978).

[a] In $g \cdot m^{-2}$. [b] In $g \cdot m^{-2} \cdot d^{-1}$.

(1974) have reported that massive transfer from the live to standing dead compartment occurred during the post-monsoon period, after maturity of herbage produced in the rainy season.

Once plant material has reached the standing dead compartment, it can remain in this compartment for varying lengths of time; eventually, however, all material in the standing dead compartment must be transferred to litter. Kelley *et al.* (1969) estimated that the mean value for the rate of transfer to the litter compartment in an *Andropogon* community is 0.98 $g \cdot m^{-2} \cdot d^{-1}$ with a range from 0.47 to 1.57 $g \cdot m^{-2} \cdot d^{-1}$ during different sampling intervals, as compared to a mean rate of 0.69 $g \cdot m^{-2} \cdot d^{-1}$ (range 0.52 to 0.87 $g \cdot m^{-2} \cdot d^{-1}$) for the *Festuca* community (Table 2.12). According to Sims & Singh (1978) the seasonlong net accumulation of litter on ungrazed areas averaged about

Table 2.12. *Monthly rates of litter fall in* Andropogon *and* Festuca *communities* $(g \cdot m^{-2} \cdot d^{-1})$

| Period | Community | |
|---|---|---|
| | *Andropogon* | *Festuca* |
| 15 March to 19 May | 0.84 | 0.52 |
| 19 May to 17 June | 0.81 | 0.70 |
| 17 June to 17 July | 0.81 | 0.80 |
| 17 July to 17 August | 0.47 | 0.87 |
| 17 August to 18 September | 1.32 | 0.83 |
| 18 September to 18 October | 1.57 | 0.59 |
| 18 October to 18 November | 0.83 | 0.53 |
| 18 November to 18 December | 1.17 | 0.66 |
| *Mean* | 0.98 | 0.69 |

From Kelly *et al.* (1969).

292 $g \cdot m^{-2}$ on the tallgrass prairie, 200 $g \cdot m^{-2}$ on the mixed-grass prairie, and about 126 $g \cdot m^{-2}$ on the shortgrass prairie. The grazed areas in these three grassland types averaged approximately 84% of the ungrazed expression. On the temperate mountain grasslands the seasonlong net accumulation of litter averaged around 108 $g \cdot m^{-2}$ compared to about 101 $g \cdot m^{-2}$ for the northwest bunchgrass type (ALE: Table 2.11). Considering all the ungrazed temperate grasslands together, the litter was produced at a rate of 0.87 $g \cdot m^{-2} \cdot d^{-1}$ compared to 0.71 $g \cdot m^{-2} \cdot d^{-1}$ for grazed temperate natural grasslands. For a tropical mixed grassland in India, Singh & Yadava (1974) have estimated that about 84% of the standing dead material is transferred to the litter compartment each year. This accounts for approximately 73% of annual aboveground net production.

### Root death and turnover

The classical root banding studies of Weaver & Zink (1946) indicated that the roots of prairie grasses live an average of 2 to 3 years. Recently Ares (1976) determined from root observation windows that 30 to 60% of the currently produced roots of blue grama die and disappear within a growing season during periods of extreme water stress. Using $^{14}CO_2$ labelling, Singh & Coleman (1974) recorded 67 to 76% by weight of functional roots in the first 0–20 cm of soil as compared with 29 to 44% in the lower depths. Of the total roots occurring to a depth of 60 cm, 62% were found to be functional. The proportion of functional roots remained fairly constant throughout the season, indicating an efficient complementarity of root growth and mortality.

Root turnover implies the rate of replacement of roots. Through an ingenious method of calculation, Dahlman & Kucera (1965) estimated that about

Table 2.13. *Turnover of root biomass on four North American ungrazed and grazed grasslands*

| Site[a] | Depth (cm) | Year | Ungrazed | Grazed |
|---|---|---|---|---|
| Cottonwood | 0–30 | 1970 | 0.29 | 0.18 |
| | | 1972 | 0.14 | 0.10 |
| Osage | 0–50 | 1970 | 0.11 | 0.12 |
| | | 1971 | 0.28 | 0.40 |
| | | 1972 | 0.68 | 0.22 |
| Pantex | 0–30 | 1970 | 0.13 | 0.26 |
| | | 1971 | 0.72 | 0.74 |
| | | 1972 | 0.65 | 0.59 |
| Pawnee | 0–30 | 1971 | 0.65 | 0.37 |
| | | 1972 | 0.29 | 0.48 |

From Sims & Singh (1978).
[a] See Table 2.11 for locations of sites.

Table 2. 14. *Turnover*[a] *of roots in certain tropical grasslands of India*

| Site | Grassland type | Turnover value | Reference |
|---|---|---|---|
| Varanasi | *Dichanthium* | 0.52 | Singh (1967) |
| Varanasi | *Dichanthium* | 0.54 | Choudhary (1967) |
| Varanasi | Mixed-grass | 0.64 | Agarwal (1970) |
| Sagar | *Heteropogon* | 0.83 | Jain & Mishra (1972) |
| Ujjain | *Dichanthium* | 0.45 | Data of C. M. Misra |
| Ratlam | *Sehima* | 0.47 | Billore (1973) |
| Kurukshetra | Mixed-grass | 0.97 | Singh & Yadava (1974) |

[a] $\text{Turnover} = \dfrac{\text{belowground net production}}{\text{maximum belowground biomass}}.$

25% of roots in a tallgrass prairie are replaced each year. Using the same method Sims & Singh (1978) have calculated the root turnover values for four North American grasslands (Table 2.13). The rates of replacement in these grasslands range from 11 to 74% annually.

Singh & Yadava (1974) reported that in the mixed grassland at Kurukshetra, India, the maximum amount of root biomass was replaced during winter (67%). Table 2.14 includes values of some turnover rates for certain tropical grasslands. In general 45 to 97% of roots are replaced each year in these grasslands. Strong seasonality in climate, more than one growth flush each year, and the high temperatures are responsible for these higher rates.

## Modelling of shoot and root mortality and turnover

Although elaborate models of photosynthesis and primary productivity have been constructed, few models to date have incorporated shoot and root mortality and turnover. Only fragmented knowledge exists as to how to attempt to model these processes. In addition, little information exists as to what driving variables are important or even how they affect these processes.

Shoot death and transfer to a standing dead category were assumed by Sauer (1978) to be related to phenology, air temperature and drought. The dependence of shoot death upon phenology in this primary producer submodel for ELM was simulated by having a linear increase in shoot death as phenology progressed from flowering to senescence. Soil water potential caused a linear increase in shoot mortality at water potentials that were less than the permanent wilting point. Cold temperatures which would cause frost damage also caused live shoot biomass to be transferred to a standing dead category in this submodel. Standing dead material became litter as a function of rainfall intensity and the water content of the standing dead. Transfer of standing dead to litter was almost totally dependent upon rainfall events. Unfortunately, wind and trampling were not considered in this transfer.

Root mortality in Sauer's (1978) submodel served as a connection between the primary producer and decomposer compartments. Root death was dependent upon soil water potential, phenology and root weight. In moist soils, a small but constant death rate was assumed. However, root death rate doubled at soil water potentials at the permanent wilting level. For annual plants, phenology was used to decrease root death of young plants and increase root death rate of older plants so that roots and shoots died simultaneously. Complete root biomass turnover varied between every 2 and every 4 years.

Ares & Singh (1974) utilized time-varying coefficients to regulate transfers of material from live roots to dead roots for a shortgrass prairie dominated by blue grama. Mortality rates had two different levels: a lower level characteristic of the growing season and a higher level corresponding to the fall and winter. The mean level of each of the coefficients was chosen to represent adequately the observed relations between live and dead carbon content of each of the compartments representing live and dead tissue of each category of roots. Sensitivity analysis indicated that the model was greatly affected by changes in mortality rates of nonsuberized roots.

In a model of simulated dynamics of blue grama production, Brown & Trlica (1977*c*) utilized soil water potential, air temperature and the amount of aboveground biomass to control the rate of shoot death. If the soil water potential was below −35 bars and the air temperature was above 39 °C, shoot death occurred at a rate proportional to the difference between the actual temperature and 39 °C. For low temperatures, shoot death occurred irrespective of soil water at the rate of 5% of the aboveground biomass for each degree Celsius below 4 °C. Root death rate in this model was based on estimates that the root biomass replacement rate of blue grama was 25% per year. Model predictions of net primary productivity of blue grama were similar to field data reported by Lauenroth (1973).

In a preliminary model of primary productivity in the Arctic tundra, Miller & Tieszen (1972) allocated photosynthate to all canopy layers. Any surplus material was used to create a new level at the top of the canopy. Canopy levels with negative production contracted; thus the canopy grew taller and leaves at the bottom of the canopy senesced when they became nonproductive. Accumulation of dead material reduced production because of shading of photosynthetic material.

In the plant production submodel developed for the Canadian IBP programme, the fate of litter and roots was not taken into consideration (Saugier *et al.*, 1974). Therefore, these producer compartments were simply allowed to accumulate biomass. Shoots used 65% of the net photosynthate and 35% was transferred to roots. Death rate of leaves was assumed to be a linear function of osmotic potential. Transfer of standing dead to litter was assumed to be a linear function of rainfall intensity.

Warembourg & Paul (1977) utilized $^{14}C$ techniques in studying carbon transfer through the soil–plant system. Their data indicated that root turnover in a Canadian grassland system followed a standard first-order reaction. The half-life of roots ($T$) was calculated to be 107 days using the following equation:

$$V = V_0 \exp\left(\frac{-0.693t}{T}\right),$$

where $V$ is the root material at time $t$ and $V_0$ is the initial root biomass.

### *Efficiency of energy capture in net production*

At low light intensities, the photosynthetic rate of individual leaves is limited by light energy only, and the leaf can fix about 12 to 15% of the incoming light energy, approaching the theoretical value of 18 to 20% expected on a photochemical basis (Cooper, 1970). As the light intensity increases, the photosynthetic rate becomes limited by other factors, such as the amount of carbon dioxide which can penetrate to the active sites in the chloroplasts or biochemical reactions, and eventually a saturation level is reached. In many temperate species the maximum photosynthetic rate is about 20 to 30 mg $CO_2 \cdot dm^{-2} \cdot h^{-1}$ which corresponds to 1.5 to 2.0 $mg \cdot cm^{-2} \cdot d^{-1}$ or an efficiency of 2 to 3%. In contrast, many tropical species have fixation values up to 70 mg $CO_2 \cdot dm^{-2} \cdot h^{-1}$ which is equivalent to 5 to 6% efficiency.

In the canopy of a grassland the lower leaves are shaded, which results in a more efficient conversion of incoming energy. Temperate grass canopies

Table 2.15. *Net primary productivity and efficiency of energy capture in some temperate US grassland communities (including the US/IBP Grassland Biome sites)*

| Grassland[a] | Productivity $(g \cdot m^{-2} \cdot d^{-1})$ | Efficiency (%) | Reference |
|---|---|---|---|
| Tallgrass prairie, Missouri | 5.5 | 1.2[b] | Kucera, Dahlman & Koelling (1967) |
| Tallgrass prairie, Colorado | 8.0[c] | 1.2[b] | Moir (1969*a*) |
| *Festuca thurberi*, Colorado | 9.0[c] | 1.2[b] | Turner & Dortignac (1954) |
| One-year weed field, New Jersey | 19.5 | 3.8[b] | Botkin & Malone (1968) |
| Shortgrass steppe, Colorado | 9.0[d] | 1.3[b] | Klipple & Costello (1960) |
| Shortgrass steppe, Colorado | 22.0[d] | 3.4[b] | Moir (1969*a*) |
| *Poa compressa*, Michigan | | 1.1[b] | Golley (1960) |
| Old-field broomsedge, South Carolina | | 0.33–0.39[e] | Golley (1965) |
| Prairie, Minnesota | | 0.61 | Ovington & Lawrence (1967) |
| Meadow steppe, West Siberia | 15.0 | 3.4[b] | Rodin & Bazilevich (1967) |
| Arid steppe, USSR | 11.0 | | Rodin & Bazilevich (1967) |
| Mixed prairie, Canada | | 0.6–0.12[e] | Redmann (1975) |
| Mixed prairie, Dickinson (U) | 4.96 | 0.73 | Sims & Singh (1971) |
| Mixed prairie, Dickinson (G) | 6.11 | 0.92 | |
| Northwest bunchgrass grassland, Bison (U) | 3.06 | 0.60 | |
| Northwest bunchgrass grassland, Bison (G) | 3.40 | 0.69 | |
| High mountain grassland, Bridger (U) | 8.89 | 0.97 | |
| High mountain grassland, Bridger (G) | 10.74 | 1.2 | |
| Shortgrass prairie, Pawnee (U) | 3.95 | 0.57 | |
| Shortgrass prairie, Pawnee (G) | 5.38 | 0.80 | |
| Shortgrass prairie, Pantex (U) | 2.06 | 0.19 | |
| Shortgrass prairie, Pantex (G) | 5.64 | 0.53 | |
| Mixed prairie, Cottonwood (U) | 3.92 | 0.47 | |
| Mixed prairie, Cottonwood (G) | 4.17 | 0.51 | |
| Mixed prairie, Hays (U) | 2.39 | 0.46 | |
| Mixed prairie, Hays (G) | 3.53 | 0.63 | |
| Tallgrass prairie, Osage (U) | 2.25 | 0.44 | |
| Tallgrass prairie, Osage (G) | 3.85 | 0.77 | |
| Desert grassland, Jornada (U) | 1.58 | 0.16 | |
| Desert grassland, Jornada (G) | 1.41 | 0.13 | |
| All sites, USA (U) | 3.18 | 0.46 | |
| All sites, USA (G) | 4.04 | 0.57 | |

From Sims & Singh (1971).

[a] U, ungrazed grassland; G, grazed grassland.
[b] Values based on 45 % of total daily solar insolation (Moir, 1969*b*).
[c] Assuming root biomass to be 0.7 total biomass (Moir, 1969*b*).
[d] Assuming root biomass to be 0.9 total biomass (Moir, 1969*b*).
[e] Based on half total annual incident solar radiation.

can achieve conversion efficiencies of 5 to 6% of the incoming energy in the summer and 12 to 15% in the lower light intensities of late autumn and early winter, and proportionately higher values for tropical species. Sims & Singh (1971) have calculated efficiencies of some US grasslands and these values (Table 2.15) are somewhat lower than other published efficiencies (Cooper, 1970).

Brown & Trlica (1977*b*) have computed efficiency of energy capture by blue grama on the basis of carbon dioxide exchange studies. They found the following efficiency values for gross photosynthesis ($P_G$) and net photosynthesis ($P_N$) using total visible irradiance: 28–29 June, $P_G = 1.97$ and $P_N = 1.72\%$; 6–7 July, $P_G = 1.70$ and $P_N = 1.58\%$; 11–12 August, $P_G = 0.56$ and $P_N = 0.24\%$; and 22–23 September, $P_G = 0.99$ and $P_N = 0.91\%$. The lowest energy conversion efficiencies occurred during the relatively high water stress conditions experienced during 11–12 August.

Recently Singh & Joshi (1976) have calculated efficiency of energy capture by tropical grasslands using harvest data. The efficiency values ranged from 0.23% for a semi-arid grassland of Pilani to 1.66% for a dry subhumid grassland at Kurukshetra, India.

### Water use

It has been commonplace in crop science for many years to predict net primary production from environmental data. The practice has been less common in ecological science, probably because of the comparative scarcity of field data against which such predictions could be tested. The results of the US/IBP Grassland Biome studies are particularly suitable for this purpose because of the subcontinental distribution of sites and the wide range of environmental conditions encountered (Table 2.16). These are discussed here as a case study example showing the dependence of primary productivity on available water.

Available water, $W_T$, as calculated for these comparisons, has two components: nongrowing season available water, $W_{NGS}$; and growing season available water, $W_{GS}$, such that

$$W_T = W_{NGS} + W_{GS}.$$

The growing season used was the thermal growing season, defined as the period of the year when the 15-day running mean air temperature exceeds 4 °C. The use of the 15-day running mean is used for computational ease, as it gives a more sharply defined beginning and end to the season than do means calculated for shorter periods. The nongrowing season is then the period between the end of one growing season and the beginning of the next.

Maximum soil water storage capacity, $W_{S(max)}$, was defined as the volume of water retained at field capacity in that depth of soil in which 90% of the plant roots (dead and alive) occur.

Table 2.16. *Comparative climatology of US/IBP Grassland Biome Sites, 1970 and 1971*

| Site[a] | Latitude and longitude | Grassland type | Year | Duration thermal growing season[b] (days) | Annual precipitation (cm) | Average annual solar radiation[c] (cal·cm$^{-2}$) |
|---|---|---|---|---|---|---|
| ALE | 46° 18′N | Northwest-bunchgrass | 1970 | 255 | 29 | 357 |
| | 119° 18′W | | 1971 | 232 | 25 | |
| Bison | 47° 19′N | Rough fescue | 1970 | 186 | 40 | 222[d] |
| | 114° 06′W | | 1971 | 207 | 30 | |
| Bridger | 45° 46′N | High mountain | 1970 | 105 | 98 | 379 |
| | 111° 03′W | | 1971 | 111 | 32 | |
| Dickinson | 46° 48′N | Mixed prairie | 1970 | 169 | 51 | 373 |
| | 102° 48′W | | 1971 | 202 | 54 | |
| Cottonwood | 43° 58′N | Mixed prairie | 1970 | 188 | 40 | 384 |
| | 101° 52′W | | 1971 | 214 | 68 | |
| Pawnee | 40° 48′N | Shortgrass prairie | 1970 | 167 | 23 | 400 |
| | 104° 46′W | | 1971 | 208 | 27 | |
| Hays | 38° 51′N | Mixed prairie | 1970 | 227 | 47 | 410 |
| | 99° 16′W | | 1971 | 246 | 60 | |
| Osage | 37° 07′N | Tallgrass prairie | 1970 | 273 | 65 | 410 |
| | 96° 10′W | | 1971 | 271 | 92 | |
| Pantex | 35° 13′N | Shortgrass prairie | 1970 | 271 | 24 | 475 |
| | 101° 42′W | | 1971 | 267 | 59 | |
| Jornada | 32° 37′N | Desert grassland | 1970 | 336 | 16 | 530 |
| | 106° 44′W | | 1971 | 281 | 20 | |

[a] See Table 2.11 for locations of sites.
[b] Period in year when 15-day running mean air temperature exceeds 4 °C.
[c] Approximate data included only for comparative purposes. Primary source: interpolation from anonymous data (1968) with refinement where site data available.
[d] Corrected for 23°, north facing slope.

$W_{NGS}$ was then taken to be the nongrowing season precipitation or $W_{S(max)}$, whichever was the smaller. Thus, $W_{GS}$ was taken to be equal to the growing season precipitation. The essentials of these calculations are presented in Table 2.17.

Total net primary production, $P_T$, has two components: net primary production of aboveground plant parts, $P_A$; and that of belowground plant parts, $P_B$, such that

$$P_T = P_A + P_B.$$

Differences in the methods of monitoring changes with time of above- and belowground plant parts, and in the respective qualities and characteristics of the data, dictated the use of different methods of estimation for $P_A$ and $P_B$.

$P_A$ was estimated from clipped plot data collected from random plots with a harvest interval of 2 to 3 weeks during the growing season. Material from the clippings is from two treatments: one, a long-term recovery from grazing by cattle or sheep; and the other, a short-term recovery from grazing (usually

Table 2. 17. *Calculation of available water on US/IBP Grassland Biome sites, 1970 and 1971*

| Site[a] | Maximum soil water storage capacity, $W_{s(max)}$ (cm) | Year | Duration nongrowing season (days) | Nongrowing season precipitation (cm) | Nongrowing season available water, $W_{NGS}$ (cm) | Growing season duration (days) | Growing season precipitation (cm) | Total available water, $W_T$ (cm) |
|---|---|---|---|---|---|---|---|---|
| ALE | 20.4 | 1970–1 | 121 | 13.4 | 13.4 | 232 | 10.3 | 23.7 |
| Bison | 4.4 | 1969–70 | 187 | 16.4 | 4.4 | 186 | 24.4 | 28.8 |
| Bridger | 13.2 | 1969–70 | 228 | 72.6 | 13.2 | 105 | 16.6 | 29.8 |
| Dickinson | 13.1 | 1969–70 | 193 | 15.6 | 13.1 | 169 | 36.8 | 49.9 |
| Cottonwood | 19.2 | 1969–70 | 180 | 13.7 | 13.7 | 188 | 26.0 | 39.7 |
| | | 1970–1 | 154 | 8.6 | 8.6 | 214 | 58.4 | 66.9 |
| Pawnee | 15.2 | 1969–70 | 193 | 10.4 | 10.4 | 167 | 14.1 | 24.5 |
| | | 1970–1 | 173 | 7.7 | 7.7 | 208 | 21.7 | 29.4 |
| Hays | 7.0 | 1969–70 | 126 | 5.1 | 5.1 | 227 | 42.8 | 47.9 |
| Osage | 19.8 | 1969–70 | 102 | 7.1 | 7.1 | 273 | 60.1 | 67.2 |
| | | 1970–1 | 74 | 7.5 | 7.5 | 271 | 78.2 | 85.7 |
| Pantex | 15.9 | 1969–70 | 112 | 6.7 | 6.7 | 271 | 19.2 | 25.9 |
| | | 1970–1 | 67 | 4.5 | 4.5 | 267 | 51.4 | 55.9 |
| Jornada | 1.7 | 1969–70 | 71 | 3.9 | 1.7 | 336 | 15.7 | 17.4 |
| | | 1970–1 | 58 | 0.4 | 0.4 | 281 | 18.4 | 18.8 |

[a] See Table 2.11 for locations of sites.

about 12 months). The clipped material was separated into individual species, for the major species, and was partitioned into categories of aboveground live, $L_A$; aboveground recent dead, $D_{A(R)}$ (material which had died during the sampling growing season); aboveground old dead; litter; etc. No use was made of the individual species information on the assumption that the technique of summation of species peaks is not entirely valid for data from destructively sampled material selected at random. Instead, the total weight for all species was used. Also, no use was made of material partitioned into categories beyond $D_{A(R)}$, on the assumption that the record of a particular season's production lies within the changes of $L_A$ and $D_{A(R)}$.

For measuring $P_A$, troughs (lowest points) and peaks (highest points) in a plot of $L_A$ against time were identified. The statistical significance of the trough–peak differences was then tested and, where a significant difference ($P < 0.05$) occurred, the trough–peak difference was noted and all such differences were summed. A plot of $D_{A(R)}$ against time was then examined, and statistically significant trough–peak differences, occurring over the same periods for which trough–peak differences had been identified in the plot of $L_A$, were summed. Thus:

$$P_A = \Sigma_1^{N'} \Delta L_A + \Sigma_1^{N} \Delta D_{A(R)},$$

where $N$ is the number of trough–peak intervals over which a significant difference occurs for $L_A$, and $\Delta D_{A(R)}$ are significant trough–peak differences in $D_{A(R)}$ within the $N$ intervals.

$P_B$ was estimated from soil core data from which living and dead plant roots ($L_B + D_B$) were extracted, the frequency of coring being generally less than for clipping, that is, varying from about every 2 to 4 weeks throughout the growing season.

$P_B$ was estimated by identifying troughs and peaks in a plot of ($L_B + D_B$) against time, testing their statistical significance and summing the significant differences. This method of estimating $P_B$ explicitly ignores decomposition and thus allows a correction to be added for simultaneous decomposition when reliable information on this process becomes available. Meanwhile, however,

$$P_B = \Sigma_1^{N'} \Delta(L_B + D_B'),$$

where $N'$ is the number of trough–peak intervals over which a significant difference occurs for ($L_B + D_B$).

Estimations of $P_T (= P_A + P_B)$, $P_A$ and $P_B$ were carried out separately for the two treatments and subsequently averaged. The values of the separate estimations are given in Table 2.18. For some sites, the absence of estimates for $P_T$ and $P_B$ is because no belowground sampling was carried out, or because the available data were inadequate for the estimation of $P_B$.

Two water-use efficiencies are defined; the first is the ratio of available water to total dry matter produced in the growing season ($W_T/P_T$) and the second is the ratio of available water to dry matter produced above ground

Table 2.18. *Aboveground ($P_A$), belowground ($P_B$) and total ($P_T$) net primary production for several grasslands in the USA, 1970 and 1971*

| Site[a] | Year | Treatment[b] | Day of first harvest of $L_A$ | g dry matter·$m^{-2}$ per growing season | | | g dry matter·$m^{-2}·d^{-1}$ | | |
|---|---|---|---|---|---|---|---|---|---|
| | | | | $P_A$ | $P_B$ | $P_T$ | $P_A$ | $P_B$ | $P_T$ |
| ALE | 1970 | A | — | — | — | — | — | — | — |
| | | B | — | — | — | — | — | — | — |
| | 1971 | A[c] | 81 | 89.8 | 245.2 | 335.0 | 0.39 | 1.06 | 1.44 |
| | | B[c] | 81 | 48.2 | 474.0 | 522.2 | 0.21 | 2.04 | 2.25 |
| Bison | 1970 | A | 135 | 177.6 | — | — | 0.95 | — | — |
| | | B | 122 | 83.8 | — | — | 0.45 | — | — |
| | 1971 | A | — | — | — | — | — | — | — |
| | | B | — | — | — | — | — | — | — |
| Bridger | 1970 | A[c] | 181 | 154.0 | — | — | 1.47 | — | — |
| | | B[c] | 180 | 135.4 | — | — | 1.29 | — | — |
| | 1971 | A | — | — | — | — | — | — | — |
| | | B | — | — | — | — | — | — | — |
| Dickinson | 1970 | A | 161 | 179.6 | 730.0[d] | 909.6 | 1.06 | 4.32 | 5.38 |
| | | B | 162 | 193.4 | 600.0[d] | 793.4 | 1.14 | 3.55 | 4.69 |
| | 1971 | A | — | — | — | — | — | — | — |
| | | B | — | — | — | — | — | — | — |
| Cottonwood | 1970 | A[c] | 126 | 256.4 | 435.9[e] | 692.3 | 1.36 | 2.32 | 3.68 |
| | | B[c] | 126 | 193.0 | 454.2 | 647.2 | 1.03 | 2.42 | 3.44 |
| | 1971 | A[c] | 96 | 208.2 | 794.7 | 1003.0 | 0.97 | 3.71 | 4.69 |
| | | B[c] | 91 | 178.4 | 733.8 | 912.3 | 0.83 | 3.43 | 4.26 |
| Pawnee | 1970 | A | 104 | 114.3 | 320.0[f] | 434.3 | 0.68 | 1.92 | 2.60 |
| | | B | 99 | 91.1 | 320.0 | 411.1 | 0.55 | 1.92 | 2.46 |
| | 1971 | A[c] | 99 | 117.1 | 334.0 | 451.1 | 0.56 | 1.61 | 2.17 |
| | | B | 121 | 128.8 | 329.4 | 458.2 | 0.62 | 1.58 | 2.20 |
| Hays | 1970 | A | 16 | 220.1 | — | — | 0.97 | — | — |
| | | B | 16 | 247.8 | — | — | 1.09 | — | — |
| | 1971 | A | — | — | — | — | — | — | — |
| | | B | — | — | — | — | — | — | — |
| Osage | 1970 | A[c] | 121 | 430.7[g] | 474.0 | 904.7 | 1.58 | 1.74 | 3.31 |
| | | B[c] | 101 | 348.2 | 474.0[f] | 822.2 | 1.28 | 1.74 | 3.01 |
| | 1971 | A | 90 | 468.1 | 477.0 | 945.1 | 1.73 | 1.76 | 3.49 |
| | | B | 90 | 359.8 | 631.5 | 991.3 | 1.33 | 2.33 | 3.66 |
| Pantex | 1970 | A | 166 | 99.6 | 240.0[h] | 339.6 | 0.37 | 0.89 | 1.25 |
| | | B | 166 | 110.0 | 240.0 | 350.0 | 0.41 | 0.98 | 1.29 |
| | 1971 | A[c] | 134 | 294.5 | 188.8 | 483.0 | 1.10 | 0.71 | 1.81 |
| | | B[c] | 134 | 246.0 | 243.2 | 489.0 | 0.92 | 0.91 | 1.83 |
| Jornada | 1970 | A | 195 | 59.7 | 57.6 | 117.3 | 0.18 | 0.17 | 0.35 |
| | | B | 196 | 59.4 | 61.4 | 120.8 | 0.18 | 0.18 | 0.36 |
| | 1971 | A | 12 | 121.0 | 58.0 | 179.0 | 0.43 | 0.21 | 0.64 |
| | | B | 83 | 32.4 | 23.7 | 56.1 | 0.12 | 0.08 | 0.20 |

[a] See Table 2.11 for locations of sites.
[b] Treatment designations: A, long-term recovery from grazing; B, short-term recovery from grazing.
[c] $L_A$ assumed to begin at 0 g·$m^{-2}$.
[d] Approximation from Sims & Singh (1971).
[e] Takes account of crown data.
[f] Treatment B belowground data deficient; $P_B$ for A substituted.
[g] Includes some retrospective accounting of $D_{A(R)}$ and *a posteriori* $\Delta(L_B + D_B)$.
[h] Treatment A belowground data deficient; $P_B$ for B substituted.

Table 2.19. *Water-use efficiencies and production ratios on US/IBP Grassland Biome sites, 1970 and 1971*

| Site[a] | Year | Treatment | $W_T/P_T$ | $W_T/P_A$ | $P_A/P_B$ | $P_A/P_T$ | $P_B/P_T$ |
|---|---|---|---|---|---|---|---|
| ALE | 1971 | A | 707.5 | 2639 | 0.366 | 0.268 | |
| | | B | 453.8 | 4917 | 0.102 | 0.092 | |
| | | mean | 580.7 | 3778 | 0.234 | 0.180 | 0.84 |
| Bison | 1970 | A | | 1622 | | | |
| | | B | | 3437 | | | |
| | | mean | | 2529 | | | |
| Bridger | 1970 | A | | 1935 | | | |
| | | B | | 2201 | | | |
| | | mean | | 2066 | | | |
| Dickinson | 1970 | A | 560.7 | 2778 | 0.246 | 0.197 | |
| | | B | 642.8 | 2580 | 0.322 | 0.244 | |
| | | mean | 601.8 | 2679 | 0.284 | 0.221 | 0.78 |
| Cottonwood | 1970 | A | 573.5 | 1548 | 0.588 | 0.370 | |
| | | B | 613.4 | 2057 | 0.425 | 0.298 | |
| | | mean | 593.5 | 1802 | 0.506 | 0.334 | 0.66 |
| | 1971 | A | 667.0 | 3213 | 0.262 | 0.208 | |
| | | B | 733.3 | 3750 | 0.243 | 0.196 | |
| | | mean | 700.1 | 3482 | 0.253 | 0.202 | 0.80 |
| Pawnee | 1970 | A | 564.1 | 2143 | 0.357 | 0.263 | |
| | | B | 596.0 | 2689 | 0.285 | 0.222 | |
| | | mean | 580.0 | 2416 | 0.321 | 0.242 | 0.76 |
| | 1971 | A | 651.7 | 2511 | 0.351 | 0.260 | |
| | | B | 641.6 | 2283 | 0.391 | 0.281 | |
| | | mean | 646.7 | 2397 | 0.371 | 0.270 | 0.74 |
| Hays | 1970 | A | | 2176 | | | |
| | | B | | 1933 | | | |
| | | mean | | 2055 | | | |
| Osage | 1970 | A | 742.8 | 1560 | 0.909 | 0.476 | |
| | | B | 817.3 | 1930 | 0.735 | 0.423 | |
| | | mean | 780.1 | 1745 | 0.822 | 0.450 | 0.55 |
| | 1971 | A | 906.8 | 1831 | 0.981 | 0.495 | |
| | | B | 864.5 | 2382 | 0.570 | 0.363 | |
| | | mean | 885.7 | 2106 | 0.775 | 0.429 | 0.57 |
| Pantex | 1970 | A | 762.7 | 2600 | 0.415 | 0.293 | |
| | | B | 740.0 | 2355 | 0.458 | 0.314 | |
| | | mean | 751.4 | 2477 | 0.437 | 0.304 | 0.73 |
| | 1971 | A | 1157.3 | 1898 | 1.560 | 0.610 | |
| | | B | 1143.1 | 2272 | 1.011 | 0.503 | |
| | | mean | 1150.2 | 2085 | 1.286 | 0.556 | 0.44 |
| Jornada | 1970 | A | 1483.4 | 2915 | 1.036 | 0.509 | |
| | | B | 1440.4 | 2929 | 0.967 | 0.492 | |
| | | mean | 1461.9 | 2922 | 1.002 | 0.500 | 0.49 |
| | 1971 | A | 1050.3 | 1554 | 2.086 | 0.676 | |
| | | B | 3351.2 | 5802 | 1.367 | 0.578 | |
| | | mean | 2200.7 | 3678 | 1.727 | 0.627 | 0.34 |

[a] See Table 2.11 for locations of sites.

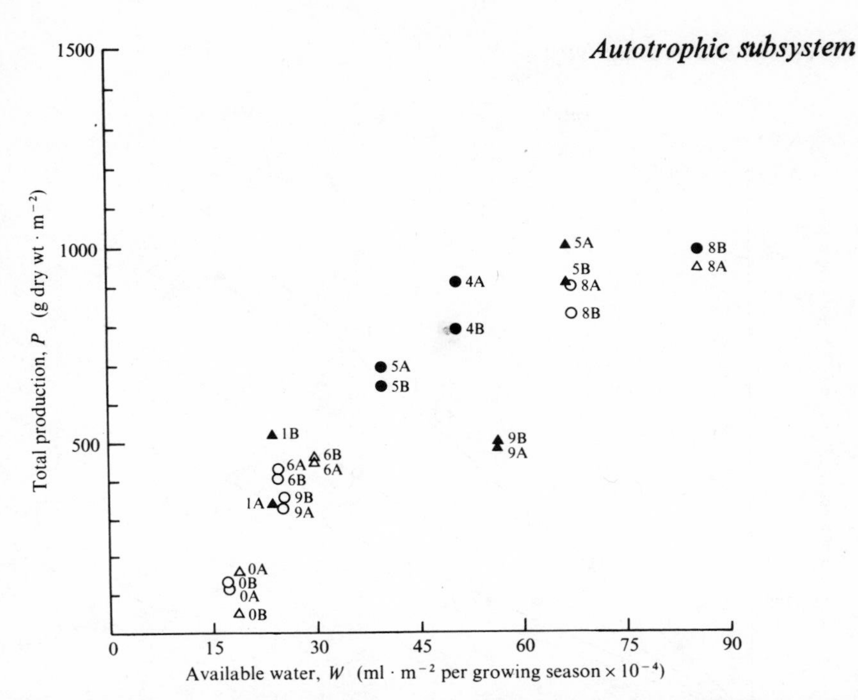


Fig. 2.26. Relationship of total production ($P_T$) to available water ($w_T$). A and B are treatments: A, long-term recovery from grazing; B, short-term recovery from grazing. Numerals refer to sites: 1, ALE; 2, Bison; 3, Bridger; 4, Dickinson; 5, Cottonwood; 6, Pawnee; 7, Hays; 8, Osage; 9, Pantex; 10, Jornada. Solid and open circles refer to 1970, solid and open triangles to 1971.

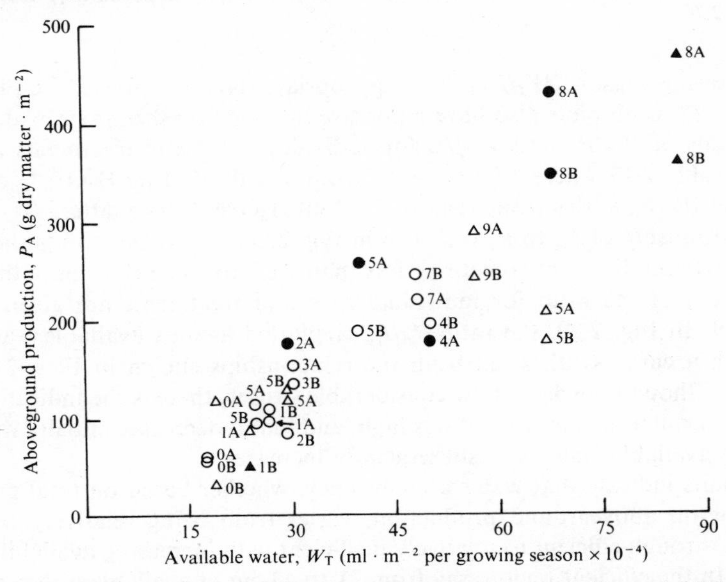


Fig. 2.27. Relationship of aboveground production ($P_A$) to available water ($W_T$). Details as for Fig. 2.26.

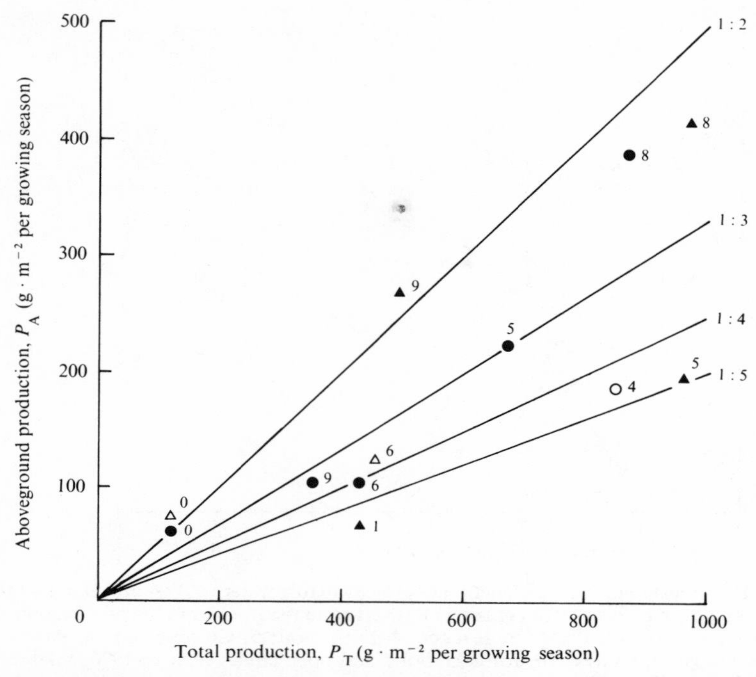


**Fig. 2.28. Relationship of aboveground production ($P_A$) to total production ($P_T$). Details as for Fig. 2.26.**

in the growing season ($W_T/P_A$). The appropriate data are plotted in Figs. 2.26 and 2.27. Both plots also have a positive intercept on the *x*-axis. Values of the ratios of $W_T/P_A$ and $W_T/P_B$ for individual sites and treatments are given in Table 2.19 where $W_T/P_T$ varies from 454 to 3351 ml $H_2O \cdot g^{-1}$ dry matter and $W_T/P_A$ varies from 1548 to 5802 ml $H_2O \cdot g^{-1}$ dry matter.

The relationship of $P_A$ to $P_T$ is shown in Fig. 2.28. There, the points show a scatter between limits of something less than 1 : 2 to somewhat more than 1 : 5. Values of the ratio for individual sites and treatments are given in Table 2.19. In Fig. 2.29, the ratio $P_A/P_B$ is plotted against available water along with a curve synthesized from the relationships shown in Figs. 2.26 and 2.27. Although the data show considerable scatter, there is the indication that at low available water the ratio is high, and that it decreases initially with increasing available water and subsequently increases.

The results indicate that water-use efficiency, whether based on total production or on aboveground production, varies from being relatively very inefficient through efficient to relatively inefficient with increasing availability of water. In the efficient region, say from 21 to 43 cm of available water, the

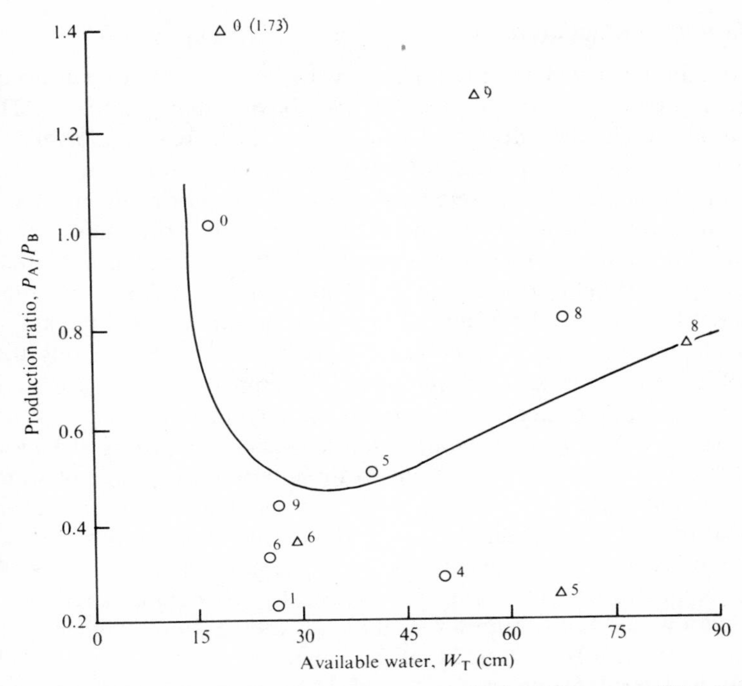


Fig. 2.29. Variation of production ratio ($P_A/P_B$) with available water ($W_T$). Details as for Fig. 2.26.

values of $W_T/P_T$ compare favourably with those found for crops. The following hypothesis is suggested to account for this observed trend in water-use efficiencies. In very sparse, desert grassland situations, it is suggested that the low efficiency of water use is caused by the inability of the vegetation to grow quickly, before surface evaporation occurs. In the moderately covered to well-covered grasslands, the high efficiency of water use is attributed to the development of these communities to make the most efficient use of available water. It is proposed that this limits growth in this intermediate situation. That is to say, there is adequate water available for the development of good cover and an efficient root system to exploit the resource. However, water is sufficiently limiting to confer distinctive evolutionary advantage on communities exploiting this resource most efficiently. In the third region, that of high available water, it is proposed that the somewhat less efficient use indicates that water is tending to become less limiting and that another factor, perhaps light, is becoming more important in limiting growth. A crude analogy is provided by the variation shown within a crop species during the growing season where a similar pattern of water use is found on going from the seedling stage to full growth.

### Variation in production ratios

The variation in the production ratio, $P_A/P_B$, with available water is suggested as, in part, representing another aspect of the above-stated hypothesis. The ratio is initially high, then drops, and finally slowly increases as available water increases. These three phases, it is suggested, correspond to three broad strategies the vegetation has developed in response to environment. These strategies, in order of response to increasing availability of water and all else that it implies, are opportunism, competition for water (below ground) and competition for light (above ground). Opportunism is the strategy most vital to desert conditions where rainfall events are infrequent and the ability to respond to these rapidly is important. Thus it would appear likely that as much of a premium is set on development of photosynthetic tissue as on root development. Admittedly the two are closely linked, and it is not unreasonable to expect greater shoot growth, which is relatively the more important in such an extreme situation. In the intermediate range of water availability, competition for available water is magnified. Hence, the strong development of root systems relative to the tops. The implication is that, no matter how much growth is put into the tops, the advantage gained by the shading of competitors would be marginal, and the concomitant reduction in root growth would fail to support such an advantage. In the higher range of water availability, it is proposed that the balance between water, as limiting growth, and available space for growth is being approached. As competition for space or light develops, so the growth of aboveground parts would be expected to become relatively more prominent. It is suggested that this is the beginning of the trend which is observed to continue into forests where aboveground production exceeds that below ground. For these grasslands, the hypothesis has been advanced that there is a general correspondence between warm- and cool-season species in the grasslands, the proportions of which vary from one extreme to the other in the grasslands studied, and $C_4$ and $C_3$ biochemical pathways of carbon fixation (Williams, 1974). Accepting, momentarily, this hypothesis to be correct, results of experimental comparisons of species of these two types under relatively simple conditions lead to the prediction that there should be a shift in the efficiency of water use with change of species type in the grasslands. The comparison illustrated in Fig. 2.30 indicates no such trend. The suggested reason is, in part, that these pathway types are more closely related to survival and microdistribution in a multi-specific community than to the collective production of all the species. Or, put in another way, on a whole-community basis, variations due to the pathway of carbon fixation are obscured by responses to relatively more important, strategic problems. This calls for more exacting experiments to determine the ecological relations in productivity of $C_3$ and $C_4$ plants.

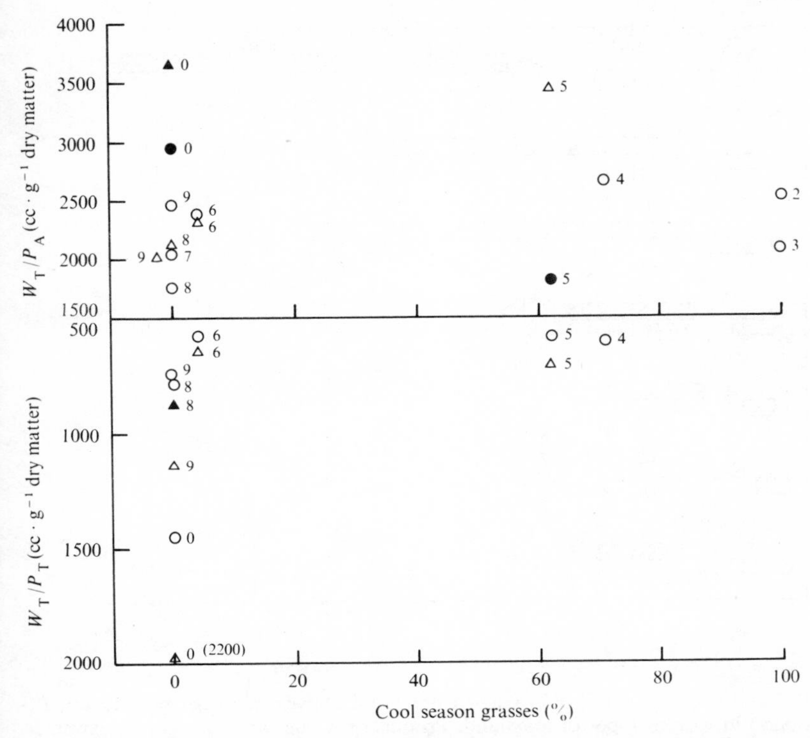


Fig. 2.30. Relationship of water-use efficiencies to proportion of cool-season grasses. Details as for Fig. 2.26.

The water-use efficiency curves in Figs. 2.26 and 2.27 can be used to identify production limits of the grassland types which the study sites were selected to represent. Thus, using rainfall extremes of 1 year in 10 for the sites, assuming no significant change in the soil water storage capacity as determined by rooting depth, and using a long-term average thermal growing season for each site, extremes of $P_A$ and $P_T$ were calculated (Fig. 2.31). The extremes for that type were compared with the sagebrush–bunchgrass type for which the annual average precipitation is similar but the variability lower. The dashed lines extending the upper limit of the shortgrass prairie well into the range predicted for tallgrass prairie arise from applying the water-use curves to the Pantex site, a situation where, because of the heavy swelling clay soils, extrapolation of the available water definition is probably unjustified. The production range for shortgrass prairie bounded by solid lines was derived from rainfall extremes for the Pawnee site.

Singh & Joshi (1976) have calculated water-use efficiency as a ratio of annual aboveground net production ($g \cdot m^{-2}$) to annual rainfall (mm) for tropical grasslands. This ratio ranged from 0.1 to 4.03. There was apparently

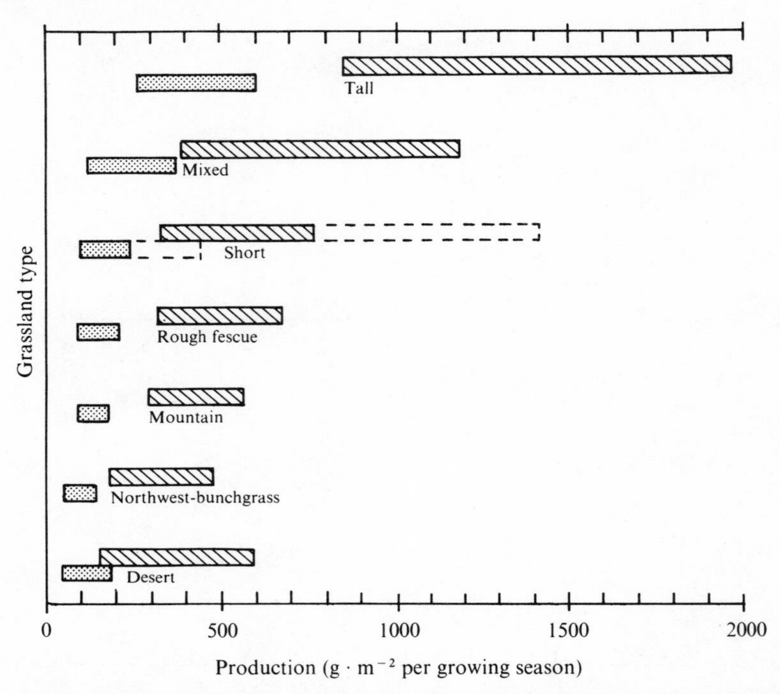


Fig. 2.31. Total net production, $P_T$ (hatched bars) and aboveground net production, $P_A$ (stippled bars) in various types of grassland. Production values are 1-in-10-year extremes (predicted).

no marked relationship between annual rainfall and the efficiency of water use, except for the fact that more communities with higher water-use efficiency tended to occur in areas receiving 200 to 900 mm of water each year. Thus there seems to be a tendency for water-use efficiency to decline in areas of high rainfall.

## Modelling water flow in the soil–plant–atmosphere continuum

Water movement through soil, through plants and into the atmosphere occurs along a catena. Liquid water is transferred from the soil to the plant root, conducted to the leaf where, after a change to the vapour phase, it moves through the stomata into the atmosphere (Fig. 2.32). Flow along this pathway is controlled by water potential gradients and resistances to water movement. Richter (1973) credits B. Huber in 1924 as the first to apply the analogue of Ohm's law to water transport through the soil–plant–atmosphere system. The simplified and widely accepted expression for water flow ($F$) is

$$F = \frac{\Delta\psi}{R},$$

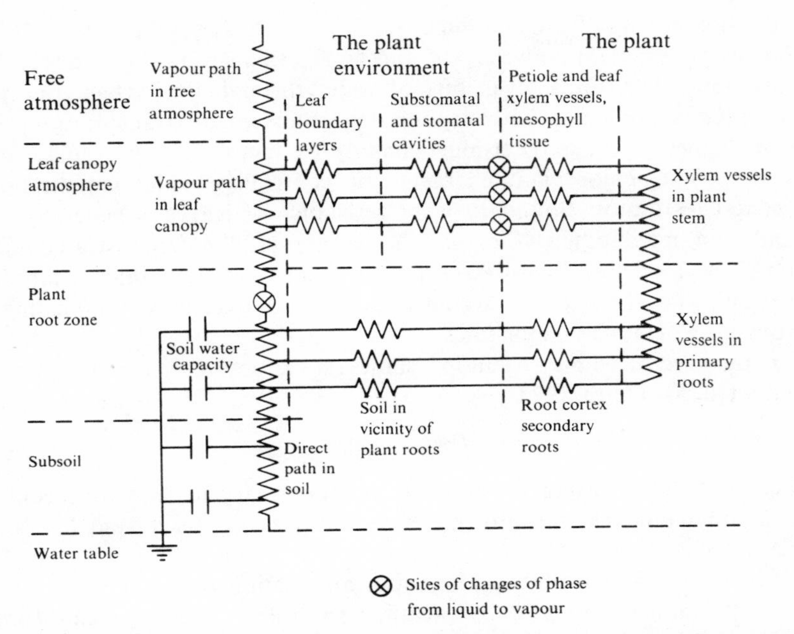


Fig. 2.32. Representation of pathways of water transport in the soil, plant and atmosphere. (After Cowan, 1965, and Cowan & Milthorpe, 1968.)

where $\Delta\psi$ is the water potential gradient and $R$ is resistance of the pathway for water flow.

Cowan (1965) and Gardner (1965) developed detailed models of water movement through the soil–plant–atmosphere system. Liquid water movement ($F$) is described in these models as:

$$F = \frac{\psi_{\text{soil}} - \psi_{\text{leaf}}}{R_{\text{soil}} + R_{\text{plant}}},$$

where $R_{\text{soil}}$ is the hydraulic resistance of the soil and $R_{\text{plant}}$ the resistance between the root surface and the leaf cells.

In the real world, this expression is complex, owing to variability in both water content and root density in the soil. Comparisons of the magnitudes of $R_{\text{soil}}$ and $R_{\text{plant}}$ are not easy, but evidence reviewed in Rutter (1975) suggests that resistance to movement through many plants exceeds by one to three orders of magnitude the resistance in soil exploited by dense roots. $R_{\text{plant}}$ probably predominates in grasslands, which are characterized by plants with dense fibrous root systems. The Cowan (1965) model demonstrates the importance of root density in controlling transpiration, but Rutter (1975) has pointed out the need for more information on the variability of resistance to water movement into and within different parts of root systems.

The potential gradient between the soil and the leaf is in the order of 10 bars, whereas the gradient between leaf and air is often several hundreds of bars. This shows that, for a given flux of water through the system, the major resistance is associated with vapour phase movement (transpiration). Flux through the system is determined mainly by factors which regulate transpiration. The major plant characteristic controlling water vapour flow is stomatal resistance. Boundary layer resistance of leaves is generally much smaller in magnitude. Waggoner & Turner (1972) demonstrated clearly using their model of a corn stand that, if a constant stomatal resistance was specified, the simulation of transpiration was inaccurate, indicating the importance of stomatal dynamics.

In the case of whole vegetation stands, a canopy resistance ($r_{can}$) can be computed (Monteith, 1973):

$$r_{can} = \frac{r_l + r_b}{LAI},$$

where $r_l$ is leaf resistance (stomatal and boundary layer resistance), $r_b$ is the aerodynamic resistance of the vegetation canopy itself, and LAI is leaf area index.

When soil water is abundant, evapotranspiration proceeds at a potential rate determined normally by weather variables. The Penman–Monteith equation (Monteith, 1973) describes potential evapotranspiration as

$$E = \frac{S_a + C_p \rho \delta e K_a}{\lambda[s + \gamma(1 + K_a/K_s)]},$$

where evapotranspiration from a stand ($E$) is a function of available radiant energy ($S_a$) and vapour pressure deficit ($\delta e$). $E$ is also a function of wind speed through its effect on boundary layer conductance ($K_a$) and on air temperature through its effect on saturation vapour pressure. $E$ also depends on stomatal conductance, $K_s$, which is influenced by both environment and plant physiology. $C_p$, $\lambda$, $\rho$ and $\gamma$ are the specific heat of air, the latent heat of vaporization of water, the density of air and the psychrometric constant, respectively.

As the soil dries, plants exhibit progressively lower potentials and in so doing they can extract water from the soil to compensate for water lost in transpiration. Potential transpiration rates can be maintained only until water stress in the plant causes stomatal closure. The dynamic relationship between transpiration and leaf water status is described by Rutter (1975): 'Although it is widely stated that leaf-water content falls when uptake is less than transpiration, it is more exact to say that the plant is constructed in such a way that it tends to a steady state in which uptake and transpiration are equal; and that the changes in the rate of uptake needed to meet the transpiration demand are met by changes in water content and water potential of the plant.'

When leaf water potential drops below some 'critical' level the stomata close, leaf resistance increases and transpiration drops. The change in $r_l$ with water potential varies among species or with environmental conditions. Diurnal fluctuations in leaf water potential and transpiration are well known (Slatyer, 1967; Rutter, 1975).

Experimental data on the relationship between whole-leaf pressure potential (turgor) and stomatal resistance are highly variable, probably because the guard cells respond independently from the remainder of the leaf cells (Turner, 1974). Meidner (1975) has recently discussed evidence that the epidermis and especially the guard cells might have their own water supply system.

When moisture, temperature, water, and nutrient status of a plant community are favourable, production processes depend only on available light and on the amount and distribution of leaves. The theory involved in the computation of radiation absorbed by a plant canopy for any leaf area, leaf angle and solar elevation appears to be well established (de Wit, 1965; Duncan *et al.*, 1967). When temperature or water status are not optimal, photosynthesis and plant growth may be reduced. A model of microclimate and water movement in the soil–plant–atmosphere system is needed to determine leaf temperature and water potential using driving variables from meteorological data. The treatment of water movement in many published models is simplified or absent (Lommen *et al.*, 1971; de Wit *et al.*, 1970; Paltridge, 1970; Curry & Chen, 1971). As Rutter (1975) has pointed out, the more mechanistic models of water flow are difficult to use in modelling real plant communities because of problems in obtaining the necessary data, especially resistances to water flow.

In recent years comprehensive models of water flow have paid particular attention to stomatal control of transpiration. Penning de Vries (1972) describes a model of stomatal functioning in which the simulated transpiration rate is determined by environmental conditions and leaf conductivity. The control mechanism of the guard cells is affected by water status, light intensity and carbon dioxide concentration in the leaf. Many parameters and functions in the model had to be estimated or calculated from other experiments, but despite this, there was good agreement between simulated and measured transpiration. According to Penning de Vries, some of the important parameters and functions about which little is known are: (1) the relative influence of subsidiary and guard cells, (2) the direct effect of light and internal carbon dioxide concentration on guard cells, (3) the constant in the photosynthesis equation, (4) the mesophyll resistance, and (5) the resistance to water flow in stem and petioles.

Van Bavel *et al.* (1973) developed a simulation model of gas and energy exchange regulation by stomatal action in plant leaves. The effects of leaf water potential, leaf temperature and leaf internal carbon dioxide level on the stomatal aperture and diffusive resistance were incorporated in the model. Simulated leaf behaviour suggests that air humidity and ambient carbon

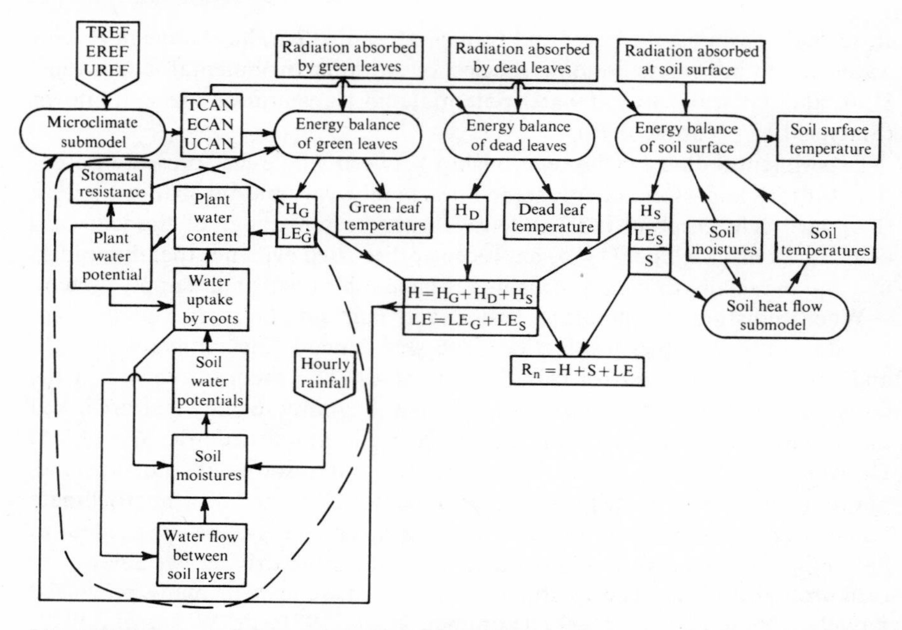


Fig. 2.33. Energy balance, water balance, microclimate and soil heat flow sections of the matador model. TREF, EREF and UREF are the air temperature, vapour pressure and wind speed at the reference level (160 cm); TCAN, ECAN and UCAN are the same parameters at mid-canopy level (7.5 cm); H, LE and S are the fluxes of sensible and latent heat into the air, and heat into the soil, respectively. Subscripts G, D and S refer to green and dead leaves, and soil. $R_n$ is the net radiation. Water flow section of the model is surrounded by dashed line. (From Saugier & Ripley, 1975.)

dioxide level are the principal environmental factors influencing water use efficiency. Light and total radiation load are secondary. The critical leaf characteristics are the compensation point and the stomatal sensitivity to water stress.

### *The Matador model*

Saugier & Ripley (1975) constructed a mechanistic model of plant growth based on physiological properties, soil physical characteristics and meteorological driving variables (see also Saugier *et al.*, 1974). Model parameters were derived largely from laboratory data, and validity was tested against field measurements of such parameters as carbon dioxide flux, leaf temperature, soil water and plant biomass.

The water flow model (Fig. 2.33) is based on the idea that plant water status is a result of the balance between water absorbed by the roots and water lost by transpiration. Plant water status acts on the stomata and regulates evaporation, which in turn affects energy balance and micro-

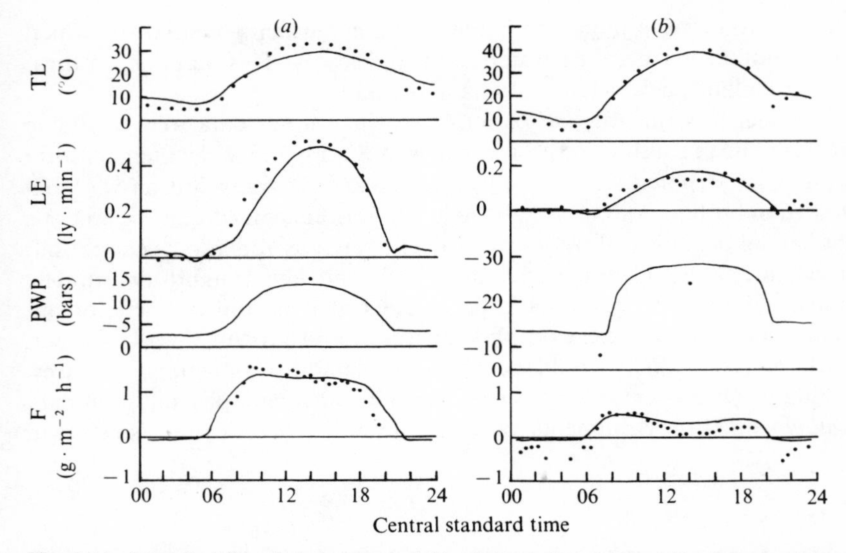


Fig. 2.34. Modelled (lines) and measured (dots) diurnal variation of leaf temperature (TL) latent heat flux (LE), plant water potential (PWP), and carbon dioxide flux (F) for 15 July 1971 (*a*) and 11 August 1971 (*b*). (From Saugier & Ripley, 1975.)

climate. Water status affects photosynthesis only by its effect on stomatal aperture, but it influences leaf expansion directly.

Certain sections of the Matador model, such as the calculation of stomatal resistance, were discussed previously and will not be repeated here. Water 'uptake' ($A$) was made proportional to the water potential drop between the soil ($\psi_s$) and the leaf ($\psi_l$), to the resistance of the bulk soil ($R_s$) and the plant resistance ($R_p$):

$$A = \frac{\psi_s - \psi_l}{R_s + R_p}.$$

Resistance of the bulk soil was considered to be very low because root density was high in the grassland ecosystem being modelled. In so doing, the authors followed Newman (1969*a*, *b*), who concluded that plant resistance was much more important than soil resistance, at least for species with well-developed root systems. In addition, since the major part of the plant resistance is located in the roots (Jarvis, 1975), it was assumed that $R_p$ was inversely proportional to the biomass of roots.

The total input of water to the plant was computed as the sum of water uptake from each of six layers of soil. Water loss through green leaf transpiration was made proportional to a water vapour density gradient between leaf and free air and to the boundary layer and stomatal resistance of the leaf:

$$T = \frac{\rho_s - \rho_a}{r_a + r_s}.$$

At each time-step the model computes the plant water potential for which there is equilibrium between water gains and losses. This value is, in turn, used to calculate a new value of stomatal resistance.

The model was run for two periods for which input data were available at the study site in southwestern Saskatchewan. Simulations of leaf temperature and plant water potential were judged adequate and transpiration was well simulated in both periods (Fig. 2.34). The authors concluded that the sections of the model dealing with physical problems such as radiation penetration, microclimate, and soil heat and water were relatively straightforward. The sections dealing with physiological parameters such as stomatal activity and the influence of water stress on photosynthesis and growth were less exact. Nonetheless, the model could be used to examine production processes, including water movement, in species for which adequate physiological data are known, and in environments for which physical properties are also known.

## References

Adegbola, A. A. & McKell, C. M. (1969*a*). Effect on nitrogen fertilization on the carbohydrate content of coastal Bermudagrass (*Cynodon dactylon* (L.) Pers.). *Agronomy Journal*, **58**, 60–4.

Adegbola, A. A. & McKell, C. M. (1966*b*). Regrowth potential of coastal Bermudagrass as related to previous nitrogen fertilization. *Agronomy Journal*, **58**, 145–6.

Agarwal, V. K. (1970). Seasonal variation in standing crop and energetics of *Dichanthium annulatum* grassland at Varanasi. PhD Thesis, Banaras Hindu University, Varanasi, India.

Alberda, T. (1957). The effects of cutting, light intensity, and night temperature on growth and soluble carbohydrate content of *Lolium perenne* L. *Plant and Soil*, **8**, 199–230.

Alberda, T. (1966). The influence of reserve substances on dry-matter production after defoliation. In *Proceedings of the X International Grassland Congress*, Helsinki, Finland, pp. 140–7. Valtioneuvoston Kirjapaino, Finland.

Albuquerque, H. E. (1968). Leaf area, and age, and carbohydrate reserves in the regrowth of tall fescue (*Festuca arundinacea* Schreb.) tillers. *Dissertation Abstracts, Section B*, **28**, 3968B.

Alcock, M. B. (1964). The physiological significance of defoliation on the subsequent regrowth of grass–clover mixtures and cereal. In *Grazing in terrestrial and marine environments*, ed. D. J. Crisp, pp. 25–41. Blackwell Scientific Publications, Oxford.

Aldous, A. E. (1930). Effect of different clipping treatments on the yield and vigor of prairie grass vegetation. *Ecology*, **11**, 752–9.

Aldous, A. E. (1935). *Management of Kansas permanent pastures. Kansas Agricultural Experimental Station Bulletin 272.*

Anderson, M. C. & Denmead, O. T. (1969). Short wave radiation on inclined surfaces in model plant communities. *Agronomy Journal*, **61**, 867–72.

Ares, J. (1976). Dynamics of the root system of blue grama. *Journal of Range Management*, **29**, 208–13.

Ares, J. & Singh, J. S. (1974). A model of the root biomass dynamics of a shortgrass prairie dominated by blue grama. *Journal of Applied Ecology*, **11**, 727–43.

Arnold, W. N. (1968). The selection of sucrose as the translocate of higher plants. *Journal of Theoretical Biology*, **21**, 13–20.

Asay, K. H., Nelson, C. J. & Horst, G. L. (1974). Genetic variability for net photosynthesis in tall fescue. *Crop Science*, **14**, 571–4.

Auda, H., Blaser, R. E. & Brown, R. H. (1966). Tillering and carbohydrate contents of orchardgrass as influenced by environmental factors. *Crop Science*, **6**, 139–43.

Baker, F. S. (1918). Aspen reproduction in relation to management. *Journal of Forestry*, **16**, 389–98.

Baker, H. K. (1963). Recent research in grassland. In *Vistas in botany*, ed. W. B. Turrill, vol. 2, pp. 36–61. Pergamon Press and Macmillan, New York.

Baker, H. K. & Garwood, E. A. (1961). Studies on the root development of herbage plants. V. Seasonal changes in fructosan and soluble-sugar contents of cocksfoot herbage, stubble, and roots under two cutting treatments. *Journal of the British Grassland Society*, **16**, 263–7.

Balasko, J. A. & Smith, D. (1973). Carbohydrates in grasses. V. Incorporation of $^{14}C$ into plant parts and nonstructural carbohydrates of timothy (*Phleum pratense* L.) at three development stages. *Crop Science*, **13**, 19–22.

Bamberg, S., Wallace, A., Kleinkopf, G. & Vollmer, A. (1972). *Gas exchange and assimilate distribution in Mojave desert shrubs. US/IBP Desert Biome Progress Report*. Logan, Utah.

Barnes, D. L. (1960). Growth and management studies on *Sabi panicum* and star grass: I. *Rhodesia Agricultural Journal*, **57**, 399–411.

Bate, G. C. & Canvin, D. T. (1971). A gas exchange system for measuring the productivity of plant populations in controlled environments. *Canadian Journal of Botany*, **49**, 601–8.

Beevers, H. (1961). *Respiratory metabolism in plants*. Row, Peterson & Co., Evanston, Illinois.

Beevers, H. (1969). Metabolic sinks. In *Physiological aspects of crop yield*, ed. J. D. Eastin, F. A. Haskins, C. Y. Sullivan & C. H. M. Van Bavel, pp. 169–84. American Society of Agronomy and Crop Science Society of America, Madison, Wisconsin.

Benedict, H. M. & Brown, G. B. (1944). The growth and carbohydrate response of *Agropyron smithii* and *Bouteloua gracilis* to changes in nitrogen supply. *Plant Physiology*, **19**, 481–94.

Biddulph, O. (1941). Diurnal migration of injected radiophosphorus from bean leaves. *American Journal of Botany*, **28**, 348–52.

Biddulph, O. & Cory, R. (1957). An analysis of translocation in the phloem of the bean plant using HTO, $P^{32}$, and $C^{14}O_2$. *Plant Physiology*, **32**, 608–19.

Biddulph, O. & Cory, R. (1960). Demonstration of two translocation mechanisms in studies of bidirectional movement. *Plant Physiology*, **35**, 689–95.

Biddulph, O. & Cory, R. (1965). Translocation of $C^{14}$ metabolites in the phloem of the bean plant. *Plant Physiology*, **40**, 119–29.

Bieleski, R. L. (1960). The physiology of sugarcane. IV. Effects of inhibitors on sugar accumulation in storage tissue slices. *Australian Journal of Biological Sciences*, **12**, 221–31.

Bielorai, H. & Mendel, K. (1969). The simultaneous measurement of apparent photosynthesis and transportation on citrus seedlings at different soil moisture levels. *Journal of the American Society for Horticultural Science*, **94**, 201–4.

Billore, S. K. (1973). Net primary production and energetics of a grassland ecosystem at Ratlam, India. PhD Thesis, Vikram University, Ujjain, India.

Biswell, H. H. & Weaver, J. E. (1933). Effect of frequent clipping on the development of roots and tops of grasses in prairie sod. *Ecology*, **14**, 368–89.

Bjorkman, O. (1968). Further studies on differentiation of photosynthetic properties in sun and shade ecotypes of *Solidago virgaurea*. *Physiologia Plantarum*, **21**, 84–99.

Bjorkman, O. (1971). Comparative photosynthetic $CO_2$ exchange in higher plants. In *Photosynthesis and photorespiration*, ed. M. D. Hatch, C. B. Osmond & R. O. Slayter, pp. 18–32. Wiley-Interscience, New York.

Bjorkman, O. & Berry, J. (1973). High efficiency photosynthesis. *Scientific American*, **229**, 80–93.

Black, C. C. & Mayne, B. C. (1970). $P_{700}$ activity and chlorophyll content of plants with different photosynthetic carbon dioxide fixation cycles. *Plant Physiology*, **45**, 738–41.

Blaisdell, J. P. & Pechanec, J. F. (1949). Effects of herbage removal at various dates on vigor of bluebunch wheatgrass and arrowleaf balsamroot. *Ecology*, **30**, 298–305.

Blaser, R. E., Brown, R. H. & Bryant, H. T. (1966). The relationship between carbohydrate accumulation and growth of grasses under different microclimates. In *Proceedings of the X International Grassland Congress*, Helsinki, Finland, pp. 147–50. Valtioneuvoston Kirjapaino, Finland.

Bommer, D. F. R. (1966). Influence of cutting frequency and nitrogen level on the carbohydrate reserves of three grass species. In *Proceedings of the X International Grassland Congress*, Helsinki, Finland, pp. 156–60. Valtioneuroston Kirjapaino, Finland.

Botkin, D. B. & Malone, C. R. (1968). Efficiency of net primary production based on light intercepted during the growing season. *Ecology*, **49**, 438–44.

Botkin, D. B., Woodwell, G. M. & Temple, N. (1970). Forest productivity estimated from carbon dioxide uptake. *Ecology*, **51**, 1057–60.

Boyer, J. S. (1970*a*). Leaf enlargement and metabolic rates of corn, soybean, and sunflower at various leaf water potentials. *Plant Physiology*, **46**, 233–5.

Boyer, J. S. (1970*b*). Differing sensitivity of photosynthesis to low leaf water potentials in corn and soybean. *Plant Physiology*, **46**, 236–9.

Brougham, R. W. (1956). The effect of intensity of defoliation on regrowth of pastures. *Australian Journal of Agricultural Research*, **7**, 377–87.

Brown, K. W. (1969). A model of the photosynthesizing leaf. *Physiologia Plantarum*, **22**, 620–37.

Brown, L. F. & Trlica, M. J. (1974). *Photosynthesis of two important grasses of the shortgrass prairie as affected by several ecological variables. US/IBP Grassland Biome Technical Report 244*. Colorado State University, Fort Collins, Colorado.

Brown, L. F. & Trlica, M. J. (1977*a*). Interacting effects of soil water, temperature and irradiance on $CO_2$ exchange rates of two dominant grasses of the shortgrass prairie. *Journal of Applied Ecology*, **14**, 197–204.

Brown, L. F. & Trlica, M. J. (1977*b*). Carbon dioxide exchange of blue grama swards as influenced by several ecological variables in the field. *Journal of Applied Ecology*, **14**, 205–13.

Brown, L. F. & Trlica, M. J. (1977*c*). Simulated dynamics of blue grama production. *Journal of Applied Ecology*, **14**, 215–24.

Brown, R. H. & Blaser, R. E. (1965). Relationship between reserve carbohydrate accumulation and growth rate in orchardgrass (*Dactylis glomerata* L.) and tall fescue (*Festuca arundinacea*). *Crop Science*, **5**, 577–82.

Brown, R. H. & Blaser, R. E. (1970). Soil moisture and temperature effects on growth and soluble carbohydrates of orchardgrass (*Dactylis glomerata*). *Crop Science*, **10**, 213–16.

Brown, W. V. (1958). Leaf anatomy in grass systematics. *Botanical Gazette*, **5**, 517–21.

Budyko, M. I. & Gandin, L. S. (1964). On using the laws of atmospheric physics in the agrometeorological research. *Meteorologiya i Gidrologiya*, **11**, 3–10.

Bukey, F. S. & Weaver, J. E. (1939). Effects of frequent clipping on the underground food reserves of certain prairie grasses. *Ecology*, **20**, 246–52.

Burr, G. O., Hartt, C. E., Tanimoto, T., Takahashi, T. & Brodie, H. W. (1958). The circulatory system of the sugarcane plant. In *International Conference on Radioisotopes in Scientific Research, Proceedings of First (UNESCO) International Conference*, ed. R. C. Extermann, pp. 351–68. Pergamon Press, Oxford.

Burton, G. W., Jackson, J. E. & Knox, F. E. (1959). The influence of light reduction upon the production, persistence, and chemical composition of coastal Bermudagrass, *Cynodon dactylon*. *Agronomy Journal*, **51**, 537–42.

Butcher, H. C. (1965). The kinetics of carbon-14 translocation in sugar beet: an effect of illumination. *Dissertation Abstracts*, **25**, 7350.

Canny, M. J. (1960*a*). The breakdown of sucrose during translocation. *Annals of Botany, New Series*, **24**, 330–4.

Canny, M. J. (1960*b*). The rate of translocation. *Biological Reviews of the Cambridge Philosophical Society*, **35**, 507–32.

Canny, M. J. (1961). Measurements of the velocity of translocation. *Annals of Botany, New Series*, **25**, 152–67.

Canny, M. J. (1962*a*). The translocation profile: sucrose and carbon dioxide. *Annals of Botany, New Series*, **26**, 181–96.

Canny, M. J. (1962*b*). The mechanism of translocation. *Annals of Botany, New Series*, **26**, 603–17.

Canny, M. J. (1971). Translocation: mechanisms and kinetics. *Annual Review of Plant Physiology*, **22**, 237–60.

Canny, M. J. (1973). *Phloem translocation*. Cambridge University Press, London.

Canny, M. J. & Askham, M. J. (1967). Physiological inferences from the evidence of translocated tracer: a caution. *Annals of Botany, New Series*, **31**, 409–16.

Canny, M. J. & Phillips, O. M. (1963). Quantitative aspects of a theory of translocation. *Annals of Botany, New Series*, **27**, 379–402.

Carr, D. J. & Wardlaw, I. F. (1965). The supply of photosynthetic assimilates to the grain from the flag leaf and ear of wheat. *Australian Journal of Biological Sciences*, **18**, 711–19.

Cartledge, O. & Connor, D. J. (1971). Structure and photosynthesis of wheat communities. *Journal of Applied Ecology*, **8**, 469–75.

Chandler, R. F. (1969). New horizons for an ancient crop. Paper presented at All-Congress Symposium on World Food Supply, 11th International Botanical Congress, 28 August 1969, Seattle, Washington. (Eka Press, Calcutta.)

Charles-Edwards, D. A. & Ludwig, L. J. (1974). A model for leaf photosynthesis by $C_3$ plant species. *Annals of Botany*, **38**, 921–30.

Chartier, P. (1967). Lumière, eau et production de matière sèche du couvert végétal. *Annales agronomiques, Paris*, **18**, 301–31.

Chartier, P. (1970). A model of $CO_2$ assimilation in the leaf. In *Prediction and meas-*

*urement of photosynthetic productivity*, ed. I. Šetlîk, pp. 307–15. Centre for Agricultural Publishing and Documentation, Wageningen, the Netherlands.

Chartier, P. (1972). Net assimilation of plants as influenced by light and carbon dioxide. In *Crop processes in controlled environments*, ed. A. R. Rees, K. E. Cockshull, D. W. Hand & R. G. Hurd, pp. 203–16. Academic Press, New York & London.

Chen, T. M., Brown, R. H. & Black, C. C. (1969). Photosynthetic activity of chloroplasts from Bermuda grass (*Cynodon dactylon* L.), a species with a high photosynthetic capacity. *Plant Physiology*, **44**, 649–54.

Choudhary, V. B. (1967). Seasonal variation in standing crop and energetics of *Dichanthium annulatum* grassland at Varanasi. PhD Thesis, Banaras Hindu University, Varanasi, India.

Colby, W. G., Drake, M., Oohara, H. & Yoshida, N. (1966). Carbohydrate reserves in orchardgrass. In *Proceedings of the X International Grassland Congress*, Helsinki, Finland, pp. 151–5. Valtioneuvoston Kirjapaino, Finland.

Cole, C. V., Innis, G. S. & Stewart, J. W. B. (1977). Simulation of phosphorus cycling in semiarid grasslands. In *Grassland simulation model*, pp. 205–30. *Springer Ecological Studies 26*, ed. G. S. Innis. Springer-Verlag, New York.

Conant, S. & Risser, P. G. (1974). Canopy structure of a tallgrass prairie. *Journal of Range Management*, **27**, 313–18.

Connor, D. J. (1973). *GROMAX: a potential productivity routine for a total grassland ecosystem model. US/IBP Grassland Biome Technical Report 208*. Colorado State University, Fort Collins, Colorado.

Connor, D. J., Brown, L. F. & Trlica, M. J. (1974). Plant cover, light interception, and photosynthesis of shortgrass prairie: a functional model. *Photosynthetica*, **8**, 18–27.

Connor, D. J. & Cartledge, O. (1970). Observed and calculated photosynthetic rates of *Chloris gayana* communities. *Journal of Applied Ecology*, **7**, 353–62.

Cook, C. W. (1966*a*). *Carbohydrate reserves in plants. Utah Agricultural Experiment Station Research Series 31*.

Cook, C. W. (1966*b*). *The role of carbohydrate reserves in managing range plants. Utah Agricultural Experiment Station Mimeograph Series 449*.

Cook, C. W. & Child, R. D. (1971). Recovery of desert plants in various states of vigor. *Journal of Range Management*, **24**, 339–43.

Cook, C. W., Stoddart, L. A. & Kinsinger, F. E. (1958). Responses of crested wheatgrass to various clipping treatments. *Ecological Monographs*, **28**, 237–72.

Cooper, C. S. & Watson, C. A. (1968). Total available carbohydrates in roots of sainfoin (*Onobrychis viciaefolia* Scop.) and alfalfa (*Medicago sativa* L.) when grown under several management regimes. *Crop Science*, **8**, 83–5.

Cooper, J. P. (1970). Potential production and energy conversion in temperate and tropical grasses. *Herbage Abstracts*, **40**, 1–15.

Coupland, R. T. (1973). *Excerpts from final reports of the Matador Project*. Canadian IBP, University of Saskatchewan, Saskatoon, Saskatchewan.

Cowan, I. R. (1965). Transport of water in the soil–plant–atmosphere system. *Journal of Applied Ecology*, **2**, 221–39.

Cowan, I. R. & Milthorpe, F. L. (1968). Plant factors influencing the water status of plant tissues. In *Water deficits and plant growth*, ed. T. T. Kozlowski, vol. 1, pp. 137–93. Academic Press, New York & London.

Coyne, P. I. (1969). Seasonal trend in total available carbohydrates with respect to phenological stage of development in eight desert range species. Unpublished PhD Thesis, Utah State University, Logan, Utah.

Coyne, P. I. & Cook, C. W. (1970). Seasonal carbohydrate reserve cycles in eight desert range species. *Journal of Range Management*, **23**, 438–44.
Crafts, A. S. (1931). Movement of organic materials in plants. *Plant Physiology*, **6**, 1–41.
Crafts, A. S. (1932). Phloem anatomy, exudation, and transport of organic nutrients in Cucurbits. *Plant Physiology*, **7**, 183–225.
Crafts, A. S. (1933). Sieve tube structure and translocation in the potato. *Plant Physiology*, **8**, 81–104.
Crafts, A. S. (1961). *Translocation in plants*. Holt, Rinehart & Winston, New York.
Crafts, A. S. & Crisp, C. E. (1971). *Phloem transport in plants*. W. H. Freeman, San Francisco, California.
Crider, F. J. (1955). *Root growth stoppage resulting from defoliation of grass. US Department of Agriculture Technical Bulletin No. 1102*. Washington, DC.
Cunningham, G. L., Balding, F. R. & Syvertsen, J. P. (1974). A net $CO_2$ exchange model for $C_4$-grasses. *Photosynthetica*, **8**, 28–33.
Curry, R. B. & Chen, L. H. (1971). Dynamic simulation of plant growth. II. Incorporation of actual daily weather and partitioning of net photosynthate. *Transactions of the American Society of Agricultural Engineers*, **14**, 1170–4.
Curtis, O. F. (1935). *The translocation of solutes in plants*. McGraw-Hill, New York.
Dahlman, R. C. & Kucera, C. L. (1965). Root productivity and turnover in native prairie. *Ecology*, **46**, 84–9.
Dahlman, R. C. & Kucera, C. L. (1968). Tagging native grassland vegetation with carbon-14. *Ecology*, **49**, 1199–203.
Dahlman, R. C. & Kucera, C. L. (1969). Carbon-14 cycling in the root and soil components of a prairie ecosystem. In *Proceedings of the II National Symposium on Radioecology*, ed. D. J. Nelson & F. C. Evans, pp. 652–60. Ann Arbor, Michigan.
Davidson, J. L. & Milthorpe, F. L. (1965). Carbohydrate reserves in the regrowth of cocksfoot (*Dactylis glomerata* L.). *Journal of the British Grassland Society*, **20**, 15–18.
Davidson, J. L. & Philip, J. R. (1958). Light and pasture growth. In *Proceedings of the Symposium on Climatology and Microclimatology*, Canberra, pp. 181–87. UNESCO Arid Zone Research, Paris: XI.
Davidson, R. L. (1969). Effects of edaphic factors on the soluble carbohydrate contents of roots of *Lolium perenne* L. and *Trifolium repens* L. *Annals of Botany*, **33**, 579–89.
DeCugnak, A. (1931). Recherches sur les glucides des graminées. *Annales des Sciences Naturelles*, **13**, 1–129.
De Vries, H. (1885). Über die Bedeutung der Circulation und der Rotation des Protoplasma für das Stofftransport in der Pflanze. *Botanische Zeitung*, **43**, 1–6, 16–26.
de Wit, C. T. (1959). Potential photosynthesis of crop surfaces. *Netherlands Journal of Agricultural Science*, **7**, 141–9.
de Wit, C. T. (1965). Photosynthesis of leaf canopies. *Verslagen van Landbouwkundige Onderzoekingen (Agricultural Research Reports) No. 663*. Centre for Agricultural Publishing and Documentation, Wageningen, the Netherlands.
de Wit, C. T. (1966). Photosynthesis of crop surfaces. *Advancement of Science*, **17**, 158–62.
de Wit, C. T. (1970). Dynamic concepts in biology. In *Prediction and measurement of photosynthetic productivity*, ed. I. Šetlîk, pp. 17–23. Centre for Agricultural Publishing and Documentation, Wageningen, the Netherlands.

de Wit, C. T., Brouwer, R. & Penning de Vries, F. W. T. (1970). The simulation of photosynthetic systems. In *Prediction and measurement of photosynthetic productivity*, ed. I. Šetlík, pp. 47–70. Centre for Agricultural Publishing and Documentation, Wageningen, the Netherlands.

Dina, S. J. & Klikoff, L. G. (1973). Effect of plant moisture stress on carbohydrate and nitrogen content in big sagebrush. *Journal of Range Management*, **26**, 207–9.

Dixon, H. H. (1923). Transport of organic substances in plants. *Notes of the Botany School of Trinity College, Dublin*, **3**, 207–15.

Dodd, J. D. & Hopkins, H. H. (1958). Yield and carbohydrate content of blue grama grass as affected by clipping. *Transactions of the Kansas Academy of Science*, **61**, 280–7.

Donart, G. B. (1969). Carbohydrate reserves of six mountain plants as related to growth. *Journal of Range Management*, **22**, 411–15.

Donart, G. B. & Cook, C. W. (1970). Carbohydrate reserve content of mountain range plants following defoliation and regrowth. *Journal of Range Management*, **23**, 15–19.

Doodson, J. K., Manners, J. G. & Myers, A. (1964). The distribution pattern of $^{14}$carbon assimilated by the third leaf of wheat. *Journal of Experimental Botany*, **15**, 96–103.

Downton, W. J. S. (1971). Adaptive and evolutionary aspects of $C_4$ photosynthesis. In *Photosynthesis and photorespiration*, ed. M. D. Hatch, C. B. Osmond & R. O. Slatyer, pp. 3–17. Wiley-Interscience, New York.

Downton, W. J. S., Berry, J. & Tregunna, E. B. (1969). Photosynthesis: temperature and tropical characteristics within a single grass genus. *Science*, **163**, 78–9.

Downton, W. J. S. & Tregunna, E. B. (1968). Photorespiration and glycolate metabolism: a reexamination and correlation of some previous studies. *Plant Physiology*, **43**, 923–8.

Duncan, W. G., Loomis, R. S., Williams, W. A. & Hanau, R. (1967). A model for simulating photosynthesis in plant communities. *Hilgardia*, **38**, 181–205.

Dye, A. J., Brown, L. F. & Trlica, M. J. (1972). *Carbon dioxide exchange of blue grama as influenced by several ecological parameters, 1971. US/IBP Grassland Biome Technical Report 181*. Colorado State University, Fort Collins, Colorado.

Eaton, F. M. & Ergle, D. R. (1948). Carbohydrate accumulation in the cotton plant at low moisture levels. *Plant Physiology*, **23**, 169–87.

El-Sharkawy, M. A. & Hesketh, J. (1965). Photosynthesis among species in relation to characteristics of leaf anatomy and $CO_2$ diffusive resistances. *Crop Science*, **5**, 517–21.

El-Sharkawy, M. A., Loomis, R. S. & Williams, W. A. (1967). Apparent reassimilation of respiratory carbon dioxide by different plant species. *Physiologia plantarum*, **20**, 171–86.

El-Sharkawy, M. A., Loomis, R. S. & Williams, W. A. (1968). Photosynthetic and respiratory exchanges of carbon dioxide by leaves in the grain amaranth. *Journal of Applied Ecology*, **5**, 243–51.

Esau, K. & Cheadle, V. I. (1961). An evaluation of studies on ultra-structure of sieve plates. *Proceedings of the National Academy of Sciences, USA*, **47**, 1716–26.

Esau, K., Engleman, E. M. & Bisalputra, T. (1963). What are transcellular strands? *Planta*, **59**, 617–23.

Evans, L. T. (1971). Evolutionary, adaptive and environmental aspects of the photosynthetic pathway assessment. In *Photosynthesis and photorespiration*, ed. M. D. Hatch, C. B. Osmond & R. O. Slatyer, pp. 130–6. Wiley-Interscience, New York.

Evans, L. T. & Dunstone, R. L. (1970). Some physiological aspects of evolution in wheat. *Australian Journal of Biological Sciences*, **23**, 725–41.

Evans, N. T. S., Epert, M. & Moorby, J. (1963). A model of the translocation of photosynthate in the soybean. *Journal of Experimental Botany*, **14**, 221–31.

Everson, A. C. (1966). Effects of frequent clipping at different stubble heights on western wheatgrass (*Agropyron smithii* Rydb.). *Agronomy Journal*, **58**, 33–5.

Evert, R. F. & Mjrmanis, L. (1965). Ultrastructure of the secondary phloem of *Tilia americana*. *American Journal of Botany*, **52**, 95–106.

Feltner, K. C. & Massengale, M. A. (1965). Influence of temperature and harvest management on growth, level of carbohydrate in roots, and survival of alfalfa (*Medicago sativa* L.). *Crop Science*, **5**, 585–8.

Fensom, D. S. (1959). The bio-electric potentials of plants and their functional significance. III. The production of continuous potentials across membranes in plant tissue by the circulation of the hydrogen ions. *Canadian Journal of Botany*, **37**, 1003–26.

Fisher, A. G. (1966). Seasonal trends of root reserves in blue grama and western wheatgrass. MS Thesis, Colorado State University, Fort Collins, Colorado.

Ford, J. & Peel, A. J. (1967). Preliminary experiments on the effect of temperature on the movement of $^{14}$C-labeled assimilates through the phloem of willow. *Journal of Experimental Botany*, **18**, 406–15.

Frame, J. & Hunt, I. V. (1971). The effects of cutting and grazing systems on herbage production from grass swards. *Journal of the British Grassland Society*, **26**, 163–71.

Fried, M. & Broeshart, H. (1967). *Soil–plant system*. Academic Press, New York & London.

Fujiwara, A. & Suzuki, M. (1961). Effects of temperature and light on the translocation of photosynthetic products. *Tohoku Journal of Agricultural Research*, **12**, 363–7.

Gaastra, P. (1959). Photosynthesis of crop plants as influenced by light, carbon dioxide, temperature, and stomatal diffusion resistance. *Mededelingen Landbouwhogeschool Wageningen*, **13**, 1–68.

Gardner, W. R. (1965). Dynamic aspects of soil water availability to plants. *Annual Review of Plant Physiology*, **16**, 323–42.

Geiger, D. R. (1966). Effect of sink region cooling on translocation rate. *Plant Physiology* (Supplement), **41**, 20.

Geiger, D. R. & Christy, A. L. (1971). Effect of sink region anoxia on translocation rate. *Plant Physiology*, **47**, 172–4.

Glasziou, K. T. (1960). Accumulation and transformation of sugars in sugarcane stalks. *Plant Physiology*, **35**, 895–901.

Golley, F. B. (1960). Energy dynamics of a food chain of an old-field. *Ecological Monographs*, **30**, 187–206.

Golley, F. B. (1965). Structure and function of an old-field broomsedge community. *Ecological Monographs*, **35**, 113–31.

Graber, L. F. (1931). Food reserves in relation to other factors limiting the growth of grasses. *Plant Physiology*, **6**, 43–72.

Greenfield, S. B. & Smith, D. (1973). Diurnal variations of nonstructural carbohydrates in each part of a shoot of switchgrass (*Panicum virgatum* L.). *Agronomy Abstracts* (1973), 25.

Greub, L. J. & Wedin, W. F. (1971). Leaf area, dry matter production, and carbohydrate reserve levels of birdsfoot trefoil as influenced by cutting height. *Crop Science*, **11**, 734–8.

Habeshaw, D. (1973). Translocation and the control of photosynthesis in sugar beet. *Planta*, **110**, 213–26.

Hanson, W. R. & Stoddart, L. A. (1940). Effects of grazing upon bunch wheatgrass. *Journal of the American Society of Agronomy*, **32**, 278–89.

Hatch, M. D. (1971). Mechanism and function of the $C_4$ pathway of photosynthesis. In *Photosynthesis and photorespiration*, ed. M. D. Hatch, C. B. Osmond & R. O. Slatyer, pp. 139–52. Wiley-Interscience, New York.

Hatch, M. D. & Slack, C. R. (1966). Photosynthesis in sugar cane leaves: a new carboxylation reaction and the pathway of sugar formation. *Biochemical Journal*, **101**, 103–11.

Hatch, M. D. & Slack, C. R. (1968). A new enzyme for the interconversion of pyruvate and phosphopyruvate and its role in $C_4$ dicarboxylic pathway of photosynthesis. *Biochemical Journal*, **10**, 141–6.

Hatch, M. D. & Slack, C. R. (1970). Photosynthetic $CO_2$ fixation pathways. *Annual Review of Plant Physiology*, **21**, 141–62.

Hedrick, D. W. (1958). Proper utilization – a problem in evaluating the physiological response of plants to grazing use: a review. *Journal of Range Management*, **11**, 34–43.

Heichel, G. H. & Musgrave, R. B. (1969). Relation of $CO_2$ compensation concentration to apparent photosynthesis in maize. *Plant Physiology*, **44**, 1724–8.

Hesketh, J. D. (1963). Limitations to photosynthesis responsible for differences among species. *Crop Science*, **3**, 493–6.

Hesketh, J. D. (1967). Enhancement of photosynthetic $CO_2$ assimilation in the absence of oxygen as dependent on species and temperature. *Planta*, **76**, 371–4.

Hesketh, J. D. & Moss, D. N. (1963). Variation in the response of photosynthesis to light. *Crop Science*, **3**, 107–10.

Hofstra, G. & Nelson, C. D. (1969). A comparative study of translocation of assimilated $^{14}C$ from leaves of different species. *Planta*, **88**, 103–12.

Holscher, C. E. (1945). The effects of clipping bluestem wheatgrass and blue grama at different heights and frequencies. *Ecology*, **26**, 148–56.

Holt, D. A. & Hilst, A. E. (1969). Daily variations in carbohydrate content of selected forage crops. *Agronomy Journal*, **61**, 239–42.

Honert, T. H. van den (1932). On the mechanism of the transport of organic materials in plants. *Proceedings of the Koninklijke Nederlandse Akademie van Wetenschappen, Series C, Biological and Medical Sciences*, **35**, 1104–11.

Horwitz, L. (1958). Some simplified mathematical treatments of translocation in plants. *Plant Physiology*, **33**, 81–93.

Hoveland, C. S. & Buchanan, G. A. (1972). Flooding tolerance of fall *Panicum* and Texas *Panicum*. *Weed Science*, **20**, 1–3.

Hsiao, T. C. (1973). Plant responses to water stress. *Annual Review of Plant Physiology*, **24**, 519–70.

Hunter, R. A., McIntyre, B. L. & McIlroy, R. J. (1970). Water-soluble carbohydrates of tropical pasture grasses and legumes. *Journal of the Science of Food and Agriculture*, **21**, 400–5.

Hyder, D. N. & Sneva, F. A. (1959). Growth and carbohydrate trends in crested wheatgrass. *Journal of Range Management*, **12**, 271–6.

Hyder, D. N. & Sneva, F. A. (1963*a*). Morphological and physiological factors affecting the grazing management of crested wheatgrass. *Crop Science*, **3**, 267–71.

Hyder, D. N. & Sneva, F. A. (1963*b*). *Studies of six grasses seeded on sagebrush–bunchgrass range: yield, palatability, carbohydrate accumulation, and developmental morphology. Oregon Agricultural Experiment Station Technical Bulletin 71.*

Ishizuka, Y. (1969). Engineering for higher yields. In *Physiological aspects of crop yield*, ed. J. D. Eastin, F. A. Haskins, C. Y. Sullivan & C. H. M. Van Bavel, pp. 15–26. American Society of Agronomy and Crop Science Society of America, Madison, Wisconsin.

Iwaki, H., Nakajima, N. & Monsi, M. (1964). Dry-matter production of *Solidago altissima* communities. I. *Proceedings of the 29th Annual Meeting of the Botanical Society of Japan*, Kanazawa, Japan, p. 24.

Jain, S. K. & Mishra, G. P. (1972). Changes in underground biomass and annual increment in an upland grassland of Sagar. *Tropical Ecology*, **13**, 131–8.

Jameson, D. A. (1963). Responses of individual plants to harvesting. *Botanical Review*, **29**, 532–94.

Jameson, D. A. (1964). Effect of defoliation on forage plant physiology. In *Forage plant physiology and soil range relationships*, pp. 67–80. *American Society of Agronomy, Special Publication 5*. Madison, Wisconsin.

Jameson, D. A. & Huss, D. L. (1959). The effect of clipping leaves and stems on number of tillers, herbage weights, root weights, and food reserves of little bluestem. *Journal of Range Management*, **12**, 122–6.

Jarvis, P. G. (1971). The estimation of resistances to carbon dioxide transfer. In *Plant photosynthetic production: manual of methods*, ed. Z. Šesták. J. Čatský & P. G. Jarvis, pp. 566–631. Dr W. Junk N. V. Publishers, The Hague, the Netherlands.

Jarvis, P. G. (1975). Water transfer in plants. In *Heat and mass transfer in the biosphere*, part I, *Transfer processes in the plant environment*, ed. D. A. de Vries & N. H. Afgan, pp. 369–94. Wiley, New York.

Jones, M. B. & Laude, H. M. (1960). Relationship between sprouting in chamise and the physiological condition of the plant. *Journal of Range Management*, **13**, 210–14.

Kamen, M. (1963). *Primary processes in photosynthesis*. Academic Press, New York & London.

Kanai, R. & Black, C. C. (1972). Biochemical basis for net $CO_2$ assimilation in $C_4$-plants. In *Net carbon dioxide assimilation in higher plants*, ed. C. C. Black, pp. 75–85. Symposium of the Southern Section of the American Society of Plant Physiologists sponsored jointly with Cotton.

Kasanaga, H. & Monsi, M. (1954). On the light-transmission of leaves, and its meaning for the production of matter in plant communities. *Japanese Journal of Botany*, **14**, 304–24.

Kelly, J. M., Opstrup, P. A., Olson, J. S., Auerbach, S. I. & Van Dyne, G. M. (1969). *Models of seasonal primary productivity in eastern Tennessee* Festuca *and* Andropogon *ecosystems. Oak Ridge National Laboratory ORNL-4310*. Oak Ridge, Tennessee.

Kennedy, R. A. & Laetsch, W. M. (1974). Plant species intermediate for $C_3$, $C_4$ photosynthesis. *Science*, **184**, 1087–9.

Kinsinger, F. E. (1953). Forage yield and carbohydrate content of underground parts of grasses as affected by clipping. MS Thesis, Kansas State College, Fort Hays, Kansas.

Kinsinger, F. E. & Hopkins, H. H. (1961). Carbohydrate content of underground parts of grasses as affected by clipping. *Journal of Range Management*, **14**, 9–12.

Klipple, G. E. & Costello, D. F. (1960). *Vegetation and cattle responses to different intensities of grazing on short-grass ranges on the Central Great Plains. US Department of Agriculture Technical Bulletin 1216*.

Knievel, D. P. & Schmer, D. A. (1971). *Preliminary results of growth characteristics of buffalograss, blue grama, and western wheatgrass, and methodology for translocation studies using $^{14}C$ as a tracer. US/IBP Grassland Biome Technical Report 86*. Colorado State University, Fort Collins, Colorado.

Knight, D. H. (1973). Leaf dynamics of a shortgrass prairie in Colorado. *Ecology*, **54**, 891–6.

Koller, D. (1970). Characteristics of the photosynthetic apparatus derived from its response to natural complexes of environmental factors. In *Prediction and measurement of photosynthetic productivity*, ed. I. Šetlík, pp. 283–94. Centre for Agricultural Publishing and Documentation, Wageningen, the Netherlands.

Kozlowski, T. T. & Keller, T. (1966). Food relations of woody plants. *Botanical Review*, **32**, 294–382.

Kramer, P. J. & Kozlowski, T. T. (1960). *Physiology of trees*. McGraw-Hill, New York.

Kucera, C. L., Dahlman, R. C. & Koelling, M. R. (1967). Total net productivity and turnover on an energy basis for a tallgrass prairie. *Ecology*, **48**, 536–41.

Kuroiwa, S. (1970). Total photosynthesis of a foliage in relation to inclination of leaves. In *Prediction and measurement of photosynthetic productivity*, ed. I. Šetlík, pp. 79–89. Centre for Agricultural Publishing and Documentation, Wageningen, the Netherlands.

Kursanov, A. L. (1961). The transport of organic substances in plants. *Endeavour*, **20**, 19–25.

Kursanov, A. L. (1963). Metabolism and the transport of organic substances in the phloem. *Advances in Botanical Research*, **1**, 209–78.

Kursanov, A. L. & Brovchenko, M. I. (1961). Effect of ATP on the entry of assimilates into the conducting system of sugar beets. *Soviet Plant Physiology*, **8**, 211–17.

Kursanov, A. L., Brovchenko, M. I. & Pariiskaya, A. N. (1959). Flow of assimilates to the conducting tissue in rhubarb (*Rheum rhaponticum* L.) leaves. *Soviet Plant Physiology*, **6**, 544–52.

Kursanov, A. L. & Turkina, M. V. (1954). In regard to forms of translocatible sugars in the conduction system of sugar beets. *Dokladȳ Akademii Nauk SSSR*, **95**, 885–8.

Laetsch, W. M. (1974). The $C_4$ syndrome: a structural analysis. *Annual Review of Plant Physiology*, **25**, 27–52.

Lake, J. V. (1967). Respiration of leaves during photosynthesis. I. Estimates from an electrical analogue. *Australian Journal of Biological Sciences*, **20**, 487–93.

Langille, A. R. & McKee, G. W. (1968). Seasonal variation in carbohydrate root reserves and crude protein and tannin in crownvetch forage, *Coronilla varia* L. *Agronomy Journal*, **60**, 415–19.

Lauenroth, W. K. (1973). Effects of water and nitrogen stresses on a shortgrass prairie ecosystem. PhD Thesis, Colorado State University, Fort Collins, Colorado.

Laycock, W. A. & Conrad, P. W. (1969). How time and intensities of clipping affect tall bluebell. *Journal of Range Management*, **22**, 299–303.

Leafe, E. L. (1972). Micro-environment, carbon dioxide exchange and growth in grass swards. In *Crop processes in controlled environments*, ed. A. R. Rees, K. E. Cockshull, D. W. Hand & R. G. Hurd, pp. 157–74. Academic Press, New York & London.

Lechtenberg, V. L., Holt, D. A. & Youngberg, H. W. (1971). Diurnal variation in nonstructural carbohydrates, in vitro digestibility, and leaf to stem ratio of alfalfa. *Agronomy Journal*, **63**, 719–24.

Lechtenberg, V. L., Holt, D. A. & Youngberg, H. W. (1972). Diurnal variation in nonstructural carbohydrates in *Festuca arundinacea* (Schreb.) with and without N fertilizer. *Agronomy Journal*, **64**, 302–5.

Lemon, E., Stewart, D. W. & Shawcroft, R. W. (1971). The sun's work in a cornfield. *Science*, **174**, 371–8.

Lommen, P. W., Schwintzer, C. R., Yocum, C. S. & Gates, D. M. (1971). A model describing photosynthesis in terms of gas diffusion and enzyme kinetics. *Planta*, **98**, 195–220.

Loomis, R. S. (1970). Dynamics of development of photosynthetic systems. In *Prediction and measurement of photosynthetic productivity*, ed. I. Šetlîk, pp. 137–41. Centre for Agricultural Publishing and Documentation, Wageningen, the Netherlands.

Loomis, R. S. & Williams, W. A. (1963). Maximum crop productivity: an estimate. *Crop Science*, **3**, 67–72.

Loomis, R. S., Williams, W. A. & Duncan, W. G. (1967). Community architecture and the productivity of terrestrial plant communities. In *Harvesting the sun*, ed. A. San Pietro, F. A. Greer & T. J. Army, pp. 291–308. Academic Press, New York & London.

Ludlow, M. M. (1971). Analysis of the difference between maximum leaf photosynthetic rates of $C_4$ grasses and $C_3$ legumes. In *Photosynthesis and photorespiration*, ed. M. D. Hatch, C. B. Osmond & R. O. Slatyer, pp. 63–7. Wiley-Interscience, New York.

Ludlow, M. M. & Wilson, G. (1971). Photosynthesis of tropical pasture plant. I. Illuminance, carbon dioxide concentration, leaf temperature, and leaf–air vapour pressure difference. *Australian Journal of Biological Sciences*, **24**, 449–70.

McCarty, E. C. (1935). Seasonal march of the carbohydrates in *Elymus ambiguus* and *Muhlenbergia gracilis* and their reaction under moderate grazing use. *Plant Physiology*, **10**, 727–38.

McCarty, E. C. (1938). *The relation of growth to the varying carbohydrate content in mountain brome. US Department of Agriculture Technical Bulletin 598.*

McCarty, E. C. & Price, R. (1942). *Growth and carbohydrate content of important mountain forage plants in central Utah as affected by clipping and grazing. US Department of Agriculture Technical Bulletin 818.*

McConnell, B. R. & Garrison, G. A. (1966). Seasonal variations of available carbohydrates in bitterbrush. *Journal of Wildlife Management*, **30**, 168–72.

McCree, K. J. (1970). An equation for the rate of respiration of white clover plants grown under controlled conditions. In *Prediction and measurement of photosynthetic productivity*, ed. I. Šetlîk, pp. 221–9. Centre for Agricultural Publishing and Documentation, Wageningen, the Netherlands.

McCree, K. J. (1972). Test of current definitions of photosynthetically active radiation against leaf photosynthesis. *Agricultural Meteorology*, **10**, 443–53.

McCree, K. J. & Troughton, J. H. (1966). Prediction of growth rate at different light levels from measured photosynthesis and respiration rates. *Plant Physiology*, **41**, 559–66.

McDonough, W. T. (1969). Carbohydrate reserves in *Mertensia arizonica* as related to growth, temperature, and clipping treatments. *Ecology*, **50**, 429–32.

McIlvanie, S. K. (1942). Carbohydrate and nitrogen trends in bluebunch wheatgrass, *Agropyron spicatum*, with special reference to grazing influences. *Plant Physiology*, **17**, 540–57.

McNaughton, S. J. & Fullem, L. W. (1970). Photosynthesis and photorespiration in *Typha latifolia*. *Plant Physiology*, **45**, 703–7.

McWilliam, J. R. (1968). The nature of the perennial response in Mediterranean grasses. II. Senescence, summer dormancy, and survival in *Phalaris*. *Australian Journal of Agricultural Research*, **19**, 397–409.

Majerus, M. E. (1975). Response of root and shoot growth of three grass species to decrease in soil water potential. *Journal of Range Management*, **28**, 473–6.

Mall, L. P., Misra, C. M. & Billore, S. K. (1973). *Primary productivity of grassland ecosystems at Ujjain and Ratlam districts of Madhya Pradesh. Progress Report, Indian National Science Academy Project*. School of Studies in Botany, Vikram University, Ujjain, India.

Mangham, S. (1917). On the mechanism of translocation in plant tissues. A hypothesis, with special reference to sugar conduction in sieve-tubes. *Annals of Botany*, **31**, 293–311.

Maranville, J. W. & Paulsen, G. M. (1970). Alteration of carbohydrate composition of corn (*Zea mays* L.) seedlings during moisture stress. *Agronomy Journal*, **62**, 605–8.

Marshall, C. (1966). Studies on the organization of the vegetative grass plant. PhD Thesis, University College of North Wales, Bangor.

Marshall, C. & Sagar, G. R. (1965). The influence of defoliation on the distribution of assimilates in *Lolium multiflorum* Lam. *Annals of Botany*, **29**, 365–72.

Mason, T. G. & Maskell, E. J. (1928*a*). Studies on the transport of carbohydrates in the cotton plant. I. A study of diurnal variation in the carbohydrates of leaf bark, and wood, and of the effects of ringing. *Annals of Botany*, **42**, 189–253.

Mason, T. G. & Maskell, E. J. (1928*b*). Studies on the transport of carbohydrates in the cotton plant. II. The factors determining the rate and the direction of movement of sugars. *Annals of Botany*, **42**, 571–636.

Mason, T. G. & Phillis, E. (1936). Further studies on transport in the cotton plant. V. Oxygen supply and the activation of diffusion. *Annals of Botany*, **50**, 455–99.

Mason, T. G. & Phillis, E. (1937). The migration of solutes. *Botanical Review*, **3**, 47–71.

Matches, A. G. (1960). The development of carbohydrate reserves by alfalfa during the initial season of establishment. *Dissertation Abstracts*, **21**, 17.

May, L. H. (1960). The utilization of carbohydrate reserves in pasture plants after defoliation. *Herbage Abstracts*, **30**, 239–45.

Meidner, H. (1975). Water supply, evaporation, and vapour diffusion in leaves. *Journal of Experimental Botany*, **26**, 666–73.

Menke, J. W. (1973). Effects of defoliation on carbohydrate reserves, vigor, and herbage yield for several important Colorado range species. PhD Thesis, Colorado State University, Fort Collins, Colorado.

Merrill, S. & Kilby, W. W. (1952). Effect of cultivation, irrigation, fertilization, and other cultural treatments on growth of newly planted tung trees. *Proceedings of the American Society for Horticultural Science*, **59**, 69–81.

Miller, P. C. & Tieszen, L. (1972). A preliminary model of processes affecting primary production in the Arctic tundra. *Arctic and Alpine Research*, **4**, 1–18.

Mitchell, K. J. (1954). Influence of light and temperature on the growth of ryegrass (*Lolium* spp.). III. Pattern and rate of tissue formation. *Physiologia Plantarum*, **7**, 51–65.

Mittler, T. E. (1953). Amino acids in phloem sap and their excretion by aphids. *Nature, London*, **172**, 207.

Mochizuki, T. & Hanada, S. (1958). The effect of nitrogen on the formation of the anisophylly on the terminal shoots of apple trees. *Soil and Plant Food*, **4**, 68–74.

Moir, W. H. (1969*a*). Steppe communities in the foothills of the Colorado Front Range and their relative productivities. *American Midland Naturalist*, **81**, 331–40.

Moir, W. H. (1969*b*). Energy fixation and the role of primary producers in energy flux of grassland ecosystems. In *The grassland ecosystem: a preliminary synthesis*, ed. R. L. Dix & R. G. Beidleman, pp. 125–47. *Range Science Department Science Series No. 2*. Colorado State University, Fort Collins, Colorado.

Monsi, M. (1968). Mathematical models of plant communities. In *Functioning of terrestrial ecosystems at the primary production level*, ed. F. E. Eckardt, pp. 131–45. UNESCO, Paris.

Monsi, M. & Murata, Y. (1970). Development of photosynthetic systems as influenced by distribution of matter. In *Prediction and measurement of photosynthetic productivity*, ed. I. Šetlîk, pp. 115–29. Centre for Agricultural Publishing and Documentation, Wageningen, the Netherlands.

Monsi, M. & Saeki, T. (1953). Über den Lichtfaktor in den Pflanzengesellschaften und seine Bedeutung für die Stoffproduktion. *Japanese Journal of Botany*, **14**, 22–52.

Monteith, J. C. (1963). Gas exchange in plant communities. In *Environmental control of plant growth*, ed. L. T. Evans, pp. 95–112. Academic Press, New York & London.

Monteith, J. C. (1972). Introduction: some partisan remarks on the virtues of field experiments and physical analogues. In *Crop processes in controlled environments*, eds. A. R. Rees, K. E. Cockshull, D. W. Hand & R. G. Hurd, pp. 107–10. Academic Press, New York & London.

Monteith, J. L. (1969). Light interception and radiative exchange in crop stands. In *Physiological aspects of crop yield*, ed. J. D. Eastin, F. A. Haskins, C. Y. Sullivan & C. H. M. Van Bavel, pp. 89–116. American Society of Agronomy and Crop Science Society of America, Madison, Wisconsin.

Monteith, J. L. (1973). *Principles of environmental physics*. Edward Arnold, London.

Mooney, H. A. (1972). The carbon balance of plants. *Annual Review of Ecology and Systematics*, **3**, 315–46.

Mooney, H. A. & Billings, W. D. (1960). The annual carbohydrate cycle of alpine plants as related to growth. *American Journal of Botany*, **47**, 594–8.

Mooney, H. A. & Billings, W. D. (1965). Effects of altitude on carbohydrate content of mountain plants. *Ecology*, **46**, 750–1.

Moore, R. M. & Biddiscombe, E. F. (1964). The effects of grazing on grasslands. In *Grasses and grasslands*, ed. C. Barnard, pp. 221–35. Macmillan, London.

Morrall, R. A. A. & Howard, R. J. (1974). *Leaf spot disease of graminoids in native grassland. Canadian Committee, IBP Technical Report 48*. University of Saskatchewan, Saskatoon, Saskatchewan.

Mothes, K. & Engelbrecht, L. (1957). Ueber Blatthonigtau. *Flora*, **145**, 132–45.

Münch, E. (1930). *Die Stoffbewegungen in der Pflanze*. Gustav Fischer, Jena.

Murata, Y. (1969). Physiological responses to nitrogen in plants. In *Physiological aspects of crop yield*, ed. J. D. Eastin, F. A. Haskins, C. Y. Sullivan & C. H. M. Van Bavel, pp. 235–59. American Society of Agronomy and Crop Science Society of America, Madison, Wisconsin.

Murata, Y. & Iyama, J. (1963). Studies on the photosynthesis of forage crops. II. Influences of air temperature upon the photosynthesis of some forage and grain crops. *Proceedings of the Crop Science Society of Japan*, **31**, 315–22.

Natr, L. & Gloser, J. (1967). Carbon dioxide absorption and dry weight increase in barley leaf segments. *Photosynthetica*, **1**, 19–27.

Neiland, B. M. & Curtis, J. T. (1956). Differential responses to clipping of six prairie grasses in Wisconsin. *Ecology*, **37**, 355–65.

Nelson, C. D. (1962). The translocation of organic compounds in plants. *Canadian Journal of Botany*, **40**, 757–70.

Nelson, C. D. & Gorham, P. R. (1957). Uptake and translocation of $C^{14}$-labelled sugars applied to primary leaves of soybean seedlings. *Canadian Journal of Botany*, **35**, 339–47.

Nelson, C. D. & Gorham, P. R. (1959). Translocation of $C^{14}$-labelled amino acids and amides in the stems of young soybean plants. *Canadian Journal of Botany*, **37**, 431–8.

Nelson, C. D., Perkins, H. J. & Gorham, P. R. (1959). Evidence for different kinds of concurrent translocation of photosynthetically assimilated $C^{14}$ in the soybean. *Canadian Journal of Botany*, **37**, 1181–9.

Newman, E. I. (1969*a*). Resistance to water flow in soil and plant. I. Soil resistance in relation to amounts of root: theoretical estimates. *Journal of Applied Ecology*, **6**, 1–12.

Newman, E. I. (1969*b*). Resistance to water flow in soil and plant. II. A review of experimental evidence on the rhizosphere resistance. *Journal of Applied Ecology*, **6**, 261–73.

Ničiporovič, A. A. (1968). Evaluation of productivity by study of photosynthesis as a function of illumination. In *Functioning of terrestrial ecosystems at the primary production level*, ed. F. E. Eckardt, pp. 261–70. UNESCO, Paris.

Ogden, P. R. & Loomis, W. E. (1972). Carbohydrate reserves of intermediate wheatgrass after clipping and etiolation treatments. *Journal of Range Management*, **25**, 29–32.

Ojima, K. & Isawa, T. (1967). Physiological studies on carbohydrates of forage plant. II. Characteristics of carbohydrate composition according to species of plant. *Journal of the Japanese Society of Grassland Science*, **13**, 39–50.

Okajima, H. & Smith, D. (1964). Available carbohydrate fractions in the stem bases and seeds of timothy (*Phleum pratensis*), smooth brome grass (*Bromus inermis*) and several other northern grasses. *Crop Science*, **4**, 317–20.

Old, S. M. (1969). Microclimates, fire, and plant production in an Illinois prairie. *Ecological Monographs*, **39**, 355–84.

Ovington, J. D. & Lawrence, D. B. (1967). Comparative chlorophyll and energy studies of prairie, savanna, oakwood, and maize field ecosystems. *Ecology*, **48**, 515–24.

Owensby, C. E., Paulsen, G. M. & McKendrick, J. D. (1970). Effects of burning and clipping on big bluestem reserve carbohydrates. *Journal of Range Management*, **23**, 358–62.

Palmquist, E. M. (1938). The simultaneous movement of carbohydrates and fluorescein in opposite directions in the phloem. *American Journal of Botany*, **25**, 97–105.

Paltridge, G. W. (1970). A model of a growing pasture. *Agricultural Meteorology*, **7**, 93–130.

Paltridge, G. W., Dilley, A. C., Garrett, J. R., Pearman, G. I., Shepherd, W. & Connor, D. J. (1972). *The Rutherglen experiment on Sherpa wheat: environmental and biological data. CSIRO Division of Atmospheric Physics, Technical Paper 22.* Australia.

Paulsen, G. M. & Smith, D. (1968). Influence of several management practices on growth characteristics and available carbohydrate content of smooth bromegrass. *Agronomy Journal*, **60**, 375–9.

Pearce, R. B., Brown, R. H. & Blaser, R. E. (1967). Net photosynthesis of barley seedlings as influenced by leaf area index. *Crop Science*, **7**, 545–6.

Peaslee, D. E. & Moss, D. N. (1968). Stomatal conductivity in K-deficient leaves of maize. *Crop Science*, **8**, 427–30.

Penman, H. L. & Schofield, R. K. (1951). Some physical aspects of assimilation and transpiration. *Symposia of the Society for Experimental Biology*, **5**, 115–29.

Penning de Vries, F. W. T. (1972). A model for simulating transpiration of leaves with special attention to stomatal functioning. *Journal of Applied Ecology*, **9**, 57–77.

Penning de Vries, F. W. T. (1974). Substrate utilization and respiration in relation to growth and maintenance in higher plants. *Netherlands Journal of Agricultural Science*, **22**, 40–4.

Penning de Vries, F. W. T. (1975). The cost of maintenance processes in plant cells. *Annals of Botany*, **39**, 77–92.

Penot, M. (1965). Etude du rôle des appels moléculaires dans la circulation libérienne. *Physiologie Végétale*, **3**, 41–89.

Peterson, C. A. & Currier, H. B. (1969). An investigation of bidirectional translocation in the phloem. *Physiologia Plantarum*, **22**, 1238–50.

Pettit, R. D. & Fagan, R. E. (1974). Influence of nitrogen and irrigation on carbohydrate reserves of buffalograss. *Journal of Range Management*, **27**, 279–82.

Plaut, Z. & Reinhold, L. (1965). The effect of water stress on $^{14}C$ sucrose transport in bean plants. *Australian Journal of Biological Sciences*, **18**, 1143–55.

Priestly, C. A. (1962). *Carbohydrate resources within the perennial plant; their utilization and conservation. Commonwealth Bureau of Horticulture and Plantation Crops, Technical Communication 27.*

Qureshi, F. A. & Spanner, D. C. (1973). The effect of nitrogen on the movement of tracers down the stolon of *Saxifraga sarmentosa*, with some observations on the influence of light. *Planta*, **110**, 131–44.

Rabideau, G. S. & Burr, G. O. (1945). The use of the $C^{13}$ isotope as a tracer for transport studies in plants. *American Journal of Botany*, **32**, 349–56.

Reardon, P. O. & Huss, D. L. (1965). Effects of fertilization on a little bluestem community. *Journal of Range Management*, **18**, 238–41.

Redmann, R. E. (1974*a*). *Photosynthesis, plant respiration, and soil respiration measured with controlled environment chambers in the field.* I. *Methods and results. Canadian Committee, IBP Technical Report 18*. University of Saskatchewan, Saskatoon, Saskatchewan.

Redmann, R. E. (1974*a*). *Photosynthesis, plant respiration, and soil respiration measured with controlled environment chambers in the field.* II. *Plant $CO_2$ exchange in relation to environment and productivity. Canadian Committee, IBP Technical Report 49.* University of Saskatchewan, Saskatoon, Saskatchewan.

Redmann, R. E. (1974*b*). *Photosynthesis, respiration, and water relations of* Agropyron dasystachyum *measured in the laboratory. Canadian Committee, IBP Technical Report* 47. University of Saskatchewan, Saskatoon, Saskatchewan.

Redmann, R. E. (1975). Production ecology of grassland plant communities in western North Dakota. *Ecological Monographs*, **45**, 83–106.

Reuss, J. O. & Innis, G. S. (1978). A grassland nitrogen flow simulation model. In *Grassland Simulation Model*, ed. G. S. Innis, pp. 186–203. *Springer Ecological Studies 26*. Springer-Verlag, New York.

Reynolds, J. H. (1962). Morphological development and trends of carbohydrate reserves in alfalfa, smooth bromegrass, and timothy under various cutting schedules. *Dissertation Abstracts*, **23**, 1855–6.

Reynolds, J. H. (1969). Carbohydrate reserve trends in orchardgrass (*Dactylis glomerata* L.) grown under different cutting frequencies and nitrogen fertilization levels. *Crop Science*, **9**, 720–3.

Richardson, M. (1968). *Translocation in plants*. St Martin's Press, New York.

Richter, H. (1973). Frictional potential losses and total water potential in plants: a re-evaluation. *Journal of Experimental Botany*, **24**, 983–94.

Ripley, E. A. (1974). *Micrometeorology*. VII. *Canopy $CO_2$ concentrations and fluxes. Canadian Committee, IBP Technical Report 38*. University of Saskatchewan, Saskatoon, Saskatchewan.

Ripley, E. A. & Redmann, R. E. (1976). Grasslands. In *Vegetation and the atmosphere*, ed. J. Monteith, Part II (Case studies), Chapt. 13. Academic Press, New York & London.

Roberts, B. R. (1964). Effects of water stress on the translocation of photosynthetically assimilated carbon-14 in yellow poplar. In *The formation of wood in forest trees*, ed. M. H. Zimmerman, pp. 273–88. Academic Press, New York & London.

Rodin, L. E. & Bazilevich, N. I. (1967). *Production and mineral cycling in terrestrial* vegetation, English edn (transl. Scripta Technica Ltd, ed. G. E. Fogg). Oliver & Boyd, London.

Rogler, G. A. & Lorenz, R. J. (1957). Nitrogen fertilization of northern Great Plains rangelands. *Journal of Range Management*, **10**, 156–60.

Rohrbaugh, L. M. & Rice, E. L. (1949). Effect of application of sugar on the translocation of sodium 2,4-dichlorophenoxyacetate by bean plants in the dark. *Botanical Gazette*, **111**, 85–9.

Rorison, I. H. (1969). *Ecological aspects of the mineral nutrition of plants*. Blackwell Scientific Publications, Oxford.

Ross, J. (1964). Mathematical models of photosynthesis in a plant stand. In *Prediction and measurement of photosynthetic productivity*, ed. I. Šetlîk, pp. 25–45. Centre for Agricultural Publishing and Documentation, Wageningen, the Netherlands.

Ross, J. & Bichele, S. (1969). Calculation of the photosynthesis of leaf canopies. II. In *Photosynthetic productivity of the plant stand, IPA*, pp. 5–43.

Ross, P. J., Heznell, E. F. & Ross, D. R. (1972). Effects of nitrogen and light in grass–legume pastures – A systems analysis approach. *Journal of applied Ecology*, **9**, 535–56.

Ruetz, W. F. (1973). The seasonal pattern of $CO_2$ exchange of *Festuca rubra* L. in a montane meadow community in northern Germany. *Oecologia*, **13**, 247–69.

Rutter, A. J. (1975). The hydrological cycle in vegetation. In *Vegetation and the atmosphere*, ed. J. L. Monteith, vol. I, pp. 111–54. Academic Press, New York & London.

Ryle, G. J. A. (1970*a*). Distribution patterns of assimilated $^{14}C$ in vegetative and reproductive shoots of *Lolium perenne* and *L. temulentum*. *Annals of Applied Biology*, **66**, 155–67.

Ryle, G. J. A. (1970*b*). Partition of assimilates in an annual and a perennial grass. *Journal of Applied Ecology*, **7**, 217–27.

Samish, Y. & Koller, D. (1968). Photorespiration in green plants during photosynthesis estimated by use of isotopic $CO_2$. *Plant Physiology*, **43**, 1129–32.

Sampson, A. W. & McCarty, E. C. (1930). The carbohydrate metabolism of *Stipa pulchra*. *Hilgardia*, **5**, 61–100.

Santarius, K. A. (1967). Das Verhalten von $CO_2$-Assimilation, NADP- und PGS-Reduktion, und ATP-Synthese intakter Blattzellen in Abhängigkeit vom Wassergehalt. *Planta*, **73**, 228–42.
Santarius, K. A. & Heber, U. (1967). Das Verhalten von Hill-Reaktion und Photophosphorylierung isolierter Chloroplasten in Abhängigkeit vom Wassergehalt. II. Wasserentzug über $CaCl_2$. *Planta*, **73**, 109–37.
Sauer, R. H. (1978). A simulation model for grassland primary producer phenology and biomass dynamics. In *Grassland simulation model*, ed. G. S. Innis, pp. 55–87. *Springer Ecological Studies 26*. Springer-Verlag, New York.
Saugier, B. & Ripley, E. A. (1975). A model of growth and water use for a natural grassland. In *Proceedings of the 1975 Summer Computer Simulation Conference*, pp. 945–53. Simulation Councils, Inc., La Jolla, California.
Saugier, B., Ripley, E. A. & Lueke, P. (1974). *A mechanistic model of plant growth and water use for the Matador grassland. Matador Project Technical Report 65.* University of Saskatchewan, Saskatoon, Saskatchewan.
Schirman, R. & Buckholtz, K. P. (1966). Influence of atrazine on control and rhizome carbohydrate reserves of quackgrass. *Weeds*, **14**, 233–6.
Sěsták, Z., Čatský, J. & Jarvis, P. G. (eds.) (1971). *Plant photosynthetic production: manual of methods.* Dr W. Junk N. V. Publishers, The Hague, the Netherlands.
Shaw, R. H. & Laing, D. R. (1966). Moisture stress and plant response. In *Plant environment and efficient water use*, ed. W. H. Pierre, D. Kirkham, J. Pesek & R. Shaw, pp. 73–94. American Society of Agronomy and Crop Science Society of America, Madison, Wisconsin.
Sheard, R. W. (1968). Relationship of carbohydrate and nitrogen compounds in the haplocorn to the growth of timothy (*Phleum pratense* L.). *Crop Science*, **8**, 658–60.
Shearman, L. L., Eastin, J. D., Sullivan, C. Y. & Kinbacher, E. J. (1972). Carbon dioxide exchange in water-stressed sorghum. *Crop Science*, **12**, 406–9.
Sims, P. L. & Singh, J. S. (1971). Herbage dynamics and net primary production in certain ungrazed and grazed grasslands in North America. In *Preliminary analysis of structure and function in grasslands*, ed. N. R. French, pp. 59–124. *Range Science Department Science Series No. 10.* Colorado State University, Fort Collins, Colorado.
Sims, P. L. & Singh, J. S. (1978). Comparative structure and function of ten western North American grasslands. IV. Compartmental transfers and system transfer functions. *Journal of Ecology*, in press.
Singh, J. S. (1967). Seasonal variation in composition, plant biomass, and net community production in the grasslands at Varanasi. PhD Thesis, Banaras Hindu University, Varanasi, India.
Singh, J. S. (1968). Net aboveground community productivity in the grasslands at Varanasi. In *Proceedings of the Symposium on 'Recent Advances in Tropical Ecology'*, ed. R. Misra & B. Gopal, part II, pp. 631–54. International Society of Tropical Ecology, India.
Singh, J. S. (1973). A compartment model of herbage dynamics for Indian tropical grasslands. *Oikos*, **24**, 367–72.
Singh, J. S. & Coleman, D. C. (1974). Distribution of photoassimilated 14-C in the root system of shortgrass prairie. *Journal of Ecology*, **62**, 359–65.
Singh, J. S. & Coleman, D. C. (1977). Evaluation of functional root biomass and translocation of photoassimilated carbon-14 in a short-grass prairie ecosystem. In *The belowground ecosystem: a synthesis of plant-associated processes*, ed.

J. K. Marshall. *Range Science Department Science Series No. 26*. Colorado State University, Fort Collins, Colorado.

Singh, J. S. & Joshi, M. C. (1976). Land use practice in semi-arid parts of India. In *Management of semiarid ecosystems*, ed. B. H. Walker. Elsevier, Amsterdam (in press).

Singh, J. S. & Yadava, P. S. (1974). Seasonal variation in composition, plant biomass, and net primary productivity of a tropical grassland at Kurukshetra, India. *Ecological Monographs*, **44**, 351–76.

Slatyer, R. O. (1967). *Plant–water relationships*. Academic Press, New York & London.

Smith, A. E. & Leinweber, C. L. (1971). Relationship of carbohydrate trend and morphological development of little bluestem tillers. *Ecology*, **52**, 1052–7.

Smith, D. (1962). Carbohydrate root reserves in alfalfa, red clover, and birdsfoot trefoil under several management schedules. *Crop Science*, **2**, 76–8.

Smith, D. (1964). Winter injury and the survival of forage plants: a review. *Herbage Abstracts*, **34**, 203–9.

Smith, D. (1968). Carbohydrates in grasses. IV. Influence of temperature on the sugar and fructosan composition of timothy plant parts at anthesis. *Crop Science*, **8**, 331–4.

Smith, D. & Graber, L. F. (1948). The influence of top growth removal on the root and vegetative development of biennial sweetclover. *Journal of the American Society of Agronomy*, **40**, 818–31.

Smith, D. & Grotelueschen, R. D. (1966). Carbohydrates in grasses. I. Sugar and fructosan composition of the stem bases of several northern-adapted grasses at seed maturity. *Crop Science*, **6**, 263–6.

Smith, D., Paulsen, G. M. & Raguse, C. A. (1964). Extraction of total available carbohydrates from grass and legume tissue. *Plant Physiology*, **39**, 960–2.

Sosebee, R. E. & Wiebe, H. H. (1971). Effect of water stress and clipping on photosynthate translocation in two grasses. *Agronomy Journal*, **63**, 14–17.

Sosebee, R. E. & Wiebe, H. H. (1973). Effect of phenological development on radiophosphorus translocation from leaves in crested wheatgrass. *Oecologia*, **13**, 102–12.

Spanner, D. C. (1958). The translocation of sugar in sieve tubes. *Journal of Experimental Botany*, **9**, 332–42.

Spanner, D. C. (1962). A note on the velocity and the energy requirements of translocation. *Annals of Botany, New Series*, **26**, 511–16.

Spanner, D. C. (1963). The mathematical pattern of tracer movement. *Progress in Biophysical Chemistry*, **13**, 246–51.

Spanner, D. C. & Prebble, J. N. (1962). The movement of tracers along the petiole of *Nymphoides peltata*. I. A preliminary study with $^{137}$Cs. *Journal of Experimental Botany*, **13**, 294–306.

Spittlehouse, D. L. & Ripley, E. A. (1974). *Micrometeorology. X. Relationships between the canopy and structure of the Matador grassland. Canadian Committee, IBP Technical Report 55*. University of Saskatchewan, Saskatoon, Saskatchewan.

Sprague, V. G. & Sullivan, J. T. (1950). Reserve carbohydrates in orchardgrass clipped periodically. *Plant Physiology*, **25**, 92–102.

Stanhill, G. (1962). The effect of environmental factors on the growth of alfalfa. *Netherlands Journal of Agricultural Science*, **10**, 247–53.

Steinke, T. D. (1969). The translocation of $^{14}$C-assimilates in *Eragrostis curvula*: an autoradiographic survey. *Proceedings of the Grassland Society of South Africa*, **4**, 19–34.

Steinke, T. D. & Booysen, P. de V. (1968). The regrowth and utilization of carbohydrate reserves of *Eragrostis curvula* after different frequencies of defoliation. *Proceedings of the Grassland Society of South South*, **3**, 105–10.

Sullivan, J. T. & Sprague, V. G. (1943). Composition of the roots and stubble of perennial ryegrass following partial defoliation. *Plant Physiology*, **18**, 656–70.

Sullivan, J. T. & Sprague, V. G. (1953). Reserve carbohydrates in orchardgrass cut for hay. *Plant Physiology*, **28**, 304–13.

Suzuki, M. (1971). Behavior of long-chain fructosan in the basal top of timothy as influenced by N, P, and K and defoliation. *Crop Science*, **11**, 632–5.

Swanson, C. A. & Böhning, R. H. (1951). The effect of petiole temperature on the translocation of carbohydrates from bean leaves. *Plant Physiology*, **26**, 751–6.

Swanson, C. A. & Geiger, D. R. (1967). Time course of low temperature inhibition of sucrose translocation in sugar beets. *Plant Physiology*, **42**, 751–6.

Tamiya, H. (1951). Some theoretical notes on the kinetics of algae growth. *Botanical Magazine, Tokyo*, **64**, 167–73.

Taylor, A. O. & Rowley, J. R. (1971). Plants under climatic stress. I. Low temperature, high light effects on photosynthesis. *Plant Physiology*, **47**, 713–18.

Taylor, B. K. (1967). Storage and mobilization of nitrogen in fruit trees: a review. *Journal of the Australian Institute of Agricultural Science*, **33**, 23–9.

Tew, R. K. (1970). Root carbohydrate reserves in vegetative reproduction of aspen. *Forest Science*, **16**, 318–20.

Thaine, R. (1954). The effect of clipping frequency on the productivity of root development of Russian wild ryegrass in the field. *Canadian Journal of Agricultural Science*, **34**, 299–320.

Thaine, R. (1961). Transcellular strands and particle movement in mature sieve tubes. *Nature, London*, **192**, 772–3.

Thaine, R. (1962). A translocation hypothesis based on the structure of plant cytoplasm. *Journal of Experimental Botany*, **13**, 152–60.

Thaine, R. (1964). The protoplasmic-streaming theory of phloem transport. *Journal of Experimental Botany*, **15**, 470–84.

Thornley, J. H. M. (1972*a*). A model to describe the partitioning of photosynthate during vegetative plant growth. *Annals of Botany, New Series*, **36**, 419–31.

Thornley, J. H. M. (1972*b*). A balanced quantitative model for root : shoot ratios in vegetative plants. *Annals of Botany, New Series*, **36**, 431–41.

Tooming, H. (1970). Mathematical description of net photosynthesis and adaption processes in the photosynthetic apparatus of plant communities. In *Prediction and measurement of photosynthetic productivity*, ed. I. Šetlîk, pp. 103–13. Centre for Agricultural Publishing and Documentation, Wageningen, the Netherlands.

Trip, P. & Gorham, P. R. (1968). Bidirectional translocation of sugars in sieve tubes of squash plants. *Plant Physiology*, **43**, 877–82.

Trip, P., Krotkov, G. & Nelson, C. D. (1963). Biosynthesis of mannitol-$C^{14}$ from $C^{14}O_2$ by detached leaves of white ash and lilac. *Canadian Journal of Botany*, **41**, 1005–10.

Trip, P., Nelson, C. D. & Krotkov, G. (1965). Selective and preferential translocation of $C^{14}$-labeled sugars in white ash and lilac. *Plant Physiology*, **40**, 740–7.

Trlica, M. J., Jr (1971). Defoliation and soil moisture regime effects on plant regrowth and carbohydrate reserve utilization and storage in eight desert range species. PhD Thesis, Utah State University, Logan, Utah.

Trlica, M. J., Jr & Cook, C. W. (1971). Defoliation effects on carbohydrate reserves of desert species. *Journal of Range Management*, **24**, 418–25.

Trlica, M. J., Jr & Cook, C. W. (1972). Carbohydrate reserves of crested wheatgrass and Russian wildrye as affected by development and defoliation. *Journal of Range Management*, **25**, 430–5.

Trlica, M. J., Jr, Buwai, M. & Menke, J. W. (1977). Effects of rest following defoliations on the recovery of several range species. *Journal of Range Management*, **30**, 21–7.

Troughton, A. (1957). *The underground organs of herbage grasses. Commonwealth Bureau of Pastures and Field Crops Bulletin 44*. Hurley, Berkshire, England.

Turner, D. J. (1969). The effects of shoot removal on the rhizome carbohydrate reserves of couch grass (*Agropyron repens* (L.) Beauv.). *Weed Research*, **9**, 27–36.

Turner, G. T. & Dortignac, E. J. (1954). Infiltration, erosion, and herbage production of some mountain grasslands in western Colorado. *Journal of Forestry*, **52**, 858–60.

Turner, N. C. (1974). Stomatal behavior and water status of maize, sorghum, and tobacco under field conditions. II. At low soil water potential. *Plant Physiology*, **53**, 360–5.

Ueno, M. & Smith, D. (1970). Growth and carbohydrate changes in the root wood and bark of different sized alfalfa plants during regrowth after cutting. *Crop Science*, **10**, 396–9.

Ullrich, W. (1961). Zur Sauerstoffabhängigkeit des Transportes in den Siebröhren. *Planta*, **57**, 402–29.

Van Bavel, C. H. M. (1975). A behavioral equation for leaf carbon dioxide assimilation and a test of its validity. *Photosynthetica*, **9**, 165–76.

Van Bavel, C. H. M., DeMichele, B. W. & Ahmed, J. (1973). *A model of gas and energy exchange regulations by stomatal action in plant leaves. Texas Agricultural Experiment Station Miscellaneous Publication 1078*.

Waggoner, P. E. (1969). Predicting the effect upon net photosynthesis of changes in leaf metabolism and physics. *Crop Science*, **9**, 315–21.

Waggoner, P. E., Moss, D. N. & Hesketh, J. D. (1963). Radiation in the plant environment and photosynthesis. *Agronomy Journal*, **55**, 36–9.

Waggoner, P. E. & Turner, N. C. (1972). Comparison of simulated and actual evaporation from maize and soil in a lysimeter. *Agricultural Meteorology*, **10**, 113–23.

Waite, R. & Boyd, J. (1953). The water-soluble carbohydrates of grasses. II. Grasses cut at grazing height several times during the growing season. *Journal of the Science of Food and Agriculture*, **4**, 257–61.

Ward, C. V. & Blaser, R. E. (1961). Carbohydrate food reserves and leaf area in regrowth of orchardgrass. *Crop Science*, **1**, 366–70.

Wardlaw, I. F. (1965). The velocity and pattern of assimilate translocation in wheat plants during grain development. *Australian Journal of Biological Sciences*, **18**, 269–81.

Wardlaw, I. F. (1967). The effect of water stress on translocation in relation to photosynthesis and growth. I. Effect during grain development in wheat. *Australian Journal of Biological Sciences*, **20**, 25–39.

Wardlaw, I. F. (1968*a*). The effect of water stress on translocation in relation to photosynthesis and growth. II. Effect during leaf development in *Lolium temulentum* L. *Australian Journal of Biological Sciences*, **22**, 1–16.

Wardlaw, I. F. (1968*b*). The control and pattern of movement of carbohydrates in plants. *Botanical Review*, **34**, 79–105.

Warembourg, F. R. & Paul, E. A. (1973). The use of $C^{14}O_2$ canopy techniques for measuring carbon transfer through the plant–soil system. *Plant and Soil*, **38**, 331–45.

Warembourg, F. R. & Paul, E. A. (1977). Seasonal transfers of assimilated $^{14}C$ in grassland: plant production and turnover, translocation and respiration. In *The belowground ecosystem: a synthesis of plant-associated processes*, ed. J. K. Marshall. *Range Science Department Science Series No. 26*. Colorado State University, Fort Collins, Colorado.

Weatherley, P. E., Peel, A. J. & Hill, G. P. (1959). The physiology of the sieve tube: preliminary experiments using aphid mouth parts. *Journal of Experimental Botany*, **10**, 1–16.

Weaver, J. E. & Zink, E. (1946). Length of life of roots of ten species of perennial range and pasture grasses. *Plant Physiology*, **21**, 201–17.

Webb, K. L. & Gorham, P. R. (1965). Radial movement of $^{14}C$-translocates from squash phloem. *Canadian Journal of Botany*, **43**, 97–103.

Weinmann, H. (1940). Seasonal changes in the roots of some South African high-veld grasses. *Journal of South African Botany*, **6**, 131–45.

Weinmann, H. (1943). Root reserves in South African highveld grasses in relation to fertilizing and frequency of clipping. *Journal of South African Botany*, **10**, 37–54.

Weinmann, H. (1948). Underground development and reserves of grasses: a review. *Journal of the British Grassland Society*, **3**, 115–40.

Weinmann, H. (1949). Productivity of Maradellas sandveld pastures in relation to frequency of cutting. *Rhodesian Agricultural Journal*, **46**, 175–89.

Weinmann, H. (1952). Carbohydrate reserves in grasses. In *Proceedings of VI International Grassland Congress*, vol. 1, pp. 655–60. Pennsylvania State College, State College, Pennsylvania.

Weinmann, H. (1961). Total available carbohydrates in grasses and legumes. *Herbage Abstracts*, **31**, 255–61.

Weinmann, H. & Reinhold, L. (1946). Reserve carbohydrates in South African grasses. *Journal of South African Botany*, **12**, 57–73.

Welch, T. G. (1968). Carbohydrate reserves of sand reedgrass under different grazing intensities. *Journal of Range Management*, **21**, 216–20.

White, L. M. (1973). Carbohydrate reserves of grasses: a review. *Journal of Range Management*, **26**, 13–18.

White, L. M., Brown, J. H. & Cooper, C. S. (1972). Nitrogen fertilization and clipping effects on green needlegrass (*Stipa viridula* Trin.). III. Carbohydrate reserves. *Agronomy Journal*, **64**, 824–8.

Whitehead, C. W., Sansing, N. G. & Loomis, W. E. (1959). Temperature coefficient of translocation in tomatoes. *Plant Physiology*, **34** (Supplement), 21.

Whitehead, D. C. (1970). *The role of nitrogen in grassland productivity. Commonwealth Bureau of Pastures and Field Crops Bulletin 48*. Hurley, Berkshire, England.

Wiebe, H. H. & Wihrheim, S. E. (1962). The influence of internal moisture stress on translocation. In *Proceedings of Symposium on Use of Radioisotopes in Soil–Plant Nutrition Studies*, pp. 279–88. Bombay, India.

Willard, E. E. & McKell, C. M. (1973). Simulated grazing management systems in relation to shrub growth responses. *Journal of Range Management*, **26**, 171–4.

Willenbrink, J. (1957). Über die Hemmung des Stofftransports in den Siebröhren lokale Inaktivierung verschiedener Atmungsenzyme. *Planta*, **48**, 269–342.

Williams, G. J., III (1974). Photosynthetic adaptation to temperature in $C_3$ and $C_4$ grasses: a possible ecological role in the shortgrass prairie. *Plant Physiology*, **54**, 709–11.

Williams, G. J. & Markley, J. L. (1973). The photosynthetic pathway type of North American shortgrass prairie species and some ecological implications. *Photosynthetica*, **7**, 262–70.

Williams, W. A., Loomis, R. S. & Lepley, C. R. (1965). Vegetative growth of corn as affected by population density. I. Productivity in relation to interception of solar radiation. *Crop Science*, **5**, 211–15.

Wilson, H. K. (1944). Control of noxious plants. *Botanical Review*, **10**, 279–326.

Woledge, J. & Jewiss, O. R. (1969). The effect of temperature during growth and subsequent rate of photosynthesis in leaves of tall fescue, *Festuca arundinacea*. *Annals of Botany, New Series*, **33**, 897–913.

Wolf, D. D. (1967). *Characteristics of stored carbohydrates in reed canarygrass as related to management, feed value, and herbage yield. Connecticut Agricultural Experiment Station Bulletin 402.*

Wright, H. A. & Stinson, K. J. (1970). Response of mesquite to season of top removal. *Journal of Range Management*, **23**, 127–8.

Wuenscher, M. L. & Gerloff, G. C. (1971). Growth of *Andropogon scoparius* (little bluestem) in phosphorus deficient soils. *New Phytologist*, **70**, 1035–42.

Younger, V. B. (1972). Physiology of defoliation and regrowth. In *The biology and utilization of grasses*, ed. V. B. Younger & C. N. McKell, pp. 305–17. Academic Press, New York & London.

Zanoni, L. J., Michelson, L. F., Colby, W. G. & Drake, M. (1969). Factors affecting carbohydrate reserves of cool season turfgrass. *Agronomy Journal*, **61**, 195–8.

Zelitch, I. (1966). Increased rate of net photosynthetic carbon dioxide uptake caused by the inhibition of glycolate oxidase. *Plant Physiology*, **41**, 1623–31.

Zelitch, I. (1971). *Photosynthesis, photorespiration, and plant productivity*. Academic Press, New York & London.

Zelitch, I. & Waggoner, P. E. (1962). Effect of chemical control on stomata on transpiration and photosynthesis. *Proceedings of the National Academy of Sciences, USA*, **48**, 1101–8.

Ziegler, H. & Vieweg, G. H. (1961). Der experimentelle Nachweiss einer Massenströmung im Phloem von *Heracleum mantegazzianum* Somm. et Lev. *Planta*, **56**, 402–8.

# 3. Small herbivore subsystem

L. ANDRZEJEWSKA & G. GYLLENBERG

**Contents**



Phytophagous animals together with saprophages, using plants as their food source, divide up among themselves the total plant production of a meadow. The number, biomass and species composition of herbivores determine the proportion of plant production of the given grassland ecosystem which they will utilize. Hence, the starting point of this study must be a review of the taxonomic groups of herbivorous animals that feed on the above- and belowground parts of plants.

Here, attention is drawn to the occurrence of herbivores in various types of meadows under different climatic conditions, viz. desert and semidesert regions, tropics, the temperate zone, tundra and mountain meadows. The biomass estimates for herbivores feeding in a given habitat usually involve only a certain number of groups, or species, which are predominant in the habitat. Simultaneous estimates of number and biomass of all animals feeding on all plant parts, i.e. both below- and aboveground parts, are rare. Analyses of herbivorous communities often omit mining insects and those feeding on nectar and pollen of flowers, also Acarina and Nematoda, especially those feeding inside the plants. (The two last groups can exceed, in terms of biomass, the insects in the aboveground plant layer.) Hence, estimates of both numbers and biomass of herbivores obtained by various authors and used here for comparison of different grasslands, usually take into account only a part of the herbivorous fauna occurring in a given habitat.

At present the majority of grassland habitats, to a greater or lesser degree, are managed and utilized by man. Meadows undergoing various treatments (fertilization, irrigation, burning) create altered nutritional and habitat conditions for the fauna grazing on them. The effect of these treatments on herbivores, their biomass and species composition is included in the second section of this chapter. The third section is devoted to a discussion of plant–herbivore interactions. In addition to estimates of consumption, as a measure of the impact of herbivores on plant vegetation, estimates of the effects of their feeding are also presented; these effects are often greater than the losses produced by consumption.

Utilization of plants consumed by herbivores is the topic of the fourth section. Parameters of herbivore population, production, and efficiencies of assimilation, ecological growth and tissue growth, are elaborated for herbivores dwelling among aboveground parts of plants in various grassland ecosystems.

The last section of the chapter is a short review of the data presented in earlier sections pertaining to the role of herbivores in the utilization and transformation of plant matter in grassland ecosystems. Three grassland ecosystems were chosen to exemplify this problem, starting from the most intensely exploited pasture, through a hay-harvested meadow to a meadow utilized only by a natural community of herbivores. General schemes depict the position of herbivores in the trophic web of the escoystem.

## 1. Herbivore biomass and population dynamics in various grassland ecosystems

When analysing the herbivore associations that occur in various grassland habitats it is impossible to exclude the features of the habitat itself – its structure and nutritional reserves. Both the species composition of the herbivores and their population dynamics must be investigated against the background of the composition of the vegetation cover and its seasonal variations.

Green plant parts and the living parts of the root system are the primary food source of phytophagous populations and overwhelmingly determine their structure, numbers and biomass. In general there is a correlation between the total biomass of phytophagous animals feeding on the grassland vegetation and its biomass or production. This relation has been shown, for example, by Ellis & French (1973) and Zlotin (1975). In both studies the authors examined the relationship between the production of aboveground plant parts and the total biomass of all phytophagous animals feeding on grasslands of the same climatic region, similar management practices and in the same vegetation season. The phrase 'all phytophagous animals' includes not only those animals feeding at particular times on various plant species but also those inhabiting various parts (levels) of the grassland vegetation. As it has been shown in numerous papers, even polyphagous phytophages reveal highly selective feeding patterns, with regard not only to the species but also the individual parts, chemical composition and physiological condition of the plants that are eaten. These are essential factors in vegetation utilization and thus influence the distribution of the plant in the habitat and the numbers and biomass of phytophagous animals.

Species diversity of phytophages also depends on the number of plant species present and the composition of the plant cover. As species diversity of phytophages increases, there is a concomitant decrease in the number of individuals belonging to each species. All management practices on grassland that lead to changes in the biomass, number of species, chemical composition or pattern of the vegetation indirectly influence the numbers, biomass and species diversity of the phytophagous fauna.

### *Food selectivity*

Both the aboveground and belowground parts of all species of the grassland vegetation can be grazed or sucked by numerous species of phytophagous animals. All plant parts can be used as food throughout the growing season because of differences in the food preferences and life cycles of the phytophages.

In diversified grasslands and when there is an abundance of food, large herbivores practice food selectivity. During the vegetation season there are also variations in the consumption of particular species or particular plant parts (Hughes, Milner & Dale, 1964; Kaufman, 1965; Voison, 1970; Gaare & Skogland, 1975; Rice & Vavra, 1971; Duffey *et al.*, 1974; Wielgolaski, 1975). Generally there are a few plant species that are preferentially consumed by any given herbivorous species, and the percentage of these preferred plants in the food is always high and relatively stable (on average over 50 %). The contribution of other species may vary over a wide range, depending on the composition of the pasture, the length of the vegetation season, etc. (Duffey *et al.*, 1974). For example, reindeer in winter may feed on only a few plant species, mainly lichens, while in spring they prefer plants with high protein content such as fungi, although lignified plant parts and lichens always constitute the main component of their diet (Bliss, 1975; Østbye *et al.*, 1975). Apart from a few essential grass species, cows will eagerly graze so-called 'weeds' to supplement their rather monotonous diet, particularly when grazing sown and cultivated pastures.

According to Duffey *et al.* (1974), grazing sheep show a regular selectivity in their utilization of plant production, a finding that also appears to be true for many other herbivorous species. It has been found that sheep prefer:

(1) leaves to stems, and green and young plant parts to dry, old ones (physiologically young tissues being preferred both because of their chemical composition and their height);

(2) plants rich in nitrogen and phosphorus (which are more valuable as an energy supply);

(3) the dominant species in the pasture if the vegetation biomass is large.

The percentage contribution of different plant species to herbivores' food changes during the season according to their palatability. The exception is if the pasture sward is poor or if the intensity of grazing is high and there is an obvious nutrient deficiency, when all plant species are grazed and palatability is of little importance.

Once a pasture has been grazed there is still a part of the plant production which is left by the feeding animals. This 'left-on-the-pasture' plant biomass averages 20 to 70 % of the total green plant biomass, the exact amount depending on grazing intensity, species composition of the sward and the species of grazing herbivore (Voisin, 1970; Duffey *et al.*, 1974; Wielgolaski, 1975). Horses and particularly sheep graze the grass close to the ground, while kangaroos consume tillering nodes and even roots (see Chapter 4, this volume). Cattle, on the other hand, do not tear the vegetation below 1.5 cm. It may be assumed that all the aboveground plant production of a grassland may, under appropriate circumstances, be utilized by vertebrates.

Small herbivorous mammals and birds consume mainly fresh plant parts, buds and seeds (Tertil, 1974; Golley, Ryszkowski & Sokur, 1975; Kallio, 1975; Østbye *et al.*, 1975). Since these constitute only a part of grassland

Table 3.1. *The ratio of available plant biomass (A) to phytophagous insect biomass (B) in meadow plots with different level of primary production*

| | Phytophagous insects feeding on: Roots A/B | Roots + tillering nodes A/B | Primary production[a] (g dry wt · m$^{-2}$) |
|---|---|---|---|
| Unfertilized meadow | 400 | 435 | 1123.9 |
| Fertilized meadow | 430 | 507 | 1709.8 |

Data from Andrzejewska (1976*a*); Plewczyńska-Kuraś (1976); Traczyk, Traczyk & Pasternak (1976*a*,*b*).

[a] As live weight of aboveground and belowground plant parts.

Table 3.2. *Ratios of herbivore biomass to available plant biomass*

| Type of grassland | Herbivore biomass (mg dry wt · m$^{-2}$) A[a] | B[b] | Plant biomass (g dry wt · m$^{-2}$) A | B | Ratio (%) | Source of data (Reference) |
|---|---|---|---|---|---|---|
| Tropics | | | | | | |
| Burned savanna | 367.8 | — | 800.0 | — | 0.05 | Africa (Lamotte, 1975) |
| Unburned savanna | 386.7 | — | ~1000.0 | — | 0.04 | |
| Reserve savanna | — | 251.9 | — | 481.0 | 0.05 | Panama (Breymeyer, 1977) |
| Grazed savanna | — | 70.0 | — | 413.7 | 0.02 | |
| Unburned savanna | 229.3 | 803.0 | 138.6 | 646.6 | A 0.2<br>B 0.1 | Panama (L. Andrzejewska & A. Myrcha, unpublished data) |
| Burned savanna | 1015.6 | 1583.0 | 143.5 | 418.0 | A 0.7<br>B 0.4 | |
| Temperate zone | | | | | | |
| Desert grassland | 1.77 | — | 53.7 | — | 0.003 | Colorado (Lewis, 1971; McDaniel, 1971) |
| Shortgrass prairie | 21.3 | — | 144.8 | — | 0.01 | |
| Mixed prairie | 30.7 | — | 94.6 | — | 0.03 | |
| True prairie | 42.3 | — | 220.8 | — | 0.02 | |
| Mountain meadows | | | | | | |
| Tussock tundra | 26.0 | 333.0 | 82.0 | 394.0 | A 0.03<br>B 0.08 | USSR, Tien Shan, (Zlotin 1975) |
| Wet meadow | 27.0 | 76.0 | 215.0 | 2160.0 | A 0.01<br>B 0.003 | |
| Steppe meadow | 172.0 | 473.0 | 140.0 | 1010.0 | A 0.1<br>B 0.05 | |
| Dry steppe | 274.0 | 351.0 | 152.0 | 425.0 | A 0.2<br>B 0.0 | |
| Desert | 19.0 | 57.0 | 67.0 | 39.0 | A 0.01<br>B 0.135 | |
| Bog | 88.0 | 33.0 | 184.0 | 2240.0 | A 0.051<br>B 0.00 | |
| Lowland meadow | 1211.0 | — | 676.0 | — | 0.2 | France (Ricou, 1976) |
| Reserve meadow | 172.4 | — | 196.4 | — | 0.2 | Central Poland (Andrzejewska, 1971) |
| Cultivated meadow | 289.5 | — | 488.0 | — | 0.09 | |
| Cultivated meadow (unfertilized) | 28.1 | 2050.0 | 117.3 | 810.2 | A 0.02<br>B 0.25 | Central Poland (Andrzejewska, 1976) |
| Cultivated meadow (fertilized) | 31.5 | 1490.0 | 374.7 | 647.6 | A 0.01<br>B 0.23 | |
| Cold zone | | | | | | |
| Peat site (bog) | 92.0 | — | 652.0 | — | 0.01 | Ireland (Moore, Dowding & Healy, 1975) |
| Wet meadow | ~300.0 | — | 306.0 | — | 0.1 | Norway (Østbye *et al.*, 1975) |

[a] Aboveground biomass. [b] Belowground biomass.

production, the size of the reserves of such foods may restrict population numbers.

Invertebrates occur in the greatest numbers and feed most intensely on underground plant parts and young leaves and stems. There is much greater selection of food by phytophagous invertebrates than is the case for vertebrates, and their food is often limited to one or a few species or even certain parts of plants. They thus utilize only a small part of the total plant production (Barnes, 1955; Ganwere, 1961; Kaufmann, 1965; Mulkern, 1967, 1970). In such cases it is that part of the plant biomass that is consumed that determines the biomass of phytophagous animals. This interdependence has been shown to exist in the case of two grassland plots with different plant production (resulting from mineral fertilization of one plot). The ratios of plant biomass to the phytophage biomass feeding on particular plant parts (roots, tillering nodes and stubble) in both habitats are very similar (Table 3.1). It means that for a biomass unit of phytophages feeding in a particular stratum in similar environmental and weather conditions there is a similar plant biomass representing their food. The food biomass considerably exceeds the phytophage biomass that feeds on it. That is true for all grasslands unless the herbivores become abnormally abundant (Table 3.2).

### *Biomass and numbers of the phytophagous fauna*

Vertebrate density under natural conditions is related to vegetation production and cannot exceed the capacity of the habitat to sustain the population. Under natural conditions animal numbers and the intensity of their grazing are regulated by free migration into, within and out of the habitat. In the case of fenced pastures grazed by domestic animals such as sheep and cows, the number of animals is fitted to plant production. The standards differ markedly between different grassland habitats and vary during the season. According to Voisin (1970) one dairy cow, mean weight 500 kg, consumes daily an average of 48 kg of fresh grass. Depending on the vegetation structure of a pasture, this ration can vary from 35 kg to over 70 kg wet weight per cow per day. When plant production is low, for example in poor habitats or where the vegetation season is short, 4 ha of pasture may feed only one cow. Under favourable conditions, when pasture production is high, 1 ha may support up to three cows (Voisin, 1970). In the Colorado prairies the average daily grass ration of a cow of mean weight 454 kg is 9–10 kg dry wt $\cdot$ $d^{-1}$ (Lewis, 1971). The average daily food consumption of a sheep (average weight 40 kg) is estimated as 2.6–4.6 kg dry wt of grass depending on general environmental and climatic conditions and the time of year (see Chapter 4, this volume). A pasture with plant production of around 135 g dry wt $\cdot$ $m^{-2}$ $\cdot$ $yr^{-1}$ can support approximately 2.5 sheep.

In some cases the amount of food available is not the decisive factor and

Table 3.3. *Periodicity of fluctuations in herbivore populations*

| Species | Period (years) | Site | Reference |
|---|---|---|---|
| *Locusta migratoria* | 3–4 | North Africa | Uvarov (1970) |
| *Locustana paradalina* | 3–4 | South Africa | Uvarov (1970) |
| *Lagopus lagopus* | 4 | Hardangervidda, Norway | Østbye *et al.* (1975) |
| *Lemmus trimucronatus* | 2–4 (typical cycle 4) | Barrow, Alaska | Bunnell, MacLean & Brown (1975) |
| *Lemmus lemmus* | 3–4 | Devon Island, Canada | Bliss (1975) |
| *Microtus arvalis* | 3–4 | Poland | French *et al.* (1975) |
| *Microtus oeconomus* | 4 | Hardangervidda, Norway | Østbye *et al.* (1975) |
| *Lepus timidus* | 10–11 | | Trojan (1975) |

Table 3.4. *Herbivore density and plant biomass in successive seasons*

| Herbivores | | Density of herbivores ($\bar{N} \cdot m^{-2}$) | Aboveground plant biomass ($g \cdot m^{-2}$) | Site of investigations | Reference |
|---|---|---|---|---|---|
| Diptera | 1972 | 29.2 | 307.6 | Cultivated meadow, Poland | Olechowicz (1976) |
| | 1973 | 214.5 | 372.8 | | |
| | 1974 | 618.2 | 374.7 | | |
| Auchenorrhyncha | 1972 | 110.2 | 307.6 | Cultivated meadow, Poland | Andrzejewska (1976) |
| | 1973 | 3.6 | 372.8 | | |
| | 1974 | 25.9 | 374.7 | | |
| Hemiptera (Psyllidae) | 1969 | 446 | 143–62 | Mountain plateau, subalpine, Hardangervidda, Norway (1100–1300 m) | Østbye *et al.* (1975) |
| | 1970 | 405 | | | |
| | 1971 | 2333 | | | |
| Auchenorrhyncha | 1968 | 220 | 191.0 | IBP Matador Project, native grassland in Saskatchewan | |
| | 1969 | 1090 | 185.8 | | |
| (peak in July) | 1970 | 560 | | | |
| Grasshoppers | 1968 | 40 | 191.0 | IBP Matador Project, native grassland in Saskatchewan | |
| | 1969 | 3 | 185.8 | | |
| (peak in July) | 1970 | 5 | | | |
| Coleoptera (*Apion loti*) | 1965 | 42.0 | | Chalk grassland, England | Morris (1967) |
| | 1966 | 196.5 | | | |
| *Dicrostonyx groenlandicus* | 1970 | 744 ($ha^{-1}$) | 80 | Truelove lowland, Devon Island, Canada | Bliss (1975) |
| | 1971 | 404 | | | |
| | 1972 | 278 | | | |
| | 1973 | 7200 | | | |
| *Clethrionomys rufocanus* | 1970 | 0 ($ha^{-1}$) | 807 | Tundra, Inari Lapland, Finland | Kallio (1975) |
| | 1971 | | | | |
| | 1972 | ~ 60 | | | |
| *Rangifer tarandus* | 1970 | ~ 4 ($km^{-2}$) | 807 | Tundra, Inari Lapland, Finland | |
| | 1971 | | | | |

does not directly influence the population size of herbivores. In most herbivore species changes in population size can be considerable from season to season, and some herbivores reach high numbers periodically (Table 3.3). These changes in herbivore numbers have no counterpart in plant biomass dynamics (Table 3.4). Lemmings in tundra show cyclic changes in their numbers every 3–4 years. Their density can reach 330 individuals per hectare, which is equivalent to 2.6 g wet wt · $m^{-2}$, the average recorded being 3 to 50 individuals per hectare or 0.024–0.4 g wet wt · $m^{-2}$ (Østbye *et al.*, 1975). The common vole *Microtus arvalis*, inhabiting mainly cultivated fields, appears in large numbers in grassland every 3–4 years, numbers reaching 1212 individuals per hectare or 3.6 g wet wt · $m^{-2}$ (French *et al.*, 1975; Myllymäki, 1975, 1976). These oscillations in numbers and periodical high densities of Microtinae have not been fully explained (Adamczewska-Andrzejewska & Nabaglo, 1977), though it is known that many factors are involved.

It is generally assumed that the density of small mammals results mainly from population mechanisms. According to Krebs, population dynamics and periodicity of vole numbers result from genetic and behavioural changes and are not related to nutrition (Krebs & Myers, 1974). Yet experiments have shown that voles provided with a considerable surplus of available food react to additional food supply by increasing in number, and there are also changes in their social behaviour and spatial distribution (Andrzejewski, 1975; Andrzejewski & Mazurkiewicz, 1976). Andrzejewski (1977) has proved that food supply, its distribution in the habitat and the size of small rodent populations and their organization are all interrelated. Cyclic changes in the numbers of small mammals occur in all climatic zones, but very large fluctuations are more common in poor habitats and in severe climatic environments (Haukioja & Hakala, 1975; Trojan, 1975).

Natural populations of large herbivores do not show big changes in numbers. While grazing, they usually move over vast areas of tundra, steppe or savanna. Their ability to utilize the food reserves of the habitat allows them to live in relatively high densities and to attain a considerably larger average biomass than is possible for domestic animals (see Chapter 4).

It seems that the density levels of vertebrates, as well as their cyclic fluctuations, are mainly the result of social intrapopulation mechanisms that regulate numbers. However, food and climate indirectly influence numbers through their effect on development and interrelations between individuals (Andrzejewski, 1975; Bunnell *et al.*, 1975; Golley *et al.*, 1975).

Except in extreme cases of population outbreaks, the biomass of phytophagous invertebrates represents only a minute fraction of the plant biomass. Figures are usually of the order of hundredths or even thousandths of a percent (Table 3.5). However, all calculated percentages should be assumed to be too low since the assessment of the biomass of phytophagous invertebrates usually includes only a part of their populations; obtaining data on species composition, number and biomass of all the phytophagous animals living in a grassland habitat is extremely difficult.

Although there are interrelations between plant biomass and the biomass of invertebrate phytophages, the constant pool of plant production per biomass unit of the phytophagous invertebrates cannot be calculated (even if food selectivity of the animals and food accessibility in the habitat are

Table 3.5. *Biomass of phytophagous invertebrates feeding on aboveground and belowground plant parts*

| | Herbivore biomass feeding on: Aboveground plant parts (A) | Belowground plant parts (B) | Ratio A/B | Source of data (Reference) |
|---|---|---|---|---|
| *Tropics* | (g fresh wt · m$^{-2}$) | | | |
| Grazing savanna: Beginning of the rainy season | 3.05 | ~5.00 | 0.61 | Grazing savanna in Panama (L. Andrzejewska & A. Myrcha, unpublished data) |
| Grazing savanna: Rainy season | 1.30 | ~22.00 | 0.06 | |
| Grazing savanna: Dry season | 0.29 | 0.40 | 0.72 | |
| *Temperate zone* | (g dry wt · m$^{-2}$) | | | |
| Meadow type | 2.5 | 3.3 | 0.75 | USSR (Zlotin, 1975) |
| Steppe meadow | 0.172 | 0.473 | 0.36 | |
| Wet meadow | 0.027 | 0.076 | 0.35 | |
| Dry steppe | 0.274 | 0.351 | 0.78 | |
| Desert | 0.019 | 0.057 | 0.33 | |
| Bogs | 0.088 | 0.033 | 2.67 | |
| Cultivated meadows | | | | |
| Fertilized | 0.096 | 5.470 | 0.017 | Poland (Andrzejewska, 1976) |
| Unfertilized | 0.084 | 6.150 | 0.013 | |
| Cold zone | | | | |
| Tundra, wet | 0.056 | 1.200 | 0.047 | Norway (Østbye *et al.*, 1975) |
| Meadow | 0.021 | 4.600 | 0.005 | |

taken into consideration). Thus it is impossible to equate a certain plant production of a grassland with a certain density of phytophages. Biomass and production of the aboveground plant parts and, to an even greater degree, those of the roots, do not usually display dramatic variations over consecutive seasons. Variability may be in the range of 15 to 50 %, but during exceptionally unfavourable weather conditions may reach 100 %. In contrast comparable figures for numbers and biomass of phytophagous insects and small mammals may vary up to twelvefold (Table 3.4). In the majority of insects there are considerable oscillations in numbers over consecutive seasons which are uncorrelated with variations in the food supply. The occasional mass appearance of a phytophagous invertebrate species does not result only from increases in available food, but can be the outcome of particularly favourable food and climatic conditions combined (Uvarov, 1931; Rafes, 1968). Such mass appearances can cause considerable damage to a grassland vegetation, especially in the tropics where that may be swarms of locusts, army worms or harvester ants (Bullen, 1970; Roffey, 1970; Weber, 1972; Ingram, 1975). But they can also be a problem in the temperate zone, where, for example, Orthoptera have become serious pests on the American prairies and in Canada (Roffey, 1970; Guseva, 1970), as have Tipulidae in tundra regions (Østbye *et al.*, 1975). As is the case with small mammals, some insects reveal periodic fluctuations in numbers, locusts, for example, appearing in large numbers every 3–4 years (Uvarov, 1970; Waloff & Conners, 1964).

Insect numbers are considerably modified by the vegetation structure of a grassland; for example Morris (1973) found a significant correlation between invertebrate numbers and plant height. Likewise, vegetation quality (i.e. its chemical composition) can influence population numbers of phytophagous insects by producing changes in their fecundity and the fitness of individuals (Harley & Thornsteinson, 1967; Rafes, 1968; Haukioja, 1974).

The densities of predators and parasites also play a considerable role in the numerical reduction of phytophages. Usually only a small proportion of phytophagous insects reaches sexual maturity. For example, at least 50% of leafhopper eggs are destroyed by parasites (Witsack, 1973; Waloff, 1975), and during the period of larval development about 96–98% of the emerging larval hoppers perish. Reduction in number of imagos depends on predator density; as the number of predators in the habitat increases, the lifespan of the adults decreases (Andrzejewska, 1971, 1976*b*).

Thus it is the quality and biomass of plant matter produced in grassland, together with climatic and weather factors, and, in higher animals, social factors in the population, that determine the changes in density and biomass of phytophages feeding on aboveground plant parts.

Phytophages feeding in the root layer are influenced relatively little by changes in the weather. Their food is also those plant parts that vary little in quality and quantity. In many grassland types and in all climatic zones the greatest biomass of phytophagous invertebrates is found in this layer, and although they are restricted to a very thin stratum of upper soil 3–5 cm in thickness, they frequently exceed by many times the biomass of the invertebrates feeding on aboveground plant parts (Table 3.6). This is not the case, however, in bogs.

When examining grassland habitats with similar plant biomass but in different climatic zones, one notices that the density and the biomass of invertebrates are largest in tropical grasslands, both above ground and in the root layers. As the climate becomes more and more harsh (progressing from tundra through mountain meadows to deserts), phytophagous invertebrates become less and less numerous. In harsh environments their biomass is usually less than a tenth of a gram per square metre (Table 3.6).

If the biomass of phytophagous vertebrates is compared with that of invertebrates, it becomes obvious that the biomass of the latter per unit area is usually considerably larger, especially in tropical and temperate zones. In tundra climates there is a marked preponderance of warm-blooded herbivores, especially in years with high numbers of lemmings (Table 3.6). In some years, though, the biomass of tipulid larvae can exceed that of reindeer.

It seems that the supply of available food in a habitat determines its potential capacity and the upper limit for the density of its phytophagous animals. However, numerous situations resulting from biocoenotic relations, habitat structure, climatic conditions and social patterns can produce

Table 3.6. *Herbivorous fauna in various climatic zones*

| | Number | Biomass | Number of species | Type of environment | Reference |
|---|---|---|---|---|---|
| *Tropics* | | | | | |
| Grass layer invertebrates | | (mg · m$^{-2}$) | | | |
| Caterpillars | 0.9 | 84.1 | — | | |
| Acrididae | 1.8 | 165.7 | 32 | | |
| Grillidae | 1.7 | 66.3 | — | | |
| Grasshoppers | 0.6 | 60.7 | — | | |
| Pentatomidae | 1.3 | 52.2 | 19 | | |
| Coreidae | 0.4 | 6.1 | — | Savanna, Lamto, Ivory Coast, Africa | D. Gillon (1971) Y. Gillon (1971) Lamotte (1975) |
| Homoptera | 2.1 | 9.6 | — | | |
| Soil invertebrates | | | | | |
| Termites | 1.3 | $\sim$ 6000.0 | 5 | | |
| Coccidae | — | — | — | | |
| Diptera (larvae) | — | — | — | | |
| Vertebrates | | | | | |
| Rodents | — | 120–160 | 12 | | |
| Antelope | — | — | — | | |
| Monkeys | — | — | — | | |
| *Temperate zone* | | | | | |
| Grass layer invertebrates | | | | | |
| Lepidoptera | — | — | — | | |
| Acrididae | 3.3 | 95.2 | 5 | | |
| Coleoptera | — | — | — | | |
| Auchenorrhyncha | 16.0 | 21.1 | 41 | | |
| Heteroptera | — | — | — | | |
| Hymenoptera | — | — | — | Native grassland *Deschampsietum* type, Central Poland | Makulec (1971) Nowak (1971) Andrzejewska (1971) |
| Diptera | 78.2 | — | 11 families | | |
| Aphididae | — | — | — | | |
| Thysanoptera | — | — | — | | |
| Acarina | — | — | — | | |
| Soil invertebrates | | | | | |
| Coleoptera (larvae) | 4.9 | — | 2 families | | |
| Diptera (larvae) | $\sim$ 1 | — | 2 families | | |
| Lepidoptera (larvae) | 3.1 | — | | | |
| Vertebrates | | | | | |
| Rodents | 0.006 | | | | |
| Birds | — | | | | |
| Grass layer invertebrates | | | | | |
| Acarina | | | | | |
| Coleoptera | 1.6 | | 6 families | | |
| Diptera | 20.0 | | 2 families | | |
| Hemiptera | 8.5 | | 5 families | | |
| Homoptera | 14.2 | 92 (mg · m$^{-2}$) | 5 families | | |
| Coccoidea | 294 | | | Mixed prairie, cottonwood, Colorado | McDaniel (1971 |
| Hymenoptera | — | | 2 families | | |
| Lepidoptera | — | | 4 families | | |
| Orthoptera | 9 | | 2 families | | |
| Gryllidae | 5 | | | | |
| Thysanoptera | 16.9 | | 2 families | | |
| Vertebrates | | | | | |
| Small mammals | | 5 (mg · m$^{-2}$) | | | |
| Large mammals (deer, antelope) | | — | | | |
| Birds | | 4.2 (mg · m$^{-2}$) | | | |
| Grass layer invertebrates | (ha$^{-1}$) | (mg wet wt · m$^{-2}$) | | | |
| Acrididae | 3.6 | 85.0 | | | |
| Cicadodea | 120.0 | 18.0 | 5 | | |
| Tipulidae | 0.8 | 1.5 | | | |
| Chrysomelidae | 0.8 | 1.5 | | | |
| Lepidoptera | 3.35 | 1.4 | | | |
| Mollusca | — | — | 2 | Wet meadow in the mountains, Tien Shan | Zlotin (1975) |
| Soil invertebrates | (m$^{-2}$) | (mg · m$^{-2}$) | | | |
| Tipulidae (larvae) | 5 | 180 | | | |
| Curculionidae (larvae) | 13 | 140 | | | |
| Scarabeidae (larvae) | 6 | 100 | | | |
| Noctuidae (larvae) | 2 | 80 | | | |
| Vertebrates | | | | | |
| Marmot | 1.5 (ha$^{-1}$) | | | | |
| Vole | — | | | | |

Table 3.6 (*continued*)

| | Number | Biomass | Number of species | Type of environment | Reference |
|---|---|---|---|---|---|
| *Tundra ecosystems* | | (mg dry wt · m⁻²) | | | |
| Aphidae | — | } ~ 26 | | | |
| Psyllidae | — | | | | |
| Cercopidae | — | | | | |
| Lepidoptera | — | | | | |
| Coleoptera | | | | | |
| Crustacea | — | } ~ 20 | | | |
| Acarina | — | | | | |
| Chironomidae | — | | | | |
| Nematoda | | ~ 13 | | | |
| Soil invertebrates | | | | Peatland site, Glenamoy, Ireland | Moore, Dowding & Healy (1975) |
| Tipulidae | | | | | |
| Vertebrates | ($km^{-2}$) | | | | |
| Red grouse (*Lagopus lagopus* L.) | 6 | | | | |
| Mountain hare (*Lepus timidus*) | 2 | ~ 33 | | | |
| Common field mouse (*Apodemus silvaticus* L.) | 20 | | | | |
| Cattle } Sheep } | ~ 10 | | | | |
| Grass layer invertebrates | ($m^{-2}$) | | | | |
| Gastropoda | 0–2 | | | | |
| Soil invertebrates | | | | | |
| Diptera (larvae) | 3–23 | | | | |
| Coleoptera (larvae) | 3–16 | | | | |
| Vertebrates | | | | | |
| Wild reindeer (*Rangifer tarandus* L.) | 9–135 (1000 ha⁻²) | 0.081–1.21 (g · m⁻²) | | | |
| Willow grouse (*Lagopus lagopus* L.) | 80–165 | 5–10 (mg · m⁻²) | | Agapa USSR, Tundra in western Taimyr | Vassilievskaja *et al.* (1975) |
| Ptarmigan (*Lagopus mutus* Montin) | 15–35 (1000 ha⁻¹) | 0.8–1.9 (mg · m⁻²) | | | |
| Anseriformes | 80–100 (1000 ha⁻¹) | | | | |
| Lemmings (*Lemmus obensis* Brants) | | up to 0.38 (g · m⁻²) | | | |
| *Clangula hyemalis* | — | | | | |
| Grass layer invertebrates | | (mg dry wt · m⁻²) | | | |
| Hemiptera | 19 | 1.0 | | | |
| Lepidoptera | 1 | 0.2 | | | |
| Coleoptera | — | | | | |
| Thysanoptera | — | | | | |
| Soil invertebrates | | | | | |
| Tipulidae (*Tipula excisa* Schum) | 20–50 | 1.2–4.60 (g wet wt · m⁻²) | | | |
| Vertebrates | | | | | |
| *Rangifer tarandus* | 4–6.4 (km⁻²) | 0.4–0.64 (g · m⁻²) | | | |
| *Lemmus lemmus* | 0–3 (ha⁻¹; in years before peak) | | | Wet meadow, Hardangervidda, Norway | Østbye *et al.* (1975) |
| *Microtus oeconomus* | | | | | |
| *Microtus agrestis* | | | | | |
| *Clethrionomys glareolus* | 25–330 (ha⁻¹; in a peak year) | | | | |
| *Clethrionomys rufucanus* | | | | | |
| *Lepus timidus* | — | | | | |
| *Lagopus lagopus* | | | | | |
| *Lagopus mutus* | | | | | |

— indicates 'no data available'.
[a] The numbers given represent the highest density recorded.
[b] Usually very few species within each group. More abundant on uncut grass and on plots sheltered from wind.

variations in the numbers of phytophages, which as a result can deviate considerably from the 'nutritional capacity' of the given habitat (Uvarov, 1970; Moore *et al.*, 1975; Trojan, 1975; Andrzejewski, 1977).

### *Diversity of the phytophagous fauna*

The faunistic diversity in a grassland habitat is influenced mainly by variety in the vegetational cover, the number of plant species and their contribution to total biomass and vegetation structure, and only to a small degree by plant biomass and production (Morris, 1971; Hurd *et al.*, 1971, 1972; Duffey *et al.*, 1974; Andrzejewska, 1976*a*). In habitats where there are only a few plant species, there are relatively few species of phytophages, though they may be represented by large numbers of individuals (Trojan, 1975). However, it is not always true that differentiation of the habitat, in terms of number of species present, determines the number of phytophagous insect species. For instance, it has been shown that concomitant with increasing insect density can be an increase in the number of species, and thus an increase in association diversity. This has been shown to be the case for hoppers (Auchenorrhyncha). Numbers of individuals and the number of hopper species on various grassland areas of one village were compared, in several cases over a 3-year period (Table 3.7). High correlations between numbers of individuals and the number of species were found, not only in the same meadows in different years but also over all meadows as a whole (Fig. 3.1). An increase in number of species with increasing total density (i.e. density of all species considered together) has been recorded for spiders (Palmgren, 1972), but the curve of species diversity against density only rises steeply if densities are high.

In poor habitats such as semideserts and tundras, as well as in cultivated meadows with little variety of plant species, the number of species of phytophagous animals is also low. In contrast, species-rich faunas of phytophages are recorded from natural habitats in tropical and temperate grasslands (Table 3.6). In the aboveground parts of tundra plants only single species of Auchenorrhyncha, Aphididae, Psyllidae, Lepidoptera, Coleoptera and Gastropoda are recorded and the phytophagous fauna of the root layer is similarly species-poor. In some types of tundra habitats high densities and biomasses of Tipulidae larvae (2–50 individuals $\cdot$ m$^{-2}$; 0.3–4.6 g wet wt $\cdot$ m$^{-2}$) and phytophagous Acarina (100–2000 individuals $\cdot$ m$^{-2}$; 0.06–0.20 g wet wt $\cdot$ m$^{-2}$) have been recorded. In temperate and tropical zones the majority of the phytophagous taxa are represented by numerous species, with Orthoptera, Hymenoptera, Auchenorrhyncha, Lepidoptera and Diptera being particularly abundant.

In climatic conditions with a long vegetation season and thus several generations of phytophagous invertebrates per year, the dominant insect groups change during the season. For instance, in Panamanian grazed

Table 3.7. *Number of Auchenorrhyncha species and log number of individuals caught in the meadows at Kuwasy, north-east Poland*

| | Type of grassland | | Σ individuals ($N$) | ln $N$ | No. of species |
|---|---|---|---|---|---|
| I.1 | Cultivated meadow (A: 10 years old) | 1953 | 1488 | 7.31 | 33 |
| 2 | | 1954 | 9037 | 9.11 | 38 |
| 3 | | 1955 | 992 | 6.90 | 25 |
| 4 | | 1955 | 3628 | 8.30 | 39 |
| II.5 | Cultivated meadow (B: 4 years old) | 1954 | 3150 | 8.06 | 37 |
| 6 | | 1955 | 678 | 6.52 | 28 |
| 7 | | 1955 | 1798 | 7.49 | 37 |
| III.8 | Natural meadow (bog) | 1953 | 1243 | 7.13 | 31 |
| 9 | | 1954 | 4421 | 8.39 | 45 |
| 10 | | 1955 | 1409 | 7.25 | 31 |
| 11 | | 1955 | 1813 | 7.50 | 38 |
| IV.12 | Wet meadow | 1953 | 295 | 5.69 | 23 |
| 13 | | 1954 | 1151 | 7.05 | 29 |
| 14 | | 1955 | 796 | 6.68 | 28 |
| V.15 | | 1955 | 1099 | 7.00 | 22 |
| VI.16 | Meadow with birch | 1954 | 573 | 6.35 | 34 |
| VII.17 | Dry meadow with *Salix repens* | 1954 | 780 | 6.66 | 37 |

Data from Andrzejewska (1966).

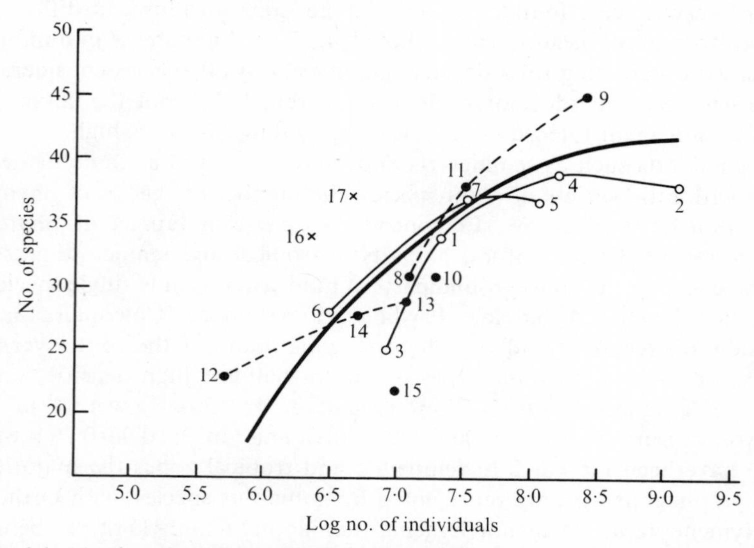


Fig. 3.1. Number of Auchenorrhyncha species as a function of their density in the meadows at Kuwasy, Poland. (The numbers refer to the left-hand column of Table 3.7.) ● – – – ●, natural meadows; ○——○, cultivated meadows; ×, meadows with shrubs. The bold line shows the relationship between number of species and log number of specimens when all data (all points on the figure) are taken into account.

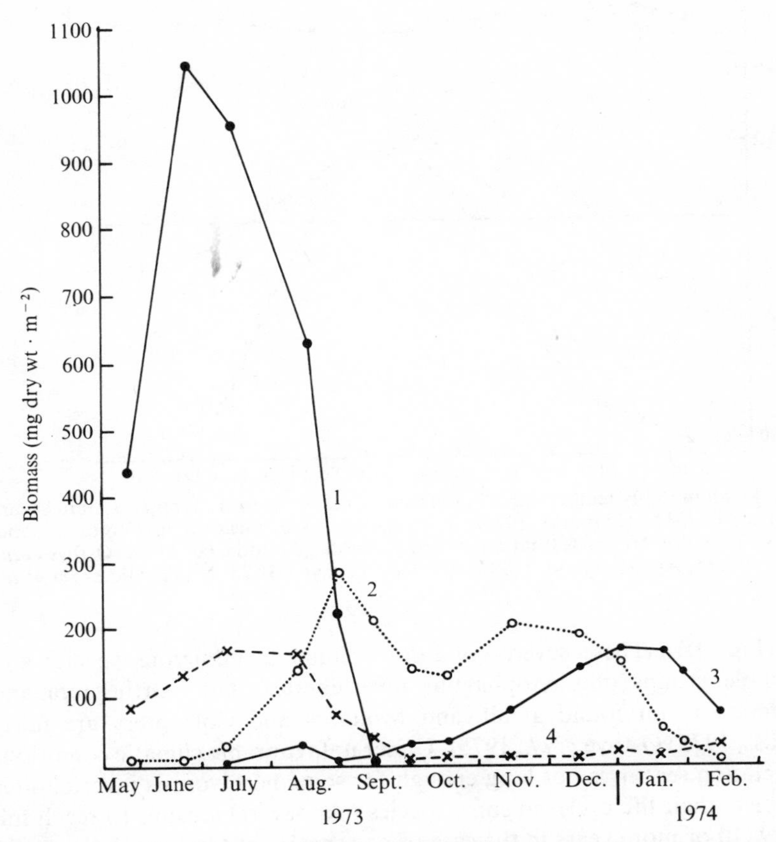


Fig. 3.2. Changes in biomass of different groups of phytophagous insects in a burned plot of pasture in Panama. 1, Lepidoptera larvae; 2, Auchenorrhyncha; 3, crickets; 4, Acrididae. (After unpublished data of Andrzejewska & Myrcha.)

savannas the young grass at the beginning of the rainy season is inhabited by numerous lepidopteran larvae (Fig. 3.2), while at the end of June and during July it is dominated by Orthopterans. Auchenorrhyncha abound during the following months of the rainy season, and the beginning of the dry season is preferred by crickets. At the same time the remains of green grass blades are intensively removed by gardening ants. These changes result in a species-rich fauna, and are possible because during the long vegetation season there is an abundance of food, and variations in the chemical composition and species composition of the meadow sward that permit numerous species to find suitable conditions for development.

In tundra and in harsh mountain climates the vegetation season (defined as the period when temperatures rise above 0 °C) lasts from a few to a dozen

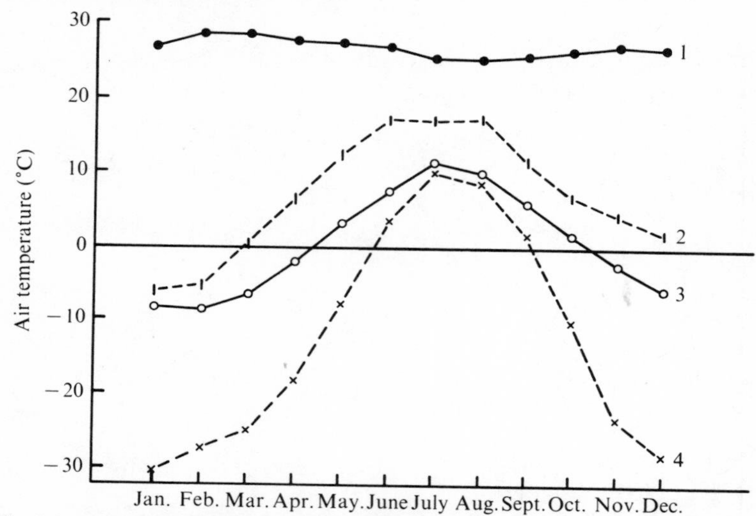


**Fig. 3.3.** Mean monthly temperatures in different climatic zones. 1, Tropics, Lamto savannah, Africa, 6° 13′ N (Lamotte, 1975); 2, temperate zone, meadow in Central Poland, 52° 12′ N; 3, tundra, Haugastøl station, Norway, Hardangervidde, 60° 31′ N (Østbye *et al.*, 1975); 4, tundra, Agapa station, USSR, Western Taimyr, 71° 25′ N (Vassiljevskaja *et al.*, 1975).

weeks (Fig. 3.3). These severe conditions eliminate numerous species and even whole groups of phytophagous invertebrates. Thus Orthoptera and Hemiptera are not found at all, and Mollusca and Coleoptera are fairly rare (Bliss, 1975; Østbye *et al.*, 1975). Under unfavourable climatic conditions the vegetation season is not long enough for some herbivorous invertebrates to complete their life cycle, so some species take several seasons to reach full maturity (10 or more years in the case of one species of Gymnaephora moth) (Bliss, 1975; Østbye *et al.*, 1975). A short vegetation season also means that all species are developing and reproducing at the same time, and thus all feeding intensively on the relatively sparse vegetation. There is a comparable situation in semideserts and in deserts, but here the limiting factor is water and not low temperature.

In summary it can be said that the diversity of the phytophagous fauna, particularly the invertebrates, is closely related to the diversity of the plant associations of the grasslands, the number of plant species, and the structure of the plant cover, but that it can be modified by other factors, such as climatic conditions, that limit the length of the growing season.

## 2. Phytophagous fauna under conditions of cultivation and the utilization of grassland habitats

Grassland habitats that are used as pastures for domestic animals or cultivated for hay production, are subjected to various management practices aimed at improving the quality and quantity of grass. The most common of these are fertilization, irrigation, and burning of the dead plant debris, all of which disrupt to some extent the natural plant cycles.

Fertilization produces an increase in aboveground plant production; at the same time root systems are reduced, there is a simplification of the species composition, and changes in the chemical composition of the plants (Mochnacka-Ławacz, 1978; Plewczyńska-Kuraś, 1976; Traczyk, 1976). Irrigating meadows and pastures not only increases plant biomass but also considerably prolongs the vegetation season. Under extreme conditions (e.g. when tropical grassland is irrigated during the dry season) the dormant period is removed allowing intensive plant growth, and its exploitation lasts all the year round. This in turn provides the phytophagous animals with an abundant food supply which, coupled with the optimal environmental conditions, gives them a practically unlimited developmental potential. Burning is used in some areas of the world to remove dry and lignified plants or plant remains and to facilitate the growth of new meadow and pasture vegetation. This practice, often regarded as a kind of fertilization, causes dramatic changes in microclimatic conditions and habitat structure (D. Gillon, 1971; Y. Gillon, 1971; Wein & Bliss, 1973; Vogl, 1973; Lamotte, 1975).

Exploitation of a meadow takes the form either of the removal of aboveground plant production by mowing, or grazing by domestic animals. Meadows mowed several times a season show cyclical changes in biomass (Fig. 3.4) and high plant growth rates (Fig. 3.5). Mowing takes place at times of maximal biomass and leaves only about 20% of the aboveground plant parts. At the same time, uncovering the litter or even the soil surface, dramatically changes the microclimatic conditions of the habitat. In pastures, biomass is dependent on the intensity of grazing. The grass is usually moderately high, and biomass changes of the aboveground plant parts during the growing season are relatively small (Fig. 3.4). However, when humidity and thermal conditions are favourable, there is a relatively constant and high growth rate (Fig. 3.5).

The intensity and frequency of cultivation practices applied to meadows also affect the production, structure and quality of the plants (Voisin, 1970; Hurd *et al.*, 1971; Sims & Singh, 1971; Traczyk, 1971, 1976; Ellis & French, 1973). These changes in the habitat and the food plants of the phytophagous fauna that result from cultivation and grassland utilization, must influence both their numbers and population structure.

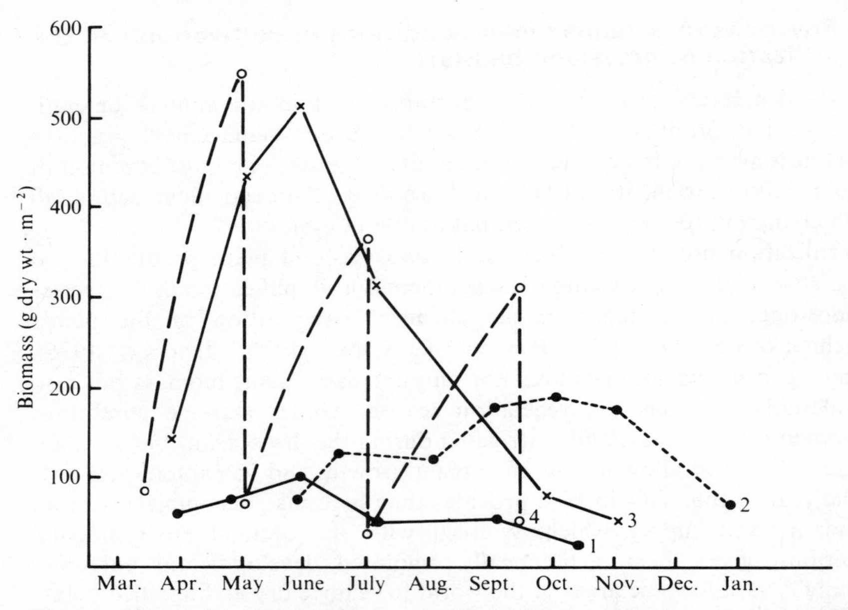


Fig. 3.4. Changes in green plant biomass in differently treated meadows. 1, sheep pasture in mountains, Poland (Andrzejewska, 1974); 2, cow pasture in Panama (L. Andrzejewska & Myrcha, unpublished data); 3, cultivated meadow, unmowed, Poland (Traczyk, 1971); 4, cultivated meadow mowed three times a year, Poland (Traczyk, 1976).

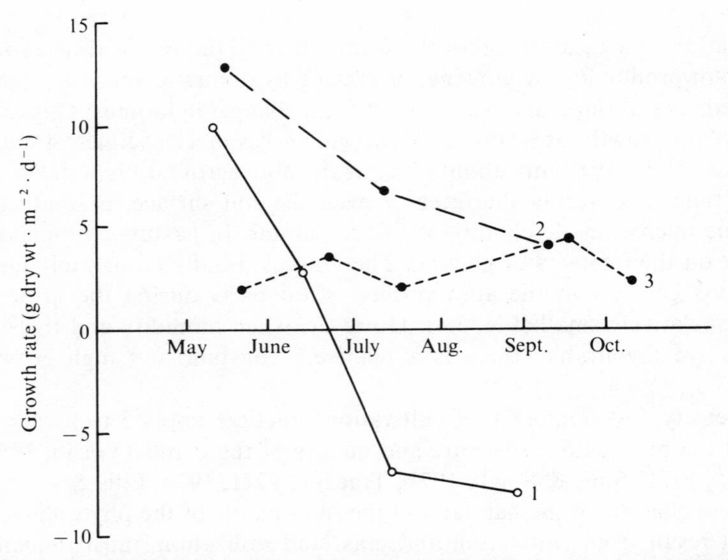


Fig. 3.5. Rates of increase of plant biomass in differently used and managed meadows in Poland. 1, managed and unmowed meadow; 2, managed and mowed meadow; 3, sheep pasture. (Data from Traczyk, 1971; Andrzejewska, 1974.)

### *Influence of grassland utilization on numbers of the phytophagous fauna*

The vegetation of grassland habitats provides not only the food but also the microenvironment for numerous species of phytophagous organisms. Different species occupy different strata of the vegetation according to their habitat requirements. The greater the differentiation of the vegetation, in terms of species composition, biomass and height, the richer and more diversified is its phytophagous fauna (Petrusewicz & Macfadyen, 1970; Hurd *et al.*, 1971, 1972; Morris, 1971).

Developmental cycles of particular phytophagous populations are not synchronized with the dynamics of the plant community as a whole, but rather with the development of a plant 'optimal' for a given species. Thus a high density of phytophages may be due to the exchange of populations or consecutive generations in each period of the season, as well as to the diversity of ecological niches that the grassland habitat provides (Ricou, 1962; (Andrzejewska, 1965; Morris, 1973). When mowing or grazing destroys the spatial structure of the vegetation, this inevitably has an effect on the phytophagous fauna. Mowing causes a drastic diminution in food supply, dramatic changes in humidity and temperature, and a reduction in living space, and always results in considerable losses of phytophages, both from mortality and migration. Larval stages are especially susceptible to changes in the habitat (Razwjazkina, 1960; Raatikainen, 1967, 1972; Andrzejewska, 1971). At the same time insects are more vulnerable to predation when crowded into the short layer of mowed grass and deprived of their plant cover.

Although the larval stages of most phytophagous insects may be found throughout the vegetation season, they occur in the greatest densities in early summer. Removal of vegetation cover at this time reduces the hoppers (Auchenorrhyncha) by approximately 80%, bugs (Heteroptera, Miridae) by about 87%, beetles (Curculionidae) by about 79–99% and Orthoptera (Acrididae) by 42% (Morris, 1967). Crowding the insects in a reduced habitat capacity may also produce or facilitate the spread of diseases. This was demonstrated in an experiment carried out in a small forest clearing (*Magnocaricetum* association) where the dominant phytophagous insect, *Cicadella viridis* L. (Cicadellidae), had a density of 148 individuals$\cdot$m$^{-2}$. Half of the clearing was then mowed, increasing the density of *C. viridis* in relation to plant biomass and space by approximately tenfold. At the same time, thermal and humidity conditions also changed. Some days later, most of the insects in the mowed area (102 individuals$\cdot$m$^{-2}$) perished from a fungal infection, whereas in the population which inhabited the unmowed area, only two individuals (0.2 m$^{-2}$) were found to be affected.

Mowing of meadow vegetation which removes about 80% of the above-ground plant parts is a 'disaster' for the phytophagous fauna and most of the

Table 3.8 *Comparison of population parameters of prey (Auchenorrhyncha) and predatory arthropod communities in two differently treated meadows*

| | Number of hoppers ($m^{-2}$) | | Mean time of ecological life | No. of predatory arthropods ($m^{-2}$) |
|---|---|---|---|---|
| | Hatched during season | Found during season | | |
| Untreated meadow | 1931 | 16 | 5.4 | 239.3 |
| Fertilized and mowed meadow | 1155 | 46 | 19.9 | 60.8 |

From Andrzejewska (1971).

insects disappear. With the growth of the new grass they soon reappear, however, frequently including migrants, sometimes from quite large distances away (Raatikainen, 1967; Emden *et al.*, 1969). The migration of phytophagous insects to habitats with favourable nutritional and environmental conditions may be illustrated by the hopper (Auchenorrhyncha) association (Table 3.8). An unmowed meadow appears to provide better conditions for development than a mowed one, almost twice as many hoppers hatching in the former than in the latter. However, the unmowed vegetation soon matures and starts to dry and many of the mature insects then migrate to cultivated and mowed meadows with lush and fresh vegetation, until insect density may reach three times the level in the unmowed habitat. The lifespan is approximately four times longer in the mowed meadow, a result of the abundance of fresh food and the considerably reduced level of predation (Kajak *et al.*, 1971; Breymeyer & Pętal, 1971).

Grazing by domestic animals (considered here as a factor that changes the environment) keeps the aboveground plant part biomass at a relatively low level through the whole season (Fig. 3.4), which, together with the mechanical damage from trampling, always diminishes the numbers and biomass of phytophagous insects (Morris, 1967; Duffey *et al.*, 1974). In sheep pastures where grass is grazed very intensively and throughout the whole season, biomass is very low and consequently phytophages in the aboveground plant parts are scarce, sheep being virtually the sole users of the available green plants (Andrzejewska, 1974). Phytophages feeding in the root layer are considerably more numerous. Less intensively grazed pastures are inhabited by greater numbers of phytophagous invertebrates, and R. Andrews (personal communication) has found that in experimental plots of shortgrass prairie in Colorado that were subjected to different grazing intensities by cattle, there was a close negative correlation between grazing intensity and biomass of phytophagous insects in the aboveground layer (Fig. 3.6).

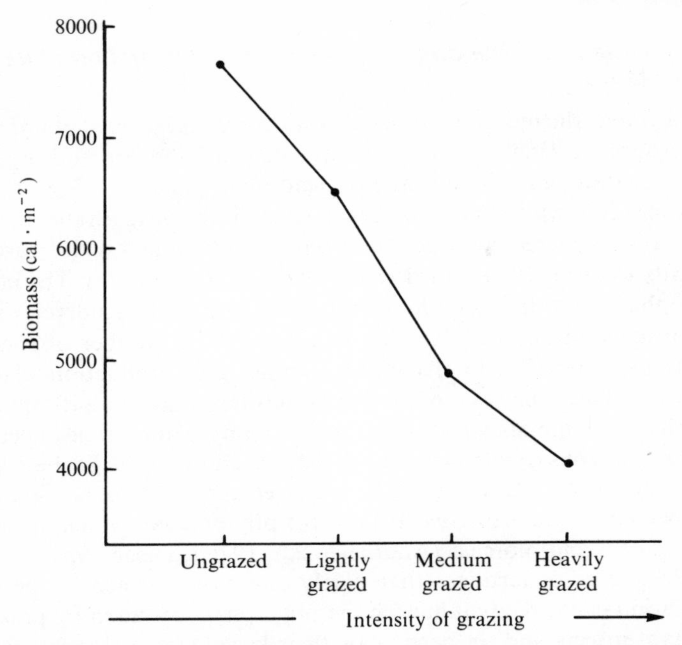


Fig. 3.6. The relation between the intensity of grazing of cow pasture and the biomass of phytophagous invertebrates in the aboveground plant layer. (Results of experiments carried out in Fort Collins, Colorado, 1972. Data from R. Andrews, personal communication.)

In summary:

(1) Grassland utilization, i.e. mowing or grazing by domestic animals, leads to a decrease in numbers and a change in the population structure of its phytophagous animals.

(2) In mowed meadows, a decrease in insect numbers occurs immediately after mowing. Soon, however, as the new vegetation grows, the fauna is renewed. Plants are fresh for a longer time and their growth rate is higher than in unmowed meadows, both of which factors favour insect life. At the same time mowing simplifies the species composition of grassland, which leads to a decrease in species diversity of phytophagous species but an increase in the number of individuals of each species.

(3) The plants in grazed meadows, although present throughout the season, are short and damaged by trampling. Under these conditions the invertebrate phytophagous fauna is species-poor and scarce. Phytophages feeding on roots, however, may be numerous. Species composition and density are dependent on grazing intensity.

### Influence of grassland cultivation on the numbers and migration of its phytophagous fauna

As mentioned above, the most common cultivation practices aimed at improving habitat resources and crop quality and quantity, are mineral and organic fertilization, irrigation, and the burning of dead plant parts.

Dressing a meadow with mineral or organic fertilizers enriches the habitat and always leads to increased production with a simultaneous decrease in species diversity (Voisin, 1970; Hurd *et al.*, 1972; Traczyk, 1976). The effect on the phytophagous animal is similar (Hurd *et al.*, 1971; Andrzejewska, 1976*a*). An increase in habitat diversity, in terms of the number of species present, leads to a decrease in the mean biomass and production of any particular species. Thus, species represented by only few individuals disappear, while phytophagous animals which can feed on many plant species become dominant and numerous (Andrzejewska, 1976*a*; Olechowicz, 1976). Fertilization, besides increasing plant biomass, enhances the nutritional value of plants: they become more attractive as food for phytophages which, in turn, increase in numbers and biomass (Andrzejewska, 1976*a*; Olechowicz, 1976).

Concentrated doses of fertilizer that partly cover the surface of the soil and the vegetation can directly influence the phytophagous fauna by poisoning less resistant insects and, in particular, their larval stages (Duffey *et al.*, 1974; Żyromska-Rudzka, 1976).

There is usually also a loosening of the turf in fertilized and exploited meadows (Plewczyńska-Kuraś, 1974; Traczyk, 1976), and the soil is covered with a thin layer of litter. Not all phytophagous animals find these conditions favourable for development, but the mass of green, fresh plants is so attractive that animals from ageing and drying meadows even some distance away are lured by it. A 'patch' of fresh green vegetation is particularly attractive during periods of low plant growth (e.g. dry season, low temperature), when most meadows and pastures are covered with a grey mass of dry vegetation from the previous year. In some parts of the world the growth of fresh plants is accelerated by burning the dry dead plants, and the quickly regrowing, fresh grass attracts large numbers of both invertebrate and vertebrate phytophages. As a result, differences in terms of both numbers and in biomass between the phytophagous animals inhabiting the burned part and those in the area covered with dead plant matter are considerable (L. Andrzejewska & A. Myrcha, unpublished data).

Fire itself, according to D. Gillon (1971), Y. Gillon (1971) and Lamotte (1975), does not cause any significant mortality of grassland fauna, for although temperatures above ground are high, they are low enough in the soil and in shelters among the debris for animals to survive short periods of fire. And most animals able to fly or run leave the burning plot. For example, the biomass of phytophagous animals before and after a savanna fire in

Table 3.9 *The biomass of a phytophagous savanna fauna 5 months after fire (Panama 1973)*

| | Burned | Unburned |
|---|---|---|
| Plant layer fauna (mg dry wt $\cdot$ m$^{-2}$) | 432.7 | 229.3 |
| Root layer fauna (g wet wt $\cdot$ m$^{-2}$) | 22352 | 8272 |

Unpublished data of L. Andrzejewska & A. Myrcha.

Lamto, Ivory Coast, was 386.7 mg $\cdot$ m$^{-2}$ and 367.8 mg $\cdot$ m$^{-2}$ respectively, and the meadow was inhabited again relatively soon after the fire. The species composition of the fauna changes somewhat but soon returns to approximately that of the unburned parts (D. Gillon, 1971; Y. Gillon, 1971). It seems, however, that the indirect influence of fire is more far-reaching in its effect on both the grassland and its fauna. The accelerated growth of new shoots after removal of the dead plant parts affects the structure, chemical composition and humidity of the habitat and attracts phytophagous animals from other areas. Even after a few months, when the green plant mass is the same on both burned and unburned areas, the density and mass of phytophages in the herb layer and the soil remain greater in the burned areas (Table 3.9).

Irrigation of grasslands also improves habitat conditions and prolongs the vegetation season of the plants. In the tropics, where the vegetation cycle is limited by the rainy season and not by temperature, irrigation in the dry season extends the period of lush growth to the whole year. Species of phytophagous insects which normally have periods of diapause determined by environmental factors such as humidity or a suitable food supply, are thus able to complete their life cycle without interruption. At the same time the areas of green irrigated grassland attract phytophagous animals from surrounding dry areas, resulting in exceptionally high densities in irrigated compared with unirrigated meadows. The effects of irrigation and burning on the density of Auchenorrhyncha in an area of Panamanian savanna are shown in Table 3.10. The density of insects was estimated on irrigated and burned plots of savanna and an untreated pasture acted as a control. The normal development cycle of plants on the Panamanian savanna lasts from the end of April (end of the dry season) to mid-December (end of the rainy season). The number of sucking phytophagous insects (Auchenorrhyncha) on the plants rises from a low level at the beginning of the season to a maximum in August and then falls again as the plants dry up. On the irrigated savanna during the same rainy season the leafhopper population had a similar species composition. During the dry season the lush vegetation of the irrigated savanna formed an oasis among the yellow grasses and leafhopper density

Table 3.10. *Density (no. of individuals·$m^{-2}$) of Auchenorrhyncha in differently treated pastures in Panama*

| | Irrigated pasture | Burned pasture | Untreated pasture |
|---|---|---|---|
| Beginning of rainy season | 208.0 | 36.9 | 13.7 |
| End of rainy season | 398.2 | 196.6 | 136.5 |
| Dry season | 430.0 | 51.4 | 37.1 |

increased to 430 individuals·$m^{-2}$ (Table 3.10) as a result both of migration from neighbouring habitats and prolonged reproductive activity of a few species. It is interesting to note that in some species diapause is not shortened regardless of environmental conditions, while others adapt easily to the changed conditions and take advantage of them to produce an increased number of generations in a season, thus enlarging their population. Such species can be classed among the invading species that are quick to exploit a new habitat (Razwjaskina, 1960).

It seems, however, that potential rates of reproduction for phytophages are considerably higher under favourable climatic and environmental conditions, i.e. appropriate humidity and temperature and a long vegetation season. In habitats where the vegetation season is short, insect populations often do not have time to complete their developmental cycle and take several seasons to reach full maturity (Østbye *et al.*, 1975; Bliss, 1975). In the tropics, on the other hand, populations of phytophagous invertebrates frequently have the chance to produce several generations a year, the only difference between seasons being the time taken to reach maturity. In general, because the reproductive potential of a species is lower in cold climates than in tropical or temperate zones, any disturbance to the fauna as the result of cultivation or utilization of the grassland would have more severe and permanent results in the colder climates.

In summary:

(1) Phytophagous animals are highly dependent on their food plants, so show changes in number, biomass, differentiation of their associations and developmental cycles with changes in these same parameters in the plant cover.

(2) Cultivation procedures indirectly disturb natural developmental cycles and can frequently change dramatically the number and structure of the phytophage populations.

(3) The destruction of the structure of a grassland habitat by mowing, burning or grazing forces phytophagous insects to migrate. However, the simultaneous preservation of physiologically fresh vegetation attracts insects, even from distant habitats.

(4) Cultivation and utilization of meadows and pastures introduces and preserves changes in the natural functioning of ecosystems. This results in the elimination of narrowly specialized species. Under these conditions, eurytopic species, often occurring as pests in plantations, are selected for and can become abundant.

## 3. The effects of interrelations between herbivores and plants

It is obvious that phytophagous animals could not exist without plants, which are their only source of food. But it is dubious whether a plant can obtain any advantage from the association, especially since it tries to rid itself of phytophages by means of special protective mechanisms in the form of chemicals (repellents) which can handicap the development of some herbivores. On the other hand, one cannot neglect the role of animals in pollination and the dispersal of seeds that ensures the spread of many plant species. It is not clear whether a plant grazed or sucked by a herbivore is only a loser or whether it may be given an opportunity for better growth. Nevertheless, the visible effects of a herbivore on a plant are that it loses biomass, becomes yellow and dries up. In addition, phytophagous insects can transport virus diseases and facilitate the infestation of plants with pathogenic fungi.

A quantitative evaluation of the action of herbivores on vegetation and the effect of the quantity and quality of plants on the herbivores are especially difficult to assess under natural conditions. In a meadow, the effect of the herbivore pressure on plants and the responses of plants are the result not of individual relationships but rather an overall interaction between the herbivorous community and the plant cover.

Data are presented below on the direct effect of the herbivorous fauna upon plants, such as consumption by herbivores and its effect (destruction or stimulation of plant growth); indirect changes of the habitat; the effect on the physiological processes in plants and plant response to grazing; and the growth rate and protective reactions of plants aimed at neutralizing or diminishing the effect of herbivores.

### *Effects of the phytophagous fauna on plant production of meadows*

The effect of herbivores on meadow plant production is usually estimated from the amount of plant biomass that is eaten by phytophages, expressed in weight or energy units. However, consumption is not a measure of the real impact of herbivores on primary production but only an indication of the amount of energy transferred from the autotrophic level to a higher one. The consumption of most phytophagous species is usually assessed experimentally in the laboratory, while consumption by grazing herbivores can be estimated from direct measurements or indirect calculations. The method most often used for the evaluation of consumption is based on measuring the difference between the weight of food supplied to the animals and the weight after they have grazed it. From the values obtained (recalculated in terms of dry weight and later in calories) the consumption rate for particular developmental stages or the whole development cycle of a given species can be estimated (e.g. Smalley, 1960; Crossley, 1966; Nakano & Monsi, 1968; Chłodny, 1969; Gyllenberg, 1970; Andrzejewska & Wójcik, 1970; Witkowski & Kosior, 1974). A similar method is based on measuring, with a planimeter, the area of plant surface that has been eaten and then converting the area measurement into weight (Gyllenberg, 1970; Stachurski & Zimka, 1976).

It is much more difficult to estimate the consumption of sucking phytophages, and few measurements exist of the amount of sap taken by sucking herbivores. Wiegert (1964) working with the spittlebug (*Philaenus spumarius*) has assumed that the volume of sap taken from the plant is almost equal to that of the excreta of the bug. However, further difficulties arise in the conversion of the amount of sap consumed into energy units, due to large changes in its calorific value with time of day, growth stage of the plant and the part of the plant it is extracted from. Another way of measuring the amount of sap intake by a sucking insect (*Cicadella viridis* L., Auchenorrhyncha) assumes that a plant under conditions of full water availability (i.e. grown hydroponically) replaces any sap deficiencies caused by the insects from the solution it is growing in. Thus measuring the difference between the volume of solution in control containers holding plants without insects and that in containers holding experimental plants with insects of a known number and biomass, allows the calculation of the sap intake per insect per unit time (Andrzejewska, 1967).

Consumption values obtained under laboratory conditions are available for a larger number of phytophagous animals. In plant sap feeders, for which the consumption rate was estimated from direct measurements, it ranges from 108 to 720 $mg \cdot mg^{-1} \cdot d^{-1}$, in plant tissue feeders from 0.15 to 3.549 $g \cdot g^{-1} \cdot d^{-1}$ (Table 3.11). These huge variations in the consumption rate between grazing and sucking phytophages result from the large difference in calorific value of the food taken. On average, the calorific value of the green parts of plants amounts to 4.5 $kcal \cdot g$ dry $wt^{-1}$ or 1.5 $kcal \cdot g$ wet $wt^{-1}$ (Golley, 1961; Lamotte, 1975), whereas that of sap is variable but always low and of the order of 2.93–9.54 $cal \cdot g^{-1}$ (Wiegert, 1964, after Bollard). Thus in order to obtain an adequate amount of food an insect has to filter a considerable volume of sap.

In complex ecosystem investigations, data referring to numbers, biomass and, sometimes, production of phytophagous animals collected from a given meadow are usually available. Division of animals into groups is very often restricted to the broader taxonomic units. Consumption is estimated by indirect methods that rely on indices of production efficiency calculated for various groups of herbivores. The range of values of these indices (i.e.

Table 3.11. *Daily food intake of different groups of herbivores*

| Herbivore species | Body weight | Consumption of herbivore per day | Developmental stage | Reference |
|---|---|---|---|---|
| **Sucking insects** | | | | |
| *Cicadella viridis* L. | 1 (mg fresh wt) | 250 (mg · mg$^{-1}$) | Larva | Andrzejewska |
| (Auchenorrhyncha) | 5 | 180 | | (1967) |
| | 10 | 120 | | |
| | 12.5 | 108 | Adult ♀♂ | |
| *Philaenus spumarius* L. | 0.39 (mg dry wt) | 720 | Larva | Wiegert (1964) |
| (Auchenorrhyncha) | 1.16 | 424 | | |
| | 3.93 | 158 | Adult ♂♀ | |
| *Neophilenus lineatus* L. | 0.089 | 704 | Larva | Hinton (1971) |
| (Auchenorrhyncha) | 0.220 | 513 | | |
| | 0.515 | 403 | | |
| | 1.125 | 314 | | |
| **Grazing animals** | | | | |
| Arthropods | | | | |
| *Tortrix viridiana* | 0.94 (mg dry wt) | 0.25 (g · g$^{-1}$ · d$^{-1}$) | Larva | Witkowski & |
| (Lepidoptera) | 1.07 | 0.45 | | Kosior (1974) |
| | 2.25 | 0.74 | | |
| | 5.37 | 0.86 | | |
| *Bombyx mori* | 0.038 (g dry wt) | 3.549 | Larva | Nakano & |
| (Lepidoptera) | 0.164 | 2.166 | | Monsi (1968) |
| | 0.69 | 2.487 | | |
| | 1.51 | 2.338 | | |
| | 2.9 | 2.575 | | |
| | 5.2 | 3.392 | | |
| *Chrysomela knabi* | 1.08 (mg dry wt) | 2.4 (mg · mg$^{-1}$) | Larva (third | Crossley (1966) |
| (Coleoptera) | 1.41 | 4.1 | instar) | |
| | 2.69 | 3.05 | | |
| | 4.31 | 2.5 | | |
| | 5.24 | 0.7 | | |
| *Oxya velox* (Acrididae) | ♀97 (mg fresh wt) | 0.137 (mg · mg$^{-1}$ | Average of all | Delvi & |
| | ♂53 | fresh wt) | stages | Pandian (1971) |
| | | 0.197 | | |
| *Mecostethus grossus* | 10–20 | 0.67 | Larva | Andrzejewska & |
| | 20–60 | 0.4 | | Wójcik (1970) |
| | 60–200 | 0.2 | | |
| | 500–700 | 0.22 | Adult | |
| *Parapleurus alliaceus* | ♀218 | 0.27 | Larva | Matsumoto |
| | ♂129 | 0.27 | | (1971) |
| | ♀264 | 0.21 | Adult young | |
| | ♂136 | 0.24 | | |
| | ♀489 | 0.20 | Adult | |
| | ♀♂474 | 0.15 | | |
| *Schistocerca gregaria* | | | | |
| (Acrididae) | 1.5–2.0 (g fresh wt) | 0.4 (g · g$^{-1}$) | Adult | Davey (1954) |
| *Conocephalus fasciatus* | 6.6 (mg dry wt) | 0.28 (mg · mg dry | Average of all | Van Hook (1971) |
| (Tettigonidae) | | wt$^{-1}$) | stages | |
| *Anaburus simplex* | 0.2 (g dry wt) | 0.45 (g · g dry wt$^{-1}$) | Larva | Cowan & |
| | | | | Shipman (1947) |
| *Anomala cuprea* | 0.235 | 0.203 | Average of all | Nakamura (1965) |
| (Scarabaeidae) | | | stages | |
| Small mammals | | | | |
| *Clethrionomys glareolus* | 19.94–21.19 (g | 0.11–0.12 | | Gębczyńska and |
| | fresh wt) | (g · g fresh wt$^{-1}$) | | Gębczyński |
| | 22.25–21.99 | 0.16–0.19 | | (1971) |
| *Arvicola terrestris* | 70.5–89.5 | 0.64–1.05 | | Drożdż, Górecki, |
| | | | | Grodziński & |
| | | | | Pelikan (1971) |
| Large mammals | | | | |
| Sheep | | | | Van Dyne *et al.* |
| early summer | 46.3 (kg) | 0.10 (g · g$^{-1}$) | | (Chapter 4) |
| mid summer | 44.5 | 0.12 | | |
| late summer | 43.6 | 0.11 | | |
| Cows | 317.3–322.3 | 0.02 | | Van Dyne *et al.* |
| | | | | (Chapter 4) |
| | 454 | 0.022 | | Lewis (1971) |
| | 500 | 0.017–0.026 | | Voisin (1970) |
| | 600 | 0.016–0.026 | | Voisin (1970) |

$P/C$, $F/C$, $R/C$) for herbivores with similar life cycles and utilizing food of similar qualities is narrow (p. 227). Respiration rate ($R$) can be calculated from a simple regression equation, $y = ax^b$, where $x$ is the fresh weight of the animal. Values for $a$ and $b$ have been calculated for various phytophages (Reichle, 1971). Kleiber (1963) has calculated that for homoiotherms $b = 0.75$; for poikilotherms Bertalanffy (1957) estimated $b = 0.67$.

Furuno & Ohmura (1971) found a high correlation between the amount of food consumed ($C$) and the quantities of the frass drop ($F$) for five species of lepidopteran larvae feeding on pine. On average, consumption exceeds frass drop by 1.3 times, the range of variation being 1.23–1.38.

The production efficiency indices ($P/C \times 100$) of phytophages show a vast range of values. It appears that the quality of food consumed is one cause of this diversity. It was found that over short periods of time the chemical composition of phytophages remained constant irrespective of the concentration of nutrients in their food (Kalinowska, 1977; Mochnacka-Ławacz, 1978). On the other hand, the amount and chemical composition of excreta varied greatly with the quality of food consumed (Kalinowska, 1977). Nakano & Monsi (1968) compared the values of $P/C$ for silkworm larvae (*Bombyx mori*) expressed in terms of either dry weight, calories or nitrogen content, and obtained values of 15, 19 and 45%, respectively. A similar diversity of results was obtained by Hiratsuka (after Nakano & Monsi, 1968). Phytophages that feed on food of a low calorific value assimilate only a minute part of the food eaten and only a small fraction of its calorific content. For example, aphids filter considerable amounts of sap and only slightly reduce its sugar content. They retain, however, the small nitrogen compounds on which their growth rate and fecundity depend (Krzywiec, 1968; Van Emden *et al.*, 1969; Van Emden & Bashford, 1969; Tulisalo, 1971). Augmenting the concentration of nitrogen compounds and sugars in plant leaves accelerates the growth and development of feeding larvae (Haukioja & Niemelä, 1976; Fischer & Andrzejewska, 1977), thus increasing utilization efficiency and decreasing maintenance costs at the same time.

In spite of the fact that calculations based on production efficiency indices allow only rough estimates of consumption, they are commonly and successfully used for constructing models of energy flow through grassland ecosystems (Macfadyen, 1967; Lewis, 1971; Coupland, Willard, Ripley & Randell, 1975; Breymeyer & Kajak, 1976). However, no matter how precise the calculations of consumption by herbivores they do not give a measure of their total effect on the plants, since the effects of feeding can not be equated with the amount of food consumed. Feeding by herbivores not only depletes the biomass of plants but it can also (1) stimulate a plant's production and alter growth, (2) disturb fundamental physiological processes in the plant, (3) change the plant's environment so as to encourage invasion by other organisms, or introduce pathogenic organisms directly.

### *Plant production in a meadow subjected to herbivore pressure*

Phytophagous animals can affect both the development and production of their food plants in a number of ways. The spot where a plant is injured is important, as are the stage of its development and the intensity of grazing. In some instances the same animal feeding on different developmental stages of the same plant species can differently affect plant production. Tertil (1974), studying the effect of a vole (*Microtus arvalis* Pall.) on the growth and crop of winter wheat, has distinguished five periods of differing sensitivity of the plant to grazing by voles, basing his results on the final quantity and quality of the grain. Grazing in winter through till spring stimulated regrowth of the plant, while if the shoots were eaten in May there were losses of 40–50% of the grain crop and grazing in May/June caused a loss of about 80%. Destruction of plants when ripening at the end of June decreased the crop by 10–15% and attacks in the pre-harvesting period by up to 40–50%. Actual consumption by the voles was only a small fraction of the crop loss caused by their grazing (Table 3.12). Grasshoppers (Acrididae) feeding on grassland vegetation destroyed 6 to 15 times more than they consumed (Andrzejewska & Wójcik, 1970; Gyllenberg, 1970). This results from the fact that Acrididae graze on leaf blades at different heights. The part of the leaf above the portion that was eaten then dries and drops off, so that the weight of plant material killed exceeds several-fold the weight actually eaten, the exact ratio depending on the shape and length of leaves attacked. Similarly the amount of sap taken by a sucking insect within 24 hours is up to 250 times the weight of the insect itself, only a minute fraction of its chemical constituents (mainly the nitrogen compounds) being used (Wiegert, 1964; Andrzejewska, 1967). The feeding activity of sap-feeders causes withering, the plant turning yellow and the affected part eventually dying. This leads to a reduction of plant biomass and, in causes of high insect densities, to a complete devastation of the plant. Although sap-feeders do not actually consume plant tissue, their feeding activity rapidly decreases both the growth rate and biomass of the plant. For example, as early as 2 weeks after infestation of plants with leafhoppers (*Cicadella viridis*) at densities typical for natural situations, the plant biomass was lowered by 25% compared to control plants (Andrzejewska, 1967).

In field studies estimates of the effect of phytophages on meadow vegetation are usually obtained by comparing various parameters of the vegetation (biomass, chemical composition, etc.) between two plots, one with and one without phytophages. Smolik (1974) has estimated using this method that the percentage loss in primary production of a pasture (shortgrass prairie) caused by nematodes is 59, 28 and 45% in three different periods of the vegetation season. In these periods the density of nematodes ranged from $2 \times 10^6$ to $6 \times 10^6$ m$^{-2}$. The removal of the invertebrate fauna from the litter

Table 3.12. *Evaluation of the effect of herbivore feeding on plants*

| Type of enviroment | Herbivores | Damage due to herbivore feeding | References |
|---|---|---|---|
| **Grasslands** | | | |
| Shortgrass prairie, USA | Nematodes (2–6 ml · m$^{-2}$) | Up to 59 % | |
| *Agropyron smithii* and *Bouteloua gracilis* shortgrass prairie, USA | Nematodes | 35–67 % | |
| Steppe, southern USSR | Grasshoppers (1–2 yard$^{-2}$) | 15.6 % (in a dry year and at 2–3 times higher density, ~ 30 %) | Serkova (1961) |
| Pastures, North America ($P_p$ ~ 600 lb · acre$^{-1}$ · yr$^{-1}$) | Grasshoppers (3 yard$^{-2}$ · y$^{-1}$) | 50 % | Anderson & Wright (1952) |
| Saltmarsh, USA | Grasshoppers | 8 % | Teal (1962) |
| Timothy grass (leys for hay), Finland | Timothy flies (*Amaurosoma* spp.) | 5.2 % (average for 45 fields) | Raatikainen & Vasarainen (1975) |
| *Echinochloa crus-galli*, Poland | Leafhoppers (*Cicadella viridis* L.) | 9–17 % | Andrzejewska (1967) |
| Pasture, in one region of Brazil | Harvester ants (*Atta carpiguara*) | Reduction of carrying capacity by equivalent of 20000 head of cattle; consumption of 10 adult colonies equivalent to consumption of 1 cow | Weber (1972) |
| Tanganyika grasslands | Large mammals | 28 % | Wiegert & Evans (1967) |
| Uganda grasslands | Large mammals | 60 % | Wiegert & Evans (1967) |
| Wet uncultivated meadow, Poland ($P_p$ = 192 dry g wt · m$^{-2}$ · yr$^{-1}$) | Arthropods | 8–14 % | Andrzejewska & Wójcik (1970) |
| Fresh, cultivated meadow, Poland ($P_p$ = 909 g dry wt · m$^{-2}$ · yr$^{-1}$) | Arthropods | 21–23 % | Andrzejewska & Wójcik (1971) |
| Tundra | Lemmings | 50–90 % | Shultz (1969) |
| Tundra (*Eriophorum* and *Carex*) | Voles (*Microtus oeconomus* and *M. middendorffi*) | Up to 30–50 voles ha$^{-1}$ stimulate plant production | Smirnov & Tokmakova (1972) |
| Temperate grasslands | Microtines, in peak densities | 1–35 % | Batzli (1975) |
| Arctic grasslands | Ground squirrel | Changes plant composition (herbs and grasses remain) | Batzli (1975) |
| Natural meadow, (max. $P_p$ ~ 41 kg dry wt · acre$^{-1}$ · yr$^{-1}$) | Rabbits | ~ 13 kg dry grass · acre$^{-1}$ · yr$^{-1}$ = ~ 32 % of $P_p$ | Vervelde (1970) |
| **Cultivated fields** | | | |
| Sugarcane, Trinidad | *Aeneolamia varia saccharina* (Cercopidae, Homoptera) | 1 ton per acre of sugar, i.e. 25–40 % of average annual yield | Fewkes (1967) |
| Oatfield, Finland | Aphids | 10–28 % | Raatikainen & Tinnila (1961) |
| Potatoes, Poland | Colorado beetles | 10–20 % | Trojan (1967) |
| Cropfield, Trinidad 1968 | Harvester ants (*Atta cephalotes* and *Acromyrmex octospinosus*) | 12 % of annual crop = 600000 Trinidad $ | Weber (1972) |
| Different crops, total for USA 1951–60 | All herbivore insects | 14 % | Bullen (1970) |
| Different crops, total for USA 1951–60 | Grasshoppers | 0.3 % | Bullen (1970) |
| All crops, Africa | Desert locusts (in invasion area) | 0.6 % | Bullen (1970) |
| All crops, Africa | All herbivore insects | ~ 13–20 % | Bullen (1970) |
| Wheat field | *Microtus arvalis* (2 individuals · m$^{-2}$ per week) | 10–80 % of crop depending on time of infestation | Tertil (1974) |

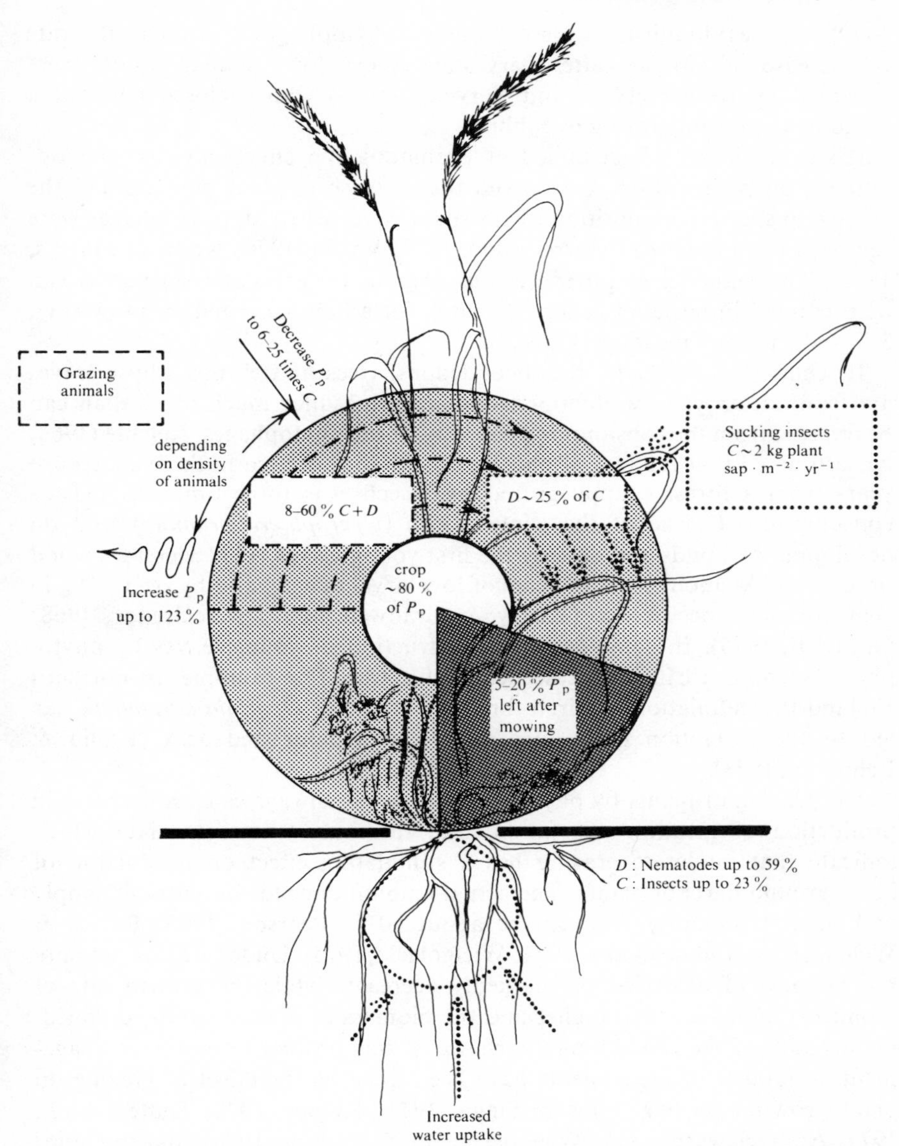


Fig. 3.7. A scheme to show the distribution of plant biomass produced on a cultivated meadow and the effect of herbivore feeding. *D*, damage to plants; *C*, consumption of herbivores; $P_p$, primary production.

layer and aboveground parts of plants in a cultivated meadow (*Arrhenatheretum medioeuropaeum*) increased plant production in subsequent months by 35, 35 and 18% (June, July and August 1966: Andrzejewska & Wójcik,

1970). Besides diminishing plant biomass, phytophages also affect the rate of increase of biomass after harvesting a meadow. Similar results were obtained by Ketner (1973) and Vervelde (1970) who enclosed parts of a meadow to prevent grazing by rabbits.

All experiments which aimed at evaluating the effect of phytophagous animals on plants attempted to find the increase in plant biomass and the change in species composition of the plant cover when all phytophages were removed from the area (Morris, 1967, 1973; Voisin, 1970; Krebs & Mayers, 1974). The influence of phytophagous animals on grassland vegetation can be presented in terms of general schemes for a hay-harvested meadow (Fig. 3.7) and a pasture meadow (Fig. 3.8).

It seems that especially the dicotyledons, trees and shrubs whose leaves are being destroyed by phytophagous animals, suffer much more than can be inferred from the consumption capacity of the phytophages. Reichle (1968) has estimated that the loss of photosynthetic surface area in three successive years from a forest canopy grazed by insects was three times as high as consumption. Larvae of the winter moth (*Operophtera brumata*) feed on developing oak buds and destroy the first young leaves. This decreases wood production by about 10% in years of low larval density and by over 43% in years of mass occurrence (Varley & Gradwell, 1962; MacGregor, 1968; Gradwell, 1975). In some cases the destruction of young leaves by phytophagous animals can cause the death of a tree. For example, in northern Finland the defoliation of birch forest by larvae of *Oporinia autumnata* has led to the formation of a huge (1350 $km^{-2}$) deforested area (Kallio & Lehtonen, 1973).

The grazing of plants by phytophages does not always cause a decrease in production of meadow vegetation, and there are numerous data which indicate that moderate grazing has a stimulating effect on production of aboveground parts of plants. The actual amount depends on nutrient supply and habitat humidity, but can be about 50% (Pearson, 1965; Rawes & Welch, 1966; Andrzejewska, 1974; Wielgolaski, 1975). Under grazing pressure the biomass of dicotyledons markedly decreases while the growth rate of monocotyledons is much higher and the biomass of grasses can be doubled. On account of the economic importance of this process to pasture management, a number of experiments have been done on the effect of grazing on plant growth rate (e.g. Sims & Singh, 1971; Ketner, 1972; Duffey *et al.*, 1974; Andrzejewska, 1974; Wielgolaski, 1975; Perkins, 1976), and the number of animals that can graze a pasture without exceeding the regenerative abilities of the plant cover ascertained. Swartzman & Singh (1974) calculated what they call 'optimal grazing strategies'. In a model for a tropical grassland they included five criteria as important measures of stability for grazing grassland: (1) the diversity of the major perennial species; (2) the utilization of the herbage; (3) the percentage of bare soil; (4) the percentage of the area

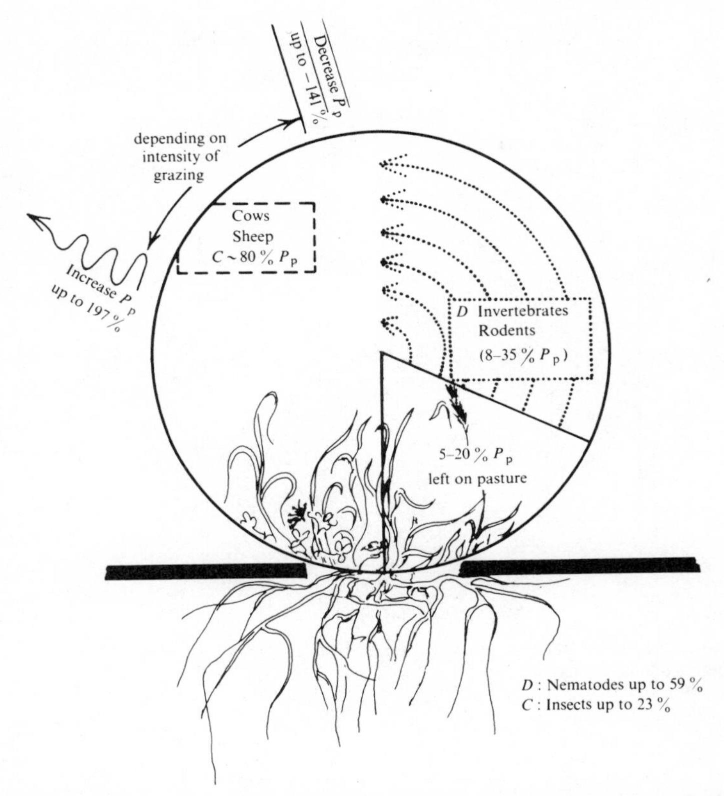


Fig. 3.8. A scheme to show the distribution of plant biomass produced on a pasture and the effect of herbivore feeding. *D*, damage to plants; *C*, consumption of herbivores; $P_p$, primary production.

occupied by legumes; (5) the percentage cover of dominant species. The overgrazing of pasture causes changes in vegetation composition, diminishing species diversity. Dicotyledons markedly decrease as does the area covered by legumes and the volume of the root system. There is an increase in the area of bare soil between plants (Fig. 3.9) which is a problem in terms of soil erosion (Voisin, 1970; Golley, 1973; Duffey *et al.*, 1974; Wielgolaski, 1975).

It seems that the effect of invertebrate and vertebrate herbivores on the plant production of a meadow is greatest in periods of intense plant growth, that is at the beginning of the season or during the regrowth of plants after mowing. Then, besides losses in biomass the growth rate of plants is decreased considerably, causing them to mature and achieve their peak biomass several weeks later than ungrazed ones (Andrzejewska, 1971). A similar delay in reaching peak biomass was caused by rabbits grazing natural grass vegetation on a salt marsh on the Dutch island of Schiermonnikoog (Vervelde, 1970;

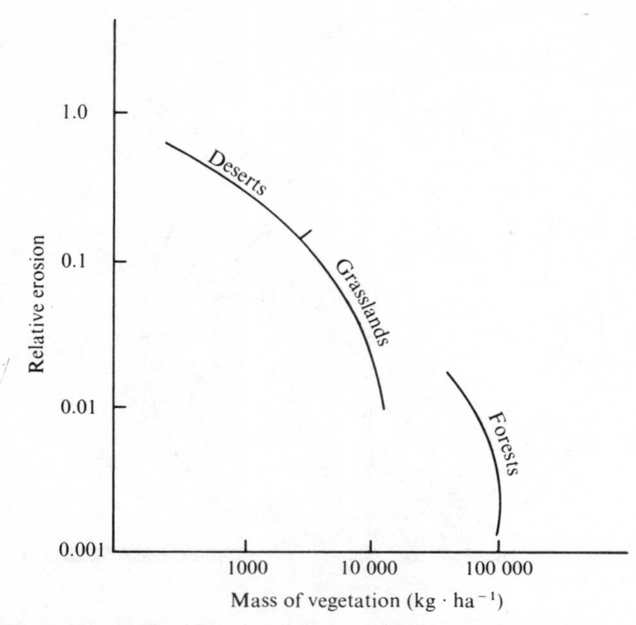


**Fig. 3.9. The relationship between mass of vegetation and relative erosion. (Golley, 1973; after Langbein & Schumm.)**

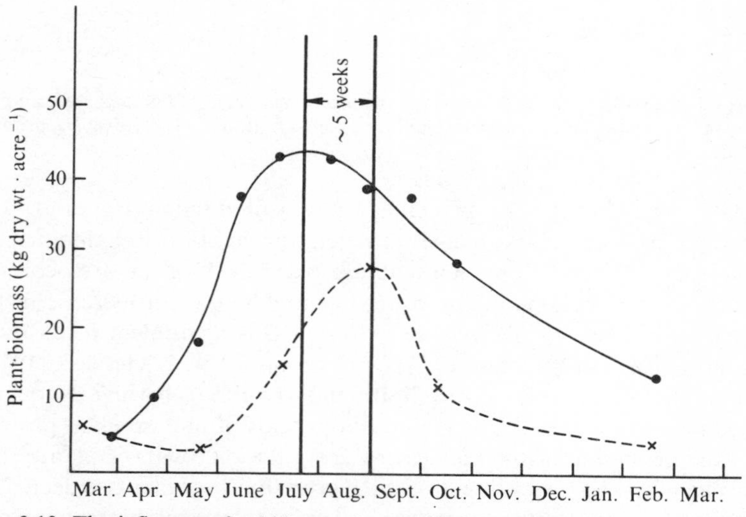


**Fig. 3.10. The influence of rabbits on the growth of vegetation. ●——●, fenced, undisturbed vegetation; ×– –×, open, grazed vegetation. Note that the peak biomass of the grazed vegetation is reached about 5 weeks later than that of the ungrazed vegetation. (After Vervelde, 1970.)**

Ketner, 1972). The production of undisturbed vegetation was not only higher, but the peak biomass was reached about 5 weeks earlier (Fig. 3.10). But the constant grazing on pastures maintains continuous growth activity. The plants do not produce seeds and, given a relatively high growth rate and favourable habitat conditions, the production of green biomass can be much higher than in an ungrazed area.

In some situations the loss in plant production resulting from phytophage grazing can be compensated for in the following season when nutrients are released from the faeces of the phytophages faster than they would be from the soil alone (Zlotin, 1971, 1975; Krebs & Mayers, 1974).

### *The effect of phytophages on the physiological processes of plants*

There are few data in the literature that indicate the degree to which the physiological processes of plants are affected by the actual feeding of herbivores. However, it is known that the disturbance and destruction of plants during feeding has a number of effects:

(1) It initiates a protective reaction by the plant which diverts its metabolic processes to the synthesis of chemical repellents. This means that extra energy is being expended by the plant (Von Rudniew, 1963; Green & Ryan, 1972; Feeny, 1975).

(2) To a greater or lesser extent respiration, transpiration and assimilation, as well as growth processes and carbohydrate and protein transformation, are all disturbed. Poskuta *et al.* (1973) have reported a reduced chlorophyll content and disturbances in plant enzyme activity.

(3) Sap-feeding insects, especially those which filter considerable amounts of plant sap, disturb the water balance of the plant. The loss of sap caused by the insect is counterbalanced by the plant augmenting its water uptake, but at high levels of infestation plants may be unable fully to compensate for the water lost and may therefore wither. This phenomenon is accentuated because transpiration is intensified during the feeding of phytophages (Casswell, Reed, Stephenson & Werner, 1973; Poskuta *et al.*, 1977).

(4) Phytophages that pierce and suck the plant can introduce toxic substances into the plant with their spittle which can affect the quality of stored reserves (Fewkes, 1967; Tulisalo, 1971).

(5) As a result of infestation by sap-feeders the direction of transport of nutrients in a plant can be altered.

In the light of the above information it seems obvious that the total impact of phytophages on plants can not be evaluated in terms of biomass or energy changes alone. Under natural conditions it is difficult to evaluate the complexity of the interactions between plant and plant-feeder, and under laboratory conditions one is limited to individual examples. The relationship should certainly not be regarded only in terms of the destruction of plants and the reduction in their biomass. Plants seem to have a certain tolerance to phytophages and have an ability to combat to some degree the feeding pressure of insects by activating appropriate physiological responses (Trojan, 1967; Zurzycki, 1975; Czarnowski, 1975).

### *Habitat alteration produced by herbivores*

Herbivores utilize only part of the food which they consume. About 10–70% is returned to the habitat as faeces which are rich in organic compounds and nutrients and are an excellent source of food for saprophagous animals. Due to the activity of saprophagous animals and bacteria, the mineralization of faeces is rapid, leading to the release of nutrients available to plants (Breymeyer, 1974; Duffey *et al.*, 1974; Olechowicz, 1974; Breymeyer, Olechowicz & Jakubczyk, 1975). Thus faeces produced by herbivores can be an important source of nutrients, especially in pastures, where about 80% of plant production is consumed by grazing animals and about 56 % of this production returns to the soil in the form of faeces (Olechowicz, 1976). In special cases, as in some Australian pastures, large saprophages are absent and the faeces deposited by sheep and cows are not mineralized. The faeces linger for months, killing vegetation and rendering as much as one-third of a pasture useless for grazing.

The faecal droplets of sucking insects contain considerable amounts of sugars and, covering the leaf surface, become an excellent substrate for the development of bacteria and fungi. Under favourable weather conditions fungal diseases can cause substantial injuries to plants or even their complete destruction.

Herbivorous mammals when grazing, making trackways or building nests, can change their habitat. Small mammals which inhabit meadows dig tunnels up to several metres long, displacing kilograms of soil. This soil is mixed with faeces and enriched with nitrogen compounds. Under such circumstances the habitat, the species composition of the plant community and quantitative relationships between species are altered (Golley, Ryszkowski & Sokur, 1975). Babińska (1972) has analysed the biomass and species composition of plants in alfalfa fields over several years in relation to their distance from the middle of a vole (*Apodemus flavicollis*) colony. In the immediate vicinity of the colony grasses replaced the alfalfa and with increasing distance from the colony the incidence of alfalfa increased again, although the overall biomass of the aboveground parts of the vegetation remained almost uniform throughout the whole area. Intense grazing by herbivores, that is removal of plants and constant treading of the plant cover, not only eliminates certain plant species but also leads to marked and often irreversible changes in the habitat. The greater the destruction of plant cover the greater is the degree of soil erosion in areas where climatic and habitat conditions favour this process (Sims & Singh, 1971; Golley 1973; Sarmiento & Monasterio, 1975; Fig. 3.9).

### *Herbivores as agents for the transmission of pathogenic organisms*

Injured areas on plants often permit the entry of bacteria and fungi. Many insect species, especially sap-feeders (aphids, leafhoppers, bugs) carry viruses which are plant pathogens. The losses caused by insects feeding are comparatively minor when compared with those which result from viruses. The intensity of virus diseases depends on weather conditions and on the resistance and condition of the host plants. The scale of plant injury may be very large and ranges from the inhibition of growth, through a drop in quantity and quality of crop, to the death of plants (Smith, 1967; Stevenson, 1970).

### *The effect of vegetation on phytophages*

As already mentioned, there is close interaction and interdependence between plants and the phytophages which feed on them. In general, the biomass and the state of plant community govern those same parameters in the animal community (discussed earlier in this chapter). Here attention will be concentrated on the effect of food quality on herbivorous populations.

Under natural conditions the number of phytophagous animals is rarely directly limited by the amount of available food. Nevertheless, the type and quality of food affect substantially the parameters governing the reproduction of phytophagous populations. Some phytophagous species can feed on several species of plants but typically on only one species, and at different developmental stages preferences are often further limited to one particular part of the plant (Thorsteinson, 1960; Ganwere, 1961; Kaufmann, 1965; Bernays, Chapman, Horsey & Leather, 1974). Palatability is one of the decisive factors determining the choice of a food species and this depends on the chemical composition of the plant being optimal for the development of the phytophage. Kaufmann (1965) carried out an experiment on the feeding behaviour of 15 species of Acridoidea which showed a definite preference for particular plant species when examined in their natural grassland habitat. Diet was found to have a great influence on mortality rate, adult longevity, body weight and number of eggs laid. The food relationships of many species of phytophagous animals were reviewed by Chapman (1974).

Caswell *et al.* (1973) put forward the hypothesis that as a general rule the type of carbon fixation pathway in plants is of major importance in determining the choice of a particular plant by a plant-feeder. All plant species use one of two pathways for fixing carbon dioxide: the $C_4$-dicarboxylic acid pathway ($C_4$ plants) or the $C_3$-Calvin cycle pathway ($C_3$ plants). Phytophages find $C_4$ plants less palatable, and if fed on them show a reduction in survival and reproductive rates, egg production, body weight and assimilation efficiency.

If a plant is damaged by the feeding of phytophages then this obviously has only harmful effects on the plant. However, plants are not completely defenceless against phytophages, and some species have evolved morphological and chemical defence mechanisms. One type of protective device is spines or hairs which cover the leaves. Or they may have tough epidermal tissues (containing lignin, silica, cork or wax) that are difficult for herbivores to graze and digest (Williams, 1954; Chapman, 1974; Bernays & Chapman, 1974). Chemical defence systems involve production of protective substances (alkaloids, phenols) as a reaction to injury, so their functioning is dependent on the intensity of grazing. The protective agent can be distributed throughout the plant within a few hours. The toxic effect of such substances on the grazing animals can hinder development, suppress fecundity and increase mortality by decreasing resistance to disease (Von Rudniew, 1963; Harley & Thorsteinson, 1967; Green & Ryan, 1972; Benz, 1974; Feeny, 1975).

Changes in the morphological, chemical and physiological states of plants during growth and ageing substantially affect herbivores, which have to adapt both ecologically and physiologically to these changes.

## 4. Bioenergetic parameters of the main groups of herbivores

The following review of herbivore energetics in grassland ecosystems deals mainly with the grazing and sucking invertebrates and largely excludes small mammals, birds and nectivores, and also omnivores such as ants. As there have been extremely few investigations into the energetics of wild populations of large mammals, some small mammal data have been included in the diagrams. It seems that the bulk of available information concerns Orthoptera and Hemiptera populations; most investigations on Lepidoptera and Coleoptera are either from woodland ecosystems or from laboratory experiments.

The following notation has been used:

$P_N$ = net primary production, comprising only aboveground vegetation unless otherwise indicated

$P$ = production, including tissue growth, reproduction and exoskeleton production, or calculated from change in biomass plus elimination (according to Petrusewicz, 1967)

$C$ = consumption, i.e. plant material actually ingested

$F$ = egestion, including defecation, secretion of urine and secretion of other products

$A$ = assimilation, the part of the food absorbed into the body tissues

$R$ = respiration, the total energy dissipated as heat, in practice measured as oxygen consumption and converted into calories

$A/C$ = assimilation efficiency (H. T. Odum, 1957; E. P. Odum, 1971), i.e. assimilation divided by consumption

$P/C$ = ecological growth efficiency (H. T. Odum, 1957; E. P. Odum, 1971), i.e. net productivity divided by consumption; equivalent to the index $NP_N/I_N$ reported by Kozlovsky (1968)

$P/A$ = tissue growth efficiency (H. T. Odum, 1957; E. P. Odum, 1971), i.e. net productivity divided by assimilation; equivalent to index $NP_N/A_N$ of Kozlovsky (1968).

### *Herbivore consumption, C*

This section deals with consumption purely from a quantitative point of view since an earlier section covered qualitative aspects (food selection, food shortage). An attempt is made to examine the relationship between total herbivore consumption and net primary production (of aboveground vegetation). The regression is shown in Fig. 3.11 on a double logarithmic scale. The equation for the line is:

$$\text{Log}\, C = 1.57 \cdot \log P_N - 4.04 \quad \text{or} \quad \log P_N = 0.39 \cdot \log C + 3.16. \quad (3.1)$$

The regression is significant ($r = 0.789$, with probability $P \sim 0.01$), but indicates quite a high variability (standard error of log $C$ for the points log $P_N$ = 2 and 3 is 0.641 and 0.62 respectively).

A list of the herbivorous groups examined and the corresponding investigation sites is given in Table 3.13. It has not been possible to deduce the total consumption of invertebrates in every investigation, and it is quite possible that this regression does not hold if consumption by other groups is significant in relation to the ones that have been investigated. The regression deals with invertebrate herbivores only (Fig. 3.11).

Even with these restrictions in mind there seems to be some general trend indicated by the slope of the line. It appears that invertebrates are more successful and remove a larger fraction of the aboveground vegetation in grasslands with a high net primary production (cf. also Wiegert & Evans, 1967). As the rate of production generally decreases as steady-state conditions are approached, it may be that a situation has evolved whereby invertebrates exploit proportionately less of the production of the grassland as succession proceeds.

### *Energy turnover in the population*

The relationship between annual production and respiration for animal populations has been investigated by Engelmann (1966), McNeill & Lawton (1970) and Funke (1972). Here we present studies of grassland herbivore

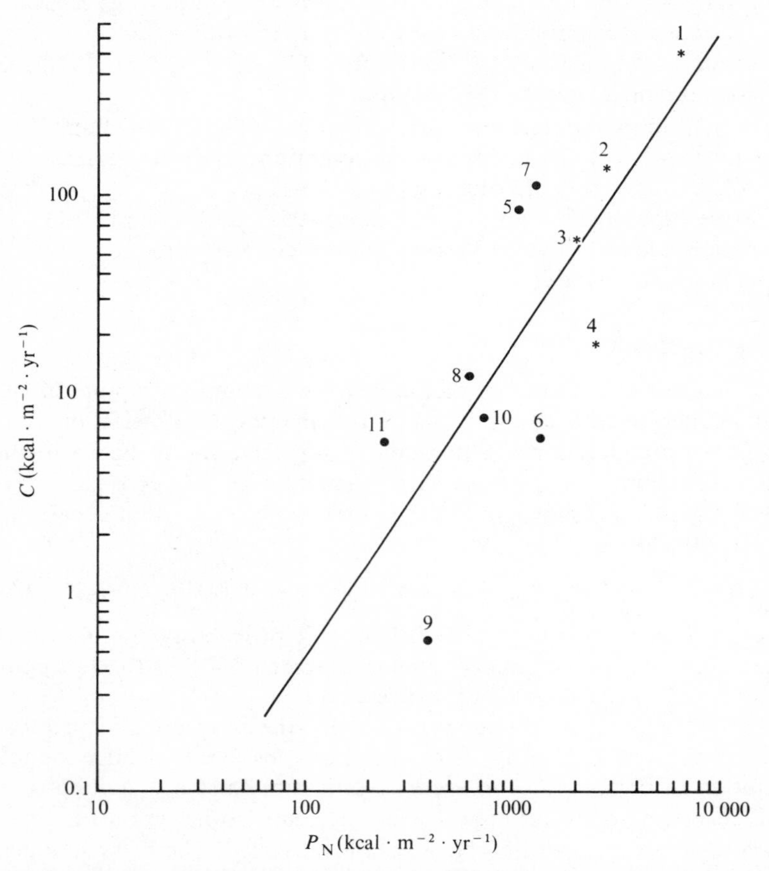


Fig. 3.11. The relationship between annual net primary production of aboveground vegetation ($P_N$) and annual consumption ($C$) of invertebrate herbivores. The numbers associated with each point correspond to the numbers in the first column of Table 3.13 Highly productive grasslands are marked with an asterisk.

populations and include estimates of annual production versus consumption, and annual assimilation versus consumption. As there are several measurements available for invertebrate consumption and assimilation, the regression for poikilotherms is probably significant even though restricted to grasslands. In the absence of adequate data for large mammal populations some data on small mammals have been included in order to make sense of the conclusions which are drawn.

A list of the data reviewed is given in Table 3.14. No data points are omitted, although this was done by McNeill & Lawton (1970) for the *Proklesia* and 'old-field' *Philaenus* populations (nos. 19 and 22). The *Philaenus* population

Table 3.13. *Annual net primary production of aboveground vegetation ($P_N$) and consumption (C) of herbivores ($kcal \cdot m^{-2}$) on grasslands*

| No. in Fig. 3.11 | Herbivore groups | Investigation area | $P_N$ | $C$ | References |
|---|---|---|---|---|---|
| 1 | Plant hoppers and grassoppers | Salt marsh, Georgia, USA | 6585 | 520.0 | Wiegert & Evans (1967) |
| 2 | Spittle bugs and grasshoppers | Alfalfa field, Michigan, USA | 2780 | 140.0 | Wiegert (1964, 1965) |
| 3 | Total herbivores (invertebrates) | Meadow, Finland | 2136 | 58.5 | Gyllenberg (1969, and unpublished data) |
| 4 | Total herbivores (invertebrates) | *Sericea lespedeza*, USA | 2530 | 17.7 | Menhinick (1967) |
| 5 | Total herbivores (invertebrates) | Old-field, South Carolina, USA | 1075 | 85.0 | Wiegert & Evans (1967) |
| 6 | Total herbivores (invertebrates) | Old-field, Michigan, USA | 1360 | 6.0 | Wiegert & Evans (1967) |
| 7 | Total herbivores (invertebrates) | Grassland, Tennessee, USA | 1280 | 112.6 | Van Hook (1971) |
| 8 | Grasshopper population | Prairie, Matador Project, Canada | 615 | 12.3 | Bailey & Riegert (1973) |
| 9 | Heteroptera population | Grassland, Silwood Park, England | 395 | 0.59 | McNeill (1971) |
| 10 | Grasshopper population | *Micanthus* area, Japan | 720 | 7.7 | Matsumoto (1971) |
| 11 | Total grasshoppers | Mountain grassland, New Zealand | 240 | 5.7 | White (1974) |

data may be inaccurate as they were influenced by immigration, but one might argue that the influence of immigration has not been considered at all in most of the investigations. Some data points are not based on independent measurements and their inclusion may be unjustified. This applies especially to the points 21, 22 and 24 in the consumption estimates. However, these points do not significantly alter the regression lines drawn in Figs. 3.12–3.14. Mean values are used where several years' energy flow data are available, otherwise the proportional weighting of populations with long investigation periods may be too high.

As McNeill & Lawton (1970) pointed out, respiration may theoretically be considered a function of production as much as production is a function of respiration. Assimilation is, strictly speaking, a function of the absorbing surface of intestine and the digestibility of food. Wiegert & Evans (1967) considered that digestibility is affected by type and quality of food, the age of the organism and the treatment the food receives when ingested. For herbivores the digestibility of food as well as the ratio of intestinal surface to body volume may be regarded as similar for all species, and thus one may assume a linear relationship between consumption and assimilation. If this relationship holds, then there should also be a linear relationship between production and consumption, though with separate regressions for homoiotherms and poikilotherms as is the case for production versus respiration (Fig. 3.11).

Table 3.14. *Annual production (P), assimilation (A) and consumption (C) (kcal · m$^{-2}$) for herbivore populations on grasslands*

| No. in Figs. 3.12–3.14 | Species | Investigation area | *P* | *A* | *C* | References |
|---|---|---|---|---|---|---|
| 1 | *Microtus pennsylvanicus* (meadow vole) | Old-field, Michigan, USA | 0.517 | 17.5 | 19.44 | Golley (1960) |
| 2 | *Peromyscus polionotus* (old-field mouse) | Old-field, Georgia, USA | 0.12 | 6.7 | 7.4 | Odum *et al.* (1962) |
| 3 | *Peromyscus* sp. (deer mouse) | Old-field, Michigan, USA | 0.01 | 0.63 | 1.05 | Wiegert & Evans (1967) |
| 4 | *Pitymys subterraneus* and *Microtus agrestis* | Alpine meadow, Poland | 0.096 | 6.1 | | Ryszkowski & Petrusewicz (1967) |
| 5 | *Cleithrionomys glareolus* (bank vole) | Beech forest, Poland | 0.09 | 3.38 | 4.08 | Grodziński *et al.* (1970) |
| 6 | *Apodemus flavicollis* (field mouse) | Beech forest, Poland | 0.07 | 2.71 | 3.15 | Grodziński *et al.* (1970) |
| 7 | *Adenota kob thomasi* (Uganda kob) | Grassland, Uganda | 0.81 | 62.5 | 74.1 | Buechner & Golley (1967) |
| 8 | *Loxodonta africana* (African elephant) | Grassland, Uganda | 0.34 | 23.3 | 71.6 | Wiegert & Evans (1967) |
| 9 | Grasshoppers | Old-field, Michigan, | 0.51 | 1.36 | 3.71 | Wiegert (1965) |
| 10 | Grasshoppers | USA (alfalfa) | 4.57 | 13.3 | 36.1 | Wiegert (1965) |
| 11 | *Orchelimum fidicinium*: Orthoptera | Salt marsh, Georgia | 10.8 | 29.4 | 107 | Smalley (1960) |
| 12 | *Chorthippus parallelus*: Orthoptera (4 years) | Meadow, Finland | 5.25 | 12.1 | 38.7 | Gyllenberg (1970) |
| 13 | *Orthochta brachycnemis*: Orthoptera | Savanna, Ivory Coast | 0.438 | 1.03 | 5.16 | Gillon (1973) |
| 14 | *Chorthippus parallelus*: Orthoptera (2 years) | Grassland, England | 0.347 | 0.611 | 2.20 | Quasrawi (1966) |
| 15 | *Encoptolophus sordidus costalis*: Orthoptera (2 years) | Prairie, Matador Project, Canada | 1.56 | 3.21 | 12.3 | Bailey & Riegert (1973) |
| 16 | *Melanoplus sanguinipes*: Orthoptera | Grassland, Tennessee | 26.1 | 51.6 | 90.4 | Van Hook (1971) |
| 17 | *Conocephalus fasciatus*: Tettigoniidae | Grassland, Tennessee | 2.34 | 7.46 | 16.3 | Van Hook (1971) |
| 18 | *Pteronemobius fasciatus*: Gryllidae (omnivorous) | Grassland, Tennessee | 1.28 | 3.97 | 9.57 | Van Hook (1971) |
| 19 | *Philaenus spumarius*: | Old-field, Michigan | 0.072 | 0.875 | 1.51 | Wiegert (1964) |
| 20 | Homoptera | (alfalfa) | 16.0 | 38.6 | 100 | Wiegert (1964) |
| 21 | *Neophilaenus lineatus*: Homoptera | Grass heath, Exeter, England | 0.348 | 0.945 | 2.27 | Hinton (1971) |
| 22 | *Proklesia* sp.: Homoptera | Salt marsh, Georgia | 70 | 275 | 413 | Wiegert & Evans (1967) |
| 23 | *Leptopterna dolabrata*: Heteroptera (5 years) | Grassland, Silwood Park, England | 0.103 | 0.187 | 0.593 | McNeill (1971) |
| 24 | *Philaenus spumarius*: Homoptera | Upper Seeds, England | 1.18 | 3.15 | 7.62 | J. B. Whittaker (personal communication) |
| 25 | *Chimabacche fagella*: Lepidoptera | Beech forest, West Germany | 0.43 | {1.49<br>2.57 | 7.40 | Winter (1971) |
| 26 | *Phragmataecia castaneae*: Lepidoptera (2-year life cycle) | Reed bed, West Germany | 2.52 | 3.44 | 13.7 | Pruscha (1973) |
| 27 | *Cicadella viridus*: Homoptera (2 years) | Grassland, Poland | 5.48 | | 64.8 | Andrzejewska (1967) |

The following regression lines have been calculated from Figs. 3.12–3.14.

### (*a*) *For homoiotherms*

$$\text{Log } P = 0.938 \cdot \log R - 1.524, \quad \text{or} \quad \log R = 1.016 \cdot \log P + 1.682. \qquad (3.2)$$

The regression is significant ($r = 0.976$ with $P < 0.001$***) with a standard error of $\log P = 0.135$ and 0.139 in $\log R_i = 0$ and 2.

$$\text{Log } P = 0.852 \cdot \log C - 1.41, \quad \text{or} \quad \log C = 1.006 \cdot \log P + 1.847. \qquad (3.3)$$

The regression is significant ($r = 0.926$ with $0.001 < P < 0.01$**) with a standard error of $\log P = 0.237$ in $\log C_i = 0$ and 2.

$$\text{Log } A = 0.93 \cdot \log C + 0.063, \quad \text{or} \quad \log C = 1.017 \cdot \log A + 0.097. \qquad (3.4)$$

The regression is significant ($r = 0.973$ with $P < 0.001$***) with a standard error of $\log P = 0.151$ in $\log C_i = 0$ and 2.

### (*b*) *For poikilotherms*

$$\text{Log } P = 0.922 \cdot \log R - 0.008, \quad \text{or} \quad \log R = 0.944 \cdot \log P + 0.321. \qquad (3.5)$$

The regression is significant ($r = 0.933$ with $P < 0.001$***) with a standard error of $\log P = 0.266$ and 0.277 in $\log R_i = 0$ and 1.

$$\text{Log } P = 1.048 \cdot \log C - 1.027, \quad \text{or} \quad \log C = 0.916 \cdot \log P + 1.066. \qquad (3.6)$$

The regression is significant ($r = 0.980$ with $P < 0.001$***) with a standard error of $\log P = 0.228$ and 0.222 in $\log C_i = 0$ and 2.

$$\text{Log } A = 1.032 \cdot \log C - 0.547, \quad \text{or} \quad \log C = 0.938 \cdot \log A + 0.613. \qquad (3.7)$$

The regression is significant ($r = 0.984$ with $P < 0.001$***) with a standard error of $\log A = 0.189$ and 0.185 in $\log C_i = 0$ and 2.

Table 3.15 shows the results of an analysis (using the *t*-test) to determine whether the separate regression lines drawn for homoiotherms and poikilotherms are statistically different. The test was done for two points on the axis of the independent variable and showed highly significant differences for the production versus respiration and production versus consumption graphs. For the assimilation versus consumption regression the two lines are not significantly different and one may be justified in using a single regression:

$$\text{Log } A = 0.996 \cdot \log C - 0.352, \quad \text{or} \quad \log C - 0.932 \cdot \log A + 0.549. \qquad (3.8)$$

The regression is significant ($r = 0.963$ with $P < 0.001$***).

It is interesting to consider the data points that show a high divergence from the regression lines. In the regression of production versus respiration, point 19 (old-field data for *Philaenus*) could be omitted for reasons stated above. Point 26 shows a high production in relation to respiration for a *Phragmataecia castaneae* (Lepidoptera) population in a reed bed. It should be pointed out that the life cycle of this population is longer than a year, the

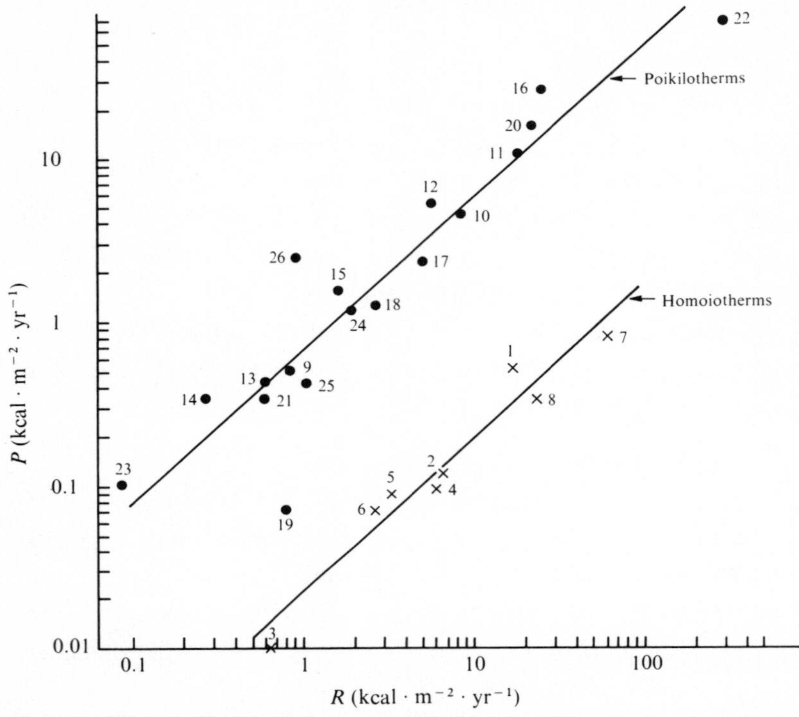


Fig. 3.12. The relationship between annual production ($P$) and annual respiration ($R$) in herbivore populations. The numbers associated with each point correspond to the numbers in the first column of Table 3.14.

Table 3.15. *t-test between the regressions for poikilotherms and homoiotherms in Figs. 3.12–3.14*

| Function | | $t$-test on points $\log X_i =$ | $t$-value | Probability ($P$) |
|---|---|---|---|---|
| $P = f(R)$ | poikilotherms | 2 | 4.467 | $P < 0.001$*** |
| | homoiotherms | 3 | 4.794 | $P < 0.001$*** |
| $P = f(C)$ | poikilotherms | 2 | 2.361 | $0.02 < P < 0.05$* |
| | homoiotherms | 4 | 3.600 | $0.001 < P < 0.01$** |
| $A = f(C)$ | poikilotherms | 2 | 1.687 | $0.1 < P < 0.2$ |
| | homoiotherms | 4 | 0.858 | $P = 0.4$ |

The $t$-test is performed for two values on the independent variable axis. d.f. = 21 on all tests.

species overwintering as larvae. Furthermore Pruscha (1971) reports that the respiration values have been corrected for activity. As this lepidopteran burrows into the stem of the reed, its habitat may protect it from large heat losses to the environment during activity. The *Melanoplus sanguinipes*

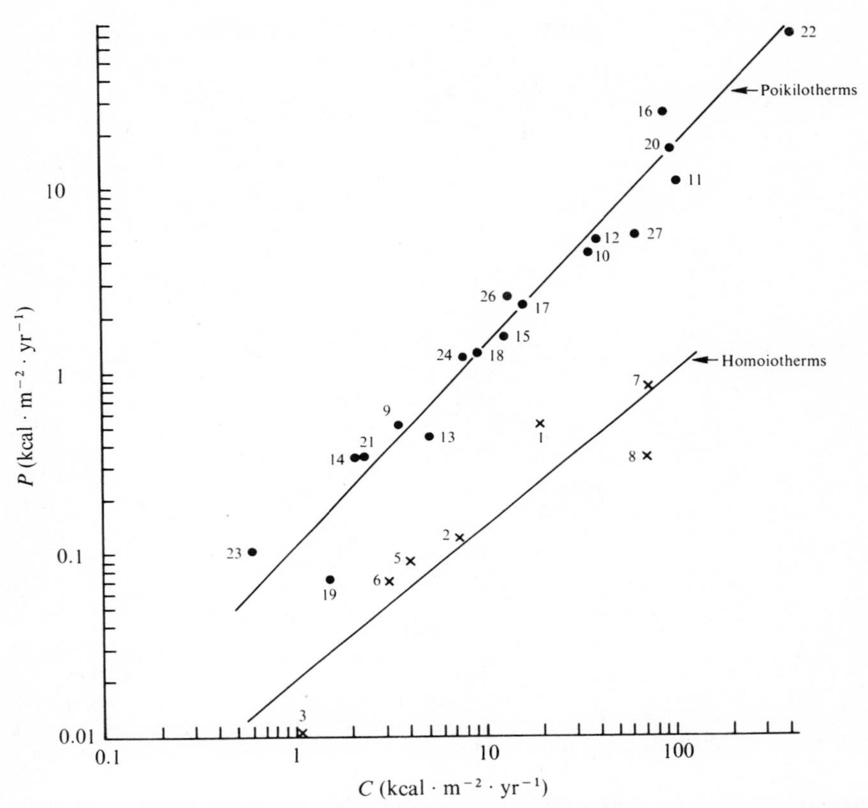


Fig. 3.13. The relationship between annual consumption (*C*) and annual production (*P*) in herbivore populations. The numbers associated with each point correspond to the numbers in the first column of Table 3.14.

population (point 16) referred to by Van Hook (1971) also appears to be highly productive with comparatively small respiration losses. As the data for this grasshopper population have been extracted from a table and one figure, the interpretation of the values given may be wrong. One possible source of error is that Van Hook used mostly adult animals for estimating the relationship between weight and oxygen consumption, and the small animals that have a high oxygen consumption in relation to weight were especially poorly represented.

In Fig. 3.14 the elephant population (point 8) shows a high divergence from other homoiotherms, with a very low assimilation relative to consumption. If this point is disregarded, the other points lie on a separate regression for homoiotherms, which differs even further from the poikilotherm regression if the *Philaenus* population data are also excluded. It

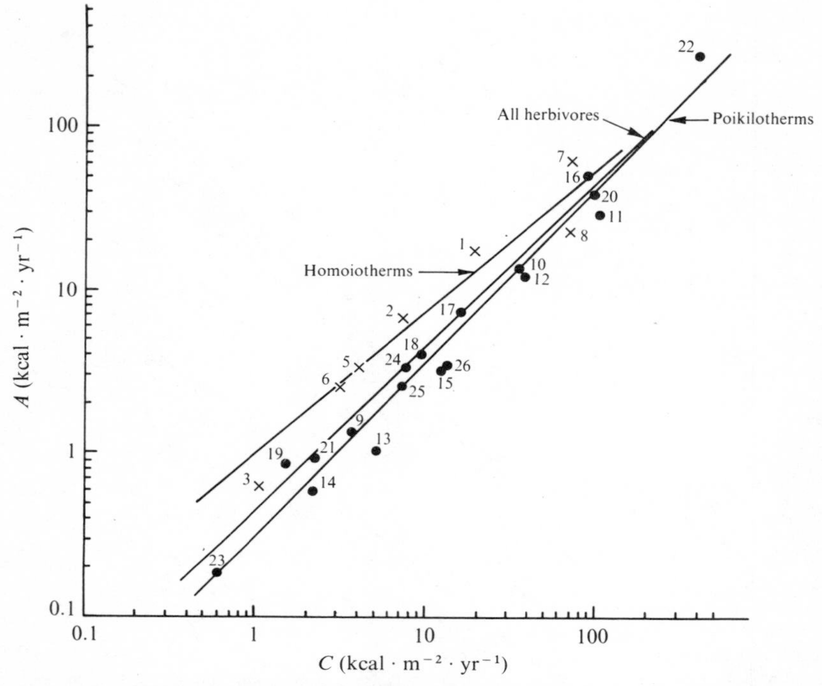


Fig. 3.14. The relationship between annual consumption ($C$) and annual assimilation ($A$) in herbivore populations. The numbers associated with each point correspond to the numbers in the first column of Table 3.14.

appears that elephants in particular are wasteful feeders with low assimilation efficiency. The regression without the elephant population would be:

$$\text{Log } A = 1.064 \cdot \log C - 0.273.$$

This regression is highly significant ($r = 0.997$, $P < 0.001$***).

These same divergences also hold for Fig. 3.13. Some reservation must be made for the *Melanoplus* population (point 16).

### *Review of efficiency values*

Table 3.16 summarizes efficiency values estimated for both laboratory and field populations from grassland ecosystems. The object was to examine mean efficiencies for groups filling different niches in the grassland ecosystem. Thus species were divided into groups comprising larger herbivores; grazing herbivores (such as grasshoppers, heteropterans and tettigonids); sucking herbivores (homopterans); and predominantly laboratory populations of caterpillars and beetles. The standard error in each group is about 5–10%

Table 3.16. *Efficiency values of herbivore populations in grassland ecosystems*

| Species | Investigation area | Efficiencies (%) A/C | P/C | P/A | References |
|---|---|---|---|---|---|
| *Microtus pennsylvanicus* (meadow vole) | Old-field, Michigan, USA | 90.0 | 2.66 | 2.95 | Golley (1960) |
| *Peromyscus polionotus* (old-field mouse) | Old-field, Georgia, USA | 86.0 | 1.62 | 1.8 | Odum *et al.* (1962) |
| *Peromyscus* sp. (deer mouse) | Old-field, Michigan, USA | 60.0 | 1.0 | 1.6 | Wiegert & Evans (1967) |
| *Clethrionomys glareolus* (bank vole) | Beech forest, Poland | 82.9 | 2.2 | 2.65 | Grodzinski *et al.* (1970) |
| *Apodemus flavicollis* (field mouse) | Beech forest, Poland | 86.1 | 2.27 | 2.64 | Grodziński *et al.* (1970) |
| *Microtus arvalis* (common vole) | Laboratory (30 days dev.) | 92.4 | 4.74 | 5.36 | Drozdz *et al.* (1972) |
| *Odocoileus virginiana* (white-tail deer) | Grassland, USA | 83.0 | | | Wiegert & Evans (1967) |
| *Adenota kob thomasi* (Uganda kob) | Grassland, Uganda | 84.3 | 1.09 | 1.29 | Buechner & Golley (1967) |
| *Loxodonta africana* (African elephant) | Grassland, Uganda | 32.5 | 0.47 | 1.46 | Wiegert & Evans (1967) |
| Average for homoiotherms (±standard error) | | 77.5 ±6.4 | 2.0 ±0.46 | 2.46 ±0.46 | |
| *Orchelimum fidicinium*: Orthoptera | Salt marsh, Georgia, USA | 27.4 | 10.0 | 36.7 | Smalley (1960) |
| *Chorthippus parallelus*: Orthoptera | Meadow, Finland (over 4 years) | 34.5 | 15.0 | 42.5 | Gyllenberg (1970) |
| *Chorthippus parallelus*: Orthoptera | Grassland, England | 27.8 | 15.8 | 57.2 | Quasrawi (1966) |
| *Chorthippus montanus*: Orthoptera | Laboratory (37 days dev.) | 63.0 | 26.0 | 41.0 | Cłodny (1969) |
| *Chorthippus dorsatus*: Orthoptera | Laboratory (37 days dev.) | 68.0 | 23.0 | 34.0 | Clhodny (1969) |
| Grasshoppers | Old-field, Michigan, USA (2 years) | 36.7 | 13.7 | 37.5 | Wiegert (1965) |
| | Alfalfa field | 36.7 | 12.7 | 34.5 | Wiegert (1965) |
| *Orthochta brachycnemis*: Orthoptera | Savanna, Lamto, Ivory Coast | 20.0 | 8.5 | 42.4 | Gillon (1973) |
| *Encoptolophus sordidus costalis*: Orthoptera | Grassland, Matador Project, Canada | 26.1 | 12.7 | 48.6 | Bailey & Riegert (1973) |
| *Melanoplus sanguinipes*: Orthoptera | Grassland, Tennessee, USA | 57.1 | 29.0 | 50.7 | Van Hook (1971) |
| *Leptopterna dolabrata*: Heteroptera | Grassland, Silwood Park, England (5 years) | 32.5 | 17.8 | 56.4 | McNeill (1971) |
| *Conocephalus fasciatus*: Tettigoniidae | Grassland, Tennessee, USA | 45.8 | 14.4 | 31.4 | Van Hook (1971) |
| Average for grazing invertebrates (±standard error) $N = 20$ | | 37.7 ±3.5 | 16.6 ±1.2 | 45.0 ±1.9 | |
| *Neophilaenus lineatus*: Homoptera | Grass heath, Exeter, England | 41.6 | 15.4 | 36.9 | Hinton (1971) |
| *Philaenus spumarius*: Homoptera | Old-field, Michigan, USA | 58.0 | 4.6 | 8.0 | Wiegert (1964) |
| | Alfalfa field | 38.5 | 15.1 | 39.2 | Wiegert (1964) |
| *Philaenus spumarius*: Homoptera | Upper Seeds, England | 41.3 | 15.5 | 37.6 | J. B. Whittaker (personal communication) |
| *Proklesia* sp.: Homoptera | Salt marsh, Georgia | 66.6 | 16.9 | 25.5 | Wiegert & Evans (1967) |
| *Cicadella viridis*: Homoptera | Laboratory (dev. period) | 47.3 | 13.3 | 28.1 | Andrzejewska (1967) |
| Average for sucking herbivores (±standard error) $N = 6$ | | 48.9 ±4.5 | 13.5 ±1.8 | 29.2 ±4.8 | |
| *Hyphantria cunea*: Lepidoptera | Laboratory (to maturity) | 30.0 | 17.0 | 55.0 | Gere (1966) |

Table 3.16 (*continued*)

| Species | Investigation area | Efficiencies (%) | | | References |
|---|---|---|---|---|---|
| | | $A/C$ | $P/C$ | $P/A$ | |
| *Leptinotarsa decemlineata*: Coleoptera | Laboratory (dev. period) | 45.4 | 27.7 | 61.0 | Chlodny *et al.* (1967) |
| *Tortrix viridiana*: Lepidoptera | Laboratory, (36 days dev.) | 67.5 | 18.5 | 27.4 | Witkowski & Kosior (1974) |
| *Pieris rapae*: Lepidoptera | Laboratory | 68.9 | 28.9 | 42.0 | Nakamura (1965) |
| *Hyalophora cecropia*: Lepidoptera | Laboratory (dev. period) | 36.6 | 19.4 | 53.1 | Schroeder (1971) |
| *Pachysphinx modesta*: Lepidoptera | Laboratory, (dev. period) | 41.4 | 19.0 | 46.0 | Schroeder (1973) |
| *Operophtera brumata*: Lepidoptera | Hazel coppice, Monks Wood, England | 40.5 | 24.0 | 57.3 | Smith (1972) |
| *Hydrionema furcata*: Lepidoptera | Hazel coppice, Monks Wood, England | 41.8 | 20.0 | 38.8 | Smith (1972) |
| *Erannis* sp.: Lepidoptera | Hazel coppice, Monks Wood, England | 43.3 | 27.0 | 71.3 | Smith (1972) |
| *Cosmia trapezina*: Lepidoptera | Hazel coppice, Monks Wood, England | 46.2 | 26.0 | 47.8 | Smith (1972) |
| Average for Lepidoptera and Coleoptera (±standard error) $N = 10$ | | 46.2 ±4.0 | 22.8 ±1.4 | 50.0 ±3.9 | |
| *Pteronemobius fasciatus*: Gryllidae (partly herbivorous) | Grassland, Tennessee, USA | 41.5 | 13.3 | 32.1 | Van Hook (1971) |
| *Phragmataecia castaneae*: Lepidoptera (borer) | Reed bed, Neusiedler See (life cycle 2 years) | 25.1 | 18.4 | 73.3 | Pruscha (1973) |
| Average for all poikilotherms (±standard error) | | 41.9 ±2.3 | 17.7 ±1.0 | 44.6 ±2.1 | |

of the mean. It should be pointed out, however, that the $A/C$ values for Homoptera populations are in most cases calculated from Wiegert's (1964) data, and thus are not independent estimates. Two laboratory populations of grasshoppers reported by Chłodny (1969) which are included in the grazer compartment should probably be placed in the laboratory population group, especially since the assimilation efficiency data given by Chłodny seem to be very high. This may be due to laboratory conditions or to the fact that the investigation period was rather short, which weights the results towards the relatively higher assimilation efficiency of the larval stage. As Gyllenberg (1969, 1970) pointed out, larval assimilation efficiency is also much higher than that of the adults in field populations of grasshoppers, an observation that holds too for the *Melanoplus sanguinipes* population data extrapolated from Van Hook (1971).

A *t*-test between the different herbivore groups is presented in Table 3.17. It is evident that all the homoiotherm efficiency values (although calculated from a small amount of data) are significantly different from corresponding values for poikilotherms. Homoiotherms seem to have high assimilation efficiencies (cf. also Engelmann, 1966), but poor production efficiencies. McNeill & Lawton (1970) concluded that animals that complete their life

Table 3.17. *A t-test between the different categories of grassland herbivores*

| | *t*-values and probability (*P*) for: | | | | | |
|---|---|---|---|---|---|---|
| Categories tested | *A*/*C* | d.f. | *P*/*C* | d.f. | *P*/*A* | d.f. |
| Grazers against suckers | 1.965<br>$0.05 < P < 0.1$ | 20 | 1.433<br>$0.1 < P < 0.2$ | 22 | 3.061<br>$0.001 < P < 0.01$** | 24 |
| Caterpillars against grazers | 1.60<br>$0.1 < P < 0.2$ | 24 | 3.362<br>$0.001 < P < 0.01$** | 26 | 1.153<br>$0.2 < P < 0.3$ | 28 |
| Caterpillars against suckers | 0.448<br>$0.6 < P < 0.7$ | 14 | 4.078<br>$0.001 < P < 0.01$** | 14 | 3.363<br>$0.001 < P < 0.01$** | 14 |
| Homoiotherms against poikilotherms | 5.235<br>$P < 0.001$*** | 41 | 14.87<br>$P < 0.001$*** | 42 | 20.06<br>$P < 0.001$*** | 44 |

cycle in one year have to produce biomass at a much faster rate than homoiotherms with a lifespan of several years which thus have longer to reach maturity. Furthermore, homoiotherms inevitably suffer heat losses, especially during the winter.

Grazing invertebrates seem to be more efficient producers than do suckers, according to the *P*/*A* values. As Wiegert (1964), Hinton (1971) and others have shown, homopterans show virtually no increase in biomass during the adult stage. The data again include the 'old-field' *Philaenus* population, but even when it is excluded the values for homopterans are significantly lower (*t*-value = 4.0, $P < 0.001$***). The *P*/*C* valuesseem to be higher for the caterpillar–beetle group as compared with both suckers and grazers. This difference is partly due to the laboratory conditions under which the caterpillar group was studied, but field estimates by Pruscha (1973) confirm that lepidopterans may be highly productive, even when the adult stage (with no production and only heat losses) is included in the estimates.

Differences in efficiency values during the life cycle of 1-year herbivores are shown in Figs. 3.15 and 3.16. Efficiency values were recalculated in terms of percentages of the efficiency of the whole (active) lifetime. Obviously this procedure is much influenced by the investigation period for each population, but if it is assumed that the investigations are actually performed from the time of hatching of the first-instar larvae to the death of the last adults, the data should be comparable. On average, transfer to the adult stage occurs when 70% of the life cycle has elapsed in Fig. 3.15 (for *P*/*A* values) and when 80% of the life cycle has passed in Fig. 3.16 (for *A*/*C* values). It should be observed that Fig. 3.16 is extrapolated from several years' data on only two field populations (one grasshopper; Gyllenberg, 1969, 1970; and one heteropteran: McNeill, 1971), whereas Fig. 3.15 is based on five independently investigated field populations of Orthoptera, Heteroptera and Homoptera.

Although the standard error of the points in Fig. 3.15 is quite high (7–20% of the mean during the larval period), some general trends may be observed. Thus it appears that the later larval stages are the most efficient producers

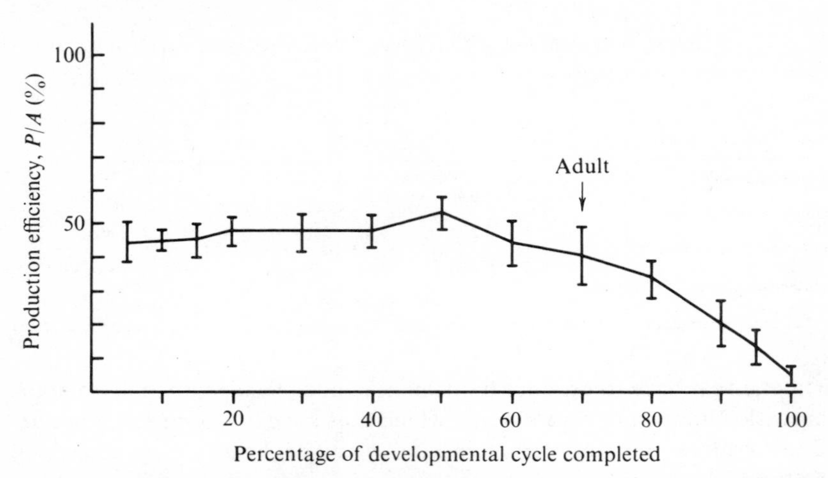


Fig. 3.15. Changes in production efficiency (in %) during the life cycle of 1-year invertebrate herbivores. The data are from five populations studied by Gyllenberg (1969; 2 years' data), Hinton (1971), McNeill (1971; 2 years' data), Smalley (1960) and Wiegert (1965); $N = 7$. Variability is given as ±S.E.

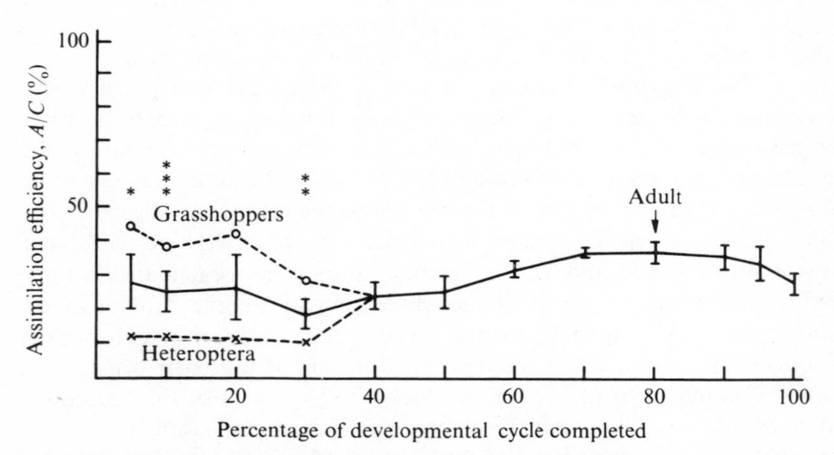


Fig. 3.16. Changes in assimilation efficiency (in %) during the life cycle of 1-year invertebrate herbivores. The data are from two populations studied by Gyllenberg (1969 and unpublished data) and McNeill (1971); $N = 6$. The asterisks indicate a significant difference between the two populations of three years (*, $P < 0.05$; **, $P < 0.01$; ***, $P < 0.001$). Variability is given as ±S.E.

and also accumulate high amounts of fat (see e.g. Gyllenberg, 1969; Hinton, 1971). The production efficiency drops drastically towards the end of the life cycle. Although female production of eggs may be high during the adult period, the males do not normally produce any biomass, and reproduction by females does not compensate for the increasing costs of maintenance in the population.

The same general trend seems to hold also for $A/C$ efficiency during the life cycle (Fig. 3.16), though the decrease in the adult stage is less pronounced. Furthermore there is a significant difference in the efficiency of the first larval stages between the two populations studied, the *Chorthippus* population having the greater efficiency. As Gyllenberg (1969) showed, this is due to the fact that young *Chorthippus* individuals have a strong preference for juicy plant species which they can assimilate efficiently.

As Slobodkin (1960) pointed out, the $P/C$ values give information about the potential yield of a species to predators. However, the actual amount removed by predators has been more or less neglected in the investigations. Theoretical speculations of potential yield reported by Wiegert (1964) and Hinton (1971) have little value as estimates of ecological efficiency, since the actual amount removed may be very small. In a Finnish meadow the amount of grasshoppers (Tettigoniidae), Asilidae and Araneae consumed by predators was estimated as only 0.5% of the amount of the energy assimilated by the grasshoppers, or 4% of their production. Other investigations on relationships between trophic levels also report a low ecological efficiency for predators on herbivores. Menhinick (1967) estimated ecological efficiency to be about 7% in a *Sericea lespedeza* stand, and Van Hook (1971) in a study of a Tennessee grassland concluded that although production efficiency ($P/C$) was 25.8%, only 4.3% was actually removed by predators. Hinton (1971) estimated the maximum potential yield as somewhere between 10 and 13%.

Slobodkin (1960) estimated that $P/C$ efficiency was between 4 and 13%, and Wiegert (1965) concluded that for populations in a steady state this efficiency cannot exceed 15–20% for any length of time. This seems to hold for all the field populations of grasslands, but laboratory populations may reach higher values (mainly because they are concerned chiefly with the highly productive larval instars – see Table 3.16).

### *Sensitivity analysis*

It is astonishing that most investigations on herbivore energetics are confined to measurements of annual production, assimilation and consumption, and the interpretation of efficiency values. As herbivore populations (especially the invertebrates) are found not to be limited by the food supply (see Engelmann, 1966; Wiegert & Evans, 1967; McNeill, 1971; Bailey & Riegert, 1972), the question immediately arises as to why they do not consume more. But as Mulkern (1967), Murdoch (1966) and White (1974) pointed out, most herbivores are selective; thus the herbivore pressure on certain plant species, or on certain localities, may reach values of 50% (White, 1974). Bernays & Chapman (1970) concluded that grasshoppers may change their food preference according to the nature of the habitat, although a homogenous food supply can interfere with normal development (Pickford, 1960, 1962).

One way to investigate the complete dynamics of a population in terms of energetics is by using simulation models. If the model includes all the factors that regulate the population, it should be capable of indicating which factor is most sensitive to fluctuations in the population. Models are based on theoretical equations, which when run on a computer do not by themselves give any conclusive answers but can indicate the controlling variable. Modelling procedures have been used by, for example, Gyllenberg (1970, 1974) and McNeill (1973).

The model constructed by Gyllenberg (1974) is derived *a priori* from statistical regression analyses between abiotic factors and energy flow parameters. These regressions and the model outputs have been tested for laboratory populations by regulating each of the weather factors within the steady-state range of the population. The energy flow derived from laboratory studies does not show significant deviations from the predictions of the model. However, certain abiotic factors, like rainfall, seem to act as catastrophic factors, which cannot bep redicted for populations in a steady state. The model does show that at northern latitudes the abiotic factors are of maximum importance in regulating the population density of invertebrates, although both density-dependent regulation and food limitation may occur in high-density populations.

McNeill's (1973) careful analysis of a *Leptopterna* population shows that the regulation of density was mediated by competition for a scarce, high-quality food source (flower and seed heads), but that actual variations between years were due to effects of weather on the survival of the later nymphs. Thus it seems that this population is actually food limited but that perturbations from year to year are caused by abiotic factors.

Another way to examine the sensitivity of factors affecting energy flow is to manipulate the system in field conditions, either by introducing artificial conditions or by paralysing a certain component of the community by the use of insecticides. Investigations of this type are referred to by Andrzejewska *et al.* (1967), Andrzejewska (1967) and Andrzejewska & Wójcik (1970), and have given promising results.

## 5. The role of herbivores in grassland ecosystems

The phytophagous community is well adapted to utilizing the plants on which it depends. In a natural, undisturbed grassland community, herbivores inhabit and feed in all vegetation levels from the roots to the aboveground parts. The total biomass of herbivores is relatively high, but in spite of this they do not cause major disturbances in the development of meadow vegetation cover. However, any environmental change which simplifies the herbivore community and reduces the number of species, thus increasing the number of individuals with the same food requirements, creates the possibility that certain plant species may be selectively destroyed. In such cases even a lower total density and biomass of herbivores can cause considerable losses of meadow vegetation. Meadow vegetation contains a variety of species and can support a biomass of invertebrate phytophages that is greater than that of the cows which would convert the meadow to pasture and reduce the aboveground plant biomass to 20% of its original value.

In the food web of grassland ecosystems herbivorous animals are obviously totally dependent on the plants which are their only source of food and at the same time the major component of their habitat. The links between phytophages and other parts of the trophic net are much less marked. Their character is of the prey–predator type, and in the case of the saprophagous level the herbivores are responsible for providing it with part of the total plant production. Throughout the whole season plant material is transferred by herbivores to saprophages in the form of faeces and fresh plant remains which fall to the ground while neighbouring parts of leaves or plants are grazed. The amount of organic plant material, its quality and the time when it becomes available to saprophages are very important influences on the development of the saprophage community (Żyromska-Rudzka, 1976).

The type of grassland environment and its management determine the nature of the herbivore community and influence both grazing and decomposition processes. Such relationships are shown in three differently used and managed grassland environments: pasture (Fig. 17*a*), fertilized and mowed meadow (Fig. 17*b*), and unused, reserve meadow (Fig. 17*c*).

In the pasture the type of plant production is determined by the grazing animals. Other herbivorous fauna is markedly reduced. Cows and sheep consume about 80% of the plant production and about 56% is returned to the habitat in the form of faeces which are acted upon by a well-developed community of coprophages. Under moderate grazing, when the soil is rich in nutrients, the production of the pasture is increased, while overgrazing decreases the production of both aboveground and belowground plant parts (Fig. 3.17*a*).

In the meadow that is fertilized, and harvested for hay two to three times a season, about 90% of aboveground plant production is removed. Only a small part of the primary production (including herbivore faeces) is transferred to saprophages. Throughout the season fresh plant biomass covers the meadow and attracts herbivores; their high densities caused marked decreases in the growth rate and production of the plants (Fig. 3.17*b*).

In the unexploited, reserve meadow practically all the plant production remains in the environment and is transferred to a well-developed saprophagous community, only a small part being utilized by the highly diversified herbivore community. The phytophagous fauna affects the structure of

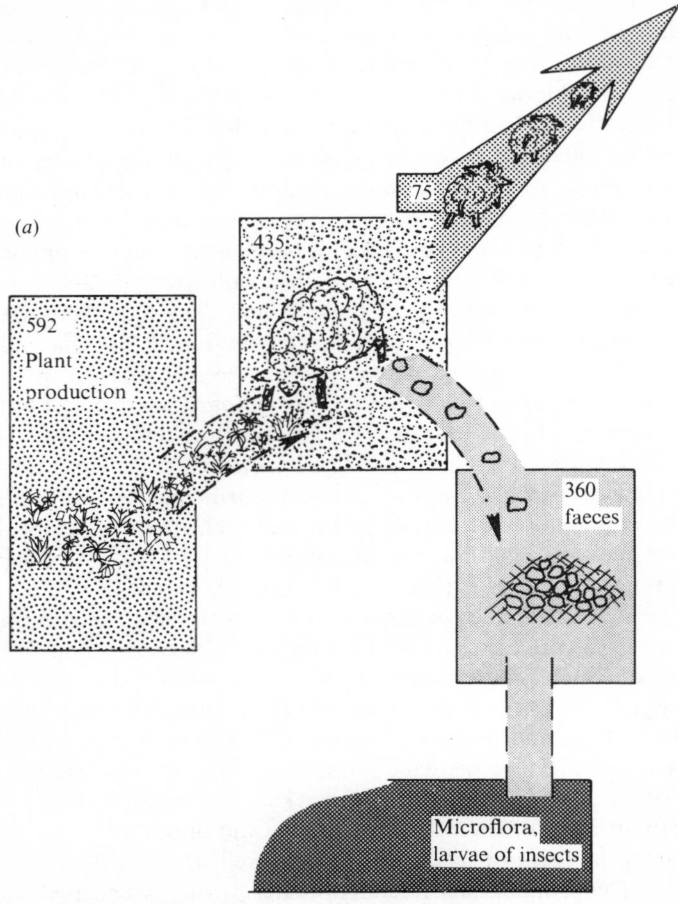


Fig. 3.17. A scheme to show production and utiization of plant biomass in the grassland ecosystem: (*a*) sheep pasture; (*b*) meadow cultivated for hay (fertilized and mowed); (*c*) unused, reserve meadow. Units are g dry wt $\cdot$ m$^{-2}$.

the whole ecosystem by slowing the growth rates of the plant parts on which it feeds, and this in turn delays the time of occurrence of their maximum biomass. This effect is compensated for to some extent by a prolongation of the period of plant growth (Fig. 3.17*c*).

The importance of herbivorous animals in the transformation of organic matter and their effect on the speed of nutrient cycling differs not only between grassland ecosystems but can also vary from season to season as the number and biomass of phytophagous animals changes.

(*b*)

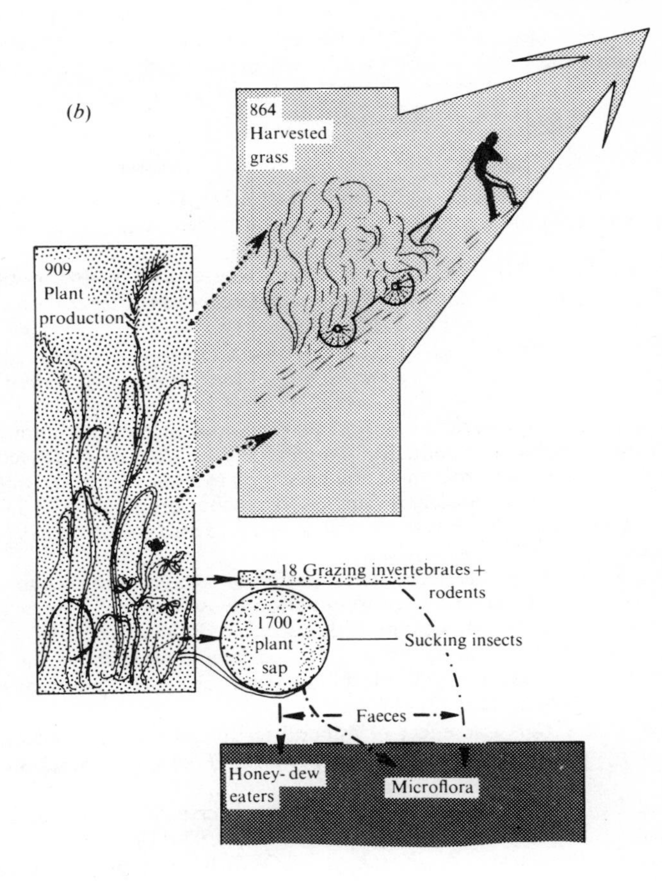
864
Harvested
grass
909
Plant
production
~ 18 Grazing invertebrates +
rodents
1700
plant
sap
Sucking insects
Faeces
Honey- dew
eaters
Microflora

(*c*)

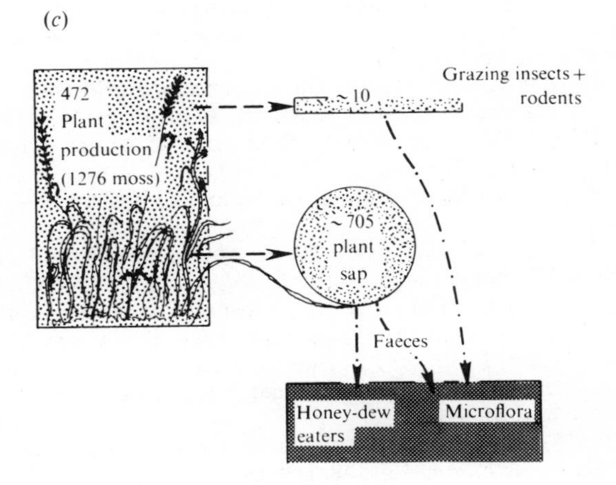
472
Plant
production
(1276 moss)
~ 10
Grazing insects +
rodents
~ 705
plant
sap
Faeces
Honey-dew
eaters
Microflora

The expense of the preparation of the typescript for this chapter was met by the Second Division of the Polish Academy of Sciences, to which I express my sincere gratitude.

## References

Adamczewska-Andrzejewska, K. & Nabagło, L. (1977). Demographic parameters and variations in numbers of the common vole. *Acta Theriologica*, **22**, 431–57.

Anderson, N. L. (1970). The assessment of range losses caused by grasshoppers. In *Proceedings of the International Study Conference on Current and Future Problems of Acridology*, pp. 173–9. London.

Anderson, N. L. & Wright, J. C. (1952). *Grasshopper investigations on Montana range lands. Technical Bulletin of the Montana Agricultural Experiment Station No. 486.*

Andrzejewska, L. (1966). Stratification and its dynamics in meadow communities of Auchenorrhyncha (Homoptera). *Ekologia Polska, Series A*, **31**, 685–715.

Andrzejewska, L. (1967) Estimation of the effects of feeding of the sucking insect *Cicadella viridis* L. (Homoptera – Auchenorrhyncha) on plants. In: *Secondary productivity of terrestrial ecosystems*, vol. 2, ed. K. Petrusewicz, pp. 791–805. Polish Academy of Sciences, Warsaw.

Andrzejewska, L. (1971). Productivity investigation of two types of meadows in the Vistula Valley. VI. Production and population density of leaf-hopper (Homoptera – Auchenorrhyncha) communities. *Ekologia Polska*, **19**, 151–72.

Andrzejewska, L. (1974). Analysis of a sheep pasture ecosystem in the Pieniny mountains (the Carpathians). V. Herbivores and their effect on plant production. *Ekologia Polska*, **222**, 527–34.

Andrzejewska, L. (1976*a*). The effect of mineral fertilization of a meadow on the Auchenorrhyncha (Homoptera) fauna. *Polish Ecological Studies*, **2** (4), 111–28.

Andrzejewska, L. (1976*b*). The influence of mineral fertilization on the meadow phytophagous fauna. *Polish Ecological Studies*, **2**(4), 93–110.

Andrzejewska, L., Breymeyer, A., Kajak, A. & Wójcik, Z. (1967). Experimental studies on trophic relationships of terrestrial invertebrates. In *Secondary productivity of terrestrial ecosystems*, vol. 2, ed. K. Petrusewicz, pp. 477–95. Polish Academy of Sciences, Warsaw.

Andrzejewska, L. & Wójcik, Z. (1970). The influence of Acridoidea on the primary production of a meadow (field experiment). *Ekologia Polska*, **18**, 89–109.

Andrzejewski, R. (1975). Supplementary food and the winter dynamics of bank vole populations. *Acta Theriologica*, **20**, 23–40.

Andrzejewski, R. (1977). Population as an ecological system. *Wiadomości Ekologiczne*, **23**, 3–33.

Andrzejewski, R. & Mazurkiewicz, M. (1976). Abundance of food supply and size of the bank vole's home range. *Acta Theriologica*, **21**, 237–53.

Babińska, J. (1972). Estimation of rodent consumption in a meadow ecosystem belonging to the community of Molinietalia order. *Ekologia Polska*, **20**, 747–61.

Bailey, C. G. & Riegert, P. W. (1973). Energy dynamics of *Encoptolophus sordidus costalis* (Scudder) (Orthoptera: Acrididae) in a grassland ecosystem. *Canadian Journal of Zoology*, **51**, 91–100.

Barnes, O. L. (1955). Effects of food plants on the lesser migratory grasshopper. *Journal of Economic Entomology*, **48**, 119–24.

Batzli, G. O. (1975). The role of small mammals in arctic ecosystems. In *Small mammals: their productivity and population dynamics*, ed. F. B. Golley, K. Petrusewicz & L. Ryszkowski, *International Biological Programme 5*, pp. 243–68. Cambridge University Press, London.

Benz, G. von (1974). Negative Rückkoppelung durch Raum- und Nahrungskonkurrenz sowie zyklische Veranderung der Nahrungsgrundlage als Regelprinzip in der Populationsdynamik des Grauen Lärchenwicklers, *Zeiraphera diniana* (Guenee) (Lep., Tortricidae). *Zeitschrift für Angewandte Entomologie*, **76**, 196–228.

Bernays, E. A. & Chapman, R. F. (1970). Food selection by *Chortippus parallelus* (Zetterstedt) (Orthoptera: Acrididae) in the field. *Journal of Animal Ecology*, **39**, 383–94.

Bernays, E. A. & Chapman, R. F. (1974). The regulation of food intake by Acridids. In *Experimental analysis of insect behaviour*, ed. B. Browne, pp. 47–59. Springer-Verlag, Berlin.

Bernays, E. A., Chapman, R. F., Horsey, J. & Leather, E. M. (1974). The inhibitory effect of seedling grasses on feeding and survival of Acridids (Orthoptera). *Bulletin of Entomological Research*, **64**, 413–20.

Bertalanffy, L. von (1957). Quantitative laws in metabolism and growth. *Quarterly Review of Biology*, **32**, 217–31.

Bliss, L. C. (1975). Truelove Lowland – a high arctic ecosystem. In *Energy flow: its biological dimensions. A summary of the IBP in Canada 1964–1974*, ed. T. W. M. Cameron & L. W. Billingsley, pp. 51–85. Royal Society of Canada, Ottawa.

Breymeyer, A. (1974). Analysis of a sheep pasture in the Pieniny mountains (the Carpathians). XI. The role of coprophagous beetles (Coleoptera, Scarabaeidae) in the utilization of sheep dung. *Ekologia Polska*, **22**, 617–34.

Breymeyer, A. (1977). Analysis of the trophic structure of some grassland ecosystems. *Polish Ecological Studies*, in press.

Breymeyer, A. & Kajak, A. (1976). Drawing models of two grassland ecosystems: a mown meadow and a pasture (IBP meeting, Dziekanów Leśny near Warsaw, 6–12 July 1973). *Polish Ecological Studies* **2** (2), 41–50.

Breymeyer, A., Olechowicz, E. & Jakubczyk, H. (1975). Influence of coprophagous arthropods on microorganisms in sheep faeces – laboratory investigation. *Bulletin de l'Académie Polonaise des Sciences, Classe II*, **23**, 257–62.

Buechner, H. K. & Golley, F. B. (1967). Preliminary estimation of energy flow in Uganda kob (*Adenota kob thomasi* Neumann). In *Secondary productivity of terrestrial Ecosystems*, vol. 1, ed. K. Petrusewicz, pp. 243–54. Polish Academy of Sciences, Warsaw.

Bullen, F. T. (1970). A review of the assessment of crop losses caused by locust and grasshoppers. *Proceedings of the International Study Conference on Current and Future Problems of Acridology*, pp. 163–71. London.

Bunnell, F. L., MacLean, S. F. & Brown, J., Jr (1975). Barrow, Alaska USA. In *Structure and function of tundra ecosystems*, ed. T. Rosswall & O. W. Heal, *Ecological Bulletins 20*, pp. 73–123. Swedish Natural Science Research Council, Stockholm.

Caswell, H., Reed, F., Stephenson, S. N. & Werner, P. A. (1973). Photosynthetic pathways and selective herbivory: a hypothesis. *American Naturalist*, **107**, 465–80.

Chapman, R. F. (1974). The chemical inhibition of feeding by phytophagous insects: a review. *Bulletin of Entomological Research*, **64**, 339–63.

Chłodny, J. (1969). The energetics of larval development of two species of grasshoppers from the genus *Chorthippus* Fieb. *Ekologia Polska, Series A*, **17**, 391–407.

Chłodny, J., Gromadzka, J. & Trojan, P. (1967). Energetic budget of development of the Colorado beetle – *Leptinotarsa decemlineata* Say (Coleoptera, Chrysomelidae). *Bulletin de l'Académie Polonaise des Sciences, classe II*, **15**, 743–7.

Coupland, R. T., Willard, J. R., Ripley, E. A. & Randell, R. L. (1975). The Matador Project. In *Energy flow: its biological dimensions. A summary of the IBP in Canada, 1964–1974*, eds. T. W. M. Cameron & L. W. Billingsley, pp. 19–50. Royal Society of Canada, Ottawa.

Cowan, F. T. & Shipman, H. J. (1947). Quantities of food consumed by Mormon crickets. *Journal of Economic Entomology*, **40**, 825–8.

Crossley, D. A. Jr (1966). Radioisotope measurement of food consumption by a leaf beetle species, *Chrysomela knabi* Brown. *Ecology*, **47**, 1–8.

Czarnowski, M. (1975). Photosynthesis and solar energy conversion. In *Polish participation in IBP, 1964–1973*, pp. 175–81. Polish Academy of Sciences/Polish National Committee for IBP, Warsaw. (Mimeograph.)

Davey, P. M. (1954). Quantities of food eaten by the desert locust *Schistocerca gregaria* (Forsk.) in relation to growth. *Bulletin of Entomological Research*, **45**, 539–51.

Delvi, M. R. & Pandian, T. J. (1971). Ecophysiological studies on the utilization of food in the paddy-field grasshopper *Oxya velox*. *Oecologia*, **8**, 267–75.

Drożdż, A., Górecki, A., Grodziński, W. & Pelikan, J. (1971). Bioenergetics of water voles (*Arvicola terrestris* L.) from Southern Moravia. (In Proceedings of the IBP meeting on secondary productivity in small mammal populations.) *Annales Zoologici Fennici*, **8**, 97–103.

Drożdż, A., Górecki, A. & Sawicka-Kapusta, K. (1972). Bioenergetics of growth in common voles. *Acta Theriologica*, **17**, 245–57.

Duffey, E., Morris, M. G., Sheail, J., Ward, Lena K., Wells, D. A. & Wells, T. C. E. (1974). *Grassland ecology and wildlife management*. Chapman & Hall, London.

Ellis, J. E. & French, N. R. (1973). Grassland consumers: structure and dynamics. In *Analysis of structure, function, and utilization of grassland ecosystems*, vol. 2, pp. 160–205. Colorado State University, Fort Collins, Colorado.

Ellison, L. (1960). Influence of grazing on plant succession of range lands. *Botanical Review*, **26**, 1–78.

Emden, H. F., Eastop, V. F., Hugens, R. D. & Way, M. J. (1969). The ecology of *Myzus persicae*. *Annual Review of Entomology*, **14**, 197–270.

Engelmann, M. D. (1966). Energetics, terrestrial field studies and animal productivity. *Advances in Ecological Research*, **3**, 73–115.

Feeny, P. (1975). Biochemical coevolution between plants and their insect herbivory. In *Coevolution of animals and plants*, ed. L. E. Gilbert & P. H. Raven, pp. 3–19. University of Texas Press, Austin & London.

Fewkes, D. W. (1967). The control of sugar cane froghoppers. *Proceedings of the British West Indies Sugar Technologists*, **6**, 2–33.

Fischer, Z. & Andrzejewska, L. (1977). Assimilation of main food components and respiration of larvae of *Arctia caja* L. (Lepidoptera). I. Experiment with food of an average nitrogen content. *Ekologia Polska* **25**, 455–65.

French, N. R., Stoddart, D. M. & Bobek, B. (1975). Patterns of demography in small mammal populations. In *Small mammals: their productivity and population dynamics*, ed. F. B. Golley, K. Petrusewicz & L. Ryszkowski, *IBP Synthesis Series 5*, pp. 73–102. Cambridge University Press, London.

Funke, W. (1972). Energieumsatz von Tierpopulation in Land-Ökosystemen. *Verhandlungen der Deutschen Zoologischen Gesellschaft* (Jahresversammlung), **65**, 95–105.

Furuno, T. & Ohmura, T. (1971). Relations between feeding and frass of leaf-eating insects, especially pine caterpillar (*Dendrolimus spectabilis* Butler) on the genus *Pinus*. *Bulletin of the Kyoto University Forests*, **42**, 27–36.

Gaare, E. & Skogland, T. (1975). Wild reindeer food habits and range use at Hardangervidda. In *Ecological Studies 17, Fennoscandian tundra ecosystems*, part 2, *Animals and systems analysis*, ed. F. E. Wielgolaski, pp. 195–215. Springer-Verlag, Berlin.

Ganwere, S. K. (1961). A monograph on food selection in Orthoptera. *Transactions of the American Entomological Society*, **87**, 67–230.

Gębczyńska, Z. & Gębczyński, M. (1971). Insulating properties of the nest and social temperature regulation in *Clethrionomys glareolus* (Schreber). (In Proceedings of the IBP meeting on secondary productivity in small mammal populations.) *Annales Zoologici Fennici*, **8**, 104–8.

Gere, G. (1956). Investigations concerning the energy turnover of *Hyphantria cunea* Drury caterpillars. *Opuscula Zoologica, Instituti Zoosystematici Universitatis Budapestensis*, **1**, 29–32.

Gillon, D. (1971). The effect of bush fire on the principal pentatomid bugs (Hemiptera) of an Ivory Coast savanna. In *Proceedings of the Annual Tall Timbers Fire Ecology Conference*, pp. 377–417.

Gillon, Y. (1971). The effect of bush fire on the principal Acridid species of an Ivory Coast savanna. In *Proceedings of the Annual Tall Timbers Fire Ecology Conference*, pp. 419–71.

Gillon, Y. (1973). Bilan énergétique de la population d'*Orthochta brachycnemis* Karsch, principale espèce acridienne de la savane de Lamto (Côte-d'Ivoire). *Annales de l'Université d'Abidjan, Series E, Ecologie*, **6**, 105–25.

Golley, F. B. (1960). Energy dynamics of a food chain of an old-field community. *Ecological Monographs*, **30**, 187–206.

Golley, F. B. (1961). Energy values of ecological materials. *Ecology*, **42**, 581–4.

Golley, F. B. (1973). Impact of small mammals on primary production. In *Ecologica energetics of homeotherms*, ed. J. A. Gessman, pp. 142–7. Utah State University Press, Logan.

Golley, F. B., Ryszkowski, L. & Sokur, J. T. (1975). The role of small mammals in temperate forests, grasslands and cultivated fields. In *Small mammals: their productivity and population dynamics*, ed. F. B. Golley, K. Petrusewicz & L. Ryszkowski, *IBP Synthesis Series 5*, pp. 223–41. Cambridge University Press, London.

Gradwell, G. R. (1975). The effect of defoliators on tree growth. In *The British oak*, pp. 182–93. London.

Green, T. R. & Ryan, C. A. (1972). Wound-induced proteinase inhibitor in plant leaves: a possible defence mechanism against insects. *Science*, **175**, 776–7.

Grodziński, W., Bobek, B., Drożdż, A. & Górecki, A. (1970). Energy flow through small rodent populations in a beech forest. *Energy flow through small mammal populations*, ed. K. Petrusewicz & L. Ryszkowski, pp. 291–8. Polish Academy of Sciences, Warsaw.

Guseva, V. S. (1970). The effect of cultivation of virgin land on grasshopper populations. In *Proceedings of the International Conference on Current and Future Problems of Acridology*, pp. 185–90. London.

Gyllenberg, G. (1969). The energy flow through a *Chorthippus parallelus* (Zett.) (Orthoptera) population on a meadow in Tvärminne, Finland. *Acta Zoologica Fennica*, **123**, 1–74.

Gyllenberg, G. (1970). Energy flow through a simple food chain of a meadow ecosystem in four years. *Annales Zoologici Fennici*, **7**, 283–9.

Gyllenberg, G. (1974). A simulation model for testing the dynamics of a grasshopper population. *Ecology*, **55**, 645–50.

Harley, K. L. S. & Thorsteinson, A. J. (1967). The influence of plant chemicals on the feeding behaviour, development and survival of the two-striped grasshopper, *Melanoplus bivittatus* (Say), Acrididae: Orthoptera. *Canadian Journal of Zoology*, **45**, 305–19.

Haukioja, E. (1974). Measuring consumption in *Eriocrania* (Eriocraniidae, Lep.) miners with reference to interaction between the leaf and the miner. *Reports of the Kevo Subarctic Research Station* **11**, 16–21.

Haukioja, E. & Hakala, T. (1975). Herbivore cycles and periodic outbreaks. Formulation of a general hypothesis. *Reports of the Kevo Subarctic Research Station*, **12**, 1–9.

Haukioja, E. & Koponen, S. (1975). Faunal structure of investigated areas at Kevo, Finland. In *Ecological Studies 17, Fennoscandian tundra ecosystems*, part 2, *Animals and systems analysis*, ed. F. E. Wielgolaski, pp. 19–28. Springer-Verlag, Berlin.

Haukioja, E. & Niemelä, P. (1976). Does birch defend itself actively against herbivores? *Reports of the Kevo Subarctic Research Station*, **13**, 44–7.

Haukioja, E. & Niemelä, P. (1977). Retarded growth of a geometrid larva after mechanical damage to leaves of its host tree. *Annales Zoologici Fennici*, **14**, 48–52.

Hinton, J. M. (1971). Energy flow in a natural population of *Neophilaenus lineatus* (Homoptera). *Oikos*, **22**, 155–71.

Hughes, R. E., Milner, C. & Dale, J. (1964). Selectivity in grazing. In *Grazing in marine and terrestrial environments*, ed. D. J. Crisp, *Symposium of the British Ecological Society*, pp. 189–202. Blackwell, Oxford.

Hurd, L. E., Mellinger, M. V., McNaughton, S. J. & Wolf, L. L. (1971). Stability and diversity at three trophic levels in terrestrial successional ecosystems. *Science*, **173**, 1134–6.

Hurd, L. E., Mellinger, M. V., Wolf, L. L. & McNaughton, S. J. (1972). Relative consumer species diversity with respect to producer diversity and net productivity. *Science*, **176**, 554–6.

Ingram, W. R. (1975). Improving control of the vegetable army-worm. *Pest Articles and News Summaries*, **21**, 162–7.

Kajak, A., Breymeyer, A. & Pętal, J. (1971). Productivity investigation of two types of meadows in the Vistula Valley. XI. Predatory arthropods. *Ekologia Polska*, **19**, 223–33.

Kalinowska, A. (1977). Rola populacji *Succinea putris* (L.) w krążeniu fosforu na dwu różnie użytkowanych łakach. *Polish Ecological Studies*, in press.

Kallio, P. (1975). Kevo, Finland. In *Structure and function of tundra ecosystem*, ed. T. Rosswall & O. W. Heal, *Ecological Bulletins 20*, pp. 193–223. Stockholm.

Kallio, P. & Lehtonen, J. (1973). Birch forest damage caused by *Oporinia autumnata* (Bkh.) in 1965–66 in Utsjoki, N. Finland. *Reports of the Kevo Subarctic Research Station*, **10**, 55–69.

Kaufman, T. (1965). Biological studies on some Bavarian Acridoidea (Orthoptera), with special references to their feeding habits. *Annals of the Entomological Society of America*, **58**, 791–801.

Ketner, P. (1972). *Primary production of salt-marsh communities on the island of Terschelling in the Netherlands*. Thoben Offset, Nijmegen.

Kleiber, M. (1961). *The fire of life*. Wiley, New York.

Kozlovsky, D. G. (1968). A critical evaluation of the trophic level concept. I. Ecological efficiencies. *Ecology*, **49**, 48–60.

Krebs, Ch. J. & Myers, J. H. (1974). Population cycles in small mammals. *Advances in Ecological Research*, **8**, 267–399.

Krzywiec, D. (1968). Biologia mszyc. In *Kurs Afidologii ogólnej*, ed. K. Berliński, pp. 53–92. Zakład, Narodowy Imienia Ossolińskich, Wrocław-Warsaw-Cracow.

Lamotte, M. (1975). The structure and function of a tropical savannah ecosystem. In *Ecological Studies 11, Tropical ecological systems: analysis and synthesis*, ed. F. B. Golley & E. Medina, pp. 179–222.

Lewis, J. K. (1971). The grassland biome: a synthesis of structure and function. In *Preliminary analysis of structure and function in grasslands*, ed. N. R. French, pp. 317–87. Colorado State University, Fort Collins, Colorado.

McDaniel, B. (1971). The role of invertebrates in the grassland biome. In *Preliminary analysis of structure and function in grasslands*, ed. N. R. French, pp. 267–315. Colorado State University, Fort Collins, Colorado.

Macfadyen, A. (1967). Methods of investigation of productivity of invertebrates in terrestrial ecosystems. In *Secondary productivity of terrestrial ecosystems*, ed. K. Petrusewicz, vol. 2, pp. 383–412. Polish Academy of Sciences, Warsaw.

MacGregor, K. A. (1968). The productivity relations between insects and oak trees. D. Phil. Thesis, Oxford University.

McNeill, S. (1971). The energetics of a population of *Leptopterna dolabrata* (Heteroptera: Miridae). *Journal of Animal Ecology*, **40**, 127–40.

McNeill, S. (1973). The dynamics of a population of *Leptopterna dolabrata* (Heteroptera: Miridae) in relation to its food resources. *Journal of Animal Ecology*, **42**, 495–507.

McNeill, S. (1977). An energy and food quality based model of individual growth and population levels in *Leptopterna dolabrata* (Heteroptera: Miridae). *Journal of Animal Ecology*, in press.

McNeill, S. & Lawton, J. H. (1970). Annual production and respiration in animal populations. *Nature, London*, **225**, 472–4.

Makulec, G. (1971). Productivity investigation of two types of meadows in the Vistula Valley, V. Introductory studies on number and energetics of Orthoptera. *Ekologia Polska*, **19**, 139–50.

Matsumoto, T. (1971). Estimation of population productivity of *Parapleurus alliaceus* Germar. (Orthoptera: Acrididae) on a *Miscanthus sinensis* Anders. grassland. II. Population productivity in terms of dry weight. *Oecologia*, **7**, 16–25.

Menhinick, E. F. (1967). Structure, stability, and energy flow in plants and arthropods in a *Serica lespedeza* stand. *Ecological Monographs*, **37**, 255–72.

Mochnacka-Ławacz, H. (1978). Mineral NPK fertilization and fluctuations of some elements in the meadow vegetation. *Polish Ecological Studies*, **4** (1), in press.

Moore, J. J., Dowding, P. & Healy, B. (1975). Glenamoy, Ireland. In *Structure and function of tundra ecosystems*, ed. T. Rosswall & O. W. Heal, *Ecological Bulletins 20*, pp. 321–43. Swedish Natural Science Research Council, Stockholm.

Morris, M. G. (1967). Differences between the invertebrate faunas of grazed and ungrazed chalk grassland: responses of some phytophagous insects to cessation of grazing. *Journal of Applied Ecology*, **4**, 459–74.

Morris, M. G. (1971). Differences between the invertebrate faunas of grazed and ungrazed chalk grassland. IV. Abundance and diversity of Homoptera – Auchenorrhyncha. *Journal of Applied Ecology*, **8**, 37–52.

Morris, M. G. (1973). The effects of seasonal grazing on the Heteroptera and Auchenorrhyncha (Hemiptera) of chalk grassland. *Journal of Applied Ecology*, **10**, 761–80.

Mulkern, G. B. (1967). Food selection by grasshoppers. *Annual Review of Entomology*, **12**, 59–78.

Mulkern, G. B. (1970). The effects of preferred food plants on distribution and numbers of grasshoppers. *Proceedings of the International Conference on Current and Future problems of Acridology*, pp. 215–18. London.

Murdoch, W. W. (1966). Community structure, population control, and competition – a critique. *American Naturalist*, **100**, 219–26.

Myllymäki, A. (1975) Control of field rodents. In *Small mammals: their productivity and population dynamics*, ed. F. B. Golley, K. Petrusewicz and L. Ryszkowski, *IBP Synthesis Series 5*, pp. 311–38. Cambridge University Press, London.

Myllymäki, A. (1976). Outbreaks and damage by the field vole, *Microtus agrestis* (L.), since World War II in Europe. In *Joint FAO/WHO/EPPO Conference on Rodents of Agricultural and Public Health Concern*, Geneva, 15–18 June, pp. 1–26. European Plant Protection Organization, Geneva.

Nakamura, M. (1965). Bio-economics of some larval populations of pleurostict Scarabaeidae on the flood plain of the River Tamagawa. *Japanese Journal of Ecology*, **15**, 1–18.

Nakano, K. & Monsi, M. (1968). An experimental approach to some quantitative aspects of grazing by silkworms (*Bombyx mori*). *Japanese Journal of Ecology*, **18**, 217–30.

Nowak, E. (1971). Productivity investigation of two types of meadows in the Vistula Valley. IV. Soil macrofauna. *Ekologia Polska*, **19**, 129–37.

Nowak, E. (1976). The effect of fertilization on earthworms and other soil macrofauna. *Polish Ecological Studies*, **2**, 1 95–207.

Odum, E. P. (1971). *Fundamentals of ecology*, 3rd edn. W. B. Saunders, Philadelphia.

Odum, E. P., Connell, C. E. & Davenport, L. B. (1962). Population energy flow of three primary consumer components of old-field ecosystems. *Ecology*, **43**, 88–69.

Odum, H. T. (1957). Trophic structure and productivity of Silver Springs, Florida. *Ecological Monographs*, **27**, 55–112.

Olechowicz, E. (1974). Analysis of sheep pasture ecosystem in the Pieniny mountains (the Carpathians). X. Sheep dung and fauna colonizing it. *Ekologia Polska*, **22**, 589–616.

Olechowicz, E. (1976). The effect of mineral fertilization on the insect community of the herbage in a meadow. *Polish Ecological Studies*, **2** (4), 129–36.

Østbye, E., Berg, A., Blehr, O., Espeland, M., Gaare, E., Hagen, A., Hesjedal, O., Hägvar, S., Kjelvik, S., Lien, L., Mysterud, I., Sandhang, A., Skar, H. J., Skartveit, A., Skre, O., Skogland, T., Solhøy, Y., Stenseth, N. C. & Wielgolaski, F. E. (1975). Hardangervidda, Norway. In *Structure and function of tundra ecosystems*, ed. T. Rosswall & O. W. Heal, *Ecological Bulletins 20*, pp. 225–64. Swedish Natural Science Research Council, Stockholm.

Palmgren, P. (1972). Studies on the spider populations of the surroundings of the Tvärminne Zoological Station, Finland. *Commentationes biologicae*, **52**, 134. Societas scientiarum fennica, Helsinki.

Pasternak, D. & Kotowska, J. (1976). Mineral fertilization and sward structure. *Polish Ecological Studies*, **2** (4), 85–92.

Pearson, L. C. (1965). Primary production in grazed and ungrazed desert communities of eastern Idaho. *Ecology*, **46**, 278–85.

Perkins, D. F. (1976). Grassland ecosystem studies in Snowdonia, Wales. (IBP Grassland Meeting, Dziekanów Leśny near Warsaw, 6–12 July 1973.) *Polish Ecological Studies*, **2** (2), 35–40.

Petrusewicz, K. (1967). Concept in studies on the secondary productivity of terrestrial ecosystems. In *Secondary productivity of terrestrial ecosystems*, vol. 1, ed. K. Petrusewicz, pp. 17–49. Polish Academy of Sciences, Warsaw.

Petrusewicz, K. & Macfadyen, A. (1970). *Productivity of terrestrial animals: principles and methods. IBP Handbook 13.* Blackwell Scientific, Oxford.

Pickford, R. (1960). Survival, fecundity and population growth of *Melanoplus bilituratus* (Wlk.) (Orthoptera: Acrididae) in relation to date of hatching. *Canadian Entomologist* **92**, 1–10.

Pickford, R. (1962). Development, survival and reproduction of *Melanoplus bilituratus* (Wlk.) (Orthoptera: Acrididae) reared on various food plants. *Canadian Entomologist*, **94**, 859–69.

Pitelka, F. A. (1972). Cyclic pattern in lemming populations near Barrow, Alaska. In *Proceedings of the 1972 Tundra Biome Symposium, University of Washington*, pp. 132–5.

Plewczyńska-Kuraś, U. (1974). Analysis of a sheep pasture ecosystem in the Pieniny Mountains (the Carpathians). IV. Biomass of the upper and underground parts of plants and of organic detritus. *Ekologia Polska*, **22**, 517–26.

Plewczyńska-Kuraś, U. (1976). Estimation of biomass of the underground parts of meadow herbage in the three variants of fertilization. *Polish Ecological Studies*, **2** (4), 63–74.

Poskuta, J., Kołodziej, A. & Kropczyńska, D. (1977). *Photosynthesis, photorespiration and respiration of strawberry plants as influenced by the infestation with* Tetranychus urticae *Koch.* Institute of Plant Protection, Agricultural Academy, Warsaw.

Pruscha, H. (1973). Biologie und Produktionsbiologie des Rohrbohrers *Phragmataecia castaneae* Hb. (Lepidoptera, Cossidae). *Sitzungsberichte der Akademie der Wissenschaften, math.-nat.*, **181**, 1–49.

Quasrawi, H. (1966). A study of the energy flow in a natural population of the grasshopper *Chorthippus parallelus* Zett. (Orthoptera, Acrididae). PhD Thesis, University of Exeter.

Raatikainen, M. (1967). Bionomics, enemies and population dynamics of *Javesella pellucida* (F.) (Hom. Delphacidae). *Annales Agriculturae Fenniae*, **6**, 1–149.

Raatikainen, M. (1972). Dispersal of leafhoppers and their enemies to oatfield. *Annales Agriculturae Fenniae*, **11**, 146–53.

Raatikainen, M. & Tinnilä, A. (1961). *Occurrence and control of aphids causing damage to cereals in Finland in 1959.* Finnish State Agricultural Board, Helsinki.

Raatikainen, M. & Vasarainen, A. (1975). Damage caused by timothy flies (*Amaurosoma* spp.) in Finland. *Biological Research Reports of the University of Jyväskylä*, **1**, 3–8.

Rafes, P. M. (1968). *Rol i znaczenje rastitielnojadnych nasjekomych w lesu.* (Role and significance of phytophagous insects in forest.) Publishing office 'Nauka', Moscow.

Rawes, M. & Welch, D. (1966). Further studies on sheep grazing in the northern Pennines. *Journal of the British Grassland Society*, **21**, 56–61.

Reichle, D. E. (1971). Energy and nutrient metabolism of soil and litter invertebrates. In *Productivity of forest ecosystems. Proceedings of the Brussels Symposium 1969*, ed. P. Duvineaud, pp. 465–77. UNESCO, Paris.

Razwjaskina, G. M. (1960). Bioekołogia sziestitocziecznych cikad roda *Macrosteles* i ich epifitołogiczeskije znaczenije. *Zoologicheskii Zhurnal*, **39**, 1855–65.

Rice, R. W. & Vavra, M. (1971). *Botanical species of plants eaten and intake of cattle and sheep grazing shortgrass prairie. US/IBP Grassland Biome Technical Report No. 103.* Colorado State University, Fort Collins.

Ricou, G. (1962). Observations recentes sur les relations entre les Cicadelles et leurs plantes-hôtes. *Revue des Sociétés Savantes de Haute-Normandie, Sciences*, **25**, 31–53.

Ricou, G. (1976). La prairie permanente du nord-ouest française le Pin-au-Haras (In IBP Grassland Meeting, Dziekanów Leśny near Warsaw, 6–12 July 1973.) *Polish Ecological Studies*, **2** (2), 51–66.

Roffey, J. (1970). The effects of changing land use on locust and grasshoppers. In *Proceedings of the International Conference on Current and Future Problems of Acridology*, pp. 199–204. London.

Ryszkowski, L. & Petrusewicz, K. (1967). Estimation of energy flow through small rodent populations. In *Secondary productivity of terrestrial ecosystems*, vol. 1, ed. K. Petrusewicz, pp. 125–46. Polish Academy of Sciences, Warsaw.

Sarmiento, G. & Monasterio, M. (1975). The sturucture and function of a tropical savannah ecosystem in tropical ecological systems. In *Trends in terrestrial and aquatic research*, ed. F. B. Golley & E. Medina, pp. 223–50. Springer-Verlag, Berlin.

Schroeder, L. A. (1971). Energy budget of larvae of *Hyalophora cecropia* (Lepidoptera) fed *Acer negundo*. *Oikos*, **22**, 256–9.

Schroeder, L. A. (1973). Energy budget of the larvae of the moth *Pachysphinx modesta*. *Oikos*, **24**, 278–81.

Schultz, A. M. (1969). The ecosystem concept in natural resource management. In *A study of an ecosystem: the Arctic tundra*, ed. G. M. Van Dyne, pp. 77–93. Academic Press, New York & London.

Serkova, L. G. (1961). On the biology and economic importance of Acridids on the summer pastures of the Sary-Arkin steppe. *Trudȳ Naučno-Issledovatelskogo Instituta Zaščity Rastenii*, **6**, 147–57.

Sims, P. L. & Singh, J. S. (1971). Herbage dynamics and net primary production in certain ungrazed and grazed grasslands in North America. *Preliminary analysis of structure and function in grasslands*, ed. N. French, *Proceedings of the Symposium of the US/IBP Grassland Biome*, pp. 59–124. Colorado State University, Fort Collins, Colorado.

Slobodkin, L. B. (1960). Ecological energy relationships at the population level. *American Naturalist*, **94**, 213–36.

Smalley, A. E. (1960). Energy flow of a salt marsh grasshopper population. *Ecology*, **41**, 672–7.

Smirnov, V. S. & Tokmakova, S. G. (1972). Influence of consumers on natural phytocenoses' production variation. In *Tundra Biome. Proceedings of the IV International Meeting on the Biological Productivity of Tundra*, ed. F. E. Wielgolaski & R. Rosswall, pp. 122–7. Academia Nauk USSR, Leningrad.

Smith, K. (1967). *Insect virology*. Academic Press, New York & London.

Smith, P. H. (1972). The energy relations of defoliating insects in a hazel coppice. *Journal of Animal Ecology*, **41**, 567–87.

Smolik, J. D. (1974). *Nematode studies at the Cottonwood site. US/IBP Grassland Biome Technical Report No. 251*. Colorado.

Stachurski, A. & Zimka, J. (1976). Methods of studying forest ecosystems: microorganism and saprophage consumption in the litter. *Ekologia Polska*, **24**, 57–67.

Stevenson, G. B. (1970). *Biology of fungi, bacteria and viruses*. Edward Arnold, London.

Swartzman, G. L. & Singh, J. S. (1974). A dynamic programming approach to optimal grazing strategies using a succession model for a tropical grassland. *Journal of Applied Ecology*, **11**, 537–48.

Teal, J. M. (1962). Energy flow in the salt march ecosystem of Georgia. *Ecology*, **43**, 614–24.

Tertil, R. (1974). Wpływ żerowania nornika polnego (*Microtus arvalis* Pall.) na plonowanie pszenicy ozimej. *Biuletyn Instytutu Ochrony Roślin*, **57**, 385–91. (The influence of common vole (*Microtus arvalis* Pall.) on the growth and yield of winter wheat.)

Thorsteinson, A. J. (1960). Acceptability of plant for phytophagous insects. In *Proceedings of the 10th International Congress of Entomology*, vol. 2, pp. 599–600. Annual Reviews, Inc., Palo Alto, California.

Traczyk, T. (1971). Productivity investigation of two types of meadows in the Vistula Valley. I. Geobotanical description and primary production. *Ekologia Polska*, **19**, 93–106.

Traczyk, T. (1976). The ecological effects of intensive mineral fertilization on mown meadow. *Polish Ecological Studies*, **2** (4), 7–14.

Traczyk, T., Traczyk, H. & Pasternak, D. (1976*a*). The influence of intensive mineral fertilization on the yield and floral composition of meadows. *Polish Ecological Studies*, **2** (4), 39–48.

Traczyk, T., Traczyk, H. & Pasternak, D. (1976*b*). Estimation of the above ground post-gathering residues in a meadow. *Polish Ecological Studies* **2** (4), 49–56.

Trojan, P. (1967). Investigation on production of cultivated fields. In *Secondary productivity of terrestrial ecosystems*, ed. K. Petrusewicz, vol. 2, pp. 545–62. Polish Academy of Sciences, Warsaw.

Trojan, P. (1975). *Ekologia ogólna*. Polish Academy of Sciences, Warsaw.

Tulisalo, U. (1971). Free and bound amino acids of three host plant species and various fertilizer treatments affecting the fecundity of the two-spotted spider mite, *Tetranychus urticae* Koch (Acarina, Tetranychidae). *Annales Entomologici Fennici*, **37**, 155–63.

Uvarov, B. P. (1931). Insects and climate. *Transactions of the Entomological Society of London*, **79**, 1–247.

Uvarov, F. E. S. (1970). A revision of the genus *Locusta* L. (= *Pachytylus* Fieb.), with a new theory as to the periodicity and migrations of locusts. *Bulletin of Entomological Research*, **12**, 135–63.

Van Emden, H. F. & Bashford, M. A. (1969). A comparison of the reproduction of *Brevicoryne brassice* and *Mysus persicae* in relation to soluble nitrogen concentration and leaf age (leaf position) in the brussel sprout plant. *Entomologia Experimentalis et Applicata*, **12**, 351–64.

Van Emden, H. F., Eastop, V. F., Hughes, R. D. & Way, M. J. (1969). The ecology of *Mysus persicae*. *Annual Review of Entomology*, **14**, 197–270.

Van Hook, R. J. Jr (1971). Energy and nutrient dynamics of spider and orthopteran populations in a grassland ecosystem. *Ecological Monographs*, **41**, 1–26.

Varley, G. C. & Gradwell, G. R. (1962). The effect of partial defoliation by caterpillars on the timber production of oak trees in England. In *Proceedings of the XI International Congress of Entomology*, vol. 2, pp. 211–14.

Vassiljevskaya, V. D., Ivanov, V. V., Bogatyrev, L. G., Pospelova, E. B., Schaleva, N. M. & Grishina, L. A. (1975). Agapa, USSR. In *Structure and function of tundra ecosystem*, ed. T. Rosswall & O. W. Heal, *Ecological Bulletins 20*, pp. 141–58. Swedish Natural Science Research Council, Stockholm.

Vervelde, G. J. (1970). *Progress report 1968–1969, Netherlands participation in the IBP. Section Productivity of Terrestrial Communities (PT)*, pp. 8–13. North-Holland, Amsterdam.

Vogl, R. J. (1973). Effects of fire on the plants and animals of a Florida Wetland. *American Midland Naturalist*, **89**, 334–47.

Voisin, A. (1970). *Produktywność pastwisk*. (*Productivité de l'herbe*.) Polish Academy of Sciences, Warsaw.

Von Rudniew, D. F. (1963). Physiologischer Zustand der Wirtspflanze und Massenvermehrung von Forstchädlingen. *Zeitschrift für Angewandte Entomologie*, **53**, 48–68.

Waloff, N. (1975). The parasitoids of the nymphal and adult stage of leafhoppers (Auchenorrhyncha: Homoptera) of acid grassland. *Transactions of the Royal Entomological Society of London*, **126**, 637–86.

Waloff, Z. & Conners, J. M. (1964). The frequencies of infestations by the desert locust in different territories. (Reprint from the FAO Plant Protection Bulletin 12, 5.)

Weber, N. A. (1972). *Gardening ants, the attines*. The American Philosophical Society, Independence Square, Philadelphia.

Wein, R. W. & Bliss, L. C. (1973). Changes in Arctic *Eriophorum* tussock communities following fire. *Ecology*, **54**, 845–52.

White, E. G. (1974). Grazing pressures of grasshoppers in an alpine tussock grassland. *New Zealand Journal of Agricultural Research*, **17**, 357–72.

Wiegert, R. G. (1964). Population energetics of meadow spittlebugs (*Philaenus spumarius* L.) as affected by migration and habitat. *Ecological Monographs*, **34**, 217–41.

Wiegert, R. C. (1965). Energy dynamics of the grasshopper in old-field ecosystems. *Oikos*, **16**, 161–76.

Wiegert, R. G. & Evans, F. C. (1967). Investigations of secondary productivity in grasslands. In *Secondary productivity of terrestrial ecosystems*, vol. 2, ed. K. Petrusewicz, pp. 499–518. Polish Academy of Sciences, Warsaw.

Wielgolaski, F. E. (1975). Comparison of plant structure on grazed and ungrazed tundra meadows. In *Ecological studies 16, Fennoscandian tundra ecosystems*, part 1, *Plants and microorganisms*, ed. F. E. Wielgolaski, pp. 86–93. Springer-Verlag. Berlin.

Williams, L. H. (1954). The feeding habits and food preferences of Acrididae and the factors which determine them. *Transactions of the Royal Entomological Society of London*, **105**, 423–54.

Winter, K. (1971). Studies in the productivity of Lepidoptera populations. In *Ecological Studies 2*, ed. H. Ellenberg, *Analysis and synthesis*, pp. 94–9. Springer-Verlag, Berlin.

Witkowski, Z. & Kosior, A. (1974). Energetics of the larval development of the oak leaf roller moth, *Tortrix viridiana* L. (Lepidoptera, Tortricidae) and an estimate of the energy budget in caterpillar development of other insects feeding on oak leaves. *Zakład Ochrony Przyrody, Series A*, **9**, 93–106.

Witsack, W. von (1973). Zur Biologie und Ökologie in Zikaden-eiern parasitierender Mymariden der Gattung *Anagrus* (Chalcidoidea, Hymenoptera). *Zoologische Jahrbücher*, **100**, 223–99.
Zlotin, R. I. (1971). Invertebrate animals as a factor of the biological turnover. In *IV Colloque International de la Faune du Sol Dijon, 1970*, pp. 455–62. Institut National de la Recherche Agronomique, Paris.
Zlotin, R. I. (1975). *Żizń w wysokogorjach*, p. 238. Izdatielstwo 'Mysl', Moscow.
Zurzycki, J. (1975). Adjustment processes of the photosynthetic apparatus to light conditions, their mechanism and biological significance. *Polish Ecological Studies*, **1** (1), 41–9.
Żyromska-Rudzka, H. (1976). The effect of mineral fertilization of a meadow on the oribatid mites and other soil mezofauna. *Polish Ecological Studies*, **2** (4), 157–82.

# 4. Large herbivore subsystem

G. M. VAN DYNE, N. R. BROCKINGTON,
Z. SZOCS, J. DUEK & C. A. RIBIC

## Contents

## Importance of large herbivores

Grazinglands throughout the world provide the majority of the nutrients for man's domestic livestock and essentially all the nutrients for numerous species of large wild herbivores. Grazinglands include natural grasslands, savannas, and shrub-steppes and man-developed pastures on land that would normally support forests if it were not for such practices as the use of fire and the re-seeding of pastures. Whether the forages consumed by grazing animals are grown in intensive pasture systems or on extensive rangeland systems, many ecological processes apply in common. Mankind uses grazinglands to produce livestock in nomadism, ranching, and mixed farming agricultural systems. Additionally, man harvests wild herbivores from many grazinglands of the world.

Demand for meat has grown throughout the world and, particularly, since the nineteenth century with the advent of refrigerated ships, meat has entered world commerce on a wide scale. Milk and meat production from grazinglands in the world have increased in the last 10 years but at the same time the area of grazinglands has decreased. Many grasslands in the high-rainfall temperate areas of the world have fallen victim to the plough and the bulldozer. Only a few humid natural grasslands remain, such as a portion of the Pampas in Argentina. The grasslands most likely to remain without cultivation or extreme modification are semi-arid grazinglands of the world, such as the shortgrass prairie in the United States. Harvesting these lands with large herbivores may be the most reasonable pattern of land use to provide food for mankind, even though the productivity is low.

In arid and semi-arid grazinglands, soil moisture limits the growing season more than temperature does. In periods when soil moisture may be available for plant growth, as many as 1 000 000 $kcal \cdot m^{-2}$ may impinge upon the surface from solar input. But only a few kilocalories per square metre may be captured in the products of large mammals used for food production and produced on these lands. In the more humid areas, there is greater efficiency of solar energy conversion into primary production and of primary production into secondary production. A general index of productivity is the ratio of animal populations to square kilometres of both crops and grasslands. In some of the more humid, western-European countries, cattle numbers are $> 100\ km^{-2}$ of crops and grass. In large arid areas, such as the Australian continent, cattle numbers are $< 10\ km^{-2}$ of crops and grasslands. In nations with both subhumid and semi-arid areas, the numbers are intermediate, generally from 25 to 75 $km^{-2}$ of cropland and grassland such as in France, UK, USA, Argentina and New Zealand.

Over recent decades, populations of large wild herbivores have decreased whereas, overall, domestic animals have increased. Yet there have been considerable shifts between sheep and goats as against cattle, with sheep and goats decreasing in many areas of the world in the last 100 years whereas cattle have increased throughout the world (Grigg, 1974). Because of the importance of the grazing animal as a harvester of the grassland crop, it is useful to review the geological history of the grasslands and grazing animals.

Grasses have been known from the Tertiary and it is assumed they emerged as a distinct class of the angiosperm complex during the late Cretaceous times or even earlier. For some 20 million years, forerunners of grasses and grazing animals have evolved together. Thus, the grazing animal and the plant which it grazes have evolved simultaneously and are therefore adapted to each other (Moir, 1968; Van Dyne, Smith, Czaplewski & Woodmansee, 1978). Within grasslands, large mammals generally do not migrate over as extensive distances as do many of the birds. Yet grazinglands are characterized by periodic intervals of harsh climate and food shortage. The large herbivores are able to survive through a combination of being able to store fat, withstand heat and cold, and subsist on low-quality roughage during the nongrowing season. All large herbivores have the advantage of having symbiotic bacteria and protozoans within their digestive tract to aid in the breakdown of cellulose. It is these cellulase-producing organisms that assist the grazing animals in converting fibrous feeds into meat.

Productivity, expressed in kilograms of meat or animal body produced per hectare in a given unit of time, is a result of many individual processes. For large herbivores, however, the major processes of concern are those involved in (i) diet selection, (ii) digestion of the food, and (iii) conversion of the digested products into fat, meat and milk. Various processes for large herbivores related to nutrition include: prehension, mastication, salivation, rumination (including regurgitation and remastication), eructation, digestion, urination, defecation, dermal excretion and absorption. The food selection process has two broad parts, that of selecting the botanical components in different proportions and secondly determining the total amount of food consumed. Other processes are important in affecting the maintenance of the population, such as immigration, emigration, reproduction and mortality.

## Objectives

Considering the importance of large herbivores in the world's grasslands and to mankind, and considering the paucity of IBP work on large herbivores, then in addition to presenting some IBP results of field and modelling studies the purposes of this chapter are (i) to review important large herbivore processes, (ii) to discuss briefly the methods by which these processes are measured, (iii) to give examples of process rates and factors influencing them, and (iv) to show how these processes are modelled and used in simulation and optimization analyses.

The animal first selects which plants it wants to eat, then consumes them, digests them, and then utilizes them in producing meat, milk, wool and other products. There are, of course, feedbacks from the content of the digestive tract and what is digested that influence the amount that is consumed. Also, what is digested and utilized has a feedback on what is selected. From the methodological standpoint, under field grazing studies, we first measure the diet, secondly the digestibility, thirdly the faecal production and fourthly the intake. Thus, in the discussions of food consumption one effectively must read ahead to the discussions of digestibility. Therefore an overall conceptual framework is needed first so that there is a larger picture in which to fit the parts.

## Conceptual framework for large herbivore processes

The large herbivore system can be viewed at various degrees of resolution or detail. For some purposes it is appropriate simply to consider the entire large herbivore system as a single entity. In other instances, such as in detailed physiological and nutritional studies, the large herbivore may be dissected to the molecular level.

### *A model diagram*

For present purposes, it is possible to conceptualize the large herbivore system and its immediate environment within the framework outlined in Fig. 4.1. This level of resolution is somewhere between considering the entire herbivore as a single entity and that of considering separately each of the individual physiological components. The large herbivore is viewed primarily

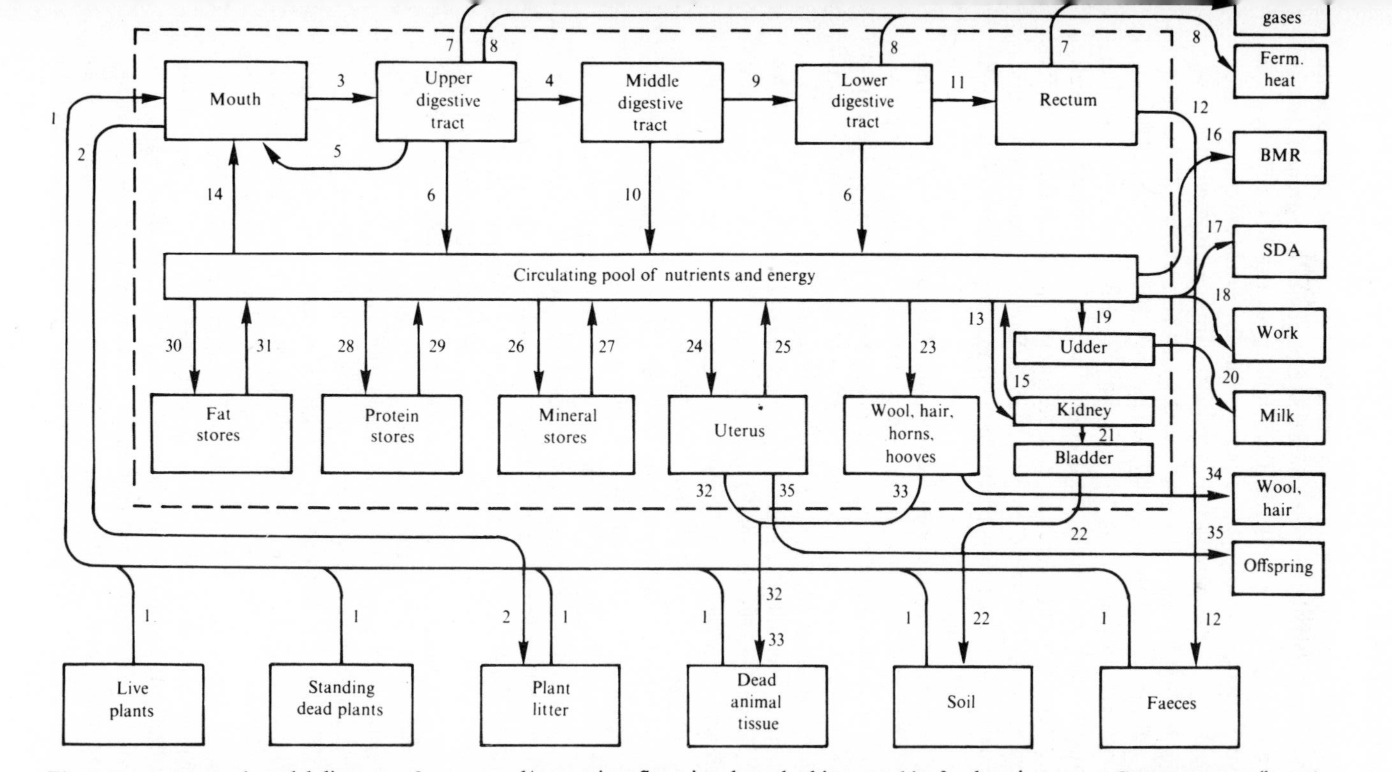


Fig. 4.1. A conceptual model diagram of energy and/or nutrient flows in a large herbivore and its food environment. Compartments (boxes) and flows (arrows) are discussed in the text. 1, Food selection from the environment; 2, food wasting, selected then rejected; 3, food swallowing; 4, ingesta transport, upper to middle digestive tract; 5, regurgitation for rumination; 6, absorption of fermentation products; 7, loss of fermentation gas, oral and anal; 8, loss of fermentation heat, upper and lower tract; 9, ingesta transport, middle to lower tract; 10, absorption of digestion products; 11, ingesta transport, lower tract to rectum; 12, defecation; 13, pool to kidney transfer; 14, pool to saliva transfer; 15, kidney to pool transfer; 16, basal metabolic rate heat loss; 17, specific dynamic effect heat loss; 18, work, activity, and combating environment loss; 19, milk formation; 20, milk secretion; 21, urine formation; 22, urination; 23, miscellaneous body tissue formation; 24, growth of foetus; 25, resorption of foetus; 26, mineral deposition; 27, mineral resorption; 28, protein deposition; 29, protein resorption; 30, fat deposition; 31, fat resorption; 32, abortion; 33, miscellaneous tissue sloughing; 34, shearing; 35, birth of young.

at the organ-system level of resolution or structural detail. The components at the bottom of the diagram represent different categories of potential foods available to the large grazing herbivore. The majority of the diet, of course, is made up of live plants or standing dead plants. But the other categories enter into the diet sometimes. The flows are noted by the lines ending with arrows.

The available food subsystem is represented by 14 compartments and outputs from the animal are represented by eight compartments. The diagrammatic model is general in that it accounts for the possibilities of both nutrient and energy flow; it is general enough to account for rumen fermenters as well as faecal fermenters. Among these 28 compartments of the animal and its immediate food environment there are about 35 different kinds of flows; and three of these occur in two of the compartments, making a total of 38 flows or processes.

Some of the compartments represent actual anatomical components of the large herbivore and some are conceptual. Some of those compartments representing anatomical features are easily identified, e.g. upper digestive tract, while others are more diffuse, e.g. protein stores. Matter or energy remains within these compartments for varying periods of time. Thus, there is a continual flux of matter through the kidney but a much slower flux through the protein stores. Several of the outputs from the animal are readily seen and measured, such as fermentation gases, milk or offspring. Other 'output compartments' are more diffuse, such as heat loss due to basal metabolic rate and specific dynamic effect, but these outputs can be calculated on the basis of total animal biomass or the quantity of one of the animal compartments.

The conceptual model is best able to account for transfer of energy and of nitrogen through the large herbivore system. At the stage food enters the mouth, separate definition would be made of the caloric, protein or mineral composition. In some instances there is an appreciable nitrogen flow but very little energy flow, such as the transfer of matter from the circulating pool into the mouth, as is the case in salivation. In other instances the compartments are oriented to heat flow rather than nitrogen flow, such as the output compartments of fermentation heat, basal metabolic rate, specific dynamic effect and work. Some compartments are purposeful aggregates of varying components such as the work compartment. This compartment could contain energy expended in grazing, in work as such, or in combating the environment as in temperature regulation. The compartment labelled 'Wool, hair' is also an aggregate accounting for various body secretions, other than milk, and includes wool, hair, horns, hooves and oil gland secretions. One of the types of miscellaneous tissues is the epithelium of the digestive tract. The amount and type of feed consumed can cause sloughing of epithelial tissue, thus leading to an output from the animal. In real life this epithelial tissue is lost with the faeces, but according to this conceptual diagram it would be accounted for with dead animal tissues.

Some of the compartments in the model diagram represent only temporary storages along one-way flows. The storage may be either under the control of the animal or under control of man. Examples here would be the bladder and rectum in the first instance and the udder in the second instance. The death of the entire animal and its transfer into the food environment complex is not shown in the diagram. Similarly, the entire animal may be harvested or emigrate from the system of concern and this would be another form of output.

In the case of ruminants, the upper digestive tract would include at least the rumen and reticulum. In the case of nonruminants, the lower digestive tract would be the point at which fermentation products would be absorbed, perhaps along with some products of intestinal digestion.

The conventional nutritional approach to energy flow initially defines gross energy as that of the food grazed. Gross energy is subdivided into components of digestible energy and faecal energy. Digestible energy is subdivided into components of metabolizable energy, urine energy and methane energy. Metabolizable energy is subdivided into components of net energy, fermentation energy and heat of nutrient metabolization. It is net energy, then, that is utilized in such processes as milk formation, fattening, growth, and reproduction. The conceptual framework in Fig. 4.1 differs in several ways. Metabolizable energy does not appear as such but is, in part, represented by the circulating pool. Methane gas loss may be either oral or anal depending upon the type of large herbivore. This is reflected in Fig. 4.1 but not in the conventional hierarchical energy flow diagrams. Our conceptual diagram allows for separate consideration of fermentation heat, i.e. from the microbes as such, and of heat from the animal itself. In practice, however, it is difficult to obtain separate measurements of those two components.

### Comparisons among herbivores

A large number of herbivores graze the world's grazinglands. We include in our tabulations data for many different species. To provide perspective, some general characteristics of these large herbivores have been summarized in Table 4.1. These data, derived from many different sources (see footnotes to table), provide a general picture of the variability in these organisms which range from the dikdik (a small ruminant of East Africa) to the capybara (a large rodent of South America) to the more common breeds of cattle and sheep. In many instances the breed of domestic animals is not specified in the literature. Values presented should be taken simply as general averages reported in the literature; they are most useful on a comparative basis. Table 4.3 (p. 288) gives names, taxonomic relationships, and major regions of occurrence for these large herbivores.

In the following several sections, results are based on analyses of data in the

Table 4.1. *General anatomical, physiological and nutritional characteristics of domestic and wild large herbivores. Unless otherwise specified (see notes) data are selected from tables compiled by Spedding (1975)*

| | Milk yield per head, typical | Period of production | Meat yield per head (kg) | Period of production | Leather yield per head (sq. ft) | Fibre yield, annual (kg) | Empty digestive tract as % of body wt | Wt of caecum as % of body wt | Rumino-reticulum capacity (l) | True stomach capacity (l) | Caecum capacity (l) | Colon capacity (l) | Relative digestive efficiencies (sheep = 100) | Fat reserves (%) | Water turnover at 37 °C (ml · $kg^{0.82}$ 24 $h^{-1}$) | Age at first breeding | Mean gestation length (d) |
|---|---|---|---|---|---|---|---|---|---|---|---|---|---|---|---|---|---|
| 1 Sheep | 150–200 | Year | 18 | 16 wk | 6 | 1–9 | — | 0.4F | 14–25 | 1.8–3.3 | 1 | 5 | 100 | 25 [5] 47 [6] | 197 | 730 | 147 |
| 2 Sheep, Masai | — | — | — | — | — | — | — | — | — | — | — | — | — | — | — | — | — |
| 3 Sheep Merino | — | — | — | — | — | — | — | — | — | — | — | — | — | — | — | — | — |
| 4 Kongoni | — | — | — | — | — | — | — | — | — | — | — | — | — | — | — | — | — |
| 5 Shorthorn | — | — | — | — | — | — | — | — | — | — | — | — | — | — | — | — | — |
| 6 Santa Gertrudis | — | — | — | — | — | — | — | — | — | — | — | — | — | — | — | — | — |
| 7 Banting | — | — | — | — | — | — | — | — | — | — | — | — | — | — | — | — | — |
| 8 Zebu cattle | 3864 | Year | 250 | 16 mo. | 35–50 | — | — | 0.6 3.3M | 92–197 | 10–20 | 10 | 28 | 100, 100, 97 | 35 [8] | — | 730 | 280 |
| 9 Ass | — | — | — | — | 15 | — | — | 3.4F | — | — | — | — | 73 [11] | — | — | — | — |
| 10 Horse | 1117 | Day | — | — | 35 | — | 6.3F | 6.2F | — | 8–18 | 33 | 80 | 102, 97, 103 | — | 37 [12] | 912–1095 | 336 |
| 11 Goat | 600 | Year | — | — | 4.5 | 0.23 [13] | — | — | — | — | — | — | 91 [11] | — | 188 | — | 150 |
| 12 Camel | 1350 | Year | — | — | 30 | — | — | 0.3 | — | — | — | — | — | 22 [13a] | 185 | — | — |
| 13 Pronghorn | — | — | — | — | — | — | — | — | — | — | — | — | — | — | — | 365 [14a] | 240 [14a] |

| | | | | | | | | | | | | | | | | | |
|---|---|---|---|---|---|---|---|---|---|---|---|---|---|---|---|---|---|
| 14 Deer | — | — | 113 [16] | 27 mo. | — | — | — | — | — | — | — | — | — | — | — | 547–730 | 214–56 |
| 15 Bighorn sheep | — | — | — | — | — | — | — | — | — | — | — | — | — | — | — | 914 [18] | 180 [18] |
| 16 Moose | — | — | — | — | — | — | — | — | — | — | — | — | 97, 103 | — | — | — | — |
| 17 Reindeer | — | — | — | — | 20 | — | — | — | — | — | — | — | 90 [11] | 25 [20] | — | — | — |
| 18 Caribou | — | — | — | — | — | — | — | — | — | — | — | — | — | — | — | — | 229 |
| 19 Musk oxen | — | — | — | — | — | 2.3–2.7 | — | — | — | — | — | — | — | — | — | — | 240 [21] |
| 20 Yak | — | — | — | — | 30 | — | — | — | — | — | — | — | — | — | — | — | — |
| 21 Alpaca | — | — | — | — | — | 1.5–5 | — | — | — | — | — | — | — | — | — | — | 343 [23] |
| 22 Llama | — | — | — | — | — | 1.5–3.5 | — | — | — | — | — | — | 112 | — | — | — | — |
| 23 Vicuña | — | — | — | — | — | 0.18 | — | — | — | — | — | — | — | — | — | 730 [25] | 330 [25] |
| 24 Huanaco | — | — | — | — | — | 1–2 | — | — | — | — | — | — | — | — | — | — | — |
| 25 Capybara | — | — | — | — | — | — | — | — | — | 0.1 [27] | — | — | — | — | — | — | 120 |
| 26 Kangaroo | — | — | — | — | 10 | — | — | — | — | — | — | — | — | — | — | 720 | 35 |
| 27 Elephant | — | — | — | — | 100 | — | — | 3.4–6.1 | — | — | — | — | 67 [28] | — | — | 6570 | 650 |
| 28 Hippopotamus | — | — | — | — | 60 | — | — | — | — | — | — | — | — | — | — | 3650 [29] 4380 | 239 [29] |
| 29 Wildebeest | — | — | — | — | 24 | — | — | — | — | — | — | — | — | 2–5 | — | 730 [29] | — |
| 30 Kudu | — | — | — | — | 20 | — | — | — | — | — | — | — | — | — | — | 730 [29] | — |
| 31 Impala | — | — | — | — | 8 | — | 3.7M | 4.6F | — | — | — | — | — | — | — | 730 [29] | 195 [29] |
| 32 Zebra | — | — | — | — | — | — | 3.3M | 0.6F | — | — | — | — | — | — | — | — | — |
| 33 Eland | 350 | — | — | — | 40 | — | 3.5F | — | — | — | — | — | — | — | — | 730 [29] | 260 [29] |
| 34 Buffalo | 900–1364 | Year | 216 [32] | 18 mo. | 35–50 | 2.3–4.5 | 3.8M | — | — | — | — | — | — | — | — | — | 310 |
| 35 Water buck | — | — | — | — | — | — | — | — | — | — | — | — | — | — | — | 930 [29] | 240 [29] |
| 36 Gazelle | — | — | — | — | — | — | — | — | — | — | — | — | — | — | — | 390 [29] | — |
| 37 Oryx | — | — | — | — | — | — | — | 0.5 | — | — | — | — | — | — | — | — | 255 |
| 38 Giraffe | — | — | — | — | — | — | — | — | — | — | — | — | — | — | — | 1095 [29] | 433 [29] |

Table 4.1 *continued overleaf*

Table 4.1 (*continued*)

| | Average lifespan (d) | Mature size, male (kg) | Mature size, female (kg) | Reproductive rate (no. of young per year) | No. of females per male for breeding | Yield, progeny carcass wt (kg) | Duration of oestrus (h) | Oestrus cycle (d) | Breeding lifespan (yr) | Min. water requirements for 40 °C day, 22 °C night (l . 100 kg body wt$^{-1}$) [*1*] | Upper limit of foraging zone (km) [2] | Milk fat % [3] | Wt (kg) [*4*] | Water turnover per 24 h (ml · kg$^{-1}$) [*4*] | Body solids (%) [*4*] |
|---|---|---|---|---|---|---|---|---|---|---|---|---|---|---|---|
| 1 Sheep | 2190 | 30–50 | 20–100 | 1–2 | 30–40 | 18–24 | 26 | 16.5 | 6 | 1.1 | — | 6.2 | 31 [7] | 107 | 32 |
| 2 Sheep, Masai | — | — | — | — | — | — | — | — | — | — | — | — | 29 | 100 | 23.1 |
| 3 Sheep, Merino | — | — | — | — | — | — | — | — | — | — | — | — | 38 | 94 | 29 |
| 4 Kongoni | — | — | — | — | — | — | — | — | — | — | — | — | 88 | 52 | 15 |
| 5 Shorthorn | — | — | — | — | — | — | — | — | — | — | — | — | 322 | 168 | 26 |
| 6 Santa Gertrudis | — | — | — | — | — | — | — | — | — | — | — | — | 523 | 144 | 37 |
| 7 Banting | — | — | — | — | — | — | — | — | — | — | — | — | 372 | 132 | 23 |
| 8 Zebu cattle | 4015 | 700–800 | 450–700 | 0.9 | 30–50 | 200–300 | 18 | 21 | 8–14 | | 1.5 | 4.4 | 532 | 121 | 38 |
| | | | | | | | | | | 3.2–6.4 | | | 417 [9] | 76 [9] | 33 [9] |
| | | | | | | | | | | | | | 197 [*10*] | 135 [*10*] | 28 [*10*] |
| 9 Ass | — | — | — | — | — | — | — | — | — | — | — | — | 160 | 126 | 21.5 |
| 10 Horse | 6935 | 1000 | 700–900 | — | 70–100 | 360 | 144 | 21 | 16–22 | — | — | 1.6 | — | — | — |
| 11 Goat | — | 48–58 | 45–54 | 1–3 | 40 | 4.3–8.4 | 28 | 19 | 6–10 | — | — | — | 40 [*14*] | 96 | 31 |
| | | | | | | | | | | | | | 70 | 52 | 33 |
| 12 Camel | — | 450–840 | 595 | 0.5 | — | 210–250 | — | — | — | — | 2.4 | — | 520 [*14*] | 61 [*14*] | 30 [*14*] |
| 13 Pronghorn | — | — | — | 1.6 [*14a*] | — | 3 [*15*] | — | — | — | — | — | 13.0 | — | — | — |
| 14 Deer | 2920 | 124 [*17*] | 75–82 [*17*] | 1 [*17*] | 2–7 | 20–64 [*17*] | — | — | — | — | 1.5–1.6 | 8.3–8.5 | — | — | — |
| 15 Bighorn sheep | 4917 | 94 | 72 | 1 [*18*] | 1.5 [*19*] | 69 | — | — | 9 | — | 1.8 | 15.9 | — | — | — |
| 16 Moose | — | — | — | — | — | — | — | — | — | — | — | 10.5 | 186 | 111 | 23 |

| | | | | | | | | | | | | | | | |
|---|---|---|---|---|---|---|---|---|---|---|---|---|---|---|---|
| 17 Reindeer | — | — | — | — | — | — | — | — | — | — | — | 22.5 | 100 | 128 | 23 |
| 18 Caribou | — | 170 | 106 | — | — | 7.8 | 48 | 11 | — | — | — | — | — | — | — |
| 19 Musk oxen | — | — | — | 0.5 | — | — | — | — | — | — | — | — | 324 | 35 | 34 |
| 20 Yak | — | 230–360 | 180–320 | — | — | — | — | — | — | — | — | — | — | — | — |
| 21 Alpaca | — | 80 [22] | 80 [22] | — | — | — | — | — | 21–36 [23] | 5.49 | — | — | — | — | — |
| 22 Llama | — | 80 – 110 [24] | 80– 110 [24] | — | — | — | — | — | — | — | — | — | — | — | — |
| 23 Vicuña | 650 [23] | — | — | 1[25] | 10 [25] | — | — | — | — | — | 5.5 [25] | — | — | — | — |
| 24 Huanaco | 10850 [26] | — | — | — | — | — | — | — | — | — | 4.0 [26] | — | — | — | — |
| 25 Capybara | — | 60 | 45 | 8.7 | — | 15.3 | — | — | — | — | — | — | — | — | — |
| 26 Kangaroo | — | 52 [15] | 26 [15] | — | — | — | — | — | — | — | — | — | — | 130 | 125 |
| 27 Elephant | 5000 [12] | 3500 | — | — | — | 120 [26] | — | — | — | 5.5 | — | — | — | — | — |
| 28 Hippopotamus | 1480 [30] | 1365 [30] | — | — | — | — | — | — | — | — | — | — | — | — | — |
| 29 Wildebeest | — | 136–227 | — | — | — | — | — | — | 4.8 | — | — | — | 175 | 53 | 27 |
| 30 Kudu | — | 101 | — | — | — | — | — | — | — | — | — | — | — | — | — |
| 31 Impala | — | 25 | — | 1 [30] | — | — | — | — | — | — | — | — | — | — | — |
| 32 Zebra | — | 365 [17] | — | — | — | — | — | — | — | — | — | 18.1 | — | — | — |
| 33 Eland | — | 261 | — | — | — | — | — | — | — | — | — | — | 247 | 78 | 20 |
| 34 Buffalo | — | 665–718 | — | — | — | 144–279 | 12 | 20–21 | 4.6 | 1.0 | — | 12.0 | 354 | 212 | 21 |
| 35 Water buck | — | 175 | — | — | — | — | — | — | 6.0 | — | — | — | — | — | — |
| 36 Gazelle | — | 15–19 [33] | — | — | — | 2.7 | — | — | 2.7–3.9 | — | — | — | — | — | — |
| 37 Oryx | — | 22.7 | — | — | — | — | — | — | — | — | — | — | 136 | 29 | 30 |
| 38 Giraffe | 1096 | — | — | — | — | 54.5 | — | — | — | 5.5 | — | — | — | — | — |

[1] Data from Louw (1970); lower cattle figure for Zebu cattle and higher for Herefords; lower gazelle figure for Thomson's and higher for Grant's.
[2] Data from Stoddart, Smith & Box (1975).
[3] Data based on Kitts *et al.* (1956).
[4] Data from Macfarlane & Howard (1974).
[5] Based on Anatolian fat-tailed sheep; data are a percentage of carcass weight.
[6] Adult female carcass weight of 32 kg of 47 % fat, with about 25 % subcutaneous, 16 % intramuscular and 6 % kidney.
[7] Ogaden.
[8] 35 % of the carcass, distributed as 11 % subcutaneous, 18 % intramuscular and 6 % kidney.
[9] Beran.
[10] Beran equatorial.
[11] Data from Maynard & Loosli (1969) based on alfalfa hay with 16 % crude protein and 27 % crude fibre.
[12] Water requirements for horse based on $kg \cdot d^{-1}$ for a resting horse; a working horse requires about double this.
[13] Based on cashmere yield.
[13a] Camel data based on Bactrian camel of live weight 460 kg; only the fat in the humps is calculated.
[14] Somali.
[14a] Data from Hoover, Till & Ogilvie (1959).
[15] Data from Dawson *et al.* (1975) on red kangaroos.
[16] Other data for deer show 82 kg at 3 years.
[17] Red deer.
[18] Data from Moser (1962).
[19] Data from Smith (1954).
[20] Reindeer data based on a buck of 103 kg live weight, for fat in two dorsal areas over rump.
[21] Based on information in Tener (1954).
[22] Alpaca mature size based on both sexes.
[23] Data from San-Martin *et al.* (1968).
[24] Llama mature size based on both sexes.
[25] Data from Koford (1957); lifespan from zoo records.
[26] Data from de la Tour (1954); lifespan is an estimate.
[27] Data taken from Gonzalez-Jimenez & Parra (1973)
[28] Data based on dry matter digestibility.
[29] Data from Talbot *et al.* (1965).
[30] Data from Longhurst (1963).
[31] Musk ox mature size is based on both sexes.
[32] Other data on buffalo show 114 kg at 14 months, 279 kg at 48 months and 250 kg for animals more than 3 years old.
[33] Data taken from graphs of Talbot & McCulloch (1965).

Appendixes (pp. 435–537) which we have compiled to provide comparative information on large herbivores. In general, we attempt to give sufficient information in the Appendixes to enable the reader to determine whether the references are of particular significance for their purposes. In most instances we have taken selected or average values from the articles and reports cited. In many places we have calculated averages, as we needed them for comparison purposes, from the tables and graphs in the literature. Since we subdivided the data into categories of botanical composition, chemical composition, intake and digestibility the same reference may be referred to in different Appendixes.

Many references are made to equations for calculating digestibility and intake. In the text, equations generally are presented without derivation. Since the notation may vary from place to place, the various equations of interest are derived in a uniform notation in Appendix 4.1.

### *A summary of herbivore diet literature*

Much of this chapter is devoted to analysis of botanical, chemical and digestibility characteristics of the diet and total intake. Results of 855 estimates of these diet characteristics of large herbivores are summarized in the Appendixes and are used in the figures and tables herein.

This literature is summarized in Table 4.2; it is restricted almost entirely to grazing situations, except in a few instances where 'simulated natural diets' were fed to wild herbivores and for which we utilized the data in our summaries. We derived these 855 estimates by taking or calculating seasonal, yearlong, or treatment means from the published literature. Within the overall literature researched on these four characteristics, the relative abundance of seasonal, yearlong and treatment data on these characteristics differed widely, being respectively 93, 137, 160 and 668 estimates for forage intake, chemical composition, digestibility and botanical composition of the diet. There were only three studies involving 13 season–treatment combinations wherein all four characteristics were investigated: (i) Smith *et al.* (1968) studied all four characteristics for cattle on early and late summer range and on winter ranges in Nevada, USA; (ii) Cook, Taylor & Harris (1962) studied the characteristics for sheep on light and heavy grazing on good and poor winter salt desert shrub ranges in Utah, USA; and (iii) Van Dyne & Heady (1965*a*, *b*), Van Dyne & Meyer (1964*a*, *b*), and Van Dyne & Lofgreen (1964) studied these characteristics for sheep and cattle grazing in common in early, middle and late summer on a California, USA, annual grass–shrub range.

## Food selection

### *Methods of evaluating diets*

Determining the food selected by large herbivores grazing under natural conditions is a first step in evaluating productivity. There are several possible methods of studying the diet of these animals, including the following:

(i) Observing free-grazing or tethered animals to note the relative abundance of different plants in their diets.
(ii) Estimating in plots the production and utilization of differing species to calculate the diet grazed.
(iii) Clipping plots before and after grazing to determine use by the difference.
(iv) Plucking plant units before and after grazing to determine chemical composition and botanical composition by difference.
(v) Using oesophageal-fistulated and rumenal-fistulated animals to collect samples of forage grazed then analysing the samples.
(vi) Killing animals to analyse ruminal or intestinal contents for botanical composition (often used with wild animals).
(vii) Examining faeces, collected on the grazingland, microscopically for botanical residue to determine indirectly dietary components.

Each of the methods has its advantages and disadvantages, as discussed by Van Dyne (1969). Lesperance, Clanton, Nelson & Theurer (1974) discuss factors affecting chemical composition of fistula samples and Theurer, Lesperance & Wallace (1976) discuss estimating dietary botanical composition by fistula methods in the field and microscopic analyses in the laboratory.

The oesophageal fistula, where it can be used, probably provides the most valid data. Paunch or stomach analysis methods have been routinely used with many large wild animals. More recently, collection of faecal material from the grazingland and analysing it microscopically has been a common methodology.

Whatever the technique used, as discussed above, the purpose is to determine what foods are selected by the animal from the grazingland. Selectivity is conditioned by two major factors – palatability and preference. Palatability refers to plant characteristics which affect or stimulate a selected response by animals (Young, 1948; Cowlishaw & Alder, 1960; Heady, 1975). Preference generally refers to the selection by the animal and thus is related to its behaviour. Palatability and preference are not synonymous, but are sometimes mistakenly referred to as such. Many factors besides the palatability of the plant influence selection. Food selection is governed by a complex mixture of factors (Heady, 1964).

Table 4.2. *Characteristics of the literature surveyed with quantitative data on diet*
*herbivores under grazingland conditions*

| | Cattle | Sheep | Camel | Goat | Horse | Burro | Bison | Pronghorn | Mule deer | White-tailed deer | Elk | Moose | Caribou | Musk oxen | Bighorn sheep | Dall sheep | Mountain goat | Deer | Barbary sheep | Red deer | Reindeer | Saiga |
|---|---|---|---|---|---|---|---|---|---|---|---|---|---|---|---|---|---|---|---|---|---|---|
| Botanical (B) | 76 | 74 | 1 | 21 | 8 | 3 | 11 | 82 | 79 | 42 | 22 | 19 | 9 | 8 | 26 | 1 | 10 | 1 | 4 | 5 | 6 | [illegible] |
| Chemical (C) | 12 | 11 | — | — | 2 | — | — | — | — | — | — | 4 | 4 | — | — | — | — | — | — | — | — | — |
| Digestibility (D) | 9 | 26 | — | — | — | — | — | 1 | 18 | 1 | — | 1 | — | — | — | — | — | — | — | — | — | — |
| Intake (I) | 9 | 13 | — | — | — | — | — | — | — | — | — | 1 | — | — | — | — | — | — | — | — | — | — |
| B+C | 19 | 4 | — | — | — | — | — | 3 | — | — | — | — | — | — | — | — | — | — | — | — | — | — |
| B+D | 3 | 4 | — | — | — | — | — | — | — | — | — | — | — | — | — | — | — | — | — | — | — | — |
| B+I | 4 | 3 | — | — | — | — | — | — | 1 | — | — | — | — | — | — | — | — | — | — | — | — | — |
| C+D | 7 | 22 | — | — | — | — | — | 1 | — | — | — | — | — | — | — | — | — | — | — | — | — | — |
| C+I | 4 | — | — | — | — | — | — | — | — | — | — | — | — | — | — | — | — | — | — | — | — | — |
| D+I | 13 | 11 | — | — | — | — | 3 | — | — | — | — | — | — | — | — | — | — | — | — | — | — | — |
| B+C+D | 9 | 14 | 6 | — | — | — | — | — | 4 | — | — | — | — | — | — | — | — | — | — | — | — | — |
| B+C+I | 2 | 4 | — | — | — | — | — | — | — | — | — | — | — | — | — | — | — | — | — | — | — | — |
| B+D+I | — | — | — | — | — | — | — | — | — | — | — | — | — | — | — | — | — | — | — | — | — | — |
| C+D+I | — | 2 | — | — | — | — | — | — | — | — | — | — | — | — | — | — | — | — | — | — | — | — |
| B+C+D+I | 6 | 7 | — | — | — | — | — | — | — | — | — | — | — | — | — | — | — | — | — | — | — | — |
| All B | 119 | 110 | 1 | 27 | 8 | 3 | 11 | 85 | 84 | 42 | 22 | 19 | 9 | 3 | 26 | 1 | 4 | 1 | 4 | 5 | 6 | |
| All C | 59 | 64 | — | 6 | 2 | — | — | 4 | 4 | — | — | 4 | 4 | — | — | | — | — | — | — | — | — |
| All D | 47 | 86 | — | 6 | — | — | 3 | 2 | 22 | 1 | — | — | 1 | — | — | — | — | — | — | — | 2 | — |
| All I | 38 | 40 | — | — | — | — | 3 | — | 1 | — | — | — | 1 | — | — | — | — | — | — | — | — | — |

## The grazing process

### Anatomical considerations

Food selection is in part determined by the anatomy of the large herbivore Cattle have no upper incisors and use their tongues as prehensile organs. The herbage is pinched between the tongue and the lower teeth and torn off Because of the structure of the lower jaw, cattle can seldom graze closer than 12 mm from the soil (Leigh, 1974). During grazing, cattle constantly sniff the herbage. Cattle generally graze from 4 to 9 $h \cdot d^{-1}$, with more time spent when on rangeland than when on dense pastures. Grazing may occur during four or five periods within each 24 h. The commencement of early-morning grazing is correlated with the season of the year. Night grazing may be more frequent during the summer and under tropical conditions. Generally, the Afrikander and Zebu breeds of cattle graze longer hours and travel greater distances than the European breeds. This may be related to the smaller capacity of the digestive system of these cattle of tropical origin. It is difficult to establish cattle preferences for herbage within a range of digestible species without taking into account the effects of early experience, local environmental variation and individual variation (Hafez, Schein & Ewbank, 1962).

*nical and botanical composition and digestibility and forage intake rate by large*

| | Mountain gazelle | Red kangaroo | Euro | Dik-dik | Warthog | Impala | Thomson's gazelle | Grant's gazelle | Hartebeest | Wildebeest | Lesser kudu | Gemsbok | Topi | Zebra | Eland | Cape buffalo | Waterbuck | Rhinoceros | Hippopotamus | Elephant | Giraffe | Total |
|---|---|---|---|---|---|---|---|---|---|---|---|---|---|---|---|---|---|---|---|---|---|---|
| ical (B) | 3 | 6 | 5 | 1 | 1 | 3 | 3 | 2 | 1 | 7 | 4 | 6 | 3 | 4 | 3 | 2 | 1 | 1 | 2 | 4 | 2 | 575 |
| ical (C) | — | — | — | — | — | — | — | — | — | — | — | — | — | — | — | — | — | — | — | — | — | 33 |
| tibility (D) | — | 2 | — | — | — | — | — | — | — | — | — | — | — | — | — | — | — | — | — | — | — | 58 |
| e (I) | — | — | — | — | — | — | — | — | — | — | — | — | — | — | — | — | — | — | — | — | — | 34 |
| | — | — | — | — | — | — | — | — | — | — | — | — | — | — | — | — | — | — | — | — | — | — |
| | — | — | — | — | — | — | — | — | — | — | — | — | — | — | — | — | — | — | — | — | — | — |
| | — | — | — | — | — | — | — | — | — | — | — | — | — | — | — | — | — | — | — | — | — | — |
| | — | — | — | — | — | — | — | — | — | — | — | — | — | — | — | — | — | — | — | — | — | — |
| | — | — | — | — | — | — | — | — | — | — | — | — | — | — | — | — | — | — | — | — | — | — |
| | — | — | — | — | — | — | — | — | — | — | — | — | — | — | — | — | — | — | — | — | — | — |
| +D | — | — | — | — | — | — | — | — | — | — | — | — | — | — | — | — | — | — | — | — | — | — |
| +I | — | — | — | — | — | — | — | — | — | — | — | — | — | — | — | — | — | — | — | — | — | — |
| +I | — | — | — | — | — | — | — | — | — | — | — | — | — | — | — | — | — | — | — | — | — | — |
| +I | — | — | — | — | — | — | — | — | — | — | — | — | — | — | — | — | — | — | — | — | — | — |
| +D+I | — | — | — | — | — | — | — | — | — | — | — | — | — | — | — | — | — | — | — | — | — | — |
| | 3 | 6 | 5 | 1 | 1 | 3 | 3 | 2 | 1 | 7 | 4 | 6 | 3 | 4 | 3 | 2 | 1 | 1 | 2 | 4 | 2 | 668 |
| | — | — | — | — | — | — | — | — | — | — | — | — | — | — | — | — | — | — | — | — | — | 137 |
| | — | — | — | — | — | — | — | — | — | — | — | — | — | — | — | — | — | — | — | — | — | 160 |
| | — | — | — | — | — | — | — | — | — | — | — | — | — | — | — | — | — | — | — | — | — | 93 |

Sheep have a cleft upper lip which permits them to graze closer to the soil surface than can cattle. The lips, the lower incisor teeth and the dental pad are used in grazing rather than the tongue as in cattle. The animal bites the forage to be grazed and jerks its head slightly forward and upward. Sheep may graze from 9 to 11 $h \cdot d^{-1}$. Thus, they graze longer than do cattle. Sheep tend to select herbage that is more easily torn than do cattle.

The goat has a mobile upper lip and a prehensile tongue and thus can graze herbage as short as can sheep. Because of its ability to climb and to stand on its hind legs, it can take browse not normally eaten by other herbivores (Maher, 1945).

Several other herbivores have unique grazing characteristics. Animals such as the kangaroo can graze very close to the ground (Ride, 1959). The kangaroo seizes food between the upper and lower incisors and detaches it with a jerk of the head. Kangaroos will also dig up roots of favoured species. Herbivores such as the camel have extremely strong jaws and thus can break large branches from trees. The white rhino with its square lip is primarily a grazer while the black and Asian rhinos are primary browsers (L. M. Talbot, personal communication). Horses have both upper and lower incisors and grasp the grass between their teeth to tear it off. The elephant feeds by plucking entire plants in large bundles of vegetation. This results in a high fibre content in the diet, particularly in the dry season. The elephant, although a nonruminant, has as fermentation chambers the colon and caecum.

In a comparison of goats and sheep grazing on a shrub range in New South Wales (Wilson, Leigh, Hindley & Mulham, 1975), goats grazed much more browse and were just as selective as were sheep, except that they took different species. Goats can reach vegetation to a height of 2 m, much out of the reach of sheep. Even at low stocking rates goats grazed more browse. The nutritive content of the diet was similar and the success of feral goats in these regions is presumably due to their ability to eat material that is unpalatable or inaccessible to the sheep rather than to a difference in quality of diet.

### *Physiological and behavioural considerations*

Detailed studies by Arnold (1966*a*, *b*) show that smell and taste are more important senses than sight and touch in the selection of food, at least by sheep. Sight allows the recognition of food but the smell and taste are more important. As has been also noted with deer (Longhurst, Oh, Jones & Kepner, 1968) the animals use smell to make initial selection and taste to continue the selection.

When grazing, sheep and cattle move in a 'horizontal plane' and select in a 'vertical plane'. It is much more difficult for sheep to graze tall, dense stands of forage than short, dense stands. The fact that animals graze leaves before stems is in part related to the plant anatomy. It is difficult to graze the stem without grazing the leaf. But it is possible to graze a leaf without grazing the stem. It is rare for the sheep to graze from the bottom of the pasture upwards in the sense of biting whole plants off at their bases (Arnold, 1964). Because sheep are so selective in their grazing for green plants, the amount of time required for grazing increases as the amount of herbage on offer decreases.

Access is much easier for grazing animals when surface soils are dry rather than when they are wet; thus some areas on heavy clay soils are generally avoided during wet weather. Strong winds prevent sheep from grazing normally and cause them to drift and be less selective in their dietary selection. Heavy concentrations of insects annoy grazing animals to a point where they seek out wind-blown ridges, etc., and heavily utilize vegetation there. Even in small pastures of fairly monotonous topography, animals will select different habitats through their innate behaviour. Thus, Schwartz, Mathews & Nagy (1976), working on shortgrass prairie, found that cattle and pronghorn tended to use heavily the lowland plots and plateaus and avoided cool slopes and ridgetops relative to their contribution to pasture area.

The grazing process for many East African large herbivores is discussed by numerous workers (e.g. Vesey-Fitzgerald, 1960; Lamprey, 1963; Talbot &

Talbot, 1962; Gwynne & Bell, 1968) and will not be elaborated fully here (see discussion on p. 311) but the general process is as follows. The basic principle is that one animal species 'prepares the area' for the next species. For example, wildebeest may eat certain vegetation species when the plants are green and other animals eat these plants when dry. Wildebeest migrate throughout the year to get the grass in the growth stage on a yearlong basis. The resident feeders such as the impala are mixed feeders, and may stay in the same area yearlong. Buffalo migrate short distances within a general area. Within the mixed herbivores some species keep the grazed grass down allowing smaller animals to come in and graze. Under pasture conditions and with domestic animals, animals don't have as much opportunity for selective grazing. However, even in pasture conditions sheep particularly have a characteristic habit as grazing proceeds. Animals such as sheep fenced on a given pasture go over and over the pasture, each time taking preferred parts until finally the whole area becomes relatively uniformly grazed. On subhumid grasslands in southern Scotland, Nicholson, Patterson & Currie (1970) consistently found that on first entering the pasture the animals were extremely selective, often grazing no more than the upper few millimetres of a leaf or shoot. Then 'centres of concentration' were grazed in which single plants or clumps of dominant species were moderately grazed, with increasing use around the edge of these clumps. This eventually resulted in a patchwork of close-grazed areas. There was repeated return to the initial centres as new shoots appeared. The close-grazed areas eventually joined and pressure over the whole area became generally uniform except for isolated areas of *Nardus* tussocks.

### *Preference by animals*

Basically, an animal must first have a desire to eat, then utilize its background and learning in selecting foods; it is then influenced by the environment in effecting that preference (Young, 1948). Various physiological states affect the hunger or the desire to eat of the animal including stage of breeding, pregnancy, lactation, fatness, fear and excitement. These factors all influence behaviour and the amount of forage intake and thus the selectivity of the animal for different plants. Part of the selectivity for food is conditioned by learned behaviour as to the particular part of a vegetation mosaic the animal will graze in, the foods it will select, the movement over different topography in different seasons and the height level it feeds on. These characteristics are perhaps best expressed in areas such as East Africa where there are large numbers of herbivores grazing the same overall region (Lamprey, 1963).

The result of all the internal animal factors and learned behaviour gives each species of grazing animal different food habits. Even within a species, each animal shows preference for certain plant species, individual plants, parts of plants, plants in certain growth stages, areas of previous use, successional stages, range sites and range types (Heady, 1975). There is also variability over a period of a few days, within the same day and among individuals (Van Dyne & Heady, 1965). Thus, studies on dietary preference should be over extended periods of time and with several animals to average out the variability noted above.

Preference for a plant species is not fixed but is conditioned by the season and the species composition of the vegetation. This is illustrated, for example, by Beck (1969) who measured estimates of diet and available herbage under morning and evening grazing by cattle on native shortgrass prairie, a pasture seeded with native species, and a successional old field. Preference ratios were calculated as the percentage of a plant in the diet divided by percentage available. The preference ratios were generally high for the forb, lambsquarter, but were higher on the seeded range than on the native range or the old field. One of the predominant native grasses, sideoats grama, had higher preferences on the native range and on the old field than in the seeded range. The forb, kochia, had higher preference values on the seeded pasture and on the native range than it did on the old field. The dominant species of the area, blue grama grass, had preference values ranging only from 0.9 to 2.0 in all years on all locations. When taken over two years and over the three grazing areas, sideoats grama and sand dropseed, both grasses, generally were higher in the afternoon diet than in the morning diet. In contrast, kochia was generally higher in morning samples than in the afternoon. For blue grama grass there was not a large difference between morning and afternoon composition of the diet and no consistent trend. In these studies the leaves of the grasses were the most important part of the plant eaten, but the entire plant of most forbs was eaten by the cattle. Any given plant species generally had higher preference value when it was young and growing than when it was mature.

### *Palatability of plants*

Various factors affect the palatability of plants to animals. The animal can recognize plants by senses of sight, touch, taste and smell. The factors causing palatability difference are not completely understood. Various chemical constituents have been shown to be related to palatability, but there is not total uniformity among research studies. Anatomical factors of the plant's morphology affect the palatability to some degree, but there are noticeable exceptions. Presence of awns, spines, hairiness, stickiness, and texture are often related to low palatability. There is no clear relationship between odour and palatability. Crude protein content is highly related to the palatability of grazingland forages to cattle and sheep. Forages high in sugars or with sugars added are correlated with high palatability to cattle and deer. High proportions of fats and ether extract are usually correlated with high palatability.

Generally, animals select cultivars high in phosphorus and potassium rather than those low in those minerals. Several chemical constituents, such as lignin content and tannins, are in general associated with low palatability. Perhaps it is the unique combination of chemical compounds rather than the individual chemical compounds that affect palatability.

The environment has an effect on both the palatability and preference. Different site influences cause variations in chemical composition, succulence and proportion of leaf material in plants. Certain plants, such as Russian thistle (*Salsola kali*), are grazed avidly by sheep when they are wet with winter snow. Many plants are consumed when 'softened' with rain but not at other times.

Because different plants vary in the proportion of stems, leaves and seeds, and because these constituents vary in chemical composition, some of the differences in palatability are due to plant part effects. Most grazing animals select leaf over stem and green material over dry material. The preference of leaves over stems may be due to the chemical composition as well as to other factors. It also may be due simply to the anatomy of the plant, because fruits and leaves can be grazed without the stems, but stems can hardly be taken without leaves.

The relative palatability of individual plants changes because plants mature at different rates. Palatability is a dynamic concept. Any treatment or site influence which changes the relative proportions and rate of maturity of plants then affects palatability. Thus, fertilization may increase the size of cells without proportionate increase in cell wall material so that the plants become more succulent and thus more palatable. Plants may be more palatable if the metabolism is changed so that the chemical composition also changes.

It is clear that palatability of a given species varies with the associated species, but the functional relationships are not clear (Westoby, 1974). The general rule is that plants with low palatability are selected to a greater degree when they compose a small rather than a large proportion of a stand (Heady, 1975). Also, the overall availability or amount of herbage affects palatability. When forage is limited, selectivity decreases and palatability is less important.

### *Preference or selection ratios*

The best way to express the dietary botanical composition for an animal is by a preference ratio or selectivity ratio. This is the ratio of the percentage by weight in the diet to the percentage of the same plant group or species in the available herbage. When expressed in this manner, most major categories of plants or plant parts will have preference ratios in the range of 0.1 up to 10.0 with only a few species or groups that are more highly selected or rejected. A preference or selectivity ratio of 1.0 means the species is selected to the same degree as it is available in the herbage. An alternative method of cal-

Table 4.3. *Geographical location of large herbivores of the world*

| Scientific name | Common name | North America | West Indies | South America | Malagasy | Africa | Europe | Asia | Southeastern Asian Is. | Philippine Is. | New Guinea | Australian Region | Antarctic Region | Arctic Region | Atlantic Ocean | Indian Ocean | Pacific Ocean |
|---|---|---|---|---|---|---|---|---|---|---|---|---|---|---|---|---|---|
| Class: Hamiphera | | | | | | | | | | | | | | | | | |
| Subclass: Prototheria (Monotrema) | | . | . | . | . | . | . | . | . | . | . | . | . | . | . | . | . |
| Subclass: Allotheria (Multituberculata) | | . | . | . | . | . | . | . | . | . | . | . | . | . | . | . | . |
| Subclass: Theria | | . | . | . | . | . | . | . | . | . | . | . | . | . | . | . | . |
| Infraclass: Pantotheria | | . | . | . | . | . | . | . | . | . | . | . | . | . | . | . | . |
| Infraclass: Metatheria | | . | . | . | . | . | . | . | . | . | . | . | . | . | . | . | . |
| Order: Marsupiala | | . | . | . | . | . | . | . | . | . | . | . | . | . | . | . | . |
| Family Macropodidae (52 spp.) | | . | . | . | . | . | . | . | . | . | . | * | . | . | . | . | . |
| Genus: *Macropus* (3 spp.) | Kangaroos | . | . | . | . | . | . | . | . | . | . | . | . | . | . | . | . |
| Infraclass: Eutheria | | . | . | . | . | . | . | . | . | . | . | . | . | . | . | . | . |
| Super-order: Carnivora | | . | . | . | . | . | . | . | . | . | . | . | . | . | . | . | . |
| Super-order: Protoungulata | | . | . | . | . | . | . | . | . | . | . | . | . | . | . | . | . |
| Super-order: Ungulata | | . | . | . | . | . | . | . | . | . | . | . | . | . | . | . | . |
| Order: Artiodactyla | | . | . | . | . | . | . | . | . | . | . | . | . | . | . | . | . |
| Sub-order: Suiformes (nonruminants) | | . | . | . | . | . | . | . | . | . | . | . | . | . | . | . | . |
| Family: Suidae (omnivorous) | Pigs or hogs | * | . | . | . | . | * | * | * | * | * | * | . | . | . | . | . |
| Family: Tayassuidae (sometimes omnivorous) | Peccaries | * | . | * | . | . | . | . | . | . | . | . | . | . | . | . | . |
| Family: Hippopotamidae | | . | . | . | . | . | . | . | . | . | . | . | . | . | . | . | . |
| Species: *Hippopotamus amphibius* | Hippopotamus | . | . | . | . | * | . | . | . | . | . | . | . | . | . | . | . |

| | | | | | | | | | | | | | | | | | |
|---|---|---|---|---|---|---|---|---|---|---|---|---|---|---|---|---|---|
| Species: *Choeropsis liberiensis* | Pigmy hippopotamus | . | . | . | . | * | . | . | . | . | . | . | . | . | . | . | . |
| Family: Camelidae | | . | . | . | . | . | . | . | . | . | . | . | . | . | . | . | . |
| Genus: *Camelus* | Camels | . | . | . | . | . | * | . | . | . | . | . | . | . | . | . | . |
| *C. dromedarius* | Dromedary | . | . | . | . | . | . | . | . | . | . | . | . | . | . | . | . |
| *C. bactrianus* | Bactrian camel | . | . | . | . | . | . | . | . | . | . | . | . | . | . | . | . |
| Genus: *Lama* | | . | . | * | . | . | . | . | . | . | . | . | . | . | . | . | . |
| *L. guanacoe* | Guanaco | . | . | . | . | . | . | . | . | . | . | . | . | . | . | . | . |
| *L. peruana* | Llama | . | . | . | . | . | . | . | . | . | . | . | . | . | . | . | . |
| *L. pacos* | Alpaca | . | . | . | . | . | . | . | . | . | . | . | . | . | . | . | . |
| Genus: *Vicugna vicugna* | Vicuña | . | . | . | . | . | . | . | . | . | . | . | . | . | . | . | . |
| Family: Tragulidae | | . | . | . | . | . | . | . | . | . | . | . | . | . | . | . | . |
| Species: *Hyemoschus aquaticus* | Water chevrotain, vion | . | . | * | . | . | . | . | . | . | . | . | . | . | . | . | . |
| Species: *Tragulus* (*c.* 6 spp.) | Asiatic mouse deer, chevrotain | . | . | * | . | . | . | . | . | . | . | . | . | . | . | . | . |
| Family: Cervidae (53 spp.) | | . | . | . | . | * | . | . | . | . | . | . | . | . | . | . | . |
| Species: *Moschus moschiferus* | Musk deer | . | . | . | . | . | . | * | * | * | . | . | . | . | . | . | . |
| Genus: *Muntiacus* (*c.* 6 spp.) | Muntjac, barking deer | . | . | . | . | . | . | * | * | . | . | . | . | . | . | . | . |
| Genus: *Dama* | | . | . | . | . | . | * | * | . | . | . | . | . | . | . | . | . |
| *D. dama* | Fallow deer | * | . | . | . | . | * | * | . | . | . | . | . | . | . | . | . |
| *D. mesopotamica* | Persian fallow deer | . | . | . | . | . | . | * | . | . | . | . | . | . | . | . | . |
| Genus: *Axis* | Axis deer, chital | . | . | . | . | . | . | * | . | . | . | . | . | . | . | . | . |
| *A. axis* | | . | . | . | . | . | . | . | . | * | . | . | . | . | . | . | . |
| *A. calamianensis* | | . | . | . | . | . | . | . | . | . | . | . | . | . | . | . | . |
| *A. porcinus* | | . | . | . | . | . | . | * | . | . | . | . | . | . | . | . | . |
| Genus: *Cervus* (*c.* 15 spp.) | | * | . | . | . | * | * | * | * | * | . | . | . | . | . | . | . |
| *C. elaphus* | Red deer | . | . | . | . | . | . | . | . | . | . | . | . | . | . | . | . |
| *C. canadensis* | Wapiti, elk | . | . | . | . | . | . | . | . | . | . | . | . | . | . | . | . |
| Species: *Elaphurus davidianus* | Père David's deer | . | . | . | . | . | . | * | . | . | . | . | . | . | . | . | . |
| Genus: *Odocoileus* | Venados, deer | * | . | * | . | . | . | . | . | . | . | . | . | . | . | . | . |
| *O. virginianus* | White-tailed deer | . | . | . | . | . | . | . | . | . | . | . | . | . | . | . | . |
| *O. hemionus* | Mule deer | . | . | . | . | . | . | . | . | . | . | . | . | . | . | . | . |
| Species: *Blastocerus dichotomus* | Venado Galheiro, swamp deer | . | . | * | . | . | . | . | . | . | . | . | . | . | . | . | . |
| Genus: *Hippocamelus* | Andean deer | . | . | * | . | . | . | . | . | . | . | . | . | . | . | . | . |
| *H. bisulcus* | Huemul | . | . | . | . | . | . | . | . | . | . | . | . | . | . | . | . |
| *H. antisensis* | Taruga | . | . | . | . | . | . | . | . | . | . | . | . | . | . | . | . |

Table 4.3 (*continued*)

| Scientific name | Common name | North America | West Indies | South America | Malagasy | Africa | Europe | Asia | Southeastern Asian Is. | Philippine Is. | New Guinea | Australian Region | Antarctic Region | Arctic Ocean | Atlantic Ocean | Indian Ocean | Pacific Ocean |
|---|---|---|---|---|---|---|---|---|---|---|---|---|---|---|---|---|---|
| Genus: *Mazama* (*c.* 10 spp.) | Brocket deer | * | · | * | · | · | · | · | · | · | · | · | · | · | · | · | · |
| Genus: *Pudu* | | · | · | * | · | · | · | · | · | · | · | · | · | · | · | · | · |
| *P. pudu* | Pudu | · | · | * | · | · | · | · | · | · | · | · | · | · | · | · | · |
| *P. mephistophiles* | | · | · | · | · | · | · | · | · | · | · | · | · | · | · | · | · |
| Species: *Alces alces* | Elk, moose | * | · | · | · | · | * | * | · | · | · | · | · | · | · | · | · |
| Species: *Rangifer tarandus* | Caribou | * | · | · | · | · | * | * | · | · | · | · | · | · | · | · | · |
| Species: *Hydropotes inermis* | Chinese water deer, gasha, kibanoru | · | · | · | · | · | · | * | · | · | · | · | · | · | · | · | · |
| Species: *Capreolus capreolus* | Rehe, roe deer | · | · | · | · | · | * | * | · | · | · | · | · | · | · | · | · |
| Family: Giraffidae | | · | · | · | · | · | · | · | · | · | · | · | · | · | · | · | · |
| Species: *Giraffa camelopardalis* | Giraffes | · | · | · | · | · | · | · | · | · | · | · | · | · | · | · | · |
| Species: *Okapia johnstoni* | Okapis | · | · | · | · | * | · | · | · | · | · | · | · | · | · | · | · |
| Family: Antilocapridae | | · | · | · | · | * | · | · | · | · | · | · | · | · | · | · | · |
| Species: *Antilocapra americana* | Pronghorn, prong bucks | * | · | · | · | · | · | · | · | · | · | · | · | · | · | · | · |
| Family: Bovidae (49 genera, *c.* 115 spp.) | | · | · | · | · | · | · | · | · | · | · | · | · | · | · | · | · |
| Genus: *Tragelaphus* (*c.* 6 spp.) | | · | · | · | · | * | · | · | · | · | · | · | · | · | · | · | · |
| *T. scriptus* | Bushbucks | · | · | · | · | · | · | · | · | · | · | · | · | · | · | · | · |
| *T.* sp. | Kudu, nyala | · | · | · | · | · | · | · | · | · | · | · | · | · | · | · | · |
| Species: *Boocercus eurycerus* | Bongo | · | · | · | · | * | · | · | · | · | · | · | · | · | · | · | · |
| Genus: *Taurotragus* | | · | · | · | · | · | · | · | · | · | · | · | · | · | · | · | · |
| *T. oryx* | Eland | · | · | · | · | * | · | · | · | · | · | · | · | · | · | · | · |
| *T. derbiamus* | Derby eland | · | · | · | · | · | · | · | · | · | · | · | · | · | · | · | · |

| | | | | | | | | | | | | | | | | | |
|---|---|---|---|---|---|---|---|---|---|---|---|---|---|---|---|---|---|
| Species: *Boselaphus tragocamelus* | Nilgai | . | . | . | . | . | . | * | . | . | . | . | . | . | . | . | . |
| Species: *Tetracerus quadricornis* | Four-horned antelope | . | . | . | . | . | . | * | . | . | . | . | . | . | . | . | . |
| Species: *Bubalus bubalis* | Water buffalo | . | . | . | . | . | . | * | * | . | . | . | . | . | . | . | . |
| Genus: *Anoa* (3 spp.) | Anoa, tamaraus | . | . | . | . | . | . | . | * | * | . | . | . | . | . | . | . |
| Genus: *Bos* (7 spp.) | | . | . | . | . | . | . | . | . | . | . | . | . | . | . | . | . |
| *B. taurus* | Cattle | * | * | * | * | * | * | * | * | * | * | . | . | . | . | . | . |
| *B. indicus* | Zebu | . | . | . | . | . | . | * | . | . | . | . | . | . | . | . | . |
| *B. grunnienus* | Yak | . | . | . | . | . | . | * | . | . | . | . | . | . | . | . | . |
| *B. banteng* | Banteng | . | . | . | . | . | . | * | . | . | . | . | . | . | . | . | . |
| Species: *Syncerus caffer* | African buffaloes, mboa | . | . | . | . | * | . | . | . | . | . | . | . | . | . | . | . |
| Genus: *Bison* | | * | . | . | . | . | * | . | . | . | . | . | . | . | . | . | . |
| *B. bison* | Bison | * | . | . | . | . | . | . | . | . | . | . | . | . | . | . | . |
| *B. bonasus* | European wisent | . | . | . | . | . | * | . | . | . | . | . | . | . | . | . | . |
| Genus: *Cephalophus* (*c.* 10 spp.) | Duikers | . | . | . | . | * | . | . | . | . | . | . | . | . | . | . | . |
| Species: *Sylvicapra grimmia* | Grayduiker | . | . | . | . | * | . | . | . | . | . | . | . | . | . | . | . |
| Genus: *Kobus* (*c.* 6 spp.) | | . | . | . | . | * | . | . | . | . | . | . | . | . | . | . | . |
| *K. lechec* | Waterbuck, lechwe antelope | . | . | . | . | . | . | . | . | . | . | . | . | . | . | . | . |
| Genus: *Redunca* (3 spp.) | Reedbucks | . | . | . | . | * | . | . | . | . | . | . | . | . | . | . | . |
| *R. fulvorufula chanleri* | Chanler's Mountain reedbuck | . | . | . | . | . | . | . | . | . | . | . | . | . | . | . | . |
| Species: *Pelea capreolus* | Reedbucks | . | . | . | . | * | . | . | . | . | . | . | . | . | . | . | . |
| Genus: *Hippotragus* (2 spp.) | Sable antelopes, roan antelopes | . | . | . | . | * | . | . | . | . | . | . | . | . | . | . | . |
| Genus: *Oryx* (4 spp.) | Oryx | . | . | . | . | * | . | * | . | . | . | . | . | . | . | . | . |
| Species: *Addax nasomaculatus* | Addax | . | . | . | . | * | . | . | . | . | . | . | . | . | . | . | . |
| Genus: *Damaliscus* (*c.* 6 spp.) | Topis, blesbok | . | . | . | . | * | . | . | . | . | . | . | . | . | . | . | . |
| Genus: *Alcelaphus* (2 spp.) | Hartebeests | . | . | . | . | * | . | . | . | . | . | . | . | . | . | . | . |
| Species: *Beatragus hunteri* | Hunter's antelope | . | . | . | . | * | . | . | . | . | . | . | . | . | . | . | . |
| Genus: *Connochaetes* (2 spp.) | Wildebeests, gnus | . | . | . | . | * | . | . | . | . | . | . | . | . | . | . | . |
| Species: *Oreotragus oreotragus* | Klipspringer | . | . | . | . | * | . | . | . | . | . | . | . | . | . | . | . |
| Species: *Ourebia ourebia* | Oribi | . | . | . | . | * | . | . | . | . | . | . | . | . | . | . | . |
| Genus: *Raphicerus* (3 spp.) | Steinboks | . | . | . | . | * | . | . | . | . | . | . | . | . | . | . | . |
| Genus: *Nesotragus* (2 spp.) | Zanzibar antelopes | . | . | . | . | * | . | . | . | . | . | . | . | . | . | . | . |
| Genus: *Neotragus* (2 spp.) | Pygmy antelopes, sanga | . | . | . | . | * | . | . | . | . | . | . | . | . | . | . | . |
| Genus: *Madoqua* (*c.* 6 spp.) | Dik-diks | . | . | . | . | * | . | . | . | . | . | . | . | . | . | . | . |

Table 4.3 (*continued*)

| Scientific name | Common name | North America | West Indies | South America | Malagasy | Africa | Europe | Asia | Southeastern Asian Is. | Philippine Is. | New Guinea | Australian Region | Antarctic Region | Arctic Region | Atlantic Ocean | Indian Ocean | Pacific Ocean |
|---|---|---|---|---|---|---|---|---|---|---|---|---|---|---|---|---|---|
| Species: *Dorcatragus megalotis* | Beira antelope | . | . | . | . | * | . | . | . | . | . | . | . | . | . | . | . |
| Species: *Antilope cervicaptra* | Blackbuck | . | . | . | . | . | . | * | . | . | . | . | . | . | . | . | . |
| Species: *Aepyceros melampus* | Impalas | . | . | . | . | * | . | . | . | . | . | . | . | . | . | . | . |
| Species: *Ammodorcas clarkei* | Clark's gazelle | . | . | . | . | * | . | . | . | . | . | . | . | . | . | . | . |
| Species: *Litocranius walleri* | Geremuks | . | . | . | . | * | . | . | . | . | . | . | . | . | . | . | . |
| Genus: *Gazella* (*c.* 12 spp.) | Gazelles | . | . | . | . | * | . | * | . | . | . | . | . | . | . | . | . |
| Species: *Antidorcas marsupialis* | Springbucks, springboks | . | . | . | . | * | . | . | . | . | . | . | . | . | . | . | . |
| Genus: *Procapra* (2 spp.) | Mongolian gazelle or zeren | . | . | . | . | . | . | * | . | . | . | . | . | . | . | . | . |
| Genus: *Pantholops hodgsoni* | Chirus, Tibetan antelope | . | . | . | . | . | . | * | . | . | . | . | . | . | . | . | . |
| Genus: *Saiga* (2 spp.) | Saiga antelope | . | . | . | . | . | * | * | . | . | . | . | . | . | . | . | . |
| Genus: *Naemorhedus* (2 spp.) | Goral, Himalayan chamois | . | . | . | . | . | . | * | . | . | . | . | . | . | . | . | . |
| Genus: *Capricornis* (2 spp.) | Serows | . | . | . | . | . | . | * | * | . | . | . | . | . | . | . | . |
| Genus: *Oreamnos americanus* | Mountain goat | * | . | . | . | . | . | . | . | . | . | . | . | . | . | . | . |
| Genus: *Rupicapra rupicapra* | Chamois | . | . | . | . | . | * | * | . | . | . | . | . | . | . | . | . |
| Genus: *Budorcas taxicolor* | Takins | . | . | . | . | . | . | * | . | . | . | . | . | . | . | . | . |
| Genus: *Ovibos moschatus* | Musk oxen | * | . | . | . | . | . | . | . | . | . | . | . | . | . | . | . |
| Genus: *Hemitragus* (3 spp.) | Tahrs | . | . | . | . | . | . | * | . | . | . | . | . | . | . | . | . |
| Genus: *Capra* (5 spp.) | Goats, cabras | . | . | . | . | * | * | * | . | . | . | . | . | . | . | . | . |
| Genus: *Pseudois nayaur* | Bluesheep, bharals | . | . | . | . | . | . | * | . | . | . | . | . | . | . | . | . |
| Genus: *Ammotragus lervia* | Barbary sheep | . | . | . | . | * | . | . | . | . | . | . | . | . | . | . | . |

| | | | | | | | | | | | | | | | | | |
|---|---|---|---|---|---|---|---|---|---|---|---|---|---|---|---|---|---|
| Genus: *Ovis* (2 spp.) | | * | . | . | . | * | . | * | . | . | . | . | . | . | . | . | . |
| *O. canadensis* | Bighorn sheep | * | . | . | . | . | . | . | . | . | . | . | . | . | . | . | . |
| *O. dalli* | Dall sheep | * | . | . | . | . | . | . | . | . | . | . | . | . | . | . | . |
| Order: Tubulidentata | | . | . | . | . | . | . | . | . | . | . | . | . | . | . | . | . |
| Family: Orycteropodiae | | . | . | . | . | . | . | . | . | . | . | . | . | . | . | . | . |
| Genus: *Orycteropus afer* (insectivore) | Aardvark | . | . | . | . | * | . | . | . | . | . | . | . | . | . | . | . |
| Order: Proboscidae | | . | . | . | . | . | . | . | . | . | . | . | . | . | . | . | . |
| Family: Elephantidae | | . | . | . | . | . | . | . | . | . | . | . | . | . | . | . | . |
| Genus: *Elephas maximus* | Asiatic elephant | . | . | . | . | . | . | * | * | . | . | . | . | . | . | . | . |
| Genus: *Loxodonta africana* | African elephant | . | . | . | . | * | . | . | . | . | . | . | . | . | . | . | . |
| Order: Perissodactyla | | . | . | . | . | . | . | . | . | . | . | . | . | . | . | . | . |
| Family: Equidae | | . | . | . | . | . | . | . | . | . | . | . | . | . | . | . | . |
| Genus: *Equus* (*c.* 8 spp.) | Horses | . | . | . | . | * | * | * | . | . | . | . | . | . | . | . | . |
| *E. burchelli* | Zebra | . | . | . | . | * | . | . | . | . | . | . | . | . | . | . | . |
| Family: Tapiridae | | . | . | . | . | . | . | . | . | . | . | . | . | . | . | . | . |
| Genus: *Tapirus* (4 spp.) | Tapirs, dantas | * | . | * | . | . | . | * | * | . | . | . | . | . | . | . | . |
| Family: Rhinocerotidae | | . | . | . | . | . | . | . | . | . | . | . | . | . | . | . | . |
| Genus: *Rhinocerus* (2 spp). | Indian rhinoceros | . | . | . | . | . | . | * | * | . | . | . | . | . | . | . | . |
| Genus: *Didermocerus sumatrensis* | Asiatic two-horned rhinoceros | . | . | . | . | . | . | * | * | . | . | . | . | . | . | . | . |
| Genus: *Diceros bicornis* | Black rhinoceros | . | . | . | . | * | . | . | . | . | . | . | . | . | . | . | . |
| Genus: *Ceratotherium simum* | Square-lipped or white rhinoceros | . | . | . | . | * | . | . | . | . | . | . | . | . | . | . | . |
| Order: Rodentia | | . | . | . | . | . | . | . | . | . | . | . | . | . | . | . | . |
| Family: Hydrochoeridae | | . | . | . | . | . | . | . | . | . | . | . | . | . | . | . | . |
| Genus: *Hydrochoerus* | Carpinchos, Capybara, Chiguire | * | . | * | . | . | . | . | . | . | . | . | . | . | . | . | . |
| *H. hydrochoerus* | | . | . | * | . | . | . | . | . | . | . | . | . | . | . | . | . |
| *H. isthmius* | | * | . | . | . | . | . | . | . | . | . | . | . | . | . | . | . |

culating a 'degree of selection' is that used by Hoefs (1974) in a detailed study of Dall sheep in Alaska, USA. His data, by plant species, for available herbage were based on botanical measurements of foliage cover, height and density and were used to calculate 'available volume'. This dietary information was based on bite counts per unit observation time. He ranked the importance of each species in the herbage and in the forage and then calculated the difference between the ranks giving an index of selection or preference, which in his study ranged from +25 to −41. This provides a qualitative rather than quantitative index.

Studies by Van Dyne & Heady (1965*b*) show that preference ratios are highest for leaves and lowest for inflorescences, with stems being intermediate. Between plant groups forb leaves generally are selected to a greater degree by sheep than by cattle. Heady (1975) recently calculated preference ratios from the literature for many plant groups in various grazing studies with cattle, sheep, goats and deer.

In almost all instances the diet of grazing animals differs from the available herbage. The degree of difference between the diet and the available herbage is in part related to the grazing intensity. We have included here data derived from IBP studies on shortgrass prairies in northeastern Colorado, USA, where cattle, sheep, bison and pronghorn were grazed in small pastures. The small pastures were grazed to a degree to simulate either heavy grazing or light grazing, as compared with large pastures nearby. Botanical measurements were made in the pastures to determine the species composition by utilizing the weight estimate method and clipped plots in a double-sampling procedure. The data on composition of the diet, composition of the pasture and preference ratios are summarized in Table 4.4 for various seasons for heavy and light grazing. Data were taken by individual plant species, but are summarized for convenience here into major plant groupings. These data show the degree of variability one can expect between large herbivores grazing the same rangeland at different intensities and in different seasons. Cattle and bison are largely grass-eaters and sheep and pronghorn are largely forb-eaters.

The seasonal variability is extremely important, as has also been noted by other workers. Rosiere, Beck & Wallace (1975) calculated preference ratios for diets of cattle in USA semidesert range. Grasses, which compose the major part of the diet, generally were, as a class, neither preferred nor rejected but occurred at about the same proportion as in the available herbage. At different seasons and in different pastures, forbs as a class were sometimes selected, but more frequently rejected. Shrubs as a group were more frequently selected than rejected. However, in individual pastures at different seasons certain species of grasses or forbs or shrubs were highly selected. Scales (1972) measured the preference index for cattle diets on sandhill range in eastern Colorado, USA, and found that highest preferences were for western wheat-

Table 4.4. *Herbage and dietary botanical composition and relative dietary preference values for cattle, bison, sheep and pronghorn under light and heavy grazing intensities on shortgrass prairie in early summer (all data in %)*

| Item | Grazing intentity | Grasses: Warm-season | Grasses: Cool-season | Forbs | Shrubs |
|---|---|---|---|---|---|
| Available herbage | Light | 53 | 22 | 10 | 15 |
| | Heavy | 49 | 27 | 7 | 17 |
| Diet | | | | | |
| Cattle | Light | 78 | 15 | 8 | <1 |
| | Heavy | 22 | 18 | 47 | 13 |
| Bison | Light | 24 | 46 | 29 | 1 |
| | Heavy | 80 | 12 | 8 | <1 |
| Sheep | Light | 22 | 19 | 52 | 3 |
| | Heavy | 23 | 4 | 72 | 1 |
| Pronghorn | Light | 10 | 3 | 87 | <1 |
| | Heavy | 17 | 2 | 81 | <1 |
| Preference | | | | | |
| Cattle | Light | 52 | 23 | 25 | <1 |
| | Heavy | 6 | 8 | 75 | 1 |
| Bison | Light | 9 | 40 | 51 | <1 |
| | Heavy | 51 | 13 | 36 | <1 |
| Sheep | Light | 8 | 7 | 85 | 1 |
| | Heavy | 4 | 1 | 94 | 1 |
| Pronghorn | Light | 2 | 1 | 97 | <1 |
| | Heavy | 3 | 1 | 96 | <1 |

grass and sandhill bluestem. However, the former was most desired in early May and in November whereas the latter was desired in June and July. These species were often as much as 10 times as abundant in the diet as in the available herbage. The dominant species on the rangeland, blue grama, was only preferred greatly in early fall. A detailed study was made of the diet of the Soay sheep (by microscopic identification of plant epidermal fragments in sheep faecal samples) and of the available herbage (by clipping quadrats) on the island of St Kilda off the coast of Scotland by Milner & Gwynne (1974). We have calculated preference ratios from their data based on percentage plant fragments in faeces and percentage weight in the herbage. These sheep often consumed two to five times as much *Festuca rubra* and *Poa* spp. as was in the available herbage, while *Agrostis* spp. and *Holcus* were generally at least twice as abundant in the herbage than they were in the diet. The striking factor here is the relative constancy of diet preference ratios with different vegetation types and seasons.

### *Other factors influencing selection*

In addition to the rather large interspecies and interseasonal differences in selection for botanical components of the diet, several other factors are important including intraseasonal influences, animal age differences and herd or flock behaviour. Some examples follow.

#### *Seasonal influences*

Sheep grazing on high mountain range in southwestern Montana, USA, consumed primarily leaves (Buchanon, Laycock & Price, 1972). On a summer-long basis, averaged over 2 years, grasses made up about 24 % of their diet (of which 18 % was in leaves) and forbs made up 76 % of their diet (of which 58 % was in leaves). Overall, 76 % of their diet was leaves. The proportion of leaves increased from early in the summer to late in the summer. However, at times the sheep would select the flowers almost exclusively. Early in the summer half of the diet was of the flowers of the forb, sticky geranium. There was even variability in leaves and stems and flowers during a sampling period of several days. Leaves decreased while stems and flowers increased in the diet from the first to the last day of each trial.

In northern Utah, USA, Cook & Harris (1950) studied sheep diets on a mountain aspen and sagebrush range by the before-and-after method of vegetation sampling. The study ran from early July to the middle of September. Overall, grasses were 42 % of the diet, forbs 30 % and shrubs 28 %; stems were 32 % of the diet and leaves 62 %. Stems at different times occupied from 16 to about 50 % of the diet, leaves varied from 47 to 82 % of the diet, and heads or inflorescences varied from less than 1 % late in the summer to a maximum of 26 % in mid-summer. There was a tendency for the sheep to use browse more as the season advanced and as grass use decreased. Forbs were used heavily and were an important part of the diet of the sheep throughout the season. Sheep made more use of the inflorescences or heads of the grasses than they did of the forbs in some seasons. In mid-summer about 25 % of the diet was grass heads while use of shrub inflorescences was negligible. Sheep made more use of grass stems than they did of forb stems or shrub stems. Shrub leaves contributed over 25 % of the diet in the second half of the summer but generally less than 10 % in the first half of the summer.

In many mountainous plant communities, killing frosts and ripening of seeds are close together in time and cause wild sheep, goat and deer populations to exploit foodstuffs of very high quality, at least temporarily (Geist, 1971). Killing frosts may convert poisonous and otherwise unpalatable plants for example, false hellebore, cow parsnip and various thistles, into highly palatable and preferred food species. Such frosts are often associated with snowfall and softening of the plants, as noted above with respect to sheep grazing Russian thistle. Severe storms, although they increase the cost of animal maintenance, sometimes provide positive side-effects. Caribou extensively utilize wind-felled lichens and lichen-covered branches in periods after storms. Mule deer have been observed to graze Douglas fir annual twigs shed in severe wind storms. Deer also feed on fallen leaves of various poplars, and such leaf fall is increased in autumn storms. Many large herbivores crave salt or other minerals and may consume plants in locations with mineral-rich soils when the same plants would not be consumed in other areas.

Although the grazing habits of wild herbivores may seem puzzling at first, under detailed study the story may be unravelled. Such a scenario is given by Val Geist (personal communication). During blizzards and storms, mule deer are concentrated in either conifer forests or pockets of still air in the foothills of western Canada. In the first instance, their food habits are characterized by eating wind-felled twigs of Douglas fir, leaf hay, some evergreen broadleafs, and frost-killed plants such as aster, senecio or false Solomonseal. In the latter case, deer feed on natural silage, some browse, thistles, leaf-fall, and the dry leaves and fermented berries of the snowberry plant. If the freeze–thaw daily cycle begins in late winter, the wild mountain sheep move to cliffs because the snow cover freezes hard. On the cliffs the sheep exploit diverse sedges and herbs growing in the cracks and crevices. When warm days make slush of the snow and green-up areas on the slope, the sheep move to those locations. They creep to higher and higher elevations as the sprouting and growth of new plants commences. When the first snow falls the animals will pick plant material above the snow and therefore eat mainly frozen and dried herbs and leaves of willows.

### *Topography, age and herding influences*

In the mixed prairie of Kansas, USA, Tomanek, Martin & Albertson (1958) found site-to-site differences in selectivity. Western wheatgrass was a preferred species on the lowland sites but not on the breaks. Buffalograss and blue grama grass were not particularly preferred or rejected on the uplands, but there was a rejection of them on the breaks and lowlands. Big and little bluestem, two climax-dominant grasses, had significantly positive selections on upland, breaks and lowland.

Doran (1943) found that in the aspen type of summer mountain range in western USA, on a percentage diet composition basis, ewes grazed about twice as much grass as did the lambs and about half as much browse. The ewes spent a larger percentage of their total feeding time grazing grasses and forbs than did the lambs, but a smaller percentage on browse. Species of *Carex* contributed heavily to the grass component in the diet of ewes but not in the lambs. Much of the *Carex* was found in the wetter places on the range.

The herding or controlling of the movement of sheep may particularly influence the diet. On summer alpine ranges in Wyoming, USA, Strasia, Thorn, Rice & Smith (1970) found that free-grazing (unherded) sheep ate 44% grass and 52% forbs whereas herded sheep grazed 32% grass and 65% forbs. The numbers of different grass and forb species in the diets were similar for herded and unherded animals but their contribution to the diet varied greatly. These data suggest that the herded sheep took the forbs as they came to them without selectivity.

These observations suggest that many wild herbivores are opportunistic. They can compromise their usual food preferences when affected by cold, mosquitoes, predators, snow crusts and other environmental variations. But with sufficient background, one can predict where the animals will be feeding and what they will take. A possible principle is that ungulates strive to diversify their intake. Perhaps they obey a 'law of least effort for maximum return'. Observation suggests, and it agrees with evolutionary theory, that during periods of growth and lactation these wild herbivores will select for the highest digestibility to maximize food intake. During periods of growth-arrest they may focus more on conserving required resources.

### *General summary of dietary botanical composition*

We have reviewed the literature for information on the botanical composition of the diet of large herbivores. These data are summarized in Appendix 4.2. Generally, we restricted our review to those situations in which the herbivores were grazing in areas with numerous plant species available. Thus, we have not summarized much information for grazing on irrigated pastures or seeded pastures where only one or a few plant species were available to the animals. We have, where possible, presented data on the basis of the grass, forb and shrub composition in the diet. Within 'grasses' we have included grass and grass-like plants, particularly the sedges. Within 'forbs' we have included the herbaceous broad-leafed plants. Within 'shrubs' we have included shrubs, half-shrubs and trees as components. We have in many instances recalculated the data of the original authors. In instances where the authors were unable to identify a small component of the diet, usually only a few per cent, we have proportioned this percentage into the identified components relative to their magnitude. In Appendix 4.2 we have listed the animal species, the general kind of grazingland, the general geographic location, methods used in determining the dietary composition, the season and/or treatment, and the overall botanical composition according to the three main plant groups. We have provided notes on species important in the diet or on other methodological considerations. In some instances we have assigned plants such as lichens, mosses, cacti and ferns to the forbs category. Of course, many other papers than those considered here are to be found in the literature, but they do not have enough components of the diet listed to enable us to calculate the grass, forb and shrub composition.

### *Triangular-coordinate graph summaries*

The data in Appendix 4.2 have been used in developing triangular-coordinate plots of the grass, forb and shrub composition at various seasons and for different animals. To read these triangular-coordinate graphs you should realize that the magnitude of the variable increases the further you go away from the axis on which the name occurs. Each component ranges from 0 to 100 %. A point on any margin means the value is 0 % for the component labelled on that side of the graph. The identification numbers for data points are keyed to those in Appendix 4.2.

In summarizing the dietary botanical composition by season in both northern and southern hemisphere temperate zones, values are grouped into appropriate monthly blocks to represent four seasons – spring, summer, autumn and winter. For the tropical situation, depending upon the description of the seasonality in the reference, we generally considered the dry season equivalent to autumn and the wet season equivalent to summer. The rationale here is that in the tropics animals are not limited in their access to food by snow cover as would be the situation in winter in some areas of the northern or southern hemisphere. Putting the wet season of the tropics equivalent to summer grazing in the northern and southern hemisphere is more arbitrary. This was done because it was felt that the abundance of herbage in the tropics, the high rainfall, and rapid plant growth rate are more equivalent to summer than to spring grazing conditions in the northern and southern hemisphere temperate zones. In the temperate zone in many of the spring-grazing studies the amount of herbage was limiting.

### *Yearlong comparisons*

Botanical composition in the diet of these large herbivores for those cases of sample size greater than 1 are summarized in Table 4.5 and Fig. 4.2. The number of samples entering into the calculation, the mean and the standard deviation are presented in Table 4.5. The data points in Fig. 4.2 are identified according to the entries in Appendix 4.2.

Data are more numerous for some domestic animals, e.g. cattle and sheep, than for most of the large wild herbivores. Among the four domestic animals, the diet of cattle and horses is dominated by grass, with forbs and shrubs each taking relatively equal secondary roles. Overall, one can say that cattle and horses consume about 70 % grass, 15 % forbs and 15 % shrubs. Sheep take, on a yearlong basis, about 50 % of their diet in grass, 30 % in forbs and 20 % in shrubs. Goats consume almost 60 % shrubs, 30 % grass and about 10 % forbs. There is, however, considerable seasonable variability (to be discussed below) as well as location effects.

Among the wild herbivores there are numerous data for the North American

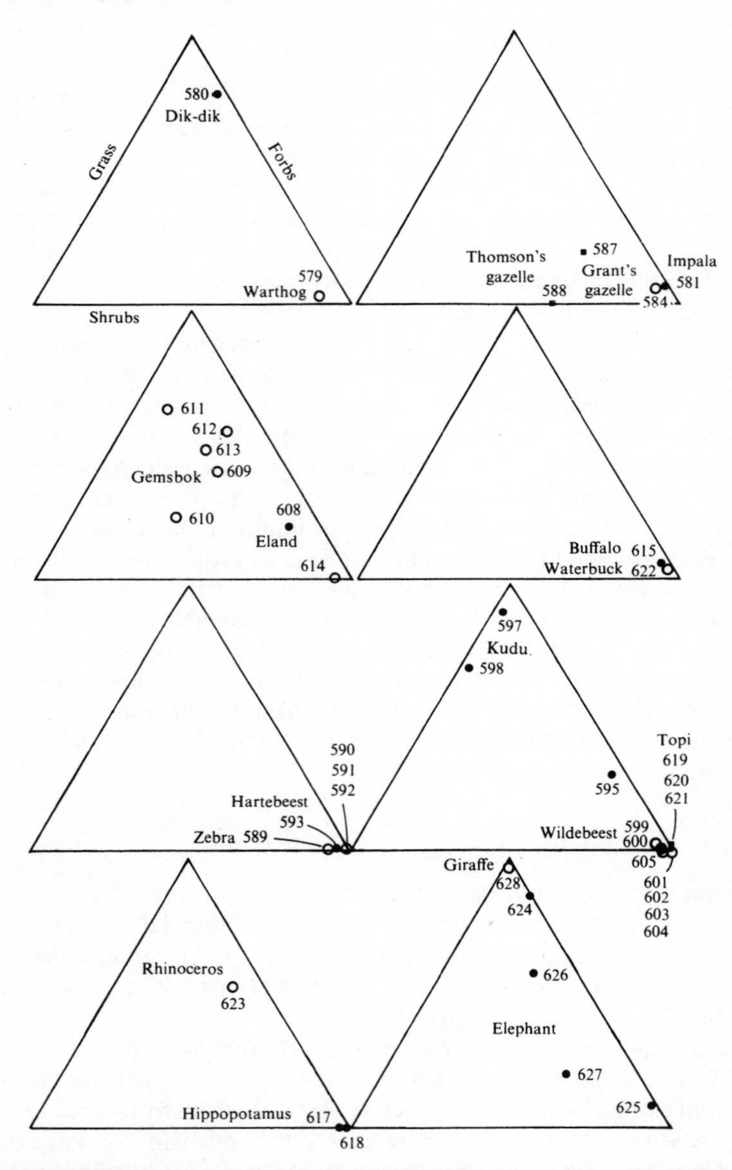


Fig. 4.2. A graphic summary of the botanical composition of the diet on a yearlong basis for important domestic and wild herbivores throughout the world. Individual data points are referred to entries in Appendix 4.2. See also Table 4.5 for means and standard deviations.

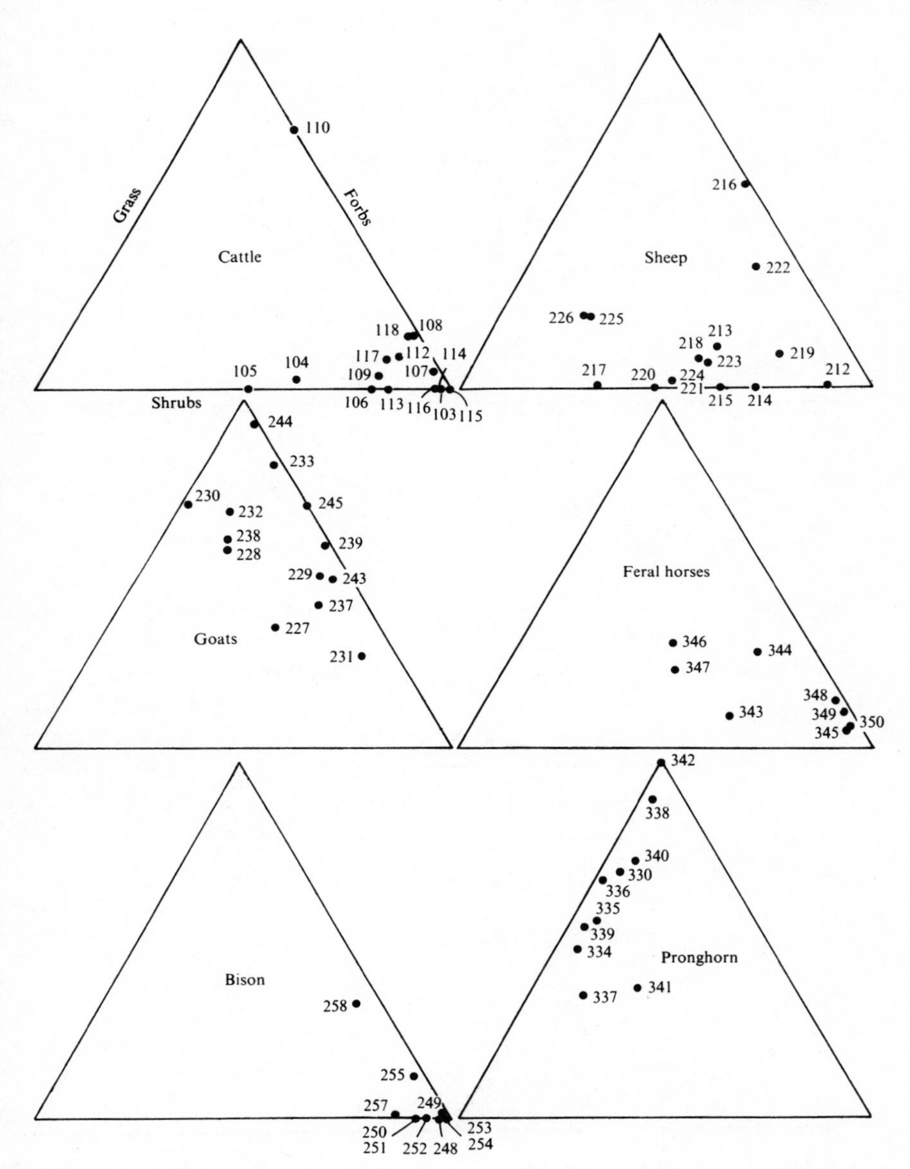


Fig. 4.2 (*cont.*). For legend see p. 300.

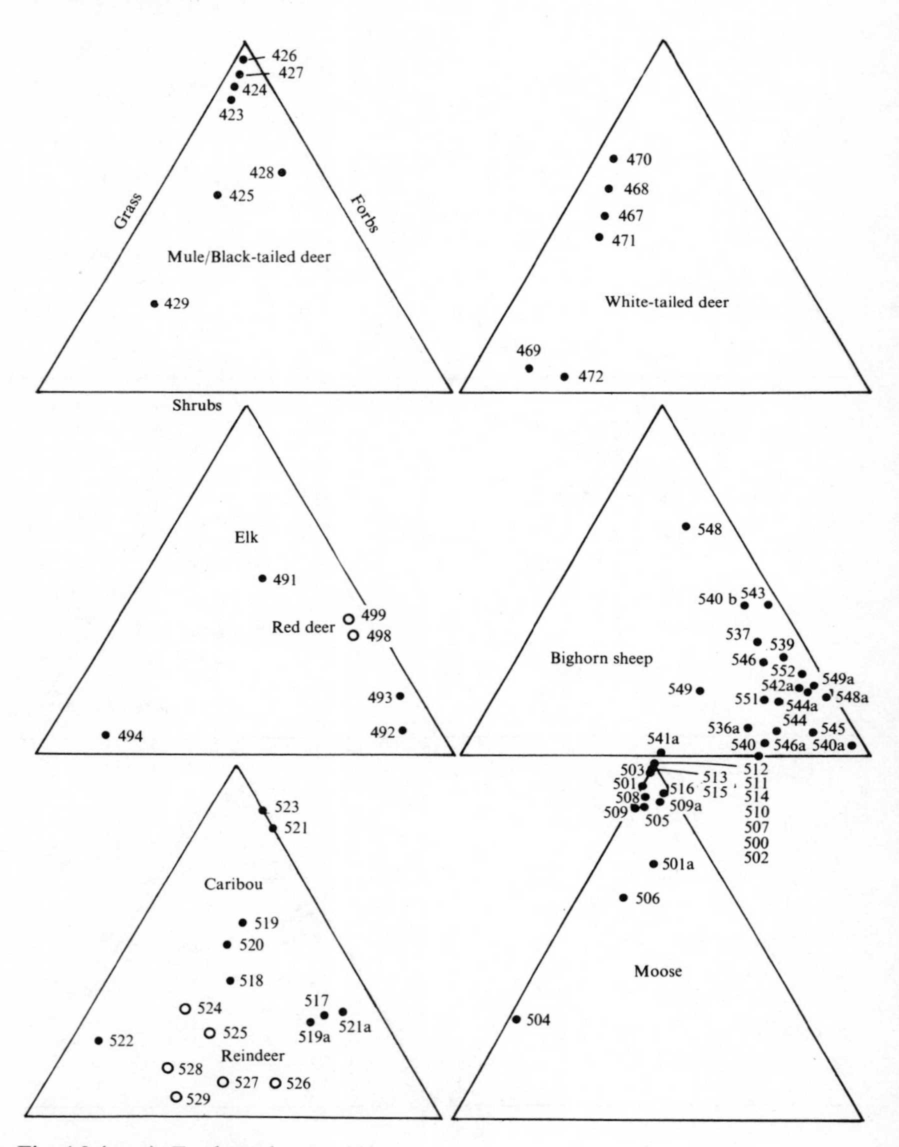


Fig. 4.2 (*cont.*). For legend see p. 300

deer and pronghorn (see Fig. 4.2 and Table 4.5). For all practical purposes, and based on gross herbage categories we summarized, there are few differences between the average diets of mule deer, black-tailed deer, and white-tailed deer. These deer consumed some 60% shrubs, 30% forbs and 10% grass on a yearlong basis. The pronghorn has somewhat more of a balanced

Table 4.5. *Summary of average dietary botanical composition for all seasons for large herbivores*

| Species | Number | Grass | Forb | Shrub |
|---|---|---|---|---|
| Cattle | 121 | 72±25 | 15±16 | 13±21 |
| Sheep | 105 | 50±23 | 30±25 | 20±21 |
| Goats | 13 | 29±19 | 12±11 | 59±18 |
| Burros | 3 | 8± 3 | 39±26 | 55±26 |
| Horses | 8 | 69±23 | 15±16 | 16±10 |
| Bison | 12 | 91±10 | 5± 4 | 4± 8 |
| Pronghorn | 82 | 15±21 | 42±28 | 43±32 |
| Mule deer + black-tailed deer | 83 | 13±17 | 28±26 | 59±30 |
| White-tailed deer | 42 | 10±11 | 30±23 | 60±27 |
| Elk | 22 | 69±26 | 14±19 | 17±22 |
| Red deer | 5 | 40±29 | 21±22 | 39±35 |
| Moose | 19 | 2± 4 | 8±16 | 90±18 |
| Caribou | 9 | 32±21 | 22±22 | 46±24 |
| Reindeer | 6 | 35±12 | 49±10 | 16±10 |
| Bighorn sheep | 21 | 65±16 | 14±12 | 21±16 |
| Mountain goat | 10 | 61±33 | 29±33 | 10±10 |
| Musk oxen | 3 | 87± 3 | 2± 1 | 11± 3 |
| Saiga | 8 | 35±24 | 49±21 | 16± 7 |
| Kangaroos | 7 | 58±17 | 10±11 | 5± 6 |
| Mountain gazelle | 3 | 61±10 | 16± 7 | 23±16 |
| Zebra | 4 | 98± 4 | 2± 2 | 0± 1 |
| Kudu | 4 | 18±22 | 21±18 | 51±36 |
| Wildebeest | 6 | 100± 1 | < 1± 1 | < 1± < 1 |
| Buffalo | 2 | 97± 4 | 1± 1 | 2± 4 |
| Gemsbok | 6 | 40±29 | 21±14 | 39±22 |
| Topi | 3 | 100± 0 | 0± 0 | 0± 0 |
| Elephant | 4 | 46±33 | 10±10 | 44±35 |
| Hippopotamus | 2 | 99± 1 | 1± 1 | 0± 0 |

Individual data are in Appendix 4.2 and means and standard deviations here are for varying sample sizes for grasses, forbs and shrubs (in %).

diet with forbs and shrubs each contributing more than 40 % and grasses 15 % on a yearlong basis.

Caribou and reindeer represent an interesting comparison when viewed together on a yearlong basis. They have perhaps more flexibility in their diet than do the other herbivores for which we have plotted data (see Fig. 4.2). Of all the herbivores summarized in Table 4.5 only the saiga and the reindeer approach 50% forbs in their diet on a yearlong basis. The values for caribou and reindeer may exceed this if we include other data (referred to later in the text) in which lichens, if classified as a forb, would increase that constituent in the diet considerably. On a yearlong basis only goats, deer, moose and kudu of the animals summarized in Table 4.5 have more than 50% shrubs in their diets. Of the large herbivores we have summarized in Table 4.5, about half of them consume more than 50% grass on a yearlong basis and half of them less than 50%. We do not have good information on the total number of grazing animals of each species in the world nor their relative intakes. But we can weight our mean species values by a factor (number of samples entering the mean ÷ standard deviation) derived from our search to give a relative importance value, and then normalize the resultant values. If we take such a weighted average of the 28 large herbivore species for which we have found suitable quantitative data, on a worldwide and yearlong basis, and then round off the data, we find they consume about 60% grass, 20% forbs and 20% shrubs.

### *Seasonal variations in diets*

Yearlong data are subdivided into seasons in Fig. 4.3 and Table 4.6. Separate plots are presented for cattle, sheep, pronghorn, mule deer and black-tailed deer, white-tailed deer, and elk and red deer.

At all seasons cattle eat considerable amounts of grass, averaging from 69% in summer to 75% in autumn. Forbs or shrubs as a component never exceed 20% on the seasonal average. The diet of sheep shows two striking features (Fig. 4.3). In the winter there is a relatively strong grass component along with either a strong forb or a strong shrub component. Our analyses show that sheep graze more grass in the autumn and less in summer than in other seasons. The other striking feature of sheep diets is that under summer grazing there is tremendous variability in the grass, forb and shrub dietary composition between the individual studies reported in the literature.

Data in Fig. 4.3 summarize the botanical composition of the diet of pronghorn. Note that in summer and autumn there is a very low percentage of grass in the diet. The only exception is in the case study discussed above for shortgrass prairie in northeastern Colorado, and otherwise, pronghorn never graze more than 25% grass at any location in summer or autumn. Surprisingly, there were several locations where pronghorn grazed more than 25% grass during the winter. Both during winter and spring there was a great variability in the dietary botanical composition for pronghorn from place to place. In a few instances the diets were almost devoid of shrubs.

Data on the dietary composition of the mule deer and white-tailed deer of North America are also summarized in Fig. 4.3. These two species of deer overlap in their ranges primarily along the east side of the Rocky Mountains in North America. Mule deer, however, live in rough breaks, the sagebrush–grass zone, foothills, and mountains, while white-tailed deer are more commonly found in river bottoms in the western USA though may be intermixed with mule deer in other areas. White-tailed deer are common on the Great Plains and further east in the USA. Two striking features show in the graphs of dietary grass, forb and shrub composition – the very low grass content in the summer, particularly, but also in the autumn, in the diet of both species

Table 4.6. *Summary of dietary botanical composition for several large herbivores on a seasonal basis*

| Species | Spring | | | | Summer | | | | Fall | | | | Winter | | | |
|---|---|---|---|---|---|---|---|---|---|---|---|---|---|---|---|---|
| | (*N*) | G | F | S | (*N*) | G | F | S | (*N*) | G | F | S | (*N*) | G | F | S |
| Cattle | 20 | 71±27 | 12±14 | 17±22 | 46 | 68±25 | 19±18 | 13±22 | 23 | 75±25 | 13±14 | 12±21 | 17 | 70±25 | 13±16 | 17±22 |
| Sheep | 11 | 50±26 | 29±24 | 21±17 | 47 | 44±26 | 38±25 | 18±20 | 9 | 62±19 | 30±23 | 8±10 | 23 | 54±19 | 14±23 | 32±26 |
| Pronghorn antelope | 16 | 29±23 | 43±22 | 28±22 | 21 | 7.0±7.0 | 70±24 | 23±27 | 17 | 18±26 | 34±21 | 48±29 | 18 | 16±24 | 21±20 | 63±33 |
| Mule deer | 17 | 24±25 | 28±19 | 48±27 | 24 | 4.0±4.0 | 42±31 | 54±33 | 16 | 9± 8 | 22±23 | 68±25 | 19 | 17±19 | 20±21 | 63±28 |
| White-tailed deer | 9 | 12±12 | 39±17 | 49±22 | 9 | 4.0±4.0 | 36±27 | 60±25 | 10 | 9±16 | 19±20 | 72±24 | 8 | 11±11 | 12±13 | 77±20 |

Individual data are in Appendix 4.2 and means and standard deviations here are for varying sample sizes (*N*) and grasses (G), forbs (F) and shrubs (S) (in %).

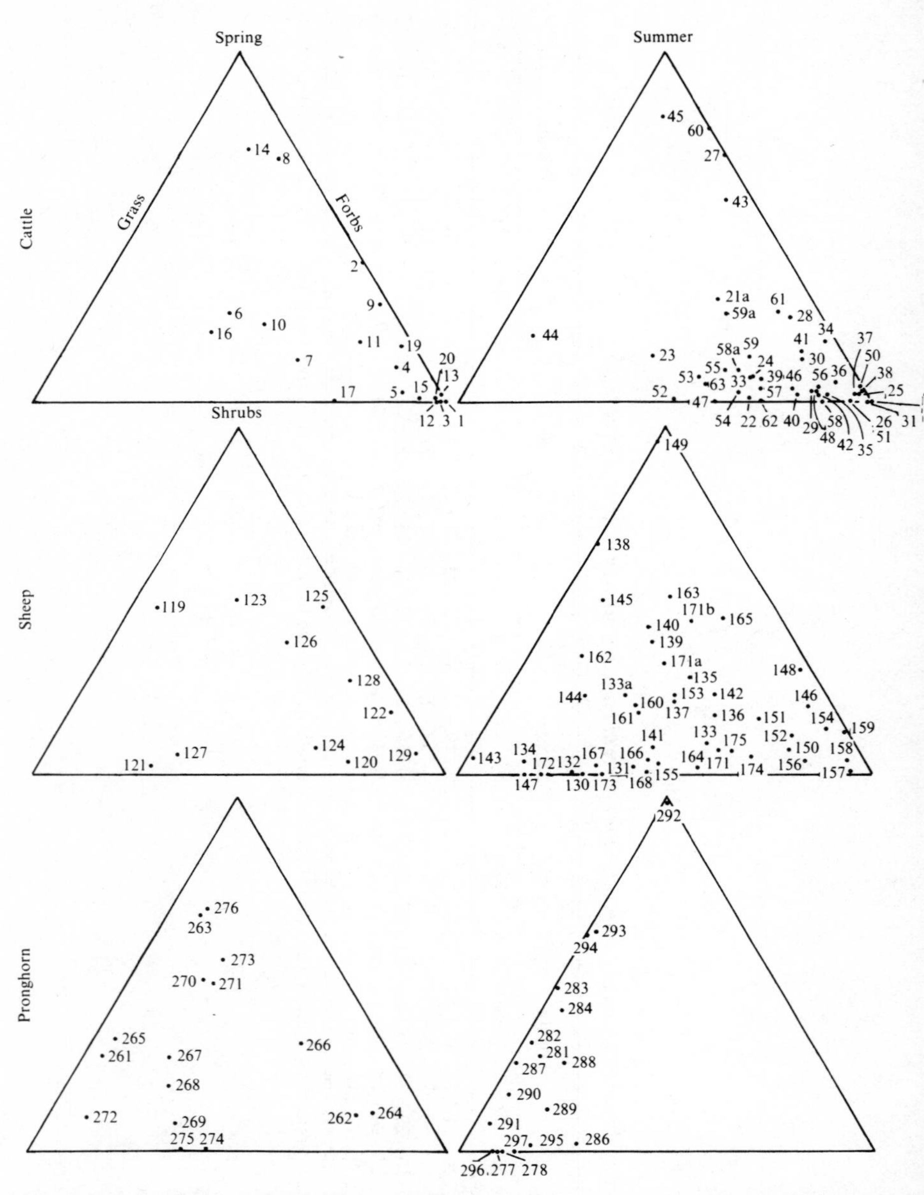


Fig. 4.3. A graphical summary of the botanical composition of the diet on a seasonal basis for several domestic and wild herbivores. See also Table 4.6 for means and standard deviations.

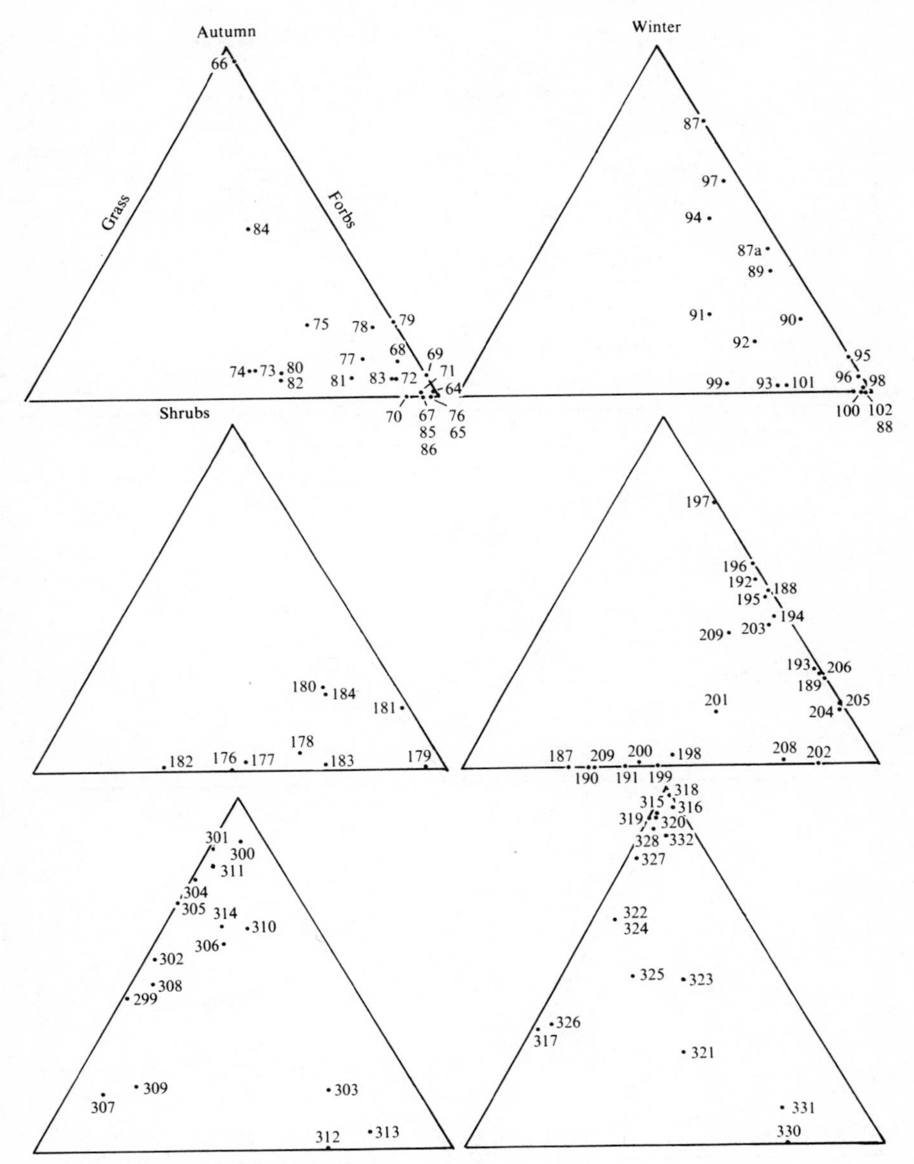


Fig. 4.3 (*cont.*). For legend see p. 306.

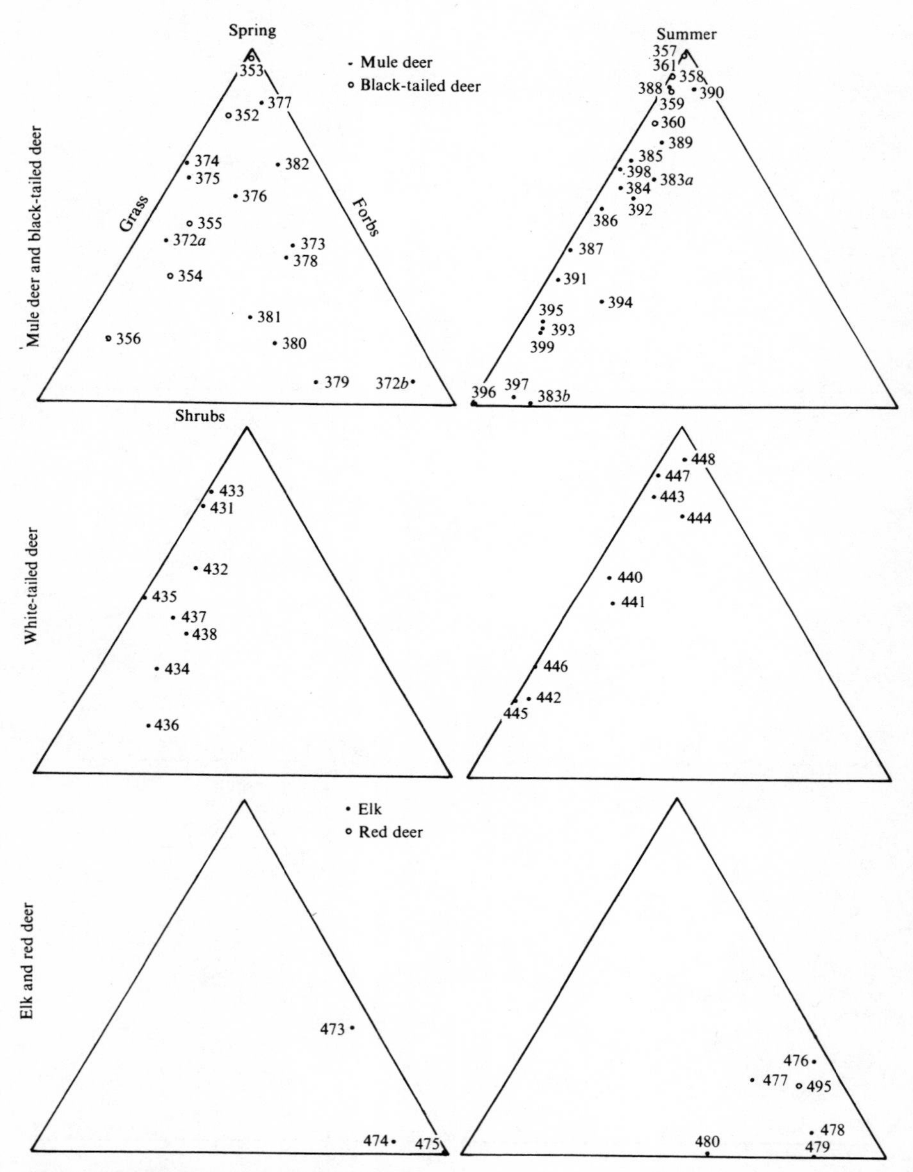


Fig. 4.3 (*cont.*). For legend see p. 306.

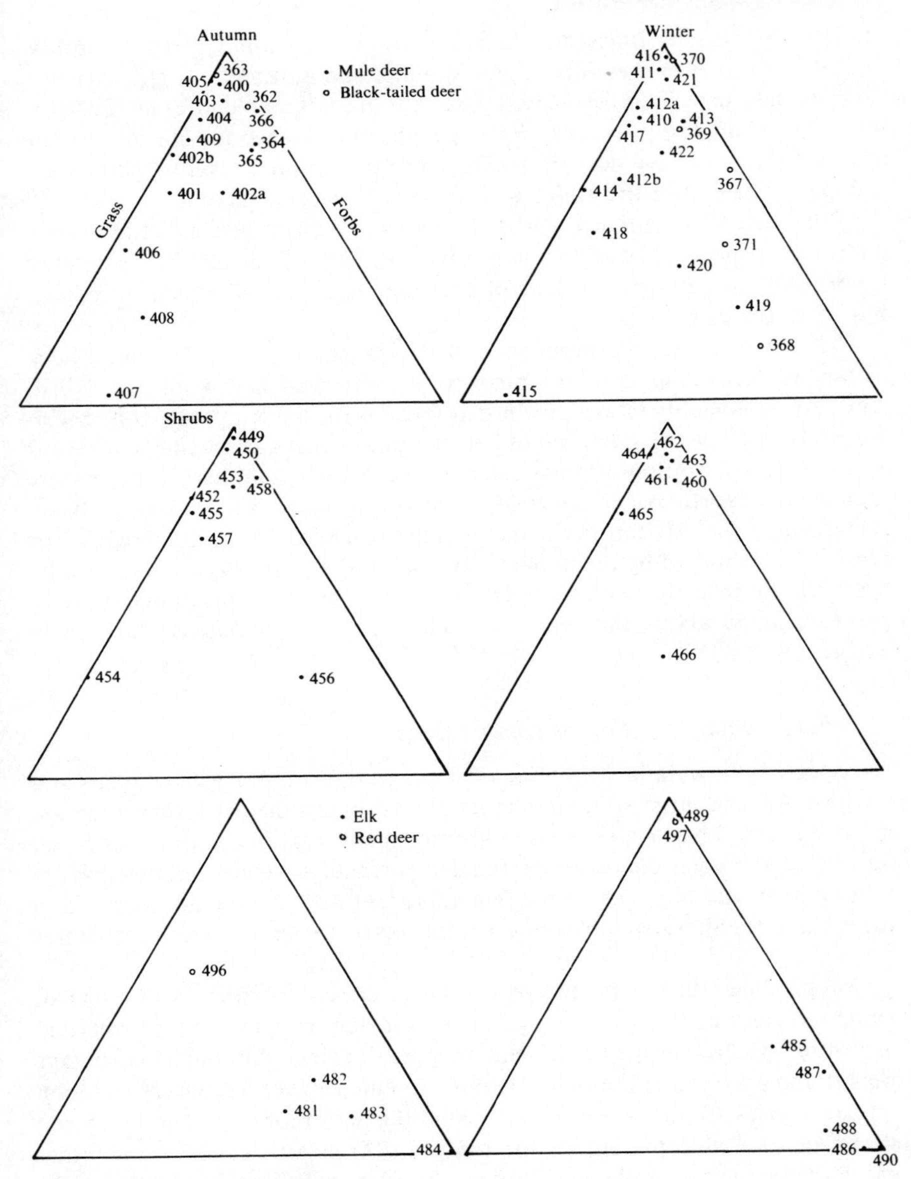


Fig. 4.3 (*cont.*) For legend see p. 306.

of deer. At various times and in various localities either forbs or shrubs dominate the diet of these deer in the summer and autumn, but grasses never do. The data presented for white-tailed deer are not as plentiful as those for mule deer, but it appears they have a higher preference for shrubs, on the average, than do mule deer in the summer and autumn. Overall, shrubs are ⩾50% of the diet of these North American deer at all seasons.

In the spring and summer, forbs are not as important in the diet of deer as one would expect. They seldom comprise over 50% of the diet in the data we report. Autumn and winter diets of deer are characterized by shrubs contributing to the diet.

The various large wild herbivores of Africa (see Fig. 4.2) in general have dietary patterns similar to the mixture of large wild herbivores for North America. Seasonality influence is not as great as for the temperate zone herbivores. They show relatively small percentages of forbs in the diets. There is a greater concentration of grass-eaters in the African wild herbivore mixture than in the North American wild herbivore mixture, the American bison excluded. In the African herbivore populations very strong preferences for grasses are exhibited by the impala, cape buffalo, waterbuck, warthog, wildebeest and hartebeest. Insufficient data are available, however, to make strong generalizations about the seasonal preferences of the African large wild herbivores.

### Further consideration of wildlife diets

As expected, the data in Appendix 4.2 focus heavily on domestic animals and on East African large wild ruminants. There is insufficient information on many large herbivore species from around the world. Because we restricted our review to papers concerning estimated percentage weight composition (or other numerical indices of use), a few important papers were not included in the tabulation. Some examples of key information for specific large herbivores follow.

Chippendale (1962) reported on the botanical composition of the diet of kangaroos and cattle based on ruminant stomach content samples collected in central Australian arid and semi-arid grazingland. Although percentage weight composition of the diet was not presented, considerable information on the species frequency in dietary sampling was reported. The kangaroo shows considerable preference for the grass *Eragrostis setifolia* and other species which grow on the gilgis on treeless flats with texture-contrast soils. This species comprised up to 90% of the diet in many samples. Other species, such as *Helipperum floribundum*, were more apparent in the kangaroo diet than that of cattle. This may be because the marsupials have an ability to graze closer to the soil surface and thus consume more prostrate plants than can cattle. Where cattle often graze dry perennial grasses, kangaroos selected mostly green leaf fragments. The cattle tended to graze more trees and shrubs than did the kangaroo.

Few data are available on the food of caribou in North America. Banfield (1951) reports on diet of caribou obtained from 14 specimens taken in the tundra in the summer in northern Canada. By comparing the plants in the stomach to those available on the range, it was found that the following plants were preferred in summer: mushrooms, lichens, willows, grasses and sedges, and glandular birch. Scotter (1967), however, has reported that in winter caribou consume much more lichen than any other plant component. Analysing 20 rumen samples collected from October through April, he reported a weight composition of: lichen 58%, grass and grass-like 3%, forbs $<1\%$, woody plants 20%, bryophytes and fungi 4%, and unidentifiable material 16%. Miller (1971) also reports some information on diets of caribou; for early and late winter, respectively, we calculate from his data approximate values of 54 and 41% lichen, 12 and 2% mushrooms, and 22 and 5% grass-like plants, thus showing considerable seasonal shift depending upon forage availability on the taiga.

The feral burros of North America are extremely adaptable in their dietary habits. It is reported that they will eat almost anything to sustain life, and there have been unconfirmed reports of burros feeding on creosote bush. This plant is very near the bottom on the scale of palatability among North American rangeland plants (McKnight, 1957). The burro is also a wasteful forager. It pulls up numerous plants, eats only a part, and drops the rest. The burro also soils water and defouls it with excreta at waterholes.

The diet of elephants in East Africa may be approximately 40% woody material and 60% herbaceous (Laws, Parker & Johnstone, 1975). But in many other situations grass comprises up to 80 to 90% of the stomach fill. For various studies of different vegetation types some data on percentage botanical composition in elephant diets are given in Table 4.7. These data suggest that elephants are widely adaptable in their diets, with respect to grass–forb–shrub combinations, dependent upon the type of habitat in which they are grazing. Debarking of trees by elephants may be a response to a nutritional need, due to the shortage of fibre in the diet. It occurs primarily at the beginning of the wet season when the protein/fibre ratio of other food may be unfavourable for the elephant's digestive process.

The great variety of grazing and browsing habits of African herbivores is illustrated by the comparison in Table 4.8 (Vesey-Fitzgerald, 1960; Talbot & Talbot, 1962; Lamprey, 1963), and further details for zebra, wildebeest, Thompson's gazelle and topi are provided by Gwynne & Bell (1968).

Table 4.7. *Botanical composition of elephant diets*

| Vegetation type | Grass | Forbs | Shrubs |
|---|---|---|---|
| Long-grass area | 45–93 | 2–49 | 1–12 |
| Shortgrass/thicket | 31–74 | 9–42 | 8–45 |
| Northeast Uganda | 32 | 28 | 40 |

Table 4.8. *Feeding habits of African herbivores*

| Animal species | Feeding habit |
|---|---|
| Warthog | Much of its sustenance comes from rooting in the ground |
| Tortoise | A distinct grazer |
| Ostrich | A grazer, but sometimes browsing |
| Black rhinoceros | Specialized diet, three species of acacia |
| Giraffe | Browser, utilizes trees at high levels |
| Steenbok | Grazes at ground level almost entirely small broadleaf herbs and euphorbs |
| Bushpig | Eats small broadleaf herbs and several species of rough grass |
| Bushbuck | Primarily a browser, occupies edges of thick brush |
| Reedbuck | Occupies tall grass, swampy areas and eats coarse grasses |
| Waterbuck | Same habitat as reedbuck but eats different grasses |
| Kudu | Feeds mostly on a series of trees, will peel and eat bark |
| Eland | A strikingly mixed feeder, eats grasses when available but survives also on browse |
| Buffalo | Largely a grazer, capable of surviving on long, dry, coarse grasses |
| Thomson's gazelle | Eats predominantly small broadleaf herbs growing beneath the grass layer |
| Wildebeest | Predominantly a grass eater, restricts grazing to grass in the vegetative stage |
| Topi | Predominantly a grass eater, red oatgrass is most important single item in diet, eats tall and dry grass |
| Zebra | Eats same grasses as the wildebeest, but when grass is at mature stage but not overly dry |

### *Relative differences of diets*

Data on the relative difference of diets of 13 of the large herbivores previously discussed, for which a reasonably large amount of data were available, are summarized in Table 4.9 and Fig. 4.4. These overall average difference values do not necessarily reflect sample size or variability within the samples. At best they are approximations but indicate the general relationships among these large herbivores based on available and relatively comparable quantitative data. As a group, the grazers having more than 50% grass in their diet include, in order of increasing importance of grass, bighorn sheep,

Table 4.9. *Relative differences of diets of 13 large herbivore species based on the yearlong averages given in Table 4.5*

| | Cattle | Sheep | Goats | Horses | Bison | Pronghorn antelope | Mule-deer + black-tailed deer | White-tailed deer | Elk | Moose | Caribou | Bighorn sheep | Mountain goat |
|---|---|---|---|---|---|---|---|---|---|---|---|---|---|
| Cattle | | 0.2 | 0.5 | <0.1 | 0.2 | 0.6 | 0.6 | 0.6 | <0.1 | 0.8 | 0.5 | 0.1 | <0.1 |
| Sheep | | | 0.4 | 0.2 | 0.4 | 0.4 | 0.4 | 0.4 | 0.2 | 0.7 | 0.3 | 0.3 | 0.2 |
| Goats | | | | 0.4 | 0.6 | 0.3 | 0.2 | 0.2 | 0.4 | 0.6 | 0.1 | 0.4 | 0.5 |
| Horses | | | | | 0.2 | 0.5 | 0.6 | 0.6 | <0.1 | 0.8 | 0.4 | 0.1 | 0.1 |
| Bison | | | | | | 0.8 | 0.8 | 0.8 | 0.2 | 0.9 | 0.7 | 0.3 | 0.2 |
| Pronghorn | | | | | | | 0.2 | 0.2 | 0.5 | 0.5 | 0.2 | 0.5 | 0.6 |
| Mule deer + black-tailed deer | | | | | | | | <0.1 | 0.6 | 0.3 | 0.1 | 0.5 | 0.6 |
| White-tailed deer | | | | | | | | | 0.6 | 0.3 | 0.1 | 0.5 | 0.6 |
| Elk | | | | | | | | | | 0.7 | 0.4 | 0.1 | 0.1 |
| Moose | | | | | | | | | | | 0.4 | 0.7 | 0.8 |
| Caribou | | | | | | | | | | | | 0.4 | 0.4 |
| Bighorn sheep | | | | | | | | | | | | | 0.1 |

Table 4.10. *Differences between diets of herbivores*

| 10 % or less | 80 % or more |
|---|---|
| Cattle and horses | Cattle and moose |
| Cattle and elk | Horses and moose |
| Cattle and mountain goat | Bison and pronghorn |
| Horses and elk | Mule deer + black-tailed deer and bison |
| Mule deer + black-tailed deer and white-tailed deer | White-tailed deer and bison |
| | Moose and bison |
| | Moose and mountain goat |

horses, elk, cattle, mountain goat and bison. The browsers, i.e. those having more than 50 % shrubs in their diet yearlong, are, in order of increasing importance, caribou, goats, mule deer plus black-tailed deer, white-tailed deer, and moose. The degree of dietary overlap among species is given in Table 4.10.

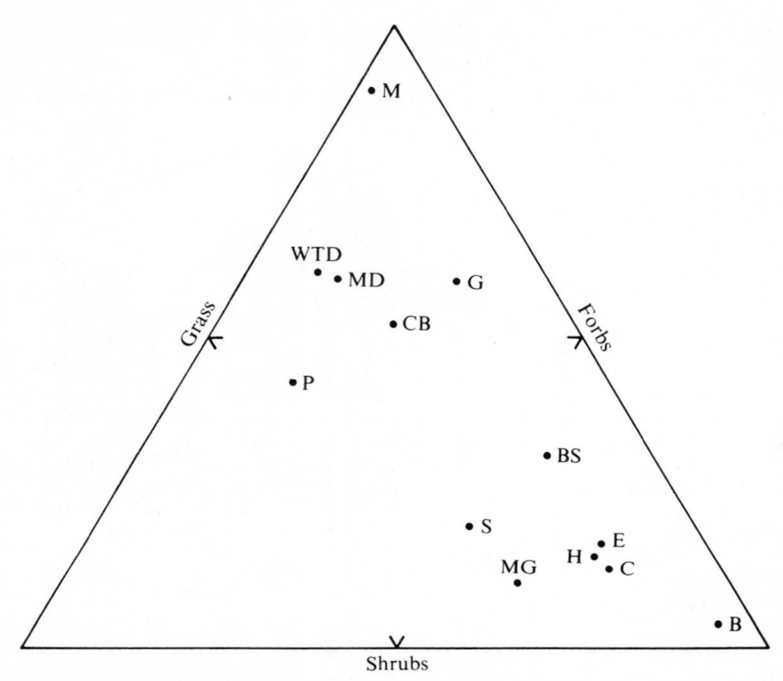


Fig. 4.4. Overall average values for dietary botanical composition for selected domestic and wild herbivores based on data of Table 4.5. See also relative differences values for species pairs in Table 4.7. B, bison; BS, bighorn sheep; C, cattle; CB, caribou; E, elk; G, goat; H, horse; M, moose; MD, mule deer; MG, mountain goat; P, pronghorn; S, sheep; WTD, white-tailed deer.

### *Relative selection: a case example*

Percentage composition of the herbage, aggregated to major botanical categories, has been presented in Table 4.4 along with the relative dietary preferences of cattle, bison, sheep and pronghorn grazing on the rangeland in various seasons (spring, early summer, autumn and winter). We have extracted these data from various published and unpublished results from the US/IBP Grassland Biome study. These animals all had access to the same herbage in the same pastures. The light grazing intensities and heavy grazing intensities imposed approximated those of long-term studies on the intensity of grazing on the USDA Central Plains Experimental Range on equivalent shortgrass prairie. The relative selectivity by these herbivores was calculated using Kuzlinski's index of the similarity and subtracting this value from 1. Our 'index of selection' compares the diet of each animal to the available herbage in the same time period. This index of selection is calculated as follows:

$$\text{Index of selection} = 1 - [(2 \sum_k W_k)/200]$$

Table 4.11. *Index of selection of four large herbivores grazing in common on shortgrass prairie in northeastern Colorado*

| Season | Cattle | Bison | Sheep | Pronghorn |
|---|---|---|---|---|
| | Light grazing | | | |
| Spring | 0.12 | 0.30 | 0.07 | 0.61 |
| Early summer | 0.21 | 0.18 | 0.46 | 0.83 |
| Late summer | 0.22 | 0.20 | 0.63 | 0.87 |
| Autumn | 0.21 | 0.28 | 0.25 | 0.26 |
| Winter | 0.29 | 0.29 | 0.27 | 0.29 |
| | Heavy grazing | | | |
| Spring | 0.08 | 0.10 | 0.02 | 0.51 |
| Early summer | 0.42 | 0.21 | 0.67 | 0.85 |
| Late summer | 0.16 | 0.30 | 0.21 | 0.82 |
| Autumn | 0.19 | 0.48 | 0.37 | 0.54 |
| Winter | 0.69 | 0.69 | 0.33 | 0.61 |

where $W_k = \min\begin{cases} H_k \\ D_{ik} \end{cases}$ if $H_k$ and $D_{ik}$ both $\geqslant 0$ and $W_k = 0$, otherwise

where $H_k$ is the percentage of plant group $k$ in available herbage and $D_{ik}$ is the percentage of plant group $k$ in the diet of animal species $i$. Our index has a range of 0 to 1. The value 1 represents maximum selectivity and 0 represents no selectivity whatsoever. We present our results in percentages. Selectivity, as defined here, means either selection or rejection in contrast to taking the plants in the diet in the same proportions as they occur in the herbage.

Overall, the four herbivore species, averaged over both grazing treatments at all seasons, expressed a selectivity of about 40% (Table 4.11). Averaged over all seasons and grazing intensities and using the selectivity value for cattle as a base, bison were about 15% more selective in their grazing, sheep 27% more, and pronghorn more than twice as selective.

This selectivity varied seasonally. Averaged over both years the greatest selectivity was either in the summer or the winter, and it was lowest in the spring. However, the selectivity pattern, averaged over all animal species, differed between heavy and light grazing. Relatively, the animals are much more selective under winter conditions under light grazing due to the selection of shrubs. Total herbage available was always sufficient in light grazing, but it became limited in heavy grazing. Thus in the winter period in heavy grazing there was less than 200 $kg \cdot ha^{-1}$ available to the animals. At the other extreme in early summer there was almost 1400 $kg \cdot ha^{-1}$ available. Yearlong, there was about one-third more herbage available for grazing under the light grazing treatment than under the heavy grazing treatment. However, selectivity was about one-quarter greater under heavy grazing than under light grazing. This is due largely to the very heavy selectivity (particularly by cattle, bison and pronghorn) in the winter grazing under heavy use. All animal species except sheep increase their selectivity under heavy grazing as compared to light grazing, and the values for sheep were about equal under both grazing treatments.

These selectivity values agree in general with the overall average from the literature (see Table 4.9), but there are relatively few studies where diets of several large herbivores have been studied simultaneously with the same methodology and to the same degree of sampling. Even among the few comparative studies there are fewer yet for which a measure of the herbage availability was obtained. The major conflict of our results with those in the literature is for pronghorn diets. Under both light grazing and heavy grazing in autumn and winter sampling, more than 70% of the diet consisted of grass and grass-like plants. The literature in general reports a much greater percentage of shrubs in the diet under autumn and winter conditions. Our data for pronghorn at other seasons agree relatively well with the literature values, giving higher percentages of forbs in the diet and relatively high percentages of grass in the early spring. However, our percentage of dietary grass in early spring averaged about 40% and this too would exceed most literature values. Our data for pronghorn were based on work of Schwartz & Nagy (1976), who used a bite-count method to determine the relative number of bites the animals made of different plants and then collected representative 'bite-units' of different species to estimate the weight eaten. These weights were used to calculate percentage weight of the individual species and these were grouped into our categories of grass, forbs and shrubs. Much of the literature on dietary botanical composition for pronghorn has been based on gross observational techniques which did not involve this weight-unit correction procedure, or values based on rumen content analysis or on faecal analysis. There is considerable opportunity for rapid digestion of succulent herbaceous materials (i.e. more than for lignified materials) in the diet of pronghorn once they reach the rumen. Thus, rumen content samples might provide estimates of forbs biased downward unless detailed microscopic examination was made. Similarly, one would expect that faecal samples might have a higher than real proportion of woody or lignified materials than occurred in the diet; perhaps the grasses and forbs would be underestimated in such samples. To our knowledge, there has never been a thorough methodological study comparing procedures for estimating dietary botanical composition of several large herbivores simultaneously and providing good data on the vegetation composition at the same time.

### *Dietary chemical composition*

There have been many fewer studies of the chemical composition of the actual diet grazed by animals than of botanical composition. It has been shown in a number of studies (see Van Dyne & Torell, 1964) that chemical composition and botanical composition of the diet may be sampled most accurately by the use of oesophageal- or ruminal-fistulated animals. Thus in Appendix 4.3 chemical compositions are included only for those studies in which oesophageal-fistulated or ruminal-fistulated animals were utilized. Furthermore, data were restricted to those cases in which, in the original author's opinion, the sampling methodology was such that samples were not biased by partial digestion in the rumen. Sampling diets with either the oesophageal or rumenal fistula has the objection of salivary contamination. We have taken care to note, where possible, whether the data are presented on a dry matter basis or on an organic matter basis. Presumably, when one of these has not been specified, particularly for hand-clipped samples representing diets, the data are on a dry matter basis. References to studies of chemical composition of the diet are primarily for domestic animals. Equivalent notes are provided in Appendix 4.3 on the kind of grazingland, geographical location, methods and seasons. See Table 4.2 for an indication of the number of season–treatment instances where both chemical and botanical composition of the diet are reported.

Crude protein is the constituent most commonly determined in dietary samples for large herbivores. In a few instances the authors have reported the nitrogen content of the sample and we have used the conventional conversion factor of 6.25 to convert this to crude protein. This is based on the percentage of nitrogen in most proteins being 16%. The second most commonly determined measures are related to the fibrous nature of forages consumed by large herbivores. Here varying constituents are determined; lignin or crude fibre determinations are most prevalent. Dietary samples obtained either with oesophageal- or ruminal-fistulated animals generally do not provide valid estimates of the calcium, phosphorus or other body minerals in the diet. Energy values have been reported in the literature in various ways, but primarily as $kcal \cdot g^{-1}$ and $kcal \cdot lb^{-1}$.

Generally, dietary chemical composition is considered to be the same morning and evening within a day. But experiments with oesophageal-fistulated sheep on *Phalaris–Trifolium* pasture in Australia have shown diurnal variation in the crude protein content of the diet (Langlands, 1965). Lowest values were found in early morning with an increase up to about midday and then a plateau of values for the remainder of the day. This occurred on different types of pastures and with sheep that were not fasted before samples were collected. If oesophageal- or ruminal-fistulated animals are fasted prior to forage one might expect even greater differences. Although they may be biased by

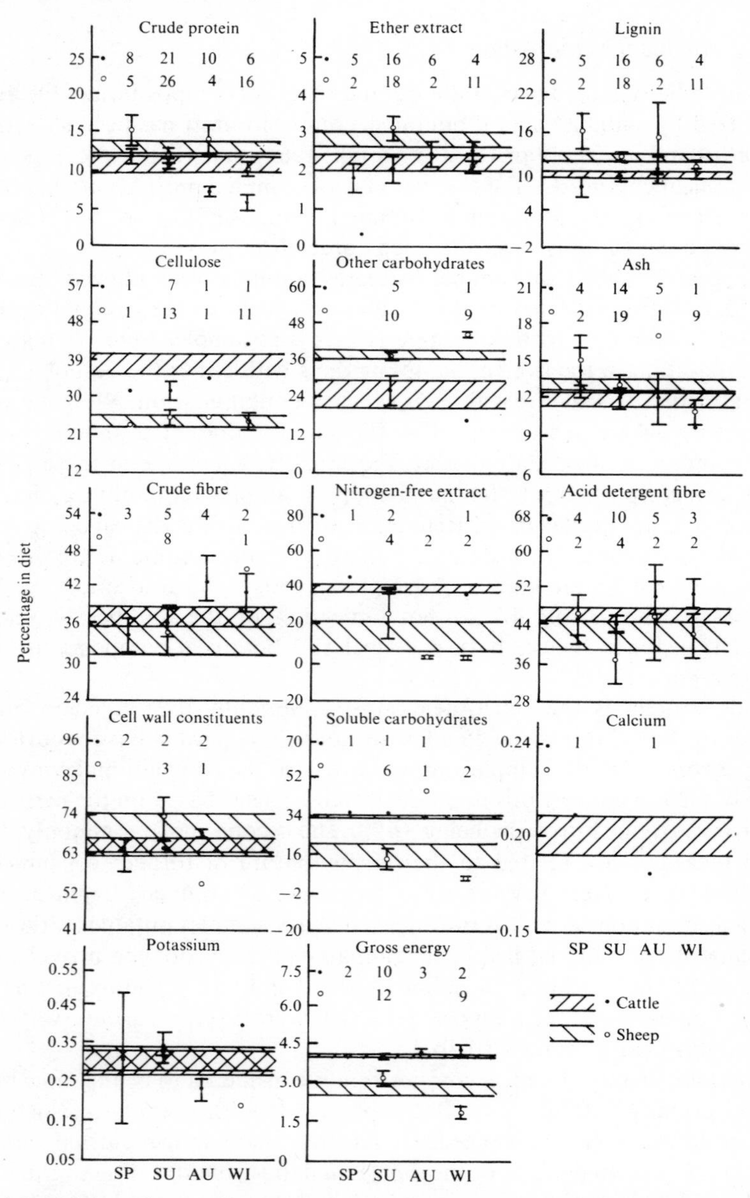


Fig. 4.5. Relative dietary chemical composition for cattle and sheep related to grazing season. Sample size values for each season or yearlong are shown at the top of each graph. Means and one standard error are plotted. The results presented here are derived from the data in Appendix 4.3. The abscissa refers to spring, summer, autumn and winter.

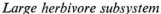

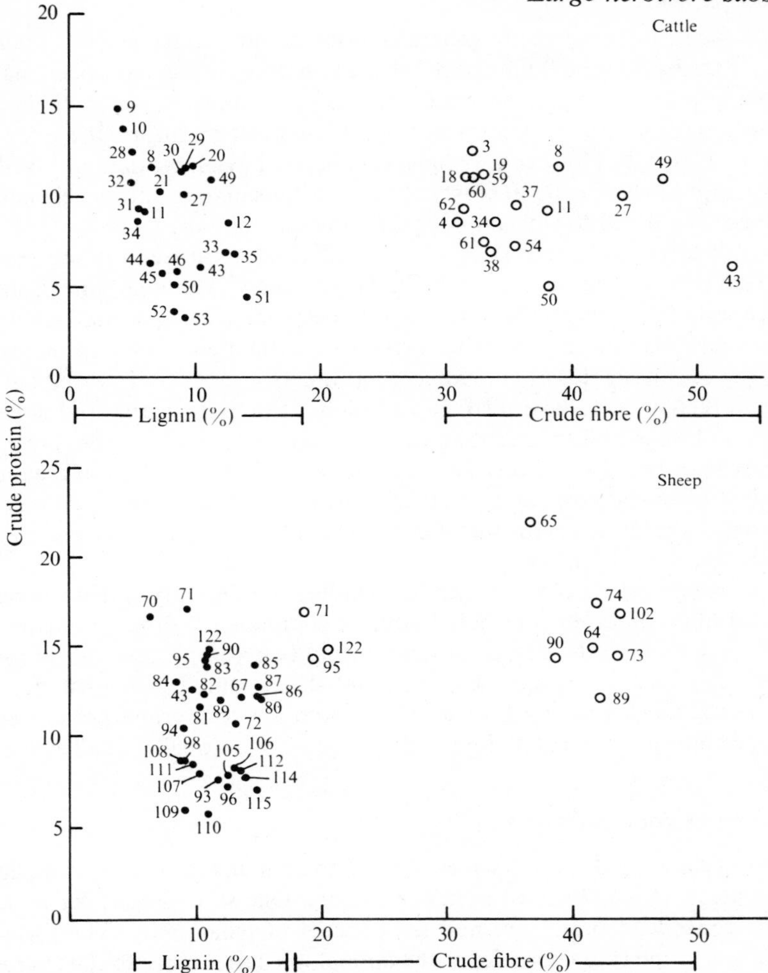


Fig. 4.6. Relationship between crude protein and fibrous components of diets consumed by cattle and sheep on the world's grazinglands. Data points are numbered according to entries in Appendix 4.3.

salivary contamination, because of their uniqueness we report here (but not in Appendix 4.3) some data on elephant diets. Average chemical composition of dietary samples taken from the oesophageal region of the stomach of elephants from East Africa, and representing November to December, January, March, and May collections, were 8% crude protein, <2% crude fat, 44% carbohydrates, 36% fibre and 11% ash (Laws *et al.*, 1975). The protein level varied from about 5 to about 12% for the dry season and the wet season, respectively.

It is difficult to make many generalizations about dietary chemical composition data based on the limited number of analyses which have been made for grazing animals. Note the relatively smaller sample sizes for chemical composition information than for botanical composition information about diets (see Table 4.2). However, we have summarized available data in Fig. 4.5 in an attempt to show general seasonal relationships among various chemical constituents analysed for in samples grazed by the animals.

For both cattle and sheep there is a distinct downward trend in the crude protein content of the diet from spring through winter. Although sample sizes and locations may vary, these data suggest sheep are grazing a diet higher in crude protein than are cattle. Ether extract in cattle diets shows an increase during the season as does lignin concentration. However, these variables in the sheep diet do not have distinct variations. For both cattle and sheep, dietary cellulose is higher later than earlier in the season. For cattle, but not for sheep, there is a distinct downward trend in ash concentration in the diet. Insufficient analyses were made of other constituents to show seasonal trends and perhaps the most useful information provided in Fig. 4.5 is the mean value for the overall season.

Crude protein generally is considered an indicator of quality of diet whereas lignin and fibre are often considered adverse components. Plots are made in Fig. 4.6 to determine the degree of relationship between protein and fibrous components of the diets grazed by cattle and sheep. Only for the instance of cattle diets is there a strong inverse relationship between crude protein and lignin percentages. In the other instances the trends were weak if at all evident.

### *A summary of food selection*

Food selection is highly variable from animal to animal, season to season, and place to place. It is influenced by many factors. One should consider separately the concept of preference and the concept of palatability. The former relates more to the characteristics of the animal, the latter more to the characteristics of the plant. But both preference and palatability may be influenced by similar factors. For example, weather conditions may influence both the animal and the plant and change the amounts of food selected. Forage preference can be expressed best as a ratio of the proportion in the diet compared to the proportion available for the animal to graze. Such selection or preference ratios generally range from 0.1 for plants being 'rejected' up to 10.0 for plants being 'selected'. For comparison of such preference ratios, they need to be normalized and expressed on a relative or percentage basis. Examples were given of relative dietary preference values for four large herbivores at several seasons of the year grazing in common on a shortgrass prairie rangeland.

Data on the botanical composition of the diet were summarized from the literature (Appendix 4.2). An attempt was made to aggregate the data into the plant groups of grasses, forbs and shrubs for comparative purposes. Plots were made of these data to illustrate the general nature of the diets of individual animal species or groups under various seasons. These data also illustrate clearly the considerable lack of information about the diets of the world's grazing herbivores. This lack of information is apparently primarily for wild herbivores, but it is also lacking for domestic herbivores under many grazingland situations.

The degree of selectivity an animal expresses can be calculated readily if one has a good measure of the herbage available. Thus, in comparative studies it is possible to show the percentage dissimilarity of the diet relevant to the herbage available, thus providing an index of selectivity. These values are provided in an example with four large herbivores.

Data on dietary chemical composition were reported from the literature (Appendix 4.3). Few general conclusions could be drawn from such data, and further comparative work on the chemical composition of diets of grazing herbivores is needed. In particular, no valid data obtained with fistulated animals seem available for mineral composition of diets or for the water content of food being grazed. This requires the use of isotopic tracers to 'tag' the mineral elements within the body of the ruminant and permits, for example, salivary contamination to be accounted for in the determination of the mineral nutrients in the fistula forage sample. It has been shown in numerous investigations that it is difficult or impossible to sample the herbage being grazed to the same degree as does the grazing animal. Thus, many botanical sampling methods or observational methods of estimating dietary botanical and chemical composition provide data that are only indicative of the actual diet.

Few general conclusions can be drawn from examining data on the dietary selectivity of a wide range of large grazing herbivores. One consideration is that the protein content of the diet generally increases with decreasing body size (Bell, 1970). It has also been shown in nutritional studies that there is a relationship between $W^{0.74}$ and endogenous nitrogen excretion. Thus, proportionately, small animals would have higher nitrogen requirements. This should be reflected in protein intake. Because the small animal also has a higher energy requirement, the diet should also have a higher proportion of soluble carbohydrates than in the case of the larger herbivores.

Most of the large herbivores reviewed are ruminants, but several important nonruminants such as the horse and the zebra occur on the world's grazinglands. These animals depend on an enlarged colon and large intestine rather than on a rumen as a site of fermentation. The rate of passage of food through the horse's alimentary canal is faster than that through the ruminant. There is some evidence that the digestibility of cellulose is greater in the ventral colon of the horse than in the rumen of cattle. If the horse could digest protein in its food at the same rate as do the ruminants then, because of faster food passage rates, it should be able to exist on a diet with a lower protein concentration. The general statement that 'zebras always appear fat' could thus be borne out by such considerations.

## Food intake rate

Food intake rate is an important phenomenon in large herbivore studies. Its measurement requires careful and sometimes ingenious methods for various herbivores. Several important physical, physiological and behavioural factors influence food intake rate. We provide here a brief summary on methodology, a discussion on factors influencing food intake rate, and then a presentation and analysis of food intake rate data.

### *Methods for determining forage intake*

About five indirect methods have been used to determine the forage intake of grazing animals from total faecal output and composition or from body weight changes: lignin and chromogen ratio procedures, faecal nitrogen index, nitrogen balance, metabolic fraction of faecal determinations and weight balances. Equations used in most of these methods are summarized in Appendix 4.1. More information on animal nutritional techniques for use under grazingland conditions is provided by Harris *et al.* (1967).

In limited situations food intake rate may be calculated from total urine collections and hence determinations of indigestible compounds therein, such as quinol and orcinol outputs from heather grazed by sheep (Martin, Milne & Moberly, 1975). Faecal collection is much more common, however. Chromic oxide ($Cr_2O_3$) is sometimes used to bypass collecting the total faecal output. Under ideal conditions chromic oxide administered to the animal is mixed well in the digestive system and is uniformly excreted. If one then takes a 'grab sample' of faecal material and determines the concentration of chromic oxide, it is possible to calculate the proportion of the total faecal output of the animal that was in the grab sample and thus calculate total faecal output. The difficulty under grazingland situations is to get uniform excretion of the faecal material. There is some evidence that there is a diurnal variation in the excretion rate.

In the nitrogen balance technique mature animals are generally used because their weight does not vary appreciably. The animal stores nitrogen only in the wool or hair, and if this can be accounted for then nitrogen intake must be equal to that plus the amount in the urine and faeces. If the percentage nitrogen in the dietary material can be determined, then the total intake can be calculated. This technique requires long-term studies to overcome the appreciable variations in excretion rate and assumptions about the amount of nitrogen stored in the animal body. These requirements preclude extensive use of the system for calculating forage intake of the grazing animal.

The water intake method requires predicting the water requirements of the animal based on body size and climatic conditions. The assumption is made that if the amount of liquid water drunk can be measured, then the remaining water that is required must come from the herbage that is grazed. Next, if the percentage water in the herbage that is grazed can be determined, then the total amount of herbage grazed can be calculated. This technique has been used reasonably successfully under grazingland conditions where the water supply may be metered to the grazing animal. With this technique care must be taken in sampling the herbage moisture content. Although herbage moisture content decreases as seasons progress, there may be as much as 3% average increase overnight as the plants regain turgor (Hyder, Bement, Norris & Morris, 1966). Rain can temporarily increase the moisture content of green grass by as much as 10% and of dry grass by as much as 20% in a semi-arid grassland.

Alldredge, Lipscomb & Whicker (1974) utilized unique data on $^{137}Cs$ concentrations in plant species collected from a Colorado, USA, mountain range area, information on the botanical composition of the diets of mule deer from rumen analysis, and $^{137}Cs$ concentrations in deer shot from the range to calculate forage intake rates of the freely grazing animals. Their paper should be referred to for a discussion of the assumptions involved in the method. Forage intake was calculated as:

$$R = Q/a \int_0^x \lambda_{(x-t)}[A\,\mathrm{e}^{-\lambda_1 t} + B\mathrm{e}^{-\lambda_2 t}]\,\mathrm{d}t,$$

where $R$ = forage intake rate
$Q$ = body burden
$a$ = assimilation value for ingested caesium
$c$ = concentration of caesium in the forage
$A, B$ = fractional retention coefficients for 'fast' and 'slow' body components (0.23 and 0.77 respectively in this study)
$\lambda_1, \lambda_2$ = elimination rate constant (0.693/half-time) of fast and slow components ($\lambda_1 = 1.2$ per day and $\lambda_2$ calculated from each carcass and weight)
$x, t$ = time values ($x = 100$ in this example)

Another unique way to measure food intake is to use the rate of depletion of an isotope in the body of an animal when it does not receive that isotope in its food or water. Benjamin *et al.* (1975) in Israel dosed sheep with tritiated water and determined the concentration in the blood periodically while the sheep were grazing a pasture but were not consuming liquid water. Dry matter intake was calculated by knowing the 24-h water turnover in litres, the dry matter percentage in the pasture herbage grazed, and the percentage water content of the pasture grazed.

It is possible to make a crude estimate of forage intake on the basis of the weight balance of the animal. The animal is weighed before the grazing and then again afterwards. It must be harnessed for faecal and urine collection and insensible weight losses have to be calculated. These techniques are not very useful under grazingland conditions.

Forage intake rates for some large herbivores, such as elephants, are difficult to measure in the field. Data are based largely on either the amount of stomach contents and the estimate of passage time or on collecting and measuring faecal output and estimating digestibility.

An 'ideal' method for determining forage intake of grazing animals should (i) be applicable to individual animals rather than to groups, (ii) be based on measurement of dietary or faecal components which can be easily and accurately analysed, (iii) not depend upon harvesting the range herbages for dry-lot digestion trials, (iv) be applicable to various large herbivores, (v) be usable on various types of grazinglands and in all seasons. Most of the methods developed to date do not meet all of these requirements. Van Dyne & Meyer (1964*b*) developed a technique of combining the use of fistulated animals and artificial rumen procedures. Dietary chemical composition was measured in samples taken from oesophageal-fistulated animals. Digestibility was measured by use of inocula obtained from ruminal-fistulated animals grazing on the same rangeland. The in-vitro system was used to digest the type of forage being grazed as well as standard forage samples. Faecal production was measured by means of total collection from the animals on the same grazinglands. Given this information, an equation was developed to calculate forage intake of the animals involved (see Appendix 4.1).

Another recent methodology is the use of oesophageal-fistulated cattle to collect large amounts of forage from a Colorado, USA, grazingland (Wallace & Denham, 1970). This forage is later fed to sheep in digestion or metabolism studies. This bypasses the use of index, indicator, or other indirect methods of estimating digestibility. It requires, however, a determination of the correlation between the abilities of cattle and sheep to digest forage.

### *Factors affecting food intake*

Food intake regulation has long been a topic of interest but reviews (Brobek, 1957; Anand, 1961; Balch & Campling, 1962; Bell, 1971) show that more information is available for monogastric animals than for ruminants.

#### *Rumen capacity and food digestibility influences*

It has long been recognized that ruminants graze to a relatively constant amount of food in the rumen. Voluntary intake is affected by rumen capacity at an upper limit. Campling & Balch (1961) found in cows that a change of 1 unit of dry matter in the reticulo-ruminal contents resulted in an inverse change in voluntary dry matter intake of 0.6 units. Thus, neither the exhaustion of the salivary glands nor the muscles of the jaw and reticulo-rumen are important in halting herbage consumption. The amount of food residues in the reticulo-rumen immediately before feeding is relatively constant over a fairly wide range of roughages for cattle (Campling, Freer & Balch, 1961). Increasing the nutritional level of the diet (protein intake) decreases retention time of residues in the alimentary tract. Some 24 h after feeding, under stall conditions, there was about the same amount of digestion in the reticulo-rumen of the cows whether or not they got a protein supplement (Campling, Freer & Balch, 1962). Thus, voluntary intake of roughage is related to the rate of disappearance from the reticulo-rumen.

Sheep fed diets of chopped alfalfa, hay and wheaten chaff had digested 42% of the material after 72 h, some 60% of which had not been excreted. Thus, about 25% of the meal remained as residues in the digestive tract at 72 h. After 10 to 14 d this amount was 0%. These data suggest that the rate of passage of fibrous foods through the digestive tract of sheep is largely determined by physical properties of the food and the dimensions of the digestive tract rather than by physiological factors (Graham & Williams, 1962).

The rate at which the rumen load is reduced depends upon the capacity of the rumen, the rate of fermentation of the feed and the rate of passage of undigested residues (Ulyatt, 1964). A high-quality forage is broken down and passed through the rumen more readily than is a low-quality forage. Thus, rumen load is reduced more rapidly with the high-quality forage. Increased levels of feed intake only slightly increased the weight of digesta in the large intestine but markedly decreased the retention time (Grovum & Hecker, 1973). With high-quality forage the animal can eat more frequently and a greater amount. The mechanism of the effect of food level or fill-controlled food intake is not clear. It is possible that there are receptors in the reticulo-rumen that are sensitive to stretch and thus control food intake. There are, however, situations in which food intake rate can be depressed, probably as the result of other factors. Intake regulation may also be correlated with the rise in metabolites in the blood after a meal. Such chemostatic regulation has been suggested for ruminants fed concentrate rations but not those feeding on grazinglands.

Rate of digestibility affects food passage in various ways. The reticulo-omasal orifice of many ruminants is relatively small and thus the roughage must be reduced in size by digestion and rumination and mastication processes before it can leave the reticulo-rumen. Generally, digestibility of a food and its time of retention in the gut are inversely correlated. Other work (Egan, 1965) suggests that mechanisms other than 'fill' of stretch-sensitive areas of the gastrointestinal tract must be employed to regulate voluntary consumption of roughage in ruminants. Thus animals fed on low-protein roughage have low forage intakes, but by supplementing the animal with nitrogen, forage intake is increased. But rate of passage of food through the alimentary tract does not alone account for greater intakes.

Increasing the protein content of the diet generally increases digestibility and thus increases herbage intake. In some instances feeding a supplement at higher levels will eventually decrease herbage intake. Thus Allden & Jennings (1962) found that in Merino wethers fed 400 g dry matter of peas there was a decrease in herbage intake of 8% for each 100 g peas fed. These animals were grazing *Phalaris*, dry annual grass, and subterranean clover pastures, with crude protein percentages varying from 4 to 16%. There was a linear decrease in herbage intake, as a percentage of metabolic weight, as the amount of supplement fed increased.

Blaxter, Wainman & Wilson (1961) have found in sheep that dry matter content of the gut content is the same for different qualities of forage fed. They indicated that the digested energy consumed per day per unit of metabolic weight could be related to voluntary intake per unit of metabolic weight by the equation: $E = 4.9(I-31)$, where $E$ is digested energy and $I$ is voluntary intake expressed as above. Thus, the voluntary intake of a forage by an animal can be used as an index of its nutrient value over a wide range of dietary quality. An increase in digestibility of a feed from 50 to 55% is not simply a relative improvement of 10% of the value of the food. This would cause an increase in voluntary intake from 62 to 72 g dry matter $(W_{kg}^{0.734})^{-1}$ and an increase in the energy digested per day from 152 to 200 kcal $(W_{kg}^{0.734})^{-1}$, or an increase of 32%. The amount of energy available above maintenance increases nearly 200%. Thus an increase in nutritive value of 10% in digestibility would increase live weight gains by nearly 100%.

In pen or stall feeding the animal usually consumes its daily diet in only one or two meals. Under grazingland conditions the animal may consume forage during five or six different periods of the day, so that there is not the clear cycle that there is under stall-fed conditions. But digestibility of forage must still influence the rate of passage and thus food intake.

In many experiments, food intake rate of ruminants on low-quality roughages is increased by supplements of concentrates containing nitrogen. Addition of starch in the diet may depress the cellulose digestibility in hay. But under grazingland conditions when the animals' diet needs to be supplemented, then energy and protein may both be beneficial.

Voluntary dry matter intake for cattle over a range of roughages from 16 down to about 3% protein varied from about 90 down to 44 g $(W_{kg}^{0.734})^{-1}$. Apparent digestibility of energy varied from about 70 down to 46% for these same feeds (Blaxter & Wilson, 1962). There was considerable variability among individual animals in food consumption rate, but those that consumed the most food digested it least efficiently. The relationship of dry matter intake, on a relative body weight basis, to apparent digestibility of energy for cattle was similar to that in sheep. Thus there was not any variation between sheep and cattle in their voluntary intake of roughages of the same quality. These animals gained about 0.35 kcal per kilocalorie of digestible intake. Detailed experiments of Greenhalgh & Reid (1971), who fed various roughages to sheep orally or via a rumen fistula, suggest that food intake rate of better quality roughages is not much influenced by palatability, but is for poor-quality roughages.

### *Animal fatness and pregnancy influences*

Several important factors of the animal's physiological status affect food intake rate by influencing capacity and body chemistry. Under both grazing and pen conditions large herbivores deposit considerable amounts of fat. Schinckel (1960) reported that the food intake of Merino wethers fed in pens declined from about 3% of the body weight for 44 kg sheep to 1.9% of the body weight for the sheep at 52 kg when they became fat. Taylor (1959) has shown that the amount of fill weight is lower in steers carrying more internal fat. More fill was found in animals later in the season than early in the season, and the increased internal fat was negatively correlated with faecal dry matter production and presumably with herbage intake rate. The results were discussed in respect of the idea of 'compensatory growth'. In such a situation, animals wintered frugally go into the spring-grazing situation with more body capacity for forage. They have higher forage intake rates, and gain back the difference in weight relative to the same type of animals wintered at a higher level of nutrition and carrying more fat. Fat cows have been reported to consume 20% less of a concentrate ration in 5 h than did the same cows when thin (Bines, 1971).

Pregnancy results in an increased demand on the animal for nutrients and thus for food intake. The pregnancy effect is positive in early stages but may be negative in late stages (Bines, 1971). There is an interesting potential feedback loop involving pregnancy, food intake rate, food digestibility and nutrient absorption. In feeding studies with pregnant and nonpregnant ewes Graham & Williams (1962) demonstrated that the passage of digesta, with feed intake held constant, became more rapid as pregnancy advanced, the mean retention time falling by 1.0 to 0.5 h per 100 g increase in estimated weight of concepta. Thus the rapid passage of digesta would depress the digestibility of some components of diets of grazing animals and might contribute to undernutrition in late pregnancy.

### *Environmental influences*

Several environmental factors affect food intake of grazing animals. Temperature is the most common of these and it generally has a positive effect. However, at extremely high temperatures, food intake is halted. Food intake rate of dairy cows increased by about 3 kg as the daily minimum ambient air temperature decreased from 4 to 1 °C. Appetite stimulation continued for at least 24 h following cold stress (MacDonald & Bell, 1958*a*, *b*). In contrast, water intake, which was roughly 1 $ml \cdot cal^{-1}$ heat produced, was reduced but its effect was overcome within 24 h. The increase in food intake in cold is to provide for heat production to maintain body temperature. Heat production for cattle in cold environments is appreciable. Equations are presented by Webster *et al.* (1975) for heat production in outdoor environments and these are used to calculate a wind chill index for cattle. For example, an air temperature of −20 °C and a wind speed of 200 $m \cdot min^{-1}$ gives the same stress as −38 °C and no wind. These calculations are further extended to determine critical temperatures (based on rectal temperature, external insulation, heat production in a thermoneutral environment and tissue insulation). Order of tolerance to cold, from greatest to least, would be a steer on full feed, a cow on maintenance-level feeding, and a young calf even if fully fed.

Digestible organic matter intake and dry matter intake for sheep are affected by shearing. Various studies have reported from 15% up to 62% increase in digestible organic matter intake following shearing (e.g. Minson & Ternouth, 1971). Values in the range of 15% to 30% occur under warm conditions, higher values under cooler conditions. Food intake would increase in the shorn sheep as the animal may need more energy to maintain body temperature. However, some workers have shown decreases in digestibility of 3 percentage units for dry matter and organic matter as a result of exposing the animal to cold typical of that in the winter on Canadian prairies (Westra, 1975; Christopherson, 1976; Kennedy, Christopherson & Milligan, 1976). This effect evidently is due to a reduced retention time of the food in the rumen. Digestion in the rumen is greatly decreased, but this is partly compensated by increased digestion in the intestines.

Another poorly understood environmental factor affecting food intake of ruminants is light. Sheep have higher voluntary food intake in summer than in winter (El-Shahat, Jones, Forbes & Boaz, 1974). At equivalent levels of feeding, animals with 16 h light per day produced significantly more than those with 8 h light and they also had 6% higher food intake rates.

Large herbivores seem to have intake limited by herbage availability in many situations. Allden (1962) has presented data showing a curvilinear relationship between intake of forage and availability of herbage dry matter from about 570 to 5700 $kg \cdot ha^{-1}$. Intake of green material by a sheep varies from less than 500 to almost 2000 $g \cdot h^{-1}$. Above about 3400 $kg \cdot ha^{-1}$ of dry matter, both grazing time and rate of intake of green material are relatively constant. But 3400 $kg \cdot ha^{-1}$ of dry matter usually is not available to the animals on arid and semi-arid grazinglands (Van Dyne *et al.*, 1977). The question becomes, for such sparsely vegetated land, 'how limiting is herbage availability?'

On desert shrub rangeland, daily forage consumption by sheep is directly related to length of grazing time and the time spent grazing is affected by snow cover, hoarfrost, rain and high winds. Normal intakes were attained with 9 h grazing. Some 5 to 6 h grazing per day resulted in only 60 to 70% of normal consumption. When the sheep were grazed or herded on pure stands of vegetation their daily intake was reduced. After a month or more on a single-species stand the reduction may have been 60% of the original intake rate. This reduction seems to be due to the lack of variety in the diet rather than the lack of available herbage (Cook, 1975). The reduction due to decreased availability of forage is of the same order of magnitude as that attained when the animals were kept on a single species for a period of time.

In some instances, under grazing conditions, large amounts of water ingested with the diet are the direct cause of low voluntary intakes (Campling, 1964). Arnold (1964) notes that high moisture content in herbage can reduce voluntary intake. There is a linear relationship between voluntary dry matter intake and the dry matter content of herbage, between 10 and 28% of herbage dry matter. Above this level there is little correlation. However, under grazingland conditions it is more often the case that high-moisture content herbage is associated with high nitrogen content and higher digestibility. We need more information on whether large herbivores on grazinglands are ever limited in dry matter intake as the result of a high water intake.

Under irrigated and subhumid pasture conditions, the amount of dung can be important in depressing palatability of herbage and, under extreme conditions, can reduce forage intake rate of grazing cattle (Marten & Donker, 1964). It is questionable, however, if this phenomenon would have a major impact on food intake rates of most species of herbivores under open range conditions.

### *General consideration of forage intake rates*

Data on forage intake rates by grazing large herbivores are summarized in Appendix 4.4. Information for each reference is provided on the animal species, vegetation type, location of the study, methods used in measuring forage intake rate, seasons or treatments, and (where given) the animal weight. In most instances, to measure forage intake rate requires an estimate of the diet, an estimate of faecal production and an estimate of digestibility. These are noted separately. Intake may be given in various units in the original publication, such as $kg \cdot animal^{-1} \cdot d^{-1}$, or in terms of animal unit equivalents as lb·AUE$^{-1}$·d$^{-1}$, percentage of body weight per day, and so forth. In not all cases has it been specified whether the intake was dry matter (which is the usual case) or organic matter. We have attempted to adjust these values to dry matter intake as a percentage of body weight per day.

Various workers have attempted to express food intake rate as a function of body weight or metabolic body weight. A rule of thumb for dry matter intake for ruminants is that they will eat daily about $0.11 W_{kg}^{0.75}$, thus intake is somewhat in excess of 10% of their metabolic body weight. There appears to be a great deal of variability in the literature, however. Early attempts to derive food intake rate values were undertaken in studies in which range forage was cut and fed to animals. Intake levels for such low-quality forages are often less than would be attained by the domestic animal on high-quality forage. For example, Guilbert (1929) fed wild oat hay to steers, which consumed 1.3% of their body weight. Guilbert & Mead (1931) fed burr clover to sheep, which consumed 2.2 to 3.3% of body weight. A mixed herbage consisting primarily of *Bromus mollis* was consumed by sheep at a rate of 1.3% of body weight; *Erodium botrys* was consumed at a rate of 1.0% of body weight in trials by Hart, Guilbert & Goss (1932). Although based on stall feeding of 11% crude protein and a ration made up largely of concentrate, burros of 164 kg consumed about 3.4 kg or 2% of their body weight per day (Knapka, Barth, Brown & Cragle, 1967). All the above values seem relatively low for grazingland conditions. Coop & Hill (1962) found that sheep on good pasture consumed about 5 to 6% of their metabolic weight per day to get their digestible organic matter intake. This amounts to about 1.4% of the body weight for a 45 kg sheep for maintenance intake of a high-quality grazed forage. Jamunapari goats in India consumed from 2.5 to 3.6% of their body weight per day while browsing (Saxena & Maheshwari, 1971).

When calculating food intake rate it should not be based just on body weight as such, but should be based also on the fat level of the animal. Thus, as will be discussed further in a later section, in some instances when modelling food intake rate, various workers have used separate values for thin animals and fat animals (Newton & Edelsten, 1976).

There is also considerable variability in the proportional food intake rate of different age and sex groups of animals. Hafez, Cairns, Hulet & Scott (1962) summarized data for sheep and indicated an average daily dry matter intake of 2.7% of body weight for young animals and 1.7% for older animals. Hafez further noted that, proportionately, forage intake is higher for lambs than for yearlings, and higher for yearlings than for adults. Similar estimates have been reported for a wide range of large herbivores. Alldredge *et al.* (1974) calculated forage intake rates of mule deer freely grazing on mountain ranges in Colorado, USA, using forage and body concentrations of fallout $^{137}Cs$. Forage intake estimates for various age groups were: 1 to 5 months, 3.4%; 6 to 11 months, 3.3%; 12 to 17 months, 3.0%; and $\geq$18 months, 1.7%.

There were no statistically different values between the sexes. Laws *et al.* (1975) estimated that daily forage intake rates for elephants would be about 4% of the body weight for immatures, mature males, and nonlactating mature females. For lactating females the value would be 6%. On a herd basis, taking into account the mixture of age and sex groups, the overall estimate was 4.4%. But this was an optimal value since it was based on the vegetation availability at the end of the rainy season; one could expect a lower intake rate under less favourable conditions. In contrast, Benedict (1936) reported laboratory studies with an elephant which consumed only 1.25% of its body weight per day of dry hay with digestibility of about 45%.

Forage intake rates under grazingland conditions may also be higher because of weather stress. Thus, the rule-of-thumb for dry matter intake of $0.11 W_{kg}^{0.75}$ would give too low a value for highly mobile animals in stressed environments, such as the case of wind stress on the northern caribou herds during the winter months. Low temperature increases heat requirements of the animals and thus increases food intake rates. This can cause decreased digestibility, as noted in a later section. Ames & Brink (1977) fed 25 kg, shorn, 4-month-old lambs a 13.5% crude protein pelleted diet at temperatures ranging from −5 to +35 °C. They found food intake rate, expressed in g $W_{kg}^{0.75})^{-1}$, could be calculated by the expression $y = 111 - 0.52T - 0.007T^2$, where $T$ is temperature (°C). Thus, considering 15 °C to be the thermoneutral temperature, and feed intake at that temperature to be 100%, dropping the temperature to 0 °C increases feed intake by about 8% and increasing temperature to 30 °C decreases it by about 14%. And these are not extreme ranges of mean daily temperature for animals on rangelands, especially in the winter when wind may also greatly increase the chill factor.

### *Summary of forage intake values*

Sufficient data were available in Appendix 4.4 to calculate seasonal and overall values of forage intake rates only for cattle and sheep (see Table 4.12). In Fig. 4.7(*a*) the numbers identifying data points refer to the case examples identified in Appendix 4.4. In Fig. 4.7(*b*) and (*c*) the pairs of numbers refer to the case examples identified in Appendixes 4.4 and 4.2, respectively. In Fig. 4.7(*d*) the pairs of numbers refer to the studies identified in Appendixes 4.4 and 4.5.

Forage intake rates for cattle are somewhat higher in the spring grazing period than in the summer. On a dry matter basis and averaged over 31 examples, cattle consumed 1.8% of their body weight per day. Surprisingly, sheep consumed as much or more forage, per unit of body weight, in the winter as in other seasons. Averaged over 41 examples, sheep consumed 2.4% of their body weight per day in dry matter.

Many factors influence forage intake rate and the data in Appendix 4.4

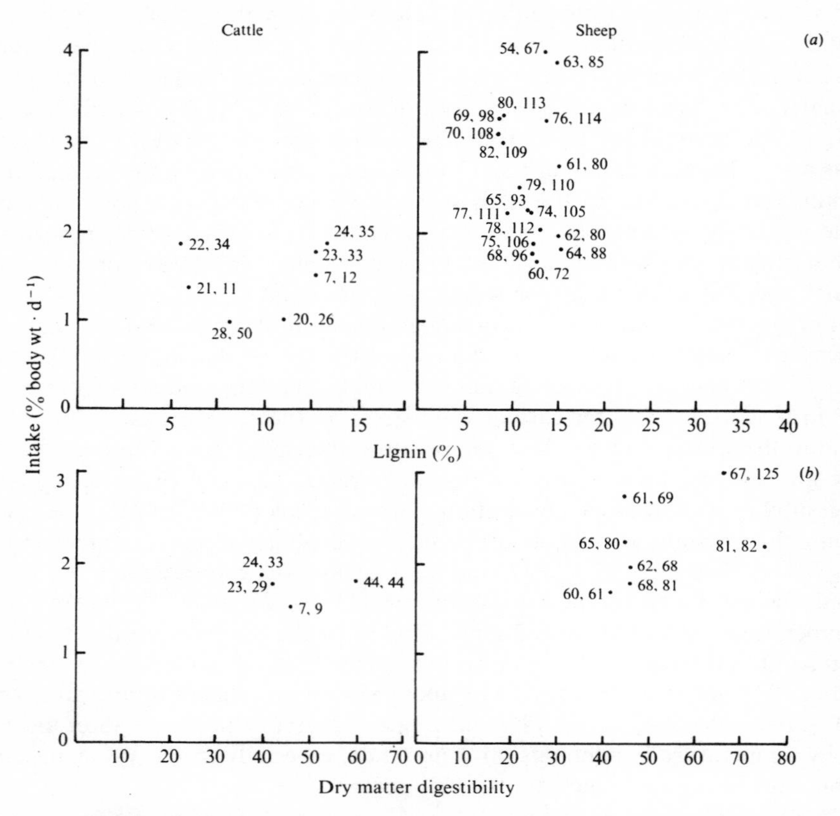


Fig. 4.7. Relationships of forage intake rate to body weight, dietary crude protein or lignin percentage, and dry matter digestibility based on data from studies noted in Appendixes 4.3, 4.4 and 4.5.

were examined for some of these factors; these are plotted in Fig. 4.7. For either cattle or sheep, for the range of body weights examined, there was no distinct relationship of weight to relative dry matter intake. However, if a few outliers are omitted in Fig. 4.7(*c*), the data suggest a slightly higher intake rate, expressed as a percentage of body weight, for smaller animals of the species involved. Relative forage intake rate is directly related to the percentage of crude protein in the diet (Fig. 4.7*d*). There was no clear relationship in the studies examined of the relative forage intake rate to the lignin concentration in the diet (Fig. 4.7*a*). Digestibility influences forage intake rate (as discussed in the next section) and intake rate could affect the digestibility. However, under the rangeland grazing conditions from which we extracted data, there was no clear relationship between relative forage intake rate and percentage dry matter digestibility of the diet (Fig. 4.7*b*).

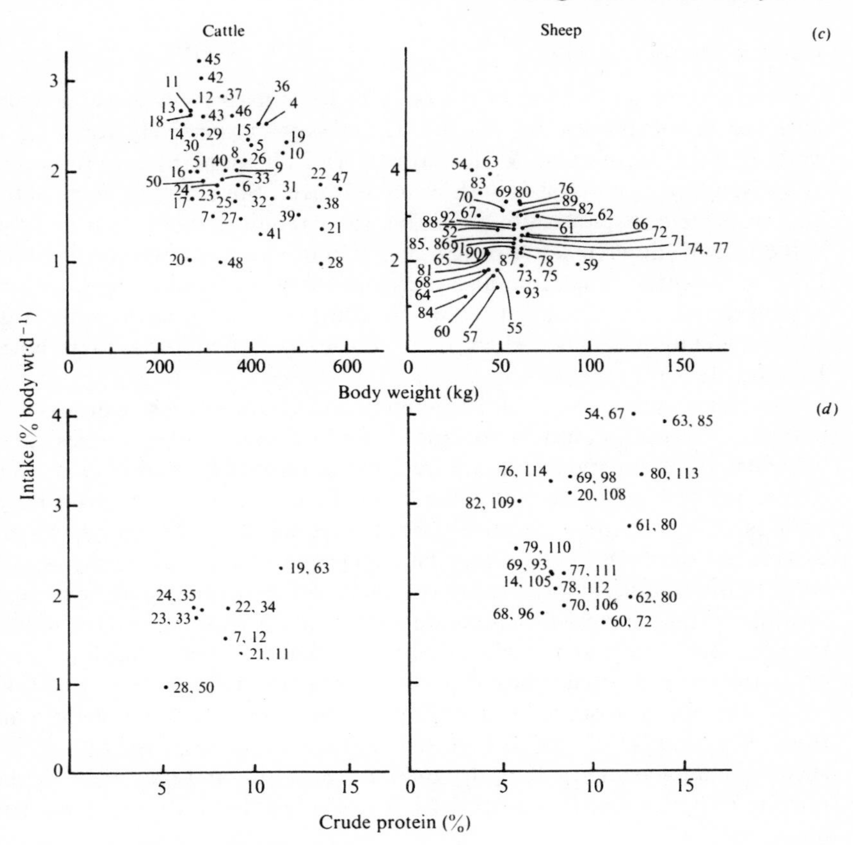


Fig. 4.7 (*cont.*). For legend see p. 332.

Table 4.12. *A summary of forage intake by grazing cattle and sheep at various seasons*

| Species | Season | Organic matter basis mean S.D. (*N*) | Dry matter basis ± mean S.D. (*N*) |
|---|---|---|---|
| Cattle | Spring | 2.4 ± 0.54 (4) | 2.1 ± 0.3 (2) |
| | Summer | 2.1 ± 0.13 (2) | 2.1 ± 0.49 (16) |
| | All seasons | 2.2 ± 0.57 (22) | 1.8 ± 0.62 (31) |
| Sheep | Spring | | 2.2 ± 0.87 (7) |
| | Summer | 2.7 ± 0.21 (2) | 2.4 ± 0.77 (8) |
| | Winter | | 2.5 ± 0.53 (11) |
| | All seasons | 2.7 ± 0.21 (2) | 2.4 ± 0.67 (40) |

Results were calculated from data in Appendix 4.4. Data are presented as daily organic matter or dry matter intake as percent of live body weight.

### *Water intake*

Generally some 60 to 75% of the body of large herbivores is water and the turnover rate extremes are 29 and 212 $ml \cdot kg^{-1}$ respectively for oryx and water buffalo (Macfarlane & Howard, 1974). In large herbivores food intake of a given type is correlated with water turnover. Some foods, such as those high in protein or potassium, alter the water required more than do others. Salt concentrations in water of less than 130 mM are required for horses and 170 mM for cattle. Lactation also increases the water requirement, and even among camels and Merino sheep in the arid tropics 44% more water is used by the lactating females than the non-lactating females (Macfarlane & Howard, 1974).

Ungulates such as oryx, addex, gerenuk and Grant's gazelle can survive for periods of several months with little or no free water; other species such as wildebeest, buffalo and zebra require water almost daily (Talbot *et al.*, 1965). The recycling of nitrogen within the body of the ruminant is an advantage in periods of water stress because then it does not take additional water to excrete the nitrogen in the urine. This recycling of nitrogen may also allow some of the wild herbivores to live on foods with an available protein content insufficient for survival of temperate-zone animals. Another factor reducing the required liquid water intake of many ruminants is that by feeding at night or in the early morning they can consume plants having from 10 to 40% water, largely hygroscopic water (Taylor, 1968), even under hot dry conditions. Benjamin *et al.* (1975) reported that the water turnover rate of 37 kg sheep grazing pasture of 18% dry matter is 3.8 $l \cdot d^{-1}$ and that of 36 kg sheep grazing 33% dry matter pasture 3.1 $l \cdot d^{-1}$. These animals did not have drinking water.

Water requirements vary considerably among species of ruminants. Under conditions of 40 °C day and 22 °C night temperatures, Taylor (1968) found Hereford cattle required about 64 g $H_2O \cdot kg$ body $wt^{-1}$ while Zebu cattle required only half that much. The oryx and Thomson's gazelle required somewhat less than did the Zebu cattle, whereas wildebeest and eland were about midway between the cattle breeds. Water intake varies widely with season. It ranged from 24 to 98 $ml \cdot kg^{-1} \cdot d^{-1}$ for oryx and 42 to 121 $ml \cdot kg^{-1} \cdot d^{-1}$ for eland (King, Kingaby, Colvin & Heath, 1975). The variation was due largely to changes in heat load expressed as solar radiation or mean ambient temperature.

In the wild, East African mammals can be divided roughly into two groups, i.e. the 'shaders' and the 'drinkers' (Vesey-Fitzgerald, 1960). The shaders (e.g. bohor reedbuck, topi and eland) may have no access to surface water for long periods in the habitats they choose. The drinkers (e.g. zebra, buffalo, puku, hippopotamus and elephant) often choose habitats with little shade but require surface water. Among domestic animals the camel, especially, and the donkey are notable in their water relations (Schmidt-Nielsen, 1959; Gauthier-Pilters, 1974). Both can lose about 25% of their body weight in dehydration, which can be more than 33% of body water for the camel. The camel's body temperature varies from 34 to 41 °C and it has, especially over the exposed back, a thick coat of hair-wool and fat insulation. The camel may go more than 2 weeks without water and the donkey 4 days. Both species may re-hydrate rapidly, regaining 25% of their body weight in as few as 10 and 2 minutes, respectively, for the camel and the donkey.

Water intake rate varies among breeds within a species, even among closely related breeds. Thus, the Bengal Zebu cattle, which are small, low-producing natives in a hot, humid area, have higher water intakes than do Hariana Zebu cattle which are from the East Punjab, India (Sharma, 1968). Dry matter intake is also higher for the Bengal cattle but they have significantly lower digestion coefficients for dry matter, crude protein, fats and total carbo-hydrates. Shorthorn and Hereford cattle drink more water than do Zebu cattle. In one comparison (Horrocks & Phillips, 1961), average water intake for Shorthorns was about 60 $mg \cdot kg$ body $wt^{-1} \cdot d^{-1}$ in contrast to 48 for Boran cattle. Both breeds averaged near 21 to 22 g $food \cdot kg$ body $wt^{-1} \cdot d^{-1}$.

## Food digestion

### *Methods for measuring digestion*

Digestibility of a forage may be determined for large herbivores by feeding them the forage, collecting the faeces, and determining the proportion digested. But under grazing conditions this cannot be done. It is impossible to feed the forage and, in many instances, impossible to make a total collection of the faecal material. Generally, faecal material is collected by putting a harness on the grazing animal and collecting the total faecal material excreted. Under grazingland conditions, frequently one group of animals is used to estimate diet composition (e.g. with oesophageal fistulas) and a different group to estimate faecal excretion rate and composition (e.g. intact animals). Determining digestibility under many grazing trials requires the use of an undigestible indicator in the forage, of which the total amount consumed will appear in the faeces. Digestibility is calculated according to the following relationship (see also Appendix 4.1):

$$\text{Digestibility of nutrient} = 1.0 - (\%\ \text{indicator in diet}/\%\ \text{indicator in faeces}) \times (\%\ \text{nutrient in faeces}/\%\ \text{nutrient in diet})$$

Various 'indicators' are used to calculate digestibility under grazingland studies. The most common indicators are lignin and silica. The lignin ratio method has given reasonable estimates of forage intake and digestion in many range studies when a good estimate was available for lignin composition of the forage grazed. But there is also a possibility of lignin being digested. This is particularly so in immature herbages. The silica ratio technique depends upon accurate sampling of the amount of silica in the diet. Even small amounts of soil contamination of herbage or faecal samples will give variable and invalid results in digestion trials using this indicator. There is also some indication that silica may be absorbed.

The basis of the faecal nitrogen index ratio procedure is that total faecal nitrogen is directly proportional to the nitrogen content of the diet. The faecal nitrogen comes from undigested forage nitrogen and metabolic nitrogen. The metabolic nitrogen is excreted in an amount proportional to the dry matter intake of the animal. This technique does not require an estimate of the diet of the grazing animals, but it does require digestion trials to develop a regression equation for the relationship between forage and faecal nitrogen for the forage under study.

Estimates of digestibility may also be obtained by microdigestion techniques utilizing in-vivo nylon bag procedures or in-vitro artificial rumen procedures. In the first case, small samples of the forage to be digested are inserted into bags and placed into the rumen for varying periods of time. On removal they are rinsed and the weight loss calculated to give the percentage digestion. A similar procedure is used in in-vitro digestion estimates. Small samples are placed in an artificial rumen system to determine digestion. Many variables affect the interpretation of these microdigestion studies. The size, grind and type of sample, the length of fermentation, and other factors are important. However, with appropriate controls and experimentation, in-vivo fermentation and in-vitro fermentation can give results within a few per cent of the per cent cellulose digestion estimates obtained conventionally. A recent approach to measuring digestibility, noted above with respect to measuring intake rate, is that of using one group of animals (e.g. oesophageal-fistulated cattle) to collect forage and then feeding it to another group of test animals (e.g. sheep) in conventional digestion trials. Details of these methods and their effect on variability are summarized by Van Dyne (1969).

### *Factors affecting digestibility*

Large herbivores may slightly increase their efficiency of digesting herbage with age (Raymond, Harris & Kemp, 1954). There is also an indication that there is a seasonal cycle in digestive efficiency – a decrease in winter and a rise during the following summer – that is independent of feed composition. The changes due to age are greater between lambs and yearlings than between yearlings and older sheep. No evidence has been presented that a similar effect occurs with cattle.

The rate of herbage intake influences digestibility to some degree when the type of feed is held constant. In fat ruminants as compared to thin ones, energy digestibility may be depressed at a lower level of increasing feed intake (Reid & Robb, 1971). When sheep are fed the same feed at a high level as compared to a low level, they have lower digestive efficiency (Raymond, Harris & Kemp, 1955; Raymond, Minson & Harris, 1959). This is due to an increased rate of passage of food through the digestive tract at the higher level of intake. This can introduce a bias in estimates of digestibility and intake phenomena derived from some methods. The faecal nitrogen index method requires cutting the herbage in the field and feeding it to animals in stalls to derive the ratio of feed intake to faecal nitrogen. Feeding at higher levels in the stalls than the animals may graze in the field would overestimate digestibility by 1.0 to 1.5%. This causes an error, an overestimate, in digestible organic matter intake of 5 to 10% (Raymond *et al.*, 1959).

Cold depresses digestibility slightly through decreasing food retention time in the rumen (as noted above in the section on factors affecting intake). This depression varies from 0.1 to 0.4% per deg C, according to several authors (Ames & Brink, 1977). This could have considerable impact on large herbivores on winter ranges. For example, let us assume that their thermal-neutral zone is 20 °C and nominal digestibility 50%. A drop in temperature to 0 °C would, assuming a 0.2% per deg C depression, result in a digestibility of 46%. Although feed intake is greater at lower temperatures, the maintenance requirement will increase greatly and the combined effect of this and decreased digestibility will be severe. Thus, when shorn 25 kg lambs were fed *ad libitum* a pelleted ration containing 13.5% crude protein and having a digestibility of 55% at 0 °C (with a reduction of 0.1% per deg C), peak gains and the highest feed efficiency and protein efficiency ratios were found near 15 °C.

The large mixture of large herbivores grazing together in East Africa has led to some interesting hypotheses regarding digestive efficiency and grazing behaviour. Many of the selectively feeding ruminants, including browsers, have a diet of relatively high nutritive value. It has low fibre and high protein content. These animals, e.g. the duiker and eland, have relatively small rumen (Arman & Hopcraft, 1975). Mixed feeders, e.g. sheep and Thomson's gazelle, take some grass plus some of the forbs from the herbage. A third group of herbivores are roughage eaters or grazers whose diet is largely grass and which is often of high fibre and low protein content, especially in the dry seasons. These animals, e.g. hartebeest, wildebeest and cattle, have large rumens. Feeding habits are more closely related to taxonomic relationships than they are to body size. The selective feeders are less dependent upon rumen fermentation than the roughage eaters for either cellulose breakdown or for protein synthesis. Rumen fermentation rates are, however, higher for small ruminants than for large ruminants (Hungate *et al.*, 1959), but in part as the result of dietary differences. The normal diet of selective feeders has a relatively low fibre content. Inefficiencies in its utilization can be compensated for by a large appetite, e.g. as in the eland. Grazers have large rumens and

Table 4.13. *Digestibility of diets grazed by large herbivores*

| Season | Small-sample microdigestion *in vitro* | | | Large-sample microdigestion | | | | | | | | | | | |
|---|---|---|---|---|---|---|---|---|---|---|---|---|---|---|---|
| | Dry matter | Organic matter | Cellulose | Dry matter | Organic matter | Protein | Ether extract | Lignin | Cellulose | Other carbohydrates | Energy | Crude fibre | Nitrogen-free extract | Cell wall constituents | Acid detergent fibre |
| **Cattle** | | | | | | | | | | | | | | | |
| All seasons | | | | | | | | | | | | | | | |
| Ave | 31 | — | 51 | 52 | 62 | 48 | 28 | 22 | 55 | 56 | 51 | 65 | 65 | 68 | 57 |
| S.D. | 32 | — | 10 | 11 | 10 | 21 | 27 | 12 | 4.4 | 5.7 | 14 | 10 | 10 | 2.7 | 4.0 |
| *N* | 4 | — | 4 | 17 | 14 | 22 | 12 | 5 | 5 | 4 | 14 | 15 | 14 | 6 | 3 |
| Spring | | | | | | | | | | | | | | | |
| Ave | — | — | 64 | 63 | 72 | 70 | — | 11 | — | — | 66 | — | — | 66 | 57 |
| S.D. | — | — | 0 | 0 | 2.8 | 0 | — | 0 | — | — | 0 | — | — | 0.71 | 0 |
| *N* | — | — | 1 | 1 | 2 | 1 | — | 1 | — | — | 1 | — | — | 2 | 1 |
| Summer | | | | | | | | | | | | | | | |
| Ave | 58 | — | 54 | 51 | 63 | 44 | 26 | 27 | 55 | 56 | 50 | 65 | 65 | 68 | 53 |
| S.D. | 3.5 | — | 0 | 12 | 12 | 22 | 27 | 14 | 4.4 | 5.7 | 14 | 11 | 11 | 3.5 | 0 |
| *N* | 2 | — | 1 | 14 | 8 | 18 | 11 | 3 | 5 | 4 | 12 | 12 | 12 | 2 | 1 |
| Autumn | | | | | | | | | | | | | | | |
| Ave | — | — | 39 | 48 | 59 | 34 | — | 20 | — | — | 46 | — | — | 69 | 61 |
| S.D. | — | — | 0 | 0 | 12 | 0 | — | 0 | — | — | 0 | — | — | 3.5 | 0 |
| *N* | — | — | 1 | 1 | 2 | 1 | — | 1 | — | — | 1 | — | — | 2 | 1 |
| Winter | | | | | | | | | | | | | | | |
| Ave | — | — | 48 | — | 17 | — | — | — | — | — | — | 60 | 64 | — | — |
| S.D. | — | — | 0 | — | 0 | — | — | — | — | — | — | 0 | 0 | — | — |
| *N* | — | — | 1 | — | 1 | — | — | — | — | — | — | 1 | 1 | — | — |
| **Sheep** | | | | | | | | | | | | | | | |
| All seasons | | | | | | | | | | | | | | | |
| Ave | 56 | 61 | — | 52 | 70 | 37 | 7.3 | 18 | 47 | 63 | 41 | — | — | 63 | 53 |
| S.D. | 5.5 | 9.6 | — | 25 | 2.8 | 7.9 | 33 | 2.6 | 5.8 | 6.4 | 5.2 | — | — | 6.8 | 2.0 |
| *N* | 7 | 25 | — | 19 | 2 | 11 | 11 | 3 | 11 | 11 | 10 | — | — | 3 | 3 |
| Spring | | | | | | | | | | | | | | | |
| Ave | 61 | 61 | — | 80 | 72 | — | — | 19 | — | — | — | — | — | 65 | 53 |
| S.D. | 0 | 8.3 | — | 0 | 0 | — | — | 0 | — | — | — | — | — | 0 | 0 |
| *N* | 1 | 8 | — | 3 | 1 | — | — | 1 | — | — | — | — | — | 1 | 1 |
| Summer | | | | | | | | | | | | | | | |
| Ave | 45 | 41 | — | 48 | 68 | 34 | 3.9 | 15 | 46 | 62 | 39 | — | — | 68 | 51 |
| S.D. | 25 | 31 | — | 6.3 | 0 | 8 | 41 | 0 | 5.8 | 4.8 | 4.6 | — | — | 0 | 0 |
| *N* | 4 | 3 | — | 10 | 1 | 7 | 7 | 1 | 7 | 7 | 7 | — | — | 1 | 1 |

| | | | | | | | | | | | | | | | |
|---|---|---|---|---|---|---|---|---|---|---|---|---|---|---|---|
| Autumn | | | | | | | | | | | | | | | |
| Ave | 51 | 56 | — | 75 | — | — | — | 20 | — | — | — | — | — | 55 | 55 |
| S.D. | 0 | 3.5 | — | 0 | — | — | — | 0 | — | — | — | — | — | 0 | |
| *N* | 1 | 3 | — | 2 | — | — | — | 1 | — | — | — | — | — | 1 | 1 |
| Winter | | | | | | | | | | | | | | | |
| Ave | 46 | 54 | — | 63 | — | 42 | 13 | — | 49 | 65 | 45 | — | — | — | — |
| S.D. | 0 | 5.1 | — | 14 | — | 3.4 | 13 | — | 5.9 | 9.0 | 5.2 | — | — | — | — |
| *N* | 1 | 7 | — | 4 | — | 4 | 4 | — | 4 | 4 | 3 | — | — | — | — |
| | | | | | | | | | Goats | | | | | | |
| All seasons | | | | | | | | | | | | | | | |
| Ave | — | 52 | — | — | — | — | — | — | — | — | — | — | — | — | — |
| S.D. | — | 3.5 | — | — | — | — | — | — | — | — | — | — | — | — | — |
| *N* | — | 6 | — | — | — | — | — | — | — | — | — | — | — | — | — |
| | | | | | | | | | Mule deer | | | | | | |
| All seasons | | | | | | | | | | | | | | | |
| Ave | — | 37 | — | — | — | — | — | — | — | — | — | — | — | — | — |
| S.D. | — | 9.9 | — | — | — | — | — | — | — | — | — | — | — | — | — |
| *N* | — | 4 | — | — | — | — | — | — | — | — | — | — | — | — | — |
| Stall feeding | | | | | | | | | | | | | | | |
| Ave | — | — | — | — | — | 40 | 45 | 40 | — | — | 40 | 28 | 54 | 58 | — |
| S.D. | — | — | — | — | — | 17 | 17 | 11 | — | — | 0 | 13 | 20 | 0 | — |
| *N* | — | — | — | — | — | 14 | 17 | 2 | — | — | 1 | 12 | 14 | 2 | — |

Results presented here are derived from analysis of values given in Appendix 4.5. All digestibility data are in percentages.

retain food for long periods of time and thus digest cellulose well and achieve high nitrogen retentions on low-protein diets. Thus cattle do relatively well on low-protein diets, as do such wild ruminants as hartebeest. Nonruminants such as the rabbit, warthog and hippopotamus when fed the same feeds, such as grass, had lower faecal nitrogen contents than most of the ruminants (Arman, Hopcraft & McDonald, 1975). Their data suggest that the nitrogen content of the faeces is not very useful for predicting the nitrogen content of the diet for a wide range of East African ruminants. A useful review of ruminant digestion and evolution is provided by Moir (1968).

### *General summary of digestibility of diet*

Data from the literature are summarized in Appendix 4.5 for examples of measures of digestibility of diets of large herbivores. These data are restricted to grazing situations, except for a few references primarily for wild herbivores in which forage was cut and fed to the animals while they were in digestion crates or metabolism crates. Information is provided on the animal species, the method, the season or treatment, and the digestibility values determined either in small sample (microdigestion procedures) or large sample (macrodigestion procedures). The method of estimating the diet, the method of estimating faecal composition, and the method of estimating digestibility are reported where possible. Thus, for example, animals equipped with oesophageal fistulas may be used to collect the forage which is analysed chemically, grab samples of faecal material may be taken for chemical determination, and the lignin ratio technique may be used to estimate digestibility.

It is difficult to make many generalizations about the digestibility data presented in Appendix 4.5 because of the limited number of examples for the same species at the same season. However, in Table 4.13 we have summarized digestibility data for cattle and sheep at various seasons, for goats and mule deer at all seasons, and a few examples for stall-feeding of harvested forages. Both 'large sample or macrodigestion estimates' and 'small sample or microdigestion estimates' are presented.

For all seasons and for a total of 36 examples noted in Appendix 4.5, mean dry matter digestibility (macrodigestion) for cattle and sheep was 52%. Not unexpectedly, dry matter digestibility values were higher in spring than later, with low values being reached in either summer or autumn. Digestibility of energy and cellulose is in the same range as that for dry matter digestibility.

Of various factors that can affect digestibility, the crude protein and lignin concentrations of the diet are often referred to. In Fig. 4.8 we have plotted values from grazingland studies around the world. We show that, in general, as crude protein concentration of the diet increases so does digestibility and, for cattle at least, as lignin concentration of the diet increases the digestibility decreases.

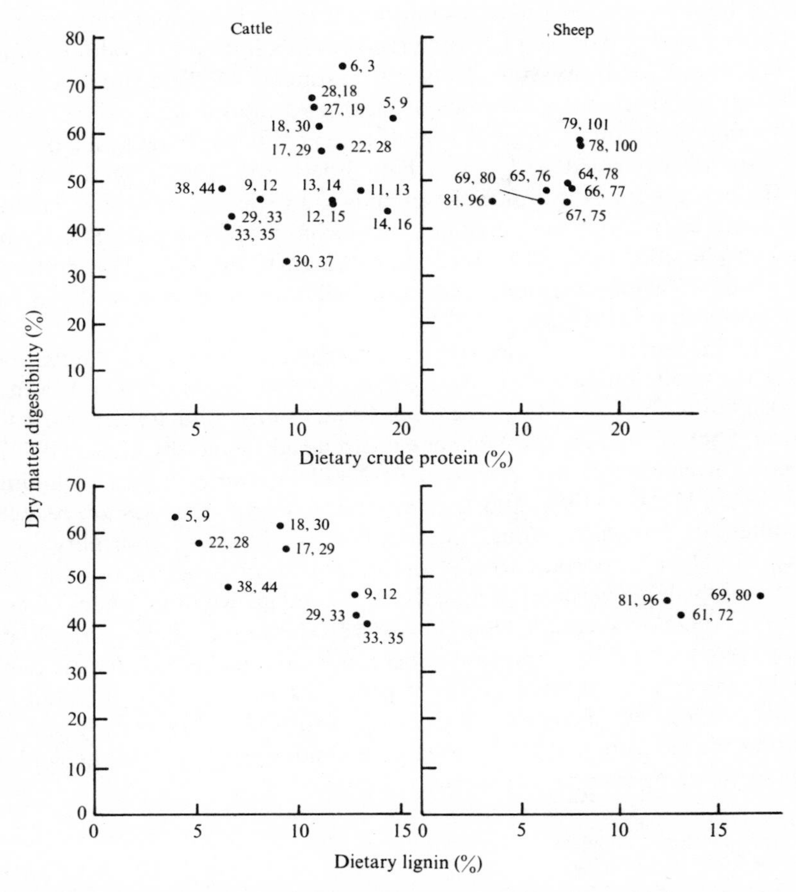


Fig. 4.8. Influence of dietary crude protein and lignin concentrations on dry matter digestibility for cattle and sheep from grazing studies. Numbers identifying the data points refer to studies noted in Appendixes 4.3 and 4.5.

It is difficult to make fully adequate comparisons between the digestive efficiencies of different species unless a cross-over experimental design is used where each animal species digests its own natural food and that of the other animal species. In numerous instances, however, comparative digestion trials use only one basic roughage, perhaps better suited naturally to only one of the animal species being tested. Thus, results of comparative studies must be interpreted judiciously.

## Other processes

There are numerous other processes important in the behaviour, physiology, nutrition and energy balance of large herbivores in addition to food selection, food intake and food digestion. Detailed treatments of these processes are unnecessary in this chapter. Methods of measuring mortality, natality, immigration and emigration of large wild herbivores are beyond the scope of this review but are discussed and covered by Eberhardt (1969). Methods for studying such processes in domestic animals are reviewed in several monographs specifically concerning methods and techniques (American Society of Animal Production, 1959; Grassland Research Institute, 1961; Agricultural Board, 1962). Various methods used with wildlife are also considered by Golley & Buechner (1969).

To examine energy exchange in large herbivores one must know the animal's metabolic rate, the evaporative loss of water (respiration and sweating), insulation (hair and fat), the body dimensions, and absorptivity to radiation. Such animal energy exchange is discussed further by Gates (1970). Energetics of ruminants are discussed in detail in several books including Brody (1945), Kleiber (1961) and Blaxter (1962). Many processes are related to population dynamics. Thus, immigration, emigration, mortality and natality would be important to discuss in an overall consideration of the dynamics of the large herbivore populations. Foetal growth and, under severe conditions, foetal resorption (Belonje & Van Niekerk, 1975; Mansell & Cringan, 1968) would be important subcomponents of the natality process, as would conception. Another group of processes not discussed in detail in the above sections are those related to energy expenditures in grazing and in combating the environment of the grazing animal, particularly in the colder regions of the temperate and subarctic climates.

### *Energetic cost of grazing*

Energy expended in the act of grazing is considerable. Wallace (1956) suggested that the maintenance energy requirements of the grazing animals should be increased by approximately 50% to account for this. Blaxter (1960) calculated that the increased energy required due to grazing activity would not exceed 10%. But both these workers were studying animals under conditions less severe than would be imposed under climatic stress on grazinglands. Joyce (1968) suggested grazing sheep require 80 to 90% more feed than sheep fed to maintenance indoors. Another estimate, that of Vercoe, Tribe & Pierce (1961), is that the energy requirement of grazing as such is less than 25% of maintenance energy. Blaxter (1964) reported estimates of maintenance requirements for sheep in indoor trials of about 1.5 $\mathrm{Mcal \cdot d^{-1}}$, whereas in various grazing experiments they averaged about 2.3 $\mathrm{Mcal \cdot d^{-1}}$. He found the maintenance energy requirement for cattle calculated from indoor studies was 11.5 Mcal·d$^{-1}$, whereas estimated from grazing studies it was 19.2 Mcal·d$^{-1}$. Thus, under grazing conditions, the estimated maintenance requirement is about 50% greater than it is indoors.

A part of this increase in energy requirement is due to the grazing activity itself. Another part is due to climatic stress. Although Blaxter (1962) stated that 'even under range conditions sheep rarely walk more than 6.4 km, and hardly ascend more than 100 m each day', Van Dyne & Van Horn (1965) found sheep on a foothill range to graze up to 13.2 km·d$^{-1}$ and climb 280 m, more than twice Blaxter's estimated extremes. The latter study was on a 1620 ha foothill rangeland in south-central Montana, USA, in which the area grazed extended from 1450 to 1730 m. The route travelled by the band of sheep was marked on maps from which the elevation climbed was estimated, and distance travelled was measured with wethers equipped with distance-measuring meters. Averaged winterlong, the sheep in the band travelled 7.6 km·d$^{-1}$, while elevation climbed varied from 85 to 280 m. Using an equivalent calculation procedure, Lynch (1974) showed that Merino sheep in Australian New South Wales open woodland arid pastures grazed from about 3 to 9.5 km·d$^{-1}$, depending on temperature and food availability. For the Montana example, energy expenditure for grazing, calculated using Blaxter's (1962) values of 0.030 and 0.0322 kcal·m$^{-1}$ respectively for horizontal and vertical movement for a 50 kg sheep, varied from 137 to 406 kcal·d$^{-1}$ per sheep. The average energy expenditure for grazing was 235 kcal·d$^{-1}$ of which less than 3% was expended for climbing. Utilizing data on digestible energy for sheep of Van Dyne & Lofgreen (1964), for equivalent protein and energy levels in the forage, the 235 kcal·d$^{-1}$ represents about 20% of the basal metabolic requirement. This value is in the same order of magnitude as would be calculated from the data of Graham (1962), who indicates that the energy cost of grazing as such is 0.6 to 0.8 kcal·h$^{-1}$·kg bodyweight$^{-1}$ of sheep, assuming a 50 kg animal with 4 h of actual grazing per day. In addition to the actual grazing time, there is considerable time spent walking between grazing areas.

We know relatively little of the energetic cost of grazing for many herbivores, but the camel may represent an extreme. The camel walks at about 0.5 km·h$^{-1}$, can cover up to 100 km·d$^{-1}$ (Gauthier-Pilters, 1974), under normal grazing covers 20 km·d$^{-1}$ taking forage in small amounts from individual plants over its route, and uses 8 to 10 h·d$^{-1}$ in grazing to get an estimated food intake of 8 to 12 kg dry matter·d$^{-1}$ on salty pastures with annuals and 5 kg dry matter·d$^{-1}$ on dried grass pastures and grazing *Acacia*. Thus, the grazing energy cost values vary from a few per cent up to 90%, and more information is needed on this component as it is sizeable. Fortunately, a recently developed method (the carbon dioxide entry-rate technique) offers considerable potential for estimating energy expenditures of grazing ruminants (Corbett, 1971; Whitelaw, 1974). Radioactive $^{14}C$ substrate is injected and the total carbon dioxide pool is thus labelled. The specific activity of carbon dioxide is used to estimate total carbon dioxide production and hence energy expenditure.

## *Efficiency of maintenance and gain*

A key problem in calculating gains of large herbivores is to determine the efficiency with which metabolizable energy can be used for maintenance and gain. This depends upon the increase in heat production with successive increments of intake. Two major components of this heat increase are fermentation heat and heat from metabolism in digestive tract tissues. Fasting metabolism is about 275 kJ ($W^{0.75}_{kg}$24 h$^{-1}$ for adult wether sheep (Webster, Osuji, White & Ingram, 1975) and about 23 % of that is due to the gut. From fasting to maintenance, this heat production was about 65 kJ·(MJ gross energy eaten)$^{-1}$; between maintenance and twice maintenance it was 120 kJ. For the sheep in the study above, when fed chopped, dried grass, the chemical energy loss from gross energy for faeces, urine and methane respectively were 39, 4 and 7%; heat losses were 1, 5, 10 and 16% for eating, fermentation, gut and remainder; the net energy was 18%.

It is generally considered that cattle are equal to sheep in energetic efficiency. However, given identical diets, digestibility may be 5% higher and voluntary food consumption per unit of maintenance requirement may be 10% lower than in sheep (Graham, 1974). Sheep tend to deposit more fat and less protein in their tissue per unit of live weight gain than do cattle. Wool growth appears to occur instead of an equivalent amount of body growth. Warwick & Cobb (1975) summarize data on nutrient requirements for maintenance for cattle which include basal metabolism and the energy expended for standing and moving about and maintaining thermal neutrality. The studies were conducted under barn-feeding and pen-feeding conditions and involved a variety of breeds generally in the range of 200 to 300 kg and approximately 1 to 2 yr age of both steers and bulls. Fasting metabolism requirements expressed in kcal $(W^{0.75}_{kg})^{-1}\cdot d^{-1}$ were as follows: Angus 72, Brahman 86, Ayrshire 91, Hereford 97, and Afrikander 102. These data illustrate the genetic variability. The efficiency of use of metabolizable energy varies widely with the age and sexual state of the herbivore.

Metabolically active animals grow rapidly but can be more seriously affected by undernutrition than less metabolically active animals. Those animals that are active metabolically have higher maintenance requirements and voluntary food consumption, both of which are proportional to metabolic rate (Kleiber, 1961). Young animals generally have higher metabolic rates than older animals. Metabolic rates between species do not necessarily correspond to size. Rattray, Garrett, East & Hinman (1973*a*, 1974) and

Rattray *et al.* (1973*b*) report the following efficiencies for use of metabolizable energy for various processes: 3- to 5-month-old lambs, maintenance 66 and gain 41%; growing ewe lambs on low-energy diets, maintenance 41, gain 18 and pregnancy 16%; for nonpregnant ewes, maintenance 66 and gain 56%; and for pregnant ewes, pregnancy 16, and total foetal development 24%. Terrill (1975) summarizes other aspects of efficiency in sheep production. He notes that genetic differences in energy requirements have not been demonstrated, but efficiency of lamb production is largely dependent upon the number of lambs marketed per ewe per year and there is great genetic variation in numbers of lambs born and reared.

The composition of the gain in an animal changes greatly as the animal matures. For example, in sheep the fat : protein : water ratio varies from 1 : 1 : 4 for lambs to 7 : 1 : 4 for sheep of 30 to 40 kg liveweight. For lambs the cost of gain is 1.5 $\text{Mcal}\cdot\text{kg}^{-1}$ and for the larger sheep it is 6 (Graham, 1974). The chemical composition and energy value of the tissue gained or lost by ruminants are variable. Gains vary from about 5 up to 9 $\text{Mcal}\cdot\text{kg}^{-1}$ and loss from about 6 to 8 $\text{Mcal}\cdot\text{kg}^{-1}$ (Reid & Robb, 1971). As dairy heifers grow, about 19 and 16% of the increase in empty body weight is in protein and fat, respectively (Reid & Robb, 1971). As the empty body weight of mature cows changes, the results are more variable, but on the average the increases are about 15 and 37% respectively for protein and fat. Dean & Rice (1975) estimated the body gain of yearling beef heifers grazing shortgrass prairie (average of about 70 kg in 150 days) to be 15% fat and 17% protein.

### *Compensatory gain*

Another important process or phenomenon to consider is that of compensatory gains of young animals deprived of optimal nutrition at early growth stages. Fig. 4.9 represents a generalization of the data of Reardon & Lambourne (1965) who kept one group of sheep on the best available improved pasture throughout the experiment (high plane), a second group on a poorer native pasture for their first 9 months and then transferred them to the improved pasture (low plane 1), and a third group on improved pasture for their first 9 months then on poorer native pasture for about 5 months before returning them to improved pasture (low plane 2). The lambs who, with their dams, had been on poor native pasture made considerable compensatory gain for almost a year after being returned to improved pasture. The lambs which were on poor native range between about 9 and 14 months of age made major compensatory gains for about half a year when returned to good pasture. By 3 years of age the differences between treatments had largely disappeared. More difference than noted here might be expected in compensatory gains under many conditions for both domestic and wild large herbivores because nutritional conditions would often vary more than they did here between pastures.

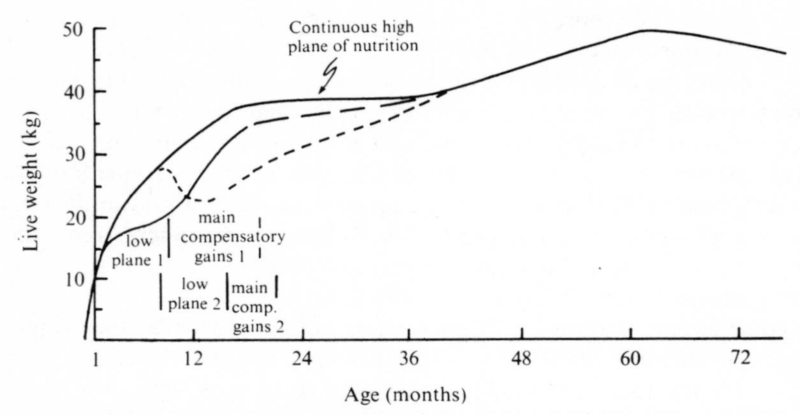


Fig. 4.9. Merino sheep on improved or native pasture, Armidale, New South Wales, Australia. (After Reardon & Lambourne, 1965.) For details see text.

Presumably, compensatory gains occur when food deprivation is rather severe and over a period of time. No difference was found in food intake rate of sheep varying from 28 to 37 kg shrunk weight (Donnelly, Davidson & Freer, 1974) when subsequently put on pasture or *ad libitum* feeding. A mature animal would have to lose more than 25 % of its weight before the factor had an impact. The gain of thin sheep, when returned to good pasture, however, may be higher in water content at first than is average.

## *Mortality and natality*

To calculate productivity of large herbivores one needs to know the natality. Some indication of the variability in natality is given below. Under grazing-land conditions on Israel semidesert rangeland there were 0.69 lambs born per ewe under a heavy stocking rate and 0.75 under a light stocking rate (Tadmor, Egal & Benjamin, 1974). On good range, embryo counts for North American deer averaged 1.71 whereas on poor range they were 1.06 (Klein, 1969). Fawn mortality was also related to range quality. From 25 to 50 % of the females would breed at one year of age on good range whereas well fewer than 10 % did so on poor range. For most large wild herbivores there is relatively low reproduction in the younger female age groups. It has been reported (references unknown) that only 15 % of the 1-year-old caribou females bred whereas 90 % of the 3-year-old females did; for North American pronghorn about 10 % of the does are barren, about 7 % of those who produce young will have one foetus and 93 % will have two, with a prenatal mortality of 5 % or less. A winter study of elk in the Rocky Mountains of North America found that very few of the calves of the year (approximately 7 months of age) carried a foetus, but that 2 % of the yearlings (19 months of age), about 73 to 95% of the 2-year-olds (31 months), and a slightly higher percentage of 3-year-olds and older (42 months or older) did (Kittams, 1953).

There is some suggestion that when does of North American deer have better nutrition there are proportionately more males born (Robinette, Gashwiles, Low & Jones, 1957). There are generally more males born than females, but there is higher infant mortality among the males. Taber (1953) comments on the higher death rate for male black-tailed deer than for females up to 19 months of age. He indicates the young bucks stray farther from their mother than do the young females. The young females also stay with their mother longer. The males grow faster and therefore eat more food. By the time they are yearlings although there may be as many as 20% more males at birth there may be 50% more females at one year of age.

### *Other factors*

Another key area of large herbivore physiology is the overall nitrogen cycle within the animal. There are major differences between breeds within a species and between species in the process of recycling nitrogen to the rumen. The role of saliva in nitrogen cycling is discussed by Kay (1965) and Kay & Hobson (1963). Those animals who do this can, of course, thrive on diets very low in nitrogen. Nitrogen recycling and water use are interrelated (as noted above in the section on water intake). Zebu cattle generally are superior to European breeds in their ability to recycle nitrogen and most wild ruminants of East Africa thus far tested are superior to cattle in this trait. In some cases, however, in periods of drought when considerable low-quality herbage is still available, dead cattle may be found near permanent water (L. M. Talbot, personal communication). The hypothesis is that cattle on a very low nitrogen diet may lose nitrogen when water is available but not lose so much as to cause death when water is more limiting.

Nonselective grazers are more likely to suffer from dietary deficiencies when the amount of pasture herbage is decreased than are more selective grazers. Selectivity in many instances is thought to be related to herbage protein composition, but in other instances an animal (such as the eland) may benefit by selecting succulent plants, thereby maintaining its water balance and thus food intake.

## Production

Mankind has domesticated only some 9 birds, 20 mammals, and 2 fish out of the many thousands of animal species available to him; and he did this probably several thousand years ago (Blaxter, 1975). Now we are beginning to look, in part because of food energy limitations, to new species such as eland, capybara, red deer, trout and others for domestication. It is important to assess carefully the production levels and production processes involved. Although it is probable that the easiest way to produce more meat is by conventional processes, there are situations where wild animals may be exploited to advantage, as in difficult habitats where physiological and behavioural characteristics may be to the wild animal's advantage (Kay, 1970).

### *Production measures and measurement*

Production can be considered as the amount of body weight gain of an animal that can be attained over a given time interval. More precisely we are interested in the portion of the animal that is utilized by man. In the case of meat we are also concerned with the quality as measured by various physical and chemical characteristics. Of course, we cannot harvest each animal so we are concerned with the 'offtake', i.e. the proportion of the herd, flock or population that can be harvested in a given interval of time.

Measurement of production of large herbivores is commonly done by measuring changes in weight over a defined period of time. In most instances with studies of domestic animals, the weights at the beginning and the end of the sampling period are taken with at least an overnight shrink. The shrink procedure is necessary to even out the influence of the content of the digestive tract, because it is such a sizeable fraction of the total weight of the organism. Obviously, this is difficult to do with wild herbivores unless they are slaughtered. More precise measures require the determination of the body composition. This may mean determining the fat content, often done by measuring the specific gravity of the organism, or by dissecting the carcass into its various components. Most of the data that are reported, however, are concerned only with the biomass change and disregard possible differences in body composition.

### *Age–weight relationships*

Production data vary widely from place to place and from year to year within the same place. However, it is useful to examine the magnitude of weight changes of animals over their life span to illustrate the growth of the individual. Data are compiled in Fig. 4.10 for age–weight curves for several wild herbivores and domestic animals. The age–weight curves for cattle are under North American grazingland conditions. These data represent, in most instances, lifetime records for large numbers of animals. A cow under the range grazing regime generally has a cyclic variation in weight throughout the year. Her weight may be least just following calving each spring. Under the stress of lactation she may not gain greatly until middle or late summer and into the autumn. During the winter foetal growth is occurring, nutritional quality is poor and, in many instances, not much herbage is available. The rather uniform weight during the winter means, because of foetal growth, that the cow herself is losing in body substance.

An example of the seasonality and magnitudes of long-term weight changes for cattle are reported by Lewis, Van Dyne, Albee & Whetzal (1956) (see also Fig. 4.10*d–f*). Heifers of about 18 months of age were started on experiments and then weighed approximately monthly for about 4 years. There were essentially no gains in weight between 18 and 24 months of age. This was followed by a rapid growth period during their second summer, some increase during their third winter, but with losses to a low point after calving. Weight then increased during the following summer until vegetation began to mature and decreased in quality and quantity while there were increasing lactation demands. Weights increased in the autumn and winter but dropped again at calving time. The pattern repeated thereafter.

Differential grazing treatments (heavy, moderate and light) began to take their toll on these animals more after several years. Animals from the light grazing treatment had superior weights after several years, which at some seasons amounted to more than 100 kg body weight difference per individual. The change in weight of the animals from their autumn weight at 18 months (the beginning of their second winter) until spring of their fifth year is particularly interesting. In the first winter there was the gain to the weight attained after calving of about 15 kg, or about 6%. In the second winter, the winter the heifers were pregnant with their first calves, there was a loss of about 1%. In the third winter there was about an 8% gain and in the fourth winter about a 6% gain. One could expect the gain to decrease then as the cows had reached their mature size.

The growth of the young animal is also of interest. For cattle under rangeland conditions the male calves are somewhat heavier than the females. An average birth weight for both sexes of calves in the herd discussed above would be about 32 kg. Average weaning weight was about 159 kg. Growth rates vary widely, of course, for any animal, but perhaps most for the young. For example, with good management and nutrition for the ewe, lambs may gain weight to weaning (slaughter) at a rate of 450 $g \cdot d^{-1}$ when suckling. If the lamb is weaned at 4 to 6 weeks and put on cold milk substitute it can gain 430–450 $g \cdot d^{-1}$; on concentrate feed it gains 250–340 $g \cdot d^{-1}$ and on pasture only 160–270 $g \cdot d^{-1}$ (Treacher, 1970). These rates for pasture are for grazing on nutritious subhumid pasture and rates are much lower on dry range. In studies in South Dakota lambs gained 157 $g \cdot d^{-1}$ (female twins) to 198 $g \cdot d^{-1}$ (male singles) in 161 days on their dams on dry mixed prairie rangeland (Bush & Lewis, 1977; Fig. 4.10*z*).

The data provided in Fig. 4.10(*a–c*, *x*) giving generalized age–weight curves for domestic animals are taken from Brody (1945); they are for animals with much more favourable nutritional regimes than grazingland conditions. The

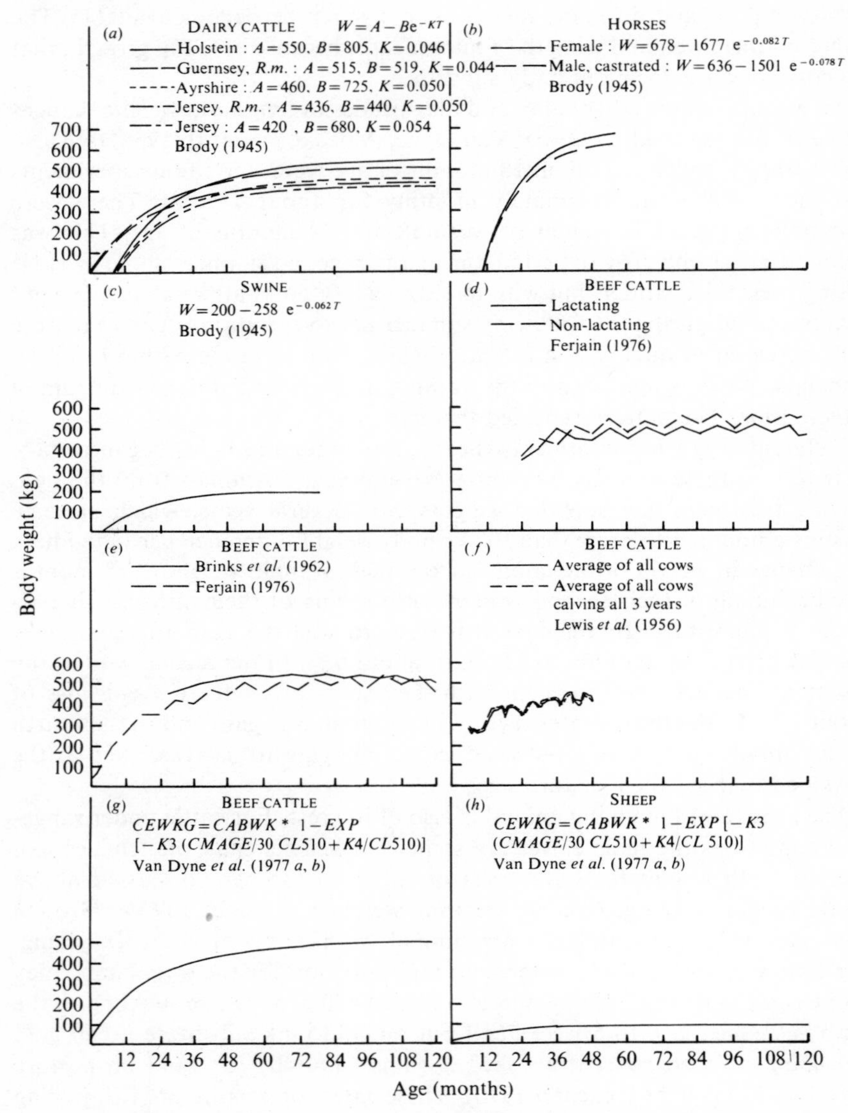


Fig. 4.10. Typical age–weight curves for large herbivores. Note the difference in scaling in the various graphs: 15 months and 120 kg, 15 months and 200 kg, and 120 months and 1200 kg.

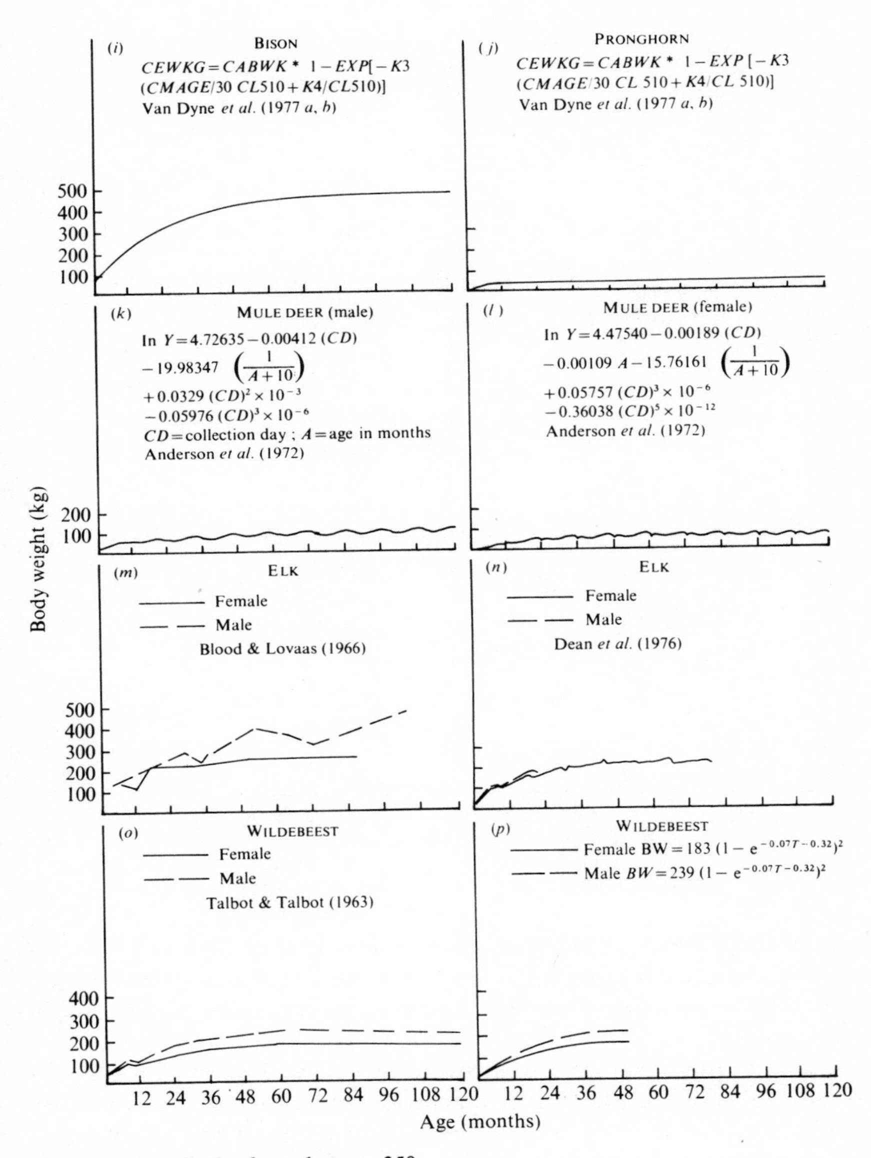


Fig. 4.10 (*cont.*). For legend see p. 350.

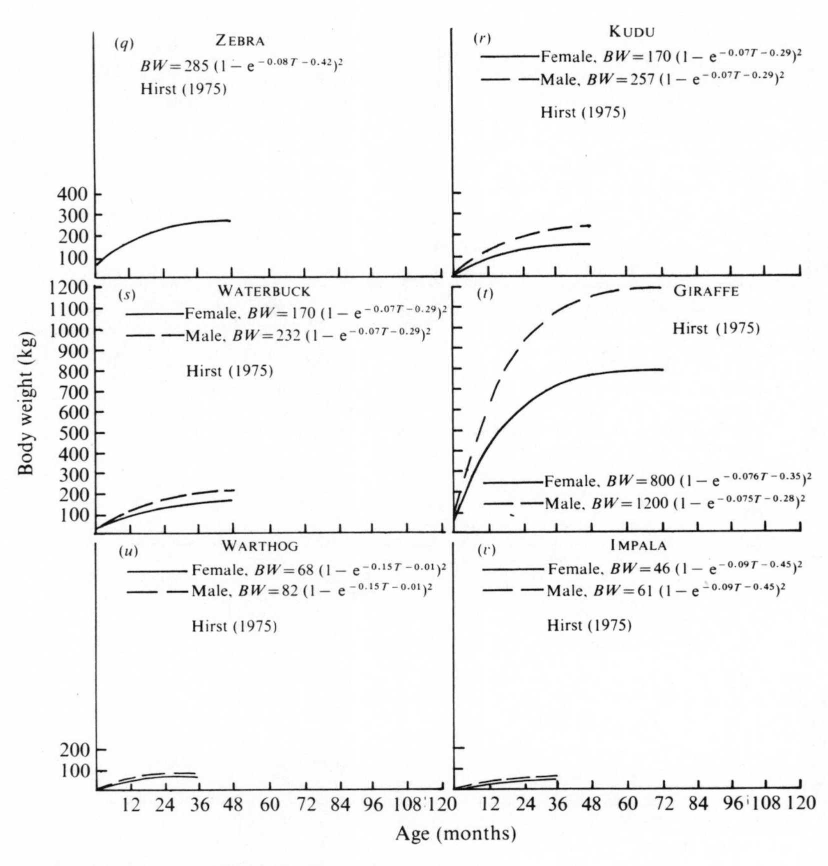


**Fig. 4.10 (*cont.*). For legend see p. 350.**

general form of Brody's growth equations is described in the graph for dairy cattle (Fig. 4.10*a*). An exponential function is used to approximate the age–weight relationship, by which the animal approaches the mature weight asymptotically.

There are many fewer data for growth rates of wild herbivores than for domestic animals. An interesting comparison and approach has been provided by Hirst (1975) who used available data to construct age–weight curves for various East African herbivores (see Fig. 4.10*g*–*v*). Then, utilizing these 'statistically fitted curves', he estimated weight increments per month for various age and sex groups. Because the females could be pregnant and the weight increases would include foetal growth, we have selected from his analyses data for the part-grown males for our comparison purposes. Weight increment per month expressed as a percentage of the body weight for part-

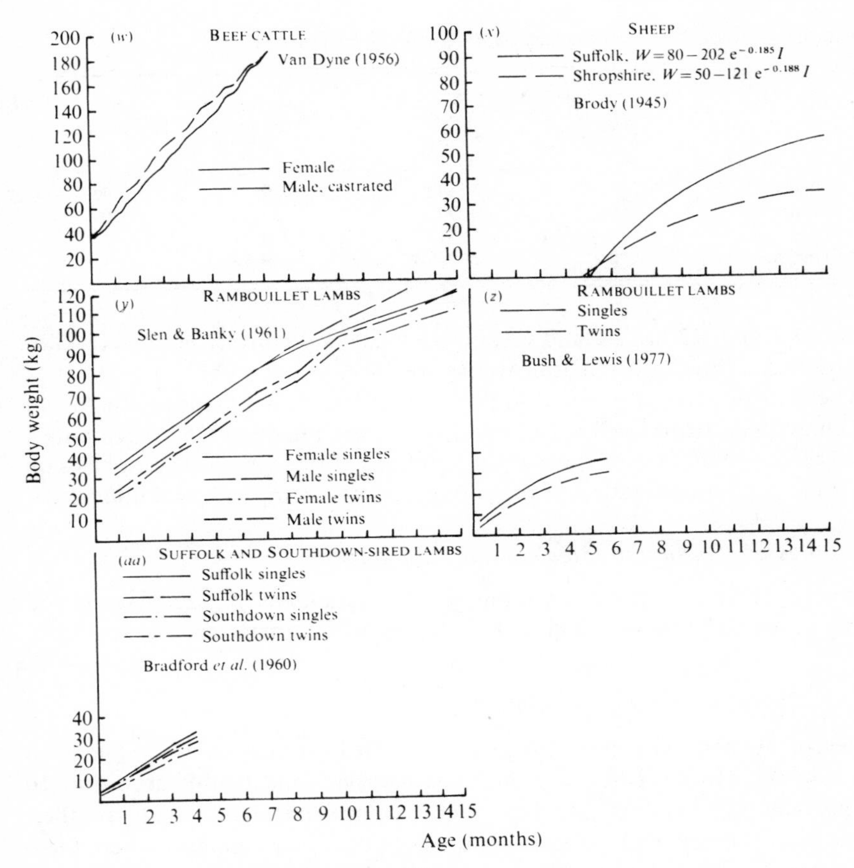


Fig. 4.10 (*cont.*). For legend see page 350.

grown males is about 3% for impala and giraffe, 5% for kudu and 6% for waterbuck. For part-grown animals of both sexes it is about 1% for warthog, 2% for wildebeest and 3% for zebra.

The growth curves for male and female elephants aged from 0 to 60 years are provided by Laws *et al.* (1975). Asymptotic weights of the male elephant over 20 years of age are about 4700 kg. For females over 20 years of age asymptotic weights are about 2700 kg.

Fig. 4.10(*g*)–(*j*) presents the overall expected growth curves for four large herbivores as used in the modelling studies of Van Dyne, Joyce & Williams (1977). The various parameters and values used in the equation in the figure are given in Table 4.14. The weight and life span values were estimated from the literature for rangeland animals; weights are in carbon biomass. Para-

Table 4.14. *Values of parameters used in Fig. 4.10 (g)–(j)*

| Species | Adult weight: CABWK (kg) | Life span: CLSID (d) | Scaling parameter: K3 | Shift parameter: K4 |
|---|---|---|---|---|
| Cattle | 58.9 | 5450 | 9 | 45 |
| Bison | 59.2 | 9490 | 14 | 52 |
| Sheep | 11.4 | 5110 | 9.5 | 15 |
| Pronghorn | 5.8 | 2190 | 9.5 | 12 |

meters K1 and K2 are 2.2 and 0.12 respectively and convert pounds to kilograms and carbon in dry matter to live weight.

The growth curves presented in Fig. 4.10 of course only represent a few of the functional forms used with large herbivores. Fitzhugh (1976) presents a good review of important age–weight functions of this type, with emphasis on the biological interpretation of the parameters and computational difficulties.

### *Examples of production rates of domestic animals*

In this section we present a few examples of weight gains of grazing cattle and sheep under different seasonal and management treatment conditions.

#### *Cattle gains on grazinglands*

In temperate zone climates cattle make their greatest seasonal weight gains in the summer. However, the size of the gain varies considerably according to intensity of grazing, year influences, and from one rangeland type to another.

Some of the better rangelands are those of the low mountains where there is frequent rainfall during the summer and lush vegetation. For example, yearling cattle grazing on lodgepole pine–pinegrass rangeland in Canada from June to September in 3 years had an average initial weight of about 250 kg, were stocked at 1.9 ha per AUM (an animal unit month is roughly equivalent to a 454 kg cow) and had a gain of about 20 $kg \cdot ha^{-1}$ (McLean, 1972). Examples of gain per head on spring pastures of *Agropyron cristatum* in Utah, USA, were reported by Frischknecht, Harris & Woodward (1953) to vary from about 1.35 through 1.49 $kg \cdot animal^{-1} \cdot d^{-1}$ in a 3-year interval under light grazing as compared to 0.92 to 1.09 $kg \cdot animal^{-1} \cdot d^{-1}$ under heavy grazing. Younger animals gain 0.85 to 0.87 $kg \cdot animal^{-1} \cdot d^{-1}$ under light grazing and 0.75 to 0.80 $kg \cdot animal^{-1} \cdot d^{-1}$ under heavy grazing.

An example of the variability within and between years in the production of the same age–sex class of animals is shown from studies on shortgrass prairie in eastern Colorado, USA. Dyck & Bement (1971, 1972) reported gains of yearling heifers for light and heavy grazing treatments respectively in

Table 4.15. *Cattle production on Texas rangeland as influenced by grazing management*

| Item | Units | Grazing treatment: Deferred rotation | Continuous light | Continuous heavy |
|---|---|---|---|---|
| Weaning weight | $kg \cdot animal^{-1}$ | 237 | 230 | 223 |
| Calf production | $kg \cdot AU^{-1} \cdot yr^{-1}$ | 221 | 211 | 200 |
| Calf crop | % | 95 | 92 | 90 |
| Calf production | $kg \cdot ha^{-1} \cdot yr^{-1}$ | 28 | 18 | 39 |

From Kothman, Mathis & Waldrip (1971).

one year to be 82 and 62 $kg \cdot animal^{-1}$ and in another year to be 90 and 54 $kg \cdot animal^{-1}$. In the two preceding years from the same experiment, Vavra, Rice & Bement (1973) reported gains for the June to August period for light and heavy grazing respectively to be 55 and 53 $kg \cdot animal^{-1}$ in one year and 59 and 52 $kg \cdot animal^{-1}$ in the second year. There is, however, a considerable variability in the gains of the animals in various months within the same year. A 3-year average gain for the 5-month season is 60 $kg \cdot animal^{-1}$ for heavy grazing and 78 for light grazing on a shortgrass prairie (Dean & Rice, 1975). Early in the growing season the vegetation is at its peak nutritive value and the animals may also be in their thinnest condition. So there is a combination of possible compensatory growth and good forage quality. Yearling heifers weighing about 210 kg may gain 1.0 $kg \cdot d^{-1}$ at this time of year. Some 5 to 6 months later they may weigh about 300 kg and be gaining only 0.15 $kg \cdot d^{-1}$. Under light grazing treatments the gains per head per day decline relatively uniformly from early summer to late summer. But under heavy grazing treatments there is much more fluctuation in the average daily gains due to periods of precipitation and 'spurts' in plant growth changing the amounts and quality of the herbage more than it does under light grazing (Hyder, Knox & Streeter, 1971). During the summer grazing interval gains are generally higher in June than in July, but on the shortgrass prairie which receives summer rainfall periodically there may be higher gains in late summer than in mid-summer. This reflects the change in amount and quality of the herbage. The herbage is of high quality in June but it is relatively limited. The animals are growing through June and July so the gains per head per month may be higher in total weight later in the summer even though the herbage quality has declined by then. Seasonlong, such cattle may convert some 4 to 7% of the gross energy consumed into gain. As a comparison, it was estimated (Davis & Golley, 1963) that white-tailed deer convert 1.2% of their food into growth.

On shortgrass and mixed prairie rangeland further south in the Great Plains of the USA than in the above experiment, but much higher rainfall conditions, several measures of animal production were obtained in an experiment involving rotation and intensity of grazing. Results are given in Table 4.15, where AU refers to an animal unit, or approximately one 454 kg cow. Deferred rotation allowed the vegetation to maintain or improve conditions even more so than under continuous light grazing.

### *Sheep gains on grazinglands*

Production of sheep varies with season in a manner similar to that of production of cattle. Generally, on arid and semi-arid grazinglands forage growth and nutritional quality are best in the summer. Thus, for example, in various studies in the summer grazing season on mountain browse, aspen and sagebrush types of vegetation, sheep on good-condition ranges in Utah, USA can gain 20% of their original body weight in a 4-month interval (Cook, Kothman & Harris, 1965). These data are for wethers which were put onto the range in fair, but not fat, condition. For ewes on similar mountain range in the summer, gains per animal are in the order of 0.6 to 0.9 $kg \cdot d^{-1}$ if the animals are nonlactating and 0.2 to 0.18 $kg \cdot d^{-1}$ if they are lactating. Under the same conditions lamb gains per animal are 0.31 to 0.35 $kg \cdot d^{-1}$ (Cook, Mattox & Harris, 1961). In many situations sheep on summer mountain ranges are herded rather than allowed to graze freely over areas in which they have a choice. In one 2-year study over about a 75-d summer-grazing period on high-mountain rangeland in Wyoming, USA, no large differences could be detected in the rate of gain of lambs under either free grazing or herding conditions. The gains of the individual lambs were about 0.14 $kg \cdot d^{-1}$ and the gains of the ewes averaged about 0.12 $kg \cdot d^{-1}$. There were more differences in the gains of animals between two years than there were between two treatments. In these studies initial weights of the individual lambs averaged 33 kg and the ewes 63 kg.

In a contrasting type of grazingland, Merino sheep were grazed on semi-desert–annual herb pasture over a 9-year interval in Israel (Tadmor *et al.*, 1974). Although there was considerable variation from year to year, ewe weights changed during the year from about 45 kg in December to about 65 kg from April through October. Lambs gained about 0.33 $kg \cdot d^{-1}$ in the green-feed season. In this study, as in many others, all the production measures evaluated were nonlinear in response to changes in stocking rate of 1 ewe on from 0.2 to 1.0 ha. A striking contrast to the carrying capacities of the arid lands is that of the humid pasturelands of Europe. But even in the higher rainfall areas there is considerable variability in carrying capacity within the same region because of soil and topographic differences. In Great Britain on poor, hill farmlands, stocking rates may be of the order of 1 ewe for 4 ha, whereas on nearby lowlands the stocking rate may be 12 ewes·ha$^{-1}$, i.e. 1/12 ha per ewe (Eadie, 1970). The resulting yield in weaned lambs is from about 15 kg·ha$^{-1}$ under poor, hill farmland up to about 600 kg·ha$^{-1}$ under the intensively utilized lowlands.

An annual cycle of sheep grazing in a semi-arid, sandhill-type pasture in Argentina was reported for 2-year-old wether sheep by Bishop, Froseth, Verettoni & Noller (1975). These wethers began the grazing trial at under 45 kg per animal in January, increased to about 50 kg in May, dropped to about 45 kg again in July, and then during the spring and summer increased to about 55 kg by November and December. The drop from May to July represented a 10% body loss. The gain from July to November represented more than a 20% gain. These data illustrate that even nonreproductive animals have considerable variability of weight yearlong under rangeland conditions.

One of the advantages of light grazing and thrifty animals comes in the ease of lambing and calving. Smoliak (1974) reported on grazing studies with sheep on shortgrass prairie in Canada. He found in an 18-year study under heavy grazing that 17% of the ewes were in poor lambing condition as compared to only 6% under the light grazing; under heavy grazing only 52% of the ewes were in good condition compared with 72% under light grazing. The mean weight of the ewes under heavy grazing was 58 kg; under light grazing they weighed an average of 62 kg.

### *Principles from grazing intensity studies*

These data illustrate a common principle observed in grazing intensity studies. Weight per animal, in this case weaning weight, is generally greater under light grazing treatments than under heavy grazing treatments. Also, reproductive performance may be better under light grazing than under heavy grazing. But, because of the greater stocking rate, production per unit area may be greater under the intensive grazing treatments rather than light grazing treatments. It appears that in arid and semi-arid grazinglands of the world a moderate grazing intensity frequently gives optimal returns at least over a short interval of time. With even longer use of these grazinglands perhaps the even lighter grazing treatments will be superior. In contrast, in the humid and subhumid grazinglands of the world the order of production per unit area is frequently heavy > moderate > light. In these situations plant species receive more water and can withstand the heavy intensity of use.

### *Examples of production rates of wild herbivores: temperate and arctic*

Wild animals throughout the world are harvested by man for food. They also provide considerable scenic resources. Because of the emotionality of the issue of conservation, the claims for production rates of wild animals are often exaggerated. In fact, very few solid data exist on which to calculate production. Therefore, in the following discussion, some of the estimates of production and some of the factors to consider are evaluated.

Game animals are important in the management of the grazinglands of the western USA because recreational hunting is a major factor on these public and private lands. Wallmo (1975) has provided estimates of the annual harvest of different species in arid and semi-arid rangelands in the western United States. Harvest rates vary from 1 to 17% of the total population size. The species, their locations, and estimated populations and harvest rates are as follows:

(i) The peccary or javelina (*Peccari tajacu*) is found primarily in the oak–juniper woodland, mountain mahogany–oak scrub, and various desert-like vegetation types including creosote bush–burr sage, palo verde–cactus shrub, creosote bush–tarbush, trans-Pecos shrub savanna, grama–tobosa shrub savanna, and riparian habitats. It occurs primarily in Arizona, New Mexico and Texas, with an estimated gross population of 20 000 animals and a 15% annual harvest.

(ii) The white-tailed deer (*Odocoileus virginianus*) occurs in the oak–juniper woodland, juniper–pinyon woodland, mountain mahogany–oak scrub and riparian habitats in most of the western states, with an estimated population at 200 000 and a 13% annual harvest.

(iii) The mule deer (*Odocoileus hemionus*) occurs in almost all the vegetation types of all of these western states. The deer population is about 1.5 million and annual harvest is 17%.

(iv) The pronghorn (*Antilocapra americana*) is most common in the Great Basin sagebrush, sagebrush steppe, saltbush–greasewood, trans-Pecos shrub savanna, and grama–tobosa shrub savannas of most of the western states. The total population is about 60 000 and the annual harvest about 17%.

(v) The desert bighorn sheep (*Ovis canadensis*) occurs primarily in the creosote bush, creosote bush–burr sage, and palo verde–cactus shrublands of California, Nevada and Arizona, with a total population of about 30 000 and a harvest rate less than 2%.

A comparison for elk (*Cervus canadensis*) in western North America is provided by Denny (1968). He notes that hunters in Colorado, USA harvest about 25% of a 60 000 animal population during a 2-week to 3-week period each autumn.

Game animal numbers vary greatly in different habitats. The mule deer is the most common western North American large wild herbivore. Mule deer populations vary widely throughout the western United States and animal numbers may vary from 6 to 39 $km^{-2}$. The high densities would be on such areas as recent burns (Longhurst, Leopold & Dasmann, 1952) where densities of animals up to 40 $km^{-2}$ with biomass 355 $kg \cdot km^{-2}$ may be found (Taber & Dasmann, 1957).

Population density and biomass estimates for wild large herbivores of the arctic and subarctic regions are more difficult to obtain. The caribou of North America and the domesticated reindeer of the tundra and taiga feed extensively on lichens. When averaged for Alaska, Canada, and Scandinavia the biomass may be in the order of 15 $kg \cdot km^{-2}$ (based on the data of Banfield, 1954, and Scheffer, 1951). The critical season in the foraging of these animals is winter. Lichens may be 50% of the winter diet. Although the winter range may have more than 3000 $kg \cdot ha^{-1}$ of standing crop of herbage, it can only be grazed periodically. There may be as much as a 5-year to 10-year cycle to allow regrowth of the lichens. Thus, animal populations are rather limited. In Fennoscandia there are $< 1 \times 10^6$ reindeer which utilize the forest–lichen ranges in the winter and the alpine tundra in the summer.

The musk ox is another large herbivore of arctic and subarctic lands. In contrast to the caribou, the musk ox does not migrate widely (Wilkinson, 1974). The musk ox is more of a browsing animal than is the caribou which feeds heavily on lichens. Mortality in the musk ox is generally not caused by predators but is due to heavy winter snow and icing conditions. In favoured portions of Devon Island, musk ox average between 300 and 400 $kg \cdot km^{-2}$ only during summer and autumn months.

### *Examples of production rates of wild herbivores: tropical*

African savannas are well known for their high populations of ungulates. Various estimates are available for their density and biomass. In the dry Mopane woods of southern Rhodesia a low population estimate of 16 species of wild ungulates combined was 30 $animals \cdot km^{-2}$ with a biomass of 900 $kg \cdot km^{-2}$ (Dasmann & Mossman, 1962). In the open savanna parkland of Albert Park in one year (1959) in the Congo, some 10 species of large wild herbivores were estimated at 35 $animals \cdot km^{-2}$ with a biomass of about 24 000 $kg \cdot km^{-2}$ (Bourlière & Verschuren, 1960). Some six species of large herbivores at about 4 $individuals \cdot km^{-2}$ had 4250 $kg \cdot km^{-2}$ in degraded *Commiphora* woodland–grassland in Tsavo National Park in Kenya (Leuthold & Leuthold, 1976).

These figures might be compared to the offtake of cattle in a tropical region in the Ankole of Uganda where cattle biomass is about $36 \times 10^3$ $kg \cdot km^{-2}$ with about a 25% annual offtake (Longhurst & Heady, 1968).

Data have been tabulated by Talbot *et al.* (1965) for various East African rangelands in units of yearlong standing crop of large herbivores. Wild ungulates on acacia savanna vary from 65 to 158 $kg \cdot ha^{-1}$. Comparable locations stocked by cattle, goats and sheep vary from only 20 to 28 $kg \cdot ha^{-1}$. The wild population exists without any management of the vegetation and without management practices commonly used with domestic animals on pasture. Otherwise the standing crop could be increased. On acacia–commiphor bushland wild ungulate biomass is in the order of 52 $kg \cdot ha^{-1}$ in contrast to sheep and goat biomasses on the equivalent rangeland of 4 to 14 $kg \cdot ha^{-1}$.

Standing crop values alone, however, are misleading and one must apply an offtake rate and also consider the utilization of the product. Harvesting of wild animals is not particularly efficient at present. Some species, such as the eland, appear to have considerable potential for domestication. There are claims that eland are easier to manage than most range cattle, can be confined by normal wire fences, have growth and reproduction rates generally superior to domestic cattle maintained in the same environment, and will even thrive under conditions of food, water and some insects in which domestic cattle cannot survive.

In calculating productivity of large wild herbivores in an area such as East Africa, one must consider their grazing area throughout the year. Consider, for example, the Serengeti–Mara Plains covering about 21 000 $km^2$ along the Kenya–Tanzania border of East Africa and which represent an area with one of the highest productivities of large herbivores in the world (Talbot & Talbot, 1962). This area is a flat to undulating plain just south of the equator. It has an average rainfall from about 50 to 75 $cm \cdot yr^{-1}$ over most of its extent. There is a distinct dry season from June to October. Much of the plain is covered by vegetation with a shortgrass aspect. Tall grasslands occur on lighter-textured soils and on the margins. Both grazing animals and fire are major influences on maintaining this vegetation in the savanna stage rather than allowing it to go to forest. The Masai tribesmen periodically set fire to the region. The area is grazed seasonally by migrating herds of wild ungulates with as many as 30 species.

The question of how so many species with seemingly similar dietary habits can coexist has long intrigued ecologists. It appears that it is their grazing selectivity and timing that allow them to occupy the same habitat. The herds may cover as many as 1600 km distance in a given year. During the wet season the open plains support a biomass of wildlife and domestic animals of about 16 000 $kg \cdot km^{-2}$, while this number drops to 0 in the dry season. In the dry season, the surrounding bush supports biomass of near 6000 $kg \cdot km^{-2}$. In the wet season areas extending some 8 to 9 km from water supply support a biomass up to 18 000 $kg \cdot km^{-2}$.

A recent study has been made to develop techniques to harvest these large wild herbivores commercially in this general area. The Kajiado district of Kenya, south of Nairobi, represents an area with one of the few remaining large migratory herds of plains game in Africa. This area of 21 000 $km^2$ is primarily semi-arid and has an estimated 200 000 plains animals. These animals can migrate into and out of Nairobi, Tsavo and Amboseli National Parks. Abundant species include the wildebeest, kongoni, zebra, and Thomson's and Grant's gazelles. Other species that are common are the giraffe, warthog, eland and impala. The annual harvest in this area amounts to about

3000 wildebeest, 2300 Grant's gazelle, 500 Thomson's gazelle and 1300 impala (Swank, Casebear, Threasher & Woodruff, 1974). These workers estimated that the meat and hide harvest of these wild animals would exceed that of domestic animals. The average dressing percentage for these animals would be about 50%. For comparison of yield by these wild animals with that of cattle, a harvest rate of 10% was used. G. W. Swank (personal communication) estimates there are about 10 000 Thomson's gazelle in the district, and that a conservative offtake of 25% could be made annually. Meat and hide harvest of domestic animals is expected to be less than that of wild animals because the wild animals use a wider variety of forage species than do the domestic animals and the wild animals are more resistant to drought and disease. The Masai, to whom these lands are being turned over in private ownership from their former trust status, have the decision as to whether to attempt to utilize their lands for wildlife recreational hunting, cropping for meat and hides, photography and tourism, commercial livestock ventures, or some combination of enterprises. This is an interesting optimization problem.

Hirst (1975) has calculated the ungulate biomass and secondary production from census data and age–weight curves for a low veld savanna region in the eastern Transveld Province in South Africa. The main ungulate species are impala, wildebeest, zebra, water buffalo and warthog (which are all grazers), and giraffe and kudu (browsers). Animal density varies from 13 to 67 $km^{-2}$. Ungulate biomass is about 40 $kg \cdot ha^{-1}$ in the wet season and 65 $kg \cdot ha^{-1}$ in the dry season. Secondary production over a 2-year study period was 0.97 $kcal \cdot m^{-2} \cdot yr^{-1}$ from a mean standing crop of about 7.5 $kcal \cdot m^{-2}$.

A careful calculation was made of the energy requirements for lambs, yearling and adult blesbock on a Rhodesian grazingland (Catto, 1976). This grazingland is burned in the spring and provides about 100 days of grazing which are critical in the yearlong nutrition of the animal. The lambs are born near the beginning of the burning and are weaned by 4 months of age; thus, there are high lactation demands on the females during the period they are on the burned grazingland. Carcass yield is 57%. Calculations were made to show that cattle produce about 3.4 times as much meat as the blesbock. Few data were available for mutton sheep, but blesbock were less favourable competitors with wool sheep than with cattle. However, if the blesbock are sold for foreign export they compete relatively well with both sheep and cattle. It was calculated that the unconfined blesbock has an energy requirement of 2.7 times the basal energy requirement; in the field situation it is 2.9 times the basal energy requirement. This contrasts to a rule-of-thumb figure of about 2 for domestic animals.

Optimal meat production in the blesbock herd was obtained with a sex ratio of five females per male, cropping of excessive lambs in December, and cropping one complete adult age segment. An optimal herd had about 1.3 lambs per female. This is considerably greater (1.7 times as much) than is the case in a field study from the same area of the natural herd composition and productivity rate.

### *Some generalizations on large herbivore production*

One of man's ultimate aims is to increase production of useful products from large herbivores. Production is the result of a balance of factors. First, one must consider those factors which affect intake, assimilation and conversion of nutrients to useful products. On the other side there are costs of maintenance and various expenditures forced on the animal. We have emphasized the 'input' factors elsewhere in this chapter, but consider here more information on the deposition of the unexpended nutrients in terms of fat and lean meat.

#### *Some domestic animal–wild animal contrasts*

The dressed carcass of many of the wild large herbivores is a higher percentage of the animal's live weight than is the case for domestic animals (Talbot *et al.*, 1965; Ledger, Sachs & Smith, 1967). For East African conditions, depending upon fatness, there is considerable variability in the dressing percentage of domestic animals – from 40 to 60% for Zebu steers. The dressing percentage for wildebeest is about 50 to 55%, for topi and kongoni about 53%, for Thomson's gazelle about 58%, for impala 58 to 60%, and for Grant's gazelle 58 to 60%. The dressing-out percentage of many of these wild ruminants is more constant than in domestic animals with regard to age, sex and condition, although in males the percentage is generally higher than in females. Most of these large, wild herbivores do not have a high fat content and tend to remain in the same general condition throughout the year. In contrast, the fat content of many domestic animals varies from a few to over 30%. Many wild animals which will give a dressing percentage of 50 or 60 will have only 2% carcass fat. Steers with a dressing percentage of 60 may have 30% carcass fat. The red deer has more 'first-quality' meat in its carcass than do sheep and cattle, the figures being about 56, 46 and 39% respectively (Blaxter, 1975).

There are striking differences among the wild mammals, though, in different environments. Deer in North America may have seasonal fluctuations amounting to 40% of their maximum live weight and much of this is due to changes in fat content.

Numerous factors can limit large herbivore production even when sufficient nutritious forage is available. Harassment may greatly influence animal requirements. Val Giest, who has had extensive experience with large wild herbivores in North America, suggests such costs may be from 20 to 200% of maintenance (personal communication). Some of such costs are due to increased energy expenditure through induced nervousness and flight. A related cost is non-use of highly productive areas with nutritious vegetation, i.e. a loss-of-opportunity cost. Even over-zealous herding of domestic animals can have significant negative impacts on production (Baskin, 1974).

The differences between standing crops of domestic animals and of wild herbivores in East Africa are due primarily to the forage composition, to water and to insects. Cattle, sheep and goats as a group have higher preferences for grasses and forbs than do many of the large wild herbivores. Even under the best range management conditions, preferred grasses may be only a small part of the total available plant food in some savannas. Heady (1960) suggests that preferred grasses for domestic animals may contribute less than 10% to the herbage in much of East Africa. Domestic animals are limited in number when compared to the incredible variety of wild herbivores in East Africa. The domestic animals evolved in a 'dietary environment' which is only poorly represented in these tropical savannas. Thus the wild animal assemblage is more adapted to, and more efficient in their use of, these highly varied grass and shrub savannas. In the dry season, watering points are so far apart that many domestic animals cannot utilize much of the area. In some locations tsetse flies prevent livestock grazing.

### *Seasonal influences on productivity*

Because of wide seasonal variations in herbage availability for wild ruminants, their fasting metabolism rate is critical because it is a large part of the maintenance requirement. There are large seasonal variations in fasting metabolism in some wild ruminants, with summer rates being up to twice those in the winter (see Topps, 1975). Rates are perhaps 50% higher for young animals than mature ones. Young eland, wildebeest and blesbock vary from around 400 to more than 500 kJ (kg $W^{0.75})^{-1} \cdot 24\ h^{-1}$ (Topps, 1975). Mature pronghorn and white-tailed deer have rates near 350 in many seasons. The interspecies mean is 293 kJ (kg $W^{0.75})^{-1} \cdot 24\ h^{-1}$, and thus the maintenance requirements of large wild herbivores are greater than that of domestic herbivores. In simulation studies, Van Dyne, Joyce & Williams (1977) found that cattle and bison produced better than did sheep and much better than pronghorn, largely as a consequence of differences in maintenance energy requirements.

Another generalized relation is that for body weight and daily gain during the grazing season within the annual cycle. This applies primarily to northern and southern hemisphere temperate zone grazinglands where there is a winter period, and to some degree to the tropical lands where there is a dry period. When vegetation growth begins the animals may be at their lowest body weight. Soon they reach their highest daily gain values, which then become successively lower as the animals grow or regain body weight. Weight losses occur in the nongrowing season and the cycle starts again.

Table 4.16. *Some examples of population and herd structures reported for large herbivores*

| Reference | Location | Animal species | Adults | | | Juveniles | | | Young | | | Notes |
|---|---|---|---|---|---|---|---|---|---|---|---|---|
| | | | Male | Female | Both | Male | Female | Both | Male | Female | Both | |
| Feist & McCulough (1975) | Western USA | Feral horse | 29 | 29 | — | 12 | 17 | — | 6 | 7 | — | |
| Lassen *et al.* (1952) | Eastern California, USA, and Western Nevada, USA | Mule deer | 9 | 51 | — | — | — | 6 | — | — | 34 | |
| Cowan (1950) | Alberta and British Columbia, Canada | Mule deer | 35 | 65 | — | — | — | — | — | — | — | |
| White (1973) | Texas, USA | White-tailed deer | 33 | 34 | — | 5 | 7 | — | 10 | 11 | — | |
| Prior (1968) | England | Roe deer | 48 | 52 | — | — | — | — | — | — | — | |
| Cowan (1950) | Alberta and British Columbia, Canada | Elk | 23 | 77 | — | — | — | — | — | — | — | |
| Gaare & Skagland (1976) | Southern Norway | Reindeer | 20 | 50 | — | — | — | 10 | — | — | 20 | |
| Bergerud (1967) | Labrador | Caribou | 20 | 64 | — | 3 | 3 | — | 2 | 8 | — | 5 % unknown distributed among the sex and age classes |
| Cowan (1950) | Alberta and British Columbia, Canada | Caribou | 33 | 67 | — | — | — | — | — | — | — | |
| Ling (1973) | Estonia | Moose | 34 | 41 | — | — | — | — | — | — | 25 | |
| Stevens (1970) | Southwest Montana, USA | Moose | 23 | 50 | — | — | — | — | — | — | 27 | |
| Houston (1968) | Wyoming, USA | Moose | 34 | 42 | — | — | — | — | — | — | 24 | |
| Knowlton (1960) | Montana, USA | Moose | 41 | 31 | — | — | — | — | — | — | 24 | |
| Pimlott (1953) | Newfoundland | Moose | 52 | 48 | — | — | — | — | — | — | — | |
| Cowan (1950) | Alberta and British Columbia, Canada | Moose | 62 | 38 | — | — | — | — | — | — | — | |
| Schultz & McDowell (1943) | Montana, USA | Moose | 45 | 27 | — | 5 | 7 | — | — | — | 16 | |
| Geist (1971) | Southern Alberta, Canada | Bighorn sheep | 59 | 31 | — | 5 | 2 | — | — | — | 3 | |
| Woolf (1968) | Wyoming, USA | Bighorn sheep | 27 | 52 | — | — | — | — | — | — | 21 | Adult female class includes yearlings of both sexes |
| Oldemeyer (1966) | Wyoming, USA | Bighorn sheep | 32 | 41 | — | — | — | 8 | — | — | 19 | 5 % unclassified distributed among the sex and age classes |
| Woodgerd (1964) | Montana, USA | Bighorn sheep | 29 | 32 | — | 8 | 4 | — | — | — | 27 | |
| Cowan (1950) | Alberta and British Columbia, Canada | Bighorn sheep | 44 | 56 | — | — | — | — | — | — | — | |
| Hjeljord (1973) | Southern Alaska, USA | Mountain goat | 67 | — | — | — | — | — | — | — | 33 | |
| de la Tour (1954) | South America | Guanaco | 13 | 87 | — | — | — | — | — | — | — | |
| Koford (1957) | Western South America | Vicuña (all male group) | 20 | — | — | 80 | — | — | — | — | — | |
| | | Vicuña (family group) | 13 | 62 | — | — | — | — | — | — | 25 | |
| Dasmann & Mossman (1962) | Southern Rhodesia | Warthog | — | — | 42 | — | — | — | — | — | 58 | |

| | | | | | | | | | | | | |
|---|---|---|---|---|---|---|---|---|---|---|---|---|
| Hvidberg-Hansen & | Kenya | Thomson's gazelle | 16 | 37 | — | 29 | 13 | — | — | — | 5 | |
| de Vos (1971) | Northwestern Tanzania | Thomson's gazelle | 31 | 44 | — | 13 | 8 | — | — | — | 4 | |
| Dasmann & Mossman (1962) | Southern Rhodesia | Impala | — | — | 31 | — | — | 43 | — | — | 26 | |
| | | Kudu | — | — | 45 | — | — | 42 | — | — | 23 | |
| | | Bushbuck | — | — | 64 | — | — | 24 | — | — | 12 | |
| Talbot & Talbot (1963) | East Africa | Wildebeest | 31 | 29 | — | 6 | 5 | — | 18 | 11 | — | |
| Dasman & Mossman (1962) | Southern Rhodesia | Wildebeest | — | — | 42 | — | — | 35 | — | — | 23 | |
| | | Zebra | — | — | 57 | — | — | 30 | — | — | 13 | |
| Heindrichs (1975) | East Africa | Reedbuck | 22 | 27 | — | 22 | 27 | — | 2 | — | — | |
| Dasmann & Mossman (1962) | Southern Rhodesia | Reedbuck | — | — | 86 | — | — | — | — | — | 14 | |
| | | Waterbuck | — | — | 26 | — | — | 42 | — | — | 32 | |
| Ross *et al.* (1976) | Uganda | Giraffe | 19 | 56 | — | 3 | 6 | 18 | — | — | 7 | 9 % unknown sex in juveniles |
| Foster (1966) | Kenya | Giraffe | 39 | 43 | — | — | — | 18 | — | — | — | Juveniles include young |
| Dasmann & Mossman (1962) | Southern Rhodesia | Giraffe | — | — | 48 | — | — | 31 | — | — | 21 | |
| Ross *et al.* (1976) | Uganda | Elephant | 17 | 29 | — | — | — | 42 | — | — | — | |
| Dagg & Foster (1976) | Nairobi Park | Giraffe | 25 | 31 | — | — | — | 30 | — | — | 14 | |

### *Age–sex composition of herds*

Data are presented in Table 4.16 on the percentage composition of males and females of different age groups within the herds of wild large herbivores. In some 30 instances where the adult population was determined separately, about two-thirds of the herd consists of adults. This leaves about one-third of the herd not sexually mature, comprising some individuals $\leqslant 1$ year of age but not having reached sexual maturity and, of course, the calves, fawns and the young of the year. In intensively hunted populations, such as North American pronghorn, elk and mule deer, around 20% of the total herd may be removed each year, and this removal comes from the sexually mature animals, primarily the males. This means that a good part of the juvenile and young animals must therefore move into the adult category each year in order to maintain herd size. Thus, losses due to predation and other environmental causes need to be minimized in order to maintain these hunting rates.

### *Comparative productivity data*

To make a rough comparison between various production data for many herbivore species we have used the approach of Wagner (1969), who compiled information on the estimated energy in standing crops and production for herbivores. We have added to his data other information and have converted all data to units of $g \cdot m^{-2}$ for standing crop and $g \cdot m^{-2} \cdot yr^{-1}$ for annual production. We summarize information on basis of weight rather than calories because of the wide variability in fat content in the herbivores and use the conversion figure of 1.5 $kcal \cdot g$ $liveweight^{-1}$ as originally used by Petrides & Swank (1966). Data are rounded to one significant digit, perhaps a level of significance appropriate to the overall result because of the roughness of estimates of field populations, production rates, weights per animal, etc. These data are summarized in Table 4.17 which shows maximal production rates of the order of 1 $g \cdot m^{-2} \cdot yr^{-1}$ and minimal production rates of 0.1 $g \cdot m^{-2} \cdot yr^{-1}$ for dense populations of the respective species. Obviously, as one goes to semi-arid and arid rangelands of the world, population densities and production rates may be even lower.

The estimates for production of white-tailed deer in Table 4.17 appear to be as high or higher than those for other North American deer populations. The data for the Rhodesian savanna with mixed large herbivores are of particular interest since the great majority of the yield comes from impala, zebra and giraffe, which have widely varying anatomy, physiology and behaviour. The data for the large herbivores of the South African savanna would include an equivalent mix. The generalization is suggested that with diverse vegetation a mixed herbivore herd will give high productivity. There is not, however, a close relationship between standing crop and annual production rate for these

Table 4.17. *Rough estimates of standing crop and annual production for various herbivore populations*

| Species | Location | Standing crop $(g \cdot m^{-2})$ | Annual production $(g \cdot m^{-2} \cdot yr^{-1})$ | Original reference |
|---|---|---|---|---|
| Bison, moose and elk | Boreal aspen forest–grassland, Alberta Canada | 6.0 | 1.1 | Telfer & Scotter (1975) |
| Savanna large herbivores | Henderson & Sons Ltd ranch, Rhodesia | 0.9 | 0.6 | Dasmann & Mossman (1964) |
| Savanna large herbivores | Lowveld savanna, South Africa | 5.0 | 0.6 | Hirst (1975) |
| Beef cattle | Hypothetical N. American grassland | 5.0 | 0.6 | Petrides & Swank (1966) |
| White-tailed deer | George Reserve, Michigan, USA | 0.9 | 0.4 | Davis & Golley (1963) |
| Elephant | Queen Elizabeth Park, Uganda | 5.0 | 0.2 | Petrides & Swank (1966) |
| Saiga | USSR steppes, Caspian area | 0.2 | 0.1 | Bannikov (1967) |
| Wildebeest | East African savanna, Kenya–Tanzania | 0.7 | 0.1 | Talbot & Talbot (1963) |

species or mixed herd populations. The data presented on production here give a very wide range of production and standing crop rates. Much more information and research on the subject is needed, including good comparative data on reproduction, growth rate and survival of the populations, as well as variability between areas and years. It is likely the values noted above are upper estimates as these are the type of data most frequently entering the literature.

## Models of large herbivores

A sizeable number of simulation and optimization models have been developed of large herbivores, or components thereof, within grazingland systems. It is beyond the scope of this chapter to review in detail the processes within all of these models. However, the following discussion will serve to highlight some of the approaches used in modelling the processes accounting for the biomass and population dynamics of large herbivores.

### *An overview of models available*

Table 4.18 summarizes models of or containing large herbivores under grazingland situations. The models are characterized as to whether they are for simulation or optimization, whether they are difference or differential or

Table 4.18. *Mathematical models containing, or of, large herbivores*

| Reference | General model | Specific model | Time stage | Computer language |
|---|---|---|---|---|
| Anway (1973) | Simulation | Difference nonlinear | 2 days | SIMCOMP |
| Armstrong (1971) | Simulation/ optimization | Algebraic/ difference | 1 day | FORTRAN SIMSCRIPT |
| Arnold & Bennett (1975) | Simulation/ optimization | Nonlinear simplex | 9 years | — |
| | Optimization | Linear programming | 1 year | Run on Cyber 72 |
| Arnold & Campbell (1972) | Simulation | Algebraic/ difference equation type | Daily | FORTRAN |
| Arnold & Galbraith (1974) | Simulation | Difference | 1 day | FORTRAN |
| Arnold & Galbraith (1974) | Simulation | Difference | 1 day | FORTRAN |
| Arnold *et al.* (1974) | Simulation | Difference | 1 day | FORTRAN |
| Baldwin (1972) | Simulation | Differential | Varies | Kinsum |
| W. A. Beyer *et al.* (unpublished data) | Simulation | Difference | 1 year | FORTRAN |
| Bledsoe (1975) | Simulation | Differential nonlinear | 0.042 day | FORTRAN IV |
| Blincoe (1975) | Simulation | Differential | 0.1 hour | FORTRAN IV |
| Bunnell *et al.* (1975) | Not specified | Not specified | Not specified | FORTRAN IV |

| Model diagram | Program code | What it was built for | Comments |
|---|---|---|---|
| Yes | No | To represent mechanistically the dynamics of any grassland mammal as well as the interconnections of that consumer with the other parts of the system | 35 state variables, 2 driving variables. Part of ELM (Innis, 1978) |
| Yes | — | Grazing management practices | 10 state variables, 2 driving variables, 13 constants, 12 flows. To compare the management strategies of set stocking and rotational grazing: model development. Plant growth and animal production. |
| No | No | Management of wether sheep producing wool, stocked yearlong on an annual range of subterranean clover. | Random sequence of climatic data used, stocking rates were varied. Constraints: size of seed pool, amount of green matter to prevent soil erosion and amount of risk (financial). |
| Yes | No | To describe crop and pasture production and utilization in a Mediterranean environment | Optimum use of 50 ha crop of lupins examined. System was a matrix of 146 activities with 75 constraints. |
| No | Available on request | To predict wool production in a ley farming system in a Mediterranean environment with nonreproductive sheep | Driving forces are rainfall, temperature and radiant energy; plant variables include seed, green vegetative mass and dead material; animal variables include numbers in different age and sex classes and animal type, mean weight and fleece weight. A component of an overall model (other submodels are crop growth, cultivation, pasture growth); each submodel can run separately. |
| Yes | No | Liveweight changes and wool production | The model is general and capable of being used in a variety of situations. More research is needed in order to improve the model results. |
| No | No | Crop and animal production | Although the models are 'first generation', they allow a wide combination of seasonal conditions, types of livestock and stocking rates to be examined. The model of lupin growth was validated against data from different sources; liveweight changes of Merino wethers and yearling cattle were simulated when grazed at 3 stocking rates. |
| No | No | Submodel of a larger model | 7 state variables, 4 driving variables, 20 processes to examine management strategy for crop use. Limited to lupin crops. In this modified version of Arnold *et al.* (1974) model, the animal group is any stock of nonbreeding sheep and cattle. Equations for planned growth on gravelly soils. |
| — | — | Energy balance at the tissue level | Energy balance and animal production at biochemical level. Deterministic nonstochastic linear equations limited only for ruminants |
| No | No | To develop a Monte-Carlo simulation of population interaction in a biome in order to provide a close fit to existing real data | The model is a denumerable state, continuous time Markov process |
| Yes | Yes | To predict time fluctuations of biological productivity fluctuation on a shortgrass prairie | 6 driving variables, 59 state variables |
| Yes | No | For iodine metabolism of lactating cattle and sheep | Code can be derived from tables and figures. Model predictions agree well with measured data. Assumes first-order processes. |
| No | No | A workshop exercise to develop model of the dynamics of a barren-ground caribou herd | Considerations in the model are population size by sex and age class, square miles of winter habitat in various successional stages, standing crops of vegetation, index of weather severity in winter and during spring calving, reproduction data, natural mortality and harvest, net production by habitat class, and rate of food consumption by animals in each habitat class. |

Table 4.18 (*continued*)

| Reference | General model | Specific model | Time stage | Computer language |
|---|---|---|---|---|
| Cale (1975) | Simulation | Differential linear | 0.01 week | CSMP |
| Christian *et al.* (1972) | Simulation/ optimization | Difference | 1 day | FORTRAN |
| Christian *et al.* (1974) | Simulation/ optimization | Difference | 1 day | FORTRAN |
| Chudleigh & Filan (1972) | Simulation | — | 1 month | — |
| Cole (1976) | Simulation | Difference | 1–2 days | SIMCOMP |
| Connolly (1974) | Optimization | Linear programming | n.a. | Unspecific |
| D'Aquino (1974) | Optimization | Linear programming | — | — |
| Davis (1967) | Simulation/ optimization | Linear programming | 20 years | IBM LP3 program |
| Donnelly & Armstrong (1968) | Simulation | Difference | — | SIMSCRIPT |
| M. I. Dyer, G. S. Innis, L. F. Paur & G. E. Ellis (unpublished manuscript) | Simulation | Algebraic | 1 week | SIMCOMP |

| Model diagram | Program code | What it was built for | Comments |
|---|---|---|---|
| No | Yes | To simulate the nominal behaviour of the Pawnee grassland and to investigate the potential of engineering analysis technique for understanding and evaluating ecosystem behaviour | 9 driving variables, 39 state variables. Low-resolution models. |
| No | No | Selecting the best systems of grazing management | 1 driving variable. Model building for management decisions. Limited to ewes during late summer. Examined a range of rotational management practices; based on Freer *et al.* (1970), compared 6 rainfall patterns, 8 stocking rates, 2 initial weights; calculated 'objective functions' based on value of liveweight at start of mating, value of food after 100 days and cost of supplementary feeding. A later version of the Freer *et al.* (1970) model. |
| Yes | No | Farm producing lambs for meat | See Christian *et al.* (1972). Management decisions. Limited to sheep and cattle; more complex than previous versions; animal management implemented through input code values; uses list-processing techniques in calculations; allow changing feed requirements for animals during the year. |
| Yes | No | A pastoral property in the Western Division of New South Wales | 18 state variables, 2 driving variables. The grazing model was developed to explore the problems involved in developing and using models based on biological relationships for decision-making studies. The paper discussed some methodological aspects of simulation with specific reference to grazing systems. |
| Yes | Yes | Objective to develop a parameterized total-system model adaptable to several semi-arid grasslands | A detailed report on ELM 1973; model contains up to 250 state variables (not all in SIMCOMP structure); discussion given of flows and rationale for their structure; 663 p. report includes following chapters: introduction, sub-models of abiotic, producer and phenology, mammalian consumer, insect consumer, decomposition, nutrient, and how to use the model; appendixes include test data deck, test run output, listing of source deck, symbolic name definition list, Pawnee site input parameter list, significant word in definition list, and definition list for preceding three appendixes. |
| n.a. | No | Objective to maximize forage utilization, animal gain and profit subject to constraints on use levels for species | Uses literature data for range cattle and sheep giving composition and utilization of vegetation. Assumes preferences are fixed and total herbage available is sufficient. |
| No | No | To help in the management of range resource system | |
| No | No | Management of deer herd and simultaneous deer and timber management | Money and labour are limitations. 10 variables, 8 equations. |
| Yes | No | To simulate the grazing of summer pasture by sheep | 10 'endogenous events', 1 driving variable, 5 state variables. To predict the response of sheep weight to grazing subdivision, rainfall, growth rate of herbage, the amount of dry food available, and the efficiency with which it is grazed. |
| No | Partially | To summarize some of the things that are known about mechanisms pertaining to predators and prey with grasslands and thereby provide a basis for deterministic predictions about these associations | |

Table 4.18 (*continued*)

| Reference | General model | Specific model | Time stage | Computer language |
|---|---|---|---|---|
| Eberhardt & Hanson (1969) | Simulation | Difference equations | 1 day | DYNAMO |
| Edelsten *et al.* (1973) | Simulation | — | 1 day | — |
| Edelsten & Newton (1977) | Simulation | Difference | 1 day | FORTRAN |
| Freer *et al.* (1970) | Simulation | Algebraic/ difference | 1 day | SIMSCRIPT |
| Fullerton (1975) | Simulation | Algebraic | 1 day | SIMCOMP |
| Garner (1967) | Simulation | Sum of exponentials | n.a. | Unspecified |
| Gilbert (1975) | Simulation | Algebraic | 1 month | SIMCOMP |
| Goodall (1969, 1970*a*, 1971, 1973) | Simulation | Difference | — | FORTRAN |
| Gross (1970) | Simulation | Algebraic | 1 year (?) | FORTRAN |
| Gross & Walters (1970) | Simulation | Algebraic | 1 week | FORTRAN IV |
| Harris & Fowler (1975) | Simulation | Differential | 1 day (est.) | FORTRAN |
| Harefield *et al.* (1974) | Simulation | Differential equations | n.a. | SAAMS |

| Model diagram | Program code | What it was built for | Comments |
|---|---|---|---|
| No | No | To represent the concentration of $^{137}Cs$ in a lichen–caribou–Eskimo food chain | Assumes exponentially decreasing fallout on lichens, seasonally varying uptake of $^{137}Cs$ by caribou, and spring and fall harvests of caribou by Eskimos. The model is primarily of value for conceptual analysis rather than for prediction |
| Yes | No | To explore different systems of lowland fat-lamb production and to define the most profitable lowland fat-lamb production | 30 state variables. To explore the effect of management variables on the growth of lambs and to explore ways of controlling these factors. Does not have an economic optimization model; it simulates a block of sheep from when the lambs are born in mid-March to when they are sold in Sept.–Oct. The initial weights of the animals are given according to probability distribution (stochastic). The rest of the model is deterministic. Simulates individual animals. |
| Yes | No | To simulate ewe and lamb growth intraseasonally allowing various grazing management and crop conservation strategies into supplemental feeding | Some initial conditions and parameters determined stochastically; thereafter model is deterministic. Individual animals simulated. Details given in Edelsten & Newton (1975). Herbage growth and herbage digestibility are inputs to model. |
| Yes | No | Grazing of summer pasture by sheep | 8 state variables, 1 driving variable, 10 parameters. To predict the response of sheep weight to changes in grazing subdivision, stocking rate and growth rate of herbage. Limited to the growing season; specific equations and diagrams and values of constants. Predictions reasonable. Plant growth and production. |
| No | No | To explore means of qualitatively assessing the performance of the mammalian consumer submodel with respect to its objective and those of the total-ecosystem model ELM | |
| Yes | No | Predict parent–daughter fission products in cows' milk | Model useful in predicting consequences of release of fission products following a criticality excursion. Compartments are rumen, plasma, udder and a 'reservoir' with kidneys and passage through the gut as sinks. |
| Yes | Yes | To link a rangeland ecosystem with the range management and range economic system | |
| No | Yes | | 7 state variables. Evolving model with spatial considerations. Sheep rangeland grazing. |
| No | Yes | To provide information on alternative decisions in managing large wild herbivores | Model requires as inputs: age and sex structure of herbivore, reproduction rates, mortality rates, growth (age–weight), population production and energy use, differential age kill, hunting effort, wounding loss, harvest rate, age thresholds for harvesting. Outputs include number of animals, quality, and weight harvested and energy required to produce yield. |
| No | Yes | To simulate the demographic and biomass dynamics of small mammals | |
| No | No | African mixed consumers on savanna | State variables include 23 animals (ages, sexes, species, dead) and 6 milk components. Driving variables include 8 plant components and precipitation. |
| Yes | No | To study kinetics of glucose in lactating cow | Paper refers to previous work using first-order constant coefficient equation system. Fit parameters to data by least squares. Uses 4-compartment model. |

Table 4.18 (*continued*)

| Reference | General model | Specific model | Time stage | Computer Language |
|---|---|---|---|---|
| Hudson (1975) | Simulation | Algebraic | 1 year | CMSP |
| K. Hutchinson (unpublished manuscript) | Simulation | Algebraic | 1 week | FORTRAN |
| Jones (1970) | Simulation | — | 108 days | DYNAMO |
| Lassiter & Hayne (1971) | Simulation | Differential | 30 years | FORTRAN IV |
| May *et al.* (1972), Till & May (1970*a*, *b*) | Simulation | Differential | Instant | Analogue |
| Mazanov & Nolan (1976) | Simulation | Difference–differential | n.a. | FORTRAN |
| McKinney (1972) | Simulation | — | — | FORTRAN CSMP |
| Milner (1974) | Simulation | Difference | 1 day | FORTRAN |
| Morley & Graham (1971) | Simulation | Algebraic/difference | 1 year | FORTRAN (KWIKTRAN) |
| Morris *et al.* (1974) | Simulation | Differential | 1 day | CSMP |
| Nelson (1972) | Simulation | Algebraic/difference | 1 month | FORTRAN |
| Newton & Edelsten (1976) | Simulation | — | 145–147 days | — |

| Model diagram | Program code | What it was built for | Comments |
|---|---|---|---|
| Yes | No | To predict the productive characteristics of a native grazing system in such an environment if managed specifically for meat production | |
| Yes | No | To show how grazing experiments and modelling can be regarded as complementary activities | 4 state variables, 4 driving variables, 12 processes. The model needs to be developed and improved with validation; example in management strategy to evaluate concept of fodder conservation levels examined; plant growth and animal production. |
| Yes | No | To model growth of lambs during the rearing period and the growth of a typical lamb during a rotational grazing period | Rate of rotation was changed in one run and stocking rate in another |
| No | Yes | To specify and thereby provide a means for quantifying some generally held ecological concepts about the dynamics of natural communities | Closed system except unlimited energy to producers, and immigration equals emigration. Biotic and abiotic factors influence a population only by changing rates of increase or death. Simulated community: 2 spp. of plants, 2 of herbivores and 2 spp. of predators and 2 abiotic factors (seasonal climatic changes and a constant environmental influence). Space limits imposed on producers. |
| No | Yes | Predictions of mineral cycle influence on sheep production | 41 state variables. Part of an experimental–analytical program; $^{35}S$ measured in various points in the system; sheep pasture sulphur cycle. |
| Yes | No | To simulate $N_2$ metabolism in whole animal and compare with experimental data | A 9-compartment model including time lags. Input: amino acids, ammonia, microbial, small intestine, caecal, rectal, body fluids and tissue. Outputs from body fluids, tissue and rectal. Based on simplifying the 25-compartment model of Nolan & Leng (1972) and using information from Nolan *et al.* (1976). Model output compared to data. Linear model. |
| Yes | No | Estimating live weight change and pasture growth during 30 days of winter | The model is similar to that published by Vickery and Hedges (1972*a*, *b*). Systems of young sheep production for meat and wool, or flock replacements, and young beef cattle grown for meat, or for herd replacement. |
| Yes | No | To visualize and interlink processes as a management tool | Model in crude form has poor predictive value. Has stimulated literature search and data collection. |
| No | No | Fodder conservation for drought | 22 'input variables'. Management decisions. The model is restricted to comparisons between two of many possible policies. |
| — | — | Understanding the dynamics of nitrogen of ruminants | 9 state variables, 4 driving variables. Teaching and experimental design. It is a compartment of a subsystem model; limited to ruminants; requires validation by experiments. Deterministic. |
| Yes | Yes | To test the feasibility of using a mix of native and domestic ruminants to harvest forage on rangeland | 1 driving variable, 75 parameters. Limitation: the model considers only green plants and ruminants (3 spp.). Far from reality; forage production is calculated on a per year basis and rain is the only driving variable. The model is a simplified abstraction of a rangeland ecosystem. Excellent documentation of derivations of flow functions; relatively few runs made with final model. |
| Yes | No | To examine two effects of nutrition on the reproduction process in sheep | One was effect of a range of ewe weights at mating on the subsequent litter size, second was different levels of nutrition during the last 8 weeks of pregnancy on growth rate, birthweight and survival. |

Table 4.18 (*continued*)

| Reference | General model | Specific model | Time stage | Computer language |
|---|---|---|---|---|
| Noble (1975) | Simulation | Algebraic equation | Variable (1 day?) | ? |
| Parton & Smith (1974*a*) | Simulation | Algebraic | 1 year | SIMCOMP |
| Parton & Smith (1974*b*) | Simulation | Algebraic | 1 year | SIMCOMP |
| Patten (1972) | Simulation | Piecewise linear and piecewise stationary | Weekly intervals (53 weeks) | CSMP |
| L. F. Paur (unpublished data) | Simulation | Difference | 1 day | — |
| Payne (1974) | Simulation | Nonlinear (model III) | 2 months | FORTRAN |
| Peden & Rice (1972) | Optimization | Linear dynamic programming | — | — |
| Redetzke & Van Dyne (1976) | Simulation | Markov matrix | 1 year | FORTRAN |
| Reichl & Baldwin (1976) | Optimization | Linear programming | n.a. | ALPSI |
| Rice *et al.* (1974) | Simulation | Difference | 1 day | CSMP |
| Seligman *et al.* (1971) | Simulation | Algebraic | 10 days | FORTRAN IV |
| Smith & Williams (1974) | Simulation/ optimization | — | 1 day | — |
| Swartzman & Van Dyne (1972) | Simulation/ optimization | Difference linear programming | Optimization 1 year; simulation 1 week | SIMCOMP |

| Model diagram | Program code | What it was built for | Comments |
|---|---|---|---|
| ? | ? | To describe a particular arid zone paddock and to validate this to the maximum extent possible | Various submodels follow energy or dry matter or elements but not through the whole model; model validation based in part upon innovative analysis of long-term photographic records of vegetation; detailed behavioural and sheep movement features; model is site-specific. |
| No | No | To explore the influence of the potential weather modification associated with the SST fleet upon a representative grassland ecosystem | |
| No | No | To explore the possible effects of the weather modification associated with cloud seeding in which the driving variables are altered according to potential climate modification | |
| Yes | Available on request | To simulate a generalized shortgrass prairie – the nominal and small perturbation behaviour of principal functional components as a total ecosystem unit | 40 state variables, 3 submodels (producer, consumer and decomposer) |
| Yes | Yes | To implement the current state of generalized mammalian consumer models | 50 state variables, 6 driving variables, 160 flows. Canonical approach. |
| Yes | Yes | To simulate population numbers and biomasses in terms of the chemical components, per unit area, for each class of animals defined in the system | Three models of varying complexity. III (most complex) simulates seasonal impact of different animal groups on the ecosystem and predicts seasonal changes in animal biomass composition and population. Simulation time in years, 16 state variables. |
| Yes | No | To permit objective handling of feeding behaviour in a mixed-species management problem | 4 constraints, 4 variables. Feeding behaviour of cattle, domestic sheep, American bison and pronghorn antelope. |
| No | No | To simulate plant cover and livestock production changes as a function of climatic variability | Transition matrices (specific to soil type, intensity of grazing and climate) developed from data analysis. Model used to predict change and validate with independent set of data. |
| n.a. | Yes | To evaluate input–output principle and test hypotheses regarding relative metabolic rates | Problem to use relative metabolic rates of 8 rumen microbial groups with 25 alternate fermentation and growth routes as constraints. |
| Yes | No | To model the grazing ruminants | 18 state variables, 2 driving variables, 7 parameters. To examine different grasslands – annual and perennial. Teaching. Does not have a good representation of grazing on herbage growth; deterministic model with unstochastic rainfall. General in form but limited to ruminants. |
| No | No | To simulate vegetation growth as a function of soil moisture and temperature | 3 driving variables. Part of NEGIV, the ecosystem model developed in Israel in 1971. |
| No | No | To postulate by computer simulation the probable importance of the length of deferment and initial plant density in relation to stocking rate | To examine the probable response of a system of animal production to deferred grazing. The combination of stocking rate and days of deferment that maximize the gross margin was approximated by use of a search routine. Because the model has not been validated against experimental data the results represent deduction from the model of biological processes. |
| No | No | To aid large-scale system decision making | 3 driving variables, 94 state variables, 243 flows. To show the interseasonal dynamics of important state variables of an arid land resource system. |

Table 4.18 (*continued*)

| Reference | General model | Specific model | Time stage | Computer language |
|---|---|---|---|---|
| G. L. Swartzman & G. M. Van Dyne (unpublished manuscript) | Optimization | Linear/ nonlinear | n.a. | FORTRAN IV |
| Trebeck (1972) | Simulation | — | — | FORTRAN |
| Van Dyne (1969) | Simulation | Algebraic (difference) | 1 day | FORTRAN |
| Van Dyne (1969) | Simulation | Difference | 1 day | FORTRAN |
| Van Dyne (1977) | Optimization | Nonlinear programming | n.a. | n.a. |
| Van Dyne *et al.* (1977*a*) | Simulation | Difference equation | 1 day | SIMCOMP |
| Van Dyne *et al.* (1977*b*) | Simulation | Difference equation | 2 days | SIMCOMP |
| Van Dyne & Rebman (1974) | Optimization | Linear programming | n.a. | n.a. |
| Vera *et al.* (1975) | Simulation | Differential | n.a. | CSMP |
| Vickery & Hedges (1972*a*, *b*) | Simulation | Difference | 1 week | FORTRAN |
| Walker (1974) | Simulation | Difference equation | 1 day illustrated | Unspecified |

| Model diagram | Program code | What it was bult for | Comments |
|---|---|---|---|
| Yes | No | To introduce ideas for using techniques of mathematical optimization in a phase of big game management decision-making | 5 elements in the objective function, 13 constraints. Includes formation of real-life complexity, a linear programming model and a linear model with stochastic elements in the objective function. |
| No | No | To assist research into extensive beef production | 2 driving variables. Stochastic rainfall, lambing percentage, wool cut per head, wool quality, clean yield of wool and feed growth. The model was validated with the same data that were used for estimating a number of the parameters of the model. |
| Yes | Yes | To present main features of dynamics of energy in a shortgrass prairie ecosystem | Simple model with one each of plant herbivore, carnivore and decomposer; simplified age-sex structure for herbivore. Includes standing dead vegetation and litter. |
| Yes | Yes | Simulate energy flow in 9 compartments of shortgrass prairie–pronghorn–blue grama grass system | Compartments are: live plants, standing dead, litter, herbivores, carnivores, dead animals. faeces, detritus and decomposers. Driving variables are solar energy, temperature and precipitation. |
| No | No | To find optimum combinations of 4 large herbivores for early-summer grazing of shortgrass prairie to maximize use of available forage | Uses dietary preference values of cattle, bison, sheep and antelope and available quantity and botanical composition of herbage to determine optimum herbivore mix for either light or heavy grazing strategies. Results in mixed-species herds whereas single species is the common practice. Analyses not based on economics. |
| Yes | No | Simulates shortgrass prairie ecosystem with varying intensity of grazing of large and small herbivores | Uses 1973 and 1974*a* versions of ELM model. Experiments for 1-year response to light, heavy and extra heavy cattle grazing with or without control of small competing herbivores. |
| Yes | No | To simulate cattle, sheep, bison and pronghorn antelope grazing on shortgrass prairie | Uses ELM model, modified from 1974*a* version, to compare 4 large hervibores under light grazing; greatest yearling animal production with cattle and least with antelope when stocked at equivalent rate based on metabolic body size. |
| No | No | To determine animal combinations to maximize economic return | Based on hypothetical grass, shrub and forb rangeland grazed by cattle, sheep and deer, also with wild turkey and catfish production available. Various constraints include biological, economic and personal preference. |
| Yes | No | Represents flow of energy between a sheep and its environment | Main compartments are fleece tip, skin, body and environment; driving variables are location, time, temperature, relative humidity, cloud cover, wind movement, animal weight. and linear measurements, and fleece weight |
| No | No | Live weight changes and wool production for sheep on *Phalaris tuberosa* pasture | 11 state variables, 3 driving variables, 11 processes. Model development. Plant growth and animal production limited in application because of the use of arbitrary assumptions; non-reproductive sheep simulated; used linear regression equations to interrelate observed and predicted data for 4 variables. |
| No | No | To provide an initial generalized structure for an African wildlife ecosystem; to understand the sensitivity of ecosystems to changes that are being imposed through man's management | Based on Lowveld of Rhodesia where plant biomass data were available. Divides area into 3 site types (permanent water area, area in which water dries up towards the end of winter, wet season grazing only) 24 functional compartments including 8 vegetation, 11 live animal, dead components and nutrients. Structure for forage intake determined according to 2 constant coefficients, one donor-controlled the second |

Table 4.18 (*continued*)

| Reference | General model | Specific model | Time stage | Computer language |
|---|---|---|---|---|
| Walters and Burnell (1971) | Simulation | Differential algebraic | 1 year | FORTRAN |
| Walters *et al.* (1975) | Simulation | Algebraic | 1 year | FORTRAN/ SIMCON |
| Walters & Gross (1972) | Simulation | Algebraic | Years | — |
| Wielgolaski (1972) | Simulation | Difference equations | 1 day | SIMCOMP |
| Wright & Van Dyne (1970) | Simulation | Difference | 1 day | FORTRAN |

| Model diagram | Program code | What it was built for | Comments |
|---|---|---|---|
| | | | receiver-controlled, depending upon plant biomass availability. Animal shift from area to area depending upon water availability. |
| Yes | No | To look at land use and big game population in British Columbia | — |
| Yes | No | To examine population regulation of caribou under hunting strategies | Population dynamics based on age structure and age-dependent survival and fecundity. Foraging submodel considers snow depth, seasonal migrations and use of winter habitat. |
| Yes | No | For management of single-species populations | Model based on interactions of density, specific natality rates and specific mortality rates |
| Of sub-models | Yes in reference document | The objective was to develop a low-resolution model of grassland tundra system generally, primarily as a learning experience in a workshop | Consumer model has a generalized herbivore; changes in population during one generation driven by an exponential survival curve; consumer intake a set percentage of body weight but reduced as a function of dietary quality and food availability; metabolic requirement primarily as a function of basal metabolism; other consumers included carnivores, omnivores and detritivores; 6 typical animals were used in tuning the model (mouse, cow/deer, grasshopper, wolf/fox, hawk/kestrel, and Oligochaetae). Controlling factors were mean temperature, percentage of nitrogen and fibre, percentage digestibility, food capacity and weight of animals; losses of methane, urine and heat production were considered. |
| No | Yes | To simulate plant cover and biomass, and cattle and jackrabbit and rodent dynamics on a semi-arid grass–shrub range | Preliminary model developed and run. Output presented indicates need for change in functions. Separate code also presented for detailed plant-physiological approach. |

algebraic equation systems, and, in some instances, whether they are linear or nonlinear. Other characteristics provided about the mathematical and computational aspects of the model include the time stage or time step, the computer language (if specified), whether a model diagram was available, and whether the program code was noted. Brief comments are given about the purpose of the model and the general biological and structural properties of the model.

We have not included in Table 4.18 the many linear programming optimization models developed for least-cost ration formulation for domestic animals. Models are not listed in this compilation unless they directly focus on large herbivores as one of the state variables or major components. Thus, many of the general predator–prey models are not included as they are not sufficiently specific for large herbivores. Many principles of such models, of course, still apply and they should be given consideration in any model-building effort related to large herbivores. There are also numerous other models of the abiotic and primary producer components of grazingland ecosystem which could be useful in building a large herbivore model (see Chapters 1 and 2, this volume, for examples). Such models were excluded, however, if they did not specifically include large herbivores.

### *A brief on models of herbivory*

Some 11 models have been selected out of those listed in Table 4.18 for further consideration. These models are generally of 'medium size' and were directly oriented to problems of large herbivore productivity. They depict intra-seasonal or interseasonal dynamics of a large herbivore. Each model contained one or more of the processes discussed above in the sections related to food selection, food intake and food digestion. The general characteristics of these selected 'intermediate models' are summarized in Table 4.19. The individual models are discussed as a series of 'briefs' or detailed abstracts in Table 4.20.

As noted in Table 4.19, ingestion is a key process occurring in all of the large herbivore models summarized. Examination of the table shows the diverse ways that forage intake or ingestion is calculated for these large grazing herbivores. Factors used and functional forms affecting forage intake rate are summarized in Table 4.21. The 11 models examined used no less than 16 different factors to control or influence forage intake rate. The functional forms varied from stepwise linear functions to exponential functions ('reflected' exponentials) to power functions to table lookups. And in some cases random factors were added, or the model was in part 'stochasticized'. Herbage availability was the most common factor used in calculating forage intake rate in these models. Animal weight was another factor frequently utilized, either directly or indirectly via rumen fill and capacity. The clear result of this

Table 4.19. *Summary of flow characteristics of selected intermediate-level simulation models of large herbivores*

| Reference | Grazing location | Food selection | Ingestion | Digestion | Food passage | Energy metabolization | Protein metabolization | Maintenance | Gain | Foetal growth | Wool growth | Lactation | Mortality | Natality |
|---|---|---|---|---|---|---|---|---|---|---|---|---|---|---|
| Goodall (1969) | × | × | × | × | . | . | . | . | × | . | . | . | . | . |
| Van Dyne (1969) | . | × | × | × | . | . | . | × | × | . | . | . | × | . |
| Freer *et al.* (1970) | . | × | × | × | . | . | . | . | . | . | . | . | . | . |
| Nelson (1970) | . | . | × | . | . | . | . | . | × | . | . | . | . | . |
| Arnold & Campbell (1972) | . | × | × | × | . | . | . | × | × | . | . | . | . | . |
| McKinney (1972) | . | . | × | × | . | × | . | × | . | . | . | . | . | . |
| Swartzman & Van Dyne (1972) | . | . | × | . | . | . | . | . | × | . | × | . | . | × |
| Vickery & Hedges (1972*a*, *b*) | . | . | × | . | . | × | . | . | × | . | . | . | . | . |
| Edelsten *et al.* (1973) | . | . | × | × | . | × | . | × | × | . | . | × | . | . |
| Rice *et al.* (1974) | . | . | × | × | × | × | × | . | × | × | . | × | × | . |
| Newton & Edelsten (1976) | . | . | × | . | . | × | . | × | × | × | . | . | . | . |

Table 4.20. *Brief summaries of selected intermediate-level simulation models of large herbivores under grazing conditions. Subheadings identify some major processes. Further information on these models is provided in Table 4.18*

Goodall (1969)

Goodall has developed several versions of a model for grazing sheep for an arid-land system. The model is one of the few that has spatial effects.

*Grazing location.* Locality is a function of the mean distance from the fence line and the mean distance from the water supply and the evaporation rate. This expression shows that there is a direct relationship of preference to distance from the fence and an inverse one for distance from water.

*Diet selection.* A relative forage preference was calculated on the basis of the amount of herbage in the different plant groups. This forage preference is:

$$F_j = \sum_i [(A_i Q_{ij}) / \sum_i A_i \sum_{i,j} Q_{ij}],$$

where $F_j$ = forage preference for area $j$

$A_i$ = relative palatability of plant $i$

$Q_{ij}$ = amount of plant $i$ in area $j$.

The effective forage is given as the product of the locality preference, the total area in that particular portion of the pasture, and the forage preference. These products are summed for each area of the pasture to give a total value which is then used to allocate time to different portions of the pasture.

*Ingestion.* Forage consumption is a product of the number of animals, the consumption rate and the time allocated in each portion of the pasture. The consumption of plant species in each area is calculated from the relative palatability and the amount of herbage. The consumption rate depends on relative availability of different herbage species.

Table 4.20 (*continued*)

Van Dyne (1969)

*Ingestion.* This generalized herbivore in the model was patterned after the pronghorn antelope on shortgrass prairie. Food intake rate is 1.4 kg herbage per day, but decreases during warm weather at 4 % per degree of temperature rise above 27 °C and increases during cool weather at 1 % per 0.21 deg C fall in maximum daily temperature below 27 °C. If precipitation rate exceeds 25 mm $d^{-1}$ as snow the herbage consumption decreases by 10 % per 25 mm increment. 25 mm of precipitation was considered equal to the snow depth. Snow was melted off the rangeland over time thus freeing herbage for grazing. Temperatures below 4.4 °C increased the nutrient requirements of the herbivore by 3 % per 0.21 deg C.

*Food selection.* The animal seeks live plant material rather than dead except when live plant material drops to a minimum level of 10 $g \cdot m^{-2}$. Standing dead is consumed for the remainder of the diet. If standing dead vegetation falls below 10 $g \cdot m^{-2}$ then herbage consumption decreases with the degree of decrease in available standing dead. When lush herbage is eaten less dry matter is consumed than with higher availability of dry herbage. Thus the animals' intake is water-limited. Under some conditions the animals may be eating 6 % more total forage if they are consuming dry as opposed to succulent plant material.

Plant digestibility was a constant 0.46. This fraction multiplied by forage intake rate gives digestible energy consumed per day and from this is subtracted the maintenance requirement calculated according to $W^{0.75}$. If the digestible energy intake exceeds the requirement then body gain is calculated at a rate of 1 kg · (1600 kcal)$^{-1}$ digestible energy. If the animal must draw upon body stores then there is a loss of 1240 kcal · $kg^{-1}$.

*Mortality.* Mortality increases when the body weight falls below 90 % of the normal weight, here 50 kg. The rate of mortality increase is 0.1 % per 0.45 kg. A random death rate varying from 0 to 0.35 % per day is added to simulate accidents, disease and losses to predators. The mortality rate of the young animals is 33 % greater than the adult mortality.

Freer *et al.* (1970)

*Ingestion.* Various aspects of this model have been described by Armstrong (1969) and Donelly *et al.* (1974). It is for a pasture grazing system. Potential forage intake was defined as the intake which would be attained if digestibility were nonlimiting. This intake was related to physiological satiation and to a sum of green and dry matter material available as follows:

$$IP = IU\{1-\exp[-k(G+D)]\},$$

where $IP$ = potential intake
$IU$ = physiological satiation
$G$ = green herbage available
$D$ = dry herbage available.

The proportion of green in the diet was calculated from the proportion of green in the pasture as follows:

$$PGE = 1-\exp\{-k[G/(G+D)]\},$$

where $PGE$ = proportion green herbage eaten.

*Diet selection.* The actual amount of any herbage class eaten was regarded as equal to its potential intake multiplied by its digestibility. Thus relative intake was related to relative digestibility.

*Digestion.* Digestibility of dry matter eaten was estimated from the digestibility coefficient for dry matter (presumably constant) and the amount of dry herbage available as follows:

$$YD = k \cdot YA/[1+k_2 \exp(-k_3 D)],$$

where $YD$ = digestibility of dry matter eaten
$YA$ = digestibility of dry matter.

Table 4.20 (*continued*)

Total herbage eaten then was calculated as follows:

$$E = IP[YD(1-PGE)+ YP \cdot PGE]$$

where $E$ = total herbage consumed.

In further developments of the model it is assumed that the sheep first eat the available green herbage and that the rate of intake at any time during the day is proportional (i) to the availability of the green herbage and (ii) to the difference between the maximum possible intake and amount eaten up to that point. This is given by the following equation:

$$\frac{dIPG}{dAG} = k \cdot AG(IU-IPG),$$

where $IPG$ = amount eaten

$AG$ = availability of herbage, which when integrated gives

$$IPG = k[1-\exp(-k \cdot AG^2)].$$

The potential dry herbage eaten is given by the availability of dry matter and the difference between the upper limit of feed intake and the potential intake of green feed:

$$IPD = k_1 [IU-IPG] [1-\exp (k_2 \cdot AD^2)],$$

where $IPD$ = potential amount of dry eaten

$IPG$ = potential amount of green eaten

$AD$ = availability of dry herbage.

These potential intakes must be multiplied by the appropriate digestibility coefficients to derive the actual amount eaten.

It should be noted, however, that under very intensive stocking, as is the case here, sheep keep the pasture grazed very short and a large proportion of the herbage is eaten during each day's grazing. In such a situation, availability of feed declines, and rate of potential intake is therefore more correctly represented as follows:

$$\frac{dIP}{dt} = k(A-IP)(IU-IP).$$

Nelson (1970)

This model predicts the dynamics of cattle, deer and sheep. Forage intake rate is in part based upon available forage. The proportion of the maximum forage intake rate that can be attained with varying levels of herbage availability was derived by fitting an exponential function of the form as follows for cattle:

$$\text{Relative forage intake} = 1-k_1 \exp(-k_2 H),$$

where $H$ = available herbage,

$$k_1 = 1.17$$
$$k_2 = 0.0024.$$

*Ingestion.* Forage intake rate was also affected by energy content of the forage. Forage intake rate per month was calculated by the following expression:

$$\text{Monthly intake in kg per } W^{0.75}_{\text{kg}} = \frac{M}{[1+k_1 \exp(k_2 D)]},$$

where $M$ = maximum intake

$D$ = digestible energy content.

Forage intake rate is also increased if the female is lactating. The intake was 26 % more than for the dry female.

*Gain.* Animal weight change is considered a function of digestible energy intake, mainten-

Table 4.20 (*continued*)

ance energy requirement and efficiency of gain. As a simplification a relationship between digestible energy intake and weight gain was defined. It was assumed that there were no associative effects of the consumption of one species on the digestible energy in another species. Weight change was given as a linear function of energy intake. Separate flow functions are used for animals gaining weight and animals losing weight. The steepness of the slope is greatest for cattle and least for deer, both for gaining and losing weight.

Arnold & Campbell (1972)

*Ingestion*. This is a model of wethers grazing annual pasture containing a grass and a forb. Digestible organic matter intake is calculated as a function of liveweight, available pasture and the digestibility of pasture on offer. The form of the equation is as follows:

$$OMI = \{k_1 \cdot \exp[-k_2(LWT-k_3)]\}\{1-\exp[k_4-k_5\,TDM]\},$$

where $OMI$ = organic matter intake; $k_1 = 55$
$LWT$ = live weight; $k_2 = 0.025$
$ADM$ = available dry matter; $k_3 = 25$
$TDM$ = total dry matter; $k_4 = 0.0005$
$ADM = TDM$ for dry matter on offer $> k_6$; $k_6 = 1000$.

If the dry matter on offer is less than $k_6$ then the available dry matter is calculated by

$$ADM = (GRM \cdot PGD) + DRM\,(1-PGD),$$

where $GRM$ = green material on offer
$PGD$ = % grass in the green herbage
$DRM$ = dry material on offer.

The sheep eat the grass and forb in proportion to the amounts of each present.

*Digestion*. Digestibility of the mixed diet is calculated from the digestibilities and nitrogen contents of the four fractions: green grass, dry grass, green forb and dry forb. These values are read into the program.

*Maintenance*. The growth of the animal is due simply to the difference between the digestible organic matter intake and the animal's maintenance requirement. The maintenance requirement was calculated as a function of liveweight and available dry matter. This function depends upon the amount of available dry matter as follows:

$$DME = \begin{cases} k_i LWT - k_2 \exp[k_3(ADM-k_4)], & ADM \leqslant 1000, \\ k_5 LWT + k_2 \exp[k_6(ADM-k_4)], & ADM \geqslant 1000, \end{cases}$$

where $DME$ = animal maintenance requirement
$k_1 = 10.35$
$k_2 = 1.15$
$k_3 = 0.0060$
$k_4 = 1050$
$k_5 = 8.05$
$k_6 = 0.0034$.

*Gain*. Animal gain is calculated by using an energy conversion figure to give liveweight change in a linear manner as follows:

$$LWC = (DOM - DME)/[k_1 + k_1 LWT)/k_3],$$

where $LWC$ = liveweight change
$DOM$ = digestible organic matter intake.
$k_1 = 0.225$
$k_2 = 0.04$
$k_3 = 0.54$.

Table 4.20 (*continued*)

Dietary selectivity is calculated by varying the percentage of green herbage in the diet as a function of the percentage green herbage (*PGP*) in the pasture and total dry matter *TDM*) as follows:

$$PGD = 100\,\{1-\exp.\,[(k_1-k_2\,TDM)\,(PGP)]\}\,(1-k_3-k_4\cdot TDM).$$

$k_1 = 0.005$
$k_2 = 0.002$
$k_3 = 1.40$
$k_4 = 0.00047.$

McKinney (1972)

*Ingestion.* Intake was calculated from available herbage by the following equation:

$$I = k_1[1-\exp\,(k_2 A)]-k_3,$$

where $I$ = dry matter intake (kg·animal$^{-1}$·d$^{-1}$)
$A$ = available herbage (kg·ha$^{-1}$)
$k_1 = 1.36$
$k_2 = 0.00108$
$k_3 = 0.03.$
An upper limit was set at 1.33 kg·animal$^{-1}$·d$^{-1}$. Maximum intake occurred at 1750 kg·ha$^{-1}$.

*Digestion.* Since the pastures are used intensively and only in winter, the herbage is always short and green and a constant digestibility of 0.72 is utilized. A constant fraction of the digestible energy, 0.82, is converted to metabolizable energy. The energy value of the herbage dry matter is 18.4 MJ·kg$^{-1}$.

*Maintenance.* The net energy requirement for maintenance was set at 0.33 MJ per unit of metabolic weight (based on a power of 0.73).

Efficiency of use of metabolizable energy for maintenance is 0.72 and for liveweight gain it is 0.51 (the energy value of liveweight gain is 20.9 MJ·kg$^{-1}$).

Swartzman & Van Dyne (1972)

*Natality and mortality.* Growth and death are calculated for 4 cattle age–sex classes (calves, weiners, cows and steers), 4 sheep age–sex classes (lambs, weiners, ewes and wethers), and kangaroos and rabbits. These animals change age according to a time function, birth rates are calculated as functions of the metabolic weight of the reproducing animals, and mortality is a function of metabolic weight. Birthweights of cattle, kangaroos and rabbits are linear functions of the metabolic weight; birthweights for sheep were 2-segment piece-wise linear functions of body weight. Cows and ewes reproduce at a given time of year; kangaroos and rabbits reproduce throughout the year.

*Gain/loss.* Animal biomass was calculated by increasing it through food intake and decreasing it through the total of the metabolic energy demands.

*Diet selection.* Food selection was calculated according to relative availabilities of and relative preferences for different plant groups – live herbs, dry herbs, live shrubs and dry shrubs. The amount of food consumed by calves also is dependent upon the amount of milk consumed. A constant coefficient represents the percentage of the total intake which is digestible and metabolizable as well as the percentage of metabolizable intake which gets converted into animal biomass.

*Food intake.* Food intake of lambs and calves was a linear function of metabolic body weight. For other animal categories food intake is a hyperbolic function of metabolic body weight, given by $q_6(q_6+q_7)$. Here $q_7$ is a threshold metabolic rate. When weight drops below $q_7$ the food intake rate increases rapidly, but when body weight increases above $q_7$ there is only a minor food intake increase.

$$A_{1i} = 1-\exp\,[-H_i^2/a],$$

where $H$ = biomass of plant group $i$.

Table 4.20 (*continued*)

$$DWC = q_5\left\{\sum_{j=1}^{4} A_{ji}\left[\prod_{k=1}^{j-1}(1-A_k)\right]\left(\frac{q_6}{q_7+q_6}\right)\right\}W_{c_i} - q_8 W_{c_i}.$$

and
$$q_6 = \left(\frac{W_{c_i}}{W_{c_i}}\right)^{0.75},$$

where $A_{ji}$ = relative availability of plant class $j$ on property $i$ and $DWC$ is the general form for food intake expressions based on animal weight ($W_{c_i}$).

Vickery & Hedges (1972*a*, *b*)

Weekly productivity and weekly rates of material removal were driven by rainfall, standard pan evaporation, and weekly mean soil drive temperature. The growth rate of a *Phalaris tuberosa–Trifolium repens* pasture grazed by sheep was described.

Dry matter intake by sheep was predicted from fleece-free liveweight and pasture digestibility, adjusted for pasture availability and partitioned into green and dead components. Herbage of 12 age classes (1 to 12 months) was maintained for both green and dead herbage, which allowed for selection by the sheep for the younger and the more digestible components of the pasture.

Dry matter intake was converted into metabolizable energy intake assuming 0.9 dry matter content which had 19.25 Mcal kg$^{-1}$ with a metabolizability of digestible energy of 0.8.

Liveweight change was calculated on the difference of metabolizable energy requirements (based on literature values) and intake. Efficiency of utilization of catabolized energy was 0.8. The efficiency of conversion of surplus energy to body reserves was calculated as a function of the digestibility of the diet.

*Ingestion.* If pasture availability is nonlimiting and if the weight of the adult dry sheep is less than 32.5 kg then a compensatory mechanism operates allowing an increase of intake up to 20 %. The potential intake of the sheep is calculated as a function of liveweight as follows:

$$DMI - k_1 + k_1 . \text{Tanh}\,[k_2(X_6 - k_3)],$$

where $DMI$ = potential intake (kg dry wt·animal$^{-1}$·wk$^{-1}$)
$X_6$ = fleece-free liveweight (and if $X_6 \geqslant 50$ then $k_3 = 50$)
$k_1 = 5.2$
$k_2 = 0.068$
$k_3 = 22.0.$

If the sheep exceeds 50 kg weight then the potential intake is reduced. Thus, fat sheep may have lower intake than thin sheep. The potential intake level may be proportionately reduced for either low availability of green herbage or low digestibility. The actual level of dry matter intake is the product of the potential dry matter intake, a factor related to compensatory mechanisms if they are operative, and a factor selected either from the effect of reduced digestibility or low availability (if they are operative).

*Metabolization.* Daily metabolizable energy requirement for maintenance is estimated from 0.132 times the metabolic weight (using 0.75 power). Digestible energy is 81 % of metabolizable. Wool growth is a linear function of the metabolizable energy requirement. But separate functions are used for sheep heavier or lighter than 20 kg.

The energy content of 1 kg liveweight is estimated by the following equations:

$$ECLC\begin{cases}\text{for } X_6 \leqslant 20 = k_1 + k_2 \exp(k_3 X_6)\\ \text{for } X_6 \geqslant 20 = (X_6 - k_4)\, k_5 + k_6\end{cases}$$

where $ECLC$ = energy content in Mcal of 1 kg liveweight change
$X_6$ = liveweight. $k_1 = 1.523$
$k_2 = 0.113$

Table 4.20 (*continued*)

$k_3 = 0.118$
$k_4 = 20.0$
$k_5 = 0.143$
$k_6 = 2.72$

Discontinuous functions are used to approximate data obtained from the literature.

*Wool production.* Each kg of wool production requires 5.4 Mcal of metabolizable energy. The efficiency of conversion of this surplus energy to body tissues such as wool is estimated for sheep in a positive energy balance by the following equation:

$$EEF = k_1\ DIG + k_2,$$

where $EEF$ = efficiency of energy conversion
$DIG$ = digestibility of total herbage dry matter.
$k_1 = 0.66$
$k_2 = 0.03.$

If the sheep is in a negative balance then only 80 % of the energy released by catabolizing body tissue can be used.

Edelsten *et al.* (1973)

This is a model of ewes with lambs, grazing on pasture. It simulates sheep from mid-March until lambs are sold in September or October.

*Ingestion, ewes.* Herbage intake for ewes is calculated from herbage available, potential intake, and concentrate eaten:

$$\text{Intake} = 1 - \exp\,(k_1\,H^2)\,PI - (CE)\,k_2,$$

where $H$ = herbage in paddock in tonne dry wt $\cdot$ ha$^{-1}$.
$CE$ = concentrate eaten
$PI$ = potential intake
$= f(t-T)\cdot FAC$
$k_1 = 0.24$
$k_2 = 0.41.$

where $T$ is time of lambing, $f(t-T)$ is a table look-up function, and $FAC$ is an intake factor = 1 for ewes.

The table adjusts intake in relation to date. Lambing ewes have a dry matter intake of 2.52 kg$\cdot$animal$^{-1}\cdot$d$^{-1}$ on the date of lambing, which increases to 2.67 at day 10, 2.82 at day 21, and then drops progressively to 1.9 at day 75 and 1.5 at day 120 and thereafter.

*Ingestion, lambs.* The table for herbage intake for lambs is the same as that for ewes except that there is a reduction for milk consumption. The potential intake for lambs is as follows:

$$PI = f(t-T)\ BW\cdot FAC^1,$$

where BW = body weight
$f(t-T)$ is given by a tabulated value for age.
$FAC^1$ = a random value (1, 0.9).

*Lactation.* The potential intake of lambs varies from 0.042 kg dm $\cdot$ kg live wt$^{-1}$ at 0 days to 0.21 at 23 and 35 days, 0.28 at 59 days and 0.34 at 68 days and thereafter. Lamb forage intake is affected by a random value whose mean is 0 and standard deviation 0.9. Lambs do not eat any herbage until they are 3 weeks old. The amount of milk that is consumed by the lamb is dependent upon the amount of milk available times a preset fixed proportion. The amount of milk available, if the ewe is not losing weight, is set by a table look-up such that at the time of lambing the lactation is 2.4 kg, between 10 and 35 days it is 3.0, at 100 days it is 1.0, and at 200 days it is 0.1. The potential milk yield is given by the above table, a random factor, and a factor adjusting for the number of lambs. The tabulated values are evidently based on two lambs per ewe. The equation for milk yield is as follows:

Table 4.20 (*continued*)

$$\text{Milk yield } (Mp) = \begin{pmatrix}\text{potential}\\ \text{milk}\\ \text{yield}\end{pmatrix} - \begin{pmatrix}\text{potential}\\ \text{metabolizable}\\ \text{energy}\\ \text{deficit}\end{pmatrix}\left((k_1)\begin{pmatrix}\text{efficiency}\\ \text{of lactation}\end{pmatrix}(k_2)\right)$$

where efficiency of lactation is 0.7, $k_1 = 0.5$, $k_2 = 1.026$ and potential metabolizable energy deficit is:

$$\text{Maintenance requirement} + Mpk_2 - \text{Metabolizable energy intake.}$$

Here the maintenance requirement in Mcal of the ewe is given by 0.1 times the metabolic weight to the power 0.73.

*Metabolization.* The metabolizable energy in the herbage intake is given by the product of the amount of herbage intake times the factor 4.4 $\text{Mcal}\cdot\text{kg}^{-1}$, times 0.82, times the digestibility.

*Digestion.* Digestibility data are obtained from a table look-up which has digestibility as a function of time with values 0.75 on day 75, 0.73 on day 121, 0.70 on day 160, 0.68 on day 250, and 0.65 on 365. The amount of energy from the concentrates eaten is calculated in a similar manner except that the digestibility remains constant at 0.82.

Lactation is calculated as a function of milk yield and energy content. Milk yield was given above in a look-up table. The energy content of milk is 1.14 $\text{Mcal}\cdot\text{kg}^{-1}$.

*Gain.* Energy not used for maintenance or lactation is kept in the body and its conversion is not dependent upon the state of the animal. The energy retained is thus an efficiency of utilization factor times the amount of energy remaining from the total metabolizable energy intake after maintenance energy and lactation energy are considered. The efficiency of utilization of metabolizable energy for maintenance is given by a function of energy content of the feed such that efficiency is 0.55 when the energy content of feed is 0 $\text{Mcal}\cdot\text{kg}^{-1}$, 0.78 at 3.4, and 0.81 at 5.4.

It is assumed that lambs utilize 95 % of the energy in the milk, 1.14 $\text{Mcal}\cdot\text{kg}^{-1}$.

The efficiency with which a lamb uses energy depends upon whether it is in a 'maintenance' or a 'growth' state. As the energy content of the feed (in $\text{Mcal}\cdot\text{kg}^{-1}$) increases from 0 the maintenance efficiency is 0.55 and fattening efficiency 0.3; with feed energy at 3.4 the values are 0.78 and 0.66; and with feed energy at 5.4 the values are 0.81 and 0.77. For a lamb in a maintenance state the requirement is 0.11 $\text{Mcal}\cdot\text{kg}^{-0.73}$ whereas if the lamb is growing the maintenance requirement is 0.075 $\text{Mcal}\cdot\text{kg}^{-0.73}$.

The conversion of retained energy into body weight for the ewe at 25 kg body wt is 3.0 $\text{Mcal}\cdot\text{kg}^{-1}$; at 50 kg or above the conversion rate is 6.6 $\text{Mcal}\cdot\text{kg}^{-1}$.

Rice *et al.* (1974)

*Ingestion.* Forage intake rate is based on availability of forage and the maximum capacity of the digestive tract. Both biomass and plant height are considered in forage availability. Green and dry availability factors were used to modify maximum intake where forage availability is limited by heavy stocking or poor plant growth. Smith & Williams' (1974) information is used to calculate plant growth in an annual grassland type. Although not presented, it was a negative exponential equation describing the relationship between available forage and intake. Green herbage intake was also influenced by height whereas dry herbage intake was not. Using data from several reports in the literature, they calculated rumen fill according to the function of 0.145 kg of rumen fill per unit metabolic weight (0.75 power). This relationship was modified in the model so that once the animals reached the mature weight, a loss in weight did not result in reduced capacity for forage consumption. Also, lactating animals were allowed to consume more feed than those not lactating.

*Digestion.* The digestibility of the food was due in part to the content of potentially digestible components, based on quality factors, and in part due to the balance between rates of digestion and rates of passage. The amount digested is the product of the amount of microbial protein present in the rumen, fermentation rate per gram of microbial protein,

Table 4.20 (*continued*)

and pool size. Food passage rate was dependent upon the size of the potentially digestible dry matter pool.

*Food passage.* Food passing the rumen is digested in the intestines and portioned into available energy, protein or faecal material depending upon the quality.

*Energy partition.* Digestible energy is used for various body functions determined by a system of priorities: (1) maintenance, (2) demands of pregnancy, (3) lactation; and (4) growth. Maintenance requirement, which includes an activity requirement, is calculated as 0.125 Mcal per unit of metabolic weight.

*Mortality.* Body energy is mobilized to provide for maintenance in pregnancy if intake of digestible nutrients is insufficient. Death occurs when adult body weight is less than 65 % of mature weight.

*Lactation.* Lactation requirements are given as a function of age, number and size of offspring, but a limit is imposed according to published lactation curves for beef cattle and sheep. Body energy can also be mobilized for lactation demands until body weight declines to 85 % of mature body size. The ability to mobilize body tissue for lactation declines to zero at 65 % of the mature weight, when death occurs.

In the nitrogen component of the model, food and microbial nitrogen are digested at different rates, 0.5 and 0.8 respectively, when they reach the lower digestive tract.

*Protein metabolization.* Digested protein is transferred into the body pool of absorbed protein and blood nitrogen. When body weight loss occurs, body nitrogen enters the labile body nitrogen pool and is available for bodily uses. If there is labile nitrogen then the excess is transferred into urinary nitrogen.

Newton & Edelsten (1976)

*Ingestion.* Dry matter intake (in $kg \cdot animal^{-1} \cdot d^{-1}$) is predicted from body size plus a random factor. The data are such that thin ewes will eat more than fat ewes. A random variable is added to predict intake based on body weight, with a standard deviation of 10 % for the intake addition or deletion. If concentrate is fed then the roughage consumed is reduced by the amount of concentrate times 0.46.

*Metabolization.* Metabolizable energy is calculated from dry matter intake times a constant 4.4 $kcal \cdot kg\ intake^{-1}$ times the digestible organic matter of the dry matter times a a fraction, 0.82, representing the remainder after losses of methane and urine energy.

The maintenance requirement of the ewe is calculated as 0.1 times the liveweight raised to the power 0.73 to give the daily requirement (in Mcal). The maintenance requirement of the foetus is 0.09 times the foetal weight. The remaining energy is used first for foetal growth and then for ewe growth.

*Energy partition.* A value of 7 kcal g foetal $wt^{-1}$ increase was used in the model. A maximum growth weight was set for each breed or cross of breeds. These values ranged from 0.0407 to 0.029 for different breeds.

Table 4.21. *Factors used and functional form affecting forage intake rate in selected large herbivore models summarized in Tables 4.19 and 4.20*

| | Factors | | | | | | | | | | | | | | | | |
|---|---|---|---|---|---|---|---|---|---|---|---|---|---|---|---|---|---|
| Reference | Grazing area available | Relative palatability of herbage spp. | Relative availability of herbage spp. or group | Nominal rate per unit weight | Temperature | Snow cover | Physiological satiation | Forage digestibility | Concentrate fed | Time since parturition | Milk available | Lactation | Digestive tract capacity | Plant height | Body condition | Random factor | Functional form |
| Goodall (1969) | × | × | × | . | . | . | . | . | . | . | . | . | . | . | . | . | Exponential reduction |
| Van Dyne (1969) | . | . | . | × | × | × | . | . | . | . | . | . | . | . | . | . | Linear, step function |
| Freer *et al.* (1970) | . | . | × | . | . | . | × | . | . | . | . | . | . | . | . | . | Exponential reduction |
| Vickery & Hedges (1972*a*, *b*) | . | . | × | × | . | . | . | × | . | . | . | . | . | . | . | . | Nonlinear, maximum reduction |
| Edelsten *et al.* (1973) | . | . | × | . | . | . | . | . | × | × | × | . | . | . | . | . | Exponential and table look-up and random factor |
| Nelson (1970) | . | . | . | × | × | . | . | × | . | . | × | . | . | . | . | . | Exponential, multiplicative |
| Arnold & Campbell (1972) | . | . | × | × | . | . | . | . | . | . | . | . | . | . | . | . | Exponential, multiplicative |
| McKinney (1972) | . | . | × | . | . | . | . | . | . | . | . | . | . | . | . | . | Exponential, step function |
| Swartzman & Van Dyne (1972) | . | × | × | × | . | . | . | . | . | . | × | . | . | . | . | . | Power function |
| Rice *et al.* (1974) | . | . | × | . | . | . | . | . | . | . | . | × | × | × | . | . | Power function, multiplicative |
| Newton & Edlesten (1976) | . | . | . | . | × | . | . | . | × | . | . | . | . | . | × | × | Negative exponential |

comparison is that there is no standard approach in presenting the calculation of forage intake rate. Generally, however, the functional forms are nonlinear and frequently some power function of body weight influences the calculations.

## *Large models*

In addition to the more empirical constant-coefficient models, and to the intermediate-level models of the same nature of size and mechanism as those shown in Table 4.20, in recent years an important series of large, complex, high-resolution models has been developed. Generally, these are total-system models dealing with abiotic, producer, consumer and decomposer segments of the ecosystem. Three such models for a shortgrass prairie ecosystem are referred to here and two for arid land ecosystems. The shortgrass prairie ecosystem models are those of Cole (1976) and Innis (1978), Bledsoe (1976) and Cale (1975). These models were built as part of the US/IBP Grassland Biome study. Preliminary discussion of the early versions of these models is provided by Innis (1975). Some features of and outputs from the ELM model are discussed in Chapter 10 of this volume. Within the US/IBP Desert Biome programme a 'general-purpose model' was structured under the direction of D. W. Goddall. The main feature of this model is that there are three more-or-less parallel models of different levels of resolution which can be combined in various ways for a particular modelling problem at hand. Generally, the flow functions in the simplest of these models are calculated by a rate constant in time. There is no specificity for the plant or animal groups. The next level of complexity has a process occurring at a constant rate in time but is specific to a plant or animal group. In the highest level there are system-dependent and species-specific rates. The animal submodel is described by Payne (1974) and has three levels of resolution all of which are time-varying, deterministic and simulate spatially homogeneous areas. The models recognize fat, protein and structural tissues within the animal. A maximum of six chemical elements are traced through the system. Carbon is divided into protein carbon, reserve carbon and structural carbon. Various animals may be included in the model by parameterization. Numbers and biomasses of the animals, in terms of chemical components, are simulated per unit area. Major processes within the animal models include consumption, predation, respiration, immigration, emigration, mortality, excretion, reproduction and growth. We are aware of no other reports or publications providing detailed descriptions or results of the runs of this overall model. The general conceptual design of the overall model is intriguing; the implementation is difficult.

Noble (1975) has developed a computer simulation model for an arid zone system. This model, in effect, is a collection of submodels with varying purposes – a paddock model containing submodels of climate, plant growth, sheep behaviour, forage consumption and animal physiology. The sheep behaviour model is probably one of the more detailed of its type available and describes the hourly movement of sheep flocks around the paddock. Heat and water balance calculations determine much of the animal behaviour. Forage consumption as a process is treated similar to a predator–prey situation in the intake model. No readily available published reports of the usage or detailed testing of this model are available to our knowledge.

The development of these large-scale models is worthy of some comment. The models of Noble, Cale and Bledsoe each represent large doctoral dissertations. They also represent modelling efforts which involved that student and other senior scientists over longer periods of time than usual, i.e. compared to most PhD dissertation projects. The Grassland Biome ELM model and the Desert Biome multicomponent model were the output of large-scale IBP programmes and teams of researchers. The full documentation of the

ELM model includes some 12 separate chapters of discussion for open-literature book publication (Innis, 1978) and a 663-page detailed report which contains a discussion and description of the code and the operation of the total ecosystem model (Cole, 1976). Some sample output from advanced versions of this model is used in Chapter 10 of this volume. About 20 professional person-years were expended in the development and documentation of the model through 1975 (Van Dyne & Anway, 1976); more time was accumulated before it was published in a detailed report and more yet before the open-literature book. Yet it can hardly be considered a 'polished, usable model'. It is large and complex and there are many omissions and inconsistencies between and within various documentations of the ELM 1973 model. And its structure leaves much to be desired. This is not a criticism of the team developing the model, for we believe it represented the state-of-the-art at its completion and it was a pioneering effort. Various modifications in the parameters (for the shortgrass prairie version), the structural form (expansions and restructuring), and the software-support system for running the model have been made leading toward an 'ELM 1977' and the improved model has been used in a variety of management hypothesis-testing experiments (Van Dyne, Joyce & Williams, 1977). The amount of output is overwhelming. About 1500 of the individual flows, driving variables, state variables and output variables which are calculated at each time step in the model are stored on magnetic tape and are now available for subsequent use in analysis. Using a 2-day time step, it takes about 7 minutes of machine time to compile and run this model on a CDC 6400 computer. Because of the memory requirements and cost of running the model it actually needs to be run in several parts. The ELM 1977 model has been run for up to 6 years and gives 'reasonable' output for most of the key state variables and flows. There is real need, however, for changes in design and operational approach to the overall model.

## *General considerations of large herbivore modelling*

### *Time trends*

The modelling of large herbivores has matured greatly within the last 10 years. The models have become larger, more mechanistic, more data based, and much more difficult to operate.

Early versions of herbivore models frequently were of the constant-coefficient form which have been reported by Patten (1972) and May, Till & Cumming (1972). More recently, detailed mechanisms and consideration of physiological and anatomical analogues have been incorporated into large herbivore models, such as those reported by Rice, Morris, Maeda & Brown (1974), Morris, Baldwin, Maeng & Maeda (1974), and Payne (1974). Early versions of grazingland system models were solved on analogue computers as systems of first-order differential equations. Almost all the recently developed large herbivore models require large, high-speed, advanced-generation digital computers for their solution.

Another trend in the recently developed large-scale models is their canonical feature with respect to mammals (e.g. see Anway in Cole, 1976, or Innis, 1978).

Another feature of the large herbivore modelling process of the last decade is that the models of domestic animals are far more advanced than are the models of wild herbivores.

Models at first were thought to be a panacea; many were developed to a low or to an 'intermediate level' of resolution, but the magnitude of effort required to carry them further has slowed their rate of development. This is useful as it allows us to examine in more detail how they predict the dynamics of large herbivore populations of various kinds. They have in numerous areas pinpointed the need for specific information (see Chapter 10, this volume). They are just now being used in testing management strategies and theory and being fully integrated with available validation data.

### *Model–information–data relationships*

The degree to which modelling parallels scientific information can be judged, for example, by examination of the parameter requirements for models of large herbivores. As an example, the AFCONS model of Harris & Francis (1972) requires a large number of parameters for six species of East African herbivores. Here the species are selected to represent a general foraging class. The parameters presented are not necessarily offered as being exact values, nor are they being overly criticized. The model is in general conceptually sound and, particularly considering it was developed rapidly by only two scientists (Harris, a zoologist, and Francis, a biostatistician), it is highly credible.

The AFCONS model for large herbivores is driven by precipitation and biomass of various vegetation categories. It includes the following herbivores – giraffe, gazelle, zebra, hartebeest, elephant and impala. The model also includes some predators and scavengers. Various parameters are needed to calculate population dynamics and examples are shown in Table 4.22. These examples were based primarily on personal experience of Harris and in part on the literature. It is interesting to compare these kinds of parameters, and their values, to those in Table 4.1 and to those discussed throughout this chapter summarizing characteristics of many large herbivores. Although the emphasis in the AFCONS model is on energetics of the population of immatures, adults and seniles, the data in Table 4.22 are only for the mature group. Harris & Fowler (1975) note that inclusion of these three age groups yields 'modest resolution in terms of monitoring demographic changes while

Table 4.22. *Some parameters used in modelling large wild herbivores*

| Parameter | Giraffe | Gazelle | Zebra | Hartebeest | Elephant | Impala |
|---|---|---|---|---|---|---|
| Offspring per female per year | 0.6 | 0.9 | 0.75 | 0.8 | 0.2 | 0.9 |
| Proportion of females | 0.62 | 0.66 | 0.45 | 0.7 | 0.6 | 0.65 |
| Julian date for parturition | 50–130 | 50–130 | 20–140 | 0–365 | 0–365 | 10–150 |
| Length of lactation (d) | 180 | 60 | 80 | 70 | 365 | 65 |
| Time span, immature (yr) | 2 | 1 | 2.5 | 1 | 11 | 1 |
| Time span, mature (yr) | 8 | 5 | 12.5 | 6 | 44 | 7 |
| Time span, old age (yr) | 15 | 4 | 7 | 5 | 20 | 4 |
| General foraging class | Leaf browser | Forb browser | Stem grazer | Leaf grazer | Tree stem browser | Browser–grazer |
| Relative dietary preference (%) of mature animals | | | | | | |
| Tree and bush leaves | 60 | 20 | — | — | 40 | 30 |
| Tree and bush stems | 40 | — | — | — | 40 | — |
| Perennial grass leaves | — | 5 | 25 | 70 | 10 | 30 |
| Perennial grass stems | — | 5 | 50 | 10 | 10 | — |
| Annual grasses | — | — | 25 | 10 | — | — |
| Forbs | — | 70 | — | 10 | — | 40 |
| Standing dead | 0 | 0 | 0 | 0 | 0 | 0 |
| Litter | 0 | 0 | 0 | 0 | 0 | 0 |
| Dietary digestible energy (%) for mature animals | | | | | | |
| Tree and bush leaves | 70 | 60 | 50 | 50 | 60 | 70 |
| Tree and bush stems | 50 | 40 | 40 | 40 | 60 | 70 |
| Perennial grass leaves | 60 | 70 | 70 | 70 | 65 | 70 |
| Perennial grass stems | 65 | 70 | 70 | 70 | 65 | 65 |
| Annual grasses | 60 | 70 | 65 | 70 | 60 | 70 |
| Forbs | 65 | 70 | 70 | 70 | 65 | 70 |
| Litter | 30 | 30 | 30 | 30 | 30 | 30 |
| Metabolizable energy in digestible energy (%) for mature animals | | | | | | |
| Tree and bush leaves | 80 | 80 | 75 | 75 | 70 | 80 |
| Tree and bush stems | 60 | 50 | 50 | 50 | 70 | 55 |
| Perennial grass leaves | 75 | 80 | 80 | 80 | 70 | 80 |
| Perennial grass stems | 75 | 75 | 75 | 75 | 70 | 75 |
| Annual grasses | 80 | 80 | 80 | 80 | 70 | 80 |
| Forbs | 80 | 80 | 75 | 80 | 70 | 80 |
| Litter | 30 | 30 | 30 | 30 | 30 | 30 |

From Harris & Francis (1972).

not being burdensome and distractive'; they are based on reproductive and mortality characteristics of Spinage (1972). Thus, in general, the parameters noted in Table 4.22 must also be determined for the immature and the senile or old-age populations.

The relative dietary preferences given in Table 4.4 can be compared with those summarized in Appendix 4.2 and in the graphs (Fig. 4.2 and 4.3) for various herbivores. Similarly, Harris & Francis (1972) estimated the proportion of the dietary gross energy which was digestible and the proportion of the digestible energy which was metabolizable. There are some discrepancies between their estimated values (see Table 4.22) and literature values.

The energy budget of the animal is calculated in AFCONS and basic energy demands such as basal requirement, activity requirement, gestation and lactation were given preference over requirements for growth and fattening. Liver, fat and protein storage compartments were calculated. Basal metabolic requirements utilized the '$W^{0.75}$ law' with a multiplier of 100 for young animals,

70 for mature animals, and 55 for old-aged animals. Daily dry matter food intake was set with limits approximating 3% of the body weight for most of the wild ungulates and 1% for the elephant (the latter based on work of Benedict, 1936, and Buss, 1961), as compared to nominal values of 2.5% for cattle and sheep. In calculating food intake rates they estimated these vertebrates to have 30% dry matter, based on the general value given by Davis & Golley (1963). Maximum growth rates allowed were 0.5% of the weight per day. In converting energy to weight they used the values of 4, 6 and 9 kcal·g dry weight$^{-1}$ for animal body, milk and fat respectively.

Various conversion efficiencies were required to transfer 'metabolized nutrients' from the liver into various products, storage or uses as well as transferring it from adipose tissue. A limited amount of literature information was available to justify the conversion efficiencies, but Harris & Francis used ARC (1965) estimates for liver-metabolizable energy-to-growth transfer and cited Flatt, Moe, Munson & Cooper (1969) to justify that remobilized fat was used more efficiently than unstructured metabolic compounds arriving from the liver. Conversion of unstructured compounds, i.e. 'net energy', from the liver was as follows for various products or activities: basal metabolism 60%, activity 60%, growth 40%, fat tissue 35%, body tissue in gestation 95%, and lactation 60%. The equivalent transfers from adipose tissue to these activities in the same order were 95, 95, 75, 35, 95 and 90%. The arbitrary rule used in partitioning 'excess energy' in young animals was 90% to growth and 10% to fattening as long as the maximum rates were not exceeded; for mature animals it was 90% to fattening and 10% to growth. The requirement for pregnancy, averaged over term, was 20% of the maintenance requirement, which included both basal maintenance and activity. Lactation rates were set at a maximum yield of 1% of the body weight per day, using 6 kcal·g$^{-1}$ for dry weight of the milk or 1.2 kcal·g$^{-1}$ for wet weight of the milk. Preferred milk consumption by the young animals was for the giraffe 20%, gazelle 30%, zebra 25%, hartebeest 30%, elephant 10% and impala 30%, as a time-weighted approximation over the entire immature period.

### *Some limitations on model complexity*

It is doubtful whether all of the above parameters are well established for cattle or sheep, much less for many wild herbivores. Yet this does not preclude, nor should it prevent, attempts at modelling. Sensitivity analysis of these models by quantitative or qualitative methods should direct attention to critical parameters and assumptions. The modelling activity thus would identify, through attempting to structure the model or to analyse it, needed information and could in effect help suggest research for many scientists for a lifetime. This process of modelling is discussed more in Chapter 10.

A problem exists in modelling large herbivores as it does with any other complicated system. And any system can be as complicated as you want to make it whether it be a cow or a county! To keep abreast with experimental efforts in a given field, modelling must be quite detailed. Thus, for example, Mazanov & Nolan (1976) simulate the dynamics of nitrogen metabolism in sheep with a nine-compartment, lagged, differential-difference equation system. Even that was a considerable simplification of a 25-pool flow model conceptualized by the same team several years earlier. Their objective was to developed a mathematical model based on an understanding of the biological nature of the processes involved. Even though they utilized first-order kinetic models (justifying this on the basis that nonlinear models become untractable) they found that time delays must be allowed for explicitly. Their philosophy calls for whole-animal models to be accompanied by a maximum number of independently measured processes within the whole animal. After studying steady-state models for a range of diets and determining coefficients in the matrices of the first-order models, they feel then that possibly a suitable variable-coefficient model can be constructed.

A similar increase in complexity of modelling and accompanying experimental work is noted in the programme of Baldwin and associates at the University of California, Davis, over the past several years (see e.g. Yang, Crist & Grichting, 1976). Their focus, now on conceptualizing hypothetical patterns of nutrient utilization in the whole animal, has reached the point of accounting for balanced metabolic equations such as in adipose tissue metabolism of the dairy cow. Increasing amounts and variety of experimental data are available on such specific topics as determining the adipose tissue energy expenditures as components of the total animal heat production (e.g. maintenance 3%, lipogenesis 1 to 5%, and lipolysis and triglyceride resynthesis $<1\%$ of total animal heat production).

The conclusion is frightening when one attempts to deal with entire ecological systems if this degree of mechanism, detail, and 'correspondence' to real-life physiology, biochemistry and anatomy is necessary. It seems we do not yet have sufficient grasp of the art of modelling large herbivore systems to allow us to simplify readily. The trend is to develop more and more detailed models requiring far more parameters as inputs (and thus experimental data) and taking us further away from the field situation. Admittedly, such detailed, high-resolution models may provide more precise answers for the limited situations to which they apply, *if* we can parameterize, set up, and solve them.

## Information and research needs

This summary and integration of data and information from IBP and related studies has pinpointed a number of research needs in the area of large herbivore process and productivity studies. A few of these needs will be discussed here briefly, but the order of presentation does not denote an order of priority.

In reviewing the literature and trying to compare data from different investigations one becomes impressed with the continued need for research on methods of studying the diets of herbivores for botanical composition. A set of statistically well-designed experiments should be made to compare some of the methods with several herbivore species, backed up with good herbage data, and conducted over a variety of seasons and vegetation types. The degree of correlation for data obtained with the oesophageal fistula, rumen fistula, bite-count, feeding-minutes, faecal examination, and botanical measurement methods needs to be established unambiguously. Much of the confusion that exists within the reported dietary botanical composition literature may be due to methodological factors.

There are very few instances of studying the diets of freely grazing herbivores under a sufficient variety of conditions to establish a sound data base from which to analyse the influence of various factors on dietary chemical and botanical composition. Most work focuses simply on the seasonal shift in diet composition without sufficient coordinated measurements to be able to determine why there are shifts. One of the factors that may affect dietary botanical composition greatly is the relative availability of different plants. To establish dietary preference ratios and to determine the range of these ratios will require measurements under a variety of situations. If such a data base were available it should be possible, at least by empirical analysis, to determine the mean values and variations in the selection ratios.

Comparative studies are relatively limited in the literature. In most instances where comparisons have been made of large herbivores, either only a few samples were taken or different methods were used on the different animal species. Furthermore, in most studies, data on the available herbage and on the grazed forage were not collected simultaneously. Further study is needed on the complementarity of diets of different herbivore species on a yearlong basis by a methodology which does not provide biased results.

Few if any data are available for changes in body composition and net energy value of the consumed herbage by freely grazing animals. Relatively little information is available on changes in body composition of the young and growing animals or of the mature animals as they go through their life.

The concept of compensatory growth has been introduced in work from several authors. What are the limits of compensation under natural grazing-land conditions with wild and domestic herbivores? How much of the compensation is due strictly to increased forage intake rates by thin animals and how much may be due to greater efficiency values for the thin animals?

Relatively little information has been tabulated on the comparative digestive efficiencies for a range of herbivores in different seasons of the year and under different levels of herbage availability. Is the digestive efficiency affected by these conditions or their combinations? Does a large herbivore grazing in the wet season with high herbage availability have a different efficiency from the same animal in the dry season when there is low herbage availability? Can different large herbivores be substituted for each other in field studies? Are there experimentally manageable equivalents to many of the large wild herbivores?

Relatively few studies have been reported where dietary botanical composition, dietary chemical composition, forage intake rate and forage digestibility have been measured simultaneously. Even fewer are those studies with an adequate number of animals on which to base estimates of means and variances. Such data would be valuable in analyses of factors affecting intake and for calculating the chemical composition of the plant part actually consumed in contrast to the total plant available.

We don't know the magnitude of importance of the relationship of the herbivore's food habits and the social system within which it grazes or browses. The subject has been poorly researched. More information is needed on the effect of herd or flock size and structure in relation to grazing performance.

We need to summarize and review literature on individual processes from a conceptual and mathematical modelling standpoint. Process rates and factors affecting the process (with functional relationships) need to be derived. This requires developing a much more detailed conceptual model, similar to that provided in Fig. 4.1, for large herbivores. Thus another research need is to develop large and detailed flow-type diagrams for energy, nitrogen and other nutrient flows within the herbivore and its immediate environment. With respect to mathematical simulation modelling of large herbivore systems, this is one of the key research needs.

There is also a need for developing detailed conceptual diagrams (noted above) of how the systems operate and then, perhaps, reducing these diagrams to more and more aggregated conceptual diagrams. The processes in each of these conceptual diagrams should be identified and a detailed, comparative study be made of the way the different processes are described in extant models. Within extant models, the parameters should be identified and a detailed, comparative literature study be made of the values for parameters and functional forms representing the processes. This would, in part, be a 'cataloguing effort' but probably would yield more value to the scientific community than an independent new-model development. In this effort, alternative functional forms could be examined in analysis experiments to test out the response, i.e. flow rate variation, according to variations in the controlling variables. This would identify the limits of operation of the various submodels. If this could be done in a uniform notation and coding scheme, the results could be utilized directly by modellers throughout the world. Unfortunately, many scientists hesitate to undertake such painstaking and extensive work. Many funding organizations are inclined to approach modelling, if not with scepticism, with a 'let's-hurry-on-with-the-job' philosophy, wanting to promote the development of a 'new piece of scientific work'. The temperament and psychology of the researcher is also relevant. Perhaps the researcher feels the rewards come in writing scientific papers of limited scope on new results and then moving on to something else rather than undertaking the large, long-term, arduous type of task discussed above.

Much of the information on wild herbivores seems difficult to secure and is often found in obscure publications. Frequently, it appears in, and remains hidden in, theses and dissertations. This part of the wildlife literature is particularly valuable as it often includes considerable behavioural information along with that on physiology and nutrition of the herbivore and frequently good summaries of environmental conditions. We do know from many of the papers in the excellent two-volume work edited by Geist & Walther (1974) that behaviour is important, but quantifying many relationships remains to be done.

We have reviewed the literature readily available to us. Thus, it has perhaps greatly under-represented the literature on large herbivores from southeast Asia and Latin America. Perhaps data and information are not as abundant for large herbivores in this part of the world and, if so, this represents an information need.

The literature appears to be lacking a thorough and reasoned summary of information available on comparative physiology of the large herbivore species throughout the world. The variety of animals in the world is great (and in fact perhaps such a review should include photographs of the animals!) and there should be genetic material within this diverse pool of germ plasm that would provide unique physiological capabilities for producing food for mankind under specific conditions that will arise in the future.

More information is needed on the influence of intensity of grazing with different species of large herbivores as to the influence on intake, digestion and production. Several studies have shown that intake does not increase linearly with increasing levels of herbage availability. When does the curve flatten out? With combinations of large herbivores can we get a different relationship of gains per animal and gains per hectare as plotted against stocking rate from that for only one large herbivore species?

These and many other challenging problems should provide a fertile field of research for scientists in decades to come. We implore they improve their standards of reporting and find ways of integrating their information, both within their study and with that available in the literature, so as to make easier the tasks of subsequent reviews.

## Summary

Grazinglands are a major habitat or biome type on all the inhabited continents of the world. Mankind's major use of these lands is to harvest products from them by means of large domestic and wild herbivores. Meat from these animals is a significant and high-quality part of our diet. With expanding human populations and with increasing use of cultivated lands for direct human food production, large herbivores and grazinglands will become even more important for the food products they can yield to mankind. Understanding this productivity and its component processes is a requisite for intensification of use. Major processes of concern are selection and digestion of food and then conversion of digested products into fat, meat and milk. Component and interrelated processes are concerned with nutrition and population dynamics. The objective of this chapter was to provide some IBP-derived information from field studies with large herbivores and from modelling of large herbivores and to interrelate this information with that in the literature. A further objective was to structure systematically this information and to provide numerical values and evaluation if possible. We accomplished this by first providing a conceptual diagram showing the relationship of processes and variables in the large herbivore subsystem of grazinglands. In subsequent sections we integrated data from field and laboratory studies on food selection, food intake, food digestion, other processes, general production information, and models. Research and information needs were pinpointed throughout these discussions and summarized in a final section.

Preference is related to animal characteristics; palatability is related to plant characteristics. Selection or preference ratios are defined as the proportion of the plant species or group in the diet compared to the proportion of the same in the available herbage. These selection ratios generally range from 0.1 for plants being rejected, up to 10.0, for plants being selected. Examples are shown, based on IBP studies, of relative dietary preference values for cattle, sheep, bison and pronghorn grazing at several seasons of the year at different intensities on shortgrass prairie rangeland.

Some 668 examples of seasonal–yearlong–treatment data for botanical composition of the diets of large herbivores on the world's grazingland were summarized from the scientific literature. Data from these studies were aggregated to show the percentage dietary composition of (i) grass and grass-like plants, (ii) forbs, and (iii) shrubs, half-shrubs and trees. Triangular-coordinate plots were made to illustrate the general nature of the diets by animal species or groups. Dietary similarity indices were calculated to show overlaps among large herbivores. Averaged over all seasons and treatment situations, cattle and horses consume about 70% grass, 15% forbs and 15% shrubs; sheep take about 50% grass, 30% forbs and 20% shrubs; goats consume about 60% shrubs, 30% grass and 10% forbs; North American deer consume about

10% grass, 30% forbs and 60% shrubs; pronghorns consume about 15% grass and somewhat more than 40% forbs and 40% shrubs. Averaged over all large herbivores studied, about half of them consume more than 50% grass on a yearlong basis and half of them less than 50%. Weighted over the 28 large herbivore species for which there are suitable quantitative data on a world-wide, yearlong basis, they consume about 60% grass, 20% forbs and 20% shrubs. The scientific literature is critiqued with respect to lack of information on mineral composition of the diet, lack of detailed data for certain animal species or groups, and inadequate consideration of sample size and methodology.

Some 137 examples of seasonal–yearlong–treatment data were analysed from the scientific literature for chemical composition of the diet. These data suggest that protein content of the diet generally decreases with increase in body size.

Some 93 examples of seasonal–yearlong–treatment data were analysed from the world's literature on forage intake rates for grazing cattle and sheep. Averaged over 31 examples and based on dry matter in forage, cattle consume 1.8% of their body weight per day; averaged over 41 examples, sheep consume 2.4%. Within each species there are slightly higher intake rates for smaller animals. Water intake rate data are less abundant and are highly variable among different species of the world's herbivores.

Some 160 examples of seasonal–yearlong–treatment data were analysed from the scientific literature for digestion of forage by herbivores on grazing-lands. Numerous factors affecting digestibility were reviewed including influences of species, cold, forage intake rate and nitrogen content. Averaged over 36 examples, mean dry matter digestibility for cattle and sheep was 52%; values were higher in the spring and lower in summer and autumn. Digestibility of energy and cellulose was in the same range as that for dry matter. There was a direct relationship between crude protein concentration of the diet and digestibility, and for cattle, at least, an inverse relationship between lignin concentration of the diet and digestibility. There is a definite lack of efficient, cross-over design trials in the scientific literature for examining digestibility influences.

Other processes were discussed but not in as much detail as those listed above. The energy cost of grazing is from 50 up to 100% of the maintenance energy requirement of the large herbivore. Efficiency of maintenance and gain is not well known for grazing herbivores; selected values from the literature of studies on large herbivores in metabolism crates, feedlot trials and physiological studies are provided. Species and breeds within species vary and example values are given. The concept of, and examples of, compensatory gain are illustrated. This is a critical phenomenon for grazing herbivores and more information is needed. Example mortality and natality rates are summarized with emphasis on temperate-climate wild herbivores.

*Processes and productivity*

Production measurements are discussed, with emphasis on offtake rates and effect of management. Comparable age–weight curves are presented for a variety of domestic and wild herbivores. The cyclic weight change, related to seasonal influences, is described. Examples are taken from arid and semi-arid grazinglands with cattle and sheep to show production rates under different intensities of grazing. These data illustrate that weight per animal and reproductive performance are generally greater under light grazing than under heavy grazing. Production per unit area may be greater under intensive grazing treatments than under light grazing treatments, at least over some interval of time. With longer use, perhaps within a century, moderate grazing treatments are superior in animal production on most semi-arid grazinglands. In the subhumid–humid grazinglands, heavier grazing intensities may be superior to light grazing intensities. More thorough evaluation is needed of the impact of grazing intensity on the plant and animal production in various grazingland systems of the world under varying management practices. Production rates of wild herbivores are estimated from data on herd size and harvest rates. Under intensive hunting, harvest rates are in the order of 15 to 20%. There appear to be some situations, such as in East Africa, where a greater meat harvest may be obtained through a mixture of a number of wild herbivores than from one or two domestic herbivores. This difference is a function of optimal utilization of vegetation, resistance to disease, resistance to heat stress, and related factors. Differences between wild herbivores and domestic animals in carcass dressing percentage and composition are summarized. A general summary of productivity shows that production rates for large herbivores on the world's grazinglands are in the order of 0.1 to 1.0 $g \cdot m^{-2} \cdot yr^{-1}$ in many instances. There is no close relationship between animal standing crop and annual production rate.

A comparative summary is made of the world's literature on mathematical simulation models of large herbivores under grazing. We indicate whether the models are difference or differential equation systems, what the time step is, if the computer language is specified, and whether a model diagram and model code are available; brief comments are provided on the purpose of the model and the general biological and structural properties. Some 11 models of medium size and resolution are further analysed and are summarized in brief with particular attention to the sections of the models related to food selection, food intake and food digestion calculations. In these 11 models some 16 different factors were used in a wide variety of mathematical calculation forms for forage intake rate. Herbage availability was the most common factor used in calculating forage intake rate, with animal weight of second importance. Large-scale, total-system, complex models including large herbivore components and developed as part of the IBP are discussed. Modelling for large herbivore systems was discussed with respect to the relationship of models and data and problems of model complexity.

The authors acknowledge A. Porges and S. H. Van Dyne for their assistance in compiling literature and information for an early draft and V. Geist and L. Talbot for their review of a draft of the manuscript and their many useful suggestions. J. E. Newton and C. R. W. Spedding also provided numerous valuable comments on the structure of the chapter. W. M. Longhurst provided manuscripts on some of his unpublished work which is much appreciated. Several other individuals alerted us to key papers and reports which were utilized in our analyses.

Barbara Carlson and her staff provided excellent services in preparing the text, tables and figures within our tight time schedule.

Part of the work in developing this chapter was supported by National Science Foundation Grants DEB73-02027 A03, GB-41233X and BMS73-02027 A02 to the Grassland Biome, US International Biological Program, for 'Analysis of Structure, Function, and Utilization of Grassland Ecosystems'. The work on and reviews of modelling large herbivore diets reported herein was supported by National Science Foundation Grant DEB76-11134.

## References

Agricultural Board, National Academy of Sciences, National Research Council, and American Society of Range Management (1962). *Basic problems and techniques in range research. National Research Council Publication 890.*

Alder, F. E. & Minson, D. J. (1963). The herbage intake of cattle grazing lucerne and cocksfoot pastures. *Journal of Agricultural Science*, **60**, 359–69.

Aldous, M. C., Craighead, F. C., Jr & Devan, J. A. (1958). Some weights and measurements of desert bighorn sheep (*Ovis canadensis nelsoni*). *Journal of Wildlife Management*, **22**, 444–5.

Allden, W. G. (1962). Rate of herbage intake and grazing time in relation to herbage availability. *Proceedings of the Australian Society of Animal Production*, **4**, 163–6.

Allden, W. G. & Jennings, A. C. (1962). Dietary supplements to sheep grazing mature herbage in relation to herbage intake. *Proceedings of the Western Society of Animal Production*, **4**, 145–53.

Alldredge, A. W., Lipscomb, J. F. & Whicker, F. W. (1974). Forage intake rates of mule deer estimated with fallout Cesium-137. *Journal of Wildlife Management*, **38**, 508–16.

Allen, E. O. (1968). Range use, foods, condition, and productivity of white-tailed deer in Montana. *Journal of Wildlife Management*, **32**, 130–41.

American Society of Animal Production (1959). *Techniques and procedures in animal production research.* American Society of Animal Production, Beltsville, Maryland.

Ames, D. R. & Brink, D. R. (1977). Effect of temperature on lamb performance and protein efficiency ratio. *Journal of Animal Science*, **44**, 136–40.

Anand, B. K. (1961). Nervous regulation of food intake. *Physiological Reviews*, **41**, 677–708.

Anderson, A. E., Snyder, W. A. & Brown, G. W. (1965). Stomach content analyses related to condition in mule deer, Guadalupe Mountains, New Mexico. *Journal of Wildlife Management*, **29**, 352–66.

Anderson, A. E., Medin, D. E. & Bowden, D. C. (1974). Growth and morphometry of the carcass, selected bones, organs, and glands of mule deer. *Wildlife Monographs*, **39**, 1–122.

Anderson, A. W. (1949). Early summer foods and movements of the mule deer (*Odocoileus hemionus*) in the Sierra Vieja Range of southwestern Texas. *Texas Journal of Science*, **1**, 45–50.

Anderson, C. C. (1953). Food habits study of game animals; elk and deer. *Wyoming Wildlife*, **17**, 26–8.

Ansotegui, R. O., Lesperance, A. L., Pudney, R. A., Papex, N. J. & Tueller, P. T. (1972). Composition of cattle and deer diets grazing in common. *Proceedings of the Western Section of the American Society of Animal Science*, **23**, 184.

Anthony, R. G. (1976). Influence of drought on diets and numbers of desert deer. *Journal of Wildlife Management*, **40**, 140–4.

Anway, J. C. (1973). Consumer simulation model: a canonical mammal. In *Proceedings of the 1973 Summer Computer Simulation Conference*, vol. 11, pp. 835–40. Simulation Councils, Inc., La Jolla, California.

Anway, J. C. (1978). A canonical mammalian model. In *Grassland simulation model – US/IBP Grassland Biome Study*, ed. G. S. Innis, vol. 26, pp. 89–125. Springer-Verlag, Berlin.

ARC (Agricultural Research Council) (1965). *The nutrient requirements of farm livestock. No. 2: Ruminants.* Agricultural Research Council, London.

Arman, P. & Hopcraft, D. (1975). Nutritional studies on the East African herbivore. I. Digestibilities of dry matter, crude fiber, and crude protein in antelope, cattle, and sheep. *British Journal of Nutrition*, **33**, 255–64.

Arman, P., Hopcraft, D. & McDonald, I. (1975). Nutritional studies on East African herbivores. II. Losses of nitrogen in the feces. *British Journal of Nutrition*, **33**, 265–76.

Arman, P., Kay, R. N. B., Goodall, E. D. & Sharman, G. A. M. (1974). The composition and yield of milk from captive red deer (*Cervus elaphus* L.). *Journal of Reproduction and Fertility*, **37**, 67–84.

Armstrong, J. S. (1969). Modelling a grazing system. *Proceedings of the Ecological Society of Australia*, **6**, 194–202.

Arnold, G. W. (1964). Factors within plant associations affecting the behavior and performance of grazing animals. In *Grazing in terrestrial and marine environments*, ed. D. J. Crisp, pp. 133–54. Blackwell Scientific Publications, Oxford.

Arnold, G. W. (1966*a*). The spatial senses in grazing animals. I. Site and dietary habits in sheep. *Australian Journal of Agricultural Research*, **17**, 521–9.

Arnold, G. W. (1966*b*). The spatial senses in grazing animals. II. Smell, taste, touch, and dietary habits in sheep. *Australian Journal of Agricultural Research*, **17**, 531–42.

Arnold, G. W., Ball, J., McManus, W. R. & Bush, I. G. (1966). Studies on the diet of the grazing animals. I. Seasonal changes in the diet of sheep grazing on pastures of different availability and composition. *Australian Journal of Agricultural Research*, **17**, 543–56.

Arnold, G. W. & Bennett, D. (1975). The problem of finding an optimum solution. In *Study of agricultural systems*, ed. G. F. Dalton, pp. 129–73. Applied Science Publishers, Barking, Essex.

Arnold, G. W. & Campbell, N. A. (1972). A model of a lay farming system with particular reference to a sub-model for animal production. *Proceedings of the Australian Society of Animal Production*, **9**, 23–30.

Arnold, G. W., Carbon, B. A., Galbraith, K. A. & Biddiscombe, E. F. (1974). Use of a simulation model to assess the effects of a grazing management on pasture and animal production. In *Proceedings of XII International Grassland Congress, Moscow*, pp. 47–52.

Arnold, G. W. & Galbraith, K. A. (1974). Predicting the value of lupins for sheep and cattle in cropping and pastoral farming systems. *Proceedings of the Australian Society of Animal Production*, **10**, 383–6.

Arnold, G. W., McManus, W. R. & Bush, I. G. (1964). Studies in the wool production of grazing sheep. I. Seasonal variation in feed intake, live weight, and wool production. *Australian Journal of Experimental Agriculture and Animal Husbandry*, **4**, 403.

Baharav, D. (1975). Energy flow and productivity in a mountain gazelle (*Gazella gazella gazella* Pallas 1766) population. PhD Thesis, Tel Aviv University, Israel.

Baker, R. D. (1976). The behavioural responses of cattle and sheep to changes in grazing conditions. Paper presented at the Association for the Study of Animal Behaviour, 14 July 1976, England.

Balch, C. C. & Campling, R. C. (1962). Regulation of voluntary food intake in ruminants. *Nutrition Abstracts and Reviews*, **32**, 669–86.

Baldwin, R. L. (1972). Tissue metabolism and energy expenditures of maintenance and production. Tenth Brody Memorial Lecture, University of Missouri, Columbia.

Baldwin, R. L., Yang, Y. T., Crist, K. & Grichting, G. (1976). Theoretical model of ruminants' adipose tissue metabolism in relation to the whole animal. *Federation Proceedings*, **35**, 2314–18.

Banfield, A. W. F. (1951). *The barren-ground caribou*. Unnumbered report. Northern Administration and Land Branch, Department of Resources Development, Ottawa, Canada.

Banfield, A. W. F. (1954). *Preliminary investigation of the barren ground caribou*. II. *Life history, ecology, and utilization. Wildlife Monograph Bulletin Series Number I, Number 10B.* Department of Northern Affairs and National Resources, National Parks Branch, Canadian Wildlife Service, Ottawa, Canada.

Bannikov, A. G. (ed.) (1967). *Biology of the saiga*. Israel Program for Scientific Translations, Jerusalem.

Barrett, R. H. (1964). Seasonal food habits of the bighorn at the Desert Game Range, Nevada. In *Desert Bighorn Council Transactions*, pp. 85–93. Desert Bighorn Council, Las Vegas, Nevada.

Baskin, L. M. (1974). Management of ungulate herds in relation to domestication. In *The behavior of ungulates and its relation to management*, ed. V. Geist & F. Walther, vol. 2, pp. 530–41. International Union for Conservation of Nature and Natural Resources, Morjes, Switzerland.

Bayless, S. R. (1969). Winter food habits, range use, and home range of antelope in Montana. *Journal of Wildlife Management*, **33**, 538–51.

Beale, D. M. & Scotter, G. W. (1968). Seasonal forage use by pronghorn antelope in western Utah. *Utah Science*, **29**, 3–6, 16.

Beale, D. M. & Smith, A. D. (1970). Forage use, water consumption and productivity of pronghorn antelope in western Utah. *Journal of Wildlife Management*, **34**, 570–82.

Beck, R. F. (1969). Diet of steers in southeastern Colorado. PhD Thesis, Colorado State University, Fort Collins.

Beck, R. F. (1975). Steer diets in southeastern Colorado. *Journal of Range Management*, **28**, 48–51.

Bedell, T. E. (1971). Nutritive value of forage and diets of sheep and cattle from Oregon subclover–grass mixtures. *Journal of Range Management*, **24**, 125–33.

Bell, F. R. (1971). Hypothalamic control of food intake. The regulation of voluntary food intake. *Proceedings of the Nutrition Society*, **30**, 103–9.

Bell, R. H. V. (1970). The use of the herb layer by grazing ungulates in the Serengeti. In *Animal populations in relation to their food resources*, ed. A. Watson, pp. 111–24. Blackwell Scientific Publications, Oxford.

Belonje, P. C. & Van Niekerk, C. H. (1975). A review of the influence of nutrition upon the oestrous cycle and early pregnancy in the mare. *Journal of Reproduction and Fertility, Supplement*, **23**, 167–9.

Benedict, F. G. (1936). *The physiology of the elephant. Carnegie Institute, Washington, Publication 474.*

Benjamin, R. W., Degen, A. A., Breighet, A., Chen, A. & Tadmor, N. H. (1975). Estimation of food intake of sheep grazing green pasture when no free water is available. *Journal of Agricultural Science*, **35**, 403–7.

Bergerud, A. T. (1967). Management of Labrador caribou. *Journal of Wildlife Management*, **31**, 621–42.

Bergerud, A. T. (1972). Food habits of the Newfoundland caribou. *Journal of Wildlife Management*, **36**, 913–23.

Bergerud, A. T. (1974). Rutting behaviour of Newfoundland caribou. In *The behavior of ungulates and its relation to management*, ed. V. Geist & F. Walther, vol. 1, pp. 395–435. International Union for Conservation of Nature and Natural Resources, Morjes, Switzerland.

Bergerud, A. T. (1975). The reproductive season of Newfoundland caribou. *Canadian Journal of Zoology*, **53**, 1213–21.

Bernt, B. (1976). Notes on a pronghorn antelope winter range. *Proceedings of the VII Biennial Pronghorn Antelope Workshop, Twin Falls, Idaho*, pp. 66–76.

Bines, J. A. (1971). Metabolic and physical control of food intake in ruminants. *Proceedings of the Nutrition Society*, **30**, 116–22.

Bishop, J. P., Froseth, J. A., Verettoni, H. N. & Noller, C. H. (1975). Diet and performance of sheep on rangelands in semi-arid Argentina. *Journal of Range Management*, **28**, 52–5.

Bissell, H. D., Harris, B., Strong, H. & James, F. (1955). The digestibility of certain natural and artificial foods eaten by deer in California. *California Fish and Game*, **41**, 572–8.

Bissell, H. D. & Weir, W. E. (1957). The digestibility of interior live oak and chemise by deer and sheep. *Journal of Animal Science*, **16**, 476–80.

Blaxter, K. L. (1960). The utilization of the energy of grassland products. In *Proceedings of the VIII International Grassland Congress*, paper no. **2B**/6. Alden Press, Oxford.

Blaxter, K. L. (1962). *Energy metabolism of ruminants*. Hutchinson, London.

Blaxter, K. L. (1964). Utilization of the metabolizable energy of grass. *Proceedings of the Nutrition Society*, **23**, 62–71.

Blaxter, K. L. (1975). Conventional and unconventional farm animals. Symposium on the Nutrition of New Farm Animals. *Proceedings of the Nutrition Society*, **34**, 51–6.

Blaxter, K. L., Wainman, F. W. & Wilson, R. S. (1961). The regulation of food intake in sheep. *Journal of Animal Production*, **3**, 51–61.

Blaxter, K. L. & Wilson, R. S. (1962). The voluntary intake of roughages by steers. *Animal Production*, **4**, 351–8.

Bledsoe, L. J. (1976). Simulation of a grassland ecosystem. PhD Dissertation, Colorado State University, Fort Collins.

Blincoe, C. (1975). Computer simulation of iodine metabolism by mammals. *Journal of Animal Science*, **40**, 342–50.

Bliss, L. C. (1975). Tundra grasslands, herblands, and shrublands and the role of

herbivores. In *Geoscience and man*, vol. 10, *Grassland ecology – a symposium*, ed. R. H. Kessel, pp. 51–79. School of Geoscience, Louisiana State University.
Blood, D. A. (1967). Food habits of the Ashnola bighorn sheep herd. *Canadian Field Naturalist*, **8**, 23–9.
Blood, D. A., Flook, D. R. & Wishart, W. O. (1970). Weights and growth of Rocky Mountain bighorn sheep in western Alberta. *Journal of Wildlife Management*, **34**, 451–5.
Blood, D. A. & Lovaas, A. L. (1966). Measurements and weight relationships in Manitoba elk. *Journal of Wildlife Management*, **30**, 135–40.
Boeker, E. L., Scott, E. F., Reynolds, H. G. & Donaldson, B. A. (1972). Seasonal food habits of mule deer in southwestern New Mexico. *Journal of Wildlife Management*, **36**, 56–63.
Borowski, S. & Kossak, S. (1972). The natural food preferences of the European bison in seasons free of snow cover. *Acta Theriologica*, **17**, 151–69.
Bourlière, F. & Verschuren, J. (1960). *Introduction a l'écologie des ongules du Parc National Albert*. Institut des Parks Nationaux du Congo Belge, Brussels.
Bradford, G. E., Weir, W. C. & Torell, D. T. (1960). Growth rate, carcass grades and net returns of Suffolk- and Southdown-sired lambs under range conditions. *Journal of Animal Science*, **19**, 493–501.
Bredon, R. M., Torell, D. T. & Marshall, B. (1967). Measurement of selective grazing of tropical pastures using esophageal fistulated steers. *Journal of Range Management*, **20**, 317–20.
Brinks, J. S., Clark, R. T., Kieffer, N. M. & Queensberry, J. R. (1962). Mature weight in Hereford range cows – heritability, repeatability, and relationship to calf performance. *Journal of Animal Science*, **21**, 501–4.
Brobek, J. R. (1957). Neural bases of hunger, appetite, and satiety. *Gastroenterology*, **32**, 169–74.
Brody, S. (1945). *Bioenergetics and growth*. Reinhold Publishing Corporation, New York.
Brown, K. W., Smith, D. D., Bernhardt, D. E. & Giles, K. R. (1975). Food habits and radionuclide concentrations of Nevada desert bighorn. In *Desert Bighorn Council Transactions*, pp. 61–8. Desert Bighorn Council, Las Vegas, Nevada.
Browning, B. (1960). Preliminary report of the food habits of the wild burro in the Death Valley National Monument. In *Desert Bighorn Council Transactions*, pp. 88–90. Desert Bighorn Council, Las Vegas, Nevada.
Buchanon, H., Laycock, W. A. & Price, D. A. (1972). Botanical and nutritive content of the summer diet of sheep on a tall forb range in southwestern Montana. *Journal of Animal Science*, **35**, 423–30.
Buechner, H. K. (1950). Life history, ecology, and range use of the pronghorn antelope in Trans-Pecos, Texas. *American Midland Naturalist*, **43**, 257–354.
Bunnell, F., Dauphine, D. C., Hilborn, R., Miller, D. R., Miller, F. L., McEwan, E. H., Parker, G. R., Peterman, R., Scotter, G. W. & Waiters, C. J. (1975). Preliminary report on computer simulation of barren ground caribou management. In *Proceedings of the I International Reindeer and Caribou Symposium (1972)*, pp. 189–93. University of Alaska.
Burton, M. (1962). *Systematic dictionary of mammals of the world*. Crowell, New York.
Bush, L. F. & Lewis, J. K. (1977). Growth patterns of range-grazed Rambouillet lambs. *Journal of Animal Science*, **45**, 953–60.
Buss, I. O. (1961). Some observations on food habits and behavior of the African elephant. *Journal of Wildlife Management*, **25**, 131–48.
Cable, D. R. & Shumway, R. P. (1966). Crude protein in rumen contents and in forage. *Journal of Range Management*, **19**, 124–8.

Cale, W. G. (1975). Simulation and systems analysis of a shortgrass prairie ecosystem. PhD Thesis, University of Georgia, Athens.
Campling, R. C. (1964). Factors affecting the voluntary intake of grass. *Proceedings of the Nutrition Society*, **23**, 80–8.
Campling, R. C. & Balch, C. C. (1961). Factors affecting the voluntary intake of food by cows. I. Preliminary observations on the effect on the voluntary intake of hay, of changes in the amount of the reticula-ruminal contents. *British Journal of Nutrition*, **15**, 523–30.
Campling, R. C., Freer, M. & Balch, C. C. (1961). Factors affecting the voluntary intake of food by cows. II. The relationship between the voluntary intake of roughages, the amount of digesta in the reticulo-rumen and the rate of disappearance of digesta from the alimentary tract. *British Journal of Nutrition*, **15**, 531–40.
Campling, R. C., Freer, M. & Balch, C. C. (1962). Factors affecting the voluntary intake of food by cows. III. The effect of urea on the voluntary intake of oat straw. *British Journal of Nutrition*, **16**, 115–24.
Casebeer, R. L. & Koss, G. G. (1970). Food habits of wildebeest, zebra, hartebeest, and cattle in Kenya, Masailand. *East African Wildlife Journal*, **8**, 25–36.
Catto, G. G. (1976). Optimal production from a blesbock herd. *Journal of Environmental Management*, **4**, 95–121.
Chamrad, A. D. & Box, T. W. (1968). Food habits of white-tailed deer in south Texas. *Journal of Range Management*, **21**, 158–64.
Chen, E. C. H., Blood, D. A. & Baker, B. E. (1965). Rocky Mountain bighorn sheep (*Ovis canadensis canadensis*) milk. I. Gross composition and fat constitution. *Canadian Journal of Zoology*, **43**, 885–8.
Chippendale, G. (1962). Botanical examination of kangaroo stomach contents and cattle rumen contents. *Australian Journal of Science*, **25**, 21–2.
Christian, K. R., Armstrong, J. S., Davidson, J. L., Donnelly, J. R. & Freer, M. (1974). A model for decision-making in grazing management system. *Proceedings of the Australian Society of Animal Production*, **9**, 124–9.
Christian, K. R., Armstrong, J. S., Donnelly, J. R., Davidson, J. L. & Freer, M. (1972). Optimization of a grazing management system. *Proceedings of the Australian Society of Animal Production*, **9**, 124–9.
Christopherson, R. J. (1976). Effects of prolonged cold and the outdoor winter environment on apparent digestibility in sheep and cattle. *Canadian Journal of Animal Science*, **56**, 201–12.
Chudleigh, P. D. & Filan, S. J. (1972). A simulation model of an arid zone sheep property. *Australian Journal of Agricultural Economics*, **16**, 183–94.
Clark, T. W. (1968). Plants used as food by mule deer in Oklahoma in relation to habitat. *Southwestern Naturalist*, **13**, 159–66.
Cole, G. F. (1956). *The pronghorn antelope: its range, use, and food habits in central Montana with special reference to alfalfa. Montana Fish and Game Department and Montana State College Agricultural Experiment Station Technical Bulletin 516.*
Cole, G. W. (ed.) (1976). *ELM: version 2.1. Range Science Department Science Series 20.* Colorado State University, Fort Collins.
Connolly, J. (1974). Linear programming and the optimum carrying capacity of range under common use. *Journal of Agricultural Science*, **83**, 259–65.
Connor, J. M., Bohman, V. R., Lesperance, A. L. & Kinsinger, I. E. (1963). Nutritive evaluation of summer range forage with cattle. *Journal of Animal Science*, **22**, 961–9.
Constan, K. J. (1973). Winter foods and range use of three species of ungulates. *Journal of Wildlife Management*, **36**, 1068–76.

Cook, C. J., Cook, C. W. & Harris, L. E. (1948). Utilization of northern Utah summer range plants by sheep. *Journal of Forestry*, **46**, 416–25.

Cook, C. W. (1970). *Energy budget of the range and range livestock. Colorado State University Experiment Station Bulletin TB 109.*

Cook, C. W. (1975). Forage utilization, daily intake, and nutrient value of desert range. In *Arid shrubland – Proceedings of the III Workshop of the US/Australia Rangelands Panel, Tucson, Arizona*, 26 March–5 April 1973, ed. D. N. Hyder, pp. 47–50. Society of Range Management, Denver, Colorado.

Cook, C. W. & Harris, L. E. (1950). *The nutritive content of the grazing sheep's diet on the summer and winter ranges of Utah. Utah State University Agricultural Experiment Station Bulletin 342.*

Cook, C. W. & Harris, L. E. (1968). *Nutritive value of seasonal ranges. Utah State University Agricultural Experiment Station Bulletin 472.*

Cook, C. W., Harris, L. E. & Young, M. C. (1967). Botanical and nutritive content of diets of cattle and sheep under single and common use on mountain range. *Journal of Animal Science*, **26**, 1169–74.

Cook, C. W., Kothman, M. & Harris, L. E. (1965). Effect of range condition and utilization on nutritive intake of sheep on summer ranges. *Journal of Range Management*, **18**, 69–73.

Cook, C. W., Mattox, J. E. & Harris, L. E. (1961). Comparative daily consumption and digestibility of summer range forage by wet and dry ewes. *Journal of Animal Science*, **20**, 866–70.

Cook, C. W., Stoddard, L. A. & Harris, L. E. (1953). Effects of grazing intensity upon the nutritive value of range forage. *Journal of Range Management*, **6**, 51–4.

Cook, C. W., Stoddard, L. A. & Harris, L. E. (1954). *The nutritive value of winter range plants in the Great Basin. Utah State University Agricultural Experiment Station Bulletin 372.*

Cook, C. W., Stoddard, L. A. & Harris, L. E. (1956). *Comparative nutritive value and palatability of some introduced and native forage plants for spring and summer grazing. Utah State University Agricultural Experiment Station Bulletin 385.*

Cook, C. W., Taylor, K. & Harris, L. E. (1962). Effect of range condition and intensity of grazing upon daily intake and nutritive value of the diet on desert ranges. *Journal of Range Management*, **15**, 1–6.

Coop, I. E. & Hill, M. K. (1962). The energy requirements of sheep for maintenance and gain. II. Grazing sheep. *Journal of Agricultural Science*, **58**, 187–99.

Corbett, J. L. (1971). Determination of the energy expenditure of grazing sheep from estimates of carbon dioxide entry rate. *British Journal of Nutrition*, **26**, 277–91.

Corbett, J. L., Langlands, J. P. & Reid, G. W. (1963). Effects of season of growth and digestibility of herbage on intake by grazing dairy cows. *Animal Production*, **5**, 119–129.

Couey, F. M. (1946). Antelope foods in southeastern Montana. *Journal of Wildlife Management*, **10**, 367.

Cowan, I. M. (1947). Range competition between mule deer, bighorn sheep, and elk in Jasper Park, Alberta. *Transactions of the North American Wildlife Conference*, **12**, 223–7.

Cowan, I. M. (1950). Some vital statistics of big game on overstocked mountain range. *Transactions of the North American Wildlife Conference*, **15**, 581–8.

Cowlishaw, S. & Alder, F. (1960). The grazing preferences of cattle and sheep. *Journal of Agricultural Science*, **54**, 257–65.

Cushwa, C. T. & Coady, J. (1976). Food habits of moose, *Alces alces*, in Alaska: a preliminary study using rumen contents analysis. *Canadian Field Naturalist*, **90**, 11–16.

Dagg, A. I. & Foster, J. B. (1976). *The giraffe*. Van Nostrand Reinhold Company, New York.

D'Aquino, S. A. (1974). A case study for optimal allocation of range resources. *Journal of Range Management*, **27**, 228–33.

Dasmann, R. F. & Mossman, A. S. (1962). Abundance and population structure of wild ungulates in some areas of Southern Rhodesia. *Journal of Wildlife Management*, **26**, 262–8.

Dasmann, R. F. & Mossman, A. S. (1964). *African game ranching*. Macmillan, New York.

Davis, D. E. & Golley, F. B. (1963). *Principles of mammalogy*. Reinhold, New York.

Davis, L. A. (1967). Dynamic programming for deer management planning. *Journal of Range Management*, **31**, 667–79.

Dawson, T. J., Denny, M. J. S., Russell, E. M. & Ellis, B. (1975). Water usage and diet preferences of free ranging kangaroos, sheep, and feral goats in the Australian arid zone during summer. *Journal of Zoology*, **77**, 11–23.

Dean, R. E. & Rice, R. W. (1975). *Energy and nitrogen flow through cattle on the shortgrass prairie. US/IBP Grassland Biome Technical Report 293*. Colorado State University, Fort Collins.

Dean, R. E., Thorne, E. T. & Yorgason, I. J. (1976). Weights of Rocky Mountain elk. *Journal of Mammalogy*, **57**, 186–9.

de la Tour, G. D. (1954). The guanaco. *Oryx*, **2**, 347–51.

Denham, A. H. (1965). *In vitro* fermentation studies on native sandhill range forage as related to cattle preference. MS Thesis, Colorado State University, Fort Collins.

De Nio, R. M. (1938). Elk and deer foods and feeding habits. *North American Wildlife Conference Transactions*, **3**, 421–7.

Denny, R. N. (1968). The case for intensive wildlife management. In *Proceedings of the Symposium on Wildlife Management and Land Use*, ed. A. de Bos & T. Jones, pp. 119–32. (*East African Agricultural Forestry Journal*, Special Issue.)

Dietz, D. R., Udall, R. H. & Yeager, L. E. (1962). Differential digestibility of nutrients in bitterbrush, mountain mahogany and big sagebush by deer. In *Proceedings of the I National White-tailed Deer Disease Symposium*, pp. 29–36, 39–50.

Dirschl, J. H. (1963). Food habits of pronghorn in Saskatchewan. *Journal of Wildlife Management*, **27**, 81–93.

Donnelly, J. R. & Armstrong, J. S. (1968). Summer grazing. In *Proceedings of the II Conference on Applied Simulation, New York*, pp. 329–32.

Donnelly, J. R., Davidson, J. L. & Freer, M. (1974). Effect of body condition on the intake of food by mature sheep. *Australian Journal of Agricultural Research*, **25**, 813–23.

Doran, C. W. (1943). Activities and grazing habits of sheep on summer range. *Journal of Forestry*, **41**, 253–8.

Dorn, R. D. (1970). Moose and cattle food habits in southwest Montana. *Journal of Wildlife Management*, **34**, 559–64.

Drawe, D. L. & Box, T. W. (1968). Forage ratings for deer and cattle on the Welder Wildlife Refuge. *Journal of Range Management*, **21**, 225–8.

Dusek, G. L. (1975). Range relations of mule deer and cattle in prairie habitat. *Journal of Wildlife Management*, **39**, 605–16.

Dyck, G. W. & Bement, R. E. (1971). *Herbage growth rate, forage intake and forage quality in 1970 on heavily and lightly grazed blue grama pastures. US/IBP Grassland Biome Technical Report 94*. Colorado State University, Fort Collins.

Dyck, G. W. & Bement, R. E. (1972). *Herbage growth rate, forage intake and forage quality in 1971 on heavily and lightly grazed blue grama pastures. US/IBP Grassland Biome Technical Report 182.* Colorado State University, Fort Collins.

Dzieciolowski, R. (1970). Foods of the red deer as determined by rumen content analysis. *Acta Theriologica*, **15**, 89–110.

Eadie, J. (1970). Sheep production and pastoral resources. In *Animal populations in relation to their food resources*, ed. A. Watson, pp. 7–24. Blackwell Scientific Publications, Oxford.

Eberhardt, L. L. (1969). Population analysis. In *Wildlife management techniques*, ed. R. H. Giles, Jr, pp. 457–75. The Wildlife Society, Washington, DC.

Eberhardt, L. L. & Hanson, W. C. (1969). A simulation model for an arctic food chain. *Health Physics*, **17**, 793–806.

Edelsten, P. R. & Newton, J. E. (1977). A simulation model of a lowland sheep system. *Journal of Agricultural Systems*, **2**, 17–32.

Edelsten, P. R., Newton, J. E. & Treacher, T. T. (1973). *A model of ewes with lambs grazing at pasture. Internal Report 260.* Grassland Research Institute, Hurley, England.

Edlefsen, J. L., Cook, C. W. & Blake, J. T. (1960). Nutrient content of diet as determined by hand plucked and esophageal fistula samples. *Journal of Animal Science*, **19**, 560–7.

Edwards, R. Y. & Richey, R. W. (1960). Foods of caribou in Wells Gray Park, British Columbia, Canada. *Canadian Field Naturalist*, **74**, 3–7.

Egan, A. R. (1965). Nutritional status and intake regulation in sheep. III. The relationship between improvement of nitrogen status and increase in voluntary intake of low-protein roughages by sheep. *Australian Journal of Agricultural Research*, **16**, 463–72.

Elliott, R. C. & Fokema, K. (1961). Herbage consumption studies on beef cattle. I. Intake studies on Afrikander and Mashona cows on veld grazing 1958–59. *Rhodesian Agricultural Journal*, **58**, 49–57.

Elliott, R. C., Fokema, K. & French, C. (1961). Herbage consumption studies on beef cattle. II. Intake studies on Afrikander and Mashona cows on veld grazing 1959–1960. *Rhodesian Agricultural Journal*, **58**, 124–130.

El-Shahat, A., Jones, R., Forbes, J. M. & Boaz, T. G. (1974). The effect of day length on the growth of lambs at two levels of feeding. *Proceedings of the Nutrition Society*, **33**, 83A–84A.

Engels, E. A. N., Van Schalkwyk, A. & Hugo, J. M. (1969). The determination of the nutritive value potential of natural pastures by means of an esophageal fistula and faecal indicator technique. *Agroanimalia*, **1**, 119–22.

Engels, E. A. N., Van Schalkwyk, A., Malan, A. & Baard, M. A. (1971). The chemical composition and *in vitro* digestibility of forage samples selected by esophageal fistulated sheep on natural pasture of the Central Orange Free State. *South African Journal of Animal Sciences*, **1**, 43–4.

Estes, R. (1972). Social organizations of the African Bovidae. In *The behaviour of ungulates and its relation to management*, ed. V. Geist & F. Walther, vol. 1, pp. 166–205. International Union for Conservation of Nature and Natural Resources, Morjes, Switzerland.

Feist, J. D. & McCullough, D. R. (1975). Reproduction in feral horses. *Journal of Reproduction and Fertility, Supplement*, **23**, 13–18.

Fels, H. E., Moir, R. J. & Rossiter, R. C. (1959). Herbage intake of grazing sheep in Southwestern Australia. *Australian Journal of Agricultural Research*, **10**, 237–47.

Ferjain, Y. (1976). Heterosis in weights of Hereford females. MS Thesis, Colorado State University, Fort Collins.

Ferjain, Y., Sutherland, T. M. & Brinks, J. S. (1976). Heterosis in weights of Hereford females. In *Twenty-seventh annual beef cattle improvement report and sale data*, ed. J. S. Brinks, pp. 14–16. Colorado State University Experiment Station, San Juan Basin Research Center General Series 954.

Field, C. R. (1970). A study of the feeding habits of the hippopotamus (*Hippopotamus amphibius* Linn.) in the Queen Elizabeth National Park, Uganda, with some management implication. *Zoologica Africana*, **5**, 71–86.

Field, C. R. & Blankenship, L. H. (1973). Nutrition and reproduction of Grant's and Thomson's gazelles, Coke's hartebeest and giraffe in Kenya. *Journal of Reproduction and Fertility, Supplement*, **19**, 287–301.

Field, C. R. & Ross, I. C. (1976). The savanna ecology of Kidepo Valley National Park – feeding ecology of elephant and giraffe. *East African Wildlife Journal*, **14**, 1–15.

Fitzhugh, H. A., Jr (1976). Analysis of growth curves and strategies for altering their shape. *Journal of Animal Science*, **42**, 1036–51.

Flatt, W. P., Moe, P. W., Munson, A. W. & Cooper, T. (1969). Energy utilization by high producing dairy cows. In *Energy metabolism of farm animals*, ed. K. L. Blaxter, p. 22. Oriel Press, London.

Foot, J. Z. & Romberg, B. (1965). The utilization of roughage by sheep and the red kangaroo, *Macrophus rufus* (Desmarest). *Australian Journal of Agricultural Research*, **16**, 429–35.

Foster, J. B. (1966). The giraffe of Nairobi National Park: home range, sex ratios, the herd and food. *East African Wildlife Journal*, **4**, 139–48.

Fox, P. H. (1971). Steer diets on three pasture types in southeastern Colorado. MS Thesis, Colorado State University, Fort Collins.

Freer, M., Davidson, J. L., Armstrong, J. S. & Donnelly, J. R. (1970). Simulation of grazing systems. In *Proceedings of the XI International Grassland Congress*, pp. 913–17. University of Queensland Press, St Lucia, Queensland.

Frischknecht, N. C., Harris, L. E. & Woodward, H. K. (1953). Cattle gains and vegetal changes as influenced by grazing treatments on crested wheatgrass. *Journal of Range Management*, **6**, 151–8.

Fullerton, S. A. (1975). *Analyses of the mammalian consumer submodel of ELM: 1973. US/IBP Grassland Biome Technical Report 294*. Colorado State University, Fort Collins.

Gaare, E. & Skogland, T. (1975). Wild reindeer food habits and range use at Hardangervidda. In *Fennoscandian tundra ecosystems*, part II; *Animals and systems analysis*, ed. F. E. Wielgolaski, pp. 195–205. Springer-Verlag, Berlin.

Galt, H. D., Theurer, B., Ehrenreick, J. H., Hale, W. H. & Martin, S. C. (1969). Botanical composition of diet of steers grazing a desert grassland range. *Journal of Range Management*, **22**, 14–19.

Garner, R. J. (1967). A mathematical analysis of the transfer of fission products to cows' milk. *Health Physics*, **13**, 205–12.

Gates, D. M. (1970). Animal climates (where animals must live). *Environmental Research*, **3**, 132–44.

Gauthier-Pilters, H. (1974). The behavior and ecology of camels in the Sahara, with special reference to nomadism and water management. In *The behavior of ungulates and its relation to management*, ed. V. Geist & F. Walther, vol. 2, pp. 542–51. International Union for Conservation of Nature and Natural Resources, Morjes, Switzerland.

Geist, V. (1971). *Mountain sheep*. University of Chicago Press, Chicago.

Geist, V. & Walther, F. (ed.) (1974). *The behavior of ungulates and its relation to management*, vols. 1 and 2. International Union for Conservation of Nature and Natural Resources, Morjes, Switzerland.

Giesecke, D. & Van Gylswyk, N. O. (1975). A study of feeding types and certain rumen functions in sex species of South African wild ruminants. *Journal of Agricultural Science, Cambridge*, **85**, 75–83.

Gihad, E. A. (1976). Intake, digestibility and nitrogen utilization of tropical natural grass hay by goats and sheep. *Journal of Animal Science*, **43**, 879–83.

Gilbert, B. J. (1975). *RANGES grassland simulation model. Range Science Department Science Series 17*. Colorado State University, Fort Collins.

Golley, F. B. & Buechner, H. K. (1969). *A practical guide to the study of the productivity of large herbivores*. Blackwell Scientific Publications, Oxford.

Gonzalez-Jimenez, E. & Parra, R. (1973). The capybara, a meat-producing animal for the flooded areas of the tropics. In Annual Report 1972–1973, *Explotación semi-domestica del Chiguire (Hydrochoerus hydrochaeris) como medio para evitar su extinción y aumentar la producción de proteina animal en zonas marginales inundables*, part 1(b), pp. 1–8. Central University of Venezuela, Institute of Animal Production, Faculty of Agronomy, Maracay, Venezuela.

Goodall, D. W. (1969). Simulating the grazing situation. In *Concepts and models of biomathematics: simulation techniques and methods*, ed. F. Heinments, vol. 1, pp. 211–36. Dekker, New York.

Goodall, D. W. (1970*a*). Simulation of grazing systems. In *Modelling and systems analysis in range science. Range Science Department Science Series 5*, ed. D. A. Jameson, pp. 51–74. Colorado State University, Fort Collins.

Goodall, D. W. (1970*b*). Studying the effects of environmental factors on ecosystems. In *Analysis of temperate forest ecosystems*, ed. D. E. Reichle, pp. 19–26. Springer-Verlag, Berlin.

Goodall, D. W. (1971). Extensive grazing systems. In *Systems analysis in agricultural management*, ed. J. B. Dent & J. R. Anderson, pp. 173–87. Wiley, New York.

Goodall, D. W. (1973). Problems of scale and detail in ecological modelling. In *Symposium of the International Institute for Applied Systems Analysis*, p. 12. International Institute for Applied Systems Analysis, Vienna.

Goodwin, G. A. (1975). *Seasonal food habits of mule deer in southeastern Wyoming. US Department of Agriculture Forest Service Research Note RM-287*. Rocky Mountain Forest and Range Experiment Station, Fort Collins, Colorado.

Graham, McC. N. (1962). Energy expenditure of grazing sheep. *Nature, London*, **196**, 289.

Graham, McC. N. (1974). Herbivore efficiency. In *Studies of the Australian arid zone*, vol. 2, *Animal production*, ed. A. D. Wilson, pp. 65–7. CSIRO, Melbourne, Australia.

Graham, McC. N. & Williams, A. J. (1962). The effects of pregnancy on the passage of food through the digestive tract of sheep. *Australian Journal of Agricultural Research*, **13**, 894–900.

Grassé, P. P. (ed.) (1955). *Traité de zoologie*, vol. 17, parts 1 and 2. Masson et Cie, Paris.

Grassland Research Institute (1961). *Research techniques in use at the Grassland Research Institute. Commonwealth Bureau of Pastures and Field Crops Bulletin 45*. Grassland Research Institute, Hurley, England.

Green, L. R., Sharp, L. A., Cook, C. W. & Harris, L. E. (1951). Utilization of winter range forage by sheep. *Journal of Range Management*, **4**, 233–41.
Green, L. R., Wagnon, K. A. & Bentley, J. R. (1958). Diet and grazing habits of steers on foothill range fertilized with sulfur. *Journal of Range Management*, **11**, 221–6.
Greenhalgh, J. F. D. & Reid, G. W. (1971). Relative palatability to sheep of straw, hay and dried grass. *British Journal of Nutrition*, **26**, 107–16.
Grigg, D. B. (1974). *The agricultural systems of the world – an evolutionary approach*. Cambridge University Press, London.
Grimes, R. C., Watkin, B. R. & Gallagher, J. R. (1966). An evaluation of pasture quality with young grazing sheep. II. Chemical composition, botanical composition and *in vitro* digestibility of herbage selected by esophageal fistulated sheep. *Journal of Agricultural Sciences*, **66**, 113–19.
Grimsdell, J. J. R. (1973). Reproduction in the African buffalo, *Syncerus caffer*, in Western Uganda. *Journal of Reproduction and Fertility, Supplement*, **19**, 303–18.
Gross, J. E. (1970). Program ANPOP: a simulation modelling exercise on the Wichita Mountain National Wildlife Refuge. Colorado Cooperative Wildlife Research Unit Program Report. Colorado State University, Fort Collins.
Gross, J. E. & Walters, C. J. (1970). Summary report on initial small herbivore mammal modelling efforts, Pawnee Site Grassland Biome. US/IBP Grassland Biome Technical Report 4. Colorado State University, Fort Collins.
Grovum, W. L. & Hecker, I. F. (1973). Rate of passage of digesta in sheep. II. The effect of level of food intake on digesta retention times and on water and electrolyte absorption in the large intestine. *British Journal of Nutrition*, **30**, 221–30.
Guilbert, H. R. (1929). *Utilization of wild oat hay for fattening yearling steers. California Agricultural Experiment Station Bulletin 481.*
Guilbert, H. R. & Mead, S. W. (1931). The digestibility of burr clover as affected by exposure to sunlight and rain. *Hilgardia*, **6**, 1.
Gwynne, M. D. & Bell, R. H. V. (1968). Selection of vegetation components by grazing ungulates in the Serengeti National Park. *Nature, London*, **220**, 390–3.
Hadjipieris, G., Jones, J. G. W. & Holmes, W. (1965). The affect of age and live weight on the feed intake of grazing wether sheep. *Animal Production*, **7**, 309–17.
Hafez, E. S. E. (ed.) (1962). *The behaviour of domestic animals*. Baillière, Tindall & Cassell, London.
Hafez, E. S. E., Cairns, R. B., Hulet, C. D. & Scott, J. P. (1962). The behavior of sheep and goats. In *The behavior of domestic animals*, ed. E. S. E. Hafez, pp. 296–348. Williams & Wilkins, Baltimore.
Hafez, E. S. E., Schein, M. W. & Ewbank, R. (1962). The behavior of cattle. In *The behavior of domestic animals*, ed. E. S. E. Hafez, pp. 235–95. Williams & Wilkins, Baltimore.
Hale, O. M., Hughes, R. H. & Knox, F. E. (1962). Forage intake by cattle grazing wiregrass range. *Journal of Range Management*, **15**, 6–9.
Halls, L. K. (1954). The approximation of cattle diet through herbage sampling. *Journal of Range Management*, **7**, 269–70.
Halls, L. K., Hale, O. M. & Knox, F. E. (1957). *Seasonal variation in grazing use, nutritive content, and digestibility of wiregrass forage. Georgia Agricultural Experiment Station Technical Bulletin 11.*
Hansen, R. M. (1976). Foods of free-roaming horses in Southern New Mexico. *Journal of Range Management*, **29**, 347.
Hansen, R. M. & Dearden, B. L. (1975). Winter foods of mule deer in Piceance basin, Colorado. *Journal of Range Management*, **28**, 298–300.

Hansen, R. M. & Reid, L. D. (1975). Diet overlap of deer, elk, and cattle in southern Colorado. *Journal of Range Management*, **28**, 43–7.

Hanson, W. C., Whicker, F. W. & Lipscomb, J. F. (1975). Lichen forage ingestion rates of free-roaming caribou estimated with fall out cesium-137. In *Proceedings of the I International Reindeer/Caribou Symposium*, pp. 71–9. University of Alaska.

Hardison, W. A., Reid, J. T., Martin, C. M. & Woolfolk, P. G. (1954). Degree of herbage selection by grazing cattle. *Journal of Dairy Science*, **37**, 89–102.

Harlow, R. F., Crawford, H. S. & Urbston, D. F. (1974). Rumen contents of white-tailed deer: comparing local with regional samples. *Proceedings of the Annual Conference of the Southeastern Association of Game and Fish Commissioners*, **28**, 562–7.

Harper, J. A., Harn, J. H., Bentley, W. W. & Yocum, C. F. (1967). The status and ecology of the Roosevelt elk in California. *Wildlife Monographs*, **16**, 1–49.

Harrington, G. N. & Pratchett, D. (1974). Stocking rate trials in Ankole, Uganda. II. Botanical analysis and esophageal fistula sampling of pastures grazed at different stocking rates. *Journal of Agricultural Science, Cambridge*, **82**, 507–16.

Harris, L. D. & Fowler, N. K. (1975). Ecosystem analysis and simulation of the Mkomazi Reserve, Tanzania. *East African Wildlife Journal*, **13**, 325–46.

Harris, L. D. & Francis, R. C. (1972). *AFCONS: a dynamic simulation model of an interactive herbivore community. US/IBP Grassland Biome Technical Report 158.* Colorado State University, Fort Collins.

Harris, L. E., Kercher, C. J., Lofgreen, G. P., Raleigh, R. J. & Bohman, V. R. (1967). *Techniques of research in range livestock nutrition. Utah State University Agricultural Experiment Station Bulletin 271.*

Harry, G. B. (1957). Winter food habits of moose in Jackson Hole, Wyoming. *Journal of Wildlife Management*, **21**, 53–7.

Hart, G. H., Guilbert, H. R. & Goss, H. (1932). *Seasonal changes in the chemical composition of range forage and their relation to nutrition of animals. California Agricultural Experiment Station Bulletin 543.*

Heady, H. F. (1960). *Range management in East Africa.* Nairobi Government Printer.

Heady, H. F. (1964). Palatability of herbage and animal preference. *Journal of Range Management*, **17**, 76–82.

Heady, H. F. (1975). *Rangeland management.* McGraw-Hill, New York.

Heindrichs, H. (1975). Observations on a population of Bohor reedbuck, *Redunca redunca* (Pallas 1767). *Zeitschrift für Tierpsychologie*, **38**, 44–54.

Herbel, C. H. & Nelson, A. B. (1966). Species preferences of Hereford and Santa Gertrudis cattle on southern New Mexico range. *Journal of Range Management*, **19**, 177–81.

Hibbs, D. L. (1966). Food habits of the mountain goat in Colorado. *Journal of Mammalogy*, **48**, 242–8.

Hirst, S. M. (1975). Ungulate–habitat relationships in a south African woodland/savanna ecosystem. *Wildlife Monographs*, **44**, 1–60.

Hjeljord, O. G. (1971). Feeding ecology and habitat preference of the mountain goat in Alaska. MS Thesis, University of Alaska, Fairbanks.

Hjeljord, O. (1973). Mountain goat forage and habitat preference in Alaska. *Journal of Wildlife Management*, **37**, 353–62.

Hlavachick, B. D. (1966). Some preferred foods of Kansas antelope. *Proceedings of Antelope States Workshop*, **2**, 60–5.

Hobson, P. N., Mann, S. O., Summers, R. & Starnes, B. W. (1976). Rumen function in red deer, hill sheep, and reindeer in the Scottish highlands. *Proceedings of the Royal Society of Edinburgh*, **75**, 181–98.

Hoefs, M. (1974). Food selection by Dall's sheep (*Ovis dalli dalli* Nelson). In *The behavior of ungulates and its relation to management*, ed. V. Geist & F. Walther, vol. 2, pp. 759–86. International Union for Conservation of Nature and Natural Resources, Morjes, Switzerland.

Hoehne, O. E., Clanton, D. C. & Streeter, C. L. (1968). Chemical composition and *in vitro* digestibility of forbs consumed by cattle grazing native range. *Journal of Range Management*, **21**, 5–7.

Hofmann, R. R. & Musangi, R. S. (1973). Comparative digestibility coefficients of domestic and game ruminants from marginal land in East Africa. *Bulletin of epizootic Diseases of Africa*, **21**, 385–8.

Holmes, W., Jones, J. G. W. & Drake-Brockman, R. M. (1961). The feed intake of grazing cattle. II. The influence of size of animal on the intake. *Animal Production*, **3**, 251–60.

Honess, R. F. & Frost, N. M. (1942). *A Wyoming bighorn sheep study. Wyoming Game and Fish Commissioners Bulletin 1.*

Hoover, J. P. (1971). Food habits of pronghorn antelope on Pawnee National Grasslands, 1970. MS Thesis, Colorado State University, Fort Collins.

Hoover, R. L. (1966). Antelope food habits and range relationships in Colorado. *Antelope States Workshop Proceedings*, **2**, 75–85.

Hoover, R. L., Till, C. E. & Ogilvie, S. (1959). *The antelope of Colorado – a research and management study. Colorado Game and Fish Department Technical Bulletin 4.*

Horrocks, D. & Phillips, G. D. (1961). Factors affecting the water and food intake of European and Zebu type cattle. *Journal of Agricultural Science*, **56**, 379–81.

Horsfield, S., Infield, J. M. & Annison, E. F. (1974). Compartmental analysis and model building in the study of glucose kinetics in the lactating cow. *Proceedings of the Nutrition Society*, **33**, 9–15.

Hosley, N. W. (1949). *The moose and its ecology. US Department of Interior Wildlife Leaflet 312.*

Houston, D. B. (1968). *The Shiras moose in Jackson Hole, Wyoming. Grand Teton National Historical Association Technical Bulletin 1.* Grand Teton National Park, Wyoming.

Hubbard, R. E. & Hansen, R. M. (1976). Diets of wild horses, cattle and mule deer in the Piceance Basin, Colorado. *Journal of Range Management*, **29**, 389–92.

Hudson, R. J. (1975). *A computer simulation study of the potential productivity of wildlife in the boreal mixed-wood zone of Alberta. 54th Annual Feeders Day Report. University of Alberta Special Issue, Agricultural Bulletin 9.*

Hughes, J. G. (1975). A study of the grazing preference of sheep on developed and undeveloped grassland at a high-country site. MS Thesis, University of Canterbury, New Zealand.

Hungate, R. E., Phillips, G. D., McGregor, A., Hungate, D. P. & Buechner, H. K. (1959). Microbial fermentation in certain mammals. *Science*, **130**, 1192–4.

Hvidberg-Hansen, H. & de Vos, A. (1971). Reproduction, population and herd structure of two Thomson's gazelle (*Gazella thomsonii* Günther) populations. *Mammalia*, **35**, 1–16.

Hyder, D. M., Bement, R. P., Norris, J. J. & Morris, M. J. (1966). Evaluating herbage species by grazing cattle. I. Food intake. In *Proceedings of the X International Grassland Congress, Helsinki*, pp. 970–4.

Hyder, D. N., Knox, K. L. & Streeter, C. L. (1971). *Metabolic components of cattle under light and heavy rates of stocking in 1970. US/IBP Grassland Biome Technical Report* 128. Colorado State University, Fort Collins.

Innis, G. S. (1975). Role of total systems models in the grassland biome study. In *Systems analysis and simulation in ecology*, ed. B. C. Patten, vol. 3, pp. 13–47. Academic Press, New York & London.

Innis, G. S. (ed.) (1978). *Grassland simulation model – US/IBP Grassland Biome Study. Springer-Verlag Ecological Studies 26.* Springer-Verlag, Berlin.

Jacobsen, E. & Skjenneberg, S. (1975). Some results from feeding experiments with reindeer. In *Proceedings of the I International Reindeer/Caribou Symposium*, pp. 95–107. University of Alaska.

Jeffries, N. W. & Rice, R. W. (1969). Nutritive value of clipped and grazed range forage samples. *Journal of Range Management*, **22**, 192–4.

Jeffries, N. W. & Rice, R. (1969). Forage intake by yearling steers on shortgrass rangelands. *Proceedings of the Western Section of the American Society of Animal Science*, **20**, 343–8.

Jensen, C. H., Smith, A. D. & Scotter, G. W. (1972). Guidelines for grazing sheep on rangelands used by big game in winter. *Journal of Range Management*, **25**, 346–52.

Jensen, P. V. (1968). Food selection of the Danish red deer (*Cervus elaphus* L.) as determined by examination of the rumen content. *Danish Review of Game Biology*, **5**, 3–44.

Jones, J. G. W. (1970). Lamb production. In *The use of models in agricultural and biological research*, ed. J. G. W. Jones, pp. 42–9. Grassland Research Institute, Hurley, England.

Joshi, D. C. (1966). Studies on high altitude Himalayan pastures (Bugiyals) of Uttar Pradesh. *Indian Veterinary Journal*, **43**, 1019–27.

Joshi, D. C., Ludri, R. S. & Kidwai, W. A. (1971). Studies on sheep nutrition. III. Comparative grazing efficiency of different breeds of sheep and the pasture intake as influenced by grazing hours. *Indian Veterinary Journal*, **48**, 282–4.

Joyce, J. P. (1968). Heat loss in sheep and cattle. *Tussock Grasslands and Mountain Lands Institute Review*, **4**, 2–10.

Julander, O. (1955). Deer and cattle range relations in Utah. *Forest Science*, **1**, 130–9.

Kay, R. M. B. (1965). The influence of saliva on digestion in ruminants. *World Review of Nutrition and Dietetics*, **6**, 292–325.

Kay, R. M. B. (1970). Meat production from wild herbivores. *Proceedings of the Nutrition Society*, **29**, 271–8.

Kay, R. M. B. & Hobson, P. N. (1963). Reviews of the progress of dairy science. *Journal of Dairy Research*, **30**, 261–313.

Kennedy, P. M., Christopherson, R. J. & Milligan, L. P. (1976). The effect of cold exposure of sheep on digestion, rumen turnover time and efficiency of microbial synthesis. *British Journal of Nutrition*, **36**, 231–42.

King, J. M., Kingaby, G. P., Colvin, J. G. & Heath, B. R. (1975). Seasonal variation in water turnover by oryx and eland on the Galana Game Ranch Research Project. *East African Wildlife Journal*, **13**, 287–96.

Kittams, W. H. (1953). Reproduction in the Yellowstone elk. *Journal of Wildlife Management*, **17**, 177–84.

Kleiber, M. (1961). *Metabolic rate and food utilization as a function of body size. Bulletin of the Missouri Agricultural Experiment Station 767.*

Kleiber, M. (1961). *The fire of life – an introduction to animal energetics.* Wiley, New York.

Klein, D. R. (1953). A reconnaissance study of the mountain goat in Alaska. MS Thesis, University of Alaska, Fairbanks.

Klein, D. R. (1969). Food selection by North American deer and their response to overutilization of preferred plant species. In *Animal populations in relation to their food resources*, ed. A. Watson, pp. 25–46. Blackwell Scientific Publications, Oxford.

Knapka, J. J., Barth, K. M., Brown, D. G. & Cragle, R. G. (1967). Evaluation of polyethylene, chromic oxide and cesium-144 as digestibility indicators in burros. *Journal of Nutrition*, **92**, 79–85.

Knowlton, F. F. (1960). Food habits, movements, and populations of moose in the Gravelly Mountains, Montana. *Journal of Wildlife Management*, **24**, 162–70.

Koford, C. B. (1957). The vicuña and the puna. *Ecological Monographs*, **27**, 153–219.

Kothman, M. M. (1966). Nutrient content of forage ingested in the morning compared to evening. *Journal of Range Management*, **19**, 95–6.

Kothman, M. M., Mathis, G. W. & Waldrip, W. J. (1971). Cow–calf response to stocking rates and grazing systems on native range. *Journal of Range Management*, **24**, 100–5.

Lamprey, H. F. (1963). Ecological separation of the large mammal species in the Tarangire Game Reserve, Tanganyika. *East African Wildlife Journal*, **1**, 63–92.

Lang, E. M. (1958). *Elk of New Mexico. New Mexico Department of Game and Fish Bulletin 8.*

Lang, E. M. (1957). *Deer of New Mexico. New Mexico Department of Game and Fish Bulletin 5.*

Langlands, J. P. (1965). Diurnal variation in the diets selected by free-grazing sheep. *Nature, London*, **207**, 666–7.

Lassen, R. W., Ferrel, C. M. & Leach, H. (1952). Food habits productivity, and condition of the Doyle mule deer herd. *California Fish and Game*, **38**, 211–24.

Lassiter, R. R. & Hayne, D. W. (1971). A finite difference model for simulation of dynamic processes in ecosystems. In *Systems analysis and simulation in ecology*, ed. B. C. Patten, pp. 368–440. Academic Press, New York & London.

Laws, R. M. (1970). Biology of African elephants. *Scientific Progress, Oxford*, **58**, 251–62.

Laws, R. M., Parker, I. S. C. & Johnstone, R. C. B. (1975). *Elephants and their habits – the ecology of elephants in North Bunyro, Uganda*. Clarendon Press, Oxford.

Laycock, W. A., Buchanan, H. & Krueger, W. C. (1972). Three methods of determining diet, utilization, and trampling damage on sheep ranges. *Journal of Range Management*, **25**, 352–6.

Ledger, H. P., Sachs, R. & Smith, N. S. (1967). Wildlife and food production. *World Review of Animal Production*, **3**, 13–36.

Leigh, J. H. (1974). Diet selection and effects of grazing on the composition and structure of arid and semiarid vegetation. In *Studies of the Australian arid zone*, vol. 2, *Animal production*, ed. A. D. Wilson, pp. 102–26. CSIRO, Melbourne, Australia.

Leigh, J. H. & Mulham, W. E. (1966). A bladder saltbush (*Atriplex vesicaria*)–cottonbush (*Kochia aphylla*) community. *Australian Journal of Experimental Agriculture and Animal Husbandry*, **6**, 460–7.

Leigh, J. H., Wilson, A. P. & Mulham, W. E. (1968). A study of merino sheep grazing a cottonbush (*Kochia aphylla*)–grassland (*Stipa variabilis–Danthonia caespitosa*) community on the Riverine Plain. *Australian Journal of Agricultural Research*, **19**, 947–61.

LeResche, R. E. & Davis, J. L. (1973). Importance of non-browse food to moose on the Kenai Peninsula, Alaska. *Journal of Wildlife Management*, **37**, 279–87.

Lesperance, A. L., Clanton, D. C., Nelson, A. B. & Theurer, C. B. (1974). *Factors affecting the apparent chemical composition of fistula samples. University of Nevada Agricultural Experiment Station Bulletin T18.*

Lesperance, A. L., Tueller, P. T. & Bohman, V. R. (1970). Symposium on pasture methods for maximum production in beef cattle; competitive use of the range forage resource. *Journal of Animal Science*, **30**, 115–21.

Leuthold, W. (1971). Studies on the food habits of lesser kudu in Tsavo National Park, Kenya. *East African Wildlife Journal*, **9**, 35–45.

Leuthold, W. & Leuthold, B. M. (1976). Density and biomass of ungulates in Tsavo East National Park, Kenya. *East African Wildlife Journal*, **14**, 49–58.

Lewis, J. K., Van Dyne, G. M., Albee, L. R. & Whetzal, F. W. (1956). *Intensity of grazing – its effects on livestock and forage production. South Dakota Agricultural Experiment Station Bulletin 459.*

Ling, Kh. I. (1973). Productivity dynamics of moose in Estonia. *Ekologiya*, **4**, 81–8.

Longhurst, D. M. & Heady, H. F. (eds.) (1968). *Report of a symposium on East African range problems.* (Symposium held at Villa Serbelloni Lake Como, Italy 24–28 June 1968.) The Rockefeller Foundation, New York.

Longhurst, W. M., Leopold, A. S. & Dasmann, R. F. (1952). *A survey of California deer herds, their ranges and management problems. California Department of Fish and Game Bulletin 6.*

Longhurst, W. M., Oh, H. W., Jones, M. B. & Kepner, R. E. (1968). A basis for the palatability for deer forage plants. *Transactions of the North American Wildlife and Natural Resource Conference*, **33**, 181–92.

Louw, G. N. (1970). Physiological adaptation as a criterion in planning production from wild ungulates. *Proceedings of the South African Society of Animal Production*, **9**, 53–6.

Lovaas, A. L. (1958). Mule deer food habits and range use, Little Belt Mountains, Montana. *Journal of Wildlife Management*, **22**, 275–83.

Lydekker, R. (1913–16). *Catalogue of the ungulate mammals in the British Museum (Natural History)*, **5** vols., vols. 3–5 with G. Balin. London. (Reprinted by Academic Press, 1966.)

Lynch, J. J. (1974). Merino sheep: some factors affecting their distribution in very large paddocks. In *The behavior of ungulates and its relation to management*, ed. V. Geist & F. Walther, vol. 2, pp. 697–707. International Union for Conservation of Nature and Natural Resources, Morjes, Switzerland.

McCaffrey, K. R., Tranetzki, J. & Piechura, J. (1974). Summer foods of deer in northern Wisconsin. *Journal of Wildlife Management*, **38**, 215–19.

McClung, G. E., Alben, R. C. & Schuster, J. L. (1976). Summer diets of steers on a deep hardland range site of the Texas high plains. *Journal of Range Management*, **29**, 387–9.

MacDonald, M. A. & Bell, J. M. (1958*a*). Effects of low fluctuating temperatures on farm animals. II. Influence of ambient air temperature on water intake of lactating Holstein–Friesian cows. *Canadian Journal of Animal Science*, **38**, 23–32.

MacDonald, M. A. & Bell, J. M. (1958*b*). Effects of low fluctuating temperatures on farm animals. III. Influences of ambient air temperatures on feed intake of lactating Holstein–Friesian cows. *Canadian Journal of Animal Science*, **38**, 148–59.

Macfarlane, W. V. & Howard, B. (1974). Ruminant water metabolism in arid areas.

In *Studies of the Australian arid zone*, vol. 2, *Animal production*, ed. A. D. Wilson, pp. 7–21. CSIRO, Melbourne, Australia.

McKinney, G. T. (1972). Simulation of winter grazing on temperature pasture. *Proceedings of the Australian Society of Animal Production*, **9**, 31.

McKnight, T. (1957). Feral burros in the American Southwest. *Journal of Geography*, **56**, 315–22.

McLean, A. (1972). Beef production on lodgepole pine–pinegrass range in the caribou region of British Columbia. *Journal of Range Management*, **25**, 10–11.

McMahan, C. A. (1964). Comparative food habits of deer and 3 classes of livestock. *Journal of Wildlife Management*, **28**, 798–808.

McMillan, J. F. (1953). Some feeding habits of moose in Yellowstone National Park. *Ecology*, **34**, 102–10.

Maher, C. (1945). The goat: friend or foe? *East African Agricultural Journal*, **11**, 115–21.

Malacheck, J. C. (1966). *Cattle diets on native and seeded ranges in the ponderosa pine zone of Colorado. US Department of Agriculture Forest Service Research Note RM-77.*

Malechek, J. C. & Leinweber, C. L. (1972). Forage selectivity by goats on lightly and heavily grazed ranges. *Journal of Range Management*, **25**, 105–11.

Mansell, D. W. & Cringan, A. T. (1968). A further instance of fetal atrophy in white-tailed deer. *Canadian Journal of Zoology*, **46**, 33–4.

Marshall, M. B. & Radloff, H. (1976). Summer range grass utilization by beef cattle. *Proceedings of the Western Section of the American Society of Animal Science*, **27**, 171–5.

Marten, G. C. & Donker, J. D. (1964). Selective grazing induced by animal excreta. I. Evidence of occurrence and superficial remedy. *Journal of Dairy Science*, **67**, 773–6.

Martin, A. K., Milne, J. A. & Moberley, P. (1975). Urinary quinol and orcinol outputs as indices of voluntary intake of heather (*Calluna vulgaris* L. (Hull)) by sheep. *Proceedings of the Nutrition Society*, **34**, 70A–71A.

Mason, E. (1952). Food habits and measurements of Hart Mountain antelope. *Journal of Wildlife Management*, **96**, 387–9.

May, P. F., Till, A. R. & Cumming, M. J. (1972). Systems analysis of sulphur kinetics in pastures grazed by sheep. *Journal of Applied Ecology*, **9**, 25–49.

Maynard, L. A. & Loosli, J. K. (1969). *Animal nutrition*, 6th edn. McGraw-Hill, New York.

Mazanov, A. & Nolan, J. V. (1976). Simulation of the dynamics of nitrogen metabolism in sheep. *British Journal of Nutrition*, **35**, 149–74.

Mendelssohn, H. (1974). The development of the populations of gazelles in Israel and their behavioral adaptations. In *The behavior of ungulates and its relation to management*, ed. V. Geist & F. Walther, vol. 2, pp. 722–43. International Union for Conservation of Nature and Natural Resources, Morjes, Switzerland.

Miller, D. R. (1971). Seasonal changes in the feeding behaviour of barren-ground caribou on the taiga winter range. In *The behaviour of ungulates and its relation to management*, vol. 2, ed. V. Geist & F. Walther, pp. 744–55. International Union for the Conservation of Nature and Natural Resources, Morjes, Switzerland.

Miller, R. F. & Krueger, W. C. (1976). Cattle use on summer foothill rangelands in northeastern Oregon. *Journal of Range Management*, **29**, 367–71.

Milner, C. (1974). Appendix: Developing a mathematical model for use in synthesis. In *Island survivors: the ecology of the Soay sheep of St Kilda*, ed. P. A. Jewell,

C. Milner & J. M. Boyd, pp. 374–8. The Athlone Press (University of London), London.

Milner, C. & Gwynne, D. (1974). The Soay sheep and their food supply. In *Island survivors: the ecology of the Soay sheep of St Kilda*, ed. P. A. Jewell, C. Milner & J. M. Boyd, pp. 273–325. The Athlone Press (University of London), London.

Minson, D. J. & Ternouth, J. H. (1971). The expected and observed changes in the intake of three hays by sheep after shearing. *British Journal of Nutrition*, **20**, 31–9.

Mitchell, G. J. & Smoliak, S. (1971). Pronghorn antelope range characteristics and food habits in Alberta. *Journal of Wildlife Management*, **35**, 238–50.

Moir, R. J. (1968). Ruminant digestion and evolution. In *Handbook of Physiology*, Section 6, *Alimentary canal*. American Physiological Society, Washington, DC.

Morley, F. M. W. & Graham, G. Y. (1971). Fodder conservation for drought. In *Systems analysis in agricultural management*, ed. J. B. Dent & J. R. Anderson, pp. 212–36. Wiley, Australia.

Morris, J. G., Baldwin, R. L., Maeng, W. J. & Maeda, B. T. (1974). Basic characteristics of a computer simulated model of nitrogen utilization in the grazing ruminant. In *Tracer studies on non-protein-nitrogen for ruminants*. International Atomic Energy Agency, Vienna.

Morris, M. S. & Schwartz, J. E. (1957). Mule deer and elk food habits on the National Bison range. *Journal of Wildlife Management*, **21**, 189–93.

Moser, C. A. (1962). *The bighorn sheep of Colorado. Colorada Game and Fish Department Technical Publication 10*.

Nellis, C. L. & Ross, R. L. (1969). Changes in mule deer food habits associated with herd reduction. *Journal of Wildlife Management*, **33**, 191–5.

Nelson, L. J. (1972). A model of competing range ruminants. MS Thesis, University of California, Davis.

Newman, D. M. R. (1973). The camel – its potential as a provider of protein in arid Australia. In *III World Conference of Animal Production, Theme I, Animal production in relation to conservation and recreation*, pp. 95–101. CSIRO, Melbourne, Australia.

Newsome, A. E. (1971). Competition between wildlife and domestic livestock. *Australian Veterinary Journal*, **47**, 577–86.

Newsome, A. E. (1975). An ecological comparison of the two arid zone kangaroos of Australia and their anomalous prosperity since the introduction of ruminant stock to their environment. *Quarterly Review of Biology*, **50**, 389–424.

Newton, J. E. & Edelsten, P. R. (1976). A model of the effect of nutrition on litter size and weight in the pregnant ewe. *Agricultural Systems*, **1**, 185–200.

Nge'the, J. C. & Box, T. W. (1976). Botanical composition of eland and goat diets on an acacia–grassland community in Kenya. *Journal of Range Management*, **29**, 290–3.

Nichols, L., Jr (1957). Forage utilization by elk and domestic sheep in the White River National Forest. MS Thesis, Colorado State University, Fort Collins.

Nicholson, I. A., Patterson, I. S. & Currie, A. (1970). A study of vegetational dynamics: selection by sheep and cattle in *Nardus* pasture. In *Animal populations in relation to their food resources*, ed. A. Watson, pp. 129–43. Blackwell Scientific Publications, Oxford.

Nixon, C. M., McClain, M. W. & Russell, K. R. (1970). Deer food habits and range characteristics in Ohio. *Journal of Wildlife Management*, **34**, 870–86.

Noble, I. R. (1975). Computer simulations of sheep grazing in the arid zone. PhD Thesis, University of Adelaide, Australia.

Nolan, J. V. & Leng, R. A. (1972). Dynamic aspects of ammonia and urea metabolism in sheep. *British Journal of Nutrition*, **27**, 177–94.

Nolan, J. V., Norton, B. W. & Leng, R. A. (1976). Further studies of the dynamics of nitrogen metabolism in sheep. *British Journal of Nutrition*, **35**, 127–47.

O'Gara, B. W. & Green, K. R. (1970). Food habits in relation to physical condition in two populations of pronghorn. *Proceedings of the Antelope States Workshop*, **4**, 131–9.

Oldemeyer, J. L. (1966). Winter ecology of bighorn sheep in Yellowstone National Park. MS Thesis, Colorado State University, Fort Collins.

Olsen, F. W. & Hansen, R. M. (1977). Food relations of wild free-roaming horses to livestock and big game in Red Desert, Wyoming. *Journal of Range Management*, **30**, 17–20.

Palmer, T. S. (1968). Index generum mammalium (a list of the genera and families of mammals). *Historiae Naturalis Classica*, vol. 58. Springer-Verlag, Berlin. (Originally published in 1904 as *North American Fauna*, vol. 23.)

Parton, W. J. & Smith, F. M. (1974*a*). Exploring some possible effects of potential SST-induced weather modification in a shortgrass prairie ecosystem. In *VI Conference on Aerospace and Aeronautical Meteorology*, pp. 255–8. American Meteorological Society, Boston, Massachusetts.

Parton, W. J. & Smith, F. M. (1974*b*). Simulating the effects of growing season rainfall enhancement and hail suppression on the production, consumption, and decomposition functions of a native shortgrass prairie ecosystem. In *IV Conference on Weather Modification*, pp. 523–8. American Meteorological Society, Boston, Massachusetts.

Patten, B. C. (1972). A simulation of the shortgrass prairie ecosystem. *Simulation*, **19**, 177–86.

Paur, L. F. (1974). A description of a generalized mammalian consumer simulation model. US/IBP Grassland Biome Study (unpublished report). Colorado State University, Fort Collins.

Payne, S. (1974). Terrestrial model: animal processes (Version III). Research Memo 74–52. In *Reports of 1973 progress*, vol. 1, *Central Office Modelling*. US/IBP Desert Biome, Utah State University, Logan.

Peden, D. G. & Rice, R. W. (1972). *A dynamic programming approach to the management of ungulate populations. US/IBP Grassland Biome Preprint 25*. Colorado State University, Fort Collins.

Peterson, R. L. (1955). *North American moose*. University of Toronto Press, Toronto, Canada.

Petrides, G. A. & Swank, W. G. (1966). Estimating the productivity and energy relations of an African elephant population. *Proceedings of the IX International Grassland Congress*, pp. 832–42. Edicoes Limitada, São Paulo, Brazil.

Phillips, G. D. (1960). The relationship between water and food intakes of European and Zebu type steers. *Journal of Agricultural Science*, **54**, 231–4.

Pickford, G. D. & Reid, E. H. (1943). Competition of elk and domestic livestock for summer range forage. *Journal of Wildlife Management*, **7**, 328–32.

Pieper, R., Cook, C. W. & Harris, L. E. (1959). Effect of intensity of grazing upon nutritive content of diet. *Journal of Animal Science*, **18**, 1031–7.

Pimlott, D. H. (1953). Newfoundland moose. *Transactions of the North American Wildlife Conference*, **18**, 563–81.

Price, D. A., Lindahl, I. L., Trederiksen, K. R., Reynolds, P. J. & Cain, C. M., Jr (1964). Nutritive quality of sheep's diet on tall forb range. *Proceedings of the Western Section of the American Society of Animal Science*, **15**, lxi.

Prior, R. (1968). *The roe deer of Cranborne Chase*. Oxford University Press, London.

Rao, M. R., Harbers, L. H. & Smith, E. F. (1974). Estimating consumption and digestibility in steers fed native Flint hills hay – methods compared. *Transactions of the Kansas Academy of Science*, **77**, 36–41.

Rattray, P. V., Garrett, W. N., East, N. E. & Hinman, N. (1973*a*). Net energy requirements of ewe lambs for maintenance, gain and pregnancy and net energy value of feedstuffs for lambs. *Journal of Animal Science*, **37**, 853–7.

Rattray, P. V., Garrett, W. N., East, N. E. & Hinman, N. (1974). Efficiency of utilization of metabolizable energy during pregnancy and the energy requirements for pregnancy in sheep. *Journal of Animal Science*, **38**, 383–93.

Rattray, P. V., Garrett, W. N., Meyer, H. H., Bradford, G. E., Hinman, N. & East, N. E. (1973*b*). Net energy requirements for growth of lambs aged 3 to 5 months. *Journal of Animal Sciences*, **37**, 1386–9.

Raymond, W. F., Harris, C. E. & Kemp, C. D. (1954). Studies in the digestibility of herbage. V. The variation, with age, of the ability of sheep to digest herbage with observations on the effect of season on digestibility. *Journal of the British Grassland Society*, **9**, 209–20.

Raymond, W. F., Harris, C. E. & Kemp, C. D. (1955). Studies in the digestibility of herbage. VI. The effect of level of herbage intake on the digestibility of herbage by sheep. *Journal of the British Grassland Society*, **10**, 19–26.

Raymond, W. F., Minson, D. J. & Harris, C. E. (1959). Studies in the digestibility of herbage. XI. Further evidence on the effect of level of intake on the digestive efficiency of sheep. *Journal of the British Grassland Society*, **14**, 75–7.

Reardon, T. F. & Lambourne, L. J. (1965). Early nutrition and lifetime reproductive performance of ewes. *Proceedings of the Australian Society of Animal Production*, **6**, 106–8.

Redetzke, K. A. & Van Dyne, G. M. (1976). A matrix model of a rangeland grazing system. *Journal of Range Management*, **29**, 425–30.

Reichl, J. R. & Baldwin, R. L. (1976). A rumen linear programming model for evaluation of concepts of rumen microbial function. *Journal of Dairy Science*, **59**, 438–54.

Reid, J. T. & Robb, J. (1971). Relationship of body composition to energy intake and energetic efficiency. *Journal of Dairy Science*, **54**, 553–64.

Reppert, J. N. (1957). Preference and consumption of native forage species by cattle at the Eastern Colorado Range Station. MS Thesis, Colorado State University, Fort Collins.

Rice, R. W., Cundy, D. R. & Weyerts, P. R. (1971). Botanical and chemical composition of esophageal and rumen fistula samples of sheep. *Journal of Range Management*, **24**, 121–4.

Rice, R. W., Dean, R. E. & Ellis, J. E. (1974). Bison, cattle, and sheep dietary quality and food intake. *Proceedings of the Western Section of the American Society of Animal Science*, **25**, 194–7.

Rice, R. W., Morris, J. G., Maeda, B. T. & Brown, R. L. (1974). Simulation of animal functions in models of production systems: ruminants on the range. *Federation Proceedings*, **33**, 188–95.

Rice, R. W. & Vavra, M. (1969). *Botanical species of plants eaten and intake of steers grazing light, medium, and heavy use shortgrass range. US/IBP Grassland Biome Technical Report 12*. Colorado State University, Fort Collins.

Ride, W. D. L. (1959). Mastication and taxonomy in the macropodine skull. In *Function and taxonomic importance*, ed. A. J. Cain, vol. 3, pp. 33–59. Publications of the Systematic Association.

Ridley, J. R., Lesperance, A. L., Jensen, E. H. & Bohman, V. R. (1963). Pasture evaluation with fistulated and intact cattle. *Proceedings of the Western Section of the American Society of Animal Science*, **14**, xxxv.

Rittenhouse, L. R., Clanton, D. C. & Streeter, C. L. (1970). Intake and digestibility of winter-range forage by cattle with and without supplements. *Journal of Animal Science*, **31**, 1215–21.

Robards, G. E., Leigh, J. H. & Mulham, W. E. (1967). Selection of diet by sheep grazing semi-arid pastures on the Riverine plain. IV. A grassland (*Danthonia caespitosa*) community. *Australian Journal of Agriculture and Animal Husbandry*, **7**, 426–33.

Robinette, W. L., Gashwiles, J. S., Low, J. B. & Jones, D. A. (1957). Differential mortality by sex and age among mule deer. *Journal of Wildlife Management*, **21**, 1–16.

Rosiere, R. E., Beck, R. F. & Wallace, J. D. (1975). Cattle diet on semidesert grassland: botanical composition. *Journal of Range Management*, **28**, 89–93.

Rosiere, R. E., Wallace, J. D. & Beck, R. F. (1975). Cattle diets on semidesert grassland: nutritive content. *Journal of Range Management*, **28**, 94–6.

Ross, I. C., Field, C. R. & Harrington, G. N. (1976). The savanna ecology of Kidepo Valley National Park, Uganda. III. Animal populations and park management recommendations. *East African Wildlife Journal*, **14**, 35–48.

Sachs, R. (1967). Liveweights and body measurements of Serengeti game animals. *East African Wildlife Journal*, **5**, 24–36.

Saiz, R. B. (1975). Ecology and behavior of the gemsbok at White Sands Missile Range, New Mexico. MS Thesis, Colorado State University, Fort Collins.

San-Martin, M., Copaira, H., Zuniga, J., Rodriguez, R., Bustinza, G. & Acosta, L. (1968). Aspects of reproduction in the alpaca. *Journal of Reproduction and Fertility*, **16**, 395–9.

Saunders, J. K., Jr (1955). Food habits and range use of the Rocky Mountain goat in the Crazy Mountains, Montana. *Journal of Wildlife Management*, **19**, 429–37.

Saxena, J. S. & Maheshwari, M. L. (1971). Studies on Jamunapari goats. II. Studies on comparative intake of nutrients by browsing goats. *Indian Veterinary Journal*, **48**, 173–5.

Scales, G. H. (1972). Nutritive value and consumption of sandhill range forage by grazing cattle. PhD Thesis, Colorado State University, Fort Collins.

Scales, G. H., Streeter, C. L. & Denham, A. H. (1971). Nutritive value and consumption of range forage. *Proceedings of the Western Section of the American Society of Animal Science*, **22**, 83–8.

Scheffer, V. B. (1951). The rise and fall of a reindeer herd. *Scientific Monthly*, **73**, 356–62.

Schinckel, P. G. (1960). Variation in feed intake as a cause of variation in wool production in grazing sheep. *Australian Journal of Agricultural Research*, **11**, 585–94.

Schmidt-Nielsen, K. (1959). The physiology of the camel. *Scientific American*, **201**, 140–51.

Schultz, W. & McDowell, L. (1943). Absaroka–Gallatin moose study. *Pittman-Robertson Quarterly*, **3**, 71–7.

Schwartz, C. C., Mathews, S. & Nagy, J. G. (1976). Foraging behavior of pronghorn and cattle on a shortgrass prairie in Colorado. *Proceedings of the VII Biennial Pronghorn Antelope Workshop*, pp. 78–9. Idaho Dept of Fish and Game, Twin Falls, Idaho.

Schwartz, C. C. & Nagy, J. G. (1976). Pronghorn diets relative to forage availability in northeastern Colorado. *Journal of Wildlife Management*, **40**, 469–78.

Scotter, G. W. (1967). The winter diet of barren ground caribou in northern Canada. *Canadian Field Naturalist*, **81**, 33–9.

Seligman, N. G., Tadmor, N. H., Noy-Meir, I. & Dovrat, A. (1971). An exercise in simulation of a semi-arid Mediterranean grassland. In *Bulletin des Recherches Agronomiques de Gembloux. Semaine d'étude des problèmes mediterranéens*, pp. 138–43. Faculté des Sciences Agronomiques de l'Etat, Gembloux, Belgium.

Severson, K. E. & May, M. (1967). Food preferences of antelope and domestic sheep in Wyoming Red Desert. *Journal of Range Management*, **20**, 21–5.

Sharma, D. C. (1968). Intake and digestion of nutrients by the Bovine under climatic stress. *Journal of Nutrition*, **94**, 317–25.

Short, H. L. (1971). Forage digestibility and diet of deer on southern upland range. *Journal of Wildlife Management*, **35**, 698–706.

Skinner, W. R. & Telfer, E. S. (1974). Spring, summer, and fall foods of deer in New Brunswick. *Journal of Wildlife Management*, **38**, 210–14.

Slade, L. M. & Hintz, H. F. (1969). Comparison of digestion of horses, ponies, rabbits, and guinea pigs. *Journal of Animal Science*, **28**, 842–3.

Slen, S. B. & Banky, E. C. (1961). Wool and body growth in lambs during the first 14 months of life. *Canadian Journal of Animal Science*, **41**, 78–88.

Smith, A. D. (1950). Sagebrush as a winter feed for deer. *Journal of Wildlife Management*, **14**, 285–9.

Smith, A. D. (1952). Digestibility of some native forages for mule deer. *Journal of Wildlife Management*, **16**, 309–12.

Smith, A. D. (1957). Nutritive value of some browse plants in winter. *Journal of Range Management*, **10**, 162–4.

Smith, A. D. & Malechek, J. C. (1974). Nutritional quality of summer diets of pronghorn antelopes in Utah. *Journal of Wildlife Management*, **38**, 792–8.

Smith, D. R. (1954). *The bighorn sheep in Idaho. Idaho Fish and Game Department Wildlife Bulletin 1.*

Smith, E. F., Young, V. A., Anderson, K. L. & Rogers, S. N. (1960). The digestibility of forage on burned and non-burned bluestem pastures as determined by grazing animals. *Journal of Animal Science*, **19**, 388–91.

Smith, J. G. & Julander, O. (1953). Deer and sheep competition in Utah. *Journal of Wildlife Management*, **17**, 101–12.

Smith, R. C. G. & Williams, W. A. (1974). Deferred grazing of Mediterranean annual pasture: a study by computer simulation. *Proceedings of the XII International Grassland Congress, Moscow*, pp. 646–55.

Smith, T. M., Lesperance, A. L., Bohman, V. R., Breckbill, R. A. & Brown, R. W. (1968). Intake and digestibility of forages grazed by cattle on a southern Nevada range. *Proceedings of the Western Section of the American Society of Animal Science*, **19**, 277–82.

Smoliak, S. (1974). Range vegetation and sheep production at three stocking rates on *Stipa–Bouteloua* prairie. *Journal of Range Management*, **27**, 23–6.

Sotala, D. J. & Kirkpatrick, C. M. (1973). Foods of white-tailed deer, *Odocoileus virginianus*, in Martin County, Indiana. *American Midland Naturalist*, **89**, 281–6.

Spedding, C. R. W. (1975). *The biology of agricultural systems*. Academic Press, New York & London.

Spinage, C. A. (1972). African ungulate life tables. *Ecology*, **53**, 645–52.

Stelfox, J. G. (1976). *Range ecology of Rocky Mountain bighorn sheep. Canadian Wildlife Service Report 39.*

Stevens, D. R. (1970). Winter ecology of moose in the Gallatin Mountains, Montana. *Journal of Wildlife Management*, **34**, 37–46.

Stoddart, L. A., Smith, A. D. & Box, T. W. (1975). *Range management*, 3rd edn. McGraw-Hill, New York.

Strasia, C. A., Thorn, M., Rice, R. W. & Smith, D. R. (1970). Grazing habits, diet and performance of sheep on alpine ranges. *Journal of Range Management*, **23**, 201–8.

Swank, G. W., Casebear, R. L., Threasher, P. B. & Woodruff, M. H. (1974). *Cropping, processing, and marketing of wildlife in the Kajiado District, Kenya. Project Working Document 6*. Kenya Government and FAO, Nairobi, Kenya.

Swartzman, G. L. & Van Dyne, G. M. (1972). An ecologically based simulation – optimization approach to natural resources planning. *Annual Review of Ecology and Systematics*, **3**, 347–98.

Taber, R. D. (1953). Studies of black-tailed deer reproduction on three chaparral cover types. *California Fish and Game*, **39**, 177–86.

Taber, R. D. & Dasmann, R. F. (1957). The dynamics of three natural populations of the deer *Odocoileus hemionus columbianus*. *Ecology*, **38**, 233–46.

Tadmor, N. H., Egal, E. & Benjamin, R. W. (1974). Plant and sheep production on semiarid annual grassland in Israel. *Journal of Range Management*, **27**, 427–32.

Talbot, L. M. & Kesel, R. H. (1975). The tropical savanna ecosystem. In *Geoscience and man*, vol. 10, *Grasslands ecology – a symposium*, ed. R. H. Kessel, pp. 15–26. School of Geoscience, Louisiana State University, Baton Rouge.

Talbot, L. M. & McCulloch, J. S. G. (1965). Weight estimation for East African mammals from body measurements. *Journal of Wildlife Management*, **29**, 84–9.

Talbot, L. M., Payne, W. J. A., Ledger, H. P., Verdcourt, L. D. & Talbot, M. H. (1965). *The meat production potential of wild animals in Africa – a review of biological knowledge. Commonwealth Agricultural Bureau Technical Communication 16*. Commonwealth Bureau of Animal Breeding and Genetics, Edinburgh.

Talbot, L. M. & Talbot, M. H. (1962). Food preferences of some East African wild ungulates. *East African Agriculture and Forestry Journal*, **27**, 131–8.

Talbot, L. M. & Talbot, M. H. (1962). The wildebeest in western Masailand, East Africa. *Wildlife Monographs*, **12**, 1–88.

Taylor, C. R. (1968). Hygroscopic food: a source of water for desert antelopes? *Nature, London*, **219**, 181–2.

Taylor, C. R. (1968). The minimum water requirements of some East African bovines. *Symposia of the Zoological Society of London*, **21**, 195–206.

Taylor, E. (1972). Food habits and feeding behavior of pronghorn antelope in the Red Desert of Wyoming. *Antelope States Workshop Proceedings*, **5**, 211–19.

Taylor, J. C. (1959). A relationship between weight of internal fat, 'fill', and the herbage intake of grazing cattle. *Nature, London*, **184**, 2021.

Telfer, E. S. & Scotter, G. W. (1975). Potential for game ranching in boreal aspen forests of Western Canada. *Journal of Range Management*, **28**, 172–80.

Tener, J. S. (1954). *A preliminary study of the musk oxen of Fosheim Peninsula, Ellesmere Island. NWT Wildlife Management Bulletin Series 1, No. 9*. Ministry of Northern Affairs and National Resources, Department of National Parks Branch, Canadian Wildlife Service, Ottawa, Canada.

Terrill, C. E. (1975). Genetic variation and nutrition of sheep. In *The effect of genetic variance on nutritional requirements of animals*, pp. 88–111. US National Academy of Sciences, Washington, DC.

Thetford, F. O., Pieper, R. D. & Nelson, A. B. (1971). Botanical and chemical composition of cattle and sheep diets on pinyon–juniper grassland range. *Journal of Range Management*, **24**, 425–31.

Theurer, C. B., Lesperance, A. L. & Wallace, J. D. (1976). *Botanical composition of the diet of livestock grazing native ranges. University of Arizona Agricultural Experiment Station Bulletin 233.*

Till, A. R. & May, P. F. (1970*a*). Nutrient cycling in grazed pastures. II. Further observations with [$^{35}$S] gypsum. *Australian Journal of Agricultural Research*, **19**, 532–43.

Till, A. R. & May, P. F. (1970*b*). Nutrient cycling in grazed pastures. III. Studies on labelling of the grazed pasture system by solid [$^{35}$S] gypsum and aqueous $Mg^{35}SO_4$. *Australian Journal of Agricultural Research*, **21**, 455–63.

Tilley, J. M. A. & Terry, R. A. (1963). A two-stage technique for the *in vitro* digestion of forage crops. *Journal of the British Grassland Society*, **18**, 104–11.

Todd, J. W. (1975). Food of Rocky Mountain bighorn sheep in southern Colorado. *Journal of Wildlife Management*, **39**, 108–11.

Tomanek, G. W., Martin, E. P. & Albertson, F. W. (1958). Grazing preference comparisons of six native grasses in the mixed prairie. *Journal of Range Management*, **11**, 191–3.

Topps, J. H. (1975). Behavioral and physiological adaptation of wild ruminants and their potential for meat production. *Proceedings of the Nutrition Society*, **34**, 85–93.

Treacher, T. T. (1970). Apparatus and milling techniques used in location studies on sheep. *Journal of Dairy Research*, **37**, 289–95.

Trebeck, D. B. (1972). Simulation as an aid to research into extensive beef production. *Proceedings of the Australian Society of Animal Production*, **9**, 94.

Trischnect, N. C., Harris, L. E. & Woodward, H. K. (1953). Cattle gains and vegetal changes as influenced by grazing treatments on crested wheatgrass. *Journal of Range Management*, **6**, 151–9.

Ullrey, D. E., Youatt, W. G., Johnson, H. E., Ku, T. K. & Fay, L. D. (1964). Digestibility of cedar and aspen browse for the white-tail deer. *Journal of Wildlife Management*, **28**, 791–7.

Ulyatt, M. J. (1964). Studies on some factors influencing food intake in sheep. *Proceedings of the New Zealand Society of Animal Production*, **24**, 43–56.

Uresk, P. W. & Rickard, W. H. (1976). Diets of steers on a shrubsteppe rangeland in south-central Washington. *Journal of Range Management*, **29**, 464–6.

Van Dyne, G. M. (1956). Effects of the intensity of grazing on range livestock production, the native vegetation, and the soil complex. MS Thesis, South Dakota State University, Brookings.

Van Dyne, G. M. (1969). *Grasslands management, research and training viewed in a systems context. Range Science Department Science Series 3.* Colorado State University, Fort Collins.

Van Dyne, G. M. (1969). Measuring quantity and quality of the diet of large herbivores. In *A practical guide to the study of the productivity of large herbivores*, ed. F. B. Golley & H. K. Buechner, pp. 59–94. Blackwell Scientific Publications, Oxford.

Van Dyne, G. M. (1977). Optimal combinations of four large herbivores for shortgrass prairie. In *XIII International Grassland Congress*, Leipzig. In press.

Van Dyne, G. M. & Anway, J. C. (1976). A research program for and the process of building and testing grassland ecosystem models. *Journal of Range Management*, **29**, 114–22.

Van Dyne, G. M. & Heady, H. F. (1965*a*). Interrelations of botanical and chemical dietary components of animals grazing dry annual range. *Journal of Animal Science*, **24**, 305–12.

Van Dyne, G. M. & Heady, H. F. (1965*b*). Botanical composition of sheep and cattle diets on a mature annual range. *Hilgardia*, **36**, 465–92.
Van Dyne, G. M., Joyce, L. A. & Williams, B. K. (1978*b*). Models and the formulation and testing of hypotheses in grazingland ecosystem management. In *Ecosystem restoration – principles and case studies*, ed. M. Holdgate & M. Woodman, pp. 41–83. Plenum Press, London.
Van Dyne, G. M. & Lofgreen, G. P. (1964). Comparative digestion of dry annual range forage by cattle and sheep. *Journal of Animal Science*, **23**, 823–32.
Van Dyne, G. M. & Meyer, J. H. (1964*a*). Forage intake by cattle and sheep on dry annual range. *Journal of Animal Science*, **23**, 1108–15.
Van Dyne, G. M. & Meyer, J. H. (1964*b*). A method for measurement of forage intake of grazing livestock using microdigestion techniques. *Journal of Range Management*, **17**, 204–8.
Van Dyne, G. M. & Rebman, K. R. (1967). Maintaining a profitable ecological balance. In *Some mathematical models in biology*, ed. R. M. Thrall, J. A. Mortimer, K. R. Rebman & R. F. Baum, pp. PL61–69. University of Michigan Report 4042-R-7 (5-Tol-GMO1457-02).
Van Dyne, G. M., Smith, F. M., Czaplewski, R. L. & Woodmansee, R. J. (1978*a*). Analyses and syntheses of grassland ecosystem dynamics. In *Glimpses of Ecology*, ed. J. S. Singh & B. Gopal. International Scientific Publ., Jaipus, India.
Van Dyne, G. M., Thomas, O. O. & Van Horn, J. L. (1964). Diet of cattle and sheep grazing on winter range. *Proceedings of the Western Section of the American Society of Animal Science*, **14**, 1–6.
Van Dyne, G. M. & Torell, D. (1964). Development and use of the esophageal fistula: a review. *Journal of Range Management*, **17**, 7–9.
Van Dyne, G. M. & Van Horn, J. L. (1965). Distance traveled by sheep on winter range. *Proceedings of the Western Section of the American Society of Animal Science*, **16**, 1–6.
Van Schalkwyk, A., Lombard, P. E. & Vorster, L. F. (1968). Evaluation of the nutritive value of a *Themada triandra* pasture in the central Orange Free State. IV. Digestibility. *South African Journal of Agricultural Science*, **11**, 679–86.
Vavra, M., Rice, R. W. & Bement, R. E. (1973). Chemical composition of the diet, intake and gain of yearling cattle on different grazing intensities. *Journal of Animal Science*, **36**, 411–14.
Vera, R. R., Koong, L. J. & Morris, J. G. (1975). A model of heat flow in the sheep exposed to high levels of solar radiation. *Computer Programs in Biomedicine*, **4**, 214–18.
Vercoe, J. E., Tribe, D. E. & Pierce, G. R. (1961). Herbage as a source of digestible organic matter and digestible nitrogen for grazing sheep. *Australian Journal of Agricultural Research*, **12**, 689–95.
Vesey-Fitzgerald, D. F. (1960). Grazing succession among East African game animals. *Journal of Mammalogy*, **41**, 161–72.
Vickery, P. J. & Hedges, D. A. (1972*a*). A productivity model of improved pasture grazed by Merino sheep. *Proceedings of the Australian Society of Animal Production*, **9**, 16–22.
Vickery, P. J. & Hedges, D. A. (1972*b*). *Mathematical relationships and computer routines for a productivity model of improved pasture grazed by Merino sheep. CSIRO Animal Research Laboratories Technical Paper 4*. CSIRO, Melbourne, Australia.

Wagner, F. H. (1969). Ecosystem concepts and fish and game management. In *The ecosystem concept in natural resource management*, ed. G. M. Van Dyne, pp. 259–307. Academic Press, New York & London.

Walker, B. M. (1974). An appraisal of the systems approach to research on and management of African wildlife ecosystems. *Journal of the South African Wildlife Management Association*, **4**, 129–35.

Walker, E. P. (1968). *Mammals of the world*, 3 vols. Johns Hopkins Press, Baltimore.

Wallace, J. D. & Denham, A. H. (1970). Digestion of range forage by sheep collected by esophageal fistulated cattle. *Journal of Animal Science*, **30**, 605–8.

Wallace, J. D., Hyder, D. N. & Van Dyne, G. M. (1972). Salivary contamination of forage selected by esophageal fistulated steers grazing sandhill grassland. *Journal of Range Management*, **25**, 184–7.

Wallace, L. R. (1956). The intake and utilization of pasture by grazing dairy cattle. In *Proceedings of the VII International Grassland Congress*, pp. 134–45.

Wallmo, O. C. (1975). Important game animals and related recreation in arid shrublands of the United States. In *Arid shrubland, Proceedings of the III Workshop of the US/Australian Rangelands Panel*, Tucson, Arizona, 26 March–5 April 1973, ed. D. N. Hyder, pp. 98–107. Society of Range Management, Denver, Colorado.

Walters, C. J. & Bunnell, F. (1971). A computer management game of land use in British Columbia. *Journal of Wildlife Management*, **35**, 644–57.

Walters, C. J. & Gross, J. E. (1972). Development of big game management plans through simulation modeling. *Journal of Wildlife Management*, **36**, 119–28.

Walters, C. J., Hilborn, R. & Peterman, R. (1975). Computer simulation of barren-ground caribou dynamics. *Ecological Modelling*, **1**, 303–15.

Warwick, E. J. & Cobb, E. H. (1975). Genetic variation in nutrition of cattle for meat production. In *The effect of genetic variance on nutritional requirements of animals*, pp. 3–18. US National Academy of Sciences, Washington, DC.

Webster, A. J. F., Osuji, P. O., White, F. & Ingram, J. F. (1975). The influence of food intake on portal blood flow and heat production in the digestive tract of sheep. *British Journal of Nutrition*, **34**, 125–39.

Weir, W. C. & Torell, D. T. (1959). Selective grazing by sheep as shown by a comparison of the chemical composition of range and pasture forage obtained by hand clipping and that collected by esophageal fistulated sheep. *Journal of Animal Science*, **18**, 641–9.

Westoby, M. (1974). An analysis of diet selection by large generalist herbivores. *American Naturalist*, **108**, 290–304.

Westra, R. (1975). The effect of temperature on digestion in sheep. MS Thesis, University of Alberta, Edmonton.

White, M. (1973). The whitetail deer of the Aransas National Wildlife Refuge. *Texas Journal of Science*, **24**, 457–89.

Whitelaw, F. G. (1974). Measurement of energy expenditure in the grazing ruminant. *Proceedings of the Nutrition Society*, **33**, 163–72.

Wielgolaski, F. E. (1972). *IBP grassland/tundra international modelling–synthesis workshop: summary of output workbook. US/IBP Grassland Biome Technical Report 197*. Colorado State University, Fort Collins.

Wielgolaski, F. E. (1975). Grazing by sheep. In *Fennoscandian tundra ecosystems*, part 2, *Animals and systems analysis*, ed. F. E. Wielgolaski, pp. 216–28. Springer-Verlag, Berlin.

Wilkins, B. T. (1957). Range use, food habits and agricultural relationships of the mule deer, Bridger Mountains, Montana. *Journal of Wildlife Management*, **21**, 159–69.

Wilkinson, P. F. (1974). Behavior and domestication of the musk ox. In *The behavior of ungulates and its relation to management*, ed. V. Geist & F. Walther. vol. 2, pp. 909–20. International Union for Conservation of Nature and Natural Resources, Morjes, Switzerland.
Wilkinson, P. F., Shank, C. C. & Penner, D. F. (1976). Muskox-caribou summer range relations on Banks Island, NWT. *Journal of Wildlife Management*, **40**. 151–62.
Wilson, A. D. (1974). Nutrition of sheep and cattle in Australian arid areas. In *Studies of the Australian arid zone*, vol. 2, *Animal production*, ed. A. D. Wilson. pp. 74–84. CSIRO, Melbourne, Australia.
Wilson, A. D., Leigh, J. H., Hindley, N. L. & Mulham, W. E. (1975). Comparison of the diets of goats and sheep on a *Casuarina crista–Heterodendrum oleifolium* woodland community in Western New South Wales. *Australian Journal of Experimental Agriculture and Animal Husbandry*, **15**, 45–51.
Wilson, A. D., Leigh, J. H. & Mulham, W. E. (1969). A study of merino sheep grazing a bladder saltbush (*Atriplex vesicaria*), cottonbush (*Kochia aphylla*) community on the Riverine Plain. *Australian Journal of Agricultural Research*. **20**, 1123–36.
Woodgerd, W. (1964). Population dynamics of bighorn sheep on Wildhorse Island. *Journal of Wildlife Management*, **28**, 381–91.
Woodward, S. L. & Ohmart, R. D. (1976). Habitat use and fecal analysis of feral burros (*Equus asinus*), Chemehuevi Mountains, California, 1974. *Journal of Range Management*, **29**, 482–5.
Woolf, A. (1968). Summer ecology of bighorn sheep in Yellowstone National Park. MS Thesis, Colorado State University, Fort Collins.
Wright, B. S. (1960). Predation on big game in East Africa. *Journal of Wildlife Management*, **24**, 1–15.
Wright, R. G. & Van Dyne, G. M. (eds.) (1970). *Simulation and analysis of a semi-desert grassland. Range Science Department Science Series 6*. Colorado State University, Fort Collins.
Yoakum, J. (1958). Seasonal food habits of the Oregon pronghorn antelope (*Antilocapra americana oregona* Bailey). *Transactions of the Interstate Antelope Conference*, **9**, 47–59.
Yoakum, J. (1964). Bighorn food habit–range relationships in the Silver Peak Range, Nevada. *Desert Bighorn Council Transactions*, pp. 95–102. Desert Bighorn Council, Las Vegas, Nevada.
Young, P. T. (1948). Appetite, palatability, and feeding habit: a critical review. *Psychological Bulletin*, **45**, 289–320.

## Appendixes

### Appendix 4.1. Equations used in calculating forage digestion and intake by grazing animals

Various equations have been referred to in the text for calculating digestion and intake by grazing animals. The derivation of the equations (after Van Dyne, 1969) are provided below.

For purposes of discussion, the following symbols are defined:

$A$ = Amount external indicator administered per day.
$A_E$ = Concentration of external indicator in a sample of faeces.
$C_i$ = Concentration, in per cent, of a digestible nutrient in the diet.
$D_i$ = Apparent digestibility coefficient, in per cent, of a nutrient.
$D_{dm}$ = Digestibility of dry matter.
$D_c$ = Digestibility of cellulose.
$E$ = Faecal excretion weight per day.
$E_s$ = Weight of a sample of faeces.
$F$ = Forage intake weight per day.
$i$ = Indicator concentration in per cent.
$I_E$ = Indicator in faeces.
$I_F$ = Indicator in forage.
$M_i$ = Microdigestion estimates for nutrient $i$ in per cent.
$N_i$ = Nutrient concentration in per cent.
$N_{iE}$ = Nutrient in faeces – then $N_{cE}$ = faecal cellulose.
$N_{iE}$ = Nutrient in forage – then $N_{cF}$ = forage cellulose.

If an indicator is truly indigestible, it follows that the amount consumed is equal to that excreted, or

$$F = E \times I_E / I_F \tag{1}$$

By definition, percentage dry matter digestion is

$$D_{dm} = (F - E)/F \times 100 \tag{2}$$

But from equation (1) $F$ may be substituted, therefore

$$D_{dm} = 100 - I_F/I_E \times 100 \tag{3}$$

Now if $D_{dm}$ is known, then $(1 - D_{dm}/100)$ is the indigestibility of a forage. Then the relation between $F$ and $E$ is

$$F = E/(100 - D_{dm}) \times 100 \tag{4}$$

The digestibility of a nutrient, $D_i$, is shown to be

$$D_i = 100 - EN_{iE}/FN_{iF} \times 100 \tag{5}$$

Using equations (1) and (5)

$$D_i = 100 - I_F N_{iE}/I_E N_{iF} \times 100 \tag{6}$$

If an external indicator of amount $A$ is administered to the animal and excreted uniformly, then faecal output $E$ can be calculated from a sample of the faeces $E_s$ by the following relation

$$E = E_s A/A_E \tag{7}$$

Therefore digestibility can be calculated by substituting equation (7) into equation (5) as follows

$$D_{dm} = 100 - E_s A/FA_E \tag{8}$$

But in range and pasture trials, feed intake ($F$) is not regulated. However, by combining

**Appendix 4.1** (*continued*)

an internal indicator and an external indicator, $F$ may be calculated by using equation (2) as follows

$$F = E_s\, A/A_E \times I_E/I_F$$

Once feed intake ($F$) and faecal output ($E$) have been calculated with equations then the apparent digestibility of any nutrient may be calculated by equation (5).

The concentration of a digestible nutrient in the diet ($C_i$) becomes

$$C_i = N_i D_i/100 \tag{10}$$

It will now be shown that microdigestion estimates for a nutrient can be used to calculate feed intake ($F$). As an example, microdigestion of cellulose ($M_c$) will be used. It is readily seen that the weight of cellulose in the diet ($F_c$) is

$$F_c = FN_{cF}/100 \tag{11}$$

Similarly, it follows that the *weight* of faecal cellulose ($E_c$) is

$$E_c = EN_{cE}/100 \tag{12}$$

Now, substituting into equation (11) by use of equation (4) and (12) it is seen that

$$F = N_{cE}\, E/100 \times N_{cF} - N_{cF} \times M_c \times 100 \tag{13}$$

If microdigestion of cellulose is expressed in a decimal form ($M_c^1$) then this simplifies to

$$F = N_{cE}\, E/N_{cF}\, (1 - M_c{}^1) \tag{14}$$

## Appendix 4.2. Botanical composition of diets of large herbivores on the world's grazinglands

| No. | Animal species | Kind of grazingland | Geographic location | Methods | Season/ treatment | Grass (%) | Forbs (%) | Shrubs (%) | Reference | Notes |
|---|---|---|---|---|---|---|---|---|---|---|
| 1 | Cattle, Zebu | *Themada–Pennisetum* grassland | Southeast, near Nairobi, Kenya | Faecal collection, microscopic analysis | Apr. | 99 | 1 | 0 | Casebeer & Koss (1970) | See notes under no. 26. |
| 2 | Cattle | Desert shrub | Southern Nevada, USA | Rumen fistula, binocularscopic analysis | Spring (Apr.–May) | 60 | 0 | 40 | Connor *et al.* (1963) | See notes under no. 87. |
| 3 | Cattle | Native sandhill range | Northeastern Colorado, USA | Direct observation | Spring | 98 | 2 | 0 | Denham (1965) | |
| 4 | Cattle Hereford and Brahman cows | Wiregrass–pine range | Coastal plains, Georgia, USA | Hand-plucked samples and observation | Spring (Apr.) | 83 | 7 | 10 | Hale *et al.* (1962) | A 1-year study showed grasses major proportion of diet with pineland threeawn and bluestems predominating during much of the growing season and curtis dropseed in the winter. |
| 5 | Cattle | Wiregrass–pine grazing | Georgia, USA | Observation and hand-plucked samples | Spring (Apr.; 3 years) | 88 | 9 | 3 | Halls *et al.* (1957) | Grass proportions increase from early spring to high in autumn and then decrease; important grasses are pine-land threeawn and curtis dropseed; bluestem is important throughout the year. No individual forbs of major importance; major shrub was saw palmetto. Cattle search for new growth in young herbage. |
| 6 | Cattle, Hereford | Desert grassland | Southern New Mexico, USA | Observation of species grazed | Spring | 35 | 40 | 25 | Herbel & Nelson (1966) | See notes under no. 73. |
| 7 | Cattle, Santa-gertrudiss | Desert grassland | Southern New Mexico, USA | Observation of species grazed | Spring | 58 | 30 | 12 | Herbel & Nelson (1966) | See notes under no. 73. |
| 8 | Cattle | Foothill rangeland; juniper–sagebrush–bitterbrush | Utah, USA | Botanical sampling | Light spring use | 25 | 6 | 69 | Julander (1955) | See notes under no. 42. |
| 9 | Cattle | Foothill rangeland; juniper–sagebrush–cliffrose | Utah, USA | Botanical sampling | Heavy spring use | 70 | 2 | 28 | Julander (1955) | See notes under no. 42. |

## Appendix 4.2 (*continued*)

| No. | Animal species | Kind of grazingland | Geographic location | Methods | Season/ treatment | Grass (%) | Forbs (%) | Shrubs (%) | Reference | Notes |
|---|---|---|---|---|---|---|---|---|---|---|
| **10** | Cattle | Foothill native bunchgrass range | Central Colorado, USA | Rumen fistula, microscopic analysis | Spring, native bunchgrass (May–June) | 45 | 33 | 22 | Malacheck (1966) | Forbs were more important during early spring and again in winter than in other periods. Fringed sagewort, a half shrub, was important in early spring and even on seeded ranges formed up to 20 % of the diet. Limited sample size. Arizona fescue and mountain muhly and western wheatgrass were main grasses; no individual forbs highly important in diet except *Potentilla pennsylvanica*. |
| **11** | Cattle | Savanna, live oak | Central Texas, USA | Observation, bite-count | Light grazing, only used in spring | 71 | 12 | 17 | McMahan (1964) | See notes under no. 124. |
| **12** | Cattle | Savanna, live oak | Central Texas, USA | Observation, bite-count | Spring, heavy use | 98 | 2 | 0 | McMahan (1964) | See notes under no. 124. |
| **13** | Cattle (heifers) | Sandhill range | Colorado, USA | Observation, before-and-after sampling | Spring (early Apr.–early June) | 98 | <1 | 2 | Reppert (1957) | See notes under no. 96. |
| **14** | Cattle | Semidesert grassland | Southern New Mexico, USA | Oesophageal fistula, microscopic analysis | Spring | 16 | 12 | 72 | Rosiere *et al.* (1975) | 5 % unidentified. Mesa dropseed main grass in diet. Any forbs species seldom more than 10 % of diet. |
| **15** | Cattle (yearlings) | Sandhill range | Northeastern Colorado, USA | Oesophageal fistula | Spring (May) | 93 | 6 | 1 | Scales (1972) | Morning and evening samples taken. Blue grama most important seasonlong with western wheatgrass second. Blue grama greatest in summer diets, less in spring and autumn; western wheatgrass highest in early spring, successively lower until early autumn; needle-and-thread highest in early spring and late autumn. Forbs never important as individual species. |
| **16** | Cattle | Piñon–juniper | South-central New Mexico, USA | Oesophageal fistula | Spring (Apr.) | 33 | 47 | 20 | Thetford *et al.* (1971) | See notes under no. 127. |
| **17** | Cattle (steers) | *Artemisia tridentata–Agropyron spicatum* pasture | South-central Washington, USA | Faecal analysis, viewed through microscope | Spring, moderate grazing | 73 | 27 | 0 | Uresk & Richard (1976) | Cusick bluegrass and hawksbeard most abundant plants grazed in early part of season, bluebunch wheatgrass mostly grazed during later part of season. |

| | | | | | | | | | | |
|---|---|---|---|---|---|---|---|---|---|---|
| **18** | Cattle | Mountain rangeland | Ruby Buttes, Nevada, USA | Rumen evacuation | Late spring (May) | 100 | <1 | 0 | Ansotegui *et al.* (1972) | |
| **19** | Cattle | Shortgrass prairie | Northeastern Colorado, USA | Oesophageal fistula, microscopic analysis | Late spring (May), light grazing | 81 | 3 | 16 | This chapter | See notes under no. 20. |
| **20** | Cattle (steers) | Shortgrass prairie | Northeastern Colorado, USA | Oesophageal fistula, microscopic analysis | Late spring (May) heavy grazing | 97 | 2 | 1 | This chapter | Herbage predominantly warm-season grasses; large amounts of prickly pear cactus in herbage, but very little in diet; shrubs more important under light grazing than under heavy grazing and as much as 50 % of the herbage at times. Herbage never too limited to prevent selection. Nonreproductive growing animals of each species used. |
| **21a** | Cattle | Mountain range | Utah, USA | Oesophageal fistula or hand-plucked samples | Early summer | 48 | 23 | 29 | Cook & Harris (1968) | |
| **21** | Cattle | Northern desert shrub | Southern Neveda, USA | Rumen fistula | Early summer | 100 | 0 | 0 | Smith *et al.* (1968) | |
| **22** | Cattle | Shortgrass prairie | Northeastern Colorado, USA | Oesophageal fistula, microscopic analysis | Early summer (June), light grazing | 70 | 29 | 1 | This chapter | See notes under no. 20. |
| **23** | Cattle | Shortgrass prairie | Northeastern Colorado, USA | Oesophageal fistula, microscopic analysis | Early summer (June), heavy grazing | 40 | 47 | 13 | This chapter | See notes under no. 20. |
| **24** | Cattle (steers) | Annual grassland | Northern California, USA | Oesophageal fistula, binocular analysis | Early summer | 68 | 25 | 7 | Van Dyne & Heady (1965) | See notes under no. 133 |
| **25** | Cattle | Mountain rangeland | Ruby Buttes, Nevada, USA | Rumen evacuation | Summer (June–July) | 96 | 2 | 2 | Ansotegui *et al.* (1972) | |
| **26** | Cattle, Zebu | *Themeda–Pennisetum* grassland | Southeast, near Nairobi, Kenya | Faecal collection, microscopic analysis | July | 95 | 5 | <1 | Casebeer & Koss (1970) | Diet remains more consistent with the combination of available grasses. Main species *Themeda triendra* (always more than 40 %), second *Pennisetum merianum* and third *Digitaria macroblephara*. |
| **27** | Cattle | Desert shrub | Southern Nevada, USA | Rumen fistula, binocularscopic analysis | Summer (July–Sept.) | 30 | <1 | 70 | Connor *et al.* (1963) | See notes under no. 87. |
| **28** | Cattle | Desert shrub | Northern Nevada, USA | Rumen fistula, binocularscopic analysis | Summer (June–Sept.) | 68 | 8 | 24 | Connor *et al.* (1963) | See notes under no. 87. |
| **29** | Cattle | Aspen range, mountain | Northern Utah, USA | Oesophageal fistula | Summer grazing (June–Sept.; 2 years) | 84 | 13 | 3 | Cook *et al.* (1967) | See notes under no. 30. |

## Appendix 4.2 (*continued*)

| No. | Animal species | Kind of grazingland | Geographic location | Methods | Season/ treatment | Grass (%) | Forbs (%) | Shrubs (%) | Reference | Notes |
|---|---|---|---|---|---|---|---|---|---|---|
| **30** | Cattle | Sagebrush–grass, mountain | Northern Utah, USA | Oesophageal fistula | Summer, grazing (June–Sept., 2 years) | 77 | 11 | 12 | Cook *et al.* (1967) | Cattle highest in grass and lowest in browse on both range types; sheep lowest in browse, equal amounts of grass and forbs on aspen but highest in grass and lowest in browse on sagebrush range. |
| **31** | Cattle | Native sandhill range | Northeastern Colorado, USA | Direct observation | Summer | 99 | 1 | 0 | Denham (1965) | 4 % other distributed among data. |
| **32** | Cattle | Willow and sedge | Southwest Montana, USA | Feeding site observations | Summer | 0 | 0 | 100 | Dorn (1970) | Cattle use of non-woody plants was not recorded. |
| **33** | Cattle | Grassland; big sagebrush and silver sagebrush | North-central Montana, USA | Feeding site examination | Summer | 67 | 26 | 7 | Dusek (1975) | |
| **34** | Cattle, Hereford and Brahman cows | Wiregrass–pine range | Coastal plains, Georgia, USA | Hand-plucked samples and observation | Summer (June) | 80 | 3 | 17 | Hale *et al.* (1962) | See notes under no. 4. |
| **35** | Cattle | Wiregrass range | South Georgia, USA | Observation and hand-plucked samples | June | 88 | 10 | 2 | Halls (1954) | Bluestems major grass component. |
| **36** | Cattle | Wiregrass–pine grazing | Georgia, USA | Observation and hand-plucked samples | Summer (June; 3 years) | 89 | 6 | 5 | Halls *et al.* (1957) | See notes under no. 5. |
| **37** | Cattle | Piñon–juniper and sagebrush | Northeastern Colorado, USA | Faecal analysis | Summer | 95 | 3 | 2 | R. M. Hansen & R. C. Clark (unpublished) | |
| **38** | Cattle | Low mountain foothill | South-central Colorado, USA | Faecal material, microscopic analysis | June–Sept. | 96 | 1 | 3 | Hansen & Reid (1975) | *Danthonia*, fescue and sedges each contributed more than 20 % seasonlong. |
| **39** | Cattle, Hereford | Desert grassland | Southern New Mexico, USA | Observation of species grazed | Summer | 72 | 22 | 6 | Herbel & Nelson (1966) | See notes under no. 73. |
| **40** | Cattle, Santa-gertrudis | Desert grassland | Southern New Mexico, USA | Observation of species grazed | Summer | 81 | 17 | 2 | Herbel & Nelson (1966) | See notes under no. 73. |
| **41** | Cattle | Foothill, rangeland, aspen–browse range | Utah, USA | Botanical sampling | Summer | 76 | 10 | 14 | Julander (1955) | See notes under no. 42. |

| | | | | | | | | | | |
|---|---|---|---|---|---|---|---|---|---|---|
| **42** | Cattle | Foothill rangeland, aspen–forb range | Utah, USA | Botanical sampling and observation | Summer | 86 | 12 | 2 | Julander (1955) | Data determined by measuring standing crop of herbage, and percentage use was used to get percentage of diet. Cattle distribution influenced mainly by steep slope and availability of water and forage. Fed mainly on grasses in the summer. |
| **43** | Cattle | Foothill rangeland, mountain-brush range | Utah, USA | Botanical sampling | Summer | 36 | 7 | 57 | Julander (1955) | See notes under no. 42. |
| **44** | Cattle | Foothill rangeland, oakbrush-type range | Utah, USA | Botanical sampling | Summer | 8 | 73 | 19 | Julander (1955) | See notes under no. 42. |
| **45** | Cattle | Foothill rangeland piñon–sagebrush–bitterbrush | Utah, USA | Botanical sampling | Summer, heavy use | 9 | 10 | 81 | Julander (1955) | See notes under no. 42. |
| **46** | Cattle | Foothill native bunchgrass range | Central Colorado, USA | Rumen fistula, microscopic analysis | Summer (July–Sept.) | 79 | 17 | 4 | Malacheck (1966) | See notes under no. 10. |
| **47** | Cattle (steers) | Shortgrass prairie | Amarillo, Texas USA | Oesophageal fistulas, point-frame microscope technique | Summer (June–Sept.), light grazing[1] | 62 | 38 | 0 | McClung *et al.* (1976) | |
| **48** | Cattle | Savanna, live oak | Central Texas, USA | Observation, bite-count | Light grazing, only used in summer | 85 | 12 | 3 | McMahan (1964) | See notes under no. 124. |
| **49** | Cattle | Savanna, live oak | Central Texas, USA | Observation, bite-count | Heavy summer use | 100 | 0 | 0 | McMahan (1964) | See notes under no. 124. |
| **50** | Cattle | Foothills, ponderosa pine–bluebunch wheatgrass | Northeastern Oregon, USA | Weight estimate method for herbage, in and out of cages, leader-length shrub measurements | Summer (Aug.–Sept.) | 95 | 1 | 4 | Miller & Krueger (1976) | Big game grazed area in spring and early summer. Grasses accounted for 52 % of the big game diet in spring and early summer. |
| **51** | Cattle (heifers) | Sandhill range | Colorado, USA | Observation, before-and-after sampling | Summer (late June–Sept.) | 95 | 5 | <1 | Reppert (1957) | See notes under no. 96. |
| **52** | Cattle (steers) | Shortgrass prairie | Northeastern Colorado, USA | Oesophageal fistula, microscopic analysis | Summer (1969), light use | 51 | 48 | 1 | Rice & Vavra (1969) | See notes under no. 54. |
| **53** | Cattle (steers) | Shortgrass prairie | Northeastern Colorado, USA | Oesophageal fistula, microscopic analysis | Summer (1969), light use | 54 | 39 | 7 | Rice & Vavra (1969) | See notes under no. 54. |
| **54** | Cattle (steers) | Shortgrass prairie | Northeastern Colorado, USA | Oesophageal fistula, microscopic analysis | Summer (1969), heavy use | 66 | 31 | 3 | Rice & Vavra (1969) | Blue grama most important grass, western wheatgrass especially in July, red threeawn only in June. Scarlet globemallow and eriogonum major forbs; main shrub fringed sagewort. |

**Appendix 4.2** (*continued*)

| No. | Animal species | Kind of grazingland | Geographic location | Methods | Season/ treatment | Grass (%) | Forbs (%) | Shrubs (%) | Reference | Notes |
|---|---|---|---|---|---|---|---|---|---|---|
| **55** | Cattle | Semidesert grassland | Southern New Mexico, USA | Oesophageal fistula, microscopic analysis | Summer | 60 | 31 | 9 | Rosiere *et al.* (1975) | Average of 1972 and 1973; distributed among data, preference changed as grazing intensity increased. |
| **56** | Cattle (yearlings) | Sandhill range | Northeastern Colorado, USA | Oesophageal fistula | Summer (June and July) | 85 | 11 | 4 | Scales (1972) | See notes under no. 15. |
| **57** | Cattle | Piñon–juniper | South-central New Mexico, USA | Oesophageal fistula | Summer (June–Aug. average) | 72 | 25 | 3 | Thetford *et al.* (1971) | See notes under no. 127. |
| **58** | Cattle (steers) | Sandhill grassland | Northeastern Colorado, USA | Oesophageal fistula, microscopic analysis | Summer (June–July) | 88 | 12 | 0 | Wallace *et al.* (1972) | Blue grama grass most important in diet overall; prairie sandreed important in July; needle-and-thread in Dec. Forbs important in early season. |
| **58a** | Cattle | Mountain range | Utah, USA | Oesophageal fistula or hand-plucked samples | Mid-summer | 63 | 28 | 9 | Cook & Harris (1968) | |
| **59** | Cattle (steers) | Annual grassland | Northern California, USA | Oesophageal fistula, binocular analysis | Mid-summer | 64 | 23 | 13 | Van Dyne & Heady (1965) | See notes under no. 133. |
| **59a** | Cattle | Mountain range | Utah, USA | Oesophageal fistula or hand-plucked samples | Late summer | 52 | 23 | 25 | Cook & Harris (1968) | |
| **60** | Cattle | Northern desert shrub | Southern Nevada, USA | Rumen fistula | Late summer | 22 | 78 | 0 | Smith *et al.* (1968) | |
| **61** | Cattle | Shortgrass prairie | Northeastern Colorado, USA | Oesophageal fistula, microscopic analysis | Late summer (Aug.), heavy grazing | 64 | 10 | 26 | This chapter | See notes under no. 20. |
| **62** | Cattle | Shortgrass prairie | Northeastern Colorado, USA | Oesophageal fistula, microscopic analysis | Late summer (Aug), light grazing | 73 | 27 | 0 | This chapter | See notes under no. 20. |
| **63** | Cattle (steers) | Annual grassland | Northern California, USA | Oesophageal fistula, binocular analysis | Late summer | 57 | 38 | 5 | Van Dyne & Heady (1965) | See notes under no. 133. |
| **64** | Cattle | Mountain rangeland | Ruby Buttes, Nevada, USA | Rumen evacuation | Early fall (Sept.) | 98 | 1 | 1 | Ansotegui *et al.* (1972) | |
| **65** | Cattle, Zebu | *Themeda–Pennisetum* grassland | Southeast, near Nairobi, Kenya | Faecal collections, microscopic analysis | Oct. | 98 | 2 | <1 | Casebeer & Koss (1970) | See notes under no. 26. |
| **66** | Cattle | Desert shrub | Southern Nevada, USA | Rumen fistula, bin-ocularscopic analysis | Autumn (Oct.–Dec.) | 4 | 0 | 96 | Connor *et al.* (1963) | See notes under no. 87. |

| | | | | | | | | | | |
|---|---|---|---|---|---|---|---|---|---|---|
| **67** | Cattle | Native sandhill range | Northeastern Colorado, USA | Direct observation | Autumn | 96 | 4 | 0 | Denham (1965) | 3 % others distributed among data. |
| **68** | Cattle | Grassland; big sagebrush and silver sagebrush | North-central Montana, USA | Feeding site examination | Autumn | 85 | 5 | 10 | Dusek (1975) | |
| **69** | Cattle (steers) | Desert grassland | Tucson, Arizona, USA | Rumen fistula | Autumn (Sept.–Dec.) | 94 | 0 | 6 | Galt *et al.* (1969) | 13 % of the diet unidentified distributed among grasses, forbs and shrubs; plains bristlegrass, lehmann lovegrass, and Arizona cottontop were major species; prickly pear cactus 5 to 15 % of the diet in winter. |
| **70** | Cattle | Wiregrass range | South Georgia, USA | Observation and hand-plucked samples | Sept. | 92 | 8 | <1 | Halls (1954) | Bluestems dominate grass; carpet grass of secondary importance. |
| **71** | Cattle | Wiregrass–pine grazing | Georgia, USA | Observation and hand-plucked samples | Autumn (Sept.; 3 years) | 95 | 4 | 1 | Halls *et al.* (1957) | See notes under no. 5. |
| **72** | Cattle, Hereford and Brahman cows | Wiregrass–pine range | Coastal plains, Georgia, USA | Hand-plucked samples and observation | Autumn (Sept.) | 87 | 8 | 5 | Hale *et al.* (1962) | See notes under no. 4. |
| **73** | Cattle, Hereford | Desert grassland | Southern New Mexico, USA | Observation of species grazed | Autumn | 51 | 41 | 8 | Herbel & Nelson (1966) | No apparent difference in total percentage of coarse plants grazed by the two breeds; Santa Gertrudis grazed more grass on yearlong basis, less forbs, and perhaps less shrubs. |
| **74** | Cattle (Santa Gertrudis) | Desert grassland | Southern New Mexico, USA | Observation of species grazed | Autumn | 50 | 42 | 8 | Herbel & Nelson (1966) | See notes under no. 73. |
| **75** | Cattle | Foothill rangeland, oak–sagebrush–grass | Utah, USA | Botanical sampling | Moderate autumn use | 58 | 21 | 21 | Julander (1955) | See notes under no. 42. |
| **76** | Cattle | Foothill native bunchgrass range | Central Colorado, USA | Rumen fistula, microscopic analysis | Autumn, hay meadow regrowth | 98 | 2 | 0 | Malacheck (1966) | See notes under no. 10. |
| **77** | Cattle | Foothill native bunchgrass range | Central Colorado, USA | Rumen fistula, microscopic analysis | Autumn, native bunchgrass (Nov.–Dec.) | 76 | 13 | 11 | Malacheck (1966) | See notes under no. 10. |
| **78** | Cattle | Savanna, live oak | Central Texas, USA | Observation, bite-count | Light grazing, only used in the autumn | 74 | 6 | 20 | McMahan (1964) | See notes under no. 124. |
| **79** | Cattle | Savanna, live oak | Central Texas, USA | Observation, bite-count | Autumn, heavy use | 79 | 0 | 21 | McMahan (1964) | See notes under no. 124. |

**Appendix 4.2** (*continued*)

| No. | Animal species | Kind of grazingland | Geographic location | Methods | Season/ treatment | Grass (%) | Forbs (%) | Shrubs (%) | Reference | Notes |
|---|---|---|---|---|---|---|---|---|---|---|
| **80** | Cattle | Semidesert grassland | Southern New Mexico, USA | Oesophageal fistula, microscopic analysis | Autumn | 58 | 35 | 7 | Rosiere *et al.* (1975) | 6 % unidentified distributed among data. Mesa dropseed main grass in diet. Any forbs species seldom more than 14 %; any shrubs species seldom more than 10 %. |
| **81** | Cattle (yearlings) | Sandhill range | Northeastern Colorado, USA | Oesophageal fistula | Autumn (Sept. and Nov.) | 78 | 18 | 4 | Scales (1972) | See notes under no. 15. |
| **82** | Cattle | Piñon–juniper | South-central New Mexico, USA | Oesophageal fistula | Autumn (Oct.) | 58 | 37 | 5 | Thetford *et al.* (1971) | See notes under no. 127. |
| **83** | Cattle | Shortgrass prairie | Northeastern Colorado, USA | Oesophageal fistula, microscopic analysis | Autumn (Oct.), light grazing | 86 | 9 | 5 | This chapter | See notes under no. 20. |
| **84** | Cattle | Shortgrass prairie | Northeastern Colorado, USA | Oesophageal fistula, microscopic analysis | Autumn (Oct.), heavy grazing | 30 | 22 | 48 | This chapter | See notes under no. 20. |
| **85** | Cattle (steers) | Sandhill grassland | Northeastern Colorado, USA | Oesophageal fistula, microscopic analysis | Sept. | 96 | 4 | 0 | Wallace *et al.* (1972) | Blue grama grass most important in diet overall; prairie sandreed important in July; needle-and-thread in Dec. Forbs important in early season. |
| **86** | Cattle, Zebu | *Themeda–Pennisetum* grassland | Southeast, near Nairobi, Kenya | Faecal collection, microscopic analysis | Autumn (Dec.) | 96 | 4 | 0 | Casebeer & Koss (1970) | See notes under no. 26. |
| **87** | Cattle | Desert shrub | Southern Nevada, USA | Rumen fistula, binocularscopic analysis | Winter (Jan.–Mar.) | 22 | 0 | 78 | Connor *et al.* (1963) | Grass composition low in autumn and winter, increased in spring and summer; cattle forced to consume primarily the browse because all other material had been grazed. In northern Nevada forbs 26 % in June; in southern Nevada the major browse *Eurotia lanata* and Brigham tea. |
| **87a** | Cattle | Desert range | Western Utah, USA | Oesophageal fistula or hand-plucked samples | Winter | 56 | 3 | 41 | Cook & Harris (1968) | |
| **88** | Cattle | Native sandhill range | Northeastern, Colorado, USA | Direct observation | Winter | 99 | 1 | 0 | Denham (1965) | 1 % others distributed among data. |
| **89** | Cattle, Hereford and Brahman cows | Wiregrass–pine range | Coastal plains, Georgia, USA | Hand-plucked samples and observation | Winter (Dec.) | 59 | 6 | 35 | Hale *et al.* (1962) | See notes under no. 4. |

| | | | | | | | | | | |
|---|---|---|---|---|---|---|---|---|---|---|
| **90** | Cattle | Wiregrass–pine grazing | Georgia, USA | Observation and hand-plucked samples | Winter (Dec.; 3 years) | 73 | 6 | 21 | Halls *et al.* (1957) | See notes under no. 5. |
| **91** | Cattle, Hereford | Desert grassland | Southern New Mexico, USA | Observation of species grazed | Winter | 50 | 27 | 23 | Herbel & Nelson (1966) | See notes under no. 73. |
| **92** | Cattle, Santa-Gertrudis | Desert grassland | Southern New Mexico, USA | Observation of species grazed | Winter | 65 | 20 | 15 | Herbel & Nelson (1966) | See notes under no. 73. |
| **93** | Cattle | Foothill native bunchgrass range | Central Colorado, USA | Rumen fistula, microscopic analysis | Winter, native meadow | 77 | 21 | 2 | Malachek (1966) | See notes under no. 10. |
| **94** | Cattle | Savanna, live oak | Central Texas, USA | Observation, bite-count | Light grazing, used in winter | 37 | 13 | 50 | McMahan (1964) | See notes under no. 124. |
| **95** | Cattle | Savanna, live oak | Central Texas, USA | Observation, bite-count | Winter, heavy use | 90 | 0 | 10 | McMahan (1964) | See notes under no. 124. |
| **96** | Cattle (heifers) | Sandhill range | Colorado, USA | Observation, before-and-after sampling | Late winter | 95 | 1 | 4 | Reppert (1957) | Samples in Feb. and Mar. Prairie sandreed, blue grama and needle-and-thread comprised at least $\frac{2}{3}$ of diet; prairie sandreed important in late winter and summer; blue grama important late summer, autumn, early winter; needle-and-thread important spring and autumn. Forbs most important in growing season, 7 %; shrubs most important winter, 14 % in mid-winter. |
| **97** | Cattle | Semidesert grassland | Southern New Mexico, USA | Oesophageal fistula, microscopic analysis | Winter | 35 | 61 | 4 | Rosiere *et al.* (1975) | 8 % inidentified. Mesa dropseed main grass in diet. Any forbs species seldom more than 14 %; any shrubs species seldom more than 10 %. |
| **98** | Cattle | Northern desert shrub | Southern Nevada, USA | Rumen fistula | Winter | 98 | 1 | 1 | Smith *et al.* (1968) | |
| **99** | Cattle | Piñon–juniper | South-central New Mexico, USA | Oesophageal fistula | Winter (Jan.) | 64 | 33 | 3 | Thetford *et al.* (1971) | See notes under no. 127. |
| **100** | Cattle | Shortgrass prairie | Northeastern Colorado, USA | Oesophageal fistula, microscopic analysis | Winter (Dec.–Jan.), light grazing | 98 | 2 | 0 | This chapter | See notes under no. 20. |
| **101** | Cattle | Shortgrass prairie | Northeastern Colorado, USA | Oesophageal fistula, microscopic analysis | Winter (Dec.–Jan.), heavy grazing | 79 | 19 | 2 | This chapter | See notes under no. 20. |
| **102** | Cattle (steers) | Sandhill grassland | Northeastern Colorado, USA | Oesophageal fistula, microscopic analysis | Dec. | 99 | 1 | 0 | Wallace *et al.* (1972) | Blue grama most important in diet overall; prairie sandreed important in July; needle-and-thread in Dec. Forbs Important in early season. |

**Appendix 4.2** (*continued*)

| No. | Animal species | Kind of grazingland | Geographic location | Methods | Season or treatment | Grass (%) | Forbs (%) | Shrubs (%) | Reference | Notes |
|---|---|---|---|---|---|---|---|---|---|---|
| **103** | Cattle | Native sandhill range | Northeastern Colorado, USA | Direct observation | Year | 98 | 2 | 0 | Denham (1965) | 3 % others distributed among data. |
| **104** | Cattle | Mesquite–buffalo grass, chap bristlegrass, and bunchgrass–annual forb communities | South Texas, USA | Botanical measurements of percentage utilization times and percentage frequency of use | Yearlong average, all soil types | 61 | 36 | 3 | Drawe & Box (1968) | See notes under no. 469. |
| **105** | Steers (yearlings) | Annual grassland | Southern California, USA | Observation and hand-plucked samples; sulphur-fertilized | Year | 51 | 49 | <1 | Green *et al.* (1958) | See notes under no. 106. |
| **106** | Steers (yearlings) | Annual grassland | Southern California, USA | Observation and hand-plucked samples; unfertilized rangeland | Year | 81 | 18 | <1 | Green *et al.* (1958) | Diet on fertilized range depended largely on relative availability of grasses and clover; grasses, grass-like plants and clovers were preferred species; foxtail fescue and softchess were major dietary components. Grasses always dominated the diet in unfertilized pasture but on fertilized pasture clover was grazed first then grasses. |
| **107** | Cattle | Mountain shrub | Piceance Basin, Colorado, USA | Microscopic analysis of faecal samples | Yearlong average | 94 | 1 | 5 | Hubbard & Hansen (1976) | See notes under no. 348. |
| **108** | Cattle | Piñon–juniper | Piceance Basin, Colorado, USA | Microscopic analysis of faecal samples | Yearlong average | 84 | 1 | 15 | Hubbard & Hansen (1976) | See notes under no. 348. |
| **109** | Cattle | Various | Western North America | Various | Yearlong average | 81 | 15 | 4 | A. Johnston & J. E. Mattox (unpublished) | Data based on review of literature of about 50 references; data are 'impressions' from the literature. |
| **110** | Cattle | Desert shrub | Nevada, USA | Rumen fistula, microscopic analysis | 2 years | 26 | 0 | 74 | Lesperance *et al.* (1970) | *Salsola kali* major forb in late summer; *Purshia glandulosa* major shrub in winter. |
| **111** | Cattle | Grassy river plain | Central Australia | Rumen samples | Year | 86 | — 14 — | | Newsome (1975) | Shrubs includes trees and forbs. |
| **112** | Cattle | Sagebrush–grass, saltbush, and rabbit brush types | Red Desert, southwestern Wyoming, USA | Microscopic analysis of faecal samples | Yearlong average | 83 | 2 | 15 | Olsen & Hansen (1977) | See notes under no. 350. |

| | | | | | | | | | |
|---|---|---|---|---|---|---|---|---|---|
| **113** | Cattle (steers) | | Southeastern Colorado, USA | Bite-counts | May–Oct. average | 85 | 15 | 0 | Beck (1975) | Nearly all plants eaten while young and growing; sand dropseed was preferred and blue grama. Grass in diet reaches low of 70 % in late summer and highest values in early summer. |
| **114** | Cattle (steers) | Seeded native pasture | Southeastern Colorado, USA | Composition–utilization measurements | May–Oct. | 97 | 3 | 0 | Fox (1971) | Blue grama and sideoats grama each about ⅓ of diet. |
| **115** | Cattle (steers) | Shortgrass prairie | Southeastern Colorado, USA | Composition–utilization measurements | May–Oct. | 100 | 0 | 0 | Fox (1971) | Three main species – blue grama, buffalograss, sand dropseed; blue grama generally more than 50 % of diet. |
| **116** | Cattle (steers) | Successional old field | Southeastern Colorado, USA | Composition–utilization measurements | May–Oct. | 97 | 3 | 0 | Fox (1971) | Sand dropseed main grass in diet, about 60 %. |
| **117** | Cattle | Northern desert shrub | Nevada, USA | Rumen fistula, microscopic analysis | May–Nov., 4 years | 80 | 11 | 9 | Lesperance *et al.* (1970) | *Salsola kali* major forb in late summer *Purshia glandulosa* major shrub in winter. |
| **118** | Cattle | Sagebrush–grass–aspen | Nevada, USA | Rumen fistula, microscopic analysis | May–Nov., 4 years | 83 | 8 | 9 | Lesperance *et al.* (1970) | *Salsola kali* major forb in late summer; *Purshia glandulosa* major shrub in winter. |
| **119** | Sheep, Merino | Bladder saltbush–pigface community | Riverine plain, Hay, New South Wales, Australia | Oesophageal fistula | Spring (Nov.) | 7 | 45 | 48 | Leigh & Mulham (1966) | About 2 and 6 % dead material unidentified in samples in spring and summer respectively. Major shrub was *Atriplex vesicaria*; important annual herbs were *Brachychome campylocarpa* and *Calocephalus sonderi*. |
| **120** | Sheep, Merino | Cotton bush–grassland community | Riverine, plains, Deniliquin, New South Wales, Australia | Oesophageal fistula | Spring | 75 | 21 | 4 | Leigh & Mulham (1966) | See notes under no. 200. |
| **121** | Sheep, Merino | Cotton bush–grassland community | Deniliquin, New South Wales, Australia | Oesophageal fistula | Spring (Sept.) | 28 | 70 | 2 | Leigh *et al.* (1968) | Average of light and heavy stocking. 24 % other distributed among data. |
| **122** | Sheep | Chaparral and annual grassland | California, USA | Rumen analysis, microscopic identification | Spring | 78 | 4 | 18 | W. M. Longhurst, G. E. Connolly, B. M. Browning & E. O. Garten (unpublished) | |
| **123** | Sheep | Savanna, live oak | Central, Texas, USA | Observation, bite-count | Light grazing, only in spring | 25 | 50 | 25 | McMahan (1964) | See notes under no. 124. |

## Appendix 4.2 (*continued*)

| No. | Animal species | Kind of grazingland | Geographic location | Methods | Season/ treatment | Grass %) | Forbs (%) | Shrubs %) | Reference | Notes |
|---|---|---|---|---|---|---|---|---|---|---|
| **124** | Sheep | Savanna, live oak | Central Texas, USA | Observation, bite-count | Spring, heavy use | 65 | 27 | 8 | McMahan (1964) | Some of the browse species that were preferred were *Quercus virginiana*, and mast from various oak species; *Bumelia lanuginosa* was a preferred species in three seasons as were *Celtis taxana*, *Euphorbia serpens* and *Hybanthus verticillatus*. *Oxalis delleni* was a preferred forb. Light use in season is average of five pastures. Only 1 cow, 1 sheep, 1 goat and 1 deer were used. |
| **125** | Sheep | Piñon–juniper | Central Utah, USA | Botanical measurements, weight-estimate methodology | Spring | 47 | 5 | 48 | Smith & Julander (1953) | See notes under no. 393. |
| **126** | Sheep | Sagebrush–oak | Central Utah, USA | Botanical measurements, weight-estimate methodology | Spring | 43 | 19 | 38 | Smith & Julander (1953) | See notes under no. 393. |
| **127** | Sheep | Piñon–juniper | South-central New Mexico, USA | Oesophageal fistula | Spring (Apr.) | 32 | 62 | 6 | Thetford *et al.* (1971) | Sheep diets generally higher in forbs and lower in grass than cattle diets, contain more grass than forbs except during summer. Cattle graze more grass than forbs except spring; blue grama most important grass followed by sideoats grama. In sheep diets the forbs sagewort, verbain and globe-mallow most important. |
| **128** | Sheep | Shortgrass prairie | Northeastern Colorado, USA | Oesophageal fistula, microscopic analysis | Late spring (May), light grazing | 64 | 9 | 27 | This chapter | See notes under no. 20. |
| **129** | Sheep | Shortgrass prairies | Northeastern Colorado, USA | Oesophageal fistula, microscopic analysis | Heavy grazing, late spring (May) | 90 | 4 | 6 | This chapter | See notes under no. 20. |
| **130** | Sheep | High mountain range | Southwestern Montana, USA | Oesophageal fistula | Early summer (2-year average) | 30 | 70 | 0 | Buchanon *et al.* (1972) | Mountain brome was major grass eaten; *Potentilla*, *Polygonum* and *Aster* were major forbs. Proportion of leaves increased from early to late summer. |
| **131** | Sheep | Shortgrass prairie | Northeastern Colorado. USA | Oesophageal fistula, microscopic analysis | Early summer (June), light grazing | 41 | 56 | 3 | This chapter | See notes under no. 20. |

| | | | | | | | | | | |
|---|---|---|---|---|---|---|---|---|---|---|
| **132** | Sheep | Shortgrass prairie | Northeastern Colorado, USA | Oesophageal fistula, microscopic analysis | Early summer (June), heavy grazing | 27 | 72 | 1 | This chapter | See notes under no. 20. |
| **133** | Sheep (wethers) | Annual grassland | Northern California, USA | Oesophageal fistula, binocularscopic analysis | Early summer | 56 | 35 | 9 | Van Dyne & Heady (1965) | Main grasses *Avena*, *Aira*, *Bromus* and *Gastridium*; main forbs *Erodium*, *Trifolium*, *Lupinus* and *Medicago*: browse was leaves and acorns from *Quercus*; perennials highly selected; Spanish moss frequent in sheep diets in late summer; forb stems grazed by sheep early, cattle later; grass leaves and total leaves highest in early summer. Afternoon samples higher in total leaves and grass than morning samples; sheep selected more leaves than cattle; sheep selected forb heads in early and mid-summer, cattle in late summer; morning samples had more genera than afternoon samples; sheep always selected as many or more genera than did cattle; sheep selected more total leaves, grass leaves, forb leaves and legumes than did cattle. |
| **133a** | Sheep | Mountain range | Utah, USA | Oesophageal fistula or hand-plucked samples | Early summer | 29 | 48 | 23 | Cook & Harris (1968) | |
| **134** | Sheep, Corridale (wethers) | Sandhills grassland | Bahia Blanca, Argentina | Oesophageal fistula | Summer grazing | 14 | 82 | 4 | Bishop *et al.* (1975) | *Cenchrus pauciflorus* was major species in diet; low values in April and May due to spiny spikelet. Other species included annuals and perennials. 10 % unknowns distributed among data. |
| **135** | Sheep | Mountain, aspen and sagebrush | Northern Utah, USA | Before-and-after vegetation sampling | Summer (mid-July–mid-Sept.) | 42 | 30 | 28 | Cook & Harris (1950) | Browse used more as season advances; main grasses *Bromus carinatus* and *Elymus glaucus*; and forb *Lathrus lucuanthus*; main shrub *Purshia tridentata*. |
| **136** | Sheep | Mountain rangeland, aspen | Cache National Forest, Utah, USA | Before-and-after vegetation sampling | Summer grazing | 54 | 29 | 17 | Cook *et al.* (1948) | See notes under no. 138. |
| **137** | Sheep (ewes) | Mountain brush-grass range | Northern Utah, USA | Herbage production and utilization measurements for diet | Summer (July–Sept.) | 42 | 37 | 21 | Cook *et al.* (1961) | Average of 4 grazing periods. Grass is more abundant in early portion of summer than in later summer. |

## Appendix 4.2 (*continued*)

| No. | Animal species | Kind of grazingland | Geographic location | Methods | Season/ treatment | Grass (%) | Forbs (%) | Shrubs (%) | Reference | Notes |
|---|---|---|---|---|---|---|---|---|---|---|
| **138** | Sheep | Mountain rangeland, sagebrush | Cache National Forest, Utah, USA | Before-and-after vegetation sampling | Summer grazing | 1 | 33 | 66 | Cook *et al.* (1948) | Estimated consumption per ewe and 1.35 lambs was 9.2 pounds per day air-dry. Leaves grazed more than stems on all types and seasons. More preference exerted in aspen-type than in sagebrush-type due to greater availability of herbage. Grass use decreased with advance in season on sagebrush-type and browse use increased. Preferred plant parts grazed 90 to 100 %, when the entire plant was utilized only 40 to 50 %. |
| **139** | Sheep | Sagebrush–grass–mountain browse–aspen | Northern Utah, USA | Oesophageal fistula | Summer (June–Sept.; 3 years), good range condition | 28 | 34 | 38 | Cook *et al.* (1965) | More browse in diet on poor range; grass is higher on good range. |
| **140** | Sheep | Sagebrush–grass–mountain browse–aspen | Northern Utah, USA | Oesophageal fistula | Summer (June–Sept.; 3 years), poor range condition | 25 | 33 | 42 | Cook *et al.* (1965) | See notes under no. 139. |
| **141** | Sheep | Mountain aspen range | Northern Utah, USA | Oesophageal fistula | Summer grazing (June–Sept.; 2 years) | 43 | 49 | 8 | Cook *et al.* (1967) | See notes under no. 30. |
| **142** | Sheep | Sagebrush–grass, mountain | Northern Utah, USA | Oesophageal fistula | Summer grazing (June–Sept.; 2 years) | 51 | 26 | 23 | Cook *et al.* (1967) | See notes under no. 30. |
| **143** | Sheep | Rocky ridges, pearl saltbush, flat country of bladder saltbush | Northwest New South Wales, Australia | Samples from junction of oesophagus and first fermentation chamber of shot animals, microscopic identification | Summer | 1 | 94 | 5 | Dawson *et al.* (1975) | |
| **144** | Sheep (ewes) | Mountain aspen range | Western Colorado, USA | Visual observation of animal | Summer (June–Sept.; 3 years) | 19 | 58 | 23 | Doran (1943) | Ewes spend larger percentage of total feeding time grazing grasses and forbs than lambs, but a smaller percentage of time on browse. |
| **145** | Sheep (lambs) | Mountain aspen range | Western Colorado, USA | Visual observation of animal | Summer (June–Sept.; 3 years) | 10 | 40 | 50 | Doran (1943) | See notes under no. 144. |

| | | | | | | | | | | |
|---|---|---|---|---|---|---|---|---|---|---|
| **146** | Sheep | Grass–heather–blueberry–lichens | Scottish highlands, Cairngorm mountains, UK | Stomach tube for rumen contents | Summer | 75 | 5 | 20 | Hobson *et al.* (1976) | See notes under no. 495. |
| **147** | Sheep | Tall forb summer range | Southwestern Montana, USA | Oesophageal fistula | Summer | 16 | 84 | 0 | Laycock *et al.* (1972) | |
| **148** | Sheep, Merino | Bladder saltbush–cotton bush | Riverine plain, Deniliquin, New South Wales, Australia | Oesophageal fistula | Summer (average of 2 years) | 68 | 2 | 30 | Leigh & Mulham (1966) | See notes under no. 200. |
| **149** | Sheep, Merino | Bladder saltbush–pigface community | Riverine plain, Hay, New South Wales, Australia | Oesophageal fistula | Summer (Feb.) | 0 | 4 | 96 | Leigh & Mulham (1966) | See notes under no. 200. |
| **150** | Sheep, Merino | Cotton bush–grassland community | Riverine plain, Deniliquin, New South Wales, Australia | Oesophageal fistula | Summer | 77 | 16 | 7 | Leigh & Mulham (1966) | See notes under no. 200. |
| **151** | Sheep Merino | Cotton bush–grassland community | Riverine plain, Deniliquin, New South Wales, Australia | Oesophageal fistula | Summer (Nov.–and Feb.) | 65 | 19 | 16 | Leigh *et al.* (1968) | Average of light and heavy stocking. 15 % other species distributed among data. |
| **152** | Sheep | Chaparral and annual grassland | California, USA | Rumen analysis, microscopic identification | Summer | 76 | 13 | 11 | W. M. Longhurst *et al.* (unpublished) | |
| **153** | Sheep | Savanna, live oak | Central Texas, USA | Observation, bite-count | Light grazing, only used in summer | 41 | 36 | 23 | McMahan (1964) | See notes under no. 124. |
| **154** | Sheep | Savanna, live oak | Central Texas, USA | Observation, bite-count | Heavy summer use | 83 | 4 | 13 | McMahan (1964) | See notes under no. 124. |
| **155** | Sheep | Open mountain meadows | Colorado, USA | Range survey | Summer | 47 | 50 | 3 | Nichols (1957) | |
| **156** | Sheep | Shortgrass prairie | Laramie, Wyoming, USA | Oesophageal fistula, microscopic analysis | June–Aug. | 82 | 14 | 4 | Rice *et al.* (1971) | Wheatgrass species 30–50 % of diet; needle-and-thread of second importance. |
| **157** | Sheep | Shortgrass prairie | Laramie, Wyoming, USA | Rumen fistula | June–Aug. | 95 | 4 | 1 | Rice *et al.* (1971) | Rumen sample values different because of possible layering of rumen contents; grass species more likely to float into rumen. Wheatgrass species 30–50 % of diet; needle-and-thread of second importance. |
| **158** | Sheep, merino | Grassland | Deniliquin, New South Wales, Australia | Oesophageal fistula | Summer | 92 | 4 | 4 | Robards *et al.* (1967) | See notes under no. 224. 46 % unidentified distributed among data. |
| **159** | Sheep | Big sagebrush–grass | South-central Wyoming, USA | Rumen samples | Summer (average 2 years) | 88 | 0 | 12 | Severson & May (1967) | Relative composition; no forb use reported. |

**Appendix 4.2** (*continued*)

| No. | Animal species | Kind of grazingland | Geographic location | Methods | Season/ treatment | Grass (%) | Forbs (%) | Shrubs (%) | Reference | Notes |
|---|---|---|---|---|---|---|---|---|---|---|
| **160** | Sheep | Currant–snowberry | Central Utah, USA | Botanical measurements, weight-estimate methodology | Summer | 33 | 47 | 20 | Smith & Julander (1953) | See notes under no. 393. |
| **161** | Sheep | Mountain range, aspen | Central Utah, USA | Botanical measurements, weight-estimate methodology | Summer | 35 | 47 | 18 | Smith & Julander (1953) | See notes under no. 393. |
| **162** | Sheep | Mountain range, spruce–fir zone | Central Utah, USA | Botanical measurements, weight-estimate methodology | Summer | 13 | 53 | 34 | Smith & Julander (1953) | See notes under no. 393. |
| **163** | Sheep | Sagebrush–chokecherry | Central Utah, USA | Botanical measurements, weight-estimate methodology | Summer | 26 | 23 | 51 | Smith & Julander (1953) | See notes under no. 393. |
| **164** | Sheep | Sagebrush–grass | Central Utah, USA | Botanical measurements, weight-estimate methodology | Summer | 57 | 41 | 2 | Smith & Julander (1953) | See notes under no. 393. |
| **165** | Sheep | Sagebrush–snowberry | Central Utah, USA | Botanical measurements, weight-estimate methodology | Summer | 42 | 13 | 45 | Smith & Julander (1953) | See notes under no. 393. |
| **166** | Sheep, Rambouillet (ewes and lambs) | Alpine mountain range | Near Cody, Wyoming, USA | Oesophageal fistula, free-grazing sheep | July–Sept. | 44 | 52 | 4 | Strasia *et al.* (1970) | Two-thirds of average diet of whiproot clover, dwarf clover, American bistort, alpine avens and fescues. Sheep more selective in choice of grass and sedge species than for individual forb species. Sheeps and red fescue most common grasses; whiproot clover most common forb. |
| **167** | Sheep, Rambouillet (ewes and lambs) | Alpine mountain range | Near Cody, Wyoming, USA | Oesophageal fistula, herded sheep | July–Sept. | 32 | 65 | 3 | Strasia *et al.* (1970) | See notes under no. 166. |
| **168** | Sheep | Piñon–juniper | South-central New Mexico, USA | Oesophageal fistula | Summer (June–Aug. average) | 45 | 54 | <1 | Thetford *et al.* (1971) | See notes under no. 127. |

| | | | | | | | | | | |
|---|---|---|---|---|---|---|---|---|---|---|
| **169** | Sheep | Belaah–rosewood community (*Casuarina–Heterodendrum*) | Ivanhoe, New South Wales, Australia | Oesophageal fistula, hand separated | Summer (Nov.), goat-grazed pasture | — 90 — | | 9 | Wilson *et al.* (1975) | See notes under no. 247. |
| **170** | Sheep | Belaah–rosewood community (*Casuarina–Heterodendrum*) | Ivanhoe, New South Wales, Australia | Oesophageal fistula, hand separated | Summer (Nov.) sheep-grazed pasture | — 80 — | | 20 | Wilson *et al.* (1975) | See notes under no. 247. |
| **171** | Sheep (wethers) | Annual grassland | Northern California, USA | Oesophageal fistula, binocular analysis | Mid-summer | 60 | 33 | 7 | Van Dyne & Heady (1965) | See notes under no. 133. |
| **171a** | Sheep | Mountain range | Utah, USA | Oesophageal fistula or hand-plucked samples | Mid-summer | 34 | 34 | 32 | Cook & Harris (1968) | |
| **171b** | Sheep | Mountain range | Utah, USA | Oesophageal fistula or hand-plucked samples | Late summer | 35 | 21 | 44 | Cook & Harris (1968) | |
| **172** | Sheep | High mountain range | Southwestern Montana, USA | Oesophageal fistula | Late summer (2-year average) | 20 | 80 | 0 | Buchanon *et al.* (1972) | See notes under no. 130. |
| **173** | Sheep | Shortgrass prairie | Northeastern Colorado, USA | Oesophageal fistula, microscopic analysis | Late summer (Aug.), light grazing | 35 | 65 | 0 | This chapter | See notes under no. 20. |
| **174** | Sheep | Shortgrass prairie | Northeastern Colorado, USA | Oesophageal fistula, microscopic analysis | Late summer (Aug.), heavy grazing | 69 | 26 | 5 | This chapter | See notes under no. 20. |
| **175** | Sheep (wethers) | Annual grassland | Northern California, USA | Oesophageal fistula, binocular analysis | Late summer | 63 | 30 | 7 | Van Dyne & Heady (1965) | See notes under no. 133. |
| **176** | Sheep | Tussock grassland | New Zealand | Microscopic faecal analysis | Autumn, fertilized and interseeded | 48 | 52 | 0 | Hughes (1975) | See notes under no. 221. |
| **177** | Sheep | Tussock grassland | New Zealand | Microscopic faecal analysis | Autumn, untreated grassland | 50 | 47 | 3 | Hughes (1975) | See notes under no. 221. |
| **178** | Sheep, Merino | Cotton bush–grassland community | Deniliquin, New South Wales, Australia | Oesophageal fistula | Autumn (Apr.) | 62 | 33 | 5 | Leigh *et al.* (1968) | Average of light and heavy stocking. 4 % other species and 27 % dead matter distributed among data. |
| **179** | Sheep | Chaparral and annual grassland | California, USA | Rumen analysis, microscopic identification | Autumn | 95 | 4 | 1 | W. M. Longhurst *et al.* (unpublished) | |
| **180** | Sheep | Savanna, live oak | Central Texas, USA | Observation, bite-count | Light grazing, only used in autumn | 59 | 17 | 24 | McMahan (1964) | See notes under no. 124. |
| **181** | Sheep | Savanna, live oak | Central Texas, USA | Observation, bite count | Autumn, heavy use | 81 | 1 | 18 | McMahan (1964) | See notes under no. 124. |
| **182** | Sheep | Piñon–juniper | South-central New Mexico, USA | Oesophageal fistula | Autumn (Oct.) | 30 | 68 | 2 | Thetford *et al.* (1971) | See notes under no. 127. |

## Appendix 4.2 (*continued*)

| No. | Animal species | Kind of grazingland | Geographic location | Methods | Season/ treatment | Grass (%) | Forbs (%) | Shrubs (%) | Reference | Notes |
|---|---|---|---|---|---|---|---|---|---|---|
| **183** | Sheep | Shortgrass prairie | Northeastern Colorado, USA | Oesophageal fistula, microscopic analysis | Autumn (Oct.), light grazing | 70 | 28 | 2 | This chapter | See notes under no. 20. |
| **184** | Sheep | Shortgrass prairie | Northeastern Colorado, USA | Oesophageal fistula, microscopic analysis | Autumn (Oct.), heavy grazing | 60 | 18 | 22 | This chapter | See notes under no. 20. |
| **185** | Sheep | Belaah–rosewood community (*Casuarina–Heterodendrum*) | Ivanhoe, New South Wales, Australia | Oesophageal fistula, hand separated | Autumn (Feb.), goat-grazed pasture | — 97 — | | 3 | Wilson *et al.* (1975) | See notes under no. 247. |
| **186** | Sheep | Belaah–rosewood community (*Casuarina–Heterodendrum*) | Ivanhoe, New South Wales, Australia | Oesophageal fistula, hand separated | Autumn (Feb.), sheep-grazed pasture | — 70 — | | 30 | Wilson *et al.* (1975) | See notes under no. 247. |
| **187** | Sheep | Saltbush–sagebrush–grass range | West-central Utah, USA | Before-and-after vegetation sampling | Winter (2-year average) | 25 | <1 | 75 | Cook & Harris (1950) | Grass and shrub utilization about 25 %. Main species winterfat and shadescale, secondary black sage and rabbitbrush; main grasses indian ricegrass, sand dropseed, and blue grama. |
| **188** | Sheep | Salt desert shrub range | West Desert, Utah, USA | Observation and hand-plucked samples | Winter (4-year average) | 50 | 0 | 50 | Cook *et al.* (1953) | Main browse species were black sage, shadescale and winterfat; main grass species were galleta, Indian ricegrass, and sand dropseed. |
| **189** | Sheep (wethers) | Salt desert shrub range | Western Utah, USA | Observation and hand-plucked samples | Winter, grass range | 76 | 0 | 24 | Cook *et al.* (1954) | Main species are black sage, chicken sage, big sage, shadescale, nuttall saltbush, rabbitbrush, winterfat; main grasses are western wheatgrass, basin wild-rye, galleta, ricegrass, squirreltail, alkali sacaton, sand dropseed, and needle-and-thread; on grass range, ricegrass, sand dropseed, and needle-and-thread were major species in diet; on saltbush range, saltbush and ricegrass were major species in diet; on sagebrush range, big sagebrush and black sagebrush major species in diet. |
| **190** | Sheep (wethers) | Salt desert shrub range | Western Utah, USA | Observation and hand-plucked samples | Winter, sagebrush range | 30 | 0 | 70 | Cook *et al.* (1954) | See notes under no. 189. |

| | | | | | | | | | | |
|---|---|---|---|---|---|---|---|---|---|---|
| **191** | Sheep (wethers) | Salt desert shrub range | Western Utah, USA | Observation and hand-plucked samples | Winter, saltbush range | 39 | 0 | 61 | Cook *et al.* (1954) | See notes under no. 189. |
| **192** | Sheep | Salt desert shrub range | Southeastern Utah, USA | Herbage production and utilization measurements to derive diet | Winter (Nov.–Mar.; 2 years) good range, light grazing | 45 | 2 | 53 | Cook *et al.* (1962) | Important grasses were Indian ricegrass, galleta grass, needle-and-thread, and threeawns and squirreltail and sand dropseed; important shrubs were yellowbrush, winterfat, big sagebrush; important forbs were globemallow and Russian thistle. Average of 2 locations, 2 5-day periods for light and heavy use. |
| **193** | Sheep | Salt desert shrub ranges | Southwestern Utah, USA | Herbage production and utilization measurements to derive diet | Winter (Nov.–Mar.; 2 years) good range, heavy grazing | 72 | 1 | 27 | Cook *et al.* (1962) | See notes under no. 192. |
| **194** | Sheep | Salt desert shrub ranges | Southwestern Utah, USA | Herbage production and utilization measurements to derive diet | Winter (Nov.–Mar.; 2 years) poor range, light grazing | 55 | 3 | 42 | Cook *et al.* (1962) | See notes under no. 192. |
| **195** | Sheep | Salt desert shrub ranges | Southwestern Utah, USA | Herbage production and utilization measurements to derive diet | Winter (Nov.–Mar.; 2 years), poor range, heavy grazing | 50 | 2 | 48 | Cooke *et al.* (1962) | See notes under no. 192. |
| **196** | Sheep | Desert shrub range | West-central Utah, USA | Hand-plucked samples, oesophageal fistula (separate methods) | Winter | 42 | 0 | 58 | Edlefsen *et al.* (1960) | Diets determined by modified methods of Cook *et al.* (1948). |
| **197** | Sheep (wethers, lambs and yearlings) | Salt desert shrub range | West-central Utah, USA | Before-and-after clipping | Winter grazing (Nov.–Apr.; 2 years) | 25 | <1 | 75 | Green *et al.* (1951) | About 25 % use on browse and grasses each. |
| **198** | Sheep | Tussock grassland | New Zealand | Microscopic faecal analysis | Winter, untreated grassland | 49 | 48 | 3 | Hughes (1975) | See notes under no. 221. |
| **199** | Sheep | Tussock grassland | New Zealand | Microscopic faecal analysis | Winter, fertilized and interseeded | 47 | 53 | 0 | Hughes (1975) | See notes under no. 221. |
| **200** | Sheep, Merino | Cotton bush–grassland community | Riverine plain, Deniliquin, New South Wales Australia | Oesophageal fistula | Winter | 42 | 57 | 1 | Leigh & Mulham (1966) | There was about 11, 1 and 6 % dead material during winter, early spring and summer respectively. Perennial grasses were most important in the spring grazing period (*Danthonia caespitosa* and *Stipa variabilis*); annual grasses were most important during the spring. Annual forbs were most important in the winter and shrubs were never a major component of the diet. |

## Appendix 4.2 (*continued*)

| No. | Animal species | Kind of grazingland | Geographic location | Methods | Season/ treatment | Grass (%) | Forbs (%) | Shrubs (%) | Reference | Notes |
|---|---|---|---|---|---|---|---|---|---|---|
| **201** | Sheep, Merino | Cotton bush–grassland community | Deniliquin, New South Wales, Australia | Oesophageal fistula | Winter (June) | 54 | 31 | 15 | Leigh *et al.* (1968) | Average of light and heavy stocking. 23 % other species and 16 % dead matter distributed among data. |
| **202** | Sheep | Chaparral and annual grassland | California, USA | Rumen analysis, microscopic identification | Winter | 86 | 14 | 0 | W. M. Longhurst *et al.* (unpublished) | Grazing sheep fed supplements. |
| **203** | Sheep | Savanna, live oak | Central Texas, USA | Observation, bite-count | Light grazing, only used in winter | 55 | 5 | 40 | McMahan (1964) | See notes under no. 124. |
| **204** | Sheep | Savanna, live oak | Central Texas, USA | Observation, bite-count | Heavy winter use | 84 | 1 | 15 | McMahan (1964) | See notes under no. 124. |
| **205** | Sheep (wethers) | Salt desert grass–shrub ranges | Millard County, Utah, USA | Hand-plucked samples and observation | Winter (Oct.–Mar.) light grazing | 83 | 0 | 17 | Pieper *et al.* (1959) | Main grasses were sand dropseed, Indian ricegrass, galleta; main shrubs were shadscale, winterfat, brigham tea, yellow brush. Diet changed from one species to another with increased use during the trial; light use 43 % of available herbage, heavy use 75 %; heavy use increased shrub component in diet. |
| **206** | Sheep (wethers) | Salt desert grass–shrub ranges | Millard County Utah, USA | Hand-plucked samples and observation | Winter (Oct.–Mar.), heavy grazing | 74 | 0 | 26 | Pieper *et al.* (1959) | See notes under no. 205. |
| **207** | Sheep | Piñon–juniper | South-central New Mexico, USA | Oesophageal fistula | Winter (Jan.) | 31 | 69 | 0 | Thetford *et al.* (1971) | See notes under no. 127. |
| **208** | Sheep | Shortgrass prairie | Northeastern Colorado, USA | Oesophageal fistula, microscopic analysis | Winter (Dec.–Jan.), light grazing | 77 | 22 | 1 | This chapter | See notes under no. 20. |
| **209** | Sheep | Shortgrass prairie | Northeastern Colorado, USA | Oesophageal fistula, microscopic analysis | Winter (Dec.–Jan.), heavy grazing | 46 | 16 | 38 | This chapter | See notes under no. 20. |
| **210** | Sheep | Belaah–rosewood community (*Casuarina–Heterodendrum*) | Ivanhoe, New South Wales, Australia | Oesophageal fistula, hand separated | Winter (May–June), goat-grazed pasture | —97— | | 3 | Wilson *et al.* (1975) | See notes under no. 247. |
| **211** | Sheep | Belaah–rosewood community (*Casuarina–Heterodendrum*) | Ivanhoe, New South Wales, Australia | Oesophageal fistula, hand separated | Winter (May–June), sheep-grazed pasture | —68— | | 32 | Wilson *et al.* (1975) | See notes under no. 247. |

| | | | | | | | | | | |
|---|---|---|---|---|---|---|---|---|---|---|
| **212** | Sheep (wethers) | Grass and grass–clover pastures | New Zealand | Oesophageal fistula | Year | 89 | 11 | 0 | Grimes *et al.* (1966) | 18 % dead matter distributed among data. |
| **213** | Sheep | Various | Western North America | Various | Yearlong average | 57 | 31 | 12 | A. Johnston & J. E. Mattox (unpublished) | See notes under no. 109. |
| **214** | Sheep | *Spinifex* steppe and arid plain | Australia | Faecal analysis, epidermal area | Year, non-drought | 72 | 28 | 0 | Newsome (1975) | Forbs includes shrubs. |
| **215** | Sheep | *Spinifex* steppe and arid plain | Australia | Faecal analysis, epidermal area | Year, drought | 63 | 37 | 0 | Newsome (1975) | Forbs includes shrubs. |
| **216** | Sheep | Sagebrush–grass, saltbush, and rabbitbrush types | Red Desert, Southwestern Wyoming, USA | Microscopic analysis of faecal samples | Yearlong average | 41 | 1 | 58 | Olsen & Hansen (1977) | See notes under no. 350. |
| **217** | Sheep | Mountain summer range | Oregon and Washington, USA | Ocular estimate by plot | Year | 33 | 66 | 1 | Pickford & Reid (1943) | |
| **218** | Sheep | Alpine range | Southern Norway | Before-and-after clipping technique | Year, dry meadow | 54 | 38 | 8 | Wielgolaski (1975) | |
| **219** | Sheep | Alpine range | Southern Norway | Before-and-after clipping technique | Year, wet meadow | 73 | 10 | 17 | Wielgolaski (1975) | |
| **220** | Sheep | Tussock grassland | New Zealand | Microscopic faecal analysis | Spring–summer, fertilized and interseeded | 47 | 53 | 0 | Hughes (1975) | See notes under no. 221. |
| **221** | Sheep | Tussock grassland | New Zealand | Microscopic faecal analysis | Spring–summer, untreated grassland | 51 | 47 | 2 | Hughes (1975) | Relative frequencies summed and converted to per cent. |
| **222** | Sheep | Grass–heather–blueberry–lichen | Scottish highlands, Cairngorm Mountains, UK | Stomach tube for rumen contents | Autumn–Spring | 55 | 10 | 35 | Hobson *et al.* (1976) | See notes under no. 495. |
| **223** | Sheep, Merino | Bladder saltbush–cotton bush | Riverine Plain, Deniliquin, New South Wales, Australia | Oesophageal fistula | Winter–spring | 57 | 36 | 7 | Leigh & Mulham (1966) | Dead material of 9 and 6 % respectively in winter–spring and summer was not separated. In winter–spring diets the dominant grass was *Hordeum leporinum*, much more annual grass than perennial; predominant forb was *Medicago polymorpha*. Under summer grazing predominant grass was the annual *Sporobolus caroli* and the predominant shrub was *Atriplex vesicaria*. |
| **224** | Sheep, Merino | Grassland | Deniliquin, New South Wales, Australia | Oesophageal fistula | Winter–spring | 51 | 47 | 2 | Robards *et al.* (1967) | About 6 % dead material unidentified. Main grass was *Danthonia caespitosa*; annual grasses and annual forbs important in winter–spring grazing. Main annual herb is *Hedybnis cretica*. |

**Appendix 4.2** (*continued*)

| No. | Animal species | Kind of grazingland | Geographic location | Methods | Season/ treatment | Grass (%) | Forbs (%) | Shrubs (%) | Reference | Notes |
|---|---|---|---|---|---|---|---|---|---|---|
| **225** | Sheep | Sagebrush–grass | Utah, USA | | Moderate grazing, May–Oct. average | 21 | 58 | 21 | Jensen *et al.* (1972) | Bitterbush was main shrub and averaged about 17 out of 21 % of total shrub use; under moderate grazing utilization varied from 22 % of the bitterbrush available in late May to as high as 72 % in late July; under heavy grazing bitterbrush utilization varied from as low as 32 % in late May to as high as 88 % in late July. |
| **226** | Sheep | Sagebrush–grass | Utah, USA | | Heavy grazing, May–Oct. average | 20 | 59 | 21 | Jensen *et al.* (1972) | See notes under no. 225. |
| **227** | Goat | Grass–shrub rangeland | Edward's Plateau, Texas, USA | Oesophageal-fistulated animals, microscopic point technique | Spring | 40 | 25 | 35 | Malechek & Leinweber (1972) | Data are average of lightly and heavily grazed pastures; on a yearlong basis diets do not differ greatly with differences in grazing intensity. Herbage availability for grass averaged over grazing treatments was about 1100, 825, 775 and 675 lb · acre$^{-1}$ in winter, spring, summer, and autumn respectively. Browse much more available on lightly grazed than on heavily grazed pastures. Plant parts in diet were: leaves 80 %; stems, 15 %; and fruits, 5 %. |
| **228** | Goat | Savanna, live oak | Central Texas, USA | Observation, bite-count | Light grazing, only used in spring | 17 | 25 | 58 | McMahan (1964) | |
| **229** | Goat | Savanna, live oak | Central Texas, USA | Observation, bite-count | Spring, heavy use | 43 | 7 | 50 | McMahan (1964) | |
| **230** | Goat | Rocky ridges, pear saltbush, flat country of bladder saltbush | Northwest New South Wales, Australia | Samples from junction of oesophagus and first fermentation chamber of shot animals, microscopic identification | Summer | 1 | 29 | 70 | Dawson *et al.* (1975) | Shrubs includes trees. |
| **231** | Goat | Grass–shrub rangeland | Edward's Plateau, Texas, USA | Oesophageal-fistulated animals, microscopic point technique | Summer | 65 | 8 | 27 | Maleckek & Leinweber (1972) | See notes under no. 227. |

| | | | | | | | | | | |
|---|---|---|---|---|---|---|---|---|---|---|
| **232** | Goat | Savanna, live oak | Central Texas, USA | Observation, bite-count | Light grazing, only used in summer | 12 | 20 | 68 | McMahan (1964) | |
| **233** | Goat | Savanna, live oak | Central Texas, USA | Observation, bite-count | Summer, heavy use | 16 | 2 | 82 | McMahan (1964) | |
| **234** | Goat | *Acacia* and tall tropical trees | Kenya | Oesophageal fistula and ocular estimation | July–Aug. | 28 | | 48 | Nge'the & Box (1976) | See notes under no. 606. |
| **235** | Goat | Belaah–rosewood community (*Casuarina–Heterodendrum*) | Ivanhoe, New South Wales, Australia | Oesophageal fistula, hand separated | Summer (Nov.), goat-grazed pasture | —52— | | 48 | Wilson *et al.* (1975) | See notes under no. 247. |
| **236** | Goat | Belaah–rosewood community (*Casuarina–Heterodendrum*) | Ivanhoe, New South Wales, Australia | Oesophageal fistula, hand separated | Summer (Nov.), sheep-grazed pasture | —13— | | 87 | Wilson *et al.* (1975) | See notes under no. 247. |
| **237** | Goat | Grass–shrub rangeland | Edward's Plateau, Texas, USA | Oesophageal-fistulated animals, microscopic point technique | Autumn | 47 | 12 | 41 | Malechek & Leinweber (1972) | See notes under no. 227. |
| **238** | Goat | Savanna, live oak | Central Texas, USA | Observation, bite-count | Light grazing only, used in autumn | 16 | 24 | 60 | McMahan (1964) | |
| **239** | Goat | Savanna, live oak | Central Texas, USA | Observation, bite-count | Autumn, heavy use | 40 | 1 | 59 | McMahan (1964) | |
| **240** | Goat | *Acacia* and tall tropical trees | Kenya | Oesophageal fistula and ocular estimation | Sept.–Oct. | 27 | | 43 | Nge'the & Box (1976) | See notes under no. 606. |
| **241** | Goat | Belaah–rosewood community (*Casuarina–Heterodendrum*) | Ivanhoe, New South Wales, Australia | Oesophageal fistula, hand separated | Autumn (Feb.), goat-grazed pasture | —51— | | 43 | Wilson *et al.* (1975) | See notes under no. 247. |
| **242** | Goat | Belaah–rosewood community (*Casuarina–Heterodendrum*) | Ivanhoe, New South Wales, Australia | Oesophageal fistula, hand separated | Autumn (Feb.), sheep-grazed pasture | —16— | | 78 | Wilson *et al.* (1975) | See notes under no. 247. |
| **243** | Goat | Grass–shrub rangeland | Edward's Plateau, Texas, USA | Oesophageal-fistulated animals, microscopic point technique | Winter | 47 | 4 | 49 | Malechek & Leinweber (1972) | See notes under no. 227. |
| **244** | Goat | Savanna, live oak | Central Texas, USA | Observation, bite-count | Light grazing, only used in winter | 6 | 1 | 93 | McMahan (1964) | |
| **245** | Goat | Savanna, live oak | Central Texas, USA | Observation, bite-count | Winter, heavy use | 30 | 0 | 70 | McMahan (1964) | |
| **246** | Goat | Belaah–rosewood community (*Casuarina–Heterodendrum*) | Ivanhoe, New South Wales, Australia | Osophageal fistula, hand separated | Winter (May–June), goat-grazed pasture | —45— | | 55 | Wilson *et al.* (1975) | See notes under no. 247. |

## Appendix 4.2 (*continued*)

| No. | Animal species | Kind of grazingland | Geographic location | Methods | Season/ treatment | Grass (%) | Forbs (%) | Shrubs (%) | Reference | Notes |
|---|---|---|---|---|---|---|---|---|---|---|
| **247** | Goat | Belaah–rosewood community (*Casuarina–Heterodendrum*) | Ivanhoe, New South Wales, Australia | Oesophogeal fistula, hand separated | Winter (May–June), sheep-grazed pasture | —31— | | 69 | Wilson *et al.* (1975) | Winter data average of low and medium stalking in two years. Major herbaceous plants are forbs. Data adjusted for unidentifiable items. Goats browse to 2 m height; they show preference for browse but no less selectivity than do sheep. Data provided for shrub and for herbaceous plants (grasses and forbs). |
| **248** | Bison | Shortgrass prairie | Northeastern Colorado, USA | Oesophageal fistula, microscopic analysis | Late spring (May), light grazing | 98 | 2 | 0 | This chapter | See notes under no. 20. |
| **249** | Bison | Shortgrass prairie | Northeastern Colorado, USA | Oesophageal fistula, microscopic analysis | Late spring, heavy grazing | 98 | 1 | <1 | This chapter | See notes under cattle, this chapter (May). |
| **250** | Bison | Shortgrass prairie | Northeastern Colorado, USA | Oesophageal fistula, microscopic analysis | Early summer (June), light grazing | 92 | 8 | 0 | This chapter | See notes under no. 20. |
| **251** | Bison | Shortgrass prairie | Northeastern Colorado, USA | Oesophageal fistula, microscopic analysis | Early summer (June), heavy grazing | 92 | 8 | 0 | This chapter | See notes under no. 20. |
| **252** | Bison | Shortgrass prairie | Northeastern Colorado, USA | Oesophageal fistula, microscopic analysis | Late summer (Aug.), light grazing | 95 | 5 | 0 | This chapter | See notes under no. 20. |
| **253** | Bison | Shortgrass prairie | Northeastern Colorado, USA | Oesophageal fistula, microscopic analysis | Late summer (Aug.), heavy grazing | 99 | 1 | 0 | This chapter | See notes under no. 20. |
| **254** | Bison | Shortgrass prairie | Northeastern Colorado, USA | Oesophageal fistula, microscopic analysis | Autumn (Oct.), light grazing | 99 | 1 | 0 | This chapter | See notes under no. 20. |
| **255** | Bison | Shortgrass prairie | Northeastern Colorado, USA | Oesophageal fistula, microscopic analysis | Autumn (Oct.), heavy grazing | 86 | 2 | 12 | This chapter | See notes under no. 20. |
| **256** | Bison | Shortgrass prairie | Northeastern Colorado, USA | Oesophageal fistula, microscopic analysis | Winter (Dec.–Jan.), light grazing | 90 | 7 | 3 | This chapter | See notes under no. 20. |
| **257** | Bison | Shortgrass prairie | Northeastern Colorado, USA | Oesophageal fistula, microscopic analysis | Winter (Dec.–Jan.), heavy grazing | 87 | 12 | 1 | This chapter | See notes under no. 20. |
| **257a** | Bison | Various grassland types | Western North America | (Unspecified) | Average all locations | 97 | 2 | 1 | A. Johnston & J. E. Mattox (unpublished) | See notes under no. 109. |

| No. | Animal | Habitat | Location | Method | Season | | | | Reference | Notes |
|---|---|---|---|---|---|---|---|---|---|---|
| **258** | European bison | Forest | Poland | Direct observation | Year | 61 | 6 | 33 | Borowski & Kossak (1972) | |
| **259** | Feral burro | Desert habitat, riparian habitat | California, USA | Microscopic faecal analysis | Yearlong | 5 | 31 | 64 | Woodward & Ohmart (1976) | Most important species in diet are desert Indianwheat, palo verde, mesquite and arrow weed. Data adjusted for 5 % unknowns. |
| **259a** | Feral burro | Creosote bush scrub | Nevada, USA | Stomach content analysis | Spring | 10 | 64 | 26 | Browning (1960) | |
| **259b** | Feral burro | Creosote bush scrub | Nevada, USA | Stomach content analysis | Autumn | 10 | 13 | 77 | Browning (1960) | |
| **260** | Camel | Desert | Australia | Faecal pellet and rumen content analysis | Yearlong | 20 | ? | ? | Newman (1973) | |
| **261** | Pronghorn | (Mixed prairie?) | Trans-pecos of Texas, USA | Observational, minute grazing proportions | Spring | 4 | 70 | 26 | Buechner (1950) | See notes under no. 281. |
| **262** | Pronghorn | Sagebrush–grassland | Petrolean County, central Montana, USA | Observation and rumen analysis | Spring | 75 | 15 | 10 | Cole (1956) | Values estimated in autumn. Big sagebrush and fringed sagewort important yearlong; alfalfa taken when animals in vicinity of fields, cudweed sagewort most important forb. |
| **263** | Pronghorn | Prairie–sagebrush | Madador, Saskatchewan Canada | Rumen analysis | Spring | 8 | 25 | 66 | Dirschl (1963) | Cactus averaged 6 % in winter diet (included in shrubs); cropland crops were 2 % in autumn (in forbs); evergreen shrubs important in spring, autunm and winter and deciduous shrubs important in summer and autumn; silver sagebrush and snowberry and ground juniper important shrubs; numerous forbs species important in summer diets, particularly cyprusels. |
| **264** | Pronghorn | Sep-montane prairie | Cyprus Hills, Saskatchewan, Canada | Rumen analysis | Spring | 78 | 11 | 11 | Dirschl (1963) | See notes under no. 263. |
| **265** | Pronghorn | Shortgrass prairie | Eastern Colorado, USA | Stomach sample analysis | Spring (Apr.–May) | 5 | 64 | 31 | Hoover (1966) | See notes under no. 322. 20 % unknown distributed among data. |
| **266** | Pronghorn | Shortgrass prairie | Northeastern Colorado, USA | Bite-count observations with tame antelope | Spring | 51 | 19 | 30 | Hoover (1971) | See notes under no. 322. |
| **267** | Pronghorn | | Colorado, USA | Stomach analysis and observations | Spring | 21 | 53 | 26 | Hoover *et al.* (1959) | 4 % cacti and miscellaneous distributed among data. |

## Appendix 4.2 (*continued*)

| No. | Animal species | Kind of grazingland | Geographic location | Methods | Season/ treatment | Grass (%) | Forbs (%) | Shrubs (%) | Reference | Notes |
|---|---|---|---|---|---|---|---|---|---|---|
| **268** | Pronghorn | Shortgrass mixed prairie | Southeastern, south-central Alberta, Canada | Rumen content analysis, volumetric determination | Spring | 25 | 57 | 18 | Mitchell & Smoliak (1971) | Autumn and winter data are averages of 2 locations. Herbage availability varied from 296 to 1422 lb · acre$^{-1}$ dry matter in 3 sampling years in 2 locations. Important foods in spring included silver sagebrush, fringed sagewort and grasses (not identified individually); important foods in summer were silver sagebrush and various forbs (*Solanum triflorum* and *Polygonum neglectum*); important foods in autumn were silver sagebrush and fringed sagewort and nipple or ball cactus; important foods in winter were silver sagebrush and fringed sagewort and snowberry. 52 forbs, 12 shrubs and an undetermined number of grasses and sedges were in the diet but most less than 1 % annually. |
| **269** | Pronghorn | Bluebunch wheatgrass steppe | National Bison Range, Moise, Montana, USA | Rumen content analysis | Spring (Mar.–May) | 32 | 61 | 7 | O'Gara & Green (1970) | See notes under no 327. |
| **270** | Pronghorn | Sagebrush–grass | Yellowstone National Park, Wyoming, USA | Rumen content analysis | Spring (Mar.–May) | 18 | 34 | 48 | O'Gara & Green (1970) | See notes under no. 327. |
| **271** | Pronghorn | Shortgrass prairie | Northeastern Colorado, USA | Bite-count method with tame pronghorn | Mar.–Apr. | 21 | 32 | 47 | Schwartz *et al.* (1976) | Data are percentage of bites. |
| **272** | Pronghorn | Shorgrass prairie | Northeastern Colorado, USA | Bite-count method with tame pronghorn | May–June | 9 | 82 | 9 | Schwartz *et al.* (1976) | Data are percentage of bites. |
| **273** | Pronghorn | Desert grass, brush | Wyoming, USA | Rumen analysis and animal-minute observation | Spring | 20 | 26 | 54 | Taylor (1972) | 23 % unknown distributed among data. |
| **274** | Pronghorn | Shortgrass prairie | Northeastern Colorado, USA | Bite-count | Late spring (May), heavy grazing | 43 | 57 | 0 | This chapter | See notes under no. 20. |
| **275** | Pronghorn | Shortgrass prairie | Northeastern Colorado, USA | Bite-count | Late spring (May), light grazing | 37 | 63 | 0 | This chapter | See notes under no. 20. |

| | | | | | | | | | | |
|---|---|---|---|---|---|---|---|---|---|---|
| **276** | Pronghorn | Sagebrush grassland primarily | California, Idaho, Nevada, and Oregon, USA | Stomach content analysis, volumetric determination | Spring | 9 | 23 | 68 | Yoakum (1958) | See notes under no. 332. |
| **277** | Pronghorn | Shortgrass prairie | Northeastern Colorado, USA | Bite-count | Early summer (June), light grazing | 9 | 91 | 0 | This chapter | See notes under no. 20. |
| **278** | Pronghorn | Shortgrass prairie | Northeastern Colorado, USA | Bite-count | Early summer (June), heavy grazing | 10 | 90 | 0 | This chapter | See notes under no. 20. |
| **279** | Pronghorn | Semidesert range | Western Utah, USA | Rumen analysis | July | 0 | 95 | 5 | Beale & Scotter (1968) | |
| **280** | Pronghorn | Semidesert range | Western Utah, USA | Rumen analysis | June | 0 | 97 | 3 | Beale & Scotter (1968) | |
| **281** | Pronghorn | (Mixed prairie?) | Trans-pecos region of Texas, USA | Observational minute grazing proportions | Summer | 6 | 67 | 27 | Buechner (1950) | Important yearlong foods were bitterweed, cutleaf daisy, sideoats and blue grama, dalea, eriogonum, gaura, deervetch, coneflower, and senecio. |
| **282** | Pronghorn | Sagebrush–grassland | Petrolean County, central Montana, USA | Observation and rumen analysis | Summer | 2 | 67 | 31 | Cole (1956) | See notes under no. 262. |
| **283** | Pronghorn | Prairie–sagebrush | Madador, Saskatchewan, Canada | Rumen analysis | Summer | 1 | 53 | 46 | Dirschl (1963) | See notes under no. 263. |
| **284** | Pronghorn | Sep-montane prairie | Cyprus Hills, Saskatchewan, Canada | Rumen analysis | Summer | 5 | 55 | 40 | Dirschl (1963) | See notes under no. 263. |
| **285** | Pronghorn | Shortgrass prairie | Eastern Colorado, USA | Stomach sample analysis | Summer (June–Sept.) | 2 | 74 | 24 | Hoover (1966) | See notes under no. 322. |
| **286** | Pronghorn | Shortgrass prairie | Northeastern Colorado, USA | Bite-count observations with tame antelope | Summer | 27 | 71 | 2 | Hoover (1971) | See notes under no. 322. |
| **287** | Pronghorn | | Colorado, USA | Stomach analysis and observations | Summer | 1 | 74 | 25 | Hoover *et al.* (1959) | 11 % cacti and miscellaneous distributed among data. |
| **288** | Pronghorn | Shortgrass mixed prairie | Southeastern, south-central Alberta, Canada | Rumen content analysis, volumetric determination | Summer | 13 | 62 | 25 | Mitchell & Smoliak (1971) | See notes under no. 268. |
| **289** | Pronghorn | Bluebunch wheatgrass steppe | National Bison Range, Moise, Montana, USA | Rumen content analysis | Summer (June–Aug.) | 15 | 73 | 12 | O'Gara & Green (1970) | See notes under no. 327. |
| **290** | Pronghorn | Sagebrush–grass | Yellowstone National Park, Wyoming, USA | Rumen content analysis | Summer (June–Aug.) | 4 | 80 | 16 | O'Gara & Green (1970) | See notes under no. 327. |

**Appendix 4.2** (*continued*)

| No. | Animal species | Kind of grazingland | Geographic location | Methods | Season or treatment | Grass (%) | Forbs (%) | Shrubs (%) | Reference | Notes |
|---|---|---|---|---|---|---|---|---|---|---|
| 291 | Pronghorn | Shortgrass prairie | Northeastern Colorado, USA | Bite-count method with tame pronghorn | July–Aug. | 3 | 89 | 8 | Schwartz *et al.* (1976) | Data are percentage of bites. |
| 292 | Pronghorn | Big sagebrush–grass | South-central Wyoming, USA | Rumen samples | Summer (average 2 years) | 2 | 0 | 98 | Severson & May (1967) | See notes under no. 159. |
| 293 | Pronghorn | Desert grass, brush | Wyoming, USA | Rumen analysis and animal-minute observations | Summer | 2 | 36 | 62 | Taylor (1972) | 11 % unknown distributed among data. |
| 294 | Pronghorn | Sagebrush–grassland primarily | California, Idaho, Nevada, and Oregon, USA | Stomach content analysis, volumetric determination | Summer | 4 | 31 | 65 | Yoakum (1958) | See notes under no. 332. |
| 295 | Pronghorn | Shortgrass prairie | Northeastern Colorado, USA | Bite-count observations with tame antelope | Mid-summer | 16 | 83 | 1 | Hoover (1971) | See notes under no. 323. |
| 296 | Pronghorn | Shortgrass prairie | Northeastern Colorado, USA | Bite-count | Late summer, (Aug.), light grazing | 8 | 92 | 0 | This chapter | See notes under no. 20. |
| 297 | Pronghorn | Shortgrass prairie | Northeastern Colorado, USA | Bite-count | Late summer (Aug.), heavy grazing | 13 | 87 | 0 | This chapter | See notes under no. 20. |
| 298 | Pronghorn | (Mixed prairie?) | Trans-pecos region of Texas, USA | Observational, minute grazing proportions | Autumn | 7 | 55 | 38 | Buechner (1950) | See notes under no. 281. |
| 299 | Pronghorn | Sagebrush–grass-land | Petrolean County, central Montana, USA | Observation and rumen analysis | Autumn | 1 | 56 | 43 | Cole (1956) | See notes under no. 262. |
| 300 | Pronghorn | Shortgrass mixed prairie | Carter County, Montana, USA | Stomach analysis | Autumn (Sept.–Oct., 2 years) | 7 | 6 | 87 | Couey (1946) | *Artemisia* most important item in diet, leaves and twigs eaten; snakeweed and snowberry also important foods; twigs, leaves and berries of the snowberry eaten. |
| 301 | Pronghorn | Prairie–sage-brush | Matador, Saskatchewan, Canada | Rumen analysis | Autumn | 1 | 14 | 85 | Dirschl (1963) | See notes under no. 263. |
| 302 | Pronghorn | Sep-montane prairie | Cyprus Hills, Saskatchewan, Canada | Rumen analysis | Autumn | 2 | 44 | 54 | Dirschl (1963) | See notes under no. 263. |

| | | | | | | | | | | |
|---|---|---|---|---|---|---|---|---|---|---|
| 303 | Pronghorn | Mixed prairie | Extreme western Kansas, USA | Observation, time records, grazing | Autumn (Sept.–Nov.) | 63 | 20 | 17 | Hlavachick (1966) | Cactus 76 % of diet distributed among data in autumn and 40 % in spring–summer (*Opuntia macrorhiza* dominant); ate considerable winter wheat in winter–early spring when in the vicinity; sand sagebrush important in diet; main forb was *Aster*; blue grama grass and sideoats grama important in diet. |
| 304 | Pronghorn | | Colorado, USA | Stomach analysis and observations | Autumn | 1 | 23 | 76 | Hoover *et al.* (1959) | 5 % cacti distributed among data. |
| 305 | Pronghorn | Shortgrass prairie | Eastern Colorado, USA | Stomach sample analysis | Autumn | 0 | 25 | 75 | Hoover (1966) | See notes under no. 323. 92 % unknown distibuted among data. |
| 306 | Pronghorn | Shortgrass mixed prairie | Southeastern, south-central Alberta, Canada | Rumen content analysis, volumetric determination | Autumn | 17 | 25 | 58 | Mitchell & Smoliak (1971) | See notes under no. 268. |
| 307 | Pronghorn | Bluebunch wheatgrass steppe | National Bison Range, Moise, Montana, USA | Rumen content analysis | Autumn (Sept.–Oct.) | 8 | 76 | 16 | O'Gara & Green (1970) | See notes under no. 327. |
| 308 | Pronghorn | Sagebrush–grass | Yellowstone National Park, Wyoming, USA | Rumen content analysis | Autumn (Sept.–Oct.) | 5 | 48 | 47 | O'Gara & Green (1970) | See notes under no. 327. |
| 309 | Pronghorn | Shortgrass prairie | Northeastern Colorado, USA | Bite-count method with tame pronghorn | Sept.–Oct. | 15 | 67 | 18 | Schwartz *et al.* (1976) | Data are percentage of bites. |
| 310 | Pronghorn | Shortgrass prairie | Northeastern Colorado, USA | Bite-count method with tame pronghorn | Nov.–Dec. | 20 | 17 | 63 | Schwartz *et al.* (1976) | Data are percentage of bites. |
| 311 | Pronghorn | Desert, grass, brush | Wyoming, USA | Rumen analysis and animal minute observations | Autumn | 4 | 16 | 80 | Taylor (1972) | 6 % unknown distributed among data. |
| 312 | Pronghorn | Shortgrass prairie | Northeastern Colorado, USA | Bite-count | Autumn (Oct.), light grazing | 71 | 29 | 0 | This chapter | See notes under no. 20. |
| 313 | Pronghorn | Shortgrass prairie | Northeastern Colorado, USA | Bite-count | Autumn (Oct.), heavy grazing | 79 | 16 | 5 | This chapter | See notes under no. 20. |
| 314 | Pronghorn | Sagebrush grassland primarily | California, Idaho, Nevada and Oregon, USA | Stomach content analysis, volumetric determination | Autumn | 14 | 23 | 63 | Yoakum (1958) | See notes under no. 332. |
| 315 | Pronghorn | Sagebrush grassland | Central Montana, USA | Rumen samples, analysed volumetrically | Winter (Dec.–Apr.) | 1 | 6 | 93 | Bayless (1969) | Big sagebrush most important species in diet; fringed sagewort relatively important; grasses eaten during green-up periods in warmer weather. |
| 316 | Pronghorn | Sagebrush–grass | Idaho, USA | Examination of feeding area for percentage diet | Winter (Jan.–June) | 4 | 1 | 95 | Bernt (1976) | |

## Appendix 4.2 (*continued*)

| No. | Animal species | Kind of grazingland | Geographic location | Methods | Season/ treatment | Grass (%) | Forbs (%) | Shrubs (%) | Reference | Notes |
|---|---|---|---|---|---|---|---|---|---|---|
| 317 | Pronghorn | (Mixed prairie?) | Trans-pecos region of Texas, USA | Observational, minute grazing proportions | Winter | 1 | 66 | 33 | Buechner (1950) | See notes under no. 281. |
| 318 | Pronghorn | Sagebrush–grassland | Petrolean County, Central Montana, USA | Observation and rumen analysis | Winter | 2 | 3 | 95 | Cole (1956) | See notes under no. 262. |
| 319 | Pronghorn | Prairie, sagebrush | Madador, Saskatchewan, Canada | Rumen analysis | Winter | 0 | 8 | 92 | Dirschl (1963) | See notes under no. 263. |
| 320 | Pronghorn | Sep-montane prairie | Cyprus Hills, Saskatchewan, Canada | Rumen analysis | Winter | 1 | 7 | 92 | Dirschl (1963) | See notes under no. 263. |
| 321 | Pronghorn | Mixed prairie | Extreme western Kansas, USA | Observation, time records, grazing | Winter (Dec.–Jan.) | 40 | 34 | 26 | Hlavachick (1966) | See notes under no. 303. |
| 322 | Pronghorn | Shortgrass prairie | Eastern Colorado, USA | Stomach sample analysis | Winter (Nov.–Mar.) | 5 | 31 | 64 | Hoover (1966) | Cactus formed as much as 6 % of the diet in winter and near 4 % in summer; higher browse diet concentrations in winter and autumn; sudden switch to forbs over browse in spring; wheat eaten readily in Nov.–Apr. when in the vicinity of wheat fields. 64 % unknown distributed among data. |
| 323 | Pronghorn | Shortgrass prairie | Northeastern Colorado, USA | Bite-count observations with tame pronghorn | Winter | 30 | 23 | 47 | Hoover (1971) | Leaves were preferred plant parts, new plants and new growth preferred where available; important species include fringed sagewort in winter and early spring and scarlet globemallow in summer. |
| **324** | Pronghorn | | Colorado, USA | Stomach sample analysis and observations | Winter | **5** | **31** | **64** | Hoover *et al.* (1959) | 15 % cacti and miscellaneous distributed among data. |
| 325 | Pronghorn | Shortgrass mixed prairie | Southeastern, south-central Alberta, Canada | Rumen content analysis, volumetric determination | Winter | 17 | 35 | 48 | Mitchell & Smoliak (1971) | See notes under no. 268. |
| 326 | Pronghorn | Bluebunch wheatgrass steppe | National Bison Range, Moise, Montana, USA | Rumen content analysis | Winter (Nov.–Feb.) | 4 | 62 | 34 | O'Gara & Green **(1970)** | See notes under no. 327. |

| No. | Animal | Vegetation | Location | Method | Season | | | | Reference | Notes |
|---|---|---|---|---|---|---|---|---|---|---|
| **327** | Pronghorn | Sagebrush–grass | Yellowstone National Park, Wyoming, USA | Rumen content analysis | Winter (Nov.–Feb.) | 2 | 17 | 81 | O'Gara & Green (1970) | Yellowstone area: forbs more than 80 % of diet in some months; grasses only important in early spring; fringed sagewort up to 20 % of the diet; big sagebrush important in autumn, winter and spring. Bison Range: forbs always important in diet; fragmented leaves and bare stems; main species are yarrow and aster, some browse throughout year with fringed sagewort and snowberry of major importance. |
| **328** | Pronghorn | Shortgrass prairie | Northeastern Colorado, USA | Bite-count method with tame pronghorn | Jan.–Feb. | 2 | 9 | 89 | Schwartz *et al.* (1976) | Data are percentage of bites. |
| **329** | Pronghorn | Desert grass–brush | Wyoming, USA | Rumen analysis and animal-minute observations | Winter | 8 | 0 | 92 | Taylor (1972) | |
| **330** | Pronghorn | Shortgrass prairie | Northeastern Colorado, USA | Bite-count | Winter (Dec.–Jan.), light grazing | 78 | 22 | 0 | This chapter | See notes under no. 20. 9 % unknown distributed among data. |
| **331** | Pronghorn | Shortgrass prairie | Northeastern Colorado, USA | Bite-count | Winter (Dec.–Jan.), heavy grazing | 72 | 18 | 10 | This chapter | See notes under no. 20. |
| **332** | Pronghorn | Sagebrush grassland primarily | California, Idaho, Nevada and Oregon, USA | Stomach content analysis, volumetric determination | Winter | 6 | 7 | 87 | Yoakum (1958) | Cheatgrass brome was most frequent grass in the diet; most common forbs were *Polygonum*, *Phlox*, and *Iva Axillaris*; browse provided bulk of diet and species in order of importance were sagebrush species, bitterbrush and rabbitbrush. |
| **333** | Pronghorn | Semidesert range | Western Utah, USA | Observation and rumen samples in hunting season | Yearlong average | 5 | 25 | 70 | Beale & Smith (1970) | Estimated data. Browse over 90 % of the diet in Nov.–Mar., primarily black sagebrush; grasses preferred in spring only; forbs in late spring and early summer, primarily globemallow and *Chaenactis*; browse back in diet by late June. |
| **334** | Pronghorn | | Colorado, USA | Stomach analysis and observations | Yearlong | 5 | 47 | 48 | Hoover *et al.* (1959) | 11 % cacti and miscellaneous distributed among data. |
| **335** | Pronghorn | Various | Western North America | Various | Yearlong average | 6 | 38 | 56 | A. Johnston & J. E. Mattox (unpublished) | See notes under no. 109. |

## Appendix 4.2 (*continued*)

| No. | Animal species | Kind of grazingland | Geographic location | Methods | Season or treatment | Grass (%) | Forbs (%) | Shrubs (%) | Reference | Notes |
|---|---|---|---|---|---|---|---|---|---|---|
| **336** | Pronghorn | Sagebrush–grass | Oregon–Nevada border, USA | Stomach analysis, volumetric determination | Yearlong average | 2 | 31 | 67 | Mason (1952) | Forbs confined to spring and summer (Apr.–July); bitterbrush utilized heavily in summer; green grass taken readily when available in late autumn and early spring; big sagebrush 61 % of yearlong diet. |
| **337** | Pronghorn | Shortgrass prairie | Newell area, Alberta, Canada | Rumen analysis | Yearlong average | 13 | 52 | 35 | Mitchell & Smoliak (1971) | Browse most important in summer, autumn and winter; forbs grazed at all seasons; grasses and sedges important in the spring; key foods were silver sagebrush and sagewort. |
| **338** | Pronghorn | Sagebrush–grass, saltbush, and rabbitbrush types | Red Desert, southwestern Wyoming, USA | Microscopic analysis of faecal samples | Yearlong average | 3 | 7 | 90 | Olsen & Hansen (1977) | See notes under no. 350. |
| **339** | Pronghorn | Various | Western North America | Various | Average values | 4 | 42 | 54 | | Of browse composition, 11 % was cactus. |
| **340** | Pronghorn | Desert grass–brush | Red Desert, Wyoming, USA | Rumen analysis and animal minute observations | Yearlong | 7 | 20 | 73 | Taylor (1972) | Big sagebrush is staple in diet. |
| **341** | Pronghorn | Mixed prairie | Extreme western Kansas, USA | Observation, time records, grazing | Spring–summer (Apr.–Aug.) | 25 | 38 | 37 | Hlavachick (1966) | See notes under no. 303. 40 % cacti distributed among diet. |
| **342** | Pronghorn | Big sagebrush–grass | South-central Wyoming, USA | Rumen samples | Autumn and winter (1 year) | 0 | 0 | 100 | Severson & May (1967) | See notes under no. 159. |
| **343** | Wild horses | Semidesert grass–shrub | Southern New Mexico, USA | Microscopic analysis of faecal samples | Spring | 61 | 30 | 9 | Hansen (1976) | See notes under no. 344. |
| **344** | Wild horses | Semidesert grass–shrub | Southern New Mexico, USA | Microscopic analysis of faecal samples | Summer | 56 | 17 | 26 | Hansen (1976) | On an annual basis, in order of relative importance in the diets of wild horses, were Russian thistle, dropseed, mesquite and junegrass. Lower percentage grass in diets here than in other regions. Data adjusted to 100 %. |
| **345** | Wild horses | Piñon–juniper and sagebrush | Northeastern Colorado, USA | Faecal analysis | Summer | 92 | 3 | 5 | R. M. Hansen & R. C. Clark (unpublished) | |
| **346** | Wild horses | Semidesert grass–shrub | Southern New Mexico, USA | Microscopic analysis of faecal samples | Autumn | 37 | 33 | 30 | Hansen (1976) | See notes under no. 344. |
| **347** | Wild horses | Semidesert grass–shrub | Southern New Mexico, USA | Microscopic analysis of faecal samples | Winter | 41 | 36 | 23 | Hansen (1976) | See notes under no. 344. 8 % unknown distributed among diet. |

| | | | | | | | | | | |
|---|---|---|---|---|---|---|---|---|---|---|
| **348** | Wild horses | Mountain shrub | Piceance Basin, Colorado, USA | Microscopic analysis of faecal samples | Yearlong average | 85 | 1 | 14 | Hubbard & Hansen (1976) | Major foods were sedges, needle-and-thread, wheatgrasses, prairie junegrass, bromes, Indian ricegrass, blue grasses, common winterfat. Serviceberry was the dominant year browse followed by piñon pine and junipers. More overlap between horse and cattle diets than between any other species compared. |
| **349** | Wild horses | Piñon–juniper | Piceance Basin, Colorado, USA | Microscopic analysis of faecal samples | Yearlong average | 89 | 1 | 10 | Hubbard & Hansen (1976) | See notes under no. 348. |
| **350** | Wild horses | Sagebrush–grass, saltbush, and rabbitbrush types | Red Desert, southwestern Wyoming, USA | Microscopic analysis of faecal samples | Yearlong average | 92 | 2 | 6 | Olsen & Hansen (1977) | Wheatgrass and needle grass constituted 11 % to 46 % for herbivores except antelope. Wild horse and cattle diets higher in grasses and sedges than those of other animals; sagebrush highest in antelope diets; saltbush consumed by each herbivore. Seasonal data available: sheep varied most seasonally. |
| **351** | Deer | Various | Western North America | Various | Yearlong average | 6 | 11 | 83 | A. Johnston & J. E. Mattox (unpublished) | See notes under no. 109. |
| **352** | Black-tailed deer | Mixed age chaparral | California, USA | Rumen analysis, microscopic identification | Spring | 4 | 15 | 81 | W. M. Longhurst *et al.* (unpublished) | |
| **353** | Black-tailed deer | Mature chaparral | California, USA | Rumen analysis, microscopic identification | Spring | 1 | 2 | 97 | W. M. Longhurst *et a'.* (unpublished) | |
| **354** | Black-tailed deer | Oak woodland | California, USA | Rumen analysis, microscopic identification | Spring | 13 | 51 | 36 | W. M. Longhurst *et al.* (unpublished) | |
| **355** | Black-. tailed deer | Chaparral and annual grassland | California, USA | Rumen analysis, microscopic identification | Spring | 10 | 39 | 51 | W. M. Longhurst *et al.* (unpublished) | |
| **356** | Black-tailed deer | Oak woodland/ chaparral ecotone | California, USA | Rumen analysis, microscopic identification | Spring | 7 | 75 | 18 | W. M. Longhurst *et al.* (unpublished) | |
| **357** | Black-tailed deer | Mixed age chaparral | California, USA | Rumen analysis, microscopic identification | Summer | 0 | 1 | 99 | W. M. Longhurst *et al.* (unpublished) | |
| **358** | Black-tailed deer | Mature chaparral | California, USA | Rumen analysis, microscopic identification | Summer | 0 | 7 | 93 | W. M. Longhurst *et al.* (unpublished) | |

**Appendix 4.2** (*continued*)

| No. | Animal species | Kind of grazingland | Geographic location | Methods | Season or treatment | Grass (%) | Forbs (%) | Shrubs (%) | Reference | Notes |
|---|---|---|---|---|---|---|---|---|---|---|
| **359** | Black-tailed deer | Chaparral and annual grassland | California, USA | Rumen analysis, microscopic identification | Summer | 2 | 9 | 89 | W. M. Longhurst *et al.* (unpublished) | |
| **360** | Black-tailed deer | Oak woodland | California, USA | Rumen analysis, microscopic identification | Summer | 3 | 17 | 80 | W. M. Longhurst *et al.* (unpublished) | |
| **361** | Black-tailed deer | Oak woodland/ chaparral ecotone | California, USA | Rumen analysis, microscopic identification | Summer | 0 | 1 | 99 | W. M. Longhurst *et al.* (unpublished) | |
| **362** | Black-tailed deer | Mixed age chaparral | California, USA | Rumen analysis, microscopic identification | Autumn | 13 | 3 | 84 | W. M. Longhurst *et al.* (unpublished) | |
| **363** | Black-tailed deer | Mature chaparral | California, USA | Rumen analysis, microscopic identification | Autumn | 1 | 6 | 93 | W. M. Longhurst *et al.* (unpublished) | |
| **364** | Black-tailed deer | Chaparral and annual grassland | California, USA | Rumen analysis, microscopic identification | Autumn | 20 | 6 | 74 | W. M. Longhurst *et al.* (unpublished) | |
| **365** | Black-tailed deer | Oak woodland | California, USA | Rumen analysis, microscopic identification | Autumn | 20 | 8 | 72 | W. M. Longhurst *et al.* (unpublished) | |
| **366** | Black-tailed deer | Oak woodland/ chaparral ecotone | California, USA | Rumen analysis, microscopic identification | Autumn | 16 | 1 | 83 | W. M. Longhurst *et al.* (unpublished) | |
| **367** | Black-tailed deer | Oak woodland/ chaparral ecotone | California, USA | Rumen analysis, microscopic identification | Winter | 33 | 2 | 65 | W. M. Longhurst *et al.* (unpublished) | |
| **368** | Black-tailed deer | Oak woodland | California, USA | Rumen analysis, microscopic identification | Winter | 65 | 20 | 15 | W. M. Longhurst *et al.* (unpublished) | |
| **369** | Black-tailed deer | Mixed age chaparral | California, USA | Rumen analysis, microscopic identification | Winter | 15 | 8 | 77 | W. M. Longhurst *et al.* (unpublished) | |
| **370** | Black-tailed deer | Mature chaparral | California, USA | Rumen analysis, microscopic identification | Winter | 3 | 0 | 97 | W. M. Longhurst *et al.* (unpublished) | |
| **371** | Black-tailed deer | Chaparral and annual grassland | California, USA | Rumen analysis, microscopic identification | Winter | 42 | 14 | 44 | W. M. Longhurst *et al.* (unpublished) | |

| | | | | | | | | | | |
|---|---|---|---|---|---|---|---|---|---|---|
| **372a** | Mule deer | Mountain range | Northeastern Colorado, USA | Rumen–reticulum content analysis | Spring | 7 | 47 | 46 | Alldredge *et al.* (1974) | |
| **372b** | Mule deer | Piñon–juniper | Central Utah, USA | Botanical measurements, weight-estimate methodology | Spring | 87 | 6 | 7 | Smith & Julander (1953) | See notes under no. 393. |
| **373** | Mule deer | Sage–oak | Central Utah, USA | Botanical measurements, weight-estimate methodology | Spring | 38 | 17 | 45 | Smith & Julander (1953) | See notes under no. 393. |
| **374** | Mule deer | Piñon–juniper, shortgrass–shrub | Guadalupe mountains, New Mexico, USA | Stomach content analysis, volumetric per cent | Spring (4 year average) | 1 | 31 | 68 | Anderson *et al.* (1965) | See notes under no. 411. |
| **375** | Mule deer | Piñon–juniper | Southwestern, New Mexico, USA | Stomach analysis, microscopic point technique | Spring | 3 | 33 | 64 | Boeker *et al.* (1972) | Sampling area 6000 to 7000 ft elevation, 16.7 in. annual precipitation. Overstorey of piñon pine and alligator juniper of about 100 trees per acre; about 60 shrubs per acre primarily birchleaf mountain mahogany, Wright silktassel, and skunkbrush sumac; perennial grass standing crop about 400 lb · acre$^{-1}$, primarily blue grama grass. Leaves and stems from shrubs and trees made up about 75 % of diet. Oak used consistently yearlong; forb species varied widely seasonally; sampling over a 5-year interval and forbs more important in diets in wet years; forb preferences not related to chemical composition. |
| **376** | Mule deer | Grassland; big sagebrush and silver sagebrush | North-central Montana, USA | Feeding site examination, rumen samples | Spring | 17 | 24 | 59 | Dusek (1975) | |
| **377** | Mule deer | Mixed shrubland, low mountain | Southern Wyoming, USA | Faecal analysis | Spring (2 years, 2 locations) | 10 | 5 | 85 | Goodwin (1975) | Mountain range 7500–9500 ft. Major browse grazed big sagebrush and bitterbrush and mountain mahogany; no single forb important, but numerous forbs in diet in summer. |
| **378** | Mule deer | Grassland to spruce fir | Central Montana, USA | Rumen samples | Spring | 38 | 20 | 42 | Lovaas (1958) | |
| **379** | Mule deer | Foothill bunchgrass rangeland | Moise, Montana, USA | Rumen analysis, volumetric percentage | Spring (Mar.–May) | 64 | 30 | 6 | Morris & Schwartz (1957) | See notes under no. 420. |

## Appendix 4.2 (*continued*)

| No. | Animal species | Kind of grazingland | Geographic location | Methods | Season/ treatment | Grass (%) | Forbs (%) | Shrubs (%) | Reference | Notes |
|---|---|---|---|---|---|---|---|---|---|---|
| **380** | Mule deer | Bunchgrass rangeland | Moise, Montana, USA | Rumen content analysis | Spring (Mar.–May) | 48 | 35 | 17 | Nellis & Ross (1969) | Data are expressed as per cent contents by volume; data are averaged for 1952 and 1963. Conifers used more in winter as part of diet than in other seasons and included in shrub component. |
| **381** | Mule deer | Mountain meadow, forest, sagebrush–bitterbrush–bunchgrass prairie | Southwestern Montana, USA | Rumen samples | Spring | 38 | 37 | 25 | Wilkins (1957) | |
| **382** | Mule deer | Mountain rangeland | Ruby Buttes, Nevada, USA | Rumen contents, slaughtered deer | Late spring (May) | 23 | 9 | 68 | Ansotegui *et al.* (1972) | |
| **383a** | Mule deer | Mountain range | Northeastern Colorado, USA | Rumen–reticulum content analysis | Summer | 11 | 25 | 64 | Alldredge *et al.* (1974) | |
| **383b** | Mule deer | Grass–shrub rangeland | Southwestern Texas, USA | Feeding minutes observations | Summer | 13 | 87 | 0 | Anderson (1949) | Grey oak and mistletoe were important browse foods but less so than was Emory's oak; grama grasses contributed most of the herbaceous material. |
| **384** | Mule deer | Piñon–juniper, shortgrass–shrub | Guadalupe Mountains, New Mexico, USA | Stomach content analysis, volumetric per cent | Summer (4 year average) | 4 | 34 | 62 | Anderson *et al.* (1965) | See notes under no. 411. |
| **385** | Mule deer | Mountain rangeland | Ruby Buttes, Nevada, USA | Rumen contents, slaughtered deer | Summer (June–July) | 2 | 28 | 70 | Ansotegui *et al.* (1972) | |
| **386** | Mule deer | Piñon–juniper | Southwestern New Mexico, USA | Stomach analysis, microscopic point technique | Summer | 2 | 42 | 56 | Boeker *et al.* (1972) | See notes under no. 375. |
| **387** | Mule deer | Grassland, big sagebrush and silver sagebrush | North-central Montana, USA | Rumen samples, feeding site examination | Summer | 1 | 55 | 44 | Dusek (1975) | 2 % unknown distributed among diet. |
| **388** | Mule deer | Low mountain rangeland | Tulumne County, California, USA | Stomach analysis, volumetric determinations | Summer range | 1 | 9 | 90 | Ferrel & Leach (unpublished) | See notes under no. 416. |
| **389** | Mule deer | Mixed shrubland, low mountain | Southern Wyoming, USA | Faecal analysis | Summer (2 years) | 7 | 18 | 75 | Goodwin (1975) | See notes under no. 377. |
| **390** | Mule deer | Piñon–juniper, sagebrush | Northeastern Colorado, USA | Faecal analysis | Summer | 7 | 3 | 90 | R. M. Hansen & R. C. Clark (unpublished) | |

| | | | | | | | | | | |
|---|---|---|---|---|---|---|---|---|---|---|
| **391** | Mule deer | Foothill bunchgrass rangeland | Moise, Montana, USA | Rumen analysis, volumetric percentage | Summer (June–Aug.) | 2 | 62 | 36 | Morris & Schwartz (1957) | See notes under no. 420. |
| **392** | Mule deer | Currant–snowberry | Central Utah, USA | Botanical measurements, weight-estimate methodology | Summer | 8 | 33 | 59 | Smith & Julander (1953) | See notes under no. 393. |
| **393** | Mule deer | Mountain range, spruce–fir zone | Central Utah, USA | Botanical measurements, weight-estimate methodology | Summer | 5 | 73 | 22 | Smith & Julander (1953) | Overall elevation from 6500 to 11 000 ft; the higher vegetation types are aspen, conifers and shrub types; the intermediate elevations are brush types. Vegetation observed prior to and after movement of a band of sheep through the area; deer use was more prolonged. Most prominent species in deer diets in the summer were *Lupinus alpestris*, *Trifolium* species and *Taraxacum officinale*; for sheep on summer range the most common species were Fendler's bluegrass, bluebells and several forbs; on spring range the most common species in the deer diet was Fendler's bluegrass followed by mountain mahogany; for sheep on spring range mountain mahogany and Fendler's bluegrass and rabbitbrush were common constituents. |
| **394** | Mule deer | Grassland to spruce–fir | Central Montana, USA | Rumen samples | Summer | 2 | 75 | 23 | Lovaas (1958) | Summer – 12 % mushrooms distributed among data. |
| **395** | Mule deer | Sage–chokecherry | Central Utah, USA | Botanical measurements, weight-estimate methodology | Summer | 15 | 55 | 30 | Smith & Julander (1953) | See notes under no. 393. |
| **396** | Mule deer | Mountain range, aspen | Central Utah, USA | Botanical measurements, weight-estimate methodology | Summer | 4 | 72 | 24 | Smith & Julander (1953) | See notes under no. 393. |
| **397** | Mule deer | Sage–grass | Central Utah, USA | Botanical measurements, weight-estimate methodology | Summer | 0 | 100 | 0 | Smith & Julander (1953) | See notes under no. 393. |
| **398** | Mule deer | Sage–snowberry | Central Utah, USA | Botanical measurements, weight-estimate methodology | Summer | 8 | 90 | 2 | Smith & Julander (1953) | See notes under no. 393. |

## Appendix 4.2 (*continued*)

| No. | Animal species | Kind of grazingland | Geographic location | Methods | Season/ treatment | Grass (%) | Forbs (%) | Shrubs (%) | Reference | Notes |
|---|---|---|---|---|---|---|---|---|---|---|
| **399** | Mule deer | Bunchgrass rangeland | Moise, Montana, USA | Rumen content analysis | Summer (June–Aug.) | 1 | 32 | 67 | Nellis & Ross (1969) | See notes under no. 380. |
| **400** | Mule deer | Mountain meadow, forest, sagebrush–bitterbrush, bunchgrass prairie | Southwestern Montana, USA | Rumen samples | Summer | 5 | 74 | 21 | Wilkins (1957) | |
| **401** | Mule deer | Mountain rangeland | Ruby Buttes, Nevada, USA | Rumen contents, slaughtered deer | Early autumn (Sept.) | 3 | 6 | 91 | Ansotegui *et al.* (1972) | |
| **402a** | Mule deer | Mountain range | Northeastern Colorado, USA | Rumen–reticulum content analysis | Autumn | 19 | 21 | 60 | Alldredge *et al.* (1974) | |
| **402b** | Mule deer | | Wyoming, USA | Stomach analysis | Autumn | 6 | 34 | 60 | Anderson (1953) | Samples taken during hunting season. |
| **403** | Mule deer | Piñon–juniper shortgrass–shrub | Guadalupe Mountains, New Mexico, USA | Stomach content analysis, volumetric percent | Autumn (4 year average) | 1 | 28 | 71 | Anderson *et al.* (1965) | See notes under no. 411. |
| **404** | Mule deer | Piñon–juniper | Southwestern New Mexico, USA | Stomach analysis, microscopic point technique | Autumn | 6 | 8 | 86 | Boeker *et al.* (1972) | See notes under no. 375. |
| **405** | Mule deer | Grassland, big sagebrush and silver sagebrush | North-central Montana, USA | Rumen samples, feeding site examination | Autumn | 3 | 16 | 81 | Dusek (1975) | |
| **406** | Mule deer | Mixed shrubland, low mountain | Southern Wyoming, USA | Faecal analysis | Autumn (2 years) | 0 | 8 | 92 | Goodwin (1975) | See notes under no. 377. |
| **407** | Mule deer | Grassland to spruce–fir | Central Montana, USA | Rumen samples | Autumn | 3 | 53 | 44 | Lovaas (1958) | |
| **408** | Mule deer | Foothill bunchgrass rangeland | Moise, Montana, USA | Rumen analysis, volumetric percentage | Autumn (Oct.–Nov.) | 19 | 78 | 3 | Morris & Schwartz (1957) | See notes under no. 420. |
| **409** | Mule deer | Bunchgrass rangeland | Moise, Montana, USA | Rumen content analysis | Autumn (Sept.–Nov.) | 17 | 58 | 25 | Nellis & Ross (1969) | See notes under no. 380. |
| **410** | Mule deer | Mountain meadow, forest, sagebrush–bitterbrush, bunchgrass prairie | Southwestern Montana, USA | Rumen samples | Autumn | 3 | 22 | 75 | Wilkins (1957) | |

| | | | | | | | | | | |
|---|---|---|---|---|---|---|---|---|---|---|
| **411** | Mule deer | Piñon–juniper shortgrass–shrub | Guadalupe Mountains, New Mexico, USA | Stomach content analysis, volumetric per cent | Winter (4 year average) | 4 | 16 | 80 | Anderson *et al.* (1965) | 5 major food items comprising 64 % of diet were wavyleaf oak, juniper, hairy cercocarpus, yucca and unidentified forbs. Juniper most important in winter; wavyleaf oak most important spring through autumn; forbs more important in years and seasons of high precipitation; more leaves than twigs eaten from oak. |
| **412a** | Mule deer | Mountain range | Northeastern Colorado, USA | Rumen–reticulum content analysis | Winter | 2 | 15 | 83 | Alldredge *et al.* (1974) | |
| **412b** | Mule deer | Piñon–juniper | Southwestern New Mexico, USA | Stomach analysis, microscopic point technique | Winter | 2 | 4 | 94 | Boeker *et al.* (1972) | See notes under no. 375. |
| **413** | Mule deer | Coniferous forest, sagebrush | Gallatin Canyon, Montana, USA | Plant use measurements | Winter (Jan.–Mar.) | 7 | 30 | 63 | Constan (1973) | |
| **414** | Mule deer | Coniferous forest, mountain rangeland | Jasper Park Alberta, Canada | | Winter | 15 | 6 | 79 | Cowan (1947) | |
| **415** | Mule deer | Grassland, big sagebrush and silver sagebrush | North-central Montana, USA | Rumen samples, feeding site examination | Winter | 0 | 40 | 60 | Dusek (1975) | |
| **416** | Mule deer | Low mountain rangeland | Tulumne County, California, USA | Stomach analysis, volumetric determinations | Winter range | 9 | 89 | 2 | Ferrel & Leach (unpublished) | Winter range samples Nov.–May, summer range June–Oct. Most important winter browse was mountain misery followed by oaks and *Ceanothus*; most grass use in winter following autumn and winter rains; *Ceanothus* most important in summer diets, oaks second. |
| **417** | Mule deer | Piñon–juniper foothills | Piceance Creek Basin, Colorado, USA | Faecal pellet microscopic analysis | Winter (Dec.–Mar.) | 2 | 1 | 97 | Hansen & Dearden (1975) | Diet essentially shrubby; values estimated for grasses and forbs. Piñon pine most important component with juniper second. |
| **418** | Mule deer | Grassland to spruce–fir | Central Montana, USA | Rumen samples | Winter-forested | 2 | 20 | 78 | Lovaas (1958) | |
| **419** | Mule deer | Grassland to spruce–fir | Central Montana, USA | Rumen samples | Winter-prairie | 8 | 44 | 48 | Lovaas (1958) | |
| **420** | Mule deer | Foothill bunchgrass rangeland | Moise, Montana, USA | Rumen analysis, volumetric percentage | Winter (Dec.–Feb.) | 54 | 20 | 26 | Morris & Schwartz (1957) | Grass diet more important for deer in early part of spring than later; only in summer was there appreciable non-grass vegetation use by elk; autumn use of grasses and forbs variable between years. |

## Appendix 4.2 (*continued*)

| No. | Animal species | Kind of grazingland | Geographic location | Methods | Season/ treatment | Grass (%) | Forbs (%) | Shrubs (%) | Reference | Notes |
|---|---|---|---|---|---|---|---|---|---|---|
| **421** | Mule deer | Bunchgrass rangeland | Moise, Montana, USA | Rumen content analysis | Winter (Dec.–Feb.) | 34 | 28 | 38 | Nellis & Ross (1969) | See notes under no. 380. |
| **422** | Mule deer | Mixed shrubland, low mountain | Southern Wyoming, USA | Faecal analysis | Winter (2 years, 2 locations) | 5 | 4 | 91 | Goodwin (1975) | See notes under no. 377. |
| **423** | Mule deer | Mountain meadow, forest sagebrush–bitterbrush, bunchgrass prairie | Southwestern Montana, USA | Rumen samples | Winter | 14 | 16 | 70 | Wilkins (1957) | |
| **424** | Mule deer | Upper desert shrub, desert grassland and oak woodland | South-central Arizona, USA | Faecal analysis | Year, drought | 5 | 12 | 83 | Anthony (1976) | See notes under no. 468. |
| **425** | Mule deer | Upper desert shrub, desert grassland and oak woodland | South-central Arizona, USA | Faecal analysis | Year, normal precipitation | 4 | 9 | 87 | Anthony (1976) | See notes under no. 468. |
| **426** | Mule deer | Forest, grassland, scrub | Montana, USA | Stomach analysis | Year | 15 | 28 | 57 | De Nio (1938) | 7 % mosses and lichens distributed among data. |
| **427** | Mule deer | Piñon–juniper | Piceance Basin, Colorado, USA | Microscopic analysis of faecal samples | Yearlong average | 2 | 3 | 95 | Hubbard & Hansen (1976) | See notes under no. 348. |
| **428** | Mule deer | Mountain shrub | Piceance Basin, Colorado, USA | Microscopic analysis of faecal samples | Yearlong average | 3 | 6 | 91 | Hubbard & Hansen (1976) | See notes under no. 348. |
| **429** | Mule deer | Sagebrush–bitterbrush | Eastern California and western Nevada, USA | Stomach analysis | Oct.–Apr. | 28 | 9 | 63 | Lassen *et al.* (1952) | |
| **430** | Mule deer | Juniper–piñon | Oklahoma, USA | Faecal analysis, rumen analysis, and direct observation | Autumn–winter | 15 | 59 | 26 | Clark (1968) | |
| **431** | White-tailed deer | Meadows, fields, bottomlands | Montana, USA | Rumen analysis | Spring (Mar.–June) | 38 | 19 | 43 | Allen (1968) | |
| **432** | White-tailed deer | Longleaf–slash pine, loblolly–shortleaf pine, and oak–gum–cypress | South Carolina, USA | Rumen samples | Spring | 1 | 22 | 77 | Harlow *et al.* (1974) | Shrubs includes fruits from shrub species. Spring 9 % fungi distributed among data. |

| | | | | | | | | | | |
|---|---|---|---|---|---|---|---|---|---|---|
| **433** | White-tailed deer | Longleaf–slash pine, loblolly–shortleaf pine and oak–gum–cypress | Southeastern coastal plain, USA | Rumen samples | Spring | 8 | 32 | 60 | Harlow *et al.* (1974) | |
| **434** | White-tailed deer | | Ohio, USA | Rumen analysis | Spring | 1 | 18 | 81 | Nixon *et al.* (1970) | 11 % other distributed among data. |
| **435** | White-tailed deer | Loblolly pine and mixed hardwood | East Texas, USA | Rumino-reticulum analysis | Spring | 13 | 56 | 31 | Short (1971) | 10 % other distributed among data. |
| **436** | White-tailed deer | Coniferous forest | New Brunswick, Canada | Rumen samples | Spring | 0 | 49 | 51 | Skinner & Telfer (1974) | 27 % other distributed among data. |
| **437** | White-tailed deer | Sumac–sassafras secondary forest succession | Southern Indiana, USA | Rumen analysis | Spring | 20 | 66 | 14 | Sotola & Kirkpatrick (1973) | 9 % other distributed among data. |
| **438** | White-tailed deer | Savanna, live oak | Central Texas, USA | Observation, bite-count | Light grazing only used in spring | 10 | 45 | 45 | McMahan (1964) | |
| **439** | White-tailed deer | Savanna, live oak | Central Texas, USA | Observation, bite-count | Spring, heavy use | 15 | 44 | 41 | McMahan (1964) | (See notes above for cattle and sheep for same reference.) |
| **440** | White-tailed deer | Meadows, fields, bottomlands | Montana, USA | Rumen analysis | Summer (June–Aug.) | 1 | 54 | 45 | Allen (1968) | |
| **441** | White-tailed deer | Savanna, live oak | Central Texas, USA | Observation, bite-count | Light grazing, used in summer | 4 | 38 | 58 | McMahan (1964) | |
| **442** | White-tailed deer | Longleaf–slash pine, loblolly–shortleaf pine and oak–gum–cypress | Southeastern coastal plain, USA | Rumen samples | Summer | 9 | 41 | 50 | Harlow *et al.* (1974) | 16 % fungi distributed among data. |
| **443** | White-tailed deer | — | Ohio, USA | Rumen analysis | Summer | 3 | 16 | 81 | Nixon *et al.* (1970) | 20 % other distributed among data. |
| **444** | White-tailed deer | Loblolly pine and mixed hardwood | East Texas, USA | Rumino-reticulum analysis | Summer | 12 | 12 | 76 | Short (1971) | 41 % other, mostly mushrooms, distributed among data. |
| **445** | White-tailed deer | Coniferous forest | New Brunswick, Canada | Rumen samples | Summer | 0 | 78 | 22 | Skinner & Telfer (1974) | Forbs includes fruits. 8 % other distributed among data. |
| **446** | White-tailed deer | Sumac–sassafras secondary forest succession | Southern Indiana, USA | Rumen analysis | Summer | 0 | 68 | 32 | Sotola & Kirkpatrick (1973) | 2 % other distributed among data. |
| **447** | White-tailed deer | Longleaf–slash pine, loblolly–shortleaf pine and oak–gum–cypress | South Carolina, USA | Rumen samples | Summer | 1 | 12 | 87 | Harlow *et a* (1974) | 8 % fungi distributed among data. Shrubs includes fruits from shrub species. |
| **448** | White-tailed deer | Savanna, live oak | Central Texas, USA | Observation, bite-count | Summer, heavy use | 5 | 4 | 91 | McMahan (1964) | |

## Appendix 4.2 (*continued*)

| No. | Animal species | Kind of grazingland | Geographic location | Methods | Season/ treatment | Grass (%) | Forbs (%) | Shrubs (%) | Reference | Notes |
|---|---|---|---|---|---|---|---|---|---|---|
| **449** | White-tailed deer | Longleaf–slash pine, loblolly–shortleaf pine and oak–gum–cypress | South Carolina, USA | Rumen samples | Autumn | 2 | 1 | 97 | Harlow *et al.* (1974) | 3 % fungi distributed among data. Shrubs includes fruits from shrub species. |
| **450** | White-tailed deer | Longleaf–slash pine, loblolly–shortleaf pine and oak–gum–cypress | Southeastern coastal plain, USA | Rumen samples | Autumn | 2 | 5 | 93 | Harlow *et al.* (1974) | 15 % fungi distributed among data. |
| **451** | White-tailed deer | Meadows, fields, bottomlands | Montana, USA | Rumen analysis | Autumn (Sept.–Nov.) | 2 | 17 | 81 | Allen (1968) | |
| **452** | White-tailed deer | | Ohio, USA | Rumen analysis | Autumn | 0 | 21 | 79 | Nixon *et al.* (1970) | 14 % other distributed among data. |
| **453** | White-tailed deer | Loblolly pine and mixed hardwood | East Texas, USA | Rumino-reticulum analysis | Autumn | 9 | 9 | 82 | Short (1971) | 64 % acorns; 3 % other distributed among data. |
| **454** | White-tailed deer | Coniferous forest | New Brunswick, Canada | Rumen samples | Autumn | 0 | 72 | 28 | Skinner & Telfer (1974) | 45 % fruit, most included in forbs. 25 % other distributed among data. |
| **455** | White-tailed deer | Sumac–sassafras secondary forest succession | Southern Indiana, USA | Rumen analysis | Autumn | 2 | 23 | 75 | Sotala & Kirkpatrick (1973) | Shrubs includes mast and fruits. |
| **456** | White-tailed deer | | Wyoming, USA | Stomach analysis | Autumn | 52 | 20 | 28 | Anderson (1953) | Grasses include wheat and corn samples taken during hunting season. |
| **457** | White-tailed deer | Savanna, live oak | Central Texas, USA | Observation, bite-count | Light grazing, only used in autumn | 8 | 24 | 68 | McMahan (1964) | |
| **458** | White-tailed deer | Savanna, live oak | Central Texas, USA | Observation bite-count | Autumn, heavy use | 13 | 2 | 85 | McMahan (1964) | |
| **459** | White-tailed deer | Meadows, fields, bottomlands | Montana, USA | Rumen analysis | Winter (Dec.–Feb.) | 6 | 29 | 65 | Allen (1968) | |
| **460** | White-tailed deer | Longleaf–slash pine, loblolly–shortleaf pine and oak–gum–cypress | Southeastern coastal plain, USA | Rumen samples | Winter | 10 | 7 | 83 | Harlow *et al.* (1974) | 3 % fungi distributed among data. |
| **461** | White-tailed deer | Longleaf–slash pine, loblolly–shortleaf pine and oak–gum–cypress | South Carolina, USA | Rumen samples | Winter | 5 | 8 | 87 | Harlow *et al.* (1974) | 3 % fungi distributed among data. Includes fruits from shrub species. |

| | | | | | | | | | | |
|---|---|---|---|---|---|---|---|---|---|---|
| **462** | White-tailed deer | Savanna, live oak | Central Texas, USA | Observation, bite-count | Light grazing, only used in winter | 4 | 5 | 91 | McMahan (1964) | |
| **463** | White-tailed deer | Savanna, live oak | Central Texas, USA | Observation, bite-count | Winter, heavy use | 6 | 5 | 89 | McMahan (1964) | |
| **464** | White-tailed deer | | Ohio, USA | Rumen analysis | Winter | 0 | 9 | 91 | Nixon *et al.* (1970) | 10 % other distributed among data. |
| **465** | White-tailed deer | Loblolly pine and mixed hardwood | East Texas, USA | Rumino-reticulum analysis | Winter | 25 | 1 | 74 | Short (1971) | Shrubs includes trees. 32 % other distributed among data. |
| **466** | White-tailed deer | Sumac–sassafras secondary forest succession | Southern Indiana, USA | Rumen analysis | Winter | 31 | 36 | 33 | Sotala & Kirkpatrick (1973) | 10 % other distributed among data. |
| **467** | White-tailed deer | Upper desert shrub, desert grassland and oak woodland | South-central Arizona, USA | Faecal analysis | Year, drought | 10 | 39 | 51 | Anthony (1976) | See notes under no. 468. |
| **468** | White-tailed deer | Upper desert shrub, desert grassland and oak woodland | South-central Arizona, USA | Faecal analysis | Year, normal precipitation | 7 | 34 | 59 | Anthony (1976) | In drought year, deer utilized evergreen and dro ught-resistant species to a greater extent than normally preferred species. |
| **469** | White-tailed deer | Mesquite–buffalo grass, chap bristlegrass, and bunchgrass annual forb communities | South Texas, USA | Botanical measurements of percentage utilization times, percentage frequency of use | Yearlong average of all soil types | 13 | 80 | 7 | Drawe & Box (1968) | Deer use greatest on browse species such as huisache, blackbrush acacia and mesquite; grass seedheads main plant portion eaten in winter. Cattle preferred seacoast bluestem, silver bluestem, spike bristlegrass and buffalo grass. |
| **470** | White-tailed deer | Live oak brush and meadow | Southern Texas, USA | Rumen analysis | Year | 4 | 29 | 67 | White (1973) | Shrubs includes 44 % mast. |
| **471** | White-tailed deer | Deciduous forest | Northern Wisconsin, USA | Rumen analysis | Apr.–Nov. | 12 | 43 | 45 | McCaffrey *et al.* (1974) | 9 % other distributed among data. |
| **472** | White-tailed deer | Grassland–brush-land complex | Southeast Texas, USA | Rumen analysis | Winter–spring | 23 | 72 | 5 | Chamrad & Box (1968) | 5 % unknown distributed among data. |
| **473** | Elk, Roosevelt | Forest–grassland–stream-sight thickets | Humboldt County, California, USA | Observation minute-feeding method | Spring | 60 | 4 | 36 | Harper *et al.* (1967) | See notes under no. 477. |
| **474** | Elk | Mountain range | 5 western mountain areas, USA | Combination of techniques including stomach analysis and observation | Spring | 86 | 10 | 4 | Howard (unpublished) | See notes under no. 483. |
| **475** | Elk | Foothill bunchgrass rangeland | Moise, Montana, USA | Rumen analysis, volumetric percentage | Spring (Mar.–May) | 100 | <1 | <1 | Morris & Schwartz (1957) | See notes under no. 420. |

## Appendix 4.2 (*continued*)

| No. | Animal species | Kind of grazingland | Geographic location | Methods | Season/ treatment | Grass (%) | Forbs (%) | Shrubs (%) | Reference | Notes |
|---|---|---|---|---|---|---|---|---|---|---|
| **476** | Elk | Piñon–juniper and sagebrush | Northeastern Colorado, USA | Faecal analysis | Summer | 70 | 3 | 27 | R. M. Hansen & R. C. Clark (unpublished) | |
| **477** | Elk, Roosevelt | Forest–grassland–stream-sight thickets | Humboldt County, California, USA | Observation minute-feeding method | Summer | 58 | 20 | 22 | Harper *et al.* (1967) | Data are for a herd mixture of cows, bulls and calves. Grasses and sedges of major importance included orchard grass, sweet vernal grass, *Agrostis alba*, California oat grass; salmonberry and trailing blackberry were main browse eaten. Calves were more selective in feeding habits than adults; elk selected some plants by odour, pulled up plant, and consumed only the root material of such species as creeping buttercup, hairy cat's ear, or narrow-leaf plantain. |
| **478** | Elk | Mountain range | 5 western mountain areas, USA | Combination of techniques including stomach analysis and observation | Summer | 80 | 13 | 7 | Howard (unpublished) | See notes under no. 483. |
| **479** | Elk | Foothill bunchgrass rangeland | Moise, Montana, USA | Rumen analysis, volumetric percentage | Summer (June–Aug.) | 85 | 15 | <1 | Morris & Schwartz (1957) | See notes under no. 420. |
| **480** | Elk | Open mountain meadows | Colorado, USA | Range survey | Summer | 58 | 41 | 1 | Nichols (1957) | |
| **481** | Elk | | Wyoming, USA | Stomach analysis | Autumn | 55 | 33 | 12 | Anderson (1953) | All samples taken during hunting season. |
| **482** | Elk, Roosevelt | Forest–grassland–stream-sight thickets | Humboldt County, California, USA | Observation, minute-feeding method | Autumn | 57 | 22 | 21 | Harper *et al.* (1967) | See notes under no. 477. |
| **483** | Elk | Mountain rangeland | 5 western mountain areas, USA | Combination of techniques including stomach analysis and observation | Autumn | 72 | 17 | 11 | Howard (unpublished) | Data are average for Little Belt Mountains, Montana, Gravely Mountains, Montana, Missouri Breaks Area, Montana, Sun River Area, Montana, and Rocky Mountain National Park; data for Boyles Prairie in California show an appreciably higher percentage of forbs in diets at all seasons, and lower grasses. |

| | | | | | | | | | | |
|---|---|---|---|---|---|---|---|---|---|---|
| **484** | Elk | Foothill bunchgrass rangeland | Moise, Montana, USA | Rumen analysis, volumetric percentage | Autumn (Oct.–Nov.) | 100 | <1 | 0 | Morris & Schwartz (1957) | See notes under no. 420. |
| **485** | Elk | Coniferous forest, sagebrush | Gallatin Canyon, Montana, USA | Plant use measurements | Winter | 60 | 10 | 30 | Constant (1973) | |
| **486** | Elk | Coniferous forest, mountain rangeland | Jasper Park, Alberta, Canada | | Winter | 97 | 0 | 3 | Cowan (1947) | |
| **487** | Elk, Roosevelt | Forest–grassland–stream-sight thickets | Humboldt County, California, USA | Observation, minute-feeding method | Winter | 76 | 2 | 22 | Harper *et al.* (1967) | See notes under no. 477. |
| **488** | Elk | Mountain rangeland | 5 western mountain areas, USA | Combination of techniques including stomach analysis and observation | Winter | 85 | 10 | 5 | Howard (unpublished) | See notes under no. 483. |
| **489** | Elk | | New Mexico, USA | Stomach analysis | Winter | 5 | 0 | 95 | Lang (1958) | |
| **490** | Elk | Foothill bunchgrass rangeland | Moise, Montana, USA | Rumen analysis, volumetric percentage | Winter (Dec.–Feb.) | 100 | <1 | <1 | Morris & Schwartz (1957) | See notes under no. 420. |
| **491** | Elk | Forest, grassland, scrub | Montana, USA | Stomach analysis | Year | 28 | 20 | 52 | De Nio (1938) | 4 % mosses and lichens distributed among data. |
| **492** | Elk | Various | Western North America | Various | Yearlong average | 84 | 8 | 8 | Johnston & Mattox (1954) | See notes under no. 109. |
| **493** | Elk | Sagebrush–grass, saltbush, and rabbitbrush types | Red Desert, southwestern Wyoming, USA | Microscopic analysis of faecal samples | Yearlong average | 79 | 3 | 18 | Olsen & Hansen (1977) | See notes under no. 348. |
| **494** | Elk | Mountain summer range | Oregon and Washington, USA | Ocular estimate by plot | Year | 13 | 81 | 6 | Pickford & Reid (1943) | |
| **495** | Red deer | Grass–heather–blueberry–lichen | Scottish highlands, Cairngorm Mountains, Scotland | Rumen content of slaughtered animals, microscopic analysis percentage of particles | Summer, 1966–70 'natural' grazing conditions | 70 | 10 | 20 | Hobson *et al.* (1976) | Lichens are considered as forbs; heather is considered a shrub. Miscellaneous items in red deer diets considered primarily herbaceous, and forbs data are averages interpolated from graph. |
| **496** | Red deer | Pine forest, coniferous forest | Poland | Rumen analysis | Autumn (Sept.–Nov.) | 13 | 36 | 51 | Dzieciolowski (1970) | 4 % other distributed among data. Shrubs includes trees. |
| **497** | Red deer | Pine forest, coniferous forest | Poland | Rumen analysis | Winter (Dec.–Feb.) | 5 | 2 | 93 | Dzieciolowski (1970) | 4 % other distributed among data. Shrubs includes trees. |
| **498** | Red deer | Beech forest and and heath | Denmark | Rumen analysis | Year | 59 | 6 | 35 | Jensen (1968) | 14 % other distributed among data. |

## Appendix 4.2 (*continued*)

| No. | Animal species | Kind of grazingland | Geographic location | Methods | Season/ treatment | Grass (%) | Forbs (%) | Shrubs (%) | Reference | Notes |
|---|---|---|---|---|---|---|---|---|---|---|
| **499** | Red deer | Grass–heather–blueberry–lichen | Scottish highlands, Cairngorm Mountains, Scotland | Rumen contents of slaughtered animals, microscopic identification | Autumn–spring | 55 | 5 | 40 | Hobson *et al.* (1976) | See notes under no. 495. |
| **500** | Moose | Shrub and deciduous trees | South-central Alaska, USA | Rumen analysis | Spring | 0 | 0 | 100 | Cushaw & Coady (1976) | Twigs of willow and birch most frequently eaten. Shrubs includes trees. See notes under no. 507. |
| **501** | Moose | Willow, aspen, Douglas fir, spruce-fir, lodgepole vegetation types | Southwest Montana, USA | Feeding site examination | Spring | 0 | 6 | 94 | Stevens (1970) | Shrubs includes trees. |
| **501a** | Moose | | Montana, USA | Faecal examination and visual estimates | Summer | 14 | 14 | 72 | Peterson (1955) | |
| **502** | Moose | Shrubs and deciduous trees | South-central Alaska, USA | Rumen analysis | Summer | 0 | 0 | 100 | Cushwa & Coady (1976) | See notes under no. 507. |
| **503** | Moose | Willow and sedge | Southwest Montana, USA | Feeding site observations | Summer | 0 | 2 | 98 | Dorn (1970) | Shrubs includes trees. |
| **504** | Moose | Willow, sagebrush, aspen, sub-alpine meadows, coniferous forest | Southwest Montana, USA | Rumen analysis | Summer | 1 | 70 | 29 | Knowlton (1960) | Shrubs includes trees. |
| **505** | Moose | Yellowstone National Park | Northwest Wyoming, USA | Direct observation | June–Sept. | 3 | 9 | 88 | McMillan (1953) | Shrubs were entirely willow. |
| **506** | Moose | | | | Summer | 11 | 26 | 63 | Slater (unpublished) | See notes under no. 514. |
| **507** | Moose | Shrubs and deciduous trees | South-central Alaska, USA | Rumen analysis | Autumn | 0 | 0 | 100 | Cushwa & Coady (1976) | Shrubs includes trees. Differences in diets between season are found in what species of shrubs and trees are eaten. |
| **508** | Moose | Willow, sagebrush, aspen, sub-alpine meadows, coniferous forests | Southwest Montana, USA | Rumen analysis | Autumn | 2 | 7 | 91 | Knowlton (1960) | Shrubs includes trees. |
| **509** | Moose | Willow, aspen, Douglas fir, spruce-fir and lodgepole vegetation types | Southwest Montana, USA | Rumen samples | Autumn | 1 | 11 | 88 | Stevens (1970) | Shrubs includes trees. |

| | | | | | | | | | | |
|---|---|---|---|---|---|---|---|---|---|---|
| **509a** | Moose | | Montana, USA | Faecal examination and visual estimates | Autumn | 5 | 4 | 91 | Peterson (1955) | |
| **510** | Moose | Shrubs and deciduous trees | South-central Alaska, USA | Rumen analysis | Winter | 0 | 0 | 100 | Cushwa & Coady (1976) | Shrubs includes trees. See notes under no. 507. |
| **511** | Moose | Willow and sedge | Southwest Montana, USA | Feeding site observations | Winter | 0 | 0 | 100 | Dorn (1970) | Shrubs includes trees. |
| **512** | Moose | Sagebrush plain | Wyoming | Stomach analysis | Winter | 0 | 0 | 100 | Harry (1957) | Shrubs includes trees. |
| **513** | Moose | Willow, sagebrush, aspen, sub-alpine meadows, coniferous forest | Southwest Montana, USA | Rumen analysis | Winter | 0 | 1 | 99 | Knowlton (1960) | Shrubs includes trees. |
| **514** | Moose | | | | Winter | 0 | 0 | 100 | Slater (unpublished) | Winter browse primarily willow; spring diet primarily grass and forbs; summer grass composed primarily of sedges and aquatics. |
| **515** | Moose | Willow, aspen, Douglas-fir, spruce-fir, lodgepole vegetation types | Southwest Montana, USA | Feeding site examination | Winter | 0 | 1 | 99 | Stevens (1970) | Shrubs includes trees. |
| **516** | Moose | Taiga | Alaska USA | Stomach content analysis | Autumn–winter | 6 | 2 | 92 | Hosley (1949) | Recalculated from identifiable portions reported by the author. Birch and willow main components of diet. |
| **517** | Caribou | Tundra | North America | Analysis of rumen samples | Spring | 57 | 13 | 30 | Beavers (unpublished) | In spring lichens were 3 % of diet, in summer 24 %, in autumn 40 %, and in winter 56 %. Fungi were important in summer and autumn, aquatic vegetation in summer, mosses in autumn and winter. 10 % unknown and lichens. distributed among data (spring). |
| **518** | Caribou | Tundra | Northwest Territories, Canada | Stomach analysis | Summer | 29 | 32 | 39 | Banfield (1954) | The forb data represent 31 % lichens, 1 % mushrooms and a trace of mosses. Major shrub in diet was *Betula glandulosa*. |
| **519** | Caribou | Tundra | North America | Analysis of rumen samples | Summer | 24 | 20 | 56 | Beavers (unpublished) | See notes under no. 517. 59 % unknown and lichen distributed among summer data. |
| **519a** | Caribou | Tundra | Banks Island, Northwest Territories, Canada | Rumen analysis | Summer | 55 | 17 | 28 | Wilkinson *et al.* (1976) | Forbs includes lichens. |
| **520** | Caribou | Tundra | North America | Analysis of rumen samples | Autumn | 23 | 28 | 49 | Beavers (unpublished) | See notes under no. 517. 57 % unknown and lichen distributed among data. |

## Appendix 4.2 (*continued*)

| No. | Animal species | Kind of grazingland | Geographic location | Methods | Season/ treatment | Grass (%) | Forbs (%) | Shrubs (%) | Reference | Notes |
|---|---|---|---|---|---|---|---|---|---|---|
| **521** | Caribou | Tundra | North America | Analysis of rumen samples | Winter | 18 | 0 | 82 | Beavers (unpublished) | See notes under no. 517. 62 % unknown and lichen distributed among data. |
| **521a** | Caribou | Tundra | Banks Island, Northwest Territories, Canada | Rumen analysis | Winter | 61 | 13 | 26 | Wilkinson *et al.* (1976) | Forbs includes lichens. |
| **522** | Caribou | Tundra and taiga | Wells Gray Park, British Columbia, Canada | Stomach analysis | Yearlong | 6 | 72 | 22 | Edwards & Richey (1960) | Forbs includes 46 % lichens and 18 % miscellaneous. |
| **523** | Caribou | Coniferous forest | Northwestern Canada | Rumen analysis | Oct.–Apr. | 13 | 0 | 87 | Scotter (1967) | 58 % lichens, 3 % others and 16 % unidentified distributed among data. |
| **524** | Reindeer | Alpine plateau | Southern Norway | Microscopic rumen analysis | Spring | 22 | 47 | 31 | Gaare & Skogland (1975) | See notes under no. 528. |
| **525** | Reindeer | Alpine plateau | Southern Norway | Microscopic rumen analysis | Summer | 32 | 44 | 24 | Gaare & Skogland (1975) | See notes under no. 528. |
| **526** | Reindeer | Grass–heather–blueberry–lichen | Scottish highlands, Cairngom Mountains, Scotland | Rumen contents of slaughtered animals, microscopic identification | Summer | 55 | 35 | 10 | Hobson *et al.* (1976) | See notes under no. 495. |
| **527** | Reindeer | Alpine plateau | Southern Norway | Microscopic rumen analysis | Autumn | 42 | 48 | 10 | Garae & Skogland (1975) | See notes under no. 528. |
| **528** | Reindeer | Alpine plateau | Southern Norway | Rumen content, binocular microscopic technique | Winter (Nov.–Mar.) | 27 | 59 | 14 | Gaare & Skogland (1975) | Litter was included with shrubs as were primarily twigs and leaves. Lichens and mosses were included with forbs. |
| **529** | Reindeer | Grass–heather–blueberry–lichen | Scottish highlands, Cairngorm Mountains, Scotland | Rumen contents of slaughtered animals, microscopic identification | Autumn–spring | 33 | 61 | 6 | Hobson *et al.* (1976) | See notes under no. 495. |
| **530** | Musk ox | Tundra grass–shrub rangeland | Ellesmere Island, Northwest Territories, Canada | Bite-count | Spring (Apr.–June) | 86 | 4 | 10 | Tener (1954) | Grass cover about 5 %, primarily *Poa*, *Alopecurus*, and *Agropyron* species; most abundant species available was *Dryas integrifolia*. *Salix arctica* was second most important. Grasses and willow highly selected as they totalled only about 6 % of the herbage cover. |
| **530a** | Musk ox | Tundra | Banks Island, Northwest Territories, Canada | Rumen analysis | Summer | 90 | 2 | 8 | Wilkinson *et al.* (1976) | Forbs includes lichens. |

| | | | | | | | | | | |
|---|---|---|---|---|---|---|---|---|---|---|
| **530b** | Musk ox | Tundra | Banks Island, Northwest Territories, Canada | Rumen analysis | Winter | 85 | 1 | 14 | Wilkinson *et al.* (1976) | Forbs includes lichens. |
| **531** | Barbary sheep | | North America | Stomach analysis | Spring | 26 | 63 | 11 | Alexander (unpublished) | See notes under no. 534. |
| **532** | Barbary sheep | | North America | Stomach analysis | Summer | 26 | 9 | 65 | Alexander (unpublished) | See notes under no. 534. |
| **533** | Barbary sheep | | North America | Stomach analysis | Autumn | 29 | 13 | 58 | Alexander (unpublished) | See notes under no. 534. |
| **534** | Barbary sheep | | North America | Stomach analysis | Winter | 86 | 3 | 11 | Alexander (unpublished) | Overall diet dominated by grass and browse; important browse species included oak and mountain mahogany. |
| **535a** | Dall's sheep | Aspen–shrub–meadow | Whitehorse, Alaska, USA | Bite-count method | Spring | 56 | 26 | 18 | Hoefs (1974) | Data are percentage bites. |
| **535b** | Dall's sheep | Aspen–shrub–meadow | Whitehorse, Alaska, USA | Bite-count method | Summer | 35 | 34 | 31 | Hoefs (1974) | Data are percentage bites. |
| **535c** | Dall's sheep | Aspen–shrub–meadow | Whitehorse, Alaska, USA | Bite-count method | Autumn | 70 | 15 | 15 | Hoefs (1974) | Data are percentage bites. |
| **535d** | Dall's sheep | Aspen–shrub–meadow | Whitehorse, Alaska, USA | Bite-count method | Winter | 63 | 7 | 30 | Hoefs (1974) | Data are percentage bites. |
| **536** | Bighorn sheep | | Idaho, USA | Feeding observation | Spring | —77— | | 22 | Smith (1954) | 1 % mosses and lichens. See notes under no. 550. |
| **536a** | Bighorn sheep | | Nevada, USA | Rumen analysis | Spring | 67 | 25 | 8 | Barrett (1964) | |
| **537** | Bighorn sheep | Mountains | Saguache County, Colorado, USA | | Spring | 57 | 10 | 33 | Todd (1975) | |
| **538** | Bighorn sheep | | Idaho, USA | Feeding observation | Summer | —86— | | 14 | Smith (1954) | See notes under no. 550. |
| **539** | Bighorn sheep | Mountains | Saguache County, Colorado, USA | | Summer | 65 | 6 | 29 | Todd (1975) | |
| **540** | Bighorn sheep | Alpine meadow | Wyoming, USA | Feeding site observations | Summer | 73 | 27 | 0 | Woolf (1968) | |
| **540a** | Bighorn sheep | | Nevada, USA | Rumen analysis | Summer | 94 | 3 | 3 | Barrett (1964) | |
| **540b** | Bighorn sheep | | Nevada, USA | Rumen analysis | Autumn | 48 | 8 | 44 | Barrett (1964) | |
| **541** | Bighorn sheep | | Colorado, USA | Observation and stomach sample analysis | Sept. | 75 | 6 | 19 | Moser (1962) | Grasslike was mostly *Poa*, *Muhlenbergia*, *Carex* and *Festuca*. Forbs mostly *Trifolium*. Shrubs mostly *Salix* and *Potentilla*. |
| **541a** | Bighorn sheep | Juniper–piñon–saltbush range | South-central Nevada, USA | Stomach analysis | Autumn | 49 | 50 | 1 | Yoakum (1964) | |
| **542** | Bighorn sheep | | Idaho, USA | Feeding observation | Autumn | —66— | | 25 | Smith (1954) | 9 % mosses and lichens, See notes under no. 550. |

## Appendix 4.2 (*continued*)

| No. | Animal species | Kind of grazingland | Geographic location | Methods | Season/ treatment | Grass (%) | Forbs (%) | Shrubs (%) | Reference | Notes |
|---|---|---|---|---|---|---|---|---|---|---|
| **542a** | Bighorn sheep | | Southern Nevada, USA | Rumen analysis | Autumn | 73 | 7 | 20 | Brown *et al.* (1975) | |
| **543** | Bighorn sheep | Mountains | Saguache County, Colorado, USA | | Autumn | 54 | 2 | 44 | Todd (1975) | |
| **544** | Bighorn sheep | Coniferous forest, sagebrush | Gallatin Canyon, Montana, USA | Plant use measurements | Winter (Jan.–Mar.) | 74 | 18 | 8 | Constan (1973) | |
| **544a** | Bighorn sheep | Juniper–piñon–saltbush range | South-central Nevada USA | Stomach analysis | Winter | 70 | 14 | 16 | Yoakum (1964) | |
| **545** | Bighorn sheep | Coniferous forest, mountain rangeland | Jasper Park, Alberta, Canada | | Winter | 83 | 10 | 7 | Cowan (1947) | |
| **546** | Bighorn sheep | Sagebrush–meadow | Wyoming, USA | Feeding site observations | Winter | 61 | 17 | 22 | Oldemeyer (1966) | |
| **546a** | Bighorn sheep | Winter ranges of fescue–sedge | Alberta and British Columbia, Canada | Range utilization transects | Winter | 73 | 23 | 4 | Stelfox (1976) | |
| **547** | Bighorn sheep | | Idaho, USA | Feeding observation | Winter | —56— | | 39 | Smith (1954) | Moss and lichens 5 %. See notes under no. 550. |
| **548** | Bighorn sheep | Mountains | Saguache County, Colorado, USA | | Winter | 23 | 11 | 66 | Todd (1975) | |
| **548a** | Bighorn sheep | | Nevada, USA | Rumen samples | Winter | 81 | 2 | 17 | Barrett (1964) | |
| **549** | Bighorn sheep | | Wyoming, USA | Stomach samples | Year | 50 | 31 | 19 | Honess & Frost (1942) | |
| **549a** | Bighorn sheep | | Nevada, USA | Rumen samples | Year | 76 | 4 | 20 | Barrett (1964) | |
| **550** | Bighorn sheep | | Idaho, USA | Feeding observation | Year | —70— | | 27 | Smith (1954) | Grass and forbs put into one category. Grass largely bluebunch wheatgrass; shrubs mostly curleaf mountain mahogany (also big sagebrush in winter). 3 % mosses and lichens. |
| **551** | Bighorn sheep | | Western North America | | Average for 5 locations | 67 | 17 | 16 | Woroselo (unpublished) | Areas include two in Colorado, one in Montana, Jasper National Park and Yellowstone National Park studies. Grasses as high as 83 % in diet in some situations; forbs as high as 45 % in diet; moss and lichens contribute to diet in winter and autumn. |

| | | | | | | | | | | |
|---|---|---|---|---|---|---|---|---|---|---|
| 552 | Bighorn sheep | Bunchgrass prairie | British Columbia, Canada | Line-point sampling, inspection of grazed area, analysis of 6 rumens | Sept.–June | 72 | 4 | 24 | Blood (1967) | |
| 553 | Mountain goat | High–low mountains | Montana, USA | Stomach analysis | Spring | 76 | 5 | 19 | Saunders (1955) | See notes under no. 554. |
| 554 | Mountain goat | High–low mountains | Montana, USA | Stomach analysis | Summer | 80 | 15 | 5 | Saunders (1955) | Conifers important in diet in winter and spring; diet comprised mostly of shrubs in winter. |
| 554a | Mountain goat | Alpine | Alaska, USA | Plot sampling with rectangular grid | Summer | 36 | 64 | 0 | Hjeljord (1971) | |
| 554b | Mountain goat | Alpine | Colorado, USA | Stomach analysis | Summer | 82 | 14 | 4 | Hibbs (1966) | |
| 555 | Mountain goat | High–low mountains | Montana, USA | | Autumn | 77 | 22 | 1 | Saunders (1955) | See notes under no. 554. |
| 556 | Mountain goat | High–low mountains | Montana, USA | | Winter | 59 | 10 | 31 | Saunders (1955) | See notes under no. 554. |
| 556a | Mountain goat | Sub-alpine | Alaska, USA | Plot sampling with rectangular grid | Winter | 0 | 90 | 10 | Hjeljord (1971) | Forbs are ferns. |
| 556b | Mountain goat | Alpine | Alaska, USA | Rumen analysis | Winter | 13 | 72 | 15 | Klein (1953) | Forbs are ferns. |
| 556c | Mountain goat | Alpine | Alaska, USA | Forage utilization | Winter | 98 | 2 | 0 | Klein (1953) | |
| 556d | Mountain goat | Sub-alpine spruce–fir | Colorado, USA | Observation of feeding animals | Winter | 88 | 0 | 12 | Hibbs (1966) | Shrubs are deciduous trees. |
| 557 | Saiga | Arid steppes and semideserts | Kalmyk, USSR | Stomach analysis with visual estimation of the percentage composition of diet | Early spring | 86 | 3 | 11 | Bannikov (1967) | Diets are corrected to be lichen-free; adult diets in winter have 10 % lichen. |
| 558 | Saiga | Arid steppes and semideserts | Kalmyk, USSR | Stomach analysis with visual estimation of percentage composition of diet | Spring | 30 | 55 | 15 | Bannikov (1967) | See notes under no. 557. |
| 559 | Saiga | Arid steppes and semideserts | Kalmyk, USSR | Stomach analysis with visual estimation of percentage composition of diet | Summer | 30 | 52 | 18 | Bannikov (1967) | See notes under no. 557. |
| 560 | Saiga | Arid steppes and semideserts | Kalmyk, USSR | Stomach analysis with visual estimation of percentage composition of diet | Adult (July) | 35 | 50 | 15 | Bannikov (1967) | See notes under no. 557. |
| 561 | Saiga | Arid steppes and semideserts | Kalmyk, USSR | Stomach analysis with visual estimation of percentage composition of diet | Young (July) | 47 | 40 | 13 | Bannikov (1967) | See notes under no. 557. |

## Appendix 4.2 (*continued*)

| No. | Animal species | Kind of grazingland | Geographic location | Methods | Season/ treatment | Grass (%) | Forbs (%) | Shrubs (%) | Reference | Notes |
|---|---|---|---|---|---|---|---|---|---|---|
| **562** | Saiga | Arid steppes and semideserts | Kalmyk, USSR | Stomach analysis with visual estimation of percentage composition of diet | Late autumn | 10 | 63 | 27 | Bannikov (1967) | See notes under no. 557. |
| **563** | Saiga | Arid steppes and semideserts | Kalmyk, USSR | Stomach analysis with visual estimation of percentage composition of diet | Winter | 8 | 74 | 18 | Bannikov (1967) | See notes under no. 557. |
| **564** | Saiga | Arid steppes and semideserts | Kalmyk, USSR | Stomach analysis with visual estimation of percentage composition of diet | Year | 32 | 48 | 20 | Bannikov (1967) | See notes under no. 557. |
| **565** | Red kangaroo | Rocky ridges, pearl saltbush, flat country of bladder saltbush | Northwest New South Wales, Australia | Samples from junction of oesophagus and first fermentation chamber of shot animals, microscopic identification | Summer | 71 | 17 | 12 | Dawson *et al.* (1975) | Shrubs includes trees. |
| **566** | Red kangaroo | Grassy river plain | Central Australia | Stomach analysis | Year | 92 | —8— | | Newsome (1975) | Shrubs includes trees and forbs. |
| **567** | Red kangaroo | Grassy river plain | Central Australia | Stomach analysis | Drought | 90 | 8 | 2 | Newsome (1975) | Shrubs includes trees. |
| **568** | Red kangaroo | *Spinifex* steppe and arid plain | Australia | Faecal analysis, epidermal area | Drought | 85 | —15— | | Newsome (1975) | Forbs includes shrubs. |
| **569** | Red kangaroo | Grassy river plain | Central Australia | Stomach analysis | Non-drought | 98 | 1 | 1 | Newsome (1975) | Shrubs includes trees. |
| **570** | Red kangaroo | *Spinifex* steppe and arid plain | Australia | Faecal analysis, epidermal area | Non-drought | 87 | —13— | | Newsome (1975) | Forbs includes shrubs. |
| **571** | Euro kangaroo | Rocky ridges, pearl saltbush, flat country of bladder saltbush | Northwest New South Wales, Australia | Samples from junction of oesophagus and first fermentation chamber of shot animals, microscopic identification | Summer | 52 | 33 | 15 | Dawson *et al.* (1975) | Shrubs includes trees. |
| **572** | Euro kangaroo | *Spinifex* steppe | Northwestern Australia | Faecal analysis, epidermal area | Non-drought | 95 | 5 | 0 | Newsome (1971) | 18 % unidentified distributed among grasses and forbs. Soft *Spinifex* 4 times more abundant in diets in drought than non-drought. |
| **573** | Euro kangaroo | *Spinifex* steppe and arid plain | Australia | Faecal analysis, epidermal area | Non-drought | 98 | 2 | 0 | Newsome (1975) | Forbs includes shrubs. |

| | | | | | | | | | | |
|---|---|---|---|---|---|---|---|---|---|---|
| **574** | Euro kangaroo | *Spinifex* steppe | Northwestern Australia | Faecal analysis, epidermal area | Drought | 95 | 5 | 0 | Newsome (1971) | 2 % unidentified material distributed to grasses and forbs. See notes under no. 572. |
| **575** | Euro kangaroo | *Spinifex* steppe and arid plain | Australia | Faecal analysis, epidermal area | Drought | 91 | —9— | | Newsome (1975) | Forbs includes shrubs. |
| **576** | Mountain gazelle | Semidesert grass–shrub | Northern Israel | Rumen analysis | Growing season (Dec.–Mar.) | 73 | 23 | 4 | Baharav (1975) | The rumen contents were used to estimate intake which was 2–3 % body weight per day; digestibility was 61–75 % dry matter. |
| **577** | Mountain gazelle | Semidesert grass–shrub | Northern Israel | Rumen analysis | Dry-off period (Apr.–June) | 53 | 15 | 32 | Baharav (1975) | See notes under no. 576. |
| **578** | Mountain gazelle | Semidesert grass–shrub | Northern Israel | Rumen analysis | Dry period (July–Nov.) | 58 | 10 | 32 | Baharav (1975) | See notes under no. 576. |
| **579** | Warthog | Open woodland | Tanzania | Visual observation | Yearlong | 88 | 9 | 3 | Lamprey (1963) | See notes under no. 584. |
| **580** | Dik-dik | Dense woodland | Tanzania | Visual observation | Yearlong | 18 | 3 | 79 | Lamprey (1963) | See notes under no. 584. |
| **581** | Impala | Open woodland | Tanzania | Visual observation | Yearlong | 93 | 1 | 6 | Lamprey (1963) | See notes under no. 584. |
| **582** | Impala | Savanna grassland | Narwok and Kajiado Districts, Kenya | Stomach analysis | Yearlong | 56 | —44— | | Talbot & Talbot (1962) | See notes under no. 600. |
| **583** | Impala | Various grass and shrub communities | South Africa | Rumen–reticulum samples | Dry season | 4 | 35 | 5 | Giesecke & Van Gylswyk (1975) | See notes under no. 616. |
| **584** | Grant's gazelle | Savanna grassland | Tanzania | Visual observation | Yearlong | 91 | 4 | 5 | Lamprey (1963) | The hartebeest grazes the savanna grassland and open woodland about equally; the elephant takes a considerable proportion of its diet from the dense woodland; the dik-dik takes a considerable portion of its diet from the open woodland. For several species of trees are important in the diet (as a component listed here as shrubs); use of trees is considerable by giraffe, elephant, eland, lesser kudu and rhinoceros. |
| **585** | Grant's gazelle | Savanna grassland | Kenya and Tanzania | Stomach analysis | Yearlong | 40 | —60— | | Talbot & Talbot (1962) | See notes under no. 600. |
| **586** | Thomson's gazelle | Savanna grassland | Kenya and Tanzania | Stomach analysis | Yearlong | 80 | —20— | | Talbot & Talbot (1962) | See notes under no. 600. |
| **587** | Thomson's gazelle | Grassland savanna | Western Serengeti, Kenya | Stomach content analysis | Dry season (Sept.) | 62 | 19 | 19 | Bell (1970) | See notes under no. 602. |

## Appendix 4.2 (*continued*)

| No. | Animal species | Kind of grazingland | Geographic location | Methods | Season/ treatment | Grass (%) | Forbs (%) | Shrubs (%) | Reference | Notes |
|---|---|---|---|---|---|---|---|---|---|---|
| **588** | Thomson's gazelle | — | Serengeti National Park, Tanzania | Collected fresh ingested material from shot animals, botanical composition by point-count method | Dry season (Sept.) | 61 | 39 | 0 | Gwynne & Bell (1968) | |
| **589** | Zebra | Savanna grassland | Tanzania | Visual observation | Yearlong | 93 | 5 | 2 | Lamprey (1963) | See notes under no. 584. |
| **590** | Zebra | Grassland savanna | Western Serengeti, Kenya | Stomach content analysis | Dry season (Sept.) | 100 | trace | | Bell (1970) | See notes under no. 602. |
| **591** | Zebra | — | Serengeti National Park, Tanzania | Collected fresh ingested material from shot animals, botanical composition by point-count method | Dry season (Sept.) | 100 | 0 | 0 | Gwynne & Bell (1968) | |
| **592** | Zebra | — | Serengeti National Park, Tanzania | Collected fresh ingested material from shot animals, botanical composition by point-count method | Wet season (May) | 100 | 0 | 0 | Gwynne & Bell (1968) | |
| **593** | Hartebeest | Savanna grassland | Tanzania | Visual observation | Yearlong | 96 | 3 | 1 | Lamprey (1963) | See notes under no. 584. |
| **594** | Springbok | Various grass and shrub communities | South Africa | Rumen–reticulum samples | Dry season | 10 | 12 | 20 | Giesecke & Van Gylswyk (1975) | See notes under no. 616. |
| **595** | Lesser kudu | Dense woodland | Tanzania | Visual observation | Yearlong | 67 | 4 | 29 | Lamprey (1963) | See notes under no. 584. |
| **596** | Kudu | Various grass and shrub communities | South Africa | Rumen–reticulum samples | Dry season | 2 | 44 | 13 | Giesecke & Van Gylswyk (1975) | See notes under no. 616. |
| **597** | Lesser kudu | Riverine bush, woodland, and shrubland | Africa | | Dry season | <1 | 10 | 90 | Leuthold (1971) | Shrubs includes trees and large shrubs; forbs includes herbs, fruits, creepers and vines, and tubers (tubers always less than 1 %). Fruits make contribution up to about 5 % in the dry season. Average of 3 habitats for the dry season and 2 habitats for the wet season. |
| **598** | Lesser kudu | Riverine bush, woodland, and shrubland | Africa | | Wet (green) season | 2 | 28 | 70 | Leuthold (1971) | See notes under no. 597. |

| 599 | Wildebeest | Savanna grassland | Tanzania | Visual observation | Yearlong | 95 | 3 | 2 | Lamprey (1963) | See notes under no. 584. |
|---|---|---|---|---|---|---|---|---|---|---|
| 600 | Wildebeest | Savanna grassland | Kenya and Tanzania | Stomach analysis | Yearlong | 98 | 1 | 1 | Talbot & Talbot (1962) | Grant's gazelle consumed more non-grass items than the other animals; selects primarily invader species in overgrazed, abused grasslands; diet complementary to those of livestock and other ruminants. Topi ate considerable dry material by choice. |
| 601 | Wildebeest | Savanna grassland | Western Masailand, East Africa | Analysis based on microscopic identification | Yearlong | 100 | trace | | Talbot & Talbot (1963) | See notes under no. 600. |
| 602 | Wildebeest | Grassland savanna | Western, Serengeti, Kenya | Stomach content analysis | Dry season (Sept.) | 100 | 0 | 0 | Bell (1970) | Forbs and shrubs analysed as total dicotyledon and split here arbitrarily for the gazelle. Grass-eating species vary in plant part (leaf, sheath and stem) taken; wildebeest eat proportionately more grass leaf than topi or zebra; zebra eat proportionately more grass stem than topi or wildebeest; all three species eat about 50 % grass sheath. |
| 603 | Blue wildebeest | Various grass and shrub communities | South Africa | Rumen–reticulum samples | Dry season | 100 | 0 | 0 | Glesecke & Van Gylswyk (1975) | See notes under no. 616. |
| 604 | Wildebeest | | Serengeti National Park, Tanzania | Collected fresh ingested material from shot animals, botanical composition by point-count method | Dry season (Sept.) | 100 | 0 | 0 | Gwynne & Bell (1968) | |
| 605 | Wildebeest | | Serengeti National Park, Tanzania | Collected fresh ingested material from shot animals, botanical composition by point-count method | Wet season (May) | 99 | 1 | 0 | Gwynne & Bell (1968) | |
| 606 | Eland | *Acacia* and tall tropical trees | Kenya | Oesophageal fistulas and ocular estimation | July–Aug. | 44 | 0 | 24 | Nge'the & Box (1976) | Data interpolated from graph; only data for most important plants given so data do not add up to 100 %. |
| 607 | Eland | *Acacia* and tall tropical trees | Kenya | Oesophageal fistulas and ocular estimation | Sept.–Oct. | 56 | 0 | 14 | Nge'the & Box (1976) | See notes under no. 606. |
| 608 | Eland | Open woodland | Tanzania | Visual observation | Yearlong | 70 | 9 | 21 | Lamprey (1963) | See notes under no. 584. |
| 609 | Gemsbok | Semidesert grassland and shrubland | White Sands Missile Range, New Mexico, USA | Faecal analysis by microscopic determination, direct observation | Spring | 37 | 22 | 41 | Saiz (1975) | Grass mostly dropseeds. |

## Appendix 4.2 (*continued*)

| No. | Animal species | Kind of grazingland | Geographic location | Methods | Season/ treatment | Grass (%) | Forbs (%) | Shrubs (%) | Reference | Notes |
|---|---|---|---|---|---|---|---|---|---|---|
| **610** | Gemsbok | Semidesert grassland and shrubland | New Mexico, USA | Faecal analysis, direct observation | Summer | 33 | 43 | 24 | Saiz (1975) | Grass mostly plains bristlegrass. |
| **611** | Gemsbok | Semidesert grassland and shrubland | New Mexico, USA | Faecal analysis, direct observation | Autumn | 10 | 26 | 64 | Saiz (1975) | |
| **612** | Gemsbok | Semidesert grassland and shrubland | New Mexico, USA | Faecal analysis, direct observation | Winter | 33 | 11 | 56 | Saiz (1975) | Grass mostly dropseeds. |
| **613** | Gemsbok | Semidesert grassland and shrubland | New Mexico, USA | Faecal analysis, direct observation | Year | 30 | 22 | 48 | Saiz (1975) | 2 % other species distributed among data. Shrubs mostly yucca and mesquite; forbs mostly Russian thistle. |
| **614** | Gemsbok | Various grass shrub communities | South Africa | Rumen–reticulum samples | Dry season | 96 | 0 | 4 | Giesecke & Van Gylswyk (1975) | See notes under no. 616. |
| **615** | Cape buffalo | Open woodland | Tanzania | Visual observation | Yearlong | 94 | 1 | 5 | Lamprey (1963) | See notes under no. 584. |
| **616** | Buffalo | Various grass and and shrub communities | South Africa | Rumen–reticulum content analysis | Dry season (July–Sept.) | 100 | 0 | 0 | Giesecke & Van Gylswyk (1975) | Percentage values not accounted for represent unidentified fine plant material. |
| **617** | Hippopotamus | Semi-deciduous forest, woodland thicket and grass savanna | Uganda | Stomach samples | Year | 98 | 2 | 0 | Field (1970) | |
| **618** | Hippopotamus | Wallows | Uganda | Stomach content analysis | Yearlong | 100 | 0 | 0 | W. M. Longhurst (unpublished) | |
| **619** | Topi | Savanna grassland | Narwok District, Kenya | Stomach analysis | Yearlong | 100 | 0 | 0 | Talbot & Talbot (1962) | See notes under no. 600. |
| **620** | Topi | Grassland savanna | Western Serengeti, Kenya | Stomach content analysis | Dry season (Sept.) | 100 | 0 | 0 | Bell (1970) | See notes under no. 602. |
| **621** | Topi | | Serengeti National Park, Tanzania | Collected fresh ingested material from shot animals, botanical composition from point-count method | Dry season (Sept.) | 100 | 0 | 0 | Gwynne & Bell (1968) | |
| **622** | Water buck | Open woodland | Tanzania | Visual observation | Yearlong | 95 | 1 | 4 | Lamprey (1963) | See notes under no. 584. |
| **623** | Rhinoceros | Open woodland | Tanzania | Visual observation | Yearlong | 38 | 9 | 53 | Lamprey (1963) | See notes under no. 584. |

| | | | | | | | | | | |
|---|---|---|---|---|---|---|---|---|---|---|
| **624** | Elephant | Open woodland | Tanzania | Visual observation | Yearlong | 13 | 1 | 86 | Lamprey (1963) | See notes under no. 584. |
| **625** | Elephant | Marsh, semi-arid savanna, and grassland | East of Lake Albert, Uganda | Direct observation and analysis of stomach contents | Dry season | 88 | 2 | 10 | Buss (1961) | Thatching grass was eaten most frequently. *Terminalia* trees appear to be damaged selectively by elephants. |
| **626** | Elephant | Grassland and *Acacia* woodland | Kidepo Valley National Park, Uganda | Feeding minute observations | Dry season (Nov.–Jan.) | 28 | 13 | 59 | Field & Ross (1976) | Shrubs also includes trees; trees always more important than shrubs as such. |
| **627** | Elephant | Grassland and *Acacia* woodland | Kidepo Valley National Park, Uganda | Feeding minute observations | Wet season | 57 | 22 | 21 | Field & Ross (1976) | See notes under no. 626. |
| **628** | Giraffe | Open woodland | Tanzania | Visual observation | Yearlong | 1 | 1 | 98 | Lamprey (1963) | See notes under no. 584. |
| **629** | Giraffe | Grassland and *Acacia* woodland | Kidepo Valley National Park, Uganda | Feeding minute observations | All times | ? | ? | 93 | Field & Ross (1976) | 93 % was minimum woody component of giraffe diet. |

## Appendix 4.3. Chemical composition of diets of large herbivores on the world's

| No. | Reference | Animal species | Season/treatment | Crude protein (%) | Ether extract (%) | Lignin (%) | Cellulose (%) | Other carbohydrates (%) | Ash (%) | Gross energy (kcal · kg$^{-1}$) |
|---|---|---|---|---|---|---|---|---|---|---|
| 1 | Cable & Shumway (1966) | Cattle | Spring (Mar.) | 16.1 | — | — | — | — | — | — |
| 2 | Cable & Shumway (1966) | Cattle | May–June | 7.8 | — | — | — | — | — | — |
| 3 | Connor *et al.* (1963) | Cattle | Spring | 12.4 | — | — | — | — | 17.2 | — |
| 4 | Halls (1954) | Cattle | June | 8.5 | 2.2 | — | — | — | 9.4 | — |
| 5 | Malachek (1966) | Cattle | Spring (May–June) native bunch-grass | 14.2 | — | — | — | — | — | — |
| 6 | Marshall & Radloff (1976) | Cattle | June | — | — | 10.3 | 32 | — | — | — |
| 7 | Rice *et al.* (1974) | Cattle | Spring (Apr.–May; 2 years) | — | — | 10.5 | — | — | 17.5 | — |
| 8 | Rosiere *et al.* (1975) | Cattle | Spring | 11.4 | — | 6.6 | — | — | — | — |
| 9 | Scales (1972) | Cattle | Spring | 14.9 | — | 3.8 | — | — | 11.7 | 4100 |
| 10 | Wallace & Denham (1970) | Cattle | June | 13.7 | 1.4 | 4.5 | — | — | — | 4100 |
| 11 | Smith *et al.* (1968) | Cattle | Early summer | 9.2 | — | 6.0 | — | 13 | — | — |
| 12 | Van Dyne & Heady (1965) | Cattle | Early summer | 8.1 | 1.7 | 12.4 | 36 | 30 | 11.5 | 4200 |
| 13 | Bedell (1971) | Cattle | Summer sub-clover–tall fescue, heavy grazing | 13.3 | — | — | — | — | 11.7 | — |
| 14 | Bedell (1971) | Cattle | Summer sub-clover–perennial ryegrass, light grazing | 11.9 | — | — | — | — | 10.8 | — |
| 15 | Bedell (1971) | Cattle | Summer sub-clover–tall fescue, light grazing | 11.9 | — | — | — | — | 10.7 | — |
| 16 | Bedell (1971) | Cattle | Summer sub-clover–perennial ryegrass, heavy grazing | 14.6 | — | — | — | — | 13.8 | — |
| 17 | Cable & Shumway (1966) | Cattle | Aug. | 21.6 | — | — | — | — | — | — |
| 18 | Connor *et al.* (1963) | Cattle | Summer (June–Sept.), northern Nevada | 10.9 | — | — | — | — | 14.0 | — |
| 19 | Connor *et al.* (1963) | Cattle | Summer | 11.0 | — | — | — | — | 19.2 | — |
| 20 | Cook *et al.* (1967) | Cattle | Summer, sage-brush–grass | 11.4 | 3.3 | 10.0 | 29 | 35 | 11.5 | 4200 |
| 21 | Cook *et al.* (1967) | Cattle | Summer, aspen rangeland | 10.2 | 3.4 | 7.0 | 31 | 37 | 11.2 | 4200 |

[a] Data have been rounded up as follows: calcium, phosphorus and silica, to the nearest 0.01 %; crude protein, ether extract, lignin and ash, to the nearest 0.1 %; cellulose, other carbohydrates, crude fibre, nitrogen-free extract, acid detergent fibre, cell wall constituents, soluble carbohydrate and total sugars, to the nearest 1.0 %; and gross energy to the nearest

**grazinglands**[a]

| No. | Crude fibre (%) | Nitrogen-free extract (%) | Acid detergent fibre (%) | Cell wall constituents (%) | Soluble carbohydrates (%) | Total sugars (%) | Calcium (%) | Phosphorus (%) | Silica (%) | Notes |
|---|---|---|---|---|---|---|---|---|---|---|
| **1** | — | — | — | — | — | — | — | — | — | Rumen fistulas used. Diet crude protein content 1.5 to 3.0 × that of clipped grass samples. Grassland with overstorey of mesquite and *Acacia*. |
| **2** | — | — | — | — | — | — | — | — | — | See notes under no. 1. |
| **3** | 32 | — | — | — | — | — | — | — | — | See notes under no. 19. |
| **4** | 31 | 46 | — | — | — | — | 0.21 | 0.14 | — | Data based on observation and hand-plucked samples. Study on wiregrass range in southern Georgia, USA. |
| **5** | — | — | — | — | — | — | — | 0.48 | — | Samples obtained by rumen evacuation technique, thus phosphorus values particularly could be biased. Most data based on 1 steer. |
| **6** | — | — | 50 | — | — | — | — | — | — | — |
| **7** | — | — | 56 | 66 | 34 | — | — | — | — | See notes under no. 129. |
| **8** | 39.4 | — | — | — | — | — | — | — | — | See notes under no. 27. |
| **9** | — | — | 32 | 58 | — | — | — | — | — | Samples collected with oesophageal-fistulated cattle, morning and evening grazing. Energy data expressed as kcal · $g^{-1}$. Sandhill grassland in Eastern Colorado. |
| **10** | — | — | 41 | — | — | — | — | — | — | Same study as Wallace *et al.* (1972) but analyses differ slightly. Gross energy expressed in kcal · $g^{-1}$. |
| **11** | 38 | 38 | 39 | — | — | — | — | — | — | Study done on moderately utilized northern desert shrub. |
| **12** | — | — | — | — | — | — | — | — | 2.8 | See notes under no. 35. |
| **13** | — | — | — | — | — | — | — | — | — | See notes under no. 75. |
| **14** | — | — | — | — | — | — | — | — | — | See notes under no. 75. |
| **15** | — | — | — | — | — | — | — | — | — | See notes under no. 75. |
| **16** | — | — | — | — | — | — | — | — | — | See notes under no. 75. |
| **17** | — | — | — | — | — | — | — | — | — | See notes under no. 1. |
| **18** | 32 | — | — | — | — | — | — | — | — | See notes under no. 19. |
| **19** | 33 | — | — | — | — | — | — | — | — | Samples with rumen fistula. Northern Nevada rangeland sagebrush–grass; southern Nevada rangeland desert shrub. |
| **20** | — | — | — | — | — | — | — | 0.30 | — | Oesophageal-fistulated animals utilized. Sheep selected higher browse diet and cattle higher grass diets; cattle took more stems and less leaves than sheep. Summer grazing from mid-June to mid-Sept. averaged over 2 years. Samples taken morning and evening by fistula. |
| **21** | — | — | — | — | — | — | — | 0.30 | — | — |

**100 kcal $kg^{-1}$. Data are expressed on a dry matter basis. This required using literature values for conversion in a few instances (see notes).**

**Appendix 4.3** (*continued*)

| No. | Reference | Animal species | Season/treatment | Crude protein (%) | Ether extract (%) | Lignin (%) | Cellulose (%) | Other carbohydrates (%) | Ash (%) | Gross energy (kcal · kg⁻¹) |
|---|---|---|---|---|---|---|---|---|---|---|
| **22** | Hoehne *et al.* (1968) | Cattle | Summer | 45.6 | — | — | — | — | 50.1 | — |
| **23** | Malachek (1966) | Cattle | Summer (July–Sept.), native bunch-grass | 13.2 | — | — | — | — | — | — |
| **24** | Marshall & Radloff (1976) | Cattle | July | — | — | 11.7 | 34 | — | — | — |
| **25** | Marshall & Radloff (1976) | Cattle | Aug. | — | — | 9.4 | 44 | — | — | — |
| **26** | Rice *et al.* (1974) | Cattle | Summer (June–Aug.; 2 years) | — | — | 11.0 | — | — | 15.3 | — |
| **27** | Rosiere *et al.* (1975) | Cattle | Summer | 10.1 | — | 9.3 | — | — | — | — |
| **28** | Scales (1972) | Cattle | Summer | 12.3 | — | 5.0 | — | — | 11.0 | 4100 |
| **29** | Vavra *et al.* (1973) | Cattle | Summer, heavy grazing | 11.3 | — | 9.2 | 25 | — | — | 3900 |
| **30** | Vavra *et al.* (1973) | Cattle | Summer, light grazing | 11.2 | — | 9.0 | 26 | — | — | 3900 |
| **31** | Wallace & Denham (1970) | Cattle | July | 9.4 | 1.6 | 5.6 | — | — | — | 4100 |
| **32** | Wallace *et al.* (1972) | Cattle | Summer | 10.6 | 1.5 | 5.0 | — | — | 7.7 | 4500 |
| **33** | Van Dyne & Heady (1965) | Cattle | Mid-summer | 6.5 | 1.3 | 12.2 | 38 | 31 | 11.7 | 4200 |
| **34** | Smith *et al.* (1968) | Cattle | Late summer moderately utilized northern desert shrub | 8.5 | — | 5.6 | — | 16 | — | — |
| **35** | Van Dyne & Heady (1965) | Cattle | Late summer (Sept.) | 6.2 | 1.0 | 12.4 | 36 | 30 | — | 4000 |
| **36** | Galt *et al.* (1969) | Cattle | Early autumn (Sept.–Oct.) | 10.0 | — | — | — | — | — | — |
| **37** | Connor *et al.* (1963) | Cattle | Autumn | 9.6 | — | — | — | — | 19.1 | — |
| **38** | Halls (1954) | Cattle | Sept. | 6.9 | 2.0 | — | — | — | 8.3 | — |
| **39** | Malachek (1966) | Cattle | Autumn (Nov.–Dec.), native bunch-grass | 6.8 | — | — | — | — | — | — |
| **40** | Malachek (1966) | Cattle | Autumn (Sept.–Oct.), hay meadow regrowth | 8.0 | — | — | — | — | — | — |
| **41** | Marshall & Radloff (1976) | Cattle | Sept. | — | — | 10.0 | 35 | — | — | — |
| **42** | Rice *et al.* (1974) | Cattle | Autumn (Oct.; 2 years) | — | — | 20.1 | — | — | 13.0 | — |
| **43** | Rosiere *et al.* (1975) | Cattle | Autumn | 6.1 | — | 10.4 | — | — | — | — |
| **44** | Scales (1972) | Cattle | Autumn | 6.4 | — | 6.4 | — | — | 10.3 | 4300 |
| **45** | Wallace & Denham (1970) | Cattle | Sept. | 5.7 | 2.1 | 7.4 | — | — | — | 4100 |
| **46** | Wallace *et al.* (1972) | Cattle | Sept. | 5.8 | 3.0 | 8.5 | — | — | 9.5 | 4500 |
| **47** | Galt *et al.* (1974) | Cattle | Late autumn (Nov.–Dec.) | 7.0 | — | — | — | — | — | — |
| **48** | Malachek (1966) | Cattle | Winter, native meadow | 7.5 | — | — | — | — | — | — |
| **49** | Rosiere *et al.* (1975) | Cattle | Winter | 10.8 | — | 11.6 | — | — | — | — |

| No. | Crude fibre (%) | Nitrogen-free extract (%) | Acid detergent fibre (%) | Cell wall constituents (%) | Soluble carbohydrates (%) | Total sugars (%) | Calcium (%) | Phosphorus (%) | Silica (%) | Notes |
|---|---|---|---|---|---|---|---|---|---|---|
| 22 | — | — | — | — | 48 | 51 | 0.59 | 0.56 | — | Hand-clipped samples of forbs analysed. |
| 23 | — | — | — | — | — | — | — | — | 0.42 | See notes under no. 5. |
| 24 | — | — | 51 | — | — | — | — | — | — | |
| 25 | — | — | 51 | — | — | — | — | — | — | — |
| 26 | — | — | 53 | 66 | 34 | — | — | — | — | See notes under no. 129. |
| 27 | 44 | — | — | — | — | — | — | — | — | Diets collected with oesophageal-fistulated cattle, thus data are presented on an organic matter basis. The per cent ash in the total sample varied from about 13 to 20 % (for conversion from o.m. to d.m. used 16.5 % ash). |
| 28 | — | — | 35 | 62 | — | — | — | — | — | See notes under no. 9. |
| 29 | — | — | 43 | — | — | — | — | — | — | Shortgrass prairie rangeland; oesophageal-fistulated animals; average of 3 months in each of 2 years. |
| 30 | — | — | 39 | — | — | — | — | — | — | — |
| 31 | — | — | 45 | — | — | — | — | — | — | See notes under no. 10. |
| 32 | — | — | 47 | — | — | — | — | — | — | See notes under no. 46. |
| 33 | — | — | — | — | — | — | — | — | 3.7 | See notes under no. 35. |
| 34 | 34 | 40 | 35 | — | — | — | — | — | — | — |
| 35 | — | — | — | — | — | — | — | — | 6.9 | Samples obtained with oesophageal-fistulated animals grazing morning and evening over several days each period. Data are expressed on a silica-free basis. Herbage availability varied from 1490 lb · acre$^{-1}$ in early summer to 420 lb · acre$^{-1}$ in late summer. No significant difference in chemical constituents between morning and afternoon diets overall; cattle afternoon diets higher in crude protein than morning diets. |
| 36 | — | — | — | — | — | — | — | — | — | Diets sampled by rumen fistula. Reconstituted diets based on hand-clipped samples of plant species as high in nutritive value as grazed diet. |
| 37 | 36 | — | — | — | — | — | — | — | — | See notes under no. 19. |
| 38 | 34 | 49 | — | — | — | — | 0.18 | 0.13 | — | See notes under no. 4. |
| 39 | — | — | — | — | — | — | — | 0.23 | — | See notes under no. 5. |
| 40 | — | — | — | — | — | — | — | 0.28 | — | See notes under no. 5. |
| 41 | — | — | 52 | — | — | — | — | — | — | — |
| 42 | — | — | 61 | 67 | 33 | — | — | — | — | See notes under no. 129. |
| 43 | 53 | — | — | — | — | — | — | — | — | See notes under no. 27. |
| 44 | — | — | 39 | 68 | — | — | — | — | — | See notes under no. 9. |
| 45 | — | — | 48 | — | — | — | — | — | — | See notes under no. 10. |
| 46 | — | — | 53 | — | — | — | — | — | — | Oesophageal-fistulated steers utilized on sandhill range. |
| 47 | — | — | — | — | — | — | — | — | — | See notes under no. 36. |
| 48 | — | — | — | — | — | — | — | 0.40 | — | See notes under no. 5. |
| 49 | 45 | — | — | — | — | — | — | — | — | See notes under no. 27. |

**Appendix 4.3** (*continued*)

| No. | Reference | Animal species | Season/treatment | Crude protein (%) | Ether extract (%) | Lignin (%) | Cellulose (%) | Other carbohydrates (%) | Ash (%) | Gross energy (kcal · kg⁻¹) |
|---|---|---|---|---|---|---|---|---|---|---|
| **50** | Smith *et al.* (1968) | Cattle | Winter, moderately utilized northern desert shrub | 5.1 | — | 8.1 | — | 16 | — | — |
| **51** | Van Dyne *et al.* (1964) | Cattle | Winter | 4.3 | 2.8 | 14.2 | 43 | — | — | — |
| **52** | Wallace & Denham (1970) | Cattle | Dec. | 3.7 | 2.3 | 8.1 | — | — | — | 4100 |
| **53** | Wallace *et al.* (1972) | Cattle | Dec. | 4.1 | 3.0 | 9.2 | — | — | 10.4 | 4700 |
| **54** | Bredon *et al.* (1967) | Cattle | Not specified | 7.0 | — | — | — | — | — | — |
| **55** | Cable & Shumway (1966) | Cattle | Nov.–Jan. | 6.8 | — | — | — | — | — | — |
| **56** | Fox (1971) | Cattle | May–Oct., short-grass prairie | 10.1 | — | — | — | — | — | 3700 |
| **57** | Fox (1971) | Cattle | May–Oct., successional old field | 10.0 | — | — | — | — | — | 3700 |
| **58** | Fox (1971) | Cattle | May–Oct., seeded native pasture | 9.6 | — | — | — | — | — | 3600 |
| **59** | Harrington & Pratchett (1974) | Cattle | Main wet period (Oct.–Dec.) | 11.0 | — | — | — | — | — | — |
| **60** | Harrington & Pratchett (1974) | Cattle | Second wet period (Apr.–May) | 10.9 | — | — | — | — | — | — |
| **61** | Harrington & Pratchett (1974) | Cattle | Main dry period | 7.4 | — | — | — | — | — | — |
| **62** | Harrington & Pratchett (1974) | Cattle | Short dry period (Feb.–Mar.) | 9.2 | — | — | — | — | — | — |
| **63** | McClung *et al.* (1976) | Cattle (steers) | June–Sept. | 11.3 | — | — | — | — | — | — |
| **64** | Arnold *et al.* (1966) | Sheep | Spring (Sept.–Oct.), native pasture | 15.0 | — | — | — | — | — | — |
| **65** | Arnold *et al.* (1966) | Sheep | Spring (Sept.–Oct.), cocksfoot pasture | 21.9 | — | — | — | — | — | — |
| **66** | Bishop *et al.* (1975) | Sheep | Spring | 18.1 | — | — | — | — | — | — |
| **67** | Engels *et al.* (1971) | Sheep | Apr.–June | 12.1 | — | 13.5 | 24 | — | — | — |
| **68** | Rice *et al.* (1974) | Sheep | Spring (Apr.–May; 2 years) | — | — | 19.1 | — | — | 13.5 | — |
| **69** | Thetford *et al.* (1971) | Sheep | Spring (Apr.) | 9.0 | — | — | — | — | 17.5 | — |
| **70** | Buchanon *et al.* (1972) | Sheep | Early summer | 16.7 | — | 5.9 | — | — | 19.0 | — |
| **71** | Price *et al.* (1964) | Sheep | Early summer, mountain range | 17.0 | 3.7 | 9.4 | — | — | — | 4300 |
| **72** | Van Dyne & Heady (1965) | Sheep | Early summer (July) | 10.3 | 1.7 | 12.5 | 33 | 33 | 12.1 | 4200 |
| **73** | Arnold *et al.* (1966) | Sheep | Summer (Dec.–Feb.) native pasture | 14.4 | — | — | — | — | — | — |
| **74** | Arnold *et al.* (1966) | Sheep | Summer (Dec.–Feb.) cocksfoot pasture | 17.5 | — | — | — | — | — | — |

| No. | Crude fibre (%) | Nitrogen-free extract (%) | Acid detergent fibre (%) | Cell wall constituents (%) | Soluble carbohydrates (%) | Total sugars (%) | Calcium (%) | Phosphorus (%) | Silica (%) | Notes |
|---|---|---|---|---|---|---|---|---|---|---|
| 50 | 38 | 38 | 46 | — | — | — | — | — | — | — |
| 51 | — | — | — | — | — | — | — | — | — | Oesophageal-fistulated animals grazed on open range; collections morning and evening; early Jan. to early Apr., 2 years. Foothill bunchgrass rangeland; limited shrubs available. More variability among days during the winter than among animals in diet composition. |
| 52 | — | — | 50 | — | — | — | — | — | — | See notes under no. 10. |
| 53 | — | — | 57 | — | — | — | — | — | — | See notes under no. 46. |
| 54 | 36 | — | — | — | — | — | — | — | — | Sanga-type Zebu cattle, Ankole area of Uganda on *Themeda* rangeland. Oesophageal fistula method used. Cattle were highly selective in their grazing. |
| 55 | — | — | — | — | — | — | — | — | — | See notes under no. 1. |
| 56 | — | — | — | — | — | — | — | 0.20 | — | Diets determined by multiplying botanical composition by utilization and summing these products to determine per cent contributed by each species. |
| 57 | — | — | — | — | — | — | — | 0.24 | — | Gross energy given in cal • $g^{-1}$. |
| 58 | — | — | — | — | — | — | — | 0.21 | — | Chemical composition of individual species determined and composited to make diets. |
| 59 | 33 | — | — | — | — | — | — | — | — | Oesophageal-fistulated Ankole bullocks. Four days per period; 4 animals; overnight fast. Animals grazed with resident bullocks; relatively heavy stocking rates. |
| 60 | 32 | — | — | — | — | — | — | — | — | — |
| 61 | 33 | — | — | — | — | — | — | — | — | — |
| 62 | 32 | — | — | — | — | — | — | — | — | — |
| 63 | — | — | — | — | — | — | — | 0.49 | — | Oesophageal fistulas used. Summer grazing (light) on shortgrass prairie. |
| 64 | 43 | — | — | — | 8 | — | — | — | — | See notes under no. 103. |
| 65 | 36 | — | — | — | 8 | — | — | — | — | See notes under no. 103. |
| 66 | — | — | — | — | — | — | — | — | — | See notes under no. 75. Used early-summer estimate of 9.0 % ash from Van Dyne & Heady (1965). |
| 67 | — | — | 41 | — | — | — | — | — | — | Used early summer estimate of 9.0 % ash from Van Dyne & Heady (1965). |
| 68 | — | — | 52 | 65 | 25 | — | — | — | — | See notes under no. 129. |
| 69 | — | — | — | 85 | — | — | — | — | — | Values read from graph. |
| 70 | — | — | 27 | — | — | — | — | — | — | Oesophageal-fistulated animals. Lignin analysis on acid detergent lignin. Southwestern Montana tall forb rangeland; average 2 years. |
| 71 | 18 | 48 | — | — | — | — | — | — | — | Oesophageal-fistulated sheep; morning grazing. High mountain forb type range; early grazing in July; late grazing in Aug. |
| 72 | — | — | — | — | — | — | — | — | 3.41 | See notes under no. 35. |
| 73 | 44 | — | — | — | 6 | — | — | — | — | See notes under no. 103. |
| 74 | 42 | — | — | — | 6 | — | — | — | — | See notes under no. 103. |

**Appendix 4.3** (*continued*)

| No. | Reference | Animal species | Season/treatment | Crude protein (%) | Ether extract (%) | Lignin (%) | Cellulose (%) | Other carbohydrates (%) | Ash (%) | Gross energy (kcal · kg⁻¹) |
|---|---|---|---|---|---|---|---|---|---|---|
| 75 | Bedell (1971) | Sheep | Summer, sub-clover–perennial ryegrass, heavy grazing | 14.8 | — | — | — | — | 12.8 | — |
| 76 | Bedell (1971) | Sheep | Summer, sub-clover–perennial ryegrass, light grazing | 12.5 | — | — | — | — | 15.0 | — |
| 77 | Bedell (1971) | Sheep | Summer, sub-clover–tall fescue, heavy grazing | 15.1 | — | — | — | — | 12.3 | — |
| 78 | Bedell (1971) | Sheep | Summer, sub-clover–tall fescue, light grazing | 14.7 | — | — | — | — | 11.6 | — |
| 79 | Bishop *et al.* (1975) | Sheep | Summer (Dec.–Feb.) | 13.3 | — | — | — | — | — | — |
| 80 | Cook *et al.* (1961) | Sheep (wet and dry ewes) | Summer grazing | 12.0 | 4.0 | 15.2 | 19 | 39 | 10.6 | 4600 |
| 81 | Cook *et al.* (1965) | Sheep | Summer, good range | 11.7 | 3.7 | 10.2 | 22 | 41 | 11.4 | 4300 |
| 82 | Cook *et al.* (1965) | Sheep | Summer (June–Sept.), poor range | 12.4 | 3.3 | 10.5 | 20 | 42 | 11.9 | 4300 |
| 83 | Cook *et al.* (1967) | Sheep | Sagebrush–grassland | 13.8 | 3.6 | 10.8 | 24 | 38 | 10.3 | 4400 |
| 84 | Cook *et al.* (1967) | Sheep | Aspen rangeland | 13.0 | 4.4 | 8.8 | 23 | 39 | 12.2 | 4400 |
| 85 | Engels *et al.* (1971) | Sheep | July–Aug. | 13.9 | — | 14.8 | 21 | — | — | — |
| 86 | Kothmann (1966) | Sheep | Summer, sagebrush–grass, morning grazing | 12.1 | 3.5 | 14.9 | 20 | 38 | 12.0 | 4300 |
| 87 | Kothmann (1966) | Sheep | Summer, sagebrush–grass, afternoon grazing | 12.7 | 3.6 | 15.0 | 20 | 38 | 11.8 | 4300 |
| 88 | Rice *et al.* (1974) | Sheep | Summer (June–Aug.; 2 years) | — | — | 15.4 | — | — | 14.6 | — |
| 89 | Strasia *et al.* (1970) | Sheep | Summer, free grazing | 11.9 | — | 11.9 | 30 | — | 14.1 | — |
| 90 | Strasia *et al.* (1970) | Sheep | Summer, herded | 14.4 | — | 11.0 | 29 | — | 13.5 | — |
| 91 | Wilson *et al.* (1975) | Sheep | Summer (Nov.), goat-grazed pasture | 10.6 | — | — | — | — | — | — |
| 92 | Wilson *et al.* (1975) | Sheep | Summer (Nov.), sheep-grazed pasture | 12.6 | — | — | — | — | — | — |
| 93 | Van Dyne & Heady (1965) | Sheep | Mid-summer (Aug.) | 7.5 | 1.5 | 11.6 | 36 | 31 | 12.7 | 4100 |
| 94 | Buchanon *et al.* (1972) | Sheep | Late summer | 10.5 | — | 9.0 | — | — | 20.0 | — |
| 95 | Price *et al.* (1964) | Sheep | Summer, mountain range – late | 14.1 | 3.3 | 11.0 | — | — | — | 4300 |
| 96 | Van Dyne & Heady (1965) | Sheep | Late summer (Sept.) | 6.9 | 1.2 | 11.9 | 37 | 34 | 8.1 | 4100 |
| 97 | Bishop *et al.* (1975) | Sheep | Autumn | 16.2 | — | — | — | — | — | — |
| 98 | Engels *et al.* (1971) | Sheep | Sept.–Dec. | 8.7 | — | 9.2 | 26 | — | — | — |
| 99 | Rice *et al.* (1974) | Sheep | Autumn (Oct.; 2 years) | — | — | 20.2 | — | — | 17.0 | — |

| No. | Crude fibre (%) | Nitrogen-free extract (%) | Acid detergent fibre (%) | Cell wall constituents (%) | Soluble carbohydrates (%) | Total sugars (%) | Calcium (%) | Phosphorus (%) | Silica (%) | Notes |
|---|---|---|---|---|---|---|---|---|---|---|
| 75 | — | — | — | — | — | — | — | — | — | Data obtained by oesophageal fistulas. Cattle were Hereford–Angus cross and Hereford; sheep were Suffolk. Samples collected from Apr. through August, averaged for 1965 and 1966. Cattle preferred grasses to sub-clover and sheep diets averaged 4 % more protein than did cattle over the entire season. |
| 76 | — | — | — | — | — | — | — | — | — | See notes under no. 75. |
| 77 | — | — | — | — | — | — | — | — | — | See notes under no. 75. |
| 78 | — | — | — | — | — | — | — | — | — | See notes under no. 75. |
| 79 | — | — | — | — | — | — | — | — | — | Oesophageal-fistulated sheep (Corridale wethers) grazing on a mixed annual–perennial grass–shrub rangeland. Used estimate of 9.23 % ash from Van Dyne & Heady (1965). |
| 80 | — | — | — | — | — | — | — | 0.35 | — | Oesophageal-fistulated ewes. |
| 81 | — | — | — | — | — | — | — | 0.3 | — | Oesophageal-fistulated animals used on sagebrush–grass range with mountain browse and aspen intermixed. Protein decreased significantly on poor range. |
| 82 | — | — | — | — | — | — | — | 0.30 | — | See notes under no. 81. |
| 83 | — | — | — | — | — | — | — | 0.34 | — | See notes under no. 20. |
| 84 | — | — | — | — | — | — | — | 0.38 | — | See notes under no. 20. |
| 85 | — | — | 37 | — | — | — | — | — | — | Used estimate of 0.23 % ash from Van Dyne & Heady (1965). |
| 86 | — | — | — | — | — | — | — | — | — | Oesophageal fistula samples; morning and evening comparison. Gross energy expressed in kcal · $lb^{-1}$. Northern Utah grazingland, sagebrush–grass type. |
| 87 | — | — | — | — | — | — | — | — | — | — |
| 88 | — | — | 51 | 68 | 32 | — | — | — | — | See notes under no. 129. |
| 89 | 42 | — | — | — | — | — | — | — | — | Oesophageal-fistulated sheep. July–mid-Sept.; 2 years. Alpine rangeland near Cody, Wyoming. |
| 90 | 38 | — | — | — | — | — | — | — | — | — |
| 91 | — | 17 | — | — | — | — | — | — | — | See notes under winter goats, Wilson *et al.* (1975). |
| 92 | — | 20 | — | — | — | — | — | — | — | See notes under goats, Wilson *et al.* (1975). |
| 93 | — | — | — | — | — | — | — | — | 2.3 | See notes under no. 35. |
| 94 | — | — | 32 | — | — | — | — | — | — | See notes under no. 70. |
| 95 | 19 | 52 | — | — | — | — | — | — | — | See notes under no. 71. |
| 96 | — | — | — | — | — | — | — | — | 4.2 | See notes under no. 35. |
| 97 | — | — | — | — | — | — | — | — | — | See notes under no. 79. Used late-summer estimate of 8.1 % ash from Van Dyne & Heady (1965). |
| 98 | — | — | 37 | — | — | — | — | — | — | Used late-summer estimate of 8.1 % ash from Van Dyne & Heady (1965). |
| 99 | — | — | 55 | 55 | 45 | — | — | — | — | See notes under no. 129. |

**Appendix 4.3** (*continued*)

| No. | Reference | Animal species | Season/treatment | Crude protein (%) | Ether extract (%) | Lignin (%) | Cellulose (%) | Other carbohydrates (%) | Ash (%) | Gross energy (kcal kg$^{-1}$) |
|---|---|---|---|---|---|---|---|---|---|---|
| **100** | Wilson *et al.* (1975) | Sheep | Autumn (Feb.), sheep-grazed pasture | 16.2 | — | — | — | — | — | — |
| **101** | Wilson *et al.* (1975) | Sheep | Autumn (Feb.), goat-grazed pasture | 16.1 | — | — | — | — | — | — |
| **102** | Arnold *et al.* (1966) | Sheep | Winter (June–July), cocksfoot pasture | 16.9 | — | — | — | — | — | — |
| **103** | Arnold *et al.* (1966) | Sheep | Winter (June–July), native pasture | 9.7 | — | — | — | — | — | — |
| **104** | Bishop *et al.* (1975) | Sheep | Winter | 20.9 | — | — | — | — | — | — |
| **105** | Cook *et al.* (1962) | Sheep | Winter, salt desert shrub range, good condition | 7.8 | 3.2 | 12.3 | 24 | 43 | 9.8 | 3900 |
| **106** | Cook *et al.* (1962) | Sheep | Winter, salt desert shrub range, poor condition | 8.3 | 2.7 | 12.6 | 22 | 43 | 11.5 | 3800 |
| **107** | Edlefsen *et al.* (1960) | Sheep | Winter | 7.8 | 1.9 | 10.2 | 22 | 43 | 15.7 | 3900 |
| **108** | Engels *et al.* (1971) | Sheep | Jan.–Mar. | 8.7 | — | 8.8 | 28 | — | — | — |
| **109** | Pieper *et al.* (1959) | Sheep | Winter, light grazing | 6.0 | 2.1 | 9.2 | 30 | 43 | 9.7 | 4100 |
| **110** | Pieper *et al.* (1959) | Sheep | Winter, heavy grazing | 5.8 | 1.6 | 11.0 | 29 | 43 | 8.6 | 4100 |
| **111** | Pieper *et al.* (1959) | Sheep | Winterfat, pure stand, light grazing | 8.5 | 2.4 | 9.8 | 24 | 46 | 9.8 | 4100 |
| **112** | Pieper *et al.* (1959) | Sheep | Winterfat, pure stand, heavy grazing | 8.0 | 2.2 | 13.3 | 21 | 44 | 11.1 | 4000 |
| **113** | Pieper *et al.* (1959) | Sheep | Four-winged saltbush, pure stand, light grazing | 12.6 | 2.3 | 9.7 | 14 | 46 | 16.2 | 4000 |
| **114** | Pieper *et al.* (1959) | Sheep | Four-winged saltbush, pure stand, heavy grazing | 7.7 | 2.3 | 13.7 | 23 | 43 | 10.7 | 4000 |
| **115** | Van Dyne *et al.* (1964) | Sheep | Winter | 7.0 | 3.1 | 14.8 | 42 | — | — | — |
| **116** | Wilson *et al.* (1975) | Sheep | Winter (May–June), sheep-grazed pasture | 13.9 | — | — | — | — | — | — |
| **117** | Wilson *et al.* (1975) | Sheep | Winter (May–June), goat-grazed pasture | 13.2 | — | — | — | — | — | — |
| **118** | Grimes *et al.* (1966) | Sheep | Year, grass pasture | 22.9 | — | — | — | 23 | — | — |
| **119** | Grimes *et al.* (1966) | Sheep | Grass and clover pasture | 25.5 | — | — | — | 22 | — | — |
| **120** | Grimes *et al.* (1966) | Sheep | Short management | 25.7 | — | — | — | 21 | — | — |
| **121** | Grimes *et al.* (1966) | Sheep | Long management | 22.9 | — | — | — | 24 | — | — |
| **122** | Weir & Torell (1959) | Sheep | Native rangeland, average several treatments in seasons | 14.8 | 2.0 | 11.2 | — | — | 9.8 | — |
| **123** | Wilson *et al.* (1975) | Goat | Winter (May–June), sheep-grazed pasture | 15.5 | — | — | — | — | — | — |
| **124** | Wilson *et al.* (1975) | Goat | Winter (May–June), goat-grazed pasture | 15.4 | — | — | — | — | — | — |

| No. | Crude fibre (%) | Nitrogen-free extract (%) | Acid detergent fibre (%) | Cell wall constituents (%) | Soluble carbohydrates (%) | Total sugars (%) | Calcium (%) | Phosphorus (%) | Silica (%) | Notes |
|---|---|---|---|---|---|---|---|---|---|---|
| **100** | — | 23 | — | — | — | — | — | — | — | See notes under no. 123. |
| **101** | — | 25 | — | — | — | — | — | — | — | See notes under no. 123. |
| **102** | 44 | — | — | — | 6 | — | — | — | — | See notes under no. 103. |
| **103** | — | — | 47 | — | 5 | — | — | — | — | Oesophageal-fistulated animals. Border Leicester × Merino ewes. Fibre analysis 'normal acid fibre'. Grazing animals eat young green leaf if possible. |
| **104** | — | — | — | — | — | — | — | — | — | See notes under no. 79. Used late-summer estimate of 8.1 % ash from Van Dyne & Heady (1965). |
| **105** | — | — | — | — | — | — | — | — | — | Oesophageal-fistulated wethers, grazed in small paddocks. 10 days of collection in each data set. |
| **106** | — | — | — | — | — | — | — | — | — | — |
| **107** | — | — | — | — | — | — | — | 0.19 | — | Results adjusted for saliva. Average of 2 grazing periods. |
| **108** | — | — | 38 | — | — | — | — | — | — | Used late-summer estimate of 8.1 % ash from Van Dyne & Heady (1965). |
| **109** | — | — | — | — | — | — | — | — | — | Samples collected by hand-plucking. Salt desert shrub range. Gross energy based on kcal · $lb^{-1}$. For mixed stands data are from 4 locations and several trials in grass and shrub rangeland. |
| **110** | — | — | — | — | — | — | — | — | — | See notes under no. 109. |
| **111** | — | — | — | — | — | — | — | — | — | See notes under no. 109. |
| **112** | — | — | — | — | — | — | — | — | — | See notes under no. 109. |
| **113** | — | — | — | — | — | — | — | — | — | See notes under no. 109. |
| **114** | — | — | — | — | — | — | — | — | — | See notes under no. 109. |
| **115** | — | — | — | — | — | — | — | — | — | See notes under no. 51. |
| **116** | — | 22 | — | — | — | — | — | — | — | See notes under no. 123. |
| **117** | — | 21 | — | — | — | — | — | — | — | See notes under no. 123. |
| **118** | — | — | — | — | — | — | — | — | — | Short management – pasture maintained at 2–4 in.; long management – unrestricted growth. Used average ash of 9.23 % from Van Dyne & Heady (1965). |
| **119** | — | — | — | — | — | — | — | — | — | — |
| **120** | — | — | — | — | — | — | — | — | — | — |
| **121** | — | — | — | — | — | — | — | — | — | — |
| **122** | 21 | — | — | — | — | — | — | — | — | Six grade Corridale wethers used. Ash are silica-free ash data. Sheep selected higher quality diet than was available by hand clipping. Data on annual grass, legume and broadleaf herb range in Mendocino County California; precise seasons unspecified; data in 1955–7. |
| **123** | — | 25 | — | — | — | — | — | — | — | Goats eat a wider range of the available vegetation than do sheep. Higher nitrogen content in goat diets are explainable in terms of browse species and the herbaceous bassias they selected. Oesophageal-fistulated animals used. |
| **124** | — | 25 | — | — | — | — | — | — | — | — |

**Appendix 4.3** (*continued*)

| No. | Reference | Animal species | Season/treatment | Crude protein (%) | Ether extract (%) | Lignin (%) | Cellulose (%) | Other carbohydrates (%) | Ash (%) | Gross energy (kcal · kg$^{-1}$) |
|---|---|---|---|---|---|---|---|---|---|---|
| **125** | Wilson *et al.* (1975) | Goat | Summer (Nov.), sheep-grazed pasture | 14.1 | — | — | — | — | — | — |
| **126** | Wilson *et al.* (1975) | Goat | Summer (Nov.), goat-grazed pasture | 13.2 | — | — | — | — | — | — |
| **127** | Wilson *et al.* (1975) | Goat | Autumn (Feb.), sheep-grazed pasture | 18.5 | — | — | — | — | — | — |
| **128** | Wilson *et al.* (1975) | Goat | Autumn (Feb.), goat-grazed pasture | 17.0 | — | — | — | — | — | — |
| **129** | Rice *et al.* (1974) | Bison | Spring (Apr.–May; 2 years) | — | — | 10.3 | — | — | 21.0 | — |
| **130** | Rice *et al.* (1974) | Bison | Summer (June–Aug.; 2 years) | — | — | 8.3 | — | — | 19.6 | — |
| **131** | Rice *et al.* (1974) | Bison | Autumn (Oct.; 2 years) | — | — | 10.6 | — | — | 21.0 | — |
| **132** | Hoover (1971) | Pronghorn | Winter | 7.2 | — | 8.6 | — | — | — | 4200 |
| **133** | Hoover (1971) | Pronghorn | Spring | 12.6 | — | 7.2 | — | — | — | 4300 |
| **134** | Hoover (1971) | Pronghorn | Summer | 12.2 | — | 7.8 | — | — | — | 4200 |
| **135** | Smith & Malachek (1974) | Pronghorn | Summer, salt desert shrub, big sagebrush rangeland | 14.7 | — | — | — | — | — | — |
| **136** | Boeker *et al.* (1972) | Mule deer | Winter | 12.0 | — | — | — | — | — | — |
| **137** | Boeker *et al.* (1972) | Mule deer | Spring | 11.0 | — | — | — | — | — | — |
| **138** | Boeker *et al.* (1972) | Mule deer | Summer | 9.0 | — | — | — | — | — | — |
| **139** | Boeker *et al.* (1972) | Mule deer | Autumn | 9.0 | — | — | — | — | — | — |
| **140** | Slade & Hintz (1969) | Horse | Feed trial, pelleted alfalfa total faecal collection | 66.7 | 5.4 | — | — | — | — | 5100 |
| **141** | Slade & Hintz (1969) | Pony | | 68.5 | 17.1 | — | — | — | — | 5200 |
| **142** | LeResche & Davis (1973) | Moose | Birch diets | 8.8 | 14.6 | — | — | 50 | 1.8 | — |
| **143** | LeResche & Davis (1973) | Moose | 75 % birch, 25 % cranberry | 8.0 | 11.5 | — | — | 54 | 1.9 | — |
| **144** | LeResche & Davis (1973) | Moose | 50 % birch, 25 % cranberry, 25 % peltigera | 10.7 | 8.2 | — | — | 58 | 3.7 | — |
| **145** | LeResche & Davis (1973) | Moose | 50 % cranberry, 50 % peltigera | 12.6 | 1.7 | — | — | 56 | 5.6 | — |
| **146** | Bergerud (1972) | Caribou | Year + spring | 15.7 | 2.8 | — | — | — | 3.6 | — |
| **147** | Bergerud (1972) | Caribou | Summer | 11.8 | 2.3 | — | — | — | 4.3 | — |
| **148** | Bergerud (1972) | Caribou | Autumn | 7.2 | 3.2 | — | — | — | 3.8 | — |
| **149** | Bergerud (1972) | Caribou | Winter | 6.1 | 2.2 | — | — | — | 1.9 | — |

| No. | Crude fibre (%) | Nitrogen-free extract (%) | Acid detergent fibre (%) | Cell wall constituents (%) | Soluble carbohydrates (%) | Total sugars (%) | Calcium (%) | Phosphorus (%) | Silica (%) | Notes |
|---|---|---|---|---|---|---|---|---|---|---|
| **125** | — | 22 | — | — | — | — | — | — | — | See notes under no. 123. |
| **126** | — | 21 | — | — | — | — | — | — | — | See notes under no. 123. |
| **127** | — | 30 | — | — | — | — | — | — | — | See notes under no. 123. |
| **128** | — | 27 | — | — | — | — | — | — | — | See notes under no. 123. |
| **129** | — | — | 64 | 76 | 24 | — | — | — | — | Oesophageal-fistulated bison (3), and cattle and sheep (6 each). 5 samples per area, 2 dates averaged for spring, 3 dates averaged for summer, 1 date for autumn; 2 grazing treatments averaged (1 cattle unit per 12 hectares per month and 3 cattle units per 12 hectares per month). Shortgrass prairie rangeland, northeastern Colorado, IBP Pawnee Site. No protein determinations made or reported. |
| **130** | — | — | 62 | 74 | 26 | — | — | — | — | — |
| **131** | — | — | 67 | 81 | 19 | — | — | — | — | — |
| **132** | — | — | 42 | 51 | — | — | — | — | — | Tame pronghorn observed; bite-count method used; diet reconstructed by hand-plucking. Leaves were preferred plant parts, new growth preferred. Winter and early spring fringed sagewort was important; in summer scarlet globemallow was important. In winter, browse 47 %; in spring, grass 51 %; in summer, forbs 70 %; year-long. browse 14 %; grass 25 %, forbs 61 %. |
| **133** | — | — | 41 | 48 | — | — | — | — | — | — |
| **134** | — | — | 32 | 44 | — | — | — | — | — | — |
| **135** | — | — | — | — | — | — | — | 0.24 | — | Data are average of 2 locations in 2 years. Samples by ocular estimate and hand collection; May–Sept. observation. Salt desert shrub, big sagebrush and black sagebrush types. Highest protein in mid-Aug.; forbs high in diet in Aug. |
| **136** | 44 | — | — | — | — | — | 1.16 | 0.19 | — | Rumen contents examined botanically; hand-clipped samples taken of equivalent species; diet composited to match that in rumen; may still be bias on plant part not matching that grazed by deer. |
| **137** | 25 | — | — | — | — | — | 0.78 | 0.26 | — | — |
| **138** | 24 | — | — | — | — | — | 0.59 | 0.17 | — | — |
| **139** | 29 | — | — | — | — | — | 0.92 | 0.18 | — | — |
| **140** | 32 | 65 | — | — | — | — | — | — | — | The alfalfa was about 20 % crude protein; digestibility 60 to 62 %. |
| **141** | 34 | 67 | — | — | — | — | — | — | — | — |
| **142** | 31 | — | — | — | — | — | — | — | — | Hand-collected forage samples in birch–spruce vegetation, Kenai Peninsula, Alaska. |
| **143** | 28 | — | — | — | — | — | — | — | — | — |
| **144** | 26 | — | — | — | — | — | — | — | — | See notes under no. 142. |
| **145** | 21 | — | — | — | — | — | — | — | — | See notes under no. 142. |
| **146** | 20 | 59 | — | — | — | — | 0.30 | 0.20 | — | Based on rumen samples taken from caribou in Newfoundland. Available foods in spring varied from 4 up to 24 % crude protein, in summer from 2 up to 16 %, in autumn from 2 up to 11 %, and in winter from 4 to 10 %. |
| **147** | 22 | 56 | — | — | — | — | 0.71 | 0.16 | — | — |
| **148** | 26 | 60 | — | — | — | — | 0.48 | 0.09 | — | — |
| **149** | 27 | 63 | — | — | — | — | 0.27 | 0.06 | — | — |

## Appendix 4.4. A summary of reported values of food intake rate for large herbivores on the world's grazinglands

| No. | Reference | Animal species | Vegetation type | Location | Methods | Season/ treatment | Animal weight (kg) | Intake (% body wt · $d^{-1}$)a | Notes |
|---|---|---|---|---|---|---|---|---|---|
| 1 | Baker (1976) | Cattle | Subhumid pasture | Southern England | Vibracorder to get bite-count, calibrated with samples taken from fistulated animals and weight | Apr.–June | Unspecified | 3 o.m. | — |
| 2 | Baker (1976) | Cattle (heifers) | Subhumid pasture | Southern England | Vibracorder to get bite-count, calibrated with samples taken from fistulated animals and weight | Apr.–June | Unspecified | 1.7 o.m. | — |
| 3 | Baker (1976) | Cattle (calves) | Subhumid pasture | Southern England | Vibracorder to get bite-count, calibrated with samples taken from fistulated animals and weight | Apr.–June | Unspecified | 2.6 o.m. | Value for growing calves; calves receiving milk consumed 1.6 % o.m. |
| 4 | Corbett *et al.* (1963) | Cattle | Ryegrass–cocksfoot–timothy–clover | Aberdeen, Scotland | Diet; faeces – chromic oxide, field collections of samples; digestibility – faecal nitrogen index | Spring | 430 | 2.5 o.m. | Spring samples calculated from 10 tria in mid-May to mid-June; summer sample in Aug. From equations, calculated herbage digestibility of 65 % would give o.m. intake of 2.5 % and herbage digestibility of 80 %, an o.m. intake of 2.7 % for a 454 kg cow, based on cut herbage. |
| 5 | Hale *et al.* (1962) | Cattle (heifers and Brahmans) | Wiregrass–pine range | South Georgia, USA | Diet – hand-plucked samples; faeces – chromic oxide and grab samples; digestibility – lignin ratio | Spring (Apr.) | 399 | 2.28 | |
| 6 | Alder & Minson (1963) | Cattle | Cocksfoot | Hurley, Berkshire, England | Diet – hand sampling; faeces – chromic oxide and grab samples; digestibility – faecal nitrogen index | Late spring | 273–454 | 1.85 o.m. | Late spring data through early June, average of 3 trials in 2 years; summer data primarily June through Aug., averaged over 5 trials in 2 years; early autumn data Sept.–Oct., average of 2 trials in 2 years. Data expressed in organic matter percentage of body weight per day. Used 364 kg on graph. |
| 7 | Van Dyne & Meyer (1964*a*) | Cattle | Annual grass–shrub | Northern California, USA | Diet – oesophageal fistula; faeces – total collection; digestibility – *in vitro* | Early summer | 318 | 1.5 | See notes under no. 60. |

| | | | | | | | | |
|---|---|---|---|---|---|---|---|---|
| **8** Alder & Minson (1963) | Cattle | Pasture grasses – legumes | Hurley, Berkshire, England | Diet – hand sampling; faeces – chromic oxide and grab samples; digestibility – faecal nitrogen index | Summer grazing period | 273–454 | 2.1 | Intake based on organic matter. Average of lucerne, cocksfoot and lucerne–cocksfoot pastures in 2 years. See notes under no. 6. |
| **9** Alder & Minson (1969) | Cattle | Cocksfoot | Hurley, Berkshire, England | Diet – hand sampling; faeces – chromic oxide and grab samples; digestibility – faecal nitrogen index | Summer | 273–454 | 2.01 o.m. | See notes under no. 6. |
| **10** Corbett *et al.* (1963) | Cattle | Ryegrass–cocksfoot–timothy–clover | Aberdeen, Scotland | Diet; faeces – chromic oxide, field collections; digestibility – faecal nitrogen index | Summer | 463 | 2.2 o.m. | See notes under no. 4. |
| **11** Dyck & Bement (1972) | Cattle (heifers) | Shortgrass prairie | Northeastern Colorado, USA | Water intake method, hand-collected samples of forage | Summer (June–Sept.), light grazing | 264.7 | 2.63 d.m. | See notes under no. 13. |
| **12** Dyck & Bement (1971) | Cattle (heifers) | Shortgrass prairie | Northeastern Colorado, USA | Water intake method, hand-collected samples of forage | Summer (June–Sept.), light grazing | 268.45 | 2.76 d.m. | See notes under no. 13. |
| **13** Dyck & Bement (1971) | Cattle (heifers) | Shortgrass prairie | Northeastern Colorado, USA | Water intake method, hand-collected samples of forage | Summer (June–Sept.), heavy grazing | 243.45 | 2.64 d.m. | Water intake measured for a group of animals; daily temperature measurements taken to use in estimating water needs. Moisture content of hand-collected forage obtained. Average weight of heifers used. |
| **14** Dyck & Bement (1972) | Cattle (heifers) | Shortgrass prairie | Northeastern Colorado, USA | Water intake method, hand-collected samples of forage | Summer (June–Sept.), heavy grazing | 251.65 | 2.50 d.m. | See notes under no. 13. |
| **15** Hale *et al.* (1962) | Cattle (heifers and Brahmans) | Wiregrass–pine range | South Georgia, USA | Diet – hand-plucked samples; faeces – chromic oxide and grab samples; digestibility – lignin ratio | Summer | 392 | 2.32 | — |
| **16** Jeffries & Rice (1969) | Cattle (steers) | Shortgrass plains | Wyoming, USA | Diet – oesophageal fistula: intake – dry weight of faeces excreted during 24 h period, digestibility by in-vitro method | June–Aug. | 268 | 2.0 d.m. | Average weight of steers used. Average of 2 years' data. |
| **17** Jeffries & Rice (1969) | Cattle (steers) | Shortgrass plains | Wyoming, USA | Diet – oesophageal fistula intake – dry weight of faeces excreted during 24 h period, digestibility by lignin method | June–Aug. | 268 | 1.7 d.m. | Average weight of steers used. Average of 2 years' data. |

[a] o.m. refers to the intake of organic matter; d.m. refers to the intake of dry matter. If neither are noted it was not clearly stated in the original work, but the data are probably based on dry matter.

**Appendix 4.4** (*continued*)

| No. | Reference | Animal species | Vegetation type | Location | Methods | Season/ treatment | Animal weight (kg) | Intake (% body wt. $d^{-1}$)[a] | Notes |
|---|---|---|---|---|---|---|---|---|---|
| **18** | Jeffries & Rice (1969) | Cattle (steers) | Shortgrass plains | Wyoming, USA | Diet – oesophageal fistula intake – dry weight of faeces excreted during 24 h period, digestibility by pepsin–HCl digestion | June–Aug. | 268 | 2.6 d.m. | Average weight of steers used. |
| **19** | McClung *et al.* (1976) | Cattle (Angus–Holstein cross steers) | Shortgrass prairie | Amarillo, Texas, USA | Clipping studies in and out of cages at bimonthly or monthly interval | Summer, light grazing | 473 | 2.3 d.m. | — |
| **20** | Rice *et al.* (1974) | Cattle | Shortgrass prairie | Northeastern Colorado, USA | Diet – oesophageal fistula; faeces – chromic oxide and grab samples; digestibility – *in vitro* | Summer grazing | 460 | 1.12 | See notes under no. 95. |
| **21** | Smith *et al.* (1968) | Cattle | Moderately utilized northern desert shrub | Southern Nevada, USA | Rumen fistula; faecal excretion – grab sample and chromic oxide indicator | Early summer | 550 | 1.35 d.m. | — |
| **22** | Smith *et al.* (1968) | Cattle | Moderately utilized northern desert shrub | Southern Nevada, USA | Rumen fistula; faecal excretion – grab sample and chromic oxide indicator | Late summer | 550 | 1.85 d.m. | — |
| **23** | Van Dyne & Meyer (1964*a*) | Cattle | Annual grass–shrub | Northern California, USA | Diet – oesophageal fistula; faeces – total collection; digestibility – *in vitro* | Mid-summer | 323 | 1.75 | See notes under no. 60. |
| **24** | Van Dyne & Meyer (1964*a*) | Cattle | Annual grass–shrub | Northern California, USA | Diet – oesophageal fistula; faeces – total collection; digestibility – *in vitro* | Late summer | 322 | 1.85 | See notes under no. 60. |
| **25** | Alder & Minson (1963) | Cattle | Cocksfoot | Hurley, Berkshire, England | Diet – hand-sampling; faeces – chromic oxide and grab samples; digestibility – faecal nitrogen index | Early autumn | 273–454 | 1.66 o.m. | See notes under no. 6. |

| | | | | | | | | |
|---|---|---|---|---|---|---|---|---|
| **26** Hale *et al.* (1962) | Cattle (heifers and Brahmans) | Wiregrass–pine range | South Georgia, USA | Diet – hand-plucked samples; faeces – chromic oxide and grab samples; digestibility – lignin ratio | Autumn | 386 | 2.12 | — |
| **27** Hale *et al.* (1962) | Cattle (heifers and Brahmans) | Wiregrass–pine range | South Georgia, USA | Diet – hand-plucked samples; faeces – chromic oxide and grab samples; digestibility – lignin ratio | Winter | 373 | 1.46 | — |
| **28** Smith *et al.* (1968) | Cattle | Moderately utilized northern desert shrub | Southern Nevada, USA | Rumen fistulas; faecal excretion – grab sample and chromic oxide indicator | Winter | 550 | 0.96 d.m. | — |
| **29** Dean & Rice (1975) | Cattle (heifers) | Shortgrass prairie | Northeastern Colorado, USA | Diet – oesophageal fistula; faeces – grab samples; intake – water intake method; digestibility – in-vitro dry matter digestibility | Year, light grazing | 292 | 2.39 | — |
| **30** Dean & Rice (1975) | Cattle (heifers) | Shortgrass prairie | Northeastern Colorado, USA | Diet – oesophageal fistula; faeces – grab samples; intake – water intake method; digestibility – in-vitro dry matter digestibility | Year, heavy grazing | 274 | 2.39 | — |
| **31** Elliott *et al.* (1961) | Cattle | Veld grazing | Rhodesia | Diet; faeces – chromic oxide and grab samples; digestibility – faecal nitrogen index | June–Oct., Afrikander pregnant cows | 477 | 1.6 o.m. | See notes under no. 38. |
| **32** Elliott & Fokema (1961) | Cattle | Veld grazing | Rhodesia | Diet; faeces – chromic oxide and grab samples; digestibility – faecal nitrogen index | June–Oct., Afrikander pregnant cows | 443 | 1.6 o.m. | See notes under no. 39. |
| **33** Elliott *et al.* (1961) | Cattle | Veld grazing | Rhodesia | Diet; faeces – chromic oxide and grab samples; digestibility – faecal nitrogen index | June–Oct., Mashona pregnant cows | 336 | 1.8 o.m. | See notes under no. 38. |
| **34** Elliott & Fokema (1961) | Cattle | Veld grazing | Rhodesia | Diet; faeces – chromic oxide and grab samples; digestibility – faecal nitrogen index | June–Oct., Mashona pregnant cows | — | 1.8 o.m. | See notes under no. 39. |

## Appendix 4.4 (*continued*)

| No. | Reference | Animal species | Vegetation type | Location | Methods | Season/ treatment | Animal weight (kg) | Intake (% body wt · $d^{-1}$)[a] | Notes |
|---|---|---|---|---|---|---|---|---|---|
| **35** | Rao *et al.* (1974) | Cattle | Tallgrass prairie | Kansas, Flint Hills, USA | Total faecal collection, faecal nitrogen index regression | June–Oct. | Unspecified | See notes | Intake rates for organic matter varied from 0.91 to 1.01 from Oct. to June per unit metabolic size. Average was 96 g per body weight (kg) to the power $\frac{3}{4}$. |
| **36** | Elliott *et al.* (1961) | Cattle | Veld grazing | Rhodesia | Diet; faeces – chromic oxide and grab samples; digestibility – faecal nitrogen index | Nov.–June, Afrikander lactating cows | 414 | 2.5 o.m. | See notes under no. 38. |
| **37** | Elliott & Fokema (1961) | Cattle | Veld grazing | Rhodesia | Diet; faeces – chromic oxide and grab samples; digestibility – faecal nitrogen index | Nov.–June, Afrikander lactating cows | 336 | 2.8 o.m. | See notes under no. 39. |
| **38** | Elliott *et al.* (1961) | Cattle | Veld grazing | Rhodesia | Diet; faeces – chromic oxide and grab samples; digestibility – faecal nitrogen index | Nov.–June, Afrikander dry cows | 545 | 1.6 o.m. | In green feed season (Nov.–June) Mashona cows grazed forage 3 % higher in digestibility than did Afrikander cows; lactation increased organic matter intake 13 %; Nov.–June herbage digestibility 60 %–43 %; in dry period herbage digestibility 42 %–50 %, June–Oct. Feed consumption proportional to digestibility but faecal production relatively constant. |
| **39** | Elliott & Fokema (1961) | Cattle | Veld grazing | Rhodesia | Diet; faeces – chromic oxide and grab samples; digestibility – faecal nitrogen index | Nov.–June, Afrikander dry cows | 500 | 1.5 o.m. | Mashona more efficient in nutrient utilization at lower levels of performance. Lactation increased . . organic matter consumption by 13 % Afrikanders consumed 29 % more organic matter than Mashona. Digestibility varied from 65 % in Nov. to 42 % in June. Afrikander cows lost 0.6 lb daily whereas Mashona cows lost 0.3 lb daily. Forage digestibility was from 45 % up to 56 % of organic matter. Data are for 1958–9. |

| | | | | | | | | |
|---|---|---|---|---|---|---|---|---|
| **40** Elliott *et al.* (1961) | Cattle | Veld grazing | Rhodesia | Diet; faeces – chromic oxide and grab samples; digestibility – faecal nitrogen index | Nov.–June, Mashona dry cows | 340 | 2.0 o.m. | See notes under no. 38. |
| **41** Elliott & Fokema (1961) | Cattle | Veld grazing | Rhodesia | Diet; faeces – chromic oxide and grab samples; digestibility – faecal nitrogen index | Nov.–June, Mashona dry cows | 418 | 1.3 o.m. | See notes under no. 39. |
| **42** Elliott *et al.* (1961) | Cattle | Veld grazing | Rhodesia | Diet; faeces – chromic oxide and grab samples; digestibility – faecal nitrogen index | Nov.–June, Mashona lactating cows | 291 | 3.1 o.m. | See notes under no. 38. |
| **43** Elliott & Fokema (1961) | Cattle | Veld grazing | Rhodesia | Diet; faeces – chromic oxide and grab samples; digestibility – faecal nitrogen index | Nov.–June, Mashona lactating cows | 295 | 2.6 o.m. | See notes under no. 39. |
| **44** Ridley *et al.* (1963) | Cattle | Tall fescue–orchard grass, irrigated pasture | Reno, Nevada, USA | Diet – rumen fistula; faeces – total collection; digestibility – chromogen indicator | Growing season | — | 1.8 d.m. | — |
| **45** Holmes *et al.* (1961) | Cattle | Ryegrass–white clover | Ashford, Kent, England | Diet; faeces – chromic oxide and grab samples; digestibility – faecal nitrogen index | Calves | 285 | 3.2 o.m. | See notes under no. 47. |
| **46** Holmes *et al.* (1961) | Cattle | Ryegrass–white clover | Ashford, Kent, England | Diet; faeces – chromic oxide and grab samples; digestibility – faecal nitrogen index | Heifers | 359 | 2.6 o.m. | See notes under no. 47. |
| **47** Holmes *et al.* (1961) | Cattle | Ryegrass–white clover | Ashford, Kent, England | Diet; faeces – chromic oxide and grab samples; digestibility – faecal nitrogen index | Cows | 591 | 1.8 o.m. | Organic matter digestibility was 74, 76 and 77 %, respectively, for calves, heifers and cows. |
| **48** Rittenhouse *et al.* (1970) | Cattle | Blue grama–needle-and-thread–western wheatgrass range | Nebraska, USA | Oesophageal fistulas; total faecal collection | No supplements, trials Jan. and Mar. | 330 | 1 | — |
| **49** Harrington & Pratchett (1974) | Cattle | *Brachiaria–Themeda–Hyparrhenia* | Ankole, Uganda | Diet – oesophageal fistula; faeces – ?; digestibility – calculated by equation | — | Unspecified | 2.1 | Calculated dry matter intake based on presence of unlimited forage of 10 % crude protein and 55 % apparent digestibility; refers to literature for dry matter intakes of 1.9–2.2 % body weight of poor quality hays of 4–8 % crude protein; refers to literature for dry matter intake up to 2.9 % of body weight with fresh grass similar to that in the wet-season forage present trial. |

## Appendix 4.4 (*continued*)

| No. | Reference | Animal species | Vegetation type | Location | Methods | Season/ treatment | Animal weight (kg) | Intake (% body wt · $d^{-1}$)[a] | Notes |
|---|---|---|---|---|---|---|---|---|---|
| **50** | Phillips (1960) | Cattle, Zebu | Star grass hay | Kenya | Diet – *ad-libitum* feeding | Controlled treatment *ad libitum* | 293 | 1.9 | See notes under no. 51. |
| **51** | Phillips (1960) | Cattle (heifers) | Star grass hay | Kenya | Diet – *ad-libitum* feeding | Controlled treatment *ad libitum* | 280 | 2 | *Ad libitum* star grass hay, crude protein content 6 %. Weights were approximately maintained. Zebu cattle at lower water intakes. Reducing water availability reduced feed intake, but less for Zebus, feed intake reduces to 1.6 % under limited water. |
| **52** | Fels *et al.* (1959) | Sheep | Clover-dominant, grass-dominant annual pasture | Nedlands, Western Australia | Diet; faeces – total collection; digestibility – faecal nitrogen index | Early spring, Sept. | 50 | 2.7 | Data calculated by average of various faecal nitrogen index regression equations. Three trials, usually with 3 methods in each case. Crude protein varied from about 4 to 25 % in different trials and organic digestion from 40 to 80 %. Feed–faeces ratio predicted from faecal nitrogen percentage by equation $y = 1.021x + 0.643$. |
| **53** | Fels *et al.* (1959) | Sheep | Clover-dominant, grass-dominant annual pasture | Nedlands, Western Australia | Diet; faeces – total collection; digestibility – faecal nitrogen index | Sept. (grass) | — | 2.0 | See notes under no. 52 |
| **54** | Engels *et al.* (1969) | Sheep | *Themeda* pasture | South Africa | Oesophageal fistula, total faecal collection | Apr.–June | 37 | 4.0 d.m. | — |
| **55** | Fels *et al.* (1959) | Sheep | Clover-dominant, grass-dominant annual pasture | Nedlands, Western Australia | Diet; faeces – total collection; digestibility – faecal nitrogen index | Oct. | 50 | 1.8 | See notes under no. 52. |
| **56** | Fels *et al.* (1959) | Sheep | Clover-dominant, grass-dominant annual pasture | Nedlands, Western Australia | Diet; faeces – total collection; digestibility – faecal nitrogen index | Oct. (grass) | — | 1.8 | See notes under no. 52. |
| **57** | Fels *et al.* (1959) | Sheep | Clover-dominant, grass-dominant annual pasture | Nedlands, Western Australia | Diet; faeces – total collection; digestibility – faecal nitrogen index | Dec. | 50 | 1.4 | See notes under no. 52. |

| | | | | | | | | |
|---|---|---|---|---|---|---|---|---|
| **58** Fels *et al.* (1959) | Sheep | Clover-dominant, grass-dominant annual pasture | Nedlands, Western Australia | Diet; faeces – total collection; digestibility – faecal nitrogen index | Dec. (grass) | — | 1.9 | See notes under no. 52. |
| **59** Hadjipieris *et al.* (1965) | Sheep | Perennial ryegrass–wild white clover | Ashford, Kent, England | Diet – hand-clipped samples; faeces – total collection; digestibility – faecal nitrogen index | Early summer | 58.7 | 2.9 o.m. | Data calculated from table 2 of Hadjipieris *et al.* 5-year-old wethers had lower relative intake than did growing sheep – 2 versus 3.2 % of live weight for organic matter. |
| **60** Van Dyne & Meyer (1964*a*) | Sheep | Annual grass–shrub | Northern California, USA | Diet – oesophageal fistula; faeces – total collection; digestibility – *in vitro* | Early summer | 46 | 1.67 | Early, middle and late summer herbage availability 1490, 1220 and 420 lb·acre$^{-1}$ respectively. Data presented are average predictions made with both in-vitro and in-vivo microdigestion of cellulose. The fistulated steers and wethers were about 2 years of age. Forage intake by cattle was less affected by lack of herbage than was forage intake by sheep; sheep were more variable than cattle in *ad libitum* intake of pelleted alfalfa under drylot feeding. Lignin and silica ratio techniques gave invalid results. |
| **61** Cook *et al.* (1961) | Sheep | Mountain brush-grass range | Northern Utah, USA | Diet – oesophageal fistula; faeces – total collection; digestibility – lignin ratio | Summer, wet lactating ewes | 64 | 2.75 | Diet protein 12 %. Average weight of animals from Cook (1970). |
| **62** Cook *et al.* (1961) | Sheep | Mountain brush-grass range | Northern Utah, USA | Diet – oesophageal fistula; faeces – total collection; digestibility – lignin ratio | Summer, dry ewes | 72 | 1.95 | See notes under no. 61. |
| **63** Engels *et al.* (1969) | Sheep | *Themeda* pasture | South Africa | Oesophageal fistula, total faecal collection | July–Sept. | 46 | 3.9 d.m. | — |
| **64** Rice *et al.* (1974) | Sheep | Shortgrass prairie | Northeastern Colorado, USA | Diet – oesophageal fistula; faeces – chromic oxide and grab sample; digestibility – *in vitro* | Summer grazing | 37 | 1.9 | See notes under no. 95. |
| **65** Van Dyne & Meyer (1964*a*) | Sheep | Annual grass–shrub | Northern California, USA | Diet – oesophageal fistula; faeces – total collection; digestibility – *in vitro* | Mid-summer | 44 | 2.24 | See notes under no. 60. |

## Appendix 4.4 (*continued*)

| No. | Reference | Animal species | Vegetation type | Location | Methods | Season/ treatment | Animal weight (kg) | Intake (% body wt · $d^{-1}$)[a] | Notes |
|---|---|---|---|---|---|---|---|---|---|
| **66** | Hadjipieris *et al.* (1965) | Sheep | Perennial ryegrass–wild white clover | Ashford, Kent, England | Diet – hand-clipped samples; faeces – total collection; digestibility – faecal nitrogen index method | Late summer | 66.8 | 2.6 o.m. | See notes under no. 59. |
| **67** | Joshi (1966) | Sheep | Alpine grassland | Uttar Pradesh, India | Total faecal collection, digestibility determined from animals fed clipped forage under stall-feeding conditions | Late summer | 40 | 3 | |
| **68** | Van Dyne & Meyer (1964*a*) | Sheep | Annual grass–shrub | Northern California, USA | Diet – oesophageal fistula; faeces – total collection; digestibility – *in vitro* | Late summer | 43 | 1.77 | See notes under no. 60. |
| **69** | Engels *et al.* (1969) | Sheep | *Themeda triandra* pasture | South Africa | Oesophageal fistula, total faecal collection | Oct.–Nov. | 55 | 3.3 d.m. | — |
| **70** | Engels *et al.* (1969) | Sheep | *Themeda triandra* pasture | South Africa | Oesophageal fistula, total faecal collection | Jan.–Mar. | 53 | 3.1 d.m. | — |
| **71** | Cook *et al.* (1953) | Sheep | Salt desert shrub range | West Desert, Utah, USA | Diet – hand-plucked samples, observation; faeces – total collection; digestibility – lignin ratio | Winter, shadescale | 63 | 2.45 | Forage intake rate decreased on each type as utilization increased; on pure stands intake dropped 10 to 20 % as utilization went from about 25 to 50 %; less change in intake due to increased utilization on mixed-grass–shrubs. Average weight of animals from Cook (1970). |
| **72** | Cook *et al.* (1953) | Sheep | Salt desert shrub range | West Desert, Utah, USA | Diet – hand-plucked samples, observation; faeces – total collection; digestibility – lignin ratio | Winter, mixed diet | 63 | 2.59 | See notes under no. 71. |
| **73** | Cook *et al.* (1953) | Sheep | Salt desert shrub range | West Desert, Utah, USA | Diet – hand-plucked samples, observation; faeces – total collection; digestibility – lignin ratio | Winter, blacksage | 63 | 1.87 | See notes under no. 71. |

|  |  |  |  |  |  |  |  |  |
|---|---|---|---|---|---|---|---|---|
| **74** Cook *et al.* (1962) | Sheep | Salt desert shrub range | Southwestern Utah, USA | Diet – oesophageal fistula; faeces – total collection; digestibility – lignin ratio | Winter, good condition | 63 | 2.23 | Diet protein under good conditions 7.8 % under poor conditions 8.3. Grass percentage under good conditions 60, under poor conditions 52. Browse percentage under good conditions 42; under poor conditions 45. Average weight of animals from Cook (1970). |
| **75** Cook *et al.* (1962) | Sheep | Salt desert shrub range | Southwestern Utah, USA | Diet – oesophageal fistula; faeces – total collection; digestibility – lignin ratio | Winter, poor condition | 63 | 1.87 | See notes under no. 74. |
| **76** Pieper *et al.* (1959) | Sheep | Salt desert shrub range | Millard County, Utah, USA | Diet – hand-plucked samples; faeces – total collection; digestibility – lignin ratio | Winter, four-winged saltbush, pure stand, heavy use | 63 | 3.25 | See notes under no. 82. |
| **77** Pieper *et al.* (1959) | Sheep | Salt desert shrub range | Millard County, Utah, USA | Diet – hand-plucked samples; faeces – total collection; digestibility – lignin ratio | Winter, winterfat, pure stand, light use | 63 | 2.21 | See notes under no. 82. |
| **78** Pieper *et al.* (1959) | Sheep | Salt desert shrub range | Millard County, Utah, USA | Diet – hand-plucked samples; faeces – total collection; digestibility – lignin ratio | Winter, winterfat, pure stand, heavy use | 63 | 2.04 | See notes under no. 82. |
| **79** Pieper *et al.* (1959) | Sheep | Salt desert shrub range | Millard County, Utah, USA | Diet – hand-plucked samples; faeces – total collection; digestibility – lignin ratio | Winter, grass–shrub, heavy grazing | 63 | 2.51 | See notes under no. 82. |
| **80** Pieper *et al.* (1959) | Sheep | Salt desert shrub range | Millard County, Utah, USA | Diet – hand-plucked samples; faeces – total collection; digestibility – lignin ratio | Winter, four-winged saltbush, pure stand, light use | 63 | 3.32 | See notes under no. 82. |
| **81** Arnold *et al.* (1964) | Sheep | *Phalaris*–subterranean clover pasture, annual grasses | Canberra, Australia | Diet; faeces – total collection; digestibility – faecal nitrogen index | Yearlong average, 2 years | 45 | 2.2 | Above figure average for 2 years. Intake varied from 500 to almost 1400 $g \cdot d^{-1}$. Body weight varied from 20 lb under mean to 30 lb over mean for 3–4-year-old wethers. Diet digestibility varied from 50 to 80 % |

## Appendix 4.4 (*continued*)

| No. | Reference | Animal species | Vegetation type | Location | Methods | Season/ treatment | Animal weight (kg) | Intake (% body wt · $d^{-1}$)[a] | Notes |
|---|---|---|---|---|---|---|---|---|---|
| **82** | Pieper *et al.* (1959) | Sheep | Salt desert shrub range | Millard County, Utah, USA | Diet – hand-plucked samples; faeces – total collection; digestibility – lignin ratio | Grass–shrub, light grazing | 63 | 3.02 | Under grass–shrub ranges, 83 % grass in diet under light use and 74 % under heavy use; protein in diet 6 %; winterfat 8 %; four-winged saltbush average 9 %. Light use areas averaged 30 % of available herbage in all trials, heavy use 72 %. Studies from late Oct. to early Mar. Average weight of animals from Cook (1970). |
| **83** | Cook & Harris (1950) | Sheep | Saltbush–grass, sagebrush–grass | West-central Utah, USA | Diet – before-and-after vegetation sampling; faeces – total collection; digestibility – lignin ratio | Nov.–Apr. average | 41 | 3.5 d.m. | Diet averaged 6 % protein, 51 % dry matter digestibility. |
| **84** | Joshi *et al.* (1971) | Sheep | Unspecified | Dehradun, India | Total faecal collection, digestibility determined in stall feeding | Winter–spring | 32 | 1.2 | — |
| **85** | Cook *et al.* (1956) | Sheep | Spring–summer grazing, seeded pasture | Foothills, Utah, USA | Diet – hand sampling; faeces – total collection; digestibility – lignin ratio | Smother weed | 59 | 2.5 d.m. | See notes under no. 92. |
| **86** | Cook *et al.* (1956) | Sheep | Spring–summer seeded pasture | Foothills, Utah, USA | Diet – hand sampling; faeces – total collection; digestibility – lignin ratio | Russian thistle | 59 | 2.5 d.m. | See notes under no. 92. |
| **87** | Cook *et al.* (1956) | Sheep | Spring–summer, seeded pasture | Foothills, Utah, USA | Diet – hand sampling; faeces – total collection; digestibility – lignin ratio | Western wheatgrass | 59 | 2.2 d.m. | See notes under no. 92. |
| **88** | Cook *et al.* (1956) | Sheep | Spring–summer, seeded pasture | Foothills Utah, USA | Diet – hand Sampling; faeces – total collection; digestibility – lignin ratio | Tall wheat-grass | 59 | 2.7 d.m. | See notes under no. 92. |
| **89** | Cook *et al.* (1956) | Sheep | Spring–summer, seeded pasture | Foothills, Utah, USA | Diet – hand sampling; faeces – total collection; digestibility – lignin ratio | Intermediate wheatgrass | 59 | 3.0 d.m. | See notes under no. 92. |

| | | | | | | | | |
|---|---|---|---|---|---|---|---|---|
| **90** Cook *et al.* (1956) | Sheep | Spring–summer, seeded pasture | Foothills, Utah, USA | Diet – hand sampling; faeces – total collection; digestibility – lignin ratio | Beardless wheatgrass | 59 | 2.3 d.m. | See notes under no. 92. |
| **91** Cook *et al.* (1956) | Sheep | Spring–summer, seeded pasture | Foothills, Utah, USA | Diet – hand sampling; faeces – total collection; digestibility – lignin ratio | Pubescent wheatgrass | 59 | 2.4 d.m. | See notes under no. 92. |
| **92** Cook *et al.* (1956) | Sheep | Spring–summer, seeded pasture | Foothills, Utah, USA | Diet – hand sampling; faeces – total collection; digestibility – lignin ratio | Crested wheatgrass | 59 | 2.8 d.m. | Data based on 130 lb wethers, which were average weight used in the trials. Confidence interval less than 10 % of the mean for most situations. Data from early May to late Aug., but most in late and early summer. |
| **93** Gihad (1976) | Sheep | *Hyparrhenia* pasture | Zambia | *Ad libitum* feeding of natural grass hay | — | 60.8 | 1.3 d.m. | — |
| **94** Gihad (1976) | Goat | *Hyparrhenia* pasture | Zambia | *Ad libitum* feeding of natural grass hay | — | 25.3 | 1.8 d.m. | — |
| **95** Rice *et al.* (1974) | Bison | Shortgrass prairie | Northeastern Colorado, USA | Diet – oesophageal fistula; faeces – chromic oxide and grab samples; digestibility – *in vitro* | Summer grazing | 348 | 1.74 | Intake data for dry matter is an average of 3 monthly sampling periods for heavy and light grazing intensities. Presumably in-vitro digestibility used in calculating intake. Average weight of the three animals given. |
| **96** Alldredge *et al.* (1974) | Mule deer | Mountain range | Northeastern Colorado, USA | Calculated from forage and body $^{137}Cs$ content (see text) | Summer, adults | 54 | 2.1 | Based on 15 animals of both sexes more than 18 months old. |
| **97** Alldredge *et al.* (1974) | Mule deer | Mountain range | Northeastern Colorado, USA | Calculated from forage and body $^{137}Cs$ content (see text) | Winter, adults | 54 | 1.7 | Based on 48 animals of both sexes, more than 18 months old. |
| **98** Hanson *et al.* (1975) | Caribou | Arctic tundra–taiga | North slope, Alaska, USA | Diet – hand-collected samples; faeces digestibility – 2-compartment, 8-parameter $^{137}Cs$ model | Winter, average of several years | 75 | 6.7 | Calculated from caesium in whole body burden, muscle, lichens, fraction of caesium injected that is absorbed, fraction body weight that is muscle, and kinetic parameters. |

## Appendix 4.5. A summary of digestibility of forage by large herbivores on the world

| No. | Reference | Animal species | Methods | Season/ treatment | Small sample microdigestion[a] | | | | | |
|---|---|---|---|---|---|---|---|---|---|---|
| | | | | | *in vitro* | | | *in vivo* | | |
| | | | | | DMD | OMD | Cel- lulose | DMD | OMD | C lul |
| 1 | Alder & Minson (1963) | Cattle | Diet – hand clipping; faeces – chromic oxide and grab samples; digestion – faecal nitrogen index | May, cocksfoot | — | — | — | — | — | — |
| 2 | Rice *et al.* (1974) | Cattle | Diet – oesophageal fistula; digestion – rumen fluid from rumen-fistulated animals | Spring (Apr. and May) | — | — | — | — | — | — |
| 3 | Rosiere *et al.* (1975) | Cattle | Oesophageal fistula, diet samples | Spring | — | 64 | — | — | — | — |
| 4 | Scales *et al.* (1971) | Cattle | Oesophageal fistulas | May–June, sandhill range | — | — | — | — | — | — |
| 5 | Scales (1972) | Cattle | Diet – oesophageal-fistulated cattle; digestion – same forage fed to sheep, total collection | Spring (May) | — | — | — | — | — | — |
| 6 | Connor *et al.* (1963) | Cattle | Diet – rumen fistula; faeces – total collection; digestion – chromogen indicator | Period 1: early summer, sagebrush–grass rangeland | — | — | — | — | — | — |
| 7 | Connor *et al.* (1963) | Cattle | Diet – rumen fistula; faeces – total collection; digestion – chromogen indicator | Period 1: early summer, desert shrub rangeland | — | — | — | — | — | — |
| 8 | Smith *et al.* (1968) | Cattle | Diet – rumen fistula; faeces – total faecal excretion by grab sample and chromic oxide indicator | Early summer, moderately utilized northern desert shrub | — | — | — | — | — | — |
| 9 | Van Dyne & Lofgreen (1964) | Cattle | Diet – oesophageal fistula; faeces – total collection; digestion – lignin ratio | Early summer | — | — | — | — | — | — |

[a] DMD refers to dry matter digestibility, OMD is organic matter digestibility, CHO refers to carbohydrate, CF is crude fib
NFE is nitrogen-free extract, CWC are cell wall constituents, and ADF is acid-detergent fibre.

## razinglands (all data in %)

| | | | | | | | | | | | | | | |
|---|---|---|---|---|---|---|---|---|---|---|---|---|---|---|
| Large sample macrodigestion | | | | | | | | | | | | | | |
| MD | OMD | Pro-tein | Ether ex. | Lignin | Cel-lulose | Other CHO | Ash | Energy | CF | NFE | CWC | ADF | Misc. | Notes |
| — | 74 | — | — | — | — | — | — | — | — | — | — | — | — | Data averaged over several days in 2 years. Data compared with forage fed to sheep in indoor digestion trials and to in-vitro digestibility. Bullocks were selecting the more digestible parts of the plants. |
| — | — | — | — | 11 | — | — | 18 | — | — | — | 66 | 57 | — | Overall average values for sheep slightly higher than cattle, and cattle somewhat higher than bison in digestibility of dry matter *in vitro*. |
| — | — | — | — | — | — | — | — | — | — | — | — | 2.4 | — | See notes under no. 20. |
| — | — | — | — | — | — | — | 14 | — | — | — | 65 | — | — | |
| 63 | 70 | 70 | — | — | — | — | — | 66 | — | — | — | — | — | Study also included chromic oxide method for field collection of total faecal data (not included here) and metabolizable nutrients in sheep digestion trial on forage grazed and collected by fistulated animals. Metabolizable energy calculated from digestible energy by equation of Blaxter (1962). Digestible energy, expressed in $kcal \cdot g^{-1}$ of digestible organic matter, varied only from 4.4 to 4.6 throughout the study. |
| 74 | 76 | 73 | 62 | — | — | — | — | 73 | 78 | 75 | — | — | 67 | Miscellaneous refers to total digestible nutrients. High apparent digestibility in period 1 in northern Nevada (sagebrush–grass rangeland) attributed to immature grasses and forbs consumed. |
| 54 | 62 | 64 | 27 | — | — | — | — | 50 | 63 | 61 | — | — | 50 | See notes under no. 7. |
| — | — | 61 | — | — | — | — | — | — | 61 | 63 | — | — | — | |
| 46 | — | 43 | 38 | — | 55 | 58 | — | 41 | — | — | — | — | — | Available herbage was 1490 $lb \cdot acre^{-1}$ in early summer and 420 $lb \cdot acre^{-1}$ in late summer. Averaged over all periods sheep had significantly higher digestion coefficients than cattle for all constituents except cellulose. Crude protein digestion coefficients varied more than for other constituents; lignin digestion was about 4 % overall. Digestibility values calculated |

**Appendix 4.5** (*continued*)

| | | | | | Small sample microdigestion[a] | | | | | |
|---|---|---|---|---|---|---|---|---|---|---|
| | | | | | *in vitro* | | | *in vivo* | | |
| No. | Reference | Animal species | Methods | Season/treatment | DMD | OMD | Cel-lulose | DMD | OMD | C lu |
| 9 | Van Dyne & Lofgreen (1964) (*cont.*) | | | | | | | | | |
| 10 | Alder & Minson (1963) | Cattle | Diet – hand clipping; faeces – chromic oxide and grab sample; digestion – faecal nitrogen index | June–Aug. | — | — | — | — | — | — |
| 11 | Bedell (1971) | Cattle | Oesophageal fistula, diet collection | Summer, sub-clover–tall fescue grazing | — | — | — | — | — | — |
| 12 | Bedell (1971) | Cattle | Oesophageal fistula, diet collection | Summer, sub-clover–tall fescue grazing | — | — | — | — | — | — |
| 13 | Bedell (1971) | Cattle | Oesophageal fistula, diet collection | Summer, sub-clover–perennial ryegrass, light grazing | — | — | — | — | — | — |
| 14 | Bedell (1971) | Cattle | Oesophageal fistula diet collection | Summer, sub-clover–perennial ryegrass, heavy grazing | — | — | — | — | — | — |
| 15 | Cook *et al.* (1967) | Cattle | Diet – oesophageal fistulas; faeces – grab samples, lignin ratio | Summer, aspen rangeland | — | — | — | — | — | — |
| 16 | Cook *et al.* (1967) | Cattle | Diet – oesophageal fistula; faeces – grab samples, lignin ratio | Summer, sagebrush–grass type | — | — | — | — | — | — |
| 17 | Rice & Vavra (1969) | Cattle | Oesophageal fistula, forage sampling | Summer, heavy grazing | 56 | — | — | — | — | — |
| 18 | Rice & Vavra (1969) | Cattle | Oesophageal fistula, forage sampling | Summer, light grazing | 61 | — | — | — | — | — |
| 19 | Rice *et al.* (1974) | Cattle | Diet – oesophageal fistula; digestion – rumen fluid from rumen fistulated animals | Summer (June, July and Aug.) | — | — | — | — | — | — |
| 20 | Rosiere *et al.* (1975) | Cattle | Oesophageal fistula, diet samples | Summer | — | 54 | — | — | — | — |

| Large sample-macrodigestion | | | | | | | | | | | | | | |
|---|---|---|---|---|---|---|---|---|---|---|---|---|---|---|
| MD | OMD | Protein | Ether ex. | Lignin | Cellulose | Other CHO | Ash | Energy | CF | NFE | CWC | ADF | Misc. | Notes |
| | | | | | | | | | | | | | | on the basis of microdigestion data and related to lignin ratio values were as follows for cattle and sheep, respectively: crude protein 108 and 118; extract 105 and 119; cellulose 101 and 111; other carbohydrates 102 and 107. 13 sheep and 11 cattle was the required number to estimate forage composition and 12 sheep and 10 cattle to estimate faecal composition for calculating digestion coefficients with the same accuracy; averaged over the summer for both species, 6 forage and 6 faecal animals would give adequate estimates of digestion coefficients for dry matter and carbohydrates but not for protein and ether extract. |
| — | 66 | — | — | — | — | — | — | — | — | — | — | — | — | See notes under no. 1. |
| 48 | — | — | — | — | — | — | — | — | — | — | — | — | — | |
| 46 | — | — | — | — | — | — | — | — | — | — | — | — | — | |
| 44 | — | — | — | — | — | — | — | — | — | — | — | — | — | |
| 44 | — | — | — | — | — | — | — | — | — | — | — | — | — | |
| — | — | 39 | 6 | 39 | 63 | 62 | — | 48 | — | — | — | — | 48 | See notes under no. 16. |
| — | — | 30 | 21 | 30 | 54 | 53 | — | 35 | — | — | — | — | 38 | Miscellaneous column is TDN. Negative ether extract digestibilities noted. Mid-June to mid-Sept.; 2 year study; northern Utah mountain rangeland. |
| — | — | — | — | — | — | — | — | — | — | — | — | — | — | Average of June–Aug. samples (may be same data as Vavra *et al.*, 1973). |
| — | — | — | — | — | — | — | — | — | — | — | — | — | — | See notes under no. 2. |
| — | — | — | 11 | — | — | 15 | — | — | — | — | 66 | 53 | — | See notes under no. 2. |
| — | — | — | — | — | — | — | — | — | — | — | — | — | 2.0 | Energy digestible in meal per kg predicted according to an equation of Rittenhouse *et al.* (1970). Averaged over 2 years for summer grazing, over 1 year for other treatments. Semidesert grassland. |

**Appendix 4.5** (*continued*)

| No. | Reference | Animal species | Methods | Season/treatment | Small sample microdigestion[a] *in vitro* DMD | *in vitro* OMD | *in vitro* Cellulose | *in vivo* DMD | *in vivo* OMD |
|---|---|---|---|---|---|---|---|---|---|
| **21** | Scales *et al.* (1971) | Cattle | Oesophageal fistulas | July, sandhill range | — | — | — | — | — |
| **22** | Scales (1972) | Cattle | Diet – oesophageal-fistulated cattle; digestion – same forage fed to sheep, total collection | Summer (June–Sept.) | — | — | — | — | — |
| **23** | Smith *et al.* (1960) | Cattle (steers) | Faeces collection, hand-picked forage samples, lignin ratio | Summer, burned bluestem pasture | — | — | — | — | — |
| **24** | Smith *et al.* (1960) | Cattle (steers) | Faeces collection, hand-picked forage samples, chromogen indicator | Summer, burned bluestem pasture | — | — | — | — | — |
| **25** | Smith *et al.* (1960) | Cattle (steers) | Faeces collection, collected forage samples by watching steers, lignin ratio | Summer, non-burned bluestem pasture | — | — | — | — | — |
| **26** | Smith *et al.* (1960) | Cattle (steers) | Faeces collection, hand-picked forage samples, chromogen indicator | Summer, non-burned bluestem pasture | — | — | — | — | — |
| **27** | Connor *et al.* (1963) | Cattle | Diet – rumen fistula; faeces – total collection; digestion – chromogen indicator | Period 2: mid-summer, desert shrub rangeland | — | — | — | — | — |
| **28** | Connor *et al.* (1963) | Cattle | Diet – rumen fistula; faeces – total collection; digestion – chromogen indicator | Period 2; mid-summer, sagebrush–grass rangeland | — | — | — | — | — |
| **29** | Van Dyne & Lofgreen (1964) | Cattle | Diet – oesophageal fistula; faeces – total collection; digestion – lignin ratio | Mid-summer | — | — | — | — | — |
| **30** | Connor *et al.* (1963) | Cattle | Diet – rumen fistula; faeces – total collection; digestion – chromogen indicator | Period 3: late summer, desert shrub rangeland | — | — | — | — | — |
| **31** | Connor *et al.* (1963) | Cattle | Diet – rumen fistula; faeces – total collection; digestion – chromogen indicator | Period 3: late summer, sagebrush–grass rangeland | — | — | — | — | — |
| **32** | Smith *et al.* (1968) | Cattle | Diet – rumen fistula; faeces – total faecal excretion, grab sample and chromic oxide indicator | Late summer, moderately utilized northern desert shrub | — | — | — | — | — |
| **33** | Van Dyne & Lofgreen (1964) | Cattle | Diet – oesophageal fistula; faeces – total collection; digestion – lignin ratio | Late summer | — | — | — | — | — |
| **34** | Alder & Minson (1963) | Cattle | Diet – hand clipping; faeces – chromic oxide and grab samples; digestion – faecal nitrogen index | Sept.–Oct. | — | — | — | — | — |
| **35** | Rice *et al.* (1974) | Cattle | Diet – oesophageal fistula; digestion – rumen fluid frøm rumen-fistulated animals | Autumn (Oct.) | — | — | — | — | — |

| MD | OMD | Protein | Ether ex. | Lignin | Cellulose | Other CHO | Ash | Energy | CF | NFE | CWC | ADF | Misc. | Notes |
|---|---|---|---|---|---|---|---|---|---|---|---|---|---|---|
| | | | | Large sample macrodigestion | | | | | | | | | | |
| — | — | — | — | — | — | — | 13 | — | — | — | 70 | — | — | |
| 57 | 62 | 60 | — | — | — | — | — | 58 | — | — | — | — | — | See notes under no. 5. |
| — | — | 27 | — | — | — | — | — | — | 71 | 63 | — | — | — | Pasture burned in mid-spring. |
| — | — | 67 | — | — | — | — | — | — | 78 | 74 | — | — | — | Pasture burned in mid-spring. |
| — | — | 11 | — | — | — | — | — | — | 68 | 60 | — | — | — | |
| — | — | 27 | — | — | — | — | — | — | 65 | 68 | — | — | — | |
| 5 | 70 | 69 | 48 | — | — | — | — | 69 | 62 | 70 | — | — | 58 | See notes under no. 6. |
| 7 | 69 | 60 | 42 | — | — | — | — | 65 | 67 | 75 | — | — | — | See notes under no. 6. |
| 2 | — | 22 | 27 | — | 53 | 49 | — | 38 | — | — | — | — | — | See notes under no. 9. |
| 3 | 35 | 54 | 19 | — | — | — | — | 31 | 35 | 34 | — | — | 18 | See notes under no. 6. |
| 0 | 62 | 60 | 53 | — | — | — | — | 59 | 60 | 64 | — | — | 56 | See notes under no. 6. |
| — | — | 71 | — | — | — | — | — | — | 74 | 71 | — | — | — | |
| 0 | — | 17 | 27 | — | 52 | — | — | 38 | — | — | — | — | — | See notes under no. 9. |
| — | 58 | — | — | — | — | — | — | — | — | — | — | — | — | See notes under no. 1. |
| — | — | — | — | 20 | — | — | 13 | — | — | — | 67 | 61 | — | See notes under no. 2. |

**Appendix 4.5** (*continued*)

| No. | Reference | Animal species | Methods | Season/treatment | Small sample microdigestion[a] *in vitro* DMD | OMD | Cel-lulose | *in vivo* DMD | OMD | C-lu |
|---|---|---|---|---|---|---|---|---|---|---|
| 36 | Rosiere *et al.* (1975) | Cattle | Oesophageal fistula, diet samples | Autumn | — | 39 | — | — | — | - |
| 37 | Scales *et al.* (1971) | Cattle | Oesophageal fistula | Sept.–Nov., sandhill range | — | — | — | — | — | - |
| 38 | Scales (1972) | Cattle | Diet – oesophageal-fistulated cattle; digestion – same forage fed to sheep, total collection | Autumn (Nov.) | — | — | — | — | — | - |
| 39 | Rosiere *et al.* (1975) | Cattle | Oesophageal fistula, diet samples | Winter | — | 48 | — | — | — | - |
| 40 | Smith *et al.* (1968) | Cattle | Diet – rumen fistula; faeces – total excretion determined by grab sample technique, chromic oxide indicator | Winter moderately utilized northern desert shrub | — | — | — | — | — | - |
| 41 | Elliott & Fokema (1961) | Cattle, Mashona | Faeces – chromic oxide indicator; digestibility – ? | Year | — | — | — | — | — | - |
| 42 | Elliott & Fokema (1961) | Cattle, Afrikander | Faeces – chromic oxide indicator; digestibility – ? | Year | | — | — | — | — | - |
| 43 | Hardison *et al.* (1954) | Cattle (steers) | Faecal collection chromogen indicator | Average of timothy, Kentucky bluegrass and meadow fescue pasture, year | — | — | — | — | — | - |
| 44 | Ridley *et al.* (1963) | Cattle | Diet – rumen fistula; digestion – chromogen indicator; faeces – total collection | Average all periods | — | — | — | — | — | - |
| 45 | Vavra *et al.* (1973) | Cattle | Diet – oesophageal fistula; faeces – total collection; digestion – *in vitro* | Light grazing year | 57 | — | — | — | — | |
| 46 | Vavra *et al.* (1973) | Cattle | Diet – oesophageal fistula; faeces – total collection; digestion *in vitro* | Heavy grazing year | 52 | — | — | — | — | |
| 47 | Arnold *et al.* (1964) | Sheep | Diet – oesophageal fistula; faeces – total collection; digestion – faecal nitrogen index | Spring, natural annual pasture | — | — | — | — | — | |
| 48 | Arnold *et al.* (1964) | Sheep | Diet – oesophageal fistula; faeces – total collection; digestion – faecal nitrogen index | Spring, natural annual pasture | — | — | — | — | — | |
| 49 | Arnold *et al.* (1964) | Sheep | Diet – oesophageal fistula; faeces – total collection; digestion – faecal nitrogen index | Spring, ryegrass–sub-clover pasture | — | — | — | — | — | |
| 50 | Arnold *et al.* (1965) | Sheep | Diet – oesophageal fistula; digestion – artificial rumen | Spring (Sept.–Oct.), native pasture | 61 | — | — | — | — | |

| | | Large sample macrodigestion | | | | | | | | | | | | |
|---|---|---|---|---|---|---|---|---|---|---|---|---|---|---|
| ID | OMD | Protein | Ether ex. | Lignin | Cellulose | Other CHO | Ash | Energy | CF | NFE | CWC | ADF | Misc. | Notes |
| – | — | — | — | — | — | — | — | — | — | — | — | — | — | 1.4 See notes under no. 20. |
| – | — | — | — | — | — | — | 11 | — | — | — | 72 | — | — | |
| 8 | 51 | 34 | — | — | — | — | — | 46 | — | — | — | — | — | See notes under no. 5. |
| – | — | — | — | — | — | — | — | — | — | — | — | — | — | 1.8 See notes under no. 20. |
| – | — | 17 | — | — | — | — | — | — | 60 | 64 | — | — | — | |
| – | 54 | — | — | — | — | — | — | — | — | — | — | — | — | |
| – | 53 | — | — | — | — | — | — | — | — | — | — | — | — | Veld grazing study 1958–69. |
| – | — | — | 43 | — | — | — | 51 | — | 62 | 69 | — | — | — | |
| 0 | 65 | 77 | — | — | — | — | 63 | — | 65 | — | — | — | 52 | Miscellaneous column is TDN. Tall fescue–orchard grass irrigated pasture mixture. Average of summer growing season. |
| – | — | — | — | — | — | — | — | — | — | — | — | — | — | See notes under no. 46. |
| – | — | — | — | — | — | — | — | — | — | — | — | — | — | Modification of Tilly & Terry (1963) technique. Average of 2 years; values higher in 1969 than in 1970. Heavy grazing significantly lower for dry matter digestibility. |
| 0 | — | — | — | — | — | — | — | — | — | — | — | — | — | See notes under no. 90. |
| 0 | — | — | — | — | — | — | — | — | — | — | — | — | — | See notes under no. 90. |
| 0 | — | — | — | — | — | — | — | — | — | — | — | — | — | See notes under no. 90. |
| – | — | — | — | — | — | — | — | — | — | — | — | — | — | See notes under no. 92. |

**Appendix 4.5** (*continued*)

| No. | Reference | Animal species | Methods | Season/treatment | Small sample microdigestion[a] *in vitro* DMD | OMD | Cellulose | *in vivo* DMD | OMD |
|---|---|---|---|---|---|---|---|---|---|
| **51** | Eadie (1970) | Sheep | Diet – hand clipped (?); digestion – faecal nitrogen index (?) | Spring–early summer (Apr.–June) | — | — | — | — | — |
| **52** | Engels *et al.* (1971) | Sheep | Diet – oesophageal fistula; digestion – *in vitro* | Spring (Sept.–Dec.) | — | 63 | — | — | — |
| **53** | Leigh *et al.* (1968) | Sheep | Diet – oesophageal fistula; digestion – *in vitro* | Spring (Sept.) | — | 75 | — | — | — |
| **54** | Rice *et al.* (1974) | Sheep | Diet – oesophageal fistula; digestion – rumen fluid from rumen-fistulated animals | Spring (Apr. and May) | — | — | — | — | — |
| **55** | Wilson *et al.* (1969) | Sheep | Diet – oesophageal fistula; digestion – artificial rumen | Spring (Dec.) | — | 50 | — | — | — |
| **56** | Wilson *et al.* (1969) | Sheep | Diet – oesophageal fistula; digestion – artificial rumen | Spring (Sept.) | — | 65 | — | — | — |
| **57** | Wilson (1974) | Sheep | Diet – oesophageal fistula; digestion – artificial rumen | Spring (Sept.) | — | 64 | — | — | — |
| **58** | Wilson (1974) | Sheep | Diet – oesophageal fistula; digestion – *in vitro* | Spring (Sept.), arid community, New South Wales, Australia | — | 64 | — | — | — |
| **59** | Wilson *et al.* (1975) | Sheep | Diet – oesophageal fistula; digestion – artificial rumen | Spring (Nov.), sheep-grazed plots | — | 54 | — | — | — |
| **60** | Wilson *et al.* (1975) | Sheep | Diet – oesophageal fistula; digestion – artificial rumen | Spring (Nov.), goat-grazed plots | — | 52 | — | — | — |
| **61** | Van Dyne & Lofgreen (1964) | Sheep | Diet – oesophageal fistula; faeces – total collection; digestion – lignin ratio | Early summer | — | — | — | — | — |
| **62** | Arnold *et al.* (1964) | Sheep | Diet – oesophageal fistula; faeces – total collection; digestion – faecal nitrogen index | Summer, ryegrass–sub-clover pasture | — | — | — | — | — |
| **63** | Arnold *et al.* (1965) | Sheep | Diet – oesophageal fistula, digestion – artificial rumen | Summer (Dec.–Feb.), native pasture | 61 | — | — | — | — |
| **64** | Bedell (1971) | Sheep | Oesophageal fistula, diet collection | Summer, sub-clover–tall fescue | — | — | — | — | — |
| **65** | Bedell (1971) | Sheep | Oesophageal fistula, diet collection | Summer, sub-clover–perennial ryegrass, light grazing | — | — | — | — | — |
| **66** | Bedell (1971) | Sheep | Oesophageal fistula, diet collection | Summer, sub-clover–tall fescue, heavy grazing | — | — | — | — | — |
| **67** | Bedell (1971) | Sheep | Oesophageal fistula, diet collection | Summer, sub-clover–perennial ryegrass, heavy grazing | — | — | — | — | — |
| **68** | Cook *et al.* (1961) | Sheep | Diet – oesophageal fistula; faeces – total and grab samples; digestion – lignin ratio | Summer, dry ewes | — | — | — | — | — |

| D | OMD | Pro-tein | Ether ex. | Lignin | Cel-lulose | Other CHO | Ash | Energy | CF | NFE | CWC | ADF | Misc. | Notes |
|---|---|---|---|---|---|---|---|---|---|---|---|---|---|---|
| | Large sample macrodigestion | | | | | | | | | | | | | |
| | 72 | — | — | — | — | — | — | — | — | — | — | — | — | Values estimated from diagram. Composite of 4-year study. Sheep set stocked on a Cheviot hill, England. |
| | — | — | — | — | — | — | — | — | — | — | — | — | — | |
| | — | — | — | — | — | — | — | — | — | — | — | — | — | See notes under no. 74. |
| | — | — | — | 19 | — | — | 14 | — | — | — | 65 | 53 | — | See notes under no. 2. |
| | — | — | — | — | — | — | — | — | — | — | — | — | — | See notes under no. 89. |
| | — | — | — | — | — | — | — | — | — | — | — | — | — | See notes under no. 89. |
| | — | — | — | — | — | — | — | — | — | — | — | — | — | See notes under no. 74. |
| | — | — | — | — | — | — | — | — | — | — | — | — | — | |
| | — | — | — | — | — | — | — | — | — | — | — | — | — | See notes under no. 131. |
| | — | — | — | — | — | — | — | — | — | — | — | — | — | See notes under no. 131. |
| | — | 36 | 30 | — | 45 | 58 | — | — | 39 | — | — | — | — | See notes under no. 9. |
| | — | — | — | — | — | — | — | — | — | — | — | — | — | See notes under no. 90. |
| | — | — | — | — | — | — | — | — | | — | — | — | — | See notes under no. 90. |
| | — | — | — | — | — | — | — | — | — | — | — | — | — | |
| | — | — | — | — | — | — | — | — | — | — | — | — | — | |
| | — | — | — | — | — | — | — | — | — | — | — | — | — | |
| | — | — | — | — | — | — | — | — | — | — | — | — | — | Average of 2 years. Willamette Valley, Oregon. Apr.–Aug. pasture. |
| | — | 44 | 40 | — | 52 | 68 | — | 44 | — | — | — | — | — | See notes under no. 69. |

**Appendix 4.5** (*continued*)

| No. | Reference | Animal species | Methods | Season/treatment | Small sample microdigestion[a] | | | | | |
|---|---|---|---|---|---|---|---|---|---|---|
| | | | | | *in vitro* | | | *in vivo* | | |
| | | | | | DMD | OMD | Cellulose | DMD | OMD | Ce lul |
| 69 | Cook *et al.* (1961) | Sheep | Diet – oesophageal fistula; faeces – total and grab sample; digestion – lignin ratio | Summer, wet ewes | — | — | — | — | — | — |
| 70 | Cook *et al.* (1967) | Sheep | Diet – oesophageal fistula; faeces – grab sample; digestion – lignin ratio | Summer, aspen rangeland | — | — | — | — | — | — |
| 71 | Cook *et al.* (1967) | Sheep | Diet – oesophageal fistula; faeces – grab sample; digestion – lignin ratio | Summer, sagebrush-grass type | — | — | — | — | — | — |
| 72 | Eadie (1970) | Sheep | Diet–hand clipped samples (?); digestion – faecal nitrogen index (?) | Summer (July–Sept.) | — | — | — | — | — | — |
| 73 | Engels *et al.* (1971) | Sheep | Diet – oesophageal fistula; digestion – *in vitro* | Summer (Jan.–Mar.) | — | 62 | — | — | — | — |
| 74 | Leigh *et al.* (1968) | Sheep | Diet – oesophageal fistula; digestion *in vitro* | Summer (Feb.) | 55 | — | — | — | — | — |
| 75 | Rice *et al.* (1974) | Sheep | Diet – oesophageal fistula; digestion – rumen fluid from rumen-fistulated animals | Summer | — | — | — | — | — | — |
| 76 | Wilson (1974) | Sheep | Diet – oesophageal fistula; digestion – *in vitro* | Summer (Feb.), arid community, Ivanhoe, New South Wales, Australia | — | 56 | — | — | — | — |
| 77 | Wilson (1974) | Sheep | Diet – oesophageal fistula; digestion – artificial rumen | Summer (Feb.) | — | 56 | — | — | — | — |
| 78 | Wilson *et al.* (1975) | Sheep | Diet – oesophageal fistula; digestion – artificial rumen | Summer (Feb.), sheep-grazed plots | 57 | — | — | — | — | — |
| 79 | Wilson *et al.* (1975) | Sheep | Diet – oesophageal fistula; digestion – artificial rumen | Summer (Feb.), goat-grazed plots | 58 | — | — | — | — | — |
| 80 | Van Dyne & Lofgreen (1964) | Sheep | Diet – oesophageal fistula; faeces – total collection; digestion – lignin ratio | Mid-summer | — | — | — | — | — | — |
| 81 | Van Dyne & Lofgreen (1964) | Sheep | Diet – oesophageal fistula; faeces – total collection; digestion – lignin ratio | Late summer | — | — | — | — | — | — |
| 82 | Arnold *et al.* (1964) | Sheep | Diet – oesophageal fistula; faeces – total collection; digestion – faecal nitrogen index method | Autumn, ryegrass-sub-clover pasture | — | — | — | — | — | — |
| 83 | Arnold *et al.* (1964) | Sheep | Diet – oesophageal fistula; faeces – total collection; digestion – faecal nitrogen index | Autumn, natural annual pasture | — | — | — | — | — | — |

| | | Large sample macrodigestion | | | | | | | | | | | | |
|---|---|---|---|---|---|---|---|---|---|---|---|---|---|---|
| MD | OMD | Protein | Ether ex. | Lignin | Cellulose | Other CHO | Ash | Energy | CF | NFE | CWC | ADF | Misc. | Notes |
| 45 | — | 43 | 34 | — | 50 | 69 | — | 43 | — | — | — | — | — | Used 45.5 DMD average for graph of DMD versus CP and lignin. Average of different periods during summer mid-July–mid-Sept. Mountain brush–grass range, northern Utah. Wet ewes consumed significantly more forage than dry ewes but digestibility did not vary appreciably. |
| — | — | 34 | 33 | — | 38 | 61 | — | 36 | — | — | — | — | 38 | See notes under no. 16. |
| — | — | 31 | 44 | — | 39 | 57 | — | 31 | — | — | — | — | 36 | See notes under no. 16. |
| — | 68 | — | — | — | — | — | — | — | — | — | — | — | — | See notes under no. 51. |
| — | — | — | — | — | — | — | — | — | — | — | — | — | — | |
| — | — | — | — | — | — | — | — | — | — | — | — | — | — | |
| — | — | — | — | — | — | — | — | — | — | — | — | — | — | *Dathonia–Stipa* grassland, Deniliquin, New South Wales, Australia. |
| — | — | — | — | 15 | — | — | 15 | — | — | — | 68 | 51 | — | See notes under no. 2. |
| — | — | — | — | — | — | — | — | — | — | — | — | — | — | Ivanhoe, New South Wales, Australia, *Stipa* community. |
| — | — | — | — | — | — | — | — | — | — | — | — | — | — | See notes under no. 131. |
| — | — | — | — | — | — | — | — | — | — | — | — | — | — | See notes under no. 131. |
| 45 | — | 25 | 41 | — | 49 | 59 | — | 42 | — | — | — | — | — | See notes under no. 9. |
| 46 | — | 23 | 41 | — | 51 | 60 | — | 42 | — | — | — | — | — | See notes under no. 9. |
| 75 | — | — | — | — | — | — | — | — | — | — | — | — | — | See notes under no. 90. |
| 75 | — | — | — | — | — | — | — | — | — | — | — | — | — | See notes under no. 90. |

**Appendix 4.5** (*continued*)

| No. | Reference | Animal species | Methods | Season/treatment | Small sample microdigestion[a] | | | | | |
|---|---|---|---|---|---|---|---|---|---|---|
| | | | | | *in vitro* | | | *in vivo* | | |
| | | | | | DMD | OMD | Cel- lulose | DMD | OMD | Ce lulo |
| **84** | Eadie (1970) | Sheep | Diet – hand-clipped samples (?); digestion – faecal nitrogen index (?) | Autumn (Oct.–Dec.) | — | — | — | — | — | — |
| **85** | Engels *et al.* (1971) | Sheep | Diet – oesophageal fistula; digestion – *in vitro* | Autumn (Apr.–June) | — | 52 | — | — | — | — |
| **86** | Leigh *et al.* (1968) | Sheep | Diet – oesophageal fistula; digestion – *in vitro* | Autumn (Apr.) | — | 56 | — | — | — | — |
| **87** | Leigh *et al.* (1968) | Sheep | Diet – oesophageal fistula; digestion – *in vitro* | Nov. | — | 59 | — | — | — | — |
| **88** | Rice *et al.* (1974) | Sheep | Diet – oesophageal fistula; digestion – rumen fluid from rumen-fistulated animals | Autumn (Oct.) | — | — | — | — | — | — |
| **89** | Wilson *et al.* (1969) | Sheep | Diet – oesophageal fistula; digestion – artificial rumen | Autumn (Mar.) | 51 | — | — | — | — | — |
| **90** | Arnold *et al.* (1964) | Sheep | Diet – oesophageal fistula; faeces – total collection; digestion – faecal nitrogen index | Winter, natural annual pasture | | | | | | |
| **91** | Arnold *et al.* (1964) | Sheep | Diet – oesophageal fistula; faeces – total collection; digestion – faecal nitrogen index | Winter ryegrass–sub-clover pasture | — | — | — | — | — | — |
| **92** | Arnold *et al.* (1965) | Sheep | Diet – oesophageal fistula; digestion – artificial rumen | Winter (June–July) native pasture | 46 | — | — | — | — | — |
| **93** | Cook *et al.* (1962) | Sheep | Diet – oesophageal fistula; faeces – total collection; digestion – lignin ratio | Winter, salt desert shrub range, poor condition | — | — | — | — | — | — |
| **94** | Cook *et al.* (1962) | Sheep | Diet – oesophageal fistula; faeces – total collection; digestion – lignin ratio | Winter, salt desert shrub range, good condition | — | — | — | — | — | — |
| **95** | Cook *et al.* (1965) | Sheep | Diet – oesophageal fistula; faeces – total collection; digestion – ratio | Winter, mountain range, poor condition | — | — | — | — | — | — |
| **96** | Cook *et al.* (1965) | Sheep | Diet – oesophageal fistula; faeces – total collection; digestion – lignin ratio | Winter, mountain range, good condition | — | — | — | — | — | — |

| | Large sample macrodigestion | | | | | | | | | | | | | |
|---|---|---|---|---|---|---|---|---|---|---|---|---|---|---|
| D | OMD | Pro-tein | Ether ex. | Lignin | Cel-lulose | Other CHO | Ash | Energy | CF | NFE | CWC | ADF | Misc. | Notes |
| - | — | — | — | — | — | — | — | — | — | — | — | — | — | See notes under no. 51. |
| - | — | — | — | — | — | — | — | — | — | — | — | — | — | |
| - | — | — | — | — | — | — | — | — | — | — | — | — | — | See notes under no. 74. |
| - | — | — | — | — | — | — | — | — | — | — | — | — | — | See notes under no. 74. |
| - | — | — | — | 20 | — | — | 17 | — | — | — | 55 | 55 | — | See notes under no. 2. |
| - | — | — | — | — | — | — | — | — | — | — | — | — | — | *Atriplex* shrub community, Wanganella, Australia. Winter rainfall zone. |
| 5 | — | — | — | — | — | — | — | — | — | — | — | — | — | Annual pastures with subterranean clover and voluntary grasses; perennial pasture received superphosphate. Little difference in digestibility among stocking rates within a pasture; data for 1958–69; wider variability between high and low in other years. |
| 5 | — | — | — | — | — | — | — | — | — | — | — | — | — | See notes under no. 90. |
| - | — | — | — | — | — | — | — | — | — | — | — | — | — | Native range compared to improved pastures; improved pastures always higher quality diets. Cockskfoot gave DMD values of 53, 74, and 65 in the respective seasons sampled. |
| - | — | 38 | 27 | — | 42 | 56 | — | — | — | — | — | — | — | See notes under no. 93. |
| - | — | 42 | 21 | — | 48 | 59 | — | 39 | — | — | — | — | — | Study Nov.–March 1957 and 1959, winter grazing season, southwestern Utah. Small paddocks grazed; data average of 2 locations; two 5-day collection periods. |
| 1 | — | 46 | 3 | — | 52 | 73 | — | 48 | — | — | — | — | 47 | See notes under no. 96. |
| 2 | — | — | 2 | — | 56. | 73 | — | 48 | — | — | — | — | 48 | Miscellaneous data are for TDN. Mixture of sagebrush–grass, mountain browse, and aspen types. Data for June–Sept. 1960–2; data are average for early and late periods within the grazing period. |

## Appendix 4.5 (*continued*)

| No. | Reference | Animal species | Methods | Season/treatment | Small sample microdigestion[a] *in vitro* DMD | OMD | Cel-lulose | *in vivo* DMD | OMD | C lul |
|---|---|---|---|---|---|---|---|---|---|---|
| **97** | Eadie (1970) | Sheep | Diet – hand-clipped samples (?); digestion – faecal nitrogen index (?) | Winter (Jan.–Mar.) | — | — | — | — | — | – |
| **98** | Engels *et al.* (1971) | Sheep | Diet – oesophageal fistula; digestion – *in vitro* | Winter (July–Aug.) | — | 50 | — | — | — | – |
| **99** | Leigh *et al.* (1968) | Sheep | Diet – oesophageal fistula; digestion – *in vitro* | Winter (June) | — | 61 | — | — | — | – |
| **100** | Wilson *et al.* (1969) | Sheep | Diet – oesophageal fistula; digestion – artificial rumen | Winter (June) | — | 52 | — | — | — | – |
| **101** | Wilson (1974) | Sheep | Diet – oesophageal fistula; digestion – artificial rumen | Winter (July) | — | 58 | — | — | — | – |
| **102** | Wilson (1974) | Sheep | Diet – oesophageal fistula; digestion – *in vitro* | Winter (July), arid community, Ivanhoe, New South Wales, Australia | — | 58 | — | — | — | – |
| **103** | Wilson *et al.* (1975) | Sheep | Diet – oesophageal fistula; digestion – artificial rumen | Winter (May–June) goat-grazed plots | — | 47 | — | — | — | – |
| **104** | Wilson *et al.* (1975) | Sheep | Diet – oesophageal fistula; digestion – artificial rumen | Winter (May–June), sheep-grazed plots | — | 51 | — | — | — | – |
| **105** | Van Schalkwyk *et al.* (1968) | Sheep | Forage fed, total faecal collection | Spring–summer | — | — | — | — | — | – |
| **106** | Van Schalkwyk *et al.* (1968) | Sheep | Forage fed, total faecal collection | Autumn–winter | — | — | — | — | — | – |
| **107** | Grimes *et al.* (1966) | Sheep | Diet – oesophageal fistula; digestion – *in vitro* | Year, short management | — | 78 | — | — | — | – |
| **108** | Grimes *et al.* (1966) | Sheep | Diet – oesophageal fistula; digestion – *in vitro* | Year, long management | — | 76 | — | — | — | – |
| **109** | Grimes *et al.* (1966) | Sheep | Diet – oesophageal fistula; digestion – *in vitro* | Year, grass pasture | — | 77 | — | — | — | – |
| **110** | Grimes *et al.* (1966) | Sheep | Diet – oesophageal fistula; digestion – *in vitro* | Year, grass and clover pasture | — | 78 | — | — | — | – |
| **111** | Cook *et al.* (1956) | Sheep | Diet – hand collection; faeces – total collection; digestion – lignin ratio | Crested wheatgrass, mature seed | — | — | — | — | — | – |
| **112** | Cook *et al.* (1956) | Sheep | Diet – hand collection; faeces – total collection; digestion – lignin ratio | Crested wheatgrass, average | — | — | — | — | — | – |
| **113** | Cook *et al.* (1956) | Sheep | Diet – hand collection; faeces – total collection; digestion – lignin ratio | Crested wheatgrass, fifth leaf | — | — | — | — | — | – |

| | | Large sample macrodigestion | | | | | | | | | | | | |
|---|---|---|---|---|---|---|---|---|---|---|---|---|---|---|
| MD | OMD | Protein | Ether ex. | Lignin | Cellulose | Other CHO | Ash | Energy | CF | NFE | CWC | ADF | Misc. | Notes |
| — | — | — | — | — | — | — | — | — | — | — | — | — | — | See notes under no. 51. |
| — | — | — | — | — | — | — | — | — | — | — | — | — | — | |
| — | — | — | — | — | — | — | — | — | — | — | — | — | — | See notes under no. 74. |
| — | — | — | — | — | — | — | — | — | — | — | — | — | — | See notes under no. 89. |
| — | — | — | — | — | — | — | — | — | — | — | — | — | — | See notes under no. 77. |
| — | — | — | — | — | — | — | — | — | — | — | — | — | — | |
| — | — | — | — | — | — | — | — | — | — | — | — | — | — | See notes under no. 131. |
| — | — | — | — | — | — | — | — | — | — | — | — | — | — | See notes under no. 131. |
| — | — | — | — | — | — | — | — | — | — | — | — | — | 42 | See notes under no. 106. |
| — | — | — | — | — | — | — | — | — | — | — | — | — | 30 | Miscellaneous data are TDN. Based on forage cut and fed to the animals. Dominant plant *Themeda triandra*. 2 year study. |
| — | — | — | — | — | — | — | — | — | — | — | — | — | — | Short management means pasture maintained at 2–4 in. |
| — | — | — | — | — | — | — | — | — | — | — | — | — | — | Long management means; unrestricted growth. |
| — | — | — | — | — | — | — | — | — | — | — | — | — | — | |
| — | — | — | — | — | — | — | — | — | — | — | — | — | — | |
| — | — | 59 | 6 | — | 58 | 72 | — | 55 | — | — | — | — | — | See notes under no. 113. |
| — | — | 61 | 18 | — | 65 | 73 | — | 58 | — | — | — | — | — | See notes under no. 113. |
| — | — | 80 | 55 | — | 80 | 89 | — | 77 | — | — | — | — | — | Plants consumed in spring and early summer. Average crop protein content: wheatgrass 18, tall wheatgrass 12, intermediate wheatgrass 11, beardless wheatgrass 10, western wheatgrass 10, Russian thistle 15, smotherweed 17. Grasses used 10 % (western wheatgrass) up to 60 % (intermediate wheatgrass). |

## Appendix 4.5 (*continued*)

| No. | Reference | Animal species | Methods | Season/treatment | Small sample microdigestion[a] *in vitro* DMD | OMD | Cel-lulose | *in vivo* DMD | OMD | C lu |
|---|---|---|---|---|---|---|---|---|---|---|
| **114** | Cook *et al.* (1956) | Sheep | Diet – hand collection; faeces – total collection; digestion – lignin ratio | Western wheatgrass, average | — | — | — | — | — | |
| **115** | Cook *et al.* (1956) | Sheep | Diet – hand collection; faeces – total collection; digestion – lignin ratio | Beardless wheatgrass, average | — | — | — | — | — | |
| **116** | Cook *et al.* (1956) | Sheep | Diet – hand collection; faeces – total collection; digestion – lignin ratio | Intermediate wheatgrass, average | — | — | — | — | — | |
| **117** | Cook *et al.* (1956) | Sheep | Diet – hand collection; faeces – total collection; digestion – lignin ratio | Russian thistle, average | — | — | — | — | — | |
| **118** | Cook *et al.* (1956) | Sheep | Diet – hand collection; faeces – total collection; digestion – lignin ratio | Tall wheatgrass, average | — | — | — | — | — | |
| **119** | Cook *et al.* (1956) | Sheep | Diet – hand collection; faeces – total collection; digestion – lignin ratio | Pubescent wheatgrass, average | — | — | — | — | — | |
| **120** | Cook *et al.* (1956) | Sheep | Diet – hand collection; faeces – total collection; digestion – lignin ratio | Smotherweed, average | — | — | — | — | — | |
| **121** | Bissell & Weir (1957) | Sheep | Cut forage fed, total faecal collection | Chemise | — | — | — | — | — | |
| **122** | Bissell & Weir (1957) | Sheep | Cut forage fed, total faecal collection | Chemise | — | — | — | — | — | |
| **123** | Foot & Romberg (1965) | Sheep | Metabolism balance trials | Oat straw | — | — | — | — | — | |
| **124** | Foot & Romberg (1965) | Sheep | Metabolism balance trials | Alfalfa hay | — | — | — | — | — | |
| **125** | Joshi (1966) | Sheep | Faeces – total collection; digestibility – in stall feeding | Late summer | — | — | — | — | — | |
| **126** | Wilson *et al.* (1975) | Goats | Diet – oesophageal fistula; digestion – artificial rumen | Spring (Nov.), sheep-grazed plots | — | 50 | — | — | — | |
| **127** | Wilson *et al.* (1975) | Goat | Diet – oesophageal fistula; digestion – artificial rumen | Spring (Nov.), goat-grazed plots | — | 53 | — | — | — | |
| **128** | Wilson *et al.* (1975) | Goat | Diet – oesophageal fistula; digestion – artificial rumen | Summer (Feb.), sheep-grazed plots | — | 57 | — | — | — | |
| **129** | Wilson *et al.* (1975) | Goat | Diet – oesophageal fistula; digestion – artificial rumen | Summer (Feb.), goat-grazed plots | — | 56 | — | — | — | |
| **130** | Wilson *et al.* (1975) | Goat | Diet – oesophageal fistula; digestion – artificial rumen | Winter (May–June), goat-grazed plots | — | 48 | — | — | — | |
| **131** | Wilson *et al.* (1975) | Goat | Diet – oesophageal fistula; digestion – artificial rumen | Winter (May–June), sheep-grazed plots | — | 51 | — | — | — | |

| Large sample macrodigestion | | | | | | | | | | | | | | |
|---|---|---|---|---|---|---|---|---|---|---|---|---|---|---|
| DMD | OMD | Protein | Ether ex. | Lignin | Cellulose | Other CHO | Ash | Energy | CF | NFE | CWC | ADF | Misc. | Notes |
| — | — | 61 | 32 | — | 73 | 75 | — | 64 | — | — | — | — | — | See notes under no. 113. |
| — | — | 54 | 30 | — | 71 | 73 | — | 61 | — | — | — | — | — | See notes under no. 113. |
| — | — | 61 | 30 | — | 74 | 74 | — | 62 | — | — | — | — | — | See notes under no. 113. |
| — | — | 79 | 34 | — | 63 | 74 | — | 68 | — | — | — | — | — | See notes under no. 113. |
| — | — | 64 | 35 | — | 72 | 70 | — | 60 | — | — | — | — | — | See notes under no. 113. |
| — | — | 62 | 23 | — | 72 | 74 | — | 62 | — | — | — | — | — | See notes under no. 113. |
| — | — | 76 | 34 | — | 55 | 74 | — | 59 | — | — | — | — | — | See notes under no. 113. |
| — | 53 | 1 | 46 | — | — | — | — | — | 20 | 56 | — | — | 41 | See notes under no. 142. |
| — | 57 | 30 | 26 | — | — | — | — | — | 13 | 63 | — | — | 48 | See notes under no. 142. |
| 39 | — | — | — | — | — | — | — | — | 42 | — | — | — | — | See notes under no. 124. |
| 53 | — | — | — | — | — | — | — | — | 46 | — | — | — | — | Body weights of sheep about 39 kg, of kangaroos, 12 kg. Dry matter intake, in g · kg metabolic wt$^{-1}$, are, for alfalfa and oat straw respectively, sheep 72 and 29 and kangaroo 53 and 40. |
| 66 | 67 | 68 | 29 | — | — | — | — | — | — | 60 | 72 | — | 68 | Miscellaneous is total carbohydrates. |
| — | — | — | — | — | — | | — | — | — | — | — | — | — | See notes under no. 131. |
| — | — | — | — | — | — | — | — | — | — | — | — | — | — | See notes under no. 131. |
| — | — | — | — | — | — | — | — | — | — | — | — | — | — | See notes under no. 131. |
| — | — | — | — | — | — | — | — | — | — | — | — | — | — | See notes under no. 131. |
| — | — | — | — | — | — | — | — | — | — | — | — | — | — | See notes under no. 131. |
| — | — | — | — | — | — | — | — | — | — | — | — | — | — | May–June samples average of various stocking rates, 5 values; other seasons and grazing |

**Appendix 4.5** (*continued*)

| No. | Reference | Animal species | Methods | Season/treatment | Small sample microdigestion[a] *in vitro* DMD | OMD | Cellulose | *in vivo* DMD | OMD | Cellulose |
|---|---|---|---|---|---|---|---|---|---|---|
| **131** | Wilson *et al.* (1975) (*continued*) | | | | | | | | | |
| **132** | C. Schwartz, J. Nagy & R. Rice (unpublished) | Pronghorn | Hand-collected samples | Year | — | — | — | 69 | — | — |
| **133** | Smith & Malachek (1974) | Pronghorn | Diet – ocular estimates, hand collection; faeces – hand-collected, digestion – lignin ratio | May–Sept., salt desert shrub, big sagebrush, black sagebrush range | — | — | — | 69 | — | — |
| **134** | Rice *et al.* (1974) | Buffalo | Diet – oesophageal fistula; digestion – rumen fluid from rumen-fistulated animals | Spring (Apr. and May) | — | — | — | — | — | — |
| **135** | Rice *et al.* (1974) | Buffalo | Diet – oesophageal fistula; digestion – rumen fluid from rumen-fistulated animals | Summer (June, July and Aug.) | — | — | — | — | — | — |
| **136** | Rice *et al.* (1974) | Buffalo | Diet – oesophageal fistula; digestion – rumen fluid from rumen-fistulated animals | Autumn (Oct.) | — | — | — | — | — | — |
| **137** | Boeker *et al.* (1972) | Mule deer | In-vitro, rumen fluid from deer collected in field | Spring | 45 | — | — | — | — | — |
| **138** | Boeker *et al.* (1972) | Mule deer | In-vitro, rumen fluid from deer collected in field | Summer | 37 | — | — | — | — | — |
| **139** | Boeker *et al.* (1972) | Mule deer | In-vitro, rumen fluid from deer collected in field | Autumn | 43 | — | — | — | — | — |
| **140** | Boeker *et al.* (1972) | Mule deer | In-vitro, rumen fluid from deer collected in field | Winter | 23 | — | — | — | — | — |
| **141** | Bissell & Weir (1957) | Deer | Cut forage fed, total faecal collection | Live oak | — | — | — | — | — | — |
| **142** | Bissell & Weir (1957) | Deer | Cut forage fed, total faecal collection | Chemise | — | — | — | — | — | — |
| **143** | Bissell *et al.* (1955) | Black-tailed deer | Cut forage fed, total faecal collection | Live oak | — | — | — | — | — | — |

**Appendix 4.5** (*continued*)

| Large sample macrodigestion | | | | | | | | | | | | | | |
|---|---|---|---|---|---|---|---|---|---|---|---|---|---|---|
| DMD | OMD | Protein | Ether ex. | Lignin | Cellulose | Other CHO | Ash | Energy | CF | NFE | CWC | ADF | Misc. | Notes |
| | | | | | | | | | | | | | | treatments from 1 value. Lower digestibility values occurred at higher stocking rates. Study in *Casuarina cristata–Heterodendron oleifolium* woodland community, western New South Wales, Australia. Comparable digestibility for goats and sheep (Ivanhoe, New South Wales study). |
| — | — | — | — | — | — | — | — | — | — | — | 76 | — | — | Number read from graph. |
| — | — | — | — | — | — | — | — | — | — | — | — | — | — | Check on method or digestion calculation and faecal collection. Data averaged over 2 years. |
| — | — | — | — | 10 | — | — | 21 | — | — | — | 76 | 64 | — | See notes under no. 2. |
| — | — | — | — | 8 | — | — | 19 | — | — | — | 74 | 62 | — | See notes under no. 2. |
| — | — | — | — | 11 | — | — | 21 | — | — | — | 81 | 67 | — | See notes under no. 2. |
| — | — | — | — | — | — | — | — | — | — | — | — | — | — | See notes under no. 140. |
| — | — | — | — | — | — | — | — | — | — | — | — | — | — | See notes under no. 140. |
| — | — | — | — | — | — | — | — | — | — | — | — | — | — | See notes under no. 140. |
| — | — | — | — | — | — | — | — | — | — | — | — | — | — | Hand-composited diet sample digested; digestibility values in winter well below general levels for most ruminants. |
| — | 43 | 11 | 33 | — | — | — | — | — | 10 | 56 | — | — | 38 | See notes under no. 141. |
| — | 54 | 37 | 48 | — | — | — | — | — | 21 | 58 | — | — | 48 | Miscellaneous is TDN. Chemise and interior live oak digestibility far less than for alfalfa hays. Protein content of chemise 7.1, of live oak 7.4. Sheep had higher digestibility for organic matter, but not for many nutrients. Forage collected May–June; new growth. Colombian black-tailed deer utilized. |
| — | — | 33 | — | — | — | — | — | — | — | — | — | — | 34 | Miscellaneous is TDN. Forage fresh daily. In 6 in. of the stem and forage, crude protein percentage was chemise 11, bitterbrush 9, sagebrush 11. Sagebrush not consumed enough |

**Appendix 4.5** (*continued*)

| No. | Reference | Animal species | Methods | Season/treatment | Small sample microdigestion[a] *in vitro* DMD | OMD | Cel-lulose | *in vivo* DMD | OMD | Cel-lulose |
|---|---|---|---|---|---|---|---|---|---|---|
| **143** | Bissell *et al.* (1955) (*continued*) | | | | | | | | | |
| **144** | Bissell *et al.* (1955) | Black-tailed deer | Cut forage fed, total faecal collection | Sagebrush | | | | | | |
| **145** | Bissell *et al* (1955) | Black-tailed deer | Cut forage fed, total faecal collection | Bitterbrush | — | — | — | — | — | — |
| **146** | Bissell *et al.* (1955) | Black-tailed deer | Cut forage fed, total faecal collection | Chemise | — | — | — | — | — | — |
| **147** | Dietz *et al.* (1962) | Mule deer | Forage fed, total faecal collection | Bitterbrush | — | — | — | — | — | — |
| **148** | Dietz *et al.* (1962) | Mule deer | Forage fed, total faecal collection | Mountain mahogany | — | — | — | — | — | — |
| **149** | Dietz *et al.* (1962) | Mule deer | Forage fed, total faecal collection | Big sagebrush, autumn collection | — | — | — | — | — | — |
| **150** | Smith (1950) | Mule deer | Cut forage fed, total faecal collection | Sagebrush digestibility | — | — | — | — | — | — |
| **151** | Smith (1952) | Mule deer | Cut forage fed, total faecal collection | Curlleaf mountain mahogany | — | — | — | — | — | — |
| **152** | Smith (1952) | Mule deer | Cut forage fed, total faecal collection | Juniper | — | — | — | — | — | — |
| **153** | Smith (1952) | Mule deer | Cut forage fed, total faecal collection | Bitterbrush | — | — | — | — | — | — |
| **154** | Smith (1957) | Mule deer | Forage fed, total faecal collection | Gamble oak | — | — | — | — | — | — |
| **155** | Smith (1957) | Mule deer | Forage fed, total faecal collection | Juniper | — | — | — | — | — | — |
| **156** | Smith (1957) | Mule deer | Forage fed, total faecal collection | Cliffrose | — | — | — | — | — | — |
| **157** | Smith (1957) | Mule deer | Forage fed, total faecal collection | Birchleaf mahogany | — | — | — | — | — | — |
| **158** | Smith (1957) | Mule deer | Forage fed, total faecal collection | Chokecherry | — | — | — | — | — | — |
| **159** | Ullrey *et al.* (1964) | White-tailed deer | Cut forage fed, total faecal collection | Cedar (*Thuja*) | — | — | — | — | — | — |
| **160** | Jacobsen & Skjenneberg (1975) | Caribou | Drylot digestion trial | Fed lichen | — | — | — | — | — | — |
| **161** | Foot & Romberg (1965) | Kangaroo | Metabolism balance trials | Alfalfa hay | — | — | — | — | — | — |
| **162** | Foot & Romberg (1965) | Kangaroo | Metabolism balance trials | Oat straw | — | — | — | — | — | — |

| | | | | | | | | | | | | | | |
|---|---|---|---|---|---|---|---|---|---|---|---|---|---|---|
| | | Large sample macrodigestion | | | | | | | | | | | | |
| DMD | OMD | Protein | Ether ex. | Lignin | Cellulose | Other CHO | Ash | Energy | CF | NFE | CWC | ADF | Misc. | Notes |
| | | | | | | | | | | | | | | to supply energy requirements. Digestibility efficiency did not vary between black-tailed deer and mule deer. |
| — | — | 42 | — | — | — | — | — | — | — | — | — | — | 56 | See notes under no. 143. |
| — | — | 57 | — | — | — | — | ... | — | — | — | — | — | 55 | See notes under no. 143. |
| — | — | * | — | — | — | — | — | — | — | — | — | — | 59 | See notes under no. 143. |
| — | — | 31 | 44 | — | — | — | — | — | 30 | 61 | — | — | 53 | See notes under no. 149. |
| — | — | — | 33 | — | — | — | — | 40 | 32 | 62 | — | — | — | Considerable weight loss when aspen was fed. Winter forage. Average of 3 trials. See notes under no. 149. |
| — | — | 53 | 77 | — | — | — | — | — | 40 | 52 | — | — | 59 | Big sagebrush determined by difference method feeding with alfalfa. All plants harvested in the autumn, less in the winter. Data for bitterbrush and mountain mahogany are for a single feed. Percentage protein: big sagebrush 10, bitterbrush 9, mountain mahogany 8. |
| — | — | 67 | 68 | — | — | — | — | — | 51 | 78 | — | — | — | Average of 6 trials. Protein composition 11 %. Daily consumption 2–6 % of body weight. |
| — | — | 54 | 43 | — | — | — | — | — | 36 | 76 | — | — | 65 | |
| — | — | 10 | 54 | — | — | — | — | — | 30 | 66 | — | — | — | |
| — | — | 36 | 53 | — | — | — | — | — | 18 | 36 | — | — | 45 | |
| — | — | 11 | 38 | — | — | — | — | — | 17 | 54 | — | — | — | See notes under no. 155. |
| — | — | 17 | 59 | — | — | — | — | — | 34 | 70 | — | — | — | Browse plants cut in winter condition. Percentage protein in forage fed varied from 5.4 to 9.9. |
| — | — | 40 | 48 | — | — | — | — | — | 4 | 59 | — | — | — | See notes under no. 155. |
| — | — | 48 | 36 | — | — | — | — | — | 32 | 60 | — | — | — | See notes under no. 155. |
| — | — | 48 | 23 | — | — | — | — | — | 9 | 56 | — | — | — | See notes under no. 155. |
| 42 | — | 10 | — | — | — | — | — | — | — | — | — | — | — | |
| — | 75 | — | 66 | — | — | — | — | — | 75 | 80 | — | — | — | *Cladonina* was fed, containing 3.1 % crude protein, 61.2 % MID, 37.6 % CF, 1.7 % ether extract and 1.4 % ash. |
| 54 | — | — | — | — | — | — | — | — | 33 | — | — | — | — | See notes under no. 124. |
| 36 | — | — | — | — | — | — | — | — | 29 | — | — | — | — | See notes under no. 124. |

* Negative

# 5. Invertebrate predator subsystem

A. KAJAK

## Contents



## Introduction

This chapter is divided into three main parts. The first part deals with the abundance of invertebrate predators in different ecosystems. The ecosystems considered are mainly grassland ecosystems, i.e. steppes, prairies, meadows, pastures, savannas, semideserts and deserts. Increasing intensification of agriculture in the world in recent years has affected not only crop fields but also grasslands. Unfortunately the data on the effects of management on the trophic structure of an ecosystem, particularly on predator abundance, are scarce and are scattered in many papers, but I have attempted to present here all the information available on the effects of different grassland treatments on predators. The majority of the data refers, however, to the temperate zone of the northern hemisphere, and reports on grassland investigations in other regions unfortunately lack data on predators.

The ecological role of the activities of predators in ecosystems is discussed in the second part of the chapter. The term 'role' is used here to cover the exploitation rate of prey production by predators and its effect on prey population density and dynamics, as well as on some processes in an ecosystem. One section deals with the effect of litter and soil predators on saprophages and, indirectly, on the rate of decomposition of organic matter. The discussion is based on the author's experimental results and on data from the literature on the variability in the proportion of predators in habitats of differing soil fertility.

The third part is devoted to the bioenergetics of common predatory species and to the efficiency of energy transfers at the individual level.

Information on the number of predators in grasslands, and even more so on their function, is limited, and here analysis is concentrated on the results of studies on energy flow through ecosystems carried out within the International Biological Programme. It is characteristic, however, at least as regards invertebrates, that the higher the trophic level the lower not only the energy input but also the number of studies on it. The most detailed studies on the effect of predation on prey numbers have been concerned with different pest species and their natural enemies, so attention has been focused mainly on prey-specific predators, in as much as they are the species generally used for biological pest control (Huffaker, Messenger & DeBach, 1971). However, in rich and stable biocoenoses, such as most grassland ecosystems, polyphagous predators that feed on many prey species are of greater importance. The diet of these predators depends largely on the actual trophic conditions in the habitat, and they prey upon the most abundant organisms available at any one time. They are relatively long-lived animals (from several months to several years). Their annual population dynamics are related not to a particular prey species but to the total abundance of the prey trophic level, and do not vary much from year to year.

Classification of animals into particular trophic levels follows that used by the authors of the papers cited. Where authors refer only to the species composition of the fauna and do not give information on the trophic level then only animals usually regarded as carnivorous are classed as predators.

The exception is ants, which were included with predators (if not already assigned to another group by the author) although many species of ants living in grasslands are omnivorous. However they feed mainly on animal food (Petal & Breymeyer, 1968; McDaniel, 1971; Athias, Josens & Lavelle, 1974), and make a considerable contribution to the reduction of arthropod populations (Petal, Andrzejewska, Breymeyer & Olechowicz, 1971). Parasitic insects (Hymenoptera, Diptera) were classed as prey-specific predators because they cause the death of their prey.

Greatest emphasis is given here to predatory polyphagous arthropods associated with the litter and top soil layers. These are mainly members of the Araneae, Chilopoda, Formicidae, Carabidae and Staphylinidae, and together form a considerable part of the total biomass of predators in grassland ecosystems.

## Predator resources and their contribution to total invertebrate biomass

### *Biomass of predators above ground*

The analysis of aboveground predators is based on data from 20 ecosystems (Table 5.1; Fig. 5.1) presented in several papers (Crossley & Howden, 1961; Zlotin, 1966; Menhinick, 1967; Olechowicz, 1970; Kajak, 1971; Lewis, 1971; Bourlière & Hadley, 1970; Gillon, 1974). Only unmanaged ecosystems are considered because management has a significant effect on the relative proportions of herbivores and predators (see below).

The biomass (dry weight) of predatory invertebrates above ground in the various types of grassland ecosystem ranges from fractions of a milligram per square metre (0.6, 0.7 $mg \cdot m^{-2}$) to several tens of milligrams per square metre (81 $mg \cdot m^{-2}$; Olechowicz, 1970; Kajak, 1971; Lewis, 1971). Most of the sampling techniques used provide underestimates of the number of animals, particularly those that are either very small (e.g. parasitic Hymenoptera), live inside plants or are very mobile. To estimate the density of these animals additional, special techniques are required. What is worse there are no standardized methods for collecting invertebrates from the field layer, and different techniques have been used at different research centres. A more detailed description of collecting methods, together with the species composition of predators and short descriptions of habitats studied, is given in Appendix 5.1 (p. 588). The best estimates of density are obtained by collecting with a suction apparatus (as used in six of the ecosystems considered), but this method is not equally efficient for all animal groups (Andrzejewska & Kajak, 1966). Despite the different sampling techniques used, discrepancies in the estimates of predator biomass between similar habitat types were not large.

The smallest predator biomass was found in the driest habitats, i.e. the desert grasslands, semidesert and dry steppes (Table 5.1). The range in these communities was 0.6 to 1.6 mg dry $wt \cdot m^{-2}$. Communities of arid prairies have predator biomasses of the order of several milligrams per square metre

Table 5.1. *Aboveground biomass of predatory invertebrates in grassland ecosystems*

| | Type of grassland | Biomass of predators (mg dry wt·m$^{-2}$) | % total invertebrate biomass | Primary production (g dry wt·m$^{-2}$·yr$^{-1}$) | References |
|---|---|---|---|---|---|
| Deserts and semideserts | 1. Desert grassland, Jornada, New Mexico | 0.6 | 27.2 | 172 | Lewis (1971); Sims & Singh (1971) |
| | 2. Desert grassland, Tien Shan | 1.57 | 10.9 | 13 | Zlotin (1966, 1975) |
| | 3. Semidesert, Tien Shan | | | | Zlotin (1966, 1975) |
| | (*a*) | 0.90 | 4.7 | 63 | |
| | (*b*) | 1.10 | | 102 | |
| Arid steppes and prairies | 4. Dry mountain steppe Tien Shan | | | | Zlotin (1966, 1975) |
| | (*a*) | 0.50–0.76 | 4.8 | 49 | |
| | (*b*) | 0.63–0.93 | | 87 | |
| | 5. Shortgrass plain, Pantex, Texas | 2.3 | 7.0 | 107 | Lewis (1971); Sims & Singh (1971) |
| | 6. Shortgrass plain, Pawnee, Colorado | 4.9 | 21.5 | 142 | Lewis (1971); Sims & Singh (1971) |
| | 7. Tropical savanna, Lamto, Ivory Coast | 7.0 | 32.3 | 831 | Bourlière & Hadley (1970); Gillon (1974); Cesar & Menaut (1974) |
| | 8. *Sericea lespedeza* stand, South Carolina | 8.0 | 13.8 | 550 | Menhinick (1967) |
| Moist prairies and meadows | 9. Moist alpine meadows Tien Shan | | | | Zlotin (1966, 1975) |
| | (*a*) | 4.1 | 8.4 | 127 | |
| | (*b*) | 15.0 | | 214 | |
| | 10. Mixed grass prairie, Cottonwood, S. Dakota | 28.6 | 35.9 | 197 | Lewis (1971); Sims & Singh (1971) |
| | 11. True prairie, Osage, Oklahoma | 17.3 | 21.9 | 337 | Lewis (1971) |
| Marshy areas | 12. Marsh, Tien Shan | 11.2 | 46.0 | 200–290 | Zlotin (1966, 1975) |
| | 13. Sedge rush lake bed, White Oak Lake | 36 | 18.2 | | Crossley & Howden (1961) |
| | 14. Marshy meadow, Kampinos Forest, Poland | 81.0 | 45.8 | 476 | Kajak (1971); Olechowicz 1970); Traczyk (1971) |

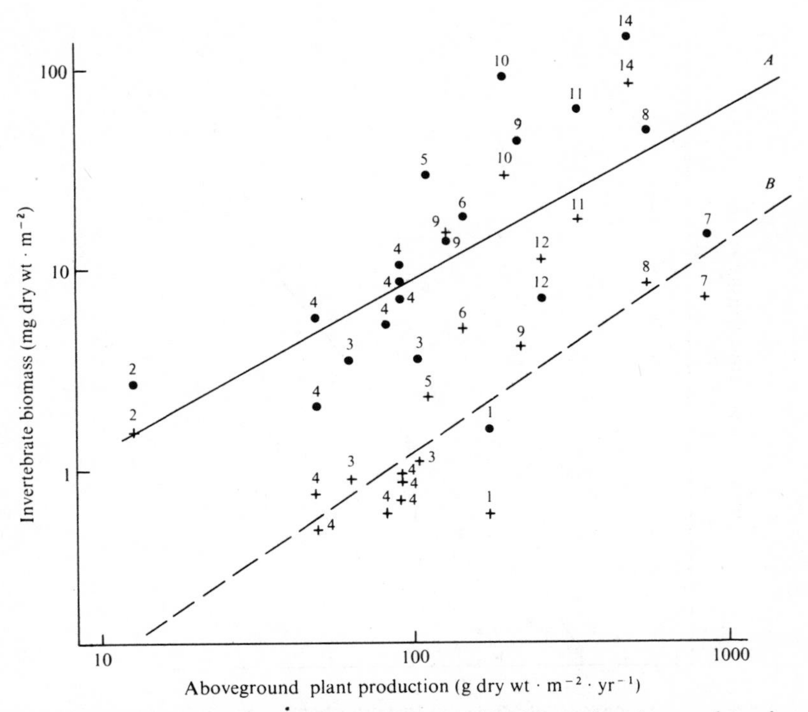


Fig. 5.1. Relationship between plant production and herbivore (●, *A*) or predator (+, *B*) biomass. Numbers represent type of grassland as in Table 5.1.

but usually below 5 mg. Fertile meadows and savanna support predator biomasses of 4 to 29 mg dry wt·m$^{-2}$. The largest biomass was found in marshy meadows (11 to 81 mg dry wt·m$^{-2}$).

On the basis of the data presented in Table 5.1 predator and herbivore biomass are found to be positively correlated with green plant production. The regression equation for herbivore biomass versus plant production is:

$$B_H = 0.18P^{0.85},$$

where $B_H$ is herbivore biomass in mg dry wt·m$^{-2}$ and $P$ is plant production in g dry wt·m$^{-2}$·yr$^{-1}$. The significance of the regression was tested by variance analysis based on $F$ statistics. The regression is significant with probability $P < 0.01$ (correlation coefficient $r = 0.64$, $F_{emp} = 19.13$, $F_{0.01} = 8.29$).

The relationship between plant production and the biomass of predators is given by the equation:

$$B_P = 0.01P^{1.05} \quad (r = 0.66, \quad F_{emp} = 18.89, \quad F_{0.01} = 8.29),$$

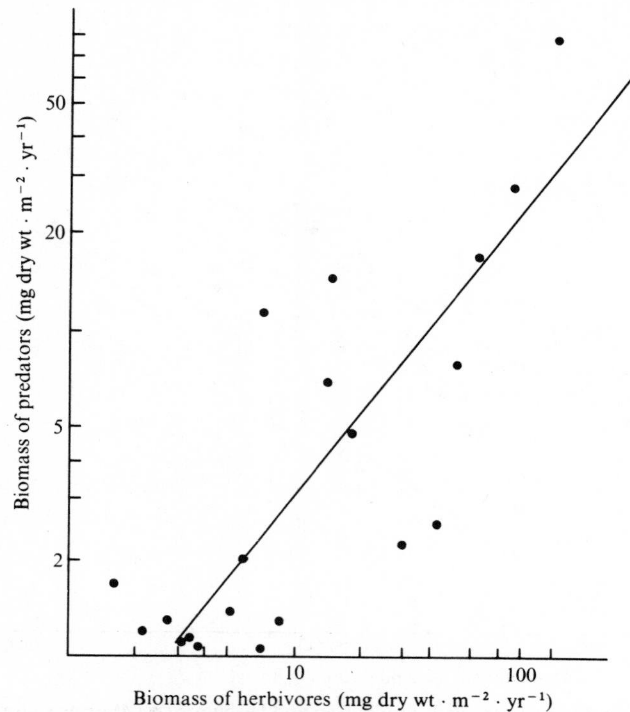


Fig. 5.2. Relationship between herbivore biomass and predator biomass.

where $B_P$ is the biomass of predators in mg dry wt$\cdot$m$^{-2}$. The regression function expressing the relation between herbivore and predator biomass has also been calculated:

$$B_P = 0.28 B_H^{0.95} \quad (r = 0.81, \quad F_{emp} = 36.00, \quad F_{0.01} = 8.29).$$

Thus, if the broad spectrum of habitats with different plant production are compared, ranging from extremely poor, arid areas to fertile meadows, it can be shown that consumer biomass is determined by the plant production (Fig. 5.1), although the biomass of predatory consumers (arthropods) is very low when expressed as a proportion of plant production (of the order of $10^{-5}$). There is a very clear relationship between herbivore and predator biomass (Fig. 5.2).

All the functions reported above are statistically significant, but the deviation of points from the theoretical curve is quite large (Fig. 5.1). Thus although they show the character of the relationship these functions cannot be used for the calculation of consumer biomass from the known plant production of a particular habitat.

A positive correlation between the biomass of Arthropoda in the field layer and plant production has also been found by Zlotin (1975) for the Tien Shan ecosystems. Gyllenberg (Chapter 3, this volume) has found a relationship between plant production and the consumption of herbivorous invertebrates in grasslands, but not, as yet, a correlation between plant production and predators.

Although predator biomass in the field layer is very small, it constitutes a considerable proportion of the total invertebrate biomass of this layer, values ranging from 5 to over 40 % (Table 5.1). The largest value has been recorded in a marshy community (46 %).

### *Biomass of soil predators*

The biomass of soil predators is considerably larger than that of predators living above ground, ranging from about 6 mg dry wt·m$^{-2}$ in arid grassland (Zlotin, 1975) to more than 3000 mg dry wt·m$^{-2}$ in limestone grassland (Macfadyen, 1963). In the majority of grassland communities, excepting the arid areas, biomass is in the range of 100 to about 600 mg dry wt·m$^{-2}$. However the contribution of predator biomass to the total invertebrate biomass is considerably lower than in the field layer and falls within a narrower range, usually between 0.4 and 12 % (Table 5.2).

Methods of sampling the soil fauna are more standardized than those applied to the aboveground fauna. The main difficulty in comparing data lies in the fact that it is not always the same groups of animals that have been studied. The assignment of species to trophic levels is difficult in some groups, for instance Acarina and Nematoda. In some cases predatory species of those groups were neglected. Detailed information on sampling and extraction techniques, sample depths and groups of animals analysed, is given in Appendix 5.2 (p. 588).

It will be readily understood that there are methodological considerations that make it difficult to compare some of the data now available. In soil habitats, as above ground, there is a relationship between trophic conditions and consumer biomass. Trophic conditions of the habitat were measured by the input of dead plant material, or by the biomass of saprophages in the case of predators. From data on consumer biomass and annual litter fall in nine grassland ecosystems regression functions expressing the relationship between saprophage or predator biomass to litter input were calculated (Fig. 5.3):

$$B_S = 0.003L^{1.75} \quad (r = 0.89, \quad F_{emp} = 26.66, \quad F_{0.01} = 11.3),$$

where $B_S$ is mean biomass of saprophages in mg dry wt·m$^{-2}$ and $L$ is annual litter input (above and below ground) in g dry wt·m$^{-2}$. And

$$B_P = 0.024L^{1.25} \quad (r = 0.82, \quad F_{emp} = 14.25, \quad F_{0.01} = 11.3),$$

where $B_P$ is mean biomass of predators in mg dry wt·m$^{-2}$.

Table 5.2. *Belowground biomass of predatory invertebrates*

| Type of grassland | Biomass of predators (mg dry wt $\cdot$ m$^{-2}$) | % total invertebrate biomass | Annual litter fall (g dry wt $\cdot$ m$^{-2}$) | References |
|---|---|---|---|---|
| 1. Desert grassland, Tien Shan | 6.3 | 12.4 | 56 | Zlotin (1975) |
| 2. Semidesert, Tien Shan | 5.7 | 3.6 | 163 | Zlotin (1975) |
| 3. Arid steppe, Tien Shan | 16.1 | 5.9 | 199 | Zlotin (1975) |
| 4. Cool steppe, Tien Shan | 13.6 | 1.8 | 205 | Zlotin (1975) |
| 5. Moist alpine meadows, Tien Shan | 81.0 | 0.5 | 767 | Zlotin (1975) |
| 6. Mixed grass prairie, Cottonwood, S. Dakota | 524.7[a] | | — | Smolik (1974) |
| 7. Steppe, central chernozem region, USSR | 120.0 | 0.4 | — | Zlotin (1969) |
| 8. Mixed grass prairie, Matador, Saskatchewan | 131.7 | 2.0 | — | Willard (1974) |
| 9. Mountain pasture (*Lolio-Cynosuretum*), Jaworki, Poland | 148.5 | 3.0 | 328 | Delchev & Kajak (1974); Pętal (1974); Wasilewska (1974) |
| 10. Marshy meadow (*Deschampsietum*), Kampinos Forest, Poland | 246.0 | 3.8 | — | Kajak, Breymeyer & Pętal (1971) |
| 11. Marshy meadow, Tien Shan | 19.5 | 0.6 | 1130 | Zlotin (1975) |
| 12. Cultivated meadow (*Arrhenatheretum*) | 566.8 | 2.0 | — | Pętal (1976); Wasilewska (1976); A. Kajak (unpublished data) |
| 13. Fertile pasture, limestone grassland | 3200.0[b] | 5.0 | — | Macfadyen (1963) |
| 14. Open shrubby savanna (*Andropogon*), burnt, Lamto, Ivory Coast | 545.0[b] | 4.6 | 1180 | Athias *et al.* (1974); Lavelle & Schaefer (1974) |
| 15. Open shrubby savanna (*Andropogon*), unburnt, Lamto, Ivory Coast | 856.0[b] | 8.9 | 2970 | Athias, Josens & Lavelle (1974); Lavelle & Schaefer (1974) |

[a] Nematodes only investigated.
[b] Recalculated from fresh weight assuming 67 % water content.

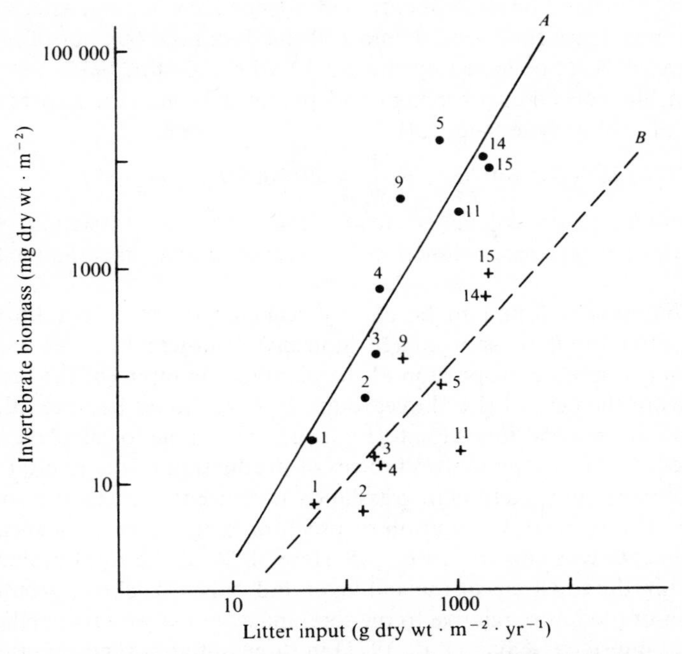


Fig. 5.3. Relationship between litter input and saprophage (●, *A*) or predator (+, *B*) biomass. Numbers represent type of grassland as in Table 5.2.

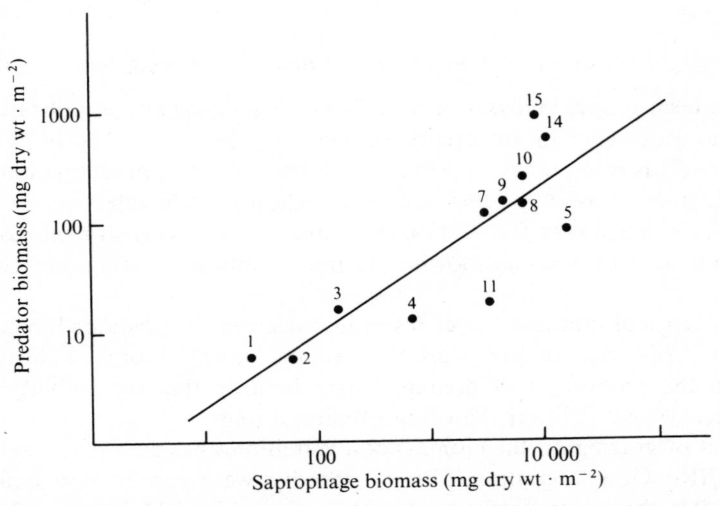


Fig. 5.4. Relationship between saprophage and predator biomass. Numbers represent type of grassland as in Table 5.2.

The differences in methodology mentioned do not have a great effect on the computation of these functions, as most of the data used (six out of nine habitats compared) were collected by the same author (Zlotin, 1975).

The relationship between saprophage and predator biomass is expressed by a function of similar type (Fig. 5.4):

$$B_P = 0.43 B_S^{0.67} \quad (r = 0.85, \quad F_{emp} = 26.10, \quad F_{0.01} = 3.82).$$

This regression has been calculated from data from 15 grasslands (see Table 5.2). All the regression functions presented above are significant ($P < 0.01$).

Consumer biomass is found to be closely related to trophic conditions (expressed as litter input or saprophage biomass) although in some cases it constitutes only a minute proportion of the plant matter input (of the order of $10^{-4}$). But as in the case of the aboveground layer it should be noted that the data are too diverse and the estimates too rough to be able to calculate the biomass of predators from the known indices of production of the habitat.

The main biomass of predators in grasslands is concentrated at the soil–litter interface. The vertical distribution of predator biomass in an *Arrhenatheretum*-type meadow is shown in Fig. 5.5. Here 38 % of the total biomass was found within the top 1 cm of the soil layer and only 1 % above ground. The proportion of predators relative to the total macrofauna was also highest at the soil–litter interface (Kajak *et al.*, 1971) in three different types of grassland. In this layer predators constituted 26–54 % of the total number of the macrofauna, compared with 0.2–3.2 % in soil and 9–14 % above ground. Thus it can be assumed that predation processes are very intense here.

### *The effect of grassland management on the biomass of predators*

There have been several analyses of the effect of management and utilization of grassland ecosystems on invertebrates, including the density and biomass of predators. This is important in so far as the proportion of predators in the trophic structure of an ecosystem can be a measure of the effectiveness of regulatory mechanisms. In this section the influence of basic treatments used in meadows, such as grazing, mowing, fertilizing and cultivating, are considered.

Generally agricultural treatments result in a decrease in predator biomass and in the proportion of predators in the total animal biomass. Such a decrease in the proportion of predators may indicate that the stability of managed ecosystems is lower than that of natural ones.

The effect of grazing on the biomass of invertebrates has been investigated in the US/IBP Grassland Biome Project. Studies were carried out at five grassland sites that were partly protected from grazing (McDaniel, 1971; Lewis, 1971; Ellstrom & Watts, 1974; Lavigne, Rogers & Chu, 1971).

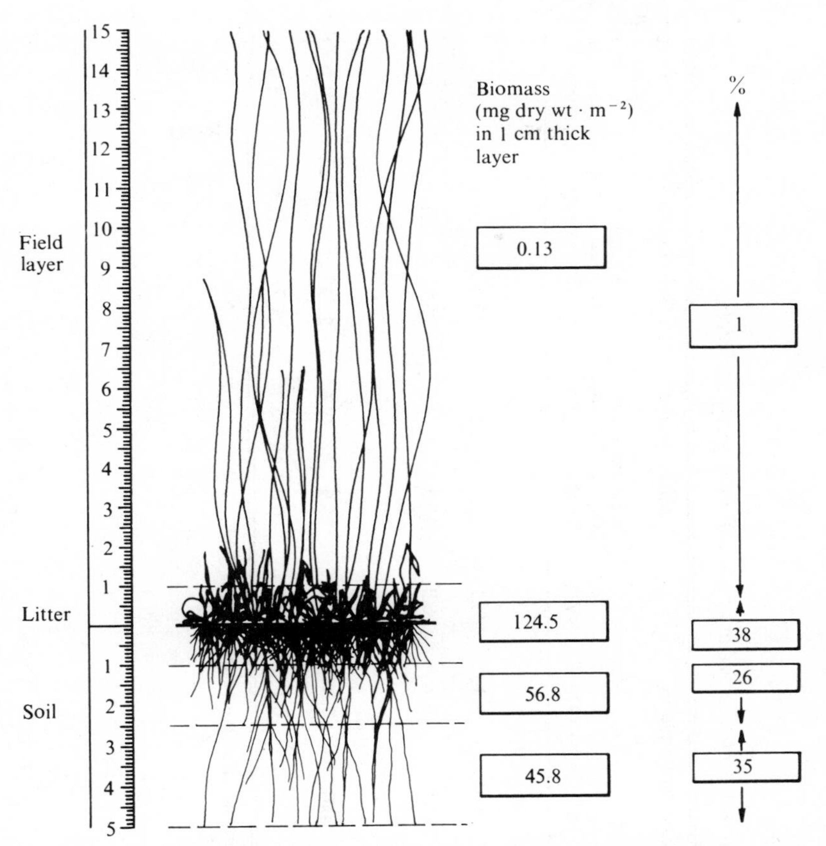


Fig. 5.5. Vertical distribution of predator biomass in an *Arrhenatheretum*-type meadow.

The average biomass of predators above ground was found to be lower in the grazed areas in both the years considered (Table 5.3). Predators are affected by grazing more markedly than are herbivores and in consequence their contribution to the total fauna biomass decreases (Table 5.3).

Detailed experiments directed mainly towards the analysis of predatory invertebrates and to changes in their activity under the influence of grazing, have been carried out at Pawnee Site (Lavigne & Rogers, 1970; Lavigne, Rogers & Chu, 1971). In a comparison between lightly, moderately and heavily grazed pastures (where cattle consumed 20, 40 and 60 % of the green plant production respectively) and a pasture grazed only in winter, it was found that the more intensive the grazing the fewer the number of ant-hills and the smaller the biomass of the dominant ant species, *Pogonomyrmex occidentalis*. This species feeds mainly on seeds but invertebrates are

Table 5.3. *Effect of grazing on the biomass of invertebrate predators in different grassland sites*

| Grassland type | Desert grassland | Shortgrass plains | | | | Mixed prairie | | True prairie | | Invertebrate biomass U/G ratio | |
|---|---|---|---|---|---|---|---|---|---|---|---|
| Site | Jornada | Pawnee | | Pantex | | Cottonwood | | Osage | | All sites | |
| Year | 1970 | 1970 | 1972 | 1970 | 1972 | 1970 | 1972 | 1970 | 1972 | 1970 | 1972 |
| Biomass (mg dry wt·m$^{-2}$) | | | | | | | | | | | |
| | | | | | | | | | | Predators | |
| Ungrazed (U) | 0.6 | 4.9 | 8.4 | 2.3 | 145.1 | 28.6 | 19.4 | 17.3 | 15.1 | 1.7 | 1.9 |
| | | | | | | | | | | Herbivores | |
| Grazed (G) | 0.7 | 6.9 | 12.1 | 7.7 | 68.3 | 10.3 | 4.3 | 6.8 | 11.3 | 0.8 | 0.3 |
| Proportion of predators in total invertebrate biomass (%) | | | | | | | | | | | |
| Ungrazed (U) | 27.2 | 21.5 | 20.0 | 7.0 | 20.1 | 35.9 | 12.6 | 21.9 | 18.1 | 21.0 | 17.0 |
| Grazed (G) | 18.4 | 37.3 | 38.9 | 16.1 | 2.4 | 6.0 | 5.4 | 11.7 | 11.5 | 11.0 | 3.4 |

After Lewis (1971), McDaniel (1971), and data from the Grassland Biome Data Bank.

[a] Saprophages were excluded.

also included in the diet. Significant differences in the number of lycosid spiders and sphecids (Hymenoptera) were also recorded. These two predatory families were most numerous in the pasture grazed only in winter and least numerous in the one which was most heavily grazed. No detrimental effects of grazing on several other predaceous groups, such as Asilidae, Mantidae, Salticidae and Cicindellidae, were found. However, among these predators only Asilidae were encountered in large numbers.

Dempster (1970) analysed in detail the intensity of predation in pasture and ungrazed areas by estimating the mortality of the cinnabar moth (*Tyria jacobaeae*), a herbivorous insect that is abundant in several environments. By testing the gut content of predatory species against anti-Cinnabar serum using the precipitin test he found that the mortality due to arthropod predators was considerably lower in the pasture (60.7 %) than in a clearing (71.2 %). The number of predators trapped was also greater in the ungrazed area.

The influence of grazing intensity on the soil macrofauna of a shortgrass plain (Pawnee Site) was also studied (Lloyd *et al.*, 1973). Number and biomass of soil invertebrates in differently utilized pastures and in a plot protected from grazing were compared. Biomass of a number of animal groups appeared, however, to be very variable, and differed more in the same area in successive years (1970 and 1971) than between plots with different utilization patterns in the same year. It is thus difficult to draw conclusions much beyond noting that the response of the total predator biomass to grazing was opposite in the two successive years. The only specific conclusion possible is that grazing did not negatively influence the biomass of Carabidae, as increasing grazing intensity increased their biomass in 1971 and did not show any tendency to decrease it in 1970.

On a chalk grassland Morris (1968) found that the invertebrates in the turf were 3.7 times as numerous in an exclosure where sheep grazing was prevented as they were in the pasture area where sheep consumed 60 % of the plant production. Carabidae and Staphylinidae larvae were among the few animal groups that were more numerous in grazed areas.

The influence of grazing and mowing on the number of spiders inhabiting the ground layer has been analysed by comparing data on the number of spiders in various types of meadows and pastures (Delchev & Kajak, 1974). Only data based on soil cores were used. The comparison revealed that the number of spiders in the turf of ungrazed meadows oscillates over a very wide range: 56–842 individuals $\cdot$ $m^{-2}$ with an average density of 250 $\cdot$ $m^{-2}$ (Table 5.4). The density of spiders in grassland which was mowed was considerably lower, having a range of 14–81 individuals $\cdot$ $m^{-2}$ and an average of approximately 50 individuals $\cdot$ $m^{-2}$. Their numbers in pastures were of a similar magnitude to those in meadows under cultivation, with the range of 33–160 individuals $\cdot$ $m^{-2}$ and an average of approximately 67 individuals $\cdot$ $m^{-2}$.

Table 5.4. *Differences in densities of spiders related to grassland management*

| | Grassland management | | |
|---|---|---|---|
| | Unused[a] | Cut[b] | Grazed[c] |
| Average density (individuals·m$^{-2}$) | 250 | 50 | 67 |
| Density range (individuals·m$^{-2}$) | 56–842 | 14–81 | 33–160 |

[a] After Duffey (1962), Cherrett (1964), Morris (1968), Van Hook (1971) and Kajak *et al.* (1971).
[b] After Dondale (1971), Kajak *et al.* (1971), and A. Kajak (unpublished data).
[c] After Wolcott (1937), Cherrett (1964), Salt, Hollick, Raw & Brian (1948), Turnbull (1966) and Morris (1968).

The data thus support the conclusion that grassland management tends to decrease the number of spiders in the meadow turf.

It has been found that the influence of mowing on field-layer spiders depends on the grass-cutting procedure as well as on the grassland type. In damp meadows where tufts of plants occur, grass cutting does not produce any significant changes in spider density because even when the grass has been removed there are still numerous places remaining that have suitable habitat features for the spiders. In meadows with a smooth surface, cutting removes most of the plant stems leaving sod only a few centimetres high and thus causing a considerable decrease in spider numbers. Particularly affected are the orb web spiders such as the Argiopidae and Tetragnathidae, and species which fasten their cocoons to grass inflorescences, such as some Thomisidae (Kajak, 1962, 1967).

These results are in agreement with those of Southwood & Emden (1967) who analysed the influence of grass cutting on the number of invertebrates. They found no negative effects of the treatment on the number of spiders if the grass was left 5–15 cm high after cutting. Likewise, no effect on the number of predatory Heteroptera was recorded. However the total number of predators in the uncut meadow was higher, as was the number of Chilopoda and the proportion of predators within Acarina and Coleoptera. Predatory Acarina accounted for 32 % of the total number of Acarina in the unmown meadow and 19 % in the mown one. The comparable figures for predatory Coleoptera were 20 % and 8.4 % respectively. Herbivorous species, however, were generally more numerous in the mown meadow.

Another common treatment of grassland is the addition of fertilizers. The influence of organic fertilizers on the pasture soil fauna was examined during three consecutive growing seasons in the mountain sheep pastures of the *Lolio-Cynosuretum* type at Jaworki, Poland (Delchev & Kajak, 1974; Pętal,

Table 5.5. *Effect of penning sheep on the biomass of soil invertebrates in a* Lolio-Cynosuretum cristati *pasture, Pieniny Mountains*

| | Biomass (mg dry wt · m$^{-2}$) | | | |
|---|---|---|---|---|
| Trophic level | Untreated pasture | | Pasture a year after penning sheep | |
| Herbivores | 277.2 | | 916.8 | |
| Elateridae | | 219 | | 850 |
| Nematoda | | 58.2 | | 66.8 |
| Decomposers | | | | |
| Lumbricidae | 4500.0 | | 20600.0 | |
| Predators | 148.5 | | 121.2 | |
| Nematoda | | 76.2 | | 21.0 |
| Araneae | | 4.2 | | 2.6 |
| Formicidae | | 52.3 | | 3.5 |
| Carabidae | | 15.8 | | 94.1 |
| Total invertebrates | 4925.7 | | 21638.0 | |
| Predatory/total invertebrates (%) | | 3.0 | | 0.6 |

Data after Delchev & Kajak (1974), Nowak (1975), Pętal (1974) and Wasilewska (1974).

Table 5.6. *Effect of penning sheep on the locomotory activity of arthropod predators in a* Lolio-Cynosuretum *pasture, Pieniny Mountains*

| | Control plot | Plots with sheep pens | | | |
|---|---|---|---|---|---|
| No. of days after penning | — | 360 | 60 | 30 | <30 |
| No. of individuals per trap per 24 hours | | | | | |
| Araneae | 3.15 ± 0.27 | 2.27 ± 0.28 | 1.11 ± 0.14 | 1.34 ± 0.17 | 1.57 ± 0.25 |
| Formicidae | 2.35 ± 0.42 | 0.54 ± 0.15 | 1.17 ± 0.36 | 0.73 ± 0.20 | 2.40 ± 0.37 |
| Carabidae | 3.70 ± 0.34 | 0.95 ± 0.12 | 4.25 ± 0.53 | 9.24 ± 0.12 | 8.50 ± 0.14 |

After Delchev & Kajak (1974), Pętal (1974) and A. Kajak (unpublished data).

1974; Wasilewska, 1974; Nowak, 1976; Zyromska-Rudzka, 1974 and unpublished data). Sheep-penning was practised there which led to the area being covered with sheep dung and the annihilation of the vegetation; the plants quickly recovered, however, and in a few months' time grew abundantly. Data obtained substantiate the conclusion that organic fertilization produces a sharp increase in the biomass of soil animals, mainly the saprophagous and coprophagous species, with a simultaneous decrease in the biomass of predators, thus leading to a considerable decrease in the contribution of predaceous animals to the total biomass of soil animals (Table 5.5).

Different predatory groups responded in different ways to the fertilization. For example, there was a significant decrease in the biomass of ants and predaceous nematodes following sheep-penning while the reduction in spider biomass lasted for 1 month only, i.e. while there was no vegetation. Differences in spider density between the manured and unmanured plots were small and statistically insignificant (Delchev & Kajak, 1974). The biomass of Carabidae increased significantly in the intensively fertilized stands (Table 5.5).

Changes in penetration of the area by Formicidae, Araneae and Carabidae following manuring were examined using pitfall traps. In the manured plots there was a significant decrease in the area penetration by the first two groups, but no regular changes in penetration by the last group were recorded (Table 5.6). The penetration by spiders increased with time after treatment. The significance of differences in the penetration of particular plots by spiders was tested by the *t*-test, which indicated that differences between areas of recent penning and those that had been penned previously were significant ($P < 0.05$). Ants showed an increase in penetration immediately following sheep-penning, probably as a result of their nest rebuilding, but approximately 1 month later this decreased again and in all intensively manured plots was significantly lower ($P < 0.05$) than in a pasture that had not been penned.

Approximately similar results were obtained in studies on the influence of mineral fertilizer application on predator biomass and activity. The application of mineral fertilizers resulted in a decrease both in the biomass of predators (Table 5.7*a*, *b*) and in the proportion of predators in the total biomass of the soil fauna. The predator biomass was lowered even when fertilization was followed by an increase in the total fauna biomass, i.e. in the potential food supply (Tables 5.5, 5.7*a*). Fertilized plots were also less intensively penetrated by arthropod predators (Pętal, 1976; A. Kajak, unpublished data). Whereas organic fertilization results in a considerable increase in the potential food supply, i.e. in both herbivore and saprophage biomass, mineral fertilization can be followed by a decrease in the number and biomass of some saprophage groups. However the response of predators to both treatments is similar, and although particular groups of predators show different reactions the zoophage trophic level as a whole shows a decrease in biomass and in intensity of penetration of the area.

The most violent management procedure on any permanent grassland is obviously its conversion into a cultivated field. This practice leads to a considerable impoverishment of the soil fauna. However, according to Willard (1974) and Zlotin (1969), the biomass of the predaceous fauna changes less drastically than that of other trophic groups, which results in an increase in its contribution to the total biomass of soil invertebrates (Table 5.8). The transformation of prairie into cultivated wheatfields leads to the complete disappearance of Formicidae, Chilopoda and pseudoscorpions, but although the total nematode biomass is diminished fourfold, there is an increase in the

Table 5.7. *The effect of mineral fertilization on the biomass of invertebrates*

(*a*) *Soil macroarthropods in shortgrass prairie, Pawnee site*[a]

| Group | Biomass (mg dry wt·m$^{-2}$) No treatment | Irrigation | Irrigation + nitrogen fertilization |
|---|---|---|---|
| Total macroarthropods | 316.5 | 439.2 | 493.8 |
| Predatory macroarthropods[b] | 184.1 | 141.5 | 171.9 |
| Carabidae | 157.3 | 125.8 | 118.6 |
| Staphylinidae | 2.6 | 1.1 | 4.0 |
| Histeridae | | | 5.4 |
| Asilidae | 22.5 | 14.6 | 43.9 |
| Therevidae | 1.7 | | |
| Predators/total macroarthropods (%) | 58.1 | 32.2 | 34.8 |

(*b*) *Soil meso- and macrofauna,* Arrhenatheretum-*type meadow*[c]

| Group | Biomass (mg dry wt·m$^{-2}$) No treatment | NPK fertilization[d] |
|---|---|---|
| Predators | 566.8 | 190.5 |
| Araneae | 34.0 | 19.7 |
| Formicidae | 241.9 | 31.2 |
| Carabidae imagos | 88.5 | 13.4 |
| Staphylinidae imagos | 20.1 | 39.9 |
| Carabidae + Staphylinidae larvae | 3.15 | 6.8 |
| Chilopoda | 179.2 | 79.5 |
| Herbivores | 2050.1 | 1490.5 |
| Saprophages | 26845.8 | 21409.6 |
| Total soil fauna | 29462.7 | 23090.6 |
| Predators/total soil fauna (%) | 2.0 | 1.0 |

[a] After Lloyd *et al.* (1973).
[b] Formicidae excluded.
[c] After Andrzejewska (1976), Nowak (1976), Żyromska-Rudzka (1976), Wasilewska (1976), Makulec (1976), Olechowicz (1976), Pętal (1976) and A. Kajak (unpublished data).
[d] Fertilizer applied at the rates of 360 kg N, 120 kg $P_2O_5$ and 200 kg $K_2O$ per hectare in a season.

proportion of predaceous species. This explains the overall rise of the predators' contribution to the total soil fauna biomass from 2.0 % in the prairie to 3.3 % in the wheatfield (Willard, 1974).

The conversion of steppe into wheatfields in the Russian lowland resulted in an enormous impoverishment of its soil fauna. Its total biomass was

Table 5.8. *Influence of cultivation on soil invertebrates*

| | Biomass (mg dry wt·m$^{-2}$) | |
|---|---|---|
| | Natural grassland | Cultivated land (wheat) |
| Total invertebrates | | |
| Steppe[a] | 32170 | 1950 |
| Mixed prairie[b] | 6479 | 1563 |
| Predators | | |
| Steppe | 63 | 20 |
| Mixed prairie | 131.7 | 50.9 |
| Biomass of predators (% of total) | | |
| Steppe | 0.2 | 1.0 |
| Mixed prairie | 2.0 | 3.3 |

[a] After Zlotin (1969). Original fresh weight data recalculated for dry weight assuming 67 % water content. [b] After Willard (1974).

diminished 16.5-fold, but predator biomass (Acarina (Gamasiidae, Trombidiidae), Chilopoda) dropped only 3.2-fold, thus again increasing the proportion of predators (Zlotin, 1969). It has also been found that ploughing the steppe causes an increase in the abundance of predatory Carabidae species (Titova & Zhavoronkova, 1965).

Detailed researches on potato and rye field ecosystems have been carried out in Poland over several years (Trojan, 1967; Dąbrowska, Karg & Ryszkowski, 1974; Ryszkowski, 1975) in areas which have been under cultivation for centuries as opposed to being newly cultivated. Here, again, a large drop in predator biomass was found. For instance predatory nematodes were completely absent (Wasilewska & Paplińska, 1975), there were no ant-hills and there was a decrease in spider biomass relative to the surrounding area Łuczak, 1975). Only the biomass of Carabidae was relatively high (Kabacik-Wasylik, 1975).

Odum (1971) provided data on the changes in numbers of phytophagous and predatory arthropods in the field layer of a millet field during the year following the time it was abandoned. Over the time of study the number of phytophagous insects was more than halved and the number of predaceous and parasitic arthropods rose considerably. At the same time the species diversity of the community increased. It is important to appreciate that the cessation of cultivation may rapidly initiate processes opposite in their effect to those which result from management.

It is a well-known fact that plant and animal communities become less complex in areas under human management. This generalization also holds for the predatory arthropods, and reveals itself as a decrease in the number of species, a lowering of diversity indices, an increased dominance of one or several species in a community and a decreasing importance of accessory species in managed grassland ecosystems (Kajak, 1962; Odum, 1971; Delchev & Kajak, 1974; Pętal, 1974).

In summary, the common occurrence of a decrease in the biomass of predators, both above and below ground, as a result of the management of an ecosystem should be emphasized. This is usually accompanied by a decrease in the contribution of predators to the total invertebrate biomass, even when the treatment (e.g. fertilization) increases the prey supply. But the influence of cultivation practices on predators requires additional elucidation, and results are not unequivocal. As a rule there is a drastic decrease in predator biomass, but in the soil of newly cultivated fields (as opposed to those that have been cultivated for some time) there may be an increase in the proportion of predators in the total biomass of soil invertebrates.

In grassland ecosystems polyphagous predatory invertebrates are, in terms of biomass, especially important, so it can be assumed that grassland management influences this group markedly. In general polyphagous predators are characterized by a lifespan longer than the time between particular grassland treatments. Because grassland management increases plant production and, in consequence, the biomass of herbivorous animals, prey-specific predators feeding on phytophages and adjusted to prey population dynamics may increase in numbers. Thus, management practices that cause a decrease in long-lived polyphagous predators may favour prey-specific species. A recent survey of the invertebrate biomass of fertilized and unfertilized plots suggests that a decrease in the biomass of Formicidae, Araneae and Carabidae after intense mineral fertilization is accompanied by an increase of prey-specific parasitic Hymenoptera (Olechowicz, 1976; Pętal, 1976; A. Kajak, unpublished data). An increase in the biomass of parasitic species was also recorded in grazed as compared with ungrazed grasslands (US/IBP Grassland Biome Data Bank). The question then arises as to how these various changes in the predator system influence regulatory processes.

## The role of predators in an ecosystem

### *Regulatory role of polyphagous predators*

Many efforts have been made to find the key factors rseponsible for the homeostasis of a system (Huffaker, 1958; Klomp, 1962; Morris, 1963; Varley & Gradwell, 1963; Breymeyer, Gałecka, Kajak & Łuczak, 1964; Rafes, 1968; Kaczmarek, 1969; Varley & Gradwell, 1971; Haukioja & Hakala, 1975). These studies have chiefly been concerned with the mechanisms of number regulation in populations of phytophagous animals – mainly forest and crop pests.

There are many studies on the effects of prey-specific predators on the population dynamics of their prey, but polyphagous predators have been less frequently studied and it has been argued that they are less effective in biological control than are prey-specific species (Huffaker *et al.*, 1971).

Table 5.9. *Predation by ants and web spiders in a* Stellario-Deschampsietum *community*

| | No. ($m^{-2} \cdot season^{-1}$) | |
|---|---|---|
| | Homoptera: Auchenorrhyncha | Diptera (imagos) |
| No. of individuals produced | 1941 | 707 |
| No. of individuals killed by: | | |
| ants | 840 | 226 |
| web spiders | | 177 |
| Elimination (%) | 43 | 57 |

Data from 1968. After Kajak *et al.* (1972) and Kajak *et al.* (1971).

But because polyphagous predators seem to be very important in grassland ecosystems and because the biomass of this group is affected by management, data on their effectiveness in such systems are presented here. In particular, attention is given to the changes which result from a reduction in the number of polyphages in an ecosystem.

It is characteristic of polyphagous predators that because of their relatively long life cycle they can not adjust their numbers rapidly to changes in the number of prey. An increase in prey density is immediately followed by the functional response, i.e. by an increase in the number of prey killed as a result of increasing individual consumption. It has been shown that the elimination rate of the prey (proportion of the total population that is killed) resulting from the functional response may follow different patterns (Holling, 1959; Dąbrowska-Prot, Łuczak & Tarwid, 1968; Reichert, 1974). The maximum elimination rate can either occur at the lowest prey densities or it can increase gradually with increasing prey density up to a maximum and then gradually decrease again. Thus, it may be concluded that polyphagous predators can effectively regulate their prey population only within certain density limits. The population dynamics of prey-specific predators, on the other hand, are closely adjusted to the population dynamics of their prey. Their reproductive rate is high if suitable food conditions exist, and thus although the response of their population is delayed it is density-dependent, i.e. percent mortality of the prey increases with increasing prey density. The fact that the response is density-dependent does not mean that it must be regulatory in character, though predation may have a regulatory effect when it follows the increase in prey density in such a way that it approaches asymptotically 100 % population production of the prey species (Varley & Gradwell, 1971; Varley, Gradwell & Hassell, 1975).

An attempt is made here to compare these statements with the results obtained under field conditions on the effectiveness of polyphagous predators.

Generally, in natural ecosystems each population is subjected not to one

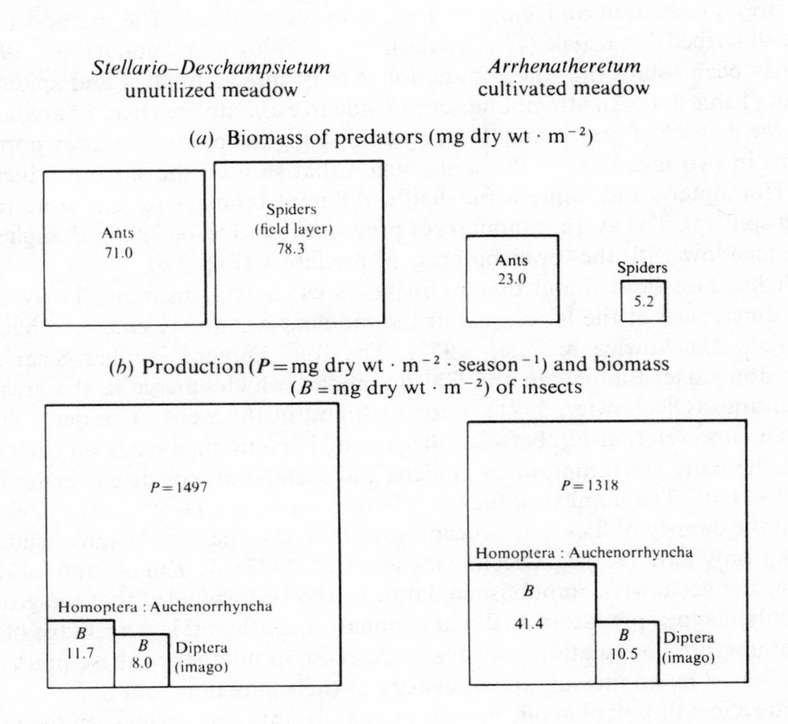


**Fig. 5.6. Influence of ants and spiders on the biomass of insects (1968). (After Andrzejewska, 1971; Kajak, Breymeyer & Pętal, 1971; Olechowicz, 1971.)**

predator only but to a system of predators. Holling (1959), when studying the effect of predatory vertebrates on the pest *Neodiprion sertifer*, found a very important coaction between predators. The maximum predation rate by each species corresponded to a different prey density, and as a result the overall effect of all prédators, i.e. the range of prey densities at which there was a high mortality rate, was considerably extended, although at very high prey densities there was a decrease in elimination rate by populations of particular predator species that was characteristic of the functional response.

The effects of predation by two important groups of polyphagous spiders and ants on Diptera and Auchenorrhyncha have been studied for several years in an unused, mid-forest meadow of the *Stellario-Deschampsietum* community (Andrzejewska, 1971; Kajak, 1971; Kajak *et al.*, 1971; Olechowicz, 1971; Pętal, 1971). Predator consumption and biomass as well as prey production and biomass have been determined. The methods used are described by Kajak (1965), Olechowicz (1970) and Pętal (1967, 1972). It has been found that the elimination rate of insects by ants and spiders is high (Table 5.9). An attempt has been made to estimate the effect of predation on the density of prey populations by comparing the biomass of prey populations in two meadows in the same region that showed the same production of Homoptera and Diptera but had a different biomass of predators (ants and spiders). The average biomass of prey was found to be 2.6 times higher in the meadow with the lower biomass of predators (Fig. 5.6).

It has also been found that web spiders catch Diptera immediately after the emergence of the latter, just at the moment when they enter the habitat (Kajak, Olechowicz & Pętal, 1972). The four Diptera families Sciaridae, Cecidomyidae, Bibionidae and Chironomidae, which emerge in the greatest quantities (Olechowicz, 1971), were all found in the webs of spiders. There was a close relationship between the rate of Diptera emergence on each day and the daily consumption by spiders, the correlation coefficient being 0.71 at $P < 0.01$. The number of flies caught in webs was, however, not correlated with the density of flies in the meadow on that day, the correlation coefficient being only 0.01 (Kajak, Olechowicz & Pętal, 1972; A. Kajak, unpublished data, E. Olechowicz, unpublished data). It may be concluded that the groups of polyphagous predators analysed eliminate a considerable percentage of the emerging prey population, and that a decrease in number of these predators is followed by an increase in the density of their prey population.

The effect of polyphagous predators, mainly ants and spiders, on the number of crop pests has been studied in Japan for many years (Kayashima, 1960; Ito, Miyashita & Sekiguchi, 1962; Kiritani, Kawahara, Sasaba & Najasuji, 1972; Sasaba, Kiritani & Urabe, 1973). These studies were stimulated by pest outbreaks that were particularly intense as a result of the general use of insecticides which severely reduced the predatory fauna while having only minor to moderate effects on phytophages.

Long-term studies of the relationship between the population of the leafhopper *Nephotettix cincticeps* (which transmits virus diseases to rice crops) and spiders, show that the highest elimination of prey larvae by spiders occurs at relatively low prey densities. The elimination of adults is density-dependent over a certain range of densities. Using empirical data and a simulation model, the ratio of spiders to prey at which the mortality rate is the most effective was determined. It has been shown that this ratio can be maintained and, as a result, pest densities can be low when selective insecticides are used. At high densities of spiders only one treatment with insecticides was needed during 10 generations of the pest (Sasaba & Kiritani, 1972, 1975; Sasaba, Kiritani & Urabe, 1973; Kiritani & Kakiya, 1975).

Another species of Homoptera, *Mogannia iwasakii*, which used not to be harmful, became a dangerous pest of sugar-cane in various regions of Japan several years ago. It is suggested that the reduction of the ant population by pesticides was responsible for this situation. In addition, an increase in the area of the crop stimulated an increase in phytophages (Nagamine, Teruya & Ito, 1975).

The suggestion that the elimination of polyphagous predators is followed by pest outbreaks is repeated in all the Japanese papers quoted (Ito, Miyashita & Sekiguchi, 1962; Kayashima, 1960; Kiritani *et al.*, 1972; Sasaba *et al.*, 1973; Sasaba & Kiritani, 1975; Nagamine *et al.*, 1975). Polyphages are thought to have a considerable effect on the number of pest insects, although the highest elimination rate by these predators occurs at low prey densities.

The winter moth *Operophtera brumata* is one of the most comprehensively studied pests (Varley & Gradwell, 1963; East, 1974; Varley *et al.*, 1975). The results of the long-term studies of this species in oak forest suggest that polyphagous ground-dwelling beetles contribute considerably to the mortality of the population. Larvae and pupae of the moth in the soil are subjected to heavy predation by Carabidae and Staphylinidae, which destroy 68 % of the initial number of pupae (East, 1974). This is a density-dependent mortality (East, 1974; Varley *et al.*, 1975).

The above data indicate that there are situations in which polyphagous predators have a considerable effect on the density of prey populations. Most authors note that the effectiveness of the predators is highest at low prey densities (Holling, 1959; Ito *et al.*, 1962; Sasaba & Kiritani, 1975; Reichert, 1974), and that their response to an increase in the prey density is immediate (Holling, 1959; Kajak, Olechowicz & Pętal, 1972). In addition, because of their long life cycle polyphagous predators are present in the ecosystem all year. The combination of these features probably accounts for the maintenance of low prey population levels, particularly in natural systems rich in biotic limiting factors.

It is of interest that the number of polyphagous predators decreases as a result of management. At the same time the number of prey-specific predators may increase. These processes can account for a transition from a state of stable equilibrium to a state of more flexible equilibrium which can be more readily disturbed.

### *The exploitation of prey populations by predators*

In a stabilized trophic chain the production rate of a population is balanced by its exploitation rate, and the latter does not affect the standing crop (Kaczmarek, 1969). Consequently the exploitation of the population is not higher than the excess of production over standing crop (in other words it depends on the *P/B* ratio of the exploited population). The results of the studies on energy flow in natural ecosystems seem to support this conclusion. The grazing of leaves in a forest does not exceed 5–10 % of production, which means that it is adjusted to the regeneration ability of trees (Kaczmarek, 1969). The exploitation of grasslands is considerably higher, the maximum recommended exploitation rate by cattle being 30–45 % of production (Wiegert & Evans, 1967), and average grazing by herbivores being up to 13–20 % (Petrusewicz & Grodziński, 1975). Seed predation is still higher (10–90 %: Chew, 1974), as is the exploitation of phytoplankton by zooplankton (3 – approx. 100 %, with an average for all lakes studied of 44 %: Hillbricht-Ilkowska, 1977).

The exploitation of prey by predators in grassland ecosystems is discussed below. Exploitation is defined here as the proportion of production of the preceding trophic level which is consumed by predators, a parameter that has also been termed 'consumption efficiency' (Kozlovsky, 1968).

Data in the literature on the production and consumption of whole trophic levels are limited, but the available sources indicate that the efficiency of predation varies within broad limits from 3 to about 100 %, the most frequent values being 18–60 % (Table 5.10). These data should be viewed with caution. It is difficult to measure precisely the relations between the trophic levels of prey and predators, and a number of physiological assumptions must be made when calculating production and consumption. Thus, the data from different sources may be based on somewhat different assumptions, and have only limited value for comparative purposes. For example, Zlotin (1975) assumes that secondary production approximately equals the mean animal biomass. Though this assumption is based on the duration of life cycles and the number of generations per year, it may be expected that the value of production is underestimated in this way and, consequently, the value of consumption efficiency overestimated. However, these data can be used as relative figures that enable one to arrange the ecosystems studied by Zlotin along a gradient of exploitation by predators.

The study carried out by Menhinick (1967) in the above-ground layer of a *Sericea lespedeza* stand suggests that the effect of predators on phytophages is remarkable, with consumption exceeding the production of phytophages. It should be noted that in this case the production of phytophages was underestimated, the author referring to it as minimum production. It was calculated as the sum of the maximal biomass of individual populations. In addition, only the food supply for phytophages was considered, not the total food available to all predators. The author concluded, however, that predators are an important factor controlling the effect of phytophages on primary production.

Van Hook (1971) calculated consumption efficiency using empirical data on production and consumption obtained by the isotope technique, but he

Table 5.10. *Consumption efficiency of the predator trophic level*

| Trophic level and habitat | $C_n/P_{n-1}$ (%)[a] | References |
|---|---|---|
| Predators/preceding trophic level (various sources) | 20–30 | Kozlovsky (1968) |
| Predators/herbivores (grassland ecosystem) | 21 | Van Hook (1971) |
| Predators/herbivores (*Sericea lespedeza* stand) | ~100 | Menhinick (1967) |
| Predators/herbivores above ground (shortgrass plain, ungrazed) | 38 | Andrews *et al.* (1974) |
| Predators/saprophages and herbivores (below ground, shortgrass plain, ungrazed) | 56 | Andrews *et al.* (1974) |
| Predators/herbivores (above ground, shortgrass plain, heavily grazed) | 28 | Andrews *et al.* (1974) |
| Predators/saprophages and herbivores (below ground, shortgrass plain, heavily grazed) | 38 | Andrews *et al.* (1974) |
| Predators/secondary production (Tien Shan) | | |
| Desert | 115 | Zlotin (1975) |
| Semidesert | 68 | Zlotin (1975) |
| Dry steppes | 92 | Zlotin (1975) |
| Cool steppes | 19 | Zlotin (1975) |
| Moist mountain meadow | 3 | Zlotin (1975) |
| Marshy meadows | 18 | Zlotin (1975) |

[a] Ratio of energy consumed by predators to the net production of the preceding trophic level.

considered a very simplified system consisting of only one group of predators (spiders of the family Lycosidae) and a few phytophagous species. According to the author these were the predominant groups; nevertheless, they do not seem to represent the whole trophic level.

As the data on the consumption efficiency of predators require so many reservations, only a few conclusions may be drawn from them. They show that the impact of predators on the trophic level of the primary consumers in grassland ecosystems is higher than the impact of primary consumers on the preceding trophic level. The average efficiency of consumption by predators is 38 % (if the highest values in Table 5.10 approximating to 100 % are ignored). The corresponding figure for the efficiency of consumption by invertebrate phytophages does not exceed 9 %, as reported by Gyllenberg (Chapter 3, this volume), or 10 % according to Breymeyer (1971); Andrzejewska (1976) reported about 5 % herbivore consumption efficiency in a fertilized meadow and 11 % in an unfertilized one. The efficiency of consumption by saprophages in grassland ecosystems also seems to be lower (10–28 % of the dead plant material input: after Zlotin & Khodashova, 1974; Pomianowska-Pilipiuk, 1976).

In poor ecosystems, with low primary production, the impact of predation is relatively higher than it is in fertile ecosystems (Zlotin, 1975). It has also been found that the impact of predators on a preceding trophic level is higher below ground than above ground (Andrews, Coleman, Ellis & Singh, 1974).

### *The role of soil predators*

The effect of predators on saprophagous animals is still less well understood than their effect on phytophages. It is easier to find data on the respiratory energy losses ($R$) of soil animals of different trophic levels than to find data on their production and consumption. The ratio of respiratory energy losses in predators to those in the lower trophic level can also be an index of the impact of predators on their prey.

The $R_n/R_{n-1}$ ratio will be called the respiratory ratio between trophic levels. This ratio has been compared for 14 ecosystems ranging from poor spruce forests, through deciduous forests and meadows to crop fields (Table 5.11). The data of a number of the authors quoted by Macfadyen (1963) are compared to more recent data (Dunger, 1968; McBrayer, Reichle & Witkamp, 1974; Andrews *et al.*, 1974; Kajak, 1974; Pętal, 1974; Delchev & Kajak, 1974; Ryszkowski, 1975; Wasilewska, 1974; Wasilewska & Paplińska, 1975; Kabacik-Wasylik, 1975; Łuczak, 1975). The habitats are arranged in the table along a gradient of increasing decomposition rate of dead plant material. Two indices of decomposition rate were used: either the C/N ratio in the aboveground litter of particular ecosystems or empirical data on the rate of litter disappearance. It is known that the decomposition rate decreases with increasing C/N ratio of a substrate. From the data for the ecosystems with known C/N in litter, a correlation was calculated between the C/N of falling leaves and the respiratory ratio of predators to saprophages (Fig. 5.7). This relationship is described by the function:

$$R_P/R_D = 0.0003x^{3.55},$$

where $x$ is the C/N ratio of litter, $R_P$ is the annual respiratory energy losses for predators, and $R_D$ is the annual respiratory energy losses for saprophages in the same area. The function was tested by the $F$-test ($r = 0.95$, $F = 8.01$, $F_{0.05} = 5.59$).

The highest proportion of predators is characteristic of coniferous forests (Table 5.11). The energy used for the respiratory metabolism of predators in these ecosystems is equal to that used by saprophages ($R_P/R_D = 80$–$126$ %). There, the C/N of falling leaves is highest, reaching 60–90 (Kazimirov & Morozova, 1973), and the complete decomposition of leaf litter takes 3–5 years.

In the majority of deciduous (beech and oak) forests, with a relatively low C/N ratio of from 47 to 51 (Wittich, 1942, 1943, after Wallwork, 1970), the respiratory ratio varies from 29 to 46 %. Next in order are meadows of different types, where the respiratory metabolism of predators amounts to

Table 5.11. *Respiratory ratio of predators to saprophages in different soil habitats*

| | Coniferous forests | | Deciduous forests | | | | |
|---|---|---|---|---|---|---|---|
| | Spruce mor[a] | Spruce mull[a] | Beech mor[a] | Beech mor[a] | Beech mull[a] | *Liliodendron* forest mull[b] | Oak mull[a] |
| Decomposition rate of leaves (yr) | 3–5 | 3–5 | 2–3 | 2–3 | 2–3 | 2 | 2–3 |
| C/N ratio (dead leaves)[g] | 59.0 | 59.0 | 51 | 51 | 51 | 48 | 37 |
| Respiration ($kcal \cdot m^{-2} \cdot yr^{-1}$) | 8.1 | 10.3 | 13.9 | 17.0 | 122.1 | — | 128.5 |
| Large decomposers ($R_{D_1}$) | — | — | — | — | — | 324.4 | — |
| Small decomposers ($R_{D_2}$) | 113.5 | 113.7 | 76.0 | 222.9 | 93.5 | — | 114.6 |
| Predators ($R_P$) | 145.9 | 99.3 | 41.5 | 93.5 | 68.7 | 94.7 | 22.2 |
| Respiratory ratio, $R_P/R_{D_1+D_2}$ (%) | 126.2 | 80.1 | 46.1 | 39.0 | 31.9 | 29.2 | 9.1 |

| | Grasslands | | | | | Forest | Arable land |
|---|---|---|---|---|---|---|---|
| | Shortgrass plains[e] | | | Mountain pasture[f] | | | |
| | Ungrazed | Grazed | Limestone grassland[a] | Untreated | Manured | *Fraxino-Ulmetum*[e] | Potato field[f] |
| Decomposition rate of leaves (yr) | 2 | 2 | — | 1 | 1 | 1 | ~1 |
| C/N ratio (dead leaves)[g] | — | — | — | 36 | 15.0 | 21–28 | — |
| Respiration ($kcal \cdot m^{-2} \cdot yr^{-1}$) | — | — | 312.7 | 19.1 | 103.0 | 55.4 | — |
| Large decomposers ($R_{D_1}$) | 16.2 | 37.0 | — | — | — | — | 69.5 |
| Small decomposers ($R_{D_2}$) | — | — | 448.0 | 60.6 | 55.0 | 29.7 | — |
| Predators ($R_P$) | 4.2 | 6.5 | 124.7 | 10.9 | 3.2 | 4.0 | 0.45 |
| Respiratory ratio, $R_P/R_{D_1+D_2}$ (%) | 25.9 | 17.5 | 16.4 | 13.6 | 2.0 | 4.7 | 0.64 |

[a] After Macfadyen (1963).
[b] After McBrayer, Reichle & Witkamp (1974).
[c] After Andrews *et al.* (1974).
[d] After Delchev & Kajak (1974), Kajak (1974), Pętal (1974) and Wasilewska (1974).
[e] After Dunger (1968).
[f] After Dąbrowska-Prot, Karg & Ryszkowski (1974), Wasilewska (1975), Łuczak (1975) and Kabacik-Wasylik (1975).
[g] C/N ratio after Wallwork (1970), McBrayer, Reichle & Witkamp (1974), Witkamp (1974), Kazimirov & Morozova 1973) and E. Olechowicz (unpublished data).

2–26 % that of saprophages. Among the grasslands under study, the highest respiratory ratio efficiency was found in ungrazed, shortgrass plains (25.9 %), the lowest in a manured pasture (2 %) (Table 5.11). A very low $R_P/R_D$ of 4.7 % was obtained in a *Fraxino-Ulmetum* forest (Dunger, 1968). The leaves of this forest have a relatively short, 1-year cycle of decomposition. Also the C/N ratio in the leaves of the predominant species of trees and shrubs was very low, varying from 21 to 28 % (Wittich, 1942, after Wallwork, 1970). The final link in this gradient of ecosystems is the crop fields, where the metabolism of predators is lower than 1 % of the metabolism of saprophages, and where the decomposition of plant material takes about a year (Ryszkowski, 1975).

Two important questions arise from the data reported. First, why the proportion of predators is higher in the soil of poor habitats, and secondly, what are the consequences of this fact in the functioning of the ecosystem.

One possible reason for predators being relatively abundant in poor habitats with a slow decomposition rate may be the accumulation of litter, which provides suitable shelter for them. It is known that habitats with acid humus of the mor type are dominated by small saprophages such as Acarina, Apterygota and Enchytraeidae, while soils with mull humus are dominated by large saprophages such as Lumbricidae, Isopoda and Diplopoda (Mac-

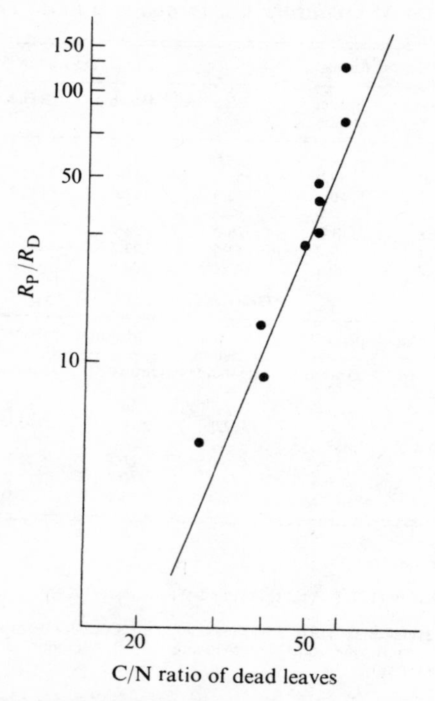


Fig. 5.7. Ratio of predator to decomposer respiration in different habitats as a function of the C/N ratio of freshly fallen leaves. $R_P$, respiration of predators ($kcal \cdot m^{-2} \cdot yr^{-1}$); $R_D$, respiration of decomposers ($kcal \cdot m^{-2} \cdot yr^{-1}$). (After Kajak & Jakubczyk, 1978.)

fadyen, 1963; Van Rhee, 1963; Dunger, 1968; Wallwork, 1970). The abundant occurrence of small saprophages is frequently accompanied by a relatively large proportion of predators. Small saprophages have shorter life cycles and higher turnover rates than large ones; their production per unit biomass is high. Hence, according to the statement from the preceding section that exploitation rate depends on the $P/B$ ratio of the population, a considerable part of the production can be exploited by predators.

The answer to the second question, i.e. the consequences of an increased proportion of predators in the soil, seems to be even more important. Can the proportion of predators in soil influence such a basic but distinct process as the rate of disappearance of dead plant material? To answer this question a field experiment was carried out in a common *Arrhenatheretum*-type meadow near Warsaw (Kajak & Jakubczyk, 1975, 1976). In this experiment a number of soil cores were taken, placed into isolating bags and replaced in their original positions in the meadow soil for 12 to 25 days. There were three series of bags. The bags of series C completely stopped the immigration

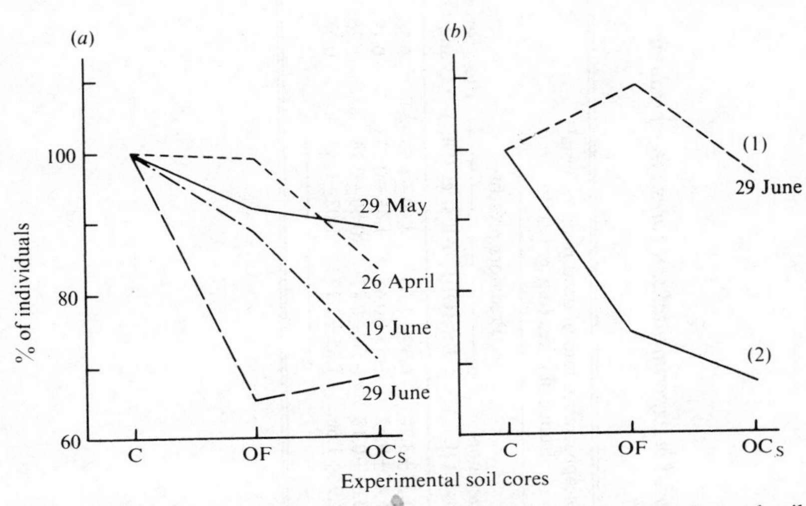


Fig. 5.8. Effect of predation on the number of Collembola in the experimental soil cores. Number of individuals in C series treated as 100 %. (*a*) Results of four experiments carried out in 1974; top 5 cm of soil analysed. (*b*) Results of variant (1) with pitfall traps partly removing predators and (2) without them; top 2.5 cm analysed. C, closed, no predators; OF, open, accessible to small (< 1 mm) predators; $OC_s$, open, accessible to all invertebrate predators.

of new predators, while those of series OF limited immigration to predators less than 1 mm in size, and those of series $OC_s$ allowed the immigration of predators of all sizes.

The subsequent analysis of the density of saprophages in the bags showed that the prevention of immigration by predators was followed by an increase in the number of saprophages, mainly Collembola (Table 5.12; Fig. 5.8). The differences in density of Collembola series C and series $OC_s$ were statistically significant ($P < 0.01$). To check that predators alone and not other factors were responsible for these differences, pitfall traps were set in an additional series of soil cores to catch some of the predators (Kajak & Jakubczyk, 1978). This procedure resulted in less marked differences in penetration by predators between series C, OF and $OC_s$. Differences in the number of Collembola between the three series were likewise considerably smaller in this experiment (Fig. 5.8). From these results it may be concluded that differences in the immigration rates of predators were really responsible for the differences in the number of Collembola recorded in the experiment conducted without pitfall traps.

In most cases the rate of disappearance of plant material was also higher in cores isolated from predators (C series) (Table 5.12). During periods of dry weather, however, the decomposition rate was primarily dependent on litter moisture and not on the accessibility to predators (Kajak & Jakubczyk, 1976).

Table 5.12. *Number of Collembola and disappearance rate of dead plant material in experimental soil cores after predator exclusion*

| No. of days since beginning of experiment | No. of individuals per 100 cm² of soil cores | | | t-test between C and $OC_s$ | Disappearance rate of dead plant material from soil cores ($mg \cdot g^{-1} \cdot d^{-1}$) | | | t-test between C and $OC_s$ |
|---|---|---|---|---|---|---|---|---|
| | | Predators present | | | | Predators present | | |
| | Predators absent (C) | < 1 mm (OF) | All sizes ($OC_s$) | | Predators absent (C) | < 1 m (OF) | All sizes ($OC_s$) | |
| 12 | 305.13 ± 13.39 | 199.80 ± 8.84 | 209.73 ± 6.25 | 6.45* | 6.53 ± 3.20 | 6.39 ± 1.65 | 4.99 ± 1.66 | 0.56 |
| 16 | 80.27 ± 5.78 | 70.90 ± 3.79 | 56.13 ± 1.91 | 3.96* | 1.42 ± 0.62 | 1.55 ± 0.45 | 1.46 ± 0.65 | −0.19 |
| 20 | 99.41 ± 2.71 | 91.50 ± 3.70 | 87.17 ± 1.87 | 4.26* | 0.62 ± 0.18 | 0.56 ± 0.17 | 0.50 ± 0.14 | 1.20 |
| 25 | 63.86 ± 2.64 | 63.72 ± 2.77 | 52.73 ± 2.05 | 3.35* | 1.94 ± 1.66 | 1.35 ± 1.19 | 1.02 ± 0.85 | 0.64 |
| | | ⟵ | | | | ⟵ | | |

[a] Data from 1974. After Kajak & Jakubczyk (1978).
* Differences significant; $P < 0.01$.
⟵, increase.

The results of these experiments show that even the short-term exclusion of predators can stimulate decomposition processes, and suggest that a decrease in predators in an ecosystem may similarly be followed by an increase in the turnover rate of organic matter. Conversely an increase in predators will decrease the turnover rate. A relationship of this type is demonstrated by the gradient of habitats shown in Table 5.11, where the higher the decomposition rate the lower the proportion of energy lost to predators in the ecosystem. On the basis of the above facts it follows that soil-litter predators are one of the factors acting in favour of the retention of matter in an ecosystem.

## Bioenergetics

Some physiological parameters are often used in ecological investigations to evaluate the energy flow through animal populations in ecosystems. An attempt has been made here to review the basic data on the metabolic rates and efficiency of energy transfers in common predatory groups.

### *Assimilation efficiency*

According to the terminology of Petrusewicz & Macfadyen (1970), assimilation ($A$) is the amount of energy ingested used for respiration ($R$) and production ($P$), or

$$A = P + R = C - FU,$$

where $FU$ are faeces plus urine (rejecta), and $C$ is consumption.

The assimilation efficiency ($A/C$) is relatively high in predators (Phillipson, 1960*a*; Turner, 1970; Reichle, 1971; Duncan & Klekowski, 1975). Reichle (1971) found a relatively large variability in assimilation efficiency, which ranged from 0.47 to 0.92 for different predators, but these values were higher than in other trophic groups. A very high efficiency of more than 0.90 is reported for predators by Kozlovsky (1968). Similar efficiencies of 0.82 to 0.93 (Table 5.13) were found for different spiders (Moulder, Reichle & Auerbach, 1970; Van Hook, 1971; Steigen, 1975*b*). A lower assimilation efficiency is characteristic of Opiliones (*Mitopus morio*: Table 5.13).

It is evident that predators feeding on food rich in protein can assimilate it efficiently, but at the same time they frequently use only a part of the prey killed and leave considerable amounts of uneaten food in the habitat. The efficiency of food utilization, or the ratio of consumption to food killed ($C/FK$), can vary remarkably according to the type of prey, degree of hunger and stage of development of predators. Moulder *et al.* (1970) found that spiders which fed on the same prey species used from 18 to 97% of the fresh weight, according to their hunger. Turnbull (1962) and Steigen (1975*b*) relate the differences in the utilization of food to the stage of development of the predator rather than to the food supply.

Table 5.13. *Assimilation efficiency (A/C) of predatory invertebrates*

| Species | $A/C$[a] | References |
|---|---|---|
| Predators (various species) | 0.47–0.92 | Reichle (1971) |
| Predators (various species) | 0.90 | Kozlovsky (1968) |
| *Pardosa* sp. | 0.93 | Van Hook (1971) |
| *Pardosa palustris* | 0.82 | Steigen (1975*b*) |
| *Pardosa rabida* adults | 0.91 | Moulder, Reichle & Auerbach (1970) |
| *Mitopus morio* | | |
| Males | 0.42 | Phillipson (1960*a*) |
| Females | 0.49 | Phillipson (1960*a*) |
| Instars | 0.47–0.74 | Phillipson (1960*b*) |

[a] $A/C$, assimilation/consumption.

A better measure of the effect of predators on prey populations is the number of prey killed, since consumption characterizes only part of their effect.

### *Respiratory metabolic rates*

In recent years much attention has been paid to measurements of energy losses in respiration. General functions have been calculated to describe the relationship between metabolic rate and body size in poikilotherms (Zeuthen, 1953; Reichle, 1971; Ryszkowski, 1975) and metabolic rates of different species have been compared (Edwards, Reichle & Crossley, 1970; Byzova, 1973). These studies show that there are no differences in oxygen consumption between predators and animals of other trophic levels.

The average metabolic rate is one of those parameters which can now be most easily found in the literature when only a general characteristic of metabolism of a population is required. On the other hand metabolic rates vary, even within the same species, depending on such factors as age, sex and physiological condition of individuals, time of day, and climatic conditions (Phillipson, 1962, 1963). Temperature is probably the only factor whose effect on metabolic rate can readily be extrapolated from laboratory measurements to field conditions, because the relationship between metabolic rate and temperature is known.

The effect of locomotory activity is very rarely considered, although it can markedly modify the metabolic rate of an animal. As the locomotory activity of predators searching for food must often be considerable, great differences can be expected between the resting metabolic rate and average daily metabolic rate of such animals. Van Hook (1971) compared the resting and the average oxygen consumption – calculated from known assimilation and production – in two herbivorous species (*Melanoplus sanguinipes* and *Conocephalus fasciatus*), one omnivorous species (*Pteronemobius fasciatus*) and one predatory species (*Pardosa rabida*). The largest ratio of average oxygen consumption to resting oxygen consumption was observed in the predatory species, the value being 3.57 compared with 1.85–2.94 in the other species. Miyashita (1969) found that carbon dioxide output was three to six times higher in active spiders (*Lycosa T-insignita*) than in resting animals. Grüm (1978) calculated the percentage of assimilated energy (*A*) used in locomotory activity in four species of adult Carabidae of the genus *Pterostichus* and four species of the genus *Carabus* and found values of 38–76% for the species which developed in autumn and 4–42% for those which developed in spring. The ratio of average energy losses for respiration to losses for resting metabolism varied from 2.64 to 6.62 in autumnal species and from 1.07 to 1.95 in spring species. Information of this kind, where the losses due to the normal activity of an animal are estimated, is extremely scarce. The data cited above show that the average metabolic rate is several times higher than resting metabolism, but they are insufficient to determine the most frequent values of these parameters.

Table 5.14 shows the metabolic rates of some common arthropod predators along with some regression functions expressing the relationship between the body weight and oxygen consumption of different species. These metabolic rates approximately correspond to the resting metabolism, because the animals were not fed or active during the experiment. To find more reliable estimates of energy loss the measurements are recorded for a 24-hour cycle (Phillipson, 1962, 1963; Moulder *et al.*, 1970; Van Hook, 1971; Manga, 1972; Steigen, 1975*b*). These estimates, however, are still low compared with the losses of energy in normal activity.

A comprehensive comparison of annual production and respiration for different animal species has been presented by McNeil & Lawton (1970). The authors show that the *P*/*R* ratio for poikilotherms depends on the duration of the life cycle. Relatively long-lived organisms (up to several years), particularly those wintering as active forms, use much of the energy assimilated for respiration and therefore have a low production efficiency. Many predatory polyphages are long-lived animals and dissipate considerable amounts of energy for respiration. This is illustrated by the *R*/*A* ratios of different predatory groups (Table 5.15). Kozlovsky (1968) found that predators dissipated more energy (0.62–0.64) than consumers of the first trophic level. Similar figures are reported for spiders: 0.72 (Edgar, 1971), 0.61 (Van Hook, 1971), 0.63 (Moulder *et al.*, 1970). Very high energy losses have been recorded in Chilopoda (Wignarajah, 1968, after McNeil & Lawton, 1970) and in ants (Nielsen, 1972; Horn-Mrozowska, 1976; Pętal, 1977). Very low values of this ratio were found, however, in the larvae of parasitic Hymenoptera (0.34 and 0.42: Chłodny, 1968), which are specific predators with a very short life cycle whose larvae develop inside the host's body, factors which contribute to an economical utilization of energy.

Table 5.14. *Mean metabolic rates of arthropod predators*

| Species | Temp. (°C) | Mean live weight (mg) | Mean metabolic rate ($\mu l\ O_2 \cdot mg^{-1} \cdot h^{-1}$) | Period of life cycle measured | References |
|---|---|---|---|---|---|
| Araneae | | | | | |
| *Pardosa rabida*: Lycosidae | 20 | 71.5 | 0.72 | Adults | Van Hook (1971) |
| | 15 | 17 ± 10.0 | 0.158 | Various stages | Moulder, Reichle & Auerbach (1969) |
| | 15 | 225 ± 17.5 | 0.150 | | |
| *Pardosa palustris* | 18 | 16.6 | 0.59 | Various stages | Steigen (1975*a*) |
| *Tarentula kochii* | 20 | 21.7 ± 7.5 | 0.21 ± 0.07 | Whole | Hagstrum (1970) |
| *Trochosa ruricola* | | 25.7 ± 8.5 | 0.378 ± 0.11 | Instars | Myrcha & Stejgwiłło-Laudańska (1973) |
| | 20 | 49.3 ± 6.1 | 0.38 ± 0.05 | Adult males | |
| | 20 | 120.6 ± 36.8 | 0.27 ± 0.09 | Females | |
| *Phidippus regius* | — | — | 0.099 | | Anderson (1970) |
| *Araneus quadratus* | 20 | 2–603 | 0.679 ± 0.179 | Whole, females | Myrcha & Stejgwiłło-Laudańska (1970) |
| Opiliones | | | | | |
| *Mitopus morio* | 16 | 3.7–56.1 | 0.91 ± 0.12–0.74 ± 0.07 | IV–VI instars, adults | Phillipson (1962) |
| *Oligolophus tridens* | 16 | 3.1–17.3 | 0.51 ± 0.05–2.74 ± 0.37 | IV–VI instars, adults | Phillipson (1962) |
| Carabidae | | | | | |
| *Nebria brevicollis* | 15 | | 0.920–0.443 | I–III instars | Manga (1972) |
| | 15 | | 0.448 | Males | |
| | 15 | | 0.391 | Females | |

| | | | | | |
|---|---|---|---|---|---|
| Staphylinidae | | | | | |
| *Oxytelus* sp. | 20 | 1.0 | 0.245 | Instars | Byzova (1973) |
| *Quedius* sp. | 20 | 6–8 | 0.201 | Instars | Byzova (1973) |
| *Philonthus* sp. | 20 | 7 | 0.150 | Instars | Byzova (1973) |
| Histeridae | | | | | |
| *Hister* sp. | 20 | 6 | 0.258 | Instars | Byzova (1973) |
| Syrphidae | | | | | |
| *Syritta pipiens* | 20 | 17 | 0.632 | Instars | Byzova (1973) |
| Formicidae | | | | | |
| *Myrmica laevinodis* | 19 | 2 | 0.50 | Workers | Brian (1973) |
| Chilopoda | | | | | |
| *Octocryptops sexspinosus* | 20 | 559 | 0.07 | — | Edwards *et al.* (1972) |

Relationship between body weight $W$ in mg live weight and oxygen consumption ($y$) per individual in $\mu l\ O_2 \cdot h^{-1}$ at 20 °C:

| | | |
|---|---|---|
| *Lycosa rabida* | $y = 0.376W^{1.236 \pm 0.238}$ | (after Van Hook, 1971) |
| Lycosidae | $y = 0.21W^{0.92}$ | (after Moulder, Reichle & Auerbach, 1969) |
| *Pardosa palustris* | $y = 1.91W^{0.75}$ | (after Steigen, 1975*b*) |
| Carabidae (14 species), adults | $y = 0.87W^{0.78}$ | (after Byzova, 1975) |

Table 5.15. *Respiration (R) to assimilation (A) efficiency of predatory invertebrates*

| | $R/A$[a] | References |
|---|---|---|
| Predators (various species) | 0.62–0.64 | Kozlovsky (1968) |
| Predatory arthropods | 0.63 | Reichle (1971) |
| Formicidae | | |
| *Lasius alienus* | 0.84 | Nielsen (1972) |
| Carabidae | | |
| *Nebria brevicollis* | 0.52 | Manga (1972) |
| Araneae | 0.63 | Moulder *et al.* (1970) |
| *Pardosa rabida* | 0.61 | Van Hook (1971) |
| *Pardosa lugubris* | 0.72 | Edgar (1971) |
| *Pardosa palustris* | 0.64 | Steigen (1975) |
| Chilopoda | | |
| *Lithobius forticatus* | 0.96 | Wignarajah (1969), after McNeil & Lawton (1970) |
| *Lithobius crassipes* | 0.95 | Wignarajah (1969), after McNeil & Lawton (1970) |
| Parasitic Hymenoptera | | |
| *Pteromalus puparum* larvae | 0.42 | Chłodny (1968) |
| *Pimpla instigator*, larvae | 0.34 | Chłodny (1968) |

[a] $R$ and $A$ in kcal·yr$^{-1}$.

### *Energy content of the body*

Calorific values of the bodies of predators are a little higher than those of other invertebrates. This has been shown by Wallwork (1975) who compared the calorific equivalents of 15 species of saprophages with the same number of predatory species. The average calorific values of the saprophages was $4698 \pm 54$ cal·g dry wt$^{-1}$, while that of predators reached $5219 \pm 81$ cal·g dry wt$^{-1}$. The difference between these two values is significant ($P < 0.01$). These results have been confirmed by other authors. Naess, Steigen & Solhoy (1975) report average calorific values for a number of invertebrate species occurring in the region of Hardangervidda. Calculations based on the average calorific values of 11 predatory species and 14 non-predatory species show that the values are higher in the predatory forms (Table 5.16).

In a study of the chemical composition of animals of 11 orders inhabiting a mown meadow, Mochnacka-Ławacz (1978) has found that the carbon content, and as a consequence the calorific value, is higher in predatory invertebrates than in herbivorous and saprophagous groups (Table 5.16).

Differences in the calorific values between representatives of different trophic levels can only be adequately demonstrated when data are drawn from the same site. This is because the calorific value of an organism varies with trophic conditions (Slobodkin & Richman, 1961) and climatic conditions (Naess *et al.*, 1975).

Table 5.16. *Body calorific content of predatory and nonpredatory invertebrates*

| Trophic level | Calorific content cal·g⁻¹ dry wt[a] | cal·g⁻¹ dry wt[b] | cal·g⁻¹ ash-free dry wt[c] |
|---|---|---|---|
| Herbivores | 4046±7.0[d] | | |
| Saprophages | 4088±9.0 | 4698±54 | 5945 |
| Predators | | 5219±81 | 6031 |
| Obligatory | 4672±15 | | |
| Non-obligatory | 4529±29 | | |

[a] From Mochnacka-Ławacz (1978).
[b] From Wallwork (1975).
[c] From Naess *et al.* (1975).
[d] Recalculated from carbon content assuming 1 g C = 9361 cal. (After Vinberg, 1972.)

## Conclusions

In this chapter the predator trophic level in grassland ecosystems of different types, including natural and managed grasslands, was analysed. Some of the more important points can be summarized as follows:

(1) In grasslands the bulk of predator biomass is concentrated at the soil–litter interface. Also the maxima of all other processes occur in this layer.

(2) In ecosystems not utilized by man predators account for 5–40% (0.5–80 mg dry wt·m$^{-2}$) of the total biomass of the aboveground fauna, while in soil and litter the proportion of predators is lower, varying from 0.4–12% at a considerably higher biomass (6–3000 mg dry wt·m$^{-2}$).

(3) The biomass of invertebrates is determined by plant production, in other words by the fertility of a habitat. Thus, not only primary consumers but also predators depend on plant production. A statistically significant regression function has been found between the production of plant biomass (live or dead) and predator biomass. It is also the case, however, that there is a higher proportion of predators in poor than rich habitats, and this is likely to mask the relationship between predators and plant production.

(4) The management of grasslands in general has a limiting effect on the biomass of predators and their proportion in a community. Even treatments which stimulate the development of the fauna, for example the addition of organic fertilizers, reduce the number of predators.

(5) It is suggested that it is mainly the polyphagous predators that are reduced in managed grasslands, while prey-specific predators may even become more numerous. Elimination of the polyphagous predators, which are capable of controlling prey populations at low densities, can result in unchecked populations of some prey species. Such species may be controlled

by prey-specific predators, but these act only after a time lag and lead to marked increases in the abundance of prey species.

(6) Data on the exploitation of prey populations by predators in grassland ecosystems are scarce and not always in agreement. They suggest that the effect of predators may vary from taking a very low percentage to almost the whole production of prey species at the lower trophic level. On the whole the role of predators is little-known, and has been neglected in many energy flow studies. Investigations on elimination rates and population size regulation are usually done in areas where there have been pest outbreaks, and very rarely in areas where the number of animals is maintained at a relatively steady level. This is probably the reason why more attention is paid to predators associated with phytophages than with saprophages.

Some evidence is presented which indicates that polyphagous predators have a considerable role in limiting the number of phytophagous insects.

(7) It has been shown that the proportion of soil-inhabiting predators is relatively high in those habitats characterized by a slow decomposition rate (of the order of several years) of dead plant material. This proportion decreases gradually as the turnover rate of the ecosystem increases. Experimental results show that litter predators can have a limiting effect on the rate of disappearance of material. This is certainly an indirect effect, through reducing the number of those organisms which feed on litter material.

I am very grateful to Dr Robin Andrews for sending me unpublished data from the Grassland Biome Data Bank, Dr Elżbieta Olechowicz for permission to use her unpublished data and Dr Teresa Wierzbowska who provided valuable help with statistical problems. This work was partially supported by the Second Faculty of the Polish Academy of Sciences and Polish Committee for IBP. I extend my gratitude to these institutions.

## References

Anderson, J. (1970). Metabolic rates of spiders. *Comparative Biochemistry and Physiology*, **33**, 51–72.

Andrews, R., Coleman, D. C., Ellis, J. E. & Singh, J. S. (1974). Energy flow relationship in a shortgrass prairie ecosystem. In *Proceedings of the First International Congress of Ecology*, The Hague, 8–14 September, pp. 22–8. Pudoc, Wageningen.

Andrzewjeska, L. (1971). Productivity investigation of two types of meadows in the Vistula Valley. VII. Production and population density of leafhopper (Homoptera – Auchenorrhyncha) communities. *Ekologia Polska*, **19**, 151–72.

Andrzejewska, L. (1976). The effect of mineral fertilization on phytophagous fauna of a meadow. *Polish Ecological Studies*, **2**, 93–109.

Andrzejewska, L. & Kajak, A. (1966). Metodyka entomologicznych badań ilościowych na łąkach – *Ekologia Polska, Series B*, **12**, 241–61.

Athias, E., Josens, G. & Lavelle, P. (1974). Analyse d'un écosystème tropical humide: la savane de Lamto (Côte d'Ivoire). Le peuplement animal des sols de la savane de Lamto. In *Les organismes endogés: Bulletin de Liaison des Chercheurs de Lamto, no. 5*, pp. 45–55.

Bourlière, F. & Hadley, M. (1970). The ecology of tropical savannas. *Annual Review of Ecology and Systematics*, **1**, 125–53.

Breymeyer, A. (1971). Productivity investigation of two types of meadows in the Vistula Valley. XIII. Some regularities in structure and function of the ecosystem. *Ekologia Polska*, **19**, 249–60.

Breymeyer, A., Gałecka, B., Kajak, A. & Łuczak, J. (1964). Różne aspekty oddziaływania drapieżcy na liczebność ofiar i warunki modyfikujące to działanie. *Polskie Pismo Entomologiczne, Series B*, **1/2**, 79–87.

Brian, M. V. (1973). Feeding and growth in the ant *Myrmica*. *Journal of Animal Ecology*, **42**, 37–53.

Byzova, Yu. B. (1973). Dykhane pochvennȳkh bezpozvonochnȳkh. In *Ekologiya pochvennȳkh bezpozvonochnȳkh*, ed. M. S. Ghilarov, pp. 3–40. Izdatel'stvo Nauka, Moscow.

Cesar, J. & Menault, J. C. (1974). Analyse d'un écosystème tropical humide: la savane de Lamto (Côte d'Ivoire). Le peuplement végétal. *Bulletin de Liaison des Chercheurs de Lamto, no. 2*, 161 pp.

Cherrett, I. M. (1964). The distribution of spiders on the Moor House National Nature Reserve, Westmorland. *Journal of Animal Ecology*, **33**, 27–48.

Chew, R. M. (1974). Consumers as regulators of ecosystems: an alternative to energetics. *Ohio Journal of Science*, **74**, 359–70.

Chłodny, J. (1968). Evaluation of some parameters of the individuals' energy budget of the larvae of *Pteromalus puparum* (L.) (Pteromalidae) and *Pimpla instigator* (Fabr.) (Ichneumonidae). *Ekologia Polska, Series A*, **16**, 505–14.

Crossley, D. A. Jr & Howden, H. F. (1961). Insect vegetation relationships in an area contaminated by radioactive wastes. *Ecology*, **42**, 302–17.

Dąbrowska-Prot, E., Karg, J. & Ryszkowski, L. (1974). An attempt to estimate the role of invertebrates in agrocenotic economics. In *Ecological effects of intensive agriculture*, ed. L. Ryszkowski, pp. 41–62. Polish Scientific Publishers, Warsaw.

Dąbrowska-Prot, E., Łuczak, J. & Tarwid, K. (1968). Prey and predator density and their reactions in the process of mosquito reduction by spiders in field experiments. *Ekologia Polska, Series A*, **16**, 773–819.

Delchev, Kh. & Kajak, A. (1974). Analysis of a sheep pasture ecosystem in the Pieniny mountains (the Carpathians). XVI. Effect of pasture management on the number and biomass of spiders (Araneae) in two climatic regions (The Pieniny and the Sredna Gora mountains). *Ekologia Polska*, **22**, 693–710.

Dempster, J. P. (1970). Some effects of grazing on the population ecology of the Cinnabar Moth. In *The scientific management of animal and plant communities for conservation*, ed. E. Duffey & A. S. Watt, *Symposium of the British Ecological Society 11*, pp. 517–26. Blackwell Scientific Publications, Oxford.

Dondale, C. D. (1971). Spiders of Heasman's field, a mown meadow near Belleville, Ontario. *Proceedings of the Entomological Society of Ontario*, **101**, 62–9.

Duffey, E. (1962). A population study of spiders in limestone grassland. Description of study area, sampling methods and population characteristics. *Journal of Animal Ecology*, **31**, 571–99.

Duncan, A. & Klekowski, R. Z. (1975). Parameters of energy budget. In *Methods for ecological bioenergetics*, ed. W. Grodziński, R. Z. Klekowski & A. Duncan, *IBP Handbook 24*, pp. 97–148. Blackwell Scientific Publications, Oxford.

Dunger, W. (1968). Produktionsbiologische Untersuchungen an der Collembolen-Fauna gestörter Böden. *Pedobiologia*, **8**, 16–22.

East, R. (1974). Predation on the soil-dwelling stages of the winter moth at Wytham Woods, Berkshire. *Journal of Animal Ecology*, **43**, 611–26.

Edgar, W. D. (1971). Aspects of ecological energetics of the wolf spider *Pardosa* (*Lycosa*) *lugubris* (Walckenaer). *Oecologia*, **7**, 136–54.

Edwards, C. A., Reichle, D. E. & Crossley, D. A. Jr (1970). The role of soil invertebrates in turnover of organic matter and nutrients. In *Analysis of temperate forest ecosystems*, ed. D. E. Reichle, pp. 147–72. Springer-Verlag, Berlin.

Ellstrom, M. A. & Watts, J. G. (1974). *Populations and trophic structure of a grassland invertebrate community. US/IBP Grassland Biome Technical Report 248.* University of Colorado, Fort Collins, Colorado.

Gillon, Y. (1974). Analyse d'un écosystème tropical humide: la savane de Lamto (Côte d'Ivoire). Conclusions. In *Les invertébrés épigés. Bulletin de liaison des Chercheurs de Lamto, no. 3*, pp. 147–50.

Gillon, Y. & Gillon, D. (1974). Analyse d'un écosystème tropical humide: la savane de Lamto (Côte d'Ivoire). Traits généraux du peuplement des Arthropodes non sociaux de la savane de Lamto. In *Les invértébres épigés. Bulletin de Liaison des Chercheurs de Lamto, no. 3*, pp. 3–23.

Grüm, L. (1978). Mortality rates of the mobile and immobile stages in the life-cycle of carabids. In *Proceedings of XV International Congress of Entomology*, Washington, DC. (In press.)

Hagstrum, D. W. (1970). Physiology of food utilization by the spider *Tarentula kochii* (Araneae: Lycosidae). *Annals of the Entomological Society of America*, **63**, 1305–8.

Haukioja, E. & Hakala, T. (1975). Herbivore cycles and periodic outbreaks. Formulation of a general hypothesis. *Report of Kevo Subarctic Research Station*, **12**, 109.

Hillbricht-Ilkowska, A. (1977). Trophic relations and energy flow in pelagic plankton. *Polish Ecological Studies*, **3**, 3–98.

Holling, C. S. (1959). The components of predation as revealed by a study of small mammal predation of a European pine sawfly. *Canadian Entomologist*, **91**, 293–320.

Horn-Mrozowska, E. (1976). Energy budget of an experimental nest of *Formica pratensis* Retzius (Hymenoptera, Formicidae). *Polish Ecological Studies*, **2**, 55–98.

Huffaker, C. B. (1958). Experimental studies on predation. II. Dispersion factors and predator–prey oscillations. *Hilgardia*, **27**, 343–83.

Huffaker, C. B., Messenger, P. S. & DeBach, P. (1971). The natural enemy component in natural control and the theory of biological control. In *Biological control*, ed. C. B. Huffaker, pp. 16–67. Plenum Press, New York.

Ito, Y., Miyashita, K. & Sekiguchi, K. (1962). Studies on the predators of rice crop insect pests using the insecticidal check method. *Japanese Journal of Ecology*, **12**, 1–11.

Kabacik-Wasylik, D. (1975). Research into the number, biomass and energy flow of Carabidae (Coleoptera) communities in rye and potato fields. *Polish Ecological Studies*, **1**, 111–21.

Kaczmarek, W. (1969). Liczebność populacji a obfitość pokarmu w zrównoważonych łańcuchach troficznych. *Ekologia Polska, Series B*, **15**, 71–6.

Kajak, A. (1962). Porównanie fauny pająków łąk sztucznych i naturalnych. *Ekologia Polska, Series A*, **20**, 1–20.

Kajak, A. (1965). An analysis of food relations between the spiders *Araneus cornutus* Clerck and *Araneus quadratus* Clerck and their prey in a meadow. *Ekologia Polska, Series A*, **13**, 717–64.

Kajak, A. (1967). Productivity of some populations of web spiders. In *Secondary productivity of terrestrial ecosystems*, ed. K. Petrusewicz, vol. 2, pp. 807–20. Polish Scientific Publishers, Warsaw.

Kajak, A. (1971). Productivity investigation of two types of meadows in the Vistula Valley. IX. Production and consumption of field layer spiders. *Ekologia Polska*, **19**, 197–211.

Kajak, A. (1974). Analysis of a sheep pasture ecosystem in the Pieniny mountains (the Carpathians). XVIII. Analysis of the transfer of carbon. *Ekologia Polska*, **22**, 711–32.

Kajak, A., Breymeyer, A. & Pętal, J. (1971). Productivity investigation of two types of meadows in the Vistula Valley. XI. Predatory arthropods. *Ekologia Polska*, **19**, 223–32.

Kajak, A., Breymeyer, A., Pętal, J. & Olechowicz, E. (1972). The influence of ants on the meadow invertebrates. *Ekologia Polska*, **17**, 163–71.

Kajak, A. & Jakubczyk, H. (1975). Experimental studies on spider predation. In *Proceedings of the VI International Arachnological Congress 1974*, pp. 82–5. Nederlandse Entomologische Verenigung, Amsterdam.

Kajak, A. & Jakubczyk, H. (1976). Experiments on the influence of predatory arthropods on the number of saprophages and disappearance rate of dead plant material. *Polish Ecological Studies*, **2**, 219–29.

Kajak, A. & Jakubczyk, H. (1978). Experimental studies on predation in soil–litter interface. In *Soil organisms as components of ecosystems*, ed. U. Lohm & T. Persson, *Proceedings of the VI International Soil Zoology Colloquium, Ecological Bulletin 25*. Swedish Natural Science Research Council, Stockholm.

Kajak, A., Olechowicz, E. & Pętal, J. (1972). The influence of ants and spiders on the elimination of Diptera on meadows. In *Proceedings of the XIII International Congress of Entomology*, vol. 3, pp. 364–7. Nauka Publishing House, Leningrad.

Kayashima, I. (1960). Studies on spiders as natural enemies of crop pests. I. Daily activities of spiders in the cabbage fields, establishment of spiders liberated in the fields and evolution of the effectiveness of spiders against crop pests. *Scientific Bulletin of the Faculty of Agriculture, Kyushu University*, **18**, 1–24.

Kazimirov, N. I. & Morozova, R. M. (1973). *Biologicheskiĭ krugovorot veshchestv v elnikakh Karelii*. Izdatel'stvo Akademii Nauk, Leningrad.

Kiritani, K. & Kakiya, N. (1975). An analysis of the predator–prey system in the paddy field. *Researches on Population Ecology*, **17**, 29–38.

Kiritani, K., Kawahara, S., Sasaba, T. & Nakasuji, F. (1972). Quantitative evaluation of predation by spiders on the green rice leafhopper, *Nephotettix cincticeps* Uhler, by a sight-count method. *Researches on Population Ecology*, **13**, 187–200.

Klomp, H. (1962). The influence of climate and weather on the mean density level, the fluctuations and the regulation of animal populations. *Archives Néerlandaises de Zoologie*, **15**, 68–109.

Kozlovsky, D. G. (1968). A critical evaluation of the trophic level concept. I. Ecological efficiencies. *Ecology*, **49**, 48–60.

Lavelle, P. & Schaefer, R. (1974). Analyse d'un écosystème tropical humide: la savane de Lamto (Côte d'Ivoire). Les sources de nourriture des organismes du sol. In *Les organismes endogés. Bulletin de Liaison des Chercheurs de Lamto, no. 5*, pp. 27–38.

Lavigne, R. & Rogers, W. E. (1970). *Effect of insect predators and parasites on grass feeding insects, Pawnee site. US/IBP Grassland Biome Technical Report 20.* Colorado State University, Fort Collins, Colorado.

Lavigne, R. I., Rogers, L. E. & Chu, I. (1971). *Data collected on the Pawnee site relating to western harvester ant and insect predators and parasites. US/IBP Grassland Biome Technical Report 107*. Colorado State University, Fort Collins, Colorado.

Lewis, J. K. (1971). The grassland biome: a synthesis of structure and function. In *Preliminary analysis of structure and function in grassland*, ed. N. R. French, *Range Science Department Science Series No. 10*, pp. 317–87. Colorado State University, Fort Collins, Colorado.

Lloyd, J. E., Kumar, R., Grow, R. R., Leetham, J. W. & Keith, V. (1973). *Abundance and biomass of soil macroinvertebrates of the Pawnee Site collected from pastures subjected to different grazing pressures, irrigation and/or nitrogen fertilization 1970–71. US/IBP Grassland Biome Technical Report 239*. Colorado State University, Fort Collins, Colorado.

Łuczak, J. (1975). Spider communities of the crop-fields. *Polish Ecological Studies*, **1**, 93–110.

McBrayer, J. F., Reichle, D. E. & Witkamp, M. (1974). *Energy flow and nutrient cycling in a cryptozoan food-web*. Oak Ridge National Laboratory, Tennessee.

McDaniel, B. (1971). The role of invertebrates in the grassland biome. In *Preliminary analysis of structure and function in grasslands*, ed. N. R. French, *Range Science Department Science Series No. 10*, pp. 267–315. Colorado State University, Fort Collins, Colorado.

Macfadyen, A. (1963). The contribution of the microfauna to total soil metabolism. In *Soil organisms*, ed. J. Doeksen & J. van der Drift, pp. 3–16. North-Holland Publishing Co., Amsterdam.

McNeil, S. & Lawton, J. H. (1970). Annual production and respiration. *Nature, London*, **225**, 472–4.

Makulec, G. (1976). The effect of NPK fertilization on the population of enchy traeid worms. *Polish Ecological Studies*, **2**, 183–93.

Manga, N. (1972). Population metabolism of *Nebria brevicollis* (F.) (Coleoptera: Carabidae). *Oecologia*, **10**, 223–42.

Menhinick, E. F. (1967). Structure, stability and energy flow in plants and arthropods in a *Sericea lespedeza* stand. *Ecological Monographs*, **37**, 255–72.

Miyashita, K. (1969). Effect of locomotory activity, temperature and hunger on the respiratory rate of *Lycosa T-insignita* Boes. et Str. (Araneae: Lycosidae). *Applied Entomology and Zoology*, **4**, 105–13.

Mochnacka-Ławacz, H. (1978). Skład chemiczny bezkręgowców na łące mineralnie nawożonej NPK. *Polish Ecological Studies*, **3**, in press.

Morris, R. F. (1963). The dynamics of epidemic budworm populations. *Memoirs of the Entomological Society of Canada*, **31**, 1–332.

Morris, M. G. (1968). Differences between the invertebrate faunas of grazed and ungrazed chalk grassland. II. The faunas of sample turves. *Journal of Applied Ecology*, **5**, 601–11.

Moulder, B. C., Reichle, D. E. & Auerbach, S. I. (1970). *Significance of spider predation in the energy dynamics of forest floor arthropod communities*. Oak Ridge National Laboratory, US Atomic Energy Commission.

Myrcha, A. & Stejgwiłło-Laudańska, B. (1970). Resting metabolism of *Araneus quadratus* (Clerck) females. *Bulletin de l'Académie Polonaise des Sciences, Classe II*, **18**, 257–9.

Myrcha, A. & Stejgwiłło-Laudańska, B. (1973). Changes in the metabolic rate of starved Lycosidae spiders. *Bulletin de l'Académie Polonaise des Sciences, Classe II*, **21**, 209–13.

Naess, S. J., Steigen, A. L. & Solhoy, T. (1975). Standing crop and calorific content in invertebrates from Hardangervidda. In *Fennoscandian tundra ecosystems*, vol. 2, *Animals and systems analysis*, ed. F. E. Wielgolaski, pp. 151–9. Springer-Verlag, Berlin.

Nagamine, M., Teruya, R. & Ito, Y. (1975). A life table of *Mogannia iwasakii* (Homoptera: Cicadiidae) in sugarcane field of Okinawa. *Researches on Population Ecology*, **17**, 39–50.

Nielsen, M. G. (1972). An attempt to estimate energy flow through a population of workers of *Lasius alienus* (Först) (Hymenoptera: Formicidae). *Natura Jutlandica*, **16**, 99–107.

Nowak, E. (1975). Population density of earthworms and some elements of their production in several grassland environments. *Ekologia Polska*, **23**, 459–91.

Nowak, E. (1976). The effect of fertilization on earthworms and soil macrofauna components. *Polish ecological Studies*, **2**, 195–207.

Odum, E. P. (1971). *Fundamentals of ecology*. W. B. Saunders, Philadelphia.

Olechowicz, E. (1970). Evaluation of number of insects emerging in meadow environment. *Bulletin de l'Académie Polonaise des Sciences, Classe II*, **18**, 389–95.

Olechowicz, E. (1971). Productivity investigation of two types of meadows in the Vistula Valley. VIII. The number of emerged Diptera and their elimination. *Ekologia Polska*, **19**, 183–95.

Olechowicz, E. (1976). The effect of mineral fertilizing on the insect community of the herbage in a mown meadow. *Polish ecological Studies*, **2**, 129–36.

Pętal, J. (1967). Productivity and consumption in a *Myrmica laevinodis* population. In *Secondary productivity of terrestrial ecosystems*, ed. K. Petrusewicz, vol. 2, pp. 841–57. Polish Academy of Sciences, Warsaw.

Pętal, J. (1972). Methods of investigating of the productivity of ants. *Ekologia Polska*, **20**, 9–22.

Pętal, J. (1974). Analysis of a sheep pasture ecosystem in the Pieniny mountains (the Carpathians). XV. The effect of pasture management on ant population. *Ekologia Polska*, **22**, 679–92.

Pętal, J. (1976). The effect of mineral fertilization on ant populations in mown meadows. *Polish Ecological Studies*, **2**, 209–18.

Pętal, J. (1977). The role of ants in ecosystems. *Production ecology of ants and termites, IBP Synthesis Series 13*, ed. M. V. Brian, pp. 293–325. Cambridge University Press, London.

Pętal, J., Andrzejewska, L., Breymeyer, A. & Olechowicz, E. (1971). Productivity investigation of two types of meadows in the Vistula Valley. X. Role of the ants as predators in a habitat. *Ekologia Polska*, **19**, 213–22.

Pętal, J. & Breymeyer, A. (1968). Reduction of wandering spiders by ants in a *Stellario-Deschampsietum* meadow. *Bulletin de l'Académie Polonaise des Sciences, Classe II*, **17**, 239–44.

Petrusewicz, K. & Grodziński, W. (1975). The role of herbivore consumers in various ecosystems. In *Productivity of world ecosystems*, ed. D. E. Reichle, I. F. Franklin & D. W. Goodall, pp. 64–70. National Academy of Sciences, Washington, DC.

Petrusewicz, K. & Macfadyen, A. (1970). *Productivity of terrestrial animals: principles and methods. IBP Handbook 13*. Blackwell Scientific Publications, Oxford.

Phillipson, J. (1960*a*). A contribution to the feeding biology of *Mitopus morio* (F.) (Phalangida). *Journal of Animal Ecology*, **29**, 35–43.

Phillipson, J. (1960*b*). The food consumption of different instars of *Mitopus morio* (F.) (Phalangida) under natural conditions. *Journal of Animal Ecology*, **29**, 299–307.

Phillipson, J. (1962). Respirometry and the study of energy turnover in natural systems with particular reference to harvest spiders (Phalangida). *Oikos*, **13**, 311–22.

Phillipson, J. (1963). The use of respiratory data in estimating annual respiratory metabolism, with particular reference to *Leiobunum rotundum* (Latr.) (Phalangida). *Oikos*, **14**, 212–23.

Pomianowska-Pilipiuk, I. (1976). The contribution of saprophages to the disappearance of plant residues on Bródno meadow. *Polish Ecological Studies*, **2**, 287–97.

Rafes, P. M. (1968). *Rol'i znachenye rastitelnoyadnykh nasekomykh v lesu.* Izdatel'stvo Nauka, Moscow.

Reichert, S. E. (1974). Thoughts on the ecological significance of spiders. *Bioscience*, **24**, 352–6.

Reichle, D. E. (1968). Relation of body size to food intake, oxygen consumption, and trace element metabolism in forest floor arthropods. *Ecology*, **49**, 538–42.

Reichle, D. E. (1971). Energy and nutrient metabolism of soil and litter invertebrates. In *Productivity of forest ecosystems*, ed. P. Duvigneaud, *Proceedings of the Brussels Symposium 1969, Paris*, pp. 465–77. UNESCO, Paris.

Ryszkowski, L. (1975). Energy and matter economy of ecosystems. In *Unifying concepts in ecology*, ed. W. H. van Dobben & R. H. Lowe-McConnell, pp. 109–26. W. Junk, Wageningen, The Netherlands.

Salt, G., Hollick, F. S. J., Raw, F. & Brian, M. V. (1948). The arthropod population of pasture soil. *Journal of Animal Ecology*, **17**, 139–50.

Sasaba, T. & Kiritani, K. (1972). Evaluation of mortality factors with special reference to parasitism of the green rice leafhoppers, *Neophotettix cincticeps* Uhler (Hemiptera: Deltocephalidae). *Applied Entomology and Zoology*, **7**, 83–93.

Sasaba, T. & Kiritani, K. (1975). A system model and computer simulation of the green rice leafhopper populations in control programmes. *Researches on Population Ecology*, **16**, 231–44.

Sasaba, T., Kiritani, K. & Urabe, T. (1973). A preliminary model to simulate the effect of insecticides on a spider–leafhopper system in the paddy field. *Researches on Population Ecology*, **15**, 9–22.

Sims, P. L. & Singh, J. S. (1971). Herbage dynamics and net primary production in certain ungrazed and grazed grasslands in North America. In *Preliminary analysis of structure and function in grasslands*, ed. N. R. French, *Range Science Department No. 10, Science Series*, pp. 59–127. Colorado State University, Fort Collins, Colorado.

Slobodkin, L. B. & Richman, S. (1961). Calories/gm in species of animals. *Nature, London*, **191**, 107.

Smolik, J. D. (1974). *Nematode studies at the Cottonwood site in Grassland Biome. US/IBP Grassland Biome. Technical Report 251.* Colorado State University, Fort Collins, Colorado.

Southwood, T. R. E. & Emden, van H. F. (1967). A comparison of the fauna of cut and uncut grasslands. *Zeitschrift für Angewandte Entomologie*, **60**, 188–98.

Steigen, A. L. (1975*a*). Respiratory rates and respiratory energy loss in terrestrial invertebrates from Hardangervidda. In *Fennoscandian tundra ecosystems*, vol. 2, *Animals and systems analysis*, ed. F. E. Wielgolaski, pp. 122–8. Springer-Verlag, Berlin.

Steigen, A. L. (1975*b*). Energetics in a population of *Pardosa palustris* (L.) (Araneae, Lycosidae) on Hardangervidda. In *Fennoscandian tundra ecosystems*, vol. 2, *Animals and systems analysis*, ed. F. E. Wielgolaski, pp. 129–44. Springer-Verlag, Berlin.

Titova, E. V. & Zhavoronkova, T. N. (1965). Vliyane raspashki celinnoï stepi na sostav i chisłennost' v populatsiyakh zhuzhelits (Carabidae). *Trudy Vsesoyuznogo Entomologicheskogo Obshchestva*, **50**, 103–20.

Traczyk, T. (1971). Productivity investigation of two types of meadows in the Vistula Valley. I. Geobotanical description and primary production. *Ekologia Polska*, **19**, 93–106.

Trojan, P. (1967). Investigations on production of cultivated fields. In *Secondary productivity of terrestrial ecosystems*, ed. K. Petrusewicz, vol. 2, pp. 545–58. Polish Academy of Sciences, Warsaw.

Turnbull, A. L. (1962). Quantitative studies of the food of *Linyphia triangularis* Clerck (Araneae, Linyphiidae). *Canadian Entomologist*, **94**, 1233–49.

Turnbull, A. L. (1966). A population of spiders and their potential prey in an overgrazed pasture in eastern Ontario. *Canadian Entomologist*, **44**, 557–83.

Turner, F. B. (1970). The ecological efficiency of consumer populations. *Ecology*, **51**, 741–2.

Van Hook, R. I. (1971). Energy and nutrient dynamics of spider and orthopteran populations in a grassland ecosystem. *Ecological Monographs*, **41**, 1–26.

Van Rhee, J. A. (1963). Earthworm activities and the breakdown of organic matter in agricultural soils. In *Soil organisms*, ed. J. Doeksen & J. van der Drift, pp. 55–9. North-Holland Publishing Co., Amsterdam.

Varley, G. C. & Gradwell, G. R. (1963). The interpretation of insect population changes. *Proceedings of the Ceylon Association of Advanced Science*, **18**, 142–56.

Varley, G. C. & Gradwell, G. R. (1971). The use of models and life tables in assessing the role of natural enemies. In *Biological control*, ed. C. B. Huffaker, pp. 93–112. Plenum Press, New York.

Varley, G. C., Gradwell, G. R. & Hassell, M. P. (1975). *Insect population ecology, an analytical approach*. Blackwell Scientific Publications, Oxford.

Vinberg, G. G. (1972). *Oboznaceniya, edinicy izmereniya i ekvivalenty vstrechaemye pri izuchenii produktivnosti presnykh vod*. Akademia Nauk USSR, Leningrad.

Wallwork, J. A. (1970). *Ecology of soil animals*. McGraw-Hill, New York.

Wallwork, J. A. (1975). Calorimetric studies on soil invertebrates and their ecological significance. In *Progress in soil zoology, Proceedings of the V International Colloquium of Soil Zoology, Prague 1973*, pp. 231–40. Akademia Praha, Prague.

Wasilewska, L. (1974). Analysis of a sheep pasture ecosystem in the Pieniny mountains (the Carpathians). XIII: Quantitative distribution, respiratory metabolism and some suggestions on predation of nematodes. *Ekologia Polska*, **22**, 651–68.

Wasilewska, L. (1976). The role of nematodes in the ecosystem of a meadow in Warsaw environs. *Polish Ecological Studies*, **2**, 137–56.

Wasilewska, L. & Paplińska, E. (1975). Energy flow through the nematode community in a rye-crop in the region of Poznań. *Polish Ecological Studies*, **1**, 75–82.

Wiegert, R. G. & Evans, F. C. (1967). Investigations of secondary productivity in grassland. In *Secondary productivity of terrestrial ecosystems*, vol. 2, ed. K. Petrusewicz, pp. 499–518. Polish Academy of Sciences, Warsaw.

Willard, J. R. (1974). *Soil invertebrates. VIII. A summary of populations and biomass. Technical Report 56, Matador Project*. University of Saskatchewan, Saskatoon.

Wolcott, G. N. (1937). An animal census of two pastures and a meadow in northern New York. *Ecological Monographs*, **7**, 1–90.

Zeuthen, R. (1953). Oxygen uptake as related to body size in organisms. *Quarterly Review of Biology*, **28**, 1–12.

Zlotin, R. I. (1966). Zonalnȳe osobennosti naseleniya nazemnȳkh zhivotnykh. In *Opyt kharakteristiki naseleniya nazemnykh chlenistonogikh na syrtakh vnutrennogo Tian-Shanya*, pp. 92–127. Izdatel'stvo Nauka, Moscow.

Zlotin, R. I. (1969). Sravnenie pochvennȳkh biotsenozov nekotorȳkh estestvennȳkh i selskochozyaïstvennȳkh ugodiĭ sredneĭ lesostepi. In *Sinantropizatsiya i domestikatsiya zhivotnogo naseleniya*, ed. Yu. A. Isakov & W. K. Rakhlin, pp. 94–7. Institut Geografii Akademii Nauk USSR, Moscow.

Zlotin, R. I. (1975). *Zhizn'v vysokogoryakh.* Izdatel'stvo Mȳsl', Moscow.

Zlotin, R. I. & Khodashova, K. S. (1974). *Rol'zhivotnykh v biologicheskom krugovorote lesostepnȳkh ekosistem.* Izdatel'stvo Nauka, Moscow.

Żyromska-Rudzka. H. (1974). Analysis of a sheep pasture ecosystem in the Pieniny mountains (the Carpathians). XIV. The occurrence of Oribatid mites, intermediate hosts of Cestodes. *Ekologia Polska*, **22**, 669–78.

Żyromska-Rudzka, H. (1976). Response of Acarina–Oribatei and other mesofauna soil components on mineral fertilizing. *Polish Ecological Studies*, **2**, 157–82.

## Appendix 5.1. Methods of sampling aboveground invertebrates, and predatory taxa analysed

(*a*) Tien Shan ecosystems

| | Deserts | Semideserts | Arid steppes | Moist alpine meadows | Marshes |
|---|---|---|---|---|---|
| Latitude (°N) | 41 | 41 | 41 | 41 | 41 |
| Altitude (m) | 3000–3200 | 3150–3300 | 3200–3500 | 3500–3800 | 3000–3900 |
| Mean annual temp. (°C) | −7.7 at 3000 m | | | | |
| Annual sum of positive temps. | 1030 | 880 | 730 | 320 | |
| Annual precipitation (mm) | 150–210 | 190–230 | 210–260 | 280–360 | |
| Dominant plant species | *Artemisia rhodantha*, *Ptilagrostis subsessiliflora* | (*a*) *A. rhodantha*, *P. subsessiliflora*, *Stipa caucasica*, *S. krylovii* (*b*) *A. rhodantha*, *Oxytropis rupifraga* | (*a*) *F. kryloviana*, *Leucopoda olgae* (*b*) *P. subsessiliflora*, *F. sulcata*, *F. kryloviana* | (*a*) *Caragana jubata*, *Cobresia capilliformis* (*b*) *C. capilliformis* | *Carex melanantha*, *Hypnum cupressiformae* |
| Methods applied | Animals counted by collecting individuals within transects 50 m long, 1–3 m wide for large animals (> 7 mm in size) or microtransects 3–30 m long, 0.25–1 m wide for smaller arthropods. The number of transects depended on the density of animals | | | | |
| Taxon analysed | Arthropods mainly | | | | |
| Dominant predatory taxa | Ar.: Lycosidae, Thomisidae, Drassidae; Ac.: Trombididae; Col.: Cicindellidae; Dip.: Asilidae | Ar.: Lycosidae, Thomisidae, Theridiidae, Drassidae; Ac.: Trombidiidae; Col.: Carabidae, Cicindellidae; Dip.: Asilidae | Ar.: Lycosidae, Thomisidae, Theridiidae; Ac.: Trombidiidae; Hym.: Braconidae, Ichneumonidae; Col.: Cantharidae, Carabidae; Dip.: Asilidae | Ar.: Lycosidae; Hem.: Reduviidae; Dip.: Syrphidae; Col.: Carabidae | Ar.: Lycosidae; Hym.: Braconidae; Chalcididae |

After Zlotin (1966, 1975).

**Appendix 5.1** (*continued*)

(*b*)

| | Tropical savanna, Lamto, Ivory Coast[a] | Strzeleckie meadows, Kampinos Forest, Poland[b] |
|---|---|---|
| Latitude (°N) | 6 | 52° 2′ |
| Altitude (m) | — | 70 |
| Mean annual temp. (°C) | 27 | 14 |
| Annual precipitation (mm) | 1182 | 470–654 |
| Dominant plant species | *Loudetia simplex* | *Carex fusca*, *C. panicea* |
| Methods applied | Individuals collected by hand in isolators 10 m² in area | Animals collected with suction apparatus from exclosures 0.64 m² in area; 10 samples taken weekly |
| Taxa analysed | Arthropods, 3 mm | Arthropods |
| Dominant predatory taxa | Ar.: Lycosidae, Salticidae, Thomisidae, Argiopidae; Ort.: Mantidae; Col.: Carabidae; Het.: Reduviidae | Ar.: Argiopidae, Thomisidae; Hym.: Formicidae, Braconidae, Cynipidae |

[a] From Cesar & Menault (1974), Gillon & Gillon (1974) and Gillon (1974).
[b] From Traczyk (1971), Olechowicz (1971), Pętal (1971), Kajak (1971) and Breymeyer (1971).

(*c*) *North American grasslands*[a]

| | Desert, Jornada, New Mexico | Shortgrass plant, Pantex, Texas | Shortgrass plain, Pawnee, Colorado | Mixed prairie, Cottonwood, S. Dakota | True prairie, Osage, Oklahoma | *Sericea lespedeza*[b] stand, S. Carolina |
|---|---|---|---|---|---|---|
| Latitude (°N) | 33 | 35 | 41 | 44 | 37 | 32 |
| Altitude (m) | 1340 | 1090 | 1430 | 850 | 380 | 82 |
| Mean annual temp (°C) | 14.9 | 15.0 | 8.4 | 8.5 | 15.0 | 18.0 |
| Annual precipitation (mm) | 235 | 528 | 324 | 360 | 953 | 1100 |

| | | | | | | |
|---|---|---|---|---|---|---|
| Dominant plant species | *Bouteloua eriopoda, Salsola kali, Tragopogon dubius, Getiervezia sarothrae* | *Bouteloua gracilis* | *Bouteloua gracilis, T. dubius* | *Buchloe dactyloidea, Bouteloua gracilis, Bromus japonicus* | *Andropogon scoparius, Stipa comata* | *Lespedeza cuneata* |
| Methods applied | Animals collected by suction apparatus from exclosures 0.5 m$^2$; later extracted in Berlese funnel or sieved | | | | | 10 sweep samples consisting of 250 strokes. An area of 225 m$^2$ was swept 20 consecutive times to estimate proportion taken in first removal |
| Taxa analysed | Arthropods | | | | | |
| Dominant predatory taxa | Hym.: Formicidae; Ar.: Lycosidae, Thomisidae, Theridiidae; Col.: Carabidae | Hym.: Formicidae; Dip.: Asilidae; Col.: Carabidae, Coccinellidae, Araneae | Hym.: Formicidae, Sphecidae; Col.: Carabidae, Cicindellidae, Dip.: Asilidae; Orth.: Mantidae; Ar.: Lycosidae, Salticidae | Hym.: Formicidae; Col.: Carabidae, Coccinellidae; Orth.: Mantidae, Araneae | Hym.: Formicidae; Col.: Carabidae, Araneae | Ar.: Attidae, Oxyopidae; Hym.: Formicidae; Hem.: Reduviidae; Dip.: Asilidae; Orth.: Mantidae |

[a] From Lewis (1971) and McDaniel (1971). [b] Menhinick (1967).

**Appendix 5.2. Methods of sampling and extracting predatory soil invertebrates, and predatory taxa analysed**

| Habitat | Invertebrate group | Sample size | Sample depth | Extraction technique | Predatory taxa analysed |
|---|---|---|---|---|---|
| Tien Shan ecosystems[a] | Macrofauna | 0.06–0.25 $m^2$ | — | Hand sorting | Material was classified to appropriate trophic levels by the author; composition of particular level not provided |
| Tien Shan ecosystems[a] | Mesofauna | 125 $cm^3$ or 1000 $cm^3$ | 30–35 cm in arid habitats; 15–20 cm in others | Extracted in photo-eclectors during 2 days | |
| Mixed prairie, Matador[b] | Arthropoda | 6.8 cm diameter | 30 cm | Flotation by Salt–Hollick procedure | Based on feeding characteristics provided by the author – Araneae, Pseudoscorpionidae and Formicidae were included in the predatory group, predatory Elateridae and Acarina were neglected |
| Mixed prairie, Matador[b] | Nematoda | Subsamples of 1/10 or 1/20 by weight of the arthropod samples | 30 cm | Centrifugal flotation | Predatory Nematoda |
| Sheep pastures (*Lolio-Cynosuretum*), Pieniny Mountains; marshy meadow (*Stellario-Deschampsietum*), Kampinos Forest[c] | Surface-dwelling Arthropoda | 900 $cm^2$ | Soil surface | Hand sorting | Araneae, Formicidae, Carabidae |

| | | | | | |
|---|---|---|---|---|---|
| Sheep pastures (*Lolio-Cynosuretum*), Pieniny Mountains; marshy meadow (*Stellario-Deschampsietum*), Kampinos Forest[c] | Nematoda | 50 $cm^3$ subsamples taken | 25 cm | Modified Batemann method | Predatory Nematoda[d] |
| *Arrhenatheretum* meadow, near Warsaw | Arthropoda | 100 $cm^2$ | 5 cm | Kempson infra-red extraction during 7 days | Araneae, Chilopoda, Carabidae, Staphylinidae, Formicidae,[e] Nematoda[f] |
| Open shrubby savanna, Lamto[g] | Macrofauna | Varies for different animals; 1–25 $m^2$ | 50 cm | Hand sorting | Araneae, Chilopoda, Formicidae, Coleoptera larvae |
| Open shrubby savanna, Lamto[g] | Mesofauna | 20 $cm^2$ | 40 cm | Berlese–Tullgren funnel | Acarina, Pseudoscorpionidae |

[a] From Zlotin (1975).
[b] From Willard (1974).
[c] From Kajak, Breymeyer & Pętal (1971), Delchev & Kajak (1974), Pętal (1974), and Wasilewska (1974).
[d] Not analysed in marshy meadow.
[e] Special technique for Formicidae was applied.
[f] Methods as in sheep pasture soil.
[g] From Athias, Josens & Lavelle (1974).

# 6. Vertebrate predator subsystem

L. D. HARRIS & G. B. BOWMAN

## Contents



## Introduction

With only minor reservations, carnivory is a contemporary euphemism for predation. But whereas predation implies more a mode of life in which food is obtained by preying upon other animals, carnivory deals more with the process of feeding on primary and higher level consumers. Reference to the concept of carnivory is generally associated with the trophic level concept of ecology and frequently implies energy transfer up the trophic pyramid. Because of the close relation between the concepts, this discussion of vertebrate carnivory necessarily draws heavily upon the predation literature.

Errington admonished his students always to distinguish between the fact of predation and the phenomenon of predation (e.g. Errington, 1964). We wish to develop this a step further and describe carnivory as a vital ecosystem process. Inasmuch as the English language is representative of the historical record, we briefly trace the attitudinal evolution of predator–prey relations. We then describe several important structural attributes of vertebrate carnivore communities and vertebrate carnivory as a process. We end with some thoughts on the role of carnivory in ecosystem function.

Predation was initially viewed in the sinister framework of predators as killers and competitors of man. This is the ugly fact of predation, and unfortunately it has dominated too much of our thought and research.

Darwin (1869) wrote that 'the stock of partridges, grouse and hares on any large estate depends chiefly on the destruction of vermin'. This vision gradually changed with the recording of more objective descriptions and the subsequent natural history works such as those of Murie (e.g. 1944, 1961), Errington (e.g. 1946, 1967), Pitelka, Tomich & Treikel (1955) and Craighead & Craighead (1956). Slightly more conceptual works (e.g. Mech, 1966; Pimlott, 1967; Schaller, 1967, 1972; Pimlott, Shannon & Kolenosky, 1969; Hornocker, 1970; Kruuk, 1972) dominated the 1960s, but still the emphasis was on the predator–prey interaction itself as distinct from the underlying trophic process. The outstanding work of Holling (e.g. 1959, 1965) during this same period exemplified the power of analytical reductionism and has given predator–prey theory a quantitative legitimacy. Yet these studies failed to relate predator–prey interactions to the energy flow or functioning of whole ecosystems in more than a qualitative manner. More recent work such as that of Golley (1960), Pearson (1964), Jordan, Botkin & Wolfe (1971) and Holmes & Sturges (1973) represents a fundamentally different philosophy of trophic dynamics and energy transfer. This along with the many feedback effects seems to represent carnivory as an ecosystem process.

Carnivory may be approached in two different ways. The first encompasses the realm of species-specific food habits, and thus demands a catalogue approach beyond the scope of this paper. The second involves the phenomenon of secondary consumption in the context of a coevolved animal community. This is the approach we shall follow.

## Time and space patterns of carnivory

The process of carnivory is directly tied to the structural characteristics of the animal community. Therefore the distribution and relative abundance of various vertebrate classes in time and space gives insight into the process itself. Of special interest are the taxonomic shifts and the ratios of various levels of carnivores to the herbivore food base in different systems.

Of all the vertebrates, the herptiles (reptiles and amphibians), mammals and birds predominate in terrestrial grassland ecosystems. Of these, the predator–prey literature greatly emphasizes the mammals while least is known about the carnivory function of herptiles. This is particularly unfortunate since carnivores overwhelmingly dominate the herptiles when compared to the other two taxa. Each of these groups shows basically different responses to temporal environmental changes.

The exothermic herptiles exhibit marked behaviour changes and functionally alter their activity by daily and seasonal quiescence. As a result, their energy demands and ecosystem function are directly related to ambient temperature and therefore season. Because of this herptiles generally become progressively more important as one proceeds towards hotter and/or drier environments. Carnivorous birds, on the other hand, exhibit dramatic numerical responses to season by migrating from one system to another. Mammals remain as the group least numerically affected by temporal changes in their environment. Even though some of them hibernate, this is relatively unimportant compared to the quiescence or migration of the herptiles and birds. A phenomenon of more importance to mammalian carnivores is a reversion to omnivory in response to stressful environmental conditions (Martin, Zim & Nelson, 1951). At least in the high latitudes, this may result from the additional energy requirements necessary for thermal regulation during the winter months. This period of higher energy demand occurs simultaneously with minimum primary production and at a time when many prey are either absent or quiescent.

Because of phenomena such as migration, temporal and spatial patterns are inseparable. Moreover, because migration is generally in response to environmental severity, gradients of temperature and moisture are also important considerations. Therefore, the degree of carnivory in birds and mammals seems to increase in response to increasing latitude and/or altitude (Morse, 1971). Coupled together, temporal and spatial phenomena imply that the relative importance of carnivory would be accentuated in the winter of high-latitude systems.

As grasslands intergrade with related biomes the size distribution, behavioural attributes and taxonomy of the vertebrate community shift noticeably. For example, with certain exceptions such as lions (*Panthera leo*) one might generalize that canid mammals are more important as the top carnivores of grasslands while felid mammals seemingly dominate the forested biomes. The rather obvious behavioural differences between the test-and-chase tactics of the canids and the ambush tactics of the felids seem to attest to the different selection pressures in woodlands versus the open grasslands. Both predator groups function in their respective biomes but one cannot help but wonder about the different selective pressure exerted on prey populations in the two different cases. Perhaps the larger group sizes of ungulates of grasslands compared to the prey of woodlands and forests results, in part, from this carnivore selection pressure (Hirth, 1977).

## Ecological displacement

Ecological displacement of organisms by functionally similar but taxonomically distinct groups is an equally compelling consideration. This is particularly apparent between different biomes (e.g. the relative importance of birds as forest carnivores versus mammals as grassland carnivores). It is also apparent within the grasslands. Obvious differences exist in the relative abundance of small (e.g. rodents) versus large (e.g. ungulates) herbivores and this in turn affects the size distribution of carnivores. Even within the small mammal community important shifts occur. North American grassland rodents are dramatically segregated at the subfamilial level (Fig. 6.1). This is not so important by itself, but the compensatory responses of ecological equivalents such as birds affect the entire community and trophic relations. In the desert grassland, for example, seed eating heteromyid rodents predominate and seemingly displace seed-eating birds. The avian fauna is completely dominated by carnivores (Harris, 1971; Weins, 1973). Conversely, on the shortgrass prairies where carnivorous small mammals were abundant, the bird species were generally dominated by omnivores (Harris, 1971; Weins, 1973). In the tallgrass a superabundance of small mammal herbivores seems to provide the food base for avian raptors, while small mammal carnivores prey on arthropods and in turn displace insectivorous birds. Ultimately this implies that neither the trophic structure nor the magnitude of carnivory can be established without considering all taxa simultaneously. Comparative examples will demonstrate the large difference to be encountered between trophic pyramids of different grasslands.

Within the small mammal group the carnivore–herbivore biomass ratio shifts from about 1:20 under tall grass prairie conditions to about 1.3:1 on the southern shortgrass (Harris, 1971). But as implied above, shifts within other taxa tend to compensate for these differences. When all important consumers are considered, the northern shortgrass prairie (Pawnee site) supports 1 g carnivore for every 146 g of herbivore (1:88 if cattle are excluded). A similar analysis for East African savanna grassland yields a ratio of 1 g to 333 g herbivore (Harris & Fowler, 1975). Finally, preliminary analysis of the trophic structure of the hydric grasslands of South Florida, USA, reveals a surprising biomass ratio of about 20 gram carnivore for each gram of herbivore (L. D. Harris, unpublished data). Clearly there is not a constant ratio of carnivores to herbivores and the relative sizes and turnover rates of the consumers must be determining factors. The East African savannas support very large herbivores (ungulates) with very slow turnover rates. As a consequence the herbivore energy turnover rate is slow and the biomass of carnivores supportable per unit weight of herbivore must be very low. Conversely, in the South Florida grasslands the primary consumers are very small with rapid turnover rates (invertebrates) and thus the abundance of carnivores can be very much greater (Fig. 6.2). Generally said, a small carnivore: herbivore ratio implies a slow energy turnover rate through the consumer pathway while a high ratio implies a rapid flow rate.

More conceptually intriguing than the simple pyramid form is the degree of trophic elaboration. The animal community of the East African savannas is overwhelmingly dominated by two trophic strata, primary consumers and primary carnivores. There is very little elaboration. Similarly, on the northern shortgrass prairie of North America 88 % of all carnivory occurs at the primary level with only 12 % occurring at the secondary level. For example,

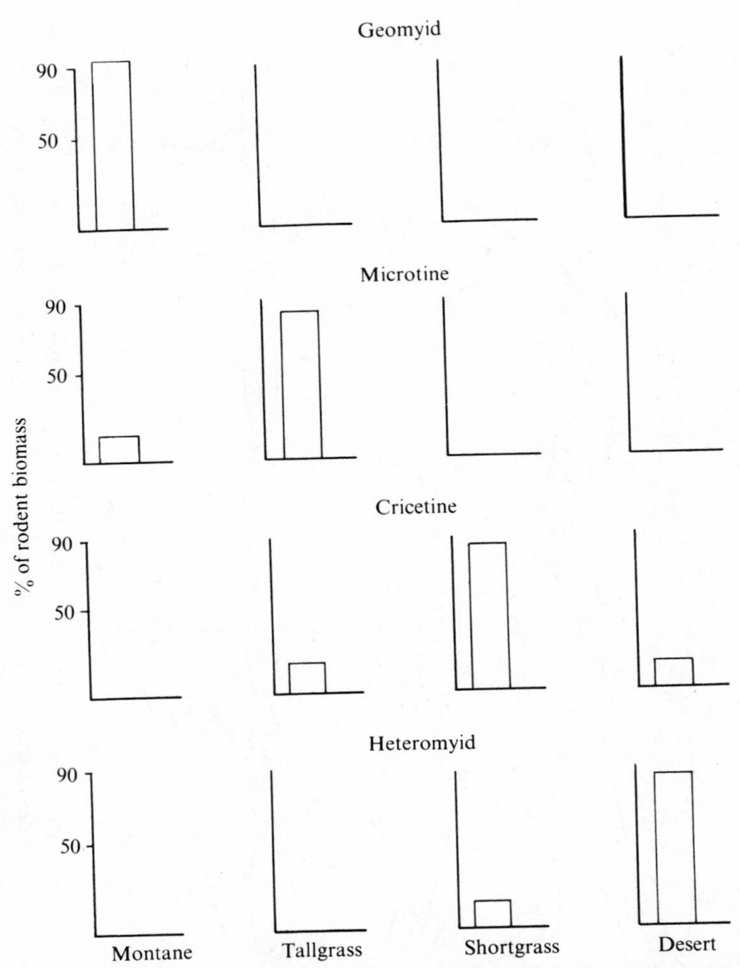


Fig. 6.1. Per cent of rodent biomass (not including sciurids) in the four major taxonomic groups occurring in four major North American grassland types during the 1971 sampling period. Whereas geomyids seem to dominate the montane grassland, microtenes dominated the tallgrass, cricetines dominated the shortgrass and heteromyids dominated the desert grassland site. (Taken from Harris, 1971.)

even top carnivores such as the golden eagle (*Aquila chrysaetos*) consume mainly herbivores (97 % of the diet) (Harris & Paur, 1972). Again, a relatively small degree of trophic elaboration exists. Contrary to the above, the overwhelming majority of vertebrates in South Florida grasslands are second, third or fourth order carnivores and the degree of trophic structure elaboration is much more pronounced (Fig. 6.2). It thus seems clear that the

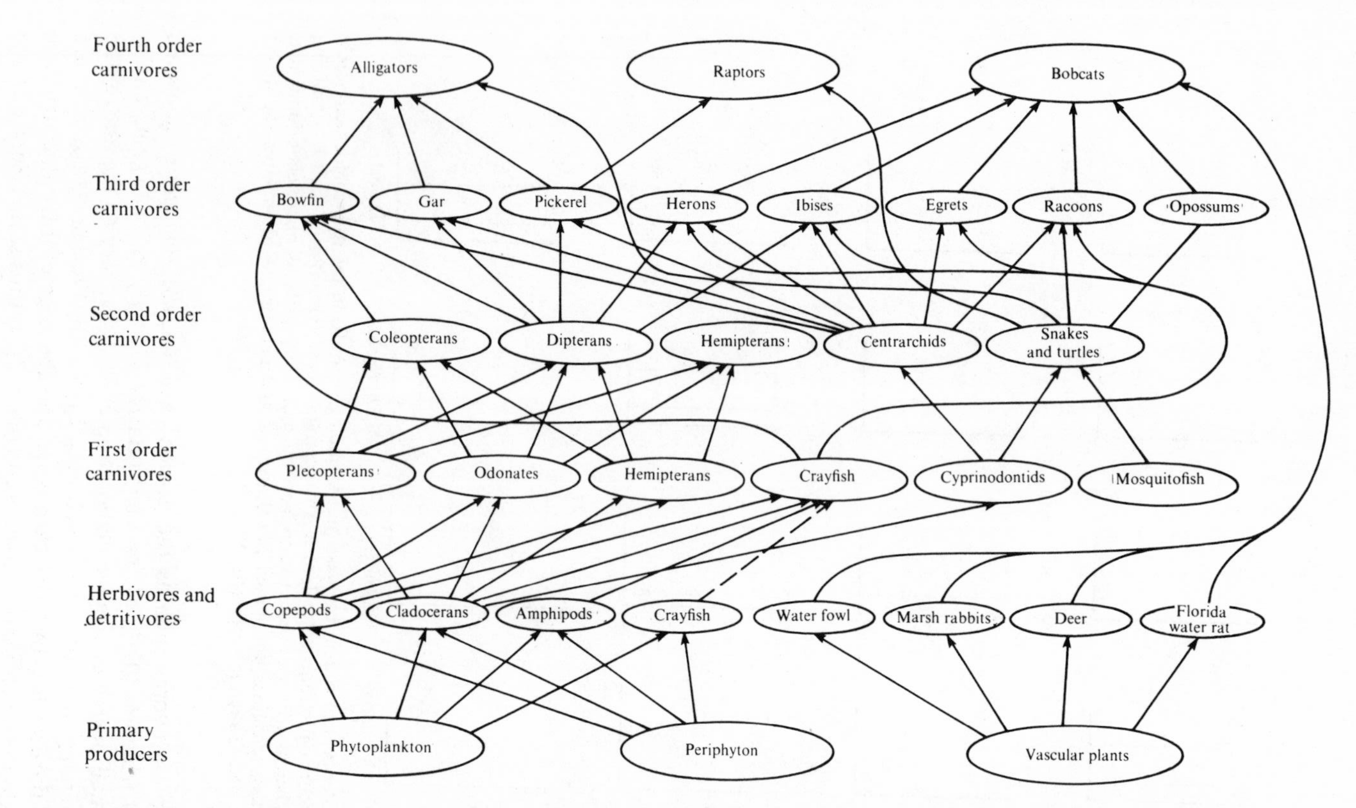


Fig. 6.2. A foodweb characteristic of the hydric grasslands of South Florida, USA, illustrating that most vertebrates (and a large proportion of the total animal biomass) occupy positions as third, fourth and even fifth order carnivores in these grassland systems.

relative abundance and function of grassland vertebrate carnivores are highly variable and highly dependent upon the characteristics of the underlying community.

In addition to the inverse relation between energy transfer per unit mass and body size, the classical population dynamics parameter, turnover rate ($\tau$), must also be considered. The alliance of these two holds inasmuch as the population turnover rate is, by and large, an inverse correlate of body size. Thus large carnivores such as lions have very slow population turnover rates while small species such as shrews turn over very rapidly. When coupled, the metabolic body size and population turnover rate dictate the form of the trophic biomass pyramid. For this reason the classical pyramids and more recent carnivore : herbivore ratios deserve special attention.

## Effects of carnivory on community structure

The effect of predation on prey community diversity is one of the more obvious structural responses. Although we know of no experimental data for grassland vertebrates, many other works serve as appropriate references. Paine (1963, 1966) demonstrated an increased diversity of prey due to predation in seashore animal communities. This increased diversity most often results from the suppression of dominants (Connell, 1961) and an increased 'evenness' (Addicott, 1974) or equitability component. The interaction between predation levels and competition is clearly involved in many instances (Cramer & May, 1972; Allan, 1974), and the co-occurrence of certain predators may even be the result of altered competition (Rosenzweig, 1966). It is also believed that under certain conditions of intense competition predation may prevent local extinctions (Slobodkin, 1964). This implies a certain utility from predation management that resource managers have at their disposal.

Table 6.1. *Comparative large herbivore diversity measures for several East African habitats and the National Bison Range, Montana, USA*

| Habitat | No. of species | Numerical diversity ($H'$)[a] | Evenness component ($e$)[b] |
|---|---|---|---|
| East Africa | | | |
| Montane | 19 | 0.74±0.10 | 0.25 |
| Bushland | 19 | 1.12±0.00 | 0.38 |
| Shortgrass | 17 | 1.46±0.22 | 0.52 |
| Savanna | 24 | 1.68±0.10 | 0.53 |
| National Bison Range | 7 | 1.63 | 0.84 |

Data from Harris & Fowler (1975).

[a] $H' = -\sum_{i=1}^{s} P_i \log P_i$, where $P_i = N_i/N$. $N_i$ is the number of individuals in the $i$th species and $N$ is the total of all large herbivores present.

[b] $e = H'/\log s$, where $H' = \sum P_i \log P_i$ as above and $s$ is the number of species.

Even though most of the native large carnivores have been extirpated or severely limited on the grasslands of North America, man now serves a similar function. For example, management of the large herbivores of the National Bison Range entails annual cropping of the dominant species. This results in a suppression of dominance within the herbivore community. Therefore, even though only seven large herbivore species inhabit this range, the overall diversity is very comparable to that of the common habitats of East Africa where approximately 20 species usually co-occur. This is almost entirely due to the evenness component of the diversity measure (Table 6.1).

## Carnivory and energy flow

The various aspects of carnivory introduced thus far deal mainly with numerical or biomass considerations. Neither of these measures is adequate in and of itself since simple numbers overemphasize the importance of the small animals while biomass overemphasizes the importance of the large. Metabolic body size is much more closely related to transfer rates (nutrient as well as energy) than any other single parameter.

The total energy and/or nutrient transfer of endothermic vertebrates can be closely estimated as a linear function of metabolic body size. Exothermic vertebrates seem to follow as closely, given that the temperature regression is known. From numerous review articles (Nice, 1938; Brody, 1945; Hemmingsen, 1950, 1960; Kleiber, 1961; Schoener, 1969; Kendeigh, 1970), we conclude that the daily energy requirement for the warm-blooded vertebrate community can be closely approximated by the equation:

$$E_R = \sum_{i=1}^{s} N_i\,(C\,\mathrm{kg}_i^{0.75}), \tag{6.1}$$

where: $E_R$ = the total community energy requirement per day;

$\sum_{i=1}^{s}$ = summation over the $s$ species present;

$N_i$ = the population number of the $i$th species;

$C$ = a constant multiplier on the metabolic weight (averaged over all age classes, seasons and activities this approximates to twice the basal metabolic rate ($2 \times 70 = 140$);

$\mathrm{kg}_i^{0.75}$ = the metabolic weight calculated as the live weight in kg raised to the power 0.75.

There are three weaknesses to this formulation. First, it requires the mean species weight to be adjusted as a function of the age or size distribution around the mean. Secondly, it does not address the difference in digestive or assimilation efficiency of different species. Thirdly, the generalized Brody–Kleiber formulation does not apply to certain species or situations and therefore as species-specific information becomes available it should be used to refine the composite estimates. Assimilation efficiency is dependent on species, foodstuff and consumer size, and therefore some care must be taken to ensure accuracy. On the other hand these estimators are available for animals ranging from lions and vampire bats (Schaller, 1972; Wimsatt, 1969) to mice and eland (*Taurotragus oryx*) (Kleiber, 1961; Rogerson, 1968). Contrary to popular belief, the possibly higher metabolic rate of vertebrate carnivores seems to counteract any increased assimilation efficiency they may possess and thus on the basis of dry weight of food, carnivores consume roughly the same proportion of their body weight per day as do herbivores (Table 6.2).

Given that assimilation efficiency values are available, then the total daily community transfer is estimated by the equation

$$E_{\mathrm{T}} = \sum_{i=1}^{s} R_i N_i \,(C\,\mathrm{kg}_i^{0.75}), \qquad (6.2)$$

where: $E_{\mathrm{T}}$ = total energy transfer of the carnivore community;
$R_i$ = the reciprocal of assimilation efficiency for the $i$th species.
(All other symbols as in equation 6.1.)

This formulation assumes that the predator community is not energy-limited and, therefore, if energy limitation is affecting predator numbers, calculations should be made at short intervals (daily). In periods of less likely limitations, calculations could span a longer period of weeks or even months.

Close consideration of equation (6.2) reveals a generally known but frequently overlooked energy transfer principle.

From the metabolic body size alone it follows that a 5 g carnivore such as a shrew is fully 10 times as metabolically active per unit weight as a 50 kg wolf (Table 6.3). From composite calculations such as the above it can be shown that a small biomass density of only 10 $\mathrm{kg \cdot ha^{-1}}$ of small rodents may transfer as much energy as massive concentrations (e.g. 175 $\mathrm{kg \cdot ha^{-1}}$) of large East African herbivores. Moreover, if small rodents have the high assimilation efficiencies reported (cf. Drożdż, 1969; Hansen & Cavender, 1970), they may well function at nearly three times the efficiency of large herbivores (Rogerson, 1968). Multiplying this threefold greater efficiency by a tenfold greater transfer rate implies an immensely greater transfer through the herbivore trophic level.

These illustrations are given to demonstrate the weakness of simple numerical or biomass interpretations of systems function. High biomass herbivore communities do not necessarily yield any greater productivity to the carnivore trophic level than low biomass communities. Indeed, the accumulated evidence to date suggests that the opposite may be true.

A final theme deserves mention. The grazing effect, or the changed productivity of vegetation in response to grazing, is well documented for the herbivore–primary producer linkage. This effect may be either positive or negative depending on the limiting factors involved. Thus if tallgrass production is light-limited because of shading, then removal by clipping or grazing serves to relieve density-dependent limitations and a positive growth response results.

Table 6.2. *Comparative daily food intake rates of comparably sized carnivores and herbivores based on dry weight of food over live weight of animal*

| Size | Representative herbivore | Representative carnivore |
|---|---|---|
| Very small | Pocket mouse ≃ 0.30[a] (*Perognathus* spp.) | Shrew ≃ 0.60[b] (*Sorex* spp.) |
| | | Shrew ≃ 0.30[b] (*Blarina brevicauda*) |
| Small | Voles ≃ 0.12[a] (*Microtus* spp.) | Vampire bat ≃ 0.08[c] (*Desmodus rotundus*) |
| | Pocket gopher ≃ 0.10[a] (*Thomomys talpoides*) | Sparrowhawk ≃ 0.07[d] (*Falco sparverius*) |
| Medium | Jackrabbit ≃ 0.04[a] (*Lepus* spp.) | Great horned owl ≃ 0.03[d] (*Bubo virginianus*) |
| | Beaver ≃ 0.03[b] (*Castor canadiensis*) | Domestic dog ≃ 0.04[e] (*Canis familiaris*) |
| Large | White-tailed deer ≃ 0.025[f] (*Odocoileus virginianus*) | Lion ≃ 0.04[g] (*Panthera leo*) |
| | Pronghorn ≃ 0.03[a] (*Antilocapra americana*) | Hyena ≃ 0.05[h] (*Crocuta crocuta*) |

[a] Hansen & Cavender (1970).
[b] Golley (1962).
[c] Wimsett (1969).
[d] Craighead & Craighead (1956).
[e] Committee on Animal Nutrition (1953).
[f] Lay (1969).
[g] Schaller (1972).
[h] Kruuk (1972).

Table 6.3. *Comparative live and metabolic weights of small versus large mammal grassland carnivores*

| Exemplary species | Approximate weight (kg) | Metabolic weight ($kg^{0.75}$) |
|---|---|---|
| Shrew | 0.005 | 0.0188 |
| Wolf | 50 | 18.80 |
| Difference between the two | $10^4$ | $10^3$ |

Since daily energy requirements are linear functions of metabolic weight it follows that the small vertebrate carnivore transfers fully 10 times as much energy per unit body weight as the large vertebrate carnivore.

On the other hand, the grazing of more density-independent, water-limited bunchgrass may well cause a suppressed growth response. We believe the analogy to be highly appropriate in understanding predation effects on secondary productivity.

Conceptual works involving the predation effect range from qualitative to mathematical (Watt, 1955, 1968; Lefkovitch, 1967; Slobodkin, 1968, 1974; Beddington & Taylor, 1973; Usher, 1973). Empirical data drawn from a single poikilothermic vertebrate and invertebrate experiments (Watt, 1955; Silliman & Gutsell, 1958; Slobodkin, 1968) generally support the premise that a reduction of prey population numbers to substantially below the saturation density causes an enhanced production response. Of course this involves mainly a release from density-dependent birthrate limitations (cf. Tanner, 1966), but there is reason to believe it also applies in density independent situations.

The mechanisms are at least threefold. In the first place, a simple reduction in the number of individuals liberates finite resources such as food, habitat or any other potentially limiting commodity. This may well result in either initial or accelerated reproduction by formerly suppressed individuals. A second mechanism involves potential shifts in the age structure of the prey population resulting from harvesting. Whether selection occurs or not, the end result of many predator–prey relations is that the harvest falls heavily on the age class extremes (young and old) and on physically impaired individuals (Mech, 1966; Pimlott, Shannon & Kolenasky, 1969; Hornocker, 1970; Schaller, 1972). This biases the remaining population toward the most productive segment, which is closely in line with the conceptual ideal. A third mechanism involves the selection of one sex over the other such that the remaining population contains a higher proportion of fecund individuals. There is little evidence that this mechanism prevails under natural conditions, but it is the obvious strategy employed by human managers ranging from cattlemen and game ranchers to legislators establishing hunting laws. Any one or all of these mechanisms may explain an increase in population turnover rate and energy transfer as a function of predation.

But the level of predation seems to stand out as the single most important consideration in this overall response. Unlike grasses and related basal-meristem plants, animal population growth is closely tied to the levels of standing crop. In this sense animal populations are very similar to apical-meristemmed plants. Thus even though the rate of change per individual ($\mathrm{d}N/N\mathrm{d}t$) in the population is maximized at low population levels, the total population productivity is maximized at a considerably higher standing crop. Predation intensities which depress the prey population below a fixed point (half of the saturation density in density-dependent, logistic growth situations) will have a detrimental effect on prey productivity. While lower levels of predation should enhance prey production, higher levels may well cause extinction. Inasmuch as prey populations occasionally increase to socially and/or ecologically intolerable levels, we may tend to conclude that vertebrate predators do not exert a significant controlling influence. This does not necessarily follow, however, since there are always bounds outside of which control mechanisms fail. The same predators that co-occur with over-populations of prey may well normally serve a control function, influence fluctuations (Pearson, 1966, 1971) and even enhance their prey population (Howard, 1974).

## Carnivory and systems function

Predator and prey populations are generally coevolved as a highly integrated community. Prey population selection, evolution and function are as much affected by predation intensity as predators are by prey population levels. The two trophic levels function together with predation serving as the selective mechanism and carnivory serving as the system's energy and material flow linkage. Thus, predator–prey interactions form a highly integrated complex of feedback mechanisms which presumably influence and possibly stabilize both population levels and the carnivory process. Vertebrate predators seem especially functional in this regard because of their greater mobility and greater ability to form distinct search images and even learned behavioural responses. This allows for complex switching between prey groups in time and space and thus the critical element of a control device, the potential for differential effect. The degree of switching and thus the cost–benefit ratio for the predator is amenable to general economic analysis (e.g. Schoener, 1971; Covich, 1972). This, along with the relative cost of searching versus handling probably dictates whether the predator is a generalist or a specialist (MacArthur & Pianka, 1966; MacArthur, 1972). This, in turn, must affect the degree of selective pressure that each population exerts on the other.

We believe the function, behaviour and well-being of any given carnivore are determined by two rather distinct sets of factors. The first set, the ultimate factors, relates to a carnivore's environmental, physiological and morphological capabilities. The second set, the proximal factors, includes factors such as hunger level, prey abundance, capture success and necessary foraging times (Fig. 6.3). It is these factors that have classically been associated with the predator–prey interaction. Holling's (1959, 1963, 1965) component analysis of these factors represents the single richest source of quantitative information available. He utilized the major variables of vertebrate predator–prey interactions originally enumerated by Leopold (1933):

(1) density of prey;
(2) density of predator;
(3) characteristics of the predator;
(4) characteristics of the prey;
(5) characteristics of the environment.

He considered these variables to be specific to the particular species and environments involved. The density variables were treated as basic to all predator–prey interactions and were separated into functional and numerical responses, each of which has several components that are subject to quantitative measurements. However, we believe his description of the functional and numerical response of predators needs to be extended to include like responses among the prey (e.g. Table 6.4). This is particularly important when considering vertebrates because of the more complex nature of the

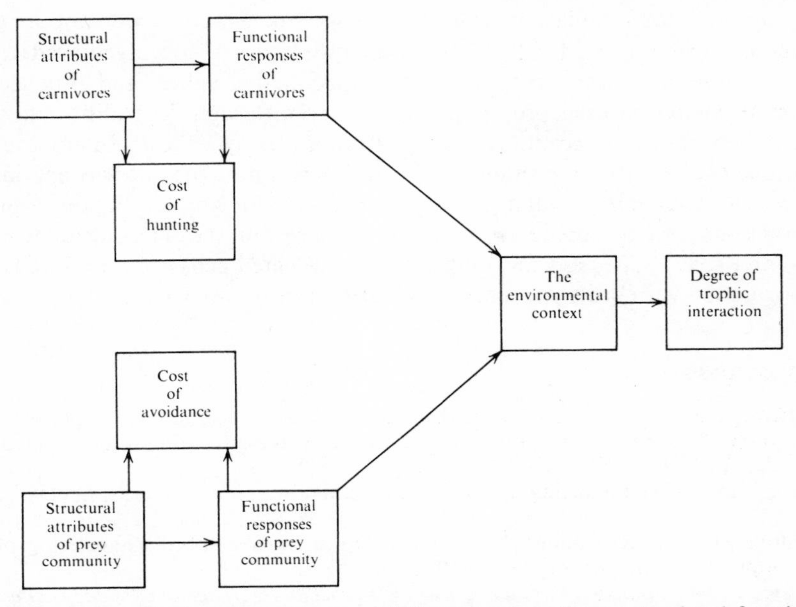


Fig. 6.3. A generalized predator–prey model illustrating how the structural and functional interactions of both predators and prey are involved in determining the degree of trophic interaction in any given system. (Modified from Rivett, 1972, p. 39.)

Table 6.4. *Components of predator and prey responses to changes in the other's abundance*

| Predator responses | Prey responses |
|---|---|
| Numerical | |
| Changes in reproduction | Change in reproduction |
| Immigration/emigration | Immigration/emigration |
| Changes in mortality | Changes in mortality |
| Functional | |
| Rate of successful search | Rate of successful avoidance |
| Time available to hunt | Time exposed to predation |
| Handling time per prey | Resistance by prey |
| Hunger level | |
| Interference between predators | Interference between prey |
| Learning by predator | Learning by prey |

predator response curves (Holling, 1959, 1965) and the prey reactions (Dill, 1974). All of these proximal factors (Table 6.4) relate back to the more general ultimate factors by means of selective pressure and dictate the coevolutionary interactions. The introduction of alien predators or prey into an otherwise stable community may perturb these interactions and cause high fluctuations in and even extinction of prey populations (Pearson, 1966) and decreases in existing predator populations (George, 1974). For example, we support the hypothesis (Slobodkin, 1968, 1974) that the predators of coevolved predator–prey communities select prey with low reproductive values and thus exert minimal impact on their prey populations – even though the extant data are insufficient to justify acceptance or rejection of this hypothesis. In any event, continued adherence to the idea that predation is simply the one way action of predators consuming available prey seems overly simple. Predator–prey interactions and the processes of carnivory are not simply coincidental in time and space. They are more likely highly selected ecosystem processes as important to vertebrate community integrity as any other natural process.

## References


Addicott, J. F. (1974). Predation and prey community structure: an experimental study of the effect of mosquito larvae on the protozoan communities of pitcher plants. *Ecology*, **55**, 475–92.

Allan, J. D. (1974). Balancing predation and competition in clacoderns. *Ecology*, **55**, 622–9.

Beddington, J. R. & Taylor, D. B. (1973). Optimum age specific harvesting of a population. *Biometrics*, **29**, 801–9.

Brody, S. (1945). *Bioenergetics and growth with special reference to efficiency complex in domestic animals.* Reinhold Publishing Co., New York.

Committee on Animal Nutrition (1953). *Nutrient requirements for dogs. National Research Council Publication 300.* National Academy of Sciences, Washington, DC.

Connell, H. H. (1961). The influence of interspecific competition and other factors on the distribution of the barnacle *Chthamalus stellatus*. *Ecology*, **42**, 710–23.

Covich, A. (1972). Ecological economics of seed consumption by *Peromyscus*. In *Growth by intussusception*, ed. E. S. Deevey, pp. 71–93. Shoe String Press, Hamden, USA.

Craighead, J. J. & Craighead, F. C. Jr (1956). *Hawks, owls and wildlife*. Stackpole, Harrisburg, Penn.

Cramer, N. F. & May, R. M. (1972). Interspecific competition, predation, and species diversity: a comment. *Journal of Theoretical Biology*, **34**, 289-93.

Darwin, C. (1869). *The origin of species*. Reprinted by The New American Library of World Literature, New York.

Dill, L. M. (1974). The escape response of the zebra danio (*Brachydani rerio*). II. The effect of experience. *Animal Behaviour*, **22**, 723–30.

Drożdż, A. (1969). Digestibility and utilization of natural food in small rodents. In *Energy flow through small mammal populations*, ed. K. Petrusewicz & L. Ryszkowski, pp. 127–9. Polish Scientific Publishers, Warsaw.

Errington, P. L. (1946). Predation and vertebrate populations. *Quarterly Review of Biology*, **21**, 144–77, 221–45.

Errington, P. L. (1964). The phenomenon of predation. *The Smithsonian Report*, **4634**, 507–19.

Errington, P. L. (1967). *Of predation and life*. Iowa State University Press, Ames, Iowa.

George, W. C. (1974). Domestic cats as predators and factors in winter shortages of raptor prey. *Wilson Bulletin*, **36**, 384–96.

Golley, F. B. (1960). Energy dynamics of a food chain of an old field community. *Ecological Monographs*, **30**, 187–206.

Golley, F. B. (1962). *Mammals of Georgia*. University of Georgia Press, Athens, Georgia.

Hansen, R. M. & Cavender, B. R. (1970). *Assimilation rates of small mammal herbivores. US/IBP Grassland Biome Technical Report 51*. Colorado State University, Fort Collins.

Harris, L. D. (1971). A précis of small mammal studies and results in the grassland biome. In *Preliminary analysis of structure and function in grasslands*, ed. N. R. French, *Colorado State University Science Series 10*, pp. 213–40. Colorado State University, Fort Collins.

Harris, L. D. & Fowler, N. K. (1975). Ecosystem analysis and simulation of the Mkomazi Reserve, Tanzania. *East African Wildlife Journal*, **13**, 325–46.

Harris, L. D. & Paur, L. (1972). *A quantitative food web analysis of a shortgrass community. US/IBP Grassland Biome Technical Report 154*. Colorado State University, Fort Collins.

Hemmingsen, A. M. (1950). The relation of standard (basal) energy metabolism to total fresh weight of living organisms. *Steno Memorial Hospital and Nordish Insulin Laboratorium* (*Niels Steensons Hospital, Copenhagen, Denmark*) Report, **4**, 58 pp.

Hemingson, A. M. (1960). Energy metabolism as related to body size and respiratory surfaces, and its evolution. *Steno Memorial Hospital and Nordisk Insulin Laboratorium* (*Niels Steensons Hospital, Copenhagen, Denmark*) *Report*, **9** (2), 1–110.

Hirth, D. H. (1977). Social behavior of white-tailed deer in relation to habitat. *Wildlife Monographs*, in press.

Holling, C. S. (1959). The components of predation as revealed by a study of small mammal predation of the European pine sawfly. *Canadian Entomologist*, **91**, 293–320.

Holling, C. S. (1963). An experimental component analysis of population processes. *Memoirs of the Entomological Society of Canada*, **32**, 22–32.

Holling, C. S. (1965). The functional response of predators to prey density and its role in mimicry and population regulation. *Memoirs of the Entomological Society of Canada*, **45**, 5–60.

Holmes, R. T. & Sturges, F. W. (1973). Annual energy expenditure by the avifauna of a northern hardwood ecosystem. *Oikos*, **24**, 24–9.

Hornocker, M. G. (1970). An analysis of mountain lion predation upon mule deer and elk in the Idaho primitive area. *Wildlife Monographs*, **21**, 39 pp.

Howard, W. E. (1974). *The biology of predator control. Modules in Biology 11*. Addison-Wesley, New York.

Jordan, P. A., Botkin, D. B. & Wolfe, M. L. (1971). Biomass dynamics in a moose population. *Ecology*, **52**, 147–52.

Kendeigh, S. C. (1970). Energy requirements for existence in relation to size of bird. *Condor*, **72**, 60–5.

Kleiber, M. (1961). *The fire of life*. Wiley, New York.

Kruuk, H. (1972). *The spotted hyena*. University of Chicago Press, Chicago.

Lay, D. W. (1969). Foods and feeding habits of white-tailed deer. In *White-tailed deer in the southern forest habitat*, pp. 8–13. *Proceedings of a Symposium of the US Forest Service, Nacogdoches, Texas*. S. Forest Experimental Station, US Dept of Agriculture.

Lefkovitch, L. P. (1967). A theoretical evaluation of population growth after removing individuals from some age groups. *Bulletin of Entomological Research*, **57**, 437–45.

Leopold, A. S. (1933). *Game management*. Charles Scribner's Sons, New York.

MacArthur, R. H. (1972). *Geographic ecology*. Harper & Row, New York.

MacArthur, R. H. & Pianka, E. R. (1966). On optimal use of a patchy environment. *American Naturalist*, **100**, 603–7.

Martin, A. C., Zim, H. S. & Nelson, A. L. (1951). *American wildlife and plants*. Dover Publications, New York.

Mech, D. L. (1966). *The wolves of Isle Royale. US National Park Service, Fauna Series 7.*

Morse, D. H. (1971). The insectivorous bird as an adaptive strategy. *Annual Review of Ecology and Systematics*, **2**, 177–200.

Murie, A. (1944). *The wolves of Mt. McKinley. US National Park Service, Fauna Series 5.*

Murie, A. (1961). *A naturalist in Alaska*. Devin-Adair Co., New York.

Nice, M. M. (1938). The biological significance of bird weights. *Bird-Banding*, **9**, 1–11.

Paine, R. T. (1963). Trophic relationships of eight sympatric predatory gastropods. *Ecology*, **44**, 63–73.

Paine, R. T. (1966). Food web complexity and species diversity. *American Naturalist*, **100**, 65–75.

Pearson, O. P. (1964). Carnivore–mouse predation: an example of its intensity and bioenergetics. *Journal of Mammalogy*, **45**, 177–88.

Pearson, O. P. (1966). The prey of carnivores during one cycle of mouse abundance. *Journal of Animal Ecology*, **35**, 217–33.

Pearson, O. P. (1971). Additional measurements of the impact of carnivores on California voles (*Microtus californicus*). *Journal of Mammalogy*, **52**, 41–9.

Pimlott, D. H. (1967). Wolf predation and ungulate populations. *American Zoologist*, **7**, 267–78.

Pimlott, D. H., Shannon, J. A. & Kolenosky, G. B. (1969). *The ecology of the timber wolf in Algonquin Provincial Park*. Dept of Lands and Forests, Ontario.

Pitelka, F. A., Tomich, P. Q. & Treikel, G. W. (1955). Ecological relations of jaegers and owls as lemming predators near Barrow, Alaska. *Ecological Monographs*, **25**, 85–117.

Rivett, P. (1972). *Principles of model building*. Wiley, New York.

Rogerson, A. (1968). Energy utilization by the eland and wildebeest. *Symposia of the Zoological Society of London*, **21**, 153–61.

Rosenzweig, M. L. (1966). Community structure in sympatric Carnivora. *Journal of Mammalogy*, **47**, 602–12.

Schaller, G. B. (1967). *The deer and the tiger*. University of Chicago Press, Chicago.

Schaller, G. B. (1972). *The Serengeti lion: a study of predator–prey relations*. University of Chicago Press, Chicago.

Schoener, T. W. (1969). Optimal size and specialization in constant and fluctuating environments: an energy–time approach. *Brookhaven Symposia in Biology*, **22**, 103–14.

Schoener, T. W. (1971). Theory of feeding strategies. *Annual Review of Ecology and Systematics*, **2**, 369–404.

Short, H. L. (1969). Physiology and nutrition of deer in southern upland forests. In *White-tailed deer in the southern forest habitat*, pp. 14–18. *Proceedings of a Symposium of the US Forest Service, Nagdoches, Texas*. S. Forest Experimental Station, US Dept of Agriculture.

Silliman, R. P. & Gutsell, J. S. (1958). Experimental exploitation of fish populations. *Fishery Bulletin, Fish and Wildlife Service, US Department of the Interior*, **58**, 215–41.

Slobodkin, L. B. (1964). Experimental populations of hydra. *Journal of Animal Ecology*, **33** (Suppl.), 131–48.

Slobodkin, L. B. (1968). How to be a predator. *American Zoologist*, **8**, 43–51.

Slobodkin, L. B. (1974). Prudent predation does not require group selection. *American Naturalist*, **108**, 665–78.

Tanner, J. T. (1966). Effects of population density on growth rates of animal populations. *Ecology*, **47**, 433–45.

Usher, M. B. (1973). *Biological management and conservation*. Chapman & Hall, London.

Watt, K. E. (1955). Studies on population productivity. I. Three approaches to the optimum yield problem in populations of *Tribalium confusum*. *Ecological Monographs*, **35**, 269–90.

Watt, K. E. (1968). *Ecology and resource management: a quantitative approach*. McGraw-Hill, New York.

Weins, J. A. (1973). Pattern and process in grassland bird communities. *Ecological Monographs*, **43**, 237–70.

Wimsatt, W. A. (1969). Transient behavior, nocturnal activity patterns and feeding efficiency of vampire bats (*Desmodus rotundus*) under natural conditions. *Journal of Mammalogy*, **50**, 233–44.

# 7. Decomposer subsystem

COORDINATORS D. C. COLEMAN & A. SASSON

Contributors A. I. Breymeyer, D. C. Coleman, M. C. Dash, Y. Dommergues, H. W. Hunt, E. A. Paul, R. Schaefer, B. Úlehlová & R. I. Zlotin

## Contents

## Introduction

It is our intention in this chapter to examine all major decomposition and soil heterotrophic processes in grasslands worldwide, in both tropical and temperate regions. We shall explore, in turn, such key processes as soil respiration and carbon turnover, organic matter decomposition, and turnover, and then discuss the use of simulation and other models in further understanding the principles behind the processes of decomposition kinetics in grasslands.

Our approach is principally twofold: (1) to examine the key processes of dissimilation of organic matter, noting the commonality of these processes worldwide, and (2) to make a comparison of the relative importance of these processes and the organisms responsible for them within the various habitats. Thus, the dry steppe regions have a decomposition pattern much different from the savannah regions of central and western Africa. In this regard, key abiotic factors, such as soil water and temperature, play an important role and account for a great part of the variability. We wish to emphasize the marked event-oriented responses of soil organisms to ambient climatic conditions and to stress the difficulty of understanding, through monthly or seasonal averages, the processes performed by fast-growing organisms.

By the late 1960s, it was understood that from 88 to 99 % of the annual net primary production entered the soil subsystems of a terrestrial ecosystem without going through the grazers (Wiegert, Coleman & Odum, 1970). In fact, numerous studies undertaken within the IBP, and covered later in this chapter, indicate that indeed grazing by large mammal herbivores in many grassland regions does not materially alter the high percentage which goes into the decomposition and mineralization pathways of energy flow. There are some exceptions, such as the African savannas; we shall explore the reasons for this.

The close interconnections between decomposition and mineral cycling are many and complex. Pertinent data on decomposition are given, and then a decomposition sub-model is presented, illustrating the importance of these processes at the subsystem and total-system levels. The interactions between our carbon-oriented chapter and the nutrient cycling chapter of Clark *et al.* (Chapter 8, this volume) should be kept in mind.

Our concern in this chapter is to characterize, in a global fashion, decomposition processes in the approximately 25 to 30 % of the earth's land area (9 % of the total area) which are considered as grassland and semi-arid and arid shrubland (Reiners, 1973). Because of constraints on length we shall discuss many processes in general and emphasize some site or area data perhaps unduly. This is inevitable, reflecting biases of editors and participants who have generously given of their time and data for what has been an arduous and frustrating task.

We have considerable incentive to emphasize this portion of global heterotrophic activity. Evolutionarily, the organisms doing the majority of the work (soil microflora) had been around for billions of years before the origin of a land flora (Devonian period), and are still responsible for the assimilation of > 90–95% of the total net primary production in terrestrial ecosystems (Wiegert *et al.*, 1970).

For global and interbiome comparisons of carbon inputs and fluxes, it is imperative to have much more information on temperate and tropical savannas and dry, grazed grasslands (Reiners, 1973). We submit that virtually all terrestrial ecosystems, but particularly grasslands, are only now receiving the total-system study which they require if adequate understanding of production biology is our intent. Thus, partitioning of primary production below ground in many grasslands exceeds 75% (Coleman, Andrews, Ellis & Singh, 1976), and the amount assimilated by cattle may be only 3 to 5% of the total net primary production.

For the purpose of this chapter, we have used not only first-hand publications, covering work mostly occurring during or sponsored by IBP, but also pre-IBP books and review articles, such as those of Clark & Paul (1970), Dommergues & Mangenot (1970), Phillipson (1970), and Sasson (1972). This chapter is as cosmopolitan as any in the volume, and, we feel, reflects the true spirit underlying the International Biological Programme effort. Several of our contributors made major compilations of not only their own, but literally dozens of other workers' efforts in various parts of the world. The contributions of Úlehlová, Zlotin, Schaefer and Dommergues are especially noteworthy in this regard. Dr Zlotin provided also a complete English review of literature printed in Russian. All of the other contributors furnished considerable amounts of materials and advice as well. The able assistance of Professor Sasson in translations of papers otherwise inaccessible to me was invaluable.

David C. Coleman

## Main microbial groups and activities

### *The microbial population in temperate grasslands*

Increased attention has been paid recently to energy flow in ecosystems, with calories (or joules) being a common denominator of their diverse structural units (Odum, 1971). Theoretical yields of bacterial cells in a given system are about 0.118 g of bacterial dry matter formed under aerobic or anaerobic conditions per 1 kcal removed from the nutrient medium (Payne, 1970). The biomass of bacterial cells that could potentially be formed from the annual production of the above- and belowground biomass in some grassland ecosystems are given in Table 7.1. The total production, e.g. of the uncut plant

Table 7.1. *Calorific values of substrates and bacterial biomass in various grasslands*

| Site | Plant community | Net primary production | | | Potential bacterial biomass | Potential calorific value |
|---|---|---|---|---|---|---|
| | | Above ground (kcal·m$^{-2}$) | Below ground (kcal·m$^{-2}$) | Total (kcal·m$^{-2}$) | (g·m$^{-2}$) | (kcal·m$^{-2}$) |
| Ojców, Poland | *Arrhenatheretum* | 1250–1360 | 1320–1760 | 3120 | 368 | 1840 |
| Ispina, Poland | *Arrhenatheretum* | 1520 | 1260–1680 | 3192 | 375 | 1875 |
| Lanžhot, Czechoslovakia | *Serratulo–Festucetum commutatae* | 1062–1912 | 1560–2080 | 3992 | 468 | 2340 |
| Lanžhot, Czechoslovakia | *Gratiola officinalis–Carex praecox-suzae* | 4847 | 1152–1536 | 4383 | 518 | 2590 |
| Lanžhot, Czechoslovakia | *Glycerietum maximae* | 4037 | 2592–3442 | 7479 | 873 | 4365 |
| Pawnee Site, USA | *Bouteloua gracilis* shortgrass prairie | 496 | 3290 | 3786 | 447 | 2235 (700)[a] |

See text for details. Ranges in aboveground and belowground values reflect error estimates and variability in climate.
[a] Calculated.

community of *Glycerietum maximae*, amounts to 1009 g of aboveground and to 765 g of belowground dry matter$\cdot m^{-2} \cdot yr^{-1}$, corresponding to 4037 and 3442 kcal, respectively. The total of 7479 kcal$\cdot m^{-2} \cdot yr^{-1}$ would theoretically yield about 870 g$\cdot m^{-2} \cdot yr^{-1}$ of dry microbial biomass. In contrast, the maximum dry weight of bacterial biomass in the upper 25 cm of soil of the same plant community is about 30 g$\cdot m^{-2}$, calculated from microbial plate counts (Úlehlová, 1973). Taking into account the short generation times of the bacterial and fungal share of the biomass, it is obvious that the decomposition of dead plant matter may be restricted to rather short periods of favourable conditions of temperature and humidity. For example, if one assumes that all organic matter produced per year is subjected to microbial action, then 2235 kcal$\cdot m^{-2}$ of microbes would be produced per year. However, a more reasonable estimate for the shortgrass prairie is 700 kcal$\cdot m^{-2} \cdot yr^{-1}$, or 9.7 generations per year for an average microbial population of 65 g$\cdot m^{-2}$ at 30 cm depth (Sparrow & Doxtader, 1973). In contrast, the mixed-grass prairie microbial biomass was higher (*c.* 200 g$\cdot m^{-2}$: Clark & Paul, 1970), with an estimated turnover rate of $< 0.001$ $h^{-1}$ (Shields, Paul, Lowe & Parkinson, 1973), or about 1800 kcal$\cdot m^{-2}$ in new microbial tissue alone. The turnover rate of microbial biomass on a yearly average is much lower than that attained during the short periods of microbial outbursts. A more meaningful estimate of microbial activity is obtained from total respiration measurements or mineralization rates (see Chapter 8, this volume), and it is our intent to focus on these major processes.

### *The microbial population in tropical grasslands*

Little information is available for various tropical grasslands. Our sole entry in this chapter concerns microbial studies in a *Borassus* palm savanna, the 'Lamto' savanna on the Ivory Coast. For edaphic, physiographic and meteorological information, the reader is referred to Lamotte (1975).

#### *Total microbial counts*

Soil samples (0–10 cm) were taken at Lamto toward the end of the dry season (February and March) and of the humid season (October to November) in the unburnt as well as in the burnt savanna. They were compared by plate counts (suspension–dilution method and growth on synthetic media). It was found that fire reduced the fungal and bacterial populations, but actinomycetes increased, resulting in a higher total microbial density in the burnt savanna soil (Pochon & Bacvarov, 1973; see Table 7.2).

The dynamics of the evolution of the various microbial groups under the influence of fire are not yet known. Although the total microflora has an apparent uniformity, there is an extreme heterogeneity of fungal species in the diverse sampling sites of every habitat as well as between the unburnt and

Table 7.2. *Microbial numbers per g dry wt soil in burnt and unburnt savanna, Lamto, Ivory Coast, March 1970*

| | Fungi | Bacteria | Actinomycetes | Total |
|---|---|---|---|---|
| Unburnt savanna | 103000 | 945200 | 1901800 | 2950000 |
| Burnt savanna | 30600 | 758400 | 2598600 | 3387000 |

After Pochon & Bacvarov (1973).

burnt savanna soils. More than 60 species of fungi have been isolated (Rambelli, 1971; Rambelli & Bartoli, 1972; Rambelli, Puppi, Bartoli & Albonetti, 1973), some of them very uncommon (*Periconia, Chaetoceratostoma, Angulimaya, Gonytrichum*). No precise relationship with soil or vegetation types has been shown; on the other hand microbial densities were relatively uniform in soil samples taken toward the end of both the dry and humid seasons.

### *Counts of functional groups of microbes*

Physiological groups of microflora have been enumerated, but results should be interpreted with caution. Thus Pochon & Bacvarov (1973) noted that the effects of delay in sample analysis and mode of sample storage are more marked than those caused by the various local pedological, microclimatic and botanical conditions.

Table 7.3 shows mean numbers of different microbial groups for two sites on two dates. A marked variability is evident. Total microflora count in October, for example, ranged from $8 \times 10^6$ to $90 \times 10^6$ for the *Hyparrhenia* savanna and *Loudetia* savanna. In addition, these differences were not consistent with the samples' organic carbon, which varied considerably between samples. Anaerobic flora is present throughout the year, mainly in the *Loudetia* sites where drainage is poor and waterlogging heavy, as indicated by the high number of anaerobic nitrogen-fixers in October.

The variations between the samples collected in October and in May probably originated, for the *Hyparrhenia* site, from a marked increase in actinomycetes and, for the *Loudetia* site, from a marked decrease in fungi and bacteria during the dry season. The latter site seems to be characterized by a very fast and frequent shift between aerobic and anaerobic phases. In contrast to these sandy soils, the more clayey ones of the *Hyparrhenia* sites on the plateau are predominately aerobic, and the water content remains near an optimal level for a longer duration.

The wide variation in total microflora counts (unburnt savanna, dry season, for example, $3 \times 10^6$ to $3 \times 10^7$) means that sites cannot thus be characterized in an absolute way, but only compared among themselves by the amplitude and direction of the seasonal numerical variations.

Table 7.3. *Functional microbial groups in two habitats,* Borassus *palm savanma*

| | Total flora | Anaerobic N-fixers | Anaerobic cellulolytic | Aerobic cellulolytic | Nitrifiers |
|---|---|---|---|---|---|
| October 1969 | | | | | |
| *Hyparrhenia* savanna | 8 | 7 | 1.3 | 5.2 | 1.5 |
| *Loudetia* savanna | 90 | 23 | 4.5 | 7.0 | 3.2 |
| May 1970 | | | | | |
| *Hyparrhenia* savanna | 30 | 7 | 0.8 | 12.4 | 2.3 |
| *Loudetia* savanna | 20 | 4 | 2.0 | 16.4 | 4.7 |

After Pochon & Bacvarov (1973).
Values for total flora are $\times 10^{-6}$ $g^{-1}$ soil, all others $\times 10^{-3}$ $g^{-1}$ soil.

Schaefer (1978) comments on humification and microbial activities in several tropical soil types at Lamto, and the reader is referred to his paper for pertinent details of these processes.

## Roles of saprophagic fauna

### *Temperate grasslands*

The role of invertebrate saprophages in grassland functioning varies widely. Several scientists have synthesized many aspects of these functional roles (Ghilarov, 1960*a*; Müller, 1965; Tischler, 1965; Phillipson, 1970). We use a somewhat modified version of Wiegert & Owen's (1971) definition of saprophagy: namely, saprophagic fauna are considered to feed on living and dead micro-organisms and/or decaying organic detritus. Striganova (1971) characterizes activities of invertebrate saprophages as follows:

(1) In the process of feeding the animals comminute small pieces of plant tissue, creating an enormous surface area and rendering them accessible to active microbiological, physical and chemical influences.

(2) By means of their own enzymes and those from symbiotic micro-organisms, the animals split some organic substances and promote their further mineralization. Additionally, they assist in the synthesis of new organic substances and the humification of plant remains.

(3) In the process of horizontal and vertical shifts within the litter and soil, the animals redistribute organic and mineral substances inside the soil profile and promote the creation of a particular soil structure favourable to active aerobic processes of transformation of organic matter (see also Kühnelt, 1957).

Undoubtedly all forms of invertebrate saprophage activity are displayed in close interaction with soil microbial complexes. The symbiotic relations of saprophages and micro-organisms have been the subject of many special studies (see Kozlovskaya & Zaguralskaya, 1966; Stebayev, 1968; Kozlovskaya, 1970; Breymeyer, Jakubczyk & Olechowicz, 1975).

Table 7.4. *Functional characteristics of various soil invertebrate saprophagic groups*

| Animal group | Mechanical destruction of litter | Mineralization (destruction of cellulose) | Primary humification |
|---|---|---|---|
| Collembola | ** | — | — |
| Oribatei | *** | — | — |
| Diplopoda | *** | *** | — |
| Oniscoidea | *** | *** | — |
| Mollusca | *** | *** | — |
| Enchytraeidae | ** | *** | — |
| Diptera (larvae) | *** | * | * |
| Lumbricidae | * | — | *** |

From Striganova (1971).
*, Participation of one or two species.
**, Participation of several species.
***, Participation of a majority or all of the species.

In the feeding process many saprophages inoculate organic materials passing through them with bacteria and fungi. Thus, it is difficult to differentiate saprophagic and microphagic feeding modes (Ghilarov, 1965). There are three size categories of invertebrate saprophages. These are: macrofauna (lumbricids, enchytraeids, millipedes, gastropods and some insect larvae); mesofauna (Acarina, Collembola, Protura and others); and microfauna (protozoans, rotifers, tardigrades and nematodes). Functional trophic classifications of these main systematic groups of soil invertebrates are presented in Table 7.4.

The Lumbricidae carry out mechanical destruction of plant litter and partial humification of plant remains. Mineralization of litter is low where the cellulolytic microflora is scarce. Dipteran larvae destroy litter fall and stimulate microbial processes and thus mineralization and humification. Enchytraeidae, molluscs, oniscoid sowbugs and diplopods both comminute and mineralize plant tissues. Oribatei and Collembola take part only in the mechanical destruction of leaf fall; many species are typical mycetophages (but see the extensive review by Luxton, 1972). Nematodes use only the contents of cells without destroying their walls (see Nielsen, 1961). The relative proportions of the three size groups of soil invertebrates change as a function of aridity. In meadow grasslands in either tundra, forest or forest-steppe zones, the macrofauna dominates the saprophagous complex, comprising $\geqslant 90\%$ of the total saprophagic fauna biomass (Table 7.5). With increasing aridity, the dominance of the macrofauna decreases and the fraction of the biomass in the mesofauna and microfauna increases. In dry steppe ecosystems, the macrofauna comprises 35% and the mesofauna and microfauna about

Table 7.5. *Biomass of invertebrate saprophages in natural grasslands of various geographic regions in the USSR* (mean values g live weight $\cdot m^{-2}$)

| Animal group | Tundra[a] | Taiga | | | Forest-steppe | | Steppe | | Semidesert | | Tien Shan Mountains | | | |
|---|---|---|---|---|---|---|---|---|---|---|---|---|---|---|
| | | Middle | | South | Meadow steppe | Steppe meadow | Typical | Dry | | | Moist meadow | Steppe meadow | Dry steppe | Semi-desert |
| | 1 | 2 | 3 | 4 | 5 | 6 | 7 | 8 | 9 | 10 | 11 | 12 | 13 | 14 |
| Lumbricidae | 60.00 | 27–34 | 111.2 | 26.9 | 82.90 | 26.50 | 18.44 | 4.02 | — | — | 49.80 | 19.00 | 0.10 | — |
| Enchytraeidae | 25.69 | — | — | — | 4.04 | 5.60 | 1.20 | — | 1.20 | 0.10 | 1.20 | 0.10 | — | — |
| Nematoda | 7.70 | — | — | — | 5.55 | 4.15 | 1.49 | 3.76 | 0.74 | 1.20 | 1.30 | 0.56 | 0.18 | 0.19 |
| Collembola | 2.22 | — | — | — | 0.41 | 0.62 | 2.03 | 1.34 | 0.11 | 0.09 | 0.32 | 0.38 | 0.16 | 0.01 |
| Acarina | 0.10 | — | — | — | 0.56 | 0.44 | 1.00 | 0.37 | 0.04 | 0.01 | 0.23 | 0.09 | 0.05 | 0.03 |
| Diplopoda | — | — | — | — | 2.72 | 7.86 | 0.38 | 0.06 | — | — | — | — | — | — |
| Others | 4.00 | — | — | — | 0.86 | 1.10 | 0.64 | 1.76 | 0.01 | 0.01 | 0.29 | 0.28 | 0.08 | 0.06 |
| *Total* | | | | | 97.04 | 46.27 | 25.18 | 11.31 | 2.1 | 1.41 | 53.14 | 20.41 | 0.57 | 0.29 |

Data of R. I. Zlotin.

1, meadow, Taimyr (Chernov, 1973); 2, bottomland meadow, Arkhangelsk region (Matveeva, 1966); 3, bottomland meadow, Moscow region (Matveeva, 1969); 4, meadow in valley, Moscow region (Matveeva, 1969); 5 and 6, Kursk region (Zlotin, 1969); 7 and 8, the Ukraine (Zlotin, 1969); 9 and 10, the western Kazakhstan (Zlotin, 1969); 11–14, Tien Shan Mountains, 3200–3700 m above sea level (Zlotin, 1975).

[a] Maximal values.

Table 7.6. *Energy flow through populations of soil invertebrate saprophages measured as active metabolism in kcal $\cdot m^{-2} \cdot yr^{-1}$*

| | Forest-steppe | | | Tien Shan Mountains | | | |
|---|---|---|---|---|---|---|---|
| Animal group | 5 | 6 | Steppe 7 | Moist meadow 11 | Steppe meadow 12 | Dry steppe 13 | Semi-desert 14 |
| Lumbricidae | 230 | 75 | 28 | 68 | 40 | 0.4 | — |
| Enchytraeidae | 30 | 50 | 8 | 10 | 2 | — | — |
| Nematoda | 180 | 120 | 220 | 22 | 16 | 6 | — |
| Collembola | 15 | 18 | 40 | 7 | 13 | 5 | 0.3 |
| Acarina | 11 | 7 | 21 | 3 | 3 | 2 | 2 |
| Others | 9 | 10 | 5 | 2 | 3 | 1 | 2 |
| *Total* | 475 | 280 | 322 | 112 | 77 | 14 | 12 |

Data of R. I. Zlotin.
For explanation of numbered ecosystems, see Table 7.5.

30% each of the total biomass. In semideserts a saprophagous macrofauna is almost totally absent and about 90% of all the saprophage biomass is due to the free-living nematodes. It should be noted that Protozoa are not included in these calculations; they represent from 0.2 to 0.3 g dry wt $\cdot m^{-2}$ in North American steppes (Elliott & Coleman, 1977) and 1 to 2 g dry wt $\cdot m^{-2}$ in mesic agro-ecosystems (Stout & Heal, 1967).

These changes in soil zoomass structure are the result of the different environmental requirements of the three groups (Ghilarov, 1965). The macrofauna requires the highest humidity and thus is abundant only in those soils having capillary-free soil water. Meso- and microfauna require less free water as they inhabit pores and natural pockets that remain constantly saturated with water, even in drier soils. The total biomass of invertebrate saprophages in grasslands ranges from $< 1$ to 100 g live wt $\cdot m^{-2}$. The largest biomass is found in meadow-steppe in the forest-steppe zone (100 g $\cdot m^{-2}$) and decreases to 1 g $\cdot m^{-2}$ in semideserts. Arid high-mountain grasslands (dry steppe and semidesert) have the lowest biomass level of all (approximately 0.3 to 0.6 g $\cdot m^{-2}$: Table 7.5).

More importantly for our estimates of ecosystem function, the energy flow through various faunal groups has been calculated (Table 7.6; see Phillipson (1970) and Zlotin (1975) for methods of calculating energy flow in these groups). The total energy flow decreased markedly with increasing aridity. Note that Lumbricidae and Enchytraeidae are very active in the most mesic sites, whereas nematodes are equally active in wet and mesic areas and continue into the drier steppes. In all cases the amount of energy flowing through the fauna (as respiration plus production) is only 10% or less of the total; as a further example, in North American shortgrass prairies the total flow through the fauna is only *c.* 0.6 % (Coleman *et al.*, 1976). This contrasts with the more faunal-dominated energy flow in tropical wet and savanna grasslands (see below).

### *Tropical grasslands*

In the Lamto Project (Ivory Coast) the saprophagic soil meso- and macrofauna were divided into two main groups:

#### (1) *Consumers of more or less decomposed dead plants and animals (i.e. saprophages)*

The main consumers of undecomposed litter are the fungi-growing termites. With only a moderate mean biomass (0.49 g dry wt·$m^{-2}$), they use not less than 1200 g·$m^{-2}$·$yr^{-1}$ of dry litter. With their symbiotic fungi and bacteria they contribute to the mineralization of organic matter (Lamotte, Barbault, Gillon & Lavelle, 1974).

The rest of the saprophages (only 10 % of the earthworm biomass, the Collembola, the pauropods, the diplopods, 52 % of the Acari, part of the ant population (Levieux, 1971) and some Coleoptera larvae make up a chain of organic matter degradation. This chain of events includes digestion, excretion and elimination and reingestion over time, which augments the continuous microbial activity.

#### (2) *Earth-eating organisms (i.e. geophages)*

These organisms ingest soil particles and assimilate the debris and the more or less humified organic matter, as well as the edaphic microflora. In terms of biomass (28 g dry wt·$m^{-2}$, emptied digestive tube), the geophages predominate among the soil fauna; they are nevertheless little diversified and include essentially the enchytraeids, the earthworms, and the 'humivorous' termites.

During their period of activity – that is when soil water is > 10–20 % by volume – earthworms ingest daily 5 to 30 times their own weight of soil, according to the species and the size of the individuals. The weight of dry soil recycled every year by these oligochaetes is thus estimated as 100 kg·$m^{-2}$ (Lavelle, 1973). Yet, because of the soil's deficiency in organic matter and because of their poor assimilating capacity, their mineralizing activity amounts only to 80 to 100 g·$m^{-2}$·$yr^{-1}$ of detrital input.

When compared with temperate grassland populations, the savanna mesofauna of Lamto contains less enchytraeids, Collembola and Acari. In contrast, the Symphyla, Protura, Diplura and scale insects are relatively abundant. But it is the termites which characterize the tropical environment at Lamto. They are relatively less abundant than in other African ecosystems (Wood, 1976), but very much diversified (more than 50 species). The same is true of the ants, of which there are more than 150 species exerting an influence at all trophic levels.

The consumers of living plants and saprophages, among which the fungi-growing termites play a prominent role, are characterized by the fact that they collect most of their food on the soil surface. They therefore have little effect on roots, even though these represent, after degradation by the microflora, a very important available source of food. We anticipate, however, that phytophagous nematodes may also be important, if their activity and numbers are similar to those in temperate grasslands (Smolik, 1974; Coleman *et al.*, 1976).

Activities of the soil fauna, particularly the soil oligochaetes, of tropical wet grasslands in India were studied by Thambi & Dash (1973), Dash, Patra & Thambi (1974), and Dash & Patra (1977). Oligochaete numbers reached a maximum of 7800 $m^{-2}$ during the rainy season (September–October), while the minimum density of around 560 $m^{-2}$ was recorded during the summer drought months of May and June. There was a significant correlation between the percentage of soil water and oligochaete numbers, and between soil temperature and oligochaete numbers. The monthly average oligochaete biomass was 8 g dry $wt \cdot m^{-2}$ with an annual secondary production of 35 $g \cdot m^{-2}$. The energy input into the grassland site from net primary production was 7680 $kcal \cdot m^{-2} \cdot yr^{-1}$. Energy utilization by the oligochaetes was chiefly through population metabolism, growth and reproduction. Oxygen consumption, about 60 $l \cdot m^{-2} \cdot yr^{-1}$, amounted to an energy equivalence of 288 $kcal \cdot m^{-2} \cdot yr^{-1}$. The energy value of mucus production (142 g dry wt $mucus \cdot m^{-2} \cdot yr^{-1}$) was 568 $kcal \cdot m^{-2} \cdot yr^{-1}$. Oligochaete tissue production was 35 g dry $wt \cdot m^{-2} \cdot yr^{-1}$, or 162 $kcal \cdot m^{-2} \cdot yr^{-1}$. All these energy values add up to 1018 $kcal \cdot m^{-2} \cdot yr^{-1}$, which is about 13.2% of the total energy input into the study site. About 15% of the total energy assimilated by earthworms is stored in their tissue, of which 95% is subsequently used in metabolism.

The belowground invertebrate biomass in different tropical grasslands in Panama amounts to 6.3 g dry $wt \cdot m^{-2}$, and this represents 78–98% of the total invertebrate biomass of the grasslands (Breymeyer, 1974). The average biomass and secondary production of oligochaetes in savannah-type grasslands in the Ivory Coast was 6.5 $g \cdot m^{-2} \cdot yr^{-1}$ and 46 $g \cdot m^{-2} \cdot yr^{-1}$, respectively (Lamotte, 1975). The biomass turnover of the population was about seven times the average biomass.

Oligochaetes dominate the invertebrate biomass in tropical grasslands, and the annual population biomass turnover (five to seven times the average biomass) indicates that they are of importance in these ecosystems. It is evident from the data of the three sites studied that the total invertebrate biomass varies greatly between sites, and that perhaps the total primary production determines the total belowground invertebrate biomass.

Wormcast production and the amount of nitrogen returned to soil by the activity of oligochaetes were studied in the Indian site (Patra & Dash, 1977). Some 77 tonnes dry wt·ha$^{-1}$·yr$^{-1}$ of wormcast were produced. There were seasonal variations in rates and amounts, with lows in the summer (May and June), and maxima during the latter part of the rainy season (September and October). This huge cast production shows the rate of soil turnover and the magnitude of the processing activity of plant material through the earthworm gut.

The nitrogen returned to the soil through earthworm mucus, dead tissue and wormcasts amounts to about 72 kg·ha$^{-1}$·yr$^{-1}$ – more than the requirement of crop plants. However, the amounts actually available to plants are not known. If one assumes no significant loss of nitrogen to the atmosphere, then the nitrogen returned to soil by earthworm activity and the inferred amount available to plants seems large.

## Loss rate of organic matter from litter

### *Wiegert–Evans paired plots*

The extent and rate of aboveground litter decomposition can be estimated by the method proposed by Wiegert & Evans (1964), or by various modifications of it (Łomnicki, Bandoła & Jankowska, 1968; Titlianova, 1973). Table 7.7 sums up the results obtained by the Wiegert–Evans method in several IBP projects on temperate zone meadows (Jankowska, 1971; Ketner, 1973; Titlianova, 1973). The Wiegert–Evans method uses 'paired plots'. Dead material is removed from one plot of a given area and weighed at time $t_0$; then the dead material is removed from a paired plot of the same area some time later, at time $t_1$. Thus the instantaneous rate of disappearance of dead material from this plot can be calculated as:

$$r = \frac{\ln (W_0/W_1)}{t_1 - t_0},$$

where $r$ is disappearance rate in g·g$^{-1}$·d$^{-1}$, $t_1 - t_0$ is the time in days, $W_0$ is the dry weight removed from the initial plot, and $W_1$ is the dry weight removed from the second plot. The amount of material decomposed per year often exceeds the maximum standing crop of the plant community. Thus, considerable amounts of plant material, such as dead leaves or litter of the moss layer, can enter the decomposition chain during the vegetative period after having contributed only temporarily to the standing crop. The seasonal patterns of decomposition often reflect the inputs of organic matter and in this sense are in some cases typical of the plant community. However, they mostly reflect changing environmental conditions (weather, floods, etc.).

Table 7.7. *Amount and rate of litter decomposition in grasslands (using the Wiegert–Evans (1964) method)*

| Plant community | Country | Year | Litter decomposition ($g \cdot m^{-1} \cdot yr^{-1}$) | Rate of decomposition ($mg \cdot g^{-1} \cdot d^{-1}$) Min–Max | Average |
|---|---|---|---|---|---|
| *Serratulo-Festucetum avennutatae* dry meadow | Czechoslovakia | 1972 | 166.4[a] | 3.4–9.3 | 6.4 |
| *Gratiola officinalis Carex praecox-suzae* moist meadow | Czechoslovakia | 1972 | 375.8[a] | 6.2–16.1 | 10.5 |
| *Junco-Caricetum extensae* | Netherlands | 1968 | 335 | 0–40 | 18 |
| | | 1969 | 205 | 2–40 | 14 |
| *Plantagini limonietum* | Netherlands | 1968 | 235 | 6–20 | 14 |
| | | 1969 | 310 | 0–16 | 9 |
| *Arrhenatheretum elatioris* association, unmown | Poland | 1965 | 742.9 | 1–16 | 9 |
| | | 1966 | 776.9 | 1–14 | 9 |
| | | 1967 | 560.0 | 1–14 | 10 |
| *Arrhenatheretum elatioris* association, mown | Poland | 1966 | 248.5 | 1–14 | 9 |
| *Thalictro-Salvietum* association | Poland | 1965 | 498 | 0.8–8 | 5 |
| | | 1966 | 517.6 | 1–13 | 5 |
| Meadow-steppe | USSR (Karachi) | 1968 | 140 | | |
| | | 1969 | 393 | | |
| | | 1971 | 90 | | |
| | | 1972 | 309 | | |
| Solonetz steppe | USSR (Karachi) | 1969 | 156 | | |
| *Calamagrostis* meadow | USSR (Karachi) | 1968 | 323 | | |
| | | 1969 | 376 | | |
| | | 1970 | 257 | | |
| | | 1971 | 303 | | |
| | | 1972 | 381 | | |
| *Puccinellia* meadow | USSR (Karachi) | 1968 | 42 | | |
| | | 1969 | 292 | | |
| Grass fen | USSR (Karachi) | 1968 | 169 | | |
| | | 1969 | 473 | | |

[a] In $g \cdot m^{-2} \cdot 240\ d^{-1}$.

### Litter mesh bags

Another method for assessing decomposition is that of the 'litter mesh bag', described by Witkamp & van der Drift (1961). The data obtained by the 'litter mesh bag' method on a *Junco-Caricetum extensae* association in Holland (Ketner, 1973) can be used for a simple comparison of litter input and decomposition rate. If the decomposition of litter continued for a month at the average rate actually measured, i.e. 14 $mg \cdot g^{-1} \cdot d^{-1}$, then 113.4 g, or 42%

of the 270 g (2-year average) of litter originally available, would be decomposed, and 53.5 g of bacterial biomass (see p. 613) would be formed. Thus more than one-third of the annual litter input can be decomposed during a single outburst of microbial activity. Similar conclusions can be drawn from the data for 1970.

None of the methods is error-free. Thus, for instance, the packing of material in the bag as well as the efficiency of its contact with the soil may vary considerably. There is also a nonlinearity between time of exposure and the litter decomposed. Estimates of decomposition using the litter mesh bag are usually lower than those obtained by the Wiegert–Evans method.

The litter mesh bag method was used in studies on the decomposability of different plant materials. In the Leningrad region, Miroshnishchenko, Pavlova & Ponyatovskaya (1972) found that in most grassland species the leaves are the part most easily decomposed, followed by the rootlets, roots and stems. The decomposition of *Alchemilla monticola* material was the fastest and that of *Alopecurus pratensis* the slowest. Ten-Chak-Mun & Fedorova (1972) showed that the decay of roots is the fastest in the tallgrass species of Sachalin, with the exception of *Fagopyrum sachalinum*. Differences between materials originating from different plant species were also observed.

Zlotin (1971, 1974) and Zlotin & Khodashova (1974) tried to differentiate various types of decomposition by using bags with different mesh sizes and by chemically excluding microbial activity with an organic microbistatic compound (toluene). They found that 24% of the litter was decomposed by abiotic processes alone, principally photochemical oxidation, 28% (an additional 4%) when micro-organisms were present, and 34% (an additional 6%) when the participation of saprophagic macrofauna was possible. Thus, about 70% of the total litter decomposition was the result of abiotic factors (principally ultraviolet and blue light at wavelengths $< 500$ nm, as determined by selective filters), and only about 30% was due to the saprophagous microbes and animals. While these data are intriguing, inferences about contributions of abiotic factors to decomposition should be viewed with caution because of the difficulty of maintaining complete sterility of litter treated with an organic microbistatic compound (toluene).

An experiment to test the roles of abiotic and biotic factors in decomposition was recently carried out on the northeastern Colorado prairie (Pawnee Site) (Vossbrinck, Coleman & Woolley, 1978). Samples of blue grama grass litter were placed in 1 mm or 53 $\mu$m nylon mesh bags. The large mesh allowed meso- and microfauna to enter, the small mesh allowed only microfloral activity, and the narrow mesh with saturated mercuric chloride and copper sulphate additions (for microbistasis) had abiotic activity only. Bags were retrieved over a 9-month period. By the end of 9 months only 8% dry weight had been lost from 'abiotic' bags, 15% from bags with microflora alone and 18% from bags with flora and fauna.

As the Kursk Station (USSR) site of Zlotin is at a lower elevation, with higher precipitation, and thicker litter layer than the Pawnee (all factors tending to favour biological activity), the contrast in experimental results is striking and as yet unexplained.

### *Cellulose materials*

More defined materials, such as cellulose, have been used as substrates in mesh bags instead of litter. Soviet workers have used cotton and linen material, while in Europe and North America an analytical grade of filter paper or cotton-wool is usually preferred. Most data on microbial activities in grassland ecosystems have been obtained with several modifications of the mesh bag method (Petrova, 1963; Yershov, 1966, 1972; Hundt & Unger, 1968; Tesarová & Úlehlová, 1968; Řehořková, Kopčanová & Řehořek, 1968; Naplekova, 1972; Zaguralskaya, Yegorova & Smantser, 1972; Zlotin & Chukanova, 1973). Hundt & Unger (1968) studied the rate of cellulose decomposition in 44 types of grassland ecosystems in East Germany. The microbial activity was generally very low in wetland plant communities such as *Caricetum fuscae* and *Cladietum marioci*, somewhat higher in *Scirpetum silvatici* and *Caricetum gracilis*, and very high in *Glycerietum maximae* and *Phalaridecum arundinaceae*. High rates of cellulose decomposition were measured in wet meadows (*Alopecuretum*), i.e. 23 to 55 $mg \cdot d^{-1}$, with low rates in *Molinietum caeruleae*. In moist meadows the decomposition was usually good (*Trisetetum*, *Arrhenatheretum*). It was low in dry plant communities growing on sand (*Corynephoretum*), while in dry communities on loess or loamy soil (*Brachypodietum*, *Stipetum*) it was relatively high. Very low cellulose decomposition rates were observed in the alpine grassland biomes in the western part of the High Tatra Mountains (Rusek, Úlehlová & Unar, 1975).

The vertical gradients of cellulose decomposition in grassland soils were studied by Zlotin & Chukanova (1973). The highest rates of cellulose losses were generally found in the upper soil layer, i.e. 0 to 10 cm (Table 7.8), but the zone of maximum cellulose decomposition may shift from the upper soil layer in the spring to the lower parts of the soil profile in summer and fall under the influence of changing moisture and, probably, temperature conditions.

In the USA, a series of litter bag decomposition studies were carried out on nine IBP Grassland Biome Sites. Burials of bags containing Whatman filter paper spanned the years 1969–73, with most data generated in 1971–2. Nearly 900 bags were buried to ensure adequate replication at each retrieval time (retrievals were designed to be after about 20% loss, in increments, across the growing season).

With few exceptions, mean decomposition rates followed climatic and production figures (Table 7.9), with Pawnee Site having the lowest and Osage Site the highest, the two values being significantly different (Tukey's *Q* test) at the

Table 7.8. *Intensity of cellulose decomposition in meadow-steppe in a watershed; expressed as losses in % dry weight per month (Kursk station)*

| Location | Season | | | Mean monthly loss in year |
|---|---|---|---|---|
| | Spring–summer | Summer–fall | Fall–winter–spring | |
| Surface | | | | |
| Litter | 2.2 | 2.0 | 0.2 | 1.3 |
| Soil | 4.4 | 4.4 | 1.3 | 2.7 |
| In soil | | | | |
| 0–10 cm | 21.1 | 8.5 | 4.1 | 8.7 |
| 10–20 cm | 11.7 | 13.0 | 1.3 | 6.2 |
| 20–30 cm | 8.2 | 10.0 | 1.0 | 4.6 |
| 0–30 cm | 13.7 | 10.5 | 2.1 | 6.5 |

5 % level. Osage, the one mesic site, differed significantly from all others at the 10 % level.

Because of extensive variations in dry matter losses, both within and between retrieval dates, differences must be very large to be detected in the analysis of variance. Thus only the desert grassland (Jornada) showed a significant ($P < 0.10$) effect of burial date, with all retrievals from 22 January averaging 0.13 g·month$^{-1}$ while those after 26 June averaged 0.31 g·month$^{-1}$. Jornada's growing season does not begin until late July.

Several sites showed significant differences in weight losses on various retrieval dates. Cellulose buried on 19 July 1972 at Pawnee Site, showed high losses (0.14 g·g$^{-1}$·month$^{-1}$) by 9 September, reflecting the effects of the mid-summer rains, while retrievals after 21 November showed monthly losses of half that size. Osage, in contrast, showed significant differences between mid-May and mid-June retrievals following mid-April burial, but these, like the more mesic Cottonwood Site had relatively large loss rates of 0.04 to 0.05 g·g$^{-1}$·month$^{-1}$.

Field experiments using cellulose strips at the Canadian Matador Station (Saskatchewan) and associated sites were of two types (Biederbeck *et al.*, 1974). The first type, termed 'single placement', was designed so that all cellulose strips were placed in the soil at the beginning of the experiment and replicates were removed at subsequent sampling dates. This type of experiment should effectively evaluate cumulative cellulose decomposition by a population of micro-organisms allowed to establish over the period of the experiment. The second type of experiment, termed 'replacement', was set up so that cellulose strips were left in the soil for 3- or 4-week periods only, at consecutive intervals during the season. This type of experiment should evaluate the initial attack of micro-organisms on new cellulose being introduced to the soil at different times during the season. Particularly in grassland, and

Table 7.9. *Abiotic, primary production and soil respiration information on North American and Eurasian grassland studies (all production values in g C·m$^{-2}$·yr$^{-1}$)*[a]

| Country, Site Reference | Vegetation type[a] | Temperature (°C) January mean | July mean | Annual mean | Precipitation (mm) Annual total | Growing total | Annual net primary production | $CO_2$-C output (g·m$^{-2}$·d$^{-1}$) Minimum | Maximum | Annual |
|---|---|---|---|---|---|---|---|---|---|---|
| USA, Tucker Kucera & Kirkham (1971) | Tallgrass | 1 | 25 | 13 | 900 | 594 | 452 | 0.10 | 2.84 | 452 |
| USA, Osage May & Risser (1973) | Tallgrass | 3 | 27 | 15 | 1000 | 600 | 398 | 0.5 | 4.15 | 593 |
| USA, Cottonwood V. Lengkeek (personal communication) | Mid-grass | −7 | 24 | 9 | 385 | 280 | 344 | 0.25 | 3.5 | 215 |
| Canada, Matador Redmann (1978) | Mid-grass | −13 | 20 | 4 | 388 | 251 | 242 | 0 | 4.3 | 120 (18 May–30 Sept.) |
| USA, Pawnee D. C. Coleman (unpublished data) | Shortgrass | −4 | 23 | 10 | 300 | 240 | 197 | 0.25 | 2.34 (1972) 3.43 (1973) | 227 |
| USA, ALE Wildung *et al.* (1975) | Bunchgrass | −2 | 24 | 13 | 158 | 34 | 150[b] | 0.18 | 1.21 | 165 |
| USA, Jornada E. E. Staffeldt (personal communication) | Desert grass | 4 | 26 | 16 | 277 | 125 | 252 | 1.0 | 5.0 | NA |

| | | | | | | | | | | |
|---|---|---|---|---|---|---|---|---|---|---|
| USA, San Joaquin J. Pigg (personal communication) | Annual | 6 | 27 | 15 | 486 | 454 | 375 | 0.85 | 3.25 | NA |
| USA, Bridger T. Weaver (personal communication) | High mountain | −8 | 15 | 1 | 960 | 125 | 136 | | | NA |
| USA, South Carolina Coleman (1973) | Old field | | | | 1010 | 700 | 420 | 0.30 | 2.28 | 409 |
| USSR, Kursk Zlotin (1974) | Forest-steppe zone meadow-steppe | −8 | 19 | 5 | 512 | 332 | 568 | | | 572+ |
| USSR, Tien Shan Zlotin (1975) | Dry steppe | −20 | 8 | −4 | 230 | 180 | 71 | | | |
| Poland, Kazuń meadow Kubicka (1973) | *Arrhenatheretum* | −5.7 | 19.5 | 7.5 | 600 | | 301 | 0.80 | 1.83 | |
| Poland, Strzeleckie meadow Breymeyer (1971), Breymeyer & Kajak (1976), Kubicka (1973) | *Stellario-Deschampsietum* | −5.7 | 19.5 | 7.5 | 600 | | 279 | 0.87 | 1.20 | |
| Czechoslovakia, Lanzhôt Tesarová & Gloser (1972) | *Gratiola-Carex* | −1.5 | 20 | 9.5 | 585 | 365 | 538 | 1.93 | 3.90 | 441+ 150 days |

[a] Tallgrass is dominated by *Andropogon gerardi* and *A. scoparius*; mid-grass by *Agropyron smithii* (Cottonwood), and *Koeleria cristata* and *Stipa* sp. (Matador); shortgrass by *Bouteloua gracilis* and *Artemisia frigida*; bunchgrass by *Agropyron spicatum*; desert grass by *Bouteloua eriopoda*; annual by *Bromus mollis*; high mountain by *Festuca idahoensis*; old field by *Andropogon virginicus* and *A. ternarius*; meadow-steppe by *Bromus riparius* and *Stipa pennata*; dry steppe by *Festuca kryloviana* and *Ptilagrostis subsessiliflora*.

[b] Total root C = 240 g.

to some extent in cultivated land, both types of cellulose decomposition are proceeding continually. New leaves, stems and roots, constantly being added to the decomposer cycle, are undergoing initial attack; decomposition of material previously attacked also continues, so that the cumulative phase of decomposition is constantly in progress.

Cumulative percentages of decomposition of the cellulose strips in natural grassland (single placement experiment) are shown in Fig. 7.1. Less than 35% of the cellulose had been decomposed during the 15-week period in the untreated area, and the rate of decomposition was similar at both the 5 and 25 cm depths. The rate of decomposition was higher in the irrigated, fertilized and irrigated+fertilized plots. The highest rates of decomposition were recorded in the area receiving irrigation only, where close to 70% of the cellulose had disappeared after 15 weeks. It is postulated that the increased plant growth on the irrigated+fertilized plots resulted in increased transpiration and dried the top soil layer, but left residual moisture at 25 cm. As a result, decomposition was limited to some extent by lack of water at 5 cm, but not at 25 cm. With irrigation alone, the drying effect of added plant growth was not as great; consequently, the rate of decomposition at the two depths was similar.

The second type of experiment (replacement) was designed so that the cellulose strips were left in the soil for 3 or 4 weeks and then replaced with new strips. This type of experiment measured the rate of initial attack on new material at intervals throughout the season. In the natural grassland treatments the rates of decomposition were low in May and early June when soil water was adequate but temperatures were too low for maximum microbial activity. The maximum rates occurred in late June and July when soil water levels were still high and temperatures were high enough to produce conditions that were probably close to the optimum for the populations of microorganisms. In August and September the rates of decomposition declined sharply again to reach minimum values in September (Table 7.10). This decline in microbial activity was apparently correlated with the drying out of the soil in mid-summer. To facilitate comparison with the single placement experiment, percentages of decomposition for the 3- or 4-week intervals were also expressed on a cumulative basis over the season. When this was done, it was found that the cumulative curves for percentages of decomposition for the two types of experiments were quite similar, but that the final cumulative percentages were greater for the replacement experiment.

In the experiments described above the weight loss of cellulose was expressed as a percentage of the initial weight before placement on an ash-free basis. This method of calculation underestimates the amount of decomposition, however, because frequently the undecomposed cellulose is impregnated with fungal hyphae and bacteria that cannot be removed by washing. A method for correcting the percentage of decomposition was developed by using the

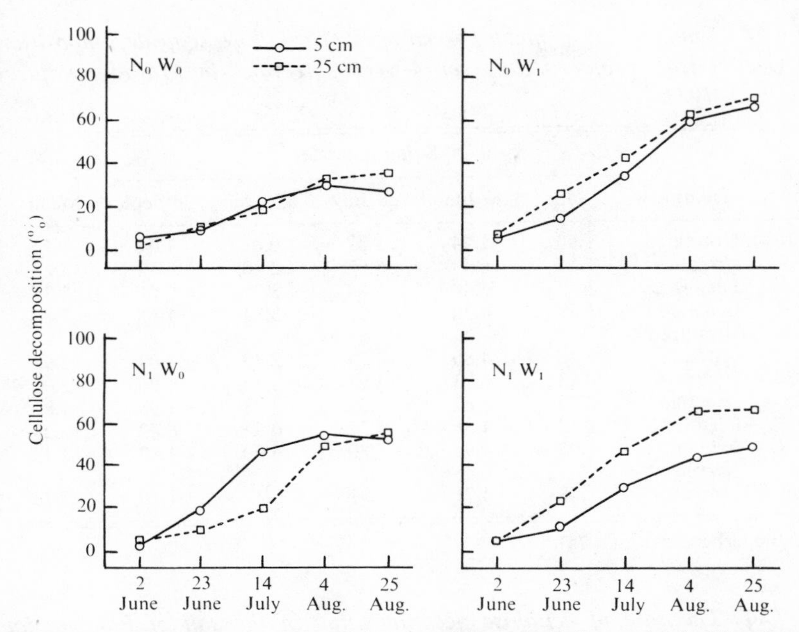


Fig. 7.1. Cumulative percentages of decomposition of cellulose strips in Matador natural grassland. All strips were put in place on 12 May 1971. $N_0W_0$, untreated; $N_0W_1$, irrigated; $N_1W_0$, fertilized; $N_1W_1$, fertilized and irrigated. (After Biederbeck *et al.*, 1974.)

phosphorus content of the recovered cellulose strips as an indicator of the amount of microbial biomass present. The phosphorus content was measured after the trace amount initially present in the cellulose itself was estimated, so it was possible to calculate (by addition) that portion of the phosphorus that originated from microbial biomass (Biederbeck *et al.*, 1974).

The same technique was also employed to estimate microbial yield coefficients (i.e. milligrams microbial biomass produced divided by milligrams substrate consumed in terms of carbon). Yield coefficients estimated from the replacement experiments averaged 0.40 for the 5 cm depth and 0.38 for the 25 cm depth (assuming a 2:1 ratio of fungal to bacterial biomass).

The loss rate of cellulose filter paper in tropical conditions was measured in the Panamanian savannah (Breymeyer, 1978). The decomposition rate was found to be relatively low in the top horizon (0 to 10 cm) of the soil (Table 7.11), being 3.76 $mg \cdot d^{-1}$ for the seed reservation and 2.62 $mg \cdot d^{-1}$ for the pasture. This corresponded to 7.84 $mg \cdot g\ cellulose^{-1} \cdot d^{-1}$ for the reservation and 5.73 $mg \cdot g\ cellulose^{-1} \cdot d^{-1}$ for the pasture.

Interesting results were also obtained in Panama in an experiment on the decomposition of standing plant material. Filter paper hung among the grasses, particularly that which was close to the ground, was decomposed at

Table 7.10. *Comparison of mean percentages of cellulose strip decomposition per week (strips replaced at 3- or 4-week intervals) in Matador natural grassland, 1971*

| Depths (cm) | Treatment | Sampling dates: 2 June | 30 June | 28 July | 25 Aug. | 24 Sept. | Means |
|---|---|---|---|---|---|---|---|
| 0–10 | Untreated | 1.50 | 1.44 | 3.37 | 0.82 | 1.46 | 1.72 |
| | Irrigated | 1.15 | 6.99 | 5.67 | 2.10 | 0.42 | 3.27 |
| | Fertilized | 1.37 | 6.48 | 5.02 | 3.21 | 0.61 | 3.34 |
| | Irrigated + fertilized | 1.89 | 4.30 | 5.01 | 3.84 | 1.65 | 3.34 |
| | *Means* | 1.48 | 4.80 | 4.77 | 2.49 | 1.03 | 2.91 |
| 20–30 | Untreated | 0.83 | 0.93 | 3.97 | 3.33 | 1.26 | 2.06 |
| | Irrigated | 0.79 | 8.62 | 9.57 | 4.28 | 1.08 | 4.87 |
| | Fertilized | 1.48 | 4.36 | 4.31 | 0.84 | 0.42 | 2.28 |
| | Irrigated + fertilized | 2.03 | 4.92 | 5.80 | 5.00 | 1.87 | 3.92 |
| | *Means* | 1.28 | 4.71 | 5.91 | 3.37 | 1.16 | 3.28 |

After Biederbeck *et al.* (1974).

Table 7.11. *The rate of cellulose decomposition in the soil of Panamanian grasslands*

| Time of exposure | Seed reservation (mean loss) | | Pasture (mean loss) | |
|---|---|---|---|---|
| | $mg \cdot d^{-1}$ | $mg \cdot g^{-1} \cdot d^{-1}$ | $mg \cdot d^{-1}$ | $mg \cdot g^{-1} \cdot d^{-1}$ |
| 17–24 August 1971 | 2.92 | 5.91 | 3.70 | 8.05 |
| 17–28 August 1971 | 5.98 | 12.44 | 2.53 | 5.23 |
| 17 August–2 September 1971 | 4.50 | 9.18 | 2.39 | 4.97 |
| 17 August–7 September 1971 | 1.64 | 3.85 | 1.87 | 3.85 |
| *Mean* | 3.76 | 7.84 | 2.62 | 5.73 |

After Breymeyer (1978).
The experiment was carried out over a period of 21 days in the rainy season. Cellulose was buried in the top 0–10 cm of soil. Each value is the mean of 5 samples.

a relatively high rate. More than 50% of the grass was decomposed above the ground. Such a high decomposition rate of standing vegetation results primarily from favourable conditions of temperature and humidity. This is in marked contrast to dry steppeland conditions in North America and Eurasia, where probably much less decomposes as 'standing dead'.

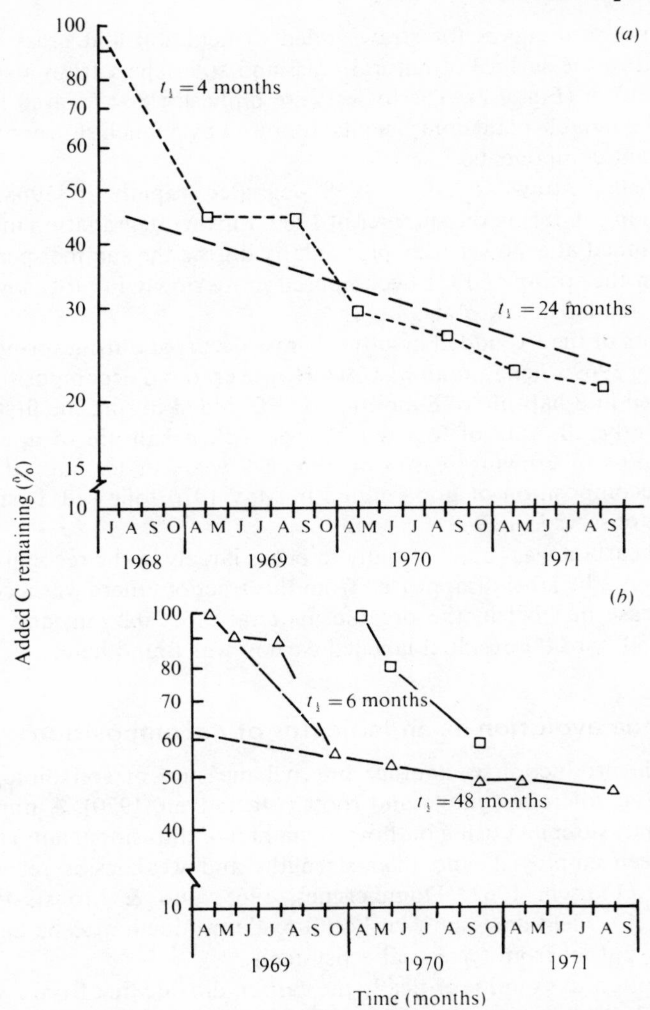


Fig. 7.2. (*a*) Decomposition of $^{14}C$-labelled wheat straw on a cultivated field. (*b*) Degradation of native vegetation on the surface of natural untreated grassland soil at Matador, Saskatchewan, Canada. Time shown does not include freeze periods. For explanations see text. (After Biederbeck *et al.*, 1974.)

### $^{14}C$-*labelled straw*

Further experiments using labelled wheat straw and labelled shoots of natural grasses were carried out to measure the rate of decomposition of plant residues. In these experiments the distribution of $^{14}C$ throughout various fractions of soil organic matter was determined at intervals as decomposition progressed (Biederbeck *et al.*, 1974).

The decomposition curves for straw added to field soil and grass shoot material added to the surface of natural grassland soil depart from a simple exponential pattern (Fig. 7.2). The losses were biphasic: first a rapid initial loss of easily degradable plant components, followed by a much slower release of more resistant components.

The $^{14}C$-labelled straw added in 1968 degraded rapidly in 1968, but remained unchanged during the summer of 1969. Further degradation in 1970 and 1971 continued at a slower rate, particularly during the summer periods. Straw added in the spring of 1971 decomposed more slowly in plots sown to wheat than in similar fallow plots.

Very little loss of the $^{14}C$ added in natural grass occurred during spring and early summer of 1969 when rainfall was low. However, rapid decomposition in autumn resulted in a half-life of 6 months for $^{14}C$ added during the first season. Subsequently, the rate of loss was slower with a half-life of approximately 48 months of growing season or almost 8 years under natural field conditions. Decomposition of grass added in May 1970 followed a pattern similar to that observed in 1969.

The labelled carbon was found initially to occur largely in the recognizable plant material. As the label disappeared from this fraction, there was a corresponding increase in label in the organic material of $> 0.2\ \mu m$, and after 2 years about 50% of the residual labelled carbon was found here.

## Carbon dioxide evolution as an indicator of decomposition

Carbon dioxide production is another internal measure of soil biological activity, including microbes, fauna and roots (Macfadyen, 1970). A number of methods of measurement using outflow (dynamic) or diffusion (static) techniques have been employed, and their strengths and weaknesses reviewed (Haber, 1958; Domsch, 1962; Dommergues, 1968; Ino & Monsi, 1969; Schlesinger, 1977). Field data show carbon dioxide production to be an important carbon output from terrestrial ecosystems.

It is our intention to examine critically the carbon dioxide flux from a wide variety of grasslands (20 or more), over a wide gradient of temperature and moisture regimes. In the temperate and tropical localities we have examples ranging from mesic to xeric conditions. Temperature regimes remain more similar in tropical than in temperate localities, however.

In the following pages we present data on net primary production (NPP), precipitation, mean air/soil temperatures and soil carbon dioxide evolution. Where possible, a comparison is made between yearlong NPP and carbon dioxide evolution. In many cases there is information only for a few months during the year. We further compare and correlate daily carbon dioxide outputs with soil water and temperature regimes to determine the proportion of total variability accounted for by these abiotic factors.

### Contributions of roots and litter to 'soil respiration'

Total soil respiration represents the activity of several biotic groupings. Even before IBP studies it was generally assumed that the soil fauna contributes < 10% of the total carbon dioxide output (Macfadyen, 1970). There is considerable disagreement, however, over the amount of carbon dioxide which is due to root respiration – in effect a 'maintenance cost' for plant nutrition and storage tissue below ground. Thus Jagnow (1958) estimated a range of 9 to 90% for roots in various plant communities; Wiant (1967) calculated root respiration in a spruce forest as 50% of the total respiration; and Mina (1960) estimated that the roots of several broad-leafed and coniferous trees accounted for 33 to 36% of the total. Recent estimates of root respiration activity in grasslands worldwide include 8–17% for *Andropogon* old fields (Coleman, 1973), 19% for mixed-grass prairie (Warembourg & Paul, 1977), 40% for tallgrass prairie (Kucera & Kirkham, 1971), and 15–51% (mean 32 ± 6%) for meadow and steppe in Europe (Zlotin, 1974). Interestingly, these values compare closely with Lundegårdh's (1924) estimate of about 30% for root respiration in a North European mesic grassland.

There is no room in this review for further exposition and commentary on the differing techniques used by various workers. However, as there is a general convergence of several findings at values of 20–30%, we will use this figure and take the remaining 70–80% as being due to heterotrophic activity in grassland soils worldwide. As will be seen, this should enable us to estimate carbon input–output for a variety of habitats, comparing season- or yearlong decomposition activities.

### Tropical savanna soil respiration

The experiments on the Lamto tree savanna soils were done using the technique of Lundegårdh (1924), with Hilger's (1963) modifications: carbon dioxide was absorbed *in situ* under a jar (after Schaefer, 1979).

Experiments were performed *in situ* in some savanna sites characterized by the dominance of either *Loudetia simplex* or Andropogonae grasses. Respiration of bare soil in the savanna was compared with that of soil covered by vegetation (Table 7.12). Cumulative evolution (over 20 days) as well as circadian rhythms of respiration and temperature (Figs. 7.3 and 7.4) were measured.

The differences observed from one year to another reflect not only different water contents in the soil at the time of the experiments, but also reflect the previous rainfall regime. There was significant respiratory activity at just above the permanent wilting point, e.g. in February when the soil humidity under *Loudetia* was near 0.5% (dry soil) or 7% of the field capacity. Evolution of carbon dioxide from ferruginous tropical soil under a cover of

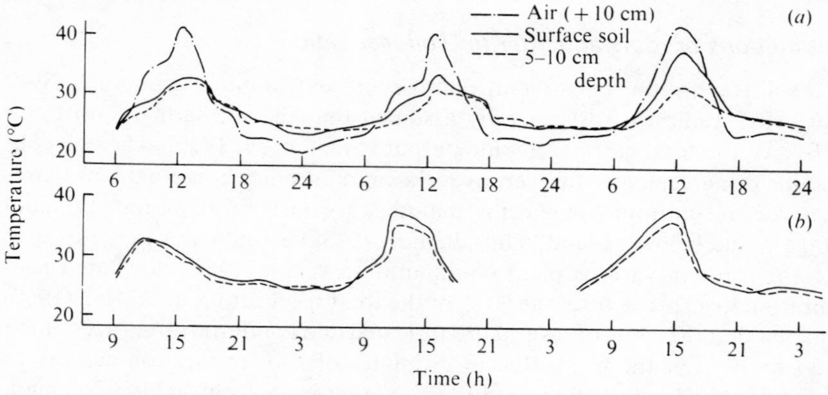


Fig. 7.3. Circadian rhythm of temperature: (*a*) *Loudetia* savanna, 9–12 October 1969, and (*b*) *Hyparrhenia* savanna, 20–23 October 1969. Station of tropical ecology, Lamto, Ivory Coast. (After Schaefer, 1979.)

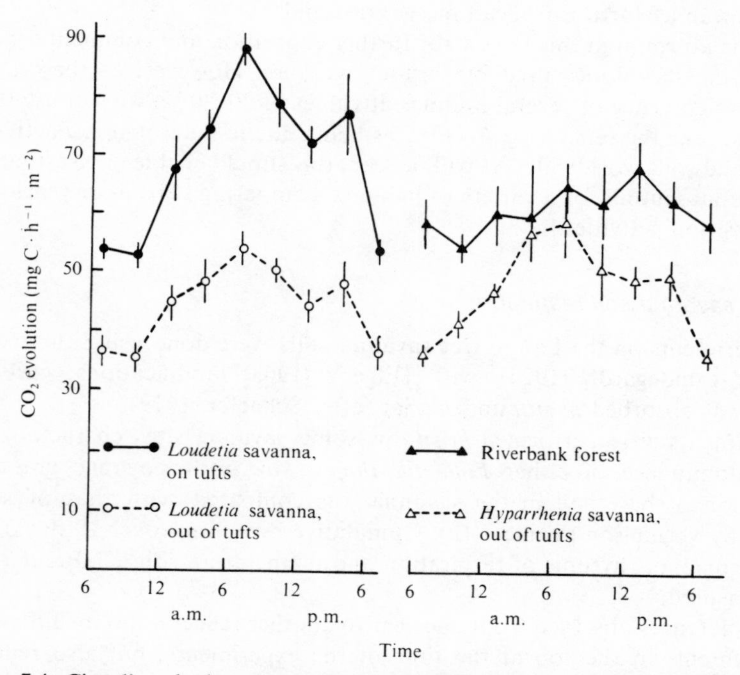


Fig. 7.4. Circadian rhythm of carbon dioxide evolution, October 1969. Type days, with standard error of the mean. Savanna and riverbank forest soils. Station of tropical ecology, Lamto, Ivory Coast. (After Schaefer, 1979.)

Table 7.12. *Soil respiration, in situ, in three habitats of the Lamto Site, Ivory Coast*

| Habitat | Date | Soil respiration | | Water content | |
|---|---|---|---|---|---|
| | | $mg\ C \cdot d^{-1} \cdot m^{-2}$ | $g\ C \cdot yr^{-1} \cdot m^{-2}$ | % dry soil | % field capacity |
| *Loudetia* (Hydromorphic sandy soil) | Oct. 1969[a] (bare soil) | 1108 | 404 | 10 | 100 |
| | Oct. 1969 (soil + vegetation) | 1723 | 629 | 10 | 100 |
| | Oct. 1973 | 601 | 216 | 10 | 100 |
| | Oct. 1973 (vertisol) | 1831 | 659 | 20 | 100 |
| | Nov. 1970 | 2100 | 767 | 7 | 100 |
| | Dec. 1970 | 1999 | 558 | 6.8 | 97 |
| | Jan. 1971 (before bushfire) | 1530 | 558 | 2.5 | 36 |
| | Feb. 1971 (after bushfire) | 1110 | 405 | 0.5 | 7 |
| Andropogonae (ferruginous tropical soil) | Oct. 1969 | 1192 | 435 | 10 | 100 |
| | Oct. 1973 | 1966 | 708 | 10 | 100 |
| Riverbank forest | Oct. 1969 | 1485 | 542 | 15 | 100 |

After Schaefer (1979).
[a] Absorbent used: 1969, sodium hydroxide; 1970–1, limed sodium hydroxide (= sodium hydroxide × 3); 1973, barium hydroxide.

*Hyparrhenia* and the vertisol under *Loudetia* were similar (Fig. 7.4). Determinations were made with bare soil (between the tufts of the grasses), soil covered with vegetation, and soil cleared of the roots.

Activity at the level of the tufts (Fig. 7.4) reflected the rhythm of the plant's physiological activity: directly from root respiration and indirectly from microbial respiration due, in part, to the energy from root exudates. The bare soil still reflected this rhythm; here a remote rhizosphere effect added to the thermal stimulation. The bare soil of the *Hyparrhenia* site had a little more activity than that of *Loudetia* vegetation, under which the thermal effect was less marked.

### *Temperate grassland soil respiration*

A wide variety of temperate grassland sites was sampled for soil respiration, ranging from moist mesic meadow tallgrass to arid shortgrass–bunchgrass steppe and desert and annual grassland. To facilitate comparison of organic matter carbon inputs and carbon dioxide carbon outputs, these data are presented (Table 7.9) along with pertinent air temperatures and precipitation information.

A general pattern of decreasing production and decomposition with decreasing rainfall holds across all the sites. Other data on Central European and Russian studies are incorporated in Table 7.9, or commented on below.

Tesarová & Gloser (1972) estimated by alkali absorption that the soil carbon dioxide production in a *Gratiola officinalis/Carex praecox-suzae* plant community was 7–14 $g \cdot m^{-2} \cdot d^{-1}$ during the vegetative period and an average of 10.8 $g \cdot m^{-2} \cdot d^{-1}$. Kubicka (1973) measured the carbon dioxide production in Kazunskie and Strzeleckie meadows, Poland (*Arrhenatheretum* and *Stellario-Deschampsietum* communities, respectively), using the method of Walter (1951) and Haber (1958), and found that during the vegetative period it varied from 2.9 to 6.7 $g \cdot m^{-2} \cdot d^{-1}$ for *Arrhenatheretum* and from 3.2 to 4.4 $g \cdot m^{-2} \cdot d^{-1}$ for *Stellario-Deschampsietum*. This technique was non-quantitative, however, as the surface: volume ratio of absorbent in the jar was too low to absorb all the carbon dioxide (Domsch, 1962). Ten-Chak-Mun & Fedorova (1972) studied the carbon dioxide production of soils under natural conditions in Sachalin by the method of Makarov (1957). The ranges encountered during the vegetative period corresponded to 4.3 to 7.3 g $CO_2 \cdot m^{-2} \cdot d^{-1}$ in the communities of tallgrasses and to 1 to 5 g $CO_2 \cdot m^{-2} \cdot d^{-1}$ in the shortgrasses. The maximum production occurred in mid-summer, coinciding with the period of the most intensive plant growth.

For simulation and regression model studies, the effects of soil water and temperature were investigated further.

## Abiotic factors affecting soil respiration

This study was undertaken to determine the feasibility of regressing carbon dioxide evolution on soil temperature and water. R. K. Steinhorst & D. C. Coleman (unpublished data) used a general-purpose multiple regression model of the form:

$$y\ (CO_2\ \text{evolution}) = a + b_1 \ln x_1 + b_2 \ln x_2,$$

where $x_1$ and $x_2$ are soil temperature and soil water, respectively, and $a$, $b_1$ and $b_2$ are parameters. The combined fit was usually good for all of the western North American sites. The coefficient of determination, multiple $R^2$, usually exceeded 0.60 (Table 7.13). The logarithm of temperature was entered first (underlined values) for Osage (tallgrass) and ALE (Pacific northwest bunchgrass) but with markedly different effects. For Osage, a mesic site, there was little effect of adding water in the equation, whereas the ALE (arid site) equation showed a strong multiplicative effect (addition of two log values) of temperature and water. This effect is discussed in detail by Wildung *et al.*

Table 7.13. *Abiotic factors affecting soil respiration in North American sites*

| Site | Year | Treatment[a] | $N$ | ln water | ln temperature |
|---|---|---|---|---|---|
| Osage (tallgrass) | 1971 | G | 30 | 0.67[b] | 0.54 |
| | | U | 30 | 0.75 | 0.71 |
| | 1972 | G | | | |
| | | U | 20 | 0.49 | 0.49 |
| Cottonwood | 1972 | U | 11 | 0.24 | 0.24 |
| Matador (mid-grass) | 1970 | U | 8 | 0.37 | 0.40 |
| | 1971 | U | 8 | 0.23 | 0.38 |
| ALE (bunchgrass) | 1971 | G | 20 | 0.72 | 0.17 |
| | | U | 20 | 0.76 | 0.15 |
| | 1972 | G+U | 116 | 0.60 | 0.21 |
| Pawnee (shortgrass) | 1971 | U | 80 | 0.64 | 0.71 |
| | 1972 | U | 80 | 0.82 | 0.82 |
| | 1972 | G | 80 | 0.78 | 0.79 |
| | 1973 | U, D | 20 | 0.31 | 0.55 |
| | 1973 | U, E | 12 | 0.80 | 0.80 |

[a] Treatments: G, grazed; U, ungrazed; D, control area; E, irrigated only. For further explanation see text.

[b] The underlined $R^2$ values are for a single independent variable – the first entered. Multiple $R^2$, with both ln water and ln temperature included in the equation, are not underlined.

(1975) who note that this 'hydrothermal' effect, recorded by earlier European workers, is well described by the statistical relationship given above.

Other drier sites, including Cottonwood, Matador & Pawnee (Table 7.9), shifted to a water response foremost (Table 7.13). The shift in response in the Pawnee irrigated treatment is of interest. Total net primary production increased 228 % to 449 g $C \cdot m^{-2} \cdot yr^{-1}$, effectively shifting from the water response of dryland to a 'wetland' response of a tallgrass prairie.

As a preliminary step in determining whether a balance exists between annual litter and root decomposition and carbon dioxide output during an annual cycle of biological turnover, the data of Kokovina (1972) were used (Table 7.14). The total respiration for the 8 months of the warm season (2100 g $CO_2 \cdot m^{-2}$) corresponds to 7350 kcal, assuming approximately 3.5 kcal per g $CO_2$ (Zlotin, 1975). In comparison, studies in the US/IBP Grassland Biome have shown an average respiratory quotient (RQ = moles $CO_2$ evolved/moles $CO_2$ consumed) of 0.7 for shortgrass prairie soil cores (Klein, 1977). Thus 0.7 mol $CO_2 \cdot$ mol $O_2^{-1}$ (oxycalorific equivalent = 4.8 kcal$\cdot l^{-1}$) gives a value of 15.7 l $CO_2$ = 107.5 kcal. This calculates to 1 g $CO_2$ = 3.49 kcal, an interesting convergence in international usage. From this quantity saprophagic organisms get about 4900 kcal and roots 2450 kcal, or 30 %. After independent estimations of the annual input of organic remains, there is a balance in the ecosystem of about 1100 g$\cdot m^{-2}$ (Afanasieva, 1966; Drozdov & Zlotin, 1974), from which roots get about 64 % and

Table 7.14. *Summed flow of carbon dioxide from soil (typical chernozem meadow-steppe watershed, Kursk) measured over 1964–70 by Kokovina (1972)*

| | Months | | | | | | | | |
|---|---|---|---|---|---|---|---|---|---|
| $CO_2$ outputs | April | May | June | July | Aug. | Sept. | Oct. | Nov. | Sum[a] |
| mg $CO_2 \cdot m^{-2} \cdot h^{-1}$ | 200.0 | 466.0 | 550.0 | 610.0 | 530.0 | 305.0 | 120.0 | 90.0 | 2100 |
| g $C \cdot m^{-2} \cdot d^{-1}$ | 1.32 | 3.08 | 3.63 | 4.03 | 3.50 | 2.01 | 0.79 | 0.59 | 572 |

[a] Sum: $g \cdot m^{-2}$ across April–Nov. period. (Unpublished data of R. I. Zlotin.)

aboveground biomass 36%. Assuming 1 g of plant litter corresponds to 4.5 kcal and annual input comprises 4950 $kcal \cdot m^{-2}$, the calculation shows very close conformity (within 50 kcal or 1%) of the annual carbon dioxide flux with the annual input of litter. Interestingly the Kursk meadow steppe respiration of 2100 g $CO_2$ = 572 g C was quite high (excluding output during the four colder months). This productivity was higher than for either tall-grass prairie site in North America (Table 7.9).

The amount of carbon dioxide efflux was higher than the organic carbon input in the range of 40 to 60%. It is probable that in soils with higher amounts of carbonates (steppes, semideserts and other semi-arid grasslands) a considerable amount of carbon dioxide is formed as a result of physico-chemical decomposition of salts such as calcium carbonate, magnesium carbonate and others (Zlotin, 1975). This abiotic carbon dioxide flux must be subtracted during calculation of the carbon dioxide from biological turnover. It is also necessary to account for a quantity of root exudates and exfoliates which may be considerable (Coleman, 1976), and the production of biomass of algae (Shtina, 1968) and autotrophic micro-organisms. Accounting for the possible variance in errors in the construction of a balance between annual carbon dioxide flux and yearly input of organic matter thus becomes a real problem.

Calculations of the balance of carbon dioxide output and litter and root decomposition were carried out for the high-mountain dry steppe and semidesert of Tien Shan (Zlotin, 1975). The value of current carbon dioxide from the soil for a year comprised 70 to 90% of the annual input of plant litter (above and below ground).

Ecosystems with large contents of calcium carbonate in the soil (Kursk) had a large release of carbon dioxide observed not only from the humus horizon, especially the upper 0 to 30 cm, but also from carbonate horizons to a depth of 200 cm (Fig. 7.5; Zlotin, 1975). Microbial numbers at that depth are extremely small (Bondarenko-Zozulina, 1955), and invertebrates are also very scarce (Ghilarov, 1960*b*; Zlotin, 1969), and as any roots were removed

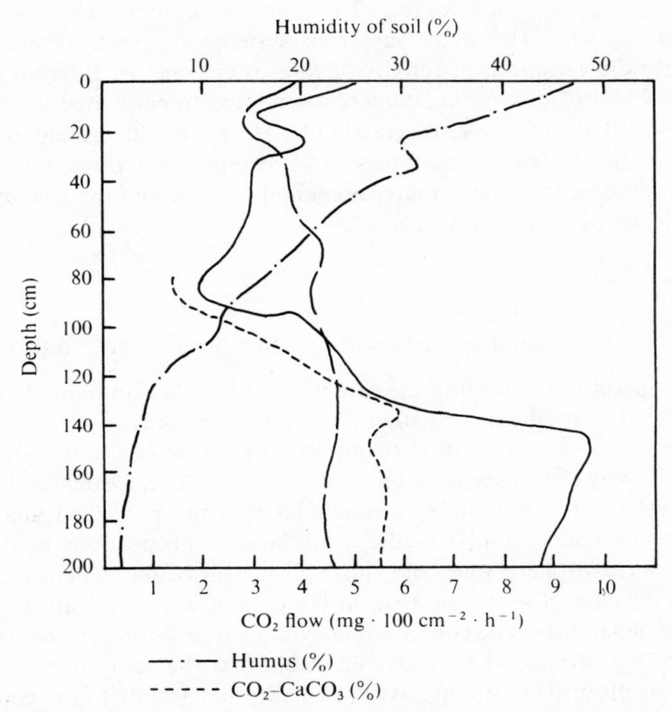


**Fig. 7.5. Carbonate contents, carbon dioxide flow, humus content and soil humidity in a typical chernozem meadow-steppe watershed (Kursk). (After Zlotin, 1975.)**

from the soil in these experiments a sizeable carbon dioxide flux from deep soil horizons may be explained only by physicochemical processes of disintegration of carbonates. Thus the abiotic portion of the carbon dioxide flux from a whole soil profile may reach 60% in carbonate chernozems and in typical chernozems about 30%.

## Soil enzymes

It is well known (Skujins, 1967) that many biological transformations occurring in soil are catalysed by enzymes found outside living soil organisms. These enzymes are often named free enzymes (Kiss, Dragan-Bularda & Radulescu, 1972), soil enzymes (Kuprevich & Shcherbakova, 1966), or sometimes extracellular enzymes (Dickinson & Pugh, 1974).

Modern investigations of soil enzymes are presently oriented toward the study of fundamental problems such as their origin and localization (e.g. Burns, Pukite & McLaren, 1972*a*; Ladd & Paul, 1973) or stabilization by inorganic and organic colloids (e.g. Burns, El Sayed & McLaren, 1972*b*), or towards discovering new enzymes (e.g. Tabatabai & Bremner, 1970; Kiss & Dragan-Bularda, 1972). However, numerous systematic investigations were concerned with the variations in activity of various enzymes in different types of ecosystem. Most of this systematic work is related to cultivated soils, but there is a small body of work on grasslands. Here we will present only a critical review, illustrated by examples, and examine the three following points: methodology, influence of environmental factors, and the role of soil enzymes in decomposition processes.

### *Methodology*

Comparison of results found in the literature is hampered by four difficulties.

(1) Lack of precision regarding the sampling site. If the sampling depth is indicated, it is often not defined whether rhizosphere or nonrhizosphere soil is being considered. Such information is important because of the rhizosphere effect on the activity of soil enzymes (Koslov, 1964; Voets & Dedeken, 1966). As this imprecision is common, one concludes that the published analyses have been done on soil samples made up of variable proportions of rhizosphere and nonrhizosphere soils, the rhizosphere soil being evidently more important in the case of grassland than in the case of weedless cultures.

(2) Storage of samples. The content of enzymes varies during the course of transportation and drying. It is clearly desirable that the analysis be run as soon as possible after the sampling, so as to correspond exactly to the content *in situ*. Unfortunately, this precaution has not always been taken by most authors.

(3) Methods of titration. These methods vary to a great extent, particularly as far as the techniques of sterilization are concerned (Skujins, 1967).

(4) Expression of results. These vary markedly, and results are often expressed differently for the same enzyme activity. Conversions from one unit into another may cause problems if certain methodological indications are lacking.

In the examples given here, the following units have been adopted:

(*a*) Saccharase (invertase) activity is expressed in $\mu$mol reducing sugars (glucose)$\cdot$g soil$^{-1}\cdot$24 h$^{-1}$.

(*b*) Amylase activity is given in $\mu$mol reducing sugars (glucose)$\cdot$g soil$^{-1}\cdot$96 h$^{-1}$.

(*c*) Urease activity is expressed in mg $NH_3$-N$\cdot$g soil$^{-1}\cdot$3 h$^{-1}$.

(*d*) Asparaginase activity is given in mg $NH_3$-N$\cdot$g soil$^{-1}\cdot$21 h$^{-1}$.

(*e*) Phosphatase activity is expressed in mg phosphate$\cdot$g soil$^{-1}\cdot$3 h$^{-1}$ (method of Hoffmann, 1967).

In conjunction with examples of enzyme activity, we have chosen to present some data on dehydrogenase activity. In this context it should be stressed that the measure of dehydrogenase activity, as it is generally used, is not a measure of enzymatic activity but a measure of overall soil biological activity, just as is the evolution of carbon dioxide or the absorption of oxygen. The dehydrogenase activity has been expressed here in $\mu$mol of TPF (formazan) formed by the reduction of TTC (triphenyl tetrazolium chloride) · g soil$^{-1}$ · 24 h$^{-1}$.

### *Enzyme content in soils of grasslands and environmental factors acting at this level*

Table 7.15 shows, for some grasslands, the activity of some enzymes usually tested, e.g. dehydrogenase (AD), invertase (SAC), amylase (AMY), urease (URE), asparaginase (ASP) and phosphatase (PHO). From this table, as from more detailed studies (e.g. Kuprevich & Shcherbakova, 1971; Pancholy & Rice, 1973), it is clear that three groups of factors govern enzyme activity in grassland soils: vegetation, climate and edaphic factors.

In the tropical environment the enzyme activity of grassland soils is consistently lower than that of forest soils. Thus, at Lamto, Ivory Coast, the invertase, amylase, urease, asparaginase and phosphatase activities of grassland (savanna) were respectively 23.0, 26.3, 0.05, 0.24 and 0.19, as against 31.8, 44.0, 0.14, 0.57 and 1.00 in forest soils (Bauzon *et al.*, 1977). In a temperate environment the situation seems, in general, to be reversed since the soils of grassland biomes are characterized by much higher enzyme activities than are those of forest biomes. This fact has been emphasized in particular by Pancholy & Rice (1973) in the framework of a comparative study between soils corresponding to three types of plant associations which recolonize formerly cultivated fields (old-field succession) in Oklahoma and Kansas.

Viewed in the framework of a given biome, it is much more difficult to show *in situ* the particular effect of each plant species on the enzyme activity of soils. Thus Ross & Roberts (1970) could not detect this specific influence when they compared grasslands showing variable proportions of grasses and legumes (essentially clover). The difficulties encountered when studying the effect of the plant species *in situ* are possibly a consequence of the interaction of numerous environmental factors.

Edaphic factors exert an influence chiefly through enzyme-protecting substances contained in the soil. That is why a positive correlation could be detected between enzyme activity (especially that of invertase) and soil organic carbon content (e.g. Ross & Roberts, 1970). The low clay content (5 to 6%) of the grassland soils at Lamto partly explains their low enzyme activity (Bauzon *et al.*, 1977). Other soil characteristics seem to be involved equally when they play the role of limiting factors. Thus low exchangeable potassium in Lamto soils caused a very low level of asparaginase activity (Bauzon *et al.*, 1977).

Table 7.15. *Biological characteristics of soils, including enzyme activities (nos. 3–8), from grassland ecosystems in tropical and temperate regions*

| Country | Type of soil | Biome ecosystem | 1 pH | 2 C/N | 3 AD | 4 SAC | 5 AMY | 6 URE | 7 ASP | 8 PHO | References |
|---|---|---|---|---|---|---|---|---|---|---|---|
| Ivory Coast (Lamto) | Ferruginous tropical soil on sandy colluvium | Grassland with *Hyparrhenia* sp. and *Loudetia* sp. | 6.4 | 15.9 | 0.28 | 23.0 | 26.3 | 0.05 | 0.24 | 0.19 | Bauzon *et al.* (1977) |
| Central African Republic | Slightly ferrallitic soil | Grass savanna | 5.8 | — | — | — | — | — | 1.14 | — | Mouraret (1965) |
| | | | 6.1 | — | — | — | — | — | 0.45 | — | |
| Tunisia | Sierozem | Artificial | 7.9 | 11.6 | 0.13 | 317.0 | — | — | — | — | Bauzon *et al.* (1968) |
| Southern France (Montpellier) | Mediterranean red soil | Grassland with *Brachypodium ramosum* | 8.8 | 10.5 | — | 214.6 | 63.4 | 0.46 | — | — | J. Cortez (personal communication) |
| New Zealand | $S_4$ Wakiwi | Mixed grassland with grasses (70–87 %) and legumes (10–20 %) | 6.0 | 11.7 | 3.17 | 184.3 | 4.7 | — | — | — | Ross & Roberts (1970) |
| | $C_4$ – 2 Ngaumu | | 5.7 | 14.3 | 2.74 | 155.5 | 31.2 | — | — | — | |
| | $C_4$ – 2 Pirinoa | | 6.0 | 10.4 | 2.30 | 169.9 | 47.2 | — | — | — | |
| | $C_8$ – 4 Levin | | 6.0 | 11.1 | 2.59 | 128.1 | 31.2 | — | — | — | |
| | $N_4$ Mangawheau | | 5.4 | 10.5 | 2.16 | 126.7 | 35.7 | — | — | — | |
| Australia (New England) | Krasnosem | Grassland with grasses and legumes | 5.2 | — | — | — | — | 0.31 | — | — | McGarity & Myers (1967) |
| | Chocolate | | 5.7 | — | — | — | — | 0.22 | — | — | |
| | Yellow podzolic | | 5.5 | — | — | — | — | 0.08 | — | — | |
| | Gley podzolic | | 5.3 | — | — | — | — | 0.07 | — | — | |
| | Red-brown earth | | 6.0 | — | — | — | — | 0.16 | — | — | |

AD, dehydrogenase; SAC, invertase; AMY, amylase; URE, urease; ASP, asparaginase; PHO, phosphatase.

In order to make the comparisons easier between data from different sources, the original units have been converted into units as defined in the text. In the case of the slightly ferrallitic soil of the Central African Republic and in the case of the Mediterranean red soil, the two extreme values of the variables are given and not the mean as in all other cases.

The influence of climate may be approached at two levels. At the level of the seasons, in a given ecosystem variations in enzyme activity can appear. But the amplitude of these variations is smaller than that caused by variations bound to climatic zonation. The theory of zonality (Mishustin, 1964) of bacterial or fungal micropopulations seems to be applicable in the case of enzymes (Table 7.15), where a striking difference between tropical biomes and temperate biomes may be seen. The enzyme activity of tropical soils is in general very low. The principal determining factor is temperature, as suggested by the *in vitro* study of Cortez, Lossaint & Billes (1972) and the comparison carried out *in situ* by Ross & Roberts (1970). The latter authors have compared enzyme activities in a sequence of five New Zealand soils (Wakiwi, Ngaumu, Pirinoa, Levin and Mangawheau) located in ecological conditions identical except for their mean annual temperatures, which were respectively 9.7, 11.4, 12.5, 12.9 and 13.4 °C. The invertase activity decreases from 184.3 to 126.7 and their dehydrogenase activity fell from 3.17 to 2.16 on going from the coldest to the warmest soils. These results indicate that the higher the temperature, the more active is the degradation of soil enzymes (in the range of temperatures remaining compatible with the activity of soil micro-organisms). Under these conditions the question arises whether enzymatic processes are less active in tropical environments than in temperate ones. It is possible but not certain that the low enzyme activity of tropical soils may be compensated for by the fact that the more favourable temperature conditions allow these enzymes to be active for longer periods during the year.

The existence of significant negative correlations between annual rainfall and the invertase activity in New Zealand soils suggests likewise the possibility that rain may also be an influencing factor (Ross & Roberts, 1970).

### *Role of soil enzymes in decomposition processes*

What are the relative roles of living micro-organisms (acting through their own enzymes) and of soil enzymes in decomposition processes? This question is important because of its implications when interpreting results concerning soil enzyme activity. Indeed, one may be aware of instances in which living micro-organisms (e.g. cellulolytic fungi) play a major role in a decomposition process (cellulolysis) whereas the free enzymes (free cellulases) are not active enough to intervene. In such cases no conclusions can readily be drawn from the enzyme analysis.

Thus far, this problem has only been approached by experimenting with soils incubated in the laboratory (Durand, 1965; Paulson & Kurtz, 1969). Using multiple regression analysis, the latter were able to show that, in the model system investigated, 79 to 89% of the variations in hydrolysis of urea could be attributed to free soil enzymes adsorbed on clays. To our knowledge, no study of this type has been conducted on grassland soils.

The contribution of root versus microbial enzymes to the total soil enzyme 'pool' remains problematic. Thus Kuprevich & Shcherbakova (1966) assert that root-elaborated enzymes are indeed much more prevalent, while other authors are less certain.

### *Conclusions*

Our present knowledge about the free soil enzymes of grassland ecosystems is essentially limited to comparisons between these and forest ecosystems and to the role of various environmental factors (plants, climate and soils) on the soil enzyme activity. This knowledge is obviously insufficient. In order to improve on it, it is necessary (1) to develop laboratory investigations on the origin of soil enzymes and on their fundamental mechanisms, mainly the processes by which they are protected, and (2) to develop a reliable field methodology with regard to sampling and analytical methods, which should further be strictly standardized.

## Conceptual and simulation models

It should be apparent by now that the impressive array of taxa involved in saprophagic heterotrophy are, in terms of their functional role in grassland ecosystems, as yet inadequately understood. Microflora are too often enumerated by plate counts, and direct counts in the literature are seldom informative on the total numbers or biomass which are active (Parkinson, Gray & Williams, 1971). Certain soil fauna have been studied in considerable detail, particularly the mesofauna (Harding & Stuttard, 1974), while the functional roles of important macroarthropods, such as termites, and microfauna, such as Protozoa, are just beginning to be elucidated.

At the current level of our understanding it is necessary to view overall processes, looking for general effects of driving variables, such as available soil water, temperature regimes and the quality of organic substrates. An important conceptualizing and integrating tool is the simulation model. One subsection of the total-system ELM grassland model (Hunt, 1977) is concerned with decomposition and it is briefly presented below.

The model (Fig. 7.6) has a series of state variables (boxes) and flows (arrows between boxes). The computer calculates flows and changes in state variables on a 1-day time step.

The three principal types of substrates in the model are humic material, faeces and dead plant and animal remains. The plant and animal tissues are further divided into labile (rapidly decomposing) and resistant (slowly decomposing) fractions. The initial nitrogen content of a given substrate predicts the proportion of rapidly decomposing material within it. Decomposition rates for each type of substrate are then predicted from temperature,

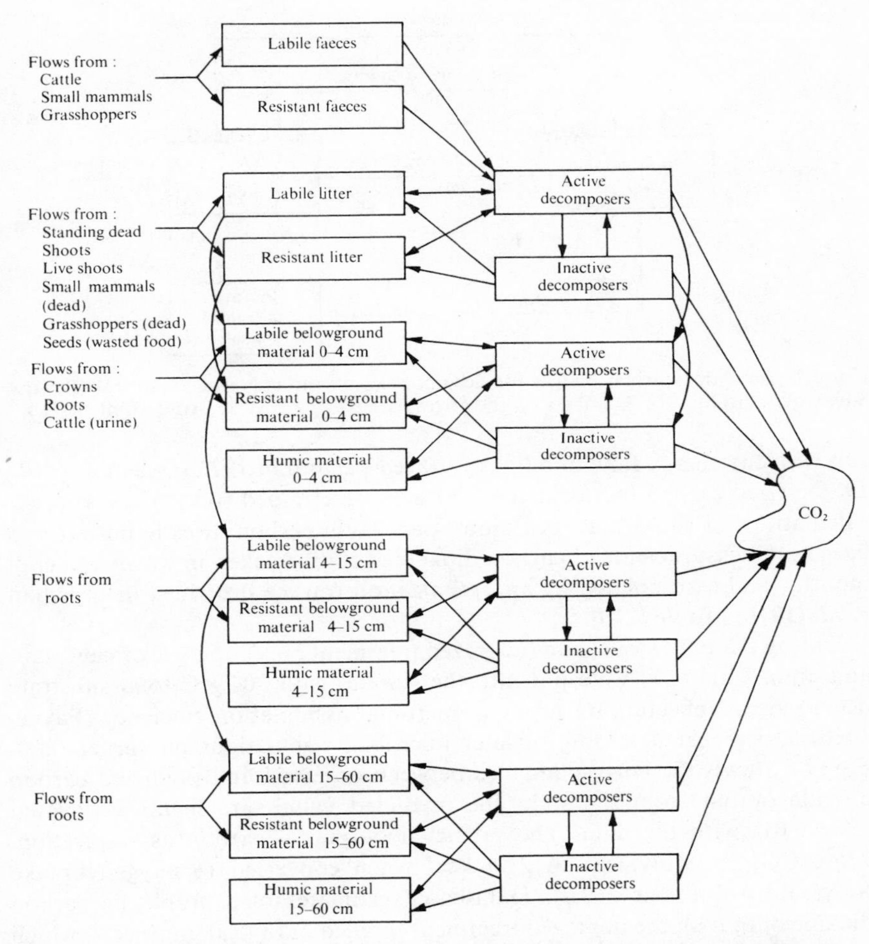


Fig. 7.6. Compartment diagram for the decomposition submodel of ELM. (After Hunt, 1977.)

water tension and inorganic nitrogen concentration (Hunt, 1977). To facilitate manageable yet meaningful population dynamics, all microbes are pooled, irrespective of taxonomic category, and only the active decomposer fraction assimilates substrate. Certain fractions of the microbes die from freezing and drying or from starvation (Fig. 7.7). The model suggests considerable recycling of material, with almost half the substrate decomposed consisting of dead microbes.

For a comparison of the model's predicted carbon and energy flow with a data-based analysis of saprophagic energy flow, we compared a 1972 model-

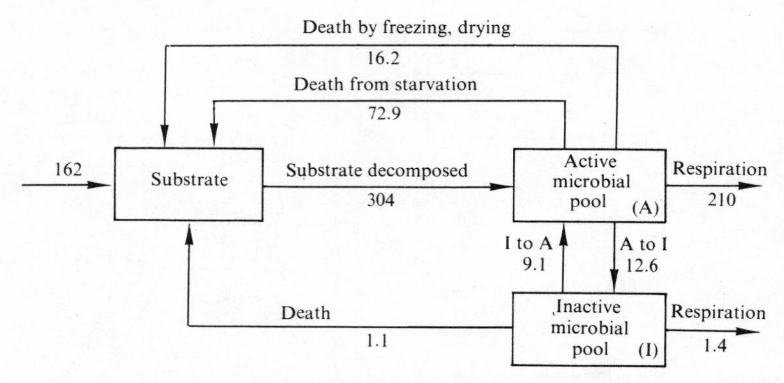


Fig. 7.7. Simulated carbon budget for decomposers on the ungrazed treatment (Pawnee Site, Colorado) in 1972. Numbers on the arrows are $g\,C \cdot m^{-2} \cdot yr^{-1}$. (After Hunt, 1977.)

run with the energy flow in a lightly grazed pasture in 1972 (Coleman *et al.*, 1976). Total carbon inputs in the field were determined from 'peak–trough' calculations of shoot and root biomasses. Total carbon dioxide output was based on measurements of carbon dioxide evolution taken in warm and cool months, and extrapolated by an optimization routine described in Coleman *et al.* (1976) (Table 7.16).

During the 1972 model-run (ungrazed treatment only), 162 g C of new plant and animal material were put into the system, with 304 g C total substrate decomposed, reflecting up to 60% microbial assimilation efficiency (Payne, 1970) and repeated feeding of later microbial populations on earlier ones. Fig. 7.8 shows the correspondence between observed and predicted carbon dioxide output from the soil. The predicted values are highly correlated ($P < 0.01$) with the data. The predicted output of carbon as respiration, 775 g $CO_2$, is equivalent to 2706 kcal when converted to oxygen uptake (respiratory quotient = 0.7). Data-based estimates of saprophagic carbon dioxide output on the ungrazed treatment totalled 2318 $kcal \cdot m^{-2} \cdot yr^{-1}$, which when compared with the predicted output show a discrepancy of 14%. This is a moderate disparity for the assumptions involved in calculating yearlong carbon dioxide output from the data could easily be out by 15 to 20%.

The model, in its interaction with other parts of the total-system model, is serving to emphasize areas where more work is needed. For example, with abundant evidence that root production and turnover events are more marked and complex than previously thought (Coleman, 1976), we have considerable incentive to develop new intertrophic approaches to the study of ecosystem dynamics. Thus microbial turnover and elemental uptake, particularly in the rhizosphere regions discussed by Dommergues (p. 640), are virtually unknown in a field context. The belated recognition of the important role of mycorrhizas in ecosystems, and certainly grasslands and shrublands, is

Table 7.16. *Energy flow in a lightly grazed pasture, Pawnee Site, 1972* (1 steer per 10.8 ha)

| Energetic parameters | kcal·m$^{-2}$ | | % of NPP | |
|---|---|---|---|---|
| Gross primary production | 5230 | | | |
| Net primary production | 3452 | | | |
| Above ground | 517 | | 15.0 | |
| Below ground (crowns and roots) | 2935 | | 85.0 | |
| Heterotrophic production (*P*) and respiration (*R*) | *P* | *R* | *P* | *R* |
| Above ground | 6.0 | 29.6 | 0.2 | 0.8 |
| Herbivores (total) | 5.8 | 27.8 | | |
| Cattle | 3.2 | 22.2 | | |
| Carnivores | 0.2 | 1.8 | | |
| Below ground | 720.9 | 2347.5 | 21.0 | 68.0 |
| Herbivores (total) | 16.4 | 25.9 | | |
| Carnivores | 1.1 | 3.6 | | |
| Saprophages (total) | 703.4 | 2318.0 | 20.5 | 67.1 |
| Microbial saprophages | 700.0 | 2302.0 | | |
| Saprophagic nematodes | 2.8 | 14.5 | 0.08 | 0.4 |
| Other saprophagic grazers (excluding Protozoa) | 0.6 | 1.6 | 0.02 | 0.05 |
| Net heterotrophic production or respiration | 727.0 | 2377.7 | 21.1 | 68.9 |

From Coleman *et al.* (1976).

All values in kcal·m$^{-2}$ per time period. For primary production this was 154 days; for poikilotherms and cattle 180 days; for saprophagic production and respiration and all other homoiotherms 366 days.

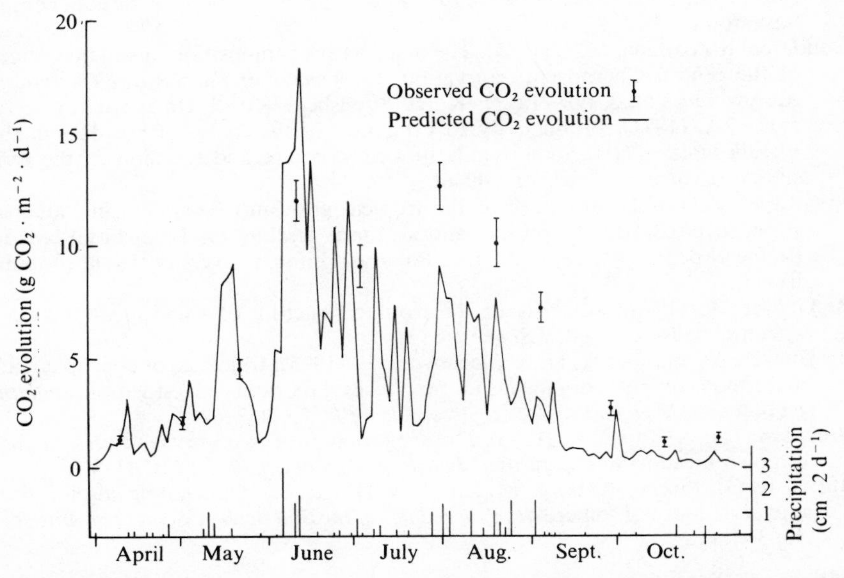


Fig. 7.8. Precipitation and carbon dioxide evolution at Pawnee Site in 1972. Bars indicate 95 % confidence limits. (After Hunt, 1977.)

important, and should lead to new interdisciplinary plant function–decomposition and elemental cycling in ecosystems (Sanders, Mosse & Tinker, 1975). Indeed, it may have been mandatory for the mycorrhizal habit to develop in Siluro-Devonian times to enable a land flora to become successfully established (Pirozynski & Malloch, 1975).

Thus our major need for future work in grasslands, indeed in all biomes, is a more unified viewpoint of ecosystem operation, including an appreciation of the importance of decomposition and other heterotrophic processes in system maintenance and function.

The preparation of the manuscript was supported in part by National Science Foundation Grant DEB73-02027 A03 to the Grassland Biome, US International Biological Programme for 'Analysis of Structure, Function, and Utilization of Grassland Ecosystems'.

## References

Afanasieva, E. A. (1966). *The Chernozem of the Middle-Russian Upland.* Nauka Publications, Moscow. (In Russian.)

Bauzon, D., Aubry, A.-M., Van den Driessche, R. & Dommergues, Y. (1977). Contribution à la connaissance de la biologie des sols de la savane de Lamto, Côte d'Ivoire. *Revue d'Ecologie et de Biologie du Sol*, **14**, 343–61.

Bauzon, D., Van den Driessche, R. & Dommergues, Y. (1968). Caractérisation respirométrique et enzymatique des horizons de surface des sols forestiers. *Science du Sol.* **2**, 55–78.

Biederbeck, V. O., Paul, E. A., Lowe, W. E., Shields, J. A. & Willard, J. R. (1974). *Soil microorganisms. II. Decomposition of cellulose and plant residues. Matador Project (Canadian Committee, IBP) Report 39.* University of Saskatchewan, Saskatoon.

Bondarenko-Zozulina, M. I. (1955). The quantitative composition of soil microflora of the central Chernozem reservation. In *Reports of the Central Chernozem Reservation*, vol. 3, pp. 212–31. Kursk Publishers, Kursk. (In Russian.)

Breymeyer, A. (1971). Productivity investigation of two types of meadow in the Vistula valley. XIII. Some regularities in structure and function of the ecosystem. *Ekologia polska*, **19**, 249–61.

Breymeyer, A. (1974). Structure of the tropical grassland ecosystem in Panama. Paper given at IBP symposium and synthesis meeting on Tropical Grassland Biome Proceedings, Dept Botany, Banaras Hindu University, India (unpublished).

Breymeyer, A. (1978). Analysis of the trophic structure of some grassland ecosystems. *Polish ecological Studies*, **4** (2), 55–128.

Breymeyer, A., Jakubczyk, H. & Olechowicz, E. (1975). Influence of coprophagous arthropods on microorganisms in sheep feces: laboratory investigation. *Bulletin de l'Académie polonaise des sciences, Classe II*, **23**, 257–62.

Breymeyer, A. & Kajak, A. (1976). Drawing models of two grassland ecosystems: a mown meadow and a pasture. *Polish ecological Studies*, **2** (2), 41–9.

Burns, R. G., Pukite, A. H. & McLaren, A. D. (1972*a*). Concerning the location and persistence of soil urease. *Proceedings of the Soil Science Society of America*, **36**, 308–11.

Burns, R. G., El Sayed, M. H. & McLaren, A. D. (1972*b*). Extraction of an urease-active organo-complex from soil. *Soil Biology and Biochemistry*, **4**, 107–8.

Chernov, Yu. I. (1973). Geozoological characteristics of Taimyr territory biogeocoenological stations. In *Taimyr biogeocoenoses and their productivity*, vol. 2, pp. 187–200. Nauka Publishing House, Moscow.

Clark, F. E. & Paul, E. A. (1970). The microflora of grassland. *Advances in Agronomy*, **22**, 375–435.

Coleman, D. C. (1973). Compartmental analysis of 'total soil respiration'. *Oikos*, **24**, 361–6.

Coleman, D. C. (1976). A review of root production processes and their influence on the soil biota of terrestrial ecosystems. In *The role of terrestrial and aquatic organisms in decomposition processes*, ed. J. M. Anderson & A. Macfadyen, pp. 417–34. Blackwell Scientific Publications, Oxford.

Coleman, D. C., Andrews, R., Ellis, J. E. & Singh, J. S. (1976). Energy flow and partitioning in selected man-managed and natural ecosystems. *Agro-Ecosystems*, **3**, 45–54.

Cortez, J., Lossaint, P. & Billes, G. (1972). L'activité biologique des sols dans les écosystèmes méditerranéens. III. Activités enzymatiques. *Revue d'Ecologie et de Biologie du Sol*, **9**, 1–19.

Dash, M. C. & Patra, U. C. (1977). Density, biomass, and energy budget of a tropical earthworm population from a grassland site in India. *Revue d'Ecologie et de Biologie du Sol*, **14**, 461–71.

Dash, M. C., Patra, U. C. & Thambi, A. V. (1974). Comparison of primary production of plant material and secondary production of Oligochaetes in a tropical grassland from southern Orissa, India. *Tropical Ecology*, **15** (172), 16–21.

Dickinson, C. H. & Pugh, G. J. F. (eds.) (1974). *Biology of plant litter decomposition*, vols. 1 and 2. Academic Press, New York & London.

Domsch, K. H. (1962). Bodenatmung: Sammelbericht über Methoden und Ergebnisse. *Zentralblatt für Bakteriologie, Parasitenkunde, Infektionskrankheiten und Hygiene*, Abt. 2, **116**, 33–78.

Dommergues, Y. (1968). Dégagement tellurique de $CO_2$. Mesure et signification. *Annales de l'Institut Pasteur*, **115**, 627–56.

Dommergues, Y. & Mangenot, F. (1970). *Ecologie microbienne du sol.* Masson et Cie, Paris.

Drozdov, A. V. & Zlotin, R. I. (1974). The dynamics of links of biological productivity with water-heat regimes in the meadow-steppe natural complexes of the middle-Russian forest-steppe. In *Contemporary state of the landscape theory*, pp. 186–90. Perm Publishers, Perm. (In Russian.)

Durand, G. (1965). Les enzymes dans le sol. *Revue d'Ecologie et de Biologie du Sol*, **2**, 141–205.

Elliott, E. T. & Coleman, D. C. (1977). Soil protozoan dynamics of a Colorado shortgrass prairie. *Soil Biology and Biochemistry*, **9**, 113–18.

Gauert, V. I. & Naplekova, N. N. (1973). Cellulolytic activity of microorganisms in chernozems of the Altai Mountains. *Izvestiya Sibirskogo Otdeleniya Akademii Nauk SSSR, Ser. biol.-Med. Nauk*, **2**, 163–6. (In Russian.)

Ghilarov, M. S. (1960*a*). The soil invertebrates as a factor of the fertility of soils. *Journal of general Biology*, **21(W2)**, 5–17. (In Russian.)

Ghilarov, M. S. (1960*b*). The soil invertebrates as an index of soil and plant cover peculiarities of forest-steppe. In Trudy *of Central Chernozem Reservation*, vol. 6, pp. 283–320. Kursk Publishers, Kursk. (In Russian.)

Ghilarov, M. S. (1965). The soil animals as components of the biocoenoses. *Journal of general Biology*, **26(W3)**, 276–89. (In Russian.)

Haber, W. (1958). Ökologische Untersuchungen der Bodenatmung. *Flora*, **146**, 109–57.

Harding, D. J. L. & Stuttard, R. A. (1974). Microarthropods. In *Biology of Plant Litter Decomposition*, vol. 2, ed. C. H. Dickinson & G. J. F. Pugh, pp. 489–532. Academic Press, New York & London.

Hilger, F. (1963). Activité respiratoire de sols équatoriaux. Application de la méthode respirométrique in situ. *Bulletin de l'Institut Agronomique Station Recherches Gembloux*, **31**(2), 154–82.

Hoffmann, G. (1967). A photometric determination of phosphatase activity in soil. *Zeitschrift für Pflanzenernaehrung Düngung Bodenkunde*, **118**, 161–72.

Hundt, R. & Unger, H. (1968). Investigations on the cellulolytic activity under grassland associations. *Tagungsbericht der deutschen Akademie der Landwirtschaftswissenschaften zu Berlin*, **98**, 263–75. (In German.)

Hunt, H. W. (1977). A simulation model for decomposition in grasslands. *Ecology*, **58**, 469–84.

Ino, Y. & Monsi, M. (1969). An experimental approach to the calculation of $CO_2$ amount evolved from several soils. *Japanese Journal of Botany*, **20**, 153–88.

Jagnow, G. (1958). Untersuchungen über Keimzahl und biologische Aktivität von Wiesenböden. *Zeitschrift für Pflanzenernaehrung Düngung Bodenkunde*, **82**, 1–50.

Jankowska, K. (1971). Net primary production during a three-year succession on an unmowed meadow of the *Arrhenatheretum elatioris* plant association. *Bulletin de l'Académie polonaise des Sciences, Classe II*, **19**, 789–94.

Ketner, P. (1973). Primary production of salt-marsh communities on the island of Terschelling in the Netherlands. *Verhandelingen Rijksinstituut Natuurbeheer*, **5**, 1–181.

Kiss, S. & Dragan-Bularda, M. (1972). Soil polysaccharidases. *Contributii Botanice*, 1972, 377–84.

Kiss, S., Dragan-Bularda, M. & Radulescu, D. (1972). Biological significance of the enzymes accumulated in soil. In *Third Symposium on Soil Biology*, pp. 19–79. Rumanian National Society of Soil Science, Bucharest.

Klein, D. A. (1977). Seasonal carbon flow and decomposer parameter relationships in a semiarid grassland soil. *Ecology*, **58**, 184–90.

Kokovina, T. P. (1972). The water regime of chernozems under agriculture and virgin herbage. In *Landscape and biogeographical study of forest-steppe*, pp. 122–42. Nauka Publications, Moscow. (In Russian.)

Koslov, K. A. (1964). Enzymatic activity of the rhizosphere and soils in the East Siberia area. *Folia microbiologica*, **9**, 145–9.

Kozlovskaya, L. S. (1971). Der Einfluss der Wirbellosen auf die Tätigkeit der Microorganismen in Torfboden. In *IVth International Colloquium on Soil Zoology*, pp. 455–62. Institut National de la Recherche Agronomique, Paris.

Kozlovskaya, L. S. & Zaguralskaya, L. M. (1966). The enchytraeids and soil microflora. In *The role of microorganisms in combatting agricultural pests*, pp. 29–71. Moscow. (In Russian.)

Kubicka, H. (1973). The evolution of $CO_2$ in two meadow communities. *Eologia polska*, **21** (5), 73–88.

Kucera, C. L. & Kirkham, D. R. (1971). Soil respiration studies on tallgrass prairie in Missouri. *Ecology*, **52**, 912–15.

Kühnelt, W. (1957). Zoogene Krümelbildung der ungestörten Böden. *Tagungsberichte der deutschen Akademie der Landwirtschaftswissenschaften zu Berlin*, **13**, 193–9.

Kuprevich, V. F. & Shcherbakova, T. A. (1966). *Soil enzymes*. Izdatel'stvo 'Nauka i Technika', Minsk. (In Russian.) (English transl. by Indian National Scientific Document Centre, New Delhi, 1971.)

Kuprevich, V. F. & Shcherbakova, T. A. (1971). Comparative enzymatic activity in diverse types of soil. In *Soil biochemistry*, vol. 2, ed. A. D. McLaren & J. Skujins, pp. 167–201. Dekker, New York.

Ladd, J. N. & Paul, E. A. (1973). Changes in enzymatic activity and distribution of acid-soluble, amino acid-nitrogen in soil during nitrogen immobilization and mineralization. *Soil Biology and Biochemistry*, **5**, 825–40.

Lamotte, M. (1975). The structure and function of a tropical savannah ecosystem. In *Tropical ecological systems*, ed. F. B. Golley & E. Medina, pp. 179–222. Springer-Verlag, New York.

Lamotte, M., Barbault, R., Gillon, Y. & Lavelle, P. (1974). Production de quelques populations animales dans une savane tropicale de Côte d'Ivoire. Paper given at IBP symposium and synthesis meeting on Tropical Grassland Biome Proceedings, Dept Botany, Banaras Hindu University, India (unpublished).

Lavelle, P. (1973). Peuplements et production des Vers de Terre dans la savane de Lamto (Côte d'Ivoire). *Annals of the University of Abidjan, Series E*, **6** (2), 79–98.

Levieux, J. (1971). Données écologiques et biologiques sur le peuplement en Fourmis terricoles d'une savane perforestière de Côte d'Ivoire. DSc Thesis, Paris.

Łomnicki, A., Bandoła, E. & Jankowska, K. (1968). Modification of the Wiegert–Evans method for estimation of net primary production. *Ecology*, **49**, 147–9.

Lundegårdh, H. (1924). *Der Kreislauf der Kohlensäure in der Natur*. G. Fischer, Jena.

Luxton, M. (1972). Studies on the oribatid mites of a Danish beechwood soil. *Pedobiology*, **12**, 434–63.

Macfadyen, A. (1970). Soil metabolism in relation to ecosystem energy flow and to primary and secondary production. In *Methods of study in soil ecology*, ed. J. Phillipson, pp. 167–72. IBP/UNESCO, Paris.

McGarity, J. W. & Myers, M. G. (1967). A survey of urease activity in soils of northern New South Wales. *Plant and Soil*, **27**, 217–38.

Makarov, B. N. (1957). A simplified method for determining soil respiration and biological activity. *Pochvovedenie*, **9**, 199–222. (In Russian.)

Matveeva, V. G. (1966). Earthworms in certain meadows of the Oneg River. In *Problems of soil zoology*, pp. 86–7. Nauka Publishing House, Moscow. (In Russian.)

Matveeva, V. G. (1969). Distribution and abundance of earthworms in certain meadows. In *Problems of soil zoology*, pp. 109–10. Nauka Publishing House, Moscow. (In Russian.)

May, S. W. & Risser, P. G. (1973). *Microbial decomposition and carbon dioxide evolution at Osage Site, 1972. US/IBP Grassland Biome Technical Report 222*. Colorado State University, Fort Collins.

Mina, V. N. (1960). Intensity of $CO_2$ evolution from forest floors and its distribution in soil in leached out chernozems under various forest stands. *Reports of the Forest Laboratory, Academy of Sciences, USSR*, **1**, 127–44. (In Russian.)

Miroshnishchenko, E. D., Pavlova, T. V. & Ponyatovskaya, V. M. (1972). Decomposition of vegetative mass in dry meadows of Leningrad district (Karelian Isthmus). *Botanicheskii Zhurnal*, **57**, 533–40. (In Russian.)

Mishustin, E. N. (1964). Different soil types and specificity of their micropopulation. *Annales de l'Institut Pasteur*, **107**, 3 (Suppl.), 63–77. (In French.)

Mouraret, M. (1965). Study on the soil enzymatic activity of asparaginase. *Mémoires*

*ORSTOM (Office de Recherche Scientifique et Technologique Outre-mer)*, **9**, 19–111.
Müller, G. (1965). *Bodenbiologie*. G. Fischer, Jena.
Naplekova, N. N. (1972). Influence of nitrogen and phosphorus on intensity of cellulose decomposition and biological activity of west Siberian zonal soils. In *Problems of abundance, biomass, and productivity of microorganisms in soil*, ed. T. V. Aristovskaya, pp. 207–15. Nauka Publications, Leningrad. (In Russian.)
Nielsen, C. O. (1961). Respiratory metabolism of some populations of enchytraeid worms and free-living nematodes. *Oikos*, **12**, 17–35.
Odum, E. P. (1971). *Fundamentals of ecology*, 3rd edn. W. B. Saunders, Philadelphia.
Pancholy, S. K. & Rice, E. L. (1973). Carbohydrases in soil as affected by successional stages of revegetation. *Proceedings of the Soil Science Society of America*, **37**, 227–9.
Parkinson, D., Gray, T. R. G. & Williams, S. T. (eds.) (1971). *Methods for studying the ecology of soil microorganisms. IBP Handbook 19*. Blackwell Scientific Publications, Oxford.
Patra, U. C. & Dash, M. C. (1978). Wormcast production and nitrogen excretion by a tropical earthworm population from a grassland site in India. *Revue d'Ecologie et de Biologie du Sol*, **15**, in press.
Paulson, K. N. & Kurtz, L. T. (1969). Locus of urease activity in soil. *Proceedings of the Soil Science Society of America*, **33**, 897–900.
Payne, W. J. (1970). Energy yields and growth of heterotrophs. *Annual Review of Microbiology*, **24**, 17–52.
Petrova, A. N. (1963). Methods for determination of biological activity of soils. In *Microorganisms in agriculture*, pp. 422–7. Moscow University Publishers, Moscow. (In Russian.)
Phillipson, J. (ed.) (1970). *Methods of study in soil ecology*. IBP/UNESCO, Paris.
Pirozynski, K. A. & Malloch, D. W. (1975). The origin of land plants: a matter of mycotrophism. *Biosystems*, **6**, 153–64.
Pochon, J. & Bacvarov, I. (1973). Données préliminaires sur l'activité microbiologique des sols de la savane de Lamto (Côte d'Ivoire). *Revue d'Ecologie et de Biologie du Sol*, **10**, 35–45.
Rambelli, A. (1971). Recherches mycologiques préliminaires dans les sols de forêt de savane en Côte d'Ivoire. *Revue d'Ecologie et de Biologie du Sol*, **8**, 210–26.
Rambelli, A. & Bartoli, A. (1972). Recherches sur la microflore fongique de terrains de Lamto en Côte d'Ivoire. *Revue d'Ecologie et du Biologie du Sol*, **9**, 41–53.
Rambelli, A., Puppi, G., Bartoli, A. & Albonetti, S. G. (1973). Recherches sur la microflore fongique de terrains de Lamto en Côte d'Ivoire. *Revue d'Ecologie et de Biologie du Sol*, **10**, 13–18.
Redmann, R. E. (1978). Soil respiration in a mixed grassland ecosystem. *Canadian Journal of Soil Science*, **58**, 119–24.
Řehořková, V., Kopčanová, L. & Řehořek, V. (1968). Influence of some ecological factors on cellulose decomposition in grassland soils. *Acta fytotechnica, Universitas Agriculturae Nitra*, **18**, 29–58. (In Slovak with English summary.)
Řehořková, V., Marendiak, D. & Kopčanová, L. (1971). Biological activity of soils of some grasslands in the alluvium of the Ipel and Slaná rivers. Internal Technical Report, Universitas Agriculturae Nitra. (In Slovak.)
Reiners, W. A. (1973). A summary of the world carbon cycle and recommendations for critical research. In *Carbon and the biosphere*, ed. G. M. Woodwell & E. V. Pecan, *AEC Symposium Series 30*, pp. 368–82. USAEC, Technical Information Center, Office of Information Services, Springfield, Virginia.

Ross, D. J. & Roberts, H. S. (1970). Enzyme activities and oxygen uptake of soils under pasture in temperature and rainfall sequences. *Journal of Soil Science*, **21**, 368–81.

Rusek, J., Úlehlová, B. & Unar, J. (1975). Soil biological features of some alpine grasslands in Czechoslovakia. In *Progress in soil zoology*, pp. 199–215. Academia, Prague.

Sanders, F. E., Mosse, B. & Tinker, P. B. (1975). *Endomycorrhizas*. Academic Press, New York & London.

Sasson, A. (1972). Microbial life in arid environments: prospects and achievements. *Annals of the Arid Zone, India*, **11**, 67–91.

Satchell, J. E. (1967). Lumbricidae. In *Soil biology*, ed. A. Burges & F. Raw, pp. 259–322. Academic Press, New York & London.

Schaefer, R. (1979). Potentiel humique et activités microbiennes dans 3 types de sol d'une savane tropicale (Lamto, Côte d'Ivoire). In *Vth Symposium International d'Ecologie Tropicale*. Kuala Lumpur, Malaya. (In press.)

Shields, J., Paul, E. A., Lowe, W. E. & Parkinson, D. (1973). Turnover of microbial tissue in soil under field conditions. *Soil Biology and Biochemistry*, **5**, 753–64.

Shtina, E. A. (1968). The algae as producers of organic matter of soil. *Soil Science*, **1**, 79–86. (In Russian.)

Skujins, J. J. (1967). Enzymes in soil. In *Soil biochemistry*, vol. 1, ed. A. D. McLaren & G. H. Peterson, pp. 371–414. Dekker, New York.

Smolik, J. D. (1974). *Nematode studies at the Cottonwood Site. US/IBP Grassland Biome Technical Report 251*. Colorado State University, Fort Collins.

Sparrow, E. B. & Doxtader, K. G. (1973). *Adenosine triphosphate (ATP) in grassland soil: its relationship to microbial biomass and activity. US/IBP Grassland Biome Technical Report 224*. Colorado State University, Fort Collins.

Stebayev, I. V. (1968). The character of aboveground and belowground zoomicrobiological complexes of steppe landscapes in west and middle Siberia. *Zoologicheskiĭ Zhurnal*, **47**, 661–75. (In Russian.)

Stout, J. & Heal, O. W. (1967). Protozoa. In *Soil biology*, ed. A. Burges & F. Raw, pp. 149–95. Academic Press, New York & London.

Striganova, B. R. (1971). The comparison of activity of different groups of soil invertebrates in the process of forest litter decomposition. *Ecology*, **4**, 36–43. (In Russian.)

Tabatabai, M. A. & Bremner, J. M. (1970). Arylsulfatase activity of soils. *Proceedings of the Soil Science Society of America*, **34**, 225–9.

Ten-Chak-Mun, A. & Fedorova, L. (1972). Yearly cycle of microbiological processes in Sakhalin soils covered with *Altherobosa*. In *Problems of abundance, biomass, and productivity of microorganisms in soil*, ed. T. V. Aristovskaya, pp. 153–68. Nauka Publishers, Leningrad. (In Russian.)

Tesarová, M. & Gloser, J. (1972). Soil respiration in a moist meadow plant community. In *Ecosystem study on grassland biome in Czechoslovakia*, pp. 91–3. *Czechoslovakia PT-PP/IBP Report 2*.

Tesarová, M. & Úlehlová, B. (1968). Cellulose decomposition under some grassland associations. *Tagungsberichte der deutschen Akademie der Landwirtschaftwissenschaften zu Berlin*, **98**, 277–87. (In German.)

Thambi, A. V. & Dash, M. C. (1973). Seasonal variation in numbers and biomass of Enchytraeidae (Olig.) populations in tropical grassland soils from India. *Tropical Ecology*, **14**, 228–37.

Tischler, W. (1965). *Agrarökologie*. G. Fischer, Jena.

Titlianova, A. A. (1971). *The study of biological turnover in biogeocenoses* (*a manual of methods*). Novosibirsk. (In Russian.)

Titlianova, A. A. (1973). *Steppe, meadow, and marsh ecosystems and agricultural crops on the divide between steppe and forest-steppe zones of the USSR*. AN USSR Publishers, Moscow. (In Russian.)

Úlehlová, B. (1973). Alluvial grassland ecosystems. Microorganisms and decay processes. *Acta Scientiarum Naturalium Academiae Scientiarum Bohemoslovacae*, **7** (5), 1–43.

Voets, J. P. & Dedeken, M. (1966). Observations sur la microflore et les enzymes dans la rhizosphere. *Annales de l'Institut Pasteur*, **111** (3) (suppl.), 197–207.

Vossbrinck, C. R., Coleman, D. C. & Woolley, T. A. (1978). Abiotic and biotic factors in litter decomposition in a semiarid grassland. *Ecology*, **59**, in press.

Walter, H. (1951). Grundlagen der Pflanzenverbreitung. In *Phytology*, vol. 3, part 1, pp. 405–12. *Stuttgart Beiträge der Naturkunde*.

Warembourg, F. & Paul, E. A. (1977). Seasonal transfers of assimilated $^{14}C$ in grassland: plant production and turnover, soil and plant respiration. *Soil Biology and Biochemistry*, **9**, 295–301.

Wiant, H. Jr (1967). Contributions of roots to forest 'soil respiration'. *Advancing Frontiers of Plant Science, India*, **18**, 163–8.

Wiegert, R. G., Coleman, D. C. & Odum, E. P. (1970). Energetics of the litter–soil subsystem. In *Methods of study in soil ecology*, ed. J. Phillipson, pp. 93–98. IBP/UNESCO, Paris.

Wiegert, R. G. & Evans, F. (1964). Primary production and the disappearance of dead vegetation on an old field in southeastern Michigan. *Ecology*, **45**, 49–63.

Wiegert, R. G. & Owen, D. F. (1971). Trophic structure, available resources, and population density in terrestrial vs. aquatic ecosystems. *Journal of theoretical Biology*, **30**, 69–81.

Wildung, R. G., Garland, T. R. & Buschbom, R. L. (1975). The interdependent effects of soil temperature and water content on soil respiration rate and plant root decomposition in arid grassland soils. *Soil Biology and Biochemistry*, **7**, 373–8.

Witkamp, M. & van der Drift, J. (1961). Breakdown of forest litter in relation to environmental factors. *Plant and Soil*, **15**, 295–311.

Wood, T. G. (1976). The role of termites (Isoptera) in decomposition processes. In *The role of terrestrial and aquatic organisms in decomposition processes*, ed. J. M. Anderson & A. Macfadyen, pp. 145–68. Blackwell Scientific Publications, Oxford.

Yershov, V. V. (1966). Influence of mineral fertilizers on microbiological processes in meadow soils of Karelia. *Mikrobiologija*, **35**, 168–73. (In Russian.)

Yershov, V. V. (1972). On the activity of microorganisms in the meadow soils of Karelia. In *Problems of abundance, biomass, and productivity of microorganisms in soils*, ed. T. V. Aristovskaya, pp. 251–8. Nauka Publishers, Leningrad. (In Russian.)

Zaguralskaya, L. M., Yegorova, R. A. & Smantser, P. G. (1972). Biological activity of swamp peat soils in southern Karelia. In *Problems of abundance, biomass, and productivity of microorganisms in soils*, ed. T. V. Aristovskaya, pp. 200–7. Nauka Publishers, Leningrad. (In Russian.)

Zlotin, R. I. (1969). Zonal peculiarities of the biomass of soil invertebrate animals in open landscapes of the Russian plain. In *The problems of soil zoology, Proceedings of the Third All-Union Conference*, pp. 75–7. Nauka Publishers, Moscow. (In Russian.)

Zlotin, R. I. (1971). Invertebrate animals as a factor of the biological turnover. In *IVth International Colloquium on Soil Zoology*, pp. 455–62. Institut National de la Recherche Agronomique, Paris.

Zlotin, R. I. (1974). The study of decomposition processes in relation to investigations of biological turnover in terrestrial ecosystems. In *Reports October 1971–June 1972, Zoology and Botany*, pp. 55–7. Moscow University Publishers, Moscow. (In Russian.)

Zlotin, R. I. (1975). *Life in high mountains*. Mysl Publishers, Moscow. (In Russian.)

Zlotin, R. I. & Chukanova, A. V. (1973). Rate of cellular tissue disintegration as an index of the intensity of microbiological processes in forest-steppe ecosystems. In *Topological aspects in the study of the behaviour of matter in geosystems*, pp. 148–50. Irkutsk.

Zlotin, R. I. & Khodashova, K. S. (1974). *The role of animals in biological turnover in forest-steppe ecosystems*. Nauka Publishers, Moscow. (In Russian: English translation, D. H. and R. Inc., Stroudsburg, Pa., USA, in press.)

# Part II. Systems synthesis

# 8. Nutrient cycling

F. E. CLARK, C. V. COLE & R. A. BOWMAN

## Contents



## Introduction

Discussions and simulations of nutrient cycles commonly employ a skeletal framework involving flow from an available nutrient pool to the live plant, thence to plant litter and thereafter to one or more reservoirs in the soil, from which there occurs replenishment of the available nutrient pool. This basic cycle may suffer diverse inputs and outputs; these as a general rule can be satisfactorily quantified.

Formulation of the internal or basic cycle is commonly dependent on a nutrient balance approach. The net annual primary production is measurable, as well as the nutrient content of that biomass. For this nutrient demand, suitable decomposition or release rates can be postulated so that the nutrient flow within the system appears credible regardless of the accuracy of the individual subflows. Formulation of the component subflows and of their compartments of origin are largely subjective, even though an aura of respectability is provided by citation of the pertinent literature. Uncertainties nevertheless remain. For nitrogen in an unfertilized grassland, precisely where does the annual supply to the plant come from? Is it primarily from last year's litter, variously from the litter of several different years, or primarily from soil humic material?

Progress in understanding the internal cycling of nutrients in the soil–plant system has been fostered by tracer studies designed to monitor the movement of carbon, nitrogen, phosphorus, sulphur or other nutrients. In Chapters 2 and 7 of this volume instances are noted in which $^{14}C$ was used to monitor carbon flow during photosynthesis and decomposition. Knowledge of carbon flow has only limited value in postulating nitrogen flow inasmuch as each nutrient has specific flows and storage compartments that must be considered individually. Flows determined for carbon are not directly transferable to nitrogen, nor are those for nitrogen duplicative of those for phosphorus.

In this chapter discussion of the nitrogen cycle will emphasize certain concepts evolved during the course of experiments using pulse application of $^{15}N$ on shortgrass prairie. In several instances these concepts differ from those heretofore commonly employed in mathematical simulations of the nitrogen cycle. In a second section, recently developed concepts concerning the phosphorus cycle are presented, again with emphasis on data compiled in recent years for North American grasslands. In closing, the flows of sulphur, potassium, calcium and magnesium in grasslands are briefly discussed.

## Nitrogen flow

Discussions of the nitrogen flow in terrestrial ecosystems commonly use box-and-arrow diagrams representing the major state variables and the directional transfers between them. Prior to the advent of simulation modelling of nutrient flows, the diagrams were relatively simple and served mainly as outlines for discussion of individual topics, such as denitrification or litter decomposition. Early attempts at mathematical simulations quickly revealed that generalized cycles of nutrient transfers were inadequate and that more specific compartmentalizations and individualized transfer rates were required. It is not our intent to review here the many variations of the nitrogen cycle that have been published in recent years, particularly those where scores of flows are shown. There will be brief discussion of the gains and losses of nitrogen in grassland, but the flows involved will not be diagrammed. More detailed discussion, supplemented by a comparatively simple box-and-arrow diagram of the compartments and flows involved, will be offered on the internal cycling of nitrogen in grassland.

### Nitrogen inputs

Nitrogen inputs to grassland, barring intensive management interventions, are primarily by symbiotic and nonsymbiotic fixation and in the ambient precipitation. These are supplemented by animal immigration and wind and surface water import. In intensively managed grasslands, fertilizers and the legumes in sown legume–grass mixtures are responsible for the major inputs.

In most grasslands the input due to nonsymbiotic dinitrogen fixation is low. Values reported are commonly in the range of 0.1 to 0.2 $g\ N \cdot m^{-2} \cdot yr^{-1}$ (Steyn & Delwiche, 1970; Reuss, 1971; Vlassak, Paul & Harris, 1973). The values cited were based on field soil cores and presumably represent total fixation by both photosynthetic and nonphotosynthetic free-living micro-organisms and with very little or no symbiotic fixation involved. Among the dozen or so genera of nonsymbiotic nitrogen-fixing bacteria, only *Azotobacter* and *Clostridium* are believed to make any significant contribution (Clark & Paul, 1970).

Photosynthetic blue-green algae have long been recognized as efficient nitrogen fixers. Short-term measurements of algal crust activity have led to extrapolations showing nitrogen fixation of from 10 to 30 $g\ N \cdot m^{-2} \cdot yr^{-1}$. Obviously annual fixation of such magnitude does not occur in grassland; otherwise, and without compensatory losses of equal magnitude that are currently unsubstantiated, the system would soon become literally flooded with nitrogen.

Of the symbiotic associations, nodulated legumes are easily the principal agents of fixation, and given suitable stand densities under favourable growing conditions their fixation can be of the order of 10 to 30 $g\ N \cdot m^{-2} \cdot yr^{-1}$. These rates are at times achieved in managed sown pastures (Williams & Andrews, 1970). In most grasslands, legumes are a minor part of the vegetation. For example, Whitman & Stevens (1952) reported that for western North Dakota grassland, legumes usually accounted for less than 10% of the total range productivity and at times for no more than 1%. Data of Sims, Uresk, Bartos & Lauenroth (1971) indicate a similar paucity of legumes in Colorado shortgrass prairie. Although the amount of nitrogen contributed annually to grassland by nodulated legumes may be low, the cumulative effect over a number of years can be important in the build-up of the organic nitrogen content of grassland soil.

Plants other than legumes may also serve as hosts for nodulating organisms (Stewart, 1966). Farnsworth & Hammond (1968) reported that sagebrush (*Artemisia ludoviciana*) and prickly pear cactus (*Opuntia fragilis*) bore root swellings that were capable of fixing atmospheric nitrogen. Because the host plants are widely distributed in many desert and semi-arid rangelands,

Table 8.1. *Some values recently reported for nitrogen in precipitation*

| N in precipitation (g $N \cdot m^{-2} \cdot yr^{-1}$) | Site | Reference |
|---|---|---|
| 0.40 | Pawnee, Colorado, USA | Reuss (1971) |
| 0.32 | Matador, Saskatchewan, Canada | Vlassak *et al.* (1973) |
| 0.30 | Curlew Valley, Utah, USA | Skujins (1975) |
| 0.35 | Kursk, USSR | Titlyanova & Bazilevich (1975) |

Farnsworth (1975) undertook to assay $N_2$ fixation by nodulated nonlegumes on a number of sites in Utah. For the 1974 season he found essentially no nodulation and concluded therefore that nodulation of sagebrush in desert, range and forest areas is a fragile phenomenon, very sensitive to very slight changes in the ambient soil temperature and moisture.

That precipitation washes down available nitrogen from the atmosphere is well known. Values recently reported for the nitrogen input to grassland via this route show surprisingly close agreement (Table 8.1).

The precipitation input is roughly of the same order of magnitude as is the input ascribed to biological fixation. A possibility heretofore commonly ignored is that the nitrogen measured in rain gauge entrapments represents only a part of the nitrogen input to the soil–plant system from the atmosphere. Studies by Hutchinson, Millington & Peters (1972) and Porter, Viets & Hutchinson (1972) have shown that plant, soil and water surfaces are a sink for atmospheric ammonia and that such direct absorption may be fully as important an input as the ammonia washed down during precipitation events. Measurements made by Denmead, Freney & Simpson (1976) within the canopy of ungrazed pasture indicated a large production of ammonia near the ground surface and almost complete absorption of it by the plant cover. The amounts absorbed appeared much too large for stomatal uptake alone.

### *Losses of nitrogen*

Nitrogen may be lost from grassland by denitrification, volatilization, leaching, wind and water erosion of particulate matter, and in harvest or animal export. The nitrogen most susceptible to these pathways of loss is mineral nitrogen in the form of nitrate and ammonia, both of which, barring their application in fertilizer, are present in very few to fractional parts per million in grassland soil. Consequently, nitrogen losses from grassland are usually far lower than those from cultivated soils. Given suitable conditions, however, some losses do occur, and therefore a few brief remarks are in order.

Leaching losses in arid and in unfertilized humid grassland are usually very low or non-existent, mainly because there is little or no soluble nitrogen in the soil to be leached. In fertilized humid grassland, leaching can remove much if not all nitrate nitrogen applied during the winter months (Woldendorp, Dilz & Kolenbrander, 1966). The same grassland suffered little or no nitrogen loss when the fertilizer was applied at conservative rates during the summer months. At that time actively growing plants quickly diminish any excess of soluble nitrogen in the soil. Power (1971) reported that semi-arid North Dakota grassland given heavy application of mineral nitrogen suffered no loss of nitrogen due to leaching in the several following years.

Denitrification can occur in soil when available nitrate is present, aeration is limited, the soil moisture, temperature and pH are favourable for microbial growth, and readily decomposable organic matter is available to the micro-organisms. This spectrum of conditions is rarely met in grassland. Denitrification therein is almost wholly limited to soils which have been heavily fertilized with nitrate and in which the plant growth is insufficient to use rapidly the applied nitrate. Woldendorp *et al.* (1966), however, observed denitrification in sandy soils low in organic matter, sown with grass, and not subject to poor aeration. They attributed the observed loss to movement of nitrate into the rhizosphere, where localized oxygen deficiency could occur because of the respiration of roots and of micro-organisms living on root exudates and exfoliates. Skujins (1975) has suggested that denitrification may occur in the decomposing algal crusts on rangelands, but confirmatory data are not given. Soulides & Clark (1958) and Broadbent & Clark (1965) reported the occurrence of denitrification in grassland soils fertilized with ammonium nitrogen. They attributed the denitrification to nitrite instability, or chemo-denitrification, following nitrification of ammonia to nitrite. Chemo-denitrification is most likely to occur in grassland given ammonium-containing or ammonium-yielding fertilizers, but it can also occur in unfertilized grassland following bursts of mineralization of organic nitrogen.

Volatilization of ammonia is known to cause substantial losses of nitrogen from either fertilized or heavily grazed grassland. Volk (1959) reported the nitrogen loss during 7 days following surface application of urea to four different grass sods to range from 20 to 30% of the urea nitrogen applied. Stewart (1970) reported that for steer urine applied to dry soil in the laboratory as much as 90% of the urea nitrogen in the urine was volatilized as ammonia. Phenomenal ammonia losses in the field have recently been reported for a grazed grass–clover pasture in Australia. For a dense stand of *Lolium rigidum* and *Trifolium subterraneum* grazed at a stocking rate of 22 sheep$\cdot$ha$^{-1}$, Denmead *et al.* (1976) observed ammonia volatilization at the rate of 30 mg N$\cdot$m$^{-2}\cdot$d$^{-1}$. For an ungrazed area of the same pasture, the volatilization was measured as 5 mg. The authors noted, however, that this high rate of volatilization in ungrazed pasture did not mean equivalent loss to the atmosphere. The ammonia released during surface litter decomposition was almost entirely absorbed by the vegetation canopy.

*Systems synthesis*

In most grasslands, ammonia losses to the atmosphere during decomposition of plant litter are probably minimal. However, Skujins (1975) has reported that as much as 10 to 25% of the nitrogen fixed in algal crusts may undergo volatilization as $NH_3$ during the course of algal crust decomposition. There is the possibility that small amounts of ammonia may be volatilized directly from higher plants or from amide-containing guttations (Martin & Ross, 1968). Porter *et al.* (1972) reported that one-tenth of the $^{15}NH_3$ introduced into polyvinyl chloride tents containing maize plants was found in the transpired water.

## *Internal cycling of nitrogen*

There is almost universal agreement that the accumulated nitrogen in grassland is mainly in the form of organic nitrogen. Barring severe perturbations, particularly those imposed by man, the nitrogen total is in steady-state equilibrium. As noted above, there are annual inputs of nitrogen, but these are not large enough to meet the annual growth requirement. Furthermore, the inputs are offset by losses of approximately equal magnitude. Consequently, the annual growth requirement of the plant for nitrogen is met largely by the cycling of nitrogen already present in the system. This internal cycling remains poorly defined.

Comprehensive empirical data on nitrogen cycling within a single ecosystem have not as yet been assembled. Early simulations of the nitrogen flows in grassland drew on the results of many different studies to formulate specific transfer rates (Dahlman & Sollins, 1970; Frere *et al.*, 1970; Reuss & Cole, 1973). Recent studies with $^{15}N$ in grassland have provided more specific data for critiquing current concepts of the routes and rates of the internal flows of nitrogen. Because our interpretation of these flows is based largely on the $^{15}N$ experiment conducted by Clark (1977) on blue grama (*Bouteloua gracilis*) grassland, a summary presentation of his data set is offered here.

The field experimentation involved mini-plots in ungrazed shortgrass prairie that were enclosed in open-ended steel cylinders driven into the soil. Following an initial $^{15}N$ fertilization in May 1971, the cylinders received no further fertilization during the following 4 years and no moisture other than that in the ambient precipitation. During the years 1971 through 1975, destructive removals of cylinders were made on 15 dates spaced over five growing seasons. Organic carbon and total nitrogen and $^{15}N$ were measured for the soil and plant materials collected. Table 8.2 shows the organic carbon and total nitrogen content in plant materials and soil. Table 8.3 shows the distribution of carbon and nitrogen in six plant material compartments.

The four plant material and two soil nitrogen compartments and the 13 nitrogen flows chosen for discussion in this chapter are shown in Fig. 8.1. They are individually listed in Table 8.4. Fig. 8.1 runs counter to the general

Table 8.2. *Organic carbon and total nitrogen contents of plant and soil materials in a blue grama grassland*

| | Organic C ($g \cdot m^{-2}$) | Total N ($g \cdot m^{-2}$) | C/N ratio |
|---|---|---|---|
| *Plant materials* | | | |
| Aboveground green | 43[a] | 1.4 | 30 |
| Aboveground dead | 64 | 2.4 | 26 |
| Crowns | 160 | 4.6 | 34 |
| Live roots | | | |
| 0–10 cm depth | 88 | 2.3 | 38 |
| 10–20 | 15 | 0.3 | 50 |
| 20–36 | 9 | 0.2 | 51 |
| 0–36 | 112 | 2.8 | 40 |
| Senescent roots | | | |
| 0–10 cm depth | 282 | 13.9 | 20 |
| 10–20 | 39 | 1.8 | 22 |
| 20–36 | 28 | 1.1 | 26 |
| 0–36 | 349 | 16.8 | 21 |
| Detrital roots | | | |
| 0–10 cm depth | 149 | 8.8 | 17 |
| 10–20 | 51 | 2.9 | 18 |
| 20–36 | 57 | 3.1 | 18 |
| 0–36 | 256 | 14.8 | 17 |
| Plant subtotal | 983 | 42.8 | 23 |
| *Soil* | | | |
| 0–10 cm depth | 1327 | 127.0 | 10 |
| 10–20 | 978 | 106.6 | 9 |
| 20–36 | 1181 | 138.9 | 8 |
| 0–36 | 3486 | 372.5 | 9 |
| *Plant–soil total* | 4469 | 415.3 | 11 |

[a] All values given are the means of values for 15 dates of sampling over 5 years. (Cited from Clark, 1977.)

Table 8.3. *Distribution of plant carbon and nitrogen in six plant material compartments for blue grama grassland*

| | Organic C (%) | Total N (%) |
|---|---|---|
| Aboveground green | 4[a] | 3 |
| Aboveground dead | 6 | 6 |
| Crowns | 16 | 11 |
| Live roots | 11 | 6 |
| Senescent roots | 35 | 39 |
| Detrital roots | 26 | 35 |

[a] Values given are the means of values for 15 dates of sampling over 5 years. (Cited from Clark, 1977.)

Table 8.4. *Identification of the compartments and nitrogen flows diagrammed in Fig. 8.1*

*Compartments*

| | |
|---|---|
| I | Live root and crown nitrogen |
| II | Green herbage nitrogen |
| III | Nitrogen in aboveground dead herbage (standing dead and litter) |
| IV | Nitrogen in belowground litter, primarily root and crown litter, but inclusive of aboveground litter that has undergone particulate transfer to belowground, plus the nitrogen in micro-organisms closely associated with belowground litter |
| V | Nitrogen in soil humus |
| VI | The available nitrogen pool, embracing the ammonium and nitrate ions available for plant uptake |

*Flows*

| | |
|---|---|
| $f_1$ | Flow from the available pool to live roots and crowns |
| $f_2$ | Return flow to pool from live roots |
| $f_3$ | Flow from live roots and crowns to green herbage to roots and crowns |
| $f_4$ | Retranslocation flow from green herbage to roots and crowns |
| $f_5$ | Return flow to available pool from green herbage |
| $f_6$ | Transfer to aboveground dead at time of death of green herbage |
| $f_7$ | Particulate transfer to belowground from aboveground litter |
| $f_8$ | Return flow to available pool from aboveground litter |
| $f_9$ | Transfer to root litter at death of live roots and crowns |
| $f_{10}$ | Return flow to available pool from belowground litter |
| $f_{11}$ | Transfer to humus during microbial decomposition of litter |
| $f_{12}$ | Flow of mineral N to micro-organisms engaged in litter decomposition |
| $f_{13}$ | Flow of N mineralized from humus to the available N pool |

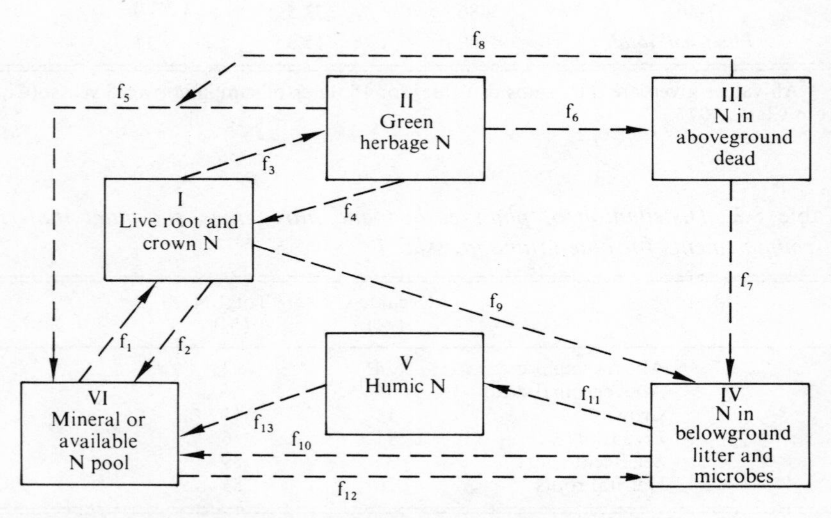


Fig. 8.1. Diagram of the internal flows of nitrogen in shortgrass prairie.

trend in recent years of showing a great number of compartments and transfers in constructions of nitrogen flow simulations. In part the relative simplicity in Fig. 8.1 stems from omission of flows showing gains and losses of nitrogen – these have been discussed separately above – and by omission of nitrogen flows to and from large herbivores. Of such flows, that of ammonia volatilization from urine has been mentioned. The within-the-system flow of nitrogen that occurs during the digestive process and in the decomposition of excreta and carrion is broadly a part of the litter decomposition cycle. We have arbitrarily chosen not to discuss nitrogen flow into and from animal excreta (see Chapter 7, this volume).

### *Nitrogen flows* ($f_{1,2}$) *between live roots and the available pool*

The forward flow of nitrogen to plant roots from the available pool is the basic flow to the growing plant and one that has been intensively studied by plant and soil scientists. Consequently, there is a wealth of information on the uptake of nitrogen by either excised roots or whole plants. Such studies commonly use a readily available form of nitrogen and are concerned with the rate control functions of the nitrogen concentrations in the soil or substrate solution and in the plant. Comprehensive reviews and discussions are available in the literature (for an excellent monographic treatment, see Epstein, 1972). Reuss & Cole (1973), in constructing a nitrogen flow simulation, modelled nitrogen uptake by roots as the sum of two processes involving a rate factor and a half-saturation constant linked to a function of soil nitrate concentration. They stated that use of the function did not imply a high degree of precision but did permit uptake to be linked satisfactorily to both low and high nitrate levels.

There appears to be no serious problem in modelling nitrogen uptake by roots from the soil solution. What is particularly needed is some definition of the time of residence of nitrogen in gramineous live roots. This time of residence will be discussed prior to any consideration of the several pathways by which nitrogen moves from live roots. Our discussion of this residence is based on $^{15}N$ data assembled by Clark (1977). In his study, the pulse fertilization given initially was at the rate of 25 $kg \cdot ha^{-1}$ for total nitrogen, of which 80% was $^{15}N$. Inorganic nitrogen analyses showed that the added nitrogen was quickly immobilized by plants and micro-organisms during the first growing season. Consequently, $^{15}N$ contents of plant materials in the four following years can safely be assumed to represent either carry-over of $^{15}N$ in live or dead material or else recycled $^{15}N$, that is, $^{15}N$ immobilized during the first year then mineralized and again immobilized in a following year. The amounts of $^{15}N$ recovered from the total soil–plant systems involved did not differ significantly over the 5 years; however, the amounts within individual compartments did differ significantly with time. In short, there were transfers of nitrogen within the system. Table 8.2 above shows that Clark treated live root and crown nitrogen as separate compartments; he also recognized separately senescent and detrital root materials. Currently we are assigning his data on live root and crown nitrogen to a single compartment and are similarly combining into one compartment his data on senescent and detrital roots.

A summary of the $^{15}N$ distribution in four plant material compartments in June of five successive years is given in Fig. 8.2. Analyses of variance of the $^{15}N$ content within compartments were made by standard procedures. For the three compartments in which significance was encountered, multiple comparisons between years were made using Tukey's Q test (Snedecor & Cochran, 1967). Differences encountered are shown by means of superscripts (the letters A to D) in Fig. 8.2.

The live roots and crowns of blue grama are an important sink for $^{15}N$ fertilizer, both in the year of application and in the several following years (Fig. 8.2). Of the total $^{15}N$ recovered in plant materials, roughly 9% was accounted for in the live roots and 20% in the crowns. The amounts of $^{15}N$ recoverable in either the crowns or live roots in the five different years did not differ significantly. How can this nearly constant $^{15}N$ content best be explained? It is unlikely that $^{15}N$ taken in by the roots and crowns in the year of application would be retained therein for the four following years without any loss whatsoever. A more likely explanation is that at least some $^{15}N$ is mineralized from one or more sources in each of the several years and that uptake of this nitrogen coupled with retention of prior-year nitrogen in long-lived roots has maintained almost an undiminished quantity of $^{15}N$ in the roots and crowns over a time span of 5 years. It is not possible to measure separately prior-year and current-year $^{15}N$ in plant roots following pulse application of $^{15}N$ in a preceding year. It is possible to estimate indirectly how much $^{15}N$ moves into the live root compartment in the current year. This is done on the basis of green herbage measurement.

Andrews, Coleman, Ellis & Singh (1974) allotted 23% of the annual net productivity in shortgrass prairie to aboveground, and 77% to belowground plant parts. The nitrogen content of the green herbage of blue grama is roughly 0.015, and that of the live roots 0.0089 (Clark & Campion, 1976). Therefore, nitrogen flow to the belowground parts should be twice that going to the green herbage ($0.015 \times 0.23$ versus $0.0089 \times 0.77$). In the Pawnee experiment, the $^{15}N$ in the green herbage in June of the years 1971 through 1975 is known; however, insofar as calculation of $^{15}N$ mineralization is concerned, values for the first 2 years should be ignored. Uptake of $^{15}N$ in 1971 is from the mineral $^{15}N$ added in 1971; herbage $^{15}N$ in 1972 may consist in part of $^{15}N$ of 1971 vintage taken into roots in late season of 1971. The $^{15}N$ contents of the green herbage in the years 1973 through 1975 can be assumed to represent nitrogen that has either undergone internal translocation or that has been

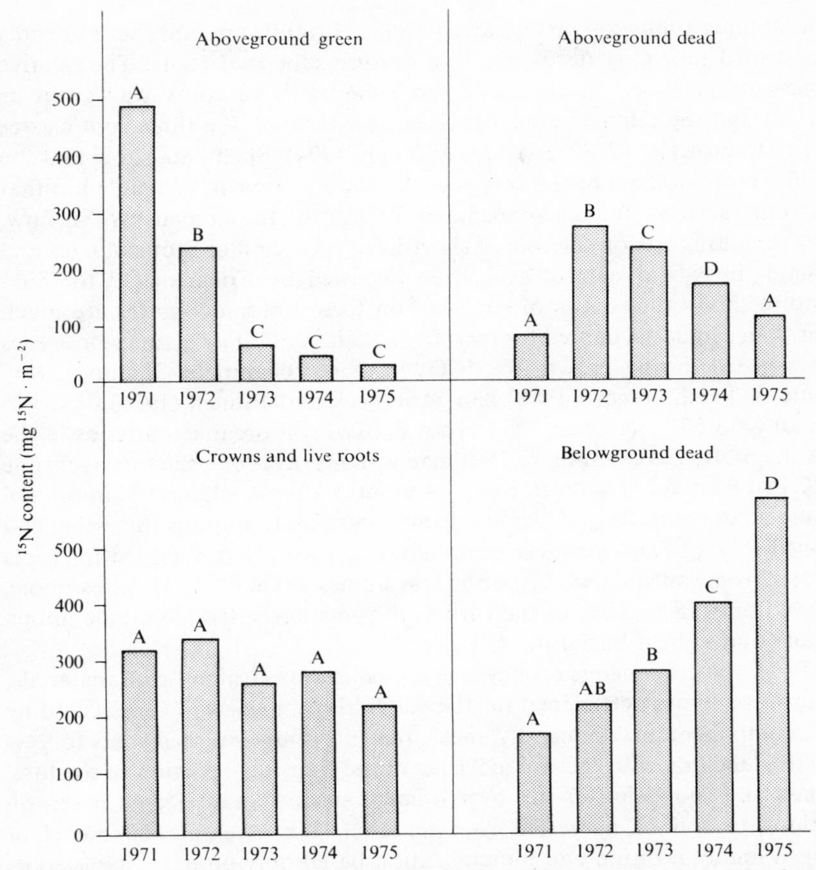


Fig. 8.2. Amounts of $^{15}N$ in four plant material compartments in June of the five years following $^{15}N$ fertilization in May 1971. Histograms show mg $^{15}N \cdot m^{-2}$ for the aboveground compartments and mg $^{15}N \cdot m^{-2}$ to a depth of 36 cm for the belowground material. Letters over bars indicate significantly different values.

immobilized and subsequently mineralized. On the premise that nitrogen flow to belowground parts is double that going to green herbage, the annual turnover rate for the nitrogen in live roots and crowns is 0.19. Is this estimate reasonable?

Pulse labelling with $^{14}C$ has indicated that root turnover in mesic prairie is accomplished in 4 years (Dahlman & Kucera, 1965; Dahlman, 1968). Data of Sims & Singh (1971) show that root turnover in xeric prairie is accomplished in 2 years. In a detailed study of roots in shortgrass prairie, Ares (1976) observed root growth and death as seen in glass-faced excavations in the field in a blue grama grassland. He observed that roughly half of the young unsuberized roots formed during the spring growth period persisted for no longer than 1 or 2 months. Very small rootlets, 'juvenile unsuberized', could quickly appear and disappear, either during the early growing season or later in the summer following precipitation events. Juvenile and unsuberized roots that escaped mortality developed into mature suberized roots. The relative biomass of juvenile, unsuberized, and suberized live roots was given as 1:68:31, and the relative annual carbon flow through the three root classes (i.e. live to dead) as 17:47:36 (Ares & Singh, 1974). Such data show that the standing crop biomass of the very juvenile roots is almost negligible but that such roots account for an appreciable fraction of the annual carbon flow. Other recent reports on the role of short-lived roots and of root exudates and exfoliates in root dynamics have been reviewed by Coleman (1976). That literature shows that losses of root carbon to soil organic matter are much larger than can be accounted for by root exudates only. For plants whose tops were exposed during growth to $^{14}CO_2$ in closed chambers, Shamoot, McDonald & Bartholomew (1968) and Sauerbeck and Johnen (1976) observed from 20% to 60% as much $^{14}C$ in root-derived soil organic matter as in the roots themselves. According to Dahlman & Kucera (1965), the turnover time of $^{14}C$ lost from roots is short. Some 14 months after labelling, $^{14}C$ in the soil organic matter was 48% of the loss from roots; by 18 months this value had dropped to 9%. Such turnover *in situ* contrasts with that obtained for roots collected from soil and used in carbon loss studies. Nyhan (1975), for example, observed only 26 to 37% of the carbon in roots harvested from blue grama was lost during field burial for 412 days.

In brief, the gramineous root system contains a spectrum of root materials, and turnover times determined for the materials *en masse* will vary according to the component mix. Generally, measurements of undisturbed roots following pulse labelling with $^{14}C$ are inclusive of the transitory portion of the total biomass and show shorter turnover times than do measurements based on additions to soil of harvested root material. Likewise, measurements of nitrogen immobilization and mineralization based only on harvestable roots can be misleading, especially if the harvest fails to include the juvenile, nitrogen-rich roots. Collections of harvestable, mature roots may contain quickly mineralizable nitrogen that escapes detection simply because mineral nitrogen measurement is made only after several weeks of incubation but not initially or after only a day or two of incubation. Power (1968) recognized this phenomenon in the course of a study on the mineralization of nitrogen in grass roots. He noted that the rate of mineralization of nitrogen was greatest at time zero; he interpreted this to mean that the root material contained considerable quantities of amides which were readily hydrolysed during analysis. Similarly, F. E. Clark & M. Conrad (unpublished data) observed that harvested roots of blue grama given soil burial showed net mineralization of nitrogen during the first 2 days of incubation, but then showed net immobilization for weeks thereafter. The quick release of nitrogen from gramineous roots during the growing season is quite likely one of the mechanisms by which nitrogen is rapidly recycled into new plant growth.

It is this recycling that is designated in Table 8.4 as flow $f_2$. It involves exuded or exfoliated nitrogen that is quickly mineralized. It bears no relation to the negative uptake or return flow from roots to soil that Reuss & Cole (1973) introduced into their simulation model. They observed that, with high soil nitrate concentration, their equation for uptake fed nitrogen into the plant roots until an improbably high root nitrogen concentration was obtained. Accordingly, they employed a computer strategy by which a return flow to the roots was triggered when the nitrate content of the roots approached a certain maximum. In this manner, a plethora of nitrogen in the roots was avoided.

Clark, Warren & O'Deen (1975) in their study of uptake of $^{15}N$ immediately following pulse fertilization with $K^{15}NO_3$ observed that $^{15}N$ did not continue to accumulate in the roots but moved rapidly to the green herbage. When nitrate is absorbed by roots, there is little or no intermingling of the newly absorbed nitrogen with the standing crop of root nitrogen but, rather, a flow-through of the nitrate to the green leaves where it encounters nitrate reductase. When ammonia nitrogen is absorbed, it is quickly built into amino acids and these move in part into newly forming roots and in part to crowns and green herbage. Thus, the total nitrogen in live roots can be viewed as transient mineral or amino nitrogen *en route* to new growth and as structural nitrogen in the existing roots. There is little or no interchange within the roots of these two forms of nitrogen.

### *Nitrogen flows* ($f_{3,4,5,6}$) *to and from green herbage*

The economic importance of the quality and quantity of the foliage in grassland has led to intensive study of the flow ($f_3$) of nitrogen to the aboveground plant parts. Yield responses to fertilizer nitrogen are extensively documented (Heady, 1975; Stoddart, Smith & Box, 1975; Valentine, 1971). For nitrogen applications not greatly in excess of the seasonal growth requirement, approximately 60% of the applied nitrogen is recoverable in the green herbage in the year of application (Woldendorp *et al.*, 1966).

In considering the internal flows of nitrogen in grassland, our concern is not so much with the first-year flow to green herbage as it is with the occurrence of the pulse nitrogen in green herbage in the several years following the initial pulse application. For the years 1971–5, the percentages of $^{15}N$ in the total nitrogen of green herbage were 58, 10, 4, 3 and 2%, respectively (Clark, 1977). At first glance, this drop from 58 to 2% over a time span of 5 years suggests a rather thorough intermingling of the applied nitrogen with the nitrogen already in the system. The percentages can be viewed differently. The initially applied $^{15}N$ constituted only 0.4% of the total nitrogen in the experimental system. A value greater than 0.4% for $^{15}N$ or, stated otherwise, a ratio lower than 250 for the total N:$^{15}N$ in any subdivision of the experimental system represents enrichment of $^{15}N$ in that subdivision. A ratio below

Table 8.5. *Total N: $^{15}N$ ratios of plant and soil compartments as determined in June of the years 1971 through 1975*

| Compartment | 1971 | 1972 | 1973 | 1974 | 1975 |
|---|---|---|---|---|---|
| Green herbage | 2 | 10 | 25 | 34 | 47 |
| Aboveground dead | 17 | 9 | 12 | 18 | 32 |
| Crowns and live roots | 22 | 21 | 26 | 22 | 28 |
| Root litter | 145 | 108 | 103 | 75 | 67 |
| Plant total | 33 | 34 | 48 | 45 | 53 |
| Soil total | 606 | 720 | 478 | 528 | 444 |
| Plant–soil system | 234 | 271 | 248 | 270 | 229 |

From Clark (1977).

250 in one or more compartments requires a ratio above 250 in one or more compartments. A summary of the total N: $^{15}N$ ratios in soil and plant compartments over a time span of 5 years is presented in Table 8.5. The ratio of 2 for green herbage after the initial pulse treatment in 1971 is a reflection of the uptake of readily available nitrogen added as fertilizer. If the rate of $^{15}N$ dilution in the green herbage (Table 8.5) is extrapolated linearly, a ratio of 250, representative of thorough intermingling of the applied nitrogen, would not be reached in the green herbage until the twenty-first year following fertilization.

In the discussion above concerning the residence of nitrogen in live roots, it was noted that the nitrogen entering green herbage in the years following fertilization is either nitrogen mineralized from organic compounds or nitrogen that has overwintered in crowns and roots prior to translocation to green tops. Whether the nitrogen moving to green herbage in a given year is newly mineralized or plant-stored is not easily determinable. The total quantity of nitrogen in plant materials aboveground, both live and dead, commonly decreases between early or mid-season and late season. In the study by Clark (1977), the quantity of aboveground $^{15}N$ in October of the years 1971 through 1975 was 54, 78, 65, 46 and 91% of that present in June of those years. Is this intraseasonal decrease, averaging 33%, attributable to late-season translocation of herbage nitrogen to belowground tissues (flow $f_4$), or is it due to other mechanisms? Inasmuch as total $^{15}N$ recovery was constant over the five years, the cumulative decrease of $^{15}N$ cannot be ascribed to loss of nitrogen from the system by such mechanisms as volatilization, wind transport of plant parts, or grazing by small herbivores. Leaching or throughfall (flow $f_5$), however, could move nitrogen from aboveground to belowground compartments.

Good evidence that such movement is substantial is lacking. The organic throughfall of vegetation is known to consist principally of sugars and fatty acids (Long, Sweet & Tukey, 1956; Carlisle, Brown & White, 1966; Malcolm

& McCracken, 1968). Leachates of gramineous leaves show negligible amounts of nitrogen in any form and only a very small amount of leachable organics; accordingly, translocation appears more plausible than throughfall as the mechanism for transport of nitrogen from aboveground to belowground in the latter part of the growing season. Wallace, Romney, Cha & Soufi (1974) cite a number of studies showing migration of nitrogen and other nutrients from leaves of trees and shrubs prior to leaf abscission. They themselves noted that for desert shrubs, often as much as 50 to 80 % of the nitrogen in the leaves appeared to be retranslocated prior to leaf abscission. Recently, Cole, Turner & Bledsoe (1977) studied annual uptake of nitrogen in red alder and Douglas fir trees in order to determine the annual requirement for growth and also the percentage of that nitrogen that came from the soil. For both trees, the annual contribution from the soil was 54 %; internal retranslocation was held responsible for the remaining 46 %. Simulations of nitrogen flow in grassland heretofore have commonly shown litter and soil organic matter as the sources of available nitrogen for new plant growth; they have largely ignored the role of translocation. If as much as 33 to 50 % of the plant nitrogen of one year can be recycled internally for use in the following year, the process of mineralization is correspondingly relieved of that much responsibility in meeting the annual growth requirement for nitrogen.

Nitrogen in live herbage not consumed by herbivores and not translocated to younger leaves or to below ground moves by mass transfer to the aboveground dead compartment upon death of the live herbage. This flow ($f_6$) to standing dead and litter is passive or mechanistic and not strictly a physiological flow, although it results from senescence and death. In grassland, death of green herbage is most routinely caused either by phenological senescence or by killing frosts or drought.

### *Nitrogen flows* ($f_{7,8}$) *from the aboveground dead compartment*

Again, our concern is primarily with the time of residence of pulse nitrogen in the litter and the availability of litter nitrogen for re-use by new plant growth. Ratios of $^{15}N$ enrichment in aboveground dead material for five successive seasons are shown in Table 8.5; total $^{15}N$ contents are shown by the histograms in Fig. 8.2. These indicate that $^{15}N$ peaks in the aboveground dead in the second year. There were intervening measurements not shown in Fig. 8.2. In October of 1971 and March of 1972 the $^{15}N$ content in the aboveground dead was 2.3 and 3.7 times that shown for June 1971, or 0.8 and 1.3 times that shown for June 1972. After June of 1972 there was an essentially linear decline in the quantity of $^{15}N$ in the aboveground dead material. The rate of loss thus indicated does not reflect the true rate of loss of $^{15}N$ from the initial litter, inasmuch as in each following year a new increment of $^{15}N$ is added in that year's green herbage. The yearly loss of $^{15}N$ from the aboveground dead compartment can be approximated from the data in Table 8.2. There were 340 mg of $^{15}N$ in this compartment in March of 1972, all of which was contributed by herbage of the 1971 growing season. There could be no admixture of $^{15}N$ herbage from 1970 or earlier inasmuch as those years pre-dated the $^{15}N$ addition. In October of 1972 the aboveground dead compartment contained 288 mg. As much as 50 to 67 % of the 240 mg of $^{15}N$ in green herbage in June of 1972 could be admixed in the 288 mg of $^{15}N$ measured in October of that year. This would leave only 108 to 168 mg as the residual from the 340 mg present in March. By such calculations, half or more of the $^{15}N$ present in the aboveground dead compartment disappeared in 188 days, that is, in slightly more than one growing season. Rapid disappearance can also be shown by a different calculation. For the years 1971 through 1975, the cumulative total of $^{15}N$ in green herbage in June of each year was 886 mg of $^{15}N$ for transfer to the aboveground dead compartment in the course of the five growing seasons. At the end of that time, in October 1975, only 100 mg of $^{15}N$ still remained in the aboveground dead compartment, a cumulative compartmental loss of 491 mg of $^{15}N$ over the five growing seasons showing a relatively quick turnover of nitrogen in the aboveground dead compartment.

There are two principal routes by which nitrogen may move from aboveground litter. One ($f_7$) concerns the quick return of litter nitrogen to the available pool in soil. This flow may involve either a simple throughfall or a more complex throughfall in which decomposition aboveground releases mineral nitrogen. This then diffuses or is leached into the soil. Included in this throughfall is the possibility of immobilization of nitrogen in micro-organisms engaged in litter decomposition followed either by quick release of such microbial nitrogen above ground or else by release below ground following water or other transport of microbial cells to below ground.

Leaching or throughfall can move appreciable quantities of inorganic nutrients (such as potassium, calcium, magnesium and manganese) from plants and surface litter (Tukey, 1970), but there are no data showing that any appreciable quantity of organic nitrogen is washed directly into soil from standing vegetation or surface litter (Clark & Paul, 1970). Particulate organic matter of aboveground origin is known to become mixed into soil. Hunt (1977), in a simulation model of decomposition in grassland, assumed that 5 % of the surface litter undergoes mass transfer into the soil annually. That rate of transfer is inadequate to account for the $^{15}N$ that, according to the different calculations given above, disappeared from the aboveground dead compartment in 1972 or cumulatively during the 5 years. Postulation of a transfer rate appreciably higher than that used by Hunt does not appear warranted. If in 1972 as much as 173 to 232 mg $^{15}N \cdot m^{-2}$ is lost from the aboveground dead compartment by particle transfer to below ground, there should have been significant accrual of $^{15}N$ in the detrital root compartment. Such accrual did not occur (Fig. 8.2).

There remains the possibility that bulk transfer of plant fragments to soil occurs and that such transfer is followed almost immediately by mineralization of the nitrogen in those fragments. If true, remnants of the aboveground herbage should be numerous in the detrital root compartment. Microscopical examination of the plant fragments in the detrital root compartment showed that, for soil depths greater than 10 cm, over 99 % of the fragments were of root origin and in the upper 10 cm over 90 %.

It can be hypothesized that decomposition takes place above ground and that the mineralized nitrogen is then leached into the soil. At first glance, this hypothesis does not appear especially plausible. Nevertheless, there appears substantial support for it in the literature. Nyhan (1975) reported that for ground-up $^{14}C$-labelled blue grama herbage buried in the top 2.5 cm of soil at two amendment levels, 54 to 57 % of the initially added carbon was lost in 412 days. Uresk, Sims & Jameson (1975) concluded that litter of blue grama decomposed during the growing season at a rate of 0.35 % per day. This indicates a half-life of 143 days, or just about one growing season. In Nyhan's study, day 1 of the 412 days was 11 February; therefore, hardly more than one growing season was involved. Observations that half the carbon in blue grama litter can be mineralized during one growing season lend support to the $^{15}N$ data showing that half the nitrogen in the aboveground dead compartment undergoes similarly early release.

This does not imply that all the litter nitrogen is rapidly mineralized. Plant litter is a chemically complex substrate containing compounds that are both easy and difficult to decompose; its decomposition is typically curvilinear, slowing with time. Hunt (1977) has treated all litter as consisting of 'soft' and 'hard' materials, each with a constant decay rate $k$, with $k_s$ equal to 30 $k_h$. Paul (1970) and Shields & Paul (1973) estimated separately the turnover time of labile and resistant fractions in litter. For immature oat leaves, the half-life of an easily decomposable fraction was 24 days; of a second and moderately resistant fraction, 325 days; and of a third fraction, 802 days. Extrapolation to field conditions indicated that the two more resistant fractions would have a half-life of approximately 10 years under the semi-arid, cool climate of Saskatchewan. The resistant materials in grass litter buried in the field showed a half-life of 96 growing season months, the equivalent of about 20 years. Such data emphasize that not all components of gramineous litter are short-lived in the field. However, insofar as nitrogen release from a given year's litter is concerned, it is the early and not the late release that is principally involved in providing nitrogen for re-use by the plant. From a given year's litter, contributions of nitrogen of a magnitude sufficient to supply more than a low percentage of the annual growth requirement would occur only during the following 1 or 2 years, that is, within the half-life of the litter.

### *Nitrogen flows* ($f_{9, 10, 12}$) *to and from root litter*

Flows to and from the root litter compartment are difficult to discuss with any degree of satisfaction. The compartment itself is multifaceted, including even some accrual from above ground, and its biomass is not easily measured. There is no routine procedure for rapidly separating live and dead roots. Direct observation, especially that using accessory optics, enables accurate characterization of individual roots or root fragments, as does also the use of vital staining techniques. Clark (1977) undertook routine compartmentalization of all the recoverable plant materials below ground other than plant crowns. Microscopic examination revealed that the materials recovered were largely roots or root fragments; therefore, the belowground biomass was classified as either 'live, senescent, or detrital' roots. These differed qualitatively (Table 8.6).

The nitrogen flux in the root biomass is commonly based on compartmentalization into live and dead compartments. The ensuing discussion treats both the senescent and detrital roots as root litter. The slight differences between senescent and detrital roots (Table 8.6) and their similar pattern of $^{15}N$ accrual strongly suggest that the senescent category represents an earlier stage of decomposition than does the detrital.

It is obvious from the histograms in Fig. 8.2 that the 5-year span of the experiment was of insufficient duration to show the pulse of $^{15}N$ into and from the root litter compartment. Each year the total amount of $^{15}N$ therein became increasingly larger; the yearly increments were 57, 58, 102 and 186 mg – a cumulative increase of 403 mg during the years 1972 to 1975. Whether or not the amount of $^{15}N$ in the root litter in 1975 was the peak amount that would have been attained remains unknown. At first glance the data suggest that if the accrual of $^{15}N$ in the root litter is due solely to death of live roots and bulk transfer of their nitrogen to the litter compartment, the turnover time of the live root compartment would be no more than 1 year and that of the detrital roots, judging from the continuing accrual over the time span of the experiment, many years. However, it is possible that the $^{15}N$ in detrital roots may accrue partly from the death of live roots and partly by some other mechanism.

In June of the fifth growing season, 62% of the $^{15}N$ in plant materials was in belowground litter. This compartment is unquestionably the major plant sink for $^{15}N$ during the 5 years following its pulse application. Accumulation of $^{15}N$ below ground is at the expense of the $^{15}N$ inventory above ground. Table 8.7 shows the distribution of $^{15}N$ in above- and belowground plant materials and soil in 1971 and 1975. From the difference column it is obvious that the major portion of the $^{15}N$ in belowground litter in 1975 must have been derived from nitrogen formerly in aboveground materials.

If the $^{15}N$ of aboveground herbage is that which moves below ground in the

Table 8.6. *Compositional dissimilarities of three categories of roots*

| Root category | Total N (%) | C/N ratio | Bound amino acids (mg·g$^{-1}$)[a] | Soluble sugars (mg·g$^{-1}$ C) |
|---|---|---|---|---|
| Live | 0.89 | 40.5 | 17.2 | 15.8 |
| Senescent | 1.90 | 20.8 | 25.7 | 5.4 |
| Detrital | 2.56 | 17.2 | 27.3 | 3.3 |

[a] Tyrosine equivalents per gram dry weight (Clark, 1977).

Table 8.7. *Distribution (%) of $^{15}N$ in plant compartments and soil in June 1971 and June 1975*

| | June 1971 | June 1975 | Difference |
|---|---|---|---|
| Aboveground plant material | 35 | 8 | −27 |
| Crowns and live roots | 20 | 13 | −7 |
| Senescent and detrital roots | 10 | 34 | +24 |
| Soil | 35 | 45 | +10 |

years 1972 to 1975, there are two possible routes for its transfer. One possibility is that the $^{15}$N mineralized annually from aboveground materials moves into the soil and then into live roots. Upon death of the live roots their $^{15}$N becomes that of dead roots. This type of transfer may be partly responsible but it can hardly be wholly responsible. Given a standing crop of 90 mg of $^{15}$N in live roots and an annual accrual in dead roots of 100 mg, even a complete turnover annually of the live root compartment would not be sufficient to account for the continuing accrual of $^{15}$N in the root litter compartment. The second possibility is that the dead roots themselves accrue $^{15}$N following its mineralization from organic materials either above or below ground. Clark & Campion (1976) reported that live, senescent and detrital roots of blue grama have mean nitrogen contents of 0.89, 1.90 and 2.56%, respectively. Table 8.6 shows that the C/N ratio in the live roots exceeds 40. In the early stages of decomposition of residues having C/N ratios greater than 25, and providing there is available nitrogen in the immediate environment, there is net inflow of nitrogen to the residues. Consequently, not only does the percentage of nitrogen in the residue increase but so too does the total nitrogen content. Strictly speaking, the inflow of nitrogen is not to the residue itself but to secondary productivity in the residue. During residue decomposition, micro-organisms become established intercellularly and intracellularly, and their nitrogen is included in the determination of the residue nitrogen.

This brings up a point that thus far has been ignored – namely, that the nitrogen associated with plant litter is not wholly plant nitrogen. It is partly microbial nitrogen. The detrital root compartment must be viewed as two compartments, the plant fragments themselves and the microbial tissues intimately associated with them.

The microbial compartment associated with the aboveground green herbage is trivial when compared to the green herbage biomass. Clark & Paul (1970) estimated the phyllosphere population as 0.001 $g \cdot m^{-2}$ in blue grama grassland in which the green herbage biomass exceeded 200 $g \cdot m^{-2}$. Likewise, the rhizosphere population is only a very minor fraction of the total live root biomass. If, however, only the transitory litter on which the rhizosphere population subsists, namely the root exudates and exfoliates, are considered, then the microbial population does approach a biomass value of one to several per cent that of the root exudates and exfoliates. In litter derived from leaves, stems, crowns and roots, and in which resistant components exist, there are successive colonizations by diverse micro-organisms. The microbial tissues, living and dead, are intimately intermingled with the plant tissue remnants. In the $^{15}N$ study, no attempt was made to measure separately the microbial nitrogen and litter nitrogen. Consequently, the $^{15}N$ recovered in the detrital root compartment does not necessarily represent structural nitrogen formerly in the live root compartment. It may represent nitrogen mineralized from aboveground or belowground litter and subsequently immobilized in microbial cells. The data in Table 8.7 suggest that $^{15}N$ released from aboveground litter has somehow entered the detrital compartment. Whether the transfer was one of mineralization followed by root uptake and subsequent death remains unknown.

McGill, Paul & Sorensen (1974), in an inventory of the nitrogen in Saskatchewan grassland, reported that the nitrogen in micro-organisms to a soil depth of 30 cm was 13 $g \cdot m^{-2}$; the nitrogen content of all the above- and belowground plant material was 23 g. There were 540 g $N \cdot m^{-2}$ to a soil depth of 30 cm in the soil organic matter. Although this nitrogen dominated the system, it is known to have an extremely slow turnover rate, whereas the turnover of nitrogen in soil micro-organisms can under favourable conditions be accomplished in hours or days. Although the soil micro-organisms intercept some freshly mineralized nitrogen that otherwise would be subject to immediate re-use by the plant, the quick release of microbially immobilized nitrogen expedites quick return of nitrogen to the plant.

### *Nitrogen flows* ($f_{11,13}$) *to and from humus*

Soil or humic nitrogen is by far the dominant form of nitrogen in the grassland system (see Table 8.2). In much of the older literature, the nitrogen of litter was channelled almost entirely to the humic compartment, from which nitrogen was mineralized and cycled back to new plant growth. In the more current literature, biomass nitrogen is channelled partly to micro-organisms and partly to quickly released mineral nitrogen; the remainder goes to humus. Frissel (1976) estimated that perhaps 40% of the litter nitrogen becomes humic nitrogen and also that perhaps 0.2% of the humic nitrogen is mineralized yearly. Do these estimates appear reasonable for grassland at the Pawnee Site?

According to data of Bokhari & Singh (1975), the annual nitrogen requirement for plant growth in this grassland is roughly 7 $g \cdot m^{-2}$. The flow diagram given by Bokhari & Singh (1975) shows that 1.8 g of this requirement is from humus and 5.2 g from litter. Table 8.2 shows this grassland to contain 372.5 g humic $N \cdot m^{-2}$ to a depth of 36 cm. If an estimated 0.2% annual rate of mineralization is applied to this grassland, then 0.745 g N should be mineralized yearly. If at the same time, 40% of the 7 g of biomass nitrogen produced yearly goes to humus, then the annual input to humus is 2.8 g N. For a grassland in a steady-state equilibrium, 2.8 g input to humus is quite incompatible with 0.74 g output from humus.

Obviously, either the 40% estimate for return of biomass nitrogen to humic nitrogen is high, or else the 0.2% rate of mineralization of humic nitrogen is low. Considering only Bokhari & Singh's 1.8 g output of humic nitrogen and state variables of 372.5 g humic N and 7 g plant growth N, and again assuming a steady-state equilibrium, the rate of mineralization of humic nitrogen would be 0.49% annually, and the amount of plant nitrogen going to humic nitrogen would be 26%. In our opinion, these values are also too high. We believe that 0.2% is a reasonable estimate of humic nitrogen mineralization in shortgrass prairie and that no more than 10% of the plant nitrogen goes to humic nitrogen. Unfortunately, 5 years proved to be too short a time span in which to monitor the flow of plant $^{15}N$ to humus. The data do show that conversion of biomass nitrogen to humic nitrogen occurs very slowly. During the 5 years, the soil $^{15}N$ increased from 35% to 45% of the total accountable $^{15}N$. However, the $^{15}N$ measured as soil nitrogen may be contained partly in soil micro-organisms. Also, it may partly be that present in finely particulate plant material, such as sloughed root hairs.

During the five experimental years, the detrital root $^{15}N$ increased from 10% to 34% (Table 8.7) without showing any evidence of peak accumulation (Fig. 8.2). Whatever the amount of actual accumulation in this compartment, it is highly improbable that the major portion of the detrital nitrogen does eventually go to humic nitrogen. It is much more probable that, during the course of further decomposition, the major portion of the detrital root nitrogen becomes mineral nitrogen.

### *Summary remarks on nitrogen flows in grassland*

In grassland in which available nitrogen is in limited supply, mineral nitrogen added as fertilizer or mineralized from organic matter will, if growing conditions are favourable, be shunted primarily to green herbage. Green herbage escaping herbivory generally dies within the year of its production. Consequently, nitrogen entering green herbage during a growing season persists in that herbage only during that season. In perennial grassland, nitrogen moves from live herbage to standing dead and litter upon death of the live herbage, to litter and the soil surface in the excreta of herbivores, and to live plant parts belowground by retranslocation. In ungrazed grassland, as much as one-third of the nitrogen in green herbage at the height of the growing season undergoes internal retranslocation to belowground parts during the latter part of the growing season. Simulations of nitrogen flow in grassland usually ignore the role of retranslocation in meeting the annual growth requirement for nitrogen, and consequently, erroneous rates are used for the mineralization of nitrogen from litter and soil organic matter.

Half or more of the nitrogen that enters the aboveground dead compartment from the green herbage of one season has moved from that compartment by the end of the next following season. The outgoing nitrogen moves below ground, either passively as organic nitrogen in litter materials that become buried in soil, or as mineral nitrogen released during decomposition of litter above ground. Tracer nitrogen experiments show that the nitrogen transfer involves both mineral and organic nitrogen, inasmuch as the litter $^{15}N$ disappears from above ground more rapidly than $^{15}N$ accrues in particulate detrital materials below ground.

The dynamics of root nitrogen remain poorly defined for several reasons. The nitrogen flows of live and dead roots have not been monitored separately or, if so, inadequately. Too much reliance has been placed on standing crop measurements of plant biomass and nitrogen and on $^{14}C$ flows as indicators of nitrogen flows. Even such $^{15}N$ data as are available are difficult to interpret. Often they are for a component mix that obscures the individual nitrogen flows of the components. The live root compartment of blue grama was observed to retain a constant amount of tracer nitrogen over a time span of five growing seasons. This does not mean that the individual roots are all extremely long-lived nor that all $^{15}N$ taken in during the first growing season persists in the same live roots for the four following seasons. Some mature suberized roots do retain their viability and at least some $^{15}N$ for several years. To the extent that root death occurs, there is depletion of the $^{15}N$ content of the live root compartment. Such depletion is balanced by the re-entry of freshly mineralized nitrogen into the persisting or into newly formed live roots.

The continuing accrual of $^{15}N$ in senescent and detrital roots over five growing seasons suggests that, additionally to the passive transfer of $^{15}N$

accompanying root death, there is also immobilization of $^{15}N$ by the microbial component of root litter. A portion of the microbially immobilized nitrogen is quickly mineralized after microbial cell death and again becomes available for re-use by the plant. A second portion of the microbial nitrogen enters heteropoly-condensates, the dark-coloured, amorphous, highly polymerized materials commonly referred to as humus, humates or humic acids. These are sufficiently resistant to decomposition that their mean residence time in soil is several hundred or even several thousand years. In blue grama grassland ungrazed by herbivores, the feedback of nitrogen from living and dead plant parts, exuded soluble organics and dead micro-organisms supplies the major portion of the annual growth requirement for nitrogen. Tracer nitrogen added to soil and immobilized in plant materials during the first growing season is either retained in or quickly recycled to the plant in the next several growing seasons and only slowly is it transferred from plant materials to the soil humus. Its release from humus is far slower than its release from primary or secondary productivity.

In brief, the annual nitrogen requirement in ungrazed blue grama grassland is met by a combination of four nitrogen-supplying mechanisms: (i) internal translocation whereby nitrogen of one season is stored over winter in belowground plant parts and then moved to new growth in the next growing season; (ii) mineralization of easily decomposable organic nitrogen compounds – among these certain herbage components, root exudates and exfoliates, and short-lived unsuberized roots; (iii) mineralization of organic nitrogen synthesized by micro-organisms subsisting on energy-rich materials, such as those mentioned in (ii) alone; and (iv) not all of the microbially synthesized organic nitrogen undergoes quick release to the available nitrogen pool; a portion of it undergoes polymerization and becomes humic nitrogen, from which there is slow feedback to the available nitrogen pool.

Discussions of the nitrogen cycle usually emphasize the movement of plant nitrogen into litter and then into the soil humus prior to its release for re-entry into the plant. Tracer nitrogen data (Clark, 1977) strongly suggest that internal translocation and quick mineralization of at least some of the organic nitrogen of plant and microbial origin are important mechanisms in shunting nitrogen back to the plant. Once a nitrogen atom makes its initial entry into the grass plant, there is greatly increased probability that the atom will again enter new herbage growth in each of several following years. This is tantamount to Odum's (1968) statement that ecosystems tend to conserve essential nutrients and that mechanisms evolve that promote quick recycling within the system.

## Phosphorus flow

The phosphorus cycle in most grasslands is closed, with no significant gains or losses even in areas of heavy precipitation. In this respect it differs significantly from the nitrogen cycle, which is open both to atmospheric fluxes and losses through leaching. Although losses of phosphorus due to leaching through the soil profile are absent or minimal, appreciable losses may occur through soil erosion following fires or other catastrophic events in which natural vegetation is disturbed. Phosphorus is held in soil largely in very slightly soluble minerals and in stable organic forms which are not biologically active.

### *Soil phosphorus forms and properties in grassland soils*

The phosphorus status of grassland soils of the United States Great Plains was examined by Haas, Grunes & Reichman (1961) by analysis of samples from dryland field stations with long-term management records. Total phosphorus, total nitrogen, organic phosphorus and labile inorganic phosphorus ($P_i$) estimates for the surface 15 cm soil layer are shown in Table 8.8. Total phosphorus generally increased from south to north and organic phosphorus content of the northern soils averaged nearly twice as high as that of the southern soils. Bicarbonate-extractable phosphorus levels provide estimates of labile inorganic phosphorus available for microbial and plant uptake. These values were highly correlated with total phosphorus in the southern soils, but showed no consistent relationship in the northern soils.

Location and description of nine additional North American grassland sites for which detailed information on phosphorus levels in soil and vegetation are available are presented in Table 8.9. Classification of the site soils has been discussed by Woodmansee (1978). The sites are widely separated geographically and the soils are formed from diverse parent materials. They are classified within the Mollisol, Entisol and Aridisol Orders. Soil textures range from loamy sand to silty clay and available moisture storage to 100 cm ranges from 45 to 200 mm. Annual precipitation ranges from 235 to 953 mm and growing season precipitation from 89 mm at the Pantex to 628 mm at the Osage Site.

Detailed measurements on the distribution of phosphorus in the various soil layers at the nine sites are shown in Table 8.10. Total phosphorus levels of these soils vary over tenfold from 94 ppm at the Jornada to 1234 ppm at the Bridger Site. Organic phosphorus averages slightly over half of the total phosphorus, but ranges widely from a high of 90 % at the Osage Site to a low of 4 % at the ALE Site. The low organic phosphorus content at the ALE site reflects the very rapid mineralization rates in soils with high temperatures during the summer months.

Resin-extractable phosphorus levels are generally considered reliable esti-

Table 8.8. *Nitrogen and phosphorus content of some United States Great Plains soils from 32° N (Big Springs, Texas) to 48° N (Havre, Montana)*

| Location | Total N (ppm) | Total P (ppm) | Organic P (ppm) | Labile $P_i$ (ppm) |
|---|---|---|---|---|
| Big Springs, Texas | 635 | 280 | 55 | 5 |
| Dalhart, Texas | 678 | 428 | 84 | 13 |
| Lawton, Oklahoma | 1550 | 262 | 128 | 4 |
| Hays, Kansas | 1810 | 570 | 174 | 6 |
| Colby, Kansas | 1530 | 622 | 158 | 28 |
| Akron, Colorado | 1340 | 539 | 115 | 18 |
| Archer, Wyoming | 1200 | 381 | 142 | 8 |
| Sheridan, Wyoming | 1470 | 586 | 120 | 7 |
| Newell, South Dakota | 1660 | 794 | 128 | 13 |
| Mandan, North Dakota | 1480 | 492 | 139 | 4 |
| Dickinson, North Dakota | 3160 | 928 | 292 | 6 |
| Mocassin, Montana | 3100 | 731 | 308 | 9 |
| Havre, Montana | 1510 | 701 | 157 | 6 |

All values given are for the 0–15 cm soil layer.

mates of the amounts of labile inorganic phosphorus. The wide range in values obtained illustrates the diversity in phosphorus behaviour in soil. In grasslands the levels of extractable organic phosphorus often exceed the inorganic phosphorus levels – in contrast to the behaviour of cultivated soils. The complexity of phosphorus solubility is shown in the last column of Table 8.10. The changes in values of the various parameters with soil depth indicates significant differences in the nature of the phosphates present in the surface layers compared to subsurface layers of the same soil. For the same amount of total phosphorus, higher proportions are extractable from the surface soils by resin or bicarbonate, and phosphorus solubilities are significantly higher at the surface in all cases.

Of the six parameters documented in Table 8.10, no one by itself is adequate to provide more than a very general estimate of the phosphorus fertility of a given grassland soil. Phosphorus solubilities must be interpreted with respect to the capacity for renewal from the labile inorganic phosphorus pools, for example. Mineralization of labile organic phosphorus also contributes significantly to the amounts of phosphorus available for plant uptake in the course of a growing season. Seasonal fluctuations may be substantial so that time of sampling should also be evaluated.

Seasonal changes in pool sizes of various forms of phosphorus at the Matador site (Halm, Stewart & Halsted, 1972) demonstrated that substantial mineralization of organic phosphate occurred. Changes in bicarbonate-extractable phosphorus from 2.3 to 5.3 ppm were accompanied by decreases in extractable organic phosphorus from 57 to 12 ppm. The balance of the organic phosphorus mineralized was accounted for by increases in the

Table 8.9. *Location and description of nine North American IBP Grassland sites*

| Site[a] | Longitude (°W) | Latitude (°N) | Soil order | Soil suborder | Texture | Clay content (%) | Normal annual temp. (°C) | Potential growing season | Normal annual precip. (mm) | 1970 Precip. in growing season (mm) | Soil water storage (mm) | Total N (%) | Total organic C (%) |
|---|---|---|---|---|---|---|---|---|---|---|---|---|---|
| ALE | 119° 33′ | 46° 24′ | Mollisol | Xeroll | Silt loam | 5 | — | — | — | — | 120 | 0.04 | — |
| Bison | 114° 16′ | 47° 19′ | Mollisol | Boroll | Silt loam | — | 7.8 | 28 Apr– 8 Oct. | 379 | 307 | 120 | — | — |
| Bridger | 110° 47′ | 45° 47′ | Mollisol | Boroll | Silt loam | 18 | — | 10 June– 10 Sept. | — | — | 120 | 0.03 | 5.1 |
| Cottonwood | 101° 52′ | 43° 57′ | Entisol | Orthent | Silty clay loam | 30 | 8.5 | 22 Apr.– 9 Oct. | 360 | 265 | 180 | 0.21 | 2.4 |
| Jornada | 106° 51′ | 32° 36′ | Aridisol | Orthid | Loamy sand | 6 | 14.9 | 22 Feb.– 15 Nov. | 235 | 132 | 45 | — | — |
| Osage | 96° 33′ | 36° 57′ | Mollisol | Ustoll | Silty clay | 29 | 15.0 | 30 Mar.– 3 Nov. | 953 | 628 | 180 | 0.20 | 3.2 |
| Pantex | 101° 32′ | 35° 18′ | Mollisol | Ustoll | Silty clay loam | 46 | 15.0 | 2 Apr.– 23 Dec. | 528 | 89 | 200 | — | 2.2 |
| Pawnee | 104° 46′ | 40° 49′ | Aridisol | Argid | Sandy loam | 18 | 8.4 | 22 Apr.– 9 Oct. | 324 | 255 | 110 | 0.10 | 0.8 |
| Matador | 107° 43′ | 50° 42′ | Mollisol | Boroll | Clay | 71 | 3.6 | 165 days | 388 | 251 | — | 0.30 | — |

[a] The Matador Site is in Canada; all others are in the USA.

Table 8.10. *Total, organic, and extractable phosphorus levels in soils of nine North American grasslands*

| Site | Depth (cm) | Total P (μg P·g⁻¹) | Organic P | Resin P | Bicarbonate-extractable Inorganic P, $P_i$ | Bicarbonate-extractable Organic P, $P_o$ | Water-soluble $P_i$ (ppm) |
|---|---|---|---|---|---|---|---|
| Cottonwood | 0–10 | 554 | 310 | 50.3 | 7.60 | 13.44 | 0.068 |
| | 10–20 | 472 | 281 | 39.7 | 2.40 | 12.95 | 0.024 |
| | 20–30 | 474 | 250 | 40.8 | 1.67 | 9.05 | 0.021 |
| | 30–40 | 474 | 232 | 41.5 | 1.18 | 7.52 | 0.041 |
| | 40–50 | 481 | 203 | 6.3 | 1.07 | 6.25 | 0.012 |
| | 50–60 | 485 | 153 | 3.6 | 0.95 | 5.69 | 0.009 |
| Bridger | 0–10 | 1234 | 675 | 37.3 | 19.77 | 76.41 | 0.219 |
| | 10–20 | 1048 | 599 | 36.2 | 7.34 | 46.67 | — |
| | 20–30 | 1048 | 500 | 37.9 | 11.61 | 41.22 | 0.049 |
| | 30–40 | 922 | 311 | 48.1 | 12.36 | 27.12 | 0.079 |
| | 40–50 | 922 | 255 | 48.3 | 16.67 | 20.13 | 0.048 |
| Osage | 0–10 | 251 | 227 | 2.6 | 3.65 | 14.03 | 0.070 |
| | 10–20 | 216 | 198 | 1.1 | 2.02 | 13.17 | 0.035 |
| | 20–30 | 187 | 175 | 0.5 | 1.07 | 12.07 | 0.016 |
| | 30–40 | 166 | 156 | 0.3 | 0.71 | 10.90 | 0.006 |
| | 40–50 | 143 | 130 | 0.2 | 0.59 | 9.25 | — |
| | 50–60 | 114 | 98 | 0.2 | 0.12 | 7.60 | — |
| ALE | 0–10 | 748 | 29 | 258.7 | 8.70 | 6.98 | 0.261 |
| | 10–20 | 736 | 26 | 270.1 | 7.32 | 3.84 | 0.154 |
| | 20–40 | 822 | 81 | 5.0 | 1.42 | 3.25 | 0.018 |
| | 30–60 | 867 | 71 | 43.2 | 2.89 | 3.66 | 0.099 |
| | 60–100 | 834 | 74 | 5.8 | 1.42 | 2.73 | 0.015 |
| Jornada | 0–5 | 94 | 37 | 1.6 | 5.43 | 2.71 | 0.348 |
| | 5–10 | 73 | 29 | 0.8 | 2.65 | 2.54 | 0.141 |
| | 10–20 | 74 | 42 | 0.4 | 0.71 | 3.18 | 0.080 |
| | 20–30 | 78 | 49 | 0.3 | 0.36 | 2.91 | 0.031 |
| | 30–40 | 82 | 53 | 0.2 | 0.24 | 2.90 | 0.026 |
| Pantex | 0–5 | 455 | 239 | 51.0 | 21.23 | 11.58 | 0.569 |
| | 5–10 | 375 | 256 | 26.8 | 5.97 | 12.39 | 0.070 |
| | 10–20 | 328 | 230 | 13.5 | 2.65 | 11.73 | 0.034 |
| | 20–30 | 251 | 140 | 12.8 | 1.67 | 7.03 | 0.018 |
| | 30–40 | 236 | 118 | 16.1 | 1.18 | 5.20 | 0.037 |
| Bison | 0–5 | 840 | 570 | 19.9 | 25.25 | 54.26 | 0.701 |
| | 5–10 | 829 | 595 | 10.1 | 17.00 | 65.84 | 0.469 |
| | 10–20 | 892 | 681 | 8.5 | 14.38 | 53.83 | 0.378 |
| | 20–30 | 989 | 673 | 10.5 | 15.35 | 41.85 | 0.285 |
| | 30–40 | 849 | 653 | 14.1 | 14.55 | 38.88 | 0.230 |
| | 40–50 | 644 | 482 | 15.6 | 12.66 | 32.19 | 0.270 |
| | 50–60 | 549 | 418 | 9.0 | 9.41 | 22.27 | 0.238 |
| | 60–70 | 407 | 308 | 7.1 | 10.13 | 18.16 | 0.370 |
| | 70–80 | 210 | 131 | 12.6 | 3.39 | 6.53 | 0.412 |
| | 80–90 | 202 | 99 | 2.8 | 1.67 | 4.47 | 0.245 |
| | 90–110 | 212 | 54 | 10.1 | 1.67 | 3.64 | 0.221 |
| Pawnee | 0–5 | 351 | 128 | 44.8 | 24.70 | 8.60 | 1.390 |
| | 5–10 | 338 | 134 | 37.3 | 18.50 | 9.40 | 1.070 |
| | 10–15 | 326 | 156 | 23.3 | 11.60 | 7.30 | 0.532 |
| | 15–20 | 285 | 142 | 9.2 | 3.70 | 8.90 | 0.179 |
| Matador | 0–15 | 593 | 332 | 4.5 | 3.80 | 57.00 | 0.003 |
| | 15–30 | 435 | 108 | 1.2 | 1.00 | 12.00 | 0.001 |

Table 8.11. *Total nitrogen and phosphorus content of plant biomass and soils to 30 cm at the peak of the growing season in* Bouteloua gracilis *dominated plots at the Pawnee Site*

| | P content ($g P \cdot m^{-2}$) | N content ($g N \cdot m^{-2}$) | Total system P (%) | Total system N (%) |
|---|---|---|---|---|
| Live tops | 0.28 | 1.34 | 0.3 | 0.3 |
| Dead tops | 0.20 | 2.13 | 0.2 | 0.5 |
| Crowns | 0.39 | 4.73 | 0.4 | 1.1 |
| Live roots | 0.55 | 9.62 | 0.6 | 2.3 |
| Root litter | 1.08 | 23.31 | 1.1 | 5.6 |
| *Total* | 2.50 | 41.13 | 2.5 | 9.8 |
| Soil organic | 34.2 | 378 | 34.0 | 90.2 |
| Soil mineral | 63.4 | — | 63.4 | — |

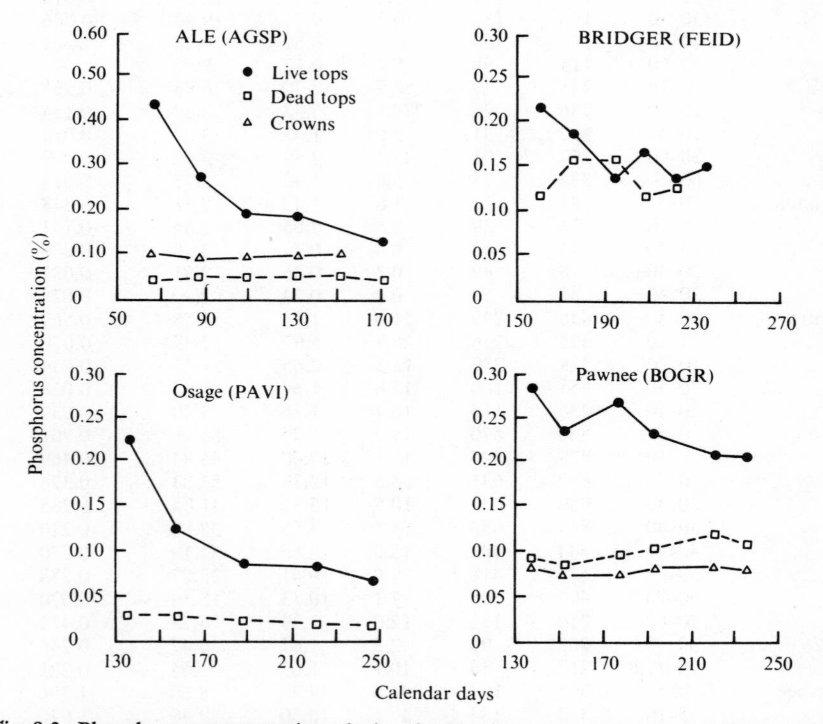


Fig. 8.3. Phosphorus concentrations during the growing season in important grasses at four US/IBP Grassland Biome study sites. Species are coded AGSP for *Agropyron spicatum*. FEID for *Festuca idahoensis*, PAVI for *Panicum virgatum*, and BOGR for *Bouteloua gracilis*,

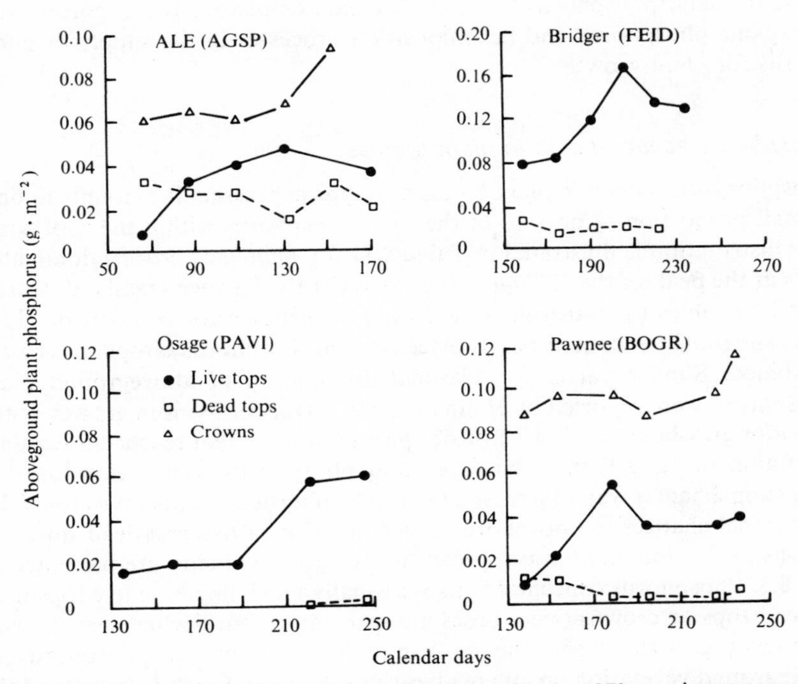


Fig. 8.4. Dynamics of aboveground plant phosphorus for the 1973 growing season at the four grassland sites for the four species noted in Fig. 8.3.

Table 8.12. *Seasonal distribution of biomass and aboveground phosphorus contents at the Matador Site in 1970*

| Date | Live tops | | | Dead tops | | | Total above ground | | |
|---|---|---|---|---|---|---|---|---|---|
| | Biomass ($g \cdot m^{-2}$) | % | Plant P ($g \cdot m^{-2}$) | Biomass ($g \cdot m^{-2}$) | % | Plant P ($g \cdot m^{-2}$) | Biomass ($g \cdot m^{-2}$) | % | Plant P ($g \cdot m^{-2}$) |
| | | | | Control | | | | | |
| 7 May | 30 | 0.30 | 0.09 | 160 | 0.07 | 0.11 | 190 | 0.10 | 0.20 |
| 7 July | 110 | 0.18 | 0.20 | 160 | 0.07 | 0.11 | 270 | 0.12 | 0.31 |
| 10 Aug. | 115 | 0.14 | 0.16 | 215 | 0.08 | 0.17 | 340 | 0.10 | 0.33 |
| 8 Sept. | 100 | 0.10 | 0.10 | 250 | 0.07 | 0.18 | 350 | 0.08 | 0.28 |
| | | | | N+P+water added previous year[a] | | | | | |
| 7 May | 120 | 0.29 | 0.35 | 455 | 0.12 | 0.55 | 575 | 0.16 | 0.89 |
| 7 July | 410 | 0.17 | 0.70 | 440 | 0.11 | 0.48 | 850 | 0.14 | 1.18 |
| 10 Aug. | 300 | 0.12 | 0.36 | 400 | 0.09 | 0.36 | 700 | 0.10 | 0.72 |
| 8 Sept. | 250 | 0.11 | 0.28 | 475 | 0.09 | 0.43 | 725 | 0.10 | 0.70 |

From Halm *et al.* (1972).
[a] 17.9 g $N \cdot m^{-2}$, 2.5 g $P \cdot m^{-2}$ and 0.8 bar suction.

calcium-bound phosphorus fraction. The data emphasize the important role of organic phosphorus and decomposition processes in the supply of phosphorus for plant growth.

### *Phosphorus accumulation in plant biomass*

Phosphorus contained in plant biomass in typical grasslands amounts to only a small proportion (2 or 3%) of the total phosphorus within the root zone. This distribution is illustrated in Table 8.11 for *Bouteloua gracilis* dominated plots at the peak of the 1973 growing season at the Pawnee grassland. Corresponding values for nitrogen contents are also shown for comparison. Live roots and root litter contain nearly twice as much phosphorus as tops and crowns combined. Similar values for seasonal distribution of aboveground plant phosphorus were reported by Halm *et al.* (1972) for cool-season grasses at the Matador grassland (Table 8.12). Root phosphorus content reached a seasonal maximum of 1.5 g $P \cdot m^{-2}$. Nitrogen and phosphorus fertilization together with supplemental water increased plant phosphorus contents up to fourfold.

Seasonal changes in phosphorus concentrations of live and dead tops and crowns of the dominant grass species at four grassland sites are presented in Fig. 8.3. Phosphorus concentrations are usually much higher in live tops than in dead tops or crowns (stem bases and rhizomes), but decline sharply with advancing growth stage. The corresponding amounts of phosphorus in aboveground vegetation on an area basis are shown in Fig. 8.4. In spite of the lower concentrations there is more aboveground phosphorus in crowns than in either live or dead tops in *Agropyron spicatum* or *Bouteloua gracilis* in the ALE and Pawnee sites respectively. Seasonal trends in phosphorus concentrations of grasses at these sites are in good agreement with the phosphorus concentrations in a number of legume, grass and sedge species at 300 sites as reported by Boawn & Allmaras (1974). Generally grasses ranging from 0.15 to 0.30% phosphorus during immature to full bloom dropped to 0.03 to 0.09 during dormant stages. Legume species and mixed forages containing legumes had significantly higher phosphorus concentrations than did grasses alone.

### *Phosphorus solubilization and diffusion in soil*

The phosphorus nutrition of plants is primarily from the very small amount of phosphorus in the solution immediately surrounding the root. Replenishment of this supply is dependent chiefly upon diffusion processes within the root zone. As the concentration of phosphorus at the root surface is reduced by plant uptake, a concentration gradient is established in the surrounding soil water films and phosphate ions flow towards the root. The extent of zones depleted in soluble phosphorus depends on the supply of solid-phase phosphorus in slightly soluble phosphate minerals and phosphate adsorbing surfaces. Phosphorus flow in the soil–plant system is thus a dynamic process with root exploration in the soil, uptake of phosphorus, and a complex pattern by which phosphorus in solution is renewed from the solid phase by diffusion and mass flow.

Chemical forms and solubility relationships of phosphorus in soil have been thoroughly reviewed by Larsen (1967) and Olsen & Flowerday (1971). Phosphorus solubility is complicated by common ion effects, ion association and pH effects, so that information about these properties of soil solutions is also necessary for interpretation of phosphorus measurements. Phosphorus solubility increases with the amounts of phosphorus adsorbed in the monolayer region on the surfaces of soil minerals. Differences in phosphorus solubility between soils of varying texture are explained by variations in surface area and adsorption capacity of the soils (Cole & Olsen, 1959). Solution phosphorus equilibrates rapidly with the labile fraction of the adsorbed phosphorus. The extent of the dependency of the rate constant for phosphorus flow from the labile to the solution phase is discussed by Olsen & Watanabe (1966). The ratio of soil labile inorganic phosphorus concentration to that of phosphorus in solution is linear over a range of phosphorus levels for most soils. The phosphorus solubility of the Ascalon soil at the Pawnee Site, for example, was highly correlated ($r = 0.84$) with labile phosphorus over a tenfold range.

A theoretical background for the evaluation of phosphorus supplying capacity and soil textural effects on phosphorus diffusion is given by Olsen & Watanabe (1963). The porous system diffusion constant, $D_p$, is calculated from the bulk solution diffusion constant by the following equation:

$$D_p = D_0 \alpha (L/L_e)^2 \theta,$$

where $D_0$ is the diffusion coefficient of phosphorus in bulk solution in $cm^2 \cdot s^{-1}$, $\alpha$ is a dimensionless coefficient accounting for diffusivity reduction due to ionic interaction and increased viscosity of water near mineral surfaces, $(L/L_e)^2$ is the tortuosity factor (dimensionless), and $\theta$ is the volumetric water content. The porous diffusion coefficient $D_p$ is experimentally evaluated by either transient or steady-state diffusion experiments (Olsen, Kemper & Van Schaik, 1965) and ranges from $0.4 \times 10^{-7}$ to $15 \times 10^{-7}$ $cm^{-2} \cdot s^{-1}$. On soils where the diffusion coefficients have not been directly determined they are estimated from the clay content (Olsen & Watanabe, 1963). Phosphorus uptake rates are restricted by lack of adequate soil moisture. Olsen, Watanabe & Danielson (1961) determined that uptake by corn roots in four experimental soils was a linear function of $\theta$ ($r = 0.91$ to $0.94$) between the limits of $\frac{1}{3}$ bar to 15 bars soil water suction. No appreciable phosphorus uptake was found at soil water suctions greater than 15 bars. This was believed due to the fact that values of $(L/L_e)^2$ extrapolate to zero at finite values of $\theta$, indicating that the water films break up.

Table 8.13. *Recovery of applied tracer phosphorus in plant and soil compartments at the Pawnee Site*

| $^{32}P$ application depth (cm) | % of applied tracer P recovered in: Tops | Roots | Plant | Soil |
|---|---|---|---|---|
| 2.5 | 0.76 | 2.42 | 3.18 | 96.8 |
| 5.0 | 0.78 | 4.16 | 4.94 | 95.1 |
| 10.0 | 1.04 | 5.08 | 6.12 | 93.9 |
| 15.0 | 0.63 | 4.55 | 5.18 | 94.8 |

Table 8.14. *Pool sizes of labile and exchangeable phosphorus following addition of tracer phosphorus to field soil*

| Soil layer depth (cm) | Exchangeable P ($\mu g \cdot g^{-1}$) Initial | Final | Increase ($g\ P \cdot m^{-2}$) |
|---|---|---|---|
| 0– 2.5 | 26.3 | 54.8 | 0.87 |
| 2.5– 7.6 | 15.7 | 50.0 | 2.09 |
| 7.6–12.7 | 9.5 | 27.5 | 1.10 |
| 12.7–17.8 | 4.1 | 22.0 | 1.09 |
| *Total* | | | 5.15 |

Isotopic tracer studies were conducted at the Pawnee site in northeastern Colorado to determine the fate of applied inorganic phosphorus. $^{32}P$ was added to four depths of the soil profile and after 30 days the distribution of $^{32}P$ in plant tops, roots and soil was determined (Table 8.13). Roughly 95% of the applied tracer was retained in the soil compartment, while the amount in the plant tops was roughly 1% or less of the amount applied. Recovery in the roots ranged from 2.4 to 5.1%. The evidence is that the applied phosphorus enters a sizeable pool of soil inorganic phosphorus. The study also enabled the calculation of the total pool of exchangeable phosphorus as shown in Table 8.14. The sizeable increase in isotope exchangeable phosphorus over the 30-day period may be accounted for by mineralization of organic phosphorus, estimated as 5.2 g $P \cdot m^{-2}$ in the top 15 cm layer. Root uptake from the solution pool may involve the daily renewal of this pool by as much as 50 times during periods of high root activity under near-ideal conditions of temperature and water. It is estimated that the solution pool is replenished by phosphorus from the labile pool at an even higher potential rate, up to 250 times per day (Cole *et al.*, 1976).

### Root uptake

The principal driving force for phosphorus uptake by plant roots is the transfer across limiting membranes at the root surface. Metabolic energy is expended to move phosphorus across the plasmalemma into the cytoplasm where the phosphorus concentration may be 50 to 100 times higher than in the surrounding solution. This flow establishes a concentration gradient in the soil solution which drives phosphorus diffusion and desorption processes. Root uptake rates in a well-stirred nutrient solution vary widely depending upon the physiological activity and nutrient status of the plant. Dependence of uptake rate on phosphorus concentration is best described by the operation of two simultaneously operating carrier mechanisms (Hagen & Hopkins, 1955; Hagen, Leggett & Jackson, 1957). An equation of the Michaelis–Menten form for two sites of uptake has been very useful in predicting the rates of phosphorus uptake over a wide range of phosphorus concentrations in solution (Carter & Lathwell, 1967):

$$\text{Uptake rates, } C_{24} = \frac{U_{\max a}}{1+\dfrac{K_{\mathrm{m}a}}{PC}} + \frac{U_{\max b}}{1+\dfrac{K_{\mathrm{m}b}}{PC}},$$

where $PC$ is the phosphorus concentration at the root surface. Values of this function derived from nutrient solution studies, in which the solution is kept well stirred to minimize the constraints of diffusion, are much higher than those observed in soil studies.

Calculations of phosphorus uptake in the field indicate a range of actual uptake rates from 0.5 to 2.0 $\mu$mol $P \cdot g\ root^{-1} \cdot d^{-1}$ ($1.5 \times 10^{-5}$ to $6.2 \times 10^{-5}$ $g\ P \cdot g\ root^{-1} \cdot d^{-1}$) (Sayre, 1948; Quirk, 1967). Growth analysis studies by Loneragan & Asher (1967) and Loneragan (1968) confirm that uptake rates of 1 and 2 $\mu$mol $P \cdot g\ root^{-1} \cdot d^{-1}$ will maintain adequate phosphorus concentration in plants growing at growth rates of 8% and 10% per day, respectively.

The major determinants of root uptake of phosphorus are the demands set up by the growth and normal functioning of plant parts as well as the external supply of phosphorus. Other nutrients affect phosphorus uptake mainly through their effects on plant growth and metabolism. Nitrogen–phosphorus interactions are the best example of this effect. The effects of nitrogen on phosphorus uptake have been observed in many plant species under a wide range of cultural conditions. Cole, Grunes, Porter & Olsen (1963) noted that phosphorus uptake by roots was highly correlated with root nitrogen content. The nitrogen stimulation of rates of phosphorus uptake was believed due to metabolic changes in the plant.

The effects of mycorrhizal fungi on phosphorus uptake have been well documented for a wide variety of plant species (Moss, 1973). Increases in phosphorus uptake by plants bearing vesicular–arbuscular mycorrhizae are commonly believed to hinge on the greater root–soil contact afforded by extensive development of infected roots and their associated hyphae (Sanders & Tinker, 1971). Uptake by the fungi is metabolically mediated similarly to root uptake, but extends to lower concentrations of phosphorus in the soil solution. This suggests that mycorrhizae benefit the phosphorus nutrition of plants mainly when diffusion rates of phosphorus through soil are slow or when there is very low phosphorus concentration in the soil solution. There are no firm data on the importance of mycorrhizae to grasses comparable to those which exist for trees. Clark (1978) has recently emphasized the paucity of information on the role of endomycorrhizae in the nutrition of grasses and forbs. What is probably far more important in phosphorus uptake than the presence or absence of mycorrhizae is the physiological vigour or growth stage of the individual roots. Cole *et al.* (1977) stated that juvenile and nonsuberized roots, accounting for roughly one-third of the total root biomass, are probably the only roots active in nutrient uptake. The abiotic factors of temperature and water are major determinants of phosphorus uptake by their effects both on plant growth and metabolism and on processes of solubilization and diffusion of soil phosphorus (Carter & Lathwell, 1967; Olsen *et al.* 1961).

### *Plant translocation*

The dynamics of phosphorus transport within plants are to a large extent controlled by the activities of meristematic tissues within newly developing organs. This results in a complex pattern of flow of nutrients into developing leaves in early stages of growth and export at later stages under the stress of demand from later developing tissues. Active recycling of phosphorus has been dramatically demonstrated using radioactive phosphorus (Koontz & Biddulph, 1957; Biddulph, Biddulph, Cory & Koontz, 1958; Tanaka, 1961; Sosebee & Wiebe, 1973). Tanaka's studies show in remarkable detail the patterns of translocation of carbohydrates and all the major nutrients throughout a developing rice plant.

Cole *et al.* (1977) concluded that the major operational constraint on phosphorus translocation was that a minimum level of phosphorus must be maintained within the exporting tissue. Thus, no translocation of phosphorus from roots below a minimum level of approximately 0.05 % phosphorus can be expected for grasses. Further, phosphorus translocation from roots is considered proportional to root phosphorus concentration between this lower limit and an upper limit (approximately 0.5 % phosphorus) above which increases in root phosphorus concentration will not further increase translocation.

The phenological index of a given plant species is the best guide available to indicate plant translocation responses to abiotic factors. There are rapid flows during juvenile stages and minimal flow when no new development is occurring. Thus, differing periods of maximum translocation can be expected in cool- and warm-season grasses. During senescence and death of plant parts, phosphorus not conserved by translocation to more active tissues is transferred passively to litter.

### *Mineralization of organic phosphorus*

The transformation of organic phosphates from plant, animal and decomposer residues into inorganic phosphate is essential to complete the cycle of phosphorus utilization in natural systems. A large proportion of the phosphorus in these residues is in the form of phosphorus esters in a wide variety of chemical forms. The stability of these compounds to enzymatic hydrolysis varies widely; the resistant compounds, especially the inositol phosphates, accumulate with long residence time in soils. Large amounts of phosphorus are contained in organic residues with $^{14}C$ residence times ranging from 350 to 2000 years (Halm, 1972). Phosphorus present as nucleic acids and phospholipids may be readily hydrolysable and hence presumably readily available to plants, whereas that present as inositol phosphorus is not.

Thompson, Black & Zoellner (1954) studied mineralization of carbon, nitrogen and phosphorus in a group of US soils developed over a wide range of vegetative and climatic conditions. Mineralization under field conditions was evaluated by comparing losses due to cultivation at paired cultivated and virgin sites for each soil. These results were compared to results of laboratory incubation over a 3-month period. The net losses of organic carbon, nitrogen and phosphorus from cultivated soils averaged 33 %, 32 % and 24 %, respectively, indicating more rapid removal of carbon and nitrogen relative to organic phosphorus under field conditions. However, relative rates of phosphorus mineralization under laboratory incubation exceeded relative rates of carbon and nitrogen mineralization. Organic phosphorus mineralization was highly correlated with organic nitrogen and carbon in the incubation experiments. The average ratio of nitrogen to phosphorus mineralized was 7.6:1 and the ratio of carbon to phosphorus mineralized was 80:1. Based on this relationship and estimates of carbon turnover, annual phosphorus mineralization rates of 3.0 and 6.4 $g\ P \cdot m^{-2}$ would be predicted at the Pawnee and Matador grassland sites, respectively (Cole *et al.*, 1977).

Organic phosphorus levels in the Olney fine sandy loam at the Pawnee Site increase from 128 ppm at the surface to a maximum of 156 ppm at 10–15 cm depth, amounting to 36 % and 48 % of the total phosphorus at these layers, respectively (Table 8.10). The Sceptre clay at the Matador Site contains a maximum organic phosphorus level of 332 ppm in the surface 15 cm (56 % of total phosphorus) which decreases to 108 in the 15–30 cm layer (25 % of total phosphorus).

Soil organic phosphorus at the Matador was differentiated into stable compounds resistant to weak acid hydrolysis and labile compounds easily hydro-

Table 8.15. *Phosphatase activities and bicarbonate-soluble organic and inorganic phosphorus levels in soils from eight North American grassland sites*

| Grassland site | Phosphatase activity ($\mu$mol nitrophenol · g soil$^{-1}$ · h$^{-1}$) | Organic P ($\mu$g · g$^{-1}$) | Inorganic P ($\mu$g · g$^{-1}$) |
|---|---|---|---|
| ALE | 7 | 7 | 9 |
| Bridger | 17 | 76 | 20 |
| Cottonwood | 12 | 13 | 8 |
| Jornada | 1 | 3 | 5 |
| Osage | 7 | 14 | 4 |
| Pantex | 8 | 12 | 21 |
| Pawnee | 3 | 9 | 15 |
| Matador | 3 | 21 | 4 |

lysed by weak acids and mostly soluble in 0.5 M sodium bicarbonate (Stewart, Halm & Cole, 1973). Seasonal increases in aluminium-bound and calcium-bound inorganic phosphorus fractions corresponded to large decreases in the soluble organic phosphorus. In these soils soluble organic phosphorus was 4 to 20 times the amount of inorganic phosphorus extracted by the same solution. The increase in organic phosphorus coincided with the period of maximum growth, highest microbial activity, and lower amounts of phosphorus associated with the litter. Significant seasonal variations in organic phosphorus levels were also observed by Dormaar (1972). He found a large build-up of organic phosphorus during the winter months. These results confirm the dynamic nature of at least a portion of the organic phosphorus reserves in soils.

Stewart *et al.* (1973) hypothesized that rapid turnover of organic phosphorus was the result of higher phosphatase activities associated with increased root activity and microbial population. Bacterial numbers and phosphatase levels were very highly correlated in a field incubation experiment at their site. Phosphatase enzymes are adsorbed and stabilized in soils, and their activity is a good indication of the past biological activity and the phosphorus mineralization potential of soils.

Phosphatase activities in grassland soils vary widely, as shown by the comparison of values for eight North American sites (Table 8.15). Plant roots are an important source of phosphatases and appear to adapt to low levels of inorganic phosphorus supply by the extracellular production of increased amounts of these important enzymes. The influence of 16 grasses and forbs on soil phosphatase activity was investigated by Neal (1973). The growth of dominant, codominant and increaser species did not significantly alter the phosphatase activity already established in the soil. However, the presence of species of plants classed as 'invaders' (because they often become the dominant species on overgrazed sites) significantly increased soil phosphatase activities. This suggests that the relative ability to mineralize organic phosphorus may be a factor to be considered in plant successional relationships.

Large amounts of phosphorus are required by decomposer micro-organisms during the decomposition of plant residues. Chang (1940) noted that the ratio of cellulose carbon decomposed to organic phosphorus formed varied from 64 to 115 depending upon conditions of incubation. Hannapel, Fuller, Bosma & Bullock (1964) studied the synthesis of organic phosphorus by $^{32}P$- and $^{14}C$-labelling experiments and found that a significant proportion of the phosphorus redistributed through the soil was of microbial origin. For every part of organic phosphorus of plant origin there were three parts of organic phosphorus derived from microbial synthesis of soil inorganic phosphorus. The uptake of phosphorus by micro-organisms is the single largest annual phosphorus flow in the ecosystem biota.

Flow rates for phosphorus in micro-organisms are calculated from estimated seasonal microbial populations, turnover rates and the phosphorus composition of representative microbial species. Johannes (1965) has indicated that the bacteria themselves release little phosphorus, for which they have a high demand, whereas bacterial grazers are important in the regeneration of bacterial phosphorus. Much of the organic matter in grassland consists of litter too poor in essential nutrients to sustain bacterial growth without uptake of nutrients from the soil. Although regeneration of nutrients from relatively nutrient-poor litter does occur, there also occurs a much faster nutrient cycling involving bacterial grazers. This microbial nutrient cycling controls decomposition processes and the availability of nutrients to plants.

Information on phosphorus budgets at the Matador and Pawnee Sites together with estimates of major flows inferred from seasonal dynamics of soil phosphorus forms, vegetation flows and estimates of microbial activity have been used to develop a simulation model of phosphorus cycling in grassland (Cole *et al.*, 1977). The state variables and the interconnecting flows for the model are shown in Fig. 8.5. The general hypotheses adopted in the formulation of model flows were:

(1) Organic phosphorus transformations play a major role in phosphorus cycling in grasslands and the micro-organisms therein play the dominant role in the cycling of organic phosphorus.

(2) Long-term effects of environmental stresses (temperature, water, nitrogen, grazing) on phosphorus cycling are mainly through effects on decomposition of plant and microbial residues.

Fig. 8.6 presents results of a simulation of phosphorus flows to plants and micro-organisms at the Pawnee Site during 1972. Cumulative seasonal decomposer uptake was five times the corresponding uptake by plant roots. Mineralization of organic phosphorus produced inputs that nearly balanced the total uptake so that levels of labile inorganic phosphorus at the end of the season were only slightly changed from the initial value.

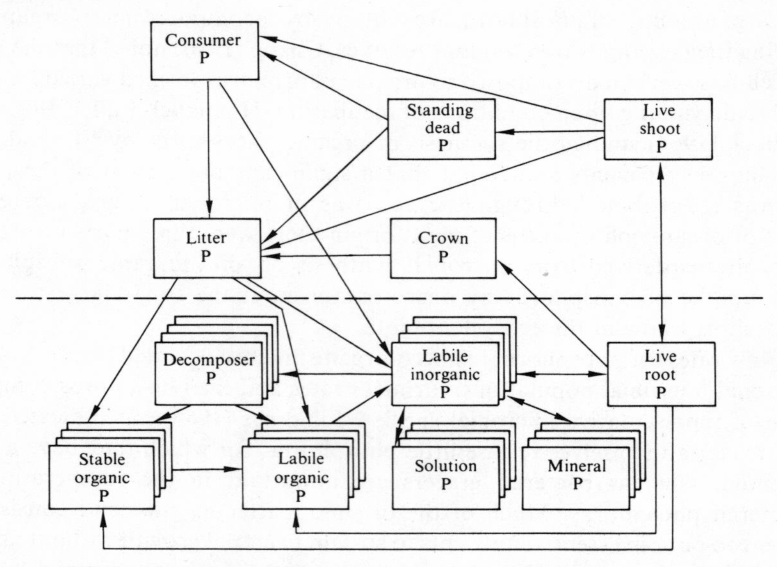


**Fig. 8.5. Diagram of a model of phosphorus cycling in grasslands showing aboveground compartments and belowground compartments at four soil depths. (After Cole *et al.*, 1977.)**

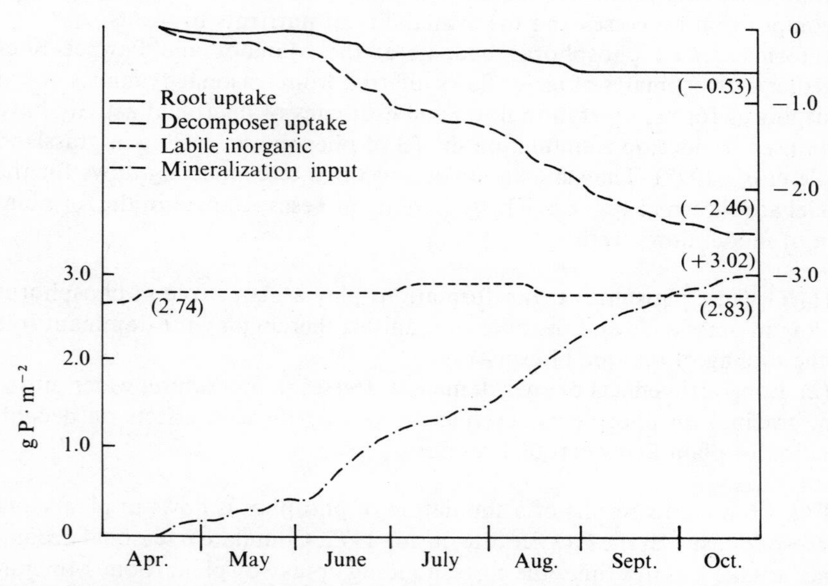


Fig. 8.6. Simulated seasonal changes in labile inorganic phosphorus (0 to 20 cm) and cumulative loss to root uptake and decomposer uptake and gain from mineralization of organic phosphorus for the 1972 season at the Pawnee Site.

## Sulphur

Sulphur ranks a close third to nitrogen and phosphorus as an essential nutrient in grassland. Many of the studies on sulphur cycling have been linked to studies on nitrogen cycling. Both elements, besides being integral parts of protein, are usually absorbed by plants as anions and both have major reserves of potentially mineralizable compounds in the soil organic matter. Sulphur, like nitrogen, is subject to atmospheric inputs into the soil, to groundwater leaching, and to a microbial reduction of the nutrient in the available form to a gaseous form. Not surprisingly, therefore, many of the conceptual cycles for sulphur in grasslands (Walker, 1957; Jordan & Ensminger, 1958; Till & May, 1970*a*; Dawson, 1974) bear striking resemblance to those for nitrogen.

### *Sulphur content of grassland soils and vegetation*

Sulphur content of grassland soils usually ranges between 0.01 and 0.05% with an average of about 0.03%. Soils in areas of heavy rainfall and sandy soils low in organic matter usually contain very little sulphur. In wet tropical areas sulphur often leaches to the subsoil where sulphate is adsorbed by iron and aluminium oxides; in arid and semi-arid areas sulphates are apt to accumulate in the form of gypsum. Grassland areas in Australia, New Zealand, South America and tropical Africa all have exhibited sulphur deficiency. Some of the grasslands in non-industrialized areas of Europe have also shown deficiency. In the United States, sulphur-deficient areas occur in northern California and in parts of Washington and Oregon. In recent years, sulphur deficiencies have been magnified by the use of sulphur-free fertilizers and by the substitution of natural gas for high-sulphur coals in industry.

Organic sulphur usually constitutes the bulk of the available sulphur reserve in grassland. Freney (1967) found this fraction to account for over 90% of the total sulphur in Australian soils (Fig. 8.7). Data of Freney show about 40 $g \cdot m^{-2}$ of labile organic sulphur, about 5 $g \cdot m^{-2}$ of inorganic sulphate sulphur and about 30 $g \cdot m^{-2}$ of resistant aromatic-type sulphur. Fractions of soil organic sulphur considered labile include the readily mineralizable sulphate esters and the large quantity of sulphate sulphur associated with the organic fraction. Little is known concerning the humic sulphur in grassland, but the soil humus undoubtedly contains the sulphur-bearing amino acids (cystine, cysteine, methionine) derived from proteins of microbial cells and plant tissues. Other organic sulphur associated with the humus include ethereal sulphates, thiourea, glucosides and alkaloids.

Bettany, Stewart & Halstead (1973) reported on carbon, nitrogen and sulphur relationships in grassland and non-grassland soils. Even though total sulphur and organic nitrogen and carbon varied considerably, the mean C:S

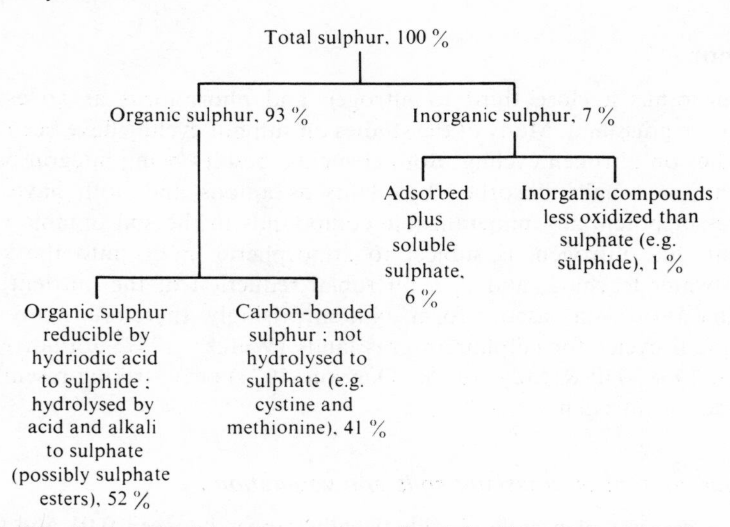


Fig. 8.7. Distribution of sulphur in some Australian soils. (After Freney, 1967.)

Table 8.16. *Concentration (%) of certain nutrients in some commonly grown forages in the hay stage*

| Forage | N | P | K | Ca | Mg | S |
|---|---|---|---|---|---|---|
| Bermuda grass | 1.42 | 0.20 | 1.47 | 0.46 | 0.17 | 0.20 |
| Kentucky bluegrass | 1.92 | 0.26 | 1.72 | 0.40 | 0.21 | 0.30 |
| Smooth bromegrass | 1.97 | 0.28 | 2.36 | 0.43 | 0.21 | — |
| Orchard grass | 1.55 | 0.37 | 2.10 | 0.45 | 0.32 | 0.28 |
| Ryegrass | 1.54 | 0.32 | 1.96 | 0.49 | 0.34 | 0.22 |
| Wheatgrass | 1.23 | 0.20 | 2.68 | 0.33 | 0.22 | — |
| Alfalfa | 2.77 | 0.26 | 1.77 | 1.64 | 0.32 | 0.35 |
| Red clover | 2.38 | 0.22 | 1.76 | 1.61 | 0.45 | 0.25 |
| Ladino clover | 3.70 | 0.40 | 2.17 | 1.38 | 0.50 | 0.35 |
| Lespedeza (Korean) | 2.35 | 0.22 | 1.04 | 1.00 | 0.31 | 0.20 |
| White clover | 3.52 | 0.39 | 2.03 | 1.24 | 0.47 | 0.35 |

After Wagner & Jones (1968).

and N:S ratios showed similar and consistent trends in 54 different soils. These workers postulated that the carbon-bonded sulphur is more likely to be incorporated into the strongly aromatic humus core by mechanisms such as quinonethiol reactions, whereas the HI-reducible inorganic and organic sulphates are more likely to be associated with active sidechain components. Neptune, Tabatabai & Hanway (1975) found that of the percentage of the total sulphur content that was carbon-bonded sulphur varied from 5 to 12 % in Brazilian soils and 7 to 18 % in Iowan soils. Generally speaking, C:N:S ratios approximate 140:10:1.3 (Williams, 1962).

Table 8.17. *Compartmental values for sulphur in a hypothetical semi-arid grassland at mid-season and to a soil depth of 20 cm*

| Compartment | Sulphur content ($g \cdot m^{-2}$) |
|---|---|
| Standing vegetation | 0.2 |
| Litter | 0.4 |
| Roots | 2.0 |
| Labile organic in soil | 40.0 |
| Resistant organic in soil | 30.0 |
| Microbial cells | 1.0 |
| Soil sulphate | 5.0 |

The sulphur content of most grasses is approximately equal to their phosphorus content. Wagner & Jones (1968) compiled concentration values of selected nutrients in some commonly grown forages in the hay stage (Table 8.16). These values are controlled by such factors as natural fertility of the soil and the application of water and fertilizers. Jones (1964) noted that a deficiency of sulphur is observed in perennial ryegrass when the sulphate sulphur in the plant tissues falls below 0.032%, a value much lower than the sulphur concentrations listed in Table 8.16. Sulphur content of blue grama at the Pawnee Site ranged from 0.14 to 0.17% during the growing season. Total sulphur in the standing green vegetation ranged from 0.05 $g \cdot m^{-2}$ in May to 0.17 $g.m^{-2}$ in midsummer; in the standing dead, the amounts were 0.08 and 0.03 $g.m^{-2}$, respectively. A compartmental inventory of sulphur in a hypothetical semi-arid grassland at mid-season and for a soil depth of 20 cm is offered in Table 8.17.

### *Sulphur cycling in grassland*

A conceptual diagram of the sulphur cycle in grassland is shown in Fig. 8.8. This diagram is patterned after that constructed by Cole *et al.* (1977) for phosphorus flow (cf. Fig. 8.5 above). As yet, we have not attempted mathematical simulation of the sulphur flows depicted, largely because we hesitate to use data piecemeal from widely separated studies. As an alternative we are summarizing site-specific data compiled by Till, May and associates (May, Till & Downes, 1968; Till & May, 1970*a*, *b*, 1971) on grazed pasture in Australia. They used $^{35}S$ in the field. Bettany *et al.* (1974) also used $^{35}S$ to study the sulphur cycle and to assess the available sulphur pool in the soil. Like $^{15}N$ and $^{32}P$, $^{35}S$ lends itself nicely to study of the integrated processes of uptake, utilization, return to the soil and re-entry into the available soil pool. As the radioactive anion is taken up, the $^{35}S$ mixes (isotopic dilution) with the rest of the cycling sulphur pool, thereby resulting in changes in the specific activity of the pool. Monitoring of this activity against time gives good

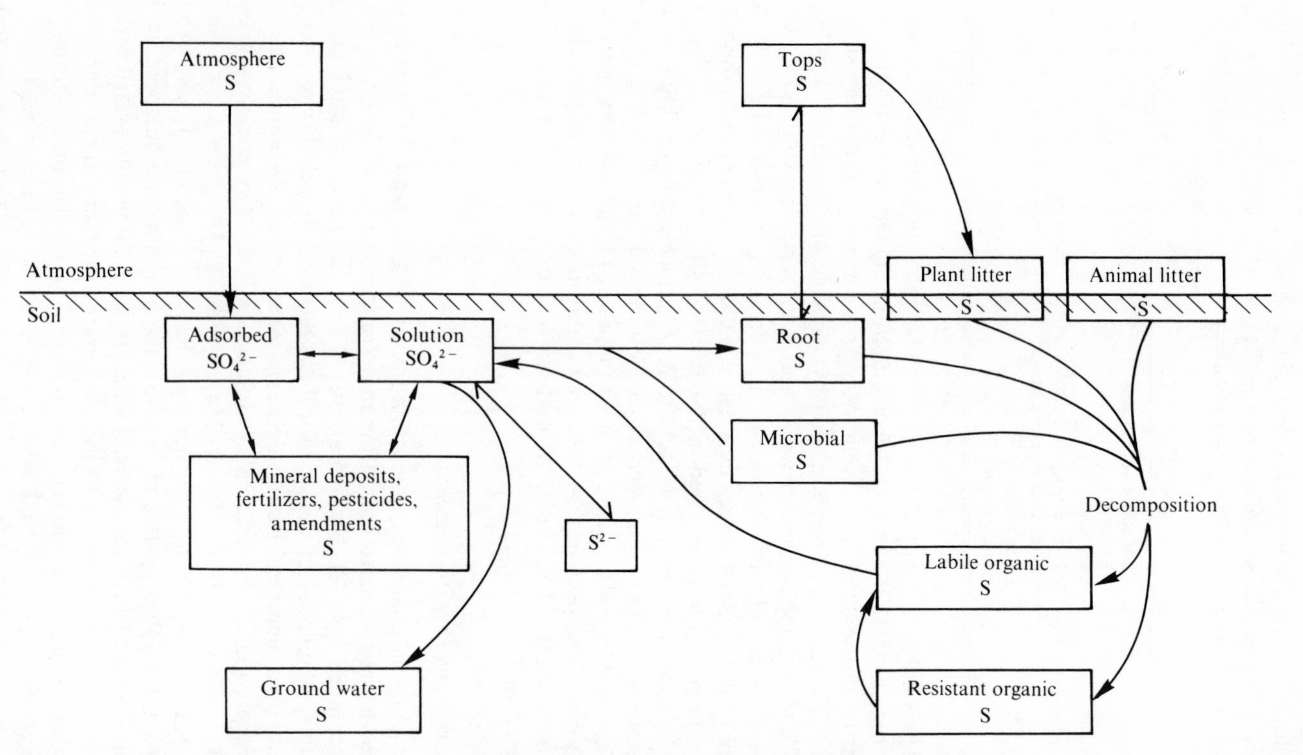


Fig. 8.8. Sulphur cycle in grassland soils.

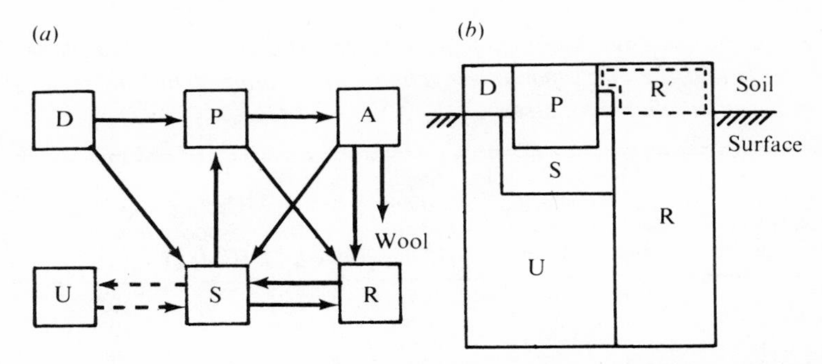


Fig. 8.9. Diagram of state variables and flows of sulphur in grazed grassland as drawn by Till & May (1970*a*). For definition of pools in (*a*) see Table 8.18. Approximate relative pool sizes are shown in (*b*).

Table 8.18. *Definitions of pools and relationships among them at equilibrium*

| Symbol | Definition | Relationships between pools at equilibrium |
|---|---|---|
| D | Fertilizer sulphur | |
| P | Plant sulphur | |
| A | Animal sulphur (in sheep) | |
| R | Organic sulphur (in dead plant matter, organisms, excreta, etc.) | |
| R′ | That portion of R not included in soil samples (e.g. faeces and plant litter) | |
| S | Soil sulphur (in forms suitable for immediate utilization by plants) | |
| C | Cycling sulphur (total cycling pool) | |
| B | Total soil sulphur | |
| G | Total sulphur in the system | |
| U | Sulphur that apparently does not enter the cycle (apparently inert) | |
| | | $D = 0$ |
| | | $C = P+A+R+S$ |
| | | $B = S+U+R-R'$ |
| | | $G = S+U+R+P+A$ |
| | | $\quad = B+R'+P+A$ |
| | | $U = G-C$ |

indication of the rates of transport or cycling and the sizes of the various compartmental pools.

May *et al.* (1968) used $^{35}S$ in the form of gypsum to measure the availability and turnover rates of fertilizer nutrients in grazed pastures. Experimental plots were 0.3 to 0.4 ha in size and contained mainly tall fescue (*Festuca elatior*) and white clover (*Trifolium repens*) with smaller amounts of perennial ryegrass (*Lolium perenne*), subterranean clover (*T. subterraneum*) and phalaris (*Phalaris tuberosa*). Three non-pregnant Merino ewes were used to graze each experimental plot.

Table 8.19. *Average wool sulphur and specific activity values over various periods and the corresponding cycling sulphur, apparently inert sulphur, and total soil sulphur pool sizes*

| Days after initial application of [$^{35}$S]gypsum[a] | Sheep group | Ave wool specific activity ($nCi \cdot g^{-1}$) | Ave S content of wool (%) | Specific activity of S ($\mu Ci \cdot g^{-1}$) | Cycling S pool (C) (kg) | S in top 25 cm of soil (B) (kg) | Apparently inert S pool (U) (kg) |
|---|---|---|---|---|---|---|---|
| 526–600[b] | 1 | 56.9 | 3.31 | 1.72 | 55 | 105 | 71 |
| 680–1100 | 4 | 42.7 | 3.32 | 1.29 | 73 | 103 | 51 |
| 1520–1660 | 4+5 | 3.21 | 3.31 | 0.097 | 62 | 100 | 58 |
| 1660–1874 | 4 | 3.03 | 3.51 | 0.086 | 70 | 90 | 38 |
| 1760–1874 | 6 | 2.93 | 3.51 | 0.084 | 71 | 114 | 66 |

[a] The second application of [$^{35}$S]gypsum was 1224 days after the initial application.
[b] Values in this row are from the data of May *et al.* (1968).

The applied sulphur that was not taken up by plants remained in the top 10 cm of soil over the 200 days of measurement. No leaching beyond the root zone occurred; neither was there any significant lateral movement. The experiment embraced a period of below-average rainfall. Specific activity values indicated that at least 70% of the soil sulphur did not exchange with the applied $^{35}$S during the first 200 days; after 600 days a large percentage of the soil sulphur still remained relatively undiluted. Changes in the specific activity of the wool indicated that 100 to 150 days were required for equilibration to occur between the sheep and the vegetation. The dilution specifics of the tracer gypsum indicated that about 47 $g \cdot m^{-2}$ of gypsum sulphur equivalent was normally available in the soil.

Till & May (1970*a*) offered a diagram of the sulphur cycle (Fig. 8.9) and defined pools and the relationships existing between them at equilibrium (Table 8.18). The average wool sulphur and specific activity values over various periods and the corresponding sulphur pool sizes are shown in Table 8.19. After 3 years the applied $^{35}$S was still turning over in the soil–plant–animal system. There was no measurable change in the total soil sulphur, indicating insignificant sulphur loss from the system; also, a large part of the soil sulphur did not exchange in the sulphur cycle. In arriving at these conclusions, Till & May made three assumptions: (*a*) that the specific activity of sulphur in wool is the same as that in plants; (*b*) that all $^{35}$S applied entered the cycle relatively quickly; and (*c*) that $^{35}$S upon entering a pool becomes mixed with sulphur already in that pool. They concluded that the amount of sulphur cycling was about 12 times that of the applied gypsum sulphur, and about five times that incorporated into plant material in 1 year. In further studies, Till & May (1970*b*, 1971) noted that after 260 days the effective specific radioactivity of the cycling sulphur had approached a steady state and that the applied sulphur remained in the available sulphur pool. They concluded that the extractable soil sulphate was a precursor of plant sulphur and that the organic sulphur pool replenished the available sulphur pool.

## Potassium, calcium and magnesium

These are also macronutrients required by plants but they differ from nitrogen, phosphorus and sulphur in that their release from organic matter is not primarily by decomposition but rather by physical weathering processes collectively called leaching. This phenomenon renders nutrients soluble and mobile within plant tissues, particularly in senescent and dying tissues. As defined by Tukey (1970), leaching is the removal of substances from plants by the action of aqueous solutions associated with rain, dew, mist and fog. The phenomenon has long been documented but was not conclusively proven until demonstration that plant-incorporated radioisotopes could be found in leachates (Long *et al.*, 1956).

Not all inorganic nutrients are leached with the same ease. Tukey, Tukey & Wittiver (1958) found that from young leaves, sodium and manganese were readily leached (25% of original content), calcium, magnesium and potassium were moderately easily leached (1 to 10%), and iron, zinc, phosphorus and chlorine were leached only with difficulty (1% or less). Tukey (1970) noted that the main factor controlling the quality and quantity of leachate is the physiological age of the leaf. Young tissues were relatively resistant to leaching of mineral nutrients, while more mature and senescent tissues were highly susceptible. Wallace (1930) reported that loss of potassium from very young leaves during 24 h of leaching was less than 5% of the initial potassium content but could exceed 80% in mature leaves. Calcium, magnesium and sodium behaved in a similar manner. Other factors affecting leaching were the health and vigour of the plants, temperature, and rain intensity and volume.

In terms of the chemistry involved, cations leach from an exchangeable cation pool in the tissue 'free space' and not from within the cells themselves. Cations in the 'free space' exchange with hydrogen ions in the leaching solution or else move directly by diffusion and mass flow when the cuticle is absent. The cations combine with dissolved carbon dioxide and when the carbonate reaches the soil they are again available for plant or microbial uptake. However, depending upon soil conditions and intensity of rainfall, some cations can be leached beyond the root zone. A classical example is the leaching of potassium ions in sandy soils.

### *Soil–plant cycling of potassium*

Simplistically speaking, potassium exists in soil in an available solution pool that is in equilibrium with the potassium exchangeable pool. Plants obtain potassium from the solution pool through direct root interception and pre-

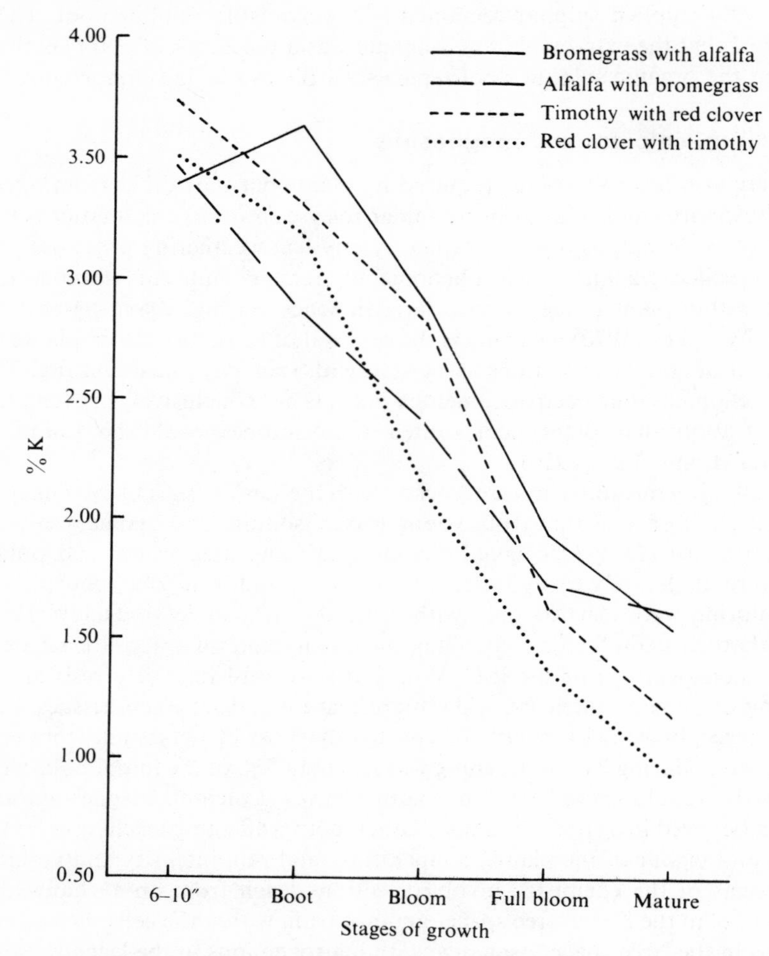


**Fig. 8.10.** Potassium content of two grasses and legumes grown in mixtures at various growth stages. (After Blaser & Kimbrough, 1968.)

dominantly through diffusion and mass flow of the potassium ion to the roots. The potassium is then absorbed actively (i.e. metabolic energy is required) and translocated to new growth. Uptake may even reach luxury limits.

The potassium content of grasses ranks a close second to, or even equals or surpasses, their nitrogen content. This is not surprising, inasmuch as soil potassium often exists in adequate supply and is absorbed with great facility (Thomas & Hipp, 1968). Potassium uptake by forages ranges from 10 to 50 $g \cdot m^{-2}$ (Wagner & Jones, 1968). Sanchez (1973) noted that potassium uptake in a tropical forest ranged from 60 to 100 $g \cdot m^{-2}$. Nelson (1968) reported that potassium uptake exceeded that of nitrogen and phosphorus in maize even after tasselling and silking. Although no attempt was made to measure the total uptake of potassium in shortgrass prairie on the Pawnee Site, total potassium in the aboveground herbage was measured at monthly intervals during the growing season. Total potassium above ground was roughly 2 $g.m^{-2}$ at mid-season. Standing green contained about 35% of this total; standing dead about 15%; and the litter about 50%.

Blaser & Kimbrough (1968) pointed out the difficulties in correlating potassium uptake with potassium dry matter yields. The potassium concentration in plants was highly dependent on growth stage (Fig. 8.10). Herbage potassium content varied from 1 to 4% and at pre-bloom it was 62% higher than at full bloom. de Wit, Dykshoom & Noggle (1963) suggested that 1% was the critical lower level for potassium at the vegetative stage of growth. MacLeod (1965) found the optimum potassium concentration of three grasses grown alone to be above 3%. Stewart & McConagher (1963) observed that the potassium content of ryegrass increased with increasing soil acidity. Dittberner (1973) noted that, in addition to pH and potassium supply, sulphate and nitrate also affected the potassium content of grasses. In legume–grass mixtures, grasses usually have a higher potassium content than do the legumes. With potassium-deficient soils, legumes tend to disappear from legume–grass mixtures (Nelson, 1968).

Although leaching is the dominant factor in returning potassium to the soil, both Sears (1950) and Barrow & Lambourne (1962) pointed out that animal urine is very high in available potassium and emphasized the importance of grazers in the cycling of potassium. Preferential grazing on young green tissues high in potassium content hastens the return of potassium to the soil prior to herbage senescence and the time that the herbage becomes especially susceptible to leaching.

### *Calcium and magnesium*

Calcium and magnesium serve as essential nutrients for plants and they are used agronomically to ameliorate the pH of acidic soils. Lime flocculates the clay fraction, improves soil structure and reduces phosphorus fixation in acid soils. An excess of lime, however, can lower the availability of trace elements (McIlroy, 1972). In Australian grasslands, liming is beneficial to pastures on acid soils (Williams & Andrews, 1970). The increase in pH fosters greater molybdenum availability and better nodulation of legumes. Magnesium was rarely deficient in Australian grasslands. Heady (1975) regarded calcium and magnesium as generally unimportant in American rangeland fertilization. Increases in yield, when obtained, were relatively minor.

Grasses usually contain 0.4 to 1.0% calcium, and 0.08 to 0.3% magnesium.

Table 8.16 shows calcium and magnesium concentrations in several forage species. Critical percentages according to Wit *et al.* (1963) are 0.1% for calcium and 0.06% for magnesium. Mean calcium and magnesium contents during the growing season of live herbage of shortgrass prairie on the Pawnee Site were 0.34 and 0.10%, respectively. Standing dead herbage contained calcium and magnesium in amounts similar to those of the live herbage. In this respect calcium and magnesium differed from potassium, whose content in live and dead herbage was 1.7%, and 0.18%, respectively. Will (1959) found similar results in forest conifers in New Zealand. Over two-thirds of the potassium was leached and calcium was only slightly leached. The meagre data at hand suggest that decomposition is more important than leaching in the recycling of calcium and magnesium, whereas the reverse is true for potassium.

This paper reports on work supported in part by National Science Foundation Grants GB-13096, GB-31862X, GB-31862X2, GB-41233X, BMS73-02027 A02 and DEB73-02027 A03 to the Grassland Biome, US International Biological Programme, for 'Analysis of Structure, Function, and Utilization of Grassland Ecosystems'.

The Pawnee Site is the field research facility of the Natural Resource Ecology Laboratory, Colorado State University, and is located on the USDA Agricultural Research Service Central Plains Experimental Range in northeastern Colorado.

## References

Andrews, R. M., Coleman, D. C., Ellis, J. E. & Singh, J. S. (1974). Energy flow relationships in a shortgrass prairie ecosystem. In *Proceedings of the I International Congress of Ecology*, pp. 22–8. Centre for Agricultural Publishing and Documentation, Wageningen, The Netherlands.

Ares, J. (1976). Dynamics of the root system of blue grama. *Journal of Range Management*, **29**, 208–13.

Ares, J. & Singh, J. S. (1974). A model of the root biomass dynamics of a shortgrass prairie dominated by blue grama (*Bouteloua gracilis*). *Journal of Applied Ecology*, **11**, 727–44.

Barrow, N. J. & Lambourne, L. J. (1962). Partition of excreted nitrogen, sulphur, and phosphorus between the faeces and urine of sheep being fed pasture. *Australian Journal of Agricultural Research*, **13**, 461–71.

Bettany, J. R., Stewart, J. W. B. & Halstead, E. H. (1973). Sulfur fractions and carbon, nitrogen and sulfur relationships in grassland, forest, and associated transitional soils. *Proceedings of the Soil Science Society of America*, **37**, 915–18.

Bettany, J. R., Stewart, J. W. B. & Halstead, E. H. (1974). Assessment of available soil sulphur in an $^{35}S$ growth chamber experiment. *Canadian Journal of Soil Science*, **54**, 309–15.

Biddulph, O., Biddulph, S., Cory, R. & Koontz, H. (1958). Circulation patterns for phosphorus, sulfur and calcium in the bean plant. *Plant Physiology*, **33**, 293–300.

Blaser, R. E. & Kimbrough, E. L. (1968). Potassium nutrition of forage crops with perennials. In *The role of potassium in agriculture*, ed. V. J. Kilmer, S. E. Younts & N. C. Brady, pp. 423–45. American Society of Agronomy, Madison, Wisconsin.

Boawn, L. C. & Allmaras, R. R. (1974). *Mineral concentrations in animal feedstuffs grown in the Columbia Plateau and adjacent valleys. College of Agriculture Research Center Bulletin 799*. Washington State University, Pullman, Washington.

Bokhari, U. G. & Singh, J. S. (1975). Standing state and cycling of nitrogen in soil–vegetation components of prairie ecosystems. *Annals of Botany*, **39**, 273–85.

Broadbent, F. E. & Clark, F. E. (1965). Denitrification. In *Soil nitrogen, Agronomy 10*, pp. 344–59. American Society of Agronomy, Madison, Wisconsin.

Carlisle, A., Brown, A. H. F. & White, J. (1966). The organic matter and nutrient elements in the precipitation beneath a sessile oak (*Quercus petraea*) canopy. *Journal of Ecology*, **54**, 87–9.

Carter, O. G. & Lathwell, D. J. (1967). Effects of temperature on orthophosphate absorption by excised corn roots. *Plant Physiology*, **42**, 1407–12.

Chang, S. C. (1940). Assimilation of phosphorus by a mixed soil population and by pure cultures of soil fungi. *Soil Science*, **49**, 197–210.

Clark, F. E. (1977). Internal cycling of $^{15}$nitrogen in shortgrass prairie. *Ecology*, **58**, 1322–33.

Clark, F. E. & Campion, M. (1976). Distribution of nitrogen in root materials of blue grama. *Journal of Range Management* **29**, 256–8.

Clark, F. E. & Paul, E. A. (1970). The microflora of grasslands. *Advances in Agronomy*, **22**, 375–435.

Clark, F. E., Warren, C. G. & O'Deen, W. A. (1975). Early uptake of $^{15}N$ in shortgrass prairie. *Geoderma*, **13**, 61–6.

Cole, C. V., Grunes, D. L., Porter, L. K. & Olsen, S. R. (1963). The effects of nitrogen on short-term phosphorus absorption and translocation in corn (*Zea mays*). *Proceedings of the Soil Science Society of America*, **27**, 671–4.

Cole, C. V., Innis, G. S. & Stewart, J. W. B. (1977). Simulation of phosphorus cycling in semiarid grasslands. *Ecology*, **58**, 1–15.

Cole, C. V. & Olsen, S. R. (1959). Phosphorus solubility in calcareous soils. II. Effects of exchange phosphorus and soil texture on phosphorus solubility. *Proceedings of the Soil Science Society of America*, **23**, 119–21.

Cole, D. W., Turner, J. & Bledsoe, C. (1977). Requirement and uptake of mineral nutrients in coniferous ecosystems. In *The belowground ecosystem – a synthesis of plant-associated processes*, ed. J. K. Marshall, *Range Science Department Science Series*. Colorado State University, Fort Collins, Colorado.

Coleman, D. C. (1976). A review of root production processes and their influence on soil biota in terrestrial ecosystems. In *The role of terrestrial and aquatic organisms in decomposition processes*, ed. J. M. Anderson & A. Macfadyen, pp. 417–34. Blackwell Scientific Publications, Oxford.

Dahlman, R. C. (1968). In *Methods of productivity studies in root systems and rhizosphere organisms*, ed. M. S. Ghilarov, pp. 11–21. Akademia Nauka, Leningrad.

Dahlman, R. C. & Kucera, C. L. (1965). Root productivity and turnover in native prairie. *Ecology*, **46**, 84–90.

Dahlman, R. C. & Sollins, P. (1970). Nitrogen cycling in grassland. In *Biological Sciences Division, Oak Ridge National Laboratory Report 4634*, pp. 71–3. Oak Ridge National Laboratory, Oak Ridge, Tennessee.

Dawson, M. D. (1974). Recycling nitrogen and sulphur in grass–clover pastures. *Sulphur Institute Journal*, **10** (2), 2–5.

Denmead, O. T., Freney, J. R. & Simpson, J. R. (1976). A closed ammonia cycle within a plant canopy. *Soil Biology and Biochemistry*, **8**, 161–4.

de Wit, C. T., Dykshoom, W. & Noggle, I. C. (1963). Ionic balance and growth of plants. *Verslagen Van Landbouwkundige Onderzoekingen*, **69**, 15–69.

Dittberner, P. L. (1973). *Soil nutrient–plant relationships at the Pawnee Site. US/IBP Grassland Biome Technical Report 122.* Colorado State University, Fort Collins, Colorado.

Dormaar, J. F. (1972). Seasonal pattern of soil organic phosphorus. *Canadian Journal of Soil Science*, **52**, 107–12.

Epstein, E. (1972). *Mineral nutrition of plants: principles and perspectives*. Wiley, New York.

Farnsworth, R. B. (1975). *Physiological and biochemical studies of nitrogen fixation in nodules of sagebrush* (Artemisia ludoviciana) *and other desert nitrogen-fixation systems. US/IBP Desert Biome Research Memorandum 75–36*. Utah State University, Logan, Utah.

Farnsworth, R. B. & Hammond, M. W. (1968). Root nodules and isolation of endophyte on *Artemisia ludoviciana*. *Proceedings of the Utah Academy of Sciences*, **45**, 182–8.

Freney, J. R. (1967). Sulfur-containing organics. In *Soil biochemistry*, ed. A. D. McLaren & G. H. Peterson, pp. 229–59. Dekker, New York.

Frere, M. H., Jensen, M. E. & Carter, J. N. (1970). Modeling water and nitrogen behavior in the soil–plant system. In *Proceedings of the 1970 summer computer simulation conference*, vol. 2, pp. 746–50. Simulation Councils, Inc., La Jolla, California.

Frissel, M. J. (1976). *Cycling of mineral nutrients in agricultural ecosystems*. Institute for Atomic Sciences in Agriculture, Wageningen, The Netherlands. (Prepublication mimeo.)

Haas, H. J., Grunes, D. L. & Reichman, G. A. (1961). Phosphorus changes in Great Plains soils as influenced by crapping and manure applications. *Proceedings of the Soil Science Society of America*, **25**, 214–18.

Hagen, C. E. & Hopkins, H. T. (1955). Ionic species in orthophosphate absorption by barley roots. *Plant Physiology*, **30**, 193–9.

Hagen, C. E., Leggett, J. E. & Jackson, D. C. (1957). The sites of orthophosphate uptake by barley roots. *Proceedings of the National Academy of Sciences, USA*, **43**, 496–506.

Halm, B. J. (1972). The phosphorus cycle in a grassland ecosystem. PhD Thesis, University of Saskatchewan.

Halm, B. J., Stewart, J. W. B. & Halstead, R. L. (1972). The phosphorus cycle in a native grassland ecosystem. In *Symposium on the use of isotopes and radiation in research on soil–plant relationships including applications in forestry*, pp. 571–89. International Atomic Energy Agency and Food and Agriculture Organization of the United Nations, Vienna.

Hannapel, R. J., Fuller, W. H., Bosma, S. & Bullock, J. S. (1964). Phosphorus movement in a calcareous soil. I. Predominance of organic forms of phosphorus in phosphorus movement. *Soil Science*, **97**, 350–7.

Heady, H. F. (1975). Rangeland fertilization. In *Rangeland management*, pp. 369–88. McGraw-Hill, New York.

Hunt, H. W. (1977). A simulation model for decomposition in grasslands. *Ecology*, **58**, 469–84.

Hutchinson, G. L., Millington, R. J. & Peters, D. B. (1972). Atmospheric ammonia: absorption by plant leaves. *Science*, **175**, 771–2.

Johannes, R. E. (1965). Influence of marine protozoa on nutrient regeneration. *Limnology and Oceanography*, **10**, 434–42.

Jones, M. B. (1964). Effect of sulfur on five annual grassland species. *California Agriculture*, **18**, 4–5.
Jordan, H. V. & Ensminger, L. E. (1958). The role of sulfur in soil fertility. *Advances in Agronomy*, **10**, 407–34.
Koontz, H. & Biddulph, O. (1957). Factors affecting absorption and translocation of foliar applied phosphorus. *Plant Physiology*, **32**, 463–70.
Larsen, S. (1967). Soil phosphorus. *Advances in Agronomy*, **19**, 151–210.
Loneragan, J. F. (1968). Nutrient concentration, nutrient flux, and plant growth. In *Transactions of the IX International Congress of Soil Science*, vol. 2, pp. 173–82.
Loneragan, J. F. & Asher, C. J. (1967). Response of plants to phosphate concentration in solution culture. II. Rate of phosphate absorption and its relation to growth. *Soil Science*, **103**, 311–18.
Long, W. G., Sweet, D. V. & Tukey, H. B. (1956). Leaching of organic compounds from leaves. *Science*, **123**, 1039–40.
McGill, W. B., Paul, E. A. & Sorensen, H. L. (1974). *The role of microbial metabolites in the dynamics of soil nitrogen. Matador Project (Canadian IBP) Technical Report 48*. University of Saskatchewan, Saskatoon.
McIlroy, R. J. (1972). *Tropical grassland husbandry*. Oxford University Press, London.
MacLeod, L. B. (1965). Effect of nitrogen and potassium on the yield and chemical composition of alfalfa, bromegrass, orchard grass, and timothy grown as pure species. *Agronomy Journal*, **57**, 261–6.
Malcolm, R. L. & McCracken, R. J. (1968). Canopy drip: a source of mobile soil organic matter for mobilization of iron and aluminium. *Proceedings of the Soil Science Society of America*, **32**, 834–8.
Martin, A. E. & Ross, P. J. (1968). A nitrogen balance study using labelled fertilizer in a gas lysimeter. *Plant and Soil*, **28**, 182–6.
May, P. F., Till, A. R. & Downes, A. M. (1968). Nutrient cycling in grazed pastures. I. A preliminary investigation of the use of [$^{35}$S]gypsum. *Australian Journal of Agricultural Research*, **19**, 531–43.
Mosse, B. (1973). Advances in the study of vesicular–arbuscular mycorrhiza. *Annual Review of Phytopathology*, **11**, 170–96.
Neal, J. L., Jr (1973). Influence of selected grasses and forbs on soil phosphatase activity. *Canadian Journal of Soil Science*, **53**, 119–21.
Nelson, W. L. (1968). Plant factors affecting potassium availability and uptake. In *The role of potassium in agriculture*, ed. V. J. Kilmer, S. E. Younts & N. C. Brady, pp. 355–83. American Society of Agronomy, Madison, Wisconsin.
Neptune, A. M. L., Tabatabai, M. A. & Hanway, J. J. (1975). Sulfur fractions and carbon–nitrogen–phosphorus–sulfur relationships in some Brazilian and Iowa soils. *Proceedings of the Soil Science Society of America*, **39**, 51–5.
Nyhan, J. W. (1975). Decomposition of carbon-14 labeled plant materials in a grassland soil under field conditions. *Proceedings of the Soil Science Society of America*, **39**, 643–8.
Odum, H. T. (1968). Work circuits and systems stress. In *Symposium on primary production and mineral cycling in natural ecosystems*, ed. H. E. Young, pp. 81–138. University of Maine Press, Orono, Maine.
Olsen, S. R. & Flowerday, A. D. (1971). Fertilizer phosphorus interactions in alkaline soils. In *Fertilizer technology and use*, 2nd edn, ed. R. A. Olson, T. J. Army, J. J. Hanway & V. J. Kilmer, pp. 153–85. Soil Science Society of America, Madison, Wisconsin.

Olsen, S. R., Kemper, W. D. & Van Schaik, J. C. (1965). Self-diffusion coefficients of phosphorus in soil measured by transient and steady-state methods. *Proceedings of the Soil Science Society of America*, **29**, 154–8.
Olsen, S. R. & Watanabe, F. S. (1963). Diffusion of phosphorus as related to soil texture and plant uptake. *Proceedings of the Soil Science Society of America*, **27**, 648–53.
Olsen, S. R. & Watanabe, F. S. (1966). Effective volume of soil around plant roots determined from phosphorus diffusion. *Proceedings of the Soil Science Society of America*, **30**, 598–602.
Olsen, S. R., Watanabe, F. S. & Danielson, R. E. (1961). Phosphorus absorption of corn roots as affected by moisture and phosphorus concentration. *Proceedings of the Soil Science Society of America*, **25**, 289–94.
Paul, E. A. (1970). Plant components and soil organic matter. *Advances in Phytochemistry*, **3**, 66–128.
Porter, L. K., Viets, F. G. & Hutchinson, G. L. (1972). Air containing nitrogen-15 ammonia: foliar absorption by corn seedlings. *Science*, **175**, 759–61.
Power, J. F. (1968). Mineralization of nitrogen in grass roots. *Proceedings of the Soil Science Society of America*, **32**, 673–4.
Power, J. F. (1971). Evaluation of water and nitrogen stress on bromegrass growth. *Agronomy Journal*, **63**, 726–8.
Quirk, J. P. (1967). *Aspects of nutrient absorption by plants from soils. Australian Society of Soil Science Publication 4*.
Reuss, J. O. (1971). Soils of Grassland Biome sites. In *Preliminary analysis of structure and function in grasslands*, ed. N. R. French, *Range Science Department Science Series 10*, pp. 35–9. Colorado State University, Fort Collins, Colorado.
Reuss, J. O. & Cole, C. V. (1973). Simulation of nitrogen flow in a grassland ecosystem. In *Proceedings of the 1973 summer computer simulation conference*, vol. 2, pp. 762–8. Simulation Councils, Inc., La Jolla, California.
Sanchez, P. A. (1973). Soil management under shifting cultivation. In *A review of soils research in tropical Latin America*, ed. P. A. Sanchez, *Technical Bulletin 219*, pp. 46–67. North Carolina Agronomy Experimental Station and US Agency for International Development, Raleigh, NC.
Sanders, F. E. & Tinker, P. B. (1971). Mechanism of absorption of phosphate from soil by endogene mycorrhizas. *Nature, London*, **233**, 278–9.
Sauerbeck, D. & Johnen, B. (1976). Der Umsatz von Pflanzenwurzeln im Laufe der Vegetationsperiode und dessen Beitrag zur 'Bodenatmung'. *Zeitschrift für Pflanzenernährung und Bodenkunde*, **3**, 315–28.
Sayre, J. D. (1948). Mineral accumulation in corn. *Plant Physiology*, **23**, 267–81.
Sears, P. D. (1950). Soil fertility and pasture growth. *Journal of the British Grassland Society*, **5**, 267–80.
Shamoot, S., McDonald, L. & Bartholomew, W. V. (1968). Rhizo-deposition of organic debris in soil. *Proceedings of the Soil Science Society of America*, **32**, 817–20.
Shields, J. A. & Paul, E. A. (1973). Decomposition of $^{14}C$-labelled plant material under field conditions. *Canadian Journal of Soil Science*, **53**, 297–306.
Sims, P. L. & Singh, J. S. (1971). Herbage dynamics and net primary production in certain ungrazed and grazed grasslands. In *Preliminary analysis of structure and function in grasslands*, ed. N. R. French, *Range Science Department Science Series 10*, pp. 59–124. Colorado State University, Fort Collins, Colorado.

Sims, P. L., Uresk, D. W., Bartos, D. L. & Lauenroth, W. K. (1971). *Herbage dynamics on the Pawnee Site: aboveground and belowground herbage dynamics on the four grazing intensity treatments; and preliminary sampling on the ecosystem stress site. US/IBP Grassland Biome Technical Report 99.* Colorado State University, Fort Collins, Colorado.

Skujins, J. (1975). *Nitrogen dynamics in stands dominated by some major cool desert shrubs. US/IBP Research Memorandum 75–33.* Utah State University, Logan, Utah.

Snedecor, G. W. & Cochran, W. G. (1967). *Statistical methods.* Iowa State University Press, Ames, Iowa.

Sosebee, R. E. & Wiebe, H. H. (1973). Effect of phenological development on radiophosphorus translocation from leaves in crested wheatgrass. *Oecologia*, **13**, 103–12.

Soulides, D. A. & Clark, F. E. (1958). Nitrification in grassland soils. *Proceedings of the Soil Science Society of America*, **22**, 308–11.

Stewart, B. A. (1970). Volatilization and nitrification of nitrogen under simulated feedlot conditions. *Journal of Environmental Sciences and Technology*, **4**, 579–82.

Stewart, J. W. B., Halm, B. J. & Cole, C. V. (1973). *Nutrient cycling. I. Phosphorus. Matador Project (Canadian IBP) Technical Report 40.* University of Saskatchewan, Saskatoon.

Stewart, J. W. B. & McConagher, S. (1963). Some effects of reaction (pH) changes in a basaltic soil on the mineral composition of growing crops. *Journal of Scientific Pd. Agriculture*, **14**, 613–21.

Stewart, W. D. P. (1966). *Nitrogen fixation in plants.* Athlone Press, London.

Steyne, P. L. & Delwiche, C. C. (1970). Nitrogen fixation by non-symbiotic microorganisms in some Southern California soils. *Environmental Science and Technology*, **4**, 1122–8.

Stoddart, L. A., Smith, A. D. & Box, T. W. (1975). *Range management.* McGraw-Hill, New York.

Tanaka, A. (1961). Studies on the nutrio-physiology of leaves of rice plants. *Journal of the Faculty of Agriculture, Hokkaido University*, **51** (3), 449–550.

Thomas, G. W. & Hipp, B. W. (1968). Soil factors affecting potassium availability. In *The role of potassium in agriculture*, ed. V. J. Kilmer, S. E. Younts & N. C. Brady, pp. 269–91. American Society of Agronomy, Madison, Wisconsin.

Thompson, L. M., Black, C. A. & Zoellner, J. A. (1954). Occurrence and mineralization of organic phosphorus in soils, with particular reference to associations with nitrogen, carbon, and pH. *Soil Science*, **77**, 185–96.

Till, A. R. & May, P. F. (1970*a*). Nutrient cycling in grazed pastures. II. Further observations with [$^{35}$S]gypsum. *Australian Journal of Agricultural Research*, **21**, 253–60.

Till, A. R. & May, P. F. (1970*b*). Nutrient cycling in grazed pastures. III. Studies on labelling of the grazed pasture system by solid ($^{35}$S) gypsum and aqueous $Mg^{35}SO_4$. *Australian Journal of Agricultural Research*, **21**, 455–63.

Till, A. R. & May, P. F. (1971). Nutrient cycling in grazed pastures. IV. The fate of sulphur-35 following its application to a small area in a grazed pasture. *Australian Journal of Agricultural Research*, **22**, 391–400.

Titlyanova, A. A. & Bazilevich, N. I. (1975). Metabolic processes and functioning of grass biogeocoenoses. In *Resources of the biosphere*, ed. N. I. Basilevich. Akademia Nauka, Leningrad.

Tukey, H. B., Jr (1970). The leaching of substances from plants. *Annual Review of Plant Physiology*, **21**, 305–32.
Tukey, H. B., Jr, Tukey, H. B. & Wittwer, S. H. (1958). Loss of nutrients by foliar leaching as determined by radioisotopes. *Proceedings of the American Society for Horticultural Science*, **71**, 496–506.
Uresk, D. W., Sims, P. L. & Jameson, D. A. (1975). Dynamics of blue grama within a shortgrass ecosystem. *Journal of Range Management*, **28**, 205–8.
Valentine, J. F. (1971). *Range development and improvements*. Brigham Young University Press, Provo, Utah.
Vlassak, K., Paul, E. A. & Harris, R. E. (1973). Assessment of biological nitrogen fixation in grassland and associated sites. *Plant and Soil*, **38**, 637–49.
Volk, G. M. (1959). Volatile loss of ammonia following surface application of urea to turf or bare soils. *Agronomy Journal*, **51**, 746–9.
Wagner, R. E. & Jones, M. B. (1968). Fertilization of high yielding forage crops. In *Changing patterns in fertilizer use*, ed. L. B. Nelson, pp. 297–326. Soil Science Society of America, Madison, Wisconsin.
Walker, T. W. (1957). The sulfur cycle in grassland soils. *Journal of the British Grassland Society*, **12**, 10–18.
Wallace, A., Romney, E. M., Cha, J. W. & Soufi, S. M. (1974). *Nitrogen transformations in rock valley and adjacent areas of the Mohave Desert. US/IBP Desert Biome Research Memorandum 74–36*. Utah State University, Logan, Utah.
Wallace, T. (1930). Experiments on the effects of leaching with cold water on the foliage of fruit trees. I. The course of leaching of dry matter, ash and potash from leaves of apple, pear, plum, blackcurrant and gooseberry. *Journal of Pomology and Horticultural Science*, **8**, 44–60.
Whitman, W. C. & Stevens, O. A. (1952). Native legumes of North Dakota grassland. *Proceedings of the North Dakota Academy of Science*, **6**, 73–8.
Will, G. M. (1959). Nutrient return in litter and rainfall under some exotic conifer stands in New Zealand. *New Zealand Journal of Agricultural Research*, **2**, 719–24.
Williams, C. H. (1962). Changes in nutrient availability in Australian soils as a result of biological activity. *Journal of the Australian Institute of Agricultural Science*, **28**, 196–205.
Williams, C. H. & Andrews, C. S. (1970). Mineral nutrition of pastures. In *Australian grasslands*, ed. R. M. Moore, pp. 321–38. Australian National University Press, Canberra.
Woldendorp, J. S., Dilz, K. & Kolenbrander, G. J. (1966). The fate of fertilizer nitrogen on permanent grassland soils. In *Nitrogen and grassland*, ed. P. F. J. Van Burg & G. H. Arnold, pp. – . Centre for Agricultural Publishing and Documentation, Wageningen, The Netherlands.
Woodmansee, R. G. (1978). Grassland ecosystem structure: soils. In *North American grasslands: ecosystem structure and function*, ed. J. E. Ellis. Dowden, Hutchinson & Ross, Inc., Stroudsburg, Pennsylvania. (In press.)

# 9. Comparative studies of ecosystem function

N. I. BAZILEVICH & A. A. TITLYANOVA

## Contents

## Introduction

The exchange of matter and energy is the principal process in ecosystems. Using the systems approach (Lyapunov, 1970; Van Dyne, 1970; Lyapunov & Titlyanova, 1971) the ecosystem is considered as consisting of compartments together with connecting fluxes that ensure matter and energy transfer between compartments. Matter and energy reserves ($Q$) are concentrated in the compartments; these reserves are measured in weight or energy units per unit area. Each flux includes several processes, each process with a certain intensity ($I$). The intensity of a flux is equal to the sum of intensities of the processes and is measured by the amount of matter and energy transported from one compartment to another per unit time and per unit area.

Since the flux intensity $I$ depends on the compartment's reserve $Q$, we can assume that $I = fQ$ or $f = I/Q$. Therefore $f$ has the dimension 1/(unit time) and shows what fraction of a given reserve is removed (or supplied) per unit time. The value $f$ will be called the specific rate of a flux (or a process) and is measured in milligrams (or grams) per day. The specific rate depends both on the external characteristics (e.g. climate) and internal characteristics (e.g. the presence of chemical compounds) of the compartment.

The compartments and connecting fluxes within the ecosystem, as well as input and output fluxes, comprise the structure of the ecosystem. The state of the ecosystem at a time $t$ is taken to be its structure at that time, with corresponding values of matter and energy reserves and flux intensities. The functioning of the ecosystem is the change in its state with time.

Grassland ecosystems vary in their functioning according to their stage of evolutionary development. It is possible to distinguish between so-called 'periodical' and 'transitional' types of functioning. The periodical type is characterized by periodic changes in flux intensities and matter reserves in compartments (e.g. from season to season) while average values remain constant over a longer time scale. This allows consideration of periodical functioning as a steady state. (Climax ecosystems are closest to a true steady state condition.) Transitional functioning is characteristic of ecosystems that are passing from one functioning regime to another (e.g. from swamp to meadow or from meadow to steppe, the climax ecosystem), and here the intensities of input and output fluxes and the matter reserves in the compartments are changing. In this type of functioning, however, it is possible that some subsystems are in a steady state while others are still at the transitional stage. This is because different processes have different characteristic rates, characteristic rate being defined as the time required to attain dynamic equilibrium with environmental conditions for any given process (or for all the exchange processes in the ecosystem as a whole).

On the basis of analysing the structure, matter reserves, intensities and specific rates of fluxes, and type of functioning, we attempt here to show the differences between various grassland ecosystems of the temperate zone of the northern hemisphere. We attempt also to show the position of grassland ecosystems in relation to other ecosystem types such as forests and deserts.

## Reserves of chemical elements in ecosystems

### *In compartments of grassland ecosystems*

Grassland ecosystems are divided here into three groups according to the degree and source of their moisture content: (1) steppes and prairies, (2) meadows, and (3) grassy swamps (only drying grassy swamps are discussed here). These are further divided into subgroups according to the same principle, and group 2 is also subdivided according to the degree of halophytism of the plant community.

The following compartments are considered: G, green plant biomass; R, live belowground plant material, called roots for short; D, standing dead; L, litter, V, dead belowground plant material; Sl, soil; and W, ground water. The amount of organic matter in compartments is given in grams of carbon.

#### *Green plant biomass reserve, $G_{max}$*

This is measured in the period of the greatest growth of the grass standing crop and depends on moisture. The $G_{max}$ values change in the following order: swamps > meadows > halophytic meadows > steppes and prairies. Below-average $G_{max}$ values are typical of dry, extracontinental, Eastern Siberian, solonetzic, psammophytic steppes and shortgrass prairies (Table 9.1; Appendixes 9.1, 9.2 and 9.3). Salinization, causing toxicity and physiological drought, acts as a moisture deficit.

The amplitude of the deviation of $G_{max}$ values from the average within a group is greatest in steppes and prairies and least in drying swamps. This is because meadow and swamp ecosystems receive moisture from surface, ground or flood water in addition to precipitation, whereas climax steppe and prairie ecosystems exist only as the result of precipitation, which varies sharply and is dependent on the spatial variation of general climatic conditions.

#### *Live root reserve, R*

This is the most stable value and changes little between the different groups of grassland ecosystems; the minimum (in halophytic meadows) differs from the maximum (in drying swamps) only by a factor of 1.5. The R value also varies relatively little within each group. The live plant biomass pool in soil must be preserved in grass communities where the aboveground biomass dies off annually, and the role of this compartment in the ecosystem is one of preservation. The R reserve is smallest in ecosystems with the best soil

Table 9.1. *Reserves of carbon (C), nitrogen (N) and mineral elements (M)*[a] *in grassy*

| Compartment | C: Steppes and prairies | C: Meadows I[b] | C: Meadows II[c] | C: Grassy swamps (drying) |
|---|---|---|---|---|
| $G_{max}$, maximum standing crop of green biomass | 95 ($n$ = 12) / 25–300 | 170 ($n$ = 13) / 80–416 | 110 ($n$ = 5) / 24–175 | 230 ($n$ = 4) / 120–400 |
| R, live belowground plant material | 570 ($n$ = 4) / 380–700 | 640 ($n$ = 6) / 330–940 | 475 ($n$ = 1) | 670 ($n$ = 4) / 320–1400 |
| V, dead belowground plant material | 610 ($n$ = 4) / 350–1100 | 660 ($n$ = 6) / 225–1720 | 215 ($n$ = 1) | 1920 ($n$ = 4) / 1030–3600 |
| R+V, live and dead belowground plant material | 820 ($n$ = 12) / 210–1480 | 1000 ($n$ = 18) / 360–2540 | 700 ($n$ = 4) / 180–1135 | 2590 ($n$ = 4) / 1350–5000 |
| D, standing dead | 104 ($n$ = 10) / 30–243 | 73 ($n$ = 4) / 18–120 | 66 ($n$ = 5) / 43–104 | 110 ($n$ = 4) / 40–230 |
| L, litter | 147 ($n$ = 10) / 60–396 | 170 ($n$ = 8) / 25–463 | 55 ($n$ = 2) / 51–60 | 560 ($n$ = 4) / 300–700 |

Numerator, average value; denominator, minimum and maximum values; $n$, number of sampling plots.
[a] Si, Ti, Ca, Mg, K, Na, Fe, Al, Mn, P, S, Cl.
[b] Meadows I comprises steppe-like, mesophytic, mesohalophytic and hygrohalophytic meadows.
[c] Meadows II comprises halophytic and hygrohalophytic meadows.

Table 9.2. *The reserves of live roots, R (g C·m*$^{-2}$ *in the 50 cm soil layer) in three series of ecosystems influenced by different limiting factors*

| Ecological series | Ecosystems | | | | |
|---|---|---|---|---|---|
| Water limitation – drying out | Flood-plain meadow,[a] plot 4.00 | Meadow steppes: Plot 1.1 | Meadow steppes: Plot 1.3 | | Extracontinental steppe, plot 1.5 |
| | 330 | 410 | 515 | | 670 |
| Oxygen limitation – excess moisture | Flood-plain meadow,[a] plot 4.00 | Flood-plain meadow,[b] plot 4.0 | Hygrophytic meadows: Plot 7.3 | Hygrophytic meadows: Plot 7.2 | Grassy swamp (drying), plot 9.1 |
| | 330 | 380 | 820 | 940 | 1400 |
| Salinization and drying out | Flood-plain meadow,[a] plot 4.00 | Mesohalophytic meadow, plot 5.2 | Mesohalophytic meadow (steppe), plot 3.4 | | Solonetzic steppe, plot 1.6 |
| | 330 | 670 | 685 | | 700 |

[a] Flooded for short periods. [b] Flooded for long periods.

conditions, and increases where there is any limitation (deficit of moisture, oxygen or nutrients) or conditioning inhibition (salinization, solonetzicity) of plant growth and development. The greater the stress on the plant, the larger is the live root reserve (Table 9.2). The increase in R biomass is limited by the quantity of hydrocarbons that is translocated from G to R, and if this input is smaller than the root respiration, R will inevitably decrease. Thus the R reserve is ultimately controlled by the G reserve and production.

The $R/G_{max}$ value is lowest in drying grassy swamps (2.9), increasing to 3.7 in steppe-like, mesophytic, mesohalophytic and hydrophytic meadows, to 4.3 in halophytic meadows, and to 6.0 in steppes and prairies.

Various types of grassland ecosystems differ little in their average reserves *ecosystems* ($g \cdot m^{-2}$)

| N | | | | M | | | |
|---|---|---|---|---|---|---|---|
| Steppes and prairies | Meadows I[b] | Meadows II[c] | Grassy swamps (drying) | Steppes and prairies | Meadows I[b] | Meadows II[c] | Grassy swamps (drying) |
| 4.0 ($n$ = 11)<br>1.6–10.0 | 5.5 ($n$ = 11)<br>3.8–8.5 | 4.6 ($n$ = 5)<br>1.5–7.7 | 8.7 ($n$ = 3)<br>3.3–16.0 | 8.5 ($n$ = 11)<br>5.3–21.2 | 15.4 ($n$ = 9)<br>10.3–22.5 | 38.9 ($n$ = 5)<br>8.5–61.3 | 22.0 ($n$ = 3)<br>12.0–39.0 |
| 10.0 ($n$ = 4)<br>8.7–12.4 | 18.4 ($n$ = 1) | 8.2 ($n$ = 1) | 30.0 ($n$ = 3)<br>22.0–40.0 | 50.0 ($n$ = 4)<br>33.0–65.0 | 95.0 ($n$ = 1) | 87.0 ($n$ = 1) | 55.5 ($n$ = 3)<br>18.5–98.0 |
| 14.0 ($n$ = 4)<br>7.0–25.0 | 7.1 ($n$ = 1) | 3.7 ($n$ = 1) | 133.9 ($n$ = 3)<br>78.0–200.0 | 71.0 ($n$ = 4)<br>30.0–162.0 | 39.0 ($n$ = 1) | 39.0 ($n$ = 1) | 159.9 ($n$ = 3)<br>75.0–257.0 |
| 20.0 ($n$ = 12)<br>8.6–34.0 | 24.8 ($n$ = 11)<br>3.0–83.0 | 6.7 ($n$ = 2)<br>1.6–11.9 | 169.4 ($n$ = 3)<br>100–227.5 | 120.0 ($n$ = 12)<br>19.0–390.0 | 91.1 ($n$ = 8)<br>24.6–225.0 | 80.1 ($n$ = 2)<br>34.2–126.0 | 218.0 ($n$ = 3)<br>98.0–355 |
| 2.5 ($n$ = 10)<br>0.5–6.2 | 3.2 ($n$ = 3)<br>1.8–5.5 | 1.5 ($n$ = 2)<br>1.2–1.7 | 1.8 ($n$ = 3)<br>1.6–2.2 | 12.1 ($n$ = 10)<br>1.6–34.6 | 12.1 ($n$ = 3)<br>8.5–16.7 | 27.5 ($n$ = 2)<br>5.0–50.0 | 6.8 ($n$ = 3)<br>3.0–13.5 |
| 3.7 ($n$ = 10)<br>0.6–12.9 | 8.8 ($n$ = 3)<br>3.8–12.2 | 1.4 ($n$ = 2)<br>1.0–1.7 | 35.5 ($n$ = 3)<br>15.6–47.0 | 16.0 ($n$ = 10)<br>4.5–43.1 | 34.4 ($n$ = 3)<br>14.3–50.6 | 31.5 ($n$ = 2)<br>4.0–59.0 | 88.5 ($n$ = 3)<br>34.5–126.0 |

of above- and belowground live biomass (G+R) – which vary from 600 to 900 $m^{-2}$ – but differ sharply in the relative proportions of the reserves. Each ecosystem has an optimal G/R ratio for its given ecological conditions, the R biomass being sufficient to provide the G biomass with water and mineral elements and the G biomass being sufficient to provide the R biomass with energy.

### *The reserve of dead belowground plant parts, V*

This varies greatly between the different groups of grassland ecosystems and also within each group. The minimum value of V (in halophytic meadows) is only one-ninth of the maximum value (in drying grassy swamps). We suggest that the V value depends above all on the degree of decomposition of root remains. This decomposition occurs fastest in saline soils where salts contribute to the breakdown of dead tissues. The decomposition slows down in cases of moisture deficit (in steppes) and becomes very slow where there is an oxygen deficit (in swamps).

The ratio V/R reflects the accumulation of dead underground parts. It has a value of 0.5 in halophytic meadows, 1.0 in steppe-like, mesophytic and mesohalophytic meadows, 1.2 in steppes and 2.8 in drying grassy swamps.

Excess dryness and excess moisture contribute to an increase in both R and V reserves. As a result the total reserve of underground organs (R+V) varies widely in different grassland ecosystems, increasing in dry grasslands (shortgrass prairies, extracontinental steppes) and in very moist ones (hygrophytic meadows and drying grassy swamps).

The ratio $(R+V)/G_{max}$, widely used by researchers, increases like the ratio $R/G_{max}$ in the sequence steppe-like, mesophytic and mesohalophytic meadows (5.9), halophytic meadows (6.4), and steppes and prairies (8.6), i.e. it parallels the deterioration of edaphic and moisture conditions. However, the largest value of this ratio (11.2) is typical of drying grassy swamps, and is primarily due to the slow decomposition of V.

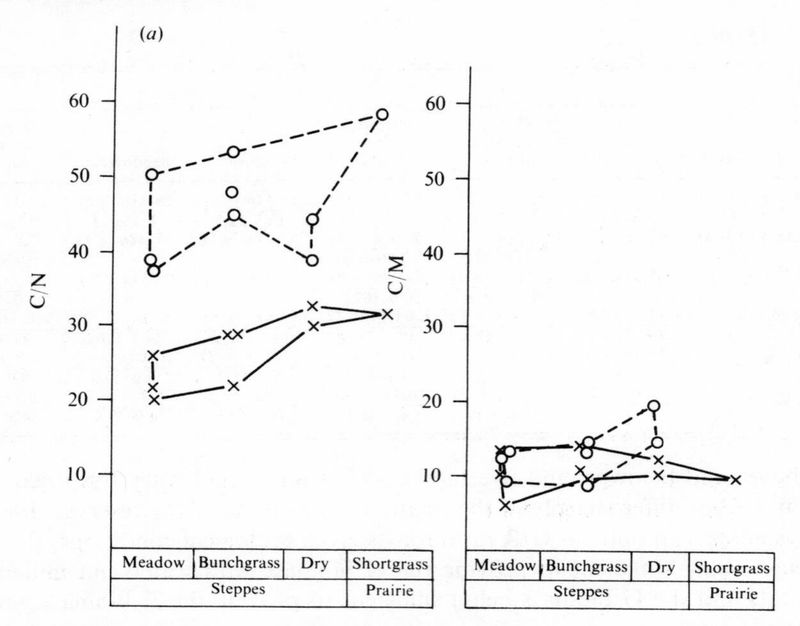


Fig. 9.1. C/N and C/M ratios in the various types of grassland ecosystems. (*a*) steppes and prairies; (*b*) meadows.

### *The reserve of standing dead, D*

This changes little either within or between the groups of grassland ecosystems. It is lowest in meadows and increases in steppes, prairies and swamps, i.e. in those ecosystems dominated by species of grasses with stiff stems and leaves that remain standing for a long time after dying off.

### *The reserve of litter, L*

This varies widely between the different groups. It is lowest in halophytic meadows and highest in grassy swamps, and depends on the rate of decomposition of aboveground plant remains. The pattern of change of L in various types of grassland ecosystems is similar to that of V.

### *The ratio of total aboveground to belowground biomass reserves, $(G+D+L)/(R+V)$*

The values for this are fairly stable: the component G+D+L in steppes and prairies is approximately 20–30 %, in meadows and drying grassy swamps 25–40 %, and up to 10–15 % in some steppe meadows. In the G+D+L subsystem G accounts for 20–40 % of the total, the lowest values being found in shortgrass prairies and the extracontinental steppes of Eastern Siberia. These ratios shift sharply in favour of G in meadows harvested for hay.

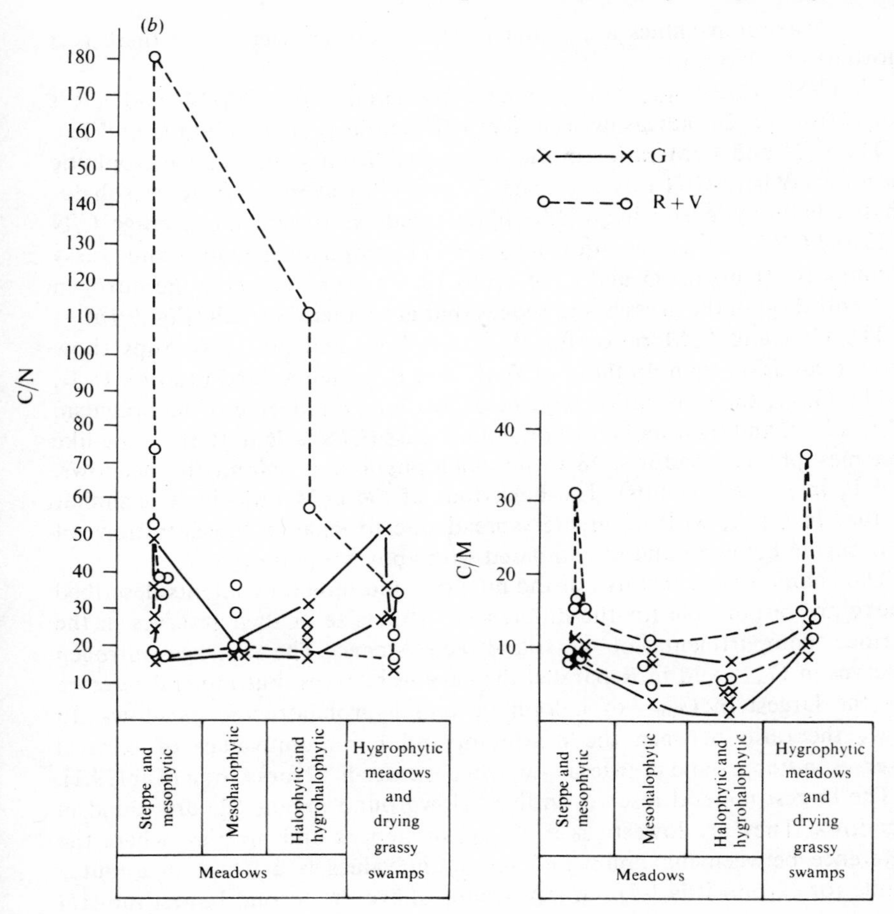


Figure 9.1. (*b*)

### *The reserves of nitrogen, N, and mineral nutrients, M, in live plant biomass*

These are generally directly proportional to the carbon (C) reserves in compartments (Table 9.1).

The C/N ratio in the $G_{max}$ of meadow-like and true steppe grasslands is 20–30, rising to 30–35 in dry steppes and shortgrass prairies (Fig. 9.1*a*; Appendixes 9.1, 9.2 and 9.3). This is due to the smaller percentage of legumes and assorted grass species in these phytocoenoses as compared with meadow-steppes and tallgrass prairies, and to the smaller biomass of grass leaves (it is known that the nitrogen and mineral content of leaves is higher than that of stems: Bazilevich, 1962). The ratio is larger in R, where it is in the range

38–58. Maximum values are found in shortgrass prairies where there is a shortage of nitrogen.

The C/M ratio in $G_{max}$ varies within narrower limits – from 6 to 14 – and in R it ranges from 7 to 20, increasing in parallel with the aridity of the habitat (Fig. 9.1*a*).

The C/N and C/M ratios in $G_{max}$ decrease from steppe and mesophytic meadows (where C/N ratios average 27 and C/M average 10) to mesohalophytic, halophytic and hygrohalophytic meadows (where the average C/N is 22 and C/M is 7). These ratios rise again in hygrophytic meadows and grassy swamps (C/N up to 35 and C/M up to 10, on average) where the nitrogen concentration in the grasses and sedges that grow there is small (Fig. 9.1*b*).

The C/N and C/M ratios for R of meadows and grassy swamps show greater variation than do those of $G_{max}$ and R in steppes and prairies. In R, unlike $G_{max}$, the C/N ratios tend to decrease as the degree of halophytism of meadows and their moisture content increase (C/N is 76 in R of steppe-like and mesophytic meadows, 48 in mesohalophytic and halophytic meadows, and 17 in grassy swamps). The behaviour of the C/M ratio in R is similar to that in $G_{max}$, with its greater spread due to smaller concentrations of nutrients in belowground as compared with aboveground organs.

The characteristic features of the nitrogen and mineral contents described above are responsible for the differences in the size of their reserves in the various compartments of grassland ecosystems. The average nitrogen reserves in $G_{max}$ and in R parallel the carbon reserves, but mineral reserves are the largest in $G_{max}$ of halophytic and hygrohalophytic meadows. In $G_{max}$ the ratio between the maximum and minimum values of mineral reserves in steppes and prairies is 4.6, while it exceeds 12 for carbon (Table 9.1).

The largest mineral reserves in live belowground organs (R) are found in meadows. They are lowest (as in $G_{max}$) in steppes and prairies, where the difference between maximum and minimum values is a factor of about 2 (while for carbon it is 1.7). Grassy swamps have the second largest mineral reserves in R, though their carbon reserves are the largest (Table 9.1).

### *The nitrogen and mineral reserves in dead aboveground plant remains (standing dead, D, and litter, L)*

These generally depend, as they do in live organs of plants, directly on carbon accumulation. However, certain characteristic features have been observed. Compartment D is short of nutrients, nitrogen in particular, compared with $G_{max}$; this is due to the leaching of chemical compounds by precipitation and their translocation to live plant organs. For example, the average carbon reserves in compartments $G_{max}$ and D of steppes and prairies are practically equal, but the average nitrogen and mineral reserves in D are smaller by a factor of about 1.5. The carbon reserve in compartment D of halophytic and hygrohalophytic meadows and grassy swamps is smaller than that in $G_{max}$ by a factor of about 2, while the difference for nitrogen is a factor of 3 and for swamps a factor of about 4 (the difference for M in the latter is about 3 times).

The same trend with respect to nitrogen is also observed for compartment L of steppes, prairies, and halophytic and hygrophytic meadows. In other types of grassy ecosystems, particularly in steppe-like, mesophytic and mesohalophytic meadows, there is a nitrogen enrichment of the litter (L) relative to the standing dead (D); this is probably due to the activity of micro-organisms. At the same time, mineral elements are also accumulated in L.

#### *Chemical elements*

The predominant chemical elements in $G_{max}$ of steppes and prairies are silicon, nitrogen and potassium. In roots potassium is replaced by calcium. In $G_{max}$ of steppe-like and mesophytic meadows nitrogen, silicon, calcium and potassium are the major elements, while in mesohalophytic meadows nitrogen, potassium and silicon are dominant and the chlorine content increases appreciably. Biohalogens, chlorine, sodium, and often sulphur, are accumulated in green aboveground organs in halophytic and hygrohalophytic meadows. Chlorine and sodium are of major importance in succulent meadows. Silicon is accumulated in large amounts in dead plant remains and calcium is stored to a certain extent in litter. Iron and aluminium are also accumulated there while potassium decreases.

In soils (Sl) of steppes and prairies, carbon and nitrogen are stored in large quantities, the content of these elements decreasing as the climate becomes more arid. The content of readily soluble salts is small, and ground water is at too great a depth to produce an effect on contemporary biogeocoenotic processes. The soils of meadows in the forest zone have smaller amounts of carbon, nitrogen and other nutrients but a larger quantity of mobile, non-salt compounds of silicon, aluminium, iron, manganese, calcium, magnesium, potassium, sodium and titanium compared with the soils of meadows of the forest-steppe and steppe zones. The content of readily soluble salts and minerals in soil and ground water increases from steppe-like and mesophytic meadows to mesohalophytic, halophytic and also hygrohalophytic meadows (Appendixes 9.1. and 9.2.)

### *In forest and desert ecosystems*

The pattern of distribution of organic matter differs sharply between grassland ecosystems and forest and desert ecosystems, where perennial lignified parts (Pr) predominate or play a significant role (accounting for up to 75 % of plant biomass in forests and 10–20 % in desert ecosystems). The reserves of live organic matter in forests and deserts are larger than those of dead organic matter (Table 9.3). Nitrogen and calcium are the predominant nutrients in taiga coniferous forest, and calcium and nitrogen in broad-

Table 9.3. *The reserves of C, N and M in forest and desert ecosystems* ($g \cdot m^{-2}$)

| Compartment | Chemical element | Forest ecosystems: Spruce forests in southern taiga, plot a | Broad-leafed forests, plot b | Tropical rain forests, plot c | Desert ecosystems: *Artemisia* dominated, plot d | *Haloxylon–Ammodendron* dominated, plot e |
|---|---|---|---|---|---|---|
| $G_{max}$, green plant biomass | C | 762.0 | 165.0 | 480.0 | 5.2 | 55.5 |
| | N | 23.5 | 8.7 | 26.5 | 0.5 | 1.9 |
| | M | 43.2 | 18.2 | 55.0 | 1.0 | 16.1 |
| Pr, perennial aboveground plant biomass | C ($\times 10^{-3}$) | 12.3 | 14.3 | 18.5 | 0.02 | 0.3 |
| | N | 39.5 | 92.5 | 125.8 | 0.7 | 4.7 |
| | M | 127.8 | 248.0 | 589.2 | 1.4 | 10.9 |
| R, live belowground plant biomass | C ($\times 10^{-3}$) | 3.6 | 4.4 | 5.0 | 0.2 | 0.9 |
| | N | 31.5 | 50.0 | 65.3 | 4.9 | 20.9 |
| | M | 72.0 | 278.0 | 224.9 | 10.0 | 81.4 |
| D, dry stems and branches | C ($\times 10^{-3}$) | 4.5 | 4.3 | 3.7 | 0.01 | 0.1 |
| | N | 14.5 | 26.6 | 44.5 | 0.2 | 1.1 |
| | M | 46.8 | 67.5 | 115.0 | 0.7 | 3.0 |
| L, litter | C ($\times 10^{-3}$) | 1.5 | 0.6 | 0.05 | 0.01 | 0.1 |
| | N | 33.4 | 17.6 | 3.0 | 0.3 | 2.1 |
| | M | 111.0 | 49.0 | 15.0 | 0.8 | 20.4 |

*Characteristics of trial plots.* a. Spruce forests in southern taiga (*P. abies*). Sod-podzolic soils. Precip. 450–780 mm. Ave yearly temp. +1.4 to +6 °C; July +16.2 to +20.1 °C, Jan. −0.9 to −15.8 °C. USSR, Russian plain, ave data (Remesov *et al.*, 1959; Rodin & Bazilevich, 1967; Kolli & Reintam, 1970; Arvisto, 1970; Smirnov, 1971). b. Broad-leafed forests – oak and beech (*Q. robur, A. platanoides, U. laevis, F. silvatica, Aegopodium podagrari, Carex pilosa*). Grey and brown forest soils. Precip. 420 – 780 mm. Ave yearly temp. +2.3 to +10.2 °C; July +16.7 to +20.2 °C, Jan. +4.8 to −15 °C. Europe, average data (Remesov *et al.*, 1959; Rodin & Bazilevich, 1967). c. Tropical rain forests (*Piptademia africana, Ceiba pentandra, Terminalia superba, Diospyros, Strombosia lucida, Nauclea dideridii*, etc.). Red-yellow ferralitic soils. Precip. 2500 to > 3000 mm. Ave yearly temp. +25 °C; min. +25 °C; max. +26 °C. Ave data (Rodin & Bazilevich, 1967; Ovington & Olson, 1970; Kira & Ogava, 1971; Dilmy, 1971; Golley, 1975). d. *Artemisia*-dominated desert (*A. kemrudica, Anabasis salsa*). Grey-brown desert soils. Precip. 100 mm. Ave yearly temp. +15.5 °C; July +26.0 °C, Jan. −6.3 °C. USSR, plateau Ustyurt. (Rodin & Bazilevich, 1967.) e. *Haloxylon*-dominated desert (*H. ammodendron, Carex physodes*). Primitive sand solonchakous soils. Precip. 100 mm. Average yearly temp. +15.9 °C; July +29.7 °C, Jan. +0.5 °C. USSR, Karakum. (Chepurko *et al.* 1972.)

leafed forests, while in moist tropical forests nitrogen, calcium, potassium with silicon, iron and aluminium all play a very significant role (Rodin & Bazilevich, 1967). In deserts nitrogen, calcium, potassium with chlorine and sodium accumulate in appreciable amounts.

The C/N and C/M ratios of forest ecosystems are larger both in the green parts (taiga 32 and 18, broad-leafed forests 19 and 9, moist tropical forests 18 and 10, respectively) and particularly in the perennial lignified aboveground organs (310 and 96, 155 and 58, 147 and 32, respectively), than they are in grassland ecosystems. The same is true of underground organs, R. In contrast, in the green annual shoots of desert ecosystems these ratios are smaller (C/N = 10 and C/M = 5 in $G_{max}$ of *Artemisia* areas; values are 29 and 3.5 respectively in $G_{max}$ of shrubby desert with *Haloxylon aphyllum*), and similar to those of halophytic and hygrohalophytic meadows.

## The intensity of exchange processes

### *Production processes in grassland ecosystems*

#### *Net primary production, P*

This is composed of aboveground plant biomass production, $P_G$, and belowground plant biomass production, $P_R$. In the sequence of grassland

Table 9.4. *Net primary production, ($P_G$) and ($P_R$), in various grassland ecosystems and the uptake of nitrogen (N) and mineral elements (M) for the production of plant biomass* ($g \cdot m^{-2} \cdot yr^{-1}$)

| Index | Chemical elements | Steppes and prairies | Meadows: steppe-like, mesophytic, mesohalophytic, hygrophytic | Meadows: halophytic and hygrohalophytic | Grassy swamps (drying) |
|---|---|---|---|---|---|
| $P_G$, net primary production of green plant biomass | C | 150 ($n$ = 12)<br>44–360 | 250 ($n$ = 14)<br>126–508 | 146 ($n$ = 5)<br>36–224 | 370 ($n$ = 4)<br>214–600 |
| $P_R$, net primary production of belowground plant biomass | C | 356 ($n$ = 7)[a]<br>200–560 | 650 ($n$ = 11)<br>126–1240 | 320 ($n$ = 5)<br>120–490 | 1880 ($n$ = 2)<br>850–2920 |
| $P_G^N$, uptake of N for $P_G$ production | N | 6.2 ($n$ = 9)<br>2.5–12.0 | 8.8 ($n$ = 12)<br>5.0–12.0 | 6.2 ($n$ = 5)<br>2.2–9.8 | 13.5 ($n$ = 3)<br>6.5–24.0 |
| $P_R^N$, uptake of N for $P_R$ production | N | 11 ($n$ = 8)<br>3.0–22.5 | 15.8 ($n$ = 12)<br>1.1–46.5 | 4.6 ($n$ = 2)<br>1.7–7.5 | 58.5 ($n$ = 1) |
| $P_G^M$, uptake of M for $P_G$ production | M | 17.1 ($n$ = 9)<br>8.5–33.0 | 23.6 ($n$ = 12)<br>10.3–40.0 | 56.0 ($n$ = 5)<br>13.6–95.0 | 34.2 ($n$ = 3)<br>18.0–58.0 |
| $P_R^M$, uptake of M for $P_R$ production | M | 62.9 ($n$ = 8)<br>6.6–147.5 | 65.0 ($n$ = 11)<br>8.2–280.0 | 50.3 ($n$ = 2)<br>21.5–79.0 | 210.0 ($n$ = 1) |

[a] Solonetzic steppe ($P_R^C$ = 1280; $P_R^N$ = 22.5; $P_R^M$ = 92) was not included in the group of steppes.
Numerator, average value; denominator, minimum and maximum values; $n$, number of measurements.

ecosystems, $\bar{P} = (\bar{P}_G + \bar{P}_R)$ ranges from 466 g $C \cdot m^{-2} \cdot yr^{-1}$ (for halophytic and hygrohalophytic meadows) to 2250 g $C \cdot m^{-2} \cdot yr^{-1}$ (for grassy swamps), i.e. values vary by a factor of 5 (Table 9.4; Appendix 9.4). Therefore the variation in $P$ values is greater than that for biomass reserves G+R, where the maximum value is only 1.5 times the minimum.

The production of meadows (excluding halophytic meadows) is 1.8 times as large as that of steppes and prairies. The carbon cycle occurs more rapidly in meadows than in steppes and prairies, the values for $(G_{max}+R)/P$ being 0.9 year and 1.3 years, respectively. The equivalent value for grassy swamps is equal to about 0.4 year and for halophytic meadows 1.2 years. Thus the rate of plant biomass renewal is fastest in grassy swamps and slowest in steppes and prairies.

### *Aboveground plant production, $P_G$*

This is about 1.2–2.2 times as large as $G_{max}$, the difference being greatest in polydominant communities. $P_G$ varies by a factor of 2.5 between the different groups of grassland ecosystems, the largest values being found in grassy swamps and the smallest in steppes, prairies and halophytic meadows. The variation of $P_G$ within a group is greatest for steppes and prairies, much smaller for meadows and smallest for grassy swamps (Table 9.4). Thus, ecosystems that receive additional moisture have a more stable regime of production of aboveground organic matter. However, the $P_G$ value depends not only on environmental conditions, but also on the average reserve of photosynthesizing plant biomass, $\bar{G}$. An estimate* shows that $\bar{G} = 0.5 \times G_{max}$. The value of the $\bar{G}/P_G$ ratio characterizes the time of the carbon cycle in the aboveground

* Estimate based on a small quantity of data: $\bar{G}$ and $G_{max}$ were measured in parallel.

sphere of the community. It is found to be 0.31 year for steppes and prairies, 0.34 year for meadows, and 0.31 year for grassy swamps. Thus various types of grassland, though differing in plant species composition and ecological conditions, can be shown to be a uniform type of ecosystem on the basis of the length of the carbon cycle in the live aboveground parts of the plant community.

### *Belowground plant production, $P_R$*

In different groups of ecosystems this varies by as much as 5 times, reaching maximum values in drying grassy swamps and minimum values in halophytic meadows (Table 9.4). The range of $P_R$ values is smallest in steppe and prairie ecosystems and largest in meadows. It is likely that control mechanisms stabilize the increment in belowground organs in steppes and prairies but that it is the increment in aboveground organs that is more strongly stabilized in meadows.

The length of the carbon cycle in the belowground plant biomass, $\bar{R}/P_R$, is equal to 1.6 years for steppe and prairie ecosystems, 1 year for meadows (excluding halophytic meadows where there is only one estimate of the R value) and 0.35 year for grass swamps. Olson (1971) also found a large rate of root renewal not only for grassy, but also for low-bush communities; this renewal is realized mainly at the expense of small roots.

The various groups of grassland ecosystems differ sharply in their values of $\bar{R}/P_R$, the cycle time increasing the greater the moisture deficit and the nearer the ecosystem is to its climax vegetation.

$P_R$ accounts for about 90–100 % of the average reserve of live underground organs, $\bar{R}$, or 60–70 % of R + V. The value of $\bar{G}$, the average reserve of photosynthesizing plant biomass, can be treated as a compartment determining the productivity of the ecosystem, for the total primary production increases as the reserve of active photosynthesizing plant biomass grows. In meadow ecosystems (excluding halophytic meadows) the average $\bar{G}$ is 1.8 times as large as it is in steppe and prairie ecosystems. The primary production, $P$, is also 1.8 times higher in meadows. The relatively high value of $P_R$ in meadows as compared with steppes and prairies is due both to the general increase in organic matter production by the photosynthesizing plant biomass and to the structure of the live plant biomass reserves. In steppes and prairies 6 g of belowground organs are available to provide water and nutrients for 1 g of photosynthesizing biomass, whereas in meadows the figure is only 3.7 g. Therefore in steppes and prairies the amount of energy used for respiration of live belowground organs is 1.6 times as large as in meadows. This larger energetic expenditure for respiration of belowground organs and the smaller photosynthesizing plant biomass are important factors in reducing the total production and $P_R$ in steppes and prairies as compared with meadows.

A pool of live belowground organs is preserved in steppes and prairies for a longer period of time than in meadows. In meadows R is renewed within 1 year, but in steppes and prairies it takes up to 1.6 years.

### *Nutrient uptake for primary production in grassland ecosystems*

Uptake of nutrients for annual primary production is directly dependent on both $P_G$ and $P_R$. Considerable differences exist between various types of grassland ecosystems (Table 9.4; Appendix 9.4). Uptake of nitrogen for formation of $P_G$ increases in parallel with production in all types of grassland ecosystems (Fig. 9.2*a*, curves 1, 3, 5 and 7); at the same time the uptake of minerals increases smoothly only in steppes, prairies and grassy swamps (curves 2 and 8). In the group comprised of steppe-like, mesophytic, meso-halophytic and hygrophytic meadows, the field of points (field 4) is above the curves for steppes, prairies and grassy swamps. This shows that the uptake of mineral elements for the formation of each unit of green plant production is more active in meadows, a result of the more important role of herbs and the better development of the leaf surface in these stands.

The field of points for halophytic meadows is located above all the others (Fig. 9.2*a*, field 6), and the C/M ratio of 3.6 here is the smallest (reaching a minimum of 0.5 in succulent halophytic meadows). This is the result of the dominance of halophyte species, whose green parts actively accumulate mineral elements, and biohalogens (chlorine, sodium and sulphur) in particular. Despite the small production of green parts, due to the scanty vegetation cover of halophytic meadows, the values of mineral uptake per unit of $P_G$ formed are the largest of all the types of grassland ecosystems.

The uptake of nutrients for the production of belowground organs is directly proportional to the amount of root production, $P_R$ (Fig. 9.2*b*). However, if the accumulation of nitrogen is subordinate to this general trend, the uptake of minerals is characterized by the same differences which were observed when we analysed the uptake of mineral elements for formation of $P_G$. In relative terms the largest mineral uptake is by the belowground organs of halophytic and hygrohalophytic meadows, but the greatest absolute values are characteristic of meadows (excluding halophytic and hygrohalophytic ones) and grassy swamps, where the $P_R$ indices are also the largest (Fig. 9.2*b*, curves 2, 4, 6).

When evaluating the reserves of nutrients (N and M) in the compartments of aboveground and belowground plant biomass, the leading role was always attributed to the belowground organs (except in succulent halophytic meadows where mineral reserves in the $G_{max}$ compartment exceed the reserves in R or R+V); however, this generalization does not always hold. Nutrient reserves in $P_G$ were found to be greater than those in $P_R$ in the following stands: steppe-like meadows of Western Siberia (plot 3.3; Appendixes 9.2

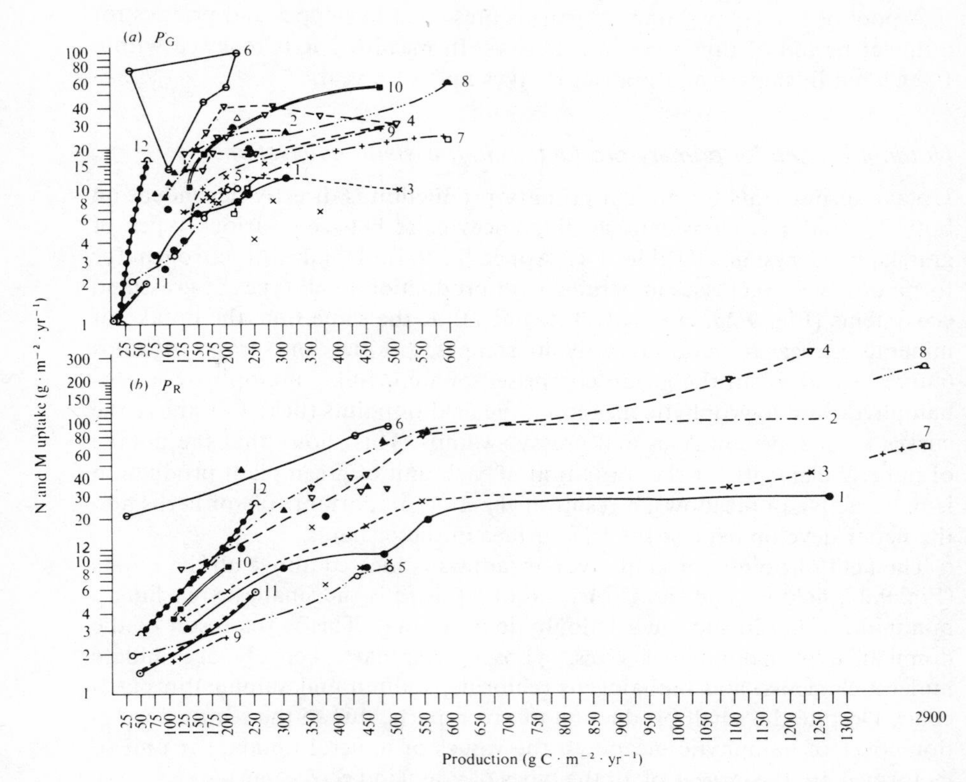


Fig. 9.2. Relationship between net primary production of (*a*) green plant biomass ($P_G$) and (*b*) belowground biomass ($P_R$) and uptake of N and M in the various types of ecosystems. 1, Steppes and prairies, N; 2, steppes and prairies, M; 3, meadows (steppe-like, mesophytic, mesohalophytic, hygrophytic), N; 4, meadows (steppe-like, mesophytic, mesohalophytic, hygrophytic), M; 5, meadows (halophytic, hygrohalophytic), N; 6, meadows (halophytic, hygrohalophytic), M; 7, grassy swamps, N; 8, grassy swamps, M; 9, forests, N; 10, forests, M; 11, deserts, N; 12, deserts, M.

and 9.4), piedmont moist meadows of Germany and Japan (plots 4.3 and 4.4), some periodically flooded meadows with high water content in Czechoslovakia (plot 7.1) and succulent halophytic meadows in Siberia (plot 6.2). This can be accounted for either by the predominance of aboveground plant production, $P_G$, over the production of belowground organs, $P_R$ (plots 3.3, 4.4 and 7.1), or by a considerably more active uptake of nutrients by aboveground organs as compared with those below ground (for instance in succulent halophytic meadows, plot 6.2).

Thus grassland ecosystems may be arranged in the following sequence according to their total production ($P_G+P_R$) and intake of nitrogen: swamps > meadows (steppe-like, mesophytic, mesohalophytic, hygrophytic) > steppes and prairies > halophytic and hygrohalophytic meadows. In terms of the uptake of minerals, halophytic meadows are transferred to second place and the sequence becomes: grassy swamps > halophytic meadows > meadows (steppe-like, mesophytic, mesohalophytic, hygrophytic) > steppes and prairies.

The main nutrients that are taken up are similar to those accumulated in the compartments of the aboveground and belowground plant biomass.

### *Production processes in forest and desert ecosystems*

In contrast to grassland ecosystems, in forests and deserts a considerable part of the annual production is accounted for by perennial lignified aboveground organs ($P_{Pr}$): 40–60 % in forests and 2–9 % in deserts (Table 9.5). However, the total production both in forest and desert ecosystems is within the same limits as the production of grasslands, ranging from 54 g $C \cdot m^{-2} \cdot yr^{-1}$ in *Artemisia* desert up to 1390 g $C \cdot m^{-2} \cdot yr^{-1}$ in wet tropical forests. Again in contrast to grasslands, the smallest share of the production in forests is that of belowground organs, $P_R$ (about 25 %). In desert ecosystems 76–88 % of the total annual production is accounted for by root production.

The value for annual production is proportional to the average reserves of actively photosynthesizing green plant biomass.* In the temperate belt average reserves of G in spruce forests are about 70 g $C \cdot m^{-2}$ and those in the broad-leafed forests about 85 $g \cdot m^{-2}$, i.e. somewhat higher than in climax steppes and prairies ecosystems where the average value is about 50 g $Cm.^{-2}$. But the total annual production in these forests is smaller (about 460 g $C \cdot m^{-2} \cdot yr^{-1}$) than that in steppes and prairies (an average of 500 g $C \cdot m^{-2} \cdot yr^{-1}$). Thus every gram of photosynthesizing green plant biomass in forests creates 4–6 $g \cdot yr^{-1}$ of new biomass as compared with 10 $g \cdot yr^{-1}$ in steppes and prairies. This is the result of both the biological characteristics of grasses and the longer growing period in the steppe zone compared with the forest zone (180 and 140 days, respectively). In desert ecosystems extreme moisture limitation means that despite a longer growing period (up to 280 days) each gram of photosynthesizing green plant biomass is able to create only in the order of 2 $g \cdot yr^{-1}$ (*Artemisia* grassland) or 1.2 $g \cdot yr^{-1}$ (shrubby desert with *Haloxylon aphyllum*). In wet tropical forest the average reserve of photosynthesizing green plant biomass is equal to $G_{max}$, i.e. 480 g $C \cdot m^{-2}$, and production is 1390 g $C \cdot m^{-2} \cdot yr^{-1}$. Thus 1 g of photosynthesizing plant biomass creates 2.9 g new biomass during a year – a smaller value than for temperate forests despite a high moisture supply in the tropics. A factor limiting annual production here is the supply of nutrients. The soils are

* It was assumed that average reserves of G are approximately $0.5G_{max}$. In coniferous forests only needles of the current year are included as actively photosynthesizing green plant biomass.

Table 9.5. *Net primary production and N and M uptake for production of green plant biomass ($P_G$), perennial plant biomass ($P_{Pr}$) and belowground plant biomass ($P_R$) ($g \cdot m^{-2} \cdot yr^{-1}$)*

| Index | Chemical element | Forest ecosystems: Spruce forests in southern taiga, plot a | Broad-leafed forests, plot b | Tropical rain forests, plot c | Desert ecosystems: *Artemisia* dominated, plot d | *Haloxylon–Ammodendron*-dominated (shrubby) plot e |
|---|---|---|---|---|---|---|
| $P_G$, net primary production of green plant biomass | C | 138.5 | 165.0 | 80.0 | 5.4 | 55.5 |
| $P_{Pr}$, net primary production of perennial plant biomass | C | 212.0 | 180.0 | 710.0 | 1.1 | 28.8 |
| $P_R$, net primary production of roots | C | 111.3 | 111.5 | 200.0 | 47.7 | 250.0 |
| $P_G^N$, uptake of N for $P_G$ production | N | 5.2 | 8.7 | 26.5 | 0.5 | 1.9 |
| $P_{Pr}^N$, uptake of N for $P_{Pr}$ production | N | 1.4 | 1.3 | 4.8 | trace | 0.5 |
| $P_R^N$, uptake of N for $P_R$ production | N | 1.7 | 1.7 | 2.5 | 1.5 | 5.6 |
| $P_G^M$, uptake of M for $P_G$ production | M | 10.2 | 18.2 | 55.0 | 1.3 | 1.1 |
| $P_{Pr}^M$, uptake of M for $P_{Pr}$ production | M | 3.3 | 2.7 | 15.3 | 0.1 | 1.5 |
| $P_R^M$, uptake of M for $P_R$ production | M | 3.6 | 4.3 | 8.8 | 2.7 | 24.9 |

heavily leached and this results in the supply of nitrogen, phosphorus and potassium being insufficient. This is probably the main cause of the small relative production capacity of these forests. Their large absolute annual production is due to a large reserve of photosynthesizing plant biomass and the long (365-day) growing period.

The rate of the carbon cycle in forest and desert ecosystems is markedly slower than it is in grassland ecosystems. For example, $(G_{max} + Pr + R)/P$ in forests totals tens of years and in deserts about 4 years, whereas in grasslands a complete carbon cycle takes only 0.4–1.3 years. The carbon cycle in the annual organs (G) in forest and desert ecosystems is also slower than in grassland. Values for broad-leafed, wet tropical and spruce forests are 0.5, 1 and 5–6 years, respectively. The longest period of time for the carbon cycle (several dozens of years) is required for the lignified aboveground organs in forests and deserts. It is somewhat shorter for R, and it can be clearly observed that the length of the carbon cycle in Pr and R in forests is directly proportional to the moisture deficit. The same is true of various desert ecosystems.

We can conclude from the above considerations that both the production capacity of each unit of photosynthesizing green plant biomass and the rate of the carbon cycle are greater in grassland ecosystems than in forest and desert ecosystems.

### *Nutrient uptake for primary production in forest and desert ecosystems*

Uptake of nutrients for $P_G$, $P_R$ and $P_{Pr}$ in forest and desert as well as in grassland ecosystems is directly proportional to production (Table 9.5). However, the concentration of nitrogen and minerals per unit of carbon in $P_R$ and $P_{Pr}$ in forests (and $P_G$ in spruce forests) is considerably smaller than in grassland ecosystems (Tables 9.3, 9.4; Fig. 9.2*a*, *b*). This is the result of less active accumulation of mineral nutrient elements by woody species. In contrast, the uptake of nitrogen and minerals in desert ecosystems is greater, a phenomenon associated with a noticeable retention of biohalogen elements in the plants (especially *Haloxylon aphyllum*). This brings deserts nearer to halophytic meadows. Broad-leafed forests of the temperate belt and wet tropical forests are more similar to grassland ecosystems in terms of the mineral and nitrogen concentrations in $P_G$. However, the content of mineral elements in the trees of wet tropical forests is somewhat peculiar, for despite a considerably greater value of $P_G$ in wet tropical forests relative to temperate broad-leafed forests, uptake of calcium, potassium, magnesium and phosphorus is similar in the two habitats. The reason for this is that tropical forests are poor in these elements, especially calcium, which is replaced in the biological cycles by iron, aluminium and manganese. These are as characteristic of tropical plant ash as silicon is (Rodin & Bazilevich, 1967).

The rates of the carbon and nutrient cycles in forest and desert ecosystems are much slower than they are in grassland ecosystems (with the exception of potassium which is actively washed out by precipitation from the tree canopy of forests and repeatedly reutilized). Every season forest trees fix in perennial lignified organs most of the carbon and nutrients they have taken up, thus excluding them for many years from the biological cycles.

## Models of exchange processes in grassland ecosystems

Processes of biomass production and nutrient uptake occur against a background of ecological and geochemical conditions on which they are dependent and with which they are interrelated. Plant growth and development, nutrient uptake, the grazing of plants by animals, the dying off of plants, transformation of plant remains and a number of other phenomena are related to biological cycles and constitute the biotic processes. Another group of processes, for example movement of chemicals in solution, their translocation by wind and their gravitational movement, etc., constitutes the abiotic processes.*

Construction of a graphical model of exchange processes connecting the compartments of one ecosystem and that ecosystem with others is the first step in mathematical modelling (Fig. 9.3). The model should reflect the structure of the ecosystem. The insertion into this graphical model of empirical

* The division of processes into these two groups is, of course, somewhat arbitrary.

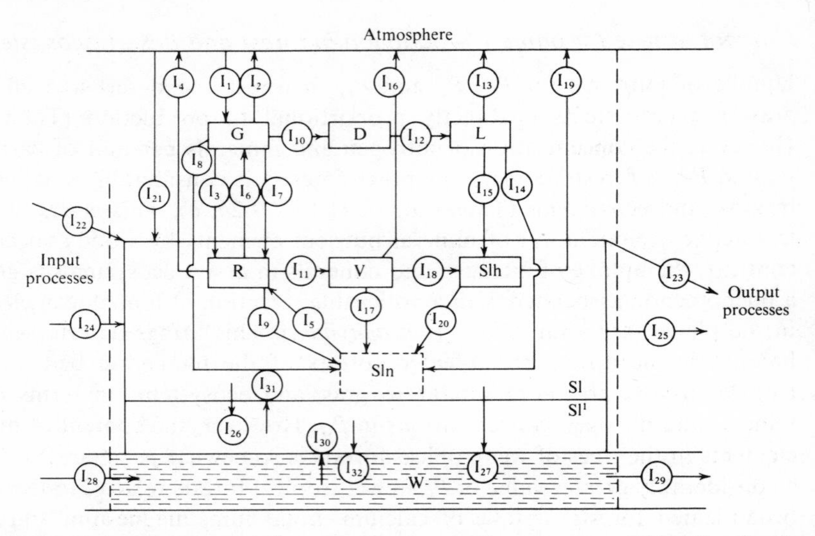


Fig. 9.3. The compartments and fluxes in the grassland ecosystem model. Sl′, subsoil; Slh, soil humus; Sln, soil nutrients; G, green plants; R, roots; D, standing dead; L, litter; W, water.

data characterizing both the reserves and fluxes of matter at a certain point in time is the second step in the modelling process. We call such a model the state of an ecosystem at a certain point in time. When we construct many such models in accordance with changes both in the reserves and in the intensities of fluxes with time, they will reflect the model of functioning of the ecosystem. A knowledge of the functioning of an ecosystem over a given period of time makes it possible to forecast its functioning over a longer time scale and to predict with some degree of certainty its new state as the result of changes in external conditions. A mathematical model of the functioning of an ecosystem should make it possible to produce more accurate predictions for longer periods of time.

A graphical model containing empirical data is called by the present authors a model of exchange processes. Such models were constructed for a number of grassland ecosystems of the USSR. In the models presented here we do not consider processes of plant biomass consumption or processes of nitrogen exchange between soil and atmosphere (the dead bodies and excrement of consumers are considered together with the litter). Zlotin & Khodashova (1973), who carried out investigations in the meadow-steppe of the Russian Plain, have established that phytophagous insects consume on average 5–6 % of the total plant biomass production (though the percentage may be considerably higher for monoculture fodder crops). The heterotrophic component of an ecosystem is considered mainly as a biological regulator, since animal and microbial populations exert predominantly an indirect influence on production processes. The direct role of animals and micro-organisms in plant destruction processes (decomposition and humification) is large. Similar conclusions have been arrived at by Petrusewicz & Grodziński (1973). However, we found it possible in the first approximation to neglect the losses of plant biomass due to consumers.

### *Description of the model*

Models incorporate the balance of the following chemical elements: carbon, nitrogen, potassium, calcium, magnesium, sodium, iron, aluminium, manganese, titanium, phosphorus, sulphur and chlorine. However, in characterizing the biotic reserves and fluxes of these elements we consider separately only carbon (C) and nitrogen (N), and group the other elements into the category 'minerals' (M). When describing the abiotic processes connected with water movements we consider carbon, nitrogen and two groups of other chemical elements:

(*a*) a group of elements found in readily soluble salts, S (chlorine, sulphur, magnesium, calcium, sodium, potassium, and carbon in carbonate and bicarbonate ions);

(*b*) a group of elements found in other mobile compounds, m (ionic, molecular, colloidal and suspensions). These are silicon, iron, aluminium, manganese, titanium, magnesium, calcium, potassium, sodium and phosphorus.

All data in the model are expressed in terms of elements.

Besides the compartments G, R, D, L, V, Sl and W described above, every model also includes the atmosphere (A), subsoil (Sl′), soil humus (Slh) and reserve of soil nutrients (Sln).

Using the data obtained in long-term investigations at the permanent research station at Karachi in Western Siberia, we have constructed models of the annual cycles of elements for the three most characteristic types of grassland ecosystems (Figs. 9.4, 9.5 and 9.6): meadow-steppe on ordinary chernozems (plot 1.3), mesohalophytic meadow on chernozem-meadow-solonetzic-solonchakous soils (plot 5.2), and drying grass swamp on peaty swampy soil (plot 9.1). The characteristics of all the plots studied and the reserves of chemical elements in the compartments of the model are presented in Appendixes 9.1 and 9.2. Data for A, Sl and Sln are not given. The values for the system of fluxes in the models are presented in Tables 9.6 and 9.7.

### *Functional models of grassland ecosystems*

In a *meadow-steppe ecosystem* biotic exchange processes occur at a much greater rate than the abiotic processes (Fig. 9.4). Production processes are accompanied by uptake of nutrients by roots from Sln, translocation from R to G (nutrients taken from soil) and translocation of synthesized substances from G to R. At the same time a certain amount of nutrients flows out from

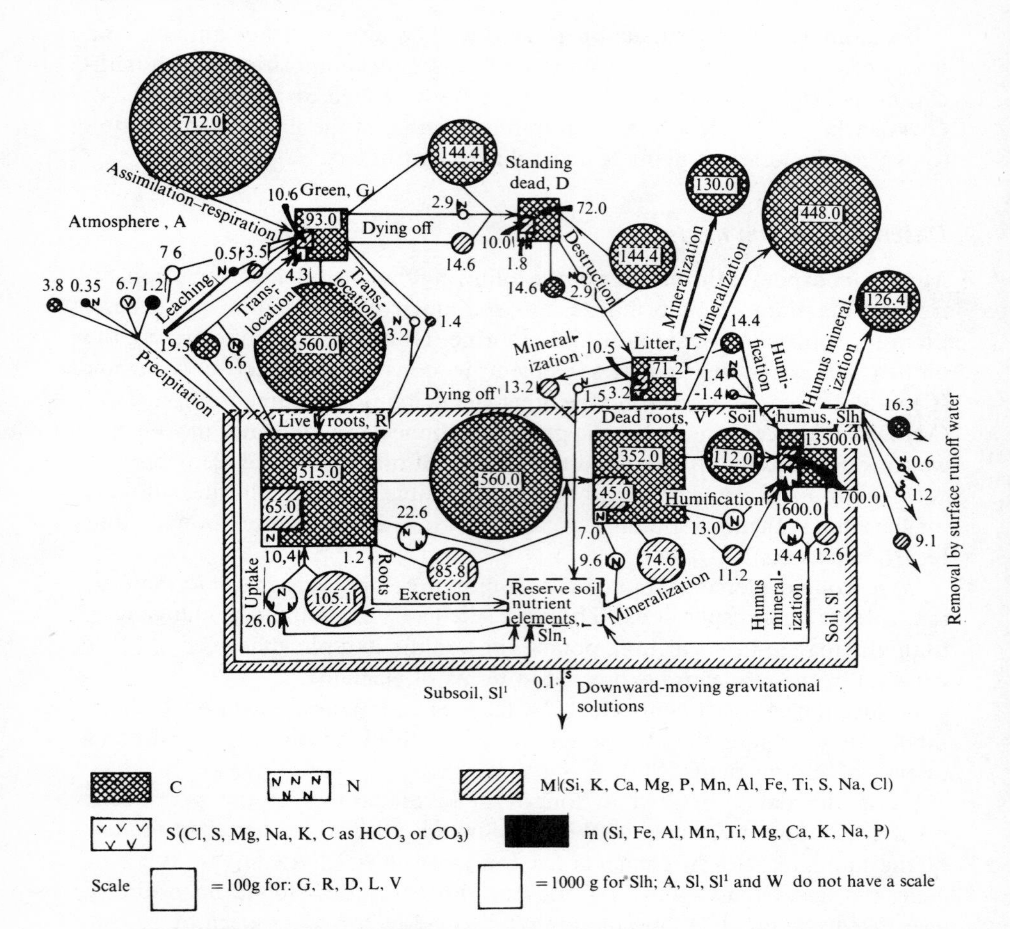


Fig. 9.4. A model of carbon (C), nitrogen (N) and mineral element (M) cycling in a meadow-steppe ecosystem. (Western Siberia, Karachi experimental station, chernozem soil, plot 1.3.) Squares represent reserves of chemical elements ($g \cdot m^{-2}$), arrows processes, and circles the rates of exchange processes ($g \cdot m^{-2} \cdot yr^{-1}$).

G to R and some are also washed out from G to Sl by precipitation. A small quantity of nitrogen and minerals is lost to the soil in root exudates. The processes of dying off of aboveground and belowground plant organs are accompanied by the storage of chemical substances in V (dead underground remains), D (standing dead) and, during the course of further degradation, in L (litter). Subsequent transformations lead to the mineralization of most (up to 85–90 %; see Resources of the Biosphere, 1975) of the dead organic matter, carbon being released to the atmosphere as carbon dioxide and nitrogen and minerals accumulating in the reserve of soil nutrients (Sln).

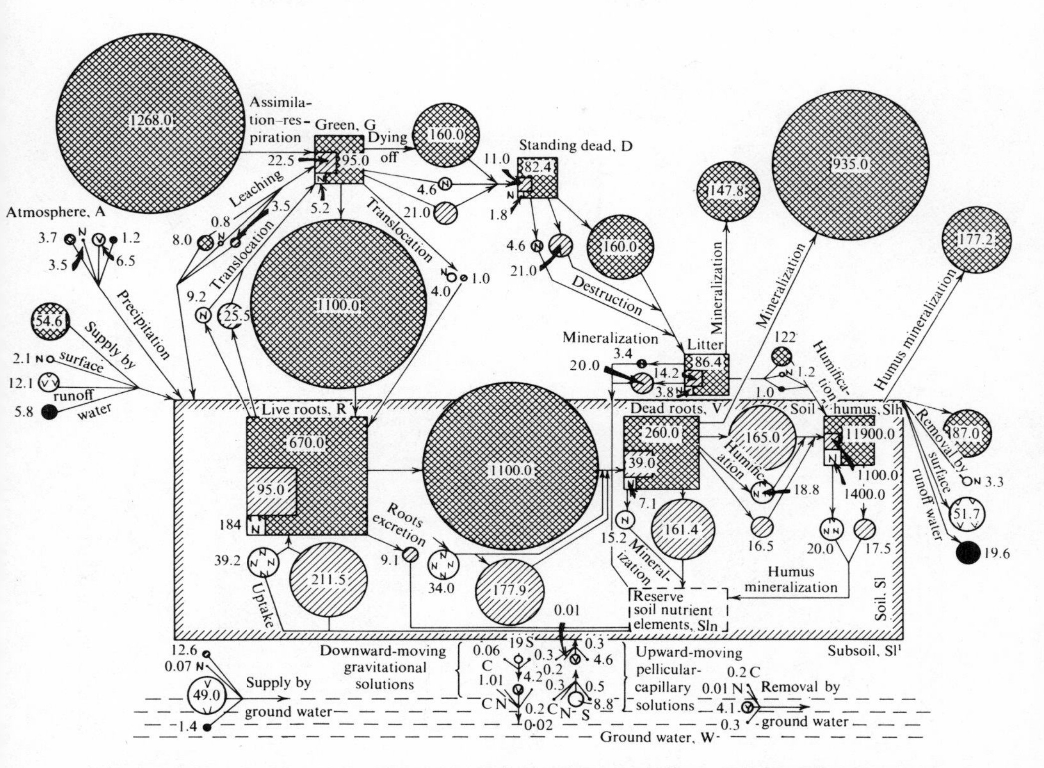


Fig. 9.5. A model of carbon (C), nitrogen (N) and mineral element (M) cycling in a meso-halophytic meadow. (Western Siberia, Karachi experimental station, chernozem meadow soil, plot 5.2.) Details and key as in Fig. 9.4.

A smaller fraction is fixed in the soil humus (Slh) in the course of the humification of aboveground and belowground plant remains. At the same time there is continuous mineralization of humic substances, releasing carbon as carbon dioxide and returning nitrogen and minerals to the soil. Thus the biological cycle of chemical elements is completed.

Both inflows and outflows of a number of chemical elements in the cycle occur against the background of abiotic exchange processes. In a meadow-steppe, precipitation ($I_{21}$) is essential in the exchange of chemical elements; surface-water flow ($I_{23}$) also plays a certain role in the transportation of elements.

An increase in salts (S) is compensated to a certain degree by removal of dead plant remains by wind; this factor is ignored in the 'balance sheet'. The loss of mobile compounds (m) reflects recent soil processes associated with

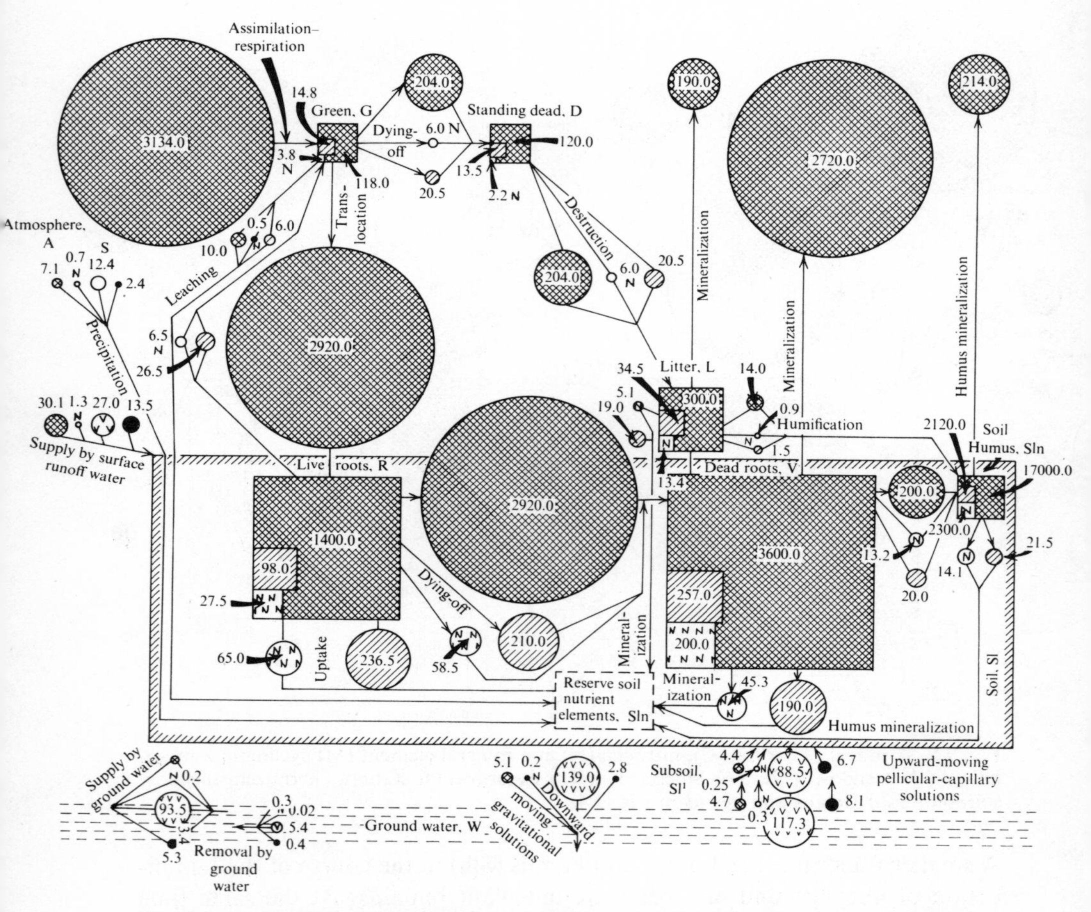


Fig. 9.6. A model of carbon (C), nitrogen (N) and mineral element (M) cycling in a grassy swamp ecosystem. (Western Siberia, Karachi experimental station, peat, low moor soil, plot 9.1.) Details and key as in Fig. 9.4.

surface denudation and occurs also in climax ecosystems, which constantly approach a steady state but practically never reach it. In the case of crop plants or hay the removal of chemical elements with yield considerably exceeds the surplus inflow of elements into the ecosystem.

In the *mesohalophytic meadow ecosystem* the biotic fluxes of carbon, nitrogen and minerals are much more intensive and the abiotic fluxes are also greater than in meadow-steppe (Fig. 9.5). Of the abiotic fluxes the inflow and outflow of chemical substances with surface runoff and their supply with ground-water flow have the greatest intensities.

A positive balance of salts, S, exists for the meadow ecosystem at the expense of influx $I_{28}$ (ground water). However, the salt balance of the soil strata is negative. Under meadows harvested for hay this negative balance is increased due to the removal of mineral elements with yield. The increasing salinization of ground water where the water level is rising increases intensity of flux $I_{31}$ (inflow of chemical elements to soil from ground water in pellicular-capillary ascending solutions) and changes the salt balance from negative to positive.

In the *grassy swamp ecosystem* the intensity of carbon and nitrogen processes associated with the biological cycle increases even more (Fig. 9.6). The intensity of abiotic exchange processes also increases.

In the annual cycle of minerals in the swamp ecosystem a positive balance of salts is formed at the expense of influx $I_{28}$, i.e. an increasing salinization of ground water occurs. However, in the soil layer at a depth of 1 m the salt balance is negative. The soil salt balance may be expected to change from negative to positive if ground-water level rises in the warm season and there are increases in the intensities of fluxes $I_{30}$ and $I_{31}$ (migration of chemical elements with ascending pellicular-capillary solutions).

## *Functioning of grassland ecosystems*

The analysis of functional models of grassland ecosystems allows us to make several general statements.

### *The biological cycle*

The biological cycle (including transport processes) consists mainly of processes of matter transformation, i.e. the synthesis of organic and organo-mineral compounds, their transformation, re-synthesis and degradation to simpler compounds. All the transformation processes together contain the movement of chemical elements within a single ecosystem.

### *Abiotic processes*

Abiotic processes (including transformation processes) are basically transport processes and play a major role in the movement of materials between ecosystems.

### *The degree of ecosystem openness*

Chemical elements are involved in both biotic and abiotic processes, and it is possible to calculate an index characterizing the relative roles of these two types of processes in the migration of any given element. The ratio of the

Table 9.6. *The intensity of some abiotic exchange processes in grassland ecosystems* ($g \cdot m^{-2} \cdot yr^{-1}$)

| Flow | Process | Experimentally measured value | Meadow-steppe, plot 1.3 | | | | Mesohalophytic meadow, plot 5.2 | | | | Grassy swamp (drying), plot 9.1 | | | |
|---|---|---|---|---|---|---|---|---|---|---|---|---|---|---|
| | | | C | N | S | m | C | N | S | m | C | N | S | m |
| A → Sl | Inflow of chemical elements with precipitation | $I_{21}$ | 3.8 | 0.4 | 6.7 | 1.2 | 3.7 | 0.4 | 6.5 | 1.2 | 7.1 | 0.7 | 12.4 | 2.4 |
| Sla → Sl | Inflow of chemical elements with surface-water flow | $I_{22}$ | 0 | 0 | 0 | 0 | 54.6 | 2.1 | 12.1 | 5.8 | 30.1 | 1.3 | 27.0 | 13.9 |
| Sl → Slb | Outflow of chemical elements with surface-water flow | $I_{23}$ | 16.3 | 0.6 | 1.2 | 9.1 | 87.0 | 3.3 | 51.7 | 19.6 | 0 | 0 | 0 | 0 |
| Sla → Sl | Inflow of chemical elements with lateral subsurface water flow | $I_{24}$ | 0 | 0 | 0 | 0 | 0.1 | trace | 0.3 | trace | 0.6 | trace | 0.3 | trace |
| Sl → Slb | Outflow of chemical elements with lateral subsurface water flow | $I_{25}$ | trace | trace | 0.1 | trace | 0.1 | trace | 0.2 | 0.1 | 0.2 | trace | 0.5 | trace |
| Sl → Sl′ | Outflow of chemical elements from soil to subsoil with downward-moving gravitational solutions | $I_{26}$ | trace | 0 | 0.1 | 0 | 0.1 | 0 | 1.9 | 0.3 | 0 | 0 | 0 | 0 |
| Sl → W | Outflow of chemical elements from soil to ground water with downward-moving gravitational solutions | $I_{27}$ | 0 | 0 | 0 | 0 | 0 | 0 | 0 | 0 | 5.1 | 0.2 | 139.0 | 2.8 |
| Wa → W | Inflow of chemical elements with flow of ground water | $I_{28}$ | 0 | 0 | 0 | 0 | 1.3 | 0.1 | 49.0 | 1.4 | 3.4 | 0.2 | 93.5 | 5.8 |
| W → Wb | Outflow of chemical elements with flow of ground water | $I_{29}$ | 0 | 0 | 0 | 0 | 0.2 | trace | 4.1 | 0.3 | 0.3 | trace | 5.4 | 0.4 |
| W → Sl′ | Inflow of chemical elements from ground water to subsoil with upward-moving pellicular-capillary solutions | $I_{30}$ | 0 | 0 | 0 | 0 | 0.3 | trace | 8.8 | 0.5 | 4.7 | 0.3 | 117.3 | 8.1 |
| Sl′ → Sl | Inflow of chemical elements from subsoil to soil with upward-moving pellicular-capillary solutions | $I_{31}$ | 0 | 0 | 0 | 0 | 0.2 | trace | 4.6 | 0.3 | 4.4 | 0.3 | 88.5 | 6.7 |
| Sl′ → W | Outflow of chemical elements from subsoil to ground water | $I_{32}$ | 0 | 0 | 0 | 0 | 0.1 | trace | 4.2 | 0.2 | 0 | 0 | 0 | 0 |

Table 9.7. *Intensity of some biotic exchange processes in the grassland ecosystems* ($g \cdot m^{-2} \cdot yr^{-1}$)

| Flow | Process | Experimentally measured value | Meadow steppe, plot 1.3 | | | Mesohalophytic meadow, plot 5.2 | | | Grassy swamp (drying), plot 9.1 | | |
|---|---|---|---|---|---|---|---|---|---|---|---|
| | | | C | N | M | C | N | M | C | N | M |
| A → G | Photosynthesis ($I_1$) | Net primary production of G | — | — | — | — | — | — | — | — | — |
| G → A | Respiration of G ($I_2$) | | — | — | — | — | — | — | — | — | — |
| G → R | Translocation of organic matter synthesized in G to belowground organs ($I_3$) | $P_G = I_1 - I_2 - I_3$ | 152.0 | — | — | 168.0 | — | — | 214.0 | — | — |
| R → A | Respiration of R ($I_4$) | Net primary production of R; $P_R = I_3 - I_4$ | 560.0 | — | — | 1100.0 | — | — | 2920.0 | — | — |
| Sl → R | Intake of chemical substances from soil by belowground organs ($I_5$) | Utilization of chemical elements by belowground organs, $P_R^{N,M} = I_5 - I_6$ | — | 19.4 | 85.6 | — | 30.0 | 186.0 | — | 58.5 | 210.0 |
| R → G | Translocation of chemical substances from belowground organs to aboveground organs ($I_6$) | Utilization of chemical substances by G ($I_6$), $P_G^{N,M}$ | — | 6.6 | 19.5 | — | 9.2 | 25.5 | — | 6.5 | 26.5 |
| G → R | Translocation of chemical substances from aboveground organs to belowground organs ($I_7$) | $I_7$ | — | 3.2 | 1.4 | — | 4.0 | 1.0 | — | — | — |
| G → Sl | Removal of chemical substances by precipitation from aboveground organs to soil | $I_8$ | 7.6 | 0.5 | 3.5 | 8.0 | 0.8 | 3.5 | 10.0 | 0.5 | 6.0 |
| R → Sl | Root excretion into soil | $I_9$ | — | — | 1.2 | — | — | — | — | — | — |
| G → D | Dying off of aboveground organs and their transformation into standing dead | $I_{10}$ | 144.4 | 2.9 | 14.6 | 160.0 | 4.6 | 21.0 | 204.0 | 6.0 | 20.5 |
| R → V | Dying off of belowground organs and their transformation into dead belowground plant remains | $I_{11}$ | 560.0 | 22.6 | 85.5 | 1100.0 | 34.0 | 177.9 | 2920.0 | 58.5 | 210.0 |
| D → L | Transformation of standing dead into litter | $I_{12}$ | 144.4 | 2.9 | 14.6 | 160.0 | 4.6 | 21.0 | 204.0 | 6.0 | 20.5 |
| L → A<br>→ Sl | Litter mineralization; transfer of C to atmosphere and N and M to soil | $I_{13}$, $I_{14}$ | 130.0 | 1.5 | 13.2 | 147.8 | 3.4 | 20.0 | 190.0 | 5.1 | 19.0 |
| L → Slh | Humification of litter | $I_{15}$ | 14.4 | 1.4 | 1.4 | 12.2 | 1.2 | 1.0 | 14.0 | 0.9 | 1.5 |
| V → A<br>→ Sl | Mineralization of dead belowground plant remains and transfer of C to atmosphere and N and M to soil | $I_{16}$, $I_{17}$ | 448.0 | 9.6 | 74.6 | 935.0 | 15.2 | 161.4 | 2720.0 | 45.3 | 190.0 |
| V → Slh | Humification of dead belowground plant remains | $I_{18}$ | 112.0 | 13.0 | 11.2 | 165.0 | 18.8 | 16.5 | 200.0 | 13.2 | 20.0 |
| Slh → A<br>→ Sl | Humus mineralization; transfer of C to atmosphere and N and M to soil | $I_{19}$ | 126.4 | 14.4 | 12.6 | 177.2 | 20.0 | 17.5 | 214.0 | 14.1 | 21.5 |

sum of the intensities of abiotic (ecosystem input or output) processes to the intensity of uptake of an element by plants during production processes may be such an index. First it is assumed that input and output of elements to and from the ecosystem take place only via water transport. The input abiotic processes that are considered (Table 9.6) are the supply of chemical elements with precipitation (A $\rightarrow$ Sl), the supply with surface water and subsurface lateral flow (Sla $\rightarrow$ Sl), and transport from subsoil to soil with ascending pellicular-capillary solutions (Sl′ $\rightarrow$ Sl). The corresponding output abiotic processes (Table 9.7) are removal of chemical elements from soil with surface water, subsurface lateral water flows (Sl $\rightarrow$ Slb) and gravitationally descending solutions (Sl $\rightarrow$ Sl′ or Sl $\rightarrow$ W). The intensity of biotic processes is taken as the net primary production – $P_G^C + P_R^C$ for carbon, $P_G^N + P_R^N$ for nitrogen, and $P_G^M + P_R^M$ for minerals (Table 9.5; Appendix 9.4).

Thus indices of the degree of ecosystem openness ($K$) are:

For carbon input: $K_{in}^C = (I_{21}^C + I_{22}^C + I_{24}^C + I_{31}^C)/(P_G^C + P_R^C)$ (cf. Fig. 9.3).
For carbon output: $K_{out}^C = (I_{23}^C + I_{25}^C + I_{26}^C)/(P_G^C + P_R^C)$ (cf. Fig. 9.3).
For nitrogen input: $K_{in}^N = (I_{21}^N + I_{22}^N + I_{24}^N + I_{31}^N)/(P_G^N + P_R^N)$.
For nitrogen output: $K_{out}^N = (I_{23}^N + I_{25}^N + I_{26}^N)/(P_G^N + P_R^N)$.
For mineral, or (S+m), input: $K_{in}^M = (I_{21}^S + I_{21}^m + I_{22}^S + I_{22}^m + I_{24}^S + I_{24}^m + I_{31}^S + I_{31}^m)/(P_G^M + P_R^M)$.
For mineral, or (S+m), output: $K_{out}^M = (I_{23}^S + I_{23}^m + I_{25}^S + I_{25}^m + I_{26}^S + I_{26}^m)/(P_G^M + P_R^M)$.

In evaluating $K_{out}^C$ and $K_{in}^C$ the denominator is the total $P_G^C + P_R^C$, because in the long-term cycle primary production is equivalent to all losses of organic matter in the process of mineralization to carbon dioxide. The same is true for nitrogen and minerals.

The values of the index for various chemical elements estimated for three grassland ecosystems are presented in Table 9.8.

Biophile elements carbon and nitrogen participating in transformation processes are retained in the ecosystem by the biological cycle. The index of openness for them is lowest in meadow-steppe. Equivalent values in meso-halophytic meadows and grassy swamps are somewhat larger due to large concentrations of water-soluble forms of humus participating in water migration. Mineral elements (M or S+m) are included in transport processes and take an active part in exchange between ecosystems; their index is higher than that for carbon or nitrogen. In meadow and swamp ecosystems where solutions circulate that are enriched both with soluble salts and other mobile mineral compounds, the index for minerals increases, and is from 2 to 20 times as large as that for carbon and nitrogen.

The nearer the ecosystem is to a steady state the more closed is its cycling of matter and the smaller is the role of abiotic processes. Of the ecosystems

Table 9.8. *Indices of the degree of ecosystem openness*

| Element | Index | Steppe meadow, plot 1.3 | Mesohalophytic meadow, plot 5.2 | Grassy swamp (drying), plot 9.1 |
|---|---|---|---|---|
| C | $K^{C}_{in}$ | 0.005 | 0.047 | 0.140 |
| | $K^{C}_{out}$ | 0.023 | 0.070 | 0[a] |
| N | $K^{N}_{in}$ | 0.013 | 0.090 | 0.035 |
| | $K^{N}_{out}$ | 0.023 | 0.080 | 0[a] |
| M | $K^{M}_{in}$ | 0.084 | 0.145 | 0.630 |
| | $K^{M}_{out}$ | 0.100 | 0.350 | 0.002 |

[a] The grassy swamps studied were typical accumulating ecosystems. Usually they lay in the lowest parts of a landscape and acted as a sink for materials in solution in water, which is a cause of the very low output values in these ecosystems.

studied the most closed cycling of matter and the smallest index of openness are found in the climax meadow-steppe.

At any given time the various subsystems of an ecosystem may not all be equally far from a steady state. For example, the biological cycle in the ecosystem may be close to a steady state (when the inflow of matter equals the outflow) while the system of exchange processes in the soil is in a transitional state because the intensities of the abiotic processes are not balanced (e.g. when soil salinization or desalinization changes). This will change the ecological conditions and the species composition will alter to bring about changes in productivity and the character of the biological cycling.

Thus the unbalanced exchange processes in an ecosystem act against the steady state of the system. This is characteristic of meadow and swamp ecosystems. The unstable state of these ecosystems increases even more under conditions of active migration of readily soluble salts.

In the light of the above, it can be seen that the simplified concept of the steady state of the whole ecosystem should be discarded and characteristic times of transformations and successions studied for every subsystem.

## Models of exchange processes in forest and desert ecosystems

Models of exchange processes in forest and desert ecosystems are constructed according to the same principle used in grassland ecosystems. Values for reserves of chemical elements in compartments are given in Table 9.3; the intensities of production processes and uptake of chemical elements are presented in Table 9.5. The intensities of other important biotic and abiotic processes are given in Table 9.9.

The intensity of biotic processes in these ecosystems is considerably greater than that of the abiotic processes. At the same time, the relative importance of some abiotic processes, particularly with respect to minerals,

Table 9.9. *The intensity of some exchange processes in forest and desert ecosystems* ($g \cdot m^{-2} \cdot yr^{-1}$)

| Flow | Process inducing flow | Value measured | Forest ecosystems: Spruce forests southern taiga, plot a C | N | M | Broad-leafed forests, plot b C | N | M | Tropical rain forests, plot c C | N | M | Desert ecosystems: *Artemisia*-dominated, plot d C | N | M | *Haloxylon–Ammodendron*-dominated, plot e C | N | M |
|---|---|---|---|---|---|---|---|---|---|---|---|---|---|---|---|---|---|
| | *Biotic processes* | | | | | | | | | | | | | | | | |
| G → Sl | Removal of chemical elements by precipitation from aboveground organs to soil | $I_8$ | 6.0 | 0.4 | 5.0 | 5.0 | 0.6 | 6.0 | 16.0 | 1.5 | 23.0 | | No data | | 0.9 | 0.1 | 10.0 |
| Pr → Sl | Removal of chemical elements by precipitation from perennial aboveground organs to soil | $I'_8$ | 0.2 | 0.01 | 0.2 | 0.25 | 0.02 | 0.3 | 15.0 | 0.2 | 5.0 | | No data | | — | — | — |
| G → D → L | Dying off of green organs and transformation into D and L | $I_{10}$, $I_{12}$ | 132.5 | 4.8 | 5.2 | 160.0 | 8.1 | 12.2 | 464.0 | 25.0 | 22.0 | 5.3 | 0.3 | 1.3 | 50.0 | 1.8 | 6.1 |
| Pr → D′ → L′ | Dying off of perennial organs (stems, branches) and their transformation into dry trees and L | $I'_{12}$ | 211.8 | 1.4 | 3.0 | 179.8 | 2.7 | 8.4 | 695.0 | 4.6 | 10.3 | 1.1 | trace | 0.3 | 28.8 | 0.5 | 1.5 |
| R → V | Dying off of roots and their transformation into dead root remains | $I_{11}$ | 111.3 | 1.7 | 3.6 | 111.5 | 1.7 | 4.3 | 200.0 | 2.5 | 8.8 | 47.7 | 1.5 | 2.7 | 250.0 | 5.6 | 24.9 |
| L → A → Sl | Litter mineralization, transfers of C to atmosphere and N and M to soil | $I_{13}$, $I_{14}$ | 119.2 | 3.4 | 4.2 | 144.0 | 6.5 | 11.0 | 417.0 | 22.5 | 20.0 | 4.7 | 0.2 | 1.2 | 45.0 | 1.3 | 5.5 |
| L′ → A → Sl | Mineralization of dead stems and L′, transfers C to atmosphere and N and M to soil | $I'_{13}$, $I'_{14}$ | 190.6 | 0 | 1.5 | 161.8 | 0.9 | 7.4 | 620.0 | 0 | 5.3 | 1.0 | trace | 0.3 | 25.8 | 0.2 | 1.8 |
| L → Slh | Humification of litter | $I_{15}$ | 13.3 | 1.4 | 1.0 | 16.0 | 1.6 | 1.2 | 47.0 | 2.5 | 2.0 | 0.6 | 0.1 | 0.1 | 5.0 | 0.5 | 0.5 |
| L′ → Slh | Humification of dry trees and branches and L′ | $I'_{15}$ | 21.2 | 1.4 | 1.5 | 18.0 | 1.8 | 1.0 | 75.0 | 4.6 | 5.0 | 0.1 | trace | trace | 3.0 | 0.3 | 0.2 |
| V → A → Sl | Mineralization of root remains, transfer of C to atmosphere and N and M to soil | $I_{16}$, $I_{17}$ | 100.2 | 0.5 | 2.6 | 100.3 | 0.5 | 3.5 | 180.0 | 0.5 | 7.3 | 42.7 | 0.9 | 2.4 | 200.0 | 1.6 | 22.9 |
| V → Slh | Humification of root remains | $I_{18}$ | 11.1 | 1.2 | 1.0 | 11.2 | 1.2 | 0.8 | 20.0 | 2.0 | 1.5 | 5.0 | 0.6 | 0.3 | 30.0 | 4.0 | 2.0 |
| Slh → A → Sl | Humus mineralization, transfer of C to atmosphere and N and M to soil | $I_{19}$, $I_{20}$ | 45.6 | 4.0 | 3.5 | 45.2 | 4.6 | 3.0 | 142.0 | 9.1 | 8.5 | 5.7 | 0.7 | 0.4 | 38.0 | 4.8 | 2.7 |
| | *Abiotic processes* | | | | | | | | | | | | | | | | |
| A → Sl | Inflow of chemical elements with precipitation | $I_{21}$ | 4.8 | 0.5 | 8.0 | 5.0 | 0.6 | 10.0 | 20.0 | 2.0 | 30.0 | 1.0 | 0.1 | 5.0 | 1.0 | 0.1 | 5.0 |
| Sl → Slb | Outflow of chemical elements with surface-water flow | $I_{23}$ | 0.8 | 0.03 | 1.0 | 0.1 | 0.01 | 0.3 | 1.5 | 0.1 | 1.0 | 0 | 0 | 0 | 0 | 0 | 0 |
| Sl → Slb | Outflow of chemical elements with lateral subsurface-water flow | $I_{25}$ | 8.0 | 0.3 | 10.0 | 2.2 | 0.3 | 5.0 | 15.5 | 1.8 | 15.5 | 0 | 0 | 0 | 0 | 0 | 0 |
| Sl → Sl′ | Outflow of chemical elements from soil to subsoil with downward-moving gravitational solutions | $I_{26}$ | 0.7 | 0.05 | 4.1 | 0.2 | 0.05 | 0.4 | 5.0 | 0.4 | 20.0 | 0 | 0 | 0 | 0 | 0 | 0 |
| W → Sl′ | Inflow of chemical elements from ground water to subsoil with upward-moving pellicular-capillary solutions | $I_{30}$ | 0 | 0 | 0 | 0 | 0 | 0 | 0 | 0 | 0 | 0 | 0 | 0 | 0 | trace | 5.0 |
| Sl′ → Sl | Inflow of chemical elements from subsoil to soil with upward-moving pellicular-capillary solutions | $I_{31}$ | 0 | 0 | 0 | 0 | 0 | 0 | 0 | 0 | 0 | 0 | 0 | 0 | trace | | 5.0 |
| Sl → A | Outflow of chemical elements from soil to atmosphere by wind | $I_{33}$ | 0 | 0 | 0 | 0 | 0 | 0 | 0 | 0 | 0 | 1.0 | 0.1 | 1.0 | 2.0 | 0.2 | 2.0 |

Table 9.10. *Indices of the degree of openness of forest and desert ecosystems*

| | | Forest ecosystems | | | Desert ecosystems | |
|---|---|---|---|---|---|---|
| Element | Index of the degree of openness | Spruce forests in southern taiga, plot a | Broad-leafed forests, plot b | Tropical rain forests, plot c | *Artemisia*-dominated, plot d | *Haloxylon ammodendron* dominated, plot e |
| C | $K_{in}^{C}$ | 0.01 | 0.01 | 0.014 | 0.019 | 0.003 |
| | $K_{out}^{C}$ | 0.02 | 0.006 | 0.016 | 0.019 | 0.006 |
| N | $K_{in}^{N}$ | 0.06 | 0.05 | 0.06 | 0.06 | 0.01 |
| | $K_{out}^{N}$ | 0.05 | 0.03 | 0.07 | 0.06 | 0.025 |
| M | $K_{in}^{M}$ | 0.47 | 0.40 | 0.38 | 1.22 | 0.24 |
| | $K_{out}^{M}$ | 0.88 | 0.22 | 0.46 | 0.24 | 0.05 |

is high. This applies, for example, to the supply of chemical elements with precipitation ($I_{21}$), in the ecosystems of southern taiga and wet tropical forests, to the removal of elements with subsurface lateral flow ($I_{25}$), and in sand desert ecosystems and woodland with *Haloxylon aphyllum* to the supply of elements with ascending pellicular-capillary solutions ($I_{31}$).

The index of ecosystem openness for the biologically important elements carbon and nitrogen is very small, as it is in climax grassland ecosystems (Table 9.10). However the index values for minerals are much larger than those for climax grassland ecosystems (Table 9.10). This indicates the relatively more important role of mineral elements supplied to or removed from the ecosystem with abiotic fluxes in the life of forest and desert ecosystems. There is an annual loss of minerals from coniferous taiga and wet tropical forest, but in small broad-leafed forests and in desert ecosystems a large accumulation of minerals is observed. These unbalanced fluxes seem to be one of the important factors shaping the evolution of ecosystems; it happens not only in ecosystems with sharply unstable functioning regimes but also in forests and deserts which are close to the climax stage.

The depletion of nutrients from the soil of forest ecosystems may result in swamping, while in contrast an excess of mineral substances in desert ecosystems can lead to soil salinization and halophytization. These problems require further investigation.

The above discussion allows us to conclude that grassland climax (or close to climax) ecosystems are characterized, relative to forest and desert ecosystems, by a much more stationary functioning regime and therefore greater stability.

## The specific rates of exchange processes in ecosystems of various types

The intensities of exchange processes depend on environmental conditions and on the initial state of the system, i.e. matter reserves in compartments (both sources and receivers). Thus the magnitude of primary production is

Table 9.11. *Specific rates of processes of production, dying off and decomposition of organic matter in various types of ecosystems* ($mg \cdot g^{-1} \cdot d^{-1}$)

| Process | Forests | | | Grasslands | | Deserts | |
|---|---|---|---|---|---|---|---|
| | Spruce forests in southern taiga | Broad-leafed forests | Tropical rain forests | Meadows | Steppe and prairie | *Artemisia*-dominated | *Haloxylon ammodendron* dominated |
| Production: | | | | | | | |
| Total production of plant biomass, $P$ | 45 | 33 | 8 | 87 | 77 | 14 | 39 |
| Green plant biomass, $P_G$ | 13 | 11 | 3 | 17 | 17 | 7 | 6 |
| Woody plant Biomass, $P_{Pr}$ | 21 | 12 | 4 | — | — | 1 | 3 |
| Roots biomass $P_R$ | 11 | 8 | 1 | 70[a] | 60[b] | 6 | 30 |
| Dying off: | | | | | | | |
| Green plant biomass, $G \rightarrow D \rightarrow L$ | 1.2 | 11.0 | 3 | 17 | 17 | 7 | 6 |
| Wood, $Pr \rightarrow L'$ | 0.10 | 0.07 | 0.1 | — | — | 0.2 | 0.3 |
| Roots, $R \rightarrow V$ | 0.20 | 0.15 | 0.1 | 6.0 | 3.4 | 0.8 | 0.9 |
| Decomposition: | | | | | | | |
| Litter, $L \rightarrow A \rightarrow Sl$ | 0.62 | 1.60 | 23.0 | 10.0 | 12.0 | 4.9 | 1.7 |
| Dead wood, $L \rightarrow A \rightarrow Sl$ | 0.32 | 0.25 | 0.46 | — | — | — | 1.4 |
| Dead roots, $V \rightarrow A \rightarrow Sl$ | No data | No data | No data | 5.6 | 3.2 | No data | No data |

[a] The specific rate of root production fluctuates from 20 to 160 $mg \cdot g^{-1} \cdot d^{-1}$.
[b] The specific rate of root production fluctuates from 20 to 200 $mg \cdot g^{-1} \cdot day^{-1}$.

proportional to the green plant biomass reserve; the intensity with which plants die off is proportional to the live matter reserve; and the intensity of decomposition of organic matter is proportional to the amount of dead matter. By dividing the value of the daily production by the green biomass reserve we can calculate the specific rate ($f$) of organic matter production. Likewise by dividing the intensity at which plants die off daily by the green biomass reserve we can calculate the specific rate at which plants die off; and by dividing the intensity of daily organic matter decomposition by the dead organic matter reserve we can calculate the specific rate of decomposition of dead organic matter. Table 9.11 shows the specific rates of the processes of production, dying off and decomposition of organic matter for ecosystems of different types. It can be seen from these data that wet tropical forests and deserts with *Artemisia* are systems where a unit of photosynthesizing plant biomass functions slowest (tropical forests, as was discussed earlier, achieve high production at the expense of large reserves of green plant biomass). In a rain forest it is insufficient light and deficiencies in mineral elements that limit the rate of production processes, while in *Artemisia* deserts moisture is the limiting factor. In shrubby desert with *Haloxylon aphyllum* the specific rate of production processes is much larger due to an additional supply of ground water.

Spruce forests have the largest specific rate of production among forest ecosystems but the values for grassland ecosystems can be larger. The specific rate of $P_G$ in grasslands ecosystems varies from 12 to 25 $mg \cdot g^{-1} \cdot d^{-1}$, with maximum values being found for grassy swamps that have sufficient moisture (Tables 9.11, 9.12). The specific rate of $P_R$ is larger than that of $P_G$ (from 10 to 340 $mg \cdot g^{-1} \cdot d^{-1}$), with maximum values for grassy swamps and solonetzic and extracontinental East Siberian steppes.

Woody parts of plants have the lowest specific rate of dying off; hence their reserves are largest in forest ecosystems. The specific rate of decomposition of dead organic matter increases in forests from north to south; this results in a reduction of the reserves of dead wood and litter in wet tropical forests (see Table 9.3). In grassland ecosystems the specific rates of dying off and decomposition of roots are approximately the same, and therefore live and dead root reserves are comparable in these ecosystems. Only in conditions that are excessively dry (dry steppes) or excessively moist (hygrophytic meadows and grassland swamps) do the V reserves (dead root remains) increase, as a result of the low specific rates of decomposition caused by lack of water or oxygen.

## The specific rates of biotic and abiotic exchange processes in grassland ecosystems of various types

A comparative analysis of various grassland ecosystems studied in detail – meadow-steppe (plot 1.3), mesohalophytic meadow (plot 5.2) and grassy swamp (plot 9.1) (Table 9.12) – shows that the specific rates at which plant biomass grows and dies off, and the intake of nutrients or their storage in dead plant remains, are highest in the swamp ecosystem and lowest in meadow-steppe. At the same time, swamps have the lowest rate of mineralization and humification of plant remains (belowground in particular) and the lowest rate of release of nutrients to the soil or the storing of them in the humus. This leads to an increase in V reserves. The specific rate of mineralization and humification of V reserves is the highest for meadows, where the soil is moist and well aerated.

The various biotic and abiotic processes in a given ecosystem show different specific rates. The rate of the processes of carbon assimilation and the rate of translocation of organic substances from G to R are the largest. Humus mineralization is characterized by very low specific rates (1000–5000 times as slow as the assimilation of carbon). In this way appreciable humus reserves are accumulated in the soil.

The same process can have a different specific rate for different chemical elements. For example, in meadow-steppe the specific rate of humification (V→Slh) is equal to 1.7 $mg \cdot g^{-1} \cdot d^{-1}$ for carbon and 10 $mg \cdot g^{-1} \cdot d^{-1}$ for nitrogen. This results in the nitrogen enrichment of the humus relative to the root remains.

The specific rates of abiotic processes are much lower than those of biotic

Table 9.12. *The specific rates of exchange processes in grassland ecosystems, Western Siberia, USSR* ($mg \cdot g^{-1} \cdot d^{-1}$)

| Process | Specific rate | C | | | N | | | M or S+m | | |
|---|---|---|---|---|---|---|---|---|---|---|
| | | Steppe, plot 1.3 | Meadow plot 5.2 | Grassy swamp, plot 9.1 | Steppe, plot 1.3 | Meadow, plot 5.2 | Grassy swamp, plot 9.1 | Steppe, plot 1.3 | Meadow, plot 5.2 | Grassy swamp, plot 9.1 |
| A → G → R | $P_G/G$ (180)[a] | 17 | 20 | 25 | — | — | — | — | — | — |
| G → R → A | $P_R/G$ (180) | 60 | 127 | 340 | — | — | — | — | — | — |
| Sl → R → G | $P_R^{N,M}/R$ (180) | — | — | — | 10.3 | 9.1 | 11.8 | 7.3 | 10.9 | 11.8 |
| R → G | $P_G^{N,M}/R$ (180) | — | — | — | 3.4 | 2.8 | 1.3 | 2.0 | 1.5 | 1.5 |
| G → R | $I_7/G$ (180) | | —No data— | | 7.4 | 8.5 | 0 | 1.4 | 0.4 | 0 |
| G → Sl | $I_8/G$ (50) | 3.0 | 3.3 | 4.2 | 4.1 | 6.1 | 7.7 | 12.0 | 6.2 | 20.0 |
| R → Sl | $I_9/G$ (180) | | —No data— | | 0 | 0 | 0 | 0.1 | 0.6 | 0 |
| G → D | $I_{10}/G$ (180) | 16.2 | 19.0 | 23.8 | 6.7 | 10.0 | 25.0 | 14.0 | 10.4 | 10.5 |
| R → V | $I_{11}/R$ (180) | 6.0 | 9.1 | 11.6 | 12.0 | 10.3 | 11.8 | 7.3 | 10.4 | 11.8 |
| D → L | $I_{12}/D$ (180) | 11.2 | 10.8 | 9.4 | 8.8 | 14.2 | 15.0 | 8.1 | 10.6 | 8.4 |
| L → A | $I_{13}/L$ (180) | 10.0 | 9.4 | 12.0 | — | — | — | — | — | — |
| L → Sl | $I_{14}/L$ (180) | — | — | — | 2.6 | 5.0 | 7.0 | 7.0 | 7.8 | 10.6 |
| L → Slh | $I_{15}/L$ (180) | 1.2 | 0.8 | 0.9 | 2.4 | 1.7 | 1.3 | 0.7 | 0.3 | 0.8 |
| V → A | $I_{16}/V$ (180) | 7.1 | 20.0 | 4.3 | — | — | — | — | — | — |
| V → Sl | $I_{17}/V$ (180) | — | — | — | 7.5 | 12.0 | 1.3 | 9.2 | 2.3 | 4.3 |

| | | | | | | | | | | |
|---|---|---|---|---|---|---|---|---|---|---|
| V → Slh | $I_{18}$/V (180) | 1.7 | 3.5 | 0.3 | 10.0 | 15.0 | 0.4 | 1.4 | 2.3 | 0.4 |
| Slh → A | $I_{19}$/Slh (180) | 0.04 | 0.08 | 0.07 | — | — | — | — | — | — |
| Slh → Sl | $I_{20}$/Slh (180) | — | — | — | 0.05 | 0.08 | 0.03 | 0.04 | 0.06 | 0.05 |
| A → Sl | $I_{21}$/Sl (50)[b] | $5 \times 10^{-3}$ | $3 \times 10^{-3}$ | 0.01 | $4 \times 10^{-3}$ | $4 \times 10^{-3}$ | 0.01 | 0.3 | 0.08 | 0.4 |
| Sla → Sl | $I_{22}$/Sl (100) | 0 | 0.02 | 0.02 | 0 | 0.01 | $6 \times 10^{-3}$ | 0 | 0.08 | 0.4 |
| Sl → Slb | $I_{23}$/Sl (50) | 0.01 | 0.04 | 0 | $4 \times 10^{-3}$ | 0.02 | 0 | 0.03 | 0.33 | 0 |
| Sla → Sl | $I_{24}$/Sl (50) | 0 | $5 \times 10^{-5}$ | $1 \times 10^{-3}$ | 0 | 0 | $1 \times 10^{-6}$ | 0 | $3 \times 10^{-3}$ | 0.01 |
| Sl → Slb | $I_{25}$/Sl (50) | $2 \times 10^{-6}$ | $7 \times 10^{-5}$ | $3 \times 10^{-4}$ | 0 | 0 | 0 | $2 \times 10^{-4}$ | $2 \times 10^{-3}$ | 0.01 |
| Sl → Sl′ | $I_{26}$/Sl (50) | $2 \times 10^{-5}$ | $1 \times 10^{-4}$ | 0 | 0 | 0 | 0 | $4 \times 10^{-3}$ | $2 \times 10^{-2}$ | 0 |
| Sl → W | $I_{27}$/Sl (50) | 0 | 0 | 0.01 | 0 | 0 | $2 \times 10^{-3}$ | 0 | 0 | 4.0 |
| Wa → W | $I_{28}$/W (365)[b] | 0 | $3 \times 10^{-4}$ | $5 \times 10^{-4}$ | 0 | $3 \times 10^{-4}$ | $6 \times 10^{-4}$ | 0 | $5 \times 10^{-4}$ | $9 \times 10^{-4}$ |
| W → Wb | $I_{29}$/W (365) | 0 | $5 \times 10^{-5}$ | 0 | 0 | $5 \times 10^{-5}$ | $6 \times 10^{-5}$ | 0 | $6 \times 10^{-5}$ | $5 \times 10^{-5}$ |
| W → Sl′ | $I_{30}$/W (100) | 0 | $3 \times 10^{-4}$ | $3 \times 10^{-3}$ | 0 | $3 \times 10^{-4}$ | $3 \times 10^{-3}$ | 0 | $4 \times 10^{-4}$ | $3 \times 10^{-3}$ |
| Sl′ → Sl | $I_{31}$/Sl (100)[b] | 0 | $1 \times 10^{-4}$ | $1 \times 10^{-3}$ | 0 | $5 \times 10^{-5}$ | $4 \times 10^{-4}$ | 0 | 0.03 | 0.42 |
| Sl′ → W | $I_{32}$/W (50)[b] | 0 | $1 \times 10^{-4}$ | 0 | 0 | $2 \times 10^{-4}$ | 0 | 0 | $2 \times 10^{-4}$ | 0 |

[a] Values in brackets give the approximate duration of the process in days.

[b] In this case the specific rate was calculated as a increment rather than a diminution of chemical elements in compartments. In all other cases it was calculated as a diminution of elements in compartments.

processes. The highest rate of migration of chemical elements is found in gravitationally descending solutions during moistening of soil in the spring and during their transport in surface-water flow. It is clear that the position of swamps in the relief of geochemically connected landscapes stipulates here the largest specific rates of practically all abiotic input processes. The lowest specific rates of abiotic processes are in the climax ecosystem of meadow-steppe, where for some processes the rate is equal to zero.

The differences between the specific rates of the same processes in various grassland ecosystems result in the variety of grassland types. Large differences in the rates of separate processes are common to all grassland ecosystems, but the relative values are similar, and as a result we observe a similar structure of matter reserves in compartments of grassland ecosystems

$$(D+L > G; \quad R > G; \quad Slh > D+L+G+R+V).$$

## Conclusions

A review of a large number of grassland ecosystems of the temperate zone indicates some of their common features. Domination of belowground organic mass over aboveground mass ($R > G$) is characteristic of all of them, both in the amounts of reserves and in the intensity of the production processes. The role of R mass increases as the environmental conditions of the ecosystem become more extreme (moisture deficit or surplus, salinization, solonetzicity, poor physical properties of soil). Similar ratios of G to R and the same types of plants in all grasslands result in the similarity of their structure; the characteristics of their functioning differ sharply from those of forest and desert ecosystems.

The annual production of the aboveground mass in grassland ecosystems, $P_G$, exceeds $G_{max}$ by 1.2–2.7 times; the production of belowground organs $P_R$, averages 0.9–1.0 of the R reserve or 0.6–0.7 of the reserve of R+V. These are much higher values than are given in the literature.

The rate of elemental cycling in grassland ecosystems is much higher than in forest or desert ecosystems.

The index of the degree of ecosystem openness is smaller in climax steppe and prairie ecosystems than it is in other grassland ecosystems (meadows and grassy swamps) or in forest and desert ecosystems.

The functioning regime of steppes and prairies is practically stationary, while climax forest and desert ecosystems are characterized by unbalanced input and output exchange processes (particularly with respect to minerals). Steppes and prairies are the most stable ecosystems on this criterion.

Meadows and grassy swamps function in the transitional or unstable regime. Halophytic and hygrohalophytic meadows are particularly unstable and are characterized by sharp annual fluctuations in the intensities of both biotic and abiotic exchange processes.

Grassland ecosystems are high-speed systems, specific rates of the processes of production, dying off of plants and intake of nutrient elements all being higher than those of other ecosystem types.

The specific rates of abiotic processes are generally much smaller than the specific rates of biotic processes for ecosystems of all types. The specific rates of abiotic processes are the smallest in climax grassland ecosystems (steppes and prairies).

## References

Arvisto, E. (1970). Content, supply and net primary production of biochemical constituents in the phytomass of spruce stands on brown forest soils. In *Estonian contributions to the IBP*, ed. L. Reintam, pp. 19–71. Academy of Sciences of the Estonian SSR, Tartu.

Bazilevich, N. J. (1962). Exchange of mineral elements in various types of steppes and meadows on chernozems, chestnut-brown soils and solonetzems. In *Problems of soil science*, ed. S. Zonn, pp. 148–207. Publishing House of the Academy of Science, Moscow. (In Russian.)

Bazilevich, N. J. (1970). *The geochemistry of soda soils*, ed. V. A. Kovda, p. 392. Israel Program for Scientific Translations, Jerusalem.

Chepurco, N. L., Bazilevich, N. J., Rodin, L. E. & Miroshnichenko, Yu. M. (1972). Biochemistry and productivity of *Haloxyloneta ammodendroni* in south-eastern Karakum desert. In *Ecophysiological foundation of ecosystem productivity in the arid zone*, ed. L. E. Rodin, *Proceedings of an International Symposium, USSR, 7–19 June 1972*, pp. 198–203. Nauka Publishing House, Leningrad.

Dahlman, R. T. & Kucera, C. L. (1965). Root productivity and turnover in native prairie. *Ecology*, **46**, 84–9.

Demin, A. P. (1970). Below-ground mass of meadow vegetation of the Oka flood plain under the influence of fertilization. *Bulletin Moskowscogo Obshchestva Ispytatelei Prirody, Biological Dept.*, **25**, 79–85. (In Russian.)

Dilmy, A. (1971). The primary productivity of equatorial tropical forest in Indonesia. In *Productivity of forest ecosystems*, ed. P. Duvigneaud, *Proceedings of the Brussels Symposium, 27–31 October 1969*, pp. 333–7, UNESCO, Paris.

Drujinina, N. P. (1973). *Phytomass of steppe communities of the south-eastern Zabaikalyl*, ed. V. B. Sochava, p. 152. Nauka Publishing House, Siberian Branch of the Academy of Science, Novosibirsk. (In Russian.)

Družina, V. D. (1972). Seasonal dynamics of ash elements of above-ground plant mass of meadow communities. *Zhurnal Rastitelnye Resursy*, 397–403. (In Russian.)

Golley, F. B. (1975). Productivity and mineral cycling in tropical forests. In *Productivity of world ecosystems*, pp. 106–15. National Academy of Sciences, Washington, DC.

Grishchenko, O. M. (1972). Below-ground plant biomass of basins and its role in biological turnover of substances and energy. In *Data on the flora and vegetation of the northern Pricaspian plain*, vol. 6, ed. V. V. Ivanov, pp. 54–63. Academy of Sciences, USSR, Leningrad. (In Russian.)

Ignatenko, J. V. & Kirillova, V. P. (1970). The change in total reserves of plant biomass under various regimes of utilization of shortgrass and forb–grass communities. In *Geobotanica*, vol. 18, *Meadow phytocoenoses and their dynamics*, ed. V. M. Ponyatovskaya, pp. 205–11. Nauka Publishing House, Leningrad. (In Russian.)

Jakrlová, J. (1972). Mineral nutrient uptake by four inundated meadow communities. In *Ecosystem study on grassland biome in Czechoslovakia*, ed. M. Rychnovska, pp. 51–5. Report No. 2, *Czechoslovak Academy of Sciences*, Brno.

Ketner, P. (1972). *Primary production of salt-marsh communities on the island of Terschelling in the Netherlands*, p. 182. Research Institute for Nature Management, Arnhem, the Netherlands.

Kira, T. L. & Ogava, H. (1971). Assessment of primary production in tropical and equatorial forests. In *Productivity of forest ecosystems, Proceedings of the Brussels Symposium, 1969*, ed. P. Duvigneaud, pp. 309–21. Unipub. Inc., New York.

Koike, K., Shoji, S. & Yoshida, S. (1973). Above-ground biomass and litter. In *Ecological studies of Japanese grassland: draft of IBP Synthesis Volume*, ed. M. Numata, pp. 45–56. University of Tokyo.

Kölli, R. & Reintam, L. (1970). Content of nitrogen and ash elements in the phytomass of spruce stands on brown forest soils. In *Estonian contributions to the IBP*, ed. L. Reintam, pp. 10–49. Academy of Sciences of the Estonian SSR, Tartu.

Kovalev, R. V. (ed.) (1974). *Structure, functioning and evolution of the Baraba biogeocoenoses*, vol. 1, p. 308. Nauka Publishing House, Siberian Branch, Novosibirsk. (In Russian.)

Kucera, C. L. & Dahlman, R. C. (1967). Total net productivity and turnover on energy basis for tallgrass prairie. *Ecology*, **48**, 536–41.

Lyapunov, A. A. (1970). What is the systems approach to studying real objects of a complicated nature? In *Upravlyayushchiye sistemy*, vol. 6, ed. A. A. Lyapunov, pp. 44–56. Academy of Sciences USSR, Siberian Branch, Novosibirsk. (In Russian.)

Lyapunov, A. A. & Titlyanova, A. A. (1971). The systems approach to studying the cycle of matter and energy flow in the biogeocoenosis. In *O nekotorych voprosach kodirovaniya i peredachi informacii v upravlyajushchikh sistemakh živoy prirody*, ed. A. A. Lyapunov & L. M. A. Belokova, pp. 99–188. Academy of Sciences USSR, Siberian Branch, Novosibirsk. (In Russian.)

Matador Project (1973). Excerpts from Final Reports Nos. 36 and 62, ed. R. Coupland. University of Saskatchewan, Canada.

Miroshnichenko, E. D. (1973). Concerning the decomposition of plant residue in meadows. *Botanicheskyi Zhurnal*, **58**, 402–12. (In Russian.)

Olson, Y. (1971). Calculation of productivity rates using different hypotheses of turnover; a sensitivity analysis. In *International Symposium on Root Systems, Eberswalde, September 1971*, ed. K. Hofman, p. 12.

Ovington, J. D. & Lawrence, D. B. (1967). Comparative chlorophyll and energy studies of prairie, savanna, oak wood and field maize ecosystems. *Ecology*, **48**, 515–24.

Ovington, J. D. & Olson, J. S. (1970). Biomass and chemical content of El Verde lower montane rain forest plants. In *A tropical rain forest*, ed. H. T. Odum & R. F. Pigeon, pp. 453–478. Office of Information Services, US Atomic Energy Commission, Oak Ridge, Tennessee.

Petrik, B. (1972). Seasonal changes in plant biomass in four inundated meadow communities. In *Ecosystem study on grassland biome in Czechoslovakia*, ed. M. Rychnovska, pp. 17–25. *Report No. 2, Czechoslovak Academy of Sciences*. Brno.

Petrusewicz, K. & Grodziński, B. (1973). The importance of phytophagous animals in ecosystems. *Soviet Ecology*, **6**, 17. (In Russian.)

Precsenyi, J. A. (1970). Study on the energy budget in *Artemisia–Festucetum Pseudovinae. Acta botanica Academiae Scientarium Hungaricae*, **16** (1–2), 179–85.

Reintam, L., Arvisto, E. & Kaupmees, K. (1970). On the formation of humus as the result of the transformation of plant residues in forest litter. In *Estonian contributions to the IBP*, ed. L. Reintam, pp. 71–81. Academy of Sciences of the Estonian SSR, Tartu.

Remesov, N. P., Bykova, L. N. & Smirnova, K. M. (1959). *Uptake and cycle of nitrogen and ash elements in the forests of European Russia*, ed. N. P. Remesov, p. 284. Moscow State University Publishing House, Moscow. (In Russian.)

Resources of the Biosphere (1975). *Productivity of steppe, meadow and grassy fen biogeocoenoses in forest-steppe*, ed. M. J. Bazilevich, pp. 56–96. *Productivity of desert communities*, ed. L. E. Rodin, pp. 128–67. Nauka Publishing House, Leningrad. (In Russian.)

Rodin, L. E. & Bazilevich, N. J. (1967). *Production and mineral cycling in terrestrial vegetation*, p. 285. Oliver & Boyd, Edinburgh & London.

Simon, T. & Kovacs-Lang, E. (1972). *Data abstracts from sites of Hungarian Lowland*, ed. T. Simon, p. 18. IBP Data Bank, Colorado State University, Fort Collins, USA.

Singh, J. S. & Coleman, D. C. (1973). Technique for evaluating functional root biomass in grassland ecosystems. *Canadian Journal of Botany*, **51**, 1868–70.

Smirnov, V. V. (1971). *Organic matter in some forest biogeocoenoses in the European part of the USSR*, ed. A. A. Molchanov, p. 360. Nauka Publishing House, Moscow.

Snitko, V. A. (1970). Geochemistry of urochishcha. In *Topology of steppe systems*, ed. V. B. Sochava, pp. 127–35. Nauka Publishing House, Leningrad.

Speidel, B. (1973). Solling Project of the Deutsche Forschungsgemeinschaft. *IBP Grassland/PT Studies*, ed. B. Speidel, p. 20. IBP Data Bank, Colorado State University, Fort Collins, USA.

Speidel, B. & Weiss, A. (1972). Zur ober- und untererdischen Stoffproduktion einer Goldhaferwiese bei verschiedener Dingung. *Der hessische Lehr- und Forschungsanstalt für Grünelandwirtschaft und Futterbau Eichhof angewandt Botanik*, **46**, 75–93.

Titlyanova, A. A. (1971). *Study of the biological cycle in biogeocenoses: methods and recommendations*, ed. R. V. Kovalev, p. 31. Academy of Sciences USSR, Siberian Branch, Novosibirsk. (In Russian.)

Vagina, T. A. & Shatokhina, N. G. (1971). Distinctive features of plant biomass accumulation in various types of grassland vegetation of the Baraba forest-steppe zone. In *Geobotanical investigations in Western and Middle Siberia*, ed. A. B. Kuminova, pp. 163–90. Nauka Publishing House, Siberian Branch, Novosibirsk. (In Russian.)

Van Dyne, G. M. (1970). Introduction to modelling and synthesis. In *Grassland ecosystems: review of research*, pp. 176–8. Colorado State University, Fort Collins, USA.

Yamane, J. & Sato, L. (1973). Seasonal change of chemical composition in *Miscanthus sinensis*. In *Ecological studies of Japanese grassland: draft of IBP Synthesis Volume*, ed. M. Numata, pp. 56–60. University of Tokyo.

Zlotin, R. J. & Khodashova, K. S. (1973). The effect of animals on the autotrophic cycle and biological turnover. In *Problems of biogeocoenology*, ed. E. M. Lavrenko & T. A. Rabotnov, pp. 105–17. Nauka Publishing House, Leningrad. (In Russian.)

Zlotin, R. J., Kazanskaya, N. S. & Khodashova, K. S. (1978). The man-made transformation of ecosystems of the subzone of true steppes. In *Proceedings of the XIV General Assembly of the International Union for the Protection of Nature*, ed. J. A. Isakov, pp. 52–67. (In Russian.)

Zolyomi, B. & Precsenyi, J. (1970). The production of the undergrowth of forest-steppe meadow in the forest at Újszentmargita. *Acta Botanica Academiae scientiarum Hungaricae*, **16** (3–4), 427–44.

Appendixes 9.1, 9.2 and 9.3 give the reserves of elements in various compartments averaged over many years.

**Appendix 9.1. Reserves of carbon (C), nitrogen (N) and mineral elements (M) in the compartments of steppe and prairie ecosystems ($g \cdot m^{-2}$)**

| Compartment | Chemical element | Steppes (USSR) | | | | | | | | Priarie (Canada) |
|---|---|---|---|---|---|---|---|---|---|---|
| | | Meadow-like | | | True | | Solonetzic | Dry | | Shortgrass |
| | | Russian plain | Western Siberia | | Western Siberia | Eastern Siberia | Western Siberia | Russian plain | Western Siberia | Saskatchewan |
| | | 1.1[a] | 1.2 | 1.3 | 1.4 | 1.5 | 1.6 | 1.7 | 1.8 | 2.1 |
| $G_{max}$, green plant biomass | C[b] | 120.0 | 300.0 | 93.0 | 160.0 | 48.0 | 70.0 | 80.0 | 56.0 | 49.0 |
| | N | 5.0 | 10.0 | 4.3 | 6.0 | 2.0 | 2.4 | 2.5 | 1.6 | 1.6 |
| | M[c] | 19.0 | 21.2 | 10.6 | 12.0 | 5.8 | 9.0 | 8.0 | 5.3 | 5.4 |
| R, live belowground organs | C | 410.0 | — | 515.0 | — | 670.0 | 700.0 | — | — | — |
| | N | 11.0 | — | 10.4 | — | 8.7 | 12.4 | — | — | — |
| | M | 33.0 | — | 65.0 | — | 56.2 | 50.0 | — | — | — |
| V, dead belowground organs | C | 600.0 | — | 352.0 | — | 1100.0 | 410.0 | — | — | — |
| | N | 16.0 | — | 7.0 | — | 25.0 | 7.3 | — | — | — |
| | M | 48.0 | — | 45.0 | — | 162.0 | 30.0 | — | — | — |
| R+V, live and dead belowground organs | C | 1010.0 | 600.0 | 867.0 | 910.0 | 1770.0 | 1100.0 | 900.0 | 380.0 | 1100.0 |
| | N | 27.0 | 16.2 | 17.4 | 17.1 | 33.7 | 19.7 | 25.0 | 8.6 | 19.2 |
| | M | 81.0 | 47.4 | 110.0 | 69.5 | 218.2 | 80.0 | 63.7 | 19.0 | 390.0 |
| D, standing dead | C | 100.0 | 243.0 | 72.0 | 126.0 | 76.0 | 73.0 | 58.5 | 30.0 | 164.0 |
| | N | 4.0 | 6.2 | 1.8 | 2.2 | 0.5 | 1.7 | 0.7 | 0.7 | 4.3 |
| | M | 17.0 | 18.6 | 10.0 | 5.2 | 7.0 | 11.5 | 3.8 | 1.6 | 34.6 |
| L, litter | C | 130.0 | 396.0 | 71.2 | 170.0 | 88.0 | 44.0 | 90.0 | 60.0 | 95.0 |
| | N | 3.0 | 12.9 | 3.2 | 5.0 | 0.6 | 1.7 | 1.6 | 1.8 | 3.6 |
| | M | 14.0 | 43.1 | 10.5 | 17.2 | 8.0 | 10.0 | 7.8 | 4.5 | 29.5 |
| Sl, soil layer | C ($10^3$) | 34.5 | 28.6 | 13.5 | 20.3 | 11.0 | 11.5 | 15.0 | 6.6 | — |
| | N ($10^3$) | 2.9 | 2.9 | 1.6 | 2.0 | 0.9 | 1.4 | 1.7 | 0.7 | — |
| | S[d] ($10^3$) | 0.3 | 0.3 | 0.5 | 0.4 | 0.5 | 2.3 | 0.4 | 0.6 | — |
| | m[e] ($10^3$) | 8.5 | 10.0 | 9.0 | 9.0 | 7.5 | 15.0 | 9.0 | 7.0 | — |

[b] The C content is assumed to equal 40 % of the dry mass of organic matter for $G_{max}$; 45 % for R, V, L and D.
[c] M is the sum of mineral elements: Si, K, Ca, Mg, Al, Fe, Mn, P, Ti, S, Na, Cl.
[d] S is the total of elements being part of readily soluble salts (Cl, S, Mg, Ca, Na, K) and C as $HCO_3$ and $CO_3$.
[e] m is the total of chemical elements being part of mobile compounds – ionic, molecular, colloidal, chemically absorbed and suspensions – Si, Fe, Al, Mn, Ti, Mg, Ca, P, K, Na.

*Characteristics of trial plots*

**1.1.** Meadow-like forb–grass steppe (*Bromus riparius*, *B. inermis*, *Koeleria gracilis*, *Stipa ioannis*, *Calamagrostis epigeios*, *Carex humilis*, *Medicago falcata*, *Filipendula hexapetala*, *Falcaria sioides* and others). Typical chernozems, clay, loamy texture. Precipitation 687 mm. Average yearly temp. +5.5 °C; July +19.5 °C, Jan. −9.1 °C. Regime of abandoned area, virgin soil. USSR, Kursk district, Streletckaya steppe (Bazilevich, 1962; Resources of the Biosphere, 1975.)

**1.2.** Meadow-like forb–grass steppe (*Stipa capillata*, *S. rubens*, *Festuca sulcata*, *Filipendula hexapetala*, *Thymus marshalianus*, *Medicago falcata* and others). Leached chernozems, loamy texture. Precipitation 480 mm. Average yearly temp. −1 °C; July +19.2 °C. Jan. −19 °C. Regime of abandoned area, virgin soil. USSR, Novosibirskii District, Priobskoe plateau. (Bazilevich, 1970.)

**1.3.** Meadow forb–grass steppe (*Phleum phleoides*, *Poa angustifolia*, *Calamagrostis epigeios*, *Filipendula hexapetala*, *Libanotis sibirica*, *Fragaria viridis*, *Medicago falcata*, *Vicia cracca*). Ordinary chernozems, loamy texture. Precipitation 438 mm. Average yearly temp. −0.5 °C; July +18.7 °C, Jan, −20.5 °C. Regime of abandoned area, old field fallow for a long time. Top of a hill. USSR, Novosibirskii District, Experimental Station Karachi, plot 12. (Vagina & Shatokhina, 1971; Kovalev, 1974; Resources of the Biosphere, 1975.)

**1.4.** Forb–bunchgrass steppe (*Stipa capillata*, *S. rubens*, *Festuca sulcata*, *Koeleria gracilis*, *Peucedanum morissonii*, *Fragaria viridis*, *Medicago falcata* and others). Ordinary chernozems, loamy texture. Precipitation 370 mm. Average yearly temp. +0.8 °C; July +19.6 °C, Jan. −18 °C. Regime of abandoned area, virgin soil. USSR, Altaiskii Krai, Priobckoe plateau. (Bazilevich, 1962.)

**1.5.** Bunchgrass steppe (*Stipa baicalensis*, *Tanacetum sibiricum*, *Clematis hexapetula*, *Serratula centauroides*, *Carex pediformis*, *Potentilla leucophylla*, *P. verticillaris*). Chernozems, sandy loamy texture. Precipitation 320 mm. Average yearly temp. −3.2 °C; July +18.5 °C, Jan. −29 °C. Regime of abandoned area, virgin soil. Depression between mountains. USSR, Chitinskii District, Zabaikalie. (Calculated by Drujinina, 1973, and Snitko, 1970.)

**1.6.** Forb–grassy solonetzic steppe (*Festuca pseudovina*, *Agropyron repens*, *Artemisia glauca*, *Galatella biflora*). Medium columnar meadow-steppe solonetzes. Clay, loamy texture. An ancient lacustrine outerbar. Precipitation 438 mm. Average yearly temp. −0.5 °C; July +18.7 °C, Jan. −20.5 °C. Regime of abandoned area, virgin soil. USSR, Novosibirskii District, Experimental Station Karachi, plot 31. (Vagina & Shatokhina, 1971; Kovalev, 1974; Resources of the Biosphere, 1975.)

**1.7.** Bunchgrass dry steppe (*Stipa ucrainica*, *Festuca sulcata*, *Galium verum*, *Falcaria sioides*, *Linosyris villosa*, *Medicago falcata*). Southern chernozems, clay, loamy texture. Precipitation 380 mm. Average yearly temp. +6 °C; July +22 °C, Jan. −1 °C. Regime of abandoned area, virgin soil. USSR, Khersonskii District, Askaniya Nova. (Bazilevich, 1962.)

**1.8.** *Artemisia*-dominated bunchgrass dry steppe (*Festuca sulcata*, *Stipa capillata*, *S. lessingiana*, *Artemisia frigida*, *A. glauca*). Precipitation 250 mm. Average yearly temp +0.5 °C; July +21 °C, Jan. −18.5 °C. Chestnut soil, sandy loamy texture. Regime of abandoned area, virgin soil. USSR, Altaiiskii Krai. (Bazilevich, 1962.)

**2.1.** Shortgrass prairie (*Agropyron dasystachyum*, *A. smithii*, *Carex eleocharis*, *Koeleria cristata*, *Artemisia frigida* and others). Weakly developed brown soil of the Sceptre Association. Precipitation 388 mm. Average yearly temp. +3.6 °C, July +18 °C, Jan. −10 °C. Regime of abandoned area, virgin soil. Canada, Saskatchewan (Matador Project, 1973.)

## Appendix 9.2. Reserves of carbon (C), nitrogen (N) and mineral elements (M)

| | | Meadows | | | | | | |
|---|---|---|---|---|---|---|---|---|
| | | Steppe | | | Mesophytic | | | |
| | | flood | dry | | flood | dry | | |
| | | | USSR | | | USSR, Russian plain | | |
| Compartment | Chemical elements | Czechoslovakia 3.1[a] | Russian plain 3.2 | Western Siberia 3.3 | Czechoslovakia 4.1 | 4.2 | 4.3 | Germany 4.4 |
| $G_{max}$, green plant biomass | C[b] | 100.0 | 132.0 | 132.0 | 160.0 | 220.0 | 94.0 | 218.0 |
| | N | 4.1 | 8.5 | 7.5 | 4.4 | 4.4 | 5.0 | 3.8 |
| | M[c] | 5.6[d] | 18.7 | 14.0 | 11.3[d] | 16.5 | 12.7 | 19.8 |
| R, live below-ground organs | C | — | — | — | — | — | — | — |
| | N | — | — | — | — | — | — | — |
| | M | — | — | — | — | — | — | — |
| V, dead below-ground remains | C | — | — | — | — | — | — | — |
| | N | — | — | — | — | — | — | — |
| | M | — | — | — | — | — | — | — |
| R + V, live and dead below-ground organs | C | 905.0 | 880.0 | 360.0 | 550.0 | 820.0 | 425.0 | 531.0 |
| | N | 16.6 | 54.0 | 10.6 | 14.5 | 10.9 | 11.2 | 2.9 |
| | M | 28.2 | 112.0 | 24.6 | 38.0 | 49.8 | — | 68.3 |
| D, standing dead | C | — | — | 72.0 | — | — | — | — |
| | N | — | — | 2.4 | — | — | — | — |
| | M | — | — | 8.5 | — | — | — | — |
| L, litter | C | — | 40.0 | 370.0 | — | — | 42.5 | 462.5 |
| | N | — | — | 12.2 | — | — | — | — |
| | M | — | — | 50.6 | — | — | — | — |
| D + L, standing dead + litter | C | 60.0 | — | 442.0 | 48.0 | 340.0 | — | — |
| | N | 1.6 | — | 14.6 | 1.7 | 13.7 | — | — |
| | M | 1.5 | — | 59.1 | 1.6 | 21.8 | — | — |
| Sl, soil layer 1 m | C ($10^3$) | — | 33.4 | 38.4 | — | 7.0 | — | — |
| | N ($10^3$) | — | 2.9 | 3.3 | — | 0.7 | — | — |
| | S[e] ($10^3$) | — | 0.8 | 0.5 | — | 0.2 | — | — |
| | m[f] ($10^3$) | — | 10.0 | 10.0 | — | 10.0 | — | — |
| W, ground water layer 1m | C ($10^3$) | — | 11.0 | 9.0 | — | 20.0 | — | — |
| | N ($10^3$) | — | 0.6 | 0.5 | — | 1.0 | — | — |
| | S ($10^3$) | — | 219.0 | 200.0 | — | 30.0 | — | — |
| | m ($10^3$) | — | 6.0 | 24.0 | — | 12.0 | — | — |

[a] Numbers in this line refer to trial plots. See below for details.
[b] The C content is assumed equal to 40 % of the dry mass of organic matter for $G_{max}$; 45 % for R, V, L and D.
[c] M is the sum of mineral elements: Si, K, Ca, Mg, Al, Fe, Mn, P, Ti, S, Na, Cl.
[d] Here and below in this column Si, Al, Fe, Mn, Ti, Cl and S were not taken into account.
[e] S is the total of elements being part of readily soluble salts (Cl, S, Mg, Ca, Na, K) and C as $HCO_3$ and $CO_3$.
[f] m is the total of chemical elements being part of mobile compounds – ionic, colloidal, chemically absorbed and suspensions – Si, Fe, Al, Mn, Ti, Mg, Ca, P, K, Na.

*Characteristics of trial plots*

**3.1.** Recently flooded meadow (*Serratula–Festuca* community; *Festuca sulcata, Anthoxanthum odoratum, Serratula tinctoria, Colchicum autumnale, Lathyrus pratensis, Fragaria vesca*). Meadow soils. Precipitation 585 mm. Average yearly temp. +9.5 °C; July +19.2 °C, Jan. −1.5 °C. Flood plain. Mowed. Czechoslovakia, Moravia (Petrik, 1972; Jakrlová, 1972.)

**3.2.** Forb–grass meadow (*Bromus riparius, Poa angustifolia, Festuca sulcata, Potentilla humifusa, Salvia pratensis*). Chernozemic meadow soils, slightly solonchakous, clay texture. Precipitation 457 mm. Average yearly temp. +5.0 °C; July +20.6 °C, Jan. −10.4 °C. Flat plain; ground water 0.5–2.2 m. Regime of abandoned area. USSR, Tambovskii District, Oksko-Donskaya lowland. (Resources of the Biosphere, 1975.)

**3.3.** Forb–grass meadow (*Calamagrostis epigeios, Poa angustifolia, P. pratensis, Agropyron repens, Koeleria gracilis, Phleum phleoides, Bromus inermis, Lathyrus pratensis, Vicia cracca, Fragaria viridis, Filipendula hexapetala*). Meadow chernozemic soils, clay loamy texture. Precipitation 380 mm. Average yearly temp. −1 °C; July +19 °C, Jan. −20 °C. Flat plain; ground water 2.5–3.5 m. Regime of abandoned area, virgin soil. USSR, Novosibirskii District. (Bazilevich, 1970.)

**in the compartments of meadow and grassy swamp ecosystems (g·m⁻²)**

| | | | | | | | | | Grassy swamp |
|---|---|---|---|---|---|---|---|---|---|
| Mesohalophytic | | | Halophytic | | | | Hygrophytic | Hygrohalo-phytic | Drying |
| dry | | | flood | dry | marsh | | flood | marsh | flood |
| | USSR | | | | | | | | |
| Japan | Russian plain | Western Siberia | USSR, Western Siberia | | Netherlands | | Czechoslovakia | Netherlands | USSR, Western Siberia |
| 5.1 | 5.2 | 5.3 | 6.1 | 6.2 | 6.3 | 6.4 | 7.1 | 8.1 | 9.1 |
| 80.0 | 95.0 | 104.0 | 69.0 | 24.0 | 175.0 | 162.0 | 416.0 | 121.0 | 118.0 |
| 4.5 | 5.2 | 5.0 | 2.1 | 1.5 | 7.3 | 7.7 | 7.8 | 4.6 | 3.3 |
| 10.3 | 22.5 | 8.8 | 8.5 | 47.7 | 61.3 | 43.2 | 24.0[d] | 33.5 | 14.8 |
| — | 670.0 | — | 475.0 | — | — | — | — | — | 1400.0 |
| — | 18.4 | — | 8.2 | — | — | — | — | — | 27.5 |
| — | 95.0 | — | 87.0 | — | — | — | — | — | 98.0 |
| — | 260.0 | — | 215.0 | — | — | — | — | — | 3600.0 |
| — | 7.1 | — | 3.7 | — | — | — | — | — | 200.0 |
| — | 39.0 | — | 39.0 | — | — | — | — | — | 257.0 |
| 132.0 | 930.0 | 126.0 | 690.0 | 180.0 | 1135.0 | — | 1310.0 | 815.0 | 5000.0 |
| 83.0 | 25.5 | 43.2 | 11.9 | 1.6 | — | — | 39.0 | — | 227.5 |
| 225.0 | 134.0 | 117.0 | 126.0 | 34.2 | — | — | 35.8 | — | 355.0 |
| — | 82.4 | 120.0 | 67.0 | 52.0 | 104.0 | 42.5 | — | 64.5 | 120.0 |
| — | 1.8 | 5.5 | 1.7 | 1.2 | — | | — | — | 2.2 |
| — | 11.0 | 16.7 | 5.0 | 50.0 | — | — | — | — | 13.5 |
| 25.0 | 86.4 | 290.0 | 51.0 | 60.0 | — | — | — | — | 300.0 |
| — | 3.8 | 10.3 | 1.7 | 1.0 | — | — | — | — | 13.4 |
| — | 14.3 | 38.1 | 4.0 | 59.0 | — | — | — | — | 34.5 |
| — | 168.8 | 410.0 | 118.0 | 112.0 | — | — | 115.0 | — | 420.0 |
| — | 5.6 | 15.8 | 3.4 | 2.2 | — | — | 3.0 | — | 15.6 |
| — | 25.3 | 54.8 | 9.0 | 109.0 | — | — | 3.9 | — | 48.0 |
| 20.6 | 11.9 | 22.6 | 10.5 | 1.1 | — | — | — | — | 17.0 |
| 2.1 | 1.4 | 1.9 | 1.2 | 0.1 | — | — | — | — | 2.3 |
| 2.3 | 1.0 | 1.6 | 2.1 | 25.0 | — | — | — | — | 0.7 |
| 15.0 | 10.0 | 15.0 | 15.0 | 12.0 | — | — | — | — | 16.0 |
| 12.0 | 12.0 | 12.0 | 24.0 | 4.0 | — | — | — | — | 18.0 |
| 0.6 | 0.6 | 0.6 | 1.2 | 0.2 | — | — | — | — | 0.9 |
| 750.0 | 300.0 | 370.0 | 2700.0 | 2700.0 | — | — | — | — | 1500.0 |
| 10.0 | 15.0 | 25.0 | 50.0 | 100.0 | — | — | — | — | 30.0 |

**4.1.** Flood plain meadow (*Gratiola officinalis–Carex praecox-suzae* association; *Alopecurus pratensis, Carex praecox-suzae, Gratiola officinalis, Sanguisorba officinalis, Lathyrus pratensis, Vicia cracca* and others.) Meadow soils; flood plain. Precipitation 585 mm. Average yearly temp. +9.5 °C; July +19.8 °C, Jan. −1.5 °C. Ground water 0–1.6 m. Mowed. Czechoslovakia, Moravia. (Petrik, 1972; Jakrlová, 1972.)

**4.I.** Forb-grass meadow (*Agrostis tenuis + Alchemilla monticola*). Sod podzolic gleyic soils. Lower part of slope; ground water 2.0 m. Precipitation 525 mm. Average yearly temp. +4.0 °C; July +17.5 °C, Jan. −6.0 °C. Regime of abandoned area. USSR, Leningradskii District. (Ignatenko & Kirillova, 1970; Družina, 1972; Miroshnichenko, 1973.)

**4.3.** Forb–grass meadow (*Festuca rubra, Anthoxanthum odoratum, Trifolium pratense*). Precipitation 1100 mm. Average yearly temp. +6.5 °C; July +16 °C, Jan. −3 °C. Mowed. Germany, Solling plateau. (Speidel & Weiss, 1972; Speidel, 1973.)

**4.4.** Grassy, secondary meadow (*Miscanthus sinensis, Pteridium aquilinum* and others). Precipitation 2335 mm. Average yearly temp. +9.8 °C; July +23 °C, Jan. −2 °C. Mowed. Foothill. Japan, Khokaido. (Koike, Shoji & Yoshida, 1973; Yamane & Sato, 1973.)

**5.1.** Solonetzic meadow (*Festuca sulcata, Artemisia monogyna, Limonium tomentellum, Atriplex litoralis*). Solonetzic meadow, clay texture. Saucer-like depression on lowland; ground water 1.7–2.3 m. Precipitation 457 mm. Average yearly temp. +5 °C; July +20.6 °C, Jan. −10.4 °C. Regime of abandoned area, virgin soil. USSR, Tambovskii District, Oksko-Donskaya lowland. (Resources of the Biosphere, 1975.)

**5.2.** Forb–grass meadow (*Calamagrostis epigeios, Vicia cracca, Artemisia rupestris, Medicago falcata*). Chernozemic meadow soils, solonchakous clay, loamy texture. Lower part of slope of ancient lacustrine outer-bar; ground water 2.0–2.5 m. Precipitation 438 mm. Average yearly temp. −0.5 °C; July +18.7 °C, Jan. −20.5 °C. Regime of abandoned area, virgin soil. USSR, Novosibirskii District, Experimental Station Karachi, plot 33. (Vagina & Shatokhina, 1971, Titlyanova, 1971; Kovalev, 1974; Resources of Biosphere, 1975.)

**5.3.** Forb–grass solonetzic meadow (*Calamagrostis epigeios, Agropyron repens, Poa pratensis, Festuca ovina,*

*Phleum phleoides, Galium verum, Galatella biflora*). Medium columnar meadow-steppe, solonetzic clay, loamy texture. Slopes of lowlands; ground water 2.5–3.0 m. Precipitation 380 mm. Average yearly temp. −1 °C, July +19 °C, Jan. −20 °C. Regime of abandoned area, virgin soil. USSR, Novosibirskii District. (Bazilevich. 1970.)

**6.1.** Solonchakous meadow (*Puccinellia tenuifolia, P. distans, Saussurea salsa*). Solonetzic meadow, clay loamy texture. Lowland by lakes; ground water 2.0–2.5 m. Precipitation 438 mm. Average yearly temp. −0.5 °C; July +18.7 °C, Jan. −20.5 °C. Regime of abandoned area, virgin soil. USSR, Novosibirskii District: Experimental Station Karachi, plot 34. (Vagina & Shatokhina, 1971; Kovalev, 1974; Resources of the Biosphere, 1975.)

**6.2.** Solonchakous meadow (*Salicornia europaea, Suaeda maritima*). Solonchakous meadow, clay. Flood plain by lakes; ground water 0.5–1.0 m. Precipitation 300 mm. Average yearly temp. 0 °C; July +19 °C, Jan. −18 °C. Regime of abandoned area. USSR, Omskii District. (Bazilevich, 1970.)

**6.3.** Solonchakous meadow (*Plantago–Limonium*-dominated community; *Plantago maritima, Limonium vulgare, Triglochin maritima*). Marshes by sea. Precipitation 760 mm. Average yearly temp. +9.5 °C; July +22.7 °C, Jan. −0.3 °C. The Netherlands. (Ketner, 1972.)

**7.1.** Flooded meadow (*Glyceria maxima*-dominated community; *Glyceria maxima, Phalaris arundinacea, Carex gracilis* and others). Meadow-swamp soils. Flood plain. Precipitation 585 mm. Average yearly temp. +9.5 °C; July +19.8 °C, Jan. −1.5 °C. Mowed from time to time. Czechoslovakia, Moravia. (Petrik, 1972; Jakrlová, 1972.)

**8.1.** Solonchakous meadow (*Juncus–Carex* community; *Juncus gerardii, Scirpus rufus, Carex extensa, Triglochin maritima*). Marshes by sea. Precipitation 760 mm. Average yearly temp. +9.5 °C; July +22.7 °C, Jan. −0.3 °C. The Netherlands. (Ketner, 1972.)

**9.1.** Grassy swamp, drying out (*Calamagrostis neglecta, Scolochloa festucacea, Carex gracilis*). Peat swamp, gley soils. Flood plain by lakes; ground water 0–3.0 m. Precipitation 438 mm. Average yearly temp. −0.5 °C; July +18 °C, Jan. −20.5 °C. Regime of abandoned area, virgin soil. USSR, Novosibirskii District, Experimental Station Karachi, plot 38. (Vagina & Shatokhina, 1971; Kovalev, 1974; Resources of the Biosphere, 1975.)

## Appendix 9.3. Some data on reserves of carbon (C), nitrogen (N) and mineral elements (M) in the compartments of grassland ecosystems ($g \cdot m^{-2}$)

| Compartment | Chemical element | Steppes | | Prairies | | | Meadows | | | | | | | Grassy swamp | | |
|---|---|---|---|---|---|---|---|---|---|---|---|---|---|---|---|---|
| | | True | Psammophytic | Shortgrass | | Tallgrass | Steppe, dry | Mesophytic, flood plain | | Mesohalophytic, dry | | Hygrophytic, flood plain | | | Drying | |
| | | 1.40[a] | 1.9 | 2.2 | 2.3 | 2.4 | 3.4 | 4.00 | 4.0 | 5.4 | 5.5 | 7.2 | 7.3 | 9.2 | 9.3 | 9.4 |
| G, green plant biomass | C[b] | 108.0 | 25.0 | — | 41.6 | 110.0 | — | — | — | 316.0 | 217.8 | 190.0 | 126.0 | 240.0 | 400.0 | 160.0 |
| | N | — | — | — | — | — | — | — | — | — | — | 9.1 | — | — | 16.0 | 6.8 |
| | M[c] | — | — | — | — | — | — | — | — | — | — | 38.1 | — | — | 39.0 | 12.0 |
| R, live belowground plant biomass | C | — | — | 550.0 | — | — | 685.0 | 330.0 | 225.0 | — | — | 940.0 | 820.0 | 320.0 | 600.0 | 360.0 |
| | N | — | — | — | — | — | — | — | — | — | — | — | — | — | 40.0 | 22.0 |
| | M | — | — | — | — | — | — | — | — | — | — | — | — | — | 50.0 | 18.0 |
| V, dead belowground remains | C | — | — | 280.0 | — | — | 310.0 | 380.0 | 230 | — | — | 1200.0 | 1720.0 | 1030.0 | 1800.0 | 1260.0 |
| | N | — | — | — | — | — | — | — | — | — | — | — | — | — | 120.0 | 78.0 |
| | M | — | — | — | — | — | — | — | — | — | — | — | — | — | 150.0 | 70.0 |
| R + V, live and dead below-ground organs | C | — | 500.0 | 830.0 | 210.0 | 780.0 | — | — | — | 952.0 | 1020.0 | 2140.0 | 2540.0 | 1350.0 | 2400.0 | 1620.0 |
| | N | — | — | — | — | — | — | — | — | — | — | — | — | — | 180.0 | 100.0 |
| | M | — | — | — | — | — | — | — | — | — | — | — | — | — | 200.0 | 9.8 |
| D, standing dead | C | — | 90.0 | — | — | — | — | — | — | 18.3 | — | — | — | 230.0 | 40.0 | 40.0 |
| | N | — | — | — | — | — | — | — | — | — | — | — | — | — | 1.6 | 1.7 |
| | M | — | — | — | — | — | — | — | — | — | — | — | — | — | 3.9 | 3.0 |
| L, litter | C | — | — | — | — | 325.0 | — | — | — | 44.0 | — | — | — | 500.0 | 700.0 | 670.0 |
| | N | — | — | — | — | — | — | — | — | — | — | — | — | — | 47.0 | 44.0 |
| | M | — | — | — | — | — | — | — | — | — | — | — | — | — | 126.0 | 100.0 |
| D + L, standing dead + litter | C | 220.0 | — | — | 150.0 | — | — | — | — | 62.3 | — | — | — | 730.0 | 740.0 | 710.0 |
| | N | — | — | — | — | — | — | — | — | — | — | — | — | — | 48.6 | 45.7 |
| | M | — | — | — | — | — | — | — | — | — | — | — | — | — | 129.9 | 103.0 |

[a] Numbers in this line refer to trial plots. See below for details.
[b] The C content is assumed to equal 40 % of the dry mass of organic matter for $G_{max}$; 45 % for R, V, L and D.
[c] M is the sum of mineral elements: Si, K, Ca, Mg, Al, Fe, Mn, P, Ti, S, Na, Cl.

## Appendix 9.3. (*continued*)

*Characteristics of trial plots*

**1.40.** Forb–grass steppe (*Festuca sulcata, Poa angustifolia, P. bulbosa, Agropyron intermedium, Stipa lessingiana, Salvia nutans, Artemisia austriaca, Medicago romanica*). Ordinary chernozem clay, loamy texture. Precipitation 380 mm. Average yearly temp. +7.5 °C; July +20 °C, Jan. −5 °C. Regime of abandoned area, virgin soil. USSR, Ukraine. (Zlotin *et al.*, 1975.)

**1.9.** Psammophytic steppe (*Festucatum vaginatum, F. wagineri* and others). Sandy steppe soils. Precipitation 650 mm. Average yearly temp. +9.0 °C; July +20 °C, Jan. −4 °C. Hungary, lowland. (Simon & Kovacs-Lang, 1973.)

**2.2.** Shortgrass prairie (*Bouteloua gracilis* and others). Precipitation 380 mm. Average yearly temp. +60 °C; July +20 °C, Jan. −3.2 °C. USA, Colorado. (Singh & Coleman, 1973.)

**2.3.** Tallgrass prairie (*Stipa spartea, Poa pratensis, Andropogon gerardi*). USA. (Ovington & Lawrence, 1967.)

**2.4.** Tallgrass prairie (*Andropogon gerardi, A. scoparius* and others.) Precipitation 520 mm. Average yearly temp. +7.8 °C; July +24 °C, Jan. −2 °C. Virgin soil, USA, Missouri. (Dahlman & Kucera, 1965; Kucera & Dahlman, 1967.)

**3.4.** Steppe-like meadow (*Festuca pseudovina, Calamagrostis epigeios, Poa angustifolia, Stipa ioannis, Medicago falcata, Filipendula hexapetala, Artemisia latifolia*). Slope of an ancient lacustrine outerbar. Deep solonetz columnar meadow, clay loamy texture; ground water 2.5 m. Precipitation 438 mm. Average yearly temp. −0.5 °C; July +18 °C, Jan. −20.5 °C Regime of abandoned area, virgin soil. USSR, Novosibirskii District, Experimental Station Karachi, plot 32. (Vagina & Shatokhina, 1971.)

**4.00.** Flooded meadow on flood plain. Precipitation 560 mm. Average yearly temp. 3.6 °C; July +19 °C, Jan. −12 °C. USSR, Gorcovskii District. (Demin, 1970.)

**4.0.** Meadow with prolonged flood, flooded plain. Precipitation 560 mm. Average yearly temp. +3.6 °C; July +19 °C, Jan. −12 °C. USSR, Gorcovskii District. (Demin, 1970.)

**5.4.** Mesohalophytic meadow (*Artemisia maritima, Festuca pseudovina*). Precipitation 620 mm. Average yearly temp. +10 °C; July +21 °C, Jan. −2 °C. Hungary. (Precsenyi, 1970; Zolyomi & Precsenyi, 1970.)

**5.5.** Mesohalophytic meadow (*Peucedanum officinale, Limonium gmelini, Aster punctatum, Alopecurus pratensis*). Precipitation 620 mm. Average yearly temp. +9.0 °C; July +20 °C, Jan. −3 °C. Hungary. (Precseny, 1970; Zolyomi & Precsenyi, 1970.)

**7.2.** Inundated meadow (*Agropyron repens, Bromus inermis, Calamagrostis epigeios* and others). Precipitation 280 mm. Average yearly temp. +9 °C; July +25 °C, Jan. −10 °C. Mowed, virgin soil. USSR. Pricaspian plain. (Grishchenko, 1972.)

**7.3.** Inundated meadow (*Agropyron repens, Poa pratensis, P. palustris, Phragmites communis, Artemisia pontica*). Peaty swamp, solodized soils, clay loamy texture. Precipitation 438 mm. Average yearly temp. −0.5 °C; July +18 °C, Jan. −20.5 °C. Swampy lowland by lake. USSR, Novosibirskii District, Experimental Station Karachi, plot 117. (Vagina & Shatokhina, 1971.)

**9.2.** Inundated grassy swamp. (*Scolochloa festucacea, Alopecurus ventricosus*). Peaty swamp, solodized soils, clay loamy texture. Precipitation 438 mm. Average yearly temp. −0.5 °C, July +18 °C, Jan. −20.5 °C. Swampy depression. USSR, Novosibirskii District, Experimental Station Karachi, plot 36. (Vagina & Shatokhina, 1971.)

**9.3.** Grassy swamp (*Phragmites communis, Carex caespitosa*). Swampy flood plain by lake. Peaty swamp soils, clay texture. Precipitation 440 mm. Average yearly temp. −1 °C; July +18.5 °C, Jan. −20.5 °C. USSR, Novosibirskii District. (Bazilevich, 1970.)

**9.4.** Grassy swamp (*Phragmites communis, Carex caespitosa*). Swampy flood plain. Peaty swamp soils. Precipitation 350 mm. Average yearly temp. −1 °C; July +19 °C, Jan. −20 °C. USSR, Novosibirskii District. (Bazilevich, 1970.)

## Appendix 9.4. Intensity of production and nutrient uptake processes in various grassland ecosystems ($g \cdot m^{-2} \cdot yr^{-1}$)

| Index | Chemical elements | Steppes: Meadow-like 1.1[a] | 1.2 | 1.3 | True 1.4 | 1.40 | 1.50 | Solonetzic 1.6 | Dry 1.7 | 1.8 | Psammo-phytic 1.9 | Prairies: Shortgrass 2.1 | 2.3 | Tall-grass 2.4 |
|---|---|---|---|---|---|---|---|---|---|---|---|---|---|---|
| $P_G$, net primary production of green plant biomass | C[b] | 224.0 | 360.0 | 152.0 | 240.0 | 195.0 | 76.0 | 108.0 | 128.0 | 90.0 | 44.0 | 100.0 | 108.0 | 184.0 |
| $P_R$, Net primary production of belowground organs | C | 456.0 | — | 560.0 | 450.0 | — | 480.0 | 1280.0 | 300.0 | 130.0 | 250 0 | 415.0 | 200.0 | 240.0 |
| $P_G^N$, uptake of N for $P_G$ production | N | 8.0 | 12.0 | 6.6 | 9.0 | — | 3.2 | 3.4 | 4.0 | 2.5 | .. | 7.3 | — | — |
| $P_G^M$, uptake of M for $P_G$ production | M[c] | 33.0 | 26.0 | 19.5 | 18.0 | — | 9.2 | 13.5 | 12.8 | 8.5 | — | 11.2 | — | — |
| $P_R^N$, uptake of N for $P_R$ production | N | 12.0 | — | 19.4 | 8.4 | — | 11.0 | 22.5 | 8.4 | 3.0 | — | 3.3 | — | — |
| $P_R^M$, uptake of M for $P_R$ production | M | 45.0 | — | 85.6 | 34.0 | — | 70.6 | 92.0 | 21.2 | 6.6 | — | — | — | — |

| Index | Chemical elements | Meadows: Steppe 3.1 | 3.2 | 3.3 | Mesophytic 4.1 | 4.2 | 4.3 | 4.4 | Mesohalophytic 5.1 | 5.2 | 5.3 | 5.4 | 5.5 |
|---|---|---|---|---|---|---|---|---|---|---|---|---|---|
| $P_G$, net primary production of green plant biomass | C | 184.0 | 270.0 | 210.0 | 288.0 | 252.0 | 126.0 | 370.0 | 180.0 | 168.0 | 167.0 | 316.0 | 267.8 |
| $P_R$, net primary production of belowground organs | C | 490.0 | 350.0 | 120.0 | 436.0 | 374.0 | 198.0 | 350.0 | 550.0 | 1100.0 | 420.0 | 1193.0 | 716.0 |
| $P_G^N$, uptake of N for $P_G$ production | N | 7.5 | 15.3 | 12.0 | 7.9 | 5.0 | 6.6 | 6.5 | 8.1 | 9.2 | 8.0 | — | — |
| $P_G^M$, uptake of M for $P_G$ production | M | 10.3[d] | 34.2 | 22.4 | 20.4[d] | 17.6 | 17.8 | 33.6 | 18.0 | 25.5 | 14.2 | — | — |
| $P_R^N$, uptake of N for $P_R$ production | N | 9.0 | 16.5 | 3.5 | 8.5 | 6.2 | 5.6 | 1.1 | 25.0 | 30.0 | 14.4 | — | — |
| $P_R^M$, uptake of M for $P_R$ production | M | 15.3[d] | 34.0 | 8.2 | 22.0[d] | 20.0 | — | 28.0 | 68.0 | 186.0 | 39.2 | — | — |

## Appendix 9.4 (*continued*)

| | | Meadows | | | | | | | | | | | |
|---|---|---|---|---|---|---|---|---|---|---|---|---|---|
| | | Halophytic | | | | Hygrophytic | | | Hygro-halo-phytic | Grassy swamps (drying) | | | |
| | | 6.1 | 6.2 | 6.3 | 6.4 | 7.1 | 7.2 | 7.3 | 8.1 | 9.1 | 9.2 | 9.3 | 9.4 |
| $P_G$, net primary production of green plant biomass | C | 108.0 | 36.0 | 224.0 | 202.0 | 508.0 | 200.0 | 215.0 | 162.0 | 214.0 | 430.0 | 600.0 | 240.0 |
| $P_R$, net primary production of belowground organs | C | 430.0 | 120.0 | 490.0 | 342.0 | 455.0 | 1240.0 | 1180.0 | 230.0 | 2920.0 | 850.0 | — | — |
| $P_G^N$, uptake of N for $P_G$ production | N | 3.3 | 2.2 | 9.8 | 9.5 | 9.5 | 9.5 | — | 6.2 | 6.5 | — | 24.0 | 10.0 |
| $P_G^M$, uptake of M for $P_G$ production | M | 13.6 | 72.1 | 95.0 | 54.0 | 29.3[d] | 40.0 | — | 45.0 | 26.5 | — | 58.0 | 18.0 |
| $P_R^N$, uptake of N for $P_R$ production | N | 7.5 | 1.7 | — | — | 15.3 | 46.5 | — | — | 58.5 | — | — | — |
| $P_R^M$, uptake of M for $P_R$ production | M | 79.0 | 21.5 | — | — | 14.0[d] | 280.0 | — | — | 210.0 | — | — | — |

Note: the calculations are tentative for plots 4.3, 4.4, 6.3, 6.4 and 8.1.

[a] Numbers in this line refer to trial plots – see notes to Appendixes 9.1, 9.2 and 9.3 for details.

[b] The C content is assumed to equal 40% of the dry mass of organic matter for $G_{max}$; 45% for R, V, L and D.

[c] M is the sum of mineral elements: Si, K, Ca, Mg, Al, Fe, Mn, P, Ti, S, Na, Cl.

[d] Here M excludes Si, Al, Fe, Mn, Cl, S and Ti.

# 10. Total-system simulation models

G. S. INNIS, I. NOY-MEIR, M. GODRON
& G. M. VAN DYNE

## Contents

## Objectives and definitions

It is useful at the outset to discuss the objectives of this chapter and to provide a number of definitions.

The objectives of this chapter are: (i) to present and discuss some examples of ecosystem-level models developed under IBP auspices, (ii) to describe the role of ecosystem models in IBP grassland projects, (iii) to describe some of the steps and problems in building such models, (iv) to describe some of the insights into systems derived from exercising grassland system models, and (v) to discuss some of the lessons learned from the development of such models.

In order to achieve these objectives, contributions from each of the authors, as well as from several other scientists modelling not only grassland ecosystems but other systems, have been included. Many of the conclusions drawn and presented here have been reached by a number of the individuals contributing to the IBP grassland modelling efforts around the world. It is also true that there are a number of substantive issues on which members of this team of authors disagree and on which we might disagree with other modelling efforts outside of this group. At the time of this writing the construction of ecosystem-level models or even process-level models is largely an intuitive or artistic (as contrasted with scientific) activity. While IBP has allowed a number of modellers to exercise this intuition, it has also given us an opportunity to separate some of the 'wheat from the chaff' and, as a result, move towards a consensus on a number of items. This convergence is incomplete and will require more time.

A number of the terms used in describing models in general, and ecosystem-level models in particular, have been given different definitions by the authors. It is worthwhile, therefore, to define some of these terms for use in this chapter.

### *Simulation models*

A model is a miniature representation of a thing. As such, it may be dynamic or static, physical or abstract, and if an abstract model the abstraction may take a number of different forms. Simulation models are dynamic models. Dynamic abstract models that have been described in mathematical terms and for which the mathematical expressions have been coded for analysis using digital computation equipment are the subject of this chapter.

### *Ecosystem-level models*

Ecosystems are recognized as being divisible into such major categories as abiotic, producer, consumer, decomposer and nutrient subsystems. An ecosystem-level model generally concerns three or more of these major subsystems. Most often one will find abiotic components, perhaps several producers, and a decomposer subsystem as being, in some sense, a minimal ecosystem-level model. The addition of consumers and of nutrient subsystems tends to make the system more inclusive. No statement is made as to the accuracy of representation of any of these subsystems. An ecosystem-level model is inclusive but not necessarily exhaustive.

### *System boundary*

The boundary of the system is an arbitrary limit determined by the objectives of the modelling effort. Things within the limit are part of the system; things outside the limit are part of the environment. Events occurring outside the system boundary are assumed to be independent of activities occurring within the system boundary during the time frame of interest (i.e. the time frame modelled). While this consideration puts some limits on the location of the system boundary, the modeller still has considerable latitude as to what he decides is inside and outside his domain of interest.

Of particular concern in this chapter is the distinction between an *ecosystem-level model* and a *process model.* Processes such as photosynthesis, birth, death, etc., are described elsewhere in this volume (see Chapters 1 to 7 inclusive) as part of the discussion of the scientific efforts of the various investigators. The ecosystem-level model presumes to put a number of these processes together in an attempt to describe the dynamics of the ecosystem.

## Some examples of grassland ecosystem models

The grassland modelling literature is large and diverse. Van Dyne & Abramsky (1975) reviewed much of it. Chapters 1, 2, 4, 12 and 13 in this volume also review many grassland or grassland component models. We shall limit our attention here to efforts conducted under the auspices of IBP. We have attempted to get contributions from each national grasslands programme, though, undoubtedly, some efforts are not adequately represented.

### *USSR*

An example of the conceptual approach to modelling ecosystems (biogeocoenoses) used as basis for mathematical models in the USSR is provided by Lyapunov & Titlyanova (1974). They conceptualize a block, substance and flow approach for carbon dynamics. Bazilevich & Titlyanova (see Chapter 9, this volume) provide a detailed description of a graphical model (diagram) for which empirical data for the exchange processes (material flows) and state variables are given. These data provide a basis from which preliminary mathematical ecosystem models could be constructed. A Russian grassland simulation model (Bazilevich, 1975) has three state variables – one for each of the three soil layers 0–50 cm, 50–100 cm and 100–150 cm. The driving variables are air temperature, precipitation and relative humidity, and the time step is 10 days. This model really does not meet the specified criteria for an ecosystem model. It is a water balance model, and the biotic components of the system are not treated explicitly.

Another modelling study of grassland ecosystems in the USSR is discussed by Gilmanov (1976). He considers the concepts of an overall system model with interconnected cycles of matter and energy flowing through macro-compartments of green plants, local atmosphere, land animals, and soil and soil organisms. Specific information is given on the plant submodel, parameterized for and validated with data from a meadow-steppe from the Karachi IBP station in the western Siberian plains. The model utilizes driving variables of air temperature, air relative humidity, precipitation, wind speed and cloudiness. Partial diagrams are presented of flows in the system, using the Forrester (1961) diagrammatic notation. Graphs are shown for simulation results and field measurements for May through September in 1971 and 1972 for aboveground live plant material, living root biomass according to depth increments, and dead and mulch material. The agreement between model results and field data appears relatively good. Preliminary examination and translation of the equations presented in the paper suggest there is considerable detail and mechanism in the model structure. For example, plant growth is calculated using a 'maximum-reduction philosophy'. A maximum growth rate is modified by factors for suboptimal conditions of soil water, temperature, phenology and competition with other plants.

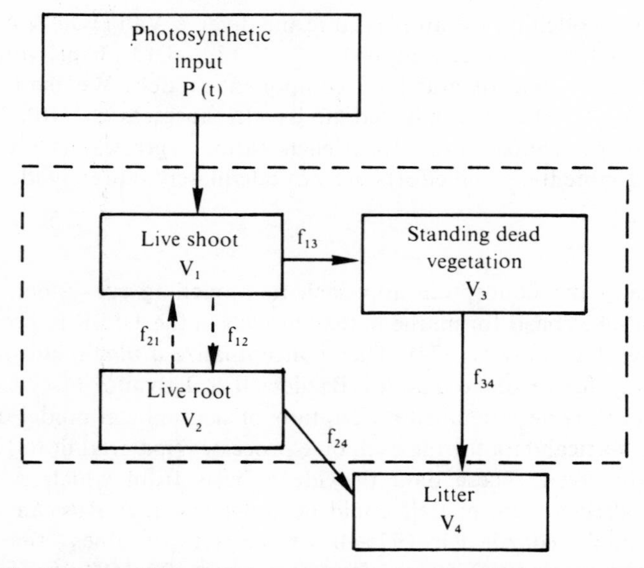


Fig. 10.1. The linear, variable-coefficient model for an ungrazed ecosystem for intraseasonal dynamics of *Miscanthus sacchariflorus* in Japan. (After Numata, 1975.)

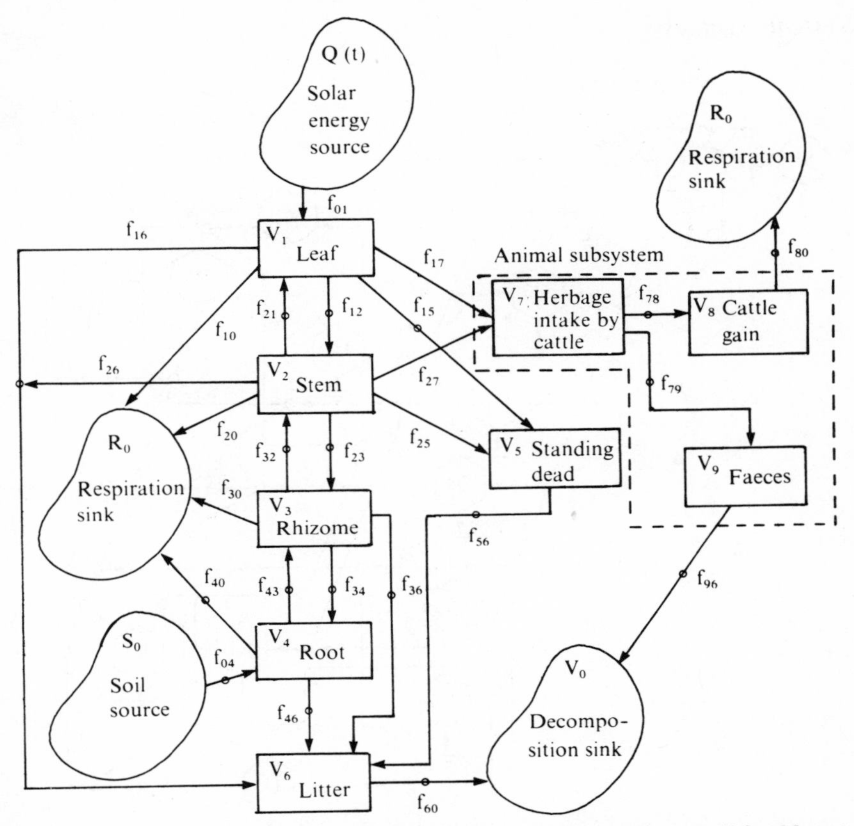


Fig. 10.2. The nonlinear, variable-coefficient model for a grazed ecosystem. (After Numata 1975.)

### *Japan*

Numata (1975, chapt. 10) reports two grassland simulation models – one for the ungrazed case, one for the grazed system. The ungrazed system model is a linear, variable-coefficient system with flow diagrams as shown in Fig. 10.1. The time dependencies are sine or exponential functions. The model has six parameters, is represented by differential equations, and has been solved via simulation. The parameters were chosen iteratively to minimize the sum of squares of the differences between observed values and model output. The model for the grazed system is considerably more complicated (Fig. 10.2). The differential equation format with variable coefficients is retained but linearity is lost because of the introduction of a number of nonlinear functions whose form depends on system state variables. The experimental procedure made comparison of model results with observations difficult. As a consequence, model results are shown and parameter values (66 of them) are given, but no observations are presented for comparison.

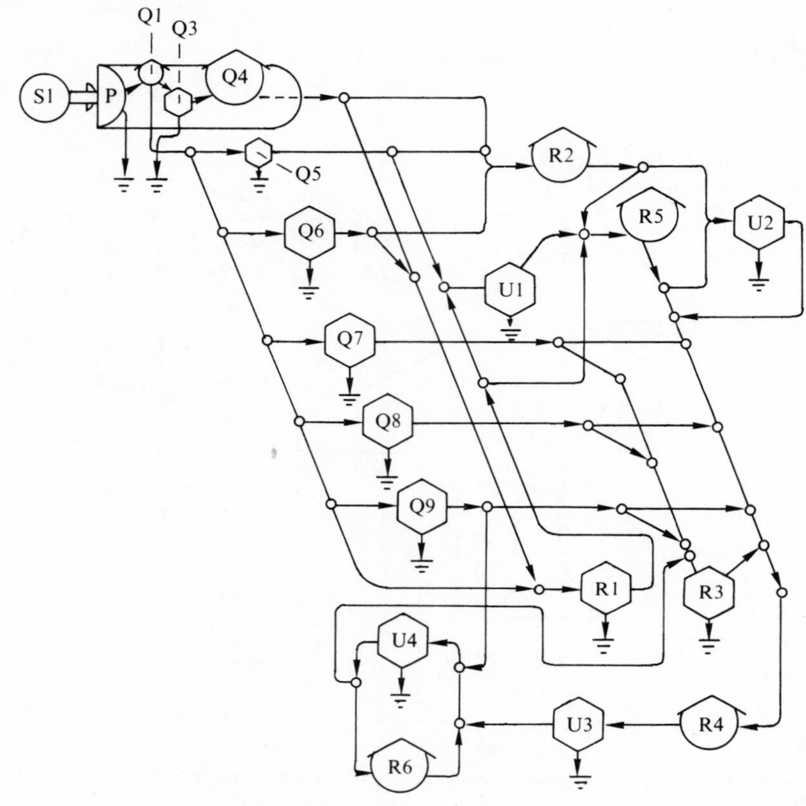


Fig. 10.3. An H. T. Odum-type diagram of the Canadian linear simulation model. P, photosynthesis; S1, solar input; Q1, labile carbohydrate; Q3, living leaf; Q4, leaf structure (cellulose); Q5, seeds; Q6, stems; Q7 shoot bases; Q8, rhizomes; Q9, roots; R1, surface invertebrates; R2, standing dead plants; R3, surface invertebrates; R4, young soil organic matter; R5, surface litter; R6, old soil organic matter; U1, birds; U2, saprophages; U3 soil, fungi; U4, soil bacteria. (From Randell & More, 1975.)

## Canada

Randell & More (1974) describe three grassland simulation models. The first of these is a linear, constant-coefficient model developed as part of a series of meetings. An H. T. Odum (1971)-type diagram of this system is shown in Fig. 10.3. Photosynthetic input (P in Fig. 10.3) is a sine function. The report does not discuss the choice of parameters or the comparison of model output with observations. Results of two model runs (each of 1100 days) with different initial conditions are given in addition to a BASIC (computer language) listing of the model. The predicted dynamics are consistent with observation and intuition. The nonlinear model (Fig. 10.4) is the product of a 3-day

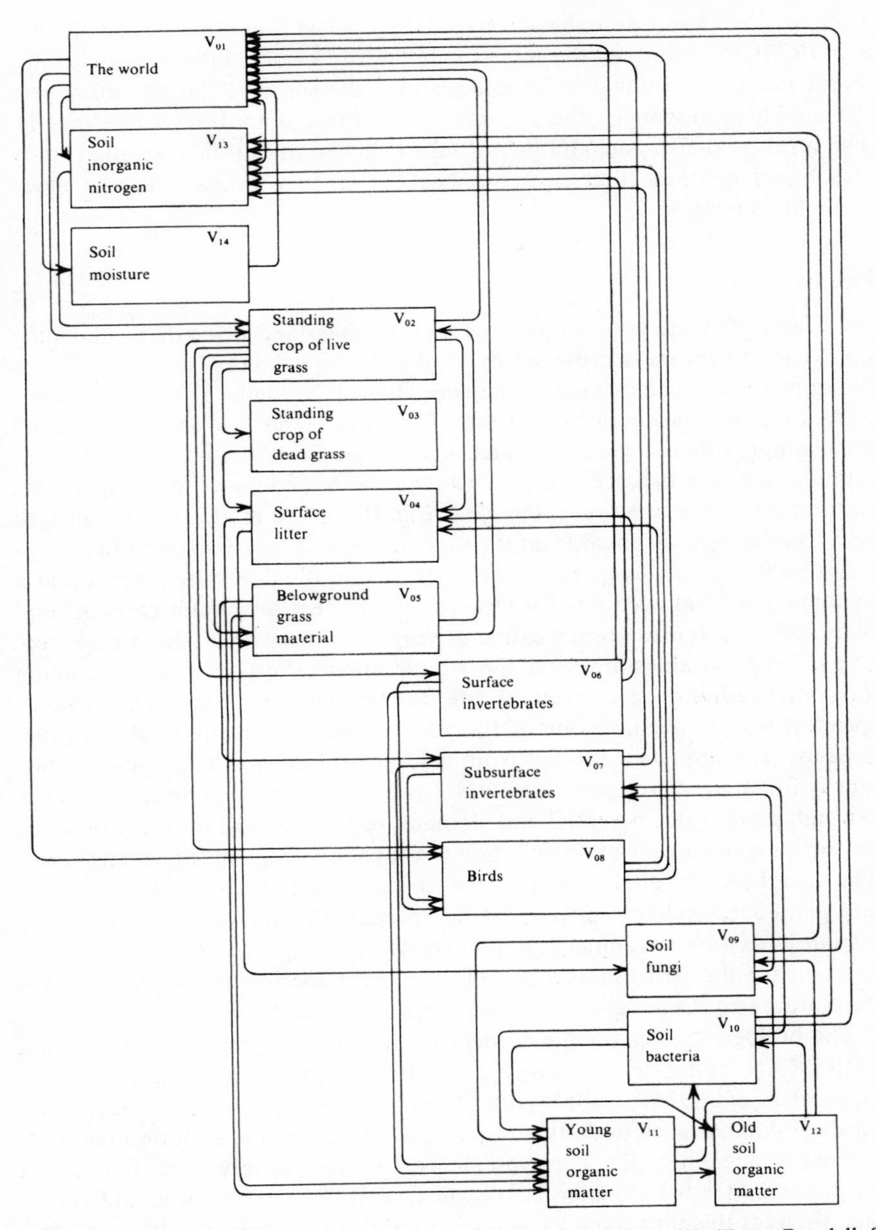


Fig. 10.4. A flow diagram of a Canadian nonlinear simulation model. (From Randell & More, 1975.)

seminar. The report describes each of the modelled flows in some detail, and a FORTRAN listing of the model is included. Two sample runs (with different initial conditions) were made, and while some of the variables were reasonably represented, others, notably the roots, did not behave realistically. The third total-ecosystem modelling effort carried out in the Canadian study was based on these first two, but because of time limitations was never brought to fruition.

### *France*

An example of the conceptual approach to grassland ecosystem modelling by French scientists is provided by Gounot & Bouché (1974). A model was developed and implemented by Saugier, Ripley & Leuke (1974). Saugier is from a research group in Montpellier, France, but this is a simulation model for plant growth and water use in a Canadian grassland ecosystem. It includes radiation, energy balance and microclimate, water, soil heat, photosynthesis, respiration and production submodels (Fig. 10.5). The model was designed to be as mechanistic as possible, at the time. It has a static component that computes microclimate, evaporation and net carbon dioxide assimilation, and a dynamic part that accounts for changes in soil heat and water contents and plant biomass. It uses seven weather driving variables with a 1-hour time step. The driving variables are total short-wave, direct short-wave and incoming long-wave radiation, precipitation rate, and temperature, humidity and wind-speed at screen height. None of the parameters used in the model is time-dependent; many were derived from field measurements. The flows are primarily functions of the state variables. Evaporation depends on soil moisture through plant water potential and stomatal resistance. Soil heat flux depends on soil temperatures. Net assimilation of the canopy depends on green shoots. The model can be run for any year for which input data are available and for any plant species whose physical structure and stomatal and photosynthetic responses to the environment are known. Thus it might be used, for example, to compare the performance of different plant species under the Matador (Saskatchewan, Canada) climate over a growing season.

The biological basis for the model, its mathematical representation and a FORTRAN computer language code listing are included in the report (Saugier *et al.*, 1974). Their chapter 'Verification of the model' describes a series of comparisons of daily and seasonal model results with observations. Generally speaking, the correspondence between observation and model output is better for the abiotic section (radiation, temperature, soil water, etc.) than for the biotic components (live and dead biomass, production, etc.). This point will be noted as a general observation on grassland models later in the chapter. The discrepancies between model output and data are attributed to unsophisticated model components (for reflected radiation), inadequate

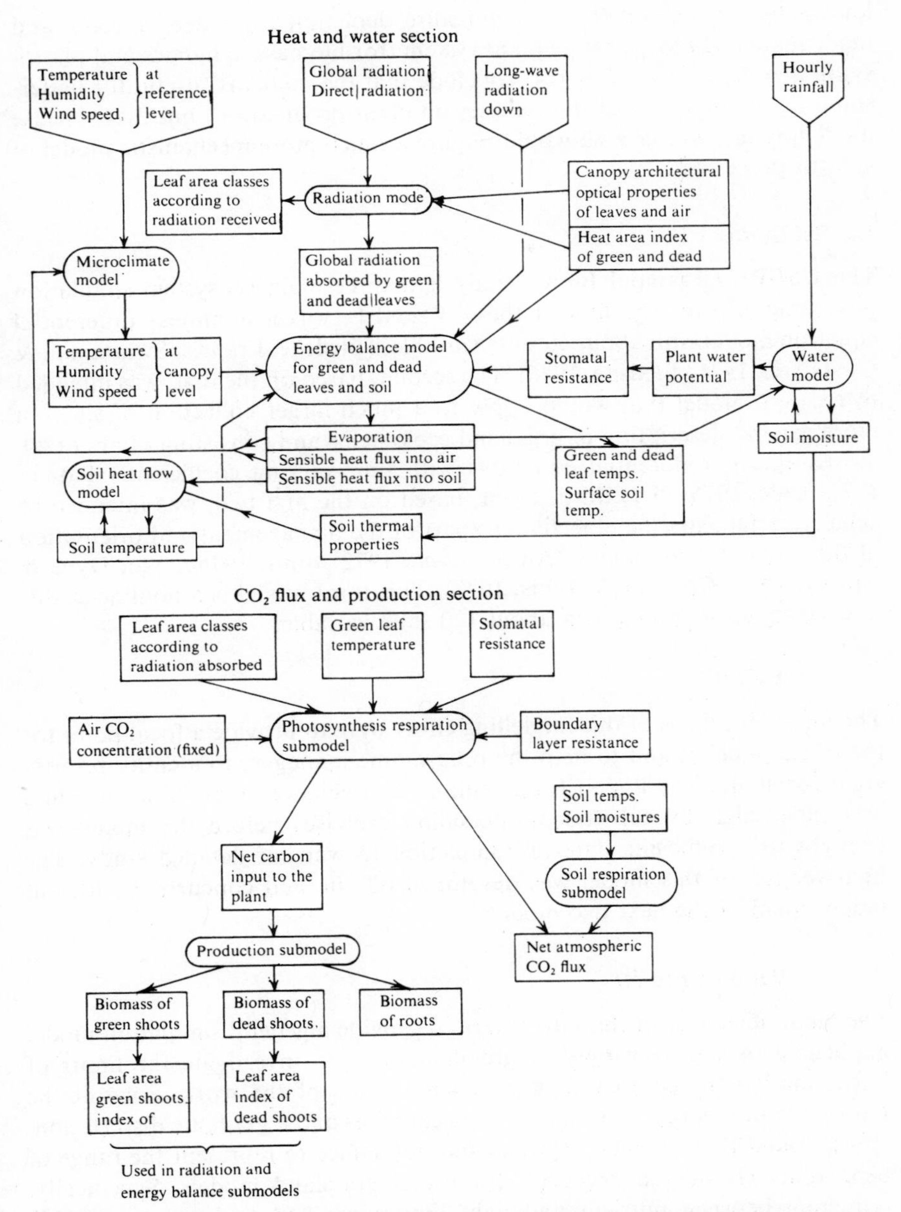


**Fig. 10.5. Structural diagrams for heat, water, carbon dioxide and plant carbon flows in a grassland system model of Saugier, Ripley & Leuke (1974).**

knowledge of the soil (for soil moisture depletion from deep layers), and inadequate knowledge of the plant system (for biomass dynamics and photosynthesis). Saugier *et al.* (1974) conclude that the model is 'useful for understanding the influence of the weather on plant production', but that 'precise modelling of plant growth would require a much more mechanistic model of growth processes'.

### *United States*

The US/IBP Grassland Biome study had three main ecosystem simulation modelling efforts. The first of these, PWNEE, was a nonlinear differential equation model with about 40 state variables (Bledsoe, Francis, Swartzman & Gustafson, 1971; Bledsoe, 1976). The second, based on the first, was intended to create a model that would apply to a much larger collection of sites – a more general description of grassland ecosystems and consisting of about 40, mostly linear, differential equations with time-varying coefficients (Patten, 1972; Cale, 1975). The third, ELM, based on the first two, was intended to achieve certain specific objectives as part of the management and orientation of the research programme (Anway *et al.*, 1972; Innis, 1975*c*; Van Dyne & Anway, 1976; Cole, 1976; Innis, 1978). This third model is a nonlinear, difference equation model with about 120 state variables.

#### *PWNEE*

The main objectives of this modelling effort were to provide a focal point for the research effort and to help the programme managers to identify priority areas for study. The PWNEE modelling effort achieved these objectives and was superseded by subsequent modelling exercises before the model was brought to a sufficient state of completion to warrant detailed study. The final version of this model was developed off site and concurrent with but independent of the next two models.

#### *Patten's model*

The main objectives of this effort were to provide a general simulation model applicable to a wide variety of grasslands and to investigate the limits of linear models in representing the system. Much of the work, done at the University of Georgia, was part of class exercises and a graduate dissertation. It was found that the linear systems did not suffice to represent the range of behaviours considered necessary for useful grassland models. Specifically, switching between nutrient and light limitations had to be state variable based, and this could not be represented linearly. Extensive analyses of this model identified equilibria, studied the dynamics near such equilibria, carried out time series analysis of model output and investigated indices of cycling

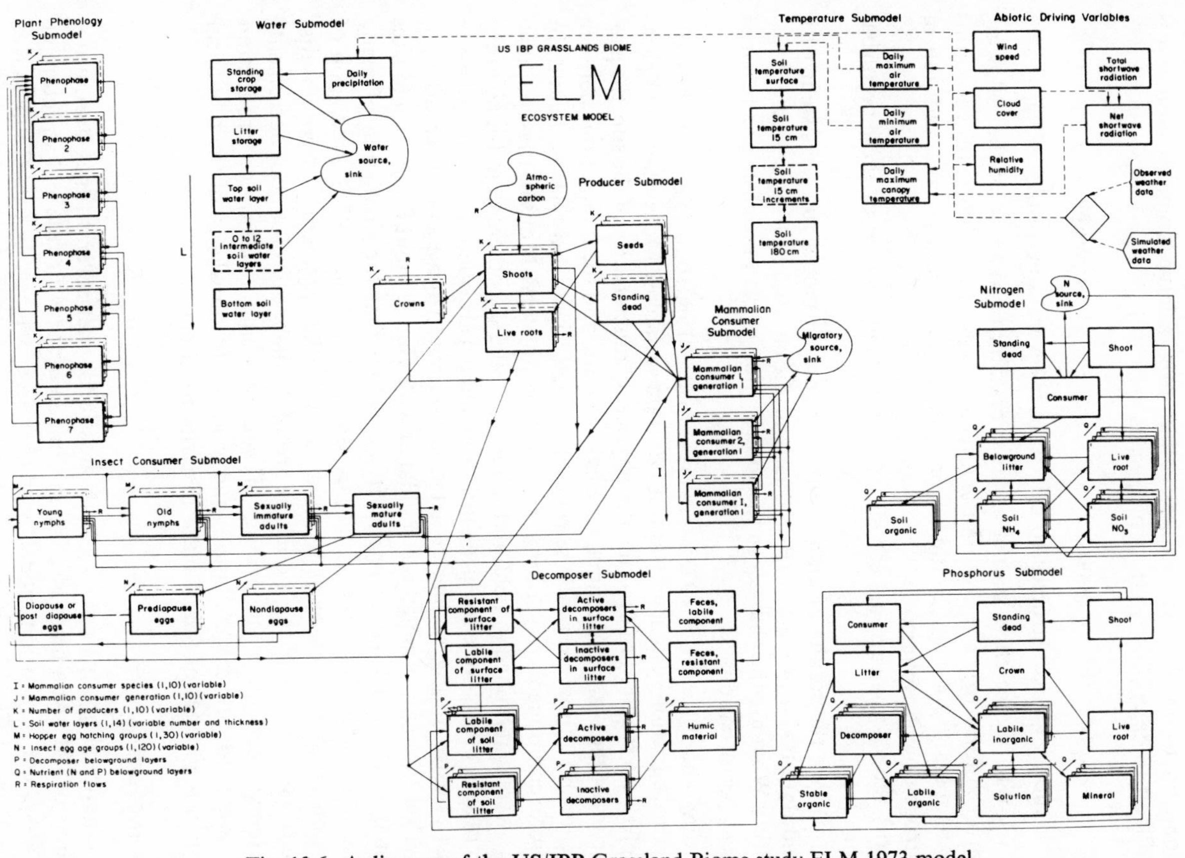


Fig. 10.6. A diagram of the US/IBP Grassland Biome study ELM 1973 model.

and connectedness in models. These results are reported in Patten (1972). One conclusion of these exercises is that given a large, complex model, it is difficult to determine what its properties are!

### *ELM*

This model was constructed to help answer several questions:

(i) What is the effect on net or gross primary production of the following perturbations: (*a*) variations in the level and type of herbivory, (*b*) variations in temperature and precipitation or applied water, and (*c*) the addition of nitrogen or phosphorus?

(ii) How is the carrying capacity of a grassland affected by these perturbations?

(iii) Are the results of an appropriately driven model run consistent with field data taken in the US/IBP Grassland Biome study? If not, why not?

(iv) What are the changes in the composition of the producers as a result of these perturbations?

A diagram of the ELM 1973 model is shown in Fig. 10.6. The model is a difference equation system with about 120 state variables. More than 1000 parameters are required to describe the physical characteristics of the site, the properties of the plant and animal species at the site, and the interactions between plants and animals (e.g. diet selection parameters) and between biota and the abiotic components (e.g. root distribution parameters). Many of these parameters were determined in field and laboratory studies, others were found in the literature, others were set using intuition and expert opinion, and a last group was adjusted during 'tuning' to get model output to agree with observation.

After tuning, the model was validated with abiotic input for years and sites not used in model development. It was found that the prediction of many abiotic variables was quite good (see Cole, 1976; Innis, 1978), prediction of certain biotic variables was reasonable, and other predictions were unsatisfactory in the view of the modelling group. Explanations for poor agreements included inadequate modelling of certain components (e.g. production responses to nutrient limitation) and inadequate knowledge of the biological system (e.g. rates of fall of plant material to litter).

### *International*

International groups working with grasslands met several times during the course of IBP. At a main meeting in Canada in 1969 the role and structure of total-system simulation models was discussed extensively (see Coupland & Van Dyne, 1969). But it was not until summer 1972 that an international group of investigators convened for 2 weeks at Fort Collins, Colorado, USA, to discuss grassland and tundra ecosystem studies and a simulation modelling effort was chosen as one focal point for discussions (Wielgoloski, Haydock &

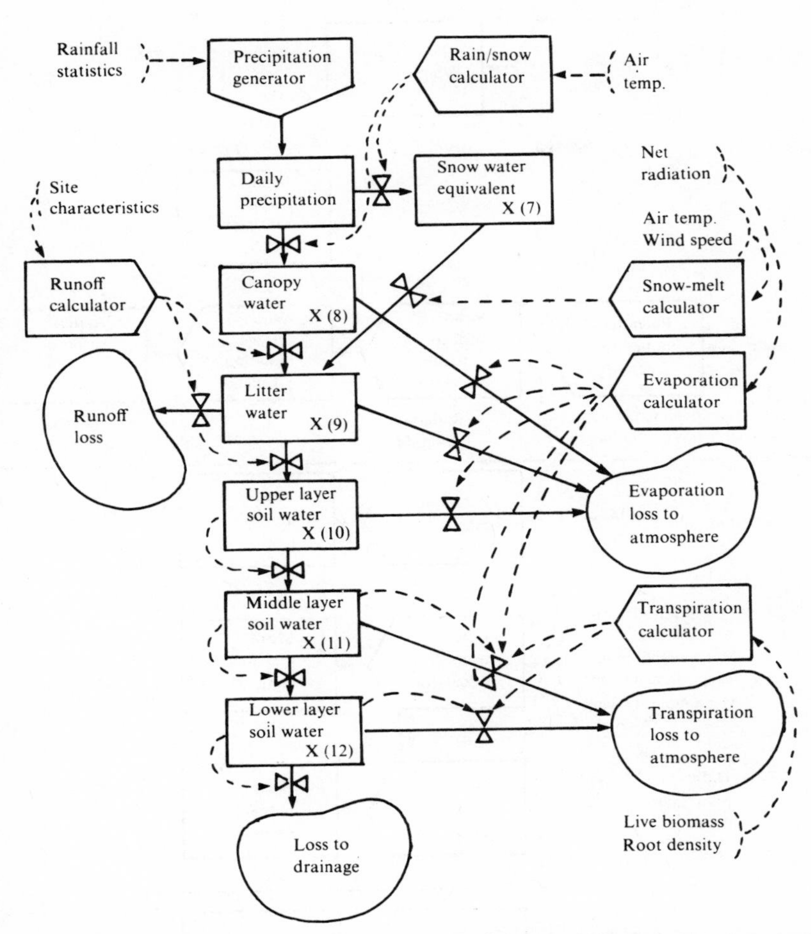


Fig. 10.7. A diagram of the model constructed at the 1972 IBP Grasslands–Tundra International Workshop.

Connor, 1972). The objective of the modelling effort was to develop a simulation model of intraseasonal dynamics of carbon and nutrient flows in the whole ecosystem which would be useful in comparing and synthesizing information from IBP sites in temperate, tropical and arid grasslands as well as in arctic, subarctic and alpine tundras. A diagram of the model is shown in Fig. 10.7. The model was coded in difference equations and parameters were estimated from data and intuition for a number of different sites. Validation was not attempted. The exercise was considered to be 'intellectually highly stimulating', but a great deal of additional work would have been needed to make the model useful for prediction and management.

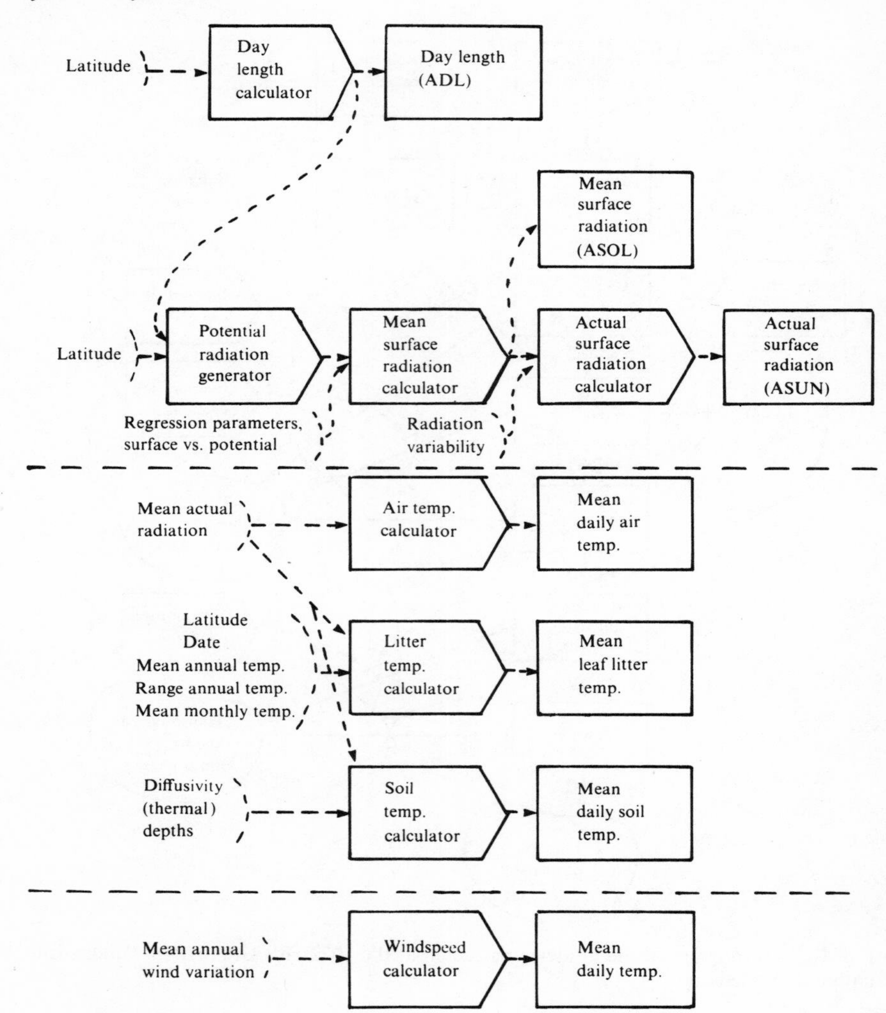


Fig. 10.7 (*cont.*)

Several other grassland ecosystem models of interest are discussed in Chapters 12 and 13 of this volume; special attention is directed to the NEGEV model noted in Chapter 13. Other important related models have been developed as part of the Arid Lands and Tundra programmes of IBP and synthesis volumes from those programmes should be consulted.

These and other total-system models of grazingland systems developed outside of IBP studies are diverse. They provide a wealth of ideas and approaches and demonstrate the possibility of modelling ecosystem-level

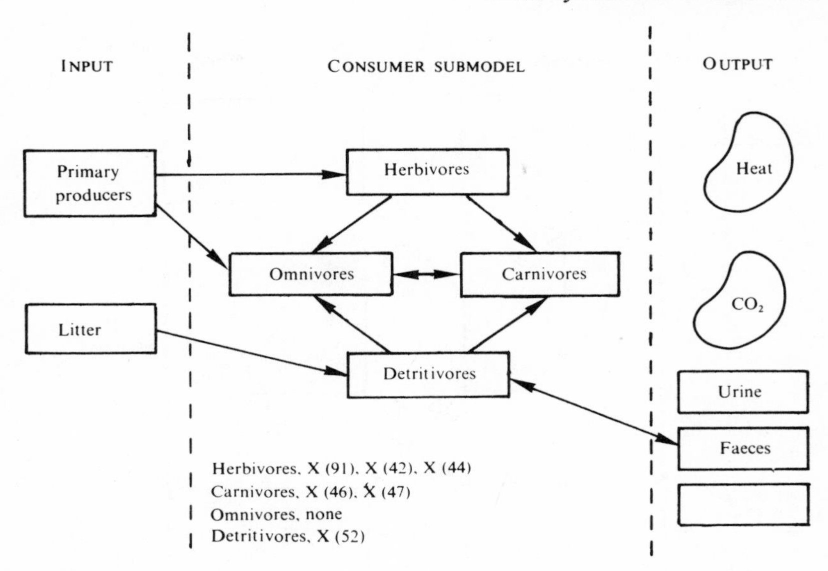


Fig. 10.7 (*cont.*)

phenomena. The existence of operational and meaningful models demonstrates that rapid advances have been made in the conduct of IBP studies. Not long ago de Wit & Goudriaan (1974) stated that we had insufficient information of quantitative and qualitative aspects of the processes to develop models that predict growth and development of single and interfering species in natural conditions. Perhaps none of the models discussed above would be considered adequate by de Wit and Goudriaan (1974). A detailed analysis and comparison of grassland models is beyond the scope of the present volume yet such an effort would yield considerable payoff for future field, laboratory and modelling studies of grasslands. Some of the main consequences of the above modelling efforts are summarized in the following sections.

## Role of ecosystem models

In the modelling exercises described above, a few common roles for the modelling effort as part of a larger study are apparent.

### *Education*

The Canadian, some of the USA, and the international modelling efforts are patently educationally oriented. They were constructed as parts of workshops or courses and served as a medium for the clarification and exchange of ideas. A careful analysis of many of the other efforts indicates that education was a big part of the job. Total-system models are interdisciplinary and as such require a significant effort in communication across disciplinary lines. This is

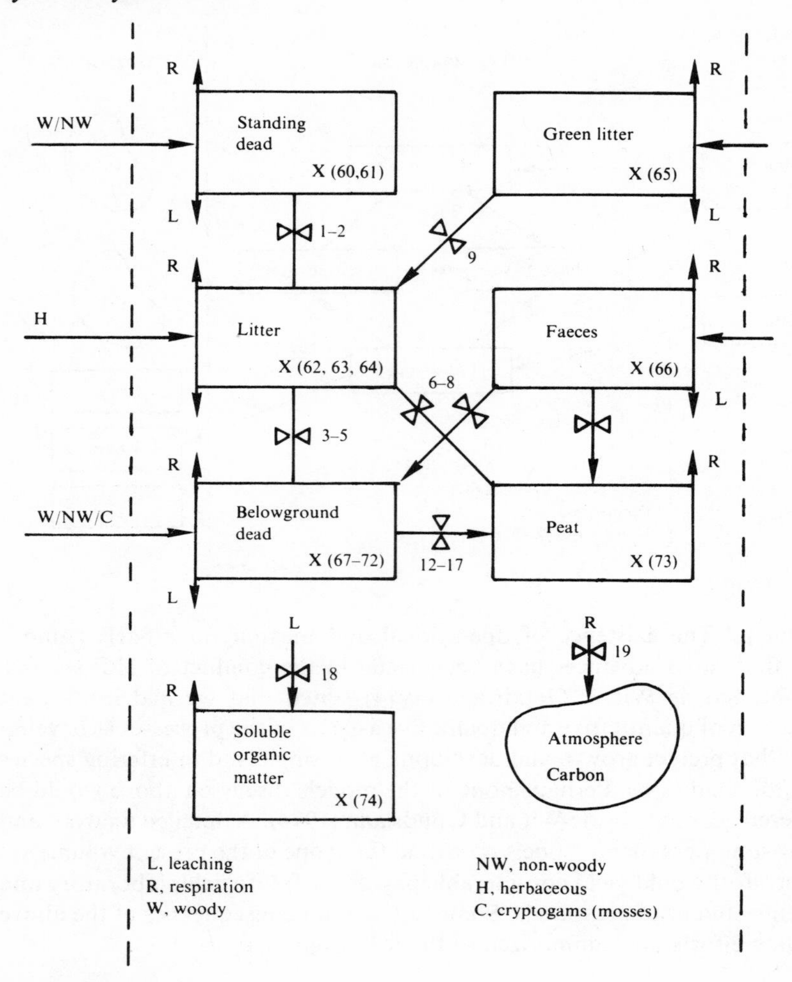


Fig. 10.7 (*cont.*)

only the first step, however, as the tools and techniques for pursuing science in the different disciplines are often distinct, and an appreciation of these differences is critical to an effective interdisciplinary effort.

### Research direction

The reports of almost all the studies discussed above wax eloquent on the role of the modelling effort in clarifying the direction of a research programme and in assigning priorities. A large part of the contribution of the worldwide IBP grassland studies to ecosystem science has been the distribution to a large audience of an appreciation of the complexity of these systems and the extent

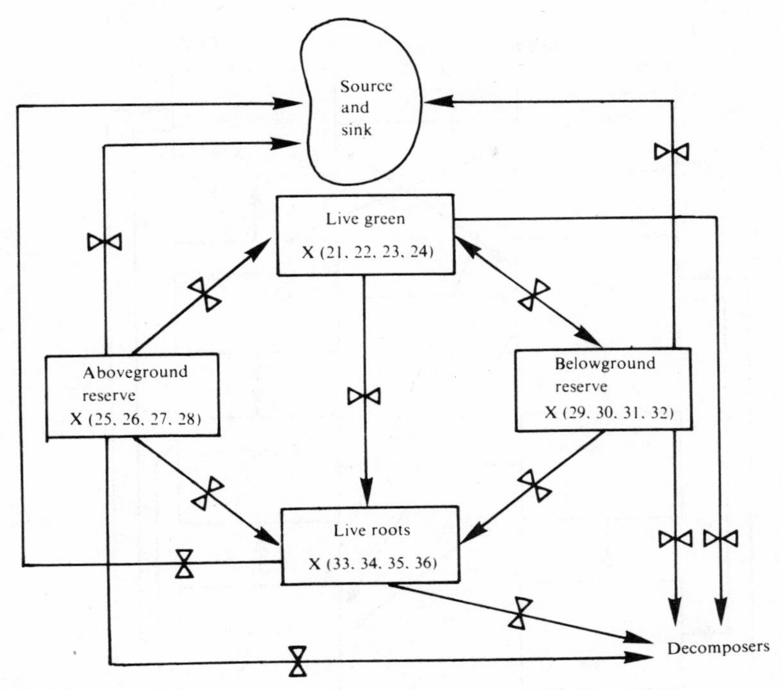


Fig. 10.7 (*cont.*)

of our ignorance of their operation. In part because of modelling efforts, research on grassland systems has changed in fundamental ways in the last decade. Areas formerly treated lightly are receiving greater attention (e.g. root studies) and research studies have shifted toward designed experiments on controls of processes with a lesser emphasis on a census of state variables. We have come to a new level of appreciation for a cybernetic system.

### *Synthesis*

The role of models in presenting an integrated picture of a research study is emphasized by several authors. Certainly simulation models are not the only way to present these results, but they have several advantages: (i) precision of statement, (ii) consequences of complexes of hypotheses can be investigated, (iii) the overall study and the interrelation of the component efforts are easily grasped in a model description while details of each piece are also available, and (iv) testing of hypotheses. The first three of these are adequately mentioned in the documents referenced above. The last, hypothesis testing, is often not given adequate attention.

Models are built of hypothesized relationships among variables (Van Dyne, Joyce & Williams, 1978). As such, they are hypotheses. Failure of a model to

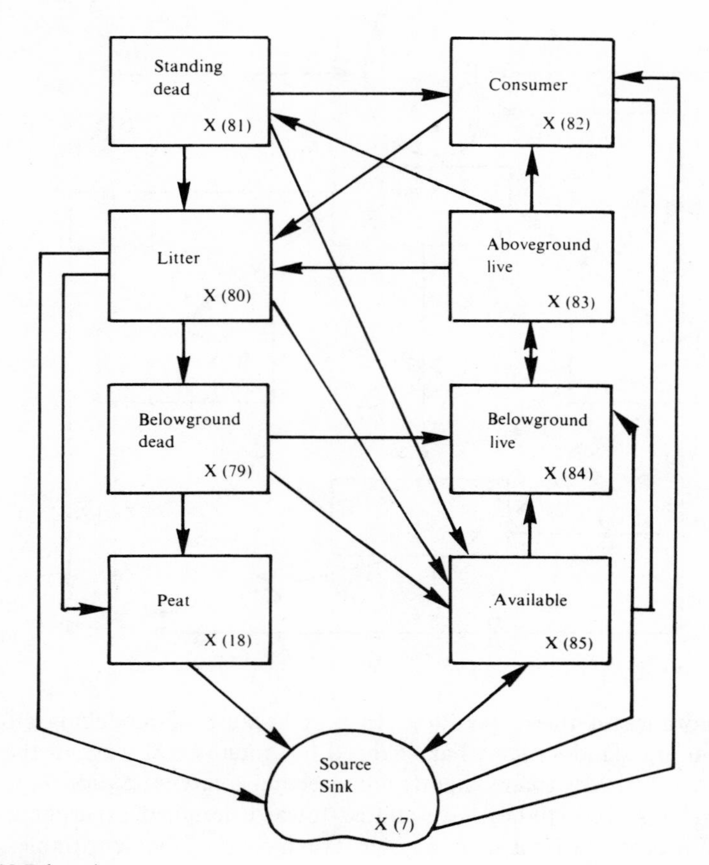


**Fig. 10.7** (*cont.*)

represent observation to a suitable degree is a matter of great scientific import – a hypothesis has been rejected. Is the model a poor representation of the hypothesis? If so, it can be improved. Is the hypothesis a poor representation of the system? If so, the investigators learn, and their hypotheses become more sophisticated and accurate. This kind of 'negative' result is so ubiquitous in modelling studies that it is ignored, or worse, treated as a failure! On the contrary, it is the very basis of the hypothetico-deductive approach to science. Investigators tend to be reluctant to evaluate 'failures' fully and to try to appreciate their basis, and often report only 'positive' results. This is unfortunate. This tendency to report only positive results is a large part of the reason why it is the modellers who gain the most from the modelling exercise and occasionally consider it more important than do other programme participants.

### *Ecosystem prediction and management*

Most of the authors of this chapter and of the models referenced above are doubtful of their ability to predict details of ecosystem dynamics and, thus, of their extensive use of these models in resource management in the near future. One author (Noy-Meir) felt that the following steps would be required to make the international model (referred to above) 'useful for predictive and management purposes': (i) increase in realism, (ii) further calibration..., (iii) tests of the sensitivity..., and (iv) formal statistical validation...'.

Indeed all of the models reported could benefit from each of the above four steps. Pursuit of these steps, however, leads quickly to considerations of diminishing returns. Noy-Meir states, 'I would venture the tentative generalization that the ratio of "total utility" (summed over all possible utilities of models) to "total effort" (or "cost") for models of this type is likely to be high at the construction stage, rapidly decreasing at the beginning of the operation stage.' Why is this? The construction stage is one of high learning and rapid appreciation of the strengths and weaknesses of one's ecosystem and modelling knowledge. In the next stage the learning is deeper (not just exchange of classical 'facts' between fields but learning what is at the frontier of one's own and others' disciplines) and, therefore, slower.

Godron notes that the choice of modelling approach must be consistent with study objectives. If the objectives are resource management and prediction, then 'empirical' models, based on accepted statistical prediction practice, may be preferred. He mentions that such an empirical model has been constructed and is being used in resource management in portions of Tunisia. On the other hand, empirical models are limited by their data bases and, in particular, can make little use of much of our qualitative knowledge of grasslands.

Each of the authors of this chapter has addressed this point of prediction and resource management and all seem to be largely in agreement. Even if we could accurately predict the response of the system to a given weather pattern, we have little hope of predicting detailed weather patterns more than a few days into the future. Forrester (1961) argues that for complex systems our objective should be the prediction of trends and properties rather than states. This can be done by considering families of model results as potential (statistically reasonable) sets of driving variables that are used to force the model. This process is expensive and has not been reported for grassland models.

## Some steps and problems in modelling

The overall programmatic role of ecosystem-level models in IBP research design and management has been discussed elsewhere for grasslands (Coupland, Zacharuk & Paul, 1969; Van Dyne, 1972, 1975; Innis, 1975*c*; Van Dyne & Anway, 1976) and for related large-scale ecological research programmes (Auerbach, 1971; Gessel, 1972; Goodall, 1972, 1974; O'Neill, 1975). These show there are some common threads involved in each of the modelling efforts described above. These threads illustrate some steps in model development that were common to the reported efforts and these, in turn, indicate some of the problems.

Each of the efforts reported was a team effort, usually involving three or more participants. These participants were drawn together to produce the model or to carry out some other activity that involved production of the model. The teams always involved people of distinct educational and scientific backgrounds with little experience in such group activities. A heavy commitment to education (mutual) was implicit or explicit.

In many of the efforts the model was explicitly a tool for bringing people together to discuss common ideas and research plans for synthesizing information collected by a number of investigators. In these situations the modelling effort achieves its purpose even though the resulting model may have little formal utility. These models are more like *Gedanken* experiments or preliminary laboratory and field experiments than more formal and careful experiments. Criticizing or evaluating them in terms of their prediction of states of the system is as inappropriate as criticizing a preliminary field experiment for not yielding statistically significant results.

### *Specifying flows versus variables*

State variables are the accumulations within the system whereas flows or processes represent the mechanisms whereby these variables are changed. Determination of the specific variables and processes that will be incorporated into the given modelling activity are determined by objectives but not in a very clear way. There are two sources of difficulty here. One is an incomplete statement of objectives, and the other is an incomplete understanding of the ecological system under consideration. In order to provide the necessary guidance, the objectives must contain accuracy criteria to be used in many decision processes, not the least of which is the determination of which state variables and flows will be incorporated. The absence of these accuracy criteria from the statement of objectives results in a wide-ranging series of decisions regarding the things to be included in the model.

On the other hand, even if we know the accuracy with which we would like to represent this system in order to achieve a specified objective, we may not know the ecology well enough to determine which state variables and which flows have to be incorporated in order to achieve that objective. As an example, an objective of representing the biomass of green plant material in a grassland ecosystem with an accuracy within one 95% confidence interval of the true mean may or may not allow one to decide on the functional groups or species of plants that have to be represented within that system in order to achieve that accuracy. This is particularly true if one is concerned not only with the nominal dynamics of the system but with the dynamics under a single stress or series of stresses and over a considerable period of time.

It is not only the flows but also the controls which operate on these flows that must be specified (see also Chapter 1, this volume). Decisions must be made as to whether or not it is sufficient to use an annual average flow rate, a seasonal average flow rate, or if the flow rates need to be controlled by other state variables in the system or by things outside the system. These decisions are difficult at best. It is important to point out, however, that the requirement for this information is not an artifact of the modelling activity. If indeed the project objectives are of biological and ecological interest and if the modelling effort addresses those objectives, then the information required to get the model to perform appropriately is the same as the information required to address the objective in any other way. Ironically, it seems that if one understands the system sufficiently thoroughly to be able to state precisely all the objectives of the model, then the modelling effort may not be of such high educational value! Modellers, experimentalists and programme managers alike must operate in an evolutionary environment. So, model objectives often evolve.

### *Process studies versus state variable measurements*

The modelling problems mentioned in the preceding subsection have an experimental counterpart associated with the conduct of studies either in the laboratory or the field. The measurement of flow rates (processes) under a variety of controls is demanding and in general requires indirect measurement techniques. Many controls operate on a given process within an ecological system. Much of the experimental work reported fails to take into consideration a number of these controls. Indeed, reports often fail to indicate the conditions under which measurements were taken in spite of the knowledge that varying those conditions would result in different experimental observations. In contrast, process studies in which the appropriate controls are measured are much less tied to a specific site and a specific time, and, therefore, are more easily generalized to other ecosystems and conditions.

The data from detailed field and laboratory studies are too voluminous to include in scientific journal articles and in some cases even in internal technical reports of the research programmes. Frequently, the measurements were taken from more than one treatment and replicate and in more than one year. The data are a valuable resource to the scientific community for future analysis and modelling efforts. Yet little has been done at present to organize the usable and reliable data sets into a standardized, documented and machine-transferable format. Such work is greatly needed, soon, to prevent irrecoverable loss of a scientific resource.

Considerable effort was expended in several IBP studies to collect sets of data from the field to be used primarily in model validation studies. Generally, these sets of data included rather continuous measurements of several driving variables, numerous periodic measurements of system state variables, and a few sporadic but sometimes intensive measurements of rate processes.

State variable measurements are easier to make in the field than are process measurements. Collecting data on state variables is the sort of thing that ecology has concerned itself with for a good part of its existence. Unfortunately, the measurement of state variables is not as easy as it appears if one wants both accuracy and precision. The state variable measurements have at least two roles in the modelling activity. First, they provide initial conditions for starting a model run, and secondly they provide validation measurements for determining the extent to which the model represents the system. The initialization of an ecosystem model, in general, requires a large number of simultaneous measurements of variables at one point in space. Thus, in theory, we are required to know the plant biomasses by species, animal biomasses by species, soil water and temperature profiles, soil and plant nutrient status, and a large number of other such things for the area being simulated all at one point in time (initial conditions). This theoretical ideal is never achieved, with the result that the selection of state variable values for initiation of the model introduces another collection of uncertainties.

State variable measurements are useful secondly in the model validation process. Measurements of state variables for validation purposes are much less demanding, except for one precaution. Driving variables must be measured at the same location as the state variables in order to have a 'true' validation. If the model and the system are assumed to be driven by variables such as precipitation, air temperature, potential evaporation, wind speed, insolation, etc., then these variables must be measured at the site or estimated from measurements near the site. This problem may be more than trivial in a situation where a significant portion of the rainfall comes in the form of highly localized, intense rain showers. Such a weather regime almost requires a meteorological station to be located on the site.

### *Levels of mechanism*

The mechanistic philosophy attempts to reduce each process to its basic physicochemical laws. Failing this, the mechanist always attempts to isolate the causal pathway associated with a process. Both the modeller and the experimentalist are continually faced with the problem of terminating this cause–effect search at an appropriate level for achieving a given set of objectives. An example of the kinds of questions that must be asked and answered follows. For the purposes of this model is the effect of phenology on photosynthesis important? If the answer is yes, then how many different phenological stages must be identified in order to achieve the required level of representation? What else affects photosynthesis, and how accurately must each of these variables be treated, etc.?

The problems of levels of mechanism should be resolved by the objectives and by the knowledge of the ecosystem. As indicated above, both the objectives and the knowledge of the ecosystem may be inadequate for the task. Choice of levels of mechanism becomes one of intuition and art and another point at which criticism may be levelled at the activity. One of the difficulties associated with this issue has to do with the tendency on the part of disciplinary-oriented individuals to want to expand the levels of mechanism in their particular area while treating the mechanisms associated with other subsystems more crudely. This point brings us back to one of the roles of system-level models in IBP, that of interdisciplinary communication. One of the most promising gains from some of these system-level modelling activities has been an appreciation on the part of some of the disciplinary scientists not only for the activities of other disciplines, but also for the importance of processes which cross disciplinary lines. The use of system-level models has allowed us to achieve a more uniform level of mechanism in some of our studies that would have been possible otherwise.

### *General classes of flow descriptions*

As the development of ecosystem-level models progresses, the modellers become aware of the extent to which a small number of processes are repeated throughout the structure. Many of the earlier modelling efforts for such things as plant simulations might have represented each plant species in a fundamentally different way. Upon rewriting a number of these processes as part of an ecosystem-level study, an appreciation for the similarity of plant processes across species is obtained. Investigation of animal processes has yielded similar recognition that a general class of flow descriptions covers a variety of biological and ecological processes. At some levels of mechanism we have even been able to represent the dynamics of different nutrients in similar ways. The recognition of these classes of flow descriptions has greatly reduced the modelling effort and has helped the ecologist to begin to look for generality as opposed to concentrating on the unique and different. He begins to see the unique and different as an exception to an interesting rule rather than as the rule itself. This kind of thinking on the part of ecologists and biologists can only accelerate studies of the ecosystem rather than its parts.

The steps taken in each model-building exercise are approximately the same but the emphases are different. Some of the modelling efforts are almost totally instructional so that all steps are taken quickly to illustrate the process. Some are managerially influenced so that prediction is more important than is process, i.e. the modellers are more concerned with an output that helps a manager make decisions about his system – be it a resource or research management activity. Some are scientifically motivated with emphasis placed on describing the biological processes at a high level of realism (mechanism). Still others are concerned more with the modelling process, its advance and limitations, than with the modelled system. Each of these emphases results in differences in the way the modelling activity is approached and in the product produced.

Critics, not acquainted with the modelling process, often fail to appreciate the above nuances. They carry the model–experiment analogy further and ask, 'Now that you have performed the experiment, what new knowledge have you produced?' This question is hard to deal with because of the number of poorly known things entering into such models. As examples, much is known about photosynthesis for certain plant species under certain conditions. Ecosystem models often address questions that require information on photosynthesis for sets of conditions and species of plants for which little is known. Are we fools to rush into this area where angels fear to tread? Or can something useful be derived from these exercises? We feel that useful results can be obtained from the construction and analysis of ecosystem-level models. Examples of results are described elsewhere in this chapter and volume.

## Ecosystem insights derived from grassland models

The grassland models described above have contributed several insights into ecosystem function. Only a few will be described.

Almost every grasslands modeller has at one time attempted to represent his system with a linear model. Such models are easy to construct and their analysis is much simpler than that of a nonlinear model – but none of them work over a useful range of inputs. It is certainly true that linearization about equilibrium points to analyse for stability is a useful exercise, but that is now generally accepted to be the limit of utility of the linear model.

In a similar way, early attempts to construct the producer section of the ELM 1973 model (described above) contained no reference to phenology of the plants. While we could achieve some of our objectives in this way, the model simply would not respond properly to certain kinds of treatments, e.g. simulated grazing or late-season rainfall. Phenology is not the only way to incorporate plant phenotype and aging into the system, but it certainly is one way. This process led us to an appreciation of the fact that something of the nature of phenology had to be incorporated into the model if we were to achieve our objective.

Ecology is at a stage where many practitioners are 'single-factor thinkers', i.e. for each phenomenon a single-factor solution is sought. This thinking has influenced a great deal of the data collection on ecological systems around the world. Participants in a modelling exercise are both impressed and frustrated by the multifactorial nature of the model they have to produce to achieve even simple objectives. These participants defend the factors included as necessary to achieve the results and become disenchanted with much of their own earlier work. Unfortunately, there is no tractable alternative, i.e. no experimental, statistical or analytical tools to deal efficiently with the number of distinct factors that come into play in ecosystem dynamics. We find ourselves trying to cope simultaneously with an advancing single-factor science (art) and a patently multifactorial system that is almost (completely?) intractable.

In order to cope with the complexity of ecosystems studied, two extreme positions are taken. One is that only single-factor studies are scientific and so even though the single-factor studies offer no hope of understanding the whole system, that is where some people put their emphasis. The other is to admit that all knowledge is incomplete and proceed, using ecosystem models as guides, to perform a series of multifactorial studies that will integrate into an ecosystem dynamic view when completed. Proponents of these extreme views are recognized among the authors of this chapter.

Exercise and analysis of completed grassland models have contributed to our understanding of ecosystem dynamics and to an appreciation of the limits of our knowledge of ecosystems. Gaining this understanding and appreciation is rapid when the models fail to perform as expected (if the reasons for failure are biological) and is much slower when the models agree with observation! Models are hypotheses and no amount of agreement between model and data will 'prove' the model, but disagreements suffice to reject. This may seem backwards if one is committed to the idea that the function of ecosystem models is to predict future states of the system. If one recognizes modelling as an integral part of the scientific process, then this view is quite consistent.

More information is available for discussion of understanding gained from the ELM model than from the other models reported. This is an artifact of funding and interest of the investigators rather than any fundamental difference in viewpoint among the authors of these models.

The 1973 version of the ELM model had a number of late-season difficulties such as excessive soil water drawdown, too much green material, too much herbivore activity. Analysis indicates that much of this was attributable to the representation of phenology – we simply do not know enough about late-season phenology and growth for native grassland plants. This insight has fostered two efforts: (i) the study of late-season phenology in native grassland plants and (ii) an attempt to replace this entire section of the model with something more fundamental, i.e. the processes which underlie phenology. Both of these efforts are underway, and we do not know at this time if significant improvements will be forthcoming.

Root death and turnover and the transfer of standing dead to litter are two other areas of the ELM producer model where information was inadequate to achieve our objective. It is surprising and educational to discover that specific tidbits of information are needed in order for the aims of the model to be achieved. Even the most experienced investigators in the field would not have expected to find that we would be limited by knowledge of these facts. When presented with these results, they were initially sceptical of the modellers' interpretation. After they studied the model and its output, however, they were convinced that these indeed were the weak links.

An analysis of the effects of grazing on the nitrogen cycle was conducted for this same model. Three different grazing intensities (light, medium and heavy) were simulated for a 6-month (growing season) grazing period. The effects of such a treatment on an ecosystem operate through a variety of pathways (Fig. 10.8). Each of these pathways responds to other factors affecting the ecosystem with the result that effects are confounded in the natural system where these other factors are not controllable. In the model, however, effects of changes in individual factors can be traced through the system.

This study showed that as grazing intensity increases, the decomposition rate and the rate of nitrogen cycling also increase. These changes are of lesser magnitude than the impact of grazing on vegetation, since cattle have a direct impact on primary producers and only indirectly affect the decomposer and nitrogen cycles. A conclusion is that many of the interactions shown in Fig. 10.8 cannot be examined in the field without total disruption of the system. Such interactions cannot be studied with models specific to only one aspect of the ecosystem. To interpret model results adequately requires examining the model structure for the underlying assumptions.

Extensive testing of the mammalian consumer portion of ELM 1973 included parameter changes, changes in model structure, addition of functions to increase realism, and combinations of the above. The investigator (Fullerton, 1975) concluded that the tests promoted 'confidence in... structure of the submodel...and its capabilities for adaptation to new, related modelling questions', i.e. further evidence is provided that the hypotheses encoded in this submodel are representative of mammalian consumers at the specified level of resolution. Thus, at this level of resolution, we can turn our attention to more pressing (less known but important) issues of consumer modelling. This is one of the most common conclusions of extensive ecosystem modelling efforts, i.e. some things are known well enough, others deserve our attention. The direction of the research changes at that point. Investigators go to the field or laboratory with a different concept of their profession and their behaviour is influenced.

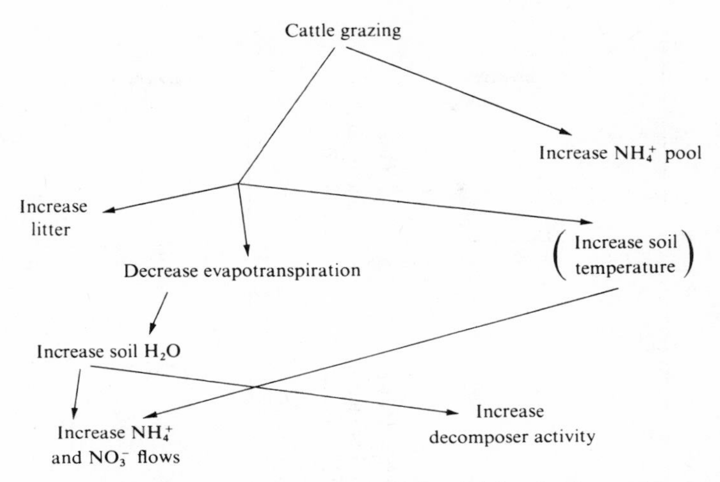


Fig. 10.8. A cause-and-effect relational diagram explaining cattle grazing impacts on various components of the grazingland system.

To produce ELM 1974 from ELM 1973, numerous modifications were made to the parameterization of ELM 1973, errors in the computer code were corrected, and a few structural changes were made (Van Dyne, Smith, Czaplewski & Woodmansee, 1978). In addition to the system state variables shown in Fig. 10.6, several output variables were coded into the model including diversity indices for plant and consumer biomass and consumer diets, several system performance ratios (e.g. photosynthesis: plant biomass, photosynthesis: system respiration and photosynthesis: intercepted insolation), and management-oriented performance indices such as protection of the soil from wind and water erosion, range condition, and projected carrying capacity for domestic animals. Model experiments were run to test intensity of grazing by cattle, with or without control of competing small mammal herbivores. Examples of year-long integrated flows plotted as a function of cattle grazing intensity in computer experiments are shown in Fig. 10.9.

Most of the system response functions obtained from these grazing intensity experiments are nonlinear. Increasing the cattle grazing intensity in these computer experiments caused decreases in intercepted insolation, water loss by transpiration, gross and net photosynthesis, primary production: respiration ratio and primary production: plant biomass ratio. Increasing intensity of cattle grazing resulted in increases in water loss by evaporation, water retained in the soil at the end of the year, grasshopper production, decomposer carbon dioxide evolution, total-system respiration and the photosynthesis: intercepted radiation ratio. Cattle production at first increased with increasing grazing intensity, but then decreased. The grazing intensity for

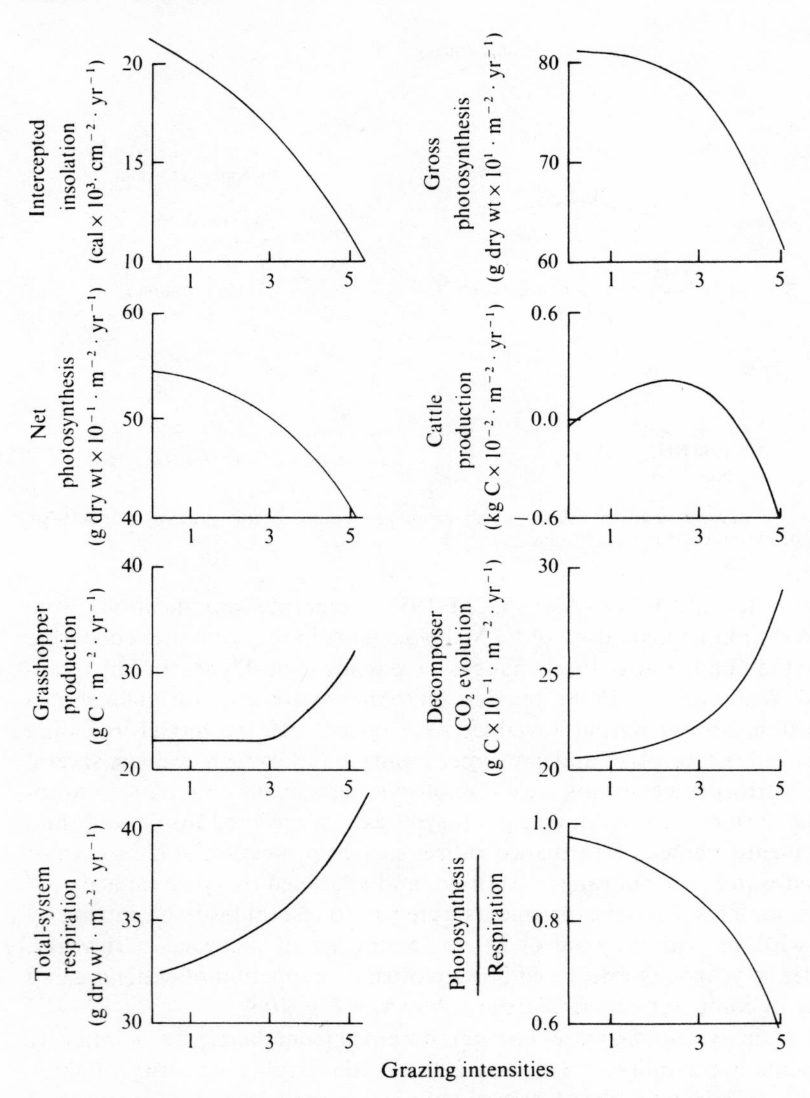


**Fig. 10.9. Impact of intensity of grazing with cattle on selected integrated system and subsystem performance measures as derived from computer simulation experiments. (Taken from Van Dyne, Smith, Czaplewski & Woodmansee, 1978.)**

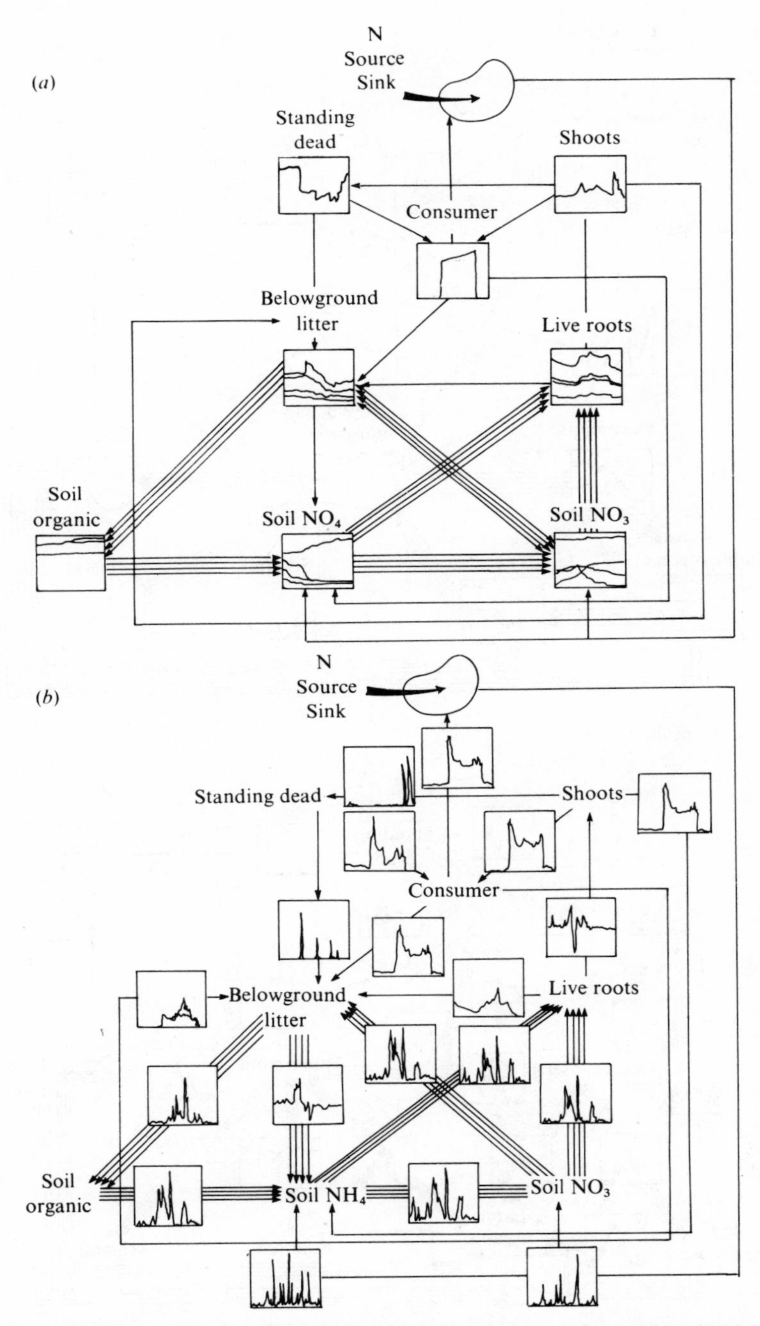


Fig. 10.10. Levels and flows of nitrogen under light grazing by cattle on a shortgrass prairie ecosystem: (*a*) levels (g N $\cdot$ m$^{-2}$) and (*b*) flows (g P $\cdot$ m$^{-2}\cdot$ 2d$^{-1}$).

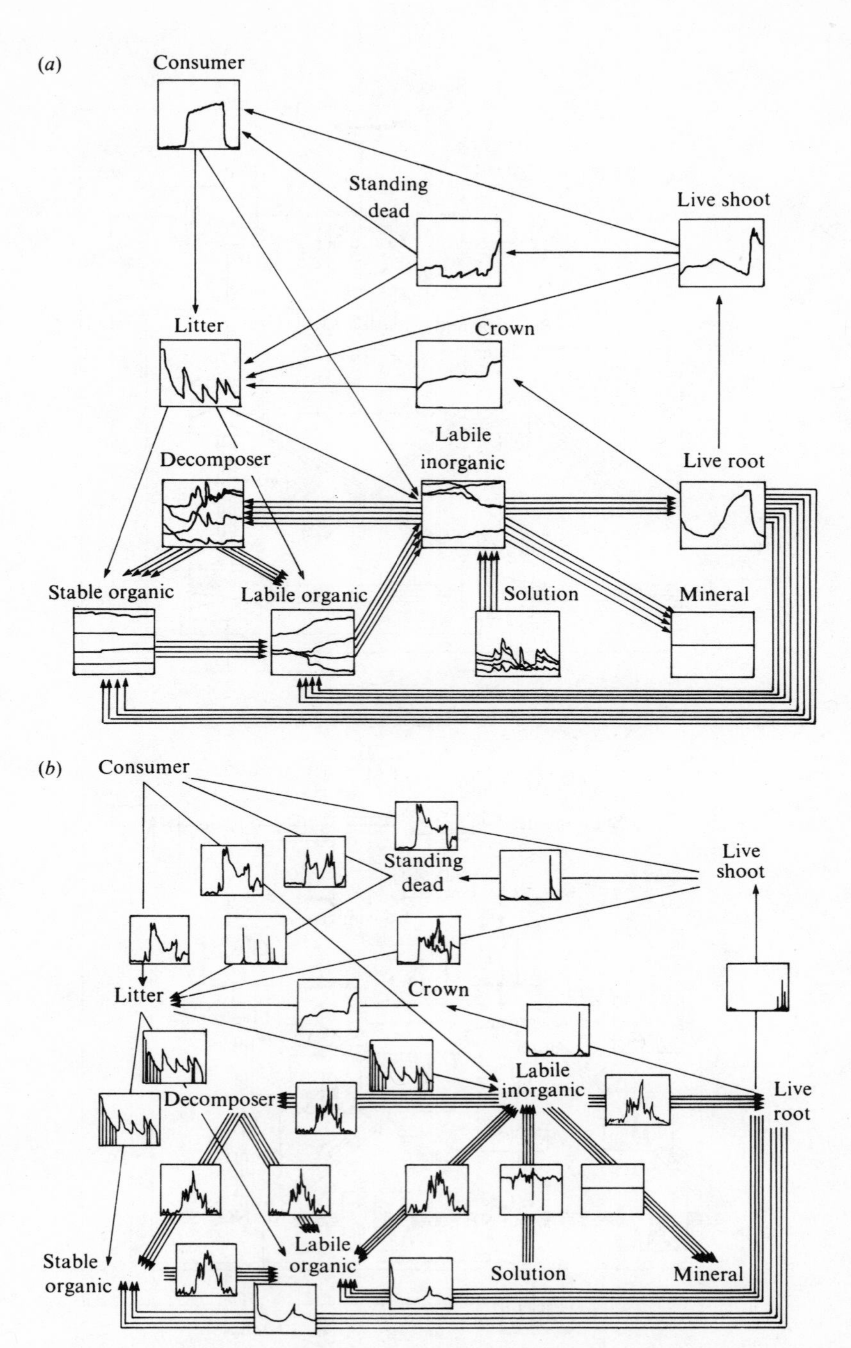


Fig. 10.11. Levels and flows of phosphorus under light grazing by cattle on a shortgrass prairie ecosystem: (*a*) levels ($g\,P \cdot m^{-2}$) and (*b*) flows ($g\,P \cdot m^{-2} \cdot 2d^{-1}$).

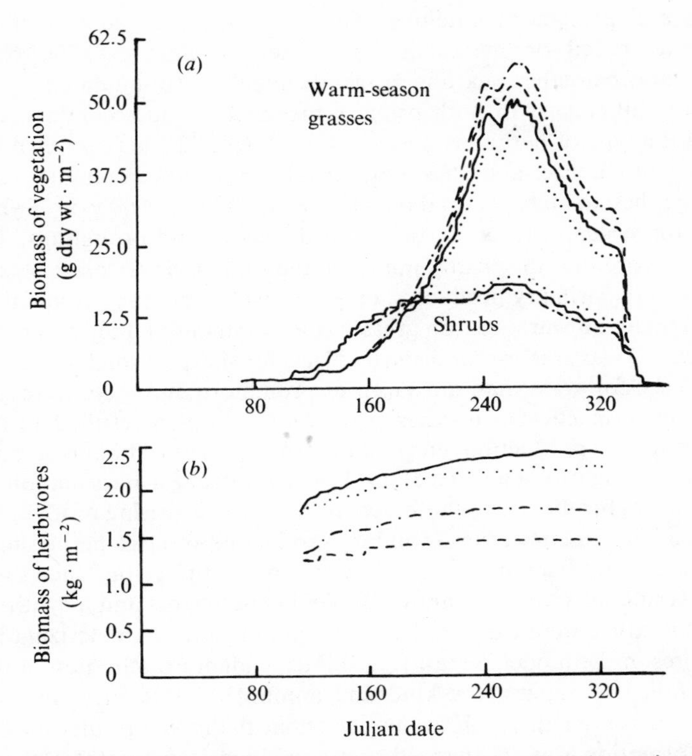


Fig. 10.12. Results of computer simulation experiments with light grazing of a shortgrass prairie with cattle (——), bison (.....), sheep (—·—·), and pronghorn antelope (- - -): (*a*) impact on standing crop of warm-season grasses; (*b*) biomass of large herbivore species in the grazing period. (From Van Dyne, Joyce & Williams, 1978.)

optimum cattle production as determined from model experiments (for a short-term simulation) was near that determined from long-term field experiments. For ELM 1974 mean nitrogen flow rates were 3.2, 4.0 and 6.4 $g \cdot m^{-2} \cdot yr^{-1}$ and mean phosphorus flow rates were 3.7, 4.0 and 4.9 $g \cdot m^{-2} \cdot yr^{-1}$, respectively, for light, heavy and extra-heavy grazing. The mean interchanges are between organic and inorganic forms of the nutrients in the soil. These preliminary modelling results suggest that extra-heavy grazing would result in a net loss of nitrogen and phosphorus from the inorganic soil pool in the system whereas under light grazing there is an approximate annual balance for these elements (Figs. 10.10 and 10.11).

Several additional large herbivores were parameterized into the ELM 1974 model for the shortgrass prairie situation where previously only cattle had been incorporated. Van Dyne, Joyce & Williams (1978) have reported a series of model experiments comparing secondary production by cattle, bison, sheep and pronghorn antelope. These animals were stocked at an equivalent rate, based on metabolic body size, for light grazing. Non-reproductive, approximately yearling growing animals were simulated. There were surprising differences in both primary production and secondary production with the type of herbivore species (Fig. 10.12). The herbivores differ in their dietary preferences and physiological characteristics. Total integrated gross photosynthesis values were about 820, 850, 850 and 860 $g \cdot m^{-2} \cdot yr^{-1}$, respectively, for systems grazed by bison, cattle, sheep and pronghorn. The species of herbivore did not greatly influence the timing of modelled phenological events in the primary producers, only the level of primary production. The relative impact on warm-season grasses (the dominant plant group in the shortgrass prairie) was greatest for bison and least for sheep. Bison had a much higher preference for grass than did sheep or pronghorn and there were also significantly different effects on other plant groups, which resulted in differences in primary production and standing crops even though grazing intensity was only light (see also Chapter 4, this volume, for information on dietary preferences for these large herbivores). Another surprising result of the model runs was the differences in secondary production of large herbivores – 8, 7, 4 and 2 $g\ C \cdot m^{-2} \cdot season^{-1}$, respectively, for cattle, bison, sheep and pronghorn. Results of several separate IBP field experiments and a synthesis of scientific literature were used to derive the parameters characterizing the large herbivores in the model. Separate and independent examination of this information failed to suggest the kind and amount of differences in large herbivore production or their relative impacts that the model results showed when the information was synthesized and combined into the total-system framework. The lower production performance of the pronghorn is probably due to their higher maintenance energy requirements and the fact that, relatively, they were older than the other herbivores in the model experiments. Data presented in Chapter 4 show that the dietary preference of pronghorn may differ considerably from the available herbage. Van Dyne (1978) has shown in nonlinear programming model analyses that optimal combinations of these herbivores, to maximize use of the available vegetation, would be to take about 67, 20, 12 and 1 % of the grazing load, respectively, with cattle, bison, sheep and pronghorn for the early-summer grazing situation.

## Some of the lessons learned

Several lessons are integrated into the discussions above, and it suffices here to highlight some of them:

(i) Models are built to achieve objectives. The model's value, or lack of it, depends in part on the relationship of the objectives to the study goals and, in part, on the extent to which the objectives are achieved. An evaluation of the model which ignores the objectives is nonsensical.

(ii) Ecosystem models are team efforts and, therefore, almost certainly require a large educational investment.

(iii) Ecosystems are multifactorial systems and as such may not be reasonably treated with a series of single-factor studies. Models and ecosystem understanding are limited by this fact.

(iv) Ecosystem models contribute greatly to the advance of ecosystem science providing a focal point for group discussions, an arena for testing complex hypotheses, a storehouse for tested and (temporarily) accepted hypotheses and a means for investigating the consequences of complex hypotheses.

(v) Ecosystem models are at present of less direct use to resource managers than to researchers because of the state of ecosystem knowledge, the relationship between models and objectives, the complexity of the models, and the time required to build a model.

Perhaps one of the reasons total-system models have not contributed more to ecology to date is that they have not been fully evaluated. Only a few large-scale, total-system models have been built, they have been documented only recently, the research teams building them have dispersed, and the scientific community has not yet had the opportunity to evaluate and interpret their results.

The evaluation of models includes both the verification and validation process. This requires either a large amount of technical expertise, or a large amount of coordinated data, or both. Perhaps not unexpectedly, the data sets necessary for model construction and development as well as model validation did not become available in condensed, uniform, readily available, computer-compatible format during the time the models were being developed. Most of the IBP data still is not available in a standardized format. It is necessary, for rapid modelling progress, to be able to mechanically access sets of initial condition values, sets of driving variables, and sets of validation values to compare with output from the model.

Many IBP modellers and experimentalists shared ideas and learned from each other in the process of the development of total-system ecological models. Yet it is difficult to document this large and valuable output from the modelling effort. This is evidenced by numerous papers by these individuals concerning the art and practice of modelling (e.g. see Goodall, 1972, 1974; Innis, 1972, 1974, 1975*a*, *b*; Van Dyne, 1972, 1975; O'Neill, 1975; Steinhorst & Garatt, 1976; Mankin, O'Neill, Shugart & Rust, 1977). Much of this gain is in the 'verification' process as the investigative team attempts to determine the 'truthfulness' of the model structure and responses.

One of the difficulties in the evaluation of total-system ecological models is the lack of a well-developed art for concise, organized, yet complete documentation. Documentation problems are more difficult for total-system models than for subsystem models primarily because of the size of the overall model and because of the different disciplinary areas which must be addressed.

A simple listing of computer codes is no substitute for, and was never intended to be, documentation. A total-system model may have 100 system state variables, some two to three flows per state variable, and from two to four variables affecting each flow. The mathematical form of flow calculations is generally nonlinear, often with lags. It is understandable that the documentation problem is difficult.

We have not yet evolved a suitable documentation approach although initial attempts have been made to structure one (Newton & Swartzman, 1973). Van Dyne & Abramsky (1975) conclude that a clear and complete statement of at least the following aspects about a model seems critical for rapid advances for modelling efforts: specify the objectives, hypotheses or assumptions; state the general mathematical form; list the driving variables and system state variables; provide a diagram of the model structure; list the equations or functional relationships for flows; list or provide access to the computer code; and discuss the model limitations. If the specific equations cannot be listed it is at least useful to provide a coupling matrix and a flow-effects matrix (see Chapter 1, this volume). For examples, Van Keulen, Louwerse, Sibma & Alberda (1975) provided a 'relational diagram' of their model which showed the box-and-arrow structure of the model and they also graphed functional relationships for some of the main flows for their model.

One of the difficulties encountered in attempting to utilize total-system models, or even their segments, with the resource management community is that the objectives of the ecological modelling effort may not be consistent with the objectives of the resource manager, even if there is no communication problem. Although large-scale, total-system models are being used in testing hypotheses related to resource management, such tests are not done in the arena of the resource manager as in a workshop mode. Instead they are done in the quiet of the researcher's office and over a longer period of time. In discussion with resource managers it is soon clear that they want to alter the structure of the model. They insist on evaluating certain objectives and discarding or discounting other portions of the model. This requires restructuring and rerunning models rapidly. It has not been possible to do so to date although Innis (1975*a*, *c*) provided some preliminary ideas on 'self-organizing modelling systems' which would be extremely useful tools in such modelling efforts. A 'software system' was envisaged that would take the long-term environment of the system being addressed, recent weather, and perturbations and objectives as input; this system would produce as output a rough model of the system appropriate to addressing the objectives specified. This requires structuring not only ecological knowledge but developing libraries of submodels and, especially, developing an understanding of how model structure relates to model objectives. He suggested that early versions of the self-organizing modelling system will use libraries of submodels as a base from which to operate.

Libraries of submodels are viewed as important and useful items in themselves. They must be structured so that they can be coupled together automatically when designing an ecosystem model. The problems of documentation and evaluation raised above appear again here. How does one decide that a certain model should be put in a library? What characteristics must it have? How does one determine that a model has these characteristics in the current state of model documentation?

One of the most useful approaches to modelling has been the long-term evolution of models coupled with experimentation and spinning off applications useful in resource management. There are a few examples of models evolving, such as the effort first reported by Goodall (1967) who developed a given model over a decade. Another example of an evolutionary grazingland model is that developed by an Australian CSIRO research team for the study of sheep on summer pastures (first reported by Donnelly & Armstrong, 1968). A third example is that of the large-scale, total-system multiple-flow ecosystem model reported here, i.e. ELM (first reported by Anway *et al.*, 1972).

To be most useful it is likely that a model (particularly a total-system model) will evolve over a relatively long period of time. Many investigators may be involved in the modelling development, verification and validation. It is then logical that some overall strategy be used in the modelling effort. Two factors are of concern here: the strategy for development and the overall structure. Two alternative tactics that have been used in model development are termed the 'top-down' and the 'bottom-up' modelling approaches (Van Dyne & Abramsky, 1975).

In the 'top-down' approach one starts with a series of objectives which dictate a model. One then continually improves the structure and function of the model in a stepwise manner. The top-down philosophy is in a sense initially mechanistic in that it subdivides the large system into a group of smaller components that can be handled adequately. But basically some units must be treated in an empirical manner. Considerable attention is needed at the outset to structure the overall model into a hierarchical form so that given subunits can be changed without having to restructure the entire effort. This approach also utilizes the fact that in the development of a given submodel, response from other submodels is needed. One approach (discussed in Chapter 4, this volume) developed by Goodall was to conceptualize the entire model as having submodels of different resolution capacities for interchange. One must start with a total-system viewpoint if one hopes to have much chance of ending up with a total-system model. Many parts of the model will be crudely represented at the outset. Part of the task of modelling management is to work continually through the various submodels to bring them closer and closer to the state-of-the-art in that subject matter area and evolve a total-system model in a balanced manner.

The 'bottom-up' approach tends to start with isolated and independent high-resolution and complex process models. The difficulty is to bring these complex submodels together into a total-system model. There is a tendency for scientists continually to increase the complexity of the individual models. The advocates of starting the development of models with high-resolution submodels of individual processes (e.g. Holling, 1966; de Wit, 1970; Holling & Ewing, 1971) argue in favour of developing and thoroughly validating individual process models and fixing the parameter values in them by experiments. They suggest *then and only then* can one progress toward a total-system model by coupling these high-resolution process models (Van Keulen, 1975). If this is done then there is high confidence in the process models and the parameters. This minimizes indeterminacy and reduces the sensitivity tests and validation that are required at the total-system level. However, at least one of us (Van Dyne) has not been able to find a large-scale, total-system model of a grazingland developed with this approach.

## Conclusions

In this chapter a number of ecosystem-level models produced in association with grassland studies, under the auspices of the International Biological Programme, are discussed. Model structures and outputs are described and some general conclusions drawn: (i) data and information for the construction of useful ecosystem models are rare, and (ii) given proper care in setting objectives, ecosystem modelling efforts contribute a great deal to ecosystem research design.

Although many diverse backgrounds were represented in the ecosystem modelling projects conducted under IBP auspices and many different objectives were pursued, the lessons learned and conclusions drawn converge: (i) Ecosystem modelling efforts play a necessary role in ecosystem studies and that role is important even if the resulting model is useless. (ii) There are diminishing returns from these efforts – or at least the gain per unit of effort does not always increase at the same rate. (iii) Biologically realistic models for resource management are more a thing of the future. Our knowledge of these systems and the time required to build a model for a specific objective are limiting factors.

## References

Anway, J. C., Brittain, E. G., Hunt, H. W., Innis, G. S., Parton, W. J., Rodell, C. F. & Sauen, R. H. (1972). *ELM: version 1.0. US/IBP Grassland Biome Technical Report 156*. Colorado State University, Fort Collins.

Auerbach, S. I. (1971). The deciduous forest biome program in the United States of America. In *Productivity of the forest ecosystems of the world, UNESCO Series in Ecological Conservation 4*, pp. 677–84. UNESCO, Paris.

Bazilevich, N. I. (ed.) (1975). Productivity of the steppe, meadow, and swamp biogeocoenoses of the forest steppe. In *Resources of the biosphere (synthesis of the Soviet studies for the International Biological Programme*), pp. 56–95. Nauka, Leningrad. (In Russian.)

Bledsoe, L. J. (1976). Models for the structure and function of a grassland ecosystem. PhD Thesis, Colorado State University, Fort Collins.

Bledsoe, L. J., Francis, R. C., Swartzman, G. L. & Gustafson, J. D. (1971). *PWNEE: a grassland ecosystem model. US/IBP Grassland Biome Technical Report 64*. Colorado State University, Fort Collins.

Cale, W. G. (1975). Simulation and systems analysis of a shortgrass prairie ecosystem. PhD Thesis, University of Georgia, Athens.

Cole, G. W. (ed.) (1976). *ELM: version 2.0. Range Science Series 20*. Range Science Department, Colorado State University, Fort Collins.

Coupland, R. T. & Van Dyne, G. M. (eds.) (1969). *Grassland ecosystems: reviews of research*. (Proceedings of the 2nd meeting, PT Grassland Working Group, International Biological Programme, at Saskatoon and Matador, Saskatchewan, Canada.) *Range Science Series 7*. Range Science Department, Colorado State University, Fort Collins.

Coupland, R. T., Zachuruk, R. Y. & Paul, E. A. (1969). Procedures for study of grassland ecosystems. In *The ecosystem concept in natural resource management*, ed. G. M. Van Dyne, pp. 25–47. Academic Press, New York & London.

de Wit, C. T. (1970). Dynamic concepts in biology. In *International Biological Programme: prediction and measurement of photosynthetic productivity*, pp. 17–23. Centre for Agricultural Publishing and Documentation, Wageningen, The Netherlands.

de Wit, C. T. & Goudriaan, J. (1974). *Simulation of ecological processes*. Centre for Agricultural Publishing and Documentation, Wageningen, The Netherlands.

Donnelly, J. R. & Armstrong, J. S. (1968). Summer grazing. In *Digest of the II conference on applications of simulation*, pp. 329–32.

Forrester, J. W. (1961). *Industrial dynamics*. MIT Press, Cambridge, Masschusetts.

Frausz, H. G. (1974). *The functional response to prey density in an acarine system*. Centre for Agricultural Publishing and Documentation, Wageningen, The Netherlands.

Fullerton, S. A. (1975). *Analyses of the mammalian consumer submodel of ELM. US/IBP Grassland Biome Technical Report 294*. Colorado State University, Fort Collins.

Gessel, S. P. (1972). Organization and research program of the Western Coniferous Forest Biome. In *Research on coniferous forest ecosystems*, ed. G. F. Franklin, L. F. Dempster & R. H. Waring, pp. 7–14. Northwest Forest and Range Experiment Station, Forest Service, US Department of Agriculture, Portland, Oregon.

Gilmanov, T. G. (1976). Plant submodel in the holistic model of a grassland ecosystem (with special reference to its belowground part). *Botanical Journal*, **61**, 1185–97. (In Russian.) (Also published in *Ecological Modelling*, **3**, 149–63.)

Goodall, D. W. (1967). Computer simulation of changes in vegetation subject to grazing. *Journal of the Indian Botanical Society*, **46**, 356–62.

Goodall, D. W. (1972). Building and testing ecosystem models. In *Mathematical models in ecology*, ed. J. N. R. Jeffers, pp. 173–94. Blackwell Scientific Publications, Oxford.

Goodall, D. W. (1974). Problems of scale and detail in ecological modelling. *Journal of Environmental Management*, **2**, 149–57.

Gounot, M. & Bouché, M. (1974). Modélisation de l'écosystème prairial: objectifs et méthodes. *Bulletin d'Ecologie*, **5**, 309–38. (In French.)

Holling, C. S. (1966). The strategy of building models of complex ecological systems. In *Systems analysis in ecology*, ed. K. E. F. Watt, pp. 195–214. Academic Press, New York & London.

Holling, C. S. & Ewing, S. (1971). Blindman's buff: exploring the response space generated by realistic ecological simulation models. In *Statistical ecology*, ed. G. P. Patil, E. C. Pielou & W. E. Waters, vol. 2, pp. 207–31. Pennsylvania University Press, University Park.

Innis, G. S. (1972). Simulation of ill-defined systems: some problems and progress. *Simulation*, **19**, 33–6.

Innis, G. S. (1974). Dynamics analysis in 'soft science' studies: in defense of different equations. In *Lecture notes in biomathematics*, ed. P. van den Driessene, vol. 2, pp. 102–22. Springer-Verlag, Berlin.

Innis, G. S. (1975*a*). One direction for improving ecosystem modelling. *Behavioral Science*, **29**, 68–74.

Innis, G. S. (1975*b*). The use of a systems approach to biological research. In *Study of agricultural systems*, ed. G. E. Dalton, pp. 369–91. Applied Science Publishers, Amsterdam.

Innis, G. S. (1975*c*). The role of total systems models in the grassland biome study. In *Systems analysis and simulation in ecology*, ed. B. C. Patten, vol. 3, pp. 13–47. Academic Press, New York & London.

Innis, G. S. (ed.) (1978). *Grassland simulation model. US/IBP Grassland Biome study. Ecological Studies Series.* Springer-Verlag, Berlin. (In press.)

Lyapunov, A. A. & Titlyanova, A. A. (1974). Systematic approach to the study of exchange processes in biogeocoenosis. *Botanical Journal* (*USSR*), **59**, 1081–92. (In Russian.)

Mankin, J. B., O'Neill, R. V., Shugart, H. H. & Rust, B. W. (1977). The importance of validation in ecosystem analysis. In *Systems ecology – where do we go from here?*, ed. G. S. Innis. Simulation Council of America, La Jolla, California.

Newton, J. F. & Swartzman, G. L. (1973). Frameworks for whole stystem study with special reference to agricultural systems of temperate grassland. In *System workshop report*, ed. J. E. Newton & G. L. Swartzman, pp. 28–35. Grassland Research Institute, Hurley, England.

Numata, N. (ed.) (1975). *Ecological studies in Japanese grasslands with special reference to the IBP area – productivity of terrestrial communities. JIBP Synthesis Volume 13.* University of Tokyo Press, Japan.

Odum, H. T. (1971). *Environment, power, and society*. Wiley-Interscience, New York.

O'Neill, R. V. (1973). Error analysis of ecological models. In *Proceedings of the III National Symposium on Radioecology*, ed. D. G. Nelson, pp. 878–90. Atomic Energy Commission, Oak Ridge, Tenessee.

O'Neill, R. V. (1975). Management of large-scale environmental modeling projects. In *Ecological modeling in a resource management framework*, pp. 251–82. Resources for the Future, Washington, DC.

Patten, B. C. (1972). A simulation of shortgrass prairie ecosystem. *Simulation*, **19**, 177–86.

Randell, R. L. & More, R. B. (1974). *Modeling. IX. Total ecosystem. Matador Project Technical Report 68.* Canadian Committee for IBP, University of Saskatchewan, Saskatoon.

Saugier, B., Ripley, E. A. & Leuke, P. (1974). *Modeling. VIII. A mechanistic model of plant growth and water use for the Matador Grassland. Matador Project Technical Report 65.* Canadian Committee for IBP, University of Saskatchewan, Saskatoon.

Steinhorst, R. K. & Garatt, M. (1976). Validation of deterministic system simulation models. In *Proceedings of the Statistical Computing Section of the American Statistical Association*, pp. 1–10.

Van Dyne, G. M. (1972). Organization and management of an integrated ecological research program – with special emphasis on systems analysis, universities, and scientific cooperation. In *Mathematical models in ecology*, ed. J. N. R. Jeffers, pp. 111–72. Blackwell Scientific Publications, Oxford.

Van Dyne, G. M. (1975). Some procedures, problems, and potentials of systems-oriented, ecosystem-level research programs. In *Procedures and examples of integrated ecosystem research*, pp. 4–58. *Barrskogslandskapets Ecolgi Technical Report 1.* Swedish Coniferous Forest Project, Uppsala, Sweden.

Van Dyne, G. M. (1978). Optimal combinations of four large herbivores for short-grass prairie. In *Proceedings of the XIII International Grassland Congress* (in press).

Van Dyne, G. M. & Abramsky, Z. (1975). Agricultural systems models and modelling: an overview. In *Study of agricultural systems*, ed. G. E. Dalton, pp. 23–106. Applied Science Publishers, Amsterdam.

Van Dyne, G. M. & Anway, J. C. (1976). A research program for and the process of building and testing grassland ecosystem models. *Journal of Range Management*, **29**, 114–22.

Van Dyne, G. M., Joyce, L. A. & Williams, B. K. (1978). Models and the formulation and testing of hypotheses in grazingland ecosystem management. In *The breakdown and restoration of ecosystems*, ed. M. Holdgate & M. Woodman, pp. 41–84. Plenum Press, London.

Van Dyne, G. M., Smith, F. M., Czaplewski, R. L. & Woodmansee, R. G. (1978). Analyses and synthesis of grassland ecosystem dynamics. In *Glimpses of ecology*, ed. J. S. Singh & B. Gopal, pp. 1–79. International Scientific Publications, Jaipur, India.

Van Keulen, H. (1975). *Simulation of water use and herbage growth in arid regions.* Centre for Agricultural Publishing and Documentation, Wageningen, The Netherlands.

Van Keulen, H., Louwerse, W., Sibma, L. & Alberda, Th. (1975). Crop simulation and experimental evaluation – a case study. In *Photosynthesis and productivity in different environments*, ed. J. E. Cooper, pp. 623–43. *International Biological Programme Synthesis Series 3.* Cambridge University Press, London.

Wielgolaski, F. E., Haydock, K. P. & Connor, D. J. (1972). *A grazinglands plant-decomposition, carbon-mineral simulation model. US/IBP Grassland Biome Technical Report 203.* Colorado State University, Fort Collins.

# 11. Trophic structure and relationships

A. I. BREYMEYER

**Contents**



## Introduction

This chapter includes some considerations of the trophic structure of grassland ecosystems and trophic relationships among components of these systems.

By trophic relationship we understand the relation between a consumer and its food. The manner in which groups of consumers are interconnected with their food is called the trophic structure of the system. Generally speaking, food sources in an ecosystem consist of live and dead organic matter and they may be either of plant or animal origin. If they are estimated by weight, the amount of organic matter of plant origin is many times larger than that derived from animals. Plant materials are readily distinguishable from animal materials when alive, or soon after the death of an organism, but it becomes more difficult to identify the origin of dead organic matter after it has been broken down.

The body of detailed knowledge of the components of an ecosystem, i.e. knowledge of the ecology of species or groups of related species living in grassland ecosystems, is large. The plant species composition is known for almost all types of grasslands from tundra to savannas, dry steppes and prairies or deserts. The phenology, production rate, habitat requirements, and frequently other detailed and specific characteristics are known for most plant species. Also, the herbivores of grassland ecosystems are relatively well known, in particular those in which man is interested either because they are a source of food or because they are competitors of his food animals, that is, other grass-consuming herbivores. In recent years much attention has been paid to decomposers which render plant nutrients available under natural conditions and are especially important in large, non-developed areas such as the tropics which have poor soils and high production rates.

## Autotrophs: some considerations on higher plants of grasslands

There are large differences in the estimates of the rate of grassland primary production. Sims & Singh (1971) report that the daily average rate of plant production varied from 1.6 to 9 $g \cdot m^{-2}$ for different North American grasslands (Table 11.1). This series of prairies includes semidesert, dry, shortgrass prairies in Colorado and New Mexico as well as wet, tallgrass, subtropical prairie in Oklahoma. Bazilevich, Rodin & Rozov (1971) found a similar range for grasslands in the USSR (Table 11.2): annual production varied from 300 $g \cdot m^{-2}$ (steppe-deserts) to 1600 $g \cdot m^{-2}$ (considered as secondary savanna wet grasslands located in the extreme south of the country). Thus, each of these continents shows a range of production values with the highest value five times the lowest. However the two series can not be directly compared on the basis of daily or annual values because the authors have not specified the length of the growing season. But if it can be assumed that the growing season lasts for about 6 months, then the Russian and American data are almost identical – the absolute values and the range of plant production in the two continents are the same. It seems, however, that the highest values reported in the two series do not characterize the maximum potential of grassland ecosystems, as the authors have no results for a wet tropical savanna where plant production is very high. Breymeyer (1978) and Singh (1976) suggest values up to 3000–4500 $g \cdot m^{-2} \cdot yr^{-1}$ under these conditions. Wrigley (1972), using data from FAO and other agricultural institutions, gives even higher values: for a pasture with 'pangola' grass located in Queensland, Australia, daily grass production reached 19.4 $g\ dry\ wt \cdot m^{-2}$. After fertilization with nitrogen this grass produced 280 $g \cdot m^{-2}$ daily. Such high values have never been recorded in the ecological literature.

The production/biomass ($P/\bar{B}$) ratio is a useful index of productivity for each trophic level. Calculated on the basis of plant production it characterizes the total potential productivity of an ecosystem. Breymeyer (1978), analysing the $P/\bar{B}$ ratio for four types of grassland ecosystems, gives the following values:

| | |
|---|---|
| North American prairie | 35% |
| Meadows in the Vistula Valley, Poland | 41–68% |
| Pasture in the Carpathian Mountains, Poland | 59% |
| Savanna in the Panama isthmus | 78–90% |

Table 11.1. *Comparison of the production rates of total (aboveground and belowground) plant biomass in different American grasslands*

| Grassland type | Production (g dry wt · $m^{-2}$ · $d^{-1}$) | Reference |
|---|---|---|
| Semidesert prairie, New Mexico | 1.58 | Sims & Singh (1971) |
| Tallgrass prairie, Oklahoma | 2.25 | |
| Mixed prairie, Kansas | 2.39 | |
| Mixed prairie, South Dakota | 3.92 | |
| Shortgrass prairie, Colorado | 3.95 | |
| Mixed grass prairie, South Dakota | 4.96 | |
| Tallgrass prairie, Missouri | 5.50 | Kucera, Dahlman & Koelling (1967) |
| High-mountain pastures, Montana | 8.89 | Sims & Singh (1971) |
| *Festuca thurberi* association, Colorado | 9.00 | Turner & Dortignae (1954) |

From Sims & Singh (1971).

Table 11.2. *Annual production of total (aboveground and belowground) plant biomass in different grasslands of the USSR*

| Ecosystem type | Production (g · $m^{-2}$ · $yr^{-1}$) |
|---|---|
| Steppe-desert[a] | 300–400 |
| Desert and semidesert steppes[a] | 500–900 |
| Mountain steppes and meadows[b] | 700–1200 |
| Arid steppes on chernozem soils[a] | 800–1300 |
| Wet prairies on chernozem soils | 1500 |
| Savanna (secondary) on black soils[c] | 1500 |
| Savanna (secondary) on red, ferruginous soils[c] | 1600 |

From Bazilevich, Radin & Rozov (1971).
[a] Sub-boreal zone.
[b] Boreal zone.
[c] Humid tropical zone.

Thus, in the savanna each gram of existing biomass produces almost another gram, while in the prairie it produces only 0.3 g and in the Polish grasslands 0.4–0.6 g. For ecosystems studied in the German Solling Project the $P/\bar{B}$ ratio computed on the basis of data from Runge (1973) is about 5% for forests and 58% for meadows. Rodin & Bazilevich (1965) present $P/\bar{B}$ ratios which enable comparison of grasslands with other terrestrial ecosystems: 10–20% for tundra, 2–5% for boreal and sub-boreal forests, 8–10% for tropical forests and 20–55% for grasslands. Thus grasslands have a very high index of production per unit of standing crop and are among the most efficient of all so-called 'natural' terrestrial ecosystems.

The vertical distribution of plant organic matter in grassland ecosystems depends mainly on water input. Breymeyer (1978) has shown a relationship between differences in the distribution of below- and aboveground plant biomass for three types of grassland ecosystems and the amount of water in these ecosystems. In humid tropical savanna the aboveground plant parts comprised more than 80% of the total biomass, while in dry prairie they constituted only 10%. The temperate meadows of the Vistula Valley showed an intermediate value, above- and belowground parts each accounting for about 50% of the total plant biomass. Similar relationships are shown by Sims & Singh (1971) for North American prairies. Tallgrass prairies of southern Oklahoma which were situated in the wettest section of the gradient studied (almost 900 mm precipitation) had a low root biomass of about 35–42% of the total plant biomass. Plant material examined by Bazilevich, Rodin & Rozow (1971) confirms this tendency – for a series of grasslands these authors found that root biomass increased in the drier habitats. In arid steppes it varied from 74 to 95% of the total biomass, in chernozem steppes and steppe-meadows it was 65–90% and in peatbogs and wetlands with mosses it dropped to 10%. For comparison, it is worthy of note that in forest ecosystems roots account for only a small part of plant biomass: for example 17 to 29% for different deciduous forests. Kira & Shidei (1967) state that in the tropical forest ecosystems a major part of the organic matter is accumulated in living aboveground parts whereas in temperate zone ecosystems organic matter is accumulated mainly in the soil. Kira & Shidei's study was concerned with forests, but their conclusions seen to be relevant when comparing grassland ecosystems of tropical and temperate zones. The information gathered by Breymeyer (1978) shows that organic carbon in the savanna is first accumulated in living aboveground parts of plants, and in the two grasslands of the temperate zone (American prairie and Polish meadow) in the belowground parts. In the prairie this organic carbon is in the form of a large biomass of roots and micro-organisms whereas in the Polish grasslands large amounts of carbon are stored as particulate soil organic matter. Thus there are different patterns of plant organic matter accumulation in different types of grassland ecosystems.

## Heterotrophs: their structure and interrelations

Consumers classified according to their food specialization, that is as phytophages, predators and decomposers, show certain characteristic features. If all animals, both vertebrates and invertebrates, are considered then the presence or absence of cattle (or sheep) will be the basic factor shaping the structure of the consumer subsystem. Simple models of grazed and ungrazed grasslands show the differences in the distribution of energy between the main pathways in these ecosystems (Fig. 11.1). If cattle or sheep are introduced into an ecosystem then not only are the proportions of different groups of consumers

(*a*) Mown meadow

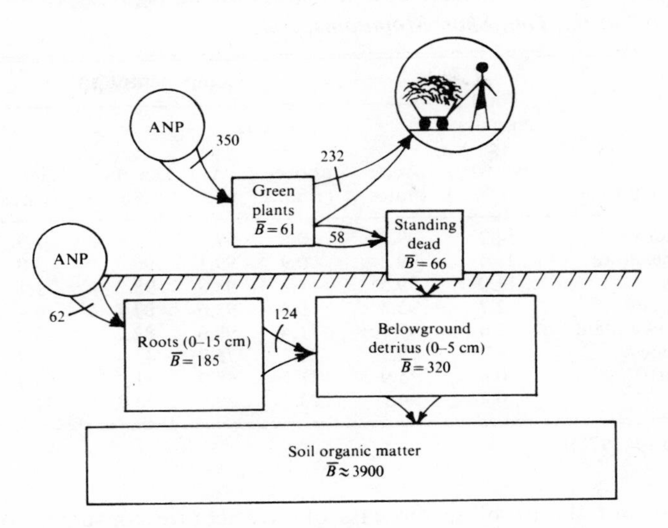


(*b*) Grazed pasture

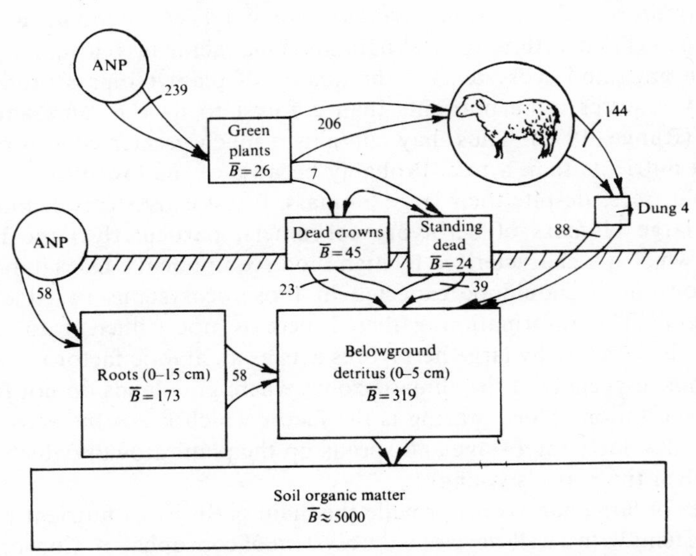


Fig. 11.1. Carbon flow and retention in two grassland ecosystems: (*a*) mown meadow and (*b*) grazed pasture. Boxes represent retention of carbon (in $g \cdot m^{-2}$); arrows represent transfer of carbon (in $g \cdot m^{-2} \cdot yr^{-1}$). ANP, annual net productivity. (After Breymeyer & Kajak, 1976.)

Table 11.3. *Animal biomass and its structure in eight types of grassland ecosystems in the Tien Shan Mountains, Asia*

| Ecosystem type | Total biomass (g live wt · $m^{-2}$) | Components (%) | | | | | |
|---|---|---|---|---|---|---|---|
| | | | Fauna | | | | |
| | | Inverte-brates | above-ground | soil | Decom-posers | Her-bivores | Predators |
| Wet meadows | 54.2 | 99.7 | 0.5 | 99.5 | 97.6 | 1.9 | 0.5 |
| Steppe-meadows | 21.5 | 99.5 | 0.9 | 99.1 | 96.7 | 3.0 | 0.3 |
| Wetlands | 11.0 | 99.5 | 1.2 | 98.8 | 98.2 | 1.1 | 0.7 |
| Cool steppes | 2.7 | 95.5 | 6.4 | 93.6 | 87.7 | 10.4 | 1.9 |
| Cushion-like plants | 2.6 | 99.7 | 1.4 | 98.6 | 82.2 | 13.9 | 3.9 |
| Arid steppes | 1.2 | 79.7 | 24.3 | 75.7 | 43.4 | 51.7 | 4.9 |
| Semideserts | 0.8 | 75.9 | 31.1 | 68.9 | 31.2 | 65.8 | 3.0 |
| Deserts | 0.2 | 96.9 | 23.4 | 76.6 | 53.8 | 34.3 | 11.8 |

After Zlotin (1970).

modified, but the absolute biomass of invertebrate consumers drops considerably (Breymeyer & Kajak, 1976). The effect of grazing as such is positive if it is maintained at a moderate level. Sims & Singh (1971) noted an increase in plant production and in the turnover rate of plant biomass for all types of grazed prairies studied. The herbivores of grassland ecosystems are among the largest associated with terrestrial habitats. One factor favouring large herbivores in grassland ecosystems is the quality of plant biomass produced. A meadow requires more nutrients than a forest to fix the same amount of energy (Runge, 1973). Thus, hay contains a much greater concentration of mineral nutrients than a tree. Probably because of the low nutritive quality of whole trees, despite their large biomass, forest ecosystems cannot maintain a large biomass of herbivore consumers, particularly large homoiotherms which are characterized by high food requirements. Does it mean that large consumers should be expected in those ecosystems based on small producers? The information gathered here is not sufficient to test this hypothesis. Grazing by large herbivores acts as an abiotic factor maintaining grasslands, especially in the climatic zones where grasslands do not form the climax vegetation. Thus, grazing is the factor which keeps the ecosystem at an early developmental stage and speeds up the plant growth (which is most rapid when the plant is young).

Faeces of large herbivores provide the main pathway of nutrient recycling in ecosystems with a well-developed subsystem of coprophages. Coprophagous invertebrates speed the decay of manure (Olechowicz, 1974, 1976); they disperse and bury pieces of manure in the soil (Breymeyer, 1974) and accelerate the activity of bacteria in the faeces (Breymeyer, Jakubczyk & Olechowicz,

Table 11.4. *Invertebrate biomass and its composition in six types of grassland ecosystems*

| Ecosystem type | Total biomass (g dry wt·m$^{-2}$) | Components (%): Macrofauna: above-ground | soil | Decomposers | Herbivores | Predators |
|---|---|---|---|---|---|---|
| Savanna, not used by man | 1.02[a]–1.49[b] | 13–31 | 87–69 | 76–60 | 10–29 | 19–11 |
| Savanna, pasture | 0.32–1.68 | 6–10 | 94–90 | 92–80 | 8–10 | 0–19 |
| Vistula Valley, meadow reservation | 6.48 | 6 | 94 | 91 | 3 | 6 |
| Vistula Valley, mown meadows | 7.43–12.33 | 6–3 | 94–96 | 94–96 | 4–2 | 2–1 |
| Mountain pasture | 6.0 | 1 | 99 | 93 | 5 | 2 |
| Prairie | 0.8 | 23 | 87 | | | |

[a] Dry season. [b] Wet season.

1975). In ecosystems where this group of decomposers is insufficient or lacking, faeces are not decomposed, they destroy the sward, and nutrients in the faeces are not recycled for long periods. Such a situation occurs in Australian pastures, where attempts have been made recently to introduce coprophagous insects. Anonymous (1972), Ferrar (1973) and Gillard (1967) discuss the advantages and ecological threats related to the introduction of coprophagous beetles of the family Scarabaeidae into Australian pastures. It seems that the coprophagous species introduced into Australia have clear-cut food specialization. They should play a positive role in the functioning of Australian grasslands, provided that quarantine techniques prevent the introduction of unknown diseases or parasites into the continent. The beetles which are being introduced come from Africa.

The invertebrate consumers of grasslands were analysed by Zlotin (1970), who studied the structure of their biomass in series of Asiatic grasslands. Some of Zlotin's (1970) results are shown in Table 11.3 and can be generalized as follows:

(1) If the spatial distribution of invertebrates is considered, the large majority (76–99 %) occurs in soil.

(2) If trophic groups are considered, decomposers have the highest biomass.

(3) The proportion of phytophages and particularly predators to the total invertebrate biomass tends to increase with decreasing moisture and plant production. Decomposers show an opposite tendency, their numbers being highest in moist habitats with larger productivity.

Breymeyer (1978) found similar relationships in the consumer biomass. Soil macrofauna accounted for 69–96 % of the total biomass of consumers, and decomposers accounted for 60–96% (Table 11.4). The invertebrate fauna in Polish meadows and pastures was many times more abundant than in the two other habitats of savanna and prairie (Table 11.4). It appears that European temperate grasslands are the most suitable habitats for the development of a soil invertebrate fauna, probably because there are no major limiting factors: food in the form of dead organic matter is abundant throughout the year and neither serious drought nor high solar radiation limit the development of the fauna.

The role of different types of consumers in energy and matter cycling in the ecosystem is different and depends on both physiological (bioenergetic) and ecological characteristics. Generally, all consumers prolong the time of nutrient cycling because they 'lock up' compounds in their tissue; on the other hand it is known that the faeces of first-order consumers contain a lot of nutrients which can be readily recycled. Therefore these consumers, excreting large amounts of rejecta (to use the terminology proposed by Petrusewicz & Macfadyen, 1970), speed up the nutrient cycling in an ecosystem. Many authors suggest that herbivores have the lowest assimilation of food and predators the highest (Kleiber, 1961; Phillipson, 1966; Reichle, 1971; Klekowski *et al.*, 1972; Heal & McLean, 1975); thus herbivores provide the largest amount of faeces, predators the smallest. In relation to saprophages and particularly coprophages, opinions differ (Myrcha, 1973; Holter, 1974).

Various consumers have different energy requirements for metabolic processes. The range is large and depends on many factors such as the type of thermoregulation, body size, and also ecological characteristics. Pętal (1967 and unpublished data) and Horn-Mrozowska (1976) show that only 2% of the food energy taken by social insects (ants) is incorporated into their bodies, the rest being used for metabolic processes. Populations of these animals can be compared to continually burning stoves. On the other hand Reichle (1971) reports that the ratio of net production to assimilation can reach almost 40% in some invertebrate predators; Klekowski *et al.* (1972) found an assimilation index of 70% for an invertebrate aquatic predator. Thus, various groups of organisms are characterized by different abilities to accumulate or dissipate energy and matter.

Some groups of consumers give the impression of living 'gratuitously', without any marked effects on the trophic structure of an ecosystem and without using its basic resources. These groups can develop, for example, on some products which are seasonally in excess and form unused sinks or deposits of matter which can accumulate at some points of the trophic web. Wiens (1972) characterizes in this way the birds inhabiting American prairies. He suggests that the prairie birds are 'frills' in these ecosystems; they live and reproduce without any significant effects on the ecosystem. It seems that in the case of migratory species they use the resources deposited in the ecosystems in such amounts that they can not be utilized by permanent components of the system. Bourlière & Hadley (1970*a*, *b*) suggest that such resources are accumulated as the result of non-rhythmical plant biomass production. In many habitats of the tropical zone with seasonal climatic changes organic matter is produced at such a high rate during the wet season that resident consumers can not utilize it. This is common in many other ecosystems; the non-rhythmical production of plant biomass accounts for its seasonal accumulation and its possible utilization by immigrants. If the immigrants are not numerous they use only a part of the accumulated matter giving the impression of living 'gratuitously', without affecting the energy flow in an ecosystem.

There is also the reverse situation when the density of consumers is higher that the level expected from the available food supply. It can occur, for example, under extremely favourable abiotic conditions. According to the author's own observations this holds for wolf spiders (Lycosidae) in moist peatbog meadows. Lycosidae reproduce there in huge numbers and they mainly feed on other members of their family. In this way almost closed loops are formed in the energy fluxes. Certainly, such loops are not self-sufficient; nevertheless, they need only small energy inputs for maintenance. Similar situations are described by Fischer (1961) and Kiran Datta (personal communication). These authors found that the predatory larvae of dragonflies fed mainly or exclusively on members of their own species showed excellent growth, and it seems that they were able to reach maturity and reproduce. Such retention of energy transfer within one species can last for a long time because the biomass of some populations is maintained at a stable level throughout the phenological cycle. When such populations are 'young' the biomass consists of a large number of small animals, whereas during the period of 'maturity' it is made up of a small number of large individuals (Breymeyer, 1967). Increases in the number of loops in a trophic web delays the transfer of matter and energy from one trophic level to another.

Micro-organisms (bacteria, fungi and actinomycetes) are the smallest consumers in grassland ecosystems and are highly sensitive to changes in moisture, aeration and temperature. Habitats where the variability of these factors exceeds the adaptive capacity of micro-organisms are characterized by irregular and insufficient decomposition of organic remains. These habitats include peatbogs, marshes, many types of cool, high-mountain meadows, desert and semidesert steppes and prairies, and also some tropical grasslands. The low efficiency of the decomposer subsystem results in a permanent or periodical accumulation of organic matter in the soil. If the productivity of the ecosystem is low, the accumulation is also low (for instance in deserts and semideserts). But if plant production is high, accumulation can also be high. It is very possible that huge resources of energy were stored in soil in past geological periods due to the high production of organic matter and the low efficiency of decomposers in those ecological systems. Was the evolution of ecosystems based on developing a more and more perfect balance between the production of organic matter and the recycling of nutrients by decomposers? It is possible, since such a balance would create conditions in which the system could exist for a long time in the same environment and it would enable the colonization of habitats deficient in nutrients.

## Ecological pyramids

The trophic structure of an ecosystem may be presented in the form of an ecological pyramid which is frequently called Elton's pyramid. Elton (1927) first described the proportions between successive consumers as a pyramid of numbers; the size of individuals becomes larger and their numbers smaller at higher tiers of the pyramid. An example of the pyramid of numbers on a small scale is furnished by a census of the organisms inhabiting the floor of a deciduous forest (Fig. 11.2). The Lindeman (1943) concept of trophic levels enabled the application of 'the pyramid model' to the whole ecosystem structure. Many authors constructed pyramids in which the size of trophic levels was measured by their biomass or production. Clarke (1954) introduced a very interesting approach. He constructed a pyramid of the matter removed from an ecosystem. At the base there is matter which is stored, and at the apex the yield taken by man (Fig. 11.3). It seems that not all ecosystems show such a distribution of the removal of matter. The increasing importance of the anthropogenic factor is followed by a shift in favour of the yield for man at the expense of deposits or delaying loops within the ecosystem.

Rigler (1973, 1975), criticizing the concept of trophic levels, emphasized one basic weakness of this notion: there are species that can not be allocated to a definite trophic level because they utilize various sources of food. This is evident for there are many multi- and omnivorous species and detailed investigations are necessary to enable us to determine their dietary composition and the proportion of different kinds of food in it. Some species of ants fall into this category. However there are typical omnivorous ants for which the proportions of plant and animal food in the diet have been adequately determined. Pętal (1967), for example, reports the dietary composition of *Myrmica laevinodis*, a common species in Polish meadows. Rogers, Lavigne & Miller (1972) have investigated the diet of *Pogonomyrmex occidentalis*, which is the most abundant species in the shortgrass prairie of Colorado. Similarly detailed data on dietary composition and its changes over the year are well known for many bird species (e.g. Moss, 1969; Wiens, 1972) and for mammals (e.g. Fleharty & Olson, 1969; Zemanek, 1972).

In the following section some examples of pyramids based on data collected during the past 5–10 years as part of IBP are considered. These data have been derived from:

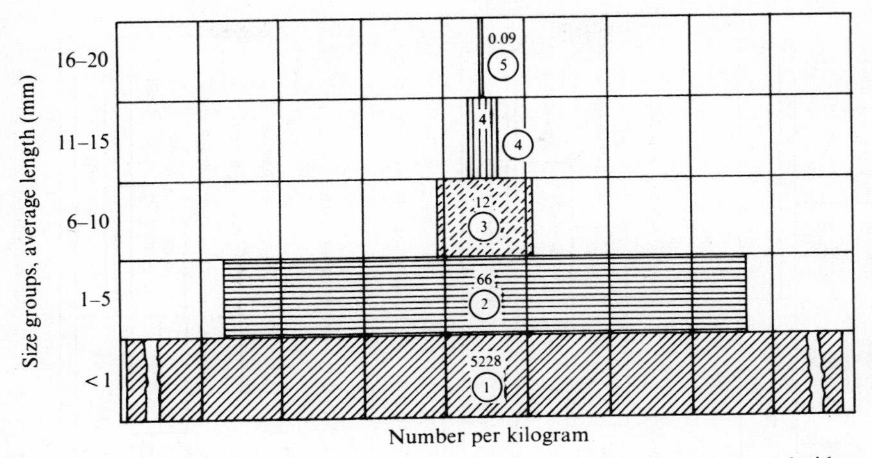


Fig. 11.2. Pyramid of numbers among the metazoans of the forest floor in a deciduous forest. (After Park, Allee & Shelford, 1939.)

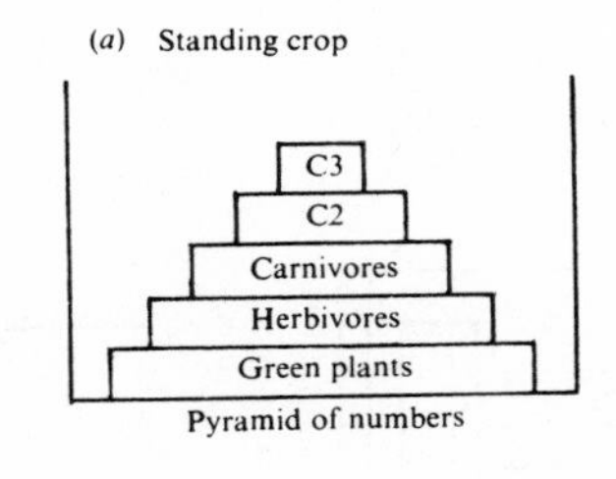


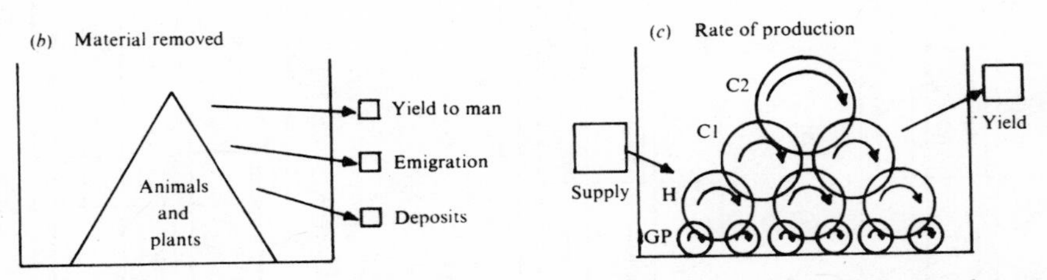


Fig. 11.3. Concepts of productivity expressed in the form of pyramids. GP, green plants; H, herbivores; C1, primary carnivores; C2, secondary carnivores; C3, tertiary carnivores. (After Clarke, 1954.)

(1) Recalculations of the results obtained by Zlotin (1970, 1975) for several types of grassland ecosystems in the Tien Shan Mountains.

(2) Recalculations of the results obtained by American ecologists for several types of prairies, based on N. R. French (personal communication).

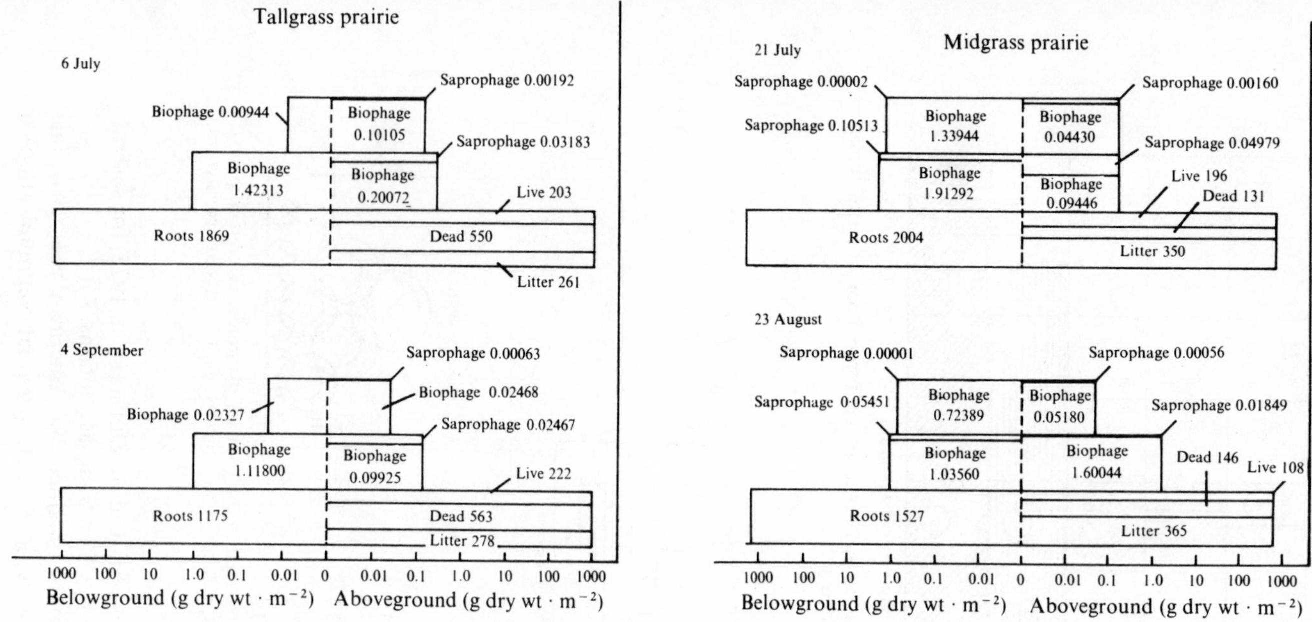


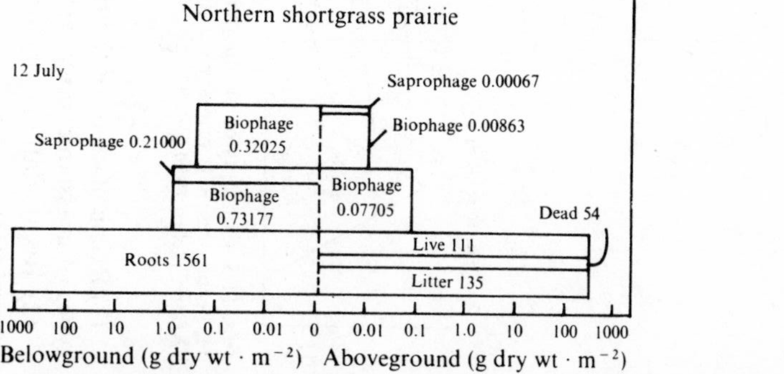


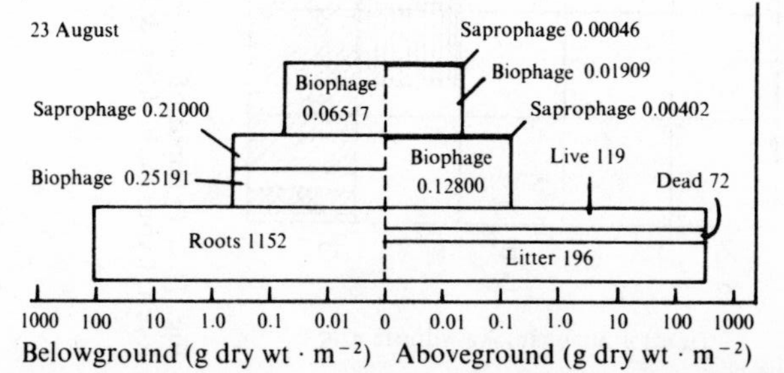


**Fig. 11.4.** Examples of trophic pyramids obtained from 1972 data of US/IBP Grassland Biome. (After N. R. French & N. K. Steinhorst.)

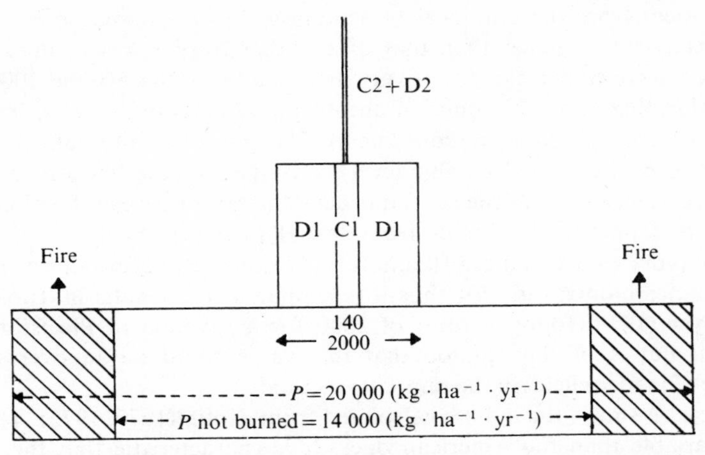


Fig. 11.5. Trophic structure of the Lamto Savanna, Ivory Coast. Pyramid of energy transfer: *P*, primary production, C1, primary consumers; C2, secondary consumers; D1, decomposers of vegetable matter; D2, decomposers of animal material. (After Lamotte, 1975.)

Table 11.5. *The ratios between biomasses of successive trophic levels calculated on the basis of 1972 data from French & Steinhorst*

| Grassland type | Producers / Consumers | Consumers I / Consumers II | Producers / Consumers I |
|---|---|---|---|
| Bunchgrass | 1375 | 2 | 1984 |
| Mountain | 748 | 4 | 922 |
| Mixed grass | 737 | 2 | 1124 |
| Tallgrass | 1306 | 18 | 1376 |
| Southern shortgrass | 1832 | 7 | 2071 |
| Northern shortgrass | 1322 | 4 | 1657 |

Their site called 'desert' is not included here because the biomass data for producers and consumers were collected from different regions.

(3) Results presented by Lamotte (1975) based on the studies of French ecologists in the African savanna of the Ivory Coast.

(4) Results obtained by the present author and colleagues in Polish meadows and pastures.

## Quantitative relations between trophic levels

Lamotte (1975) and French & Steinhorst (personal communication) present biomass pyramids graphically (Figs. 11.4, 11.5). The quantitative relations between the biomasses of different trophic levels are not shown in the figures but are given in tables. Table 11.5 shows the well-established feature (also evident from the graphical presentation) of the biomass of producers being many times higher than that of all other trophic levels; the ratio of producer to consumer biomass is very high and fluctuates around 1000. This means that there are 1000 units of autotroph biomass (producers) for every unit of heterotroph biomass (consumers). This index falls dramatically at the transfer from the second trophic level (consumers of the first order) to the third level (consumers of the second order), the ratio between the biomass of these two trophic levels ranging from 2 to 18 (Table 11.5).

In the Ivory Coast studies of Lamotte (1975) the ratio of producer biomass to consumer biomass is 750; thus it is similar to the results in Table 11.5. Breymeyer (1978) found a ratio of 2000 for a savanna in the isthmus of Panama, but is of the opinion that this value could be an overestimate because of inadequacies in the sampling methods.

Table 11.6 shows the indices calculated from Zlotin (1970, 1975). They are more variable than the American ones. It is characteristic that the largest differences between producer and consumer biomass occur in extreme habitats. For the group of very dry habitats the ratio varies from 1375 to 3785; for wetlands and cushion-like plants it is about 2000. The most efficient transformation of producer biomass into consumer biomass is recorded in ecosystems under more temperate, optimum environmental conditions. This is shown by the data for wet and steppe-meadows, which have an index of 221–313. Even lower values of the producer biomass/consumer biomass ratio are reported by Breymeyer (1978) for Polish meadows and pastures, where they range from 77 to 113 (Table 11.7). The other feature of the data shown in Table 11.7 is that the amount of plant biomass per gram of fauna biomass is many times higher above ground than it is below ground. Invertebrate biomass below ground is considerably larger, both in absolute units and in relation to the biomass of belowground plant parts.

The ratios of producer to consumer biomass are rough indexes of efficiency; the ratios based on the productivities of whole trophic levels are probably burdened with large errors in the productivity estimates. However, some attempts have been made in the literature to calculate ecological efficiencies based on productivity estimates of whole trophic levels of terrestrial ecosystems. The results obtained by Reichle (1971) are of special interest. According to this author the efficiency of production calculated as $A_n/NP_{n-1}$ (where $A_n$ is assimilation at the level $n$, and $NP_{n-1}$ is net production at the level $n-1$) for different trophic levels is:

| | |
|---|---|
| Saprophages | 0.11–0.17 |
| Phytophages | 0.02–0.07 |
| Predators | 0.02 |

Thus the efficiency of production (more precisely assimilation, although the author calls it 'trophic level production efficiency') varies from 2 to 17%, being lowest for predators and highest for saprophages.

Table 11.6. *The ratios of consumer to producer biomass for eight high-mountain ecosystems in the Tien Shan, recalculated from Zlotin (1971)*

| Grassland type | Producer biomass / Consumer biomass |
|---|---|
| Wetlands | 1917 |
| Cushion-like plants | 2034 |
| Wet meadows | 221 |
| Steppe-meadows | 313 |
| Cool steppes | 786 |
| Dry steppes | 1375 |
| Semideserts | 2115 |
| Deserts | 3785 |

Data on plant biomass are average values; dry biomass is calculated as one-third of the live biomass.

Table 11.7. *Ratios of the biomass of producers to invertebrate consumers in four grassland ecosystems*

| Ecosystem type | Aboveground vegetation / Herb-layer fauna | Belowground vegetation / Soil fauna | Σ vegetation / Σ fauna |
|---|---|---|---|
| Panamanian savanna, not used by man[a] | 5718 | 481 | 2227 |
| Poland, Vistula Valley, not used by man reservation meadow[b] | 325 | 79 | 89 |
| Poland, Vistula Valley, mown meadow[c] | 650 | 44 | 77 |
| Poland, pasture in Pieniny Mountains[d] | 297 | 100 | 113 |

After Breymeyer (1978).
[a] Calculated for the wet season.
[b] Vegetation without mosses.
[c] Calculated for mown plots.
[d] Calculated for grazed plots.

Odum (1971) quoted anonymous values of 10–20% for production efficiency between trophic levels ($NP_n/NP_{n-1}$) and he stated a large variability in assimilation efficiency, which ranged from 10 to 50% or more.

A wide range of efficiency indices suggests that it is not possible to predict the secondary production of ecosystems from their primary production. 'Present data do not support the premise that heterotrophs are simply "parasitic" on the autotrophs', as O'Neil (1976) says. Also the index of assimilation efficiency varies widely. Respiration efficiency can be still more surprising (Reichle *et al.*, 1973). Therefore, it seems that it is not at present possible to estimate biomass, productivity and the related bioenergetic indices for consumers from estimates of the biomass and productivity of producers. The same refers to groups of the heterotrophs that feed on each other. For this reason many attempts to calculate the budget of energy flowing between trophic levels have failed. Such authors as Trojan (1967, 1968), Ryszkowski (1972) and Andrews, Coleman, Ellis & Singh (1974) assumed that the productivity of heterotrophs was proportional to the productivity of autotrophs. It seems, however, that these relationships are much more complicated. O'Neil (1976), analysing six different ecosystems, stated that a relationship between heterotrophic standing crop and net primary production does not exist, but he found a distinct relationship between the ratio of heterotroph biomass to autotroph biomass and the system turnover time. This means that the heterotroph/autotroph ratio should be treated as an ecosystem characteristic related to the other features of the system.

## Concluding remarks

The trophic structure of grassland ecosystems shows some features which suggest the possible evolution and strategy of these systems.

Decomposition of organic matter seems to be the process most sensitive to environmental changes. In particular, micro-organisms, which are the last component of this trophic chain, exhibit an immediate and sharp response to changes in environmental conditions. The deactivation of micro-organisms (in very dry or very wet soils, by freezing, etc.) slows down the rate of decomposition and should lead to an increase in the accumulation of organic debris. However, in the highly productive tropical grasslands where there is a scanty soil fauna and a reasonably low decomposition rate, there is also a lack of organic matter accumulation. This suggests that the other mechanisms responsible for the disappearance of dead organic matter have been developed in the ecosystem. Odum, Lugo & Burns (1970), who analysed the ecology of humid equatorial forest, found that mycorrhiza can function as a living system, collecting nutrients dissolved in rainfall. Went & Stark (1968) presented the following theory for the direct cycling of nutrients in the tropics. Mycelium of fungi, which is present on the soil surface and in the humus layer, can probably 'digest' dead litter and transfer nutrients directly to living cells of roots. In this way nutrients are recycled before they reach the soil where they are liable to be leached. At the same time Zlotin (1971) emphasized the role of abiotic decomposition, that is, decomposition occurring without the activity of living organisms. According to this author the abiotic decomposition in a steppe accounted for more than 50% of the total losses of dead organic matter – which is a very high estimate. It is probable that abiotic decomposition is even higher in the tropics, where the leaves of grasses are burnt by the sun and leached during heavy rainfall. So, although the role of decomposers is well known and fully appreciated, ecosystems have also developed parallel mechanisms for recycling nutrients. These shorter pathways of recycling are particularly valuable in regions with critical deficiencies in nutrients, where the ecosystems have to develop defence mechanisms against nutrient losses.

The quality and diversity of consumers vary between ecosystems and this influences the rate of inter-level fluxes. Recent papers stress the regulatory role that consumers, in spite of their comparatively small biomass, play. Lee & Inman (1975) and O'Neil (1976) prove mathematically that consumers play a rate regulation role and increase the ability of an ecosystem to recover from perturbations. Only small changes in heterotroph biomass are necessary (theoretically) to re-establish equilibrium in the system. How do ecological mechanisms lead to this equilibrium? Some possibilities are discussed below. Ecosystems with a well-developed trophic structure, made up of a large number of competing species, which are always ready to utilize the food resources of the system, are characterized by slow inter-level energy flow and still slower nutrient cycling, as large amounts of nutrients are immobilized in the bodies of numerous and highly differentiated consumers (Fig. 11.6). The Strzeleckie Meadows in the Vistula Valley are an example of such an ecosystem. These meadows form a nature reserve which is fully protected from human interference. Natural succession is occurring at a slow rate, while a more and more developed niche system is gradually filled by increasing numbers of inhabitants. The number of species in a single taxonomic group of consumers exceeds the total number of species of producers. Turnover of nutrients in the soil of this habitat is rather slow; numerous organisms involved in energy and matter transfer retain nutrients as long as they are alive. It can be expected that the populations of large, long-lived animals have the most delaying effect on nutrient cycling. The most rapid and complete nutrient transfer occurs where nutrients pass through small organisms with a short lifespan and high turnover rate. The evolution of species has resulted in the creation of larger, long-lived forms, which are more successful in the struggle for existence and more efficient in retaining nutrients. Does the evolution of ecosystems consist in the regulation of proportions between the populations of long-lived animals and those of rapid 'transmitters'? Do ecosystems accept definite numbers of these two kinds of populations depending on the energetic and material possibilities of the place? This seems probable under circumstances where resources of energy and nutrients are limited and ecosystems which can be maintained in a given place are under selection pressure.

Considering the role of consumers from another viewpoint, it should be stated that the storage of nutrients in their bodies acts as a conservation mechanism. Only the organism's death releases all the elements bound in its body and the frequency of release will depend in the main on the size of the organism and the climatic conditions. Southwood (1976) has illustrated the general relationship between the size of an organism and its generation time;

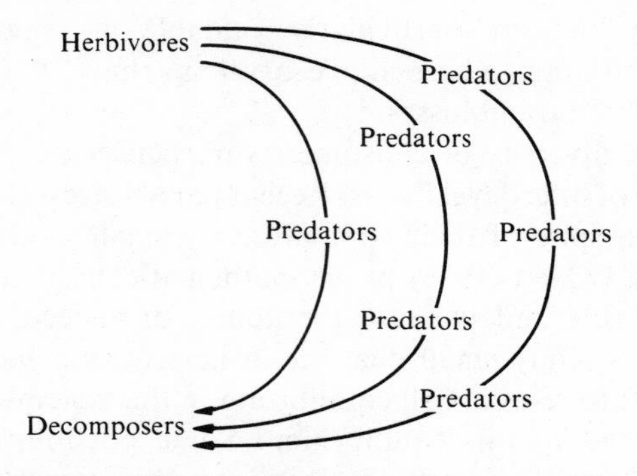


Fig. 11.6. The prolongation of food chains leads to the storage of organic matter in the bodies of consecutive consumers.

Lamotte & Meyer (1977) present data on the speeding-up of turnover among animals of tropical savanna. Their findings enable one to suggest that the rate and time governing the release of nutrients bound in the bodies of consumers can be seen as dependent on general ecosystem needs and strategy.

Another group of questions arises when one considers the rate of development of particular trophic levels. In all grassland ecosystems the production of organic matter is non-rhythmical and this seems to be one of the most important factors affecting the functioning of the trophic web of these systems and one of the reasons for the retention of unused organic matter. For example, as mentioned above, the very favourable conditions above ground in the humid tropics result in an extremely high production rate of green plant biomass. The consumers of these ecosystems are not adapted to such a high production and, as a result, plant biomass is not fully utilized. Thus, considering the annual budget, it may be stated that an acceleration in production rate in one subsystem cannot always be used for development of other subsystems within the ecosystem. The utilization of food resources by different subsystems is, in a sense, independent. What is the degree of this independence? Various parts of the ecosystem reach a climax balance with the environment at different rates (Margalef, 1968; Bazilevich & Titlyanova, Chapter 9, this volume). This should imply that within an ecosystem there can coexist subsystems already in a steady state with the carrying capacity of the environment and other subsystems which have not yet reached a state of balance with the environment. Nowadays, when man interferes in natural ecosystem succession, some subsystems are permanently out of balance.

## References

Andrews, R., Coleman, D. C., Ellis, J. E. & Singh, J. S. (1974). Energy flow relationships in a shortgrass prairie ecosystem. In *Proceedings of the I International Congress of Ecology, The Hague*, pp. 22–8. International Scholarly Book Services, Forest Grove, Oregon.

Anonymous (1972). *Dung beetles on the move. Rural Research in CSIRO,* **75**, 2.

Bazilevich, N. I., Rodin, L. Ye. & Rozov, N. N. (1971). Geographical aspects of biological productivity. *Soviet Geography, Review and Translation*, May 1971, 293–317.

Bourlière, F. & Hadley, M. (1970*a*). The ecology of tropical savannas. *Annual Review of Ecology and Systematics*, **1**, 125–52.

Bourlière, F. & Hadley, M. (1970*b*). Combination of qualitative and quantitative approaches. In *Analysis of temperate forest ecosystems*, ed. D. Reichle, pp. 1–6. Springer-Verlag, Berlin.

Breymeyer, A. (1967). Preliminary data for estimating the biological production of wandering spiders. In *Secondary productivity of terrestrial ecosystems*, vol. 2, ed. K. Petrusewicz, pp. 821–35. Polish Academy of Sciences, Warsaw.

Breymeyer, A. (1974). Analysis of a sheep pasture ecosystem in the Pieniny Mountains. XI. The role of coprophagous beetles (Coleoptera: Scarabaeidae) in the utilization of sheep dung. *Ekologia Polska*, **22**, 617–34.

Breymeyer, A. (1978). Analysis of trophic structure of some grassland ecosystems. *Polish Ecological Studies*, **4**, 55–128.

Breymeyer, A., Jakubczyk, H. & Olechowicz, E. (1975). Influence of coprophagous arthropods on microorganisms in sheep faeces: laboratory investigation. *Bulletin de l'Académie polonaise des Sciences, Classe II*, **23**, 257–62.

Breymeyer, A. & Kajak, A. (1976). Drawing models of two grassland ecosystems. a mown meadow and a pasture. *Polish Ecological Studies*, **2**, 41–9.

Clarke, G. L. (1954). *Elements of ecology*. Wiley, New York.

Elton, C. S. (1927). *Animal ecology*. Sidgwick & Jackson, London. (Reprinted 1966 by Chapman & Hall, London.)

Ferrar, P. (1973). CSIRO dung beetle project – *Wool Technology and Sheep Breeding*, **20** (1), 73–5.

Fischer, Z. (1961). Cannibalism among the larvae of the dragonfly *Lestes nympha* Selys. *Ekologia Polska, Series B*, **7**, 33–9.

Fleharty, E. D. & Olson, L. E. (1969). Summer food habits of *Microtus ochrogaster* and *Sigmodon hispidus*. *Journal of Mammalogy*, **50**, 475–86.

Gillard, O. (1967). Coprophagous beetles in pasture ecosystems. *Journal of the Australian Institute of Agricultural Science*, **33**, 30.

Heal, O. W. & McLean, S. F. (1975). Comparative productivity in ecosystems: secondary productivity. In *Unifying concepts of ecology*. ed. W. H. Van Dobben & R. H. Love, pp. 89–109. McConnell, The Hague.

Holter, P. (1974). Food utilization of dung-eating *Aphodius* larvae (Scarabaeidae). *Oikos*, **25**, 75–9.

Horn-Mrozowska, E. (1976). Energy budget elements of an experimental nest of *Formica pratensis* Retzius 1783 (Hymenoptera, Formicidae). *Polish Ecological Studies*, **2** (3), 55–98.

Kira, T. & Shidei, T. (1967). Primary production and turnover of organic matter in different forest ecosystems of the western Pacific. *Japanese Journal of Ecology*, **17**, 70–87.

Kleiber, M. (1961). *The fire of life: an introduction to animal energetics*. Wiley, New York.

Klekowski, R. Z., Fischer, E., Fischer, T., Ivanova, M. B., Prus, T., Shushkina, E. A., Stachurska, T., Stepien, Z. & Zyromska-Rudzka, H. (1972). Energy budgets and energy transformation efficiencies of several animal species of different feeding types. In *Productivity problems of fresh waters*, ed. Z. Kajak & A. Hillbricht-Ilkowska, pp. 749–63. Warsaw.

Kucera, C. L., Dahlman, L. C. & Koelling, N. R. (1967). Total net productivity and turnover on an energy basis for tallgrass prairie. *Ecology*, **48**, 536–41.

Lamotte, M. (1975). The structure and function of a tropical savannah ecosystem. In *Trends in tropical ecology*, pp. 179–222. Springer-Verlag, Berlin.

Lamotte, M. & Meyer, J. M. (1977). Use of the turn-over rate *P/B* in the analysis of the energetic functioning of ecosystems. In *Resumenes recibidos para el IV Simpósium Internacional de Ecología Tropical*, ed. Henk Wolda, pp. 65–7. Editora de la Nación, Panama.

Lee, J. J. & Inman, D. L. (1975). The ecological role of consumers: an aggregated systems view. *Ecology*, **56**, 1455–8.

Lindeman, R. L. (1942). The trophic–dynamic aspect of ecology. *Ecology*, **23**, 399–418.

Margalef, R. (1968). *Perspectives in ecological theory*. University of Chicago Press.

Moss, R. (1969). A comparison of red grouse (*Lagopus L. scoticus* Stochs) with the production and nutritive value of heather (*Calluna vulgaris*). *Journal of Animal Ecology*, **28**, 103–12.

Myrcha, A. (1973). Bioenergetics of the development period of *Copris lunaris* L. *Ekologia Polska*, **21**, 13–35.

Odum, E. P. (1971). *Fundamentals of ecology*. Saunders, Philadelphia.

Odum, H. T., Lugo, A. & Burns, L. (1970). Metabolism of forest floor microcosm. In *A tropical rain forest*, ed. H. T. Odum & R. F. Pigeon, pp. 35–56. National Technical Information Service, Springfield, Virginia.

Olechowicz, E. (1974). Analysis of a sheep pasture ecosystem in the Pieniny Mountains. X. Sheep dung and the fauna colonizing it. *Ekologia Polska*, **22**, 589–616.

Olechowicz, E. (1976). The role of coprophagous dipterans in a mountain pasture ecosystem. *Ekologia Polska*, **24**, 125–65.

O'Neil, R. W. (1976). Ecosystem persistence and heterotrophic regulation. *Ecology*, **57**, 1244–53.

Park, O. W. C., Allee, V. E. & Shelford, W. (1939). *A laboratory introduction to animal ecology and taxonomy*. University of Chicago Press.

Pętal, J. (1967). Productivity and the consumption of food in the *Myrmica laevinodis* Nyl. population. In *Secondary productivity of terrestrial ecosystems*, vol. 2, ed. K. Petrusewicz, pp. 841–57. Polish Academy of Sciences, Warsaw.

Petrusewicz, K. & McFadyen, A.) 1970). *Productivity of terrestrial animals: principles and methods. IBP Handbook 13*. Blackwell Scientific Publications, Oxford.

Phillipson, J. (1966). *Ecological energetics*. Edward Arnold, London.

Reichle, D. E. (1971). Energy and nutrient metabolism of soil and litter invertebrates. In *Productivity of forest ecosystems: proceedings of the Brussels symposium*, ed. P. Duvigneaud, pp. 465–77. Unipub. Inc., New York.

Reichle, D. E., Dinger, B. E., Edwards, N. T., Harris, W. F. & Sollins, P. (1973). Carbon flow and storage in a forest ecosystem. In *Carbon and the biosphere*, ed. G. W. Woodwell & E. V. Pecan, pp. 354–65. London.

Rigler, F. H. (1973). Adequacies and shortcomings of the energetic concept and the mineral cycling concept. *Wiadomosci Ekologiczne*, **19**, 194–204. (In Polish.)

Rigler, F. H. (1975). The concept of energy flow and nutrient flow between trophic levels. In *Unifying concepts of ecology*, ed. W. H. Van Dobben & R. H. Love, pp. 15–26. McConnell, The Hague.

Rodin, L. Ye. & Bazilevich, N. I. (1965). *Production and mineral cycling in terrestrial vegetation*, transl. G. E. Fogg. Oliver & Boyd, Edinburgh.

Rogers, L., Lavigne, R. & Miller, J. N. (1972). Bioenergetics of the western Harvester Ant in the shortgrass plain ecosystem. *Environmental Entomology*, **1**, 763–8.

Runge, M. (1973). Energieumsätze in den Biozönosen terrestrischer Ökosysteme. *Scripta Geobotanica, Göttingen*, **4**, 1–71.

Ryszkowski, L. (ed.) (1972). *Ecological effects of intensive agriculture*. Zeszyty Nauk, Inst. Ekol. PAN, 5, 340 pp.

Sims, P. L. & Singh, J. S. (1971). Herbage dynamics and net primary production in certain grazed and ungrazed grasslands in North America. In *Preliminary analysis of structure and function in grasslands*, ed. N. R. French, pp. 59–124. *Range Science Department Science Series No. 10*. Colorado State University, Fort Collins, Colorado.

Singh, J. S. (1976). Structure and function of tropical grassland vegetation in India. *Polish Ecological Studies*, **2**, 17–34.

Southwood, T. R. E. (1976). Bionomic strategies and population parameters. In *Theoretical ecology*, ed. R. May, pp. 26–49. Blackwell Scientific Publications, Oxford.

Trojan, P. (1967). Investigation on production of cultivated fields. In *Secondary productivity of terrestrial ecosystems*, vol. 1, ed. K. Petrusewicz, pp. 545–61. Polish Academy of Sciences, Warsaw.

Trojan, P. (1968). Agrocoenosis as a biological productive system. *Polskie Pismo Entomologiczne*, **28**, 647–55. (In Polish.)

Turner, G. T. & Dortignae, E. J. (1954). Infiltration, erosion and herbage production of some mountain grasslands in Western Colorado. *Journal of Forestry*, **52**, 838–60.

Went, P. W. & Stark, N. (1968). Mycorrhizae. *Bioscience*, **18**, 1035–9.

Wiens, J. A. (1972). Pattern and process in grassland bird communities. *Ecological Monographs*, **43**, 237–70.

Wrigley, R. (1972). *Tropical agriculture*. Faber & Faber, London.

Zemanek, M. (1972). Food and feeding habits of rodents in a deciduous forest. *Acta Theriologica*, **17**, 315–25.

Zlotin, R. I. (1970). Structure and productivity of biogeocoenoses of the Tien Shan mountains. Nauka, Moscow. (In Russian.)

Zlotin, R. I. (1971). Invertebrate animals as a factor in biological turnover. In *IV Colloque International de la Faune du Sol, Dijon*, pp. 455–62.

Zlotin, R. I. (1975). *Life in high mountains*. Nauka, Moscow. (In Russian.)

# Part III. System utilization

# 12. Management impacts on structure and function of sown grasslands

K. J. HUTCHINSON & K. L. KING

## Contents



## Introduction

The maximization of economic productivity is the primary aim of management in sown and fertilized grasslands. Radical management practices have been used to increase productivity where the climatic potential exists. These have involved the replacement of native vegetation with sown species, the use of fertilizers, and an increase in domestic flocks and herds to provide products for man. Control of grazing intensity is the most powerful single factor influencing animal production from sown and fertilized pastures. The effects of this practice on the structure and function of these grasslands is the major theme of this chapter. Other management options including seasonal and rotational grazing (Heady, 1975), fodder conservation (Hutchinson, 1974) and the timing of reproduction of domestic animals (Davies, 1962), are not discussed. While there have been a number of studies of the effects of such practices on domestic animal production there is little information on their impact on the total grassland community.

Data from 'the Armidale sites' are used throughout this chapter to illustrate changes in the structure and function of sown grasslands that are grazed, year-long, with sheep at three rates of stocking. Details of sampling methods for herbage and invertebrates have been reported by King & Hutchinson (1976). Descriptions of the climate of Armidale and some other research activities of the Pastoral Research Laboratory are given in Coupland (1977). Comparisons are drawn between Armidale data and published observations from other intensively grazed grassland sites. They include botanical trends, under grazing, in the cool temperate pastures of the UK and changes in grazed annual-type 'Mediterranean' pastures which are used widely for domestic animal production (Rossiter, 1966). Calculations of energy expended by the invertebrates of a limestone grassland at Moor House, UK (Cragg, 1961) also provide valuable comparisons.

Grassland development programmes have been the preserve of the agronomist and field studies have been based on a restricted set of the full range of biological processes that occur in all grassland communities, either perturbed or in their natural states. Shoot growth has received most attention since this process represents the quantitative and qualitative potential for increasing the production of domestic herbivores. Field programmes have been supported by a wide range of studies, including varietal comparisons, nutrient and rhizobial responses, and by grazing experiments, the more valuable ones being based on experimental ecosystems. While dramatic responses in primary production have been achieved, there is a lack of basic knowledge of the effects of management on the full range of community processes and there have been few attempts to apply ecosystem theories to intensively utilized grasslands.

Floristic changes are characteristic in grasslands that are grazed intensively and these may have substantial effects on primary and secondary production. Accompanying changes occur in the invertebrate fauna and in ecosystem function as measured by biological energy flow and the cycling rates of nutrients. Knowledge of these responses provides a basis for the rational management decisions that are required to sustain increased productivity without environmental deterioration.

## Floristic changes in intensively grazed grasslands

The application of husbandry and management practices has largely freed domestic herbivore numbers from control by natural mechanisms of population regulation. Hence it is not surprising that domestic flocks and herds can exert powerful effects on the botanical composition of intensively utilized pasture communities. This situation is in contrast to natural grasslands, where the numbers and biomass of herbivores are generally small and vegetational changes are controlled mainly by factors such as climate and fire.

Five groups of mechanisms control floristic changes in intensively grazed grasslands:

(i) Plants may possess attributes that increase their resistance to grazing; low palatability and accessibility (Norman, 1957), including the ability to develop a prostrate growth form (Kydd, 1966), are examples of these.

(ii) Animals differ in their grazing behaviour with some exercising greater dietary selection than others.

(iii) Defoliation frequency and timing can affect the production and dispersal of seed in some species (Rossiter, 1961; Walton, 1975).

(iv) Floristic composition is influenced by the levels of plant nutrients (Rossiter, 1966; Wolfe & Lazenby, 1973*a*, *b*) and their spatial distribution (Hilder, 1964).

(v) Climatic events can exercise powerful direct and interactive effects.

### *Temperate perennial-type grasslands*

There has been some notable experimentation in intensively grazed, cool temperate, perennial grasslands, particularly of the UK. Harper (1969) classified these experiments on the basis of whether animals are withdrawn, added or varied in number to impose different seasonal patterns of grazing intensity. The last approach was employed in the classical experiments of Martin Jones (1933*a*, *b*) which produced such dramatic evidence of the power of grazing management to control botanical composition that Harper (1969) wrote that 'so impressive was this achievement in agronomic practice that for years afterwards many farmers seemed to be managing their livestock to create the image of the perfectly composed pasture rather than using their pastures to grow the most perfect livestock'. This raises the important question as to the extent to which the primary and secondary production in managed swards may be reduced by botanical changes that are considered to be agronomically undesirable. A similar situation exists for North American grasslands where several plant species are considered to be indicators of rangeland condition and potential productivity. Species 'desirability' appears to have been based largely on traditional ecological views on the key importance of climax vegetation.

The exclusion of the rabbit, *Oryctolagus cuniculus*, from the chalk flora of the South Downs of England resulted in marked reductions in the number of sward species, notably dicotyledons, and a reduced number of vegetative shoots. The production of fertile tillers from a reduced number of grass species was associated with taller and self-shading vegetation (Tansley & Adamson, 1925; Hope-Simpson, 1940). The series of papers by Jones (1933*a*,*b*) described botanical changes in *Lolium perenne*/*Trifolium repens* pasture subjected to a range of grazing managements using sheep. These two pasture species are highly palatable to sheep and overgrazing reduced their numbers but increased the total number of species present. This finding is in contrast to experiments of Milton (1940, 1947) on hill pastures in Wales. Here the dominants comprised the native species *Festuca rubra*, *Agrostis tenuis* and *Molinia* *caerulea*, which are considered to have a low palatability to sheep. Floristic richness did not increase, even with applied fertilizer, until the grazing intensity was sufficiently high to compel the sheep to graze the unpalatable dominants.

These UK experiments confirmed the operation of competitive exclusion at low grazing intensities. This results in the emergence, from a narrow spectrum of species, of dominants that successfully compete for niche space. Grass dominance is favoured by light stocking along with an increased number of fertile tillers and an erect habit that shades out shorter species (Stern & Donald, 1962). As the stocking intensity increases there is an increase in the number and proportion of vegetative tillers; a prostrate habit is favoured and there is an increase in the number of species of annual grasses and dicotyledons. Where the dominant species is unpalatable, then the response in richness is delayed until grazing pressure overrides food preferences. Domestic herbivores also modify the edaphic environment. Grazing increases the rate of nutrient cycling through excretal return and it also affects the spatial distribution of nutrients by urine patches, dung pats and the 'camping behaviour' of some domestic herbivore species (Hilder, 1966).

Botanical changes in the UK experiments would appear to have been determined mainly by biotic mechanisms since the sites were generally well favoured climatically. However, climate exercises the major control on vegetation and the effects of climatic events, such as drought, interact strongly with biotic factors such as numbers of herbivores. Botanical data from the Armidale sites are presented to provide an example of the importance of these interactions.

The Armidale experimental area was sown with *Phalaris tuberosa* and *Trifolium repens* in 1958 and received 250 $kg \cdot ha^{-1}$ of superphosphate and 125 $kg \cdot ha^{-1}$ of potassium chloride fertilizer each autumn. The area was lightly grazed until 1963 when replicated sites were grazed with sheep at 10 $sheep \cdot ha^{-1}$ (LSR), 20 $sheep \cdot ha^{-1}$ (MSR) and 30 $sheep \cdot ha^{-1}$ (HSR). Grazing at these stocking rates was continuous for 11 years except for a short period (October 1965 to July 1966) which followed heavy sheep losses in a drought year. The Armidale climate is cool temperate; the major difference from the UK sites that were discussed earlier is the high variability of annual rainfall (see Coupland, 1979). Botanical changes were recorded early each spring as percentage of basal cover using a 10-pin vertical quadrat on 80 locations per site.

Floristic changes are shown in Fig. 12.1 together with annual rainfalls for the period September to September. Because of the high levels of input of phosphorus, potassium and sulphur in all treatments, and assuming adequate nitrogen fixation by *T. repens*, at least in the LSR and MSR treatments, it is unlikely that the levels or the spatial distributions of nutrients played a major role in determining these botanical changes. The wide range of sheep numbers

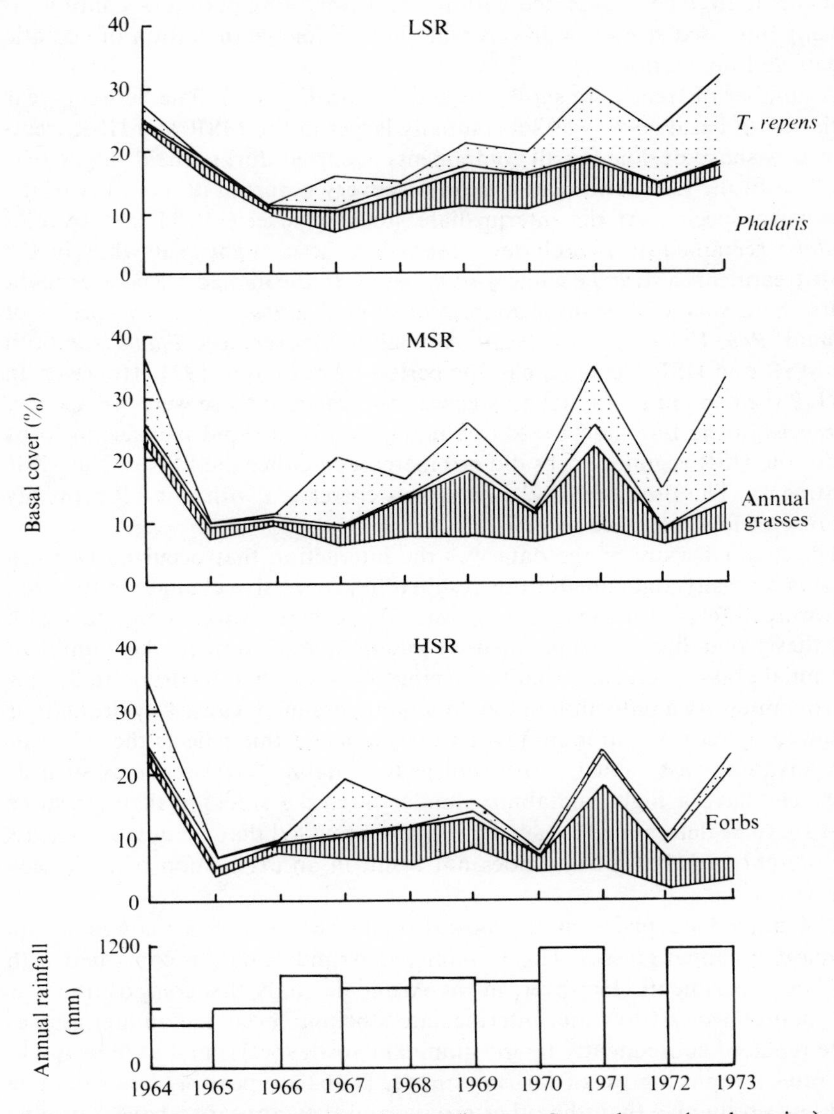


Fig. 12.1. Changes in percentage basal cover (between September and early spring) over 10 years for sown pastures that were grazed continuously with 10 sheep $\cdot$ ha$^{-1}$ (LSR) 20 sheep $\cdot$ ha$^{-1}$ (MSR) or 30 sheep $\cdot$ ha$^{-1}$ (HSR). Data, including annual rainfall from September to September, are from the Armidale sites (see text).

that was used, when combined with the relatively long period of continuous grazing, imposed stress conditions and allowed for the operation of climatic effects and interactions.

A number of trends are shown by the data in Fig. 12.1. The between-year variation in basal cover was substantially larger in the MSR and HSR treatments. A sharp decline for all components occurred during the drought year of 1965. In the LSR treatment a steady recovery occurred in the cover of the two sown species. At the intermediate stocking level (MSR) the cover of *Phalaris* remained at a stable level following the drought year while in the HSR treatments a steady decline was observed. In the immediate post-drought years there was a substantial ingress of annual grasses, notably species of *Bromus*, *Poa*, *Vulpia* and *Hordeum*. Annual grasses replaced *Phalaris* in both the MSR and HSR treatments in the period from 1966 to 1971. However, in 1971–2 the content of annual grasses was reduced and these were replaced by *T. repens* in the LSR and MSR treatments and by a rapid increase in forbs under the HSR regime. Forbs did not increase in either the LSR or the MSR treatments. The presence of *T. repens* was correlated with rainfall probably because of low tolerance to moisture stress.

The major feature of the data was the interaction that occurred between grazing intensity and climate; this resulted in progressive changes in the flora that were different for each stocking rate. These changes were consistent with the thesis that high grazing intensity causes a reduction in the dominant perennial grass species and that this change is accelerated by drought. At low grazing intensity a reduction in the dominant perennial, caused by drought, is followed by recovery in more favourable years and this reflects the competitive advantage associated with a tall growth habit. Sown legumes such as *T. repens* have a high palatability; under intensive stocking their presence appears to be determined largely by climate, provided that adequate nutrients are present and that grazing does not result in an exhaustion of their seed reserves.

The immediate phase of the post-drought recovery was marked by an increase in annual grasses. This trend in the Armidale data is consistent with the UK experiments. However, in the Armidale study this composition was not maintained. At low and intermediate stocking levels the annual grasses were replaced subsequently by the dominant sown species and at high stocking rates a rapid ingress of forbs occurred. Forbs are possibly less palatable and less productive than the other groups and they appear to have a prolific seed production. A study of the dynamics of seed reserves is clearly needed. Factors such as seed consumption by invertebrates, changed microclimate and the physical removal of seeds where the surface cover is low, also need further investigation.

### *Mediterranean annual-type grasslands*

The climate for these grasslands is characterized by mild wet winters and hot dry summers; the grasslands are distributed widely through the Mediterranean countries, California, Chile and the southern part of South Africa and Australia. They are quite commonly fertilized and grazed intensively by domestic herbivores. The dominant annual grasses are species of *Bromus*, *Hordeum*, *Lolium* and *Vulpia*; legumes are represented by species of *Medicago* and *Trifolium* with *T. subterraneum* being of particular importance.

The primary production of these annual grasslands increases substantially with fertilizer application (Ofer & Seligman, 1969). With increased rates of sheep stocking the content of grasses is reduced and there are increases in forbs and clovers (Rossiter, 1966). Young pastures tend towards clover dominance with intensive stocking, presumably because the level of soil nitrogen resulting from nitrogen fixation by legumes is moderate and the seed supplies of vigorous forbs, such as *Cryptostemma* and *Erodium* spp., may not have built up. At a later stage of grassland development forbs may become dominant. The pattern of seasonal rainfall is also important. Legumes are less tolerant to moisture stress and early autumn rains followed by repeated dry periods result in loss of seedlings and exhaustion of seed reserves. Grasses and forbs are more drought tolerant and tend to become dominant under these climatic conditions.

The persistence of all annual species is obviously dependent upon adequate seed production. Defoliation prior to flowering can increase seed production of *T. subterraneum* (Rossiter, 1961). Severe defoliation treatments can triple the amount of seed that is buried by this species (Walton, 1975) presumably by stimulating the development of stronger peduncles which provides the mechanism for seed burial (Barley & England, 1970).

### *Simulation*

Simulation models for predicting botanical changes in intensively grazed swards can be based on the mechanisms that have been discussed. Given a set of initial conditions, then the annual change in total cover could be stated as functions of stocking intensity and climatic events. Change in the sown perennial fraction might be evaluated in terms of palatability status and grazing intensity. Changes in the annual grasses could be based on the status of the dominant perennial grass and their level of seed reserves. The legume component could be predicted from soil moisture status and seed reserves. The forb fraction might be represented in niches that are vacated by other groups which fail to persist. The more vigorous forbs with high seed productions may have competitive advantages over legumes.

A more comprehensive model would need to account for species requirements for nutrients, the effects of domestic animal behaviour on the spatial distribution of nutrients, dietary preferences and plant reproductive processes. The coupling of a suitable model for simulating floristic change under intensive grazing with the productivity models, discussed in later parts of this chapter, would be a desirable research objective. Swartzman & Singh (1974) have employed a transfer matrix approach to predict botanical changes under different grazing intensities. A background of long-term data from experimental ecosystems is required to validate predictions of successional changes and to substantiate the changes in productivity that are commonly assumed.

## Effects of management on invertebrates

Grassland management practices have profound effects on the habitats of pasture and soil invertebrates and consequently on their distribution, abundance and species richness. Many species are considered to play useful roles in organic matter decomposition, nutrient cycling and the improvement of soil physical characteristics. Some species, however, notably of the orders Coleoptera, Lepidoptera and Diptera, have been given the status of pests. Little research has been done to determine relationships between high livestock numbers and invertebrates. This is surprising since domestic herbivores graze most grasslands of the world and human needs, economic forces and climatic failures commonly result in overgrazing. Information on the effects of domestic animals on the abundance and biomass of invertebrates is needed to assess the extent to which invertebrate consumers may compete with domestic animals. There is also the allied question of the effects of domestic herbivore numbers on the spatial distribution and species diversity of invertebrates.

Morris (1967, 1968, 1969) found that the numbers of phytophagous insects and some soil fauna increased in abundance following the cessation of sheep grazing in a chalk grassland in Bedfordshire, UK. Rogers, Lavigne & Miller (1972) showed that the western harvester ant (*Pogonomyrmex occidentalis*) had its highest density in moderately grazed areas and its lowest density in heavily grazed areas in shortgrass prairie in North America. Hutchinson & King (1970) reported large reductions in the abundance of mesofauna as the numbers of grazing sheep increased at the Armidale sites. Further data for invertebrates are given in Table 12.1.

Trends in invertebrate biomass with increased sheep numbers were similar to responses in abundance (Table 12.1), with the exception of Scarabeidae. Live biomass estimates for the scarab larvae were 221, 201 and 39 kg·ha$^{-1}$ for the three stocking treatments LSR, MSR and HSR respectively. The adult beetles of the main species at these sites show a preference for bare ground for ovipositing (Hilditch, 1974) and the high numbers of small larvae in the HSR treatments may have resulted from the combination of high egg numbers and poor larval nutrition.

Table 12.1. *The effects of sheep numbers on the mean abundance (no. $m^{-2}$) of the major groups of invertebrates recorded at the Armidale sites between 1970 and 1974*

| | Sheep numbers ($ha^{-1}$) | | |
|---|---|---|---|
| Invertebrate groups[a] | 10 (LSR) | 20 (MSR) | 30 (HSR) |
| Coleoptera larvae | | | |
| Scarabeidae | 56 | 64 | 54 |
| Other | 46 | 29 | 12 |
| Oligochaeta (large) | 71 | 83 | 50 |
| Diplopoda | 277 | 150 | 60 |
| Enchytraeidae | 6870 | 3810 | 2760 |
| Collembola | 64440 | 36090 | 4780 |
| Acarina | 38030 | 32720 | 9160 |
| Nematoda | 298000 | 215000 | 112000 |

[a] Mesofauna were collected over 19 samplings and macrofauna over 32 samplings.

The data given in Table 12.1 suggest that there is a negative association between domestic herbivore numbers and invertebrate abundance. Increased stocking rate modifies the whole range of environmental conditions that constitutes the habitat of invertebrates. The major changes that occur are in type and amount of food available, living space and the microclimate. These factors interact in a complex way; they are also correlated and it is difficult to isolate the causal factors of any change in population size. Excluding immigration and emigration, the abundance of animals depends upon the balance between natality and mortality. This balance can be studied by constructing life tables which can give important leads on the management factors that cause changes in abundance. New Zealand research workers (East & Pottinger, 1975) used this approach for defining potential management controls for the grass grub, *Costelytra zealandica* (Scarabeidae).

Invertebrates are frequently clumped in their spatial distribution and these populations are represented by two parameters of biological interest, viz. the mean ($m$) and an exponent ($k$); the inverse ($1/k$) is an index of aggregation. Active aggregation results from the organization of social insects, e.g. Isoptera and Hymenoptera, but many others are clumped because of the operation of other factors including reproduction and the restriction of tolerable habitats representing essential needs for food, shelter and microclimate. For these factors, aggregation ($1/k$) represents the potential for a species to survive under harsh conditions and to increase its abundance by spreading out from residual foci when conditions improve. Hence it is not uncommon to find negative relationships between $1/k$ and $m$. Data for Collembola and Acarina from the Armidale sites showed an increase in aggregation in those seasons where abundance was low (King & Hutchinson, 1976). However, of greater interest

Table 12.2. *The influence of grazing intensity on mesofaunal numbers, m (thousands per m²) and aggregation (1/k) for some Armidale samplings*

| | Collembola | | Acarina | |
|---|---|---|---|---|
| Sheep ($ha^{-1}$) | *m* | $1/k$[a] | *m* | $1/k$ |
| 10 | 126 | 0.57 | 53 | 0.52 |
| 20 | 72 | 1.82 | 39 | 0.55 |
| 30 | 6.5 | 2.62 | 13 | 1.16 |

From King & Hutchinson (1976).
[a] Values estimated by maximum likelihood method (Bliss & Fisher, 1953).

here is the negative correlation between abundance and aggregation of mesofauna associated with increased numbers of grazing sheep (Table 12.2).

Mesofauna appear to have an important role in organic matter decomposition and the cycling of nutrients in grasslands. Physical, chemical and behavioural explanations have been proposed for some important functional interactions between mesofauna and micro-organisms. These include the comminution and mixing of plant litter (Berthet, 1964; Kevan, 1968), and the dispersal of micro-organisms through the litter and soil (Macfadyen, 1961; Hale, 1967). In addition to the effects on abundance (Table 12.2) the possible influence of grazing intensity on species diversity may also be relevant. Changes in species diversity have been studied in relation to grazing (Morris, 1971), dryness (Janzen & Schoener, 1968), mowing and burning (Bulan & Barrett, 1971) and radiation (McMahan & Sollins, 1970).

Twenty-eight species of Collembola have been recorded at the Armidale sites (King, Hutchinson & Greenslade, 1976). Collembola, collected in 1971 and 1972, were classified into species and cumulative richness–abundance curves are shown in Fig. 12.2. Richness, or the number of species present ($S$) is one component of species diversity. High grazing intensity reduced the richness of the Collembola fauna from 25 to 19 species (replicated sites pooled). The second component of diversity is evenness or equitability ($E$). Values for $E$ were calculated from Brillouin's (1960) index ($H$) and are included in Fig. 12.2. These values show a compensating increase in $E$ for the reductions observed in $S$. The abundances of the hemiedaphic collembolan species were the most severely reduced of any group. However, the major species of this group, *Hypogastrura communis*, *Brachystomella parvula* and *Cryptopygus thermophilus*, were still present, in reduced numbers, at the highest stocking intensity. It would appear that reductions in any functional role of Collembola may follow from reductions in abundance rather than from the elimination of important species.

Increased numbers of large herbivores strongly influence the vegetation and the edaphic environment. The levels of live and dead herbage and of roots are

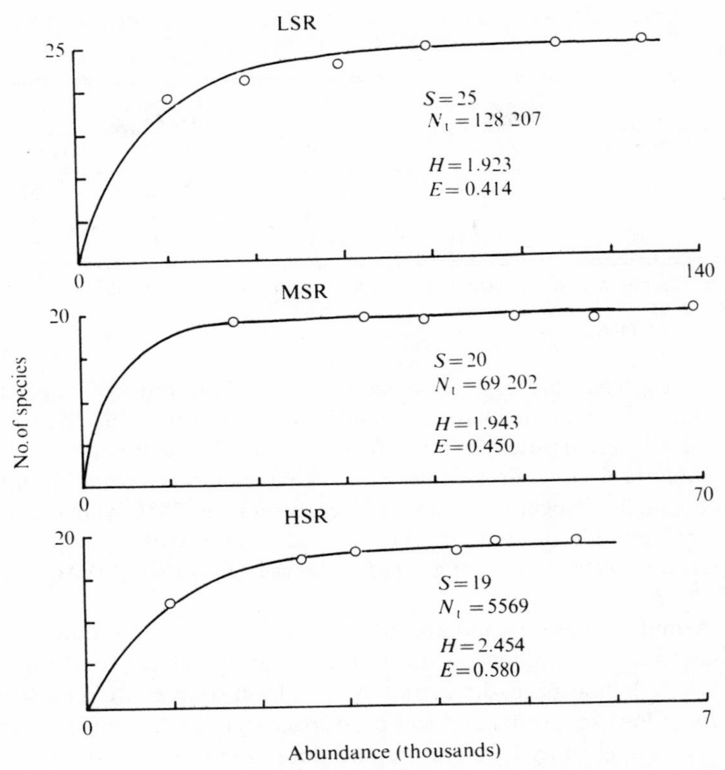


Fig. 12.2. Collembolan species–abundance curves for Armidale sites grazed at different stocking intensities (details as in Fig. 12.1). Exponential curves of the form $y = a(1-e^{-bx})$, have been fitted. The values for species richness ($S$), Brillouin's (1960) diversity index ($H$) and evenness ($E$) determined for the total numbers of animals classified ($N_t$) are shown. (Data are derived from King, Hutchinson & Greenslade, 1976.)

reduced as stocking intensity increases; primary productivity may increase, however. Floristic changes have already been discussed. The trampling effect of increased numbers of domestic herbivores increases the bulk density of the surface soil (Langlands & Bennett, 1973) and this effect, which is exacerbated when the cover of vegetation is low, reduces water infiltration rate and increases runoff and soil loss. However, there may be a compensating reduction in evapotranspiration losses so that the balance of soil water content may not change. Soil compaction also reduces pore space, causing a reduction in living space for non-burrowing soil animals (e.g. Collembola and Acarina); it may also increase the energy expenditure of burrowing animals (e.g. millipedes, oligochaetes and Coleoptera larvae) if they have to tunnel greater distances in search of food and a more suitable microclimate. Surface soil

Table 12.3. *Effect of sheep numbers on invertebrate habitat at the Armidale sites*

| Sheep ($ha^{-1}$) | Green herbage | Dead herbage (kg dry wt·$ha^{-1}$) | Roots | Soil temp.,[a] annual range (°C) | Soil porosity (vol. %) |
|---|---|---|---|---|---|
| 10 | 3125 | 4225 | 6800 | 7–18 | 52 |
| 20 | 1959 | 1238 | 4263 | 7–19 | 49 |
| 30 | 620 | 104 | 2516 | 7–21 | 44 |

The values for roots and herbage are seasonal means for 3 years, 1970–2 (King & Hutchinson, 1976).
[a] Readings at 09.00 h.

temperature responses are also significantly affected by reduced vegetative cover. Herbage and edaphic changes associated with the stocking treatments imposed at the Armidale sites are given in Table 12.3. The temperature effects shown in Table 12.3 are conservative. The diurnal range of temperature was greatest at the highest stocking rate and temperatures up to 35 °C were recorded for short time periods in summer. This would impose stress conditions on some important invertebrates, e.g. scarab larvae (Davidson, Wiseman & Wolfe, 1972*a*, *b*).

For the Armidale sites it would appear that the effects of sheep numbers on the distribution and abundance of invertebrates can be related to changes in habitat. Shelter, living space, direct and indirect food sources are represented by the levels of herbage, roots and soil pore space that are available. Physiological stresses imposed by temperature and poor aeration are also mediated through these factors. The balance of community energy transactions should also be considered. McMeekan (1956) has drawn attention to the powerful influence of stocking rate on domestic animal production per hectare and this is largely achieved by increasing energy flow through domestic animals. Hence on general thermodynamic grounds a reduction in the productivity of other consumers and decomposers should be expected to accompany increases in numbers of domestic animals.

## Management and ecosystem productivity

Management practices aimed at increasing the productivity and utilization of intensively grazed grasslands fall into two groups. The introduction of new pasture species, fertilizers and irrigation are directed towards increasing primary productivity. Increases in domestic animal numbers and practices such as the conservation of excess herbage as hay or silage, are aimed at increasing the utilization of this primary production by the animals of man's choice. Simulation models for studying the effects of these practices on domestic animal production are the subject of the section that follows. However, the

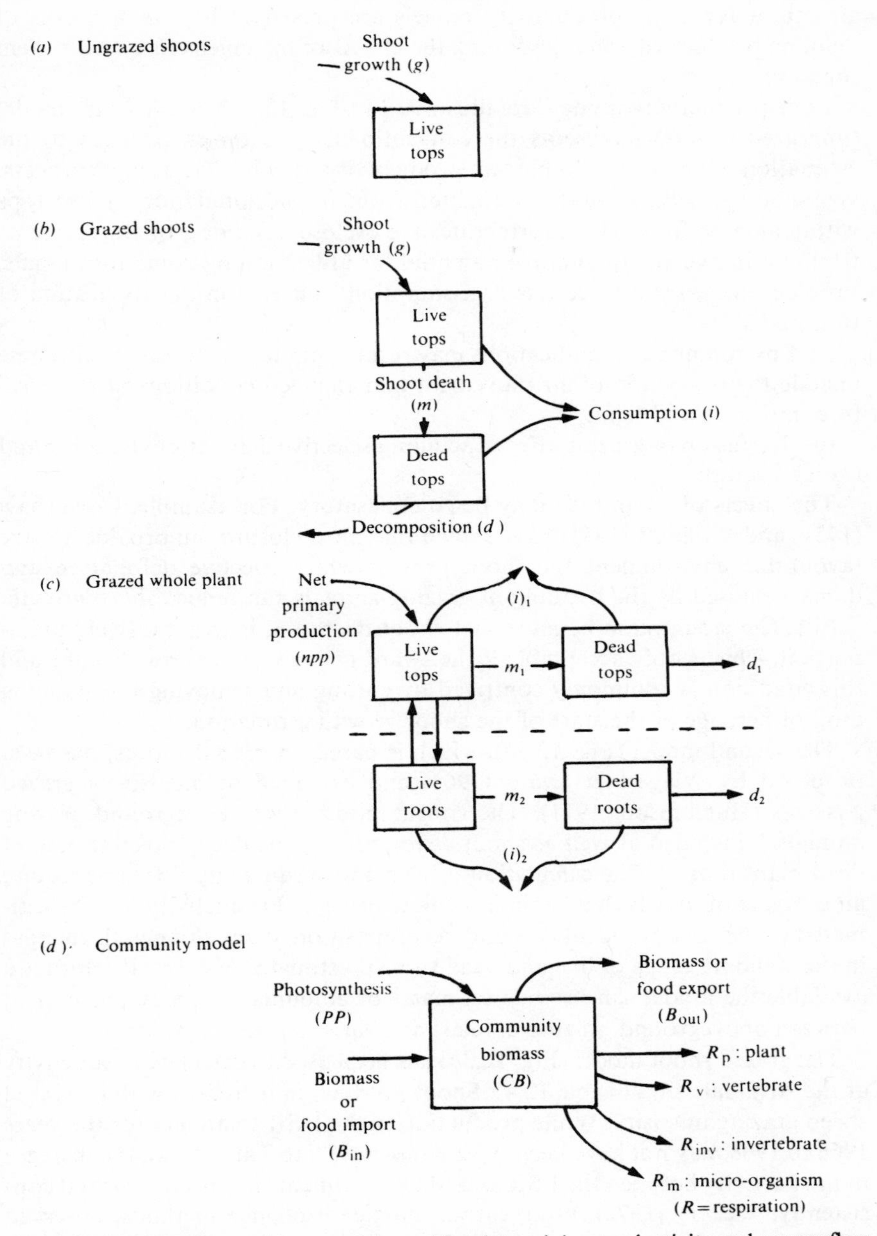


Fig. 12.3. The structures of four models used for determining productivity and energy flow in grazed grasslands.

structures for some productivity models are presented here as a means of defining productivity and describing the effects of management on ecosystem functions.

Four productivity models are illustrated in Fig. 12.3. The exclosure model (ungrazed shoots) represents the conventional agronomic approach to the estimation of aboveground pasture productivity. It considers a single process, viz. shoot growth, *g*, that is estimated from the accumulation of live tops within an area from which vertebrates are exclosed. Shoot growth represents the quantitative and qualitative potential for utilization by domestic animals. Problems arise from three sets of assumptions inherent in the application of this model:

(i) Environmental modifications may result from the enclosure itself. These include the restriction of air movement and changed conditions of temperature, moisture and light.

(ii) Exclusion of grazing effects including selective defoliation, treading and excretal return.

The effects of (i) and (ii) may be compensatory. For example, Cowlishaw (1951) and Williams (1951) have shown that an enclosure can provide a more favourable environment for shoot growth while selective defoliation and damage caused by the treading of grazing animals can reduce shoot growth.

(iii) The assumption is made that shoot death, *m*, is quantitatively unimportant. This is only acceptable if the sward is in an active growth stage and this condition is commonly contrived by cutting and removing the standing crop of herbage at the start of the shoot growth estimation.

The second model (Fig. 12.3*b*), which is based on grazed shoots, has been proposed by Wiegert & Evans (1964) and extended to intensively grazed pastures (Hutchinson, 1971). The consumption, *i*, of aboveground grazing animals is included as well as shoot death, *m*, and the decomposition rate of dead plant tops, *d*. The calculation of primary productivity does not require an estimate of *m*, which is an intermediate process. Productivity may be estimated as the sum of the intakes and decomposition losses along with changes in the standing crops of live and dead tops. If estimates of excretal return are available the model can provide estimates of amounts of energy partitioned between aboveground grazing animals and the litter decomposers.

The grazed shoot model (Fig. 12.3*b*) has been used to calculate productivity at the Armidale sites (Table 12.4). Shoot production increased with increased sheep grazing intensity. While production in the HSR treatment for the years 1966 to 1968 may not have been typical (see note b to Table 12.4), the increase in productivity between the LSR and MSR treatments has been observed consistently. Vickery (1972), using carbon dioxide exchange methods, observed a 40% increase in net primary production (*npp*) between the LSR and MSR treatments for the 17 months from August 1968 to July 1970. There are a number of mechanisms that could explain an increase in *npp* with increases in

Table 12.4. *Annual shoot production and its utilization* ($MJ \times 1000\ ha^{-1}$) *for the Armidale sites calculated by the grazed shoot model* (*Fig. 12.3b*) *for the years 1966–8*

| | Avenues of utilization | | | | |
|---|---|---|---|---|---|
| | | Decomposers | | | Annual clean wool |
| Sheep ($ha^{-1}$) | Assimilated by sheep | Plant litter | Sheep excreta | Annual shoot production[a] | production ($kg \cdot ha^{-1}$) |
| 10 (LSR) | 43 | 51 | 25 | 115 | 33 |
| 20 (MSR) | 83 | 25 | 50 | 158 | 59 |
| 30 (HSR) | 97 | 9 | 60 | 166[b] | 60 |

[a] Estimated from the sum of columns 2, 3 and 4, plus the difference between the initial and final energy states of the plant pools.

[b] Rainfall for the years 1966–8 was above the mean and a substantial amount of white clover was present in the HSR sites (see Fig. 12.1). In other years shoot production was depressed at this stocking level.

grazing intensity. These include the removal of shading effects leading to more efficient utilization of light, an increase in the number of young shoots that are more efficient photosynthetically, and an increased rate of nutrient cycling.

The calculations shown in Table 12.4 also show the effect of sheep numbers on the avenues of energy flow. At a stocking rate of 10 sheep$\cdot$ha$^{-1}$ the sheep assimilated 37% of the shoot growth; at 30 sheep$\cdot$ha$^{-1}$ this proportion increased to 58% and this would appear to be close to the maximum since the return of dead tops to the decomposers was reduced to about 5%. The total energy return to the decomposers did not vary greatly between the different levels of grazing intensity. However, there was a substantial change in the avenues of energy return. As the sheep numbers increased the relative importance of excretal return increased. These results emphasize the dominant role of the domestic herbivore in the utilization of organic matter in intensively grazed grasslands. A reduced role for invertebrate consumers and decomposers appears to follow the introduction of greater numbers of domestic herbivores.

A grazed pasture model at the whole plant level is shown in Fig. 12.3(*d*). The processes of grazing, death of plant tissue and decomposition occur below ground as well as above ground. A large and varied fauna of invertebrates is present belowground with a biomass that can exceed that of the aboveground animals. Moreover, a number of important consumers of roots, e.g. Scarabeidae, are indiscriminate feeders and losses from the severance of living roots may be at least equal to consumption. The large domestic herbivore is more discriminating in its feeding behaviour and ingests most of the herbage that it prehends during grazing; flocks and herds can, however, cause shoot death by trampling (Edmond, 1966). The input for this model is net primary production (*npp*), which represents the balance between photosynthesis and whole plant respiration. The whole plant model includes belowground processes of quantitative importance; however the model is limited in the experimental sense since the measurement of many of these processes is very difficult.

The productivity of intensively grazed grasslands can also be represented by an open-system thermodynamic model (Fig. 12.3*d*). Mass, or its energy equivalent, can move into or out of the system. A typical input is the introduction, from an external source, of supplementary feed for domestic animals. The energy content of livestock products, used or consumed by man, provides an example of the output of mass. Net primary production (*npp*) is equal to the net input of radiation energy entering the system as plant chemical energy. Thermal balance is maintained in the system by the passive, two-way transfer processes of radiation, conduction, convection and evaporation. Over a long time period the sum of these passive transfers is zero since the system as a whole may be assumed to be in thermal balance. In such cases the energy budget for the total system can be represented, in energy units, by the following equation, the symbols for which are given in Fig. 12.3(*d*):

$$npp = pp - R_{\mathrm{p}} = R_{\mathrm{v}} + R_{\mathrm{inv}} + R_{\mathrm{m}} + (B_{\mathrm{out}} - B_{\mathrm{in}}) \pm \Delta CB.$$

This equation shows that community productivity is equal to the total respiration energy expended by fauna and micro-organisms, adjusted for changes in biomass expressed in energy terms. The total respiration of the fauna and micro-organisms is equal to *npp* for a balanced system which is closed to the import and export of food and biomass. A number of workers have used this approach for estimating community productivity (Cragg, 1961; Macfadyen, 1963). The community may be divided into trophic levels (Lindeman, 1942) and the study of energy flow between these levels can provide a basis for comparing community structures.

Calculations of vertebrate and invertebrate respiration for the Armidale sites are given in Table 12.5 for the period 1970 to 1973. Non-breeding, adult sheep were used for grazing and the energy content of the wool exported from the system was relatively small. The respiration estimates for the sheep were based on numbers, mean biomass per animal and the application of Young & Corbett's (1968) equation for maintenance requirements as assimilated or metabolizable energy. Calculations for invertebrates were based on data for abundance, mean biomass per individual and the application of Hemmingsen's (1960) equation. Invertebrate metabolism was adjusted for temperature using a $Q_{10}$ of 2 and soil and air temperature data were available from the sites. All calculations were based on a 42-day interval and were integrated over a 3-year period to give estimates of mean annual respiration.

As sheep numbers increased from 10 to 20 $ha^{-1}$ the respiration energy cost of the flock approximately doubled. However, a slight decline occurred with further increase in numbers; this was due to low liveweights per animal associated with undernutrition. However, the energy expenditure of the sheep

Table 12.5. *Mean annual respiration energy expenditure ($MJ \cdot ha^{-1}$) calculated for soil invertebrates versus sheep grazed at three rates on the Armidale sites*

| | Sheep numbers ($ha^{-1}$) | | |
|---|---|---|---|
| | 10 (LSR) | 20 (MSR) | 30 (HSR) |
| Invertebrates | | | |
| Coleoptera | | | |
| Scarabeidae larvae | 5250 | 4780 | 1070 |
| Other larvae | 670 | 270 | 40 |
| Adults | 380 | 300 | 200 |
| Orthoptera | 60 | 60 | 40 |
| Lepidoptera larvae | 80 | 20 | 10 |
| Diplopoda | 2080 | 1300 | 420 |
| Oligochaeta (large) | 2720 | 3600 | 1970 |
| Enchytraeidae | 1430 | 810 | 570 |
| Nematoda | 1020 | 730 | 370 |
| Acarina | 2120 | 1970 | 560 |
| Collembola | 5580 | 3730 | 500 |
| Araneae | 110 | 50 | 20 |
| *Invertebrate total* | *21500* | *17620* | *5770* |
| Sheep | 39160 | 74390 | 66540 |
| *Total* | *60660* | *92010* | *72310* |

remained high in the HSR treatment; the sheep compensated in part for low herbage availability by increasing their grazing time. In contrast, the respiration energy expended by the invertebrates declined rapidly as sheep numbers increased (Fig. 12.4). These comparisons suggest that sheep can compete with invertebrates but invertebrates cannot tolerate the effects of high grazing intensity by sheep. This raises the question of the need for applying chemicals to reduce the numbers of invertebrates in pastures. At low levels of utilization there may be sufficient plant material available for all fauna and the invertebrates may perform a useful function in the breakdown of organic matter. At high levels of utilization the invertebrates cannot compete; the system appears to become less diverse with the functions of energy flow and nutrient cycling being associated increasingly with the activities of large herbivores and micro-organisms.

The scarabeid larvae and the Collembola at the Armidale sites were the most important invertebrates in terms of energy expenditure. Every group (Table 12.5) showed a depressed level of activity in the HSR treatment. Data have been grouped in Fig. 12.5 to show the relative amounts of energy expended by invertebrate herbivores, large decomposers, small decomposers and predators. This comparison has been extended to include the data compiled for limestone grassland at Moor House in the north of England (Cragg, 1961). The series of sector diagrams exhibits a surprising degree of uniformity in their structure. The Moor House sites, studied by Cragg, were lightly grazed so that the data on total respiration fit into the sheep stocking sequence. Herbivores

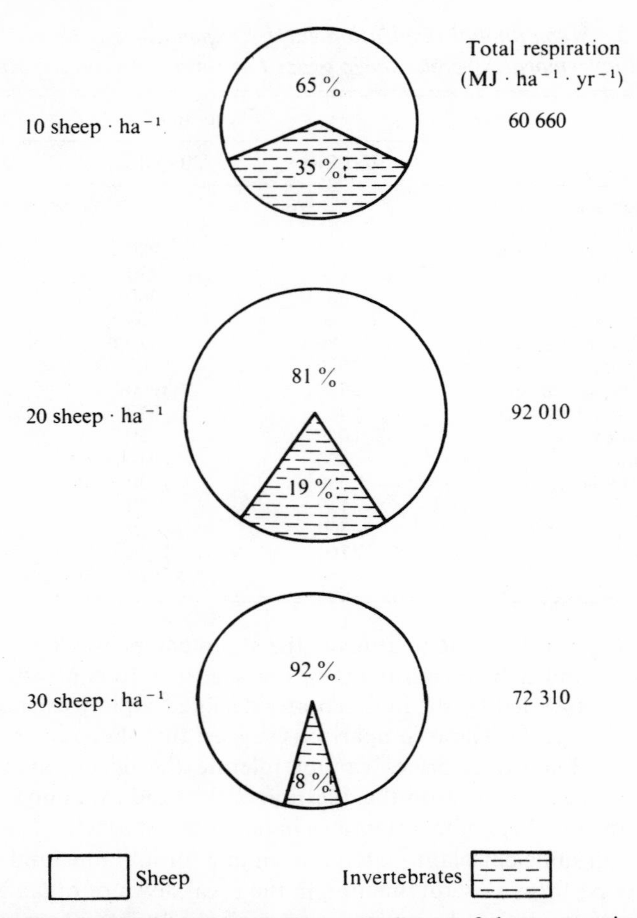


Fig. 12.4. Calculated annual respiration energy expenditures of sheep versus invertebrates for the Armidale sites grazed at different stocking intensities.

account for 25 to 31 % of the invertebrate total expenditure, and predators a constant 1 %. However, the dominant fauna for each group vary between Moor House and the Armidale sites. Tipulids were the major herbivores at Moor House and scarabeids at Armidale. Oligochaetes were the dominant large decomposer group at both locations. At Moor House, however, the large decomposer group was quantitatively more important than the small decomposers, in which enchytraeids were dominant; at Armidale the reverse was apparent with the small decomposer group playing the more important role and Collembola being the dominant group.

Finally there are the effects of sheep numbers on the relative energy expenditures of the invertebrate groups at Armidale. Sheep stocking had the

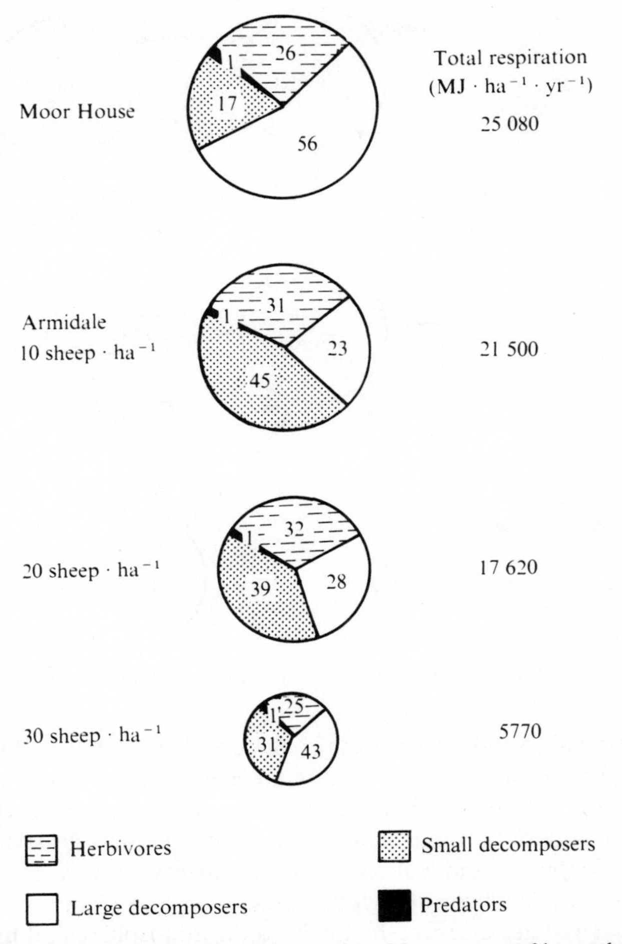


Fig. 12.5. Annual respiration energy expenditure (%) of four groups of invertebrates from lightly grazed chalk grasslands at Moor House, UK (Cragg, 1961) and from the Armidale sites grazed at different stocking intensities.

greatest effect on the activity of the small decomposers, notably Collembola and Acarina. The impact of sheep on habitat modification for these animals has been discussed earlier. The larger, and mainly burrowing, decomposers were less affected by domestic herbivore numbers.

## Nutrient cycling in intensively grazed pastures

The transfer of materials by cycling is an important part of ecosystem function. A simplified diagram for the cycling of nutrients is given in Fig. 12.6. Global aspects are represented as two 'source–sinks' from which nutrients

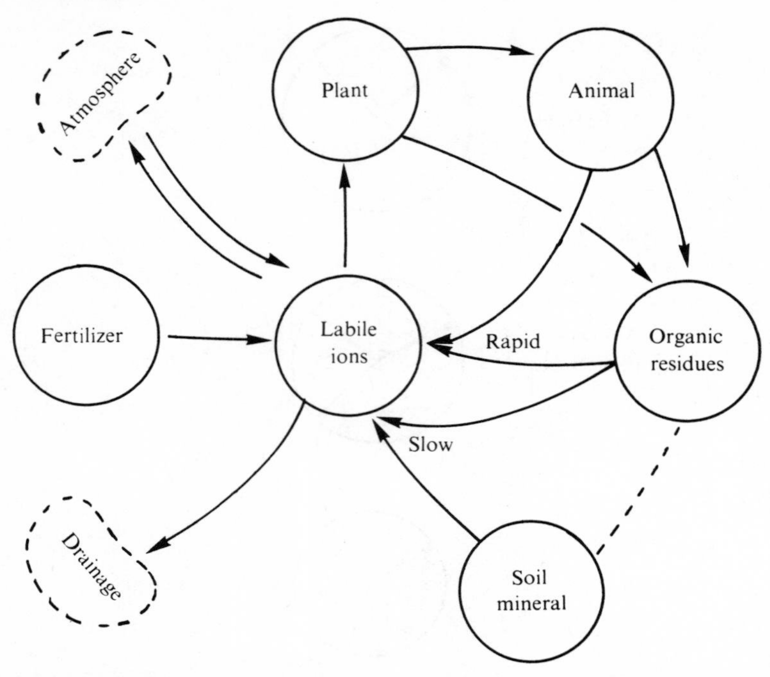


Fig. 12.6. Generalized network diagram for the net flows of nutrients in a grazed grassland.

are transferred to and from the atmosphere and are lost by surface runoff and leaching. Labile and available ions are taken up by plants while exudates from tops and roots, dead plant material and animal excreta are recycled at varying rates, generalized as 'rapid' and 'slow'. For a natural grassland, where no fertilizer is applied and nutrients are approximately in a steady state, the inputs of materials from the global circulation and from the weathering of parent material are balanced by losses to the atmosphere and by streamflow. For natural grasslands the amounts of nutrients involved in these transactions are usually small compared with the inputs of fertilizers used in intensively managed swards. Great increases in the use of artificial fertilizers have occurred in this century and an increasing proportion of these have been applied to permanent and temporary grasslands, e.g. in the UK (Church & Webber, 1971). However, fertilizer use has been associated with an increased concentration of nutrients in streams (Green, 1973) and intensive grazing can result in substantial losses of nitrogen to the atmosphere (Denmead, Simpson & Freney, 1974).

The application of fertilizers to responsive grassland species has resulted in large increases in productivity in many regions of the world. Phosphatic fertilizers and the introduction of *Trifolium* spp. have resulted in multiple

Table 12.6. *A comparison of the aboveground annual net primary production of two communities of native grasses and a fertilized pasture based on introduced species*

| Type of pasture | Annual production (kg dry wt·ha$^{-1}$·yr$^{-1}$) |
|---|---|
| Native: *Bothriochloa* sp. | 2840 |
| Native: *Chloris/Danthonia* sp. | 1120 |
| Introduced species with fertilizer: *Phalaris tuberosa/Trifolium repens/T. subterraneum* | 11230 |

Data from Armidale, NSW, over 2 years, have been taken from Begg (1959).

increases in primary production and livestock numbers in the higher rainfall areas of Australia and New Zealand. Data that illustrate these responses are given in Table 12.6.

Where the grazing intensity is low the principal route for nutrient transfer is through the breakdown of dead plant material. This transfer route is relatively slow (Fig. 12.6). Increased grazing intensity increases the overall rate of cycling by augmenting the transfer through excreta. The return of urine is particularly effective since the nutrients contained in this part of the excreta are very labile. However, local concentrations of excreta increase the potential losses of nutrients by volatilization and leaching. The 'camping' behaviour of gregarious breeds of domestic livestock, e.g. the Merino sheep, can result in heavy local concentration of excreta (Hilder, 1966).

Radiotracer methods have been used in nutrient transfer studies (Patten & Witkamp, 1967). However, this approach has rarely been used for estimating transfer rates in whole grassland systems where interactions between climate, soil, plant and animal communities can operate freely. One example has been the cycle for $^{35}S$ for which the basic model and the methods for estimating the transfer parameters have been described by May, Till & Cumming (1972). One data set from this project has been obtained for the Armidale sites (Till & Blair, 1974) and these data are given to illustrate the effect of stocking intensity on the cycling of a nutrient (Fig. 12.7).

Each site received an annual input of 30 kg S·ha$^{-1}$·yr$^{-1}$ from fertilizer. An increase in sheep numbers from 10 to 20 ha$^{-1}$ resulted in an increased uptake of sulphur by the grazed plants from 19 to 25 kg·ha$^{-1}$·yr$^{-1}$. A response in primary productivity of a similar order can be inferred because the sulphur contents of the plant material from both treatments did not differ markedly. Vickery (1972) reported a similar increase in net photosynthesis for the same sites. Primary production responses to grazing would involve a number of mechanisms apart from sulphur transfer; changes in light utilization and photosynthesis, floristic composition and sward structure and the cycling of other nutrients would all be involved.

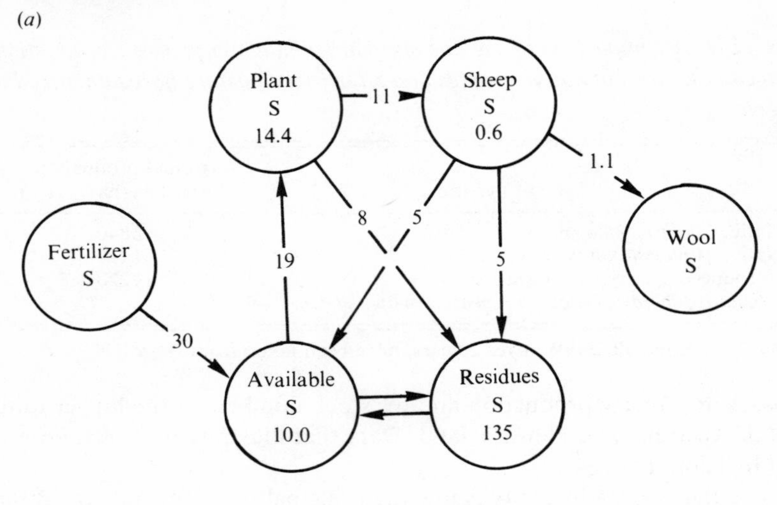


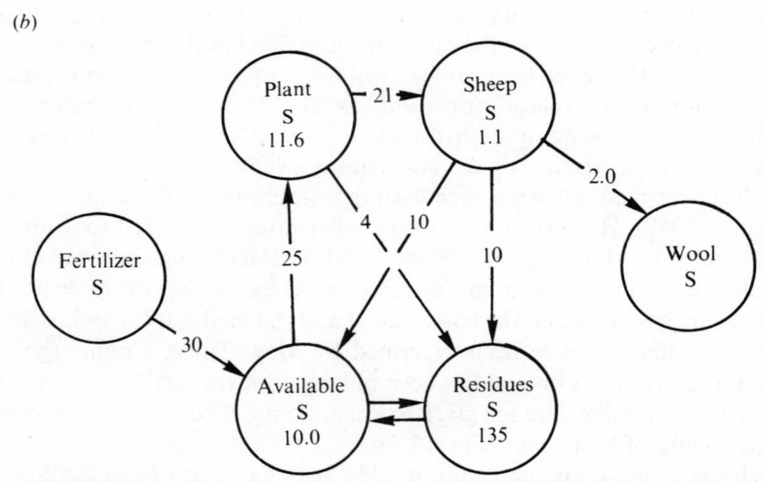


Fig. 12.7. Annual flow rates (kg S · ha$^{-1}$ · yr$^{-1}$) and mean pool sizes (kg S · ha$^{-1}$) for sulphur in the Armidale sites grazed at two stocking rates of sheep: (*a*) 10 sheep · ha$^{-1}$; (*b*) 20 sheep · ha$^{-1}$. (Data are derived from Till & Blair, 1974.)

The mean levels of sulphur present in the herbage were 14.4 and 11.6 kg · ha$^{-1}$ and from these pool sizes and the corresponding uptake rates, the mean residence times can be calculated to be 280 and 170 days for the two stocking treatments. The increase in cycling rate for sulphur can be attributed mainly to increased consumption and excretal return by the increased numbers of sheep; the excretal route provides for more rapid cycling than return through the decomposition of dead plant material.

Fertilized grasslands are subsidized ecosystems and from the viewpoint of the use of global resources there is an urgent need to evaluate minimum requirements for maintaining high productivity levels. Knowledge of the nutrient transfer rates that operate within grasslands and that respond to management practices, provide the only sound basis for solving this problem.

## Discussion

The introduction of new plant species, fertilizers and increases in stocking rates can produce notable responses in domestic animal production in areas where the climatic potential exists for grassland improvement. However, continuing advances will require an understanding of the effects of these and other management practices on the structure and function of the total communities of these systems. Ecological concepts such as energy budgets and nutrient cycling provide useful frameworks for studies at this level. There is a need to recognize also that both the floristic and faunal composition of intensively utilized grasslands respond both to management and to climatic variation.

Control of the numbers of domestic grazing animals has a powerful influence on the yield of products for man and this is not simply a consequence of increasing the percentage of plant shoot growth that is utilized for this purpose. Stocking rate affects the net primary production of grasslands; this has been demonstrated in the Armidale sites by carbon dioxide exchange (Vickery, 1972) and by calculations based on intakes and litter decomposition rates (Table 12.4). Such responses can be inferred from calculations of the respiration energy expenditure of consumer and decomposer communities, excluding micro-organisms (Table 12.5), and from estimates of $^{35}$S uptake of plants grazed at different intensities (Fig. 12.7).

Increased numbers of domestic herbivores modify the whole range of environmental conditions that constitute the habitat of invertebrates. Southwood (1971) compiled data on invertebrates of grazed pastures in the UK and proposed a negative relationship between the extent of management and the biomass of invertebrates. Increased numbers of grazing sheep were associated with a reduction in the abundance (Table 12.1) and biomass of invertebrates. Some phytophagous species can be regarded as potential competitors of domestic herbivores; however, competition needs to be expressed in some measurable way before invertebrate species can be classified as grassland 'pests'. Energy flow comparisons (Table 12.5) provide one basis for defining a competitive situation. The annual respiration energy expenditure by sheep approximately doubled when numbers were increased from 10 ha$^{-1}$ to 20 ha$^{-1}$; the corresponding values for the invertebrates show an overall decline of 18% in respiration. A further increase in sheep numbers from

20 $ha^{-1}$ to 30 $ha^{-1}$ resulted in a further decline in invertebrate respiration of 68 % while the energy expended by the sheep flock was reduced by 11 % only. Thus, grazing sheep show some competitive advantage over invertebrates in their capacity to derive energy for maintenance. Domestic herbivores are protected by man from predation and disease, they are highly mobile in their grazing activity and their numbers can affect substantially the microclimate and living space for invertebrates.

Roberts (1973) has sampled invertebrates from another set of experimental sites at Armidale. The influence of the breed of sheep and time of lambing on invertebrate abundance has been demonstrated in addition to the effects of sheep numbers. Curvilinear relationships were reported between stocking rate and the abundance of phytophagous species with the peak invertebrate numbers occurring at an intermediate level of stocking where the net primary production is near the maximum. Consistent relationships of this kind would support the possibility of the control of pests by manipulating grazing intensity (Roberts, 1973).

The conservation of predatory species may be desirable if any decline in their species richness were to be associated with reduced predatory activity (Southwood, 1971). However, the activity of predatory invertebrates, expressed as a percentage of the total respiration energy expenditure, may remain relatively constant when managed grasslands are subjected to a range of grazing intensities (Fig. 12.5). The species richness of some other 'useful' invertebrates, such as Collembola, may be reduced (Fig. 12.3), but a reduction in any functional role would be more seriously affected by a decline in their abundance (Table 12.1). It is difficult also to distinguish between 'useful' and 'essential' roles; a decrease in either richness or abundance may not be of great consequence if the function in question e.g. decomposition, can be efficiently performed by other organisms, particularly micro-organisms, which may be able to tolerate the environmental changes associated with intensive management.

The effects of domestic herbivores on the floristic composition of sown grasslands have been widely recognized. However, the evidence for an accompanying decline in net primary production is less convincing. There is a need for further study of the amount of floristic change that can be tolerated in sown grasslands without producing a significant decline in plant or animal production. The general theory underlying botanical change in grazed pastures is far from complete – for example there is little quantitative knowledge of the effects of management on plant reproductive processes and the status of seed reserves. There is also the possibility, little explored as yet, of allelochemical reactions operating between plant species (Whittaker & Feeny, 1971; Newman & Rovira, 1975).

## Research and information needs

Management research in intensively used grasslands should aim to achieve and sustain a high level of grassland production using minimum inputs and avoiding deterioration of the environment. The evaluation of the effectiveness and costs of different management inputs, used singly or in combination, continues to be a high priority area for applied research. The concept of energy flow can provide a framework for comparing different management strategies at the farm level. New species, irrigation and fertilizers are introduced either to increase the input of plant chemical energy into the grassland system or to change the seasonal distribution of this energy and its quality to provide for the demands of domestic animals. Other management practices are aimed at increasing the percentage flow of energy assimilated by domestic flocks and herds versus other consumers and decomposers. Choice of stocking level is the most powerful control of this type. Grassland hay and silage is another practice that may increase the proportion of primary production that is utilized ultimately by domestic herbivores. This practice, within the restriction of a unit area of grassland supporting a resident flock or herd, can be viewed as providing an additional energy flow loop which can operate to divert energy from other consumers and decomposers (Hutchinson, 1971). Feed supplement from outside the system has the effect of stabilizing the productivity of domestic herbivores and its interactive effects on behaviour and net primary production can be evaluated in energy terms. More emphasis is needed on the environmental impact of management in intensively used grasslands and this will require a widening of the traditional areas of interest of the grassland agronomist.

Throughout this chapter attention has been drawn to some of the gaps in our knowledge of the structure and function of the grassland ecosystem and particularly the impact of stocking management on the processes of production. High priority in basic research should be given to the retention and cycling of native and introduced mineral nutrients. Little quantitative information is available on the impact of management on the activities of the biota involved in decomposition and mineralization. Functional interrelationships between invertebrate decomposers and micro-organisms have yet to be studied rigorously (Satchell, 1974). There may be gains to be made from a better understanding of the interface between plant roots and particular micro-organisms such as the endomycorrhizas (Crush, 1975), and despite the large biomass of total micro-organisms reported in grassland soils (Clark & Paul, 1970) and their established role in mineralization there have been few studies of the influence of management practices in this area.

Long-term studies, at the agro-ecosystem level, are essential for understanding the impact of management on grazing communities. Their major advantage is that they allow the operation of important interactions between climate, soil and biota. The operation of individual processes can be studied within the experimental ecosystems and analytical solutions of functional parameters can be obtained using system models (e.g. May *et al.*, 1972). However, the range of management options and their combinations is so wide that it is impracticable to solve all management problems by 'real' system studies. The complementary role of simulation modelling provides a more flexible approach which can seek optimum solutions for management; this is the main topic for the chapter that follows.

## Summary

Fertilizers and the control of domestic herbivore numbers have important effects on the productivity of intensively used grasslands. However there is little information on the effects of these practices on ecosystem structure and function. Data are presented from temperate, sown perennial grasslands (Armidale, Australia) and comparisons are drawn with published observations from other intensively grazed sites.

Floristic changes under grazing are described and the importance of interactions between climatic and biotic mechanisms is shown. There is a need for more comprehensive simulation models, which incorporate plant species/nutrient effects, plant reproduction responses and relevant aspects of animal behaviour. A desirable objective would be the coupling of models for predicting botanical change under grazing with models of production and utilization. Long-term data from experimental ecosystems are in short supply.

Domestic animal numbers are shown to have substantial effects on the abundance, biomass, aggregation and species richness of grassland invertebrates. These result from changes in habitat which include levels of plant litter, root biomass and soil pore space. Productivity and utilization models are discussed and calculations are presented for the Armidale sites. Sheep numbers have a large influence on the partitioning of respiration energy expenditure between domestic animals and invertebrates; such comparisons provide a basis for assessing competition between animal groups and hence the rationale for applying control measures.

Energy budgets constructed from data from experimental ecosystems are valuable for comparing the efficiency of different management practices. At the basic research level there is a great need for information on the retention and cycling rates of native and fertilizer minerals and the possible role of management to increase and sustain high recovery rates through the mineralization of organic residues.

## References

Barley, K. P. & England, P. J. (1970). Mechanisms of burr burial by subterranean clover. In *Proceedings of the XI International Grasslands Congress, Surfers Paradise*, sect. 2, pp. 103–7. University of Queensland Press, St. Lucia, Queensland.

Begg, J. E. (1959). Annual pattern of soil moisture stress under sown and native pastures. *Australian Journal of Agricultural Research*, **10**, 518–29.

Berthet, P. (1964). L'activité des oribatides d'une chênaie. *Mémoires de l'Institut Royale des Sciences Naturelles de Belgique*, **152**, 1–152.

Bliss, C. I. & Fisher, R. A. (1953). Fitting the negative binomial distribution to biological data. *Biometrics*, **9**, 176–200.

Brillouin, L. (1960). *Science and information theory*, 2nd edn. Academic Press, New York & London.

Bulan, C. A. & Barrett, G. W. (1971). The effects of two acute stresses on the arthropod component of an experimental grassland ecosystem. *Ecology*, **52**, 597–605.

Church, B. M. & Webber, J. (1971). Fertilizer practice in England and Wales. *Journal of the Science of Food and Agriculture*, **22**, 1–7.

Clark, F. E. & Paul, E. A. (1970). The microflora of grassland. *Advances in Agronomy*, **22**, 375–435.

Coupland, R. T. (ed.) (1979). *Grassland ecosystems of the world: analysis of grasslands and their uses. IBP Synthesis Series 18*. Cambridge University Press.

Cowlishaw, S. J. (1951). The effect of sampling cages on the yields of herbage. *Journal of the British Grassland Society*, **6**, 179–82.

Cragg, J. B. (1961). Some aspects of the ecology of moorland animals. *Journal of Animal Ecology*, **30**, 205–34.

Crush, J. R. (1975). Occurrence of endomycorrhizas in soils of the Mackenzie Basin, Canterbury, New Zealand. *New Zealand Journal of Agricultural Research*. **18**, 361–4.

Davidson, R. L., Wiseman, J. R. & Wolfe, V. J. (1972*a*). Environmental stress in the pasture scarab, *Sericesthis nigrolineata* Boisd. I. Mortality in larvae caused by high temperature. *Journal of Applied Ecology*, **9**, 783–97.

Davidson, R. L., Wiseman, J. R. & Wolfe, V. J. (1972*b*). Environmental stress in the pasture scarab, *Sericesthis nigrolineata* Boisd. II. Effects of soil moisture and temperature on survival of first instar larvae. *Journal of Applied Ecology*' **9**, 799–806.

Davies, H. L. (1962). Studies on time of lambing in relation to stocking rate in South Western Australia. *Proceedings of the Australian Society of Animal Production*, **4**, 113–20.

Denmead, O. T., Simpson, J. R. & Freney, J. R. (1974). Ammonia flux into the atmosphere from a grazed pasture. *Science*, **185**, 609–10.

East, R. & Pottinger, R. B. (1975). Starling (*Sturnus vulgaris* L.) predation on grass grub (*Costelytra zealandica* (White), Melononthinae) populations in Canterbury. *New Zealand Journal of Agricultural Research*, **18**, 417–52.

Edmond, D. B. (1966). The influence of animal treading on pasture growth. In *Proceedings of the X International Grassland Congress, Finland*, pp. 453–8. Valtioneuvoston Kirjapaino, Helsinki.

Green, F. H. W. (1973). Aspects of the changing environment: some factors affecting the aquatic environment in recent years. *Journal of Environmental Management*, **1**, 377–91.

Hale, W. G. (1967). Collembola. In *Soil biology*, ed. A. Burges & F. Raw, pp. 397–411. Academic Press, New York & London.

Harper, J. L. (1969). The role of predation in vegetational diversity. *Brookhaven Symposia in Biology*, **22**, Upton, 48–62.

Heady, H. F. (1975). Management of seasonal grazing. In *Rangeland management*, pp. 189–235. McGraw-Hill, New York.

Hemmingsen, A. M. (1960). *Energy metabolism as related to body size and respiratory surfaces, and its evolution. Report of the Steno Memorial Hospital and the Nordisk Insulinlaboratorium 9*. Copenhagen.

Hilder, E. J. (1964). The distribution of plant nutrients by sheep at pasture. *Proceedings of the Australian Society of Animal Production*, **5**, 241–8.

Hilder, E. J. (1966). Distribution of excreta by sheep at pasture. In *Proceedings of the X International Grassland Congress, Helsinki, Finland*, pp. 977–81. Valtioneuvoston Kirjapaino, Helsinki.

Hilditch, J. A. (1974). Effects of temperature and soil moisture on the abundance and distribution of pasture scarabs (Coleoptera: Scarabeidae). PhD Thesis, The University of New England, Armidale, NSW.

Hope-Simpson, J. (1940). Studies of the vegetation of the English chalk. VI. Late stages in succession leading to chalk grassland. *Journal of Ecology*, **28**, 386–402.

Hutchinson, K. J. & King, K. L. (1970). Sheep numbers and soil arthropods. *Search*, **1**, 42–3.

Hutchinson, K. J. (1971). Productivity and energy flow in grazing fodder conservation systems. *Herbage Abstracts and Reviews*, **41**, 1–10.

Hutchinson, K. J. (1974). Fodder conservation in grazing systems. *Proceedings of the New Zealand Grassland Association*, **35** (1), 142–8.

Janzen, D. H. & Schoener, T. W. (1968). Differences in insect abundance and diversity between wetter and drier sites during a tropical dry season. *Ecology*, **49**, 98–110.

Jones, M. G. (1933*a*). Grassland management and its influence on the sward. II. The management of a clovery sward and its effects. *Empire Journal of Experimental Agriculture*, **1**, 122–8.

Jones, M. G. (1933*b*). Grassland management and its influence on the sward. III. The management of a grassy sward and its effects. *Empire Journal of Experimental Agriculture*, **1**, 223–34.

Kevan, D. K. McE. (1968). Soil fauna and humus formation. In *Transactions of the IX International Congress of Soil Science, Adelaide*, vol. 2, pp. 1–10. International Society of Soil Science/Angus & Robertson, Sydney.

King, K. L. & Hutchinson, K. J. (1976). The effects of sheep stocking intensity on the abundance and distribution of mesofauna in pastures. *Journal of Applied Ecology*, **13**, 41–55.

King, K. L., Hutchinson, K. J. & Greenslade, P. (1976). The effects of sheep numbers on associations of Collembola in sown pastures. *Journal of Applied Ecology*, **13**, 731–9.

Kydd, D. D. (1966). The effect of intensive sheep stocking over a five-year period on the development and production of the sward. I. Sward structure and botanical composition. *Journal of the British Grassland Society*, **21**, 284–8.

Langlands, J. P. & Bennett, I. L. (1973). Stocking intensity and pastoral production. I. Changes in the soil and vegetation of a sown pasture grazed by sheep at different stocking rates. *Journal of Agricultural Science, Cambridge*, **81**, 193–204.

Lindeman, R. L. (1942). The trophic–dynamic aspect of ecology. *Ecology*, **23**, 399–418.

Macfadyen, A. (1961). Metabolism of soil invertebrates in relation to soil fertility. *Annals of Applied Biology*, **49**, 216–19.

Macfadyen, A. (1963). *Animal ecology: aims and methods*, 2nd edn. Pitman, London.

McMahan, E. A. & Sollins, N. F. (1970). Diversity of microarthropods after irradiation. In *A tropical rain forest: a study of irradiation and ecology at El Verde, Puerto Rico*, ed. H. T. Odum & R. F. Pigeon, book 2, E151-8. US Atomic Energy Commission, Washington, DC.

McMeekan, C. P. (1956). Grazing management and animal production. In *Proceedings of the VIII International Grassland Congress, Palmerston North, New Zealand*, pp. 146–56. Wright & Carman Ltd, Wellington, New Zealand.

May, P. F., Till, A. R. & Cumming, M. J. (1972). Systems analysis of 35-sulphur kinetics in pastures grazed by sheep. *Journal of Applied Ecology*, **9**, 25–49.

Milton, W. E. J. (1940). The effect of manuring, grazing and cutting on the yield, botanical and chemical composition of natural hill pastures. I. Yield and botanical section. *Journal of Ecology*, **28**, 326–56.

Milton, W. E. J. (1947). The yield, botanical and chemical composition of natural hill herbage under manuring, controlled grazing and hay conditions. I. Yield and botanical section. *Journal of Ecology*, **35**, 65–89.

Morris, M. G. (1967). Differences between the invertebrate faunas of grazed and ungrazed chalk grassland. I. Responses of some phytophagous insects to cessation of grazing. *Journal of Applied Ecology*, **4**, 459–74.

Morris, M. G. (1968). Differences between the invertebrate faunas of grazed and ungrazed chalk grassland. II. The faunas of sample turves. *Journal of Applied Ecology*, **5**, 601–11.

Morris, M. G. (1969). Differences between the invertebrate faunas of grazed and ungrazed chalk grassland. III. The Heteropterous fauna. *Journal of Applied Ecology*, **6**, 475–87.

Morris, M. G. (1971). Differences between the invertebrate faunas of grazed and ungrazed chalk grassland. IV. Abundance and diversity of Homoptera–Auchenorrhyncha. *Journal of Applied Ecology*, **8**, 37–52.

Newman, E. I. & Rovira, A. D. (1975). Allelopathy among some British grassland species. *Journal of Ecology*, **63**, 727–37.

Norman, M. J. T. (1957). The influence of various grazing treatments upon the botanical composition of a Downland permanent pasture. *Journal of the British Grassland Society*, **12**, 244–56.

Ofer, Y. & Seligman, N. G. (1969). Fertilization of annual range in Northern Israel. *Journal of Range Management*, **22**, 337–41.

Patten, B. C. & Witkamp, M. (1967). Systems analysis of 134-caesium kinetics in terrestrial microcosms. *Ecology*, **48**, 813–24.

Roberts, R. J. (1973). Some effects of grazing management on population of invertebrates in pastures. Thesis: Diploma of Membership of Imperial College (Zoology), London.

Rogers, L., Lavigne, R. & Miller, J. L. (1972). Bioenergetics of the western harvester ant in the shortgrass plains ecosystem. *Environmental Entomology*, **1**, 763–8.

Rossiter, R. C. (1961). The influence of defoliation on the components of seed yield in swards of subterranean clover (*Trifolium subterraneum* L.). *Australian Journal of Agricultural Research*, **12**, 821–33.

Rossiter, R. C. (1966). Ecology of the Mediterranean annual pasture type. *Advances in Agronomy*, **18**, 1–56.

Satchell, J. E. (1974). Introduction: litter-interface of animate/inanimate matter. In *Biology of plant litter decomposition*, ed. C. H. Dickinson & G. J. F. Pugh, vol. 1, pp. xiii–xliv. Academic Press, New York & London.

Southwood, T. R. E. (1971). Farm management in Britain and its effect on animal population. In *Proceedings of the Tall Timbers Conference on Ecological Animal Control by Habitat Management*, vol. 3, pp. 29–51.

Stern, W. R. & Donald, C. M. (1962). Light relationships in grass–clover swards. *Australian Journal of Agricultural Research*, **13**, 599–614.

Swartzman, G. L. & Singh, J. S. (1974). A dynamic programming approach to optimal grazing strategies using a succession model for a tropical grassland. *Journal of Applied Ecology*, **11**, 537–48.

Tansley, A. G. & Adamson, R. S. (1925). Studies of the vegetation of the English chalk. III. The chalk grasslands of the Hampshire–Sussex border. *Journal of Ecology*, **13**, 177–223.

Till, A. R. & Blair, G. J. (1974). Sulphur and phosphorus cycling in grazed pastures. In *Transactions of the X International Congress of Soil Science, Moscow*, vol. 2, pp. 153–7. Nauka Publishing House, Moscow.

Vickery, P. J. (1972). Grazing and net primary production of a temperate grassland. *Journal of Applied Ecology*, **9**, 307–14.

Walton, G. H. (1975). Response of burr burial in subterranean clover (*Trifolium subterraneum*) to defoliation. *Australian Journal of Experimental Agriculture and Animal Husbandry*, **15**, 69–73.

Whittaker, R. H. & Feeny, P. P. (1971). Allelochemics: chemical interactions between species. *Science*, **171**, 757–70.

Wiegert, R. G. & Evans, F. C. (1964). Primary production and the disappearance of dead vegetation on an old field in south eastern Michigan. *Ecology*, **45**, 49–62.

Williams, S. S. (1951). Microenvironment in relation to experimental techiques. *Journal of the British Grassland Society*, **6**, 207–17.

Wolfe, B. C. & Lazenby, A. (1973*a*). Grass–white clover relationships during pasture development. I. Effects of superphosphate. *Australian Journal of Experimental Agriculture and Animal Husbandry*, **13**, 567–74.

Wolfe, E. C. & Lazenby, A. (1973*b*). Grass–white clover relationships during pasture development. II. Effect of nitrogen fertilizer with superphosphate. *Australian Journal of Experimental Agriculture and Animal Husbandry*, **13**, 575–80.

Young, B. A. & Corbett, J. L. (1968). Energy requirements for maintenance of grazing sheep measured by calorimetric techniques. *Proceedings of the Australian Society of Animal Production*, **7**, 327–34.

# 13. Simulation of intensively managed grazing systems

N. G. SELIGMAN & G. W. ARNOLD

## Contents



In recent years systems analysis with simulation models has been used increasingly for studying the ecology and management of agricultural and pastoral systems. This approach provides a means whereby much accumulated knowledge about the system can be integrated into a consistent, dynamic format and applied to actual management problems. At present it is easier to point out the inadequacies of simulation models and analyses so far published than their creative achievements (Van Dyne & Abramsky, 1975; van Keulen, 1976; Seligman, 1975; Chapter 10, this volume). Clearly, when dealing with such complex situations as biological systems, the need for a well-defined approach and sound conceptions of the possibilities and limitations are essential guides to useful simulation. Some basic principles have been proposed (de Wit, 1970; Swartzman, 1972; Maynard Smith, 1974; van Keulen, 1976), but the 'state of the art', as applied to simulation of agropastoral systems, has not yet reached maturity.

In this chapter some of the dynamic simulation models that have been used for the study of intensively managed grazing systems will be examined in some detail. Such systems are generally man-made pastures on relatively productive land and under fairly reliable climatic conditions. Production, compared to natural or semi-natural pastoral systems, is fairly high and is maintained by agrotechnical means, mainly cultivation, sowing of improved pasture species and chemical fertilization (Jones & Brockington, 1971). These systems are generally more homogeneous and more easily manipulated than natural pasture ecosystems. They are not unique ecosystems in the sense that what holds for one sample of the system holds, within acceptable bounds of confidence, for other samples of the same system. As a result these systems are more amenable to conventional experimentation. Consequently, processes have been studied in considerable detail and this information is available for use in modelling. Simulation of such systems can proceed from a fairly solid base and verification of the simulation results is feasible within reasonable bounds of time and facilities. The need for a powerful analysis technique may be greater in the heterogeneous, extensively managed grazing systems, but the possible applications in analysis and management of the simpler, but still complex, agropastoral situations are numerous and promising.

Management of agropastoral systems aims at maximizing various benefits. These may include satisfaction of subjective inclinations like low risk, a uniformly coloured beef herd, a good-looking pasture or even a tax-deductable loss; but it is often assumed that the benefit to be maximized is long-term profit. This involves a combination of knowledge about system responses and business acumen. The latter qualification includes acquaintance with market prices of materials, feeds and livestock combined with an ability to judge future fluctuation accurately or, alternatively, a lot of luck! On the whole individual farms probably vary economically more than they do biologically, as the talents and the tastes of the manager add greatly to the complexity of the system. A study of the biological and business aspects requires different disciplines and approaches. The interface between these aspects involves additional problems, mainly of scale and resolution, and raises the need for yet another approach (Dillon, 1971; Swartzman & Van Dyne, 1972; Smith, 1975; Arnold & Bennett, 1975).

In a survey of the application of programming techniques in resource management, only three out of the 26 models surveyed were based on simulation models (Swartzman, 1972). Simulation models have been used mainly to provide input for separate routines that optimize various management parameters. Actual input/output functions from the system itself would probably have been preferable had they been available. In most cases, however, they are not available but can be constructed by simulation from available data and knowledge of system behaviour. Optimization of management parameters within the simulation model itself is often done for individual parameters or a relatively small number of combinations of a small number of parameters (e.g. Morley, 1974; Edelsten & Newton, 1975). Simultaneous optimization of more than three or four parameters in a moderately complex simulation model can become an expensive exercise.

Simulation models as such are at present often being used as part of a wider research programme that includes field and laboratory experiments. They are being used to study behaviour of the system and to identify areas of greater sensitivity to management. The discussion in this chapter is concerned mainly with some of the intensive agropastoral systems that have been studied in this way and with some of the problems associated with this approach. As criteria for judging the practical or conceptual validity of such simulation models depend on the expectations and objectives of both judge and modeller, it is difficult to put them all on a common base. Our approach will be to assess the models in terms of what they set out to achieve and how they resolved (or ignored!) some common dilemmas like generality versus realism or precision versus low resolution. We recognize also that where models are part of a research programme based on systems analysis, their utility or success cannot be fully evaluated out of this context. Nevertheless, at this stage where many different approaches to simulation are being adopted, there is a case for reviewing some of the models on their own merits.

## Grazing system models and conventional grazing experiments

Our understanding of grazed ecosystems has till recently been based on conventional grazing experiments and experience. But progress in developing a comprehensive pasture science with enough generality and flexibility to apply universally has been slow, mainly because of the naturally cumbersome nature of most grazing trials and the slow turnover of results. Any more than a few treatments and a minimum number of replications make these experiments expensive. Large areas of pasture are needed to feed and maintain experimental flocks for long periods, generally yearlong. Fencing, water supply, access, supplemental feeding and management of the livestock make heavy demands on funds and manpower. Then, once an experiment has been set up, it must run for a number of years so as to evaluate interactions between treatments and years with different climatic conditions as well as the effects of one year on subsequent years. A large amount of data on a number of processes must be accumulated if the reasons for changes in the system are to be analysed and understood. The experiment can include only a few treatments from the very large number of possibilities, so that it is very difficult to construct continuous response surfaces over the whole valid range of variation. Three and higher dimensional response surfaces are extremely rare in published grazing experiments even though they are essential for fuller understanding of system behaviour. Finally, the results of grazing experiments may not be directly applicable to other areas and other years, so that repetition in a number of locations is necessary to assess the generality of a particular management practice.

*System utilization*

The hope held out by simulation of grazing systems is that they will at least extend the relevance of conventional grazing experiments beyond the site and years to which they are specific. The maximalist aim of completely replacing field experiments is probably premature if not misguided. The reason is that a model of a biological system is always a simplification of the real system. The simplification means that many processes have been omitted or lumped together into a low-resolution empirical relationship, or a 'black box'. The simplification is based on the assumption that what has been omitted is of negligible importance to the objective of the study, an assumption that can be made only if it is based on knowledge of the system being studied. Thus modelling, to be relevant, must be combined with experimentation. The interaction between the models and experiments will certainly alter the type and emphasis of the field experiments. As insight into the dynamics of the system deepens, more detailed studies on the component processes will be needed, together with experiments to validate the model's predictions about the behaviour of the system under various changes in management. The model will then become not only a means of integrating knowledge and evaluating new concepts and hypotheses, but also a dynamic framework for storing and communicating information about the system. In time, management, i.e. manipulation of the system to attain desired objectives, should be able to lean increasingly on analysis of models as confidence in their reliability grows.

## Big models, small models and biological complexity

The biological part of a grazing system can be defined as a set of processes interacting under certain environmental and management constraints. A process is well-defined when represented in chemical or physical terms (Thornley, 1976). Only restricted biological systems or parts of more comprehensive systems (e.g. the micrometeorological sections of advanced crop canopy models) can be so represented. This is as much due to the extreme complexity involved in such a representation as to the fact that so little is understood of biological systems in such basic terms. The common solution, for those who believe that a solution exists, is to aggregate or lump the myriad basic processes into a much smaller number of empirical relationships. These are often physiological relationships that have been well studied and have achieved general acceptance, e.g. the photosynthesis rate/radiation intensity functions, or energy requirements for maintenance of livestock. More often they are *ad hoc* relationships that have been constructed for the occasion, e.g. vegetation growth rate/soil moisture relationships or physiological age of a canopy as a function of time of germination and flowering. All of these empirical relationships may be valid in themselves but whether their validity holds when they interact in a system, where there is feedback onto the parameters (or coefficients) of the function itself, needs to be verified by experimenting with the system. Clearly, as the *ad hoc* empirical relationships proliferate in a model, so does its *a priori* validity decrease, and the need for detailed verification increases. There are many facets to grassland grazing systems and the temptation to simulate all the details is strong. From what has been said above there appears to be a good case for circumscribing the scope of such models drastically so that the untested interactions between the *ad hoc* empirical 'black boxes' are kept to a minimum. 'Since the size of this problem is a power function of the number of variables, this in itself is almost sufficient justification for small models' (Arnold & Bennett, 1975). One way of concentrating the modelling effort on the most relevant aspects of the system is the objective-centred hierarchical approach described by Spedding (1975). According to this approach the objective is defined in terms of its immediate components. The components are then defined in greater or lesser detail depending on their importance in defining the objective. Thus the objective is the basis of the model and an integral part of it. This approach is apparently not as obvious as it seems, as most pastoral modellers have implicitly assumed an objective but have defined the system 'from the bottom up', entering all components assumed to be relevant but without direct reference to the objective. Defining the system from the central objective outwards is easily done with simulation languages like CSMP, whereas the structural logic of FORTRAN puts the objective at the end. In this way the language seems to have influenced the structure and complexity of some models.

## Some grazing management models

The history of grazing management models probably starts with Arcus (1963) who outlined how simulation might be used in the study of grazing management problems. The first model outlined in detail was PASTOR, a model of a semi-arid pasture paddock grazed by sheep. The paddock is divided into land units and zones preferred by sheep; five plant species with different characteristics are included. The management problem is to determine the effect of watering points on efficiency of pasture utilization and livestock production (Goodall, 1967). The aim of this model was to see whether whole-system synthesis was feasible and it is presented without any attempt at verification. Subsequent to PASTOR, other grazing management models were developed: LEYFARM (Arnold & Campbell, 1972; Arnold, Carbon, Galbraith & Biddiscombe, 1974), the Armidale model (Vickery & Hedges, 1972*a*, *b*, 1974) and others in Australia. In England pasture management models have been constructed at Hurley (Brockington, 1971; Edelsten & Newton, 1975) and at other centres more are being developed. In the USA, most of the grazingland modelling effort has been concentrated in the framework of the Grassland Biome and Desert Biome projects. These are mainly total-ecosystem models or models of extensively utilized natural grazing lands that fall outside the

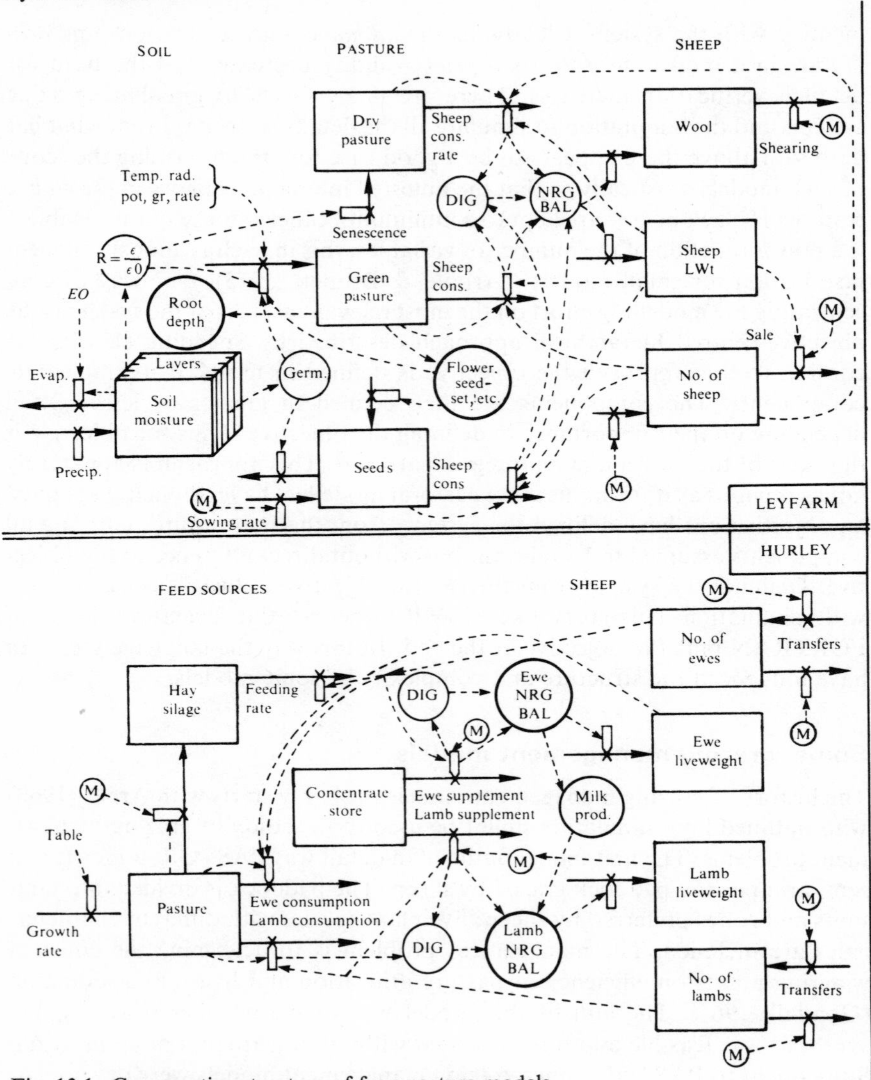


Fig. 13.1. Comparative structure of four pasture models.

scope of the present discussion. Examples are discussed in Chapter 10 of this volume. In Israel a grassland grazing model, NEGEV, was developed but, like PASTOR, with no serious attempt at detailed verification (Seligman, Tadmor, Noy-Meir & Dovrat, 1972). More recently some work on generalized grazing systems has yielded some interesting results (Noy-Meir, 1974*a*, *b*, 1975).

The models that will be discussed in this section have been selected to illustrate a number of different approaches (Fig. 13.1). They differ mainly in the

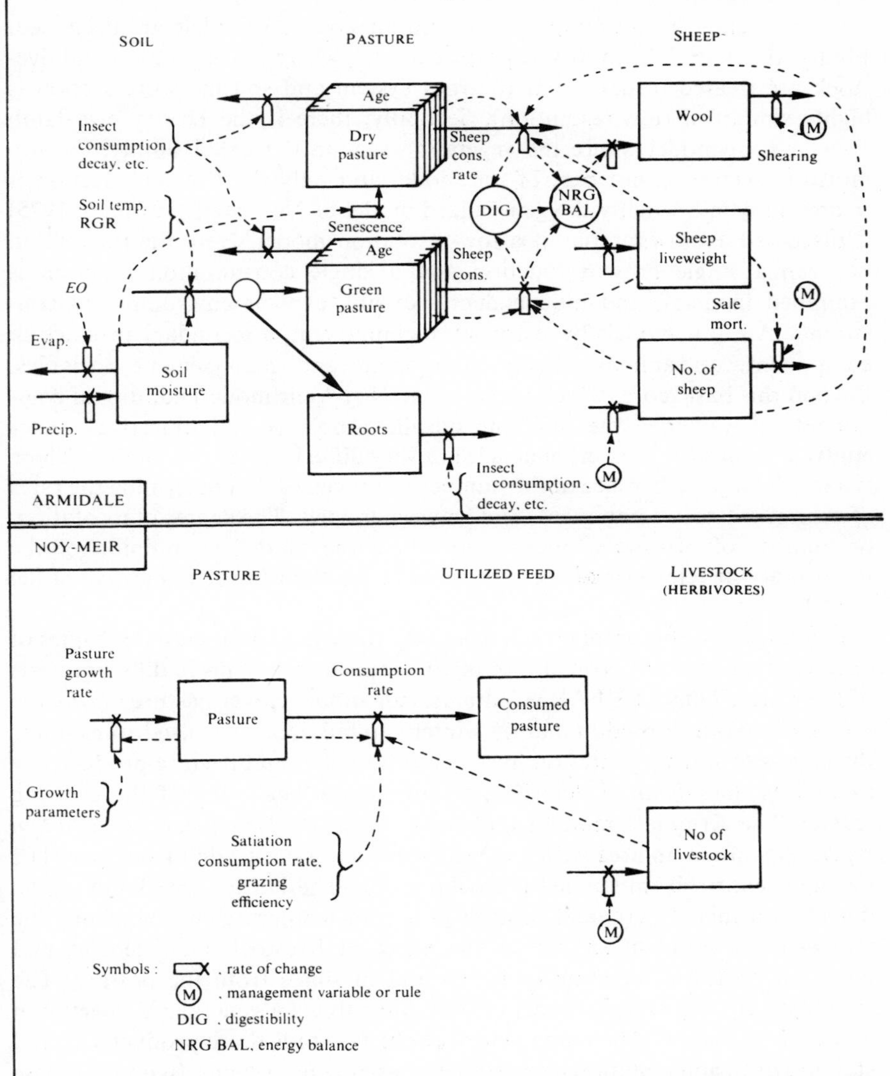


distribution of 'resolution' among the various components of the system. By resolution is meant the number of processes that comprise a given system component. The lowest resolution for a component would be a single empirical relationship between a rate and a state. The highest resolution grazing models are those with the largest number of processes per system component. These processes are generally described as empirical functions.

An example of a high-resolution model in this sense is LEYFARM, where the components (vegetation, soil moisture and livestock) are treated in considerable detail. Two examples of medium-resolution models are discussed. Firstly, the Armidale improved-pasture model, where the vegetation and livestock are treated in detail but the root system and soil moisture section is highly simplified (low-resolution). Secondly, there is the Hurley ewe–lamb management model where the vegetation section is treated simply, the soil moisture section is not part of the model and only the livestock section is treated in detail. Finally the generalized model of Noy-Meir (1974*a*, *b*, 1975) is discussed as an example of a low-resolution model. Here the interaction between a single growth function and a single consumption function is examined in detail and independently of any explicit environmental constraints. All four models have a pasture component, a secondary production component, an animal density component and management variables. Beyond this basic core (which is the whole Noy-Meir model) additional components characterize the different specific problems and objectives under study. The number of components is not very different between models. There is a much larger difference in the number of processes that determine the rates of change between components and between models. The degree of resolution, or number of processes, appears to reflect the modellers' opinion of the importance or relevance of these processes with regard to the objective of the model.

In addition to this gradient of resolution, these models are also examples of different management problems of intensive pasture systems in different parts of the world. Thus LEYFARM simulates an annual clover pasture (*Trifolium subterraneum*) in a predominantly winter rainfall area. It is used to examine the effects of stocking rates on livestock production and pasture productivity as well as the effects of deferring grazing at the beginning of the growing season. The Armidale model simulates a mixed *Phalaris tuberosa–Trifolium repens* pasture in an area with a considerable amount of rain in summer. The management problem studied is stocking rate. The Hurley ewe–lamb model describes a mixed perennial pasture in a cool temperate environment. The management problem studied is the effect of livestock management and nutrition on the production of lambs and of silage from the pasture. The simplified grazing system model of Noy-Meir treats a generalized vegetation and grazing animal. The management problems studied are productivity and stability of various rotational and continuous grazing systems. In none of these models has plant nutrition been considered.

### *LEYFARM (Arnold & Campbell, 1972; Arnold* et al., *1974)*

This is a moderately complex, comprehensive model that in its present form simulates sheep grazing on annual legume pasture. The pasture is described from seed germination through growth, flowering, seed formation, death, decay and consumption through grazing to the softening of hard seeds for the

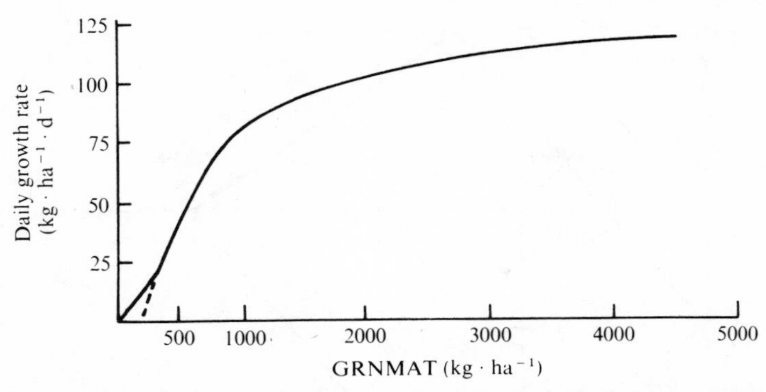


Fig. 13.2. LEYFARM: potential daily growth rate as a function of amount of live biomass (GRNMAT). (Derived from Arnold *et al.*, 1974.)

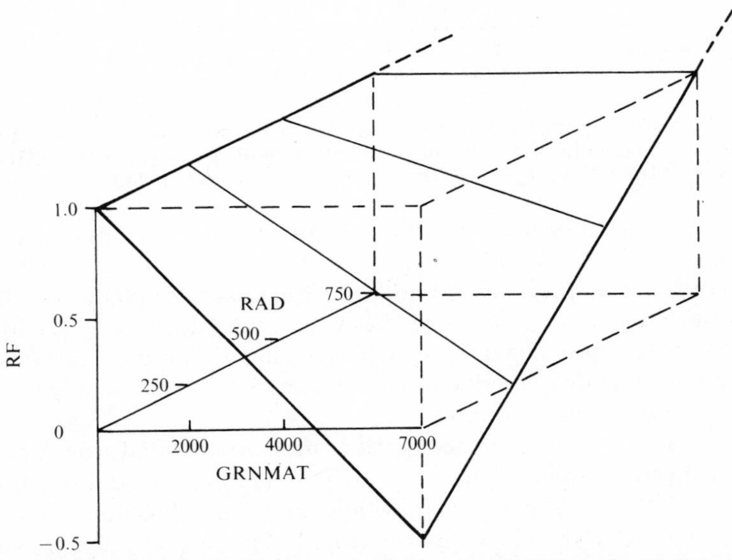


Fig. 13.3. LEYFARM: dependence of radiation factor (RF) on live biomass (GRNMAT, in $kg \cdot ha^{-1}$) and radiation intensity (RAD, in $cal \cdot cm^{-2} \cdot d^{-1}$). (Derived from Arnold *et al.*, 1974.)

next season's germination (Fig. 13.1). The moisture balance in the soil and the grazing animal are simulated in detail (Carbon & Galbraith, 1975). Considerable attention is given to the seeds which are formed as hard seeds and have to soften before they can germinate. Germination takes place in waves and the separate waves are monitored to form the basis for calculating a mean emergence day and a mean flowering day. These days are used to determine the weight of individual seeds, the yield of seeds and the rate of ageing of the

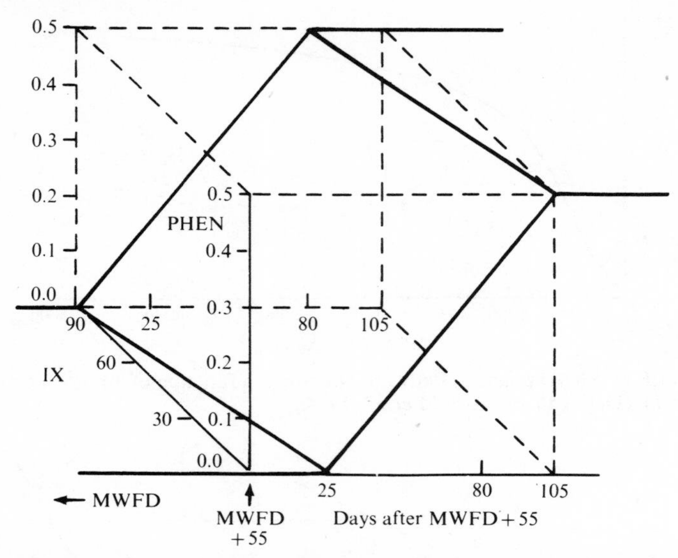


Fig. 13.4. LEYFARM: dependence of herbage death due to maturity (PHEN) on time after flowering (MWFD) and on IX, a function of mean time of seedling emergence (MERGD), where IX = 0 < (MERGD − 138) < 90. (Derived from Arnold *et al.*, 1974.)

herbage. After germination and establishment, growth is initiated by converting germinated seed into live biomass (weight of seedlings = 0.5 × weight of germinated seed). From then on and until the live biomass eventually dies, the pasture grows with a daily growth rate, dies with a daily death rate. The live and dead material is consumed by sheep selectively and the dead material also decays. The functions that describe these processes are derived mainly from field data and partly from literature.

The daily growth rate is determined by calculating the potential growth rate when light, temperature and soil moisture are non-limiting and adjusting this value for these factors. The potential growth is entered as a function of green matter dry weight (GRNMAT) in $kg \cdot ha^{-1}$ (Fig. 13.2). The maximum growth rate approaches 124 $kg \cdot ha^{-1} \cdot d^{-1}$.

The radiation factor is a function of radiation ($cal \cdot cm^{-2} \cdot d^{-1}$) and GRNMAT (Fig. 13.3). The dependence on radiation when the canopy is fully developed (7000 $kg \cdot ha^{-1}$) is linear.

The effect of soil moisture on growth is mediated by R = E/EO, where E is actual transpiration and EO is potential transpiration when the canopy is closed (GRNMAT > 1000 $kg \cdot ha^{-1}$). When the canopy is not closed, account is taken of evaporation from bare soil (Carbon & Galbraith, 1975). R serves as a soil moisture factor on potential growth.

The growth rate decreases after flowering and this decrease is set to begin 55 days after the mean date of flowering but at a different rate depending on the time

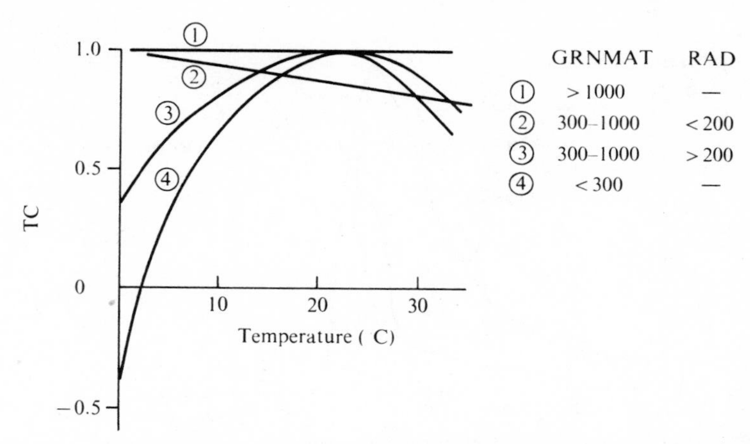


Fig. 13.5. LEYFARM: functions used for temperature factor (TC). (Derived from Arnold *et al.*, 1974.)

of year when mean flowering date is reached. Again, the construction is a three-dimensional surface based on direct interpretation of available data (Fig. 13.4).

The temperature factor is composed of a set of curves, the choice of which is dependent on green matter dry weight and radiation. When GRNMAT is greater than 1000 kg · ha$^{-1}$, temperature ceases to influence growth (Fig. 13.5).

The daily death rate of the vegetation is calculated from an initial death rate based on the amount of live biomass, which is increased by 50% as the plant matures and by a further 25% as the soil dries out. The time at which death rate increases is related to mean flowering date (Fig. 13.4).

The effect of the grazing animal on the growth rate of the pasture is assumed to operate solely through the reduction of GRNMAT caused by forage consumption. This assumption is substantiated by experimental data which indicate that potential growth rate is very closely related to the amount of green pasture whether it is being grazed or not (Arnold, 1975). In fact Greenwood & Arnold (1968) have shown that when green matter dry weight is greater than 100 kg · ha$^{-1}$ the amount of herbage removed by the sheep per day at normal stocking densities is generally much less than 10% of what is available. The effect on an actively growing canopy is thus quantitatively weak. However, at the beginning of the season when there is very little herbage available, and what there is is concentrated in clumps, as much as 30% of what is available can be removed in one day. Arnold (1975) maintains that even at such intensities of defoliation the major quantitative effect on growth is due directly to the reduced biomass. This is probably true for relatively low biomass values and when the vegetation is not sufficiently differentiated to allow for strong selection effects due to animal preference. However, where there is much stemmy vegetation and the animals are selecting

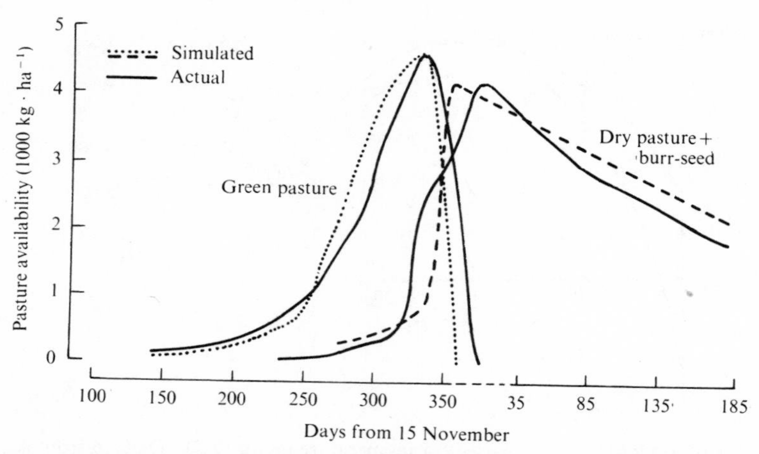


Fig. 13.6. LEYFARM: measured and simulated values for pasture availability when stocking rate is 10 sheep$\cdot$ha$^{-1}$. (Derived from Arnold *et al.*, 1974.)

actively growing leaves, the simple mediation of growth through total plant biomass only could be the source of deviations between model results and observations. An aspect of this problem is treated by the Armidale model (see below). An additional effect of grazing on the pasture is effected by the consumption of seeds. This has a critical effect on the year-to-year dynamics of the pasture and is represented in fair detail in LEYFARM.

The digestibility and nitrogen content of the green and dry herbage on offer are functions of the amounts of green and dry herbage, but values for dry herbage are adjusted for leaching by rain. Intake of organic matter by the sheep is calculated as a potential based on the liveweight of the sheep which is adjusted according to the selectivity of the sheep, herbage availability and digestibility of herbage eaten. The conversion of feed consumed to energy for maintenance, liveweight change and wool growth is based on metabolic energy values.

The model produced satisfactory results (Fig. 13.6), despite the arbitrary functions and awkward conceptualizations in certain parts.

The system can be manipulated by changing stocking rates, grazing systems (continuous, deferred) and initial amounts of clover seed and herbage available. When used to investigate the effect of grazing at the beginning of the growing season on subsequent productivity it showed that in some years deferment was crucial to maintaining production whereas in others the effect was small. This result is confirmed by experience in the field and is determined by the relationship between the growth rate of a unit area of pasture and the intake rate by the herd from the same unit area over the season. A theoretical stability analysis of this type of phenomenon has been presented by Noy-Meir (1974*b*) and illustrates the potential contribution of a simpler, but more rigorous approach to the understanding of growth/grazing dynamics.

### *The Armidale model (Vickery & Hedges, 1972a, b, 1974)*

This is of an improved *P. tuberosa–T. repens* pasture grazed by Merino sheep at different stocking densities. The emphasis in the vegetation section is on the growth, forage value and consumption of the pasture. As it is a perennial pasture, germination and seed formation are neglected. The soil moisture balance is also treated simply: a 75 mm soil moisture capacity is filled by rain and emptied by evapotranspiration. Potential evapotranspiration (0.8 × pan evaporation) is modified by a reduction factor dependent on soil moisture (Fig. 13.7). Root distribution is assumed always to be adequate for moisture extraction – probably a reasonable assumption given the specific conditions, but a possible source of error. The time unit in which rates are expressed is 1 week and the time step used in updating the state variables is also 1 week. This results in considerable saving of computer time but may be rather coarse for periods when rapid change is taking place. It may also cause considerable inaccuracy when unstable equilibrium points (or turning points) are encountered in the system (see Noy-Meir, 1974*b*).

The relative growth rate of the herbage is calculated as a potential rate, dependent on soil temperature at 4 cm depth. This value is adjusted by a series of reduction factors for soil moisture, leaf area, age and a dry soil factor. In the temperature-dependent potential growth curve (Fig. 13.8), there is no growth below 4 °C and a maximum relative growth rate of 0.5 $week^{-1}$. The soil moisture growth reduction factor is nearly linear between 20 mm and 75 mm soil moisture. Below 20 mm growth rate is severely reduced.

The leaf area reduction factor starts to operate when the green herbage exceeds 2500 kg dry wt · $ha^{-1}$ and reduces growth to 0.1 of potential when it exceeds 5000 kg · $ha^{-1}$.

The ageing factor is dependent on stocking density and time of the year and takes values of 0.75 to 1.5. It is entered as a table and is based on tiller weight data of Hutchinson (1969). Thus, the growth can be raised above the temperature-dependent 'potential' rate. The dry soil factor has a value 0.2 and operates for higher stocking densities only during weeks 4 and 5 (end January/beginning February). It represents dormancy in *P. tuberosa* induced by low soil moisture conditions in mid-summer. Total weekly growth is then calculated by multiplying relative growth rate by the live shoot biomass plus a fraction of the root biomass. The fraction is given in a table, dependent on time of the year, and has a range from 0.25 to 0.65. The total growth is limited to a maximum of 1500 kg · $ha^{-1}$ · $week^{-1}$ and is partitioned between roots and tops according to a function dependent on stocking density and time of year. This function is also entered as a table.

The shoot biomass dies at a rate which ranges from 0.001 to 0.075 $week^{-1}$, depending on the time of year. The current value is increased by a factor of 3.5 when soil moisture is less than 20 mm concurrent with a soil temperature

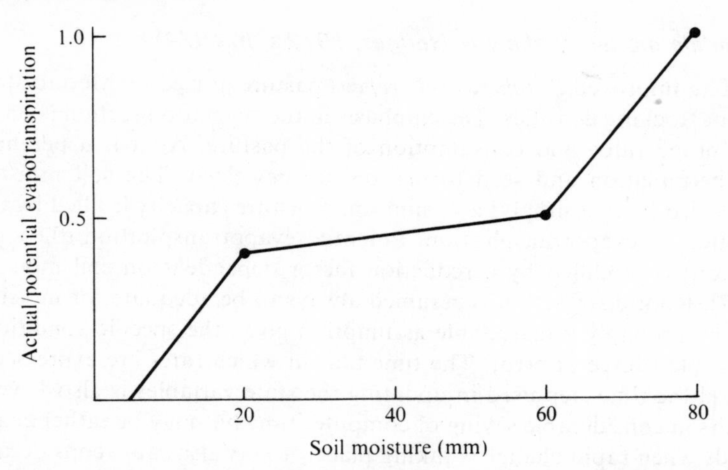


**Fig. 13.7. Armidale: dependence of evapotranspiration on soil moisture. (From Vickery & Hedges, 1972*a*.)**

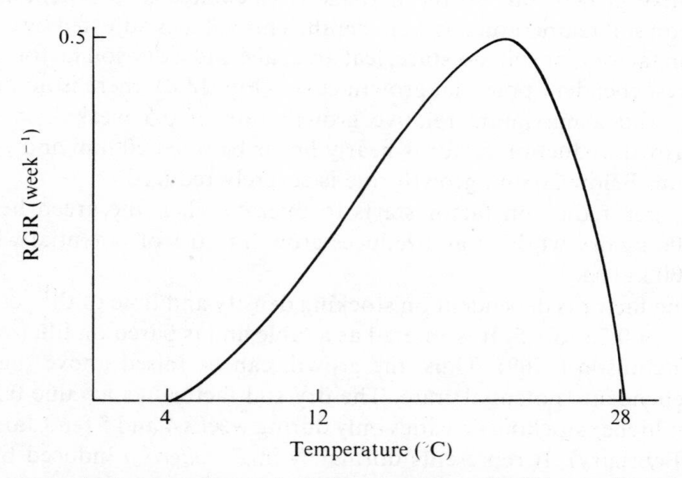


**Fig. 13.8. Armidale: dependence of relative growth rate (RGR) on soil temperature measured at 4 cm depth. (From Vickery & Hedges, 1972*a*.)**

above 19.5 °C. In addition to removal of biomass by grazing there is also consumption by insects. Root biomass dies with a time-dependent mortality factor and some of it is consumed by soil fauna.

The growth and death of the pasture vegetation is thus a highly site-specific process, fairly rigidly determined by empirical time-dependent functions. Here, as in LEYFARM, grazing affects the current pasture growth only by reducing the live and dead biomass by consumption. This has been shown to

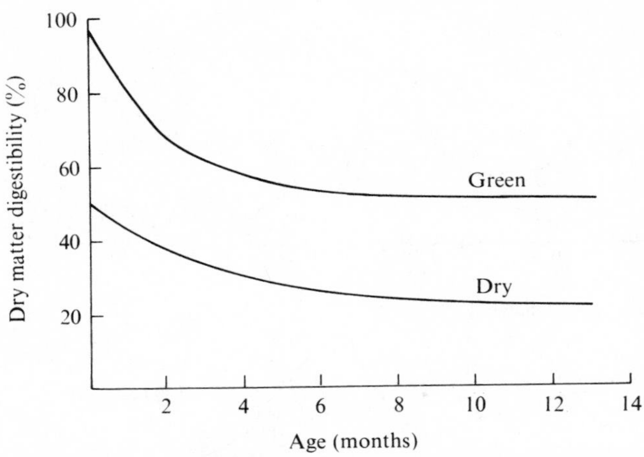


Fig. 13.9. Armidale: relationship between age and digestibility of green and dead herbage. (From Vickery & Hedges, 1972*a*.)

be a reasonable simplification for moderate and low stocking densities. For high stocking densities it may be less reasonable, as indicated by the discrepancies that occur when the model is run under such conditions (Fig. 13.10).

The most distinctive part of the Armidale model is probably the approach to determining the forage value of the pasture and the herbage consumption by the sheep. The emphasis is on the ageing of the live and dead biomass. Both are divided into thirteen 4-week age classes. New growth enters the youngest age class and progresses through the classes, residing in each class for 4 weeks. Live biomass that dies is decremented from the oldest classes and entered into the youngest dead biomass class. This procedure is adopted so as to determine the digestibility of the pasture which is here age-dependent according to the functions in Fig. 13.9. It is then used to determine the forage value of the herbage actually consumed by the sheep, as the method allows for selection of young green herbage in preference to old. This is done by accumulating the amount of herbage in the successive age classes and reading off the proportion consumed from each successive class from a curve that relates the proportion of green herbage in the diet to the total amount of green herbage available. As this is a negative exponential saturation curve the proportion consumed from younger herbage is higher than that from the older herbage classes. In this way younger herbage is selected preferentially. This is admittedly an arbitrary approximation but is used for want of better data on herbage selection.

Despite the fact that herbage is divided into 4-weekly age classes, it ages continuously, i.e. in weekly steps. Thus some herbage is moved from one class to the next every week. The actual amount that remains to be aged is determined by the proportion grazed in the current week. It is assumed that the

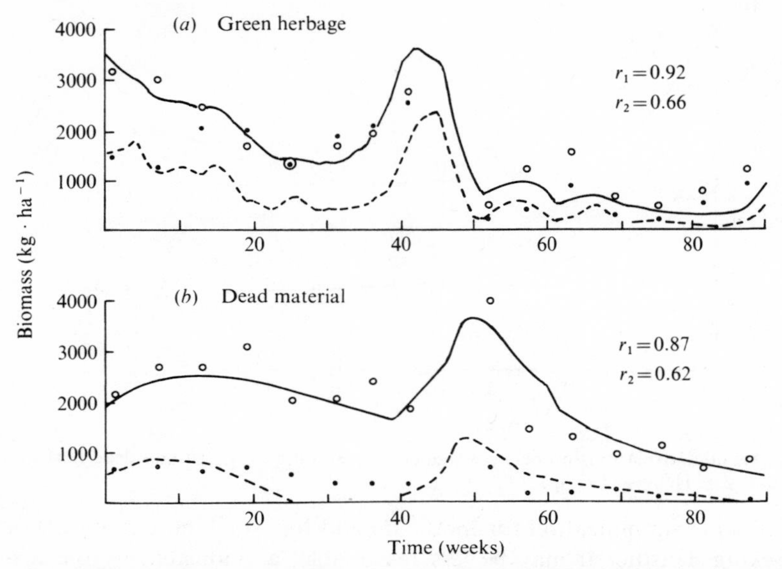


**Fig. 13.10. Armidale: comparisons of predicted values obtained from the sheep production model with actual data of Hutchinson (1969). Actual data noted by points or circles. (*a*) Green herbage (kg·ha$^{-1}$); (*b*) dead material (kg·ha$^{-1}$). Stocking densities 9.9 (——) and 29.7 (- - - - - - - -) wethers·ha$^{-1}$. $r_1$ and $r_2$ are correlation coefficients for the high and low stocking densities respectively.**

same proportion was grazed in the previous 3 weeks. Thus grazing affects the average 'physiological age' of the pasture.

The livestock section can be divided into two parts: one determines the forage intake and its feed value and the other converts the intake into weight change and wool growth. The intake section is different from LEYFARM in that selection between young and old herbage is included for both green and dry pasture. The total amount eaten is dependent on the amount of pasture on offer and its forage value is then calculated according to the age composition of the green and dry herbage.

Conversion of intake into animal weight change and wool growth is fairly straightforward and similar procedures and functions are used in other models too. Wool growth is hardly affected by nutrition and depends mainly on the weight of the animal.

The output of the model was checked against field observations and some of the results for herbage growth (Vickery & Hedges, 1972*b*) are given in Fig. 13.10 for two stocking densities, 9.9 and 29.7 wethers·ha$^{-1}$. The model results for the lower stocking density are much better than those for the higher stocking density. This would indicate that the interaction between the grazing animal and the growing pasture is not represented well enough to account for the actual growth of the pasture under conditions where this interaction is important, as it is at high stocking densities.

The model has been used to study the behaviour of the grazing system and to identify areas of greatest sensitivity.

### *The Hurley ewe–lamb model (Edelsten & Newton, 1975)*

This is designed *a priori* as a management model with particular emphasis on the effect of nutrition (supplementary feeding), grazing rotation and stocking rates on the production of lambs and silage. Management algorithms are defined and the conversion of feed consumed to milk, lamb and ewe live-weight change is treated in some detail. On the other hand, the growth of the pasture is entered as a table dependent only on whether it is being grazed or not (Fig. 13.11). When not being grazed, growth is not even season-specific. It is thus assumed that there is no feedback of grazing on herbage growth rates and that growth undisturbed by grazing is independent of different climatic conditions from year to year. Whatever effect grazing had on growth, it was regarded as constant and independent of the management or stocking densities defined in the model. Such an assumption obviously limits the scope of the model to cases where this condition holds. This would be the case where the pasture is a closed canopy with a growth rate independent of the amount of biomass in the canopy – a situation that would indicate that respiration is dependent on growth and not on total biomass. It is also implied that grazing is never heavy enough to cause exposure of bare ground. Animal feed intake and growth is not identical for all animals of a class. Stochastic variation is allowed for, even in milk intake between twin lambs.

Some results of this model are given in Table 13.1. The growth rates of the lambs are close to those observed, especially at the higher stocking densities. This may not be surprising as the management of the flocks is set to attain acceptable growth rates. Management included concentrate feeding which is influenced by the amount of feed grazed off the pasture. The simulated amounts of silage harvested were substantially different from those observed. The discrepancy could arise from an oversimplified representation of plant growth which ignores feedback from grazing. The authors also feel that their growth input data may have been faulty.

The use of a low-resolution herbage growth component is justified when the main interest is in another part of the system. The validity of such use would depend on the sensitivity of the growth component to feedback from other parts of the system. If the source of the deviation between the model results and observation is because this component is sensitive to feedback, especially when heavier stocking density reduces biomass to the extent that the canopy is not continuously closed any more, it may be possible to remove the discrepancy by increasing resolution slightly to account for it.

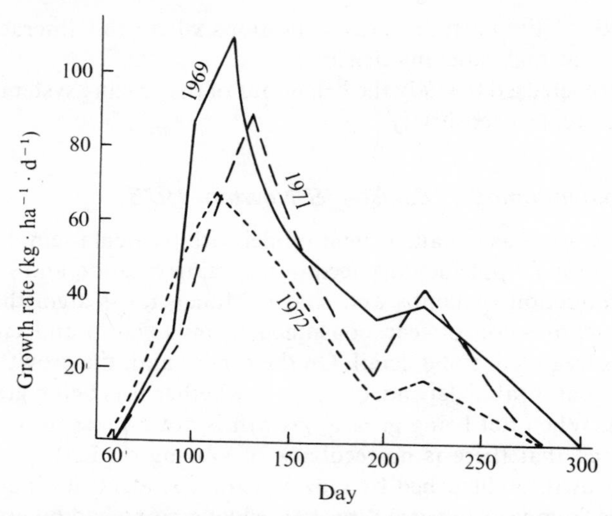


**Fig. 13.11. Growth rate functions of grazed pasture used in the Hurley ewe–lamb model. (After Edelsten *et al.*, 1973.)**

Table 13.1. *Some results of the Hurley ewe–lamb model*

| Item | Experimental results | | | Model results | | |
|---|---|---|---|---|---|---|
| Stocking density (ewes·ha$^{-1}$) | 14 | 17 | 20 | 14 | 17 | 20 |
| Average growth rate of lambs (g·d$^{-1}$) | 220 | 192 | 198 | 193 | 208 | 200 |
| Silage produced (tonnes DM·ha$^{-1}$) | 2.3 | 2.4 | 1.0 | 4.9 | 3.8 | 3.3 |
| Concentrates fed (kg·ewe$^{-1}$) | 16 | 20 | 20 | 19 | 19 | 19 |

After validation, the model has been used to evaluate the effect of a set of different management strategies on the profitability of a ewe–lamb production system.

The three models discussed so far have much in common in their animal consumption and energetics routines and indeed these have greatest precision and generality. The plant production routines vary greatly but none have generality. Modelling plant growth in a grazing system is certainly a much more difficult problem.

## *Simple, generalized models (Noy-Meir, 1974a, b, 1975)*

The complexity of 'realistic' models and their specific nature often make their conclusions limited in application. Some generalization is possible but it is usually a secondary consideration. A different approach has been used by

Noy-Meir (1974*a*, *b*, 1975). He defined a deliberately simplified model as the basis for an analytical approach to the problem of pasture stability under different stocking densities and different grazing methods. This model consists essentially of a vegetative growth rate function dependent only on 'quantity' of vegetation in the pasture and a family of consumption functions dependent on stocking density and on the quantity of pasture on offer. He shows that many properties of such a system under continuous grazing can be determined by simple graphic means (Noy-Meir, 1974*a*, *b*). However, to analyse the system for rotational grazing it is necessary to define an explicit mathematical model (Noy-Meir, 1975). The model simulates net change in vegetative biomass, $V$:

$$\frac{\mathrm{d}V}{\mathrm{d}t} = G - C,$$

where $G$ is the growth rate of the pasture and $C$ the rate of consumption by the grazing animal.

$$G = gC\left(1 - \frac{V}{V_{\mathrm{m}}}\right),$$

a logistic growth curve, and

$$C = c_{\mathrm{m}}H\frac{V - V_{\mathrm{r}}}{(V - V_{\mathrm{r}}) + (V_{\mathrm{k}} - V_{\mathrm{r}})} \quad (\text{if} \quad V < V_{\mathrm{r}}, \quad C = 0),$$

a Michaelis saturation function. Here $g$ is maximum relative growth rate; $V_{\mathrm{m}}$ is maximum plant biomass; $c_{\mathrm{m}}$ is maximum consumption rate per animal; $H$ is stocking density; $V_{\mathrm{r}}$ is 'residual' ungrazable plant biomass; and $V_{\mathrm{k}}$ is plant biomass at which consumption is half that at satiation (Michaelis constant). $V_{\mathrm{k}}$ serves as a measure of 'grazing efficiency'. The term $V_{\mathrm{k}} - V_{\mathrm{r}}$ should, in fact, be $V_{\mathrm{k}}$ only. The 'Michaelis constant' as defined here would then be $V_{\mathrm{k}} + V_{\mathrm{r}}$. The Michaelis function may not be the most appropriate function for describing consumption by grazing animals, but it is a fair approximation for studies on the qualitative behaviour of plant growth/animal consumption systems.

When $\mathrm{d}V/\mathrm{d}t = 0$ the system is in equilibrium and stability conditions can then be defined.

An example of the graphic stability analysis of continuous grazing is given in Fig. 13.12. Five different stability situations are defined: (*a*) undergrazed steady state; (*b*) overgrazing to extinction; (*c*) overgrazing to a low biomass steady state; (*d*) steady state and turning point to extinction; (*e*) two steady states.

Noy-Meir concludes that 'this approach has yielded a series of general conclusions about stability and productivity of the system. These appear to be relevant at least to some classes of real-world pastoral systems and to some problems in their practical management' (Noy-Meir, 1974*b*). It can be added

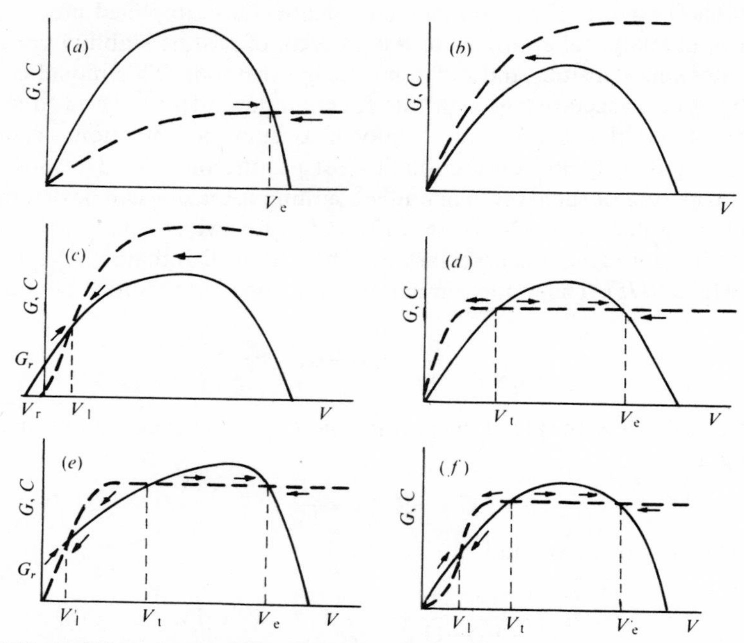


Fig. 13.12. Stability of grazing systems: application of predator–prey graphs. Possible stability combinations of $G$ and $C$ curves at given $H$. (*a*) Undergrazed, stable steady state ($V_e$). (*b*) Overgrazed to extinction. (*c*) Overgrazed to a low biomass steady state $V_l$. $V_r$, reserve (ungrazable) biomass; $G_r$, residual growth potential. (*d*) Steady state and unstable turning point ($V_t$) to extinction. (*e*) Two steady states ($V_e$, $V_l$) separated by a turnin gpoint. (*f*) As for (*e*), but caused by a sigmoid $C$-curve, not by ungrazable plant reserve. (From Noy-Meir, 1974*b*.)

that this approach allows for the stepwise development of a comprehensive theory of grassland dynamics. It also links grazing problems to existing theory of predator–prey systems thus increasing, hopefully, the fruitful contact between grassland management and ecological theory.

## Resolution of model components

On the whole, resolution is highest in those components of the system that are of greatest interest to the modeller. Generally, they reflect his belief that these are the important components that have the greatest effect on the performance, accuracy and predictive power of the model. Thus, in the Armidale model the ageing of the pasture is treated in considerable detail whereas the soil moisture balance is treated very simply. In the Hurley model the growth of the pasture is a table but the livestock management component is given much closer attention. In LEYFARM, seed dynamics and the soil water balance are treated in much greater detail than the growth/radiation relationship; nutrient supply and nutrient regime in the plant have not been treated at all. In Noy-Meir's model all exogenous variables have been ignored and even the two processes included are of low resolution. But these are clearly defined conceptually and as a result the study of their interactions seems to have practical relevance of a general nature.

Thus, by deciding on different levels of resolution for different parts of the model, it is possible to direct the major effort of the study to those aspects of of the system that are of central interest. However, there are limitations associated with the different levels of resolution. Low-resolution elements mainly include parameters, tables or empirical functions ('black boxes'). The constants of these elements are not widely used physical or chemical constants; they are generally site-specific and are often derived from a small set of data and sometimes simply approximated (or guessed) from unmeasured observation. These constants are sometimes used to adjust, initialize or 'tune' a model to perform reasonably. They are insensitive to feedback from other components of the model. If the feedback on these elements in the real system is negligible or irrelevant in the context of the study, then it could reasonably be neglected in the model. Thus, in the Hurley model the growth rate of the pasture is a table dependent only on time and on whether or not the pasture has been grazed. It is independent of grazing pressure or of the amount of standing green canopy. As has been pointed out previously, this simplification could be the source of some serious discrepancies between simulated and observed results. Nevertheless, low-resolution elements are essential in cutting down time needed for model development and in simplifying the structure of a model. They should be used to define peripheral components that are relatively independent of feedback from other elements in the context of the system being studied.

High-resolution elements are open to feedback from other elements in the model but the accuracy of the model is not automatically increased as the resolution increases. This is because every element should be open to both positive and negative feedback. If this is not fully covered by the model, the element will be 'out of balance' and the output may be less accurate than that of a low-resolution model. An example is given by Whitfield (1972*a*, *b*). He presents two models that describe the growth of a tundra shrub, *Dryas integrifolia*. One model, of low resolution, is called 'an empirical model'. Here the growth function of the shrub is given as a series of four tables, one for each of four phenological phases. Growth is a function of radiation and the amount of standing biomass, both of which are entered as driving variables. The total annual growth is predicted within an accuracy of 3%. The second model, called a 'process model', is a much more detailed canopy photosynthesis model, including simulation of the microclimate in and around the shrub. The prediction of growth by the model is fair for the first stages of growth, but deviates widely for the latter part of the growth period. Clearly some processes have been omitted or have been incorrectly represented.

Precision and high resolution do not necessarily go together. Neither do precision and low resolution necessarily go together. Accuracy would seem to depend more on the conceptual validity (or structural balance) of the model than on its resolution.

## Optimization of grazing management decisions

Although most grassland models have been used mainly to simulate the behaviour of systems as measured in conventional grazing experiments, increasingly they are being used as a means for finding optimal management combinations of variables that can be manipulated. Here, as with the purely biological models, solutions can be sought for specific situations or for more general cases. The two approaches cannot be completely separated but certainly the difference in emphasis can be distinguished. An example of an optimization model that is rather specific is that of Smith & Williams (1973, 1974). An example of a more general approach is that of Morley (1974).

In the Smith & Williams model the effect of deferring grazing of a Mediterranean annual pasture at the beginning of the growing season is evaluated by an objective function that calculates the gross margin. The effects on the gross margin of length of deferment and initial plant density were examined and plotted out. The results 'indicated that optimal combinations of stocking rate and length of deferment could vary widely at one site due to variation in initial plant density and economic weights given to the supplementary feed input and animal output' (Smith & Williams, 1974). Thus the results stop short of generalization. Instead, the authors feel that 'experimentation will be needed to test the postulates made by the model and to incorporate more fully factors such as climate and soil fertility'. It is probable that such a conclusion springs from the conceptual basis of the model, which assumes that, with increasing knowledge of the system gained through experimentation, a model can evolve from a formal approximation of system behaviour towards the ultimate 'isomorphic and unique (unattainable) model' which describes exactly how the system behaves (Smith, 1975).

The 'isomorphic model' will presumably be as complete as the system itself and, granting that evolution has produced fairly efficient systems, it is doubtful whether the computer copy will have any advantage over the real thing. Reproducing rapid physical, chemical and even physiological reactions on the computer sometimes takes considerably longer than the actual reactions themselves! The aim of approaching isomorphic models of agropastoral models seems to lead into a labyrinth fraught with dead-ends. In the case in point, an alternative could be to analyse the dynamics of prehension of a young sward in terms of the relationship between growth rates of the vegetation and defoliation rates by the animals. An analysis like that proposed by Noy-Meir (1974*b*) would indicate that the system has two levels of stability, a low one which is maintained when the rate of prehension exceeds the growth rate from the start of the relevant grazing period and a high one which is approached when growth exceeds prehension from the beginning. An intermediate state of unstable equilibrium is indicated too. Such a conclusion could direct the subsequent research and modelling effort to a study of growth rates at the beginning of the season and interactions with grazing, rather than to an attempt to incorporate more fully factors such as climate and soil fertility, as proposed.

This study does show that the optimal management combinations are highly sensitive to the economic weights attached to the inputs and outputs. Thus, the higher the value of the animal product compared to the supplementary feed, the longer is the optimal deferment and the higher the optimal stocking rate. With prices as unstable as they are, such specific models could be helpful in analysing real management options.

The models used by Morley (1974) in his study are apparently of moderate complexity, similar to that of the Smith & Williams (1973) model. The objective is to determine the effect of a 20% increase in pasture growth rate, due to fertilization, on the animal response and economic benefit. Four grazing systems are studied: (*a*) sheep meat and wool; (*b*) flock ewe replacements; (*c*) beef production; and (*d*) heifer replacements. The results of this study lead to a rather generalized conclusion: 'A 20 percent increase in pasture growth rate increased all gross margins... provided the stocking rates were approximately those giving maximum gross margins from the unfertilized pasture. Most of the benefits could have been obtained without increasing stocking rate. At lower stocking rates the gross margins (due to fertilization) were less and even negative. Therefore if stocking rate is below the "optimum", it should be increased before fertilizer is applied' (Morley, 1974).

This study has arrived at a fairly definite conclusion even though the models are only a rough approximation of reality. Yet the conclusions appear reasonable. In this case there appears to be little point in further developing the model, unless new situations are encountered (Dillon, 1971; Morley & Graham, 1971).

## Applications to the individual farm

Simulation models can be used to develop guidelines for management. They may even be used for continuous control and management of an individual enterprise. However, that would be the exception, as data for initialization of the pasture, soil and animal variables in even moderately complex simulation models of agropastoral systems are generally unavailable. At present, most farmers would have to apply a generalized guideline to their own specific situation. Aspects of the system may be included in a computerized advisory service that provides a recommendation for each field. An example is the

'DECIDE' superphosphate fertilizer advisory program developed by the Western Australian Department of Agriculture and CSIRO (Bennett & Ozanne, 1973). It is based on a series of equations relating response to soil and plant characteristics. But such a service for the pastoral enterprise as a whole seems to be a long way off. It may be questionable whether the problem can be 'translated into precise and unequivocal numerical procedures' (Christian *et al.*, 1974). The experience, local knowledge and intuition of the farm manager are probably part of the vital network of negative feedback pathways that act to dampen tendencies of the system to deviate widely from a desired objective. Thus, *a priori* 'optimal' solutions are probably suboptimal if they ignore the active function of the intelligent manager reacting sympathetically to the running of the system. Anything that increases his understanding of the dynamics of the system and widens his basis of theory will probably be of direct practical use. Techniques for optimizing management decisions under conditions of operational uncertainty generally require the probability distribution of various benefits (Dillon, 1971). Simulation with pasture models, relevant to the managed system, has been used to provide such input (Wright, 1970). Using simulation models of pastoral systems as a management game is also gaining popularity as a means of developing managerial skills.

Because of the limitations of dynamic simulation models, management analyses of farm situations are usually done with static models like linear programming, with input from simulation models (Swartzman & Van Dyne, 1972: Arnold & Bennett, 1975) or from the system itself. An example of the latter is a project presently being conducted in Colorado where 40 individual ranches are being programmed for optimal production. A standard linear programming package is being used. 'The project investigators, in cooperation with the Extension Service and the Soil Conservation Service, are going into the kitchens of the ranchers, sitting down with coffee and remote terminal, and inputting specific ranch characteristics, looking at results, making changes, drinking more coffee...In this way the local knowledge and intuition of the manager are put into the analysis.' (G. M. Van Dyne, personal communication.)

At the other end of the specific/general continuum, simulation models and analysis should contribute towards the development of a comprehensive theory of grassland dynamics. It is difficult to say whether progress in this direction will be faster through generalized models or through the more common quasi-realistic site-specific models. Some convergence between the two approaches will probably occur. When the behaviour of many systems has been clarified, some general principles may begin to emerge. Arnold & Bennett (1975) have used the LEYFARM model to study the dynamics and economics of one system. Noy-Meir (1974*a*, *b*, 1975), Smith (1975), Morley (1974), Christian *et al.* (1974) and others have used models to study the dynamics of other systems. Maybe this is a hesitant beginning. A developing theory will help structure more realistically lower-resolution models in which the objective is to produce generalizable management strategies and tactics or to provide realistic inputs for economic models. If the present and future grassland models cannot, in the long term, aid better management decisions by those using the world's grasslands then the resources used to produce them should be diverted into approaches that can. It is a crucial challenge to the modellers that will probably be decided in the next decade.

## Summary

Intensive grazing systems are generally managed to maximize long-term profit. This general objective needs to be translated into short-term biological objectives like higher lambing percentages, earlier pasture readiness or lower concentrations of feed inputs. Overall long-term strategy is commonly expressed as general guidelines based mainly on economic considerations and on the production potential of the system. For the day-to-day operation of the system, managers must also decide on short-term tactics based on their knowledge of the biological behaviour of the system and on its short- to medium-term responses to manipulation. These tactics determine the managers' response to expected but unpredictable events like market changes, climatic vagaries or disease and pest depredation. Their application is, of course, also dependent on economic constraints. Dynamic, continuous simulation has been used mainly to study the behaviour of intensive grazing systems under various operating (management) conditions. These studies have increased understanding of the response patterns of the system and have facilitated the identification of sensitive system components. They have also made it possible to extrapolate from available data and provide fairly comprehensive input for economic optimization routines. In some cases the simulation programme itself has been used to find optimum economic management strategies (Morley, 1974; Edelsten & Newman, 1975), but the more flexible approach seems to be to use separate programs for simulation and economic optimization. Swartzman & Van Dyne (1972) have described a combination of simulation and economic optimization with linear programming, the simulation running continuously but the optimization being conducted twice a year. Arnold & Bennett (1975) have separated the simulation and optimization, but have used a non-linear optimizing routine that made possible the use of functions that preserve some of the dynamic aspects of the system. Economic optimization of agropastoral systems can, however, also be done without simulation at all, the input functions coming directly from the system or from previous managerial experience.

The role of simulation in the management of intensive grazing systems has not yet been clearly established, but most of the studies have increased the modellers' understanding of the system being studied. Transfer of this understanding to better management is possible in principle and some methods presented here have been shown to give useful results. But these are only beginnings. The increase in simulation studies in recent years indicates that the association between simulation and management of intensive grazing systems will become closer in the future. With a little luck this development may yet fulfil the promise it seems to hold. It could well be that only after many simulation studies have been conducted will there be sufficient experience and confidence to develop effective and efficient bridges to wide-scale practical application.

In the long run, the indirect role of simulation models could well have the greatest impact on intensive grazing management by influencing research and guiding experimental effort more clearly to the most sensitive components of the systems being studied.

We would like to thank Dr P. Vickery for permission to use diagrams from his publications.

## References

Arcus, P. L. (1963). An introduction to the use of simulation in the study of grazing management problems. *Proceedings of the New Zealand Society of Animal Production*, **23**, 159–68.

Arnold, G. W. (1975). Effects of herbivores on arid and semi-arid rangelands: defoliation and growth of forage plants. In *Proceedings of the II US/Australian Rangelands Panel, Adelaide, 1972*. Society of Range Management, Denver.

Arnold, G. W. & Bennett, D. (1975). The problem of finding an optimum solution. In *Study of agricultural systems*, ed. G. E. Dalton, pp. 129–73. Applied Science Publishers, Amsterdam.

Arnold, G. W. & Campbell, N. A. (1972). A model of a ley-farming system, with particular reference to a sub-model for animal production. *Proceedings of the Australian Society of Animal Production*, **9**, 23–30.

Arnold, G. W., Carbon, B. A., Galbraith, K. A. & Biddiscombe, E. F. (1974). Use of a simulation model to assess the effects of a grazing management on pasture and annual production. In *Proceedings of the XII International Grasslands Congress, Moscow*, pp. 47–52.

Bennett, D. & Ozanne, P. G. (1973). *Deciding how much superphosphate to use. Annual Report of the Division of Plant Industry*, 1972. CSIRO, Melbourne.

Brockington, N. R. (1971). Using models in agricultural research. *Span*, **14**, 1–4.

Carbon, B. A. & Galbraith, K. A. (1975). Simulation of the water balance for plants growing on coarse textured soils. *Australian Journal of Soil Research*, **13**, 21–31.

Christian, K. R., Armstrong, J. D., Davidson, J. L., Donnelly, J. R. & Freer, M. (1974). A model for decision making in grazing management. In *Proceedings of the XII International Grasslands Congress, Moscow*, pp. 126–31.

Dillon, J. L. (1971). Interpreting systems simulation output for managerial decision making. In *Systems analysis in agricultural management*, ed. J. B. Dent & J. R. Anderson, pp. 188–211. Wiley & Sons (Australasia), Sydney.

Edelsten, P. R. & Newton, J. E. (1975). *A simulation model of intensive lamb production from grass. Technical Report 17.* Grassland Research Institute, Hurley, England.

Edelsten, P. R., Newton, J. E. & Treacher, T. T. (1973). *A model of ewes with lambs grazing on pasture. Internal Report 260.* Grassland Research Institute, Hurley, England.

Goodall, D. W. (1967). Computer simulation of changes in vegetation subject to grazing. *Journal of the Indian Botanical Society*, **46**, 356–62.

Greenwood, E. A. N. & Arnold, G. W. (1968). The quantity and frequency of removal of herbage from an emerging annual grass sward by sheep in a set stocked system of grazing. *Journal of the British Grassland Society*, **23**, 144–8.

Hutchinson, K. J. (1969). Fodder conservation and energy flow in grazing systems. PhD Thesis, University of New England, Australia. (Quoted by Vickery & Hedges, 1972*b*.)

Jones, J. G. W. & Brockington, N. R. (1971). Intensive grazing systems. In *Systems analysis in agricultural management*, ed. J. B. Dent & F. R. Anderson, pp. 85–120. Wiley & Sons (Australasia), Sydney.

Keulen, H. van (1976). Evaluation of models. In *Critical evaluation of systems analysis in ecosystems research and management*, ed. C. T. de Wit & G. W. Arnold, pp. 22–9. PUDOC, Wageningen, The Netherlands.

Maynard Smith, J. (1974). *Models in ecology*. Cambridge University Press, London.

Morley, F. H. W. (1974). Evaluation by animal production of increases in pasture growth; using computer simulation. In *Proceedings of the XII International Grasslands Congress, Moscow*, pp. 416–20.

Morley, F. H. W. & Graham, G. Y. (1971). Fodder conservation for drought. In *Systems analysis in agricultural management*, ed. J. B. Dent & J. R. Anderson, pp. 212–36. Wiley & Sons (Australasia), Sydney.

Noy-Meir, I. (1975*a*). The use of models in the study of grazing ecosystems. In *Ecological research on development of arid zones with winter precipitation*, pp. 65–78. *Agricultural Research Organisation Special Publication 39*. Bet Dagan, Israel.

Noy-Meir, I. (1975*b*). Stability of grazing systems: an application of predator–prey graphs. *Journal of Ecology*, **63**, 459–81.

Noy-Meir, I. (1976). Rotational grazing in a continuously growing pasture: a simple model. *Agricultural Systems*, **1**, 87–112.

Seligman, N. G. (1976). A critical appraisal of some grassland models. In *Critical evaluation of systems analysis in ecosystems research and management*, ed. C. T. de Wit & G. W. Arnold, pp. 60–97. PUDOC, Wageningen, The Netherlands.

Seligman, N. G., Tadmor, N. H., Noy-Meir, I. & Dovrat, A. (1971). An exercise in simulation of a semi-arid mediterranean grassland. In *Bulletin des Recherches Agronomiques de Gembloux. Semaine d'étude des problèmes méditerranéens*, pp. 138–43. Faculté des Sciences Agronomiques de l'Etat, Gembloux, Belgium.

Smith, R. C. G. (1975). Computer simulation and the evaluation of grazing systems with an example of the deferred grazing of a Mediterranean annual type pasture. In *Proceedings of the III World Conference on Animal Production*, ed. R. L. Reid, pp. 196–202. Sydney University Press.

Smith, R. C. G. & Williams, W. A. (1973). Model development for a deferred grazing system. *Journal of Range Management*, **26**, 454–60.

Smith, R. C. G. & Williams, W. A. (1974). Deferred grazing of Mediterranean annual pasture: a study by computer simulation. In *Proceedings of the XII International Grasslands Congress, Moscow*, pp. 646–55.
Spedding, C. R. W. (1975). *The biology of agricultural systems*. Academic Press, New York & London.
Swartzman, G. L. (ed.) (1972). *Optimization techniques in ecosystem and land-use planning. US/IBP Grassland Biome, Technical Report 143*. Colorado State University, Fort Collins.
Swartzman, G. L. & Van Dyne, G. M. (1972). An ecologically based simulation–optimization approach to natural resource planning. *Annual Review of Ecology and Systematics*, **3**, 347–98.
Thornley, J. H. M. (1976). *Mathematical models in plant physiology*. Academic Press, New York & London.
Van Dyne, G. M. & Abramsky, Z. (1975). Agricultural systems models and modelling: an overview. In *Study of agricultural systems*, ed. G. E. Dalton, pp. 23–106. Applied Science Publishers, Amsterdam.
Vickery, P. J. & Hedges, D. A. (1972*a*). *Mathematical relationships and computer routines for a productivity model of improved pasture grazed by Merino sheep. Technical Paper 4, Animal Research Laboratories*. CSIRO, Melbourne.
Vickery, P. J. & Hedges, D. A. (1972*b*). A productivity model of improved pasture grazed by Merino Sheep. *Proceedings of the Australian Society for Animal Production*, **9**, 16–22.
Vickery, P. J. & Hedges, D. A. (1974). Simulation in animal–pasture ecosystem research. *Simulation Today*, **23**, 89–92.
Whitfield, D. W. A. (1972*a*). Empirical model of *Dryas integrifolia* net $CO_2$ assimilation. In *Measurement and modelling of photosynthesis in relation to productivity. Proceedings of the CCIBP/PP-Ps Workshop*, pp. 53–9. University of Guelph.
Whitfield, D. W. A. (1972*b*). Process model of *Dryas integrifolia*. In *Measurement and modelling of photosynthesis in relation to productivity. Proceedings of the CCIBP/PP-Ps workshop*, pp. 59–70. University of Guelph.
Wit, C. T. de (1970). Dynamic concepts in biology. In *Prediction and measurement of photosynthetic activity, Proceedings of the IBP/PP Technical meeting, Trebon, Sept. 1969*, pp. 17–23. PUDOC, Wageningen, The Netherlands.
Wit, C. T. de & Goudriaan, J. (1974). *Simulation of ecological processes. Simulation Monographs 3*. PUDOC, Wageningen, The Netherlands.
Wright, A. (1970). *Systems research and grazing systems: management-oriented simulation. Farm Management Bulletin 4*. University of New England, Australia.

# 14. Reflections and projections

G. M. VAN DYNE

## Contents



## Introduction

This volume is one of two major books beginning the description, analysis, synthesis, and integration with the scientific literature of results from the IBP grassland studies. The two international volumes are supplemented with several national volumes already published, now in press, or in preparation.

It will be some time before all of these volumes are available for detailed reading and interpretation.

The purpose of this chapter is not to integrate the scientific information so much as to reflect on and project about the organization, conduct and value of large-scale integrated ecosystem studies. Special emphasis will be placed on grassland studies and in particular my own experience in the United States IBP Grassland Biome Study.

### *Objectives*

Specific objectives of this chapter are (i) to review evolution in the conceptualization of ecological studies, (ii) to review general considerations of modelling and modelling needs, and (iii) to discuss implications for future programmes.

My bias is clear, for I have been involved in grassland studies as part of the IBP for more than a decade. However, as noted previously (Van Dyne, 1972), in a review of the research programme the objective is not to whitewash our failures, nor to chastise our critics, nor to platitudinize to our supporters – instead, it is to record information on the organization and conduct of such ecological research and to provide some constructive criticism so that any future programmes might benefit.

### *Definitions*

Terminology changes with time and varies between individuals and countries. To make the following discussion more coherent, my usage of terminology is defined:

*Driving variables* are those variables which affect the system but are not affected by the system (e.g. precipitation).

A *system state variable* is a variable which changes over time, is a component of the system, and is usually expressed in measures of mass or energy concentration units (e.g. aboveground live plant biomass).

A *rate process* is a phenomenon which accounts for the transfer of matter or energy from one part of the system to another by physiological, physical or sociological mechanisms (e.g. infiltration of water is a physical process, and ingestion of food or photosynthesis are physiological processes).

*Parameters* are coefficients or terms in mathematical expressions of rate processes.

*Dynamic models* have time as at least one independent variable; examples include simulation models.

In most *static models* time is not an independent variable; examples include most least squares statistical models.

A *field validation study* provides results for evaluating ecological system simulation models. Such a study includes simultaneous collections of periodic or continuous measurements of driving variables, systems state variables, and sometimes rate processes.

A *model* is a representation of a thing. Usually the model is a miniature version of the item being modelled, but it is not an exact representation, only a simplification.

A *synthesis* is a combination of separate elements or parts into a whole.

A *system*, for present purposes, may be defined as any organization with a function.

An *ecosystem* is a result of the integration of living and non-living factors of the environment. In this sense the term ecosystem implies a concept and not a specific unit of landscape or seascape.

*Systems ecology* can be considered to be the growing science of studying the development, dynamics and disruption of ecosystems. It utilizes a large number of mathematical and analytical tools from systems analysis and operations research.

*Systems analysis*, for present purposes, may be considered the orderly and logical organization of data and information into models.

A *systems approach* requires (i) combining, condensing, and synthesizing a great amount of information concerning components of the system; (ii) examining in detail the structure of the system; (iii) translating this knowledge of system components, functions and structures into models of the system; and (iv) using the models to derive new insights into the operation, management and utilization of the system.

*Grasslands*, as interpreted here, includes planted pastures as well as native rangelands, and the latter include a wide variety of conditions from semi-arid to subhumid.

*Grazinglands* are grasslands, deserts, tundras and savannas which are used by man for domestic and wild large herbivores (see Chapter 4).

## Evolution of conceptualization of ecological studies

### *Component-oriented studies*

In many instances it has been possible to compare grassland phenomena between nations by examining only one or two components or segments of the ecosystem at a time. For example, data have been assembled comparing standing crops of live vegetation across geographic regions. Generally the focus in a component-oriented study is examination of state variable values, such as the peak standing crop of the vegetation above ground. Less emphasis is given to the rate processes. In many of the component-oriented studies there is good replication in time and in space, but seldom is there information about treatments or manipulations of the entire ecosystem as in the concept of 'stress ecology' (Barrett, Van Dyne & Odum, 1976).

*System utilization*

Examples of component-oriented comparisons are provided by Van Dyne, Smith, Czaplewski & Woodmansee (1978), based on data examined in an IBP Grassland and Tundra Working Groups workshop in August 1972. Data from different sites were organized in a standard manner, and analyses made by statistical methods to determine differences in interrelationships of variables. Generally data were available on vegetation and on climate. For example, vegetation data were summarized for aboveground live, aboveground standing dead, aboveground litter and belowground total live plus dead. Values were given for the annual minimum, maximum and mean of the total plant community. Where possible, the flow of this vegetation into consumers of various types was summarized. Graphs were provided to show the intraseasonal dynamics of each of these components or compartments over the year.

Several generalities can be shown from analysis of these data:

(i) There is a weak (in the sense of the degree of correlation) exponential decreasing relationship between maximum live biomass of vegetation and degrees of latitude from the equator.

(ii) There is a weak concave curvilinear relationship with a maximum of about 2000 $g \cdot m^{-2}$ of live vegetation biomass with respect to precipitation over a range of near 100 up to 2400 mm mean annual precipitation.

(iii) There is a moderately strong exponential increasing relationship between maximum live biomass and mean annual temperature over the range of 0 to more than 24 °C.

(iv) There is a weak exponential increasing relationship between the ratio of belowground (total live plus dead) to aboveground (live only) herbage and latitude from the equator. These ratios vary from about 5 at locations less than 10° latitude, up to 15 to 30 for locations 60° from the equator.

(v) There is a relatively strong exponentially decreasing relationship between the ratio of belowground (total live plus dead) to aboveground (live only) herbage biomass and mean annual precipitation. Ratio values of 20 to 30 are found at locations with less than 300 mm precipitation in contrast to ratios near 2 for locations of 1200 mm precipitation or more.

(vi) There is a moderately strong exponentially decreasing relationship between the ratio of belowground (live plus dead) to aboveground (live only) herbage and mean annual temperature.

(vii) There is a weak exponentially decreasing relationship between aboveground net primary productivity and latitude. From 15° to 30° latitude the productivity is in the range of 1 to 3 $kg \cdot m^{-2}$. From 45° to 60° latitude the productivity is in the range of 0.1 to 1.0 $kg \cdot m^{-2}$.

These analyses show only the broad geographic relationships between major components of grassland ecosystems. The correlations generally are not very high because of differences in methodology in sampling the vegetation and year-to-year effects. Detailed study must be made of descriptions of sampling methodology, intensity of sampling, data presentation and variables to be collected. Without such in-depth analysis it is unlikely that much further information will be provided by this type of study. Such an analysis has been made by the various chapter authors in Coupland (1979).

### *Systems-oriented studies*

The systems-oriented studies rally around the ecosystem concept. The ecosystem concept was the higher-order or level of concept required to provide the integration and interdisciplinarity. Effectively this was a 'conceptual quantum jump', although the term ecosystem was introduced in 1935 by Tansley and the concept has a much longer history (Major, 1969).

The adoption of the ecosystem concept provided the theoretical and philosophical framework that was missing in the largely single-factor studies of classical ecology. Systems ecology is a 'new ecology' that has been very much developed in the IBP. McIntosh (1974) believes that the major difference between the 'old and the new ecology' is not the ecosystem concept or the recognition of its complexity and wholeness, but is instead the accumulation of the techniques, instruments and wherewithal for analysis of ecosystems as a whole.

Key concepts that were strengthened or developed in the new ecology, but prior to the IBP, were productivity, nutrient cycling and energetics of ecosystems. It was only within the time span of the IBP, however, that these phenomena were sufficiently quantified to enable their modelling and simulation. Models of one form or another have been around for many years in ecology, particularly in grassland studies. Generally, however, these were 'static' rather than dynamic.

Another phenomenon of general interest arising out of systems-oriented studies is that of plant–animal interactions. Chemical interactions between plants and animals have been of considerable concern, as have studies of food preference, palatability and nutritional value. Much of the information being presented in plant–animal interaction studies ties in directly with the evolution and functioning of the ecosystems in the long term.

Large-scale, long-term ecological research that is characteristic of systems-oriented studies has some characteristics of programmatic research. Notterman & Mintz (1965) note that programmatic support of research permits, but does not ensure, the gradual development of experimental technique, the corroboration of puzzling findings, and the eventual emergence of data-based theories. However, the systems-oriented studies in IBP grassland ecology investigations had a maximum duration of 10 years of major funding. This in some cases led to the acquisition of a large amount of data which was never thoroughly examined, analysed and synthesized within the time span of IBP. The large-scale programmes did, however, afford the favourable

Table 14.1. *Examples of syntheses of ecosystem studies arising from IBP efforts and related programmes in grazinglands of the world*

| Reference | Location | Type of grazingland | Remarks |
|---|---|---|---|
| Breymeyer (1971) | Vistula Valley, Poland | *Stellario-deschampsietum* and *Arrhenatheretum medioeuropaeum* associations | 14 papers; geobotanical and soil descriptions; decomposition rate, microbiological and soil macrofauna studies; Orthoptera, Homoptera, Diptera, spider, ants, predatory arthropods and shrike (*Lanius collurio*) investigations; impact of insect grazing on plant biomass; general structure–function analyses. |
| Bliss (1972) | Devon Island, Northwest Territories, Canada | *Carex* and *Eriophorum* meadows | About 24 papers; abiotic studies of geology, soils, permafrost, micrometeorology, energy budgets and $CO_2$ flux; vegetation studies on pollen, plant production, mycorrhizal survey, $CO_2$ assimilation and nitrogen fixation; consumer studies of musk ox productivity, bird bioenergetics and population dynamics, invertebrates, mammalian carnivores and lemming; decomposer and dung decomposition studies; traffic, oil spill and scraping impacts; energy flow and heat flow modelling. |
| Ellenberg (1971) | German Solling Project | *Festuca rubra* facies of *Trisetetum flavescentis* meadow | Primary production of meadow; green leaf area indices of grassland communities; methods of study of functional and non-functional roots; grassland mapping in the High Solling; preliminary reports. |
| Ketner (1972) | Terschelling, Netherlands | Salt marsh communities | 1966–70 field study; above- and belowground vegetation standing crop measurements; production as a function of years and community; relationship of production to abiotic factors; decomposition rate of dead plant material; solar energy use efficiency; grazing impact analysis. |
| Coupland, Willard & Ripley (1974) | Canada | Mixed prairie | Summary of detailed 1967–74 study; site description; abiotic components; primary production; secondary production; activities of micro-organisms; modelling; lists of reports; budgetary expenditures. |
| Petrusewicz (1974) | Carpathian Mountains, Poland | *Lolio-cynosuretum* and *Hieracio-nardetum* associations | 17 papers; geobotanical characteristics; primary production above and below ground; herbivore impact on plants, soils and microflora; microflora development and decomposition; coprophagous beetle, earthworm, nematode, oribatid mite, ant and spider studies; detritus versus grazing foodchain carbon flows and balances; 3-year field study. |

Table 14.1 (*cont.*)

| Reference | Location | Type of grazingland | Remarks |
|---|---|---|---|
| Numata (1975) | Japan | *Zoysia* and *Miscanthus* grasslands | About 50 short papers; emphasis on abiotic and primary producer components; three-compartment seasonally varying simulation of biomass flow for *Miscanthus* and *Zoysia* plant communities; nine-compartment grass–litter–cattle simulation model for a *Zoysia* grazed grassland. |
| Rosswall & Heal (1975) | Arctic, alpine, subarctic and subalpine tundras world-wide | Tundra to mountain grassland to mires | Word models for 15 national projects; simulation models for primary production, herbivory and decomposition processes at Point Barrow, Alaska; a carbon flow, 14-compartment model of a pure stand of *Dupontia fischeri*. |
| Rychnovska & Blanca (1975) | Czechoslovakia | *Stipa* spp. and *Bromus erectus* on xerothermic habitats | Mineral composition of plants; soil physical, chemical and biological characteristics; growth, water relations, photosynthesis and respiration of *Stipa* spp.; geographic distributional analyses of *Stipa* spp. |
| Soriano (1975) | Buenos Aires Province, Argentina | Humid pastures of saline depressions and semi-desert pastures of east central Patagonia | Physiographic description; soils; plant communities. |
| USSR/IBP (1975) | USSR | Grassland, tundra, desert | Terrestrial productivity and production processes; tundra, meadow, steppe, swamp, desert and mountain grassland communities; six chapters of concern regarding vegetation with some data for standing crop, production, litterfall or death above and below ground, water balance, photosynthesis, chemical composition and nutrient budgets; general survey on wild animal population structure and dynamics; numbers and biomass of various groups of soil fauna for different communities. |
| Wiegolaski (1975*a*, *b*) | Hardangervidda, Norway; Abisko, Sweden; Kabo, Finland | Alpine and arctic tundra and related grasslands | More than 70 papers, two volumes; site descriptions, abiotic variable measurements, primary producer structure and processes, decomposers, nutrient cycling, animal community structure, dynamics, bioenergetics and herbivory; integrative and predictive models; carbon flow simulation mode for Hardangervidda of about 11 compartments, deterministic, 1-day time step; energy-flow simulation for invertebrates; 10-compartment simulation model focusing on Hardangervidda rodents; energy calculation model for acquisition and utilization of energy by root voles; integrative word model of functioning of Fennoscandian tundra ecosystems. |

Table 14.1 (*cont.*)

| Reference | Location | Type of grazingland | Remarks |
|---|---|---|---|
| Ayyad (1976) | Northern Egypt | Dune ecosystems: *Ammophilia*, *Euphorbia*, *Pancratum*, *Ononis*, *Elymus*, *Echinops* and *Thymelaea* spp. | Mathematical models. |
| Ayyad (1975) | Northern Egypt, Sandene | Dune ecosystems: *Ammophila*, *Euphorbia*, *Pancratum*, *Ononis*, *Elymus* *Echinops* and *Thymelaea* spp. | General biogeographic description; autecology of *Thymelaea hirsuta*; wate flow and productivity; vertebrates and invertebrates including soil fauna and flora; chemical composition of plants and animals; soil algae and bacteria and fungi and litter studies. |
| Traczyk (1976) | Poland | *Arrhenatheretum medioeuropaeum* artificial meadow | 24 papers; mineral fertilization studies: impact on yield and composition of vegetation, above and below ground, structure and succession; studies of phytophagous fauna, Homoptera, nematodes, oribatid mites, other mesofauna, enchytraeid worms, earthworms, ants, predators and soil microflora; litter decomposition in relation to fertilizers, environmental factors, saprophagous fauna, microflora and chemical composition. |
| Ayyad (1977) | Northern Egypt | Dune ecosystems: *Ammophila*, *Euphorbia*, *Pancratum*, *Ononis*, *Elymus*, *Echinops* and *Thymelaea* spp. | Primary producer phenology, standing crop and growth; waterflow; nutrient accumulation; vertebrate and invertebrate biomass; chemical composition of plants and animals; so decomposers. |
| Duvigneaud & Kestemont (1977) | Belgium | *Festucetum*, *Filipenduletum*, seed grassland | Primary production of fescue alluvial prairie productivity; biogeochemical cycles of *Bistorto-filipenduletum*. |
| Pandeya *et al.* (1977) | Western India | *Cenchrus ciliaria* complex grazinglands | Geography, climate and water resource population differences in *Cenchrus* spp state of knowledge summary of Indian grazinglands regarding primary production and dynamics; soil–plant–climate relationships; ecotypes, seed and germination studies; systems analysis of a village ecosystem in Rajkot; time-series dynamics of drivin variables and abiotic and biotic state variables; compartmental model diagrams of energy and water flow. |
| Rodin (1977) | Arid zone, USSR | Desert shrub and shrub-steppe; *Ammodendron*, *Artemisia*, *Kochia*, *Agropyron*, | |

Table 14.1 (*cont.*)

| Reference | Location | Type of grazingland | Remarks |
|---|---|---|---|
| | | *Astragalus*, *Haloxylon* and *Carex* spp. | |
| Heal & Perkins (1978) | Great Britain | Moors and montane grasslands | Synthesis of detailed studies, primary and secondary production. |

opportunity for investigators or teams of investigators to publish their work as a whole, rather than piecemeal. The use of models as a synthetic medium was particularly important here as it compelled researchers to compare the logic and interpretation of findings reported in one place with those appearing in another. Large-scale programmes generally also examined a greater variety of experimental situations with a common methodology, thus increasing the likelihood of generalizations eventually being made from the data.

Synthesis or attempts at synthesis were a key component of the large-scale grassland studies in various countries around the world. I think synthesis and systems descriptions were a result of the large number of ecological researchers working in an integrated way in the IBP. Examples of such syntheses of grazinglands, summarized in Table 14.1, further illustrate, however, the newness of such information.

The brief remarks on the volumes reviewed in Table 14.1 show that their contents are similar to some of the information presented in Chapters 9 and 11 of this volume. Likewise, the models presented in the volumes listed in Table 14.1 should be compared with the models presented throughout this volume of individual processes or groups of processes at a given trophic level, of a subsystem, or of total-system dynamics (Chapters 1, 2, 10, 12 and 13). The concepts and processes of synthesis are considered further in the following section.

### *Concepts of and approaches to synthesis*

A question that arises in conducting and synthesizing systems-oriented studies, is how much detail to include in the system. A facetious definition I heard once indicated that 'in instances where you have from one to three variables you have a science, when you have from four to seven variables in a study you have an art, and when you have more than seven variables you have a system'! The art of synthesizing a system is practised in different ways.

The first level of synthesis of many of these studies comes in the development of 'word models'. This is a methodical and detailed way of describing what was there, what happened to what and why. A second step in the synthesis includes the development of detailed box-and-arrow diagrams showing flow of matter and energy through the system. We outlined the steps of this approach in Chapter 1: (i) development of a diagram, a coupling matrix and a flow-effects matrix; (ii) implementation of equations to calculate the flows, combining these flows into difference or differential equations; and (iii) the running of the model by setting initial conditions then driving the model with records of external or driving variables, and simulating what happens to the state variables over time. In this process one is always fitting information into a higher-order organization, i.e. the system model. Thus, studies on individual plants fit into the overall function and structure of the vegetation community which is in turn fitted into the ecosystem.

### *Interest in mathematical approaches*

One of the major changes in ecological science dramatized by projects in the IBP, and in which grassland studies have been a central part, has been the greatly increased awareness of ecologists of models and the modelling process. For many individuals this has been a fascinating experience; for others it has been discouraging. Models of various types have been used by ecologists for some time, but simulation modelling, in addition to being relatively new to ecologists, is also new in many other areas of environmental biology including natural resource management and many aspects of agriculture. Most new fields rapidly gain their protagonists and antagonists. The antagonists said early, and continue to say, that it is unrealistic at present to simulate ecological systems (see further discussion in Chapter 10). They say such systems are far too complex and our knowledge of the processes operating in these systems is far too fragmentary, both quantitatively and qualitatively (see Van Dyne, 1975, for further discussions of this topic). Most protagonists of the modelling concept applied to ecology recognize that models are always imperfections of reality, only abstractions of the real world.

Much progress has been made in modelling of ecological systems largely because many programmes included in their planning the concept of models and modellers. This approach has drawn in more mathematically trained persons than was characteristic of the grassland ecology of the past.

For more than 40 years theoretical population ecologists have used the early model formats of Volterra, Lotka & Gause, so one might ask 'Are simulation models new?' A quick review of the types of models reviewed in several chapters in the present volume (especially Chapter 10) will show that a conceptual quantum jump has been made. The models now include dynamics of the entire system and follow flow of energy, water, carbon, numbers and other currencies. Thus, a start has been made in thinking of the complexity of ecosystems in an organized, quantitative and synthetic way.

## General consideration of modelling and modelling needs

Examples of grassland and grazingland system and subsystem models for research or for management purposes have been introduced throughout this volume (see Chapters 1, 2, 4, 10, 12 and 13). The purpose of this section is not so much to discuss the models as such as to consider the modelling process and its needs and how these were or were not met in some of the IBP integrated research programmes. Some of the information presented here was derived from a workshop on the state of the art in grassland modelling held at the National Academy of Sciences in Washington, DC, in February 1978. I thank the 25 participants, most of whom were or still are participants in large-scale integrated research programmes, for helping me to crystallize my ideas and for sharing with me their experiences. They should not, of course, be held responsible for the comments in this section.

### *Modelling organizational problems*

There are few reports on the modelling process and there are not many experts available to discuss the management of ecological research. References to scientific management problems in large-scale ecological research programmes seem to be restricted to IBP studies (e.g. Coupland, Zacharuk & Paul, 1969; Auerbach, 1971; Gessel, 1972; Goodall, 1972; Van Dyne, 1972). There are also relatively few accounts of organizational problems with respect to mathematical modelling. The study by Mar (1974) of problems encountered in the development of multidisciplinary resource and environmental simulation models is of particular importance here. He notes several problems that should be discussed, including those of documentation, funding of modelling efforts, review and critique of models, and steps of modelling.

### *Model documentation, review and analysis*

Surveys of modelling efforts show that documentation is seldom developed along with the model. If documentation is done, it is usually after the fact. There are at least two reasons for this. First, modellers are preoccupied with the construction of the model. Secondly, modellers, or the organizations they work for, may have proprietary interests in the model and want to use it for gaining further funding. With respect to the funding problem, Shubik & Brewer (1972) found in many defence projects that some research groups have maintained families of models for many years by separately proposing components of these models as discrete models to be funded, used and evaluated independently. This dangerous trend could appear in ecological modelling.

### *System utilization*

Numerous investigators (e.g. Mar, 1974; Van Dyne & Abramsky, 1975) show that most models do not receive detailed professional review nor are the usual scientific standards of evaluation applied when the models are examined. Large, complex models may thus be of little practical use, because problems in the areas of documentation, communication and conceptualization are exaggerated with size. Relatively little information on the cost of modelling has been summarized but Van Dyne (1978) provides some such data for models developed within and around the US/IBP Grassland Biome study. Large-scale models of this type may require up to 20 professional person-years of effort.

### *Key problems in modelling management*

Mar (1974) identifies three major issues in an environmental modelling effort that must be resolved if it is to be productive: (i) a clear definition of the system being analysed and a clear understanding of the theory, experience, intuition and data that already exist for the system components; (ii) an 'orchestration' of individuals and management of project resources in such a way as to integrate information on each of the system's components in the overall analysis; and (iii) the ability to document and transfer model results to potential users. This last point is perhaps not so critical because it is difficult to identify the user of basic research, if modelling research can be considered basic.

### *Modelling strategies with respect to management*

It is worthwhile here briefly to revisit ideas introduced in Chapter 10. Consider here the two alternative strategies, i.e. a top-down or a bottom-up approach, with respect to the management task.

As regards the hierarchical organization of the model it was suggested early that one should try to keep the number of levels of organization within the model to two or three (deWit, 1970). Consider the process of starting with an overall system model and continually improving it by changing its structure. Let this improvement be made not by increasing the number of state variables, but instead by increasing the mechanism in flow descriptions. It is possible to improve it uniformly so that all the state variables stay at one or two levels of organization, and the mechanisms to describe the changes in state variables are perhaps at one or two levels of organization lower. In a second round of development it may be necessary to divide some of the state variables, thus introducing new ones. But it is still possible to use this approach to model development with a total-system framework. It is, in effect, the 'top-down' modelling approach discussed in Chapter 10.

The contrasting approach in model development, i.e. the 'bottom-up' strategy, starts with isolated, independent, high-resolution, complex process models. This results in considerable difficulty. The difficulty is that, given a sector of a model to construct, individual scientists will prefer to explore the frontiers of this topic rather than develop a representation that is adequate within the framework of the whole model (Mar, 1974). In my experience, scientfic groups tend to make models progressively more complex and mechanistic with higher levels of resolution. Thus, over a period of time, the models tend toward describing less and less complex structural levels of organization in the flow functions in order to explain changes in the state variables in the higher level of organization.

### *The modelling leader*

To be able to work towards the original goals of the modelling effort the modelling team must devote considerable time to learning the 'languages' of two or even three disciplines other than their own, and adding new disciplines requires an additional year or so in the life of the team. As with an overall research team leader, the modelling team leader who lacks technical expertise and academic standing will probably not be effective in a university research environment. Such a leader might have more opportunity in an industrial environment. The reason is that the major function of an interdisciplinary modeller is to arbitrate disputes between disciplines and catalyse cooperative modelling efforts (Mar, 1974).

### *Some modelling needs*

An evaluation of the models discussed in several chapters of this volume leads me to the following conclusion. The models of grazingland systems developed to date are only the forerunners of those that are needed if they are to be adopted by those responsible for managing renewable resources and result in new insights and practical applications. As is the case in agricultural management in general (Anderson & Dent, 1975), the resource management arena has two key features: complexity of the managerial situation, as a result of the numerous dynamic interrelationships which must be considered, and the pervasive influence of stochastic factors, which adds complexity to the decision-making process.

The complexity of the resource being considered means that several segments must be simulated simultaneously in a management-oriented model. To date, almost all of the grazingland simulation models describe a given hypothetical point in space rather than having spatial components. If it is necessary to introduce models with multiple spatial components then size or complexity increases very rapidly. If variables in the different spatial components have varying relaxation times (which Van Keulen, Louwerse,

Sibma & Alberda (1975) defined as the time taken to recover from small changes) then it may be wasteful to run the entire model at a single time step and multiple time steps should be used. Such a feature has only recently become available in simulation languages.

The problem relating to the pervasive influence of stochastic factors is perhaps more serious. Considerable uncertainty may exist in information about the driving variables, which are predominantly climatic variables. DeMichele (1975) notes that it is generally accepted that random events do not occur – other than perhaps at the level of subatomic particles – and states that apparent randomness observed in systems we model is due to a lack of detailed knowledge about the mechanism in the system. Consider the implications.

An ecosystem model contains many parameters. Each parameter may not be known with certainty, but instead may be estimated as a mean value with some variance. We know little about the distribution of these parameters or of the variance–covariance matrix for them. But assume that we have knowledge of the means, variances and covariances for (i) the parameters in the flow functions, (ii) the initial conditions, and (iii) the driving variables. Then, what would the statistical properties of the output from these calculations be? I don't know.

Although the problem of validation is widely argued, we can generally obtain some time series for the purposes of evaluating a model. Normally we obtain means and variances for different points in time for selected state variables. We want to compare the output from a model, for which at present we have little idea of the distribution properties, with the validation points. It has been suggested by some that we should use a Monte Carlo approach and derive the distribution properties of the predicted state variable output time series. To do this one would select at random the values of the parameters of the flow function and run the model many times for a given set of initial conditions and driving variable records. One would obtain many outputs from the model, i.e. prediction curves for each state variable. Then one would calculate the variances for each state variable at each point in time for which we have data points. Even disregarding the assumptions about the distributions, the task is terribly large. Consider that the model may have 500 parameters for an ecosystem-level simulator. Consider, too, that we may have to run the model several hundred times in the Monte Carlo approach. We would soon be stopped by costs, caution or criticism.

There are different philosophical viewpoints on the strategy for validation. Anderson & Dent (1975), Van Dyne & Abramsky (1975), and Steinhorst (1973) focus on these problems. Essentially, the purists suggest that all of the components of the system being simulated must be validated by appropriate real-world experiments. The pragmatists feel that this is not necessary and use the analogy that decisions in real-world management must be made even when some elements of the system are poorly known.

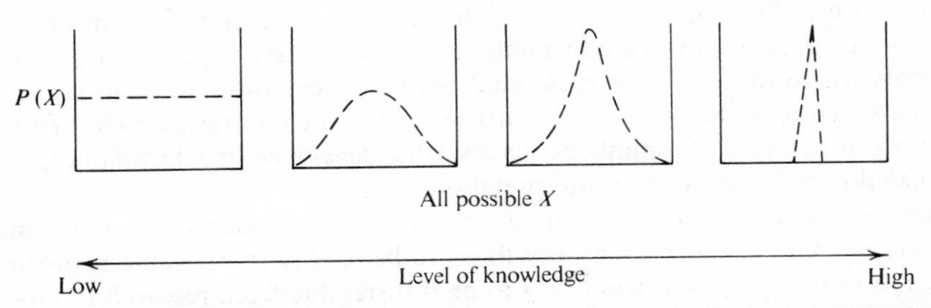


Fig. 14.1. An example of the impact of variability in parameters in the flow functions, i.e. level of knowledge, on the distribution of the output of stochastic models.

The same data should not be used both in construction and evaluation of the model. Fortunately, most of the information used in constructing the model is concerned with flow functions. Hypotheses about flow functions are generally derived from different information than the field data consisting of time series for different state variables and driving variables. If, however, it is possible to derive parameter values or functional forms for flow expressions from analysis of field data sets, then separate data sets should be used to derive these values and to evaluate model outputs.

Yet, we should be concerned with the outcome predictability from a simulation model run. A good example is outlined by DeMichele (1975). I have adapted his concept in Fig. 14.1. If we know nothing about our system, i.e. we have no mechanisms in the flow functions, we must expect a response that can be treated as completely random. As we acquire information through data and theories, we develop an output that for practical purposes will be normally distributed. But all outputs are not equally likely. It is probable that a non-uniform distribution function will describe the likelihood of any output. If, however, the entire system is built on solidly developed theory and the inputs to the system are known, then the output should be predictable and could be described by a delta function. At the present state of knowledge for grazingland models we are probably just to the left of centre on the arbitrary scale in Fig. 14.1.

A fortuitous situation exists with respect to the model documentation and usage which is of interest here. Consider the situation where one models a phenomenon as deterministic when it really is stochastic. If this error is made, then it may be a good thing that models are so poorly documented and nontransferable. If they were not, we might reach the level of misuse that has appeared in recent years for the use of complex statistical methodologies. Jeffers (1973) notes that it is now extremely easy to employ a statistical technique without understanding any of the basic assumptions of the technique or its limitations. The misuse occurs frequently now because so many ‘computer packages’, or advanced methods of analysis are available to any interested user from many major computer centres. A second problem exists with use of these sophisticated techniques: there are few papers in scientific journals where the results of these analyses have been interpreted in detail. It seems that it is the underlying mathematics of the analysis, which, being of more interest to the mathematicians who developed the techniques, is discussed in detail instead of interpretations.

Another modelling need is in the development of submodels. Total-system models are, for various reasons, not likely to be readily transferable between research groups and even less likely to be transferable from research groups to users. Thus, some have suggested the possibility of establishing autonomous modules representing relationships, that apply with acceptable accuracy over a broad range of situations (Anderson & Dent, 1975). It is presumed that such modules could be linked together so as to form the core of a model or at least the starting point for constructing a total-system model. The unique characteristics of the system would then specify the addition. This approach has been proposed by Innis (1975*b*) but not yet fully implemented or recorded in the grazingland literature.

### *Model utility*

Simulation models of grazinglands have had relatively little use to date in the field of managing renewable resources. This is in part related to their complexity (discussed above) in contrast, for example, to linear programming algorithms. The complexity of the simulation models, and the resultant problems in documentation, yield a product that is difficult to transfer to the user. The collection of articles published on simulation modelling (see for example Patten, 1971, 1972, 1975), contributes little to resource management decision-making because of the technical writing in the papers. For models to be of use in decision-making processes one should be able to test potential management policies in a short period of time and identify those management strategies which yield optimal economic gains. Notice that the term optimization is introduced here. It has been noted that most real-life problems concerning the management of renewable natural resources require predictions of responses to various possible management strategies. Few examples of simulation–optimization models exist (Van Dyne & Abramsky, 1975).

Defining efficient experiments to conduct with simulation models is another new problem. Anderson & Dent (1975) have indicated that further study on the specific problems associated with design and analysis of computer-based experiments will be highly rewarding. Van Dyne and co-workers (Van Dyne, Joyce, Williams & Kautz, 1977; Van Dyne, Joyce & Williams, 1978) conducted a series of such experiments with grassland simulation models. First a series of hypotheses was formulated, with the assistance of a group of knowledgeable scientists who were not involved in the construction of the model, describing how the system should respond to management. The existing model was then modified in such a way as to 'do' a series of experiments. The output was compared qualitatively with the hypotheses of how the system might respond. Disagreement between model output and the *a priori* experience contained in the hypotheses led to an awareness of errors in the construction or parameterization of the model and also to some counterintuitive ideas about the real system.

### *The interrelationship of model and experiment*

In Chapter 10 we noted how a large-scale model which evolved alongside experimental results helped influence the nature of new experiments. However, as one reviews the literature and the IBP experience, it is clear that most models develop at the end of, or at best lag behind, the experimentation. Thus models did not have a full opportunity to influence the design of integrated research programmes and specify some of the tasks they should include. In this section I suggest a procedure by which one can relate objectives, programme management and design, and individual tasks, and the role of models therein. First I will address the problem of objectives and heirarchies, then the relation to experimental design, and finally the relationship to an energy flow model of the grazinglands.

There are many points of view on the necessity and role of objectives in a programme of basic research. Some say it is impossible to narrowly define objectives in such research. Others argue that without objectives no research should be done. One of the problems of considerable importance in programmatic strategy concerns dispersion of effort. With limited resources with which to conduct research, objectives must be defined more closely than where there are unlimited resources. Most pure scientists will agree that some objectives provide for some efficiency in research. Yet, there is a special problem in defining objectives for large-scale, long-term ecological research programmes.

In any large-scale programme the overall objective may be very broad and sometimes vaguely stated. In a programme oriented to technology (but which still may contain many pure research objectives) it is often easier to state a precise, simple and readily understood objective. For example, consider the objective of putting a man on the moon by a certain date. It is understood by all. Any given research project can be evaluated fairly readily as to whether or not it contributes to that objective. On the other hand, consider an objective for an ecological research programme of 'understanding the structure, function and utilization of certain ecosystems'. Such an objective, while inclusive, may not satisfy some scientists as being precise enough. Thus, it is necessary to break down the long-term, global objective into smaller and shorter-term objectives.

Theoretically, it should be possible to derive objectives at various levels of resolution, each set of which will nest in the objective in the level above. In classical experimental research, and in a statistical sense, objectives are often stated as null hypotheses. In many instances it is possible to design experiments based on such precisely stated objectives. But in a broader sense, and at a lower (and yet perhaps more common and more useful) level of resolution objectives are stated in such terms as 'to formulate', 'to extend', 'to enhance', 'to understand', 'to analyse', 'to estimate', 'to establish', and so on *ad infinitum*! But most research is done in a context of limited funds. Given limited funds, decisions must be made concerning which studies to support and which studies to omit. If these decisions are based on intuitive feelings, then disagreement is more likely to arise. Yet in our present state of theoretical ecological knowledge many intuitive decisions must be taken regarding research design and direction. No completely objective procedure is available at present on which to base all of our decisions. I feel the 'single objective approach' is more applicable to applied than to basic research.

A logical approach to defining objectives would be first to subdivide them into component parts. For example, within the area of the 'structure of the ecosystem' some specific objectives might be: (1) to determine the presence and relative abundance of the biotic components of the ecosystem; (2) to characterize the abiotic elements of the system; (3) to study the structural interrelationship between the biotic and abiotic components. The objective of 'understanding function' of the ecosystem might be subdivided as follows: (1) to study the energy flow through the system; (2) to study the cycling of nutrients in the system; (3) to study the cycling of water in the system. Analysis of 'utilization of the ecosystem' (implying utilization by man) might be subdivided as follows: (1) to study the effects of harvesting autotrophs; (2) to study the effects of harvesting heterotrophs; (3) to study water utilization; (4) to determine and examine aesthetic factors.

At the next level of the resolution each of the above objectives may again be divided. For example, in the study of energy flow as a component of function one might consider the development of a mathematical model to simulate the energy flow in ecological systems. This becomes much more specific and leads to the statement of objectives about the model. This may be particularly important in interdisciplinary researches because many analytically trained and oriented people have grown accustomed to working with precise formulations of problems.

Consider, for example, the objective of developing a simulation model for energy flow of a given ecosystem. First, one must define the context in which the word ecosystem is used. Does it include all components, such as abiotic, producer, consumer and decomposer? How many state variables are needed for each component? Should the model be able to predict the dynamics of energy in the system over one week, one year, one century, or what? What sort of precision is required in the prediction? Does one simply want to predict the direction and order of magnitude of the system's response to naturally occurring variations in the driving variables? Or does one want to be able to predict also the direction and order of magnitude of the response to certain stresses, perturbations or management manipulations imposed upon the system? If so, what management manipulations will be applied and with what degree of stress?

Simply predicting the direction and order of magnitude of the response of the system to natural or man-imposed perturbations may be inadequate. Perhaps some more precise criteria of performance of the model should be considered. The proof of the pudding is in its eating, and the proof of a model is in its ability to predict real life. The validation of a model, as discussed earlier, is a complex process, though most field biologists feel that they can measure enough components of real life to provide points over time against which to check model output. However, the criteria of comparison must be quantified. One might say, for example, that the model is adequate (at least initially) if the output predictions are within one experimentally measured standard deviation of the measured mean values of state variables for 80% of the time (for those variables which are common to the field experiment and the model). Additionally, validation points may be taken from field-determined rate processes. If the validation points are carefully selected, if the standard deviations are reasonably small, and if a reasonable amount of biological and physical knowledge is incorporated in the rate process representations within the model, then this may provide an adequate test. Note, however, there are several 'ifs' in the preceding sentence. This means a great deal of interpretation and wisdom must be put into the design and collection of field measurements and into the design and operation of the model.

Consider the objective of being able to test the energy flow model and its response to a given range of natural or man-made experimental conditions. One soon realizes that not only must one have driving variables, state variables, and field rate process validation measurements, but one must also have them from a variety of conditions. In order to test hypotheses about system performance, one must perform experiments on the system. If one simply studies the system in great detail, one is merely conducting an intensive survey. This provides a certain, limited amount of information about the interrelations of structure, function and utilization and allows the development of hypotheses about the system, but only the inclusion in the research of experimental stresses on the system will allow the testing of such hypotheses.

In the process of running and testing the model it is likely that many changes will have to be made to improve the model's performance. This identifies additional objectives over time, i.e. specifies new rate process studies to be examined. In some ways, however, to start an ecosystem study by measuring all the rate processes would be self-defeating, for there are just too many of them. Some procedure is needed to help select, from the large number of rate processes that could be studied, those which are most needed for the model's objectives.

## The present level of understanding of grazingland ecosystems

Although the main purpose of this chapter is not to evaluate in detail scientific findings of IBP researches, it is useful to reflect on the general level of understanding attained. I will focus on only a few points, drawing examples from the US/IBP Grassland Biome study. Many scientists provided the raw data for the compilations from which I extracted data to prepare figures used here. Thus, Smith, Singh, Coleman, Ellis & Andrews (1974) summarized data of several North American grasslands, as did Sims & Singh (1978*a*, *b* *c*, *d*), and this work has been further elaborated on by Dyer, Dodd, Leetham and others in the US/IBP Grassland Biome study. I have used these data in comparing standing crops and productivities for selected North American sites. Andrews, Coleman, Ellis & Singh (1974) and Coleman, Andrews, Ellis & Singh (1976) have summarized energy flow in some of our grassland ecosystems and I have drawn on those data to illustrate our level of knowledge for shortgrass prairie. Clark *et al.* (Chapter 8) and Woodmansee, Dodd, Bowman, Clark & Dickinson (1978) have summarized information on nutrient cycling, and I have used some of these data to compile a diagram on nitrogen cycling for shortgrass prairie.

### *Some comparisons of North American grasslands*

A major investigation in the US/IBP Grassland Biome study was the comparative determination of aboveground and belowground standing crops and productivities in different types of grassland systems. Some of these data have been reviewed by Coupland (1979). The data summarized here illustrate the general state of our knowledge for autotrophs and heterotrophs.

Data are summarized in Fig. 14.2 for aboveground and belowground standing crop and productivity for ten North American grassland sites. The sites studied in our programme, the type of grassland, and the dominant plants are noted in Table 14.2. In general, moving westward across the Great Plains, we find tallgrass prairie, mixed-grass prairie, and shortgrass prairie. In the northwest are shrub-steppe bunchgrass types, there are annual grasslands in the interior valleys of California, and desert grasslands are found in the southwest. Scattered throughout the mountains of the western United States are smaller segments of grasslands, some of which are seral, and some of which, for all practical purposes, are climax. The rainfall pattern varies widely, from some 200 mm in the desert grassland up to 1000 mm in the

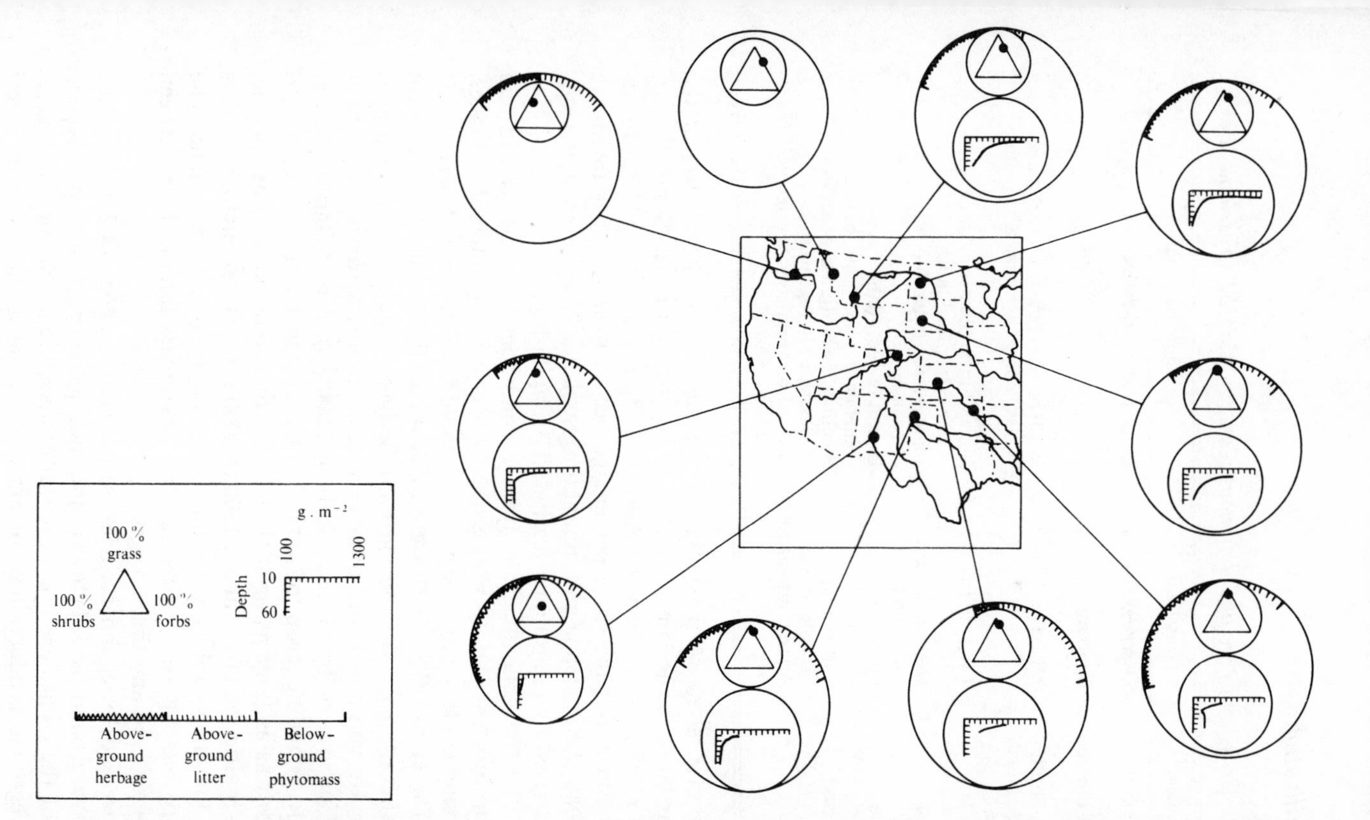


Fig. 14.2. Comparisons of vegetation production and standing crop of grasslands. The areas of the outer circles are logarithmically proportional to annual average standing crop. The areas of the upper and lower inner circles, respectively, are logarithmically proportional to aboveground and belowground annual production. The triangles show the relative contribution of grasses and grasslike plants, forbs, and shrubs to the annual production. The graph in the lower circle shows the depth distribution of live plus dead root biomass.

Table 14.2. *Grassland type and dominant plants of US/IBP grassland sites*

| Location | Grassland type | Dominant plants |
|---|---|---|
| New Mexico | Desert grassland | Black grama grass, mesa dropseed grass, soaptree yucca |
| Texas, Colorado | Shortgrass prairie | Blue grama grass, buffalograss, cactus |
| Kansas, South Dakota, North Dakota | Mixed prairie | Western wheatgrass, little bluestem grass, sideoats grama, blue grama |
| Oklahoma | Tallgrass prairie | Big and little bluestem, Indiangrass, prairie dropseed grass |
| Montana | High mountain grassland | Idaho fescue grass, lupine, wheatgrass |
| Montana | Low mountain grassland | Rough fescue, Idaho fescue, bluebunch wheatgrass, lupine |
| Washington | Shrub-steppe grassland | Bluebunch wheatgrass, Sandberg bluegrass, big sagebrush |
| California | Annual grassland | Annual bromegrasses and fescue grasses |

tallgrass prairie, some of the mountain grasslands and some parts of the annual grasslands in the interior valleys along the west coast. Some 75 % of the annual rainfall occurs between April and September for the Great Plains grasslands and about the same percentage occurs from October to March in the annual, shrub-steppe and mountain grasslands. The desert grasslands are characterized by late summer/early fall precipitation.

In Fig. 14.2 the data illustrate average conditions for both grazed and ungrazed locations at various sites generally over a 3-year interval or longer. The large outer circles denote the average annual standing crop of herbaceous material. The areas of the circles are logarithmically scaled. When average annual standing crops of aboveground herbage, litter and belowground phytomass are combined, the total vegetation biomass varies from about 300 $g \cdot m^{-2}$ on desert grasslands to about 2500 $g \cdot m^{-2}$ in tallgrass prairie. Surprisingly, standing crop of total herbaceous material on our mixed prairie sites was as large as or slightly larger than that on tallgrass prairie. Shortgrass prairies had from about 1500 to 2000 $g \cdot m^{-2}$ total phytomass. This was about the same level as was found in the high mountain grassland. The shrub-steppe bunchgrass type had a total phytomass intermediate between the desert grassland and the shortgrass prairie.

The upper and lower inner circles, respectively, in Fig. 14.2 refer to the annual aboveground and belowground net primary production. Again, the areas of the circles are logarithmically proportional to annual production. Aboveground production of plants varies from about 100 $g \cdot m^{-2} \cdot yr^{-1}$ in desert grasslands, shortgrass prairie and shrub-steppe bunchgrass, to more than 300 $g \cdot m^{-2} \cdot yr^{-1}$ in mixed and tallgrass prairies. The shortgrass prairie in

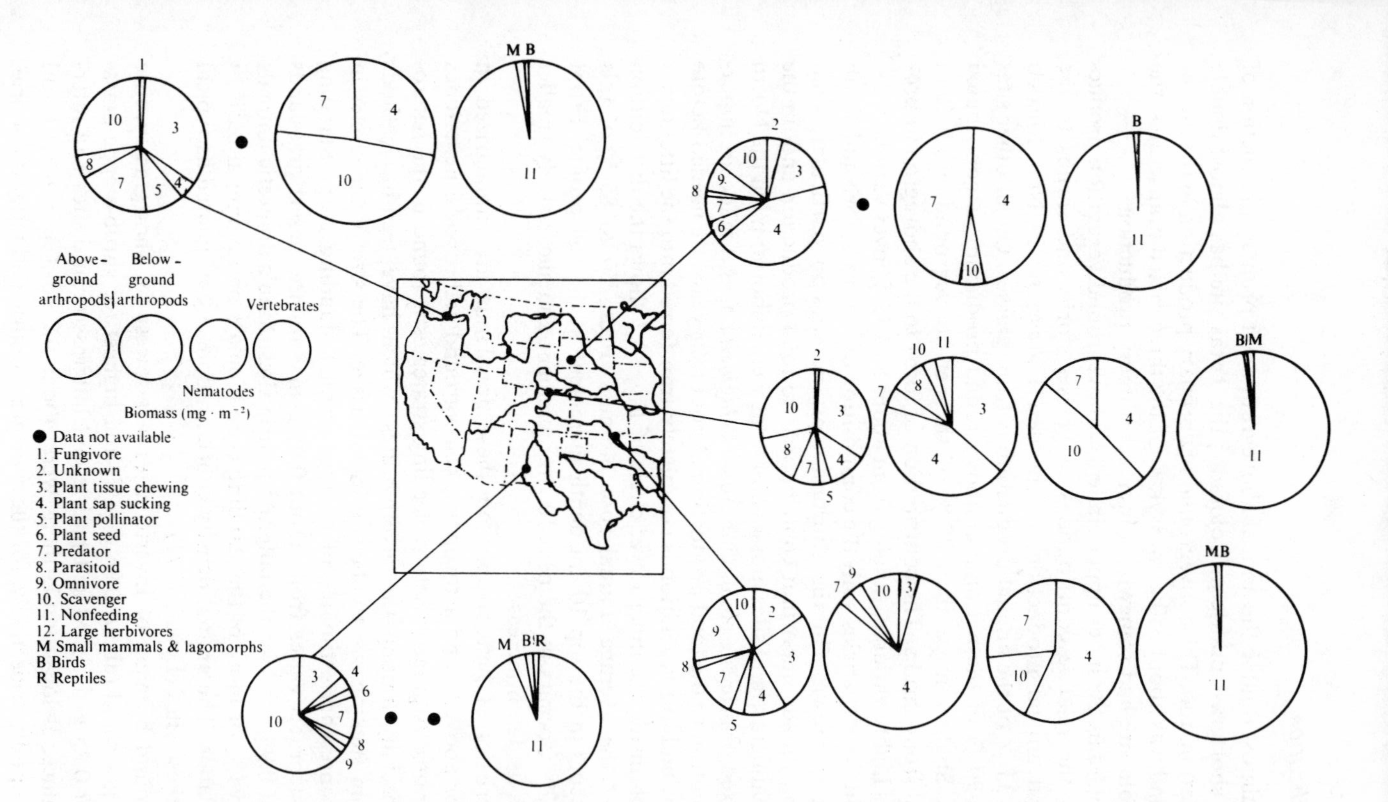


Fig. 14.3. Comparisons of animal biomass on selected grasslands. Arthropods, aboveground and belowground, and nematodes are identified by functional groups. Vertebrates are grouped into largely taxonomic categories. Areas of circles are logarithmically proportional to average annual standing crop biomass.

Texas has almost double the annual aboveground net primary production of that of the shortgrass prairie in Colorado; the Texas site has almost double the precipitation, too. The aboveground net primary production in the mountain grassland was about 250 $g \cdot m^{-2} \cdot yr^{-1}$. Of the three mixed prairie sites, the South Dakota site had somewhat lower net primary production.

The triangles in the inner upper circle for aboveground vegetation productivity show the relative contributions of grasses, forbs and shrubs to the aboveground annual production. In the desert grassland in New Mexico only about 33 % of the annual production is from grasses. On all other sites from 70 to 90 % of the net annual aboveground production is contributed by grasses. Slightly more than 25 % of the net aboveground production was derived from shrubs. The contribution of forbs to the aboveground production was highly variable, ranging from about 5 to 35 % over various sites. Note that, in these triangles, only the composition of the grass, forb and shrub component is shown. On the shortgrass prairie sites in particular, cacti (*Opuntia* spp.) made important contributions to total production and on the high mountain grassland sites mosses and ferns were important. On the North Dakota mixed prairie site small clubmoss (*Selaginella densa*) made an important contribution to the total production. About four years are needed before the aboveground live vegetation is completely transferred into the litter layer.

The graph inside the inner lower circle of Fig. 14.2 shows the distribution by depth of the average annual root biomass. From 50 to 85 % of this biomass occurs in the top 10 cm depth increment in the soil profile. From about 5 to 20% occurs in the next 10 cm increment and successively smaller proportions as depth increases.

Characteristics of heterotrophs on these grasslands are summarized in Fig. 14.3 for aboveground arthropods, belowground arthropods, nematodes and vertebrates. A great variety of feeding strategies is found in the heterotrophs, especially among the invertebrates, which have feeding methods ranging from sucking sap to chewing plant tissue. The size of the circles in Fig. 14.3 represents estimates of average annual standing crop. Standing crop of vertebrates varies from about 0.05 $g \cdot m^{-2}$ for the desert grasslands up to about 0.9 $g \cdot m^{-2}$ for the tallgrass prairie sites. Man's domestic animals contribute 94% or more of the standing crop biomass on an annual basis in these grasslands. The animal density concentrations are estimated as typical values for these sites (Lewis, 1971).

Aboveground biomass of invertebrates was lowest for shortgrass prairie and desert grassland (about 0.06 $g \cdot m^{-2}$) and highest for shrub-steppe grassland (about 0.25 $g \cdot m^{-2}$). Mixed prairie and tallgrass prairie sites had intermediate values. Belowground, there were about 0.2 to 1.3 $g \cdot m^{-2}$ of soil macroarthropods, respectively, for the shortgrass prairie and tallgrass prairie. Our preliminary studies for nematodes suggest their biomass may exceed that of the belowground arthropods.

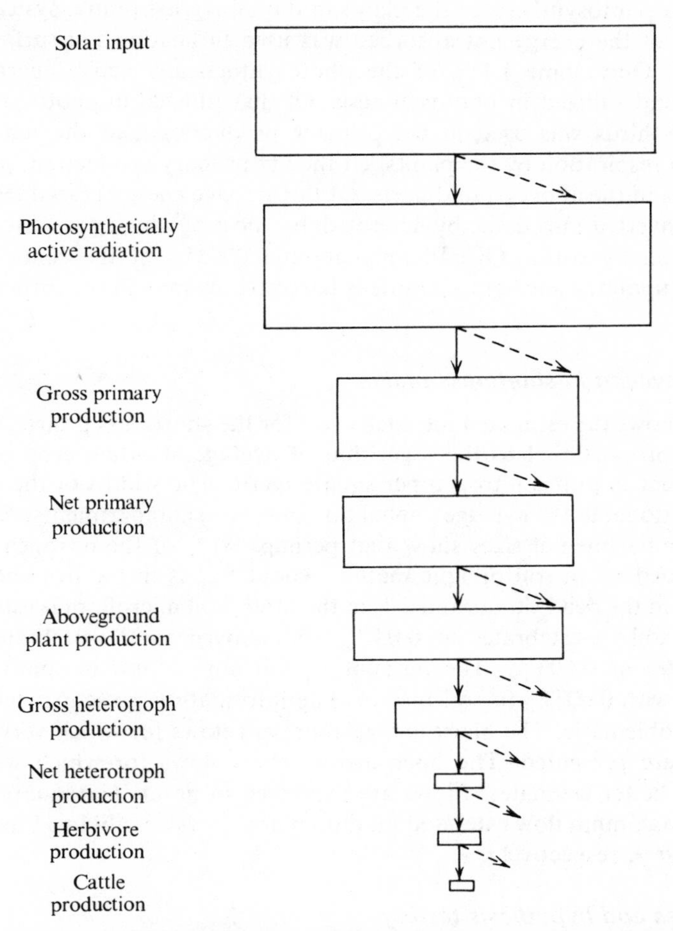


**Fig. 14.4.** Flow of energy through a shortgrass prairie grassland in a 5-month growing season. Boxes are logarithmically proportional to kilocalories in each compartment. © 1978 by *Frontiers*, published by the Academy of Natural Sciences of Philadelphia.

### *Energy flow in shortgrass prairie*

Analysis of energy flow has intrigued ecologists for more than 30 years, but relatively few detailed energy budgets are available for grasslands. I have summarized in Fig. 14.4 data for the shortgrass prairie in Colorado, where the boxes are scaled logarithmically proportional to kilocalories. The data represent energy flow for a 154-day growing season, defined by the period when soil water was sufficient for plant growth rather than on the basis of temperature. During this period about 1000000 kcal$\cdot$m$^{-2}$ of energy was inputed to the prairie's surface. Less than half of this energy is of wavelengths suitable for photosynthesis of the plants in the shortgrass prairie system. The remainder of the energy not absorbed was used in heating soil surface and vegetation. Only some 1.1 % of the photosynthetically active energy was absorbed and utilized in photosynthesis. Of that utilized in photosynthesis, about two-thirds was used in net primary production and the remainder was lost in respiration by the plants. Of the net primary production, perhaps one-sixth is in the aboveground parts. Of the herbage energy grazed less than 10 % is converted into tissue by heterotrophs, most of which goes into herbivores, especially cattle. Overall only about 0.0003 % of the solar energy impinging upon the shortgrass prairie is harvested by man in the form of beef cattle.

### *Nitrogen cycling in shortgrass prairie*

Fig. 14.5 shows the estimated nitrogen cycle for the shortgrass prairie. Circles again are proportional to the logarithm of average standing crop of each compartment in grams nitrogen per square metre. The widths of the arrows are proportional to the average annual flows between compartments. Average annual compartmental sizes show that perhaps 90 % of the nitrogen in the system is tied up in soil organic matter. About 5 % is in the live and dead roots, 1 % in the dead shoots and 2 % in the litter. Soil microflora is estimated at 0.8 %, soil invertebrates at 0.03 %, and aboveground vertebrates and invertebrates at 0.003 %. The amount in soil ammonium is about 0.2 % compared with 0.003 % in soil nitrate. The information presented on flows is more problematic. The black arrows represent flows for which very crude estimates are presented. The open arrows show flows for which we have somewhat better estimates. Flows are expressed in grams per square metre per year. Maximum flow rates and minimum flow rates are about 11 and 0.01 $g\ N \cdot m^{-2} \cdot yr^{-1}$, respectively.

### *Hypotheses and hypothesis testing*

As a contribution towards eventually developing a theory for grasslands, we have begun to formulate hypotheses about their structure, function and utilization. They are formulated in order that we can test the logical and empirical consequences of our assumptions. Hypotheses about grassland ecosystems can essentially be categorized into three overlapping levels (Van Dyne, Joyce & Williams, 1978): (i) those concerning organs or species; (ii) those at the population or community level; and (iii) those at the total-system level. The ecosystem processes are organized according to abiotic, producer, consumer or decomposer categories in Table 14.3. A name is provided for each process within each category, together with the source and the destination of the material involved in the process. Within the abiotic processes, separate categories are shown for heat, water and nutrient

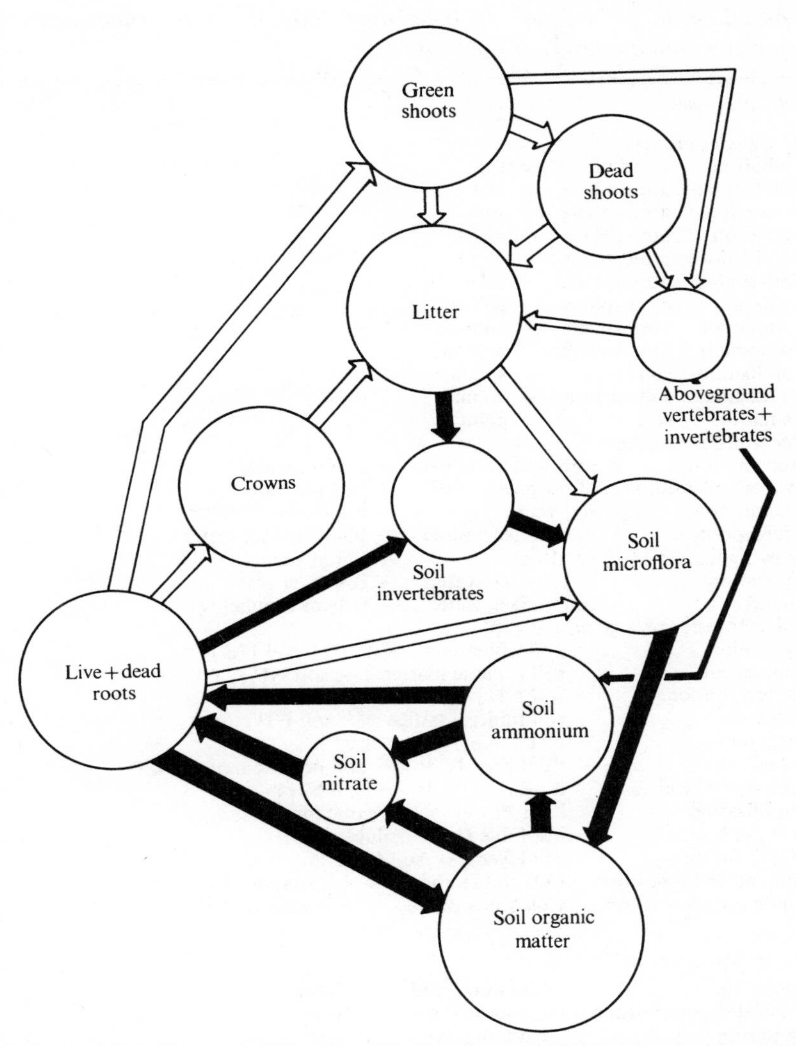


Fig. 14.5. Nitrogen cycling in a shortgrass prairie grassland. Circles are proportional to the logarithm of the average standing crop of each compartment in grams nitrogen per square metre. The widths of the arrows are proportional to the average annual flows between compartments. The black arrows represent rough estimates and illustrate our paucity of knowledge regarding this cycle. © 1978 by *Frontiers*, published by the Academy of Natural Sciences of Philadelphia.

transfers. Within the producer, decomposer and consumer categories carbon would generally be the material involved.

Hypotheses about ecosystems have been illustrated by Van Dyne, Joyce & Williams (1978) and need not be discussed in detail here. Examples of these hypotheses are shown in Fig. 14.6. A first step in the quantification of a

Table 14.3. *Example of processes for showing hierarchical organization at the physiological, organismal or population level of biotic components and equivalent abiotic levels of the ecosystem*

| Process | From | | To |
|---|---|---|---|
| *Abiotic processes* | | | |
| Heat transfer process | | | |
| Radiation | atmosphere | → | soil |
| Conduction | atmosphere | → | soil |
| Convection | atmosphere | → | soil |
| Radiation | atmosphere | → | plant |
| Conduction | atmosphere | → | plant |
| Convection | atmosphere | → | plant |
| Radiation | atmosphere | → | animals |
| Conduction | atmosphere | → | animals |
| Convection | atmosphere | → | animals |
| Conduction | soil layer | → | soil layer |
| Conduction | soil surface | → | animals |
| Radiation | soil | → | plant |
| Water transfer processes | | | |
| Precipitation | atmospheric water | → | surface water |
| Evapocondensation | atmospheric water | → | surface water |
| Evaporation | soil water | → | atmospheric water |
| Interception | atmospheric water | → | plant surface water |
| Snow melt | solid water | → | liquid water |
| Infiltration | soil surface water | → | soil layer water |
| Runoff | soil surface water | → | stream channel water |
| Nutrient transfer processes | | | |
| $N_2$ fixation | atmosphere | → | soil $NH_4^+$ and $NO_3^-$ |
| Mineralization | soil organic matter | → | soil $NH_4^+$, P |
| Immobilization | soil $NH_4^+$ | → | soil organic matter |
| Exchange | soil mineral lattice | → | soil $NH_4^+$ food |
| Oxidation | $NH_4^+$ | → | $NO_3^-$ |
| Humification | dead root N, P | → | soil organic matter N, P |
| Ammonification | dead root N, P | → | soil $NH_4^+$ |
| Humification | litter N | → | soil organic matter |
| Transformation | insoluble P | → | soluble P |
| $CO_2$ diffusion | soil layer | → | soil layer |
| Mineral decomposition | $CO_2$ soil storage | → | soil solution $CO_2$ |
| Buffer transformation | $CO_2$ soil storage | → | soil solution $CO_2$ |
| *Producer processes* | | | |
| Shattering | standing dead | → | litter |
| Physical decomposition | standing dead | → | litter |
| Shattering | standing live | → | litter |
| Expiration | standing live | → | standing dead |
| Expiration | live root | → | dead root |
| $CO_2$ diffusion | external atmosphere | → | stomatal atmosphere |
| Carbon fixation | stomatal atmosphere | → | organic compound |
| Translocation | leaf | → | stem |
| Translocation | stem | → | seed |
| Translocation | stem | → | crown storage |
| Translocation | crown | → | root |
| Respiration | stem | → | atmosphere |
| Respiration | root | → | soil $CO_2$ |
| Chemical transformation | labile | → | nonlabile compounds |

Table 14.3 (*cont.*)

| Process | From | | To |
|---|---|---|---|
| Absorption | soil solution nutrient | → | live root nutrient |
| Exudation | live root nutrient | → | soil solution nutrient |
| *Decomposer processes* | | | |
| Consumption | aboveground litter (mulch) | → | microbiota |
| Consumption | belowground litter, humic compounds, animal residues, dead roots | → | microbiota |
| Expiration | live active microbiota | → | dead microbiota |
| State transformation | live active microbiota | → | inactive microbiota |
| Expiration | inactive microbiota | → | dead microbiota |
| Humification | dead microbiota | → | humic compounds |
| Extracellular decomposition | belowground litter, humic compounds, animal residues, dead roots | → | degraded compounds |
| *Consumer processes* | | | |
| Emigration | in system | → | out of system |
| Immigration | out of system | → | in system |
| Diet selection | food available | → | food handled |
| Ingestion | food selected | → | food consumed |
| 'Wastetation' | food selected | → | food wasted |
| Storage | food selected | → | cache |
| Transportation | food selected | → | carried to young |
| Digestion | food consumed | → | food digested |
| Defecation | food consumed | → | faeces |
| Metabolization | food digested | → | food metabolized |
| Urination | food digested | → | urine |
| Eructation | food digested | → | atmosphere |
| Transformation | food metabolized | → | basal metabolism |
| Transformation | food metabolized | → | heat maintenance |
| Transformation | food metabolized | → | reproductive tissue |
| Transformation | food metabolized | → | milk |
| Transformation | food metabolized | → | nonfat |
| Transformation | food metabolized | → | fat |
| Transformation | food metabolized | → | integument |
| Transition | age/sex state$_i$ | → | age/sex state (see below) |

| | State$_i$ | | State$_j$ | |
|---|---|---|---|---|
| 1 | Hibernating or dipausing | → | 2–4, 8–10, 16–18 | 21 |
| 2 | Preproductive feeding | → | 1, 3, 4, 5, 8, 9 | 21 |
| 3 | Preproductive nonfeeding | → | 1, 2, 4, 5, 8, 9 | 21 |
| 4 | Male sexually active | → | 1, 5, 6 | 21 |
| 5 | Male sexually inactive | → | 1, 4 | 21 |
| 6 | Male incubating | → | 7 | 21 |
| 7 | Male nonincubating | → | 1, 4–6 | 21 |
| 8 | Female sexually active | → | 1, 9, 12, 17 | 21 |
| 9 | Female sexually inactive | → | 1, 8 | 21 |
| 10 | Female lactating | → | 11, 16 | 21 |
| 11 | Female nonlactating | → | 1, 8, 9 | 21 |
| 12 | Female gravid | → | 1, 10, 13, 14, 16–18 | 21 |
| 13 | Female nongravid | → | 8, 9 | 21 |
| 14 | Female incubating | → | 15, 18 | 21 |
| 15 | Female nonincubating | → | 1, 8, 9 | 21 |
| 16 | Female sexually active lactating | → | 8, 10 | 21 |
| 17 | Female sexually active gravid | → | 16 | 21 |
| 18 | Female sexually active incubating | → | 8, 14 | 21 |
| 19 | Female lactatinging gravid | → | 10, 12 | 21 |
| 20 | Female gravid incubating | → | 12, 13 | 21 |
| 21 | Dead | | | |

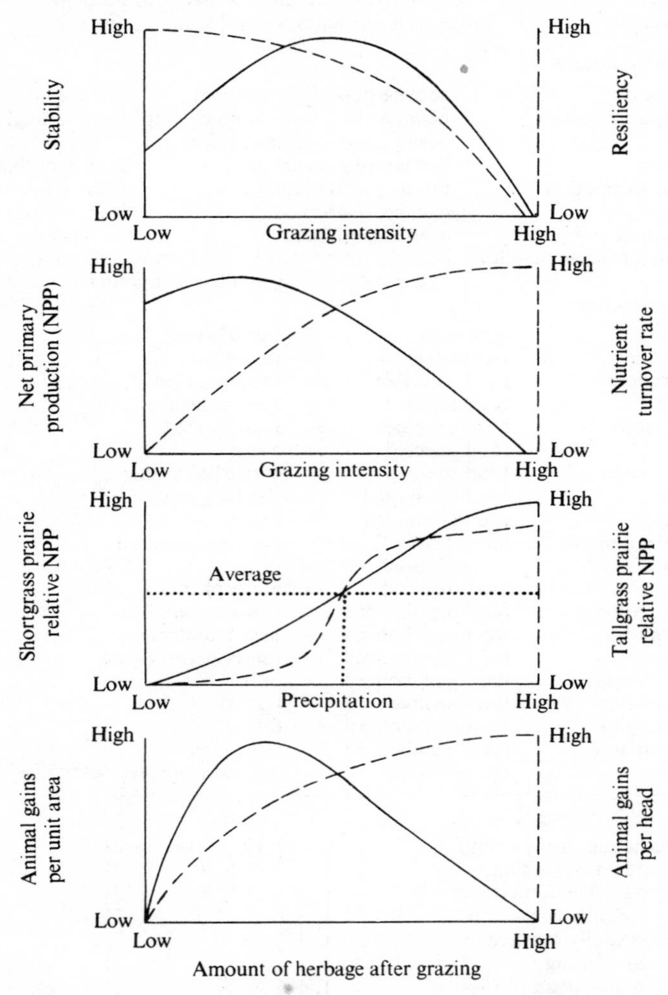


**Fig. 14.6. Some general hypotheses concerning natural semi-arid grassland responses to grazing and precipitation. (After Van Dyne, Joyce & Williams, 1978.)**

hypothesis is to put it into graphic form as illustrated. For example, in the lower graph of Fig. 14.6 the hypothesis is that as grazing intensity increases, gains per individual animal (of domestic animals) will continually decrease whereas gains per unit area will at first increase and then decrease.

Another set of hypotheses relates to the expected response of the system to stress. For example, in intensive renewable resource management, as compared with extensive management, the questions the managers ask are concerned with the sorts of responses that result from combinations of treatments such

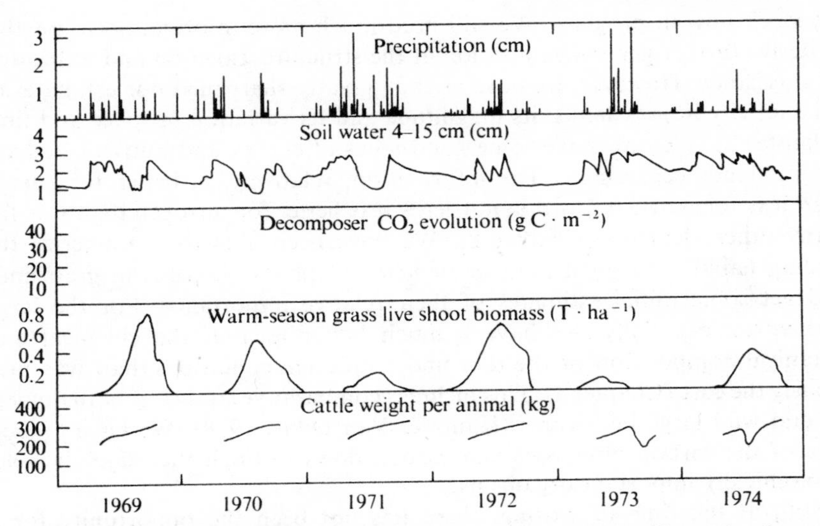


**Fig. 14.7. An example output from a grassland simulation model run illustrating dynamics of a shortgrass prairie over 6 years.**

as grazing, herbicides and fire. For the shortgrass prairie we recently developed such a set of 50 formal hypotheses derived from field experience and from general theory about ecosystems (Van Dyne *et al.*, 1977). We have conducted a set of computer experiments to test many of these hypotheses. To test a given hypothesis requires several computer runs, each run being specified by the model structure, the parameter values used in the flow expressions, the initial conditions and the driving variable records. Output obtained from one such run can be illustrated by the time-series dynamics in Fig. 14.7. Thus, in the example shown, precipitation is one of the driving variables, soil water is an abiotic state variable, warm-season grass live shoot biomass and cattle weight are biotic state variables, and decomposer carbon dioxide evolution represents a flow.

Only part of the information to be derived from model experiments can be gained from examination of such time-series graphs. Because we may run the model under several conditions to test a hypothesis, we frequently integrate flows and plot these integrated values against the level of a treatment (see Chapter 10, Fig. 10.9). Then we can fit empirical functions relating the integrated flow rate to the level of the treatment imposed. In effect, at this stage, we are making 'analyses of syntheses', where the analysis represents the empirical analysis made of the output of a series of simulation model runs; each simulation model is a synthesis of information about processes operative in the ecosystem.

The types of examples described in this section are only an illustration of the work now in progress. We still need much more information to be able to derive further generalizations about the structure, function and utilization of grasslands. However, we have already greatly sharpened our estimates of productivity in grasslands, its magnitude and its variation in space and time (Chapter 2). We now have some good ideas of energy and nutrient budgets in grasslands ecosystems. The information, relatively, is better for energy than it is for nutrients; within nutrients it is better for nitrogen than it is for many other elements (Chapter 8). We have been able to characterize the feeding habits of a great number of heterotrophic organisms in grasslands and get some initial estimates of their relative importance. For the large herbivores, especially, we have a much better idea of the chemical and botanical composition of the diet under different conditions than was previously the case (Chapter 4). This includes comparative studies of both domestic and wild large herbivores (Kautz & Van Dyne, 1978). We have a good idea of the carbon biomasses and carbon flows through the major, visible, economically important organisms.

Still, at the time of writing, there has not been the opportunity for a thorough examination of the data from many of the US/IBP Grassland Biome programmes. I am sure that when this is done we can look forward to finding many unexpected properties of and counterintuitive results about grassland ecosystems.

## Applications and implications for future efforts

In this section I will make projections about the needs and approaches for future efforts. Special attention will be given to the problems of managing renewable natural resources and the potential role of systems analysis, with emphasis on simulation modelling.

### *The nature of management-oriented modelling needs*

Cooper (1976) has provided a good discussion of ecosystem models and environmental policy. For practical purposes, we can consider environmental policy here in the sense of renewable natural resource management. The policy-maker's aim in this field is to improve man-dominated ecosystems, or at least to inhibit their degradation. Generally there is insufficient time and money to be able to test by experiment the consequences of alternative policy choices, and the manager then uses a mental construct or model of the system in his decision-making. This intuitive image of the world can often be improved upon and many ecologists contend that models based on ecosystem theory and principles will lead to sounder management decisions.

The manager does not have an easy task, however. Many of the problems he has to deal with are extremely complex and involve simultaneous consideration of many variables. Management of these systems with the usual tools of human judgement and intuition may lead to wrong decisions. Understanding the dynamics and utilization of resource systems requires being able to simulate the system under varying conditions and then being able to manipulate variables within the system for optimal return. Therefore, in the future we will see a new class of models developing which will have both simulation and optimization components. Not only does management of ecosystems lead to the need for dynamic optimization techniques, but specifically it must deal with the problems of long-range planning objectives. Some mathematical techniques are available, e.g. control theory, dynamic programming and calculus of variations, which can be used on problems of this type, but these techniques are inadequate for all but the simplest of real-life problems. Most resource system management models to date couple a simulation model, such as a system of difference equations, with an optimization model, such as linear programming (e.g. Swartzman & Van Dyne, 1972). The main use of a resource system management model will be to give managers better insight into the potential effects of their manipulations of resource systems under a variety of conditions. The use of such models, of course, should be an aid, not a prescriber, in making management decisions. Frequently the use of such models will be undertaken in a 'gaming' context.

Perhaps scientists have stayed in their ivory tower rather than delving into real-world problems. But, as Jeffers (1973) notes, the scientist must now risk his reputation in the application of his discipline to real problems of resource management. He can no longer retreat with honour to an academic application of methods to abstract problems. Likewise, the resource manager must modify if not abandon the now outdated mode of management decision-making based on subjective weighting of undefined assumptions.

As we move into this bold new world, we should proceed cautiously. Holling (1974) warns us against applying any integrated set of modelling techniques and procedures unless they have been designed from the start with the management policy questions as a first consideration. Thus, it will require structuring new models rather than simply transferring old models to the new problem.

Another caution concerns the mass of information that might be generated from applying models. The key results from modelling experiments must be extracted, condensed and communicated simply and clearly. Again, the problem is twofold. Attending any scientific meeting, particularly where model output is presented, is an experience in unreadable slides, indecipherable computer output and incomprehensible jargon (Cooper, 1976). The decision-maker may also build up resistance if he is forced to face massive amounts of computer output and to interpret it alone. On the other hand, the resource managers must realize that the new tool is powerful, but not infallible. To assess its true worth will require considerable effort on their part in examining different kinds and amounts of output from computer models.

The unfamiliarity of the decision-makers, and sometimes even the heads of research organizations, with ecological modelling is further emphasized by SCOPE (1976) and Frenkiel & Goodall (1978). Frequently the decision-makers are unwilling to trust to the results of models when those results run counter to their expectations. Thus it is especially important that they understand the inputs to and the outputs from models. Frequently resource management agencies do not have many employees who are familiar with the modelling process. One technique to achieve involvement, training and interpretation is the use of workshops, where representatives of management agencies, industries, local interest groups and scientists can be brought together. In some instances models are constructed, operated and interpreted within the workshop itself, but there is a possibility here that the harm done by producing erroneous results may outweigh the good done in generating enthusiasm and commitment among the participants. Generally, the results should not be presented to laymen until the model has been carefully evaluated and accepted by knowledgeable scientists. Perhaps a compromise can be reached by using in the workshop 'building blocks' of submodels or subprogrammes which are already well-documented and discussed. Secondly, it is necessary to caution all participants of the preliminary nature of any findings derived in a 2-week workshop.

Some further needs for such management models can be derived from Cooper's (1976) article. When policy-oriented models are designed, an attempt should be made to consider their future use. It may be practical to design the model so that it can be applied to related problems with a minimum of reprogramming. This leads to the art of model structuring, which is beyond the scope of this section, but which is referred to in several other places in this volume (Chapters 1 and 10).

There are some key problems in budgeting modelling activities in the resource management decision-making field. Generally the budget for a programme is developed by a scientist who then finds that the funds available are less than those requested. This requires negotiation between the agency and the investigator. Usually it is items of documentation and external evaluation that get cut from the budget. A model may be developed eventually but is difficult or impossible to transfer to the potential users because of inadequate structure or documentation. This may be one of the reasons why SCOPE (1976) concludes that the most successful users of models tend to be the modellers themselves. They infer that this is due to their familiarity with and confidence in the effort. But the purpose of modelling in resource management is to develop models usable by decision-makers. It may be, however, that we need more decision-makers who were once modellers or who at least have a fairly sophisticated knowledge of technical modelling matters and systems analysis methodology. This implies some change in the training of managers and scientists.

If a model is going to be useful to a decision-maker it must be run under a wide variety of conditions and it must be subjected to formal sensitivity analysis. Although the art of sensitivity analysis is not well developed, various ingenious methods for conducting such analyses can be derived and applied to particular problems.

### *Organization and design of long-term studies*

One of the self-criticisms of the IBP was that 'we were too often trying to do too much in too short a time' (Bourlière, 1975). The need for long-term studies was repeated again and again in both early and late phases of IBP. For example, in a discussion of long-term ecological measurements a group of scientists noted that 'too much of our research seems to appear out of a temporal and spatial vacuum' (Anonymous, 1977).

One of the difficulties for long-term ecological research in the USA comes from our method of funding research, which seems to obstruct rather than promote continued, long-term studies by good scientists on good projects. Mechanisms to promote long-term collection of high-quality data have been put forward. Selected projects of a long-term nature should be funded, perhaps providing younger scientists relatively small sums to do inexpensive studies involving infrequent measurements. Even if this is done, however, it must be realized that many records must be kept for longer than one person's career. Thus, some of the projects must be institutionalized. Government agencies seem to be the logical choice but there is real need for many scientists – those needing the data as well as those collecting it – to be involved in the planning of long-term government projects.

Many of the above recommendations are not just needed in the USA but also in other nations. For example, in Canada recommendations have been put forward for integrated research, management and development specifically directed at long-term ecosystem studies. The Science Council of Canada (1970) notes that IBP studies have clearly indicated the consequences of resource uses on ecosystems and the potential for intentional large-scale ecosystem manipulations. Even though we don't yet fully know how ecosystems work, a principle derived from IBP studies is that a management treatment in one part of the system may set off a chain of repercussions. We also know that there are limits to the resilience and stability of natural associations which man modifies and stresses.

The research required to solve the problems discussed above must be large-scale and long-term. The large-scale aspect is worthy of further consideration. Consider the interactive effects of multiple-purpose use of resources. To understand and document the influence of mechanized logging operations on runoff, stream quality, nutrient leaching, ecological succession, and a host of effects on fish and wildlife, will require projects on a landscape scale (Science Council of Canada, 1970). If such studies of stressing ecosystems are done on a large scale, it is more likely that the responses produced will be equivalent to those resulting when the practices are incorporated into management. It is difficult to overemphasize the dual needs of long duration and large area for ecological studies if they are to be directly relevant to resource management.

## A personal reflection on international scientific cooperation

Writing this chapter and examining the materials that have accumulated causes me to stop and reflect on the IBP experience. As I sort through the mass of assorted correspondence accumulated during the development and editing of this volume, I cannot help but reflect on the comments of Jim Cragg in the Preface and Alicja Breymeyer in the Introduction. More than 10 years have passed since my introduction to the international aspects of grassland studies. I have had a good opportunity to meet a large number of grassland scientists from around the world. They have been extremely tolerant of my lack of understanding of their language and have helped me understand more of their research philosophy and the way they approach ecological problems in their countries. In the US/IBP Grassland Biome study we were extremely fortunate in having access to good library materials, support staffs and the means of generating manuscripts and reports readily. We have attempted to disseminate many of our materials to groups throughout the world. Undoubtedly this burdened them with a great deal of paper. But they did examine it and I feel there was a great deal of vigour gained from the international cross-fertilization of ideas.

We have greatly benefited in our grassland project from the participation of scientists from overseas. I would like to comment particularly on the outstanding contributions of J. S. Singh of India, who not only published a great deal during his 3-year stay with us but also gave much inspiration to several participants in our programme. Similar contributions and accomplishments could be noted for other scientists.

I am particularly interested in the way many foreign scientists adapted to our 'systems approach' and made important contributions in this area (see for example, some of the modelling papers in the Fennoscandian synthesis publications as well as individual papers of Wielgolaski and Framsted of Norway, Connor and Haydock of Australia, and Morris of South Africa).

Van Wyk of South Africa made a major contribution by developing special devices for breaking up root cores in order to rinse out root materials (Van Wyk, 1972, 1974) and an air flotation system for separating live matter from dead. The last device, affectionately called the 'giraffe' to denote the South African origin, saved us hundreds of hours of tedious separation of the live from dead shoots of the warm-season shortgrasses from hundreds of plots in our Pawnee Site study.

Breymeyer and Grodzinski brought us much knowledge about detailed approaches to population dynamics and energetics of invertebrates and small mammals as developed in Polish studies.

Last, but not least, we had a great deal of beneficial contact with Canadian investigators. A group of 38 of our scientists went to visit their field site and met and discussed with their scientists. Coupland and others involved in the administration of the Matador programme gave freely of their time in helping us review some of our proposals as we continued to change our organization and approach during the 10 years.

These are just a few of the examples of how research in our programme was enriched by international cooperation. It would be most interesting to ask the many IBP grassland scientists in 10 years time what impact the programme has had subsequently on their careers and research. To me, the international aspects of the IBP experience were enlightening and rewarding.

It is with great pleasure that I acknowledge the participants of the US/IBP Grassland Biome Study with whom I have been associated in various degrees over the past 10 years. I have drawn freely upon information generated by or summarized by these participants in my presentation.

Thanks are due to Sallie Van Dyne for her assistance and for the tabulation of information required for the graphs and tables of this chapter, and to Ellen Gorenzel and Lindsay Nimrod for their dedication to the task of helping me produce this chapter to a stiff deadline and in competition with many other activities.

Lastly, I want to express my thanks to Cambridge University Press for their cooperation in many ways in the development of this volume. Particularly, I want to single out Jane Farrell who has done a most painstaking job on detecting inconsistencies, illogical statements, and poor organization in the manuscripts. Her assistance has greatly improved the chapters of the volume.

## References

Anderson, J. R. & Dent, J. B. (1975). Agricultural systems analysis: retrospect and prospect. In *Systems analysis in agricultural management*, ed. J. B. Dent & J. R. Anderson, pp. 383–8. Wiley, New York.

Andrews, R. M., Coleman, D. C., Ellis, J. E. & Singh, J. S. (1974). Energy flow relationships in a shortgrass prairie ecosystem. In *Structure, functioning and and management of ecosystems: Proceedings of the First International Congress of Ecology*, pp. 22–8, Centre for Agricultural Publishing and Documentation, Wageningen, The Netherlands.

Anonymous (1977). *Long-term ecological measurements.* Report of a Conference, Woods Hole, Massachusetts, 16–18 March 1977.

Auerbach, S. I. (1971). The deciduous forest biome program in the United States of America. In *Productivity of forest ecosystems: Proceedings of the Brussels Symposium, 1969*, pp. 677–84. HMSO, London.

Ayyad, M. (Principal Investigator) (1975). *Systems analysis of Mediterranean desert ecosystems of Northern Egypt 'Samdene'*. EPA Foreign Currency Grant PR 3-54-1, Progress Report 2, University of Alexandria, Egypt.

Ayyad, M. (Principal Investigator) (1976). *Systems analysis of Mediterranean desert ecosystems of Northern Egypt 'Samdene'*. EPA Foreign Currency Grant PR 3-54-1, Progress Report 3, University of Alexandria, Egypt.

Barrett, G. W., Van Dyne, G. M. & Odum, E. P. (1976). Stress ecology. *BioScience*, **26**, 192–4.

Bliss, L. C. (ed.) (1977). *Devon Island IBP project: high arctic ecosystem.* Project Report, Department of Botany, University of Alberta.

Bourlière, F. (1975). The meaning of IBP for the future. In *The evolution of IBP*, ed. E. B. Worthington, *IBP Synthesis Series 1*, pp. 131–9. Cambridge University Press, London.

Breymeyer, A. (ed.) (1972). Productivity investigation of two types of meadows in the Vistula Valley. *Ekologia Polska*, **19**, nos. 7–19.

Coleman, D. C., Andrews, R. M., Ellis, J. E. & Singh, J. S. (1976). Energy flow and partitioning in selected man-managed and natural ecosystems. *Agro-Ecosystems*, **3**, 45–54.

Cooper, C. F. (1976). Ecosystem models and environmental policy. *Simulation*, **23**, 133–8.

Coupland, R. T. (ed.) (1979). *Grassland ecosystems of the world: analysis of grasslands and their uses. IBP Synthesis Series 18*. Cambridge University Press, London.

Coupland, R. T., Willard, J. R. & Ripley, E. A. (1974). *Summary of activities, 1967–74. Matador Project Technical Report 69*. University of Saskatchewan, Saskatoon, Saskatchewan, Canada.

Coupland, R. T., Zacharuk, R. Y. & Paul, E. A. (1969). Procedures for study of grassland ecosystems. In *The ecosystem concept in natural resource management*, ed. G. M. Van Dyne, pp. 25–47. Academic Press, New York & London.

DeMichele, D. W. (1975). An evaluation of modelling, systems analysis, and operations research in defining agricultural research needs and priorities in pest management. *Iowa State Journal of Research*, **49**, 597–621.

deWit, C. T. (1970). Dynamic concepts in biology. In *Prediction and measurement of photosynthetic productivity*, pp. 17–23. Centre for Agricultural Publishing and Documentation, Wageningen, The Netherlands.

Duvigneaud, P. & Kestemont, P. (ed.) (1977). *Productivité biologique en Belgique*. Duculot, Paris & Gembloux.

Ellenberg, H. (ed.) (1971). *Integrated experimental ecology: methods and results of ecosystem research in the German Solling project*. Springer-Verlag, Berlin.

Frenkiel, F. N. & Goodall, D. W. (ed.) (1978). *Simulation modelling of environmental problems*. Wiley, New York.

Gessel, S. G. (1972). Organization and research program of the western coniferous forest biome. In *Research on coniferous ecosystems: first year's progress in the coniferous forest biome, US/IBP*, ed. G. F. Franklin, L. F. Dempster & R. H. Waring, pp. 7–14. Pacific Northwest Forest and Range Experiment Station, Forest Service, USDA, Portland, Oregon.

Goodall, D. W. (1972). Building and testing ecosystem models. In *Mathematical models in ecology*, ed. J. N. R. Jeffers, pp. 173–94. Blackwell Scientific Publications, Oxford.

Heal, O. W. & Perkins, D. F. (1978). *The ecology of some British moors and montane grasslands. Ecological Studies 27*. Springer-Verlag, New York.

Holling, C. S. (1974). *Modelling and simulation for environmental impact analysis.* International Institute for Applied Systems Analysis, Research Memorandum RM-74-4. Schloss Laxenburg, Austria.

Innis, G. S. (1975*a*). Role of total system models in the grassland biome study. In *Systems analysis and simulation in ecology*, vol. 3, ed. V. C. Patten, pp. 13–47. Academic Press, New York & London.

Innis, G. S. (1975*b*). A self-organizing approach to ecosystem modelling. In *New directions in the analysis of ecological systems*, part 2, ed. G. S. Innis, pp. 179–87. Simulation Council Proceeding Series, Society for Computer Simulation, La Jolla, California.

Jeffers, J. N. R. (1973). Systems modelling and analysis in resource management. *Journal of Environmental Management*, **1**, 13–28.

Kautz, J. E. & Van Dyne, G. M. (1978). Comparative analyses of diets of bison, cattle, sheep, and pronghorn antelope on shortgrass prairies in northeastern Colorado, USA. In *Proceedings of the First International Rangeland Congress*, pp. 438–43. Society of Range Management, Denver, Colorado.

Ketner, P. (1972). *Primary production of salt-marsh communities on the island of Terschelling in the Netherlands*. Thobsen Offset, Nijmengen.

Lewis, J. K. (1971). A synthesis of structure and function, 1970. In *Preliminary analysis of structure and function in grasslands*, ed. N. R. French, *Range Science Department*, *Science Series 10*, pp. 317–87. Colorado State University, Fort Collins.

McIntosh, R. P. (1974). Plant ecology 1947–1972. *Annals of the Missouri Botanical Gardens*, **61**, 132–65.

Major, J. (1969). Historical development of the ecosystem concept. In *The ecosystem concept in natural resource management*, ed. G. M. Van Dyne, pp. 9–22. Academic Press, New York & London.

Mar, B. W. (1974). Problems encountered in multidisciplinary resources and environmental simulation models development. *Journal of Environmental Management*, **2**, 83–100.

Notterman, J. M. & Mintz, D. E. (1965). *Dynamics of response*. Wiley, New York.

Numata, M. (ed.) (1975). Ecological studies in Japanese grasslands with special reference to IBP area. In *JIBP synthesis*, *volume 13*.

Pandeya, S. C., Sharma, S. C., Jain, H. K., Pathak, S. T., Palwal, K. C. & Bhanot, V. M. (1977). *The environment and Cenchrus grazing-lands in Western India*: *an ecological assessment*. Dept of Biosciences, Saurashtra University, Rajkot, India.

Patten, B. C. (ed.) (1971). *Systems analysis and simulation in ecology*, vol. 1. Academic Press, New York & London.

Patten, B. C. (ed.) (1972). *Systems analysis and simulation in ecology*, vol. 2. Academic Press, New York & London.

Patten, B. C. (ed.) (1975). *Systems analysis and simulation in ecology*, vol. 3. Academic Press, New York & London.

Petrusewicz, K. (ed.) (1974). Analysis of a sheep pasture ecosystem in the Pieniny Mountains. *Ekologia Polska*, **22** (3/4), 732pp.

Rodell, C. F. (1977). A grasshopper model for a grassland ecosystem. *Ecology*, **58**, 227–45.

Rosswall, T. & Heal, O. W. (ed.). Structure and function of tundra ecosystems. Papers presented at IBP Tundra Biome Fifth International Meeting on Biological Productivity of Tundra, Abisko, Sweden, April 1974.

Rychnovska, M. & Ulehlova, B. (1975). *Autökologische Studie der Tschechoslowakischen Stipa-Arten*. Czechoslovakian Academy of Sciences, Prague.

Science Council of Canada (1970). *This land is their land: a report on fisheries and wildlife research in Canada*. Queens Printer for Canada, Ottawa, Ontario.

SCOPE (1976). *Environmental modelling and decision making: the United States experience*. Praeger, New York.

Shubik, M. & Brewer, G. (1972). *Models, simulation, and games: a survey*. Rand Corporation Report R1060-ARPA/RC.

Sims, P. L. & Singh, J. S. (1978*a*). The structure and function of ten western North American grasslands. I. Abiotic and vegetational characteristics. *Journal of Ecology*, **66** (1), 251–85.

Sims, P. L. & Singh, J. S. (1978*b*). The structure and function of ten western North American Grasslands. II. Intraseasonal dynamics in primary producer compartments. *Journal of Ecology*, **66** (2), 547–72.

Sims, P. L. & Singh, J. S. (1978*c*). The structure and function of ten western North American grasslands. III. Net primary production, turnover, and efficiencies of energy capture and water use. *Journal of Ecology*, **66** (2), 513–97.

Sims, P. L. & Singh, J. S. (1978*d*). The structure and function of ten western North American grasslands. IV. Compartmental transfers in system transfer functions. *Journal of Ecology*, in press.

Smith, F. M., Singh, J. S., Coleman, D. C., Ellis, J. E. & Andrews, R. M. (1974). Total system studies of seven grassland types of North America. In *Improvement of natural and production of seeded meadows and pastures*, compiled by V. G. Iglovikow, A. P. Movssissyants, F. F. Sarganov & A. A. Stepanenko. XII International Grassland Congress, Moscow, pp. 383–7.

Soriano, A. (ed.) (1975). *Productividad primaria neta de sistemas herbaceosos: descripción del area de estudio en los pastizales de la depresión del Salado, Provincia de Buenos Aires, Argentina*. Commission de Investigaciones Científicas de la Provincia de Buenos Aires, Argentina.

Steinhorst, R. K. (1973). Validation tests for ecosystem simulation models. (Manuscript unpublished.)

Swartzman, G. L. & Van Dyne, G. M. (1972). An ecologically based simulation–optimization approach to natural resource planning. *Annual Review of Ecology and Systematics*, **3**, 347–98.

Tansley, A. G. (1935). The use and abuse of vegetation concepts and terms. *Ecology*, **16**, 284–307.

Traczyk, T. (ed.) (1976). The effect of intensive fertilization on the structure and productivity of meadow ecosystems. *Polish Ecological Studies*, **2** (4), 332pp.

USSR/IBP (1975). *Resources of the biosphere*, *Synthesis of the Soviet Studies for the International Biological Programme 1*. Academy of Sciences of the USSR.

Van Dyne, G. M. (1972). Organization and management of an integrated ecological research program – with special emphasis on systems analysis, universities, and scientific cooperation. In *Mathematical models in ecology*, ed. J. N. R. Jeffers, pp. 111–72. Blackwell Scientific Publications, Oxford.

Van Dyne, G. M. (1975). Some procedures, problems, and potentials of systems-oriented, ecosystem-level, research programs. In *Procedures and examples of integrated ecosystem research*, pp. 4–58. *Technical Report No. 1, Swedish Coniferous Forest Project*. Uppsala, Sweden.

Van Dyne, G. M. (1978). Foreword: perspectives on the ELM model and modelling effort. In *Grassland simulation model*, ed. G. S. Innis, *Ecological Studies 26*, pp. v–xxii. Springer-Verlag, New York.

Van Dyne, G. M. & Abramsky, Z. Z. (1975). Agricultural systems models and modelling: an overview. In *Study of agricultural systems*, ed. G. E. Dalton, pp. 23–106. Applied Science Publishers Ltd, London.

Van Dyne, G. M., Joyce, L. R. & Williams, B. K. (1978). Models and the formulation and testing of hypotheses in grazing land ecosystem management. In *The breakdown and restoration of ecosystems*, ed. M. W. Holdgate & M. J. Woodman, pp. 41–84. Plenum Press, New York.

Van Dyne, G. M., Joyce, L. A., Williams, B. K. & Kautz, J. E. (1977). *Data hypotheses, simulation and optimization models, and model experiments for shortgrass prairie grazinglands. A summary of recent research.* Report to the Council on Environmental Quality, Executive Office of the President, Washington, DC (NTIS Accession PB273 738).

Van Dyne, G. M., Smith, F. M., Czaplewski, R. L. & Woodmansee, R. G. (1978). Analyses and syntheses of grassland ecosystem dynamics. In *Glimpses of Ecology*, ed. J. S. Singh & B. Gopal. (In press.)

Van Keulen, H., Louwerse, W., Sibma, L. & Alberda, T. H. (1975). Crop simulation and experimental evaluation: a case study. In *Photosynthesis and productivity in different environments*, ed. J. P. Cooper, *IBP Synthesis Series 3*, pp. 623–43. Cambridge University Press, London.

Van Wyk, J. J. P. (1972). *A preliminary report on new separation techniques for live–dead aboveground grass, herbage, and roots from dry soil cores. US/IBP Grassland Biome Study Technical Report 144.* Colorado State University, Fort Collins, Colorado.

Van Wyk, J. J. P. (1974). A mechanical root washer. In *Proceedings of the Grassland Society of South Africa*, vol. 9, pp. 165–7.

Wielgolaski, F. E. (ed.) (1975*a*). *Fennoscandian tundra ecosystems*, part 1, *Plants and microorganisms*. Springer-Verlag, New York.

Wieglolaski, F. E. (ed.) (1975*b*). *Fennoscandian tundra ecosystems*, part 2, *Animals and systems analysis*. Springer-Verlag, New York.

Woodmansee, R. G., Dodd, J. L., Bowman, R. A., Clark, F. E. & Dickinson, C. E. (1978). Nitrogen budget of a shortgrass prairie ecosystem. *Oecologia*, **34**, 363–76.

# Index

Large herbivores are generally indexed by common name rather than Latin name. Table 4.3, pp. 288–93, gives Latin and common names for selected large herbivores.

# Index

# PROFESSIONAL CONDUCT CASEBOOK

## SECOND EDITION

# PROFESSIONAL CONDUCT CASEBOOK

SECOND EDITION

Kenneth Hamer

*Barrister of the Inner Temple*
*Recorder of the Crown Court*

UNIVERSITY PRESS

Great Clarendon Street, Oxford, OX2 6DP,
United Kingdom

Oxford University Press is a department of the University of Oxford.
It furthers the University's objective of excellence in research, scholarship,
and education by publishing worldwide. Oxford is a registered trade mark of
Oxford University Press in the UK and in certain other countries



First Edition published in 2013
Second Edition published in 2015

Impression: 1



Published in the United States of America by Oxford University Press
198 Madison Avenue, New York, NY 10016, United States of America

British Library Cataloguing in Publication Data
Data available

Library of Congress Control Number: Data Available

ISBN 978–0–19–872859–7

Printed and bound by
Lightning Source UK Ltd

# FOREWORD

The publication of the first edition of Kenneth Hamer's *Professional Conduct Casebook* in 2013 was welcomed by barristers and solicitors working in this field as an invaluable means of keeping on top of the ever-increasing flow of reports of disciplinary cases. As Roger Henderson QC observed in his foreword to the first edition, this is an area where the sheer volume of cases makes it extremely difficult for practitioners to keep up to date and to identify those decisions which raise key issues of law. As a result, Kenneth's comprehensive work in a readily accessible format has met a real need.

The first edition was published at a time when the Law Commission was engaged with the Scottish Law Commission and the Law Commission of Northern Ireland on a project on the Regulation of Health Care Professionals and the Regulation of Social Care Professionals in England. The lawyers working on the project found it of the greatest assistance. They made extensive use of the book when drafting the final report and draft bill (Law Com No 345 / Scot Law Com No 237 / NILC 18 (2014)) and only wished that it had been available while they had been working on the consultation paper.

This book takes a helpfully broad view of the relevant fields of professional conduct, extending far beyond health care professionals and legal professionals to include chartered accountants, veterinary surgeons, footballers and even the clergy. The second edition includes an extremely important new section on the concept of deficient professional performance as a ground for regulatory intervention, a matter on which the Law Commissions placed particular emphasis in their report, where they argued that it should be given greater prominence in a new legal framework. It also includes a revised chapter on "no case to answer" which is a timely reminder that due legal process is essential in such cases.

The publication of this second edition so soon after the first is an indication of the speed with which this area of the law is developing. I hope I may be forgiven for expressing the hope that the implementation of the Law Commissions' recommendations on Regulation of Health Care Professionals and the Regulation of Social Care Professionals in England may soon necessitate the publication of a third edition.

For his labours in producing this second edition, Kenneth deserves the warm thanks and congratulations of a grateful profession.

Rt. Hon. Lord Justice Lloyd Jones
Chairman of the Law Commission
1 May 2015

# PREFACE

The number of professional conduct cases coming before the Courts has continued to increase, and is now running at about 150 cases a year, mainly in the Administrative Court in London and Manchester, or in the Court of Appeal. Additionally cases involving important issues in the field of regulatory and disciplinary law are being heard by the Court of Session or the High Court in Northern Ireland. I have sought to include in this edition of *Professional Conduct Casebook* all significant judgments handed down by the senior courts up to 1 March 2015.

The structure of the second edition remains the same. There are 71 chapters listed in alphabetical order ranging from Absence of the Practitioner to Witnesses, with a chapter on Words and Phrases. The aim is for each chapter to identify the relevant disciplinary cases relating to that topic, whilst at the same time guiding the reader to other relevant chapters and cases that may assist relating to the subject matter. New to this edition are key words included in the margin beside each case, enabling the reader to see at a glance the critical features of the case in question. The book seeks to cover an extensive range of professions from doctors to other healthcare professionals, lawyers to surveyors, accountants to financial service providers, teachers, the police, the prison service and sports law.

There have been significant developments since the first edition was published in March 2013. In April 2014 the Law Commission, the Scottish Law Commission and the Northern Ireland Law Commission published their final report *Regulation of Health Care Professionals: Regulation of Social Care Professionals in England* setting out a new single legal framework for the regulation of all health and social care professionals, including doctors, dentists, nurses and pharmacists. The final report included a draft Bill and we await to see what developments take place in the next Parliament. The General Medical Council (Fitness to Practise) Rules 2004 were amended in 2009, 2013 and most recently on 25 June 2014. Further amendments are currently out for consultation to be added to the 2004 rules. I have referred to the present position for convenience as the 2014 rules although this should be understood as meaning the 2004 rules incorporating current amendments. The General Medical Council (Fitness to Practise and Over-arching Objective) and the Professional Standards Authority for Health and Social Care (References to Court) Order 2015, made on 19 March 2015, establishes the Medical Practitioners Tribunal Service as a statutory committee of the GMC to strengthen the separation between its investigation and adjudication arms, and makes provision for there to be an overriding objective for the making of procedural rules for securing that cases are dealt with fairly and justly. Other regulators have made amendments to their fitness to practise rules since the last edition, including the General Optical Council, the Bar Standards Board, the Chartered Institute of Public Finance & Accountancy, and the Police.

The case studies remain at the heart of the book. There are over 1,000 cases reported in the text, with some 250 new cases since the last edition. The book contains extensive summaries of the cases, outlining the facts and decision reached by the court in each case, often citing key passages from the judgments or summarising the reasons given by the judge or judges concerned. All cases are fact sensitive and the cases in the book are merely examples in which the subject matter at hand has been judicially considered by the courts.

Kenneth Hamer<br>Henderson Chambers<br>2 Harcourt Buildings<br>Temple<br>London EC4Y 9DB
25 April 2015

# ACKNOWLEDGEMENTS

This edition would not have been possible without the support and encouragement I have received from professional colleagues and friends, and to whom I am indebted and remain grateful. I should like to express my appreciation to Sir David Lloyd Jones, Chairman of the Law Commission and a Lord Justice of Appeal, for writing the Foreword to the second edition. I am grateful to James Leonard of the Professional Discipline Lawyers Group on LinkedIn, and to Nigel Parry and Peter Gribble for their invaluable help in drawing my attention to cases on the Internet. I should also like to acknowledge the help I have received from Central Law Training and IBC Legal Conferences who regularly host excellent conferences on the latest developments in regulatory and disciplinary proceedings, and to the General Medical Council for its training days for legal assessors, all of which enable recent cases in the field of regulation and disciplinary law to be discussed. I owe a debt of gratitude to Oxford University Press, who have as always been helpful, courteous and supportive throughout. I should like especially to thank Andy Redman, Editorial Director, Faye Mousley, Commissioning Editor, Practitioner Law, and Amy Jones, former Assistant Commissioning Editor. I am extremely grateful to Barath Rajasekaran and Gnanambigai Jayakumar who have overseen the production of the second edition through to publication and have been my main contacts throughout the production process. The book would not have been brought to fruition without the support of Vanessa Plaister, Editorial freelancer and an Advanced Member of the Society for Editors and Proofreaders, Debbie Harris for compiling the tables of cases and legislation, and Kim Harris for compiling the index. My deepest thanks must also go, yet again, to Mrs Janice Alexander, who has typed and re-typed this edition like the earlier one with dedication and charm, and without whose assistance it would not have been possible to complete the work. Finally, my wife Victoria for proofreading the book and showing patience and forbearance.

# TABLE OF CONTENTS

Amendment
Integrity
Proportionality

# TABLE OF CASES


7722656 Canada Inc. v. Financial Conduct Authority [2013] EWCA Civ 1662; [2013] LR (FC) 207 . . . 28.21, 43.18

A&B Solicitors v Law Society of Hong Kong CACV 269/2004 . . . 35.11
A Health Authority v X [2001] EWCA Civ 2014 . . . 18.25
A v General Medical Council [2004] EWHC 880 (Admin); [2004] A.C.D. 54 . . . 18.22
AT v University of Leicester *See* R (AT) v University of Leicester
AWG Group Ltd v Morrison [2006] EWCA Civ 6; [2006] 1 W.L.R. 1163; [2006] 1 All E.R. 967; Independent, January 25, 2006 . . . 6.07
AXM v Bar Standards Board [2011] EWHC 2990 (QB) . . . 18.15
Aaron v Law Society [2003] EWHC 2271 (Admin); [2003] N.P.C. 115 . . . 17.23, 62.08
Abdalla v Health Professions Council [2008] EWHC 3498 (Admin) . . . 1.07, 45.14
Abdullah v General Medical Council [2012] EWHC 2506 (Admin) . . . 38.30
Abiodun v Nursing and Midwifery Council [2015] EWHC 434 (Admin) . . . 19.80
Abrahaem v General Medical Council [2008] EWHC 183 (Admin) . . . 11.10, 25.36, 25.38, 36.08, 59.06, 59.11, 59.18
Abusheikha v. General Medical Council [2003] UKPC 80 . . . 22.11
Adams v Law Society of England and Wales [2012] EWHC 980 (QB) . . . 49.10, 52.13, 52.14
Addis v Crocker [1961] 1 Q.B. 11; [1960] 3 W.L.R. 339; [1960] 2 All E.R. 629; (1960) 104 S.J. 584 . . . 52.12
Adediwura v Nursing and Midwifery Council [2013] EWHC 2238 (Admin) . . . 19.70
Adeeko v Solicitors Regulation Authority [2012] EWHC 841 (Admin) . . . 15.15
Adegboyega v Nursing and Midwifery Council [2013] EWHC 2647 (Admin) . . . 19.71
Adegbulugbe v Nursing and Midwifery Council [2013] EWHC 3301 (Admin); (2014) 135 B.M.L.R. 171 . . . 4.41
Adelakun v Solicitors Regulation Authority *See* R (Adelakun) v Solicitors Regulation Authority
Adesemowo and Alafe-Aluko v Solicitors Regulation Authority [2013] EWHC 2020 (Admin) . . . 4.26, 28.19
Adesemowo v Solicitors Regulation Authority [2013] EWHC 2015 (Admin) . . . 56.20
Adeyemi v General Medical Council *See* R (Adeyemi) v General Medical Council
Adu v General Medical Council [2014] EWHC 1946 (Admin) . . . 6.15, 35.21
Adu v General Medical Council [2014] EWHC 4080 (Admin) . . . 6.16
Afolabi v Solicitors Regulation Authority [2011] EWHC 2122 . . . 25.13
Afolabi v Solicitors Regulation Authority [2012] EWHC 3502 (Admin) . . . 60.49
Aga v General Medical Council *See* R (Aga) v General Medical Council
Agassi v Robinson [2005] EWCA Civ 1507; [2006] 1 W.L.R. 2126; [2006] 1 All E.R. 900; [2006] S.T.C. 580; [2006] 2 Costs L.R. 283; [2006] B.T.C. 3; [2005] S.T.I. 1994; (2005) 155 N.L.J. 1885; (2006) 150 S.J.L.B. 28; [2005] N.P.C. 140; Times, December 22, 2005; Independent, December 7, 2005 . . . 47.45
Agyeman v Solicitors Regulation Authority [2012] EWHC 3472 (Admin); [2012] All ER (D) 63 (Nov) . . . 15.17
Ahmed v Bar Standards Board, 16 May 2013 (unreported) . . . 56.23
Ahmed v General Optical Council *See* R (Ahmed) v General Optical Council
Ahmed v The Inns' Conduct Committee [2012] EWHC 3270 (QB) . . . 56.22
Airey v Ireland (1979) 2 EHRR 305 . . . 32.11, 45.09
Ajali v Nursing and Midwifery Council [2012] EWHC 2976 . . . 14.26
Akewushola v Secretary of State for the Home Department [2000] 1 W.L.R. 2295; [2000] 2 All E.R. 148; [1999] Imm. A.R. 594; [1999] I.N.L.R. 433; Times, November 3, 1999 . . . 14.07, 43.16
Akodu v Solicitors Regulation Authority [2009] EWHC 3588 (Admin) . . . 47.39
Al-Daraji v General Medical Council [2012] EWHC 1835 (Admin) . . . 1.17

Al-Fallouji v General Medical Council [2003] UKPC 30 . . . . . . . . . . . . . . . . . . . . . . . 13.04
Al-Khawaja v UK [2011] 32 BHRC 1 . . . . . . . . . . . . . . . . . . . . . . . . . . . . . . . . . . . . 25.02
Al-Mehdawi v Secretary of State for the Home Department [1990] 1 A.C. 876 . . . . . . . . 25.39
Al-Mishlab v Milton Keynes Hospital NHS Foundation Trust [2015]
EWHC 191 (QB) . . . . . . . . . . . . . . . . . . . . . . . . . . . . . . . . . . . . . . . . . . . . . . 16.18
Al-Zayyat v General Medical Council *See* R (Al-Zayyat) v General Medical Council
Alacakanat v General Medical Council *See* R (Alacakanat) v General Medical Council
Alami and Botmeh v Health and Care Professions Council and Young
(Interested Party) [2013] EWHC 1895 (Admin) . . . . . . . . . . . . . . . . . . . . . . . . . . 39.29
Albert and Le Compte v Belgium (1983) 5 EHRR 533 . . . . . . . . . . . . . . . 17.23, 32.04, 32.10, 32.12, 32.16, 53.02
Alexander v Rayson [1936] 1 KB 169 . . . . . . . . . . . . . . . . . . . . . . . . . . . . . . . . . . . . 50.05
Algar v Spain [1988] 30 EHRR 827 . . . . . . . . . . . . . . . . . . . . . . . . . . . . . . . . . . . . . 6.17
Alhy v General Medical Council *See* R (Alhy) v General Medical Council
Ali v General Medical Council *See* R (Ali) v General Medical Council
Allatt v Chief Constable of Nottinghamshire Police and IPCC [2011]
EWHC 3908 (Admin) . . . . . . . . . . . . . . . . . . . . . . . . . . . . . . . . . . . . . . . . . . . . . 39.30
Allender v Royal College of Veterinary Surgeons [1951] 2 All E.R. 859; [1951]
2 T.L.R. 872; [1951] W.N. 527; (1951) 95 S.J. 698 . . . . . . . . . . . . . . . . . . . . 4.02, 56.02
Allinson v General Council of Medical Education and Registration [1894]
1 Q.B. 750 . . . . . . . . . . . . . . . . . . . . . . . . . . . . . . . . . . . . . . . . . . . 47.02, 47.19, 47.22
Almond v Council of the Royal College of Veterinary Surgeons, 1951, unreported . . . . . 56.02
Amao v Nursing and Midwifery Council [2014] EWHC 147 (Admin) . . . . 37.17, 60.62, 67.10
American Cyanamid Co. v Ethicon Ltd [1975] A.C. 396; [1975] 2 W.L.R. 316;
[1975] 1 All E.R. 504; [1975] F.S.R. 101; [1975] R.P.C. 513; (1975)
119 S.J. 136 . . . . . . . . . . . . . . . . . . . . . . . . . . . . . . . . . . . . . . . . . . . . . . 10.15, 38.70
Amro International SA v Financial Services Authority *See* R (Amro
International SA) v Financial Services Authority
Anderson v HM Advocate 1996 JC 29 . . . . . . . . . . . . . . . . . . . . . . . . . . . . . . . . . . . 67.12
Andersons Solicitors v Solicitors Regulation Authority [2012] EWHC 3659 (Admin) . . . . . . 51.07
Andreou v Institute of Chartered Accountants in England and Wales [1998]
1 All ER 14 . . . . . . . . . . . . . . . . . . . . . . . . . . . . . . . . . . . . . . . . . . . . . . . . . . . 42.08
Anoom v Bar Standards Board [2015] EWHC 439 (Admin) . . . . . . . . . . . . . . . . . . . . . 58.16
Ansari v General Pharmaceutical Council [2012] EWHC 1159 (Admin) . . . . . . . . 4.31, 4.48, 14.25, 25.37
App. no. 33958/96 . . . . . . . . . . . . . . . . . . . . . . . . . . . . . . . . . . . . . . . . . . . . . . . . 6.01
Arch Financial Products LLP v Financial Services Authority [2012]
Lexis Citation 114 . . . . . . . . . . . . . . . . . . . . . . . . . . . . . . . . . . . . . . . . . . . . . . . 51.08
Archbold v Royal College of Veterinary Surgeons [2004] UKPC 1 . . . . . . . . . . . . . . . . 19.42
Archer v South-West Thames Regional Health Authority (1985)
The Times, 10 August . . . . . . . . . . . . . . . . . . . . . . . . . . . . . . . . . . . . . . . . . . . . 10.03
Arora v General Medical Council [2008] EWHC 1596 (Admin) . . . . . . . . . . . . . . . . . 19.09
Arora v General Medical Council [2012] EWHC 1560 (Admin) . . . . . . . . . . . . . . . . . . 59.15
Arunkalaivanan v General Medical Council [2014]
EWHC 873 (Admin) . . . . . . . . . . . . . . . . . . . . . . . . . . . . . . . . . . 27.17, 30.07, 66.07
Asare-Konadu v Nursing and Midwifery Council [2014] EWHC 4385
(Admin) . . . . . . . . . . . . . . . . . . . . . . . . . . . . . . . . . . . . . . . . . . . . . . . . . . . . . . 19.80
Ashiq v Bar Standards Board, PC 2010/0185/A, 27 March 2013A . . . . . . . . . . . . . . . . . 7.03
Ashraf v General Dental Council [2014] EWHC 2618 (Admin) . . . . . . . . . . . . . . . . . . 21.11
Ashton v General Medical Council [2013] EWHC 943 (Admin) . . . . . . . . . . . . . . . . . 38.68
Assicurazioni v Generali SpA v Arab Insurance Group [2002] EWCA
Civ 1642; [2003] 1 W.L.R. 577 . . . . . . . . . . . . . . . . . . . . . . . . . . . . . . . . . . . . . . 4.31
Associated Provincial Picture House Ltd v Wednesbury Corporation [1948]
1 KB 223 . . . . . . . . . . . . . . . . . . . . . . . . . . . . . . . . . . . . . . . . . . . . . . . . 42.25, 45.12
Atkinson v General Medical Council [2009] EWHC 3636 (Admin) . . . . . . . . . . 19.55, 36.09
Atlantic Law LLP v Financial Services Authority, FIN/2009/0007 . . . . . . . . . . . . . . . . 19.21
Atlantic Law LLP v Financial Services Authority [2010] UKUT B7(FS) . . . . . . . 28.05, 28.12

Attorney-General's Reference (No. 1 of 1990) [1992] Q.B. 630; [1992] 3 W.L.R. 9; [1992] 3 All E.R. 169; (1992) 95 Cr. App. R. 296; (1992) 156 J.P. 593; [1993] Crim. L.R. 37; (1992) 156 J.P.N. 476; (1992) 89(21) L.S.G. 28; (1992) 142 N.L.J. 563; Times, April 16, 1992; Independent, May 1, 1992 . . . . . . . . . . . . . . . . . . . . . . . . 2.02, 2.05, 17.02, 17.17, 17.23, 17.24, 50.04
Attorney-General's Reference (No. 2 of 2001) [2003] UKHL 68; [2004] 2 A.C. 72; [2004] 2 W.L.R. 1; [2004] 1 All E.R. 1049; [2004] 1 Cr. App. R. 25; [2004] H.R.L.R. 16; [2004] U.K.H.R.R. 193; 15 B.H.R.C. 472; [2004] Crim. L.R. 574; (2004) 101(4) L.S.G. 30; (2004) 148 S.J.L.B. 25; Times, December 12, 2003; Independent, December 16, 2003 . . . . . 17.04, 17.23, 17.24, 17.27, 17.29
Attorney-General's Reference (No. 52 of 2003) (Webb) [2003] EWCA Crim 3731; [2004] Crim. L.R. 306; Times, December 12, 2003 . . . . . . . . . . . . . . . . . . . . . . . 36.07
Au Wing Lun v Solicitors Disciplinary Tribunal CACV 4154/2001 . . . . . . . . . . . . . . . . 35.11
Augustine v Nursing and Midwifery Council [2009] EWHC 517 (Admin) . . . . . . . . . . 60.37
Austin v Teaching Agency *See* R (Austin) v Teaching Agency
Autofocus Ltd v Accident Exchange Ltd [2010] EWHC Civ 788 . . . . . . . . . . . . . . . . . 52.13
Awan v Law Society [2003] EWCA Civ 1969 . . . . . . . . . . . . . . . . . . . . . . . 3.08, 3.16, 45.10
Aziz v General Medical Council *See* R (Aziz) v General Medical Council
Azzam v General Medical Council [2008] EWHC 2711 (Admin); [2009] LS Law Medical 28; (2009) 105 B.M.L.R. 142 . . . . . . . . . . . . . . . . . . . . . . . 33.06, 33.11

B (Children) (Care Proceedings: Standard of Proof) (CAFCASS intervening) [2008] UKHL 35; [2009] 1 A.C. 11; [2008] 3 W.L.R. 1; [2008] 4 All E.R. 1; [2008] 2 F.L.R. 141; [2008] 2 F.C.R. 339; [2008] Fam. Law 619; [2008] Fam. Law 837; Times, June 12, 2008 . . . . . . . . 54.33, 62.04, 62.06, 62.12, 62.20, 62.21
B v Auckland District Law Society [2003] UKPC 38; [2003] 2 A.C. 736; [2003] 3 W.L.R. 859; [2004] 4 All E.R. 269; (2003) 100(26) L.S.G. 38; (2003) 147 S.J.L.B. 627; Times, May 21, 2003 . . . . . . . . . . . . . . . . . . . . . . . . . . . . 52.05, 52.08
B v Chief Constable of Avon and Somerset Constabulary [2001] 1 W.L.R. 340; [2001] 1 All E.R. 562; [2000] Po. L.R. 98 . . . . . . . . . . . . . . . . . . . . . . . . . . . . . . 62.04
B v Director of Public Prosecutions *See* R (B) v Director of Public Prosecutions
B v Hampshire County Council *See* R (B) v Hampshire County Council
B v Independent Safeguarding Authority (Royal College of Nursing intervening) [2012] EWCA Civ 977; [2013] 1 W.L.R. 308; Times, September 21, 2012 . . . . . . . . 32.21, 43.14
B v Nursing and Midwifery Council *See* R (B) v Nursing and Midwifery Council
B v Secretary of State for the Home Department *See* R (B) v Secretary of State for the Home Department
Baba v General Medical Council (2001) 62 BMLR 34 . . . . . . . . . . . . . . . . . . . . . . . . . . 3.07
Bains v Solicitors Regulation Authority [2015] EWHC 506 (Admin) . . . . . . . . . . 19.34, 67.13
Baker v British Boxing Board of Control [2014] EWHC 2074 (QB) . . . . . . . . . . . . . . . 38.70
Baker v Police Appeals Tribunal [2013] EWHC 718 (Admin) . . . . . . . . . . . . . . . . . . . . 43.13
Balamoody v Nursing and Midwifery Council [2009] EWHC 3235 (Admin) . . . . . . . . . . . . . . . . . . . . . . . . . . . . . . . . . . . . . . . . . . 37.08, 58.07, 58.08
Balamoody v Nursing and Midwifery Council (No. 2) [2010] EWHC 2256 (Admin) . . . . . . . . . . . . . . . . . . . . . . . . . . . . . . . . . . . . . . . 58.08
Balani v Spain (1994) 19 EHRR 566 . . . . . . . . . . . . . . . . . . . . . . . . . . . . . . . . . . . . . 50.07
Baldock v Webster [2004] EWCA Civ 1869; [2006] Q.B. 315; [2006] 2 W.L.R. 1; [2005] 3 All E.R. 655; (2005) 102(7) L.S.G. 27; Times, January 13, 2005 . . . . . . . . . . . . . . . . . . . . . . . . . . . . . . . . . . . . . . . . . . . . 13.13, 13.15
Balfour v Occupational Therapists Board (2000) 51 B.M.L.R. 69; Times, July 24, 1999 . . . . . . . . . . . . . . . . . . . . . . . . . . . . . . . . . . . . . . . . . . 47.23
Ballesteros v Nursing and Midwifery Council [2011] EWHC 1289 (Admin) . . . . 19.62, 37.14
Bamgbelu v General Dental Council [2013] EWHC 1169 (Admin) . . . . . . . . . . . . . . . . 59.17
Bank of Scotland v Investment Management Regulatory Organisation Ltd 1989 S.C. 107; 1989 S.L.T. 432; 1989 S.C.L.R. 386 . . . . . . . . . . . . . . . . . . . . . . . . . . . 42.06
Bar Standards Board v L, 24 January 2007 (unreported) . . . . . . . . . . . . . . . . . . . . . . 17.27
Bar Standards Board v O'Riordan [2014] Lexis Citation 24 . . . . . . . . . . . . . . . . . . . . 19.74

Bar Standards Board v Sivanandan, 13 January 2012 (unreported) . . . . . . . . . . . . . . . . 47.45
Barakat v General Medical Council [2013] EWHC 3427 (Admin) . . . . . . . . . . . . . . . . 54.27
Barlow Clowes International Ltd v Eurotrust International Ltd [2005] UKPC 37; [2006] 1 W.L.R. 1476; [2006] 1 All E.R. 333; [2006] 1 All E.R. (Comm) 478; [2006] 1 Lloyd's Rep. 225; [2005] W.T.L.R. 1453; (2005-06) 8 I.T.E.L.R. 347; (2005) 102(44) L.S.G. 32; [2006] 1 P. & C.R. DG16 . . . . . . . . . 19.12
Bartosik v General Medical Council [2011] EWHC 3906 (Admin) . . . . . . . . . . . . . . . . 4.34
Bass and Ward v Solicitors Regulation Authority [2012] EWHC 2012 (Admin) . . . . . . . . . . . . . . . . 15.27, 15.29, 28.09
Basser v Medical Board of Victoria [1981] VR 953 . . . . . . . . . . . . . . . . 24.03
Bates v District Judge Zani (Independent Adjudicator) *See* R (Bates) v District Judge Zani (Independent Adjudicator)
Batra v Financial Conduct Authority [2014] UKUT 0214 (TCC) . . . . . . . . . . . . . . . . 19.22
Baxendale-Walker v Law Society [2006] 3 All E.R. 675; [2008] 1 W.L.R. 426, CA . . . . . . . . . . . . 15.05, 15.07, 15.14, 15.20, 15.23, 15.25, 15.26, 15.27, 15.28, 15.29, 35.11, 52.14
Baxendale-Walker v Middleton [2011] EWHC 998 (QB) . . . . . . . . . . . . 49.09, 52.12, 52.13
Beard v Bar Standards Board, 17 July 2014 (unreported) . . . . . . . . . . . . . . . . 60.60
Begum v Solicitors Regulation Authority [2007] WL 511 6865 . . . . . . . . . . . . . . . . 56.05
Belal v General Medical Council [2011] EWHC 2859 (Admin) . . . . . . . . . . . . . . 11.11, 25.26
Belgian Linguistic Case (No. 2) (1968) 1 EHRR 252 . . . . . . . . . . . . . . . . 56.25
Belilos v Switzerland (1988) 10 EHRR 466 . . . . . . . . . . . . . . . . 6.01
Benham Ltd v Kythira Investments Ltd [2003] EWCA Civ 1794 . . . . . . . . . . . . . . . . 50.05
Benmax v Austin Motor Co Limited [1955] A.C. 370; [1955] 2 W.L.R. 418; [1955] 1 All E.R. 326; (1955) 72 R.P.C. 39; (1955) 99 S.J. 129 . . . . . . . . . . . . . . . . 27.05
Bentley v. Jones Harris & Co. [2001] EWCA Civ 1724 . . . . . . . . . . . . . . . . 50.05
Beresford v Solicitors Regulatory Authority [2009] EWHC 3155 (Admin) . . . . . . . . . . . . . . . . 15.14, 15.20, 54.15
Bevan v General Medical Council *See* R (Bevan) v General Medical Council
Bhadra v General Medical Council [2002] UKPC 55; [2003] 1 W.L.R. 162 . . . . . . . . . . . 4.10
Bhamjee v Forsdick [2003] EWCA Civ 1113; [2004] 1 W.L.R. 88; [2003] C.P. Rep. 67; [2003] B.P.I.R. 1252; (2003) 100(36) L.S.G. 41; Times, July 31, 2003; Independent July 29 2003 . . . . . . . . . . . . . . . . 9.04
Bhandari v Advocates Committee [1956] 1 W.L.R. 1442; [1956] 3 All E.R. 742; (1956) 100 S.J. 836 . . . . . . . . . . . . . . . . 62.09
Bhatnagar v General Medical Council [2013] EWHC 3412 (Admin) . . . . . . . . . . . . . . 38.16
Bhatt v General Medical Council *See* R (Bhatt) v General Medical Council
Bhattacharya v General Medical Council [1967] 2 A.C. 259; [1967] 3 W.L.R. 498; (1967) 111 S.J. 316 . . . . . . . . . . . . . . . . 39.02, 44.08, 47.10
Bijl v General Medical Council [2001] UKPC 42; [2002] Lloyd's Rep. Med. 60; (2002) 65 B.M.L.R. 10; Times, October 24, 2001 . . . . . . . . . . . . . . . . 4.16, 60.22
Biogen Inc. v Medeva Plc [1997] RPC 1;(1997) 38 B.M.L.R. 149; (1997) 20(1) I.P.D. 20001; Times, November 1, 1996 . . . . . . . . . . . . . . . . 4.27
Birks v Commissioner of Police of the Metropolis *See* R (Birks) v Commissioner of Police of the Metropolis, Independent Police Complaints Commission (Interested Party)
Black v Professional Conduct Committee of the General Dental Council [2013] HCJAC 39 . . . . . . . . . . . . . . . . 33.20
Boddington v British Transport Police [1992] 2 A.C. 143 . . . . . . . . . . . . . . . . 42.25
Bolam v Friern Hospital Management Committee [1957] 1 W.L.R. 582; [1957] 2 All E.R. 118; [1955-95] P.N.L.R. 7; (1957) 101 S.J. 357 . . . . . . . . . 26.09, 47.43, 49.01, 49.02, 49.03, 62.13
Bolitho v City and Hackney Health Authority [1998] A.C. 232; [1997] 3 W.L.R. 1151; [1997] 4 All E.R. 771; [1998] P.I.Q.R. P10; [1998] Lloyd's Rep. Med. 26; (1998) 39 B.M.L.R. 1; [1998] P.N.L.R. 1; (1997) 94(47) L.S.G. 30; (1997) 141 S.J.L.B. 238; Times, November 27, 1997 . . . . . . . . . . . . . . 49.03
Bolt v Chief Constable of Merseyside Police *See* R (Bolt) v Chief Constable of Merseyside Police

Bolton v Law Society [1994] 1 W.L.R. 512; [1994] 2 All E.R. 486;
[1994] C.O.D. 295; Times, December 8, 1993 .................... 4.15, 4.16, 4.17, 15.05, 19.36, 19.40, 19.41, 19.42, 19.49, 19.51, 19.54, 19.60, 19.64, 56.05, 60.02, 60.07, 60.12, 60.16, 60.29, 60.36, 60.43
Bonhoeffer v General Medical Council *See* R (Bonhoeffer) v General Medical Council
Boodoo v General Medical Council [2004] EWHC 2712 (Admin) ................ 46.03
Booth v General Dental Council [2015] EWHC 381 (Admin) .................... 47.63
Borgers v Belgium (1991) 15 EHRR 92 ........................................ 46.04
Bose v General Medical Council [2003] UKPC 52; (2004) 79 B.M.L.R. 1 ........... 44.15
Bourgass and Hussain v Secretary of State for Justice *See* R (Bourgass and Hussain) v Secretary of State for Justice
Boyce v Wyatt Engineering [2001] EWCA Civ 692 ............................ 50.05
Brabazon-Drenning v United Kingdom Central Council for Nursing, Midwifery and Health Visiting (2001) HRLR 6 .................... 2.07, 3.06, 54.04, 54.06
Bradley v Jockey Club [2004] EWHC 2164 (QB) ............................ 60.09
Bradley v Jockey Club [2005] EWCA Civ 1056; Times, July 14, 2005 .............. 60.09
Bradshaw v General Medical Council [2010] EWHC 1296 (Admin) .......... 38.27, 38.39
Brennan v Health Professions Council [2011] EWHC 41 (Admin); (2011) 119 B.M.L.R. 1 ........................................... 54.20, 60.46
Brett v Solicitors Regulation Authority [2014] EWHC 2974 (Admin) .............. 55.07
Brew v General Medical Council [2014] EWHC 2927 (Admin) .............. 37.18, 45.28
Briggs v Law Society [2005] EWHC 1830 (Admin) ..................... 60.35, 66.03
Brocklebank v General Medical Council [2003] UKPC 57 ...................... 31.06
Brooks v Director of Public Prosecutions [1994] 1 A.C. 568; [1994] 2 W.L.R. 381; [1994] 2 All E.R. 231; (1994) 138 S.J.L.B. 34 .................. 21.04
Broomhead v Solicitors Regulation Authority [2012] EWHC 509 (Admin) .......... 11.13
Broomhead v Solicitors Regulation Authority [2014] EWHC 2772 (Admin) ......... 15.20
Brown v United Kingdom (1998) 28 EHRR CD 233 .................... 28.02, 32.12
Bryan v United Kingdom (1995) 21 EHRR 342 ............................... 44.02
Bryan v United Kingdom (1996) 21 EHRR 342 ............................... 32.07
Bryant and Bench v Law Society [2007] EWHC 3043 (Admin); [2009] 1 W.L.R. 163; (2008) 105(3) L.S.G. 28; (2008) 158 N.L.J. 66 ................ 11.14
Bryant v Law Society [2007] EWHC 3043 (Admin); [2009] 1 W.L.R. 163; (2008) 105(3) L.S.G. 28; (2008) 158 N.L.J. 66 ................ 15.24, 19.12, 30.03
Bryant v Solicitors Regulation Authority [2012] EWHC 1475 (Admin); (2012) 156(23) S.J.L.B. 35 ........................................... 11.14
Bultitude v Law Society [2004] EWCA Civ 1853; (2005) 102(9) L.S.G. 29; Times, January 14, 2005 .................... 4.17, 19.06, 19.08, 19.12, 19.46, 55.01
Burke v General Teaching Council [2009] EWHC 3138 (Admin) ................ 60.40
Burke v Independent Police Complaints Commission *See* R (Burke) v Independent Police Complaints Commission
Burnham-Slipper v General Medical Council [2001] EWHC Admin 656 ........... 38.66
Burns v Financial Conduct Authority [2013] Lexis Citation 34 .................... 51.09
Burrowes v Law Society [2002] EWHC 2900 (Admin) .......................... 4.17
Butt v Solicitors Regulation Authority [2010] EWHC 1381 (Admin) ............... 56.11

Calhaem v General Medical Council [2007] EWHC 2606 (Admin); [2008] LS Law Medical 96 ........................... 16.06, 16.11, 47.37, 47.47, 49.04
Calver v Adjudication Panel for Wales *See* R (Calver) v Adjudication Panel for Wales
Calvin v Carr [1980] A.C. 574; [1979] 2 W.L.R. 755; [1979] 2 All E.R. 440; (1979) 123 S.J. 112 ............................. 4.04, 38.70, 48.05, 48.10
Camacho v Law Society (No. 1) [2004] EWHC 1042 (Admin) ............. 11.05, 11.17
Camacho v Law Society (No. 2) [2004] EWHC 1675 (Admin); [2004] 1 W.L.R. 3037; [2004] 4 All E.R. 126; (2004) 101(33) L.S.G. 37; Times, October 5, 2004 ........................................... 11.05, 11.17
Camancho v Law Society (No. 2) [2004] EWHC 1675 (Admin); [2004] 1 W.L.R. 3037; [2004] 4 All E.R. 126; (2004) 101(33) L.S.G. 37; Times, October 5, 2004 ........................................ 15.12, 15.13

Cambridge City Council v Alex Nestling Limited *See* R (Cambridge City Council) v Alex Nestling Limited
Campbell and Fell v United Kingdom (1984) 7 EHRR 165 . . . . . . . . . . . . . . . . . . 45.07, 45.08
Campbell v General Medical Council *See* R (Campbell) v General Medical Council
Campbell v Hamlet [2005] UKPC 19; [2005] 3 All E.R. 1116 . . . . 17.26, 62.09, 62.11, 62.15
Caparo Industries Plc v Dickman [1992] A.C. 605 . . . . . . . . . . . . . . . . . . . . . . . . . . . 49.13
Carmichael v General Dental Council [1990] 1 W.L.R. 134; (1990) 87(4) L.S.G. 34; (1989) 134 S.J. 22 . . . . . . . . . . . . . . . . . . . . . . . . . . . . . . . . . . . . . . . . . . . . . . . 60.20
Carruthers v General Medical Council [2003] UKPC 42; [2004] Lloyd's Rep. Med. 514; (2004) 75 B.M.L.R. 59 . . . . . . . . . . . . . . . . . . . . . . . . . . . . . . . . . . . . 11.04
Carter v Ahsan [2005] ICR 1817 . . . . . . . . . . . . . . . . . . . . . . . . . . . . . . . . . . . . . . . . 43.15
Carter v Northmore Hale Daley & Leake (1995) 183 CLR 121 . . . . . . . . . . . . . . . . . . . . 52.05
Casey v General Medical Council [2011] NIQB 95 . . . . . . . . . . . . . . . . . . . . . . . . . . . 27.08
Cathcart v Law Society of Scotland [2013] CSIH 104 . . . . . . . . . . . . . . . . . . . . . . . . . 60.55
Cela v General Pharamceutical Council *See* R (Cela) v General Pharmaceutical Council
Central London Community Healthcare NHS Trust v Information Commissioner [2013] UKUT 551 (AAC) . . . . . . . . . . . . . . . . . . . . . . . . . . . . . . . 18.33
Chhabra v West London Mental Health NHS Trust [2013] UKSC 80; [2014] 1 All E.R. 943; [2014] ICR 194. . . . . . . . . . . . . . . . . . . . . . . . 18.28, 39.36, 39.37, 48.16
Chakrabarty v Ipswich Hospital NHS Trust and the National Clinical Assessment Service (Interested Party) [2014] EWHC 2735 (QB) . . . . . . . . . . . . . . . . . . 16.10, 16.17
Chaligne, Sejean and Diallo v Financial Services Authority [2012] FS/2011/0001, FS/2011/0002, FS/2011/0005 . . . . . . . . . . . . . . . . . . . . . . . . 28.12, 28.14
Chamba v Law Society [2009] EWHC 190 (Admin) . . . . . . . . . . . . . . . . . . . . . . . . . . 19.53
Chan Hei Ling Helen v Medical Council of Hong Kong [2009] 4 HKLRD 174 . . . . . . 44.20
Chandler v Alberta Association of Architects [1989] 2 SCR 848 . . . . . . . . . . . . . . . . . 24.03
Chandrasekera v Nursing and Midwifery Council [2009] EWHC 144 (Admin) . . . . . . 14.24
Chaudhari v General Pharmaceutical Council [2011] EWHC 3433 (Admin) . . . . 3.16, 25.14, 25.20
Chaudhari v Royal Pharmaceutical Society of Great Britain *See* R (Chaudhari) v Royal Pharmaceutical Society of Great Britain
Chaudhary v Specialist Training Authority Appeal Panel [2005] ICR 1086, CA . . . . . . 39.39
Chauhan v General Medical Council [2010] EWHC 2093 (Admin) . . . . . . . . . . 22.25, 22.34
Cheatle v General Medical Council [2009] EWHC 645 (Admin); [2009] LS Law Medical 299 . . . . . . . . . . . . . . . . . . 4.18, 4.19, 11.07, 32.08, 33.07, 33.10, 60.40
Chegwyn v Ethical Standards Officer of the Standards Board for England [2010] EWHC 471 (Admin); [2011] PTSR 165 . . . . . . . . . . . . . . . . . . . . . . . . . 54.18
Chester v Afshar [2004] UKHL 41; [2005] 1 A.C. 134; [2004] 3 W.L.R. 927; [2004] 4 All E.R. 587; [2005] P.I.Q.R. P12; [2005] Lloyd's Rep. Med. 109; (2005) 81 B.M.L.R. 1; [2005] P.N.L.R. 14; (2004) 101(43) L.S.G. 34; (2004) 154 N.L.J. 1589; (2004) 148 S.J.L.B. 1215; Times, October 19, 2004 . . . . . . . . . . . 26.07
Chief Constable of Avon and Somerset v Police Appeals Tribunal *See* R (Chief Constable of Avon and Somerset) v Police Appeals Tribunal
Chief Constable of British Transport Police v Police Appeals Tribunal and Whaley *See* R (Chief Constable of British Transport Police) v Police Appeals Tribunal and Whaley
Chief Constable of Derbyshire v Goodman, 2 April 1998, DC (unreported) . . . . . . . . . . 15.03
Chief Constable of Dorset v Police Appeals Tribunal and Salter *See* R (Chief Constable of Dorset) v Police Appeals Tribunal and Salter
Chief Constable of Durham v Police Appeals Tribunal *See* R (Chief Constable of Durham) v Police Appeals Tribunal and Cooper
Chief Constable of Hampshire Constabulary v Police Appeal Tribunal and McClean [2012] EWHC 746 (Admin) . . . . . . . . . . . . . . . . . . . . . . . . 4.25, 70.06
Chief Constable of the North Wales Police v Evans [1982] 1 W.L.R. 1155; [1982] 3 All E.R. 141; (1983) 147 J.P. 6; (1982) 79 L.S.G. 1257; (1982) 126 S.J. 549 . . . . . . 48.08, 58.09
Chief Constable of Wiltshire Police v Police Appeals Tribunal and Andrews [2012] EWHC 1798 (Admin) . . . . . . . . . . . . . . . . . . . . . . . . . . . . . . . . . . . . . . . . 4.24
Chief Justice of Gibraltar, Referral under Section 4 of the Judicial Committee Act 1833, Re [2009] UKPC 43 . . . . . . . . . . . . . . . . . . . . . . . . . . . . . . . . . . . . . 41.02

Chien Singh-Shou, In re [1967] 1 W.L.R. 1155, PC .......... 44.06
Choudhury v Solicitors Regulation Authority [2014] EWHC 809 (Admin) .......... 19.28
Christian v Nursing and Midwifery Council [2010] EWHC 803 (Admin) .......... 45.21, 67.12
Christou v Haringey London Borough Council [2013] EWCA Civ 178; [2014] Q.B. 131; [2013] 3 W.L.R. 796; [2014] 1 All E.R. 135; [2013] I.C.R. 1007; [2013] I.R.L.R. 379; (2013) 163 N.L.J. 323 .......... 21.10, 24.05
Chuah v Nursing and Midwifery Council [2013] EWHC 894 (Admin); [2013] EWCA 1799 .......... 14.30
Chyc v General Medical Council [2008] EWHC 1025 (Admin) .......... 27.03
City of Bradford Metropolitan District Council v Booth (2000) 164 J.P. 485; (2001) 3 L.G.L.R. 8; [2000] C.O.D. 338; (2000) 164 J.P.N. 801; Times, May 31, 2000 .......... 15.02, 15.03, 15.05, 15.07, 15.23
Clark v Kelly [2003] UKPC D 1; [2004] 1 A.C. 681; [2003] 2 W.L.R. 1586; [2003] 1 All E.R. 1106; 2003 S.C. (P.C.) 77; 2003 S.L.T. 308; 2003 S.C.C.R. 194; [2003] H.R.L.R. 17; [2003] U.K.H.R.R. 1167; 14 B.H.R.C. 369; (2003) 100(17) L.S.G. 27; (2003) 147 S.J.L.B. 234; 2003 G.W.D. 7-164; Times, February 12, 2003; Independent, March 17, 2003 .......... 44.02
Clark v Kelly [2003] UKPC D 1; [2004] 1 A.C. 681; [2003] 2 W.L.R. 1586; [2003] 1 All E.R. 1106; 2003 S.C. (P.C.) 77; 2003 S.L.T. 308; 2003 S.C.C.R. 194; [2003] H.R.L.R. 17; [2003] U.K.H.R.R. 1167; 14 B.H.R.C. 369; (2003) 100(17) L.S.G. 27; (2003) 147 S.J.L.B. 234; 2003 G.W.D. 7-164; Times, February 12, 2003; Independent, March 17, 2003 .......... 44.02
Clark v Vanstone [2004] FCA 1105, (2004) 81 ALD 21 .......... 41.02
Clery v Health and Care Professions Council [2014] EWHC 951 (Admin) .......... 47.59
Cohen v General Medical Council [2008] EWHC 581 (Admin); [2008] LS Law Medical 246; (2008) 105(15) L.S.G. 27 .......... 22.25, 26.05, 33.04, 33.05, 33.06, 33.10, 33.11, 33.15
Colgan v The Kennel Club, Case No.01/TLQ/0673, 26 October 2001 .......... 14.21, 60.24
Collins v Office for the Supervision of Solicitors [2002] EWCA Civ 1002 .......... 49.06
Colman v General Medical Council *See* R (Colman) v General Medical Council
Commissioner of Inland Revenue v West, Walker [1954] NZLR 191 .......... 52.01
Commissioner of Police for the Metropolis v Police Appeals Tribunal and Naulls *See* R (Commissioner of Police for the Metropolis) v Police Appeals Tribunal and Naulls [2013] EWHC 1684 (Admin)
Commissioner of Police of the Metropolis v Police Appeals Tribunal and Peart *See* R (Commissioner of Police of the Metropolis) v Police Appeals Tribunal and Peart [2011] EWHC 3421 (Admin)
Compton v General Medical Council [2008] EWHC 2868 (Admin) .......... 44.19, 67.03
Compton v Wiltshire Primary Care Trust (No.2) *See* R (Compton) v Wiltshire Primary Care Trust (No.2)
Conlon, Gordon and Williams v Bar Standards Board Case no. D2007/098, 4 July 2014 .......... 47.62
Conlon v Bar Standards Board [2014] Lexis Citation 136 .......... 43.20
Connelly v Director of Public Prosecutions [1964] A.C. 1254; [1964] 2 W.L.R. 1145; [1964] 2 All E.R. 401; (1964) 48 Cr. App. R. 183; (1964) 128 J.P. 418; (1964) 108 S.J. 356 .......... 2.02
Connelly v DPP [1964] A.C. 1254; [1964] 2 W.L.R. 1145; [1964] 2 All E.R. 401; (1964) 48 Cr. App. R. 183; (1964) 128 J.P. 418; (1964) 108 S.J. 356 .......... 21.01
Connolly v Law Society [2007] EWHC 1175 (Admin) .......... 22.18
Constantinides v Law Society [2006] EWHC 725; (2006) 156 N.L.J. 680 .......... 25.09, 25.11, 25.19, 25.20
Conteh v Onslow Fane, Transcript No. 291 of 1975, CA, Times, June 6, 1975 .......... 10.06, 45.05
Coppard v Customs & Excise [2003] EWCA Civ 511; [2003] Q.B. 1428; [2003] 2 W.L.R. 1618; [2003] 3 All E.R. 351; (2003) 100(24) L.S.G. 36; (2003) 147 S.J.L.B. 475; Times, April 11, 2003 .......... 13.13, 13.15
Cordle v Financial Services Authority [2013] Lexis Citation 01 .......... 19.26
Cordle v Financial Services Authority, FS/2012/0006, 2 January 2013 .......... 56.21
Cornish v General Medical Council [2012] EWHC 1196 (Admin) .......... 27.10

Corporation of Accountants v Society of Accountants in Edinburgh (1903) 11 SLT 424 . . . . . 56.18
Council for Health Care Regulatory Excellence v Nursing and Midwifery Council [2012] CSIH 24; 2012 G.W.D. 11-210 . . . . . 43.11
Council for the Regulation of Health Care Professionals v General Dental Council and Fleischmann [2005] EWHC 87; Times, February 8, 2005 . . . . . 14.22, 14.32, 56.05
Council for the Regulation of Health Care Professionals v General Medical Council and Basiouny [2005] EWHC 68 (Admin); (2005) 102(13) L.S.G. 28; Times, February 7, 2005 . . . . . 4.15, 26.03
Council for the Regulation of Health Care Professionals v General Medical Council and Biswas [2006] EWHC 464 (Admin) . . . . . 33.03, 47.30, 62.10
Council for the Regulation of Health Care Professionals v General Medical Council and Grant [2011] EWHC 927 (Admin) . . . . . 33.15, 33.20, 33.21, 66.05
Council for the Regulation of Health Care Professionals v General Medical Council and Leeper [2004] EWHC 1850 (Admin); Times, September 1, 2004 . . . . . 36.03, 36.06
Council for the Regulation of Health Care Professionals) v General Medical Council and Rajeshwar *See* R (Council for the Regulation of Health Care Professionals) v General Medical Council and Rajeshwar
Council for the Regulation of Health Care Professionals v General Medical Council and Ruscillo [2004] EWCA Civ 1356; [2005] 1 W.L.R. 717; [2005] Lloyd's Rep. Med. 65; [2005] A.C.D. 69; [2005] A.C.D. 99; (2004) 148 S.J.L.B. 1248; Times, October 27, 2004; Independent, October 28, 2004 . . . . . 8.03, 21.05
Council for the Regulation of Health Care Professionals v General Medical Council and Saluja [2006] EWHC 2784 (Admin); [2007] 1 W.L.R. 3094; [2007] 2 All E.R. 905; [2007] LS Law Medical 237; (2006) 92 B.M.L.R. 153; [2007] A.C.D. 29; (2006) 156 N.L.J. 1767 . . . . . 2.08, 17.27
Council for the Regulation of Health Care Professionals v General Medical Council and Solanke [2004] EWHC 944 (Admin); [2004] 1 W.L.R. 2432; [2004] Lloyd's Rep. Med. 377; [2004] A.C.D. 55; Times, May 10, 2004 . . . . . 4.12, 4.13, 19.45, 36.05
Council for the Regulation of Health Care Professionals v Health Professions Council and Jellett [2005] EWHC 93 (Admin) . . . . . 58.05
Council for the Regulation of Health Care Professionals v Nursing and Midwifery Council [2004] EWHC 585 (Admin) . . . . . 4.12
Council for the Regulation of Health Care Professionals) v Nursing and Midwifery Council *See* R (Council for the Regulation of Health Care Professionals) v Nursing and Midwifery Council
Council for the Regulation of Health Care Professionals v Nursing and Midwifery Council and Trustcott [2004] EWCA Civ 1356; [2005] 1 W.L.R. 717; [2005] Lloyd's Rep. Med. 65; [2005] A.C.D. 69; [2005] A.C.D. 99; (2004) 148 S.J.L.B. 1248; Times, October 27, 2004; Independent, October 28, 2004 . . . . . 8.03
Council for the Regulation of Health Care Professionals v General Medical Council and Khanna [2009] EWHC 596 (Admin); [2010] Med. L.R. 157 . . . . . 60.38
Countryside Alliance v Attorney General *See* R (Countryside Alliance) v Attorney General
County Council, A v W (Disclosure) [1997] 1 F.L.R. 574; [1996] 3 F.C.R. 728; [1997] Fam. Law 318 . . . . . 47.22
Cowan v Hardey [2014] CSIH 11 . . . . . 47.57
Crabbie v General Medical Council [2002] UKPC 45; [2002] 1 W.L.R. 3104; [2002] Lloyd's Rep. Med. 509; (2003) 71 B.M.L.R. 9; Times, October 10, 2002 . . . . . 31.04, 31.05, 31.07, 60.06
Crane v Director of Public Prosecutions [1921] 2 A.C. 299 . . . . . 4.04
Craven v Bar Standards Board, 30 January 2014 (unreported) . . . . . 47.58
Crawford v Legal Ombudsman [2014] EWHC 182 (Admin) . . . . . 42.25
Crompton v General Medical Council [1981] 1 W.L.R. 1435; [1982] 1 All E.R. 35 . . . 18.02, 31.02, 48.07
Crompton v General Medical Council (No.2) [1985] 1 W.L.R. 885; (1985) 82 L.S.G. 1864; (1985) 129 S.J. 468 . . . . . 31.02
Cronin v Greyhound Board of Great Britain Ltd [2013] EWCA Civ 668 . . . . . 54.26

Crosby v IPCC *See* R (Crosby) v IPCC
Crowley, In re [1964] IR 106 . . . . . 11.20
Cullen v General Medical Council [2005] EWHC 353 (Admin) . . . . . 26.02, 31.08
Currie v Chief Constable of Surrey [1982] 1 W.L.R. 215; [1982] 1 All E.R. 89; (1981) 125 S.J. 121 . . . . . 70.02

D (Secretary of State for Northern Ireland intervening), In re [2008] UKHL 33; [2008] 1 W.L.R. 1499; [2008] 4 All E.R. 992; [2008] N.I. 292; [2009] 1 F.L.R. 700; [2008] Prison L.R. 219; [2009] Fam. Law 192; (2008) 105(29) L.S.G. 36; (2008) 152(25) S.J.L.B. 32; Times, June 24, 2008 . . . . . 62.05, 62.12
D v Conduct and Competence Committee of Nursing and Midwifery Council [2014] SLT 1069 . . . . . 27.18, 45.29, 67.12
D v Law Society [2003] EWHC 408 (Admin) . . . . . 19.05
D v Nursing and Midwifery Council [2014] EWHC 1298 (Admin) . . . . . 14.33
DH v Czech Republic (2008) 47 EHRR 3 . . . . . 49.12
DPP v Nasrulla [1967] 2 A.C. 238; [1967] 3 W.L.R. 13; [1967] 2 All E.R. 161; (1967) 111 S.J. 193 . . . . . 21.01
DV v General Medical Council [2010] EWHC 2873 (QB) . . . . . 9.02, 30.04
Dad v General Dental Council [2000] 1 W.L.R. 1538; [2000] Lloyd's Rep. Med. 299; (2000) 56 B.M.L.R. 130; Times, April 19, 2000 . . . . . 14.25, 19.58, 60.03, 60.06
Dad v General Dental Council [2010] CSIH 75; 2010 G.W.D. 30-620 . . . . . 19.58
Dally v General Medical Council , Privy Council Appeal No. 7 of 1987 . . . . . 13.03
Daly v General Medical Council [1952] 2 All E.R. 666; [1952] 2 T.L.R. 331; [1952] W.N. 413; (1952) 96 S.J. 546 . . . . . 44.03
Daniels v Nursing and Midwifery Council [2014] EWHC 3287 (Admin) . . . . . 4.46
Daraghmeh v General Medical Council [2011] EWHC 2080 (Admin) . . . . . 11.10, 59.11
Darker v Chief Constable of West Midlands Police [2001] 1 A.C. 435 . . . . . 52.12, 52.13
Dart Harbour and Navigational Authority v Secretary of State for Transport, Local Government and the Regions *See* R (Dart Harbour and Navigational Authority) v Secretary of State for Transport, Local Government and the Regions
Datta v General Medical Council , Privy Council Appeal No. 34 of 1985 . . . . . 6.02, 6.17
David v General Medical Council [2004] EWHC 2977 (Admin); (2005) 84 B.M.L.R. 30; Times, January 12, 2005 . . . . . 39.09
Davidson v Scottish Ministers [2004] UKHL 34; 2005 1 S.C. (H.L.) 7; 2004 S.L.T. 895; 2004 S.C.L.R. 991; [2004] H.R.L.R. 34; [2004] U.K.H.R.R. 1079; [2005] A.C.D. 19; 2004 G.W.D. 27-572; Times, July 16, 2004 . . . . . 35.12, 35.20
Davies v Financial Services Authority *See* R (Davies) v Financial Services Authority
Davis v Solicitors Regulation Authority [2011] EWHC 3645 (Admin) . . . . . 56.16
De Gregory v General Medical Council [1961] A.C. 957; [1961] 3 W.L.R. 645; (1961) 105 S.J. 681 . . . . . 47.07
Delezuch v Chief Constable of Leicestershire Constabulary *See* R (Delezuch) v Chief Constable of Leicestershire Constabulary
Dellow's Will Trusts, In re [1964] 1 W.L.R. 451; [1964] 1 All E.R. 771; (1964) 108 S.J. 156 . . . . . 62.03
Dennis v IPCC [2008] EWHC 1158 (Admin) . . . . . 39.31
Dental Council of New Zealand v Gibson [2010] NZHC 912 . . . . . 24.03
Depner v General Medical Council [2012] EWHC 1705 (Admin) . . . . . 16.13, 39.39, 66.06
Dey v General Medical Council [2001] UKPC 44; [2002] Lloyd's Rep. Med. 68; (2002) 64 B.M.L.R. 43 . . . . . 14.20, 19.37
Dhorajiwala v General Pharmaceutical Council [2013] EWHC 3821 (Admin) . . . . . 25.17
Diennet v France [1995] ECHR 28 . . . . . 53.07
Dimes v Proprietors of Grand Junction Canal (1852) 3 HL 759 . . . . . 35.12
Disclosure and Barring Service (formally Independent Safeguarding) v Harvey [2013] EWCA Civ 180 . . . . . 32.21
Dixon, Re [2001] EWHC Admin 645 . . . . . 17.12
Donkin v Law Society [2007] EWCA Civ 414 (Admin); (2007) 157 N.L.J. 402 . . . . . 19.08, 25.31, 30.02, 30.03

Doshi v Southend-on-Sea Primary Care Trust *See* R (Doshi) v Southend-on-Sea Primary Care Trust
Doughty v General Dental Council [1988] A.C. 164; [1987] 3 W.L.R. 769; [1987] 3 All E.R. 843; (1987) 84 L.S.G. 2767; (1987) 131 S.J. 1357 . . . . . . . . . . . . 47.18, 47.20, 47.22
Dowland v Architects Registration Board [2013] EWHC 893 (Admin) . . . . . . . . . . . . . 58.12
Doyle v Canada (Restrictive Trade Practices Commission) (1985) 21 DLR (4th) 366 . . . . . . 43.15
Dr X v General Medical Council [2001] EWHC Admin 447 . . . . . 10.09, 10.16, 38.21, 38.42
D'Souza v Law Society [2009] EWHC 2193 (Admin); (2009) 153(30) S.J.L.B. 30 . . . . . . . . . . . . . . . . . . . . . . . . . . . . . . . . 15.12, 15.14, 15.17, 28.04, 28.17
Dublas v Nursing and Midwifery Council [2012] EWHC 214 (Admin) . . . . . . . . . . . . . 14.27
Duggan v Association of Chief Police Officers [2014] EWCA Civ 1635 . . . . . . . . . . . . . 39.40
Dunn v Murray (1829) 9 B&C 780 . . . . . . . . . . . . . . . . . . . . . . . . . . . . . . . . . . . . . . . . . 24.03
Durant v FSA [2003] EWCA Civ 1746 . . . . . . . . . . . . . . . . . . . . . . . . . . . . . . . . . . . . . . 18.30
Duthie v Nursing and Midwifery Council [2012] EWHC 3021 (Admin) . . . . . . . . . . . . . . . . . . . . . . . . . . . . . . . . . . . . . . . . 22.32, 27.15, 54.24
Dutt v General Medical Council *See* R (Dutt) v General Medical Council
Dutta v General Medical Council [2013] EWHC 132 (Admin) . . . . . . . . . . . . . . . . . . . 65.08
Dyer v Watson [2002] UKPC D 1; [2004] 1 A.C. 379; [2002] 3 W.L.R. 1488; [2002] 4 All E.R. 1; 2002 S.C. (P.C.) 89; 2002 S.L.T. 229; 2002 S.C.C.R. 220; [2002] H.R.L.R. 21; [2002] U.K.H.R.R. 542; 2002 G.W.D. 5-153; Times, February 4, 2002 . . . . . . . . . . . . . . . . . . . . . . . . . . . . . . . . . . . . . 17.03, 17.05, 17.29
Dzikowski v General Medical Council [2006] EWHC 2648 (Admin) . . . . . . . . . 13.05, 47.32
Dzikowski v General Medical Council [2009] EWHC 1090 (Admin) . . . . . . . . . 44.21, 59.07

E v T 1949 SLT 411 . . . . . . . . . . . . . . . . . . . . . . . . . . . . . . . . . . . . . . . . . . . . . . . 47.03
Eagil Trust Co Ltd v Pigott-Brown [1985] 3 All E.R. 119 . . . . . . . . . . . . . . . . . . . . . . 54.10
Ebhogiaye v Solicitors Regulation Authority [2013] EWHC 2445 (Admin) . . . . . . . . . . 11.17
Edward Wong Finance Co Ltd v Johnson Stokes & Master [1984] A.C. 296; [1984] 2 W.L.R. 1; (1983) 80 L.S.G. 3163; (1983) 127 S.J. 784 . . . . . . . . . . . . . . . . 49.03
Eguakhide v Governor of HMP Gartree *See* R (Eguakhide) v Governor of HMP Gartree
E I DuPont de Nemours & Company v ST DuPont [2006] 1 W.L.R. 2793; [2006] C.P. Rep. 25; [2004] F.S.R. 15, CA . . . . . . . . . . . . . . . . . . . . . . . . . . . . . . . . . . . . 4.26
El Farargy v El Farargy [2007] EWCA Civ 1149; [2007] 3 F.C.R. 711; (2007) 104(46) L.S.G. 26; (2007) 151 S.J.L.B. 1500; Times, November 23, 2007 . . . . . . . . 6.17
Elliott v Financial Services Authority [2005] UKFSM 019 . . . . . . . . . . . . . . . . . . . . . 25.16
Elliott v Solicitors Disciplinary Tribunal *See* R (Elliott) v Solictors Disciplinary Tribunal
Ellis-Carr v Solicitors Regulation Authority [2014] EWHC 2411 (Admin) . . . . . . . . . . . 58.15
Enderby Town Football Club Ltd v Football Association Ltd [1971] Ch 591 . . . . . . . . . . 45.03
Engel v The Netherlands (No. 1) (1976) 1 EHRR 647 . . . . . . . . . . . . . . . . . . . . . 32.12, 45.08
English v Emery Reinbold and Strick Ltd [2002] EWCA Civ 605; [2002] 1 W.L.R. 2409; [2002] 3 All E.R. 385; [2002] C.P.L.R. 520; [2003] I.R.L.R. 710; [2002] U.K.H.R.R. 957; (2002) 99(22) L.S.G. 34; (2002) 152 N.L.J. 758; (2002) 146 S.J.L.B. 123; Times, May 10, 2002; Independent, May 7, 2002 . . . . . . . . . . 27.09, 54.10, 54.13, 54.14, 54.15, 54.30
Erenbilge v IPCC *See* R (Erenbilge) v IPCC
Ettl v Austria (1987) 10 EHRR 255 . . . . . . . . . . . . . . . . . . . . . . . . . . . . . . . . . . . . . . 44.02
Evans v General Medical Council, unreported, 19 November 1984, PC . . . 4.08, 19.44, 60.19
Evans v Solicitors Regulation Authority [2007] WL 511 6865 . . . . . . . . . . . . . . . . . . . . 56.05

F (Children), In the matter of [2012] EWCA Civ 828 . . . . . . . . . . . . . . . . . . . . . . . . . 54.30
F Hoffman La Roche & Co AG v Secretary of State for Trade and Industry [1975] A.C. 295 . . . . . 39.34
Fabiyi v Nursing and Midwifery Council [2012] EWHC 1441 (Admin) . . . . . . . . . . . . . . 1.13
Facey v Midas Retail Security Limited [2001] I.C.R. 287; [2000] I.R.L.R. 812 . . . . . . . 48.12
Fajemisin v General Dental Council [2013] EWHC 3501 (Admin); [2014] 1 W.L.R. 1169 . . . . . . . . . . . . . . . . . . . . . . . . . . . . . . . . . . . . . . . . . . . 43.16, 56.29

Faniyi v Solicitors Regulation Authority [2012] EWHC 2965 (Admin) .......... 1.14, 4.47
Farag v General Medical Council [2009] EWHC 2667 (Admin) ............. 19.13, 27.07
Farah v General Medical Council *See* R (Farah) v General Medical Council
Faridian v General Medical Council [1971] A.C. 995; [1970] 3 W.L.R. 1065;
[1971] 1 All E.R. 144; (1970) 114 S.J. 934 ............................... 47.11
Fatnami v General Medical Council [2007] EWCA Civ 46; [2007]
1 W.L.R. 1460; [2007] I.C.R. 811; (2007) 104(5) L.S.G. 29; (2007)
151 S.J.L.B. 127 .................................... 4.16, 4.18, 14.25, 60.10
Faulkner v Nursing and Midwifery Council [2009] EWHC 3349 (Admin) ........... 1.09
Fawdry v Murfitt [2002] EWCA Civ 643; [2003] Q.B. 104; [2002] 3 W.L.R.
1354; [2003] 4 All E.R. 60; [2002] C.P. Rep. 62; [2002] C.P.L.R. 593;
Independent, May 23, 2002 ................................... 13.13, 13.15
Felix v General Dental Council [1960] A.C. 704; [1960] 2 W.L.R. 934; [1960]
2 All E.R. 391; (1960) 104 S.J. 446 ..................... 44.08, 47.06, 47.08, 47.17,
47.18, 47.19, 47.20
Ferguson v Scottish Football Association, 1 February 1996 (unreported) ............ 43.06
Fernando v General Medical Council [2014] EWHC 1664 (Admin) ............... 19.75
Fidelitas Shipping Co Ltd v V/O Exportchleb [1966] 1 Q.B. 630; [1965] 2 W.L.R.
1059; [1965] 2 All E.R. 4; [1965] 1 Lloyd's Rep. 223; (1965) 109 S.J. 191 ........ 24.03
Financial Ombudsman Service v Heather Moor & Edgecombe Ltd [2008]
EWCA Civ 643 ................................................ 15.11
Financial Services Authority v Fox Hayes [2009] EWCA Civ 76; [2009]
1 B.C.L.C. 603; (2009) 106(9) L.S.G. 17; [2009] Bus. L.R. D109 .............. 28.03
Financial Services Authority v Information Commissioner [2009] EWHC 1548
(Admin); [2009] Bus. L.R. 1287; [2010] 1 B.C.L.C. 53; [2009] A.C.D. 72 ....... 18.24
Findlay v United Kingdom (1997) 24 EHRR 221 ............................. 44.02
Finegan v General Medical Council [1987] 1 W.L.R. 121; (1987) 84 L.S.G. 577;
(1987) 131 S.J. 196 ............................................ 11.02
First Financial Advisers Ltd v Financial Services Authority, FS/2010/0038 ........... 19.21
Fish v General Medical Council [2012] EWHC 1269 (Admin); [2012]
Med. L.R. 512 ..................................... 19.18, 22.31, 44.23
Fisher, In Re (2009) 202 P 3d 1186, 1199 (Supreme Court, Colorado) .............. 24.03
Flannery v Halifax Estate Agencies Ltd [2000] 1 W.L.R. 377; [2000] 1 All E.R. 373;
[2000] C.P. Rep. 18; [1999] B.L.R. 107; (1999) 11 Admin. L.R. 465; (1999) 15
Const. L.J. 313; (1999) 96(13) L.S.G. 32; (1999) 149 N.L.J. 284; [1999] N.P.C. 22;
Times, March 4, 1999; Independent, February 26, 1999 .................... 54.30
Fleurose v Securities and Futures Authority Ltd [2001] EWCA Civ 2015; [2002] I.R.L.R.
297; Times, January 15, 2002; Daily Telegraph, January 17, 2002 ......... 25.06, 32.12
Florida Bar v Rodriguez (2007) 967 So 2d 150 ............................. 24.03
Florida Bar v St Louis (2007) 967 So 2d 108 .............................. 24.03
Fotheringham, petitioner [2008] CSOH 170 ................................ 6.09
Fox Hayes v Financial Services Authority, 12 May 2011, Upper Tribunal (Tax and
Chancery Chamber) Financial Services ................................ 28.03
Fox v General Medical Council [1960] 1 W.L.R. 1017; [1960] 3 All E.R. 225;
(1960) 124 J.P. 467; (1960) 104 S.J. 725 ............ 4.03, 44.04, 44.05, 44.08, 44.09
Frankowicz v Poland, App No 53025/99, Judgment of 16 December 2008,
ECtHR .................................................... 32.19
Fraser v Mudge [1975] 1 W.L.R. 1132; [1975] 3 All E.R. 78; (1975) 119 S.J. 508 ..... 45.04, 45.06
French v Chief Constable of West Yorkshire Police *See* R (French) v Chief
Constable of West Yorkshire Police
Fuglers LLP v Solicitors Regulation Authority [2014] EWHC 179 (Admin) .......... 28.22
Fuyane v Nursing and Midwifery Council [2012] EWHC 3229 (Admin) ............ 14.29
Fynes v St George's Hospital NHS Trust [2014] EWHC 756 (QB) ................ 39.37

G (A Minor) (Child Abuse: Standard of Proof), In re [1987] 1 W.L.R. 1461;
[1988] 1 F.L.R. 314; [1988] Fam. Law 129; (1987) 84 L.S.G. 3415; (1987) 131
S.J. 1550; Independent, June 6, 1997 ................................ 62.03

G v Governors of X School (Secretary of State for the Home Department intervening) *See* R (G) v Governors of X School (Secretary of State for the Home Department intervening)
Gadd v Solicitors Regulation Authority [2013] EWCA Civ 837 . . . . . . . . . . . . . . . . . . . 17.31
Gage v General Chiropractic Council [2004] EWHC 2762 (Admin). . . . . 18.08, 48.13, 62.11
Gainer v Bar Standards Board [2013] Lexis Citation 81 . . . . . . . . . . . . . . . . . . . . . . . . 65.11
Gangar v General Medical Council [2003] UKPC 28; [2003] H.R.L.R. 24 . . . . 19.04, 22.09
Garfoot v General Medical Council [2002] UKPC 35 . . . . . . . . . . . . . . . . . . . . 60.25, 60.26
Gatawa v Nursing and Midwifery Council [2013] EWHC 3435 (Admin) . . . . . . . . . . . . . 3.19
Gaunt v Office of Communications *See* R (Gaunt) v Office of Communications
Gautrin v France (1999) 28 EHRR 196. . . . . . . . . . . . . . . . . . . . . . . . . . . . . . . . . . . . 32.09
Gbidi v Nursing and Midwifery Council [2015] EWHC 237 (Admin) . . . . . . . . . . . . . . 19.80
Gee v General Medical Council [1987] 1 W.L.R. 564; [1987] 2 All E.R. 193; (1987) 131 S.J. 626. . . . . . . . . . . . . . . . . . . . . . . . . . . . . . . . . . . . . 22.05, 22.08, 40.02, 40.03
General Dental Council v Jamous [2013] EWHC 1428 (Admin). . . . . . . . . . . . . . . . . . 14.10
General Dental Council v Rimmer [2010] EWHC 1049 (Admin) . . . . . . . . . . . . . . . . . 18.25
General Dental Council v Savery [2011] EWHC 3011 (Admin); [2012] Med. L.R. 204; [2012] A.C.D. 11 . . . . . . . . . . . . . . . . . . . . . . . . . . . . . . . . . . . . . . . . . . . . . . . 18.26
General Medical Council v Adeosun [2013] EWHC 2367 (Admin). . . . . . . . . . . . . . . . 38.54
General Medical Council v Agrawal [2014] EWHC 1669 (Admin) . . . . . . . . . . . . . . . . 38.63
General Medical Council v British Broadcasting Corpn [1998] 1 W.L.R. 1573; [1998] 3 All E.R. 426; [1998] E.M.L.R. 833; (1998) 43 B.M.L.R. 143; (1998) 95(25) L.S.G. 32; (1998) 148 N.L.J. 942; (1998) 142 S.J.L.B. 182; Times, June 11, 1998; Independent, June 17, 1998 . . . . . . . . . . . . . . . . . . . . . . . . . . . . . . . . . . . . . . . . 51.02
General Medical Council v Dr E [2013] EWHC 3425 (Admin) . . . . . . . . . . . . . . 38.56, 38.57
General Medical Council v Gill [2012] EWHC 2069 (Admin) . . . . . . . . . . . . . . . . . . . . 38.50
General Medical Council v Hiew [2007] EWCA Civ 369; [2007] 1 W.L.R. 2007; [2007] 4 All E.R. 473; [2007] LS Law Medical 309; Times, June 15, 2007 . . . . . . . 38.07, 38.19, 38.43, 38.57
General Medical Council v Homaei [2013] EWHC 1676 (Admin) . . . . . . . . . . . . . . . . 61.06
General Medical Council v Jooste [2013] EWHC 1751 (Admin) . . . . . . . . . . . . . . . . . 68.06
General Medical Council v Kaluba [2013] EWHC 4212 (Admin) . . . . . . . . . . . . . . . . 38.20
General Medical Council v Kor [2011] EWHC 2825 (Admin) . . . . . . . . . . . . . . . . . . . 38.10
General Medical Council v Lauffer [2009] EWHC 3497 (Admin) . . . . . . . . . . . . . . . . 38.44
General Medical Council v Makki [2014] EWHC 769 (Admin) . . . . . . . . . . . . . . . . . . 38.61
General Medical Council v Malik [2013] EWHC 122 (Admin) . . . . . . . . . . . . . . . . . . 38.53
General Medical Council v Michalak [2014] UKEAT/0213/14/RN . . . . . . . . . . . . . . . 39.39
General Medical Council v Nakhla [2014] EWCA Civ 1522; *Times*, January 8, 2015 . . . . . . . . . . . . . . . . . . . . . . . . . . . . . . . . . . . . . . . . . . 4.31, 56.32
General Medical Council v Paterson [2014] EWHC 201 (Admin) . . . . . . . . . . . . . . . . 38.18
General Medical Council v Qureshi [2014] EWHC 775 (Admin) . . . . . . . . . . . . . . . . 38.60
General Medical Council v Razoq [2012] EWHC 2135 (Admin) . . . . . . . . . . . . . 10.14, 38.48
General Medical Council v Shalaby [2013] EWHC 4211 (Admin) . . . . . . . . . . . . . . . . 38.59
General Medical Council v Sheill [2006] EWHC 3025 (Admin) . . . . . . . . . . . . . . . . . 38.05
General Medical Council v Sondhi [2013] EWHC 4233 (Admin) . . . . . . . . . . . . . . . . 38.58
General Medical Council v Spackman [1943] A.C. 627; [1943] 2 All E.R. 337; (1943) 59 T.L.R. 412; (1943) 169 L.T. 226; (1943) 87 S.J. 298 . . . . . . . . . . . . . . . . . . . . . . . . . . . . . . . . 25.03, 25.14, 25.20, 39.02
General Medical Council v Srinivas [2012] EWHC 670 (Admin) . . . . . . . . . . . . . . . . . 10.12
General Medical Council v Srinivas [2012] EWHC 2513 (Admin) . . . . . . . . . . . . . . . . 38.49
General Medical Council v Uruakpa [2007] EWHC 1454 (Admin) . . . . . . . . . . . . . . . 38.06
General Osteopathic Council v Sobande [2011] CSOH 39 . . . . . . . . . . . . . . . . . . . . . 56.18
George v General Medical Council *See* R (George) v General Medical Council
Gerstenkorn v General Medical Council [2009] EWHC 2682 (Admin) . . . . . . . . . 4.20, 8.07
Ghosh v General Medical Council [2001] UKPC 29; [2001] 1 W.L.R. 1915; [2001] U.K.H.R.R. 987; [2001] Lloyd's Rep. Med. 433; Times, June 25, 2001 . . . . . . . 4.08, 4.09, 4.14, 4.16, 4.17, 4.28, 17.19, 19.44, 19.49, 27.02, 60.04

Ghosh v Knowsley NHS Primary Care Trust *See* R (Ghosh) v Knowsley NHS Primary Care Trust
Giambrone v Solicitors Regulation Authority [2014] EWHC 1421 (Admin) .... 56.30, 60.58
Gibson v General Medical Council *See* R (Gibson) v General Medical Council
Giele v General Medical Council [2005] EWHC 2143 (Admin); [2006] 1 W.L.R. 942; [2005] 4 All E.R. 1242; (2006) 87 B.M.L.R. 34 ............ 8.05, 37.12, 60.08, 65.04
Gillies v Secretary of State for Work and Pensions [2006] UKHL 2; [2006] 1 W.L.R. 781; [2006] 1 All E.R. 731; 2006 S.C. (H.L.) 71; 2006 S.L.T. 77; 2006 S.C.L.R. 276; [2006] I.C.R. 267; (2006) 9 C.C.L. Rep. 404; (2006) 103(9) L.S.G. 33; (2006) 150 S.J.L.B. 127; 2006 G.W.D. 3-66; Times, January 30, 2006 ............ 35.08
Gilthorpe v General Medical Council [2012] EWHC 672 (Admin) ............ 4.36, 4.38
Ginikanowa v United Lingdom (1988) 55 DR 252 ............ 17.23
Gitau v Nursing and Midwifery Council [2014] EWHC 2581 (Admin) ............ 14.35
Giwa-Osagie v General Medical Council [2013] EWHC 1514 (Admin) ............ 47.49
Gleadall v Huddersfield Magistrates' Court [2005] EWHC 2283 (Admin) ............ 5.03
Goodchild-Simpson v General Medical Council [2014] EWHC 1343 (Admin) ....... 31.11
Goodwin v Health and Care Professions Council [2014] EWHC 1897 (Admin) ...... 17.34
Goose v Wilson, Sandford & Co [1998] TLR 85 ............ 17.26
Gopakumar v General Medical Council [2006] EWHC 729 (Admin) ............ 30.01
Gopakumar v General Medical Council [2008] EWCA 309; (2008) 101 B.M.L.R. 121 ............ 44.17, 44.18, 44.23, 45.16, 45.18
Gorlov v Institute of Chartered Accountants in England and Wales *See* R (Gorlov) v Institute of Chartered Accountants in England and Wales
Gosai v General Medical Council , Privy Council Appeal No. 20 of 2002 ........... 58.04
Graham v Chorley Borough Council [2006] EWCA Civ 92 ............ 50.06
Gray v Avadis [2003] EWHC 1830 (QB); (2003) 100(36) L.S.G. 43; Times, August 19, 2003 ............ 52.12
Greaves v Newham London Borough Council Employment Appeals Tribunal, 16 May 1983 ............ 62.21
Green v Police Complaints Authority *See* R (Green) v Police Complaints Authority
Green and Valentine, Case no.8606/1, Report of Lloyd's Disiciplinary and Appellate Proceedings ............ 47.17
Guest v General Medical Council [2013] EWHC 3121 (Admin) ............ 1.18
Gulamhusein v General Pharmaceutical Council [2014] EWHC 2591 (Admin) ..... 10.16, 38.42
Gupta v General Medical Council [2001] EWHC Admin 612; [2001] EWHC Admin 631 ............ 38.65, 38.66
Gupta v General Medical Council [2001] UKPC 61; [2002] 1 W.L.R. 1691; [2002] I.C.R. 785; [2002] Lloyd's Rep. Med. 82; (2002) 64 B.M.L.R. 56; Times, January 9, 2002 ............ 4.16, 4.27, 4.28, 19.40, 19.41, 27.02, 27.05, 54.05, 54.06, 54.10, 54.11, 54.12, 54.15, 54.17, 54.30
Gupta v General Medical Council [2007] EWHC 2918 (Admin) ............ 10.07, 16.07
Gurpinar v Solicitors Regulation Authority [2012] EWHC 192 (Admin) ............ 1.12
Gyurkovits v General Dental Council *See* R (Gyurkovits) v General Dental Council

H (A Barrister), In re [1981] 1 W.L.R. 1257; [1981] 3 All E.R. 205; (1981) 125 S.J. 609 ............ 57.02
H (Minors) (Sexual Abuse: Standard of Proof), In re [1996] A.C. 563; [1996] 2 W.L.R. 8; [1996] 1 All E.R. 1; [1996] 1 F.L.R. 80; [1996] 1 F.C.R. 509; [1996] Fam. Law 74; (1995) 145 N.L.J. 1887; (1996) 140 S.J.L.B. 24; Times, December 15, 1995; Independent, January 17, 1996 ..... 16.04, 44.25, 62.03, 62.04, 62.05
H (Minors) (Sexual Abuse: Standard of Proof), re [1996] A.C. 563; [1996] 2 W.L.R. 8; [1996] 1 All E.R. 1; [1996] 1 F.L.R. 80; [1996] 1 F.C.R. 509; [1996] Fam. Law 74; (1995) 145 N.L.J. 1887; (1996) 140 S.J.L.B. 24; Times, December 15 ........... 19.29, 19.33, 54.33, 62.06, 62.21
H v Belgium (1988) 10 EHRR 339 ............ 32.06, 32.07

H v H (Minors) (Child Abuse: Evidence) [1990] Fam. 86; [1989] 1 F.L.R. 212; [1989] F.C.R. 257; 87 L.G.R. 166; [1989] Fam. Law 148; (1989) 153 J.P.N. 289; (1989) 86(40) L.S.G. 44 .......... 62.03
H v Nursing and Midwifery Council *See* R (H) v Nursing and Midwifery Council
Haase v District Judge Nuttall *See* R (Haase) v District Judge Nuttall
Haikel v General Medical Council [2002] UKPC 37; [2002] Lloyd's Rep. Med. 415 .......... 17.19, 44.12
Hameed v Central Manchester University Hospitals NHS Foundation Trust [2010] EWHC 2009 (QB); [2010] Med. L.R. 412 .......... 32.23, 32.28, 35.15, 45.22
Hamilton v Al Fayed (No. 2) [2001] EMLR 15 .......... 25.38
Hamilton v Al Fayed, unreported, Court of Appeal, 21 December 2000 .......... 4.11, 25.35
Han and Yau v Commissioners of Customs and Excise [2001] EWCA Civ 1048; [2001] 1 W.L.R. 2253; [2001] 4 All E.R. 687; [2001] S.T.C. 1188; [2001] H.R.L.R. 54; [2001] U.K.H.R.R. 1341; [2001] B.V.C. 415; 3 I.T.L. Rep. 873; [2001] S.T.I. 1015; (2001) 98(32) L.S.G. 37; (2001) 151 N.L.J. 1033; (2001) 145 S.J.L.B. 174; Times, August 3, 2001; Independent, July 5, 2001 .......... 32.12
Hannam v Financial Conduct Authority [2014] UKUT 0233 (TCC) .......... 62.21
Harford v Nursing and Midwifery Council [2013] EWHC 696 (Admin) .......... 47.51
Harris v Appeal Committee of the Institute of Chartered Accountants of Scotland 2005 SLT 487 .......... 25.07
Harris v Registrar of Approved Driving Instructors [2010] EWCA Civ 808; [2011] R.T.R. 1; Times, July 26, 2010 .......... 56.14
Harris v Solicitors Regulation Authority [2011] EWHC 2173 (Admin) .......... 55.04
Harrison v General Medical Council *See* R (Harrison) v General Medical Council
Harrison v General Medical Council [2011] EWHC 1741 (Admin) .......... 4.33
Harry v General Medical Council *See* R (Harry) v General Medical Council
Harry v General Medical Council [2012] EWHC 2762 (QB) .......... 38.31, 38.33
Hart v Standards Committee (No.1) of the New Zealand Law Society [2012] NZSC 4 .......... 51.06
Hasan v General Medical Council [2003] UKPC 5 .......... 19.39
Hassan v General Optical Council [2013] EWHC 1887 (Admin) .......... 19.65, 19.72, 44.24, 56.24
Haward v General Medical Council *See* R (Haward) v General Medical Council
Haworth v Nursing and Midwifery Council *See* R (Haworth) v Nursing and Midwifery Council
Hay v Institute of Chartered Accountants in Scotland 2003 SLT 612 .......... 52.04
Hayat v General Medical Council [2014] EWHC 1477 (Admin) .......... 38.41
Haye v General Teaching Council for England (2013) 157 Sol Jo 35 .......... 37.16
Hazelhurst v Solicitors Regulation Authority [2011] EWHC 462 (Admin); [2011] N.P.C. 29 .......... 28.06, 54.22
Headdon v Institute of Actuaries *See* R (Ranson, Headdon and Nash) v Institute of Actuaries
Health Professions Council v Information Commissioner, EA/2007/01116, 14 March 2007 .......... 18.32
Heath v Commissioner of Police of the Metropolis [2004] EWCA Civ 943; [2005] I.C.R. 329; [2005] I.R.L.R. 270; [2004] Po. L.R. 259; (2004) 148 S.J.L.B. 913; Times, July 22, 2004 .......... 52.07, 52.16
Heath v Home Office Policy and Advisory Board for Forensic Pathology *See* R (Heath) v Home Office Policy and Advisory Board for Forensic Pathology
Heather v Leonard Cheshire Foundation *See* (R) Heather v Leonard Cheshire Foundation
Heatley v Tasmanian Racing and Gaming Commission (1977) 137 CLR 487 .......... 42.06
Hedley Byrne & Co. Ltd v Heller & Partners [1964] A.C. 465 .......... 49.13
Heesom v Public Service Ombudsman for Wales [2014] EWHC 1504 (Admin); [2014] 4 All E.R. 269 .......... 60.57
Hefferon v Professional Conduct Committee of the United Kingdom Central Council for Nursing Midwifery and Health Visiting (1988) 10 BML 1 .......... 4.31, 4.48, 25.34, 48.09

Helow v Secretary of State for the Home Department [2008] UKHL 62; [2008] 1 W.L.R. 2416; [2009] 2 All E.R. 1031; 2009 S.C. (H.L.) 1; 2008 S.L.T. 967; 2008 S.C.L.R. 830; (2008) 152(41) S.J.L.B. 29; 2008 G.W.D. 35-520; Times, November 5, 2008 .......... 6.06, 6.17, 35.13
Henderson v General Teaching Council for England *See* R (Henderson) v General Teaching Council for England
Henshall v General Medical Council [2005] EWCA Civ 1520; [2006] Lloyd's Rep. Med. 103; (2006) 88 B.M.L.R. 146; Times, January 9, 2006 .......... 18.09, 39.11
Hertfordshire Investments Ltd v Bubb [2000] 1 W.L.R. 2318; [2001] C.P. Rep. 38; [2000] C.P.L.R. 588; Times, August 31, 2000 .......... 4.11, 25.35
Heydon-Burke v Nursing and Midwifery Council [2012] EWHC 2435 (Admin) .......... 19.65
Hickey v General Medical Council *See* R (Hickey) v General Medical Council
Hill v Institute of Chartered Accountants in England and Wales *See* R (Hill) v Institute of Chartered Accountants in England and Wales
Hill v Institute of Chartered Accountants in England and Wales [2012] EWHC 1731 (QB) .......... 13.17
HK v General Pharmaceutical Council [2014] CSIH 61 .......... 65.15
Hobbs v Financial Conduct Authority (formerly Financial Services Authority) [2013] Bus LR 1290, CA .......... 43.19
Holden v Solicitors Regulation Authority [2012] EWHC 2067 (Admin) .......... 62.17
Holder v College of Physicians and Surgeons of Manitoba [2003] 1 WWR 19 .......... 24.03
Holder v Law Society [2005] EWHC 2023 (Admin); [2006] P.N.L.R. 10 .......... 35.07, 52.09
Hollington v F Hewthorn & Co. Ltd [1943] K.B. 587 .......... 25.20
Hollis v Association of Chartered Certified Accountants *See* R (Hollis) v Association of Chartered Certified Accountants
Holloway v Solicitors Regulation Authority [2012] EWHC 3393 (Admin) .......... 62.19
Holmes v Royal College of Veterinary Surgeons [2011] UKPC 48 .......... 35.13
Holtin v General Medical Council [2006] EWHC 2960 (Admin); (2007) 93 B.M.L.R. 74; [2007] A.C.D. 66 .......... 16.05, 18.10
Holy v Law Society [2006] EWHC 1034 (Admin) .......... 55.02
Homburg Houtinport BV v Agrosin Private Ltd [2003] UKHL 12; [2004] 1 A.C. 715; [2003] 2 W.L.R. 711; [2003] 2 All E.R. 785; [2003] 1 All E.R. (Comm) 625; [2003] 1 Lloyd's Rep. 571; [2003] 1 C.L.C. 921; 2003 A.M.C. 913; (2003) 100(19) L.S.G. 31; Times, March 17, 2003 .......... 56.03
Homes v General Medical Council *See* R (Homes) v General Medical Council
Hoodless and Blackwell v Financial Services Authority, 3 October 2003 .......... 19.21
Hornal v Neuberger Products Ltd [1957] 1 Q.B. 247; [1956] 3 W.L.R. 1034; [1956] 3 All E.R. 970; (1956) 100 S.J. 915 .......... 62.03, 62.21
Hornsey v General Medical Council [2011] EWHC 3779 (Admin) .......... 19.16
Horrocks v Lowe [1975] A.C. 135; [1974] 2 W.L.R. 282; [1974] 1 All E.R. 662; 72 L.G.R. 251; (1974) 118 S.J. 149 .......... 52.06
Hosny v General Medical Council [2011] EWHC 1355 (Admin); (2012) 123 B.M.L.R. 20 .......... 19.59
Hossack v General Dental Council [1998] 40 BMLR 97 .......... 47.21
Hossain v General Medical Council [2001] UKPC 40 .......... 60.23
Hounslow LBC v School Appeal Panel *See* R (Hounslow LBC) v School Appeal Panel
Houshian v General Medical Council [2012] EWHC 3458 (QB) .......... 38.33, 38.37
Howard v Stanton [2011] EWCA Civ 1481 .......... 1.21
Hucks v Cole [1993] 4 Med. L.R. 393 .......... 49.03
Hughes v Architects' Registration Council of the United Kingdom [1957] 2 Q.B. 550; [1957] 3 W.L.R. 119; [1957] 2 All E.R. 436; (1957) 101 S.J. 517 .......... 4.02, 47.05, 47.19
Hughes v Singh Times, April 21, 1989; Independent, April 24, 1989 .......... 25.37
Hume [2007] CSIH 53; 2007 S.C. 644 .......... 4.32
Hunt v Commissioners for Her Majesty's Revenue & Customs [2014] UKFTT 1084 (TC) .......... 56.33

Hunter v Chief Constable of the West Midland Police [1982] A.C. 529; [1981] 3 W.L.R. 906; [1981] 3 All E.R. 727; (1981) 125 S.J. 829; Times, November 26, 1981 . . . . . 2.06, 14.04, 17.24
Hunter v Hanley 1955 S.C. 200; 1955 S.L.T. 213; [1955-95] P.N.L.R. 1 . . . . . 49.02
Hurnam v Bar Standards Board, 6 February 2012 (unreported) . . . . . 14.08
Hussain v General Medical Council [2012] EWHC 2991 (Admin) . . . . . 38.14
Hussain v General Medical Council [2013] EWHC 3865 (Admin) . . . . . 22.35
Hussain v General Medical Council [2014] EWCA Civ 2246 . . . . . 19.30
Hussein v General Medical Council [2013] EWHC 3535 (Admin) . . . . . 26.12, 50.12
Hutchinson v General Dental Council [2008] EWHC 2896 (Admin) . . . . . 17.28, 62.14

Ibrahim v Feltham Magistrates' Court *See* R (Ibrahim) v Feltham Magistrates' Court
Idenburg v General Medical Council (2000) 55 BMLR 101 . . . . . 25.21
Ige v Nursing and Midwifery Council [2011] EWHC 3721 (Admin) . . . . . 19.61, 33.17, 37.13
Ighalo v Solicitors Regulation Authority [2013] EWHC 661 (Admin) . . . . . 35.20
Ikarian Reefer *See* National Justice Compania Naviera SA v Prudential Assurance Co Ltd (The Ikarian Reefer) (No.1)
Imam v General Medical Council [2011] EWHC 3000 (Admin) . . . . . 60.47
Inayatullah v General Medical Council [2014] EWHC 3751 (Admin) . . . . . 30.09
Independent Police Complaints Commission v Assistant Commissioner Hayman *See* R (Independent Police Complaints Commission) v Assistant Commissioner Hayman
Iqbal v Solicitors Regulation Authority [2012] EWHC 3251 (Admin); [2012] All E.R. (D) 217 (July) . . . . . 25.33, 49.11
Iqbal v Solicitors Regulation Authority [2012] EWHC 4097 (QB) . . . . . 52.14
Iran v Secretary of State for the Home Department *See* R (Iran) v Secretary of State for the Home Department
Isaghehi v Nursing and Midwifery Council [2014] EWHC 127 (Admin) . . . . . 14.31
Islam v Bar Standards Board, 1 August 2012 (unreported) . . . . . 56.19
Izuchukwu v Solicitors Regulation Authority [2013] EWHC 2106 (Admin) . . . . . 55.06
Izzet and Cazaly v Law Society [2009] EWHC 3590 (Admin) . . . . . 60.43

J v Upper Tribunal (Immigration Asylum Chamber) *See* R (J) v Upper Tribunal (Immigration Asylum Chamber)
JB v General Teaching Council for Scotland [2013] CSIH 114 . . . . . 56.28
Jago v District Council of New South Wales (1989) 168 CLR 23; (1989) 63 ALJ 640 . . . . . 2.02
Jain v Trent Strategic Health Authority [2009] UKHL 4; [2009] 1 A.C. 853; [2009] 2 W.L.R. 248; [2009] P.T.S.R. 382; [2009] 1 All E.R. 957; [2009] H.R.L.R. 14; (2009) 12 C.C.L. Rep. 194; [2009] LS Law Medical 112; (2009) 106 B.M.L.R. 88; (2009) 106(5) L.S.G. 14; (2009) 153(4) S.J.L.B. 27; Times, January 22, 2009 . . . . . 49.08
Jalloh v Nursing and Midwifery Council *See* R (Jalloh) v Nursing and Midwifery Council
Jasinarachchi v General Medical Council [2014] EWHC 3570 (Admin) . . . . . 4.48, 25.42, 65.14
Jatta v Nursing and Midwifery Council [2009] EWCA Civ 828 . . . . . 61.03
Javed v Solicitors Regulation Authority [2012] EWHC 114 (Admin) . . . . . 19.17
Jeffery v Financial Conduct Authority [2013] Lexis Citation 49 . . . . . 17.32, 19.21
Jeffery v Financial Services Authority [2012] EWCA Civ 178 . . . . . 70.04
Jeffery v Financial Services Authority, FS 2010/0039, 5 December 2012 . . . . . 6.13
Jenkinson v Nursing and Midwifery Council *See* R (Jenkinson) v Nursing and Midwifery Council
Jespers v Belgium (1983) 5 EHRR 305 . . . . . 18.12
Jesudason v Alder Hey Children's NHS Foundation Trust [2012] EWHC 4265 (QB) . . . . . 69.02

Jeyaretnam v Law Society of Singapore [1989] A.C. 608; [1989] 2 W.L.R. 207; [1989] 2 All E.R. 193; (1989) 86(1) L.S.G. 37; (1989) 133 S.J. 81 . . . . . . . . . . . . . . . . . 14.03, 14.06
Jideofo v Law Society [2007] EW Misc 3 . . . . . . . . . . . . . . . . . . . . 56.05, 56.07, 56.11, 56.16
John v Rees [1970] Ch 345 . . . . . . . . . . . . . . . . . . . . . . . . . . . . . . . . . . . . . . . . . . . . . 18.17
Johnson and Maggs v Nursing and Midwifery Council (No. 2) *See* R (Johnson and Maggs) v Nursing and Midwifery Council
Johnson v Johnson (2000) 201 CLR 488 . . . . . . . . . . . . . . . . . . . . . . . . . . . . . . . . 6.01, 6.06
Johnson and Maggs v Professional Conduct Committee of the Nursing and Midwifery Council [2008] EWHC 885 (Admin) . . . . . . . . . . . . . . . . . 6.07, 18.12, 22.21
Johnson v Professional Conduct Committee of the Nursing and Midwifery Council *See* R (Johnson and Maggs) v Professional Conduct Committee of the Nursing and Midwifery Council
Jones v Bar Standards Board, 12 March 2014 (unreported) . . . . . . . . . . . . . . . . . . . . . . 65.12
Jones v Commission for Social Care Inspection [2004] EWCA Civ 1713; [2005] 1 W.L.R. 2461; (2005) 149 S.J.L.B. 59; Times, January 4, 2005 . . . . . . . . . . . . . . . . . 56.04
Jones v Judicial Appointments Commission [2014] EWHC 1680 (Admin) . . . . . . . . . . . 41.05
Jones v University of Warwick [2003] EWCA Civ 151; [2003] 1 W.L.R. 954; [2003] 3 All E.R. 760; [2003] C.P. Rep. 36; [2003] P.I.Q.R. P23; (2003) 72 B.M.L.R. 119; (2003) 100(11) L.S.G. 32; (2003) 153 N.L.J. 231; (2003) 147 S.J.L.B. 179; Times, February 7 . . . . . . . . . . . . . . . . . . . . . . . . . . . . . . . . . . . . . . . . . . . . . . . . . . 2.08
Jooste v General Medical Council [2010] EWHC 2558 (Admin) . . . . . . . . . . . . . . . . . 38.28
Jooste v General Medical Council [2013] EWHC 2463 (Admin) . . . . . . . . . . . . . . . . . . 9.06
Joyce v Secretary of State for Health [2008] EWHC 1891 (Admin); [2009] PTSR 266; [2009] 1 All E.R. 1025; (2008) 11 C.C.L. Rep. 761; [2008] A.C.D. 89; Times, August 29, 2008 . . . . . . . . . . . . . . . . . . . . . . . . . . . . . . . . . . . . . . . . . . 47.38
Jurkowska v Hlmad Ltd [2008] EWCA Civ 231 . . . . . . . . . . . . . . . . . . . . . . . . . . . . 4.40

K v HM Advocate [2002] UKPC D 1; [2004] 1 A.C. 379; [2002] 3 W.L.R. 1488; [2002] 4 All E.R. 1; 2002 S.C. (P.C.) 89; 2002 S.L.T. 229; 2002 S.C.C.R. 220; [2002] H.R.L.R. 21; [2002] U.K.H.R.R. 542; 2002 G.W.D. 5-153; Times, February 4, 2002 . . . . . . . . . . . . . . . . . . . . . . . . . . . . . . . . . . . . . . . . . . . . . . . . . 17.03
Kaftan v General Medical Council *See* R (Kaftan) v General Medical Council
Kakade v General Medical Council [2013] EWHC 2110 (Admin) . . . . . . . . . . . . . . . . 38.15
Kamel v General Medical Council [2007] EWHC 313 (Admin) . . . . . . . . . . . . . . . . . . 59.04
Kanda v Government of the Federation of Malaya [1962] A.C. 322; [1962] 2 W.L.R. 1153; (1962) 106 S.J. 305 . . . . . . . . . . . . . . . . . . . . . . . . . . . . . . . . . . . . . 48.02
Karan v Financial Services Authority, Ref No FS/2010/0025, 15 May 2012 . . . . . . . . . . 28.08
Karpe v Financial Services Authority, Ref No.FS/2010/0019, 15 May 2012 . . . . . . . . . . . 28.07
Karwal v General Medical Council [2011] EWHC 826 (Admin); (2011) 120 B.M.L.R. 71 . . . . . . . . . . . . . . . . . . . . . . . . . . . . . . . . . . . . . . . . . . . . . 37.11, 59.10
Kashif v General Medical Council [2013] EWHC 4185 (Admin) . . . . . . . . . . . . . . . . . 38.17
Kataria v Essex Strategic Health Authority [2004] EWHC 641 (Admin) . . . . . . . . . . . . 59.03
Katti v General Medical Council *See* R (Katti) v General Medical Council
Kay v Chief Constable of Northumbria Police *See* R (Kay) v Chief Constable of Northumbria Police
Keane v Law Society [2009] EWHC 783 (Admin) . . . . . . . . . . . . . . . . . . . . . . . . . . . 54.16
Kearns v General Council of the Bar [2002] EWHC 1681 (QB); [2002] 4 All E.R. 1075; Independent, November 4, 2002 . . . . . . . . . . . . . . . . . . . . . . . . . 52.06
Kearns v General Council of the Bar [2003] EWCA Civ 331; [2003] 1 W.L.R. 1357; [2003] 2 All E.R. 534; [2003] E.M.L.R. 27; (2003) 153 N.L.J. 482; (2003) 147 S.J.L.B. 476 . . . . . . . . . . . . . . . . . . . . . . . . . 52.06
Keazor v Law Society [2009] EWHC 267 (Admin) . . . . . . . . . . . . . . . . . . . . . . . . . . . 55.03
Kehinde Bimbe Adegbulugbe v NMC [2001] EWHC Admin 3301 . . . . . . . . . . . . . . . . 4.44
Kelly v Edward *See* R (Kelly) v Edward
Kerman & Co v. Legal Ombudsman *See* R (Kerman & Co) v. Legal Ombudsman
Khan v Chief Constable of Lancashire *See* R (Khan) v Chief Constable of Lancashire
Khan v Financial Conduct Authority FS/2013/002, 8 April 2014 . . . . . . . . . . . . . . . . . 28.23

Khan v General Medical Council *See* R (Khan) v General Medical Council
Khan v General Medical Council [1996] ICR 1032 . . . . . . . . . . 39.39
Khan v General Medical Council [2009] EWHC 535 (Admin) . . . . . . . . . . 59.08
Khan v General Medical Council [2013] EWHC 2187(Admin) . . . . . . . . . . 27.16
Khan v General Medical Council [2015] EWHC 301 (Admin) . . . . . . . . . . 15.30, 19.79
Khan v General Medical Council (unreported) . . . . . . . . . . 4.05
Khan v General Teaching Council for England [2010] EWHC 3404 (Admin) . . . . . . . . . . 3.14
Khan v Solicitors Regulation Authority [2010] EWHC 1555 (Admin) . . . . . . . . . . 11.13, 56.13
Khokhar v Health Professions Council [2006] EWHC 2484 (Admin) . . . . . . . . . . 60.33
Kibe v Nursing and Midwifery Council *See* R (Kibe) v Nursing and Midwifery Council
Kidd v Scottish Legal Complaints Commission [2011] CSIH 75 . . . . . . . . . . 39.27
Kilshaw v Office of the Supervision of Solicitors [2005] EWHC 1484 (Admin) . . . . . . . . . . 3.09
King (Ch Ct of York), In re [2009] 1 W.L.R. 873 . . . . . . . . . . 60.36
King v Secretary of State for Justice *See* R (King) v Secretary of State for Justice
Kirk v Royal College of Veterinary Surgeons [2003] UKPC 3; [2003] UKPC 47 . . . . . . . . . . 18.06
Kirk v Royal College of Veterinary Surgeons, Privy Council Appeal No. 51 of 2002 . . . . . . . . . . 14.05
Kituma v Nursing and Midwifery Council [2010] EWCA Civ 154 . . . . . . . . . . 33.12
König v Federal Republic of Germany (1978) 2 EHRR 170 . . . . . . . . . . 32.02
Kress v France, App No. 39594/98, ECHR 2001-V1, 43 . . . . . . . . . . 46.04
Krippendorf v General Medical Council [2001] 1 W.L.R. 1054; [2001] Lloyd's Rep. Med. 9; (2001) 59 B.M.L.R. 81; (2001) 145 S.J.L.B. 5; Times, November 29, 2000 . . . . . . . . . . 16.02, 16.03, 16.04
Kudrath v Ministry of Defence, 26 April 1999, unreported . . . . . . . . . . 48.12
Kulkarni v Milton Keynes Hospital NHS Foundtion Trust [2009] EWCA Civ 789; [2010] I.C.R. 101; [2009] I.R.L.R. 829; [2009] LS Law Medical 465; (2009) 109 B.M.L.R. 133; Times, August 06, 2009 . . . . . . . . . . 32.22, 32.24, 32.28, 35.14, 45.20
Kumar v General Medical Council [2012] EWHC 2688 (Admin) . . . . . . . . . . 26.11, 47.46, 55.05
Kumar v General Medical Council [2013] EWHC 452 (Admin) . . . . . . . . . . 38.36, 61.05

L & B (Children), Re [2013] UKSC 8; [2013] 1 WLR 634 . . . . . . . . . . 38.16
L v Health and Care Professions Council [2014] EWHC 994 (Admin) . . . . . . . . . . 4.30
L v Law Society [2008] EWCA Civ 811 . . . . . . . . . . 53.04
L v The Health and Care Professions Council [2014] EWHC 994 (Admin) . . . . . . . . . . 54.32
Ladd v Marshall [1954] 1 W.L.R. 1489 . . . . . . . . . . 25.38, 65.14
Lake v British Transport Police [2007] ICR 1293 . . . . . . . . . . 52.16
Lamey v Belfast Health and Social Care Trust [2013] NIQB 91 . . . . . . . . . . 10.15
Land v Executive Counsel of the Accountants' Joint Disciplinary Scheme *See* R (Land) v Executive Counsel of the Accountants' Joint Disciplinary Scheme
Lanford v General Medical Council [1990] 1 A.C. 13; [1989] 3 W.L.R. 665; [1989] 2 All E.R. 921; (1989) 133 S.J. 1202 . . . . . . . . . . 62.11
Langford v Law Society [2002] EWHC 2802 (Admin); (2003) 153 N.L.J. 176 . . . . . . . . . . 4.17, 17.35
Laud v General Medical Council, Privy Council Appeal No. 18 of 1979 . . . . . . . . . . 4.05
Lavis v Nursing and Midwifery Council [2014] EWHC 4083 (Admin) . . . . . . . . . . 19.31
Law Liat Meng v Disciplinary Committee [1968] A.C. 391; [1967] 3 W.L.R. 877; (1967) 111 S.J. 619 . . . . . . . . . . 22.02
Law Society of England and Wales v Shah [2014] EWHC 4382 (Ch) . . . . . . . . . . 43.22
Law Society of England and Wales v Society of Lawyers [1996] FSR 739 . . . . . . . . . . 56.18
Law Society of Scotland v Scottish Legal Complaints Commission 2011 SC 94 . . . . . . . . . . 39.21
Law Society (Solicitors Regulation Authority) v Emeana [2013] EWHC 2130 (Admin) . . . . . . . . . . 28.18, 60.16
Law Society (Solicitors Regulation Authority) v Henry, 29 January 2015 (unreported) . . . . . . . . . . 18.19
Law Society v Adcock [2006] EWHC 3212 (Admin); [2007] 1 W.L.R. 1096; [2007] A.C.D. 63; (2007) 104(3) L.S.G. 29; (2007) 157 N.L.J. 66; Times, January 16, 2007 . . . . . . . . . . 15.06, 15.09, 47.34
Law Society v Master of the Rolls *See* R (Law Society) v Master of the Rolls

Law Society v Waddingham [2012] EWHC 1519 (Admin); [2012] 6 Costs L.O. 832; (2012) 156(24) S.J.L.B. 35 . . . . . . . . . . . . . . . . . . . . . . . . 28.10, 62.18
Law v National Greyhound Racing Club Ltd [1983] 1 W.L.R. 1302; [1983] 3 All E.R. 300; (1983) 80 L.S.G. 2367; (1983) 127 S.J. 619 . . . . . . . . . . . . . 42.03, 42.06
Lawal v Northern Spirit Ltd [2003] ICR 856 . . . . . . . . . . . . . . . . . . . . . . . . . . . . . . 6.01, 6.17
Lawrence v Attorney General of Grenada [2007] UKPC 18, [2007] 1 W.L.R. 1474 . . . . . . . . . . . . . . . . . . . . . . . . . . . . . . . . . . . . . . . . . . . . . . . . . . . 41.02
Lawrence v General Medical Council [2012] EWHC 464 (Admin) . . . . . 13.12, 25.27, 26.10, 27.09, 70.05
Lawther v Council of the Royal College of Veterinary Surgeons [1968] 1 W.L.R. 1441; (1968) 112 S.J. 625 . . . . . . . . . . . . . . . . . . . . . . . . . . . . . . . . . . . . . . 44.07
Le Compte, Van Leuven and De Meyere v Belgium (1982) 4 EHRR 1 . . . . . . . . 32.03, 32.07, 32.10, 32.12, 32.28
Le Scroog v General Optical Council [1982] 1 W.L.R. 1238; [1982] 3 All E.R. 257; (1982) 79 L.S.G. 1138; (1982) 126 S.J. 610 . . . . . . . . . . . . . . . . . . . 47.14
Leathley v Visitors of the Inns of Court *See* R (Leathley) v Visitors of the Inns of Court
Lebow v Law Society [2008] EWCA Civ 411 . . . . . . . . . . . . . . . . . . . . . . . . . . . 11.14, 11.18
Legal Ombudsman v Young [2011] EWHC 2923 (Admin); [2012] 1 W.L.R. 3227 . . . . 18.14
Levinge v Health Professions Council [2012] EWHC 135 (Admin) . . . . . . . . . . . 11.12, 26.08
Levy v Solicitors Regulation Authority [2011] EWHC 740 (Admin) . . . . . . . . . . . 15.14, 60.14
Libman v General Medical Council [1972] A.C. 217; [1972] 2 W.L.R. 272; [1972] 1 All E.R. 798; (1971) 116 S.J. 123 . . . . . . . . . . . . . . . . . . . . . 4.09, 19.72, 27.05, 30.01, 44.08, 44.24, 54.03
Lim v Law Society [2009] EWHC 1706 (Admin) . . . . . . . . . . . . . . . . . . . . . . . . . . . . . 19.54
Lim v Royal Wolverhampton Hospitals NHS Trust [2011] EWHC 2178 (QB); [2012] Med. L.R. 146; (2011) 122 B.M.L.R. 43 . . . . . . . . . . . . . . . . . 16.10, 16.17, 39.24
Lincoln v Daniels [1962] 1 Q.B. 237; [1961] 3 W.L.R. 866; [1961] 3 All E.R. 740; (1961) 105 S.J. 647 . . . . . . . . . . . . . . . . . . . . . . . . . . . . . . . . . . . . . 4.07, 52.15, 70.02
Livingstone v Adjudication Panel for England [2006] EWHC 2533 (Admin); [2006] H.R.L.R. 45; [2006] B.L.G.R. 799; [2007] A.C.D. 22; (2006) 156 N.L.J. 1650; Times, November 9, 2006 . . . . . . . . . . . . . . . . . . . . . . . . . . . . . . . . . . . . . . . . 47.31
Lloyd v John Lewis Partnership [2001] EWCA Civ 1529 . . . . . . . . . . . . . . . . . . . . . . . . 50.05
Locabail (UK) Ltd v Bayfield Properties Ltd [2000] Q.B. 451; [2000] 2 W.L.R. 870; [2000] 1 All E.R. 65; [2000] I.R.L.R. 96; [2000] H.R.L.R. 290; [2000] U.K.H.R.R. 300; 7 B.H.R.C. 583; (1999) 149 N.L.J. 1793; [1999] N.P.C. 143; Times, November 19, 1999; Independent, November 23, 1999 . . . . . . . . 6.07, 35.20
Lonnie v National College for Teaching and Leadership *See* R (Lonnie) v National College for Teaching and Leadership
Loufti v General Medical Council [2010] EWHC 1762 (Admin) . . . . . . . . . . . . . . . . . 61.04
Low v General Osteopathic Council *See* R (Low) v General Osteopathic Council
Lucas v Millman [2002] EWHC 2470 (Admin); [2003] 1 W.L.R. 271; (2003) 100(1) L.S.G. 25; Times, November 22, 2002 . . . . . . . . . . . . . . . . . . . . . . . . . . . . . . . . 50.07
Luksch v Austria (2002) 35 EHRR 17 . . . . . . . . . . . . . . . . . . . . . . . . . . . . . . . . . 17.21, 32.14
Luthra v General Dental Council *See* R (Luthra) v General Dental Council
Luthra v General Medical Council [2013] EWHC 240 (Admin) . . . . . . . . . . . . . . . 8.09, 16.15
Lutton v General Dental Council [2011] CSIH 62; 2011 G.W.D. 33-696 . . . . . . . . . . . . 66.04
Lutton v General Dental Council [2011] CSOH 96; 2011 S.L.T. 671; 2011 G.W.D. 19-444 . . . . . . . . . . . . . . . . . . . . . . . . . . . . . . . . . . . . . . . . . . . . . . . . . . 66.04
Lynch v General Medical Council [2013] EWHC 3521 (Admin) . . . . . . . . . . . . . . . . . 11.19

M (A minor) (Appeal) (No 2), Re [1994] 1 F.L.R. 419; [1994] 2 F.C.R. 759; [1994] Fam. Law 427; Times, December 1, 1993 . . . . . . . . . . . . . . . . . . . . . . . . . . . 62.03
M v Worcestershire County Council [2002] EWHC 1292 . . . . . . . . . . . . . . . . . . . . . . . 27.09
MS v Sweden (1999) 28 E.H.R.R. 313; 3 B.H.R.C. 248; (1999) 45 B.M.L.R. 133 . . . . . . . . . . . . . . . . . . . . . . . . . . . . . . . . . . . . . . . . . . . . . 18.18, 18.26
McAllister v General Medical Council [1993] A.C. 388 . . . . . . . . . . . . . . . . . . . . 43.04, 62.11

McCandless v General Medical Council [1996] 1 W.L.R. 167; [1996] 7 Med. L.R. 379; (1996) 30 B.M.L.R. 53; (1996) 93(5) L.S.G. 31; (1996) 140 S.J.L.B. 28; Times, December 12, 1995 .......... 47.20, 49.04
McCoan v General Medical Council [1964] 1 W.L.R. 1107; [1964] 3 All E.R. 143; (1964) 108 S.J. 560 .......... 4.03, 4.06, 47.08, 47.10
McConnell v United Kingdom (2000) 30 EHRR 289 .......... 48.12
McDaid v Nursing and Midwifery Council [2013] EWHC 586 (Admin); (2013) 132 B.M.L.R. 190 .......... 1.15, 19.76
McEniff v General Dental Council [1980] 1 W.L.R. 328; [1980] 1 All E.R. 461; (1980) 124 S.J. 220; Times, December 1, 1979 .......... 47.17
McGraddie v McGraddie [2013] UKSC 58; [2013] 1 W.L.R. 2477 .......... 4.31
McInnes v Her Majesty's Advocate [2010] UKSC 7 .......... 18.17
McInnes v Onslow-Fane [1978] 1 W.L.R. 1520; [1978] 3 All E.R. 211; (1978) 122 S.J. 844 .......... 48.04, 54.26
Mackaill v Independent Police Complaints Commission *See* R (Mackaill) v Independent Police Complaints Commission
McKeown v British Horse Racing Authority [2010] EWHC 508 .......... 27.09
McKernan's Application, Re (1985) 135 N.L.J. 1164; The Times, 26 October, 1985 .......... 43.05
McInnes v Onslow-Fane [1978] 1 W.L.R. 1520; [1978] 3 All E.R. 211; (1978) 122 S.J. 844 .......... 48.04, 54.26
Maclean v Workers' Union [1929] 1 Ch 602 .......... 45.02
Macleod v Royal College of Veterninary Surgeons [2006] UKPC 39; (2006) 150 S.J.L.B. 1023 .......... 19.58, 57.04
McMillan v General Medical Council , Privy Council Appeal No. 62 of 1992 .......... 11.03
McNally v Secretary of State for Education and Employment *See* R (McNally) v Secretary of State for Education and Employment
McNicholas v Nursing and Midwifery Council *See* R (McNicholas) v Nursing and Midwifery Council
McNicholls v Judicial and Legal Service Commission [2010] UKPC 6 .......... 41.03
Madam Justice Levers (Judge of the Grand Court of the Cayman Islands) Referral under Section 4 of the Judicial Committee Act 1833, Re [2010] UKPC 24 .......... 41.04
Madan v General Medical Council (No. 2) [2001] EWHC Admin 577; [2001] Lloyds LR Med 539 .......... 38.03, 38.19, 38.33, 42.10
Maggs v Professional Conduct Committee of the Nursing and Midwifery Council *See* R (Johnson and Maggs) v Professional Conduct Committee of the Nursing and Midwifery Council
Mahfouz v Professional Conduct Committe of the General Medical Council *See* R (Mahfouz) v Professional Conduct Committee of the General Medical Council
Mahmood v General Medical Council *See* R (Mahmood) v General Medical Council
Mahon v Rahn (No.2) [2000] 1 W.L.R. 2150; [2000] 4 All E.R. 41; [2000] 2 All E.R. (Comm) 1; [2000] E.M.L.R. 873; [2000] Po. L.R. 210; (2000) 97(26) L.S.G. 38; (2000) 150 N.L.J. 899; Times, June 14, 2000 .......... 52.03, 52.12, 52.15
Maistry v Solicitors Regulation Authority [2012] EWHC 3041 (Admin); (2012) 156(39) S.J.L.B. 31 .......... 28.11
Makhdum v Norfolk and Suffolk NHS Foundation Trust [2012] All E.R. (D) 278 (Oct) .......... 17.30
Makki v General Medical Council [2009] EWHC 3180 (Admin) .......... 19.14
Malik v General Medical Council [2013] EWHC 2902 (Admin) .......... 38.17, 38.38
Malik v General Medical Council [2014] EWHC 2408 (Admin) .......... 1.21, 4.47
Malik v Waltham Forest NHS Primary Care Trust *See* R (Malik) v Waltham Forest NHS Primary Care Trust
Mallon v General Medical Council [2007] CSIH 17 .......... 47.35
Manzur v General Medical Council [2001] UKPC 55; (2002) 64 B.M.L.R. 68 .......... 19.38
Marcus v Nursing and Midwifery Council [2011] EWHC 3505 (Admin) .......... 22.29
Marinovich v General Medical Council [2002] UKPC 36 .......... 4.16, 60.05

Mark Anthony Financial Management v Financial Services Authority [2012] Lexis Citation 113 .......... 28.16
Marrinan v Vibert [1963] 1 Q.B. 528 .......... 52.12
Marten v Royal College of Veterinary Surgeons' Disciplinary Committee [1966] 1 Q.B. 1; [1965] 2 W.L.R. 1228; [1965] 1 All E.R. 949; (1965) 109 S.J. 876 .......... 47.09, 47.17, 47.19, 47.22
Martin v General Medical Council [2011] EWHC 3204 (Admin) .......... 13.11, 33.16, 50.10
Martin v General Medical Council [2014] EWHC 1269 (Admin) .......... 38.40
Martin v Tauranga District Ciuncil [1995] 2 NZLR 419 .......... 17.04
Martin-Artajo v Financial Conduct Authority [2014] UKUT 0340 (TCC) .......... 4.45
Mateu-Lopez v General Medical Council [2003] UKPC 44 .......... 60.26
Matthews v Solicitors Regulation Authority [2013] EWHC 1525 (Admin) .......... 28.17
Mattu v University Hospitals of Coventry and Warwickshire NHS Trust [2012] EWCA Civ 641; [2012] 4 All E.R. 359; [2012] I.R.L.R. 661 .......... 32.28, 35.18, 39.37, 45.25
May v Chartered Institute of Management Accountants *See* R (May) v Chartered Institute of Management Accountants
Mayer v Hoare [2012] EWHC 1805 (QB) .......... 52.15
Maynard v Osmond [1977] Q.B. 240; [1976] 3 W.L.R. 711; [1977] 1 All E.R. 64; [1976] Crim. L.R. 633; (1976) 120 S.J. 604 .......... 45.05
Maynard v West Midland Regional Health Authority [1984] 1 W.L.R. 634; [1985] 1 All E.R. 635; (1984) 81 L.S.G. 1926; (1983) 133 N.L.J. 641; (1984) 128 S.J. 317; Times, May 9, 1983 .......... 49.02, 49.03
Meadow v General Medical Council [2006] EWCA Civ 1390; [2007] Q.B. 462; [2007] 2 W.L.R. 286; [2007] 1 All E.R. 1; [2007] I.C.R. 701; [2007] 1 F.L.R. 1398; [2006] 3 F.C.R. 447; [2007] LS Law Medical 1; (2006) 92 B.M.L.R. 51; [2007] Fam. Law 214; [2006] 44 E.G. 196 (C.S.); (2006) 103(43) L.S.G. 28; (2006) 156 N.L.J. 1686; Times, October 31, 2006; Independent, October 31, 2006 .......... 4.18, 4.21, 4.28, 4.29, 25.20, 26.04, 27.05, 27.06, 32.08, 33.06, 33.07, 33.11, 39.29, 47.01, 47.32, 47.35, 47.46, 47.47, 49.04
Medicaments and Related Classes of Goods (No.2), Re [2001] 1 W.L.R. 700; [2001] U.K.C.L.R. 550; [2001] I.C.R. 564; [2001] H.R.L.R. 17; [2001] U.K.H.R.R. 429; (2001) 3 L.G.L.R. 32; (2001) 98(7) L.S.G. 40; (2001) 151 N.L.J. 17; (2001) 145 S.J.L.B. 29; Times, February 2, 2001; Independent, January 12, 2001 .......... 6.01, 6.17, 35.05, 35.12
Mehey v Visitors of the Inns of Court *See* R (Mehey) v Visitors of the Inns of Court
Mercer v Pharmacy Board of Victoria [1968] VR 72 .......... 47.17
Merelie v General Dental Council [2009] EWHC 1165 (QB) .......... 49.07
Merrick v Law Society [2007] EWHC 2997 (Admin) .......... 15.10, 15.12, 15.13, 15.16, 28.04
Meyers v Casey (1913) 17 CLR 90 .......... 24.05
Micalizzi v Financial Conduct Authority [2014] UKUT 0335 (TCC) .......... 28.24
Michalak v General Medical Council *See* R (Michalak) v General Medical Council
Michel van der Wiele NV v Pensions Regulator [2011] PLR 109 .......... 25.16
Millar v Dickson [2002] 1 W.L.R. 1615; 2002 S.C.(P.C.) 30; 2001 S.L.T. 988; [2001] H.R.L.R. 59 .......... 6.17
Miller (t/a Waterloo Plant) v Margaret Cawley [2002] EWCA Civ 1100 .......... 50.05
Miller v General Medical Council *See* R (Miller) v General Medical Council
Miller v Minister of Pensions [1947] 2 All E.R. 372; 63 T.L.R. 474; [1947] W.N. 241; [1948] L.J.R. 203; 177 L.T. 536; (1947) 91 S.J. 484 .......... 7.02, 62.02
Mills v General Dental Council [2014] EWHC 89 (Admin) .......... 19.73
Mireskandari v Solicitors Regulation Authority [2013] EWHC 907 (Admin) .......... 15.18
Misra v General Medical Council [2003] UKPC 7; (2003) 72 B.M.L.R. 108 .......... 22.10, 33.08
Mitchell v Nursing and Midwifery Council [2009] EWHC 1045 (Admin) .......... 4.32, 4.33
Modahl v British Athletic Federation Ltd [2001] EWCA Civ 1447; [2002] 1 W.L.R. 1192; (2001) 98(43) L.S.G. 34; (2001) 145 S.J.L.B. 238 .......... 48.10
Modohal v British Athletic Federation [2001] EWCA Civ 1447; [2002] 1 W.L.R. 1192; (2001) 98(43) L.S.G. 34; (2001) 145 S.J.L.B. 238 .......... 38.70

Mond v Association of Chartered Certified Accountants [2005]
EWHC 1414 (Admin); [2006] B.P.I.R. 94 . . . . . . . . 22.16
Moneim v General Medical Council [2011] EWHC 327 (Admin) . . . . . . . . 19.15
Monger v Chief Constable of Cumbria Police *See* R (Monger) v Chief
Constable of Cumbria Police
Monibi v General Dental Council [2014] EWHC 1911 (Admin) . . . . . . . . 30.08
Monji v General Pharmaceutical Council [2014] EWHC 3128 (Admin) . . . . . . . . 60.62
Montgomery v HM Advocate [2003] 1 A.C. 641; [2001] 2 W.L.R. 779;
2001 S.C. (P.C.) 1; 2001 S.L.T. 37; 2000 S.C.C.R. 1044; [2001] U.K.H.R.R. 124;
9 B.H.R.C. 641; 2000 G.W.D. 40-1487; Times, December 6, 2000 . . . . . . . . 51.04
Montgomery v Police Appeals Tribunal and the Commissioner of Police of the
Metropolis *See* R (Montgomery) v Police Appeals Tribunal and the Commissioner
of Police of the Metropolis
Moody v General Osteopathic Council [2004] EWHC Admin 967 . . . . . . . . 19.58
Moody v General Osteopathic Council [2008] EWCA Civ 513 . . . . . . . . 60.33
Moody v General Osteopathic Council [2009] 1 W.L.R. 526 . . . . . . . . 38.67
Moore v Nursing and Midwifery Council [2013] EWHC 4620 (Admin) . . . . . . . . 54.31
Mort v United Kingdom Application no. 44564/98, 6 September 2001, ECtHR . . . . . . 44.02
Moseka v Nursing and Midwifery Council [2014] EWHC 846 (Admin) . . . . . . . 4.28, 19.25,
67.11, 67.12
Moseley v Solicitors Regulation Authority [2013] EWHC 2108 (Admin) . . . . . . . . 60.54
Mould v General Dental Council [2012] EWHC 3114 (Admin) . . . . . . . . 11.15, 65.07
Mubarak v General Medical Council [2008] EWHC 2830 (Admin) . . . . . 27.14, 37.07, 54.30
Mucelli v Government of Albania [2009] UKHL 2; [2009] 1 W.L.R. 276;
[2009] 3 All E.R. 1035; [2009] Extradition L.R. 122; (2009) 153(4)
S.J.L.B. 28; Times, January 27, 2009 . . . . . . . . 4.32, 4.33
Muldoon v IPCC [2009] EWHC 3633 (Admin) . . . . . . . . 39.31, 39.32
Mulla v Solicitors Regulation Authority [2010] EWHC 3077 (Admin) . . . . . . . . 56.15
Mullins v Appeal Board of the Jockey Club *See* R (Mullins) v Appeal Board
of the Jockey Club
Murnin v Scottish Legal Complaints Commission [2012] CSIH 34; 2012
S.L.T. 685; 2012 G.W.D. 14-292 . . . . . . . . 17.11
Murphy v General Teaching Council of Scotland 1997 SLT 1152 . . . . . . . . 25.05
Murtagh v Solicitors Regulation Authority [2013] EWHC 2024 (Admin) . . . . . . . . 15.09
Muscat v Health Professions Council [2008] EWHC 2798 (Admin) . . . . . . . . 6.08
Muscat v Health Professions Council [2009] EWCA Civ 1090 . . . . . . . . 60.41
Musonza v Nursing and Midwifery Council [2012] EWHC 1440 (Admin) . . . . . . . . 67.07
Mvenge v General Medical Council [2010] EWHC 3529 . . . . . . . . 60.45
Myers v Elman [1940] A.C. 282; [1939] 4 All E.R. 484 . . . . . . . . 43.05

Nadarajah v Secretary of State for the Home Department [2005]
EWCA Civ 1363 . . . . . . . . 2.09, 32.24
Nagiub v General Medical Council [2011] EWHC 366 (Admin) . . . . . . . . 67.06
Nagiub v General Medical Council [2013] EWHC 1766 (Admin) . . . . . . . . 4.39
Nagle v Feilden [1966] 2 Q.B. 633 . . . . . . . . 42.06
Nahal v Law Society [2003] EWHC 2186 (Admin) . . . . . . . . 4.17, 60.29
Naheed v General Medical Council [2011] EWHC 702 (Admin) . . . . . . . . 19.63
Nandi v General Medical Council [2004] EWHC 2317 (Admin) . . . . . . . 47.27, 47.33, 47.35,
49.04, 66.02
Narayan v General Medical Council [1982] 1 W.L.R. 1227; (1982) 79
L.S.G. 988; (1982) 126 S.J. 483 . . . . . . . . 60.18
Nash v Chelsea College of Art and Design *See* R (Nash) v Chelsea College of
Art and Design
Nash v Institute of Actuaries *See* R (Ranson, Headdon and Nash) v Institute of Actuaries
National Justice Compania Naviera SA v Prudential Assurance Co Ltd
(The Ikarian Reefer) (No.1) [1993] 2 Lloyd's Rep. 68; [1993] F.S.R. 563; [1993]
37 E.G. 158; Times, March 5, 1993 . . . . . . . . 26.01, 27.05

Needham v Nursing and Midwifery Council [2003] EWHC 1141 (Admin) . . . . . . . . . . 54.06
Newfield v Law Society [2005] EWHC 765 (Admin) . . . . . . . . . . . . . . . . . . . . . . . . . . . . 4.17
Nicholas-Pillai v General Medical Council [2009] EWHC
1048 (Admin) . . . . . . . . . . . . . . . . . . . . . . . . . . . . . . . . . . . 19.79, 33.08, 33.15, 60.62
Nicholas-Pillai v General Medical Council [2012] EWHC 2238 (Admin) . . . . . . . . . . . . 8.08
Nicholas-Pillai v General Medical Council [2015] EWHC 305 (Admin) . . . . . . . 45.30, 70.11
Nicolaides v General Medical Council *See* R (Nicolaides) v General Medical Council
Norton v Bar Standards Board [2014] EWHC 2681 (Admin) . . . . . . . . . . . . . . . . . . . . . 3.20
Nowak v Nursing and Midwifery Council *See* R (Nowak) v Nursing and Midwifery Council
Nowak v Nursing and Midwifery Council and Guy's & St Thomas' NHS
Foundation Trust [2013] EWHC 1932 (QB) . . . . . . . . . . . . . . . . . . . . . . . . . . . . . . 9.04
Nursing and Midwifery Council v Farah [2014] EWHC 1655 (Admin) . . . . . . . . . . . . . 14.34
Nursing and Midwifery Council v Gatawa [2014] EWHC 3153 (Admin) . . . . . . . . . . . . 38.71
Nursing and Midwifery Council v Hitchenor [2012] EWHC 3565 (Admin) . . . . . . . . . 38.51
Nursing and Midwifery Council v Kidd and De'filippis [2014] EWHC 847 (Admin) . . . . . 38.62
Nursing and Midwifery Council v Maceda [2011] EWHC 3004 (Admin) . . . . . . . . . . . 38.46
Nursing and Midwifery Council v Miller [2011] EWHC 2601 (Admin) . . . . . . . . . . . . . 38.45
Nursing and Midwifery Council v Nowak [2013] EWHC 4330 (QB); [2014]
EWHC 336 (QB) . . . . . . . . . . . . . . . . . . . . . . . . . . . . . . . . . . . . . . . . . . . . . . . . 9.05
Nursing and Midwifery Council v Ogbonna [2010] EWCA Civ 1216 . . . . . . . . . . 25.23, 70.03
Nursing and Midwifery Council v Okafor [2013] EWHC 1045 (Admin) . . . . . . . 38.43, 38.52
Nursing and Midwifery Council v Pratt [2011] EWHC 3626 (Admin) . . . . . . . . . . . . . . 38.47
Nwabueza v General Medical Council [2000] 1 W.L.R. 1760; (2000)
56 B.M.L.R. 106; Times, April 11, 2000 . . . . . . . . . . . . . . . . 6.04, 22.07, 34.04, 44.10
Nwogbo v General Medical Council [2012] EWHC 2666 (Admin) . . . . . . . . . . . 14.09, 54.32

O v Secretary of State for Education and National College for Teaching and
Leadership (Interested Parties) [2014] EWHC 22 (Admin) . . . . . . . . . . . . . . . . . . 54.30
Obi v Solicitors Regulation Authority [2012] EWHC 3142 (Admin) . . . . . . . . . . . . . . . 60.48
Obukofe v General Medical Council [2014] EWHC 408 (Admin) . . . . . . . . . . . . 14.32, 59.18
O'Connor v Bar Standards Board, 17 August 2012 (unreported) . . . . . . . . . . . . . . 47.45, 47.53
O'Connor v Bar Standards Board [2014] EWHC 4324 (QB) . . . . . . . . . . . . . . . 32.30, 49.12
Odes v General Medical Council [2010] EWHC 552 (Admin); (2010)
113 B.M.L.R. 139 . . . . . . . . . . . . . . . . . . . . . . . . . . . . . . . . . . . . . . . . . . . . . . . 37.09
Odoi-Asare v Nursing and Midwifery Council [2014] EWHC 1151 (Admin) . . . . . . . . . 33.21
Official Receiver v Stern [2000] 1 W.L.R. 2230; [2001] 1 All E.R. 633; [2001]
B.C.C. 121; [2000] 2 B.C.L.C. 396; [2000] U.K.H.R.R. 332; Independent,
February 10, 2000 . . . . . . . . . . . . . . . . . . . . . . . . . . . . . . . . . . . . . . . . . . . . . . . 32.12
Ogango v Nursing and Midwifery Council [2008] EWHC 3115 (Admin) . . . . . . . . . . . 54.12
Ogbonna-Jacob v Nursing and Midwifery Council [2013] EWHC 1595 (Admin) . . . . . 14.35,
25.30, 59.16
Ogudele v Nursing and Midwifery Council [2013] EWHC 2748 (Admin) . . . . . . . . . . . 25.29
Ogundele v Nursing and Midwifery Council [2013] EWHC 2748 (Admin) . . . . . . . . . . 22.33
Ogunnowo v Solicitors Regulation Authority [2013] EWHC 1882 (Admin) . . . . . . . . . 11.16
Okee v Nursing and Midwifery Council [2014] EWHC 1763 (Admin); [2014]
All E.R. (D) 07 (Mar) . . . . . . . . . . . . . . . . . . . . . . . . . . . . . . . . . . . . . . . . . . . . 19.77
Okeke v Nursing and Midwifery Council [2013] EWHC 714 (Admin) . . . . . . . . 16.16, 17.38
Okorji v Bar Standards Board, 30 April 2014 (unreported) . . . . . . . . . . . . . . . . . . . . . 47.60
Oluyemi v Nursing and Midwifery Council [2015 EWHC 487 (Admin) . . . . . . . . . . . . 19.80
Ong Bak Hin v General Medical Council [1956] 1 W.L.R. 515; [1956]
2 All E.R. 257; (1956) 100 S.J. 358 . . . . . . . . . . . . . . . . . . . . . . . . . . . . . . . . . . . 14.02
Osmani v Camden LBC [2005] HLR 325 . . . . . . . . . . . . . . . . . . . . . . . . . . . . . . . . . 42.25
Otote v General Medical Council [2003] UKPC 71 . . . . . . . . . . . . . . . . . . . . . . . . . . 59.02

P, a barrister v General Council of the Bar [2005] 1 W.L.R. 3019 . . . . . . . 13.16, 15.04, 35.06
P v Commissioner of Police of the Metropolis [2014] UKEAT/0449/13/JOJ . . . . . . . . . . 52.16

Paponette v Attorney General of Trinidad and Tobago [2010] UKPC 32; [2012] 1 A.C. 1; [2011] 3 W.L.R. 219; Times, December 15, 2010 . . . . . . . . . 2.09, 32.24, 65.13
Parkin v Nursing and Midwifery Council [2014] EWHC 519 (Admin) . . . . . . . . . . . . . . 4.44
Parkinson v Nursing and Midwifery Council [2010] EWHC 1898 (Admin) . . . . . 1.10, 19.56, 19.65, 44.24
Parry-Jones v Law Society [1969] 1 Ch. 1; [1968] 2 W.L.R. 397; [1968] 1 All E.R. 177; (1967) 111 S.J. 910 . . . . . . . . . . . . . . . . . . . . . . . . . . . . . . . 52.01, 52.08
Patel v General Medical Council *See* R (Patel) v General Medical Council
Patel v General Medical Council [2003] UKPC 16; [2003] I.R.L.R. 316 . . . . . . . 19.40, 19.41
Patel v General Medical Council [2012] EWHC 3688 (Admin) . . . . . . . . . . . . . 38.34, 38.35
Patel v Solicitors Regulation Authority [2012] EWHC 3373 (Admin) . . . . . . . . . . . . . . . 47.50
Pattar v General Medical Council *See* R (Pattar) v General Medical Council
Peatfield v General Medical Council [1986] 1 W.L.R. 243; [1987] 1 All E.R. 1197; (1986) 83 L.S.G. 200; (1985) 129 S.J. 894 . . . . . . . . . . . . . . . . . . . . . . . . 22.04, 54.03
Pembrey v General Medical Council [2003] UKPC 60; (2003) 147 S.J.L.B. 1084 . . . . . 60.27
Pepper v Pepper . . . . . . . . . . . . . . . . . . . . . . . . . . . . . . . . . . . . . . . . . . . . . . . . . . . . . 25.03
Perry v Nursing and Midwifery Council [2012] EWHC 2275 (Admin) . . . . . . . . . . . . . 38.13
Perry v Nursing and Midwifery Council [2013] EWCA Civ 145; [2013] 1 W.L.R. 3423; [2013] Med. L.R. 129; (2014) 135 B.M.L.R. 61 . . . . . . . . . . . . . . . 38.13
Pett v Greyhound Racing Association Ltd (No.1) [1969] 1 Q.B. 125; [1968] 2 W.L.R. 1471; [1968] 2 All E.R. 545; (1968) 112 S.J. 463 . . . . . . . . . . 45.02, 45.03, 45.04, 45.05
Pett v Greyhound Racing Association Ltd (No.2) [1970] 1 Q.B. 67 (Note); [1970] 2 W.L.R. 256; [1970] 1 All E.R. 243 . . . . . . . . . . . . . . . . . . . . . . . . . . . . . . . . . . . . . 45.02
Phillips v General Medical Council *See* R (Phillips) v General Medical Council
Phipps v General Medical Council [2006] EWCA Civ 397; [2006] Lloyd's Rep. Med. 345 . . . . . . . . . . . . . . . . . . . . . 4.14, 27.09, 54.10, 54.13, 54.14, 54.15, 54.30
Piersack v Belgium [1983] 5 EHRR 169 . . . . . . . . . . . . . . . . . . . . . . . . . . . . . . . . . . . . 35.20
Pine v Law Society, 25th October 2001, case number C/2000/35/61 and 35/61A . . . . . . 32.17
Pine v Law Society [2001] EWCA Civ 1574; [2002] U.K.H.R.R. 81; (2002) 99(1) L.S.G. 19 . . . . . . . . . . . . . . . . . . . . . . . . . . . . . . . . . . . . . . 32.11, 45.09, 45.10
Pinto v Nursing and Midwifery Council [2014] EWHC 403 (Admin) . . . . . . . . . . . . . . . 4.43
Pitt, ex parte (1833) 2 Dowl 439 . . . . . . . . . . . . . . . . . . . . . . . . . . . . . . . . . . . . . . . . 43.05
Plenderleith v Royal College of Veterinary Surgeons [1996] 1 W.L.R. 224; (1996) 29 B.M.L.R. 1; (1996) 140 S.J.L.B. 26; Times, December 19, 1995 . . . . . . . . . . . . 47.19
Polanski v CondéNast Publications Ltd [2005] UKHL 10; [2005] 1 W.L.R. 637; [2005] 1 All E.R. 945; [2005] C.P. Rep. 22; [2005] E.M.L.R. 14; [2005] H.R.L.R. 11; [2005] U.K.H.R.R. 277; (2005) 102(10) L.S.G. 29; (2005) 155 N.L.J. 245; Times, February 11, 2005; Independent, February 16, 2005 . . . . . . 70.07
Pomiechowski v Poland [2012] UKSC 20; [2012] 1 W.L.R. 1604; [2012] 4 All E.R. 667; [2012] H.R.L.R. 22; [2013] Crim. L.R. 147; (2012) 162 N.L.J. 749; *Times* June 7, 2012 . . . . . . . . . . . . . . . . . . . . . . . . . . . . . . . . . . . . . . . . . . . . . 4.40
Pool v General Medical Council [2014] EWHC 3791 (Admin) . . . . . . . . . . . . . . . . . . . 65.17
Pope v General Dental Council [2015] EWHC 278 (Admin) . . . . . . . . . . . . . . . . 7.04, 19.33
Porteous v West Dorset District Council [2004] EWCA Civ 244; [2004] H.L.R. 30; [2004] B.L.G.R. 577; (2004) 148 S.J.L.B. 300; Independent, March 12, 2004 . . . . . . . . . . . . . . . . . . . . . . . . . . . . . . . . . . . . . . . . . . . . . . . . . 43.16
Porter v Magill [2001] UKHL 67; [2002] 2 A.C. 357; [2002] 2 W.L.R. 37; [2002] 1 All E.R. 465; [2002] H.R.L.R. 16; [2002] H.L.R. 16; [2002] B.L.G.R. 51; (2001) 151 N.L.J. 1886; [2001] N.P.C. 184; Times, December 14, 2001; Independent, February 4, 2002; Daily Telegraph, December 20, 2001 . . . . . . 6.01, 6.06, 6.09, 6.15, 6.17, 17.23, 35.12, 35.13, 35.20, 51.04
Potato Marketing Board v Merricks [1958] 2 Q.B. 316; [1958] 3 W.L.R. 135; [1958] 2 All E.R. 538; (1958) 102 S.J. 510 . . . . . . . . . . . . . . . . . . . . . . . . . . . 35.02
Pottage v Financial Services Authority [2013] Lloyd's Rep FC 16; (2012) 109(35) L.S.G. 20 . . . . . . . . . . . . . . . . . . . . . . . . . . . . . . . . . . . . . . . . . . . . . . 47.48
Powell v Streatham Manor Nursing Home [1935] A.C. 243 . . . . . . . . . . . . . . . . . . . . . . 4.07
Practice Direction (Justices: Clerk to Court) [2000] 1 W.L.R. 1886 . . . . . . . . . . . . . . . . 44.02

Prasad v General Medical Council [2015] EWHC 338 (Admin) . . . . . . . . . . . . . . . . . . . . 6.18
Prashad v General Medical Council [1987] 1 W.L.R. 1697 . . . . . . . . . . . . . . . . . . . . . . 43.03
Preiss v General Dental Council [2001] UKPC 36; [2001] 1 W.L.R. 1926; [2001] I.R.L.R. 696; [2001] H.R.L.R. 56; [2001] Lloyd's Rep. Med. 491; (2001) 98(33) L.S.G. 31; Times, August 14, 2001 . . . . . . . . 4.09, 4.17, 4.28, 27.02, 35.04, 35.13, 47.24, 47.33, 47.35, 47.46, 47.47, 57.03
President of the Republic of South Africa v South African Rugby Football Union [1999] 4 SA 147 . . . . . . . . . . . . . . . . . . . . . . . . . . . . . . . . . . . . . . . . . . . . . 6.17
Procter & Gamble UK v HMRC [2009] EWCA Civ 407; [2009] STC 1990 . . . . . . . . . . 4.31
Professional Standards Authority for Health and Social Care v General Chiropractic Council and Briggs [2014] EWHC 2190 (Admin) . . . . . . . . . . . . . . . 22.37
Professional Standards Authority for Health and Social Care v General Dental Council [2014] EWHC 2280 (Admin) . . . . . . . . . . . . . . . . . . . . . . . . . . . . . . . . . . 19.27
Professional Standards Authority for Health and Social Care v General Medical Council [2014] EWHC 1903 (Admin) . . . . . . . . . . . . . . . . . . . . . . . . . . . . . . . . . . 47.61
Professional Standards Authority for Health and Social Care v General Pharmaceutical Council and Onwughalu [2014] EWHC 2521 (Admin) . . . . . . . . . . . . . . . . . . . . . . . . . . . . . . . . . . . . . . . . . . . . . 15.26, 60.59
Professional Standards Authority for Health and Social Care v Health and Care Professions Council and Elizabeth David [2014] EWHC 4657 (Admin) . . . . . . . . . 19.32
Professional Standards Authority for Health and Social Care v Nursing and Midwifery Council [2013] EWHC 4369 (Admin) . . . . . . . . . . . . . . . . . . . . 43.17, 56.27
Professional Standards Authority for Health and Social Care v Nursing and Midwifery Council and Macleod [2014] EWHC 4354 (Admin) . . . . . . . . . . . . . . . 22.38
Professional Standards Authority v Health and Care Professions Council and Ghaffar [2014] EWHC 2708 (Admin); [2014] EWHC 2723 (Admin) . . . . . . . . . . 19.78
Professional Standards Authority v Health and Care Professions Council and Ghaffar [2014] EWHC 2723 (Admin) . . . . . . . . . . . . . . . . . . . . . . . . . . . . . . . . . . 33.22
Pugsley v General Medical Council [2010] EWHC 2247 (Admin) . . . . . . . . . . . . . . . . 13.07
Pullum v Crown Prosecution Service [2000] COD 206 . . . . . . . . . . . . . . . . . . . . . . . . 54.09
Puri v Bradford Teaching Hospitals NHS Foundation Trust *See* R (Puri) v Bradford Teaching Hospitals NHS Foundation Trust

Quarters Trustees Ltd v Pensions Regulator [2012] Pens LR 415 . . . . . . . . . . . . . . . 2.11, 25.16
Quinn Direct Insurance Ltd v Law Society [2010] EWCA Civ 805; [2011] 1 W.L.R. 308; [2010] Lloyd's Rep. I.R. 655; (2010) 160 N.L.J. 1044; (2010) 154(29) S.J.L.B. 33; [2010] N.P.C. 80 . . . . . . . . . . . . . . . . . . . . . . . . . . . . 52.10
Quinn v Bar Standards Board, 25 February 2013 (unreported) . . . . . . . . . . . . . . . 15.28, 54.25
Qureshi v General Medical Council [2003] UKPC 56 . . . . . . . . . . . . . . . . . . . . . . . . . 16.03

R (Abrahaem) v General Medical Council [2004] EWHC 279 (Admin); [2004] A.C.D. 37 . . . . . . . . . . . . . . . . . . . . . . . . . . . . . . . . . . . . . . . . . . . . . . . . 36.02, 36.05, 36.08, 37.02
R (Adelakun) v Solicitors Regulation Authority [2014] EWHC 198 (Admin) . . . . . . . . . 11.18
R (Adesina) v Nursing and Midwifery Council [2004] EWHC 2410 (Admin) . . . . . . . . 37.03
R (Adesina) v Nursing and Midwifery Council [2013] EWCA Civ 818; [2013] 1 WLR 3156 . . . . . . . . . . . . . . . . . . . . . . . . . . . . . . . . . . . . . . . . . . . . . . . 4.40, 4.46
R (Adeyemi) v General Medical Council [2012] EWHC 425 (Admin) . . . . . . . . . . . . . . 59.13
R (Aga) v General Medical Council [2012] EWHC 782 (Admin) . . . . . . . . . . . . . 42.18, 47.43
R (Ahmed) v General Optical Council [2012] EWHC 3699 (Admin) . . . . . . . . . . . . . . 14.28
R (Al-Zayyat) v General Medical Council [2010] EWHC 3213 (Admin) . . . . . . 13.09, 68.03
R (Alacakanat) v General Medical Council [2013] EWHC 1866 (Admin) . . . . . . . . . . . 56.26
R (Alhy) v General Medical Council [2011] EWHC 2277 (Admin); [2011] A.C.D. 114 . . . . . . . . . . . . . . . . . . . . . . . . . . . . . . . . . . . . . . . . . . . . . . . 54.21
R (Ali) v General Medical Council [2008] EWHC 1630 (Admin) . . . . . . 38.19, 38.24, 38.28
R (Allatt) v Chief Constable of Nottinghamshire Police and IPCC [2011] EWHC 3908 (Admin) . . . . . . . . . . . . . . . . . . . . . . . . . . . . . . . . . . . . . . . . . 39.30

R (Amro International SA) v Financial Services Authority [2010] EWCA Civ 123; [2010] Bus. L.R. 1541; [2010] 3 All E.R. 723; [2010] 2 B.C.L.C. 40; (2010) 107(10) L.S.G. 17; Times, May 3, 2010 .......... 18.13
R (Aston) v Nursing and Midwifery Council [2004] EWHC 2368 (Admin) .......... 45.12, 45.21, 45.30, 67.12
R (AT) v University of Leicester [2014] EWHC 4593 (Admin) .......... 70.10
R (Attwood) v Health Service Commissioner [2008] EWHC 2315 (Admin); [2009] P.T.S.R. 1330; [2009] 1 All E.R. 415; [2009] A.C.D. 6 .......... 62.13
R (Austin) v Teaching Agency [2013] EWHC 254 (Admin) .......... 6.14, 60.53
R (Aziz) v General Medical Council [2004] EWHC 3325 (Admin) .......... 17.07, 17.25
R (B) v Director of Public Prosecutions [2009] EWHC 106 (Admin); [2009] 1 W.L.R. 2072; [2009] 1 Cr. App. R. 38; [2009] U.K.H.R.R. 669; (2009) 106 B.M.L.R. 152; [2009] M.H.L.R. 61; [2009] Crim. L.R. 652; [2009] A.C.D. 19; (2009) 153(5) S.J.L.B. 29; Times, March 24, 2009 .......... 39.18
R (B) v Hampshire County Council [2004] EWHC 3193 (Admin) .......... 45.30
R (B) v Nursing and Midwifery Council [2012] EWHC 1264 (Admin) .......... 42.17, 43.12, 43.16
R (B) v Secretary of State for the Home Department [2012] EWHC 3770 (Admin) .......... 39.26
R (Balasubramanian) v General Medical Council [2008] EWHC 639 (Admin) .......... 60.34
R (Bar Standards Board) v Disciplinary Tribunal of the Council of the Inns of Court [2014] EWHC 1570 (Admin); [2014] 4 All ER 759 .......... 15.31
R (Bates) v District Judge Zani (Independent Adjudicator) [2011] EWHC 3236 (Admin) .......... 48.14
R (Begley) v Chief Constable of West Midlands Police [2001] EWCA Civ 1571 .......... 56.10
R (Bevan) v General Medical Council [2005] EWHC 174 (Admin); [2005] Lloyd's Rep. Med. 321 .......... 22.14, 36.06, 37.04, 37.05, 60.07
R (Bhatt) v General Medical Council [2011] EWHC 783 (Admin) .......... 2.10, 4.21, 4.29, 14.25, 21.09, 21.11, 25.12, 27.06, 27.11
R (Birks) v Commissioner of Police of the Metropolis, Independent Police Complaints Commission (Interested Party) [2014] EWHC 3041 (Admin); [2015] ICR 204 .......... 65.13
R (Bolt) v Chief Constable of Merseyside Police [2007] EWHC 2607 (QB); [2007] Po. L.R. 212 .......... 58.09
R (Bonhoeffer) v General Medical Council [2011] EWHC 1585 (Admin); [2012] I.R.L.R. 37; [2011] Med. L.R. 519; [2011] A.C.D. 104 .......... 25.25, 25.28, 25.30, 70.06
R (Bourgass and Hussain) v Secretary of State for Justice [2012] EWCA Civ 376; [2012] 1 W.L.R. 3602; [2012] 4 All E.R. 44; [2012] H.R.L.R. 17; Times, June 27, 2012 .......... 32.29
R (Burke) v Independent Police Complaints Commission [2011] EWHC Civ 1665 .......... 4.35
R (Calver) v Adjudication Panel for Wales [2012] EWHC 1172 (Admin); [2012] A.C.D. 81 .......... 32.27, 47.42
R (Cambridge City Council) v Alex Nestling Limited [2006] EWHC 1374 (Admin); (2006) 170 J.P. 539; [2006] L.L.R. 397; (2006) 170 J.P.N. 975; Times, July 11, 2006 .......... 15.07
R (Campbell) v General Medical Council [2005] EWCA Civ 250; [2005] 1 W.L.R. 3488; [2005] 2 All E.R. 970; [2005] Lloyd's Rep. Med. 353; (2005) 83 B.M.L.R. 30; Times, April 18, 2005; Independent, April 12, 2005 .......... 27.05, 30.01, 30.02, 30.03, 33.03, 47.29, 47.35
R (Cela) v General Pharmaceutical Council [2012] EWHC 2785 (Admin) .......... 4.38
R (Chaudhari) v Royal Pharmaceutical Society of Great Britain [2008] EWHC 3190 (Admin) .......... 39.12, 53.03
R (Chief Constable of Avon and Somerset) v Police Appeals Tribunal [2004] EWHC 220 (Admin); [2004] Po. L.R. 116; Times, February 11, 2004 .......... 4.11
R (Chief Constable of British Transport Police) v Police Appeals Tribunal and Whaley [2013] EWHC 539 (Admin) .......... 4.48, 25.39
R (Chief Constable of the Derbyshire Constabulary) v Police Appeals Tribunal [2012] EWHC 2280 (Admin); [2012] A.C.D. 126 .......... 4.25, 22.30

R (Chief Constable of Dorset) v Police Appeals Tribunal and Salter [2011] EWHC 3366 (Admin); aff'd [2012] EWCA Civ 1047 . . . . . . . . . . 19.60
R (Chief Constable of Durham) v Police Appeals Tribunal and Cooper [2012] EWHC 2733 (Admin) . . . . . . . . . . 4.22, 4.25
R (Chief Constable of West Yorkshire Police) v Independent Police Complaints Commission and Armstrong [2013] EWHC 2698 (Admin); [2014] EWCA Civ 1367; [2015] ICR 184, CA . . . . . . . . . . 39.30
R (Chief Constable of Wiltshire Police) v Police Appeals Tribunal and Woollard (Interested Party) [2012] EWHC 3288 (Admin), 4752 . . . . . . . . . . 4.25
R (Clarke) v United Kingdom Central Council for Nursing, Midwifery and Health Visiting [2004] EWHC 1350 (Admin) . . . . . . . . . . 37.03
R (Coke-Wallis) v Institute of Chartered Accountants in England and Wales [2011] UKSC 1; [2011] 2 A.C. 146; [2011] 2 W.L.R. 103; [2011] 2 All E.R. 1; [2011] I.C.R. 224; (2011) 108(5) L.S.G. 18; Times, January 31, 2011 . . . . . . . . . . 24.03, 24.05
R (Colman) v General Medical Council [2010] EWHC 1608 (QB); [2011] A.C.D. 38 . . . . . . . . . . 6.11
R (Commissioner of Police for the Metropolis) v Police Appeals Tribunal and Naulls [2013] EWHC 1684 (Admin) . . . . . . . . . . 4.25, 15.08, 15.25, 47.52
R (Commissioner of Police of the Metropolis) v Police Appeals Tribunal and Peart [2011] EWHC 3421 (Admin); [2012] A.C.D. 43 . . . . . . . . . . 42.16
R (Compton) v Wiltshire Primary Care Trust (No.2) [2009] EWHC 1824 (Admin); [2010] PTSR (CS) . . . . . . . . . . 6.15
R (Council for the Regulation of Health Care Professionals) v General Medical Council and Rajeshwar [2005] EWHC 2973 (Admin) . . . . . . . . . . 22.15, 22.19
R (Council for the Regulation of Health Care Professionals) v Nursing and Midwifery Council [2007] EWHC 106 (Admin) . . . . . . . . . . 22.19
R (Countryside Alliance) v Attorney General [2006] EWCA Civ 817; [2007] Q.B. 305; [2006] 3 W.L.R. 1017; [2007] Eu. L.R. 139; [2006] H.R.L.R. 33; [2006] U.K.H.R.R. 927; (2006) 150 S.J.L.B. 886; [2006] N.P.C. 73; Times, June 30, 2006 . . . . . . . . . . 32.18
R (Crawford) v Legal Ombudsman [2014] EWHC 182 (Admin) . . . . . . . . . . 42.25
R (Crosby) v IPCC [2009] EWHC 2515 (Admin) . . . . . . . . . . 39.31, 39.32
R (D) v General Medical Council [2013] EWHC 2839 (Admin) . . . . . . . . . . 39.33
R (D) v Independent Police Complaints Commission [2011] EWHC 1595 (Admin) . . . . . . . . . . 39.23
R (D) v Secretary of State for Health [2006] Lloyd's Rep Med 457 . . . . . . . . . . 38.07
R (Dart Harbour and Navigational Authority) v Secretary of State for Transport, Local Government and the Regions [2003] EWHC 1494 (Admin); [2003] 2 Lloyd's Rep. 607 . . . . . . . . . . 4.22
R (Davies) v Financial Services Authority [2003] EWCA Civ 1128; [2004] 1 W.L.R. 185; [2003] 4 All E.R. 1196; [2004] 1 All E.R. (Comm) 88; [2005] 1 B.C.L.C. 286; [2003] A.C.D. 83; (2003) 100(38) L.S.G. 34; Times, October 6, 2003 . . . . . . . . . . 42.11
R (Delezuch) v Chief Constable of Leicestershire Constabulary [2014] EWCA Civ 1635 . . . . . . . . . . 39.40
R (Dennis) v IPCC [2008] EWHC 1158 (Admin) . . . . . . . . . . 39.31, 39.32
R (Doshi) v Southend-on-Sea Primary Care Trust [2007] EWHC 1361 (Admin); [2007] LS Law Medical 418; [2007] A.C.D. 70 . . . . . . . . . . 62.11
R (Duggan)v. Association of Chief Police Officers [2014] EWCA Civ 1635 . . . . . . . . . . 39.40
R (Dutt) v General Medical Council [2009] EWHC 3613 (Admin), 4.19done . . . . . . . . . . 45.19
R (Eguakhide) v Governor of HMP Gartree [2014] EWHC 1328 (Admin) . . . . . . . . . . 42.24
R (El-Baroudy) v General Medical Council [2013] EWHC 2894 (Admin) . . . . . . . . . . 22.34
R (Elliott) v Solicitors Disciplinary Tribunal [2004] EWHC 1176 (Admin) . . . . . . . . . . 1.04, 1.14
R (Erenbilge) v IPCC [2013] EWHC 1397 (Admin) . . . . . . . . . . 39.31, 39.32
R (Farah) v General Medical Council [2008] EWHC 731 (Admin) . . . . . . . . . . 19.79
R (French) v Chief Constable of West Yorkshire Police [2011] EWHC 546 (Admin) . . . . . . . . . . 3.17, 6.12

R (G) v Governors of X School (Secretary of State for the Home Department intervening) [2011] UKSC 30; [2012] 1 A.C. 167; [2011] 3 W.L.R. 237; [2011] P.T.S.R. 1230; [2011] 4 All E.R. 625; [2011] I.C.R. 1033; [2011] I.R.L.R. 756; [2011] H.R.L.R. 34; [2011] U.K.H.R.R. 1012; [2011] B.L.G.R. 849; [2011] E.L.R. 310; [2011] Med. L.R. 473; (2011) 161 N.L.J. 953; (2011) 155(26) S.J.L.B. 27; Times, July 4, 2011 . . . . . . . . . . . . . . . . . . 32.25, 32.28, 35.17, 45.24
R (Gaunt) v Office of Communications [2011] EWCA Civ 692; [2011] 1 W.L.R. 2355; [2011] E.M.L.R. 28; [2011] H.R.L.R. 33 . . . . . . . . . . . . . . . 32.26, 47.41
R (George) v General Medical Council [2003] EWHC 1124 (Admin); [2004] Lloyd's Rep. Med. 33; Independent, June 30, 2003 . . . . . . . . . . . . . . . . . . . 38.04
R (Ghosh) v Knowsley NHS Primary Care Trust [2006] EWHC 26 (Admin); [2006] Lloyd's Rep. Med. 123; [2006] A.C.D. 60; Times, February 2, 2006 . . . . . 45.13
R (Gibson) v General Medical Council [2004] EWHC 2781 (Admin) . . . . . . . . . 17.24, 17.27, 17.29, 68.02
R (Gorlov) v Institute of Chartered Accountants in England and Wales [2001] EWHC Admin 220; [2001] A.C.D. 73 . . . . . . . . . . . . . . . . . . . . . . 15.03, 15.05, 15.23, 43.07
R (Grabinar) v General Medical Council [2013] EWHC 4480 (Admin) . . . . . . . . . . . . . 39.35
R (Green) v Police Complaints Authority [2004] UKHL 6; [2004] 1 W.L.R. 725; [2004] 2 All E.R. 209; [2004] H.R.L.R. 19; [2004] U.K.H.R.R. 939; [2004] Po. L.R. 148; [2004] Inquest L.R. 1; (2004) 101(12) L.S.G. 36; (2004) 148 S.J.L.B. 268; Times, February 27, 2004 . . . . . . . . . . . . . . . . . . . . 18.07, 39.15
R (Gwynn) v General Medical Council [2007] EWHC 3145 (Admin); [2008] LS Law Medical 112 . . . . . . . . . . . . . . . . . . . . . . . . . . . . . . . . . . . . . . . . . . . . . . . 17.10
R (Gyurkovits) v General Dental Council [2013] EWHC 4507 (Admin) . . . . . . . . . . . . . 4.42
R (H) v Ashworth Hospital Trust [2002] EWCA Civ 923; [2002] EWCA Civ 923; [2003] 1 W.L.R. 127; (2002) 5 C.C.L. Rep. 390; (2003) 70 B.M.L.R. 40; [2002] M.H.L.R. 314; [2002] A.C.D. 102; (2002) 99(34) L.S.G. 29; (2002) 146 S.J.L.B. 198; Times, July 10, 2002; Daily Telegraph, July 11, 2002 . . . . . . . . . . . . . . . . . . . 26.02
R (H) v Nursing and Midwifery Council [2013] EWHC 4258 (Admin) . . . . . . . . 5.04, 14.11, 25.18, 54.28
R (Haase) v District Judge Nuttall [2007] EWHC 3079 (Admin); [2008] 1 W.L.R. 1401; [2008] H.R.L.R. 19; [2008] Prison L.R. 378; [2008] A.C.D. 26; Times, February 11, 2008 . . . . . . . . . . . . . . . . . . . . . . . . . . . . . . . . . . . . . . . . . . . 35.09
R (Haase) v District Judge Nuttall [2008] EWCA Civ 1089; [2009] Q.B. 550; [2009] 2 W.L.R. 1004; [2009] H.R.L.R. 2; [2008] U.K.H.R.R. 1260; [2008] Prison L.R. 391; [2009] Prison L.R. 241; [2009] A.C.D. 14; (2008) 105(41) L.S.G. 21; Times, October 28, 2008 . . . . . . . . . . . . . . . . . . . . . . . . . . . . . . . . . . . . . . 35.09
R (Harrison) v General Medical Council [2011] EWHC 1741 (Admin) . . . . . . . . . . . . . . 4.36
R (Harry) v General Medical Council [2006] EWHC 3050 (Admin) . . . . . . . . . . 33.15, 60.31
R (Haward) v General Medical Council [2007] EWHC 2236 (Admin); [2007] A.C.D. 102 . . . . . . . . . . . . . . . . . . . . . . . . . . . . . . . . . . . . . . . . . . . . . . . . . . 17.09
R (Health Professions Council) v Disciplinary Committee of the Chiropodists Board [2002] EWHC 2662 (Admin) . . . . . . . . . . . . . . . . . . . . . . . . . . . . . . . . . . . 43.09
R (Heath) v Home Office Policy and Advisory Board for Forensic Pathology [2005] EWHC 1793; Times, October 18, 2005 . . . . . . . . . . . . . . . . . . . . . . . . 8.04, 39.10
R (Heather) v Leonard Cheshire Foundation [2002] EWCA Civ 366; [2002] 2 All E.R. 936; [2002] H.R.L.R. 30; [2002] U.K.H.R.R. 883; [2002] H.L.R. 49; (2002) 5 C.C.L. Rep. 317; (2003) 69 B.M.L.R. 22; [2002] A.C.D. 43; Times, April 8, 2002 . . . . . . . . . . . . . . . . . . . . . . . . . . . . . . . . . . . . . 42.13
R (Henderson) v General Teaching Council for England [2012] EWHC 1505 (Admin) . . . . . . . . . . . . . . . . . . . . . . . . . . . . . . . . . . . . . . . . . . . . . . . . . 37.15, 60.56
R (Hibbert) v General Medical Council [2013] EWHC 3596 (Admin) . . . . . . . . . . . . . . 39.34
R (Hickey) v General Medical Council [2010] EWHC 1608 (QB); [2011] A.C.D. 38 . . . . . . . . . . . . . . . . . . . . . . . . . . . . . . . . . . . . . . . . . . . . . . . . . . . . 6.11
R (Hill) v Institute of Chartered Accountants in England and Wales [2012] EWHC 1731 (QB) . . . . . . . . . . . . . . . . . . . . . . . . . . . . . . . . . . . . . . . . . . . . . . . 48.15

R (Hill) v Institute of Chartered Accountants in England and Wales [2013] EWCA Civ 555; [2014] 1 W.L.R. 86, CA . . . . 43.15
R (Holden) v Solicitors Regulation Authority [2012] EWHC 2067 (Admin) . . . . 62.17
R (Hollis) v Association of Chartered Certified Accountants [2014] EWHC 2572 (Admin) . . . . 25.19, 25.20
R (Holmes) v General Medical Council [2001] EWHC Admin 321; [2001] Lloyd's Rep. Med. 366; (2002) 63 B.M.L.R. 131; [2002] EWCA Civ 1838 . . . . 18.08, 39.06
R (Hounslow LBC) v School Appeal Panel [2002] EWCA Civ 900; [2002] 1 W.L.R. 3147; [2002] 3 F.C.R. 142; [2002] B.L.G.R. 501; [2002] E.L.R. 602; (2002) 99(32) L.S.G. 33; Times, October 3, 2002 . . . . 42.12
R (Howlett) v Health Professions Council [2009] EWHC 3617 . . . . 60.44
R (Ibrahim) v Feltham Magistrates' Court [2001] 1 W.L.R. 1293 . . . . 18.08
R (Independent Police Complaints Commission) v Assistant Commissioner Hayman [2008] EWHC 2191 (Admin) . . . . 62.12
R (Independent Police Complaints Commission) v Chief Constable of West Midlands Police [2007] EWHC 2715 (Admin); [2007] Po. L.R. 205; Times, December 17, 2007 . . . . 59.05
R (Independent Police Complaints Commission) v Commissioner of Police of the Metropolis [2009] EWHC 1566 . . . . 39.15
R (Iran) v Secretary of State for the Home Department [2005] EWCA Civ 982; [2005] Imm. A.R. 535; [2005] I.N.L.R. 633; Times, August 19, 2005 . . . . 25.37
R (J) v Upper Tribunal (Immigration Asylum Chamber) [2012] EWHC 3770 (Admin) . . . . 39.26
R (Jackson) v General Medical Council [2013] EWHC 2595 (Admin) . . . . 68.07
R (Jalloh) v Nursing and Midwifery Council [2009] EWHC 1697 (Admin) . . . . 11.08, 33.09
R (Jenkinson) v Nursing and Midwifery Council [2009] EWHC 1111 (Admin) . . . . 14.07, 42.15, 43.12, 43.16
R (Johnson and Maggs) v Nursing and Midwifery Council (No. 2) [2013] EWHC 2140 (Admin) . . . . 17.33, 42.19, 47.54
R (Jones) v Judicial Appointments Commission [2014] EWHC 1680 (Admin) . . . . 41.05
R (K) v Chief Constable of Lancashire Police [2009] EWCA Civ 1197 . . . . 56.09
R (Kaftan) v General Medical Council [2009] EWHC 3585 (Admin) . . . . 54.13, 65.05
R (Kashyap) v General Medical Council [2010] EWHC 603 (Admin) . . . . 42.14
R (Katti) v General Medical Council [2004] EWHC 238 (Admin) . . . . 67.02
R (Kaur) v Institute of Legal Executives Appeal Tribunal [2011] EWCA Civ 1168; [2012] 1 All E.R. 1435; [2012] 1 Costs L.O. 23; [2011] E.L.R. 614; (2011) 108(45) L.S.G. 20; (2011) 155(40) S.J.L.B. 31; [2011] N.P.C. 106; [2012] P.T.S.R. D1 . . . . 35.12
R (Kay) v Chief Constable of Northumbria Police [2009] EWHC 1835 (Admin); [2010] I.C.R. 962 . . . . 56.10, 58.09
R (Kerman & Co) v. Legal Ombudsman [2014] EWHC 3726 (Admin); [2015] 1 WLR 2081 . . . . 39.41, 43.23
R (Khan) v Chief Constable of Lancashire [2009] EWHC 472 (Admin) . . . . 56.09, 56.10
R (Khan) v General Medical Council [2008] EWHC 3509 (Admin) . . . . 19.10, 19.47
R (Khan) v General Medical Council [2014] EWHC 404 (Admin) . . . . 4.48, 25.40
R (Khan) v Independent Police Complaints Commission [2010] EWHC 2339 (Admin) . . . . 39.18
R (Kibe) v Nursing and Midwifery Council [2013] EWHC 1402 (Admin) . . . . 19.68
R (King) v Secretary of State for Justice [2010] EWHC 2522 (Admin); [2011] 1 W.L.R. 2667; [2011] 3 All E.R. 776; [2010] U.K.H.R.R. 1245; [2011] A.C.D. 13; Times, March 8, 2011 . . . . 35.19, 48.13
R (King) v Secretary of State for Justice [2012] EWCA Civ 376; [2012] 1 W.L.R. 3602; [2012] 4 All E.R. 44; [2012] H.R.L.R. 17; Times, June 27, 2012, 3519 . . . . 32.27, 32.29, 45.26
R (L) v Executive Counsel of the Accountants' Joint Disciplinary Scheme [2002] EWHC 2086 (Admin); [2002] Pens. L.R. 545; (2002) 152 N.L.J. 1617 . . . . 10.10

R (L) v London Borough of Waltham Forest Special Educational Needs and Disability Tribunal [2003] EWHC 2907 (Admin); [2004] E.L.R. 161 .................... 27.09
R (Law Society) v Master of the Rolls [2005] 1 W.L.R. 2033 .................... 56.03
R (Leathley) v Visitors of the Inns of Court [2013] EWHC 3097 (Admin) .................... 13.15, 13.16, 43.20
R (Levett) v Health and Care Professions Council [2013] EWHC 3330 (Admin) ...... 70.08
R (Levy) v General Medical Council [2011] EWHC 2351 (Admin) ................ 59.12
R (LI) v General Medical Council [2013] EWHC 522 (Admin) .................... 68.05
R (Lonnie) v National College for Teaching and Leadership [2014] EWHC 4351 (Admin) .................... 60.63
R (Lonsdale) v Bar Standards Board 8 October 2014 (unreported) ................ 15.21
R (Low) v General Osteopathic Council [2007] EWHC 2839 (Admin) ............. 14.23
R (Luthra) v General Dental Council [2004] EWHC 458 (Admin) ........... 54.07, 54.13
R (M) v Independent Police Complaints Commission [2012] EWHC 2071 (Admin) .................... 25.15
R (McCann) v Crown Court at Manchester [2002] UKHL 39; [2003] 1 A.C. 787; [2002] 3 W.L.R. 1313; [2002] 4 All E.R. 593; [2003] 1 Cr. App. R. 27; (2002) 166 J.P. 657; [2002] U.K.H.R.R. 1286; 13 B.H.R.C. 482; [2003] H.L.R. 17; [2003] B.L.G.R. 57; [2003] Crim. L.R. 269; (2002) 166 J.P.N. 850; (2002) 146 S.J.L.B. 239; Times, October 21, 2002; Independent, October 23, 2002 .................... 62.04
R (McCarthy) v Visitors to the Inns of Court [2013] EWHC 3253 (Admin); [2015] EWCA Civ 12 .................... 18.17, 48.16, 48.18
R (Mackaill) v Independent Police Complaints Commission [2014] EWHC 3170 (Admin) .................... 39.38
R (McNally) v Secretary of State for Education and Employment [2001] EWCA Civ 332; [2002] I.C.R. 15; [2001] 2 F.C.R. 11; (2001) 3 L.G.L.R. 47; [2002] B.L.G.R. 584; [2001] E.L.R. 773; (2001) 98(19) L.S.G. 36; (2001) 145 S.J.L.B. 91; Times, March 23, 2001 .................... 48.11
R (McNicholas) v Nursing and Midwifery Council [2009] EWHC 627 (Admin) .................... 17.14, 39.13
R (McNicholas) v Nursing and Midwifery Council (No.2), 23 March 2010 (unreported) .................... 17.14, 39.14
R (Madan) v General Medical Council (No. 1) [2001] EWHC Admin 322 .......... 38.02
R (Mafico) v Nursing and Midwifery Council [2014] EWHC 363 (Admin) .... 19.72, 44.24
R (Mahfouz) v Professional Conduct Committee of the General Medical Council [2004] EWCA Civ 233; [2004] Lloyd's Rep. Med. 377; (2004) 80 B.M.L.R. 113; (2004) 101(13) L.S.G. 35; Times, March 19, 2004 ......... 6.05, 42.12, 42.23, 42.27, 44.16, 50.11, 51.04
R (Mahmood) v General Medical Council [2007] EWHC 474 (Admin); (2007) 95 B.M.L.R. 229 .................... 1.05
R (Malik) v Waltham Forest NHS Primary Care Trust [2007] EWCA Civ 265; [2007] 1 W.L.R. 2092; [2007] 4 All E.R. 832; [2007] I.C.R. 1101; [2007] I.R.L.R. 529; [2007] H.R.L.R. 24; [2007] U.K.H.R.R. 1105; [2007] LS Law Medical 335; Times, April 10, 2007 .................... 32.18
R (May) v Chartered Institute of Management Accountants [2013] EWHC 1574 (Admin) .................... 18.27, 19.20
R (Mehey) v Visitors of the Inns of Court [2014] EWCA Civ 1630 ................ 13.16
R (Michalak) v General Medical Council [2011] EWHC 2307 (Admin) ............ 13.10
R (Miller) v Chief Constable of Merseyside Police [2014] EWHC 400 (Admin) ....... 53.08
R (Miller) v General Medical Council [2013] EWHC 1934 (Admin) ............... 53.07
R (Monger) v Chief Constable of Cumbria Police [2013] EWHC 455 (Admin) ....... 42.21
R (Montgomery) v Police Appeals Tribunal and the Commissioner of Police of the Metropolis [2012] EWHC 936 (Admin) .................... 4.22, 4.25, 10.13
R (Morgan Grenfell & Co Ltd) v Special Commissioner of Income Tax [2002] UKHL 21; [2003] 1 A.C. 563; [2002] 2 W.L.R. 1299; [2002] 3 All E.R. 1; [2002] S.T.C. 786; [2002] H.R.L.R. 42; 74 T.C. 511; [2002] B.T.C. 223; 4 I.T.L. Rep. 809;

[2002] S.T.I. 806; (2002) 99(25) L.S.G. 35; (2002) 146 S.J.L.B. 126; [2002] N.P.C. 70; Times, May 20, 2002; Independent, May 21, 2002 . . . . . . . . . . . 18.19, 52.01, 52.08
R (Mullins) v Appeal Board of the Jockey Club [2005] EWHC 2197 (Admin); [2006] A.C.D. 2; Times, October 24, 2005 . . . . . . . . . . 42.13
R (N) v Mental Health Review Tribunal (Northern Region) [2005] EWCA Civ 1605; [2006] Q.B. 468; [2006] 2 W.L.R. 850; [2006] 4 All E.R. 194; (2006) 88 B.M.L.R. 59; [2006] M.H.L.R. 59; Times, January 12, 2006 . . . . . . . . . . 62.05
R (Nakash) v Metropolitan Police Service and General Medical Council (Interested Party) [2014] EWHC 3810 (Admin) . . . . . . . . . . 18.18, 33.23
R (Nash) v Chelsea College of Art and Design [2001] EWHC Admin 538; Times, July 25, 2001 . . . . . . . . . . 54.16
R (Nicholds) V Security Industry Authority [2006] EWHC 1792 (Admin); [2007] 1 W.L.R. 2067; [2007] I.C.R. 1076 . . . . . . . . . . 32.18
R (Nicolaides) v General Medical Council [2001] EWHC Admin 625; [2001] Lloyd's Rep. Med. 525 . . . . . . . . . . 35.05
R (North Yorkshire Police Authority) v Independent Police Complaints Commission [2010] EWHC 1690 (Admin); [2012] P.T.S.R. 268; [2011] 3 All E.R. 106 . . . . . . . 39.17
R (Nowak) v Nursing and Midwifery Council [2013] EWHC 4341 (Admin) . . . . . . . . . 38.39
R (O'Brien) v General Medical Council [2006] EWHC 51 (Admin) . . . . . . . . . . 40.04
R (Onwuelo) v General Medical Council [2006] EWHC 2739 (Admin) . . . . . . . 19.48, 37.05
R (Pamplin) v Law Society [2001] EWHC Admin 300 . . . . . . . . . . 18.21
R (Patel) v General Medical Council [2013] EWCA Civ 327; [2013] 1 W.L.R. 2801; (2013) 133 B.M.L.R. 14; Times, June 2, 2013 . . . . . . . . . . 42.22, 56.25
R (Pattar) v General Medical Council [2010] EWHC 3078 (Admin) . . . . . . . . . . 59.09
R (Peacock) v General Medical Council [2007] EWHC 585 (Admin); [2007] LS Law Medical 284 . . . . . . . . . . 17.08
R (Perinpanathan) v City of Wetminster Magistrates' Court [2010] EWCA Civ 40; [2010] 1 W.L.R. 1508; [2010] 4 All E.R. 680; [2010] 4 Costs L.R. 481; [2010] L.L.R. 514; (2010) 160 N.L.J. 217; Times, March 2, 2010 . . . . . . 15.07
R (Peters) v Chief Constable of West Yorkshire Police [2014] EWHC 1458 (Admin) . . . . . . . . . . 39.32
R (Phillips) v General Medical Council [2004] EWHC 1858 (Admin); (2005) 82 B.M.L.R. 135 . . . . . . . . . . 17.13, 21.11
R (Puri) v Bradford Teaching Hospitals NHS Foundation Trust [2011] EWHC 970 (Admin); [2011] I.R.L.R. 582; [2011] Med. L.R. 280 . . . . . . . 32.24, 32.28, 35.16, 45.23
R (Raheem) v Nursing and Midwifery Council [2010] EWHC 2549 (Admin) . . . . . . . . . 1.11
R (Ramsden) v Independent Police Complaints Commission, Chief Constable of West Yorkshire Police (Interested Party) [2013] EWHC 3969 (Admin) . . . . . . . . . . 39.31, 39.32, 70.09
R (Ranson, Headdon and Nash) v Institute of Actuaries [2004] EWHC 3087 (Admin) . . . . . . . . . . 10.11
R (Redgrave) v Commissioner of Police of the Metropolis [2003] EWCA Civ 4; [2003] 1 W.L.R. 1136; [2003] Po. L.R. 25; (2003) 100(11) L.S.G. 34; (2003) 147 S.J.L.B. 116; Times, January 30, 2003 . . . . . . . . . . 21.04, 21.09, 21.11
R (Remedy UK Ltd) v General Medical Council [2010] EWHC 1245 (Admin); [2010] Med. L.R. 330; [2010] A.C.D. 72 . . . . . . . . . . 16.08, 39.16, 47.40
R (Rengarajaperunal) v General Medical Council [2008] EWHC 1953 (Admin) . . . . . . . 3.13
R (Rhodes) v Police and Crime Commissioner for Lincolnshire [2013] EWHC 1009 (Admin) . . . . . . . . . . 65.09
R (Rich) v Bar Standards Board [2012] EWCA Civ 320 . . . . . . . . . . 39.25
R (Richards) v General Medical Council [2001] Lloyd's Rep. Med. 47; Times, January 24, 2001 . . . . . . . . . . 18.08, 39.04, 39.06
R (RJM) v Secretary of State for Work and Pensions [2008] UKHL 63; [2009] 1 A.C. 311; [2008] 3 W.L.R. 1023; [2009] 2 All E.R. 556; [2009] P.T.S.R. 336; [2009] H.R.L.R. 5; [2009] U.K.H.R.R. 117; 26 B.H.R.C. 587; Times, October 27, 2008 . . . . . . . . . . 25.02

R (Rogers) v General Medical Council [2004] EWHC 424 (Admin); [2004] A.C.D. 41 . . . . . 19.44, 19.79, 31.07
R (Roomi) v General Medical Council [2009] EWHC 2188 (Admin) . . . . . 22.23, 22.34, 44.22
R (Rosen) v Solicitors Disciplinary Tribunal [2002] EWHC 1323 (Admin) . . . . . 18.05
R (Royal College of Nursing) v Secretary of State for the Home Department [2010] EWHC 2761 (Admin); [2011] P.T.S.R. 1193; [2011] 2 F.L.R. 1399; [2011] U.K.H.R.R. 309; (2011) 117 B.M.L.R. 10; [2011] Fam. Law 231; (2010) 154(46) S.J.L.B. 30 . . . . . 32.21, 43.14
R (Russell) v General Medical Council [2008] EWHC 2546 (Admin) . . . . . 4.31, 4.48, 25.36, 25.38
R (Rycroft) v Royal Pharmaceutical Society of Great Britain [2010] EWHC 2832 (Admin); [2011] Med. L.R. 23 . . . . . 39.20
R (Shaikh) v General Pharmaceutical Council [2013] EWHC 1844 (Admin) . . . . . 1.16, 4.48, 25.28, 25.38
R (Shanker) v General Medical Council [2004] EWHC 565 (Admin) . . . . . 61.02
R (Shanker) v General Medical Council [2010] EWHC 3689 (Admin) . . . . . 9.03
R (Sharaf) v General Medical Council [2013] EWHC 3332 (Admin) . . . . . 42.23, 50.11
R (Sharma) v General Dental Council [2010] EWHC 3184 (Admin) . . . . . 37.10
R (Sheikh) v General Dental Council [2007] EWHC 2972 (Admin) . . . . . 38.23, 38.26, 38.39, 38.42
R (Siborurema) v Office for the Independent Adjudicator [2007] EWCA Civ 1365 . . . . 42.25
R (Sivills) v General Social Care Council [2007] EWHC 2576 (Admin) . . . . . 56.25
R (Sinha) v General Medical Council [2008] EWHC 1732 (Admin) . . . . . 21.08, 25.39, 44.18, 45.18
R (Smith) v General Teaching Council for England [2007] EWHC 1675 (Admin) . . . . 60.32
R (Smith) v Parole Board (No. 2) [2003] EWCA Civ 1269; [2004] 1 W.L.R. 421; [2004] Prison L.R. 31; [2004] Prison L.R. 216; (2003) 100(38) L.S.G. 34; (2003) 153 N.L.J. 1427; Times, September 2, 2003 . . . . . 20.04
R (Soar) v Secretary of State for Justice [2015] EWHC 392 (Admin) . . . . . 48.18
R (Solicitors Regulation Authority) v Solicitors Disciplinary Tribunal [2013] EWHC 2584 (Admin) . . . . . 58.14
R (Sosanya) v General Medical Council [2009] EWHC 2814 (Admin); [2010] Med. L.R. 62 . . . . . 38.09, 38.25, 38.42
R (Squier) v General Medical Council [2015] EWHC 299 (Admin) . . . . . 22.39, 25.20, 42.27
R (SS) v Knowsley NHS Primary Care Trust [2006] EWHC 26 (Admin); [2006] Lloyd's Rep. Med. 123; [2006] A.C.D. 60; Times, February 2, 2006 . . . . . 45.13
R (Subner) v Health Professions Council [2009] EWHC 2815 (Admin) . . . . 17.29, 37.06, 60.42
R (Thompson) v Law Society [2004] EWCA Civ 167; [2004] 1 W.L.R. 2522; [2004] 2 All E.R. 113; (2004) 101(13) L.S.G. 35; (2004) 154 N.L.J. 307; (2004) 148 S.J.L.B. 265; Times, April 1, 2004; Independent, March 29, 2004 . . . . . 20.04, 32.15, 32.16, 32.28, 45.11
R (Thompson) v Professional Conduct Committee of the General Chiropractic Council [2008] EWHC 2499 (Admin) . . . . . 3.11, 8.06, 8.07
R (Tinsa) v General Medical Council [2008] EWHC 1284 (Admin); (2008) 103 B.M.L.R. 41 . . . . . 3.12, 45.17
R (Tutin) v General Medical Council [2009] EWHC 553 (Admin) . . . . . 50.03, 50.09, 50.11
R (Uruakpa) v General Medical Council [2012] EWHC 1960 (Admin) . . . . . 59.14
R (Vali) v General Optical Council [2011] EWHC 310 (Admin) . . . . . 16.11, 25.24, 33.14
R (Van Vuuren) v General Medical Council [2002] EWHC 2661 (Admin) . . . . . 17.22
R (Veizi) v General Dental Council [2013] EWHC 832 (Admin) . . . . . 1.17, 4.47
R (Walker) v General Medical Council [2003] EWHC 2308 (Admin) . . . . . 38.19, 38.22
R (Wheeler) v Assistant Commissioner House of the Metropolitan Police [2008] EWHC 439 (Admin) . . . . . 22.22
R (Willford) v Financial Services Authority [2013] EWCA Civ 677 . . . . . 42.20, 53.06
R (Wilson, Chief Constable of Fife Constabulary) v Police Appeals Tribunal [2008] CSOH 96; 2008 S.L.T. 753; 2008 S.C.L.R. 598; 2008 G.W.D. . . . . . 28.04

R (Wood) v Secretary of State for Education [2011] EWHC 3256 (Admin) [2011] EWHC 3256 (Admin); [2012] E.L.R. 172; [2012] A.C.D. 24 . . . . . . . 2.09, 24.04, 32.24
R (Woods and Gordon) v Chief Constable of Merseyside Police [2014] EWHC 2784 (Admin); [2015] I.C.R. 125 . . . . . . . . . . . . . . . . . . . . 16.19, 42.26, 66.08
R (Wright) v Secretary of State for Health [2009] UKHL 3; [2009] 1 A.C. 739; [2009] 2 W.L.R. 267; [2009] P.T.S.R. 401; [2009] 2 All E.R. 129; [2009] H.R.L.R. 13; [2009] U.K.H.R.R. 763; 26 B.H.R.C. 269; (2009) 12 C.C.L. Rep. 181; (2009) 106 B.M.L.R. 71; (2009) 106(5) L.S.G. 15; (2009) 153(4) S.J.L.B. 30; Times, January 23, 2009 . . . . . . . . . . . . . . . . . . . . . . . . . . . . 32.20, 38.13
R (X) v General Medical Council [2011] EWHC 3271 (Admin); [2012] Med. L.R. 139 . . . . . . . . . . . . . . . . . . . . . . . . . . . . . . . . . . . . . . . . . . . . . . . 68.04
R v Adams [2007] EWCA Crim 1; [2007] 1 Cr. App. R. 430; [2007] Crim. L.R. 559; (2007) 151 S.J.L.B. 123; Times, January 30, 2007 . . . . . . . . . . . . . 34.03
R v Advertising Standards Authority Ltd, ex parte Insurance Service Plc (1990) 2 Admin. L.R. 77; (1990) 9 Tr. L.R. 169; [1990] C.O.D. 42; (1989) 133 S.J. 1545 . . . . . . . . . . . . . . . . . . . . . . . . . . . . . . . . . . . . . . . . . . . . . . . . . . . 42.06
R v Army Board of the Defence Council, ex parte Anderson [1992] Q.B. 169 . . . . . . . . . 20.02
R v Ashton [2006] EWCA Crim 794; [2007] 1 W.L.R. 181; [2006] 2 Cr. App. R. 15; [2006] Crim. L.R. 1004; Times, April 18, 2006 . . . . . . . . . . . . . . . . . . . . . . . . . 38.08
R v Assessment Committee of St Mary Abbots, Kensington [1891] 1 Q.B. 378, CA . . . . 45.02
R v Association of British Travel Agents, ex parte Sunspell Ltd (t/a Superlative Travel), CO/4295/1999, 12 October 2000 . . . . . . . . . . . . . . . . . . . . . . . . . . . . . . . . . . . . . 42.09
R v Board of Visitors of Frankland Prison, ex parte Lewis [1986] 1 W.L.R. 130; [1986] 1 All E.R. 272; [1986] Crim. L.R. 336; (1986) 83 L.S.G. 125; (1986) 130 S.J. 52; Times, November 7, 1985 . . . . . . . . . . . . . . . . . . . . . . . . . . . . . . . . . . . . . . . . . 6.03
R v Board of Visitors of HM Prison, The Maze, ex parte Hone [1988] A.C. 379; [1988] 2 W.L.R. 177; [1988] 1 All E.R. 321; (1988) 132 S.J. 158; Times, January 22, 1988; Independent, January 22, 1988; Guardian, January 22, 1988 . . . . . . . . . . . . . . . . . . . . . . . . . . . . . . . . . . . . . . . . . . 45.08
R v Board of Visitors of Hull Prison, ex parte St Germain (No.1) [1979] Q.B. 425; [1979] 2 W.L.R. 42; [1979] 1 All E.R. 701; (1979) 68 Cr. App. R. 212; (1978) 122 S.J. 697; Times, October 4, 1978 . . . . . . . . . . . . . . . . . . . . . . . . . . . . . . . . . 48.06
R v Board of Visitors of Hull Prison, ex parte St Germain (No.2) [1979] 1 W.L.R. 1401; [1979] 3 All E.R. 545; [1979] Crim. L.R. 726; (1979) 123 S.J. 768 . . . . . . . . . . . . . 48.06
R v Bolivar [2003] EWCA Crim 1167 . . . . . . . . . . . . . . . . . . . . . . . . . . . . . . . . . . . . . 45.12
R v Bow Street Metropolitan Stipendiary Magistrate, ex parte Pinochet Ugarte (No. 2) [2000] 1 A.C. 119; [1999] 2 W.L.R. 272; [1999] 1 All E.R. 577; 6 B.H.R.C. 1; (1999) 11 Admin. L.R. 57; (1999) 96(6) L.S.G. 33; (1999) 149 N.L.J. 88; Times, January 18, 1999; Independent, January 19, 1999 . . . . . . . . . . . . . . . . . . . . 35.06, 35.12
R v Brentford Justices, ex parte Wong [1981] Q.B. 445; [1981] 2 W.L.R. 203; [1981] 1 All E.R. 884; (1981) 73 Cr. App. R. 67; [1981] R.T.R. 206; [1981] Crim. L.R. 336 . . . . . . . . . . . . . . . . . . . . . . . . . . . . . . . . . . . . . . . . . . . . . . . . . . . . . 2.02
R v British Broadcasting Corporation, ex parte Lavelle [1983] 1 W.L.R. 23; [1983] 1 All E.R. 241; [1983] I.C.R. 99; [1982] I.R.L.R. 404; (1983) 133 N.L.J. 133; (1982) 126 S.J.L.B. 836 . . . . . . . . . . . . . . . . . . . . . . . . . . . . . . . . . . . . . . . . . . . . . . 10.02
R v Chaaban [2003] EWCA Crim 1012; [2003] Crim. L.R. 658; Times May 9 2003; Independent, May 6, 2003 . . . . . . . . . . . . . . . . . . . . . . . . . . . . . . . . . . . . . . . . . . 8.02
R v Chance, ex parte Smith [1995] B.C.C. 1095; (1995) 7 Admin. L.R. 821; Times, January 28, 1995 . . . . . . . . . . . . . . . . . . . . . . . . . . . . . . . . . . . . . . 10.07, 10.08
R v Charnley [2007] EWCA Crim 1354; [2007] 2 Cr. App. R. 33; [2007] Crim. L.R. 984 . . . . . . . . . . . . . . . . . . . . . . . . . . . . . . . . . . . . . . . . . . . . . . . . . . . . . 34.03
R v Chief Constable of British Transport Police, ex parte Farmer [1999] COD 518 . . . . . . . . . . . . . . . . . . . . . . . . . . . . . . . . . . . . . . . . . . . . . . . . . . 56.09, 56.10
R v Chief Constable of Devon and Cornwall Constabulary, ex parte Hay [1996] 2 All E.R. 711; Times, February 19, 1996 . . . . . . . . . . . . . . . . . . . . . . . . . . . . . . . . 17.17
R v Chief Constable of Devon and Cornwall Constabulary, ex parte Police Complaints Authority [1996] 2 All E.R. 711; Times, February 19, 1996 . . . . . . . . . 17.17

R v Chief Constable of Merseyside Police, ex parte Calveley [1986] Q.B. 424 ....... 17.15, 17.16
R v Chief Constable of Merseyside Police, ex parte Merrill [1989] 1 W.L.R. 1077; [1990] C.O.D. 61; (1989) 153 L.G. Rev. 1010; Times, May 19, 1989 ........ 17.16
R v Chief Constable of West Midlands Police, ex parte Carroll (1995) 7 Admin. L.R. 45; Times, May 20, 1994; Independent, June 6, 1994 .............. 56.09, 56.10
R v Chief Rabbi of the United Hebrew Congregation of Great Britain and the Commonwealth, ex parte Wachmann [1992] 1 W.L.R. 1036; [1993] 2 All E.R. 249; (1991) 3 Admin. L.R. 721; [1991] C.O.D. 309; Times, February 7, 1991 ........ 42.05, 42.06
R v City of London Corporation, ex parte Matson [1997] 1 W.L.R. 765; 94 L.G.R. 443; (1996) 8 Admin. L.R. 49; [1996] C.O.D. 161; Times, October 20, 1995; Independent, September 27, 1995 ........ 31.03, 54.02
R v Clarke (1931) 22 Crim App Rep 58 ........ 18.17
R v Connor and Rollock [2004] 1 A.C. 1118 ........ 34.03
R v Council for Licensed Conveyancers, ex parte Watson, CO/02/98, 21 December 1999 ........ 21.02
R v Council for Licensed Conveyancers, ex parte West, CO 4901/99, 7 June 2000 (unreported) ........ 17.18
R v Criminal Injuries Compensation Baord, ex parte Lain [1967] 2 Q.B. 864; [1967] 3 W.L.R. 348; [1967] 2 All E.R. 770; (1967) 111 S.J. 331 .... 42.03, 42.06, 42.09
R v D [2010] EWCA Crim 1213 ........ 25.02
R v Day [2003] EWCA Crim 1060 ........ 45.12
R v Derby Magistrates' Court, ex parte B [1996] A.C. 487; [1995] 3 W.L.R. 681; [1995] 4 All E.R. 526; [1996] 1 Cr. App. R. 385; (1995) 159 J.P. 785; [1996] 1 F.L.R. 513; [1996] Fam. Law 210; (1995) 159 J.P.N. 778; (1995) 145 N.L.J. 1575; [1995] 139 S.J.L.B. 219; Times, October 25, 1995; Independent, October 27, 1995 ........ 52.05
R v Director of Public Prosecutions, ex parte Manning [2000] 3 W.L.R. 603 ......... 39.18
R v Disciplinary Board of the Metropolitan Police, ex parte Borland, 20 July 1982 (unreported) ........ 21.01
R v Disciplinary Committee of the Jockey Club, ex parte Aga Khan [1993] 1 W.L.R. 909; [1993] 2 All E.R. 853; [1993] C.O.D. 234; (1993) 143 N.L.J. 163; Times, December 9, 1992; Independent, December 22, 1992 ........... 42.06, 42.13
R v Disciplinary Committee of the Jockey Club, ex parte Massingberd-Mundy [1993] 2 All E.R. 207; (1990) 2 Admin. L.R. 609; [1990] C.O.D. 260 .......... 42.06
R v Executive Counsel Joint Disciplinary Scheme, ex parte Hipps [1996] CLY 5 ...... 10.08
R v F (S) [2011] EWCA Crim 1844; [2012] Q.B. 703; [2012] 2 W.L.R. 1038; [2012] 1 All E.R. 565; [2011] 2 Cr. App. R. 28; [2012] Crim. L.R. 282; Times, July 25, 2011 ........ 2.05, 17.06, 50.04
R v Football Association Ltd, ex parte Football League Ltd [1993] 2 All E.R. 833; (1992) 4 Admin. L.R. 623; [1992] C.O.D. 52; Times, August 22, 1991 ........ 42.06
R v G [2003] UKHL 50; [2004] 1 A.C. 1034; [2003] 3 W.L.R. 1060; [2003] 4 All E.R. 765; [2004] 1 Cr. App. R. 21; (2003) 167 J.P. 621; [2004] Crim. L.R. 369; (2003) 167 J.P.N. 955; (2003) 100(43) L.S.G. 31; Times, October 17, 2003 ........ 47.46
R v Galbraith [1981] 1 W.L.R. 1039; [1981] 2 All E.R. 1060; (1981) 73 Cr. App. R. 124; [1981] Crim. L.R. 648; (1981) 125 S.J. 442 .... 2.05, 50.02, 50.04, 50.09, 50.11
R v General Council of the Bar, ex parte Percival [1991] 1 Q.B. 212; [1990] 3 W.L.R. 323; [1990] 3 All E.R. 137; (1990) 2 Admin. L.R. 711; (1990) 87(24) L.S.G. 44 ........ 42.04
R v General Council of the Bar, ex parte Percival [1991] Q.B. 212 ................ 39.02
R v General Council of Medical Education and Registration in the United Kingdom [1930] 1 KB 562 ........ 39.02
R v General Medical Council ex parte El Shanawany, 2 December 1991 (unreported) .... 35.05

R v General Medical Council , ex parte Toth [2000] 1 W.L.R. 2209; [2000] Lloyd's Rep. Med. 368; (2001) 61 B.M.L.R. 149; (2000) 97(27) L.S.G. 39; Times, June 29, 2000 ..... 17.24, 18.04, 18.08, 39.03, 39.05, 39.06, 39.07, 39.11, 39.13
R v Ghosh [1982] Q.B. 1053; [1982] 3 W.L.R. 110; [1982] 2 All E.R. 689; (1982) 75 Cr. App. R. 154; [1982] Crim. L.R. 608; (1982) 126 S.J. 429 ......... 19.01, 19.04, 19.05, 19.12, 19.19, 19.31
R v Gough [1993] A.C. 646; [1993] 2 W.L.R. 883; [1993] 2 All E.R. 724; (1993) 97 Cr. App. R. 188; (1993) 157 J.P. 612; [1993] Crim. L.R. 886; (1993) 157 J.P.N. 394; (1993) 143 N.L.J. 775; (1993) 137 S.J.L.B. 168; Times, May 241993; Independent, May 26, 1993; Guardian, May 22, 1993 ............ 35.12
R v Governor of Brixton Prison, ex parte Armah [1968] A.C. 192; [1966] 3 W.L.R. 828; [1966] 3 All E.R. 177; (1967) 131 J.P. 43; (1966) 110 S.J. 890 ..... 39.05
R v Greenaway 115 E.R. 436; (1845) 7 Q.B. 126 ............................ 70.02
R v Hayward [2001] EWCA Crim 168; [2001] Q.B. 862; [2001] 3 W.L.R. 125; [2001] 2 Cr. App. R. 11; (2001) 165 J.P. 281; [2001] Crim. L.R. 502; (2001) 165 J.P.N. 665; (2001) 98(9) L.S.G. 38; (2001) 145 S.J.L.B. 53; Times, February 14, 2001; Independent, February 8, 2001 ........................ 67.03
R v Hayward, Jones and Purvis [2001] EWCA Crim 168; [2001] Q.B. 862; [2001] 3 W.L.R. 125; [2001] 2 Cr. App. R. 11; (2001) 165 J.P. 281; [2001] Crim. L.R. 502; (2001) 165 J.P.N. 665; (2001) 98(9) L.S.G. 38; (2001) 145 S.J.L.B. 53; Times, February 14, 2001 .................. 1.02, 1.08, 3.16, 44.19
R v Heston-Francois [1984] Q.B. 278; [1984] 2 W.L.R. 309; [1984] 1 All E.R. 795; (1984) 78 Cr. App. R. 209; [1984] Crim. L.R. 227 .............. 2.02
R v Horncastle [2009] UKSC 14; [2010] 2 A.C. 373; [2010] 2 W.L.R. 47; [2010] 2 All E.R. 359; [2010] 1 Cr. App. R. 17; [2010] H.R.L.R. 12; [2010] U.K.H.R.R. 1; [2010] Crim. L.R. 496; (2009) 153(48) S.J.L.B. 32; Times, December 10, 2009 ..... 25.02
R v Horseferry Road Magistrates' Court, ex parte Bennett [1994] 1 A.C. 42; [1993] 3 W.L.R. 90; [1993] 3 All E.R. 138; (1994) 98 Cr. App. R. 114; [1994] C.O.D. 123; (1993) 157 J.P.N. 506; (1993) 143 N.L.J. 955; (1993) 137 S.J.L.B. 159; Times, June 25, 1993; Independent, July 1, 1993 .... 2.03, 2.04, 17.04
R v Huddersfield Justices, ex parte D [1997] COD 27 .......................... 42.12
R v Hull University Visitor, ex parte Page [1993] A.C. 682 ...................... 42.07
R v Hurle-Hobbs, ex parte Simmons [1945] K.B. 165; [1945] 1 All E.R. 573 ......... 70.02
R v Ibrahim [2012] EWCA Crim 837; [2012] 4 All E.R. 225; [2012] 2 Cr. App. R. 32; (2012) 176 J.P. 470; [2012] Crim. L.R. 793 ................. 25.02
R v Institute of Chartered Accountants of England and Wales, ex parte Taher Nawaz [1997] PNLR 433; [1997] CLY 1 ...................... 52.02, 52.09
R v Institute of Chartered Accountants, ex parte Brindle [1994] B.C.C. 297; Times, January 12, 1994 ............................... 10.06, 10.07, 10.08
R v Insurance Ombudsman, ex parte Aegon Life Insurance Ltd [1995] L.R.L.R. 101; [1994] C.L.C. 88; [1994] C.O.D. 426; Times, January 7, 1994; Independent, January 11, 1994 ....................................... 42.09
R v Jisl, Tekin and Konakli [2004] EWCA Crim 696; Times, April 19, 2004 .......... 8.02
R v Jockey Club, ex parte R.A.M. Racecourses Ltd [1993] 2 All E.R. 225; (1991) 5 Admin. L.R. 265; [1990] C.O.D. 346 ................................. 42.06
R v Jones [2002] UKHL 5; [2003] 1 A.C. 1; [2002] 2 W.L.R. 524; [2002] 2 All E.R. 113; [2002] 2 Cr. App. R. 9; (2002) 166 J.P. 333; [2002] H.R.L.R. 23; (2002) 166 J.P.N. 431; (2002) 99(13) L.S.G. 26; (2002) 146 S.J.L.B. 61; Times, February 21, 2002; Independent, February 27, 2002 ................ 1.02, 1.05, 1.08, 3.10, 3.16, 3.20
R v Kelly [2000] Q.B. 198; [1999] 2 W.L.R. 1100; [1999] 2 All E.R. 13; [1999] 2 Cr. App. R. 36; [1999] 2 Cr. App. R. (S.) 176; [1999] 2 Cr. App. R. (S.) 185; [1999] Crim. L.R. 240; Times, December 29, 1998 ........................ 17.11
R v Khan [1997] A.C. 558 ............................................ 25.21
R v Latif [1996] 1 W.L.R. 104; [1996] 1 All E.R. 353; [1996] 2 Cr. App. R. 92; [1996] Crim. L.R. 414; (1996) 93(5) L.S.G. 30; (1996) 146 N.L.J. 121; (1996) 140 S.J.L.B. 39; Times, January 23, 1996; Independent, January 23, 1996 ........................................... 2.03, 2.04

R v Legal Aid Board, ex parte Kaim Todner[1999] Q.B. 966; [1998] 3 W.L.R. 925; [1998] 3 All E.R. 541; (1998) 95(26) L.S.G. 31; (1998) 148 N.L.J. 941; (1998) 142 S.J.L.B. 189; Times, June 15, 1998; Independent, June 12, 1998 . . . . 51.05, 53.03
R v Legal Services Commission, ex parte Jarrett [2001] EWHC Admin 389 . . . . 32.11, 45.09
R v Liddell [1995] 1 NZLR 538 . . . . 51.06
R v Liverpool City Justices, ex parte Topping [1983] 1 W.L.R. 119; [1983] 1 All E.R. 490; (1983) 76 Cr. App. R. 170; (1983) 147 J.P. 154; [1983] Crim. L.R. 181; (1983) 127 S.J. 51; Times, November 16, 1982 . . . . 6.03
R v Lloyd's of London, ex parte Briggs [1993] 1 Lloyd's Rep. 176; [1993] C.O.D. 66; Times, July 30, 1992; Independent, September 16, 1992; Financial Times, July 29, 1992 . . . . 42.09
R v London Metal Exchange Ltd, ex parte Albatross Warehousing BV, 30 March 2000 (unreported) . . . . 42.09
R v Longworth [2006] UKHL 1; [2006] 1 W.L.R. 313; [2006] 1 All E.R. 887; [2006] 2 Cr. App. R. (S.) 62; [2006] Crim. L.R. 553; (2006) 103(6) L.S.G. 35; (2006) 150 S.J.L.B. 132; Times, February 1, 2006; Independent, February 2, 2006 . . . . 14.16
R v Looseley [2001] UKHL 53; [2001] 1 W.L.R. 2060; [2001] 4 All E.R. 897; [2002] 1 Cr. App. R. 29; [2002] H.R.L.R. 8; [2002] U.K.H.R.R. 333; [2002] Crim. L.R. 301; (2001) 98(45) L.S.G. 25; (2001) 145 S.J.L.B. 245; Times, October 29, 2001 . . . . 2.08
R v Lucas [1981] Q.B. 720; (1981) 73 Crim App Rep 159 . . . . 19.31, 44.23
R v Maguire [1992] Q.B. 936; [1992] 2 W.L.R. 767; [1992] 2 All E.R. 433; (1992) 94 Cr. App. R. 133; Times, June 28, 1991; Independent, June 27, 1991; Guardian, June 27, 1991 . . . . 18.03
R v Manchester City Stipendiary Magistrates, ex parte Snelson [1977] 1 W.L.R. 911; (1978) 66 Cr. App. R. 44; [1977] Crim. L.R. 423; (1977) 121 S.J. 442 . . . . 21.04
R v Marquis [2009] UKSC 14; [2010] 2 A.C. 373; [2010] 2 W.L.R. 47; [2010] 2 All E.R. 359; [2010] 1 Cr. App. R. 17; [2010] H.R.L.R. 12; [2010] U.K.H.R.R. 1; [2010] Crim. L.R. 496; (2009) 153(48) S.J.L.B. 32; Times, December 10, 2009 . . . . 25.02
R v Master of the Rolls, ex parte McKinnell [1993] 1 W.L.R. 88; [1993] 1 All E.R. 193; [1993] C.O.D. 34; (1992) 142 N.L.J. 1231; (1992) 136 S.J.L.B. 259; Times, August 6, 1992; Independent, August 12, 1992 . . . . 58.02
R v Maxwell [2010] UKSC 48; [2011] 1 W.L.R. 1837; [2011] 4 All E.R. 941; [2011] 2 Cr. App. R. 31; (2011) 108(31) L.S.G. 17; Times, July 29, 2011 . . . . 2.03
R v Metropolitan Police Disciplinary Tribunal, ex parte Police Complaints Authority, CO/1780/91, 20 March 1992 (unreported) . . . . 2.06
R v Metropolitan Stipendiary Magistrate, ex parte Gallagher (1972) 136 JP 80 . . . . 6.03
R v Miah [1997] 2 Cr. App. R. 12; [1997] Crim. L.R. 351; Times, December 18, 1996 . . . . 34.03
R v Middleton [2000] TLR 293 . . . . 19.31
R v Mullen [2000] Q.B. 520; [2000] Crim. L.R. 873 . . . . 2.04
R v Newton (1982) 4 Cr App R(S) 388 . . . . 60.14
R v North and East Devon Health Authority, ex parte Coughlan [2001] Q.B. 213 . . . . 65.13
R v P [2002] 1 A.C. 146; [2001] 2 W.L.R. 463; [2001] 2 All E.R. 58; [2001] 2 Cr. App. R. 8; (2001) 98(8) L.S.G. 43; (2001) 145 S.J.L.B. 28; Times, December 19, 2000; Independent, December 20, 2000 . . . . 2.08
R v Panel on Takeovers and Mergers, ex parte Datafin Plc [1987] Q.B. 815; [1987] 2 W.L.R. 699; [1987] 1 All E.R. 564; (1987) 3 B.C.C. 10; [1987] B.C.L.C. 104; [1987] 1 F.T.L.R. 181; (1987) 131 S.J. 23 . . . . 42.06, 42.09
R v Panel on Takeovers and Mergers , ex parte Fayed [1992] B.C.C. 524; [1992] B.C.L.C. 938; (1993) 5 Admin. L.R. 337; Times, April 15, 1992; Financial Times, April 14, 1992 . . . . 10.05, 10.07, 10.08
R v Patel [2006] EWCA Crim 2689; [2007] 1 Cr. App. R. 12; [2007] I.C.R. 571; [2007] Crim. L.R. 476 . . . . 14.17
R v Police Complaints Board, ex parte Madden [1983] 1 W.L.R. 447; [1983] 2 All E.R. 353; [1983] Crim. L.R. 263; (1983) 127 S.J. 85 . . . . 21.01

R v Price (1990) 90 Cr. App. R. 409; [1990] Crim. L.R. 200 ...................... 19.02
R v Professional Conduct Committee of the General Medical Council, ex parte Jaffe, 25 March 1988 (unreported) ............................ 44.09
R v Provincial Court of the Church in Wales, ex parte Williams (1998) CO/2880/98 ................................................ 62.21
R v Riat [2012] EWCA Crim 1509; [2013] 1 W.L.R. 2592; [2013] 1 All E.R. 349; [2013] 1 Cr. App. R. 2; [2013] Crim. L.R. 60 ................. 25.02
R v Roberts (1987) 84 Cr. App. R. 117; [1986] Crim. L.R. 188 ............... 19.02, 19.04
R v Royal Borough of Kensington and Chelsea, ex parte Grillo (1995) 94 LGR 144 ................................................ 54.02
R v Secretary of State for the Home Department, ex parte Doody [1994] 1 A.C. 531; [1993] 3 W.L.R. 154; [1993] 3 All E.R. 92; (1995) 7 Admin. L.R. 1; (1993) 143 N.L.J. 991; Times, June 29, 1993; Independent, June 25, 1993 ............................................ 31.03, 54.02
R v Secretary of State for the Home Department, ex parte Tarrant [1985] Q.B. 251; [1984] 2 W.L.R. 613; [1984] 1 All E.R. 799; (1984) 81 L.S.G. 1045; (1984) S.J. 223; Times, November 9, 1983 .............................. 45.06, 45.08
R v Secretary of State for the Home Department, ex parte Thornton [1987] Q.B. 36; [1986] 3 W.L.R. 158; [1986] 2 All E.R. 641; (1986) 83 L.S.G. 2493; (1986) 130 S.J. 246 ............................................ 14.15
R v Servite Houses and the London Borough of Wandsworth Council, ex parte Goldsmith (2001) 33 H.L.R. 35; (2000) 2 L.G.L.R. 997; [2001] B.L.G.R. 55; (2000) 3 C.C.L. Rep. 325; [2001] A.C.D. 4 ................... 42.09
R v Shippey [1988] Crim LR 767 ........................................ 50.03
R v Solicitors Complaints Bureau, ex parte Curtin (1994) 6 Admin. L.R. 657; [1993] C.O.D. 467; [1994] C.O.D. 390; (1993) 137 S.J.L.B. 167; Times, December 3, 1993; Independent, December 13, 1993 ...................... 20.03
R v Solicitors Disciplinary Tribunal, ex parte Gallagher, 30 September 1991 (unreported) ................................................ 10.04
R v Solicitors Disciplinary Tribunal, ex parte Toth [2001] EWHC Admin 240; [2001] 3 All E.R. 180; (2001) 151 N.L.J. 502; Times, May 3, 2001 ........................................ 39.05, 39.11, 39.13
R v Statutory Committee of Pharmaceutical Society of Great Britain, ex parte Pharmaceutical Society of Great Britain [1981] 1 W.L.R. 886; [1981] 2 All E.R. 805; (1981) 125 S.J. 428 ..................... 14.14, 17.36, 21.04, 42.02
R v Statutory Committee of the Royal Pharmaceutical Society of Great Britain, ex parte Sokoh, 3 December 1986 (unreported) .......................... 47.16
R v Stow [2005] EWCA Crim 1157; [2005] U.K.H.R.R. 754 ................... 35.09
R v Takeovers and Mergers Panel, ex parte Guiness [1990] 1 Q.B. 146; [1989] 2 W.L.R. 863; [1989] 1 All E.R. 509; (1988) 4 B.C.C. 714; [1989] B.C.L.C. 255; (1988) 138 N.L.J. Rep. 244; (1989) 133 S.J. 660 ...................... 10.08
R v Thakrar [2001] EWCA Crim 1906 .................................... 45.12
R v Totnes Licensing Justices, ex parte Chief Constable of Devon and Cornwall (1992) 156 J.P. 587; [1990] C.O.D. 404; (1992) 156 J.P.N. 538; (1992) 156 L.G. Rev. 904; Times, May 28, 1990 ................................ 15.03
R v Uljee [1982] 1 NZLR 561 ............................................ 52.05
R v United Kingdom Central Council for Nursing Midwifery and Health Visiting, ex parte Thompson, Machin and Wood [1991] COD 275 .............. 3.04
R v Visitors to the Inns of Court, ex parte Calder [1994] Q.B. 1; [1993] 3 W.L.R. 287; [1993] 2 All E.R. 876; [1993] C.O.D. 242; (1993) 143 N.L.J. 164; Times, January 26, 1993; Independent, January 29, 1993 ...... 4.07, 42.07, 43.20
R v Visitors to the Inns of Court, ex parte Persaud [1994] Q.B. 1; [1993] 3 W.L.R. 287; [1993] 2 All E.R. 876; [1993] C.O.D. 242; (1993) 143 N.L.J. 164; Times, January 26, 1993; Independent, January 29, 1993 ........... 4.07, 42.07
R v Ward [1993] 1 W.L.R. 619; [1993] 2 All E.R. 577; (1993) 96 Cr. App. R. 1; [1993] Crim. L.R. 312; (1992) 89(27) L.S.G. 34; (1992) 142 N.L.J. 859; (1992) 136 S.J.L.B. 191; Times, June 8, 1992 ............................ 18.03

R v Westminster City Council, ex parte Ermakov [1996] 2 All E.R. 302; [1996] 2 F.C.R. 208; (1996) 28 H.L.R. 819; (1996) 8 Admin. L.R. 389; [1996] C.O.D. 391; (1996) 160 J.P. Rep. 814; (1996) 140 S.J.L.B. 23; Times, November 20, 1995 . . . . . 54.16
R v Wiltshire Appeal Tribunal, ex parte Thatcher (1916) 86 LJ KB 121 . . . . . 70.02
R & T Thew Ltd v Reeves (No. 2) [1982] Q.B. 1283 . . . . . 43.05
Raheem v Nursing and Midwifery Council *See* R (Raheem) v Nursing and Midwifery Council
Rahman v Bar Standards Board [2013] EWHC 4202 (QB) . . . . . 17.39, 65.10
Rai v General Medical Council, Privy Council Appeal No. 54 of 1983 . . . . 4.06, 13.02, 54.03
Rajan v General Medical Council [2000] Lloyd's Rep. Med. 153 . . . . . 18.03
Rajasooria v Disciplinary Committee [1955] 1 W.L.R. 405; (1955) 99 S.J. 256 . . . . . 39.02, 47.04
Raji v General Medical Council [2003] UKPC 24; [2003] 1 W.L.R. 1052; [2003] Lloyd's Rep. Med. 280; [2003] A.C.D. 63; (2003) 100(21) L.S.G. 30; Times, March 31, 2003 . . . . . 58.03
Ramsden v Independent Police Complaints Commission *See* R (Ramsden) v Independent Police Complaints Commission
Ranson v Institute of Actuaries *See* R (Ranson, Headdon and Nash) v Institute of Actuaries
Rao v General Medical Council [2002] UKPC 65; [2003] Lloyd's Rep. Med. 62; (2003) 147 S.J.L.B. 113 . . . . . 44.13, 47.25, 47.29
Raschid v General Medical Council; Fatnani v General Medical Council [2007] EWCA Civ 46; [2007] 1 W.L.R. 1460; [2007] I.C.R. 811; (2007) 104(5) L.S.G. 29; (2007) 151 S.J.L.B. 127 . . . . . 4.16, 4.18, 14.25, 37.07, 60.10
Rashid v General Medical Council [2012] EWHC 2862 (Admin) . . . . . 38.11, 38.32
Rasool v General Pharmaceutical Council [2015] EWHC 217 (Admin) . . . . . 6.17, 60.64
Ratnam v Law Society of Singapore [1976] 1 MLJ 195 . . . . . 14.03
Rauniar v General Medical Council [2011] EWHC 782 . . . . . 46.05
Raza v General Medical Council [2011] EWHC 790 (Admin) . . . . . 32.14
Razak v General Medical Council [2004] EWHC 205 (Admin) . . . . . 36.04, 50.08
Razeen v Law Society [2008] EWCA Civ 1220 . . . . . 11.14
Razzaq v General Medical Council [2006] EWHC 1300 (Admin); (2006) 91 B.M.L.R. 108 . . . . . 25.22
Red Bank Manufacturing Co Limited v Meadows [1992] ICR 204 . . . . . 48.12
Reddy v General Medical Council [2012] EWCA Civ 310; [2012] C.P. Rep. 27 . . . . . 4.37
Rees v Crane [1994] 2 A.C. 173; [1994] 2 W.L.R. 476; [1994] 1 All E.R. 833; (1994) 138 S.J.L.B. 71 . . . . . 41.04
Rehman v Bar Standards Board, PC2008/0235/A and PC2010/0012/A, 19 July 2013 . . . . . 25.32, 47.53, 67.09
Rengarajaperunal v General Medical Council *See* R (Rengarajaperunal) v General Medical Council
Rey v Government of Switzerland [1999] 1 A.C. 54; [1998] 3 W.L.R. 1; (1998) 142 S.J.L.B. 167 . . . . . 54.05
Reza v General Medical Council [1991] 2 A.C. 182; [1991] 2 W.L.R. 939; [1991] 2 All E.R. 796; [1991] 2 Med. L.R. 255; (1991) 135 S.J. 383; Times, March 5, 1991; Daily Telegraph, March 25, 1991 . . . . . 22.06, 40.03
Rice v Health Professions Council [2011] EWHC 1649 (Admin) . . . . . 4.30
Rich v Bar Standards Board *See* R (Rich) v Bar Standards Board
Rich v Bar Standards Board [2011] EWHC 1099 (Admin) . . . . . 51.05
Rich v General Medical Council [2013] EWHC 1673 (Admin) . . . . . 3.18, 45.27, 60.52
Richards v General Medical Council *See* R (Richards) v General Medical Council
Richards v Law Society (Solicitors Regulation Authority) [2009] EWHC 2087 (Admin) . . . . . 22.24, 62.15
Richardson v Redpath Brown & Co Ltd [1944] A.C. 62; [1944] 1 All E.R. 110 . . . . . 46.02
Richardson v Solihull Metropolitan Borough Council [1998] EWCA Civ 335 . . . . . 27.09

Ridge v Baldwin [1964] A.C. 40; [1963] 2 W.L.R. 935; [1963] 2 All E.R. 66; (1963) 127 J.P. 295; (1963) 127 J.P. 251; 61 L.G.R. 369; 37 A.L.J. 140; 234 L.T. 423; 113 L.J. 716; (1963) 107 S.J. 313 . . . . . . . . . . . . . . . . . 48.03, 48.04, 48.08
Rimmer v General Dental Council [2011] EWHC 3438 (Admin); (2012) 124 B.M.L.R. 40 . . . . . . . . . . . . . . . . . . . . . . . . . . . . . . . . . . . . . . . . . . . . . 26.07
Roberts v Hook [2013] EWHC 1349 (Admin) . . . . . . . . . . . . . . . . . . . . . . . . . . . . . 39.28
Robinson v Solicitors Regulation Authority [2012] EWHC 2690 (Admin); (2012) 156(40) S.J.L.B. 31 . . . . . . . . . . . . . . . . . . . . . . . . . . . . . . . . . . . . . . . . . . . . . 65.06
Rogan v Nursing and Midwifery Council [2011] NIQB 12 . . . . . . . . . . . . . . . . . . . . . . 38.12
Rogers v General Medical Council *See* R (Rogers) v General Medical Council
Rogers v General Medical Council [2008] EWHC 2741 (Admin) . . . . . . . . . . . . . . . . . . 17.19
Rogers v Hoyle [2014] EWCA Civ 257; [2015] Q.B. 265; [2014] 3 W.L.R. 148; [2014] 3 All E.R. 550; [2014] C.P. Rep. 30; [2014] 1 C.L.C. 316 . . . . . . . . . . . . . . . 25.20
Royal Aquarium and Summer and Winter Garden Society v Parkinson [1892] 1 Q.B. 431 . . . . . . . . . . . . . . . . . . . . . . . . . . . . . . . . . . . . . . . . . . . . . . . . . . . 52.12
Royal College of Nursing v Secretary of State for the Home Department *See* R (Royal College of Nursing) v Secretary of State for the Home Department
Roylance v General Medical Council, The Times, 27 January 1999 . . . . . . . . . . . . . . . . 34.02
Roylance v General Medical Council (No. 2) [2000] 1 A.C. 311; [1999] 3 W.L.R. 541; [1999] Lloyd's Rep. Med. 139; (1999) 47 B.M.L.R. 63; (1999) 143 S.J.L.B. 183; Times, March 26, 1999 . . . . . . . . . . . 34.03, 47.01, 47.22, 47.27, 47.33, 47.35, 49.04
Ruscillo v Council for the Regulation of Health Care Professionals and General Medical Council [2004] EWCA Civ 1356; [2005] 1 W.L.R. 717; [2005] Lloyd's Rep. Med. 65; [2005] A.C.D. 69; [2005] A.C.D. 99; (2004) 148 S.J.L.B. 1248; Times, October 27, 2004; Independent, October 28, 2004 . . . . 4.13, 6.08
Russell v Bar Standards Board, 12 July 2012 (unreported) . . . . . . . . . . . . . . . . . . 13.13, 13.14
Russell v General Medical Council *See* R (Russell) v General Medical Council
Rycroft v Royal Pharmaceutical Society of Great Britain *See* R (Rycroft) v Royal Pharmaceutical Society of Great Britain
Ryell v Health Professions Council [2005] EWHC 2797 (Admin) . . . . . . . . . . . . . . . . 54.09

S (A Barrister), In re [1970] 1 Q.B. 160; [1969] 2 W.L.R. 708; [1969] 1 All E.R. 949; (1969) 113 S.J. 145 . . . . . . . . . . . . . . . . . . . . . . . . . . . . . . . 3.03, 43.02
S-B (Children) (Care Proceedings: Standard of Proof), In re [2010] 1 A.C. 678 . . . . . . . 62.06
SS v Knowsley NHS Primary Care Trust *See* R (SS) v Knowsley NHS Primary Care Trust
Sacha v General Medical Council [2009] EWHC 302 (Admin) . . . . . . . . . . 4.21, 21.11, 27.06
Sadighi v General Dental Council [2009] EWHC 1278 (Admin) . . . . . . . . . . . . . . . . . 35.10
Sadler v General Medical Council [2003] UKPC 59; [2003] 1 W.L.R. 2259; [2004] H.R.L.R. 8; [2004] Lloyd's Rep. Med. 44; Times, September 29, 2003 . . . . . . . . . . . . . . . . . . . . . . . . 16.04, 16.11, 46.02, 46.05, 62.11
Saeed v Inner London Education Authority [1985] I.C.R. 637; [1986] I.R.L.R. 23; (1985) 82 L.S.G. 2911 . . . . . . . . . . . . . . . . . . . . . . . . . . . . . . . . . . . 21.04
Saeed v Royal Wolverhampton Hospitals NHS Trust [2001] I.C.R. 903; [2001] Lloyd's Rep. Med. 111; Times, January 17, 2001 . . . . . . . . . . . . . . . . . . . . . . . . . 47.28
Saha v General Medical Council [2009] EWHC 1907 (Admin); [2009] LS Law Medical 551 . . . . . . . . . . . . . . . . . . . . . . . . . . . . . . . . . . . . . . . . . . 18.23, 33.10
Salha v General Medical Council [2003] UKPC 80; [2004] E.C.D.R. 12; (2004) 80 B.M.L.R. 169 . . . . . . . . . . . . . . . . . . . . . . . . . . . . . . . 19.43, 22.11, 22.27
Salsbury v Law Society [2008] EWCA Civ 1285; [2009] 1 W.L.R. 1286; [2009] 2 All E.R. 487; (2008) 105(47) L.S.G. 18; (2008) 158 N.L.J. 1720; (2008) 152(46) S.J.L.B. 30; Times, January 15, 2009 . . . . . . . . . . 4.17, 4.26, 19.51, 19.54, 19.60, 19.64, 60.12, 60.45
Samuel v Royal College of Veterinary Surgeons [2014] UKPC 13 . . . . . . . . . . . . 14.12, 62.20
Sanders v Kingston [2005] EWHC 1145 (Admin); [2005] B.L.G.R. 719; Times, June 16, 2005 . . . . . . . . . . . . . . . . . . . . . . . . . . . . . . . . . . . . . . . . . . . . 32.27

Sanders v Kingston (No.2) [2005] EWHC 2132 (Admin); [2006] B.L.G.R. 111; (2005) 102(44) L.S.G. 33; Times, November 14, 2005 ........... 36.07
Sandler v General Medical Council [2010] EWHC 1029 (Admin); [2010] Med. L.R. 491; (2010) 114 B.M.L.R. 141 ................ 38.09, 38.14, 38.26, 38.28
Sarkar v General Medical Council [2012] EWHC 4008 (Admin) .................. 16.14
Sarker v General Medical Council [2012] EWHC 4008 (Admin) .................. 60.50
Sarkodie-Gyan v Nursing and Midwifery Council [2009] EWHC 2131 (Admin) ..... 31.10
Saunders v United Kingdom [1997] B.C.C. 872; [1998] 1 B.C.L.C. 362; (1997) 23 E.H.R.R. 313; 2 B.H.R.C. 358; Times, December 18, 1996; Independent, January 14, 1997 .............................................. 52.09
Saverymuttu v General Medical Council [2011] EWHC 1139 (Admin) ............. 22.27
Saville-Smith v Scottish Legal Complaints Commission [2012] CSIH 99 ............ 39.27
Schodlok v General Medical Council [2013] EWHC 2280 (Admin) ........... 8.10, 70.07
Scholten v General Medical Council [2013] EWHC 173 (Admin) ................. 38.33
Schubert Murphy (a firm) v Law Society [2014] EWHC 4561 (QB) ................ 49.13
Scott v Scott [1913] A.C. 417 ................................................ 53.07
Secretary of State for Education and Skills v Mairs [2005] EWHC 996 (Admin); [2005] I.C.R. 1714; [2005] A.C.D. 93; Times, June 15, 2005 .......... 25.08
Selvanathan v General Medical Council [2001] Lloyd's Rep. Med. 1; (2001) 59 B.M.L.R. 95; Times, October 26, 2000 ....................... 54.03, 54.05
Selvarajan v General Medical Council [2008] EWHC 182 (Admin) ................ 17.36
Shah v General Dental Council [2011] EWHC 3003 (Admin) .................... 37.12
Shah v General Medical Council [2011] EWHC 3003 (Admin) .................. 37.12
Shah v General Pharmaceutical Council (formerly Royal Pharmaceutical Society of Great Britain) [2011] EWHC 73 (Admin) ....................... 60.13
Shaikh v General Pharmaceutical Council *See* R (Shaikh) v General Pharmaceutical Council
Shaikh v National Co-operative Chemists Ltd and Royal Pharmaceutical Society of Great Britain [2010] EWHC 2602 (QB) ........................ 39.19
Shamsian v General Medical Council [2011] EWHC 2885 (Admin) ............... 19.28
Shanker v General Medical Council *See* R (Shanker) v General Medical Council
Sharaf v General Medical Council [2013] EWHC 3332 (Admin) ................. 42.23
Sharma v General Dental Council [2010] EWHC 3184 (Admin) ................. 11.09
Sharma v General Dental Council *See* R (Sharma) v General Dental Council
Sharma v General Medical Council [2014] EWHC 1471 (Admin) ...... 19.29, 19.33, 44.25
Sharma v Solicitors Regulation Authority [2012] EWHC 3176 (Admin); [2012] All E.R. (D) 289 (Oct) ................................................ 15.16
Sharp v Law Society of Scotland 1984 S.C. 129; 1984 S.L.T. 313 ................. 47.15
Sharp v Nursing and Midwifery Council [2011] EWHC 2174 (Admin) ............. 22.28
Shaw and Turnbull v Logue [2014] EWHC 5 (Admin) ..... 4.27, 19.24, 30.06, 38.69, 54.29
Sheikh v General Dental Council *See* R (Sheikh) v General Dental Council
Sheill v General Medical Council [2008] EWHC 2967 (Admin) ....... 19.11, 22.20, 22.31
Shepherd v Governor of HMP Whatton [2010] EWHC 2474 (Admin) ............. 54.19
Shepherd v Law Society [1996] EWCA Civ 977 ......................... 14.04, 14.06
Shiekh v General Dental Council *See* R (Shiekh) v General Dental Council
Shiekh v General Dental Council [2009] EWHC 186 (Admin) .............. 17.37, 19.52
Shrimpton v General Council of the Bar [2005] EWHC 2472 .... 15.04, 15.22, 15.29, 22.17
Siborurema v Office of the Independent Adjudicator *See* R (Siborurema) v Office of the Independent Adjudicator
Siddiqui v General Medical Council [2013] EWHC 1083 (Admin) ................ 19.67
Siddiqui v Health Professions Council [2012] EWHC 2863 (Admin) .......... 27.11, 62.16
Silver v General Medical Council ; [2003] Lloyd's Rep. Med. 333; Times, May 9, 2003 ............................................ 47.26, 47.29, 47.47
Simmons v British Steel Plc 2004 SC (HL) 109 ............................. 27.04
Simms v Law Society [2005] EWCA Civ 749 ................................. 52.08
Simms v Law Society [2005] EWHC 408 (Admin) ...................... 22.13, 52.08
Simpson v General Medical Council , Times, November 9, 1955 .................. 14.13

Singh v General Medical Council [2003] UKPC 15 .......... 60.28
Singh v General Medical Council, Privy Council Appeal No. 73 of 1997 .......... 60.21
Singh v Reading Borough Council [2013] ICR 1158 .......... 52.16
Singleton v Law Society [2005] EWHC 2915 (Admin) .......... 19.07, 22.12, 22.27
Sinha v General Medical Council *See* R (Sinha) v General Medical Council
Sivanandan v Bar Standards Board, 17 August 2012 (unreported) .......... 47.53
Sivarajah v General Medical Council [1964] 1 W.L.R. 112 .......... 44.05, 44.08, 44.09
Sivills v General Social Care Council *See* R (Sivills) v General Social Care Council
Skidmore v Dartford and Gravesham NHS Trust [2003] UKHL 27; [2003] 3 All E.R. 292; [2003] I.C.R. 721; [2003] I.R.L.R. 445; [2003] Lloyd's Rep. Med. 369; (2003) 73 B.M.L.R. 209; (2003) 100(27) L.S.G. 35; (2003) 153 N.L.J. 881; (2003) 147 S.J.L.B. 624; Times, May 23, 2003 .......... 47.28
Slater v Solicitors Regulation Authority [2012] EWHC 3256 (Admin) .......... 27.13
Sloan v General Medical Council [1970] 1 W.L.R. 1130; [1970] 2 All E.R. 686; (1970) 114 S.J. 514 .......... 22.03, 36.07
Smith v General Teaching Council for England *See* R (Smith) v General Teaching Council for England
Smith v Linskill [1996] 1 W.L.R. 763; [1996] 2 All E.R. 353; (1996) 146 N.L.J. 209; (1996) 140 S.J.L.B. 49; Times, February 7, 1996 .......... 14.04, 14.06
Society of Accountants in Edinburgh v Corporation of Accountants (1843) 20R 750 .......... 56.18
Sohal v Solicitors Regulation Authority [2014] EWHC 1613 (Admin) .......... 4.48, 25.41
Sokunbi v Health Professions Council [2013] EWHC 672 (Admin) .......... 4.30
Solicitor, ex parte Incorporated Law Society, Re a [1903] 1 KB 857; [1903] 2 KB 205 .......... 43.05
Solicitor, ex parte Law Society, In re a [1912] 1 KB 302 .......... 47.04
Solicitor, In the Matter of a, CO/2504/2000, 20 November 2000 .......... 24.01
Solicitor, In re a [1945] 1 All E.R. 445 .......... 62.08
Solicitor, In re a [1956] 1 W.L.R. 1312; [1956] 3 All E.R. 516; (1956) 100 S.J. 787 .......... 14.18
Solicitor, In re a [1972] 1 W.L.R. 869; [1972] 2 All E.R. 811; (1972) 116 S.J. 275 .......... 47.12, 49.04
Solicitor, In re a [1975] Q.B. 475; [1975] 2 W.L.R. 105; [1974] 3 All E.R. 853; (1974) 118 S.J. 737 .......... 47.13
Solicitor, In re a [1993] Q.B. 69; [1992] 2 W.L.R. 552; [1992] 2 All E.R. 335; (1991) 141 N.L.J. 1447; Times, September 24, 1991; Independent, July 12, 1991 .......... 25.04, 25.19, 62.07, 62.08, 62.09, 62.15
Solicitor, No 20. of 2008, Odunlami, Re a [2008] EWCA Civ 1598 .......... 11.06
Solicitor, No 4 of 2009, Afsar [2009] EWCA Civ 842 .......... 56.07
Solicitor, re a [2001] NIQB 52 .......... 11.20
Solicitor, re a [2014] NIQB 46 .......... 11.20
Solicitor v Law Society of New Brunswick [2004] NBQB 95 .......... 24.03
Solicitors, ex parte Peasegood, Re [1994] 1 All E.R. 298; (1993) 143 N.L.J. 778; [1993] N.P.C. 76; Times, May 6, 1993; Independent, May 6, 1993 .......... 43.05
Solicitors Nos 21 and 22 of 2007, Ali and Naeem, Re [2008] EWCA Civ 769 .......... 56.06
Solicitors Regulation Authority v Ali [2013] EWHC 2584 (Admin) .......... 58.14
Solicitors Regulation Authority v Anderson Solicitors [2013] EWHC 4021 (Admin) .......... 15.19, 22.36, 28.20
Solicitors Regulation Authority v Davis and McGlinchey [2011] EWHC 232 (Admin) .......... 15.13, 15.17, 15.20, 28.17
Solicitors Regulation Authority v Dennison [2012] EWCA Civ 421; (2012) 162 N.L.J. 542 .......... 19.64
Solicitors Regulation Authority v Kaberry, 30 October 2012 (unreported) .......... 58.11
Solicitors Regulation Authority v Sharma [2010] EWHC 2022 (Admin) .......... 19.57, 19.69, 19.72, 44.24
Solicitors Regulation Authority v Solicitors Disciplinary Tribunal *See* R (Solicitors Regulation Authority) v Solicitors Disciplinary Tribunal

Solicitors Regulation Authority v Spence [2012] EWHC 2977 (Admin) . . . . . . . . . . . . . 19.66
Solicitors Regulation Authority v Uddin [2014] EWHC 4553 (Admin) . . . . . . . . . . . . . 65.16
Soni v General Medical Council [2015] EWHC 364 (Admin) . . . . . . . . . . . . . . . 19.81, 50.13
Sosanya v General Medical Council *See* R (Sosanya) v General Medical Council
South Bucks District Council v Porter (No. 2) [2004] 1 W.L.R. 1953 . . . . . . . . . . . . . . . 54.30
Southall v General Medical Council [2010] EWCA Civ 407; [2010] Med.
L.R. 252; (2010) 154(18) S.J.L.B. 30 . . . . . . . . . . . . . . . . 4.21, 4.29, 13.06, 13.07, 26.08,
27.05, 27.06, 27.09, 54.17, 54.32
Spencer v General Osteopathic Council [2012] EWC 3146 (Admin) . . . . . . . . . . . . . . . 47.47
Spencer v Maryland State Board of Pharmacy (2003) 846 A 2d 341, 352
(Maryland Court of Appeal) . . . . . . . . . . . . . . . . . . . . . . . . . . . . . . . . . . . . . . . . . . . 24.03
Spiers (Procurator Fiscal) v Ruddy [2007] UKPC D2; [2008] 1 A.C. 873; [2008]
2 W.L.R. 608; 2009 S.C. (P.C.) 1; 2008 S.L.T. 39; 2008 S.C.L. 424; 2008
S.C.C.R. 131; [2008] H.R.L.R. 14; 26 B.H.R.C. 567; 2007 G.W.D. 40-700;
Times, December 31, 2007 . . . . . . . . . . . . . . . . . . . . . . . . . . . . . . . . . . . . . . . . . . 17.05
Spofforth v General Dental Council, Privy Council Appeal No.5 of 1999 . . . . . . . . . . . . . 3.05
Squier v General Medical Council *See* R (Squier) v General Medical Council
Sreenath v General Medical Council [2002] UKPC 56 . . . . . . . . . . . . . . . . . . . . . . . . . 31.05
Srirangalingham v General Medical Council [2002] UKPC 77 . . . . . . . . . . . . . . . . . . 59.15
Stannard v General Council of the Bar, 24 January 2006 (unreported) . . . . . . . . . . . . . . 14.06
Stansbury v Datapulse Plc [2003] EWCA Civ 1951; [2004] I.C.R. 523; [2004]
I.R.L.R. 466; [2004] U.K.H.R.R. 340; (2004) 101(6) L.S.G. 32; (2004)
148 S.J.L.B. 145; Times, January 28, 2004 . . . . . . . . . . . . . . . . . . . . . . . . . . . . . . . 48.12
Stefan v General Medical Council [1999] 1 W.L.R. 1293; [2000] H.R.L.R. 1;
6 B.H.R.C. 487; [1999] Lloyd's Rep. Med. 90; (1999) 49 B.M.L.R. 161;
(1999) 143 S.J.L.B. 112; Times, March 11, 1999 . . . . . . . . . . . . . . . . 31.03, 54.02, 54.04
Stefan v United Kingdom (1998) 25 EHRR 130 . . . . . . . . . . . . . . . . . . . . . . . . . . . . . 32.07
Stewart v Secretary of State for Scotland 1998 SC (HL) 81 . . . . . . . . . . . . . . . . . . . . . 41.02
Stock v Central Midwives Board [1915] 3 KB 756 . . . . . . . . . . . . . . . . . . . . . . . . . . . 25.34
Stokes v Law Society [2001] EWHC Admin 1101 . . . . . . . . . . . . . . . . . . . . . . . . . . . 18.05
Straker v Graham 150 E.R. 1612; (1839) 4 M&W 721 . . . . . . . . . . . . . . . . . . . . . . . . 34.03
Strouthos v London Underground Ltd [2004] EWCA Civ 402 . . . . . . . . . . . . . . . . . . . 22.34
Subner v Health Professions Council *See* R (Subner) v Health Professions Council
Subramanian v General Medical Council [2002] UKPC 64 . . . . . . . . . . . 22.08, 51.03, 51.04
Sukul v Bar Standards Board [2014] EWHC 3532 (Admin) . . . . . . . . . . . . . . . . . . . . . 60.61
Sultan v Bar Standards Board, 6 November 2013 (unreported) . . . . . . . . . . . . . . . . . . . 47.56
Sultan v General Medical Council [2013] EWHC 1518 (Admin) . . . . . . . . . . . . . . . . . 60.51
Sumukan Ltd v Commonwealth Secretariat (No. 2) [2007] EWCA Civ 1148 . . . . . . . . . 13.15
Surrey Police Authority v Beckett [2001] EWCA Civ 1253; [2002] I.C.R. 257;
[2001] Emp. L.R. 1157; [2001] Po. L.R. 305; (2001) 151 N.L.J. 1408; Times,
August 8, 2001 . . . . . . . . . . . . . . . . . . . . . . . . . . . . . . . . . . . . . . . . . . . . . . . . . . . 43.08
Sutherland-Fisher v Law Society of Scotland 2003 SC 562 . . . . . . . . . . . . . . . . . . . . . 67.12
Swain v The Law Society [1983] 1 A.C. 598; [1982] 3 W.L.R. 261; [1982]
2 All E.R. 827; (1982) 79 L.S.G. 887; (1982) 126 S.J. 464 . . . . . . . . . . . . . . . . . . . 42.06
Swanney v General Medical Council 2008 SC 592 . . . . . . . . . . . . . 21.07, 24.02, 25.10, 43.10

Tait v Royal College of Veterinary Surgeons [2003] UKPC 34; (2003)
147 S.J.L.B. 536 . . . . . . . . . . . . . . . . . . . . . . . . . . . . . . . 1.03, 1.08, 19.41, 19.42, 19.79
Tariquez-Zaman v General Medical Council [2006] UKEAT 0292/06
and 0517/06 . . . . . . . . . . . . . . . . . . . . . . . . . . . . . . . . . . . . . . . . . . . . . . . . . . . . 39.39
Tarnesby v General Medical Council, Privy Council Appeal No. 21 of 1969 . . . . . . . . . . 4.03
Taylor v Director of the Serious Fraud Office [1999] 2 A.C. 177; [1998] 3 W.L.R.
1040; [1998] 4 All E.R. 801; [1999] E.M.L.R. 1; Times, November 4, 1998;
Independent, November 3, 1998 . . . . . . . . . . . . . . . . . . . . . . . . . . . . . . . . . . . . . . 52.12
Taylor v General Chiropractic Council [2009] EWHC 301 (Admin) . . . . . . . . . . . . . . . 56.08
Taylor v General Medical Council [1990] 2 A.C. 539; [1990] 2 W.L.R. 1423; [1990] 2 All
E.R. 263; [1990] 2 Med. L.R. 45; (1990) 87(20) L.S.G. 36; (1990) 134 S.J. 757 . . . . . 65.02

Taylor v General Medical Council [2010] EWHC 984 (QB) . . . . . . . . . . . . . . . . . . . . . 52.11
Tehrani v United Kingdom Central Council for Nursing, Midwifery and Health Visiting 2001 S.C. 581; 2001 S.L.T. 879; [2001] I.R.L.R. 208; 2001 G.W.D. 4-165 . . . . . . . . . . . . . . . . . . . . . . . . . . 32.10, 32.16, 32.28, 35.03, 35.13, 44.11
Teinaz v Wandsworth London Borough Council [2002] EWCA Civ 1040; [2002] ICR 1471; [2002] I.R.L.R. 721; [2002] Emp. L.R. 1107; (2002) 99(36) L.S.G. 38; Times, August 21, 2002 . . . . . . . . . . . . . . . . . . . . . . . . . . . . . . . . . . . 3.02
Temblett v Bar Standards Board, 26 July 2012 (unreported) . . . . . . . . . . . . . . . . . . . . 13.14
Thaker v Solicitors Regulation Authority [2011] EWHC 660 (Admin) . . . . 3.15, 22.26, 25.11
Therrien v Canada (Minister for Justice) [2001] 2 SCR 3 . . . . . . . . . . . . . . . . . . . . . . . . 41.04
Thiruvengadam v General Medical Council , 3 November 2010, GIL 7983 (NI) . . . . . . 67.05
Thobani v Solicitors Regulation Authority [2011] EWHC 3783 (Admin) . . . . . . . 58.10, 58.11
Thomas v Council of the Law Society of Scotland 2006 SLT 183 . . . . . . . . . . . . . . . . . 21.06
Thomas v Thomas [1947] A.C. 484; [1947] 1 All E.R. 582; 1947 S.C. (H.L.) 45; 1948 S.L.T. 2; 1947 S.L.T. (Notes) 53; 63 T.L.R. 314; [1948] L.J.R. 515; 176 L.T. 498 . . . . . . . . . . . . . . . . . . . . . . . . . . . . . . . . . . . . . . . . . . 4.07, 27.02, 27.04
Thompson v Law Society *See* R (Thompson) v Law Society
Thompson v Professional Conduct Committee of the General Chiropractic Council *See* R (Thompson) v Professional Conduct Committee of the General Chiropractic Council
Thomson v Kvaerner Govan Ltd 2004 SC (HL) 6 . . . . . . . . . . . . . . . . . . . . . . . . . . . . 27.04
Thorneycroft v Nursing and Midwifery Council [2014] EWHC 1565 (Admin) . . . . . . . 25.31
Thrasyvoulou v Secretary of State for the Environment [1990] 2 A.C. 273; [1990] 2 W.L.R. 1; [1990] 1 All E.R. 65; 88 L.G.R. 217; (1990) 2 Admin. L.R. 289; (1990) 59 P. & C.R. 326; [1990] 1 P.L.R. 69; [1990] 13 E.G. 69; (1990) 154 L.G. Rev. 192; [1989] E.G. 178 (C.S.) . . . . . . . . . . . . . . . . . . . . . . . . 24.03, 24.05
Three Rivers District Council v Governor and Company of the Bank of England [2006] EWHC 816 (Comm) . . . . . . . . . . . . . . . . . . . . . . . . . . . . . . . . . . . 15.30, 49.11
Threlfall v General Optical Council [2004] EWHC 2683 (Admin); [2005] Lloyd's Rep. Med. 250; [2005] A.C.D. 70; (2004) 101(48) L.S.G. 25; Times, December 2, 2004 . . . . . . . . . . . . . . . . . . . . . . . . . . . 4.31, 4.48, 25.35, 32.13, 54.08
Tinkler v Solicitors Regulation Authority [2012] EWHC 3645 (Admin) . . . . . . . . . . . . 28.15
Tinsa v General Medical Council *See* R (Tinsa) v General Medical Council
Tosounides v General Medical Council [2012] All E.R. (D) 206 (Jul) . . . . . . . . . . . . . . 38.29
Townrow v Financial Services Authority, FS/2012/0007, 10 January 2013 . . . . . . 58.13, 59.19
Turner v Nursing and Midwifery Council [2014] EWHC 520 (Admin); [2014] Med LR 205 . . . . . . . . . . . . . . . . . . . . . . . . . . . . . . . . . . . . . . . . . . . . . . . . . . 25.30
Tutin v General Medical Council *See* R (Tutin) v General Medical Council
Twinsectra Ltd v Yardley [2002] UKHL 12; [2002] 2 A.C. 164; [2002] 2 W.L.R. 802; [2002] 2 All E.R. 377; [2002] P.N.L.R. 30; [2002] W.T.L.R. 423; [2002] 38 E.G. 204 (C.S.); (2002) 99(19) L.S.G. 32; (2002) 152 N.L.J. 469; (2002) 146 S.J.L.B. 84; [2002] N.P.C. 47; Times, March 25, 2002 . . . . . . . 19.03, 19.05, 19.08, 19.12, 54.29, 55.01, 62.18
Two Solicitors, Re [1938] 1 KB 616 . . . . . . . . . . . . . . . . . . . . . . . . . . . . . . . . . . . . . . 43.05

U (A Child) (Department for Education and Skills intervening) [2004] EWCA Civ 567; [2005] Fam. 134; [2004] 3 W.L.R. 753; [2004] 2 F.L.R. 263; [2004] 2 F.C.R. 257; [2004] Fam. Law 565; (2004) 101(22) L.S.G. 31; (2004) 154 N.L.J. 824; Times, May 27, 2004; Independent, May 18, 2004 . . . . . . . . . . . . . . . . . . . . . . . . . . . . . 62.04
Uddin v General Medical Council [2012] EWHC 1763 (Admin) . . . . . . . . . . . . . . . . . 33.18
Uddin v General Medical Council [2012] EWHC 2669 (Admin) . . . 4.28, 19.19, 19.25, 19.31
Uddin v General Medical Council [2013] ICR 793 . . . . . . . . . . . . . . . . . . . . . . . . . . . 39.39
Udom v General Medical Council [2009] EWHC 3242 (Admin); [2010] Med. L.R. 37; (2010) 112 B.M.L.R. 47 . . . . . . . . . . . . . . . . . . 11.07, 11.10, 46.05, 59.11
Ujam v General Medical Council [2012] EWHC 580 (Admin) . . . . . . . . . . . . . . . . . . 33.19
Ujam v General Medical Council [2012] EWHC 683 (Admin) . . . . . . . . . . . . . . . 32.18, 60.15
United Arab Emirates v Abdelghafar [1995] ICR 65 . . . . . . . . . . . . . . . . . . . . . . . 4.40, 4.44

Uruakpa v General Medical Council *See* R (Uruakpa) v General Medical Council
Uruakpa v General Medical Council [2010] EWHC 1302 (Admin); (2010) 115 B.M.L.R. 52, 16.09done . . . . . . 59.14

Vaghela v General Medical Council [2013] EWHC 1594 (Admin) . . . . . . 67.08
Vaidya v General Medical Council [2010] EWHC 984 (QB) . . . . . . 49.08, 52.11
Vali v General Optical Council *See* R (Vali) v General Optical Council
Van Marle v Netherlands (1986) 8 EHRR 486 . . . . . . 32.07, 32.18
Vane, No 1687/2002 . . . . . . 58.10
Varley v General Osteopathic Council [2009] EWHC 1703 (Admin) . . . . . . 60.39
Varma v General Medical Council [2008] EWHC 753 (Admin); [2008] LS Law Medical 313; (2008) 102 B.M.L.R. 84; (2008) 152(18) S.J.L.B. 30 . . . . . . 1.06
Veen v Bar Standards Board [2001] All ER (D) 223 (Oct) . . . . . . 18.16
Veizi v General Dental Council *See* R (Veizi) v General Dental Council
Venton v Solicitors Regulation Authority [2010] EWHC 1377 (Admin) . . . . . . 56.12
Vermeulen v Belgium (1996) 32 EHRR 313 . . . . . . 46.04
Vidya v General Medical Council [2007] EWHC 1497 (Admin) . . . . . . 3.10
Virdee v General Pharmaceutical Council [2015] EWHC 169 (Admin) . . . . . . 54.33, 70.12
Virdi v Law Society (Solicitors Disciplinary Tribunal intervening) [2010] EWCA Civ 100; [2010] 1 W.L.R. 2840; [2010] 3 All E.R. 653; [2010] A.C.D. 38; (2010) 107(9) L.S.G. 14 . . . . . . 6.10, 35.11, 43.15
Visser v Association of Professional Engineers and Geoscientists 2005 BCSC 1402 . . . . . . 24.03
Visser and Fagbulu v Financial Services Authority [2011] Lloyd's Rep. F.C. 551; [2011] All E.R. (D) 57 (Oct) . . . . . . 28.05, 28.12, 28.24
Vranicki v Architects Registration Board [2007] EWHC 506 (Admin); [2007] 13 E.G. 254 (C.S.) . . . . . . 47.36
Vukelic v Financial Services Authority, 13 March 2009 . . . . . . 19.21

W (Minors) (Sexual Abuse: Standard of Proof) [1994] 1 FLR 419; [1994] 2 F.C.R. 759; [1994] Fam. Law 427; Times, December 1, 1993 . . . . . . 62.03
WR v Austria (2001) 31 EHRR 43 . . . . . . 17.20, 32.13
Waghorn v General Medical Council [2014] EWHC 1214 (Admin) . . . . . . 1.19, 38.19
Wagner v Financial Services Authority [2012] FS/2011/0015 . . . . . . 28.13
Wakefield v Channel 4 Television Corpn [2006] EWHC 3289 (QB); (2007) 94 B.M.L.R. 1; (2007) 104(5) L.S.G. 28 . . . . . . 18.11, 18.18
Walker v Bar Standards Board, 19 September 2013 (unreported) . . . . . . 47.55
Walker v General Medical Council *See* R (Walker) v General Medical Council
Walker v General Medical Council [2002] UKPC 57; (2003) 71 B.M.L.R. 53; Times, December 16, 2002; Independent, November 14, 2002 . . . . . . 44.14
Walker v General Medical Council [2010] EWHC 3849 (Admin) . . . . . . 53.05
Walker v Royal College of Veterinary Surgeons [2007] UKPC 64 . . . . . . 19.49
Walker v Royal College of Veterinary Surgeons [2008] UKPC 20 . . . . . . 15.23
Walker v Secretary of State for Education [2014] EWHC 267 (Admin) . . . . . . 60.56
Walker-Smith v General Medical Council [2012] EWHC 503 (Admin); (2012) 126 B.M.L.R. 1; (2012) 109(12) L.S.G. 20 . . . . . . 26.09, 47.45
Wallace v The Queen [1997] 1 Cr. App. R. 396; [1997] Crim. L.R. 356; (1997) 141 S.J.L.B. 12; Times, December 31, 1996 . . . . . . 54.05
Ward v Nursing and Midwifery Council [2014] EWHC 1158 (Admin) . . . . . . 1.20, 4.47
Warnik v Townend [1979] AC 731 . . . . . . 56.18
Warren v Attorney General for Jersey [2011] UKPC 10; [2012] 1 A.C. 22; [2011] 3 W.L.R. 464; [2011] 2 All E.R. 513; [2011] 2 Cr. App. R. 29; Times, March 29, 2011 . . . . . . 2.04
Wasu v General Dental Council [2013] EWHC 3782 (Admin) . . . . . . 4.29
Watson v General Medical Council [2005] EWHC 1896 (Admin); [2006] I.C.R. 113; [2005] Lloyd's Rep. Med. 435; (2005) 86 B.M.L.R. 152; (2005) 155 N.L.J. 1356; Time, October 7, 2005 . . . . . . 46.04

Watson v General Medical Council [2006] EWHC 18 (Admin); (2006) 91 B.M.L.R. 162; [2006] A.C.D. 42 . . . . . . . . 25.36, 27.09, 54.11
Webb v Solicitors Regulation Authority [2013] EWHC 2078 (Admin) . . . . . . . . 19.23
Webster v General Teaching Council [2012] EWHC 2928 (Admin) . . . . . . . . 4.23, 27.12
Wendenburg v Germany (2003) 2 Reports of Judgments and Decisions 347 . . . . . . . . 32.18
Wentzel v General Medical Council [2004] EWHC 381 (Admin); (2005) 82 B.M.L.R. 127 . . . . . . . . 60.30
Westcott Financial Services Ltd v Financial Ombudsman Service [2014] EWHC 3972 (Admin) . . . . . . . . 10.17
Westcott v Westcott [2008] EWCA Civ 818; [2009] Q.B. 407; [2009] 2 W.L.R. 838; [2009] 1 All E.R. 727; [2009] E.M.L.R. 2; (2008) 105(30) L.S.G. 17; Times, August 27, 2008 . . . . . . . . 52.15
White v Nursing and Midwifery Council [2014] EWHC 520 (Admin); [2014] Med LR 205 . . . . . . . . 25.30
White v Southampton University Hospitals NHS Trust [2011] EWHC 825 (QB); [2011] Med. L.R. 296; (2011) 120 B.M.L.R. 81 . . . . . . . . 52.11
Whitehart v Raymond Thompson Limited, 11 September 1984, unreported . . . . . . . . 48.12
Whitehouse v Jordan [1981] 1 W.L.R. 246; [1981] 1 All E.R. 267; (1981) 125 S.J. 167 . . . . . . . . 49.02
Wickramsinghe v United Kingdom (1998) EHRR 338 . . . . . . . . 32.08, 32.12, 54.05, 54.06
Willford v Financial Services Authority *See* R (Willford) v Financial Services Authority
Williams v Bar Standards Board, 31 October 2014 (unreported) . . . . . . . . 15.29
Williams v Royal College of Veterinary Surgeons [2008] UKPC 39 . . . . . . . . 19.50
Wisson v Health Professions Council [2013] EWHC 1036 (Admin) . . . . . . . . 30.05, 40.05
Wood v Law Society [1995] N.P.C. 39; Times, March 2, 1995; Independent, March 1, 1995 . . . . . . . . 49.05
Wood v Secretary of State for Education *See* R (Wood) v Secretary of State for Education
Woodman-Smith v Architects Registration Board [2014] EWHC 3639 (Admin) . . . . . . . . 43.21, 56.31
Woods and Gordon v Chief Constable of Merseyside Police *See* R (Woods and Gordon) v Chief Constable of Merseyside Police [2014] EWHC 2784 (Admin)
Woods v General Medical Council [2002] EWHC 1484 (Admin) . . . . . . . . 39.07, 39.12
Woolgar v Chief Constable of Sussex Police [2000] 1 W.L.R. 25; [1999] 3 All E.R. 604; (2000) 2 L.G.L.R. 340; [1999] Lloyd's Rep. Med. 335; (1999) 50 B.M.L.R. 296; (1999) 96(23) L.S.G. 33; (1999) 149 N.L.J. 857; (1999) 143 S.J.L.B. 195; Times, May 28, 1999 . . . . . . . . 18.18, 18.20, 18.21
Woolmington v The Director of Public Prosecutions [1935] A.C. 462; (1936) 25 Cr. App. R. 72 . . . . . . . . 7.04
Wright v Secretary of State for Health *See* R (Wright) v Secretary of State for Health

X v General Medical Council *See* R (X) v General Medical Council
X v United Kingdom (1984) 6 EHRR 136 . . . . . . . . 32.11, 32.16, 45.09

Y v General Medical Council [2013] EWHC 860 (Admin) . . . . . . . . 38.35
Yaacoub v General Medical Council [2012] EWHC 2279 (Admin) . . . . . . . . 27.14, 54.23
Yanah v General Medical Council [2006] EWHC 3843 (Admin); [2007] LS Law Medical 143 . . . . . . . . 31.09
Yeong v General Medical Council [2009] EWHC 1923 (Admin); [2010] 1 W.L.R. 548; [2009] LS Law Medical 582; (2009) 110 B.M.L.R. 125; Times, August 25, 2009 . . . . . . . . 13.08, 14.35, 26.06, 32.11, 32.18, 33.11, 33.15, 33.19, 54.13, 54.14
Yerolemou v Law Society [2008] EWHC 682 (Admin); (2008) 152(12) S.J.L.B. 31 . . . . 60.35
Yiacoub v The Queen [2014] UKPC 22; [2014] 1 W.L.R. 2996; Times, July 23, 2014 . . . . . . 6.17
Yousef v Solicitors Regulation Authority, 23 January 2015 (unreported) . . . . . . . . 56.34

Yuill v Yuill [1945] P. 15; [1945] 1 All E.R. 183 .................................. 4.07
Yusuf v Royal Pharmaceutical Society of Great Britain [2009] EWHC 867
(Admin) .................................................... 1.08, 67.04

Zia v General Medical Council [2011] EWCA Civ 743; [2012] 1 W.L.R. 504;
[2012] I.C.R. 146 .......................................... 16.12, 39.22
Ziderman v General Dental Council [1976] 1 W.L.R. 330; [1976] 2 All E.R. 334;
(1975) 120 S.J. 48; Times, December 10, 1975 ......... 5.02, 14.19, 19.35, 21.04, 21.11
Zygmunt v General Medical Council [2008] EWHC 2643 (Admin); [2009]
LS Law Medical 219 .................................. 33.04, 33.05, 33.10

# TABLE OF LEGISLATION

# TABLE OF STATUTORY INSTRUMENTS

Care Homes Regulations 2001, SI 2001/3965 . . . . . . . . . . . . . . . . 56.04
Chartered Certified Accountants Complaints and Disciplinary Regulations 2014, reg 11(2)(d). . . . . . . . . . . . . . . . . . .25.19
Civil Procedure Rules, SI 1998/3132 . . . . . . . . . . . 15.07, 50.05
r.3.1 . . . . . . . . . . . . . . . . . . . . . . . . 4.33
r.3.1(2)(a) . . . . . . . . . . . . . . . . . . . . . 4.32
r.3.9 . . . . . . . . . . . . . . . . . . . . . . . . 4.38
r.3.9(a) . . . . . . . . . . . . . . . . . . . . . . 4.38
r.3.9(e) . . . . . . . . . . . . . . . . . . . . . . 4.38
r.3.11 . . . . . . . . . . . . . . . . . . . . . . . 9.01
pt.6 . . . . . . . . . . . . . . . . . . . . . . . . 38.20
PD 6B . . . . . . . . . . . . . . . . . . . . . . 61.06
pt.8 . . . . . . . . . . . . . . .18.25, 38.03, 38.13
r.8.5(1) . . . . . . . . . . . . . . . . . . . . . . 38.03
r.8.6. . . . . . . . . . . . . . . . . . . . . . . . 38.09
pt.22 . . . . . . . . . . . . . . . . . . . . . . .47.45
r.25.15 . . . . . . . . . . . . . . . . . . . . . .15.18
r.31.4 . . . . . . . . . . . . . . . . . . . . . . . 70.01
r.34.4. . . . . . . . . . . . . . . . . . 18.10, 70.02
pt.35 . . . . . . . . . . . . . . . . . . . . . . . 26.01
r.39.2(3). . . . . . . . . . . . . . . . . . . . . 53.03
pt.52. . . . . . . . . . . . . . . .4.01, 25.37, 38.03
r.52.9 . . . . . . . . . . . . . . . . . . . . . . . 1.21
r.52.9(1)(c) . . . . . . . . . . . . . . . . . . . .15.18
r.52.11 . . . .4.12, 4.21, 25.37, 54.27, 60.63
r.52.11(1) . . . . . . . . 4.15, 4.26, 4.31, 19.51
r.52.11(3) . . . . . . . . . . . . .4.20, 6.17, 25.37
r.52.13 . . . . . . . . . . . . . . . . . . . . . . 4.14
pt 52 PD . . . . . . . . . . . . . . . . . . . . . 4.01
pt 52 PD 22.3(3) . . . . . . . . . . . . . . . . 4.26
pt 52 PD 116(2) . . . . . . . . . . . . . . . . . 4.18
r.53.11(3) . . . . . . . . . . . . . . . . . . . . 4.31
pt.54 . . . . . . . . . . . . . . . . . . 42.01, 42.13
r.54.1 . . . . . . . . . . . . . . . . . . . . . . 42.13
r.54.7 . . . . . . . . . . . . . . . . . . . . . . 38.20
r.54.52 . . . . . . . . . . . . . . . . . . . . . 42.14
Criminal Procedure Rules 2014, SI 2014/2610
Pt. 33 . . . . . . . . . . . . . . . . . . . . . . 26.01
Family Procedure Rules 2010, SI 2010/2955
Pt. 25 . . . . . . . . . . . . . . . . . . . . . . 26.01
Financial Services and Markets Act 2000 (Service of Notices) Regulations 2001 (SI 2001/1420) . . . . . . . . . . . . . . 43.19
General Medical Council (Constitution of Panels and Investigation Committee) Rules Order of Council, SI 2004/2611. . . . 13.05, 13.11
General Medical Council Professional Conduct Committee (Procedure) Rules 1998 . . . . . . . . . . . . . . . . . . 47.29
General Medical Council (Professional Performance) Rules 1997, SI 1997/1529. . . . . . . 16.05
r 12(3) . . . . . . . . . . . . . . . . . . . . . . 61.02
Gibraltar Constitution Order 2007
Art 64(2) . . . . . . . . . . . . . . . . . . . . 41.02
Art 64(3) . . . . . . . . . . . . . . . . . . . . 41.02
Art 64(4) . . . . . . . . . . . . . . . . . . . . 41.02
Health and Social Work Professions Order 2001 . . . . . . . . . . . . . . . . . 43.09
Art 5 . . . . . . . . . . . . . . . . . . . . . . . 58.05
Art 22 . . . . . . . . . . . . . . . . . . 31.01, 33.01
Art 29 . . . . . . . . . . . . . . . . . . 11.01, 60.01
Art 31 . . . . . . . . . . . . . . . . . . . . . . 38.01
Art 33 . . . . . . . . . . . . . . . . . . . . . . 58.01
Art 35 . . . . . . . . . . . . . . . . . . . . . . 46.01
Health and Social Work Professions Order 2001 (Legal Assessors) Order of Council 2003, SI 2003/1578 . . . . . . . . . . . . . . . . 44.01
Judicial Committee (Medical Rules) (No.2) Order 1971. . . . . . . . . . . . 44.08
Judicial Discipline (Prescribed Procedures) Regulations 2014, SI 2014/1919 . . . 41.01
Medical Act 1983 (Amendment) Order 2002, SI 2002/3135
Art 13 . . . . . . . . . . . . . . . . . . . . . . 39.07
National Health Service (Performers Lists) Regulations 2004, SI 2004/585 . . . . . . . . . . . . . . . . . 32.18
reg 10. . . . . . . . . . . . . . . . . . . . . . . .45.13
Nursing and Midwifery Order 2001, SI 2001/253 . . . . . . . . . . . . . 9.04, 43.12
Art 22 . . . . . . . . . . . . 16.01, 31.01, 33.01
Art 25 . . . . . . . . . . . . . . . . . . . . . . 18.01
Art 27 . . . . . . . . . . . . . . . . . . . . . . 38.13
Art 29 . . . . . . . . .4.41, 11.01, 60.01, 65.01
Art 29(6) . . . . . . . . . . . . . . . . . . . . .16.16
Art 29(9) . . . . . . . . . . . . . . . . . . . . . 4.40
Art 29(10) . . . . . . . . . . . . . . . . .4.32, 4.40
Art 31 . . . . . . . . . . . . . . . . . . . . . . 38.01
Art 31(1)(c). . . . . . . . . . . . . . . . . . . 38.71
Art 31(12) . . . . . . . . . . . . . . . . . . . 38.13
Art 33 . . . . . . . . . . . . . . . . . . . . . . 58.01
Art 35 . . . . . . . . . . . . . . . . . . 16.01, 46.01
Art 38 . . . . . . . . . . . . . . . . . . . . . . 58.08

# TABLE OF INTERNATIONAL TREATIES AND CONVENTIONS



# TABLE OF HOME OFFICE CIRCULARS

# TABLE OF PROFESSIONAL RULES AND GUIDANCE

# LIST OF ABBREVIATIONS

| | |
|---|---|
| A&E | accident and emergency |
| ABTA | Association of British Travel Agents |
| ACCA | Association of Chartered Certified Accountants |
| APER | (Financial Services Authority Handbook) Statements of Principle and Code of Practice for Approved Persons |
| ARB | Architects Registration Board |
| ARDL | Association of Regulatory and Disciplinary Lawyers |
| BBC | British Broadcasting Corporation |
| BPS | British Psychological Society |
| BSB | Bar Standards Board |
| BSc | Bachelor of Science |
| BSE | bovine spongiform encephalopathy |
| CAA | Civil Aviation Authority |
| CBI | Confederation of British Industry |
| CCC | Competence and Conduct Committee |
| CEO | chief executive officer |
| CFO | chief finance officer |
| CHRE | Council for Healthcare Regulatory Excellence |
| CILEX | Chartered Institute of Legal Executives |
| CIMA | Chartered Institute of Management Accountants |
| CIPFA | Chartered Institute of Public Finance and Accountancy |
| CNEP | continuous negative extrathoracic pressure ventilation |
| CPD | continuous professional development |
| CPP | (GMC) Committee on Professional Performance |
| CPR | Civil Procedure Rules |
| CPS | Crown Prosecution Service |
| CQC | Care Quality Commission |
| CRB | Criminal Records Bureau |
| CVA | company voluntary arrangement |
| Defra | Department of the Environment, Food and Rural Affairs |
| DEPP | (Financial Services Authority Handbook) Decision Procedure and Penalties Manual |
| DPP | Director of Public Prosecutions |
| EAT | Employment Appeal Tribunal |
| ECHR | European Convention on Human Rights |
| ECT | electroconvulsive therapy |
| ECtHR | European Court of Human Rights |
| ENT | ear, nose, and throat |
| ERC | European Rugby Cup |
| FA | The Football Association |
| FCA | Financial Conduct Authority |
| FHSAA | Family Health Services Appeal Authority |
| FOIA | Freedom of Information Act 2000 |
| FOS | Financial Ombudsman Service |
| FRC | Financial Reporting Council |
| FSA | Financial Services Authority |
| GBGB | Greyhound Board of Great Britain |
| GCC | General Chiropractic Council |

| | |
|---|---|
| GDC | General Dental Council |
| GHB | gamma-Hydroxybutyric acid |
| GMC | General Medical Council |
| GOC | General Optical Council |
| GOsC | General Osteopathic Council |
| GP | general practitioner |
| GPhC | General Pharmaceutical Council |
| HCPC | Health and Care Professions Council |
| HMRC | Her Majesty's Revenue and Customs |
| HPC | Health Professions Council |
| IAJLJ | International Association of Jewish Lawyers and Jurists |
| ICAS | Institute of Chartered Accountants in Scotland |
| ICAEW | Institute of Chartered Accountants in England and Wales |
| ICD | (WHO) International Classification of Diseases |
| IPCC | Independent Police Complaints Commission |
| ISA | Independent Safeguarding Authority |
| JDS | Joint Disciplinary Scheme |
| JLSC | (Trinidad and Tobago) Judicial and Legal Service Commission |
| JP | Justice of the Peace |
| LLP | limited liability partnership |
| LME | London Metal Exchange |
| LSE | London Stock Exchange |
| LVI | local veterinary inspector |
| MBBS | Bachelor of Medicine, Bachelor of Surgery |
| MMR | measles, mumps, and rubella |
| MPTS | (GMC) Medical Practitioners Tribunal Service |
| MRCS | Membership of the Royal College of Surgeons |
| MRI | magnetic resonance imaging |
| NCAS | National Clinical Assessment Service |
| NHS | National Health Service |
| NMC | Nursing and Midwifery Council |
| OIA | Office of the Independent Adjudicator |
| OSCE | objective structured clinical examination |
| OSS | Office for the Supervision of Solicitors |
| PCA | Police Complaints Authority |
| PCC | (GMC) Professional Conduct Committee |
| PCT | primary care trust |
| PLAB | (GMC) Professional and Linguistics Board |
| PNC | Police National Computer |
| POVA | Protection of Vulnerable Adults |
| PPC | (GMC) Preliminary Proceedings Committee |
| PSA | Professional Standards Authority for Health and Social Care |
| QC | Queen's Counsel |
| RCVS | Royal College of Veterinary Surgeons |
| RFC | Rugby Football Club |
| RFU | Rugby Football Union |
| RICS | Royal Institute of Chartered Surveyors |
| RSC | Rules of the Supreme Court |
| SDT | Solicitors Disciplinary Tribunal |
| SEC | (US) Securities and Exchange Commission |
| SFA | Securities and Futures Authority |
| SFO | Serious Fraud Office |

| | |
|---|---|
| SHO | senior house officer |
| SIDS | sudden infant death syndrome |
| SLCC | Scottish Legal Complaints Commission |
| SOP | (NMC) Standard Operating Procedure |
| SRA | Solicitors Regulation Authority |
| SSDT | Scottish Solicitors' Discipline Tribunal |
| UCH | University College Hospital |
| UCL | University College London |
| UKCC | United Kingdom Central Council for Nursing, Midwifery and Health Visiting |
| VAT | value added tax |
| WHO | World Health Organization |
| YOI | young offender institute |

# ABOUT THE AUTHOR

**Kenneth Hamer** is a practising Barrister at Henderson Chambers, Temple, London (http://www.hendersonchambers.co.uk), and is a Recorder of the Crown Court. He has substantial experience in all areas of professional discipline and regulation. He has, for many years, been on the prosecuting panel for the Bar Standards Board (formerly the Bar Council), and he prosecutes and defends in disciplinary cases involving barristers, solicitors, and accountants. From 2006 to 2012, he was chairman of the Appeal Committee of the Chartered Institute of Management Accountants. He is a legal assessor/legal adviser to the General Medical Council, the Nursing and Midwifery Council, and the General Dental Council. In 2008, he spent six months at the Financial Services Authority, and he regularly appears on behalf of the General Pharmaceutical Council (formerly the Royal Pharmaceutical Society of Great Britain) and advises on its fitness to practise processes.

Kenneth Hamer is joint editor of the Association of Regulatory and Disciplinary Lawyers (ARDL) *Quarterly Bulletin* and a former member of the committee of ARDL. Since its inception in 2002, he has written the 'Legal Update' section for the *Quarterly Bulletin*, and in 2012 he chaired ARDL's response to the Law Commission Consultation Paper *Regulation of Health Care Professionals*.

# 1

# ABSENCE OF PRACTITIONER



## A. Legal Framework

**1.01** Examples include:

Nursing and Midwifery Council (Fitness to Practise) Rule 2004, rule 21

Bar Standards Board Disciplinary Tribunals Regulations 2014, rule E148 (relevant procedure has been complied with, and tribunal considers it just to proceed to hear and determine the charge or charges)

General Medical Council (Fitness to Practise) Rules 2014, rule 31 (panel may nevertheless proceed to consider and determine the allegation if satisfied that all reasonable efforts have been made to serve the practitioner with notice of the hearing in accordance with the Rules)

Institute of Chartered Accountants in England and Wales Disciplinary Committee Regulations, regulation 22 (tribunal may proceed in the defendant's absence where it is satisfied that regulations as to notice of hearing and service of documents have been observed)

## B. General Principles

***R v. Hayward, Jones and Purvis*** **[2001] QB 862, CA;** ***R v. Jones (Anthony)*** **[2003] 1 AC 1, HL**

**1.02** Principles to be applied when proceeding in absence of defendant

Rose LJ, handing down the judgment of the Court of Appeal, Criminal Division (Rose LJ, and Hooper and Goldring JJ), said:

> 22. In our judgment, in the light of the submissions which we have heard and the English and European authorities to which we have referred, the principles which should guide the English courts in relation to the trial of a defendant in his absence are these.
>
> (1) A defendant has, in general, a right to be present at his trial and a right to be legally represented.
>
> (2) Those rights can be waived, separately or together, wholly or in part, by the defendant himself. They may be wholly waived if, knowing, or having the means of knowledge as to, when and where his trial is to take place, he deliberately and voluntarily absents himself and/or withdraws instructions from those representing him. They may be waived in part if, being present and represented at the outset, the defendant, during the course of the trial, behaves in such a way as to obstruct the proper course of the proceedings and/or withdraws his instructions from those representing him.
>
> (3) The trial judge has a discretion as to whether a trial should take place or continue in the absence of a defendant and/or his legal representatives.
>
> (4) That discretion must be exercised with great care and it is only in rare and exceptional cases that it should be exercised in favour of a trial taking place or continuing, particularly if the defendant is unrepresented.
>
> (5) In exercising that discretion, fairness to the defence is of prime importance but fairness to the prosecution must also be taken into account. The judge must have regard to all the circumstances of the case including, in particular: (i) the nature and circumstances of the defendant's behaviour in absenting himself from the trial or disrupting it, as the case may be and, in particular, whether his behaviour was deliberate, voluntary and such as plainly waived his right to appear; (ii) whether an adjournment might result in the defendant being caught or attending voluntarily and/or not disrupting the proceedings; (iii) the likely length of such an adjournment; (iv) whether the defendant, though absent is, or wishes to be, legally represented at the trial or has, by his conduct, waived his right to representation; (v) whether an absent defendant's legal representatives are able to receive instructions from him during the trial and the extent to which they are able to present his defence; (vi) the extent of the disadvantage to the defendant in not being able to give his account of events, having regard to the nature of the evidence against him; (vii) the risk of the jury reaching an improper conclusion about the absence of the defendant; (viii) the seriousness of the offence, which affects defendant, victim and public; (ix) the general public interest and the particular interest of victims and witnesses that a trial should take place within a reasonable time of the events to which it relates; (x) the effect of delay on the memories of witnesses; (xi) where there is more than one defendant and not all have absconded, the undesirability of separate trials, and the prospects of a fair trial for the defendants who are present.
>
> (6) If the judge decides that a trial should take place or continue in the absence of an unrepresented defendant, he must ensure that the trial is as fair as the circumstances permit. He must, in particular, take reasonable steps, both during the giving of evidence and in the summing up, to expose weaknesses in the prosecution

> case and to make such points on behalf of the defendant as the evidence permits. In summing up he must warn the jury that absence is not an admission of guilt and adds nothing to the prosecution case.

The appellant's appeal against conviction was dismissed by the House of Lords. Lord Bingham of Cornhill, at [6], said that the Court has:

> ... a discretion, to be exercised in all the particular circumstances of the case, whether to continue a trial or to order that the jury be discharged with a view to a further trial being held at a later date. The existence of such a discretion is well-established,... But it is of course a discretion to be exercised with great caution and with close regard to the overall fairness of the proceedings; a defendant afflicted by involuntary illness or incapacity will have much stronger grounds for resisting the continuance of the trial than one who has voluntarily chosen to abscond.

See further:

- Lord Bingham, at [13]–[15]:

  > 13. ... The Court of Appeal's check-list of matters relevant to exercise of the discretion (see paragraph 22(5)) is not of course intended to be comprehensive or exhaustive but provides an invaluable guide. I would add two observations only.
  >
  > 14. First, I do not think that "the seriousness of the offence, which affects defendant, victim and public", listed in paragraph 22(5)(viii) as a matter relevant to the exercise of discretion, is a matter which should be considered....
  >
  > 15. Secondly, it is generally desirable that a defendant be represented even if he has voluntarily absconded....

- Lord Nolan, at [18]:

  > In the case of an absconding defendant the critical question for the judge is whether the defendant has deliberately and consciously chosen to absent himself from the court.

- Lord Hoffmann, at [20]:

  > 20. But I do not read the European cases as laying down that a trial may proceed in the absence of the accused only if there has been a waiver of the right to a fair trial. The question in my opinion is not whether the defendants waived the right to a fair trial but whether in all the circumstances they got one...

- Lord Hutton, at [38]:

  > The discretion of a judge to proceed with a trial in the absence of the defendant is one to be exercised with great care, but in my opinion there can be circumstances where in the interests of justice a judge is entitled to proceed, particularly where the defendant has deliberately absconded to avoid trial.

- Lord Rodger of Earlsferry, at [54]–[68] (prosecuting counsel looked to see whether there are any areas that ought to be highlighted in view of the fact that the defendants were not present or represented, and the judge's summing up was obviously of even greater importance than usual; the jury were told specifically that they must not speculate as to the reasons for the defendants' absence and that they should not assume that the defendants' failure to attend court in any way at all established that either or both of them were guilty; the jury should carefully assess the evidence as they would have done had the defendants been present and had he been represented by counsel).

## C. Disciplinary Cases

### *Tait v. Royal College of Veterinary Surgeons* [2003] UKPC 34

**1.03** Discretion to proceed—advice inadequate—decision to proceed quashed

T appealed under section 17 of the Veterinary Surgeons Act 1966 against decisions made by the Disciplinary Committee of the Royal College of Veterinary Surgeons (RCVS) on 27–28 June 2002 and 31 July 2002, when the Committee found two charges of disgraceful conduct in a professional respect proved and directed that T's name be removed from the register. The hearing of the inquiry was due to take place on 2 May 2002. T did not attend and the hearing was adjourned to 27 June 2002. Again, T did not attend. He applied in writing for an adjournment on the ground of ill health, viz. alleged hypertension. He produced no medical evidence. The legal assessor directed the Committee that 'at all times, the committee will remember that it has absolute discretion'. The Committee refused the application for an adjournment and proceeded to hear the charges against T in his absence. The Privy Council (Lord Steyn, Lord Slynn of Hadley, and Lord Walker of Gestingthorpe) held that the legal assessor's advice was wrong. In delivering its reasons in allowing T's appeal and quashing the removal of his name from the register, Lord Steyn said, at [8]:

> This direction does not comply with the requirements of the decision in Jones. It was not correct to say that the committee had an absolute discretion. The discretion was a severely constrained one. The direction did not mention that the appellant was at serious risk of removal from the register. It did not draw attention to the risk of a wrong decision about the appellant's alleged ill-health. It did not sufficiently highlight the risk of the committee coming to a wrong conclusion on merits without the account of the appellant. The guidance placed before the committee was inadequate. Normally, one would expect the committee to rely on the guidance of the legal assessor. *Prima facie* therefore the committee appears to have misdirected itself. If the committee relied on the direction, it was based on materially incorrect guidance.

### *R (Elliott) v. Solicitors Disciplinary Tribunal* [2004] EWHC 1176 (Admin)

**1.04** Defendant attended at start of hearing and left before evidence—no grounds to order rehearing

The claimant, E, sought to quash a decision of the Solicitors Disciplinary Tribunal (SDT) which, on 23 October 2003, dismissed his application pursuant to rule 25 of the Solicitors Disciplinary Proceedings Rules 1994 for a rehearing of charges brought against him by the Law Society heard on 3 and 4 December 2001, and which led to his being struck off the roll. Rule 25 provides that, at any time before the filing of the Tribunal's findings and order with the Law Society, or within one calendar month of such filing, if the defendant has neither attended in person nor been represented at the hearing and the Tribunal has proceeded in his absence, he may apply to the Tribunal for a rehearing. E attended at the start of the hearing on 3 December 2001, but absented himself after his application for an adjournment was refused and before the investigation of the merits of the complaint. E said that he had been advised that he must withdraw, that he did so, and that the Tribunal continued in his absence. Leveson J held that rule 25 did not bite in these circumstances. On its true construction, E did not fall within the description of a respondent who had neither attended in person nor been represented at the hearing. In any event, E could not rely on his own deliberate and informed decision not to attend the substantive hearing as providing a basis for the exercise of the Tribunal's discretionary remedy of a rehearing.

### *R (Mahmood) v. General Medical Council* [2007] EWHC 474 (Admin)

**1.05** Practitioner admitted to hospital on morning of hearing—involuntary illness—procedural irregularity to continue in his absence

On 27 March 2006, the first day of the fitness to practise hearing, M attended at the premises of the General Medical Council (GMC), but shortly after doing so became ill and was admitted to hospital. On 24 March 2006, he wrote to the GMC stating that he intended to attend in relation to allegations that, whilst he was a treating doctor to Miss A, he had conducted a personal and sexual relationship with her, that he had borrowed money from her that he did not repay, and that he had prescribed medication for her outside a clinical setting. M was seen by Miss A shortly before 9 a.m. outside and near to the GMC's premises. At 2.55 p.m. on 27 March, M telephoned the GMC's offices to tell the Council that he had fallen in the street and had been admitted to the coronary care unit in University College Hospital (UCH). Mitting J allowed M's appeal and remitted the case to a fresh panel. Mitting J said that the test that the panel should have applied was clearly and uncontroversially set out in *R v. Jones* [2003] 1 AC 1, at [6], where Lord Bingham had observed that a defendant afflicted by involuntary illness or incapacity will have much stronger grounds for resisting the continuance of a trial than one who has voluntarily chosen to abscond. At [13], Lord Bingham observed that if the absence of the defendant is attributable to involuntary illness or incapacity, it would very rarely, if ever, be right to exercise the discretion in favour of commencing a trial, at any rate unless the defendant is represented and asks that the trial should begin. If, in the instant case, the panel had decided on proper grounds that M had deliberately absented himself from the hearing and had sought admission to UCH simply as a ploy, its decision to continue in his absence would have been unchallengeable. On the information that it had and on the facts that it found, the panel could not properly have concluded that his absence from the hearing was deliberate. It was a clear and serious procedural irregularity to continue with the hearing.

### *Varma v. General Medical Council* [2008] EWHC 753 (Admin)

**1.06** Medical evidence—practitioner capable of participating in proceedings—voluntary election not to be present

Forbes J held that the panel had properly borne in mind that it had to balance the appellant's private rights against the public interest of having serious allegations properly investigated. The appellant had been granted a couple of short stays on medical grounds, but an indefinite stay had been refused on the basis that there had already been inordinate delay and the medical evidence indicated that the appellant was capable of participating in the proceedings. The panel had justifiably treated the appellant's failure to attend the hearing as a voluntary election not to be present. There was no basis on which the Court could interfere with the panel's exercise of its discretion. At [29], Forbes J said:

> It is also important to note that, having dismissed the application for a stay, the Panel gave separate consideration to whether it should proceed with the substantive hearing in the belief that [the appellant] would neither be present nor represented—a belief that turned out to be only intermittently true. As [Counsel for the GMC] observed, the Panel had a broad power to do so, as provided in Rule 31 [of the GMC Fitness to Practise Rules 2004], which provides: "*Where the practitioner is neither present nor represented at a hearing, the... Panel may nevertheless proceed to consider and determine the allegation if they are satisfied that all reasonable efforts have been made to serve the practitioner with notice of the hearing in accordance with these rules.*" In my judgment, the Panel was clearly entitled, at that stage, to conclude that [the appellant] had "*voluntarily chosen to waive his right to be present and give evidence and be represented*" and [was] right to decide to proceed with the hearing in his absence for the reasons that it gave.

### *Abdalla v. Health Professions Council* [2008] EWHC 3498 (Admin)

**1.07** No breach of Article 6 ECHR— lack of funds for legal representation— insufficient reason for not attending hearing

The appellant, a radiographer, claimed that a committee's decision to proceed in her absence was in breach of her rights under Article 6(1) of the European Convention on Human Rights (ECHR); alternatively, it was in breach of natural justice for the committee to proceed in her absence. Sullivan J rejected this argument. The appellant did not produce any medical evidence to support the claim that she was unable to attend, and the committee had satisfied itself that she had received the relevant papers and knew of the hearing date. Sullivan J observed, at [6]:

> The appellant was certainly entitled—whether under Article 6(1) or as a matter of fairness or natural justice—to an opportunity to attend the hearing. But if she chose not to avail herself of that opportunity then the committee was entitled to proceed in her absence.

The Court held that lack of funds to pay for legal representation would not justify any further adjournment where there was no realistic prospect that the position would change in the reasonably near future and where the appellant could have appeared herself (her legal representatives having refused to attend as a result of lack of funds). Sullivan J said, at [13]–[15]:

> Thus it will be seen that the primary reason put forward by the solicitors for their not attending the hearing is the lack of funds. That would not have prevented the appellant herself from attending the hearing and explaining the position to the committee.... In summary, there was simply no good reason for her failing to attend the hearing.

The Court went on to state that it was in the public interest, and the interests of both parties, to resolve the allegations as soon as reasonably possible. At [16], Sullivan J said:

> Moreover the specific allegations against the appellant related, in the main, to a period between September 2005 and October 2006. It was clearly desirable that allegations, some of which went back to some two-and-a-half years, should be investigated without further delay. There is a clear public interest in allegations against members of the health professions being resolved as soon as reasonably possible. In addition, of course, speedy resolution of such allegations will be very much in the interests of the individual practitioners themselves who will be able to clear their names if it is concluded that the allegations against them are not well-founded.

### *Yusuf v. Royal Pharmaceutical Society of Great Britain* [2009] EWHC 867 (Admin)

**1.08** Key issue whether Y had voluntarily chosen not to attend— discretion to proceed exercised with utmost caution— decision fair, measured, and unassailable

After reviewing the decision of the Court of Appeal in *R v. Hayward* [2001] QB 862, approved on appeal by the House of Lords, sub nom *R v. Jones (Anthony)* [2003] 1 AC 1, and applied to the disciplinary proceedings of professional bodies by the Privy Council in *Tait v. The Royal College of Veterinary Surgeons* [2003] UKPC 34, Munby J held that the chairman of the Disciplinary Committee of the Royal Pharmaceutical Society of Great Britain had correctly identified the key issue as being, in this particular case, whether the appellant had voluntarily chosen not to attend. At [39], Munby J said:

> [Counsel for the Society] says that the Chairman correctly identified the key issue as being, in this particular case, whether the appellant had voluntarily chosen not to attend. I agree. Moreover, as [counsel] points out, the Committee was properly alert to the need to exercise its discretion to proceed in the appellant's absence with the utmost caution. He submits that the Committee's decision was fair, measured and unassailable. I agree. The Committee was entitled to find, as it did, that the appellant had voluntarily chosen neither to appear nor to be represented. It then proceeded to exercise its discretion, having, in my judgment, directed itself impeccably in law. And, to repeat,

it was entitled, proceeding as it said with the "utmost caution", to conclude, and for the reasons it gave, that it was proper for it to proceed in the absence of the appellant.

In short, the Committee was properly alert to the need to exercise its discretion to proceed in the appellant's absence with the utmost caution. The Committee's decision was fair, measured, and unassailable, and the Committee was entitled to find, as it did, that the appellant had voluntarily chosen neither to appear nor to be represented. As to taking proper and sufficient steps to ensure that the appellant's case is put, see paragraph 67.04.

### *Faulkner v. Nursing and Midwifery Council* [2009] EWHC 3349 (Admin)

**1.09**

During retirement to consider sanction F surfaced and addressed committee—F failed to make prior application for adjournment—voluntarily chose not to attend

The Administrative Court (Cranston J) held, in dismissing the instant statutory appeal, that the Conduct and Competence Committee (CCC) of the Nursing and Midwifery Council (NMC) had not fallen into error in exercising its discretion to hear allegations made against the appellant nurse in his absence, and then to find those allegations proved and to order a striking off. The appellant was made the subject of six allegations relating to his time as a staff nurse at a general-practitioner-led community hospital, which provided rehabilitative and remedial care for adults over the age of 50. Each of the allegations related to incidents of inappropriate behaviour: two concerned causing stress to a patient; one concerned the abuse of a patient; the other three allege that the appellant engaged in inappropriate sexually related activity. The hearing before the Committee had been adjourned on a previous occasion owing to the appellant's non-appearance. After finding the allegations proved and that the appellant's fitness to practise was impaired, and during the Committee's retirement to consider sanction, the appellant surfaced to address the Committee. Notwithstanding this, the Committee ordered the striking off of the appellant from the register. On appeal, the appellant submitted that the exercise of discretion to proceed in his absence at the fact finding stage was flawed on two counts: first, that too much weight had been given to the fact that witnesses were present and ready for the hearing; and second, that, on the previous occasion, the Committee had taken the view that the appellant would not be able to cross-examine one of the witnesses. Cranston J dismissed the appeal, holding that the appellant had not adduced any evidence either as to his medical condition or as to the reason for his not having obtained legal assistance. The appellant was aware of the hearing date, but failed to make any application for an adjournment. He had voluntarily chosen not to attend.

### *Parkinson v. Nursing and Midwifery Council* [2010] EWHC 1898 (Admin)

**1.10**

See paragraph 19.56.

### *R (Raheem) v. Nursing and Midwifery Council* [2010] EWHC 2549 (Admin)

**1.11**

No guidance by legal assessor as to how committee should exercise discretion—significant omission—need to demonstrate discretion to proceed exercised with care and caution

The Administrative Court held that there had been no proper exercise of discretion by the NMC's CCC as to whether it had been appropriate to continue in the absence of the appellant in respect of a hearing conducted to decide whether she had been guilty of misconduct. Holman J recognized that notice of the hearing had been properly served under the 2004 Rules even though the appellant had not received the recorded delivery envelope and was unaware of the hearing. However, the legal assessor did not give any sufficient advice or direction to the Committee as to the separate exercise of discretion to proceed in the absence of the practitioner under rule 21. He gave no guidance as to how the Committee should approach that exercise of discretion. Since it was clear on authority that a continuation in the absence of a practitioner must be exercised with the 'utmost caution', the learned judge said that this was a significant omission from the advice given by the legal assessor. Further, there was no indication from the transcript that the Committee broke off or adjourned to consider the matter, and it seemed to the judge that the chairman moved very rapidly to its ruling. Holman J said that it was extremely important that a committee or tribunal in question demonstrates by its language that it appreciates that the discretion to proceed in the absence of the practitioner that it is exercising is one that must be exercised with care and caution. In the instant case, there was a fatal procedural defect in the approach of the Committee. The appeal must be allowed and the whole matter reheard by the CCC from scratch.

***Gurpinar v. Solicitors Regulation Authority* [2012] EWHC 192 (Admin)**

**1.12** Co-defendants—one present, one absent—tribunal could not be criticized for proceeding in absence of co-defendant

The primary ground of appeal by G was that the SDT wrongly proceeded to hear the allegations in his absence, despite the fact that it had been told that he was unable to be present owing to a combination of ill health and the disruption to air travel caused by the Icelandic volcano in April 2010. Moore-Bick LJ, in dismissing G's appeal, said that he was not persuaded that certain letters or emails purportedly sent by G ever reached their intended recipients and that the only explanation for that was that they were not sent. At the hearing on 22 April 2010, the Tribunal was faced with a difficult and frustrating situation. One co-defendant in partnership with G, N, was present, but G was not, nor was anyone there to represent him. He had failed to serve his defence bundle. The Tribunal thought that G was 'playing fast and loose', and Moore-Bick LJ said it had good reason to think so. In the circumstances, the Tribunal could not be criticized for proceeding in G's absence.

***Fabiyi v. Nursing and Midwifery Council* [2012] EWHC 1441 (Admin)**

**1.13** Witness statement containing arguable defence lodged on morning of hearing—no evidence that she wanted hearing to proceed in absence—decision to proceed after committee adjourning for 8 minutes—procedurally unfair

F was charged with dishonesty in working as a midwife in a non-supernumerary role. She was neither present nor represented at the hearing and the NMC's CCC found the charge of dishonesty to be proved. On the morning of the hearing, a representative from a firm of solicitors attended the offices of the NMC with copies of a witness statement on behalf of F, stated that she would not be attending today, and asked that the witness statement—which was signed, but not dated—be placed before the Committee. When advising the Committee in deciding whether or not to exercise its discretion to proceed in the absence of F and any representative on her behalf, the legal assessor correctly reminded the Committee that F had a right to be present and to attest the NMC's case, and that such rights could be waived if she deliberately and voluntarily absented herself. The legal assessor then, again correctly, advised the Committee that it should address the question of why F was not present. However, he did not point out that F's witness statement, just delivered, clearly stated that she did not admit that she had been dishonest and that this was the only contested issue of fact left for determination since the witness statement admitted all of the other allegations. He stated that F had not put forward any reason for her non-attendance, but then invited the Committee to infer that her legal representative was effectively taking the position that F wanted the hearing to continue. Neither the person attending nor her solicitors, as a firm, had taken or were taking any position about whether F wanted or did not want the hearing to proceed in her absence. Nothing had been said to give rise to the inference that F wanted the hearing to proceed in her absence. Judge Thornton QC said that, in the context of this case, the Committee should have been advised to take into account the fact that the allegation of dishonesty was inadequately particularized, and that the evidence relied on had not been summarized and was not set out in the hearing bundle in a readily intelligible order. As a result, without that being done, there was a serious danger of procedural unfairness, particularly in the absence of both F and her legal representative. F appeared to have an arguable defence (on dishonesty) that the principal witnesses had previously accepted. Since that defence involved a consideration of her state of mind at the time during which she worked unsupervised, it would be very difficult for the Committee to address that defence fairly in her absence and without hearing her give evidence. The Committee retired for eight minutes to consider whether to proceed. The decision to proceed in F's absence was procedurally unfair and she was greatly prejudiced by the hearing proceeding without either herself or her legal representative being present. The Committee could not fairly reach any conclusion as to why F was not present or as to whether her non-appearance and non-representation had arisen on the basis of her informed consent.

### *Faniyi v. Solicitors Regulation Authority* [2012] EWHC 2965 (Admin)

**1.14** Failure to attend—whether adjournment before sentence

The appellant did not appear before the SDT to face allegations of providing misleading statements to prospective professional indemnity insurers and matters relating to a firm of which he was the senior partner. There were nine separate allegations, supported by a rule 5 statement and a supplementary rule 7 statement. No witnesses were called by the Solicitors Regulation Authority (SRA), which invited the Tribunal to consider the allegations against the appellant on the basis of the documentary evidence that had been served and of which notice to admit had been given under rule 13. The Tribunal agreed to receive the rule 5 and rule 7 statements as evidence, and to treat the documents exhibited to each of them as authentic. In dismissing F's appeal, the Court said that he did not attend the hearing (deliberately) nor did he make submissions as to why it was necessary for the various witnesses to be called and documents to be proved. He did not engage constructively in the process at all. Further, the Tribunal did not act unreasonably in not affording the appellant an opportunity to make submissions by way of mitigation after it had made its adverse findings against him. A solicitor who deliberately absents himself cannot argue that he is entitled to a rehearing: see *R (Elliott) v. Solicitors Disciplinary Tribunal* [2004] EWHC 1176 (Admin). A second tribunal dismissed F's application for a rehearing. The burden of establishing that justice required a rehearing lies on the parties seeking it—in particular where, as in this case, there was clearly credible evidence before the original tribunal that F had deliberately chosen not to attend.

### *McDaid v. Nursing and Midwifery Council* [2013] EWHC 586 (Admin)

**1.15** Matters on behalf of defendant—extent of tribunal's duty

The appellant did not attend the hearing before the NMC's CCC. The Committee is required, during the giving of evidence by the NMC witnesses, to take reasonable steps to expose weaknesses in the NMC case and to make such points on behalf of the defendant as the evidence permits. However, it has no duty to cross-examine witnesses in the same way as might a litigant in person or any legal representative. The appellant had sent in a letter from the Newham University Hospital Trust, which, on the face of it, undermined the Committee's decision in relation to one charge of misconduct, and (arguably) the Committee's conclusion in relation to that charge 'infected' its conclusions in relation to the other charges. Despite the NMC arguing that the letter was not 'genuine', the Court quashed the decision and remitted the case for rehearing by a new committee.

### *R (Shaikh) v. General Pharmaceutical Council* [2013] EWHC 1844 (Admin)

**1.16** Failure to attend—extent of tribunal's duty to test evidence

The Fitness to Practise Committee, chaired by Mr Christopher Gibson QC, found two charges against S proved in his absence. In dismissing S's appeal, Supperstone J said, at [23], that the Committee had regard to the fact that the appellant deliberately absented himself so that he would not have been in a position to cross-examine witnesses. The Committee in its determination, however, added: 'Nevertheless, we consider that it is our responsibility to test the evidence in the absence of [S] to the extent that we reasonably can.'

### *R (Veizi) v. General Dental Council* [2013] EWHC 832 (Admin)

**1.17** Absence of appellant at hearing of appeal

The appellant did not appear at the hearing of his appeal against the direction that his name be erased from the dental register. He had not made contact with the Court and did not respond to an invitation from the respondent's solicitors to withdraw his appeal. Mr Philip Mott QC (sitting as a deputy High Court judge) said that, in these circumstances, the appropriate course is to hear the appeal in the appellant's absence and determine it—a course that was adopted by Wyn Williams J in similar circumstances in the case of *Al-Daraji v. General Medical Council* [2012] EWHC 1835 (Admin).

### *Guest v. General Medical Council* [2013] EWHC 3121 (Admin)

**1.18** Appellant suffering relapse of multiple sclerosis—no evidence that her capacity or ability adversely affected—clear desire to proceed

G was found guilty of dishonestly making claims for expenses totalling £12,000. G sent emails to the GMC, stating that she was unable to secure affordable legal representation because she was not covered by insurance, and that she had suffered a relapse of multiple sclerosis and would not be attending the hearing. Dismissing G's appeal and sanction of erasure, the Court said that it could not properly have been regarded by the fitness to practise panel as necessarily determinative that G had said that she was not seeking an adjournment and was not claiming in terms that she was unable to participate owing to ill health, and indeed that she was saying that she simply wanted the panel to get on with it. However, there was no evidence before the panel that G was operating under any medical or other condition such as would—or even might have—adversely affected her capacity or her ability to make an informed decision as to how to respond to the forthcoming hearing. She had clearly communicated, not once, but on four occasions, her clear desire that the hearing should proceed in her absence and that, on its face, was a perfectly rational decision for her to make. Further, it was not a case in which the panel could realistically have considered that matters could be dealt with by way of a short adjournment. It could not seriously be said that the panel should, of its own volition, have adjourned to allow the appellant to obtain further medical evidence.

### *Waghorn v. General Medical Council* [2014] EWHC 1214 (Admin)

**1.19** Panel to keep wary eye open to ensure no unfairness

In dismissing W's appeal, Charles J observed, at [22], that:

> [A]s [W] decided to absent himself from the process, the Panel had to keep a wary eye open to see whether or not at some point there was some unfairness being inflicted upon or caused to [W] because he simply was not there to raise points that he wished to.... [W] was perfectly able and certainly in a position to identify the points he wished to put as to [the expert witness] providing the wrong standard by reference to the innovative of procedure that he was carrying

> out. The reason that was not put to [the expert witness] was either that [W] had not formulated it in that way at the time or, perhaps more correctly, he decided to absent himself from the process and simply deprive himself of the opportunity of putting those points to the expert relied on by the GMC and the other witnesses. That, to my mind, is a decision he made with his eyes wide open and did not give rise at any point during the hearing to evidence being given which the fitness to practise panel should have considered warranted a revisit to the application for an adjournment.

### *Ward v. Nursing and Midwifery Council* [2014] EWHC 1158 (Admin)

**1.20** Absence of appellant at hearing of appeal

W appealed against the panel's findings of fact, impairment, and sanction. In refusing her application that the hearing be postponed, Hickinbottom J said, at [7]:

> (i) It is open to the court to proceed in an appellant's absence, if it is in the interests of justice to do so.
> (ii) The Appellant is clearly on notice of the hearing, and on notice that her solicitors were no longer acting for her. She does not appear to have made any attempt to contact the court at all.
> (iii) No formal application to adjourn has been made, and no evidence has been submitted in support of an adjournment in the form of medical evidence concerning either the Appellant's father or, if appropriate, the Appellant herself.
> (iv) The events underlying this appeal occurred over three years ago. The NMC are entitled to finality, and it is in the best interests of all parties, including the Appellant herself, that this appeal is determined promptly.
> (v) No communication has been received from the Appellant, her solicitor's letter was very late, the NMC have appeared, and an adjournment would result in costs being thrown away.
> (vi) I have had the benefit of very full grounds of appeal lodged on the Appellant's behalf, which I have taken fully into account.
> (vii) The Appellant did not appear at the hearing before the Panel, who found that she had voluntarily absented herself. In my view, it is likely that, even if the appeal hearing were adjourned, the Appellant would not attend any hearing fixed in the future. If there are good grounds for her non-attendance today, it is in any event open to her to apply to set aside any order made on that basis.

In the circumstances, the Court would proceed to hear the appeal, on the basis that it would not result in any unfairness to the appellant and that it was in the interests of justice to do so.

### *Malik v. General Medical Council* [2014] EWHC 2408 (Admin)

**1.21** Absence of appellant at hearing of appeal

M appealed against the findings of fact made by the fitness to practise panel and its decision to order erasure. He did not attend; no explanation was forthcoming and there was no application for an adjournment. In the circumstances, the Court addressed the possibility of striking out the appeal or hearing it on its merits, so far as those could be gleaned from M's paperwork. After referring to *Howard v. Stanton* [2011] EWCA Civ 1481, Sir David Eady said that an appellate court, whether the Court of Appeal or the administrative court hearing a statutory appeal under section 40 of the General Medical Act 1983, must have power to control the proceedings and, in appropriate circumstances, to strike out an appeal. The answer may be contained in CPR 52.9, which provides the court with the power to strike out an appeal notice where there is a compelling reason to do so. Since the appeal had not been pursued, it was appropriate that the appeal notice be struck out.

## D. Other Relevant Chapters

**1.22** Adjournment, Chapter 3
Appeals, Chapter 4
Legal Assessors, Chapter 44
Unrepresented Practitioner, Chapter 67

# 2

# ABUSE OF PROCESS



## A. Legal Framework

**2.01** European Convention on Human Rights (ECHR), Article 6 (see paragraph 32.01)

Most regulators provide that the procedure before a fitness to practise panel or committee shall begin with the panel or committee hearing and considering any preliminary legal arguments or objection to the charge on a point of law. For example, see:

Nursing and Midwifery Council (Fitness to Practise) Rules 2004, rule 24(2)
General Pharmaceutical Council (Fitness to Practise and Disqualification etc) Rules 2010, rule 31(2)
Bar Standards Board Disciplinary Tribunals Regulations 2014, rules E107–123A (appointment of a directions judge to include giving directions for separate hearings, applications to sever charges, and applications to strike out charges)
General Medical Council (Fitness to Practise) Rules 2014, rule 17(2)

## B. General Principles

### *Attorney-General's Reference (No. 1 of 1990)* [1992] QB 630

**2.02** Permanent stay should be exception, rather than rule—effect of delay—prejudice

Lord Lane CJ, reading the opinion of the Court of Appeal, Criminal Division, said that abuse of process may arise in many different forms. It may involve complaints against the methods used to investigate the offence: see *R v. Heston-Francois* [1984] QB 278. It may be based, as *Connelly v. Director of Public Prosecutions* [1964] AC 1254 itself was, on the allegation that the defendant is being prosecuted more than once for what is, in effect, the same offence. It may be a misuse of the process of the court to escape statutory time limits: see *R v. Brentford Justices, ex parte Wong* [1981] QB 445. However, the most usual ground is that based on delay—that is, the lapse of time between the commission of the offence and the start of the trial. At 643G–644C, Lord Lane said:

> Stays imposed on the grounds of delay or for any other reason should only be employed in exceptional circumstances. If they were to become a matter of routine, it would be only a short time before the public, understandably, viewed the process with suspicion and mistrust. We respectfully adopt the reasoning of Brennan J in *Jago v. District Council of New South Wales* (1989) 168 CLR 23. In principle, therefore, even where the delay can be said to be unjustifiable, the imposition of a permanent stay should be the exception rather than the rule. Still more rare should be cases where a stay can properly be imposed in the absence of any fault on the part of the complainant or prosecution. Delay due merely to the complexity of the case or contributed to by the actions of the defendant himself should never be the foundation for a stay. In answer to the second question posed by the Attorney-General, no stay should be imposed unless the defendant shows on the balance of probabilities that owing to the delay he will suffer serious prejudice to the extent that no fair trial can be held: in other words, that the continuance of the prosecution amounts to a misuse of the process of the court. In assessing whether there is likely to be prejudice and if so whether it can properly be described as serious, the following matters should be borne in mind: first, the power of the judge at common law and under the Police and Criminal Evidence Act 1984 to regulate the admissibility of evidence; secondly, the trial process itself, which should ensure that all relevant factual issues arising from delay will be placed before the jury as part of the evidence for their consideration, together with the powers of the judge to give appropriate directions to the jury before they consider their verdict.

### *R v. Maxwell* [2011] 1 WLR 1837

**2.03** Two categories: (1) accused cannot receive fair trial; (2) to try accused would offend court's sense of justice and propriety

Lord Dyson JSC said:

> 13. It is well established that the court has the power to stay proceedings in two categories, namely (i) where it will be impossible to give the accused a fair trial, and (ii) where it offends the court's sense of justice and propriety to be asked to try the accused in the particular circumstances of the case. In the first category of case, if the court concludes that an accused cannot receive a fair trial, it will stay the proceedings without law. No question of the balancing of competing interests arises. In the second category of case, the court is concerned to protect the integrity of the criminal justice system. Here a stay will be granted where the court concludes that in all the circumstances a trial will offend the court's sense of justice and propriety (per Lord Lowry in *R v. Horseferry Road Magistrates' Court, ex parte Bennett* [1994] 1 AC 42, 74G) or will undermine public confidence in the criminal justice system and bring it into disrepute (per Lord Steyn in *R v. Latif* [1996] 1 WLR 104, 112F).

### *Warren v. Attorney General for Jersey* [2012] 1 AC 22

**2.04** Prosecutorial misconduct—balance between competing public interests

The appellants were convicted of conspiracy to import into Jersey 180 kg of cannabis with a street value in excess of £1 million. They applied for a stay of proceedings on the grounds of abuse of process. The basis of the application was that crucial evidence on which the prosecution wished to rely had been obtained as a result of serious prosecutorial misconduct in installing audio device equipment in the knowledge that the Attorney General and the Chief of Jersey Police had not given authority, and that foreign authorities had refused their consent. Sir Richard Tucker (sitting as a commissioner) dismissed the appellants' application for a stay, whilst describing the prosecution conduct as 'most reprehensible'. The Court of Appeal of Jersey dismissed a renewed application for leave to appeal and the defendants were subsequently convicted of the offence. On their appeal against the refusal of the stay, the Privy Council affirmed the decisions of Sir Richard Tucker and the Court of Appeal of Jersey. Lord Dyson JSC said, at [26]:

> The Board recognises that, at any rate in abduction and entrapment cases, the court will generally conclude that the balance favours a stay. But rigid classifications are undesirable. It is clear from *R v. Latif* [1996] 1 WLR 104 and *Mullen's case* [*R v. Mullen*] [2000] QB 520, that the balance must always be struck between the public interest in ensuring that those who are accused of serious crimes should be tried and the competing public interest in ensuring that executive misconduct does not undermine public confidence in the criminal justice system and bring it into disrepute. It is true that in *Ex parte Bennett* [1994] 1 AC 42 the need for a balancing exercise was not mentioned, but that is no doubt because the House of Lords considered that the balance obviously came down in favour of a stay on the facts of that case (the kidnapping of a New Zealand citizen to face trial in England).

### *R v. F* (S) [2012] 1 All ER 565

**2.05** Delay—historic sex allegations—distinction between stay and no case to answer

The Court of Appeal, Criminal Division (Lord Judge CJ, Hughes and Goldring LJJ, and Ouseley and Dobbs JJ) said, at [36], that the authority of *R v. Galbraith* [1981] 1 WLR 1039, with its emphasis on the responsibilities of the jury as the fact-finding body responsible for delivering the verdicts, is undiminished. The principles have neither been modified nor extended for the purposes of addressing trials that involve historic unreported sexual crimes. In accordance with the second limb of *R v. Galbraith*, there will continue to be cases in which the state of the evidence called by the prosecution, and taken as a whole, is so unsatisfactory, contradictory, or so transparently unreliable that no jury, properly directed, could convict. The Court continued:

> 38. When, by contrast with a submission of "no case to answer", the court is considering an application to stay proceedings on the grounds of prejudice resulting from delay in the institution of proceedings, the appropriate test was directly addressed and answered in *A-G's Reference (No 1 of 1990)*. It continues to provide the benchmark....
>
> 39. Accordingly in our judgment any suggestion that, on the basis of delay, the judge may be responsible for assessing whether in advance of a conviction, the conviction would be unsafe, is based on a misunderstanding of the principles in both *R v. Galbraith* and *A-G's Reference (No 1 of 1990)*. The judge, of course, is responsible for the conduct of the trial. That responsibility extends to deciding whether the trial should be stayed because it would constitute an abuse of process, applying the principles relevant to that question, and the distinct, separate question, whether at the end of the prosecution case, the jury should be directed to return a "not guilty" verdict or verdicts on *R v. Galbraith* principles. These are distinct features of the trial process and neither of these separate responsibilities of the judge should be elided with each other, or with the equally distinct responsibilities of the jury.
>
> 40. In the overwhelming majority of historic sex allegations the reasons for the delayed complaint, and whether and how the delay is explained or justified, bear directly on the credibility of the complainant. They therefore form an essential part of the factual matrix on which the jury must make its decision. That is the principle, and in the overwhelming majority of cases, the only relevance of the evidence on these issues. When, in the authorities to which we have referred, it is clearly stated that an abuse of process argument cannot succeed unless prejudice has been caused to the defendant, the principles do not normally encompass the explanation for the delay, nor do they extend to the explanation or explanations which the judge himself or herself may regard as inadequate or unsatisfactory or inconsistent. Indeed features like these are revealed by and become apparent through the ordinary processes of trial, and these questions remain pre-eminently for the jury. Although therefore they may be relevant

to submissions that there is no case to answer (carefully considered in the context of the limitations imposed by *R v. Galbraith*) it is difficult to conceive of circumstances in which they have any relevance to an abuse of process argument, unless in some manner they impact on the question whether there can be a fair trial. The explanations for delay are relevant to an application to stay only if they bear on how readily the fact of prejudice may be shown. Unjustified delay in the making of the complaint, and even more so institutional prosecutor misconduct leading to delay (which is what the court was considering in *A-G's Reference (No 1 of 1990)*), may make the judge more certain of prejudice, which may even have been the aim of the delay. That is the import of the references in the cases to the reasons for the delay. That is, however, a long way from the proposition that unjustified delay is by itself a sufficient reason for a stay. It is not.

## C. Disciplinary Cases

***R v. Metropolitan Police Disciplinary Tribunal, ex parte Police Complaints Authority*, [1992] COD 405**

**2.06** Police officers charged with providing false evidence in criminal proceedings—defendant in criminal trial acquitted—subsequent disciplinary proceedings—no collateral attack on verdict

Police Constable Spittles and Superintendent Bean appeared before a disciplinary tribunal convened under section 94 of the Police and Criminal Evidence Act 1984 to face charges brought against them under the Police Disciplinary Code at the direction of the Police Complaints Authority (PCA). Both officers were charged with knowingly making false statements in their original witness statements and again when they gave evidence at Knightsbridge Crown Court, at which time the defendant in criminal proceedings, Mrs A-S, was convicted on a charge of keeping a disorderly house. Mrs A-S made a complaint about the conduct of the officers. Before the disciplinary charges were put, it was submitted on behalf of the two officers that the proceedings should be stayed as an abuse of the process of the tribunal. The members of the tribunal upheld that submission. The PCA sought an order of certiorari to quash the decision of the tribunal. The principal submissions on behalf of the two officers were based on the decision of the House of Lords in *Hunter v. Chief Constable of the West Midlands Police and ors* [1982] AC 529—namely, where a final decision has been made by a criminal court of competent jurisdiction, it is a general rule of public policy that the use of a civil action to initiate a collateral attack on that decision is an abuse of the process of the court. The two officers argued that a final decision against Mrs A-S had been made by a criminal court of competent jurisdiction, that she had made no appeal against that decision, and that the purpose of her complaint against the two officers was plainly to initiate a collateral attack upon that decision by seeking to discredit their evidence, on the basis of which she had been convicted, and thus to prepare the ground for her to seek leave to appeal out of time to the Court of Appeal, Criminal Division. Nolan LJ and Jowitt J said that the possibility of a successful appeal against conviction by Mrs A-S was a matter of indifference to the PCA. No matter how great the sense of grievance felt by the two officers at being subjected to disciplinary proceedings, the fact remained that if the conduct alleged fell within the terms of the Disciplinary Code, then it was a proper subject for disciplinary proceedings. It was not suggested that the PCA had acted otherwise than in strict accordance with the Act and the Code in directing that the proceedings should be brought. The disciplinary proceedings were not being brought by Mrs A-S. The PCA, at whose direction they were brought, was not concerned in the criminal trial. The PCA cannot be accused of pursuing a collateral attack. Mrs A-S's motives in making her complaint may have bearing on her credibility, but that is another matter entirely. In granting the PCA's application, the Court said that it must clearly be recognized that *Hunter* provides no authority for any general proposition that false statements made at the trial of a convicted defendant cannot form the subject of disciplinary proceedings so long as that conviction stands.

***Brabazon-Drenning v. United Kingdom Central Council for Nursing, Midwifery and Health Visiting* [2001] HRLR 6**

**2.07** Decision by regulator not to bring charges—legitimate expectation—subsequent charge based on same facts quashed

The registrant was the owner of a nursing and residential home in Redruth, Cornwall, and, between August 1996 and April 1997, she allowed her registration with the UK Central Council for Nursing, Midwifery and Health Visiting (UKCC) to lapse. This was potentially a disciplinary offence, but on 27 June 1997 she was notified that the Preliminary Proceedings Committee (PPC) had considered the allegation of non-registration and decided not to act upon it. The letter also indicated that all documentation relating to the matter would be destroyed. Subsequently, six charges of complaint were made, which included a charge that she had failed to be registered. In quashing the charge, Elias J (with whom Rose LJ agreed) said that, quite independently of the question of legitimate expectation, once the Committee had made its ruling and had determined that there should be no further action taken in respect of the charge, then, unless there was some misrepresentation or unless it was acting under some fundamental misconception of the true position, the Committee was bound by that determination. It was not open to it to resurrect the matter at will, or because it had discovered other charges and wished to strengthen the case in some way against the registrant. In any event, it would be unfair for the matter to be resurrected given the unambiguous and unequivocal way in which the decision not to pursue it had been notified to the appellant. The appellant did have a substantive legitimate expectation that the matter would not be reopened and there was no countervailing public interest that justified the Committee frustrating that expectation.

***Council for the Regulation of Health Care Professionals v. General Medical Council and Saluja* [2006] EWHC 2784 (Admin)**

**2.08** Undercover journalist posing as patient—panel erred in granting stay

On 19 January 2006, the fitness to practise panel stayed proceedings against Dr S as an abuse of process on the basis of entrapment. An undercover journalist posing as a patient asked Dr S to provide her with a sickness certificate, telling him that she wanted the certificate to enable her to take time off work and have a holiday. In holding that the panel had manifestly erred in granting a stay, Goldring J derived the following from the authorities:

> 79. First, to impose a stay is exceptional.
>
> 80. Second, the principle behind it is the court's repugnance in permitting its process to be used in the fact of the executive's misuse of state power by its agents. To involve the court in convicting a defendant who has been the victim of such misuse of state power would compromise the integrity of the judicial system.
>
> 81. Third, as both domestic and European authority make plain, the position as far as misconduct of non-state agents is concerned, is wholly different. By definition no question arises in such a case of the state seeking to rely upon evidence which by its own misuse of power it has effectively created. The rationale of the doctrine of abuse of process is therefore absent. However, the authorities leave open the possibility of a successful application for a stay on the basis of entrapment by non-state agents. The reasoning I take to be this: given sufficiently gross misconduct by the non-state agent, it would be an abuse of the court's process (and a breach of Article 6) for the state to seek to rely on the resulting evidence. In other words, so serious would the conduct of the non-state agent have to be that reliance upon it in the court's proceedings would compromise the court's integrity. There has been no reported case of the higher courts, domestic or European, in which such "commercial lawlessness" has founded

a successful application for a stay. That is not surprising. The situations in which that might arise must be very rare indeed.

82. As will become apparent, I do not accept that for a journalist to go into a doctor's surgery and pretend to be a patient in circumstances such as the present is similar to abuse of power by an agent of the state.

83. Fourth, in the present disciplinary hearing there is no state involvement in the proceedings being brought. These are proceedings brought against a doctor by his regulator in order to protect the public, uphold professional standards and maintain confidence in the profession. These are to a significant degree different considerations from those that apply to a criminal prosecution and misuse of executive powers by the state's agents.

84. Fifth, it would an error of law in considering any application for abuse of process for the tribunal not to have well in mind the differences to which I have referred. It would not be appropriate for an FPP to approach the conduct of journalists as though they were agents of the state.

85. Sixth, "commercial lawlessness" can be a factor in an application to exclude evidence under section 78, although again different considerations apply as between state and non-state agents.

86. Seventh, when deciding in any given case whether there has been an abuse of process, the tribunal, here the FPP, is exercising a discretion. In doing so, it must consider all the facts of the case as well as the factors to which I have referred. While guidance can be obtained from such aspects as were referred to in *R v. Looseley* [2001] 1 WLR 2060, no one aspect is determinative and the aspects there set out are not exhaustive.

87. Eighth, if the defendant's Article 8 rights have been infringed that is merely a matter to be taken into account when deciding whether there has been an abuse of process or, (and it amounts to the same thing), his Article 6 rights have been infringed: see for example *R v. P* [2002] 1 AC 146 and *Jones v. University of Warwick* [2003] 1 WLR 954.

88. Ninth, section 78 is concerned with the admissibility of evidence. As Lord Nicholls said in paragraph 12 of *Looseley* (above), it is directed primarily at matters going to the fairness of the conduct of the trial; the reliability of the evidence, how the defendant might test it and so on. Entrapment does not mean the evidence must be excluded. It is a factor to take into account. In considering broader matters going to fairness it is necessary to bear in mind the features referred to above (among others).

### *R (Wood) v. Secretary of State for Education* [2011] EWHC 3256 (Admin)

**2.09**

Letter sent by Secretary of State informing W of no further action to bar him from working with children—legitimate expectation—later decision to bar him—public interest in protecting children

In this claim for judicial review, W challenged the decision of the Secretary of State under section 142 of the Education Act 2002 to bar him from working with children. W's entire professional life had been spent as a teacher. In essence, the claimant's complaint was that his case had been investigated by the Department of Education between 2003 and 2005, but that, at that stage, it had been decided that no action would be taken to bar him. In particular, he had been sent a letter dated 15 April 2005, which, W submitted, made it clear that no further action would be taken against him in the absence of further misconduct coming to the Department's attention. It was common ground in the instant claim that there was no evidence or allegation of any misconduct since that time. In those circumstances, the claimant submitted, the decision to bar him in October 2009 was unlawful. His main ground was that the decision was an abuse of process because it was taken in breach of a substantive legitimate expectation.

Singh J accepted that the Department's letter of 15 April 2005 created a legitimate expectation that the claimant would not have further action taken against him unless further misconduct were to come to the Department's attention. In his view, the letter did contain a representation to that effect, which was clear, unambiguous, and devoid of relevant qualification. The main point taken on behalf of the Secretary of State was whether the defendant was entitled to change his mind because there was an overriding reason in the public interest to do so. It was common ground that the test for determining whether the Secretary of State has acted lawfully when acting in a way that is inconsistent with a substantive legitimate expectation is whether: (a) there was a legitimate aim in the public interest; and (b) his conduct satisfied the principle of proportionality (*Nadarajah v. Secretary of State for the Home Department* [2005] EWCA Civ 1363; *Paponette v. Attorney General of Trinidad and Tobago* [2011] 3 WLR 219). As to (a), there was, and could be, no real dispute. There clearly was a legitimate aim: the public interest in protecting children—in particular, in protecting them from the risk of sexual abuse—is manifest and present. As to (b), the principle of proportionality, the learned judge said that it did not mean that every decision to reconsider a case will always be proportionate; everything depends on the facts of each case. Whilst conscious of the impact on the claimant of the Secretary of State's decision to reconsider W's case and to make the barring order, the learned judge said that he was not persuaded that there was sufficient to outweigh the overriding public interest that justified the decision. The Secretary of State's decision satisfied the principle of proportionality. There was no breach of the claimant's rights under article 6 or 8 of the Human Rights Act 1998.

***Bhatt v. General Medical Council*** **[2011] EWHC 783 (Admin)**

**2.10** See paragraph 21.09.

***Quarters Trustees Ltd and ors v. Pensions Regulator and anor*** **[2012] Pens LR 415**

**2.11** See paragraph 25.16.

## D. Other Relevant Chapters

**2.12**

Appeals, Chapter 4
Bias, Chapter 6
Case Management, Chapter 8
Delay, Chapter 17
Double Jeopardy, Chapter 21
Estoppel, Chapter 24
Evidence, Chapter 25
Investigation of Allegations, Chapter 39
Natural Justice, Chapter 48

# 3

# ADJOURNMENT



## A. Legal Framework

**3.01** Examples include:

Nursing and Midwifery (Fitness to Practise) Rules 2004, rule 32 (paragraph (5) provides that, before adjourning the proceedings, the practice committee shall consider whether or not to make an interim order)

General Pharmaceutical Council (Fitness to Practise and Disqualification etc) Rules 2010, rule 37 (where a person concerned applies for a postponement or adjournment on grounds of ill health, the person concerned must adduce appropriate medical certification and the chair may require the person concerned to submit to be examined by a registered medical practitioner approved by the Council)

General Medical Council (Fitness to Practise) Rules 2014, rule 29 (postponements and adjournments)

Solicitors Disciplinary Tribunal Policy/Practice Note on adjournments (the existence of other proceedings, lack of readiness, ill health, and inability to secure representation will not generally be regarded as providing justification for an adjournment; the tribunal will expect the respondent to support a late application for adjournment with a statement of truth as to the reasons for the sought adjournment)

## B. General Principles

### *Teinaz v. Wandsworth London Borough Council* [2002] ICR 1471, CA

**3.02** Onus on applicant for adjournment—medical grounds—fair trial

Peter Gibson LJ (with whom Arden LJ and Buckley J agreed) said:

> 21. A litigant whose presence is needed for the fair trial of a case, but who is unable to be present through no fault of his own, will usually have to be granted an adjournment, however inconvenient it may be to the tribunal or court and to the other parties. The litigant's right to a fair trial under article 6 of the European Convention on Human Rights demands nothing less. But the tribunal or court is entitled to be satisfied that the inability of the litigant to be present is genuine, and the onus is on the applicant for an adjournment to prove the need for such an adjournment.
>
> 22. If there is some evidence that a litigant is unfit to attend, in particular if there is evidence that on medical grounds the litigant has been advised by a qualified person not to attend, but the tribunal or court has doubts as to whether the evidence is genuine or sufficient, the tribunal or court has a discretion whether or not to give a direction such as would enable the doubts to be resolved. Thus, one possibility is to direct that further evidence be provided promptly. Another is that the party seeking the adjournment should be invited to authorize the legal representatives for the other side to have access to the doctor giving the advice in question. The advocates on both sides can do their part in assisting the tribunal faced with such a problem to achieve a just result. I do not say that a tribunal or court necessarily makes any error of law in not taking such steps. All much depends on the particular circumstances of the case. I make these comments in recognition of the fact that applications for an adjournment on the basis of a medical certificate may present difficult problems requiring practical solutions if justice is to be achieved.

## C. Disciplinary Cases

### *In re S (A Barrister)* [1970] 1 QB 160

**3.03** Defendant taken by surprise—no opportunity to answer charge

The Visitors (Paull, Lloyd-Jones, Stamp, James, and Blain JJ) said, at 176D–176E, that the standing orders of the Senate of the Four Inns of Court provide for the chairman giving such directions regarding the conduct of the hearing as he considers necessary for securing a proper opportunity for the barrister concerned to answer any charges. An adjournment ought, of course, to be granted if the barrister is taken by surprise at any of the evidence and has no proper opportunity of meeting it.

### *R v. United Kingdom Central Council for Nursing Midwifery and Health Visiting, ex parte Thompson, Machin and Wood* [1991] COD 275

**3.04** Lengthy adjournment between hearing dates—whether breach of natural justice

Disciplinary proceedings against the applicants were started on 30 July 1990. The hearing resumed on 29 and 30 October 1990, when the complainant gave her evidence and was cross-examined, and submissions were begun on behalf of one of the defendants. It became clear that the Professional Conduct Committee (PCC) would not have time to hear submissions on behalf of all of the registrants on that date and the proceedings were then adjourned until 4 March 1991, being the earliest date on which the Committee could reconvene because of the difficulty of reassembling the same Committee and the number of cases coming before the Committee as a whole. In proceedings for judicial review, Potts J said that the Court assumed that it had jurisdiction to interfere with the decision of the Committee and to consider whether there had been prejudice by reason of a breach of the principles of natural justice or an abuse of process. Whether or not there was prejudice amounting to a breach of natural justice, or an abuse of process, was a matter of degree depending on all of the circumstances of the case. Given the nature of the charges and the issues, the state of the evidence, the timescale, and the fact that the respondent had no alternative but to adjourn the matter as it did, the applicants had failed to establish prejudice to such a degree as to enable the Court to grant the relief sought. There was no breach of the principles of natural justice and no abuse of process.

***Spofforth v. General Dental Council*, Privy Council Appeal No. 5 of 1999**

**3.05** Registrant being treated by consultant psychiatrist—undue weight given to public interest—insufficient weight given to mitigating circumstances—counsel not instructed if matter not adjourned

The Privy Council allowed an appeal from the decision of the PCC to refuse an adjournment. S pleaded guilty before Liverpool Crown Court of seven counts of forgery and false accounting in relation to two grants from the Department of Trade and Industry. The sentencing judge observed that the Department would probably have given S the money anyway had he gone about his application in a different way; accordingly, it was not a case for a prison sentence. S's solicitors instructed counsel to attend the disciplinary hearing before the respondent's PCC on 25 January 1999 to make an application for an adjournment. Counsel advanced two reasons: one was the mental condition of S, who was being treated by a consultant psychiatrist; the other was that S was in the process of moving house as a result of bankruptcy proceedings. Counsel for S informed the Committee that she was not instructed to represent him if the matter was not adjourned and proceeded to a hearing. In allowing S's appeal, the Privy Council (Lord Hoffmann, Lord Hutton, and Sir Andrew Leggatt) said that it was concerned only to consider the lawfulness of the Committee's decision to refuse an adjournment. However, in order to decide this issue, their Lordships had to express some views on the nature of the case, although the expression of such views was not intended to influence the decision to which a differently constituted Committee might ultimately come. In the opinion of their Lordships, the criminal offences that S committed, although grave, were unusual, in that he was entitled to the monies that he claimed and could have obtained those monies properly had he supported his claims with valid documents. There was no suggestion that S's patients had suffered in any way because of the crimes that he had committed and there was no suggestion that he had made improper claims for payment from the National Health Service. Their Lordships considered that S had a case to make before the Committee that his conduct had not been so grave as to show that he was unfit to continue to practise his profession. There were reasons of considerable weight for granting an adjournment in order to give S an opportunity to instruct counsel, when he was in a fit mental condition to do so, or to present his case in person to the Committee. Their Lordships considered that there was little weight in the points that the public interest required the matter to proceed on 25 January 1999, or that there would have been public concern or a lack of public confidence if the Committee had granted an adjournment because of S's mental ill health. The Committee misdirected itself in giving undue weight to what it regarded as the public interest, and in giving inadequate weight to the unusual and mitigating circumstances of the criminal offences, to the previously unblemished reputation of S, and to the mental ill health from which he suffered.

*Brabazon-Drenning v. United Kingdom Central Council for Nursing, Midwifery and Health Visiting* [2001] HRLR 6

**3.06** Adjournment on medical grounds sought for short period—appellant not practising

On 1 December 1999, the PCC found the appellant guilty of each of six charges and removed her from the register. The appellant challenged the decision of the Committee on a number of grounds, first among which was that the Committee should have adjourned the hearing so as to permit her to be represented and to appear. Council submitted that, in view in particular of the medical evidence, it was an infringement of the principles of natural justice to deny the appellant the right to an adjournment. A doctor's letter before the Committee stated that she was suffering from mixed anxiety and depression, and in her present state was not able to withstand the rigours of a disciplinary hearing, although, in the doctor's opinion, her eventual recovery would depend on her getting the inquiry out of the way and therefore a new date, reasonably early in the new year, would be appropriate. Counsel noted that this was her first application for an adjournment, that it was only for a relatively short period, and that both the doctor and the appellant's mother indicated that a new hearing could take place in the new year. The appellant had no intention of practising as a nurse until a decision had been taken. Elias J (with whom Rose LJ agreed) said that the hearing plainly should have been adjourned. Save in very exceptional cases in which the public interest points strongly to the contrary, it must be wrong for a PCC, which has the livelihood and reputation of a professional individual in the palm of its hands, to go on with a hearing when there is unchallenged medical evidence that the individual is simply not fit to withstand the rigours of the disciplinary process.

*Baba v. General Medical Council* (2001) 62 BMLR 34

**3.07** Medical evidence before Committee inadequate to support adjournment—fresh evidence raising real question whether doctor fit to give instructions

The Privy Council (Lord Bingham of Cornhill, Lord Hope of Craighead, and Sir Philip Otton) said that the Board could not criticize the decision of the PCC of the General Medical Council (GMC) to refuse B's application for an adjournment on 27 January 2000. The medical certificate signed by B's general practitioner (GP), which was dated 26 January 2000, which the Committee was well placed to evaluate, fell far short of what was needed to support such an application at such a stage. The Committee was bound to have regard to the undesirability of requiring one particular witness to attend on another occasion. The decision was one for the Committee and, on the material before it, it was fully entitled to reach the decision that it did. However, the medical evidence now before the Board did raise a real question whether B was, at the time of the hearing, in a fit state to give coherent instructions. The GP's notes described him on 26 January 2000 as 'incoherent' and on the following day as 'unable to hold proper conversations'. Reports obtained from various sources suggested that B may have been hypoglycaemic on 27 January 2000, with consequent detriment to his powers of comprehension and concentration. The evidence caused the Board to doubt whether, despite appearances, B was able on that date to apply his mind to the issues, to appreciate the effect of the advice that he was receiving, and to give instructions accordingly. In the opinion of the Board, there was some risk of injustice and it was appropriate that the matter should be remitted to a differently constituted committee to consider the matter afresh.

*Awan v. Law Society* [2003] EWCA Civ 1969

**3.08** Tribunal entitled to consider whether medical evidence properly supports adjournment—Tribunal not obliged to grant adjournment

On 7 August 2001, the Solicitors Disciplinary Tribunal (SDT) struck A from the roll of solicitors. He appealed to the Administrative Court, which, on 12 December 2001, dismissed his appeal. He appealed against the decision of the Administrative Court with the permission of Schiemann LJ to the Court of Appeal (Lord Phillips MR, and May and Carnworth LJJ). Prior to the hearing before the SDT, A obtained a certificate from his doctor stating that he should refrain from work on account of anxiety and depression; he submitted this certificate to the SDT. The Tribunal replied, emphasizing that it required a doctor's certificate and prognosis. No additional certificate was submitted and the Court of Appeal said that the original certificate did not show that A was not fit to attend the hearing. It made no mention of his having had a stroke or having high blood pressure—the conditions that he put forward to the SDT over the telephone when he said he was not fit to attend. The Court of Appeal said that, having regard to the past history, the Tribunal was entitled to proceed on the basis that A had failed to demonstrate that his health prevented him from attending the hearing. The requirement for evidence could not have been made clearer, as equally was true of the warning as to the consequences that might follow if he did not produce the necessary evidence. His conduct had all of the hallmarks of prevarication. May LJ said that if medical reasons are advanced, the Tribunal may well require production of a medical report or certificate in support of the application. If a report or certificate is produced, the Tribunal is entitled to consider whether it sufficiently supports the reason of the adjournment upon which it is relying. It is not obliged to grant the application to adjourn simply because a medical certificate is produced, whatever its content.

*Kilshaw v. Office of the Supervision of Solicitors* [2005] EWHC 1484 (Admin)

**3.09** Medical evidence—insufficient to show appellant unable to attend, prepare, or conduct his case

On 18 July 2002, some five days before the hearing before the SDT on 23 July 2002, the appellant first raised the question of his health, and sought an adjournment on the basis that he was unwell and had been so for some time. He said he could not travel from Chester, where he lived, to London the following week because this would be against medical advice. He enclosed a note from his GP, which stated that he had symptoms of anxiety and depression. This was followed up by a medical note, which stated that the appellant had difficulty preparing his case at the present time and that there were some elements of depressive illness. In refusing the application for an adjournment, the Tribunal said that the medical evidence was insufficient to convince it that the appellant was not well enough to attend, or to prepare or conduct his case. In dismissing his appeal to the Divisional Court against an order of striking-off, Rose LJ said:

> 28. As it seems to me, so far as the question of the SDT granting an adjournment is concerned, that decision has to be assessed in the light of the material which was before the tribunal on the date on which the hearing was refused. It is apparent from the history of events which I have set out that the application to adjourn was made at a very late stage, some five days before the hearing. It is also apparent that, whatever the reason why [the appellant] had only very recently sought medical evidence, the fact was, so far as the tribunal were concerned, that he had only sought medical assistance a few days before the hearing was due to take place. The shortcomings of the medical evidence provided in support of the application are apparent on the face of those documents as is in part conceded by [the appellant's representative].
>
> 29. All of those very recent events, the tribunal were entitled to look at, as it seems to me, in the context of the earlier dealings which [the appellant] had had with the tribunal, up to and including, in particular, the letter to which I have referred, written by him on 22nd June.

30. In the light of all those matters, it seems to me that there was ample material entitling the tribunal to conclude that the application for an adjournment was a manoeuvre by [the appellant] to put off the evil day.

### *Vaidya v. General Medical Council* [2007] EWHC 1497 (Admin)

**3.10** Hearing refixed for convenience of V—delay by V seeking legal representation—no oral application to adjourn by solicitor, counsel, or V

V appealed from the decision of the fitness to practise panel given on 14 September 2006, whereby it found charges of serious professional misconduct proved and directed that his name be erased from the medical register. His grounds of appeal included that he was not given a fair hearing, contrary to Article 6 of the European Convention on Human Rights (ECHR), on the basis that the panel wrongly failed to grant him the adjournment he sought and instead conducted the hearing in his absence. It was common ground that Article 6 ECHR was engaged, and it was also common ground that the notice of proceedings was duly and properly served on V. What was in dispute was whether the panel was entitled, in the circumstances of the case, to exercise its admitted discretion to refuse to adjourn the hearing and/or to proceed in the absence of V. After referring to *R v. Jones* [2003] 1 AC 1, Bennett J said, at [47], that there were a number of important matters in this case in considering the question of whether the panel failed to exercise its discretion by refusing the request for an adjournment. The GMC was anxious to fix the hearing to a date when V could attend and had vacated several dates, at the request of V, prior to fixing the dates in April 2006. The date of 11 September and the following four days were convenient to V. At no time did he, or his solicitor, contend that V was unable to attend on these dates. V delayed seeking legal representation for a period of three months. Unknown to the panel, V's solicitor had tried to instruct a particular counsel, but, when he learned of her unavailability, had apparently done nothing to find anyone else to represent V. No counsel or solicitor was ever instructed to attend before the panel to make an oral application for an adjournment despite being informed that it would be open to V to do just that. Bennett J held that, to adopt the words of Lord Hutton at [36] in *Jones*, where no defence was put forward by V at the hearing in consequence of his deliberate decision not to be present, there was no violation of Article 6(3)(c) ECHR. In the circumstances, the panel was entitled to exercise its discretion to refuse the application for a postponement and to continue in V's absence.

### *R (Thompson) v. Professional Conduct Committee of the General Chiropractic Council* [2008] EWHC 2499 (Admin)

**3.11** Defence expert unavailable—procedural fairness—evidence potentially relevant

The claimant challenged the decision of the PCC to hold a substantive hearing when a particular expert witness, whom the claimant proposed to call, would not be available. The Committee had been given a list of dates of the expert's availability. The Committee had chosen the hearing date, after considering the interests of all parties and the need to avoid unnecessary delay. It took the view that the expert's evidence was not necessary because it did not go to the heart of the case. The Court granted judicial review. Lloyd Jones J analysed the issue as being one of procedural fairness of the proceedings before the Committee. Whilst courts would be slow to interfere with case management decisions taken by a professional disciplinary body, this was subject to the supervisory jurisdiction of the Court to ensure that the parties were given a fair hearing. The evidence of the claimant's expert witness was crucial to his defence. The Court made clear that it was expressing no view as to whether the proposed defence may or may not be valid, but it seemed to the Court that it was a matter that was potentially relevant and which the claimant should be permitted to ventilate at the hearing before the Committee. The Court rejected the Committee's claim that a further short adjournment (which would have permitted the expert to attend) would have resulted in any substantial prejudice to its obligations to hear cases promptly.

### *R (Tinsa) v. General Medical Council* [2008] EWHC 1284 (Admin)

**3.12** Subsequent medical evidence—whether appellant not in fit state to defend himself

The appellant (who had been convicted of two offences of public order and failing to surrender to bail) claimed that the proceedings should have been adjourned after the finding of serious professional misconduct because he was suffering from mental illness, which prevented him from representing himself properly, and that the panel should have obtained a psychiatric report to assist with sanction. The Court (Underhill J) found that no medical evidence was brought before the panel to support any claim that the appellant was suffering at the time from any mental ill health or felt under any difficulty in defending himself, still less that he wanted an adjournment. The Court acknowledged that an appeal against a decision of the panel may, however, be allowed in circumstances in which subsequent evidence establishes that an appellant was not in a fit state to defend himself at a hearing, even though the evidence in question was not before the panel at the time. However, the Court held that the evidence before it did not prove that the claimant had not been in a fit state to conduct his own defence. There was no basis whatsoever on which the panel could be said to have erred in law in failing on its own initiative to propose an adjournment to obtain medical evidence.

### *R (Rengarajaperunal) v. General Medical Council* [2008] EWHC 1953 (Admin)

**3.13** Impairment—whether to adjourn for up-to-date information—performance assessment to confirm or displace current impairment

R appealed against the panel's finding that his fitness to practise was impaired because of misconduct and deficient professional performance. He was made subject to conditions. The panel found a number of allegations proved in relation to R's practice as a consultant orthopaedic surgeon in relation to three patients. One allegation was that R had acted beyond the limits of his professional competence and another was that he lacked insight into the limits of his professional competence. There was no challenge to the findings, and counsel accepted that there would have been no challenge to the conclusion that those findings showed serious misconduct and deficient professional performance. After finding the facts proved, the panel heard submissions as to whether R's fitness to practise was impaired. It decided that it had insufficient information to reach a decision, and determined to adjourn and to direct that an assessment of R's professional performance be carried out. This led to some consternation on counsel's part, who submitted that if the panel was satisfied that there was insufficient evidence to enable it to find that R's fitness to practise was impaired, it had to find that there was no impairment. The panel decided to set aside its decision on the adjournment and, following further submissions, it returned with its final conclusion that R's fitness to practise was indeed impaired. Ouseley J dismissed R's appeal based on an alleged error of law through perversity. The learned judge rejected the submission of counsel on behalf of R that the gravity of the facts found were not sufficiently serious or persistent to warrant a finding of misconduct or deficient professional performance. In its decision on impairment, the panel had not changed its mind. The most logical explanation was that the panel, in the light of submissions made to it, thought that it should not only reach a conclusion based on the occasions on which it had made adverse findings of fact. If the panel members had asked themselves whether they had up-to-date and full information with which to look forward, they would have said that they did not. To that end, a performance assessment was sought to help to protect the public. That course was consistent with the conclusion that there was no inconsistency between the decisions. More up-to-date information could be useful in confirming or displacing the position that the panel had arrived at in relation to fitness to practise, based on its findings of fact.

### *Khan v. General Teaching Council for England* [2010] EWHC 3404 (Admin)

**3.14** Adjournment to examine original statements—no basis for adjournment

The Administrative Court (Ouseley J) dismissed the appellant's appeal against the decision of the respondent Council's PCC not to adjourn a hearing into her conduct, finding that, on the facts of the instant case, the Committee had not erred in refusing to grant the adjournment. The appellant allegedly made derogatory and inappropriate remarks whilst teaching at a school. The appellant's request for an adjournment of the hearing before the Committee was refused and the hearing went ahead in her absence. A prohibition order for two years was imposed. The essence of the appellant's case was that the adjournment that she sought should have been granted and that the hearing was unfair because it relied upon the transcribed words of the students concerned, whilst she did not see the original statements, nor did the Committee, nor the General Teaching Council. As to the adjournment, Ouseley J said that it was necessary for the appellant to show that the refusal of an adjournment was outside the proper exercise by the Committee of its discretion, that it erred in principle, that it was based on irrelevant considerations or ignored relevant ones, or that it was a thoroughly unreasonable or unfair decision. The learned judge held that the Committee was right to proceed with the hearing in the absence of the appellant and that even if there was a proper case for an adjournment, the Committee did not reach a decision that could in any way be described as wrong in principle or erroneous in the exercise of its discretion. In reality, there simply was no justification for an adjournment. On the merits of the appeal, it was for the appellant to show that the decision on the facts by the PCC was wrong. If the transcripts were accurate—and there was no reason the Committee had to doubt that, especially as the Committee knew that the appellant's representatives had had the opportunity to check the accuracy of the statements—sight of the original statements would not reveal whether there had been collusion or not, nor would it reveal whether there had been manipulation or not. That could be addressed only by examining the circumstances in which the statements came to be taken, the period of time between the allegations being made and the statements being taken, and the nature and contents of the statements themselves. The learned judge held that both bases of appeal—namely, wrongful refusal of an adjournment and a wrong decision on the facts—were not made out at all.

### *Thaker v. Solicitors Regulation Authority* [2011] EWHC 660 (Admin)

**3.15** Case opened beyond identified transactions—T and counsel not expecting to meet case beyond pleaded transactions

T appealed against a decision of the SDT that he be struck off the roll of solicitors. The principal ground was that the Tribunal erred in failing to grant an adjournment at the start of the hearing. A second ground was that the Tribunal erred in allowing the Solicitors Regulation Authority (SRA) to make submissions and to call evidence that went beyond the twelve pleaded transactions. The Administrative Court (Jackson LJ and Sweeney J) said that the pleadings were in a state of some chaos and that T had known for only ten days which transactions were relied upon by the SRA. Further, when the solicitor advocate opened the case before the Tribunal, he sought to pursue allegations that went well beyond the twelve relevant transactions that had been identified. It became clear that the SRA was presenting a case that went far beyond the twelve relevant transactions and it was not a case that T or his counsel was expecting to meet. T was being subjected to a process that was simply unfair. Furthermore, this was a process in which his livelihood was at stake.

*Chaudhari v. General Pharmaceutical Council* **[2011] EWHC 3433 (Admin)**

**3.16** Medical certificate—Committee entitled to scrutinize medical evidence supplied—party to demonstrate that health prevents attendance at hearing

King J said that the Disciplinary Committee of the General Pharmaceutical Council (the Royal Pharmaceutical Society at the time of the disciplinary hearing) had given detailed rulings refusing the appellant's applications for adjournment on the grounds of ill health made during the course of the fitness to practise hearing. It was clear that the Committee had a discretion under regulation 19 of the Pharmaceutical Society (Statutory Committee) Regulations 1978 to proceed with the inquiry in the registrant's absence or to adjourn the hearing. It was not bound to grant any adjournment simply because of medical certificates supplied by the registrant and the Committee was entitled fully to scrutinize the medical evidence supplied. A party who claims not to be fit to attend a professional disciplinary inquiry has to demonstrate that his health prevents attendance at the hearing: see, in particular, *Awan v. Law Society* [2003] EWHC 1969, [13]–[14] *per* Lord Phillips MR and [60] *per* May LJ. King J considered that the Committee, in each of its rulings, gave careful reasons, including by reference to the previous history of applications for adjournments, why it was not satisfied that the appellant had discharged the burden upon her to demonstrate the requisite degree of ill health and why, in effect, it concluded that the applications to adjourn were no more than a delaying tactic by the appellant to put off the evil day. The Committee had considered a number of authorities, including *R v. Jones* [2002] UKHL 5 and *R v. Haywood and ors* [2001] QB 862. King J said:

> 84. In this case, the Committee gave detailed rulings on 10th February 2009, 11th February 2009 and 14th February 2009 refusing the appellant's applications for adjournment on the grounds of ill-health. It is clear that the Committee had a discretion under [regulation 19 of the Pharmaceutical Society (Statutory Committee) Regulations 1978] and were not bound to grant any adjournment simply because of the medical certificates supplied by the registrant and they were entitled carefully to scrutinise the medical evidence supplied. I accept the proposition submitted by the respondent that a party who claims to be not fit to attend a professional disciplinary inquiry has to demonstrate that his health prevents attendance at the hearing; see in particular Lord Phillips MR in *Awan v. Law Society* [2003] EWHC 1969 at paragraphs 13 and 40 and May LJ at paragraph 60, who said this: "Courts and tribunals have sometimes to consider applications to adjourn which look as if they may be advanced for insubstantial reasons in order to put off a hearing which the Applicant would rather not face up to. If medical reasons are advanced the Tribunal may well require production of a medical report or certificate in support of the application. If a report or certificate is produced the Tribunal is entitled to consider whether it sufficiently supports the reason of the adjournment which is relied upon. It is not obliged, in my judgment, to grant the application to adjourn simply because a medical certificate is produced, whatever its content."

*R (French) v. Chief Constable of West Yorkshire Police* **[2011] EWHC 546 (Admin)**

**3.17** Officer in case charged with misconduct—socializing with drug dealer—defence that drink spiked—forensic report produced on morning of hearing—panel right to conclude that report did not exclude possible spiking

On three separate occasions during 2009 and 2010, PC F (a police constable with the West Yorkshire police force) faced misconduct proceedings brought under the Police (Conduct) Regulations 2008. On 24 August 2009, a misconduct meeting took place concerning the drunken behaviour of PC F on 28 October 2008. The allegation was admitted. PC F was given a final warning for a period of eighteen months. On 17 December 2009, a misconduct hearing took place concerning the failure of PC F to attend court on 19 June 2009 and to give evidence on behalf of the prosecution in relation to a driving offence. The misconduct panel consisted of three members, including an independent member Mr F. The allegation was admitted by PC F. After hearing mitigation including evidence of good performance, the panel decided not to dismiss F, but to extend the final warning by a further eighteen months. On 25 August 2010, a misconduct hearing took place concerning the behaviour of PC F on the night of 4–5

March 2010. It was alleged that, during a night out in Wakefield, she socialized with a man (H) charged with a drugs offence in respect of which she was the officer in the case. She became drunk, provided information to H about the case against him, and behaved in a sexually inappropriate way. PC F's defence was that her drink had been spiked by some form of drug and that her behaviour was therefore involuntary. The misconduct panel consisted of three members including the same independent member, Mr F. PC F was not made aware that Mr F was a member of the panel until the day of the hearing. Further, during the course of the hearing, a report from a forensic scientist, Dr P, was introduced in evidence. The report contained an analysis of a hair sample that had been provided by PC F in May 2010. Faced with this report and the presence of Mr F, an application was made for an adjournment on behalf of PC F. The application was refused and the hearing continued. The allegations were found proved and PC F was summarily dismissed without notice.

In judicial review proceedings, His Honour Judge Behrens rejected the claim that an adjournment should have been granted because of the presence of Mr F. Mr F was a relatively experienced member of the panel. He was one of three people on the police authority list. He became an independent member in April 2004; he had sat on six panels prior to 25 August 2010. He had undergone training in 2004 and 2008. Even if the decision rationale document that contained detailed information about the circumstances had been before the panel at the misconduct hearing on 17 December 2009 (of which there was no direct evidence), the Court was satisfied that Mr F and the panel were entitled to conclude that the proceedings on 25 August 2010 were not only fair, but could also be seen to be fair. The issue to be decided related to the question of spiking PC F's drink. It was totally different from the issue in relation to the earlier hearing. Much of the material from the earlier decision rationale was, in any event, contained in the material before the panel on 25 August 2010. A fair-minded observer would not regard it as unfair for Mr F to continue. As to the expert forensic evidence, the question was whether a reasonable panel, properly directed as to the law and in the circumstances of this case, could have properly refused to grant the adjournment requested on behalf of PC F. Dr P had provided a report of a blood sample analysis that had been served prior to the hearing and which was, to a large extent, unfavourable to PC F. Hair samples had been taken and it was this report that was produced only at the hearing. The Court agreed with counsel for the claimant that Dr P's second report was an important piece of evidence and that it would have been better if PC F had sight of it before the hearing. It was important because it provided no support for the suggestion that any sedative drugs, such as gamma-Hydroxybutyric acid (GHB), had been administered on the night of 4–5 March 2010. It also eliminated a number of other drugs. Taken with the first report on the blood sample, a large number of drugs were eliminated. The panel was, however, right to comment that the report did not exclude possible spiking. Since the hearing before the panel, PC F had obtained a forensic scientific report. This stated that there was no clear analytical evidence to show that the claimant was administered GHB or any other compound around the relevant time, but the possibility that she could have ingested the drug on the night in question remained. Overall, the findings did not assist in determining whether or not PC F had been exposed to exogenous GHB and therefore this remained a possibility. The Court, in dismissing the claim for judicial review, said that, in reality, PC F's expert agreed with Dr P. Neither expert provided any support for the proposition that PC F's drink was spiked. Equally neither was able to exclude the possibility that it was. Thus the panel's reasoning in refusing the adjournment on the grounds that there was no evidence to state categorically whether or not PC F took a sedative or date-rape drug on the night in question was proved accurate. The decision to adjourn was essentially discretionary and the Court would interfere only if very clear grounds were shown. The decision had to be shown to be unreasonable, and the panel's overall decision that it would neither be fair nor in accordance with the principles of natural justice to delay the proceedings by adjourning them could not be said to be so unreasonable that the Court was entitled to interfere.

***Rich v. General Medical Council* [2013] EWHC 1673 (Admin)**

**3.18** Adjournment to seek legal representation

An application was made by R for adjournment of his appeal hearing on the grounds that he needed time to seek legal representation and that he had no savings with which to pay any lawyers whom he might instruct. The application was refused by Males J. The effect of the challenge by the applicant was that the erasure remained in abeyance pending the outcome of the appeal, and at time of writing it had already been almost a year since it was imposed. It was in the public interest that the position should be resolved one way or the other. It was also impossible to imagine what difference an adjournment and legal representation would have made to the appellant's case.

***Gatawa v. Nursing and Midwifery Council* [2013] EWHC 3435 (Admin)**

**3.19** Doctor's letter—discretion to adjourn—procedural fairness—letter did not indicate G was unfit to attend or instruct someone

At the disciplinary hearing, G's representative applied for an adjournment on the basis of a doctor's letter stating that G was suffering from depression and anxiety, and that the doctor believed that G's ability to give coherent answers in cross-examination could be affected. Dismissing G's appeal, Andrews J said that the case turned on its own facts and that one has to look at the facts of the particular case. The distinguishing features in this case were, first of all, that the doctor's letter, although unchallenged medical evidence, did not indicate that G was unfit to instruct anybody to represent her. It did not say that she was going to be medically unfit to attend the hearing and it did not say that she lacked mental capacity. What it did say was that she may be incapable of producing coherent statements or instructions in evidence. The Conduct and Competence Committee (CCC) of the Nursing and Midwifery Council (NMC) had a discretion whether to grant an adjournment. Where someone suffers from mental illness, the Committee has to make a value judgment based on the evidence before it. The doctor's letter had not indicated that G was unfit to instruct anyone to represent her. The issue was one of procedural unfairness. The Committee could not be criticized for refusing an adjournment.

***Norton v. Bar Standards Board* [2014] EWHC 2681 (Admin)**

**3.20** Failure to apply *R v. Jones* test—no finding that defendant avoided attending tribunal hearing

N was charged with disciplinary offences of failure to declare on his call to the Bar two criminal convictions of unlawful possession of a CS gas spray and an offence of wilfully obstructing the police (charge 1); and wrongly declaring that he had degrees from Stafford University and Harvard University (charge 2). N claimed that he became aware of relevant letters from the Bar Standards Board only shortly before the hearing on 7 February 2014, which had been received on an old computer system, and asked for an eight-week adjournment in order to prepare for the hearing. N said that he also had to arrange suitable child care and that the train fare would be 'highly excessive', and that this was his first proposed adjournment of the case. In refusing an adjournment, a disciplinary tribunal said that there was a public interest in matters being concluded without undue delay, that N had not made the tribunal aware of his intended defence or any evidence on which he would rely in support of it, and that it was therefore doubtful whether an adjournment would achieve anything. In allowing N's appeal, quashing the order of disbarment, and remitting the case for a rehearing before a fresh disciplinary tribunal, the Divisional Court (Fulford LJ and Stewart J), hearing the appeal under section 24 of the Crime and Courts Act 2013, said that the attention of the tribunal was not drawn to *R v. (Anthony) Jones* [2013] 1 AC 1, that the submissions made to the tribunal tended to suggest that the discretion to proceed in the absence of the person charged was general and unfettered, and that prejudice was not a relevant issue. The procedures of the tribunal do not require the accused—at any stage—to indicate whether he intends to contest the charges or to provide a description of the defence that he proposes to advance. Although the extent of the information provided by the accused is a relevant consideration when ruling on an application on his behalf to adjourn, this must be accorded proportional treatment. The tribunal is obliged to focus on the *Jones* criteria, amongst which is the need for the tribunal to bear in mind the extent of the disadvantage to the defendant in not being able to give his or her account of events, having regard to the nature of the case against him or her. There were fundamental problems with the central matters referred to by the tribunal when set alongside the non-exhaustive list of factors that the Court of Appeal in *Jones* indicated need to be taken into account.

(1) Although the tribunal decided that N had been served with the documents substantially in advance of the hearing, it made no findings as to whether he had deliberately avoided attending the hearing, thereby waiving his right to appear, or had voluntarily chosen to be absent. Accordingly, the tribunal failed to address whether the reasons advanced by N justified his absence, regardless of when he had received the documentation.
(2) The tribunal failed to consider whether an adjournment might result in N attending at the next hearing.
(3) Although the tribunal expressed the conclusion that delaying the hearing was unlikely to achieve anything, this was based solely on N's failure to set out his defence in the application to adjourn as opposed to a more general review of the issues and evidence in the case.
(4) Although the tribunal correctly expressed the view that it is in the public interest for proceedings of this kind to be concluded timeously, there was no consideration of the lack of any victims or witnesses who would be prejudiced by a delay, and this was not a case in which the memories of witnesses would be adversely affected.

In summary, in order to exercise the 'severely constrained' discretion to conduct a trial in the accused's absence, the tribunal must proceed with the 'utmost care and caution', and it must apply those parts of the *Jones* criteria that are relevant. There was a real risk that the tribunal did not properly apply the factors that are of 'prime importance' in order to secure fairness to the defence, whilst also taking into account the need to be fair to the prosecution.

## D. Other Relevant Chapters

**3.21**

Absence of Practitioner, Chapter 1
Case Management, Chapter 8
Unrepresented Practitioner, Chapter 67

# 4

# APPEALS

*R (Cela) v. General Pharmaceutical Council* [2012] EWHC 2785 (Admin) 4.38
*Nagiub v. General Medical Council* [2013] EWHC 1766 (Admin) 4.39
*R (Adesina) v. Nursing and Midwifery Council; R (Baines) v. Nursing and Midwifery Council* [2013] EWCA Civ 818; [2013] 1 WLR 3156, CA 4.40
*Adegbulugbe v. Nursing and Midwifery Council* [2013] EWHC 3301 (Admin) 4.41
*R (Gyurkovits) v. General Dental Council* [2013] EWHC 4507 (Admin) 4.42
*Pinto v. Nursing and Midwifery Council* [2014] EWHC 403 (Admin) 4.43
*Parkin v. Nursing and Midwifery Council* [2014] EWHC 519 (Admin) 4.44
*Martin-Artajo v. Financial Conduct Authority* [2014] UKUT 0340 (TCC) 4.45
*Daniels v. Nursing and Midwifery Council* [2014] EWHC 3287 (Admin) 4.46
**D. Non-Attendance** 4.47
**E. Fresh Evidence on Appeal** 4.48
**F. Other Relevant Chapters** 4.49

Lincoln Crawford

## A. Legal Framework

**4.01** Appeals to the High Court are governed by Civil Procedure Rules, Part 52 (see also the non-exclusive list of statutory and special appeals in Practice Direction 52D).

The Hearings before the Visitors Rules 2010 cease to have effect against relevant decisions made on or after 7 January 2014.

Appeals relating to the regulation of the Bar now generally lie to the High Court, under section 24 of the Crime and Courts Act 2013.

Appeals under section 17 of the Veterinary Surgeons Act 1996 against a direction by the Disciplinary Committee of the Royal College of Veterinary Surgeons lie to the Privy Council.

Some regulatory bodies have internal appeal procedures, for example:

- the Institute of Chartered Accountants in England and Wales (ICAEW);
- the Chartered Institute of Management Accountants (CIMA);
- the Chartered Institute of Legal Executives (CILEX);
- the Jockey Club;
- the Police Appeals Tribunals Rules 2012; and
- the Bar Standards Board Fitness to Practise Regulations 2014, rule E317 (appeal against a decision to impose, extend, vary, or replace a period of restriction).

## B. General Principles

### *Hughes v. Architects' Registration Council of the United Kingdom* [1957] 2 QB 550

**4.02** Wide power of appeal—similar from judge to Court of Appeal

Lord Goddard CJ, at 557–8, said that the first and, to his mind, perhaps the most important question in the case was to determine the powers of the court on an appeal. Section 9 of the Architects (Registration) Act 1931 provides that any person aggrieved by the removal of his name from the register or by a determination of the Council that he be disqualified for registration during any period may appeal to the High Court, and on any such appeal 'the court may give such directions in the matter as they think proper, and the order of the court shall be final'. Lord Goddard said that a section in these terms confers a right of appeal as wide as one from a judge to the Court of Appeal:

see *Allender v. Royal College of Veterinary Surgeons* [1951] 2 All ER 859. Whilst an appellate court will always attach great importance to the findings of a lower court, especially on findings of fact, if, in its opinion, the decision below is wrong, the appellate court must give effect to its opinion and reverse that decision.

### *Tarnesby v. General Medical Council*, Privy Council Appeal No. 21 of 1969

**4.03** Rehearing—weight to be attached to decision of committee

The appellant was found guilty of advertising himself and his practice in breach of provisions under the Abortion Act 1967; he appealed against the finding of guilt and against the direction for erasure. In delivering the judgment of the Privy Council, Lord Pearson said that the question of whether a doctor's conduct amounts to advertising for his own professional advantage and thus to an improper attempt to profit at the expense of professional colleagues is entrusted to the decision of the disciplinary committee, subject to appeal to their Lordships. In a matter of this kind, the decision of the disciplinary committee, composed of members of the same profession, must carry weight. As Lord Upjohn said in *McCoan v. General Medical Council* [1964] 1 WLR 1107, 1112 (although in relation to misconduct of a different character): 'The Medical Acts have always entrusted the supervision of the medical advisers' conduct to a committee of the profession, for they know and appreciate better than anyone else the standards which responsible medical opinion demands of its own profession.' Lord Pearson continued that although the decision of the disciplinary committee on the eminently professional questions involved in this case must carry great weight and although, in relation to questions of credibility, the committee has the advantage of seeing and hearing the appellant as he gave his evidence and so of observing his demeanour, nevertheless the appeal is by way of rehearing and puts in issue matters of fact, as well as matters of law. Lord Radcliffe said in *Fox v. General Medical Council* [1960] 3 All ER 225, 226:

> The appeal in this case lies as of right and by statute—see section 56 of the Medical Act 1956. The terms of the statute that confers the right do not limit or qualify the appeal in any way, so that an appellant is entitled to claim that it is in a general sense nothing less than a re-hearing of his case and a review of the decision. Nevertheless, an appellate court works under certain limitations which are inherent in any appeal that does not take the form, as this does not, of starting the case all over again and hearing the witnesses afresh.

### *Calvin v. Carr* [1980] AC 574

**4.04** Decision void—breach of natural justice—whether 'cured' by appeal proceedings

The appellant was the part-owner of Count Mayo, a horse that ran in a handicap at Randwick Racecourse in New South Wales, Australia. Following an inquiry by the stewards, the jockey was found guilty of an offence against rule 135(a) ('every horse shall run on its merits'); the appellant was a party to the breach and the stewards disqualified him for one year. The appellant appealed to the Committee of the Australian Jockey Club alleging a breach of natural justice. The Committee had the transcript of the proceedings before the stewards. The chairman of the stewards and the other stewards, with one exception, gave evidence. All persons who gave evidence before the stewards were in attendance before the Australian Jockey Club and were available for cross-examination; several of them were called. Other witnesses also gave evidence, including the appellant, the trainer's foreman, and the jockey. The form of the race and forms of the horse's performance in New Zealand were shown in the presence of the parties. The Committee dismissed the appeals of the appellant and the jockey, and the appellant brought an action against the chairman of the Committee of the Australian Jockey Club as representing the Club, the members of the Committee, and the stewards holding office under the rules of racing. The action was dismissed. The judgment of the Privy Council was delivered by Lord Wilberforce, who said that the first issue was whether the Committee had any jurisdiction to enter upon the appeal. The appellant's proposition was that the stewards' decision was void. The Privy Council held that where the question is whether an appeal lies, the impugned decision cannot be considered as totally void, in the sense of being legally non-existent. An analogous situation in the law exists with regard to criminal proceedings. In *Crane v. Director of Public Prosecutions* [1921] 2 AC 299, there were irregularities at the trial that had the effect of rendering the trial a 'nullity'. Nevertheless, an appeal was held to lie to the Court of Criminal Appeal. Lord Wilberforce said that a decision of an administrative or domestic tribunal reached in breach of natural justice, although it may be called—indeed may be for certain purposes—'void', is nevertheless susceptible of an appeal. The appellant's second argument was that such defects of natural justice as may have existed as regards the proceedings before the stewards were not capable of being cured by the appeal proceedings before the Committee, even though, as was not contested before the Board, these were correctly and fairly conducted. The Privy Council said that, upon the assumed basis that there had been a failure to observe natural justice in the proceedings before the stewards, no clear and absolute rule can be laid down on the question of whether defects in natural justice arising at an original hearing, whether administrative or quasi-judicial, can be 'cured' through appeal proceedings. The situations in which this issue arises are too diverse, and the rules by which they are governed so various, that this must be so. There are, however, a number of typical situations in relation to which some general principle can be stated. First, there are cases in which the rules provide for a rehearing by the original body, or some fuller or enlarged form of it. This is the situation in relation to social clubs. It is not difficult in such cases to reach the conclusion that the first hearing is superseded by the second, or, to put it in contractual terms, that the parties are taken to have agreed to accept a decision of the hearing body, whether original or adjourned. At the other extreme are cases in which, after examination of the whole hearing structure, in the context of the particular activity to which it relates (trade union membership, planning, employment, etc.) the conclusion is reached that a complainant has the right to nothing less than a fair hearing at both the original and the appeal stages. Intermediate cases exist in which it is for the court, in the light of the agreements made and in addition having regard to the course of proceedings, to decide whether, at the end of the day, there has been a fair result, reached by fair methods, such as the parties should fairly be taken to have accepted when they joined the association. Naturally, there may be incidences in which the defect is so flagrant, the consequences so severe, that the most perfect of appeals or rehearings will not be sufficient to produce a just result. Many rules anticipate that such a situation may arise by giving power to remit for a new hearing. There may also be cases in which the appeal process is itself less than perfect and may be vitiated by the same defect as the original proceedings.

### *Laud v. General Medical Council*, Privy Council Appeal No. 18 of 1979

**4.05** Sanction—appropriate approach on appeal

L was convicted of dishonesty consisting of cloning and receiving from the National Health Service salaries and expenses that he falsely stated he had paid to his staff. He appeared before a disciplinary committee of the General Medical Council, which suspended his registration for a period of twelve months. His appeal to the Privy Council was dismissed and their Lordships (Lord Salmon, Lord Russell of Killowen, and Lord Keith of Kinkel) adopted the following passage from the judgment delivered by Lord Wilberforce on 5 July 1979 in *Safi Ullah Khan v. General Medical Council* (unreported):

> Their Lordships have affirmed on many occasions that they will not interfere with the decision of professional disciplinary bodies as to the proper sentences to be imposed

in cases of professional misconduct. This principle applies equally when the matters complained of also constitute criminal offences which have been sanctioned by the criminal law. If, as well as being criminal, offences also constitute professional misconduct, it is for the appropriate disciplinary body to deal with them on that aspect. There can be no doubt that these offences were capable of being so regarded. This is the general rule from which their Lordships do not wish to depart.

### *Rai v. General Medical Council*, Privy Council Appeal No. 54 of 1983

**4.06** Sanction—appropriate approach on appeal

R was found guilty of abusing his professional position as a medical practitioner by issuing numerous prescriptions for dipipanone hydrochloride with cyclizine, methylphenidate, and other drugs in return for fees, otherwise than in the course of bona fide treatment; the Professional Conduct Committee (PCC) directed that his name should be erased from the register. In dismissing R's appeal, the Privy Council (Lord Scarman, Lord Brandon of Oakbrook, and Lord Brightman) said that punishment is for the professional judgment of the Committee unless the penalty imposed can be demonstrated to be so severe that it was not merited even upon the view of the appellant's conduct that the Committee had felt obliged to take: *McCoan v. General Medical Council* [1964] 1 WLR 1107. In the present case, if the Committee was right to take the view that it did of the appellant's conduct, erasure from the register was plainly the appropriate penalty, notwithstanding the previous good record of the doctor. In their Lordships' opinion, there was evidence upon which the Committee could properly reach its conclusion that the appellant was guilty of serious professional misconduct.

### *R v. Visitors to the Inns of Court, ex parte Calder; R v. Visitors to the Inns of Court, ex parte Persaud* [1994] QB 1

**4.07** Appellate rehearing—not review

C and P sought judicial review to quash the findings by the Visitors to the Inns of Court. The appeals of C and P were allowed by the Court of Appeal. Sir Donald Nicholls V-C, at 42D–42F, said:

> There remains [C]'s fourth ground of appeal: that the visitors misunderstood their role. She contends that the visitors were sitting as an appellate tribunal, not (as they seem to have thought), as a reviewing tribunal, and hence they failed fully and properly to carry out their duties as visitors. As to this, first, I can see no reason to doubt that an appeal to the judges as visitors is precisely that: an appeal. It is so described in the authorities. In *Lincoln v. Daniels* [1962] 1 QB 237, 256, Devlin LJ referred to it as "a hearing on appeal". Thus the visitors will look afresh at the matters in dispute and form their own views. The procedure followed in the conduct of such an appeal is a matter for the visitors. The current visitors' rules provide that fresh evidence will be admissible only in exceptional circumstances. In the absence of fresh evidence the appeal will be comparable to an appeal in the Civil Division of the Court of Appeal. Regarding sentence, it will be for the visitors to exercise their own discretion and judgment.

Stuart-Smith LJ, at 61H–62D, said:

> I come then to the final ground of appeal, namely, that the visitors misdirected themselves as to the nature of their jurisdiction in that they treated the matter as one of review rather than appeal by way of rehearing on the merits. It was not contested before us that the proper approach was that of an appellate court rehearing the case on its merits, such as is the position of the Court of Appeal on appeal in a civil case from the decision of a judge alone. Although the point has never fallen to be decided, I agree that this is the correct approach. All the cases dealing with judges' jurisdiction as visitors refer to it as an appeal to the visitors. There is no warrant for thinking that they limited themselves to the circumstances in which the prerogative writs of

prohibition, mandamus or certiorari would lie, that being the foundation of the judicial review jurisdiction. The language of the Hearings before the Visitors Rules 1991 is appropriate for an appeal and not a review only. Thus the appellant is referred to as such and not an applicant: rule 2(2). The grounds of appeal are against the finding and the petition should refer to the evidence relied upon: rules 5 and 7(2)(e). The visitors may either allow the appeal or order a rehearing: rule 11(3). They are not limited to quashing the order. Like any other appellate court, the visitors do not as a rule hear evidence from witnesses unless they give leave under rule 10(6) and (7). Accordingly they should adopt the same approach to findings of fact made by the tribunal as the Court of Appeal do to findings of the trial judge: see *Yuill v. Yuill* [1945] P 15; *Watt or Thomas v. Thomas* [1947] AC 484 and *Powell v. Streatham Manor Nursing Home* [1935] AC 243.

### *Ghosh v. General Medical Council* [2001] 1 WLR 1915

**4.08** Appellate hearing on facts and sanction—appropriate approach

G appealed against the decision of the PCC, which erased her name from the register for failing to comply with arrangements made for retraining during a period of conditional registration imposed following a finding that she had been guilty of serious professional misconduct. In dismissing her appeal, the Privy Council (Lord Bingham of Cornhill, Lord Cooke of Thorndon, and Lord Millett), in a judgment delivered by Lord Millett, said:

> 33. Practitioners have a statutory right of appeal to the Board under section 40 of the Medical Act 1983, which does not limit or qualify the right of the appeal or the jurisdiction of the Board in any respect. The Board's jurisdiction is appellate, not supervisory. The appeal is by way of a rehearing in which the Board is fully entitled to substitute its own decision for that of the committee. The fact that the appeal is on paper and that witnesses are not recalled makes it incumbent upon the appellant to demonstrate some error has occurred in the proceedings before the committee or in its decision, but this is true of most appellate processes.
>
> 34. It is true that the Board's powers of intervention may be circumscribed by the circumstances in which they are invoked, particularly in the case of appeals against sentence. But their Lordships wish to emphasise that their powers are not as limited as may be suggested by some of the observations which have been made in the past. In *Evans v. General Medical Council* (Unreported) 19 November 1984 the Board said:
>
> > 'The principles upon which this Board acts in reviewing sentences passed by the Professional Conduct Committee are well settled. It has been said time and again that a disciplinary committee are the best possible people for weighing the seriousness of professional misconduct, and that the Board will be very slow to interfere with the exercise of the discretion of such a committee... The committee are familiar with the whole gradation of seriousness of the cases of various types which come before them, and are peculiarly well qualified to say at what point on that gradation erasure becomes the appropriate sentence. This Board does not have that advantage nor can it have the same capacity for judging what measures are from time to time required for the purpose of maintaining professional standards.'
>
> For these reasons the Board will accord an appropriate measure of respect to the judgment of the committee whether the practitioner's failings amount to serious professional misconduct and on the measures necessary to maintain professional standards and provide adequate protection to the public. But the Board will not defer to the committee's judgment more than is warranted by the circumstances. The council conceded, and their Lordships accept, that it is open to them to consider all the matters raised by [G] in her appeal; to decide whether the sanction of erasure was appropriate and necessary in the public interest or was excessive and disproportionate; and in the latter event either to substitute some other penalty or to remit the case to the committee for reconsideration.

*Preiss v. General Dental Council* [2001] 1 WLR 1926

**4.09** Appellate hearing—appropriate approach

In delivering the judgment of the Privy Council (Lord Bingham of Cornhill, Lord Cooke of Thorndon, and Lord Millett), Lord Cooke said:

> 26. As indicated at the outset of this judgment, the provisions of section 29(1)(a) and (3) of the Dentists Act 1984 appear manifestly designed to give a full right of appeal to Her Majesty in Council, extending to the questions of fact as well as law and not limited even as to matters of degree or discretion, though as with most such general appeals the Judicial Committee would have to be satisfied before allowing an appeal that the decision of the PCC has been shown to have been wrong. It would be unusual for the Board to hear oral evidence, and allowance must be made for any advantages that the PCC has derived from seeing and hearing the witnesses; but this does not mean that for the purposes of article 6(1) the Board lacks full jurisdiction over the case.
>
> 27. Since the coming into operation of the Human Rights Act 1998, with its adjuration in section 3 to read and give effect to legislation, so far as it is possible to do so, in a way compatible with the Convention rights, any tendency to read down rights of appeal in disciplinary cases is to be resisted. In *Ghosh v. General Medical Council* [2001] 1 WLR 1915, 1923F–H the Board has recently emphasized that the powers are not as limited as may be suggested by some of the observations which have been made in the past. An instance, on which some reliance was placed by the General Dental Council in the argument of the present appeal, is the observation in *Libman v. General Medical Council* [1972] AC 217, 221, suggesting that findings of a professional disciplinary committee should not be disturbed unless sufficiently out of tune with the evidence to indicate with reasonable certainty that the evidence was misread. That observation has been applied from time to time in the past, but in their Lordships' view it can no longer be taken as definitive. This does not mean that respect will not be accorded to the opinion of a professional tribunal on technical matters. But, as indicated in *Ghosh*, the appropriate degree of deference will depend on the circumstances. In the instance case the weaknesses already identified in the dental disciplinary structure and the failure to comply with rule 11(2) go to diminish any reluctance that the Board might otherwise have in differing from the PCC. Against this background the Board now gives its own opinion on the case.

*Bhadra v. General Medical Council* [2003] 1 WLR 162

**4.10** Delay—prosecution—appeal dismissed

The Judicial Committee of the Privy Council has jurisdiction to dismiss an appeal for non-prosecution. Lord Hope of Craighead, delivering the judgment of their Lordships, said that while over a year had passed since the records and the bundle of documents were lodged, the onus lies on an appellant who has failed to prosecute his appeal with due diligence for such a long period to explain his default and to show cause why his appeal should not be dismissed. No explanation or excuse had been offered in this case for the appellant's failure to take any steps to prosecute his appeal. The Board might have been prepared to allow the appeal to proceed if it had been persuaded that the failure was the result of a mistake, oversight, or some other excusable cause, or that, for some other reason, the appellant would suffer an injustice if the appeal were to be dismissed—but this was not the position in this case. The respondent petition seeking to dismiss the appeal was granted.

*R (Chief Constable of Avon and Somerset) v. Police Appeals Tribunal* [2004] EWHC 220 (Admin)

**4.11** Appeal before Police Appeals Tribunal—proper approach

The test applicable to appeals before the Police Appeals Tribunal is not one of review of the decision below. On an appeal by a police constable against a decision of the chief constable requiring him to resign from the force, the Tribunal was entitled to substitute the decision for what amounted to a financial penalty in lieu. A claim by the chief constable for judicial review of the decision of the Police Appeals Tribunal was dismissed by Collins J, who, at [26], said that there was nothing in the statutory framework that indicated the way in which the Tribunal must approach its task. It is, in terms, an appeal. It is an appeal to an expert tribunal. In those circumstances, one would expect that that Tribunal, particularly where it can hear fresh evidence and consider all matters that are put before it (and it has the power to substitute, in the case of sanctions, any that could have been imposed by the tribunal from which it is hearing the appeal), has a full power to reconsider and exercise its own judgment as to what the appropriate outcome should be.

### *Council for the Regulation of Health Care Professionals v. General Medical Council and Solanke* [2004] 1 WLR 2432

**4.12** Appeal by council of 'unduly lenient' decision—test to be applied by court

On 15 December 2003, the PCC of the General Medical Council (GMC) determined that S was guilty of serious professional misconduct in relation both to his involvement in a sexual relationship with a patient, and to his alteration of his birth certificate and curriculum vitae so as to reduce his apparent age. Bearing in mind that he had already been suspended from general practice for six months, the Committee ordered that he be suspended from the medical register for a period of three months. Pursuant to section 29 of the National Health Service Reform and Health Care Professions Act 2002, the Council for the Regulation of Health Care Professionals (the Council) appealed against the sentence on the grounds that it was unduly lenient. On 5 April 2004, Leveson J heard the appeal and, at the conclusion of the argument, dismissed it. The learned judge, at [14]–[15], said that he agreed that there was a very real difference between the criminal and disciplinary jurisdictions, because criminal sentencing is concerned with punishment and disciplinary sanction, with protection of the public. But that difference is not to the point in relation to the proper construction of section 29(4) of the 2002 Act. The context in which undue lenience must be considered may be different for criminal cases (concerned with retribution, deterrents, and rehabilitation) and disciplinary cases (concerned with the protection of the public and the reputation of the profession), but the relevant question remains the same—namely, having regard to the purposes of the particular sanction being imposed (whether criminal or disciplinary), is this particular sanction outside the range of sanctions that the sentencing tribunal, applying its mind to all of the factors relevant to its jurisdiction, could reasonably consider appropriate? In the circumstances, the learned judge said that he rejected the submission that the court is not required to consider whether a decision is unduly lenient. Furthermore, 'wrong' in CPR 52.11 itself means more than 'less than the court would impose'; it means outside the range of sanctions that the relevant disciplinary panel, applying its mind to all of the factors relevant to its jurisdiction, could reasonably consider appropriate. At [16], the learned judge said that he entirely endorsed the conclusion of Collins J in *Council for the Regulation of Health Care Professionals v. Nursing and Midwifery Council* [2004] EWHC 585 (Admin), shortly expressed in these terms:

> 10. ... I see no reason to doubt that the true construction of section 29 requires that the court will only allow the appeal if satisfied that undue leniency and desirability for the protection of the public is made out. If undue leniency is established, it will only be in the rarest of cases that a different view from that of the council is likely to be appropriate in respect of desirability.
>
> 11. I see no reason not to apply mutatis mutandis the same test as the Court of Appeal applies in deciding whether a sentence in a criminal case is unduly lenient.

***Ruscillo v. Council for the Regulation of Health Care Professionals and General Medical Council; Council for the Regulation of Health Care Professionals v. Nursing and Midwifery Council and Trustcott* [2005] 1 WLR 717**

**4.13** Expertise of disciplinary tribunal—degree of deference

On an appeal by the Council for the Regulation of Health Care Professionals under section 29 of the National Health Service Reform and Health Care Professions Act 2002, the Court of Appeal (Lord Phillips of Worth Matravers MR, and Chadwick and Hooper LJJ), at [76], after referring to *Solanke's case* (see paragraph 4.12 above), said that it considered the test of whether a penalty was unduly lenient in the context of section 29 to be whether it is one that a disciplinary tribunal, having regard to the relevant facts and to the object of the disciplinary proceedings, could reasonably have imposed. At [78], the Court considered the extent to which the Council and the Court should defer to the expertise of the disciplinary tribunal. That expertise is one of the most cogent arguments for self-regulation. At the same time, Part 2 of the 1992 Act had been introduced because of concern as to the reliability of self-regulation. Where all material evidence has been placed before a disciplinary tribunal and it has given due consideration to the relevant factors, the Council and the Court should place weight on the expertise brought to bear in evaluating how best the needs of the public and the profession should be protected. Where, however, there has been a failure of process, or evidence is taken into account on appeal that was not placed before the disciplinary tribunal, the decision reached by that tribunal will inevitably need to be reassessed.

***Phipps v. General Medical Council* [2006] EWCA Civ 397**

**4.14** Second appeal to Court of Appeal

P, a consultant general surgeon, was found guilty of serious professional misconduct before the GMC's PCC and was suspended from practice for a period of twelve months. His appeal was heard in the Administrative Court by Newman J and was dismissed. He sought the permission of the Court of Appeal to appeal against Newman J's decision. Laws LJ considered the application on paper and adjourned it to an oral hearing on notice to the GMC. In refusing permission to appeal, Wall LJ (with whom Sir Mark Potter P and Arden LJ agreed) said that it was apparent that P's application for permission to appeal related to a second appeal and that section 55(1) of the Access to Justice Act 1999 applied. Section 55(1) provides that no appeal may be made to the Court of Appeal from a decision of the High Court unless the Court of Appeal considers that the appeal would raise an important point of principle or practice, or there is some other compelling reason for the Court of Appeal to hear it. CPR 52.13 follows the statutory language in the Act. P's application passed neither of the tests identified by section 55(1) of the Act, although the Court, having allowed counsel to develop his arguments set out in the appellant's notice and skeleton argument, and noting that success or failure was a matter of considerable moment for P, nonetheless examined P's application on its merits.

***Council for the Regulation of Health Care Professionals v. General Medical Council and Basiouny* [2005] EWHC 68 (Admin)**

**4.15** See paragraph 26.03.

***Raschid v. General Medical Council; Fatnani v. General Medical Council* [2007] 1 WLR 1460**

**4.16** Sanction—appropriate approach on appeal

In *Raschid*, the fitness to practise panel of the GMC ordered the suspension of the practitioner from the register for twelve months and directed a further hearing or review of his position. Collins J allowed an appeal, substituted the twelve-month period of suspension with one month's suspension, and quashed the order for further hearing or review. In *Fatnani*, the fitness to practise panel directed that the name of the practitioner be erased from the medical register for her admitted criminal conviction. Collins J allowed her appeal and substituted for the erasure order a suspension of twelve months. On appeal by the GMC, the Court of Appeal, having granted permission to appeal in both cases, reversed the decisions of the learned judge. Laws LJ (with whom Chadwick LJ and Sir Peter Gibson agreed) said:

> 16. In these circumstances it seems to me to be clear that we should follow the guidance given in the cases decided before the change in the appeal system effected on 1st April 2003. First, the Privy Council is of course a source of high authority; but, secondly, we are in any event considering an effectively identical statutory regime. As it seems to me there are in particular two strands in the relevant learning before 1st April 2003. One differentiates the function of the panel or committee in imposing sanctions from that of a court imposing retributive punishment. The other emphasises the special expertise of the panel or committee to make the required judgment.
>
> 17. The first of these strands may be gleaned from the Privy Council decision in *Gupta v. General Medical Council* [2002] 1 WLR 1691, para 21, in the judgment of their Lordships delivered by Lord Rodger of Earlsferry:
>
> > 'It has frequently been observed that, where professional discipline is at stake, the relevant committee is not concerned exclusively, or even primarily, with the punishment of the practitioner concerned. Their Lordships refer, for instance, to the judgment of Sir Thomas Bingham MR in *Bolton v. Law Society* [1994] 1 WLR 512, 517–519 where his Lordship set out the general approach that has to be adopted. In particular he pointed out that, since the professional body is not primarily concerned with matters of punishment, considerations which would normally weigh in mitigation of punishment have less effect on the exercise of this kind of jurisdiction. And he observed that it can never be an objection to an order for suspension that the practitioner may be unable to re-establish his practice when the period has passed. That consequence may be deeply unfortunate for the individual concerned but it does not make the order for suspension wrong if it is otherwise right. Sir Thomas Bingham MR concluded, at p519: "The reputation of the profession is more important than the fortunes of any individual member. Membership of a profession brings many benefits, but that is a part of the price." Mutatis mutandis the same approach falls to be applied in considering the sanction of erasure imposed by the committee in this case.'
>
> 18. The panel then is centrally concerned with the reputation or standing of the profession rather than the punishment of the doctor. This, as it seems to me, engages the second strand to which I have referred. In *Marinovich v. General Medical Council* [2002] UKPC 36 Lord Hope of Craighead, giving the judgment of the Board, said:
>
> > '28. . . . in the appellant's case the effect of the committee's order is that his erasure is for life. But it has been said many times that the Professional Conduct Committee is the body which is best equipped to determine questions as to the sanction that should be imposed in the public interest for serious professional misconduct. This is because the assessment of the seriousness of the misconduct is essentially a matter for the committee in the light of its experience. It is the body which is best qualified to judge what measures are required to maintain the standards and reputation of the profession.
> >
> > 29. That is not to say that their Lordships may not intervene if there are good grounds for doing so. But in this case their Lordships are satisfied that there are no such grounds. This is a case of such a grave nature that a finding that the appellant was unfit to practise was inevitable. The committee was entitled to give greater weight to the public interest and to the need to maintain public confidence in the profession than to the consequences to the appellant of the imposition of the penalty. Their Lordships are quite unable to say that the sanction of erasure which the committee decided to impose in this case, while undoubtedly severe, was wrong or unjustified.'

19. There is, I should note, no tension between this approach and the human rights jurisdiction. This is because of what was said by Lord Hoffmann giving the judgment of the Board in *Bijl v. General Medical Council* [2002] Lloyd's Rep Med 60, paras 2 and 3, which with great respect I need not set out. As it seems to me the fact that a principal purpose of the panel's jurisdiction in relation to sanctions is the preservation and maintenance of public confidence in the profession rather than the administration of retributive justice, particular force is given to the need to accord special respect to the judgment of the professional decision-making body in the shape of the panel. That I think is reflected in the last citation I need give. It consists in Lord Millett's observations in *Ghosh v. General Medical Council* [2001] 1 WLR 1915, 1923, para 34: "the Board will afford an appropriate measure of respect to the judgment of the committee whether the practitioner's failings amount to serious professional misconduct and on the measures necessary to maintain professional standards and provide adequate protection to the public. But the Board will not defer to the committee's judgment more than is warranted by the circumstances."

20. These strands in the learning then, as it seems to me, constitute the essential approach to be applied by the High Court on a section 40 appeal. The approach they commend does not emasculate the High Court's role in section 40 appeals: the High Court will correct material errors of fact and of course of law and it will exercise a judgment, though distinctly and firmly a secondary judgment, as to the application of the principles to the facts of the case.

[...]

26. I acknowledge without cavil that Collins J's judgments are careful and humane. But I have to say that they do not in my view remotely offer sufficient recognition of the two principles which are especially important in this jurisdiction: the preservation of public confidence in the profession and the need in consequence to give special place to the judgment of the specialist tribunal. Applying these principles I am driven to conclude that there was not in either of these cases any proper basis established for overturning the sanctions set by the Fitness to Practise Panel.

### *Salsbury v. Law Society* [2009] 1 WLR 1286

**4.17** Sanction—appropriate approach on appeal

S was admitted as a solicitor in 1984. In 1999, he was appointed clerk to the trustees of a school in Sussex. On 15 November 2000, he altered the amount of a cheque in his favour for £862.50, so as to read £1,862.50. He was convicted at Croydon Crown Court of one count of obtaining a money transfer by deception contrary to section 15A of the Theft Act 1968. The sentence imposed by the Court was a conditional discharge for twelve months, which the sentencing judge described as an exceptional course given all of the circumstances. The Solicitors Disciplinary Tribunal (SDT) ordered that S be struck off the roll of solicitors. On appeal, the Divisional Court allowed S's appeal and substituted an order for three years' suspension. On appeal, the Court of Appeal reviewed the authorities including: *Bolton v. Law Society* [1994] 1 WLR 512; *Ghosh v. General Medical Council* [2001] 1 WLR 1915; *Preiss v. General Dental Council* [2001] 1 WLR 1926; *Langford v. Law Society* [2002] EWHC 2802 (Admin); *Burrowes v. Law Society* [2002] EWHC 2900 (Admin); *Nahal v. Law Society* [2003] EWHC 2186 (Admin); *Bultitude v. Law Society* [2004] EWCA Civ 1853; and *Newfield v. Law Society* [2005] EWHC 765 (Admin). Jackson LJ then said:

30. From this review of authority I conclude that the statements of principle set out by Sir Thomas Bingham MR in *Bolton v. Law Society* [1994] 1 WLR 512 remain good law, subject to this qualification. In applying the *Bolton* principles the Solicitors Disciplinary Tribunal must also take into account the rights of the solicitor under articles 6 and 8 of the Convention. It is now an overstatement to say that "a very strong case" is required before the court will interfere with the sentence imposed by the

Solicitors Disciplinary Tribunal. The correct analysis is that the Solicitors Disciplinary Tribunal comprises an expert and informed tribunal, which is particularly well placed in any case to assess what measures are required to deal with defaulting solicitors and to protect the public interest. Absent any error of law, the High Court must pay considerable respect to the sentencing decisions of the tribunal. Nevertheless if the High Court, despite paying such respect, is satisfied that the sentencing decision was clearly inappropriate, then the court will interfere. It should also be noted that an appeal from the Solicitors Disciplinary Tribunal to the High Court normally proceeds by way of review: see CPR rule 52.11(1).

### *Cheatle v. General Medical Council* [2009] EWHC 645 (Admin)

**4.18** Test—whether decision wrong

Cranston J, at [12]–[15] under the heading 'Appeals from a Fitness to Practise Panel', said that an appeal to the High Court under section 40 of the Medical Act 1983 is by way of a rehearing. The relevant Practice Direction offers no guidance as to what this means: see CPR 52 PD 116(2). Clearly, it is not an appeal confined to a point of law, but neither at the other end of the spectrum is it a *de novo* hearing, in which the Court hears the witnesses giving evidence again. The basis of intervention appears to be broader than that for judicial review. After referring to *Meadow v. General Medical Council* [2007] QB 462 and *Raschid and Fatnani v. General Medical Council* [2007] 1 WLR 1460, the learned judge said that, in his view, the approaches in *Meadow* and *Raschid* are readily reconcilable. The test on appeal is whether the decision of the fitness to practise panel can be said to be wrong. That follows because it is an appeal by way of rehearing, not review. In any event, grave issues are at stake and it is not sufficient for intervention to turn on the more confined grounds of public law review such as irrationality. However, in considering whether the decision of a fitness to practise panel is wrong, the focus must be calibrated to the matters under consideration. With professional disciplinary tribunals, issues of professional judgment may be at the heart of the case. *Raschid* was an appeal on sanction and professional judgment is especially important in that type of case. As to findings of fact, as with any appellate body, there will be reluctance to characterize findings of fact as wrong. That follows because findings of fact may turn on the credibility or reliability of a witness, an assessment of which may be derived from his or her demeanour and from the subtleties of expression that are evident only to someone at the hearing. Decisions on fitness to practise, such as assessing the seriousness of any misconduct, may turn on an exercise of professional judgment. In this regard, respect must be accorded to a professional disciplinary tribunal such as a fitness to practise panel. However, the degree of deference will depend on the circumstances. One factor may be the composition of the tribunal. One cannot be completely blind to the current composition of fitness to practise panels.

### *Dutt v. General Medical Council* [2009] EWHC 3613 (Admin)

**4.19** Findings of fact

Cranston J said that, in relation to an appeal, it is necessary to focus on the matter under consideration. With professional disciplinary tribunals, issues of professional judgment may be at the heart of the case. On the other hand, there may be important findings of fact that found their conclusion. In relation to the latter, as the learned judge explained in *Cheatle*, his view is that there is no difference between the Court's role in relation to a panel's findings of fact and findings of fact in other appellate contexts. With any appellate body, there is a reluctance to characterize findings of fact as wrong. That follows because findings of fact may turn on the credibility or reliability of the witness. An assessment of those factors may be derived from a witness's demeanour and from the subtleties of expression that are evident only to someone at the hearing. It may be that, in some cases, an appellate body can conclude that there was insufficient evidence or no evidence to support a finding of fact. But, in general, appellate bodies are reluctant to interfere with findings of fact by tribunals entrusted with that function.

***Gerstenkorn v. General Medical Council* [2009] EWHC 2682 (Admin)**

**4.20** Test—wrong or unjust

At [39], Kenneth Parker J said that the Court should allow an appeal only where it is satisfied that the decision of the fitness to practise panel was (a) wrong or (b) unjust, because of a serious procedural or other irregularity in the proceedings of the lower court: see CPR 52.11(3).

***Bhatt v. General Medical Council* [2011] EWHC 783 (Admin)**

**4.21** Test—proper approach

Langstaff J, after referring to CPR 52.11 and the cases of *Sacha v. General Medical Council* [2009] EWHC 302 (Admin), [8], *Southall v. General Medical Council* [2010] EWCA Civ 407, [47], and *Meadow v. General Medical Council* [2007] QB 462, [128], said:

> 9. I accept and adopt the approach outlined in these authorities, in particular that although the court will correct errors of fact or approach:
>
> (i) it will give appropriate weight to the fact that the Panel is a specialist tribunal, whose understanding of what the medical profession expects of its members in matters of medical practice deserves respect;
> (ii) that the tribunal has had the advantage of hearing the evidence from live witnesses;
> (iii) the court should accordingly be slow to interfere with the decisions on matters of fact taken by the first instance body;
> (iv) findings of primary fact, particularly if found upon an assessment of the credibility of witnesses, are close to being unassailable, and must be shown with reasonable certainty to be wrong if they are to be departed from;
> (v) but that where what is concerned is a matter of judgment and evaluation of evidence which relates to police practice, or other areas outside the immediate focus of interest and professional expertise of the [fitness to practise panel], the court will moderate the degree of deference it will be prepared to accord, and will be more willing to conclude that an error has, or may have been, made, such that a conclusion to which the Panel has come is or may be "wrong" or procedurally unfair. To this extent I accept and adopt the submissions of [counsel for the appellant].

***R (Chief Constable of Durham) v. Police Appeals Tribunal and Cooper (Interested Party)* [2012] EWHC 2733 (Admin)**

**4.22** Appeal to Police Appeals Tribunal—appropriate test

The Chief Constable of Durham moved, by way of judicial review, to challenge the Police Appeals Tribunal's determination that allowed the appeal of C, a police constable, against a finding of a misconduct panel that he was guilty of gross misconduct and should be dismissed. The underlying facts related to an incident in which M had telephoned the police to say that she was wanted on a warrant for failure to appear in court. She was, at the time, at her grandmother's house. There was no dispute that it was C who called on his own at the grandmother's house to arrest M. It was not in dispute that, for a period of over 20 minutes or so, the police officer was alone with M in her bedroom. The evidence showed that M had a blood alcohol level of what would have been four-and-a-half times the legal limit had she been driving. The allegation was, effectively, either of oral rape or attempted oral rape. The misconduct panel concluded that M had been telling the truth, despite certain inconsistencies in her account and the effect of drink. The Police Appeals Tribunal, which was chaired by an experienced Queen's Counsel, concluded that the panel had reached an unreasonable conclusion and had adopted an unreasonable approach to the evidence. In dismissing the chief constable's application for judicial review, the Administrative Court (Moses LJ and Hickinbottom J) said that relevant to the question of the appropriate test adopted by the Police Appeals Tribunal are the rules made by the Secretary of State under the Police Act 1996. Rule 4 of the Police Tribunal Rules 2008 provides the circumstances in which a police officer may appeal to a tribunal. Rule 4(4) provides that the grounds of appeal under the rule are:

(a) that the finding or disciplinary action imposed was unreasonable; or
(b) that there is evidence that could not reasonably have been considered at the original hearing which could have materially affected the finding or decision on disciplinary action [not a rule relevant for this appeal]; or
(c) that there was a breach of the procedures set out in the Conduct Regulations...or other unfairness which could have materially affected the finding or decision on disciplinary action.

Moses LJ, giving the judgment of the Court, said that:

> 6. The imposition of a test which asks whether the decision of the misconduct panel was unreasonable has led some to take the view that that imported a test of *Wednesbury* unreasonableness, a test appropriate to that applied by this court in questions of public law. That, in my view, is erroneous.... The test is not one of *Wednesbury* unreasonableness. Firstly, the test must be seen in its correct statutory context, namely that of a specialist appeal tribunal considering the decision of a misconduct panel. A *Wednesbury* unreasonableness test is that test which is conventionally adopted where courts review decisions of the executive or expert panels....
>
> 7. ...[T]he test imposed by the [Police Appeals Tribunal Rules] is not the *Wednesbury* test but is something less. That does not mean that the appeal tribunal is entitled to substitute its own view for that of the misconduct hearing panel, unless and until it has already reached the view, for example, that the finding was unreasonable. Nor...is the Police Appeals Tribunal entitled, unless it has already found that the previous decision was unreasonable, to substitute its own approach.... The Police Appeals Tribunal is only allowed and permitted to substitute its own views once it has concluded either that the approach was unreasonable, or that the conclusions of fact were unreasonable...

Moses LJ pointed to *R (Dart Harbour and Navigational Authority) v. Secretary of State for Transport, Local Government and the Regions* [2003] EWHC 1494 (Admin) and *R (Montgomery) v. Police Appeals Tribunal and the Commissioner of Police of the Metropolis* [2012] EWHC 936 (Admin) as authority in a different statutory context, and continued:

> 8. ...In the latter case, Collins J had to consider whether it was permissible for an appeal tribunal to look at fresh documents. He decided that it was...
>
> 9. This case turns on the question whether the Police Appeals Tribunal restricted itself to the permissible approach under rule 4(4)(a) or on the contrary, as the Chief Constable for Durham contends, was guilty of substituting its own findings for those of the misconduct panel without any justifiable basis for reaching the conclusion that the misconduct hearing panel's decision and conclusion was unreasonable.

Moses LJ went on to confirm that, in his view, the appeal panel had adopted a decision that was carefully reasoned and fully set out the basis upon which it concluded that the misconduct panel had reached an unreasonable conclusion. There were substantial inconsistencies in M's evidence, and whilst the misconduct panel was entitled to reach a conclusion that M was a credible and genuine witness, it could reach that view only if it fairly grappled with the inconsistencies. The Police Appeals Tribunal was entitled to take the view that the panel did not fairly grapple with those inconsistencies and did not accord them the weight to which the cumulative effect of those inconsistencies were entitled. The Tribunal also found a number of other reasons on which it concluded that the decision of the misconduct panel was unreasonable. Moses LJ confirmed that the Court agreed that, from time to time, the Tribunal appeared to be substituting its own approach for that of the misconduct panel without proper justification for doing so—but it did identify an unreasonable approach and an unreasonable conclusion in a number of respects that were sufficient for it to substitute its own conclusions on the material before it for those of the misconduct panel. The Court found that the Police Appeals Tribunal properly set out all of its reasons for doing so.

### *Webster v. General Teaching Council* [2012] EWHC 2928 (Admin)

**4.23** Teacher—appeal against findings

W, a young, newly qualified teacher, appealed against the decision of the PCC of the General Teaching Council, which found her guilty of unacceptable professional conduct contravening section 8(1) of Schedule II to the Teaching and Higher Education Act 1998. W faced an allegation that, between February 2008 and the spring term of 2009, she engaged in an inappropriate relationship with Pupil A, a 17-year-old girl pupil at Abbey School, Faversham, Kent. The PCC found four of ten particulars of the allegations proved and six unproved. The proved particulars included allegations that, in March 2008, W walked hand-in-hand with and kissed Pupil A in Whitstable, went into a public house with Pupil A in Whitstable, and allowed herself to be photographed with Pupil A in a manner that suggested that they were in a relationship with each other. In dismissing W's appeal against the PCC's findings of fact, Irwin J, after reviewing the evidence, said, at [20]:

> The function of this court is not to rehear proceedings before a Professional Disciplinary Committee such as this. As all parties have recognised, the function of this court is to quash findings and substitute other findings, or remit the matter, only if the conclusions of the Committee are wrong. I find no basis in this case for concluding that the conclusions of the Committee are wrong. It seems to me very clear that they intended to find a sexual or intimate relationship between Pupil A and the appellant. It seems clear to me that there was sufficient basis for them to do so, taking all of these particulars together. Sensibly, it is not submitted by [counsel for the appellant] if this was indeed a conclusion that there was an inappropriate relationship, in the sense of a sexual relationship between the appellant and Pupil A, that the sanction passed of two years' suspension was in any way excessive.

### *Chief Constable of Wiltshire Police v. Police Appeals Tribunal and Andrews (Interested Party)* [2012] EWHC 1798 (Admin)

**4.24** Decision of Police Appeals Tribunal—whether irrational

The Chief Constable of Wiltshire Police challenged the decision of the Police Appeals Tribunal, which had allowed in part the appeal of A, a police sergeant, against the decision of a police disciplinary panel that had found A guilty of three charges of misconduct and had imposed a sanction of dismissal. The chief constable challenged the decision of the Tribunal on the grounds that it had come to an irrational conclusion in its findings. A faced three charges before the panel arising out of an incident that took place on 4 July 2008, when, at about 10 a.m., Ms S was brought into the custody suite at Melksham police station, where A was on duty as the custody sergeant. Charge 1 alleged that, throughout the period of her detention, A had failed to treat Ms S with courtesy and respect, contrary to paragraph 3 of the Code of Conduct set out in Schedule 1 to the Police (Conduct) Regulations 2004. Charge 2 contained two allegations of misconduct by the use of excessive force and abuse of authority by A. The charge concerned the manner in which A removed Ms S to the cells when she was being uncooperative with the booking-in procedure, and the manner in which A took her back to the cells when she refused to undertake a breath test and had become abusive to an officer. Charge 3 alleged that A acted in a way likely to bring discredit upon the police service. Charge 3 was treated by the Tribunal as covering much of the same ground as charge 1.

In allowing A's appeal on charge 2 from the decision of the panel, the Tribunal found that A was entitled to conclude that Ms S was not going to cooperate and, from what he had been told and had observed himself, that she was likely to become violent and need to be removed to a cell. The Tribunal found that the manner in which Ms S was removed by A was undignified, but it could not be overlooked that A had not created the problem; Ms S had. The Tribunal quashed charge 2, substituted for the sanction of dismissal imposed by the disciplinary panel a fine amounting to thirteen days' pay, and ordered A's reinstatement to the Wiltshire Constabulary as a sergeant. In dismissing the chief constable's claim, Ouseley J said that the grounds of challenge were better suited to a merits appeal than to a rationality challenge. The Tribunal's conclusion was based in part on A's evidence to it, which it accepted. It was not irrational to accept that evidence and, on that basis, it was not remotely arguable that the conclusion that the Tribunal reached in relation to charge 2 was irrational. Any appeal on the merits was unarguable. The decision by A to remove Ms S was a proper one; there was a reasonable explanation given, which the Tribunal accepted, as to why she was removed without prior communication with other officers. The Tribunal found that the method used by A in removing Ms S to her cell was unremarkable until she deliberately sat down to try to avoid being put back in her cell. He did not then use excessive force to move her from that point. However, the language that he used whilst dragging her—'I'll drag you to your cell . . . it doesn't bother me, love'—and the tone of voice rendered that part of the episode unacceptable and a breach of the Code of Conduct. Permission to apply for judicial review would be refused.

***R (Chief Constable of Wiltshire Police) v. Police Appeals Tribunal and Woollard (Interested Party)* [2012] EWHC 3288 (Admin)**

**4.25** Appeal to Police Appeals Tribunal—appropriate test

W, a police constable, admitted or was found guilty of ten allegations of inappropriate behaviour arising out of a police investigation of sexual offences by M, a friend of W. A misconduct panel found that the allegations cumulatively constituted gross misconduct and that W should be dismissed without notice by reason of that misconduct. On appeal to the Police Appeals Tribunal, the Tribunal decided that two of the allegations that had been made against W (the only two allegations that he contested) should not have been found proved and it also decided that the misconduct panel should not have imposed the sanction of dismissal without notice; rather, it should have issued a final written warning. As a consequence of its decision, the Tribunal ordered W's reinstatement as a police constable. In judicial review proceedings, the Chief Constable of Wiltshire Police challenged this decision. In granting judicial review and quashing the decision of the Tribunal, Wyn Williams J said that there had been a number of recent decisions in which the courts have grappled with what is meant by the word 'reasonable' in rule 4(4)(a) of the Police Appeals Tribunal Rules 2008 (the grounds of appeal are that the finding or disciplinary action imposed was unreasonable). The learned judge referred to *R (Montgomery) v. Police Appeals Tribunal* [2002] EWHC 936 (Admin) (Collins J), *R (Chief Constable of Hampshire) v. Police Appeals Tribunal* [2012] EWHC 746 (Admin) (Mitting J), *R (Chief Constable of the Derbyshire Constabulary) v. Police Appeals* Tribunal [2012] EWHC 2280 (Admin) (Beatson J), and *R (The Chief Constable of Durham) v. Police Appeals Tribunal* [2012] EWHC 2733 (Admin) (a Divisional Court consisting of Moses LJ and Hickinbottom J). In his decision in the *Derbyshire* case, Beatson J expressed the view that the issue of whether a finding or sanction was unreasonable should be determined by asking whether the panel in question had made a finding or imposed a sanction that was within the range of reasonable findings or sanctions based upon the material before it. Beatson J clearly considered that his view was consistent with the views expressed in the earlier decisions in the *Montgomery* and *Hampshire* cases. The approach of Beatson J was echoed in the approach adopted in the *Durham* case by Moses LJ (with whom Hickinbottom J agreed). During the course of his judgment, Moses LJ considered whether or not the use of the word 'unreasonable' within rule 4(4)(a) mandated the Tribunal to apply what is familiarly known as the '*Wednesbury* test' when determining whether or not a finding or sanction is to be categorized as unreasonable. His conclusion was that the test imposed by the Police Appeals Tribunal Rules is not the *Wednesbury* test, but something less (see paragraph 4.22 above). Wyn Williams J said that he proposed to follow the same approach to the word 'unreasonable' as that which had been adopted by Beatson J in the *Derbyshire* case and Moses LJ and Hickinbottom J in the *Durham* case.

See further *R (Commissioner of Police for the Metropolis) v. Police Appeals Tribunal and Naulls* [2013] EWHC 1684 (Admin): paragraph 47.52.

### *Adesemowo and Alafe-Aluko v. Solicitors Regulation Authority* [2013] EWHC 2020 (Admin)

**4.26** Solicitors' appeal—flexible review rather than rehearing

Section 49(1) of the Solicitors Act 1974 confers a statutory right of appeal from decisions of the SDT to the High Court. Neither the Solicitors Act 1974 nor Practice Direction 52 specifies whether an appeal is to be by way of rehearing or review. Practice Direction 52, paragraph 22.3(3), lists statutory appeals from other professional regulatory bodies that are held to be by way of rehearing, but appeals from the Law Society are not included in the list. Therefore, by virtue of rule 52.11(1), the starting point is that the appeal is normally by way of review rather than rehearing: *Law Society v. Salsbury* [2009] 2 All ER 487, [30]. The nature of the review is flexible, as explained in *El Dupont de Nemours & Company v. ST Dupont* [2006] 1 WLR 2793 CA, and the scope of the review that the Court can carry out in an appeal means that it is unnecessary to hold a rehearing.

### *Shaw and Turnbull v. Logue* [2014] EWHC 5 (Admin)

**4.27** Solicitors' appeal—'wrong' means 'plainly wrong'

S and T were found guilty of dishonesty and were struck off the roll of solicitors by the SDT. The parties were agreed that the Court could intervene only if satisfied that the SDT's decision was 'wrong', which meant in reality 'plainly wrong': see *Gupta v. General Medical Council* [2002] 1 WLR 1691; *Biogen Inc. v. Medeva Plc* [1997] RPC 1; *McGraddie v. McGraddie and anor* [2013] 1 WLR 2477. The Court's function is to review the evidence and to apply to it a strict yardstick; it is not to second-guess the SDT's findings, nor to substitute its own views even if, for example, it were of the opinion that the conclusions were *probably* wrong. 'Plainly wrong' imports a higher onus of persuasion—and for good reason: the reviewing court does not see and hear the witnesses.

### *Moseka v. Nursing and Midwifery Council* [2014] EWHC 846 (Admin)

**4.28** Findings of fact—appropriate approach on appeal

Green J said, at [21], that the appellate court may draw any inference of fact that it considers justified on the evidence. The Court considered *Meadow v. General Medical Council* [2000] QB 462 (in particular at [197]), *Preiss v. General Dental Council* [2001] 1 WLR 1926 (in particular at [26]), *Ghosh v. General Medical Council* [2001] 1 WLR 1915 (in particular at [33]), *Gupta v. General Medical Council* [2001] UKPC 61, [10], and *Uddin v. General Medical Council* [2012] EWHC 2669, [4] *et seq.* The Court said, at [22], that, for present purposes, it sufficed to summarize the principal points emerging from the case law in the following way:

i) The appellate body has full jurisdiction on the appeal but this does not mean that it will, necessarily, not pay some deference to the fact finding function of the lower court or body (e.g. *Preiss* paragraph [26]);
ii) It will be unusual for the appellate body to hear oral evidence (e.g. *Preiss* paragraph [26]);
iii) In assessing the findings of fact of the lower court or tribunal the appellate court will have regard to the fact that the lower body is a specialist tribunal whose membership is selected for its experience in the subject matter of the matters before it (*Meadow* paragraph [197] *per* Auld LJ);
iv) The appellate tribunal will recognise that the lower court or tribunal has had the benefit of hearing and seeing the witnesses and is therefore in a better position to judge their credibility and reliability than is the appellate court... ([*Gupta*] paragraph [10C]).
v) The same degree of the judicial deference will not arise where the decision of the court or tribunal below is not based upon an assessment of the credibility or reliability of witnesses. For example where the appellate body is in materially the same position as the lower court or tribunal then it will be more inclined to form its own view on the facts or matters in issue.
vi) Accordingly to succeed upon an appeal the appellant will *normally* be required to demonstrate a procedural error before the NMC or in its decision.

### *Wasu v. General Dental Council* [2013] EWHC 3782 (Admin)

**4.29** General Dental Council—appeal—principles applicable

In dismissing L's appeal against an order striking her off the Health and Care Professions Council's register of practitioner psychologists, Haddon-Cave J said, at [17], that the general principles applicable to an appeal against a decision of a professional disciplinary committee can be summarized as follows:

(1) The Court will give appropriate weight to the fact that the Panel is a specialist tribunal, whose understanding of what the health care professional expects of its members in matters of medical practice deserves respect.
(2) The Court will have regard to the fact that the tribunal has had the advantage of hearing the evidence from live witnesses.
(3) The Court should accordingly be slow to interfere with decisions on matters of fact taken by the first instance body.
(4) Findings of primary fact of the first instance body, particularly if founded upon an assessment of the credibility of witnesses, are close to being unassailable, and must be shown with reasonable certainty to be wrong if they are to be departed from.
(5) Where what is concerned is a matter of judgment and evaluation of evidence which relates to areas outside the immediate focus of interest and professional experience of the body, the Court will moderate the degree of deference it will be prepared to accord, and will be more willing to conclude that an error has, or may have been, made, such that a conclusion to which the Panel has come is or may be "wrong" or procedurally unfair.

(See the helpful summary of the authorities by Langstaff J in *Bhatt v. General Dental Council* [2011] EWHC 783 (Admin), in particular at [9]. In particular, see also *Meadow v. General Medical Council* [2007] QB 462 and *Southall v. General Medical Council* [2010] EWCA Civ 407.)

### *L v. Health and Care Professions Council* [2014] EWHC 994 (Admin)

**4.30** Similar principles were applicable here in relation to the Health and Care Professions Council. (See also *Rice v. Health Professions Council* [2011] EWHC 1649 (Admin), [11]–[17], and *Sokunbi v. Health Professions Council* [2013] EWHC 672 (Admin).)

### *General Medical Council v. Nakhla* [2014] EWCA Civ 1522

**4.31** Doctor—Registration Appeal Panel—appeal to county court—second appeal to Court of Appeal

N, who practised as a surgeon in Egypt, applied to the GMC to be registered as a specialist in trauma and orthopaedic surgery. His application was rejected by the registrar, whose decision was upheld by the Registration Appeal Panel (RAP). N appealed to the county court against the decision of the RAP and, on 30 May 2014, Her Honour Judge Faber allowed his appeal. The GMC appealed to the Court of Appeal. In allowing the appeal in part and ordering that M's appeal to the RAP should be redetermined in the light of more recent procedure-based assessments that M said he had obtained, Lewison LJ (with whom Longmore and Burnett LJJ agreed) said:

> 11. Unlike many appeals to the court from specialist tribunals, an appeal to the county court from the RAP is not limited to an appeal on a point of law.
>
> 12. The county court, acting in is appellate role, will normally conduct a review of the decision of the RAP rather than a rehearing: CPR Part 52.11(1). It will allow an appeal under CPR Part 53.11(3) [*sic*] where the decision under appeal was (a) wrong or (b) unjust because of a serious or other procedural irregularity.
>
> 13. The first general point that I wish to make arises out of the fact that an appeal to this court is a second appeal. In appeals of that kind the real question for this court is whether the intermediate appeal court (here the county court) was entitled to interfere with the decision of the first tier tribunal (here the RAP). In other words the focus of this appeal is whether the RAP were wrong: *Procter & Gamble UK v. HMRC* [2009] EWCA Civ 407; [2009] STC 1990.
>
> 14. The second general point is about appeals on fact. An appeal from the RAP is not limited to an appeal on a question of law, but that does not dilute the reluctance which an appeal court should have about overturning factual conclusions. Factual conclusions for this purpose include not only findings of primary fact but also evaluations of those facts, value judgments based on them, and the application of the facts to legal standards. The appeal court should not retry the case on the transcript. All this is contained in the recent judgment of the Supreme Court in *McGraddie v. McGraddie* [2013] UKSC 58; [2013] 1 WLR 2477 as well as in *Procter & Gamble UK* at [9] (Jacob LJ), [60] (Toulson LJ) and *Assicurazioni v. Generali SpA v. Arab Insurance Group* [2002] EWCA Civ 1642, [2003] 1 WLR 577 at [16] (Clarke LJ). Moreover where, as here, the appeal is an appeal from a specialist tribunal an appeal court's reluctance to intervene on questions of expertise should be all the greater: *Procter & Gamble UK* at [11] (Jacob LJ), [48] (Toulson LJ). This tribunal included a medically qualified member, and received advice from a specialist adviser.

## C. Late Appeals

### *Mitchell v. Nursing and Midwifery Council* [2009] EWHC 1045 (Admin)

**4.32** Nursing and Midwifery Order 2001, art 29(10)—extension of time—lack of merits

A decision letter dated 15 December 2006 was sent by the Nursing and Midwifery Council (NMC) to the appellant notifying her of the decision of the Conduct and Competence Committee (CCC) that she should be struck off the register of nurses. By article 29(10) of the Nursing and Midwifery Order 2001, any appeal against a striking-off order must be brought before the end of the period of twenty-eight days beginning with the date on which notice of the order or decision appealed against is served on the person concerned. On 16 January 2007, the appellant issued a claim form for permission to apply for judicial review, which would have been in time or possibly one day late. When the application for permission was considered by Ouseley J on the papers, he said that the claimant had failed to show any basis upon which the NMC's decision was arguably flawed and that the complete absence of any merit discernible in her case should militate against the exercise of any discretion in her favour. A renewed application in open court was dismissed by Stanley Burnton J on 30 January 2008, who said that it was unarguable that the Committee did not apply the law correctly and that the claimant's case was not arguable. On 30 September 2008, the appellant filed a notice of appeal out of time. When it was lodged, Bean J said that the appellant was able to refer to a decision of the Inner House of the Court of Session in the case of *Hume* [2007] CSIH 53, in which the Inner House (the Lord President, Lord MacFadyen, and Sir David Edward) held that article 29(10) did not deprive the court of jurisdiction to extend time if it saw fit, notwithstanding that the statutory time limit had not been complied with. That decision, however, had plainly been overruled by the recent decision of the House of Lords in *Mucelli v. Government of Albania* [2009] 1 WLR 276, which held that if an appeal is out of time, there is no power under the Civil Procedure Rules to extend. Bean J said that the hurdle that was placed in the appellant's path by article 29(10) is neither a rule of court, a Practice Direction, nor a court order, and so CPR 3.1(2)(a) did not assist. The learned judge added that even had he had a discretion to grant an extension of time, he would not have done so in this case. This was primarily because the merits of the appellant's appeal or application had already been considered twice—once by Ouseley J and once by Stanley Burnton J—and in each case the Court had held that there was no arguable basis of challenge to the NMC's decision on its merits.

***Harrison v. General Medical Council* [2011] EWHC 1741 (Admin)**

**4.33** Medical Act 1983, section 40(4)—CPR 3.1

Blake J held that the Court had no power to extend time for the lodging of an appeal under section 40(4) of the Medical Act 1983, as amended, against the determination of a fitness to practise panel and, in the alternative that, even if the Court had the power under CPR 3.1, it could not conclude on the facts that the interests of justice required an extension of time. On 26 February 2010, the GMC sent notice of the panel's decision that Dr H's registration be erased on grounds of a finding of impairment of fitness to practise based upon a serious criminal conviction. Dr H was a serving prisoner who did not attend the fitness to practise hearing. His appeal was lodged on 29 April 2010 and long after the period of twenty-eight days provided for by the 1983 Act. Blake J agreed with the decision of Bean J in the case of *Mitchell v. Nursing and Midwifery Council* [2009] EWHC 1045 (Admin), in which the subject matter was a question of whether the Court had power to extend time in a nursing and midwifery disciplinary appeal. Following the decision of the House of Lords in *Mucelli v. Government of Albania* [2009] 1 WLR 276, Bean J held that where a primary statute lays down the time limit for an appeal, the Court does not have power to extend the time beyond the period set out in the statute unless the statute gives the Court power to extend.

***Bartosik v. General Medical Council* [2011] EWHC 3906 (Admin)**

**4.34** Time limits to be strictly observed

His Honour Judge Jeremy Richardson QC (sitting as a deputy judge of the High Court) held that there was a strict time limit of twenty-eight days for an appeal. The period may not be extended by order of the court, because it is a statutory appeal. The time limits are to be strictly observed.

***R (Burke) v. Independent Police Complaints Commission* [2011] EWCA Civ 1665**

**4.35** Police (Complaints and Misconduct) Regulations 2004, regulation 10(1)

B made a complaint about the conduct of four officers of the Metropolitan Police involved in his arrest. The complaint was referred to the Independent Police Complaints Commission (IPCC) on 1 August 2006. The IPCC decided that the matter should be investigated by the Metropolitan Police, with a right of appeal to the IPCC if the appellant was not satisfied with the outcome. On 19 September 2007, the Metropolitan Police sent a decision letter addressed to B at his home, informing him that his allegations against the police officers in his arrest could not be proved. B said that he did not receive such a letter and that he learned of the decision only in early February 2009; he sought judicial review proceedings against the refusal of the IPCC to extend time for any appeal. Regulation 10(1) of the Police (Complaints and Misconduct) Regulations 2004 provides that any appeal shall be made within twenty-eight days of the date on which the appropriate authority sends a notification to the complainant of its determination. Regulation 10(8) provides that the IPCC may extend time in any case in which it is satisfied that, by reason of special circumstances, it is just to do so. The Court of Appeal (Richards, Hallett, and Stanley Burnton LJJ) dismissed B's appeal from the decision of Wynn Williams J. The Court held that the time for appealing runs from the date on which the appropriate authority sends a notification of its determination to the complainant. The fact that an appeal is out of time because the complainant does not receive notification of the decision within the time limit for appealing would be a very weighty factor in favour of an extension of time. If the complainant were to act promptly once notification was received, the Court would expect it normally to be a decisive factor, sufficient to establish that, in the special circumstances of the case, it was just to extend time. But the Court did not accept that it must always, as a matter of law, lead to a grant of the extension of time. There may be other relevant considerations to which it was both legitimate and necessary to have regard when deciding whether to exercise the discretion to extend time. In the present case, the overall circumstances of the delay and the effect of the delay were given careful and rational consideration by the IPCC in reaching its decision whether to extend time, and when the circumstances as a whole were taken into account, as they were, the result was not unreasonable.

### *Gilthorpe v. General Medical Council* [2012] EWHC 672 (Admin)

**4.36** Medical Act 1983, section 40(4)

The appellant sought to appeal against the decision of the GMC's fitness to practise panel made on 11 October 2011 to suspend his registration for four months. On 12 October 2011, the GMC wrote to the appellant setting out the panel's decision in full and drawing to his attention that any appeal must be lodged on or before 8 November 2011. On 15 November 2011, the appellant filed a notice of appeal out of time. Section 40(4) of the Medical Act 1983 provides that a person in respect of whom an appealable decision has been taken may, before the end of the period of twenty-eight days beginning with the date on which notification of the decision was served, appeal against the decision to the relevant court. Supperstone J held, following the decision in *R (Harrison) v. General Medical Council* [2011] EWHC 1741 (Admin), that the Court had no power to extend time. The learned judge said that he would not have extended time if the Court had jurisdiction to do so in the interests of justice. The appellant had not given a satisfactory explanation as to why the appeal, whether by statutory notice as required or by judicial review claim if that was permissible, was not made in time. There did not appear to be any merit in the grounds of appeal.

### *Reddy v. General Medical Council* [2012] EWCA Civ 310

**4.37** Medical Act 1983, section 40(4)

R sought to appeal against an order made by His Honour Judge Hand QC dismissing his appeal against the refusal of the GMC of his application for entry on the specialist medical register as a general surgeon. Judge Hand held that the Court had no jurisdiction to entertain R's appeal because it had not been brought within the prescribed time, being twenty-eight days beginning with the date on which notice of the determination had been given to him. The Court of Appeal (Mummery, Moore-Bick, and Black LJJ) held that the judge was right to hold that the Court had no power to extend time in favour of R. The judge was of the view that even if the Court had power to extend time, it would be inappropriate to do so in this case.

***R (Cela) v. General Pharmaceutical Council* [2012] EWHC 2785 (Admin)**

**4.38** Pharmacy Order 2010, article 58(3)—CPR 3.9—interest and justice

The respondent, the General Pharmaceutical Council, applied to strike out C's notice of appeal on the grounds that it was lodged out of time and there were no justifiable grounds to extend time. Article 58(3) of the Pharmacy Order 2010 provides that any notice of appeal must be filed at the High Court and served on the Council 'within 28 days beginning with the day on which the written notice for the reasons for the decision was sent or within such longer period as the High Court may, in accordance with the rules of court, allow'. On 12 December 2011, C was informed in writing of the decision of the Council that he be suspended from practice for twelve months and he was informed that the expiry of his period for appeal would be 9 January 2012. C's notice of appeal was issued on 24 January 2012 and not served on the Council until 2 March 2012. No explanation was ever furnished to the Court and no evidence was ever served on the Court as to why the notice of appeal was issued late. His Honour Judge Seys Llewellyn QC (sitting as a High Court judge) said that he turned first to consider whether C's appeal was one that may have merit, because ultimately it is the interests of justice that are likely to determine whether time should be extended: see *Gilthorpe v. General Medical Council* [2012] EWHC 672 (Admin). In the instant case, C was a pharmacist who was convicted in the magistrates' court on 11 January 2010 of theft by an employee, to which he was sentenced to a community order. The offence was carried out in breach of trust and when employed in the practice of a pharmacist, and involved taking money from the hands of customers and not putting it into the till or registering the relevant order or prescription for which he had received the relevant monies. Before the fitness to practise panel, C admitted the allegations. The decision of the panel as to impairment was unappealable and its decision on sanction, to impose a period of suspension of twelve months, was in some respects a humane approach, in that many cases of dishonesty may lead to erasure and this was a breach of trust to a high degree because it was committed during C's practice as a pharmacist. Under CPR 3.9, the Court is directed to consider all matters, including the merits. Since in the instant case there was a sanction, if the appeal were not brought in time and it were not extended, this would be analogous to the situation in which a litigant needs to apply for relief from sanction under the Civil Procedure Rules. As to rule 3.9(a), the interests of the administration of justice, in the instant case the interests of justice did not require an extension of time to be granted to the appellant because the appeal was without merit. As to (b), whether the application for relief has been made promptly, it does not follow even where an application is made promptly, that the Court must always grant an extension of time. There may be other relevant considerations. Here, in any event, application had not been made, let alone promptly. As to (c), whether the failure to comply was intentional, it was right to say that C had never furnished positive evidence or explanation that failure to comply was caused by mistake or inadvertence. As to (d), whether there is a good explanation for the failure, the answer in this case was emphatically not. As to compliance with other court orders, rule 3.9(e), it was to some extent striking that C was required to serve with his appellant's notice a skeleton argument and grounds of appeal, and that he had never done so, nor had there ever been an attempt to remedy that. For these reasons, to extend time would be of prejudice to the respondent Council, the panel of which had made a decision that appeared to the Court to be impeccable and which would be denuded of effect by an extension of time to present an appeal. For these reasons, the Court refused to extend time beyond that set out in article 58(3) of the Pharmacy Order 2010.

### *Nagiub v. General Medical Council* [2013] EWHC 1766 (Admin)

**4.39** Incorrect court fee paid

An appeal notice was lodged in time, but the incorrect fee paid. Therefore the appeal notice was not issued. It was held that there was no duty on court staff to give advice and no power to extend the time allowed.

### *R (Adesina) v. Nursing and Midwifery Council; R (Baines) v. Nursing and Midwifery Council* [2013] EWCA Civ 818; [2013] 1 WLR 3156, CA

**4.40** Principles to be applied following *Pomiechowski v. Poland* [2012] 1 WLR 1604 (Sc)

Two appeals were lodged by registrants after the end of the period of twenty-eight days provided for under article 29(9) of the Nursing and Midwifery Order 2001. There is no express provision permitting the court to extend time on a discretionary or any other basis. After judgment in the Administrative Court dismissing the appellants' appeals—[2012] EWHC 2615 (Admin)—the Supreme Court decided *Pomiechowski v. Poland* [2012] 1 WLR 1604, in which it held that apparently absolute time limits may, in some circumstances, have to yield to the requirements of Article 6 of the European Convention on Human Rights (ECHR). The Court of Appeal (Maurice Kay, Patten, and Floyd LJJ) held that the decision in *Pomiechowski v. Poland* requires adoption of a discretion 'in exceptional circumstances' and where the appellant 'personally has done all he can to bring [the appeal] timeously'. The discretion would arise in only very few cases. Although the absolute approach can no longer be said to be invariable, the scope for departure from the twenty-eight-day time limit is extremely narrow. On the facts of the present appeals, it was held that there were no exceptional circumstances to extend time. In giving judgment, Maurice Kay LJ said that:

> 14. ... [E]xclusion from a profession, is still one of great importance to an appellant. There is good reason for there to be time limits with a high degree of strictness. However, one only has to consider hypothetical cases to appreciate that, without some margin for discretion, circumstances may cause absolute time limits to impair "the very essence" of the right of appeal conferred by statute. Take, for example, a case in which a person, having received a decision removing him or her from the register, immediately succumbs to serious illness and remains in intensive care; or a case in which notice of the disciplinary decision has been sent by post but never arrives and time begins to run by reason of deemed service on the day after it was sent (Nursing and Midwifery Council (Fitness to Practise) Rules 2004, rule 34(4)). In such cases, the nurse or midwife in question might remain in blameless ignorance of the fact that time was running for the whole of the 28 day period. It seems to me that to take the absolute approach in such circumstances would be to allow the time limit to impair the very essence of the statutory right of appeal.
>
> 15. The real difficulty is where to draw the line. [Counsel], on behalf of the appellants, does not contend for a general discretion to extend time. Parliament is used to providing such discretions, often circumscribed by conditions (see, for example, Employment Rights Act 1996, section 111(2), in relation to unfair dismissal). The omission to do so on this occasion was no doubt deliberate. If Article 6 and section 3 of the Human Rights Act require article 29(10) of the Order to be read down, it must be to the minimum extent necessary to secure ECHR compliance. In my judgment, this requires adoption of the same approach as that of Lord Mance JSC in *Pomiechowski*. A discretion must only arise "in exceptional circumstances" and where the appellant "personally has done all he can to bring [the appeal] timeously" (paragraph 39). I do not believe that the discretion would arise save in a

very small number of cases. Courts are experienced in exercising discretion on a basis of exceptionality. See, for example, the strictness with which the discretion is approached in relation to the 42 day time limit and the discretion to extend in connection with appeals from Employment Tribunals to the Employment Appeal Tribunal: *United Arab Emirates v. Abdelghafar* [1995] ICR 65; *Jurkowska v. Hlmad Ltd* [2008] EWCA Civ 231.

### *Adegbulugbe v. Nursing and Midwifery Council* [2013] EWHC 3301 (Admin)

**4.41** One day late—prospects of success minimal

A's appeal under article 29 of the Nursing and Midwifery Order 2001 was filed one day late. In dismissing A's appeal, Andrews J said that all professional lawyers who represent clients before the courts owe duties to their clients and to the court. In this case, it seemed that A had not been well served by either her solicitors or counsel. The appeal was shoddily prepared and its presentation fell below the professional standards that are expected in any court. Wherever the blame lay for the failure to lodge the documents on time, albeit only a day too late, the appeal was doomed to fail in the absence of strong and persuasive evidence that the essence of the right to appeal would be impaired if the Court did not allow the merits to be heard. After receipt of the NMC's skeleton argument, the appellant prepared a last-minute witness statement, but it did not address the point. Moreover, on the basis of the materials before the Court, A's prospects of success on the merits were minimal.

### *R (Gyurkovits) v. General Dental Council* [2013] EWHC 4507 (Admin)

**4.42** In this case, it was held that there were no exceptional circumstances to justify an appeal out of time.

### *Pinto v. Nursing and Midwifery Council* [2014] EWHC 403 (Admin)

**4.43** In this case, too, there were no exceptional circumstances to justify an appeal out of time.

### *Parkin v. Nursing and Midwifery Council* [2014] EWHC 519 (Admin)

**4.44** Merits of appeal—extent to which merits may be relevant

P's notice of appeal was lodged one day late. In considering whether or not to grant an extension, the appellant submitted, it was relevant to consider the nature of the appeal, including the facts that the case involved the ability of P to conduct his profession, that the grounds of appeal included serious procedural irregularities, and that this was not a marginal or unmeritorious appeal. Eder J observed:

> 16. The question of principle that then arises is how should the court proceed when those points are raised. [Counsel on behalf of the NMC] initially submitted that the merits, the underlying merits of the case, are not relevant to the exercise of the discretion. As it seems to me, that submission may well be right: it may be supported, for example, by the views expressed by Andrews J in *Kehinde Bimbe Adegbulugbe v. NMC* [2013] EWHC 3301 (Admin), in particular at paragraph 28. However, I am prepared to assume at least in this case in [Counsel on behalf of the appellant]'s favour that the merits of the appeal may be relevant. I should make it plain that I do not decide that this is indeed the case. But I am prepared to assume in [Counsel on behalf of the appellant]'s favour that it is the case here. That appears at least from a decision in a slightly different context in *United Arab Emirates v. Abdelghafar* [1995] ICR 65 where Mummery J referred to the exercise of the judicial discretion in that type of case. That was in the [Employment Appeal Tribunal]. He said, in particular, as follows: "The merits of the case may be relevant but they are usually of little weight. It is not appropriate on an application for leave to extend time for the appeal tribunal to be asked to investigate in detail the strength of the appeal, otherwise

> there is a danger that an application for leave will be turned into a mini hearing of the substantive appeal."
>
> 17. So I reject [Counsel on behalf of the NMC]'s submission that the merits are not relevant at least to this extent, that I am prepared to assume, as I have said, in favour of [Counsel on behalf of the appellant]'s submissions that they are or may be relevant at least in part, although the court should be very wary about not turning this into a mini hearing of the substantive appeal.

Having heard submissions, the judge held that the underlying merits of the appeal were weak at best and that the appeal was inevitably doomed, even if the appeal had been brought in time.

*Martin-Artajo v. Financial Conduct Authority* [2014] UKUT 0340 (TCC)

**4.45** Reference lodged with Upper Tribunal outside statutory time limit—public interest—prejudice to appellant

The applicant, M-A, was head of Europe, the Middle East, and Africa (EMEA) Credit and Equity Trading for JP Morgan Chase Bank NA (JPM) in London between 2007 and July 2012, and managed the synthetic credit portfolio in JPM's chief investment office. As a result of large trading losses within the portfolio, its activities were investigated in both the United Kingdom and the United States. JPM agreed to a settlement with the Financial Conduct Authority (FCA) following the conclusion of the Authority's investigation, which resulted in JPM accepting a financial penalty of £137,610,000. A final notice addressed to JPM imposing this penalty, and setting out in full the facts and matters relied on and the reasons for the decision, was published on 19 September 2013. In the meantime, the applicant had been made the subject of an investigation by the Authority as to his own personal conduct. On 15 October 2013, just before the expiry of the twenty-eight-day time limit for filing a reference with the Upper Tribunal, the applicant's lawyers wrote to the Authority stating that, in the light of the FCA's final notice to JPM, they would be grateful for an update in relation to 'your proposed timetable for the progression of the FCA's investigation with respect to [M-A]'. Subsequently, on 16 December 2013, the Authority replied, stating that, in the light of proceedings brought against the applicant in the United States by the US Department of Justice and the US Securities and Exchange Commission, the FCA had decided to discontinue its investigation against the applicant—although this decision did not mean that the FCA had concluded that the applicant's conduct complied with the standards expected of him under the law and applicable regulatory regime. The applicant, taking the view that he had been identified in the Authority's final notice against JPM, filed a reference with the Tribunal. The reference was filed four months and twenty-two days after the expiry of the twenty-eight-day statutory time limit. In extending time, Judge Timothy Herrington said that whilst the importance of the time limit was recognized, it was lessened in this case because of the lack of significant prejudice to the Authority. Whilst the delay was significant, it would make no material difference to disposing of the litigation efficiently because of an ongoing reference by another party, and it was appropriate to stay the applicant's reference until the Court of Appeal's decision in relation to the other party. Bearing in mind the public interest considerations and the fact that the applicant had not previously had the opportunity to make representations, the prejudice to the applicant in not extending time would clearly outweigh the prejudice to the Authority if time were not extended. The Tribunal also took into account the fact that, at all times, the applicant acted in good faith, upon professional advice, and with a degree of hindsight that matters might have been handled differently. It would be too severe a consequence in all of the circumstances and unfair to the applicant to refuse to extend time.

***Daniels v. Nursing and Midwifery Council* [2014] EWHC 3287 (Admin)**

**4.46** Attempts to raise funds and lodge appeal in time—appellant's circumstances—good reason to extend time

D applied to extend time to appeal against the sanction imposed by the NMC made on 8 February 2014, when the NMC determined that a caution order for a period of three years was appropriate. The allegations were not of clinical misconduct, but were of a failure to comply with the conditions of supervised practice on three occasions and a failure to comply with specific instructions on a further date. Pursuant to the rules, D had twenty-eight days in which to appeal. Time expired on 8 March 2013; in fact, grounds of appeal were filed on 11 March 2013. In granting an extension of time, Nicola Davies J said that she was satisfied that there were exceptional circumstances and a good reason why this particular appellant was unable to lodge her grounds of appeal in time. At the date on which D received notice of the sanction of the NMC, she had been unemployed for three years. She was living on state benefits and, at the time of the hearing, she had at different times been legally represented. Nurses who appeared before the NMC do not have the advantage of such defence organizations as are available to doctors and dentists in respect of their regulatory proceedings, and as a result can encounter difficulties in funding legal representation. The appellant obtained funding from family and friends, but this was very limited in nature. Her solicitors did not specialize in this field and worked for her either at a reduced cost, or, on occasion, on a pro bono basis in respect of the NMC proceedings. Such was the limit of their representation that there was an occasion on which the appellant herself had to cross-examine NMC witnesses. The appellant had no money to put her solicitors in funds for an appeal and did not have easy recourse to funds, because her family and friends had helped her already. On the day before time expired, she approached counsel who had acted for her, but did not have the money to pay the court fee of £235 to lodge an appeal. When the monies for the court fee were found, the fee was paid and grounds were lodged. The learned judge said that she had considered very carefully the authority of *R (Adesina) v. Nursing and Midwifery Council* [2013] EWCA Civ 818. In this case, there was good reason why the appeal could not be lodged in time: the appellant could not find £235 with which to pay the court fee. That was unsurprising, given that she was living on benefits, had been dependent on family and friends to help her through the NMC proceedings, and had been unemployed for a period of three years. The time by which she was outside the relevant twenty-eight days was a matter of some two or three days, and it could not be said that any particular prejudice was thereby suffered by the NMC in dealing with her case.

## D. Non-Attendance

**4.47** See paragraphs 1.14, 1.17, 1.20, and 1.21, and the following cases:

- *Faniyi v. Solicitors Regulation Authority* [2012] EWHC 2965 (Admin);
- *R (Veizi) v. General Dental Council* [2013] EWHC 832(Admin);
- *Ward v. Nursing and Midwifery Council* [2014] EWHC 1158 (Admin); and
- *Malik v. General Medical Council* [2014] EWHC 2408 (Admin).

## E. Fresh Evidence on Appeal

**4.48** See paragraphs 25.34–25.42 and the following cases:

- *Hefferon v. Professional Conduct Committee of the United Kingdom Central Council for Nursing Midwifery and Health Visiting* (1988) 10 BML 1;

- *Threlfall v. General Optical Council* [2004] EWHC 2683 (Admin);
- *R (Russell) v. General Medical Council* [2008] EWHC 2546 (Admin);
- *Ansari v. General Pharmaceutical Council* [2012] EWHC 1563 (Admin);
- *R (Shaikh) v. General Pharmaceutical Council* [2013] EWHC 1844 (Admin);
- *R (Chief Constable of British Transport Police) v. Police Appeals Tribunal and Whaley* [2013] EWHC 539 (Admin);
- *R (Khan) v. General Medical Council* [2014] EWHC 404 (Admin);
- *Sohal v. Solicitors Regulation Authority* [2014] EWHC 1613 (Admin); and
- *Jasinarachchi v. General Medical Council* [2014] EWHC 3570 (Admin).

## F. Other Relevant Chapters

**4.49** Bias, Chapter 6
Dishonesty, Chapter 19
Drafting of Charges, Chapter 22
Evidence, Chapter 25
Findings of Fact, Chapter 27
Human Rights, Chapter 32
Impairment of Fitness to Practise, Chapter 33
Misconduct, Chapter 47
Natural Justice, Chapter 48
Sanction, Chapter 60
Service, Chapter 61

# 5

# BAD CHARACTER

A. Legal Framework 5.01
B. Disciplinary Cases 5.02
*Ziderman v. General Dental Council* [1976] 1 WLR 330 5.02
*Gleadall v. Huddersfield Magistrates' Court* [2005] EWHC 2283 (Admin) 5.03
*R (H) v. Nursing and Midwifery Council* [2013] EWHC 4258 (Admin) 5.04
C. Other Relevant Chapters 5.05

## A. Legal Framework

**5.01** Criminal Justice Act 2003, section 143(2):

> In considering the seriousness of an offence ("the current offence") committed by an offender who has one or more previous convictions, the court must treat each previous conviction as an aggravating factor if (in the case of that conviction) the court considers that it can reasonably be so treated having regard, in particular to—
> (a) the nature of the offence to which the conviction relates and its relevance to the current office, and
> (b) the time that has elapsed since the conviction.

## B. Disciplinary Cases

### *Ziderman v. General Dental Council* [1976] 1 WLR 330

**5.02** *Previous offences and erasure—relevance to sanction*

In June 1971, Z's name had been erased from the dentists' register following his conviction of ten offences of obtaining money by false pretences from the Executive Council of the National Health Service and sixty-two other similar offences being taken into consideration. In November 1972, his name was restored to the register. On 31 December 1974, Z was convicted of a charge of shoplifting, to which he pleaded guilty and for which he was fined. Disciplinary proceedings were brought against him in respect of the shoplifting conviction and, on 14 May 1974, it was again ordered that his name be erased from the register. From that order, he appealed to the Privy Council and relied on section 25(2)(a) of the Dentists Act 1957, which provides that a person's name shall not be erased on account of a conviction for an offence that does not, either from the trivial nature of the offence or from the circumstances under which it was committed, disqualify a person from practising dentistry. Lord Diplock, in giving the judgment of the Board, said that section 25(2)(a) does no more than state expressly and in negative form the obverse of the positive duty of a committee that is implicit in the nature of its disciplinary jurisdiction and the consequences of the only penalty that it has power to impose. The reference to 'the trivial nature of the offence' and 'the circumstances under which it was committed' is wide enough to cover, with one exception, all of those matters that the committee ought to take into consideration in deciding whether the practitioner committing the offence justified the conclusion that he was unfit to continue in practice as a member of the dental profession. The one exception is his conduct after the conviction. This may show that he has so rehabilitated himself by the time the committee pronounces sentence that he has ceased to be unfit to continue in practice as a dentist. Lord Diplock continued, at 333G–334F:

> Convictions previous to that on which the charge before the committee is based do not fall within the category of subsequent conduct. In their Lordships' view, the previous convictions of a person charged with an offence are relevant matters to be taken into account by a sentencing tribunal in assessing the gravity of the offence with which he is charged. They are thus relevant in determining whether or not it is an offence of too trivial a nature to justify the infliction of a particular penalty, particularly where that penalty is severe. Justice requires that the gravity of an offence should be reflected in the severity of the sentence imposed; and, as is manifest not only from the universal sentencing practice of judges but also from statutory provisions limiting the severity of penalties that may be imposed upon first offenders, the national sense of what constitutes justice in the field of sentencing recognises that an offence which is committed by a person who has offended before is graver than a similar offence committed by a person who offends for the first time. Indeed in Scotland, to which the Dentists Act 1957 also applies, an indictment contains a final paragraph (not disclosed to the jury) setting out the previous convictions of the accused as constituting an "aggravation" of the offence with which he is charged. The weight to be attached to previous convictions in assessing the gravity (or the trivial nature) of the offence charged is a matter for the tribunal charged with the duty of determining what the penalty shall be. No doubt previous convictions for technical offences which involve no moral turpitude will be given little weight in the assessment of the appropriate penalty for an offence which does. The weight to be attached to previous convictions for offences which involve some moral turpitude, such as dishonesty, will no doubt vary with their similarity or otherwise to the offence committed. But questions of weight are not questions of law on which the legal assessor to the committee is required or entitled to advise them. They are left to the collective good judgment of the members of the committee in whom the sentencing function is vested by the Dentists Act 1957. In their Lordships' view there was no need for the legal assessor to give to the committee any elaborate advice upon their legal duty in exercising their sentencing power in the instant case; and, in particular, he would have been wrong had he advised them as a matter of law that they were not entitled to take into account the appellant's previous convictions for offences of dishonesty in determining whether the offence of shoplifting, which he had committed and which was the subject of the charge before them, was in all the circumstances so grave as to justify the conclusion that the appellant was unfitted to continue in practice as a member of the dental profession. What weight was to be attached to those previous convictions was a question for the members of the committee themselves—not for their legal assessor.

### *Gleadall v. Huddersfield Magistrates' Court* [2005] EWHC 2283 (Admin)

**5.03** No requirement generally to investigate bad character of prosecution witness

G was charged with an offence of common assault. It was alleged that he had punched V in the street. There were four eyewitnesses to the assault, all of whom supported the contention that G had punched V. G entered a 'not guilty' plea. As primary disclosure, the prosecution informed the defence that none of its lay witnesses had any previous convictions. G served a questionnaire, which asked whether the witnesses had ever been the subject of criminal proceedings or a disciplinary investigation or hearing (regardless of the outcome). The Crown Prosecution Service (CPS) replied, stating that it would not answer the questionnaire and that it would apply the usual rules in relation to unused material. In dismissing G's application for a stay of the prosecution and judicial review, the Divisional Court (Smith LJ and Simon J) said that the real question was whether the kind of investigation called for by G was necessary in the interests of justice. It was not. What was necessary in the interests of justice was investigation that is reasonable in the circumstances of the individual case. The gravity of the offence charged would be an important factor. So far as an investigation of character is concerned, the centrality of a particular witness's evidence to the prosecution case will be important. But it cannot be said that the interests of justice require that, in every case, comprehensive inquiries are made about the character of every prosecution witness whose evidence is to be challenged.

***R (H) v. Nursing and Midwifery Council* [2013] EWHC 4258 (Admin)**

**5.04** Cross-examination of bad character—no requirement for leave—relevance

An application was made on behalf of the registrant nurse before the Conduct and Competence Committee (CCC) of the Nursing and Midwifery Council (NMC) to adduce evidence of the criminal convictions of the complainant and service user, and to cross-examine him about them. In allowing H's appeal on different grounds, His Honour Judge Pelling QC (sitting as a judge of the High Court) said, at [18], that it was unclear why it was thought necessary to make this application at all. Rule 31 of the Nursing and Midwifery Council (Fitness to Practise) Rules 2004 dealing with evidence, under which the application was made, is not concerned with the admissibility of particular facts or matters, but the nature of the evidence relied on as proving the factual allegation or assertion made. It suggests, as might be expected, that a panel will take a rather more liberal view of the methods by which a fact or allegation can be proved than would a court of law. Thus it was not appropriate to make the application at all. The only basis on which it could be contended that the fact of a criminal conviction ought to be excluded was on grounds of relevancy or inadmissibility by operation of statute, such as the rehabilitation of offenders legislation. Convictions for offences of dishonesty, or convictions for any offence following a plea of 'not guilty' in which the person concerned gave evidence that was disbelieved or, for that matter, evidence that a person's evidence was disbelieved in a civil case, are material that is at least potentially relevant to an assessment of credibility. On appeal the registrant did not challenge the Committee's decision to exclude the material, although the learned judge said, at [20], that, for his part, he did not consider fairness to be material to the assessment: either the evidence is admissible or it is not. The Committee concluded that the convictions were not relevant (although the basis for that conclusion was not spelt out) and also that they would have little probative value. This last point is inconsistent with the conclusion that the convictions are inadmissible or irrelevant. If they were relevant, then the appellant should have been permitted to cross-examine about them and their weight was something for the Committee to evaluate, having heard all of the evidence.

## C. Other Relevant Chapters

**5.05**

Good Character, Chapter 30
Insight, Chapter 37
Investigation of Allegations, Chapter 39

# 6

# BIAS



## A. Legal Framework/General Principles

**6.01** European Convention on Human Rights (ECHR), Article 6 (see paragraph 31.01)

In *Porter v. Magill* [2002] 2 AC 357, 494H, Lord Hope of Craighead said that '[t]he question is whether the fair-minded and informed observer, having considered the facts, would conclude that there was a real possibility that the tribunal was biased'.

In *Lawal v. Northern Spirit Ltd* [2003] ICR 856, HL, [14], Lord Steyn, giving the judgment of the House of Lords (Lords Bingham of Cornhill, Nicholls of Birkenhead, Steyn, Millett, and Rodger of Earlsferry) said that, in *Porter v. Magill*, the House of Lords approved a modification of the common law test of bias enunciated in *R v. Gough* [1993] AC 646. This modification was first put forward in *In re Medicaments and Related Classes of Goods (No. 2)* [2001] ICR 564. The purpose and effect of the modification was to bring the common law rule into line with the Strasbourg jurisprudence. As a result, there is now no difference between the common law test of bias and the requirements under Article 6 ECHR of an independent and impartial tribunal. The small, but important, shift approved in *Porter v. Magill* has at its core the need for 'the confidence which must be inspired by the courts in a democratic society': *Belilos v. Switzerland* (1988) 10 EHRR 466, 489, [67]; *Wettstein v. Switzerland*, App. no. 33958/96, [44]; *In re Medicaments and Related Classes of Goods (No. 2)* [2001] ICR 564, 591, [83]. Public perception of the possibility of unconscious bias is the key. It is unnecessary to delve into the characteristics to be attributed to the fair-minded and informed observer. What can confidently be said is that one is entitled to conclude that such an observer will adopt a balanced approach. This idea was succinctly expressed in *Johnson v. Johnson* (2000) 201 CLR 488, 509, [53], by Kirby J, when he stated that '[a] reasonable member of the public is neither complacent nor unduly sensitive or suspicious'. Lord Steyn continued, at [21], by stating: 'The principle to be applied is that stated in *Porter v. Magill*, namely whether a fair-minded and informed observer, having considered the given facts, would conclude that there was a real possibility that the tribunal was biased.'

## B. Disciplinary Cases

### *Datta v. General Medical Council*, Privy Council Appeal No. 34 of 1985

**6.02** Chairman involved in earlier proceedings against registrant

On 16 July 1985, the Professional Conduct Committee (PCC) of the General Medical Council (GMC) found D guilty of serious professional misconduct and ordered his name to be erased from the register. The charge against D was that he had failed to visit and to provide or arrange treatment for a patient for whose general medical care he was responsible at the material time, despite the patient's condition and despite requests to visit the patient who was in need of urgent medical attention. D submitted that the same chairman presided over earlier proceedings before the Committee, which, on 29 November 1984, had found D guilty of serious professional misconduct in that he had abused his position as a medical practitioner by borrowing substantial sums of money from a patient and that a cheque for repayment had been dishonoured on presentation. Those proceedings resulted in an admonition. Lord Griffiths, in delivering the judgment of the Privy Council in dismissing D's appeal, said at 6, that there was no substance to the submission:

> It is inevitable that those who sit in a judicial or a quasi-judicial capacity will, from time to time, have to hear cases against accused persons who have appeared before them on previous occasions. As a general rule there can be objection to the practice. Those entrusted with judicial or quasi-judicial functions must and can be trusted to try the case on the evidence before them and to put out of their minds knowledge, arising out of any earlier appearance before them by the same accused person. If in the particular circumstances of an individual case it is thought to be undesirable that the case should be heard by the same person then objection to the tribunal should be taken at the outset of the proceedings so that consideration can be given to the objection. In the absence of any such objection in the present case, the general rule must prevail and this ground of challenge fails.

### *R v. Board of Visitors of Frankland Prison, ex parte Lewis* [1986] 1 WLR 130

**6.03** Chairman of Board of Visitors also JP member of Parole Board

At the hearing of disciplinary proceedings, the applicant thought that he recognized the chairman of the Board of Visitors, but was unable to place him. It was only after the hearing that he realized that the chairman was a member of the local review committee that had seen him for the purpose of the Parole Board considering, in due course, whether he should be released on licence. The Divisional Court (Woolf J) dismissed an application for judicial review to quash the finding made by the Board of Visitors. The chairman deposed in an affidavit that he had considerable experience as a member of the Board and he was also a Justice of the Peace (JP). After reviewing *R v. Metropolitan Stipendiary Magistrate, ex parte Gallagher* (1972) 136 JPJ 80 and *R v. Liverpool City Justices, ex parte Topping* [1983] 1 WLR 119, Woolf J said that a distinction can and has to be drawn between the situation of a member of the Board of Visitors and a JP. It is inevitable in relation to the functions of a Board of Visitors, apart from those of hearing charges of a disciplinary nature, that the members of the Board must know more about those who appear before them than would normally be the case with JPs. In considering whether or not there has been a departure from the rules of natural justice, two principles should be borne in mind. First, a Board of Visitors, like Justices of the Peace, undoubtedly has a discretion not to proceed with the hearing of disciplinary proceedings and to require a different panel to be constituted if it considers that it is not possible to proceed fairly and justly, having regard to its background knowledge. Second, when exercising that discretion, the Board should take into account the functions that it has to perform, and bear in mind that it is required to adjudicate and to deal with disciplinary offences. It is important that those who adjudicate upon them should have knowledge of the workings of prisons in general and the prison of which the prisoner concerned was an inmate in particular. The Board of Visitors should not be too ready to regard general background knowledge of a particular prisoner as something that makes it desirable for it not to continue with the adjudication on a particular disciplinary charge. In deciding whether or not it is possible for a particular member of a Board to adjudicate fairly upon disciplinary offences, the reasonable and fair-minded bystander would have to take into account the nature of the proceedings and the nature of the duties that the Board has to perform.

### *Nwabueze v. General Medical Council* [2000] 1 WLR 1760

**6.04** Lay member of PCC with connections to locality in which doctor practised

The allegations against N all related to the period between September 1993 and November 1996, during which he was in practice as a doctor at a surgery in Prestatyn, Clwyd. Between August 1992 and September 1993, N had been working at a medical centre in Wrexham. A lay member of the PCC, W, had undisclosed local connections with Wrexham. In a statement, W said that she did not know anyone working at the practice in Wrexham and that the only occasion prior to the commencement of the hearing on which she had heard of the doctor was when she was telephoned by the GMC about her availability to sit. She replied that she did not know of N or anything about him. The Privy Council said that W was eminently qualified to sit on the Committee as one of its lay members. She brought to that membership an extensive knowledge of the health service in Wales, as a result of having worked there for many years as a nurse and midwife, and her period of service as director of the South East Wales Institute. It was in the public interest that those who serve as lay members on disciplinary bodies of this kind should be well informed and have experience of working in the area within which cases are likely to arise on which they may be called upon to adjudicate. It could not possibly be suggested that anything could give rise to the danger or possibility of bias on her part.

### *R (Mahfouz) v. Professional Conduct Committee of the General Medical Council* [2004] EWCA Civ 233

**6.05** Newspaper report of doctor's prior involvement with GMC

M practised in London as a cosmetic surgeon specializing in laser surgery. During the course of proceedings before the PCC, the *Evening Standard* (circulating in London) published an article that included reference to the fact that M had previously been struck off when working as a general practitioner's (GP's) assistant. Davis J dismissed a challenge to the decision of the Committee not to discharge itself and its decision to refuse an adjournment for the purposes of a High Court application. The Court of Appeal dismissed an appeal on the first point, but allowed it on the second.

See paragraph 51.04.

### *Helow v. Secretary of State for the Home Department* [2008] 1 WLR 2416

**6.06**
Judge—member of International Association of Jewish Lawyers and Jurists—asylum claim brought by supporter of Palestinian Liberation Organization

The appellant, H, was a Palestinian by birth and a supporter of the Palestinian Liberation Organization (PLO). She arrived in the United Kingdom in August 2001 and claimed asylum here on 4 September 2001. Her application was refused by the Home Secretary and, on appeal, by an adjudicator. The appellant was refused leave to appeal to the Immigration Appeal Tribunal. She then lodged a petition in the Court of Session seeking a review of that refusal. The petition was considered by Lady Cosgrove, who dismissed it. H made no criticism of Lady Cosgrove's reasons for dismissing her petition. Instead, in a petition to the nobile officium, she craved the Court to set aside Lady Cosgrove's interlocutor on the ground that it was vitiated from 'apparent bias and want of objective impartiality'. H contended that Lady Cosgrove's membership of the International Association of Jewish Lawyers and Jurists (IAJLJ) suggested that she might be the kind of supporter of Israel who could not be expected to take an impartial view of a petition for review concerning a claim for asylum made by a PLO supporter involved in legal proceedings against Israel. The Court of Session and the Inner House refused the petition. In dismissing H's appeal, the House of Lords said that the legal test to be applied in cases of apparent bias is to be found in the speech of Lord Hope of Craighead in *Porter v. Magill* [2002] 2 AC 357, 494H: 'The question is whether the fair-minded and informed observer, having considered the facts, would conclude that there was a real possibility that the tribunal was biased.' It was equally well established that the fair-minded observer is not unduly sensitive or suspicious: *Johnson v. Johnson* (2000) 201 CLR 488, 509, [53] *per* Kirby J. In the instant case, it was not suggested that Lady Cosgrove had ever said or expressed support for any extreme views such as those expressed by the president of the IAJLJ, and a fair-minded and informed observer would not impute to Lady Cosgrove the published views of other members of the IAJLJ, by reason only of her membership of that organization. The alternative argument that there was a real possibility of bias by reason of Lady Cosgrove having been influenced by the views expressed in articles published by the IAJLJ involved the inherent unlikelihood that Lady Cosgrove, despite her training and experience as a judge, would not have been able to put aside what she had read.

### *R (Johnson and Maggs) v. Professional Conduct Committee of the Nursing and Midwifery Council* [2008] EWHC 885 (Admin)

**6.07**
Failure by Committee to criticize NMC—structure of professional regulation—allegations to be particularized and supported by evidence

Beatson J dismissed a rolled-up application for permission with immediate determination of the substantive claim in respect of objections made by two registrants to a number of charges, and the refusal of the Nursing and Midwifery Council (NMC) to stay charges. The claimants alleged that there was apparent bias by the PCC in favour of a prosecution, and an undue willingness to overlook defaults in the NMC as prosecutor in respect of late disclosure and non-disclosure of correspondence about the investigation. At [115]–[123], the learned judge turned to the challenge based on bias. The failure of the Committee to criticize the NMC in respect of these matters (and the NMC may well have been open to criticism) did not arguably give rise to a real possibility that it was biased under the familiar tests in, for example, *AWG Group Ltd v. Morrison* [2006] 1 WLR 1163 and *Locabail (UK) Ltd v. Bayfield Properties Ltd* [2000] QB 451. The structure of professional regulation in this country, for better or for worse, is one that gives a large role in regulation to members of the profession. To that extent, it is always possible for a practitioner facing serious charges to feel that the prosecutor and the tribunal are somehow close. It is, however, significant that, when there are examinations of the rigour of professional regulation of this sort (for example, the criticisms of the GMC's approach in the *Shipman* inquiry), the criticism is not of excessive rigour and bias against practitioners, but quite the reverse. Where allegations of bias are made, it is important that they should be particularized and supported by evidence. This did not happen in this case. The challenge on the bias ground was utterly unarguable.

### *Muscat v. Health Professions Council* [2008] EWHC 2798 (Admin)

**6.08** Panel member's cross-examination of witness

M, a radiographer, appealed against the decision of the Health Professions Council's Conduct and Competence Committee (CCC), in which it was found that his fitness to practise was impaired and an order was made striking him off the register. The allegations made against M, which he strongly disputed, were that he conducted an X-ray of a female patient by lifting her gown to just below her chest and conducted a magnetic resonance imaging (MRI) scan of another female patient during which he required her to undress completely. On appeal, an allegation of bias or misconduct was made against one member of the Committee. It was alleged that the Committee member, W, misconducted herself during the hearing in that she proceeded to cross-examine M in a very hostile way and asked him questions by shaking her finger at him in an aggressive manner. Silber J offered both counsel the opportunity to apply to cross-examine M or any of the other makers of the witness statements, but they declined. Silber J said that there was no reason why he could reject the evidence of the members of the Committee that they did not remember any aggressive behaviour on the part of W. At the end of the day, the learned judge concluded that, in the absence of cross-examination, he could not be satisfied to even the lowest civil standard of proof that W either wagged her finger in an aggressive way or moved her chair towards the appellant. The questioning by W did not reach the point at which she had descended into the arena so as to impair her ability properly to evaluate and weigh the evidence. In conclusion, the learned judge held, at [59]–[67], that the challenge based on actual or apparent bias had to be rejected for four reasons.

(1) Decision-makers should be more proactive and interventionist: *Ruscillo v. Council for Regulation of Healthcare Professionals* [2005] 1 WLR 717, [80].
(2) A member of a disciplinary tribunal is entitled to express a provisional view of some evidence based on his or her understanding of it, even if that understanding is wrong, without there being any possibility of that member being regarded as being biased.
(3) The cases show that cases of bias or apparent bias depend on showing either (a) some pre-existing relationship between the person said to be actually or apparently biased and the victim of the bias, or (b) at least behaviour during the entire proceedings that establishes actual or apparent bias.
(4) If one ignores the allegations made by M, then the case for rejecting the appellant's claims of bias and apparent bias becomes even stronger.

### *Fotheringham, petitioner* [2008] CSOH 170

**6.09** Chairman—secretary of regional football association and chairman of victim's club

In this case, the petitioner challenged the decision of a disciplinary committee of the Scottish Football Association (FA) on the grounds of bias because its chairman was the secretary of a regional football association comprising several clubs, including that to which the alleged victim belonged. It was also claimed that the appeal hearing before the Scottish FA Disciplinary Tribunal was tainted by bias because the disciplinary committee appeared as a party and was represented by counsel, and the president of the Scottish FA was a member of the Disciplinary Tribunal. The claims of bias were rejected. Applying the test in *Porter v. Magill*, the Court of Session (Lord Pentland) found that a fair-minded and informed observer, having considered all of the facts, would not conclude that there was a real possibility that the original committee was biased because its chairman was the elected secretary of the regional football association. The link between the chairman and the alleged victim's club was too weak and insubstantial to give rise to any serious doubt about his ability to act independently and objectively as chairman. The mere fact that the president of the Scottish FA was a member of the Disciplinary Tribunal in circumstances in which the disciplinary committee was represented would be insufficient to cause a fair-minded and informed observer to conclude that there was a real possibility of bias.

### *Virdi v. Law Society (Solicitors Disciplinary Tribunal intervening)* [2010] 1 WLR 2840

**6.10** See paragraph 35.11.

### *R (Colman) v. General Medical Council; R (Hickey) v. General Medical Council* [2010] EWHC 1608 (Admin)

**6.11** Deputy registrars preparing draft determination—retiring with Committee

In each of these applications for judicial review, the claimants challenged the decision of the PCC of the GMC that they had been guilty of serious professional misconduct and that their respective names be erased from the register. Both applications gave rise to the same issue—namely, whether the determination of the Committee was, in each case, rendered unlawful by virtue of apparent bias. Both claimants argued that that was the effect of the involvement of the GMC's deputy registrars in the preparation of the proceedings, in preparing a draft determination in advance of the hearing, and in retiring with the Committee when it considered its determination. Owen J refused both applications, holding that the role of the deputy registrar was not that of a prosecutor, but was essentially to assemble the material on which a screening decision would be made. The draft determinations had not influenced the Committee. Whilst the deputy registrars were present throughout its deliberations, they played no part in the deliberations and, on the evidence, the cases could not be identified as cases in which an outsider had dealings with the Committee or could have influenced it. The deputy registrars were present as secretaries and the panels were made up of a substantial number of independent medical practitioners; the integrity of the proceedings were subject to the further safeguard of the presence of the legal assessor, who was under a specific duty to inform the Committee of any irregularity in the proceedings. A fair-minded and informed observer would not consider that there was a real possibility that either panel was biased.

### *R (French) v. Chief Constable of West Yorkshire Police* [2011] EWHC 546

**6.12** See paragraph 3.17.

### *Jeffery v. Financial Services Authority*, FS 2010/0039, 5 December 2012

**6.13** Judge sharing office with another judge

In this case, a judge of the Upper Tribunal and former chairman of the Regulatory Decisions Committee of the Financial Services Authority (FSA) shared an office with the judge hearing the appeal. There was no allegation of actual bias and no ground for recusal of the judge.

### *R (Austin) v. Teaching Agency* [2013] EWHC 254 (Admin)

**6.14** Gender of panel members

Two members of a three-member panel were female. The appellant was seeking the case to be heard by heterosexual males because of the delicate nature of some documentary material in his defence. His application was refused on the basis that the case fell 'a long way short of constituting a case of apparent bias'.

### *Adu v. General Medical Council* [2014] EWHC 1946 (Admin)

**6.15** Judge—professional association with legal assessor

The medical speciality in which A practised was paediatrics. At the commencement of the fitness to practise hearing in which it was alleged that A's fitness to practise was impaired by reason of deficient professional performance, the legal assessor disclosed to the panel and the parties some matters about health problems encountered by one of his children and questions or concerns that arose about the medical care that child had received. A argued that the legal assessor should recuse himself because the case involved issues of competence regarding paediatric care. The panel gave a reasoned ruling and determined that the circumstances were not such that the legal assessor should recuse himself. The panel referred to *R (Compton) v. Wiltshire Primary Care Trust (No. 2)* [2009] EWHC 1824 (Admin); [2010] PTSR (CS), in which Cranston J held that the bias of advisers was capable of vitiating a decision when there was a real possibility that it had adversely infected the views of the decision-maker. The fitness to practise panel then proceeded to hear the case over a total of thirteen days in May and December 2013, and found A's fitness to practise to be impaired by reason of deficient professional performance. On appeal, A's case included the complaints that this decision of the panel was wrong and that the legal assessor should have stood down, and an assertion that the fact that he did not do so led to a hearing that was, or appeared to be, biased against A. At the start of the appeal before Warby J, the learned judge declared that the legal assessor was a person well known to him as a fellow member of the set of barristers' chambers of which he was, until very recently, a member. The legal assessor and the learned judge had been fellow members of those chambers for some twenty years, and the professional relationship had ended only recently; the learned judge and the legal assessor had a continuing professional relationship to the extent that the judge was editing a book to which the legal assessor was a contributor and, in that connection, they remained in contact. A submitted that the learned judge should recuse himself because of the length of professional association between him and the legal assessor, coupled with the importance to A's case of the complaint that he made about the legal assessor's role in the case. The learned judge concluded, with reluctance, that he should indeed recuse himself for the following reasons:

> 13. The *Porter* test [*Porter v. Magill* [2002] 2AC 357] is satisfied if the informed fair-minded observer would see a real possibility of bias. Whilst fanciful or tenuous objections must be disregarded, the threshold is not an especially high one. It is not necessary that you show a likelihood of bias or a real danger, for instance. Here, the issues for decision involved complaints of actual and/or apparent partiality against an individual with whom I have a long, reasonably close professional association, and an association which is continuing at present albeit in a limited context. Those complaints form a significant plank of the appellant's case, in an appeal which is concerned with his right to work as a doctor and the public interest in ensuring that, if unfit to do so, he does not so work. If the complaint against the Assessor is, as it may be, one of actual bias then it seems clear to me that the links between me and the Assessor mean that I should not sit in judgment on the case. Even if the allegation made by [A] is no graver than apparent bias in the [fitness to practise panel] proceedings I have concluded that the fair-minded observer knowing the relevant facts would see a possibility that, in applying what should be an objective test, I would bring to bear my subjective impressions of the individual's character and personality.
>
> 14. In the end, it was the combination of all the circumstances I have mentioned rather than any one of them by itself that led me, in this case, to recuse myself. If, however, there was one factor that weighed with me more than others it was the nature of the issues raised by this aspect of [A]'s appeal. If I had proceeded to hear and determine this appeal that would have involved me in ruling upon the propriety, or at least the

apparent propriety, of the conduct as Legal Assessor of a person who is well-known to me in the ways and contexts I have described above. The fair-minded observer, knowing those facts, would in my judgment conclude that there was a real possibility that I would approach my task with a pre-determined view of whether a long-term associate would be biased, or could reasonably be seen as possibly biased.

### *Adu v. General Medical Council* [2014] EWHC 4080 (Admin)

**6.16** On 16 October 2014, McGowan J dismissed A's appeal on the grounds that a fair-minded and informed observer would not consider that an incident some nine years earlier concerning a failure in the treatment of the legal assessor's daughter had had any influence. Any actual or apparent bias on the part of the legal assessor could be discounted.

### *Rasool v. General Pharmaceutical Council* [2015] EWHC 217 (Admin)

**6.17** Alleged apparent bias—chairman presiding at interim orders hearing relating to pharmacist working in the same pharmacy as registrant—relevant considerations

The disciplinary proceedings against R arose out of an undercover investigation by the British Broadcasting Corporation (BBC) into the allegedly unlawful supply of prescription-only medicines by a number of pharmacies in central London. One such pharmacy was Al Farabi Pharmacy at 39 Edgeware Road, London, W2, where R was the superintendent pharmacist. The allegations were broadcast on television on 17 December 2012. On appeal against an order removing R's name from the register, R alleged apparent bias on the part of the chairman on the grounds that he ought to have recused himself, having heard matters relating to R in the course of an interim orders hearing against another pharmacist and employee at the pharmacy, S. In dismissing R's appeal, Carr J said:

> 24. An appeal will be allowed where the decision of the court or tribunal was wrong or unjust because of a serious procedural or other irregularity in the proceedings below (see CPR 52.11(3))....
>
> 25. The test for bias can be stated un-controversially as follows: would a fair-minded observer, having considered the relevant facts, conclude that there was a real possibility that the tribunal was (consciously or subconsciously) biased: see *Porter v Magill* [2002] 2 AC 357 and *Lawal v Northern Spirit Limited* [2003] ICR 856 (HL) (at paragraph 21). The court must first ascertain all the circumstances which have a bearing on the suggestion that the judge was biased (see *In Re Medicaments and Related Classes of Goods (no 2)* [2001] 1 WLR 700 (at paragraph 85)). The appearance of independence and impartiality is just as important as the question of whether these qualities exist in fact. Justice must not only be done, it must be seen to be done (see for example Lord Hope's statement in *Millar v Dickson* [2002] 1 WLR 1615 at paragraph 63, as endorsed by the Privy Council in *Yiacoub and another v. The Queen* [2014] 1 WLR 2996).
>
> 26. Both parties have referred me to *Castillo Algar v Spain* [1988] 30 EHRR 827 (at paragraphs 45 and 46). The proposition was there stated that any judge in respect of whom there was a legitimate reason to fear a lack of impartiality must withdraw. In assessing whether there was such a legitimate reason, the standpoint of the accused was important but not decisive. The decisive test was an objective one. The mere fact that a judge had already taken decisions before the trial could not in itself be regarded as justifying anxieties as to his impartiality. The issue of bias was highly fact-sensitive.

After reviewing further *Datta v. General Medical Council* (Privy Council Appeal No. 34 of 1985), *Helow v. Secretary of State for the Home Department* [2008] 1 WLR 2416 (HL) (Sc), and *El Farargy v. El Farargy and ors* [2007] EWCA Civ 1149, quoting the observations of the Constitutional Court of South Africa in *President of the Republic of South Africa v. South African Rugby Football Union* [1999] 4 SA 147, 177, her Ladyship concluded at [30] that '[u]ltimately, each case will turn on its own facts'. In the present case, at [31], it was said in essence that the chairman was not in a position to put out of his mind the earlier hearing and matters relating to R arising out of S's proceedings, that there should have been full disclosure of the material before the committee presiding over the interim orders review hearing for S, and that the chairman misdirected himself as to the law and as to the appropriate standard necessary for recusal. In rejecting these grounds, the learned judge said, at [32], that the relevant circumstances could be identified as follows:

> a) although [S] and [R] were both working at the pharmacy, the incident relating to [S in her proceedings] was a different one to the occasions relied on against [R]. Each occasion was treated separately;
> b) the material from [S] amounted to no more than an out of court assertion in different proceedings in respect of which no finding was made;
> c) that material was not being relied on in any way in relation to [R] and was not before the Committee in evidence at the hearing against him. It was not referred to at any stage in the Committee's findings (or reasons);
> d) indeed the material does not appear to have been of any relevance even to the issues facing [S], as evidenced by the fact that the Chairman said in terms [at the interim order hearing] that it was of no assistance;
> e) the Chairman could not remember what was said about [R] at the interim order hearing. He took no notice of it;
> f) at no stage was it suggested that the Chairman had formed any view, let alone an adverse one, in respect of [R] during [S's] proceedings;
> g) the Chairman was an experienced QC and chairman who regularly presided over disciplinary hearings involving the GPhC. He could be assumed to be well aware of his duties and obligations to try hearings fairly and by reference only to the evidence and materials before him, and assumed to be capable of discounting anything that had been asserted by [S]. Whilst not determinative without more, this last factor is of considerable weight.
>
> 33. Against this background and the relevant authorities, it cannot in my judgment be said that the Chairman was wrong to conclude that a fair-minded and informed observer would not conclude that there was a real possibility that the Chairman was biased and to rule that he should not recuse himself.

### *Prasad v. General Medical Council* [2015] EWHC 338 (Admin)

**6.18** Chairman (medical member)—knowledge of witnesses interviewed by appraisers—knowledge of practice manager

A fitness to practise panel of the GMC made findings of misconduct and deficient professional performance against the appellant, and found that his fitness to practise was currently impaired and that his registration should be subject to conditions. Dr C, the chairman of the panel, was a medical member, and the appellant contended on appeal that the chairman was conflicted or apparently biased by reason of his professional knowledge of two medical witnesses whom he had himself appraised as medical director of their region, and also by reason of his professional knowledge of a factual witness, in consequence of which the chairman should have recused himself. The two medical witnesses were not witnesses at the hearing, but were interviewed by the GMC's appraisers in the course of the allegation of deficient professional performance. In dismissing the appellant's appeal, His Honour Judge Simon Barker QC (sitting as a judge of the High Court) said that it was unfortunate that the chairman had professional knowledge of the two witnesses, having been their appraisers on an earlier occasion. However, the record of the hearing made clear that the chairman disclosed this prior relationship and that the parties were afforded a proper opportunity to consider whether to ask the chairman to recuse himself. The upshot was that the appellant's counsel expressed the appellant's willingness to proceed. The panel also had another qualified doctor in addition to the chairman. As to the lay witness, the basis for this submission was that she was a long-serving practice manager at a surgery in the area of which the chairman was director and that he must have or should be taken to have known her. The learned judge said that may be correct, but that the circumstances would have been or should have been just as apparent to the appellant and his legal team at the time of the hearing as they were at the time of the appeal. Had there been any concern, it could and should have been raised with the chairman then. In any event, the lay witness's evidence was favourable to and complimentary of the appellant.

## C. Other Relevant Chapters

**6.19**

Abuse of Process, Chapter 2
Appeals, Chapter 4
Human Rights, Chapter 32
Independent and Impartial Tribunal, Chapter 35
Natural Justice, Chapter 48

# 7

# BURDEN OF PROOF



## A. Legal Framework

**7.01** Although, in all cases, the burden of proving any allegation or charge will rest on the regulator, some regulatory bodies expressly provide that the burden of proof in relation to allegations before the fitness to practise committee or disciplinary committee shall rest on the regulator, for example:

General Pharmaceutical Council (Fitness to Practise and Disqualification etc) Rules 2010, rule 42(1)

Chartered Institute of Management Accountants Disciplinary Committee Rules 2011, rule 13(1)

## B. General Principles

### *Miller v. Minister of Pensions* [1947] 2 All ER 372

**7.02** Denning J said, at 374A–B:

Reasonable degree of probability sufficient in civil cases—if more probable than not, burden is discharged, but if probabilities are equal, it is not

> [T]he case must be decided according to the preponderance of probability. If at the end of the case the evidence turns the scale definitely one way or the other, the tribunal must decide accordingly, but if the evidence is so evenly balanced that the tribunal is unable to come to a determinate conclusion one way or the other, then the man must be given the benefit of the doubt. This means that the case must be decided in favour of the man unless the evidence against him reaches the same degree of cogency as is required to discharge a burden in a civil case. That degree is well settled. It must carry a reasonable degree of probability, but not so high as is required in a criminal case. If the evidence is such that the tribunal can say: "We think it more probable than not," the burden is discharged, but, if the probabilities are equal, it is not.

### *Ashiq v. Bar Standards Board*, PC 2010/0185/A, 27 March 2013

**7.03** Duty to respond not a reversal of the burden of proof

A was charged with professional misconduct for failure to pay a judgment debt against him. A claimed to have sent a cheque in settlement. A gave evidence, but refused to produce bank documents. The primary submission on appeal was that a disciplinary tribunal had, in effect, reversed the burden of proof by requiring A to prove his innocence. A's duty under the code of conduct was to respond promptly to any request from the Bar Standards Board. 'Respond' plainly means to 'respond positively', not by way of a refusal. The evidence before the tribunal was thus an unjustified refusal. It was for the tribunal to consider what inference it should draw from the refusal. That was not a reversal of the burden of proof. The appeal was dismissed.

### *Pope v. General Dental Council* [2015] EWHC 278 (Admin)

**7.04** Burden of proof remains throughout on prosecution—if evenly balanced, case not proved

P, a dentist, faced a broad range of allegations of misconduct at a hearing before the Professional Conduct Committee (PCC) of the General Dental Council (GDC). The charges were directed not only towards P's clinical performance, but also towards his probity. In particular, it was contended that he had repeatedly made dishonest claims for remuneration to which he knew he was not entitled under his contract with the primary care trust. The PCC found twenty out of thirty-one allegations of dishonesty to have been made out and the sanction that it imposed was erasure from the register. In dismissing P's appeal, Turner J said:

> 13. There are in English law two categories of burden of proof: the persuasive (or legal) and the evidential. We are concerned here only with the persuasive burden so nothing more need be said about the evidential burden.
>
> 14. In a civil case, the burden of proof is fixed at the beginning of the trial by the state of the pleadings and remains there never shifting. Similarly, in criminal cases, the burden of proof lies on the same party (in practice usually, but not always, the prosecution) throughout. In this case, the burden of proof on the issue of dishonesty in respect of each relevant charge undoubtedly rested upon the shoulders of the GDC throughout the proceedings.
>
> 15. The general practical impact of this is that at no stage of the proceedings is the fact finder entitled to say that the evidence on any given issue has accreted to the extent that the persuasive burden of proof has, as a result, been effectively shifted from one party onto the other party.
>
> 16. This point was authoritatively and emphatically made by Viscount Sankey LC in *Woolmington v. The Director of Public Prosecutions* [1935] AC 462 at page 481.
>
> 17. The PCC in this case could not, therefore, legitimately have said at any intermediate stage in the proceedings that, in the light of the evidence against Mr Pope (however compelling) with respect to any issue, that the burden of proof on that issue had thereby shifted onto him. The benefit of the burden of proof is preserved intact and the consequences of its application fall to be adjudicated upon at the end of the case when all the evidence is in.
>
> [...]
>
> 19. The time at which the burden of proof must be applied is, therefore, at the end of the case without either attenuation or reversal as a result of the ebb and tide of evidential fortunes in the hearing which has preceded it.
>
> [...]
>
> 24. The evidence in favour of one party is put in one pan of the scales and that of the other in the other pan. As the case progresses, one pan may rise as the other falls and vice versa. When the evidence had concluded, the scales will have tipped in one direction or another or will have ended up evenly balanced. The fact that one party bears the burden of proof means that he will lose not only if the pan has fallen in favour of the other party but also if the scales end up evenly balanced.
>
> 25. The application of the burden of proof does not, however, involve putting some unspecified weight into the pan of the party that does not bear the burden of proof before any evidence is called. The burden of proof and the standard of proof comprise the criteria which are to be applied to all of the evidence after it is complete in order to determine how any given issue is to be resolved. As such, the burden of proof has no

"weight" either in the scale analogy or, literally, in the context of a contested issue as a piece of evidence in itself. To say that the burden remains on one party throughout is merely to make the point that, however imbalanced the scales may appear to be at any given stage in the proceedings, the test to be applied remains unchanged throughout.

## C. Other Relevant Chapters

**7.05** Restoration to Register, Chapter 58
Review Hearings, Chapter 59
Standard of Proof, Chapter 62

# 8

# CASE MANAGEMENT



## A. Legal Framework

**8.01** Examples include:

Solicitors (Disciplinary Proceedings) Rules 2007, rule 11 (directions)
General Pharmaceutical Council (Fitness to Practise and Disqualification etc) Rules 2010, rules 20 and 21 (case management meetings and directions)
Bar Standards Board Disciplinary Tribunals Regulations 2014, rules E107–116 (standard directions—directions judge—matters that are expedient for the efficient conduct of the hearing)

Examples of provision of documents and/or other material to the committee in advance of the substantive hearing of fitness to practise or disciplinary proceedings include:

Solicitors (Disciplinary Proceedings) Rules 2007
General Optical Council (Fitness to Practise) Rules 2013
Institute of Chartered Accountants in England and Wales (ICAEW) Disciplinary Committee Regulations

## B. General Principles

### *R v. Jisl, Tekin and Konakli* [2004] EWCA Crim 696

**8.02** *Principles for judicial management*

After a trial that lasted nearly seventy working days and which was itself a retrial, Tekin and Jisl were convicted on 5 October 2001 at the Central Criminal Court, before His Honour Judge Dunn QC and a jury, of fraudulent evasion of a prohibition on the importation of goods. Konakli pleaded guilty on rearraignment. The goods in question were 139 brown packages of diamorphine, with a street value in excess of £7 million. Tekin was sentenced to twenty-one years' imprisonment, Konakli was sentenced to fifteen years' imprisonment, and Jisl to fourteen years. The Court of Appeal, Criminal Division (Judge LJ, and Nelson and McCombe JJ) made observations at [113]–[120], under the heading 'Case Management':

> 114. The starting point is simple. Justice must be done. The defendant is entitled to a fair trial: and, which is sometimes overlooked, the prosecution is equally entitled to a reasonable opportunity to present the evidence against the defendant. It is not however a concomitant of the entitlement to a fair trial that either or both sides are further entitled to take as much time as they like, or for that matter, as long as counsel and solicitors or the defendants themselves think appropriate. Resources are limited... Time itself is a resource... It follows that the sensible use of time requires judicial management and control.

The Court referred to *R v. Chaaban* [2003] EWCA Crim 1012, in which the Court endeavoured to explain that, in principle, the trial judge should exercise firm control over the timetable, where necessary, making clear in advance and throughout the trial that the timetable will be subject to appropriate constraints. The Court continued:

> 116. The principle therefore, is not in doubt. This appeal enables us to re-emphasise that its practical application depends on the determination of trial judges and the co-operation of the legal profession. Active, hands on, case management, both pre-trial and throughout the trial itself, is now regarded as an essential part of the judge's duty. The profession must understand that this has become and will remain part of the normal trial process, and that cases must be prepared and conducted accordingly.
>
> 117. The issues in this particular trial were identified at a very early stage, indeed during the course of the previous trial itself. In relation to each of the defendants, in a single word, the issue was knowledge. And indeed, the issue in most trials is equally readily identified.
>
> 118. Once the issue has been identified, in a case of any substance at all, (and this particular case was undoubtedly a case of substance and difficulty) the judge should consider whether to direct a timetable to cover pre-trial steps, and eventually the conduct of the trial itself, not rigid, not immutable, and fully recognising that during the trial at any rate the unexpected must be treated as normal, and making due allowance for it in the interests of justice. To enable the trial judge to manage the case in a way which is fair to every participant, pre-trial, the potential problems as well as the possible areas for time saving, should be canvassed. In short, a sensible informed discussion about the future management of the case and the most convenient way to present the evidence, whether disputed or not, and where appropriate, with admissions by one or other or both sides, should enable the judge to make a fully informed analysis of the future timetable, and the proper conduct of the trial. The objective is not haste and rush, but greater efficiency and better use of limited resources by closer identification of and focus on critical rather than peripheral issues. When trial judges act in accordance with these principles, the directions they give, and where appropriate, the timetables they prescribe in the exercise of their case management responsibilities, will be supported by this court. Criticism is more likely to be addressed to those who ignore them.
>
> 119. If these principles had been applied to this trial, it seems to us inconceivable that it would have taken 70 days before the jury reached its verdict, or given the issue in Tekin's case, that his evidence would have lasted four days in chief, and part of a further ten days in cross-examination by counsel for the Crown, or that Tekin himself would have been permitted without warning, to produce a bundle of documents some 400 pages long and seek to adduce it in evidence.

## C. Disciplinary Cases

### *Council for the Regulation of Health Care Professionals v. General Medical Council and Ruscillo; Council for the Regulation of Health Care Professionals v. Nursing and Midwifery Council and Truscott* [2005] 1 WLR 717

**8.03** Procedures for disciplinary proceedings

Under the heading 'Procedural shortcomings and fresh evidence', the Court of Appeal (Lord Phillips of Worth Matravers MR, and Chadwick and Hooper LJJ) said:

> 79. Where a defendant is prosecuted for a crime the precise charge or charges and the evidence relied upon to support them are matters for the prosecution. The defence can make admissions. Prosecution and defence can agree facts. The procedure is adversarial and the judge plays a passive role in the factual inquiry. If the defendant is convicted the judge sentences him for the offences of which he has been convicted in the light of the evidence that has been given at the trial.
>
> 80. The procedures for disciplinary proceedings under the various statutes referred to in section 29(1) of the National Health Service Reform and Health Care Professions Act 2002 are not identical. In general they involve a preliminary investigation of conduct of the practitioner of which complaint has been made. If it is decided to bring disciplinary proceedings, a charge will be proffered which alleges the facts relied upon as demonstrating professional misconduct. Admissions may be made by the practitioner, facts may be agreed and evidence may be called. The disciplinary tribunal will be faced with an act or omission, or more typically a course of conduct, which it is alleged constitutes professional misconduct. The disciplinary tribunal should play a more proactive role than a judge presiding over a criminal trial in making sure that the case is properly presented and that the relevant evidence is placed before it.

### *R (Heath) v. Home Office Policy and Advisory Board for Forensic Pathology* [2005] EWHC 1793 (Admin)

**8.04** Intrinsic powers of tribunal—procedural gaps—fair, impartial, and just hearing

The Home Office Policy Advisory Board for Forensic Pathology was established in 1991, its purpose being to maintain the quality and standards of the profession of forensic pathologists. The claimant, H, a well-known and senior forensic pathologist, accredited in that capacity by the Home Office, sought judicial review of a decision to refer to a disciplinary tribunal complaints made against him in respect of his conduct as a forensic pathologist and expert witness in relation to two murder trials. Newman J rejected H's argument that the Home Secretary, in setting up the Board, had acted ultra vires and that the decision to refer the complaints to a disciplinary tribunal was made without jurisdiction. As to the disciplinary procedures and the position before the tribunal, Newman J said, at [47], that he should like to endorse what the chairman of the tribunal convened in the claimant's case for the directions hearing, Mr Andrew Pugh QC, said:

> [T]he Tribunal does have intrinsic powers, simply by virtue of being a tribunal. It has the obligation to observe the rules of natural justice and to conduct its proceedings fairly and to decide procedural matters which are not expressly dealt with in the rules... It may well be that a tribunal acting fairly can fill in the procedural gaps...

The learned judge concluded, at [51], that, in his judgment, it was in the interests of the claimant and in the public interest that the issues raised by the complaints should be dealt with by a tribunal without any further delay. As regards the procedure to be adopted at the hearing, that was a matter for the tribunal in so far as it required to make specific directions. Any issue in connection with the applicability of rules other than the 1997 Rules (the constitution of the Board) could be raised with the tribunal. The overriding consideration was that, at that hearing, the claimant should be accorded a fair, impartial, and just hearing.

***Giele v. General Medical Council* [2006] 1 WLR 942**

**8.05** Delays and length of hearings

The hearing lasted twenty-nine days, spread over fifteen months between December 2003 and March 2005. Collins J said that hearings of these lengths, which have to be conducted with substantial gaps owing largely to the difficulties in bringing panel members together, are clearly undesirable. The learned judge said, at [2]:

> One of the problems appears to be the lack of any means whereby the defence can be properly identified in advance. It is apparent that the GMC should seriously consider amendments to its rules to ensure that there is power, which should be exercised robustly but fairly to avoid unnecessary delays and length of hearings, to ensure that issues are properly identified and, so far as possible, evidence on both sides served in advance. I am not in a position to nor do I seek to cast any blame in this case, but the present system can undoubtedly result in a substantial and unnecessary expenditure of money which the profession through its contributions to the GMC will have to bear. The sooner steps are taken to avoid these altogether too lengthy hearings the better.

***R (Thompson) v. Professional Conduct Committee of the General Chiropractic Council* [2008] EWHC 2499 (Admin)**

**8.06** Procedural fairness—case management decisions—supervisory jurisdiction by court

T, a registered chiropractor, sought permission to challenge a decision of the defendant's Professional Conduct Committee (PCC) taken on 11 August 2008, to hold the substantive hearing of proceedings against him on 13 October 2008. The essential ground of the proposed challenge was that, as the Committee was made aware, Dr H, the claimant's expert witness, would not be available to give evidence on the date on which Dr H was to be called. Nevertheless, the Committee decided that the hearing should go ahead and eight days had been allocated. T faced sixteen allegations, which related to complaints made by two patients. The allegations included taking X-rays without clinical justification, and excessive recommendation and provision of multiple treatment sessions. If proved, then it was certainly possible that T might face suspension from practice as a chiropractor. The complaint made by patient A was not made until 26 February 2007, although T's treatment of that patient had ended in December 2005. T was notified of the complaint to the investigating committee on 21 June 2007; on 25 January 2008, he was notified that the complaint had been referred to the PCC. On 8 April 2008, without any prior consultation with T, the Committee notified him that the full hearing would take place over five days beginning 11 August 2008. That hearing became a directions hearing. Despite the fact that a complete list of dates of the availability of Dr H was provided to the Committee in advance of the directions hearing, the substantive hearing was fixed for 13 October. Lloyd Jones J, in granting relief, said that it was not an appeal from a case management decision and that what was really in issue was the procedural fairness of the proceedings before the Committee. The Court will normally be slow to interfere with case management decisions taken by a professional disciplinary body. However, this must be subject to the supervisory jurisdiction of the Court to ensure that the parties before such a tribunal are given a fair hearing. Counsel for the Committee submitted that Dr H's evidence could not assist T and that the PCC will be required to judge the case by standards of general application. However, the learned judge said that it seemed to him that the Committee, in considering the case and in considering the defence that T proposed to run, would have to address the nature and efficacy of the therapy, and whether it justified the course of treatment prescribed by T and the taking of X-rays that he had recommended. The learned judge accepted without hesitation the defendant's submission that the Committee had a public duty to safeguard the public by hearing its proceedings as promptly as possible. However, the matter could be heard before the end of 2008 and the learned judge did not see that the short further adjournment could result in any substantial prejudice in that regard. The learned judge concluded that the Committee did act unfairly, when faced with the need to adjourn the August hearing and refix it, in failing to refix it for a date on which T's proposed expert could be available.

### *Gerstenkorn v. General Medical Council* [2009] EWHC 2682 (Admin)

**8.07** Adjournment—case management decision—fair hearing

In dismissing G's appeal brought under section 40 of the Medical Act 1983, Kenneth Parker J said, at [99], that the panel gave a full and carefully reasoned determination running to nine pages of transcript when refusing G's application for an adjournment of the hearing to enable him to present his evidence. The panel's decision was a case management one taken by a professional disciplinary body, with which the Court should be slow to interfere, provided that it is satisfied that the appellant had a fair hearing: see *R (Thompson) v. General Chiropractic Council* [2008] EWHC 2499 (Admin).

### *Nicholas-Pillai v. General Medical Council* [2012] EWHC 2238 (Admin)

**8.08** Hearing venue—ill health of doctor—transfer from Manchester to London

The doctor, who was subject to an interim suspension order, applied for an order transferring the substantive disciplinary proceedings to London, on the basis that he was suffering from ill health and would be unable to travel to Manchester. A medical report in relation to the practitioner was provided in support of the transfer application. The report stated that the practitioner had suffered an acute heart attack resulting from stenosis in the left main coronary artery and had a stent. The report stated that he needed to prevent stressful situations and long-distance travel from London, and that, in view of his current health status, it was recommended that the case with the General Medical Council (GMC) should be transferred from Manchester to London. His Honour Judge Pelling QC (sitting as a judge of the High Court) said it seemed to him that, in principle, it would be appropriate for the fitness to practise hearing, which was listed for seven working days commencing 14 January 2013, to be transferred to London. As to the GMC's application to extend the period of the interim suspension order, which was due to expire next week, the mere fact that it would be a more expensive process if the case were transferred to London was not a material consideration when the evidence established that the practitioner was unfit to travel to Manchester. Accordingly, the Court would make an order continuing the interim suspension for a period of nine months to cover the substantive hearing listed in January 2013, subject to any application on the part of the practitioner to apply or discharge the order on forty-eight hours' notice to the GMC, such application to be made in the Royal Courts of Justice in London.

### *Luthra v. General Medical Council* [2013] EWHC 240 (Admin)

**8.09**

The proceedings, which lasted five days, were exhaustive and 'no stone was left unturned', said Mostyn J, at [24]:

> ...Of course when a man's professional reputation, and ability to practice [*sic*] in the future, are on the line the disciplinary inquiry must be thorough and scrupulously fair. But I cannot help observing that an inquiry with this degree of elaboration, and at inevitably enormous expense to the profession, is perhaps a counsel of perfection, particularly in this age of austerity in which we live.

***Schodlok v. General Medical Council* [2013] EWHC 2280 (Admin)**

**8.10** As to the discretion whether to admit evidence via video link, see paragraph 70.07.

## D. Other Relevant Chapters

**8.11** Adjournment, Chapter 3

Disclosures, Confidentiality, Data Protection, and Freedom of Information, Chapter 18

# 9

# CIVIL RESTRAINT ORDERS



## A. Legal Framework

**9.01** Senior Courts Act 1981, section 42 (application by the Attorney General to restrict vexatious legal proceedings)

Civil Procedure Rules, rule 3.11 (power of the court to make civil restraint orders)

## B. Disciplinary Cases

### *Vaidya v. General Medical Council* [2010] EWHC 2873 (QB)

**9.02** Persistent claims and applications found to be totally without merit—general civil restraint order necessary

In this case, the Queen's Bench Division dismissed Dr V's application to set aside a general civil restraint order made against him. In her judgment, Mrs Justice Nicola Davies DBE referred to the background of Dr V's numerous claims against the General Medical Council (GMC) and others. In summary, since 2007, Dr V had issued proceedings in the Queen's Bench Division and the Administrative Court, in county courts, and an employment tribunal against the GMC, named individuals who had been involved in the GMC disciplinary process against him, NHS trusts, named doctors employed by those trusts, the Crown Prosecution Service (CPS), and Her Majesty's Court Service. The causes of action were many and varied. They included: complaints of racial and sexual discrimination and harassment; harassment contrary to the Protection from Harassment Act 1997; breaches of Articles 3 and 6 of the European Convention on Human Rights (ECHR); claims for professional negligence, libel, malicious falsehood, negligent misstatement, and conspiracy to injure; and claims pursuant to the Data Protection Act 1998. Damages claimed in some proceedings exceeded £1 million and it was of note that such damages were claimed in actions against individually named doctors. The learned judge stated that she was satisfied that Dr V persisted in issuing claims and applications that had been found to be totally without merit. It was clear that he had been warned by judges of the consequences of repeatedly issuing proceedings, but had demonstrated that he was not deterred by the threat of an extended civil restraint order. In her judgment, the learned judge said that, based upon the evidence and given the history of the matter and the continuing conduct of Dr V, she did not believe that an extended civil restraint order would provide adequate protection to prospective defendants, and that a general civil restraint order was not only justified, but was also necessary.

### *R (Shanker) v. General Medical Council* [2010] EWHC 3689 (Admin)

**9.03** Extended civil restraint order for two years granted

Ouseley J, in making an extended civil restraint order for two years, said that it cannot be said that everything that Dr S had done had been wholly without merit. Following the sanction by the GMC, he was entitled to appeal and the bringing of such proceedings cannot fall into the category of being totally without merit, but it was difficult to see that any of the other cases brought by Dr S had any merit. Certainly, two with which Ouseley J had dealt and two with which the Court of Appeal had dealt were totally without merit, and the general manner in which the others had been conducted and the comments made by the judges in relation to them made this a process that ought to be controlled through the making of an extended civil restraint order.

### *Nowak v. Nursing and Midwifery Council and Guy's & St Thomas' NHS Foundation Trust* [2013] EWHC 1932 (QB)

**9.04** Likelihood of further applications totally without merit unless restrained—strong public interest in protecting court system from abuse—scope of order

Leggatt J said, at [36], that N had made a number of applications, which had been dismissed as totally without merit. He continued to make such applications despite his claim being struck out, and it appeared likely that, unless restrained, N would make further applications that he had no right to make or which were otherwise totally without merit. Leggatt J continued:

> 58. As explained by the Court of Appeal in the leading case of *Bhamjee v Forsdick* [2004] 1 WLR 88, the rationale for the regime of civil restraint orders is that a litigant who makes claims or applications that have absolutely no merit harms the administration of justice by wasting the limited time and resources of the courts. Such claims and applications consume public funds and divert the courts from dealing with cases that have real merit. Litigants who repeatedly make hopeless claims or applications impose costs on others for no good purpose and usually at little or no cost to themselves.... In these circumstances, there is a strong public interest in protecting the court system from abuse by imposing an additional restraint on their use of the court's resources.

In the instant case, in the space of five months, N made no fewer than eight applications that were found to be totally without merit. The learned judge held that the appropriate order was one that was wider than the default form of an extended civil restraint order in that the order should require N to obtain permission before issuing any claim or making any application, in the High Court or any county court, against either defendant to the present proceedings or against any other party in relation to any matter involving, or relating to, or touching upon, or leading to the proceedings in which the present order was made. Any extended civil restraint order was not to affect N's right to seek to set aside earlier orders or any right of appeal, or right of appeal under the Nursing and Midwifery Order 2001, if the disciplinary proceedings against him were to result in the imposition of a sanction.

### *Nursing and Midwifery Council and anor v. Nowak and ors* [2013] EWHC 4330 (QB); [2014] EWHC 336 (QB)

**9.05** Anti-harassment injunction granted

The Nursing and Midwifery Council (NMC) and another party brought an application for an anti-harassment injunction, to restrain the defendants, pending trial, from pursuing various courses of conduct at various locations. An interim application was granted by McDuff J and continued on the return date by King J against N, a nurse currently facing a hearing before the NMC's Conduct and Competence Committee (CCC), his representative before the Committee, and B, his supporter.

***Jooste v. General Medical Council* [2013] EWHC 2464 (Admin)**

**9.06** Extended civil restraint order for one year proportionate

In this case, it was held that the proportionate civil restraint order that would be appropriate in the circumstances was, in principle, an extended civil restraint order that sought to restrain J from issuing claims or applications concerning any matter involving, relating to, or touching upon, or leading to the proceedings in which the order was made. The order was expressed to last for one year.

# 10

# CONCURRENT PROCEEDINGS



## A. Legal Framework

**10.01** Senior Courts Act 1981, section 49(3) (power of the Court of Appeal or the High Court to stay any proceedings before it, where it thinks fit to do so)

(See further *Civil Procedure, vol. 2* (London: Sweet and Maxwell, 2015), paragraphs 9A.183–9A.184)

For examples of the regulator's power to postpone or adjourn disciplinary proceedings, see paragraph 3.01.

Police (Conduct) Regulations 2012, regulation 9(3) ('For any period during which the appropriate authority considers any misconduct proceedings or special case proceedings would prejudice any criminal proceedings, no such misconduct or special case proceedings shall take place')

## B. Disciplinary Cases

### *R v. British Broadcasting Corporation, ex parte Lavelle* [1983] 1 WLR 23

**10.02** Charge of theft—disciplinary proceedings—decision not to adjourn—supervisory role of court

The applicant, L, was charged on 4 February 1982 with theft of certain tapes, the property of the British Broadcasting Corporation (BBC), which were found at her home. She intended to plead not guilty and it was anticipated that the hearing of the charge would be in October 1982. The BBC alleged that the removal of the tapes was, in itself, misconduct justifying dismissal irrespective of whether or not that removal amounted to theft. It decided to conduct disciplinary proceedings and, on 6 April 1982, L was dismissed. In judicial review proceedings, Woolf J said that if an employee makes an application to a domestic tribunal to adjourn its proceedings until after the conclusion of criminal proceedings on the basis that the continuation of the disciplinary proceedings would prejudice criminal proceedings, that application should be sympathetically considered by the tribunal. If it comes to the conclusion that the employee will suffer real prejudice if the domestic proceedings continue, then, unless there is good reason for not doing so, the disciplinary proceedings should be adjourned (at 36A–36C). However, if the disciplinary tribunal does not adjourn the proceedings, there should be no automatic intervention by the Court. While the Court must have jurisdiction to intervene to prevent a serious injustice occurring, it will do so only in very clear cases in which the applicant can show that there is a real danger, and not merely a notional danger, that there would be a miscarriage of justice in the criminal proceedings if the Court were not to intervene (at 39A–39C). In the instant case, in refusing to adjourn the disciplinary proceedings on 6 April 1982, the position of the applicant, but also the position of the BBC, was to be taken into account. The BBC would have to continue to pay the applicant until the appeal procedure was exhausted and, if the trial were not to take place until the early part of October, any adjournment would have to be for a substantial time. Furthermore, the disciplinary proceedings were to be in private, there was no risk of a real injustice to the applicant, and the matters to be proved in the criminal proceedings were much more extensive than those in the disciplinary proceedings. In the disciplinary proceedings, the removal of the tapes would be sufficient to establish a disciplinary offence.

### *Archer v. South-West Thames Regional Health Authority* (1985) The Times, 10 August

**10.03** 'Until the result of the police action is known'—meaning

The plaintiff, who faced both disciplinary proceedings and criminal proceedings arising from the same events, sought an interlocutory injunction restraining the defendant from holding or concluding the disciplinary proceedings until the final determination of the criminal proceedings. The plaintiff was employed as a higher clerical officer at the South London Blood Transfusion Centre. An audit revealed a loss of many thousands of pounds by reason of systematic overpayment to two companies that supplied certain materials to the Centre. The defendant's code of disciplinary procedure stated that: 'In cases where the police have been called to make investigations, disciplinary proceedings should not be involved until the result of the police action is known.' Steyn J said that the real question was whether the paragraph availed the plaintiff in the present case: did it relate only to the police investigation in the narrow sense (as the defendants contended) or did it also embrace the criminal trial and any appeal (as the plaintiff contended)? The provision is far from clear. On balance, however, the natural meaning of the words 'until the result of the police action is known' seems to refer to the decision to bring charges or not. Thereafter, the role of the police recedes into the background. The words 'police action' do not seem apt to describe a criminal trial by a judge and jury or appellant proceedings. The narrower meaning is also more sensible, and if the provision is capable of both meanings, it is therefore to be preferred. After all, it would be somewhat odd if the defendants were not entitled to bring disciplinary proceedings before a criminal trial or an appeal would be concluded in a case in which, for example, an employee admitted his guilt to the defendants, or even intimated an intention to plead guilty in due course. Accordingly, the plaintiff's application was dismissed.

***R v. Solicitors Disciplinary Tribunal, ex parte Gallagher*, 30 September 1991 (unreported)**

**10.04** Charge of mortgage fraud—effect on disciplinary proceedings

G's renewed application for judicial review of a decision by the Solicitors Disciplinary Tribunal (SDT) declining to adjourn disciplinary proceedings was dismissed by the Court of Appeal (Parker, Stuart-Smith, and Mann LJJ). G faced a large number of charges before the Tribunal associated with mortgage frauds and which had become the subject of criminal charges. The charges before the Tribunal did not allege dishonesty, whereas all of the criminal charges did so. However, G said that, in desiring to give an explanation of his conduct and to advance explanations that he was incompetent or ill-advised, he may have to limit his mitigation or reveal his defence to the prosecution or to a co-defendant in the criminal proceedings. The Court considered that, because the criminal trial was unlikely to be heard for at least a year, these matters were very remote. If, when the trial arrives, a local newspaper publishes the outcome of the disciplinary proceedings so that the jury might see it and possibly be influenced, that was a matter that would no doubt be dealt with by the judge at the criminal trial.

***R v. Panel on Takeovers and Mergers, ex parte Fayed* [1992] BCC 524**

**10.05** Civil proceedings—investigation by Panel on Takeovers and Mergers—prejudice—procedure

The applicant applied for judicial review to prevent disciplinary proceedings until the trial of proceedings brought by Lonrho plc against the applicant and others in the High Court, and until the Panel on Takeovers and Mergers had conducted its own investigation into the material that formed the basis of its allegations in the disciplinary proceedings. The Court of Appeal (Neill, Scott, and Steyn LJJ) dismissed the renewed application for permission. The continuation of the disciplinary proceedings would not prejudice the applicant's prospects in the High Court action. If there were any real risk of prejudice, the risk would provide a strong—and perhaps in the present case a conclusive—reason why the disciplinary proceedings should be stayed pending the trial of the action. It was important to bear in mind that the action would not be a jury action, but would be tried by a judge alone. Should the Panel come to a conclusion adverse to the applicant, the judge would not treat that conclusion as evidence against the applicant or any relevant issue in the action. The judge would decide matters regardless of any opinions that the Panel may have expressed. There was no procedure whereby an appellate court, exercising its limited supervisory powers by way of judicial review, could prevent an inferior court or tribunal from embarking on a hearing on the ground that the 'prosecution' had not assembled sufficient evidence or had not carried out a proper investigation before the hearing. If the evidence is inadequate or inadmissible, that is a matter, in the first instance, for the court or tribunal conducting the hearing. In its annual report, the Panel said that it would normally consider it inappropriate to pursue an investigation into the conduct of parties to a takeover or merger at a time when that conduct was the subject of legal proceedings that had already been set in train. Scott LJ said that he was prepared to accept that, as a general proposition, a tribunal carrying out a quasi-judicial function ought not to depart in an individual case from the general rules of procedure that it has prescribed for itself unless it has good reason for doing so. He did not think that this proposition had to be based on 'legitimate expectation'; it was based on the principle that a tribunal ought, in general, to apply the same procedural rules to similar cases with which it deals. The tribunal may have power to change its general practice. But if, while maintaining its general practice, it adopts a different procedure for a particular case, the differentiation must be justified by some good reason. Another way of putting the point is that an individual should not, without good reason, be subjected to less favourable treatment than others in a similar position. This is part of what jurists in the US mean by 'due process'. In the instant case, the Panel executive had given reason for not following the normal procedure and had stated that it was difficult to see any possible prejudice to the civil proceedings resulting from any action taken by the Panel.

### *R v. Institute of Chartered Accountants, ex parte Brindle and ors* [1994] BCC 297

**10.06** Civil proceedings by liquidator—virtual overlap with disciplinary proceedings

The appellant partners in Price Waterhouse coordinated the worldwide audits of the accounts of Bank of Credit and Commerce International Holdings (Luxembourg) SA and its subsidiaries (BCCI Group). The collapse of the BCCI Group on 5 July 1991 led to civil proceedings against the appellants by the liquidators of BCCI and by, or on behalf of, a number of creditors, and the setting up of disciplinary proceedings under the provision of the Joint Disciplinary Scheme (JDS) operated by the accountancy profession. The appellants requested the committee of inquiry to suspend the operation of the disciplinary proceedings until, at any rate, the first stage of the civil litigation in this country had been completed. The request was refused and the appellants' judicial review proceedings at first instance were dismissed. On appeal to the Court of Appeal, Nolan LJ said that the critical question was whether the continuation of the disciplinary proceedings would delay, impede, and prejudice the appellants in the conduct of their defence in the civil proceedings to an extent that could not be justified in the public interest. There seemed to be a positive advantage in the committee having the benefit of findings of fact, arrived at by a judge in a fair trial after the issues with which the committee was principally concerned had been extensively canvassed, before it proceeds to a decision in the disciplinary inquiry. Hirst LJ said that the degree of overlap between the issues raised in the disciplinary proceedings and those raised in the liquidators' action were so complete as to amount to a virtual total eclipse. In both proceedings, the same facts were in issue, and the basic professional standards invoked were identical and non-controversial. The fact that, in the action, some more controversial embellishments were added did not affect the comparison of the basic standards relied upon. It would be inherently unfair that two tribunals should contemporaneously be considering the same issue: *Conteh v. Onslow Fane*, Transcript No. 291 of 1975, CA. Accordingly, the committee should be prohibited from continuing the disciplinary proceedings until the first-instance trial of the liquidators' action against the appellants had been concluded. The respondents were at liberty to apply to vary or set aside the order for prohibition, so as to cover the possibility of unforeseen developments in the proceedings.

### *R v. Chance, ex parte Smith and ors* [1995] BCC 1095

**10.07** Civil proceedings—fraudulent use of public funds—disciplinary proceedings—public interest

The applicants, partners in the firm Coopers & Lybrand, challenged the decision of the executive counsel appointed for the purposes of the JDS of the Institute of Chartered Accountants in England and Wales (ICAEW) refusing the applicant's request for a stay of the scheme enquiry and disciplinary proceedings against the firm pending the disposal of civil proceedings arising from the death of Robert Maxwell. On 5 November 1991, Robert Maxwell, chairman of Mirror Group Newspapers and the effective controller of a web of private family companies, died. Following his death, it became apparent that there had been large-scale misappropriation of assets belonging to the pension funds of various public companies with which he was concerned. Coopers & Lybrand was the auditor of the pension funds and the companies that were the trustees and managers of relevant assets. The Court (Henry LJ and Kay J) said that the test to be applied was clear. It was enunciated by Neill LJ in *R v. Panel on Takeovers and Mergers, ex parte Fayed* [1992] BCC 524, 531E:

> It is clear that the court has power to intervene to prevent injustice where the continuation of one set of proceedings may prejudice the fairness of the trial of other proceedings . . . but it is a power which should be exercised with great care, and only where there is a real risk of serious prejudice which may lead to injustice.

The Court concluded that *R v. ICAEW, ex parte Brindle* [1994] BCC 297 was decided on its own facts and was distinguishable. (Amongst the distinguishing features were that, in *Brindle*, the open-ended nature and scale of the disciplinary proceedings were such that to pursue them contemporaneously with the civil proceedings would prejudice the defence of the civil proceedings, whereas in the instant case the disciplinary proceedings were specific and would not jeopardize the proper defence of the civil proceedings. In *Brindle*, there was no allegation of misconduct or special consideration of urgency, whereas in the instant case serious allegations were made and the disciplinary proceedings clearly could be completed before the civil proceedings. In *Brindle*, the Court took the view that, on the facts, there was little public interest in the pursuit of the disciplinary proceedings and a real risk of serious prejudice to the civil proceedings were they not stayed, whereas in the instant case the Court took a diametrically opposite view.) The Court said that disciplinary proceedings are necessary both in order to ensure that regulatory questions arousing public concern are addressed within a reasonable timescale, and in order that they be addressed by a body with power to deal with them by way of fine and disqualification. Both the community and Parliament have seen the need for such proceedings, and they would not be properly efficacious if they had to await the resolution—whether by compromise or otherwise—of civil proceedings, which might happen long after the event. Moreover, given the nature of the JDS, there was no warrant for assuming that such proceedings would arrive at an erroneous conclusion. The Court accepted without hesitation the strains that are placed on named individuals having to fight a battle on two fronts. But these strains must be greatly reduced by the fact that they have the support of their employers (and of their considerable resources). Additionally, the disciplinary allegations were specific; they were not general and wide-ranging. The extent of the overlap would mean that the proper preparation for defence on the one front would also serve as the defence on the other front. As to the contention that the JDS enquiry might prejudice the criminal trials, the parties and the court in those trials would be alive to any potential risk of prejudice to them. The responsibility for preservation of the integrity of those trials rests with the court in those trials. In the unlikely event of their integrity being threatened by the hearing (in private, in the first instance) of the disciplinary proceedings or the publication of the findings thereof, then the criminal court has ample powers to obtain an order restraining publication until after the conclusion of the criminal trials. Accordingly, there were no grounds for granting any stay of the disciplinary proceedings in this case.

### *R v. Executive Counsel Joint Disciplinary Scheme, ex parte Hipps and ors* [1996] CLY 5

**10.08** Civil proceedings—principles to be applied—balancing exercise

The applicants were partners in Stoy, the auditor for Polly Peck International, which collapsed in the autumn of 1990. Stoy requested the defendant to adjourn investigations pending the outcome of civil proceedings commenced by the administrator of Polly Peck, in which it was alleged that Stoy had been negligent and a sum in excess of £225 million in damages and interest was claimed. In dismissing the application, Dyson J said that he derived the following principles from the authorities:

(1) The court is not concerned with a *Wednesbury* review of (the defendant's decision not to adjourn the proceedings). Rather I am required to exercise an original jurisdiction whether to grant a stay: see *R v. Takeovers and Mergers Panel, ex parte Guinness* [1990] 1 QB 146, 178G–H, 184C–E, and *R v. Chance, ex parte Smith* [1995] BCC 1095 and 1100G.
(2) The jurisdiction to stay one of two concurrent sets of proceedings must be exercised sparingly and with great care: see *R v. Panel on Takeovers and Mergers, ex parte Fayed & Ors* [1992] BCC 524, 531E, and *R v. ICAEW, ex parte Brindle* [1994] BCC 297, 310D–E.
(3) Unless a party seeking a stay can show that if a stay is refused there is a real risk of serious prejudice which may lead to injustice in one or both of the proceedings, a stay must be refused: see *ex parte Fayed* at 531 *ex parte Brindle* at 316G–H.
(4) If the court is satisfied that, absent a stay, there is a real risk of such prejudice then the court has to balance that risk against the countervailing considerations. Those considerations will almost always include the strong public interest in seeing that the disciplinary process is not impeded: see *ex parte Brindle* at 310E–G, *ex parte Smith* at 1100G, 1103B–D.
(5) In a case where the balancing exercise is carried out, the court will give great weight to the view of the person or body responsible for the decision as to the factors militating against the stay and the weight to be given to them, but the court is the ultimate arbiter for what is fair: see *ex parte Smith* at 1101F–G, 1102H to 1103F, and *ex parte Guinness* at 184D–E.
(6) Each case turns on its own facts. Accordingly, only limited assistance can be derived when comparing the facts of a particular case with those of other cases where a stay was granted (as in *ex parte Brindle*) or where a stay was refused (as in *ex parte Smith*).

The learned judge did not accept on the evidence that the existence of a parallel inquiry, in so far as it was concerned with matters that were also the subject of the litigation, would impair Stoy's ability to defend itself in the civil proceedings. The question of delay per se was irrelevant to the threshold test. The only question was whether, in the events that had happened, there was a real risk of serious prejudice leading to possible injustice. There will be cases in which there is such a risk, even where there has been no delay, and the converse is also true. The issue is whether the concurrency of the two sets of proceedings give rise to the risk. If the answer is that it does and the concurrency has been the result of delay on the part of the ICAEW, then, in a sense, the delay has caused the risk. It does not follow that delay is a factor to be taken into account in deciding whether the risk of serious prejudice arises. Since there was no real risk of serious prejudice to the civil proceedings or the inquiry, and since there was a substantial public interest in the continuation of the disciplinary proceedings, the balancing exercise in this case inevitably brought the scales down heavily against a stay.

### *Dr X v. General Medical Council* [2001] EWHC Admin 447

**10.09** Interim order—criminal proceedings awaiting trial

On 28 February 2001, Dr X, a general practitioner, was charged by the police with six counts of indecent assault on two of his nieces (then aged 15 and 13). On 2 March 2001, the Interim Orders Committee (IOC) of the General Medical Council (GMC) ordered that Dr X's registration as a medical practitioner should be suspended with immediate effect for a period of 18 months. Dr X applied to the High Court under section 41A(10) of the Medical Act 1983 to terminate the interim suspension order imposed on his registration. In dismissing his application, Pill LJ (with whom Silber J agreed) said, at [24]:

> I have referred to the limited nature of the material which was before the IOC. It was for them to examine the material before them with care. It is plainly a worrying situation when a professional man may be suspended on the basis of allegations of criminal conduct which, as yet, are untested in a court of law. I cannot, however, accept that the power to suspend by way of interim order, provided in section 41A, must not be exercised because the allegations are untested in a court. Nor, in my judgment, can it be said that the exercise of the power to suspend was inappropriate because the conduct alleged was not towards patients of the claimant.

### *R (Land and ors) v. Executive Counsel of the Accountants' Joint Disciplinary Scheme* [2002] EWHC 2086 (Admin)

**10.10** Civil proceedings—no real risk of serious prejudice

In these proceedings, the claimants, partners in Ernst & Young, sought an order staying the investigation by the Executive Counsel of the Accountants' Joint Disciplinary Scheme into their work as auditors of the Equitable Life Assurance Society until after the conclusion of civil proceedings instituted by Equitable Life against them. It was contended that the continuation of the investigation would give rise to a real risk of serious prejudice to the claimants. The prejudice alleged by the claimants was prejudice in that litigation, in the JDS investigation itself, and in any subsequent disciplinary tribunal proceedings, and to the private and professional lives of the partners and personnel who were involved in the audits and regulatory work of the firm for Equitable Life. The claimants contended that this prejudice outweighed the public interest in the present continuation of the investigation. Stanley Burnton J concluded that the claimants had not established the precondition for a stay—namely, that there was now a real risk of serious prejudice in any of the proceedings if the JDS inquiry were to continue—and it was unnecessary for the Court to carry out a balancing exercise weighing the risk of that prejudice against the countervailing considerations.

### *R (Ranson, Headdon and Nash) v. Institute of Actuaries* [2004] EWHC 3087 (Admin)

**10.11** Civil proceedings—fairness—prejudice—public interest

The three claimants were actuaries subject to the discipline of the Institute of Actuaries. The claimants had been involved in the management of the Equitable Life Assurance Society and were three among a number of defendants in two parallel actions due to start in April 2005 and expected to last six months. On 26 June 2004, the defendant's investigating committee notified the claimants that various charges alleging misconduct would be referred to a tribunal. The claimants, who were unrepresented in the disciplinary proceedings, asked the tribunal for an adjournment until after the civil proceedings had been completed. The tribunal refused an adjournment and proposed to start a one-month disciplinary hearing on 2 November 2004. On 28 October 2004, Moses J granted, as a matter of urgency following a full hearing, a stay of the disciplinary hearing. The learned judge accepted and wished to emphasize the importance of the disciplinary proceedings: they are central to the role of the Institute. The actuarial profession has an obligation in the public interest to provide the best possible service advice. Tribunals should be astute to refuse adjournments unless fairness demands it. It would be unfair now to have required these three defendants to face charges before the civil proceedings at first instance were determined. First, they were litigants in person, unable to deploy anything other than very limited legal assistance in the civil proceedings; they had to represent themselves in the disciplinary hearing. That fact is a serious impediment to the ability to be engaged at the same time in both disciplinary and civil proceedings. Second, the fact that the claimants were familiar with the background, and that they had assistance from the investigating committee with the documents and the statements, was not sufficient to make it fair to require them to handle both the disciplinary and the civil proceedings. Facing the disciplinary proceedings would inevitably require considerable time to be spent in preparation and, in any event, there would be one month for the disciplinary hearing. That alone would prejudice the preparation of the civil proceedings. They could not conduct both the disciplinary hearings and, at the same time, properly prepare for the civil litigation, and it would be unfair for them to be forced to drop their preparation of the civil proceedings at this stage in order to prepare for the disciplinary charges, however familiar they were with the facts. Third, whilst accepting the importance of the public interest in the disciplinary proceedings, one further feature is important: the Institute never sought to impose, and had never imposed, temporary suspension upon the claimants. The claimants offered to undertake not to practise as actuaries until the resolution of the disciplinary charges. Whilst that did not fully meet the public interest, it did at least afford protection of the public should that prove necessary.

### *General Medical Council v. Srinivas* [2012] EWHC 670 (Admin)

**10.12** Interim order—extension

The GMC sought an extension for twelve months of an interim order previously imposed by the IOC. In granting the extension for five months, His Honour Judge Pelling QC (sitting as a judge of the High Court) said, at [12]:

> I accept the submission that once the defendant became the subject of criminal prosecution proceedings the ability of the GMC to actively investigate on its own terms the allegations which are being made was severely hampered and for all practical purposes prevented because I accept the submission that any approach they made to witnesses who were to be called by the prosecution – or might be – could be misconstrued for fairly obvious reasons. I also accept that the practice of the GMC is to ensure as far as possible that the totality of allegations made against the doctor are investigated in one go and that all material evidence is made available for one hearing. The reasons for this are obvious. It is in the public interest and in the interests of the doctor that all relevant information be considered at one hearing and that a multiplicity of disciplinary proceedings is avoided if at all possible. In addition, such an approach is likely to be cost effective for a heavily pressed regulatory body.

### *R (Montgomery) v. Police Appeals Tribunal* [2012] EWHC 936 (Admin)

**10.13** Criminal conviction subject to appeal—whether to adjourn disciplinary proceedings pending appeal of conviction—extent to which findings or outcome of criminal court may be relevant in appropriate case

The claimant was a detective inspector with the Metropolitan Police. Following a drunken incident at a London hotel when the claimant was off duty, he was convicted on 27 July 2009 at Westminster Magistrates' Court of one charge of common assault following a contested trial; thereafter, the claimant was served with a notice pursuant to regulation 21 of the Police Conduct Regulations 2008. He asked for an adjournment of the disciplinary proceedings on the basis that he was appealing to the Crown Court against his conviction and it would not be appropriate for the panel to proceed to hear the matter before the Crown Court decision was reached. The application for an adjournment was not granted and the hearing went ahead; the allegation was proved and the claimant was dismissed from the Metropolitan Police. In the event, his appeal to the Crown Court, heard after the disciplinary proceedings, was dismissed. Collins J said that, in the circumstances of this case, the panel was entitled in the exercise of its discretion to refuse an adjournment on the basis that it could look, and could properly decide to look, at all of the material that was before it. The learned judge did not think in the circumstances that it was necessarily to be regarded as a wrong decision and an unfair decision not to await the hearing before the Crown Court. But he would introduce a caveat here. While the Court was prepared to uphold the decision in the circumstances of this case, it was dangerous for a panel to refuse to adjourn a matter pending a hearing before a criminal court when there is a real issue as to whether a particular offence on which the case relies has been committed. While the standard of proof is lower in the disciplinary process, where facts are in issue, it is likely to be important to be able to consider findings in a criminal court. While rule 9 of the Police Conduct Regulations 2008 provided that proceedings should proceed without delay, the Court said that speed cannot override fairness. All that rule 9 does is indicate the need for there to be no delay; it says nothing about whether there should be a deferment, and so an adjournment, where the officer in question is the subject of criminal proceedings. It is possible that to pursue misconduct proceedings before the officer is able to be heard in his defence in the criminal proceedings may, ipso facto, be prejudicial to him and one can imagine circumstances in which that might well be the case. The Court was very far from saying that it can never be appropriate to carry on before the hearing before the Crown Court, but care must be taken and it must be recognized that if there is a significant dispute of fact that can go to whether there has been gross misconduct, it may well be unfair to refuse to adjourn pending the hearing before the Crown Court, or indeed perhaps any criminal court, and that is something that must be taken into account because rule 9 does not in any way prevent such an adjournment in appropriate cases.

***General Medical Council v. Razoq* [2012] EWHC 2135 (Admin)**

**10.14** Interim order—extension—complaint by doctor of delay

R was arrested for fraud and charged in relation to failure to disclose full Criminal Records Bureau (CRB) check results to Ipswich NHS Trust and a locum agency through which he was employed, failures to disclose his exclusion from work to locum agencies, and misrepresentations made in respect of qualifications as a doctor that he purported to have. The frauds concerned approximately £100,000 over a period of twenty-one months. On 27 May 2010, R was convicted on seven counts and was subsequently sentenced to two years' imprisonment on each count concurrently. His appeals against conviction and sentence were dismissed by the Court of Appeal, Criminal Division, on 29 February 2012. In seeking to oppose the extension of an interim suspension order imposed by the General Medical Council's (GMC's) Interim Orders Panel (IOP), R submitted that the GMC erred in staying its hand in respect of disciplinary proceedings pending the criminal charges. Hickinbottom J rejected this argument, saying that it was appropriate for the GMC to stay its hand in respect of the disciplinary investigation and procedures against R whilst the criminal proceedings were running their course. The frauds that were the subject matter of the criminal proceedings concerned R's conduct as a doctor—and those frauds were clearly going to be crucial to any disciplinary proceedings against him before the GMC. Since the conclusion of the criminal proceedings in February 2012, the GMC investigators had sought disclosure of the police investigation file into the frauds, to determine the full basis on which the charges were brought and upon which he was found guilty, with a view to ensuring that the charges before the fitness to practise panel were comprehensive. The GMC understandably wished all matters in respect of R to be dealt with at one fitness to practise hearing. No criticism of the GMC could be made as to the manner in which it had dealt with the allegations against R, or the time that had been taken, or the decision not to proceed whilst the criminal charges were pending. The interim suspension order would be extended for nine months to expire on 25 March 2013.

### *Lamey v. Belfast Health and Social Care Trust* [2013] NIQB 91

**10.15** Applicant excluded from work—hearing whether to terminate employment—GDC disciplinary proceedings ongoing

L, a senior oral medicine consultant in a regional specialist dental hospital, sought an injunction restraining the defendant from holding a meeting to determine the termination of his employment with the hospital in advance of the receipt of the conclusions of the Professional Conduct Committee (PCC) currently before the General Dental Council (GDC). Deeney J held that the principles in *American Cyanamid Co. v. Ethicon Ltd* [1975] AC 396 applied. The defendant was a public body discharging important public duties. A special factor and relevant to the balance of convenience was that the plaintiff had been excluded from work now for two years and eight months, and there was no clear date for completion of the proceedings before the GDC. The plaintiff was both a doctor and a dentist, and disciplinary proceedings before the GMC had been concluded on undertakings given by the plaintiff. The injunction was refused.

### *Gulamhusein v. General Pharmaceutical Council* [2014] EWHC 2591 (Admin)

**10.16** Interim order—extension—application to terminate

G, a pharmacist, was involved in running a pharmacy in Maida Vale. During an undercover operation, G was approached while working in the pharmacy by a journalist and employees of the BBC posing as members of the public. It was alleged that G supplied medication without prescription, including sixty Cytotec 200mg Misoprostol tablets, in the knowledge that they were going to be used for the purpose of an illegal termination of a pregnancy. On 16 January 2013, G's registration was suspended for eighteen months by the Council's Fitness to Practise Committee. An attempt by G to challenge that suspension was rejected by the Committee in a determination dated 27 May 2014 and G appealed pursuant to article 58 of the Pharmacy Order 2010. By the time of the hearing on 27 May 2014, G had been charged with supplying Misoprostol without a prescription and had elected for trial in the Crown Court. At the time of the hearing in the High Court, he had not entered a plea to this count. There were no criminal charges in relation to the other supplies by G. In dismissing G's appeal and granting the Council's application for an extension for six months, Sir Stephen Silber said:

> 45. I agree that there was no need for the allegations made against the appellant to be proved before they could be considered relevant for disciplinary proceedings as was explained by Pill LJ when giving judgment in the Divisional Court in *Dr X v. General Medical Council* [2001] EWHC Admin 447, [24]. The fact that the allegations have not been proved is clearly a matter that the disciplinary Committee was bound to take into account. What is important is that it must scrutinise the unproven allegations with great care to test their cogency. That is what this Committee did after taking account of the previous good record of the appellant and the matters put forward in his favour.
>
> 46. The Committee concluded correctly, in my view, that if the allegations are found proved, the appellant would be shown to be prepared to supply dangerous prescription only drugs improperly when asked, and where, in the case of Misoprostol (Cytotec) the risk to the public is that of procuring an illegal abortion in relation to which no prescription had been obtained and in relation to which the appellant had no knowledge.
>
> 47. It is true that the Committee stated that a reasonable onlooker with knowledge of the facts would be appalled if no order was made and the allegations against the appellant were eventually proved. The Committee was nonetheless very clear in stressing that, in its judgment, the order was necessary for the protection of the public and in the public interest.

***Westcott Financial Services Ltd and ors v. Financial Ombudsman Service* [2014] EWHC 3972 (Admin)**

**10.17** Civil proceedings—Commercial Court—whether appropriate to stay complaint to Ombudsman

The claimants were independent financial advisers (IFAs) who were the subject of a complaint to the Financial Ombudsman Service (FOS) by former clients. Each claimant sought a stay of the determination of the complaint, pending the resolution of litigation in the Commercial Court. In each case, the stay was refused, the Ombudsman upholding the complaint and directing the claimants to pay compensation. The claimants sought judicial review of the decision not to stay the determination of the complaints. A very large number of investors had suffered significant losses allegedly arising out of the claimants' recommendation to clients of particular financial products marketed and distributed between July 2005 and April 2007. Many had made claims on the Financial Services Compensation Scheme (FSCS), which had so far paid out compensation in excess of £300 million. The investors' claims against their IFAs had been assigned to the FSCS, which began proceedings against a large number of advisers in 2011. At a case management conference in March 2013, the Court held that there should be test cases against the lead defendants. Those cases and the defendants were identified in August 2013; all other cases were stayed. There was no trial date set at time of writing. If there is to be a trial in this case, it will not take place before the end of 2015 or beginning of 2016. In dismissing the claimants' claims for judicial review, Thirlwall J, in a judgment handed down on 2 December 2014, said at [32], after a review of the Financial Services and Markets Act 2000 and the Ombudman's rules, that the scheme is designed to permit disputes to be resolved quickly and informally by people who have appropriate qualifications and experience. There is no requirement that processes (or indeed decisions) should mirror those of the courts; on the contrary, this is an alternative method of resolving disputes. Provided a person is eligible to complain he is entitled to have his complaint considered. Complaints which are also causes of action are not excluded—save where the complainant has brought proceedings. Complaints are to be determined by reference to what is, in the opinion of the Ombudsman, fair and reasonable in all of the circumstances of the case. A decision by a judge in the Commercial Court in respect of the alleged professional negligence of an IFA is reached by the application of the relevant law to the facts found. A decision of fact by the Commercial Court in respect of that risk would be one of the matters to be taken into account by the Ombudsman dealing with a complaint in respect of the product. The weight to be given to it would be a matter for the Ombudsman.

## C. Other Relevant Chapters

**10.18**

Adjournment, Chapter 3
Conviction and Caution Cases, Chapter 14
Double Jeopardy, Chapter 21
Evidence, Chapter 25
Interim Orders, Chapter 38
Natural Justice, Chapter 48

# 11

# CONDITIONS OF PRACTICE ORDERS



## A. Legal Framework

**11.01** In relation to healthcare professionals, examples include:

Medical Act 1983, section 35A
Health and Social Work Professions Order 2001, article 29
Nursing and Midwifery Order 2001, article 29
Pharmacy Order 2010, article 54

In relation to other professional bodies, see for example:

Solicitors Regulation Authority Practising Regulations 2011, regulation 7 (conditions on practising certificate)
Chartered Institute of Management Accountants Disciplinary Committee Rules 2013, rule 24(6) (conditions on membership)

Guidance on conditional registration is often contained in the regulator's indicative sanctions guidance. For example, see:

General Medical Council Indicative Sanctions Guidance for the Fitness to Practise Panel, April 2009 (with April 2014 revisions), paragraphs 56–68 (conditions are likely to be appropriate where the concerns about the doctor's practice are such that a period of retraining and/or supervision is likely to be the most appropriate way of addressing them; conditions might be appropriate in cases involving the doctor's health, performance, or following a single clinical incident, or where there is evidence of shortcomings in a specific area or areas of the doctor's practice; the purpose of conditions is to enable the doctor to deal with his or her health issues and/or to remedy any deficiencies in his or her practice, whilst in the meantime protecting patients from harm)

## B. Disciplinary Cases

***Finegan v. General Medical Council* [1987] 1 WLR 121**

**11.02** Condition precluding any prescribing of medicines—whether wider than necessary

The charge against the doctor was that he had repeatedly and at frequent intervals issued prescriptions for large quantities of controlled drugs and other euphoriants, most of which were used for purposes other than the bona fide treatment of his National Health Service (NHS) patients. The doctor's wife, a registered nurse, had prevailed upon him to prescribe controlled drugs for her in large quantities following her giving birth to a child, who had died a few years later. The doctor had stopped prescribing drugs, but was prevailed upon to resume doing so in response to threats from his wife to himself and their young child. All of the facts alleged in the charge were admitted and the doctor did not dispute that the facts disclosed serious professional misconduct. The Professional Conduct Committee (PCC) of the General Medical Council (GMC), rejecting conditions on his registration such as would prevent him from prescribing for his wife or dealing with her as a patient, directed that, for a period of three years, he should not prescribe or possess any controlled drugs or any prescription-only medicines. On appeal, it was maintained that Mrs F was the only member of the public with whose protection the Committee had occasion to be concerned, that the conditions imposed went far wider than was necessary for her protection, and that they were not necessary for the doctor's interests and would prevent him for all practical purposes from carrying on his practice of a general practitioner (GP), the only medical field for which he had the requisite training and experience. In dismissing the appeal, Lord Keith of Kinkel, delivering the judgment of their Lordships, said that the Committee was entitled to take the view that the wider public interest required that the doctor should have no access at all, over a lengthy period, to drugs that required a prescription and that this was in the best long-term interests of the doctor himself. Restriction of practice is not the same thing as suspension of registration. Suspension is a much more serious matter. In response to the submission that the doctor was, for practical purposes, in a worse position than if his registration had been suspended for the maximum of twelve months or his name had been erased from the register, their Lordships said that the Committee had power under section 36(4)(b) of the Medical Act 1983, not subject to any limitation as to time of its exercise, to revoke a direction for conditional registration or to revoke or vary any of the conditions imposed by the direction.

***McMillan v. General Medical Council*, Privy Council Appeal No. 62 of 1992**

**11.03** Conditions preventing any private work—whether necessary—public interest

M, a consultant obstetrician and gynaecologist, held an NHS post at Whipps Cross Hospital, Leytonstone, London E11. He became involved in a private clinic that was concerned with cosmetic surgery in the form of removal of tattoos by means of a laser. It was in relation to his involvement with this clinic that he was charged with serious professional misconduct. The PCC directed that, for a period of eighteen months, his registration should be conditional upon his compliance with the requirements that he should not engage in any form of professional practice except under the NHS and that he should not undertake any professional commitment in relation to private practice. On appeal against sanction, it was argued that there was no evidence to justify the direction for the protection of members of the public or in M's interests, and that, in breach of the rules of natural justice, no warning had been given that the Committee was minded to impose any such condition. In dismissing M's appeal, Lord Goff of Chieveley, delivering the judgment of their Lordships, said that not only sentences of erasure and suspension, but also the imposition of conditions, may have a severe impact not only upon the doctor himself, but also upon innocent persons who may be affected. Such innocent persons may include not only patients, but also others who are dependent on the doctor for their employment. However, from time to time, it is nevertheless necessary to impose such penalties, in the public interest, for the purpose of registering disapproval of unprofessional conduct and for maintaining high standards of conduct in the medical profession. In the present case, the Committee was entitled to require that, for a limited period of time, the appellant's position should be confined to the NHS, with all of the structural safeguards built into that service to prevent conduct of the present kind from occurring. The appellant was represented before the Committee by very experienced solicitors and counsel. It must have been obvious that either suspension or conditions would be the likely outcome of the proceedings; in either event, it would have been material to draw to the attention of the Committee the impact upon others of any penalty that precluded the appellant from continuing his private practice in obstetrics and gynaecology.

*Carruthers v. General Medical Council* [2003] UKPC 42

**11.04** Terms and content of conditions—review hearing

In allowing C's appeal to the extent of deleting two conditions that the GMC's PCC had imposed on his registration, the Board said that it was open to the parties in the course of their submissions before the Committee to make suggestions as to the content or wording of conditions that might meet the circumstances of the case. Neither side did so, beyond acknowledging in general terms that the imposition of conditions was a possibility. Their Lordships appreciate that to descend to this level of detail might well have been thought by the appellant's counsel to be inconsistent with his primary argument, which was that a finding of serious professional misconduct was not justified by the evidence. If the Committee has decided to impose a condition that is impracticable or unworkable, the matter can be dealt with at an early or expedited resumed hearing in the light of the evidence. If the only issue that the registered person wishes to raise relates to the content of the conditions that the Committee has decided to impose on him, the appropriate way of dealing with the matter is to ask for the matter to be dealt with at a resumed hearing by the Committee. But that was not this case. The appellant had challenged the determination that was made against him on the merits, as well as on grounds that related to the appropriate penalty and the conditions.

*Camacho v. Law Society* [2004] 1 WLR 3037

**11.05** Solicitors Disciplinary Tribunal—power of Tribunal to impose conditions

On 1 April 2004, the Court allowed an appeal by C, a solicitor, against a decision of the Solicitors Disciplinary Tribunal (SDT) by which he was suspended indefinitely and reduced the period of suspension to eighteen months. In the course of its judgment—[2004] EWHC 1042 (Admin)—the Court considered that certain conditions should be imposed upon C in relation to his practice. At a further hearing to determine the powers of the Tribunal to impose conditions, the Court (Thomas LJ, and Silber and Goldring JJ) held that the terms of section 47(2) of the Solicitors Act 1974 were sufficiently wide to give the Tribunal itself power to impose conditions under which a solicitor might practise as distinct from making a recommendation to the Law Society so that the Law Society could implement the recommendations by imposing conditions on the solicitor's practising certificate. The Court did not accept the two principal reasons submitted as to why the Tribunal should not exercise its power to impose terms—namely, reasons of subsequent policing and enforcement, and the role of practising certificates. It was the duty of the Tribunal in each case in which it considered that restrictions were required to consider imposing those restrictions itself. The powers conferred by section 47 were very wide. For example, there was no impediment to the Tribunal imposing, if it thought fit, conditions for an indefinite period with permission to apply to it, so that it could reconsider the conditions and vary them.

### *Re a Solicitor, No. 20 of 2008, Odunlami* [2008] EWCA Civ 1598

**11.06** Appeal from SRA adjudicator

O was admitted as a solicitor on 2 April 2006. He had practised before that as a registered foreign lawyer. On 2 April 2008, an authorized officer of the Solicitors Regulation Authority (SRA) imposed conditions on O's practising certificate. The conditions imposed were that he may act as a solicitor only in employment or partnership, or as a member, officer holder, or shareholder of an incorporated solicitors' practice, the arrangements for which have been first approved by the SRA. O's appeal to an SRA adjudicator was dismissed on 30 May 2008 and he appealed to the Master of the Rolls (Sir Anthony Clarke) on 9 December 2008. In dismissing O's appeal, Sir Anthony Clarke said:

> 23. The issue in this appeal is whether the imposition of practising certificate conditions was justified in this case and, if so, whether the conditions imposed were reasonable and proportionate. In my judgment the conditions were both justified, and reasonable and proportionate when they were imposed. Indeed this seems to me to be a straightforward case. [O]'s practising history leaves a lot to be desired, and it is fair to him to note that he recognises in his two statements to which I have referred that that was the case. The thrust of his submissions today is that that is all in the past and that he should be allowed to practise in an unfettered way in the future. He undoubtedly had professional difficulties while practising with [his former partner]. He admitted those errors, however, before the SDT in 2007. It seems to me that when the matter was presented to the Adjudicator and the appeal panel earlier this year, the picture was such that they were entirely justified in imposing the conditions that they did. He is a solicitor whose practice certainly needed to be supervised, and they were correct to say that he is a solicitor who needs to demonstrate over a period of time that he is practising consistently with his professional obligations. In my judgment there is no basis for complaint about the way in which the SRA had dealt with those matters.
>
> 24. What then of the present position? . . .
>
> 25. So the position at present is, as I understand it, and this is accepted by [counsel instructed by the SRA], that the SRA will now have to consider whether in the light of the current position to grant a practising certificate for 2008/2009 without conditions, which is no doubt what [O] has applied for. It is not for me to prejudge how they will view that application. They will, I am sure, as is their duty, have regard not only to the history but, importantly, to the current position.

### *Udom v. General Medical Council* [2009] EWHC 3242 (Admin)

**11.07** Primary object of conditions—effect on ability to practise—whether incompatible with registration

The panel found that U's fitness to practise was impaired by reason of both deficient professional performance and his physical or mental health, and directed that his registration be subject to conditions for a period of nine months with immediate effect. U was a consultant anaesthetist, and, in May 2006, there was an allegation made against him that he had failed to monitor a patient properly post-operatively and that the patient had suffered a bleed. At a GMC performance assessment, U was well below the majority of his peers, ten objective structured clinical examinations were rated as unacceptable, and he fell below the minimum standard of the Royal College of Anaesthetists in eight out of ten examinations. On U's appeal on sanction, Hickinbottom J said that the hallmark of suspension is identified in section 35D(2)(b) of the Medical Act 1983: for a determined period, the registration of the relevant doctor is divested of all effect. Suspension prevents the doctor from practising during the period of suspension. Conditions on registration under section 35D(2)(c) presuppose that the doctor will continue to be effectively registered and will continue to practise, but subject to conditions that enable him to deal with his health or deficiency issues whilst protecting patients, as well as the integrity of the medical profession, from harm. The primary object of imposing a sanction is to maintain the standing of the profession and the confidence of the public in the profession, although the need to protect individual patients is also a purpose. The impact of the sanction on the practitioner is also relevant, because the panel can impose only a sanction that is proportionate. But because the primary objectives concern the wider public interest, the impact of a sanction on a practitioner has been said not to be 'a primary consideration': *Cheatle v. General Medical Council* [2009] EWHC 645 (Admin), [38], [40]. Subsections (b) and (c) of section 35D(2) of the 1983 Act are based on different and mutually exclusively premises: section 35D(2)(a) (suspension) is based upon the premise that, in substance, registration—and hence the ability to practise—cease, at least temporarily; section 35D(2)(b) (the imposition of conditions) is based upon the premise that, in substance, registration—and hence the ability to practise—continue, but are restrictive. They are alternatives—hence the word 'or' that joins them—precisely because they are mutually exclusive. It is logically impossible to have conditions limiting the ability to practise medicine attaching to a registration that is of no effect. If conditions are such that they divest registration of all effect, then they cannot properly be called a 'conditional registration'. In substance, such a registration is no registration at all. A panel cannot undermine this statutory structure by bringing to an end the ability to practise through the imposition of conditions under subsection (c), rather than suspension under subsection (b). The panel in this case erred in law in applying conditions that resulted in the emasculation of the appellant's registration, with further conditions as to his conduct towards his rehabilitation as a doctor. As a result of that error, the sanctions imposed by the panel were clearly wrong. It was not appropriate for the Court to substitute its own conditions for those of the panel and the case was to be remitted for consideration of sanction by a differently constituted panel, on the basis of the undertaking offered by the appellant not to practise in the meantime on terms to be agreed between the parties.

### *R (Jalloh) v. Nursing and Midwifery Council* [2009] EWHC 1697 (Admin)

**11.08** Proportionate sanction—concern of future ability to respond

The charges against the appellant, a psychiatric nurse, were that, whilst employed by the South London and Maudsley NHS Trust as a grade E nurse, she gave patient B medicine that had not been prescribed for him and failed to respond adequately when patient B was found hanging from the ceiling following an apparent suicide attempt, in that she failed to ensure that basic life support measures were taken in respect of patient B prior to the arrival of the crash team. The facts alleged were admitted and the Conduct and Competence Committee (CCC) of the Nursing and Midwifery Council (NMC) went on to find that the appellant's fitness to practise was impaired; it then made, against the appellant, conditions of practice orders for eighteen months. On appeal, Silber J dismissed J's appeal on impairment on the grounds that the Committee had failed to consider properly the mitigating factors relating to the particular event that led to the charge. The position was that the appellant, as a trained nurse, should have known the great urgency that was indeed required and that the effect of delaying treatment could lead to the patient suffering brain damage resulting from lack of oxygen reaching the brain. On the appeal against conditions, the thrust of the appellant's submissions was that there was inadequate consideration given as to whether a caution would have been an appropriate sanction having regard to the relevant factors in the indicative sanctions guidance. The learned judge said, at [42], that, at the end of the day, the conclusion that conditions should be imposed was a proportionate response to the proven failure of the appellant. The Committee was entitled to have concern about her ability in the future to respond to a similar incident in an acute setting. It must not be forgotten that the consequences in the future of a failure to act, and to act correctly, could be very serious. The imposition of conditions could not be impugned.

***Sharma v. General Dental Council* [2010] EWHC 3184 (Admin)**

**11.09** In this case, the appeal of S in relation to the conditions of practice order was allowed to the extent that it related to the period covered by the order, in the light of the length of time that had elapsed since the order had been made with immediate effect by the fitness to practise panel.

***Daraghmeh v. General Medical Council* [2011] EWHC 2080 (Admin)**

**11.10** Doctor not practising since 2005—stringent conditions imposed—whether necessary for public protection

King J dismissed an appeal from findings of a review hearing held to investigate whether the appellant's fitness to practise was still impaired. In November 2005, the appellant had been suspended from his post as a specialist registrar in medicine for the elderly at a number of hospitals in Scotland. He had not worked in any clinical way since 2005. At a hearing before the fitness to practise panel in February 2009, the panel had determined that the appellant's fitness to practise was impaired by reason of deficient professional performance. The panel imposed a sanction of twelve months' suspension. At a review in February 2010, the panel found that the appellant's fitness to practise remained impaired by reason of his deficient professional performance and imposed a number of conditions on his registration. On appeal, the appellant challenged a number of the conditions and contended that:

- the conditions imposed by the review panel were irrational, disproportionate, and impracticable;
- their cumulative effect was to defeat the purpose of conditions as expressed in the GMC's Guidance—namely, to enable the doctor to remedy any deficiencies in his practice; and
- the conditions made it a practical impossibility for the appellant to obtain a post of employment in the United Kingdom.

King J recognized that *Udom v. General Medical Council* [2010] Med LR 37 established that it would be an error of law for a panel to impose conditions on registration that were, in effect, incompatible with registration. However, the panel was faced with having to balance the interests of the appellant against the need to protect patients. In the instant case, the panel was concerned with the ability of the GMC to monitor the performance of a doctor who had not practised clinically since 2005 and in the context of evidence that the assessment team itself had debated whether the appellant was capable of returning to any sort of work. In such circumstances, it was entirely open to the panel to impose stringent conditions, and to include within the conditions that they concluded were necessary for the protection of the public a restriction confining the appellant in the first instance to the care of the elderly and a restriction providing that his work should be supervised by a named consultant. The learned judge agreed with the observations of Blake J in the case of *Abrahaem v. General Medical Council* [2008] EWHC 183 (Admin), at [34], point (b), where it was said that the best evidence of the impact of the conditions would have been the appellant's attempts to find British employment or to seek advice from a postgraduate dean or other qualified person.

***Belal v. General Medical Council* [2011] EWHC 2859 (Admin)**

**11.11** Condition to notify third parties—whether notice to recruitment agency sufficient—whether contract concluded

On 27 April 2007, the GMC's Interim Orders Panel (IOP) placed conditions on the registration of B, a general practitioner. Condition 9 provided that he must inform any organization or person employing or contracting with him to undertake medical work that his registration was subject to conditions. In February 2008, International Medical Recruitment forwarded B's details to Dr U, a general practitioner in New Zealand, who was seeking to appoint a general practitioner to her practice in Hamilton. Dr U contacted B and it was common ground that she was not told that B's registration was subject to conditions. Lloyd Jones J said that whatever B may have told the agency, the IOP was clearly correct in its conclusion that it was his responsibility to notify Dr U of the true position in relation to his practising status. Accordingly, any disclosure to the agency could not relieve B of liability for breach of the condition if he were to fail to tell Dr U. However, B also contended that the obligation under condition 9 was to disclose the conditions to an organization or person 'employing or contracting' with him to undertake medical work. The GMC accepted that there had to be a contract before condition 9 could apply and did not seek to argue that there could be a breach of condition 9 in the absence of a binding contract. Having reviewed the evidence, the learned judge came to the conclusion that there was no binding contract, and accordingly, in view of the basis on which the GMC put its case below and on appeal, the finding of the IOP that B was in breach of condition 9 was wrong and must be set aside.

***Levinge v. Health Professions Council* [2012] EWHC 135 (Admin)**

**11.12** Complaint that registrant went beyond teaching techniques—condition imposed requiring her to undergo clinical supervision—no criticism of registrant as clinician—condition unreasonable and disproportionate

L was a music therapist and employed with the Royal Welsh College of Music and Drama as programme leader for music therapy. The respondent received an anonymous complaint about L's conduct during the course of her work at the College and, following a hearing before a panel of the Health Profession Council's CCC, certain allegations of misconduct and/or lack of competence were proved, and the panel found that L's fitness to practise was impaired. As a consequence of such findings, the Committee imposed a conditions of practice order for three years. The conditions included a requirement that L should undergo clinical supervision by an arts therapist registered with the respondent (within the music mode), who was to submit a report every nine months to confirm whether L's practice conformed to the respondent's Standards of Proficiency for Arts Therapists and Standards of Conduct, Performance and Ethics. In allowing L's appeal, Wyn Williams J held that no explanation for this condition, in particular, was to be found in the reasoning of the panel's decision. There had never been any criticism of L as a clinician and, accordingly, the imposition of the condition relating to clinical practice was, on the facts of the case, unreasonable and disproportionate. There were two reasons for this conclusion. First, the condition had a direct impact upon L as a clinician and, in her long and distinguished career as a practising therapist, there had never been a complaint about L's conduct towards her patients. The thrust of the complaint made against L was that, in effect, she had treated students, on occasions, as if they were her patients; rather than restricting herself to teaching students about psychodynamic therapeutic techniques, she had practised such techniques upon them. Second, the condition imposed upon L a financial burden the extent of which was never investigated.

***Broomhead v. Solicitors Regulation Authority* [2012] EWHC 509 (Admin)**

**11.13**
Test—whether condition on practising certificate is necessary and proportionate for protection of public interest

B appealed against the decision of the SRA to impose conditions upon his practising certificate, including a requirement that he inform prospective employers of the conditions. Conditions had first been placed on B's practising certificate in 1994 and, after the conditions were relaxed in 2004, further conditions were attached in 2009. The parties were agreed that the relevant test in respect of the imposition of a condition on a practising certificate is whether the condition is necessary and proportionate for the protection of the public interest. Sales J said that, as in *Khan v. Solicitors Regulation Authority* [2010] EWHC 1555 (Admin), he did not find it necessary to reach a concluded view on the significance, if any, of the fact that the SRA itself concluded that the imposition of the condition was necessary and proportionate for the protection of the public interest. Whether the approach of the Court was one of a complete substitution for the views of the SRA or an approach involving some element of deference to the views of the SRA, the appeal should be dismissed in the instant case. It was necessary and proportionate to ensure that B's practising arrangements were subject to an appropriate level of support and supervision in order to safeguard the public interest. It was necessary and proportionate for the protection of the public interest that B should inform any actual or prospective employer of the various conditions and the reasons for their imposition by the SRA. This was to ensure that if B were to practise as a solicitor with a firm, the firm would be properly on notice of the concerns that had arisen in the past in relation to his conduct, and would therefore be likely to take proper and fully informed steps to ensure that procedures and supervision were put in place to safeguard members of the public dealing with him and with the firm.

***Bryant v. Solicitors Regulation Authority* [2012] EWHC 1475 (Admin)**

**11.14**
Test—need to protect the public and reputation of the profession—necessary and proportionate

In October 2007, B, a solicitor, appealed against a finding of dishonesty made by the SDT, which ordered him to be struck off the roll of solicitors. The finding of dishonesty was quashed by the Divisional Court in December 2007 and a two-year suspension from 17 October 2006 was substituted for the striking-off order: *Bryant and Bench v. Law Society* [2007] EWHC 3043 (Admin), [2009] 1 WLR 163. The Court had concluded that B was guilty of conduct unbefitting a solicitor, but to a much lesser extent than that found by the Tribunal. Once the two-year suspension period ended, in October 2008, B was granted a series of practising certificates subject to conditions. The conditions were directed towards protecting and reassuring the public. An appeal committee of the SRA dismissed B's appeal against the conditions on 30 March 2011. It was against that decision that B appealed to the Administrative Court. In dismissing B's appeal, Eady J said that there was no dispute as to the appropriate test to be applied on such an appeal. In *Lebow v. Law Society* [2008] EWCA Civ 411, Sir Anthony Clarke MR explained that such an appeal was by way of rehearing, although it was important to have in mind that the imposition of conditions on a practising certificate is a regulatory decision, based on the need to protect the public and the reputation of the profession. Conditions, however, if they are to be imposed, must be both necessary and proportionate; see also *Razeen v. Law Society* [2008] EWCA Civ 1220. The Court recognized that previously the imposition of conditions on a practising certificate was compatible with continuing professional life and that, subject to a period of suspension, the solicitor would be able to resume practice thereafter. According to the evidence before the Court, the position has fundamentally changed: the imposition of conditions is now, in practical terms, recognized to be 'the kiss of death'. To all intents and purposes, they render the prospect of further practice impossible. In a sense, therefore, it may be said that what were originally intended to be temporary and precautionary measures have, in reality, become permanent and punitive. However, it is unthinkable that B would apply to any prospective employer without revealing his disciplinary record and that what would trouble an employer is not so much the mere fact of the conditions, but rather the disciplinary hinterland that they represent. In these circumstances, even if B were to succeed fully in his appeal, in the sense of having the conditions removed from his certificate, he would still face the practical difficulties in obtaining employment purely by reason of his disciplinary record. Considering the five conditions in the instant case, it was quite impossible to conclude that the imposition of the conditions was irrational, illogical, unnecessary, or disproportionate. They were sensible and were directed towards proper objectives. Furthermore, there was no reason for bypassing the well-established statutory regime and seeking to address the legitimate concerns of the SRA by way of undertakings. Either they would make no difference or they would be less effective in serving the SRA purposes of protecting the public and maintaining confidence in the profession.

### *Mould v. General Dental Council* [2012] EWHC 3114 (Admin)

**11.15** For this case on failure to comply with conditions, see paragraph 65.07.

### *Ogunnowo v. Solicitors Regulation Authority* [2013] EWHC 1882 (Admin)

**11.16** Regulatory history—relevant factor

Conditions imposed on O's practising certificate included: (i) that he could act as a solicitor only in employment approved by the SRA; (ii) that he could not operate as a sole practitioner and was prohibited from being a manager or owner of a firm or practice; and (iii) that he had to inform prospective employers of the conditions and reasons for the imposition of the conditions. In dismissing O's appeal, Collins J said, at [16]: 'It is important to look at the record of the appellant, because that is a matter which it is entirely proper for the regulatory authority to take into account and will indicate what is appropriate by way of condition.' The learned judge also said, at [38], that the way in which this appellant had dealt with his obligations as a solicitor to comply with the regulatory regime was a matter that could be properly taken into account and which did underline the need for him to be subject to proper supervision in any employment that he seeks.

### *Ebhogiaye v. Solicitors Regulation Authority* [2013] EWHC 2445 (Admin)

**11.17** Conditions imposed by SDT exercising disciplinary sanction—compare Law Society regulatory powers

On 15 November 2012, the appellant appeared before the SDT with her daughter, and they jointly admitted ten breaches of the Solicitors Accounts Rules 1998 and the Solicitors Code of Conduct 2004. In addition, the appellant admitted four further breaches on her own account. A six-month suspension was imposed by the SDT (not challenged on appeal), together with '*Camacho* conditions'—namely, that, at the expiration of the suspension, the appellant could work as a solicitor only in employment approved by the SRA and could not work as a sole practitioner or partner. Haddon-Cave J held that the SDT had power under section 47(2) of the Solicitors Act 1974 to impose conditions on a practice certificate to run after a period of suspension: see *Camacho v. Law Society (No. 1)* [2004] EWHC 1042 (Admin) and *Camacho v. Law Society (No. 2)* [2004] 1 WLR 3037. The powers of the SDT under section 47(2) are very wide: the Tribunal has a range of sanctions available to it, including imposing conditions on future practice, and it is not generally appropriate that the SDT should delegate its disciplinary functions, or part of those functions, to the Law Society. If appropriate, the SDT should itself impose conditions as part of the sanction in the particular case; the SDT can impose conditions indefinitely and appellants should be given liberty to apply to vary such conditions in the future. The powers that the SDT exercises are disciplinary powers, as distinct from the Law Society's general regulatory powers under sections 9–18 of the Solicitors Act 1974 to impose conditions on a practising certificate, which are intended to be essentially non-punitive in nature. The imposition of conditions by the SDT is about both risk and reputation: it is about both the risk to the public *and* protecting the reputation of the solicitors' profession. There was no procedural unfairness by the SDT in the instant case and the penalty imposed was not inappropriate.

### *R (Adelakun) v. Solicitors Regulation Authority* [2014] EWHC 198 (Admin)

**11.18** Appeal from regulatory decision—approach of court

A appealed under section 13 of the Solicitors Act 1974 against the decision of the SRA's appeal panel, which upheld an adjudicator's decision to impose conditions on his practising certificate—in particular, that he could not practise as a sole practitioner, but must act in an employed capacity. Dismissing A's appeal, Singh J said, at [54], that it was important to recall that, in this context, the Court is considering an appeal from the decision of a professional regulatory body. The correct approach on such an appeal was set out by the then Master of the Rolls in *Lebow v. Solicitors Regulation Authority* [2007] EWCA Civ 411, [23], that the imposition of conditions on a practising certificate is a regulatory decision, and one based on the need to protect the public and the reputation of the profession—although conditions, if they are to be imposed, must be both necessary and proportionate.

### *Lynch v. General Medical Council* [2013] EWHC 3521 (Admin)

**11.19** Condition to notify third parties—breach

L, a consultant obstetric and gynaecological surgeon, was subjected to conditions in September 2011, which included that he must inform any organization or person employing or contracting with him to undertake medical work. In breach of this condition, L failed to inform Highgate Hospital that his registration was subject to conditions when undertaking private work at the hospital. L's case was that, because he was under no obligation to accept work from the hospital, the nature of the agreement or arrangement was not such as to fall within the requirements to notify the hospital of the condition. It was suggested on his behalf that if such were intended, it could have been achieved by the use of more extensive wording by the IOP. Dismissing L's appeal, the Court said that the arrangement between the appellant and the hospital constituted an arrangement for undertaking medical work, and that the IOP was entitled to so conclude. The case was entirely different from *Belal v. General Medical Council* [2011] EWHC 2859 (Admin), which did not assist the appellant.

### *Re a Solicitor* [2014] NIQB 46

**11.20** Long unblemished record—no loss to clients—decision of SDT correct

Despite the unblemished long history of practice by the appellant solicitor until 2010 and the absence of any loss to clients, the conclusion of the Solicitors Disciplinary Tribunal of Northern Ireland ordering the appellant to be restricted from practising on his own account and permitting him to work in partnership only with a solicitor of at least seven years' post-qualification experience was correct in the light of the persistent failure of the appellant to successfully address book-keeping issues. The restriction was postponed for three months to allow the appellant to put his affairs in order. Following an inspection of the appellant's accounts, it was discovered that his books of account were not up to date and that there was an unreconciled deficit in client accounts. In August 2012, the appellant had been admonished in relation to similar matters. The Tribunal had regard to the facts that the appellant was a solicitor of thirty years' standing, that there was no suggestion of any dishonesty on his part, and that he rectified the deficits identified by the Law Society's accountant at the earliest opportunity. Morgan LCJ said that the task upon which the court was engaged in cases in which the appeal is from a refusal to issue a practising certificate was different from the determination of an appeal from the Tribunal. In the first case, the power to refuse a practising certificate was purely protective: see *In re Crowley* [1964] IR 106. The factors to be taken into account are the interests of the public, the interests of the profession, the interests of the clients of the solicitor in question, and the interests of the solicitor himself. There is no presumption that the Society's view was correct: *Re a Solicitor* [2001] NIQB 52. In an appeal from the Tribunal, issues of protection remain relevant considerations; to these must be added the punishment of misconduct and the deterrence of repetition by others. The court should also give weight to the Tribunal's decision Appeal dismissed.

## C. Other Relevant Chapters

**11.21**

Experts, Chapter 26
Health (Adverse Physical or Mental Health), Chapter 31
Review Hearings, Chapter 59
Suspension, Chapter 65
Undertakings and Warnings, Chapter 66

# 12

# CONSENSUAL DISPOSAL



## A. Legal Framework

**12.01** Examples include:

Nursing and Midwifery Council (Fitness to Practise) Rules 2004, rules 9–11A (allegation to be considered at a meeting or a hearing)

General Pharmaceutical Council (Fitness to Practise and Disqualification etc.) Rules 2010, rules 10 and 26 (agreement of undertakings by the investigating committee, and agreement of undertakings and giving of advice and warnings by the fitness to practise committee)

Chartered Institute of Public Finance and Accountancy Disciplinary Regulations 2013, rule 4.5 (where the investigations committee is of the opinion that there is a prima facie breach of the CIPFA's bye-law and the gravity of the breach does not warrant a more severe penalty than an entry on record or a reprimand, and the respondent so consents)

Bar Standards Board Complaints Regulations 2014, rules E67–83 (Determination by Consent) (A complaint that the Professional Conduct Committee is otherwise intending to refer to the Disciplinary Tribunal may, with the consent of the relevant person against whom the complaint is made, be determined by the PCC having regard to the regulatory objective. It must be in the public interest to resolve the complaint under the determination by consent procedure; the potential professional misconduct, if proved, must not appear to be such as to warrant a period of suspension or disbarment; and there must be no substantial disputes of fact that can only be resolved by oral evidence. If the relevant person accepts a determination by consent, no one may appeal against it; the sanctions available are—fine, imposition of conditions, reprimand, advice, and an order to complete continuing professional development, or CPD)

General Medical Council (Fitness to Practise Rules) 2014, rule 10 (If, after considering the allegation, it appears to the case examiners that (a) the practitioner's fitness to practise is impaired, or (b) the practitioner suffers from a continuing or episodic physical or mental condition that, although in remission at the time of the assessment, may be expected to cause a recurrence of impairment of the practitioner's fitness to practise, the case examiners may recommend that the practitioner be invited to comply with such undertakings as they think fit—including any limitations on the practitioner's practice: rule 10(2); the registrar shall not invite the practitioner to comply with any such undertakings where there is a realistic prospect that if the allegation were referred to a fitness to practise panel, his or her name would be erased from the register: rule 10(5))

Chartered Institute of Management Accountants Investigation Committee Guidance Notes (disposal by consent order with one or more of the following sanctions—admonishment, reprimand, severe reprimand, fine, and costs; right of appeal to the complainant against the decision to the appeal committee)
Institute of Chartered Accountants in England and Wales Disciplinary Bye-laws, bye-law 16 (consent orders)

## B. Other Relevant Chapters

**12.02**

Disposal without an Oral Hearing, Chapter 20
Undertakings and Warnings, Chapter 66
Voluntary Erasure, Chapter 68

# 13

# CONSTITUTION OF PANEL



## A. Legal Framework

**13.01** Examples include:

[2011] EWHC 2307 (Admin) | 13.10 |

[2011] EWHC 3204 (Admin) | 13.11 |
| Dally
| Southall v. General Medical Council [2010] EWCA Civ 407 | 13.06 | R (Leathley and ors) v. Visitors of the Inns of Court [2013] EWHC 3097 (Admin) | 13.15 |
| Pugsley v. General Medical Council [2010] EWHC 2247 (Admin) | 13.07 | R (Mehey and ors) v. Visitors of the Inns of Court [2014] EWCA Civ 1630 | 13.16 |
| Yeong v. General Medical Council [2010] 1 WLR 548 | 13.08 | R (Hill) v. Institute of Chartered Accountants in England and Wales [2014] 1 WLR 86, CA | 13.17 |
| R (Al-Zayyat) v. General Medical Council [2010] EWHC 3213 (Admin) | 13.09 | C. Other Relevant Chapters | 13.18 |

General Medical Council (Constitution of Panels and Investigation Committees) Rules 2004
General Medical Council (Constitution of Panels and Investigation Committees) (Amendment) Rules 2005, 2006, 2010, and 2013
Bar Standards Board Disciplinary Tribunals Regulations 2014, rules E131–141 (composition of disciplinary tribunals)
Chartered Institute of Management Accountants Regulations, Part II ('Discipline')
Institute of Chartered Accountants in England and Wales Disciplinary Bye-Laws, bye-laws 19 and 27 (disciplinary tribunals and appeal panels)

## B. Disciplinary Cases

### *Rai v. General Medical Council*, Privy Council Appeal No. 54 of 1983

**13.02** Member of panel leaving—quorum

The appellant appeared before the Professional Conduct Committee (PCC) of the General Medical Council (GMC), charged with an allegation that he had abused his professional position as a medical practitioner by issuing numerous prescriptions for dipipanone hydrochloride with cyclizine, methylphenidate, and other drugs, in return for fees, otherwise than in the course of bona fide treatment. The appellant was judged to have been guilty of serious professional misconduct and the Committee directed that his name should be erased from the register. On appeal, which was dismissed, counsel for the appellant submitted that natural justice had been infringed. Counsel submitted that it was improper that a member of the Committee should have withdrawn from the hearing after the first day. In delivering the judgment of the Board, Lord Scarman said that a quorum, however, remained and the member of the Committee who had withdrawn did not return. There was nothing in the point.

### *Dally v. General Medical Council*, Privy Council Appeal No. 7 of 1987

**13.03** Members ceasing to take part—hearing quorate

D complained of the fact that four of the nine members of the PCC at the start of the disciplinary hearing did not complete the hearing. The inquiry began on 9 December 1986. Of the ten members of the Committee invited to attend, one was ill and took no part, and another withdrew either at or shortly after the commencement of the hearing for the same reason. When the hearing was resumed in the New Year, on 26 January 1987, two of the members of the Committee who had previously attended had not been re-elected and so had ceased to be members; of the six who remained, one immediately withdrew in protest against a ruling on an issue relating to confidentiality. The remaining five members completed the hearing, the Committee being quorate throughout the inquiry. Lord Goff of Chieveley, giving the judgment of the Privy Council in dismissing D's appeal, said that there was no suggestion that any member of the Committee who withdrew after the inquiry had begun took any part in the deliberations of the Committee. The only suggestion is that those members of the Committee who made the final decision may have had their minds and opinions affected by views previously expressed by the members who had withdrawn. Their Lordships were able, however, to see no substance in this objection.

### *Al-Fallouji v. General Medical Council* [2003] UKPC 30

**13.04** Allegations against surgeon—no surgeon on panel—qualifications of panel members

The PCC of the GMC heard evidence and submissions in relation to charges against the appellant, an experienced and highly qualified surgeon, brought under nine heads. In summary, the charges alleged: irresponsible and inappropriate remarks made to female nurses and medical students; irresponsible and inappropriate physical conduct towards female nurses, support workers, and medical students; unacceptable conduct to a young and vulnerable female patient; unnecessary public criticism of and rudeness to junior doctors, nurses, and patients; and making criticisms of junior doctors that were misleading. None of the charges alleged any failure of technical skill as a surgeon. The Committee found most of the charges to have been proved and directed the appellant's name to be erased from the register. On appeal before the Board, the appellant challenged the findings of the Committee, and alleged bias on the part of the Committee and unfairness in its handling of his case. Among the points made, the appellant complained of the constitution of the Committee because it comprised (as well as three lay members) two general practitioners (GPs) and the chairman, Mrs W, whose training and background were in nursing. There were no surgeons on the Committee. In delivering the judgment of the Board, Lord Walker of Gestingthorpe said that those facts fell far short of raising (in the mind of a fair-minded and informed observer) the perception that the Committee (and Mrs W in particular) would be in some way biased against surgeons, or in favour of nurses. Several of the most important witnesses were doctors, not nurses, and no technical issues arose as to surgical skills or techniques. No application was made at the time for Mrs W to recuse herself.

### *Dzikowski v. General Medical Council* [2006] EWHC 2468 (Admin)

**13.05** Doctor specializing in drug addiction—no specialist on panel

D, a medical practitioner, ran a clinic specializing in the problems of drug addiction and its prevention. It was alleged that, in prescribing methadone mixture to patient A, he had acted in an inappropriate and irresponsible way, and not in the best interests of his patient. The allegations included failure to assess patient A prior to issuing her with a prescription, prescribing a daily dose of methadone that was too high, and that patient A should have been prescribed with daily-collection instalments rather than three times per week. The fitness to practise panel found D guilty of serious professional misconduct and imposed conditions on his registration, and hence his method of practising in his clinic. On appeal, D complained on grounds, amongst other things, that the hearing was unfair because the members of the panel were not specialists in the field of addiction. D said that it was not fair that there were a limited number of doctors on the panel that heard and judged him, and that he should not only have been judged by his peers, but also by peers who were specialists in the same area of work in which he had been practising for a number of years. In dismissing this argument and D's appeal, Hodge J said that the panel was established under the rules set out in the General Medical Council (Constitution of Panels and Investigation Committee) Rules Order 2004 made by the Privy Council. The rules were binding on the GMC and the Court. Members of the panel are appointed from a list of medical and lay persons who are regarded as appropriate to be on the panel. Panels that hear cases have to comprise medical and lay panellists. There are restrictions on who can be a member. The members of individual panels are selected by the registrar of the GMC. There were five members of this particular panel; two were doctors. The submission of the GMC was that both the lay and medical members of panels are experienced to deal with medical issues that arise in determining the fitness to practise of practitioners, and that no particular medical specialization is required. Hodge J accepted those submissions.

*Southall v. General Medical Council* [2010] EWCA Civ 407

**13.06** Panel—medical members—expertise of panel

The panel initially comprised two medical practitioners and three lay members. When the hearing resumed in 2007, only one of the medical practitioners (an orthopaedic surgeon) remained. Counsel, on appeal, suggested that this limited and called into question the expertise that the panel brought to the decision, which, as a result, required the Court to pay less deference to its conclusions. Rejecting this argument, Leveson LJ (with whose judgment Dyson and Waller LJJ agreed) noted that there was no rule that there should be a medical panellist whose speciality matches that of the practitioner. Leveson LJ continued:

> 67. Far from it being appropriate to have an expert from the same field, I consider the converse to be the case:... any issues requiring particular specialist knowledge should be dealt with through the calling of expert evidence; neither the GMC nor the doctor would be in a position to challenge the opinion of a member of the panel and, if a professional in the same field, the risk would be that a decision would be made on the basis of an expert view that had not been the subject of evidence or argument.

*Pugsley v. General Medical Council* [2010] EWHC 2247 (Admin)

**13.07** No requirement for same specialty in deficient professional performance case

P, a consultant cardiothoracic surgeon, appealed against the decision of the fitness to practise panel that his fitness to practise was impaired by reason of deficient professional performance. The panel imposed conditions on P's registration with immediate effect. Mr Michael Supperstone QC (sitting as a deputy judge of the High Court) said that the Court of Appeal in *Southall v. General Medical Council* [2010] EWCA Civ 407 had made it clear that there was no rule that there should be a medical panellist whose speciality matched that of the practitioner. The learned deputy judge said: 'It must follow that deference to the views of the Panel does not depend on the composition of the Panel.'

### *Yeong v. General Medical Council* [2010] 1 WLR 548

**13.08** Experience and expertise of panel

In dismissing Y's appeal, Sales J said, at [37]–[38], that he accepted the points made by the GMC—namely, that the panel was entitled to draw on its own experience and judgment, and was not obliged to accept the assessment of Dr K, the practitioner's expert psychiatrist; that the panel had in fact, in substance, given reasons for its own view where that view differed from Dr K's; and that there was no realistic basis on which there could be any expectation or possibility that if the impairment decision were quashed and the matter remitted to the panel, it would arrive at any different conclusion on the question of impairment. So far as concerns the panel's assessment of Dr K's evidence, the learned judge said that the panel was well entitled to draw upon its own experience and judgment in forming a view on whether Y presented a present risk to patients. Dr K's evidence was to the effect that Y did not suffer from any psychological disorder that underlay his misconduct. In the light of that assessment, Dr K's expression of opinion as to the risk posed by Y carried little weight attributable to any special expertise on the part of Dr K. The question of the possibility of a recurrence of such misconduct by Y was a matter of the ordinary assessment of likely human behaviour, in relation to which a psychiatrist's expertise conferred no special privileged insight. The assessment of risk of any particular form of future behaviour is the sort of task that courts and tribunals regularly perform without needing to refer to expert psychiatric evidence.

### *R (Al-Zayyat) v. General Medical Council* [2010] EWHC 3213 (Admin)

**13.09** Experience and expertise of panel

In quashing the panel's refusal to accede to the practitioner's application for voluntary erasure, Mitting J said, at [21], that one of the members of the panel was a consultant psychiatrist. He, like the other members of the panel, had witnessed the doctor's interview in private with the practitioner, with the panel and representatives listening in. Clearly, he was permitted to bring his own learning and experience to bear upon the critical question, identified by the legal assessor as being whether the alleged incapacity was genuine or fabricated, and whether the practitioner was unable to attend and/or effectively to participate in the proceedings.

### *R (Michalak) v. General Medical Council* [2011] EWHC 2307 (Admin)

**13.10** Substitution of panel member—effect on evidence—potential for injustice

In June 2010, disciplinary proceedings brought against M started before the fitness to practise panel. The panel comprised three members: the chairman, a medical member (Professor H) and a lay member. The hearing was lengthy. After thirty-four days and after twenty-eight witnesses had been called by the GMC in support of the allegations against M, the medical member, Professor H, became ill and it became clear that he would not be able to resume participation at any resumed hearing in the foreseeable future. The registrar decided to substitute another medical member so as to return the panel to three members and make it quorate. Kenneth Parker J, in judicial review proceedings brought by M to challenge the substitution and the panel's proposed procedure to rehear the evidence with the substituted medical member, held that there was a power under the rules to make the substitution in the interests of justice: General Medical Council (Constitution of Panels and Investigation Committees) Rules 2004, rule 4; and General Medical Council (Constitution of Panels and Investigation Committees) (Amendment) Rules 2010, rule 7. In principle, there was power to appoint panel members at any time and it was of considerable importance that, if the panel was potentially liable to be inquorate, either because of unavailability or ineligibility, the registrar should have the power to do so if it was in the interests of justice. What gave the learned judge great trouble, however, was the decision reached by the panel that the reconstituted panel should hear all of the witnesses again. The panel took the view that it would not be conducive to the promotion of justice for the new member simply to read the transcripts to bring himself up to speed. The learned judge said, at [17]–[18], that it is generally acknowledged that that is not a perfect solution because, in some cases, the way in which the evidence has been given can be significant and therefore a person who has not heard the witness may be at a disadvantage to those who have. Appellate courts time and again refer to the advantage that a first-instance judge has had over an appellate court in the consideration of evidence because the first-instance judge has heard the witness, and has observed the demeanour and the manner in which that evidence has been given. However, in some cases, having regard to the nature of the case and the evidence that has been given, the potential for injustice is relatively slight and the advantages for the administration of justice in general countervail such slight disadvantage that has been perceived. The learned judge considered that if two members of the panel were essentially to hear the prosecution evidence a second time, it would give rise to an impression of unfairness. But, more importantly, the third member would be put in an entirely different position: the third member would not have had the advantage of the first two in seeing those witnesses on the previous occasion. At [19], the judge said:

> In a case where the panel has already concluded that the nature of the evidence and the nature of the case is such that it is important that the panel members should have heard and seen the witnesses, it seems to me most unsatisfactory that one of them then will have seen the witnesses on one occasion and the other two will have seen the same witnesses on two occasions. There is a risk in such circumstances that different impressions of the weight of the evidence that has been given could be formed and that therefore there would be real potential for tension between the two original members and the third who is placed in an entirely different position.

The alternative was a fresh panel entirely and the matter was accordingly remitted to the defendant to consider beginning the case again before a fresh panel.

### *Martin v. General Medical Council* [2011] EWHC 3204 (Admin)

**13.11** Lay members—degree of deference

In dismissing M's appeal against the fitness to practise panel's decision on misconduct and sanction that M's name be erased from the medical register, Lang J said that, in her view, in considering whether a practitioner's fitness to practise is impaired by reason of misconduct, it was helpful for a panel to consider the GMC guide *Good Medical Practice* to identify what standards of behaviour are expected of registered doctors. In the instant case, the panel found that M had violated many of the duties in *Good Medical Practice* and the supplementary guidance *Maintaining Boundaries.* The panel concluded that M's conduct amounted to 'a serious departure from the standards of behaviour required of doctors'. The learned judge agreed. M also argued that since the three-person panel comprised two lay members, the degree of deference or respect afforded to the panel should be less, because the members were not medical specialists and a judge was just as well equipped as a lay member to make judgments about the public interest. Lang J did not accept this submission. A fitness to practise panel is a statutory committee established under the Medical Act 1983 to perform statutory functions in relation to the fitness to practise of registered doctors. It was Parliament's intention that the primary decision-making body in relation to fitness to practise was the panel and that the courts should have only an appellate function. The current fitness to practise scheme was introduced by amendment in 2004, by which time Parliament would have been aware of the requirements of Article 6 of the European Convention on Human Rights (ECHR) in relation to disciplinary hearings where erasure or suspension was an issue, because the Human Rights Act 1998 was by then in force. The introduction of lay panel members had been approved by the Privy Council and by Parliament: see the General Medical Council (Constitution of Panels and Investigation Committees) Rules Order 2004. The learned judge said that, in her view, the introduction of lay members into GMC panels did not justify a more interventionist role by judges, nor did it justify a departure from long-established principles on the proper approach to appeals from professional disciplinary bodies.

### *Lawrence v. General Medical Council* [2012] EWHC 464 (Admin)

**13.12** Experience and expertise of panel

The fitness to practise panel comprised three members, of whom one was lay and two were medical. The chairman of the panel was a former consultant psychiatrist, part-time judicial member of the Mental Health Review Tribunal, and former part-time psychiatrist member of the Parole Board. The other medical member was a general practitioner (see [150]). The practitioner, a consultant psychiatrist, challenged the panel's findings of fact, in particular on the grounds of its determination 'from its own experience'. Stadlen J said, at [229], that the rules of natural justice preclude members of a specialist tribunal including expert members from giving evidence to themselves that the parties have no opportunity to challenge, and that where a specialist tribunal drawing on its own knowledge and experience independently identifies an important fact or matter that may influence its decision, but which has not been the subject of evidence adduced by or submissions advanced by the parties, it should state this openly and give the parties an opportunity to seek to adduce evidence and/or to make submissions on it.

See further paragraph 27.09.

### *Russell v. Bar Standards Board*, 12 July 2012 (unreported)

**13.13** Arrangements for appointment of panel members—not part of Disciplinary Tribunals Regulations—de facto authority

The appellant, a non-practising barrister who had been a solicitor, appealed against the decision of a disciplinary tribunal of the Council of the Inns of Court dated 21 June 2010. The appellant appealed against the tribunal's decision finding a single charge proved on the grounds that the tribunal was not properly constituted. On 15 April 2010, the president of the Council of the Inns of Court issued a convening order, pursuant to the Disciplinary Tribunals Regulations 2009. By that convening order, the president nominated five persons to constitute the tribunal—namely, a judge, two lay members, and two barristers. Both barristers satisfied the requirements in the 2009 Regulations of being not less than seven years' standing. S, one of the barrister members, had been added on 2 May 2001 as a panel member for disciplinary tribunals. He had remained on the list ever since. On 10 May 2006, the Council of the Inns of Court adopted new arrangements for establishing tribunal members, which provided that, in relation to barristers, existing panel members were to be permitted to remain on the panel for up to three years. Consequently, if the new arrangements were to be applied to S, he would, as an existing panel member, have remained on the list of barrister volunteers only until 10 May 2009. The tribunal in the instant case sat on 17–20 May 2010 and 21 June 2010. The appellant contended that because S was no longer a person who, under the current arrangements, was eligible for inclusion on the list of barrister members, his nomination by the president was ultra vires the 2009 Regulations. In delivering the decision of the Visitors, Sir Rabinder Singh said that they did not accept the appellant's argument that the 2009 Regulations and the new arrangements for vetting and appointing tribunal members should be read together as constituting a 'code' governing the composition of disciplinary tribunals. In the view of the Visitors, it is the Regulations that govern the powers of the president in nominating members to sit on a tribunal. There is no further limitation or restriction on his powers over and above the Regulations. There is not even any cross-reference to the new terms of appointment, so it cannot be said that they have been incorporated by reference into the requirements of the Regulations. In order to be validly constituted as a matter of law, the tribunal members had to meet the requirements of the Regulations, which included the requirement that a barrister member should be one of seven years' standing, as S was. S was also nominated by the president of the Council of the Inns of Court in accordance with the Regulations. The fact that he was no longer eligible under the new arrangements for inclusion in the list of barristers volunteering for hearings did not render his nomination by the President invalid. The Regulations do not require the persons nominated to be on the list of barristers vetted and approved by the current arrangements. Accordingly, the tribunal was properly constituted in accordance with the Regulations. Further, and in any event, S had de facto authority to sit as a member of the tribunal. He had colourable authority in accordance with well-established cases such as *Fawdry v. Murfitt* [2003] QB 104, *Coppard v. Customs & Excise* [2003] QB 1428, and *Baldock v. Webster* [2006] QB 315. S had been included on a list of volunteers in 2001 and met the criteria. He had received guidance on sentencing. He believed himself to be validly appointed, which is why he sat. He sat with four other members, including a judge, all of whom must have assumed that he was validly sitting or they would not have continued the hearing with him. The appellant had been served with the convening order, which named S as a volunteer. All of these factors demonstrated that he was not a 'usurper' exercising an authority that he knew he did not possess or as to which he was wilfully blind.

***Temblett v. Bar Standards Board*, 26 July 2012 (unreported)**

**13.14** Appointment of tribunal members—whether validly appointed

T, a barrister, applied to adjourn the hearing of his appeal against a decision of a disciplinary tribunal in relation to findings against him of professional misconduct. T submitted that the decision of the Visitors in *Russell v. Bar Standards Board* was distinguishable, because it involved one member of the tribunal, whereas the present appeal concerned four members; also, the instant case was concerned with the Disciplinary Tribunals Regulations 2005, whereas *Russell* was concerned with the Disciplinary Tribunals Regulations 2009. Sir Andrew Nicol, sitting as Visitor to the Inns of Court, refused T's application on the grounds that *Russell* was not distinguishable.

***R (Leathley and ors) v. Visitors of the Inns of Court* [2013] EWHC 3097 (Admin)**

**13.15** Tribunal members—term expired—whether tribunal established by law

The claimant barristers sought permission to challenge, by way of judicial review, findings in relation to professional misconduct. The most significant issue related to the constitutions of the disciplinary tribunals convened to hear the charges against them and of the Visitors of the Inns of Court who heard their appeals. The claimants submitted that some of the members of the disciplinary tribunals, and of the Visitors, were not qualified to sit because the limited duration of their eligibility to sit had expired. Accordingly, they were not tried by a tribunal established by law, within the meaning of Article 6 ECHR; the tribunals had no power to try them and the Visitors had no power to uphold the findings. In dismissing the applications, Moses LJ (with whom Kenneth Parker J agreed) said that the sole requirements for appointment to any particular disciplinary tribunal were those contained in the Disciplinary Tribunals Regulations 2009 themselves and in the Hearings before the Visitors Rules 2005. The legal authority to sit is derived from those Regulations and Rules, and not from the arrangements of the Council of the Inns of Court (COIC) constituting the pool of those who have been appointed by the COIC. The Court reached this conclusion with regret. The Bar must surely be at the forefront of setting standards as to how institutions should regulate themselves.

Anyone reading the COIC's constitution, the relevant guidance, and the memorandum of understanding would gain the clearest impression that the qualifications devised by the COIC for eligibility to the pool were themselves the qualifications for appointment to disciplinary tribunals or Visitors' hearings. In the light of this conclusion, it was unnecessary to consider the alternative argument advanced by the Bar Standards Board—that the consequences of a defective judicial appointment can be avoided by the doctrine of de facto authority. After reviewing *Fawdry v. Murfitt* [2003] QB 104, *Coppard v. Customs & Excise* [2003] QB 1428, *Baldock v. Webster* [2006] QB 315, and *Sumukan Ltd v. Commonwealth Secretariat (No. 2)* [2007] EWCA Civ 1148, the learned judge said that the doctrine is not designed to *ratify* a retrospective action, but rather to *recognize* authority, where, by virtue of the doctrine, such authority exists at the time that the impugned tribunal is appointed. It is recognition of an authority that exists at the time that the tribunal whose authority is impugned is formed or established. Further, there is no reason to conclude that the doctrine offended Article 47 of the Charter of Fundamental Rights of the European Union ('Everyone is entitled to a fair and public hearing within a reasonable time by an independent and impartial tribunal *previously* established by law'). The doctrine recognizes an authority that existed at the time the tribunal in question was appointed. In the instant case, the nomination of the members whose authority was impugned derived from the fact that all believed them to be qualified and that it was on this basis that they were nominated. Had it been necessary, the Court would have upheld their authority to make the decisions in the instant cases.

### *R (Mehey and ors) v. Visitors of the Inns of Court* [2014] EWCA Civ 1630

**13.16** Appointment of tribunal members—whether validly appointed

Two of the claimants in *R (Leathley and ors) v. Visitors of the Inns of Court* [2013] EWHC 3097 (Admin) sought permission to appeal, together with another barrister, challenging the validity of disciplinary proceedings brought against them on the ground that some of the individuals who heard those proceedings or appeals therefrom were disqualified from sitting. In refusing permission to appeal, Jackson LJ (with whom Ryder and Sharp LJJ agreed) said that the appellants had no prospect of successfully appealing on the grounds that certain disciplinary tribunal members or Visitors were not members of the COIC pool. Their membership of the Bar Standard Board's PCC was invalid. The scheme set up by the COIC created a pool of barristers, lay representatives, and others from which nominations could safely and properly be made for the purposes of disciplinary hearings and appeals. There was no absolute ban on appointing persons from outside the pool as members of disciplinary tribunals or as Visitors. On occasion, it may be appropriate to appoint a barrister or lay representative with particular expertise that was not available within the COIC pool. The court of the Visitors would step in to protect a defendant barrister if an ineligible person were appointed to sit. This is precisely what happened in the case of *P (a barrister) v. General Council of the Bar* [2005] 1 WLR 3019. His Lordship did not accept the proposition that a disciplinary tribunal or a panel of Visitors appointed from barristers or lay representatives outside the pool would not be independent or would not be guaranteed to be free from outside pressure.

### *R (Hill) v. Institute of Chartered Accountants in England and Wales* [2014] 1 WLR 86, CA

**13.17**

See paragraph 43.15.

## C. Other Relevant Chapters

**13.18** Bias, Chapter 6
Independent and Impartial Tribunal, Chapter 35
Jurisdiction, Chapter 43
Natural Justice, Chapter 48

# 14

# CONVICTION AND CAUTION CASES

## A. Legal Framework

**14.01** Examples include:

Rehabilitation of Offenders Act 1974 (Exceptions) Order 1975 (excepts most regulatory bodies from the operation of the 1974 Act)
Medical Act 1983, section 35C(2)(c) (a person's fitness to practise shall be regarded as impaired by reason of a conviction or caution)
Powers of Criminal Courts (Sentencing) Act 2000, section 14 (an absolute or conditional discharge shall not been deemed to be a conviction for any purpose other than the purposes of the proceedings in which the order is made)
Home Office Circular 16/2008, *Cautioning of Adult Offenders*, paragraph 34 ('A Simple Caution is not a form of sentence (which only a court can impose), nor is it a criminal conviction. It is, however an admission of guilt... [and] forms part of an offender's criminal record...')
General Pharmaceutical Council (Fitness to Practise and Disqualification etc) Rules 2010, rule 24(4) and (5)
Pharmacy Order 2010, article 51(1)(e)
Bar Standards Board Disciplinary Tribunals Regulations 2014, rule E146
General Medical Council (Fitness to Practise) Rules 2014, rule 34(3) (production of a certificate confirming that a person has been convicted shall be conclusive evidence of the offence committed, and the only evidence that may be adduced in rebuttal is evidence for the purposes of proving that the practitioner is not the person referred to in the conviction)

## B. Disciplinary Cases: Principles

### *Ong Bak Hin v. General Medical Council* [1956] 1 WLR 515

**14.02** Admission of criminal record before committee—consent—no necessity for witnesses to give oral evidence

The appellant had been convicted and sentenced by the Supreme Court of Malacca on 14 August 1953, for performing an operation with intent to cause a miscarriage, which caused the death of a woman contrary to section 314 of the Penal Code for the Federation of Malaya. The formal charge before a disciplinary committee of the General Medical Council (GMC) alleged that, on 18 May 1953, the appellant, being registered under the Medical Acts, unlawfully performed an operation of abortion with intent to cause the miscarriage of G; that the operation caused her death, and thereby the appellant committed an offence under section 314 of the Penal Code of the Federation of Malaya of which he was convicted in the High Court of Malacca on 14 August 1953 and sentenced to five years' imprisonment (reduced on appeal to two years), and that, in relation to the facts alleged, he was guilty of infamous conduct in a professional respect. The evidence sought to be adduced in support of the charge consisted of a statutory declaration by the registrar of the Court of Appeal for the Federation of Malaya, to which was exhibited (a) a certificate by the assistant registrar of the Supreme Court of Malacca of the conviction, and (b) the record of the hearing in the High Court consisting of the charge, notes of evidence made by the trial judge, the judge's reasons for sentence, the list of exhibits at the trial, a statement of the deceased woman, extracts from medical textbooks put in at the trial, copies of notice and petition of appeal to the Court of Appeal, and certified copies of the order and judgment of the Court of Appeal. At the inquiry before the GMC, the appellant's solicitor submitted at the outset that justice could not be done to his client unless the witnesses who had given evidence at his trial were called to give oral evidence before the committee. He agreed that the documentary evidence was admissible, but contended that the committee, in its discretion, should exclude it in the interests of justice. He expressly stated that he did not ask for an adjournment. The legal assessor pointed out that the committee had no power to compel the attendance of witnesses from Malaya and advised the committee to proceed on the documentary evidence, although certain passages were deleted or covered up at the request of the appellant's solicitor. The summing up of the judge at the trial was also omitted. Lord Tucker, giving the reasons for dismissing the appeal, said that, in the present case, the record with certain excisions was put in by consent as soon as the committee had quite properly decided that the attendance of the witnesses from Malaya was not practicable and that their absence did not vitiate the proceedings. The consent to the admission of the record was fatal to any objection on that score now being put forward before the Board.

### *Jeyaretnam v. Law Society of Singapore* [1989] AC 608

**14.03** Criminal conviction bad in law—no opportunity to appeal conviction—exceptional circumstances

On 19 October 1987, the High Court of Singapore ordered that J, a solicitor, be struck off the roll of advocates and solicitors of the Supreme Court of Singapore. The Attorney-General of Singapore had reported J to the Law Society following his conviction of three offences under section 421 and one offence under section 199 of the Singapore Penal Code. The section 421 offences related to the dishonest disposal of three cheques allegedly the property of the Workers' Party of Singapore of which J was the secretary-general and one charge relating to a purported false statutory declaration. Realistically, the only chance that J had of avoiding being struck off was if the Court could be persuaded to go behind the convictions. At the outset of the hearing, objection was taken to Wee Chong Jin CJ sitting, on the ground that he had been involved in the earlier criminal proceedings, and the Court in the disciplinary proceedings was to be invited to go behind the convictions, and condemn Wee Chong Jin CJ's decision to refuse to reserve questions of law to the Court of Criminal Appeal. In *Ratnam v. Law Society of Singapore* [1976] 1 MLJ 195, Lord Simon of Glaisdale, delivering the judgment of the Board, said that although it was open to go behind a conviction, this would be justified only in exceptional circumstances. For example, if a plea of guilty had been made under a misunderstanding and there was no opportunity of rectifying it on appeal, justice would demand that the conviction should not be conclusive against the accused in the course of disciplinary proceedings, the object of which themselves is, after all, to promote justice. In the instant case, their Lordships considered the circumstances sufficiently exceptional at least to warrant examination on the grounds that the convictions being attacked were bad in law. The conviction of J on the accounts charge depended on a construction of section 199 of the Penal Code first propounded by Wee Chong Jin CJ, which was attacked as bad in law; J had had no opportunity to test any of the questions of law by appeal to the Court of Criminal Appeal of Singapore because Wee Chong Jin CJ refused to reserve any questions of law and, in the absence of such reservation, neither the Court of Criminal Appeal nor the Board had any jurisdiction to entertain any appeal. In the event, the Board held that the charge under section 199 was misconceived in law and J's conviction was fatally flawed. In relation to the cheque charges under section 421, no offence by J was committed. The Workers' Party never had more than a defeasible title to the proceeds of the cheques and, before the title was perfected, the cheque was in each case lawfully disposed of in accordance with the donor's instructions. J was innocent of any offence under section 421 and had a good defence on the merits to the charge under section 199. He had suffered a grievous injustice, had been fined, imprisoned, and publicly disgraced for offences of which he was not guilty, and had been disqualified for a year from practising his profession.

### *Shepherd v. Law Society* [1996] EWCA Civ 977

**14.04** Criminal conviction—collateral attack—abuse of process

S sought leave to appeal against the rejection by the Divisional Court of his appeal against the findings and order of the Solicitors Disciplinary Tribunal (SDT) when he had been struck off following his conviction of fifteen offences of dishonesty and sentence of three years' imprisonment in respect thereof. Before the Tribunal, a certificate of conviction evidencing the convictions and sentence had been admitted without objection. The ground of appeal advanced related to the Tribunal's refusal to allow S to adduce evidence in support of his assertion that he was not, in fact, guilty of the offences of which he had been convicted. The Divisional Court, in a reserved judgment delivered by Lord Taylor of Gosforth CJ, dismissed the appeal and refused leave to appeal. The Divisional Court accepted that the effect of the Solicitors Disciplinary Proceedings Rules 1994, which incorporated section 11 of the Civil Evidence Act 1968, was that a certificate of conviction was admissible to prove the fact of the convictions and as prima facie evidence that S was guilty of the offence. It was prima facie evidence only because it was subject to the qualification in section 11(2)(a) 'unless the contrary is proved'. However, the charge that brought discredit on S and the profession, and which rendered him guilty of conduct unbefitting a solicitor, was the very fact that he had been convicted and sentenced to imprisonment, and the practice of the Tribunal not to go behind a conviction unless there were exceptional circumstances was lawful and justified. The cases of *Hunter v. Chief Constable of the West Midlands Police and ors* [1982] AC 529 and *Smith v. Linskill (A Firm)* [1996] 2 All ER 353 showed that it was an abuse of process to challenge in a civil action a previous decision of a court of competent jurisdiction. The Court of Appeal (Leggatt and Hutchison LJJ), in dismissing S's application for leave to appeal, said that a particularly important part of the judgment of Lord Taylor of Gosforth CJ was as follows:

> Public policy requires that, save in exceptional circumstances, a challenge to a criminal conviction should not be entertained by a Disciplinary Tribunal for the reasons quoted above from the Master of the Rolls' judgment. If this appellant's argument were right, he should have been allowed to challenge his conviction before the Tribunal even if he had appealed unsuccessfully to the Court of Appeal, Criminal Division. That could, in theory, have led after a conviction by a jury on the criminal burden of proof, upheld by three Appeal Court Judges, to exoneration by a Disciplinary Tribunal on the civil burden of proof. Moreover, to achieve it, the witnesses from the criminal case would have had to undergo the trauma of a rehearing. In the absence of some significant fresh evidence or other exceptional circumstances such an outcome could not be in the public interest. Here the appellant had not even applied for leave to appeal. There were no exceptional circumstances. What he wished to do was to have a rehearing of the criminal trial in which he could conduct his own case, as he submitted to us, better than his leading counsel. We are in no doubt that the Tribunal were right to refuse an adjournment and to refuse the appellant an opportunity to mount such an operation.

### *Kirk v. Royal College of Veterinary Surgeons* [2004] UKPC 4

**14.05** Appellant cannot go behind conviction—evidence of underlying facts admissible if not inconsistent with conviction

K was convicted of eleven criminal offences that the Disciplinary Committee of the Royal College of Veterinary Surgeons (RCVS) found rendered him unfit to practise veterinary surgery and the Committee directed that K's name be removed from the register. Delivering the judgment of the Board, Lord Hoffmann said, at [6]–[8], that the legal effect of a statute such as section 16(1)(a) of the Veterinary Surgeons Act 1966, which entitles the Disciplinary Committee to find that a conviction for a criminal offence renders a registered veterinary surgeon unfit to practise is to preclude the practitioner from denying the truth of any facts necessarily implied in the conviction. As Viscount Simon LC said in *General Medical Council v. Spackman* [1943] AC 627,

634–5: 'The decision of the council is properly based on the fact of the conviction, and the practitioner cannot go behind it and endeavour to show that he was innocent of the charge and should have been acquitted.' On the other hand, rule 8 of the Veterinary Surgeons and Veterinary Practitioners (Disciplinary Committee) (Procedure and Evidence) Rules 1967 permits the College and the practitioner to adduce evidence about the underlying facts upon which the conviction is based, provided that the facts which such evidence is relevant to prove are not inconsistent with the finding that the practitioner was guilty of the offence. What the practitioner cannot do is to relitigate the conviction before the Committee. In the instant case, the convictions were for common assault, threatening behaviour, resisting a constable in the execution of his duty, and an offence under section 4 of the Public Order Act 1986. In addition, K had been convicted of an offence of failing to prevent the deposit of controlled waste, a planning application matter, and road traffic offences. In dismissing K's appeal, their Lordships said, at [33], that they were very conscious that deprivation of K's services as a veterinary surgeon would be a loss to the animal-owning public in South Wales and that this could be said to be contrary to the public interest:

> On the other hand, veterinary surgeons as professionals have wider duties than the care of animals. They are expected to conduct themselves generally in accordance with the standards of professional men and women and failure to do so may reflect upon the reputation of the profession as a whole. If for example, Mr Kirk had been found guilty of serious dishonesty, there can be no doubt that the Committee would have been entitled to take the view that he was unfit to be a member of the profession.

Their Lordships said that they found it 'difficult to say that violent or anti-social behaviour of the kind involved in Mr Kirk's convictions cannot in principle be a ground for a finding that he is unfit to practise as a member of the profession'.

### *Stannard v. General Council of the Bar*, 24 January 2006 (unreported)

**14.06** Evidence to go behind conviction—need for exceptional circumstances

S was found guilty before a disciplinary tribunal of a charge of engaging in dishonest conduct contrary to paragraph 201(a)(i) of the Code of Conduct of the Bar of England and Wales (5th and 6th edns). The particulars of the offence were that, between 1 January 1993 and 31 October 1997, with intent to defraud, S cheated Her Majesty The Queen and the Commissioners of Inland Revenue of public revenue—namely, corporation tax—for two offences of which he was convicted at Southwark Crown Court before Judge Fingret and a jury. He was sentenced to four-and-a-half years' imprisonment and nine years' disqualification under the Company Directors Disqualification Act 1986. A central issue on appeal before the Visitors to the Inns of Court (Hart J, chairman) was the extent to which it was open to the Visitors to go behind the conviction. S's submission was that the question of his honesty or dishonesty in relation to the events in question should be approached by the Visitors with an open mind and that, while they could pay regard to the jury's verdict, they should nevertheless come to an independent judgment on the whole of the evidence, including specialist knowledge of the workings of tax schemes and developments in the law since the date of his trial. In dismissing S's appeal, the Visitors said that they saw no significance in the fact that, in *Shepherd v. The Law Society* [1996] EWCA Civ 977, the solicitors' disciplinary rules had a statutory origin, whereas the Bar's do not. Precisely the same considerations of policy apply to the approach to be taken by the disciplinary tribunal of the Bar as were held by the Court of Appeal in *Shepherd* to apply in the case of the solicitors' disciplinary tribunals. Accordingly, as a matter of law, the tribunal was entitled to refuse to hear evidence that sought to go behind the conviction unless there were exceptional circumstances. The Visitors accepted that even if they were to reject S's primary submission (as they indeed did), they were still entitled to look afresh at the evidence underlying his conviction if he could demonstrate exceptional circumstances: this was consistent with *Smith v. Linskill* [1996] 2 All ER 353, *Shepherd*, and the course taken by the Privy Council in *Jeyaretnam v. Law Society of Singapore* [1989] 2 All ER 193. However, there were no such exceptional circumstances present here. This was not a case in which S was seeking to adduce any new evidence such as might 'entirely change the aspect of the case' (see *Smith v. Linskill*, at 360F). He was seeking to say that, on the evidence before the jury, he ought not to have been convicted of dishonesty, or that the Court of Appeal had been wrong to find that the jury had been properly directed. Developments in the law since the date of S's trial did not help and, in any event, the Court of Appeal in its judgment assumed the facts in S's favour. The decision of the jury, confirmed on appeal to the Court of Appeal, was that S had been guilty of offences of dishonesty. 'Dishonesty' in the Bar's code of conduct does not bear some special meaning peculiar to barristers, whether they be specialist tax experts or otherwise.

### *R (Jenkinson) v. Nursing and Midwifery Council* [2009] EWHC 1111 (Admin)

**14.07** Conviction quashed on appeal—effect on decision by disciplinary committee

Following her conviction in 1996 for causing grievous bodily harm with intent, the claimant had been found guilty in 1998 of misconduct by the Professional Conduct Committee (PCC) of the Nursing and Midwifery Council (NMC) and had been struck off the nursing register. In 2005, on referral from the Criminal Cases Review Commission, the Court of Appeal, Criminal Division, quashed the conviction when it became clear that the expert evidence founding the conviction—namely, how the ventilator of a patient in her charge operated—was erroneous. Thereafter, the claimant sought to have the Committee's decision to strike her off the nursing register set aside. The subsequent Committee accepted the advice of its legal assessor that it had no jurisdiction to set its original decision aside and declined to do so. The NMC ultimately supported the claimant's judicial review application and sought guidance as to how it should deal with situations such as this. Cranston J, in granting the application and quashing the original decision to strike the claimant off the nursing register, said that it was unwise for the Court to provide specific guidelines. However, it was plain from *Akewushola v. Secretary of State for the Home Department* [2000] 1 WLR 2295, CA, and *Wade and Forsyth on Administrative Law*, 9th edn (Oxford: Oxford University Press, 2004), p. 262, that the powers of the NMC were not stillborn and that, in cases of accidental slips, mistakes, flaws, or miscarriage of justice, it had the power to act and rectify a mistake. It was clear on the facts of the instant case that the original decision that the claimant was guilty of misconduct, so that it was appropriate to strike her off the nursing register, was based on a mistake—namely, that she was guilty of a criminal offence. Once that conviction was quashed, the subsequent finding of misconduct and sanction fell away. Accordingly, the original decision amounted to a miscarriage of justice based upon a mistake.

### *Hurnam v. Bar Standards Board*, 6 February 2012 (unreported)

**14.08** Appellant alleged wrongly convicted—no exceptional circumstances

H, a barrister who had practised for many years in Mauritius, appealed against a decision of the Disciplinary Committee of the Council of the Inns of Court at which the Committee found him guilty of professional misconduct contrary to paragraph 301(a)(i) of the Code of Conduct of the Bar of England and Wales (8th edn). The Committee found that, on 30 January 2008, H had been convicted by the Intermediate Court of Mauritius of the offence of conspiracy—namely, to hinder police in an enquiry regarding a larceny committed at the Grand Bois State Bank by fabricating an alibi for his client—and his name was ordered by the Supreme Court of Mauritius to be erased from the roll of barristers. The Committee disbarred H. On appeal to the Visitors, H's main point was that he was wrongly convicted by the Intermediate Court and many of his arguments related to subsequent proceedings in which he attempted unsuccessfully, either directly or collaterally, to challenge that conviction. Rule 14(8) of the Hearings before the Visitors Rules 2010, which reflect the position derived from the case law, provides that: 'An appellant or defendant (as the case may be) may only challenge before the Visitors a decision of a court of law on which the relevant decision was based in exceptional circumstances and with the consent of the Visitors.' There were no exceptional circumstances to justify going behind H's conviction. The Supreme Court in Mauritius validly ordered that his name be erased from the roll of barristers admitted to practise in Mauritius. The Disciplinary Committee was therefore fully entitled to find H guilty of professional misconduct.

### *Nwogbo v. General Medical Council* [2012] EWHC 2666 (Admin)

**14.09** Oral evidence of circumstances surrounding offence—purpose of evidence

On 21 April 2010, N was charged by Greater Manchester Police with an offence of assaulting his wife at their home. The incident was alleged to have taken place on 31 March 2010. N admitted that there had been an incident, but he always denied and continued to deny that he was guilty of any assault. At the time, N was employed by the Clatterbridge Centre for Oncology National Health Service (NHS) Foundation Trust, where he held a staff-grade position in oncology. On 4 August 2010, the criminal charge was tried in the magistrates' court and N was convicted of an offence under section 39 of the Criminal Justice Act 1988. Following a pre-sentence report, N was sentenced on 22 September 2010 to a term of imprisonment of two months, suspended for eighteen months, with a requirement that he be under the supervision of the Probation Service and that he attend a domestic violence programme. N's appeal to the Crown Court was dismissed on 21 January 2011. His Honour Judge Stephen Davies (sitting as a judge of the High Court) said that, under rule 34(3) of the General Medical Council (Fitness to Practise) Rules 2004, the conviction for the criminal offence of assault by battery was conclusive evidence that N committed that offence. It was not therefore open to him as a matter of law to challenge the allegation that he committed the offence, even though he sought to do so before the fitness to practise panel, and nor would it be open to him to challenge the panel's decision on that point before the Court.

> 14.... There is, however, a question which the appellant has raised as to whether or not the panel were entitled to go on to make adverse findings against him in relation to the circumstances surrounding the assault and his response to his conviction. In his written supplementary submissions the appellant has complained that the panel placed undue weight on the evidence of the attending officers, whose evidence was inconsistent, and that the panel unfairly interrupted his cross-examination of those officers. He further submitted that the panel was wrong to have regard to evidence in relation to the surrounding circumstances going beyond the fact of the conviction itself.
>
> 15. In considering those complaints, I have reminded myself of what counsel then instructed by the GMC said to the panel in her opening address about why the attending officers were being called. She said that the purpose of so doing was to ensure that the panel had the fullest picture of the circumstances of and around the index offence and the information it needed to deal with the matter fully and fairly. That, as it seems to me, was an approach which the GMC was fully entitled to take in a case such as this....
>
> 16. It is apparent, therefore, that the panel [in its decision in relation to impairment] accepted the evidence of the police officers that the appellant's attitude towards them was obstructive and difficult. That is clearly, in my judgment, a conclusion which

> they were entitled to reach.... the panel was fully entitled to accept [the evidence of the police officers] as credible, notwithstanding that the appellant was able, as he has done, to point to some inconsistencies in their evidence....

The Court continued to affirm that the panel was obviously entitled to place reliance on the findings of the magistrates' court, expressed in a memorandum of conviction, in relation to the circumstances of the conviction and that it was also entitled to place reliance upon the contents of the pre-sentence report, which had been placed before the magistrates in relation to, as they said, the appellant's continuing unwillingness to accept the full facts of the assault and his conviction for the offence. There was some issue as to whether a letter written by a probation officer, which dealt with the supervision element of the order made by the magistrates, and said that the appellant's response to this had always been good and that he expressed some regret for his behaviour, was put before the panel. Even if it was, the Court considered that all that it demonstrated was that the appellant had completed the course, that he was informed as to the relevant issues, and that he had expressed some regret for his own antisocial behaviour (not his responsibility for the assault). None of that was seen to detract in any way from the panel's conclusion that the appellant was still unable to accept the full facts of the assault and his conviction. Accordingly, the Court rejected the appeal in so far as it complained either of the panel's approach and findings in relation to the circumstances of the assault, or to the appellant's attitude to that assault and subsequent conviction.

### *General Dental Council v. Jamous* [2013] EWHC 1428 (Admin)

**14.10** Unlawful provision of dentistry—breach of Dentists Act 1984

The General Dental Council (GDC) appealed, by way of case stated, a district judge's dismissal of criminal proceedings. The defendant, who offered to the public the service of tooth-whitening, was not a registered dentist. The Divisional Court, in reversing the decision of the district judge, held that the all-important statutory question under section 37 of the Dentists Act 1984 was whether tooth-whitening treatment is usually performed by a dentist. Based on the expert evidence and guidance before the court, the evidence did prove that the tooth-whitening treatment given by the defendant was the practice of dentistry and therefore that the defendant was guilty of a breach of sections 38 of the Dentists Act 1984, which prohibits the practice of dentistry by laymen, and 40, which prohibits anyone being treated in the course of the business of dentistry.

### *R (H) v. Nursing and Midwifery Council* [2013] EWHC 4258 (Admin)

**14.11** Criminal conviction—admissibility—cross-examination

The NMC alleged that H had engaged in a sexual relationship with A whilst he was under her care and she was treating him. H alleged that the relationship began subsequently. The charge hinged on whether the panel preferred one version of events over the other. There was a straight conflict between the evidence of A and H. Their accounts were fundamentally irreconcilable. In these circumstances, where there is a conflict of evidence of the sort identified by the panel, the conventional approach to determining issues of fact between the parties involves primarily testing the relevant contentions of the parties by reference to such contemporaneous documentation as is available. An application was made to the panel on behalf of H to adduce evidence of the criminal convictions of A and to cross-examine him about them. His Honour Judge Pelling QC (sitting as a judge of the High Court) said that it was unclear why it was thought necessary to make this application at all. Rule 31 of the NMC (Fitness to Practise) Rules 2004, the rule under which the application was made, is not concerned with the admissibility of particular facts or matters, but the nature of the evidence relied on as proving the factual allegation or assertion made. In any event, the only basis on which it could be contended that the fact of a criminal conviction ought to be excluded was on grounds of relevancy or inadmissibility by operation of statute, for example the rehabilitation of offenders legislation. Convictions for the offences of dishonesty or convictions for any offence following a plea of not guilty, in which the person concerned gave evidence that was disbelieved or, for that matter, evidence that a person's evidence was disbelieved in the civil case, are material that is at least potentially relevant to an assessment of credibility.

### *Samuel v. Royal College of Veterinary Surgeons* [2014] UKPC 13

**14.12** Conviction—underlying facts—finding of unfitness to practise quashed

S, a registered veterinary surgeon, pleaded guilty at Cardiff Magistrates' Court on 22 November 2011 to theft of a camera and memory card contrary to sections 1 and 7 of the Theft Act 1968, common assault contrary to section 29 of the Criminal Justice Act 1988, and using threatening, abusive, or insulting words or behaviour contrary to section 4 of the Public Order Act 1986. He was sentenced to concurrent terms of twenty-eight days' imprisonment for theft and common assault, and twelve weeks' imprisonment for the public order offence, all suspended for twelve months. He was also ordered to carry out 140 hours' unpaid work and to pay £75 compensation to the victim. On 18 February 2013, the Disciplinary Committee of the RCVS found S unfit to practise and, the following day, directed that his name be removed from the register. From the outset of the disciplinary proceedings, S had made his case plain that he had been provoked by racial abuse and had responded in a way that he regretted. The Board concluded that the Committee's finding of unfitness to practise could not fairly stand. On any view, what happened was a spontaneous outburst in the course of an angry quarrel between neighbours. It was a nasty occurrence, but not one that would be expected to cause a profound effect on the victim. The snatching of the camera, which led to the conviction of theft, was done with the aim of removing the SIM card recording photos that the victim had taken; it was not an act carried out for financial gain. The SIM card was recovered from S when he was searched at the police station and the camera itself was returned. S was a man of good character. The Board added that if it had upheld the determination of unfitness to practise, it would have concluded that the sanction of removing S's name from the register was disproportionately severe.

## C. Absolute or Conditional Discharge

### *Simpson v. General Medical Council* (1955) The Times, 9 November

**14.13** Probation order—effect of conviction and sentence

The appellant medical practitioner, at Chelmsford Assizes in November 1954, was charged with, and pleaded guilty to, a number of very grave offences against his female patients. Having pleaded guilty, he was duly convicted, but the learned judge, having heard medical evidence, did not pass any sentence upon him, but placed him on probation, the probation order being that he should submit to treatment as a resident patient at a mental hospital. The result of these proceedings was that, by virtue of section 12 of the Criminal Justice Act 1948, the appellant's conviction could not be deemed to be a conviction for any purpose other than for the purposes of the criminal proceedings and must be disregarded for the purposes of any enactment that imposed or required the imposition of any disqualification (provisions now contained in section 14(1) and 14(2) of the Powers of Criminal Courts (Sentencing) Act 2000). Section 29 of the Medical Act 1858 (the then relevant statutory provision) provided that if any registered medical practitioner shall be convicted of any crime or offence, or shall, after due enquiry, be judged by the GMC to have been guilty of infamous conduct in any professional respect, the GMC may, if it sees fit, direct the registrar to erase the name of such medical practitioner from the register. Viscount Simonds, delivering the judgment of the

Privy Council and giving the advice of the Board that the appeal of the practitioner should be dismissed against an order of erasure, said:

> [B]y virtue of section 12 of the Criminal Justice Act 1948 ... it was therefore not open to the Committee whose duty it was to review the conduct of the appellant to proceed upon the footing that he had been convicted of a crime. It was for them to determine after due enquiry whether he had been guilty of infamous conduct in any professional respect and, if they so determined, then, if they saw fit, to direct the Registrar to erase his name from the Register. It was agreed by the appellant before the hearing by the Committee that the depositions of certain witnesses taken at the Magistrates' Court should be put in as evidence and at the hearing the facts alleged in the charge were agreed and admitted on his behalf... The Medical Acts are designed at the same time to protect the public and to maintain the high professional and ethical standard of an honourable calling. If a practitioner, having committed grave offences of which the appellant has been guilty, call upon such a plea successfully to resist the charge of infamous conduct and the erasure of his name from the Register, the public will lack their proper protection and the honour of the profession may be endangered by the continued practices of one who can still claim to be of their number.

### *R v. Statutory Committee of Pharmaceutical Society of Great Britain, ex parte Pharmaceutical Society of Great Britain* [1981] 1 WLR 886

**14.14** Conviction for unlawful wounding—conditional discharge—effect of sentence—tribunal not prevented from acting on facts leading to conviction

The respondents, three students at the London School of Pharmacy who were student members of the Pharmaceutical Society, were involved in a fracas in which another student was seriously injured. The respondents were convicted at the Central Criminal Court of unlawful wounding, contrary to section 20 of the Offences against the Person Act 1861, but each was conditionally discharged. The Statutory Committee sat on 12 March 1979 and the legally qualified chairman ruled that the Committee did not have jurisdiction to hear the complaints because, under section 13(3) of the Powers of Criminal Courts Act 1973 (now section 14(3) of the 2000 Act), there was no jurisdiction to allow evidence of the facts that led to the conviction to be adduced before the Committee. On an application for judicial review by the Pharmaceutical Society of Great Britain, Lord Lane CJ said that he assumed, for the purposes of argument, that the removing of a name from the register was a disqualification or disability within the provisions of the 1973 Act, although it was perhaps a somewhat tortuous way of describing it, 'but it seems to us clear that whatever Section 13 may do, it does not purport to prevent a tribunal such as this one from acting upon the facts which underly the finding of guilt'. The Court held that a tribunal such as the Society's Statutory Committee was not prevented by section 13(3) of the 1973 Act from acting on facts leading to a conviction in a criminal court, since the Committee was merely prohibited by section 13(3) from relying on a conviction itself as evidence of misconduct. Lord Lane, at 811C, said that if it had been intended that not only the conviction, but also the facts underlying the conviction, should be disregarded in any future proceedings, then the 1973 Act should have said so and it did not.

### *R v. Secretary of State for the Home Department, ex parte Thornton* [1987] QB 36

**14.15** Conditional discharge of criminal offence—effect on subsequent police disciplinary proceedings

The applicant was a serving police officer and was subject to the Police (Discipline) Regulations 1977. He was convicted of an offence contrary to section 5(2) of the Criminal Law Act 1967 and given a conditional discharge. He was then charged with an offence against police discipline based on having been 'found guilty by a court of law of a criminal offence'. The Police Conduct Committee found the allegation proved and the applicant was required by the chief constable to resign from the force as an alternative to dismissal. In proceedings for judicial review, the applicant submitted that the effect of section 13(1) and (3) of the 1973 Act was that the imposition of a conditional discharge was not to be deemed to be a conviction for any purposes other than the criminal proceedings and that, therefore, there had been no breach of the Police (Discipline) Regulations 1977. The application for judicial review was refused by the judge and the appeal to the Court of Appeal was dismissed. The Court of Appeal (Purchas, Ralph Gibson, and Nicholls LJJ) dealt with the submissions under section 13(1) and (3) separately. As to section 13(1) (deeming the conviction not to be a conviction for any purpose other than the purposes of criminal proceedings), the Court held that the word 'conviction' in the subsection should be given its strict meaning of 'finding and sentence'—that is, the ultimate disposal of the case—whereas the police disciplinary proceedings were concerned only with the applicant having been 'found guilty by a court of law of a criminal offence'. It followed that section 13(1) did not prevent the fact that the applicant had been found guilty of a criminal offence from being taken into consideration when the police disciplinary proceedings were heard. Purchas LJ said, at 646D–E, that the words in the Regulations were synonymous with the informal meaning of the word 'conviction'—that is, 'finding of guilt'—and were to be distinguished from the formal meaning of the word 'conviction' used in section 13(1) of the 1973 Act. Ralph Gibson LJ, in his judgment at 647, referred to the fact that a probation order was also to be deemed not to be a conviction for any purpose other than the purposes of the proceedings in which it was made, and said that 'a probation order may be imposed in a case in which the offence is of some gravity, such as would require in the public interest the removal from the force of the officer convicted'. And Nicholls LJ, at 650D–E, said that the clear impression he obtained from looking at section 13(1) of the 1973 Act was that the word 'conviction' used in the subsection was being used in the sense of the final disposal of a case, such as the person being found guilty, but also being dealt with by a conditional or absolute discharge or a probation order. Thus 'conviction' in section 13 did not have the same meaning as 'found guilty by a court of law of a criminal offence' in the Police (Discipline) Regulations. As to section 13(3), the Court of Appeal held that whilst the Regulations were an 'enactment or instrument' within the meaning of section 13(5) of the 1973 Act, the punishments open to a police authority, including dismissal or requiring an officer to resign, did not amount to the imposition of a disqualification or disability. A police officer, although dismissed and ceasing to be a member of the police force, is not as a matter of law prevented from being re-engaged and remains eligible for re-engagement at any time. Nicholls LJ, at 649D–E, contrasted the position under, for example, the Representation of People Act 1983, whereby a person convicted of certain offences is 'incapable' from being elected to or sitting in the House of Commons, or of holding any public or judicial office. Similarly, the imposition of a disqualification of a director under the Companies Acts disqualifies the person from being a director or taking part in the management of a company.

***R v. Longworth* [2006] 1 WLR 313**

**14.16** Conditional discharge for offence of making indecent photographs—effect on registration under Sex Offenders Act 1997

The House of Lords considered the provisions of section 14(1) of the 2000 Act in the context of the trial judge passing a sentence of conditional discharge in relation to an offence of making indecent photographs of children contrary to the Protection of Children Act 1978, and directing the appellant to be subject to the notification and registration requirements of the Sex Offenders Act 1997. The House held that the appellant having been conditionally discharged for an offence under Protection of Children Act 1978, the judge's ruling of registration under the Sex Offenders Act 1997 was a nullity because the effect of section 14(1) of the 2000 Act deemed there to be no conviction for the purposes of the Sex Offenders Act 1997. Lord Mance, in his speech at [23], observed that the statute book contains a number of enactments in which Parliament has excluded the effect of subsections (1) and/or (3) of section 14, as contemplated by section 14(6) of the 2000 Act. For example, under section 46(1) of the Road Traffic Offenders Act 1988, a court, on convicting a person of an offence involving obligatory or discretionary disqualification and making an order discharging him absolutely or conditionally, may, notwithstanding section 14(3), also disqualify the person from driving.

*R v. Patel (Rupal)* [2006] EWCA Crim 2689

**14.17** Conditional discharge for shoplifting—failure to disclose on job application—effect on charge of obtaining pecuniary advantage by deception

The defendant was indicted for an offence of obtaining a pecuniary advantage by deception. In May 2003, she had applied for a job on the civil staff of the Metropolitan Police and, on her application form, had ticked the box marked 'No' in response to the question: 'Have you ever been convicted of an offence?' In fact, nine years earlier, when she was 17 years old, the defendant had appeared at a magistrates' court for an offence of shoplifting. The court had made an order for her conditional discharge of twelve months and ordered her to pay £30 in costs. In dismissing the Crown's application for leave to appeal against the ruling of the trial judge dismissing the proceedings, the Court of Appeal held that the word 'conviction' in section 14(1) was no different from the word 'convicted' on the application form and therefore section 14(1) was a bar to the subsequent prosecution.

## D. Sanction

*In re a Solicitor* [1956] 1 WLR 1312

**14.18** Indecent assault—three months' imprisonment—whether striking off roll proportionate—approach of court in criminal cases

The appellant, a solicitor, was found guilty on two charges of indecent assault and was sentenced to three months' imprisonment. On appeal from the decision of the Disciplinary Committee striking him off the roll of solicitors, Lord Goddard CJ (with whom Hilbery and Ashworth JJ agreed) said, at 1314, that the Court is, and always has been, very loath to interfere with the findings of the Disciplinary Committee either on a matter of fact or with regard to the penalty that it imposes:

> If it is a matter of professional misconduct, it would take a very strong case to induce this court to interfere with the sentence passed by the Disciplinary Committee, because obviously the Disciplinary Committee are the best possible people for weighing the seriousness of professional misconduct. There is no suggestion of professional misconduct in this case. That being so, I think this court is bound to consider, as the Court of Criminal Appeal would have to do, whether or not the sentence is in proportion or out of proportion to the misconduct which has been proved.

The penalty was varied to two years' suspension.

*Ziderman v. General Dental Council* [1976] 1 WLR 330

**14.19** Previous convictions—relevant to sanction

In July 1970, Z had been convicted of ten offences of obtaining money by false pretences from the Executive Council of the National Health Service (NHS). He asked for sixty-two other similar offences to be taken into consideration. The convictions had been the subject of disciplinary proceedings before the Disciplinary Committee of the GDC, as a result of which his name was erased from the dentists' register on 12 June 1971. In November 1972, Z's name was restored to the register. On 21 December 1974, Z was convicted at Marlborough Street Magistrates' Court on a charge of theft, viz. shoplifting. He had stolen a tie and a car leather, worth £3.40, at Selfridges on the previous day. To this offence, he pleaded guilty. He was fined £250 and ordered to pay £50 costs. Disciplinary proceedings were brought against Z in respect of the shoplifting conviction and the Committee again ordered that his name be erased from the register. In dismissing Z's appeal, Lord Diplock, giving the judgment of the Board, said that previous convictions of a person charged with an offence are relevant matters to be taken into account by a sentencing tribunal in assessing the gravity of the offence with which he is charged.

See further paragraph 5.02.

### *Dey v. General Medical Council* [2001] UKPC 44

**14.20** Previous good character—whether erasure excessive

D was a married man of previously good character. He was convicted of fourteen counts of furnishing false information submitted to the local health authority for payment of health screening tests. The GMC's PCC directed that D's name be erased from the register. On appeal, the material placed before the Privy Council by way of mitigation included a petition signed by some 300 patients and twenty or so letters in D's support. They testified to his excellence as a doctor and the care with which he had provided a vital service to local residents. D's clinical performance was not in issue. Their Lordships said at [11]:

> The object of disciplinary proceedings against a medical practitioner who has been convicted of a criminal offence is twofold. It is to protect members of the public who may come to him as patients and to maintain the high standards and reputation of the profession. It is not to punish him a second time for the same offence. Nevertheless the same conduct which constitutes the offence for which he has been convicted may also demonstrate that the need to maintain the standards and reputation of the profession or to protect the public or both requires the erasure of his name from the Register. There is no clear line of demarcation: the difference lies not in the facts themselves but in the perspective from which they are viewed.

D not only submitted fraudulent claims for payment, but also sought to substantiate the claims by deliberately falsifying patients' records so that they showed records of medical examinations that had not, in fact, taken place. The Committee understandably took an extremely serious view of D's conduct in this regard, which obviously placed patients at risk. Other practitioners must be able to rely on the accuracy of patients' records and the integrity of the practitioner who made them. The sentence of erasure was not excessive and the Committee was entitled to conclude that such irresponsible behaviour cannot be tolerated in a registered medical practitioner, and that the need to maintain high standards and to protect the public, taken together, required the erasure of D's name from the register.

### *Colgan v. The Kennel Club*, Case no. 01/TLQ/0673, 26 October 2001

**14.21** Absolute discharge by magistrates' court—disqualification by Committee for five years—disproportionate

The claimant, C, was an experienced and well-regarded breeder of Newfoundland dogs. On 19 January 1999, C was convicted at Leicester Magistrates' Court of sixteen offences under section 1(1)(b) of the Protection of Animals Act 1911, in as much as she permitted to be conveyed or carried in a vehicle an animal in such a manner as to cause that animal unnecessary suffering. On 15 May 1998, C set off from kennels in Bury St Edmunds on a hot day in two vehicles containing thirty-two pedigree Newfoundland dogs. She intended to take the dogs to kennels in Derbyshire, some 110 miles away. Whilst the vehicle was stuck in slow-moving traffic on the M1, despite every effort to revive the dogs, ten died or had to be put down because of their parlous condition as a result of the build-up of temperature inside one of the vehicles. In their sentencing remarks, the magistrates said that they were fully aware of the distress, trauma, and financial loss suffered by C, and that they were confident that nothing like this incident would ever happen again. The magistrates gave C an absolute discharge and ordered her to pay £2,000 towards costs. Subsequently, the Kennel Club brought disciplinary proceedings against C. On 21 April 1999, the Kennel Club suspended C from exhibiting at, taking part in, attending, and/or having any connection with any event licensed by the Club and disqualified her from taking part in the management of any event licensed by the Club for a period of five years. C's solicitors sought to appeal against the ruling, but the Kennel Club said that there was no scope for an appeal, but if there were fresh evidence, the Club would consider it. In June 2000, nearly one year later, C's solicitors wrote a long letter to the Kennel Club seeking a review of its ruling. The Kennel Club responded saying that the penalties were not disproportionate and declined to remove or modify the penalties imposed. C issued proceedings. Cooke J said that it was common ground between the parties that the disciplinary rules were part of a contractual relationship between C and the Kennel Club; it was also common ground that the Kennel Club was obliged to exercise its disciplinary powers in a manner that was reasonable and proportionate, and consistent with common law and relevant statutory provisions. The penalties imposed were manifestly excessive and disproportionate to the objectives to be achieved, given the limited culpability of C, the inherent unlikelihood of any repetition of the offence by her, the huge publicity already given to the dangers to dogs in transporting or leaving them in vehicles exposed to the sun, and the financial loss and trauma already suffered by C in the loss of valuable pedigree dogs to whose welfare she was devoted. The disciplinary rulings made by the Kennel Club were not unlawful by reason of the Rehabilitation of Offenders Act 1974, nor did the review or failure to review fall foul of that statute. The penalties imposed, however, were disproportionate and C was entitled to a declaration to that effect.

***Council for the Regulation of Health Care Professionals v. General Dental Council and Fleischmann*** **[2005] EWHC 87 (Admin)**

**14.22** Criminal sentence not completed

F pleaded guilty to twelve offences of downloading child pornography from the Internet. He received a community rehabilitation order for three years. In allowing an appeal by the Council for the Regulation of Health Care Professionals and substituting a penalty of erasure in place of twelve months' suspension, Newman J said, at [54]:

> I am satisfied that, as a general principle, where a practitioner has been convicted of a serious criminal offence or offences he should not be permitted to resume his practice until he has satisfactorily completed his sentence. Only circumstances which plainly justify a different course should permit otherwise. Such circumstances could arise in connection with a period of disqualification from driving or time allowed by the court for the payment of a fine. The rationale for the principle is not that it can serve to punish the practitioner whilst serving his sentence, but that good standing in a profession must be earned if the reputation of the profession is to be maintained.

***R (Low) v. General Osteopathic Council*** **[2007] EWHC 2839 (Admin)**

**14.23** Criminal sentence—not necessarily reliable guide—maintenance of public confidence

L, a registered osteopath, pleaded guilty to ten specimen counts of possession of indecent photographs of children. There were 842 level 1 photographs and six level 2 photographs. The judge in the Crown Court imposed fines totalling £5,200 and an order for costs. L appealed under section 31 of the Osteopaths Act 1993 against the decision of the respondent's PCC that he should be removed from the register. In dismissing the appeal, Sullivan J said, at [20]:

> Because of these considerations the seriousness of the criminal offence, as measured by the sentence imposed by the Crown Court, is not necessarily a reliable guide to its gravity in terms of maintaining public confidence in a particular profession. Thus a relatively minor offence of financial dishonesty may well be considered to be of the utmost gravity by the Law Society when dealing with a solicitor who has care of his

clients' funds. While the offences to which the appellant pleaded guilty in this case would have been serious offences whatever the profession or occupation of the defendant, the committee were entitled to regard them as being of grave concern in terms of damaging public confidence in a profession where the professional inevitably has a one to one "hands-on" relationship with a partially undressed patient.

### *Chandrasekera v. Nursing and Midwifery Council* [2009] EWHC 144 (Admin)

**14.24** Appellant guilty of manslaughter on grounds of diminished responsibility—appeal against striking off dismissed

The appellant, aged 61, had been a nurse for all of her adult life. It was not suggested that her work as a nurse had been anything less than satisfactory. Whilst suffering from a depressive disorder and anxiety brought on by the conduct of her husband, she killed her husband with an axe. Having spent nine months in custody on remand, she was given a three-year community punishment, having pleaded guilty to manslaughter on the grounds of diminished responsibility. The NMC's Conduct and Competence Committee (CCC) subsequently struck the appellant off the nursing register. On appeal, the deputy judge said that he approached the matter on the basis that the appellant was not a risk to the public generally or her patients. While expressing some sympathy for the appellant, the judge held that the appeal must be dismissed.

### *Ansari v. General Pharmaceutical Council* [2012] EWHC 1159 (Admin)

**14.25** Alcohol-related driving offences—bringing profession into disrepute—removal—degree of deference to Committee's decision

A faced two allegations—the first, that he had been convicted of a total of six criminal offences involving two alcohol-related driving matters, two offences of driving whilst disqualified, and two offences of driving without insurance. In March 2007, he had been sentenced to thirty days' imprisonment. The second allegation was misconduct, the particulars being that, when applying for registration as a pharmacist, A had dishonestly failed to disclose his first conviction of failing to provide a specimen of breath for analysis, for which he was disqualified from driving for fifteen months. The Disciplinary Committee found dishonesty proved and that A's fitness to practise was currently impaired at the date of the hearing in March 2011; it directed that A's name be removed from the register of pharmacists pursuant to article 52(3) of the Pharmacists and Pharmacy Technicians Order 2007. On appeal, Sales J, in dismissing A's appeal, said that it was an established principle that a decision on whether a conviction was likely to bring a profession into disrepute was one primarily for the PCC and one to which the Court should exercise a degree of deference before deciding that the decision was wrong. In the instant case, the Committee had had the advantage of hearing evidence from the appellant and was particularly well placed to make an evaluative assessment in respect of the degree of insight into what he had done, which was an important foundation for its reasoning on sanction. After reviewing *Dad v. General Dental Council* [2000] 1 WLR 1538, *Raschid v. General Medical Council; Fatnani v. General Medical Council* [2007] 1 WLR 1460, and *Bhatt v. General Medical Council* [2011] EWHC 783 (Admin), the Court was not in a position to say that the Committee was wrong in the required sense in imposing the sanction that it had. The Committee had clearly had in mind that the appellant was a competent pharmacist and that the public would be deprived of the benefit of his services if the sanction of erasure were imposed, but it had been entitled to conclude that the pattern of conduct and the offending were so serious as to warrant the imposition of the sanction of erasure.

### *Ajala v. Nursing and Midwifery Council* [2012] EWHC 2976 (Admin)

**14.26** Community order of unpaid work—suspension by Committee

The appellant, a registered nurse, pleaded guilty in the Crown Court to two offences of dishonesty: the first, being possession of a document with the intention of using it for establishing a registerable fact about herself; the second, being possession of documents for use in the course of or in connection with a fraud. The appellant was found in possession of a driving licence, a library card, a bank card, a bank statement, and a utility bill in the name of another person. On 1 October 2010, Mr Recorder Malins passed a prison sentence of sixteen weeks concurrent for both offences, suspended for a period of twelve months. He also ordered the appellant to do 100 hours of unpaid work. On 24 August 2011, the NMC's CCC suspended the appellant's registration for six months. In its determination, the Committee said that 'the convictions involving dishonesty were too serious' for a caution order and that 'to make a caution order would not be sufficient to protect the wider public interest in the reputation of the profession'. In dismissing the appellant's appeal against the suspension order of six months, Haddon-Cave J said:

> 34. [Counsel for the appellant] subjected those words, and some others [in] the ruling, to something of forensic analysis and submitted that the CCC were taking too broad a brush view that all convictions for dishonesty were too serious. This, in my judgment, was an incorrect reading of the CCC's reasoning. It is quite clear that the CCC determined "the convictions involving dishonesty" *in this case* were too serious. That indeed was the view taken by Recorder Malins who determined that the custody threshold had been passed.
>
> 35. There is a further point on the wording of the rulings by the CCC taken by [counsel for the appellant], namely that the CCC, she submits, did not give full weight to the finding by Recorder Malins that the appellant had been subject to coercion by another man in the commission of these offences. [Counsel for the appellant] said that the CCC had used the phrase "may have been influenced by a partner" in relation to the fraud matters.
>
> 36. In my judgment however, again this is an unfair reading of the CCC's reasons. The Panel quite clearly accepted the premis that the appellant may have been influenced by a partner and in using the word "influenced" immediately before the passage in question on page 37 of the transcript quoted from the decision of Recorder Malins, they clearly would have well in mind his finding about coercion.
>
> 37. It is not always helpful to subject findings of such Panels to a minute forensic examination, as if it is an exercise in statutory construction. A fair reading of the CCC's reasons leads me to conclude that the hearing and the reasoning cannot be faulted.

### *Dublas v. Nursing and Midwifery Council* [2012] EWHC 4214 (Admin)

**14.27** Wilful neglect of patient—removal

D appealed against the decision to remove her name from the nursing register following her criminal conviction for wilful neglect of an elderly female dementia patient. D was sentenced to nine months' imprisonment, suspended for twelve months. It was not for the Court to substitute its professional judgment or assessment for that of the CCC. It is evident that the committee gave anxious, careful, and thorough consideration to the totality of the history of the case.

### *R (Ahmed) v. General Optical Council* [2012] EWHC 3699 (Admin)

**14.28** Driving offences—twelve months' suspension—reduced to four months

A, a practising optometrist, appealed against the sanction of twelve months' suspension imposed by the Fitness to Practise Committee of the General Optical Council. A was convicted in the magistrates' court of driving in excess of the speed limit and wilfully obstructing a police constable in the execution of his duty. A was stopped in a 40 mph zone for driving at 54 mph and initially gave a false name. Within minutes, he acknowledged that he had given false details, and provided his true name and date of birth. The Court was concerned that the maximum period of suspension was imposed by the Committee. It may be said that it was considering erasure and so suspension, even for the maximum term, reflected all of the mitigating areas and the necessary protection for the public that was required and the need to maintain confidence in the general reputation of the profession as a whole. However, the length of the suspension was disproportionate to the finding of impairment and it was appropriate to reduce the period of suspension to four months.

### *Fuyane v. Nursing and Midwifery Council* [2012] EWHC 3229 (Admin)

**14.29**
Dishonesty—failed asylum seeker—suspension substituted for removal—approach of court

The appellant, a national of Zimbabwe, pleaded guilty at Birmingham Crown Court to two offences of dishonesty: one relating to the funding of her course for which the NHS had paid £21,000; the other, to a bursary that she had received of £20,500. She received a suspended sentence of imprisonment and a community order of 180 hours' unpaid work. On 31 May 2012, the CCC of the NMC ordered that her name be removed from the register. In allowing F's appeal and substituting an order suspending her registration for one year from the hearing of the appeal before the panel, His Honour Judge Langan QC (sitting as a judge of the High Court) recognized that an appeal against a penalty imposed by a professional body was subject to severe constraints. In particular:

(1) a strong case must be put forward for interfering with the decision of the panel;
(2) the personal circumstances of an appellant must often give way to the public interest in maintaining confidence in the integrity of the profession; and
(3) proved or admitted dishonesty frequently—perhaps almost always—led to removal from a professional register, particularly where, as in this case, the amounts involved are substantial.

In the instant case, the offences were committed whilst the appellant was in limbo as a failed asylum seeker; she immediately admitted her wrong-doing; academic, personal, and professional references demonstrated that, these offences apart, she was a woman of impeccable character; she had lost her primary career as a lawyer; and if struck off the nursing register, the training and experience that she had gained over the past few years would be lost to the community.

### *Chuah v. Nursing and Midwifery Council* [2013] EWHC 894 (Admin); [2013] EWCA Civ 1799

**14.30**
Assault on hospital premises—removal—approach of court—whether excessive or disproportionate—public interest

The appellant was convicted of driving with excess alcohol and of an offence of assault by beating. The latter offence was committed on hospital premises and involved an assault on a member of the hospital staff. The offence was committed after the appellant had been drinking heavily. The appellant's appeal against the decision to remove his name from the register was dismissed by the Administrative Court (Simon J). Although the Court would accord an appropriate measure of respect to the judgment of the panel, it should not defer to the judgment more than was warranted by the circumstances. The Court should decide whether the sanction was appropriate and necessary in the public interest, or excessive or disproportionate. The Court was entitled to interfere if it concluded that the sentencing decision was 'clearly inappropriate'. The test is not dissimilar to that which is applied by the Court of Appeal, Criminal Division, on an appeal against sentence in the Crown Court. The sentence will be quashed where the Court considers the sentence to be either manifestly excessive or wrong in principle: *per* Simon J, at [51]. Leave to appeal was refused by Lloyd Jones LJ: [2013] EWCA Civ 1799. The sanction imposed was a severe one, but not disproportionate or wrong in principle. The panel was faced with previous conduct involving an assault by an off-duty nurse in a drunken state in a hospital on a member of the hospital security staff. The panel was clearly entitled to conclude that the only appropriate sanction was a striking-off order. The panel was entitled to conclude that a striking-off order was necessary in these circumstances.

***Isaghehi v. Nursing and Midwifery Council* [2014] EWHC 127 (Admin)**

**14.31** Road rage—conviction following trial—removal—twelve months' suspension substituted

I, a psychiatric nurse, following trial, was convicted of dangerous driving and sentenced to fifteen months' imprisonment. He had grown frustrated by another vehicle in the outside lane of the A11 road, driven too close, and driven into the inside lane to overtake; when the vehicles were parallel, he swore and gesticulated at the other driver, and thereafter steered twice in front of the other driver's vehicle. His vehicle made contact and both drivers lost control. By good fortune, no serious physical injuries were sustained, but the incident was terrifying for the other driver and her passenger. The car in which they were travelling turned over and the passenger had to be cut out. She remained distressed for a considerable period in the aftermath of the accident. I appealed against the CCC's decision to strike him off the register. In allowing I's appeal and substituting a sanction of twelve months' suspension, Turner J, on reviewing the NMC's Indicative Sanctions Guidance, said that this was a single incident of misconduct; however, it was not fundamentally incompatible with I's continuing to be a registered nurse and the public interest could be satisfied by a less severe outcome than permanent removal from the register. There was no evidence of harmful deep-seated personality or attitudinal problems and there was no record of concerns relating to his abilities as a nurse, in which capacity he was unfailingly patient and calm with his patients at all times; nor was there evidence that I had been involved in any similar behaviour since the incident. On the evidence, I had insight and did not pose a significant risk of repeating his bad behaviour. Public confidence in the profession could be sustained if I were not to be removed from the register and the sanction of striking off was disproportionately high.

***Obukofe v. General Medical Council* [2014] EWHC 408 (Admin)**

**14.32** Suspended sentence complete—sex offenders' register requirement still running

In April 2011, O was convicted following a trial at Leicester Crown Court of three counts of sexual assault. He was sentenced on each count to six months' imprisonment, suspended for twelve months, the two periods to run concurrently, and a sex offenders' register requirement for a period of seven years. In June 2013, a fitness to practise panel imposed a sanction on O's registration of twelve months' suspension. The legal assessor drew attention to *Council for the Regulation of Health Care Professionals v. General Dental Council and Fleishmann* [2005] EWHC 87 (Admin), in which Newman J said that, as a general principle, where a practitioner has been convicted of a serious criminal offence or offences, he should not be permitted to resume his practice until he has satisfactorily completed his sentence. Dismissing O's appeal, Popplewell J said, at [58], that the important element of Newman J's reasoning related to the sex offender's treatment programme to which Fleishmann was subject for three years, rather than his subjection to the notification requirements of the sex offenders' register for five years. If, in this case, the panel had thought that the *Fleishmann* case required it to continue the suspension until O's criminal sentence had been completed (including the period during which he remained subject to the notification requirements of the sex offenders' register), that would have been an error of law. But it was clear that this was not the way in which the panel relied on the *Fleishmann* decision. It was clear that the panel exercised its judgment on the sanction that ought to be imposed independently of that element of the sentence on O which related to the sex offenders' register.

***D v. Nursing and Midwifery Council* [2014] EWHC 1298 (Admin)**

**14.33** Battery—twins aged 2½ and husband—serious departure from standards expected of nurse

On 8 September 2011, the appellant was convicted on a plea of guilty of three counts of battery, contrary to section 39 of the Criminal Justice Act 1988, and sentenced to eighteen weeks' imprisonment on each count, to be served concurrently. The appellant was a mother of twins aged 2½. She used a cane to beat the first child for a long period of time. She grabbed her husband by the shoulders and threw him on the floor, and then beat the other child with a cane also. The CCC ordered that the appellant's name be struck off the register. The Committee fully took into account that there had been no repetition of the behaviour and that the appellant had an unblemished nursing career, and that she had undertaken remedial work and had gained some insight into her conduct. In dismissing the appellant's appeal, Her Honour Judge Alice Robinson (sitting as a judge of the High Court) said that while it might be thought that a conviction for battery was not the most serious criminal offence, the facts of these particular offences disclosed very serious conduct: the attack was out of all proportion to the minor incident that preceded it, and the panel was entirely right to conclude that it was a serious departure from the standards expected of a nurse and that a striking-off order was the only sanction that was sufficient to protect the public interest.

### *Nursing and Midwifery Council v. Farah* [2014] EWHC 1655 (Admin)

**14.34** Fraud—abuse of trust—removal

F was convicted at Blackfriars Crown Court on 23 May 2012 of an offence of making a false representation to make a gain for himself, contrary to sections 1(2)(a) and 2 of the Fraud Act 2006, and sentenced to a suspended sentence order, with various community requirements. Between June and August 2009, F was working at King's College Hospital whilst being off sick and receiving sick pay from his employment as a nurse at the University College London Hospital, the loss to University College London being £2,517.86. In his sentencing remarks, the judge observed that there were a number of aggravating features of the case including the abuse of the appellant's position of trust with his employer, the deceit of his general practitioner, his attempts to lay blame for the offence on one of his work colleagues, and the preplanned nature of the offence, which was dishonest from its inception and took place over a period of time. F's appeal against the sanction by the CCC of striking off was dismissed.

### *Gitau v. Nursing and Midwifery Council* [2014] EWHC 2581 (Admin)

**14.35** Fraud—sentence still running—removal—limited insight

On 31 October 2012, G pleaded guilty at Croydon Magistrates' Court to two charges of fraud and was committed to Croydon Crown Court for sentencing. The first charge related to use of a forged British birth certificate; the second related to a false 'leave to remain' stamp. On 7 December 2012, G was sentenced to six months' imprisonment, concurrent for each offence and suspended for two years, and 100 hours' unpaid work. At the hearing before the CCC on 23 August 2013, G admitted the convictions and gave oral evidence. The Committee noted that the suspension period of the sentence was still running. It thought that the two offences were very serious. G's evidence demonstrated only limited insight. She was a single mother with two young children, one of whom had medical problems requiring him to attend hospital four times a year. In dismissing G's appeal, Nichol J said that the Court's task was to decide whether the Committee's decision was wrong: see further *Ogbonna-Jacob v. Nursing and Midwifery Council* [2013] EWHC 1595 (Admin), [22], *per* Sales J. The matters were indeed serious. The Committee was clearly entitled to say that they brought the nursing profession into disrepute and were contrary to the professional tenets that the appellant was expected to maintain. This was so whether or not there was a likelihood of repetition. The absence of any risk of repetition was more likely to be relevant where disciplinary proceedings are taken because of the registrant's clinical failings than where a sanction is required to make a firm declaration of professional standards: see, for example, *Yeong v. General Medical Council* [2009] EWHC 1923 (Admin), [51]. The appellant did have difficult personal circumstances; she had expressed remorse; she had engaged in the regulatory process: this was all in her favour and these were all points that were taken into account by the Committee. Nonetheless, the CCC was entitled to conclude that no sanction short of striking off would be appropriate. It was not wrong to come to that conclusion.

## E. Other Relevant Chapters

**14.36** Dishonesty, Chapter 19
Drafting of Charges, Chapter 22
Evidence, Chapter 25

# 15

# COSTS
|     *Shrimpton v. General Council of the Bar* [2005] EWHC 2472 | 15.04 |
|     *Baxendale-Walker v. Law Society* [2006] 3 All ER 675; [2008] 1 WLR 426, CA | 15.05 |
|     *Law Society v. Adcock* [2007] 1 WLR 1096 | 15.06 |
|     *R (Perinpanathan) v. City of Westminster Magistrates' Court* [2010] 1 WLR 1508, CA | 15.07 |
|     *R (Commissioner of Police for the Metropolis) v. Police Appeals Tribunal and Naulls* [2013] EWHC 1684 (Admin) | 15.08 |
|     *Murtagh v. Solicitors Regulation Authority* [2013] EWHC 2024 (Admin) | 15.09 |
| C. Costs against the Practitioner | 15.10 |
|     *Merrick v. Law Society* [2007] EWHC 2997 (Admin) | 15.10 |
|     *Financial Ombudsman Service v. Heather Moor & Edgecombe Ltd* [2008] EWCA Civ 643 | 15.11 |
|     *D'Souza v. Law Society* [2009] EWHC 2193 (Admin) | 15.12 |
|     *Solicitors Regulation Authority v. Davis and McGlinchey* [2011] EWHC 232 (Admin) | 15.13 |
|     *Levy v. Solicitors Regulation Authority* [2011] EWHC 740 (Admin) | 15.14 |
|     *Adeeko v. Solicitors Regulation Authority* [2012] EWHC 841 (Admin) | 15.15 |
|     *Sharma v. Solicitors Regulation Authority* [2012] EWHC 3176 (Admin) | 15.16 |
|     *Agyeman v. Solicitors Regulation Authority* [2012] EWHC 3472 (Admin) | 15.17 |
|     *Mireskandari v. Solicitors Regulation Authority* [2013] EWHC 907 (Admin) | 15.18 |
|     *Solicitors Regulation Authority v. Anderson Solicitors* [2013] EWHC 4021 (Admin) | 15.19 |
|     *Broomhead v. Solicitors Regulation Authority* [2014] EWHC 2772 (Admin) | 15.20 |
|     *R (Lonsdale) v. Bar Standards Board*, 8 October 2014 (unreported) | 15.21 |
| D. Costs on Appeal | 15.22 |
|     *Shrimpton v. General Council of the Bar* [2005] EWHC 2472 | 15.22 |
|     *Walker v. Royal College of Veterinary Surgeons* [2008] UKPC 20 | 15.23 |
|     *Bryant v. Law Society* [2009] 1 WLR 163 | 15.24 |
|     *R (Commissioner of Police for the Metropolis) v. Police Appeals Tribunal and Naulls* [2013] EWHC 1684 (Admin) | 15.25 |
|     *Professional Standards Authority for Health and Social Care v. General Pharmaceutical Council and Onwughalu* [2014] EWHC 2521 (Admin) | 15.26 |
|     *Bass and Ward v. Solicitors Regulation Authority* [2013] 5 Costs LO 651 | 15.27 |
|     *Quinn v. Bar Standards Board*, 25 February 2013 (unreported) | 15.28 |
|     *Williams v. Bar Standards Board*, 31 October 2014 (unreported) | 15.29 |
|     *Khan v. General Medical Council* [2015] EWHC 301 (Admin) | 15.30 |
|     *R (Bar Standards Board) v. Disciplinary Tribunal of the Council of the Inns of Court* [2014] 4 All ER 759 | 15.31 |

## A. Legal Framework

**15.01** Examples include:

Solicitors (Disciplinary Proceedings) Rules 2007, rule 18 (tribunal may make such order as to costs as the tribunal shall think fit)
Chartered Institute of Management Accountants Disciplinary Committee Rules 2011, rule 25 (panel may decide that the respondent be required to pay all or part of the costs of the proceedings against him)
General Optical Council (Fitness to Practise) Rules 2013, rule 52 (committee may summarily assess the costs of any party to the proceedings and order any party to pay all or part of the costs or expenses of any other party)
Institute of Chartered Accountants in England and Wales Appeal Committee Regulations, regulations 35–41
Institute of Chartered Accountants in England and Wales Disciplinary Bye-laws, bye-law 33 (powers of tribunals and panels as to costs)
Institute of Chartered Accountants in England and Wales Disciplinary Committee Regulations, regulations 40–44

## B. Costs against the Regulator

### *Bradford Metropolitan District Council v. Booth* (2000) 164 JP 485, DC

**15.02** General principles

The City of Bradford Metropolitan District Council appealed, by way of case stated, from a decision of justices to award costs against the local authority. The respondent was successful on a complaint against a vehicle licensing decision of the local authority, but when the local authority had not, in making the decision appealed against, acted unreasonably or in bad faith. In allowing the local authority's appeal, Lord Bingham CJ (with whose judgment Silber J agreed) said:

> 23. I would accordingly hold that the proper approach to questions of this kind can for convenience be summarised in three propositions. (1) Section 64(1) [of the Local Government (Miscellaneous Provisions) Act 1976] confers a discretion upon a magistrates' court to make such order as to costs as it thinks just and reasonable. That provision applies both to the quantum of the costs (if any) to be paid, but also as to the party (if any) which should pay them. (2) What the court will think just and reasonable will depend on all the relevant facts and circumstances of the case before the court. The court may think it just and reasonable that costs should follow the event, but need not think so in all cases covered by the subsection. (3) Where a complainant has successfully challenged before justices an administrative decision made by a police or regulatory authority acting honestly, reasonably, properly and on grounds that reasonably appeared to be sound, in exercise of its public duty, the court should consider, in addition to any other relevant fact or circumstances, both (i) the financial prejudice to the particular complainant in the particular circumstances if an order for costs is not made in its favour; and (ii) the need to encourage public authorities to make and stand by honest, reasonable and apparently sound administrative decisions made in the public interest without fear of exposure to undue financial prejudice if the decision is successfully challenged.

### *R (Gorlov) v. Institute of Chartered Accountants in England and Wales* [2001] EWHC 220 (Admin)

**15.03** Misguided and unreasonable conduct

G, a chartered accountant, challenged the refusal of the Appeal Committee of the Institute of Chartered Accountants in England and Wales (ICAEW) to award him the costs of disciplinary proceedings in which he was ultimately successful. The secretariat laid two complaints before the Investigation Committee, which resolved to prefer a single formal complaint to the Disciplinary Committee. The wording of the formal complaint did not make sense and the professional standards office of the Institute sent the claimant an entirely different complaint. The claimant's solicitors objected and reserved their right to draw attention to the breaches in the Institute's bye-laws. The hearing duly went ahead and the Disciplinary Committee found the complaint proved; it ordered that G be reprimanded and that he pay a fine of £2,500 and costs of £5,500. The Appeal Committee held that the Disciplinary Committee had no jurisdiction to deal with the complaint before it and that the hearing had been a nullity. It refused G's application for costs. Jackson J quashed the Appeal Committee's decision on costs. After reviewing *R v. Totnes Licensing Justices, ex parte Chief Constable of Devon and Cornwall* (1990) 156 JP 587, *Chief Constable of Derbyshire v. Goodman*, 2 April 1998, DC (unreported), and *City of Bradford Metropolitan District Council v. Booth* [2000] COD 338, Jackson J said that it was clear from the wording of the Institute's bye-laws and from the authorities cited that costs do not necessarily follow the event. The fact that the claimant won his appeal before the Appeal Committee was a factor in his favour; it was not, however, decisive. One then turns to wider matters. The ICAEW is a professional body, which, acting in the public interest, brings disciplinary proceedings against accountants. That is a factor that points against any automatic award of costs in disciplinary proceedings that fail. The present case, however, had special features. The disciplinary proceedings brought by the Institute were 'a shambles from start to finish'. The Disciplinary Committee proceeded to make an order that was a nullity; the claimant appealed against that order; the Institute resisted that appeal. In due course, the Appeal Committee set aside as a nullity the order made by the Disciplinary Committee. The conduct of the Institute throughout the disciplinary proceedings was, of course, honest and well intended, but the conduct was misguided and unreasonable.

### *Shrimpton v. General Council of the Bar* [2005] EWHC 2472

**15.04**
Defect in constitution of panel

In allowing S's appeal from the Disciplinary Tribunal of the General Council of the Bar, Lindsay J said that although the judgment of the Visitors to the Inns of Court ranged over other issues, crucial to S's success was the acceptance by the General Council of the Bar, only shortly before the hearing before the Visitors began, that the decision of the Tribunal from which S appealed was a decision without legal effect. That was so because, in the light of *P, a barrister v. General Council of the Bar* [2005] 1 WLR 3019, the Tribunal had been constituted in a way that failed to satisfy Article 6 of the European Convention on Human Rights (ECHR). The defect lay in the constitution of the panel that made up the Tribunal. Accordingly, there was no reason not to order S's costs before the Disciplinary Tribunal and before the Visitors to follow the event, and to be paid by the Bar Council.

### *Baxendale-Walker v. Law Society* [2006] 3 All ER 675; [2008] 1 WLR 426, CA

**15.05**
Principles—good reason

The Law Society complained that the conduct of the appellant solicitor was 'unbefitting a solicitor' in two separate respects: first, that he gave evidence in High Court proceedings that the Court found to be 'manifestly untrue'; and second, that he provided a reference in circumstances that he knew or ought to have known were improper and/or unprofessional. The first allegation was dismissed by the Solicitors Disciplinary Tribunal (SDT); the second allegation was admitted, but the precise ambit of the admission was not admitted. The Tribunal found the solicitor guilty of conduct unbefitting a solicitor and suspended him from practice for three years. The Tribunal ordered that the Law Society should pay 30 per cent of the solicitor's costs of the proceedings. The Divisional

Court (Moses LJ and Stanley Burnton J) [2006] 3 All ER 675 dismissed the solicitor's appeal against the Tribunal's decision to suspend him from practice and allowed the Law Society's cross-appeal on costs, subsequently ordering the solicitor to pay 60 per cent of the Law Society's costs of the disciplinary proceedings. Moses LJ said:

> [43] The question thus arises as to whether the order that the Law Society should pay a proportion of the appellant's costs and that no costs should be paid by the appellant was correct, as a matter of law. The principles in relation to an award of costs against a disciplinary body were not in dispute. A regulator brings proceedings in the public interest in the exercise of a public function which it is required to perform. In those circumstances the principles applicable to an award of costs differ from those in relation to private civil litigation. Absent dishonesty or a lack of good faith, a costs order should not be made against such a regulator unless there is good reason to do so. That reason must be more than that the other party has succeeded. In considering an award of costs against a public regulator the court must consider on the one hand the financial prejudice to the particular complainant, weighed against the need to encourage public bodies to exercise their public function of making reasonable and sound decisions without fear of exposure to undue financial prejudice, if the decision is successfully challenged.
>
> [44] Those principles can be derived from a number of cases summarised by Jackson J in *R (on the application of Gorlov) v. Institute of Chartered Accountants in England and Wales* [2001] EWHC 220 (Admin) at [30]–[35], [2001] ACD 73 (and see, in particular, the three principles distilled by Lord Bingham of Cornhill CJ in *Booth v. Bradford MDC* [2000] COD 338).
>
> [45] In the instant appeal, in my view there was no basis for the order made by the tribunal. The only ground on which it relied was that the appellant had been successful in his defence of the first allegation. That was not a sufficient ground to order the Law Society to pay any of his costs. There was no finding that the allegation was misconceived, without foundation or born of malice or some other improper motive. In those circumstances the order was without foundation. The mere fact that resistance to the first allegation required greater expenditure in time and money was not a basis for making the Law Society pay any of the appellant's costs.

The Court of Appeal (Sir Igor Judge P, and Laws and Scott Baker LJJ) dismissed the solicitor's appeal against the costs order. Sir Igor Judge P, in handing down the judgment of the Court at [34], said:

> Our analysis must begin with the Solicitors Disciplinary Tribunal itself. This statutory tribunal is entrusted with wide and important disciplinary responsibilities for the profession, and when deciding any application or complaint made to it, section 47(2) of the Solicitors Act 1974 undoubtedly vests it with a very wide costs discretion. An order that the Law Society itself should pay the costs of another party to disciplinary proceedings is neither prohibited nor expressly discouraged by section 47(2)(i). That said, however, it is self-evident that when the Law Society is addressing the question whether to investigate possible professional misconduct, or whether there is sufficient evidence to justify a formal complaint to the tribunal, the ambit of its responsibility is far greater than it would be for a litigant deciding whether to bring civil proceedings. Disciplinary proceedings supervise the proper discharge by solicitors of their professional obligations, and guard the public interest, as the judgment in *Bolton v. Law Society* [1994] 1 WLR 512 makes clear, by ensuring that high professional standards are maintained, and, when necessary, vindicated. Although, as [counsel for the appellant] maintained, it is true that the Law Society is not obliged to bring disciplinary proceedings, if it is to perform these functions and safeguard standards, the tribunal is dependent on the Law Society to bring properly justified complaints of professional misconduct to its attention. Accordingly, the Law Society has an independent obligation of its own to ensure that the tribunal is enabled to

fulfil its statutory responsibilities. The exercise of this regulatory function places the Law Society in a wholly different position to that of a party to ordinary litigation. The normal approach to costs decisions in such litigation—dealing with it very broadly, that properly incurred costs should follow the "event" and be paid by the unsuccessful party—would appear to have no direct application to disciplinary proceedings against a solicitor.

### *Law Society v. Adcock* [2007] 1 WLR 1096

**15.06** Appeal against costs order—approach of court

A and M were partners in the firm of Adcock, and faced disciplinary proceedings that they had provided misleading information to clients regarding the cost of local searches and had delivered inaccurate bills to clients. The SDT dismissed the case against each solicitor and, on appeal, the Divisional Court (Waller LJ and Treacy J) allowed the Law Society's appeal in part. By a respondent's notice, A and M cross-appealed against the refusal of the Tribunal to order that the whole of the solicitors' costs be paid by the Law Society. Waller LJ said, at [41]:

> This Court should only disturb an order for costs in rare circumstances and only if, in the exercise of its discretion, the tribunal had misdirected itself or reached a conclusion that the Court could not have reached, and where the solution preferred by the tribunal had exceeded the general ambit within which a reasonable disagreement is possible.

### *R (Perinpanathan) v. City of Westminster Magistrates' Court* [2010] 1 WLR 1508, CA

**15.07** Principles applicable to costs against regulatory body

On 25 April 2006, P's daughter, who was aged 15, was stopped at Heathrow Airport. She was carrying some £150,000 in cash. The cash was detained by the police on the basis that there were reasonable grounds to suspect that it was intended for use in unlawful conduct—namely, terrorism. On 13 December 2007, the City of Westminster Magistrates' Court dismissed an application by the Commissioner of Police of the Metropolis under section 298 of the Proceeds of Crime Act 2002 for forfeiture of the cash, but refused to award costs to P upon dismissing the application. On 10 March 2008, the Divisional Court dismissed P's claim for judicial review of the costs order. In dismissing the claimant's appeal to the Court of Appeal (Lord Neuberger of Abbotsbury MR, and Maurice Kay and Stanley Burnton LJJ), Stanley Burnton LJ said:

> 40. I derive the following propositions from the authorities to which I have referred. (1) As a result of the decision of the Court of Appeal in *Baxendale-Walker v. Law Society* [2008] 1 WLR 426, the principle in the *Bradford* case 164 JP 485 is binding on this court. Quite apart from authority, however, for the reasons given by Lord Bingham CJ I would respectfully endorse its application in licensing proceedings in the magistrates' court and the Crown Court. (2) For the same reasons, the principle is applicable to disciplinary proceedings before tribunals at first instance brought by public authorities acting in the public interest: see *Baxendale-Walker v. Law Society.* (3) Whether the principle should be applied in other contexts will depend on the substantive legislative framework and the applicable procedural provisions. (4) The principle does not apply in proceedings to which the [Civil Procedure Rules] CPR apply. (5) Where the principle applies, and the party opposing the order sought by the public authority has been successful, in relation to costs the starting point and default position is that no order should be made. (6) A successful private party to proceedings to which the principle applies may nonetheless be awarded all or part of his costs, if the conduct of the public authority in question justifies it. (7) Other facts relevant to the exercise of the discretion conferred by the applicable procedural rules may also justify an order for costs. It would not be sensible to try exhaustively to define such matters, and I do not propose to do so.

> 41. Lord Bingham CJ stated that financial prejudice to the private party may justify an order for costs in his favour. I think it clear that the financial prejudice necessarily involved in litigation would not normally justify an order. If that were not so, an order would be made in every case in which the successful private party incurred legal costs. Lord Bingham CJ had in mind a case in which the successful private party would suffer substantial hardship if no order for costs was made in his favour. I respectfully agree with what Toulson J (with whom Richards LJ agreed) said in *R (Cambridge City Council) v. Alex Nestling Limited* (2006) 170 JP 539, para 12: "As to the financial loss suffered by the successful appellant, a successful appellant who has to bear his own costs will necessarily be out of pocket, and that is the reason in ordinary civil litigation for the principle that costs follow the event. But that principle does not apply in this type of case. When Lord Bingham CJ referred to the need to consider the financial prejudice to a particular complainant in the particular circumstances, he was not... implying that an award for costs should routinely follow in favour of a successful appellant; quite to the contrary."

Lord Neuberger of Abbotsbury MR said, at [76]:

> In a case where regulatory or disciplinary bodies, or the police, carrying out regulatory functions have acted reasonably in opposing the grant of relief, or in pursuing a claim, it seems appropriate that there should not be a presumption that they should pay the other party's costs. It is not as if the other party would have no right to recover costs in such a case: as Lord Bingham CJ made clear, at paras 25 and 26, one must take into account "all the relevant facts and circumstances of the case", and in particular "the financial prejudice to the particularly complainant... if an order for costs is not made in his favour".

***R (Commissioner of Police for the Metropolis) v. Police Appeals Tribunal and Naulls* [2013] EWHC 1684 (Admin)**

**15.08** See paragraph 15.25.

***Murtagh v. Solicitors Regulation Authority* [2013] EWHC 2024 (Admin)**

**15.09** Disciplinary proceedings properly brought

On 17 April 2012, the SDT gave oral reasons for finding that the case brought by the Solicitors Regulation Authority (SRA) had not been proved against M, but the Tribunal also made no order for costs. M appealed, seeking an order that the SRA pay his costs before the Tribunal. In dismissing M's appeal, Dingemans J said that applying the test in *Law Society v. Adcock* [2007] 1 WLR 1096, [41], he did not consider that the Tribunal misdirected itself or that it had reached a conclusion that the Court would not have done, which solution exceeded the general ambit within which a reasonable disagreement is possible.

## C. Costs against the Practitioner

***Merrick v. Law Society* [2007] EWHC 2997 (Admin)**

**15.10** Effect of suspension or striking off—means of defendant

The most serious allegation faced by M involved a charge that he had improperly utilized client money for his own benefit by making an improper transfer of £10,000 from client account to office account. However, no allegation of dishonesty was made by the Law Society against M. In the event, the SDT concluded that the transfer was improper and constituted a serious regulatory breach, amounting to conduct unbefitting a solicitor. Conduct could be improper even if it were not dishonest. M was suspended from practice for twelve months and ordered to pay the Law Society's costs, summarily assessed by the SDT, in the sum of £45,000. The Administrative Court (Thomas LJ and

Gross J) dismissed M's appeal against the order for suspension from practice, but allowed M's appeal against the order for payment of costs. In considering whether the order for costs was 'manifestly excessive or otherwise inappropriate', Gross J said that the order for suspension would very likely bring to an end M's professional career, who was aged 67 years—his means of livelihood. The order for costs was not justified and was manifestly excessive or inappropriate for the following reasons:

> 61. First, there can be no general rule that the SDT should *not* impose an order for costs in addition to an order of suspension or an order striking off a solicitor. Were it otherwise, the more serious the misconduct, the less likely that the Law Society could recoup the costs to which it had been put in dealing with it. That cannot be right.
>
> 62. Secondly, whether in any individual case it is appropriate to add an order for costs to an order suspending a solicitor from practice or striking him off must depend on the facts. In some cases, the order for suspension or striking off will be sufficient punishment. In others, it may not.
>
> 63. Thirdly, when an order is made, effectively depriving a solicitor of his livelihood, the question necessarily arises as to how any order for costs would be paid. An analogous issue arises when in a criminal case a defendant is given a custodial sentence. In my view, if an order for costs is being considered, the right course is to inquire into the means of the solicitor before coming to a decision on the question of costs.
>
> 64. Fourthly, no inquiry into the means of [M] was conducted by the SDT in this case. This Court has, however, caused such an inquiry to be made. It is clear from the materials now available that [M] would not be in a position to satisfy the order for costs, on the assumption that he is now suspended from practice.
>
> 65. Fifthly, in the circumstances that no inquiry as to means was hitherto made, I am less reticent than I might otherwise have been to intervene on a matter of costs. Looking at this matter afresh and in the light of the material now available as to [M]'s means, it is plain that the order in the amount of £45,000 cannot stand.
>
> 66. Sixthly, there remains the question of whether any order for costs should survive. Mr Williams QC submitted that *an* order for costs was justified. With respect and on balance, I am unable to accept that submission. In all the circumstances of this case, having regarding to [M]'s hitherto clean record and the likely consequences of the order for suspension, to my mind that order is punishment enough. The order for costs should be set aside.

### *Financial Ombudsman Service v. Heather Moor & Edgecombe Ltd* [2008] EWCA Civ 643

**15.11** Compensation scheme—standard case fee—recovery

The Financial Ombudsman Service (FOS) appealed the decision of District Judge Rutherford, sitting in the Trowbridge County Court, dismissing the FOS's claim against the respondent for £1,440, being the FOS's standard case fee of £360 in relation to four complaints rejected by the Ombudsman. The scheme rules relating to the funding of the FOS provided that a firm must pay the standard case fee in respect of each chargeable case relating to that firm that is closed by the FOS, unless a special case fee is payable or has been paid in respect of the case. The four complaints in the instant case had been investigated by the FOS, but rejected on their merits, although each of the complaints was a chargeable case in respect of which the standard fee was payable by the respondent. The Court of Appeal (Laws, Rix, and Stanley Burnton LJJ), in allowing the FOS's appeal, said that the rule requiring payment of a standard case fee was not unreasonable or unlawful. Section 234 of the Financial Services and Markets Act 2000 restricts the recovery of costs from complainants. A system under which firms make a payment of a fee in respect of the services of the Ombudsman in investigating and deciding complaints against them is a perfectly rational response to the need to fund the scheme. The Financial Services Authority (FSA) rules require firms to deal with complaints fairly and efficiently, and a complaint to the FOS is likely to be made only when a firm's response to a complaint is regarded by the complainant as unsatisfactory. The standard case fee is not payable if there are no more than two complaints in a year against a firm. Wholly unmeritorious complaints do not generate a case fee if they are dismissed. The practicality of a case fee payable only in respect of successful complaints was considered in a discussion paper entitled *Financial Ombudsman Service Compulsory Jurisdiction: Funding Review* in May 2006. A scheme under which the decision-maker's decision on the merits of a complaint affects the income of the decision-maker was undesirable.

***D'Souza v. Law Society* [2009] EWHC 2193 (Admin)**

**15.12** Means of defendant

The appellant appealed against the decision of the SDT that he should be subject to a financial penalty of £1,500 and pay costs of just under £8,000. Thomas LJ and Coulson J, in allowing the appellant's appeal, quashed the sanctions imposed by the Tribunal and instead ordered that the appellant pay a total of £2,000 by way of fine and costs. The Court said that *Camancho v. Law Society* [2004] EWHC 1675 (Admin) and *Merrick v. Law Society* [2007] EWHC 2997 (Admin) were authority for the proposition that the means of a defendant to tribunal proceedings may be a relevant consideration in calculating the appropriate sanction as to the level of fines and costs. This would usually arise when a solicitor is being suspended from practice or struck off, but there will be exceptional cases in which, even though a solicitor is allowed to continue in practice, his income may be a relevant consideration both as to any costs sanction and in respect of any financial penalty that might be imposed.

***Solicitors Regulation Authority v. Davis and McGlinchey* [2011] EWHC 232 (Admin)**

**15.13** Disciplinary proceedings—costs principles

The SRA appealed against the decision of the SDT to make no order for costs against D and M after findings that both should be removed from the roll of solicitors. D was bankrupt and had no funds, and M was on Jobseeker's Allowance. In refusing to make an order for costs, the chairman of the Tribunal said that, 'in view of both the respondents' financial position', there should be no order for costs. On appeal, the SRA submitted that proceedings before the SDT were civil proceedings, so that the ordinary civil costs rule should apply as a matter of principle, that the Solicitors Disciplinary Rules made no express provision that means should be taken into account, and that the possibility that a solicitor's means should be taken into account would impose upon the remainder of the profession the financial burden of disciplinary proceedings brought against an aberrant solicitor. In rejecting these arguments, Mitting J said that disciplinary proceedings are not ordinary civil proceedings. A solicitor does not participate in the disciplinary proceedings voluntarily; he is subjected to them. Cases of *Camancho* and *Merrick* had been applied subsequently in *D'Souza*, and were authority for the proposition that the means of a defendant to tribunal proceedings may be a relevant consideration in calculating the appropriate sanction as to the level of fines and costs.

***Levy v. Solicitors Regulation Authority* [2011] EWHC 740 (Admin)**

**15.14** Defendant partly successful—whether discount appropriate

L was suspended from practice for a period of nine months and ordered to pay costs of £26,000. He was successful at the SDT in defeating the allegation of dishonesty. On appeal, a sanction of suspension was dismissed. As to costs, L submitted that the Tribunal gave no discount when awarding costs against him when he was successful in defeating the allegation of dishonesty. Cranston J (with whom Jackson LJ agreed) said that the starting point in considering costs was that rule 18 of the Solicitors (Disciplinary Proceedings) Rules 2007 conferred a wide discretion on the Tribunal as to awarding costs, even when no allegation of misconduct is proved against a respondent. It was clear that the ordinary rules of costs following the event do not apply in disciplinary proceedings. A disciplinary body is not in the same position as a party in ordinary civil litigation. It is acting in the public interest and should not be dissuaded from properly brought proceedings by an adverse costs order. However, it may be that it is appropriate to give a discount on the costs awarded to the disciplinary body when its success is only partial. Whether and to what extent the Tribunal should order a discount on costs awarded to the SRA, where one or more of its allegations is defeated, is very much a matter within its discretion. As Moses LJ observed in *Baxendale-Walker*, at [49], the Court has none of the advantages that the Tribunal has of assessing the extent to which the appellant should bear the costs of the hearing in the circumstances of the particular case. In *Beresford v. Solicitors Regulatory Authority* [2009] EWHC 3155 (Admin), [118], it was said that appeal courts reopen costs decisions only if plainly wrong. In the instant case, the Tribunal could not be said to be plainly wrong. It had heard the evidence and decided that it was reasonable for the appellant's honesty to be investigated. Although ultimately it found that that allegation of dishonesty was not proved, it was entirely appropriate for the Tribunal, seized of all of the circumstances, to decide that the appellant should bear all of the costs.

### *Adeeko v. Solicitors Regulation Authority* [2012] EWHC 841 (Admin)

**15.15** Enforcement of order for costs—stay

A challenged adverse findings made by the SDT that he had failed to keep books of accounts properly written up, contrary to rule 32 of the Solicitors Accounts Rules 1998. There were a number of allegations against A. Some of the matters included allegations of dishonesty. As matters proceeded, some allegations were withdrawn after the late provision of information by A. But a number of the allegations were denied and were tried by the Tribunal. In the event, the Tribunal convicted A of a single allegation only—namely, the failure to keep books properly contrary to rule 32. Having found a breach of rule 32, A was suspended from practice for a period of eighteen months with immediate effect and ordered to pay costs of £10,000, not to be enforced without leave of the SDT. Dismissing A's appeal against the suspension order, the Court of Appeal (Stanley Burnton and McFarlane LJJ, and Treacy J) turned to the question of costs. A submitted that the sum of £10,000 was unrepresentative of the relative success of the parties in the disciplinary proceedings: at the conclusion of the proceedings, only a single breach had been established. Moreover, it was contended that the Tribunal failed to consider A's financial circumstances before making the award of costs: at the time of the hearing, A had been suspended for over a year and was said to be without income. He owed £800,000 on a judgment debt and was otherwise heavily indebted. In dismissing A's appeal on costs, the Court said that A's financial circumstances were put forward before the Tribunal. Before the Tribunal, the sum of £12,000 was agreed as reasonable in principle for costs, but it was argued that no order should be made in view of A's financial circumstances. The Tribunal apportioned the costs between the appellant as to £10,000 and FK £2,000, who had not participated in the proceedings. The investigation was properly brought and the Tribunal was not in error in making the order for costs in the sum of £10,000; by further ordering that the costs should not be enforceable without further order, the Tribunal appropriately paid regard to the difficult financial circumstances faced by the appellant. In three months following the appeal, the appellant may be free to practise once again as a solicitor.

### *Sharma v. Solicitors Regulation Authority* [2012] EWHC 3176 (Admin)

**15.16** Means—defendant to adduce evidence

S, a solicitor, admitted allegations of dishonesty. The SDT ordered that S be struck off and pay costs of £49,000. S appealed against the costs order. The Court (Elias LJ and Singh J) held that there could be no general rule that the Tribunal should not impose an order for costs in addition to an order striking off a solicitor or an order of suspension. In *Merrick v. Law Society* [2007] EWHC 2997 (Admin), the Tribunal had not conducted a means inquiry in relation to the order of £45,000 in that case. When an order for costs is made, effectively depriving a solicitor of his livelihood, the question necessarily arises as to how any order for costs will be paid. In the instant case, S had been alive to the need to address the Tribunal as to his means. He owned three properties and received rental income from two of them. He had produced schedules detailing his income, expenditure, assets, and liabilities, but had not adduced any independent valuations of the properties. There had been no error of law in regard to the issue of costs, and S had presented matters to the Tribunal both orally and on paper through counsel. Accordingly, the appeal would be dismissed.

### *Agyeman v. Solicitors Regulation Authority* [2012] EWHC 3472 (Admin)

**15.17** Defendant absent at hearing—need to warn of costs consequences

The appellant solicitor did not attend the hearing before the SDT. The hearing went ahead in his absence and the Tribunal found the allegations proved, and ordered him to be suspended indefinitely and to pay costs of £35,824. In allowing the appellant's appeal in part, on the issue of costs, the Divisional Court (Elias LJ and Singh J) said that, when ordering a solicitor to pay costs, it was relevant for the Tribunal to consider the solicitor's means: see *D'Souza v. Law Society* [2009] EWHC 2193 (Admin). After the Tribunal has determined whether disputed charges were made out, it might be necessary to hear evidence and submissions on costs for that matter to be fairly addressed: see *Solicitors Regulation Authority v. Davis* [2011] EWHC 232. Although the appellant in the instant case had chosen not to appear, it seemed that he had not been warned of the potential costs consequences. Accordingly, the issue of costs should be remitted to the Tribunal and, to that extent only, the appeal succeeded.

### *Mireskandari v. Solicitors Regulation Authority* [2013] EWHC 907 (Admin)

**15.18** Security for costs

The SRA applied for security for costs under CPR 25.15 and costs as a condition of appeal under CPR 52.9(1)(c), in relation to the appellant's statutory appeal under section 49 of the Solicitors Act 1974. The appellant was in the United States and had been declared bankrupt. The Court ordered the appellant to provide security in the sum of £150,000. In addition, the SRA sought an order for payment of £206,598, representing the balance due on outstanding costs orders to be paid as a condition for pursuing the appeal. The Court considered that such an order was open to be made, even though there may be a statutory right of appeal. However, on the facts, to oblige the appellant to make a further payment in addition to the £150,000 for security for costs would be to deny him a right of appeal.

### *Solicitors Regulation Authority v. Anderson Solicitors* [2013] EWHC 4021 (Admin)

**15.19** Costs schedule—award—calculation unclear

The respondent firm and five solicitors faced eight allegations under the Solicitors Code of Conduct 2007. The charges were mainly in relation to conveyancing matters: failing to provide adequate costs information and/or providing clients with misleading costs information; overcharging conveyancing clients; and making charges in excess of those set out in the firm's costs documentation. The respondents were acquitted of lack of integrity and taking unfair advantage of clients. The Administrative Court (Treacy LJ and King J) said that taking advantage of another must involve an element of consciously taking unfair advantage; it was not sufficient for the solicitor to have taken deliberate action that had the consequence of taking unfair advantage to the client. No penalty was imposed on the firm, but the penalty of £1,000 imposed by the SDT on the sole equity partner would be increased to £15,000, to mark the gravity of the offending alongside such mitigation as was available, and the penalties on each of the remaining respondents should be increased from £1,000 to £5,000. What occurred in relation to costs was unsatisfactory. The SRA produced a costs schedule for £142,277 at the end of a five-day hearing on liability. The Tribunal summarily assessed costs in the sum of £80,000 on a joint and several basis. However, the reasons given were wholly inadequate and gave no clear indication as to how the sum of £80,000 was arrived at. An award of costs of this magnitude could not be sustained given the sparsity of reasoning and the level of costs awarded were somewhat surprising, given the relatively nominal penalties imposed by the Tribunal. The question of costs was to be remitted for further consideration by the Tribunal, together with any mitigation in relation to the penalties.

***Broomhead v. Solicitors Regulation Authority* [2014] EWHC 2772 (Admin)**

**15.20** Order—means—whether to grant adjournment—reduction for partial success

The SDT found proved two allegations against B, a solicitor. It rejected a third. It reprimanded B and ordered that he pay the costs of the SRA, which had brought the proceedings. Those costs were to be subject to detailed assessment. In dealing with the costs order, Nicol J (with whom Elias LJ agreed) said that Mitting J in *Solicitors Regulation Authority v. Davis* [2011] EWHC 232 (Admin), [21], proposed that the means of an individual against whom a costs order is proposed should be investigated by the Tribunal. If the solicitor were to assert that he or she was impecunious and that, for that reason, no order should be made, it would be incumbent on him or her to lay sufficient information before the Tribunal to persuade it that that is indeed the case. Mitting J was saying that a solicitor who contested charges could not reasonably be expected to make disclosure of his or her means in advance of the hearing. He was not saying that the solicitor, in the event of a finding of guilt, would be entitled to an adjournment to put his or her financial evidence before the Tribunal. Whether to grant an adjournment would be a matter for the Tribunal to consider in its discretion. The usual assumption is that the solicitor must come to the Tribunal together with all of the material he or she wishes to put before it. In the instant case, the advice that the SRA gave to B in its letters of 22 and 27 March 2012 was to that effect, and it was correct. This ground of appeal failed. Before the Tribunal, the SRA had submitted that there should be no discount. First, the allegation that failed had been unsuccessful in its specific facts and the investigation had formed a limited part of the whole case. Second, in any event, the SRA submitted, all of the charges had been properly brought. The Tribunal accepted the second of these arguments; it is not clear whether it accepted the first. Whilst the Court would not take issue with the finding that all three charges were properly brought, it would allow the appeal to the extent of varying the Tribunal's costs order so that B was required to pay 80 per cent of the SRA's costs before the Tribunal, to be subject to detailed assessment if not agreed. Nicol J said:

> 40. The SRA had proved two of the three charges they brought. Undoubtedly the charges on which the SRA succeeded had occupied the principal part of the time before the Tribunal. The argument as to whether the Appellant was subject to the regime in Schedule 1A of the 1974 Act was also common to the first two charges. Nonetheless, [a client of the appellant] had been called to give evidence exclusively on the charge which the SRA had failed to substantiate and [another witness] and the

Appellant, in part, had had to give evidence on that matter as well. The part of the proceedings on which the SRA had failed could not be dismissed as trivial. The position in this case is therefore different from *Beresford v. Solicitors Regulation Authority* [2009] EWHC 3155 (Admin) where no discount was made because "the allegations on which the Appellant succeeded were but a small fraction of a very serious whole" – see [120]. Had the Tribunal made a similar finding in this case the position may well have been different.

41. I would not take issue with the finding that all three charges were properly brought. The propriety of bringing unsuccessful charges is a good reason why the SRA should not have to bear the costs of the solicitor. The SRA is, after all, a regulator and should not be dissuaded from carrying out its task fearlessly because of a concern that it would have to pay costs if unsuccessful – see *Baxendale-Walker v. The Law Society* [2008] 1 WLR 426 at [39] (always assuming that the charges are properly brought).

42. However, while the propriety of bringing charges is a good reason why the SRA should not have to pay the solicitor's costs, it does not follow that the solicitor who has successfully defended himself against those charges should have to pay the SRA's costs. Of course there may be something about the way the solicitor has conducted the proceedings or behaved in other ways which would justify a different conclusion. Even if the charges were properly brought it seems to me that in the normal case the SRA should have to shoulder its own costs where it has not been able to persuade the Tribunal that its case is made out. I do not see that this would constitute an unreasonable disincentive to take appropriate regulatory action.

### *R (Lonsdale) v. Bar Standards Board*, 8 October 2014 (unreported)

**15.21** Order for costs to include expenses

In awarding costs against the appellant, Patterson J held that the word 'costs' in regulation 31(1) of the Disciplinary Tribunals Regulations 2009 ('a Disciplinary Tribunal shall have power to make such Orders for costs, whether against or in favour of a defendant, as it shall think fit') was apt to embrace all costs and expenses incurred by the Bar Standards Board, including those sums that the Board was liable to pay for the running costs of the Disciplinary Tribunal. The regulatory system was operated in the public interest and it was fair for the costs of running the system to be met in part by those who had used it.

## D. Costs on Appeal

### *Shrimpton v. General Council of the Bar* [2005] EWHC 2472

**15.22** See paragraph 15.04.

### *Walker v. Royal College of Veterinary Surgeons* [2008] UKPC 20

**15.23** Successful appeal—costs against regulator

In allowing W's appeal against an order removing his name from the register and substituting an order for his suspension for six months, the Board ordered the Royal College of Veterinary Surgeons (RCVS) to pay W's costs of the appeal. The authorities of *City of Bradford Metropolitan District Council v. Booth* [2000] COD 338, *Gorlov v. Institute of Chartered Accountants* [2001] EWHC 220 (Admin), and *Baxendale-Walker v. The Law Society* [2007] EWCA Civ 233 concerned the different position of costs before disciplinary tribunals or before a court upon a first appeal against an administrative decision by a body such as a police or regulatory authority. In the present case, the Disciplinary Committee of the RCVS made no order for costs in respect of the proceedings before it and no one had challenged that. The appeal before the Board was under section 17 of the Veterinary Surgeons Act 1966, and the Board has in practice made costs orders against the College when an appeal has succeeded and in its favour in cases of unsuccessful appeals. A similar position has applied with appeals from other disciplinary committees. No order for costs was made in two cases in which the appeal failed on liability, but succeeded on penalty.

### *Bryant v. Law Society* [2009] 1 WLR 163

**15.24** Successful appeal—solicitors substantial winners—normal CPR applies

On the successful appeal of solicitors to the Divisional Court, counsel submitted that the Law Society should pay the costs of the appeal and that the solicitors should have to bear only 50 per cent of the costs of the tribunal hearing. Richards LJ and Aikens J said they accepted that the allegations against the solicitors before the tribunal could have been expressed in clearer and more precise language, and that the tribunal hearing might have been shorter as a result. But the fact remained that the solicitors, by their professional misconduct, brought the proceedings upon themselves. Accordingly, the proper way in which to deal with costs was to leave the order for the costs of the tribunal hearing undisturbed. As for the appeal, the solicitors should have their costs. The Court said, at [251]:

> As for the appeal, we see no reason why the normal approach under the CPR should not apply. We consider that the solicitors are the substantial winners, even though they have not succeeded in displacing all the findings that were challenged. We therefore order that the Law Society should pay the costs of both solicitors on this appeal. We will also order that those costs be set off against the costs that the solicitors were ordered to pay in respect of the tribunal hearing.

### *R (Commissioner of Police for the Metropolis) v. Police Appeals Tribunal and Naulls* [2013] EWHC 1684 (Admin)

**15.25** Costs before internal appeal body

On 12 October 2010, at the conclusion of misconduct proceedings, Inspector Naulls was dismissed without notice by a misconduct hearing panel. Inspector Naulls appealed against that dismissal and, following a hearing on 21 October 2011, the Police Appeals Tribunal decided that the appeal should be allowed as to the disciplinary action taken, so that Inspector Naulls was reinstated and given a final written warning to last for eighteen months. The Tribunal also decided that Inspector Naulls' costs should be paid by the police authority. In dismissing the Commissioner's challenge to the decision to reinstate Inspector Naulls, Dingemans J quashed the decision of the Tribunal as to costs. The learned judge said that the reasoning by the Tribunal in allowing Inspector Naulls' costs on his successful appeal from the original misconduct hearing panel did not take account of *Baxendale-Walker v. Law Society* [2006] 3 All ER 675, [44], which was affirmed on appeal: [2008] 1 WLR 426. The principle relating to costs in regulatory proceedings applies to any internal appeal before the Police Appeals Tribunal. The principles of law were not considered by the Tribunal, and the failure to take such fundamental principles into account meant that its decision was not lawful.

### *Professional Standards Authority for Health and Social Care v. General Pharmaceutical Council and Onwughalu* [2014] EWHC 2521 (Admin)

**15.26** Successful appeal by PSA

On 25 February 2013, O pleaded guilty to two counts of cruelty to a child; on 12 April 2013, she was sentenced to a term of fourteen months' imprisonment concurrent on each count. After a hearing on 29 January 2014, the Fitness to Practise Committee of the General Pharmaceutical Council found that the registrant's fitness to practise was impaired by reason of the convictions. However, in relation to sanction, the Committee declined to order her removal from the Council's register and decided instead to suspend her from the register for a period of twelve months. The Professional Standards Authority (PSA) appealed that decision on the grounds that it was unduly lenient. The

Council did not resist the appeal, and supported the PSA's request for the Committee's decision on sanction to be quashed and for the Court to substitute an order for the registrant's removal from the register. Cox J said that, in her judgment, the ordinary rule should apply—namely, that costs should follow the event. The PSA's statutory function is to bring an appeal in those cases in which it is considered that the decision of the relevant regulatory body has been unduly lenient. The fact that, as in this case, the regulator has drawn the PSA's attention to a perceived deficiency in a decision does not mean that costs should not follow the event if the appeal succeeds. The Council was under an obligation to bring the Committee's error to the attention of the PSA and it could not seek a discount in relation to costs as a consequence of fulfilling that obligation. *Baxendale-Walker v. The Law Society* [2007] EWCA Civ 233 does not assist in these circumstances. Further, the appeal had succeeded as a result of a failing on the part of the regulator, through one of its committees, and the costs incurred were therefore an unavoidable consequence of that failing. There was no basis for the suggestion that such a committee is somehow independent of the Council. In relation to quantum, the PSA accepted that the Council effectively, and entirely properly, conceded the appeal and that it had not sought to escalate costs in this case. Notwithstanding the Council's early decision to concede the appeal, O's right to be heard rendered it necessary for there to be full preparation for a hearing, at which she may have attended and taken part. The learned judge ordered that costs should be divided equally between the respondents, the Council and O therefore being responsible for 50 per cent of the assessed costs.

***Bass and Ward v. Solicitors Regulation Authority* [2013] 5 Costs LO 651**

**15.27** Successful appellant—costs awarded

In allowing the appeals of B and W, Bean J reduced the sanction to a reprimand and reduced the financial penalty by half. As to costs below, the SDT made orders for costs against each respondent corresponding to the amount of the penalty. The learned judge said that he proposed to make the same reduction in the order for costs below, not simply to track the amount of penalty, but because it was a sensible and rough-and-ready reflection of the fact that had one of the allegations not been pursued, the hearing would have been likely to be shorter—probably one day, rather than two. As to costs of the appeal, the learned judge said that, in his judgment, the decision in *Baxendale-Walker v. The Law Society* [2007] EWCA Civ 233 does not govern costs on appeal, and therefore there was no reason in principle why B and W should not recover costs on the appeal. As to the facts, whilst the appellants did not do as well as they were arguing for, W and B had to come to the court to get the finding on liability on one issue set aside. They succeeded. They had to come to the court to get the sanction reduced. They succeeded. The arguments on which the appellants were unsuccessful did not significantly add to the length of the hearing. Accordingly, the appellants were to have their costs of the appeal against the SRA.

***Quinn v. Bar Standards Board*, 25 February 2013 (unreported)**

**15.28** Successful appeal before Visitors—costs awarded

In 2012, Q was convicted by a disciplinary tribunal of three charges of professional misconduct and the tribunal imposed concurrent sentences of suspension from practice for two years. On 25 February 2013, the Visitors to the Inns of Court (Sir Wyn Williams, chairman) allowed an appeal and quashed the convictions. Q applied for costs, initially for an order that the Bar Standards Board should pay her costs both of the appeal and below, but ultimately confined to seeking an order for costs in relation to the appeal. Sir Wyn Williams said that the Visitors had no doubt that the approach derived from a series of cases including *Baxendale-Walker v. The Law Society* is justified in first-instance proceedings—that is, in this context before the tribunal. The Visitors doubted, however, whether the principle is applicable when the decision of the first-instance body is under appeal. He continued:

> 46. Be that as it may, in this case there are factors which pointed clearly in favour of making an order for costs. Almost as soon as the appeal was launched the Appellant's legal advisers invited the Respondent to concede the appeal on the ground which has succeeded. That offer was rejected, most recently, shortly before the hearing. In our judgment, there was always a very good prospect that the appeal would succeed...yet the respondent decided to contest the appeal with vigour.
>
> 47. There can be little doubt from the information available that funding the appeal has been difficult for the Appellant. In the context of this appeal we took that consideration into account.
>
> 48. Unhesitatingly,...justice demanded that the Respondent should pay the Appellant's costs of and incidental to the appeal.

### *Williams v. Bar Standards Board*, 31 October 2014 (unreported)

**15.29** Successful appellant—unfettered discretion by Visitors to award costs

Following W's successful appeal to the Visitors (see paragraph 47.62), the Visitors (Asplin J, chairman) stated that, in the light of the broad wording of the Hearings before the Visitors Rules 2010, and of the decision in *Bass and Ward v. Solicitors Regulation Authority* [2013] 5 Costs LO 651, with which they agreed, they did not consider themselves bound to follow the *Baxendale-Walker* principle in relation to the costs of the appeal. As a result, the Visitors' discretion is unfettered. Although finely balanced, the Visitors considered by a majority that the correct order in this case was that there be no order as to costs on the grounds that the facts at issue in the appeal and the circumstances in which W was involved were complex and to some extent unclear. Although W was ultimately successful, this was not an appeal in which there were necessarily 'very good prospects of success' from the outset, as in the *Quinn* case (above); and unlike *Shrimpton v. General Council of the Bar* [2005] EWHC 2472, this was not a case in which the Bar Standards Board could properly be criticized for bringing the original proceedings or defending the appeal, or for the way in which the proceedings were conducted. The authorities that do not relate to Bar disciplinary proceedings are not particularly helpful as to how the Visitors should exercise their discretion. Whilst W was successful in the appeal, this was a factor in his favour, but not determinative, and there were no other factors that would justify the exercise of discretion to award costs against the Bar Standards Board or in favour of W, save for his success on the appeal.

### *Khan v. General Medical Council* [2015] EWHC 301 (Admin)

**15.30** Indemnity costs—appellant guilty of conduct out of norm

The appellant was convicted following trial of two offences of fraud for financial gain and two offences of theft, and was sentenced to nine months' imprisonment. The appellant retained some blank prescription forms when leaving a medical practice where he had been employed as a doctor, and wrote two fake prescriptions for diazepam and temazepam in favour of someone known as 'SK'. The appellant did not attend the fitness to practise proceedings and, on 20 February 2014, a fitness to practise panel of the GMC erased his name from the register. In dismissing his appeal, Mostyn J said, at [28], that: 'Very surprisingly the appellant appealed to this court on 30 March 2014. I say very surprisingly as it is arguably an abuse of process to appeal when you have not troubled to engage anything other than minimally with the FTPP proceedings and have not attended that hearing.' On 9 January 2015, the appellant sent an email to the GMC in which he stated that he would be willing to consider withdrawing his appeal if the GMC were to apologize and admit that he had been treated unfairly, as compared to other doctors who had been in similar situations, and if the GMC would restore his licence to practise. The GMC applied for indemnity costs. In granting indemnity costs, the learned judge referred to *Three Rivers District Council and ors v. Governor and Company of the Bank of England* [2006] EWHC 816 (Comm), in which Tomlinson J, as he then was, set out the relevant principles that apply when a claim for indemnity costs is made including, for example, 'where a claimant pursues a claim which is, to put it most charitably, thin and in some respects far-fetched'. In the instant case, the appellant had been guilty of conduct that took the case out of the norm.

### *R (Bar Standards Board v. Disciplinary Tribunal of the Council of the Inns of Court* [2014] 4 All ER 759*)*

**15.31** Award of costs—assessment

When awarding costs against the Bar Standards Board the Civil Procedure Rule 1998 do not apply, and the correct basis for assessing such costs is in accordance with the BSB's own rules under reg 31(1) of the Disciplinary Tribunals Regulations 2009, namely to award such costs as the Tribunal thinks fit. A reasonable figure would be £60 per hour, taking into account that the successful barrister, who had represented herself during the disciplinary proceedings determined in her favour, was not practising at the time.

# 16

# DEFICIENT PROFESSIONAL PERFORMANCE



## A. Legal Framework

**16.01** *R (Vali) v. General Optical Council* | |
| B. **Disciplinary Cases** | 16.02 | [2011] EWHC 310 (Admin) | 16.11 |
| *Krippendorf v. General Medical* | | *Zia v. General Medical Council* [2012] | |
| *Council* [2001] 1 WLR 1054 | 16.02 | 1 WLR 504, CA | 16.12 |
| *Qureshi v. General Medical Council* | | *Depner v. General Medical Council* | |
| [2003] UKPC 56 | 16.03 | [2012] EWHC 1705 (Admin) | 16.13 |
| *Sadler v. General Medical Council* | | *Sarkar v. General Medical Council* | |
| [2003] 1 WLR 2259 | 16.04 | [2012] EWHC 4008 (Admin) | 16.14 |
| *Holton v. General Medical Council* | | *Luthra v. General Medical Council* | |
| [2006] EWHC 2960 (Admin) | 16.05 | [2013] EWHC 240 (Admin) | 16.15 |
| *Calhaem v. General Medical Council* | | *Okeke v. Nursing and Midwifery Council* | |
| [2007] EWHC 2606 (Admin) | 16.06 | [2013] EWHC 714 (Admin) | 16.16 |
| *Gupta v. General Medical Council* | | *Chakrabarty v. Ipswich Hospital NHS* | |
| [2007] EWHC 2918 (Admin) | 16.07 | *Trust and the National Clinical* | |
| *R (Remedy UK Ltd) v. General Medical* | | *Assessment Service (Interested Party)* | |
| *Council* [2010] EWHC 1245 | | [2014] EWHC 2735 (QB) | 16.17 |
| (Admin) | 16.08 | *Al-Mishlab v. Milton Keynes Hospital* | |
| *Uruakpa v. General Medical Council* | | *NHS Foundation Trust* [2015] | |
| [2010] EWHC 1302 (Admin) | 16.09 | EWHC 191 (QB) | 16.18 |
| *Lim v. Royal Wolverhampton* | | *R (Woods and Gordon) v. Chief* | |
| *Hospitals NHS Trust* [2011] | | *Constable of Merseyside Police* | |
| EWHC 2178 (QB) | 16.10 | [2014] EWHC 2784 (Admin); | |
| | | [2015] ICR 125 | 16.19 |

Examples include:

Medical Act 1983, sections 35C and 35D (allegation that a practitioner's fitness to practise is impaired by reason of deficient professional performance)

Nursing and Midwifery Order 2001:

- article 22 (allegation that a registrant's fitness to practise is impaired by reason of lack of competence)
- article 35 (medical assessors to advise screeners and practice committees on matters within their professional competence)

Pharmacy Order 2010:

- article 51 (allegation that the fitness to practise of a registered pharmacist or registered pharmacy technician is impaired by reason of deficient professional performance, which includes competence, or of failure to comply with a reasonable requirement imposed by an individual assessor or an assessment team in connection with carrying out a professional performance assessment)
- article 55 (professional performance assessment)
- article 64 (clinical and other specialist advisers)

General Medical Council (Fitness to Practise) Rules 2014:
rule 3 (appointment of panels of advisers, assessors, and examiners)
rule 7 (investigation of allegations, including an assessment of the practitioner's performance or health to be carried out in accordance with Schedule 1 or 2)
rule 10 (undertakings)
rule 14 (appointment of specialist advisers)
rule 17 (procedure before a fitness to practise panel)

## B. Disciplinary Cases

***Krippendorf v. General Medical Council* [2001] 1 WLR 1054**

**16.02** Nature of assessment—paediatrician—assessment of work practitioner doing—no findings made concerning complaint

K had extensive expertise in many parts of the world in training and in health promotions in various aspects of public health, especially maternal and child health. She was internationally recognized as a specialist in child health. Whilst employed as a locum consultant community paediatrician by the Shetland Health Board, a formal written complaint was made to the General Medical Council (GMC) concerning K's professional competence. Specifically, it was alleged that she had carried out a schools immunization programme and had administered vaccine to children and had used the incorrect injection technique, leading to an unusually high incidence of side effects. This complaint triggered a screening process by the GMC, the appointment of an assessment panel, and an eventual hearing before the Committee on Professional Performance (CPP). The Committee determined that the standard of K's professional performance was seriously deficient and directed that her registration should be suspended for a period of twelve months. K appealed. In quashing the determination of the Committee, the Privy Council agreed with the definition of 'serious deficient performance' as being 'a departure from good professional practice, whether or not it is covered by specific GMC guidance, sufficiently serious to call into question a doctor's registration'. However, the assessment report, although long and careful, demonstrated a basic error in law in the panel's approach to its functions. The wording of section 36A of the Medical Act 1983 showed that it is the past professional *performance*, not the professional *competence*, of the practitioner in the work that he had actually been doing to which the Committee (and likewise the assessment panel) must direct its attention. Everything in the rules suggests that it is the duty of the assessment panel and the Committee to have regard to the track record of the practitioner in the work that he has actually been doing. It is not the function of the assessment panel and the Committee to conduct an examination equivalent to that of a student's examination. In the present case, instead of focusing its attention primarily on K's track record and the serious complaint in Shetland that had led to the assessment, the panel directed its attention primarily to assessing K's professional competence in a number of areas of work falling within what it perceived to have been her various job descriptions. The assessment was apparently conducted almost entirely without reference to K's performance in the work that she had actually been doing and the panel did not investigate or make any findings in relation to the complaint, even though it triggered the whole of the proceedings and was directly related to K's professional performance. As to the performance hearing, it was clear from the Committee's determination that it had relied primarily on the reports of the assessment panel. That report was flawed, and the failure of both the panel and the Committee to investigate the complaints reflected their erroneous concentration on K's professional competence rather than her actual professional performance. Accordingly, in the opinion of their Lordships, the Committee misdirected itself in law in reaching its determination.

### *Qureshi v. General Medical Council* [2003] UKPC 56

**16.03** Assessment in anaesthetics—no criticism or assessment in ophthalmology—whether assessment useful to Committee

A complaint was made arising out of Q's work as a locum staff grade doctor in anaesthetics at Doncaster Royal Infirmary. Q was willing to give up anaesthetics and specialize in ophthalmology, in relation to which no criticisms of his professional competence had been made. In delivering the judgment of the Privy Council, Lord Hoffmann said that, in the present case, ophthalmology was work that Dr Q had actually been doing and there seemed to their Lordships no reason in principle why he should not have been assessed in that field if the CPP had thought it would be helpful for this to be done. The complaint related only to performance in anaesthetics, but some skills were essential to the practice of virtually any branch of medicine and evidence of deficiencies in these respects in the practice of one form of medicine may suggest that there were similar deficiencies in the practice of other forms as well. Thus the nature of the complaint and the other materials considered by the screener may suggest to him that the practitioner's actual past performance in another branch of medicine may also have been seriously deficient. Their Lordships did not think that *Krippendorf's* case prevents an assessment of performance in that other branch. Such an assessment may be particularly useful to the Committee in a case such as the present in which the practitioner declares an intention to specialize in that other branch. Dr Q's appeal was dismissed.

### *Sadler v. General Medical Council* [2003] 1 WLR 2259

**16.04** Purpose of assessment—competence and performance—not negligence—burden and standard of proof

Lord Walker of Gestingthorpe, in delivering the judgment of the Board, said, at [17], that, without casting any doubt on the decision in *Krippendorf*, their Lordships felt that the distinction between competence and performance drawn in that case should not be taken too far:

> It is important that any assessment panel should have proper regard to the complaint or other information which originally set the assessment in motion. But in most cases there is an obvious correlation between competence and performance. Moreover the assessment panel is concerned, not only with assessing past professional performance, but also with what needs to be done to improve a practitioner's performance, both in the public interest and in the practitioner's own best interests.... the purpose of assessment is not to punish a practitioner whose standards of professional performance have been seriously defective, but to improve those standards, if possible, by a process of supervision and retraining, for the protection and benefit of the public. The process of assessment must include forming a view as to the standard of past performance, but if it is to achieve its objectives the process must not be restricted to that sort of backward-looking exercise.

At [63], the Board stated that it did not consider negligence to be a relevant or useful concept at a performance hearing. Negligence is concerned with compensating loss proved to have been caused by a breach of a practitioner's duty of care. Seriously deficient performance is a much wider concept, since it can extend to such matters as poor record-keeping, poor maintenance of professional obligations of confidentiality, or even deficiencies (if serious and persistent) in consideration and courtesy towards patients. It does not depend on proof of causation of actionable loss. On the other hand, one isolated error of judgment by a surgeon might give rise to liability in negligence, but would be unlikely, unless very serious indeed, to amount by itself to seriously deficient performance. The burden of proving sufficiently deficient performance rests on the regulator throughout and the standard should be the ordinary civil standard of proof. There may be exceptional cases in which a heightened civil standard might be appropriate as in *Re H* [1996] AC 563. S's appeal was dismissed.

***Holton v. General Medical Council* [2006] EWHC 2960 (Admin)**

**16.05** Assessment panel found no deficient performance—committee not bound by finding—standard applicable to post to which practitioner was appointed and to work being carried out

H's employer, the University Hospitals of Leicester National Health Service (NHS) Trust, suspended him in May 2001 as a result of concerns regarding his clinical performance in respect of the diagnosis and treatment of epilepsy. There ensued considerable public disquiet as to what was said to be his overdiagnosing of epilepsy in young children and his overprescribing of drugs. The Royal College of Paediatrics and Child Health produced an independent performance review in November 2001, which concluded that there were significant areas of concern in H's practice. It strongly recommended a period of further professional development. An assessment panel appointed under the General Medical Council (Professional Performance) Rules 1997 produced a lengthy report in January 2004. It graded H's professional performance under fifteen heads and concluded that his performance under all but two heads was acceptable; they graded those other two as cause for concern. They concluded that H's professional performance had not been seriously deficient, and the lead assessor (a consultant paediatrician) and the second medical assessor (a consultant paediatric neurologist) concluded that no remedial action was required. However, the lay examiner did not agree that no further action should be taken and, in consequence, the case was referred to the fitness to practise panel of the GMC. The panel found that the standard of H's professional performance had been seriously deficient; it departed from four of the assessors' gradings and directed that H's registration should be subject to conditions for a period of three years. Stanley Burnton J said that the case was unusual because of the contradiction between the finding of the fitness to practise panel and the assessors who had been appointed to consider the standard of H's professional performance. The assessors had concluded that his professional performance had not been seriously deficient. The case raised important issues relating to the matters to be taken into account by the fitness to practise panel in deciding whether a doctor's performance had been seriously deficient. In concluding that H's appeal against the findings of seriously deficient professional performance should be dismissed, the judge said that the assessors took the view, broadly, that the appropriate standard was that for which H had been trained; the panel took the view that the appropriate standard was that applicable to the post to which he was appointed and the work he was carrying out. There was no doubt that the panel was correct. Deficiency is to be judged against the standard of the doctor's professional work that is reasonably to be expected of the practitioner. In the instant case, there was no complaint relating to H's professional performance in general paediatrics. The question therefore arises whether deficiency only in part of a doctor's work can lead to a finding of seriously deficient professional performance. The judge said that he had no doubt that it can. It is certainly sufficient that that performance is in a significant part of a doctor's work. In the present case, the alleged seriously deficient professional performance was, on any basis, in a very substantial part of H's work. It follows that the panel was right to depart from the standard applied by the assessors. The panel had to do its best to identify the standard to be expected of a practitioner. The learned judge continued:

> 74. Consistently with this approach, it is irrelevant that the practitioner has not had sufficient training. Professional performance is no less deficient because the practitioner has not been sufficiently trained or educated to be able to render adequate performance. In my judgment, facts and factors personal to the doctor whose performance is being assessed are irrelevant to the question whether it is deficient. The test is objective in that sense. Thus his education, training and personality are irrelevant. Those are matters that may be addressed, if his performance is seriously deficient, by the conditions imposed by the panel, as it rightly said.

> 75. On the other hand, factors external to and independent of the doctor, such as the pressure of work, any lack of resources, and professional isolation due to the lack or absence of colleagues are relevant factors. As I remarked during argument, no one can sensibly expect, for example, a doctor working in Accident and Emergency at a time of crisis (for example after a road or rail crash when many serious casualties arrive at the same time) to be able to give the same time to patients as he would if he were not under pressure. His performance should be that which is to be expected of a competent practitioner in the circumstances. Professional isolation due to a doctor's personality or behaviour, on the other hand, is not a factor to be taken into account in assessing the adequacy of professional performance.

The fact that a doctor accepts that his performance has been seriously deficient and undertakes to remedy it cannot prevent a finding that it has been so. The learned judge went on to say that the panel must take the assessors' findings into account and will normally give them great weight, but is not bound by them. It is certainly not bound by them if it finds, correctly, that the assessors reached their conclusions on an incorrect basis. In that situation, the panel has no choice but to reach its own conclusions on the facts. If the panel does not have the benefit of the relevant expert evidence, it should normally ask its expert adviser to express his views on the facts on the basis that the panel considers to be applicable. The conditions imposed by a panel must relate to the specific deficient professional performance found by it; they must be necessary for the protection of the public and/or for remedying the deficiencies found, and they must be proportionate—that is, such conditions should not operate disproportionately in their interference with the practitioner's right to practise. In the instant case, the learned judge allowed H's appeal against two of the conditions, which were replaced; the remaining conditions were to remain.

### *Calhaem v. General Medical Council* [2007] EWHC 2606 (Admin)

**16.06** Test—level of performance indicating fitness to practise impaired—single episode exceptional

C was an experienced consultant anaesthetist practising at the North Staffordshire Nuffield Hospital in Newcastle-under-Lyme. He admitted that there were significant failures in relation to an operation on a patient. The fitness to practise panel found that C's fitness to practise was impaired because of misconduct and deficient professional performance, and suspended his registration for three months. On appeal, Jackson J held that the finding of impairment because of misconduct was upheld, but the finding of impairment because of deficient professional performance should be quashed and the order for suspension, together with the accompanying conditions, should in part, because of the time that had elapsed, be quashed. The learned judge said that, despite recent changes in the statutory regime and that allegations of deficient professional performance were previously dealt with by the GMC's CPP, the earlier authorities were still relevant. The word 'misconduct' in section 35C(2)(a) of the Medical Act 1983 does not mean any breach of the duty owed by a doctor to his patient; it connotes a serious breach that indicates that the doctor's fitness to practise is impaired. The phrase 'deficient professional performance' does not mean any instance of substandard work; it connotes a level of professional performance which indicates that the doctor's fitness to practise is impaired. Although the language in section 35C(2)(a) and (b), in its present form, differs from the language used in the old sections 36 and 36A, the meaning remains the same. In the instant case, the judge accepted that C's breaches of duty all concerned one patient. He would hesitate to call any of those breaches 'isolated'. Over a period of some five hours, C made a series of serious mistakes that put his patient at risk. Depending upon the circumstances, even one act or omission, if sufficiently serious, can amount to misconduct. In the present case, the series of serious mistakes that were made, when taken together, readily crossed the dividing line between (a) mere negligence and (b) serious professional misconduct, or misconduct. As to deficient professional performance, although it is possible for a finding of deficient professional performance to be based upon a single episode of treatment on one day, such a finding will be very rare and will arise from exceptional circumstances. The legal assessor failed in his advice to the panel to explain properly the respects in which deficient professional performance differed from misconduct. At [39] Jackson J derived five principles from a review of the authorities:

(1) Mere negligence does not constitute "misconduct" within the meaning of section 35C(2)(a) of the Medical Act 1983. Nevertheless, and depending upon the circumstances, negligent acts or omissions which are particularly serious may amount to "misconduct".
(2) A single negligent act or omission is less likely to cross the threshold of "misconduct" than multiple acts or omissions. Nevertheless, and depending upon the circumstances, a single negligent act or omission, if particularly grave, could be characterized as "misconduct".
(3) "Deficient professional performance" within the meaning of section 35C(2)(b) is conceptually separate both from negligence and from misconduct. It connotes a standard of professional performance which is unacceptably low and which (save in exceptional circumstances) has been demonstrated by reference to a fair sample of the doctor's work.
(4) A single instance of negligent treatment, unless very serious indeed, would be unlikely to constitute "deficient professional performance".
(5) It is neither necessary nor appropriate to extend the interpretation of "deficient professional performance" in order to encompass matters which constitute "misconduct".

### *Gupta v. General Medical Council* [2007] EWHC 2918 (Admin)

**16.07** Assessors' Handbook—cause for concern

G complained that the assessors' report did not follow the Assessors' Handbook and did not contain a full quotation on the meaning of 'cause for concern' given in the guidance to assessors. Burton J, in dismissing G's appeal against the panel's finding of deficient professional performance, said that 'cause for concern' was intended to be a category in which the assessment team was unable to make the clear finding of 'unacceptable'. The words set out in the report were not dissimilar to those in the Assessors' Handbook, although it would have been much better had the report said 'there is insufficient evidence to classify it as unacceptable'. It is plain that a doctor can be criticized in respect of those categories in which there are findings of 'cause for concern'. A category in respect of which there is a conclusion that the conduct is 'unacceptable' is plainly a matter for criticism as to deficiency. But it is still a serious matter of criticism for a doctor if he cannot prove competence, even if a conclusion of incompetence, or lack of proficiency, cannot be proved to the respect required.

### *R (Remedy UK Ltd) v. General Medical Council* [2010] EWHC 1245 (Admin)

**16.08** See paragraph 47.40.

### *Uruakpa v. General Medical Council* [2010] EWHC 1302 (Admin)

**16.09** Refusal by doctor to undertake assessment—need for assessment to ensure public safety

In this case, Saunders J held that it was for the GMC and not the High Court to decide what was the appropriate test for medical competence. It was not for a doctor to refuse to take an assessment because he did not like its structure. Where a doctor had continually refused to complete an assessment to ascertain his professional performance, the GMC's fitness to practise panel had been entitled to find his fitness to practise impaired and to impose the sanction of suspending his name from the medical register, given its obligation to ensure the safety of the public. The appellant doctor appealed against a decision of the panel suspending his name from the medical register for twelve months. He had qualified as a doctor in Nigeria and Australia, had come to the United Kingdom, and had been granted full registration by the GMC in 2003. He practised in the field of obstetrics and gynaecology, and worked in a number of hospitals. In 2005, his work was referred to the GMC in respect of the issues concerning the conduct of operations, performance on call, poor communication with patients, and missed diagnoses. The GMC's fitness to practise panel made an interim order restricting his practice. Before the final hearing of the issue of fitness to practise, the GMC asked the appellant to undergo assessment of his professional performance. Despite having twice agreed to do so, he ultimately declined to undertake any assessment, criticizing the structure of the proposed tests. The panel found that his fitness to practise was impaired and imposed the suspension. Saunders J held that where a charge against a doctor concerned clinical work, an appellate court had to accord deference to the decision of a panel of doctors. In the instant case, the panel had been concerned about the appellant's failure to undertake an assessment and, because of the length of time since he had practised, it was not possible to decide the extent of his deficiencies. Proper assessment was needed to ascertain his skills and, given his refusal to complete the assessment process and in the light of the panel's obligation to ensure the safety of the public, the sanction imposed was the correct one.

***Lim v. Royal Wolverhampton Hospitals NHS Trust***
**[2011] EWHC 2178 (QB)**

**16.10**
**NHS capability hearing—NCAS assessment**

Slade J held that the defendant NHS Trust would be in breach of a doctor's contract of employment by conducting a hearing about his capability without first referring the matter to a National Clinical Assessment Service (NCAS) panel for assessment and the panel advising that the doctor's performance was so flawed that no action plan would have a realistic chance of success. The claimant was a consultant anaesthetist employed by the Royal Wolverhampton Hospitals NHS Trust, and the claim concerned a proposed capability and conduct hearing to be held by the defendant. The Court held that a capability hearing should not be held until an NCAS assessment panel had determined that his professional performance was so fundamentally flawed that no educational and/or organizational action plan had a realistic prospect of success. Slade J held that the defendant would not be in breach of any contractual obligation in pursuing allegations of misconduct against the claimant, but that it would be in breach of contract in failing to comply with the procedure promulgated for dealing with issues of capability.

See also paragraph 16.17 below and the contrasting decision in *Chakrabarty v. Ipswich Hospital NHS Trust and the National Clinical Assessment Service (Interested Party)* [2014] EWHC 2735 (QB).

***R (Vali) v. General Optical Council* [2011] EWHC 310 (Admin)**

**16.11**
**Distinction between deficient professional performance and misconduct**

See paragraph 33.13 as to the facts. The Fitness to Practise Committee of the General Optical Council (GOC) found V's conduct amounted to misconduct and deficient professional performance. Although V was guilty of a single negligent act, Ouseley J said that the Committee was entitled to take the serious view that it did of a failure to follow a simple step in relation to a patient who exhibited three or, more probably, four indicators of glaucoma, a serious eye disease. Accordingly, the Committee's finding of misconduct would not be disturbed (the learned judge subsequently found no current impairment of V's fitness to practise). The Committee went on to consider the question of deficient professional performance. Ouseley J said, at [27], that although he had reservations about the third of Jackson J's principles in *Calhaem v. General Medical Council* [2007] EWHC 2606 (Admin)—because it does not entirely accurately reflect the judgment of Lord Walker in *Sadler v. General Medical Council* [2003] 1 WLR 2259, [62], which makes it clear that deficient professional performance may readily arise on a single event depending on the nature of the medicine being practised and the allegation—the learned judge had no hesitation at all about the fifth of Jackson J's principles, which cautions, entirely appropriately, against extending 'deficient professional performance' to encompass misconduct and, he would add, vice versa. The learned judge went on to draw the distinction between deficient professional performance and misconduct:

> 28. It is important that "deficient professional performance" should not be contorted so that it is a mere synonym for "misconduct" in practice, and that the essence of "deficient professional performance" is more in contrast to than coterminous with misconduct. It is intended at least to be different in that one would often, if not normally, expect to find a pattern of conduct underlying the allegation of deficient professional performance.
>
> 29. I do not intend to lay down any particular principle in those observations but to draw attention to the need for care over charging exactly the same conduct under both heads. Of course there may be different facets which justify the same facts, because viewed from different prospectives, being charged under both heads, but that is not this case. There may also be circumstances in which an allegation of misconduct may be made but be insufficiently strong in the end to warrant a finding, but the underlying facts should nonetheless not fall out of the picture for the purposes of a deficient professional performance charge.
>
> 30. Bearing in mind those observations, I conclude that the Committee seems wrong to find that this was deficient professional performance, because it had found that it was misconduct. It is necessary to keep, so far as possible, some distinction between the two. I accept that what happened here could have been found to be deficient professional performance, but it would in those circumstances have been wrong simply to found a misconduct charge proved.

### *Zia v. General Medical Council* [2012] 1 WLR 504, CA

**16.12** Refusal of assessment—registrar's decision to refer direct to panel—no inflexible rule to refer to case examiners

A hospital trust that had employed Z complained about his performance and competence to the GMC. The GMC's registrar referred Z for a performance assessment pursuant to rule 7(3) of the General Medical Council (Fitness to Practise) Rules 2004. Z refused to submit to the assessment and the registrar referred the allegations to the GMC's fitness to practise panel. The panel found the allegations proved and ordered Z's suspension from the register. Z appealed against that decision and argued, as a preliminary issue in the High Court, that the registrar's referral to the fitness to practise panel was impermissible as contrary to the obligation to refer any allegation in the first instance to case examiners. The Court of Appeal allowed the GMC's appeal against the decision of the High Court in favour of Z on this point. The Court of Appeal held that the stated objective of the Medical Act 1983, as set out in section 1, was for the GMC to protect, promote, and maintain the health and safety of the public, and the purpose of the 2004 Rules was to achieve a balance between meeting that aim and regulating the procedure for investigation and resolution of allegations against medical practitioners. Whilst the majority of cases were considered by both the registrar and case examiners, there was no inflexible requirement that a matter had always to be considered by the case examiners before proceeding further. The rules enabled the registrar to carry out his own investigation and to direct a performance assessment;

because of Z's non-compliance, the registrar's decision to refer the allegations directly to the panel was lawful.

***Depner v. General Medical Council* [2012] EWHC 1705 (Admin)**

**16.13** See paragraph 66.06.

***Sarkar v. General Medical Council* [2012] EWHC 4008 (Admin)**

**16.14** Whether performance assessment required before sanction

S was a general practitioner in Grimsby. He was a sole practitioner and had worked for almost twenty years until February 2009, when he was suspended by the Northeast Lincolnshire Care Trust as a result of concerns that had been raised by his practice manager. The concerns and the investigation that was subsequently undertaken related to: failure to provide or arrange adequate cover; inappropriate delegation, and thus inadequate provision of clinical management and patient care; inadequate record-keeping; non-compliance with the Quality Outcomes Framework; failure to provide management and leadership of his practice and the team working there; and submitting claims for payment in respect of extended hours when such hours had not, in fact, been performed. Charges against S of deficient professional performance and misconduct were found proved, and the panel directed that S's name be erased from the medical register. Rule 17(9) of the GMC (Fitness to Practise) Rules 2004 provides that, at any stage before making its decision as to sanction, the panel 'may adjourn for further information or reports to be obtained in order to assist it in exercising its functions'. The panel recorded that the legal assessor advised it that the option of a performance assessment would have been appropriate for the panel at the impairment stage under rule 17(4), rather than immediately prior to sanctions under rule 17(9), and that the panel accepted the advice of the legal assessor in this regard. S contended that the panel erred in accepting the advice of its legal assessor that a performance assessment was not appropriate at the sanction stage. In rejecting this ground of appeal, Beatson J said that, first, the panel did not state that it had no power to adjourn for a performance assessment; the legal assessor's advice was in effect that this would not have been "appropriate". Secondly, the panel was aware of the general power under rule 17(9) because it had been discussed only just before its ruling.

***Luthra v. General Medical Council* [2013] EWHC 240 (Admin)**

**16.15** Whether sanction of erasure disproportionate

The appellant, aged 70, faced an allegation that his fitness to practise was impaired by reason of deficient professional performance. No allegation of misconduct was made. The allegations were found proved and the panel determined that erasure was the only appropriate sanction. In dismissing L's appeal, Mostyn J said, at [38], that the fact that, in some respects, the appellant performed positively does not at all inform an assessment as to whether he could recognize his deficiencies and respond to retraining in relation to those fields in which he had failed. The learned judge said that, in that regard, the key failings must be considered and the appraisal of the assessors, and the panel, was that no amount of further opportunity for retraining would cure the appellant's core deficiencies in knowledge and defects in competence so as to eliminate the risk, irrespective of his positive performance in other fields.

***Okeke v. Nursing and Midwifery Council* [2013] EWHC 714 (Admin)**

**16.16** Whether power to strike off—Nursing and Midwifery Order 2001

The appellant faced an allegation of lack of competence, arising out of her employment as a band 6 midwife at Queen Mary's Hospital, and a separate allegation of misconduct. The legal assessor advised that a striking-off order was an available sanction in relation to the charge of lack of competence. Leggatt J disagreed, saying that the clear scheme of the regulations and their overall effect is that, by reason of article 29(6) of the Nursing and Midwifery Order 2001, a first finding of lack of competence against a person cannot result in an order that he or she be struck off the register; the person concerned may, at the most, be suspended. Only if the overall period of suspension or conditions exceeded two years can the person concerned be struck off.

***Chakrabarty v. Ipswich Hospital NHS Trust and the National Clinical Assessment Service (Interested Party)* [2014] EWHC 2735 (QB)**

**16.17** NHS capability hearing—NCAS assessment

The claimant, who was employed by the Ipswich Hospital NHS Trust, sought an injunction restraining the Trust from referring his case to a capability hearing panel to consider the extent of any deficiencies in his capability or competence to work as a consultant in cardiology and general medicine under the Trust's internal competence/capability procedures and the national procedure known as *Maintaining High Professional Standards in the Modern NHS* (MHPS). The primary basis for the injunction sought was that there had been no assessment of the prospects of success of remediation in the claimant's case by the NCAS and, accordingly, the Trust could not lawfully proceed to a capability hearing at all. Simler J concluded, at [128]–[146], that, having considered the national MHPS procedure, the Trust could proceed to a capability hearing despite the NCAS not having carried out an assessment of the claimant's case. The learned judge said that she respectfully disagreed with the conclusions of Slade J in *Lim v. Royal Wolverhampton Hospitals NHS Trust* [2011] EWHC 2178 (QB). Where the NCAS considers that no assessment is appropriate, that can conclude its involvement and leave open to the employer the possibility of achieving a resolution of any unresolved capability concerns by means of a capability hearing.

***Al-Mishlab v. Milton Keynes Hospital NHS Foundation Trust* [2015] EWHC 191 (QB)**

**16.18** Deficiencies found in surgeon's clinical practice and technical ability—exclusion by trust for four years—whether justified

The claimant, a colorectal, laparoscopic, and general surgeon, sought a declaration that he had been excluded from his employment with the defendant Trust in breach of contract and an injunction requiring the Trust to permit him to return to clinical practice. He had been excluded from all clinical work since March 2011 and before that had practised subject to restrictions from July 2010. An investigation by the Royal College of Surgeons had found deficiencies in the claimant's clinical practice and his technical ability. In dismissing the claimant's action, Laing J noted that the document entitled *Maintaining High Professional Standards in the Modern NHS* (MHPS) provided that the express purposes of MHPS include that NHS employers should use exclusion in only the most exceptional circumstances, when it is strictly necessary, for the shortest possible period, and that alternatives to exclusion must be considered. Under paragraph 35 of MHPS, there should normally be a maximum limit of six months' exclusion, except for those cases involving criminal investigations of the practitioner concerned. In the instant case, the Court concluded that the trust had not breached the terms of MHPS either in initially excluding the claimant, or in maintaining that exclusion, latterly described as a restriction, to date. The claimant's own representative suggested in September 2011, in a meeting to discuss his exclusion, that it would be preferable to the claimant to return to supervised practice at another trust, although he recognized that it would be premature to develop this before the internal investigation had finished. This supported the Trust's view that any return to clinical practice should, at any rate, be started elsewhere. Moreover, given the history of the matter, and the fact that the relationship between the claimant and key members of the multidisciplinary team at the Trust had broken down, and there was no suitable and willing consultant at Milton Keynes Hospital to supervise his work, the claimant's exclusion or restriction was necessary, and the Trust was reasonably entitled, under the terms of MHPS, to see it as necessary. Even if the Court had decided that the claimant's initial and/or continued exclusion had been in breach of MHPS, it would not have ordered the Trust to take him back to work. The Court would have made a declaration only and ordered an assessment of damages.

***R (Woods and Gordon) v. Chief Constable of Merseyside Police* [2014] EWHC 2784 (Admin); [2015] ICR 125**

**16.19** See paragraph 66.08.

# 17

# DELAY

## A. Legal Framework

**17.01** Examples include:

European Convention on Human Rights (ECHR), Article 6(1) (see paragraph 32.01)
General Pharmaceutical Council (Fitness to Practise and Disqualification etc) Rules 2010, rule 6 (registrar must not refer an allegation where more than five years have elapsed unless he considers it necessary for the protection of the public, or otherwise in the public interest, for the allegation to be referred)
General Medical Council (Fitness to Practise) Rules 2014, rule 4(5):

> No allegation shall proceed further if, at the time it is first made or first comes to the attention of the General Council, more than five years have elapsed since the most recent events giving rise to the allegation, unless the Registrar considers that it is in the public interest, in the exceptional circumstances of the case, for it to proceed.

## B. General Principles

### *Attorney General's Reference (No. 1 of 1990)* [1992] QB 630

**17.02** See paragraph 2.02.

### *Dyer v. Watson and anor; K v. HM Advocate* [2004] 1 AC 379

**17.03** Delay between charge and hearing—explanation—investigation

In the first case, W and B were police officers who gave evidence at a trial in the sheriff court at Linlithgow. When the trial ended, the sheriff, in open court, expressed the opinion that the officers had committed perjury. This statement received wide publicity in the press at the time. Both were charged with perjury and proceedings against them were commenced by way of summary complaint. Before the pleading diet, each of the officers gave notice of a devolution issue and claimed that there had been such delay in bringing proceedings against them as to breach the reasonable time requirement. The House of Lords allowed an appeal by the Crown from the decision of the High Court of Justiciary, confirming the decision of the sheriff to uphold pleas in bar of trial. At [56]–[57], Lord Bingham of Cornhill said that the period of twenty months between the charging of the officers and their trial was not, on its face and without more, such as to suggest that a basic human right of the officers may have been infringed. They were not in custody. Police officers are, by reason of their occupation, peculiarly susceptible to accusations of misconduct, many or most of which are found upon examination to be malicious, fabricated, or self-serving. The credibility of the accusers is often very suspect. The strong public interest in the integrity of the police and the interest of individual officers in vindicating their reputations require that accusations of misconduct against officers, when made, should be carefully and independently investigated by a body other than the police themselves. Even when, as here, strong criticisms of a police officer are voiced from the bench, the need for careful and independent investigation remains before any proceedings, whether criminal or disciplinary, are initiated. This inevitably takes time.

In the second case, which concerned a minor, then aged 13, charged with serious sexual offences against other minors, aged between 3 and 8, the House of Lords said that the reasonable time requirement in the European Convention on Human Rights (ECHR) must, when dealing with children, be read in the light of the UN Convention on the Rights of the Child and the Beijing Rules, both of which applied to K and both of which highlighted the need for criminal proceedings, if brought at all, to be prosecuted with all due expedition. A period of twenty-seven months had elapsed between the charging of K on 31 October 1998 and the service of the indictment upon him on 29 January 2001. On 29 March 2001, K, now aged 16, lodged a devolution issue. No satisfactory explanation had been given for a lapse of time that was inordinate and inexcusable, and which breached the reasonable time requirement. A trial in January 2002 (the date of decision by the Privy Council), some three-and-a-half years or more after the date of charge, would not be acceptable.

*Attorney General's Reference (No. 2 of 2001)* [2004] 2 AC 72

**17.04** Test—no longer fair trial or otherwise unfair to try defendant—remedy—time running

In February 2000, informations were laid against seven prisoners following a serious disturbance in April 1998 in an English prison. In June 2000, the defendants were committed for trial in the Crown Court, where they were charged in an indictment containing a single count of violent disorder contrary to section 2(1) of the Public Order Act 1986. The trial in the Crown Court was fixed to begin on 29 January 2001. When the matter came before the court, counsel for the defendants submitted to the trial judge that the delay in bringing the charge against the defendants had been such that to proceed with the trial would be to act in a way incompatible with Article 6 ECHR. The judge accepted this argument and ordered that the proceedings be stayed. Subsequently, the prosecution offered no evidence and the defendants were acquitted. On a reference by the Attorney-General to the Court of Appeal and referral for determination by the House of Lords, Lord Bingham of Cornhill said:

> 24. If, through the action or inaction of a public authority, a criminal charge is not determined at a hearing within a reasonable time, there is necessarily a breach of the defendant's Convention right under article 6(1). For such breach there must be afforded such remedy as may (section 8(1)) be just and appropriate or (in Convention terms) effective, just and proportionate. The appropriate remedy will depend on the nature of the breach and all the circumstances, including particularly the stage of the proceedings at which the breach is established. If the breach is established before the trial, the appropriate remedy may be a public acknowledgement of the breach, action to expedite the hearing to the greatest extent practicable and perhaps, if the defendant is in custody, his release on bail. It will not be appropriate to stay or dismiss the proceedings unless (a) there can no longer be a fair hearing or (b) it would otherwise be unfair to try the defendant. The public interest in the final determination of criminal charges requires that such a charge should not be stayed or dismissed if any lesser remedy will be just and proportionate in all the circumstances.... If the breach of the reasonable time requirement is established retrospectively, after there has been a hearing, the appropriate remedy may be a public acknowledgement of the breach, a reduction in the penalty imposed on a convicted defendant or the payment of compensation to an acquitted defendant. Unless (a) the hearing was unfair or (b) it was unfair to try the defendant at all, it will not be appropriate to quash the conviction ...
>
> 25. The category of cases in which it may be unfair to try a defendant of course includes cases of bad faith, unlawfulness and executive manipulation of the kind classically referred to in *R v. Horseferry Road Magistrates' Court, ex parte Bennett* [1994] 1 AC 42, but [counsel for the acquitted person] contended that the category should not be confined to such cases. That principle may be broadly accepted. There may be cases (of which *Darmalingum v. The State* [2000] 1 WLR 2303 is an example) where the delay is of such an order, or where a prosecutor's breach of professional duty is such (*Martin v. Tauranga District Council* [1995] 2 NZLR 419 may be an example), as to make it unfair that the proceedings against a defendant should continue. It would be unwise to attempt to describe such cases in advance. They will be recognisable when they appear. Such cases will however be very exceptional, and a stay will never be an appropriate remedy if any lesser remedy would adequately vindicate the defendant's Convention right.

[...]

27. As a general rule, the relevant period [for the purposes of article 6(1)] will begin at the earliest time at which a person is officially alerted to the likelihood of criminal proceedings against him. This formulation gives effect to the Strasbourg jurisprudence but may (it is hoped) prove easier to apply in this country. In applying it, regard must be had to the purposes of the reasonable time requirement: to ensure that criminal proceedings, once initiated, are prosecuted without undue delay; and to preserve defendants from the trauma of awaiting trial for inordinate periods. The Court of Appeal correctly held that the period will ordinarily begin when a defendant is formally charged or served with a summons, but it wisely forbore to lay down any inflexible rule.

***Spiers (Procurator Fiscal) v. Ruddy* [2008] 1 AC 873**

**17.05** Facts—determination—delay but fair trial still possible

R lodged a devolution issue with the House of Lords, contending that his right under Article 6(1) ECHR to be tried within a reasonable time in relation to various road traffic offences had been breached and that the Lord Advocate accordingly had no power to continue the proceedings against him. On the reference, the House of Lords held, first, that only where the facts had been found by the Scottish courts (as they were in *Dyer v. Watson* [2004] 1 AC 379) can the Board properly entertain a devolution issue relating to the question of unreasonable delay. Second, it is axiomatic that if an accused cannot be tried fairly, he or she should not be tried at all, and where either of the conditions in *Attorney General's Reference (No. 2 of 2001)* [2004] 2 AC 72 apply, the proceedings must be brought to an end: *per* Lord Bingham of Cornhill, at [15]. Where the fairness of the trial has not been or will not be compromised, such delay does not give rise to a continuing breach that cannot be cured save by discontinuance of proceedings. It gives rise to a breach that can be cured, even where it cannot be prevented, by expedition, reduction of sentence, or compensation, provided always that the breach, where it occurs, is publicly acknowledged and addressed: *per* Lord Bingham of Cornhill, at [16].

***R v. F (S)* [2012] 1 All ER 565**

**17.06**

See paragraph 2.05.

## C. Allegations First Made More than Five Years after the Event

***R (Aziz) v. General Medical Council* [2004] EWHC 3325 (Admin)**

**17.07** Event—application for hospital post—prior employment

By a letter dated 31 August 2000, the Postgraduate Dean and Director at the University of Newcastle wrote to the General Medical Council (GMC) following the dismissal of A as a senior registrar in clinical pharmacology, on the ground that, when applying for the post of senior registrar in 1996, he deceived the appointments panel that he had previously been suspended and then dismissed from a post at a hospital in Ireland. Accordingly, less than five years had elapsed since the events giving rise to the allegation. Stanley Burnton J held that, for the purposes of rule 6(7) of the GMC Preliminary Proceedings Committee and Professional Conduct Committee (Procedure) Rules, the allegation of misconduct did not concern anything that had happened in Ireland, where A was employed between 1994 and 1996; it concerned what had been said by A or what had been written by him at the time he applied in 1996 for his position in Newcastle. On the face of it, therefore, rule 6(7) would not prevent the case relating to the conduct that was the subject of that complaint being referred to the Preliminary Proceedings Committee (PPC) by a screener. If there were a delay after the first making of the complaint, then different considerations would arise and it would be open to A at the opening of the proceedings before the Professional Conduct Committee (PCC) to object that the proceedings against him were an abuse. It was far better that questions of delay and their effect were considered by the PCC, which is an expert body and better able to address and assess complaints of the kind raised by A.

*R (Peacock) v. General Medical Council* [2007] EWHC 585 (Admin)

**17.08** Letter of complaint by solicitors—post-settlement of claim—seven years after treatment—three years after medical evidence

On 16 May 1998, the complainant, W, was admitted to Bedford General Hospital, where she suffered a series of cardiac arrests. In 1998 or 1999, W consulted solicitors to investigate a potential claim against the claimant, her general medical practitioner, and Bedford General Hospital in relation to events that led to her sustaining the injuries. In December 1999, through her solicitors, W received a positive medical report from a professor of pharmacology; in March 2000, she received a positive medical report from a professor of general practice. In January 2001, civil proceedings were issued on behalf of W against the claimant and Bedford General Hospital. In April 2002, liability was compromised, the claimant accepting 25 per cent, and damages were eventually agreed in the sum of £3.5 million in 2005. Meanwhile, W remained registered as a patient at the claimant's practice and, on 9 May 2005, solicitors acting on her behalf first wrote to the GMC with a letter of complaint from their client dated 4 May 2005. The letter made a complaint in relation to prescriptions by the claimant in February–May 1998. In December 2005, the registrar decided to waive the five-year rule—that is, rule 4(5) of the 2004 Rules. The decision letter made no reference to exceptional circumstances. By a subsequent letter dated 10 May 2006, the GMC wrote a further letter setting out the full text of rule 4(5) and the registrar's reasoning. Gibbs J held that the claimant's application for judicial review to quash the decision of the registrar was not premature and the Court should entertain it. There were no circumstances that were capable of being found exceptional so as to make it in the public interest to take the complaint further. Despite the tragic circumstances, there was no potential basis for finding exceptional circumstances. The claimant had practised as a doctor for many years before and since the events relied upon, and had done so without any other complaint or blemish on his record. At its gravest, the allegation was one of serious negligence in the case of one patient in relation to one example of a series of prescriptions of drugs. Given the history of the case, it could not reasonably be said that there was a sensible explanation for the delay. All of the medical evidence relating to the claimant's negligence was available to the complainant's solicitors by 2002 at the latest. There was no reason to delay making the complaint for a further three years.

*R (Haward) v. General Medical Council* [2007] EWHC 2236 (Admin)

**17.09** Judicial review—delay by claimant

In March 2005, the GMC decided to waive the five-year rule in relation to two complaints, notwithstanding the age of the allegations. However, the challenge by judicial review was not made until April 2006, some thirteen months later. Burton J, in refusing to grant an extension of time, observed that there was obviously a powerful case to be made by the claimant that the decision of the registrar to waive the five-year rule was irrational when the events took place some fourteen years or more (and in some cases nineteen years) ago. The learned judge said: 'What the strength would have been of that case had an application been made within 3 months (of the decision) I need not conclude.'

*R (Gwynn) v. General Medical Council* [2007] EWHC 3145 (Admin)

**17.10** Exceptional circumstances—reasons—decision by registrar—case examiners

Sullivan J ordered a stay of proceedings against the claimant, a vascular and breast surgeon, in respect of five patients. The allegations against the claimant were that his fitness to practise was impaired by reason of deficient professional performance in respect of his conduct of breast surgery on ten patients over a period from September 1997 to December 2003. In the case of five patients, all of the complaints were made after the five-year period had elapsed. Counsel for the GMC submitted that the GMC's aide-mémoire was for guidance. The learned judge said, at [46], that it was not appropriate to either water down or reword the rule 4(5) test: the registrar must be satisfied that there are circumstances of the case that can fairly be described as 'exceptional circumstances' and that proceeding with the case is in the public interest, in those exceptional circumstances. The exceptional circumstances may, but need not necessarily, relate to the lapse of time. Whilst the 2004 Rules do not impose a duty to give reasons for a decision to allow, or not to allow, proceedings to continue under rule 4(5), fairness to both the practitioner and the patient requires that reasons are given by the registrar. The need to give adequate reasons for a rule 4(5) decision is all the more important if that decision is a belated decision in an attempt by the GMC to cure an earlier procedural irregularity. The practitioner must be able to identify the 'exceptional circumstances of the case' that had led the registrar to conclude that it is in the public interest, in those circumstances, that it should proceed; otherwise, he or she will be left with a very real doubt as to whether the underlying reason for the registrar's decision was not the exceptionality of the case, but a corporate desire on the part of the GMC to avoid the inconvenient or embarrassing consequences of its own procedural errors. He continued, at [59], to note that while the allegations, taken with others, formed a pattern of substandard performance raising serious concerns about the claimant's fitness to practise, that is not, of itself, an exceptional circumstance, since a case will not come before a fitness to practise panel unless there are serious concerns about the practitioner's fitness to practise. Under the 2004 Rules, it was for the case examiners, not the registrar to decide whether the fitness to practise panel should hear the allegation. The registrar's only function under rule 4(5) was to decide whether there were any 'exceptional circumstances of the case' and, if so, whether it was in the public interest, in those circumstances, that the allegation should proceed further—that is, that it should be referred to the case examiners (see [60] and [68]).

### *Murnin v. Scottish Legal Complaints Commission and anor* [2012] CSIH 34

**17.11** Solicitor—financial inspection—exceptional circumstances—meaning

The appellant was a solicitor and the first respondent was responsible for the sifting of complaints about solicitors' conduct. Rule 4(6) of the first respondent's rules provides that it will not accept a complaint where it is made after the expiry of one year from the date of the misconduct complained of 'unless the [first respondents consider] that the circumstances are exceptional'. On 14 June 2011, a complaint was submitted to the first respondent in relation to events occurring before 18 May 2010, when a financial compliance inspection of the appellant's firm by the Law Society of Scotland (the second respondent) revealed a potential deficit on the firm's client account of £232,000. The complaint was that the appellant may be guilty of breaches of rules 4 and 6 of the Solicitors (Scotland) Accounts etc. Rules 2001. In dismissing the appellant's appeal, the Court said that the first issue to determine was whether the complaint was time-barred. The language of rule 4(6) was expressed in the context of a preliminary process designed to sift out unmeritorious complaints because of the delay in their presentation. Ordinary or common complaints will fall foul of the provision if they are tardily made. But all that the wording requires, before the first respondents can accept a late complaint, is a determination that there are circumstances that are exceptional. In *R v. Kelly (Edward)* [2000] QB 198, Lord Bingham CJ said, at 208, that the word 'exceptional' was to be construed:

> ...as an ordinary, familiar English adjective, and not as a term of art. It describes a circumstance which is such as to form an exception, which is out of the ordinary

> course, or unusual, or special, or uncommon. To be exceptional a circumstance need not be unique, or unprecedented, or very rare; but it cannot be one that is regularly, or routinely, or normally encountered.

Thus, in the context of rule 4(6), the first respondents would look for unusual, special, or uncommon features of a nature that ought to be regarded as capable of justifying accepting a complaint outwith the one-year time limit. The Court has little difficulty in holding that the gravity of any alleged misconduct can be an exceptional circumstance in this context. Although, in many cases, the feature identified may relate to some event, such as a mistake or oversight, which results in a complaint not being timeous, there must be situations in which the misconduct is so grave that, even if the reason for the lateness of the complaint were wholly inexcusable, nevertheless the public interest would demand that the complaint be investigated and, if well founded, the solicitor dealt with according to the profession's disciplinary rules and procedures.

## D. Decisions of Preliminary Proceedings Committee or Investigation Committee

*Re Dixon* [2001] EWHC 645 (Admin)

**17.12** Application to stay—appropriate forum—PPC or PCC—whether clear case

D, a gynaecologist, applied to quash the decision of the PPC to refer a complaint to the PCC by a patient because of delay. In 1993, D was commencing an operation on the patient to carry out a hysterectomy, when he discovered that she was pregnant. He completed the hysterectomy. Following a lengthy trial, D was acquitted in 1995. In 1998, the patient made a complaint to the GMC. On 15 November 2000, the PPC referred the matter to the PCC and, in a witness statement, the chairman stated that it was the view of the PPC that the appropriate forum in which to decide whether there was an abuse of process should be the PCC, because at the PCC, in contrast to the PPC, there would be the opportunity for representations by counsel, legal argument, and advice from a legal assessor on the merits of the points raised on behalf of D. The hearing would also be in public and would afford the complainant an opportunity to make representations as to why, in her view, the complaint should be made the subject of a full inquiry. In dismissing D's application for judicial review, Collins J said that it seemed to him in a case such as this that there clearly was going to have to be an explanation by the patient as to why she had delayed. If the matter were dealt with in private on the papers, there would be no opportunity for any questions to be asked of the complainant by the doctor nor could any matters be argued that might have troubled the PPC. The Committee was entitled to have regard to the question of public confidence in the system as a whole, and public concern that inevitably would result if, in a case such as this, which clearly gave rise to important issues, the matter was disposed of in a private hearing without the complainant being able to be present or to know precisely what had gone on. Collins J continued:

> 22. In reality, in my view, the PPC should only decide such matters for themselves in a very clear case. If there are or may be factual issues and if there are or are likely to be arguments raised on either side whether indeed if (sic) is a case which should be struck out for an abuse of process (bearing in mind that there is a case to answer as to whether there has been serious professional misconduct by a doctor) then normally those matters should be dealt with in the forum which has the ability to obtain the evidence, to listen to the arguments on behalf of the parties, and to deal with it openly

and in public. Of course, there may be cases where it is plain that on any view the case will have to be stopped because it would be an abuse of process. In such a case the PPC should itself decide.

### *R (Phillips) v. General Medical Council* [2004] EWHC 1858 (Admin)

**17.13** Decision of PPC—jurisdiction to quash—sparingly exercised

P applied to quash the decision of the PPC to refer complaints to the PCC. P had been acquitted of eight counts of indecent assault at the Crown Court, the allegations being that he had indecently assaulted female patients in the course of intimate medical examinations. A further twenty-four cases were not dealt with in the criminal proceedings. Newman J, in refusing to quash the decision of the PPC, said that, in his judgment, the jurisdiction of the Administrative Court to pre-emptively restrain the process would be sparingly exercised. That the jurisdiction exists is undoubtedly correct, but where the application relates to an independent tribunal such as here, established by rules, governed by its own procedures, and having a specialized expertise to bring to play within its jurisdiction, the responsibility for deciding whether its procedures have been abused should, unless weighty circumstances point to another conclusion, be decided by it.

### *R (McNicholas) v. Nursing and Midwifery Council* [2009] EWHC 627 (Admin); *R (McNicholas) v. Nursing and Midwifery Council (No. 2)*, 23 March 2010 (unreported)

**17.14**

See paragraphs 39.13 and 39.14.

## E. Delay in the Hearing of Substantive Allegation

### *R v. Chief Constable of Merseyside Police, ex parte Calveley and ors* [1986] QB 424

**17.15** Notice of complaint—breach of regulations—destruction of documents—prejudice

The applicants were five police officers. They had been found guilty of disciplinary offences by the Chief Constable of Merseyside and dismissed by the force or required to retire. On 30 June 1981, following an incident in the early hours of 21 June 1981 that led to formal complaints being made against the conduct of the police officers concerned, an investigating officer was appointed, but it was not until some two-and-a-half years later, in November–December 1983, that the applicants were officially informed of the fact that complaints had been made or that they were being investigated. They applied for judicial review based on the failure to inform them of the complaints, in breach of regulation 7 of the Police (Discipline) Regulations 1977. In allowing the applicants' appeal, Sir John Donaldson MR said that regulation 7 provided an essential protection for police officers facing disciplinary charges and that, save in the rare case in which an investigation of the complaint or a related investigation would be prejudicial by giving notice, or where the nature of the complaint was unclear or it was clearly frivolous, it would be difficult to justify any appreciable delay in giving the officer concerned notice of the complaint. The words 'as soon as is practicable' should be read with all of the emphasis on the word 'soon'. The primary purpose of the regulation is to put the officer on notice that a complaint has been made and to give him a very early opportunity to put forward a denial, which in some cases might even take the form of an alibi or an explanation, and to collect evidence in support of that denial or explanation. The Court went on to state that the normal rule in cases such as this is that an applicant for judicial review should first exhaust whatever other rights he has by way of appeal. To the normal rule, there will, of course, be exceptions. In the instant case, abuse was shown. Apart from the failure to give the notices timeously under regulation 7, over three years passed between the alleged offences and the hearing before the chief constable. In that time, the radio log sheets and the parade sheets, and other documents that would have shown what other officers were on duty at the relevant time, had been destroyed. Further, the investigating officer did nothing between the end of 1981 and July 1983, and, in addition, the chief constable did not appear to have had the question of possible prejudice sufficiently in mind when he rejected the preliminary point on the regulation 7 notices taken at the start of the disciplinary hearing.

### *R v. Chief Constable of Merseyside Police, ex parte Merrill* [1989] 1 WLR 1077

**17.16** Criminal trial—notice of complaint—served post-trial

The applicant, a police officer, was involved in an incident on 28 August 1984 also involving a car driven by E. E was subsequently charged with reckless driving and, following a retrial, was acquitted in October 1985. In January 1986, an investigating officer interviewed the applicant and he was served with a regulation 7 notice of an investigation into an allegation of complaint. The Court of Appeal, in allowing the applicant's application for judicial review and quashing the disciplinary proceedings, held that *Calveley* did *not* decide that a regulation 7 notice was to be served as soon as reasonably possible after the conclusion of any criminal proceedings. The chief constable and the Divisional Court had not given any weight, or indeed consideration, to the prejudice inherent in depriving a police officer for fifteen months of the information to which he was entitled under the regulation—namely, notice that his conduct was formally under investigation. As to whether to remit the matter to the chief constable for reconsideration, the Court noted that the situation had arisen not through any fault of the applicant. The public interest in complaints against police officers being fully investigated and adjudicated was undoubted, but it must be done speedily. Regulation 7 requires the investigating officer to give the member subject to investigation written information 'as soon as is practicable'. 'Practicable' is a common English word the nearest synonym of which is probably 'feasible' and the word is so defined in the *Shorter Oxford Dictionary*.

### *R v. Chief Constable of Devon and Cornwall Constabulary, ex parte Hay; R v. Chief Constable of Devon and Cornwall Constabulary, ex parte Police Complaints Authority* [1996] 2 All ER 711

**17.17** Dismissal of disciplinary charges—judicial review to quash—no prejudice to officers

In October 1993, H was shot dead under siege conditions by officers of the Devon and Cornwall Constabulary. The deceased man's brother, H, and the Police Complaints Authority, brought judicial review proceedings against the decision to dismiss the disciplinary proceedings in May 1995, at which one of the officers successfully applied to dismiss the proceedings as an abuse of process, on the grounds of late notice of the amended charges combined with delay in the disciplinary hearing, which had prejudiced his defence. Sedley J, in granting judicial review, said that *A-G's Reference (No. 1 of 1990)*, presided over by Lord Lane CJ, was a definitive decision and that the headnote at [1992] QB 630, 631, sufficiently summarized the principles. In the instant case, there was no serious or culpable delay; all of the time that had elapsed could be well accounted for and there was no evidence that the accused officer had suffered any identifiable prejudice by reason of the delay, or of the late amendment of the charge and late service of some further evidence in support of it. The amended charge lay within the four corners of the regulation 7 notice, which had been served shortly after the incident itself.

### *R v. Council for Licensed Conveyancers, ex parte West*, CO 4901/99, 7 June 2000 (unreported)

**17.18**
'As soon as possible'—context of particular case—reliance on Trading Standards Department

W was, at all material times, a licensed conveyancer. In July 1997, two clients made a complaint to the Trading Standards Office of the Solihull Metropolitan Borough Council. On 21 August 1997, the complaint was also put in writing by the two clients to the Council for Licensed Conveyancers. The essence of the complaint was that the conveyancing clients were dissatisfied with the appellant's charges. One feature of those charges was a bill received for local authority searches. The clients wanted a copy of the search for which they had paid and alleged that the applicant was evasive. On 10 October 1997, the investigating committee of the Council made a number of resolutions, including a resolution to intervene in the practice of the applicant, and to institute disciplinary proceedings against him and to suspend his licence with immediate effect. Criminal investigations were carried out by the Trading Standards Department and, in February 1999, the applicant pleaded guilty to two offences under the Trade Descriptions Act 1968. On 8 June 1999, notice of hearing was given by the council of disciplinary proceedings. Against that background, the applicant contended that there had been considerable delay from 10 October 1997 to 8 June 1999 before sending a notice of hearing setting out the charges against him. Further, paragraph 4(1) of the Licensed Conveyancers Discipline and Appeals Committee (Procedure) Rules 1987 provides that 'as soon as possible' after a case has been referred to the investigating committee, the Council shall send to the respondent a notice of hearing. Gibbs J, in refusing to grant an order to quash the decision of the Disciplinary Appeals Committee, said that the words 'as soon as possible' in relation to the sending of a notice of hearing amount, on any view, to a more stringent test than that which would have had to be met by the Council for the purposes of natural justice or for the purposes of Article 6 ECHR, important though those requirements are. The real question was whether the Disciplinary Appeals Committee was entitled to hold that paragraph 4(1) was complied with. The chairman of the disciplinary tribunal did not find the decision easy. Twenty months in which to formulate disciplinary charges and to give notice of them might seem, on the face of it, not as prompt as it could have been. The Court was inclined to the view that the provision in the instant case was mandatory. However, that by no means provides a full answer to the question in this case. The phrase 'as soon as possible' must be seen in the circumstances of the particular case. The question posed is therefore whether the notice is given as soon as possible having regard to the seriousness of the allegations against the applicant, the way in which these expanded following investigation, the Council's statutory duty under section 12(2) of the Administration of Justice Act 1985 to ensure that the standards of competence and professional conduct among persons who practise as licensed conveyancers are sufficient to secure adequate protection for consumers, and the various arguments advanced on either side by counsel. The Disciplinary Appeals Committee did not err in law in finding that notice was given as soon as possible. It was entitled to conclude, as it did, that, having received the files from the Trading Standards Department, the Council needed to review them and collate them carefully. It was entitled to conclude that, whilst the Council was pointed in the direction of the eventual charges in October 1997 by the original complaint, what started as an apparently small infringement of the rules grew in the period that followed as more and more files were examined. It was also not unreasonable for the Council, in the circumstances of this case, to rely on the Trading Standards Department to conduct the enquiries necessary on the ground to establish the factual position. These enquiries were considerable and needed to be carried out.

### *Haikel v. General Medical Council*, Privy Council Appeal No. 69 of 2001

**17.19** Historic allegations—period of delay—GMC letter notifying allegations and start of hearing—not excessive

H, a general practitioner (GP), faced charges of improper and unprofessional conduct in relation to consultations and examinations of six female patients. His name was ordered to be erased from the register. The allegations covered the period 1988–97. The hearing before the PCC took place in July 2001. On appeal to the Privy Council, counsel for H referred to a PPC decision in *Rogers v. General Medical Council* [2008] EWHC 2741 (Admin), in which a doctor's right to a fair trial within a reasonable time was recognized where the delay was two years and ten months from the date on which the GMC had informed the doctor that his case was being referred to the PPC. The PCC subsequently held that the timescale was unreasonable and that there had been a breach of Article 6(1) ECHR. In dismissing H's appeal, Sir Philip Otton, delivering the judgment on behalf of the Board, said:

> 13. In reaching their conclusion, their Lordships recognise that the PCC is a public authority for the purposes of the Human Rights Act and as such must act in a way which is not incompatible with a convention right. The proceedings against a registered practitioner for professional misconduct are hybrid, having the criminal burden and standard of proof and largely, although not exclusively, criminal rules of evidence and procedure, but they also involve a determination of the practitioner's civil rights and obligations. Accordingly they attract the protection of article 6(1) (see *Ghosh v. General Medical Council* [2001] 1 WLR 1915).
>
> 14. In order to determine whether there has been a breach of the right to a hearing within a reasonable time, it is first necessary to establish the period of time over which the protection of article 6(1) applies. It is well established in the jurisprudence of the European Court of Justice and the Human Rights Act that article 6(1) is concerned with procedural delay in the course of proceedings. It is not concerned with the delay between the commission of the allegedly wrongful actions and the commencement of the proceedings. Hence the material period for consideration is the time during which the administrative and judicial authorities were dealing with the case. (Counsel for the GMC) realistically conceded that the relevant period in the instant case was the time that elapsed between the 23rd September 1999 letter from the GMC to the appellant outlining in general terms the nature of the allegations and the 19th July 2001 when the hearing commenced. Their Lordships consider that an explanation for this lapse of time might have been illuminating and helpful to the PCC. However in the absence of an explanation it was still open to the Committee to determine whether or not the delay was such as to amount to an abuse of process. There was no evidence or suggestion by the appellant that he had suffered any material prejudice in addition to the lapse of time. There was no complaint that witnesses were no longer available who would have been but for the delay. The records of the patients were available and full and even those of Ms A were eventually produced, albeit incomplete. In the absence of any such assertions it was open to the Committee to reach the decision it did and to proceed to hear the complaints.
>
> 15. Their Lordships have not been persuaded that on the facts of this case it can be said that the reasonable time requirement has been violated. To stay proceedings on the ground of abuse of process is a rare step only to be taken in exceptional cases. There was undoubtedly an important countervailing public interest that allegations of such a serious nature should be heard and determined rather than stayed.

### *WR v. Austria* (2001) 31 EHRR 43

**17.20** Article 6(1) ECHR—effect on disciplinary proceedings

The applicant was a practising lawyer by profession. On 30 April and 15 May 1985, respectively, the president of the Mauerkirchen District Court laid a disciplinary information against the applicant. Subsequently, preliminary investigations were carried out. On 15 June 1987, the Disciplinary Council of the Upper Austrian Bar Chamber decided to open disciplinary proceedings against the applicant. He was charged with several counts of professional misconduct. On 16 May 1988, the Disciplinary Council joined a further set of disciplinary proceedings against the applicant. On 18 January 1989, the Disciplinary Council convicted the applicant on three counts and acquitted him on three other counts. It ordered him to pay a fine of 5,000 Austrian Schillings (Sch). The written version of the Disciplinary Council's decision was served on the applicant on 4 April 1990. The Appeals Board received his appeal in May 1990. On 25 January 1993, the Appeals Board dismissed the applicant's appeal. On the appeal of the Bar Chamber, the Appeals Board found the applicant guilty of two disciplinary offences of which he had been acquitted by the Disciplinary Council. In the same decision, the Appeals Board decided two further sets of disciplinary proceedings. In all, the applicant was found guilty of seven counts and the Appeals Board imposed a fine of Sch25,000 on him. On 17 May 1993, the applicant lodged a complaint with the Constitutional Court. He complained in particular about the length of the disciplinary proceedings brought against him. On 12 October 1994, the Constitutional Court dismissed the applicant's complaint. The applicant applied to the Commission on 13 February 1995. He alleged, inter alia, a violation of Article 6(1) ECHR on account of the length of the disciplinary proceedings against him. The European Court of Human Rights confirmed that the applicant's right to practise as a lawyer was a 'civil right' within the meaning of Article 6(1). In the present case, the possible penalties for disciplinary offences included a suspension of the right to practise as a lawyer for up to one year. Thus the applicant ran the risk of a temporary suspension of his right to practise his profession. It follows that his right to continue to practise as a lawyer was at stake in the disciplinary proceedings against him. Accordingly, Article 6(1) was applicable. The government accepted that the disciplinary proceedings lasted from 15 June 1987, when the disciplinary proceedings against the applicant started, until 12 October 1994, when the Constitutional Court gave judgment. The Court saw no reason to disagree with the conclusions reached by the Commission that there had been a violation of Article 6(1). The overall duration of the disciplinary proceedings of seven years and four months, at three levels of jurisdiction, cannot be considered 'reasonable' within the meaning of Article 6(1). The applicant claimed Sch30,000 as compensation for non-pecuniary damages resulting from distress caused by the disciplinary proceedings against him. The government did not comment on this claim, and the Court found the claim reasonable and therefore allowed it in full.

### *Luksch v. Austria* (2002) 35 EHRR 17

**17.21** Article 6(1) ECHR—effect on disciplinary proceedings

The applicant was an accountant by profession. On 2 April 1986, the Chamber of Accountants instituted disciplinary proceedings against him. Subsequently, the proceedings were adjourned, having regard to criminal proceedings pending against the applicant. The applicant was subsequently convicted and, by the end of 1990, the file relating to the criminal proceedings reached the Disciplinary Court; on 8 November 1991, the hearing date was set for 8 May 1992. On 8 May 1992, the Disciplinary Court of the Chamber of Accountants, having regard to the applicant's conviction, found that he had infringed the profession's reputation and ordered the suspension of the applicant's right to practise for one year. On 7 May 1993, the Appeals Board dismissed his appeal. On 28 February 1997, his appeal was dismissed by the Administrative Court and its decision was served on 22 March 1997. In finding a breach of Article 6(1) ECHR, the European Court of Human Rights said that the proceedings at issue with the present case started on 2 April 1986 and were terminated on 22 March 1997. Thus they lasted for a little less than eleven years. They were not particularly complex and there were no delays attributable to the applicant; however, considerable periods of inactivity were imputable to the competent authorities. In particular, there was no explanation as to why almost three-and-a-half years elapsed between the termination of related criminal proceedings on 3 October 1988 and the hearing of the applicant's disciplinary case on 8 March 1992; nor was there an explanation for the lapse of one year and more than eight months between the decision of the Appeals Board on 7 May 1993 and the service of this decision on 30 January 1995. The duration of the proceedings before the Administrative Court of two years was also substantial. The Court therefore found that the length of the proceedings exceeded a 'reasonable time'. The applicant claimed Sch6 million for loss of earnings and related pension claims, as well as for loss of reputation and distress. As to the pecuniary damage, the Court agreed with the government that there was no causal link between the breach of the Convention found and the alleged loss of earnings and pension claims. Consequently, it made no award under this head. In so far as the applicant could be understood to be claiming compensation for non-pecuniary damage, the Court, having regard to its case law and making an assessment on an equitable basis, awarded the applicant Sch130,000.

***R (Van Vuuren) v. General Medical Council* [2002] EWHC 2661 (Admin)**

**17.22** Complexity of case—delay not unreasonable—whether fair trial can be held

In March 1998, there was a formal complaint made to the GMC in respect of the conduct of the claimant. The complaint was in relation to 'interprofessional relationships and dealings' rather than the way in which the claimant dealt with his patients. In August 2000, more than two years later, the GMC sent a rule 6 letter informing the claimant that the complaint had been passed to the PPC. In November 2000, the claimant was notified that a charge would be formulated and, in May 2002, the claimant was given notice of an inquiry setting out the charge, which was in identical terms to those used in the rule 6 letter. Field J, in dismissing the claimant's renewed application for permission to apply for judicial review, said that, in his judgment, the Committee was entitled to have regard to the complexity of the case. It was open to it to have concluded that the period of delay in itself was not unreasonable. The whole question raised by Article 6 ECHR was essentially one of fairness, and the Committee was entitled to express its reasoning and its conclusion as to whether or not there had been an unreasonable delay, in terms of whether a fair trial could be held.

***Aaron v. Law Society* [2003] EWHC 2271 (Admin)**

**17.23** Application of Article 6 ECHR—proof of prejudice not required

A's complaint before the Solicitors Disciplinary Tribunal (SDT) of delay and/or abuse of process was in connection with an allegation made in October 1996 of non-payment or delay in payment of counsel's fees in 1987–91. The matter was disposed of by the Tribunal in July 2002. In dismissing A's appeal, Auld LJ (with whom Goldring J agreed) said, at [25] as to the law, that it is established that Article 6 ECHR applies to professional disciplinary proceedings: *Albert and Le Compte v. Belgium* [1983] 5 EHRR 533, [28] and [29]; *Ginikanowa v. United Kingdom* (1988) 55 DR 252, 257–8. It was also established that failure to determine disciplinary proceedings within a reasonable time may violate Article 6 without proof of prejudice to the accused: *Porter v. Magill* [2002] 2 AC 357, *per* Lord Hope at [106], [108], and [109]. But there was uncertainty as to whether the courts may nevertheless, in their discretion, still refuse the relief of a stay where to proceed would not amount to an abuse of process—that is, the traditional common law approach, as stated by Lord Lane CJ in *Attorney General's Reference (No. 1 of 1990)* [1992] QB 630, 643. In *Attorney General's Reference (No. 2 of 2001)* [2001] 1 WLR 1869, the Court of Appeal had effectively kept to that approach in the Article 6 context as to the exercise of a court's discretion in determining what, if any, relief to grant for its violation.

The decision of the Court of Appeal in *Attorney General's Reference (No. 2 of 2001)* was affirmed by the House of Lords: [2004] 2 AC 72.

### *R (Gibson) v. General Medical Council* [2004] EWHC 2781 (Admin)

**17.24** Unacceptable delays—no special rule in disciplinary proceedings—public interest—no manifest unfairness

G, a retired medical practitioner, was a consultant pathologist at Kent and Canterbury Hospital during the period 1990–95. In November 1997, he was notified that the GMC was conducting an investigation into his professional practice and conduct in the light of an inquiry report. In September 2000, solicitors for eleven women complainants submitted complaints to the GMC. In February 2003, the PPC decided to refer the case to the PCC. G sought judicial review of the decision of the PCC, in the course of a preliminary hearing in September 2004, not to stay the proceedings against him and to refuse his application of a voluntary erasure. In dismissing both applications, Elias J said that, in relation to delay, the legal principles were not fundamentally in dispute. The case was put both on the basis that it was an abuse of process at common law for the proceedings to continue and also on the ground that it would be an infringement of Article 6 ECHR. After reviewing *Attorney General's Reference (No. 1 of 1990)*, *Attorney General's Reference (No. 2 of 2001)*, and *Hunter v. Chief Constable of the West Midlands Police* [1982] AC 529, Elias J said that he rejected the notion that there was a special rule in disciplinary proceedings, giving unique protection to the defendant. In *R v. General Medical Council, ex parte Toth* [2000] 1 WLR 2209, Lightman J said that both the legitimate expectation of complainants and the public confidence in the regulation of the medical profession required that the complaint should, in the absence of some special and sufficient reason, be publicly investigated. It would undermine that important principle if mere unreasonable delay, absent prejudice, were to require a stay to be granted. In the instant case, the PCC was fully entitled to conclude that a fair hearing could be given to G, notwithstanding the unacceptable delays. This was not a case in which it was being said that crucial documents had been lost, or important witnesses were no longer available. As to whether it was fair to continue the hearing, there were both a public interest and the interest of the complainants to consider, and there should be a very plain case of manifest unfairness in the conduct of proceedings before this principle came into play. One can, of course, sympathize with the position in which G found himself, facing serious charges while in retirement and so long after the events have occurred. But the Court could not say that the Committee was compelled to hold that it would be unfair for the proceedings to continue and it was entitled to conclude that there were no sufficiently exceptional circumstances to warrant that very unusual step.

### *R (Aziz) v. General Medical Council* [2005] EWHC 2695 (Admin)

**17.25** Decision of panel quashed—whether rehearing—ten years after events and five-and-a-half years after complaint

On 11 March 2005, the fitness to practise panel found that A was guilty of serious professional misconduct and ordered that he be suspended from the register for a period of two months. The hearing took place in the absence of A and he was not represented. The allegation arose out of his application for the post of senior registrar in clinical pharmacology at a hospital in Newcastle upon Tyne. In his application form and his CV, he did not mention that he had been dismissed from employment at a hospital in Ireland. Shortly after the decision of the fitness to practise panel, the GMC appreciated that there had been procedural improprieties in relation to the hearing, and that the decision could not stand and A's appeal would have to be allowed. The GMC suggested a consent order to that effect, but that the matter should then be remitted by the Court to the GMC for the proper steps to be taken and for the matter to be reheard by a different panel. A challenged this, claiming that it would be wrong for the matter to go back and that he should not be retried. Collins J, in holding that it would be unfair and an abuse for the matter to proceed, said that the key point was whether it would be right for the matter to go back to be reconsidered on the basis that there had been such a delay since the complaint was made. The relevant period of delay was the delay since the complaint was made to the GMC. Delay between the time when the alleged offence was committed and the complaint was made is not directly material when one is considering whether there has been an abuse of the process by reason of delay. It is, of course, relevant, because the longer the time that has elapsed since the matter on which the complaint is based happened, the more important it is that the complaint is dealt with as speedily as possible. It was now unlikely, if the matter were to go back, assuming that the decision to continue with disciplinary proceedings was against A, that the matter could be heard before ten years after the events in question and some five-and-a-half years after the complaint was originally made to the GMC. That seemed, in the circumstances, a delay that was altogether too long; despite the importance of the proceedings, it would be wholly unfair and wrong and disproportionate for the matter to proceed any further.

### *Campbell v. Hamlet* [2005] 3 All ER 1116

**17.26** Delay in delivering judgment—no adverse effect on determination

The appellant, C, an attorney-at-law practising in Trinidad and Tobago, appeared before the Attorneys-at-Law Disciplinary Committee in 1988 to face a complaint that he had received US$29,400 to purchase two parcels of land, but had neither conveyed the land nor returned the purchase price. The hearing lasted, in all, eleven days stretching over a ten-month period from February to December 1988. It was not until 29 October 1996 that the Committee produced its findings and orders, finding the allegation of professional misconduct substantiated and ordering C to pay the respondent compensation of $29,400, together with interest from the date of the order until payment and costs. In dismissing C's appeal, the Privy Council said that the delay of eight years in giving judgment was highly reprehensible—indeed, unforgivable. But it did not follow that the appellant was in any way prejudiced by this delay, nor did it afford him any substantial ground of appeal. No sanction was imposed upon the appellant besides being ordered to repay the money and costs. There are, of course, cases in which delay may adversely affect the determination: *Goose v. Wilson, Sandford & Co* [1998] TLR 85 was one such case. The present case, however, was very different on its facts. Whereas, in *Goose*, the judge's ability to decide the case was clearly compromised, here, it was not.

### *Bar Standards Board v. L*, 24 January 2007 (unreported)

**17.27** Defendant to establish grounds—stay in disciplinary proceedings exceptional

L applied to strike out certain disciplinary charges brought against him by the Bar Council. The events went back to 2000 and related to claims made by L for publicly funded fees, which, according to the Bar Council, L claimed recklessly not caring whether they were excessive and unjustifiable or not. Warren J, in dismissing L's application, said that much of the law had been developed in the criminal context and, broadly speaking, the same principles apply for the purposes of disciplinary proceedings. It is for the defendant seeking a stay on the grounds of abuse, whether based on delay or any other factor, to show that (a) he would not receive a fair trial or (b) that it would be unfair for him to be tried: *Attorney General's Reference (No. 2 of 2001)* [2004] 2 AC 72, *per* Lord Bingham at [17]. To impose a stay in disciplinary proceedings is exceptional: *Council for the Regulation of Health Care Professionals v. General Medical Council and Saluj; R (Gibson) v. General Medical Council*. There was not any real dispute about the applicable principles, but only about which side of the line the facts fall and how the Court should exercise the discretion whether to stay proceedings. The Court was not persuaded that it would not be possible for L to have a fair hearing and this was not a case in which abuse was established.

### *Hutchinson v. General Dental Council* [2008] EWHC 2896 (Admin)

**17.28** Tribunal to consider prejudice to practitioner—analogy with criminal trial

In April 2007, the PCC found various hygiene-related charges against H proved, his fitness to practise was found to be impaired, and an order for erasure was made. The events went back over many years and alleged, amongst other things, that: between 1978 to 2006 H did not wear gloves or wash his hands when treating patients; between 1990 and 2006 he used dental instruments to clean his fingernails and ears; and between 1983 and 1999 he did not wear gloves when treating Patient A and her family unless specifically asked, and did not wash his hands when seeing her and her family. Blair J said that, however vague the charges were in terms of timing, they were explicit in terms of the behaviour alleged. It was obvious that even occasional incidences of the kind alleged would be completely unacceptable. It did not render a fair hearing impossible. Given that a stay on grounds of abuse will be rare, the important thing in cases like this is that the tribunal should reach its findings on the evidence, with possible prejudice to the practitioner caused by factors such as delay and lack of specificity firmly in mind. To take the analogy of a criminal trial, on facts like these, an express direction to the jury as to the necessity to guard against potential prejudice caused by delay would be essential. In the present circumstances, the Committee was right to reject the abuse of process application and to go on to consider the charges on the merits: see 62.14.

### *R (Lloyd Subner) v. Health Professions Council* [2009] EWHC 2815 (Admin)

**17.29** Period of delay between complaint and hearing—no unfairness

The appellant was struck off the register of operating department practitioners following a hearing in October 2008. The allegation was that, in December 2006 at Kings College Hospital, where the appellant was employed as an anaesthetic practitioner, he physically assaulted and verbally abused another member of staff, and left work early without permission. The complaint was not made until 26 March 2007, and the appellant submitted that the delay between the date when the complaint was received and the hearing in October 2008 was not reasonable. Kenneth Parker J, in dismissing the appeal, said that the leading authority on reasonableness of delay under Article 6 ECHR in this jurisdiction is *Dyer v. Watson* [2004] 1 AC 379. The learned judge said that, given that any case takes some period to prepare witness statements and otherwise to make the case ready for trial or hearing, he was not able to find that a period from March 2007 to October 2008—that is, about sixteen-and-a-half months—comes anywhere within the range that would cause real concern that a basic human right of the appellant had been infringed. The period is significantly shorter than that in *Dyer* and shorter by far than the periods that have caused concern to the European Court of Human Rights, to which Lord Bingham referred. In any event, the appellant faced a further insurmountable obstacle. It is now well established by the authority that breach of Article 6 is not in itself a ground for striking out proceedings, nor for setting them aside or quashing them after the event. It is also necessary to show that the unreasonable delay would not permit a fair trial to take place or, after the event, that a fair trial had not been possible: *Attorney General's Reference (No. 2 of 2001)* [2004] 2 AC 72, [24]. The statement of law in that paragraph was applied in the context of professional disciplinary proceedings in *R (Gibson) v. General Medical Council* [2004] EWHC 2781 (Admin), [26], [36], and [37].

### *Makhdum v. Norfolk and Suffolk NHS Foundation Trust* [2012] EWHC 4015 (QB)

**17.30** Disciplinary proceedings by Trust—injunction—delay by claimant

M, a consultant psychiatrist employed by the defendant Trust, sought an injunction preventing the defendant from continuing disciplinary proceedings against him until the conclusion of his claim that the defendant's investigation and disciplinary process was in breach of his contract of employment. In early 2012, following the publication of a report, six allegations of misconduct were made against the claimant by the defendant. The investigation was delayed for six months because the claimant was signed off sick. In September 2012, a disciplinary meeting was held and adjourned to a date in November 2012. The claimant commenced proceedings and sought an injunction preventing the continuation of the disciplinary process. Beatson J, in dismissing M's application, said that, on the facts, the claimant had known the scope and content of the investigation from a very early stage. He waited until after the disciplinary meeting had commenced in September 2012 before commencing proceedings. M was seeking to invoke an equitable remedy that relied on promptness.

### *Gadd v. Solicitors Regulation Authority* [2013] EWCA Civ 837

**17.31** Law Society intervention—application to withdraw—extension of time

On 7 December 2009, the Law Society intervened in G's practice on the grounds of suspected dishonesty and failure to comply with the account rules. Paragraph 6(4) of Schedule 1 to the Solicitors Act 1974 provides that a person served with an intervention notice must, within eight days of service, apply to the court to withdraw the notice. G took proceedings against the Law Society more than twelve months after the intervention. Sharp J upheld an application by the Law Society for summary judgment. In dismissing G's renewed application for permission to appeal, Elias LJ (with whose judgment Beatson LJ agreed) said, at [8] and [11]–[12], that even if the Court were prepared to read down paragraph 6(4) so as to allow for applications out of time in exceptional cases, nonetheless impecuniosity was not a justification for applying that principle. There may certainly be circumstances in which an applicant is prejudiced without seeing the basis on which the intervention is made, but even allowing for the possibility that this would justify, under Convention principles (and in particular Article 6), some departure from the eight-day period, it was plainly critical for G to act very speedily thereafter. The reason for a need for a speedy challenge to the intervention notice is to protect the interests of creditors and clients of the firm. But it is also to protect the solicitor's own position, because the longer the delay and the longer the period for which assets are frozen, the more damaging it is to the standing and goodwill of the firm. Whatever the merits of the contention that there may be exceptional cases in which the claim can be pursued outside the eight-day limit, it was not justified in this case.

### *Jeffery v. Financial Conduct Authority* [2013] Lexis Citation 49

**17.32** Knowledge to justify investigation—limitation defence

J, an insurance broker, disputed allegations made by the Financial Conduct Authority (formerly the Financial Services Authority) and, in addition, raised a limitation defence under section 66(4) of the Financial Services and Markets Act 2000 (two years, now three years, from the date on which the Authority knew of the misconduct) in respect of the imposition of a penalty. It was held that, for time to start running it is not necessary that the Authority has the full picture that would justify the issue of a warning notice; the limitation period starts to run from the date on which the Authority knows of the misconduct or has information from which misconduct can reasonably be inferred. The Authority must, however, have sufficient knowledge of the particular misconduct or knowledge sufficiently capable of giving rise to reasonable inference of misconduct, to justify an investigation. Mere suspicion is not enough, nor is any general impression that misconduct may have taken place: at [337]. On the facts, the Authority did not know of any of the misconduct prior to May 2008 and there was no limitation impediment to any of the actions taken by the Authority under section 66 of the 2000 Act.

### *Johnson and Maggs v. Nursing and Midwifery Council (No. 2)* [2013] EWHC 2140 (Admin)

**17.33** Finding of misconduct by committee—no further action—finding quashed on appeal

The hearing lasted eighty-six days, spread over a period of two years and nine months. By the time the PCC of the Nursing and Midwifery Council (NMC) made its decisions, the events that were the subject of the charges were between thirteen and nine years old. The total time that elapsed between the date on which the registrants were notified of the allegations and the conclusion of the disciplinary proceedings was more than eight years. The NMC accepted—as, in the circumstances, it was bound to do—that the delays in the conduct of the proceedings violated the rights of the registrants, under Article 6 ECHR, to a hearing within a reasonable time. When every allowance was made for the extent to which the conduct of the defence contributed to the delay, the length of time that the disciplinary proceedings took remained disgraceful. At the conclusion of the proceedings, the Committee found misconduct, but decided to take no further action. The finding of misconduct was to be quashed: see paragraph 47.54.

### *Goodwin v. Health and Care Professions Council* [2014] EWHC 1897 (Admin)

**17.34** Self-referral August 2011—hearing July 2013—no excessive delay

G was a biomedical scientist. In October 2013, the Conduct and Competence Committee (CCC) of the Health and Care Professions Council (HCPC) found allegations against G to be proved. These were that he had not demonstrated the required competency levels in nine specific areas, had not demonstrated appropriate time management skills, had not prioritized tasks appropriately, and had not demonstrated an appropriate awareness of health and safety issues. The Committee found that G had fallen significantly below the basic level of biomedical scientist practice across all areas of clinical benchwork in the laboratory, time management, and health and safety over a considerable period of time. The errors had the potential to put many patients at risk. The Committee imposed a sanction of suspension for twelve months. G had referred himself to the HCPC in August 2011. He was subsequently dismissed five months later by his employers. Because he was no longer employed when the hearing began in July 2013, he could not afford legal representation. In addition to a complaint of delay in holding the hearing, G argued that whilst one member of the Committee was registered on the same part of the register as G, neither that member, nor any other member of the Committee, had expertise in the specific fields in which G's competence was in issue. In dismissing G's appeal, Nicol J said that the Committee was right to reject his submission that the passage of time meant that the proceedings should be stayed as an abuse of process. The period of time between 2011 to the start of the hearing in July 2013 had not been excessive, particularly given the complex nature of the investigations that had to be completed and the need too to give G an opportunity to respond to them. There was no procedure for funding legal assistance for those facing allegations before the HCPC or any other of the regulatory bodies. G's further complaint about the composition of the Committee was not well founded.

## F. Effect of Delay on Sanction

***Langford v. Law Society* [2002] EWHC 2802 (Admin)**

**17.35** Substantial deficiency—no reduction appropriate

In dismissing the appellant's solicitor's appeal against an order of striking off, Rose LJ said that it was difficult to see how, in the circumstances of the present case, the penalty imposed by the SDT could properly be reduced by reason of any delay in the proceedings. The deficiency on client account was in the region of £150,000–£230,000. Rose LJ said, at [26]:

> Of course there are cases in which, if there has been inappropriate and unreasonable delay, it is possible and appropriate to reflect this in the penalty imposed: that may be a means whereby satisfaction can be given for breach of an Article 6 right.

***Selvarajan v. General Medical Council* [2008] EWHC 182 (Admin)**

**17.36** Criminal trial—not guilty verdict entered—delay relevant to sanction—erasure upheld

In this case, the practitioner was acquitted of a criminal charge of conspiracy to defraud arising out of dishonest conduct that resulted in a loss to the primary care trust of some £150,000. The jury failed to reach a verdict and, by the time of the retrial, the practitioner's co-defendant had become ill; 'not guilty' verdicts were entered against the practitioner and the co-defendant, because a trial against the practitioner alone was thought to be unfair. In subsequent disciplinary proceedings, the practitioner admitted a charge of professional misconduct, but appealed against the sanction of erasure in part owing to the passage of time. Blake J observed that the absence of any common law double jeopardy rule in professional misconduct proceedings was specifically noted in *R v. Statutory Committee of the Pharmaceutical Society of Great Britain* [1981] 2 All ER 805, in which disciplinary proceedings had been brought in respect of conduct for which there had been an acquittal. Blake J held (contrary to the views of the GMC and the legal assessor) that delay was relevant to sanction, but nonetheless this was a very serious and sustained dishonest conduct in a professional capacity, totally undermining the trust and respect that should be accorded to medical professionals. It therefore demanded a severe sanction to vindicate the standing of the profession in public esteem. Accordingly, the practitioner's appeal was dismissed.

***Shiekh v. General Dental Council* [2009] EWHC 186 (Admin)**

**17.37** Period since commission of offences—properly considered by Committee

S appealed against the decision of the General Dental Council (GDC), made on 1 February 2008, ordering the erasure of his name from the register of practitioners on the grounds, amongst other things, that the PCC had wrongly disregarded the length of time since the offence for which he was convicted. About a month after the start of his trial, the appellant offered a plea of guilty to conspiracy to defraud the Dental Practice Board by dishonestly making claims for repayment within the National Health Service (NHS) between 19 December 1994 and 4 August 2000. The extent of the fraud, in financial terms, was unquantifiable. The appellant accepted that his criminal activity had spanned about four years between December 1994 and December 1998. He was sentenced to twelve months' imprisonment, suspended for eighteen months, and ordered to pay a fine of £50,000. The Court of Appeal, Criminal Division, in upholding the sentence, said that the sentence of imprisonment was appropriate and necessary. Of the judge's decision to suspend the sentence of imprisonment, the Court said that it may be thought that it was a merciful course. Civil proceedings were settled in which the appellant repaid the Dental Health Board the sum of £1.1 million. In dismissing S's appeal, Lloyd Jones J said that the Committee had taken full account of the conduct of the appellant, what he had done, and what he had achieved since the commission of the offence, and that those matters were properly considered by the Committee in the context of their bearing on the maintenance of public confidence in the profession.

### *Okeke v. Nursing and Midwifery Council* [2013] EWHC 714 (Admin)

**17.38** On 13 January 2012, the NMC's CCC found two allegations of lack of competence against the registrant midwife proved. The registrant was given notice of the allegations against her by the NMC in March 2007. The period for consideration of delay under Article 6 ECHR was nearly five years. On reviewing the steps taken during that period, Leggatt J concluded that, looking at the matter overall, the delay was unreasonable and it followed that there had been a breach of Article 6. Whilst rejecting the primary submission of the appellant that she did not receive a fair hearing, the learned judge said that the interests of the registrant are one of the matters that the Committee must consider when deciding what is the appropriate sanction. It is relevant to take into account the fact that, as a result of the delay in bringing the case to a hearing, the appellant had already been subject to suspension from practice under interim orders for a very significant period. The Committee would also need to consider whether the delay may have affected the attitude that the appellant took to her defence of the allegations and whether, by way of remedy for what must be acknowledged to have been a breach of the appellant's right under Article 6, it is appropriate to make a reduction in the sentence that would otherwise have been imposed. These were matters that were best reconsidered by the Committee itself.

Breach of Article 6 ECHR—relevant to sanction

### *Rahman v. Bar Standards Board* [2013] EWHC 4202 (QB)

**17.39** R, a barrister, pleaded guilty in June 2010 to two charges of professional misconduct and received a suspension of eighteen months. He filed notice of appeal in relation to the sentence in July 2010, but the appeal did not come to hearing until February 2013. The Visitors (Eady J, chairman) considered the effect of the delay on the barrister. The original decision of the panel was quite understandable, but had the sentence taken effect straight away, the appellant would have been back at work in January or February 2012. Instead, it had cast a shadow over him for the past two-and-a-half years, during which time the suspension had affected him not only personally, but in relation to his practice. Because of continuing publicity of the suspension on the Bar Standards Board's website, in the interests of openness and keeping the public and the profession informed, R's practice had been significantly affected. The delay was held to be in no sense the fault of either the appellant or the respondent, but that it would be wholly unfair to ignore it. Accordingly, the sentence was to be reduced to three months' suspension.

Effect of delay on appeal—effect on practice

## G. Other Relevant Chapters

**17.40**

Abuse of Process, Chapter 2
Human Rights, Chapter 32
Interim Orders, Chapter 38
Investigation of Allegations, Chapter 39

# 18

18.29 |
|     Data Protection Act 1998 | 18.29 |
|     Durant v. Financial Services Authority [2003] EWCA Civ 1746 | 18.30 |
|     Freedom of Information Act 2000 | 18.31 |
|     Health Professions Council v. Information Commissioner, EA/2007/0116, 14 March 2007 | 18.32 |
|     Central London Community Healthcare NHS Trust v. Information Commissioner [2013] UKUT 551 (AAC) | 18.33 |
| E. Other Relevant Chapters | 18.34 |

# DISCLOSURE, CONFIDENTIALITY, DATA PROTECTION, AND FREEDOM OF INFORMATION
|     Rajan v. General Medical Council [2000] Lloyd's Rep Med 153 | 18.03 |
|     R v. General Medical Council, ex parte Toth [2000] 1 WLR 2209 | 18.04 |
|     R (Rosen and anor) v. Solicitors Disciplinary Tribunal [2002] EWHC 1323 (Admin) | 18.05 |
|     Kirk v. Royal College of Veterinary Surgeons [2003] UKPC 3 and [2003] UKPC 47 | 18.06 |
|     R (Green) v. Police Complaints Authority [2004] 1 WLR 725 | 18.07 |
|     Gage v. General Chiropractic Council [2004] EWHC 2762 (Admin) | 18.08 |
|     Henshall v. General Medical Council [2005] EWCA Civ 1520 | 18.09 |
|     Holton v. General Medical Council [2006] EWHC 2960 (Admin) | 18.10 |
|     Wakefield v. Channel Four Television Corporation [2006] EWHC 3289 (QB) | 18.11 |
|     R (Johnson and Maggs) v. Professional Conduct Committee of the Nursing and Midwifery Council [2008] EWHC 885 (Admin) | 18.12 |
|     R (Amro International SA) v. Financial Services Authority [2010] EWCA Civ 123, [2010] 3 All ER 723 | 18.13 |
|     Legal Ombudsman v. Young [2011] EWHC 2923 (Admin) | 18.14 |
|     AXM v. Bar Standards Board [2011] EWHC 2990 (QB) | 18.15 |
|     Veen v. Bar Standards Board [2011] All ER(D) 223 (Oct) | 18.16 |
|     R (McCarthy) v. Visitors to the Inns of Court [2013] EWHC 3253 (Admin); [2015] EWCA Civ 12 | 18.17 |
|     R (Nakash) v. Metropolitan Police Service and General Medical Council (Interested Party) [2014] EWHC 3810 (Admin) | 18.18 |
|     Law Society (Solicitors Regulation Authority) v. Charles Henry, 29 January 2015 (unreported) | 18.19 |
| C. Confidential Information | 18.20 |
|     Woolgar v. Chief Constable of Sussex Police [2000] 1 WLR 25 | 18.20 |
|     R (Pamplin) v. Law Society [2001] EWHC 300 (Admin) | 18.21 |
|     A v. General Medical Council [2004] EWHC 880 (Admin) | 18.22 |
|     Saha v. General Medical Council [2009] EWHC 1907 (Admin) | 18.23 |
|     Financial Services Authority v. Information Commissioner [2009] EWHC 1548 (Admin) | 18.24 |
|     General Dental Council v. Rimmer [2010] EWHC 1049 (Admin) | 18.25 |
|     General Dental Council v. Savery and ors [2011] EWHC 3011 (Admin) | 18.26 |
|     R (May) v. Chartered Institute of Management Accountants [2013] EWHC 1574 (Admin) | 18.27 |
|     Chhabra v. West London Mental Health NHS Trust [2014] ICR 194, SC | 18.28 |
| D.

## A. Legal Framework

18.01

Examples of disclosure include:

Medical Act 1983, section 35A (power to require disclosure of information)
Dentists Act 1984, section 33B (power to require disclosure of information)
Nursing and Midwifery Order 2001, article 25
Royal College of Veterinary Surgeons and Veterinary Practitioners (Disciplinary Committee) (Procedure and Evidence) Rules 2004, rule 8 (College shall send to the respondent any evidence that may assist the respondent's case or harm the College's case)
Pharmacy Order 2010, article 49 (disclosure of information)
Police (Conduct) Regulations 2012, regulation 4 (information in documents that are stated to be subject to 'the harm test' shall not be supplied to the officer concerned if preventing disclosure is necessary in the interests of national security, the prevention or detection of crime, or the apprehension or prosecution of offenders, or for the purposes of prevention or detection of misconduct by other police officers)
Architects Registration Board Investigations Rules, rule 17 (registrar shall have power to call upon any registered person to produce such information, books, papers, records, and plans as considered necessary for discharging functions under the Rules, save that this requirement shall not apply to any information in relation to which the registered person is entitled to legal professional privilege)
Solicitors Disciplinary Tribunal Practice Direction:

> Where directions are sought as to disclosure or discovery of documents, the Tribunal will adopt the view that material should be disclosed which could be seen on a sensible appraisal by the Applicant:
> (i) To be relevant or possibly relevant to an issue in the case;
> (ii) To raise or possibly raise a new issue whose existence is not apparent from the evidence the Applicant proposes to use, and which would or might assist the Respondent in fully testing the Applicant's case or in adducing evidence in rebuttal;
> (iii) To hold out a real (as opposed to a fanciful) prospect of providing a lead on evidence which goes to (i) or (ii).
> There may be exceptional circumstances in which the Tribunal, balancing the interest in disclosure of a document against a competing public interest such as a specific or compelling need for confidentiality, may decide not to order disclosure of a document which falls within (i), (ii) or (iii) above.

## B. Disclosure of Documents

### *Crompton v. General Medical Council* [1981] 1 WLR 1435

18.02

Expert reports given in confidence to regulator—natural justice

In allowing the doctor's appeal to the Judicial Committee, their Lordships said that the observance of the rules of natural justice would have demanded that the psychiatric medical evidence upon which the Committee proposed to act should be disclosed to the doctor, and an opportunity given to him to answer it and to adduce, if he so wished, expert psychiatric evidence on his own behalf to contradict it. At a review hearing in December 1980, the Committee decided to extend the doctor's suspension and told the doctor that, shortly before the date of resumption, he would be asked to furnish the names of professional colleagues, including two consultants in adult psychiatry, to whom the General Medical Council (GMC) could apply for information *to be given in confidence* on his fitness to resume medical practice. Pursuant to that requirement, C furnished the names of two consultant psychiatrists and their reports were before the Committee at the resumed hearing, but the Committee refused to tell C what these reports said about what the chairman of the Committee described as his 'condition'. The Judicial Committee said that such expert medical evidence does not fall within the exception of information to be given confidentially by professional colleagues and other persons of standing nominated by the doctor. Challenging the Committee's direction to erase his name from the register, C relied upon the Committee's failure to observe the rules of natural justice by its refusal to let him see the reports by the psychiatric consultants upon which it based its decision, or even to inform him of the general nature of those reports. On this ground, their Lordships felt reluctantly compelled to recommend that the appeal be allowed, with the consequence that—the last period of suspension validly fixed by the Committee having now expired—C's name remained upon the register. Their Lordships did not think it right to read themselves the psychiatric reports relied on by the Committee, since they could not in fairness to C do so without showing them to him, and they could not show them to C without breaking the promise of confidentiality upon which the reports were furnished by the consultants.

***Rajan v. General Medical Council* [2000] Lloyd's Rep Med 153**

**18.03** Patient's diary entries—disclosure at hearing—evidence on appeal

On 25 July 1997, Ms B first consulted the appellant. The appellant was charged with an inappropriate and improper examination of Ms B, alleged to have taken place at a consultation on 20 August 1997. On the first morning of the hearing, a photocopy of Ms B's diary for the entry of 20 August was produced, which referred to a consultation with the appellant at 3.40 p.m. The appellant's case was that his surgery did not open until 4.30 p.m. on that day, and his first appointment was at 5 p.m. At the hearing before the Professional Conduct Committee (PCC), the charge was proved and the appellant's name was ordered to be erased from the medical register. On appeal to the Privy Council, it was submitted that Ms B's diary should have been disclosed to the appellant prior to the date of the hearing. In particular, if the entry for 20 August had been disclosed earlier, the appellant would have been in a position to make the positive case, which he now wished to make, that the alleged consultation had never taken place and was a fantasy. Their Lordships permitted the admission into evidence before them of further affidavit evidence from two former receptionists of the appellant, who averred that the evening surgery did not begin until 4.30 p.m. and that, after 11.45 a.m. (when the morning surgery ended), the surgery closed until 4.30 p.m. In allowing the appellant's appeal, their Lordships stated that the entry in Ms B's diary was clearly relevant and material, and that, under the principle established in cases such as *R v. Macguire and ors* [1992] 2 All ER 433 and *R v. Ward* [1993] 2 All ER 577, it should have been disclosed to the appellant's legal advisers on a date well before the date of the hearing, and that the failure to disclose in proper time rendered the finding of the Committee in respect of Ms B's complaint unsafe, because that failure denied to the appellant and his advisers a proper opportunity to advance a case that might have succeeded—namely, that the examination of which Ms B complained had never taken place. In the interests of justice, the complaint should be reheard. The appellant also applied at the hearing that their Lordships should make an order that Ms B produce her 1997 diary for inspection by the appellant's legal advisers, so that they could examine it for other entries that might appear relevant to the matters arising in the case. Their Lordships decided that they would not make such an order because they were of opinion that it was unnecessary to do so for the determination of the appeal. However, rule 21 of the General Medical Council Preliminary Proceedings Committee and Professional Conduct

Committee (Procedure) Rules 1988 provided that any party to an inquiry may, at any time, give to any other party notice to produce any document relevant to the inquiry alleged to be in the possession of that party. On the rehearing of the charges, it was open to the appellant under rule 21 to give notice to produce the 1997 diary. Ms B, although requested by the solicitors for the GMC to do so, had refused to produce her entire diary for inspection on the ground that it was a very personal and private document. That attitude was understandable, but their Lordships were of opinion that, in the particular circumstances of this case, on the rehearing of the charge the demands of fairness require that not the entire diary, but the entries in the diary from 25 July 1997 to 26 August 1997 (being the date of Ms B's letter of complaint to the GMC), should be produced for inspection by the appellant's legal advisers a reasonable time before the date of the rehearing and that, if they were not produced, Ms B's complaint against the appellant should be dismissed by the Committee.

### *R v. General Medical Council, ex parte Toth* [2000] 1 WLR 2209

**18.04** Disclosure—transparency—confidentiality—undertaking in respect of any confidential material disclosed

On 9 October 1993, T's 5-year-old son W, who suffered from glycogen storage disease, became hypoglycaemic. T called his doctor, Dr J, who made a home visit. The central complaint of T was that he and his partner told Dr J of W's condition and of his urgent need for intravenous glucose, but that Dr J failed promptly to realize (as he should have done) that this was the case and instead treated W with sedative drugs. W died on 16 October 1993. The applicant sought judicial review to quash two decisions of the GMC not to proceed with his complaint, and also sought disclosure of material concerning Dr J's health disclosed to the GMC in response to the complaint. Lightman J, in holding that the GMC was entitled to require of T an undertaking of confidentiality in respect of any confidential medical evidence adduced by Dr J, said that he was not concerned to consider whether, and if so, how far, the material could be relevant to any decision to be made by the GMC in considering the complaint. The learned judge observed:

> 15. Under the [General Medical Council Preliminary Proceedings Committee and Professional Conduct Committee (Procedure) Rules 1988] it is apparent that neither before the screener nor before the [Preliminary Proceedings Committee] is the complainant entitled to see the material made available to the screener or the PCC. But today, with the imminent coming into force of the Human Rights Act 1998, the GMC properly acknowledges the responsibility that its practices and procedures should, so far as possible, be transparent and to this end it has decided to adopt a new form of practice as from 1 July 2000....
>
> 16. Whilst the GMC is not bound to make such disclosure to a complainant of material put before the screener, it is not precluded by the Rules from doing so and accordingly it is free to do so, at any rate unless precluded from doing so by a confidentiality obligation owed to the parties supplying the material. The issue raised is whether, as a condition of voluntarily making disclosure to [T] of confidential medical evidence relating to the health of [Dr J], and accordingly of material which [Dr J] has every reasonable ground to wish should remain confidential, the GMC can insist on [T] providing an undertaking of confidentiality. [T] submits that, as the GMC has no statutory power to exact such an undertaking, it cannot require such an undertaking as a condition of making voluntary disclosure. I reject this submission. The statutory power to require an undertaking is only relevant where there is a statutory obligation to make disclosure. There is no such obligation. In the absence of such statutory obligation, if the GMC voluntarily in accordance with the principles of fairness decides that in principle disclosure should be made, it is entirely free to impose conditions which likewise accord with the principles of fairness. In my view, in insisting on respect being afforded by [T] for the confidentiality of the medical evidence relating

> to [Dr J]'s health, the GMC is acting entirely properly. To do otherwise would be calculated to discourage practitioners from submitting relevant, but confidential, material to the GMC for consideration by the screener. [T] has no legitimate ground for objecting to furnishing the undertaking: he can only legitimately require to see and use the material for the purposes of the hearings before the screener and, if the matter proceeds further, before the PPC. I accordingly hold that the GMC is entitled as a condition of making the material available to him to require [T] to give an undertaking of confidentiality limiting the use and disclosure to use for the purposes which I have referred to.

### *R (Rosen and anor) v. Solicitors Disciplinary Tribunal* [2002] EWHC 1323 (Admin)

**18.05** Employment of disqualified solicitor—file of other cases—relevant disclosure—fair hearing

The applicants were partners in a firm that acquired the practice of another firm that had employed a struck-off solicitor without permission of the Law Society. The applicants were charged with a breach of section 41 of the Solicitors Act 1974, which provides that no solicitor shall, except in accordance with the written permission granted by the Law Society, employ or remunerate a solicitor who, to his knowledge, is disqualified from practising as a solicitor. The applicants' case was that, as soon as they became aware of the problem, they immediately acted and contacted the Law Society. The applicants sought disclosure of certain documents, including the files in relation to earlier cases in which the Law Society had allowed former solicitors to be employed in breach of section 41. In quashing the order of the Solicitors Disciplinary Tribunal (SDT) refusing to order disclosure, Sullivan J noted that the challenge was concerned with an interlocutory decision of the Tribunal and that doubt had been expressed, for example in *Stokes v. Law Society* [2001] EWHC 1101 (Admin), as to whether it was appropriate for judicial review proceedings to be instituted in respect of interim decisions of the Tribunal. However, the blanket refusal of the Tribunal to order disclosure of internal files could not be justified. The request for disclosure was too wide, but the answer was for the Tribunal to exercise its discretion, so as to cut down the breadth of any unnecessarily wide request for disclosure, to identify what kinds of document, within a particular file, were in truth being sought, to focus its attention upon the relevance, or otherwise, of such documents, and, if relevant, to consider whether, on a proper balancing exercise, bearing in mind such matters as confidentiality, they should be disclosed in the interests of a fair trial.

### *Kirk v. Royal College of Veterinary Surgeons* [2003] UKPC 3 and [2003] UKPC 47

**18.06** Documents in possession of regulator—to support case or undermine regulator's case—record of complaints

The applicant, a veterinary surgeon, sought disclosure of documents pending his appeal against a finding by the Disciplinary Committee of the Royal College of Veterinary Surgeons (RCVS) of disgraceful conduct in a professional respect. The finding was largely based upon a number of convictions recorded against the appellant, mainly concerning motoring offences, and also offences of assault and using threatening behaviour, and a specific incident on the beach near Cardiff. Lord Hutton, giving the judgment of the Board on 14 January 2003, said:

> 2. It is clear that [the applicant] is entitled to disclosure of relevant documents. That is accepted by [counsel for the Royal College], who has also appeared before the Board today. Rule 6 of the Veterinary Surgeons and Veterinary Practitioners Disciplinary Committee Procedure Evidence Rules Order In Council 1967 provides: "Upon application by any party to the inquiry and on payment of proper charges the solicitor shall send to that party a copy of any statutory declaration, complaint, answer, admission explanation or other similar document sent to the College by any party to the inquiry."

3. In addition to that provision it is clear under general principles that [the applicant] was entitled and is entitled to the disclosure of any document which is relevant. That is accepted by the College. The main thrust of [the applicant]'s submission to the Board today is that there are additional documents in the possession of the College which have not been disclosed to him but which should have been disclosed. It is clear that he is only entitled to disclosure of documents if they are relevant and the Board has questioned [the applicant] and has heard him at some length as to what those documents are. But [the applicant] has failed to satisfy the Board that there has been a failure to disclose relevant documents to him.

At a subsequent hearing on 17 June 2003, Lord Hoffmann, delivering the judgment of the Board, said:

3. Now there are two bases upon which [the applicant] is entitled to documents from the College. One is his entitlement simply as a litigant in relation to the charges on which he was convicted.

He was entitled in the ordinary way to any documents which might support his case or tend to undermine the case of the College. Under that heading there have been disclosed to him certain unused materials from witnesses who were not called at the disciplinary hearing. In addition to that and quite apart from this litigation he is entitled under the Data Protection Act to electronic records kept by the College which refer to him and a number of documents have been disclosed to him under that head although the College say that to the best of their belief there were others which have been lost and which they are unable to disclose. [The applicant] says now that what he wishes to see particularly are complaints which were made about him by two local veterinary surgeons who are in competition with him. One of them did show up under the Data Protection Act disclosure and he says that there are others. The one which showed up under the Data Protection Act was not regarded as being relevant under the first heading to the particular charges because it was simply a response by the veterinary surgeon concerned to a complaint which had been made by the owner of an animal which he had seen after it had been treated by [the applicant]. It is the case, however, that the writer of the letter took the opportunity not only to respond to that particular inquiry but to make a very large number of other allegations against [the applicant]. Those however do not appear to have been pursued and the letter was not treated within the College as a separate complaint requiring formal investigation. The same may be true of other letters in which these two veterinary surgeons expressed adverse views about [the applicant]'s conduct.

4. The position now is that the College say that they have now been through their records several times and to the best of their knowledge and belief they have produced all the records to which [the applicant] is entitled either as a litigant or under the Data Protection Act [1998].

### *R (Green) v. Police Complaints Authority* [2004] 1 WLR 725

**18.07** Police Act 1996, section 80—whether disclosure necessary

G lodged a complaint against the police, alleging that he had been deliberately knocked down by a police car. Proceedings against the officer for driving without due care and attention were brought, to which the officer pleaded guilty and was fined £250 with five penalty points. In reviewing whether to recommend disciplinary proceedings against the officer, G sought disclosure from the Police Complaints Authority of all witness statements and documents available to the Authority. The House of Lords, in dismissing G's appeal, held that section 80 of the Police Act 1996 contained a general ban on the Authority disclosing any information received except in certain circumstances. The relevant question was whether the disclosure of the witness statements sought by G was necessary for the proper discharge of the particular function of the Authority. A complainant, such as G, was in a position to make an effective contribution to the process of reaching the final decision on the complaint without seeing the witness statements and other primary material. G could also draw attention to any legal or other flaw in the reasoning of the proposed decision.

***Gage v. General Chiropractic Council*** **[2004] EWHC 2762 (Admin)**

**18.08** Expert report obtained by regulator—seen by investigation committee—fair trial

G, a chiropractor, appealed against the decision of the General Chiropractic Council's PCC that he was guilty of unacceptable conduct. In 2003, G distributed copies of a booklet to patients and other members of the public that gave information about the clinic, but also set out the stories of certain patients. A formal complaint was received from an osteopath and the General Chiropractic Council referred the complaint to its investigation committee. Unbeknown to G, the Council sought a report from an expert registered chiropractor and asked him to comment on G's booklet. The report was thirty-seven pages long and accompanied by thirty pages of appendices. The gist of the report was that, by distributing the booklet, G had acted in breach of the Chiropractors' Code in a number of respects. The report was seen by the investigation committee, but was not disclosed to G at that stage. Jackson J said that the Council's power to obtain expert evidence must denote expert evidence that is disclosed to the person under investigation. A silver thread that runs through the investigation rules is that any significant evidence that the committee receives must be copied to the chiropractor, so that he or she may comment. Accordingly, the Committee was wrong to rule that the investigating committee was under no duty to disclose the report. Further, this did not prevent a fair trial. Even if the report had been disclosed to G, there was nothing that he could have said or done to prevent the case going forward to the PCC. G had a full opportunity to consider the report before the hearing. Whilst G's rights must be protected, so also there is a strong public interest in ensuring that serious allegations of misconduct against health professionals proceed to a full and proper investigation: *R v. General Medical Council, ex parte Toth* [2000] 1 WLR 2209, 2218–20; *R (Richards) v. General Medical Council* [2001] Lloyd's Rep Med 47, [58]–[62]; *R (Holmes) v. General Medical Council* [2001] EWHC 321 (Admin), [25]. In the context of criminal proceedings, a procedural irregularity of the kind that had occurred in the present case would not lead to the proceedings being stayed or dismissed: see *R (Ibrahim) v. Feltham Magistrates' Court* [2001] 1 WLR 1293.

***Henshall v. General Medical Council*** **[2005] EWCA Civ 1520**

**18.09** Response by practitioner—approach to disclosure—complainant

H challenged the decision of the GMC's Preliminary Proceedings Committee (PPC) not to refer her complaints against three registered medical practitioners to the PCC. In particular, she had not had disclosed to her the response of one of the medical practitioners, Professor S. The Court of Appeal (Sedley and Jonathan Parker LJJ, and Auld LJ dissenting) allowed H's appeal and said that, of course, not every document that came to the PPC required disclosure. If that had been the case, every complaint would risk becoming an endless war of words. But it was unacceptable to derive, by implication from rule 16 of the General Medical Council Preliminary Proceedings Committee and Professional Conduct Committee (Procedure) Rules 1988 (no complainant shall have any right of access to any documents where the Committee had decided not to refer a case for inquiry), a general inhibition. Enabling a practitioner to put in potentially contentious material in response and to deny sight of it to the complainant was capable of stifling the individual's right to bring a tenable complaint to the attention of the PCC: *per* Sedley LJ, at [82]. The solution was to consider in each case what the practitioner had put in, to decide whether, in fairness, it was something to which the complainant should be able to respond, and, if it was, to tell the practitioner that, unless he or she agreed to the disclosure of the material, it would be ignored by the PPC. Non-disclosure of a response might be permitted, which, for instance, placed the practitioner at risk of violence if it were to be disclosed. If the practitioner tenders evidential material to the PPC on terms that all or part of it is not to be disclosed to the complainant, or that disclosure is subject to conditions that the complainant is not willing to accept, then the PPC has an inherent discretion whether or not to take such material into account in reaching its decision. In the instant case, the complainant was willing to give an undertaking not to use Professor S's responses in any other context, but he did not consider that such an undertaking would afford him sufficient protection. Accordingly, he maintained an absolute prohibition on disclosure. That was a matter for him. However, the question then for the PPC was whether, in all of the circumstances and given Professor S's absolute prohibition on disclosure, it was appropriate for the PPC to take his responses into account in reaching its decision.

### *Holton v. General Medical Council* [2006] EWHC 2960 (Admin)

**18.10** Medical notes—availability

Counsel for the appellant doctor submitted that the panel had erred in law in accepting and acting on the evidence of the lay complainants, whom she had not cross-examined because the medical notes and hospital records of the patient to whom the complaint related were not available at the hearing. Stanley Burnton J said, at [92], that the appropriate course for those representing the appellant was to require the GMC to make available for the hearing any contemporaneous documents on which it wished to rely, so as to enable them to be deployed at the hearing. If the documents in question were unavailable to the GMC, it could have sought the issue of a witness summons under Civil Procedure Rules Part 34.4. If the GMC were to refuse to make the documents available, a direction should be sought from the panel and, if necessary, an adjournment of the hearing until they are produced. In the instant case, no application had been made on behalf of the appellant for the medical notes.

### *Wakefield v. Channel Four Television Corporation* [2006] EWHC 3289 (QB)

**18.11** Libel proceedings—documents held by regulator—use of—defence

The defendants in libel proceedings commenced by W, a gastroenterologist, sought disclosure of documents received by him from the GMC during the course of its investigation into his conduct. The background to the litigation was the controversy surrounding the measles, mumps, and rubella (MMR) vaccine, following the broadcast of a television programme. W commenced proceedings against the defendant. The GMC objected to W disclosing documents that fell essentially into three categories: (a) documents disclosed to the GMC by the Legal Services Commission; (b) witness statements and exhibits for the disciplinary proceedings; and (c) documents disclosed to the GMC by University College London. Some of the documents derived from patients to whom undertakings had been given of confidentiality; other documents had been obtained under statutory powers given to the GMC. In ordering disclosure and inspection, Eady J said that one background factor was that the parties seeking inspection were not intending to use the information to launch proceedings for defamation; the defendants wished to refer to the documents for the purpose of defending themselves against such proceedings. The claimant was seeking to vindicate his reputation and it was inherently undesirable that the defendants should be precluded from access to relevant information. That is not only a matter of protecting their own interests and reputations, but also of seeking to protect the public interest. There was also a need for the Court to have regard to the defendants' rights to a fair trial in accordance with Article 6 of the European Convention on Human Rights (ECHR). Another factor was that it was no part of the defendant's purpose to reveal any of the confidential information that was contained in the documents in question. The defendants accepted that provisions must be put in place to ensure that confidentiality was protected.

### *R (Johnson and Maggs) v. Professional Conduct Committee of the Nursing and Midwifery Council* [2008] EWHC 885 (Admin)

**18.12** Evidence—whether duty to obtain

A second challenge to the decision of the PCC in this case was that the Nursing and Midwifery Council (NMC) had not complied with its duty to gather evidence in favour of the claimants (as well as against them), and that this also breached their Article 6 ECHR rights. The claimants faced a number of charges in disciplinary proceedings relating to their management of a nursing home. The Court (Beatson J) held that there was no free-standing positive duty on those bringing disciplinary proceedings to gather evidence in favour of registrants, as well as evidence against them. Beatson J said, at [64], that the decision in *Jespers v. Belgium* (1983) 5 EHRR 305 was not authority for a free-standing positive duty on those bringing disciplinary proceedings to gather evidence for the following reasons: first, the decision was in the factual context of a criminal prosecution; second, it concerned non-disclosure of material in the hands of the prosecutor, not material that it was said the prosecutor was under a duty to obtain; third, the Commission was concerned with equality of arms; fourth, the Commission was concerned solely with disclosure of material that the prosecutor had; and fifth, what the Commission stated was fundamentally inconsistent with the submission that there was a free-standing duty on prosecutors to obtain evidence and give it to the subjects of disciplinary proceedings. In the instant case, the Court held that, whilst there was a duty to prevent inequality of arms, there was no inequality of arms on the facts of this case. The claimants had not been at a disadvantage in obtaining evidence from documents. Whether they had been afforded adequate time and facilities for the preparation of their defence was fact-specific.

### *R (Amro International SA) v. Financial Services Authority* [2010] EWCA Civ 123, [2010] 3 All ER 723

**18.13** Foreign regulatory proceedings—notice to produce—use of investigative powers

The Court of Appeal (Sir Anthony May P, Stanley Burnton, and Jackson LJJ) held that the Financial Services Authority (FSA) had power to issue notices to require the production of documents from a firm of accountants at the request of the US Securities and Exchange Commission (SEC). It had no duty to investigate or verify the information being sought before exercising its powers under the Financial Services and Markets Act 2000. The Court observed that financial transactions were increasingly international and that it was of the greatest importance that national financial regulators cooperated, particularly where financial fraud or misconduct was suspected. There was nothing in the Financial Services and Markets Act 2000 that required the FSA to second-guess a foreign regulator as to its own laws and procedures, or as to the genuineness or validity of its requirement or information or documents. The FSA had to, and did, consider the request when deciding whether to exercise its statutory discretion by the exercise of its investigative powers. It was clear that the FSA decided to exercise its investigative powers, having considered the matters required by the Act of 2000. Accordingly, there was no error of law or principle in the FSA's decision to appoint the investigators who were to assist the SEC with its ongoing civil action in New York.

### *Legal Ombudsman v. Young* [2011] EWHC 2923 (Admin)

**18.14** Ombudsman—obligation to produce documents

Y failed to respond to a complaint about legal services provided by him to former clients. The Legal Ombudsman required Y to produce specific documents, files, and bills that the Ombudsman considered were necessary to enable him to determine the complaint. Y failed to comply with that requirement. The Ombudsman submitted a report to the Solicitors Regulation Authority (SRA) under sections 143, 145, and 148(2) of the Legal Services Act 2007, stating that, among other things, he believed that Y had failed to comply with the requirements imposed on him by a section 147 notice. The Ombudsman requested the Court, under section 149 of the Act, to enquire into the case and to 'deal with' Y for his alleged default on the requirements of the notice issued by the Ombudsman. Lindblom J held that the Ombudsman's application succeeded and that a financial penalty was justified. Fixing this at £5,000, the learned judge had in mind that Y was also facing a substantial award of costs against him.

### *AXM v. Bar Standards Board* [2011] EWHC 2990 (QB)

**18.15** Report to Conduct Committee—disclosure

The applicant, a barrister, faced a number of disciplinary charges alleging professional misconduct arising out of his handling of the defence case in two criminal matters. He sought disclosure of 'the sponsor's report' prepared by a nominated member of the PCC in advance of the Committee's hearing to decide whether or not to instigate disciplinary proceedings. The sponsor's report is designed to avoid the need for each and every member of the Committee to go through every page of the papers making up the material that is before them. What the sponsor does is to prepare, using the material, a summary of the facts alleged against the barrister concerned, the nature of the evidence, and a summary of the responses of the barrister concerned to the allegations made. A separate report is prepared containing recommendations of the sponsor. King J, in ordering disclosure of the report other than the recommendations of the sponsor, said that everything in the sponsor's report going to the summary of the facts and the summary of the defendant barrister's response to the complaints made against him should be disclosed. The Bar Standards Board need not disclose the recommendations made by the sponsor. In the context of the particular case, the particular issues raised, for which disclosure was sought, go to the nature of the decision-making process of the Complaints Committee. In 2007, a review by Professor Behrens recommended that there should be more transparency in the decision-making process as to whether or not to prosecute, and that whereas the sponsor's reports were not discloseable in the practice then followed, that practice should be altered and sponsor's notes should be disclosed, although taking into account any data protection issues or matters of sensitivity. In the instant case, in the context of an application that the charges should be dismissed and the proceedings struck out because of a systematic failure in the process of the Committee to consider properly and fairly the responses of the barrister concerned, fairness required disclosure of the report.

### *Veen v. Bar Standards Board* [2011] All ER(D) 223 (Oct)

**18.16** File notes—disclosure

V, a barrister, appealed to the Visitors against a finding of a disciplinary tribunal of the Council of the Inns of Court of professional misconduct in, having accepted a brief to represent clients in a licensing application, refusing to represent them unless his fees were increased. The Visitors (Thirlwell J, Mr Jonanthan Barnes, and Mr William Henderson), in dismissing the appeal, referred to failures of disclosure by the Bar Standards Board. At the hearing before the disciplinary tribunal, the Board had disclosed part of an email sent to it by a trainee from V's instructing solicitors and, following the hearing, the Board served a file note of a conversation between it and V's instructing solicitors. Thirlwell J said that no reason had been given for the failure to disclose the file notes at the proper time. They plainly should have been disclosed before the hearing before the tribunal.

### *R (McCarthy) v. Visitors to the Inns of Court* [2013] EWHC 3253 (Admin); [2015] EWCA Civ 12

**18.17** Draft witness statements—disclosure—inconsistency—fairness

The claimant barrister was found guilty by the Bar Disciplinary Tribunal of producing fabricated or forged letters in relation to a direct access case and was disbarred. The Bar Standards Board served the complainant's witness statement after receipt of M's witness statement, and a number of paragraphs responded in detail to matters referred to by M in his witness statement. A draft earlier statement had been prepared for the complainant that had never been served and came to light only on appeal before the Visitors. In granting permission to apply for judicial review by a majority, but refusing to quash the Visitors' decision and dismissing the claimant's appeal, the Divisional Court (Moses LJ) said, at [17], that it was beyond question that, in disciplinary proceedings with the potential for such grave consequences, draft statements capable of being used to discredit a witness should be disclosed. The criminal process had, for nearly eighty years, recognized the obligation to disclose statements that were potentially of use for the purpose of cross-examining witnesses: *R v. Clarke* (1931) 22 Crim App Rep 58. It was well understood that, in criminal courts, draft statements of witnesses that might reasonably be considered capable of undermining the case for the prosecution or of assisting the case for the accused are discloseable: see Criminal Procedures Investigations Act 1996, section 3(1)(a), and the Attorney General's Guidelines. If there was no inconsistency, the prosecution has nothing to lose, and if there was inconsistency, there is no warrant for concealing. There was no basis why the position should be any different in relation to disciplinary proceedings brought on behalf of the Bar Standards Board. The failure to serve the complainant's draft earlier statement was a breach of rule 7 of the Disciplinary Tribunals' Regulations 2009 and, moreover, a breach of natural justice. Where unfairness is alleged, it is necessary to consider both the nature and degree of the unfairness, and the extent to which it might have made a difference: *John v. Rees* [1970] Ch 345, 402C–E.

In allowing the claimant's appeal, the Court of Appeal (Burnett LJ, with whom Newey J and Dame Janet Smith agreed) said that what had happened was extraordinary. A conscious decision was taken by an official at the Board, which had the effect of subverting the rules that provide for disclosure, and furthermore suggested that he was blind to any sense of fairness in the conduct of a disciplinary prosecution. The matter was compounded by inviting the complainant to assume the role of surrogate prosecutor. The witness's witness statement of 158 paragraphs was an amalgam of evidence properly so-called, comment, and argument intended to demolish the claimant's defence to the charges, rather than to provide unvarnished evidence. It was this document that stood as evidence in chief. The first statement remained undisclosed. At [25], Burnett LJ said that the ultimate question is whether the proceedings as a whole were fair. In criminal cases involving Article 6 ECHR, the test was settled by the Supreme Court in *McInnes v. Her Majesty's Advocate* [2010] UKSC 7, in which Lord Hope identified two questions that fall to be considered in disclosure cases. The first is whether the material under consideration should have been disclosed. The second is the consequences of any failure to disclose. His Lordship continued:

> 26. There is no reason to apply a different test in disciplinary proceedings when it is established that there has been material non-disclosure. Is there a real possibility that

the Tribunal would have come to a different conclusion had the disclosure been made? That involves a consideration not only of the content of the undisclosed material but also an evaluation of the various ways in which its disclosure might have affected the course of the proceedings. In substance Moses LJ applied this test. The question is whether his conclusion was correct.

The standard of proof required in the Tribunal proceedings was the criminal standard. Its members had to be sure before finding the charges proved. In the light of the central place that the complainant's credibility occupied in the Tribunal hearing, and that cross-examination on the first statement was capable of undermining the witnesses' credibility given the differences between the two statements, there was a real possibility that the Tribunal would have come to a different conclusion had disclosure been made. Accordingly, the appeal should be allowed.

### *R (Nakash) v. Metropolitan Police Service and General Medical Council (Interested Party)* [2014] EWHC 3810 (Admin)

**18.18** Disclosure of information to regulator—police interview—documents unlawfully seized—whether outweigh legitimate aim of disclosure under Medical Act 1983, section 35A

In June 2011, the claimant was working as a specialist registrar in the obstetrics and gynaecology department of the Royal Free Hospital in London, where he had been employed for about four years. On the morning of 5 June 2011, he carried out a trans-vaginal scan upon a young female patient referred to as 'M'. Although he initially asked a female nurse to accompany him as chaperone, in accordance with hospital policy, it was not in dispute that the claimant invited the nurse to leave at some point during the procedure, leaving him alone in the room with M. Later on that day, M made a complaint of sexual assault by the claimant in the course of this procedure and the police were contacted. In the afternoon of 6 June, three police officers arrived at the claimant's home. It was not in dispute that the police action on 6 June was unlawful. The claimant appears to have been arrested, although no contemporaneous record of the reasons for his arrest was made. The claimant was not informed that he was being arrested, or of the grounds for his arrest, and no consideration was given as to the necessity for his arrest. The claimant's home was searched and computers containing private correspondence were seized, in purported exercise under section 18 of the Police and Criminal Evidence Act 1984. The claimant agreed to be interviewed and, in the early evening of 6 June, he gave full and detailed answers to a wide-ranging series of questions about M's complaint and the surrounding circumstances. The police examination of the claimant's computer, seized from his home at the time of the search, revealed a significant amount of private correspondence during the months before M's complaint of sexual assault. Much of it was of a personal and intimate nature, but it contained, on one page, via the Internet telecommunication facility 'Skype', an exchange between the claimant and W, another doctor, occurring some three weeks before the alleged assault on M. On that one page, the claimant made references to his own sexual arousal during a previous conversation with a different patient. The officer in charge of the investigation was subsequently suspended and later dismissed by the Metropolitan Police Service. The claimant was eventually charged with two offences—namely, sexual assault and assault by penetration. He denied both charges and the matter proceeded to trial. M and the claimant both gave evidence at the trial at Southwark Crown Court, which concluded on 2 July 2013, when the claimant was acquitted on both charges. Tragically, the GMC was notified that M had committed suicide on 8 August 2013. On 3 July 2013, the police informed the GMC that they had material in their possession relating to the claimant that was not adduced at his trial—namely, the police interview of 6 June 2011 and the Skype communication. On 8 July 2013, the GMC, relying on section 35A of the Medical Act 1983, requested that the police provide a copy of the police file for the case, including the material not adduced at trial. Section 35A of the Act provides that the Council, or any of its committees, or the registrar, in carrying out functions in respect of a practitioner's fitness to practise, may require the practitioner or any other person to supply information or produce any document that appears to be relevant to the discharge of any such function. In dismissing the claimant's claim for judicial review to prevent disclosure of the police interview and the Skype communication, Cox J said that the power given to the GMC to request disclosure under section 35A was broad in scope and served an important purpose. There is a strong public interest in the GMC's performance of its statutory functions, as they relate to a doctor's fitness to practise. The disclosure of material to the GMC by other agencies, including the police, has an important role to play in the exercise of those functions. After reviewing *Wakefield v. Channel 4 Television Corpn and ors* [2006] EWHC 3289 (QB), *Woolgar v. Chief Constable of Sussex Police and UKCC* [2000] 1 WLR 25, CA, and a decision of the European Court of Human Rights in *MS v. Sweden* [1997] 3 BHRC 248, and Article 8(1) and (2) ECHR, her Ladyship said that she acknowledged that the unlawful seizure of the Skype extract represented a serious intrusion into the claimant's private communications, within the confines of his own home, but added:

> 56. I cannot, however, accept [counsel for the claimant's] submission that this extract cannot be said to have any relevance to the GMC's investigation. While the extract may not, in itself, be probative of M's complaint, or of any other complaint against the claimant, or of there having been any specific act of misbehaviour by him, or even of any intended act, it is not necessary for the material to be directly probative in this sense in order for it to have relevance. The inquiry being carried out by the GMC, as the claimant's regulatory body, is broader than the focus inquiry of the [Crown Prosecution Service] and the Crown Court. In my view, this conversation cannot be said to be irrelevant to the GMC's investigation into this claimant's fitness to practise as a doctor.
>
> 57. The GMC will be concerned not only with M's specific allegation of criminal conduct, but with all the circumstances surrounding that allegation, including the circumstances in which M was examined, what the claimant was doing and thinking at the time, and how the claimant was behaving generally, both towards her and towards the nurse acting as "chaperone". The claimant's acquittal on the criminal charges does not mean that the GMC, acting entirely independently, cannot entertain professional concerns as to his fitness to practise. The fact that the Skype evidence was considered inadmissible at his criminal trial does not render it immune from disclosure under section 35A...
>
> [...]
>
> 61. Given my clear view as to the relevance of this Skype extract to the regulator's inquiry, I accept [counsel for the GMC's] submission that the circumstances in which it was obtained will carry little weight in the balancing exercise required under Article 8. The scope and purpose of the GMC's investigation remain exactly the same. The fact that the material was obtained unlawfully does not, in my judgment, outweigh the legitimate aim served by its disclosure under section 35A, namely to enable the GMC, in the exercise of their statutory functions, to protect public health and safety and to protect the rights and freedoms of others. For these reasons disclosure of the Skype extract does not violate the claimant's rights under Article 8. Disclosure is a proportionate response to that important and legitimate aim.

Her Ladyship's conclusion in relation to the disclosure of the police interview was the same—namely, that its disclosure to the GMC was reasonable and justified. Both the Skype extract and the interview were documents that were relevant to the GMC's investigation and, notwithstanding the circumstances in which they were obtained, their disclosure was justified under Article 8(2) ECHR: at [68].

***Law Society (Solicitors Regulation Authority) v. Charles Henry*, 29 January 2015 (unreported)**

**18.19** Third-party disclosure order

The SRA applied for an order for disclosure of documents from the employer of three solicitors who were the subject of investigations of misconduct. The employer was a charity, and it and its practice manager were subject to general civil restraint orders. The charity was not regulated by the SRA, but had engaged the three solicitors who were practising at its office. In granting an order requiring the charity to disclose certain documents, Laing J held that the test for making the order against the charity was satisfied. The practice manager had been involved in unmeritorious actions in respect of which the SRA was seeking disclosure of documents. Under section 44BB of the Solicitors Act 1974, the SRA had a general concern that the three solicitors had not supervised the litigation conducted by the charity and the practice manager. Such interference with the practice manager's rights under Article 8 ECHR had to be balanced against the public interest in the investigation of solicitors' misconduct. It was not disproportionate to require disclosure and it was established that the statutory regime did not breach legal professional privilege: see *R (Morgan Grenfell & Co. Ltd) v. Special Commissioners of Income Tax* [2003] 1 AC 563.

## C. Confidential Information

***Woolgar v. Chief Constable of Sussex Police* [2000] 1 WLR 25**

**18.20** Police interview—information given under caution—disclosure to regulatory body—public interest

The plaintiff, a registered nurse and the matron of a nursing home in Worthing, was arrested and interviewed by the police following the death of a patient in her care. The allegation that particularly concerned the police was one of overadministration of diamorphine, but there were also other allegations of concern to the registration and inspection of the home. The plaintiff was informed that the evidence did not meet the evidential test required for criminal charges and the matter was then referred to the United Kingdom Central Council for Nursing Midwifery and Health Visiting (UKCC), the regulatory body for nursing, midwifery, and health visiting. The plaintiff appealed to the Court of Appeal against the order of Astill J, who dismissed her application for an order that the chief constable be restrained from disclosing to the UKCC the contents of her police interview. On her behalf, it was submitted that, when she answered questions, she did so in the reasonable belief that what she said to the police would go no further unless it was used for the purposes of criminal proceedings. The caution administered to her so indicated and, in order to safeguard the free flow of information to the police, it was essential that those who gave information should be able to have confidence that what they say will not be used for some collateral purpose. In dismissing the plaintiff's appeal, Kennedy LJ (with whom Otton and Waller LJJ agreed) said at 37D–G:

> However, in my judgment, where a regulatory body such as UKCC, operating in the field of public health and safety, seeks access to confidential material in the possession of the police, being material which the police are reasonably persuaded is of some relevance to the subject matter of an inquiry being conducted by the regulatory body, then a countervailing public interest is shown to exist which, as in this case, entitles the police to release the material to the regulatory body on the basis that, save insofar as it may be used by the regulatory body for the purposes of its own inquiry, the confidentiality which already attaches to the material will be maintained. As [counsel for the chief constable] said in his skeleton argument: "A properly and efficiently regulated nursing profession is necessary in the interest of the medical welfare of the country, to keep the public safe, and to protect the rights and freedoms of those vulnerable individuals in need of nursing care. A necessary part of such regulation is the ensuring of the free flow of the best available information to those charged by statute with the responsibility to regulate."

*R (Pamplin) v. Law Society* [2001] EWHC 300 (Admin)

**18.21**
Police file disclosed to Law Society—public interest

P was a solicitor's clerk. Following his arrest in September 1998 and the decision by the Crown Prosecution Service to take no action in respect of the matter, the Law Society received a police file from the Chief Constable of Lancashire. P issued judicial review proceedings challenging the decision of the Office for the Supervision of Solicitors to consider whether to make an order against him under section 43(2) of the Solicitors Act 1974 that no solicitor should employ him in connection with his or her practice without written consent. In the light of *Woolgar v. Chief Constable of Sussex Police* [2000] 1 WLR 25, the applicant amended his statement of grounds to allege that the police did not exercise their power to disclose the material lawfully, because they did not give prior notice to P and did not provide him with an opportunity to comment on the proposal to send the file to the Law Society, and that this was a breach of Article 8 ECHR. In dismissing P's application, Newman J said that the rationale of the case of *Woolgar* was against the proposition that it is a precondition to disclosure—in the circumstances under scrutiny, being legal—that notice should be given to the person affected. The *ratio* of the Court of Appeal's decision was that even if there is no request from the regulatory body, if the police come into possession of confidential information that, in their reasonable view, in the interests of public health or safety should be considered by the professional or regulatory body, then the police are free to pass that information to the relevant body for its consideration. In the instant case, the interests of the public are the interests in the proper administration of justice, the integrity of the solicitors' profession, and the maintenance and regulation of those involved in the legal profession—all of which makes it necessary, for the better administration of justice, that there be disciplinary control of matters coming to the notice of either the police or the Law Society that may have a bearing on and put at risk those matters which it is in the public interest to uphold. The decision of the Office of Solicitors to investigate was a proper and reasonable response to the material that it had been given. The Office received the file from the police on 20 May 1999 and it did not write to the claimant until 2 August 2000. That was a long time for a decision to be made as to whether to start an investigation. Having regard to the delay that occurred, the notice—namely, that comments should be returned by the claimant within fourteen days—was an insufficient period of time and, on the face of it, unnecessarily pressing. Thereafter, discussion did take place about an extension of time. As to Article 8 ECHR, the learned judge said that, so far as confidentiality was concerned, it could extend no further than the interview that took place between P and the police, and which formed but part of the documentation disclosed by the police. However, the matter did not get under way. It is to confuse matters to assume that just because something is confidential or capable of being confidential, then that which occurs gives rise to the engagement of Article 8. That which is confidential is not necessarily connected with a person's private life. What took place was in accordance with law, and centred upon the employment of the claimant and was part of his public life—his public life in the important profession of providing legal services. There was nothing that gave rise to any Convention point.

*A v. General Medical Council* [2004] EWHC 880 (Admin)

**18.22**
Child—welfare—paediatrician—use of confidential information—public interest

Miss A, suing by her litigation friend and father, sought judicial review of the decision of the PCC that Dr C was not guilty of serious professional misconduct. The allegation against Dr C, a consultant paediatrician, was that, despite a letter from Miss A's parents that they no longer wished Dr C to be involved in the treatment of their daughter, he wrote letters to a consultant paediatrician and a senior clinical medical officer, and sought to influence the course of Miss A's clinical management, and sought and obtained access to confidential information about her ongoing care to which he was not entitled. In dismissing the judicial review proceedings, Charles J referred, at [36]–[44], to the competing public interests between (a) promoting doctor/patient confidentiality, and (b) promoting the welfare of children and thus their protection from harm. It was necessary to determine where the overall public interest lies if a doctor wishes to, or has, used or disclosed information provided to him, or obtained by him, in connection with the treatment of a child for purposes relating to the welfare of that child if that use or disclosure has not been authorized, or has been forbidden, by his or her parents. General factors that are likely to be relevant in the balancing exercising include:

(1) the reasons that underlie the competing public interest;
(2) the roles of the courts and the local authority in respect of child protection, and issues concerning the medical treatment of children;
(3) the role of the parents of a non *Gillick* competent child and the respect that should be accorded to their views; and
(4) the roles of others involved in the relevant use and disclosure, and the need for there to be cooperation and communication between people who acquire information that may impact on issues relating to the welfare of children (such as doctors, health visitors, and teachers) and local authorities, which have statutory duties relating to the promotion of the welfare of children.

Another important general factor is that there is an important distinction between (i) disclosure to a person who is, or persons who are, aware of the confidentiality of the information and who, in connection with the purpose for which the information is disclosed, have a role to play in its consideration or evaluation, and (ii) more general disclosure or publication. When considering statements concerning the nature and importance of duties of confidence owed by a doctor to a patient, and the use or disclosure of information covered by those duties, it needs to be remembered that the positions of an adult and a child patient are different. The difference flows from the autonomy of an adult patient of sound mind and his or her right to refuse treatment, which is different from the position of a child and his or her parents, because of the ability of the courts to make decisions relating to the treatment of a child under the inherent jurisdiction of the Children Act. Thus, in the case of a child patient, there are public bodies who can override the decision of the parents.

### *Saha v. General Medical Council* [2009] EWHC 1907 (Admin)

**18.23** Medical records—disclosed by GP to GMC—confidentiality—public interest

This was an appeal by S against a finding of misconduct based on his failure to cooperate fully and to provide relevant information, in breach of paragraph 30 of *Good Medical Practice*, in connection with an investigation by the GMC into his conduct. The investigation concerned the fact that S was infected with hepatitis B, and he refused to disclose to the GMC details relating to his employment and previous employers. S's medical records were disclosed to the GMC by his general practitioner (GP). The appellant's case was that this was a breach of his right to confidentiality in his own medical records, and, accordingly, the GMC was not lawfully entitled to ask him for information about his current and past employers; for that reason, there could be no breach of paragraph 30 of *Good Medical Practice* nor of section 35A(2) of the Medical Act 1983. In dismissing S's appeal, the deputy judge (Mr Stephen Morris QC) said, at [67], that medical confidentiality is not an absolute right, but necessarily involves a balancing of competing public interest. The public interest in patient safety and welfare is an extremely important consideration. A further highly relevant consideration is the persons to whom the disclosure has taken place or is envisaged: disclosure to a person who is aware of the confidentiality and who has a role in its consideration or evaluation is to be distinguished from general disclosure or publication. In the instant case, the disclosure by the three doctors of the appellant's records, ultimately, to the GMC was justified by reference to the concerns that each of them had at the relevant time, and indeed each was under a duty to do so.

***Financial Services Authority v. Information Commissioner* [2009] EWHC 1548 (Admin)**

**18.24** Disclosure of confidential information held by FSA—Financial Services and Markets Act 2000, section 348—disclosure under Freedom of Information Act 2000

This appeal by the FSA raised a short, but important, question of construction as to the true meaning and effect of section 348 of the Financial Services and Markets Act 2000. Section 348 of 2000 Act provides that confidential material relating to the business or other affairs of a person must not be disclosed by a primary recipient, or by any person obtaining the information directly or indirectly from a primary recipient, without the consent of a person concerned to whom the information relates. The purpose of section 348 is to protect confidential information that has found its way into the FSA's hands. Munby J held that the names and identities of certain firms involved in the provision of selling endowment mortgages for which higher charges had been applied after the policies had been taken out could have come to the knowledge of the FSA only as a result of information supplied to it by the relevant firm; accordingly, the information tribunal was wrong to order disclosure sought under the Freedom of Information Act 2000. Similarly, disclosure of the lists of firms investigated in relation to equity release schemes, and found to have failed to explain to customers the advantages and disadvantages of such schemes, would have meant disclosure of matters relating to the business or other affairs of the firms. However, disclosure of the lists of firms used for the exercise generally was quite different and did not involve disclosure of confidential information contrary to the statute.

***General Dental Council v. Rimmer* [2010] EWHC 1049 (Admin)**

**18.25** Dental and clinical records—patient computer records—confidentiality—access to regulator

This was an application by the General Dental Council (GDC) for a declaration authorizing the granting of access to, and the inspection and copying of, the defendant's computerized records by the claimant's forensic computer expert. R, a general dental practitioner, was subject to a fitness to practise investigation by the GDC. The allegation was that he retrospectively amended some, or all, of the computer records of sixteen patients, an allegation he denied. Not all patients consented to their records being used. Lloyd Jones J said that the first question that arose was one of the jurisdiction of the Court to hear an application under Civil Procedure Rules Part 8. He was entirely satisfied that CPR Part 8 was the appropriate means by which to bring the application before the Court. The learned judge said that the Court was concerned essentially with questions of confidentiality at common law and with rights arising under Article 8 ECHR. A patient's dental or clinical records are protected by obligations of confidentiality. That confidentiality does not apply to anonymized records, provided that those records can no longer be used to identify individual patients. However, that did not assist in the present case, because the hard drives in question could not be accessed without identifying the individual patients. The Court may properly be asked to sanction the disclosure of material that is confidential if sufficient justification exists. On such an application, it is for the Court to assess the competing public interests involved in such an application: the interests of the individual patients and the public at large in the confidentiality of such records; and the public interest in the proper and effective pursuit of professional disciplinary proceedings and, as in this case, proceedings as to the fitness of professional persons to practise. It is for the Court to satisfy itself, on the balance of these considerations, that the disclosure is appropriate. The learned judge referred to *A Health Authority v. X and ors*

[2001] EWCA Civ 2014, which was concerned with disclosure of medical records in connection with proceedings under the Children Act 1989. In that case, Thorpe LJ said at [19] that there is obviously a high public interest, analogous to the public interest in the due administration of criminal justice, in the proper administration of professional disciplinary hearings, particularly in the field of medicine. At [20], he observed that a balance still had to be struck between competing interests. In *A Health Authority v. X*, the balance came down in favour of production, as it invariably does, save in exceptional circumstances. Lloyd Jones J said that he was satisfied that, in this case, the public interest enabling effective disciplinary proceedings to be considered and conducted by a regulator against a dental practitioner is a proper and lawful purpose, and within the meaning of Article 8 ECHR. The objective falls within Article 8(2), both for the reason that it is for the protection of health and more generally because it is for the protection of the rights and freedoms of others. The public interest is fostered by the proper administration of such disciplinary proceedings. The learned judge was satisfied that the order sought was necessary to enable the investigation to be carried out.

### *General Dental Council v. Savery and ors* [2011] EWHC 3011 (Admin)

**18.26** Dental records—use of patient records for disciplinary proceedings—absence of consent—confidentiality of Committee members

In this case, the GDC applied for a declaration that it may use and disclose the dental records of fourteen patients and former patients of the fifteenth interested party, Dr A-N, who was a registered dentist. The GDC wished to be able to use the patient records for the purposes of professional disciplinary proceedings, which had been commenced against Dr A-N. The GDC wished to establish that the registrar of the GDC (and those working in his office) who already had copies of the relevant patient records in their possession, may pass those records to an investigating committee of the GDC to enable that committee to conduct an investigation into the allegations of professional misconduct and impairment of fitness to practise against Dr A-N. Efforts had been made by the GDC to contact all of the patients to obtain their consent to the use of their records for this purpose. Of the fourteen patients, ten had expressly refused to give their consent and four had not replied to the GDC's enquiries. The net effect, therefore, was that, in relation to all fourteen of the patients, there was an absence of consent to the use of their records by the GDC for the purposes of the professional misconduct proceedings against Dr A-N. The fourteen patients each took out dental health insurance with an insurance company and, in doing so, the patients authorized the release of their medical records to the insurance company to inform the GDC about its concerns in regard to claims for reimbursement of monies paid to Dr A-N for treatment. Sales J said that section 33B(2) of the Dentists Act 1984 gave power to the GDC to require the insurance company to provide information and patient records as it did. The obligation on the registrar under section 27(5)(a) to refer an allegation to the investigating committee includes an obligation to refer all evidential material relevant to the allegation to the committee as well. This included the patient records in this case. There is no requirement under this regime for the registrar first to obtain an order of the court before passing such records on to the investigating committee. The members of the committee who receive the patient records will also take them subject to common law duties of confidentiality owed to the patients and dentist in question. The investigating committee will, if it determines that the allegation ought to be considered by a practice committee, be under an obligation to refer the allegation to such committee under section 27A(4)(a) of the Act. Again, it is clear that, by implication, the referral includes an obligation to refer all evidential material relevant to the allegation to the practice committee and there is no requirement to obtain an order from the court. The members of the practice committee will receive the patient records subject to common law obligations of confidentiality in relation to them. Sales J held that the provisions in the 1984 Act fall to be read as qualified by Article 8 Convention rights. The learned judge said that the leading Strasbourg authority regarding one public authority transmitting confidential patient records to another public authority to enable the second authority to carry out functions in the public interest was *MS v. Sweden* (1999) 28 EHRR 313. The learned judge concluded that the proposed disclosure in the instant case fell within Article 8(2), as being in the interests of public safety, for the protection of health and morals, and for the protection of the rights and freedoms of others. The proposed disclosure was also in accordance with the law, since it was made pursuant to the clear statutory regime in the Dentists Act 1984. The proposed disclosure of the patient records also satisfied the requirements of being necessary in a democratic society. Accordingly, the Court confirmed that the GDC was permitted to use and disclose patient dental records for the purpose of fitness to practise proceedings even though the patients had not consented to their use and disclosure.

### *R (May) v. Chartered Institute of Management Accountants* [2013] EWHC 1574 (Admin)

**18.27** Letter marked 'Strictly Private & Confidential'—disclosure—whether information confidential

M, a member of the governing council of the Chartered Institute of Management Accountants (CIMA), received a letter from the chief executive headed 'Strictly Private & Confidential' in response to an email circulated by M to members of the CIMA council. The chief executive's letter recorded that M's actions were wholly inappropriate and asked her to desist forthwith from sending emails of this nature. M was found guilty of professional misconduct and a breach of CIMA's code of ethics in failing to respect confidentiality in relation to the dissemination of the chief executive's letter, which M sent on to others. In allowing M's appeal, Stadlen J said, at [118], that the language of the code meant that it was a necessary precondition of a breach of duty thereunder that there should be identified confidential information acquired by M as a result of her professional relationship with the chief executive. Even if it was right to characterize the chief executive's letter as 'information', the fact that, by forwarding his letter, M disclosed those views to those recipients did not, in the circumstances, necessarily render the fact or expression of those views confidential.

### *Chhabra v. West London Mental Health NHS Trust* [2014] ICR 194, SC

**18.28** Patient confidentiality—central to trust between patients and doctors

C, a consultant forensic psychiatrist at Broadmoor Hospital, faced allegations under her employer's disciplinary procedures for doctors and dentists in the National Health Service (NHS). The allegations included breach of patient confidentiality during a train journey when, in a busy carriage, C discussed an incident involving a patient in the secure unit at the hospital, and was reading a medical report on a patient whose name and personal details could be clearly identified, and an allegation that C had dictated patient reports when travelling on a train. The Supreme Court agreed with the Court of Appeal—[2013] EWCA Civ 11—that the Trust's medical director and case manager in relation to the concerns raised about C would have been entitled to take the view that there was evidence in the case investigator's report that could amount to serious misconduct, and that he could properly have convened a conduct panel on that basis rather than refer the allegations to a capability hearing under the framework document *Maintaining High Professional Standards in the Modern NHS* and the Trust's performance and disciplinary policies. There is no doubt that patient confidentiality is an overriding principle and is central to trust between patients and doctors—General Medical Council, *Good Medical Practice* (2006), page 5, paragraphs 21 and 37; *Guidance on Confidentiality* (2009), paragraph 6: *per* Lord Hodge, at [33]. Accordingly, the evidence in the case investigator's report was capable of supporting a complaint of serious misconduct, although, because of a number of irregularities in the proceedings against C, the convening of the panel was unlawful.

## D. Data Protection and Freedom of Information

### Data Protection Act 1998

**18.29** Section 7 provides a right of access to personal data held by a professional regulator. A regulator can avoid having to provide such data where that disclosure would be likely to prejudice the discharge of functions relating to the protection of the public against dishonesty, malpractice, or other seriously improper conduct by, or the unfitness or incompetence of, persons authorized to carry on any profession or other activity (section 31).

The prohibition on disclosing personal data to third parties without the data subject's consent is disapplied where the disclosure is required by any rule of law or order of a court, and therefore the Data Protection Act does not prevent the disclosure of personal data to a regulator or professional body, which has powers to require disclosure (see paragraph 18.01), nor does it prevent the use of personal data in tribunal/committee proceedings (see section 35).

### *Durant v. Financial Services Authority* [2003] EWCA Civ 1746

**18.30** Information held by FSA—claim against bank—purpose of section 7—whether information infringes privacy—not to obtain disclosure to support litigation

D had sued his bank—Barclays—in 1993 and lost. He thought that the FSA had information that would help him to reopen his claim against Barclays, so in 2001 he made a section 7 subject access request under the Data Protection Act 1998 for all data that the FSA held on him. The FSA made partial disclosure of some material to D. D started a claim under section 7(9) to enforce his subject access request rights to the undisclosed data. Eventually, various questions came before the Court of Appeal and, in resolving those questions, the Court found that the purpose of section 7 was to 'enable him to check whether the data controller's processing of it [his data] unlawfully infringes his privacy and, if so, to take such steps as the Act provides...to protect it' (*per* Auld LJ, at [27]). Auld LJ went on to say that the section was not there to 'assist him, for example, to obtain discovery of documents that may assist him in litigation or complaints against third parties'. Speaking on behalf of the Court at [30], Auld J held that:

> I do not consider that the information of which [D] seeks further disclosure—whether about his complaint to the FSA about the conduct of Barclays Bank or about the FSA's own conduct in investigating that complaint—is "personal" within the meaning of the Act. Just because the FSA's investigation of the matter emanated from a complaint by him does not, it seems to me, render information obtained or generated by that investigation, without more, his personal data.... His claim is a misguided attempt to use the machinery of the Act as a proxy for third party discovery with a view to litigation or further investigation....

### Freedom of Information Act 2000

**18.31** Generally, it is very hard to use the Freedom of Information Act 2000 to obtain information from regulators about an ongoing investigation or proceedings. Sections 30, 31, 41, and 42 all provide exemptions to the duty to disclosure information by a regulator. Section 31, in particular, exempts disclosure where it would, or would be likely to, prejudice the exercise of a public body of any of its functions for the purposes of ascertaining a person's fitness or competence in relation to any profession that a person is, or is seeking to become, authorized to carry on.

### *Health Professions Council v. Information Commissioner*, EA/2007/0116, 14 March 2007

**18.32**

Complaint—regulator finding no case to answer—FOIA request—regulator to forewarn registrants of its statutory duties

Ms Lee made an allegation to the Health Professions Council (HPC) that a particular registrant's fitness to practise was impaired. This was investigated and papers duly put before an investigating committee. The committee decided that there was no case to answer. Ms Lee, who was unhappy with this decision, then made a Freedom of Information Act (FOIA) request in relation to the papers that were put to the investigating committee. This was refused by the HPC on the grounds that certain exemptions applied—namely, sections 30 (investigations and proceedings conducted by public authorities), 40 (personal data), and 41 (confidentiality). Ms Lee then made an application under section 50 of the 2000 Act for a decision by the Information Commissioner as to whether her request had been dealt with in accordance with law. As part of his investigation, the Information Commissioner sought to be provided with the disputed information that had not been provided to Ms Lee. Initially, he sought this without recourse to the issuing of an information notice. The HPC resisted on the basis that, absent a formal information notice, it did not have the power to disclose the information. The HPC set out its concerns in correspondence over the months of July–September 2007 as to the potential damage to the process were registrants to be aware that information had been passed to the Information Commissioner. The information notice was issued on 1 October 2007 and the HPC filed a notification of appeal against the notice on 29 October 2007. Counsel for the HPC at the hearing set out the appeal on the grounds that the Information Commissioner had either (a) erred in exercising his discretion to issue an information notice, and/or (b) the information notice was not in accordance with the law on the basis that the Commissioner had failed to have regard to relevant matters—namely, the way in which the process would be undermined as a result of the notice. It was argued that, because registrants believed that the information they provided to the HPC at the 'case to answer' stage was confidential, the effect of the information notice would be to discourage the provision of information at this early stage, for fear of disclosure to the Commissioner and onward disclosure by the Commissioner to the public. On testing the evidence, the Information Tribunal concluded, however, that the damage to the process anticipated by the HPC would not be as significant as feared. Registrants had a self-interest in disclosing a broad range of information at the early stage—this was, after all, the best way in which to keep the matter out of the public domain (that is, by virtue of a 'no case to answer' finding). The Tribunal considered that the HPC was overstating the confidentiality expected by registrants. Not all registrants received the assurances as to confidentiality and, even then, the content of their observations were subject to partial disclosure to complainants. The Tribunal noted, moreover, that registrants were unaware of the fact that the HPC, on occasion, made disclosures in the public interest, such as in child protection and police matters. The reality was that the 'aura' of confidence, as counsel for the HPC put it, attached to the information provided by registrants was not as clear or as bright as contended. The Tribunal was of the view that the HPC would be able to revise its procedures to ensure that those providing information were accurately forewarned that the HPC's dealing with this information would be subject to its duties under the 2000 and 1998 Acts.

***Central London Community Healthcare NHS Trust v. Information Commissioner* [2013] UKUT 551 (AAC)**

**18.33** Disclosure of sensitive patient data to hospice—inadvertent disclosure by fax to third party—penalty

The appellant Trust regularly faxed patient-sensitive data to a hospice and, under its relevant protocol, the hospice would acknowledge receipt by telephone. In error, the Trust's administrator sent the data to an additional fax number. The recipient shredded the documents and the Trust informed the Information Commissioner, who imposed a penalty of £90,000 on the basis of a serious breach of data protection principles and in accordance with the Information Commissioner's guidance on monetary penalties. The Trust's appeal to the Upper Tribunal, upholding the penalty notice issued by the Information Commissioner under section 55A of the Data Protection Act 1998, was dismissed.

## E. Other Relevant Chapters

**18.34**

Abuse of Process, Chapter 2
Case Management, Chapter 8
Natural Justice, Chapter 48
Privilege, Chapter 52
Whistleblowing, Chapter 69
Witnesses, Chapter 70

# 19

# DISHONESTY
| **A. General Principles** | 19.01 |
| *R v. Ghosh* [1982] QB 1053 | 19.01 |
| *Twinsectra Ltd v. Yardley and ors* [2002] 2 AC 164 | 19.03 |
| **B. Disciplinary Cases: Principles** | 19.04 |
| *Gangar v. General Medical Council* [2003] UKPC 28 | 19.04 |
| *D v. Law Society* [2003] EWHC 408 (Admin) | 19.05 |
| *Bultitude v. Law Society* [2004] EWCA Civ 1853 | 19.06 |
| *Singleton v. Law Society* [2005] EWHC 2915 (Admin) | 19.07 |
| *Donkin v. Law Society* [2007] EWHC 414 (Admin) | 19.08 |
| *Arora v. General Medical Council* [2008] EWHC 1596 (Admin) | 19.09 |
| *R (Khan) v. General Medical Council* [2008] EWHC 3509 (Admin) | 19.10 |
| *Sheill v. General Medical Council* [2008] EWHC 2967 (Admin) | 19.11 |
| *Bryant and anor v. Law Society* [2007] EWHC 3043 (Admin) | 19.12 |
| *Farag v. General Medical Council* [2009] EWHC 2667 (Admin) | 19.13 |
| *Makki v. General Medical Council* [2009] EWHC 3180 (Admin) | 19.14 |
| *Moneim v. General Medical Council* [2011] EWHC 327 (Admin) | 19.15 |
| *Hornsey v. General Medical Council* [2011] EWHC 3779 (Admin) | 19.16 |
| *Javed v. Solicitors Regulation Authority* [2012] EWHC 114 (Admin) | 19.17 |
| *Fish v. General Medical Council* [2012] EWHC 1269 (Admin) | 19.18 |
| *Uddin v. General Medical Council* [2012] EWHC 2669 (Admin) | 19.19 |
| *R (May) v. Chartered Institute of Management Accountants* [2013] EWHC 1574 (Admin) | 19.20 |
| *Jeffery v. Financial Conduct Authority* [2013] Lexis Citation 49 | 19.21 |
| *Batra v. Financial Conduct Authority* [2014] UKUT 0214 (TCC) | 19.22 |
| *Webb v. Solicitors Regulation Authority* [2013] EWHC 2078 (Admin) | 19.23 |
| *Shaw and Turnbull v. Logue* [2014] EWHC 5 (Admin) | 19.24 |
| *Moseka v. Nursing and Midwifery Council* [2014] EWHC 846 (Admin) | 19.25 |
| *Cordle and anor v. Financial Services Authority* [2013] Lexis Citation 01 | 19.26 |
| *Professional Standards Authority for Health and Social Care v. General Dental Council* [2014] EWHC 2280 (Admin) | 19.27 |
| *Choudhury v. Solicitors Regulation Authority* [2014] EWHC 809 (Admin) | 19.28 |
| *Sharma v. General Medical Council* [2014] EWHC 1471 (Admin) | 19.29 |
| *Hussain v. General Medical Council* [2014] EWCA Civ 2246 | 19.30 |
| *Lavis v. Nursing and Midwifery Council* [2014] EWHC 4083 (Admin) | 19.31 |
| *Professional Standards Authority for Health and Social Care v. Health and Care Professions Council and Elizabeth David* [2014] EWHC 4657 (Admin) | 19.32 |
| *Pope v. General Dental Council* [2015] EWHC 278 (Admin) | 19.33 |
| *Bains v. Solicitors Regulation Authority* [2015] EWHC 506 (Admin) | 19.34 |
| **C. Sanction in Dishonesty Cases** | 19.35 |
| *Ziderman v. General Dental Council* [1976] 1 WLR 330 | 19.35 |
| *Bolton v. Law Society* [1994] 1 WLR 512, CA | 19.36 |
| *Dey v. General Medical Council* [2001] UKPC 44 | 19.37 |
| *Manzur v. General Medical Council* [2001] UKPC 55 | 19.38 |
| *Hasan v. General Medical Council* [2003] UKPC 5 | 19.39 |
| *Patel v. General Medical Council* [2003] UKPC 16 | 19.40 |
| *Tait v. Royal College of Veterinary Surgeons* [2003] UKPC 34 | 19.41 |
| *Archbold v. Royal College of Veterinary Surgeons* [2004] UKPC 1 | 19.42 |
| *Salha v. General Medical Council* [2003] UKPC 80 | 19.43 |
| *R (Rogers) v. General Medical Council* [2004] EWHC 424 (Admin) | 19.44 |
| *Council for the Regulation of Health Care Professionals v. General Medical Council and Solanke* [2004] 1 WLR 2432 | 19.45 |

*Bultitude v. Law Society* [2004] EWHC 1370 (Admin) 19.46
*R (Khan) v. General Medical Council* [2004] EWHC 292 (Admin) 19.47
*R (Onwuelo) v. General Medical Council* [2006] EWHC 2739 (Admin) 19.48
*Walker v. Royal College of Veterinary Surgeons* [2007] UKPC 64 19.49
*Williams v. Royal College of Veterinary Surgeons* [2008] UKPC 39 19.50
*Salsbury v. Law Society* [2009] 1 WLR 1286 19.51
*Shiekh v. General Dental Council* [2009] EWHC 186 (Admin) 19.52
*Chamba v. Law Society* [2009] EWHC 190 (Admin) 19.53
*Lim v. Law Society* [2009] EWHC 1706 (Admin) 19.54
*Atkinson v. General Medical Council* [2009] EWHC 3636 (Admin) 19.55
*Parkinson v. Nursing and Midwifery Council* [2010] EWHC 1898 (Admin) 19.56
*Solicitors Regulation Authority v. Sharma* [2010] EWHC 2022 (Admin) 19.57
*Dad v. General Dental Council* [2010] CSIH 75 19.58
*Hosny v. General Medical Council* [2011] EWHC 1355 (Admin) 19.59
*R (Chief Constable of Dorset) v. Police Appeals Tribunal and Salter* [2011] EWHC 3366 (Admin); [2012] EWCA Civ 1047 19.60
*Ige v. Nursing and Midwifery Council* [2011] EWHC 3721 (Admin) 19.61
*Ballesteros v. Nursing and Midwifery Council* [2011] EWHC 1289 (Admin) 19.62
*Naheed v. General Medical Council* [2011] EWHC 702 (Admin) 19.63
*Solicitors Regulation Authority v. Dennison* [2012] EWCA Civ 421 19.64
*Heydon-Burke v. Nursing and Midwifery Council* [2012] EWHC 2435 (Admin) 19.65
*Solicitors Regulation Authority v. Spence* [2012] EWHC 2977 (Admin) 19.66
*Siddiqui v. General Medical Council* [2013] EWHC 1083 (Admin) 19.67
*R (Kibe) v. Nursing and Midwifery Council* [2013] EWHC 1402 (Admin) 19.68
*Hassan v. General Optical Council* [2013] EWHC 1887 (Admin) 19.69
*Adediwura v. Nursing and Midwifery Council* [2013] EWHC 2238 (Admin) 19.70
*Adegboyega v. Nursing and Midwifery Council* [2013] EWHC 2647 (Admin) 19.71
*R (Mafico) v. Nursing and Midwifery Council* [2014] EWHC 363 (Admin) 19.72
*Mills v. General Dental Council* [2014] EWHC 89 (Admin) 19.73
*Bar Standards Board v. O'Riordan* [2014] Lexis Citation 24 19.74
*Fernando v. General Medical Council* [2014] EWHC 1664 (Admin) 19.75
*McDaid v. Nursing and Midwifery Council* [2014] EWHC 1862 (Admin) 19.76
*Okee v. Nursing and Midwifery Council* [2014] EWHC 1763 (Admin) 19.77
*Professional Standards Authority v. Health and Care Professions Council and Ghaffar* [2014] EWHC 2708 (Admin); [2014] EWHC 2723 (Admin) 19.78
*Khan v. General Medical Council* [2015] EWHC 301 (Admin) 19.79
*Soni v. General Medical Council* [2015] EWHC 364 (Admin) 19.81
**D. Other Relevant Chapters** 19.82


## A. General Principles

### *R v. Ghosh* [1982] QB 1053

**19.01** Criminal proceedings—*Ghosh* test

G was a surgeon acting as a locum tenens consultant at a hospital. He was convicted on four counts of an indictment under the Theft Act 1968, which alleged that he had falsely represented that he had himself carried out a surgical operation to terminate pregnancy or that money was due to himself or an anaesthetist for such an operation, when in fact the operation had been carried out by someone else and/or under the National Health Service (NHS) provisions. His defence was that there was no deception and that the sums paid to him were due for consultation fees that were legitimately payable under the regulations, or else were the balance of fees properly payable—in other words, that there was nothing dishonest about his behaviour on any of the counts. The effect of the jury's verdict was that he had made false representations and had acted dishonestly. In dismissing G's appeal against conviction, Lord Lane CJ, reading the judgment of the Court of Appeal, Criminal Division, said at 1064D–1064G:

> In determining whether the prosecution has proved that the defendant was acting dishonestly, a jury must first of all decide whether according to the ordinary standards of reasonable and honest people what was done was dishonest. If it was not dishonest by those standards, that is the end of the matter and the prosecution fails.
>
> If it was dishonest by those standards, then the jury must consider whether the defendant himself must have realised that what he was doing was by those standards dishonest. In most cases, where the actions are obviously dishonest by ordinary standards, there will be no doubt about it. It will be obvious that the defendant himself knew that he was acting dishonestly. It is dishonest for a defendant to act in a way which he knows ordinary people consider to be dishonest, even if he asserts or genuinely believes that he is morally justified in acting as he did. For example, Robin Hood or those ardent anti-vivisectionists who remove animals from vivisection laboratories are acting dishonestly, even though they may consider themselves to be morally justified in doing what they do, because they know that ordinary people will consider these actions to be dishonest.

**19.02** *Archbold Criminal Pleading Evidence and Practice* (London: Sweet & Maxwell, 2015), paragraph 21-2d, states that there is no need to give a *Ghosh*-based direction unless the defendant has raised the issue that he did not know that anybody would regard what he did as dishonest: *R v. Roberts (W)* (1987) 84 Cr App R 117, CA. See also *R v. Price (RW)* (1990) 90 Cr App R 409, CA (*Ghosh* direction need be given only where the defendant might have believed that what he was alleged to have done was in accordance with the ordinary person's idea of honesty).

### *Twinsectra Ltd v. Yardley and ors* [2002] 2 AC 164

**19.03** Criminal proceedings—application of *Ghosh* principles

L was a solicitor. He acted for Y in a transaction that included the negotiation of a loan of £1 million from T. L did not take steps to ensure that the money was utilized solely for the acquisition of property on behalf of Y; he simply paid it out upon Y's instructions. The result was that £350,000 was used by Y for purposes other than the acquisition of property. In an action by T against Y and L, the trial judge found that L was 'misguided', but not dishonest. The House of Lords agreed. Lord Hoffmann said, at [20], that a dishonest state of mind requires a 'consciousness that one is transgressing ordinary standards of honest behaviour'. Lord Hutton said, at [27]:

> ... [B]efore there can be a finding of dishonesty it must be established that the defendant's conduct was dishonest by the ordinary standards of reasonable and honest people and that he himself realised that by those standards his conduct was dishonest.

Lord Hutton added, at [36]:

> ... [D]ishonesty requires knowledge by the defendant that what he was doing would be regarded as dishonest by honest people, although he should not escape a finding of dishonesty because he sets his own standards of honesty and does not regard as dishonest what he knows would offend the normally accepted standards of honest conduct.

## B. Disciplinary Cases: Principles

### *Gangar v. General Medical Council* [2003] UKPC 28

**19.04** Separate charges—intention to mislead—reckless dishonesty—no *Ghosh* direction

G was a consultant in the obstetrics and gynaecology department of Ashford Hospital. He admitted that, in applying to the Royal College of Obstetricians and Gynaecologists for a preceptorship in laparoscopic surgery, he had made a number of statements about his experience in carrying out hysterectomies and associated pathology that were wrong. It was alleged that, in submitting those statements, he had intended to mislead and to act dishonestly. The Professional Conduct Committee (PCC) found that G had acted dishonestly, but had not intended to mislead the Royal College. G's appeal to the Privy Council, on the grounds that the findings were inconsistent and that the legal assessor had failed to give a *Ghosh* direction, was dismissed. Their Lordships held that the two heads of charge were alternatives, that, on the facts, the more serious allegation was the intent to mislead (an intent deliberately to mislead a professional body), and that dishonesty was a lesser allegation (recklessly submitting information without caring whether it was true or false, or whether the body would be misled). The appellant had admitted that he was 'utterly foolish and utterly careless' in making a 'wild guess' on his application form as to the number of hysterectomies carried out. Accordingly, their Lordships were satisfied that was no need for the PCC to have been advised by the legal assessor that there were two aspects to dishonesty (subjective and objective) in line with *Ghosh*. Consistent with *R v. Roberts (W)* (1987) 84 Cr App R 117, there was no need to give a *Ghosh* direction unless the accused had raised the issue that he did not know that anybody would regard what he did as dishonest. The defence did not raise this issue.

### *D v. Law Society* [2003] EWHC 408 (Admin)

**19.05** Subjective test—'wilfully ignored'—'grossly reckless'

D was found guilty of conduct unbefitting a solicitor and was ordered to be struck off the roll. At the heart of the allegations against him was said to be his involvement in bank instrument transactions admitted to be dubious or fraudulent, to which he was introduced by a man called S. The Solicitors Disciplinary Tribunal (SDT) found that none of the transactions would be regarded by an honest and competent solicitor as other than spurious. D appealed on the grounds that the Tribunal had applied an incorrect, purely objective test for dishonesty contrary to *Ghosh* and *Twinsectra*. The Divisional Court (Rose LJ and Mackay J), in dismissing D's appeal, said that it was impossible to contend that the Tribunal had misdirected itself as to the legal test for dishonesty. The words 'wilfully ignored', as used by the Tribunal in its decision, did not connote an objective test. The words 'recklessly obtuse' do, although it was to be noted that those words were used in the specific context of D continuing to deal with S, rather than as part of a consideration of the legal elements of dishonesty. The words 'grossly reckless', in the Tribunal's determination, must likewise be read in their context. When the whole of the Tribunal's reasons were read, it was apparent that it had applied the right test.

### *Bultitude v. Law Society* [2004] EWCA Civ 1853

**19.06** No intention to deprive permanently—reckless dishonesty—conscious impropriety

B, a solicitor, was found guilty of reckless dishonesty in transferring a composite sum of client balances of £50,099 to office account without seeing any supporting documentation, and when his partner R had devised a system of false debit notes that were made out to look like bills of costs to get round the firm's computerized accounts package, whereby it was impossible to transfer money from client to office accounts without a bill having been drawn. B was guilty of dishonesty in that he neither knew nor cared whether his firm was entitled to be paid the funds. Kennedy LJ (with whom

Laws and Arden LJJ agreed) said, at [35]–[36], that proof of dishonesty in this context was not dependent upon proving an intention permanently to deprive clients of their funds. B signed a cheque for £50,000 transferring his clients' funds to his office account without any supporting documentation and thus, it must be inferred, without knowing or caring whether his firm was entitled to be paid those funds. That satisfied both legs of the *Twinsectra* test. At some stage, B became aware of the debit notes and, once he saw those bogus documents, it must have been clear to him what had been done to clear the credit balances, but he did nothing to backtrack. As the SDT found, he was guilty of conscious impropriety amounting to dishonesty in endorsing what had been done. See further paragraph 55.01.

### *Singleton v. Law Society* [2005] EWHC 2915 (Admin)

**19.07** Allegation of dishonesty to be pleaded

The SDT found dishonesty in relation to two allegations—namely, that the appellant had made false entries in books of accounts and had made a secret profit as a result of improper charges for telegraphic transfer fees. Dishonesty was not pleaded in the allegation. The Divisional Court (Maurice Kay LJ and Penry-Davey J), in quashing the finding of dishonesty, said that it was axiomatic that an allegation of dishonesty made against a solicitor is particularly serious; if proved, the consequences are inevitably severe. At [13], the Court said:

> [W]e conclude that it is unacceptable for the Tribunal to make findings of dishonesty when there has been no documentary pleading of such an allegation in a clear and timeous way. It may be that, with proper notice of such an allegation, the appellant would have chosen to be professionally represented, as he was in 2001. It may be that, if the Tribunal had reminded him of his options when the allegation of dishonesty was being pursued on 14th December, he would have asked for an adjournment or elected to give sworn evidence, on the basis that this would be likely to carry more weight than his unsworn submission. It may be that he would have adduced further evidence, such evidence which he seeks to adduce on this appeal to the effect that his clients were notified of the £20 charge for telegraphic transfers. We simply do not know whether matters would have proceeded in these different ways. In our judgment, this is not a case in which this court should simply say that compliance with procedural fairness would not have made a difference. That is not the point. We conclude that in bringing disciplinary charges against a solicitor, the Law Society is under an obligation to give timely notice of an allegation of dishonesty with relevant particulars when appropriate unless it is obvious from the nature of the charge (for example, where the allegation is that the solicitor has committed or been convicted of a criminal offence which necessarily involves dishonesty).

### *Donkin v. Law Society* [2007] EWHC 414 (Admin)

**19.08** Dishonesty—subjective test

D had practised for almost thirty years without attracting any concern about his integrity. He was a solicitor of high repute. His difficulties began when he and his firm were defrauded by a dishonest partner. An allegation against D was made of dishonest misappropriation of client's funds relating to three clients in respect of whom D had taken money from the client account to discharge the liabilities of the practice. In each case, the transaction was recorded in the accounts of the firm and the money was repaid with interest at the client account rate. The case for the appellant was that, in each case, the withdrawals were by way of authorized loans and that he was not dishonest. In opening the case before the SDT on dishonesty, the solicitor-advocate for the Law Society said that D knew, or ought to have known, that what he was doing was wrong. In its written reasons, the Tribunal stated that it was inconceivable that a solicitor of the appellant's experience could be unaware of the light in which his fellow solicitors and members of the public would regard his actions. In allowing D's appeal and remitting the case to a differently constituted Tribunal for rehearing, the Divisional Court (Maurice Kay LJ and Goldring J) said that, in the course of submissions before the Divisional Court, reference was made to *Bultitude v. Law Society* [2004] EWHC Civ 1853, which illustrates that dishonesty does not require proof of an intention permanently to deprive the clients of their funds and that the subjective element in the two-stage *Twinsectra* test can be satisfied by a reckless disregard that negatives honest belief. Counsel for the Law Society (which did not appear in the Tribunal) properly acknowledged that, by using the words 'ought to have known', the advocate for the Law Society was misleading the Tribunal. The Court said that the references in the Tribunal's decision to 'inconceivable' and 'a solicitor of the appellant's experience' bore a hint of the objective. The conclusion was that, in the instant case, the reasons stated by the Tribunal did not demonstrate a clear and consistent understanding, nor application of the subjective stage of the *Twinsectra* test, and the finding of dishonesty could not be sustained.

***Arora v. General Medical Council* [2008] EWHC 1596 (Admin)**

**19.09** Performers list—name—failure to ensure accuracy—not dishonest

Between April 2005 and April 2006, the appellant worked as a general practitioner (GP) doing locum work in Birmingham. Following regulations enacted in 2004, from 1 April 2004 every practitioner is required to be on a 'performers list' maintained by a primary care trust (PCT). In May 2005, the appellant applied to be on the Wandsworth PCT list, but the application was never processed. In April 2006, the Eastern Birmingham PCT wrote to the appellant stating that it had come to its attention that he was working as a GP in Birmingham, but was not on any performers list. The appellant replied, stating that he was on the Wandsworth PCT performers list. Blair J set aside the finding of dishonesty made by the fitness to practise panel of the General Medical Council (GMC), stating that a failure to take steps to ensure the accuracy of a statement was different from making a statement dishonestly. The findings of the panel were insufficient to ground such a serious charge and the finding of dishonesty against the appellant could not stand.

***R (Khan) v. General Medical Council* [2008] EWHC 3509 (Admin)**

**19.10** Failure to take steps to verify—whether dishonesty

K made an application for a job in Merseyside as an orthopaedic surgeon. His application referred to a published article of which he was the author, giving a reference date of 10 August 2003 as the date of publication. The article was not published on that date and the fitness to practise panel found that K knew when he wrote the application in June 2003 that it would not be published then. The panel found that he presented this as a published article with a view to buttressing the academic weight of his application. In its decision and in argument following the handing down of its decision, the panel referred to the GMC guide *Good Medical Practice*, which commented on doctors being honest and trustworthy when writing reports, or completing or signing forms, and stated that this meant that doctors must take reasonable steps to verify any statement before they sign a document. It was accepted that the proposition that dishonesty can be established by a failure to take reasonable steps to verify a statement before a document is signed is bad law and the test of dishonesty in *Good Medical Practice* did not represent an incorrect legal test of dishonesty. Stadlen J, in quashing the finding of dishonesty, said that the chairman was plainly labouring under a misapprehension as to the test of dishonesty.

***Sheill v. General Medical Council* [2008] EWHC 2967 (Admin)**

**19.11**

See paragraph 22.20.

***Bryant and anor v. Law Society* [2007] EWHC 3043 (Admin)**

**19.12** Far beyond the standards expected—'standards laid down by the profession'—no finding that solicitor was aware that, by those standards, he was acting dishonestly—no subjective test

B was a 56-year-old solicitor with positive good character. The second appellant was aged 67 and formerly a legal executive, but was admitted as a solicitor in 2001. They faced seven charges of professional misconduct, which the SDT found largely substantiated. The allegations were that they acted for, and continued to act for, clients who were involved in dubious or fraudulent transactions that bore the hallmarks of fraudulent investment schemes. The Tribunal made findings of dishonesty against B and ordered him to be struck off. It found the second appellant's involvement to have been much more limited and not to have been dishonest. In allowing B's appeal on dishonesty, the Administrative Court (Richards LJ and Aikens J) said, at [134], that a striking feature of the Tribunal's reasoning was the extent to which an objective test was applied. The Tribunal concluded that B had acted dishonestly because his conduct 'was so far beyond the standards to be expected of an honest and competent solicitor as to justify condemnation' and that it amounted to dishonesty 'by the standards laid down for the profession'. The case advanced on appeal was that the Tribunal thereby fell into error. It applied a purely objective test and found B guilty of dishonesty notwithstanding that, on the evidence accepted by the Tribunal, his state of mind was not dishonest. The Court said that the decision of the Court of Appeal in *Bultitude* stood as binding authority that the test to be applied in the context of solicitors' disciplinary proceedings was the *Twinsectra* test, because it was widely understood before the decision of the Privy Council in *Barlow Clowes International Ltd v. Eurotrust International Ltd* [2006] 1 WLR 1476 that it is a test that includes the separate subjective element. The Administrative Court continued:

> [154] In any event there are strong reasons for adopting such a test in the disciplinary context and for declining to follow in that context the approach in *Barlow Clowes.* As we have observed earlier, the test corresponds closely to that laid down in the criminal context by *R v. Ghosh*; and in our view it is more appropriate that the test for dishonesty in the context of solicitors' disciplinary proceedings should be aligned with the criminal test than with the test for determining civil liability for assisting in a breach of trust. It is true as [counsel for the Law Society] submitted that disciplinary proceedings are not themselves criminal in character and that they may involve issues of dishonesty that could not give rise to any criminal liability (e.g. lying to a client as to whether a step had been taken on his behalf) but the tribunal's finding of dishonesty against a solicitor is likely to have extremely serious consequences for him both professionally (it will normally lead to an order striking him off) and personally. It is just as appropriate to require that the defendant had a subjectively dishonest state of mind in this context as the court in *R v. Ghosh* considered it to be in the criminal context....
>
> [155] Accordingly, the tribunal in the present case should, in our judgment, have asked itself two questions when deciding the issue of dishonesty: first, whether B acted dishonestly by the ordinary standards of reasonable and honest people; and, secondly, whether he was aware that by those standards he was acting dishonestly.
>
> [156] There is nothing to show that the tribunal asked itself the second of those questions. At no point did it articulate with any clarity the test that it was applying, and the test applied cannot be derived from the authorities cited, since the passages selected for quotation do not lay down any single test. Most pertinently, although the tribunal found that B acted dishonestly by the standards of an honest and competent solicitor, it did not make any finding or even any suggestion that B was aware that by those standards he was acting dishonestly.
>
> [157] It follows that in our judgment the tribunal's finding of dishonesty is vitiated by a serious legal error.

### *Farag v. General Medical Council* [2009] EWHC 2667 (Admin)

**19.13**

Sick pay—allegations of faking illness—not alleged dishonesty by concealing other employment

F was a staff-grade doctor in obstetrics and gynaecology, employed by the Penine Acute Hospitals NHS Trust. Proceedings were brought by the GMC against F, alleging that he had performed work at the Marie Stopes International Clinics in Leeds and Manchester while he was certified sick and unable to work for the Trust. The GMC panel found that F had dishonestly concealed from the Trust that, whilst off sick and in receipt of full sick pay, he had been working part-time for Marie Stopes International. The panel's findings of dishonesty did not extend to a finding that F had deliberately submitted sickness certificates to the Trust when he was well. In allowing F's appeal and remitting the matter for rehearing, Standlen J said that 'the case, as put before the panel, was that the dishonesty alleged against F consisted of faking or exaggerating his illness so as to obtain sick pay'. It was not opened by counsel for the GMC or put to F in cross-examination that, even if he was genuinely ill, and even if he was not dishonestly exaggerating or faking his illness and submitting fraudulent certificates, he was still dishonest in concealing his other work for fear of prompting an enquiry that might lead to the Trust disallowing or reducing his sick pay.

### *Makki v. General Medical Council* [2009] EWHC 3180 (Admin)

**19.14**

Disciplinary proceedings in Ireland—finding of dishonesty—proceedings in England

M appeared before the Irish Medical Council, which found that he had presented false information when applying for a post in Ireland. By the time of the hearing, M was no longer on the Irish medical register and so the disposal recommended by the Irish Fitness to Practise Committee to the Medical Council was that the finding of professional misconduct against M should be taken into account by the Council should he apply for any further period of registration in Ireland. In proceedings before the GMC's fitness to practise panel based on the Irish allegations, M contended that dishonesty was not explicit on the charge. Irwin J said, at [7], that the charge as formulated before the Irish Medical Council contained the clearest possible allegation of a false declaration in order to achieve a paid medical post, and that any reader of the charge from the Irish Medical Council would have understood that dishonesty was implicit and clearly involved in the allegation. The GMC panel had focused heavily on M's subjective state of mind and his own correspondence conceded the objective implication of dishonesty. It seemed to the learned judge that the panel was not in the position of a criminal trial in which the objective appearance of dishonesty was in question. It was doubted whether such a legalistic approach would ever arise in a fitness to practise hearing turning on dishonesty, save in specific circumstances; certainly, it would not arise in the circumstance in which false particulars are advanced to obtain employment, since it could hardly rationally be suggested that sensible-thinking members of the public would regard that as not being objectively dishonest or that a doctor, who must be taken to have a relatively high level of intelligence and perception, would think that that could be the public view.

### *Moneim v. General Medical Council* [2011] EWHC 327 (Admin)

**19.15**

Patient records—dishonest alteration

Between April 2003 and early 2007, M was a partner in a general practice in Reading. Late in 2006, two of his partners in the practice became concerned about his clinical performance. On 9 January 2007, they told M that they were considering referring him to the relevant PCT. On 20 January 2007, the partners sent him a formal email, copied to the PCT, attaching a list of cases of concern. On further investigation, they discovered that M had made a number of retrospective alterations in patients' records after he had received their email. M was removed from the practice and a complaint to the GMC was made. M was found to have acted dishonestly in amending the records of six patients. In dismissing M's appeal and the sanction of twelve months' suspension, Lloyd Jones J said, at [22]:

> Dishonesty has many facets and manifestations. In the context of this case, concerned with altering a record, an important distinction emerged between conduct which may have been dishonest because the defendant acted for an improper motive and conduct which may have been dishonest because the content of the amended statement was to the knowledge of the maker itself false. Examples of the first category would include cases where a person with a dishonest motive alters a record to make it look better or to cast his conduct in a more favourable light or to divert attention from his role in a given matter but where the resulting amended record is not a false statement of what actually occurred. The amendment of a medical record by a doctor to include a true fact which he had failed to record earlier might be such a case if made with a dishonest motive. Despite the fact that the content of the resulting statement is not false, the alteration may still be dishonest if the author had acted from a dishonest motive. Examples of the second category might include cases where a doctor alters a medical record with the result of including a diagnosis he did not make or advice he did not give. At the hearing before me these situations were referred to respectively as cases of "dishonesty because of improper motive" and "dishonesty because of falsity"....

### *Hornsey v. General Medical Council* [2011] EWHC 3779 (Admin)

**19.16** Allegations of dishonesty—doctor's explanation not disputed—implausible that doctor would have acted dishonestly

H appealed against a finding that she had acted dishonestly in directing a member of staff at her surgery to shred documents, which included patient letters and results. A complaint was made by an employee of the practice concerning the maintenance of proper patient records, which had not been entered on the computer or Lloyd George records. The complaint resulted from the discovery, in the course of a decoration of the surgery, of a large quantity of documents in a storeroom adjacent to H's consulting room at the surgery. The complaint was passed to the relevant PCT, which appointed an external investigator. During the investigation, H brought two bags of documents from her car containing hospital records that were potentially relevant to the investigation and asked an employee of the surgery to shred them. In setting aside the finding of dishonesty by the GMC's fitness to practise panel, the Court said that the onus of proving dishonesty rested fairly and squarely on the GMC from first to last, on the balance of probabilities. The evidence of H to the effect that she had examined the documents and had come to the conclusion that those that she had directed should be shredded were redundant was not effectively challenged by any positive evidence to the contrary, or any inferential material, or indeed by rigorous questioning, and the point remained unanswered and unaddressed in the reasoning of the panel. That, in combination with it being inherently implausible that a dishonest doctor would bring documents to the surgery and direct that they be destroyed other than in circumstances in which she considered that they were genuinely redundant, led the Court to conclude that the panel was wrong to find that H acted dishonestly in the way alleged by the GMC and that the panel's conclusion must therefore be quashed.

### *Javed v. Solicitors Regulation Authority* [2012] EWHC 114 (Admin)

**19.17** Solicitor—mortgage fraud—objective and subjective approach

J faced an allegation that he had behaved in a way that was likely to diminish the trust that the public placed in the legal profession, in that he had been involved in mortgage fraud transactions. The Solicitors Regulation Authority (SRA) alleged, and the SDT found, that J had acted dishonestly. In its decision, the Tribunal said that, having considered all of the evidence, both written and oral, it was satisfied, so that it was sure, that the allegation had been proved to the higher standard. Moreover, the Tribunal was satisfied that, in conducting the sale of the property in question in total disregard of all of the various hallmarks of mortgage fraud, J was aware that his conduct had been dishonest by the standards of reasonable and honest people, and that he himself had realized that, by those standards, his conduct was dishonest. Nicol J, in dismissing J's appeal, said that J's complaint that the Tribunal applied a wholly objective criterion of dishonesty ignored the way in which the Tribunal phrased itself in its decision. The ground of appeal was wholly unmeritorious.

### *Fish v. General Medical Council* [2012] EWHC 1269 (Admin)

**19.18** Allegation of dishonesty—need for solid grounds—clearly particularized and put

The appellant was a consultant anaesthetist who was regularly employed by an agency and worked as a locum. It was alleged that he had been dishonest in relation to declarations made when submitting time sheets. In relation to the allegation of dishonesty, Foskett J said:

> 67. What, however, seems to be a proposition of common sense and common fairness is this: an allegation of dishonesty should be not be found to be established against anyone, particularly someone who has not been shown to have acted dishonestly previously, except on solid grounds. Given the consequences of such a finding for an otherwise responsible and competent medical practitioner, any Panel will almost certainly (without express reminder) approach such an allegation in that way.
>
> 68. An allegation of dishonesty against a professional person is one of the allegations that he or she fears most. It is often easily made, sometimes not easily defended and, if it sticks, can be career-threatening or even career-ending. Who would want to employ or otherwise deal with someone against whom a finding of dishonesty in a professional context has been made? I am, of course, dealing with the issue of dishonesty in a professional person simply because that is the issue before me. It is, however, a finding that no-one, whatever their walk in life, wishes to have recorded against his or her name.
>
> 69. I do not think that I state anything novel or controversial by saying that it is an allegation (a) that should not be made without good reason, (b) when it is made it should be clearly particularised so that the person against whom it is made knows how the allegation is put and (c) that when a hearing takes place at which the allegation is tested, the person against whom it is made should have the allegation fairly and squarely put to him so that he can seek to answer it. It is often uncomfortable for an advocate to suggest that someone has been deliberately dishonest, but it is not fair to shy away from it if the same advocate will be inviting the tribunal at the conclusion of the hearing to conclude that the person being cross-examined was dishonest...
>
> 70. At the end of the day, no-one should be found to have been dishonest on a side wind or by some kind of default setting in the mechanism of the inquiry. It is an allegation that must be articulated, addressed and adjudged head-on.

### *Uddin v. General Medical Council* [2012] EWHC 2669 (Admin)

**19.19** *Ghosh* test—relevance to disciplinary proceedings

U, a GP, together with her husband, owned a care home. The manager of the care home referred U to the GMC after U referred the manager to the Department of Health under the Protection of Vulnerable Adults (POVA) scheme, under which H was placed on a list barring her from working with vulnerable adults. The decision was later reversed. In its determination, the GMC's fitness to practise panel, applying the test in *R v. Ghosh*, made five findings of dishonesty against U. The panel concluded that U's conduct in referring H under the POVA scheme was misleading and irresponsible, and that witness statements submitted by U had been falsified and that the material sent to POVA to substantiate the referral was inaccurate. The panel included that erasure from the medical register was the only appropriate sanction. In allowing U's appeal against the panel's findings of dishonesty, Singh J held that the findings were flawed and could not stand. The Court directed that the case should be remitted to a fresh panel for reconsideration. In future, in considering issues of dishonesty, the GMC should take care in applying the *Ghosh* test, which was devised in the criminal law context. The standard of proof in proceedings before the GMC was the civil, and not the criminal, standard. Even in the criminal context, it was not the general practice to give a *Ghosh* direction in all cases and the advice given by the Judicial College to Crown Court judges was that no direction was generally required on the meaning of dishonesty. A *Ghosh* direction was given where an issue was raised as to whether a relevant charge was dishonest by the standards of ordinary people. The real issue in the instant case was whether the conduct took place and whether it was known that it was false, or that it was innocent or negligent. Singh J said:

> 30. In the appeal before me, the advice of the legal assessor, and the panel's clear acceptance of it, were not in themselves the subject of any challenge. Nevertheless, it was submitted that it might be helpful for this court to make any observations which may be helpful for future consideration by the respondent in its approach to cases that may raise issues of dishonesty. I would tentatively accept that invitation, although it is not necessary for a final decision in this particular case and I have not heard full argument on the points. Nevertheless, in the hope that it may be helpful, I would wish to make two brief observations which I would hope will be considered by the General Medical Council in the future. The first is that care needs to be taken about applying a test which was devised in the context of criminal law. As is clear from the standard of proof which is relevant in the present context, the standard of proof is the ordinary civil standard of a balance of probabilities and not the criminal standard. The question of dishonesty can arise in civil contexts as well as criminal ones. There can be, for example, torts or other civil disputes which raise the question of whether a person did or said what they did or said dishonestly, for example fraudulently as distinct from negligently.
>
> 31. The second observation to bear in mind is that even in the criminal context it is not general practice to give the so-called *Ghosh* two-part direction. In many cases, the advice which is given now by the Judicial College to judges who sit in the Crown Court is that no direction is required on the meaning of dishonesty. One context in which the twofold *Ghosh* direction may be required is where, on behalf of a defendant in criminal proceedings, an issue is raised whether he or she realised that the conduct charged was dishonest by the standards of reasonable and honest people. In many cases, there will be no such issue of fact raised. It will be perfectly apparent that if the conduct alleged did take place then it clearly was dishonest. The real issue in many cases may be whether the conduct took place and with what state of mind. For example, was a false representation made? But even if it was, was it done knowing that it was false or may have been, for example, innocent or even a negligent mistake?

### *R (May) v. Chartered Institute of Management Accountants* [2013] EWHC 1574 (Admin)

**19.20** Integrity—dishonesty

It was alleged that the practitioner had acted contrary to the Chartered Institute of Management Accountants (CIMA) code of ethics, which was applicable to all members and required them to act with integrity: see paragraph 18.24. Section 100.4(a) of the code, under the heading 'Integrity', provided that a professional accountant 'should be straightforward and honest' in all professional business relationships. Section 110.1, also headed 'Integrity', provided that the principle of integrity imposed an obligation on all professional accountants to be straightforward and honest in professional and business relationships. It added that 'integrity also implies fair dealing and truthfulness'. Stadlen J, in allowing M's appeal, said that 'integrity', when used in the sense of a human quality, is broadly synonymous with 'honesty'. Accordingly, in the absence of an allegation of dishonesty, there could be no finding of acting contrary to the fundamental principle of integrity.

### *Jeffery v. Financial Conduct Authority* [2013] Lexis Citation 49

**19.21** Integrity—dishonesty

The practitioner was charged with a breach of Principle 1 of the Financial Services Authority (FSA) Statements of Principle and Code of Conduct for Approved Persons, concerned with integrity and which provides that an approved person must act with integrity in carrying out his or her controlled function. The Upper Tribunal (Judge Roger Berner, chairman) stated that the expression has been considered in a number of cases before the Tribunal and its predecessor, the Financial Services and Markets Tribunal: see *Hoodless and Blackwell v. Financial Services Authority*, 3 October 2003; *Atlantic Law LLP v. Financial Services Authority*, FIN/2009/0007; *Vukelic v. Financial Services Authority*, 13 March 2009; and *First Financial Advisers Ltd v. Financial Services Authority*, FS/2010/0038. A person may lack integrity even though it is not established that he or she has been dishonest. Even though a person might not have been dishonest, that person will lack integrity if he or she either lacks an ethical compass or his or her ethical compass points him or her in the wrong direction.

### *Batra v. Financial Conduct Authority* [2014] UKUT 0214 (TCC)

**19.22** Integrity—dishonesty

A lack of integrity does not necessarily equate to dishonesty. While a person who acts dishonestly is obviously also acting without integrity, a person may lack integrity without being dishonest. One example of a lack of integrity not involving dishonesty is recklessness as to the truth of statements made to others who will, or may, rely on them, or wilful disregard of information contradicting the truth of such statements.

### *Webb v. Solicitors Regulation Authority* [2013] EWHC 2078 (Admin)

**19.23** Solicitor's undertaking—breach—explanation lacking credibility—adverse inference

W deliberately breached an undertaking, as opposed to having made an innocent mistake in doing so. Having undertaken to withdraw an application for registration at HM Land Registry, W subsequently wrote to the Land Registry renewing the application for registration, which was duly completed. In dismissing W's appeal against the SDT's finding of dishonesty, Jeremy Baker J said that if the renewed application had been an innocent mistake, one would have anticipated W to have taken some steps to seek to rectify the position, or at the very least to have explained why he could not do so, when the matter was brought to his attention. W's explanation lacked credibility and the Tribunal was entitled to draw an adverse inference. Whilst undoubtedly the absence of a motive involving a benefit to W or another arising out of his application to register the transfer was a powerful argument to deploy, it was no part of the task of the SRA to establish such a motive in these proceedings, nor was it a prerequisite to an adverse finding being made by the Tribunal in relation to the question of dishonesty. The fact that no such motive could be discerned was made clear by the Tribunal in its findings, just as the lack of such a finding was not a bar to a finding of dishonesty. The application for registration was made six weeks after the undertaking and it was held to be almost inconceivable that W would not have been fully aware that he was breaching his undertaking when he made the application to the Land Registry.

### *Shaw and Turnbull v. Logue* [2014] EWHC 5 (Admin)

**19.24**

See paragraph 54.29.

### *Moseka v. Nursing and Midwifery Council* [2014] EWHC 846 (Admin)

**19.25** *Ghosh* test—approach

In dismissing M's appeal on charges of dishonesty, Green J said that it was not necessary for him to delve into the exact definition and meaning of 'dishonesty' in the particular case. However, the learned judge said that he agreed with the observations of Singh J in *Uddin v. General Medical Council* [2012] EWHC 2669 (Admin) that (i) these are not criminal proceedings, and (ii) the standard of proof is different being the civil standard of a balance of probabilities. Green J said, at [24], that he would make one additional observation:

> The consequences for an Appellant of being struck off are particularly severe. They entail the loss of livelihood. This can be a much more severe sanction than many a low level criminal punishment. The severity of the consequences is a factor that a regulatory authority will need to bear in mind when applying the civil standard of proof. It is trite that the standard of proof can adjust to the context. In a regulatory case where dishonesty is the issue even on the civil standard the [Conduct and Competence Committee] CCC should apply particular care. If dishonesty is found then the likelihood of a severe sanction being imposed (including striking off) is quite high.

### *Cordle and anor v. Financial Services Authority* [2013] Lexis Citation 01

**19.26** Previous investigation—findings—application for authorization—non-disclosure

C and SBL, a company through which C operated, referred decision notices issued by the FSA rejecting his and SBL's application for authorization to carry on regulated financial activities on account of C's lack of honesty and integrity. In applying for authorization, C and SBL made no mention of allegations of misconduct previously made against them whilst acting as the appointed representative of a financial services network, which had terminated the appointment. The FSA's Fit and Proper Test for Approved Persons provides that, in determining a person's honesty, integrity, and reputation, the FSA will have regard to all relevant factors including, but not limited to, whether the person is or has been the subject of any investigation that might lead to disciplinary proceedings, or has been candid and truthful in all of his or her dealings. In upholding the decision notices, Sir Stephen Oliver QC, giving the judgment of the Upper Tribunal, said that the FSA had sound reasons for rejecting the applications for authorization and approval. When assessing fitness and propriety of a person to perform a particular controlled function, the most important considerations include the person's honesty, integrity, and reputation. The correct approach is to have regard to all of those factors. C drew attention to his long and virtually unblemished career in financial services, and to the reputation for competence and honesty that he enjoyed among his clientele. A number of good references were sent to the Tribunal. However, honesty and integrity cannot be ignored. C knowingly misled the FSA by not disclosing the earlier investigation of his appointment as a representative of a financial services network and its outcome, and he persisted in misleading the FSA at least until he owned up to the truth in the course of the Regulatory Decisions Committee hearing. Also relevant to the assessment will be any positive steps towards rehabilitation that have been taken by the individual. In that connection, C had conducted industry training in order to increase his knowledge in matters with which he was not previously familiar. Also the FSA was not seeking to question C's competence and capability.

### *Professional Standards Authority for Health and Social Care v. General Dental Council* [2014] EWHC 2280 (Admin)

**19.27** Email—deliberately misleading, but not dishonest—irrational

D, a dentist, in response to an email from a patient who complained that her fillings appeared rough, replied that the work was not yet complete and that the fitted fillings were only provisional. The PCC of the General Dental Council (GDC) found that D's email had been misleading and had been intended to mislead, but that his conduct had not been dishonest. The Committee referred to the fact that D had been willing to carry out further dental work for the patient, that he had apologized for not making things clearer about the status of her treatment, and that the email did not reach the high threshold for a finding of dishonesty. In allowing the appeal of the Professional Standards Authority, King J held that the Committee's finding that dishonesty had not been established was manifestly wrong and irrational: its reasons were either irrelevant or tended to support a finding of dishonesty. The fact that D had been willing to carry out further dental work was irrelevant as to whether his conduct was deliberately misleading. D's apology formed part of the deceit alleged. Having found that D had knowingly misled the patient, his good character was irrelevant to the objective test for dishonesty in *Ghosh*.

### *Choudhury v. Solicitors Regulation Authority* [2014] EWHC 809 (Admin)

**19.28** Finding of dishonesty—appeal

In dismissing C's appeal against findings of dishonesty made by the SDT, Bean J said, at [7], that he was prepared to assume for present purposes that an appellant could seek to overturn a finding that he was dishonest despite the Tribunal having seen and heard witnesses. In *Shamsian v. General Medical Council* [2011] EWHC 2885 (Admin), the Court had allowed an appeal against findings of dishonesty made against a professional where the Tribunal's findings had been made on an illogical basis.

### *Sharma v. General Medical Council* [2014] EWHC 1471 (Admin)

**19.29** Failure to disclose GMC warning to employers—whether event likely to have occurred—advice to panel

On 10 January 2014, the GMC's fitness to practise panel found three allegations of dishonesty against the appellant proved:

(1) that he had dishonestly failed to inform his employers of a warning given to him by the GMC in 2007;
(2) that he had dishonestly failed in 2010 to complete accurately an employer details form by omitting details of his previous hospital employers; and
(3) that he had dishonestly failed to inform his hospital employers that he had been the subject of a conditions order imposed by the Interim Orders Panel (IOP) in October 2010.

In allowing the appellant's appeal in relation to the 2007 warning, but dismissing the 2010 employer details form and the failure to disclose the decision of the IOP, His Honour Judge Pelling QC (sitting as a judge of the High Court) said that it was common ground between the parties that, for a finding of dishonesty to be made, the GMC must prove (a) that the act or omission concerned was dishonest by the standards of reasonable and honest people, and (b) that the practitioner must have realized that what he or she was doing was dishonest if applying those standards. The onus of proof rests throughout on the GMC and the applicable standard of proof is the civil standard—that is, the balance of probabilities. In *Re H (Minors) (Sexual Abuse: Standard of Proof)* [1996] AC 563, 586, Lord Nicholls observed that, in assessing the balance of probabilities, the more serious the allegation, the less likely it is that the event occurred and hence the stronger should be the evidence before a court concludes that the allegation is established on the balance of probabilities. His Lordship agreed that the principle identified by Lord Nicholls of Birkenhead in *Re H* applies in relation to allegations of dishonesty such as those that gave rise to the findings of dishonesty in this case. It was noteworthy that the legal assessor did not draw the attention of the panel to this point, which is one to which all fact-finding tribunals applying the civil standard of proof are bound to have regard when considering an allegation of dishonesty. He ought to have done so. However, on the facts of this case, it might have made a difference only in relation to the 2007 warning issue and it was immaterial to the two other findings of dishonesty.

### *Hussain v. General Medical Council* [2014] EWCA Civ 2246

**19.30** *Ghosh* direction—application to disciplinary proceedings

In dismissing H's appeal from the decision of His Honour Judge Bird (sitting as a deputy judge of the Administrative Court at Manchester), dismissing his appeal against a decision of a fitness to practise panel on 26 June 2013 that his name should be erased from the register, Longmore LJ said, at [51]:

> I would only add that I am a little troubled about the *Ghosh* direction given by the legal assessor in this case. It would have been standard in a criminal case. But this was a professional disciplinary hearing and it seems to me that in future it would be right and proper for the first part of the direction to be adapted to read that the panel should decide "whether according to the standard of reasonable and honest *doctors* [not people] what was done was dishonest". There may be a not unimportant difference between the two as shown by the decision of the judge in this very case.

### *Lavis v. Nursing and Midwifery Council* [2014] EWHC 4083 (Admin)

**19.31** Midwife—finding of dishonesty relating to record-keeping—whether panel applied appropriate test—alternative reasons for erroneous notes

L, a registered midwife, appealed against the determination by a panel of the CCC of the Nursing and Midwifery Council (NMC), dated 27 May 2014, and its determination to suspend her from practice for a period of four months. In summary, it was alleged and found proved by the panel that L failed, in June 2011, to provide adequate care during labour to Ms A and/or her unborn baby. The panel found that L failed to maintain accurate records in relation to the care given to Ms A, in that she recorded in the antenatal notes an alleged discussion about procedures that she had not in fact had, recorded that Ms A had requested the artificial rupture of membranes procedure when in fact she had not, recorded in the labour notes that she had changed Ms A's pad when in fact she had not, and recorded that she had noticed the umbilical cord when in fact she had been informed of this by Ms A's husband. The panel found that L's conduct in relation to these specific matters relating to record keeping was dishonest. In allowing L's appeal, and quashing the finding of dishonesty and remitting it back to the panel to consider the issue of dishonesty again, Cobb J held that the panel had applied the wrong legal test in relation to dishonesty. The panel had purported to follow the two-stage *Ghosh* test, but had improperly imported into the objective stage of the test the concept of 'deliberate misleading' (a subjective element), which would have had the effect of distorting (and probably fortifying) the objective test. Having erred in its application of the 'first stage' of the test, it would, in the circumstances, have been more reassuring if the 'second stage' test had been expressed in a manner that was more faithful to Lord Lane LCJ's original phrase from *Ghosh*: '[T]he defendant must have realised that what he was doing *was by those standards* dishonest' became '[The Appellant] must have realized that what [she was] doing *would be regarded* as dishonest'. Moreover, the panel did not seem to consider any explanation for the entries that did not involve dishonesty. While dishonesty was plainly one of the possible explanations, it was not the only explanation: it would have been appropriate and proper for the panel to have explicitly considered, in respect of each of the entries, whether L had acted in an unthinking way, out of habit, in a 'slapdash' manner, or while distracted. It would have been appropriate for there to be some consideration of whether the record (or any of the records) had been made carelessly, or even automatically recording the normal practice without proper attention to whether the normal practice had actually been observed on this occasion. It is also possible that the sense of each recording, indeed the accuracy, could have been distorted by being inappropriately abbreviated or written in a form of shorthand: at [56]–[58]. The learned judge said that the panel may have been better advised to consider the approach of Singh J in *Uddin v. General Medical Council* [2012] EWHC 2669 (Admin), [31], where it was pointed out that: 'The real issue in many cases may be whether the conduct took place and with what state of mind. For example, was a false representation made? But even if it was, was it done knowing that it was false or may it have been, for example, innocent or even a negligent mistake?' Cobb J said that although neither counsel referred the Court to the decision of *R v. Lucas (Ruth)* [1981] QB 720 and *R v. Middleton* [2000] TLR 293 explicitly, the guidance from those cases was surely relevant, for, as appears from these authorities, a conclusion that a person is lying or telling the truth about point A does not mean that she is lying or telling the truth about point B. A party or witness may lie for many reasons, for example shame, panic, duress or distress; the fact that a witness has lied in respect of one matter does not mean that he or she has lied in respect of everything. In the instant case, it did not appear that the panel had considered the recordings of the events individually when examining the reasons for the entries and, in the light of the guidance in *Lucas*, the Court considered that it should have done so—and should have done so carefully. The panel ought to have considered the second 'subjective' test of *Ghosh* in relation to the charges taken individually; it was possible that, in relation to one or more, it may have concluded that the entry was not dishonest.

### *Professional Standards Authority for Health and Social Care v. Health and Care Professions Council and Elizabeth David* [2014] EWHC 4657 (Admin)

**19.32** *Ghosh* test—subjective limb

The second respondent, D, was a social worker and childminder who faced disciplinary proceedings before the Health and Care Professions Council's CCC concerning her running of a childminding business from home and in respect of lies that she told to Ofsted inspectors when they made an unannounced visit. The Committee found that whilst D had acted dishonestly in relation to misleading the inspectors, she had not acted dishonestly in breaching the conditions of her childminding registration. The Committee said that, on the balance of probabilities, D would 'not necessarily' have understood that other people would find her actions dishonest. In allowing an appeal by the Professional Standards Authority for Health and Social Care under section 29 of the National Health Service Reform and Health Care Professions Act 2002, and remitting the issue of dishonesty to a freshly constituted Committee, Popplewell J said that D would not necessarily have understood that other people would have found her actions dishonest if they were applying the criminal standard of proof contained in the subjective limb of the *Ghosh* test: *R v. Ghosh* [1982] QB 1053. The Committee purported to be addressing the question on the basis of the civil standard of proof of a balance of probabilities. It followed that the Committee did apply the wrong test by applying the subjective limb of *Ghosh* in a way that applied a criminal standard of proof. In disciplinary proceedings before the Health and Care Professions Council, the standard of proof is the civil standard of proof of the balance of probabilities. The Committee, while purporting to address the question on the basis of the civil standard of proof, applied the wrong test by applying the subjective limb of *Ghosh* in a way that applied a criminal standard of proof.

### *Pope v. General Dental Council* [2015] EWHC 278 (Admin)

**19.33** Legal adviser—advice to Committee on likelihood of registrant behaving dishonestly

In dismissing P's appeal against findings of dishonesty made by the GDC's PCC, Turner J said that the extent to which it may be necessary in any given case for a legal adviser specifically to make explicit reference to the point that the more serious an allegation, the less likely it is to be true, or that a person with a good reputation in a position of trust would not be likely to risk his or her good name and livelihood by behaving dishonestly, will depend on the facts of the case and how it has been presented. It is to be noted that the Court in *Sharma v. General Medical Council* [2014] EWHC 1471

(Admin) did not, however, consider that the legal adviser's omission was fatal to all of the findings of dishonesty in that case. In the instant case, although the written legal advice did not make reference to the point in *Re H (Minors) (Sexual Abuse: Standard of Proof)* [1996] AC 563, it did set out clearly the effect of good character on the issue of dishonesty. Furthermore, during the course of the hearing, counsel for the GDC had carefully, expressly, and accurately conceded that the appellant's positive good character was to be taken into account both on the issues of propensity and credibility. Whilst it would have been open to the legal adviser to reinforce the point further with reference to the inherently greater likelihood that someone of good character will act mistakenly rather than dishonestly, the omission was not fatal to the safety of the finding of the Committee in the circumstances of this case. The Committee made detailed findings of dishonesty and described the appellant's acts of dishonesty as 'blatant'.

***Bains v. Solicitors Regulation Authority* [2015] EWHC 506 (Admin)**

**19.34** Solicitor—mental health—impact on dishonesty

B, a solicitor with many years' unblemished service, faced two charges of breaches of the Solicitors Accounts Rules 1998. Charge 1 was that he had withdrawn monies from the client account in respect of four clients that exceeded the funds held for those clients, giving rise to a shortage. Charge 2 was that he had improperly utilized client monies for the purposes of other clients. The SDT found both charges proved and that B had been reckless in respect of the first charge and dishonest in respect of the second charge, and that he should be struck off the solicitor's roll. B adduced evidence to the Tribunal that his partner in the practice had become ill and had died rapidly, as a result of which B had ended up acting as a sole practitioner—but that his own mother was seriously ill and had died, and that he had himself been treated for depression and had some physical problems. On appeal, B submitted that the Tribunal had not sufficiently taken his medical history into account. In dismissing B's appeal, Holman J said that the Tribunal had had B's medical evidence well in mind, as well as his personal and professional pressures, and was satisfied that he was not disadvantaged in conducting his own case because of his ill-health. The Tribunal had addressed separately the question of whether B's mental health was such that he did not know that what he was doing was wrong. B's impairment was not such that he could not recognize that his behaviour was dishonest. B's ill-health could not impact on the ultimate findings of the Tribunal and its conclusion of dishonesty. B was intelligent and had conducted his case with skill, and the Tribunal had taken care to look at the disadvantages that B had in acting in person or being slightly unfit.

## C. Sanction in Dishonesty Cases

***Ziderman v. General Dental Council* [1976] 1 WLR 330**

**19.35** Previous disciplinary findings—relevance

In July 1970, Z, a registered dentist, had been convicted of ten offences of obtaining money by false pretences from the Executive Council of the NHS. Sixty-two other similar offences were taken into consideration. Those convictions were the subject of disciplinary proceedings and, in June 1971, Z's name was erased from the dentists' register. In November 1972, his name was restored to the register. On 31 December 1974, Z was again convicted of a criminal offence—shoplifting—having stolen a tie and other goods worth £3.40 from Selfridges on the previous day. Disciplinary proceedings were brought against him in respect of the shoplifting conviction and, on 14 May 1975, his name was again ordered to be erased from the register. From that order, he appealed to the Privy Council. In dismissing his appeal, Lord Diplock, giving the judgment of the Board, said that the previous convictions of a person charged with an offence are relevant matters to be taken into account by a sentencing tribunal in assessing the gravity of the offence with which he is charged. The weight to be attached to previous convictions in assessing the gravity (or the trivial nature) of the offence charged is a matter for the tribunal charged with the duty of determining what the penalty shall be. No doubt previous convictions for technical offences which involve no moral turpitude will be given little weight in the assessment of the appropriate penalty for an offence which does. The weight to be attached to previous convictions for offences which involve some moral turpitude, such as dishonesty, will no doubt vary with their similarity or otherwise to the offence committed.

### *Bolton v. Law Society* [1994] 1 WLR 512, CA

**19.36**

See paragraph 60.02.

### *Dey v. General Medical Council* [2001] UKPC 44

**19.37** Criminal conviction—sanction imposed by disciplinary committee—purpose—reputation of profession

Following D's conviction of fourteen counts of furnishing false information, the PCC directed that his name be erased from the register. D had submitted to the local health authority 1,085 requests for payment for health screening tests. He pleaded guilty on the third day of trial and prior to calling any evidence. In dismissing D's appeal against sanction, Lord Millett, giving the judgment of the Board, said that:

> 11. The object of disciplinary proceedings against a medical practitioner who has been convicted of a criminal offence is twofold. It is to protect members of the public who may come to him as patients and to maintain the high standards and reputation of the profession. It is not to punish a second time for the same offence. Nevertheless the same conduct which constitutes the offence for which he has been convicted may also demonstrate that the need to maintain the standards and reputation of the profession or to protect the public or both requires the erasure of his name from the Register. There is no clear line of demarcation: the difference lies not in the facts themselves but in the perspective from which they are viewed.
>
> [...]
>
> 17. Their Lordships are satisfied that the sentence of erasure was not excessive. The Committee was entitled to conclude that such irresponsible behaviour cannot be tolerated in a registered medical practitioner and that the need to maintain high standards and to protect the public, taken together, required the erasure of [D]'s name from the Register.

### *Manzur v. General Medical Council* [2001] UKPC 55

**19.38** Criminal conviction—erasure—disparity with co-defendants—lesser role

M pleaded guilty to five charges of false accounting and five further similar offences to be taken into consideration. M, an ophthalmic surgeon, was one of three persons who were investigated for dishonestly obtaining public money from a health authority. The other two persons were a dispensing optician and another ophthalmic surgeon. All three were charged with false accounting. The dispensing optician pleaded guilty, but was not exposed to any form of disciplinary proceedings. The criminal proceedings against the other ophthalmic surgeon were abandoned and he was not brought before the PCC. The Privy Council, in allowing M's appeal against an order of erasure from the medical register, noted that the amount attributed to him was not great, and substantially less than attributed to the dispensing optician and the other doctor. He had had a long career practising in the United Kingdom and in Bangladesh. His ability as a doctor had never been questioned, and he was engaged in alleviating the suffering of the disadvantaged in Bangladesh and setting up a centre providing socio-cultural activities for the poor and illiterate. He was approaching the age of 69 and, in the light of all of these matters, erasure was unduly harsh and disproportionate. In substitution, M's registration was suspended for a period of three months.

*Hasan v. General Medical Council* [2003] UKPC 5

**19.39** Bogus medical reports—financial gain—erasure—appeal dismissed—public trust

The appellant provided bogus medical reports to undercover journalists. For a fee of £1,000 and in the belief that the report would be submitted to the immigration authorities, the appellant produced a report in which he claimed to have examined M (an undercover journalist) and to have given a diagnosis of acute chronic renal failure; he stated that investigations were under way to confirm the diagnosis and that the appellant had made arrangements for M to be admitted to a clinic. None of this was true. In dismissing the appellant's appeal against a direction that his name be erased from the medical register, Lord Hope of Craighead, in giving the judgment of the Board, said that the gravity of the appellant's conduct could not be overemphasized. The relationship between the medical profession and those liable to be affected by such conduct is one of trust. The work of the immigration authorities and of insurers in personal injury cases would be seriously impeded if investigations had to be conducted in every case about the probity of all of the medical reports that are put in front of them. It is a tribute to the high regard in which the profession is held that these reports are normally accepted and acted on without any questions being asked as to whether or not they are genuine. The appellant showed by his conduct that he was prepared to abuse that relationship for personal reward in the most cynical manner without any regard to the consequences. His response to the findings that were made against him revealed a disturbing lack of insight into their gravity. If public trust in the profession is to be maintained, conduct of this kind must be rooted out wherever it is found. Erasure, coupled with immediate suspension, was the only appropriate sanction.

*Patel v. General Medical Council* [2003] UKPC 16

**19.40** Criminal conviction—false information—locum fees—erasure

P was a GP working at an army base, employed by the Ministry of Defence. Following a three-week trial, he was convicted of eight counts of dishonestly furnishing false information contrary to section 17(1)(b) of the Theft Act 1968. The criminal conduct consisted in claiming for locums fees in relation to doctors and one nurse, when, to his knowledge, no such locum work had been undertaken. The counts spanned an eight-month period and involved a sum of £6,300. P was given an immediate custodial sentence of six months' imprisonment. His name was erased from the medical register. Dismissing P's appeal, Lord Steyn, giving the judgment of the Board, said that the PCC was right to be guided by the judgment in *Bolton v. Law Society* [1994] 1 WLR 512. It is true that, in that case, misconduct by a solicitor was at stake—but the approach in that case applies to all professional persons. There can be no lower standard applied to doctors: *Gupta v. General Medical Council* [2002] 1 WLR 1691, *per* Lord Rodger of Earlsferry at [21]. For all professional persons, including doctors, a finding of dishonesty lies at the top end in the spectrum of gravity of misconduct.

*Tait v. Royal College of Veterinary Surgeons* [2003] UKPC 34

**19.41** Medical treatment—false representations—erasure—public interest

T was convicted before the Disciplinary Committee of the Royal College of Veterinary Surgeons (RCVS) of making false representations in relation to the treatment of an English bull terrier bitch and the Committee directed the removal of T's name from the register. The Privy Council allowed T's appeal against the decision of the Committee to refuse an adjournment. In giving judgment, the Board said that, strictly, it was unnecessary to consider T's appeal that the removal of his name from the register was manifestly excessive. However, having heard full argument from counsel for the appellant, the Privy Council considered it appropriate to comment on the point. For all professional men, a finding of dishonesty lies at the top end of the spectrum of gravity of misconduct: *Bolton v. Law Society* [1994] 1 WLR 512; *Gupta v. General Medical Council* [2002] 1 WLR 1691, [21]; *Patel v. General Medical Council*, Privy Council Appeal No. 48 of 2002, [10]. If the first charge were well-founded, it would fall into this category. Had it been necessary to do so, their Lordships would have ruled that removal from the register was necessary to protect the public interest.

### *Archbold v. Royal College of Veterinary Surgeons* [2004] UKPC 1

**19.42** Veterinary certificates—false information—consumer protection—reputation of profession

A, a veterinary surgeon, appealed against an order removing his name from the register. The charges were based on two certificates that A had completed and signed in February 2000, certifying that he had attended and verified the identity of two cows, and had administered lethal injections to them. The certificates were part of the 'over 30 months' scheme that was introduced to prevent bovine animals over 30 months of age from entering the food chain as part of a package of measures to control bovine spongiform encephalopathy (BSE). In the words of the Disciplinary Committee of the RCVS, the role that the veterinary surgeon plays in signing the certificates is crucial to the operation of the scheme, which provides for consumer protection, allows proper monitoring of disease control, and helps to ensure animal welfare. The public is entitled to trust the signature on such a certificate. A admitted the facts alleged and the charges, and admitted that he was guilty of disgraceful conduct in a professional respect. The Committee concluded that the truth was that A had not attended and verified the identity of the two animals on the farm, and he had not administered the lethal injections. Further, he had passed barbiturates in hypodermic needles and syringes to the farmer, and had never identified the animals. He had failed in his fundamental responsibilities as a veterinary surgeon. In dismissing A's appeal, Sir Kenneth Keith, giving the judgment of the Privy Council, said that the case was one of dishonesty in a professional respect. As was said in *Bolton v. Law Society* [1994] 1 WLR 512 and in *Tait v. Royal College of Veterinary Surgeons* [2003] UKPC 34, proven dishonesty comes at the top end of the spectrum of gravity for misconduct. The gravity of the matter arose in this case not only from the dishonesty, but also from the possible consequences of the false declaration. The declaration in the certificates that A signed stated that, in his opinion after carrying out a clinical examination and making due inquiries, the animal showed no signs of disease or pathological condition that would render the carcase unfit for human consumption. Their Lordships agreed that the imposing of the sanction of removal in the circumstances of this case properly fulfilled the Committee's stated duties to protect animals and the public, and to uphold the honour and reputation of the veterinary profession. The penalty was just, fair, and proportionate in all of the circumstances.

### *Salha v. General Medical Council* [2003] UKPC 80

**19.43** Plagiarism—dishonesty—not alleged—lesser sanction substituted

S was found to have submitted for publication in a journal a review paper and case report that included seven plagiarized paragraphs from a previous paper written by another doctor. The allegation in the charge was that his actions were unethical and unprofessional. The PCC found the charge of serious professional misconduct proved. On sanction, the Committee said that S's actions were unacceptable and dishonest. The Privy Council considered that there was no justification for the finding of dishonesty. Dishonesty was not alleged, and it was a fundamental principle of fairness that a charge of dishonesty should be unambiguously formulated and adequately pleaded. The question then was what their Lordships would do about the sentence. It was not disputed that the failure to prevent plagiarism of substantial parts of the paper was serious professional misconduct, even though in this case the plagiarism related to the mode of expression rather than the scientific content. The Committee regarded a three-month suspension as appropriate for dishonest copying and it must follow that it would have regarded a lesser sentence as appropriate for a negligent failure to prevent copying. The appeal was to be allowed, and the order for suspension set aside and a reprimand substituted.

### *R (Rogers) v. General Medical Council* [2004] EWHC 424 (Admin)

**19.44** Criminal conviction—controlled drugs—personal use—use of prescription forms—repeated dishonesty

R, a GP working in recent years as a locum, was of previous good character. In November 2001, he pleaded guilty at Cardiff Magistrates' Court to twelve offences of obtaining by deception prescription drugs, mainly opiates, and eleven offences of possession of controlled drugs, all but two of class A. All were for R's own use and he harmed no one other than himself. He was committed for sentence to Cardiff Crown Court. He was sentenced on 14 December 2001 to twelve months' imprisonment, suspended for two years. On 1 August 2003, the PCC directed that R's name should be erased from the register. On appeal, it was contended that erasure was a disproportionate and excessive sanction. Mitting J—after referring to *Ghosh v. General Medical Council* [2001] 1 WLR 1915, [33] and [34], and *Evans v. General Medical Council*, 19 November 1984, PC, [34]—said that the offences were not merely offences of dishonesty. One can imagine circumstances in which, owing to a momentary lapse of character, a doctor may commit an isolated offence of dishonesty such as shoplifting, which would not require the sanction of erasure from the register to be imposed. But these offences were not isolated offences of dishonesty; rather, they were repeated. There were twelve of them in all. Further, they involved dishonesty in the context of class A controlled drugs. The dishonest use of prescription forms to obtain such drugs is a serious matter and the Committee was entitled, and was right, to treat that factor as a serious and aggravating feature. Further, the dishonesty was not peripheral to the doctor's discharge of his duties as a doctor. The dishonesty was in the performance of one of his duties: the completion of prescription forms. That, too, was a serious and aggravating feature of the case. The Committee was clearly right to state that the offences demonstrated a clear abuse of the trust that is placed in any practitioner and a breach of the principles of good medical practice, and that such conduct inevitably undermined the confidence that members of the public place in the profession. Of course, the Committee had to consider not only the aggravating features of the offences, but also the mitigating features: the previous lack of any disciplinary record of R; his illness; and the medical evidence that the offences were secondary to the illness. The Committee had to perform a balancing exercise and to decide whether the aggravating features demanded the sanction of erasure notwithstanding the presence of those mitigating features. Having performed the same exercise itself, the Court asked whether such grave misconduct demanded, as necessary for the maintenance of public confidence in the medical profession, the imposition of the most severe sanction of erasure. It concluded that it did and that, notwithstanding the sympathy that anyone reviewing R's case might have for his personal circumstances, the Court would unhesitatingly uphold the decision of the Committee and reject his appeal.

### *Council for the Regulation of Health Care Professionals v. General Medical Council and Solanke* [2004] 1 WLR 2432

**19.45** Dishonest alteration of birth certificate and CV—disclosure to employer—competent doctor

On 15 December 2003, the GMC's PCC found that Dr S was guilty of serious professional misconduct in relation both to his involvement in a sexual relationship with Ms X, a patient, and as to his dishonest alteration of his birth certificate and CV, so as to reduce his apparent age. Bearing in mind that he had already been suspended from general practice for six months, the Committee ordered that he be suspended from the medical register for a period of three months. As to the birth certificate and CV, S made it clear in evidence before the Committee that he had himself disclosed the falsifications to his trainer at the general practice where he was employed. He said that he had falsified the documents because he had had difficulty obtaining employment as a doctor and believed that his age was against him. He said that he knew it was wrong to falsify the documents. He went on to add that he would not repeat his conduct again. At the time of his relationship with Ms X, he had separated from his wife, and he explained that the relationship, which was 'mutual', lasted about six months. He said that he knew the relationship was wrong and that he had undertaken counselling subsequently. In its determination, the Committee said that it took a serious view of the charges, but it believed that S would not repeat this behaviour: he was remorseful and had readily admitted his misdemeanours. At the time of the events, he was in a stressful state owing to unfortunate circumstances in his personal life. The submissions made on his behalf from professional colleagues indicated that S was a competent doctor, who was well regarded by colleagues and patients. In dismissing the appeal on leniency by the Council for the Regulation of Healthcare Professionals, Leveson J said that there was no challenge to S's medical competence. On the basis of the assessment made by the Committee about S, and given the additional evidence from the counsellor and five other medical practitioners with whom S had come into contact about his behaviour and his reaction, the conclusion was that, although not as informed as it might have been, the decision of the Committee could not be characterized as unduly lenient.

***Bultitude v. Law Society*** **[2004] EWHC 1370 (Admin)**

**19.46** Reckless dishonesty—no allegation that defendant intended to retain funds regardless of entitlement—suspension substituted for striking off

The allegation of dishonesty against the appellant was that he had signed a cheque transferring client funds to his office account without knowing or caring whether or not his firm was entitled to be paid those funds. Such conduct would be dishonest, even if there was an intention to repay the clients, if it were found on a subsequent investigation that the firm was not in fact entitled to the sums transferred under the cover of the debit notes. B was guilty of the dishonesty alleged against him—namely, his transfer of funds from his client to his office account, not knowing at the time whether he was entitled to make those transfers. There was no allegation against him of dishonesty in the additional sense of an intention by him to retain such funds in the office account regardless of any such entitlement. Accordingly, the SDT wrongly made a finding that he did not intend to carry out an analysis of the individual clients' ledgers at a later time. The Court would not wish to cast doubt on the proposition that striking off will usually follow if there has been a finding of dishonesty, but the imposition of a lesser penalty may be justified in exceptional circumstances. In the instant case, the dishonesty arose suddenly over the span of a weekend, under considerable and somewhat questionable pressure by the firm's reporting accountant to 'clear' the clients' accounts by the Monday. In the very unusual circumstances of the case, it was open to the Court, without violation of the importance of scrupulous attention by solicitors to the Solicitors' Accounting Rules, to substitute a penalty of suspension from the roll for two years for that of striking off imposed by the Tribunal. The order would operate from the date of the strike-off.

***R (Khan) v. General Medical Council*** **[2004] EWHC 292 (Admin)**

**19.47** Conviction—fraudulent receipt of monies—health authority—suspension

The appellant, a GP with surgery premises at Ashvale Road, London SW17, received £35,850 in rent from Merton, Sutton, and Wandsworth Health Authority for a flat contained within his surgery. The flat was occupied by a tenant who paid the doctor rent in respect of that tenancy and, accordingly, he received payments from the Health Authority to which he was not entitled. He was charged with, and found guilty of, dishonesty before the PCC and the penalty was a suspension from practice order for a period of twelve months. Pitchford J, in dismissing K's appeal, said that there was no basis for the complaint that the Committee failed to consider the appropriate penalty on a step-by-step basis. The Court did not accept that the suspension for twelve months would stop K's medical career. As the Committee concluded in its remarks, the opportunity was available, which it urged K to take, to keep himself in touch so that his career would not be terminated. The fact that K was aged 63 did not put him beyond the prospect of a return to practice.

### *R (Onwuelo) v. General Medical Council* [2006] EWHC 2739 (Admin)

**19.48** Theft—failure to disclose—suspension

O was convicted of theft on 25 May 2000. The charges before the fitness to practise panel arose out of his failure to disclose this conviction when applying for a position with the Surrey and Sussex Healthcare NHS Trust in January 2001 and his subsequent failure to consent to a Criminal Records Bureau (CRB) check or to disclose this conviction on one of a number of occasions. O had completed and signed an employment status form dated 15 January 2001, which he submitted to the Trust. In answer to the question 'Have you ever been convicted of a criminal offence?' on the form, the answer 'No' was circled. In May 2003, O failed to consent to a CRB check and answered 'No comment' when interviewed on 4 July 2003 by the Trust as to whether he had previous convictions of any kind. In ordering O's name to be erased from the register, the panel stated that his conduct throughout was dishonest, intended to mislead the Trust, and was neither in the public interest nor of the standard to be expected of a medical practitioner. In allowing O's appeal and substituting a sanction of twelve months' suspension, Walker J said that O did not, in 2003, compound the lie of 2001. True, he did not make a clean breast of matters, but he did not cross over the line into dishonesty in 2003. The primary contention on behalf of O in relation to sentence was that the panel, when sentencing, proceeded on the footing that there had also been dishonest and misleading conduct in 2003. That complaint, on the part of the appellant, appears to have been well founded. It must have had a strong influence on the panel when it came to conclude that erasure, rather than suspension, was the appropriate penalty. The panel plainly erred in law when regarding O as guilty of misleading conduct and of dishonesty in 2003, for the purposes of sentence.

### *Walker v. Royal College of Veterinary Surgeons* [2007] UKPC 64

**19.49** False certificates—misguided conduct—suspension

In allowing W's appeal against an order removing his name from the register following a finding of providing false certifications on two separate and similar occasions, the Board reminded itself of the approach indicated by the Privy Council in *Ghosh v. General Medical Council* [2001] 1 WLR 1915, [33] and [34]. The principles apply equally to an appeal relating to the RCVS. The Board also reminded itself of the guidance given by Sir Thomas Bingham MR in *Bolton v. Law Society* [1994] 1 WLR 512. The correctness of veterinary certificates is a matter of importance and can, in some contexts, bear on animal, and indeed human, health. False certification is an extremely serious matter, and the choice lay between suspension and removal. Although W issued false certificates calculated to mislead the Jockey Club, he did so misguidedly, in order to be helpful and to avoid restarting a primary course of injections that had no medical purpose and would have entailed some degree of extra risk. The circumstances were quite distinct from those of cases in which there was a deliberate misleading of insurers, purchasers, or export agencies about the physical status or condition of an animal, or in which there was a risk to animal or human health. It was a case of relatively unthinking ante-dating on two isolated occasions, in the course of a long, and otherwise unblemished and excellent, career. W made no financial gain. He was frank and remorseful throughout, and the likelihood of any reoccurrence of such conduct was remote. The appeal was allowed and resulted in suspension for six months in lieu of removal from the register.

### *Williams v. Royal College of Veterinary Surgeons* [2008] UKPC 39

**19.50** False certificates—previous misconduct—removal

W, a vet of long experience, was asked to arrange to certify, for export purposes, three horses. He signed three certificates, stating that appropriate tests had been undertaken at a Department of the Environment, Food and Rural Affairs (Defra)-approved laboratory with a negative result in each case. No Defra laboratory tests accompanied the certificates and, in the case of one horse, the relevant swabs had not been taken and Defra refused to countersign the certificate. Defra removed W from its panel of local veterinary inspectors (LVIs). There was a history of three previous occasions on which W had been suspended by Defra or its predecessor ministry from LVI duties. In removing W's name from the register, the Disciplinary Committee of the RCVS said that his attitude to certification was irresponsible, or cavalier, or both. In order to maintain public confidence in veterinary certification and to reinforce to the profession the importance of accurate certification, and having regard to the fact that W had issued inaccurate export certificates on several previous occasions, which were followed by clear warnings to take the utmost care when issuing such certificates, the Committee considered that it had no alternative but to remove his name from the register. In dismissing W's appeal, the Privy Council accepted that, although there was no intention to deceive, W's attitude to export certification created at least some risk that an unfit animal might be exported and brought the integrity of export certification into question.

### *Salsbury v. Law Society* [2009] 1 WLR 1286

**19.51** Principles—conviction—receiving money by deception—striking off—approach on appeal

S, a solicitor practising in Sussex for many years, was appointed clerk to the trustees of a local school. S received payment for his services as clerk to the trustees and further payment for any legal work that he undertook on behalf of the trustees. S asked for, and was given, a cheque for £862.50 on account of fees and he altered the amount so as to read £1,862.50. The cheque was presented to his bank and was duly cleared. He was subsequently convicted at Croydon Crown Court of one count of obtaining money by deception contrary to section 15A of the Theft Act 1968. The sentence imposed was a conditional discharge for twelve months. The SDT concluded that it was both appropriate and proportionate to order that S be struck off the roll of solicitors. S's appeal was allowed by the Divisional Court, which substituted an order for three years' suspension. The Court of Appeal allowed an appeal by the Law Society to reinstate the order for striking off. After a review of the authorities, Jackson LJ (with whom Arden LJ and Sir Mark Potter P agreed) said:

> 30. From this review of authority I conclude that the statements of principle set out by Sir Thomas Bingham MR in *Bolton v. Law Society* [1994] 1 WLR 512 remain good law, subject to this qualification. In applying the *Bolton* principles the solicitors Disciplinary Tribunal must also take into account the rights of the solicitor under articles 6 and 8 of the Convention. It is now an overstatement to say that "a very strong case" is required before the court will interfere with the sentence imposed by the Solicitors Disciplinary Tribunal. The correct analysis is that the Solicitors Disciplinary Tribunal comprises an expert and informed tribunal, which is particularly well placed in any case to assess what measures are required to deal with defaulting solicitors and to protect the public interest. Absent any error of law, the High Court must pay considerable respect to the sentencing decisions of the tribunal. Nevertheless if the High Court, despite paying such respect, is satisfied that the sentencing decision was clearly inappropriate, then the court will interfere. It should also be noted that an appeal from the Solicitors Disciplinary Tribunal to the High Court normally proceeds by way of review: see CPR rule 52.11(1).

***Shiekh v. General Dental Council* [2009] EWHC 186 (Admin)**

**19.52** Fraud—NHS—travel expenses—erasure

S appealed against a decision of the GDC's PCC, made on 1 February 2008, ordering the erasure of his name from the register of practitioners. The case arose out of the following facts. S owned some thirteen dental practices in Nottingham and Derby. Between 1994 and 2000, dental practitioners at those surgeries provided out-of-hours emergency dental services to thousands of patients. There was no question in this case that the patients received treatment, which was at least adequate. Payments were claimed by or on behalf of dental practitioners who were practising in S's practices on the basis that they had travelled a distance to reopen the surgery, whereas they had not in fact travelled the circuit. Of the payments received, 50 per cent would be retained by S and 50 per cent would be paid to the dental practitioner who claimed to have been recalled. The reality was that this was a fraud that was being perpetrated against the NHS. It was said that, under the influence of S, others were induced to assist in the fraud and that S employed a large number of foreign dentists whom he was able to manipulate. S pleaded guilty at Northampton Crown Court to an indictment of conspiracy to defraud the Dental Practice Board between December 1994 and August 2000 by dishonestly making claims for repayment within the NHS that were, and which he knew to be, false. He was sentenced to twelve months' imprisonment, suspended for eighteen months, and ordered to pay a fine of £50,000. The Court of Appeal, Criminal Division, in upholding the sentence, said that the judge's decision to suspend the sentence of imprisonment was a merciful course. In civil proceedings, which were settled before the criminal case was heard, S had repaid to the Dental Practice Board the sum of £1.15 million. However, that sum did not relate solely to the conduct that formed the subject of the proceedings, which was considerably less. In dismissing S's appeal, Lloyd Jones J said that he was entirely satisfied that the Committee did have regard to the particular circumstances of this case in deciding whether it was necessary to impose the sanction of erasure. All relevant matters were fully considered and the conclusion was one that it was clearly entitled to reach.

***Chamba v. Law Society* [2009] EWHC 190 (Admin)**

**19.53** Use of client monies—repayment—whether striking off appropriate rather than exceptional

C faced two allegations of dishonesty. First, he drew a cheque on a client account for £30,000 to settle a liability to a former partner. The payment was allocated to a client ledger, which showed that there were no monies in the client account at the time to cover the payment. C agreed that he had knowingly made the payment and did not have funds available at the time that the payment was due. The second allegation arose when C and his wife were in the process of purchasing a residential property. He knowingly utilized client monies to complete his own purchase, although later repaying the sums in three tranches. C made candid admissions, expressed his apologies, and was able to advance cogent personal mitigation and produce some forty testimonials from professional colleagues and others who spoke highly of his competence and integrity, referring also to his generosity in the community. Nevertheless, the SDT proceeded to order that C be struck off. The Divisional Court accepted that the correct approach was for the SDT to decide whether striking off was appropriate having regard to all of the circumstances of the individual case, rather than to adopt a test of 'a most exceptional case', which would impose a burden on the solicitor to establish exceptional grounds for not imposing a striking-off order. However, the SDT correctly identified the need to give proper consideration to the imposition of a striking-off order and gave careful consideration to the submission that it need not, in the circumstances, impose the ultimate sanction. Appeal dismissed.

### *Lim v. Law Society* [2009] EWHC 1706 (Admin)

**19.54** False payslip

L created a false payslip purporting to be issued by his current employer and delivered the payslip to a prospective employer for use in salary negotiations. The SDT found that he dishonestly misrepresented the true position and ordered that L be struck off the roll of solicitors. On appeal, the Administrative Court said it was now well established that, where dishonesty was proved, a solicitor will be struck off unless there are exceptional circumstances: see *Bolton v. Law Society* [1994] 1 WLR 518 and *Salsbury v. Law Society* [2009] 1 WLR 1286. In the instant case, this was 'a hopeless appeal'.

### *Atkinson v. General Medical Council* [2009] EWHC 3636 (Admin)

**19.55** False statement of qualifications and experience

A, a Ukrainian national, had committed some twenty-two dishonest acts on twelve occasions over a period of five-and-a-half years. Overwhelmingly, these were dishonest statements as to her qualifications and medical experience, made for the purpose of obtaining employment and/or training posts in the United Kingdom. She had also made false statements in the course of enquiries made by employers and others as to whether she had mis-stated her position. The fitness to practise panel concluded that A's fitness to practise was impaired by reason of misconduct and directed the sanction of erasure. The first ground of appeal was that even if the panel were entitled to conclude that the appellant was unreliable in her evidence, it should not have gone further and concluded that she had lied. However, Blake J said that it was difficult to see how the panel could have avoided the conclusion that the appellant had been evasive and dishonest with it in her evidence. The centrepiece of the disciplinary inquiry was precisely her qualifications: which qualifications were false and whether any qualifications were genuine. She had already been interviewed about these matters. She could hardly have been surprised to have been asked questions about them again before the panel. The panel was fully entitled to reach the conclusion that it did. On the second ground of appeal—namely, whether the sanction of erasure was wrong—Blake J concluded that even if, contrary to the decision that he had reached on the first question, there were room for sufficient doubt caused by confusion, incomprehension, and language difficulties, the panel would be fully entitled to conclude that the only appropriate sanction that was proportionate and appropriate was erasure. This was not a case in which, as against admitted or proven dishonesty, there was a body of reliable material relating to the rehabilitation of the appellant, although it had never been suggested that her clinical competence was itself a basis for concern. That is frequently a feature of cases in which dishonesty is concerned. Nevertheless, the interests of maintenance of the reputation of the profession required nothing less than erasure by way of sanction.

### *Parkinson v. Nursing and Midwifery Council* [2010] EWHC 1898 (Admin)

**19.56** Employment—dishonest receipt of sick pay—second job—non-appearance before tribunal

P, a registered mental health nurse, appealed against the decision of the NMC's fitness to practise panel of 4 August 2009 directing that his name be erased from the register. He was employed for nineteen years at Mersey Care NHS Trust. Unknown at the time to the Trust, he also had a second job as a nurse at Blair House nursing home. Mitting J observed that there was nothing wrong in that, although P might have been well advised to have told his principal employers about his second employment. His employers at Blair House knew about his employment for the Trust. It occurred to them, when the Working Time Directive came into force, that, by carrying out two jobs, he might be infringing the Directive. So they contacted the Trust to find out the hours that he worked for the Trust. This, unfortunately for him, revealed a set of facts that led to the proceedings that led to his erasure. During 2005, P had extensive periods of sickness. He was paid sick leave by the Trust in June and September 2005, and, during the period

31 October 2005 to 3 January 2006, he worked at Blair House. In consequence, as is not now and indeed never was disputed, he received sick pay from the Trust to which he was not entitled for at least some of that time. The charges against him that led to his erasure recited these facts and alleged that P's conduct was dishonest. Although P responded in writing to the charges, he did not attend the hearing. Indeed, he sent an email on the first day of the hearing saying that he was not going to attend. The hearing proceeded in his absence. In dismissing P's appeal, Mitting J said that, on the information that the panel had, he could not say that the decision of the panel was wrong. Certainly, it was stern, but it was properly stern, because, as the panel noted, one of its tasks is to maintain public confidence in the profession. A nurse found to have acted dishonestly is always going to be at severe risk of having his or her name erased from the register. A nurse who has acted dishonestly, who does not appear before the panel either personally or by solicitor or counsel to demonstrate remorse, a realization that the conduct criticized was dishonest, and an undertaking that there will be no repetition, effectively forfeits the small chance of persuading the panel to adopt a lenient or merciful outcome and to suspend for a period, rather than to direct erasure. Unhappily, as P acknowledged, he did not take that course. Had he done so, it is possible, said Mitting J, that he might have persuaded the panel to exercise leniency in his favour.

***Solicitors Regulation Authority v. Sharma* [2010] EWHC 2022 (Admin)**

**19.57** Dishonesty—appeal by SRA

S, a partner in a well-known firm of city solicitors, improperly signed a number of documents purporting to be signed by another and gave a misleading representation in a letter about them. On 14 May 2009, the SDT found the two allegations proved and that S had acted dishonestly. He was suspended from practice for three years. S had set up an investment company incorporated in the Isle of Man, of which he was the sole beneficial owner and the shares of which were held by two directors as nominees, pursuant to a bare trust of which S was the sole beneficiary. The directors provided corporate services and S's company was indebted to them for outstanding fees. S decided that, instead of paying the fees due for corporate services, he would forge the signatures of the two directors on various company forms that needed their signature for a reverse takeover of an English company. In September 2007, S forged the directors' signatures on stock transfer forms, a sale and purchase agreement, a power of attorney, and a letter; on 9 October 2007, he wrote on the headed notepaper of his firm a letter to Capita Registrars, enclosing the stock transfer document and stating that it had been 'duly signed by two directors'. That was, of course, untrue. In allowing the SRA's appeal and substituting a sanction of striking off, the Divisional Court (Laws LJ and Coulson J) said that the question was whether the Tribunal's decision could be described as excessively lenient. If it could, then the Court would substitute for the Tribunal's sentence the sentence that it considered to be commensurate with the offences. If the sentence could not be regarded as excessively lenient, even if it were not necessarily the sentence that the Court would itself have imposed, the sentence should remain unchanged. The Tribunal was quite right to approach the question of sentencing on the basis that, unless exceptional circumstances could be shown, the respondent should be struck off the roll of solicitors. However, the Tribunal was plainly wrong to conclude that the circumstances of this case were exceptional such that the usual sanction should not be applied. This was a serious case, which simply could not justify the leniency shown by the Tribunal: it involved repeated acts of forgery; the dishonesty was not a one-off event; the letter of 9 October 2007 was a serious aggravating factor; the respondent earned a benefit from the forgeries; and—although this is a lesser point—the respondent denied dishonesty before the Tribunal. The denial acted as a further bar to any real litigation.

### *Dad v. General Dental Council* [2010] CSIH 75

**19.58** Fraud on NHS—non-disclosure of police investigation on re-registration application—non-disclosure of previous disciplinary process

On 8 April 2004, D, a dentist, defrauded the NHS of £8,336.60, representing sums that he claimed he had paid to Glasgow District Council as non-domestic rates for two properties that he claimed were dental surgeries. Although he had sent a cheque for that sum to Glasgow District Council, he knew that the cheque would be dishonoured. Moreover, in respect of one of the properties, D had represented to the NHS that it was being used as a dental surgery when, in fact, it was not. On 5 October 2007, following a plea of guilty, D was convicted of two charges of fraud and was subsequently sentenced to a community service order for 240 hours. His registration with the GDC had ceased on 24 January 2006 as a result of his non-payment of the annual retention fee. On 2 February 2007, three days before appearing in Glasgow sheriff court in respect of charges of dishonesty, D applied to the GDC to restore his name to the dentists' register. He did not disclose that he was currently the subject of police investigations that might lead to a conviction and he also failed to disclose that he had been the subject of previous disciplinary proceedings in 1999: *Dad v. General Dental Council* [2000] 1 WLR 1538. On 5 and 6 May 2010, D attended a hearing before the PCC, which found allegations proved including dishonesty, and the Committee ordered that D's name be erased from the register. In dismissing D's appeal, Lord Hardie, giving the judgment of the Inner House (Lord Hardie, Lord Mackay of Drumadoon, and Lord Dummond Young), said that it was apparent that D did not learn his lesson from his previous involvement with the GDC or with the criminal law. Instead, about four years after the decision of the Privy Council, he committed two offences of fraud. He submitted an application form to his professional body that he acknowledged to be misleading, intended to mislead, and dishonest. Unlike his previous convictions for offences under the road traffic legislation, the convictions for fraud were associated with his practice as a dentist. Similarly, the intentionally misleading and dishonest statements submitted to the GDC were made in a successful attempt to restore his name to the register to enable him to practise as a dentist. The guidance issued by the GDC, *Guidance for the Professional Conduct Committee*, recognizes that dishonesty is highly damaging to a registrant's fitness to practise and to public confidence in the dental profession, particularly where that dishonesty is associated with professional practice. In the context of professional misconduct, it does not appear to be significant that the appellant repaid the money, which he had obtained by fraud. Nor is it significant that, in his application to be included in the NHS list of practitioners, he disclosed the criminal proceedings, although he had not done so in his application to his professional body. In short, the PCC was entitled to conclude that the issues raised on behalf of D were not of sufficient weight to justify the unusual course of not ordering the erasure of his name from the register when he had committed two offences of fraud and had subsequently knowingly submitted a dishonest application to his professional body. The decision of the Committee was not plainly wrong, the test being enunciated by Collins J in *Moody v. General Osteopathic Council* [2004] EWHC 967 (Admin) and approved in *Macleod v. Royal College of Veterinary Surgeons* [2006] UKPC 39, [23].

### *Hosny v. General Medical Council* [2011] EWHC 1355 (Admin)

**19.59** False references—failure to disclose interim suspension—exceptional mitigation

H appealed against the decision of the fitness to practise panel of the GMC, which imposed a twelve-month suspension on her right to practise. H was an anaesthetist, who faced a number of allegations, including clinical matters and two allegations of dishonesty. The allegations in relation to the clinical matters were either found not proved or did not result in a finding of impairment. The two allegations of dishonesty were found proved and resulted in a finding that her fitness to practise was impaired. The first of the allegations of dishonesty related to the sending of a false reference to five locum agencies. The false reference purported to be from the lead clinician in anaesthetics at Lincoln County Hospital. The second allegation of dishonesty related to H's failure to disclose on Internet applications for employment to two separate bodies that her registration had been suspended by the GMC's IOP. It is plain, from paragraph 82 of the GMC's Indicative Sanctions Guidance, that erasure may well be the appropriate sanction when dishonesty is involved, especially when persistent or covered up. This was a case in which two separate acts of dishonesty were proved. However, it was also a case in which there were a number of mitigating factors. In the light of those factors, the fitness to practise panel felt able to describe H's case as 'exceptional'. The sanction of suspension was well within the bracket of possible sanctions open to the panel. The panel took into account all of the mitigating factors and the decision on sanction could not begin to be described as wrong.

### *R (Chief Constable of Dorset) v. Police Appeals Tribunal and Salter* [2011] EWHC 3366 (Admin); [2012] EWCA Civ 1047

**19.60** Road traffic accident—destruction of evidence—coroner's inquest—approach—honesty and integrity in investigations

In the early hours of Sunday, 26 October 2008, M, a serving police constable, was killed in a road traffic accident. No other vehicles were involved. S, a sergeant, was appointed as deputy senior investigating officer. It became apparent that M had a long-term partner and was also involved with a member of another force with whom he had spent the night before his death. His partner was unaware of the relationship. Two mobile telephones were recovered from the crashed vehicle and it became known to S that one of those telephones contained stored text messages that evidenced the relationship. S instructed another officer to go to the vehicle recovery centre, find the telephone, and destroy it. The death was the subject of a coroner's investigation. The officer was not prepared to destroy evidence and raised the matter with senior colleagues. S appeared before a misconduct panel of the Dorset police, which required him to resign from the force. S's appeal to the Police Appeals Tribunal resulted in him being reduced in rank from sergeant to police constable. The chief constable brought proceedings for judicial review of the decision of the Tribunal. Burnett J, in quashing the decision of the Tribunal, said that the correct approach to the question of sanction of a finding of serious impropriety by a police officer in the course of his duty is reflected in the principles articulated in *Bolton* and *Salsbury*. The reasons that underpin the strict approach applied to solicitors and barristers apply with equal force to police officers. Honesty and integrity in the conduct of police officers in any investigation are fundamental to the proper workings of the criminal justice system. They are no less important for the purposes of other investigations carried out by police forces, including those on behalf of coroners. The public should be able unquestionably to accept the honesty and integrity of a police officer. S's appeal was dismissed by the Court of Appeal (Maurice Kay, Stanley Burnton, and Gross LJJ): *Salter v. Chief Constable of Dorset* [2012] EWCA Civ 1047. Maurice Kay LJ said that, as to personal mitigation, just as an unexpectedly errant solicitor can usually refer to an unblemished past and the esteem of his colleagues, so will a police officer often be able to do so. However, because of the importance of public confidence, the potential of such mitigation is necessarily limited. Gross LJ added that a central role of the police involves the gathering and preservation of evidence. The destruction of evidence is inimical to the office of constable, and all the more so when it entailed an instruction to a junior officer to do so.

### *Ige v. Nursing and Midwifery Council* [2011] EWHC 3721 (Admin)

**19.61** Mortgage frauds—insight—public confidence

I was convicted following trial at Harrow Crown Court on two interrelated counts alleging offences of dishonesty arising out of a mortgage fraud. She was sentenced to nine months' imprisonment on each count, concurrent and suspended for two years, and was made the subject of a confiscation order in the sum of £18,685. By reason of her conviction, the NMC's CCC found that I's fitness to practise was impaired and imposed a sanction of striking off. At the time of the offending, I was a student nurse at Hertford University; at the date of her trial and conviction, she had qualified and was a registered nurse. In his sentencing remarks, the judge referred to 'a deliberate, carefully planned, brazen mortgage fraud'. In dismissing I's appeal, King J said that he was quite satisfied that the panel was fully justified in concluding that I had never accepted or recognized the dishonesty of which she had been found guilty. There was no suggestion that her clinical abilities were impaired. However, the panel clearly found impairment on a public interest basis. It is well established that, when considering whether a practitioner's fitness to practise is impaired by reason of misconduct, a panel is not confined to considering simply whether the practitioner continues to present a risk to members of the public in his or her current role. The panel is fully entitled, in an appropriate case involving past conduct amounting to a criminal offence of dishonesty, to determine whether public confidence in the profession would be undermined if a finding of impairment were not made in the particular circumstances. The panel is fully entitled to reach a finding of impairment on this basis by reference to the need to uphold professional standards. Lack of insight and failure to demonstrate remorse as regards matters of dishonesty are relevant not only to the impairment stage, but also at the sanction stage. The appeal was therefore dismissed.

### *Ballesteros v. Nursing and Midwifery Council* [2011] EWHC 1289 (Admin)

**19.62** Alleged sickness—other employment—second dishonesty—false timesheets

B appealed against a finding of impairment and the imposition of a striking-off order. B, while employed full-time at University College London (UCL) Hospital, presented herself as being unfit for work, but then worked at a number of other hospitals for which she was paid. Her conduct over a significant period of time was dishonest, despite considerable extenuating circumstances—namely, that her mother was very seriously sick indeed, and that her mother later died from her condition, at which time the appellant was under financial pressure. Not long after having been dismissed by UCL Hospital, B began yet another course of dishonest conduct: she was then employed by Mayday Healthcare and she created a false time sheet which she then herself signed in the name of other nurses who were working at the relevant hospital or hospitals in order to support false claims for payment. In dismissing B's appeal, Kenneth Parker J said that the panel was justified in its conclusions on impairment and sanction. This was not an isolated incident, but deliberate dishonesty over a period of many months. Taking account of other features—namely, the worrying concerns about lack of insight into her conduct—the panel concluded that the risk was too great, and that there was indeed a risk that conduct of this kind or similar in nature could be committed by B again in the future and that would be contrary to the public interest. The second set of dishonest behaviour caused real concern about this particular appellant. Had her dishonesty stopped at UCL Hospital, in all probability a lesser sanction would have been imposed. What shifted the panel was the further serious misconduct—dishonesty against the background described.

### *Naheed v. General Medical Council* [2011] EWHC 702 (Admin)

**19.63**
False application—use of another's career history and experiences—dishonest explanation

N applied for speciality training at Yorkshire and the Humberside Postgraduate Deanery by dishonestly cutting and pasting someone else's career history into her own application, and claiming that that other person's experiences were, in fact, her own. Her application was thus, to her knowledge, fundamentally misleading. When challenged by the associate postgraduate dean, N gave a dishonest explanation of the circumstances behind the application and told further lies. N appeared before the IOP and maintained that the application form was accurate in material respects. To her credit, when she came ultimately before the fitness to practise panel at the substantive hearing, she did admit in large part the charges against her. The panel, having found impairment by reason of misconduct, concluded that erasure was the appropriate sanction. On appeal, it was argued, amongst other points, that, in a parallel case of Dr E, who had also misconducted himself in a similar fashion by making use of a plagiarized document, a separate panel had dealt with the matter differently. Parker J, in dismissing N's appeal, accepted that there were striking similarities between the two cases. However, Dr E did not seek to deceive the IOP, and Dr E recognized his wrongdoing and had genuine insight into it. On a challenge in relation to proportionality, the Court has to ask itself where a panel has made a finding of fundamental dishonesty and has concluded that the doctor in question has not demonstrated insight into wrongdoing, and whether there was material before the panel upon which such a conclusion could have been based. In that respect, certain of N's answers given in cross-examination were of some relevance. Also, N had maintained that she had not acted in a misleading or dishonest way in relation to her clinical experience, but it did appear that the account upon which she was ultimately relying differed materially from the one that she put forward in her application form. Therefore, particularly taking account of the special position that the panel enjoys, there was here material to support the finding by the panel that there had not been a full recognition of the dishonest behaviour. That was a matter eminently for the panel to evaluate.

### *Solicitors Regulation Authority v. Dennison* [2012] EWCA Civ 421

**19.64**
Interest in company—deceit of partners—no misuse of client monies

D, a solicitor, appealed against the order of the Divisional Court that he be struck off the roll of solicitors. The order of the Divisional Court was made on an appeal by the SRA against the order of the SDT that D be fined a sum of £23,500 and pay the costs of the proceedings in relation to allegations of professional misconduct that it found had been proved against him. The Tribunal found that D had deliberately failed to disclose to his partners or to clients of the firm his interest in a company that provided expert reports for the sourcing, funding, and representation of claimants in personal injury cases. The Tribunal found that D had deliberately deceived his partners because he wanted to retain the whole of the benefit of his interest in the company for himself. The Tribunal found that D had acted dishonestly by the ordinary standards of reasonable and honest people, and, moreover, that he had been aware that, by those standards, he had been acting dishonestly. The Court of Appeal held that, although the case did not involve the dishonest use of clients' money, it was a case that could be expected to have resulted in D being struck off for the reasons given in *Bolton v. Law Society*. Despite the Tribunal's view that it was not appropriate or necessary for D to be struck off or suspended, the Court was unable to accept that his dishonesty was so trivial as to fall into what has been described in *Salsbury v. Law Society* as a residual category of cases for which striking off is not an appropriate penalty.

### *Heydon-Burke v. Nursing and Midwifery Council* [2012] EWHC 2435 (Admin)

**19.65** Police nurse—false medical record—non-attendance before tribunal

The appellant, a nurse for thirty-four years with an unblemished record, appealed against the decision of the NMC striking her off the nursing register. She was employed as a custody nurse to undertake clinical assessments of persons detained at police stations. She was charged that, on 17 December 2009 at West Didsbury police station, she breached her employer's policy in that she administered to a detained person methadone without a forensic medical examiner being present, and dishonestly completed the detainee's medical record stating that the methadone had been self-administered by the detainee at home prior to arrest. The incorrect completion of the medical record was of the utmost significance and, in the end, ultimately led to the appellant being struck off. In dismissing her appeal, His Honour Judge Raynor QC (sitting as a judge of the High Court) said that the panel concluded, inevitably, that she had breached relevant parts of the Nursing Code. As regards sanction, the panel considered, correctly, whether or not lesser sanctions would be appropriate. It took account of the fact that it was an isolated incident that had not been repeated and the appellant had an otherwise unblemished nursing career. Unfortunately, the appellant did not attend the disciplinary hearing and the panel noted that, while she had admitted what she had done, she had not demonstrated remorse and had not shown insight into her actions, had provided no references or testimonials, nor any evidence of rehabilitative steps. Dishonesty was not admitted. The Court said that, as to sanction, the appellant's absence deprived the panel of the opportunity of assessing her in person. As Mitting J observed, it is perfectly plain that if there is a finding of dishonesty and a failure to appear before the panel either personally, or represented by solicitors or counsel, the person effectively forfeits the small chance of persuading the panel to adopt a lenient and merciful outcome: see *Parkinson v. Nursing and Midwifery Council* [2010] EWHC 1898 (Admin). Whilst the Court had the greatest sympathy with the appellant, in the end she chose not to attend. There was no procedural irregularity and the decision was not wrong. Given that there was no acceptance of dishonesty until right at the end of the hearing of the appeal, the decision was not even then a wrong decision, although it was a very harsh one.

### *Solicitors Regulation Authority v. Spence* [2012] EWHC 2977 (Admin)

**19.66** Dishonest information supplied to SRA—insurance cover and practising certificate—suspension—appeal by SRA

The SRA appealed against the sanction of three years' suspension imposed on S, contending that it was too lenient in the circumstances. S faced six allegations, the first containing allegations of dishonest and misleading statements made to the SRA during the period December 2007 to February 2008. The dishonest and misleading statements were that:

(1) on 12 December 2007, he submitted an application for a practising certificate in which he asserted he had professional indemnity insurance, which was untrue—there was no professional indemnity insurance in place;
(2) when the SRA began an inspection on 14 February 2008, S was not in his office and, on being telephoned, gave an untrue statement as to where he was; and
(3) on 13 February 2008, S spoke on the telephone to a case worker at the SRA and said that a lady who was a solicitor with her own practice had been dealing with all of his live client matters, and that he had not practised since he and the case worker had last spoken, which statement was untrue—the lady solicitor had had nothing to do with his practice.

At the hearing before the SDT, S accepted that he had taken a conscious decision to present a misleading picture to the SRA, including trying to mislead it over the role of the lady solicitor. The Tribunal in those circumstances understandably came to the conclusion that he had been dishonest in connection with those matters. In allowing the SRA's appeal and substituting an order for striking off, Pitchford LJ (sitting with Foskett J) said that the Tribunal erred in law in two respects. First, it appeared to treat repeated lies to the SRA as a less culpable form of dishonesty than dishonesty in relation to clients' money. This was a mistaken approach. The purpose and effect of S's dishonesty was to enable him to continue to practise without insurance, or a practising certificate. It is difficult to think of a more serious risk to the interest of those clients than the absence of professional indemnity insurance. It would, in addition, be deeply misleading for practitioners to think that lying to their regulator was somehow not as culpable a form of professional misconduct as the misapplication of client's funds. Second, the Tribunal treated S's personal mitigation as sufficient to justify an exceptional course. In this aspect too, the Tribunal was mistaken. Unhappily, none of the pressures of S's personal and professional life were exceptional, and Pitchford LJ said that he could not have justified taking the course that the Tribunal selected.

***Siddiqui v. General Medical Council*** **[2013] EWHC 1083 (Admin)**

**19.67** Amendment of patient's records—single incident—long career

S, a GP, admitted dishonestly amending a patient's records and was suspended for six months. Her appeal against sanction was dismissed. The panel accepted that the dishonesty was a one-off episode in a long career spanning more than forty years and that the appellant did not have a harmful or deep-seated attitudinal problem. Whilst she said, in essence, that a period of six months' suspension would, in practical terms, preclude her from practising medicine again and deprive her patients of her care, the appeal nevertheless was dismissed. The case law requires that if the appeal is to succeed, it must be demonstrated that the decision reached was wrong.

***R (Kibe) v. Nursing and Midwifery Council*** **[2013] EWHC 1402 (Admin)**

**19.68** A nurse failed to mention on a job application form and at interview a previous dismissal. The appeal against a resulting striking off order was dismissed.

***Hassan v. General Optical Council*** **[2013] EWHC 1887 (Admin)**

**19.69** Police caution—student registration—subsequent disclosure—advice to Committee

On 23 January 2009, H, a student at City University studying optometry, received a police caution for fraud in seeking to make an insurance claim arising from a road traffic accident. In June 2009, H completed a form to apply to remain on the Council's register of student optometrists, but did not disclose the police caution. He subsequently supplied details to the General Optical Council (GOC) in April 2011 at a time when he needed to register as a full member. In March 2012, the defendant's Fitness to Practise Committee found that H's fitness to undertake training as a student registrant was impaired by misconduct and by the caution, and directed that his name be erased from the register of student optometrists. In allowing H's appeal and remitting the case to be reconsidered by the Committee, Leggatt J said that the legal adviser misdirected the Committee; the learned judge also had real misgivings about the Committee's perception of the facts and the approach taken by the Committee. By referring to *Solicitors Regulation Authority v. Sharma* [2010] EWHC 2022 (Admin), the legal adviser, whilst stating that it was in a different regulatory arena, introduced a qualification or restriction on the guidance issued by the GOC in its Indicative Sanctions Guidance. It was a misdirection for the legal adviser to say that, save in exceptional circumstances, findings of dishonesty would lead to a striking off. The principles stated in the *Sharma* case were intended to apply only to solicitors.

### *Adediwura v. Nursing and Midwifery Council* [2013] EWHC 2238 (Admin)

**19.70** Midwife—practising also as nurse—non-registration

The registrant was qualified and registered as a midwife in the United Kingdom, but not as a nurse, despite being qualified in Nigeria both as a midwife and a nurse. She was found guilty of dishonestly acting as a practice nurse between November 2004 and July 2008 without being registered as a nurse. The panel imposed a suspension order for six months as sufficient to mark the seriousness with which the conduct was regarded, and to maintain public confidence in the profession and the regulatory process. The appellant had placed patients at risk, and although there was no demonstrable patient harm, the potential for serious harm was evident. The appellant had held herself out as being a registered nurse and as possessing the relevant expertise to carry out that role. In dismissing the appellant's appeal, Stuart-Smith J said, at [28], that a prolonged period of dishonesty involving practising without registration as a nurse for a period of years is very serious precisely because it raises the public interest concerns that the panel addressed.

### *Adegboyega v. Nursing and Midwifery Council* [2013] EWHC 2647 (Admin)

**19.71** Dishonest entry made in hospital records—consequences devastating—striking off

In this case, the appeal against a striking off order for dishonesty was dismissed. Allegations had been made that the appellant had signed a patient's observation chart to confirm that she had completed observations when, in fact, she had not done so, and that she acted dishonestly. The patient died, and that death involved a succession of acts and omissions over some hours. The consequences were devastating.

### *R (Mafico) v. Nursing and Midwifery Council* [2014] EWHC 363 (Admin)

**19.72** Police caution—theft of drugs from hospital—advice to Committee

In November 2011, M, a band 5 nurse, accepted a police caution in respect of the theft of unspecified quantities of amitriptyline and tramadol, which she admitted taking from the ward at the hospital where she was employed. On 5 September 2013, the NMC's CCC imposed a sanction of striking off. M's health had deteriorated during the period of the offences and, in August 2011, she had attempted suicide by an overdose of quinine sulphate tablets, which she had purchased online. M appealed, claiming that the legal assessor erred when commending to the Committee the application of the decision in *Solicitors Regulation Authority v. Sharma* [2010] EWHC 2022, which gave the Committee the erroneous impression that striking off would be appropriate in all cases of dishonesty absent exceptional circumstances, and that the present case was on all fours with *Hassan v. General Optical Council* [2013] EWHC 1887. In dismissing M's appeal, His Honour Judge Gore QC (sitting as a judge of the High Court) said, at [18], that, in alleged misdirection cases, two quite distinct questions must be addressed based on *Libman v. General Medical Council* [1972] AC 217—namely, first, was there a material misdirection, and if so, secondly, was it of sufficient significance to the result to invalidate the decision? The learned judge said, at [31]–[34], that whilst *Hassan v. General Optical Council* was persuasive and weight should be given to it, Leggatt J did not ask or then answer the second question posed in *Libman*—namely, was any misdirection of sufficient significance to the result to invalidate the decision? Moreover, there were several points of distinction between *Hassan* and this case that were highly relevant. The matters taken into account by the Committee were much more carefully and extensively set out in the decision letter than they were in *Hassan*. This was a decision-making process undertaken by the relevant panel in exercising its professional judgment to which respect should be accorded. Whereas the reasoning of the panel in *Hassan* ran to but four short paragraphs over half a page, in this case the reasons given by the Committee were much more extensive, running to some four single-spaced pages of detailed analysis. If reference to *Sharma* constituted a misdirection, it was not material, and it certainly could not be demonstrated that it had such a significant effect on the outcome as to invalidate the decision.

### *Mills v. General Dental Council* [2014] EWHC 89 (Admin)

**19.73** NHS—fraudulent claim for remuneration

M had been in practice as a dentist in Truro since 2003. The GDC's PCC found proved that, between 2009 and 2011, M had dishonestly submitted to the NHS inappropriate claims for remuneration in respect of forty-three patients. He did so under circumstances in which he was facing the prospect of having to repay to the primary care trust an overpayment of £4,153. The Committee's decision that M's name should be erased from the register, on the basis of the need to uphold proper professional standards and public confidence in the profession, was entirely justified.

### *Bar Standards Board v. O'Riordan* [2014] Lexis Citation 24

**19.74** False CV—sentence—appeal—fresh evidence

The Visitors allowed the Bar Standards Board's appeal of three years' suspension and substituted an order of disbarment. The respondent had submitted false particulars in a curriculum vitae, which were dishonest. The Disciplinary Tribunal of the Inns of Court, by a majority of three to two, concluded that the appropriate sentence should be three years' suspension, the minority favouring disbarment. Prompted by publicity consequent upon the hearing, a barrister in independent practice wrote to the Board, drawing attention to an expert report by the respondent that also falsely set out his qualifications. The Visitors said that they were satisfied that this evidence would not have been apparent upon an investigation of the work diaries of the respondent and other material filed by him for the purposes of the hearing before the Tribunal. They were further satisfied that, had the Tribunal known of this evidence, the majority view as to the appropriate sentence in this case would undoubtedly have been different.

### *Fernando v. General Medical Council* [2014] EWHC 1664 (Admin)

**19.75** Fraudulent prescriptions—use of prescription pads—obtaining drugs for own use—police caution

The appellant faced fourteen allegations. Most related to, or followed on from, the appellant presenting a fraudulent prescription to a Sainsbury's store in August 2012, in that, although made out by the appellant, it was for a fictitious patient and purported to be signed by a different doctor. The appellant intended to obtain and use the tablets subject of the prescription for himself. He was cautioned for a criminal offence of fraud by dishonest false representation contrary to sections 1 and 2 of the Fraud Act 2006 in relation to the presentation of the prescription. After his arrest, his house and car were searched, and prescription pads found from seven other medical practices. The appellant had, on one occasion, written out a prescription knowing that there was no patient of that name and signed it with a false name. Again, he had intended to use the prescription to obtain listed drugs for himself. The other prescription pads had not been returned to the relevant medical practices. When the appellant told his fellow partners at his medical practice of his arrest, he failed to inform them of the reason for it and falsely led them to believe that his arrest related to a road traffic offence. He also failed to disclose his lack of professional indemnity insurance. Dismissing the appellant's appeal against a sanction of erasure, Patterson J said that, in considering the appropriate sanction, the assessment on the seriousness of the misconduct was essentially a matter for the panel in the light of its experience. The committee comprised a chair and two medical members. The medical members were able to question the appellant when he gave his evidence as to his underlying condition and its implications. They would be well qualified also to understand what measures were required to maintain the standards and reputation of the profession. It was clear from their determination that the issue of proportionality was expressly part of their judgment and that, having set that out, erasure was the appropriate sanction.

### *McDaid v. Nursing and Midwifery Council* [2014] EWHC 1862 (Admin)

**19.76** Forged letter sent to patient—use of name of another employee

Following a successful appeal, a second panel of the CCC of the NMC found that M had dishonestly written a letter addressed to a patient in the name of the trust's head of employment. In dismissing M's appeal against the decision of the second panel striking her name off the register of midwives and nurses, Blake J said that a finding that M had forged documents, and hence was dishonest, would of itself make her striking off inevitable. M was the author of the letter.

### *Okee v. Nursing and Midwifery Council* [2014] EWHC 1763 (Admin)

**19.77** Police caution—non-disclosure

In October 2005, F was convicted of failing to provide a specimen of breath for analysis; in January 2006, she received a second conviction of driving with excess alcohol. F was admitted to the register as a nurse in October 2010. In December 2010, she successfully applied for a nursing post, but did not disclose the convictions. Subsequently, a CRB check revealed the convictions. At a meeting with the trust, F falsely said that her sister had committed the offences using her car and driving licence. This was untrue. F's appeal against the striking off order was dismissed.

### *Professional Standards Authority v. Health and Care Professions Council and Ghaffar* [2014] EWHC 2708 (Admin); [2014] EWHC 2723 (Admin)

**19.78** Caution—false representation for gain—suspension

On 11 October 2012, G, a biomedical scientist, was convicted on a guilty plea at Leeds Crown Court of dishonestly making false representations for gain, contrary to section 2 of the Fraud Act 2006. He was sentenced to six months' imprisonment, suspended for twelve months, and ordered to fulfil 200 hours of unpaid work. The conviction arose out of the production by the registrant of a forged Master of Science (MSc) degree certificate, which he used for the purpose of applying for a job for which, as he well knew at the time, he was not qualified. The circumstances surrounding the offence included repeated acts of dishonesty and a subsequent attempt to conceal. On 9 January 2014, the panel of the Health and Care Professions Council (HCPC) found that G's fitness to practise was not impaired by reason of the conviction. Carr J, having given judgment quashing the decision of the panel and imposing a finding of impairment of fitness to practise, with the agreement of the parties, proceeded to consider what sanction, if any, should be imposed in the light of that finding. After reference to the Council's Indicative Sanctions Policy and mitigation on behalf of the registrant, the learned judge said that the only two realistic options were striking off or suspension. After careful consideration and having had the benefit of submissions on behalf of the registrant, the learned judge reached the conclusion that striking off was not merited. Essentially, although this was a very serious and troubling dishonesty, there were powerful testimonials. The registrant had taken active steps to enrol on courses. He had shown undoubted remorse. There were question marks over his insight, but, in the grand scheme of things, there was at least a substantial degree of insight on the material before the Court. The appropriate and proportionate sanction should be suspension for a period of six months and, at the conclusion of that period, there should be a review.

***Khan v. General Medical Council* [2015] EWHC 301 (Admin)**

**19.79** Guidance—dishonesty with lack of insight

In dismissing K's appeal in relation to the dishonesty, theft, and forgery of prescription forms, Mostyn J said:

> 6. The decisions from this court have demonstrated that a very strict line has been taken in relation to findings of dishonesty. This court and its predecessor, the Privy Council, has repeatedly recognised that for all professional men and women, a finding of dishonesty lies at the top end of the spectrum of gravity of misconduct; see *Tait v. Royal College of Veterinary Surgeons* [2003] UKPC 34 at paragraph 13.
>
> 7. Dishonesty will be particularly serious where it occurs in the performance by a doctor of his or her duties and/or involves a breach of trust placed in the doctor by the community. Both elements are serious and aggravating features and both are present in a case of dishonestly using prescription forms to obtain drugs. See *R (Rogers) v. General Medical Council* [2004] EWHC 424 (Admin) per Mitting J at [28]–[30].
>
> 8. In cases of proven dishonesty, the balance can be expected to fall down on the side of maintaining public confidence in the profession by a severe sanction against the doctor concerned. See *Nicholas-Pillai v. General Medical Council* [2009] EWHC 1048 (Admin) per Mitting J at [27] where he stated: "That sanction will often and perfectly properly be the sanction of erasure, even in the case of a one-off instance of dishonesty."
>
> 9. Where proven dishonesty is combined with a lack of insight (or is covered up) the authorities show that nothing short of erasure is likely to be appropriate. As Sullivan J put it in *R (Farah) v. General Medical Council* [2008] EWHC 731 (Admin), a case which involved the theft and forgery of prescription forms in order to obtain drugs, at paragraph 21: "...given the nature of the appellant's dishonesty and given the Panel's finding that there had been a persistent lack of insight into that dishonesty, whatever the mitigating factors were, the inevitable consequence was that erasure from the register was an entirely proportionate response to the appellant's conduct. The Panel was entitled to come to the view that where a doctor had engaged in deliberate dishonesty and abused his position as a doctor and then had shown a persistent lack of insight into that conduct, he simply could not continue to practice in the medical profession."

**19.80** Further recent examples of sanction in dishonesty cases include:

- *Asare-Konadu v. Nursing and Midwifery Council* [2014] EWHC 4385 (Admin) (dishonestly failing to disclose previous employment details and working whilst on special leave; appeal against striking-off order dismissed);
- *Gbidi v. Nursing and Midwifery Council* [2015] EWHC 237 (Admin) (dishonest failure to disclose previous employment; appeal against three-year caution dismissed; caution was an exceptional sanction in that a more serious sanction could readily have been visited had G's circumstances differed, *per* Knowles J at [21]);
- *Abiodun v. Nursing and Midwifery Council* [2015] EWHC 434 (Admin) (false references given by appellant to gain a nursing role; appeal against striking-off order dismissed); and
- *Oluyemi v. Nursing and Midwifery Council* [2015] EWHC 487 (Admin) (false passport obtained by appellant purporting to show that she had leave to remain in United Kingdom and so was eligible for work; appeal against striking-off order dismissed).

***Soni v. General Medical Council* [2015] EWHC 364 (Admin)**

**19.81** See paragraph 50.13.

## D. Other Relevant Chapters

**19.82**

Absence of Practitioner, Chapter 1
Drafting of Charges, Chapter 22
Good Character, Chapter 30
Insight, Chapter 37
Sanction, Chapter 60

# 20

# DISPOSAL WITHOUT ORAL HEARING



## A. Legal Framework

**20.01** Examples include:

General Pharmaceutical Council (Fitness to Practise and Disqualification etc) Rules 2010, rule 38 (disposal of allegations without hearings)
Chartered Institute of Management Accountants Regulations, Part II 'Discipline' (consent order procedure)
Institute of Chartered Accountants in England and Wales Disciplinary Bye-Laws, bye-law 16 (consent orders)

## B. Disciplinary Cases

### *R v. Army Board of the Defence Council, ex parte Anderson* [1992] QB 169

**20.02** Army Board—investigation—fairness

The applicant, A, applied for judicial review of the decision of the Army Board of the Defence Council denying redress of his grievance that, while a serving soldier, he was discriminated against on racial grounds by being subjected to abuse, harassment, and assault. In granting A's application for judicial review, the Divisional Court (Taylor LJ and Morland J) said that the Army Board was the forum of last resort and that, dealing with an individual's fundamental statutory rights, it must, by its procedures, achieve a high standard of fairness. The principles were as follows.

(1) There must be a proper hearing of the complaint, in the sense that the Board must consider, as a single adjudicating body, all of the relevant evidence and contentions before reaching its conclusions.
(2) The hearing does not necessarily have to be an oral hearing in all cases. Provided that the Board achieves the degree of fairness appropriate to its task, it is for the Board to decide how it proceeds and there is no rule that fairness always requires an oral hearing.
(3) The opportunity to have the evidence tested by cross-examination is again within the Army Board's discretion.

(4) Whether oral or not, there must be what amounts to a hearing of any complaint. This means that the Army Board must have such a complaint investigated, consider all of the material gathered in the investigation, give the complainant an opportunity to respond to it, and consider his response.

***R v. Solicitors Complaints Bureau, ex parte Curtin* [1994] COD 390**

**20.03** Solicitor—conditions—delegation of powers

C, a practising solicitor, sought judicial review of the decision by an assistant director of the Law Society to impose conditions on his practising certificate. The grounds were that the decision by the assistant director was void, because the delegation to him of the decision-making process by the Law Society through its Council was ultra vires, and that the procedure adopted was in breach of the rules of natural justice in not offering the applicant an oral hearing. The Court of Appeal (Russell, McCowan, and Steyn LJJ), in dismissing C's appeal against the decision of the Divisional Court, said that there was nothing unlawful or ultra vires in the way in which the Council delegated its functions under section 79 of the Solicitors Act 1974 and that the absence of an oral hearing did not vitiate the decision, because C was not prejudiced. The conditions were not unusual, C having been found guilty of disciplinary offences that included deliberately misleading clients and which resulted in him being suspended for three months.

***R (Thompson) v. Law Society* [2004] 1 WLR 2522**

**20.04** Inadequate professional service—adjudicator—fairness

T applied for permission to bring judicial review proceedings to quash the decisions of the Law Society of findings of providing inadequate professional service on the ground that the decisions had been reached without an oral hearing. In the first case, T did not seek an oral hearing; in the second case, his application was dismissed by the adjudicator on the ground that the matter was not of sufficient complexity to warrant an oral hearing and that the written material was sufficiently detailed to enable the adjudicator to reach a fair conclusion. In dismissing T's appeal, Clarke LJ (with whom Kennedy and Jacob LJJ agreed) said that the adjudicator (or equivalent) had a duty to act fairly. What is 'fair' depends upon the circumstances of the particular case. One can imagine circumstances in which an adjudicator or appeal panel might think it appropriate to hold an oral hearing and there may even be cases in which the Court would intervene to quash a decision refusing to do so. The relevant principles were considered in *R (Smith) v. Parole Board (No. 2)* [2004] 1 WLR 421, in which the Court of Appeal considered and rejected a submission that the Parole Board should have held an oral hearing. In that context, at [37], Kennedy LJ approved the following test proposed by counsel. An oral hearing should be ordered where there is a disputed issue of fact that is central to the Board's assessment and which cannot fairly be resolved without hearing oral evidence. Clarke LJ said that, to succeed, the claimant would have to show that the procedure adopted was unfair. The application was refused upon the principles of the common law. As to Article 6(1) of the European Convention on Human Rights (ECHR), Clarke LJ said, at [80], that there was a distinction between the kind of case in which there is, say, a reprimand and the kind of case in which there is a suspension of the right to practise. It is only in the latter class of case that it can be said that the right to continue to practise is at stake, so as to cross the line between a disciplinary process and a process that determines civil rights and obligations. In the instant case, the decision of the Law Society to reprimand the claimant severely was not a determination of his civil rights.

## C. Other Relevant Chapters

**20.05** Consensual Disposal, Chapter 12
Human Rights, Chapter 32
Natural Justice, Chapter 48
Registration, Chapter 56
Undertakings and Warnings, Chapter 66
Voluntary Erasure, Chapter 68

# 21

# DOUBLE JEOPARDY



## A. Disciplinary Cases

### *R v. Police Complaints Board, ex parte Madden* [1983] 1 WLR 447

**21.01** Criminal proceedings—insufficient evidence to prosecute—whether double jeopardy—meaning

The applicant made a complaint about the conduct of police officers. The available evidence was referred to the Director of Public Prosecutions (DPP), who concluded that there was insufficient evidence to justify the institution of criminal proceedings against any police officer. The applicant brought proceedings for judicial review against the decision of the Police Complaints Board that when the DPP had determined not to bring criminal proceedings on the same or similar evidence, it would not be appropriate to bring a corresponding disciplinary charge. In quashing the Board's decision, McNeill J said that there was no question that the background to the legislation and the guidance operated by the Board is that a police officer should not be put in double jeopardy. Double jeopardy, properly understood, is best described by the phrase 'No man should be tried twice for the same offence': see *Connelly v. DPP* [1964] AC 1254, 1305, and *DPP v. Nasrulla* [1967] 2 AC 238, 249. McNeill J referred to *R v. Disciplinary Board of the Metropolitan Police, ex parte Borland*, 20 July 1982 (unreported). In that case, the applicant, a sergeant in the Metropolitan police force, sought judicial review of a decision of a disciplinary board finding him guilty of two disciplinary offences. It was a case in which the DPP had decided not to institute criminal proceedings. Ormrod LJ, giving the first judgment of the Divisional Court, with which McCullough J agreed, observed that it was not a 'double jeopardy case'. The Commissioner of Police exercised his discretion to proceed with disciplinary charges notwithstanding the decision of the DPP not to prosecute. Until the exercise of that discretion is attacked in appropriate proceedings (and it would have to act on the *Wednesbury* principles), the matter had no further relevance.

***R v. Council for Licensed Conveyancers, ex parte Watson,* CO/2022/98, 21 December 1999; The Times June 16, 2000**

**21.02** Civil proceedings—dismissal—statutory role of Council

W, a licensed conveyancer, brought proceedings for judicial review against the decision of the investigating committee of the Council for Licensed Conveyancers. W had acted for the purchasers in a conveyancing transaction. The purchasers subsequently issued proceedings in the county court alleging negligence in breach of contract. The claim was dismissed and there was no appeal. The purchasers then complained to the respondent Council, which directed W to pay compensation. W's main argument was that the investigating committee was not entitled, where a county court had dismissed the claim in negligence, to consider an allegation that a licensed conveyancer had not provided a service that it was reasonable to expect if the basis of that allegation was that the conveyance was negligent. Owen J, in dismissing the claim for judicial review, said that whilst it was conceded that there may be overlapping of issues, the Council's general answer was that a claim in negligence was not to be equated with a complaint that a service fell below the quality that it was reasonable to expect. The Council was established pursuant to the Administration of Justice Act 1985 and the Act must be construed in a purposive way: the objects of its provisions are to protect the public by ensuring that the services provided are such as a client is entitled to expect. Claims in court may fail for many reasons, including the passage of time, and whilst no doubt, as happened, the result of proceedings must be taken into account, there was no statutory objection to the investigating committee being satisfied that the conveyancer should have obtained a copy of the relevant conveyance in the instant case, as it was, in the register and, being so satisfied, determining that the conveyancer should pay compensation.

***Brabazon-Drenning v. United Kingdom Central Council for Nursing, Midwifery and Health Visiting* [2001] HRLR 6**

**21.03** See paragraph 2.07.

***R (Redgrave) v. Commissioner of Police of the Metropolis* [2003] 1 WLR 1136**

**21.04** Criminal proceedings—defendant discharged—identical evidence—whether double jeopardy

Simon Brown LJ asked, at [1]:

> Is a police disciplinary board entitled to proceed with disciplinary proceedings against a police officer notwithstanding that an examining magistrate has previously discharged that officer on the ground that there was no sufficient evidence to warrant putting him on trial for substantially the same offence?

The claimant, a police officer, claimed that the double jeopardy rule and/or the doctrine of res judicata precluded the bringing or continuation of disciplinary proceedings in such circumstances. On 14 December 1998, the claimant was charged with conspiracy to pervert the course of justice. At a committal hearing on 17 May 1999, the district judge at Bow Street Magistrates' Court declined to commit the claimant (and his co-accused) for trial and accordingly discharged them. On 8 June 1999, Eady J refused the Crown's application for a voluntary bill of indictment on the same charge. The claimant was subsequently charged with an offence of discreditable conduct pursuant to the Police (Discipline) Regulations 1985. On 8 March 2001, the Police Disciplinary Board decided, at a preliminary hearing, that it would not be prejudicial or unfair to conduct a substantive hearing into the charge on account of delay and that no issue of double jeopardy arose from the claimant's discharge at the committal hearing, notwithstanding that it was accepted by the Board that the evidence on which the presenting officer in the disciplinary proceedings relied was, in effect, identical to that relied upon by the Crown in the earlier committal proceedings. In dismissing the claimant's application for judicial review, the Court of Appeal said that the role of examining justices was to determine whether or not there was sufficient evidence to warrant a jury trial. In such circumstances, the double jeopardy rule has no application: see *R v. Manchester City Stipendiary Magistrate, ex parte Snelson* [1977] 1 WLR 911 and *Brooks v. Director of Public Prosecutions* [1994] 1 AC 568. The authorities in relation to disciplinary proceedings—*Ziderman v. General Dental Council* [1976] 1 WLR 330; *R v. Statutory Committee of Pharmaceutical Society of Great Britain, ex parte Pharmaceutical Society of Great Britain* [1981] 1 WLR 886; *Saeed v. Inner London Education Authority* [1985] ICR 637—established that, even assuming that there has been an acquittal by a criminal court, the double jeopardy rule has no application save to other courts of competent jurisdiction and there is therefore no bar to the bringing of disciplinary proceedings in respect of the same charge. The character and purpose of the proceedings is entirely different—the central point made by Lord Diplock in the *Ziderman* case. No less importantly, the material before the tribunal is likely to be different, in part because different rules of evidence are likely to apply and in part because judicial discretions may well be differently exercised—generally, less strictly in the disciplinary context, where at least the accused's liberty is not at stake. Simon Brown LJ put the matter in this way:

> 37. These authorities, to my mind, establish that, even assuming there has been an acquittal by a criminal court, the double jeopardy rule has no application save to other courts of competent jurisdiction, and there is therefore no bar to the bringing of disciplinary proceedings in respect of the same charge. And it is surely right that this should be so. Plainly it is so when the standard of proof is different: even the passage from *Friedland, Double Jeopardy* (1969), p319 quoted by Popplewell in *Saeed's* case [*Saeed v. Inner London Education Authority* [1985] ICR 637] recognises that. But, in my judgment, it is right also even where the standard of proof is the same, i.e., where the disciplinary charge too has to be proved beyond reasonable doubt...
>
> 38. There are two main reasons why the double jeopardy rule should not apply to tribunals even where they apply the criminal standard of proof. In the first place, it must be recognized that the character and purpose of the proceedings is entirely different – the central point made by Lord Diplock in the *Ziderman* case. Secondly, however, and no less importantly, the material before the tribunal is likely to be different, in part because different rules of evidence are likely to apply and in part because judicial discretions may well be differently exercised – generally, less strictly in the disciplinary context where at least the Accused's liberty is not at stake. It may also be that on occasions, as [counsel for the defendant] suggests, witnesses will be readier to give evidence at disciplinary proceedings held in private than in the full glare of open court proceedings.

### *Council for the Regulation of Health Care Professionals v. General Medical Council and Ruscillo* [2005] 1 WLR 717

**21.05** Decision by Council to refer case to High Court—protection of public

Dr R appeared before the Professional Conduct Committee (PCC) of the General Medical Council (GMC) charged with serious professional misconduct and was acquitted. The Council for the Regulation of Health Care Professionals referred the decision to the High Court under section 29 of the National Health Service Reform and Health Care Professions Act 2002. On the hearing of a preliminary issue, Leveson J held that the Council had power to refer an acquittal to the Court. The Court of Appeal said, at [42], that the intervention of the Council under section 29, whether to put in issue an acquittal or the adequacy of a sentence, clearly places a practitioner under the stress of having his case reopened when it would otherwise be closed. This element of double jeopardy is, however, necessarily inherent in the scheme of review under section 29. The object of that scheme is the protection of the public and the Council can refer a decision to the High Court only when it considers that this is necessary for the protection of the public. Where this requirement is satisfied, considerations of double jeopardy should take second place.

### *Thomas v. Council of the Law Society of Scotland* 2006 SLT 183

**21.06** Criminal proceeding—no case to answer—subsequent disciplinary proceedings

T was tried on an indictment at Glasgow Sheriff Court of effecting two transactions knowing, or having reasonable grounds to believe, that the properties concerned represented the proceeds of drug trafficking. T had acted for a client, B, who was concerned in the smuggling of cannabis and, on B's instructions, T had effected the gratuitous conveyance of B's interests in two properties to B's brother. T was acquitted of both charges, the sheriff having upheld a submission of no case to answer. The Law Society of Scotland thereafter brought a complaint against T before the Scottish Solicitors' Discipline Tribunal (SSDT). It alleged that T had been guilty of professional misconduct by acting dishonestly as an accomplice of B in carrying out the conveyances in order to avoid a confiscation order. The Law Society relied on the same factual allegations as those on which the Crown relied in the prosecution. T took a preliminary plea of res judicata, which the Tribunal dismissed. In dismissing T's appeal, the Lord Justice Clerk (Gill), giving the opinion of the Court, said that the decision of the sheriff cannot bar disciplinary proceedings before a domestic tribunal arising out of the same circumstances. Where the prior proceedings are a straightforward prosecution brought by the Lord Advocate, the outcome cannot found a plea of res judicata in civil proceedings relating to the same facts. The Law Society was not a party to those proceedings and, in bringing this complaint, it represented the public interest only in relation to the professional conduct of solicitors. Furthermore, the subject matter of the two sets of proceedings was not the same. In the criminal proceedings, the issue was whether the petitioner was innocent or guilty of the charges. In the present proceedings, the issue was whether he was innocent or guilty of professional misconduct. The disciplinary proceedings would be conducted and determined by different evidential and procedural rules. Finally, the proceedings brought before a criminal court are to obtain a conviction and sentence; the proceedings brought before a domestic tribunal are to obtain a finding of misconduct and the imposition of sanctions under the Solicitors (Scotland) Act 1980.

### *Swanney v. General Medical Council* 2008 SC 592

**21.07** See paragraph 43.10.

### *R (Sinha) v. General Medical Council* [2008] EWHC 1732 (Admin); [2009] EWCA Civ 80

**21.08** Criminal proceedings—*voir dire*—subsequent disciplinary proceedings

In this case, criminal proceedings of inappropriate sexual behaviour towards female patients had been dropped following an extensive *voir dire* during the trial. The *voir dire* focused on the conduct of the investigation by the police and the suggestion that the evidence of the complainants may have been contaminated or that they may have colluded, meaning that their evidence against Dr S was unreliable. In subsequent disciplinary proceedings based on the same allegations, the practitioner claimed that no subsequent investigation by the GMC could remedy the failures of, retrieve the material that had been lost by, or undo the damage caused by the flawed police investigation. The panel found the allegations proved. On appeal against the finding of misconduct and the sanction of erasure, the claimant claimed that the panel failed to take into account the collusion of witnesses and contamination of evidence resulting in the dismissal of the criminal proceedings. The Court (Irwin J), rejecting this claim, held that there is no strict rule of double jeopardy in relation to the dismissal of criminal proceedings in subsequent disciplinary proceedings. Dr S's renewed application to the Court of Appeal for permission to appeal was dismissed by Wall LJ on 18 February 2009. In a reserved judgment, Wall LJ said:

> 7. ... criminal proceedings are designed to establish guilt or innocence by a member of the public with a view to punishment by society if the verdict is guilty, and acquittal if the verdict is not guilty. Proceedings before a professional body are designed to establish whether or not professional men and women have fallen below the standards expected of their profession; whether or not the professionals concerned should remain members of the profession concerned, and if so, on what terms.
>
> 8. A moment's thought will suffice to demonstrate that the mere fact of an acquittal in criminal proceedings cannot be the be all and end all of the matter for other purposes. Supposing, for example, that a professional man is acquitted of murder or grievous bodily harm by a jury on the direction of the judge on a purely technical and unmeritorious point. He is not guilty in the eyes of the criminal law. But that would not stop—nor should it stop—his professional body re-investigating the matter and deciding both that he had been guilty of serious professional misconduct, and that he should be disciplined according to the rule of the profession concerned. A professional body is, after all, charged with the duty to protect the public from members of the profession which fall below its standards.
>
> 9. In the present application, as I have already indicated, a judge in the crown court has found that, because of the manner in which the investigation into the applicant's conduct was handled by the police, a fair trial in the criminal court was not possible. After an extensive *voir dire* in the absence of the jury, the prosecution offered no evidence and the judge directed the jury to bring in a verdict of not guilty. So the applicant asks: "if the Crown Court judge thought I could not have a fair trial, how can my professional body conduct such a trial and find me guilty?"
>
> 10. The answer, of course, as I have already stated, is that the functions of the Crown Court and the GMC are different. The hearing before the FPP was not a second criminal trial. It was an investigation by the FPP into the applicant's professional conduct. The fact that the applicant had been acquitted in the criminal proceedings was plainly a factor in the matters they had to consider. But it was not conclusive in the applicant's favour.

### *R (Bhatt) v. General Medical Council* [2011] EWHC 783 (Admin)

**21.09** Criminal proceedings—serious flaws—contaminated evidence—defendant acquitted—subsequent disciplinary proceedings

On 12 August 2010, a GMC fitness to practise panel found B guilty of charges against him relating to six female patients. In respect of four of the patients, the panel found that B had been sexually, rather than medically, motivated when he intimately examined them. Langstaff J observed that a central feature was that B was tried before the Crown Court on seven counts alleging unlawful sexual interference with the same six patients and was acquitted by verdict of the jury on all. On appeal from the findings by the fitness to practise panel, it was alleged on behalf of B that he should not have been exposed to double jeopardy before the GMC. It was also alleged that the criminal investigation was so flawed or contaminated that the evidence arising from it should not have been admitted before the panel. On the issue of abuse of process, reliance was placed by B on the judgment of Simon Brown LJ at [46] in *R (Redgrave) v. Commissioner of Police for the Metropolis* [2003] 1 WLR 1136. Simon Brown LJ commended to disciplinary boards generally two particular paragraphs in the 1999 Home Office *Guidance on Police Unsatisfactory Performance, Complaints and Misconduct Procedures*, which state that the degree of proof required increases with the gravity of what is alleged and its potential consequences, and that where criminal proceedings have been taken for an offence arising out of the matter under investigation and those proceedings have resulted in the acquittal of the officer, that determination will be relevant to a decision on whether to discipline the officer. Langstaff J, in dismissing B's appeal, said that a stay for abuse is an exceptional course. It should be granted either where a doctor cannot receive a fair hearing or where it would be unfair for the doctor to face a hearing. *Redgrave* was concerned with the latter. However, it is guidance and, even if it is still currently applicable, falls short of obliging a disciplinary panel to regard it as abusive for matters upon which a professional has been acquitted in the criminal court to be revisited in the course of professional regulation. All the more does it not constitute an abuse, given that the purpose of disciplinary proceedings (regulation to maintain proper standards in the profession in the best interests of the public and the profession) is different from the purpose served by the criminal courts. The learned judge went on to state that section 107 of the Criminal Justice Act 2003 (stopping the case where evidence is contaminated) is applicable only to criminal trials before a jury, and then only in respect of evidence of bad character, as defined in the 2003 Act. It is thus not obviously applicable to disciplinary proceedings.

***Christou v. Haringey London Borough Council* [2014] QB 131, CA**

**21.10** See paragraph 24.05.

***Ashraf v. General Dental Council* [2014] EWHC 2618 (Admin)**

**21.11** Criminal proceedings—defendant acquitted—whether disciplinary proceedings abuse of process

On 21 October 2013, a PCC of the General Dental Council (GDC) adjudged that Dr A, a dental practitioner, had made inappropriate and dishonest claims to the National Health Service (NHS) and, in addition, had interfered with witnesses who had been approached in the investigation. On the following day, it made consequential findings of misconduct and impairment. In consequence, the PCC determined that Dr A's name should be erased from the register. Central to the appeal was the fact that, prior to the PCC hearing, between 5 and 21 March 2012, Dr A had been prosecuted on an indictment containing three counts of fraud and theft in relation to the same NHS claims, and had been acquitted by the jury on all counts. That led to an application to stay the disciplinary proceedings as an abuse, which, on 9 September 2013, the PCC rejected. The PCC noted that the GDC charges were both wider and narrower than those in the criminal trial: wider, because the allegations were that the claims made to the NHS were inappropriate as well as dishonest; and narrower, in that they concerned fewer patients. The panel accepted the advice of its legal adviser that the doctrine of double jeopardy had no application to disciplinary proceedings and that to try a matter in disciplinary proceedings where there has previously been an acquittal was not inherently abusive. On appeal, pursuant to section 29 of the Dentists Act 1984, Dr A submitted that it would be unfair to pursue misconduct charges against him when he had been acquitted of substantially the same charges. It was submitted on his behalf that it is unfair to routinely subject professionals to both criminal and disciplinary proceedings in relation to the same subject matter, and that there is a discretion that falls to be exercised, which was wrongly exercised by PCC in this case. After reviewing *R (Redgrave) v. Commissioner of Police of the Metropolis* [2003] 1 WLR 1136, *Ziderman v. General Dental Council* [1976] 1 WLR 330, *Phillips v. General Medical Council* [2004] EWHC 1858 (Admin), *Sacha v. General Medical Council* [2009] EWHC 302 (Admin), and *Bhatt v. General Medical Council* [2011] EWHC 783, Sir Brian Leveson (president of the Queen's Bench Division) said, at [33], that it is essential that regulators are confident in exercising their discretion in these matters and that the continued anxious citation of this line of authority ought to be discouraged. The approach in [37] and [38] of *Redgrave*, confirmed in *Phillips*, *Sacha*, and *Bhatt*, is clearly correct. Simon

Brown LJ (at [46] of *Redgrave*), referring to the 1999 Home Office Guidance on *Police Unsatisfactory Performance, Complaints and Misconduct Procedures*, was not enunciating the law, but merely commending the approach set out in the Guidance (which has since been altered). The latest Home Office Guidance is dated November 2012. It is concerned with misconduct action following criminal proceedings at paragraphs 2.35–2.40, and rather than suggesting that it would usually be unfair to proceed following an acquittal, it states in paragraph 2.36 that an acquittal is simply 'a relevant factor which should be taken into account in deciding whether to continue with those proceedings'. In addition, professional conduct panels do not require the criminal standard of proof, but apply the civil standard. These two factors are crucial in demonstrating that [46] of *Redgrave* cannot be considered authority—if it ever was—for the proposition that it would necessarily be an abuse of process to bring disciplinary proceedings against a person on substantially the same subject matter as had been the subject of failed criminal proceedings. At [35], Sir Brian Leveson said that although it is not inherently unfair to bring misconduct charges against a professional who has already been acquitted in the criminal courts, this does not mean that there will not be circumstances in which it may well be unfair to proceed. Allegations of crime (which, if leading to a conviction, would justify erasure) may, in some circumstances, not justify further investigation by a regulator. Without seeking to be determinative, it might be that no further investigation by the regulator is justified, because the allegations do not, in any way, touch upon professional responsibilities either to patients or (as here) to the NHS (which is required to invest trust in the integrity of the professional to fulfil the terms of the funding contract honestly). This elaboration, however, is not intended to be definitive guidance: regulators must each determine how they go about achieving their regulatory objectives and, bearing those objectives in mind, faithfully apply the well-known principles engaged with the concept of abuse of process. Turning to the facts of this case, there was no error of law and the appeal was to be dismissed. Cranston J agreed.

## B. Other Relevant Chapters

**21.12** Abuse of Process, Chapter 2
Conviction and Caution Cases, Chapter 14
Estoppel, Chapter 24
Evidence, Chapter 25
Jurisdiction, Chapter 43

# 22

# DRAFTING OF CHARGES

## A. Legal Framework

**22.01** Examples include:

Nursing and Midwifery Council (Fitness to Practise) Rules 2004, rules 3, 9, and 11 (particulars of the allegation and the alleged facts in support of the allegation)
Solicitors (Disciplinary Proceedings) Rules 2007, rule 5(2) (application to be supported by a statement setting out the allegations, and the facts and matters supporting the application, and each allegation contained in it)
General Medical Council (Fitness to Practise) Rules 2014:
rule 7 (as soon as reasonably practicable after referral of an allegation, the registrar shall write to the practitioner informing him of the allegation and stating the matters that appear to raise a question as to whether his fitness to practise is impaired)
rule 15 (the notice of hearing shall particularize the allegation against the practitioner and the facts upon which it is based)
Institute of Chartered Accountants in England and Wales Disciplinary Committee Regulations, rule 3 (notice of the terms of the formal complaint within forty-two days after referral to the Disciplinary Committee)

As to general principles when drafting charges brought by the General Medical Council, see:

General Medical Council, *Guidance: Drafting Charges* (2014)

As to amendment of allegations see, for example:

Nursing and Midwifery Council (Fitness to Practise) Rules 2004, rule 28
Solicitors Disciplinary Proceedings Rules 2009, rule 7(1) (supplementary statements)
General Optical Council (Fitness to Practise) Rules 2013, rule 46
General Medical Council (Fitness to Practise) Rules 2014, rule 17(3)
Chartered Institute of Management Accountants Disciplinary Committee Rules, rule 16

## B. Disciplinary Cases

### *Lau Liat Meng v. Disciplinary Committee* [1968] AC 391

**22.02** Finding of Committee—no charge—no opportunity to challenge—natural justice

The appellant was a member of a firm of advocates and solicitors in Singapore. He was charged with a disciplinary offence of accepting $700 by entering into a champertous agreement to take 25 per cent of damages arising from a road traffic accident. The charge was found proved. Additionally, the appellant had received $500. No charge was proffered in respect of that sum. No application was made for amendment of the charges despite the appellant making an admission. The Disciplinary Committee held that the appellant had received the additional sum of $500. On appeal, the Privy Council said that, while acknowledging the gravity of the admission made by the appellant as to the $500 that he put into his own pocket without disclosure to his client, he was not charged either with having made excessive charges for professional work or having committed any specific fraudulent act. Formal amendment of the charges might have been dispensed with provided that adequate notice of the charge was given, but natural justice requires adequate notice of charges and also the provision of an opportunity to meet them. This requirement was not met. It would be unjust to allow the finding with regard to the $500 to stand. If disciplinary proceedings were thereafter at any time to be taken against the appellant in respect of this sum, no conviction or acquittal would stand in their way, for no charge relating to this matter had ever been made. The case was remitted to the High Court of Singapore to reconsider the sentence passed upon the appellant.

### *Sloan v. General Medical Council* [1970] 1 WLR 1130

**22.03** Alternative charges—charge ambiguous

The appellant was charged with prescribing tablets and administering injections, and with representation that the purpose was to procure a miscarriage. The charges were so framed that whether it turned out in evidence that the representations were true or false, the appellant would equally be guilty of professional misconduct if he had supplied the tablets and administered the injections. There was evidence that the pills and injections were not intended to procure a miscarriage. The Privy Council stated that if it were desired to prefer alternative charges, then they should be preferred in the alternative in the recognized form, leaving the Disciplinary Committee of the General Medical Council (GMC) to decide on the evidence which alternative had been established. In their Lordships' views, it is embarrassing to the doctor to prefer a charge that, on the face of it, is ambiguous and presents two alternatives for the Committee's consideration. It was a 'trap charge', so that whichever explanation was given by the doctor, he could not fail to be convicted. The charge ought to have been one of making false representations that the pills and injections were given with the intention of procuring a miscarriage. However, the appellant was not in any way prejudiced by the form of the charges.

### *Peatfield v. General Medical Council* [1986] 1 WLR 243

**22.04** Treatment of 'individual patients'—whether duplicitous—course of conduct

The appellant, a medical practitioner, was charged with supplying 'individual patients' with drugs between June 1980 and November 1984 without first adequately examining each patient, or consulting the patients' general practitioners (GPs), or making adequate enquiries as to the patients' health. Their Lordships took the view that, in the present case, the charge can fairly be read as alleging a course of conduct by the doctor in the conduct of his practice over the period stated in which patients were treated and that it was this course of conduct that amounted to serious professional misconduct. The use of the words 'individual patients' was inappropriate, but it was apparent from the record of the hearing that both parties before the GMC's Professional Conduct Committee (PCC) treated the charge as a complaint with regard to a course of conduct. The point taken that the charge was bad for duplicity was not a good reason for allowing the appeal.

### *Gee v. General Medical Council* [1987] 1 WLR 564, HL

**22.05** Duplicity—course of conduct—whether capable of amounting to misconduct

G was charged with abusing his professional position as a medical practitioner by supplying to individual patients quantities of drugs over extensive periods between May 1982 and August 1983 without first adequately examining the patients, without first consulting the patients' GPs, without making adequate enquiries about the patients' health, and without offering appropriate advice. G submitted that the charge was duplicitous in that it alleged distinct instances of professional misconduct in respect of individual patients and that to proceed with a charge in this form was not in accordance with the principles of fairness. The GMC submitted that the charge was not bad or duplicitous, since it alleged a course of conduct capable of constituting serious professional misconduct. The House of Lords allowed the charge to proceed.

Lord Mackay of Clashfern, giving the lead speech, said at 575 that the rule against duplicity would be impossible to apply universally to cases relating to conduct before the PCC. The rule is necessary in the interests of fairness where the only answer that can be returned in respect of a particular charge is guilty or not guilty to the whole charge. It is not necessary in order to obtain fairness in a procedure that the PCC be required to make a determination that distinguishes between the facts alleged that are found proved and those that are not before moving forward to considering a determination as to guilt of serious professional misconduct based on its determination of facts found proved. In any event, the charge in the present case was one that, properly construed, narrated a course of conduct in relation to the practice carried on by G and it was in relation to this course of conduct that it was alleged that he was guilty of serious professional misconduct. It was open to the PCC to hold that the facts of the alleged course of conduct had not been proved at all, or had been proved only to some extent more limited than indicated in the charge, or that the facts had been wholly established as charged. Lord Mackay went on to say that, in his opinion, there is no unfairness in a procedure in which a number of allegations of fact are set out in one charge and it is alleged against a medical practitioner that these matters of fact, if established, render him guilty of serious professional misconduct, provided that he has fair notice in time to prepare his defence of the nature of the evidence to be led in support of these allegations. Also, there is no unfairness provided that the PCC, charged to adjudicate upon the matter, makes plain which of the allegations of fact, if any, it has found proved in time for the practitioner to make appropriate submissions and to lead any further relevant evidence available to him before a determination is made whether he is guilty of serious professional misconduct. The procedures set out in the rules meet these provisos, and if they were properly followed, no unfairness would result to G.

***Reza v. General Medical Council* [1991] 2 AC 182**

**22.06** Several facts in one charge—course of conduct

The appellant was charged with making improper and indecent remarks to four named employees at his surgery premises, and making improper and indecent remarks to four named patients. In dismissing the appeal, their Lordships said, at 197C–198F, that the inference from the rules is that several facts may be presented as constituting one charge, such that if some facts should fall by the wayside, other facts may remain in relation to which it is open to a committee to find the practitioner guilty of serious professional misconduct. The rules appear to contemplate an inquiry by one committee into every matter put before it, whether relating to convictions or conduct or both. The whole picture is of a committee that is to be informed of all of the facts alleged and all of the background that could help it to determine in the interests of the public and the profession what, if anything, is to be done by way of erasure or suspension or the imposition of conditions. The procedure is clearly and strictly laid down, so that it would be impossible, and not merely undesirable, to wait until two or more separate committees had made interim findings and then have a final adjudication. In *Gee v. General Medical Council* [1987] 1 WLR 564, HL, it was held that in a case relating to conduct, where two distinct types of misconduct are alleged and where the determination that one type of misconduct was established could not reasonably aggravate the seriousness of the other misconduct, it is preferable and in the interests of clarity for two separate charges to be alleged. As in *Gee's case*, their Lordships were, in this case, of the opinion that the charge was one that, properly construed, narrated a course of conduct in relation to the practice carried on by the appellant.

*Nwabueze v. General Medical Council* [2000] 1 WLR 1760

**22.07** Irrelevant charge—delete from notice of inquiry

N faced allegations of sexual misconduct with patients and two allegations of dishonesty. During the course of evidence, it emerged that one of the named persons had ceased to be a patient prior to any relationship with N. The professional relationship had ceased more than one year before the alleged relationship took place. In quashing the finding in respect of this head of charge, their Lordships said that a charge, or part of a charge, which contains an allegation that had no bearing on the practitioner's conduct as a medical practitioner is irrelevant to a charge that he is guilty of serious professional misconduct. As such, it is objectionable on grounds of law and it should be deleted from the notice of inquiry.

*Subramanian v. General Medical Council* [2002] UKPC 64

**22.08** Different types of misconduct—whether need for separate charges

The charge against S concerned his misdiagnosis of the patient and his failure to cooperate with the subsequent inquiry. On appeal, S objected to the form of the charge relying on *Gee v. General Medical Council* [1987] 1 WLR 564, 575–6, where Lord Mackay of Clashfern said that, in a case relating to conduct in which two distinct types of misconduct are alleged and in which the determination that one type of misconduct was established could not reasonably aggravate the seriousness of the other misconduct, it would be better and in the interests of clarity for two separate charges to be alleged. In dismissing S's appeal, their Lordships said that those reasons did not apply in the present case, in which the second limb could, and did, aggravate the seriousness of the appellant's misconduct and in which there was no need in the interests of clarity for there to be two separate charges. Further, no application was made before or at the hearing for the two limbs of the case to be heard separately or to be treated independently as separate limbs. It has never been suggested that the GMC cannot charge an entire course of conduct: S retained responsibility to the patient and her mother until the complaints process was over. That was one course of conduct and not two.

*Gangar v. General Medical Council* [2003] UKPC 28

**22.09** Different patients and events—single charge—no unfairness

G submitted that it was wrong and unfair for the GMC to have proceeded against him in respect of a single charge of misconduct when the substance of the allegations making up that charge related to different patients and/or different matters, at different times. There could be no complaint of unfairness when a number of allegations of fact are set out in one charge provided that there was a connection or nexus between the allegations. The Board was unable to accept this submission, and accepted the submission of the GMC that the allegations comprised a series of examples of high-handed, inappropriate, or dishonest behaviour by G towards patients and professional colleagues, and to his own professional body. Thus there was nothing wrong with the PCC proceeding on the basis of an omnibus charge containing various heads.

*Misra v. General Medical Council* [2003] UKPC 7

**22.10** Treatment of patient—allegations of dishonesty—unnecessary and oppressive—allegations of alcoholism—prejudice

M was an overworked sole practitioner in Corby, a socially deprived area with a shortage of GPs. The charges related to his treatment of a 65-year-old patient of his, Mrs B. The substantive allegations against M were that he had been informed on four occasions of telephone calls and requests for home visits. M had admitted being informed of only two of them. The factual issues of the case were, in reality, simply whether M had or had not been informed of the other two telephone calls. Included in the charge were allegations that M gave information to Mrs B's son that he knew to be untrue. Their Lordships, in allowing M's appeal, said that the addition of the allegations of dishonesty in the present case were unnecessary and oppressive. They added nothing to what would have been shown by the degree of culpability of M if the substantive allegations that he had declined to admit were found proved against him. Further, a matter of background, not referred to in the charge, but which loomed large at the hearing before the GMC's Preliminary Proceedings Committee (PPC), related to M's history of alcoholism. It was not relevant to the charge that M was facing. He was not charged with having failed to visit Mrs B because he was under the influence of alcohol, nor was he charged with being regularly under the influence of alcohol while on duty as a doctor. His problems with alcohol, capable of being highly prejudicial to the view of him taken by the Committee, ought to have been kept in the background and disclosed to the Committee only in the event that it found the charge of serious professional conduct to be proved.

***Salha v. General Medical Council; Abusheikha v. General Medical Council* [2003] UKPC 80**

**22.11** Dishonesty—need to allege and particularize

The GMC formulated charges against A and S that they submitted a review paper and case report for publication in the journal Human Reproduction Update (2000), which included seven plagiarized paragraphs from a previous paper written by others and published in 1994. The allegation against A and S was that they each failed to adequately review the 2000 paper before its publication, and that their actions were unethical and unprofessional. Each claimed that he had nothing to do with the composition of the paper and, in the light of their explanations, the charge against A and S was a failure 'adequately to review the paper before publication' and not conscious copying. In its determination of serious professional misconduct against both, the GMC's PCC said that to accept the benefits of authorship while evading the responsibilities for any deficiencies in the paper was 'unacceptable and dishonest'. The Committee went on to say that the conduct of A and S reflected 'poorly on the integrity of the medical profession' and that to impose conditions would not address 'the fundamental lack of honesty and integrity displayed in this case'. The Committee accordingly decided to suspend the doctors from the register for three months. In allowing the doctors' appeals and substituting a reprimand, Lord Hoffmann, giving the judgment of the Privy Council, said that their Lordships did not consider there to be any justification for the findings of dishonesty or lack of integrity against the doctors. Dishonesty was not alleged against either of them and they gave no evidence because they admitted the whole of the facts alleged by the GMC. Those facts might be regarded as indicating a high degree of negligence, but could not, without more, amount to dishonesty. Lord Hoffmann said, at [14]: 'It is a fundamental principle of fairness that a charge of dishonesty should be unambiguously formulated and adequately particularized.'

***Singleton v. Law Society* [2005] EWHC 2915 (Admin)**

**22.12** Allegation of dishonesty—obligation to give timely notice, with particulars

S was charged with conduct unbefitting a solicitor. The allegation was particularized under nine paragraphs. At the outset of the hearing before the Solicitors Disciplinary Tribunal (SDT), the allegations were admitted. It became apparent that there was an issue as to whether S had acted dishonestly in relation to one of the allegations or any of them. He maintained that he had not. In the event, the SDT found dishonesty in relation to two allegations. In allowing S's appeal, the Divisional Court (Maurice Kay LJ and Penry-Davey J) said that it was axiomatic that an allegation of dishonesty made against a solicitor is particularly serious. If proved, the consequences are inevitably severe. The essential complaint was that the appellant was not provided with adequate notice or particulars of the allegation of dishonesty. At no point in the rule 4 statement or in any other document was there an unequivocal reference to dishonesty. At the end of the opening, the chairman appeared to raise the question as to whether it was a dishonesty case, to which the solicitor for the Law Society replied that it was. The appellant, who was representing himself, then proceeded to address the Tribunal. In the Court's view, the failure expressly to allege or to particularize dishonesty in a document in advance of the hearing constituted a procedural flaw. It may be that, with proper notice of such an allegation, the appellant would have chosen to be professionally represented, as he was on a previous occasion. However, that is not the point. In bringing disciplinary charges against a solicitor, the Law Society is under an obligation to give timely notice of an allegation of dishonesty with relevant particulars when appropriate, unless it is obvious from the nature of the charge (for example where the allegation is that the solicitor has committed or been convicted of a criminal offence that necessarily involves dishonesty).

### *Simms v. Law Society* [2005] EWHC 408 (Admin)

**22.13** Rule 5 statement—need to identify case against solicitor—ability to defend

In dismissing S's appeal, the Divisional Court (Latham LJ, and Curtis and Newman JJ) said that the purpose of a rule 4(2) statement under the Solicitors (Disciplinary Proceedings) Rules 1994 (now rule 5 of the Solicitors (Disciplinary Proceedings) Rules 2007) is clear: it is to identify for the solicitor the nature of the charges that he has to face and to provide enough information about the nature of the case that the Law Society seeks to make against him for him to be able to defend himself. It was not particularly helpful to describe it as a 'pleading', or an 'indictment' or 'charge sheet', save in so far as it shared the same characteristic—that is, to enable the person who is charged, or sued, to know what the case against him or her is. In the present case, the rule 4(2) statement properly identified the nature of the allegations that were being made. The first paragraph made clear the general nature of each of the allegations and identified what matters were to be relied upon in support of each of those allegations. The remainder of the statement included within it reference to the material and the evidence upon which the Law Society intended to rely. It did so in a way that could have left the solicitor in no doubt as to the case that he had to meet.

### *R (Bevan) v. General Medical Council* [2005] EWHC 174 (Admin)

**22.14** Heads of charge—conduct

The misconduct in question against Dr B was constituted by an adulterous sexual relationship with a patient, who was referred to as 'Mrs A', together with improprieties in the provision of prescription medicines to her. Paragraph 10 of the charge alleged that Dr B's relationship with Mrs A was 'inappropriate', 'unprofessional', 'an abuse of his position of trust', 'not in the best interests of the patient', and 'liable to bring the profession into disrepute'. Paragraph 14, the allegation of the provision of medicines, was said to be 'irresponsible', 'inappropriate', and 'not in the best interests of Mrs A'. Collins J said that headings 10 and 14 were perhaps more matters of judgment than issues of fact. The purpose of the charge was to set out the matters that the GMC proposed to establish and, from that, to seek to persuade the PCC that serious professional misconduct had been made out. No doubt it is helpful to the practitioner to know what conclusions are sought to be drawn, but to require admissions to that extent does not seem to be necessarily appropriate. However, it is always open to a practitioner to admit the facts, or as many of them as are appropriate, and to accept that he has been guilty of serious professional misconduct.

### *R (Council for the Regulation of Health Care Professionals) v. General Medical Council and Rajeshwar* [2005] EWHC 2973 (Admin)

**22.15** Undercharging—no allegation of conduct being sexually motivated/indecent

This was an appeal under section 29 of the National Health Service Reform and Health Care Professions Act 2002 against a decision of the GMC's fitness to practise panel in respect of a charge brought against the second respondent, R. The charge against R related to his examination of two patients, Miss S and Miss B. The charge alleged that he examined their breasts and that, inter alia, his examinations were not medically justified, were inappropriately carried out, and were inadequately recorded. The charge did not allege that R's conduct was sexually motivated and/or indecent. On behalf of the Council for the Regulation of Health Care Professionals, it was alleged that the GMC 'undercharged' the case against R. It should have been alleged that R's conduct was sexually motivated and/or indecent, and not simply incompetent and/or inappropriate. Newman J held that there had been a serious procedural irregularity, and that there was no doubt that the GMC should have amended the charge so as to allege indecency and/or a sexual motivation. As a result, that issue was not considered by the panel, and the matter was to be remitted to amend the charge and to be reheard by a differently constituted panel.

### *Mond v. Association of Chartered Certified Accountants* [2005] EWHC 1414 (Admin)

**22.16** Bye-laws—overlap between charges—drafting

M was appointed supervisor of a company voluntary arrangement (CVA). It was alleged that M was liable to disciplinary action 'under bye-law 8(a)(i) and/or (ii)' in that he failed to act with due skill, care, diligence, and expedition, and with regard for the technical and professional standards expected in relation to the CVA monies. Bye-law 8(a)(i) provided that a member shall be liable to disciplinary action if he has been guilty of misconduct. Bye-law 8(a)(ii) related to professional work that was improper, inefficient, or incompetent. Collins J, in allowing M's appeal, said that the charge was bad. While there may be a considerable overlap between bye-law 8(a)(i) and (ii), there can be no doubt that (i) is regarded as more serious than (ii). While misconduct is not to be equated with dishonesty, it is more serious than negligence or incompetence. It carries a stigma and that, coupled with the penalty of a severe reprimand and a substantial fine, would convey to the profession that the claimant's conduct was regarded as having a high degree of culpability. This does not mean that a contravention of bye-law 8(a)(ii) may not, on its facts, be serious enough to justify a severe penalty even though there was no deliberate or reckless conduct. The charges were also badly drafted: they should have reflected the language of the relevant paragraphs of bye-law 8. Thus, if a contravention of bye-law 8(a)(i) were being charged, it should have been specifically alleged that what the claimant had done amounted to misconduct. As it was, he and his advisers were misled by the absence of any specific allegation of misconduct or impropriety. It may be said that the fact that bye-law 8(a)(i) and (ii) were referred to in the preamble to the charges should have made it clear to them that those allegations were included. However, it is apparent that they were misled, and they approached the hearing on the basis that any finding of misconduct must be based on more than negligence and incompetence; they no doubt felt that they had a powerful case that a breach of bye-law 8(a)(i) could not be established.

### *Shrimpton v. General Council of the Bar* [2005] EWHC 844

**22.17** Discreditable conduct

S, a barrister, faced charges that he had engaged in discreditable conduct in what he had said or claimed to others about the complainant. The charges alleged only what S had said or claimed. In allowing S's appeal, the Visitors said that, in their judgment, the charges themselves should generally outline what it is that makes discreditable that which is not necessarily so. Should the charges themselves not make that clear, then it is incumbent on the prosecutor to make clear in good time before the hearing what, at least in outline, the nature of the prosecutor's case is, so as clearly and succinctly to show why it is that that which need not have been discreditable was discreditable. There was force in the analogy with cases of civil contempt.

***Connolly v. Law Society* [2007] EWHC 1175 (Admin)**

**22.18** Use of 'and/or' in allegations

The Divisional Court (Laws LJ and Stanley Burnton J) observed that the Law Society should avoid, where possible, formulating charges that include 'and/or' allegations, such as in the present case, which comprised numerous alternatives. Where such a charge is laid, the tribunal should make specific findings as to which allegation has been proved.

***R (Council for the Regulation of Health Care Professionals) v. Nursing and Midwifery Council and Kingdom* [2007] EWHC 1806 (Admin)**

**22.19** Dishonesty—not alleged—serious procedural error

The charges brought against K concerned false documentation regarding her qualifications. At the end of the hearing, the legal adviser to the Conduct and Competence Committee (CCC) of the Nursing and Midwifery Council (NMC) raised the fact that, whereas much of the evidence went to the issue of K's dishonesty and whether she was dishonest in what she had done with the documents, the charge did not specify dishonesty. The charge alleged misconduct by K in producing a false document. The Committee decided that it was not appropriate for it to consider whether K had been dishonest because dishonesty had not been expressly alleged. Beatson J, in allowing an appeal by the Council for the Regulation of Health Care Professionals, concluded that the failure to charge K properly was a serious procedural error. As to the consequences, if, applying the test in *R (Council for the Regulation of Health Care Professionals) v. General Medical Council and Rajeshwar* [2005] EWHC 2973 (Admin), the issue of dishonesty had been on the charge sheet, had been considered by the Committee, and had been resolved against K, the finding in this case that there had been no misconduct would undoubtedly have been unduly lenient. The learned judge said that he rejected the submission that the Committee could not reconsider or amend the charge once the facts were proved. One of the purposes of this jurisdiction is to deal with a perception that professional regulatory bodies may sometimes undercharge or impose lenient penalties. The matter would be remitted to a differently constituted Committee, with a direction that the NMC amend the charges to include an allegation of dishonesty so that the matter could be considered afresh.

***Sheill v. General Medical Council* [2008] EWHC 2967 (Admin)**

**22.20** Dishonesty—particulars—source of allegation

Foskett J was critical of the phraseology of a head of charge alleging dishonesty. He said that the head of charge was non-specific about the circumstances in which it was alleged that the medical practitioner had made a false claim. When a false or dishonest claim was made, for example in a document, it would be usual for the document to be identified in the charge—perhaps by date, but certainly by description, which showed clearly the source of the allegation. That was not done in this case and no request for particulars of the charge appears to have been made, either in writing before the hearing or in some application to the panel at the outset of the hearing. His Lordship quashed the panel's findings on the dishonesty charge.

### *R (Johnson and Maggs) v. Professional Conduct Committee of the Nursing and Midwifery Council* [2008] EWHC 885 (Admin)

**22.21** Particulars—clarity to identify case to meet and prepare defence

The grounds of challenge in this case were that the charges were insufficiently particularized, which was unfair and contrary to Article 6 of the European Convention on Human Rights (ECHR). Specifically, the claimants relied on Article 6(3)(b), which provides that those charged with a criminal offence have the right to adequate time and facilities for the preparation of their defence. The Court held that the charges must be sufficiently particularized to enable those charged to know, with reasonable clarity, the case that they have to meet and to prepare a defence. The PCC had not erred in this case. Whilst a challenge based on lack of particularity was not necessarily premature when made at the opening of a hearing, a stay would be an exceptional remedy, to be used only when it was clear that a fair trial was not possible.

### *R (Wheeler) v. Assistant Commissioner House of the Metropolitan Police* [2008] EWHC 439 (Admin)

**22.22** Vagueness of charges—need to particularize before hearing

In this case, the applicant applied for judicial review of the decision of Assistant Commissioner House to uphold the earlier decision of a disciplinary panel that found the applicant to be in breach of the code of conduct applicable to police officers. There were two breaches of the code of conduct alleged against the applicant: one, that he failed to ensure that another officer carried out 'to an acceptable standard his duty with respect to the management and supervision of investigations into allegations of child abuse'; the second, that the applicant 'failed to ensure that investigations were carried out by the team to an acceptable standard'. In quashing the decision of the Assistant Commissioner, Stanley Burnton J criticized the vagueness of the charges, and stated that the hearings before the panel and the Assistant Commissioner would have been better focused had both charges not been in the vague terms that they were. His Lordship stated, at [6]:

> Vagueness is a ground for judicial review if it leads to unfairness in the proceedings, and the danger with a vague charge is that the parties, and in particular the respondent, do not know with some precision what is alleged against them, and therefore are not fully able to address those matters in the course of a hearing.

His Lordship stated that it is sufficient if a charge is particularized subsequent to it being first formulated, but certainly it should be sufficiently particularized well before the hearing so that the respondent to disciplinary charges knows not only what it is alleged he failed to do, but also in what respects he failed, so that he can see whether or not, consistent with his other duties, he could or should have done that which it is alleged he should have done.

### *R (Roomi) v. General Medical Council* [2009] EWHC 2188 (Admin)

**22.23** Findings by Committee—allegations not contained in notice

In this case, Collins J allowed an appeal by the practitioner against the finding of impairment by reason of deficient professional performance, on the grounds that the finding by the panel went beyond the allegations contained in the notice of hearing. The appellant was able to call evidence to show that he had taken steps to improve his skills and that evidence led the panel to decide that the deficiencies identified in the notice of allegation had been remedied. But the panel justified its finding of continuing deficiency on the basis of a failure by the practitioner to carry out regular or systematic medical and clinical audits. The judge criticized the legal assessor and said that the panel ought to have been advised that it could not properly rely on this matter unless it formed part of the allegations made against the practitioner. The whole hearing was on the basis that what was in issue was the practitioner's skill, nothing else.

### *Richards v. Law Society (Solicitors Regulation Authority)* [2009] EWHC 2087 (Admin)

**22.24** Rule 5 statement—need to identify case against solicitor—ability to defend

The Divisional Court (Sir Anthony May P and Saunders J) held that the purpose of a rule 4(2) statement under the Solicitors (Disciplinary Proceedings) Rules 1994 (now rule 5(2) under the Solicitors (Disciplinary Proceedings) Rules 2007) is to inform the solicitor fairly and in advance of the case that he has to meet. The rule 4(2) statement in this case did not put the case against R on the basis found by the SDT—namely, that the solicitor should have seen to it that the client had financial advice. However, the case was made clear in correspondence well before the hearing, and, accordingly, the solicitor was given fair and due notice of the financial advice point in advance of the hearing.

### *Chauhan v. General Medical Council* [2010] EWHC 2093 (Admin)

**22.25** Matters not formulated by charges—findings of Committee—findings used to support allegations in charges

The Administrative Court (King J) allowed the appellant consultant's appeal against findings of dishonesty and impairment on the basis of the fitness to practise panel's failure to confine itself to the proper ambit of the disciplinary charges and for importing into its conclusions prejudicial factual matters that had not been stated in the notice of hearing. The appellant applied for the post of consultant in trauma and orthopaedic surgery at a National Health Service (NHS) trust. He faced charges of dishonesty in relation to his experience in revision surgery and hip resurfacings, and his experience to undertake a technique known as 'Birmingham hip resurfacing'. King J said that, in so far as the panel, at stage one of its decision process, made material findings of fact adverse to the practitioner that could themselves have been the subject of a charge of professional misconduct, which, however, was not within the charges as formulated, then those findings could not properly or fairly be used by the panel to support its findings; in so far as the panel so used them, then the charges as formulated and found were liable to be vitiated and set aside. In *Cohen v. General Medical Council* [2008] EWHC 581 (Admin), Silber J, at [48], said that findings in relation to any particular charge at stage one 'must be focused solely on the heads of the charges themselves'. King J said that there were examples of the panel unfairly introducing into their considerations in determining whether dishonesty had been established evidence directed at behaviour that was not the subject matter of any charge. The learned judge rejected a submission by the respondent that it was entitled to introduce such evidence even if strictly outside the ambit of the charges as propensity evidence—that is, evidence of the appellant's propensity to exaggerate dishonestly the true extent of his medical experience—or that it should have been made clear to the appellant during the course of the hearing that the respondent was inviting the panel to make findings wider than those strictly related to his experience.

### *Thaker v. Solicitors Regulation Authority* [2011] EWHC 660 (Admin)

**22.26** Rule 4 statement—case presentation—allegations beyond identified transactions—fair process

T appealed against the decision of the SDT that he be struck off the roll of solicitors. The issue was the way in which the charges had been drafted and the case opened by the Solicitors Regulation Authority (SRA). The Administrative Court (Jackson LJ and Sweeney J) allowed T's appeal and ordered a rehearing before a new panel. The first ground was that the Tribunal erred in failing to grant an adjournment at the start of the hearing. The Court readily acknowledged that the question whether or not to adjourn was a matter of the discretion of the Tribunal, but said that, as the opening proceeded, it became clear that the case presented by the SRA went far beyond the allegations in the rule 4 statement containing details of the allegations and that, at the very least, the Tribunal should have granted a period of adjournment for T to consider matters. Further, by allowing the proceedings to range far and wide, and allowing submissions and evidence beyond the identified relevant transactions, the Tribunal unwittingly caused injustice to T. Additionally, there were erroneous findings of fact made by the SDT. Jackson LJ said that, drawing the threads together, if a solicitor is going to be struck off the roll for acts of dishonesty and gross recklessness, he is entitled to a fair process and a fair hearing before that decision is reached. In this case, T did not receive either a fair process or a fair hearing. This occurred because of the manner in which the case against him was pleaded and presented to the Tribunal. If the rule 4 statement had alleged that T knew or ought to have known certain matters, the facts giving rise to that actual or constructive knowledge should have been set out, and in a complex case the Tribunal needs to have a coherent and intelligible rule 4 statement, in order to do justice between the parties.

### *Saverymuttu v. General Medical Council* [2011] EWHC 1139 (Admin)

**22.27** Private health insurance payments—dishonesty—letter—opportunity to recall witnesses—adjournment—fairness

S, the appellant, a consultant physician with an interest in gastroenterology, undertook private work in respect of which he commonly sought payment from the patient's private health insurer. The insurer used a code to identify diagnostic or therapeutic oesophago-gastro duodenoscopies (OGDs). The insurer paid more for therapeutic treatment, which was a different code number. Before the fitness to practise panel, the GMC said that the appellant had misled the insurer and had acted dishonestly. The panel found that the appellant did know that he was misleading the insurer after he received a letter from AXA PPP dated 17 March 2005. The letter was not pleaded as part of the allegation against the practitioner and, on day 19 of the hearing before the panel and in the course of closing submissions, an issue arose as to whether the ambit of the allegations against the doctor was wide enough for the panel to take account of the letter. The panel allowed the appellant to recall any witnesses that he wished. Counsel representing the appellant suggested a two-week adjournment, to which the panel agreed. In the event, counsel recalled only one witness (the appellant's expert witness) and, despite being invited to do so, did not recall the appellant. The further evidence took just half a day. Nichol J, after being referred to *Salha v. General Medical Council* [2003] UKPC 80 and *Singleton v. Law Society* [2005] EWHC 2915, agreed that there was some force in counsel's submission that reliance on the letter of 17 March 2005 was a new variant of the case of dishonesty against the appellant. However, counsel's argument that no application was made to amend, whilst true, was a formulistic response. The appellant was in no doubt as to what was proposed. Whether the proposal was put as an amendment or dealt with in the way in which it was, the issue was the same—namely, whether this could be done consistently with the panel's obligation to act fairly. The panel's view was that any potential unfairness to the appellant could be dealt with by an adjournment, with the doctor having the opportunity to recall witnesses and to make further submissions. Nichol J said that, in his view, the panel was not only entitled to take that decision, but was also right that these measures were sufficient to cure any potential unfairness to the appellant. It allowed counsel to recall any witnesses and to ask the further questions that he considered relevant in view of the reliance being placed on the letter of 17 March 2005.

*Sharp v. Nursing and Midwifery Council* [2011] EWHC 2174 (Admin)

**22.28** Dishonesty—allegations not part of charges—whether Committee influenced by material

Holman J allowed the claimant's appeal against the determination of the NMC's CCC on grounds that the Committee had taken into account allegations of dishonesty amongst the papers that did not form part of the charge and had not been raised during the two-day hearing. The appellant was a registered nurse, who claimed to be sick and unable to work for periods during which it was subsequently discovered he had worked shifts as a bank nurse, for which he was paid. The appellant was charged with two substantive charges. The first was to the effect that he had acted dishonestly by working shifts as a bank nurse, while claiming to be off sick on three nights in question in December 2006. The second charge was to the effect that, in various stated ways, he had failed to adhere to his employer's sickness absence policy during the period November 2006 to March 2007. In its decision on fitness to practise, the Committee stated that it noted that the registrant had worked four additional shifts as a bank nurse in January 2007 and that this contradicted his assertion that he had worked only the three shifts in December 2006. Holman J, in allowing the appeal, said that he was left with the clear impression that the Committee was, or may have been, considerably influenced by taking into account the additional known occasions on which the registrant had repeated his dishonest behaviour. This was a serious error by the Committee in a case that had not been raised with the case presenter or the legal assessor.

*Marcus v. Nursing and Midwifery Council* [2011] EWHC 3505 (Admin)

**22.29** Allegation by complainant—different from charges—issues refined—no new allegation

M challenged the decision of the NMC's CCC to find that her fitness to practise was impaired by reason of misconduct. She said that the allegations of fact, as refined and set out in the decision of the Committee, were not the same as those originally put forward by the University College Hospital NHS Trust when the matter was first investigated against her. At that stage, the allegations were that she had been grossly negligent whilst caring for a ventilated baby in a neonatal unit. It was also alleged that she was uncooperative in meetings when asked about an incident that occurred whilst caring for the baby. The appellant claimed that the allegations had somehow been 'titrated' into a narrow set of allegations. Judge Mackie QC (sitting as a deputy High Court judge) said that this may be right, but it seemed to him that there was nothing untoward about it. It was right for the Committee to narrow what were broad and serious allegations. The appellant knew precisely what was going to be alleged against her well ahead of the hearing. It would have been of assistance to her that the broader allegations were dropped, so that consideration concentrated on the matters that were of real relevance. The appellant suggested that there had been a certain amount of shifting of ground by the NMC. However, the appellant accepted that there were no new allegations being put forward at the hearing; it was simply a refining of a broader case as had been put before.

*R (Chief Constable of the Derbyshire Constabulary) v. Police Appeals Tribunal* [2012] EWHC 2280 (Admin)

**22.30** Police disciplinary notice—standards of professional behaviour—need to specify breaches—factual findings

The chief constable challenged the decision of the Police Appeals Tribunal overturning the findings of gross misconduct made against two police constables, G and S, by a misconduct panel acting pursuant to the Police (Conduct) Regulations 2008. The panel had ordered that the officers be dismissed without notice, having been guilty of 'gross misconduct' (that is, a breach of standards so serious that dismissal would be justified). Notices pursuant to regulation 21 of the 2008 Conduct Regulations were served, which alleged: that G and S had failed to take any or any adequate steps to secure as evidence at the scene of an inquiry a plastic bag containing a white powder, despite having reasonable grounds to believe that it consisted of or contained a controlled substance; that they failed to take any, or any adequate, steps to prevent the girlfriend of the suspect from disposing of the plastic bag and white powder; that, in the case of G, he advised her to do so; and that, in the case of S, he failed to challenge, prevent or report G for so doing. The allegations were strongly denied, but the misconduct panel found the factual allegations proved. The Police Appeals Tribunal concluded that a regulation 21 notice, in identifying how the conduct is alleged to amount to 'misconduct', must identify the specific breaches of standards of professional behaviour alleged. It therefore concluded that there was a breach of the procedures set out in the Regulations. As to whether the breach could have materially affected the finding or decision, the Police Appeals Tribunal concluded that the answer depended on whether the factual findings made by the misconduct panel as to what occurred would in any event have led to a finding of gross misconduct based on a breach of the standard of professional behaviour of honesty and integrity. However, the misconduct panel's factual findings would not, in any event, have led to a finding of gross misconduct. This was because, while the misconduct panel made explicit findings of fact, nowhere in those findings did it explain how they led to any finding of breach of the standard of professional behaviour of honesty and integrity. The only reference to the officers' state of mind concerned the use of the term 'misleading'. Nowhere did the misconduct panel use an expression such as 'deliberately false' or suggest a lack of honesty and integrity by the officers. On its own examination of the evidence, the Police Appeals Tribunal could not conclude that a breach of the standard of honesty and integrity had occurred. In dismissing the chief constable's appeal, Beatson J said that regulation 21 of the 2008 Conduct Regulations expressly requires written notice of how the conduct that is the subject matter of the case 'is alleged to amount to misconduct or gross misconduct'. In the Regulations, 'misconduct' and 'gross misconduct' are not free-standing concepts: 'misconduct' is a breach of the standards; 'gross misconduct' is a breach of those standards so serious that dismissal would be justified. They can be understood only by reference to the 2008 Conduct Regulations. The regulations set out ten standards of professional behaviour in the Schedule—and they do so for a reason. There are differences in the nature of the specified type of conduct and the extent of the culpability involved in breaches of the different standards. There is also overlap between the number of standards. In order to set out how the conduct alleged amounts to misconduct or gross misconduct, it is necessary to set out which standard(s) of professional behaviour are alleged to have been breached. The Home Office Guidance also supports this. The Guidance states that the particulars of the behaviour of the officer, and also the reasons why it is thought that behaviour amounts to misconduct or gross misconduct, must be specified. The Guidance also states that 'clear particulars' should describe the actual behaviour that is alleged to have fallen below the standard expected.

### *Fish v. General Medical Council* [2012] EWHC 1269 (Admin)

**22.31** Dishonesty—allegation—need to fully particularize

In a case involving an allegation of dishonesty against a consultant anaesthetist arising out of time sheets submitted by the doctor as a locum and the removal of the declaration contained in the time sheets, Foskett J, repeating what he said previously in *Sheill v. General Medical Council* [2008] EWHC 2967 (Admin), said:

> 69. I do not think that I state anything novel or controversial by saying that [dishonesty] is an allegation (a) that should not be made without good reason, (b) when it is made it should be clearly particularised so that the person against whom it is made

knows how the allegation is put and (c) that when a hearing takes place at which the allegation is tested, the person against whom it is made should have the allegation fairly and squarely put to him so that he can seek to answer it. It is often uncomfortable for an advocate to suggest that someone has been deliberately dishonest, but it is not fair to shy away from it if the same advocate will be inviting the tribunal at the conclusion of the hearing to conclude that the person being cross-examined was dishonest…

70. At the end of the day, no-one should be found to have been dishonest on a side wind or by some kind of default setting in the mechanism of the inquiry. It is an issue that must be articulated, addressed and adjudged head-on.

### *Duthie v. Nursing and Midwifery Council* [2012] EWHC 3021 (Admin)

**22.32** Notes—alteration—need for specific charge

At [182], Irwin J said:

> It is absolutely clear from the conduct of this hearing that the Presenter knew from the outset that she intended to make the suggestion that some of the Appellant's notes were manufactured after the event. In my judgment, if the case was to be advanced in that way, it would have been proper to lay charges encapsulating that.

See further paragraph 27.15.

### *Ogundele v. Nursing and Midwifery Council* [2013] EWHC 2748 (Admin)

**22.33** Pattern of incompetence—whether sufficiently particularized

The appellant, a band 5 staff nurse at Royal Liverpool and Broadgreen Hospitals, challenged the adverse findings of the CCC and the sanction of conditions imposed for twelve months. Following serious concerns about the appellant's professional performance, she was given formal warnings by the management of the hospital in October 2008. She was suspended in December 2008 and, following a disciplinary meeting on 16 April 2009, she was dismissed. The NMC's charges were divided into three categories: (a) failure, between January 2007 and December 2008, to demonstrate the standards of knowledge, skill, and judgment required to practise without supervision; (b) failure, between April and September 2008, whilst undergoing a period of performance review, to demonstrate that she was capable of working safely on the department without supervision; and (c) failure, between October 2008 and January 2009, whilst undergoing a further period of performance review, to demonstrate that she was capable of working safely on the department without supervision. A total of eleven subheadings against the appellant were found proved. A further nine subheadings were found not proved. In dismissing the appellant's appeal (including the appeal on conditions), His Honour Judge Jeremy Richardson QC (sitting as a judge of the High Court) said that the first ground of appeal averred that it was impossible to meet certain of the charges owing to their imprecise, vague, and unclear nature. The point was argued before the panel more than once as the case unfolded. The basis of the case against the appellant was that there was a pattern of conduct and that the evidence amply revealed it. The panel accepted that and declined to uphold the appellant's objection. The learned judge said:

> 15. There can be no doubt whatever that a person brought before a Professional Conduct Committee must know what is alleged against them and have disclosure of the evidence that forms the foundation of the charge or charges. That is a matter of elementary fairness. There was no application by the appellant made in advance of the hearing for further particulars of the charges. It seems to me that there is force in the submission of the NMC that in a case where it is asserted there is a long-standing pattern of lack of confidence, and not simply a few isolated episodes, there inevitably will be a volume of incidents revealing the pattern. The material before the Panel plainly

> revealed this, and in my judgment, there cannot be a shred of misunderstanding as to the thrust and detail of the alleged incompetence.
>
> 16. The evidence of the NMC (as revealed in the few extracts I have selected) reveals a plain pattern of incompetence. The appellant had to meet this case... there was sufficient detail in the charges. The evidence relating to each is as plain as plain could be to reveal a pattern of utter incompetence.

### *R (El-Baroudy) v. General Medical Council* [2013] EWHC 2894 (Admin)

**22.34** Death in police custody—no allegation of defendant causing death—finding by panel of neglect and 'opportunity to survive was lost'

The appellant, a forensic medical examiner, was called to Chelsea police station to carry out an examination on AR, who had been arrested and was in a cell. The custody records noted that AR suffered from epilepsy and schizophrenia. The appellant was told that AR was the worse for drink and that, in a struggle to control him, he had banged his head on the floor. The custody sergeant asked the appellant whether or not AR should go to hospital. The appellant entered the cell and was in attendance for about a minute. It was crucial to perform an adequate medical assessment. It was not challenged that the appellant failed even to make basic assessments of AR's condition, with the result that the appellant did not realize that AR was unconscious rather than asleep and that he needed an immediate transfer to hospital. Within about three hours, or a little more, of the appellant having informed the police that AR was fit to be detained and did not need to be transferred to hospital, AR was found dead in his cell. The appellant faced a charge that he had failed to provide good clinical care to AR. There was no allegation that the appellant's conduct either caused the death of AR, or even caused AR to lose a substantial or significant or real chance of survival. Following *Roomi v. General Medical Council* [2009] EWHC 2188 (Admin), *Strouthos v. London Underground Ltd* [2004] EWCA Civ 402, and *Chauhan v. General Medical Council* [2010] EWHC 2093 (Admin), it was held that if the GMC had wished to pursue allegations that the misconduct either caused death or caused the loss of any realistic chance of survival, which would have been highly material, then such allegations should have been clearly stated in the charges; in the absence of such allegations being stated, evidence directed to those issues should not have been led and the panel should not in any way have based a judgment as to whether the appellant's fitness to practise was impaired or as to sanction on any question of causation. Given that the panel, in finding impairment, had referred to 'the tragic outcome for AR' and, on sanction, had said that 'through your neglect of duty and care an opportunity for AR to survive was lost', it was reasonably clear that the panel had taken account of the fact that the appellant's lack of care did have a causative effect in relation to AR's death. Accordingly, the matter was to be remitted to a fresh panel to look at the allegations afresh on the basis that it was not alleged that the appellant's misconduct either caused AR's death or caused the loss of any real chance of survival. Evidence available directed to those issues was to be disregarded.

### *Hussain v. General Medical Council* [2013] EWHC 3865 (Admin)

**22.35** Separate allegations of 'false' and 'dishonest'—duplication

The allegations against H, in paragraphs 2 and 4, alleged that he had 'falsely' stated that he had been awarded certain degrees in pharmacy and pharmacology, and had submitted 'false' information on a feedback form. The allegation went on to allege separately, in paragraph 9, that H's conduct, as set out above, was 'dishonest'. No point was taken before the panel on the phrasing of the allegations. In dismissing H's appeal, His Honour Judge Bird (sitting as a judge of the High Court) said:

> 50. ...It does however appear to me that the addition of ['dishonesty' in] allegation 9 to ['falsity' in] allegations [2 and 4] may give rise to difficulties. It is difficult to see

how, in the context of the present allegations, making an untrue statement intending that it should be relied upon could be anything other than dishonest. If that be right then allegation 9 appears to add nothing to these allegations.

[...]

52. It may be that the GMC would in the future wish to ensure that there is no doubling up of allegations of "falsity"...and "dishonesty". The difficulty would disappear if the word "falsely" in allegation 2 was replaced by "inaccurately" and the word "false" in allegation 4(c) was replaced by the word "untrue".

### *Solicitors Regulation Authority v. Anderson Solicitors* [2013] EWHC 4021 (Admin)

**22.36** Form of charges—complicated structure

In giving judgment, Treacy LJ (with whom King J agreed) said, at [3], that during the course of the hearing, the Court commented on the highly complicated structure of the charges, involving multiple allegations of breaches of rules, either cumulatively or in the alternative. Counsel for the SRA acknowledged the force of this criticism. The Court would hope that, in the future, consideration would be given to a significantly clearer method of framing charges.

### *Professional Standards Authority for Health and Social Care v. General Chiropractic Council and Briggs* [2014] EWHC 2190 (Admin)

**22.37** Dishonesty—not alleged—appeal by PSA—case remitted

B was charged with providing chiropractic treatment to one or more patients when he was registered with the General Chiropractic Council as a non-practising chiropractor, and providing chiropractic treatment without having appropriate professional indemnity insurance in place. Dishonesty was not alleged and the PCC correctly said that an allegation of dishonesty ought to be pleaded if it is to be relied upon. The Committee found the charges proved and imposed a six-month suspension order, without provision for any review. In allowing the Professional Standards Authority (PSA) to appeal under section 29 of the National Health Service Reform and Health Care Professions Act 2002, Lang J said that, on carefully considering the material, there was plainly evidence that supported an allegation of dishonesty. Under section 20(12) of the Chiropractors Act 1994, the investigating committee *must* refer an allegation to the PCC where it concludes that there is a case to answer. It has no discretion. It was a serious error not to refer an additional allegation of dishonesty. It arose out of the episode that formed the basis of the principle allegations and was directly connected to the allegations that were referred. Therefore the case was to be remitted to a fresh panel for a rehearing.

### *Professional Standards Authority for Health and Social Care v. Nursing and Midwifery Council and Macleod* [2014] EWHC 4354 (Admin)

**22.38** Undercharging—charges not adequately reflecting the seriousness of registrant's conduct

The second respondent, M, was a band 5 registered mental health nurse working on a specialist acute psychiatric ward caring for female patients suffering from serious mental illness. The charges arose out of an incident that took place on 22 October 2011. On that date, M was working an early shift on the ward. Charge nurse X was also working on that shift; she was a band 6 nurse and the most senior nurse on the shift. M was the next most senior nurse on the shift. During the shift, patient A struck charge nurse X with a hairbrush, causing a laceration above her eye that necessitated treatment. X called an ambulance herself, was taken to accident and emergency, and remained off work for approximately five months. M did not complete a serious incident report and it was only in August 2011 that he revealed to his superiors that he was unable to support charge nurse X's version of events. M stated that he had witnessed charge nurse X standing on a sofa, jumping onto the back of patient A and holding her in a headlock. He also said that charge nurse X had been verbally abusive to the patient. The relevant charge, as laid against M, was that he failed to complete a serious incident report in relation to the allegation and failed immediately to report that he had witnessed charge nurse X acting inappropriately towards the patient. The NMC's CCC found the charges proved and imposed a conditions of practice order upon M's registration for a period of nine months. The PSA appealed and contended that there was a serious procedural irregularity in that the charges of professional misconduct laid against M did not sufficiently reflect the gravity of his conduct. Andrews J said that, in her judgment, this was not a case in which it was obvious that dishonesty should have been alleged. M did not provide a positively false account of events on the day in question at any stage. However, it was a matter of serious concern that nowhere in the decision of the CCC did it address the reason why it took M five months to report X's inappropriate behaviour. There was a failure by the NMC to adequately reflect the seriousness of M's conduct in the charges. This was therefore a case of undercharging, and there was ample material to justify an allegation that M failed to escalate his concerns because of a wish to support or protect charge nurse X. The matter would be remitted for consideration by a differently constituted CCC, with a direction to the NMC that it must amend the charges so as to clearly assert that the reason (or motive) for M's failure immediately to escalate his concerns was to support or protect charge nurse X.

***R (Squier) v. General Medical Council* [2015] EWHC 299 (Admin)**

**22.39** Particulars of allegation—schedule of evidence and draft opening statement

Paragraph 3 of the allegation against S was that, whilst providing expert opinion evidence, she had failed to present her report and the research material upon which she relied in a way that was as complete and accurate as possible. Paragraph 5 alleged that S's actions and omissions were misleading, irresponsible, deliberately misleading, dishonest, and likely to bring the reputation of the medical profession into disrepute. On an application for judicial review, Ouseley J said that paragraph 3 was hopelessly inadequately particularized. No fair trial was possible on this allegation without further particulars. These could be short, and would identify the omissions and inaccuracies that should have been avoided. It should not be difficult to state what the omissions and inaccuracies relied on were, in a schedule if need be. Neither the existing schedule of evidence nor the GMC's draft opening statement remedied the deficiency. Moreover, the focus of particularizing an allegation is different from the focus of the evidence that supports it. The learned judge said that he did not wish to be overly prescriptive about how the GMC should present the charges, since the most important aspect was that S should know what was alleged against her, so that she could prepare her defence. It may be legitimate to refer to a schedule of evidence. The Court was reluctant to endorse the notion that a draft opening statement can supply the deficiency in the particulars of the allegation. In principle, there was no reason why the necessary task of providing particulars should not be done in the list of allegations themselves or, to avoid burdening them, in a schedule of particulars.

See further paragraph 42.27.

## C. Other Relevant Chapters

**22.40** Case Management, Chapter 8
Dishonesty, Chapter 19
Evidence, Chapter 25
Findings of Fact, Chapter 27
Investigation of Allegations, Chapter 39
Joinder, Chapter 40
Natural Justice, Chapter 48

# 23

# ERASURE/STRIKING OFF



## A. Other Relevant Chapters

Sanction, Chapter 60 **23.01**

# 24

# ESTOPPEL



## A. Disciplinary Cases

### *In the Matter of a Solicitor*, CO/2504/2000, 20 November 2000

**24.01** Disciplinary proceedings—subsequent criminal conviction—further disciplinary hearing—whether separate penalty

The respondent solicitor appeared before the Solicitors Disciplinary Tribunal (SDT), and admitted allegations involving utilizing clients' funds for his own purposes and misappropriating clients' funds. Because of very strong mitigating circumstances, the Tribunal came to the conclusion that, for the offences that the respondent admitted, he should be suspended from practice for three years. Subsequently, the respondent was convicted in the Crown Court of six offences of theft. The offences were charged on the basis of essentially the same facts as those that had been considered by the Tribunal. The respondent was sentenced to a term of one year's imprisonment. A differently composed SDT heard an application against the solicitor involving an allegation that he had been guilty of conduct unbefitting a solicitor by reason of the criminal conviction. In regard to the second set of disciplinary proceedings, the Tribunal ruled that, because the conviction arose from the same facts as those that had been considered by the earlier Tribunal (which was not in dispute), it was an abuse of process for the second application to be made. In allowing the Law Society's appeal, the Divisional Court (Lord Woolf CJ, and Hallett and Rafferty JJ) said that it was very important not to forget the interests of the public and the importance of the standing of the profession. In the Court's judgment, the situation that was before the first SDT was significantly different from that which was before the second. The important fact of the conviction could not be brought before the first Tribunal. It was therefore not dealing with that conviction, although it was dealing with the underlying facts. There may be situations in which the second Tribunal would inevitably come to the same conclusion as the first as to the appropriate punishment. However, whether the Tribunal would come to such a conclusion is a matter for the decision of the second Tribunal. It was perfectly in order for the Law Society to decide that the situation had changed as a result of the conviction to such an extent that the position was sufficiently different to justify the commencement of the second disciplinary proceedings. However, taking into account all of the matters, it would not be appropriate to add any further order to that made by the first Tribunal.

### *Swanney v. General Medical Council* 2008 SC 592

**24.02** See paragraph 43.10.

### *R (Coke-Wallis) v. Institute of Chartered Accountants in England and Wales* [2011] 2 AC 146

**24.03** Successive disciplinary complaints—dismissal of first complaint—final determination—res judicata

The Supreme Court held that the doctrine of cause of action estoppel applied to successive complaints before a professional disciplinary body, that disciplinary proceedings were civil in nature and that therefore the principles of res judicata applied, and that there was no reason why cause of action estoppel should not apply to successive sets of proceedings before the Disciplinary Committee of the Institute of Chartered Accountants in England and Wales (ICAEW). The Supreme Court so held in allowing an appeal by the claimant, Piers Coke-Wallis, a chartered accountant, against the Court of Appeal, which had upheld the dismissal of his application for judicial view of the decision by the Committee to refuse to dismiss a second complaint based on the same facts of a first complaint that had been dismissed on the merits. The first complaint had been dismissed by the Committee in April 2005 on the basis that the claimant's conviction in Jersey of an offence under the Financial Services (Jersey) Law 1998 did not correspond to one that was indictable in England and Wales. The ICAEW then preferred a second complaint in March 2006 based on the facts that gave rise to the original conviction. The Supreme Court held that what had been called 'the conviction complaint' and 'the conduct complaint' were not accurate descriptions. The dismissal of the first complaint against the claimant was a final decision on the merits and, in all of the circumstances, all of the constituent elements of cause of action estoppel were established. Since the first and second complaints relied upon the same conduct, once the first complaint was dismissed, it was contrary to the principles of res judicata to allow the ICAEW to proceed with the second complaint. In giving the leading judgment in the Supreme Court, Lord Clarke of Stone-cum-Ebony JSC said:

> [25] It is important to note that this appeal is concerned only with the case where there have been two successive sets of disciplinary proceedings. It is not concerned with a case in which either set of proceedings was either criminal or civil proceedings. In *Spencer Bower & Handley, Res Judicata*, 4th ed (2009), para 1.05 it is stated that res judicata can either give rise to a cause of action estoppel or to an issue estoppel. In this case the claimant relies upon cause of action estoppel, which is concisely defined in para 1.06 in this way: "If the earlier action fails on the merits a cause of action estoppel will bar another."
>
> [...]
>
> [29] In these circumstances I see no reason why the principles of cause of action estoppel should not apply to proceedings before a disciplinary tribunal set up under the byelaws. The provisions of the charter and supplemental charter are akin to statutory provisions and it seems to me that similar principles to those identified by Lord Bridge in *Thrasyvoulou v. Secretary of State for the Environment* [1990] 2 AC 273 apply to them. It was not suggested in the course of the argument that there was anything in the charter or supplemental charter to lead to the conclusion that the principles of cause of action estoppel should not apply to successive sets of disciplinary proceedings.

Lord Collins of Mapesbury JSC said:

> [57] It has been held or assumed in a number of decisions in other common law jurisdictions that res judicata principles apply to successive complaints before professional disciplinary bodies. Many professional disciplinary bodies are established or regulated by legislation, but the principles apply equally irrespective of the status of the disciplinary body. The reason is that from the earliest times it has been recognised that the

principle of finality or res judicata applies to tribunals established by the parties, such as an arbitral tribunal: *Dunn v. Murray* (1829) 9 B&C 780; *Fidelitas Shipping Co Ltd v. V/O Exportchleb* [1966] 1 QB 630, 643, per Diplock LJ.

[58] For example, in Canada it was accepted by the Manitoba Court of Appeal that principles of res judicata applied to a complaint by the College of Physicians against a doctor. On the facts it was held that the college could take proceedings against the doctor for sexual misconduct notwithstanding that four years previously the college had rejected the complaint, but that was because the earlier decision was not regarded as a final decision: *Holder v. College of Physicians and Surgeons of Manitoba* [2003] 1 WWR 19. In *Solicitor v. Law Society of New Brunswick* [2004] NBQB 95 the Law Society was held to be barred from bringing a complaint based on alleged fraudulent billing, when the solicitor had already been reprimanded for billing irregularities arising out of the same matters; and in *Visser v. Association of Professional Engineers and Geoscientists* 2005 BCSC 1402 it was held that the association was not entitled to bring successive disciplinary proceedings for different offences based on the same conduct. In Australia it was held that a doctor who had been censured by a medical board could not subsequently be the object of a second inquiry into alleged infamous conduct: *Basser v. Medical Board of Victoria* [1981] VR 953. See also in New Zealand *Dental Council of New Zealand v. Gibson* [2010] NZHC 912 (dentist bound by findings of disciplinary tribunal). In some cases the same result has been achieved by finding that the disciplinary tribunal is functus officio after the first decision: *Chandler v. Alberta Association of Architects* [1989] 2 SCR 848 (Canadian Supreme Court). In the United States, in *Florida Bar v. St Louis* (2007) 967 So 2d 108 and *Florida Bar v. Rodriguez* (2007) 967 So 2d 150 the Supreme Court of Florida accepted that res judicata principles applied to successive complaints brought by the Bar, but held that on the facts the causes of action were different. But it has also been said that res judicata or double jeopardy principles may not apply to disciplinary bodies because their "disciplinary requirements serve purposes essential to the protection of the public, which are deemed remedial, rather than punitive": *Spencer v. Maryland State Board of Pharmacy* (2003) 846 A 2d 341, 352 (Maryland Court of Appeals); cf *In Re Fisher* (2009) 202 P 3d 1186, 1199 (Supreme Court, Colorado).

### *R (Wood) v. Secretary of State for Education* [2011] EWHC 3256 (Admin)

**24.04** See paragraph 2.09.

### *Christou v. Haringey London Borough Council* [2014] QB 131, CA

**24.05** Employer's disciplinary procedure—procedural errors—written warning—subsequent disciplinary charges of misconduct

The first claimant was the team leader and the second claimant, the social worker, responsible for 'Baby P'. Prior to the mother pleading guilty to causing or allowing Baby P's death and the commissioning of a report into safeguarding arrangements for children at Haringey Borough Council, the claimants had been disciplined under the Council's simplified disciplinary procedure and given a written warning. Fresh disciplinary proceedings were instituted against both claimants, which resulted in their dismissal for gross misconduct. The Court of Appeal (Laws, Elias, and McCombe LJJ) dismissed the claimants' appeals on the grounds that the disciplinary process resulting in their dismissals had been barred by the doctrine of res judicata, or constituted an abuse of process, on account of the previous disciplinary proceedings relating to the same matters. Elias LJ said, at [47], that it was wrong to describe the exercise of disciplinary power by an employer as a form of adjudication. The purpose of the procedure is not 'a determination of any issue which establishes the existence of a legal right', as Lord Bridge put it in *Thrasyvoulou v. Secretary of State for the Environment* [1990] 2 AC 273, nor is it properly regarded as 'determining a dispute'. In a loose sense, a disciplinary body set up by an employer can be described as a domestic tribunal; it is far removed from the professional or sporting bodies of the kind found in *Meyers v. Casey* (1913) 17 CLR 90 or *R (Coke-Wallis) v. Institute of Chartered Accountants in England and Wales* [2011] 2 AC 146, in which the disciplinary body regulates the conduct of members of the profession or sport concerned, who stand on an equal footing with each other: at [52]. It was true that the factual substratum was the same, but the particular focus of complaint in the second proceedings was very different. The first proceedings focused on procedural errors, and the second concentrated much more firmly on substantive errors of judgment and breaches of the care plan for Baby P: at [55].

## B. Other Relevant Chapters

**24.06** Abuse of Process, Chapter 2
Double Jeopardy, Chapter 21
Investigation of Allegations, Chapter 39
Jurisdiction, Chapter 43
Natural Justice, Chapter 48

# 25

# EVIDENCE


A. **Legal Framework** 25.01

B. **Criminal Proceedings: Hearsay Evidence** 25.02

*R v. Riat* [2013] 1 All ER 349 25.02

C. **Previous Civil and Other Proceedings** 25.03

*General Medical Council v. Spackman* [1943] AC 627 25.03

*In re a Solicitor* [1993] QB 69 25.04

*Murphy v. General Teaching Council of Scotland* 1997 SLT 1152 25.05

*Fleurose v. Securities and Futures Authority Ltd* [2001] EWCA Civ 2015 25.06

*Harris v. Appeal Committee of the Institute of Chartered Accountants of Scotland* 2005 SLT 487 25.07

*Secretary of State for Education and Skills v. Mairs* [2005] ICR 1714 25.08

*Constantinides v. Law Society* [2006] EWHC 725 (Admin) 25.09

*Swanney v. General Medical Council* 2008 SC 592 25.10

*Thaker v. Solicitors Regulation Authority* [2011] EWHC 660 (Admin) 25.11

*Bhatt v. General Medical Council* [2011] EWHC 783 (Admin) 25.12

*Afolabi v. Solicitors Regulation Authority* [2011] EWHC 2122 (Admin) 25.13

*Chaudhari v. General Pharmaceutical Council* [2011] EWHC 3433 (Admin) 25.14

*R (M) v Independent Police Complaints Commission* [2012] EWHC 2071 (Admin) 25.15

*Quarters Trustees Ltd and ors v. Pension Regulator and anor* [2012] Pens LR 415 25.16

*Dhorajiwala v. General Pharmaceutical Council* [2013] EWHC 3821 (Admin) 25.17

*R (H) v. Nursing and Midwifery Council* [2013] EWHC 4258 (Admin) 25.18

*R (Hollis) v. Association of Chartered Certified Accountants* [2014] EWHC 2572 (Admin) 25.19

*R (Squier) v. General Medical Council* [2015] EWHC 299 (Admin) 25.20

D. **Hearsay Evidence** 25.21

*Idenburg v. General Medical Council* (2000) 55 BMLR 101 25.21

*Razzaq v. General Medical Council* [2006] EWHC 1300 (Admin) 25.22

*Nursing and Midwifery Council v. Ogbonna* [2010] EWCA Civ 1216 25.23

*R (Vali) v. General Optical Council* [2011] EWHC 310 (Admin) 25.24

*R (Bonhoeffer) v. General Medical Council* [2011] EWHC 1585 (Admin) 25.25

*Belal v. General Medical Council* [2011] EWHC 2859 (Admin) 25.26

*Lawrence v. General Medical Council* [2012] EWHC 464 (Admin) 25.27

*Shaikh v. General Pharmaceutical Council* [2013] EWHC 1844 (Admin) 25.28

*Ogudele v. Nursing and Midwifery Council* [2013] EWHC 2748 (Admin) 25.29

*White v. Nursing and Midwifery Council; Turner v. Nursing and Midwifery Council* [2014] Med LR 205 25.30

*Thorneycroft v. Nursing and Midwifery Council* [2014] EWHC 1565 (Admin) 25.31

E. **Lies before the Tribunal** 25.32

*Rehman v. Bar Standards Board,* PC2008/0235/A and PC2010/0012/A, 19 July 2013 25.32

F. **Failure to Give Oral Evidence** 25.33

*Iqbal v. Solicitors Regulation Authority* [2012] EWHC 3251 (Admin) 25.33

G. **Fresh Evidence on Appeal** 25.34

*Heffron v. Professional Conduct Committee of the United Kingdom Central Council for Nursing Midwifery and Health Visiting* [1988] 10 BMLR 1 25.34

*Threlfall v. General Optical Council* [2004] EWHC 2683 (Admin) 25.35

*R (Russell) v. General Medical Council* [2008] EWHC 2546 (Admin) 25.36

## A. Legal Framework

**25.01** Examples include:

Nursing and Midwifery Council (Fitness to Practise) Rules 2004, rule 31
Solicitors (Disciplinary Proceedings) Rules 2007, rule 13 ('the Civil Evidence Act 1968 and the Civil Evidence Act 1995 shall apply in relation to proceedings before the Tribunal in the same manner as they apply in relation to civil proceedings')
General Pharmaceutical Council (Fitness to Practise and Disqualification etc) Rules 2010, rule 24
Bar Standards Board Disciplinary Tribunals Regulations 2014, rule E144 (the rules of natural justice apply and, subject to these, the tribunal may admit any evidence, whether oral or written, whether direct or hearsay, and whether or not the same would be admissible in a court of law; the tribunal may exclude any hearsay evidence if it is not satisfied that reasonable steps have been taken to obtain direct evidence of the facts sought to be proved by the hearsay evidence)
General Medical Council (Fitness to Practise) Rules 2014, rule 34 (the panel may admit any evidence that it considers fair and relevant to the case before it, whether or not such evidence would be admissible in a court of law, save that where the evidence would not be admissible in criminal proceedings, the panel shall not admit it unless satisfied that its duty of making due inquiry makes its admission desirable)
Institute of Chartered Accountants in England and Wales Disciplinary Committee Regulations, regulation 28 (the hearing shall be informal and the strict rules of evidence shall not apply)

## B. Criminal Proceedings: Hearsay Evidence

### *R v. Riat* [2013] 1 All ER 349

**25.02** General principles—admission of hearsay evidence in criminal proceedings—guidance

The Court of Appeal, Criminal Division (Hughes LJ, and Dobbs and Globe JJ) heard consecutively five cases that involved the admission of hearsay evidence. Giving guidance as to the correct approach in English law to such cases, the Court said, at [2]:

> ...For the purposes of a Crown Court in England and Wales dealing day to day with cases of this kind, five propositions are central:
> i) the law is, and must be accepted to be, as stated in United Kingdom statute, viz the Criminal Justice Act 2003 ("CJA 03");
> ii) if there be any difference, on close analysis, between the judgment of the Supreme Court in [*R v. Horncastle, R v. Marquis* [2010] 2 AC 373] and that of the ECtHR in [*Al-Khawaja v. UK* [2011] 32 BHRC 1], the obligation of a domestic court is to follow the former: see *R (RJM) v. Secretary of State for Work and Pensions* [2009] 1 AC 311 at [64] and [*R v. Ibrahim* [2012] 4 All ER 225] at [87];

iii) there are indeed differences in the way in which principle is stated, but these may well be more of form than of substance; in particular, the importance of the hearsay evidence to the case is undoubtedly a vital consideration when deciding upon its admissibility and treatment, but there is no over-arching rule, either in the ECtHR or in English law, that a piece of hearsay evidence which is "sole or decisive" is for that reason automatically inadmissible;

iv) therefore,... the Crown Court judge need not ordinarily concern himself any further with close analysis of the relationship between the two strands of jurisprudence and need generally look no further than the statute and *Horncastle*...;

v) however, neither under the statute nor under *Horncastle*, can hearsay simply be treated as if it were first hand evidence and automatically admissible.

The Court of Appeal added that:

> 3. ... The common law prohibition on the admission of hearsay evidence remains the default rule but the categories of hearsay which may be admitted are widened. It is essential to remember that although hearsay is thereby made admissible in more circumstances than it previously was, this does not make it the same as first-hand evidence. It is not. It is necessarily second-hand and for that reason very often second-best. Because it is second-hand, it is that much more difficult to test and assess....

At [5], the Court noted that the issue in *Horncastle* was whether English law knew an overarching general rule that hearsay which could be described as the sole or decisive evidence was not to be admitted, or would inevitably result in an unfair trial if it was. The answer was 'no'. At [7]:

> The statutory framework provided for hearsay evidence by the CJA 03 can usefully be considered in these successive steps.
> i) Is there is a specific statutory justification (or "gateway") permitting the admission of hearsay evidence (s[s] 116–118)?
> ii) What material is there which can help to test or assess the hearsay (s 124)?
> iii) Is there a specific "interests of justice" test at the admissibility stage?
> iv) If there is no other justification or gateway, should the evidence nevertheless be considered for admission on the grounds that admission is, despite the difficulties, in the interests of justice (s 114(1)(d))?
> v) Even if prima facie admissible, ought the evidence to be ruled inadmissible (s 76 [of the Police and Criminal Evidence Act 1984, or PACE] and/or s 126 CJA)?
> vi) If the evidence is admitted, then should the case subsequently be stopped under section 125?

In summary, the Court of Appeal said that admissibility of hearsay evidence must be justified under one or other of the statutory exceptions. Most of the common law exceptions are preserved by section 118, and business and similar records are admissible under section 117. The more controversial cases are likely to be those covered by section 116(2)—death, illness, absence abroad, the lost witness, and fear—or by the additional possible gateway in section 114(1)(d):

> 15. The general principle underlying [the gateways] in s 116 cases is clearly that the necessity for resort to second-hand evidence must be demonstrated. Illness or, *a fortiori* death, may demonstrate such necessity. Absence abroad will do so only if it is not reasonably practicable to bring the witness to court, either in person or by video link. If the witness is lost, all reasonable practicable steps must have been taken to get him before the court: this will include not only looking for him if he disappears but also keeping in touch with him to avoid him disappearing: see for example *R v. Adams* [2008] 1 Cr App R 430 and *R v. Kamahuza* [[2008] EWCA Crim 3060].

At [17], the Court said that, in the case of fear, section 116 is to be widely construed; it includes fear for others and may include fear of financial loss. However, the court must be satisfied that it is in the interests of justice for the statement to be admitted. In deciding whether it is or is not in the interests of justice to admit it, every relevant circumstance is to be considered, including the risk of unfairness to any party that would ensue from its admission. The availability of special measures should be considered to enable a fearful witness to give evidence notwithstanding his or her apprehension. Finally, section 114(1)(d) contains a general residual power to admit hearsay evidence that does not otherwise pass a statutory gateway if the judge is satisfied that it is in the interests of justice for it to be admitted. It must not become a route by which all or any hearsay evidence is routinely admitted without proper scrutiny. That would be to subvert the express provisions that follow in sections 116–118. The Court emphasized, in *R v. D* [2010] EWCA Crim 1213, that section 114(1)(d) cannot be used routinely to avoid the statutory conditions for the admission of evidence that properly falls to be considered under sections 116–118. Whichever is the statutory power under consideration, it is clear that hearsay must not simply be 'nodded through'. A focused decision must be made whether it is to be admitted or not.

## C. Previous Civil and Other Proceedings

### *General Medical Council v. Spackman* [1943] AC 627

**25.03** Findings in previous civil proceedings—challenge—prima facie evidence—compare findings in previous criminal proceedings

In 1941, Langton J, sitting in the divorce court in the case of *Pepper v. Pepper (Spackman cited)*, found that the party cited, S, had committed adultery with the petitioner, Mrs P. The original petition was that of the wife asking for her marriage to be dissolved on the grounds of her husband's cruelty. The wife's allegations were supported by the evidence of S, a doctor who was attending her. On the issue of cruelty, Langton disbelieved S and found against the wife, and awarded a decree nisi to the husband on the grounds of the wife's adultery with S. The General Medical Council (GMC) subsequently brought disciplinary proceedings on the grounds that, by reason of the finding of adultery, S was guilty of infamous conduct in a professional respect. The solicitor to the Council opened the case and placed before the Council a certified copy of the decree absolute. Each member of the Council was provided with extracts from the shorthand notes of the proceedings in the divorce court, including a copy of the judgment of the learned judge. S's solicitor admitted that his client stood in a professional relationship with Mrs P, but submitted that he should be allowed to call evidence that was not before the divorce court, with a view to challenging the correctness of the judge's conclusion on the issue of adultery. The Council was not prepared to hear fresh evidence and, relying on the facts alleged in the charge, directed that S's name should be erased from the medical register. The House of Lords, affirming the decision of the Court of Appeal, held that there had been no 'due inquiry' within section 29 of the Medical Act 1958. Viscount Simon LC drew a distinction between a case in which a practitioner had been convicted of a criminal offence and a case in which the allegation of misconduct arose from adverse conclusions reached in a civil court of law. In the former case, the practitioner cannot go behind the conviction, and endeavour to show that he was innocent of the charge and should have been acquitted. In the latter case, whilst an adverse conclusion reached in a court of law might lead to a charge of professional misconduct, the conclusion reached in the courts would be prima facie proof of the matter alleged. Viscount Simon said, at 635–6:

> There is no question of estoppel or of *res judicata*. In such cases the decision of the courts may provide the council with adequate material for its own conclusion if the facts are not challenged before it, but if they are, the council should hear the challenge and give such weight to it as the council thinks fit. The same view must, I think,

> be taken if the practitioner challenges the correctness of a finding of adultery by the Divorce Court. The decree provides a strong prima facie case which throws a heavy burden on him who seeks to deny the charge, but the charge is not irrebuttable... What matters is that the accused should not be condemned without being first given a fair chance of exculpation. This does not mean that the council has to rehear the whole case by endeavouring to get the previous witnesses to appear before it, though in special circumstances the recalling of a particular witness, in the light of what the accused or his witnesses assert, may, if feasible, be desirable. The council will primarily rely on the sworn evidence already given at the trial. It is not required to conduct itself as a court.

### *In re a Solicitor* [1993] QB 69

**25.04** Findings of disciplinary proceedings in Australia—approach and weight to be attached

The appellant was admitted as solicitor in England and Wales in 1976; in 1981, she qualified in Australia. On 15 November 1988, the Supreme Court of Western Australia ordered that the appellant's name be struck off the roll of practitioners there. The allegations found proved against her were that her application for dissolution of marriage in 1982, supported by an affidavit, had contained two material falsehoods: first, that she and her husband had lived separately and apart for the twelve months prior to the application; and second, that there were no children of the marriage. On 1 November 1990, the Solicitors Disciplinary Tribunal (SDT) ordered that the appellant be struck off the roll of solicitors. Lord Lane CJ, giving the judgment of the Divisional Court (Lord Lane CJ, and Simon Brown and Jowitt JJ), said that a tribunal can, subject to its own rules, use evidence that might in strict law be inadmissible. It was open to the Tribunal to make such use of the findings of the Supreme Court of Western Australia and the Barristers' Board of Western Australia as was proper in the circumstances of the case. He continued, at 80F–H:

> It seems to us that the task of the tribunal in a case such as the present is to have regard to all the evidence which is adduced before them, including the board's findings, the course of events in Western Australia, the evidence of the appellant and the contents of the affidavits adduced on her behalf and to ask themselves whether or not they are satisfied to the requisite standard of proof that the charges are made out. In deciding the weight to be attached to the board's findings i.e. the board's actual conclusions upon the allegations of misconduct, the tribunal will clearly bear in mind a variety of considerations and not least (a) the evidence adduced before the board, (b) the apparent fairness or otherwise of the proceedings before the board, (c) the standard of proof adopted by the board, (d) the absence of any right of appeal from the board's findings. That there is no such right of appeal is apparent from *Bercove v. Barristers' Board* [1986] WAR 50, although this was clearly not appreciated by the tribunal.

### *Murphy v. General Teaching Council of Scotland* 1997 SLT 1152

**25.05** Police warning given to defendant—no reference in charge—lack of legal advice to Committee

The main ground of appeal to the Court of Session was that information was supplied to the Disciplinary Committee about a warning given by the police to the appellant, but which was not referred to in the notice of allegation, and that the appellant had not been called upon to admit or deny it. The Committee received no advice to disregard the police warning and the Court observed that, even if it had, it would have been almost impossible for the Committee to put it out of its mind. The Inner House upheld the appellant's appeal and quashed the decision of the Disciplinary Committee, which had directed that his name be removed from the register; the Inner House directed that the proceedings be terminated. The basis of the Court's judgment was that justice was not seen to be done because the members of the Committee had before them material that was irrelevant and which was prejudicial to the appellant. In giving the judgment of the Court, Lord Justice-Clerk (Ross) said, at 1156B–E:

> We recognise that in the decision letter there is nothing to suggest that the Disciplinary Committee were in fact influenced by their knowledge of the earlier police warning, and that they were given adequate directions as to the test they should apply in determining this complaint. However, it is axiomatic that, in proceedings of this nature, justice must not merely be done, but must be seen to be done. In our opinion, even though the Disciplinary Committee may have applied the correct test, justice was not seen to be done because the members of the Disciplinary Committee had before them material which was irrelevant and which was prejudicial to the appellant. They were never told to disregard the material and, accordingly, one cannot exclude the possibility that some if not all of the members of the Disciplinary Committee were influenced to some extent by that objectionable material. For all these reasons we are satisfied that the decision of the Disciplinary Committee must be quashed.

### *Fleurose v. Securities and Futures Authority Ltd* [2001] EWCA Civ 2015

**25.06** Evidence given before Stock Exchange inquiry—no objection before disciplinary tribunal—admissible

F, a senior trader in securities employed by JP Morgan, was found guilty by the Disciplinary Tribunal of the Securities and Futures Authority (SFA) of improper conduct. He was suspended for two years and ordered to pay £175,000 towards the SFA's costs. Following a sudden fall in the FTSE 100 index, an immediate investigation was carried out by the London Stock Exchange (LSE), which concluded that the fall had stemmed from substantial sales by JP Morgan conducted by the appellant. JP Morgan's own investigation concluded that F had breached paragraph 2.10 of the LSE Rules, which prohibits a member firm from doing an act or engaging in a course of conduct the sole intention of which is to move the index value. F was not, at the material time, a member of the LSE, and therefore not within its disciplinary reach. On appeal, F contended that the answers given by him under questioning by the LSE and before the Disciplinary Tribunal were improperly admitted. The Court of Appeal (Schiemann and Clarke LJJ, and Wall J), in dismissing F's appeal, held that these were not criminal proceedings. In the circumstances, neither under domestic law nor under any Convention law was there any question of the evidence being inadmissible in the absence of objection. In civil proceedings, the question of admissibility is one of discretion in the tribunal of first instance. F, although under undoubted practical pressures, was under no legal compulsion to answer the questions in the LSE interview, which was being conducted not by his regulator, but by his firm's regulator; nor was he under any compulsion to answer the questions in cross-examination before the Disciplinary Tribunal. The Tribunal rightly, in the absence of any objection, allowed it to be included. The allegedly compelled evidence was but a part of the total evidence against F. It is not inconceivable that the Disciplinary Tribunal would have come to the same conclusion even had it never seen that evidence.

### *Harris v. Appeal Committee of the Institute of Chartered Accountants of Scotland* 2005 SLT 487

**25.07** Summaries of conclusions reached by investigation committee—affidavit containing complainant's views—prejudicial and irrelevant material

The petitioner faced thirty-six separate allegations before the Disciplinary Committee of the Institute of Chartered Accountants of Scotland, the most serious being that he had conspired with the directors of a company in liquidation and a Mr C to liquidate the company's assets so as to advance their interests. Before the Disciplinary Committee was an affidavit of Mr C, much of which was prejudicial to the petitioner, including an expression of the view that he in fact knew what was going on. Also before the Disciplinary Committee were two summaries that contained detailed narratives of the findings and conclusions reached by the investigation committee from which the allegations in the charges had been drawn. Despite the Disciplinary Committee stating that it had formed the view that the petition was not a part of any criminal or fraudulent conspiracy to defraud creditors, and it was not prepared to rely upon the affidavit of 'a convicted fraudster', the Outer House allowed the appeal because of the prejudicial irrelevant material being adduced at the hearing.

### *Secretary of State for Education and Skills v. Mairs* [2005] ICR 1714

**25.08** Statutory inquiry—witness—critical findings—report admissible—findings not binding on subsequent tribunal

M managed the team of social workers responsible for Victoria Climbié, whose death on 25 February 2000 led to comprehensive investigations being undertaken in relation to the failure of child protection agencies to prevent the tragedy. A statutory inquiry conducted by Lord Laming found M guilty of misconduct and, following a disciplinary hearing, she was dismissed by the social services department for the London Borough of Haringey and was placed on a list kept by the Secretary of State of persons unsuitable to work with children. M appealed to the Care Standards Tribunal under the Protection of Children Act 1999 against her inclusion on the list. The Tribunal ordered her name to be removed from the list. In dismissing the Secretary of State's appeal, Leveson J pointed out that M was not a party to Lord Laming's inquiry; she was only a witness—albeit a witness who had the right and opportunity to make representations as to the findings that ought to be made. She had no power to cross-examine any witnesses or to determine who should give evidence; in that regard, she was not a participant at all. As to the status of Lord Laming's report, the learned judge said that the Tribunal was not bound by the rules of evidence, and the report was clearly admissible both as to the facts and as to the conclusions. However, the conclusions on the facts were Lord Laming's opinion, based on his assessment of the evidence, and could not possibly bind the Tribunal, which was required by statute to reach its own conclusions on the matters in issue. Common sense dictates that the views expressed following a statutory inquiry undertaken with a wide-ranging remit will be highly persuasive and will require careful consideration. That is not the same, however, as saying that the Tribunal is bound to follow them or is inhibited in attaching greater weight to certain features of the evidence than did the Tribunal. The Tribunal was entitled to exercise its own judgment in accordance with its obligations under the 1999 Act, albeit paying appropriate attention to the views of the inquiry, and doubtless explaining, for reasons of clarity, where and why it took a different view.

### *Constantinides v. Law Society* [2006] EWHC 725 (Admin)

**25.09** High Court judgment—findings of dishonesty against solicitor—background—tribunal to reach its own findings

The appellant appealed from a decision of the SDT striking him off the roll of solicitors. He challenged the Tribunal's findings of dishonesty and contended that, in reaching its conclusion, it was inappropriately influenced by a judgment of Peter Smith J adduced in evidence by the Law Society in which the appellant had been a party. A preliminary issue arose at the outset. The Law Society had given notice that it proposed to rely upon the judgment of Peter Smith J under rule 30 of the Solicitors (Disciplinary Proceedings) Rules 1994. This provides, amongst other things, that:

(a) the judgment of any civil court may be proved by producing a certified copy of the judgment;
(b) the findings of fact upon which a judgment is based shall be admissible as prima facie proof of those facts; and
(c) the Tribunal has a discretion not to apply the strict rules of evidence at any hearing.

The appellant's representative objected to the Tribunal reading the judgment. It was said that the judgment was so overwhelmingly and comprehensively prejudicial to the appellant that no Tribunal that had read it could thereafter perform its task fairly and properly. The Tribunal resolved the preliminary issue in favour of the Law

Society. The Court (Moses LJ, and Holland and Walker JJ), in dismissing the solicitor's appeal, said:

> [28] The judge's views as to the appellant's dishonesty and lack of integrity were not admissible to prove the Law Society's case against this appellant in these disciplinary proceedings. We are far from ruling that a judge's conclusions as to dishonesty cannot amount to findings of fact within the meaning of Rule 30. There will be cases when a finding of fact, be it in a civil or criminal case, of dishonesty will be prima facie evidence of that dishonesty. But in the instant case the judge's conclusions were far more wide-ranging than the allegations made against the appellant in the disciplinary proceedings. They were not relied upon by the Law Society as proof of dishonesty. At paragraph 21 it was recorded that the Law Society only intended to rely upon the judge's description of the appellant's behavior and to limit the references to the allegations made in the disciplinary proceedings. Furthermore, the Tribunal directed itself that it was an expert and experienced Tribunal which was bound to apply a different standard of proof to that applied by the judge.
>
> [...]
>
> [30] ... [I]t is plain to us that the tribunal disregarded the judgment when reaching their own conclusion as to the Appellant's dishonesty. The chairman said so in terms after the tribunal retired to consider its decision as to dishonesty (see p77 of the transcript of the second day of the hearing). Further, the tribunal in its written decision stated: "The tribunal has reached this decision upon the facts placed before it by the Applicant, which the Respondent did not dispute, and without taking into account any part of the judgment relating to the civil trial".
>
> [...]
>
> [33] We ought, however, to record that we do not see why it was necessary to refer to the judgment at all. The background facts were not in dispute. Provided they were clearly set out within the Rule 4 statement there was no need to rely upon it save insofar as it emerged that the appellant disputed those primary facts. However, the Rule 4 statement was itself a mixture of assertion of fact and argument. We would suggest that had a simple account of the facts been set out with a reference to the relevant paragraphs in the judgment there would have been no further need to refer to it.

### *Swanney v. General Medical Council* 2008 SC 592

**25.10** See paragraph 43.10.

### *Thaker v. Solicitors Regulation Authority* [2011] EWHC 660 (Admin)

**25.11** High Court judgment—admissible to prove background facts

In allowing T's appeal against a decision of the SDT that he be struck off the roll of solicitors, Jackson LJ said that, in *Constantinides v. Law Society*, the Divisional Court held that there could be no reasonable objection to the SDT reading a civil or criminal judgment in which the judge had made findings as to the dishonesty of a solicitor appearing before the Tribunal, provided that the Tribunal was clear and rigorous in its approach to that judgment. The judgment would be admissible to prove background facts, but not to prove the Law Society's case against the solicitor in the disciplinary proceedings. If a solicitor is going to be struck off the roll for acts of dishonesty and gross recklessness, he is entitled to a fair process and a fair hearing before the decision is reached. In this case, T did not receive either a fair process or a fair hearing. This occurred because of the manner in which the case against him was pleaded and presented to the Tribunal.

***Bhatt v. General Medical Council* [2011] EWHC 783 (Admin)**

**25.12** See paragraph 21.09.

***Afolabi v. Solicitors Regulation Authority* [2011] EWHC 2122 (Admin)**

**25.13** Employment tribunal—evidence—findings—not conclusive before SDT

A qualified and was admitted as a solicitor in 2007, and the Law Society was informed that she had immediately become a partner in a firm. One of the allegations against A was that she gave evidence to an employment tribunal that the tribunal considered not to be honest. The SDT found the allegation proved. It arose from a claim by an employee against the firm in relation to unlawful deduction from wages. On an application by the firm to set aside a default judgment, Mrs A gave evidence on oath on behalf of the firm. The chairman of the employment tribunal, in written reasons, stated that her evidence was not honest before the tribunal. In quashing the decision of the SDT, which had ordered A's name to be removed from the roll of solicitors, Holman J observed that rule 15(2) of the Solicitors (Disciplinary Proceedings) Rules 2007 provided that the findings of fact upon which a judgment was based were admissible as proof, but not as conclusive proof of those facts. The standard of proof adopted by the chairman of the employment tribunal was the civil standard of proof. The finding of the chairman was little more than the trigger for the allegation and, under rule 15(4), it was admissible as proof, but not conclusive proof; because the required standard before the SDT was the criminal standard, it was vital that the Tribunal reached a final conclusion of its own to that standard upon the critical question. An order for a rehearing by a differently constituted tribunal would be made.

***Chaudhari v. General Pharmaceutical Council* [2011] EWHC 3433 (Admin)**

**25.14** Injunction and orders granted by county court and High Court—evidential status of findings—prima facie evidence

C, a registered pharmacist, appealed against an order striking her off the register made under section 10 of the Pharmacy Act 1954. The charges faced by the appellant arose out of the way in which she, with others, including her sister, pursued a public campaign against, in particular, five Great Ormond Street Hospital doctors and a pathologist who carried out a post mortem of her sister's baby, accusing the doctors of murder and the pathologist of being involved in a cover-up. C had been the subject of various court orders and civil judgments against her in which it was said that she had been involved in, amongst other things, publishing material on the Internet, demonstrating in hospital grounds, and publishing leaflets, accusing the doctors of murder. In dismissing C's appeal, King J, at [71]–[80], considered the evidential status of the court findings and orders. The learned judge said that the Disciplinary Committee of the General Pharmaceutical Council demonstrated that it was alive to the correct evidential status of court findings—namely, that 'they were strong prima facie evidence of their contents, but were not conclusive of the matter'. This was in accordance with the decision of the House of Lords in *Spackman v. General Medical Council* [1943] AC 627, which is to the effect that a finding in civil proceedings is not conclusive evidence in subsequent disciplinary proceedings, but does provide prima facie evidence of the facts found, and that the practitioner should be given a fair chance to explain himself, but a disciplinary tribunal is not required to conduct itself as a court of law rehearing all of the evidence underlying those findings. In the instant case, the Committee did properly apply the law as set out in *Spackman*. The Committee was not concerned with the truth or otherwise of the appellant's beliefs as to what had caused the death of the baby, but with the appropriateness of the conduct alleged against her in the charges that she faced, even accepting, which the Committee did, that she believed the various allegations that she made.

### *R (M) v. Independent Police Complaints Commission* [2012] EWHC 2071 (Admin)

**25.15** Police local resolution procedure—statements—whether admissible in subsequent disciplinary proceedings

In March 2006, the claimant, Ms M, attended the accident and emergency department of Wycombe General Hospital, alleging a serious sexual assault. Two police constables attended at the hospital. The claimant made a complaint against the Police Service regarding the officers and subsequently agreed to the matter being locally resolved. The complainant was dissatisfied with the outcome of the local resolution procedure and appealed against it to the Independent Police Complaints Commission (IPCC). That appeal was rejected on 22 June 2006. In the course of subsequent county court proceedings, the police disclosed the witness statements of the two constables who attended at the hospital. The result was a further complaint made by Ms M against the officer who conducted the local resolution procedure in 2006. Ms M alleged that the officer had misrepresented statements from the two police constables to her in the course of conducting the local resolution procedure. The force's professional standards department found that there was no case to answer against the officer and the IPCC rejected the complainant's appeal. In dismissing Ms M's judicial review claim, Bean J considered to what extent the statements made in the course of the local resolution procedure were admissible. Paragraph 8(3) of Schedule 3 to the Police Reform Act 2002, dealing with the local resolution of complaints, provides that a statement made by any person for the purposes of the local resolution of any complaint shall not be admissible in any subsequent criminal, civil, or disciplinary proceedings (except to the extent that it consists of an admission relating to a matter that has not been subjected to local resolution). The question is what is meant by the provision that a statement made by any person for the purposes of the local resolution procedure is inadmissible in any subsequent criminal, civil, or disciplinary proceedings. The rationale of paragraph 8(3) was not in dispute and was obvious. There is an analogy between the local resolution procedure envisaged by Schedule 3 and mediation in civil or family proceedings. In a police disciplinary investigation, regulations require that an officer under suspicion is given a warning—essentially similar to that given to suspects in criminal proceedings—that he is not obliged to answer any question, although if he refuses to answer, inferences may be drawn from his silence. No such warning is given under the local resolution procedure, because its terms make clear that it does not lead to disciplinary proceedings. The purpose of paragraph 8(3) is therefore to encourage frankness by all concerned in the investigation. Parliament could have enacted that a statement made in the course of a local resolution procedure shall not be admissible against the maker of the statement, but that is not what the paragraph says. It is a clear and absolute prohibition. It says quite clearly that such statements shall not be admissible in *any* subsequent criminal, civil, or disciplinary proceedings. It is not limited to admissibility as against the maker of the statement. It is a prohibition on admissibility, not on discoverability, in civil proceedings. The parties to subsequent civil proceedings such as this case cannot dispense with it, even by consent.

### *Quarters Trustees Ltd and ors v. Pension Regulator and anor* [2012] Pens LR 415

**25.16** Reference to Upper Tribunal—applications to strike out by Pensions Regulator and FSA—decision and findings of SDT—admissible evidence

The Pensions Regulator applied to strike out the applicants' reference notice served on 13 April 2011 by which the applicants sought to refer to the Upper Tribunal an order made by the determinations panel of the Pensions Regulator, prohibiting the applicants from acting as trustees of any occupational pension scheme. The second respondent, the Financial Services Authority (FSA), also sought to strike out reference notices served by Q and B, the second and third applicants. By decision notices of the FSA, those applicants' approval for performing any function in relation to regulated activities had been withdrawn. The applications of the Pensions Regulator and of the FSA were linked, and were heard together as an application under rule 8(3)(c) of the Tribunal Procedure (Upper Tribunal) Rules 2008, which provides that the Upper Tribunal may strike out the whole or a part of proceedings if it considers there is no reasonable prospect of the appellant's case succeeding. Before granting a strike-out application, the Tribunal needs to be satisfied that there is a real, and not simply a fanciful, prospect of success on the part of the applicants. 'Success' means that the applicant in question must have a real prospect of securing from the Tribunal a determination as to the appropriate action that is more favourable to him or her than that contained in the notice of the regulator in question. The fact that the application to strike out the reference may raise difficult issues does not mean that those issues should be addressed only at the final hearing; a difficult issue may, on analysis, admit of a clear answer in favour of striking out in which case an order should be made: *Michel van der Wiele NV v. Pensions Regulator* [2011] PLR 109. The Pensions Regulator's grounds for prohibition of the applicants in the instant case had been their failure to meet the standards of (a) fitness to be trustees, including competence and capability, and (b) financial soundness. Q and B, as directors of QTL, bore the responsibility for the acts and omissions of QTL. As professional trustees, they were to be held to a higher standard of competence than unpaid lay trustees, and as solicitors, Q and B ought to have been particularly aware of the importance of high standards of probity and efficiency required when dealing with clients' money. The FSA relied on both the fact of the Pensions Regulator's determination and on the FSA's own primary investigation into Q's and B's alleged misconduct. On 14 October 2011, the SDT made determinations against Q and B. The conclusion of the SDT was that, overall, Q and B had shown a wholesale disregard of the Solicitors' Accounts Rules and the way in which a proper solicitor's practice should be run. The SDT concluded that, even without the allegations of dishonesty, Q's conduct had been so bad that a striking-off order would have been warranted. B was prohibited from being re-registered as a registered foreign lawyer. In directing that the applicants' reference notices should be struck out, Sir Stephen Oliver QC considered to what extent it was permissible for the Upper Tribunal to take into account the outcome of the judgment of the SDT. In this regard, both regulators relied on the preliminary decision of the Financial Services and Markets Tribunal in *Elliott v. Financial Services Authority* [2005] UKFSM 019 as authority for the proposition that findings of the SDT are admissible evidence before the Upper Tribunal as to a person's fitness and propriety, and that the FSA may rely upon them without needing to reprove each and every allegation that the SDT found to be proved. In *Elliott*, the Tribunal had observed that its task had been to decide whether the applicant in question had been a fit and proper person. Paragraph 36 of the Tribunal's decision contains these words:

> We conclude that it will be an abuse of process to permit the Applicant to mount a collateral attack on the findings and order of the Solicitors' Disciplinary Tribunal because he had a full opportunity of contesting the decision by way of appeal to the Divisional Court and also in his judicial review proceedings. Accordingly, we are of the view that we are free to make such use of the Solicitors' Disciplinary Tribunal findings as is proper in the circumstances.

Sir Stephen Oliver QC concluded that, on the strength of that decision and of the authorities on which it was based, it was permissible to take account of the conclusions of the SDT in the present case. Moreover, no attempt had been made to challenge or lodge an appeal against the SDT's judgment; indeed, the facts upon which its conclusions were based were not challenged before him.

### *Dhorajiwala v. General Pharmaceutical Council* [2013] EWHC 3821 (Admin)

**25.17** Police interview—criminal trial—conviction quashed—whether interview admissible in disciplinary proceedings

D, a pharmacist with an unblemished record for thirty years, was convicted at Reading Crown Court of theft from her employer. Her defence that she had been acting on instructions from her employer to remove monies from the till so that they did not pass through the books of the business was rejected by the jury. Subsequently, D's conviction was quashed by the Court of Appeal, Criminal Division, on the grounds that there had been a procedural irregularity on the part of the trial judge in admitting a documentary record of an interview in which D was recorded as making a series of damaging admissions. The Court of Appeal did not hold that the record of interview was intrinsically inadmissible; the appeal succeeded on the basis that there had been a failure to implement procedural safeguards. It was held that the General Pharmaceutical Council's Fitness to Practise Committee had not erred in admitting the record of interview at the hearing of D's subsequent disciplinary proceedings, even though one of the two inspectors had since died. The Committee had correctly considered the possibility of breaches of the PACE Code and the question of oppression.

### *R (H) v. Nursing and Midwifery Council* [2013] EWHC 4258 (Admin)

**25.18**

See paragraph 14.11.

### *R (Hollis) v. Association of Chartered Certified Accountants* [2014] EWHC 2572 (Admin)

**25.19** High Court judgment—findings by judge—case management hearing to admit material—judicial review

H, a licensed insolvency practitioner and a member of the Association of Chartered Certified Accountants (ACCA), applied by way of judicial review to quash a decision of ACCA's Disciplinary Committee at a case management hearing to admit material into evidence to stand as prima facie evidence of misconduct against him. H was a party to proceedings in Companies House in which Henderson J was very critical of H's conduct as administrator of a company in promoting a company voluntary arrangement. The learned judge said, in his judgment, that he was driven to conclude that there was a deliberate misrepresentation of the true position and that H was unwilling to disclose the truth. The Committee decided to admit the judgment, under regulation 11(2)(d) of the Disciplinary Regulations 2014, to stand 'as prima facie evidence in the proceedings' against H and on the basis that the judgment comprised a 'finding of fact' by the judge, within the meaning of the regulation. H contended that the relevant parts of the judgment did not qualify as relevant 'findings of fact' for the purposes of regulation 11(2)(d) and, accordingly, could not be treated as admissible prima facie evidence in the proceedings. H accepted that the judgment could be read as background, but without attaching prima facie evidential weight to it. In rejecting H's submission and dismissing the application for judicial review, Sales J said, at [41], that the meaning of the phrase 'finding of fact' is straightforward and accords with the ordinary meaning attached to those words. It covers any matter in a judgment in civil proceedings that, as a matter of ordinary language, is properly described as a finding of fact made by the court in the course of giving its judgment. The mention of a 'finding of fact', when used in reference to a judgment, is one that is very familiar, and there is no indication that the drafter of regulation 11 intended the phrase to bear any strained or unusual meaning when used in that provision. In *Constantinides v. The Law Society* [2006] EWHC 725 (Admin), the Divisional Court, when considering the Law Society's disciplinary code, appeared to have no difficulty in giving the term its ordinary meaning. The weight to be given to the findings of fact made by Henderson J in the judgment was a matter at large for determination at the substantive hearing, on the basis of evidence and submissions at that stage (cf. *In re a Solicitor* [1993] QB 69, DC, 80F–H): at [49]. It would be unhelpful to describe the effect of regulation 11(2)(d) as creating, in formal terms, an evidential burden upon any person wishing to disprove a finding of fact in a judgment that falls within the provision. The regulation simply says what it says. It makes a relevant finding of fact in a judgment admissible as prima facie evidence in the proceedings, but no more. Something can be prima facie evidence in proceedings and yet ultimately be found to have no weight at all, such as a witness's evidence in a witness statement or a document that, on full examination of the case, is shown to be wholly mistaken or irrelevant. However, the more a finding of fact in a judgment represents the considered view of a judge after hearing detailed argument and evidence, with full notice to the ACCA member involved and an opportunity for him or her to participate in the hearing, the greater the weight it is likely to carry. This may impose a practical onus on the member to be willing to come forward before the Disciplinary Committee to give his or her own, exculpatory, version of events. In this way, the weight to be given to a finding of fact will be modulated by, amongst other things, issues regarding the fairness of the underlying proceedings to the member and a nuanced assessment of the probative force in the context of the disciplinary proceedings against the member of the finding made by the court in a different context: [51]–[52].

### *R (Squier) v. General Medical Council* [2015] EWHC 299 (Admin)

**25.20** Expert witness—judgments of High Court and Court of Appeal, Criminal Division—whether judgments admissible in disciplinary proceedings—relevance and fairness

S, a consultant paediatric neuropathologist, sought judicial review to quash the decision of a fitness to practise panel of the GMC's Medical Practitioner's Tribunal Service to adduce in evidence five judgments of the High Court and one of the Court of Appeal, Criminal Division, in six cases of alleged non-accidental head injury to infants—'shaken baby syndrome'—in which S had given evidence as an expert witness. The GMC agreed that redactions should be made of the adverse comments and findings in the judgments. The issue before the panel concerned the relevance and fairness of admitting the judgments as evidence. After reviewing *General Medical Council v. Spackman* [1943] AC 627, *Constantinides v. Law Society* [2006] EWHC 725 (Admin), *Chaudhari v. General Pharmaceutical Council* [2011] EWHC 3433 (Admin), *R (Hollis) v. Association of Chartered Certified Accountants* [2014] EWHC 2572 (Admin), and *General Medical Council v. Meadow* [2006] EWCA Civ 1390, Ouseley J, in rejecting S's contention that the decision on admissibility was unlawful, said:

> 39. In the light of those authorities, the FTPP did not act unreasonably in concluding that the judgments would be relevant in providing an insight into the background to the cases and the forensic context in which [S] prepared and gave her evidence, and in providing prima facie evidence of facts about the circumstances of the deaths, the post-mortems, what the parents said, and the medical issues faced at trials to which [S]'s evidence was relevant. The judgments are relevant to the scope of the medical issues and why particular factual bases were relevant for consideration, and to the potential effect on the outcome of the cases. They are relevant to show the nature of the issue about the cause of death. The gravity and nature of the issues may be relevant to the care and precision required in understanding what reports say and the importance of a full description of their limitations and nuances.
>
> 40. The FTPP did not decide that the judgments were to be admitted to prove the cause of death; that is not an issue for the FTPP. They are not relevant to prove [S]'s evidence was not accepted or was found to be lacking in certain qualities. The issue before the FTPP is not whether [S] was right or wrong which was the issue before the judges, but concerns the basis upon which she gave her evidence, its scope and her use of the underlying research papers. That is the crucial issue for the FTPP. The actual outcome of the trials, and any finding in or inferred from the redacted judgments that [S]'s evidence was rejected, is not relevant to these allegations of misconduct. The fact

> that the issues which were before the judges and the issues now before the FTPP are different does not mean that the judgments are irrelevant to the background to her giving evidence or to the forensic context in which the evidence was given, even if before the judge that context was highly contentious.

The learned judge said that the modern expression of *Hollington v. F Hewthorn & Co. Ltd* [1943] KB 587 was found in *Hoyle v. Rogers* [2014] EWCA Civ 257, [39], in the judgment of Clarke LJ:

> The trial judge must decide the case for himself on the evidence that he receives, and in the light of the submissions on the evidence made to him. To admit evidence of the findings of fact of another person . . . or the opinion of someone who is not the trial judge is, as a matter of law, irrelevant and not one to which the trial judge ought to have regard.

Ouseley J went on to say that:

> 43. There is not so great a divide between *Hoyle v Rogers* and *Spackman* and the more recent disciplinary cases . . . The crucial point about the role of the disciplinary tribunal is that it should be the decision maker on the issues and evidence before it; it should not adopt the decision of another body, even of several judges, as a substitute for reaching its own decision on the evidence before it, on the different issues before it. None of that precludes the GMC under its Fitness to Practise Rules considering the judgments . . . [But] it is the FTPP's statutory duty to decide the issues before it. . . . It may also be unfair for the judgments to be a significant influence on the mind of the tribunal on the crucial issues before it.
>
> [. . .]
>
> 45. . . . The FTPP should be very careful to avoid any actual or inferred findings of the judges on the quality of [S]'s evidence being used as evidence of the truth of the allegations, because that risks substituting another body for its functions. . . .

Moreover, noted the learned judge, the specific findings had been redacted. It was therefore not unfair for the judgments to be admitted.

See further paragraph 42.27.

## D. Hearsay Evidence

### *Idenburg v. General Medical Council* (2000) 55 BMLR 101

**25.21** Medical notes—whether legally obtained—admissibility

The Professional Conduct Committee (PCC) of the GMC found proved an allegation that I did not give any, or any adequate, prior notice of her intention to leave her post at a hospital in Scotland, and that certain statements that she later made to the GMC in letters were misleading and false. I had claimed that she had left her job in Scotland for medical reasons and complications with her pregnancy, which necessitated rest to try to prevent a miscarriage. The GMC asked I to provide evidence of the dates of consultations, and for permission to approach her general practitioner and gynaecologist for the information. I gave consent, but took strong objection to the doctor disclosing, without her permission, detailed medical information, which she claimed was confidential. The doctor confirmed that she first attended after her contract with the hospital expired. In dismissing I's appeal and suspension for three months, Lord Clyde, giving the judgment of the Privy Council, said that their Lordships were satisfied that the admissibility of the medical evidence did not depend upon it having been legally obtained. The principle in *R v. Khan (Sultan)* [1997] AC 558 (the test of admissibility is relevant, and that relevant evidence, even if illegally obtained, was admissible) was not restricted to criminal cases and it applied to disciplinary proceedings, such as that with which the present appeal was concerned. Indeed, rule 50 of the General Medical Council Preliminary Proceedings Committee and Professional Conduct Committee (Procedure) Rules 1988 enables an even wider scope to admissibility than would be permitted under criminal proceedings. What is essential is that it be relevant to the issue to be determined—and that was very plainly the case here. That the appellant had not seen the doctor until after her contract expired pointed strongly to the conclusion that she had not been unable to return to work as a result of a problematic pregnancy.

### *Razzaq v. General Medical Council* [2006] EWHC 1300 (Admin)

**25.22** Documentary evidence—signature—evidence—whether challenge well founded

R appealed against two charges of serious professional misconduct. The misconduct alleged was the signing of National Health Service (NHS) optical vouchers certifying that he had tested the sight of patients when, in fact, he had not done so. It was alleged, and found by the panel, that R's conduct was dishonest, an abuse of his professional position, and intended to defraud the NHS. The panel ordered that his name be erased from the register. In dismissing R's appeal, Stanley Burnton J said that the unchallenged documentary evidence in the case was evidence of the truth of its contents. Where a document contains a statement and has been signed by a person, the normal inference is that the signatory intended to make that statement. Thus the signature of the patient on the form, if hers, was evidence that she had received a pair of glasses under the NHS optical voucher scheme. However, if it was true, as she said, that she had not read the document before signing it and had signed it at the insistence of the optician, the document was not evidence of any statement by her apparently contained in it. The panel had to weigh the fact that she had signed the document and the evidence of R against her evidence, and the other evidence before it, and had to decide whether it was sure that her evidence was accurate. Section 117 of the Criminal Justice Act 2003 takes the matter no further. The section is not specifically incorporated into the procedural rules of the panel, and is more appropriate to a trial before a judge and jury than to proceedings before the panel. Essentially, once the truth of the contents of a document is challenged by admissible evidence, as in the present case, the tribunal of fact must decide whether that factual challenge is well founded. If it is, the document ceases to be evidence of the facts stated. There was nothing in the present case to lead the Court to conclude that the panel did not deal with the documents appropriately.

### *Nursing and Midwifery Council v. Ogbonna* [2010] EWCA Civ 1216

**25.23** Witness—non-attendance—evidence challenged—arrangements for cross-examination

In relation to the first charge, X was the sole witness. At the hearing before the Conduct and Competence Committee (CCC) of the Nursing and Midwifery Council (NMC), the registrant opposed the reading of X's witness statement on the grounds that the evidence was roundly disputed. Rimer LJ (with whom Pill and Black LJJ agreed) said that it was obvious that, in the circumstances, fairness to the registrant demanded that, in principle, the witness statement ought only to be admitted if the registrant had an opportunity of cross-examining the witness upon it. It should have been obvious to the NMC that it could and should have sought to make arrangements to enable such cross-examination to take place—either by flying the witness to the United Kingdom at its expense, or else by setting up a video link. The NMC had given no thought to anything like that, and the CCC proceeded instead on the groundless, and mistaken, assumption that the witness had said she was unable to come to the UK. If, despite reasonable efforts, the NMC could not have arranged for the witness to be available for cross-examination, then the case for admitting her hearsay statement might well have been strong. But the NMC made no such efforts at all. The Court of Appeal disagreed with the NMC's submission that, having admitted the statement, then it was for the Committee to make a careful assessment of its weight. That submission overlooked the point that the criterion of fairness was whether the statement should be admitted at all. As to the other charges, because the Committee's findings on these charges may be tainted by its finding on the first charge, the Court of Appeal directed that the second and third charges alone should be reheard afresh by a new panel.

### *R (Vali) v. General Optical Council* [2011] EWHC 310 (Admin)

**25.24** Complainant—non-attendance—son present at examination and available to give evidence—issue of what happened—whether unfair to admit father's statement

The registrant qualified as an optometrist in 2005 and, whilst employed with Specsavers Optical Group Ltd, dealt in 2005 with patient A. The allegation against the registrant was that, at a consultation, she did not repeat the recording of intraocular pressure measurements, did not perform a visual field test, and did not make a referral of the patient to a medical practitioner concerning his abnormal intraocular pressure measurements. The patient was a black African from Ethiopia in his early 60s. He was accompanied at the consultation by his son, who spoke good English, while the father's English was distinctly more limited. When the allegations came before the Fitness to Practise Committee of the General Optical Council (GOC), one of the first issues concerned the admissibility of patient A's statement. Patient A was in Ethiopia and had refused for at least a year to engage with the GOC. He had not responded to letters. His son was present to give evidence, but the statements of the father and son were taken together in November 2007 in England at the GOC premises. Both statements were in English, which was not a language the father spoke well. Looking at the fairness of the proceedings as a whole, Ouseley J said that he did not consider that it had been shown that the admissibility of the father's witness statement led to unfairness. The son had been present at the appointment and was able to give evidence covering—and could be subject to cross-examination on—all of the issues about what happened. The narrow issue was whether there had been oral advice by the appellant to patient A to see a doctor about his eye pressure. However, the appeal would be allowed on other grounds. A finding of impairment in 2010 was not justified. This was a one-off, albeit serious, oversight. The error had not been repeated, nor had any like errors been repeated. The level and number of testimonials suggested that the error had not been repeated in circumstances falling short of those that might give rise to a complaint. Previous employees and employers had all supported the appellant, and she had progressed in her supervisory roles and in her teaching roles.

### *R (Bonhoeffer) v. General Medical Council* [2011] EWHC 1585 (Admin)

**25.25** Witness—sole evidence—non-attendance—decision not to call—unfair to admit statement

The Administrative Court (Laws LJ and Stadlen J) observed that the case raised important issues relating to the circumstances in which hearsay evidence may be admitted in disciplinary proceedings. The claimant, an eminent consultant paediatric cardiologist, applied for judicial review of the decision of the GMC's fitness to practise panel to admit hearsay evidence of witness A in fitness to practise proceedings against the claimant. It was alleged by the GMC that the claimant was guilty of serious sexual misconduct while undertaking work in Kenya. The evidence against the claimant in respect of the majority of the charges that he faced came from a single source, witness A, whose identity was disguised. Although witness A indicated that he was willing and able to travel to the UK to give evidence in person in support of the allegations, the GMC decided not to call him as a witness. The GMC contended that calling witness A would place him at a risk of reprisals from homophobic elements in Kenya were he to be identified as having engaged in sexual activity with the claimant, and that he could be exposed to a risk of harm from those who were loyal to the claimant and who might wish to exact revenge for witness A's participation in the fitness to practise proceedings. The High Court quashed the decision of the fitness to practise panel to allow witness A's evidence to be read, holding that the panel's decision was irrational and a breach of the registrant's right to a fair hearing. The panel should have concluded that the general obligation of fairness imposed by the common law and Article 6 of the European Convention on Human Rights (ECHR), and rule 34 of the GMC's Fitness to Practise Rules, meant that it should not admit the evidence of witness A under the hearsay provisions. Witness A was the sole witness in relation to most of the allegations and he was willing to give live evidence. The arguments for affording the claimant the opportunity to cross-examine witness A were formidable. The claimant was an extremely eminent consultant cardiologist of international repute and the allegations against him could hardly have been more serious. If proved, they would have a potentially devastating effect on his career, reputation, and financial position, and, in so far as the allegations involved alleged misconduct towards other victims, those victims denied that the allegations were true. At [108], Stadlen J said that, in disciplinary proceedings that raise serious charges amounting in effect to criminal offences, there would, if the evidence were to constitute a critical part of the evidence against the accused party and if there were no problems associated with securing the attendance of the accuser, need to be compelling reasons why the requirement of fairness and the right to a fair hearing did not entitle the accused party to cross-examine the accuser. At [116], Stadlen J noted that of relevance also was the fact that the standard of proof in fitness to practise hearings was changed in 2008, in response to the *Shipman* case, from the criminal standard of being satisfied 'beyond reasonable doubt' to the civil standard of the 'balance of probabilities'; nor is there a requirement that the fitness to practise panel must be satisfied beyond reasonable doubt that the regulatory body was not able to adduce the evidence by calling the witness.

***Belal v. General Medical Council* [2011] EWHC 2859 (Admin)**

**25.26** Personal data form—Criminal Justice Act 2003, section 117—admissibility in disciplinary proceedings

Dr B faced various charges, which included an allegation that, in April 2009, he did not inform the Health Authority of Abu Dhabi that his UK registration was suspended. In April 2009, an agency acting on behalf of Dr B submitted a personal data form signed by Dr B to the Health Professionals Licensing Department of the Health Authority of Abu Dhabi. Under the heading 'Relevant Licence(s)', an entry referred to Dr B's annual registration in the United Kingdom and under the subheading 'Status' appeared the word 'Active'. The panel found the charge proved. On appeal, Dr B submitted that the personal data form was not admissible as evidence before the panel. Dr B submitted that the document would not be admissible in criminal proceedings in England. In particular, its provenance was unknown, it was not produced by a witness, and there was no evidence from the maker of the document. The document was, on its face, incomplete and it contained a manuscript passage that apparently had been added some time after the form was signed by Dr B. Lloyd Jones J, in dismissing the ground of appeal, said that the GMC's concession that the document could not be admissible in criminal proceedings under section 117 of the Criminal Justice Act 2003 was correctly made. There was no witness or witness statement to adduce the document. It would not have been admissible because there would have been no evidence of the matters set out in section 117(2). However, the panel's decision to admit the document under rule 34(2) of the General Medical Council (Fitness to Practise) Rules 2004 was clearly correct. The GMC did not seek to adduce the personal data form as evidence of the truth of its contents; rather, it was intended to adduce it to prove that a statement had been made in the document that the annual registration of Dr B with the GMC was active. The form was relevant to the allegation in the charge. The panel was entitled to conclude that its duty of inquiry and its overriding duty to protect the public made it desirable to admit the document in evidence. Furthermore, Dr B was not prejudiced in any way by any uncertainty as to the provenance of the document. In the event, Dr B accepted in his evidence that he had signed the form after it had been completed by the agency. Its admission did not cause any unfairness to Dr B.

### *Lawrence v. General Medical Council* [2012] EWHC 464 (Admin)

**25.27** Case notes—original destroyed—typed summary—whether unfair to admit typed notes

L, a consultant psychiatrist, faced allegations that he had an inappropriate relationship with a patient that was sexually motivated. The GMC called as a witness B, a psychotherapist, who had prepared case notes of psychotherapy sessions that she had with the patient. B had destroyed her original notes when she prepared her typed summary. Stadlen J held that the fact that B's original notes had been destroyed did not make it unfair to admit the typed notes. It was not suggested in cross-examination that B had destroyed them in bad faith and it was clear from the transcript that it did not occur to her that the original handwritten notes would be of interest to anyone. B's evidence was that the typed notes contained no factual information other than what had been contained in the original handwritten notes, and that where the original notes recorded the actual words used by the patient to her, that was shown in the typed notes by the use of italics. The fitness to practise panel was entitled to admit B's typed notes.

### *Shaikh v. General Pharmaceutical Council* [2013] EWHC 1844 (Admin)

**25.28** Witness statements—witness unavailable through age, illness, or reluctance to attend—admissibility of evidence—relevance and fairness

The Fitness to Practise Committee of the General Pharmaceutical Council found two charges against S, a pharmacist, proved. The first was an allegation that he was dishonest in dealing with National Co-operative Chemists Ltd, for whom he used to work, in respect of £26,000 that was overpaid in error and which he then refused to pay back. The second was that, as the superintendent pharmacist of a pharmacy, he was dishonest in dispensing repeat prescriptions to patients without their consent and, in one case, presenting a prescription for payment to the Prescription Pricing Authority for reimbursement, notwithstanding that the medication had not been supplied and was refused by the patient. The Council had sought to rely on witness statements of six patients at the hearing before the Committee. The witnesses did not attend through age, illness, or a reluctance to give oral evidence, and the Committee admitted the evidence under rule 24 of the General Pharmaceutical Council (Fitness to Practise and Disqualification) Rules 2010 ('subject to the requirements of relevance and fairness, the Committee may receive any documentary evidence whether or not such evidence would be admissible in any subsequent civil proceedings if the decision of the Committee were appealed to the relevant court'). On appeal, S submitted that the rules did not allow the use of hearsay; in the alternative, the use of the particular hearsay evidence was unfair and outwith the guidance in *R (Bonhoeffer) v. General Medical Council* [2011] EWHC 1585 (Admin). In dismissing S's appeal, Supperstone J said that rule 24 gives the Committee the power to decide all questions of admissibility of evidence subject only to the requirements of the relevance and fairness. It is quite clear that the Committee can receive any evidence under rule 24 so long as it meets the requirements of relevance and fairness and, in the case of a witness statement, is in the form required by the rules. It was clear that the Committee gave careful consideration to the admissibility of the evidence. It considered individually the circumstances of each witness and considered in each case the reasons for their non-attendance. The Committee had regard to the importance of their evidence in the case and to the fact that the appellant had deliberately absented himself so that he would not have been in a position to cross-examine witnesses, but added that it would nevertheless seek to test the evidence in the absence of S to the extent that it reasonably could. The evidence to support the charge of dishonest dispensing of repeat prescriptions was not confined to the evidence of the patients. One patient gave oral evidence and six other witnesses also gave evidence: a practice manager, a pharmacy manager, a general practitioner, a pharmacist, the former owner of a pharmacy, and an inspector with the Council. The evidence from the absent patients was contained in contemporaneous witness statements and the evidence from live witnesses; doctors and pharmacists corroborated the evidence of the absent patients. It is also clear that the Committee considered the repeat prescription service allegations and the Co-op overpayment allegations separately. The decision of the Committee to admit the hearsay evidence could not be faulted.

### *Ogudele v. Nursing and Midwifery Council* [2013] EWHC 2748 (Admin)

**25.29** Witness—non-attendance—other evidence—not fatal

Patient A attended outpatients in October 2008 to have a dressing removed. It was alleged that the appellant undertook the task very poorly. There were four witnesses to the event: patient A and the appellant, a ward sister, and a consultant orthopaedic surgeon. Patient A was ready willing and available to give evidence at the hearing in August 2012. Because of delays, the case was adjourned part-heard to February 2013. Unfortunately, patient A was then on holiday in Thailand. His statement was admitted by the panel. There was no application to adjourn his evidence until after his return in a few days. The panel heard evidence from the ward sister. It was held that there was a need for the panel to proceed with caution in the absence of cross-examination of patient A, who was an important factual witness. It was important for the panel to view his evidence to see if there was any inherent or other weakness. Additionally, the panel needed to ascertain if there was evidence supporting or buttressing the untested evidence. In the result, the panel was requested to make a fair assessment in all of the circumstances. The panel made that assessment. The appellant was acquitted on one charge and this was not a case of blanket acceptance of the NMC case against the appellant. Additionally, the panel gave cogent reasons for its findings and expressly stated that it took into account the disadvantage to the appellant by the absence of cross-examination of patient A. Whilst regrettable that patient A could not be cross-examined, it was not fatal to the fairness of the proceedings, as the detail of the findings of the panel demonstrated.

### *White v. Nursing and Midwifery Council; Turner v. Nursing and Midwifery Council* [2014] Med LR 205

**25.30** Anonymous letters—admissibility—fairness—other admissible evidence

W and T, two nursing sisters working in an accident and emergency department, were subject to disciplinary proceedings, the main allegation being that they had falsified patients' records to show that they had spent less time in the department in order to meet departmental targets. Before the CCC, the NMC adduced in evidence three anonymous letters. Dismissing W and T's appeals, Mitting J said that it was settled law that Article 6(1) ECHR applied to disciplinary proceedings that might result in the removal of professional status. However, the protections of Article 6(3) in criminal proceedings, including the right to examine witnesses did not apply. Rule 31 of the NMC Fitness to Practise Rules 2004 provided that, upon receiving advice of the legal assessor and subject only to the requirements of relevance and fairness, a practice committee considering an allegation could admit oral, documentary, or other evidence, whether or not such evidence would be admissible in civil proceedings. Rule 31 was, however, subject to the requirements of fairness and an opportunity to test the evidence: *Ogbonna v. Nursing and Midwifery Council* [2010] EWCA Civ 1216 and *R (Bonhoeffer) v. General Medical Council* [2011] EWHC 1585 (Admin). The absence of any means of testing the anonymous hearsay evidence meant that the test of fairness could not be satisfied and the CCC had erred in admitting the anonymous letters. However, the Committee's conclusions were not expressly based on the anonymous evidence and there was other admissible evidence to support the charges. There may be different categories of evidence that, although anonymous, could be fair to admit, such as hospital records as a contemporaneous note even if it were not possible to identify the author.

### *Thorneycroft v. Nursing and Midwifery Council* [2014] EWHC 1565 (Admin)

**25.31** Witnesses—non-attendance—sole or decisive evidence—approach

T was a senior nurse (modern matron) at a residential care home. He had twenty-five years' experience and worked primarily with clients with learning disabilities or neurological impairment. The allegation against T was that he had used derogatory language about patients. It was not suggested that any of the words had been used directly towards patients. The alleged incidents took place between 2002 and 2007. The allegations did not come to light until an investigation in 2011. Four witness statements were disclosed in support of the NMC's case: from a support worker, two nurses, and a member of the human resources team who had investigated the allegations. T, in a written statement to the panel, emphatically denied the allegations and said that they were based on 'rumours, distortions and lies', and that the investigation had been part of a 'witch-hunt' instigated by senior executives against him. He produced testimonials from other colleagues who had never heard him use unprofessional language of the type alleged. T put forward specific reasons why the support worker and one of the nurses might have personal antipathy towards him. The second nurse's evidence related to a single conversation in 2002 and the member of the human resources team had no first-hand knowledge of any of the incidents. The evidence of the support worker and the first nurse was the sole and decisive evidence on all but one of the charges. T did not attend the fitness to practise hearing and his decision not to attend was 'very unwise'. However, the appeal would be allowed. The support worker and the first nurse refused to attend the hearing. The second nurse was the only witness present. The member of the human resources team had emigrated, but would give evidence over a telephone link. The panel was led into error in its approach to the evidence of the two missing witnesses. The decision to admit the witness statements despite their absence required the panel to perform a careful balancing exercise. It was essential, in the context of the present case, to take the following matters into account:

(1) whether the statements were the sole or decisive evidence in support of the charges;
(2) the nature and extent of the challenge to the contents of the statements;
(3) whether there was any suggestion that the witnesses had reasons to fabricate their allegations;
(4) the seriousness of the charge, taking into account the impact that adverse findings might have on T's career;
(5) whether there was a good reason for the non-attendance of the witnesses;
(6) whether the NMC had taken reasonable steps to secure their attendance; and
(7) the fact that T did not have prior notice that the witness statements were to be read.

The panel was not given the opportunity to read T's witness statement before it took its decision to admit the witness statements of the support worker and the first nurse, and would not have been aware of the potential motives that T had suggested these two witnesses might have to lie. The issues raised by T were not outlined to the panel. Insufficient consideration was given to the fact that, despite some reference to ill health, the witnesses did not appear to have good reasons for their failure to attend. Their reluctance to attend in itself was capable of undermining the credibility of their evidence. There was no consideration of the further steps that could have been taken to compel the attendance of the witnesses. There was no reference to the serious consequences for T if the evidence were to be admitted and accepted. There was a second material error. The panel should have been provided, at the fact-finding stage, with all of the documents that T had submitted. As *Donkin v. Law Society* [2007] EWHC 414 (Admin) shows, there are cases in which character evidence goes to the credibility of the allegation itself. The statements submitted by T were relevant and admissible for that purpose. The error in this case was more fundamental. The decision on admissibility was a judgment for the panel to make, not the legal assessor or the case presenter. In the absence of T, the only proper way for the panel to judge the relevance and admissibility of the statements was to read them for themselves. That was not done. The transcript shows that, on the advice of the legal assessor, the panel left it to the case presenter to decide whether or not it should read the statements. The determination of the panel was to be quashed and a rehearing was undesirable. The allegations dated back many years, the last incident being approximately seven years previous. The appellant had already served long periods of suspension from the trust before his dismissal and by virtue of an interim order.

## E. Lies before the Tribunal

### *Rehman v. Bar Standards Board*, PC2008/0235/A and PC2010/0012/A, 19 July 2013

**25.32** Defendant's evidence—lies—separate charge

R faced a charge of discreditable conduct in that he failed to comply with a court order—namely, a judgment made against him on 10 July 2009 by the Clerkenwell & Shoreditch County Court, whereby he was ordered to pay £2,751.67 in instalments of £200 per month, payable from 10 August 2009. Prior to the hearing, R told the judgment creditor that he had sent a cheque for £2,000, but had later found that the cheque he had in mind was to the Inland Revenue. R later stated in an email to the Bar Standards Board that he then sent a cheque to the judgment creditor, which was never cashed and must have been lost in the post. R repeated these matters in evidence before the Disciplinary Tribunal of the Inns of Court. The Tribunal found that R had lied about sending the cheque and had also lied in his evidence to the Tribunal to the same effect. In dismissing R's appeal, the Visitors (Sir Raymond Jack, chairman) said that the Tribunal did not fall into error in finding that R had lied to the judgment creditor and to the Bar Standards Board. The emphasis of its decision was on the lies to the judgment creditor and in the email to the Board. However, the Visitors had no hesitation in rejecting the submission of the Board that the Tribunal was entitled to take into account R's lies to the Tribunal in deciding whether his failure to comply with the court order was discreditable conduct. What R said to the Tribunal was not part of the circumstances relating to the non-payment. If a defendant lies to a disciplinary tribunal, that may be a serious matter that can be met with a subsequent charge of dishonest conduct, but it cannot be part of the original offence: at [16].

## F. Failure to Give Oral Evidence

### *Iqbal v. Solicitors Regulation Authority* [2012] EWHC 3251 (Admin)

**25.33** Failure to give evidence—matter to be taken into account

The appellant, I, a solicitor, appealed against the decision of the SDT, which, whilst making no finding of dishonesty, nevertheless ordered that the appellant be struck off the roll of solicitors. The appellant appealed against that sanction. In dismissing B's appeal, the Divisional Court (Sir John Thomas P and Silber J) said that the appellant did not give evidence before the SDT despite the seriousness of the allegations. Of course, he was perfectly entitled to decline to give evidence, but it has been a principle in civil courts from time immemorial to take into account a person's failure to give evidence when reaching a conclusion in respect of him. Similarly, in the criminal courts, since the change in the law at the end of the last century, the court, when considering the evidence, is entitled to take into account the position that the defendant has taken as regards the giving of evidence. The practice of the SDT is not to take into account the failure to give evidence by a solicitor. However, the ordinary public would expect a professional person to give account of his or her actions and it would be appropriate for the SDT to review this practice. It can only be in the public interest that this practice, which might have been justified by analogy to the law as it used to be in the last century, might be brought up to date for the present century.

## G. Fresh Evidence on Appeal

### *Heffron v. Professional Conduct Committee of the United Kingdom Central Council for Nursing Midwifery and Health Visiting* [1988] 10 BMLR 1

**25.34** Fresh expert evidence—epidemiologist—evidence before tribunal—discretion

H, the appellant, applied to introduce fresh evidence on appeal. It took the form of an affidavit from an epidemiologist and a well-qualified expert on public health. The expert's expertise was the surveillance of the efficacy and safety of vaccines in children. What she had stated in her affidavit served to counter evidence given before the PCC by the senior nurse at the clinic at the time as to what were claimed by her to be the deleterious effects of giving the particular vaccine, especially to a premature baby, which this child apparently was. Watkins LJ (with whom Nolan J agreed) said that, in appropriate circumstances, there is power to admit fresh evidence into an appeal of this kind: see *Stock v. Central Midwives Board* [1915] 3 KB 756, 761. It has to be acknowledged, of course, that this is not a rehearing in the sense that the Crown Court rehears a case on an appeal from a magistrates' court; what this court is called upon to do is, among other things, scrutinize the evidence given at the hearing from which the appeal emanates. Where there has been an opinion expressed or an assertion of relevant fact made in the course of that hearing, as revealed in the transcript of evidence given, it is competent for this court to receive in its discretion fresh evidence going to that expression of opinion or that assertion of fact. It was upon this basis that the court came to the conclusion that it was proper to admit the evidence by affidavit of the expert in this case. The effect of the evidence is to completely destroy the opinion expressed by the senior nurse at the clinic at the time. She expressed an opinion to the Committee that it should not have accepted in the absence of confirmation from elsewhere. If the Committee wished to be informed of the effects of the inoculation carried out by the appellant, it was necessary for the prosecution to produce an expert upon such a matter to inform the Committee adequately of what could possibly amount to a serious aspect of this affair.

### *Threlfall v. General Optical Council* [2004] EWHC 2683 (Admin)

**25.35** Explanation for practitioner's earlier evidence—whether special grounds

The appellant applied to put her witness statement in evidence concerning her account of the events at the hearing before the GOC's Disciplinary Committee on 12 July 2002. The basis of the application was that the appellant had experienced such an abnormal degree of anxiety and panic when she gave her oral evidence to the Committee that she was not able to give a coherent account of events, and that it was not in the interests of justice for the appeal court to determine the appeal without her having an opportunity to give her account of events. Stanley Burnton J refused the application. The general rule is that an appeal court will not receive evidence that was not before the lower court without special grounds justifying doing so: see CPR 52.11(2), *Hertfordshire Investments Ltd v. Bubb* [2000] 1 WLR 2318, and *Hamilton v. Al-Fayed (Joined Party)* [2001] EMLR 15. The witness statement in the present case was not new evidence in the sense of evidence that was unavailable to the appellant at the date of the hearing before the Committee; rather, it was her evidence in a new form. To accept a witness statement in place of, or in addition to, her oral evidence before the lower court would allow a witness to improve her evidence as compared with that before the lower court. In addition, the appellant's witness statement would not be the subject of cross-examination, whereas the evidence before the Committee was so subject. Moreover, the appeal court is not in a position to assess the reliability and credibility of that evidence in comparison to that given by the patient, since it has not heard, and would not hear, either of them give evidence orally. Different considerations might arise if it were to be shown that the appellant had been treated unfairly or oppressively by or before the Committee. In that event, in most cases the appropriate remedy would not be to receive a witness statement as evidence, but to quash the decision of the Committee and to order a rehearing before a differently constituted Committee. However, in the present case, unfairness or oppression was not, and could not be, suggested. There were no special reasons that could justify the reception of the witness statement and there were good reasons why it should not be received in evidence.

### *R (Russell) v. General Medical Council* [2008] EWHC 2546 (Admin)

**25.36** Order for suspension—effect—fresh evidence—whether to admit

R, a registered medical practitioner practising as a consultant in child and adolescent psychiatry, appeared before a fitness to practise panel of the GMC in August 2007. After hearing evidence and giving a decision on fitness to practise, the hearing was adjourned until 21 November 2007 for sanction. The panel found that R's fitness to practise was impaired by reason of her adverse mental or physical health—namely, by reason of a bipolar affective disorder. There was no challenge to that finding. At the resumed hearing on 21 November 2007, the panel suspended R's registration for two months under section 35D of the Medical Act 1983. On appeal, it was alleged that R was the victim of procedural unfairness in that the panel reached its decision without indicating to her that it was considering a suspension. R sought to adduce fresh evidence on appeal, comprising a witness statement from her treating psychiatrist, Dr W, who gave evidence at the hearing, and a letter from the medical director of the Tees, Esk and Wear Valley NHS Trust. The former included a statement that it was impossible to predict how R would respond to the stress of suspension, that it was fairly clear that she would find any threat to her career extremely distressing and stressful, and that suspension could lead to a relapse of R's illness, thus exacerbating any potential threat to her very future in medical employment. R's treatment psychiatrist said that had he been asked to express an opinion on these issues at the hearing, he would have done so in these terms. The letter from the medical director of the NHS Trust referred to the adverse effect on the delivery of health care in the specialist field in which R practised—namely, child learning disability—should the suspension continue. Dyson LJ, sitting in the Administrative Court, said that he was not prepared to admit this evidence. First, it could, with reasonable diligence, have been obtained before the August 2007 hearing, and certainly before the November 2007 hearing. The possibility of a suspension was, or should have been, obvious in any event. The Court endorsed what Collins J said in *Watson v. General Medical Council* [2006] EWHC 18 (Admin) at [13] (the fact that the appellant had not been advised to obtain a report from a psychiatrist cannot justify an attempt now to fill the gap) and what Blake J said in *Abrahaem v. General Medical Council* [2008] EWHC 183 (Admin) at [34] (the admission of fresh contested evidence is very much the exception in section 40 [Medical Act 1983] appeals and would require compelling justification).

### *Ansari v. General Pharmaceutical Council* [2012] EWHC 1563 (Admin)

**25.37** Fresh character evidence—CPR 52—guidelines

An application was made on this appeal for additional evidential material to be admitted in the form of a further character reference from Mr N, who was the managing director of a small chain of community-based pharmacists that employed the appellant. Put shortly, the further letter from Mr N described the professional capabilities of the appellant in glowing terms, and ventured opinions about the reaction of the appellant to his mistakes and the outcome of the hearing before the PCC. In ruling that it would not be right to admit the further character evidence, Sales J said that the appeal took the form of a rehearing on the merits. It was covered by CPR Part 52 and, in particular, CPR 52.11. The guidance in the White Book in relation to admission of evidence concerning matters that occurred after the date of the trial or hearing under appeal says:

> Under the former rules the express but exceptional power to admit further evidence as to matters occurring after the date of the trial or hearing will be exercised sparingly and with due regard to the need for finality in litigation: see *Hughes v. Singh* The Times April 21st 1989. The position under the CPR was summarised in *R (Iran) v. Secretary of State for the Home Department* [2005] EWCA Civ 982 at paragraphs 34 to 37. It remains the case that evidence of change to circumstances since the date of the original decision should only be sparingly admitted. Examples of cases in which it was appropriate to admit such evidence were given by Brooke LJ at paragraph 34.

In the instant case, N provided a character testimonial in respect of the appellant, which was considered at the time of the disciplinary hearing against him and was taken into account by the Committee. There was no question mark against the professional capabilities of the appellant, who was accepted to be a good pharmacist in his professional practice. Therefore the further evidence in the letter from Mr N would not assist the Court in relation to the appeal, so far as that matter is concerned. Evaluation of the appellant's evidence and the degree of insight that he has acquired into his offending behaviour is essentially a matter for assessment, in the first instance, by the Committee and then on appeal in the Court. The focus for this Court must, in accordance with CPR Part 52.11(3), be upon whether the lower court was wrong in the judgment to which it came. The Committee considered a range of testimonials in relation to the appellant, including one from Mr N, when it reached its decision. It was able to make its own assessment of the attitude of the appellant in the light of all of the available evidence, including, importantly, oral evidence given by him. This new testimonial from Mr N, giving his opinion about the appellant's attitude, could not be taken to override the Committee's assessment and did not show that the Committee's assessment was wrong. Evidence of change of circumstances since the date of an original decision should only be sparingly admitted.

### *R (Shaikh) v. General Pharmaceutical Council* [2013] EWHC 1844 (Admin)

**25.38** Fresh factual evidence—late application

In the week prior to S's appeal, his solicitors issued an application notice seeking an order that the fresh evidence be admitted. The application was supported by three witness statements from the appellant, his brother, and the manager of a pharmacy. The evidence also exhibited a new witness statement containing new matters not in her previous statement. In refusing S's application to admit the fresh evidence, Supperstone J said that the legal principles relating to the receipt of fresh evidence by an appeal court are well known. Before May 2000, the fresh evidence could be received upon appeal only where there were special grounds, as set out in *Ladd v. Marshall* [1954] 1 WLR 1489:

(1) that the evidence could not have been obtained with reasonable diligence for use at the trial;
(2) that the evidence must be such that, if given, it would probably have an important influence on the result of the case, although it need not be decisive; and
(3) that the evidence must be such as is presumably to be believed—that is, it must be apparently credible, although it need not be incontrovertible.

After the introduction of the Civil Procedure Rules, the principles reflected in *Ladd v. Marshall* are matters that an appeal court must consider in the exercise of its discretion, and it must do so in the light of the overriding objective of the Rules (see *Hamilton v. Al Fayed (No. 2)* [2001] EMLR 15, [1], *per* Lord Phillips MR). In the instant case, the evidence that the appellant now wished to introduce by way of fresh evidence could plainly have been obtained with reasonable diligence before trial. Not only had the appellant, the panel found, deliberately chosen not to attend the hearing, but the day before the hearing the respondents had been informed for the first time that the appellant did not, in fact, go to Saudi Arabia as he had said he intended to do when requesting an adjournment of the hearing. It was not until the day before the appeal that the statements on which the appellant now relied were produced to the Council. The learned judge said that he was not satisfied as to the credibility of the material parts of the witness statements of the appellant, his brother, and the witness who had previously given a statement. Much of the documentation now produced pre-dated the disciplinary hearing and could have been produced at the hearing. Further, the Court was not satisfied that the evidence that the appellant wished to adduce would probably have had an important influence on the result of the case. The admission of fresh contested evidence is very much the exception in appeals, such as this, and would require compelling justification: see *Abrahaem v. General Medical Council* [2008] EWHC 183 (Admin), [34], *per* Blake J, and *R (Russell) v. General Medical Council* [2008] EWHC 2546 (Admin), [28]–[30], *per* Dyson LJ, as he then was.

### *R (Chief Constable of British Transport Police) v. Police Appeals Tribunal and Whaley* [2013] EWHC 539 (Admin)

**25.39** Police Appeals Tribunal—medical evidence—fresh evidence—whether unfair—Police Appeal Tribunal Rules

W, a sergeant with British Transport Police (BTP), was convicted of harassment and was sentenced by way of a conditional discharge for a period of two years. The decision of a disciplinary panel was that W should be dismissed from the force as a result of gross misconduct. In 2009, W was diagnosed with bipolar disorder. A report from a psychiatrist, Dr W, was before the panel and the report concluded that W had bipolar 1 affective disorder. However, the report did not explicitly deal with whether W's mental condition had been causative of the misconduct in question. On appeal before the Police Appeals Tribunal, W sought to rely on a second report from Dr W, which did state clearly that there was a direct causation between the mental illness and the misconduct that was alleged. Rule 4(4) of the Police Appeal Tribunal Rules 2008 (which are applied to the BTP under the British Transport Police Appeals Tribunal Regulations 2008) provide that an appellant is entitled to appeal on the grounds:

(a) that the finding or disciplinary action imposed was unreasonable; or
(b) that there is evidence that could not reasonably have been considered at the original hearing which could have materially affected the finding or decision on disciplinary action; or
(c) there is . . . other unfairness which could have materially affected the finding or decision on disciplinary action.

The Police Appeals Tribunal dismissed W's appeal based on rule 4(4)(a) or (b). Dr W's subsequent report was new or fresh evidence and was not admissible under rule 4(4)(b). However, the Tribunal considered that the failure to prepare and serve Dr W's second report resulted in an unfairness in the proceedings both to W and to the panel itself, and allowed the appeal under rule 4(4)(c). By way of judicial review proceedings, the Chief Constable of BTP said that the Police Appeals Tribunal was wrong to consider that a failure to adduce evidence—which could and should have been produced at the original hearing—amounted to unfairness, even though the result could have been that the wrong decision was made when one looks at the subsequent evidence. Collins J agreed. At [19], the learned judge said that, in his view, there is no question but that, applying *Al-Mehdawi v. Secretary of State for the Home Department* [1990] 1 AC 876 and *R (Sinha) v. General Medical Council* [2008] EWHC 1732 (Admin), 'unfairness' in this context means unfairness to the individual police officer that results from something done or not done, either by the panel or Tribunal in question, or by the police constable, or those representing him or her, who bring the charges against the police officer. Such failures can produce unfairness within the meaning of rule 4(4)(c). The learned judge considered whether, in the exercise of discretion, he should refuse relief to the claimant on the basis that the Court now knew that there was a direct causal link between the mental condition and the conduct; he held that doing so would presuppose that the decision based upon that causal link covered the whole of the conduct and excused that conduct. Furthermore, it was by no means entirely clear, even from Dr W's second report, that the whole of the conduct in question could be put down to the mental condition. It did not seem right in those circumstances for the Court to exercise discretion against granting the relief claimed to quash the decision of the panel.

***R (Khan) v. General Medical Council* [2014] EWHC 404 (Admin)**

**25.40** Medical evidence—whether relevant to sanction

K applied to admit fresh evidence in the form of three letters from various medical practitioners. The application arose in the context of an appeal by K against a sanction of four months' suspension imposed by the panel because of certain criminal convictions that K had accrued. In admitting the evidence, notwithstanding the fact that it was not before the panel, Lewis J said, first, that there was no prejudice here to the GMC. The Council accepted that. Secondly, it was important that the Court understood what the evidence was about the position in relation to K when it came to evaluate the misconduct and the consequences. Thirdly, it was important to K, who had engaged in criminal activity, and that had to be taken into account and weighted against the sanction.

***Sohal v. Solicitors Regulation Authority* [2014] EWHC 1613 (Admin)**

**25.41** Medical evidence—earlier road traffic accident—psychological injuries

On appeal to the Administrative Court, S, against whom an order of striking off had been made by the SDT, sought to rely on fresh evidence that had not been provided to the Tribunal—namely, further medical evidence relating to both the physical and psychological effects of a road traffic accident that he sustained in May 2010. The reports were from a consultant surgeon, a consultant psychological professor, and a consultant clinical pharmacologist. The reports showed that S was suffering from psychological difficulties at the time of the events in question, which adversely affected his judgment. Baker J considered the evidence *de bene esse*. It was quite clear that the more recent expert evidence was credible, but it was equally clear that it could have been obtained with reasonable diligence for use before the Tribunal. No criticism could be made of the appellant's previous legal team. The reason why such further evidence had not been obtained by those representing the appellant at the time was that they had considered that such relevant evidence that could be provided had already been obtained and presented to the Tribunal from Dr H, consultant psychiatrist. This reflected the reality of the situation and the Court was not satisfied that the contents of the more recent reports, either individually or collectively, added any significant material over and above that contained in Dr H's earlier opinion. The Court was not satisfied that the more recent reports would have had an important, or indeed any, influence on the result of the appeal over and above that evidence which was already available from Dr H.

***Jasinarachchi v. General Medical Council* [2014] EWHC 3570 (Admin)**

**25.42** See paragraph 65.14.

## H. Other Relevant Chapters

**25.43** Appeals, Chapter 4
Conviction and Caution Cases, Chapter 14
Disclosure, Confidentiality, Data Protection, and Freedom of Information, Chapter 18
Findings of Fact, Chapter 27
Investigation of Allegations, Chapter 39

# 26

# EXPERTS



## A. Legal Framework

**26.01** *Rimmer v. General Dental Council* [2011] EWHC 3438 (Admin) | 26.07 |
| B. Disciplinary Cases | 26.02 | *Levinge v. Health Professions Council* [2012] EWHC 135 (Admin) | 26.08 |

[2012] EWHC 503 (Admin) | 26.09 |
| *Council for the Regulation of Health Care Professionals

[2012] EWHC 464 (Admin) | 26.10 |
| *Meadow

[2012] EWHC 2688 (Admin) | 26.11 |
| *Cohen

EWHC 581 (Admin) | 26.05 | *Hussein

[2013] EWHC 3535 (Admin) | 26.12 |
| *Yeong v. General Medical Council* [2009] EWHC 1923 (Admin) | 26.06 | C. Other Relevant Chapters | 26.13 |

Civil Procedure Rules, Part 35
Criminal Procedure Rules, Part 33
Family Procedure Rules, Part 25
*The Ikarian Reefer* [1993] 2 Lloyds Rep 68, 81–2 (key principles relevant to the expert's duties)

## B. Disciplinary Cases

### *Cullen v. General Medical Council* [2005] EWHC 353 (Admin)

**26.02** Cognitive disorder—expert opinion—rejection of evidence—reasons—expert's presence during evidence

C appealed against the determination of the Health Committee, which found that his fitness to practise was seriously impaired by reason of mild cognitive disorder. He contended that the Committee acted irrationally in preferring the evidence of Dr K to the other expert evidence before it, and that the Committee failed to give adequate reasons for its acceptance of the evidence of Dr K and its rejection of the other expert evidence before it. C, aged 72, was a general practitioner (GP) and, following his retirement, he sought work as a locum. C took up locum posts at RAF Digby and RAF Halton, where questions arose as to his treatment of a number of patients, and his behaviour towards colleagues and members of staff. The preponderance of medical opinion prior to the hearing before the Health Committee was that C was fit to practise. Substantial doubt had been cast on the validity of Dr K's conclusion of generalized cognitive impairment. In quashing the Committee's decision and remitting the matter for hearing to a differently constituted Committee, Stanley Burnton J said that there was a preponderance of medical opinion before the Committee that there was no evidence of any cognitive disorder of C. That evidence was of two eminent doctors, Professors P and T: the former was one of the General Medical Council's (GMC's) nominated medical examiners; the latter was described as an eminent neuropsychiatrist. Their evidence was ultimately unequivocal, logically sustained, and cogent. Professor T's criticism of Dr K's logic and conclusion were unshaken in the course of his questioning by the Committee, and the GMC did not seek to undermine them by cross-examination. The Committee rejected that evidence. It gave no reason for doing so. Its reference to Dr K's credibility was not a reason to reject what may have been the equally, or more, credible opinions of Professors P and T. The consistency between the test results had been taken into account by Professors P and T. Furthermore, the Committee accepted that the neuropsychological tests were not diagnostic. The judge said, at 53:

> In my judgment, it was particularly important in the present case for the committee to explain why they felt able to reject the conclusions of Professors P and T. Those conclusions had not been challenged by the GMC during the hearing. Dr K's conclusion that there had been a decline in C's functioning was unsustainable, and confirmed to be so by Professor T, in the absence of any results of tests performed on C at any time before she saw him.

Referring to *R (H) v. Ashworth Special Health Authority* [2002] EWCA Civ 923, a decision of a mental health review tribunal (in which Dyson LJ said it is important that the tribunal should state which expert evidence, if any, it accepts and which it rejects, giving reasons), Stanley Burnton J said that this applies with even greater force where the tribunal decides to reject most of the expert evidence and adopt the minority view. Stanley Burnton J went on to say that the proceedings of the Health Committee are not criminal proceedings; they are regulatory. There is much to be said for the Committee adopting what is now the normal course in civil proceedings, in which the factual evidence is heard first, in the presence of the experts if they are available, and both parties then call their oral expert evidence. This would have the considerable advantage that the medical experts would be able to comment on the behaviour of the practitioner when he gave his evidence and be able to respond to the Committee's questions concerning his evidence. There is no good reason why medical expert witnesses, whose good faith is not in issue, should be excluded from the hearing. It was unfortunate that, in the present case, neither Professor T, nor Professor P, nor Dr K, were able to comment on C's evidence, on which the Committee placed reliance.

### *Council for the Regulation of Health Care Professionals v. General Medical Council and Basiouny* [2005] EWHC 68 (Admin)

**26.03** Finding—expert evidence rejected—reasons

B, a doctor in the ear, nose, and throat (ENT) department of a hospital, faced a charge that, when he examined a 23-year-old Swedish woman, he inappropriately examined her groin and breasts. The Professional Conduct Committee (PCC), having found him guilty of serious professional misconduct, determined that he be suspended from the medical register for a period of six months. Richards J allowed an appeal by the Council for the Regulation of Health Care Professionals on the grounds that the decision was unduly lenient and remitted the matter to the PCC to reach a fresh decision as to penalty. The Committee's finding under head 5(c) that B's examination of the patient was not inappropriate and was clinically indicated was untenable. The finding was inconsistent with the only expert evidence before the PCC—namely, that of E, a highly qualified and experienced ENT consultant. Richards J said that whilst the Committee was a specialist tribunal with medical expertise of its own, there would have to be clear and compelling reasons, which he could not find, for it to reject expert medical evidence of the kind given by E. Second, the finding was inconsistent with parts of B's own evidence.

***Meadow v. General Medical Council* [2007] QB 462, CA**

**26.04** See paragraph 47.33.

***Cohen v. General Medical Council* [2008] EWHC 581 (Admin)**

**26.05** Finding—expert evidence rejected—reasons

The GMC's expert anaesthetist, in her report in response to a question of whether C's conduct amounted to misconduct, stated that: 'C's management of this case can be criticised on a number of grounds. However, I do not consider that these were so serious as to amount to misconduct, such that his registration might be called into question.' When giving evidence, the expert explained that: 'I think that the core anaesthetic was carried out to a standard entirely in keeping with what might be expected of a consultant anaesthetist, the heart of the matter.' In allowing C's appeal, Silber J said, at [68], that the panel was obliged to explain why it did not accept the opinion of the expert called by the GMC. The panel failed to give any reasons, or any cogent reasons, why it disagreed with this opinion.

See further paragraph 33.04.

***Yeong v. General Medical Council* [2009] EWHC 1923 (Admin)**

**26.06** Finding—misconduct—risk to patients—whether expert evidence

Sales J, in dismissing Y's appeal in relation to a finding of misconduct arising from a sexual relationship between the doctor and a former patient, said that the fitness to practise panel was 'well entitled to draw upon its own experience and judgment in forming a view whether Y presented a present risk to his patients'. The effect of the expert evidence called on behalf of Y was that he did not suffer from any psychological disorder that underlay his misconduct. In the light of that assessment, Sales J said that the expert's expression of opinion as to the risk posed by Y carried little weight attributable to any special expertise on the part of the expert. The question of the possibility of a reoccurrence of such misconduct was a matter of the ordinary assessment of likely human behaviour, in relation to which a psychiatrist's expertise confers no special privileged insight. The assessment of risk of any particular form of future behaviour is the sort of task that courts and tribunals regularly perform without needing to refer to expert psychiatric evidence.

See further paragraph 33.11.

***Rimmer v. General Dental Council* [2011] EWHC 3438 (Admin)**

**26.07** Body of expert opinion—risk to patients—conflicting views—no finding by Committee

The appellant, a general dental practitioner, administered to fifteen of his patients, who were children, intravenous polypharmacy sedation to calm them so as to permit him better to undertake dental procedures upon them. A complaint was made by the local primary care trust (PCT) about his sedation practices. The charge was that the appellant did not provide the parents or guardians of the patient with an explanation that intravenous conscious sedation was not the standard method of providing sedation to children and, in the circumstances, he carried out treatment under conscious sedation without the informed consent of the parents or guardians of the patients. Evidence was given to the PCC about intravenous polypharmacy by reputable experts, including Professor S for the General Dental Council (GDC) and Dr R for the appellant. Their views differed significantly. Underlying the heads of charge was a debate within the profession of dentistry about the risk or relative lack of risk of sedation by intravenous polypharmacy. Mitting J, in quashing the decision of the Committee, said that it was implicit from the speech of Lord Steyn in *Chester v. Afshar* [2005] 1 AC 134, 143 (at [14] and [16]), that the law does not impose on the medical or dental practitioner an obligation to warn a patient about the risks involved in a procedure if those risks are non-existent or not serious, or are not a risk, however small, of serious injury or harm. In the instant case, the evidence of Dr R would appear to have been that the risk of any harm was very small, and the risk of serious harm negligible, such that there was no requirement to inform the patient or the parent of a patient before intravenous polypharmacy sedation was attempted. Professor S's view was clearly to the contrary. The Committee made no express finding about which view it preferred. It founded its decision simply on the wording of the head of charge that intravenous polypharmacy sedation 'was not the standard method of providing sedation to children'. The Court was left with the position that it did not know what, if any, view the Committee formed about the difference of opinion between Professor S and Dr R. This was unfortunate, because if Professor S's view were right, then, applying the law as stated by Lord Steyn in *Chester v. Afshar*, intravenous polypharmacy sedation would have been unlawful without the informed consent of a child's parent or guardian. It is difficult to conceive that an unlawful procedure would not amount to misconduct.

### *Levinge v. Health Professions Council* [2012] EWHC 135 (Admin)

**26.08** Music teacher—teaching techniques—relevance of expert evidence

L was a music therapist employed at the Royal Welsh College of Music and Drama. The first allegation asserted that the appellant had utilized psychodynamic therapeutic techniques inappropriately in a higher education setting. It was common ground that music therapists commonly deploy psychodynamic therapeutic techniques in a clinical setting—that is, when a therapist is treating a patient. Equally clearly, it was a necessary part of the course of which the appellant was the leader that students should be taught to apply psychodynamic therapeutic techniques. The thrust of the complaint made against the appellant was that, in effect, she had treated students, on occasions, as if they were her patients; rather than restricting herself to teaching students about psychodynamic therapeutic techniques, she had practised such techniques upon them. The appellant denied vehemently that she had behaved in this fashion. The thrust of her case was that she had never crossed the line between teaching therapeutic techniques to students and practising such techniques upon them. She relied upon expert evidence from S, the head of music therapy at the Guildhall School of Music and Drama, in support of that contention. Wyn Williams J, in holding the finding of the Conduct and Competence Committee (CCC) in relation to the first allegation to be flawed, said that the issue of whether or not the appellant had strayed over the line between teaching psychodynamic therapeutic techniques and practising the same on her students was a difficult issue to resolve. It was, of course, permissible for the Committee to attach weight to the views of the students who gave evidence and upon whom such techniques had allegedly been practised. It was incumbent upon the Committee to scrutinize carefully the evidence that the appellant gave in relation to the first allegation. The evidence of S was capable of giving significant support to the appellant's case. Her evidence could not be dismissed simply on the basis that she had not visited the college. S's evidence was that there was nothing within the written statements of the students who complained about the appellant to support the allegation that the appellant was going beyond the role of a tutor/supervisor. S's evidence was that the appellant's approach was consistent with that used widely in psychodynamic supervision, within music therapy training. In those circumstances, it was incumbent upon the Committee to explain why it was that S's evidence was not accepted. The first allegation demanded that there be a careful appraisal of the appellant's teaching techniques and, just as importantly, an appraisal of whether those techniques were acceptable judged by reputable professional standards. S's evidence was important to both aspects of the appraisal. This was the sort of case that was exceptional in accordance with the principles laid down in *Southall v. General Medical Council* [2010] EWCA Civ 407, [56].

### *Walker-Smith v. General Medical Council* [2012] EWHC 503 (Admin)

**26.09** Clinical treatment—whether appropriate medical practice or research—conflicting expert opinion—*Bolam* test—lack of finding by Committee

Professor W-S, together with Dr W and Professor M, faced very lengthy fitness to practise proceedings arising from concerns expressed in *The Lancet*, which triggered a growth in public concern about a possible link between the triple vaccine for measles, mumps, and rubella (MMR) and the occurrence of developmental disorders in young children, often diagnosed as autism. In a written decision handed down on 24 May 2010, the GMC's fitness to practise panel concluded that Dr W and Professor W-S were guilty of serious professional misconduct, but that Professor M was not. It ordered that the names of Dr W and Professor W-S be erased from the register. Both initially appealed, but Dr W subsequently abandoned his appeal. The central allegation against Professor W-S was that, in relation to eleven children treated at the Royal Free Hospital, he subjected them to a programme of investigation for research purposes without ethics committee approval, that he caused some of them to undergo treatment that was not clinically indicated, and that, in relation to all of the children, his conduct was contrary to their clinical interests. The case histories of the eleven children were summarized in a paper published in *The Lancet* on 28 February 1998. On appeal, Mitting J said, at [9]–[18], that at the heart of the GMC's case against Professor W-S were two simple propositions:

(1) that the investigations undertaken under his authority on the children were not done as part of a research project, which required, but did not have, approval from the ethics committee of the Royal Free Hospital; and
(2) that the treatment was clinically inappropriate.

Professor W-S's case was that the investigations were clinically appropriate attempts at diagnosis of bowel and behavioural disorders in children with broadly similar symptoms, and that he was conducting medical practice, which did not require approval, as distinct from conducting research, which required ethics committee approval. Both sides accepted that the starting point for the distinction between medical practice and research was the January 1990 guidance of the Royal College of Physicians in a report entitled *Research Involving Patients*. The premise upon which the GMC's case was founded was an objective approach, and a subjective test would significantly water down the obvious requirements for medical research to be approved and monitored by ethics committees: if the intention of the practitioner were the sole, or even principal, determinant, an undesirable result may occur. The Court held that the purely objective test proposed by the GMC was not a reasonable interpretation of the Royal College's guidance. The panel made no express finding on this issue, and if it had founded its determination on the GMC's interpretation of the guidance, it would have been wrong to do so. Professor W-S's defence was that he intended and genuinely believed that what he was doing was solely or primarily for the clinical benefit of the children. The Court said that, when such an issue arises, a panel will almost always have to determine the honesty or otherwise of the practitioner. The issue arose starkly in this case. The panel was invited by the GMC not to determine Professor W-S's intention and was also invited not to determine his truthfulness in his dealings with the ethics committee. The GMC's approach to the fundamental issues in the case led it to believe that that was not necessary—an error from which many of the subsequent weaknesses in the panel's determination flowed. On the question of whether the treatment for the children was clinically inappropriate, the Court noted, at [23], that expert evidence was given on both sides on this issue. It was a striking feature of the panel's decision that it expressed no view about the expertise and objectivity of the experts, and even more striking that, when their views were in conflict, it expressed no conclusion about which of them it preferred. This was a serious weakness in its reasoning. In the case of each child, the experts were asked to consider whether the investigations undertaken were clinically indicated and, if not, contrary to the clinical interests of the child. It was common ground that the test in *Bolam v. Friern Hospital Management Committee* [1957] 1 WLR 582 applied to both issues. When, as was in fact the case, two experts expressed the view, respectively, that a colonoscopy (and, if appropriate, barium meal and follow-through) or lumbar puncture were clinically indicated and were not contrary to the clinical interests of the child, a finding that their view was not one held by a responsible body of medical opinion would have been an essential prerequisite to the dismissal of their evidence in respect of that child. The panel made no such finding. In conclusion, the learned judge said, at [186], that the panel's overall conclusion that Professor W-S was guilty of serious professional misconduct was flawed in two respects: inadequate and superficial reasoning; and, in a number of incidences, a wrong conclusion.

***Lawrence v. General Medical Council* [2012] EWHC 464 (Admin)**

**26.10** See paragraph 27.09.

***Kumar v. General Medical Council* [2012] EWHC 2688 (Admin)**

**26.11** See paragraph 47.46.

***Hussein v. General Medical Council* [2013] EWHC 3535 (Admin)**

**26.12** See paragraph 50.12.

## C. Other Relevant Chapters

**26.13**

Findings of Fact, Chapter 27
Impairment of Fitness to Practise, Chapter 33
Misconduct, Chapter 47
Reasons, Chapter 54

# 27

# FINDINGS OF FACT



## A. Legal Framework

**27.01** Most regulatory bodies provide for fitness to practise and disciplinary hearings to be conducted in stages, and for findings of fact to be considered and announced by the panel or committee before any consideration of misconduct, impairment, or sanction. Examples include:

General Dental Council (Fitness to Practise) Rules 2006, rule 19
General Medical Council (Fitness to Practise) Rules 2014, rule 17
Chartered Institute of Management Accountants Disciplinary Committee Rules, rule 10

## B. Disciplinary Cases

### *Gupta v. General Medical Council* [2002] 1 WLR 1691

**27.02** Appeal—principles

In delivering the judgment of the Privy Council, Lord Rodger of Earlsferry said, at [10]:

> The decisions in *Ghosh* [*Ghosh v. General Medical Council* [2001] 1 WLR 1915] and *Preiss* [*Preiss v. General Dental Council* [2001] 1 WLR 1926] are a reminder of the scope of the jurisdiction of this Board in appeals from professional conduct committees. They do indeed emphasise that the Board's role is truly appellate, but they also draw attention

> to the obvious fact that the appeals are conducted on the basis of the transcript of the hearing and that, unless exceptionally, witnesses are not recalled. In this respect these appeals are similar to many other appeals in both civil and criminal cases from a judge, jury or other body who has seen and heard the witnesses. In all such cases the appeal court readily acknowledges that the first instance body enjoys an advantage which the appeal court does not have, precisely because that body is in a better position to judge the credibility and reliability of the evidence given by the witnesses. In some appeals that advantage may not be significant since the witnesses' credibility and reliability are not in issue. But in many cases the advantage is very significant and the appeal court recognises that it should accordingly be slow to interfere with the decisions on matters of fact taken by the first instance body. This reluctance to interfere is not due to any lack of jurisdiction to do so. Rather, in exercising its full jurisdiction, the appeal court acknowledges that, if the first instance body has observed the witnesses and weighed their evidence, its decision on such matters is more likely to be correct than any decision of a court which cannot deploy those factors when assessing the position. In considering appeals on matters of fact from the various professional conduct committees, the Board must inevitably follow the same general approach. Which means that, when acute issues arise as to the credibility or reliability of the evidence given before such a committee, the board, duly exercising its appellate function, will tend to be unable properly to differ from the decisions as to fact reached by the committee except in the kinds of situation described by Lord Thankerton in the well-known passage in *Watt or Thomas v. Thomas* [1947] AC 484, 487–8.

In *Watt*, Lord Thankerton had said that the principle is a simple one and may be stated thus:

> I. Where a question of fact has been tried by a judge without a jury, and there is no question of misdirection of himself by the judge, an appellate court which is disposed to come to a different conclusion on the printed evidence, should not do so unless it is satisfied that any advantage enjoyed by the trial judge by reason of having seen and heard the witnesses, could not be sufficient to explain or justify the trial judge's conclusion;
>
> II. The appellate court may take the view that, without having seen or heard the witness, it is not in a position to come to any satisfactory conclusion on the printed evidence;
>
> III. The appellate court, either because the reasons given by the trial judge are not satisfactory, or because it is unmistakably so appears from the evidence, may be satisfied that he has not taken proper advantage of his having seen and heard the witnesses, and the matter will then become at large for the appellate court. It is obvious that the value and importance of having seen and heard the witnesses will vary according to the class of case, and, it may be, the individual case in question.

### *Chyc v. General Medical Council* [2008] EWHC 1025 (Admin)

**27.03** Allegation of sexual misconduct—complainant—reliability and credibility

C was found guilty of serious professional misconduct and his name was erased from the medical register. The allegations found proved by the panel included an allegation that he had behaved inappropriately towards a patient, and abused his position of trust by fondling her breast during an examination and asking if he could kiss her. It was conceded that if the Court were to reject the appeal concerning the patient, C would not be able successfully to submit that the finding of serious professional misconduct was inappropriate or that the penalty of erasure was disproportionate. In dismissing C's appeal, Foskett J said that there were questions about the account given by the patient. As a fitness to practise panel will inevitably know, a doctor can become the subject of a false or exaggerated allegation in circumstances in which raising an answer to the allegation is extremely difficult. A doctor can, of course, simply be the subject of a misunderstanding in the way in which an examination has been carried out. Matters such as those raised in relation to the patient in the instant case are legitimate areas for investigation before a tribunal of fact determining the truth or otherwise of what is alleged. But it is almost always—and emphasizing the word 'almost', given the parameters within which an appeal from a tribunal of fact can be considered by the Court—how the witness against whom these kind of matters are raised deals with them in evidence that will shape the decision of the tribunal. How they are dealt with will, of course, be seen in the context of the rest of the evidence. Any tribunal of fact would need to consider with care the issues going to the credibility and reliability of a complainant such as P in the present case, and would wish to observe closely how such a complainant dealt with questioning (whether in the form of evidence-in-chief, cross-examination, or questions from the panel) on those issues. In addition to observing P's evidence, the transcript also revealed that all three members took advantage of the opportunity (not, incidentally, offered to a jury trying a criminal case) to ask P questions at the conclusion of her questioning by counsel. The questions ranged over the delay in making the complaint (which she explained at least partly on the basis that she had tried to 'block out' the incident), the circumstances in which she came to continue to see the appellant, and the circumstances of the incident itself. At the conclusion of that, and having had the benefit of detailed submissions on both sides and the advice of the legal assessor, the panel reached the conclusion that the allegation was proved. Although not strictly obliged to give reasons (since the issue was essentially one of credibility and reliability), the panel gave reasons and said that, despite the time that had elapsed since the incident, P's account of the salient features of the assault was cogent, credible, and consistent throughout.

### *R (Wilson, Chief Constable of Fife Constabulary) v. Police Appeals Tribunal* [2008] CSOH 96

**27.04**

Inappropriate conduct—police misconduct hearing—no deficiency in reasoning—Police Appeals Tribunal—whether entitled to substitute its opinion of evidence

At a police misconduct hearing held over a number of days and concluded on 16 December 2004, C, a police constable, was found to have conducted himself in an oppressive or improper way towards a female, contrary to the Police (Scotland) Regulations 1996. The hearing took place before the chairman, a superintendent, who sat with two assessors, a solicitor and a chief superintendent. The disposal was the dismissal of C from Fife Constabulary with immediate effect. C appealed against the decision to the Police Appeals Tribunal under section 30 of the Police (Scotland) Act 1967. The Tribunal, consisting of four persons, allowed C's appeal on the casting vote of the chairman. The chief constable sought judicial review of the decision of the Police Appeals Tribunal on the grounds that it had misconducted itself in law by failing to give adequate reasons for its decision and fell into error when it came to deal with the evidence. In granting the claim and remitting the case for a fresh hearing before a differently constituted Police Appeals Tribunal, Lord Uist said that the misconduct hearing did not err in its approach to the standard of proof and that the Police Appeals Tribunal was wrong in so finding. The Tribunal was not invited to rehear the evidence. It recognized, correctly, that, that being the case, the only basis upon which it could interfere with a finding of fact made by the misconduct hearing was that set out by the House of Lords in *Watt or Thomas v. Thomas* [1947] AC 484, particularly Lord Thankerton at 487–8. The Police Appeals Tribunal was correct to decide that the requirements of *Watt or Thomas v. Thomas* had to be met before it was entitled to interfere with the misconduct hearing's conclusion on the facts, and whether it was entitled to interfere with the conclusion of the hearing and reach its own conclusion solely on the basis of the transcript of the evidence. The Tribunal was not correct in law in so holding. It is to be noted that the Tribunal did not say that the misconduct hearing was not entitled to believe the complainer. What the Tribunal had done in this case was to have succumbed to the same temptation as the Inner House did in *Thomson v. Kvaerner Govan Ltd* 2004 SC (HL), 1, and *Simmons v. British Steel Plc* 2004 SC (HL), 94, of retrying the case on the printed evidence. It substituted its own opinion of the evidence for that of the misconduct hearing. It did not seem to have had regard to what Lord du Parcq said in *Watt or Thomas v. Thomas*, at 493, about 'the undesirability of deciding a case on the written record against the view of the judge who heard the witnesses'. To apply the words of Lord Simonds in the same case at 492 to the present case, the misconduct hearing came to certain conclusions of fact and the Tribunal was entitled and bound, unless there was compelling reason to the contrary, to assume that it had taken the whole of the evidence into consideration. The misconduct hearing explicitly said that, having listened to and watched the witnesses give evidence, it considered the account of the complainer to be credible. The Police Appeals Tribunal's criticism of the misconduct hearing's decision on the facts can be described as fundamentally unsound criticism. The misconduct hearing was not required to comment on every point of conflict in the evidence and to engage in a close analysis of it. It cannot be inferred from the fact that it did not do so that it had forgotten or ignored such evidence. There was no deficiency in the reasoning of the misconduct hearing.

***Southall v. General Medical Council* [2010] EWCA Civ 407**

**27.05** Principles—findings of primary fact—appropriate measure of respect

Leveson LJ, giving guidance on findings of fact, said, at [47], that, as a matter of general law, it is well established that findings of primary fact, particularly if founded upon an assessment of the credibility of witnesses, are virtually unassailable: see *Benmax v. Austin Motor Co. Ltd* [1955] AC 370 and *National Justice Compania Naviera SA v. Prudential Assurance Co. Ltd (The Ikarian Reefer)* [1995] 1 Lloyd's Rep 455, 458. Further, the Court should reverse a finding on the facts only if it can be shown that the findings were sufficiently out of tune with the evidence to indicate with reasonable certainty that the evidence had been misread: *Libman v. General Medical Council* [1972] AC 217, *per* Lord Hailsham of St Marylebone LC at 221F; *R (Campbell) v. General Medical Council* [2005] 1 WLR 3488, *per* Judge LJ at [23]; and *Gupta v. General Medical Council* [2002] 1 WLR 1691, *per* Lord Rodger at [10]. All that said, it nonetheless remains for a court—in 'appropriate cases' and if 'necessary'—to come to its own view and substitute that for the decision of a disciplinary body: *Meadow v. General Medical Council* [2007] QB 462, *per* Auld LJ at [120], albeit that he too recognized that the courts should accord disciplinary bodies assessing evidence of professional practice in their respective fields an appropriate measure of respect.

***Bhatt v. General Medical Council* [2011] EWHC 783 (Admin)**

**27.06** Principles

Langstaff J, after referring to CPR 52.11 and the cases of *Sacha v. General Medical Council* [2009] EWHC 302 (Admin), [8], *Southall v. General Medical Council* [2010] EWCA Civ 407, [47], and *Meadow v. General Medical Council* [2007] QB 462, [128], said:

> 9. I accept and adopt the approach outlined in these authorities, in particular that although the court will correct errors of fact or approach:
> (i) it will give appropriate weight to the fact that the Panel is a specialist tribunal, whose understanding of what the medical profession expects of its members in matters of medical practice deserves respect;
> (ii) that the tribunal has had the advantage of hearing the evidence from live witnesses;
> (iii) the court should accordingly be slow to interfere with the decisions on matters of fact taken by the first instance body;

(iv) findings of primary fact, particularly if found upon an assessment of the credibility of witnesses, are close to being unassailable, and must be shown with reasonable certainty to be wrong if they are to be departed from;

(v) but that where what is concerned is a matter of judgment and evaluation of evidence which relates to police practice, or other areas outside the immediate focus of interest and professional expertise of the [fitness to practise panel] FTPP, the court will moderate the degree of deference it will be prepared to accord, and will be more willing to conclude that an error has, or may have been, made, such that a conclusion to which the Panel has come is or may be "wrong" or procedurally unfair. To this extent I accept and adopt the submissions of [counsel for the appellant].

### *Farag v. General Medical Council* [2011] EWHC 1212 (Admin)

**27.07** Appellant's evidence—rejected by panel—whether proper basis—documentary evidence

The challenge in this appeal was to the factual findings of the panel. In summary, what was contended by counsel for the appellant was that the panel fell into error in rejecting the appellant's oral evidence in circumstances in which, according to him, either there was no evidence to counter the appellant's oral evidence, or the oral evidence to the contrary was either contained in documents that had not been properly established by oral or other evidence to be reliable and were contradicted by the appellant, or in oral evidence that was introduced only at a late stage in the case and was unreliable. Counsel submitted that the panel had provided no proper or demonstrated basis for rejecting the appellant's evidence, and that even though the panel in its decision recognized that the burden was on the General Medical Council (GMC) to prove its case, in fact the panel appeared to have reversed the burden and put it on to the appellant. In dismissing the appellant's appeal, His Honour Judge Davies said that the evidence given by the appellant as to the relevant circumstances were all circumstances that the panel would have been entitled to view as unsatisfactory. The panel was entitled not to accept the appellant's evidence and to prefer the contemporaneous documentary evidence, backed up by its own assessment of the position. It was a conclusion that the panel was fully entitled to reach. Although it was clear from the transcript that, initially, counsel for the appellant objected to the admission of documentary evidence, it is also clear that, following discussion, it was agreed that all documents, both the GMC's and the appellant's documents produced for the hearing, should be admitted into evidence, but that the weight to be given to them should be a matter for submission before the panel in due course. In any event, it was quite clear that the panel would have had jurisdiction to admit the documentation into the proceedings, on the basis that they would have been admissible in criminal proceedings as admissible hearsay and in any event on the basis that it was desirable to do so.

### *Casey v. General Medical Council* [2011] NIQB 95

**27.08** Sexual misconduct—complainant—inconsistencies—reliability and credibility

C, a general medical practitioner, sought to set aside the determination of the panel in relation to an allegation of misconduct brought against the doctor by a patient. The patient alleged that C carried out an examination of her chest and abdomen in a sexually inappropriate manner. The panel, in its determination of the facts, found the patient's account to be consistent, reliable, and credible, and without embellishment. C challenged the decision on a number of grounds: that the panel was in serious error in finding that the patient was a consistent, reliable, and credible witness; that the panel was wrong to rely on the evidence of the patient's employer when it concluded that the employer's evidence corroborated the patient's evidence; that the panel failed to address the inconsistencies in the evidence of the patient's subsequent partner to whom the patient made the allegations about her treatment by the doctor;

and that the panel failed to explain why it rejected the appellant's defence, and why it found him to be incredible and unbelievable. In setting aside the decision of the panel, Girvan LJ observed that the evidence, as given by the patient to the panel, differed in a number of significant respects from that made by the patient to the police in her police statement. The inconsistencies in her evidence in relation to the chest examination were followed by inconsistencies in relation to evidence in respect of the abdominal examination. It was evident that the patient had, over time, presented a number of quite inconsistent versions of the same events. This should have been obvious to the panel as the decider of facts. The inconsistences were serious ones. The earlier versions of events (which themselves had inconsistencies between them) were quite different from the evidence as presented to the panel. The finding by the panel that the patient presented as a consistent, reliable, and credible witness is one that no tribunal properly directing itself on the evidence could have made in the circumstances. In a case that turns on which of two contradictory witnesses a tribunal should believe, a careful examination of important inconsistencies is necessary in evaluating reliability and credibility. While a witness who has presented contradictory evidence may ultimately be accepted as telling the truth on some one or more issues, a tribunal faced with such a witness should, in fairness to the party whose evidence is rejected, explain why the evidence of the witness who has given seriously conflicting and inconsistent evidence is to be preferred to the other witness. The evidential difficulty arising from serious inconsistencies, and from making serious and ultimately unfounded allegations, is one with which the tribunal must demonstrably appreciate and rationally deal with. The unavoidable inference from the panel's decision in the present instance must be that it failed to take properly into account the importance and significance of the inconsistencies of the patient's evidence when looked at as a whole in its entire context: [11].

### *Lawrence v. General Medical Council* [2012] EWHC 464 (Admin)

**27.09** Panel's 'own expertise'—expert evidence—rejection—reasons

Stadlen J, in a lengthy judgment, said that L was a consultant psychiatrist and, in the course of his psychiatric and psychotherapeutic treatment of a female married patient, allegedly acted in various ways that were said to have been inappropriate, not in the best interests of his patient, and an abuse of his professional position. The most serious allegations were that he attempted to pursue an emotional relationship with the patient, encouraged her to believe that he wanted to pursue an emotional relationship with her, revealed to her sexual fantasies that he had about her, and told her personal information about himself. It was alleged that his conduct was sexually motivated. The hearing before the GMC's fitness to practise panel lasted fifteen days, at the end of which the panel found the allegations proved and directed that L's name be erased from the medical register. On appeal, the findings of fact were challenged 'root and branch', and in particular two findings—namely, the panel's determination 'from its own expertise' that a woman with low self-esteem, unless encouraged, would be unlikely to fantasize that she was attractive to another person and that, whilst mindful of the evidence of the GMC expert, the panel did not find that the patient would be likely, on her own, to believe that L had feelings for her. The panel comprised three members, of whom one was lay and two of whom were medical. Of the two medical members, one was a general practitioner (GP) and the other, the chairman of the panel, a former consultant psychiatrist, part-time judicial member of the Mental Health Review Tribunal, and a former part-time psychiatrist member of the Parole Board. In relation to the panel's finding based on 'its own expertise', Stadlen J said that a specialist tribunal, including a GMC fitness to practise panel, remains bound by the rules of natural justice, that the rules of natural justice preclude members of a specialist tribunal including expert members from giving evidence to themselves that the parties have no opportunity to challenge, and that where a specialist tribunal, drawing on its own knowledge and experience, independently identifies an important fact or matter that may influence its decision, but which has not been the subject of evidence adduced by or submissions advanced by the parties, it should state this openly and give the parties an opportunity to seek to adduce evidence and/or to make submissions on it: *Watson v. General Medical Council* [2005] EWHC 1896 (Admin); *Richardson v. Solihull Metropolitan Borough Council* [1998] EWCA Civ 335; *Lucie M v. Worcestershire County Council* [2002] EWHC 1292 (Admin); *R (L) v. London Borough of Waltham Forest Special Educational Needs and Disability Tribunal* [2003] EWHC 2907 (Admin); *McKeown v. British Horse Racing Authority* [2010] EWHC 508; *Southall v. General Medical Council* [2010] EWCA Civ 407. Stadlen J went on to state that one of the problems in the instant case was that the panel did not explain what it meant by 'its own expertise'. The learned judge agreed with L's submission that the natural and obvious interpretation is that the panel intended to refer to the professional expertise of the chairman in his capacity as a former consultant psychiatrist. Turning to the second finding of the panel, the learned judge accepted counsel's submission that the panel's conclusion would require to be based on expert evidence. Whilst the panel was entitled in reaching a conclusion on expert evidence to rely on its individual and/or collective expertise and/or experience, and it was open to the panel to reject expert evidence, none of the expert witnesses had expressed the view that the panel found as a fact. The extent and nature of the duty to give reasons was authoritatively stated by the Court of Appeal in *English v. Emery Reinbold and Strick Ltd* [2002] 1 WLR 2409. There are two essential requirements: it must be apparent to the parties why one side has won and the other has lost; and the judgment must enable an appellate court to understand why the judge reached his decision. In *Phipps v. General Medical Council* [2006] EWCA Civ 397, the Court of Appeal held that although the decision in *English v. Emery Reinbold* was primarily addressed to the professional judiciary, the application of the principles set out in that judgment are universal and its conclusions are applicable to any tribunal charged with the duty to reach a judicial or quasi-judicial conclusion. The learned judge concluded that the interests of justice required that the panel's findings on the disputed facts and the findings of misconduct against L must be quashed.

### *Cornish v. General Medical Council* [2012] EWHC 1196 (Admin)

**27.10**

Misuse of drugs—hospital premises—evidence before panel—circumstantial evidence

The appellant, a consultant anaesthetist at Yeovil District Hospital, did not contest the panel's finding that his fitness to practise was impaired on the grounds of his conviction for theft of drugs and misconduct. He did, however, appeal against the panel's determination that he self-administered drugs within the hospital buildings. In dismissing the appellant's appeal, the Court said that the panel's reasons were fully and clearly stated, and met the standard required of a lay panel of a professional disciplinary body. The evidence demonstrated that the appellant persistently lied about his drug abuse. The panel was entitled, on the evidence, to conclude that the appellant's drug abuse had become so extensive, 'out of control', and 'chaotic' that he would not have had the self-control not to self-administer within the hospital. It was telling that the appellant confessed that he sometimes could not even wait for the 20 minutes it took to drive out of Yeovil before he needed to inject. Whilst he would have faced practical difficulties in absenting himself for periods of time, up to an hour, without being detected, plainly the conditions at the hospital enabled him to do so. His evidence was that self-administration could take any time between 10 minutes and an hour. The panel was entitled to reject the appellant's explanation for the drugs paraphernalia in his hospital locker as simply a holding place for drug detritus brought from his car or home. It was circumstantial evidence supporting the GMC's submission that he was injecting in the hospital buildings.

### *Siddiqui v. Health Professions Council* [2012] EWHC 2863 (Admin)

**27.11** Sexual misconduct—complainant—credibility—circumstantial evidence

S, a radiologist, sought to challenge the conclusions of the Health Professions Council that he should be struck off from the relevant register on the basis of findings made by the respondent in relation to three complaints. No challenge to the findings was made in relation to two of the complaints, which concerned serious allegations of incompetence and malpractice. The third complaint concerned an examination carried out by S on a patient, identified as KD, on 27 June 2006 on his own authorization, without clinical justification, without a chaperone; in doing so, it was alleged, S was sexually motivated to carry out the procedure. In dismissing S's appeal, His Honour Judge Pelling QC (sitting as a judge of the High Court) said that, tested against the test identified by Langstaff J in *Bhatt v. General Medical Council* [2011] EWHC 783 (Admin), [9], the question was whether it can be said with reasonable certainty that the conclusion of the panel was wrong in relation to its assessment of the credibility of KD. The Court was unable to reach that conclusion.

> 21. ... It is perfectly true to say that there were points that could be made, and indeed points that were made concerning the credibility of KD. However, it was primarily for the tribunal, who had the benefit of seeing the witness, to assess her credibility. I accept that in an appropriate case it would be open to the court to overturn such a finding, but the circumstances when that will be so inevitably are rare, as was noted by Langstaff J, and the material that is available to me in this case does not come near being that which would allow me to conclude that the panel's assessment of the credibility of KD as a witness should be overturned.
>
> 22. Furthermore, there was in my judgment more than an adequate amount of material to enable the panel to conclude that the appellant was sexually motivated. That this is so can be inferred from at least the following facts and matters. Firstly, the procedures that were adopted from beginning to end were wholly outwith conventional procedural activity. It is not conventional or appropriate for medical professionals to contact patients directly by email or any other mechanism. It was contrary to the policy of the relevant health authority for this to be done. It is not conventional practice to invite patients to tell untruths to their GP for the purpose of obtaining appointments. It is not conventional practice to engineer matters so as to ensure that a patient ostensibly needing a particular procedure should be referred to a particular radiographer over any other when there is no justification for so doing. The content of the emails, when read collectively, and particularly when the email of 27 June is considered, is very strongly consistent with that motivation being the relevant motivation.
>
> 23. There is a further point... If the appellant had considered that it was appropriate to carry out the internal examination that ultimately was carried out, then there was no reason why that could not have been carried out at the time of the first appointment.... Furthermore, it is, to put it no higher, extremely unusual that a radiographer should make decisions as to what examination should be carried out and should not be carried out without the input of a clinician.... what is surprising, and open to criticism, is the notion that a radiographer should contact a patient and suggest further examinations, particularly ones of an intimate and intrusive nature, other than via the treating clinician. All of this was material from which it was open to the panel to infer a sexual motivation...

The Court dismissed the appeal.

### *Webster v. General Teaching Council* [2012] EWHC 2928 (Admin)

**27.12** See paragraph 4.23.

### *Slater v. Solicitors Regulation Authority* [2012] EWHC 3256 (Admin)

**27.13** *Newton* hearing

The Solicitors Regulation Authority (SRA) was unwilling to accept the basis of plea of the solicitor in proceedings before the Solicitors Disciplinary Tribunal (SDT). The Tribunal held a *Newton* hearing—after *R v. Newton* (1983) 77 Cr App R 13, CA—in which it found as a fact that the solicitor was evasive. The Divisional Court (Elias LJ and Singh J) held that it would have been unrealistic for the Tribunal to ignore its own finding of fact during the *Newton* hearing when deciding on the appropriate sanction for the admitted offences of misappropriating client money and breaching the conditions of S's practising certificate.

### *Yaacoub v. General Medical Council* [2012] EWHC 2779 (Admin)

**27.14** Sexual misconduct—complainant—evidence

Y, a GP, appealed against a decision of a fitness to practise panel of the GMC, dated 13 October 2011, that his fitness to practise was impaired by reason of sexual misconduct in relation to a female patient and that his name should be erased from the register. At the heart of Y's appeal was the contention that, contrary to the finding of the panel, the patient was plainly not a credible witness and no reasonable panel could have found otherwise. That contention in turn was founded on a critical examination of how her complaint of sexual misconduct on the part of Y had evolved over time. Kenneth Parker J said that there was sufficient evidence, arising from what the patient had said in examination-in-chief and under cross-examination, to support the essential findings of the panel that were adverse to Y. The patient had consistently maintained from the outset that Y had used inappropriate sexual language during both visits and, on the basis that he had done so, it was a justifiable inference that his conduct was an attempt at sexual grooming of the patient. The task for a panel is to consider whether the core allegations are true. It is commonplace for there to be inconsistency and confusion about details of varying importance: see *Mubarak v. General Medical Council* [2008] EWHC 2830 (Admin), [20]. However, this was a case in which the patient, over time, had developed a fundamentally different description of what she said had happened to her. When she reported the matter at first to NHS Direct, her sole concern was that, in future, only female doctors should be sent to her. She herself did not want to involve the police. There was no hint that Y had performed an act of gross indecency in her presence and certainly no suggestion whatsoever that he had tried to rape her, let alone that he had in fact raped her. Then, following a lengthy interview with a female police officer, she alleged that both a sexual assault and an act of gross indecency had occurred on the second visit—an account essentially different from what she had told NHS Direct. In allowing Y's appeal from the decision of the panel to find him guilty of sexual misconduct, the learned judge said that, in his view, the treatment of the patient's evidence by the panel did not indicate with sufficient clarity and precision that the panel truly recognized the difficulty posed by such a fundamental shift of evidence. The reasons had to address that difficulty squarely and to explain, even if briefly, the basis upon which the panel deemed it safe, notwithstanding the radical shift, to accept the account as developed by the patient and as given in her evidence to the panel. On the latter aspect, the panel simply stated that it was 'not surprising' that 'the detail of Patient A's complaint was made on a piecemeal basis and that she did not provide exactly the same account' on each occasion when interviewed. Again, that would suggest that the central core of Patient A's account remained basically consistent, and that only the 'detail' was developed and varied over time. However, that was not the position here.

### *Duthie v. Nursing and Midwifery Council* [2012] EWHC 3021 (Admin)

**27.15** Midwife—evidence of complainants—conflict with contemporaneous records—panel's failure to resolve

D, a midwife, appealed against the decision of the Conduct and Competence Committee (CCC) of the Nursing and Midwifery Council (NMC) finding her fitness to practise to be impaired by reason of misconduct and striking her off the register. The matter in issue was D's conduct during her care of Mrs A, before and during her labour at the beginning of September 2007. Baby L was stillborn at home on 3 September 2007. Mother and baby were transferred to hospital as an emergency in the evening of 3 September, but resuscitation failed to revive Baby L. The most important point in issue was whether D effectively dissuaded the parents from going into hospital earlier. In a detailed judgment comprising 191 paragraphs, Irwin J said that the essential route by which the panel reached its conclusions could be not unfairly characterized as the following series of steps.

(1) It found that D altered and expanded her notes, dishonestly, after the event, in an attempt to protect herself.
(2) It found this to damage her credibility.
(3) It found that this conclusion supported the credibility of Mr and Mrs A.
(4) It accepted the evidence of Mr and Mrs A, where it was in conflict with the notes made or the evidence given by D.

Irwin J said that, in his judgment, there was a flaw in that approach. The finding that D altered or manufactured notes could, of course, properly undermine her credibility, but it was a slender basis in this case for supporting the credibility of Mr and Mrs A. The important problem, which was never properly addressed by the panel, was the conflict between their evidence and the contemporaneous records—specifically, the records created by professionals other than D and those records created by D that were unchallenged. There was a very great problem with the credibility of Mr and Mrs A. In particular, the evidence of Mrs A as to her willingness—or even, as she would have it, desire—to have a caesarean section from early on was in strong contrast with the content of almost all of the unimpeached contemporaneous records in the case. One feature that emerged from all of the evidence was the degree to which Mrs A and her husband asserted their own desires and their own position at every point along the way. There was no evidence that the panel grappled with these difficulties. That did not imply an obligation on the part of the panel to write a full reasoned judgment or an essay justifying its conclusions. However, the conclusions here simply did not seem consistent with the reliable contemporaneous material, coming into being before and after the tragic stillbirth of Baby L. In such circumstances, it will be wise for a panel to make its reasoning clear enough to render the unexpected conclusion explicable. Mere reference to the demeanour of the witness is unlikely to carry the matter home. The panel's explanation of its reasons here was therefore found to be inadequate.

### *Khan v. General Medical Council* [2013] EWHC 2187 (Admin)

**27.16** Clinic—cosmetic surgery—financial discount—overall management of patient

K, a GP, was, with his wife, a director and owner of a company operating a cosmetic surgery clinic in Harley Street. Three female friends approached K at his National Health Service (NHS) practice in Birmingham to have cosmetic surgery carried out on them. K was charged with unreasonably inducing a patient to accept surgery by agreeing to undertake financially discounted cosmetic surgery if three patients booked together. The arrangements were made at the clinic by G, who was not medically qualified, and in evidence K accepted responsibility for G's actions. In dismissing K's appeal against the factual findings of the allegation, Dingemans J held that these were proceedings directed solely to the issue of K's professional conduct; they were not private law proceedings in tort or for breach of contract, K was a director and shareholder of the clinic, K had seen the medical notes, which contained references to deals and a discount if all three patients were to book together, and K had professional responsibilities for delegation and continuing responsibilities for the overall management of the patient. Paragraph 54 of *Good Medical Practice* states that delegation involves asking a colleague to provide treatment or care, and although a doctor will not be accountable for the decisions and actions of those to whom he or she delegates, he or she will still be responsible for the overall management of a patient and accountable for the decision to delegate. Properly analysed, K was rightly accepting professional responsibility for the fact that an inappropriate inducement had been offered by G.

***Arunkalaivanan v. General Medical Council*** **[2014] EWHC 873 (Admin)**

**27.17** See paragraph 30.07.

***D v. Conduct and Competence Committee of Nursing and Midwifery Council*** **2014 SLT 1069**

**27.18** See paragraph 67.12.

## C. Other Relevant Chapters

**27.19**

Appeals, Chapter 4
Misconduct, Chapter 47
Reasons, Chapter 54

# 28

## FINES



| D'Souza v. Law Society [2009] EWHC 2193 (Admin) | 28.04 | Matthews v. Solicitors Regulation Authority [2013] EWHC 1525 (Admin) | 28.17 |
| Visser and Fagbulu v. Financial Services Authority [2011] LR (FC) 551 | 28.05 | Law Society (Solicitors Regulation Authority) v. Emeana and ors [2013] EWHC 2130 (Admin) | 28.18 |
| Hazelhurst and ors v. Solicitors Regulation Authority [2011] EWHC 462 (Admin) | 28.06 | Adesemowo and Alafe-Aluko v. Solicitors Regulation Authority [2013] EWHC 2020 (Admin) | 28.19 |
| Karpe v. Financial Services Authority, FS/2010/0019, 15 May 2012 | 28.07 | Solicitors Regulation Authority v. Anderson Solicitors [2013] EWHC 4021 (Admin) | 28.20 |
| Bass and Ward v. Solicitors Regulation Authority [2012] EWHC 2012 (Admin) | 28.09 | Fuglers LLP and ors v. Solicitors Regulation Authority [2014] EWHC 179 (Admin) | 28.22 |
| Law Society v. Waddingham and ors [2012] EWHC 1519 (Admin) | 28.10 | Khan v. Financial Conduct Authority, FS/2013/002, 8 April 2014 | 28.23 |
| Maistry v. Solicitors Regulation Authority [2012] EWHC 3041 (Admin) | 28.11 | Micalizzi v. Financial Conduct Authority [2014] UKUT 0335 (TCC) | 28.24 |
| Chaligne, Sejean and Diallo v. Financial Services Authority, FS/2011/0001, FS/2011/0002, FS/2011/0005, 25 April 2012 | 28.12 | C. Other Relevant Chapters | 28.25 |

## A. Legal Framework

**28.01** Examples include:

Solicitors Act 1974, section 47(2) (power of the Solicitors Disciplinary Tribunal to make an order for the payment by a solicitor or former solicitor of an unlimited penalty, which shall be forfeit to Her Majesty)

Chartered Institute of Management Accountants Disciplinary Committee Rules 2011, rule 24(6) (in the case of registered students, limited to £2,000)

Solicitors Regulation Authority Disciplinary Procedure Rules 2011 (powers of an adjudicator and an adjudication panel)

Institute of Chartered Accountants in England and Wales Disciplinary Bye-Laws, bye-law 22

## B. Disciplinary Cases

### *Brown v. United Kingdom* (1998) 28 EHRR 233

**28.02** Maximum penalty—time of offence and decision—Article 7 ECHR

On 12 March 1996, at the conclusion of a hearing before the Solicitors Disciplinary Tribunal (SDT), the charges against the applicant were found to be proved and he was fined a total of £10,000. The applicant faced three allegations of conduct unbefitting a solicitor—namely: (1) that he established and operated a solicitors practice principally managed by an unqualified clerk in circumstances that were improper; (2) that, being in possession of information indicating that said clerk was dishonest, having been convicted of serious offences of dishonesty and sentenced to a term of imprisonment, he failed to take any reasonable or sufficient steps to protect the profession or the public; and (3) that he failed to comply with the Solicitors Practice Rules 1987 and 1988 (fee sharing) in that he established and operated a solicitors practice on the basis that said unqualified clerk would be entitled to receive one half of the gross fees thereof. The applicant's appeal was dismissed by the Divisional Court. The applicant complained of a violation of Article 7 of the European Convention on Human Rights (ECHR) (no heavier penalty shall be imposed than that which was applicable at the time that the criminal offence was committed) in that the maximum penalty available to the Tribunal at the time of his offence was £3,000, whereas he was fined £10,000. In dismissing the complaint, the European Court of Human Rights (ECtHR) held that, in order for the applicant to avail himself of the protection of Article 7, the proceedings had to amount to the determination of a criminal charge, that the charges against the applicant were classified under domestic law as disciplinary offences, being examined in a tribunal without any involvement by the police or prosecuting authorities, and that disciplinary proceedings involved the determination of civil rights and obligations, and that the charges faced by the applicant related to matters of professional behaviour and organization, emphasis being given to the standards of conduct befitting a solicitor. The Court considered whether, notwithstanding the non-criminal character of the proscribed misconduct, the nature and degree of severity of the penalty that the person concerned risked incurring may bring the matter into the 'criminal' sphere. However, having regard in particular to the essential disciplinary context of the charges, the Court found that the severity of the penalty was not, of itself, such as to render the charges 'criminal' in nature.

### *Financial Services Authority v. Fox Hayes* [2009] EWCA Civ 76; *Fox Hayes v. Financial Services Authority*, 12 May 2011, Upper Tribunal (Tax and Chancery Chamber) Financial Services

**28.03** Penalty—seriousness and recklessness of misconduct—value of secret profits obtained

At all material times, Fox Hayes was a firm of solicitors in Leeds and an authorized person for the purposes of section 21 of the Financial Services and Markets Act 2000 (engaged in investment activity). On 29 September 2006, the Financial Services Authority (FSA) served a decision notice on Fox Hayes imposing a penalty of £150,000 for breach of the Conduct of Business Rules. Fox Hayes referred that decision notice to the Financial Services and Markets Tribunal. In the lead-up to the hearing, it was revealed that M, the senior partner of Fox Hayes and the partner responsible for this part of the firm's business, had personally received substantial sums by way of commission from unauthorized overseas companies, quite apart from Fox Hayes' receipt of its ordinary fee for its services. The commissions totalling £454,770 were not revealed by M to the British shareholders who bought shares recommended by the companies, nor, at the time that he received them, to his fellow partners. Most of the shares offered to the public pursuant to promotions approved by Fox Hayes were 'high-risk, illiquid shares' in small companies not quoted in the United Kingdom. As a result of the promotions approved by Fox Hayes, 670 investors altogether invested the sum of US$20,350,986.83, most of which was lost to them. Following a hearing, the Tribunal decided that there should be a reduction in the total penalty from £150,000 to £146,000. On appeal by the FSA to the Court of Appeal (Longmore, Wilson, and Lawrence Collins LJJ), the Court said that the penalty should reflect both the seriousness and the recklessness of the misconduct. In a case of this kind in which ordinary investors have lost many millions, it would be inappropriate, other things being equal, to fix a penalty less than £750,000. In the instant case, in the light of the history, the penalty should be reduced to £500,000, added to which it was appropriate that Fox Hayes should pay, by way of penalty, the commissions received by M, making a total therefore of £954,770. The partners should have an opportunity to take independent advice before a final figure was arrived at with respect to the penalty. At a subsequent hearing before Sir Stephen Oliver QC, Colin Senior, and Christopher Burbidge, the Upper Tribunal considered that the right amount of the penalty should not be less than the secret profits obtained by M, which were to be regarded as assets of the partnership. To impose a penalty that failed to reflect what the partnership obtained would be wrong. The breach was so severe that it would be inappropriate to reduce the penalty to below the figure of £454,770.

*D'Souza v. Law Society* **[2009] EWHC 2193 (Admin)**

**28.04** Means

The appellant appealed against the decision of the SDT that he should be subject to a financial penalty of £1,500 and pay costs of £8,000. The appellant claimed that the Tribunal failed to take into account his means at arriving at its decision on sanction and costs. The Divisional Court (Thomas LJ and Coulson J) said that *Camacho v. Law Society* [2004] EWHC 1675 (Admin) and *Merrick v. Law Society* [2007] EWHC 2997 (Admin) are authority for the proposition that the means of a defendant to tribunal proceedings may be a relevant consideration in calculating the appropriate sanction as to the level of fines and costs. This would usually arise when a solicitor is being suspended from practice or struck off, but there will be exceptional cases in which, even though a solicitor is allowed to continue in practice, his income may be a relevant consideration both as to any costs sanction and in respect of any financial penalty that might be imposed. In the instant case, the Court capped the overall amount at £2,000, divided up as £500 fine and £1,500 costs.

See further paragraph 15.12.

*Visser and Fagbulu v. Financial Services Authority* **[2011] LR (FC) 551**

**28.05** Systematic market abuse—penalties manual—additional factors—€35 million loss—£2 million fine for CEO—lesser sum for CFO

On 15 March 2010, the FSA issued decision notices arising from the conduct of the applicants in the running of Mercurius Capital Management Ltd, a UK company authorized by the FSA, which acted as an investment manager for a Cayman Islands based hedge fund called Mercurius International Fund Ltd. V was the chief executive officer (CEO) and F was employed as its chief finance officer (CFO). The Fund collapsed, with a loss of approximately €35 million investors' funds. The essence of the FSA's case was that the applicants committed market abuse, deliberately and repeatedly breached the investment restrictions placed on the Fund, undertook transactions designed to give a false impression of the value of the Fund's assets, and concealed from investors important information, such as the true nature of the Fund's investments, the resignation of its prime brokers and the consequential loss of margin, and its increasingly precarious financial position. The FSA imposed a penalty on V of £2 million for breach of Principle 1 of the FSA's Statements of Principle for Approved

Persons and for engaging in market abuse, and a financial penalty on F of £500,000. On appeal, the Financial Services and Markets Tribunal observed that the penalties that it was required to consider were imposed in accordance with section 66(3)(a) of the Financial Services and Markets Act 2000 in respect of the applicants' conduct as a whole, and additionally in accordance with section 123(1), which is directed specifically at market abuse. The Act of 2000 does not contain any criteria governing the imposition of a penalty, nor the amount of any penalty decided upon. It requires no more than that the FSA is satisfied that it is appropriate to impose a penalty, the amount of which is left to its discretion, subject only to reference to the Tribunal. The FSA has, however, published some guidance about its policy, as it is required to do by sections 69 and 124 of the Act. The guidance is to be found in Chapter 6 of the Decision Procedure and Penalties Manual (DEPP 6) within the FSA Handbook. The Tribunal is not bound by either the policy or the guidance, but they are a convenient and useful starting point. DEPP 6.1.2G states that the principal purpose of imposing a financial penalty is to promote high standards of conduct by deterring those who have committed breaches from reoffending and by deterring others from committing breaches. DEPP 6.3 relates specifically to penalties for market abuse. DEPP 6.5 describes the factors that the FSA will consider when determining the appropriate level of financial penalty. The non-exhaustive list includes: the nature, seriousness, and impact of the breach in question; the extent to which it was deliberate or reckless; whether the person on whom the penalty is to be imposed is an individual; the financial resources and other circumstances of the person on whom the penalty is to be imposed; the amount of benefit gained or loss avoided; the difficulty of detecting the breach; the person's conduct following the breach; and his or her disciplinary record and compliance history. In the Tribunal's view, those are all pertinent considerations. We would add to them repetition, persistence, concealment, and whether or not the person concerned has attempted to make amends. DEPP 6.5.2G sets out factors that may be relevant to determine the appropriate level of financial penalty to be imposed on a person under the Act: the size, financial resources, and other circumstances of the person on whom the penalty is to be imposed; whether there is verifiable evidence that it would cause serious financial hardship or financial difficulties if the person were to pay the level of penalty appropriate for the particular breach; and that the purpose of a penalty is not to render a person insolvent or to threaten the person's solvency. In the Tribunal's view, the policy constitutes proper guidance to the determination of the level of a penalty, in conjunction with a close consideration of an individual applicant's own circumstances. In its decision at [114], the Tribunal quoted *Atlantic Law LLP v. Financial Services Authority* [2010] UKUT B7(FS), in which it was said at [110]:

> The fact that the purpose of imposing a financial penalty is not to bring about insolvency does not mean that the Tribunal cannot and should not fix a penalty which may have that unfortunate result. Victims of boiler room schemes [the conduct found to have been perpetrated in that case] have to take the financial consequences of the losses perpetrated upon them. Those who help cause those losses do not deserve special protection.

In the instant case, the failings were numerous, repeated, and sustained. Neither applicant paid any regard to the terms upon which Mercurius was required to manage the Fund. Each engaged in market abuse in a way that caused significant increases in the price of illiquid securities. Each systematically and repeatedly deceived the Fund's investors, and their respective acts and omissions contributed to the liquidation of the Fund. Faced with so little evidence and nothing of substance in mitigation beyond the limited acceptance of culpability, the Tribunal was not persuaded that the starting point for V of £2 million should be adjusted. The penalty in F's case should be £100,000.

***Hazelhurst and ors v. Solicitors Regulation Authority* [2011] EWHC 462 (Admin)**

**28.06** Theft by employee—repayment by firm—no client loss—fine quashed—reprimand

This was an appeal by four partners in a firm of solicitors against orders of financial penalty made against each partner by the SDT. Between February 2003 and May 2006, an employee of the firm, M, stole £101,826 from the firm's client account. The thefts were discovered by chance. Subsequent enquiries by the firm revealed widespread misuse of private funds. The firm immediately self-reported the matter to the Solicitors Regulation Authority (SRA), and appointed a locum to examine its files in order to identify the full nature and extent of the misuse of funds by M. All funds were repaid to the accounts by the partners within the firm. Indemnity insurance was insufficient for this purpose and, as a consequence, the partners of the firm paid between £80,000 and £90,000 in order to meet the liabilities. No client of the firm suffered any loss. There was no suggestion that any appellant was involved in dishonest practice. M was dismissed from employment by the firm. The case against the appellants was founded upon breaches of the Solicitors Accounts Rules and a breach of the Solicitors Practice Rules—namely, a failure to supervise an employee who was discovered to be dishonest. The appellants pleaded guilty to all charges. The sanction imposed was that each appellant was to pay a penalty of £4,000 and costs. No appeal was raised as to the costs order, but the appellant relied upon two facts: that, during the three years of the thefts, the firm's accounts were independently audited by accountants in accordance with the guidance contained in the Solicitors Accounts Rules and the auditors discovered nothing untoward; and that M was a trusted employee of the firm, who had worked there for eight years and, until the discovery of the fraud, there had been no reason to believe that she was anything other than trustworthy. In allowing the appeal, and quashing the orders of the financial penalties made by the SDT and substituting a reprimand, Nicola Davies J said that, in short, the acceptance of a failure to supervise, which was made primarily on the basis of a fraud carried out over a period of three years, had to be balanced against the fact that others had failed to identify any wrongdoing and that such wrongdoing was perpetuated by a member of staff whose conduct had given no cause to question her honesty. In the judge's view, the SDT had failed in their written reasons to address adequately the submissions of the appellants as to why the thefts went undiscovered for a period of three years. In failing to address the appellant's submissions, the SDT did not provide adequate reasons for its findings as to the breaches of the Rules, and specifically the lack of supervision.

***Karpe v. Financial Services Authority*, FS/2010/0019, 15 May 2012**

**28.07** Unauthorized trading—client losses US$42 million—transfers to disguise losses—£1.25 million fine

K worked for UBS in its international wealth management business, which focused on providing wealth management services to individuals. He had worked in the financial services industry since 1990. He was the desk head of the Asia II Desk, which provided wealth and management services to customers resident in India, or of Indian origin. He managed a team of twenty-five UBS employees. The FSA imposed a financial penalty on K of £1,250,000, pursuant to section 66 of the Financial Services and Markets Act 2000, and made an order prohibiting him from performing any function in relation to any regulated activity, on the grounds that he was not a fit and proper person, because his conduct demonstrated a lack of honesty and integrity. The statement of case in summary alleged that K had, amongst other things, carried out unauthorized trading on various customer accounts, which resulted in substantial losses; he had carried out unauthorized trading transfers between unconnected customer accounts to disguise losses that had arisen as a result of the unauthorized trading; he had orchestrated the movement of funds from one customer account to others to disguise losses; and he had arranged the movement of funds to be effected through various methods, which included using purported loans and transfers between unconnected accounts. Investigations showed that he had undertaken unauthorized trading across thirty-nine client accounts. This had resulted in substantial losses to twenty-one of those and had prompted compensation payments by UBS in the sum of US$42.4 million. Deloitte LLP and others carried out investigations, which concluded that there had been widespread unauthorized trading that had caused losses of many millions of dollars across a large number of client accounts. K referred the decision notice to the Upper Tribunal (Tax and Chancery Chamber) Financial Services. In dismissing K's appeal and directing the FSA to impose a financial penalty of £1.25 million, Sir Stephen Oliver QC, giving the judgment of the Upper Tribunal, said that K was responsible for unauthorized transactions and transfers between clients of a substantial aggregate amount. The scale of the unauthorized operations attributable to him was large enough and continued over a long enough period to justify a substantial penalty. In reaching the conclusion that the fine imposed of £1.25 million was appropriate to the circumstances, the Tribunal took into account:

> 171. ...the scale of the losses that resulted from [K]'s carrying out of unauthorised transactions and transfers between clients over many years. We think that [K] induced others serving on his desk to participate in what was an obvious dishonest course of conduct. He has not cooperated with the FSA. We recognize that this is not a case for fixing the penalty at an amount designed to produce "disgorgement" of ill-gotten gains.... And we infer that the whole motivation was to benefit him indirectly and in the long term by obtaining new clients through his apparent prestige, increasing funds under management and thereby advancing his career and increasing his bonuses.

### *Karan v. Financial Services Authority*, FS/2010/0025, 15 May 2012

**28.08** Client adviser—see *Karpe* above—experienced banker—participation—£80,000 fine

Ms K was a client adviser, working for UBS wealth management within the London branch. She worked on the Asia II Desk and reported directly to K throughout the relevant period January 2006 to January 2008: see paragraph 28.07 above. By a decision notice, the FSA informed Ms K of its decision to impose a financial penalty of £90,000 and to make an order prohibiting her from performing any function in relation to any regulated activity, on the grounds that she was not a fit and proper person, because her conduct demonstrated a lack of honesty and integrity. Ms K referred the decision notice to the Upper Tribunal (Tax and Chancery Chamber) Financial Services and it was heard at the same time as that of K's reference. In its judgment delivered by Sir Stephen Oliver QC, the Tribunal said that, assuming in favour of Ms K that it had been K and not her who had initiated the various acts of wrongdoing, the Tribunal was nonetheless satisfied that the duty of integrity can be breached where someone else initiates wrongdoing and where the person in question is put in a position of choosing whether to go along with it. Ms K allowed her name to figure as the client adviser in relation to numerous transactions.

> 90. We recognise that [Ms K] had been placed in an extremely awkward situation through the manipulation of [K]. The fact, however, is that over and over against she chose to go along with and, on occasions, to facilitate [K]'s wrongdoing.
>
> 91. ...We do not, however, accept that [Ms K] was putty in his hands. She herself was an experienced banker with a substantial client base. She knew how client accounts and relationships ought to be run.... She presented an apparently solid front when cross-examined and persisted with her explanations notwithstanding their evident incredulity.... [S]he was fully alerted to the transactions that [the customers] participated in

> and the movements of money. The clear inference must be that she knew about many, if not all, of these transactions and payments. In that connection [Ms K] never suggested that she had been under any form of duress...
>
> 92. ...The fact is...that she actually participated in transactions that enabled [K] to continue making unauthorised transactions and unauthorised payments....
>
> [...]
>
> 95. ...This is not a case of embezzlement by her. Nonetheless, she was prepared, over and over again, to lend herself to the misconduct initiated by [K] because it suited her career and, we infer, her remuneration package.

The Tribunal was satisfied that the prohibition against Ms K was appropriate. As to the level of the appropriate penalty, £75,000 was deemed an appropriate amount of penalty in the present case: Ms K's conduct showed a lack of integrity over a relatively long period of time and her basic pay for 2007, which covered the relevant period, was £80,000.

### *Bass and Ward v. Solicitors Regulation Authority* [2012] EWHC 2012 (Admin)

**28.09** Unauthorized client account withdrawals—£3,500 shortage—no dishonesty—admissions—equity partners—fines reduced to level of principal defendant

B and W were, at all material times, the two equity partners in a firm of solicitors in Leeds. P, another solicitor, was a 'fixed share partner' for the period May 2007 to October 2007, when she resigned. P had made improper withdrawals and/or transfers from the client bank account of residential balances held on behalf of clients without client authority and without justification. In total, forty-four transfers were made in this way. All were under £500; some were very small amounts indeed. The total was £4,296.31. After making adjustments for sums that the appellants considered were or might have been due to the firm, there remained a client account shortage of £3,513.73. Dishonesty was not alleged 'in any shape or form'. At the hearing before the SDT, all three solicitors admitted the factual allegations. The sanctions imposed by the Tribunal were that B and W each be fined £10,000 and P £5,000. Bean J considered that justice would be done by reducing the penalty imposed on each of the appellants to the same amount as that imposed on P—namely, £5,000.

### *Law Society v. Waddingham and ors* [2012] EWHC 1519 (Admin)

**28.10** Trust funds—loans to solicitors—conflict of interest—no dishonesty—fines quashed—suspension

The Law Society, acting through the SRA, appealed against findings and orders made by the SDT in proceedings against three solicitors W, S, and P. The solicitors were partners in a firm in Preston. The main allegations concerned the use of trust funds being loaned to S and P in conflict of interests, and a loan of trust funds to a company by W and S, which amounted to a conflict of interests. During the period July 2002 to March 2003, S and P received £86,000 and £73,000 respectively from the estate of a client of which W was the trustee. In the event, W ultimately repaid the total sums personally. In August 2000, £10,000 was transferred from the trust fund to a client ledger in the name of a company, as a loan to be repaid with interest. The £10,000 was repaid in due course, but without any interest on the loan. The Tribunal found that the allegations that W, S, and P had acted dishonestly had not been made out. W and S were each fined £12,000, and P, £7,000. In addition, W, S, and P were each ordered to pay costs of £6,800. On appeal, the Divisional Court (Richards LJ and Maddison J) held that, having regard to the criminal standard of proof, the Tribunal was right to conclude that the allegations of dishonesty had not been made out. However, on any view, the solicitors were parties to the removal of tens of thousands of pounds from what they knew was a trust fund administered by their partner W. W had since undertaken to cause his name to be removed from the roll and not to seek readmission at any future time. In these circumstances and by consent of the parties, the Court ordered that the appeal relating to W be withdrawn. The financial penalties imposed by the Tribunal were wholly inappropriate. The Court said that, after anxious consideration, this was not a case in which a striking-off order would be appropriate. However, it would be appropriate and proportionate to make orders quashing the fines imposed by the Tribunal, but suspending S from practice as a solicitor for eighteen months and P for twelve months.

### *Maistry v. Solicitors Regulation Authority* [2012] EWHC 3041 (Admin)

**28.11** Failure to supervise—fine and costs—late admissions

M, a solicitor and the managing partner of a firm, admitted a failure to exercise appropriate supervision over unqualified staff in breach of rule 5.01 and/or rule 5.03 of the Solicitors Code of Conduct 2011, and a failure to ensure that material facts were disclosed to mortgagee clients of the firm, in breach of rule 1 of the Code. The SDT ordered M to pay a fine of £2,000 and costs of £10,000 in view of her late plea. M's appeal was dismissed by Cranston J. After careful consideration, the Tribunal had conducted a fair hearing and made no error of law. M had the benefit of expert representation in the form of leading counsel, who advanced mitigation before the Tribunal, which the Tribunal considered. The fine was appropriate given the misconduct involved and the standards of conduct required of solicitors. The Tribunal approached the question of costs in a proper manner. It was entitled to take into account the lateness of the admissions and the earlier denials of responsibilities by M in setting the figure of £10,000.

### *Chaligne, Sejean and Diallo v. Financial Services Authority*, FS/2011/0001, FS/2011/0002, FS/2011/0005, 25 April 2012

**28.12** Market abuse—fines, plus disgorgement of benefits

On 22 December 2010, the FSA's Regulatory Decisions Committee issued decision notices addressed to each of the applicants, C, S, and D. The notices contained the Committee's conclusion that each of the applicants had committed market abuse in breach of section 118 of the Financial Services and Markets Act 2000. In essence, the complaint was that they had deliberately increased the closing prices of various quoted stocks by artificial, manipulative trading. S and D had their approved person status withdrawn (C was not an approved person). C was made subject to a financial penalty of £900,000 plus the disgorgement of personal benefits of £266,924. S was subjected to a penalty of £550,000. The decision notice addressed to D stated that the Committee would have subjected him to a financial penalty of £100,000 but for its conclusion that its doing so would inflict serious financial hardship upon him. C, S, and D referred the decision notices to the Upper Tribunal, Tax and Chancery Chamber, and the references were heard together. In upholding the prohibition notices, Judge Colin Bishopp, giving the judgment of the Upper Tribunal, said that the FSA was correct to withdraw S's and D's approvals. In relation to the penalties, Judge Bishopp said:

> 108. There is relatively little precedent in respect of penalties for market abuse, and such precedent as there is covers several different types of conduct. It offers some guidance, but we shall ultimately have to reach a conclusion of our own. Precedents to which we were referred include (among others) *Jabre* (insider dealing, and therefore conduct of a very different character; *Atlantic Law LLP and Greystoke* ("Boiler room" selling, and again rather different conduct); *Visser and Fagbulu* (price manipulation) and *Eagle* (share ramping). The latter two are helpful, although the conduct in each case was conspicuously worse than that here.
>
> [...]
>
> 110. We start, in what follows, from the position that any form of market manipulation is a serious matter. As we have already pointed out, users of markets are entitled

> to assume that other participants are operating within the rules of the market, that prices (whether of stocks, commodities or derivatives, for example) reflect supply and demand faithfully, and that bids or offers are not being placed in such a manner as to confuse or mislead not only human traders, but also the computers which are increasingly commonly utilised, not least in the uncrossing stage of the auction we have described.

The Upper Tribunal concluded that the penalty imposed on C in respect of the disgorgement element should be £362,950, plus the £900,000 fine. The penalty imposed on S should be £650,000, subject to adjustment for his personal circumstances in such matter as the Tribunal shall determine at a later hearing. D was not subject to any monetary penalty.

***Wagner v. Financial Services Authority*, FS/2011/0015, 15 October 2012**

**28.13** False mortgage applications—observations on level of penalty

W referred a decision notice issued by the FSA to the Upper Tribunal, Tax and Chancery Chamber. The decision notice informed W of the FSA's decision to make a prohibition order on him and to impose a financial penalty of £100,000. The FSA found that W, an authorized person, had knowingly inflated his income to obtain a residential mortgage, had deliberately allowed false and inflated income figures to be submitted on his behalf in order to obtain four buy-to-let mortgages, and had failed to have in place proper systems and controls, as a result of which two employees had been able to submit mortgage applications containing false information. As a preliminary point, W challenged the FSA's power to impose any penalty. The alleged misconduct, he noted, took place in May 2005. Until June 2008, however, it had not been FSA policy to impose financial penalties on individuals for knowing involvement in mortgage fraud; instead, the individual in question had been prohibited pursuant to section 56 of the Financial Services and Markets Act 2000. W also relied on Article 7(1) ECHR, which provides that a heavier penalty shall not be imposed than that which was applicable at the time the criminal offence was committed. Sir Stephen Oliver QC, giving the decision of the Upper Tribunal, said that the preliminary point was misplaced. It is correct that Part XIV of the 2000 Act is headed 'Disciplinary Measures' and that section 206(2), in the form it took until a change in the law in 2010, excluded the FSA from both requiring an authorized person to pay a penalty and withdrawing his authorization. However, the present penalty was imposed in pursuance of the 'Disciplinary Powers' provisions in section 66, which is directed to 'approved persons'. There was no equivalent to section 206(2) in section 66 that could apply to approved persons such as W. Assuming in W's favour that the relevant penalty provisions are of a criminal nature for the purposes of Article 7(1), a £100,000 penalty could lawfully have been imposed when the violation took place. On the substantive issue of W's involvement, following observations from the Tribunal as to the state of the evidence available to it, the parties were agreed that: the prohibition order should stand on the basis that W did not deliberately mislead the lender, but, as an approved person, he failed to act with integrity; no financial penalty should be imposed on W; and no order for costs should be made.

***Sejean v. Financial Services Authority*, FS/2011/0002, 4 December 2012**

**28.14** Published guidance—DEPP

The starting point for considering whether or not adjustment is appropriate is the FSA's Decision Procedure and Penalties Manual (DEPP). Those subject to disciplinary measures should be entitled to rely on published guidance, and a tribunal consequently should depart from that guidance adversely to an individual only when it is plain that the guidance is wrong and there is compelling reason. No such limitation is appropriate or necessary when the tribunal differs from the published guidance, but in the individual's favour—but even here there is a public interest in consistency of application. The penalty of £650,000 for S would remain.

### *Tinkler v. Solicitors Regulation Authority* [2012] EWHC 3645 (Admin)

**28.15** Solicitors' Account Rules

A fine of £40,000 was reduced to £20,000 in respect of failure to comply with the Solicitors' Account Rules and other allegations, and misleading clients as to the nature of the firm's partnership. The fine was to be paid by instalments.

### *Mark Anthony Financial Management v. Financial Services Authority* [2012] Lexis Citation 113

**28.16** False mortgage applications—£150,000 fine

A penalty of £150,000 was levied in respect of mortgage applications that contained knowingly false and misleading information. A sum of £50,000 was to be paid within one month and £100,000 over the course of the subsequent twelve months.

### *Matthews v. Solicitors Regulation Authority* [2013] EWHC 1525 (Admin)

**28.17** Means

M pleaded guilty to two charges: one of failing to act in the best interests of his firm's lender clients, and the other of failing to make arrangements so as to ensure adequate supervision of an employee of his practice. The result of these matters was that M's practice closed down in September 2009, after which time he had found it impossible to find a position as a solicitor. M was nearly 63 years old and also suffered from periods of ill health. The SDT ordered M to pay a fine of £5,000 and costs of £16,000. Means is regarded as material by the Court both as to the amount of costs and in respect of the quantum of any fine: see *D'Sousa v. Law Society* [2009] EWHC 2193 (Admin); *Solicitors Regulation Authority v. Davis and McGlinchey* [2011] EWHC 232 (Admin). See also paragraphs 15.12 and 15.13. The overall financial liability is a relevant consideration that must be taken into account in deciding on the appropriate orders, both of costs and of fine, if fine is considered to be necessary. The Tribunal was entitled to take the view that the case merited a fine rather than a reprimand—but the level of fine, coupled with the costs, was excessive. A total liability of fine and costs was to be limited to £5,000. Accordingly, the fine would be reduced to £500 and the balance of £4,500 would be by way of costs.

### *Law Society (Solicitors Regulation Authority) v. Emeana and ors* [2013] EWHC 2130 (Admin)

**28.18** Sham partnership agreement—fines quashed—strike-off order

Fines were imposed in relation to solicitors who had entered into a sham partnership agreement that was held to be inappropriate. Participation in a sham partnership undermines the profession and the protection to which the public are entitled to derive from regulation and the proper working of partnerships. E had been able to practise as a de facto sole principal, when he was in fact not qualified to do so. That alone would have justified the solicitors responsible for this being struck off. Further, the SDT underestimated the gravity of other offences.

### *Adesemowo and Alafe-Aluko v. Solicitors Regulation Authority* [2013] EWHC 2020 (Admin)

**28.19** Unpaid insurance

The appellants owed a total of £105,061 in respect of unpaid insurance premiums for professional cover. It was held that, notwithstanding that the second appellant was a salaried partner, a fine of £5,000 and costs of £3,879 were reasonable.

***Solicitors Regulation Authority v. Anderson Solicitors* [2013] EWHC 4021 (Admin)**

**28.20** See paragraph 15.19.

***7722656 Canada Inc. v. Financial Conduct Authority* [2013] EWCA Civ 1662; [2014] LR (FC) 207**

**28.21** Market abuse—cynical course of intensive manipulation of LSE—wholesale disregard of regulatory requirements

The Regulatory Decisions Committee of the Financial Conduct Authority (FCA) imposed a penalty on the appellant of £8 million for market abuse. In upholding the penalty as appropriate, the Upper Tribunal (Judge Colin Bishopp, chairman) said that the appellant was guilty of a cynical course of intensive manipulation of the London Stock Exchange (LSE) for the benefit of itself and dealers within its network. The conduct was carried out over a period of a little over a year, attempts were made to conceal it, and, far from demonstrating remorse, those controlling the appellant had done everything possible to escape the consequences of their actions. Of its kind, said the Tribunal, it was as serious a case as might be imagined and there was nothing that could possibly be said by way of mitigation. The Upper Tribunal said that it did not approach determination of the penalty from the viewpoint of ability to pay, nor by reference to the profits made by the appellant or the losses sustained by others. What was clear is that this was a prolonged, cynical course of market abuse committed by a company that exhibited a wholesale disregard of regulatory requirements. There was little precedent about the level of penalty appropriate to conduct of this kind. Left to itself, the Upper Tribunal might well have concluded that £8 million was insufficient—but it was a substantial sum and not so obviously too little that the Tribunal felt that it should take the step, which may operate as a disincentive to other, meritorious, applicants of increasing it. There was, however, no basis on which a lesser penalty could properly be imposed: at [139]–[143]. The appellant's appeal on separate grounds of jurisdiction and whether the company's behaviour occurred within section 118 of the Financial Services and Markets Act 2000 was subsequently dismissed.

***Fuglers LLP and ors v. Solicitors Regulation Authority* [2014] EWHC 179 (Admin)**

**28.22** Client account—use as banking facility for client

Fines of £50,000 against a West End firm of solicitors, and £20,000 and £5,000 respectively against its two equity partners, were upheld. Four charges against the appellants were upheld—namely, making use of the firm's client account by using it as a banking facility for a client; operating the client account contrary to rule 15 of the Solicitors' Accounts Rules 1998; providing services to a client other than those that a recognized body is permitted to provide, contrary to rule 14 of the Solicitors' Code of Conduct 2007; and acting in a way that was likely to diminish the trust that the public placed in them and the profession, contrary to rule 1.06 of the Solicitors' Code of Conduct 2007. All of the charges arose in relation to the same conduct. The gravamen of the charges was that the appellants had allowed the firms' client account to be used by a client of the firm, Portsmouth City Football Club Ltd, as a banking facility between 5 October 2009 and 8 February 2010. A total of about £10 million passed through the account over the four-month period, throughout which the Club was in a perilous financial state and subject to negotiations for its purchase by a consortium of potential buyers, for whom the firm was also acting.

***Khan v. Financial Conduct Authority*, FS/2013/002, 8 April 2014**

**28.23** False mortgage application—£80,000 fine

K was an approved person, and the sole director and employee of Sovereign Worldwide Ltd, which had permission to carry out regulated mortgage and general insurance activities. In October 2009, K knowingly submitted a personal mortgage application through Sovereign that contained false and misleading information about his income, and which he supported with falsified payslips. In May 2010, K knowingly confirmed that the false information supporting the application was correct when submitting a substituted mortgage application. The FCA contended that this was not the first time that K had supplied false details: he had also done so in 2007. In dismissing K's appeal against a financial penalty of £80,000, the Upper Tribunal (Tax and Chancery Chamber) said that the correct approach was to assess K's means to pay the financial penalty by reference to his verified assets and outstanding liabilities at the time that the assessment falls to be made. It was not for the Tribunal to direct that any particular asset should be realized to meet the liability for the financial penalty, but it anticipated that property realizations would be necessary. In those circumstances, it would be appropriate to give K a period of twelve months in which to pay the financial penalty.

***Micalizzi v. Financial Conduct Authority* [2014] UKUT 0335 (TCC)**

**28.24** Fraudulent conduct by trader—financial losses to investors and lenders of US$41.8 million—personal circumstances of trader—financial penalty £2.7 million

M was the CEO of Dynamic Decisions Capital Management Ltd (DDCM). DDCM was the manager of a fund, the DD Growth Premium Master Fund. The Fund was marketed by DDCM as a low-risk, highly liquid, market-neutral fund, which pursued a pairs strategy using equities and outperformance options. In the final quarter of 2008, the Fund sustained catastrophic losses to investors and lenders. M abused that position to conceal the losses. Nomura Bank International Plc was a provider of finance to the Fund and invested US$41.8 million following M's representations to it. The Upper Tribunal (Judge Roger Berner, chairman) said that although it was not possible to assess a particular figure for loss, from the evidence of investors and Nomura, considerable losses were incurred that were likely not to have been incurred had M presented to investors and lenders the true position of the Fund's losses. M was dishonest in a number of respects, and his dishonesty adversely affected the interests of investors and lenders to the Fund. Such misconduct merits a substantial penalty, both to reflect the seriousness of the case and as a deterrent to others. It was accepted by the FCA that M was 'balance sheet insolvent', and that any level of penalty would cause him serious financial hardship and may cause his bankruptcy. The learned judge said that whilst, in suitable cases, it may be appropriate to reflect an individual's personal circumstances in the level of penalty, the Upper Tribunal accepted the Authority's submission that this was not appropriate in this case, given the seriousness of the findings. It would not be right, in a case such as this, for the penalty to be reduced to reflect M's own financial circumstances. The financial penalty proposed by the Authority was £3 million, influenced, at least to some extent, by the remarks of the Tribunal in *Visser v. Financial Services Authority* [2011] All ER (D) 57 (Oct), [120]. Taking all of the circumstances into account, M's misconduct was at the most serious end of the scale and a penalty of £2.7 million was appropriate in this case.

## C. Other Relevant Chapters

**28.25** Costs, Chapter 15

# 29

# *GALBRAITH* SUBMISSIONS



## A. Other Relevant Chapters

**29.01** No Case to Answer, Chapter 50

# 30

# GOOD CHARACTER



## A. Disciplinary Cases

### *Gopakumar v. General Medical Council* [2006] EWHC 729 (Admin)

**30.01** Indecent conduct—advice to panel—adequacy

G had served as a doctor in the National Health Service (NHS) without complaint for more than thirty years and was a partner in a busy practice. He was the subject of two complaints by female patients, known as Mrs A and Miss B, at his surgery. His examination of the patients was alleged to have been inappropriate and indecent. The panel found the charges relating to Miss B proved, but did not find that he had acted indecently in his examination of Mrs A. In his advice to the panel, the legal assessor said that all witnesses are of good character unless the panel hears to the contrary; good character was, of course, relevant, G was of good character, his good character supported his credibility, and the fact that he was of good character may mean he was less likely than otherwise might be the case to have acted as was alleged he had acted in this particular case. In dismissing G's appeal and the sanction of erasure, Underhill J agreed that, viewed as a character direction in a criminal trial, what the legal assessor said was unsatisfactory in that saying all witnesses—that is, including the complainants—were of good character undermined the force of the direction. However, the role of a legal assessor assisting a disciplinary panel is not analogous to that of a judge summing up in a criminal trial: *R (Campbell) v. General Medical Council* [2005] 1 WLR 3488, [23], citing *Libman v. General Medical Council* [1972] AC 217. The learned judge said that he did not believe that the way in which the assessor dealt with the issue of good character invalidated the panel's decision. The assessor's advice was clearly adequate to remind it—if a reminder were needed—to consider that G had no criminal convictions or findings of misconduct against him. As regards the submission that an enhanced good character direction was required, even in a criminal trial it would be very much a matter for the judge's discretion, and the learned judge said he did not believe that fairness required that the panel here should have been explicitly reminded of the length of G's (unblemished) record in the NHS. See further paragraph 44.17.

### *Donkin v. Law Society* [2007] EWHC 414 (Admin)

**30.02** Dishonesty—credibility and propensity—positive good character—relevance to state of mind

D was a solicitor who had practised for almost thirty years without attracting any concern about his integrity. He and his firm were defrauded by a dishonest partner. An allegation against D of dishonest misappropriation of clients' funds related to three clients in respect of whom D had taken money from client account in order to discharge liabilities of the practice. The case for D was that, in each case, the withdrawals were by way of authorized loans. At the commencement of the hearing, counsel for D sought to place before the Solicitors Disciplinary Tribunal (SDT) a large number of references speaking to D's professional and social reputation. They were forty-seven in number. They came from circuit judges, a magistrate, Queen's Counsel, junior counsel, police officers, and others of high standing. The Tribunal chair agreed to receive the material, although she commented that the weight to be attached to it might be less in view of the fact that the material was in writing and its authors (with one exception) had not attended for cross-examination. The Tribunal made no references whatsoever to the cogent evidence of positive good character when setting out its findings against D on dishonesty. In allowing D's appeal, Maurice Kay LJ (with whom Goldring J agreed) said:

> 22. Whilst it is true that, in some professional disciplinary cases, evidence of character is only relevant at the second stage, there are other cases where it has potential relevance at the earlier stage. An example of irrelevance at the first stage would be a case where the alleged misconduct does not require proof of a guilty state of mind. Once the conduct has been proved or admitted, it cannot avail the person charged to say that his previous exemplary character prevents the conduct from being misconduct. These issues were authoritatively canvassed in *The Queen (Campbell) v. General Medical Council* [2005] EWCA Civ 250. Giving the judgment of the Court of Appeal Judge LJ referred to issues of culpability and mitigation as being distinct, with a need for them to be addressed and decided sequentially (paragraph 43). However, the court did not suggest that material relevant to the discrete issues is always mutually exclusive. The same passage continues: "The fact that in some cases there will be an overlap, or that the same material may be relevant to both issues, if they arise, does not justify treating evidence which is exclusively relevant to personal mitigation as relevant to the prior question, whether serious professional misconduct has been established."
>
> 23. In other words, it is the context which determines whether material which would be relevant to personal mitigation is also relevant to "the prior question". The mischief which was the concern of the court in *Campbell* was the situation where personal mitigation might be misused to downgrade what would otherwise amount to serious professional misconduct to some lesser form of misconduct (see paragraph 46(iii)).
>
> 24. On behalf of the appellant, [counsel] submits that where the issue is dishonesty, evidence of good character, particularly evidence as reliable and extensive as was produced in this case, is relevant to credibility and to propensity just as it would be in a criminal trial. She further suggests that it is also relevant to an examination of the circumstances in which the misconduct took place although, ultimately, they may add little to propensity in the sense that that word surely denotes propensity to commit the offence in the circumstances which are established.
>
> 25. In my judgment the evidence of good character in this case was relevant to the issue of dishonesty. As in a criminal trial, it cannot afford a defence in itself. Moreover, the weight to be attached to it is in the last resort a matter for the tribunal. In the present case, the reasons stated by the tribunal do not disclose that it gave any consideration at all to this evidence in this context. I am not satisfied from the text of the stated Reasons that it played any part in its consideration of dishonesty. I find that to be a significant legal error.

### *Bryant and anor v. Law Society* [2007] EWHC 3043 (Admin); [2009] 1 WLR 163

**30.03** Cogent evidence of good character—relevant to dishonesty

The appellants were solicitors practising in partnership together in London. They faced seven charges of professional misconduct, which the SDT found largely substantiated. The Tribunal also made a finding of dishonesty in respect of one of the solicitors and ordered him to be struck off the roll of solicitors. On appeal, it was contended that there was an error in the Tribunal's approach to the issue of dishonesty. The Tribunal was asked to consider, prior to its decision on liability, a number of written character references adduced on the solicitor's behalf. They were unusually impressive and came from a retired High Court judge, a former deputy chairman of the Board of Inland Revenue, a former president of the Confederation of British Industry (CBI), a former chairman of the Competition Commission, a senior solicitor, and the rector of the solicitor's parish. They testified to the solicitor's honesty and integrity. It was submitted to the Tribunal that the evidence was relevant to the question of propensity to be dishonest and also to the solicitor's credibility when giving evidence. The Tribunal rejected the application, relying on *R (Campbell) v. General Medical Council*. Accordingly, it left the character evidence out of account when reaching its finding of dishonesty. In allowing the solicitor's appeal, the Divisional Court (Richards LJ and Aikens J), after referring to the decision of the Divisional Court in *Donkin* at [22]–[25], said, at [162]:

> We are in full agreement with that reasoning, which applies with equal force in the circumstances of the present case. The character references in support of Mr Bryant were cogent evidence of positive good character and were of direct relevance to the issue of dishonesty. The tribunal's refusal to take that evidence into account when deciding the question of dishonesty was a significant legal error.

### *DV v. General Medical Council* [2007] EWHC 1497 (Admin)

**30.04** Testimonials—practitioner absent—not disclosed to panel

V, a staff-grade physician whose speciality was gastroenterology, faced a series of allegations arising out of his employment at Grantham Hospital. He did not attend, nor was he represented at the hearing before the fitness to practise panel in September 2006. However, shortly prior to the hearing, he sent by fax to the GMC and to Eversheds, the GMC's solicitors, five testimonials. Owing to an error of the GMC, the panel was shown only two of the five testimonials. Bennett J said that this was a mistake. The GMC said that the three testimonials not placed before the panel would not have changed the panel's mind. In allowing V's appeal and remitting sanction to the panel for reconsideration, Bennett J said that, be that as it may, V was entitled to expect the GMC to put before the panel all of the five testimonials, particularly if there was a possibility, as occurred, of his name being erased from the register and of such a sanction being imposed in his absence. Justice was not seen to be done in this respect, although it occurred through no fault of the panel.

### *Wisson v. Health Professions Council* [2013] EWHC 1036 (Admin)

**30.05** Practitioner—credibility—weight to be attached to good character

Good character is obviously a matter that can go to mitigation, but good character can be material, and is material, in many cases when considering the credibility of the individual in question. Where there is a dispute between the registrant's evidence in relation to a particular allegation and that of a complainant, or indeed any other witness, the credibility of the registrant is obviously a matter of some importance. Good character is a factor that can be taken into account; obviously the weight to be attached to it is a matter for the tribunal and in assessing the defendant's evidence whether he or she is to be believed. It may be that, in many instances, it does not take the matter that much further. The obvious case is one in which dishonesty is being alleged, but good character is not limited to dishonesty.

### *Shaw and Turnbull v. Logue* [2014] EWHC 5 (Admin)

**30.06** 'Good character direction'— not essential

The SDT is an expert, professional jury that does not need the sort of 'good character direction' that one sees in criminal trials (and neither counsel sought to give one), and does not need to have demonstrated that it took the appellants' good character into account by express reference to these trite principles in the body of its judgment: *per* Jay J, at [182]. It is obvious to anyone experienced in this line of work that the SDT would be very slow to find subjective dishonesty unless driven by the evidence to do so.

### *Arunkalaivanan v. General Medical Council* [2014] EWHC 873 (Admin)

**30.07** Patient examination—whether sexually motivated—inference from primary facts—good character—whether alleged conduct inherently unlikely

The appellant, A, was a consultant obstetrician and urogynaecologist with an impressive curriculum vitae. He appealed against a decision of the fitness to practise panel of the GMC, which found that he had conducted a breast examination of a patient in the absence of a chaperone and in an inappropriate manner, and that his conduct was sexually motivated. The disputed allegations turned on the evidence of the patient. The issue in relation to the appropriateness of A's handling of the patient's breasts was a straightforward factual one to be determined by the panel, having considered the evidence of the only two people who were present when it happened—that is, the patient and A. This was a classic case for the Tribunal to hear oral evidence and, having done so, to decide which version of events it preferred. What was called for was a comparative evaluation of each party's evidence. It is clear that the panel undertook that exercise before reaching its decision and its primary findings were unassailable. The position was somewhat different in relation to the finding that A's actions were sexually motivated. Although a finding of fact, it depended not on direct evidence, but on the inference to be drawn from the primary facts as found by the panel and the surrounding circumstances. The panel made no reference at all to A's good character in its reasoning on the issue of sexual motivation. Having considered its reasons, it appeared to have taken the character evidence into account in the first stage when determining the factual dispute between A and the patient. However, having decided that the patient was the more credible and reliable witness (which it was perfectly entitled to do), the panel apparently did not weigh up the extent to which the evidence of A's character might be relevant to the final issue of whether he was sexually motivated. This was a material omission. The panel's reasoning in relation to sexual motivation was very closely tied to its decision that the patient was the more credible and reliable witness. The choice for the panel was effectively between, as the defence submitted, a clumsy, inappropriate (and the Court added) insensitive examination or a sexual assault. There was no real justification within the panel's reasons for finding one rather than the other. Matters suggesting an absence of sexual intent included the lack of any evidence of sexual gratification—A's demeanour remained entirely normal throughout—and the inherent improbability of a man with an unblemished history and a long record of working in gynaecology suddenly committing an indecent assault on a day on which he was in a hurry to pick up his child and, in fact, did hurry away at the end of the appointment. It was unlikely that A's actions were sexually motivated and far more likely that he carried out an inappropriate examination because he was rushing, probably distracted and thus clumsy and insensitive to the patient. As a result, he left her feeling violated, even though he did not intend to touch her sexually. Accordingly, while the Court would decline to interfere with the panel's primary findings of fact as to how the patient's breasts were handled by A, it was appropriate to reverse the finding that his conduct was sexually motivated. That part of the decision was wrong and could not be supported by the evidence. The GMC, in such circumstances, did not contend that A's fitness to practise was impaired on the basis of the remaining findings, but the matter was to be remitted to a differently constituted panel to consider whether, in the light of the primary findings, it would be appropriate to impose a warning under rule 17(2)(l) of the 2004 Rules.

### *Monibi v. General Dental Council* [2014] EWHC 1911 (Admin)

**30.08** Reasons—Committee's determination—likelihood of dishonestly altering records

Between 2007 and 2009, M provided dental services to a patient. The patient was dissatisfied with her treatment and the matter was referred to the General Dental Council (GDC), which instituted disciplinary proceedings. The Professional Conduct Committee (PCC) determined that key facts alleged against M were proved, including an allegation that he had dishonestly altered entries in his notes for the patient. M's challenge to the Committee's determination on the factual allegations was rejected, but the challenge to the Committee's determination on dishonesty succeeded and the sanction of four month's suspension was quashed. In dealing with the Committee's failure to give sufficient weight to M's previous good character, Stuart-Smith J said, at [59]:

> The Committee said that it had regard to the Appellant's previous good character but did not otherwise identify what weight it had attached to it. The evidence from [the appellant's] Receptionist, his Practice Manager and a colleague who gave evidence on his behalf was that [the appellant] is highly competent and respected in his field as a clinician and that he is a person of the highest integrity. That evidence goes directly to the likelihood of [the appellant] acting dishonestly when altering his records, and I have taken it into account. Of course, the Committee had the advantage of seeing [the appellant] giving evidence whereas I have only been able to read it. It is also fair to point out that [the appellant] was singularly unable to explain how he came to make the original mistakes or the precise circumstances in which he made the alterations. However, there is nothing in his evidence that suggests that his character witnesses were wrong in their assessment; to the contrary, he showed himself ready on many occasions when giving evidence to accept responsibility from misunderstandings that may have arisen between him and Patient A. Despite his inability to give satisfactory explanations, the evidence of his general integrity raises questions about the likelihood that he would make two dishonest alterations while casually making six altogether. His general integrity is not on its own a sufficient reason to depart from the findings of the Committee, but it goes into the balance when reviewing them.

Once it was decided that five alterations were honestly made, that fact and M's general integrity weighed heavily in the balance against a finding that the sixth alteration was dishonest.

### *Inayatullah v. General Medical Council* [2014] EWHC 3751 (Admin)

**30.09** Good character direction—whether appropriate

The appellant was the subject of a sting operation undertaken by investigative journalists, who arranged for actors to pose as patients and present themselves to his general practice clinic, making complaints of various symptoms. A number of charges made against the appellant were not proved. The charges that were proved related to two patients in particular and the conduct of their consultations. In relation to one patient, it was found proved: that the appellant had failed to obtain details of weight change and rectal bleeding, including the nature of that bleeding, for the purposes of understanding the patient's complaints; that the appellant failed to undertake a rectal examination and failed to refer the patient for further assessment; that the patient had not been diagnosed by the appellant; and that the appellant had failed to make an adequate record of the consultation and manage it appropriately. At a subsequent consultation with the same patient, the appellant had failed to take a further medical history and failed to assess or diagnose the patient, or to properly manage the consultation, and inaccurately recorded in the patient's notes that he had examined him when, in fact, he had not. Turning to the second patient, it was found that the appellant had failed to ask what his presenting symptoms were or to take an adequate history, to assess or diagnose the patient, to provide treatment, or to properly manage the consultation. The panel concluded that the only appropriate sanction was erasure. On appeal, it was submitted on behalf of the appellant that, by analogy with the position in the criminal law, a good character direction was essential in this case. In dismissing the appellant's appeal, Dove J said, at [46], that he was unable to accept that submission: whilst it might not have been inappropriate to give such a direction, the real question in this case was whether fairness required it. The learned judge continued:

> [46] ...[T]he analogy with criminal proceedings is not one which is close-fitting. These are not proceedings which are identical and raise the same issues as to legal soundness which a judge's direction to a jury may. There is, as the authorities make clear, a distinction to be drawn between experienced panel members – and here there were two lay members and one who was medically qualified – and a jury and the instruction or advice which either might need will differ as a result of that context. Whilst it would be wrong to say that good character would not be an issue in all cases that come before the Respondent, it is a reasonable observation to say that in far more cases than might be the case in a criminal context, those who appear before the Respondent will have a good and unblemished character which is being impeached by those proceedings.
>
> [47] In my view, the fact that the Appellant presented to the Panel as a recognised medical practitioner and, subject to the charges, fit to practise, is an obvious part of the backdrop of the proceedings in this case. The two limbs of the criminal good character direction, mainly the impact of good character on the creditability of his evidence and its impact on his propensity as a qualified doctor to commit the offence in this case, namely to lie, would be both clear and obvious to the panel without the formality or necessity of a direction to render the proceedings fair.
>
> [48] I am therefore satisfied that the absence of such a good character direction was not a legal error in this case. I do not consider there was any obligation on the Legal Assessor to give one. I have reached that conclusion without needing to consider the further submissions which were made by [counsel for the GMC] in this respect, namely the absence of a request from the Appellant's representative that a good character direction should be given...

## B. Other Relevant Chapters

**30.10** Bad Character, Chapter 5
Impairment of Fitness to Practise, Chapter 33
Misconduct, Chapter 47
Registration, Chapter 56

# 31

# HEALTH (ADVERSE PHYSICAL OR MENTAL HEALTH)



## A. Legal Framework

**31.01** Examples include:

Medical Act 1983, section 35C (fitness to practise impaired by reason of adverse physical or mental health)
Opticians Act 1989, section 13D
Dentists Act 1994, section 27
Health and Social Work Professions Order 2001, article 22
Nursing and Midwifery Order 2001, article 22
Pharmacy Order 2010, article 51 (adverse physical or mental health that impairs the ability to practise safely and effectively, or which otherwise impairs the ability to carry out the duties of a pharmacist or pharmacy technician in a safe and effective manner)

## B. Disciplinary Cases

### *Crompton v. General Medical Council (No. 2)* [1985] 1 WLR 885

**31.02** Findings of Health Committee—expert evidence—conditions

C appealed from a decision of the Health Committee of the General Medical Council (GMC) that his fitness to practise was seriously impaired by reason of his mental condition and from a direction that his registration as a fully registered medical practitioner should be conditional for twelve months on his compliance with three requirements: (a) that he should practise only in a laboratory post; (b) that he should consult a psychiatrist; and (c) that he should also consult a neurologist. C was convicted in 1975, 1977, and 1978 of certain criminal offences, and appeared before the GMC's Disciplinary

Committee. Two consultants in adult psychiatry, Drs H and F, supplied reports to the GMC, but these were not shown to C. On that ground alone, the Privy Council allowed C's appeal: *Crompton v. General Medical Council* [1981] 1 WLR 1435 (see further paragraph 48.07). In September 1981, the registrar wrote to C, notifying him that the GMC had received information that appeared to raise a question whether his fitness to practise was seriously impaired by reason of mental disorder. In June 1982, the Health Committee found that C's fitness to practise was seriously impaired by reason of his mental condition and directed that his registration should be suspended for twelve months. C's registration was suspended for two further periods of twelve months in June 1983 and July 1984. Oral evidence was given by Drs H and F, as well as a Dr E. C thereafter lodged an appeal to the Privy Council. His primary submission was that there was not sufficient evidence to support a finding that his fitness to practise was seriously impaired by reason of his mental condition. Their Lordships said that they had no hesitation in saying that, in their opinion, the Health Committee had before them ample evidence to justify such a finding. They continued at 894G–H:

> Secondly, only in an exceptional case would this Board feel justified in rejecting a finding of the Health Committee, assisted by medical assessors, on an issue which is a matter of medical judgment. Thirdly, Dr Crompton, throughout the long history of this case since September 1981, has persistently declined to submit to any medical examination in relation to his mental condition.

C's second submission of substance was that the conditions imposed on him were unworkable and unfair. As regards condition (a), it did not confine C to a laboratory post that had to be filled by a fully registered medical practitioner. As regards conditions (b) and (c), it was implicit that the treatment or tests should be reasonable in the circumstances. The conditions did not deprive C of his common law right to refuse to undergo any treatment or tests that he may have wished to decline. If he failed to comply with a condition, the Committee had a discretion under section 37(2) of the Medical Act 1983 whether or not to suspend his registration. Clearly, the Committee would not seek to suspend his registration if his refusal to undergo a particular treatment or test were reasonable. Additionally, the president had power to expedite any resumed hearing to enable the Committee to vary the conditions. If a particular condition were to be shown by C to be occasioning hardship, it would be open to him to request and expedite his resumed hearing, and in an appropriate case the president would no doubt so direct.

### *Stefan v. General Medical Council* [1999] 1 WLR 1293

**31.03** Suspension—reasons—fairness

S, a registered medical practitioner, appealed from a determination of the Health Committee of the GMC on 23 February 1998 that her fitness to practise was seriously impaired and that her registration should be suspended indefinitely. After an initial hearing on 2 June 1998 before a panel of three, their Lordships directed that the case should be adjourned to be heard by a board of five and that an amicus curiae should be appointed. Lord Clyde, giving the judgment of their Lordships on 8 March 1999, said that there was no express statutory duty on the Health Committee to state reasons for its decisions. Their Lordships are unable to spell out an implied obligation to state reasons. Detailed as the procedural provisions are, it cannot be concluded that there is an implied statutory duty to give reasons. But, correspondingly, their Lordships were not persuaded that the Medical Act 1983 or the rules were to be read as excluding an obligation to give reasons where the common law would require reasons to be given. It is certainly within the power of the Committee to state its reasons even though the rule does not itself imply an obligation to do so.

> 22. The trend of the law has been towards an increased recognition of the duty upon decision-makers of many kinds to give reasons. This trend is consistent with current developments towards an increased openness in matters of government and administration. But the trend is proceeding on a case-by-case basis... and has not lost sight of the established position of the common law that there is no general duty, universally imposed on all decision-makers. It was reaffirmed in *R v. Secretary of State for the Home Department, ex parte Doody* [1994] 1 AC 531, 564, that the law does not at present recognise a general duty to give reasons for administrative decisions. But it is well established that there are exceptions where the giving of reasons will be required as a matter of fairness and openness....
>
> [...]
>
> 25. Turning to the particular circumstances of the present case their Lordships are persuaded that there was a duty at common law upon the Committee in the present case to state the reasons for their decision. In the first place there is the consideration that the decision was one which was open to appeal under the statute.... Secondly, a consideration of the whole procedure and function of the Committee prompts the conclusion that the procedures which it follows and the function which it performs are akin to those of a court where the giving of reason would be expected....

The Court continued, at [26]: 'Thirdly, the issue was one of considerable importance to the practitioner.' At [27], 'Fourthly, [S] has repeatedly asked for an explanation of the Committee's view and for the diagnosis which they have reached of her condition.' Fifthly, it was not clear why, in the light of the only expert witness to give evidence before the Committee, Dr A, the Committee reached the decision that it did. At [29]: 'Sixthly, this was the first time that an indefinite suspension was decided upon.' In allowing the appeal and remitting the case to a freshly constituted Health Committee, their Lordships said:

> 32. The extent and substance of the reasons must depend upon the circumstances. They need not be elaborate nor lengthy. But they should be such as to tell the parties in broad terms why the decision was reached. In many cases, as has already been indicated in the context of Article 6(1) of the European Convention for the Protection of Human Rights and Fundamental Freedoms, a very few sentences should suffice to give such explanation as is appropriate to the particular situation. Their Lordships do not anticipate that the recording of a generally agreed statement of their reasoning would add to the burden of the decision-making process. While the decision involves the application of some medical expertise in the assessment of fitness, the articulation of the reasons for a value judgment should not give rise to difficulty (*R v. City of London Corporation, ex parte Matson* [1997] 1 WLR 765, 783). Their Lordships have observed that in certain other appeals from the Health Committee which have come before them succinct but adequate reasons have been stated in the decision. Unfortunately such a course was not adopted in the present case.

### *Crabbie v. General Medical Council* [2002] UKPC 45; [2002] 1 WLR 3104

**31.04** Sanction—possible erasure—PCC—Health Committee

Following a serious road traffic accident, C, a registered medical practitioner, was charged with causing death by dangerous driving and driving with excess alcohol. She pleaded guilty to both charges, was sentenced to five years' imprisonment, and was disqualified from driving for ten years. The convictions were reported to the GMC, which instituted disciplinary proceedings. Medical evidence before the Professional Conduct Committee (PCC) indicated, that at the time of the accident, C had been exhibiting features of an alcoholic dependency, and counsel for the doctor asked for the case to be remitted to the Health Committee to be dealt with by that Committee rather than by the PCC. Unlike the PCC, the Health Committee has no power to direct erasure. The PCC accepted that C's fitness to practise might be seriously impaired by reason of alcohol dependency, but, in view of the serious nature of the criminal convictions, refused the request for the case to be referred to the Health Committee. The PCC directed that C's name be erased as the only appropriate sentence, adding that, in its view, erasure was proportional to the nature and gravity of the offence. In dismissing C's appeal, Lord Scott of Foscote, in giving the judgment of the Board, said:

> 18. Accordingly, in their Lordships' view, if the case is one in which erasure is a serious possibility, neither the medical screener nor the Preliminary Proceedings Committee should refer the case to the Health Committee notwithstanding that it may be one where the fitness to practise of the practitioner in question appears to be seriously impaired by reason of his or her physical or mental condition.
>
> 19. As to the PCC, the question whether the PCC should exercise its rule 51 power to refer the case to the Health Committee should be considered in conjunction with the question whether the case is or may be one which calls for a direction of erasure. The PCC should not, in their Lordships' view, refer a case to the Health Committee unless and until satisfied that a direction of erasure would not be the right direction to make. And once the PCC has decided that a direction of erasure is the right direction to make, the question whether the case should be referred to the Health Committee has received its answer.
>
> 20. It follows that in the present case the first and main question is whether the direction of erasure was justified. If it was, there is nothing left in the submission that the PCC should have referred the case to the Health Committee.

### *Sreenath v. General Medical Council* [2002] UKPC 56

**31.05** PCC—Health Committee

S appealed from a direction given by the GMC's PCC on 15 March 2002, under section 36 of the Medical Act 1983, as amended, that his name be erased from the register. The direction followed the Committee's refusal to accede to an application made on S's behalf that his fitness to practise should be referred to the Health Committee, or alternatively that the enquiry should be adjourned to allow him to obtain definitive medical evidence of his medical condition before proceeding further. The appeal was effectively an appeal from that refusal. S was charged with serious professional misconduct—that is, behaving inappropriately and indecently towards two of his female patients on two separate occasions while acting as a locum general practitioner (GP). He was also charged with having committed breaches of conditions imposed on his registration, which had been imposed by the Interim Orders Committee (IOC). The PCC having found the charges relating to S's conduct towards his patients proved, his counsel sought and was granted a short adjournment to enable S to be examined by a consultant psychiatrist. At the resumed hearing, Dr C, a consultant psychiatrist, gave oral evidence in camera. Dr C concluded that S was suffering from an acute anxiety state and that it was possible he might be suffering from organic brain damage in the form of dysexecutive syndrome (DES), affecting his understanding and judgment. In evidence, Dr C said that he was not in a position to make a definitive diagnosis, but that DES was an organic personality change resulting from significant brain insult. To confirm the diagnosis, it would be necessary to undertake a battery of neuropsychological tests, with a much more detailed history and probably brain imaging. The PCC took the view that S's conduct amounted to serious professional misconduct and that the protection of the public required the erasure of his name from the register whether or not he was suffering from a psychiatric disorder at the relevant time. Following the conclusion of the inquiry, S obtained and placed before the Board reports from a neuropsychologist and a psychiatrist. These showed that S was currently suffering from bipolar affective disorder associated with disinhibition and socially inappropriate behaviour. The consultants were of the opinion that S's mental condition affected his ability to practise effectively and may have contributed to his condition at the time of the offences in question. Lord Millett, in giving the judgment of the Board in dismissing S's appeal, said that the question that the PCC had to consider was whether the material sought to be obtained would be relevant to any question that it had to decide. The Committee was entitled to take the view that further evidence corroborating Dr C's report would not be of assistance and to refuse an adjournment sought for the purpose of obtaining such evidence. The possibility of erasure could not be excluded at the time of the application and was still open if the PCC were to be directed to reconsider the question of penalty. Lord Millett said:

> 17. The functions of the Professional Conduct Committee and the Health Committee, though complementary, are distinct. The Professional Conduct Committee is concerned to maintain professional standards of integrity and competence and the reputation of the medical profession. Its function is disciplinary. If it finds that a practitioner has been guilty of serious professional misconduct, it must consider whether the safety of the public and the reputation of the profession require that his name be erased from the register so that he cannot carry on a medical practice even if medically fit to do so; or whether something lesser penalty such as conditional registration or suspension for a limited period would be sufficient. The functions of the Health Committee are not disciplinary. It is concerned to protect the public (and the practitioner himself) from the dangers of a practitioner being allowed to carry on practice while he is medically unfit to do so, whether or not he has committed a disciplinary offence. Its powers are limited to enable it to achieve this purpose. It cannot order the practitioner's name to be erased from the register. The most it can do in an appropriate case is to order indefinite suspension, and such an order is reviewable at any time. If the Health Committee is satisfied that the practitioner has recovered sufficiently to be allowed to resume practice, it will terminate his suspension.
>
> 18. But these safeguards are unnecessary if the practitioner's name is to be erased from the register so that he cannot practice even if medically fit to do so. They are necessary only where his name is not to be erased from the register, either because the Professional Conduct Committee has found the charges to be unfounded or because it considers that some lesser penalty is appropriate. That is why in *Crabbie v. General Medical Council* [2002] 1 WLR 3104 their Lordships held that the Professional Conduct Committee should not refer a case to the Health Committee unless it was satisfied that an order directing the erasure of the practitioner's name from the register was not appropriate.

***Brocklebank v. General Medical Council*** **[2003] UKPC 57**

**31.06** Drug and alcohol dependency—recurrence—serious impairment

B appealed against the finding of the Health Committee that her fitness to practise was seriously impaired by reason of her physical or mental condition. The Committee found the causes of impairment to be a drug dependency and a tendency to harmful use of alcohol, both being currently in remission. Rule 24(2) of the General Medical Council Health Committee (Procedure) Rules 1987 provides:

> In reaching their judgment the Committee shall be entitled to regard as current serious impairment either the practitioner's current physical or mental condition, or a continuing and episodic condition, or a condition which, although currently in remission, may be expected to cause recurrence of serious impairment.

Dr B, a consultant psychiatrist, said that she regarded B's drug dependency syndrome as a diagnosis that should be maintained even after years of abstinence. She had seen patients relapse after many years and some authorities treated it as a lifetime diagnosis. Whilst the chance of a return to opiate dependency was very low—'practically nil'—nevertheless a return to stressful conditions could precipitate a relapse. The vulnerability for alcohol relapse would remain elevated over the long term if B were under severe stress for any reason. In dismissing B's appeal, Lord Hoffmann, giving the judgment of the Board, said that the evidence provided ample support for the finding of the Committee that a serious impairment existed, but was currently in remission. As to the words 'may be expected to cause recurrence of serious impairment' in rule 24(2):

> 11. Their Lordships do not think that the jurisdiction of the Committee depends upon an assessment of the chances of the impairment recurring. It is sufficient that the condition, if it recurs, may be expected to cause serious impairment. Of course, the Committee will have regard to what it considers to be the likelihood of recurrence in deciding whether to impose conditions upon the registration and for how long. But the existence of an underlying condition capable of causing serious impairment if it should recur is sufficient to found jurisdiction.

***R (Rogers) v. General Medical Council* [2004] EWHC 424 (Admin)**

**31.07** Criminal conviction—erasure—PCC—Health Committee

R practised as a GP and, in recent years, as a locum. In November 2001, he pleaded guilty at Cardiff Magistrates' Court to twelve offences of obtaining prescription drugs—namely, opiates—by deception and eleven offences of possession of controlled drugs, all but two of Class A. He was committed for sentence to Cardiff Crown Court. He was sentenced in December 2001 to twelve months' imprisonment, suspended for two years. The convictions were admitted before the PCC, which directed that R's name should be erased from the register. In rejecting R's submission to refer the case to the Health Committee, the PCC said that 'from the evidence presented, the Committee is not satisfied that your current state of health is sufficiently impaired to warrant a referral to the Health Committee'. In dismissing R's appeal, the Privy Council accepted the criticism of R's counsel that the observations of the Committee concerning the possibility of a reference to the Health Committee were inapt. The Committee should, in fact, only have gone on to consider a reference to the Health Committee if it had first rejected erasure as a sanction, following the decision of the Privy Council in *Crabbie v. General Medical Council* [2002] 1 WLR 3104, [19].

***Cullen v. General Medical Council* [2005] EWHC 353 (Admin)**

**31.08** See paragraph 26.02.

***Yanah v. General Medical Council* [2006] EWHC 3843 (Admin)**

**31.09** Medical assessor—specialist advice—consideration by Committee

Y appealed against a finding that his fitness to practise was seriously impaired by reason of cognitive impairment. In allowing Y's appeal and remitting the matter for redetermination to a differently constituted panel, Sullivan J said that, in normal circumstances, he would have considered that the degree of deference to be accorded to a panel of the GMC that is considering fitness to practise on health grounds should be accorded a considerable degree of respect. That is because, in matters relating to health, perhaps rather more than in matters relating to professional conduct or sanctions, the views of a tribunal with a medical membership command particular respect. However, the other side of that coin is that particular respect should also be accorded to the advice given to the panel by its own specialist health adviser. While the ultimate decision is for the panel, and it is entitled to disagree with its specialist health adviser and can reach a different conclusion, one would expect to see very careful reasoning to justify such a conclusion. The panel's reference to the specialist advice that it had received in the present case was brief in the extreme, and it was not entirely clear whether the panel was saying that its specialist health adviser's evidence was inconsistent or that the evidence about cognitive impairment generally was inconsistent. There were two questions for the panel to consider. First, was there mild-to-moderate cognitive impairment as opposed to no cognitive impairment whatsoever? Second, how did this mild-to-moderate cognitive impairment affect Y's performance as a surgeon? That second question was particularly important, given that it was accepted that there was no evidence of deterioration and given that this was a case in which the chain of events had been commenced not by a complaint by any patient that he or she had suffered, but by a concern expressed by a professional colleague. This is a case in which no concerns had been expressed about the appellant's performance or functioning as a surgeon. The specialist adviser said in terms to the panel that it should look at 'his performance and his functioning', and said 'that is most important here'. Sullivan J said that it seemed to him that the panel had in essence sidestepped the latter question.

### *Sarkodie-Gyan v. Nursing and Midwifery Council* [2009] EWHC 2131 (Admin)

**31.10** Procedure

The registrant appealed against the decision of the Health Committee of the Nursing and Midwifery Council (NMC), finding that her fitness to practise was impaired. It was common ground that her appeal had to be allowed because of procedural irregularities, in that the Committee had wrongly combined the fact-finding stage with the impairment stage. The Court emphasized that, in a health case under the rules, it was still necessary for the Committee to apply the three-stage process of fact-finding, impairment, and, if appropriate, sanction.

### *Goodchild-Simpson v. General Medical Council* [2014] EWHC 1343 (Admin)

**31.11** Suspension—risk of relapse—patient safety—whether conditions adequate

In November 2006, the appellant's fitness to practise was found to be impaired by reason of adverse mental health and he was subjected to eighteen-month conditions. Further fitness to practise hearings were convened in 2008, 2009, and 2011, which found that the appellant's fitness to practise continued to be impaired by reason of adverse mental health. In the course of 2011, the GMC received negative feedback regarding the appellant's abilities and performance, which appeared to be unrelated to his health. A performance assessment report concluded that the appellant's performance was acceptable in certain areas, but unacceptable in others, and that he could work as a junior doctor, which would, by definition, be a closely supervised training post in which his work would be reviewed by a senior professional. At a hearing in August 2013, the panel found that the appellant's fitness to practise was impaired by reason of adverse mental health and deficient professional performance, and imposed a sanction to suspend the appellant's registration for nine months. The panel considered that the imposition of conditions would be inadequate. In dismissing the appellant's appeal, Green J said, at [12], that the fitness to practise panel was required to ask itself whether the conduct of the appellant gave rise to a risk of harm to patients. The panel concluded that the appellant suffered from a serious condition, which was at risk of relapse. The learned judge said, at [40], that there was no question but that the task of the panel was to address the issue of sanction for itself. It could not delegate the task to the assessors and, by the same token, it was clearly not bound to accept the evidence or the recommendations of any of the individual experts who gave evidence before it. The panel had before it a range of experts, who gave evidence on different aspects of the case. Some experts gave evidence on health matters; others, on performance. The assessment report addressed only performance, not health. The key matters that led the panel to choose the option of suspension, and not the imposition of conditions, were the oral evidence of certain medical experts to the effect that the appellant had a 'serious mental illness' and that residual symptoms remained. The panel was swayed by the medical evidence that linked the appellant's mental illness to problems associated with the exercise of good medical judgment. The panel was also swayed by the evidence that suggested that patient safety was at risk of being compromised. It also took account of the medical evidence of Dr D that, in her view, the appellant needed a complete break. The decision of the panel fell squarely within the centre of its area of expertise. It was not credible that the High Court should interfere in that decision.

## C. Other Relevant Chapters

**31.12** Experts, Chapter 26
Human Rights, Chapter 32
Impairment of Fitness to Practise, Chapter 33
Medical Assessors, Chapter 46
Reasons, Chapter 54

# 32

# HUMAN RIGHTS



## A. Legal Framework

**32.01** *Frankowicz v. Poland* App. no. 53025/99, 16 December 2008, ECtHR | 32.19 |
| B. **Disciplinary Cases** | 32.02 | | |
| *König v. Federal Republic of Germany* (1979–80) 2 EHRR 170 | 32.02 | *R (Wright and ors) v. Secretary of State for Health* [2009] 1 AC 739 | 32.20 |
| *Le Compte, Van Leuven and De Meyere v. Belgium* (1982) 4 EHRR 1 | 32.03 | *R (Royal College of Nursing and ors) v. Secretary of State for the Home Department* [2010] EWHC 2761 (Admin); [2011] PTSR 1193 | 32.21 |
| *Albert and Le Compte v. Belgium* (1983) 5 EHRR 533 | 32.04 | | |
| *Van Marle and ors v. Netherlands* (1986) 8 EHRR 483 | 32.05 | *Kulkarni v. Milton Keynes Hospital NHS Foundation Trust* [2010] ICR 101 | 32.22 |
| *H v. Belgium* (1988) 10 EHRR 339 | 32.06 | *Hameed v. Central Manchester University Hospitals NHS Foundation Trust* [2010] EWHC 2009 (QB) | 32.23 |
| *Stefan v. United Kingdom* (1998) 25 EHRR 130 | 32.07 | | |
| *Wickramsinghe v. United Kingdom* (1998) EHRLR 338 | 32.08 | *R (Puri) v. Bradford Teaching Hospitals NHS Foundation Trust* [2011] EWHC 970 (Admin) | 32.24 |
| *Gautrin and ors v. France* (1999) 28 EHHR 196 | 32.09 | | |
| *Tehrani v. United Kingdom Central Council for Nursing, Midwifery and Health Visiting* 2001 SC 581 | 32.10 | *R (G) v. Governors of X School (Secretary of State for the Home Department and anor intervening)* [2012] 1 AC 167 | 32.25 |
Human Rights Act 1998:
| *Pine v. Law Society* [2001] EWCA Civ 1574 | 32.11 | *R (Gaunt) v. The Office of Communications* [2011] 1 WLR 2355, CA | 32.26 |
| *Fleurose v. Securities and Futures Authority Ltd* [2001] EWCA Civ 2015 | 32.12 | *R (Calver) v. Adjudication Panel for Wales* [2012] EWHC 1172 (Admin) | 32.27 |
| *WR v. Austria* (2001) 31 EHRR 43 | 32.13 | | |
| *Luksch v. Austria* (2002) 35 EHRR 17 | 32.14 | *Mattu v. University Hospitals of Coventry and Warwickshire NHS Trust* [2012] 4 All ER 359, CA | 32.28 |
| *R (Thompson) v. Law Society* [2004] 1 WLR 2522 | 32.15 | | |
| *Threlfall v. General Optical Council* [2004] EWHC 2683 (Admin) | 32.16 | *R (King) v. Secretary
| *Macpherson v. Law Society* [2005] EWHC 2837 (Admin) | 32.17 | | |
| *R (Malik) v. Waltham Forest NHS Primary Care Trust* [2007] 1 WLR 2092 | 32.18 | *O'Connor v. Bar Standards Board* [2014] EWHC 4324 (QB) | 32.30 |
| | | C. **Other Relevant Chapters** | 32.31 |

section 3 ('So far as is possible to do so, primary legislation and subordinate legislation must be read and given effect in a way which is compatible with Convention rights')

section 4 (declaration of incompatibility may be granted only by the High Court, Court of Appeal, or Supreme Court) section 6(1) ('It is unlawful for a public authority to act in a way which is incompatible with a Convention right')

European Convention on Human Rights (ECHR):

Article 6 (right to a fair trial):

6(1) In the determination of his civil rights and obligations or of any criminal charge against him, everyone is entitled to a fair and public hearing within a reasonable time by an independent and impartial tribunal established by law. Judgment shall be pronounced publicly but the press and public may be excluded from all or part of the trial in the interest of morals, public order or national security in a democratic society, where the interests of juveniles or the protection of the private lives of the parties so require, or to the extent strictly necessary in the opinion of the court in special circumstances where publicity would prejudice the interests of justice.

(2) Everyone charged with a criminal offence shall be presumed innocent until proved guilty according to law.

(3) Everyone charged with a criminal offence has the following minimum rights:

- (a) to be informed promptly, in a language which he understands and in detail, of the nature and cause of the accusation against him;
- (b) to have adequate time and facilities for the preparation of his defence;
- (c) to defend himself in person or through legal assistance of his own choosing or, if he has not sufficient means to pay for legal assistance, to be given it free when the interests of justice so require;
- (d) to examine or have examined witnesses against him and to obtain the attendance and examination of witnesses on his behalf under the same conditions as witnesses against him;
- (e) to have the free assistance of an interpreter if he cannot understand or speak the language used in court.

Article 8 (right to respect for private and family life):

8(1) Everyone has the right to respect for his private and family life, his home and his correspondence.

(2) There shall be no interference by a public authority with the exercise of this right except such as is in accordance with the law and is necessary in a democratic society in the interests of national security, public safety or the economic well-being of the country, for the prevention of disorder or crime, for the protection of health or morals, or for the protection of the rights and freedoms of others.

Article 9 (freedom of thought, conscience, and religion):

9(1) Everyone has the right to freedom of thought, conscience and religion; this right includes freedom to change his religion or belief and freedom, either alone or in community with others and in public or private, to manifest his religion or belief, in worship, teaching, practice and observance.

(2) Freedom to manifest one's religion or beliefs shall be subject only to such limitations as are prescribed by law and are necessary in a democratic society in the interests of public safety, for the protection of public order, health or morals, or for the protection of the rights and freedoms of others.

Article 10 (freedom of expression):

10(1) Everyone has the right to freedom of expression. This right shall include freedom to hold opinions and to receive and impart information and ideas without interference by public authority and regardless of frontiers. This Article shall not prevent States from requiring the licensing of broadcasting, television or cinema enterprises.

(2) The exercise of these freedoms, since it carries with it duties and responsibilities, may be subject to such formalities, conditions, restrictions or penalties as are prescribed by law and are necessary in a democratic society, in the interests of national security, territorial integrity or public safety, for the prevention of disorder or crime, for the protection of health or morals, for the protection of the reputation or rights of others, for preventing the disclosure of information received in confidence, or for maintaining the authority and impartiality of the judiciary.

Article 14 (prohibition of discrimination):

14 The enjoyment of the rights and freedoms set forth in this Convention shall be secured without discrimination on any ground such as sex, race, colour, language, religion, political or other opinion, national or social origin, association with a national minority, property, birth or other status.

## B. Disciplinary Cases

### *König v. Federal Republic of Germany* (1979–80) 2 EHRR 170

**32.02**

Article 6(1) ECHR—ENT specialist—withdrawal of authorization—duration of proceedings

In October 1962, proceedings against the applicant, an ear, nose, and throat (ENT) specialist, for unprofessional conduct were instituted by the Regional Medical Society before the Tribunal for the Medical Profession attached to the Frankfurt Administrative Court, and he was declared unfit to practise in July 1964. The accusations against the applicant included payment for the introduction of clients, persuading a patient to have treatment not covered by social security by assurances that he would be able to use more effective methods, refusing to make out for one of his clients an account corresponding to the fee actually paid, performing an operation not within the field in which he specialized, and widely publicizing his practice in the press and using nameplates, notepaper, and prescription forms containing wording contrary to the rules of the medical profession. In 1967, the applicant had his authorization to run his clinic withdrawn and then, in 1971, his authorization to practise. Criminal proceedings were taken against him in 1972 for the illegal practice of medicine. Actions brought by the applicant to challenge both of these withdrawals had been in progress before the Frankfurt Administrative Court since November 1967 and October 1971, respectively. The applicant did not challenge the length of the disciplinary proceedings or the criminal proceedings instituted against him, but he did complain of the length of the proceedings taken by him against the withdrawals of the authorizations. The applicant complained of the length of the actions in the Frankfurt Administrative Court, and alleged violation of Article 6(1) of the European Convention on Human Rights (ECHR). In its judgment, a majority of the European Commission of Human Rights was of the opinion that Article 6(1) was applicable to the rights claimed by the applicant before the Frankfurt Administrative Court—namely, the right to run his clinic and the right to exercise his profession of medical practitioner; it considered these rights to be 'civil'. The Commission continued:

99. The reasonableness of the duration of proceedings covered by Article 6 para 1 of the Convention must be assessed in each case according to its circumstances. When enquiring into the reasonableness of the duration of criminal proceedings, the Court has had regard, inter alia, to the complexity of the case, to the applicant's conduct and to the manner in which the matter was dealt with by the administrative and judicial authorities. The Court... considers that the same criteria must serve in the present case as the basis for its examination of the question whether the duration of the proceedings before the administrative courts exceeded the reasonable time stipulated by Article 6 para 1.

There had been a violation of Article 6(1) both as regards the duration of the proceedings relative to the withdrawal of the applicant's authorization to run his clinic and the withdrawal of his authorization to practise. In an overall assessment of the various factors and taking into account what was at stake in the proceedings—namely, the applicant's whole professional livelihood—the Commission considered, at [111], that, notwithstanding the delays attributable to the applicant's behaviour, the investigation of the case was not conducted with the necessary expedition.

***Le Compte, Van Leuven and De Meyere v. Belgium*** **(1982) 4 EHRR 1**

**32.03** Article 6(1) ECHR—right to practise medicine—interference—public hearing

In October 1970, the West Flanders Provincial Council of the *Ordre des Médecins* ordered that C's right to practise medicine be suspended for six weeks and, on 30 June 1971, the Provincial Council of the *Ordre des Médecins* ordered another suspension for three months of the applicant's right to practise. The grounds were that he had given interviews and had publicized in the press matters relating to the disciplinary organs of the *Ordre* incompatible with the dignity and reputation of the profession. C's appeal to the Appeals Council of the *Ordre* was dismissed, as was his appeal to the Court of Cassation. In January 1973, thirteen medical practitioners filed a complaint with the East Flanders Provincial Council of the *Ordre des Médecins* regarding L and M. In October 1973, the Provincial Council directed that their right to practise medicine be suspended for a period of one month for having charged fees limited to the amounts reimbursed by social security, and for having contributed to a magazine and having made therein public utterances judged offensive to their colleagues. L and M's appeals were dismissed by the Appeals Council and the Court of Cassation. On applications before the European Court of Human Rights (ECtHR), the Court held first that, as regards the applicability of Article 6(1) ECHR, disciplinary proceedings as such cannot be characterized as 'criminal'; nevertheless this may not hold good for certain specific cases. Again, disciplinary proceedings do not normally lead to a *contestation* (dispute) over 'civil rights and obligations'. However, this does not mean that the position may not be different in certain circumstances. The suspensions ordered by the Provincial Council of the *Ordre* were to deprive the three applicants temporarily of their right to practise. That right was directly in issue before the Appeals Council and the Court of Cassation, which bodies had to examine the applicants' complaints against the decisions affecting them. The Court thus concluded that Article 6(1) was applicable. Unlike certain other disciplinary sanctions that might have been imposed on the applicants (warning, censure, and reprimand), the suspension of which they complained undoubtedly constituted a direct and material interference with the right to continue to exercise the medical profession. The fact that the suspension was temporary did not prevent its impairing that right. Since the dispute over the decisions taken against the applicants has to be regarded as a dispute relating to 'civil rights and obligations', it follows that they were entitled to have their case heard by 'a tribunal' satisfying the conditions laid down in Article 6(1). However, whilst Article 6(1) embodies the 'right to a court', it nevertheless does not oblige the contracting states to submit '*contestations*' over 'civil rights and obligations' to a procedure conducted at each of its stages before 'tribunals' meeting the Article's various requirements. Demands of flexibility and efficiency, which are fully compatible with the protection of human rights, may justify the prior intervention of administrative or professional bodies and, a fortiori, of judicial bodies that do not satisfy the requirements in every respect. As to compliance with the requirements of Article 6(1) in the instant case, there can be no doubt as to the independence of the Court of Cassation and this also applies to the Appeals Council. It is composed of exactly the same number of medical practitioners and members of the judiciary, and one of the latter, designated by the Crown, always acts as chairman and has a casting vote. Besides, the duration of a Council member's term of office (six years) provides a further guarantee in this respect. Under the Royal Decree of 6 February 1970, all publicity before the Appeals Council is excluded in a general and absolute manner, both for the hearings and for the pronouncement of the decision. The applicants were entitled to have the proceedings conducted in public. Conducting disciplinary proceedings of this kind in private does not contravene the Convention, provided that the person concerned consents. In the present case, however, the applicants clearly wanted and claimed a public hearing. To refuse them such a hearing was not permissible under Article 6(1) and the public character of the proceedings before the Court of Cassation cannot suffice to remedy this defect. The applicants' case was not heard publicly by a tribunal competent to determine all of the aspects of the matter and in this respect there was, in the particular circumstances, a breach of Article 6(1).

### *Albert and Le Compte v. Belgium* (1983) 5 EHRR 533

**32.04** Article 6(1)–(3) ECHR—application to professional disciplinary proceedings

In June 1974, the Brabant Provincial Council of the *Ordre des Médecins* suspended A's right to practise medicine for a period of two years. He was accused of having issued spurious certificates, and was found to have carried out no medical examinations to warrant issuing certificates of a state of unfitness to work. A's appeals to the Appeals Council of the *Ordre* and to the Court of Cassation were rejected. In March 1974, the West Flanders Provincial Council of the *Ordre des Médecins* suspended C's right to practise medicine for a period of two years following further complaints regarding 'improper publicity' and 'contempt of the *Ordre*', relating to interviews that he had given to magazines and a letter sent to the president of the Provincial Council. In October 1974, the Appeals Council rejected C's appeal and changed the suspension into striking his name from the register of the *Ordre*. Ce's subsequent appeal to the Court of Cassation was dismissed in November 1975 and the striking-off order took effect in December 1975. On applications to the ECtHR for alleged breaches of Article 6(1) ECHR, the Court said that it saw no cause to depart from its earlier judgment in the case of *Le Compte, Van Leuven and De Meyere*, especially since C, the government, and the Commission each referred back to their respective arguments in that case. The effect of the disciplinary sanctions in question was to divest the applicants temporarily (A) or permanently (C) of their civil right, within the meaning of Article 6(1), to practise medicine, which they had duly acquired and which allowed them to pursue the goals of their professional life. The right to continue to practise constituted, in the case of the applicants, a private right and thus a civil right within the meaning of Article 6(1), notwithstanding the specific character of the medical profession—a profession that is exercised in the general interest—and the special duties incumbent on its members. In many member states of the Council of Europe, the duty of adjudicating on disciplinary offences is conferred on jurisdictional organs of professional associations. Even in incidences in which Article 6(1) is applicable, conferring powers in this manner does not in itself infringe the Convention. Nonetheless, in such circumstances, the Convention calls at least for one of the two following systems: either the jurisdictional organs themselves comply with the requirements of Article 6(1); or they do not so comply, but are subject to subsequent control by a judicial body that has full jurisdiction and does provide the guarantees of Article 6(1). In the present case, the applicants' cases were dealt with by three bodies—namely, the Provincial Council, the Appeals Council, and the Court of Cassation. The Court did not consider it necessary to decide whether, in the specific circumstances, there was a 'criminal charge'. A relied on Article 6(2) and on 6(3)(a), (b), and (d), but, in the opinion of the Court, the principles enshrined therein are, for the present purposes, already contained in the notion of a fair trial as embodied in Article 6(1); the Court would therefore take these principles into account in the context of Article 6(1). As to compliance with Article 6(1), the manner of appointment of the medical practitioners sitting on the Appeals Councils provides no cause for treating those individuals as biased: although elected by the Provincial Councils, they act not as representatives of the *Ordre des Médecins*, but, like the legal members nominated by the Crown, in a personal capacity. No issue can be taken as to the impartiality of the Court of Cassation. However, the cases of A and C were not heard publicly by the Appeals Council, and the public character of the Cassation proceedings does not suffice to remedy the defect found to exist at the stage of the disciplinary proceedings. In the opinion of the Court, the principles set out in Article 6(2) and in the provisions of Article 6(3)(a), (b), and (d) are applicable, *mutatis mutandis*, to disciplinary proceedings subject to Article 6(1) in the same way as in the case of a person charged with a criminal offence. The Court considered that, on the facts in this case, there was no violation of Article 6.

### *Van Marle and ors v. Netherlands* (1986) 8 EHRR 483

**32.05** Article 6(1) ECHR—accountant's registration

The applicant and others practised as accountants. The Dutch Parliament passed two Acts designed to regulate and delimit the profession of accountant, which until then had not been subject to any statutory control. The Acts laid down standards of professional competence required of practising accountants. Each of the applicants applied to be registered as a certified accountant in accordance with the transitional provisions contained in the Acts. Their applications to the Board of Admission, and thereafter appeals to the Board of Appeal, were dismissed on the ground that the applicants' statements had been unsatisfactory on certain essential points and that their replies to questions did not show sufficient professional competence. In holding by eleven votes to seven that Article 6(1) ECHR was not applicable, the ECtHR said that the Board of Appeal re-examined the applicants, calling them to interview, at which they had the opportunity to comment on balance sheets that they had drawn up and to answer questions on accountancy theory and practice. An assessment of this kind, evaluating knowledge and experience for carrying on a profession under a particular title, is akin to a school or university examination and is so far removed from the exercise of the normal judicial function that the safeguards in Article 6 cannot be taken as covering resultant disagreements. There was thus no 'contestation' (dispute) within the meaning of Article 6, which therefore was not applicable in the present case. The fact that, in domestic law, the Board of Appeal is considered to be a tribunal does not alter this conclusion.

### *H v. Belgium* (1988) 10 EHRR 339

**32.06** Article 6(1) ECHR—barrister—*Ordre des Avocats*—judicial function—procedural safeguards

The applicant was struck off the Antwerp Bar roll in June 1963 and had twice applied unsuccessfully to be reinstated. The Council of the *Ordre des Avocats* of Antwerp was satisfied that H had wrongly persuaded a client that he (the client) risked arrest if he did not immediately pay a sum of BFR20,000. H had no criminal convictions. Restoration to the role of an *avocat* who has been struck off is governed by Article 471 of the Judicial Code, which provides that no *avocat* who has been disbarred may be entered on a roll of the *Ordre* until ten years have elapsed from the date on which the decision to strike off becomes final and unless exceptional circumstances warrant it. The ECtHR held, by twelve votes to six, that, in relation to H's application for restoration to the role after disbarment, Article 6(1) ECHR was applicable and had been violated. Having examined the various aspects of the profession of *avocat* in Belgium, the Court found that they confer on the asserted right to practise the character of a civil right within the meaning of Article 6(1), which was thus applicable. As to compliance with Article 6(1), no appeal lay against the decisions of the Council of the *Ordre des Avocats* of Antwerp.

Moreover, the Council of the *Ordre*, when taking its decisions on the application for readmission made by the applicant (who had been struck off in 1963), was performing a judicial function further to its disciplinary responsibilities. It took action in the case on occasions that were separated by a considerable lapse of time (in 1963, and then in 1979–81) and in different contexts (disbarment and application for readmission). Additionally, the relevant procedure of the Council of the *Ordre des Avocats* of Antwerp was open to criticism in two respects. In the first place, it was very difficult for the applicant to adduce appropriate evidence of the 'exceptional circumstances' that might, in law, have brought about his restoration to the role and, more generally, to argue his case with the requisite effectiveness; in particular, neither the applicable provisions nor the previous decisions of the councils of the *Ordre* gave any indication of what could amount to 'exceptional circumstances'. At the same time, he had cause to fear that there was some risk of being dealt with arbitrarily, especially because there was no provision allowing him a right to challenge and because the Antwerp Bar did not have any internal rules of procedure. The procedural safeguards thus appear unduly limited. Their inadequacy is of especial importance in view of the seriousness of what is at stake when a disbarred *avocat* seeks restoration to the role and the imprecise nature of the statutory concept of 'exceptional circumstances'.

### *Stefan v. United Kingdom* (1998) 25 EHRR 130

**32.07** Article 6(1) ECHR—doctor—health—fitness to practise—procedural guarantees

At a hearing on 21 June 1993, the Health Committee of the General Medical Council (GMC) found that S's fitness to practise was seriously impaired and directed that her registration should be conditional on compliance with certain conditions. On 20 June 1994, the Health Committee, after consideration of a further medical report that diagnosed S as suffering from a paranoid disorder, again imposed conditions on her registration, this time for a period of eight months. On 23 February 1995, at the resumed hearing, S did not agree to certain conditions relating to her remaining under medical supervision and having limitations placed on the scope of her professional practice. Her registration was again suspended for a further period of eight months. The Privy Council dismissed an appeal on 17 July 1995. S complained to the European Commission of Human Rights of a violation of Article 6 ECHR, contending that the proceedings did not afford her a hearing by an independent and impartial tribunal. The government submitted that the evaluation of the fitness of an individual to practise on medical grounds did not constitute a determination of the applicant's civil rights and obligations within the meaning of Article 6(1). It relied on *Van Marle v. Netherlands* (1986) 8 EHRR 483, in which a majority of the Court held that Article 6 was not applicable to a dispute concerning individuals' registration as certified accountants. The Commission considered that the present case was distinguishable from the case of *Van Marle v. Netherlands*. In that case, which concerned a complaint by applicants who had been denied registration as accountants, the Court commented that an assessment evaluating knowledge or experience for carrying on a profession under a particular title was akin to a school or university examination, and was far removed from the exercise of the normal judicial function that the safeguards in Article 6 were covering. The present case did not concern an evaluation as to whether S had sufficient education and experience to hold the title of doctor; rather, it concerned an investigation into whether a provisionally registered doctor had the mental fitness to continue practising. It cannot be said the Health Committee hearing was akin to a school or university examination. The proceedings therefore determined S's civil rights and obligations, and thus fell within the ambit of Article 6(1): see *Le Compte, Van Leuven and De Meyere v. Belgium* (1982) 4 EHRR 1, [44]–[50]. There is no indication in the case law of the ECtHR that the mere fact that disciplinary proceedings against professional persons are determined by members of the profession amounts to a lack of 'independence', even when the professional body concerned regulates a number of functions of the profession (as was the case in *H v. Belgium* (1988) 10 EHRR 339, [50], [51]). There was no allegation of bias or lack of impartiality on the part of the members of the Health Committee in the present case. As to procedural guarantees, there remained, however, areas in which the independence of the Health Committee could be seen to be open to doubt. However, even where an adjudicatory body determining disputes over civil rights and obligations does not comply with Article 6(1) in some respect, no violation of the Convention can be found if the proceedings before the body are subject to subsequent control by a judicial body that has full jurisdiction and does provide the guarantees of Article 6(1): *Bryan v. United Kingdom* (1996) 21 EHRR 342. It follows that this part of the application was manifestly ill-founded.

### *Wickramsinghe v. United Kingdom* (1998) EHRLR 338

**32.08** Doctor—indecent behaviour—disciplinary proceedings—not criminal

The applicant, a medical doctor, was removed from the register following a hearing before the Professional Conduct Committee (PCC) of the GMC in December 1994, which found that he was guilty of indecent behaviour towards a patient in relation to an incident in June 1992. W's appeal to the Privy Council was dismissed in July 1995. W's application to the European Commission of Human Rights was declared inadmissible, as being manifestly ill-founded. The allegations faced by W were not of a 'criminal nature' and, despite the seriousness of the charge and the nature of the sanction, the charge should not be treated as criminal. The Commission held that professional disciplinary matters were essentially matters that concerned the relationship between professional associations and individuals, rather than a law of general applicability.

### *Gautrin and ors v. France* (1999) 28 EHHR 196

**32.09** Article 6(1) ECHR—public hearing

The applicants—all members of an association called 'SOS Médecins', the object of which is to provide emergency medical services on call to patients—were found guilty of a breach of article 23 of the Code of Professional Conduct by displaying the name 'SOS Médecins' on their vehicles and prescriptions. Some of the applicants were suspended from practising medicine for periods of one or two months; the remaining applicants received a reprimand. Before the National Council of the *Ordre des Médecins*, the decision of the Regional Council as to the finding that there had been a breach of the Code of Professional Conduct was upheld, but the penalties were reduced. The suspension for two of the doctors was reduced to fifteen days, the doctors who had received a month's suspension were given a reprimand, and those who had been reprimanded were given a warning. They did not appeal to the *Conseil d'État*. The ECtHR held unanimously that there had been a violation of Article 6(1) ECHR in that the applicants' case was not heard in public. The judgment constituted sufficient just satisfaction for any alleged non-pecuniary damage.

### *Tehrani v. United Kingdom Central Council for Nursing, Midwifery and Health Visiting* 2001 SC 581

**32.10** Article 6(1) ECHR—nurse—disciplinary proceedings—independent and impartial tribunal—right of appeal to court

T, the petitioner, formerly the matron of a nursing home, was the subject of disciplinary proceedings in relation to her non-attendance and alleged failure to maintain adequate staffing levels at the nursing home. T brought judicial review proceedings and sought a declaration against the respondents that the decision to hold a meeting of the PCC to determine the charge of misconduct against the petitioner was unlawful because the decision was incompatible with her Convention right under Article 6(1) to a hearing before an independent and impartial tribunal. In dismissing T's petition, Lord Mackay of Drumadoon said that, in his opinion, having regard to the consequences for the petitioner that would follow, a decision to remove her name from the register would constitute a direct interference with her right to practise her chosen profession as a nurse. However, the respondents' decision to institute disciplinary proceedings would only fall to be categorized as incompatible with T's Convention right to a hearing before an independent and impartial tribunal if the requirements of Article 6(1) would not be met either in the proceedings before the PCC or in the course of any appeal under section 12 of the Nurses, Midwives and Health Visitors Act 1997. The petitioner's Convention rights do not require that the PCC itself be an independent and impartial tribunal, within the meaning of Article 6(1). At [52], the learned judge said:

> Under the Strasbourg jurisprudence the position in relation to tribunals charged with responsibility for disciplinary or administrative matters is quite clear. Where the decision of such a tribunal involves the determination of a dispute over civil rights and obligations, no violation of the Convention can be found in the proceedings before the tribunal, if the tribunal's decision is subject to subsequent control by a court that has full jurisdiction and does provide the guarantees required by art 6(1) (see eg *Le Compte, Van Leuven and De Meye v. Belgium* (1982) 4 EHRR 1, paras 50–51, *Albert and Le Compte v. Belgium* (1983) 5 EHRR 533, para 29, *Obermeier v. Austria* (1991) 13 EHRR 290 at p306, para 70, and *Bryan v. United Kingdom* (1996) 21 EHRR 342 at p349, para 40).

Having regard to the relevance of a statutory right of appeal, under section 12 of the Nurses, Midwives and Health Visitors Act 1997, the learned judge said that the question of whether the PCC meets all of the requirements of an independent and impartial tribunal becomes academic. As to the membership of the panel and the role of the respondents' officials, the learned judge said:

> 81. In the United Kingdom, it is not unusual for the disciplinary bodies of independent professions to contain "lay members". They are so described because they do not have any qualifications of the profession in question. That is not to say that such individuals are without relevant qualifications and experience. On the contrary, it is because of their qualifications and experience in other fields, legally separate from, but in many incidences not unconnected with, the profession in question, that they are selected to serve in the capacity that they do. Far from rendering such individuals unsuitable, their professional qualifications and experience in other areas are why such persons are appointed to serve on the disciplinary tribunals of professions, other than their own. Their qualifications and experience, when linked to their independence from the profession in question, are one reason why such individuals make a very valuable contribution to self-regulation of the profession. In general, there is no reason to believe that such lay members of disciplinary bodies discharge their duties in anything other than a responsible manner—as individuals, independent of the profession in question, but also acting independently of any professional body, employment or other organisation in which they may be involved.
>
> 82. Against that background, the fact that a non-council member of the PCC may be a member of another profession, whose members practise in areas closely related to those of the nursing profession, does not in my opinion provide any objective basis for concern as to the impartiality of the PCC. Nor would any such concern arise merely because such an individual might be associated with a consumer organisation in the fields of health and social care. No doubt, on occasion, such organisations find themselves involved in controversial issues or political campaigns about health and care issues, where their views may be at odds with those of the Government of the day, professional bodies and other organisations. In the event that was to occur, it is conceivable that a particular disciplinary case might arise, in which it would be inappropriate for a non-council member, associated with a particular organisation, to serve. In such a situation, the non-council member would be required to disqualify herself from the PCC, for the case in question....

83. Equally, in my opinion, there is no substance in any challenge based on what was said to be the limited role for the legal assessor.... The legal assessor to the PCC has a clearly defined role to play. That role is set out in the 1983 Order.... His role is to advise the decision-makers, not to usurp or improperly influence their functions. In the circumstances, I consider that the legal assessor plays as full a role as one could expect or wish any legal assessor to undertake.

84. Likewise, I am not persuaded that there is any substance in the criticism founded upon the role of the respondents' officials. It is not immediately apparent why correspondence relating to the affairs of the [Preliminary Proceedings Committee] PPC is handled by the office of the Director of Professional Conduct, whose officials also handle correspondence on behalf of the PCC. Nevertheless, such officials play no part in the hearing before the PCC nor, more importantly, any part in the deliberations of the PCC. As from 3 November 2000, none of the respondents' officials have retired with the members of the PPC and been present during their deliberations. If the hearings before the PCC are public, then the respondents' officials have as much right to be present as anybody else. In such circumstance I do not consider that the role played by the officials of the respondents in the conduct of disciplinary proceedings, has any bearing on whether the PCC complies with the requirements of Article 6(1).

85. ... The fact that the same individuals sit on both the PPC and the PCC is the factor of great significance.... I consider that any objective observer would consider it unusual that those involved, from time to time, in the taking of decisions to initiate disciplinary proceedings against members of a profession, are also involved, at other times, in adjudication upon such proceedings. The fact that the same individuals can move backwards and forwards between these two roles, throughout their terms of office, is of particular significance....

### *Pine v. Law Society* [2001] EWCA Civ 1574

**32.11** Article 6 ECHR—solicitor—disciplinary proceedings—lack of legal advice or representation—whether unfair hearing

The Solicitors Disciplinary Tribunal (SDT) found all the allegations against P to have been substantiated and ordered that P be struck off the roll of solicitors. The Tribunal said that it had before it a catalogue of dishonest behaviour on the part of the respondent. This included deposing to affidavits that were false and misleading, giving false and misleading information to clients, and serious breaches of the Solicitors Accounts Rules. The Tribunal did not accept that P was unaware of the very clear obligations with regard to the Solicitors Accounts Rules. He had appeared before the Tribunal previously when it was clear that his books of account were in a 'deplorable state'. P's appeal to the High Court pursuant to section 49 of the Solicitors Act 1974 was dismissed. P applied to the Court of Appeal for permission to appeal on the ground that the lack of any provision for legal advice or representation in connection with the proceedings before the disciplinary tribunal was a breach of Article 6 ECHR. In dismissing P's appeal, the Court of Appeal said that the issue was whether or not P had obtained a fair hearing before the Tribunal. It was common ground that the allegations made against P were not charges of a criminal offence. Thus Article 6(2) and (3) have no direct application, although they may be said to reflect aspects of a fair trial under Article 6(1): see *Airey v. Ireland* (1979) 2 EHRR 305. Even in the case of a charge of a criminal offence, the right to free representation is not absolute, but arises only when the interests of justice so require. The requirements of Article 6 with respect to legal advice and representation depend on the facts of any given case. Thus if P could show on the facts of his case that legal advice and representation for the purpose of the disciplinary proceedings before the Tribunal were required by Article 6, then he did not need to rely on any more abstract principle. Counsel for P (Mr Nicholas Blake QC) accepted that the relevant principle was that extracted by the Commission in *X v. United Kingdom* (1984) 6

EHRR 136 from the judgment of the Court in *Airey* and applied by Burton J in *R v. Legal Services Commission, ex parte Jarrett* [2001] EWHC 389 (Admin)—namely, whether 'the withholding of legal aid would make the assertion of a civil claim practically impossible, or where it would lead to obvious unfairness of the proceedings'. In the instant case, counsel for P did not claim that the withholding of legal aid made his defence to the allegations practically impossible. He contended that it led to obvious unfairness, not because of any complexity in the procedure or the facts of the case, but because of the nature of the charges, the severity of the possible consequences for P, and the inhibiting effect on the prosecution of his defence of P's emotional involvement. In the Court's view, the procedure was not complex. The relevant facts were within the knowledge of P. P was a solicitor experienced in commercial litigation. P had ample opportunity to indicate any defences that he might wish to advance. Neither the seriousness of the likely consequences nor the emotional involvement of P, when considered in the light of the absence of any legal advice or representation, gave rise to any unfairness.

***Fleurose v. Securities and Futures Authority Ltd* [2001] EWCA Civ 2015**

**32.12** Securities trader—suspension—whether determination of criminal charge

F, a trader in securities, appealed against a judgment of Morison J, dismissing his application for judicial review of a decision of the Disciplinary Appeals Tribunal of the Securities and Futures Authority (SFA). F had been suspended from acting as a 'registered person' for two years and ordered to pay £175,000 towards the SFA's costs. As to whether the hearing before the Tribunal constituted the determination of a criminal charge, the judge answered this question negatively after consideration of the Strasbourg jurisprudence including: *Engle v. The Netherlands (No. 1)* (1976) 1 EHRR 647; *Le Compte, Van Leuven and De Meyere v. Belgium* (1982) 4 EHRR 1; *Albert and Le Compte v. Belgium* (1983) 5 EHRR 533; *Bendenoun v. France* (1994) 18 EHRR 54; *Wickramsinghe v. United Kingdom* (1998) EHRR 338; and *Irving Brown v. United Kingdom* (1998) 28 EHRR CD 233. Since the decision of the judge, there had been two decisions of the Court of Appeal: *Han and Yau v. Commissioners of Customs and Excise* [2001] EWCA Civ 1048 and *Official Receiver v. Stern* [2001] 1 WLR 2230. In dismissing F's appeal, the Court of Appeal (Schiemann and Clarke LJJ, and Wall J) said that to be debarred from gaining one's livelihood in an activity in which one has done so much of one's life is a serious matter. However, applying the principles set out in *Han and Yau*, the Court was not persuaded that the proceedings instituted by the SFA against F were properly to be regarded as involving a criminal charge or offence. Nor was there any ground for suggesting that the hearing was in any way unfair.

***WR v. Austria* (2001) 31 EHRR 43**

**32.13** See paragraph 17.20.

***Luksch v. Austria* (2002) 35 EHRR 17**

**32.14** See paragraph 17.21.

***R (Thompson) v. Law Society* [2004] 1 WLR 2522**

**32.15** See paragraph 20.04.

***Threlfall v. General Optical Council* [2004] EWHC 2683 (Admin)**

**32.16** Sanction—powers of disciplinary tribunal—application of Article 6 ECHR

T, a registered ophthalmic optician, was found guilty of serious professional misconduct in relation to a charge of adequately failing to examine the right eye of a patient in July 2002. The Disciplinary Committee imposed a penalty order of £500. As to the applicability of Article 6 ECHR, Stanley Burnton J said, at [33]–[35], that the authorities established that the decision by a disciplinary tribunal to suspend or to disqualify a professional person is a determination of civil rights and obligations: *Albert and Le Compte v. Belgium* (1983) 5 EHRR 533. On the other hand, there are authorities that establish that the decision of a disciplinary tribunal to admonish a professional person is not such a determination and which suggest that Article 6 did not apply to the tribunal's proceedings, even if the tribunal had power to suspend or to disqualify him: for example, *X v. United Kingdom* (1983) 6 EHRR 583. However, it seems to be obvious that the applicability of Article 6 must be determined on the basis of the jurisdiction and powers of the tribunal rather than its ultimate decision. The adjectival law applicable to its proceedings must be determined before the proceedings begin rather than after they have been completed. Thus the question of whether a person subject to disciplinary proceedings is entitled to a 'fair and public hearing by an independent and impartial tribunal' must be determined before the hearing and before its result is known. *Tehrani v. United Kingdom Central Council for Nursing, Midwifery & Health Visiting* [2001] IRLR 208, [33], was cited with approval by the Court of Appeal in *R (Thompson) v. Law Society* [2004] 1 WLR 2522, [83]. The cases in which it is has been held that a professional person has no remedy under Article 6 in respect of a decision to reprimand him must be explained on the basis that his civil rights and obligations have not been affected, and he cannot therefore complain of a breach of Article 6, in the same way as a defendant in criminal proceedings who has been acquitted cannot complain of a procedural irregularity in those proceedings. However, it follows from the fact that the disciplinary proceedings in the present case might have resulted in a decision to suspend or to disqualify T that Article 6 applied to those proceedings. In addition, a financial penalty imposed by a disciplinary tribunal creates an obligation to pay it, and for this reason there may have been an actual determination of T's civil rights and obligations. This point was not argued and the judge said that he expressed no concluded view on it.

### *Macpherson v. Law Society* [2005] EWHC 2837 (Admin)

**32.17** Proceedings before the SDT not criminal in nature

In dismissing M's appeal from the SDT pursuant to section 49 of the Solicitors Act 1974, Maurice Kay LJ said, at [7], that: 'It is simply not the case that the proceedings before the SDT are criminal in nature. That was accepted by the Court of Appeal in *Pine v. The Law Society* 25th October 2001, case number C/2000/35/61 and 35/61A.'

### *R (Malik) v. Waltham Forest NHS Primary Care Trust* [2007] 1 WLR 2092

**32.18** Doctor—inclusion on NHS list—whether a 'possession' under Article 1 to First Protocol

Waltham Forest National Health Service (NHS) Primary Care Trust (PCT) and the Secretary of State for Health appealed from the ruling of Collins J in favour of the respondent, M. In his ruling, Collins J held that M had suffered an interference with his 'possessions' contrary to Article 1 of Protocol No. 1 ECHR when the PCT unlawfully suspended him from the medical performers list maintained by it pursuant to the National Health Service (Performers Lists) Regulations 2004. The central issue is whether a general practitioner (GP) providing NHS services for a PCT under the Performers Lists Regulations has a possessory right under Article 1 of Protocol No. 1, by virtue of his inclusion on the list, to perform such services. If so, every GP included on such a list has a potential claim under Article 1, Protocol No. 1, in the event of unlawful removal from it in addition to whatever other claims might flow from it. In allowing the appeal by the PCT and the Secretary of State, the Court of Appeal (Auld, Rix, and Moses LJJ) held that the questions of law raised by the appeal included the following.

- In what respects may future income be an Article 1, Protocol No. 1, 'possession'—goodwill and legitimate expectation?
- Is a personal permission, in the form of inclusion on a professional list or a licence, an Article 1, Protocol No. 1, 'possession'?

Auld LJ said that, in summary on the issues of goodwill and legitimate expectation, there is clear Strasbourg authority, in *Wendenburg v. Germany* (2003) 2 Reports of Judgments and Decisions 347 and other cases, and domestic authority, in *R (Countryside Alliance) v. Attorney General* [2007] QB 305, that the assets of a business may include possessions for the purpose of Article 1, Protocol No. 1, in the form of 'clientele' or goodwill of the business. Where such clientele/goodwill exists, measures that diminish its value, such as interference with professional practice in *Van Marle v. The Netherlands* (1986) 8 EHRR 483, may engage Article 1. But where it does not exist, as it did not in the present case, the Court of Appeal's decision in *Countryside* upholding the reasoning of the Divisional Court is clear authority for the proposition that, without it, mere prospective loss of future income cannot amount to a possession for the purpose. Equally, any consideration of a further category of Article 1 possession based on a notion of legitimate expectation in this context would unacceptably blur that distinction of principle. It would lead to great difficulties of practical application in the next stages of the Article 1 exercise of identifying precisely what legitimately expected 'possession' had been interfered with and to what extent. On the strength of the Strasbourg and domestic jurisprudence, in particular *Countryside* and *R (Nicholds) v. Security Industry Authority* [2007] 1 WLR 2067, it would be wrong to conclude that the personal right of M to practise in the NHS flowing from his inclusion in the performers list was a 'possession' within Article 1, Protocol No. 1.

### *Frankowicz v. Poland* App. no. 53025/99, 16 December 2008, ECtHR

**32.19** Article 10 ECHR—doctor—medical report—criticism of another doctor

The applicant, a gynaecologist, made critical remarks of another doctor in a medical report. He was charged with, and found guilty of, unethical conduct, in breach of the principle of professional solidarity, contrary to the Polish Code of Medical Ethics. He claimed that there had been an interference with his right to freedom of expression, contrary to Article 10 ECHR, in that he should have the right to state his opinion on the treatment received by his patient. The Court held that the applicant's Article 10 rights had been violated. An absolute prohibition of any criticism between doctors was likely to discourage doctors from providing patients with an objective opinion on their health and treatment, which could compromise the very purpose of the medical profession. The interference with the applicant's Article 10 rights was disproportionate.

### *R (Wright and ors) v. Secretary of State for Health* [2009] 1 AC 739

**32.20** Children—list of barred persons—Article 6(1) ECHR—whether process fair

The claimants, all registered nurses, appealed against decisions to include each of them provisionally on the Protection of Vulnerable Adults (POVA) list kept by the Secretary of State for Health pursuant to the Care Standards Act 2000 and on a list of persons unsuitable to work with children maintained under the provisions of the Protection of Children Act 1999. In allowing the claimants' appeals, the House of Lords said that the key provision with which it was concerned was the process for provisional inclusion on the list under section 82(4) of the Care Standards Act 2000. At [27], Baroness Hale of Richmond said that no one can be in any doubt of the need for some scheme to protect children and vulnerable adults from being harmed by the people who regularly come into contact with them in the course of work. The most practicable way of providing such a scheme may well be to have a list of banned individuals maintained administratively and in relation to which the initial decisions are made by officials. If the process is working as it should, many people will accept that they should indeed be on the list. However, the scheme, as enacted in the Care Standards Act, does not comply with Article 6(1) ECHR. The process does not begin fairly, by offering the care worker an opportunity to answer the allegations made against her, before imposing upon her possibly irreparable damage to her employment or prospects of employment. The care worker suffers possibly irreparable damage without being heard, whatever the nature of the allegations made against her. The care worker may have a good answer to the allegations, no matter how serious they are. There may well be cases in which the need to protect the vulnerable is so urgent that an *ex parte* procedure can be justified. But one would then expect there to be a swift method of hearing both sides of the story and of doing so before irreparable harm is done. Nor is there any method provided of assessing the true urgency of the case. As it happens, no great urgency was felt in the four cases in the appeal, in which there was a gap of four–six months between the referral and the provisional listing.

### *R (Royal College of Nursing and ors) v. Secretary of State for the Home Department* [2010] EWHC 2761 (Admin); [2011] PTSR 1193

**32.21** Barred list—automatic inclusion—oral hearing

The claimants, the Royal College of Nursing and certain aggrieved nurses, challenged the lawfulness of the scheme established under the Safeguarding Vulnerable Groups Act 2006, which prohibits those placed on lists established under the scheme from working with children and/or vulnerable adults. A person convicted of, or cautioned in respect of, a sexual offence against persons with a mental disorder is placed upon the relevant list and has no right to make representations with a view to seeking his or her removal from the list. A person convicted of, or cautioned in relation to, other sexual offences, offences involving violence, and offences relating to mistreatment of children is also placed upon the list, but the person has the right to make representations to the effect that he or she should be removed. An individual who is included in a barred list may appeal to the Upper Tribunal under Schedule 3 to the 2006 Act. In allowing the claimant's claim for judicial review in part, Wyn Williams J held that the inclusion of a person's name upon one or more of the barred lists is an act that is concerned with that person's civil right to remain in the employment currently enjoyed or, if the person is unemployed, to engage in the whole of the wide variety of jobs available in the nursing sector: at [47]. The inclusion of the claimants' names on the barred lists prior to receipt of representations was not a legitimate and proportionate holding exercise: at [63]. Automatic inclusion of all persons convicted of, or cautioned with, specified offences cannot be justified simply to cater for what must be a very small number of truly urgent cases: at [66]. The issues as to whether Article 6 requires an oral hearing in every case before the Independent Safeguarding Authority (ISA), established by the Home Secretary under the 2006 Act to administer the scheme, must be judged by the flexible principles that generally prevail when assessing the level of procedural protection that Article 6 affords in any given case. The scheme does not preclude the ISA convening an oral hearing if it thinks it appropriate. Section 4 of the 2006 Act confers a right of appeal. Although permission to appeal is required, the scheme envisages that an appeal may be brought when, at least arguably, the ISA has made a mistake on any point of law or in any finding of fact upon which its decision is based. In the event that the appeal succeeds, the Upper Tribunal may either direct the ISA to remove the person's name from the barred list or remit the matter to the ISA for a new decision.

As to appeals relating to inclusion of individuals on a barred list, see, for example, *B v. Independent Safeguarding Authority (Royal College of Nursing intervening)* [2013] 1

WLR 308, CA, and *Disclosure and Barring Service (formally Independent Safeguarding) v. Harvey* [2013] EWCA Civ 180.

***Kulkarni v. Milton Keynes Hospital NHS Foundation Trust* [2010] ICR 101**

**32.22** See paragraph 45.20.

***Hameed v. Central Manchester University Hospitals NHS Foundation Trust* [2010] EWHC 2009 (QB)**

**32.23** NHS Trust—employment—disciplinary hearing—Article 6 ECHR

The claimant, a staff-grade ophthalmologist at the Manchester Royal Eye Hospital, was dismissed for gross misconduct. In an action, she sought a declaration that her dismissal was in breach of contract, and that the investigation and subsequent disciplinary hearing were unlawful. In relation to Article 6 ECHR, it was argued that the panel hearing the case was not independent or impartial: only one member of the panel was not employed by the defendant Trust. In dismissing the claim, Swift J said that the claimant's case did not fall within the type of exceptional circumstances envisaged by Smith LJ in *Kulkarni v. Milton Keynes Hospital NHS Foundation Trust* [2010] ICR 101 and that Article 6 did not apply. Neither can there have been any implied term in the contract of employment that the disciplinary hearing would be conducted in accordance with Article 6. Such an implied term would have been completely inconsistent with the terms of *Maintaining High Professional Standards in the Modern NHS*, published by the Department of Health in 2003, and the Trust Procedure, which envisage that disciplinary hearings will be dealt with locally by employing authorities, rather than by independent panels as previously.

***R (Puri) v. Bradford Teaching Hospitals NHS Foundation Trust* [2011] EWHC 970 (Admin)**

**32.24** NHS Trust—employment—disciplinary hearing—Article 6 ECHR

The claimant, P, was employed by the defendant Trust as a consultant neurologist, until his dismissal following an internal disciplinary hearing on grounds of misconduct. Essentially, the issue for the High Court on judicial review was whether Article 6 ECHR was engaged in the disciplinary proceedings that led to P's dismissal, and if so, whether the disciplinary panel of the Trust that decided to dismiss him was independent and impartial, so as to comply with Article 6. Blair J concluded that this was not a case in which the effect of the disciplinary proceedings had been to deprive P of the right to practise his profession, within or outside the NHS. Further, there were important distinctions between the present case and *Kulkarni v. Milton Keynes Hospital NHS Trust* [2009] EWCA Civ 789. The charges against K, if proved, would have constituted a criminal offence; no such issue remotely arose in the present case. The instant case was not a case in which an NHS doctor faced charges that were of such gravity that, in the event that they were found proved, he would be effectively barred from employment in the NHS. Blair J went on to say that, had he held that Article 6 was engaged, he would not have held that compliance required a disciplinary panel comprising persons external to the Trust. Although the disciplinary panel was chaired by the chairman of the Trust, and predominantly comprised Trust members and employees, the Court concluded that the panel was not non-compliant by reason of its composition.

***R (G) v. Governors of X School (Secretary of State for the Home Department and anor intervening)* [2012] 1 AC 167**

**32.25** See paragraph 45.24.

*R (Gaunt) v. The Office of Communications* **[2011] 1 WLR 2355, CA**

**32.26** Broadcast—offensive insults—breach of Ofcom Code—no interference with Article 10 ECHR rights

The claimant, a talk show host, interviewed a local authority councillor responsible for children's services about a controversial proposal to ban smokers from becoming foster parents on the grounds that passive smoking was likely to harm children. During the course of the interview, the claimant became aggressive and called the councillor, among other things, 'a Nazi' 'a health Nazi', and 'you ignorant pig'. The Office of Communications (Ofcom), the relevant regulatory authority for complaints from members of the public, found that the interview breached Ofcom's Broadcasting Code, and rejected the claimant's representations that its finding would involve a disproportionate interference with the claimant's right to freedom of expression under Article 10 ECHR. Dismissing the claim for judicial review, the Administrative Court (Sir Anthony May P and Blair J)—[2011] 1 WLR 663—held that the Court's task was to determine whether Ofcom's finding constituted a disproportionate interference with the claimant's Article 10 right to freedom of expression, that since the interview was political and controversial, the claimant's freedom of expression had to be accorded a high degree of protection, but that such protection, although capable of extending to offensive expression, did not extend to gratuitously offensive insults or abuse that had no contextual content or justification, or to repeated abusive shouting, which served to express no real content. In dismissing the claimant's appeal, the Court of Appeal (Lord Neuberger MR, and Toulson and Etherton LJJ) said that, when considering whether the interview broke Ofcom's Broadcasting Code, the interview had to be considered as a whole and in its context.

*R (Calver) v. Adjudication Panel for Wales* **[2012] EWHC 1172 (Admin)**

**32.27** Councillor—code of conduct—alleged breaches—Article 10 ECHR

C was a member of a community council and was required to comply with its code of conduct pursuant to the Local Government Act 2000. On 5 November 2010, Pembrokeshire County Council's Standards Committee found breaches of the code by C in respect of thirteen comments. It found that C had failed to show respect and consideration to others in breach of paragraph 4(b), and had brought the community council into disrepute in breach of paragraph 6(1)(a) of the code. C's appeal to the Adjudication Panel for Wales was dismissed on 25 May 2011. In judicial review proceedings before Beatson J, it was common ground that the questions to be answered were those formulated by Wilkie J in *Sanders v. Kingston* [2005] EWHC 1145 (Admin), [72], which, adapting the questions to reflect the facts of the present case, were as follows.

(1) Were the Standards Committee and the Panel entitled as a matter of fact to conclude that C's conduct in respect of the thirteen comments was in breach of paragraphs 4(b) and/or 6(1)(a) of the code of conduct?
(2) If so, was the finding in itself or the imposition of a sanction prima facie a breach of Article 10?
(3) If so, was the restriction involved one that was justified by reason of the requirements of Article 10(2)?

Before turning to the application of the questions to the circumstances of the present case, Beatson J made five observations about the underlying principles. These concerned the common law, the approach of the Court to the first of Wilkie J's questions, the relevant legal principles, the process of balancing, and, as a general proposition, that freedom of expression includes the right to say things that 'right-thinking people' consider dangerous or irresponsible or which shock or disturb. In the instant case, the Standards Committee and the Panel were entitled to conclude that the thirteen comments by C breached the code. The second and third questions identified in *Sanders v. Kingston* focused on the position under the Convention. It is not arguable that the legislative scheme making provision for codes of conduct for councillors or the codes of conduct made under the 2000 Act are too uncertain to qualify as being prescribed by law. Accordingly, the real issue concerns the third question: whether the restriction was one that was justified by reason of the requirements of (and the application of the factors in) Article 10(2). In the instant case, the Panel took an overly narrow view of what amounts to 'political expression'. Not all of C's comments were political expression even in the broad sense in which the term has been used. Some of the comments were sarcastic and mocking, and some sought to undermine another councillor in an unattractive way. However, notwithstanding their tone, the majority related to the way in which the Council meetings were run and recorded. Some of them were about the competence of councillor G, who, albeit in a voluntary capacity in the absence of a Council official, was taking the minutes and no doubt trying to do her best. Others were about the provision of minutes to councillors or the approach of councillors to declarations of interest. The comments were in no sense 'high' manifestations of political expression. But they (or many of them) were comments about the inadequate performance of councillors in their public duties. As such, they fell within the term 'political expression' in the broader sense in which the term has been applied in the Strasbourg jurisprudence. Accordingly, C's application was granted and the Panel's decision had to be set aside.

***Mattu v. University Hospitals of Coventry and Warwickshire NHS Trust***
**[2012] 4 All ER 359, CA**

**32.28**
NHS Trust—employment—disciplinary hearing—Article 6 ECHR

M was a consultant in non-invasive cardiology and general medicine, employed by the defendant. He was dismissed by the chief executive. In proceedings for alleged breach of contract, it was alleged that M was entitled to a hearing before an independent and impartial tribunal, and that his dismissal by a public body such as the defendant Trust, which was, in turn, part of the NHS, was equivalent to being struck off the medical register and barred from practice in the NHS. Consequently, it was argued that M's dismissal by the chief executive could not be said to be by someone independent of the Trust. This argument was rejected by the deputy judge: [2011] EWHC 2068 (QB). On appeal to the Court of Appeal (Stanley Burnton, Elias LJJ, and Sir Stephen Sedley), Stanley Burnton LJ said that the appeal raised an important and fundamental issue as to the application of Article 6 ECHR to disciplinary proceedings of public authority employees. The real question is: does a decision by an employer whether to dismiss an employee under a contract of employment determine a civil right of the employee within the meaning of Article 6? After reviewing *Le Compte, Van Leuven and De Meyere v. Belgium* (1982) 4 EHRR 1, *Tehrani v. United Kingdom Central Council for Nursing, Midwifery & Health Visiting* 2001 SC 581, *R (Thompson) v. Law Society* [2004] 1 WLR 2522, *Kulkarni v. Milton Keynes Hospital NHS Foundation Trust* [2010] ICR 101, *Hameed v. Central Manchester University Hospitals NHS Foundation Trust* [2010] EWHC 2009 (QB), *R (Puri) v. Bradford Teaching Hospital NHS Trust* [2011] EWHC 970 (Admin), and *R (G) v. Governors of X School* [2011] 3 WLR 237, Stanley Burnton LJ said that the judgments in *G* confirmed that the Trust's disciplinary decision, dismissing M in the instant case under his contract of employment, did not determine any civil right of his: the Trust's disciplinary proceedings were proceedings in which an individual's civil rights or obligations were not being explicitly determined (proceedings 'A' in Lord Dyson's formulation in *G*). The disciplinary proceedings did not engage Article 6. They did not determine any civil right of M. His dismissal was the exercise, or purported exercise, of a contractual right, not the determination of a civil right.

***R (King) v. Secretary of State for Justice; R (Bourgass and Hussain) v. Secretary of State for Justice*** **[2012] 1 WLR 3602, CA**

**32.29** See paragraph 35.19.

***O'Connor v. Bar Standards Board*** **[2014] EWHC 4324 (QB)**

**32.30** See paragraph 49.12.

## C. Other Relevant Chapters

**32.31** Delay, Chapter 17
Independent and Impartial Tribunal, Chapter 35
Legal Representation, Chapter 45
Misconduct, Chapter 47
Public or Private Hearing, Chapter 53
Restoration to Register, Chapter 58

# 33

# IMPAIRMENT OF FITNESS TO PRACTISE



## A. Legal Framework 33.01 | *Raza v. General Medical Council* [2011] EWHC 790 (Admin) | 33.14 |
| B. | General Principles | 33.02 | *Council for Healthcare Regulatory Excellence v. Nursing and Midwifery Council and Grant* [2011] EWHC 927 (Admin) | 33.15 |
| C. | Disciplinary Cases | 33.03 | | |
| | *Council for the Regulation of Health Care Professionals v. General Medical Council and Biswas* [2006] EWHC 464 (Admin) | 33.03 | *Martin v. General Medical Council* [2011] EWHC 3204 (Admin) | 33.16 |
| | *Cohen v. General Medical Council* [2008] EWHC 581 (Admin) | 33.04 | *Ige v. Nursing and Midwifery Council* [2011] EWHC 3721 (Admin) | 33.17 |
| | *Zygmunt v. General Medical Council* [2008] EWHC 2643 (Admin) | 33.05 | *Uddin v. General Medical Council* [2012] EWHC 1763 (Admin) | 33.18 |
| | *Azzam v. General Medical Council* [2008] EWHC 2711 (Admin) | 33.06 | *Ujam v. General Medical Council* [2012] EWHC 580 (Admin); [2012] EWHC 683 (Admin) | 33.19 |
| | *Cheatle v. General Medical Council* [2009] EWHC 645 (Admin) | 33.07 | *Black v. Professional Conduct Committee of the General Dental Council* [2013] HCJAC 39 | 33.20 |
| | *Nicholas-Pillai v. General Medical Council* [2009] EWHC 1048 (Admin) | 33.08 | *Odoi-Asare v. Nursing and Midwifery Council* [2014] EWHC 1151 (Admin) | 33.21 |
| | *Jalloh v. Nursing and Midwifery Council* [2009] EWHC 1697 (Admin) | 33.09 | *Professional Standards Authority v. Health and Care Professions Council and Ghaffar* [2014] EWHC 2723 (Admin) | 33.22 |
| | *Saha v. General Medical Council* [2009] EWHC 1907 (Admin) | 33.10 | | |
| | *Yeong v. General Medical Council* [2009] EWHC 1923 (Admin) | 33.11 | *R (Nakash) v. Metropolitan Police Service and General Medical Council (Interested Party)* [2014] EWHC 3810 (Admin) | 33.23 |
| | *Kituma v. Nursing and Midwifery Council* [2010] EWCA Civ 154 | 33.12 | | |
| | *R (Vali) v. General Optical Council* [2011] EWHC 310 (Admin) | 33.13 | D. Other Relevant Chapters | 33.24 |

Examples include: **33.01**

Medical Act 1983, section 35C
Dentists Act 1984, section 27
Opticians Act 1989, section 13D
Nursing and Midwifery Order 2001, article 22
Health and Social Work Professions Order 2001, article 22
Pharmacy Order 2010, article 51

## B. General Principles

**33.02** In *Safeguarding Patients: Lessons from the Past—Proposals for the Future* (Cmd 6394, London: HMSO, 2004), the fifth report of the Shipman Inquiry, the Rt Hon Dame Janet Smith DBE said, at paragraph 25.50:

> I think it will be helpful, in the resolution of the problems that I am about to outline, if I analyse the reasons why a decision-maker might conclude that a doctor is unfit to practise or that his/her fitness to practise is impaired. In the examples I discussed above, four reasons for unfitness recurred. They were (a) that the doctor presented a risk to patients, (b) that the doctor had brought the profession into disrepute, (c) that the doctor had breached one of the fundamental tenets of the profession and (d) that the doctor's integrity could not be relied upon. Lack of integrity might or might not involve a risk to patients. It might or might not bring the profession into disrepute. It might be regarded as a fundamental tenet of the profession. I think it right to include it as a separate reason why a doctor might be regarded as unfit to practise, because it is relevant even when it arises in a way that is quite unrelated to the doctor's work as a doctor.

## C. Disciplinary Cases

### *Council for the Regulation of Health Care Professionals v. General Medical Council and Biswas* [2006] EWHC 464 (Admin)

**33.03** Impairment—matter of judgment rather than proof

At the conclusion of the misconduct stage, the legal assessor advised the panel as to a finding of serious professional misconduct, saying that the panel, as with matters of fact had to be 'sure' before making a finding of serious professional misconduct. Jackson J agreed with the Council for the Regulation of Health Care Professionals that that was an erroneous direction because it substituted a test of proof for the correct approach, which is one of judgment. In its deliberations on impairment, the panel was required to carry out an exercise of judgment or assessment. The learned judge said:

> 40. In relation to these issues, I turn for guidance to the wide-ranging and detailed reports prepared by Dame Janet Smith following the Shipman Inquiry. In her fifth report, at paragraphs 21.31 to 21.32, Dame Janet Smith stated as follows:
>
> > '21.31 Rule 28 of the 1988 Professional Conduct Rules provided that, where a PCC panel found the facts, or some of the facts, alleged in a charge proved or admitted (and, presumably, that they were not insufficient to support a finding of SPM), it should then invite the GMC's representative (or the complainant) to address it: "... as to the circumstances leading to those facts, the extent to which such facts are indicative of serious professional misconduct on the part of the practitioner, and as to character and previous history of the practitioner. The Solicitor or the complainant may adduce oral or documentary evidence to support an address under this rule."
> >
> > 22.32 The doctor (or his/her representative) was then invited to address the PCC panel in mitigation and adduce evidence in support if desired. After that, the PCC panel would deliberate again. It would consider whether the facts proved did amount to SPM and, if so, what sanction should be imposed. In my view, these were both matters of judgment for the PCC panel, rather than a matter of proof. However, there are indications in the GMC documents that some people were of the view that SPM must be proved "beyond reasonable doubt". In my view, only the facts were a matter for "proof", the other issues were matters of judgment. In a Consultation Paper in March 2001, the GMC said that opinions differed on whether the criminal standard of proof should apply to the decision whether the facts proved amounted

to SPM and to sanction. I understand that the GMC now takes the view that only the facts need to be proved to the criminal standard, and that whether the facts which have been proved amount to SPM is a matter of judgment.'

41. I entirely agree with the views expressed by Dame Janet Smith in paragraph 21.32 of her fifth report. The Court of Appeal appear to have taken a similar view in *R (Campbell) v. General Medical Council* [2005] EWCA Civ 250 (see paragraphs 5 and 43 of that judgment).

### *Cohen v. General Medical Council* [2008] EWHC 581 (Admin)

**33.04** Consultant anaesthetist—operation—misconduct—expert evidence—current impairment

In this case, Dr C, a consultant anaesthetist, appealed against the decision of the General Medical Council's (GMC's) fitness to practise panel that his fitness to practise was impaired and to impose conditions on his registration. The complaint that led to the practitioner's appearance before the panel was made by Mr B, who underwent surgery for suspected cancer of the colon. Apart from Mr B's case, the practitioner was of good character and had been a consultant anaesthetist since 1980, with no previous adverse findings made against him, and with many references extolling his skills and expertise. Significantly, the GMC called an expert consultant anaesthetist who, whilst critical of the way in which the practitioner had treated Mr B in relation to his pre- and post-operative care and assessment, and of his note-taking, nevertheless did not consider that these matters were so serious as to amount to misconduct such that the practitioner's registration might be called into question; the expert said that the core anaesthetic treatment of Mr B was carried out to a standard entirely in keeping with what might be expected of a consultant anaesthetist. Silber J, in allowing the appeal and setting aside the conditions, held that whether the practitioner's fitness to practise was impaired was a relevant factor at the impairment stage, rather than at the sanctions stage. On the facts, the errors of the practitioner were easily remediable and the panel should have concluded that his fitness to practise was not currently impaired. The learned judge said:

62. Any approach to the issue of whether a doctor's fitness to practise should be regarded as "impaired" must take account of "the need to protect the individual patient, and the collected need to maintain confidence in the profession as well as declaring and upholding proper standards of conduct and behaviour of the public in their doctors and that public interest includes amongst other things the protection of patients, maintenance of public confidence in the profession". In my view, at stage 2 when fitness to practise is being considered, the task of the Panel is to take account of the misconduct of the practitioner and then to consider it in the light of all the other relevant factors known to them in answering whether by reason of the doctor's misconduct, his or her fitness to practise [has been*] impaired. It must not be forgotten that a finding in respect of fitness to practise determines whether sanctions can be imposed: section 35D of the Act.

63. I must stress that the fact that the stage 2 is separate from stage 1 shows that it was not intended that every case of misconduct found at stage 1 must automatically mean that the practitioner's fitness to practise is impaired.

64. There must always be situations in which a Panel can properly conclude that the act of misconduct was an isolated error on the part of a medical practitioner and that the chance of it being repeated in the future is so remote that his or her fitness to practise has not been impaired. Indeed the Rules have been grafted on the basis that once the Panel has found misconduct, it has to consider as a separate and discrete exercise whether the practitioner's fitness to practise has been impaired. Indeed section 35D(3) of the Act states that where the Panel finds that the practitioner's fitness to practise is not impaired, "they may nevertheless give him a warning regarding his future conduct or performance".

* See *Zygmunt v. General Medical Council* [2008] EWHC 2643 (Admin), below.

### *Zygmunt v. General Medical Council* [2008] EWHC 2643 (Admin)

**33.05** Neurosurgeon—wrong diagnosis—current impairment

The appellant, a neurosurgeon, challenged the panel's finding that his fitness to practise was impaired by reason of misconduct and the imposition of a two-month suspension. The allegation arose out of a wrong diagnosis made by Professor Z that the patient suffered from a tumour and not an infected abscess. The Court (Mitting J) noted that even if a panel properly finds that a practitioner has been guilty of misconduct, it may nonetheless conclude that his or her fitness to practise is not impaired. The learned judge said, at [17], that in many—perhaps the great majority of—cases, the issue will not be live, but in cases in which it is, it must be separately and appropriately addressed by the panel. As to the meaning of 'fitness to practise', Mitting J adopted the summary of potential causes of impairment offered by Dame Janet Smith in the Fifth Shipman Inquiry Report (see paragraph 33.03 above). Mitting J noted, at [30], that Dame Janet Smith recognized that present impairment of fitness to practise can be founded on past matters, that a doctor's current fitness to practise must be gauged partly by his or her past conduct or performance, and that it must also be judged by reference to how he or she is likely to behave or perform in the future. At [31], Mitting J said:

> In a misconduct or deficient performance case, the task of the panel is to determine whether the fitness to practise is impaired by reason of misconduct or deficient performance. It may well be, especially in circumstances in which the practitioner does acknowledge his deficiencies and takes prompt and sufficient steps to remedy them, that there will be cases in which a practitioner is no longer any less fit to practise than colleagues with an unblemished record.

Mitting J went on to say that he agreed with Silber J in *Cohen* that, when fitness to practise was being considered, the task of the panel is to take account of the misconduct of the practitioner and then to consider in the light of all of the other relevant factors known to the panel whether his or her fitness to practise is (rather than has been) impaired. Accordingly, the judge quashed the decision of the panel on the question of fitness to practise being impaired and remitted it to the panel to redetermine in the light of the guidance given in the judgment. He added that if the panel did determine that fitness to practise is not impaired, it could, of course, give Professor Z a warning as to his future conduct or performance, which would not be free of effect.

### *Azzam v. General Medical Council* [2008] EWHC 2711 (Admin)

**33.06** Medical errors—evidence of remediation—admissibility and weight—whether current impairment

A, a specialist registrar in obstetrics and gynaecology, appealed against a finding that his fitness to practise was impaired by reason of misconduct, and that he be suspended for one month. The panel found that A did not interpret or recognize the signs of foetal distress shown by a cardiotocograph reading and failed to act upon the signs of that distress, nor did he attach a foetal scalp electrode. On delivery, the umbilical cord was wrapped around the baby's neck and he was not breathing. Attempts at resuscitation were made, but tragically the child died on the same day. The cause of death was asphyxia developing in the second stage of labour. At the conclusion of the panel's findings on the facts, counsel for A applied to the panel to admit evidence on A's behalf in three broad categories, as follows.

(1) Testimonial evidence
(2) Evidence as to A's training following the incident in this case
(3) Evidence from Doctor P as to A's current performance, particularly in relation to his cardiotocograph reading ability at that time

The application was opposed by the GMC, but the panel decided that it might be assisted by the additional evidence in determining whether A's fitness to practise was impaired; it therefore decided to accede to the application. However, in its finding on impairment, the panel said that it had 'borne in mind the testimonial evidence, but gave it little weight at this stage since the facts found proved represent such a departure from the acceptable standard of care that these issues do not affect the panel's decision'. In allowing A's appeal and quashing the decision of the panel, McCombe J said:

> 44. It seems to me that, in the light of the authorities cited [*Meadow v. General Medical Council* [2007] QB 462 and *Cohen v. General Medical Council* [2007] EWHC 581 (Admin)], it must behove a [fitness to practise] FTP panel to consider facts material to the practitioner's fitness to practise looking forward and for that purpose to take into account evidence as to his present skills or lack of them and any steps taken, since the conduct criticised, to remedy any defects in skill. I accept [counsel for the GMC's] submission that some elements of reputation and character may well be matters of pure mitigation, not to be taken into account at the "impairment" stage. However, the line is a fine one and it is clear to me that evidence of a doctor's overall ability is relevant to the question of fitness to practise. Even if [counsel for the GMC] is correct as to the construction of rule 17(2)(j) [of the General Medical Council (Fitness to Practise) Rules 2004] (which I doubt, but do not have to decide) the rule clearly envisages the admission of relevant further evidence at stage 2. The panel must consider that evidence (in the same manner as any other evidence received) and weigh it up, decide whether to accept it and then to determine whether, in the light of the further evidence that it does accept *and* the facts found proved at stage 1, the practitioner's fitness to practise is impaired.
>
> 45. The evidence adduced consisted of a 38 page bundle of testimonials setting out in very complimentary terms the writer's assessments of [A]'s skill as an obstetrician.

McCombe J went on to state, at [51], that the panel was concerned with an error, or errors, which it considered to have been irresponsible. Whether it is to be seen as a single error with inevitable consequences or a series of errors between 16:15 and 16:35 on that one day, it was not a rape or misconduct of that kind. The panel was concerned with a serious error of professional judgment over a limited period. The decisions in *Meadow* and *Cohen* required the panel to look forward, and to consider in the light of that, and of the evidence as to the doctor's conduct and ability demonstrated in the intervening period, whether his fitness to practise was impaired by the events over 20 minutes on 1 September 2003.

> 52. In my judgment, it was quite impossible for the panel to afford only "little weight" to the evidence that it heard and read at stage 2. Giving all respect to the experience of the panel in hearing and determining medical conduct matters, the evidence of [A]'s rehabilitation was outstanding and uncontested. It required to be given substantial weight in deciding whether the doctor's fitness to practise was truly impaired for the future. In failing to do that I consider the panel's decision at stage 2 was flawed.
>
> 53. If proper weight had been given to the evidence of [A]'s actions to remedy his deficiencies and his then current ability and skill, I consider that the panel could not have found that his fitness to practise was still impaired as at October 2007.

### *Cheatle v. General Medical Council* [2009] EWHC 645 (Admin)

**33.07** Two-step process—misconduct and impairment

In *Cheatle*, Cranston J said, at [17], that impairment of fitness to practise was a somewhat elusive concept. However, he considered that the four examples given by Dame Janet Smith in her Fifth Shipman Report helpfully set out the reasons why a decision-maker might conclude that a registrant was unfit to practise or that his fitness to practise was impaired. The four examples were (a) that the practitioner presented a risk to patients, (b) that the practitioner had brought the profession into disrepute,

(c) that the practitioner had breached one of the fundamental tenets of the profession, and (d) that the practitioner's integrity could not be relied upon. Cranston J continued:

> 19. Whatever the meaning of impairment of fitness to practise, it is clear from the design of section 35C [of the Medical Act 1968, as amended] that a panel must engage in a two-step process. First, it must decide whether there has been misconduct, deficient professional performance or whether the other circumstances set out in the section are present. Then it must go on to determine whether, as a result, fitness to practise is impaired. Thus it may be that despite a doctor having been guilty of misconduct, for example, a fitness to practise panel may decide that his or her fitness to practise is not impaired.
>
> [...]
>
> 21. There is clear authority that in determining impairment of fitness to practise at the time of the hearing regard must be had to the way the person has acted or failed to act in the past. As Sir Anthony Clarke MR put it in *Meadow v. General Medical Council* [2007] 1 WB 462:
>
> > "In short, the purpose of [fitness to practise] proceedings is not to punish the practitioner for past misdoings but to protect the public against the acts and omissions of those who are not fit to practise. The FPP thus looks forward not back. However, in order to form a view as to the fitness of a person to practise today, it is evident that it will have to take account of the way in which the person concerned has acted or failed to act in the past" (para 32).
>
> 22. In my judgment this means that the context of the doctor's behaviour must be examined. In circumstances where there is misconduct at a particular time, the issue becomes whether that misconduct, in the context of the doctor's behaviour both before the misconduct and to the present time, is such as to mean that his or her fitness to practise is impaired. The doctor's misconduct at a particular time may be so egregious that, looking forward, a panel is persuaded that the doctor is simply not fit to practise medicine without restrictions, or maybe at all. On the other hand, the doctor's misconduct may be such that, seen within the context of an otherwise unblemished record, a fitness to practise panel could conclude that, looking forward, his or her fitness to practise is not impaired, despite the misconduct.

### *Nicholas-Pillai v. General Medical Council* [2009] EWHC 1048 (Admin)

**33.08** Evidence during hearing—impairment—relevant to determination

The panel found that a note prepared by the appellant practitioner following an operation was both inaccurate and not prepared contemporaneously, or as soon as possible after the matters to which it related, was unprofessional, was prepared with a view to misleading Patient A's solicitors, and was dishonest. In dismissing the appellant's appeal, Mitting J said that, on the evidence, the panel was clearly entitled to reach those findings and its findings were not challenged on appeal. However, the panel, in its determination on impairment, said that it considered the practitioner's dishonest conduct to have been 'compounded by the fact that [he had] given inconsistent and unreliable evidence at this hearing'. Later, when considering sanction, the panel observed that, in addition, the practitioner had given misleading instructions to his solicitors about the note. Mitting J considered the extent to which the practitioner's misleading instructions to his solicitors were relevant to impairment, and said:

> [16] ... It is rightly common ground that sanctions are imposed for the public interest, which includes protection of patients, the maintenance of public confidence in the profession and declaring and upholding proper standards of conduct and behaviour. Given that purpose, the panel are, in my view, clearly entitled to take into account, at the stage at which they determine whether fitness to practise is impaired, material

other than the allegations which they have considered which suggest that it is either not impaired or that it is impaired.

[17] To take an instance not far removed from this case, this was an isolated act of professional dishonesty. If [the appellant] had acknowledged that he had made up the notes after the event, or had inserted a date that he had no reason to believe was right after the event, and had accepted that, in so doing, he intended to mislead the patient's solicitors, then hard though it may have been to make those admissions, they would have stood to his credit, and might have tended to suggest that his fitness to practise was not as impaired as otherwise it would ordinarily be found to have been. But he did not do that.

[18] In the view of the panel, which is not disputed, he contested the critical allegations of dishonesty and intention to mislead. That was a fact which the panel were entitled to take into account in determining whether or not his fitness to practise was impaired, even though it did not form a separate allegation against him. Indeed, it is hard to see how it could have done. One can envisage circumstances in which lying to a disciplinary panel may itself amount to professional misconduct such as to lead to a finding that fitness to practise is impaired and a severe sanction. In a case, for example, of alleged clinical error, where a doctor had given false evidence to the panel about it, the panel would not be entitled to treat that as a freestanding ground of impairment of fitness to practise leading to a sanction. If it found that the original clinical error which founded the allegation did not impair his fitness to practise and it was only the lies told to the panel, then that would have to be pursued in separate proceedings, with the charge made the subject of a separate allegation. But that set of circumstances is likely to be highly unusual.

[19] In the ordinary case such as this, the attitude of the practitioner to the events which give rise to the specific allegations against him is, in principle, something which can be taken into account either in his favour or against him by the panel, both at the stage when it considers whether his fitness to practise is impaired, and at the stage of determining what sanctions should be imposed upon him.

[20] I would reach that conclusion without the benefit of authority, but in fact it appears to have been assumed to be the case in *Misra v. General Medical Council* [2003] UKPC 7, at paragraph 17, when the Privy Council observed in relation to entirely different facts: "If his denial were to be disbelieved then the Committee would have to consider his conduct regarding [Mrs B] on the footing that he had received four requests to visit her but had failed to do so and on the footing also that he had lied on oath about two of the telephone calls."

[21] The Privy Council there clearly accepted that lying on oath under the old procedure would be a factor relevant to the determination of the panel. If it was relevant then, so it seems to me it is relevant now.

In refusing permission to appeal to the Court of Appeal at [2009] EWCA Civ 1516, Hooper LJ said that the fact the practitioner had given dishonest evidence must compound the original dishonesty and be a factor that a panel is entitled to take into account.

### *Jalloh v. Nursing and Midwifery Council* [2009] EWHC 1697 (Admin)

**33.09**

Nurse—procedures and treatment—risk of repetition—mitigation

In this case, Silber J. dismissed the registrant's appeal against the finding of impairment and a conditions of practice order for eighteen months. The registrant was an experienced mental health worker. The overall picture was of a series of serious mistakes by the registrant, a failure to comply on a number of occasions with proper procedures, and a disregard of the interests of a vulnerable patient. This was not the case of one error, but a series of errors. Even after taking account of the mitigating factors, there was a great deal of evidence that showed that, because of the appellant's repeated failures, her fitness to practise was impaired. The Conduct and Competence Committee (CCC) of the Nursing and Midwifery Council (NMC) considered there was a risk of repetition despite the mitigating factors, which included that the patient was hostile and angry, the event in question was very traumatic and was likely to cause a degree of panic, lack of practical training, the presence of a lot of equipment on the trolley that impeded resuscitation, and the registrant's unblemished record and excellent conduct both before and after the event. Silber J said, at [36]: 'I agree with the committee in reaching the decision which they did, especially as there was a risk of repetition.' At [37], Silber J went on to say that the judgment of the Committee deserved respect as the body best qualified to judge what the profession expects of its members, and the measures necessary to maintain high standards of professional practice and treatment. The Committee consisted of three members, two of whom were nurses, one of whom had psychiatric experience, and that would be a relevant factor.

### *Saha v. General Medical Council* [2009] EWHC 1907 (Admin)

**33.10** Two-stage process—misconduct and impairment

In this case, the panel held that the fitness to practise of the appellant was impaired by reason of misconduct and directed his registration to be erased. The relevant misconduct found was a failure by the appellant to cooperate fully and to provide relevant information, in breach of paragraph 30 of *Good Medical Practice*, in connection with an investigation by the GMC into the appellant's conduct. The investigation concerned the fact that the appellant was, or had been, a healthcare worker who was infected with hepatitis B. One of the issues that fell for determination was the question of separate consideration of 'misconduct' and 'impairment' at stage 2 of the proceedings. Mr Stephen Morris QC (sitting as a deputy High Court judge) held that there was no requirement in all cases for there to be a formal 'two-stage process' in considering the issues of misconduct and impairment, and no requirement that, in all cases, the reasons for a finding of impairment had to be distinct from the reasons of a finding of misconduct. The panel was required to consider whether there had been misconduct and, further, whether that misconduct was such as to impair fitness to practise; often, a finding of impairment would follow from one of misconduct. In the instant case, the panel had considered both issues and found, broadly, that one and the same facts gave rise to the misconduct and the impairment. That approach was not erroneous as a matter of law. The learned judge said:

> 94. [Counsel for the appellant] submitted that the Panel erred in law in not applying a "two-stage process" to the issues of "misconduct" and "impairment". The decisions of this court in *Cohen*, *Zygmunt* and more recently, *Cheatle* [2009] EWHC 645 (Admin), impose a requirement upon an FTP panel to consider and decide separately these two issues. By contrast, in the present case (counsel for the appellant) contended that there was no such "two-stage process", that the panel never actually determined that the appellant's actions amounted to misconduct, and that the panel applied exactly the same reasoning in respect of "misconduct" and "impairment".
>
> 95. This is an argument of some substance. In the present case, the panel did not expressly identify (a) findings on "misconduct" and (b) findings on "impairment"; and the delineation between the two is not easy to identify. Moreover, the panel gave almost the same reasons for its finding of misconduct and its finding of impairment, namely, breach of *Good Medical Practice*, not in the best interests of patients and undermining public confidence in the medical profession. In the case of impairment, the panel, additionally, characterised the breach as a breach of "fundamental principles".
>
> 96. In my judgment, it would certainly have been better, particularly, in the light of this Court's observations in *Zygmunt* and *Cohen*, if the Panel in the present case, had clearly indicated distinct consideration of the two issues of "misconduct" and "impairment".

> 97. However, I accept (Counsel for the GMC's) submission that, as a matter of law, there is no requirement in *all* cases for there to be a formal "two-stage" process. The requirement under the Act is that there are two "steps"; the panel must consider whether there has been misconduct and further whether that misconduct is such as to impair fitness to practise. As pointed out by Cranston J in *Cheatle* whilst misconduct is about the past, impairment is an assessment addressed to the future, albeit made in the context of the past misconduct.
>
> [...]
>
> 99. Nor, as a matter of law, is there a requirement that, in all cases, the reasons for a finding of impairment must be distinct from the reasons for the finding of misconduct. Often a finding of impairment will follow from past misconduct, but that is not necessarily the case. As Mitting J put it in *Zygmunt* "even though the panel... finds... misconduct, it *may* conclude that fitness to practise is not impaired." After saying that in perhaps the majority of cases, the issue will not be live (i.e. in such cases, a finding of impairment will follow from the finding of misconduct), Mitting J continued, in contrast, by stating that in cases in which the issue is live, then impairment "must be separately and appropriately" addressed. It is thus necessary to distinguish between cases where misconduct is, of itself, likely to lead to a finding of impairment and cases where misconduct does not necessarily lead to a finding of impairment, because of other factors to be taken into account. Such factors usually comprise events between the date of misconduct and the date of the panel hearing, such as a one-off event of misconduct followed by the passage of substantial time, [and] otherwise unblemished record, or subsequent retraining. In each of *Zygmunt*, *Cohen* and *Cheatle*, the panel had failed to take into account what had happened in the period between a one-off incident of past clinical misconduct and the date of the assessment of fitness at the panel hearing.

### *Yeong v. General Medical Council* [2009] EWHC 1923 (Admin)

**33.11** Doctor—sexual relationship with patient—impairment—whether remediable—public confidence

In this case, Y's registration with the GMC was suspended for twelve months by reason of misconduct following a sexual relationship with a former patient. Y obtained an expert report from an experienced psychiatrist, who assessed that he did not have a psychological disposition to engage in sexual relationships with patients, the likelihood of recurrence was extremely low, and that Y did not pose a risk to patients in his capacity practising as an obstetrician and gynaecologist. On appeal, Y contended (amongst other grounds) that the panel applied an incorrect test of impairment of fitness to practise. Sales J, in his judgment at [31], said that the panel, in its impairment decision, plainly considered that Y did present a heightened risk of improper conduct in relation to his patients in future and that this was treated by the panel as a relevant consideration weighing in favour of the decision that it took on impairment. Sales J went on to state at [38] that:

> The question of the possibility of a recurrence of such misconduct by [Y] was a matter of the ordinary assessment of likely human behaviour, in relation to which a psychiatrist's expertise confers no special privileged insight. The assessment of risk of any particular form of future behaviour is the sort of task which courts and tribunals regularly perform without needing to refer to expert psychiatric evidence.

In dismissing Y's appeal, Sales J said, at [40], that:

> [I]mportantly, the panel's view was that the general public interest in clearly marking proper standards of behaviour for doctors in respect of relationships with their patients so as to uphold public confidence in the medical profession was by far the weightiest factor pointing in favour of the finding of impairment of fitness to practise and the sanction which was imposed.

As to whether Y's current fitness to practise was impaired, Sales J considered that *Cohen*, *Meadow*, and *Azzam* were to be distinguished from the present case on the basis that each of *Cohen*, *Meadow*, and *Azzam* was concerned with misconduct by a doctor in the form of clinical errors and incompetence. Sales J accepted the submission of counsel for the GMC that:

> Where a [fitness to practise panel] FTPP considers that the case is one where the misconduct consists of violating such a fundamental rule of the professional relationship between medical practitioner and patient and thereby undermining public confidence in the medical profession, a finding of impairment of fitness to practise may be justified on the grounds that it is necessary to reaffirm clear standards of professional conduct so as to maintain public confidence in the practitioner and in the profession. In such a case, the efforts made by the medical practitioner in question to address his behaviour for the future may carry very much less weight than in a case where the misconduct consists of clinical errors or incompetence.

In relation to such types of misconduct, the question of remedial action taken by the doctor to address his areas of weakness may be highly relevant to the question of whether his fitness to practise is currently (that is, at the date of consideration by the panel) impaired. But the position in relation to Y's case—that is, improperly crossing the patient–doctor boundary by entering into a sexual relationship with a patient—was different. As Sales J made clear, in the latter type of case, the efforts made by the medical practitioner in question to address his behaviour for the future may carry very much less weight. Sales J referred to the overarching function of the GMC, as set out in section 1(1A) of the Medical Act 1983, to have regard to the public interest in the form of maintaining public confidence in the medical professional generally and in the individual medical practitioner when determining whether particular misconduct on the part of that medical practitioner qualifies as misconduct that currently impairs the fitness to practise of that practitioner. The public's confidence in engaging with him and with other medical practitioners may be undermined if there is a sense that misconduct that violates a fundamental rule governing the doctor–patient relationship may be engaged in with impunity. Second, a firm declaration of professional standards so as to promote public confidence may be required and efforts made by the practitioner to reduce the risk of recurrence may be of less significance than in other cases, such as those involving clinical errors or incompetence.

### *Kituma v. Nursing and Midwifery Council* [2010] EWCA Civ 154

**33.12** Midwife—misconduct—good record—no subsequent complaint—weight to be attached to particular facts

K, a registered midwife, appealed against an order of King J, by which he dismissed her appeal against a decision of the NMC's CCC, which had found misconduct proved and had imposed the sanction of a striking-off order. K was employed as an agency midwife at Watford General Hospital. In January 2004, she undertook the care of patient D, who was in labour with her first child. Four of the charges of misconduct that K faced and which were found proved all related to her clinical management of the labour and delivery. The second head of charges related to events after complaint was made by patient D, and to the appellant's custody of photocopies of patient D's medical and midwife notes, and of the letter of complaint that the appellant had been given. When dealing with the issue of misconduct, the judge said that the Committee did expressly take into account K's general good record both before and since the events under investigation. Nothing in the cases cited is authority for the proposition that professional misconduct arising out of one isolated set of events (even if this be a proper characterization of the present case, which may be doubtful given the separate charge under head two) can never give rise to an adverse finding on fitness to practise. Every case must be determined on its own facts. In dismissing K's appeal, the Court of Appeal (Mummery, Richards, and Rimer LJJ) said that the Committee stated in terms that it had considered very carefully the written evidence that K had submitted in relation to her clinical work. The question of weight was pre-eminently a matter for the specialist Committee. The Committee was entitled to reach the conclusion that it did, notwithstanding the evidence concerning K's performance since the events that were the subject of the misconduct allegations. The judge's analysis and conclusions were correct, and there is no conflict at all between what he said and the decision in *Azzam*. Different results were reached in the two cases on the application of the same principles to different facts. The appeal against sanction was dismissed.

### *R (Vali) v. General Optical Council* [2011] EWHC 310 (Admin)

**33.13** Single examination—misconduct—subsequent steps—whether impairment

The allegation against the appellant, V, was that her fitness to practise was impaired because, at a consultation with a patient in July 2005, the appellant recorded intraocular pressure measurements that were abnormal, but allegedly did not repeat the recording either immediately and/or at a different time of day, did not perform a visual field test, and did not make a referral of the patient to a medical practitioner concerning the abnormal intraocular pressure measurements. Ouseley J rejected V's arguments on the admissibility of the patient's witness statement (the patient being in Ethiopia) and the appellant's arguments on the findings of misconduct, but concluded that the finding by the fitness to practise panel on impairment could not be justified and the panel's decision should be quashed. The consultation occurred four-and-a-half years before the hearing. The panel had positive evidence that there had been no complaints either to the General Optical Council (GOC) or by those for whom the appellant cared about the way in which she dealt with patients over the period since July 2005. It also had, from a variety of sources, testimonials and support that were not consistent with somebody having so basic a failing in knowledge as the panel supposed. The appellant's career was on a good upwards trajectory, including supervision and teaching. There was also evidence of recent acquisition of relevant skill base. The Court said that it was unfortunate that the panel used the language that it did about the appellant's lack of insight. Absence of insight, if it means no more than that the appellant's evidence was not accepted, was an inappropriate use of the concept as a basis for a finding of impairment. The Court concluded that it was hard to conceive that somebody who is a continued risk to public safety or to public confidence in the competence and standing of the profession would have progressed as the appellant had done.

### *Raza v. General Medical Council* [2011] EWHC 790 (Admin)

**33.14** Isolated incident—steps taken to prevent recurrence—reasons for impairment

R appealed against the decision of the GMC's fitness to practise panel that his fitness to practise was impaired by his misconduct and the decision to impose upon him a sanction of twelve months' suspension. It was alleged that R conducted himself in the course of a consultation with a patient in a way that was inappropriate, sexually motivated, and an abuse of his professional position. The submissions on appeal relevant to the issue of impairment were (a) that this was an entirely innocent incident, (b) that R's record was an unblemished one, and (c) that there was no pattern of predatory behaviour that had been identified or established by the evidence adduced before the panel. In allowing R's appeal, the Administrative Court (His Honour Judge Pelling QC, sitting as a judge of the High Court) criticized the panel's finding that the matters complained of stemmed from R's 'underlying attitude'. The reasons given by the panel did not define what that attitude was alleged to have been, and, more fundamentally, failed to explain the basis for the conclusion and how it was consistent with the points made on behalf of the doctor. This was an isolated incident by a doctor with an otherwise unblemished record. The judge considered that the panel was wrong to dismiss the steps taken by the doctor now always to have a chaperone present, when examining women patients, as being to protect himself rather than to protect the patients. The Court concluded that the panel's obligation to decide whether, and then to explain why, the doctor's current fitness was impaired by reason of his past misconduct had not been discharged correctly. In the Court's view, the panel did not explain, even to the modest standards imposed on tribunals in these circumstances, why it reached the conclusion that it did. That, of itself, would justify quashing its decision.

### *Council for Healthcare Regulatory Excellence v. Nursing and Midwifery Council and Grant* [2011] EWHC 927 (Admin)

**33.15** Nurse/midwife—bullying behaviour—inappropriate patient care—misconduct—impairment—public confidence

The Council for Healthcare Regulatory Excellence (CHRE) appealed against a decision of the NMC's CCC that the second respondent, G, a registered nurse and midwife, was guilty of misconduct, but that her fitness to practise was not impaired. G worked as a midwifery sister in a hospital. The charges against her included that she had, over a period of some twenty months, failed to provide assistance to a junior colleague, and subjected that colleague to bullying and harassment for reporting her, had failed to provide appropriate care to a patient admitted for delivery of her baby, who had died in utero, and had failed properly to record that a baby born at 20 weeks' gestation had been born alive. The Committee found that the charges were proved and amounted to misconduct. However, they found that G's attitude had improved and that she had addressed her poor performance, so that her fitness to practise was not currently impaired. In allowing CHRE's appeal, supported by the NMC, but opposed by G, the Administrative Court (Cox J) said that it was essential, when deciding whether fitness to practise was currently impaired, not to lose sight of the need to protect the public, and the need to declare and uphold proper standards of conduct and behaviour so as to maintain public confidence in the profession. A panel should consider not only whether the practitioner continued to present a risk to members of the public in his or her current role, but also whether the need to uphold professional standards and public confidence in the profession would be undermined if a finding of impairment were not made. In the course of an extensive review of the authorities, including *Cohen*, *Nicholas-Pillai*, and *Yeong*, the learned judge said:

> 65. The term "impairment of fitness to practise" has not been defined in [Nursing and Midwifery Council (Fitness to Practise) Rules 2004], and this is also the position in relation to those schemes which apply to other medical practitioners. Thus, as Dame Janet Smith pointed out in her Fifth Report from *The Shipman Inquiry* (9th December 2004), the concept has the advantage of flexibility, being capable of embracing a multiplicity of problems, but also the disadvantages that flow from a lack of clarity and definition. Further, recognizing impaired fitness to practise inevitably involves making a value judgment (see paragraph 25.42 et seq).
>
> [...]
>
> 69. It is clear, ... that the question is always whether [fitness to practise] is impaired as at the date of the hearing, looking forward in the manner indicated by Silber J in his judgment [in *R (Cohen) v. General Medical Council* [2008] EWHC 581 (Admin)]. The question for this Committee as at 21 April 2010 [the hearing before the CCC in *Grant*] was therefore "is this registrant's current fitness to practise impaired?"
>
> 70. An assessment of current fitness to practise will nevertheless involve consideration of past misconduct and of any steps taken subsequently by the practitioner to remedy

it. Silber J recognized this when referring, at paragraph 65, to the necessity to determine whether the misconduct is easily remediable, whether it has in fact been remedied and whether it is highly unlikely to be repeated.

71. However it is essential, when deciding whether fitness to practise is impaired, not to lose sight of the fundamental considerations emphasised at the outset of this section of his judgment at paragraph 62, namely the need to protect the public and the need to declare and uphold proper standards of conduct and behavior so as to maintain public confidence in the profession.

72. This need to have regard to the wider public interest in determining questions of impairment of fitness to practise was also referred to by Goldring J in *R (Harry) v. General Medical Council* [2006] EWHC 3050 (Admin) and by Mitting J in *Nicholas-Pillai*, where he held that the panel were entitled to take into account the fact that the practitioner had contested critical allegations of dishonest note-keeping.

73. Sales J also referred to the importance of the wider public interest in assessing fitness to practise in *Yeong v. GMC* [2009] EWHC 1923 (Admin), a case involving a doctor's sexual relationship with a patient....

74. I agree with that analysis and would add this. In determining whether a practitioner's fitness to practise is impaired by reason of misconduct, the relevant panel should generally consider not only whether the practitioner continues to present a risk to members of the public in his or her current role, but also whether the need to uphold proper professional standards and public confidence in the profession would be undermined if a finding of impairment were not made in the particular circumstances.

75. I regard that as an important consideration in cases involving fitness to practise proceedings before the NMC where, unlike such proceedings before the General Medical Council, there is no power under the rules to issue a warning, if the committee finds that fitness to practise is not impaired. As [the NMC] observes, such a finding amounts to a complete acquittal, because there is no mechanism to mark cases where findings of misconduct have been made, even where that misconduct is serious and has persisted over a substantial period of time. In such circumstances the relevant panel should scrutinise the case with particular care before determining the issue of impairment.

76. ... [A]t paragraph 25.67 [Dame Janet Smith's Fifth Report from *Shipman*] she referred the following as an appropriate test for panels considering impairment of a doctor's fitness to practise, but in my view the test would be equally applicable to other practitioners governed by different regulatory schemes:

> "Do our findings of fact in respect of the doctor's misconduct, deficient professional performance, adverse health, conviction, caution or determination show that his/her fitness to practise is impaired in the sense that s/he: (a) has in the past acted and/or is liable in the future to act so as to put a patient or patients at unwarranted risk of harm; and/or (b) has in the past brought and/or is liable in the future to bring the medical profession into disrepute; and/or (c) has in the past breached and/or is liable in the future to breach one of the fundamental tenents of the medical profession; and/or (d) has in the past acted dishonestly and/or is liable to act dishonestly in the future."

The value of this test, in my view, is threefold: it identifies the various types of activity which will arise for consideration in any case where fitness to practise is in issue; it requires an examination of both the past and the future; and it distils and reflects, for ease of application, the principles of interpretation which appear in the authorities. It is, as it seems to me, entirely consistent with the judicial guidance to which I have already referred, but is concisely expressed in a way which is readily accessible and readily applicable for all panels called upon to determine this question.

The Committee in the instant case had not referred in its reasons to the importance of wider public interest considerations or to the need for substantial weight to be given to the protection of the public, the maintenance of public confidence in the profession, and to the upholding of proper standards of conduct and behaviour. Nor was there anything in the reasons to suggest that it had in fact had regard to these wider considerations without making any express reference to them.

### *Martin v. General Medical Council* [2011] EWHC 3204 (Admin)

**33.16** Good Medical Practice

In considering whether a practitioner's fitness to practise is impaired by reason of misconduct, Lang J said, at [76], that it is helpful for a fitness to practise panel to consider the GMC guide *Good Medical Practice* to identify what standards of behaviour are expected of registered doctors—that is, what constitutes 'fitness to practise'. In this case, the panel found that M had violated many of the duties in *Good Medical Practice* and the supplementary guidance *Maintaining Boundaries*. The panel concluded that M's conduct amounted to 'a serious departure from the standards of behaviour required of doctors'.

### *Ige v. Nursing and Midwifery Council* [2011] EWHC 3721 (Admin)

**33.17** See paragraph 19.61.

### *Uddin v. General Medical Council* [2012] EWHC 1763 (Admin)

**33.18** Procedure—admitted impairment—evidence—findings of impairment—whether any unfairness

U was employed between 2007 and 2009 as a trainee general practitioner (GP) in the Wales Deanery. The charges against him alleged that, between October 2007 and January 2009, he made entries in his e-portfolio that were false and that, at a meeting with his programme directors, he claimed that he had not falsified any e-portfolio records. At the hearing before the fitness to practise panel, U admitted misconduct and that his fitness to practise was impaired. The panel decided to hear evidence at the impairment stage and made findings in reaching its conclusion on impairment—namely: (i) that U had acted with a view to obtaining a professional qualification by deception; (ii) that he had shown a lack of insight into the effect of his misconduct; (iii) that the panel was unable to conclude that the conduct was unlikely to be repeated; (iv) that U posed a risk to patient safety; (v) that the panel was unable to accept that, although falsified, the entries in his e-portfolio were accurate; and (vi) that U had made a mockery of the assessment process. U appealed against the sanction of erasure, and submitted that the panel had erred procedurally and substantively by taking into account, in reaching its conclusion on sanction, the findings that it had made at the impairment stage. U had admitted that his fitness to practise was impaired, so there should not have been a factual enquiry into that issue and the appellant should not have been required to give evidence at the impairment stage. It was further submitted that the panel made findings beyond the charges that had been made against the appellant. Stadlen J dismissed the appeal. He said that the panel had not gone beyond the original allegations. Had the evidence not been given at the impairment stage, it was inevitable that it would have been given at the sanction stage. It was to be inferred that the same evidence would have been given at the sanction stage as had in fact been given at the impairment stage. There was no basis for saying that the panel had not followed the correct procedure. The findings had been made on the evidence and it had been open to the panel to make them. Stadlen J said:

> 20. In my judgment, so far as the first criticism of the FTPP is concerned, namely that it was wrong for it to admit any evidence on impairment in the light of the indication on behalf of [U] at the outset that he was going to admit impairment, it is met with the insuperable obstacle that (counsel) on [U's] behalf expressly indicated that he was content with the procedure of there being evidence on impairment after the admissions had been made and after the evidence relied on by the GMC in support

of the allegations had been led, and that he was content that not only should there be evidence but in particular that he should call [U]. In those circumstances it seems to me that there is no scope for any complaint at this appeal stage for an argument that it was either procedurally unfair or beyond the jurisdiction of the tribunal (the FTPP) to permit that evidence to be adduced. It was adduced with the consent of [U] with the benefit of legal advice.

[...]

33. ... [U] gave evidence voluntarily at the impairment stage as indeed he was bound to do as a matter of common sense in order to set in context the evidence which had been prayed-in-aid against him in respect of the allegations which he had admitted. And, in any event, had those matters not been raised at the impairment stage because of his admission, in my judgment, it is inevitable that he would have given evidence at the sanction stage in order, if for no other reason, to indicate that he had sufficient insight into his behaviour to justify suspension rather than erasure as a sanction, in which event it is to be inferred that precisely the same evidence would have been given.

### *Ujam v. General Medical Council* [2012] EWHC 580 (Admin); [2012] EWHC 683 (Admin)

**33.19** Findings of fact—overlap with impairment—remediation—interim order

U was employed between August 2007 and November 2008 at a hospital on a general practitioner (GP) training programme. The charge included allegations of inappropriate and sexually motivated conduct towards two female doctors and a female nurse at the hospital. A fitness to practise panel found the allegations proved and suspended U's registration for a period of six months. The panel concluded that he failed to recognize the proper boundaries of professional and social relationships between colleagues, and that his misconduct had brought the medical professional into disrepute. The panel's determination had included a general passage on the credibility of the various witnesses, specifically indicating why they found the relevant complainants credible. On appeal, it was suggested by counsel for U that, having come to these general conclusions, the panel then appeared to treat them rather mechanistically as a basis for going on to make findings adverse to the appellant. Counsel submitted that even if his client was a less-than-impressive witness on recounting what had taken place, it did not follow that any of the transgressions should therefore be taken as impairing his fitness to practise as a doctor. Eady J, in dismissing U's appeal, said that, as a general proposition, of course, that is unassailable. However, there was no convincing evidence that the panel did allow its general view on credibility to prejudice the separate task of judging impairment of fitness to practise. There is, however, an element of overlap and specifically in the context of U's insight or lack of it. If the panel were of the view that U presented himself as being somewhat insensitive to the effect of his behaviour on other people, it might also take the view, legitimately, that such an attitude would be relevant to the issues of impairment and, if relevant, sanction. There was no possible basis for ruling that the panel's assessment of U's misconduct, as having been serious and such as to bring the profession into disrepute, was wrong. On impairment, it was suggested that the allegations of sexual harassment could properly be regarded as a series of events 'never to be repeated' and should be judged in isolation. There was nothing before April 2007 to suggest that U's record had been other than exemplary, or since. Eady J said that the conduct discussed in *Yeong v. General Medical Council* [2009] EWHC 1923 (Admin) was far more serious than that in the present case, but the distinction remains important, as between clinical incompetence and misconduct by way of 'crossing boundaries' in relation to the relative significance of remedial action. The panel was right to give relatively little weight to the 'remedial' factors in this case and, in particular, to statements from fellow practitioners to the effect that U had effectively 'turned over a new leaf', or was being more guarded in his conduct. Because of the importance attached in this context to maintaining the confidence of the public, the panel was entitled to come to the conclusion that the appellant's fitness to practise remained impaired as at the date of the hearing in December 2010. On sanction, between July 2009 and February 2010, U's registration had been suspended by the GMC's Interim Orders Panel (IOP). It would be undoubtedly right that the suspension the IOP imposed should be borne in mind as part of the background circumstances, but it would certainly be inappropriate to regard it as analogous to a period of imprisonment served while on remand (which would normally be deducted from any custodial term imposed by a sentencing court). In the GMC Indicative Sanctions Guidance, paragraph 22 states:

> ... [I]n making their decision on the appropriate sanction, panels need to be mindful that they do not give undue weight to whether or not a doctor has previously been subject to an interim order for conditions or suspension imposed by the Interim Orders Panel, or the period for which that order was effected... an interim order and the length of that order are unlikely to be of much significance for panels...

There is no alternative to dismissing the appeal on the sanction imposed.

***Black v. Professional Conduct Committee of the General Dental Council* [2013] HCJAC 39**

**33.20** Professional indemnity insurance—misconduct—whether impairment

The charged faced by the appellant was that, between May 1993 (alternatively November 1997) and January 2012, he was not indemnified or insured in respect of claims from patients arising out of his practice as a dental practitioner. On appeal, B accepted that his conduct amounted to misconduct, but asserted that the Professional Conduct Committee (PCC) had erred in finding that his fitness to practise was impaired, and that the decision to erase his name from the register was excessive and disproportionate. In dismissing B's appeal on impairment and sanction, Lord Bracadale, giving the judgment of the Inner House (Lady Paton, Lady Smith, and Lord Bracadale), said that, from 1993 onwards, it became increasingly clear that insurance cover was a requirement for practice. The General Dental Council (GDC) had issued guidance in 1993, 1997, and 2005, and the Dentists Act 1984 had been amended to include a new section 26A, which introduced a statutory requirement for insurance cover. The learned judge continued:

> 24. It seems clear to us that, at the latest by 1997, it was plain that the requirement for insurance cover was mandatory. In our opinion, the committee was entitled to conclude that the General Dental Council had been clear in its guidance since 1997 that membership of a fence organisation or indemnity insurance was obligatory. Having regard to the specialist nature of the committee we are unable to criticise its conclusion that the requirement was a mandatory ethical requirement and that a majority of the profession would find such a lack of membership or indemnity insurance deplorable. The committee was entitled to conclude that as the appellant had treated patients while disregarding the respondents' standards on this point his actions amounted to misconduct. Before us that was accepted on behalf of the appellant.
>
> 25. In relation to the first two grounds of appeal, we reject the submission made by [counsel for the appellant] that because the misconduct had been remedied by the taking out a policy of indemnity insurance and that, as a result, patients, both past and future, were now protected, and because there was no prospect of the misconduct being repeated, the committee was not entitled to find that the current fitness to practise of the appellant was impaired. It was necessary for the committee in deciding whether the fitness to practise was impaired to have regard to wider considerations than the consideration that the misconduct had been remedied and there was little likelihood of repetition. The committee [was] required to have regard to the need to protect the

public and the need to declare and uphold proper standards of conduct and behaviour so as to maintain public confidence in the profession. We agree with the analysis of these issues by Cox J in *Council for Healthcare Regulatory Excellence v. Nursing and Midwifery Council and Grant* [2011] EWHC 927 (Admin) . . . The committee was entitled to consider all the information which had been placed before it on behalf of both the appellant and the respondents. The appellant had made a deliberate decision not to renew his insurance after he had been refused a no claims bonus in 1982. As a result, he practised for many years without professional indemnity insurance. . . . [W]e note that the absence of insurance cover emerged as a side-wind from an unrelated complaint. It is clear that but for that discovery, the appellant would never have taken out professional indemnity insurance. He did so only because he had been found out. In these circumstances it seems to us that the fact that the absence of insurance cover had been remedied by the time of the hearing before the committee was of little significance.

26. . . . [T]he Committee was entitled to have regard to history of the conduct of the appellant, the nature of the alternative arrangement put in place by him, his denial that his conduct amounted to misconduct, and the seriousness of the misconduct and to draw from that history the inference that the appellant had shown and continued to show a lack of insight and appreciation of the risks of the possible detriment of his patients which he ran for many years.

27. In our opinion the committee was entitled on the basis of the information before it to come to the view that the appellant's fitness to practise was impaired.

### *Odoi-Asare v. Nursing and Midwifery Council* [2014] EWHC 1151 (Admin)

**33.21** Two-stage process—misconduct and impairment

The appellant was employed as a band 6 health visitor for Ealing Hospital National Health Service (NHS) Trust. Her role involved conducting child and maternal health needs assessments, developmental reviews, healthy child clinics, group work, and safeguarding and child protection work. The appellant admitted and/or the panel found proved allegations concerning Child A, including failing to notify his social worker of threats made by the mother, failing to put an alert on his record of an incident of domestic violence, failing to record observations during the home visit, failing to conduct a health review, and failing to notify other professionals that Child A was subject to a child protection plan. The appellant maintained that the panel failed properly to apply the test for impairment, as set out in various decisions of the Administrative Court. In dismissing the appellant's appeal against a finding of impairment, Supperstone J said, at [46], that the panel followed the correct approach in relation to impairment by adopting the required two-stage process: first, to consider whether or not the findings of fact amount to misconduct; and then, if so, to proceed to consider whether or not the appellant's fitness to practise is currently impaired, having regard to all of the information available to it. The learned judge said that, in assessing the issue of current impairment, the panel considered the guidance provided by Dame Janet Smith in her Fifth Report from the Shipman Inquiry and asked itself whether the appellant had acted so as to put a patient, or patients, at unwarranted risk of harm, had brought the nursing profession into disrepute, and/or had breached one of the fundamental tenets of the nursing profession. Supperstone J continued:

46. I accept [the NMC's] submission that the Panel's reasons make plain that it viewed the Appellant's clinical failings to be so serious that even the accepted remediation was not enough to enable it to uphold the public interest without a finding of impairment.

47. In [*Council for Healthcare Regulatory Excellence v. Nursing and Midwifery Council and Grant* [2011] EWHC 927 (Admin)] Cox J is giving general guidance (see para 76). The approach to be applied in each case is the same. It is the act of misconduct which has to be considered, not whether the misconduct is clinical or non-clinical. The wider public interest is engaged in each case.

***Professional Standards Authority v. Health and Care Professions Council and Ghaffar*** **[2014] EWHC 2723 (Admin)**

**33.22** Dishonesty—impairment

In allowing the appeal of the Professional Standards Authority (PSA) on impairment, Carr J said, at [51]–[53], that a finding of impairment does not, of course, necessarily follow upon a finding of dishonesty, although it was accepted by the panel that it will be a frequent one. While each case will turn on its own facts, it will be an unusual case in which dishonesty is not found to impair fitness to practise. In the instant case, the dishonesty was in the health and care environment, and gave rise to real public health risks. The panel also erred in finding that the registrant's conviction and sentence were sufficient to meet the public interest. The criminal and regulatory strands are different and perform different functions. The need to uphold proper professional standards and to uphold public confidence required a finding of impairment quite separate and distinct from the criminal sanctions imposed on the registrant: at [56]–[57].

See paragraph 19.78.

***R (Nakash) v. Metropolitan Police Service and General Medical Council (Interested Party)*** **[2014] EWHC 3810 (Admin)**

**33.23** Impairment to be judged at date of hearing—totality of evidence about practitioner

In judicial review proceedings relating to disclosure by the Metropolitan Police Service of material requested by the GMC that was unlawfully obtained by the police, Cox J said, at [36]:

> It is now well established that, in deciding whether a doctor's fitness to practise is impaired, the Panel will take account not only of any past act of misconduct or deficient performance by the practitioner, but also whether the act is likely to be repeated, and of the need to maintain public confidence in the profession and to uphold proper standards of conduct and behaviour generally. The Panel looks to the future and not to the past. The question is always whether, in the light of all the relevant factors known to them as at the date of the hearing, the doctor's fitness to practise is impaired. The Guidance refers (at paragraph 55) to the fact that two or more allegations of impairment, of any kind, may be considered at the hearing and that the hearings will be "holistic". Allegations will be brought forward based on the totality of the evidence obtained at the investigation stage and may comprise a combination of allegations relating to a practitioner's health, performance or conduct.

## D. Other Relevant Chapters

**33.24**

Insight, Chapter 37
Misconduct, Chapter 47
Review Hearings, Chapter 59
Sanction, Chapter 60

# 34

# IN CAMERA DISCUSSIONS



## A. Legal Framework

**34.01** Some bodies include a requirement that the panel 'shall deliberate in private'. Examples include:

Nursing and Midwifery Council (Fitness to Practise) Rules 2004:

rule 24(11)–(14) (initial hearing)
rule 25 (review or restoration hearing)
rule 26 (interim orders hearing)

General Dental Council (Fitness to Practise) Rules 2006, rule 21
Institute of Chartered Accountants in England and Wales Disciplinary Committee Regulations, rule 29 (tribunal 'may deliberate in camera' at any time)

Some bodies have legal assessors rules. Examples include:

Veterinary Surgeons (Disciplinary Proceedings) Legal Assessor Rules 1967, rule 5
General Medical Council (Legal Assessors) Rules 2004 (any advice tendered after the committee has begun to deliberate and where the committee considers that it would be prejudicial to the discharge of its duties for the advice to be tendered in the presence of the parties or their representatives, the legal assessor shall, as soon as possible, inform the committee of his advice and the information so given by him shall be recorded)

## B. Disciplinary Cases

### *Roylance v. General Medical Council* (1999) The Times, 27 January

**34.02** Disclosure of notes—public interest immunity

From 1989 to 1995, R was the chief executive officer (CEO) of the United Bristol Healthcare National Health Service (NHS) Trust. On 18 June 1998, the Professional Conduct Committee (PCC) of the General Medical Council (GMC) found R guilty of serious professional misconduct and directed that his name be erased from the register. The Committee found that R had failed to take action over the years during which concerns were being raised about the excessive mortality of infants following a number of cardiac operations on very young children at the Bristol Royal Infirmary. Various matters were raised by the doctor on appeal, including an allegation of bias on the part of the chairman of the Committee, which was advanced in the course of the inquiry, and renewed and developed before the Board. Prior to the hearing of the appeal, R sought from the Board an interlocutory order to have the shorthand notes of the in camera deliberations of the Committee disclosed. On 19 January 1999, their Lordships refused the application. In reaching their decision, Lord Steyn said:

> Their Lordships are satisfied that such an order would be inappropriate. It is acknowledged to be an unprecedented attempt to probe into in camera discussions. Counsel submits that the exceptional circumstances of the case warrants such an order. Their Lordships are wholly unpersuaded that this case can be so categorised. If the submission were to be accepted it would seriously inhibit freedom of discussion during in camera sessions. It is ruled out in the present case by public interest immunity attaching to the in camera discussions of the Professional Conduct Committee. Their Lordships are not satisfied that there are good or sufficient reasons for overriding that immunity.

### *Roylance v. General Medical Council (No. 2)* [2000] 1 AC 311

**34.03** Evidence—affidavit—admissibility—legal assessor

On the hearing of the substantive appeal in February 1999, R sought to present a further argument levelled at the stage of the deliberations by the PCC on its final determination. R tendered an affidavit by a Professor D, which the doctor wished to use to support an argument that there had been some unfairness in the course of the Committee's deliberations. Professor D had written an article that was published in the *British Medical Journal* expressing a variety of concerns about the handling of the matter before the Committee. That had evidently prompted a telephone call from someone who was understood by Professor D to have reliable information about what had passed during the in camera deliberations by the Committee. Their Lordships accepted the affidavit de bene esse in order to understand more fully the substance of its contents, but, having done so and heard argument upon its admission, came to a clear view that it should not be admitted. Some assistance can be obtained from the consideration of the corresponding problem that has on occasion arisen in connection with juries. In that context, the refusal of the Court to inquire into the processes of a jury's determination was affirmed long ago in *Straker v. Graham* (1839) 4 M&W 721, and is now regarded as a settled rule of long standing: *R v. Miah* [1997] 2 Cr App R 12. [*R v. Connor and Rollock* [2004] 1 AC 1118; but see now *R v. Adams* [2007] 1 Cr App R 34 and *R v. Charnley* [2007] 2 Cr App R 33.] Not only is the doctor's application contrary to that general principle of confidentiality, but it is the more inappropriate in the particular context of the PCC. Unlike a jury, the Committee has the benefit of a legal assessor who, under the Procedure Rules, may retire with it. The assessor has an express duty under the Legal Assessors Rules not only to inform the Committee of any irregularity in the conduct of proceedings before it that comes to his knowledge, but also to advise it of his own motion where it appears to him that there is a possibility of the mistake of law being made. That provision appears wide enough to cover improprieties in the process of discussion and determination of the issues before the Committee. Furthermore, the parties are entitled to be informed of any advice tendered after the Committee has begun to deliberate, so that, in the circumstances of the present case, the parties not only have the reassurance that the Committee has the opportunity for advice from the assessor, but also even have the right to know what advice on matters of law has been given to the Committee.

### *Nwabueze v. General Medical Council* [2000] 1 WLR 1760

**34.04**
Panel member—confidentiality—Legal Assessor's Rules—possible breach

N faced four allegations of sexual misconduct and two allegations of dishonesty. He was found guilty of serious professional misconduct and his name was directed to be erased from the register. Shortly after the hearing before the PCC was concluded on 30 April 1999, one of the members of the Committee, Dr C, contacted the doctor's representative, indicating that he had concerns about the way in which the case had been handled by the Committee. The GMC objected to the doctor's counsel and solicitors entering into discussions with Dr C about what had occurred while the Committee was deliberating in camera. On 3 February 2000, the GMC obtained an injunction ordering that, subject as might be directed by the Privy Council, Dr C was to maintain the confidentiality of, and was not to disclose to anyone, any information concerning what was said or had occurred in the course of the in camera deliberations. At the hearing of the substantive appeal on 9 February 2000, their Lordships decided that, except in regard to one matter, insufficient grounds had been made out for outweighing the very strong presumption that exists in favour of preserving the confidentiality of the deliberations of the PCC when it is considering its decision in camera. They held that the doctor was not entitled to any information as to what took place during those deliberations either by way of interviewing Dr C, or by way of seeing a transcript or the full judgment of the hearing on 3 February 2000. The one matter on which their Lordships were of the view that information as to what took place during the in camera deliberations should be made available to N related to an alleged breach of the General Medical Council (Legal Assessors) Rules 1980. Their Lordships were satisfied that the doctor was entitled to the basic factual material that he needed to advance the argument that advice that the legal assessor gave to the Committee while it was deliberating in camera was given in breach of the Legal Assessors Rules. However, in regard to the procedural aspects, counsel accepted the assurance of counsel for the GMC that the legal assessor's advice was not tendered in response to any question from the Committee, but was proffered by him spontaneously of his own motion in the course of the discussions. Their Lordships were not persuaded that there was a breach of the Legal Assessors Rules. Because the advice was given by the legal assessor of his own motion and not in response to a question from the Committee, what occurred in this case fits more easily with the provisions of rule 3 rather than those of rule 4 of the 1984 Rules (now rule 4(a) and (b) of the 1984 Rules). Their Lordships continued, at 1775C–F:

> That having been said, their Lordships consider that the principle which lies behind the requirement that the parties should be informed of the assessor's advice to the committee is that of fairness, and that fairness requires that the parties should be afforded an opportunity to comment on that advice and that the committee should have an opportunity to consider their comments before announcing their determination. The transcript of the proceedings indicates that the chairman regarded the legal assessor's statement about the legal advice which he had tendered to the committee while they were deliberating in camera as a mere formality, as the committee had already arrived at their determination which he was about to announce. This was a misconception, as the reason why the legal assessor's advice to the committee must be given or made known to the parties afterwards in public is so that the parties may have an opportunity of correcting it or of asking for it to be supplemented as the circumstances may require. In this respect the requirements of the common law would appear to be at one with those of [A]rticle 6 of the [European] Convention [on Human Rights].

*Walker v. General Medical Council* [2002] UKPC 57

**34.05** Legal assessor—record of interventions

In quashing the sanction of erasure and directing that the matter of appropriate sanction be remitted to the PCC, differently constituted and advised by a different legal assessor, the Privy Council (Lord Hope of Craighhead, Lord Walker of Gestingthorpe, Sir Murray Stuart-Smith, the Rt Hon Justice Gault, and Sir Philip Otton) said that where, as here, there arose a question of non-compliance with the General Medical Council (Legal Assessors) Rules, some limited intrusion into the confidentiality of in camera discussions of the PCC may be necessary. It was unfortunate that there was no agreement on what occurred in relation to the advice and comments of the legal assessor. To avoid such a situation in the future, consideration should be given to the adoption of a practice of making a record of interventions and responses by legal assessors in the course of in camera deliberations of the PCC and Health Committee.

See further paragraph 44.14.

## C. Other Relevant Chapters

**34.06** Legal Assessors, Chapter 44

# 35

# INDEPENDENT AND IMPARTIAL TRIBUNAL

A. Legal Framework 35.01
B. Disciplinary Cases 35.02
*Potato Marketing Board v. Merricks* [1958] 2 QB 316 35.02
*Tehrani v. United Kingdom Central Council for Nursing, Midwifery and Health Visiting* 2001 SC 581 35.03
*Preiss v. General Dental Council* [2001] 1 WLR 1926 35.04
*R (Nicolaides) v. General Medical Council* [2001] EWHC 625 (Admin) 35.05
*In re P (A Barrister)* [2005] 1 WLR 3019 35.06
*Holder v. Law Society* [2005] EWHC 2023 (Admin) 35.07
*Gillies v. Secretary of State for Work and Pensions* [2006] 1 WLR 781 35.08
*R (Haase) v. District Judge Nuttall* [2009] QB 550 35.09
*Sadighi v. General Dental Council* [2009] EWHC 1278 (Admin) 35.10
*Virdi v. Law Society (Solicitors Disciplinary Tribunal intervening)* [2010] 1 WLR 2840 35.11
*R (Kaur) v. Institute of Legal Executives Appeal Tribunal* [2011] EWCA Civ 1168 35.12
*Holmes v. Royal College of Veterinary Surgeons* [2011] UKPC 48 35.13
*Kulkarni v. Milton Keynes Hospital NHS Foundation Trust* [2010] ICR 101 35.14
*Hameed v. Central Manchester University Hospitals NHS Foundation Trust* [2010] EWHC 2009 (QB) 35.15
*R (Puri) v. Bradford Teaching Hospitals NHS Foundation Trust* [2011] EWHC 970 (Admin) 35.16
*R (G) v. Governors of X School (Secretary of State for the Home Department and anor intervening)* [2012] 1 AC 167 35.17
*Mattu v. University Hospitals of Coventry and Warwickshire NHS Trust* [2012] 4 All ER 359, CA 35.18
*R (King) v. Secretary of State for Justice; R (Bourgass and Hussain) v. Secretary of State for Justice* [2012] 4 All ER 44, CA 35.19
*Ighalo v. Solicitors Regulation Authority* [2013] EWHC 661 (Admin) 35.20
*Adu v. General Medical Council* [2014] EWHC 1946 (Admin) 35.21
C. Other Relevant Chapters 35.22

## A. Legal Framework

**35.01** European Convention on Human Rights (ECHR), Article 6 (see paragraph 32.01)

## B. Disciplinary Cases

### *Potato Marketing Board v. Merricks* [1958] 2 QB 316

**35.02** Barrister—chairman of disciplinary committee—fee

M was charged with obstructing officers of the Potato Marketing Board in their attempts to check the acreage of potatoes grown at his farm, and failing to comply with a demand requiring him to furnish to the Board information relating to potatoes planted in 1956 and 1957. Section 5(1) of the Agricultural Marketing Act 1949 provided for a disciplinary committee of the Board to be constituted with 'a chairman, who is not a member of the board but is an independent person who is a barrister of not less than seven years' standing'. Devlin J held that what is contemplated is that the chairman should be a person who will be able to bring an entirely independent mind to bear on any case that is presented to the committee and who, because of his legal experience, will be able to guide the committee and ensure that it approaches the cases that it has to decide in a true judicial spirit. Accordingly, notwithstanding that its chairman received a fee from the Board, the committee was properly constituted under the provisions of the scheme.

***Tehrani v. United Kingdom Central Council for Nursing, Midwifery and Health Visiting* 2001 SC 581**

**35.03** See paragraph 32.10.

***Preiss v. General Dental Council* [2001] 1 WLR 1926**

**35.04** Preliminary screener—chair of PCC

The disciplinary charges against the appellant all related to his treatment of one patient, Mrs H, a lady in her 60s. Following a hearing before the Professional Conduct Committee (PCC), the appellant's registration was suspended for twelve months. The appellant challenged various aspects of the disciplinary procedure. The Board saw no substance in the contentions that the PCC went beyond the charges or gave insufficient reasons for its decision. However, the argument founded on natural justice and Article 6(1) ECHR required further consideration. By paragraph 11 of the Schedule to the Dentists Act 1984, the PCC consisted of the president and ten other members of the General Dental Council (GDC). The president did preside in the instant case. He had also acted as preliminary screener, whose function is to set in train proceedings before the Preliminary Proceedings Committee (PPC) unless he or she considers that the matter need not proceed further. Lord Cooke of Thorndon, delivering the judgment of the Board, said that the contention for the appellant that the role of the preliminary screener is prosecutorial cannot be accepted. It is more akin to the role of examining justices or a judge ruling on a submission of no case to answer. But that by no means disposes of the appellant's points under Article 6(1). In the opinion of the Board, when the participation of the president both as preliminary screener and as chairman of the PCC is seen in conjunction with the predominance of Council members in both the PPC and the PCC, and in conjunction moreover with the fact that the disciplinary charge is brought on behalf of the Council, the cumulative result is an appearance and a real danger that the PCC lacked the necessary independence and impartiality. Only the ultimate right of appeal to Her Majesty in Council saves the day.

***R (Nicolaides) v. General Medical Council* [2001] EWHC 625 (Admin)**

**35.05** Second misconduct hearing—no real danger of bias

N, a professor and consultant in obstetrics of international repute, appeared before the PCC in May 1998. The charge was dismissed, but, in the course of N's evidence on oath, it was alleged that he lied to the Committee. The matter was referred to the Crown Prosecution Service (CPS), which decided that there was insufficient evidence for a realistic prospect of conviction of perjury. However, a second disciplinary hearing took place with a different panel in November/December 2000. The second panel found that N had given a patently incorrect answer under oath and that he was guilty of serious professional misconduct; the panel issued what it described as the severest of reprimands. In subsequent judicial review proceedings, N submitted that the second disciplinary hearing was an abuse of process because of lack of independence and impartiality. In dismissing N's claim, Sir Richard Tucker said that, in his judgment, adopting the test laid down by Lord Phillips MR in *In Re Medicaments and Related Classes of Goods (No. 2)* [2001] 1 WLR 700, no fair-minded and informed observer would conclude that there was a real possibility or danger that the second PCC was biased. There were several factors that led to this conclusion: first, the absence of any connection between the two panels; second, there was nothing to suggest any loyalty on the part of the second panel to the first; third, the functions of the PCC, importantly, at the second disciplinary hearing are separate from those of the General Medical Council (GMC) as a whole; and fourth, the character of the issue—that is, whether the claimant knowingly told a lie in order to deceive the first PCC. The Court was not influenced one way or the other by the decision of the CPS not to prosecute N for perjury. That decision did not bind the GMC: see the decision of Rose J in *R v. General Medical Council, ex parte El Shanawany*, 2 December 1991 (unreported). So far as Article 6 ECHR is concerned, it adds nothing to the common law requirements of natural justice.

### *In re P (A Barrister)* [2005] 1 WLR 3019

**35.06** Visitors—lay member—member of Committee—recusal

P, a barrister was convicted of conduct that was dishonest or otherwise discreditable contrary to paras 301(a)(i) and 901 of the Code of Conduct of the Bar, and suspended from practice for a period of three months. Two days before the hearing by the Visitors of P's appeal, counsel for the appellant served submissions that raised objection to the lay member participating as a member of the panel of Visitors. The basic ground for the objection was that the lay member was a member of the Professional Conduct and Complaints Committee of the Bar Council, which was the body responsible under the Code of Conduct for deciding whether to prosecute a member of the Bar against whom a complaint had been raised. Accordingly, the lay member would be a judge in her own cause, although it was accepted that she had taken no part in the particular decision to prosecute the appellant. The Visitors held as a preliminary issue that the lay representative should be recused, applying *R v. Bow Street Metropolitan Stipendiary Magistrate, ex parte Pinochet Ugarte (No. 2)* [2000] 1 AC 119. In the instant case, amongst the functions of the Committee was the taking of decisions as to whether the facts of complaints presented to it by the Complaints Commissioner justified the prosecution of charges against members of the Bar. If the Committee concluded that there was a prima facie case of professional misconduct, its duty was to nominate a representative from its members to be responsible for the conduct of the proceedings on its behalf. Thus the decision of the Committee to institute proceedings against a barrister imposes upon the Committee as agent for the Bar Council a duty to prosecute that person and, consistently with the applicable procedure, to present the case against the barrister in a manner designed to procure conviction. The Visitors' conclusion was that, were the hearing to proceed before it with the lay member as a panel member, the Committee could not be said to be an independent tribunal for the purposes of Article 6(1) ECHR.

### *Holder v. Law Society* [2005] EWHC 2023 (Admin)

**35.07** SDT—Law Society—nature of Tribunal

H appealed against a decision and order of the Solicitors Disciplinary Tribunal (SDT) whereby, following dishonesty in utilizing clients' money for his own purposes, he was struck off the roll of solicitors. On appeal, it was argued that the SDT was not a sufficiently independent and impartial tribunal from the Law Society for the purposes of common law or Article 6 ECHR. The Divisional Court (Smith LJ and McCombe J) dismissed this argument. There was material to show that the members of the Tribunal are appointed by the Master of the Rolls. There is an open and transparent system of selection of members, after advertisement, sifting of written applications, and interview. The processes were administered by the Tribunal with an overseer to represent the Master of the Rolls. The solicitor members serve without remuneration. All receive reimbursement of their expenses. The Law Society has no hand in the appointments system and members of the Society's Council are not eligible to serve. Removal from office of any Tribunal member would be a matter for the Master of the Rolls. Taking into account all of the circumstances, the nature of the Tribunal is entirely adequately independent and impartial for the purposes for which it is constituted. The reasonable bystander, properly informed of the facts, could not consider otherwise.

### *Gillies v. Secretary of State for Work and Pensions* [2006] 1 WLR 781

**35.08** Doctor—medical reports—government agency

There was no basis for holding that the medical member of a disciplinary appeal tribunal was biased because she provided reports on behalf of the Benefits Agency in Disability Living Allowance and Incapacity Benefit cases as an examining medical practitioner. The critical issue was whether the fair-minded and informed observer would conclude, having considered the facts, that there was a real possibility that she would not evaluate reports by other doctors objectively and impartially against the other evidence. One of the strengths of the tribunal system as it has been developed is the breadth of relevant experience that can be built into it by the use of lay members to sit with members who are legally qualified.

### *R (Haase) v. District Judge Nuttall* [2009] QB 550

**35.09** Prisoner—prison disciplinary proceedings—adjudicator—prosecutor

The claimant was a serving prisoner, who challenged the determination made by an independent adjudicator exercising jurisdiction in disciplinary proceedings in prison. He contended that the proceedings before the independent adjudicator were unfair and in breach of his Convention rights under Article 6, because the prosecution lacked sufficient independence. The facts were that, in October 2004, the claimant was sentenced to fourteen years' imprisonment. In January 2006, a prison officer at HM Prison Full Sutton, acting under the Prison Governor's authority and in accordance with Prison Rules, required the claimant to provide a sample of urine for the purpose of testing for the presence of a controlled drug. He refused, and was charged with disobeying a lawful order. At the subsequent disciplinary proceedings, the prosecution of the claimant was essentially conducted by the reporting prison officer who had requested the original sample. The charge was found proved and the claimant was sentenced to serve twenty-one additional days. In the Administrative Court ([2007] EWHC 3079 (Admin)), Stanley Burnton J held, when dismissing the application by the claimant for judicial review, that prison disciplinary proceedings chaired by an independent adjudicator in which the prosecution case was presented by a prison officer who might also be a witness were not incompatible with a right to a fair trial guaranteed by Article 6 ECHR. His Lordship said that there was nothing in the Strasbourg authorities to indicate that the prosecutor was required to be independent. The Court of Appeal affirmed the decision of Stanley Burnton J and dismissed the claimant's appeal. Richards LJ (with whom Sir Anthony Clarke MR and Scott Baker LJ agreed) said:

> 18. I am not persuaded that article 6(1) imposes any general requirement as to the independence and impartiality of the prosecutor. If, on the particular facts, a lack of independence or impartiality on the part of the prosecutor has resulted in unfairness to the defendant, then of course there may be a violation of article 6(1); but a lack of prosecutorial independence or impartiality does not in itself render the proceedings unfair or give rise to an automatic violation. Thus I would reject [counsel for the appellant's] contention that the system of prison disciplinary hearings as currently operated is institutionally incompatible with article 6(1).
>
> 19. It is highly significant that the express requirement of independence and impartiality laid down by article 6(1) relates only to the *tribunal*, i.e. to the decision-makers. As

> the court said in *R v. Stow* [2005] UKHRR 754, "it is the independence and impartiality of those involved in the decision-making process which is fundamental to a fair trial": para 31 of the judgment. Unless the tribunal is independent and impartial there can be no assurance of fairness. But if the tribunal is independent and impartial, then it seems to me that fairness can in principle be achieved without imposing an additional general requirement as to the independence and impartiality of the prosecutor. There is insufficient justification for implying such an additional requirement into article 6(1).

The Court of Appeal went on to say that the decision in *R v. Stow* should be treated with caution. The judgment did not lay down a principle of general application as to the independence and impartiality of prosecutors. The Court's focus was on the specific context of a court martial, and what was said about prosecutorial independence and impartiality should not be treated as governing cases arising in other contexts.

### *Sadighi v. General Dental Council* [2009] EWHC 1278 (Admin)

**35.10** PCC—members appointed by GDC

S was suspended from practice by the GDC for six months. The PCC found proved that S, together with his sister, rewrote part of the dental records of a patient who made a complaint against his dental practice and presented the partially rewritten records to the GDC as though they were original records made in respect of the patient. S's appeal on the grounds that the members of the PCC were appointed by the Council and the chairman was a former elected member of the Council was dismissed by Plender J. The chairman had ceased to be an elected member of the Council no less than five years before the institution of the proceedings. At the opening of the hearing, the chairman asked if there were any objections; none were made. The members of the PCC were appointed by an appointments committee of the Council. Plender J said that he very much doubted that the imposition of the appointments committee was insufficient to resolve any doubt that an objective bystander might have in the impartiality of the PCC. Professional self-regulation is not prohibited by the European Convention. To say that a body lacks impartiality when its members are appointed by an appointments committee set up in accordance with a statutory instrument, because the appointments committee is itself based on individuals nominated by the professional body, is very close to advancing the proposition that professional self-regulation is no longer permitted.

### *Virdi v. Law Society (Solicitors Disciplinary Tribunal intervening)* [2010] 1 WLR 2840

**35.11** Tribunal clerk—retirement with Tribunal—drafting written record

In this case, the SDT found five allegations relating to money laundering and dishonest transactions proved against the appellant solicitor. The Tribunal gave an extempore judgment finding the appellant grossly reckless and ordering suspension from practice. Its written findings were not delivered until almost one year after conclusion of the hearing. It also emerged that the Tribunal's clerk had assisted substantially in drafting the reasons and the appeal raised issues as to the lawfulness of the part played by the clerk to the Tribunal. The Divisional Court (Scott Baker LJ and David Clarke J) considered that the delay in delivering the Tribunal's written reasons was both inordinate and inexcusable. However, the Court was not persuaded that the delay had caused any injustice to the appellant, because he knew the decision of the Tribunal when it was announced and the basic reasons for it, and his suspension still had another twenty months to run. The appellant's main ground of appeal related to the Tribunal clerk. The Court heard that it was customary for the clerk to retire with the Tribunal, to hear its discussions and decisions, and to take a note. It was also customary for the clerk to have the initial

responsibility for producing the written record, because he or she was the best person to ensure that the record captured the Tribunal's decisions and reasons accurately, and that nothing had been overlooked. In dismissing the appeal, both the Divisional Court and the Court of Appeal (Jacob, Lloyd, and Stanley Burnton LJJ) held that the important basic fact was that the Tribunal gave its decision orally and outlined its reasons for it, and that, on the evidence, the clerk took no part in the decision-making process. The order was drawn up immediately following the hearing and the appellant's suspension began to run at that point. Thereafter the Tribunal was *functus officio*.
Stanley Burnton LJ said:

> 33. In my judgment, the procedure of the Tribunal included their withdrawing to consider their decision in private with their clerk and her role in this case. (Counsel for the appellant) submitted that the procedure of the Tribunal within the meaning of rule 31(a) is confined to the trial process. There is no basis for so limiting the rule. The procedure of the Tribunal did not come to an end when they retired to consider their decision. As was held in *Baxendale-Walker* [2006] EWHC 643, once they had announced their decisions, both on whether the appellant had been guilty of serious professional misconduct and on sanction, they were functus officio in that they could not reconsider or change those decisions; but they retained the power and the duty to provide adequate written findings. The provision of formal written findings is as much part of the procedure of the Tribunal as the trial process and the announcement of their decisions. But if I am wrong about this, I have no doubt that the Tribunal had implied power, if power was required, to permit or to invite their clerk to retire with them and to assist them in the manner she did in this case.
>
> 34. The assistance of the clerk in drafting the formal written Findings of the Tribunal occurred and occurs after the decision of the Tribunal has been given orally and its formal order filed with the Law Society. At that point the decision is effective, and the Tribunal has no power to reconsider it: *Baxendale-Walker* at paragraphs 23 to 28. It follows that what occurs subsequently cannot in general give rise to a ground of appeal against the decision.

The Court distinguished two Hong Kong Court of Appeal cases—*Au Wing Lun v. Solicitors Disciplinary Tribunal* CACV 4154/2001 and *A&B Solicitors v. Law Society of Hong Kong* CACV 269/2004—in which the clerk had retired with the Tribunal and drafted its findings before it had given its decision and made its order. In those cases, the Court considered that there was a grave suspicion that justice had not been done, in that it was unclear whether the reasons for the decision of the Tribunal were, in fact, its reasons rather than those of the clerk. In *Virdi*, Stanley Burnton LJ said, at [39], that the facts of those cases differed from those of the present case, in which it was conceded that, on the basis of the facts as now known, nothing untoward occurred. The findings of the Tribunal in *Virdi* were 'clearly their Findings', and 'in these circumstances, it is inappropriate to consider issues that could conceivably arise in other cases, particularly since the rules of the Tribunal now make express and specific provision on the role of the clerk'.

### *R (Kaur) v. Institute of Legal Executives Appeal Tribunal* [2011] EWCA Civ 1168

**35.12** Tribunal and appeal panels—members of Council—whether judge in own cause

K was a student member of the Institute of Legal Executives (ILEX). In May 2007, she and other student members sat certain law and practice examinations. Following investigations and a decision by an ILEX disciplinary committee, K and five other student members were charged with various disciplinary offences including cheating in two examinations. K and four others were excluded from ILEX following a hearing before the disciplinary tribunal. K appealed to ILEX's appeal tribunal and raised as a preliminary matter an objection to the presence on the panel of ILEX's vice-president.

Judicial review was refused by Foskett J. K appealed to the Court of Appeal, and the short issue was whether the presence of an ILEX council member and director of ILEX on the disciplinary tribunal, and of the council's vice-president on the appeal tribunal, were in breach of the doctrines that no one may be a judge in his own cause and/or of apparent bias, requiring those decisions to be quashed. Rix LJ (with whom Sullivan and Black LJJ agreed)—after reviewing the leading cases on apparent bias of *R v. Bow Street Metropolitan Stipendiary Magistrate, ex parte Pinochet Ugarte (No. 2)* [2000] 1 AC 119, *Dimes v. Proprietors of Grand Junction Canal* (1852) 3 HL 759, *Porter v. Magill* [2001] UKHL 67, in which the modern law of apparent bias was definitively stated in the speech of Lord Hope of Craighead, building on *R v. Gough* [1993] AC 646 and *In re Medicaments and Related Classes of Goods (No. 2)* [2001] 1 WLR 700, and the observations of Lord Bingham of Cornhill in *Davidson v. Scottish Ministers* [2004] UKHL 34—observed that the cases are concerned with ensuring that governing members of an organization are barred from involvement in the investigation or decision-making process and ensuring the proper separation of disciplinary panels from those concerned with the overall governance of the organization. The appeal was allowed.

***Holmes v. Royal College of Veterinary Surgeons* [2011] UKPC 48**

**35.13** Disciplinary procedures—fairness—Veterinary Surgeons Act 1966

H contended that features of the general system operated by the Royal College of Veterinary Surgeons (RCVS) for the determination of disciplinary complaints and features of its operation specific to the instant proceedings represented deficiencies that combined to give rise to an appearance of bias against him on the part of the Disciplinary Committee of the RCVS. The Privy Council, in a judgment delivered by Lord Wilson, said that the Board was satisfied that the RCVS had made strenuous attempts to ensure that its disciplinary procedures were fair and, since the coming into force of the Human Rights Act 1998, were in accordance with its obligations to the registrant under Article 6 ECHR. In particular, the College had made elaborate efforts to separate what one might regard, however loosely, as the parts of its system that contribute to the prosecution of the charge against a registrant from the parts that determine it. However, the College had found itself hamstrung by the Veterinary Surgeons Act 1966. The preference of the RCVS, publicly announced, is that its members should not be drawn from the Council and it had lobbied the government, so far in vain, to support an amendment of the Act so as to preclude members of the Council from being members of either of the investigation or disciplinary committees. Lord Wilson observed that the features complained of in *Tehrani v. UK Central Council for Nursing, Midwifery and Health Visiting* 2001 SC 581 and *Preiss v. General Dental Council* [2001] 1 WLR 1926 were absent from the present case. The question was whether a fair-minded and informed observer, having considered the facts, would conclude that there was a real possibility that the Committee in the instant case was biased: *Porter v. Magill* [2002] 2 AC 357; *Helow v. Secretary of State for the Home Department* [2009] SC (HL) 1. The Privy Council concluded that the fair-minded and informed observer would not conclude in the instant case that there was a real possibility that the College's Disciplinary Committee was biased against H.

***Kulkarni v. Milton Keynes Hospital NHS Foundation Trust* [2010] ICR 101**

**35.14** See paragraph 45.20.

***Hameed v. Central Manchester University Hospitals NHS Foundation Trust* [2010] EWHC 2009 (QB)**

**35.15** See paragraph 32.23.

***R (Puri) v. Bradford Teaching Hospitals NHS Foundation Trust*** **[2011] EWHC 970 (Admin)**

**35.16** See paragraph 32.24.

***R (G) v. Governors of X School (Secretary of State for the Home Department and anor intervening)*** **[2012] 1 AC 167**

**35.17** See paragraph 45.24.

***Mattu v. University Hospitals of Coventry and Warwickshire NHS Trust*** **[2012] 4 All ER 359, CA**

**35.18** See paragraph 32.28.

***R (King) v. Secretary of State for Justice; R (Bourgass and Hussain) v. Secretary of State for Justice*** **[2012] 4 All ER 44, CA**

**35.19** Prison discipline—governor

In April 2009, K was serving a seven-year sentence in a young offender institute (YOI) for causing death by dangerous driving. He was detained at HM Prison Portland. Between 11 and 13 April, he was subjected to confinement to cell following a disciplinary charge of failing to comply with a lawful order. B was serving a sentence of life imprisonment for the murder of a police officer, and seventeen years' imprisonment for conspiracy to commit public nuisance by the use of poisons and/or explosives. In March 2010, whilst detained at HM Prison Whitemoor, he was subjected to segregation for reasons of good order and discipline pursuant to rule 45 of the Prison Rules 1999. He was segregated from 10 March until 22 April and again from 23 April until October or November of that year. H was detained at HM Prison Frankland and was serving a long sentence for terrorism-related offences. On 24 April 2010, he was found to have carried out a serious attack on another prisoner. He was subjected to segregation until October 2010. Each appellant claimed that the decisions to place and/or keep them in confinement to cell or segregation were unlawful, principally by reference to Article 6 ECHR, on the grounds that the adjudication of confinement to cell or segregation by the governor was not taken by an 'independent and impartial tribunal'. K's application for judicial review was dismissed by Pitchford LJ and Maddison J on 13 October 2010: [2010] EWHC 2522 (Admin), [2011] 3 All ER 776, [2011] 1 WLR 2667. The applications on behalf of B and H were dismissed by Irwin J on 18 February 2011: [2011] EWHC 286 (Admin). The claimants appealed to the Court of Appeal. In dismissing their appeals, Maurice Kay LJ, with whom Lloyd LJ agreed, said that:

> 44. ... Prison or YOI Governors have the responsibility of maintaining good order and discipline in a complex and potentially combustible setting. They have to make urgent decisions about such matters as segregation based on their experience, expertise and judgment. They do so not just in a binary mode as between themselves and an individual prisoner. They are acting in the interests of the security of the institution as a whole. Sometimes they may have to make a decision which has an immediate restricting effect on the whole or a large part of the institution—for example, the immediate "lockdown" of an entire wing on receipt of apparently credible information about a planned breakout. Such urgent matters are not susceptible to a judicialisation of the decision-making process. In the present cases, it would be quite unrealistic to require the initial decision to segregate to be taken by "an independent and impartial tribunal established by law"—presumably an independent adjudicator... The need for action is often immediate. The cases of [B] and [H] are paradigm examples. It is true that when the [Segregation Review Board] SRB stage is reached, whilst initially the timeline is short, the urgency is less extreme. However, it seems to me that the review is

> one best entrusted to those with the necessary experience and expertise as an exercise of collective professional discretion, with built-in safeguards, albeit falling short of Article 6 standards....

Maurice Kay and Lloyd LJJ considered that Article 6 was not engaged at the stages of the governor's decision or the SRB decision. Elias LJ, in agreeing with the result, said at [94] that the prison context is a very special one and that he would be very reluctant to require key areas of prison discipline to be subjected to external determination unless compelled by authority to do so.

> 95. ...First, as a matter of policy, there are powerful reasons for permitting matters of prison discipline to be handled internally. The prison governor is particularly well placed to fix the appropriate sanction by reference to the particular individual and the needs and problems of the prison itself....Second, whilst I see force in the concern about the perception of bias, there is in general no reason why prison governors ought not in fact to be able to make fair and independent assessments, notwithstanding that they are not institutionally independent of the prison....Third, the internal procedures include the essence of a fair procedure, and the initial decision is reviewed within 72 hours and periodically thereafter by the SRB which includes a wide range of individuals. Fourth, it would greatly increase the cost and time for these decisions to be given to judges; and it would be particularly unsatisfactory where, as is almost always the case, a speedy decision is required. Fifth, it is in my view highly relevant that in all of the prison cases where it has been held that civil rights are engaged, nobody appears to have suggested that there is anything inappropriate about the initial decision being taken internally. The concern has been with the alleged lack of any effective review.

***Ighalo v. Solicitors Regulation Authority* [2013] EWHC 661 (Admin)**

**35.20** SDT—Tribunal member—previously member of SRA panel

The appellant, a solicitor, challenged the decision of the SDT on the grounds that the composition of the Tribunal was such that it could not be said to have been independent or impartial. H, a solicitor member of the Tribunal, had previously held an appointment with the Solicitors Regulation Authority (SRA) as a member of its Adjudication Panel. H's appointment as an adjudicator ended more than two years before the hearing before the SDT. The Divisional Court (Laws LJ and Swift J) dismissed the appellant's appeal. After reviewing *Porter v. McGill* [2002] 2 AC 357, *Davidson v. Scottish Ministers (No. 2)* [2004] UKHL 34, *Locabail (UK) Ltd v. Bayfield Properties Ltd* [2000] QB 451, and *Piersack v. Belgium* [1983] 5 EHRR 169, Swift J, in giving the judgment of the Court, said, at [41], that H's previous appointment could not of itself give rise to a reasonable apprehension of bias. There are, of course, good reasons why an individual should not hold an appointment with the SRA at the same time as sitting on tribunals hearing disciplinary cases, since that would involve simultaneous involvement in both the investigatory/prosecutorial and decision-making arms of the disciplinary process. In this case, however, there was no question of H having acted as an adjudicator for the SRA at the same time as sitting as a member of a tribunal at a disciplinary hearing. At the time of the appellant's case, H had not carried out any work for the SRA for more than two years. Further, it is clear that there is no requirement for an individual sitting in a judicial capacity to disclose every previous activity or association that he or she may have had, whether or not the activity or association is capable of forming the basis for a reasonable apprehension of bias. The duty extends only to activities or associations that would or might provide the basis for such a reasonable apprehension: at [43].

***Adu v. General Medical Council* [2014] EWHC 1946 (Admin)**

**35.21**

See paragraph 6.15.

## C. Other Relevant Chapters

**35.22** Bias, Chapter 6
Constitution of Panel, Chapter 13
Human Rights, Chapter 32
In Camera Discussions, Chapter 34
Investigation of Allegations, Chapter 39
Legal Representation, Chapter 45
Natural Justice, Chapter 48

# 36

# INDICATIVE SANCTIONS GUIDANCE

A. **Legal Framework** 36.01

B. **Disciplinary Cases** 36.02

*R (Abrahaem) v. General Medical Council* [2004] EWHC 279 (Admin) 36.02

*Council for the Regulation of Health Care Professionals v. General Medical Council and Leeper* [2004] EWHC 1850 (Admin) 36.03

*Razak v. General Medical Council* [2004] EWHC 205 (Admin) 36.04

*Council for the Regulation of Health Care Professionals v. General Medical Council and Solanke* [2004] 1 WLR 2432 36.05

*Bevan v. General Medical Council* [2005] EWHC 174 (Admin) 36.06

*Sanders v. Kingston (Ethical Standards Officer)* [2005] EWHC 2132 (Admin) 36.07

*Abrahaem v. General Medical Council* [2008] EWHC 183 (Admin) 36.08

*Atkinson v. General Medical Council* [2009] EWHC 3636 (Admin) 36.09

C. **Other Relevant Chapters** 36.10

## A. Legal Framework

**36.01** Most regulators have developed indicative sanctions guidance for fitness to practise panel members or disciplinary committee members in the interests of openness and transparency, and in line with the regulator's duty to uphold public confidence in the profession.

Examples in the healthcare professions include:

- the General Medical Council;
- the Nursing and Midwifery Council;
- the General Pharmaceutical Council;
- the General Optical Council;
- the General Dental Council; and
- the General Chiropractic Council.

Other examples include:

- the Chartered Institute of Management Accountants;
- the Association of Chartered Certified Accountants; and
- the Council of the Inns of Court.

Matters usually covered include: the regulator's statutory powers; the sanctions available; the threefold purpose of sanctions—namely, the protection of the public, the maintenance of public confidence in the profession, and the maintenance of proper standards of conduct and behaviour; and general principles of fairness and proportionality.

## B. Disciplinary Cases

***R (Abrahaem) v. General Medical Council* [2004] EWHC 279 (Admin)**

**36.02** Indicative Sanctions Guidance—accurate description of purpose of sanctions

Newman J, in allowing the doctor's appeal against sanction, said that the approach of the Professional Conduct Committee (PCC) in its most material part is set out in the *Indicative Sanctions Guidance for the Professional Conduct Committee* produced by the General Medical Council (GMC) dated July 2003. With great clarity and accuracy, the purpose of the PCC sanctions is described as follows:

> 9. The purpose of the sanctions is not to be punitive, but to protect the public interest, although they may have a punitive effect.
>
> *The Public Interest*
>
> 10. There is clear judicial authority that the public interest includes:
>    a. The protection of members of the public.
>    b. The maintenance of public confidence and the profession.
>    c. Declaring and upholding proper standards of conduct.
>
> 11. The public interest may also include the doctor's return to work if he or she possesses certain skills, competencies or knowledge, for example expertise in a particular area, or language skills.

As to what the guidance says in relation to erasure and immediate suspension, Newman J said, at [36]:

> Those are very useful guidelines and they form a framework which enables any tribunal, including this court, to focus its attention on the relevant issues. But one has to come back to the essential exercise which the law now requires in what lies behind the purpose of sanctions, which, as I have already pointed out, is not to be punitive but to protect the public interest; public interest is a label which gives rise to separate areas of consideration.

***Council for the Regulation of Health Care Professionals v. General Medical Council and Leeper* [2004] EWHC 1850 (Admin)**

**36.03** Indicative Sanctions Guidance—consistent approach to imposition of penalties

Collins J said, at [24]:

> The GMC's Indicative Sanctions Guidance for the Professional Conduct Committee is the equivalent to a sentencing guide. It helps to achieve a consistent approach to the imposition of penalties where serious professional misconduct is established. The PCC must have regard to it although obviously each case will depend on its own facts and guidance is what it says and must not be regarded as laying down a rigid tariff. It points out that the purpose of the sanctions is not to be punitive, but to protect the public interest, which includes the maintenance of public confidence in its profession and upholding proper standards of conduct.

***Razak v. General Medical Council* [2004] EWHC 205 (Admin)**

**36.04** Penalty in line with Indicative Sanctions Guidance

On 30 July 2003, the PCC of the GMC directed that R's name be erased from the medical register. R was a hospital practitioner in psychiatry and the allegation against him, which the Committee found proved, was that whilst treating a female patient for her psychiatric condition, he had had an improper emotional relationship with her. In dismissing R's appeal, Stanley Burnton J said, at [44], that the penalty was in line with the Indicative Sanctions Guidance for the PCC published by the GMC.

### *Council for the Regulation of Health Care Professionals v. General Medical Council and Solanke* [2004] 1 WLR 2432

**36.05** Indicative Sanctions Guidance indicative only—not legally binding on committee—guidance provides consistency in the interests of the public, doctors, and GMC

In dismissing an appeal by the Council for the Regulation of Health Care Professionals under section 29 of the National Health Service Reform and Health Care Professions Act 2002 on the grounds of undue leniency, Leveson J observed that the GMC's Indicative Sanctions Guidance provides specifically at paragraph 9: 'The purpose of the sanctions is not to be punitive, but to protect the public interest, although they may have a punitive effect.' The learned judge continued:

> 29. . . . I entirely agree with [counsel for the GMC] that neither the committee nor the court is bound to reach its decision within the framework of the guidance. It is clearly and so stated to be indicative only; it is not legally binding on the committee, let alone the council or the court. Having said that, I further agree with [counsel for the GMC] that, together with the body of relevant case law, the guidance assists the committee to reach consistent decisions while at the same time taking account of the particular circumstances of each case. Such consistency is in the interests of the public, doctors and the GMC alike. In that regard I also endorse the observations of Newman J in *R (Abrahaem) v. General Medical Council* [2004] EWHC 279 (Admin) that the description of the purpose of sanctions and the public interest reveals "great clarity and accuracy" (para 7) and that the section of the guidance dealing with suspension, with which I am also concerned in this case, consists of "very useful guidelines" which form "a framework which enables any tribunal, including [the] court, to focus its attention on the relevant issues": see para 36.
>
> 30. Neither do I consider that it is the function of the court to intervene in the formulation of modifications to the guidance although its decisions will doubtless inform the GMC in relation to areas that may require reconsideration. The council, on the other hand, is in an admirable position to take part in the process of revising this "living instrument". Section 26(2)(c) of the 2002 Act specifically permits the council to recommend to a regulatory body changes to the way in which it performs any of its functions. Its regulatory experience, across the range of health care professionals, will doubtless be of great value in the process of reconsidering guidelines to ensure that the right decisions are taken in relation to misconduct of differing gravity in a constant fashion both in different cases within the medical profession and across different health care professionals.

### *Bevan v. General Medical Council* [2005] EWHC 174 (Admin)

**36.06** Use of Indicative Sanctions Guidance by legal assessor

Collins J said, at [40]:

> Despite [counsel for the practitioner]'s somewhat muted concerns that the guidance is not an appropriate document to be relied upon before the Committee because it is produced and drafted by the GMC, I remain of the view, which I expressed in *Leeper's* case, and which has been confirmed by others of my brethren, that they are indeed most helpful and I note that the legal assessor in this case specifically referred to them and relied upon them in the assistance that he gave to the Committee. It does not seem to me to matter who drafted them, provided their contents are sensible and helpful, and quite clearly their contents are indeed, in my judgment, sensible and helpful.

### *Sanders v. Kingston (Ethical Standards Officer)* [2005] EWHC 2132 (Admin)

**36.07** No reference to guidance in decision on sanction—guidance appropriate starting point—failure to engage with guidance was an error of principle

S appealed under section 79(15) of the Local Government Act 2000 against a decision of a tribunal of the Adjudication Panel for England disqualifying him from being a local councillor for eighteen months. S was a long-standing member of a council and its leader. The tribunal found him guilty of intimidation and bullying officers, and of bringing both his office and his authority into disrepute. In allowing S's appeal against penalty and substituting a shorter period of suspension of six months, Sullivan J noted that there appeared to have been no reference in the tribunal's decision on penalty to the guidance to tribunals issued by the president of the Adjudication Panel for England under section 75(9)(b) of the 2000 Act. The guidance is issued on action to be taken by a case tribunal where a respondent has been found to have failed to comply with a code of conduct. Sullivan J said, at [27], that he appreciated that the guidance is just that—guidance, not a list of mandatory requirements—but it will almost invariably be the appropriate starting point from which a departure may, or may not, be justified to a greater or lesser extent, depending on the facts of the particular case. Making express reference to the guidance will ensure that, at the very least, the parties and the tribunal consider whether the guidance is of any assistance in the circumstances of the particular case, and if not, why not. The new practice of making express reference to the guidance is in line with the approach towards the duties of prosecuting counsel in a criminal trial: see *Attorney-General's Reference No. 52 of 2003 (Ian David Webb)* [2004] Crim Lr 306, 307. Reminding the tribunal of the sanctions available, and any guidance from the president as to their imposition, is not to be equated with arguing for a particular penalty, which would be inappropriate. Sullivan J continued:

> 30. I do not suggest that a failure to refer to the Guidance will necessarily mean that there has been an error of principle. In *Sloam v. Standards Board for England* [2005] EWHC 124 (Admin), Bennett J was prepared to assume that the Tribunal, as a specialist tribunal, would have had the Guidance in mind even though it was not specifically referred to in the decision (see para 16). That assumption will be justified in those cases where the Appellant's conduct and the Tribunal's response by way of penalty, fall clearly within the Guidelines. In such cases it would be most unfortunate if reference to the Guidelines in the Tribunal's reasons was required as a kind of mantra. However, in cases such as the present appeal, where there is (to put it at its lowest) scope for debate as to where the Appellant's conduct should be placed on the scale of seriousness, and what the appropriate response should be by way of penalty, a brief explanation of whether, and if so how, the Guidelines have been applied will be necessary, if only to ensure that the Tribunal's reasoning is adequate. On the particular facts of this case, the Tribunal's failure to engage with the Guidance in its decision was an error of principle. If it had engaged with the Guidance, it is difficult to see how it could reasonably have concluded that the appellant's conduct could be equated with the kinds of conduct described in paragraph 6 of the Guidance.

### *Abrahaem v. General Medical Council* [2008] EWHC 183 (Admin)

**36.08** Indicative Sanctions Guidance—relevance at review hearings

In January 2004, Newman J concluded that the penalty of erasure in this case was disproportionately severe, and the learned judge substituted a sanction of one year's suspension: see [2004] EWHC 279 (Admin). In January 2005, the twelve-month period of suspension came to an end, and a fitness to practise panel had to review the latter and decide what course to adopt. Blake J said that, in his judgment, consistent with the statutory scheme contained in the Medical Act 1983, as amended, and the 2004 Rules, the Indicative Sanctions Guidance is intended to apply at reviews and will therefore have relevance as to the panel's finding whether fitness to practise remains impaired following a period of suspension.

***Atkinson v. General Medical Council*** **[2009] EWHC 3636 (Admin)**

**36.09** Indicative Sanctions Guidance is just that: guidance—need for panel to take into consideration—departure from guidance may need explanation

In dismissing A's appeal against the sanction of erasure, Blake J said:

> 11. All those factors were before the panel and taken into consideration when it gave its reasons for its decision for the sanction of erasure. Understandably, the panel indicated that in reaching its decision it took account of the GMC's Indicative Sanctions Guidance (April 2009) and bore in mind that any sanction must be proportionate and that its purpose was not to be punitive, although it may have a punitive effect. The panel was conscious that it has to balance the appellant's interests with the wider public, including not only the protection of patients but the maintenance of public confidence in their profession and the keeping and upholding of proper standards of conduct and behaviour. None of that is legally controversial and is indeed an appropriate self-direction on the task to be achieved in the question of sanctions.
>
> 12. The Indicative Sanctions Guidance is just that: guidance. It is something which every panel must take into consideration, and departures from it may need some explanation, but it is not the source of legal obligation.

## C. Other Relevant Chapters

**36.10**

Review Hearings, Chapter 59
Sanction, Chapter 60

# 37

# INSIGHT

A. Legal Framework 37.01
B. Disciplinary Cases 37.02
*R (Abrahaem) v. General Medical Council* [2004] EWHC 279 (Admin) 37.02
*R (Adesina) v. Nursing and Midwifery Council* [2004] EWHC 2410 (Admin) 37.03
*R (Bevan) v. General Medical Council* [2005] EWHC 174 (Admin) 37.04
*R (Onwuelo) v. General Medical Council* [2006] EWHC 2739 (Admin) 37.05
*R (Lloyd Subner) v. Health Professions Council* [2009] EWHC 2815 (Admin) 37.06
*Mubarak v. General Medical Council* [2008] EWHC 2830 (Admin) 37.07
*Balamoody v. Nursing and Midwifery Council* [2009] EWHC 3235 (Admin) 37.08
*Odes v. General Medical Council* [2010] EWHC 552 (Admin) 37.09
*R (Sharma) v. General Dental Council* [2010] EWHC 3184 (Admin) 37.10
*Karwal v. General Medical Council* [2011] EWHC 826 (Admin) 37.11
*Shah v. General Dental Council* [2011] EWHC 3003 (Admin) 37.12
*Ige v. Nursing and Midwifery Council* [2011] EWHC 3721 (Admin) 37.13
*Ballesteros v. Nursing and Midwifery Council* [2011] EWHC 1289 (Admin) 37.14
*Henderson v. General Teaching Council for England* [2012] EWHC 1505 (Admin) 37.15
*Haye v. General Teaching Council for England* (2013) 157 Sol Jo 35 37.16
*Amao v. Nursing and Midwifery Council* [2014] EWHC 147 (Admin) 37.17
*Brew v. General Medical Council* [2014] EWHC 2927 (Admin) 37.18
C. Other Relevant Chapters 37.19

## A. Legal Framework

**37.01** Insight is frequently identified as a factor for consideration by fitness to practise panels or disciplinary committees when considering impairment and sanction. For example, see:

General Medical Council, *Indicative Sanctions Guidance for the Fitness to Practise Panel* (April 2009, with revisions), paragraph 34 (Insight—that is, the expectation that a doctor will be able to stand back and accept that, with hindsight, he or she should have behaved differently, and that it is expected that he or she will take steps to prevent a reoccurrence—is an important factor in a hearing; when assessing whether a doctor has insight, the panel will need to take into account whether he or she has demonstrated insight consistently throughout the hearing, for example has not given any untruthful evidence to the panel or falsified documents; the panel should be aware of cultural differences in the way in which insight is expressed)

## B. Disciplinary Cases

### *R (Abrahaem) v. General Medical Council* [2004] EWHC 279 (Admin)

**37.02** Penalty—lack of insight—proportionality

In allowing A's appeal and substituting a sentence of twelve months' suspension in place of a direction for erasure, Newman J said:

> 39. The problem that [the] doctor poses is that because of what might be described as his less than penitential approach to the gravity of what he did, one has a lingering consideration that his insight is less than that which it should be to qualify for consideration within the frame of a lesser penalty. As to that, in my judgment, I have to ask: how far does the ultimate sanction of erasure remain in play when the real position is that what is at risk is that the individual has just not got sufficient insight into the gravity of that which he did. It seems to me that one is entering into areas of excessive and disproportionate penalty if what one does is, in effect, assume that this doctor is quite incapable of having sufficient insight into the gravity of that which he did so as to lead to the risk that he would commit an offence such as this again. In my judgment, it is asking or inferring too much, from what I regard as his less than satisfactory evidence to the court in his statement, to conclude that he will not develop from this experience, and from what I would regard as an appropriate continuing penalty proportionate to the situation, sufficient and overall insight. If, as may be the case, he has not now learnt, in the period of time which he has had since these troubles broke out, that he offended a core aspect of professional conduct in this country as a doctor, then he should have some further time to think about it.

### *R (Adesina) v. Nursing and Midwifery Council* [2004] EWHC 2410 (Admin)

**37.03** Sanction—protection of public—patient confidentiality

A, a registered midwife, appealed against an order that her name be erased from the register. The finding against her was that she had accessed patient notes for a non-clinical purpose, that she had made inappropriate contact with a former patient in circumstances distressing to that patient, and that she had spoken to that patient inappropriately. In dismissing A's appeal, Munby J said:

> 8. It is highly relevant, as it seems to me, in a case such as this, to bear in mind the character of the particular profession concerned and the nature of the complaint which has been found established. The appellant is a member of the midwifery profession. Fundamental to that profession, as to the nursing profession and the medical profession, are the principle that the interests of the patient come first and principles of patient confidentiality. Those were the very principles which, in this particular case, formed the basis of the findings of the Committee.
>
> 9. The appellant, as I have said, was found to have accessed patient notes for a non-clinical, that is to say for a non-professional, purpose, and to have made inappropriate contact with and spoken inappropriately to a former patient. It seems to me highly relevant where the findings were of that nature, for the Committee to assess the extent to which the appellant did, or as they found did not, recognise the seriousness of what she had done.
>
> 10. Equally it was, it seems to me, highly relevant for the Committee to assess the extent to which the appellant did, or did not, as in the event they found, have insight into what had happened. The Committee was not, as I read their decision, applying any general principle that a denial of itself justified any particular penalty. What the Committee was saying was that in this particular case, bearing in mind the nature of the complaints which had been found established against the appellant, it was highly relevant, in their view, when they came to consider the

protection of the public, to assess such matters as insight, remorse, recognition of the seriousness of what the appellant had done, and the other matters to which they made reference.

11. There was, in my judgment, no error of law and no error of principle or approach in the course which the Committee adopted. They were entitled to take into account, in my judgment, each of the matters to which they drew attention in their reasons. They were entitled, in my judgment, to treat each of these matters as being, in the particular circumstances of this case, aggravating features, which made this, as they recognised, one-off matter, as they put it, "more serious" than it might otherwise have been.

12. ...Counsel...has helpfully drawn my attention to the judgment of Richards J in the case of *R (Clarke) v. United Kingdom Central Council for Nursing, Midwifery and Health Visiting* [2004] EWHC 1350 (Admin) paragraph 35. There complaint was made, in a very similar type of case, of the fact that the Committee had, in part, justified the penalty imposed because the appellant had not, in that case, demonstrated that he was aware of the seriousness of his actions.

13. Commenting on that complaint, Richards J said: "In my judgment, however, it was highly relevant for the Committee to consider whether the appellant had shown insight into the misconduct found proved against him, and the assessment made was plainly one reasonably open to the Committee on the evidence before it." That, in my judgment, applies as much to the present case as to the case with which Richards J was concerned.

### *R (Bevan) v. General Medical Council* [2005] EWHC 174 (Admin)

**37.04** Recognition of wrongdoing—likelihood of reoccurrence

The Court commented upon a paragraph in the Indicative Sanctions Guidance of the General Medical Council (GMC) that reads:

> 34. This 'insight'—the expectation that a doctor will be able to stand back and accept that, with hindsight, they should have behaved differently, and that it is expected that he/she will take steps to prevent reoccurrence—is an important factor in a hearing.

At [39], Collins J said:

> That paragraph makes clear what in my view is in fact implicit in dealing with questions of insight: that insight is most material to ensure that the doctor has realised that he has indeed gone wrong and therefore will not do anything similar in the future. That is the purpose behind a need to recognise insight. Insight does not seem to me to be really an appropriate way of looking at a situation where there is no danger of any recurrence but there is a concern that there has not been necessarily a full acceptance of the facts which have been alleged against the doctor.

### *R (Onwuelo) v. General Medical Council* [2006] EWHC 2739 (Admin)

**37.05** Dishonesty—no risk of recurrence—dedicated and competent doctor—effect of insight on penalty

In November 2005, O was found guilty of serious professional misconduct and his name was ordered to be erased from the medical register. The charges before the panel arose out of his failure to disclose a conviction for theft in May 2000 when applying for a position with the Surrey and Sussex Healthcare National Health Service (NHS) Trust in January 2001, and his subsequent failures to consent to a Criminal Records Bureau (CRB) check or to disclose the conviction on one of a number of occasions. O's case before the fitness to practise panel, in summary, was that he had made a mistake in forms he had originally filled out for the Trust. The panel, having seen and heard evidence over three days, decided, in key respects, that it did not believe him. Walker J said that he had no difficulty in accepting both propositions that it was not an aggravating factor with regard to sentence that the doctor had offered at the hearing an account that was found to be untrue and that an unsuccessful plea of not guilty should not be a predominant factor when sentencing. The panel was not guilty of any such error of law. However, Walker J said that the panel's concern was that the way in which O had presented his case demonstrated a profound lack of insight and evidenced a deep-seated attitudinal problem. The question, which the judge said troubled him, is whether the degree of lack of insight, or the other matters that concern the panel in this regard, could reasonably justify the draconian sentence of erasure. The observation of Collins J in *Bevan v. General Medical Council* (above)—'Insight does not seem to me to be really an appropriate way of looking at a situation where there is no danger of any recurrence but there is a concern that there has not been necessarily a full acceptance of the facts which have been alleged against the doctor'—was directly in point in the present case. It was difficult to see that there was any risk of any recurrence. The panel had not found a general or systematic dishonesty. It had been concerned with one discrete event. The panel acknowledged that, at the stage at which this event came to light, O was experiencing a distressing time in his personal life. It had before it outstanding testimonials describing O as a dedicated and well-respected medical practitioner. There was a complete absence of any criticism of O's clinical competence. It was certainly necessary for a doctor to be completely open and honest in his dealings with people, including employers. Where there is material on the primary facts that warrants some inference that there is a real risk of the doctor not being completely open and honest in the future, it may well be that the severe sanction of erasure will be called for. Taking account of all matters, however, the Court substituted a sentence of suspension for twelve months.

### *R (Lloyd Subner) v. Health Professions Council* [2009] EWHC 2815 (Admin)

**37.06** Anaesthetist—assault and abuse of staff—non-attendance at hearing—no insight of remedial measures

The allegations against the appellant, an operating department practitioner employed as an anaesthetic practitioner by King's College Hospital, were that, in December 2006, during the course of his employment, he (a) physically assaulted and (b) verbally abused another member of staff, and that (c), following these episodes, he left work early and without permission. The Conduct and Competence Committee (CCC) of the Health Professions Council (HPC) found the allegations well founded and struck him off the register. The appellant did not attend and the Committee decided to proceed in his absence. Kenneth Parker J said that physical assault—by throwing a box of ten 1 litre units of intravenous fluids—and verbal abuse are plainly very serious matters indeed. The Committee provided coherent and sensible reasons as to why the options of taking no further action, a caution, a conditions of practice order, or suspension were not considered to be appropriate. The Committee noted in particular that the appellant had not attended to put forward any explanation as to his conduct, any assurances about future conduct, or any evidence that he had reflected on what had occurred, and that he had not accepted the seriousness of his conduct. As the HPC's Indicative Sanctions Policy notes, a key factor in many cases will be the extent to which a health professional recognizes his or her failings and is willing to address them, and in deciding what, if any, sanction is required, the issue that the panel needs to determine is whether the health professional has genuinely recognized his or her failings and the steps needed to address them. In the instant case, the Committee was plainly correct to give considerable weight to the fact that the appellant had not attended, had shown no insight whatsoever into the seriousness of the findings against him, and had given no indication whatsoever that he was minded in any way to embark upon remedial measures, in so far as they plainly would have been appropriate in this case. The appeal was consequently dismissed.

### *Mubarak v. General Medical Council* [2008] EWHC 2830 (Admin)

**37.07** Sanction—failure to appreciate behaviour wrong—lack of remorse

M, a general practitioner (GP), was found guilty of improper sexual conduct towards a female patient. In dismissing M's appeal and the sanction of erasure, Burnett J said:

> 38. The repeated reference to "insight" in the context of sanction arose from the need to consider the guidance on sanction promoted by the General Medical Council. Professional misconduct not infrequently involves behaviour which is judged inappropriate, but which the professional himself may not appreciate is inappropriate. In the medical context dealings with patients and colleagues may give rise to such complaints. Insight into such conduct is potentially an important factor in considering sanction. A lack of insight would suggest the doctor concerned continued to fail to appreciate that his conduct was wrong. If someone has insight into a failing it is more likely to be controlled. In the case before the panel there was some behaviour complained of, in the nature of inappropriate comment, to which ordinary concepts of "insight" might readily apply. However, it would be a strange doctor indeed who failed to appreciate that any of the three aspects of sexualised conduct were profoundly wrong. In this case the appellant readily recognised that if the unnecessary digital penetration, cupping of the navel piercing or breast examination took place they were wrong.
>
> 39. It seems to me that the panel was using lack of insight, in this context, as a synonym for lack of remorse. There was, of course, no remorse because the appellant denied the allegations and continued to do so. [Counsel for the doctor] submits that the panel's treatment of this question shows that they penalised the appellant for having contested the proceedings. It would be wrong to impose an additional sanction on that account just as it is wrong in principle to increase what would otherwise be an appropriate sentence because someone has denied a criminal charge. That is so even though the exercise of imposing a sanction and punishment are not identical, as Laws LJ explained in *Raschid*. Acceptance of responsibility may properly lead to a less severe sanction, in an appropriate case, just as it may result in a less harsh sentence. However, I can detect nothing in the careful approach to sanction of this panel which suggests that they fell into that suggested error.

### *Balamoody v. Nursing and Midwifery Council* [2009] EWHC 3235 (Admin)

**37.08** Restoration to register—whether safe to practise—not insight

B appealed against the refusal of the Nursing and Midwifery Council (NMC) to restore him to the register. In allowing B's appeal and ordering that his application for restoration be reheard by another panel, Langstaff J said:

> 31. Underlying all these complaints were two questions. The first: what was it, asked the appellant, that he had to prove in order to satisfy the committee that he was a proper person to return to the register as a nurse? Secondly, a general complaint that the result of the proceedings in the context which I have set out was desperately unfair to him. The first of those he raised at the meeting itself. He asked the committee what was the meaning of "insight" in the context? It does plainly concern him and for good reason. The "lack of insight" had become familiar short hand for being unsafe to practise. The definition given by the nursing member of the panel, the other two being a chairman and a lay person, was that she could not argue with a dictionary definition but would comment that "I would want you to be thinking not only how your actions have affected others and the full ramifications of that. That is what I would add for that".
>
> 32. That was not obviously a helpful definition. It may be thought to be compounded by the observations in the decision that the panel would expect an applicant for restoration to have enough insight to appreciate what evidence would be helpful to his application. This is "insight" used in a different context and sense. It is, I think, to suggest that it is entirely up to an individual to determine what matters he should bring to prove his case to the tribunal. That does not seem to me to fit easily with the scheme set up by the Order. In the context of someone who is a caring person, whose offences

have caused no actual harm that could be established, for whom they represent an isolated occasion though repeated twice over 10 days in the course of a long career otherwise unmarked by complaint, they were singularly unhelpful, bearing in mind the consequences of a refusal of registration to this individual.

[...]

40. ... It seems to me that the panel have focused, given the history presented to them, upon whether or not [the appellant] had "insight" into that which had happened in the past as a surrogate for asking the question which their minds should really and centrally have addressed; that is, whether he was potentially safe in the future.

[...]

42. ... Given the fact that the convictions rose out of a failure of management as a nursing home proprietor, it would seem to cry out for [the appellant] not being put in a position in which he supervised the actions of others. It is in this context in which a failure of insight might be important even though he appreciates fully the mechanics which have to be operated.

### *Odes v. General Medical Council* [2010] EWHC 552 (Admin)

**37.09** Consultant physician—operations—lack of acknowledgement of mistakes—whether conditional registration appropriate

O was a locum consultant physician in general (internal) medicine and cardiology. The charges against O before the GMC arose out of the treatment of two patients in 2004 and 2006. The 2004 patient had been admitted to the hospital with a spontaneous pneumothorax (air between the lung and chest wall) for which, under the care of the appellant and his team, he was treated with a series of chest drains. The allegation found proved was that the removal of the patient's drain was inappropriate and contrary to the advice of the cardiothoracic specialist. The 2006 patient was admitted complaining of pain in her right hip and lower right abdomen. The panel found proved a lack of adequate investigation and diagnosis. In dismissing O's appeal on findings of fact, impairment, and sanction of four months suspension, Hickinbottom J said that the panel's approach to sanction could not be criticized. The panel found that the seriousness of the misconduct, in all of the circumstances, was such that taking no action, relying upon the appellant's own undertakings as he offered, or a period of conditional registration would be inadequate, given the pattern of behaviour demonstrated, and the appellant's misjudgments and lack of insight, as the panel found them to be. The panel noted that the appellant had still not acknowledged that he had made mistakes that were his fault—and he had failed to supervise adequately his junior staff. It found that this exhibited an attitude that could not properly be addressed by a conditional registration, which would also not have reflected the seriousness of the misconduct in the panel's view.

### *R (Sharma) v. General Dental Council* [2010] EWHC 3184 (Admin)

**37.10** Impairment—change in dental practices—no objective evidence to address deficiencies

S, a registered dentist, admitted the allegations made against him, which related to widespread deficiencies in the treatment of a single patient over a period of five months and S being dismissive of the patient's concerns to the point of being unprofessional. The Committee concluded that it would be proportionate and appropriate to impose conditions intended to address the deficiencies in S's practice. Ouseley J said, in relation to impairment, that it was rightly a matter of concern that the facts were not admitted until the hearing. The value of the responses of S, such as reducing his workload and altering his practices, was, as counsel for the General Dental Council (GDC) pointed out, rather dependent upon the extent to which S had analysed correctly why he had failed in those respects and had analysed correctly his ability through those means to address them. The information about those changes was conveyed to the Professional Conduct Committee by S's counsel, and not through any statement or oral evidence from S enabling the PCC to test how far he had shown true insight into what the Committee saw as the cause of the problems and to test how far what he was doing genuinely addressed those deficiencies. There was no objective, supportive written material demonstrating his workload management for the future, or how the written treatment and consent system now operated was operating satisfactorily to redress the informed consent deficiency, and there was no written or objective material to explain why he had behaved in the way that he had towards this patient to show that what he was now doing addressed his cavalier and disrespectful attitude towards a wholly justifiable complaint. The concerns that the PCC had about S's lack of insight into his behaviour went to the heart of the adequacy of his analysis of his problems and self-selected remedies.

### *Karwal v. General Medical Council* [2011] EWHC 826 (Admin)

**37.11** Dishonesty—review hearing—whether continuing lack of insight

This was an appeal against a further nine months' suspension imposed by the GMC's fitness to practise panel at a review hearing. In June 2008, an earlier panel found that the appellant doctor, K, had knowingly made false representations to a professional colleague about an investment scheme so as fraudulently to reassure him that £188,000 he had invested would be repaid. The panel suspended K from the medical register for twelve months. At a review hearing in March 2010, the panel found that K's fitness to practise was still impaired and she was further suspended for nine months from the expiry of the current suspension. A key question was whether K had sufficient insight into or had fully appreciated the gravity of the original offence. The review panel had determined that K's behaviour continued to demonstrate lack of insight and no real acceptance of the original findings of dishonesty. In the Administrative Court, Rafferty J said that, in her judgment, the panel was not only entitled, but also obliged to address K's dishonesty, but also her lack of insight. The GMC's Indicative Sanctions Guidance provides that a review panel will need to satisfy itself that the doctor has fully appreciated the gravity of the offence, has not reoffended, and has maintained his or her skills or knowledge. Rafferty J said that insight—in the sense of determining whether the doctor had appreciated gravity—was inevitably an issue at a review. Dishonesty by a doctor, albeit unconnected with the practice of medicine, undermines the profession's reputation and public confidence. The appellant had always maintained her innocence of the original findings. The Court was not persuaded that equating maintenance of innocence with lack of insight was the same. The panel was scrupulous to make clear that it did not see acceptance of culpability as a condition precedent for insight. The findings of the panel demonstrated its justifiable view that K had not fully appreciated the gravity of her offence, rather than that she sought to minimize it.

### *Shah v. General Dental Council* [2011] EWHC 3003 (Admin)

**37.12** Sanction—lack of remorse—danger of reoccurrence

S appealed against the sanction of erasure imposed by the GDC. In April 2009, a patient attended his surgery for an emergency appointment, complaining of pain. He advised her that root canal treatment was required, and misleadingly and dishonestly stated that it was not available on the NHS. The appellant knew that he should have offered to provide NHS treatment to the patient and that she could probably ill afford to pay for this treatment privately. In finding that the appellant's fitness to practise was impaired, the PCC took into account that he had alleged in the course of the hearing that the patient had fabricated her accounts, that he had not told the truth to the Committee, and that he had shown no remorse or insight, and the Committee was not satisfied that the appellant's failings that underlay his course of conduct had been remedied nor that they were easily remediable. On appeal, it was submitted on behalf of the appellant that the Committee fell to the error of eliding lack of remorse with lack of insight. In rejecting this argument and dismissing the appellant's appeal, Supperstone J said that, in his view, the Committee had proper regard to the mitigation put forward on the appellant's behalf and bore in mind the principle of proportionality. Agreeing with the analysis of Collins J in *Giele v. General Medical Council* [2005] EWHC 2144 (Admin), [31], Supperstone J said, at [24]:

> The fact that the appellant alleged Patient A fabricated the allegations, this forming part of his defence, did not justify the imposition of a more severe sanction than the misconduct deserved. However, the fabrication allegation, together with the lies that the appellant told the committee, do disclose a lack of remorse and, in my view, in the present case, indicate a lack of insight so that it can be said that there is a danger of recurrence.

### *Ige v. Nursing and Midwifery Council* [2011] EWHC 3721 (Admin)

**37.13** See paragraph 19.61.

### *Ballesteros v. Nursing and Midwifery Council* [2011] EWHC 1289 (Admin)

**37.14** See paragraph 19.62.

### *Henderson v. General Teaching Council for England* [2012] EWHC 1505 (Admin)

**37.15** Denial of allegations over three years—absence of practitioner at hearing—represented by counsel—lack of evidence to judge insight

The appellant, H, a former headteacher of a school in Islington, appealed against a prohibition order imposed by the PCC of the General Teaching Council for England for unacceptable professional conduct. The order struck the appellant off the teaching register. Its effect was that he was not entitled to work as a teacher and could not apply for permission to re-register for a period of two years. A school disciplinary hearing on 1 May 2009 found that the appellant was seen masturbating in his office in front of his computer by two members of the public, on two occasions, that he had used the computer equipment to access, download, and to view pornographic images and films during school hours when children were present in the building, and that he had been negligent in the recruitment and vetting process of the afterschool drama leader. He was summarily dismissed. His appeal was dismissed on 3 July 2009 and the local authority informed the General Teaching Council of its investigation. It was only after notice of the hearing before the PCC on 17 August 2011 that the appellant instructed counsel and admitted all of the allegations. Previously, he had admitted the facts relating to the downloading of the pornography and the failure to carry out the appropriate procedure when recruiting the after-hours leader, although he maintained that the position and procedures in relation to the latter were unclear. On appeal to the High Court, the appellant claimed that the Committee erred in its approach to insight. It wrongly gave weight to the appellant's denial of the masturbation over the period and his history of denial in respect of that, and equated that denial with lack of insight. Beatson J, in dismissing the appellant's appeal, said that the Committee did not equate denial with lack of insight. It took into account that, over three years, the appellant had many formal opportunities to admit the masturbation charge, but did not. It was entitled to take that into account. Whilst it is correct that two sections of the Committee's decision did not sit comfortably with each other, they were not necessarily inconsistent. It is important that the appellant did not attend the hearing and that the Committee was unable to ask questions or to assess the questions designed to assess the depth of his insight. The fact that an admission is made and an apology offered through counsel does not in itself show appropriate insight, particularly when the admission came, as this did, only after the individual received robust advice from counsel. Obviously, there is some credit to be given for that. But it would not be real to regard that sort of admission and apology as showing the same insight as an early unpressured one, unless there were some further explanation to show that that was so. All that the Committee had were counsel's submissions. Counsel's speech to the Committee relied solely on the admission by the appellant that he had done wrong. There was no other evidence before the Committee as to the appellant's insight. There was his statement, and a statement from a friend and fellow headmaster. There was no medical evidence. In those circumstances, it was not clear to the Committee, and the Committee was entitled to regard it as not clear, what the basis was upon which the appellant was motivated to make the admission that he made. The criticisms of the Committee's treatment of lack of insight were rejected.

*Haye v. General Teaching Council for England* (2013) 157 Sol Jo 35

**37.16** Teacher—views expressed at school—lack of insight

H, a teacher, appealed against a prohibition order with a two-year minimum review arising out of comments made at school on issues of homosexuality and religion. The panel stated that it was not H's beliefs that were inappropriate, but his lack of insight of the inappropriateness of the comments, and his lack of insight meant that it was likely that the event would be repeated. In dismissing H's appeal, it was held that the panel had been right when it said that if H were to wish to return to the teaching profession in the future, he would have to demonstrate genuine insight into the public perception of teachers as promoters of tolerance and respect for the rights and beliefs of others.

*Amao v. Nursing and Midwifery Council* [2014] EWHC 147 (Admin)

**37.17** Impairment and sanction—single incident—lack of focus on recurrence—inappropriate to ask registrant if she agreed with panel's findings

A, a practice nurse, faced a single allegation that, on 13 July 2009, she shouted at the practice manager at the clinic where both worked and, in the course of losing her temper, made threats to the practice manager. At the hearing, A did not have legal assistance. She attended, gave evidence, asked questions of witnesses, and made submissions. The panel found the allegations proved an impairment, and imposed a striking-off order. Walker J observed that where a person facing disciplinary proceedings is unrepresented, the tasks arising at the hearing for the unrepresented person, and for the legal assessor, are unlikely to be easy. The Court concluded that this was a case in which:

(1) procedural mishaps occurred during the first part of the hearing, concerned with factual matters; but
(2) the effect of those mishaps was not such as to require the panel's findings of fact to be quashed in the interests of justice; although
(3) when the panel moved on to consider misconduct, impairment, and sanction, A was not given a fair opportunity to address the panel's concern that she had no insight with regard to her future professional obligations, and it followed that the decision to strike off could not stand; and
(4) the appropriate order to have been made by the panel would have been an order that A's registration be suspended for a period of one year.

A had now, in effect, been subjected not only to a one-year suspension, but also to a substantial extension of that suspension. In dealing with impairment and sanction, Walker J said:

> 155. ... [I]n her recent submissions, [A] comments that the word "impairment" is suggestive of a "medical condition that affects my ability to look after patients." I agree that to a layperson the word "impairment" is indeed suggestive of something along these lines. Of course, it was not suggested at the hearing that "impairment"

has this meaning – and [A] does not suggest otherwise. However I agree with her that the transcript demonstrates confusion on her part as to what "impairment" really involved.

156. The feature of "impairment" which, to my mind was never properly addressed in evidence and submissions, concerned the distinction between a failure to have insight into the misconduct which had occurred, and a failure to have insight into the need in future to avoid being in a position where such conduct was likely to recur. The relentless focus of the prosecution was upon what happened on 13th July 2009. It is apparent from the transcript that the proceedings on day 7 left [A] confused. It is difficult to see how, without legal representation, she could have been anything other than confused. Before [A] gave evidence on impairment, the legal assessor had advised that because the charge sheet had alleged impairment by reference only to factual allegations (a) to (f), it might not be proper when considering impairment to consider the first and second incidents referred to in the evidence of (the practice manager). However he then went on to say that the position might be different if [A]'s evidence on impairment were to make any suggestion as to her conduct before 13th July 2009 or her future conduct....

157. Here, it seems to me, there was a very important missed opportunity. There was a need to stand back from the emotionally charged incident of 13th July 2009. As the NMC acknowledges, the panel's findings of fact meant that [A] faced a substantial risk of being struck off. In determining whether that was the appropriate penalty, a key question for the panel would be whether [A] had insight into the desirability of taking steps to ensure that in future no misconduct occurred. There was nothing to suggest that [A] had any understanding of how important it was for her to give evidence and make submissions about her ability to take steps to ensure that in future there would be no incident similar to what happened on the 13th July 2009. She had thus far proceeded, and continued thereafter to proceed, on the basis that the crucial question was whether she had said the things alleged against her on 13th July 2009 and whether she was at fault in doing so.

158. Far from indicating that it would be highly desirable for [A] to stand back from events on 13th July 2009, and consider what she could say to assuage concerns about the future, [A] was in effect told that this was a topic she should avoid.

It was both inappropriate and unnecessary for the prosecution to seek to question A about whether she agreed with the panel's findings on each of the factual allegations. The legal assessor rightly intervened. However, it would have been desirable to go further, for here was another opportunity to shift the focus. The distinction drawn between agreeing with the panel on the one hand, and accepting its findings on the other, was a subtle one. It was a distinction that was first suggested while A was in the course of cross-examination. It was a distinction that A, understandably, found impossible to grasp. Moreover, she was perfectly entitled to say that she did not accept the findings of the panel: she had a right of appeal, which she was entitled to exercise. In the circumstances, it was thoroughly inappropriate—almost Kafka-esque—to cross-examine her in a way that implied that she would be acting improperly if she did not 'accept the findings of her regulator'. She was at a loss, and confused, as to how to deal with this line of questioning. The remainder of the proceedings did nothing to make her any less confused. The learned judge continued:

163. The reality was that [A] did not have any appreciation of the real nature of the case that she had to meet in relation to impairment, namely that it was not just her insight into what happened on the 13th July which the panel would wish to consider, but also her insight into what could be done in the future to avoid an incident of that kind recurring. In these circumstances, the panel's finding that there was a high risk of repetition was vitiated by an unfair procedure.

***Brew v. General Medical Council* [2014] EWHC 2927 (Admin)**

**37.18** See paragraph 45.28.

## C. Other Relevant Chapters

**37.19** Impairment of Fitness to Practise, Chapter 33
Sanction, Chapter 60

# 38

# INTERIM ORDERS
| A. **Legal Framework** | | 38.01 | *R (Ali) v. General Medical Council* [2008] EWHC 1630 (Admin) | 38.24 |
| | *Madan v. General Medical Council (No. 2)* [2001] EWHC Admin 577, [2001] Lloyd's LR Med 539 | 38.03 | *Bradshaw v. General Medical Council* [2010] EWHC 1296 (Admin) | 38.27 |
| | *R v. Ashton* [2007] 1 WLR 181 | 38.08 | *Rashid v. General Medical Council* [2012] EWHC 2862 (Admin) | 38.32 |
| | *Waghorn v. General Medical Council* [2012] EWHC 3427 (Admin); [2013] 1 CMLR 45 | 38.19 | D. **Extension of Interim Orders** | 38.43 |
| | *General Medical Council v. Kaluba* [2013] EWHC 4212 (Admin) | 38.20 | *General Medical Council v. Hiew* [2007] 1 WLR 2007, CA | 38.43 |
| C. **Necessity, etc., for Interim Order** | | 38.21 | *General Medical Council v. Lauffer* [2009] EWHC 3497 (Admin) | 38.44 |
| | *Dr X v. General Medical Council* [2001] EWHC Admin 447 | 38.21 | *Nursing and Midwifery Council v. Miller* [2011] EWHC 2601 (Admin) | 38.45 |
| | *R (Walker) v. General Medical Council* [2003] EWHC 2308 (Admin) | 38.22 | *Nursing and Midwifery Council v. Maceda* [2011] EWHC 3004 (Admin) | 38.46 |
| | *R (Shiekh) v. General Dental Council* [2007] EWHC 2972 (Admin) | 38.23 | | |

*Nursing and Midwifery Council v. Pratt* [2011] EWHC 3626 (Admin) 38.47
*General Medical Council v. Razoq* [2012] EWHC 2135 (Admin) 38.48
*General Medical Council v. Srinivas* [2012] EWHC 2513 (Admin) 38.49
*General Medical Council v. Gill* [2012] EWHC 2069 (Admin) 38.50
*Nursing and Midwifery Council v. Hitchenor* [2012] EWHC 3565 (Admin) 38.51
*Nursing and Midwifery Council v. Okafor* [2013] EWHC 1045 (Admin) 38.52
*General Medical Council v. Malik* [2013] EWHC 122 (Admin) 38.53
*General Medical Council v. Adeosun* [2013] EWHC 2367 (Admin) 38.54
*Jooste v. General Medical Council* [2013] EWHC 2463 (Admin) 38.55
*General Medical Council v. Dr E* [2013] EWHC 3425 (Admin) 38.56
*General Medical Council v. Dr E* [2014] EWHC 1620 (Admin) 38.57
*General Medical Council v. Sondhi* [2013] EWHC 4233 (Admin) 38.58
*General Medical Council v. Shalaby* [2013] EWHC 4211 (Admin) 38.59
*General Medical Council v. Qureshi* [2014] EWHC 775 (Admin) 38.60
*General Medical Council v. Makki* [2014] EWHC 769 (Admin) 38.61
*Nursing and Midwifery Council v. Kidd and De'filippis* [2014] EWHC 847 (Admin) 38.62
*General Medical Council v. Agrawal* [2014] EWHC 1669 (Admin) 38.63
**E. Measures Pending Appeal** 38.64
*Gupta v. General Medical Council* [2001] EWHC Admin 612, [2001] EWHC Admin 631 38.65
*Burnham-Slipper v. General Medical Council* [2001] EWHC 656 (Admin) 38.66
*Moody v. General Osteopathic Council* [2009] 1 WLR 526 38.67
*Ashton v. General Medical Council* [2013] EWHC 943 (Admin) 38.68
*Shaw v. Logue* [2013] All ER (D) 33 (Sep) 38.69
*Baker v. British Boxing Board of Control* [2014] EWHC 2074 (QB) 38.70
*Nursing and Midwifery Council v. Gatawa* [2014] EWHC 3153 (Admin) 38.71

## A. Legal Framework

**38.01** Examples include:

Medical Act 1983, section 41A (where the panel is satisfied that it is necessary for the protection of members of the public or is otherwise in the public interest or is in the interests of the practitioner for his or her registration to be suspended or be made subject to conditions for a period not exceeding eighteen months, the panel may make such an order; any interim order is to be subject to review, when it may be revoked or varied, and the period of eighteen months may be extended by the court)
Dentists Act 1984, section 32
Opticians Act 1989, section 13L
Chiropractors Act 1994, sections 21 and 24
Health and Social Work Professions Order 2001, article 31
Nursing and Midwifery Order 2001, article 31
Pharmacy Order 2010, article 56

See also the relevant regulator's fitness to practise rules for interim orders.

## B. General Principles and Procedure

### *R (Madan) v. General Medical Council* [2001] EWHC Admin 322

**38.02** Appeal—statutory appeal—judicial review

M sought to challenge the decision of the Interim Orders Committee (IOC) of the General Medical Council (GMC) given in November 2000 to suspend her registration for eighteen months pending a full hearing by the Professional Conduct Committee (PCC) of allegations of inappropriate and irresponsible prescribing of appetite suppressants by her. M initially made an application for permission to apply for judicial review. After objection was taken by the GMC to the appropriateness of the judicial review route, M made a statutory application to the Court under section 41A(10) of the Medical Act 1983:

> (10) ... [T]he court may—
> (a) in the case of an interim suspension order, terminate the suspension
> (b) in the case of an order for interim conditional registration, revoke or vary any condition imposed by the order;
> (c) in either case, substitute for the period specified in the order... some other period which could have been specified in the order...

Richards J said that the two routes are in substance the same, but, given the availability of a statutory remedy under section 41A(10) by way of an application to the Court, the defendant's objections to the bringing of judicial review proceedings were well founded and judicial review was inappropriate in this case. Having reviewed the merits, the reasons for the IOC's decision were adequate and intelligible reasons, sufficient to comply with the duty, be it a statutory duty or a duty at common law, to give reasons for the decision. The basis upon which the Committee concluded that a suspension was necessary was clearly articulated and M can have been left in no doubt as to that basis.

### *Madan v. General Medical Council (No. 2)* [2001] EWHC Admin 577, [2001] Lloyd's LR Med 539

**38.03** Claim form CPR Part 8—not CPR Part 52—Article 6 ECHR—proportionality—powers of court

In May 2001, the IOC, exercising its powers under section 41A(2) of the Medical Act 1983, reviewed the initial order for suspension previously made in November 2000, and decided to continue the suspension and not to substitute for it an order by way of conditional registration. On appeal, the Administrative Court (Brooke LJ and Newman J) said that it is highly likely that interim suspension hearings engage Article 6 of the European Convention on Human Rights (ECHR) and that the fundamental position in this case comes down to the following.

(1) The Committee had to make a choice between conditional registration and interim suspension, and had to consider whether the need to protect the public interest required suspension rather than conditional registration. That is what section 41 of the Act requires. But—
(2) It was incumbent upon the Committee to balance the need so addressed under (1) against the consequences that an order for suspension would have upon the applicant and to satisfy itself that the consequences of the remedy on the applicant were not disproportionate to the risk from which it was seeking to protect the public.

An application under section 41A(10) may be made by issuing a claim form under CPR Part 8, as opposed to a notice of appeal under CPR Part 52. Under the Part 8 procedure, the claimant's written evidence must be filed with her claim form (CPR 8.5(1)) and the defendant's written evidence must be filed with its acknowledgement of service, which is due not more than fourteen days after the service of the claim form. The powers of the Court, by reason of the very limited terms of section 41(10), are limited. The Court's powers are restricted to determining the suspension or substituting a different period of suspension for the period specified.

### *R (George) v. General Medical Council* [2003] EWHC 1124 (Admin)

**38.04** IOC—role of Committee—judgment and experience—review

In this case, Collins J said:

> 42. Now I should make it plain that the Committee did not, and was not required to make any findings as to whether the allegations were or were not established. It was sufficient for them to act, if they took the view that there was a prima facie case and that the prima facie case, having regard to such material as was before them by the medical practitioner, required that the public be protected by a suspension order.

> 43. They were not making any final decision because, as I say, they were not reaching any conclusions of fact. That is important, because it must not be taken that I have made any conclusions of fact. I have not. It has not been my task in the context of this application to do so.
>
> 44. I merely record that before me [G] has put forward a very lengthy and detailed statement in which he refutes most of the allegations that are made against him and explains some of the others, indicating that what may seem to be of significance, in reality is not. Those are matters which he has not yet had an opportunity of airing before the committee, the Interim Orders Committee, or any other committee of the GMC. It is submitted, on his behalf, that if all that material is taken into account, it would be wrong for any interim suspension to be made.
>
> 45. The problem I faced was, as it seemed to me, that that was a matter which was peculiarly within the judgment and expertise of fellow medical practitioners, and it would be only in an extreme case that this court would feel able to interfere with that judgment. Indeed, it is perhaps in some ways a somewhat unsatisfactory procedure that is before the court, because there is no specific power given, as there is in almost all other cases where an appeal lies to this court, to remit for reconsideration. That is the normal practice of this court, and indeed of the Privy Council, because it is very rare for this court and, I think, virtually unknown for the Privy Council, to hear any evidence. If it is felt that something may have gone wrong, the proper course would normally be to remit for reconsideration. But, as I say, there is no express power in section 41A(10) for the court to do that.
>
> 46. However, counsel have agreed, and in particular [counsel for the GMC] has recognised, that the court can achieve that, and it can achieve it in this way. It has power to vary the length of any suspension, and if it feels that the matter ought to be reconsidered and there ought to be a review, it can indicate as much in the judgment and can give effect to that by reducing the period of suspension with a view to there being a review of the matter within a short period of time.

Before Collins J was an application under section 41A(10), but also an application for permission for judicial review of the decision of the IOC. The point made by counsel for the applicant, essentially, was that the remedy provided by section 41A(10) was insufficient, because the Court was unable to quash the original decision to suspend, which would remain on the doctor's record and be a matter that might adversely affect his future. On reviewing the authorities in which an alternative remedy is available, the learned judge said that the position here was that Parliament has introduced a particular right of, or equivalent to, appeal in a situation in which there would have been a right of judicial review of the decision of the Committee, albeit that it is not in formal terms an appeal. Accordingly, counsel's concerns did not justify the application for judicial review and, in the exercise of discretion, the Court would not grant judicial review.

*General Medical Council v. Sheill* **[2006] EWHC 3025 (Admin)**

**38.05** Interim suspension—reviews—application to terminate original suspension after reviews

S was suspended from practice on 14 October 2005 by the GMC's Interim Orders Panel (IOP). The suspension was the subject of review hearings on 5 January, 16 June, and 15 September 2006. The suspension was due to expire on 16 December 2006. The GMC applied for an extension of the order for suspension pursuant to section 41A(6) of the Medical Act 1983, as amended. S applied for termination of the suspension under section 41A(10)(a). In granting the GMC's application for extension for twelve months, Crane J said, at [30], that he concluded that in principle a doctor may submit that the original suspension was wrong, even if reviews have taken place. However, the Court will not necessarily reconsider the correctness of the original order. The Court has a discretion: subsection (1) uses the word 'may'. If the application is made after several reviews, for example, there may be little value in such a reconsideration. If the doctor has fully participated in reviews without any application to the Court, the Court may decline to embark on a lengthy reconsideration. The Court might also decline to embark on a lengthy reconsideration of an order that had repeatedly been reviewed, if the definitive disciplinary hearing was imminent and particularly if such a hearing might be delayed or its preparation interfered with.

### *General Medical Council v. Uruakpa* [2007] EWHC 1454 (Admin)

**38.06** Application for extension—expiry of interim order—jurisdiction—power of panel to make fresh order

U, a medical practitioner with a specialty in obstetrics and gynaecology, was the subject of a reference to the GMC from the National Clinical Assessment Service (NCAS) in June 2005 following allegations that his clinical performance had fallen below proper standards in a number of respects. On 28 July 2005, U's registration was made subject to conditions by the IOP for a period of eighteen months. In January 2007, before the eighteen months expired, an application was made by the GMC to the Court for an extension of twelve months. The matter came before Mitting J on 24 January 2007, who ordered that the application for an extension of the conditional registration order be 'adjourned' on condition that the conditions 'remain in place as an order of the court'. On the hearing of the adjourned application on 19 April 2007, Collins J observed that the order was apparently not an extension of the interim order, but was a separate court order to preserve the position pending the hearing of the adjourned application. However, the Court was concerned here with the situation in which the application for an extension has been made before the expiry of the original order. Accordingly, there was no doubt that the Court had jurisdiction to make the necessary order, or to make such order as appropriate in this respect. It is difficult to imagine circumstances in which the GMC would think it right to let an order expire and then, at some later stage, make an application for an extension. Whether or not the Court would accept such an application is a matter that would have to be decided if that situation were to arise:

> 30. [Counsel for the GMC] submits that if there is a situation whereby the GMC considers it desirable that notwithstanding that an order has expired there should be a fresh order made, then that application should be made to the [IOP]. It would undoubtedly be wrong for the [IOP] to entertain such an application unless there were good reason to do so, and such good reason might arise because of the production of fresh material which had not been and had not been able to be considered before and which might justify the imposition of a fresh order. But it is obvious that the [IOP] would have to tread very carefully, because otherwise it would no doubt be suggested that it was trying to get round the time limit of 18 months imposed by Parliament in entertaining a fresh application once that 18 months had expired. But that there is power in appropriate circumstances for a fresh order to be made I have no doubt. I suspect that those circumstances would be somewhat rare. As I say, I am not deciding, nor am I in the least encouraging applications to be made beyond the time that the original order has expired. In fairness, in my experience I am not aware that the GMC has ever sought to extend an interim order that has expired.

### *General Medical Council v. Hiew* [2007] 1 WLR 2007, CA

**38.07** Criteria for extension—similar to original interim order

H appealed from the order of Bean J—[2006] EWHC 2699 (Admin)—extending, for a period of six months, an order for the suspension of H's registration as a medical practitioner previously made by the GMC's IOP. The judge's order was made under section 41A(7) of the Medical Act 1983 ('the relevant court may extend or further extend for up to twelve months the period for which the order has effect'). The original order of the IOP for eighteen months was an interim suspension order, later substituted by an order for interim conditional registration. In dismissing H's appeal to the Court of Appeal, Arden LJ (with whom Tuckey and Lawrence Collins LJJ agreed) said:

> 28. Section 41A(7) does not set out the criteria for the exercise by the court of its powers under that subsection in any given case. In my judgment, the criteria must be the same as for the original interim order under section 41A(1), namely the protection of the public, the public interest or the practitioner's own interests. This means, as [counsel] for the GMC submits, that the court can take into account such matters as the gravity of the allegations, the nature of the evidence, the seriousness of the risk of harm to patients, the reasons why the case has not been concluded and the prejudice to the practitioner if an interim order is continued. The onus of satisfying the court that the criteria are met falls on the GMC as the applicant for the extension under section 41A(7).
>
> 29. The judge must, however, reach his decision as to whether to grant an extension on the basis of the evidence on the application. He will need to examine that evidence with care. One of the difficulties in this case was that the witness statement in support of the application was relatively perfunctory with respect of the narrative of events and moreover set out the reasons for the application in summary form only: see paragraph 32 of the witness statement in support of the application. The evidence did not explain, to the requisite level of detail, the reasons why the GMC made the application and the judge had to elicit that information from the submissions of counsel and the large volume of contemporary documents with which he was provided. In my judgment, the witness statement should fairly explain, in summary, but as a self-standing document, the GMC's reasons for the application for an extension.
>
> [...]
>
> 32. The evidence on the application will include evidence as to the opinion of the GMC, and the [IOP] or Fitness to Practise Panel, as to the need for an interim order. It is for the court to decide what weight to give to that opinion. It is certainly not bound to follow that opinion. Nor should it defer to that opinion. All that is required is that the court should give that opinion such weight as in the circumstances of the case it thinks fit. Weighing up the opinion of a body that has special statutory responsibilities and relevant experience and expertise is again part of the ordinary task of judicial decision-making.
>
> 33. [Counsel for the medical practitioner] relies on the proposition, regarded as axiomatic by Laws LJ in *R (D) v. Secretary of State for Health* [2006] Lloyd's Rep Med 457, para 26, that the more serious a public authority's interference with an individual's interest, the more substantial will be the justification which the court will require if the interference is to be permitted. But, in this case, the decision of the court is simply that there should be an extension of the period of suspension. The court is not expressing any view on the merits of the case against the medical practitioner. In those circumstances, the function of the court is to ascertain whether the allegations made against the medical practitioner, rather than their truth or falsity, justify the prolongation of the suspension. In general, it need not look beyond the allegations. If the medical practitioner contends that the allegations are unfounded, the medical practitioner should challenge by judicial review the original order for suspension or the failure to review it and make some other decision in accordance with section 41A(2). On such an application, the decision of the [IOP] or Fitness to Practise Panel will then be examined on well-established judicial review grounds. I do not consider that a judge is bound to treat a medical practitioner's opposition to an application under section 41A(7) as if it were an application for judicial review on the grounds that the allegations are without foundation, and there is a danger, if he does so, other than in a plain and obvious case (which I have already observed will be rare) that the wrong test will be applied. If the judge proceeds in the manner described above, he is unlikely to be diverted by the task of having to consider the seriousness of the risk to the public on the evidence provided by the GMC by contentions that the allegations are unfounded.

### *R v. Ashton* [2007] 1 WLR 181

**38.08** Substantive period of—review

*R v. Ashton* is usually relied on as authority that if there has been a procedural failure such as a slippage in the time for a review hearing, the interim order is not nullified, provided that any review shall take place within the substantive period of eighteen months (or such lesser period) specified in the order, or such period as extended by the court, and that it is in the interests of justice to continue the interim order and there is no prejudice to the registrant caused by any procedural failure.

### *Sandler v. General Medical Council* [2010] EWHC 1029 (Admin)

**38.09** Evidence on appeal—not before committee

Nicol J, in relation to adducing fresh evidence, accepted that CPR 8.6 gives the Court a discretion as to whether to receive evidence that was not filed with the claim form. Section 41A(10) of the Medical Act 1983 would allow the Court to admit evidence that was not before the panel. An example of that appears to have occurred in *R (Sosanya) v. General Medical Council* [2009] EWHC 2814 (Admin), [28]. In this case, however, the Court would not admit the evidence. It is not only the GMC that is prejudiced by the late service of evidence; the Court is also deprived of the views of the panel of the impact of the additional material. Absent any evidence explaining why the evidence was not served earlier than it was, the Court was not in a position to conclude that there was some good reason for the delay in serving additional witness statements. Further, S would have the opportunity of persuading the panel at the review hearing to take a different view of his suspension in the light of the evidence.

### *General Medical Council v. Kor* [2011] EWHC 2825 (Admin)

**38.10** Application for extension—order expiring prior to hearing—jurisdiction—position where expired before application

On 30 March 2010, the IOP made an interim suspension order against the doctor for eighteen months. The order expired on 29 September 2011. On 1 September 2011, before the order expired, the GMC issued an application for an extension, but before the hearing of the application on 3 October 2011, the original order had expired. In finding that the Court had jurisdiction to grant an extension, His Honour Judge Pelling QC (sitting as a judge of the High Court) said:

> 12. In my judgment if an order made by an Interim Order Panel expires before an application to extend is made, then (a) the GMC has no power to make a new order, or at any rate in relation to the same facts and matters that caused it to make the initial order; and (b) the court has no power to extend in such circumstances because the concept of extension cannot apply to an order that has already expired. In my judgment, to conclude otherwise would be to defeat the statutory scheme, which imposes a limit on the period that a doctor can be suspended by the GMC (see section 41A(1)(a)). If Parliament had intended that the GMC could impose a new order in respect of the same facts and matters after expiry of an earlier order, then Parliament would not have inserted the maximum duration provision contained in section 41A(1)(a), nor would it have inserted the power of the court to grant an extension of the order contained in section 41A(6) and (7), for such provisions would have been plainly unnecessary. Further, whilst for reasons I will come to in a moment I am satisfied that a court has power to extend an order that was extant when the application to extend was issued, I do not consider that to be so in relation to an order that has expired before the application is issued. To conclude otherwise would not merely defeat the statutory scheme but would also be inconsistent with the language of the statute, which refers to a power to extend which is consistent only with there being an order that is in existence but which (but for the extension order) would expire by effluxion of time.
>
> 13. In my judgment, the position where an order expired before the issue of an application and the determination of the application is different. In such circumstances the GMC does not have jurisdiction to make a fresh order for the reasons already given.

However, I am satisfied that in those circumstances, on a true construction of section 41A(6) and (7) together, the court retains the power to extend.

***Rashid v. General Medical Council* [2012] EWHC 2862 (Admin)**

**38.11** Further evidence

On 24 April 2012, the GMC's IOP suspended R's registration for eighteen months. He had only three working days' notice of the hearing, although he was able to instruct counsel to represent him. The Court permitted him to file further evidence.

***Rogan v. Nursing and Midwifery Council* [2011] NIQB 12**

**38.12** Interim suspension order during adjournment of substantive hearing

The hearing of R's case before the Conduct and Competence Committee (CCC) of the Nursing and Midwifery Council (NMC) commenced on 11 January 2010. It was adjourned part-heard and resumed again on 27 April 2010 for a further five days, when the Committee found proved a number of allegations against R and dismissed other allegations (stage one of the hearing). The case was adjourned to 6 September 2010, when the Committee's decision on misconduct and impairment (the second stage of the hearing) was handed down. The Committee then further retired to consider the appropriate sanction, thereafter making a striking-off order. At the conclusion of stage one, the Committee suspended R's registration. In dismissing R's appeal, Gillen J said that the decision to suspend the appellant's registration during the lengthy adjournment was a perfectly logical step to take pending a final outcome of the case in order to ensure that the public was protected in the interim. It predicated no final determination of the matter and was an occurrence that is regularly reflected in court proceedings elsewhere, when interim measures are often adopted pending a final determination.

***Perry v. Nursing and Midwifery Council* [2013] 1 WLR 3423, CA**

**38.13** Nurse—inappropriate conduct towards patient—Articles 6 and 8 ECHR

The applicant was a 53-year-old registered mental health nurse. He qualified in March 2001. In October 2010, a patient made a complaint that he had acted inappropriately towards her. He accepted that he had overstepped professional boundaries and had sent text messages in response to texts that she had sent him. He did not accept any sexual touching and contended that the allegations were fabricated to punish him for seeking to distance himself from the complainant. Although the police were fleetingly involved, it was not apparent that there was any allegation of criminal conduct. The police took no action. The applicant was summarily dismissed by the Health Board, which referred the matter to the NMC. At a hearing before the IOP on 29 November 2011, the applicant accepted some of the allegations against him and offered four undertakings to the Panel. The Panel suspended the applicant from practice as a nurse for a period of eighteen months. The applicant applied to terminate the interim suspension order pursuant to article 31(12) of the Nursing and Midwifery Order 2001, and also made a CPR Part 8 claim that the hearing before the IOP and the order that it made were contrary to Articles 6 and 8 ECHR. Thirlwall J held that an interim suspension order was not necessary in this case. Because the Court had no power to terminate the suspension and to substitute conditions of practice, the peroration of the suspension was to terminate after twenty-eight days or at such earlier date as the NMC may convene a panel to consider the imposition of suitable conditions of practice: [2012] EWHC 2275 (Admin). The applicant appealed the judgment of Thirlwall J rejecting that the interim orders hearing infringed his rights under Articles 6 and 8. There was no appeal against that part of the judge's order terminating the applicant's suspension from practice. In dismissing the applicant's appeal against the dismissal of his Part 8 claim, Sir Stanley Burnton (with whom Davis and Hughes LJJ agreed) said that the appeal to the Court of Appeal raised a question of general importance as to the procedure of an investigating committee of the NMC (and similar committees of other professional regulatory bodies) when considering whether to make interim orders pending the substantive hearing of a complaint against a member of the profession. The learned judge said that he was content to proceed on the basis that Articles 6 and 8 were engaged at the hearing of the IOP. It was not suggested that the Panel was not impartial or independent, or that the applicant was prevented from making submissions to it. The only question was whether fairness required the applicant to have the opportunity to give evidence for the Panel to consider on the truth of the allegations made against him. He was prevented from giving such evidence:

> 19. What is required by fairness depends on the nature of the enquiry being conducted by the tribunal in question. The statutory function of the committee relevant in this appeal is its duty to determine whether to make an interim order, and the statutory right of the registrant under article 26 of the 2004 Rules to give "any relevant evidence in this regard" refers to evidence relevant to that question. For this purpose the committee must decide whether, on the basis of the allegation and evidence against the registrant, including any admission by him, it is satisfied that an order is necessary for the protection of the public, or otherwise in the public interest or in the interests of the registrant himself. The committee must of course permit both parties to make their submissions on the need for an interim order and, if one is to be made, its nature and terms. For that purpose it must consider the nature of the evidence on which the allegation made against the registrant is based. It is entitled to discount evidence that is inconsistent with objective or disputed evidence or which is manifestly unreliable. The committee may receive and assess evidence on the effect of an interim order on the registrant, and the registrant is entitled to give evidence on this. The registrant may also give evidence, if he can, to establish that the allegation is manifestly unfounded or manifestly exaggerated; but the committee is not otherwise required to hear his evidence as to whether or not the substantive allegation against him is or is not well founded: that is not the issue on the application for an interim order.
>
> 20. What the committee cannot do, and should not do, is to seek to decide the credibility or merits of a disputed allegation: that is a matter for the substantive hearing of the allegation by the conduct and competence committee, pursuant to article 27 of the 2001 Order. Necessarily, at the interim stage, the committee must not and cannot decide disputed issues of fact in relation to the substantive allegations. The committee must also be extremely cautious about rejecting or discounting evidence on the basis that it is incredible or implausible.
>
> [...]
>
> 31. In my judgment, *Wright's* case [*R (Wright) v. Secretary of State for Health* [2009] AC 739] is not authority for the proposition that fairness requires that a respondent to an allegation of unfit to practise his profession must be given an opportunity to give evidence as to the substance of that allegation before a tribunal considering whether to make an interim suspension or other interim order under a legislative scheme such as the present.
>
> 32. Furthermore, I accept [counsel for the NMC]'s submission that [counsel for the applicant]'s submissions are inconsistent with the statutory scheme. Under that scheme, it is the conduct and competence committee, and not the investigation committee that decides on the merits of the allegations against the registrant. Moreover, the committee may make an interim order when it has not yet reached a decision that there is a case to answer, and it did so in the present case.
>
> 33. In my judgment, if a registrant is to be given an opportunity to give evidence to an investigation committee on the substance of the allegations against him, with a view to the committee rejecting those allegations on the merits, fairness would require that the NMC should have the opportunity to call the complainant and any other evidence in support of its allegations. The result would be a trial before the trial. That is not what the statutory scheme envisages or what fairness requires at the interim stage.

*Hussain v. General Medical Council* [2012] EWHC 2991 (Admin)

**38.14** Interim orders panel hearing—subsequent rule 7 letter

On 24 September 2012, a rule 7 letter was sent, subsequent to an IOP hearing on 26 June 2012, setting out allegations on which the GMC now intended to rely. His Honour Judge Pelling QC (sitting as a judge of the High Court) held that the Court was entitled to consider subsequent developments: see *Sandler v. General Medical Council* [2010] EWHC 1029 (Admin), [12]. Allegations relied on before the IOP were not included in the rule 7 letter; the only issues featuring in the rule 7 letter were probity issues, which were apparently not considered appropriate for interim relief by an earlier IOP. The learned judge said that the question that had to be considered was whether, in the aggregate, the low-grade probity issues justified the imposition of any of the conditions that had been imposed. It was relevant to take account of the fact that the claimant had worked for most of his medical career in accident and emergency medicine. The material available suggested that he had been a highly valued practitioner in that specialism and at that grade. There was no material available to suggest that the claimant had any patient safety issues in relation to his work, and it was difficult to see how the low-level probity issues to which reference had been made could, of themselves, impact indirectly on current patient safety issues. Interim conditions were terminated.

*Kakade v. General Medical Council* [2013] EWHC 2110 (Admin)

**38.15** Second interim orders panel hearing—fuller evidence

A complaint was made against a general practitioner (GP) by solicitors on behalf of relatives four years after the patient had died. No order was made by the initial IOP, because the incident had occurred more than five years prior to the hearing, no further clinical concerns had been identified, and there was nothing to suggest a repetition. A subsequent hearing was based on a second medical report, fuller than that obtained before the first hearing, and a subsequent allegation concerning an erroneous answer given by the practitioner in a performance list application form. The Panel in its determination found 'wide-ranging concerns' and imposed an interim suspension order. The suspension was terminated on appeal by the Court. The initial Panel had decided to take no action in relation to the treatment of the patient and the subsequent probity issue identified was of a low-level nature.

*Bhatnagar v. General Medical Council* [2013] EWHC 3412 (Admin)

**38.16** Application to terminate suspension—hearing—judgment not handed down—subsequent panel decision to impose conditions—effect on Court's judgment

On 17 January 2013, the GMC's IOP ordered that the registration of B, a consultant ophthalmologist in private practice, should be suspended for twelve months. This followed inspections by the Care Quality Commission (CQC) of B's clinic in Rotherham and allegations received by the GMC from the former manager of a clinic in Liverpool. Following a hearing in the Administrative Court, but before handing down judgment, the Court notified the parties on 2 July 2013 that the judge, Edwards-Stuart J, had decided to dismiss B's application. On 16 July 2013, the IOP met to review the order made on 17 January 2013 and replaced the suspension with conditions. On 11 October 2013, a further IOP carried out a review and decided to continue the order of conditions on the same terms. B applied for a review of the decision notified to the parties on 2 July 2013. In dismissing B's application to adduce further evidence, Edwards-Stuart J accepted that it has now been held by the Supreme Court in *Re L & B (Children)* [2013] UKSC 8 that an exceptionality test is not the correct approach to be adopted when a court is deciding whether or not to alter its decision at any time before the order giving effect to it is drawn up. However, the decision to dismiss the appeal should stand.

(1) The decision had been communicated to the parties, and whilst it is open to the court to revoke or vary its decision, in this type of case there would have to be compelling circumstances before such a course would be appropriate.

(2) The decision was based on the situation in January 2013, not upon the situation in July 2013. The continuing uncertainty about when the substantive hearing would take place was a factor that could properly have carried more weight in July 2013 than it did in January 2013.
(3) The determination of the IOP in July 2013 suggests, on its face, that it took into account further information over and above that which was before the Panel in January 2013.
(4) Even if the Panel (albeit differently constituted) had simply changed its mind as to the sanction that would have been appropriate in January 2013, that would not of itself be a reason why the Court's appraisal of the situation—reached independently—should change.

Accordingly, the two decisions of the IOP did not provide any sound justification for altering the conclusion reached by the Court following the hearing.

***Kashif v. General Medical Council* [2013] EWHC 4185 (Admin)**

**38.17** Panel hearing—subsequent events

On 18 December 2012, the IOP suspended K's registration for eighteen months; on 29 May 2013, the order was continued. K challenged the decision to maintain the suspension. In August 2013, the GMC prepared its rule 8 letter. The allegations were significantly narrowed and, by the time of the hearing of K's application before Cranston J in October 2013, the situation was different from that which was before the Panel in May 2013. The learned judge noted that, in the course of his decision in *Malik v. General Medical Council* [2013] EWHC 2902 (Admin), His Honour Judge Raynor QC (sitting as a judge of the High Court) said, at [3], that he had to consider the position on the evidence before him and that the evidence was not limited to that which was before the IOP. Cranston J said that the Court needed to consider the matter as it now was. In all of the circumstances, the continuation of the interim suspension would be disproportionate and it would be discharged.

***General Medical Council v. Paterson* [2014] EWHC 201 (Admin)**

**38.18** Doctor lacking capacity—provision to enable extension to be varied or set aside

The GMC applied, pursuant to section 41A of the Medical Act 1983, for an order extending an interim order of suspension for a period of twelve months. His Honour Judge Pelling QC (sitting as a judge of the High Court) said that the complicating factor in this case was that, by a letter dated 10 December 2013, the solicitors who were acting for the defendant, R, informed those who were acting for the GMC and informed the Court that they did not consider the defendant to have capacity to give instructions in relation to the hearing and, in consequence, they were not in attendance. The allegations against the defendant were serious, involving a number of operative procedures in relation to female patients concerning breast cancer. It was clear from the correspondence from the defendant's solicitors that there was an ongoing discussion with the Official Solicitor as to whether or not the Official Solicitor was prepared to act as the litigation friend of the defendant. The issue was whether any, and if so what, provision ought to be made in any order extending the period of suspension to take account of the defendant's capacity issue. The GMC's submission was that it was not strictly necessary to make provision within any order extending the period of supervision given that there are statutory mechanisms by which interim orders can be challenged or varied where additional evidence is available. However, the Court considered that this did not quite meet the problem, which is that if someone lacks capacity and an interim order is made against him or her, then there ought to be provision within the order that enables such a person, once properly represented, to apply to vary or set aside the order. Accordingly, to guard against the possibility that applications might be made using any residual power included in the order, in a way that was not contemplated, the order should be subject to a power enabling the defendant, by his litigation friend, if so appointed, to apply to vary or set aside the order.

### *Waghorn v. General Medical Council* [2012] EWHC 3427 (Admin); [2013] 1 CMLR 45

**38.19** Section 41A Medical Act 1983—provisions explained—reference to ECJ refused

In dismissing W's application to refer questions to the European Court of Justice as to whether the power to impose restrictions of up to eighteen months on a practitioner's registration, without provision for the practitioner to call evidence, was contrary to EU law, Stuart-Smith J said that the effect of the provisions in section 41A of the Medical Act 1983 as amended can be summarized as follows:

> 5. The [IOP] may make an order that is stated to last up to 18 months. But there are four separate safeguards which provide the practitioner with protection. First, where an interim order has been made that is stated to last for more than six months it must be reviewed within six months of the date on which the order was made and must thereafter be reviewed within six months of the immediately preceding review or at any time more than three months after the immediately preceding review if the practitioner requests such a review – see subs (2)(a). Secondly if the General Medical Council think that an order should be extended beyond 18 months, it must apply to the court which has the power to extend the order for a further period of 12 months – see subs (6) and (7). Thirdly, the practitioner may apply at any stage to the court, for the order, including an order for interim suspension to be revoked or varied – see subs (10). Fourthly, in addition the order may be reviewed at any stage if fresh evidence comes to light – see subs (2)(b).
>
> 6. The task of an [IOP] under subs (41A)(1) is to determine whether or not it is necessary for the protection of members of the public or is otherwise in the public interest or in the interests of the doctor to impose interim restrictions on the doctor's registration. It is not the function of the [IOP] to investigate or make findings of fact regarding allegations against the doctor in question: see *R (on the application of Ali) v. General Medical Council* [2008] EWHC 1630 (Admin) at [31], [35] and [38].
>
> 7. At a review hearing, such as is mentioned in section 41A, the [IOP] has power to revoke the order, revoke or vary any condition imposed by the order, replace an order for interim conditional registration with an interim suspension order or vice versa: see section 41A(3). The function of the [IOP] on review is to undertake "a comprehensive reconsideration of the initial order in the light of all the circumstances which are then before the [IOP]": see *Madan v. General Medical Council* [2001] EWHC Admin 577 at [73].
>
> 8. On an application to the High Court under section 41A(6) the High Court can extend the order for up to 12 months: see section 41A(7). The criteria for granting an extension are the same as for an original interim order under section 41A(1), namely the protection of the public, the public interest or the practitioner's own interest – see *General Medical Council v. Hiew* [2007] EWCA Civ 369 at [26]–[28].
>
> 9. An application under section 41A(10) is a "full appeal" by way of rehearing and not simply a review of the [IOP]'s decision. The court does not interfere on a review ground but itself decides what order is appropriate. The task of the court on a section 41A(10) application is to decide whether or not the decision made by the [IOP] is correct and should be extended, if not, what decision to substitute for it, in terms of either termination of the suspension, revocation or variation of conditions or abbreviation of the period specified in [the] interim order: see *R (on the application of Walker) v. General Medical Council* [2003] EWHC 2308 (Admin) at [3] and [8].

The learned judge said that nothing had been placed before the Court to give it any reason to think that there was, in fact, a conflict between the statutory structure that is in place in the United Kingdom and any relevant provisions of European law. He was satisfied that it was unnecessary for any reference to be made.

### *General Medical Council v. Kaluba* [2013] EWHC 4212 (Admin)

**38.20** Application for extension of interim order—service—CPR Part 6

In granting an extension of the current period of suspension of the defendant doctor, His Honour Judge Gore QC (sitting as a judge of the High Court) said that since the defendant was not present or represented, he must first be satisfied that he had validly and effectively been served with notice of the GMC's application. Service, as required by CPR 54.7, must be on the defendant, but does not have to be personal service. Therefore CPR Part 6 applies and permits service by a number of different means, including, if so ordered, service by alternative means. Pursuant to an application previously made, the learned judge made an order for service by alternative means in this case—namely, by email. The GMC and the defendant, in the course of investigations, had corresponded with each other via an email address and there was recent correspondence with that email address. Transmission of the GMC's application by email to that address did occur and there was no evidence of that email bouncing back or of any delivery failure in respect of it. Accordingly, the Court was satisfied that the defendant had been validly and effectively served with notice of the proceedings.

## C. Necessity, etc., for Interim Order

### *Dr X v. General Medical Council* [2001] EWHC Admin 447

**38.21** Allegations of indecent assault—criminal proceedings— no determination—powers of court

X, a GP, faced allegations of indecent assault on two of his nieces (now aged 15 and 13). X applied to the Court by virtue of section 41A(10) of the Medical Act 1983 to quash an order of the IOC to suspend his registration for eighteen months. Criminal charges against the claimant had not yet proceeded to trial in the Crown Court. In dismissing X's application, Pill LJ (with whom Silber J agreed) said:

> [24] I have referred to the limited nature of the material which was before the IOC. It was for them to examine the material before them with care. It is plainly a worrying situation when a professional man may be suspended on the basis of allegations of criminal conduct which, as yet, are untested in a court of law. I cannot, however, accept that the power to suspend by way of interim order, provided in section 41A, must not be exercised because the allegations are untested in a court. Nor, in my judgment, can it be said that the exercise of the power to suspend was inappropriate because the conduct alleged was not towards patients of the claimant.
>
> [...]
>
> [26] The three grounds [in section 41A(1)] overlap, reflecting different aspects of the duties of the IOC as a professional body concerned with the protection of the public and with the professional standards of its members. Each of the grounds must nevertheless be considered specifically. In my judgment on each of the grounds there was material upon which the IOC were entitled to reach the conclusion they did. They were also entitled to reach it as a general conclusion.

It was common ground that it is not open to the Court to substitute an interim conditional registration or an interim suspension order upon an application under section 41A(10). The power of the Court, subject to its power under section 41A(10)(c) is either to quash or to uphold the order of the Committee.

### *R (Walker) v. General Medical Council* [2003] EWHC 2308 (Admin)

**38.22** Criminal proceedings—suspension—reasons—public interest—prejudice to practitioner

In dismissing W's application under section 41A(10) of the Medical Act 1983 on the basis that the IOC was entitled on review to continue an interim suspension order, Stanley Burnton J said that the reasons given by the Committee were inadequate. They did not expressly identify the public interests concerned. The Committee did not, in terms, refer to a need to suspend W for the protection of members of the public and there was no analysis of the consequences of the order for W. A second source of discomfort was that the question of prejudice was not adequately investigated before the Committee. There were two aspects of prejudice to be considered. One was the general effect of any suspension, which precludes a doctor from working. It is the most serious interim measure and is to be applied only in serious cases. The second aspect that fell for consideration was the immediate effect of the suspension on the doctor's remuneration and his office (compare section 47(3) of the Medical Act 1983). In the instant case, in which there were multiple manslaughter charges against a doctor, very serious consideration had to be given to an interim suspension, although, of course, such a measure cannot be applied as a matter of course. The decision and the jurisdiction are those of the IOC and not the Crown Prosecution Service (CPS). But, when one takes into account that such charges are not prosecuted except on substantial grounds, an IOC will, in cases in which there has been no previous consideration of the relevant facts, give the greatest and most anxious consideration to an interim suspension of the doctor in question. Confidence in the medical profession will be at risk if doctors who face charges of that level of seriousness are free to practise during the dependency of the criminal proceedings. The instant case differed from the generality in that the PCC had previously reached a view as to the seriousness of the appropriate action before criminal proceedings were instituted, but the new charges cast another light on that.

### *R (Shiekh) v. General Dental Council* [2007] EWHC 2972 (Admin)

**38.23** Criminal conviction—false travel claims—suspension on public interest grounds—appropriate test

On 21 November 2006, the IOC of the General Dental Council (GDC) suspended S's registration for eighteen months. S had pleaded guilty at Northampton Crown Court to one count of conspiracy to defraud in relation to false claims for travelling expenses made by associates at his dental practice. S stood to receive a percentage of the travel claims made by those associates. On 19 December 2006, S issued a claim form under CPR rule 8 to challenge the decision of the Committee under section 31(12) of the Dentists Act 1984. It was common ground that, for the purposes of section 32(4), the only relevant statutory test for an interim order that applies is that which relates to the public interest. It was agreed that interim suspension was neither sought, nor could be justified by reference to considerations of what is necessary for the protection of the public or what was in the interests of the practitioner concerned. In terminating the suspension, Davis J observed:

> 15. As a matter of strict language, no grammatical interpolation of the word "necessary" falls to be applied to the phrase "or is otherwise in the public interest". But that is not the end of the matter because it does seem to me that if "the public interest" is to be invoked in this context, under the statute, then that, to my mind, does at least carry some implication of necessity; and certainly it at least carries with it the implication of desirability...
>
> 16. At all events, in the context of imposing an interim suspension order, on this particular basis, it does seem to me, adopting the words of [counsel for the appellant], that the bar is set high; and I think that, in the ordinary case at least, necessity is an appropriate yardstick. That is so because of reasons of proportionality. It is a very serious thing indeed for a dentist or a doctor to be suspended. It is serious in many cases just because of the impact on that person's right to earn a living. It is serious in

> all cases because of the detriment to him in reputational terms. Accordingly, it is, in my view, likely to be a relatively rare case where a suspension order will be made on an interim basis on the ground that it is in the public interest. I do not use the words "an exceptional case" because such language is easily capable of being twisted and exploited in subsequent cases; but I do think, as I say, it is likely to be a relatively rare case. Ultimately, of course, all these things have to be decided on the facts of each particular case.

The difficulty in the present case was trying to get a purchase on why it was that the Committee thought that interim suspension was needed. This was not a case of ongoing or future risk, or anything like that. Many of the points that the Committee made might well have been apposite if one were considering at the final hearing why suspension was needed. But what was difficult to grasp, certainly from the reasons that were given by each of the panels so far as there were reasons, was their assessment that suspension was needed on an interim basis. If it was to be the case that the justification was that of public perception and public confidence, then one might have thought that that could be reflected by an appropriate decision by the panel, if so minded, at the final hearing when all of the facts had been fully explored, all of the mitigation fully advanced, and the position finally assessed at that stage. It was not clear from the reasons of the panel why *interim* suspension, on public interest grounds, was called for in this particular case. Whether suspension is called for at the *final* hearing is a different matter which should be decided by the panel then hearing the case in the light of all of the arguments and evidence put before it.

### *R (Ali) v. General Medical Council* [2008] EWHC 1630 (Admin)

**38.24** Protection of public—no requirement to find impairment—nor findings of fact

On 19 April 2007, the IOP determined to impose conditions on A's registration that confined him to working in senior house officer (SHO) posts in anaesthesia for periods of not less than three months' duration under the supervision of a named consultant. On 11 October 2007, the Panel replaced the previous order attaching conditions with an order suspending A's registration for the remainder of the eighteen-month period. On 9 January 2008, the order for suspension was maintained, and thereafter A commenced a Part 8 claim pursuant to section 41A(10)(a) of the Medical Act 1983 to terminate the interim order of suspension. In dismissing A's application, Stadlen J said, at [31], that, by contrast with section 35D of the Medical Act 1983 (no direction can be made unless the fitness to practise panel makes a finding that the person's fitness to practise is impaired), in the case of section 41A (the making of a suspension or conditional direction), the test is that the IOP or the fitness to practise panel must be satisfied that it is necessary to make such a direction 'for the protection of members of the public or is otherwise in the public interest, or is in the interests of a fully registered person'. It is not there specified as a condition precedent that the panel, before making such a direction, must make a finding of fact that his fitness to practise is impaired. That is not surprising, because the permanent erasure of a name is a more draconian consequence, and one that can be brought about only as a result of a full factual investigation. The same is not the position in the case of an interim suspension or imposition of conditions because it is in the nature of things that there may be circumstances in which, before it is possible fully to examine the facts with all of the procedures and safety net provided by the rules, the allegations that are made against a person are of such seriousness as to require a temporary suspension for the safety of the public.

> [38] It is not, in my judgment, the function of this court in determining an application under section 41A(10)(a) to terminate the suspension order, to reach findings of fact as to whether the allegations that were before the Interim Orders Panel are true. It was not, in my judgment, as a matter of law, the function of the Interim Orders Panel itself

on 9 January 2008 to make findings as to whether, on the balance of probabilities, the allegations against Dr Ali, which will be determined by the Fitness to Practise Panel, are true.

[39] What that Panel was concerned with was whether it was satisfied that it was necessary for the protection of members of the public that Dr Ali's suspension should continue. In reaching that decision, it was bound to take account, and it plainly did take account, of the nature of the allegations that had been made against him. I would add that, as appears from the transcript of the hearing, the Panel had very much in mind the contents of the report of the assessment team.

### *R (Sosanya) v. General Medical Council* [2009] EWHC 2814 (Admin)

**38.25** Davis J summarized the facts as follows.

Criminal proceedings—money laundering—public interest—need for cogent reasons—proportionality

7. The background leading to [S] coming before the Interim Orders Panel is, shortly put, this. [S] is married. Her husband is apparently an accountant. Relatively recently her husband was charged with what has been described as an advance fees fraud. In due course he pleaded guilty. He was sentenced to a term of two years and nine months' imprisonment. [S] herself was arrested in the latter part of 2008. In March 2009 she was herself charged with an offence, being an offence of money laundering. The details of the offence with which she is charged are very sparsely set out in the papers before me. They were very sparsely set out in the papers placed before the panel.... The trial is due to take place, I gather, in the earlier part of next year.

8. [S] denies the charge or charges made against her and has pleaded not guilty. Neither I nor the panel have been shown any indictment or defence statement for example, nor is it revealed what sums of money are at stake. Given the fact that her husband has on his own plea been convicted of an offence of fraud, it may be that it is alleged that moneys which apparently are said to have gone into [S]'s bank account may have come from a fraudulent source connected with her husband: although that in itself is not revealed at the moment....

9. At all events, it seems clear... that what prompted the matter to be put before the Interim Orders Panel of the GMC was the fact that [S] had been charged with money laundering and was facing trial in the Crown Court.

The hearing before the IOP was relatively short. There was no oral evidence given and no cross-examination. In making a suspension order for eighteen months, the Panel, in its decision, stated that it was satisfied that there may be impairment of S's fitness to practise that posed a real risk to members of the public, or might adversely affect the public interests or her own interests. Davis J, in terminating the suspension, said at [23] that it was very unclear as to why the Panel thought that all three limbs of section 47A(1) could be invoked in this context.

21. ... [S]elf-evidently there was no evidence to suggest that the conduct of [S] relied upon had any bearing on clinical issues. There was no suggestion that the charge of money laundering had any relevance in itself to a clinical issue; indeed the evidence was to the contrary, that S had shown herself a good doctor.

22. It is also very hard to understand why the panel concluded that it was in [S]'s own interest that she be suspended.... Her husband could not earn because by now he was in prison. She had two young children to look after. Above all, she had her reputation to think of.

As to the public interest:

27. ... [S] was facing charges, which can be described as serious charges... but which were not of the gravest kind and which were charges which she was denying[.] She had been convicted of nothing. No risk in her continuing to practice [*sic*] in the interim

> was identified.... It may well be that if she was convicted at the end of the day a fitness to practise panel at that stage could impose the appropriate sanction as it thought necessary. But that does not meet the point as to whether it was necessary to suspend her on an *interim* basis.

In this case, the IOP had not given any cogent reasons as to why it is necessary in the public interest or otherwise for S to be suspended as an interim measure. Interim suspension was not necessary or proportionate given the nature of the charges that she faced (and which she denied).

### *Sandler v. General Medical Council* [2010] EWHC 1029 (Admin)

**38.26** Criminal proceedings—cremation certificates—allegation in course of clinical duties—public interest grounds

S was employed by the Heart of England National Health Service (NHS) Foundation Trust. In December 2008, the Trust held a disciplinary hearing relating to forms that S had completed under the Cremation Act 1901. The allegations against S were that he received payments for the period 2007–08 for cremation duties, whilst failing to examine the bodies of patients and failing to meet his professional duties. Subsequently, S was charged with offences under section 8(2) of the 1901 Act of wilfully signing false certificates. The GMC's IOP suspended S's registration for eighteen months. It decided that it was not necessary to make an interim order for the protection of the public or in S's own interests. However, it did think that:

> ... [T]hese are serious matters which demonstrate that there may be impairment of your fitness to practise which adversely affects the reputation of the profession in the eyes of the public, and, after balancing your interests against those of the public, an interim order is necessary.

As to the merits, Nicol J, at [23], said the charges brought against S were serious. One incident might have been regarded as an aberration, but here the wilful signing of false certificates is alleged to have taken place on at least 116 occasions over a number of years. It is significant that this lack of probity is alleged to have occurred in the course of the doctor's clinical duties (a distinction, incidentally, from the frauds that were alleged in the *Sheikh* case and which Davis J did not think justified interim suspension). Whilst the Trust thought that a final warning and conditions would suffice, it referred the matter to the GMC and the police. The GMC has a wider responsibility and the scale of S's alleged offending had enlarged somewhat by the time the Panel came to consider the matter. The application to terminate the suspension was dismissed.

### *Bradshaw v. General Medical Council* [2010] EWHC 1296 (Admin)

**38.27** Employee—CAA—issues of probity and integrity—public interest grounds

The claimant doctor applied to terminate an order of the GMC's IOP to suspend his registration. The claimant, whilst employed as a medical officer by the Civil Aviation Authority (CAA), had been suspended on full pay pending an investigation into a number of allegations of misconduct arising out of an alleged affair between him and another employee, a doctor. The CAA's disciplinary hearing held that, had he not resigned during the investigation, he would have been dismissed without notice. The IOP subsequently suspended his registration on the grounds that there could be impairment of his fitness to practise that posed a real risk to members of the public or could adversely affect the public interest. His Honour Judge Roger Kaye QC (sitting as a judge of the High Court), in dismissing the application, held that the Panel had been correct to order the claimant's suspension. The allegations did not involve a criticism of his clinical competence, and note was to be taken of his impressive academic record and positive testimonials, the potential financial and career consequences of suspension, the fact that the full hearing might not take place for some time, and the claimant's denial of the charges. An interim suspension would usually be viewed as disproportionate where allegations arose out of an alleged personal intimate relationship and there was absent any suggestion or criticism of clinical performance or abuse of patient safety. In addition, the making of interim suspension orders on public interest grounds in cases of non-clinical allegations would ordinarily expect something that would impinge more directly on members of the public, such as murder, rape, or abuse of children. However, in the instant case, the matters were serious, with serious implications as to the appellant's probity and integrity. The allegations went much further than accusation and counter-accusation against and by persons involved in an intimate relationship, to include allegations of false accusations, fabricating and altering documents, and lying to the investigator. Although the allegations involved a colleague and not a patient, a member of the public could ask whether the appellant would seek to cover up or lie or to make false accusations to defend himself if a complaint were made against him by a patient. Such factors would likely undermine public confidence in a doctor's core duties and responsibilities of honesty and integrity.

### *Jooste v. General Medical Council* [2010] EWHC 2558 (Admin)

**38.28** Doctor's premises—not registered with CQC—risk to members of public

J applied for an order terminating an interim suspension order made in his absence by the GMC's IOP. The CQC referred J to the Fitness to Practise Directorate following an inspection of his premises and concerns. In refusing J's application, Nicola Davies J confirmed that the case law—*R (Ali) v. General Medical Council* [2008] EWHC 1630 (Admin) and *Sandler v. General Medical Council* [2010] EWHC 1029 (Admin)—reflects that the function of an IOP is not to make findings of fact as to whether allegations before it are true. The task of the Panel is to determine whether it is necessary for the protection of members of the public, in the public interest or the doctor's interest, that an interim order should be made. In the instant case, before the Panel was undisputed evidence that prescriptions written by J were found in unregistered premises where it was said no doctor or pharmacist was employed. There was evidence before the Panel that raised concerns as to where J was working, what he was doing, and whether such premises were registered with the CQC. Such concerns were relevant to the issue of J's fitness to practise and any risk that he might pose to members of the public. On the documentary evidence before it, it was open to the Panel to determine that an interim order of suspension was necessary.

### *Tosounides v. General Medical Council* [2012] All ER (D) 206 (Jul)

**38.29** Application for post—form—advice sought from GMC—interim suspension terminated

T, a consultant histopathologist, applied to set aside an order of interim suspension made by the IOP in March 2012. An NHS Trust had referred T to the GMC regarding differing accounts provided by him on his application form for a post as to the reasons for leaving his previous employment and the account given by his previous employers in a letter of reference. T had sought advice from the GMC as to whether he was required to disclose the fact that his previous employer had written to the GMC. On the form to his new prospective employer, he had stated, correctly, that he had resigned and indicated that he was under investigation by the GMC, but gave no further details. The IOP, in directing that T's registration be suspended for eighteen months, found that there were concerns over his probity, and that no conditions could be imposed that would adequately protect the public and the profession. His Honour Judge Mackie QC (sitting as a judge of the High Court) granted T's application on the grounds that the Panel had fallen into error. All allegations of a want of probity were serious, but the allegations in the instant case were not grave compared to examples given in the guidance and in other cases. Further, the material relied on was thin given the serious consequences for T compared with what was seen in other cases.

### *Abdullah v. General Medical Council* [2012] EWHC 2506 (Admin)

**38.30** Allegations of sexual misconduct—patient—no criminal proceedings—interim suspension—twelve months

Lindblom J held that the GMC's IOP was entitled to suspend A's registration pending allegations of serious sexual misconduct against him being investigated. A had an unblemished professional record and denied the claims. No criminal proceedings had been brought and his local primary care trust (PCT) had decided not to suspend him. However, there was a need to maintain public confidence in the medical profession, and the allegations against A were very serious and were not so vague or inconsistent as to justify no action being taken. However, an interim suspension order of eighteen months was excessive and any order longer than twelve months would be disproportionate. Lindblom J said:

> 87. The scope of the court's jurisdiction under section 41A(10) of the 1983 Act is well established. The relevant jurisprudence is clear. In a case such as this the court is not constrained by the principles of public law that govern a claim for judicial review. I must decide whether the [IOP] were right to suspend the claimant while the allegations he faces are investigated. I must judge whether their decision was, and is, both justified and proportionate. Suspension will have been justified if it was necessary as a means of protecting members of the public, or if it was otherwise in the public interest, or if it was in the interests of the claimant himself as a registered person. I must look at the [Panel]'s determination and consider what weight I should give to it, remembering that Parliament has entrusted to them the power in the first instance to make decisions on a doctor's freedom to practise while his or her fitness to do so is investigated, that they bring to bear on their task their own experience and expertise and their own knowledge of the public's expectations of the medical profession, and that it is not their responsibility—or the court's—to make findings of fact or to resolve factual disputes.
>
> [...]
>
> 89. In this case, the complaints that have been made about the claimant's conduct are not trivial; they are, indisputably, very serious. They are complaints of sexual misconduct with a patient. Such allegations are treated as a category of their own in the GMC's guidance. Where sexual misconduct has been alleged, as paragraph 23 of the guidance makes clear, the "impact on public confidence if the doctor were to continue working unrestricted in the meantime" calls for "particular consideration". This is necessary not only when the doctor is being investigated by the police for a "sexual criminal offence", but generally when "sexually inappropriate behaviour towards patients" has been alleged. In paragraph 31 of its guidance the GMC acknowledges that, when an allegation of this kind is made, there may be "a significant risk to patient safety and public confidence in the profession if decisions at the interim stage are not seen to reflect the seriousness of the individual case". An important consideration here, as the guidance recognises (in paragraph 34), is the need "to maintain public confidence in the medical profession or the medical regulator". Action will sometimes have to be taken "to protect public confidence even where there is no immediate risk to patients".
>
> [...]
>
> 96. This is not to say that in every case where an allegation of sexual misconduct is made against a doctor, he or she must automatically be suspended while the investigation of the complaint runs its course. That notion would be misconceived. It would find no support in relevant authority or in the GMC's guidance. Had it come into the [IOP]'s thinking when they made their decision to suspend the claimant, they would have misdirected themselves. But there is no hint of it in their determination. Nor, as I understood him, did (counsel for the claimant) suggest there was. If this was his submission, I would reject it.

*Harry v. General Medical Council* [2012] EWHC 2762 (QB)

**38.31** Blood samples—breach of regulations governing transportation—interim suspension terminated

On 4 April 2012, H's registration was suspended by the GMC's IOP for eighteen months. H was a consultant in genito-urinary medicine. The principal allegations that primarily featured in the Panel's decision to suspend were that H had transported human blood samples in breach of regulations governing their transportation and that he had requested a member of staff to amend the records of a patient. The Court observed that the transportation of human blood is governed by strict regulations. On a return journey from Nigeria in December 2010, H accepted that he carried two samples of blood, one infected with HIV, in his hand luggage. He said that they were appropriately packed in accordance with the regulations, but agreed that the regulations required them to go in the hold. He said that he was unaware of this. On return to the United Kingdom, H opened the package in which the samples had been transported at home. He then took them to the laboratory. The regulations require that samples be opened in a laboratory. The rationale behind these strictures is not difficult to divine: should an incident occur in which a phial of infected blood is broken, there is a risk that people might inadvertently come into contact with it and be exposed to infection. The allegation of requesting a member of staff to amend the record of a patient was admitted by H and came about in this way. When he was confronted with the suggestion that he was using NHS resources for private patients, he asked a member of staff to amend the paperwork relating to a patient. But, within an hour and without any intervention by another person, he realized the folly of that course and reversed his request. In terminating the suspension of eighteen months imposed by the IOP, Burnett J said that the Panel was clearly concerned about the transportation of samples from Nigeria and the short-lived and admitted attempt by H to cover his tracks. H's case raised no concerns about 'patient safety' in the sense so often encountered in GMC cases. All too often the alleged conduct is such as to give rise to a real risk that if the doctor concerned were to continue to practise, patients would be put at risk. The risk in this case concerned the possibility that, as a result of the inappropriate transportation of blood samples in December 2010 and in the event of damage, a member of the public might have come into contact with infected blood. Suspension is concerned to protect against a real continuing risk. It is looking to the future, albeit in the light of what is alleged to have occurred in the past. Any continuing risk to the public could arise only if H were to repeat the error that he had made in the past with regard to transporting blood samples. Nothing in the material before the Panel, or in the papers before the Court, supported the proposition that, in the face of all that had occurred, H realistically might again transport blood contrary to the regulations in place. There was no risk to members of the public in H's continuing to practise. The Court was not satisfied that the inappropriate transportation of blood samples, accepted by H, coupled with the improper request to his colleague, also accepted, could justify his suspension in the public interest pending the resolution of the disciplinary proceedings. In considering whether the public interest calls for suspension (even before looking at proportionality), it is relevant to consider what the reaction of the public would be when the disciplinary process is complete and it becomes apparent that suspension did not occur. The information currently available suggested that H, a senior consultant of unblemished record, carried packed samples in his hand luggage when they should have been in the hold and opened them in his home rather than the laboratory. There was no doubt that H requested a member of staff to amend a record, but that he countermanded his request very quickly. That was unquestionably a serious matter—but public confidence would not be damaged were H able to continue to practise pending the resolution of the disciplinary proceedings. As to proportionality, the financial consequences of suspension are often likely to be serious. Even if the Court had been persuaded that the public interest called for suspension to be considered, it was very doubtful whether it would have been proportionate in this case.

***Rashid v. General Medical Council* [2012] EWHC 2862 (Admin)**

**38.32** Doctor arrested as part of alleged conspiracy—dispute with partner and PCT—no clinical concerns—suspension terminated

The GMC had received a letter from the deputy medical director of the PCT to which R was contracted. The GMC was informed that R had been arrested by West Yorkshire Police, who were investigating an alleged significant conspiracy to defraud. The allegation involved part of a 'cash for crash' fraud, and referred to 'a doctor' engaged in handling examinations and subsequent reports. R had also been involved in a recent dispute with his former partner and, since the split, he had been in dispute with the PCT over his contractual arrangements. R had secretly recorded a conversation with the head of primary care contracting at the PCT and part of the conversation was published on YouTube. A further allegation was that patient notes were strewn around the doctor's house and car, and that eight packets of medication prescribed to five different people other than the doctor and his family were seized from his bedroom. There were, however, no clinical concerns concerning R and he produced a number of positive references to the IOP. Counsel for the GMC at the hearing before the Panel contended that both protection of members of the public and the public interest applied, but on appeal accepted that only the public interest limb could be justified. In terminating the interim order for suspension, His Honour Judge Gosnell (sitting as a judge of the High Court) said that the Panel in this case had not given sufficiently cogent reasons why R should be suspended as an interim measure. There were no concerns about R's clinical fitness. The IOP relied on the seriousness of the matters before it, without going into any assessment of which were the more serious and why it took that view. No risk in R continuing to practise on an interim basis was identified, save for serious public concerns that were not otherwise defined. The allegations, individually or cumulatively, were not sufficiently serious to justify his suspension.

***Scholten v. General Medical Council* [2013] EWHC 173 (Admin)**

**38.33** Length of suspension—proportionality

Dr S was a consultant plastic and cosmetic surgeon practising in female genital plastic and cosmetic surgery. On 27 February 2012, whilst practising from Fitzwilliam Hospital in Peterborough, he took a photograph of a patient's external female genitalia without her consent. He used the camera on his iPhone. On 2 March 2012, the Fitzwilliam Hospital suspended Dr S's practising privileges and referred him to the GMC. On 10 April 2012, the GMC's Interim Orders Panel suspended Dr S's registration for eighteen months. The Panel reviewed the order on 8 October 2012 and maintained it. Dr S applied under section 41A(10) of the Medical Act 1983 to terminate the suspension. In granting Dr S's application, Supperstone J said, at [34]–[35], that, in the decision of 8 October 2012 maintaining the suspension, there was no sufficient balancing of the risk to members of the public (protection of patients) and the public interest, and the impact of a suspension order upon Dr S. The Panel did not identify the risk posed by Dr S remaining in practice pending the resolution of the allegations against him or, more importantly, the degree of that risk (see *Houshian v. General Medical Council* [2012] EWHC 3458 (QB), [34]). The learned judge went on to say, at [42], that, in its determination, the Panel did not indicate what, if any, consideration it gave to the consequences of the suspension for Dr S when concluding that the order of suspension was a proportionate response.

> 43. Further, there is no reference in the Panel's decision to the length of the suspension order in the context of considering whether maintaining the order is a proportionate response. In *Harry v. General Medical Council* [2012] EWHC 2762 (QB) at para 18 Burnett J said, in response to an indication from counsel that if an interim suspension

order is made it is likely to be for 18 months: "It should not be overlooked that Parliament has provided that 18 months is the maximum period of suspension that the Panel can impose. There will be many cases in which suspension is proportionate for a short period but not for as long as 18 months, given the very serious consequences it has upon the doctor concerned. 18 months should not become a default position." [Counsel for the GMC] makes the point that the Panel conducting the review in October 2012 did not have the power to impose a shorter period of suspension than that imposed by the IOP in April 2012. That is correct. Nevertheless the length of the suspension order that has been imposed is, in my view, a factor to be taken into account when considering whether the order of suspension is a proportionate response.

44. In *Madan* [*Madan v. General Medical Council* [2001] EWHC Admin 577] at para 68 Newman J observed: "This case demonstrates the length of the delay which can take place and how the period of suspension from practice can be very long. The suspension is capable of giving rise to serious and grave consequences for the future professional career of a doctor, as well as creating immediate consequences of hardship." The facts of the present case are relatively straight-forward. It concerns a single incident in respect of which Dr S has admitted the conduct which is the subject matter of the investigation by the GMC. He has been suspended for nine months and the Rule 7 stage in the GMC's procedures has not yet been reached. [Counsel for Dr S] does not expect any hearing to take place before a FTP Panel before autumn 2013. [Counsel for the GMC] does not suggest that it will be before the summer of 2013. As Burnett J noted in *Harry v. GMC* at para 18, "the pressure on the Fitness to Practise Panel of the GMC is well known".

45. For these reasons I have given, in my view, the Panel conducted no sufficient balancing of the risk to members of the public (protection of patients) and public interest, and the impact of a suspension order upon Dr S.

### *Patel v. General Medical Council* [2012] EWHC 3688 (Admin)

**38.34** Criminal proceedings—suspension—whether confidence in profession undermined—suspension terminated

P was a 73-year-old GP, who was referred to the GMC by the Metropolitan Police on 7 April 2011. He had been arrested in connection with the discharge of his duties as a governor of a school and the authorization of substantial payments to members of staff to which they were allegedly not entitled. More than a year later, on 7 June 2012, P was charged with conspiracy to defraud and with committing fraud by abuse of position. It was said that the unauthorized payments totalled £1.8 million and that the wrongdoing extended over the period 2003–09. On 4 July 2012, the GMC's IOP made an order of suspension for a period of eighteen months. P made an application to the Court under section 41A(10) of the Medical Act 1983 to terminate the interim order. Essentially, the ground relied upon was that P was a person of exemplary and unblemished character, and that the interim order was neither necessary, nor proportionate. Specifically, it was said that the public interest did not require interim suspension given that the allegations did not have any bearing upon clinical matters. The trial was due to take place in September 2013. In terminating the suspension, Eady J said that it seemed, from the Panel's determination, that the suspension was imposed simply as being 'otherwise in the public interest'; it was not suggested that the suspension was necessary 'for the protection of members of the public'. There was no reason to suppose, even if the charges brought against P were ultimately made out, that such conduct would reflect on his clinical competence or in any way have endangered the health of patients. The learned judge said that he had in mind, of course, that the total sum involved was very large and that the alleged wrongdoing went on for some six years. But what was critical was the nature of the wrongdoing: it would not be right to suspend purely because of the sum involved if otherwise, in principle, it would not be appropriate. The learned judge said, at [28]:

> I have to ask what a reasonable onlooker would think, in the event of this applicant ultimately being convicted of the conspiracy charge, about his being allowed to go

> on practising in the meantime. Would confidence in the profession be undermined? I would add that such a reasonable onlooker needs to have attributed to him knowledge of the relevant facts. Otherwise, the danger is of proceeding on a superficial analysis or even mere prejudice.

In terminating the suspension, the learned judge said that, in all of the circumstances, no reasonable and properly informed member of the public would be offended or surprised to learn, even following a hypothetical conviction at some point in 2013, that the applicant had been permitted to go on serving his patients in the interim. There was no evidence of any threat at all to their welfare. The decision would not undermine confidence in the medical profession.

### *Y v. General Medical Council* [2013] EWHC 860 (Admin)

**38.35** Allegations of sexual misconduct—conditions—chaperone

Following the decision of a fitness to practise panel being quashed because of inadequate reasoning—[2012] EWHC 2770—an IOP imposed conditions on Y's registration. The allegations of serious sexual misconduct made against Y involved a vulnerable patient. They were denied, and the patient and complainant had not given a consistent account. On 29 November 2012, the Panel imposed conditions on Y's registration, including the use of chaperones for a period of nine months. In dismissing Y's appeal, His Honour Judge Gilbart QC (sitting as a judge of the High Court) said that he did not accept that the Panel must apply a 'reasonable onlooker' test, derived from *Patel v. General Medical Council* [2012] EWHC 3688 (Admin). In that case, Eady J unsurprisingly held that it was unnecessary to suspend a doctor because of a rather technical fraud alleged against him in his role as a school governor and expressed the view that, given the facts of the case in question (and they were quite specific and particular), no member of the public would be offended or surprised to learn that he had been permitted to continue in practice. Public confidence in the profession, and the varying effect upon it of knowing the facts in issue, may be very different when the allegation is of sexual misconduct against a patient, as opposed to a fraud that had nothing to do with practice as a doctor. The IOP (and the Court, under section 41A) must take a view of the effect on public confidence of the information it has before it. The 'reasonable onlooker' test may be one that it chooses to adopt, but it is not the only one, nor need it be deployed in any and every case: [38]. The learned judge said that the Court's sympathy for him must be tempered by the need to guard against possible risks to patients, to the public interest, and to the public's confidence in the medical profession.

### *Kumar v. General Medical Council* [2013] EWHC 452 (Admin)

**38.36** Conditions—effect on ability to work—proportionality

Allegations were made against K, a locum hospital doctor, regarding his attitude and his clinical management of patients. An IOP imposed conditions, including that K confine his medical practice no higher than CT1 trust doctor or senior house officer (SHO) level in NHS hospital posts, where his work must be closely supervised by a consultant, or in general practice with no fewer than three partners, where his work will be supervised by a GP principal. The appellant claimed that the effect of the conditions had been greatly to restrict his employability and that he had, in fact, not found it possible to obtain employment, except possibly for one short period. What is more, he had lost the opportunity to become an Army doctor, had got into financial difficulties, and had lost his house. In dismissing K's appeal, Underhill J said that those were very serious impacts of the conditions. However, once the IOP had reached the view that there was a real risk that his clinical competence to work unsupervised was in serious doubt, of which there was evidence, it became impossible to argue that the imposition of the condition was disproportionate: at [37]. The learned judge said, at [39], that it was inevitable that the need to protect the public during the period before a competent investigation can be properly completed will often cause hardship and that, sometimes, that hardship will, in the end, prove to have been unjustified because the allegations will not be proved. That is a great misfortune in the cases in which it happens. But it is inevitable in any system in which a form of interim protection is required. (It can, of course, happen in other circumstances as well, including criminal proceedings in cases in which bail is denied.) The best that can be done is for the body making the interim decision to do its best to hold the balance fairly, but there was no reason to believe that that had not been done in this case.

### *Houshian v. General Medical Council* [2012] EWHC 3458 (QB)

**38.37** Finding of forgery in employment proceedings—interim suspension—proportionality—whether serious damage to public confidence

H, a consultant orthopaedic surgeon, sought the termination of an interim suspension order. The central allegation arose from a finding made against him by an employment tribunal that he had forged documents in relation to his claim for unfair dismissal against Lewisham NHS Trust. An interim order was made to protect the reputation of the profession and to maintain public confidence. In allowing H's appeal, King J said, at [13], that the importance of proportionality in determining whether an interim order should be made pending the resolution of as yet unproved allegations faced by the practitioner cannot be overstated. A suspension has potentially three important consequences for a practitioner: its impact upon the person's right to earn a living (in this case, the applicant's pre-suspension salary was in the region of £150,000); the obvious detriment to him or her in terms of reputation; and it deprives the practitioner of showing that, during the relevant period, he or she has conducted himself or herself well and competently. At [34], the learned judge said that it is necessary to consider the degree of risk and, in this context, the likelihood of *serious* damage to public confidence in the profession, and hence to the reputation of the profession, if the applicant were allowed to continue to work with patients pending the resolution of the unproven allegations against him. The learned judge said, at [35], that the tribunal did not, however, expressly focus upon the likelihood of *serious* damage to the public interest: it cannot be sufficient simply to reiterate the seriousness of the allegations faced by the applicant. Coupled with the delay, the court terminated the interim suspension order.

### *Malik v. General Medical Council* [2013] EWHC 2902 (Admin)

**38.38** Allegation of dishonesty—competent doctor—whether risk of repetition pending substantive hearing

M, a doctor working in the obstetrics and gynaecology department of Warwick Hospital, was alleged to have falsified assessments between December 2010 and April 2011. M asserted that he had acted honestly. There was no suggestion about M's clinical competence and M was regarded as a good doctor. The Court held that interim suspension could not be justified as being necessary for the protection of members of the public. As to whether an interim order of suspension ought to be continued to ensure public confidence, there was no evidence of any lack of probity or misconduct after April 2011, and it would not be right to proceed upon the basis that, during the period of investigation, there was likely to be serious risk of any repetition of the alleged dishonest conduct. The interim order for suspensions was quashed.

### *R (Nowak) v. Nursing and Midwifery Council* [2013] EWHC 4341 (Admin)

**38.39** Public interest—necessity for order—impact on registrant—balance

In dismissing N's Part 8 claim challenging the decision of the NMC's IOP imposing an interim suspension order of eighteen months on public interest grounds only, Carr J said, at [24], that, in relation to public interest, the bar is set high. In an ordinary case, necessity is the appropriate yardstick: see *R (on the application of Shiekh) v. General Dental Council* [2007] EWHC 2972 (Admin), and *Bradshaw v. General Medical Council* [2010] EWHC 1296 (Admin). Additionally, the Panel must take into account the impact that an order may have on the registrant, including financially and reputationally. The need for an order needs to be balanced against the consequences.

### *Martin v. General Medical Council* [2014] EWHC 1269 (Admin)

**38.40** Hospital appointment—forged references—no risk to patients at appropriate grade and supervision—doctor not suitable for middle-grade post

M asked the court to terminate an eighteen-month interim suspension order, or alternatively to substitute a shorter period, on the grounds that there was no evidence of risk to patients or risk to the public confidence that would have justified an order of suspension. Between August 2012 and January 2013, M worked in a middle-grade post in the accident and emergency (A&E) department of the Leighton Hospital Foundation Trust in Crewe. Subsequently, he obtained a placement through Medilink at Lancaster A&E department at Morecambe Bay at specialist registrar (SPR) grade. A letter of reference from Medilink was a forgery, as was a letter of reference purportedly written by a consultant doctor at Leighton Hospital. M denied that he had forged any document, although there was sufficient cogency in the evidence implicating the claimant in the forged references to justify an interim suspension order, if such an order were otherwise justified by reference to relevant risk. Authentic references and assessments from Leighton Hospital could be relied on to support the position that, as at January 2013, M did not pose a risk to patients, provided that he was appointed at the appropriate level and with the appropriate degree of supervision. However, those who had observed and supervised M over a sustained and extended period of time had reached the clear conclusion that he was not suitable for a middle-grade post in A&E, nor close to being suitable for such a post. If the claimant were involved in the production of the two documents that were accepted to be fraudulent, he would have been deliberately putting himself forward as competent and suitable for a middle-grade post within an A&E department. He would have been putting himself forward by reference to views that he was fabricating and attributing such fabrications to those whose considered view was precisely the opposite. He would have been putting forward a glowingly positive picture emanating from sources that he knew would not have supported it and who would have said something very different. A doctor who is engaged in this kind of fraudulent activity would directly be placing the public at a real tangible identifiable and present risk. Securing a medium-grade post, on the basis of that picture of reassurance, would mean knowingly and from the day on which the doctor walked into the A&E department discharging functions when the assessment was known to have been that the doctor lacked the competence to be entrusted with those functions. A doctor prepared to act in that way would be exhibiting one of two things: either a reckless self-confidence in his or her own competence, or a reckless indifference. Each of these constitutes 'real, tangible, identifiable and present risks'. As to the duration of suspension, this was a case in which the matter of interim suspension had been dealt with promptly, and therefore the investigation was necessarily only at an early stage. The claim would be dismissed.

### *Hayat v. General Medical Council* [2014] EWHC 1477 (Admin)

**38.41** GP—inadequate patient records—suspension—effect on doctor—period reduced from eighteen to twelve months

H's registration was suspended for eighteen months following a review of his GP practice records. The review concluded that there was a consistent and persistent lack of evidence that he had documented relevant and important findings in patient records that would allow reasonable safeguarding of patients and would contribute towards their continuity of care. It was evident that the standards of medical care provided by H were significantly deficient in terms of the entire core principles expected of a GP applying *Good Medical Practice*. Charles J said that the report made troubling reading. However, the Court was concerned with the length of the suspension, which was the maximum period of eighteen months. The learned judge, in shortening the period of suspension to a further six months, said:

> 32. As soon as you stop to think of what you are doing to a doctor who is suspended, and when you identify what is being done to [H]; given the mortgage that he has taken out to buy the premises for his practice and the expenditure he has, the considerable significance of depriving somebody of being able to make his living by exercising his professional qualifications as a doctor is very apparent. To find it follows that there is a real need to identify what the next steps are, and how long they are going to take and thereby to seek to minimize the period of any suspension.
>
> [...]
>
> 34. ... [A]ny panel charged with a regulatory exercise such as this should be asking itself those sorts of questions when exercising its discretion, because it needs to properly inform itself.

***Gulamhusein v. General Pharmaceutical Council* [2014] EWHC 2591 (Admin)**

**38.42** Pharmacist—undercover operation—supply of prescription-only drugs without prescription—suspension necessary for public safety and public interest

G, a 55-year-old pharmacist, was involved in running a pharmacy in Maida Vale, which he purchased in about 1987. During an undercover operation in September 2012, carried out by journalists from News International and the British Broadcasting Corporation (BBC), there was evidence that G had been supplying products without a prescription for them, even though such products could be supplied only with a prescription. On 27 September 2012, at the pharmacy, G was alleged to have provided to an undercover reporter a pack of sixty Cytotec 200mg Misoprostol tablets without prescription and in the knowledge that they were going to be used for the purpose of an illegal termination of pregnancy. G was interviewed by the police and, apart from confirming his name, he made no comment to the questions that were put to him. On 16 January 2013, the Fitness to Practise Committee of the General Pharmaceutical Council imposed a suspension for eighteen months as being necessary for both the public safety and the public interest. The suspension was confirmed at reviews in January 2013 and on 11 December 2013. At a further review on 27 May 2014, the order was continued. In granting the Council's application for the period of suspension for six months and rejecting G's application for the imposition of conditions, Sir Stephen Silber said that, in its determination on 27 May 2014, the Committee first found it to be necessary to continue to suspend G's registration on the grounds of protection of the public and the public interest. In other words, the Committee did not consider this to be a case to be determined solely on a public interest basis. In *R (Shiekh) v. General Dental Council* [2007] EWHC 2972 (Admin) and in *R (Sosanya) v. General Medical Council* [2009] EWHC 2814 (Admin), the allegations did not relate to the way in which the professional person performed his or her duties. This is important, bearing in mind that the threshold for imposing an order for suspension is necessity. It is therefore a remedy of the last resort where there is no other proportionate and appropriate remedy available. The learned judge continued:

> 30. A corollary of the test of necessity, which is set out in rule 56 of the Pharmacy Order 2010, is that an order for suspension should only be imposed if no other remedy would be appropriate and proportionate. Therefore a Disciplinary Committee should first consider whether any other remedy, such as the imposition of conditions, would ensure that the interests of the public and the profession were satisfied because an order for suspension would not be necessary if another remedy, such as the imposition of conditions, would be appropriate.
>
> 31. Whether conditions can be imposed as a proportionate remedy will depend on the nature of the complaint against the practitioner concerned. It might be easier for a

> Disciplinary Committee to conclude that imposing conditions would be proportionate where the complaint related to lack of capability rather than where the complaint was one of serious misconduct. If, for example, the breach related to capability and showed a lack of training or experience, then conditions relating to supervision might be appropriate.
>
> 32. The position might therefore be different if the allegations made related to misconduct such as theft. The question for the Disciplinary Committee would therefore be to see whether the complaints that had been made relating to the pharmacist could be adequately addressed by conditions if they would, among other things, ensure matters complained of would not be repeated.

The learned judge said that the Committee did not regard the present case as a public interest only case. It was right to regard the instant case as one concerning public safety, because the allegations against G involved unlawfully supplying to members of the public prescription-only medicine without being supplied with a prescription. There was evidence that medication was supplied in this manner by G. By the time of the hearing on 27 May 2010, G had been charged with supplying prescription-only medicine—namely, Misoprostol—without a prescription and he had elected trial in the Crown Court. There were no criminal charges in relation to the other supplies. The learned judge said that there was no need for the allegation made against G to be proved before it could be considered relevant for disciplinary proceedings, as was explained by Pill LJ when giving judgment in the Divisional Court in *Dr X v. General Medical Council* [2001] EWHC Admin 447, [24]. The fact that the allegations had not been proved was clearly a matter that the Committee was bound to take into account. What is important is that it must scrutinize the unproven allegations with great care to test their cogency. That is what this Committee did after taking account of the previous good record of G and the matters put forward in his favour. The Committee concluded correctly that if the allegations were found proved, G would be shown to be prepared to supply dangerous prescription-only drugs improperly when asked and where, as in the case of Misoprostol (Cytotec), the risk to the public is that of procuring an illegal abortion in relation to which no prescription had been obtained and in relation to which G had no knowledge. The Committee said it would take into account the reputational damage to G caused by the imposition and the continuation of the suspension. In addition, the Committee was aware of G's personal position as a pharmacist of twenty-eight years' standing and that if his registration were suspended, it would cost him some £80,000 a year. Nevertheless, the learned judge said that, in the light of the seriousness of G's alleged conduct, the Committee was right to reject the contention that the imposition of conditions would be an appropriate and proportionate order. Conditions might have satisfied the requirement of being appropriate and proportionate had G's alleged offence been committed because he was unaware of the rules, or because he had made an honest mistake, which he had acknowledged. That was not the case. What is clear is that conditions would not be appropriate where, as in this case, the conduct complained of was apparently a blatant breach of the statutory provisions for which no excuse has been suggested:

> 50. In conclusion, it is important not to lose sight of the very important role played by pharmacists in ensuring that drugs for which prescriptions are required are not supplied to those who do not have prescriptions for them. This role is significant in many ways as the pharmacists are the gatekeepers, ensuring that drugs only go to people approved to use them and requiring them by their medical practitioners.

## D. Extension of Interim Orders

*General Medical Council v. Hiew* [2007] 1 WLR 2007, CA

**38.43** General principles

In terms of making an interim order, this case sets out the general principles, and the factors to consider in deciding to extend an interim order include the gravity of the allegations, the seriousness of the risk of harm to patients, the reason why the case has not been concluded, and the prejudice to the practitioner if an interim order is continued. [The burden of showing that all of those criteria are met is on the regulatory body: *Nursing and Midwifery Council v. Okafor* [2013] EWHC 1045 (Admin), [2], *per* Clive Lewis QC (sitting as a deputy High Court judge).]

See paragraph 38.07.

*General Medical Council v. Lauffer* [2009] EWHC 3497 (Admin)

**38.44** Whether allegations justify prolongation of suspension—minimum period necessary

On 9 July 2008, L's case was first considered by an interim orders panel, which decided to suspend his registration for eighteen months. The grounds were to protect the public and in his own interests. The complaint consisted of allegations that a small number of patients had suffered injury as a result of negligent surgery. Some of them, sadly, had died, allegedly as a result of negligent surgery. This was coupled with an allegation that L had failed to comply with restrictions imposed by his employing NHS trust pending further investigation. In extending the order for six months, His Honour Judge Pelling QC (sitting as a judge of the High Court) said, at [8], that the Court is required to approach an application of this sort by asking itself two questions: (a) whether the allegations justify the prolongation of the suspension; and (b) whether, in the circumstances, any suspension ought to be extended for the period sought by the GMC or for some lesser period. The GMC needs anxiously to consider the period required, bearing in mind the obvious point that the longer the time for which a medical practitioner, and in particular a surgeon, is suspended, the more deskilled he or she is likely to become, and thus if care is not taken, the resulting restriction being imposed on the medical practitioner concerned irrespective of the outcome in relation to the original allegations. It should never be the default position that the maximum period should be routinely applied or applied for. Each case has to be considered on its merits, and, particularly where a medical practitioner had been suspended for the maximum period at the outset, any extension must be the minimum necessary and requires justification by the GMC. At [9], the learned judge added:

> [A]dministrative difficulties faced by the GMC in arranging hearings can never, at any rate in my judgment, be a sufficient reason for prolonging the period of suspension of a practitioner's ability to practice, which in some cases, though happily not so far in this, can result in the medical practitioner concerned being deprived of his ability to earn a living without any of the allegations made against him having been determined.

*Nursing and Midwifery Council v. Miller* [2011] EWHC 2601 (Admin)

**38.45** Failure to prosecute with reasonable expedition—extension refused

In May 2009, an interim suspension order was imposed for eighteen months. The allegations related to events occurring in the summer of 2008 concerning allegations of aggressive conduct and a dishonest claim that the registrant had worked a shift, when he had not. In November 2010, Mitting J granted an extension of nine months, saying that he did so with a little disquiet because it seemed to him that the relatively minor matters could be investigated a little more quickly. In the summer of 2011, a date for the substantive hearing was fixed for February 2012. On 30 September 2011, Blake J dismissed the NMC's application for a further extension. The learned judge said that, in his judgment, there had been a failure to progress the matter with reasonable expedition: it was an uncomplicated case and it was inconceivable that it would take three years to list it for a hearing.

### *Nursing and Midwifery Council v. Maceda* [2011] EWHC 3004 (Admin)

**38.46** Extension not indefinite—application refused

In February 2011, an interim suspension order was made for eighteen months in relation to a nurse at the Oxford Radcliffe Hospital who had been dismissed in the summer of 2009. The NMC appointed a solicitor to carry out an external investigation. In September 2010, the solicitor reported that there was insufficient evidence of misconduct or lack of competence. The NMC was dissatisfied with the solicitor's report and commissioned a supplemental report. In August 2011, His Honour Judge Birtles (sitting as a deputy judge of the High Court) granted an extension for three months only. In November 2011, Bean J said that, on the evidence before him, he was not satisfied that it was right to renew the interim order and, accordingly, the NMC's application was dismissed. The learned judge said that he took into account that, in February 2010, the investigating committee made an order because it considered it necessary to protect the public from harm. But that cannot be decisive for all time. If it were, Parliament would have given the investigating committee of the NMC, and the corresponding bodies in the GMC and other regulators, the power to renew such orders for an indefinite period. Similarly, it is not conclusive on an application to renew made to the High Court, otherwise every application would be granted.

### *Nursing and Midwifery Council v. Pratt* [2011] EWHC 3626 (Admin)

**38.47** Systematic delays—extension agreed by registrant

The defendant came to the attention of the NMC in May 2009 following a referral from the Winchester and Eastleigh Healthcare NHS Trust. The allegations were that, between November 2008 to March 2009, the registrant failed to notice that medicine was incorrectly labelled, gave incorrect instructions to a student, failed to follow instructions concerning a health visitor, and made an error in administering fluid using a nasogastric tube. On 30 June 2010, a suspension order was imposed for eighteen months. Simon J said that this was a case that had not been disposed of three years after the first reference. It was marked by casual administrative incompetence and by systematic delays, which were the consequence of a greatly overloaded system. It was not the most serious or obvious case for an interim order. But for the fact that the respondent had agreed to an extension for eight months, the learned judge said that he would have refused the application.

### *General Medical Council v. Razoq* [2012] EWHC 2135 (Admin)

**38.48** Criminal proceedings—extension pending determination

As to extending an interim suspension order pending the investigation and prosecution of criminal proceedings, see paragraph 10.12. Hickinbottom J said that it was appropriate for the GMC to stay its hand in respect of the disciplinary investigation and procedures against R whilst the criminal proceedings were running their course. The frauds that were the subject matter of the criminal proceedings concerned R's conduct as a doctor—and those frauds were clearly going to be crucial to any disciplinary proceedings against him before the GMC.

### *General Medical Council v. Srinivas* [2012] EWHC 2513 (Admin)

**38.49** Criminal proceedings—doctor acquitted—police/CPS material—investigation

On 2 June 2010, S, a locum GP, was charged with seven counts of misconduct contrary to section 3 of the Sexual Offences Act 2003 and eleven counts of misconduct contrary to section 2 of the Sexual Offences Act 2003. On 10 June 2010, the IOP decided to convert a previous interim order of conditions to one of suspension. In March 2011, a further three counts were added. In May 2011, S was acquitted of seven of the counts that he faced and, following a retrial in October 2011, he was found not guilty on all of the remaining counts. The GMC applied for a twelve-month extension of the interim order of suspension. His Honour Judge Pelling QC (sitting as a judge of the High Court) granted an extension of five months to enable the GMC's investigation to be completed, bearing in mind that the practitioner had been suspended or subject to conditions for a period of slightly in excess of two-and-a-half years. The learned judge said:

> 12. I accept the submission [of the GMC] that once the defendant became the subject of criminal prosecution proceedings the ability of the GMC to actively investigate on its own terms the allegations which were being made was severely hampered and for all practical purposes prevented because I accept the submission that any approach they made to witnesses who were to be called by the prosecution – or might be – could be misconstrued for fairly obvious reasons. I also accept that the practice of the GMC is to ensure as far as possible that the totality of allegations made against the doctor are investigated in one go and that all material evidence is made available for one hearing. The reasons for this are obvious. It is in the public interest and in the interests of the doctor that all relevant information be considered at one hearing and that a multiplicity of disciplinary proceedings is avoided if at all possible. In addition, such an approach is likely to be cost effective for a heavily pressed regulatory body.

***General Medical Council v. Gill* [2012] EWHC 2069 (Admin)**

**38.50** Duty to court—disclosure of relevant information—principles similar to without-notice applications

On 13 September 2011, the Court extended, under section 49A(6) and (7) of the Medical Act 1983, an interim order of suspension for a further period of nine months. On that occasion, the Court was critical of the way in which the GMC had conducted its investigation in relation to the doctor and was particularly critical of the leisurely approach that had been adopted at various stages, which, in the aggregate, had resulted in a very extensive period of interim suspension when, on the fact of it, the case could have been placed before a fitness to practise panel much more quickly. Shortly before the nine-month extension period expired, the GMC applied for a further extension. It was supported by a witness statement from the head of investigations, who stated merely that 'An extension of nine months was granted by the High Court on 14th September 2011'. His Honour Judge Pelling QC (sitting as a judge of the High Court) said that it was surprising that there was no reference to the judgment that was delivered on that occasion. It is particularly so since a transcript was available with a neutral citation reference attached to it. Given that these applications are, if heard at all, almost invariably heard without attendances from the doctor concerned, the principles that apply to the hearing of without-notice applications apply in substantially the same way to applications of this sort—that is, there is a duty of frankness that applies to the GMC when applying for extensions to ensure that all relevant information is available to the Court, irrespective of whether it supports the GMC's case or not. It was surprising therefore that, in the witness statement, there was no reference to the earlier judgment or to the criticisms of the progress that had been made to that date, or any reference to the reasons why the Court was prepared to extend the period of interim suspension on that occasion only for a period of nine months and, more pertinently given the circumstances, no attempt to explain why it was that the steps that were anticipated as being capable of being taken over the nine-month period of extension had not already been taken. The Court ordered that the period of extension be extended for six months, but, in order to maintain a degree of active case management, any further applications for extension of time were to be reserved to Judge Pelling QC.

### *Nursing and Midwifery Council v. Hitchenor* [2012] EWHC 3565 (Admin)

**38.51** Extension refused

H was dismissed by an NHS trust for removing and self-administering an unknown quantity of tramadol whilst on duty between July 2008 and January 2009. An eighteen-month interim suspension order was made in November 2010. Subsequently reviewed on six occasions, it was extended by the High Court in May 2012 to expire on 24 November 2012. On 23 November 2012, the Court refused to extend time for a further six months until the scheduled hearing in February 2013. On the previous occasion, the Court was told that the NMC expected to conclude matters within the six-month extension sought. There was no evidence of current or recent substance abuse by H. He had turned his life around and was presently clean, and had been clean since 2010. An extension of the present order was not shown to be necessary either for the protection of patients or in the public interest.

### *Nursing and Midwifery Council v. Okafor* [2013] EWHC 1045 (Admin)

**38.52** Extension for short period to conclude case

The learned judge, Clive Lewis QC (sitting as a deputy High Court judge), said that the allegations against O, a midwife, were serious. They related to misconduct and lack of competence over a period of time, and two of the allegations involved allegations of dishonesty. There was a real serious risk of harm to patients, and therefore it was appropriate in the interests of public protection and the public to grant an extension of the present interim order. What concerned him, however, was the length of time that the matter had taken to come to a final conclusion. Some of the allegations related to March 2009; some, to July 2011. Some delays were because of problems with getting witnesses. The final hearing was now scheduled for 22–26 April 2013. The Court would extend the order not for the original period sought, which was five months, but for a shorter period of time, to 28 May 2013.

### *General Medical Council v. Malik* [2013] EWHC 122 (Admin)

**38.53** Extension refused

Allegations against M concerned inappropriate examination of a female patient in November 2010. An interim suspension order had been made in June 2011 and replaced by conditions in December 2012. Whilst serious allegation requires investigating seriously, it also includes investigating expeditiously. His Honour Judge Gilbart QC (sitting as a judge of the High Court) deemed it unacceptable that a straightforward allegation should still be unresolved more than three years after the matter complained of, and which matter had been brought to the attention of the GMC in early 2011. The GMC was to accept that if it acts in a dilatory way, then it will not succeed in obtaining an extension: [9]. Extension of the imposition of conditions was refused.

### *General Medical Council v. Adeosun* [2013] EWHC 2367 (Admin)

**38.54** Extension granted

In this case, extension of time was granted, but for a significantly shorter period than the GMC had sought. The case was reserved to a named judge for any further application for extension, so as to ensure that a degree of continuity would be maintained in managing the case through to a conclusion.

### *Jooste v. General Medical Council* [2013] EWHC 2463 (Admin)

**38.55** Extension granted—appeal to Court of Appeal—extension pending appeal

On 20 March 2013, Stuart-Smith J extended an interim suspension order for three months from 7 April to 6 July 2013: [2013] EWHC 1751 (Admin). The doctor lodged an application for permission to appeal to the Court of Appeal. On the GMC's further application for an extension from 6 July 2013, His Honour Judge Pelling QC (sitting as a judge of the High Court) said that if, as was the case, he were satisfied that the interim suspension ought to be continued, and if, as appeared to be the case, there were a pending appeal by which the doctor sought to challenge the conclusions of Stuart-Smith J, the appropriate course would be to continue the interim suspension. If the appeal succeeds, the Court of Appeal may be able to discharge the order. The interim order should be extended with liberty to apply, to vary, or to discharge after final determination by the Court of Appeal of the doctor's appeal.

### *General Medical Council v. Dr E* [2013] EWHC 3425 (Admin)

**38.56** General practitioner—inappropriate relationship with patient—conditions—chaperone—form of conditions

In April 2012, the GMC's IOP imposed conditions on E's registration for eighteen months, which included a condition that, except in life-threatening emergencies, he was not to undertake consultations with female patients without a chaperone being present, and that the chaperone was to be a fully registered medical practitioner or a fully registered nurse or midwife. E claimed that the effect of the order was, for all practical purposes, the equivalent of suspension, because E had worked for most of his professional life as a GP and an out-of-hours GP. In October 2013, the GMC applied to the High Court for an extension of the order for twelve months. His Honour Judge Pelling QC (sitting as a judge of the High Court) said that the single, albeit very serious, allegation against E was of a serial rape of a former vulnerable patient KA over a period of years. The police decided to take no further action in relation to the allegations and, on a review of the material available to the Court, the evidence of KA appeared to be sufficiently unreliable that it must be discounted, save where it was agreed or was supported by other evidence from an otherwise unimpeachable source. Thus the evidence available established that E entered into a relationship with KA and that he had a child by her, who was now living with him by an order of the Family Court. The allegation that the relationship started whilst KA was E's patient was denied and was not supported by KA in the statement that she gave at the time of the first investigation. It remained the case, however, that E had commenced a relationship with someone who was, by his own admission, a former patient and who he knew or must have known was a vulnerable adult by reason of her mental health issues. It also remained the case that he was credibly alleged to have actively misled his former partners on the question of whether he admitted having a relationship with a vulnerable person. Despite there being no other allegations in twenty years of practice, looking at the allegation rather than arriving at any judgment concerning its truthful accuracy did suggest a risk that E might attempt to develop a relationship with a vulnerable patient, and, accordingly, interim conditions were necessary for the protection of both patients and E. However, a condition not to undertake consultations, except in life-threatening emergencies, with female patients without a chaperone being present was disproportionate and was to be replaced with the following conditions, at [30]:

(a) not to treat or have any personal contact with any female patient other than in a consultation fixed by prior appointment by the patient concerned to a GP's surgery or other NHS unit.
(b) not to have telephone contact with any female patient other than in response to calls from such patients to the surgery or other NHS unit at which the claimant is employed
(c) Not to make any contact with any patient by e-mail or other electronic means other than to respond to emails received from a female patient sent to the surgery or other NHS unit of which the claimant is employed.
(d) except in life threatening emergencies not to treat or undertake a consultation with any female patient more than three times in any three month period.
(e) save to the extent provided by (a) to (c) above, not to make contact with any female patient or visit or meet with such patients.
(f) on every occasion when he treats or conducts a consultation with a female patient, he must offer that patient a chaperon [*sic*].

> 31. The conditions ought to enable E to practice [*sic*] as a locum GP during normal surgery hours but at the same time eliminates to the maximum possible extent the opportunities for developing inappropriate relationships with female patients in the way that is alleged by the GMC in this case....

The order would be extended for a further period of six months. By then, E would have been subject to an interim order of conditions for two years and the GMC would have, or ought reasonably to have, decided whether it wished to proceed, and if so, on what basis.

*General Medical Council v. Dr E* [2014] EWHC 1620 (Admin)

**38.57** Complainant ceasing to cooperate with GMC—further extension refused

Following the hearing on 17 October 2013—[2013] EWHC 3425 (Admin)—the complainant stopped cooperating with the GMC in or about November 2013. On 11 March 2014, the GMC served a rule 7 letter limited to allegations that Dr E's relationship with KA was inappropriate owing to his previous professional relationship with her and her known vulnerabilities, and his failure to inform colleagues that he had conducted an extramarital relationship with a former patient whose identity he refused to reveal. The allegations of rape and pursuing an inappropriate relationship with KA while she was his patient were no longer pursued. In refusing to grant any further extension, Stuart-Smith J said that the GMC's witness statement did not address the seriousness of the remaining allegations, the evidence that was available, the reasons why the case had not been concluded, the prejudice to Dr E by the order continuing and the question of public interest, and that the allegation of rape was no longer pursued. Two courses of action were available to the Court. The first would be simply to refuse any extension at all; the second would be to grant an extension of one or two weeks, to enable the GMC to make good the deficiencies in its evidence on the next application, if it could. Notwithstanding that the residual allegations were serious, it was not self-evident that a prolonged extension of the current conditions was necessary or any conditions justifiable. Applying the principles in *General Medical Council v. Hiew* [2007] EWCA Civ 369, there had been a total failure by the GMC to address the matters on which the Court must be satisfied if it is to grant an extension. In the circumstances, it would be wrong in principle to grant any extension at all.

*General Medical Council v. Sondhi* [2013] EWHC 4233 (Admin)

**38.58** Substantive hearing part-heard—allegations of financial impropriety—no impact on clinical competence—further extension refused

In January 2010, the GMC's IOP suspended S's registration for eighteen months. Further extensions were granted to September 2013. The present application for an extension arose at a point in the proceedings when the hearing before the fitness to practise panel had commenced, but those proceedings had been adjourned part-heard. At the date of his suspension, S was a GP working in Croydon, who had practised for twenty-nine years without any adverse findings concerning his clinical skills or practice having been made against him. The main issues before the fitness to practise panel concerned S's financial and operational management of an out-of-hours GP service carried on by a private company called Croydoc Ltd, of which S was chairman and operational medical director. The allegations concerned S's financial dealings as a director of Croydoc, and failure to provide good clinical care to patients in organizing the Croydoc rota and whilst on duty on the Croydoc rota. It was not alleged by the GMC that any matter causally adversely affected any patient at any stage. Any money that was allegedly dishonestly or wrongly obtained by S was obtained from Croydoc and was money that belonged to that company, and was not in any relevant sense public money. In refusing to extend the interim order, the Court said that the real issue was whether suspension ought to be continued in the public interest in the light of the allegations of financial impropriety that had been made for the purpose of protecting the reputation of the medical professional generally. The issue was whether public confidence in the profession would be damaged if the public were to learn, following a hypothetical finding in favour of the GMC by the fitness to practise panel, that S had been working with patients while matters were being investigated and determined. It was highly unlikely that reasonable people would regard the allegations made, even if subsequently established, as impacting on S's clinical competence in a way that endangered the health of patients. Even if this were wrong, the imposition of conditions that precluded S from working in a financial management or administrative role would have provided proportionate protection. To suspend S's registration based on the allegations of dishonesty was not necessary for the protection of the public, and was unnecessary and disproportionate. No patients had been put at risk. Additionally, the length of time that had elapsed since S was first suspended was a factor that was material when balancing the interests of the practitioner against those of the public. The adverse effect on a practitioner of suspension is the loss of his or her ability to earn a living, in combination with the risk of progressive de-skilling. Clearly, the longer the interim suspension continues, the more significant will become each of these effects. The delay in the instant case was plainly unacceptable when viewed in the aggregate.

***General Medical Council v. Shalaby* [2013] EWHC 4211 (Admin)**

**38.59** Criminal conviction—extension granted—costs of application

Following a trial in the magistrates' court, S, a consultant orthopaedic practitioner, was convicted of assault contrary to section 39 of the Criminal Justice Act 1988 and, contrary to his professional obligations, failed to report the conviction to the GMC. It was not in the public interest to refuse to grant the GMC's application to extend the period of suspension of S's registration until the determination of the fitness to practise proceedings in May 2014. However, bearing in mind the delay, there should be no order for costs against S. Costs, ultimately, are a matter for the general discretion of the Court, and it was not appropriate—notwithstanding that S had failed to fully and properly engage with the fitness to practise process—that the GMC should recover its costs. Accordingly, there would be no order as to costs. The suspension would be continued until 30 May 2014 unless revoked by a fitness to practise panel at an earlier date.

***General Medical Council v. Qureshi* [2014] EWHC 775 (Admin)**

**38.60** Health—doctor suffering mild cognitive impairment—further extension granted

In August 2012, the GMC's IOP suspended Q's registration for eighteen months following two incidents in July 2012. In the first incident, he insulted the patient's husband and sought to attack him; in the second incident, he punched the patient's partner in the face. The police decided not to take any action. However, both patients reported the matters to the GMC. In April 2013, Males J dismissed Q's appeal against the decision of the Panel to continue an order for suspension of the registration. Males J said that there was clearly material to conclude that there may be impairment affecting Q's ability to practise safely, and that it was necessary for the protection of patients—and indeed in Q's own interests—that his practice should be suspended. In February 2014, the GMC applied for an extension of the period of suspension. Q had been the subject of neuropsychological assessment and it was common ground that he suffered from mild cognitive impairment. The incidents involving patients, if established in due course, would be matters that would affect the public interest to the extent that they would establish significant inappropriate behaviour by a professional medical doctor towards two of his patients. Although appreciative that some delays can take place, both in the ability to gain witness evidence from the complainants and seeking the assistance of the police in the investigations that they have made, Jeremy Baker J said that he would not, if this were the only matter, be satisfied that an extension of the interim suspension order would be justified. However, it was now clear on the medical evidence on both sides that Q did indeed suffer from some cognitive impairment. It is of the essence of being a medical practitioner of whatever field, and in particular the onerous role of a general practitioner, that he or she has sufficient mental acuity in order to carry out that task appropriately and safely. The impact of somebody not having sufficient mental acuity to be able to make appropriate decisions dealing with a patient's affairs is a serious matter, giving rise to serious potential risk. However, expedition was now required. A rule 7 letter had been provided to Q and, allowing twenty-eight days in which to respond, the papers would then go before a medical and lay examiner under rule 8 to decide whether or not the matter should proceed further, and in particular as to whether it should proceed to a fitness to practise panel of the Medical Practitioner's Tribunal Service. The present interim order for suspension was therefore for a period of three months. It was then to be a matter for the GMC, in the light of further evidence, to decide whether to seek a further extension to cover any period leading up to any hearing.

### *General Medical Council v. Makki* [2014] EWHC 769 (Admin)

**38.61** Conditions—chaperone—extension continued—condition deleted

Following a complaint by a female patient, the GMC's IOP imposed conditions on M's registration, which included a condition restraining M from consulting with female patients between the ages of 16 and 75 except either in life-threatening emergencies or without a chaperone present, such chaperone being a fully registered medical practitioner or fully registered nurse or midwife. His Honour Judge Stephen Davies (sitting as a judge of the High Court) said that he was not satisfied that it was appropriate to continue this condition. There does come a time when the Court has to say that, on the evidence before it, it is simply not appropriate for an intrusive condition to be continued for a period of time beyond that which is reasonably required for an investigation to be brought to a conclusion. The evidence showed that there would be no significant risk to patient safety if that condition were not continued, nor would public confidence be shaken if that condition were removed, particularly bearing in mind that the essential allegation in this case was one of one incident of inappropriate behaviour involving the doctor and a female patient, in circumstances in which the doctor and patient were already known to each other outside the doctor–patient relationship, in circumstances in which the conduct was, on any view, consensual and short-lived, and in which there had been no repetition and, indeed, a degree of contrition. It was held to be a significant intrusive condition not only for the doctor, but also to the operation of the activity in which he is involved. The interim order would be continued for a further period of seven months, but the condition would be deleted from that order.

### *Nursing and Midwifery Council v. Kidd and De'filippis* [2014] EWHC 847 (Admin)

**38.62** Applications for extension—service and lodging of papers—opportunity for respondent to respond—draft order—effect on other court users

The application to extend was lodged with the Administrative Court Office on 4 March 2014. It related to an order that expired on 11 March 2014. The hearing was scheduled for 7 March 2014. The respondent, K, was served with a hard-copy application by means of recorded delivery and first-class post on 5 March 2014. Accordingly, pursuant to CPR 6.14, since the claim is deemed to be served on the second business day, service was not effected until 7 March 2014. With regard to the electronic service, even if this were made at some point before 16:30 on 4 March 2014, it would still be deemed served on 6 March. This was in relation to a hearing that occurred on 7 March 2014. At some point on 5 March 2014, the NMC sent to K and his union representative a draft consent order. The consent order sought complete capitulation. K was unhappy about this state of affairs. Because the interim order expired imminently, the Administrative

Court Office felt compelled to list the matter urgently. Green J said that there were a number of serious problems relating to this course of events:

> 11. First, and foremost, it is the fact that the respondent was given insufficient time to prepare any sort of a case to put before the court. Invariably respondents are litigants in person who have no legal representatives to advise them or act on their behalf during hearings. Sometimes they appear in court with a McKenzie Friend or a union representative or with some other professional association advisor. Experience tells one that this ability to digest a well prepared application and to respond is hindered not only by a lack of legal advice but by the pressure of time. There is a serious and elementary issue of fairness which arises. An interim order strips a person of the right to practice a profession prior to a final hearing on fitness to practise. It is a significant incursion into that person's civil liberties. At the very least when the regulators seek court orders to extend interim orders the respondent must be given an adequate time to prepare... The CPR provides rules as to notice of applications. However in cases of this sort the courts will wish to look closely at how applications are made. It will be rare indeed that short notice should ever be given. Good practice might entail giving litigants in person longer notice than the bare minimum. Having had the opportunity to discuss this matter with both the Administrative Court Office and with colleagues, in circumstances such as these, good practice suggests that a respondent should ordinarily receive a minimum of seven calendar days' notice of an application to extend. In many instances a respondent given proper time will agree to the extension and will in such a period have had an opportunity to consider the regulator's evidence and seek advice or assistance. If the order is on the cusp of expiring and a respondent decides to challenge the application the court might then have to decide whether to make directions for a contested hearing which might include an extension of the order until further order of the court. The short but critical point is that respondents to proceedings of this nature must be given a fair chance to be heard and the courts will be astute to ensure that regulators act fairly.
>
> 12. Secondly, these cases are *not* intrinsically urgent. They are artificially made so by the delay on the part of the regulator in bringing them to court such that by the time the application is made the order in issue is very close to expiry. The Administrative Court Office has felt compelled to place such applications before a Judge urgently but it is not, as I have stated, for any good reason. Even in a case where a respondent has made clear that upon expiry of the order he or she intends to obtain work or otherwise conduct themselves in a manner causing a real risk to the public, there would still be no urgency. This would simply be evidence likely to lead the court to extend the order to prevent such eventuality occurring. Nothing in the present case prevents the NMC from giving the respondent adequate notice and then liaising with the court to find an appropriate time for the case to be heard.
>
> 13. Thirdly, the practice of the NMC in tendering a consent order to the respondent just days before the scheduled hearing risks using the imminence of the hearing as a means of pressurising a respondent into agreeing to a consent order in the onerous terms that I have referred to. One factor that risks pressurising a nurse or midwife into consenting is the alternative to not consenting, namely the potentially very intimidating prospect of being required to turn up in person before a High Court Judge at very short notice not having had any sort of a realistic chance to prepare. I am not suggesting that the proffering of draft consent orders is necessarily inappropriate; but it has the potential to be unfair if tendered at the 11th hour to a litigant in person shortly before an oral hearing.
>
> 14. Fourthly, the *modus operandi* adopted by the NMC in the present case also imposes pressure upon the court whose lists are always crowded. To make room for urgent applications of this type other matters might have to be deferred and/or Judges diverted from other duties. In circumstances where, as I have observed, the lateness of the application is a problem of the regulator's own making it is not right that this

should be allowed to have adverse consequences for the efficient administration of the court service.

Extensions would be granted in each case subject to the respondents being at liberty to seek to set aside and/or vary the orders.

### *General Medical Council v. Agrawal* [2014] EWHC 1669 (Admin)

**38.63** Clinical work—evidence—expert report not before panel—strength of case—delay—order no longer necessary or otherwise in public interest

In October 2011, A was suspended by his employers as a result of clinical allegations. On 1 May 2012, the IOP imposed conditions on A's registration. Reports from eminent consultants instructed on behalf of A were prepared, which showed that his clinical work had been exonerated by experts of undoubted repute. A rule 7 letter was sent on 2 April 2014 and was responded to by A on 11 April 2014. In refusing to extend the interim order further, His Honour Judge Raynor QC (sitting as a judge of the High Court) considered the allegations in the rule 7 letter, together with the latest opinion of the GMC's expert, whose further report was not before the last panel. The case against A was weak, and this was a matter that the Court was entitled to take into account when considering whether it justified to continue the conditions. Other relevant matters were delay. The GMC's conduct of the investigation had been deplorable. Had the investigation been properly conducted, the decision as to whether to proceed to a hearing would have been taken at least twelve months ago and so the prejudice to A by such delay was manifest. This obviously was a relevant matter that the Court was entitled to take into account. A had undergone further training with a leading surgeon, who had provided a reference. It was for the GMC to satisfy the Court, on the balance of probabilities, that the continuation of the conditions was 'necessary' for the protection of members of the public or was otherwise in the public interest. The GMC had failed to do so on the particular facts of this case.

## E. Measures Pending Appeal

**38.64** In this regard, the legal framework includes:

Medical Act 1983, section 38 (power to order interim suspension after a finding of unfitness to practise)
Osteopaths Act 1993, section 24
Pharmacy Order 2010, article 58

### *Gupta v. General Medical Council* [2001] EWHC Admin 612, [2001] EWHC Admin 631

**38.65** No opportunity to make representations—inadequate reasons—procedural unfairness

On 25 July 2001, the Administrative Court (Brooke LJ and Newman J) announced its decision to allow G's application pursuant to section 38(6) of the Medical Act 1983 and to give its reasons in due course. The brief reasons were that the procedure by the PCC to order immediate suspension of G's registration at the conclusion of the substantive hearing on 1 June 2001 was faulty in two respects: first, that specific representations were not sought on the question of whether the erasure should be suspended pending an appeal; and second, that, although the Privy Council has said that brief reasons will suffice, it is not sufficient simply to quote the words of the statute in the reasons and then to leave it to people to look at the submissions made to the Committee and try to work out from that what the Committee's reasons probably were. The Committee found proved charges to the effect that G had allowed her husband and former medical partner, who earlier had been struck off the register, to hold consultations with patients at two surgeries, and that she well knew that he was holding consultations at the practice surgeries. The Committee directed that her registration be erased from the register and that it would be in her 'best interests that your registration in the Register should be suspended with immediate effect'. Giving the judgment of the Court on 9 August 2001 ([2001] EWHC 631 (Admin)), Newman J said that no notice was given to the applicant that the Committee contemplated an order for immediate suspension of her registration, and it was obvious that the only reason given by the Committee for the immediate suspension of registration was that it would be in the applicant's 'best interests':

> 10. In our judgment [immediate suspension] conclusively determines the position pending appeal, just as an interim suspension order under section 41(A) determines the position pending the hearing of reviews, in a manner which prejudges the outcome of the future hearing. Immediate suspension has a character of finality which the review procedure in connection with interim orders mitigates. The period pending appeal to the Privy Council can be significant. The financial consequences of being unable to practise are highly likely to be serious, invariably the professional consequences will be very serious. Where a body exercising penal powers is contemplating imposing a penalty which is more severe than that which might normally be expected to flow from a finding of guilt, in our judgment, common principles of fairness require that the person who will be affected should have an opportunity of making representations against the making of such a special order.

***Burnham-Slipper v. General Medical Council* [2001] EWHC 656 (Admin)**

**38.66** No opportunity to make representations—whether order wrong on merits—whether need to protect public outweighs impact upon doctor

On 31 July 2001, Moses J dismissed the appellant's application against the order of the PCC that he be suspended forthwith following its order of erasure. The sole, but important, purpose of the appeal was to preserve the appellant's entitlement to practise pending his appeal to the Privy Council, which it was thought could take several months to be heard. The appellant was aged 62 and had been a qualified doctor for thirty-seven years. The allegations found proved related to a patient, L, to the appellant's removal of medical records from the surgery where he practised, and to his treatment of five patients in relation to the removal of records. The Committee found that the appellant's actions placed at risk the treatment and safety of the relevant patients, in that it prevented other doctors and healthcare professionals from having access to a patient's documented medical history. In relation to the patient cases, the evidence revealed that, over a period of time, the appellant seriously disregarded his professional responsibilities and that, in the case of four of the patients, there was a real risk of cancer, although it proved not to be the case. The appellant's essential plea was to be permitted to work at a medical centre for two hours twice a week as an occupational health physician. In dismissing the application, Moses J said that the case of *Gupta v. General Medical Council* had been drawn to his attention, but that there was as yet no written decision. In so far as the Court was likely to accept the proposition that a proper opportunity must be given to those appearing before the Committee to address the question of immediate suspension, Moses J accepted that proposition. In the instant case, the Committee gave no advance warning of its proposal to order immediate suspension. The question of immediate suspension ought to have been considered separately and the doctor ought to have been given an opportunity to respond to any proposal for immediate suspension. To that extent, the procedure was defective. But the difficulty in such an instance is that the jurisdiction of the Court hearing an appeal is limited. It has no power to order a rehearing on that point. It must consider merely whether the failure in that procedure led to a wrongful order of immediate suspension, as apparently it did in the case of *Gupta*. Considering the merits of the appeal in the instant case, Moses J continued:

> 34. ...If this appellant is not immediately suspended he will continue to have all the responsibilities of a doctor to a patient wherever he works. He owes those responsibilities to those visiting the occupational health centre just as much as he would owe them to patients of a general practitioner....
>
> 35. The appellant's failures went to the heart of his responsibility as a doctor, namely the responsibility for safety and welfare of those for whom he was responsible. He demonstrated a lack of care for the safety of five patients and for the safety of many more in relation to his management of records. It matters not that there will also be others responsible for those visiting the [occupational health centre]....

At [38] and [39], Moses J found that the Committee did not set out the effect upon the doctor of an order of immediate suspension and that it was incumbent upon the Committee to consider whether the need to protect the public was outweighed by the impact upon this doctor of not being able to work at all. However, the Committee could hardly have been unaware of that impact and there was no basis for saying that the Committee did not conduct that balancing exercise.

### *Moody v. General Osteopathic Council* [2009] 1 WLR 526

**38.67** Appeal against interim measures—merits of substantive appeal—expedited hearing of substantive appeal

Following the PCC's announcement of sanction that M's name should be removed from the register of osteopaths, counsel representing the General Osteopathic Council (GOsC) applied for an interim suspension order. The making of such an order was resisted. Nonetheless, the Committee made the order stating it was 'satisfied that it is necessary to make such an order in the light of all the evidence it has heard concerning the deficiencies in [M]'s practice'. Before imposing the sanction of removing M's name from the register, it gave a number of reasons for this, including that he posed a significant risk to patient safety. M appealed against the finding of professional incompetence and also against the sanction imposed upon him. On M's appeal against the interim suspension order, Wyn Williams J said that he was prepared to accept that it is permissible to look at the merits of the substantive appeal in order to assess whether or not it was wrong of the Committee to have imposed an interim suspension order. However, save in fairly unusual cases, it will not be possible to form a definitive view upon the merits of the substantive appeal, and unless the Court, exercising its appellate jurisdiction, is strongly of the view that the substantive appeal is bound to succeed, it should be slow to categorize as wrong a decision that has at its heart the protection of the public. In the instant case, there were aspects of the Committee's decision that made this appeal truly arguable, but unless the Court was strongly of the view that a finding of professional incompetence should be set aside or that the penalty would not even involve suspension measured in months, the Court should not categorize a decision to impose an interim suspension as wrong. In the instant case, the Court could not take either of those views of this appeal. Accordingly, the appeal was dismissed, but the Court would direct that there should be an expeditious hearing of the substantive appeal.

### *Ashton v. General Medical Council* [2013] EWHC 943 (Admin)

**38.68** Six months' suspension—doctor not working—immediate suspension not necessary

On 20 November 2012, A's registration was suspended for six months by a fitness to practise panel, which ordered the suspension to be with immediate effect. Stuart-Smith J dismissed A's appeal against the findings of misconduct and impairment, and the sanction of six months' suspension, but allowed A's appeal against the immediate order of suspension. The learned judge said that the GMC's power to impose a sanction of immediate suspension derives from section 38(1) of the Medical Act 1983, and must be challenged by an application for termination of the order pursuant to section 38(8) of the Act and not by an appeal under section 40. A was a registered medical practitioner who had ceased to practise as a GP and, after a period practising emergency medicine in a hospital, most recently had been employed carrying out cosmetic surgical procedures. He had been in practice as a GP for more than thirty years without any previous complaint having been made about his conduct or competence. His registration was, however, already subject to conditions imposed on health grounds relating to the consumption of alcohol. The reasons given by the panel for an immediate order for suspension were 'very thin' and not materially supported by reference to section 38 of the Medical Act 1983 or to the terms of paragraphs 121–6 of the GMC's Indicative Sanctions Guidance. A was not working at all and his employment, should he return to it, was in carrying out cosmetic surgery procedures. In general terms, the evidence was that his work in that field was satisfactory. It was highly unlikely that he would return to general practice. An interim measures order was not *necessary* to protect the public. The upholding of proper standards of conduct and behaviour may reasonably be said to have been achieved by the panel's decision to impose a six-month suspension, without the need for an immediate suspension as well.

***Shaw v. Logue*** **[2013] All ER (D) 33 (Sep)**

**38.69** Solicitor—striking-off order—application to stay

The SDT found allegations of dishonesty proved against S, a solicitor with considerable experience, and ordered that he be struck off the roll of solicitors. S's appeal was to be heard in December 2013. He applied on 4 September 2013 for a stay of the strike-off decision to allow him to continue to practise prior to the appeal hearing. His application was refused by Turner J, in the Administrative Court. A balance had to be struck between, on the one hand, prejudice to the applicant and, on the other, the public interest in upholding high standards in the profession. Although the Court accepted that there was prejudice to the applicant, the time between the hearing and the substantive appeal was relatively short.

***Baker v. British Boxing Board of Control*** **[2014] EWHC 2074 (QB)**

**38.70** Boxing manager's licence—withdrawn—application to restore pending appeal

On 19 November 2013, the British Boxing Board of Control, in accordance with its regulatory procedures, found B guilty of misconduct. B was the chairman and managing director of the Professional Boxers Promoters Association. On 11 February 2014, the Board ordered that B's licence be withdrawn. In accordance with the applicable rules, B issued an appeal to the Stewards, which was likely to be heard in July 2014. Sir David Eady said that the matter for immediate determination was whether B was entitled to an interim order to have his licence restored, pending the hearing of his appeal in July 2014. In dismissing B's application, the learned judge said that he was persuaded that the broadly based challenge to the relevant regulations did not give rise to a serious question to be tried, and that B was unable to fulfil the requirement for obtaining an injunction in accordance with *American Cyanamid Co. v. Ethicon Ltd* [1975] AC 396, whether in relation to his challenge to the validity of the regulations or to the decision-making process of the Board. In so far as the withdrawal of B's licence may cause B financial damage pending the outcome of the appeal, that could be adequately compensated, should it prove necessary, by an award of damages. Before any legal challenge is mounted to the disciplinary procedure under the regulations, that procedure should be allowed to come to its natural conclusion, so that any such attack can be made on the procedure as a whole: see *Modahl v. British Athletic Federation* [2002] 1 WLR 1192; *Calvin v. Carr* [1980] AC 574.

### *Nursing and Midwifery Council v. Gatawa* [2014] EWHC 3153 (Admin)

**38.71** Interim measures continued to cover time between High Court and hearing in Court of Appeal

On 30 January 2013, G was struck off the NMC register of nurses and midwives, and an interim suspension order for eighteen months was made under article 31(1)(c) of the Nursing and Midwifery Order 2001. The charges laid against G were serious charges involving alcohol and dereliction of duty as a nurse. G appealed to the High Court against the decision of the NMC's CCC. The appeal was heard initially on 22 October 2013 by Andrews J, who dismissed the appeal. The matter returned to Andrews J on 8 November 2013 by way of a rehearing. Save for finding one charge not made out, the learned judge upheld the other findings. Following G lodging an application for permission to appeal to the Court of Appeal, the NMC applied for an extension of the interim suspension order to cover the period until the Court of Appeal ruled on the question of permission to appeal, and if necessary, the appeal itself, if such permission were granted. In granting an extension of twelve months, Haddon-Cave J said that normally the NMC asks for extensions prior to the hearing of a disciplinary committee. The charges in this case had not only been laid, but also been found made out. A serious sanction had been imposed and that sanction had been upheld by the High Court. G had chosen to seek a further appeal against her striking off from the register. That was her choice and right. However, it meant that there was an unanswerable case by the NMC that the interim suspension order that was put in place be continued until such time as the proceedings in the Court of Appeal come to an end. Giving judgment on 23 July 2014, the learned judge said that the current order was due to expire on 29 July and therefore that the twelve-month extension would run until 28 July 2015.

# 39

# INVESTIGATION OF ALLEGATIONS


**A. Legal Framework** 39.01

**B. Disciplinary Cases** 39.02

*R v. General Council of the Bar, ex parte Percival* [1991] 1 QB 212 39.02

*R v. General Medical Council, ex parte Toth* [2000] 1 WLR 2209 39.03

*R (Richards) v. General Medical Council* [2001] Lloyd's Rep Med 47 39.04

*R v. Solicitors Disciplinary Tribunal, ex parte Toth* [2001] EWHC Admin 240 39.05

*R (Holmes) v. General Medical Council* [2001] EWHC Admin 321; [2002] EWCA Civ 1838 39.06

*Woods v. General Medical Council* [2002] EWHC 1484 (Admin) 39.07

*R (Aurangzeb) v. Law Society of England and Wales* [2003] EWHC 1286 (Admin) 39.08

*David v. General Medical Council* [2004] EWHC 2977 (Admin) 39.09

*Heath v. Home Office Policy and Advisory Board for Forensic Pathology* [2005] EWHC 1793 (Admin) 39.10

*Henshall v. General Medical Council* [2005] EWCA Civ 1520 39.11

*R (Chaudhari) v. Royal Pharmaceutical Society of Great Britain* [2008] EWHC 3464 (Admin) 39.12

*R (McNicholas) v. Nursing and Midwifery Council* [2009] EWHC 627 (Admin) 39.13

*R (McNicholas) v. Nursing and Midwifery Council (No. 2)*, 23 March 2010 (unreported) 39.14

*R (Independent Police Complaints Commission) v. Commissioner of Police of the Metropolis* [2009] EWHC 1566 (Admin) 39.15

*R (Remedy UK Ltd) v. General Medical Council* [2010] EWHC 1245 (Admin) 39.16

*R (North Yorkshire Police Authority) v. Independent Police Complaints Commission* [2010] EWHC 1690 (Admin) 39.17

*R (Khan) v. Independent Police Complaints Commission* [2010] EWHC 2339 (Admin) 39.18

*Shaikh v. National Co-operative Chemists Ltd and Royal Pharmaceutical Society of Great Britain* [2010] EWHC 2602 (QB) 39.19

*R (Rycroft) v. Royal Pharmaceutical Society of Great Britain* [2010] EWHC 2832 (Admin) 39.20

*Law Society of Scotland v. Scottish Legal Complaints Commission* 2011 SC 94 39.21

*Zia v. General Medical Council* [2011] EWCA Civ 743; [2012] 1 WLR 504 39.22

*R (D) v. Independent Police Complaints Commission* [2011] EWHC 1595 (Admin) 39.23

*Lim v. Royal Wolverhampton Hospitals NHS Trust* [2011] EWHC 2178 (QB) 39.24

*R (Rich) v. Bar Standards Board* [2012] EWCA Civ 320 39.25

*R (B) v. Secretary of State for the Home Department; R (J) v. Upper Tribunal (Immigration Asylum Chamber)* [2012] EWHC 3770 (Admin) 39.26

*Saville-Smith v. Scottish Legal Complaints Commission* [2012] CSIH 99 39.27

*Roberts v. Hook* [2013] EWHC 1349 (Admin) 39.28

*Alami and Botmeh v. Health and Care Professions Council and Young (Interested Party)* [2013] EWHC 1895 (Admin) 39.29

*R (Chief Constable of West Yorkshire Police) v. Independent Police Complaints Commission and Armstrong* [2013] EWHC 2698 (Admin); [2015] ICR 184, CA 39.30

*R (Ramsden) v. Independent Police Complaints Commission, Chief Constable of West Yorkshire Police (Interested Party)* [2013] EWHC 3969 (Admin) 39.31

*R (Peters) v. Chief Constable of West Yorkshire Police* [2014] EWHC 1458 (Admin) 39.32

*R (D) v. General Medical Council* [2013] EWHC 2839 (Admin) 39.33
*R (Hibbert) v. General Medical Council* [2013] EWHC 3596 (Admin) 39.34
*R (Grabinar) v. General Medical Council* [2013] EWHC 4480 (Admin) 39.35
*Chhabra v. West London Mental Health NHS Trust* [2014] ICR 194, SC 39.36
*Fynes v. St George's Hospital NHS Trust* [2014] EWHC 756 (QB) 39.37
*R (Mackaill) v. Independent Police Complaints Commission* [2014] EWHC 3170 (Admin) 39.38
*General Medical Council v. Michalak* [2014] UKEAT/0213/14/RN 39.39
*R (Delezuch) v. Chief Constable of Leicestershire Constabulary; R (Duggan) v. Association of Chief Police Officers* [2014] EWCA Civ 1635 39.40
*R (Kerman & Co) v. Legal Ombudsman* [2014] EWHC 3726 (Admin); [2015] 1 WLR 2081 39.41
**C. Other Relevant Chapters** 39.42

## A. Legal Framework

**39.01** Most, if not all, regulatory bodies have detailed rules for the investigation of allegations and complaints. These are frequently found in the relevant statute or in rules. Examples include:

Medical Act 1983, section 35C
General Medical Council (Fitness to Practise) Rules 2014, Part 2
Police (Conduct) Regulations 2012
Association of Chartered Certified Accountants Complaints and Disciplinary Regulations 2014
Architects Registration Board Investigations Rules (third-party review under rule 16 where complainant or architect is dissatisfied with the process)
Chartered Institute of Legal Executives Investigation Rules
Institute of Chartered Accountants in England and Wales Disciplinary Bye-Laws, bye-laws 9–16A

## B. Disciplinary Cases

### *R v. General Council of the Bar, ex parte Percival* [1991] 1 QB 212

**39.02** *Barrister—decision to prosecute—challenge by complainant as to form of charges—judicial review—whether decision of PCC flawed*

The applicant, a former Solicitor-General and a Queen's Counsel (QC) in active practice, accused another QC in active practice, S, of acting dishonestly whilst they both belonged to the same chambers and of engaging in conduct that might bring the profession of barristers into disrepute. The role of a prosecutor exercising discretion in the sifting and assessment of complaints, and empowered to prosecute complaints before the disciplinary tribunal, was delegated to the Professional Conduct Committee (PCC). Following investigation and obtaining counsels' advice, the PCC decided to forward to the disciplinary tribunal charges against S only of breaches of professional standards, and not to charge S with the more serious charge of professional misconduct. On the applicant's application for judicial review, the Divisional Court (Watkins LJ and Garland J) held that the acts and omissions of the PCC could be challenged by way of judicial review, that the applicant had and retained locus standi, and that the decision whether or not to prosecute was reviewable and there was no ground of public policy that should inhibit the Court from reviewing a decision of the PCC. The mere fact that professional standards are called in question does not mean that the Court should exercise its discretion not to interfere. Authorities such as *R v. General Council of Medical Education and Registration in the United Kingdom* [1930] 1 KB 562, *General Medical Council v. Spackman* [1943] AC 627, 639–40, *Rajasooria v. Disciplinary Committee*

[1955] 1 WLR 405, and *Bhattacharya v. General Medical Council* [1967] 2 AC 259 were concerned with the decisions of disciplinary tribunals as to whether or not, in particular instances, the rules or standards of professional conduct had been transgressed. However, as to the decision in the instant case, the Court could not find anything irregular or out of conformity with the rules in the conduct of the PCC. It was not bound to refer the applicant's complaint to the disciplinary tribunal. It was acting well within its discretion in receiving counsels' opinions and allowing them to influence its judgment. The test that the Court had to apply in deciding whether or not the decision of the PCC was flawed was obviously not whether the Court, in the Committee's position, would have come to the same decision as it did. There may very well be room on the special facts for divergent views as to the proper charge that S should face before the disciplinary tribunal. What could not be controverted was that the PCC acted within a broad discretion, which undoubtedly it had, upon correct principles, and with impartiality and fairness. Even if the Committee did in fact change its mind following the receipt of a second opinion, the Court would not regard it, in the particular circumstances, as having acted unreasonably or perversely, or otherwise wrongly, in the exercise of its discretion in that respect. The PCC relied on the proper test of a prima facie case. Accordingly, the application was dismissed.

***R v. General Medical Council, ex parte Toth* [2000] 1 WLR 2209**

**39.03** Complainant—filtering process—screener—PPC—legitimate expectation—public hearing before PCC

T made a complaint to the General Medical Council (GMC) against his general practitioner (GP), Dr J, concerning the treatment of his 5-year-old son. The central complaint of T was that he and his partner told J of the need for intravenous glucose, but that J wrongly treated the child with sedative drugs, and that his untreated condition and lack of intravenous glucose led to the death of the child. T sought judicial review of two decisions made by the GMC not to refer the complaint to the Preliminary Proceedings Committee (PPC). Under the General Medical Council Preliminary Proceedings Committee and Professional Conduct Committee (Procedure) Rules 1988, complaints against registered medical practitioners have to go through two filters or processes of examination before they are heard by the PCC. The first involves an examination by a member of the GMC, colloquially referred to as a 'screener', to decide whether the complaint need not proceed further; the second, by the PPC deciding whether the complaint ought to be referred for inquiry to the PCC. In granting T's application for judicial review, Lightman J said, at [11], that the provisions in the Medical Act 1983 and rules made thereunder are designed to protect the public from the risk of practice by practitioners who for any reason (whether competence, integrity, or health) are incompetent or unfit to practise, and to maintain and sustain the reputation of, and public confidence in, the medical profession. The Act and rules set out to provide a just balance between the legitimate expectation of the complainant that a complaint of serious professional misconduct will be fully investigated and the need for legitimate safeguards for the practitioner, who, as a professional person, may be considered particularly vulnerable to and damaged by unwarranted charges against him. The learned judge said, at [14], that his conclusions were as follows:

(1) The general principles underlying the Act and Rules are that: (a) the public have an interest in the maintenance of standards and the investigation of complaints of serious professional misconduct against practitioners; (b) public confidence in the GMC and the medical profession requires, and complainants have a legitimate expectation, that such complaints (in the absence of some special and sufficient reason) will be publicly investigated by the PCC; and (c) justice should, in such cases, be seen to be done. This must be most particularly the case where the practitioner continues to be registered and to practise.

(2) There are a series of processes designed to filter out complaints that need not, or ought not, to proceed further.
(3) The registrar's role is merely to ensure that the complainant has complied with the formal requirements laid down for investigation of a complaint.
(4) The role of the screener is a narrow one: it is to filter out from complaints that have complied with the formal requirements laid down for investigation, not those that, in his view, ought not to proceed further, but those that he is satisfied (for some sufficient and substantial reason) need not proceed further. For this purpose, he must be satisfied of a negative—namely, that the normal course of the complaint proceeding to the PPC need not be followed. The assumed starting point is (1) above and the need referred to is the need to honour the legitimate expectation that complaints (in the absence of some special and sufficient reason) will proceed through the PPC to the PCC.
(5) The PPC's role is to decide whether the complaint "ought to proceed". This language must be read in the context of a scheme under which the complainant has no right to the practitioner's comments on the complaint or other material put before the PPC, and a scheme of which the central feature is the investigation of complaints by the PCC before whom alone there is full disclosure of documents and evidence and a form of hearing where the complainant (and public) can see, and be reassured by seeing, the proper examination of the merits of the complaint. The PPC may examine whether the complaint has any real prospect of being established, and may themselves conduct an investigation into its prospects, and may refuse to refer if satisfied that the real prospect is not present, but they must do so with the utmost caution bearing in mind the one-sided nature of their procedures under the Rules, which provide that, whilst the practitioner is afforded access to the complaint and able to respond to it, the complainant has no right of access to or to make an informed reply to that response, and the limited material likely to be available before the PPC compared to that available before the PCC. It is not their role to resolve conflicts of evidence.
(6) In the exercise of their respective jurisdictions, the screener and PPC should be particularly slow in halting a complaint against a practitioner who continues to practise, as opposed to one who has since retired, for the paramount consideration must be the public's protection in respect of those continuing to practise.

### *R (Richards) v. General Medical Council* [2001] Lloyd's Rep Med 47

**39.04** Clinical care—patient—allegations of negligent treatment and fabricating records—test—whether case to answer

R sought judicial review of a decision of the PPC contained in a decision letter dated 7 April 2000 not to refer her complaint against two GPs to the PCC, because the PPC was 'not of the opinion that the matters before them appeared to raise a question whether serious professional misconduct had been committed'. The GPs (who were husband and wife) had treated the claimant's sister, who died of a pulmonary embolism, aged 36. In her complaint to the GMC, the claimant alleged that one of the doctors had been negligent in the treatment given, that he had fabricated the patient's records to cover up that negligence, and that the other doctor had colluded with her husband in that fabrication. In granting judicial review, quashing the decision letter, and remitting the complaint to a new PPC, Sullivan J said that it was necessary to stand back and consider the decision letter as a whole. The Court accepted that the PPC believed that it was applying the test set out in the rules, but perhaps precisely because of its expertise, it set out to answer the matters in dispute rather than to decide whether there was a question for the PCC to answer. Whilst understanding the basis on which the members of the PPC reached their own conclusions as to clinical care, in the light of all of the evidence the Court did not understand how a body with a screening function, even one as expert as this PPC, could possibly have formed the opinion that the material before it did not even 'raise a question whether serious professional misconduct had been committed'. It followed that the decision must be quashed.

***R v. Solicitors Disciplinary Tribunal, ex parte Toth* [2001] EWHC Admin 240**

**39.05** Complaint—misleading Court—referral of complaint by SDT to Law Society—whether appropriate—whether prima facie case

T considered that the solicitor acting for two doctors (in respect of whom he had made a complaint to the GMC: [2000] 1 WLR 2209) had misled the Court in T's claim for damages and had otherwise acted improperly. In February 2000, T made a complaint to the Solicitors Disciplinary Tribunal (SDT) pursuant to rule 4 of the Solicitors (Disciplinary Proceedings) Rules 1994. Rule 4 provides that the application shall be considered by a solicitor member of the Tribunal, who shall certify whether a prima facie case is established and, after the finding of a prima facie case, the Tribunal shall fix a date for the hearing. If no prima facie case is established, the Tribunal may dismiss the application. By rule 28, the Tribunal may, at any stage in which an application is not made on behalf of the Law Society, refer the case to the Society for consideration. By letter dated 13 March 2000, the Tribunal informed T that two solicitor members and a lay member of the Tribunal had 'declined to reach a conclusion as to whether or not a prima facie case is established' and had decided that the appropriate body to conduct an investigation was the Office for the Supervision of Solicitors (OSS). Stanley Burnton J said that, as a matter of construction, the powers conferred by rule 28 to refer a case to the OSS and to adjourn the application applied both before and after certification under rule 4. However, the power to refer an application to the OSS is qualified: it must be for the purpose of it enabling the OSS to decide whether to lodge a further application against the respondent solicitor or to undertake the prosecution of the original application. There was no justification for the decision of the Tribunal to 'decline' to reach a conclusion as to whether or not a prima facie case was established. If it was because it delegated that decision to the OSS, it was clearly wrong for it to do so. If the evidence before it establishes a prima facie case, the Tribunal should not, without good reason, defer certification in order to refer the case to the OSS under rule 28. The cases must be rare in which it is not possible to determine whether a prima facie case exists but it is appropriate to refer to the OSS. The establishment of a prima facie case is not to be regarded as a high hurdle: see *R v. Governor of Brixton Prison ex parte Armah* [1968] AC 192, in which the House of Lords distinguished between the relatively low requirement of a prima facie case and the requirement, in an extradition case, of a 'strong or probable presumption' that the defendant committed the offence charged.

***R (Holmes) v. General Medical Council* [2001] EWHC Admin 321; [2002] EWCA Civ 1838**

**39.06** Complaint—delay—retirement of practitioner—factor for screener/PPC—caution before closing case

The claimants were the cohabiting partner and the parents of D, who died aged 34 on 26 July 1995 from a colloid cyst on his brain. On 24 July 1995, D had been seen by Dr R in his surgery. The next day, he had been seen by Dr S at his home in the evening. In the case of Dr R, three screeners, who included a lay member, decided that the complaint should not proceed to the stage of consideration by the PPC. In the case of Dr S, the complaint was considered by the PPC, but it decided that the complaint should not proceed to the PCC. The claimants lodged judicial review proceedings challenging the decisions. In granting relief at first instance—[2001] EWHC 321 (Admin)—Ouseley J said that there was clearly undue delay in this case by the claimants. Dr R was by that time retired and Dr S's retirement was imminent. What would be the value in those circumstances of these proceedings being continued, with the associated stress and health problems, and with the background of delay? However, if the Court were to take the view that the actual and impending retirements were a sufficient basis in this case not to quash the decisions, the Court would be engaged in what would be an inappropriate value judgment. It is for the screening committee or the PPC to weigh those factors against the concern that could arise that retirement might eliminate the regulatory and investigatory process in circumstances in which it might otherwise be warranted. If the position were that the case was as weak as counsel for Drs R and S had asserted, the delay and prejudice would not be that significant. The greater the delay, the stronger the case against them must have become, and the stronger the public interest in ensuring the proper regulation of the medical profession and the examination of the circumstances of their case. The Court of Appeal (Laws, Jonathan Parker and Keene LJJ), in dismissing the appeal, said that generally decisions taken by bodies such as the PPC ought to stand or fall by the terms in which those decisions are promulgated. Of course, further explanations may sometimes be required to clarify the meaning of a particular expression, or to resolve an ambiguity appearing on the face of the decision—but, in general, those affected by such a decision ought to be able to take it at face value. The Court endorsed the observations of Hooper J in *Edwards v. Cornwall County Council* [2001] All ER (D) 395 (Jul) and Richards J in *R (Molloy) v. Powys County Council* [2001] EWHC Admin 332 as to the need for great caution in considering whether to admit evidence from a tribunal commenting on its decision. In general, as Richards J said, decisions should speak for themselves.

### *Woods v. General Medical Council* [2002] EWHC 1484 (Admin)

**39.07** Inquiry—removal of organs—referral to GMC—principles—aide-mémoire to be applied by PPC

The claimant, Christine Woods, was the mother of a baby boy, who was born on 30 September 1989 and died the following day. His body was transferred to Alder Hey Children's Hospital on 4 October 1989. A post mortem was performed on 9 October 1989. Ten years later, the claimant was informed, as many other parents similarly learned, that a number of her baby's organs had been removed and retained without her consent. In December 1999, the Parliamentary Under-Secretary of State for Health established an independent inquiry, the report of which was published on 30 January 2001. A copy of it was passed to the GMC for consideration of disciplinary proceedings against doctors named in the report. Of the thirteen doctors considered by the GMC, two, including Professor V, the main target of the report's criticism and condemnation, were referred to the PCC. Of the remaining eleven, two (Dr C and Dr K) were screened out by the screeners, and nine cases were screened out by the PPC and not referred on to the PCC. On hearing the claimant's application for judicial review of the decisions of the screeners and the PPC, Burton J said that, save in the case of Dr K, which he directed should be reconsidered by the screeners, there was no ground in law for interfering with the reasoned conclusions. The principles that underlined the provisions could be summarized as follows.

(1) They constitute a fine balance between three competing desirables:
    (a) the protection of the public from the risk of practice by practitioners who, for any reason (whether competence, integrity, or health), are incompetent or unfit to practise, and the maintenance of standards;
    (b) the maintenance of the reputation of, and public confidence in, the medical profession, and the legitimate expectation of the public, and of complainants in particular, that complaints of serious professional misconduct would be fully and fairly investigated; and
    (c) the need for legitimate safeguards for the practitioner, who, as a professional person, may be considered particularly vulnerable to, and damaged by, unwarranted charges against him.
(2) The filtering exercise is especially required in pursuit of the last of these three principles. It is necessary in order to ensure, given the sensitive and high-profile role of doctors and the ease of, and the understandable, but often misguided resort to, making complaints against them, with all of the time-consuming and damaging

consequences for the doctor of such an investigation, that only those cases are taken forward in which there is a real prospect of the complaint succeeding. On the other hand, because of the importance of the other two principles, it is necessary for these filtering exercises not to be ratcheted to too high a level, and that caution should be exercised before filtering out a complaint, so that, if there is doubt, it must be resolved in favour of referring the matter on for investigation.

The GMC took the opportunity, after Lightman J's judgment in *Toth*, to seek advice from two senior barristers, Robert Englehart QC and Mark Shaw, who formulated an aide-mémoire, to which the GMC now works. With one alteration, to sub-paragraph 3(4), in which the Court would substitute the word 'a' for the word 'the', the learned judge approved and commended the aide-mémoire in its entirety, as follows.

*The Approach to be Applied by the PPC in Conduct Cases*

***Aide Memoire***

1. In conduct cases the PPC's task is to decide whether, in its opinion, there is a real prospect of serious professional misconduct being established before the PCC. Serious professional misconduct must be considered in the context of conduct so grave as potentially to call into question a practitioner's registration whether indefinitely, temporarily or conditionally.

2. The "real prospect" test applies to both the factual allegations and the question whether, if established, the facts would amount to serious professional misconduct. It reflects not a probability but rather a genuine (not remote or fanciful) possibility. It is in no-one's interest for cases to be referred to the PCC when they are bound to fail, and the PPC may properly decline to refer such cases. On the other hand, cases which raise a genuine issue of serious professional misconduct are for the PCC to decide.

3. The following does not purport to be an exhaustive list, but in performing its task the PPC:

(1) should bear in mind that the standard of proof before the PCC will be the criminal standard (beyond reasonable doubt);
(2) is entitled to assess the weight of the evidence;
(3) should not, however, normally seek to resolve substantial conflicts of evidence;
(4) should proceed with caution given that, among other considerations, it is working from documents alone and does not generally have the benefit of [a] complainant's response to any reply to the complaint submitted on behalf of the practitioner;
(5) should proceed with particular caution in reaching a decision to halt a complaint when the decision may be perceived as inconsistent with a decision made by another public body with medical personnel or input (for example, an NHS body, a Coroner or an Ombudsman) in relation to the same or substantially the same facts and, if it does reach such a decision, should give reasons for any apparent inconsistency;
(6) it should be slower to halt a complaint against a practitioner who continues to practise than one who does not;
(7) if in doubt, should consider invoking rule 13 of the procedure rules and in any event should lean in favour of allowing the complaint to proceed to the PCC; and
(8) should bear in mind that, whilst there is a public interest in medical practitioners not being harassed by unfounded complaints, there is also a public interest in the ventilation before the PCC in public of complaints which do have a real prospect of establishing serious professional misconduct.

[Serious professional misconduct has been replaced by the categories of impairment of fitness to practise provided by section 35C of the Medical Act 1983, as substituted by article 13 of the Medical Act 1983 (Amendment) Order 2002. The criminal standard of proof was replaced by the civil standard of proof for new hearings on 31 May 2008.]

### *R (Aurangzeb) v. Law Society of England and Wales* [2003] EWHC 1286 (Admin)

**39.08** Solicitor—decision by Law Society—referral to SDT—whether appropriate for Court to intervene

The claimant, a solicitor, sought judicial review challenging the decision of the Law Society to refer allegations to the SDT. The allegations were of practising in breach of conditions that, on an earlier occasion, had been applied to his right to practise and a failure to notify the Law Society that he had been made bankrupt. The applicant claimed that the decision to refer the allegations to the SDT was flawed in two significant respects. In refusing relief, Newman J said that, in his view, the question is whether, as a matter of principle, it is right for the Court to intervene by way of judicial review when the ambit and reach of the decision under challenge goes no further than to place the allegations in question before a disciplinary tribunal. The Court would not conclude that it had no power to intervene at this stage of the disciplinary process in connection with the Law Society or any other disciplinary body having similar powers. But the Court must, in accordance with basic elementary principles, have considerable reservations about the desirability of intervening in the manner suggested, save in circumstances in which the facts call out for that intervention, such as where irreparable harm or unfairness is likely to occur, or justice could be met only by intervention. Where what has occurred is that there has been a procedural failure and the matter is before the tribunal, the proper conclusion is that the tribunal will have ample opportunity to cure any of the failures that to date it is said have occurred. There will be an occasion for justice then to be done. The decision of the Tribunal was therefore to be subject to review.

### *David v. General Medical Council* [2004] EWHC 2977 (Admin)

**39.09** Anaesthetist—referral of complaint to PCC—caution before concluding not to refer—decision of PPC not determinative of any civil right—reasons

D, a consultant anaesthetist, sought an order quashing the decision of the PPC to refer for public inquiry by the PCC the question of whether or not she committed serious professional misconduct in relation to the treatment of a patient at Basildon Hospital in Essex. Stanley Burnton J said that the authorities establish that the PPC is entitled (and in general obliged) to refer a case to the PCC if it concludes that there is a real prospect that it will be established that the practitioner in question was guilty of serious professional misconduct. If the PPC concludes that there is no such prospect, it should not refer the case to the PCC, but it must exercise caution before so concluding and must take into account the limitations of the material available to it. The learned judge accepted that, just as the PPC should exercise caution when deciding not to refer a case to the PCC, so should the Court exercise caution when considering whether to quash a decision of the PPC to refer a case to the PCC. That caution is particularly appropriate if the Court is asked (as it was in the present case) to assess evidence of medical practice, or expert medical reports, with a view to determining whether the PPC could legitimately conclude that there is a real prospect of a charge or charges being proved. The Court, unlike the PPC, is not medically qualified and it may intervene only if it concludes that the view taken by the medically qualified committee was perverse. The degree of caution that is appropriate is, however, significantly less if the issue before the Court is whether the PPC correctly interpreted or applied the Procedure Rules. It is appreciated that a hearing before the PCC involves stress, adverse publicity, and cost for the practitioner. However, where the decision in question is that a case should be referred to the PCC, the Court must take into account that the decision of the PPC is not determinative of any civil (or other) right or obligation of the practitioner. The practitioner will have a fair trial before the PCC and, if innocent of serious professional misconduct, can expect to be vindicated. When it decides to refer a case to the PCC, the PPC is not obliged to give extensive reasons for its decision and the scrutiny of the Court on judicial review is not required to be, and should not be, intensive. Only in the clearest of cases could it be appropriate for the Court, on an application for judicial review, to scrutinize the expert medical evidence before the PPC with a view to deciding whether it could reasonably have referred a case to the PCC. In dismissing the application, the learned judge said, however, that he had considerable sympathy with the claimant's submission that, in general, the PCC is not the appropriate forum to investigate genuine ethical questions. Serious professional misconduct normally involves conduct going beyond established ethical rules. To read the decision letter in the instant case as stating that the PPC referred the doctor's case to the PCC because of the public interest in the resolution of a controversial ethical question was too literal a reading. No such ethical question is to be seen in the charges, or in the expert report, or in the submissions made on behalf of the doctor to the PPC.

### *Heath v. Home Office Policy and Advisory Board for Forensic Pathology* [2005] EWHC 1793 (Admin)

**39.10** Pathologist—referral to disciplinary tribunal—judicial review refused

The claimant, H, a well-known and senior forensic pathologist accredited by the Home Office, sought judicial review of a decision to refer to a disciplinary tribunal complaints that were made against him in respect of his conduct as a forensic pathologist and expert witness in relation to two murder trials. The Home Office Policy Advisory Board for Forensic Pathology was established in 1991 to monitor standards and to advise the Home Secretary. In dismissing H's application, Newman J observed that:

> 42. The application for permission to apply for judicial review has reached the court before any substantive disciplinary hearing has commenced. In exceptional cases, a need may arise for action to be taken to avoid an obvious miscarriage of justice. In the normal course, [the Court's] supervisory jurisdiction is concerned with the legality of a decision made by a disciplinary tribunal not to monitor the procedural process governed by a wide discretion which is exercised by disciplinary tribunals to rule their own affairs. In the usual course, the roles and duties of a screening committee are not appropriately made the subject of application for judicial review. The stage of the process under the Constitution of this Board, which is determinative of rights or obligations or which affects a substantive interest, is the decision of the tribunal. That is not to say that prejudice will not arise from the giving of notice that a referral to a tribunal has been made, but it is inescapable and lasts only so long as the tribunal decision is delayed.
>
> 43. The way to avoid such prejudice being allowed to continue is for the hearing before the tribunal to take place promptly.... The events...give the strong impression that, from the outset, the claimant set out to block the process as best he could.... he has perpetuated delay by raising as many points as he possibly could which might serve to avoid or delay a decision to refer to a tribunal. His demand, for example, for the supply of transcripts of the criminal proceedings before matters proceed, was misguided and excessive. The transcripts may or may not be necessary at the substantive hearing, if there is to be one.... The documentation which had been served upon him enabled him to make all necessary representations...However firmly it might have been believed he was entitled to more time, he should have responded to the best of his ability on the basis of the information which he obviously had in his possession.

### *Henshall v. General Medical Council* [2005] EWCA Civ 1520

**39.11** Complainant—clinical trials for treatment of premature babies—decision not to refer to PCC—test—'real prospect of success'

H challenged a decision of the PPC not to refer to the PCC her complaints against three registered medical practitioners, Professor S, Dr Sa, and Dr Sp. In the early 1990s, those doctors had been involved in a clinical trial of a treatment for premature babies with breathing difficulties at the North Staffordshire Maternity Hospital in Stoke. The treatment was known as 'continuous negative extrathoracic pressure ventilation' (CNEP). H had two premature babies, both of whom were included in the CNEP trial: the first died 60 hours after her birth; the second was subsequently found to have cerebral palsy. H complained about the integrity of the CNEP trial, which was designed and established by Professor S and Dr Sa, and about the supervision and conduct of it at the hospital, for which Dr Sp was responsible. In allowing the appeal (Sedley and Jonathan Parker LJJ, and Auld LJ dissenting in the conclusion), Auld LJ said, at [31]–[33], that Lightman J's and the GMC's test of a 'real prospect' of establishment of a complaint has been approved and applied in a series of High Court judgments—namely, *R (Richards) v. General Medical Council* [2001] Lloyds Med Rep 47, [58], *per* Sullivan J, subject to two qualifications not affecting the fundamental nature of the test; *R v. General Medical Council, ex parte McNicholas* [2001] EWHC 279 (Admin), [12], *per* Sullivan J; and *R (Woods) v. General Medical Council* [2002] EWHC 1484 (Admin), [14(iii)], *per* Burton J. It was also seemingly approved by Jonathan Parker LJ, with whom Laws and Keene LJ agreed, in *R (Holmes) v. General Medical Council* [2002] EWCA Civ 1838, [74]. It is only at the PCC stage that the matter assumes the character of a traditional forensic process, with sequential mutual disclosure of documents, a public hearing that begins by the reading and putting of the 'charges' to the accused, a hearing in which the GMC and the accused doctors can be represented by lawyers, at which evidence is prepared and given in traditional form, orally, and/or in writing, and at which the evidence of witnesses on each side can be tested in cross-examination. The learned judge went on to say, at [47], that the PPC's conclusion that a complaint had a real prospect of success, as distinct from a probability or a real probability of success, was just another way of saying that it appeared to raise a question that ought to be referred to the PCC for determination. The test, 'a realistic prospect of conviction', in the Code for Crown Prosecutors is a markedly different test, taking into account the criminal standard of proof for the realization of such prospect before the PCC and the need to proceed with caution as the GMC's aide-mémoire advises. In this context, it is interesting to note how Jonathan Parker LJ, in his seeming acceptance in *Holmes*, at [74], of the 'real prospect of success' test in *Toth* and *Richards*, equated it with 'an arguable case' of serious professional misconduct. Continuing at [49]–[51], the learned judge said that it is necessary to keep in mind the different functions and procedures of the PPC and PCC. Whilst the former was to act as a 'filter' for the latter before referring to it complaints of serious professional misconduct and the latter was to decide on evidence put before it whether such complaints were established, the filtering was clearly intended to take a more rigorous form than that conducted by the screener, although not so rigorous as the determinative and forensic role of the PCC. It is not the job of the PPC to conduct an inquiry in the full or evidential sense. Its role was not to consider 'evidence'; that was for the PCC's consideration if the matter reached it, in relation to evidence that could be forensically tested and with the benefit of mutual disclosure. The PPC's role was to consider whether material put before it on paper raised a question as to serious professional misconduct that 'ought to be' the subject for evidential presentation to an inquiry by the PCC. Jonathan Parker LJ, in agreeing with Sedley LJ that the appeal should be allowed, said that the PPC in the instant case had identified the correct legal test as laid down in the rules, and as explained by Lightman J in *Toth* and by the GMC's aide-mémoire. However, having identified the correct test, the PPC failed to apply it properly to the material before it. The PPC went beyond the limits of its function. It is one thing to evaluate the available evidential material in order to determine whether, in its opinion, such material appears to raise a question of whether the practitioner has committed serious professional misconduct, but quite another to purport to resolve disputed factual issues. In making the findings recorded in the decision letter, and in particular the finding that the CNEP trial was properly conducted, the PPC went further than was necessary for the purpose of deciding whether the material before it satisfied the test. In so doing, it trespassed on an area that was properly the province of the PCC, should the case be referred to it.

### *R (Chaudhari) v. Royal Pharmaceutical Society of Great Britain* [2008] EWHC 3464 (Admin)

**39.12** Complaint—prescription of drugs—identity of pharmacist at pharmacy—whether hospital pharmacist responsible for dispensing relevant prescription

The mother of a child who died aged 5 months brought proceedings against the Royal Pharmaceutical Society of Great Britain on the grounds of its failure to determine her complaint against its members properly as regards prescription of drugs for the baby. The complaint in question related to two matters: first, the dispensation of drugs by a retail pharmacy; and second, the dispensing of drugs by a hospital pharmacist when the child was an in-patient. In dismissing the complainant's renewed application for permission to apply for judicial review, Blair J considered the statutory basis upon which the Society considers complaints made against pharmacists. These are dealt with under statutory rules and guidance, which is issued to the Disciplinary Committee and sets out the principles to be applied to ensure that the only cases that are referred to the Committee are those in which there is a real prospect of establishing misconduct that would render the person concerned unfit to remain on the register. The Court had to consider both the legal framework and the factual framework. So far as the factual framework was concerned, in the first complaint, it was not possible to identify which of the two pharmacists on duty at the pharmacy dispensed the drug concerned; so far as the hospital complaint was concerned, it could not be concluded that the named pharmacist was responsible for dispensing the relevant prescription. A similar factual situation was considered in *Woods v. General Medical Council* [2002] EWHC 1484 (Admin), in which Burton J referred to the balancing that has to take place on these occasions and referred further to the 'real prospect of success' test. He held that, on that basis, a decision of the GMC was not susceptible to challenge and dismissed the claim. Blair J said that the same principle applied in this case.

### *R (McNicholas) v. Nursing and Midwifery Council* [2009] EWHC 627 (Admin)

**39.13** PPC—Nurses, Midwives and Health Visitors (Professional Conduct) Rules 1993, rule 9(1) and (3)—two-stage process

The claimant, the father of the late Sarah Jane McNicholas, who died on 31 July 1997 at Halton Hospital, Cheshire sought to review judicially the decision of the PPC to refer his complaint against nine registered nurses who at the relevant time were employed or worked at the hospital to the solicitor of the Nursing and Midwifery Council (NMC) for further investigation. Rule 9(1) of the Nurses, Midwives and Health Visitors (Professional Conduct) Rules 1993 provides that the PPC shall consider allegations of misconduct and shall, subject to any determination under rule 8(3) (adjourned consideration of the matter), and where it considers that the allegations may lead to removal from the register, direct the registrar to send to the practitioner a notice of proceedings. Rule 9(3) provides that where a notice of proceedings has been sent to a practitioner, the PPC shall consider any written response by the practitioner and, subject to any determination under rule 8(3), shall refer to the PCC a case that it considers justifies a hearing before the Committee with a view to removal from the register. In granting the claimant's application, Stadlen J held that rule 9 contemplated a two-stage process. The PPC may only at the rule 9(1) stage be able to form a preliminary view as to whether, subject to any answer that the practitioner may come up with in any response under rule 9(3), there is a reasonable prospect of success on the basis of the material before it. This may well be a less onerous task than that involved at the rule 9(3) stage. By that stage, the PPC (if a matter has reached that stage) has the benefit of any response received from the practitioner in response to the request that the PPC will have directed the registrar to make under rule 9(1) to the practitioner. Thus, at the rule 9(3) stage of these rules, the PPC is at a stage similar to, and in a similar position to, that of the PPC in the rules considered by Lightman J in *Toth*, when deciding whether the case shall be referred to the PCC for inquiry. The PPC's function under rule 9(1) is a very limited one: of deciding whether the allegations may lead to removal and thus to require directing the registrar to send to the practitioner a notice of proceedings. It should remind itself that, for the purpose of making that decision, it will have an opportunity to consider definitively whether there is a real prospect of success on receipt of any response from the practitioner should the matter proceed to that stage. The power of the PPC under rule 8(3) to require further investigations to be conducted is not an at large power to initiate fact finding investigations for their own sake. It arises only in the context of the PPC's consideration of existing allegations made against a practitioner. In issuing a direction to the registrar to send the practitioner a notice of proceedings, it is perfectly proper for a PPC to seek the assistance and advice of a solicitor for the purpose of examining the material placed before the PPC, which has been presented by the claimant and/or obtained as part of the NMC's rule 8(2) investigation, and formulating the potential wording of allegations for the purpose of inclusion in a notice of proceedings if the PPC reasonably considers that to be necessary to enable it to discharge its functions under rule 9(1). There may be many cases in which, for a variety of reasons, the complaint received or the material supporting it are not in a sufficiently advanced or coherent state to enable a judgment to be reached on whether they might lead to removal, or to enable a coherent set of allegations to be included in the notice of proceedings, or both. On the facts of this case, there was a mountain of material before the PPC, and whilst it would not have been illegitimate to receive help from a solicitor in terms of marshalling the material to enable the Committee to discharge its function under rule 9(1), there is nothing to indicate that it took that step. Accordingly, the matter would be remitted back to the PPC.

### *R (McNicholas) v. Nursing and Midwifery Council (No. 2)*, 23 March 2010 (unreported)

**39.14** Nurses, Midwives and Health Visitors (Professional Conduct) Rules 1993, rule 9(1)

At a subsequent hearing, Collins J observed that the rule 9(1) stage of the Nurses, Midwives and Health Visitors (Professional Conduct) Rules 1993 was, as Stadlen J indicated in [2009] EWHC 627 (Admin), a relatively straightforward exercise. It did not involve any assessment of the strength of the evidence. In argument, Collins J said that, obviously, if it is material which on any view cannot amount to a row of beans, that is one thing, but otherwise it is not at the rule 9(1) stage for the PPC to go into questions of investigation or at least not beyond a fairly superficial manner. The Committee needs to consider whether the material on the various allegations is such as needs to be answered. It obviously has to be persuaded that it is a serious enough allegation that could amount to removal. Is there evidence that could lead to removal? That is the test under rule 9(1). It is not for the PPC to do more than decide whether the evidence looks as if it needs to be answered, or is such as needs to be answered under rule 9(3). It is not for the PPC to go into deciding whether it is correct or not.

### *R (Independent Police Complaints Commission) v. Commissioner of Police of the Metropolis* [2009] EWHC 1566 (Admin)

**39.15** Provisional letter concluding complaint—subsequent reopening of investigation—delay—discretionary relief

On 1 March 2005, a case worker with the Independent Police Complaints Commission (IPCC) wrote to the Directorate of Professional Standards of the Metropolitan Police Service, enclosing a draft letter to a complainant stating that the IPCC was 'minded' to conclude that misconduct proceedings against two officers could not be justified. Subsequently, after further investigation, the IPCC reopened the investigation and sought further information from the Metropolitan Police Service. In judicial review proceedings, the IPCC challenged the defendant's decision, by letter dated 28 April 2008, to refuse to respond to the claimant's request for further information, on the grounds that the complaint and investigation had previously been concluded. Stanley Burnton J concluded that the letter of 1 March 2005 was not a final decision on the part of the IPCC not to recommend disciplinary proceedings against the two officers. The letter stated that the IPCC was 'minded' to conclude that misconduct proceedings cannot be justified. That was the language of provisional decision, of a mind that might be changed, and not that of a final decision. In *R (Green) v. Police Complaints Authority* [2004] 1 WLR 725, [85], Lord Carswell approved Simon Brown LJ's statement in the Court of Appeal on the ability of the Police Complaints Authority to reopen an investigation if it thinks it necessary in the light of representations made or evidence supplied, following the issue of a provisional decision letter. In the end, however, in the instant case, the events in question occurred over six years ago. When the events took place so long ago, it is simply not sensible or practicable now to ask the officers to remember what they knew, or considered, or did, and when, which is what is sought by a request for further information. The impact of the delay was compounded by the not unreasonable belief on the part of the Directorate of Professional Standards that the matter had been concluded and the fact that, through no fault of their own, the officers were given to believe that indeed it had. For these reasons, no relief was to be granted.

### *R (Remedy UK Ltd) v. General Medical Council* [2010] EWHC 1245 (Admin)

**39.16** Chief Medical Officer—role in appointment of training posts—appointments system flawed—whether subject to fitness to practise proceedings

In this case, the claimant company, which was founded to represent doctors and campaigned on a wide range of medical and professional issues affecting doctors, especially appointments for junior doctors' training posts, sought to subject the Chief Medical Officer for England and the chair of the Department of Health's recruitment and selection steering group, to the GMC's disciplinary processes. The claimant company applied for judicial review of the decision of the GMC not to refer allegations of misconduct to case examiners arising from a number of public investigations that had recognized that the appointments system was a deeply flawed scheme. The Divisional Court (Elias LJ and Keith J), in dismissing the application for judicial review on the grounds that the allegations against the Chief Medical Officer and the chair of the steering group did not fall within section 35C(2) of the Medical Act 1983, held that the concept of fitness to practise was not limited to clinical practice alone and could extend to other aspects of a doctor's calling. There was no reason why a doctor who was seriously deficient in research, or who engaged in teaching students in an incompetent manner, could not properly be subject to the GMC's fitness to practise procedures for those failings. However, the administrative functions being exercised by the Chief Medical Officer and the chair of the steering group could not be described as exercising functions that were part of their medical calling, or sufficiently closely linked to the practice of medicine. Their essential skills were not medical. The making and implementation of government health policy was not a medical function, even where the policies in issue directly related to doctors and closely affected the medical profession. The functions being exercised by the Chief Medical Officer and the chair of the steering group were too remote from the profession of medicine to bring them within the scope of section 35C(2) of the Medical Act 1993. To fall within section 35C, the conduct had to be of a kind that justified some kind of moral censure, or had to involve conduct that would be considered disreputable for a doctor.

### *R (North Yorkshire Police Authority) v. Independent Police Complaints Commission* [2010] EWHC 1690 (Admin)

**39.17** Police investigation into care home—complaint—whether 'conduct'—whether direction and control of police force

North Yorkshire Police Authority applied for judicial review of the decision of the IPCC upholding a complaint, and determining that it related to police conduct rather than to direction and control of a police force. The police authority refused to record a complaint about the 'conduct' of the chief constable relating to the investigation of treatment to a patient in a care home prior to her death, on the grounds that the complaint related to the direction and control of a police force and was outside the scope of the IPCC. His Honour Judge Langan QC (sitting as a judge of the High Court), in dismissing judicial review proceedings, held that the word 'conduct' did not carry with it the notion that the behaviour must be of a particular quality, whether good or bad, and that the IPCC was right to treat the complaint as one that related to the conduct of the chief constable. The concept of direction and control was essentially concerned with matters that are of a general nature, and, on this basis, a decision by a chief officer that is confined to a particular subject falls outside the scope of direction and control. The judge rejected a 'flood-gates argument' that persons dissatisfied with a decision not to commence an investigation, or with a decision after investigation that there should be no prosecution, might overload the system by making pointless requests to chief officers to have the matter reconsidered. The instant case was concerned with the recording of a complaint that was, in essence, a matter of registration. If a complaint is repetitious or an abuse of the complaints procedure, it can be disposed of on an application for dispensation to the IPCC and the availability of the dispensation procedure mitigates any fear that the system may become clogged up.

### *R (Khan) v. Independent Police Complaints Commission* [2010] EWHC 2339 (Admin)

**39.18** Police officers—alleged police violence—criminal proceedings—officers acquitted—decision not to institute disciplinary proceedings—judicial review

This was an application for permission to apply for judicial review of a decision of the IPCC when it decided not to recommend the institution of misconduct proceedings against a number of police officers. In relation to an incident in June 2007, it was alleged that the claimant and others were stopped and put in the back of a police van, where they were subjected to violence and racial abuse before arriving at the police station. After a trial lasting almost a month, all of the police officers were acquitted. In dismissing the claimant's application for judicial review, Sir Michael Harrison said that there was no dispute that the IPCC's decision was susceptible to judicial review, although, as Lord Bingham said in *R v. Director of Public Prosecutions, ex parte Manning* [2001] QB 330, the power to review a decision not to prosecute (which is analogous to a decision not to recommend misconduct proceedings) is one to be exercised sparingly, albeit that the standard of review should not be set too high so as to deny a citizen an effective remedy. In *R (B) v. Director of Public Prosecutions* [2009] 1 WLR 2072, Toulson LJ described judicial review of a prosecutorial decision as a highly exceptional remedy. In effect, he stated that a prosecutor can ordinarily be expected to have properly informed himself and to have asked himself the right questions before arriving at a decision whether or not to prosecute.

### *Shaikh v. National Co-operative Chemists Ltd and Royal Pharmaceutical Society of Great Britain* [2010] EWHC 2602 (QB)

**39.19** Pharmacist—complaint by employer to Society—claim by pharmacist against Society for harassment—protection from Harassment Act 1997, section 3(b)—Pharmacy Technicians Order 2007—action struck out

Nicola Davies J dismissed the claimant's appeal from the judgment of the master ordering that the claim form as against the second defendant, Royal Pharmaceutical Society of Great Britain, be struck out and the action be dismissed. Following a complaint from the first defendant, the claimant's former employer, the Society investigated the complaint and referred it to its investigating committee. In essence, the claimant contended that the first defendant was wrong to refer what was a contractual matter to the Society and that the Society was wrong to investigate the complaint. He began proceedings against both defendants for damages under the Protection from Harassment Act 1997. The learned judge held that there was no reasonable prospect of the claimant succeeding on his claim, and that the actions of the registrar and the Society were at all times such that they came within the exception provided by section 3(b) of the Protection from Harassment Act 1997. Section 3(b) provides exclusion to any alleged course of conduct if the person who has pursued it shows that it was pursued under an enactment or rule of law. The Pharmacists and Pharmacy Technicians Order 2007 gave the registrar of the Society power to consider allegations, and the

2007 Rules included the carrying out of any appropriate investigations and the decision to refer an allegation to the investigating committee. Accordingly, there could not be a reasonable prospect of the claimant succeeding upon any claim pursuant to the 1997 Act as against the Society.

***R (Rycroft) v. Royal Pharmaceutical Society of Great Britain* [2010] EWHC 2832 (Admin)**

**39.20** Superintendent pharmacist—complaint—referral by registrar—obligation to refer complaints within five years—reasonable time for consideration

The Administrative Court refused the claimant's application for judicial review on the basis that the registrar had not acted unlawfully when he considered whether to refer fitness to practise allegations to the investigating committee. The claimant was employed as the superintendent pharmacist for a chain of pharmacies owned by a company. The Society received a complaint alleging the company was involved in resupplying patient-returned medication to customers. Wyn Williams J, in refusing to quash the registrar's referral of the allegation, said that the episode occurred within five years of the referral and consequently the registrar was bound to refer the allegation under the Rules. Rule 9 of the 2007 Rules provides that the registrar shall not refer an allegation to a fitness to practise panel if more than five years have elapsed since the circumstances giving rise to the allegation, unless the registrar considers that it is necessary for the protection of the public, or otherwise in the public interest, for the allegation to be referred. The learned judge said that he had no difficulty in proceeding on the basis that the registrar is under an implicit obligation to make a referral within a reasonable time. However, he did not accept that a failure to make a referral within a reasonable time amounts to a reason to quash the referral and to stay the proceedings, unless it is also established that the failure to act within a reasonable time has caused prejudice to such an extent that no fair disciplinary process is possible or that it is unfair for the process to continue. In the instant case, there was no such prejudice and the referral was within five years on any view. Whether the investigating committee considered it ought to refer the allegation to the Disciplinary Committee remained to be seen, but the decision of the registrar was lawful.

***Law Society of Scotland v. Scottish Legal Complaints Commission* 2011 SC 94**

**39.21** Complaint—solicitor—referral by SLCC to Law Society of Scotland—misunderstanding of role and duties of solicitor concerned

On 6 July 2009, D, a solicitor in Edinburgh, wrote on instructions to Mr and Mrs M as a result of their trespass on his client's land. D stated that if there were a repeat by Mr and Mrs M of their accessing the land in question, his clients had instructed him to raise court proceedings, in the form of an interdict action, without further notice. Mr and Mrs M lodged a complaint with the Scottish Legal Complaints Commission (SLCC) alleging that D's letter was overly aggressive, intimidating, and threatening. They admitted to accessing a field, but said that the majority of the letter was based on inaccurate facts and untruths, and it was 'an absolute outrage that solicitors are sending this type of correspondence to people'. Thereafter, the SLCC prepared a summary of complaint and recommendation in which the author stated that he did not believe that D's letter was overly aggressive, intimidating, and threatening, that the solicitor was detailing the points that he was requested by his client and stating facts relating to this, that the letter was a standard letter sent out by a solicitor on instructions from his client and in good faith, and that the complaint against D should be considered to be totally without merit. Notwithstanding these recommendations, the SLCC proceeded thereafter to determine that the complaint should not be rejected on the grounds that it was frivolous, vexatious, or totally without merit, and that it should be remitted to the appellant, the Law Society of Scotland, for investigation and determination as a complaint in accordance with section 6 of the Legal Profession and Legal Aid

(Scotland) Act 2007. The Court (Lord Kingarth and Lord Reed; Lord Malcolm dissenting) allowed an appeal by the Council of the Law Society of Scotland under section 21 of the 2007 Act against the decision of the SLCC. Lord Kingarth (with whom Lord Reed agreed) said that the relevant letter was sent in standard terms by an agent reporting his client's concerns in relation to certain apparent and past continuing actions, and requesting that these actions cease, and intimating that, if they did not, legal proceedings could be raised. The solicitor was representing one party against a prospective opponent in litigation. In these circumstances, the language used could only be regarded as measured. The solicitor's duty was to report his client's concerns and in no sense could the solicitor, in these circumstances, be said to warrant, or to be personally responsible for, the accuracy of what he was told. The solicitor had no obligation personally to carry out some kind of investigation in relation to where the truth lay between the competing claims. Accordingly, the SLCC erred in law: the determination proceedings were reached on a misunderstanding of the role and duty of the solicitor in the circumstances, and being one that, on the information available, was not reasonably open. How the SLCC performs its important preliminary sifting duty will no doubt depend on the facts of any case and will vary from case to case. In these circumstances, it would be inappropriate, in this appeal, to attempt to give any detailed guidance. It does, however, appear that the SLCC will be required in every case to obtain at least basic information as to the basis upon which the complaint is being made. To take an example suggested by the appellant: if the respondents receive a complaint that solicitor A is a thief, the respondents would, it seems, at least have to ask upon what basis that allegation is being made before they could assess whether there was any merit in it. If, for example, on questioning, the complainer were to answer that it was because solicitor A had red hair, the lack of merit of the complaint would be obvious. Further, section 17 of the 2007 Act gives the respondents power to examine documents and to demand explanations in connection with conduct or services or complaints where the Commission is satisfied that it is necessary for it to do so. This was wholly inconsistent with their current approach. Lord Malcolm, whilst dissenting on the present appeal, said that the 2007 Act expressly allows for complaints of 'unsatisfactory professional conduct', a much broader, easier to establish concept than the more serious matter of professional misconduct. When considering its task under section 2(4), the SLCC should ask itself whether there is any possibility that the relevant professional body would consider that the complaint merits consideration, with or without further enquiries. In other words, and putting the matter colloquially: 'Might there be something in it?' If the answer is 'Possibly, yes', the hurdle is cleared and the SLCC must refer the complaint to the professional body. An example, discussed at the hearing, to which the answer to this question would be 'Clearly, no', concerned a complainer criticizing his advocate for referring the judge to a recent Supreme Court authority adverse to his client's interests, which had been overlooked by his opponent's counsel. It is wholly understandable that this might irritate the client and leave him feeling aggrieved. Such a complaint could not be described as frivolous, vexatious, or an abuse of process. However, it is beyond doubt that counsel would simply be fulfilling his duty to the court and no amount of further investigation could alter that fact. This would be an example of a complaint that was totally without merit within the meaning of section 2(4). Where the SLCC is of the view that further investigation is needed, it does not follow that it must carry out that investigation before it can determine the section 2(4) issues. On the contrary, a need for investigation is likely to demonstrate that the complaint is not totally without merit. This remains so even if the likelihood is that further investigation will exonerate the solicitor.

***Zia v. General Medical Council* [2011] EWCA Civ 743; [2012] 1 WLR 504**

**39.22** Referral by registrar directly to fitness to practise panel—no referral to case examiners—no inflexible requirement

A hospital trust that had employed Z complained about his performance and competence to the GMC. The GMC's registrar referred Z for a performance assessment pursuant to rule 7(3) of the General Medical Council (Fitness to Practise) Rules 2004. Z refused to submit to the assessment and the registrar referred the allegations to the GMC's fitness to practise panel. The panel found the allegations proved and ordered Z's suspension from the register. Z appealed against that decision and argued in the High Court, as a preliminary issue, that the registrar's referral to the fitness to practise panel was impermissible as contrary to the obligation to refer any allegation in the first instance to case examiners. The Court of Appeal allowed the GMC's appeal against the decision of the High Court in favour of Z on this point. The Court of Appeal held that the stated objective of the Medical Act 1983, as set out in section 1, was for the GMC to protect, promote, and maintain the health and safety of the public, and that the purpose of the 2004 Rules was to achieve a balance between meeting that aim and regulating the procedure for investigation and resolution of allegations against medical practitioners. Whilst the majority of cases were considered by both the registrar and case examiners, there was no inflexible requirement that a matter always had to be considered by the case examiners before proceeding further. The rules enabled the registrar to carry out his own investigation and to direct a performance assessment, and because of Z's non-compliance, the registrar's decision to refer the allegations directly to the panel was lawful.

***R (D) v. Independent Police Complaints Commission* [2011] EWHC 1595 (Admin)**

**39.23** Police officer—allegation of rape—conduct of investigation—determination by IPCC

D was raped in January 2005, when she was aged 15. F, a police officer, was assigned to the case as the sexual offences investigation trained officer. In the months leading up to the trial of D's alleged assailant, D's mother asked F many times about mobile telephone records that would corroborate D's account. F confirmed each time that the call data was available. In fact, it was never obtained by the police and D's alleged assailant was, in consequence, acquitted. Following an internal police inquiry, the IPCC undertook its own investigation into D's complaint that F had not given her the right information about the call data. The IPCC concluded that it was not possible to prove on the balance of probabilities that F had been dishonest or had failed in her duties, and there was insufficient evidence to conclude that there had been misconduct. The Administrative Court (Collins J) granted judicial review, and held that F gave D and her mother inaccurate information and that she should have known it was inaccurate. F failed in her duty to D and her family, and it was an important factor that F knew that D was vulnerable and was nervous about giving evidence, and that she might not do so if she knew that there was no corroborating telephone call evidence. There was a prima facie case that F's conduct fell well below what was required by the Police (Conduct) Regulations 2004 and the IPCC's decision not to request disciplinary action against her was bad in law. However, because of the substantial lapse of time since the events in question, there was no longer any point in ordering disciplinary proceedings. Instead, a declaration that the IPCC's decision was unlawful was adequate.

***Lim v. Royal Wolverhampton Hospitals NHS Trust* [2011] EWHC 2178 (QB)**

**39.24** Anaesthetist—contract of employment—capability hearing—NCAS assessment

Slade J held that the defendant National Health Service (NHS) Trust would be in breach of a doctor's contract of employment by conducting a hearing about his capability without first referring the matter to a National Clinical Assessment Service (NCAS) panel for assessment, and the panel advising that the doctor's performance was so flawed that no action plan would have a realistic chance of success. The claimant was a consultant anaesthetist employed by the Royal Wolverhampton Hospitals NHS Trust, and the claim concerned a proposed capability and conduct hearing to be held by the defendant. The Court held that a capability hearing should not be held until an NCAS assessment panel had determined that his professional performance was so fundamentally flawed that no educational and/or organizational action plan had a realistic prospect of success. Slade J said that the defendant would not be in breach of any contractual obligation in pursuing allegations of misconduct against the claimant, but that it would be in breach of contract in failing to comply with the procedure promulgated for dealing with issues of capability.

### *R (Rich) v. Bar Standards Board* [2012] EWCA Civ 320

**39.25** Barrister—report to complaints committee—whether fair and accurate report—alternative remedies available

R, a barrister, challenged the decision of the Bar Standards Board to lay a disciplinary charge against him, on the grounds that the 'sponsor's' report to the Complaints Committee was not a fair and accurate summary of the detailed and lengthy rebuttal to the claim that R had made in correspondence to the Board. Buckinghamshire County Council, as prosecuting authority, had lodged a complaint arising out of a conversation that its representative had with R concerning the prosecution of a farmer. Davis J considered that there was no material basis for complaining about the fairness and accuracy of the sponsor's report, and in consequence dismissed R's application for permission in judicial review proceedings. On a renewed application, Rix LJ considered the sponsor's report against the background of the complaint. It was clear that the conversation was an advocate-to-advocate conversation, and the report dealt with the fact that it was a conversation between lawyers and that the two men were discussing the charges, which R was seeking to describe as insignificant and unsupported by the evidence. The ground upon which R asked the Bar Standards Board to review its Complaints Committee decision was adequately dealt with in summary in the sponsor's report. The summary was fair and accurate, and there was no real prospect of the Court granting judicial review. Further, there was force in the judge's second ground for refusing permission, in that there is a remedy for striking out summarily a complaint available from the directions judge. There was also an alternative remedy under rule 5(9) of the Bar Standards Board Rules, which involves a decision to reconsider a complaint that has been referred to a disciplinary tribunal leading to instructions to counsel for the Board to offer no evidence.

### *R (B) v. Secretary of State for the Home Department; R (J) v. Upper Tribunal (Immigration Asylum Chamber)* [2012] EWHC 3770 (Admin)

**39.26** Advocates—duty to court

Advocates and litigators owe a duty to the Court that is a long-standing part of the law of England and Wales, but is also reflected in section 188 of the Legal Services Act 2007. In asylum and immigration matters, lawyers should undertake only cases in which they have a proper knowledge of the law to be able to put forward competent arguments and should bear in mind always their paramount duty to the Court. In the light of the apologies given and the explanations tendered, the Court would take no action in referring the solicitors' firms, the solicitor advocate or counsel, to the relevant regulatory authorities.

### *Saville-Smith v. Scottish Legal Complaints Commission* [2012] CSIH 99

**39.27** Complaint—SLCC—sifting function—complaint without merit

The appellant complained to the SLCC about the actions of the solicitor representing his former employers against whom he had a dispute. In dismissing the appellant's claim, an Extra Division of the Inner House (Lords Clarke, Drummond Young, and Wheatley) said that, generally speaking, the Commission is empowered to consider complaints about the conduct of services provided by solicitors practising in Scotland and to take a range of actions in order to bring about the investigation or other resolution of such complaints. At an initial stage, the Commission undertakes a sifting function and determines whether complaints against solicitors are frivolous, vexatious, or 'totally without merit': Legal Professions and Legal Aid (Scotland) Act 2007, section 2(4). It is not for the Commission to determine the substantive merits of any conduct complaint nor to undertake any investigation beyond that necessary to discharge its sifting function. In particular, it is not for it to resolve any material dispute of fact that may have a bearing on the complaint: *Kidd v. Scottish Legal Complaints Commission* [2011] CSIH 75, [10]. Whilst the test in section 2(4) of the 2007 Act is clearly a low threshold, when properly analysed, the complaint here was totally without merit and the Commission's decision was one that such a body could properly make.

***Roberts v. Hook* [2013] EWHC 1349 (Admin)**

**39.28** SDT—complaint against solicitor—no case to answer—dismissed on paper—reasons

R appealed under section 49 of the Solicitors Act 1974 against the decision of the SDT that his application against the defendant solicitors should be dismissed on consideration of the papers by a panel of three members of the Tribunal. A solicitor member, who has power to screen out applications where there is no case to answer, referred R's application to a panel of three members to certify whether there was a case to answer. The present practice of a panel making a rule 5 determination was not to give reasons. The Court considered that even if Article 6 of the European Convention on Human Rights (ECHR) were not applicable, the common law requirement of fairness would usually require brief reasons for the result arrived at, and it was to be hoped that the Tribunal would look again at its practice when dismissing an application on the grounds of no case to answer. On the merits, the Court concluded that it would not be fair, and would be unjust, if the proceedings were allowed to continue.

***Alami and Botmeh v. Health and Care Professions Council and Young (Interested Party)* [2013] EWHC 1895 (Admin)**

**39.29** Parole Board—expert report—allegations of misconduct and lack of competence—claim dismissed

Y, a forensic neuropsychologist, carried out a risk assessment for the Parole Board on A and B, two violent offenders who sought release from prison on parole. Y was supervised in her assessments. A and B's complaints of misconduct or lack of competence were dismissed by the Investigatory Committee of the Health and Care Professions Council on the ground of no case to answer. In dismissing A and B's claim for judicial review, Collins J said that despite errors in Y's report, absent bad faith or recklessness, it would be only a very rare case in which a finding of serious professional misconduct should be made: *Meadow v. General Medical Council* [2007] 462, [194], [211], and [278]. In the instant case, the reports were produced more than seven years previous. Y was and remained a respected professional in dealing with violent offenders generally. She acted in good faith and took advice from an apparent expert. Justice did not require a reconsideration, particularly as a different conclusion would not be reached and any misconduct did not reach a sufficient level to justify any sanction.

***R (Chief Constable of West Yorkshire Police) v. Independent Police Complaints Commission and Armstrong* [2013] EWHC 2698 (Admin); [2015] ICR 184, CA**

**39.30** Report by IPCC—Police Reform Act 2002, section 10

In quashing the report of the IPCC upon a complaint against a police officer, His Honour Judge Jeremy Richardson QC (sitting as a judge of the High Court) held that the IPCC's function under section 10 of the Police Reform Act 2002 was to *record matters* and not to make findings or rulings upon such matters. The scope of any report is to fully investigate facts and to record them, and for the report to contain (if appropriate) an opinion as to whether there is a case to answer in respect of misconduct or gross misconduct. The language used in the instant case amounted to a suggestion of determination rather than opinion. An appeal by the IPCC to the Court of Appeal was dismissed. Sir Colin Rimer (with whom Beatson and Gloster LJJ agreed) said that the investigators' report in this case exceeded their powers. The case of *R (Allatt) v. Chief Constable of Nottinghamshire Police and IPCC* [2011] EWHC 3908 (Admin) raised a challenge to the lawfulness of an IPCC investigators' report on the ground that, as in this case, it had made findings that were properly within the remit of a disciplinary tribunal. Langstaff J, in the Administrative Court, held that the criticism of the report in this respect was not established, whereas the Court agreed with the judge that a different view deserved to be taken of the investigators' report in this case. Langstaff J recognized that it was not for an IPCC investigator to make final decisions on matters arising in a case that he or she finds calls for an answer.

### *R (Ramsden) v. Independent Police Complaints Commission, Chief Constable of West Yorkshire Police (Interested Party)* [2013] EWHC 3969 (Admin)

**39.31** Allegations of assault following crowd disturbance at football match—investigation—IPCC—witness statements

The claimant, his daughter, and their friends were caught up in an unpleasant and frightening situation of serious crowd disturbance following a football match between Leeds United and Manchester United in September 2011. The claimant alleged that, in the course of crowd disturbances following that match, he was subject to two separate incidents of assault by police officers: first, being pushed and shoved with a police officer's baton; and secondly, that a mounted police officer rode his horse at him against mesh fencing and that he was kicked by that officer. The claimant's complaint was investigated by the police, who found that his allegations were not proven. He appealed to the IPCC. That appeal was upheld. As a result, the police investigated again and declined to take action. The claimant appealed to the IPCC a second time. By its decision dated 29 October 2012, that appeal was rejected. The claimant was granted permission to apply for judicial review to seek a declaration that the decision not to direct statements to be obtained from his daughter and the son of his friend was unreasonable. Stephen Morris QC (sitting as a deputy High Court judge) said, at [21], that a number of principles could be derived from *R (Dennis) v. IPCC* [2008] EWHC 1158 (Admin), *R (Crosby) v. IPCC* [2009] EWHC 2515 (Admin) (in particular [5], [39]–[42]), *Muldoon v. IPCC* [2009] EWHC 3633 (Admin) (in particular [18], [19], [24] and [40]), and *R (Erenbilge) v. IPCC* [2013] EWHC 1397 (Admin):

(1) The question for the police investigation is whether the allegations made in the complaints have been established on the balance of probabilities, taking account of proportionality: *Muldoon* at para 18 and *Crosby* (cited in *Muldoon*) at para 41.
(2) The IPCC's appeal procedure is by way of review; in considering the question under paragraph 25(5)(b) of Schedule 3 to the Police Reform Act 2002, the IPCC's task is to ensure that, *following a proportionate investigation*, an appropriate conclusion has been reached by the police investigation: *Muldoon* at paras 18, 24. Was the conclusion on the police investigation one which was fair and reasonable?
(3) An IPCC appeal decision is not expected to be "tightly argued" – nevertheless the conclusion should be clear and the reasons readily understandable: *Dennis* at para 20.
(4) The function of the Court on an application for judicial review of an IPCC appeal decision is confined to the question whether the IPCC has reached a decision which was fairly and reasonably open to it, even if the court might have reached a different conclusion. IPCC decisions involve matters of judgment and the court will allow the IPCC a discretionary area of judgment: *Muldoon* at paras 19, 40.
(5) Where the IPCC upholds the decision of the police investigation, the question for the court involves an element of "double rationality": was the decision of

> the IPCC that the decision of the police investigation was fair and reasonable itself fair and reasonable? The question is not whether the court would necessarily have reached the same conclusions as the police or the IPCC, nor whether it can be seen with hindsight that an error may have been made: *Muldoon* at paras 24, 34.

The learned judge said, at [92]–[98], that the decision in the instant case of the police and the IPCC not to seek witness statements from the claimant's daughter and his friend's son was one that was reasonably and fairly open to them, taking into account considerations of proportionality.

(1) There was said to be no duty, upon an investigator, nor upon the IPCC, to interview all relevant witnesses (let alone all police officers). There are no hard-and-fast rules as to the conduct of an investigation. The decision as to the evidence to be gathered in an investigation is a matter for the discretion of the investigating officer, taking account of proportionality in the particular circumstances of the case: see *Statutory Guidance to the Police Service and Police Authorities on the Handling of Complaints 2010*, paragraphs 303, 308, and 310.
(2) The claimant had sustained little, if any, injury as a result of the relevant events, and the police had conducted a reasonably detailed investigation, covering evidence from at least six police officers, an expert, and a detailed review of a very substantial amount of CCTV footage, site visit evidence, and consideration of still photographs.
(3) The IPCC's decision was, by and large, a clear, fair, and accurate reflection and assessment of the evidence that had been gathered. The claimant himself did not raise the issue of his two witnesses in his first appeal.
(4) The IPCC expressly concluded that, even assuming that the claimant's account of the facts was correct, there was no sufficient case of misconduct against the officers and, on that basis, further investigation would not be proportionate.

***R (Peters) v. Chief Constable of West Yorkshire Police* [2014] EWHC 1458 (Admin)**

**39.32** Public order offence—complainant acquitted after trial—allegations against police officers—investigation by another force

At about 2.30 a.m. on the morning of 2 March 2012, the claimant was arrested at a lap-dancing club in Leeds following a call from staff to the police. The claimant was arrested for a racially aggravated public order offence. Between 3 and 6 December 2012, the claimant was tried for the alleged offence at Leeds Crown Court and was acquitted. The claimant thereafter made a complaint to the IPCC against a number of police officers, including the arresting officers. Under its standing procedures, the IPCC referred the complaints to the force. The complaint was not upheld and the Chief Constable of West Yorkshire Police refused an appeal by the claimant against the decision, rejecting the complaint against the police officers. In dismissing the claimant's claim for judicial review, Blair J said that there were a number of authorities on judicial review as regards complaints of misconduct against police officers: *R (Dennis) v. IPCC* [2008] EWHC 1158 (Admin); *R (Crosby) v. IPCC* [2009] EWHC 2515 (Admin); *Muldoon v. IPCC* [2009] EWHC 3633 (Admin); *R (Erenbilge) v. IPCC* [2013] EWHC 1397 (Admin); and *R (Ramsden) v. IPCC* [2013] EWHC 3969 (Admin). The learned judge said that the last of these, at [21], contained a useful summary: see paragraph 39.31. These cases involved the IPCC, whereas in the present case, under the statutory scheme, the investigation and subsequent appeal was carried out within the police force itself. The learned judge said, at [38], that whilst it is true that the IPCC is established as an independent body and that independence is an important attribute in this regard, the element of independent consideration can and should also be present when the complaint is dealt with within the police force in question. The Court did not consider that merely because it is so dealt with, the process should be accorded less respect than that of the IPCC.

### *R (D) v. General Medical Council* [2013] EWHC 2839 (Admin)

**39.33** 'Five-year rule'—GMC Rules 2004, rule 4(5)—consultant obstetrician and gynaecologist—allegation of sexual misconduct—twenty-one years since alleged acts

D, a consultant obstetrician and gynaecologist, challenged the decision of the assistant registrar of the GMC to waive the 'five-year rule' under rule 4(5) of the General Medical Council (Fitness to Practise) Rules 2004 and to refer allegations involving sexual misconduct by D with a child to the GMC's case examiners, notwithstanding that they related to matters more than five years old. In 1990, D was the subject of a complaint to the police lodged by his wife concerning their 15-year-old stepdaughter. The police recorded that the child was unable or unwilling to elaborate any specific acts or inappropriate touching. In December 1990, a child protection conference concluded that the allegations were not substantiated and that no further action could be taken. In January 1991, the police decided that no further police action would be taken. In 1995, D's employer decided not to refer the 1990 allegations to the GMC. More than twenty-one years later, in April 2011, D was the subject of a fresh allegation, again made by D's ex-wife, concerning her 2-year-old granddaughter. The allegation was subject to a multi-agency investigation by police and social services, and the outcome of the joint investigation was that the allegation was not substantiated. Following its own investigation, the GMC notified D in September 2012 that a decision had been made to refer the 1990 allegations to a fitness to practise panel, but that the 2011 allegation would not be pursued owing to insufficient evidence. The GMC's case examiner subsequently determined that there was a realistic prospect of a finding that D's fitness to practise was impaired. Haddon-Cave J, quashing the decision of the assistant registrar, held that there were no 'exceptional circumstances' justifying waiving the five-year rule. On the contrary, the present case represented a paradigm case for the application of the five-year rule, involving, as it did, stale twenty-one-year-old allegations that had been thoroughly investigated by the police and social services at the time, and found to be without foundation, and the absence of any further fresh allegations that had been found to be of substance.

### *R (Hibbert) v. General Medical Council* [2013] EWHC 3596 (Admin)

**39.34** 'Five-year rule'—GMC Rules 2004, rule 4(5)—expert report in Family Court proceedings—registrar's decision flawed—procedural irregularity—case remitted to GMC to reconsider

In 2006, H was appointed by a local authority to provide an expert report in Family Court proceedings concerning Miss B and her child. Miss B had a history of mental illness and the local authority instructed H to carry out a residential assessment of her. She was accordingly admitted to a residential placement for a period of three months in 2006 and, during the period of residential placement, H provided detailed reports based on his assessment of her. More than six years later, Miss B made a complaint to the GMC alleging misconduct by H in the care and treatment that he provided to her and her child. Without reference to H, who was given no opportunity whatsoever to make representations, an assistant registrar waived the five-year rule under rule 4(5) of the General Medical Council (Fitness to Practise) Rules 2004, which provides that if more than five years have elapsed since the most recent events giving rise to the allegation, the investigation should not proceed unless the registrar considers that it is in the public interest, in the exceptional circumstances of the case, for it to proceed. At the hearing of H's judicial review proceedings, the GMC acceded to an order quashing the decision and sought an order permitting the complaint to be reconsidered by a different registrar, pursuant to the 2004 Rules. H's response was that the GMC had no jurisdiction to consider the allegation afresh following any quashing order. Rejecting H's argument, Simler J said that the general rule is that a decision successfully challenged on judicial review and found to be unlawful is treated as if it never had any legal effect at all. The nullification of such a decision is ordinarily retrospective and not merely prospective: *F Hoffman-La Roche & Co AG v. Secretary of State for Trade and Industry* [1975] AC 295, 365–6, *per* Lord Diplock. Accordingly, the effect of a quashing order in this case will be that no decision will have been taken under the 2004 Rules and there will accordingly be a duty on the GMC under rule 4(1) to consider the allegation as required by those Rules. The policy of the 2004 Rules is to protect the public and ensure that the GMC can properly exercise its statutory powers, the main objectives of which are to protect the public: see, for example, section 1(1A) of the Medical Act 1983. That purpose would not be served were the GMC to be estopped from exercising its statutory powers in accordance with the Rules. The GMC was not only to be permitted to consider the allegation, but would also be required to do so by rule 4(1), subject, of course, to consideration under rule 4(5) in the circumstances of the case.

***R (Grabinar) v. General Medical Council* [2013] EWHC 4480 (Admin)**

**39.35** 'Five-year rule'—GMC Rules 2004, rule 4(5)—general practitioner—registrar's decision flawed—procedural irregularity—whether to remit case to GMC to reconsider

G, a GP in south-west London, applied to quash the decision of the GMC's assistant registrar to waive the five-year rule on an investigation into his fitness to practise. From 1996 to 2010, allegations had been made at the practice of inappropriate behaviour by R, a doctor in the practice. Between 1996 and 1999, the claimant appeared to be the senior practice partner or at least one of them. In 2003, he left and severed all connections with the practice. In March 2008, R was acquitted of sexual assault following a criminal prosecution, but in April 2010, he was convicted of a sequence of nine further sexual offences and received a sentence of two years' imprisonment with ancillary orders. In December 2010, a report was commissioned by the primary care trust (PCT). In August 2012, the GMC wrote to the claimant, informing him that the professional colleagues of R, including the claimant, may have failed in their professional obligations to act quickly to protect patients from risk. In November 2012, the deputy registrar waived the five-year rule without sending the claimant a copy of the report commissioned by the local PCT, and no particulars were given of the dates of failure by the claimant personally that were of concern. The GMC agreed that the deputy registrar's decision was procedurally flawed and should be quashed, but said that it should be remitted for reconsideration by the GMC. Blake J said, at [22], that it might have been open to the Court in an appropriate case either to refuse to remit to the registrar or to issue mandamus directing a particular outcome of a decision, but such an exercise of jurisdiction could be made only where there was a single outcome available to a properly self-directing registrar. In a case in which, in particular, the core facts were not yet fixed, it would be inconceivable that the Court should exercise such a jurisdiction. Doubtless, a fresh decision would take into account the aide-mémoire issued by the GMC on rule 4(5) issues and the consistent learning of the orders that 'exceptional' means something other than routine and truly out of the ordinary. But the balance between the weight to be attached to the public interest and whether, on the facts properly construed, there is a sufficient case to meet that stringent test is one for the GMC properly directing itself after a fair inquiry. The learned judge continued, at [23], that fairness requires this claimant to have all of the information upon which the deputy registrar may act when applying the structured test commended in the aide-mémoire. It may be that all of the material held by the GMC had now been supplied, but the PCT may hold material that it may be able to supply to the GMC, and if it were to do so, the GMC should make any such material available to the claimant. Further, the claimant should have an opportunity to make representations to respond to all such material before a decision is taken.

### *Chhabra v. West London Mental Health NHS Trust* [2014] ICR 194, SC

**39.36** NHS disciplinary procedures—concerns about doctor's performance—roles of case investigator and case manager

B, the Trust's medical director and case manager, appointed T, a consultant forensic psychiatrist, from another Trust as case investigator to investigate concerns raised about C, a consultant forensic psychiatrist at Broadmoor Hospital. B instructed T to investigate four matters about C—namely, an allegation of breach of patient confidentiality during a train journey, an allegation that C had dictated patient reports when travelling on the train, concerns about C's working relationship with her clinical team, and a solicitors' complaint. T completed and signed her report, and found evidence supporting the first three allegations, but concluded that the fourth issue, the solicitor's complaint, did not have merit. In allowing C's appeal and quashing B's decision to refer the first two complaints to a disciplinary panel, the Supreme Court said that the case raised an important question about the roles of the case investigator and the case manager when handling concerns about a doctor's performance in the NHS under the framework document called *Maintaining High Professional Standards in the Modern NHS* (MHPS). The aim of MHPS is to have someone, who can act in an objective and impartial way, investigate the complaints identified by the case manager to discover if there is a prima facie case of a capability issue and/or misconduct. The case investigator gathers relevant information by interviewing people and reading documents. The testimony of the interviewees is not tested by the practitioner or his or her representative. In many cases, the case investigator will not be able to resolve disputed issues of fact. If the case investigator were to conclude that there was no prima facie case of misconduct, there would normally be no basis for the case manager to decide to convene a conduct panel. But if the report were to record evidence that was perverse, the case manager would not be bound by that conclusion. The case manager can make his or her own assessment of the evidence that the case investigator records in the report. It would introduce an unhelpful inflexibility into the procedures if the case investigator were not able to report evidence of misconduct that was closely related to, but not precisely within, the terms of reference, or if the case manager were to be limited to considering only the case investigator's findings of fact when deciding on further procedure. In the instant case, the case manager's decision to convene a conduct panel in relation to the first and second allegations was unlawful. The findings of fact and evidence that the case investigator recorded were not sufficiently capable, when taken at the highest, of supporting a charge of gross misconduct as to potentially make any further relationship and trust between the Trust and C impossible. The breaches were not wilful in the sense that they were deliberate breaches of confidentiality; they were qualitatively different from a deliberate breach of confidentiality, such as speaking to the media about a patient. Moreover, T had made amendments to her draft report, which had the effect of stiffening the criticism of C, following sending a draft of her report to the Trust's associate human resources director. There would generally be no impropriety in a case investigator seeking advice from an employer's human resources department, for example on questions of procedure. Nor would it be illegitimate for an employer, through its human resources department or a similar function, to assist a case investigator in the presentation of a report, for example to ensure that all necessary matters have been addressed and achieve clarity. But, in this case, T's report was altered in ways that went beyond clarifying its conclusions. The report had to be the product of the case investigator. It was not.

### *Fynes v. St George's Hospital NHS Trust* [2014] EWHC 756 (QB)

**39.37** NHS disciplinary procedures—classification of allegations—conduct hearing or capability hearing—role of court

The claimant, a consultant in urogynaecology, was involved in a number of work-related incidents that formed six specific allegations of misconduct. She was invited by the defendant trust to a disciplinary hearing to consider those allegations. The claimant's case was that each of the allegations had wrongly been classified as a conduct allegation under the defendant's disciplinary policy and the Secretary of State for Health's guidance contained in MHPS. The claimant contended that the allegations should be classified as health matters, or alternatively as mixed professional capability and health allegations, which would require a capability hearing and procedural protections guaranteed. In dismissing the claimant's claim, His Honour Judge Birtles (sitting as a judge of the High Court) said that the case of *Mattu v. University Hospitals Coventry and Warwickshire NHS Trust* [2013] ICR 270 is authority for the proposition that the Court decides the classification, rather than the case manager: see the judgments of Stanley Burnton LJ at [31]–[34], Elias LJ at [81]–[88], and Sir Stephen Sedley at [134]–[155], especially at [137]. However, the learned judge would also agree with the proposition that the Court ought to be slow to readily interfere with the conclusion of an experienced case manager on classification (in that case, an experienced and independent panel): *per* Stanley Burton LJ at [34] and Elias LJ at [88]. The Court did not accept the suggestion that *Mattu* has been undermined by the *obiter* comments of Lord Hodge JSC in *West London Mental Health NHS Trust v. Chhabra* [2013] UKSC 80 (QB), [41]. There is no reference to *Mattu* in the judgment of Lord Hodge JSC (with whom the other justices of the Supreme Court agreed). However, the six allegations in the instant case fell within the ambit of Part 1, paragraph 3.1, of the defendant Trust's medical staff procedure, modelled on the MHPS, and were properly classified by the case manager as misconduct, and there was evidence in support of the allegations. Having examined all of the material, the Court was firmly of the view that each one could be said to properly raise a case of misconduct such as to justify the claimant appearing before a conduct enquiry.

### *R (Mackaill) v. Independent Police Complaints Commission* [2014] EWHC 3170 (Admin)

**39.38** Police force carrying out investigation—flawed report—IPCC dissatisfied with report—whether IPCC has power to carry out further investigation

The three claimants, serving police officers, were the subject of an investigation as to their conduct after they had made statements to the media arising out of what Andrew Mitchell MP had said at a meeting in his constituency regarding the 'Plebgate' affair. On the evening of 19 September 2012, an incident occurred at the gates of Downing Street in which Mr Mitchell MP was involved. On 12 October 2012, the IPCC directed an investigation by relevant police forces into what the claimants had said at a meeting at Mr Mitchell's Sutton Coldfield constituency office. In due course, determinations were made by the police forces that there was no case to answer by each of the three police officers. On 30 October 2013, the IPCC purported to redetermine the mode of investigation into the conduct of the claimants pursuant to paragraph 15(5) to Schedule 3 of the Police Reform Act 2002, which provides that the Commission may, at any time, make a further determination to replace an earlier one. The claimants sought to quash the decision of the IPCC on the grounds that it had no power to redetermine, or justification in redetermining, the matter. As part of its answer to the claim, the IPCC in turn challenged the validity of the purported prior determinations of the appropriate police authorities. The Divisional Court (Davis LJ and Wilkie J) declared that, on the IPCC's cross-application, the original reports had been flawed as a consequence of irregularities. Significant and material procedural irregularities had arisen. Various versions had been produced, contrary to the statutory scheme. Looking at matters overall, the resulting position was that no report compliant with the statutory scheme was ever produced. The matter would be remitted to the IPCC for fresh consideration; a new decision-maker was to decide whether or not to exercise the power available under paragraph 15(5), and if so, to determine what form that investigation would take.

### *General Medical Council v. Michalak* [2014] UKEAT/0213/14/RN

**39.39** Disciplinary proceedings—investigation of allegation—discrimination—no claim before employment tribunal

The claimant complained to an employment tribunal that she had been discriminated against by the GMC in the course of its consideration of her case on referral by her employer to the Council. The GMC contended that section 120(7) of the Equality Act 2010 precluded jurisdiction. It was common ground that the matters complained of by the claimant were matters that could be considered in an application for judicial review. After reviewing *Khan v. General Medical Council* [1996] ICR 1032, *Chaudhary v. Specialist Training Authority Appeal Panel and ors* [2005] ICR 1086, CA, *Tariquez-Zaman v. General Medical Council* [2006] UKEAT 0292/06 and 0517/06, *Depner v. General Medical Council* [2012] EWHC 1705 (Admin), *Jooste v. General Medical Council* [2012] EQLR 1048, and *Uddin v. General Medical Council* [2013] ICR 793, Langstaff J held that *Jooste* should be followed and was authority that, centrally and alone in terms, established as a matter of principle and *ratio* that where an application for judicial review may be brought in respect of the act complained of, the effect of section 120(7) of the Equality Act 2010 is to deny an employment tribunal jurisdiction.

### *R (Delezuch) v. Chief Constable of Leicestershire Constabulary; R (Duggan) v. Association of Chief Police Officers* [2014] EWCA Civ 1635

**39.40** Deaths following alleged use of force by police officers—investigation—guidance

In this case, the claimants challenged the lawfulness of guidance issued by the College of Policing in September 2014 relating to the post-incident management of investigations into deaths that follow the use of force by police officers. The first claimant was the father of RD, a young man who suffered a cardiac arrest and died in hospital in August 2012 soon after he had been forcibly restrained and detained by police officers. The second claimant was the mother of MD, who died in August 2012 in circumstances that had been the subject of extensive publicity: the vehicle in which he was travelling was stopped by police officers, one of whom shot MD twice, fatally injuring him. In each case, the main argument was that, in failing to require the immediate separation of the officers who either used force or witnessed its use, the College of Policing guidance was unlawful, because of the risk of deliberate collusion or innocent contamination of evidence given that officers had an opportunity to confer with one another before they made statements for the purposes of the investigation. In dismissing both applications for judicial review, Richards LJ (with whom Moore-Bick and Tomlinson LJJ agreed) said that, overall, the guidance left open a risk of collusion, but that, in the light of safeguards that the guidance did provide and bearing in mind that the adequacy of an investigation for the purposes of Article 2 ECHR would have to be assessed by reference to all of the features of the investigation, the risk of a breach of Article 2 was a relatively low one. It was not an unacceptable risk such as would justify a finding that the guidance was unlawful.

***R (Kerman & Co) v. Legal Ombudsman*** **[2014] EWHC 3726 (Admin); [2015] 1 WLR 2081**

**39.41** Solicitor—successor firm—jurisdiction of legal ombudsman

The claimant firm of solicitors sought judicial review of the decision of the defendant that as the successor firm of L, a sole practitioner, it was responsible for dealing with a complaint made by a former client of L to the legal ombudsman. Dismissing the firm's claim, Patterson J said that there was one issue in the case, namely, whether the ombudsman had jurisdiction to deal with the complaint. The complaint arose from service given by L as trustee when he was a solicitor in his own firm prior to joining the claimant as a consultant. The learned judge held that the claimant had succeeded to the whole of L's practice and was an "authorised person" for the purposes of the Legal Services Act 2007 and the Legal Ombudsman Scheme Rules 2013, and accordingly the defendant had jurisdiction in which to proceed to investigate the complaint in which the firm was the appropriately named respondent.

## C. Other Relevant Chapters

**39.42**

Abuse of Process, Chapter 2
Consensual Disposal, Chapter 12
Delay, Chapter 17
Disclosure, Confidentiality, Data Protection, and Freedom of Information, Chapter 18
Drafting of Charges, Chapter 22
Estoppel, Chapter 24
Judicial Review, Chapter 42
Privilege, Chapter 52

# 40

# JOINDER

| | | | |
|---|---|---|---|
| A. **Legal Framework** | 40.01 | *R (O'Brien) v. General Medical Council* [2006] EWHC 51 (Admin) | 40.04 |
| B. **Disciplinary Cases** | 40.02 | *Wisson v. Health Professions Council* [2013] EWHC 1036 (Admin) | 40.05 |
| *Gee v. General Medical Council* [1987] 1 WLR 564 | 40.02 | C. **Other Relevant Chapters** | 40.06 |
| *Reza v. General Medical Council* [1991] 2 AC 182 | 40.03 | | |

## A. Legal Framework

**40.01** Examples include:

Nursing and Midwifery Council (Fitness to Practise) Rules 2004, rule 29
General Pharmaceutical Council (Fitness to Practise and Disqualification etc) Rules 2010:

rule 27 (joinder of allegations for a joint hearing)
rule 28 (if the particulars of the allegation relate to more than one category of impairment of fitness to practise and those particulars include a conviction or caution, the chair must ensure that the committee makes its findings of facts in relation to the allegations that do not relate to the conviction or caution before it hears and makes its findings of fact in relation to the conviction or caution)

General Medical Council (Fitness to Practise) Rules 2014, rule 32 (committee or panel may consider and determine together (a) two or more allegations against the same practitioner, or (b) allegations against two or more practitioners, where it would be just to do so)

## B. Disciplinary Cases

### *Gee v. General Medical Council* [1987] 1 WLR 564

**40.02** See paragraph 22.05.

### *Reza v. General Medical Council* [1991] 2 AC 182

**40.03** Medical Act 1983—Rules—single inquiry—course of conduct

By a notice of inquiry the doctor was charged with making improper and indecent remarks to four separate employees, and making improper and indecent remarks to three patients. In dismissing the doctor's appeal, Lord Lowry, giving the judgment of the Board, said, at 197C–H, that the inference from the General Medical Council Preliminary Proceedings Committee and Professional Conduct Committee (Procedure) Rules 1988 is that several facts may be presented as constituting one charge, such that if some facts should fall by the wayside, other facts may remain in relation to which it is open to the committee to find the practitioner guilty of serious professional misconduct. The Rules appear to contemplate an inquiry by one committee into every matter put before it, whether relating to convictions or conduct or both, including any new charge or charges introduced at a resumed hearing. The whole picture is of a committee that is to be informed of all of the facts alleged and all of the background that could help it to determine, in the interests of the public and the profession, what, if anything, is to be done by way of erasure or suspension or the imposition of conditions. A provision for the separate hearing of charges or parts of charges would match ill with the scheme of the Rules and with the Medical Act 1983. Not only would segregation impede the objects of the Rules, which contemplate that the committee will fully inform itself before reaching a decision, but if a practitioner is immediately suspended after the hearing of one matter, while several other matters are pending, there will be no possibility of inquiring into the other matters, because the practitioner will have ceased to be registered. A further difficulty is that the committee, with a view to giving a suitable direction (which might be one of erasure), will not be able to consider the practitioner's conduct as a whole in the way that is clearly intended. Lord Lowry continued, at 198A–F, by saying that an important case is *Gee v. General Medical Council* [1987] 1 WLR 564, in which the House of Lords held that the rule against duplicity was inapplicable in a case before the PCC. As in *Gee* (at 575C), their Lordships were, in this case, of the opinion that the charge was one that, properly construed, narrates a course of conduct in relation to the practice carried on by the doctor. As Lord Mackay said in *Gee* at 577G, there is no definition of the word 'case', but it appears that it can contain a number of different allegations and result in a number of different charges. It appeared to their Lordships from the Rules that the *charge*, and not the separate incidents designated by the heads or particulars of the charge, is the unit of accusation and determination. If any of the particulars of a charge survive, serious professional misconduct may be found in respect of the charge of which they are particulars. But if there is a finding of not guilty on one *charge*, that charge and the conduct relating to it are no longer before the committee. This leads to the view that separate charges are suitable where two distinct types of misconduct are alleged *and* where the determination of one type of misconduct, if established, could not reasonably aggravate the seriousness of the other misconduct—as Lord Mackay said in *Gee*, at 575H:

> In a case relating to conduct where two distinct types of misconduct are alleged and where the determination that one type of misconduct was established could not reasonably aggravate the seriousness of the other misconduct I should think it would be better and in the interests of clarity for two separate charges to be alleged.

### *R (O'Brien) v. General Medical Council* [2006] EWHC 51 (Admin)

**40.04** Doctors—sick notes—investigative journalists—severance—whether simply to hear cases together

Charges were brought by the General Medical Council (GMC) against five separate general practitioners (GPs) for providing sick notes to patients in circumstances in which the individual was not ill and wanted a day off, or more than one day off, work. The case arose as a result of an operation undertaken by two journalists with the *Sunday Times*. None of the GPs knew each other before the events in question and there was no suggestion of any joint enterprise in so far as they were concerned. The issues raised on behalf of each doctor were not the same. An issue of entrapment was raised by some, but not the claimant. The claimant's defence essentially was that he did not issue any sick note and he wished to investigate the circumstances in which the journalist played her part. The claimant and some of the other doctors wished to raise arguments under Article 8 of the European Convention on Human Rights (ECHR). The decision made by the GMC was that all five cases ought to be heard together. The justification for that was that the circumstances arose from an operation that was set up by the *Sunday Times* and thus had a common source, that the allegations were substantially identical in the sense that, in each case, it was alleged that a claim was made for a sick note by a journalist masquerading as a patient, and in circumstances in which the doctor was informed, and clearly informed, that a sick note was not to be for genuine reasons of sickness. The panel ruled against severance. In refusing the claimant's application for judicial review, Collins J said that the analogy in which criminal charges can be heard jointly is not exact because the rules applicable to criminal cases, although obviously material in the sense that they give some indication as to what might be regarded as fair and just, do not apply in all their force to disciplinary proceedings such as this. The test is, and always has been, that which is now explicitly set out in rule 32 of the 2004 Rules—namely, whether it is just for the charges to be heard together. It is obvious that there must normally be some nexus between the various defendants or the circumstances alleged if there is to be a joint hearing. It is to be noted that the other four doctors were being heard together and that three of them made no objection to that. The fourth did support the claimant's argument, but there was no attempt to seek any sort of redress from the Court. Thus the case against the other four would, by their agreement, or by their failures to disagree in a positive way, be heard together. There was though, on the face of it, a distinction being drawn in the present case in as much as the defence was somewhat narrower and a different journalist was involved. The Article 8 argument was clearly common to all and, in the circumstances, it was highly undesirable, the hearing having started, that that should be dealt with in isolation. As to concerns that the claimant's part is a small one and if there were a continuing joint hearing someone would have to be there to represent the interests of the claimant throughout, the panel may think that there is merit in a sensible use of its powers to deal with procedural matters in a way that is fair to all. The Court was not persuaded that it was arguable that the decision to hold a joint hearing was unlawful.

***Wisson v. Health Professions Council* [2013] EWHC 1036 (Admin)**

**40.05** Linked allegations—approach

Two complaints against a podiatrist were heard together. The first concerned a number of inappropriate remarks made by the registrant to another employee when employed at a particular place and the registrant showing his mobile telephone, which contained pictures of a pornographic nature. The second complaint arose out of a patient attending surgery where the registrant was employed as a locum and when the registrant was alleged to have acted inappropriately, with a sexual motivation behind what he did. It was held that the same panel could hear both charges under paragraph 54(b) of the Health Professions Council Conduct and Competence Procedure Rules 2003, which provides that the committee may consider and determine together: (a) two or more allegations against the same registrant; or (b) allegations against two or more health professionals where it would be just to do so. A practice note dated October 2009 stated that joining allegations against a single registrant would be appropriate only where the allegations are linked in nature, time, or by other factors, for example several allegations based on the same acts, events, or course of dealing, or based on connected or related acts, events, or courses of dealing. In dismissing W's appeal, Collins J stated:

> 14. Although the criminal law does not have a direct application it clearly has an analogy when one is considering whether it would be in the interests of justice that the matters be heard together. It is of course to be noted that so far as the panel itself is concerned it has to consider whether a particular course of conduct is such that merits any particular sanction because it amounts to conduct which is not to be accepted.
>
> 15. Accordingly, it is always necessary that the totality of any alleged conduct is decided where there are issues and where there are disputes before any sanction is to

be imposed. That does not of itself necessarily mean that the same panel must deal with all issues but it is a pointer in that direction, and the other point that must be borne in mind is that the criminal provisions have been put into effect and devised as a safeguard for a particular defendant, knowing that he was to be tried, certainly at the Crown Court level, by a jury who cannot be expected necessarily to have the expertise to be able to differentiate between conduct on one occasion and another; and they might well be adversely affected if there is a joinder of charges against the individual where there is no proper link and no proper basis for that joinder.

16. The situation is somewhat different when one is dealing with a panel of specialists as is the case with panels such as the one in question and the same point has arisen in relation to other disciplinary bodies such as, for example, the General Medical Council. Albeit the rules of the GMC may be slightly different, the approach is similar to that which is applicable in cases involving the HPC.

## C. Other Relevant Chapters

**40.06** Drafting of Charges, Chapter 22

# 41

# JUDICIAL CONDUCT



## A. Legal Framework

**41.01** The Judicial Discipline (Prescribed Procedures) Regulations 2014 (which came into force on 18 August 2014, replacing earlier regulations) make provision for the investigation and determination of complaints of misconduct by a judicial officer holder. The term 'judicial office holder' is defined in section 109(4) of the Constitutional Reform Act 2005. The Judicial Conduct Investigations Office replaced the Office for Judicial Complaints. The Regulations make provision for rules to be made by the Lord Chief Justice, with the agreement of the Lord Chancellor dealing with the investigation process, the nomination of judges, and the constitution of disciplinary panels.

## B. Disciplinary Cases

### *Re Chief Justice of Gibraltar, Referral under Section 4 of the Judicial Committee Act 1833* [2009] UKPC 43

**41.02** Chief Justice—removal from office—'inability' or 'misbehaviour'—meaning—bringing office into disrepute

Following a complaint by all of the Queen's Counsel (QCs) in Gibraltar, with the exception of the Speaker in the House of Assembly, stating that they had lost confidence in the ability of the Chief Justice, The Hon Mr Justice Schofield, to discharge the functions of his office, the Governor appointed a tribunal pursuant to article 64(4) of the Gibraltar Constitution Order 2007. Article 64(2) provides that the Chief Justice may be removed from office 'only for inability to discharge the functions of his office (whether arising from infirmity of body or mind or any other cause) or for misbehaviour'. The tribunal found proved all but one of twenty-three episodes between 1999 and 2007 in which the conduct of the Chief Justice had been the subject of criticism. The tribunal commented that, in a number of instances, the conduct of the Chief Justice amounted to impropriety, giving four examples of this, but added that no single instance of misbehaviour showed that the Chief Justice was unfit to hold office. Pursuant to article 64(3) of the 2006 Order, the tribunal's report was referred to the Privy Council under section 4 of the Judicial Committee Act 1833. By a majority of four to three (Lord Phillips, Lord Brown, Lord Judge, and

Lord Clarke; Lord Hope, Lord Rodger, and Lady Hale dissenting), the Privy Council concluded that the conduct of the Chief Justice had brought him and his office into disrepute, that his actions had rendered his position as Chief Justice untenable, and that although there were a number of incidents that qualified as misbehaviour, it would follow the example of the tribunal in finding that these were incidents in a course of conduct that had resulted in an inability on the part of the Chief Justice to discharge the functions of his office. Accordingly, he should be removed from office. In considering article 64(2) of the Order, Lord Phillips said, at [201]–[206], that there is considerable jurisprudence on the test of both 'misbehaviour' and 'inability' in the context of the removal from office of a judge or public official. This demonstrates that there is a degree of overlap between the two. The most recent authoritative guidance on the meaning of 'misbehaviour' is to be found in the advice of the Privy Council in *Lawrence v. Attorney General of Grenada* [2007] UKPC 18; [2007] 1 WLR 1474. Giving the advice of the majority of the Judicial Committee, Lord Scott of Foscote remarked that 'misbehaviour' was a word that drew its meaning from the context in which it was used. In the instant case, and applying the decision of Gray J in *Clark v. Vanstone* [2004] FCA 1105, (2004) 81 ALD 21, the following four questions arose.

(i) Has the Chief Justice's conduct affected directly his ability to carry out the duties and discharge the functions of his office?
(ii) Has that conduct adversely affected the perception of others as to his ability to carry out those duties and discharge those functions?
(iii) Would it be perceived to be inimical to the due administration of justice in Gibraltar if the Chief Justice remains in office?
(iv) Has the office of Chief Justice been brought into disrepute by the Chief Justice's conduct?

So far as 'inability' is concerned, assistance can be derived from the decision of the House of Lords in *Stewart v. Secretary of State for Scotland* 1998 SC (HL) 81. That appeal arose out of a report by the Lord President and the Lord Justice-Clerk to the Secretary of State for Scotland pursuant to section 12(1) of the Sheriff Courts (Scotland) Act 1971 that the appellant was 'unfit for office by reason of inability, neglect of duty or misbehaviour'. The conduct that was the subject of investigation consisted of a consistent pattern of bizarre behaviour both on and off the bench. The finding was that this did not constitute 'misbehaviour', but was attributable to a character flaw, which amounted to 'inability'. The House upheld the finding of the Court of Session that 'inability' was not to be restricted to unfitness through illness, but extended to unfitness through a defect in character. In the leading speech, Lord Jauncey of Tullichettle rejected the submission that the importance of judicial independence required that 'inability' be accorded a narrow meaning. He held that the fact that the decision as to a sheriff's unfitness lay with two senior judges was the bulwark standing between the sheriff and any undue interference by the executive. A similar point could be made in the present case. There is good reason to give 'inability' in article 64(2) of the Order the wide meaning that the word naturally bears. If, for whatever reason, a judge becomes unable properly to perform his judicial function, it is desirable in the public interest that there should be power to remove him, provided always that the decision is taken by an appropriate and impartial tribunal. In the instant case, it was open to the tribunal to proceed on the basis that defect of character and the effects of conduct reflecting that defect, including incidents of misbehaviour, were cumulatively capable of amounting to 'inability to discharge the functions of his office' within article 64(2).

### *McNicholls v. Judicial and Legal Service Commission* [2010] UKPC 6

**41.03** Chief Magistrate—disciplinary charges—JLSC—whether acting ultra vires

The appellant, the Chief Magistrate of Trinidad and Tobago, appealed to the Privy Council against the decision of the Court of Appeal in Trinidad and Tobago refusing to quash the decision of the Judicial and Legal Service Commission (JLSC) to prefer disciplinary charges against him and to suspend him. The JLSC is the body constituted under section 111 of the Constitution of Trinidad and Tobago to appoint, remove, and exercise disciplinary control over all judges and judicial officers except the Chief Justice. The disciplinary charges arose out of what was said to be a refusal by the Chief Magistrate to give evidence for the prosecution in committal proceedings against the then Chief Justice of Trinidad and Tobago. The allegations against the Chief Justice arose out of a complaint made by the appellant to the Prime Minister that the Chief Justice had attempted to influence him in a decision that he was about to take in a criminal trial of the Leader of the Opposition. In dismissing the appellant's appeal, the Privy Council said that there were four issues before the Board as follows.

(i) Was the JLSC acting ultra vires in preferring the charges?
(ii) Was the JLSC acting unfairly and/or contrary to the rules of natural justice in preferring the charges?
(iii) Are the charges unsustainable in fact and law?
(iv) Was the JLSC's conduct of the disciplinary process fundamentally unfair?

The Board found that all the grounds of appeal failed. The JLSC did not act ultra vires, the appellant was not treated in any relevant respect unfairly, and the Board had every confidence that the disciplinary proceedings before an appropriate tribunal would be fair. The appellant had a case to answer, but what decision the tribunal was to reach would be a matter for it and not for the Board.

### *Re Madam Justice Levers (Judge of the Grand Court of the Cayman Islands), Referral under Section 4 of the Judicial Committee Act 1833* [2010] UKPC 24

**41.04** Judge of Grand Court—standard of behaviour—removal—whether public confidence would be undermined

In March 2007, the Chief Justice of the Cayman Islands was approached with a complaint about the manner in which Levers J conducted criminal trials. The complainant alleged, amongst other matters, that the judge had a practice of issuing arrest warrants against jurors who failed to attend court, regardless of the circumstances, and demonstrated bias against women and in favour of male defendants. The Chief Justice received separate complaints from two unsuccessful women litigants in relation to the manner in which they had been treated by the judge in family proceedings and, over the next two months, considered a number of transcripts of trials in which the judge had presided. Based in part on these, the Chief Justice prepared a schedule of seventeen incidents that demonstrated summary arrest of jurors, discourtesy to counsel, unfavourable treatment of female complaints, lack of sensitivity, and injudicious use of language and criticism of fellow judges. A tribunal was appointed by the Governor of the Cayman Islands. In a lengthy report dated 12 August 2009, the tribunal advised the Governor to request that the question of the removal of the judge be referred to the Judicial Committee of the Privy Council. The tribunal expressed the view in strong terms that the judge was guilty of misbehaviour that justified her removal from office. In ordering removal, the Privy Council (Lord Phillips, Lord Saville, Lady Hale, Lord Manse, Lord Judge, Lord Carr, and Dame Janet Smith) said that whilst it was implicit that the tribunal, after investigating the facts, would recommend a reference to the Privy Council only if it were to consider that the judge's conduct amounted to misbehaviour justifying removal, the Board considered that it was not appropriate for the tribunal to castigate the judge's conduct in the extreme terms adopted in its Executive Summary. It was one thing for an investigating tribunal to identify conduct that it considered amounted to misbehaviour justifying removal; it was quite another to do so in terms that may irreparably damage the reputation of a judge before his or her conduct has been appraised by the Judicial Committee. After reviewing *Rees v. Crane* [1994] 2 AC 173 (in which the Privy Council held that the Chief Justice of Trinidad and Tobago had acted beyond his powers in suspending a judge of the High Court and that the investigation by the JLSC had been in breach of a duty to treat the judge fairly before recommending the appointment of a tribunal), the Board dismissed Levers J's complaints against the Chief Justice and the Governor. Lord Phillips, giving the judgment of the Board, said:

> 50. The public rightly expects the highest standard of behaviour from a judge, but the protection of judicial independence demands that a judge shall not be removed from misbehaviour unless the judge has fallen so far short of that standard of behaviour as to demonstrate that he or she is not fit to remain in office. The test is whether the confidence in the justice system of those appearing before the judge or the public in general, with knowledge of the material circumstances, will be undermined if the judge continues to sit – see *Therrien v. Canada (Minister for Justice)* [2001] 2 SCR 3. If a judge by a course of conduct, demonstrates an inability to behave with due propriety misbehaviour can merge into incapacity.

In conclusion, the Board said that it was time to stand back and look at the overall picture. Whilst the judge had many admirable qualities, the Chief Justice did not refer her conduct immediately for consideration by the Governor, but instead, following his memorandum for her consideration, the judge continued to behave in a manner that was unacceptable in the performance of her judicial duties. The Board was most concerned with those occasions on which the judge had been guilty in court of completely inexcusable conduct that had given the appearance of racism, bias against foreigners, and bias in favour of the defence in criminal cases. They had been fatal flaws in a judicial career that had many admirable features. The Board did not endorse the unqualified terms in which the tribunal saw fit to condemn the judge. The Board, however, was satisfied that, by her conduct, the judge showed that she was not fit to continue to serve as a judge of the Grand Court and should be removed from that office on the ground of misbehaviour.

### *R (Jones) v. Judicial Appointments Commission* [2014] EWHC 1680 (Admin)

**41.05** Appointment—driving offences—good character—public confidence

The claimant, a solicitor and deputy district judge, entered the June 2013 competition for full-time appointment as a district judge. His application was rejected because he was not of good character for the purposes of section 63(3) of the Constitutional Reform Act 2005. The reason was that he currently had seven penalty points on his driving licence. The relevant Guidance states that if the total number of penalty points currently endorsed on an individual's licence exceeds six, then this will normally prevent that individual from being selected for judicial appointment. Dismissing the claimant's claim, Sir Brian Leveson P (with whom Supperstone J agreed) said that any policy enunciated by the Judicial Appointments Commission (JAC) is obviously subject to review on public law grounds, and this particular policy is hedged with a discretion (reflected by the word 'normally'). It is sufficient to conclude that the JAC is entitled to take the view that public confidence in the standards of the judiciary would not be maintained if persons appointed to judicial office were to have committed motoring offences resulting in penalty points at the level identified in the Guidance. In the case of those who already hold judicial office, whether full-time or part-time, the *Guide to Judicial Conduct* makes clear that members of the judiciary must report to the Lord Chief Justice motoring offences that result in disqualification, the award of six points (for a single offence), or (if a lesser number of points are awarded) where total points endorsed exceed six. It is then a matter for the Lord Chief Justice and the Lord Chancellor to decide what action, if any, to take. In other words, there is a comparable level of offending that, for those who hold judicial office, triggers the requirement of reporting. Given the outstanding success that the claimant otherwise had in the district judge competition, the Court concluded by hoping that, because the first of his motoring convictions was to fall away later that year, he would consider reapplying during the next competition.

# 42

# JUDICIAL REVIEW



## A. Legal Framework

**42.01** Senior Courts Act 1981:

section 31 (application for judicial review)
section 31A (transfer to Upper Tribunal)

Tribunals, Courts and Enforcement Act 2007, sections 15–18 (Upper Tribunal's judicial review jurisdiction)
Civil Procedure Rules, Part 54 (judicial review and statutory review)

## B. Disciplinary Cases

### *R v. Statutory Committee of the Pharmaceutical Society of Great Britain, ex parte Pharmaceutical Society of Great Britain* [1981] 1 WLR 886

**42.02** The Pharmaceutical Society of Great Britain applied for judicial review by way of orders of certiorari and mandamus to direct the Statutory Committee of the Pharmaceutical Society of Great Britain to hear complaints under the Pharmacy Act 1994, section 8, against two student pharmacists.

See paragraph 14.14.

### *Law v. National Greyhound Racing Club Ltd* [1983] 1 WLR 1302

**42.03** Trainer—licence—suspension—stewards' jurisdiction—contractual arrangements

On 9 December 1982, the stewards of the National Greyhound Racing Club, a company limited by guarantee, held an inquiry that the plaintiff attended, where it was decided that he had had in his charge a greyhound that, on examination, showed presence in its tissues of substances that would affect its performance. The stewards suspended the plaintiff's trainer's licence for six months. It was this decision that the plaintiff challenged in his originating summons for judicial relief. In refusing relief, the Court of Appeal (Lawton, Fox, and Slade LJJ) said that such powers as the stewards had to suspend the plaintiff's licence were derived from a contract between him and the defendants. This was so for all who took part in greyhound racing in stadia licenced by the defendant. A stewards' inquiry under the defendant's Rules of Racing concerned only those who voluntarily submitted themselves to the stewards' jurisdiction. There was no public element in the jurisdiction itself. Its exercise, however, could have consequences from which the public benefited, such as the stamping out of malpractices, and from which individuals might have their rights restricted, such as being prevented from employing a trainer whose licence has been suspended. In the past, the courts have always refused to use the orders of certiorari to review the decisions of domestic tribunals: *R v. Criminal Injuries Compensation Board, ex parte Lain* [1967] 2 QB 864, 882.

### *R v. General Council of the Bar, ex parte Percival* [1991] 1 QB 212

**42.04** See paragraph 39.02.

### *R v. Chief Rabbi of the United Hebrew Congregation of Great Britain and the Commonwealth, ex parte Wachmann* [1992] 1 WLR 1036

**42.05** Rabbi—inappropriate conduct—contract of employment—role of Chief Rabbi—non-regulatory system

In 1990, allegations were raised against the applicant, an Orthodox rabbi, essentially of adultery with members of his congregation. Following a report by a commission to investigate the allegations, the Chief Rabbi considered the applicant's conduct to be incompatible with his rabbinical standing and activities, and on 24 August 1990 the executive and council of the Congregation passed a resolution to terminate the applicant's employment. By notice of application, the applicant sought leave to apply for judicial review. The issue before the Court was whether the Chief Rabbi is subject to judicial review in respect of the discharge of his essential functions as the spiritual head of the United Hebrew Congregation of Great Britain and the Commonwealth. In dismissing the applicant's claim, Simon Brown J said, at 1041D–1043A, that:

> [Counsel for the Chief Rabbi] invites my attention to certain passages in the judgments of the Court of Appeal both in *Law v. National Greyhound Racing Club Ltd* [1983] 1 WLR 1302 and in *R v. Panel on Takeovers and Mergers, ex parte Datafin Plc* [1987] QB 815. I need not recite them. Their effect is clear enough. To say of decisions of a given body that they are public law decisions with public law consequences means something more than that they are decisions which may be of great interest or concern to the public or, indeed, which may have consequences for the public.
>
> To attract the court's supervisory jurisdiction there must be not merely a public but potentially a governmental, interest in the decision-making power in question. And, indeed, generally speaking the exercise of the power in question involves not merely the voluntary regulation of some important area of public life but also what [counsel for the Chief Rabbi] calls a 'twin track system of control'. In other words, where non-governmental bodies have hitherto been held reviewable, they have generally been operating as an integral part of a regulatory system which, although itself non-statutory, is nevertheless supported by statutory powers and penalties clearly indicative of government concern.
>
> [...]
>
> It cannot be suggested, [counsel for the Chief Rabbi] submits and I accept, that the Chief Rabbi performs public functions in the sense that he is regulating a field of public law and but for his offices the government would impose a statutory regime. On the contrary, his functions are essentially intimate, spiritual and religious functions which the government could not and would not seek to discharge in his place were he to abdicate his regulatory responsibility.
>
> [...]
>
> [As to public policy], the court is hardly in a position to regulate what is essentially a religious function – the determination whether someone is morally and religiously fit to carry out the spiritual and pastoral duties of his office. The court must inevitably be wary of entering so self-evidently sensitive an area, straying across the well-recognised divide between church and state.

### *R v. Disciplinary Committee of the Jockey Club, ex parte Aga Khan* [1993] 1 WLR 909

**42.06** Jockey Club—Disciplinary Committee—Rules of Racing—private rights

On 10 June 1989, the filly Aliysa, owned by the applicant, His Highness the Aga Khan, won the Oaks at Epsom. In a routine examine after the race, a metabolite of camphor was said to be found in a sample of the filly's urine. Under the Jockey Club's Rules of Racing, camphor was a prohibited substance, and the Disciplinary Committee of the Jockey Club held an inquiry. On 20 November 1990, the Committee ruled that the urine contained a metabolite of camphor, that the source of the metabolite was camphor, that the filly should be disqualified for the race in question, and the filly's trainer should be fined £200. On 3 July 1991, the Divisional Court (Woolf LJ and Leonard J) held that the decision was not susceptible to judicial review and dismissed the notice of motion. The Court of Appeal (Sir Thomas Bingham MR, and Farquharson and Hoffmann LJJ) reviewed a considerable body of authority relied on as relevant in determining the scope of judicial review, including: *R v. Criminal Injuries Compensation Board, ex parte Lain* [1967] 2 QB 864; *Law v. National Greyhound Racing Club Ltd* [1983] 1 WLR 1302; *R v. Disciplinary Committee of the Jockey Club, ex parte Massingberd-Mundy* [1993] 2 All ER 207 (in which the applicant sought judicial review of a decision that his name be removed from the list of those qualified to act as chairman of a panel of local stewards); *R v. Jockey Club, ex parte RAM Racecourses Ltd* [1993] 2 All ER 225 (in which the applicant for judicial review was a racecourse management company that sought to challenge the Jockey Club's allocation of racing fixtures); *R v. Panel on Takeovers and Mergers, ex parte Datafin Plc* [1987] QB 815; and *R v. Football Association Ltd, ex parte Football League Ltd* (1991) The Times, 22 August. Sir Thomas Bingham MR concluded, at 923–4:

> I have little hesitation in accepting the applicant's contention that the Jockey Club effectively regulates a significant national activity, exercising powers which affect the public and are exercised in the interest of the public. I am willing to accept that if the Jockey Club did not regulate this activity the Government would probably be driven to create a public body to do so.
>
> But the Jockey Club is not in its origin, its history, its constitution or (least of all) its membership a public body. While the grant of a Royal Charter was no doubt a mark of official approval, this did not in any way alter its essential nature, functions or standing. Statute provides for its representation on the Horserace Betting Levy Board, no doubt as a body with an obvious interest in racing, but it has otherwise escaped mention in the statute books. It has not been woven into any system of governmental control of horseracing, perhaps because it has itself controlled horseracing so successfully that there has been no need for any such governmental system and such does not therefore exist. This has the result that while the Jockey Club's powers may be described as, in many ways, public they are in no sense governmental...
>
> I would accept that those who agree to be bound by the Rules of Racing have no effective alternative to doing so if they want to take part in racing in this country. It also seems likely to me that if, instead of Rules of Racing administered by the Jockey Club, there were a statutory code administered by a public body, the rights and obligations conferred and imposed by the code would probably approximate to those conferred and imposed by the Rules of Racing. But this does not, as it seems to me, alter the fact, however anomalous it may be, that the powers which the Jockey Club exercises over those who (like the applicant) agree to be bound by the Rules of Racing derive from the agreement of the parties and give rise to private rights on which effective action for a declaration, an injunction and damages can be based without resort to judicial review. It would in my opinion be contrary to sound and long-standing principle to extend the remedy of judicial review to such a case.
>
> It is unnecessary for purposes of this appeal to decide whether decisions of the Jockey Club may ever in any circumstances be challenged by judicial review and I do not do so. Cases where the applicant or plaintiff has no contract on which to rely may raise different considerations and the existence or non-existence of alternative remedies may then be material. I think it better that this court should defer detailed consideration of such a case until it arises. I am, however, satisfied on the facts of this case the appeal should be dismissed.

In concurring, Hoffmann LJ said, at 931–3:

> There is no reason why a private club should not also exercise public powers. The Law Society is essentially a club, incorporated by Royal Charter, perhaps less exclusive than the Jockey Club, but private nonetheless. Not all solicitors choose to belong. But the Law Society also exercises public powers, conferred by statute in the public interest. In exercising these powers, the Law Society operates in the realm of public law: see *Swain v. The Law Society* [1983] 1 AC 598. In the case of the Jockey Club, however, there is no public source for any of its powers....
>
> *R v. Panel on Takeovers and Mergers, ex parte Datafin Plc* [1987] QB 815 shows that the absence of a formal public source of power, such as statute or prerogative, is not conclusive. Governmental power may be exercised de facto as well as de jure. But the power needs to be identified as governmental in nature. In *ex parte Datafin Plc* Sir John Donaldson MR explained how the panel had come to occupy the position it did....

What one has here is a privatisation of the business of government itself. The same has been held to be true of the Advertising Standards Authority (*R v. Advertising Standards Authority Ltd, ex parte Insurance Service Plc* (1989) 2 Admin LR 77) and the Investment Management Regulatory Organisation ('IMRO'): *Bank of Scotland v. Investment Management Regulatory Organisation Ltd* 1989 SLT 432. Both are private bodies established by the industry but integrated into a system of statutory regulation. There is in my judgment nothing comparable in the position of the Jockey Club....

It is true that in some countries there are statutory bodies which exercise at least some control over racing: e.g. *Heatley v. Tasmanian Racing and Gaming Commission* (1977) 137 CLR 487... The fact that certain functions of the Jockey Club could be exercised by a statutory body and that they are so exercised in some other countries does not make them governmental functions in England. The attitude of the English legislator to racing is much more akin to his attitude to religion (see *R v. Chief Rabbi of the United Hebrew Congregations of Great Britain and the Commonwealth, ex parte Wachmann* [1992] 1 WLR 1036): it is something to be encouraged but not the business of government.

[...]

It may be that in some cases the remedies available in private law are inadequate. For example, in cases in which power is exercised unfairly against persons who have no contractual relationship with the private decision-making body, the court may find it easy to fashion a cause of action to provide a remedy. In *Nagle v. Feilden* [1966] 2 QB 633, for example, the court had to consider the Jockey Club's refusal on grounds of sex to grant a trainer's licence to a woman....

[...]

In the present case, however, the remedies available in private law to the Aga Khan seem to me entirely adequate. He has a contract with the Jockey Club, both as a registered owner and by virtue of having entered his horse in the Oaks. The club has an implied obligation under the contract to conduct its disciplinary proceedings fairly. If it is has not done so, the Aga Khan can obtain a declaration that the decision was ineffective (I avoid the slippery word void) and, if necessary, an injunction to restrain the club from doing anything to implement it. No injustice is therefore likely to be caused in the present case by the denial of a public law remedy.

### *R v. Visitors to the Inns of Court, ex parte Calder; R v. Visitors to the Inns of Court, ex parte Persaud* [1994] QB 1

**42.07** Judges—Visitors

Decisions of judges acting as visitors are judicially reviewable, the scope of judicial review being limited in the way described in *R v. Hull University Visitor, ex parte Page* [1993] AC 682, *per* Sir Donald Nicholls V-C at 40D–41A, Stuart-Smith LJ at 47F–51D, and Staughton LJ at 66D–66G.

### *Andreou v. Institute of Chartered Accountants in England and Wales* [1998] 1 All ER 14

**42.08** Accountant—disciplinary hearing—exclusion—time for appeal under bye-law expired—agreement to be bound by Royal Charter and bye-laws

On 14 December 1993, the Disciplinary Committee of the Institute of Chartered Accountants in England and Wales (ICAEW) found the plaintiff, A, guilty of serious disciplinary charges and, as a result, he was excluded from membership. He failed to bring an internal appeal within twenty-eight days, the time prescribed by the Institute's bye-law 85(c). Being unable to appeal, A commenced judicial review proceedings to challenge the vires of bye-law 85(c) and the decision of the Institute that it had no power to extend time. A's judicial review claim was refused on grounds of delay, the Court ordering, however, that the claim be continued as if begun by writ. In doing so, it expressed no opinion and made no finding as to whether A had a right that could be pursued in this way. Subsequently, the Institute applied to strike out the claim as an abuse of process under Rules of the Supreme Court (RSC) Order 18, rule 19, and/or the inherent jurisdiction of the Court. In striking out A's claim, the Court of Appeal (Lord Woolf MR, Brooke LJ, and Sir Brian Neill) said that the validity of bye-law 85(c), which limits the right of appeal to twenty-eight days, was at the heart of his application for judicial review and it had been dismissed. This meant that A would be acting in a manner that was an abuse of process if he were, as part of his remaining private law proceedings, to rely on public law claims on which he was previously relying as part of his judicial review proceedings. The fact that the Institute is a public body does not prevent it entering into contractual relations giving its members private rights. However, any valid claim that A had must exist in private law independently of those public law claims. Given that a member undertakes to be bound by the Institute's royal charter and its bye-laws, there can be no room for a general implied term that would seek to make the express provisions in bye-law 85(c) subject to a qualification that it should be exercised fairly and reasonably. Any private law protection that is needed is intended to be provided by the requirement of a two-thirds majority, in accordance with the royal charter.

### *R v. Association of British Travel Agents, ex parte Sunspell Ltd (t/a Superlative Travel)*, CO/4295/1999, 12 October 2000

**42.09** Travel agent—ABTA Code of Conduct—trade association—contractual arrangements

The applicant, trading as Superlative Travel, was a member of the Association of British Travel Agents (ABTA). In early 1998, it offered holidays in France for the World Cup 1998 including tickets for the matches themselves, and it took bookings from a number of members of the public on that basis. However, the official agents to the French Football Association failed to produce the tickets despite having agreed to do so. As a result, the applicant cancelled the holidays booked with it and refunded the would-be travellers. However, the respondent received complaints and, as a result, it took disciplinary action under ABTA's Code of Conduct. The applicant was eventually fined £12,000. In dismissing its claim for judicial review, Keene J said that the evidence showed that ABTA was established in 1950 as a trade association for travel agents and tour operators. It is a company limited by guarantee. The members of it are bound by their contracts with ABTA, and hence by the terms and conditions of membership set out in the articles of association. The codes of conduct of ABTA are binding on members. The travel and holiday industry is the subject of various public controls. Certain aspects are subject to powers possessed by the Director General of Fair Trading, the Advertising Standards Authority, and local trading officers, together with a raft of statutory regulations. The right to a fair trial under Article 6 of the European Convention on Human Rights (ECHR) operates in private law proceedings as it does in public law hearings. Even if it applies to ABTA's disciplinary proceedings, it does not, because of that, render them amenable to judicial review. The fact that ABTA's disciplinary powers derive from contracts between itself and its members, and not from any statute, regulation, prerogative, or other public source of power, is to be seen as very important. That emerges from *R v. Panel on Takeovers and Mergers, ex parte Datafin Plc* [1987] QB 815, *R v. Insurance Ombudsman, ex parte Aegon Life Insurance Ltd* [1994] COD 426, and *R v. Servite Houses and the London Borough of Wandsworth Council, ex parte Goldsmith*, 12 May 2000 (unreported). The reason for this being a very important factor is that judicial review is concerned with the performance of public functions and duties (see the classic case of *R v. Criminal Injuries Compensation Board, ex parte Lain* [1967] 2 QB 864, 882B). Where the power derives from a truly voluntary submission to the authority of a body, it is difficult to see that the decision of that body can be regarded as public or governmental. It may sometimes be different where the contractual relationship is not wholly voluntary, but is a necessary condition for pursuing a particular activity, as in *R v. London Metal Exchange Ltd, ex parte Albatross Warehousing BV*, 30 March 2000 (unreported). In such cases, the function of the body in question may be seen to be sufficiently woven into a system of governmental control, which was the test applied by Richards J in the *London Metal Exchange* case. It is also crucial to bear in mind that the ultimate issue is whether the particular function or decision being questioned is a matter of public law. Not all decisions of patently public bodies are susceptible to judicial review: the decision in *R v. Lloyd's of London, ex parte Briggs and ors* [1993] 1 LL R 176 was not a matter of public law.

***Madan v. General Medical Council (No. 2)*** **[2001] EWHC Admin 477, [2001] Lloyd's LR Med 539**

**42.10** See paragraph 38.03.

***R (Davies and ors) v. Financial Services Authority*** **[2004] 1 WLR 185**

**42.11** Financial services—statutory scheme—alternative remedy

The Court of Appeal (Kennedy, Mummery, and Carnwath LJJ) held that judicial review is not apt where an appeal exists. Following complaints, the Financial Services Authority (FSA) issued warning notices against the claimants, former employees of a member of the London Metal Exchange (LME), under section 57 of the Financial Services and Markets Act 2000. The FSA warned them that it proposed to make prohibition orders against them under section 56 of the 2000 Act on the grounds that they were not fit and proper persons to perform certain functions relating to regulated activities. The claimants appealed against the judge's dismissal of their application for judicial review. In dismissing the appeal, the Court held that the claimants had an alternative remedy. Mummery LJ said, at 193B–F, that the application would, if granted, bypass the comprehensive statutory scheme. It was specifically set up by the 2000 Act. It enabled persons against whom decisions are made and actions taken to refer the matter to a specialist tribunal, with a right of appeal on points of law direct from the tribunal to the Court of Appeal. Even if the only point raised were a point of law on the lawfulness of the decision or action of the authority, that could be dealt with by the tribunal. It was unnecessary to apply for judicial review to resolve it. There are no exceptional circumstances in this case, in which the grounds of challenge to the warning notice are not restricted to purely jurisdictional grounds. The legislative purpose evident from the detailed statutory scheme was that those aggrieved by the decisions and actions of the FSA should have recourse to the special procedures and to the specialist tribunal rather than to the general jurisdiction of the Administrative Court. Only in the most exceptional cases should the Administrative Court entertain applications for judicial review of the actions and decisions of the FSA, which are amenable to the procedures for making representations to the FSA, for referring matters to the tribunal, and for appealing directly from the tribunal to the Court of Appeal. No exceptional circumstances existed in this case that would justify following the judicial review route rather than leaving the claimants to invoke the statutory procedure.

***R (Mahfouz) v. Professional Conduct Committee of the General Medical Council*** **[2004] EWCA Civ 233**

**42.12** Doctor—disciplinary proceedings—adjournment during hearing—judicial review

M, a cosmetic surgeon facing charges relating to advice and treatment given to patients, sought judicial review in the course of the fitness to practise proceedings to challenge the panel's ruling and its failure to discharge itself from further hearing the case. Carnwath LJ (with whom Sedley and Waller LJJ agreed) said that he agreed with Mr Robert Englehart QC, counsel for the General Medical Council (GMC), that in general it is preferable for proceedings to be allowed to take their course and a challenge to their validity to be taken by way of appeal: see *R (Hounslow LBC) v. School Appeal Panel* [2002] 1 WLR 3147 (which concerned a hearing before a panel relating to admission of children to particular schools). May LJ expressed concern that the proceedings had got 'bogged down with questions of legality and the possibility of judicial review'. He made clear that applications for judicial review in the course of an appeal to an appeal panel were to be discouraged. Those remarks were made in the context of a statutory scheme in which speed was essential to enable school lists to be finalized as quickly as possible. Similarly, in *R v. Huddersfield Justices ex parte D* [1997] COD 27, it was emphasized that magistrates should in general seek to avoid adjourning cases part-heard for applications to be made to the Divisional Court. The present case was rather different, and there should have been an adjournment to enable the application to be made based on the particular facts of this case. The issue (four members of the panel having seen a newspaper report that stated that M had previously been struck off the medical register) had arisen on the second day of a programmed eight-day hearing; it had been treated as an important issue, requiring detailed legal argument, and there had been an apparent difference of view between the panel and its legal assessor as to the correct test. The GMC might well have wanted its own counsel to attend the application and to be heard, particularly on the question of a stay. In those special circumstances, it seems that justice and the appearance of justice required at least an opportunity to be given for that matter to be raised before a High Court judge.

### *R (Mullins) v. Appeal Board of the Jockey Club* [2005] EWHC 2197 (Admin)

**42.13** Jockey Club—Appeal Board—private law

Stanley Burnton J held: that the decision of the Court of Appeal in *R v. Disciplinary Committee of the Jockey Club, ex parte Aga Khan* [1993] 1 WLR 909 (decisions of the Disciplinary Committee of the Jockey Club are not amenable to judicial review) is applicable to the decisions of the Appeal Board of the Jockey Club; that review of the disciplinary decisions of the Jockey Club and its organs, including the Appeal Board, is a matter of private law, not public law; and that the enactment of the Human Rights Act 1998, the amendment to Part 54 of the Civil Procedure Rules (CPR), and the decision of the House of Lords in *Aston Cantlow and Wilmcote with Billesley Parochial Church Council v. Wallbank* [2004] 1 AC 546 have not changed the law. On 30 November 2002, Be My Royal, a horse of which the claimant was the trainer, was first past the post in the Hennessy Gold Cup at Newbury. A urine sample was taken from Be My Royal. It was found to contain morphine. On 2 May 2003, the claimant was informed that he was required to attend an inquiry before the Disciplinary Committee of the Jockey Club as to whether, in the light of the finding of morphine, there had been a breach of rule 53 of the Rules of Racing. The Committee found that there had been a breach of rule 53 and disqualified Be My Royal. The claimant appealed to the Appeal Board established by the Jockey Club. It upheld the decision of the Disciplinary Committee. In the present proceedings, the claimant sought judicial review of that decision of the Appeal Board of the Jockey Club, with whom the claimant has no contract. At [28], Stanley Burnton J said that:

> The Appeal Board's jurisdiction, like that of the Disciplinary Committee in *Aga Khan*, was derived from and entirely dependent upon the Rules of Racing of the Jockey Club. In the *Aga Khan* case the decision of the Disciplinary Committee was treated as that of the Jockey Club, since it was made by an organ of and with the authority of the Club. The only difference between the Appeal Board and the Disciplinary Committee whose decision was the subject of the judgment of the Court of Appeal is that the former is an appellate body the members of which are independent of the Jockey Club. Both the Aga Khan and [the claimant] had contracts with the Jockey Club that incorporated the Rules of Racing. The nature of the decisions of both was the same, and it is the nature of the decision—whether it was in the exercise of a public function—on which CPR Part 54.1 focuses. Moreover a private body (which on the assumption that *Aga Khan* was correctly decided the Jockey Club was before it created the Appeal Board)

> cannot by itself either create a public body or convert a private function (the exercise of its domestic disciplinary powers) into a public function. In other words, it could not, by creating an Appeal Board, convert the private function exercised by its Disciplinary Committee into a public function exercised by its Appeal Board. Whether, when it created the Appeal Board, the Jockey Club considered that it was or might be a public authority within the meaning of the 1998 Act is, as a matter of law, irrelevant.

At [36], the learned judge said that it seemed to him that changes in the law were the only possible basis for refusing to follow the decision of the Court of Appeal in *Aga Khan*. The procedural rules applicable to judicial review have changed since that case was decided. They are now contained in CPR Part 54. They were previously contained in RSC Order 53. In *R (Heather and ors) v. The Leonard Cheshire Foundation* [2002] EWCA Civ 366, Lord Woolf CJ, giving the judgment of the Court, referred to the distinction between the approach of RSC Order 53 and CPR Part 54 (the former focused on the nature of the application, and whether it was an application for an order of mandamus, prohibition, or certiorari, whereas Part 54 has changed the focus of the test so that it is also partly functions-based). The provisions of Part 54 came into force on 2 October 2000—that is, on the same date that the Human Rights Act 1998 came into force. Part 54.1 uses similar language to section 6 of the Act. Part 54, like section 6, falls to be interpreted by taking into account the jurisprudence of the European Court of Human Rights (ECtHR). The essential question that arose in the present case was whether the test to be applied under Part 54.1 differs from that applied by the Court of Appeal in *Aga Khan*. The test applied in that case was whether the functions of the Jockey Club were governmental. That test is, in substance, the test applied by the House of Lords in the *Aston Cantlow* case: see Lord Nicholls at [7]–[10]; Lord Hope at [47], [49], and [59]; Lord Hobhouse at [88]; and Lord Rodger at [163]. It follows that the Court of Appeal decision in *Aga Khan* is authority for the proposition that the Jockey Club is not a public authority for the purposes of section 6 and that the particular function exercised in that case, which was identical to the function exercised in the present case, was not a function of a public—that is, governmental—nature. And it also follows that the decision of the Appeal Board was not made in the exercise of a public function for the purposes of CPR Part 54.

### *R (Kashyap) v. General Medical Council* [2010] EWHC 603 (Admin)

**42.14** Doctor—referral to fitness to practise panel—challenge—delay—choice of remedy—judicial review or contest allegation on merits

The claimant, a consultant orthopaedic surgeon specializing in hip and knee surgery, was the subject of a performance assessment by an assessment panel appointed by the GMC. On 16 October 2007, the GMC case examiners notified the claimant of their decision to refer his case to a fitness to practise panel as a result of concerns raised about him by the assessors. A panel was convened and, following a hearing lasting twenty-eight days, it found, on 3 February 2009, that the claimant's fitness to practise was, as at the date of its decision, impaired by reason of deficient professional performance. The hearing was adjourned until 3 August 2009 to consider what, if any, sanctions should be imposed upon the claimant in consequence of the panel's findings. On 15 April 2009, the claimant lodged an application for judicial review with a view to quashing the decision of the case examiners to refer the case to the fitness to practise panel and the decision of the panel that his fitness to practise was impaired by reason of deficient professional performance. In refusing permission to apply for judicial review, Mitting J said that whether or not there was merit in the contention that the registrar should not have referred the case to the panel at all, if it was to be raised as a challenge in judicial review proceedings, it should have been raised within three months of its occurrence, or, at the very latest, within three months of the date on which the decision was confirmed, on 16 October 2007. All of the facts on which the challenge could be made were known before the panel began to hear the case on 28 April 2008. The claimant submitted to the panel that the referral was improper and unlawful, and not in accordance with the General Medical Council Fitness to Practise Rules 2004. The panel decided, correctly, that it did not have jurisdiction to entertain such a challenge. Such a challenge could be made only by a claim for judicial review. If it had been and if it had succeeded, a twenty-eight-day hearing, with more to come if this challenge did not succeed, would have been avoided. The claimant had a choice of remedy: to seek to apply for judicial review or to contest the allegations on their merits. He chose the latter. He had to abide by his choice. It was far too late now to bring a claim for judicial review on this ground. Under section 40 of the Medical Act 1983, the claimant would have a right of appeal only if the panel were to give a direction for erasure, suspension, or conditional registration. Judicial review is required as a judicial remedy only if no sanction on registration is imposed. On express instructions of the GMC, counsel accepted that if a judicial review challenge is brought as soon as reasonably practicable, and in any event within three months, after a decision not to impose one of the three sanctions identified in section 40 is made, such a claim would be brought in time. The GMC's formal concession, and its acceptance by the Court, should give every confidence to practitioners in future cases that, in the event that they delay bringing a challenge on judicial review grounds to the decision of the fitness to practise panel that does not result in the imposition of one of the sanctions identified in section 40, they can do so in the certain knowledge that their claim will not be rejected simply on the grounds that it is brought too late.

***R (Jenkinson) v. Nursing and Midwifery Council* [2010] EWHC 1111 (Admin)**

**42.15** See paragraph 14.07.

***R (Commissioner of Police of the Metropolis) v. Police Appeals Tribunal and Peart* [2011] EWHC 3421 (Admin)**

**42.16** Police officer—assault—misconduct hearing—conviction quashed—transcript—Police Appeals Tribunal

Ordinarily, judicial review proceedings will not be appropriate as a means of challenging interim or procedural decisions under the Police Appeals Tribunal Rules 2008. If there is a challenge to an interim or procedural decision under the Rules that is capable of correction by the Police Appeals Tribunal, that should be the first port of call. Exceptionally, in this case, matters had developed outside the ordinary run of such cases to the extent that there was now a purported redetermination of an interim decision by the chairman that was of an entirely opposite effect to the chair's earlier decision. In these exceptional circumstances, the Court would consider the impugned decisions made. In December 2008, the interested party, a serving police officer, was involved in an incident in Regent Street in the centre of London. In September 2009, he was convicted of common assault at South Western Magistrates' Court arising out of the incident. At a hearing held under the Police Conduct Regulations 2008, the panel made a finding of gross misconduct and he was dismissed from the Police Service. He gave notice of appeal and, in the meantime, his appeal against conviction was allowed at Kingston Crown Court. The chair of the Police Appeals Tribunal initially ordered that the transcript of the hearing in the Crown Court should be admitted, but was later persuaded to rescind his determination. The Court quashed both determinations and said that there was no reason why the parties cannot make submissions about the admissibility of the Crown Court transcript to the Police Appeals Tribunal.

***R (B) v. Nursing and Midwifery Council* [2012] EWHC 1264 (Admin)**

**42.17** See paragraph 43.12.

***R (Aga) v. General Medical Council* [2012] EWHC 782 (Admin)**

**42.18** See paragraph 47.43.

***Johnson and Maggs v. Nursing and Midwifery Council (No. 2)* [2013] EWHC 2140 (Admin)**

**42.19** See paragraph 47.54.

***R (Willford) v. Financial Services Authority* [2013] EWCA Civ 677**

**42.20** Decision notice—reasons—whether adequate reasons given to refer to Upper Tribunal

W brought a claim for judicial review seeking to quash a decision notice given by the (then) Financial Services Authority (FSA) on the grounds that W needed the Regulatory Decisions Committee to state more fully its reasons, to enable him to make an informed decision whether to refer the matter to the Upper Tribunal. On 14 May 2010, the Committee issued a formal warning notice to W, proposing to impose a penalty of £150,000 on him for failing to comply with Principle 6 (failure to exercise due skill, care, and diligence in managing the business of a firm). In due course, on 27 October 2010, following an oral hearing and representation, the Committee issued a decision notice imposing on W a financial penalty of £100,000. The Committee found that W, the former group finance director of Bradford & Bingley Plc, had failed to ensure that the finance department brought potentially important financial information to his personal attention promptly. The decision notice itself ran to twenty-three pages and included a section headed 'Reasons for the action', containing a summary of the conduct on which the decision was based. It contained further sections headed: 'Facts and matters relied on', running to five pages; 'Representations', summarizing W's submissions; and 'Conclusions', running to five pages, setting out in some detail the Committee's conclusions. W maintained that the Committee had failed to give adequate reasons for its decision because it had not specifically addressed each of the individual submissions that he had made. He therefore brought a claim for judicial review, seeking to have the decision notice quashed. The FSA said that the reasons given in the decision notice were sufficient to comply with the requirements of section 388 of the Financial Services and Markets Act 2000, and that even if they were not, the appropriate course would be for W to refer the matter to the Upper Tribunal, which could consider the whole matter afresh. There was therefore an alternative remedy available to W. The Court of Appeal (Pill, Moore-Bick, and Black LJJ) held that judicial review was not appropriate and that it was not necessary for the Regulatory Decisions Committee to respond in detail to every point made by W, provided that it gave sufficient reasons to enable him to understand the basis of its decision. A challenge to a decision notice involves a full rehearing before the tribunal, at which the reasons given by the Committee would play little or no part. Moreover, the Court was not persuaded that W needed the Committee's reasons to be stated more fully in order for him to make an informed decision whether or not to refer the matter to the Upper Tribunal.

***R (Monger) v. Chief Constable of Cumbria Police* [2013] EWHC 455 (Admin)**

**42.21** Special constable—misconduct—dismissal—correct regulations not followed

The claimant, a special constable aged 20, was dismissed from the Cumbria Constabulary on the strength of misconduct allegations that he denied. He was not dealt with under the misconduct procedures set out in the Police (Conduct) Regulations 2008, but instead under different regulations, which offered an alternative route in the case of a special constable. Allowing M's claim for judicial review, Supperstone J held that Parliament could not have intended that, in the case of misconduct, a police force can choose to bypass the 2008 Regulations, which specifically laid down appropriate procedures and safeguards for police officers, including special constables, in cases of misconduct. Accordingly, M's dismissal was unlawful.

***R (Patel) v. General Medical Council* [2013] 1 WLR 2801, CA**

**42.22** See paragraph 56.25.

### *R (Sharaf) v. General Medical Council* [2013] EWHC 3332 (Admin)

**42.23** Doctor—disciplinary proceedings—application for judicial review during hearing

S sought permission to apply for judicial review during the course of disciplinary proceedings before the fitness to practise panel. S sought to challenge the panel's findings of fact. Permission was refused. Judicial review was a remedy of last resort and, in the event of an adverse finding, S would be able to appeal. The general rule was that there should be no challenge during the course of disciplinary proceedings: *R (Mahfouz) v. General Medical Council* [2004] Lloyd's Rep Med 377. There could be no inflexible rule, but in general it was preferable for proceedings to take their course and any challenge to be dealt with on appeal. There was a public interest in respecting the integrity of the process, and the right time to challenge was at the conclusion of the hearing.

### *R (Eguakhide) v. Governor of HMP Gartree* [2014] EWHC 1328 (Admin)

**42.24** Prison governor—drugs testing in prison

The claimant was a prisoner at HM Prison Gartree, serving a life sentence for robbery that was imposed in 2004. On 24 November 2012, he was ordered to attend the mandatory drug-testing suite in the prison to provide a sample, but refused to do so. He was charged with disobeying a lawful order. On 10 January 2013, a governor at the prison, having carried out an adjudication, found the charge proved and imposed a penalty of five days' cellar confinement and loss of privileges. An application for review within the National Offender Management Service (NOMS) was unsuccessful. The claimant sought judicial review of the governor's adjudication. The claimant was a practising Muslim. The date 24 November 2012 was the Muslim festival of Yaum Ashurah. This festival involves devout Muslims fasting on three consecutive days from sunrise to sunset. The obligation to fast includes, as it does during the lunar month of Ramadan, an obligation not to drink water. In dismissing the claimant's application, Bean J said that, of course, it would be possible never to test a Muslim prisoner on a day when he was observing a fast. It would be possible to avoid the three days of fasting involved in the observance of Yaum Ashurah even more obviously than it would be possible to avoid the entire lunar month of Ramadan. But the defendant's argument, which the Court accepted, was that to do so would be to undermine the random nature of the drug testing programme. Paragraph 3.14 of Prison Service Order 3601 emphasizes this. It is notorious to anyone with any knowledge of the prison system that the possession and use of illegal drugs in British prisons is rife. It is a very serious problem to which section 16A of the Prison Act 1952, enforced since 1995 ('any prison officer may, at the prison, in accordance with prison rules, require any prisoner who is confined in the prison to provide a sample of urine for the purpose of ascertaining whether he has any drug in his body'), and provisions in the Prison Service Order and Prison Discipline Manual are directed. To have any safe periods, even of as short a length as three days, during which Muslim prisoners, or any other prisoners, could count on not being selected for mandatory drug testing would be to undermine the random nature of the drug testing programme, which is essential to its effectiveness.

### *R (Crawford) v. Legal Ombudsman* [2014] EWHC 182 (Admin)

**42.25** Barrister—allegation of inadequate service—Legal Ombudsman Scheme

N complained to the Legal Ombudsman about the service provided by C, a barrister, at a conference that took place in September 2012. N had paid £780 (including VAT) in advance for advice at the conference. The Ombudsman held that only limited advice had been given, as a result of which half the fee should be repaid. Popplewell J said that at the heart of the challenge to the decision was the submission that the Ombudsman's process of reasoning was in three stages:

(1) that C had not provided any note of the conference or advice thereafter that evidenced what advice was given;

(2) that the inference to be drawn from that failure was that he had provided only limited advice; and
(3) that the provision of only limited advice constituted poor service, for which C should be deprived of half the fee.

The learned judge, having referred to the Legal Ombudsman Scheme created by Part 6 of the Legal Services Act 2007 and the Scheme Rules, said that the provisions illustrated two important aspects: the scheme was intended to resolve complaints swiftly and informally; and the Ombudsman was afforded a considerable latitude of discretion. In exercising powers of review, the Court does not put itself in the position of the Ombudsman and does not review the merits of the decision as if it were exercising the statutory powers itself. To do so would be to subvert the intention of Parliament. The Ombudsman's decision may be overturned only as unreasonable in the *Wednesbury* sense: *Associated Provincial Picture House Ltd v. Wednesbury Corporation* [1948] 1 KB 223. A common modern formulation is that the decision must be outside the range of reasonable responses open to the decision-maker: see, for example, *Boddington v. British Transport Police* [1992] 2 AC 143, 175H, *per* Lord Steyn. This is a high threshold, particularly in the context of a scheme intended to resolve complaints swiftly and informally, in which the decision-maker is afforded a wide discretion to do what he or she thinks is fair and reasonable in all of the circumstances. Decisions of the Legal Ombudsman are to be read with a degree of benevolence—see *R (Siborurema) v. Office for the Independent Adjudicator* [2007] EWCA Civ 1365—and should not be construed as if they were statutes or judgments, nor subjected to pedantic exegesis: see *Osmani v. Camden LBC* [2005] HLR 325, [38(9)], *per* Auld LJ. Nevertheless, it was impossible to read the decision in any other way than as adopting an illogical process of reasoning as the sole basis for its conclusion. Counsel for C submitted that a barrister in C's position would not be expected to have a note of the conference (a) because if he took a note, it would record what he was being told, not what he was saying, and (b) because such notes as would be taken would be for the barrister's use and benefit, not for the use and benefit of the client; a failure to provide the Ombudsman with a note of the advice allegedly given therefore could not support an inference as to whether any, and if so, what, advice was given. The Court agreed that the decision of the Ombudsman was irrational and was to be quashed, and the matter was remitted for further consideration.

**_R (Woods and Gordon) v. Chief Constable of Merseyside Police_ [2014] EWHC 2784 (Admin), [2015] ICR 125**

**42.26** See paragraph 66.08.

**_R (Squier) v. General Medical Council_ [2015] EWHC 299 (Admin)**

**42.27** Decisions of panel prior to conclusion of disciplinary proceedings—principles

On a rolled-up hearing of an application for judicial review of two decisions of a fitness to practise panel of the GMC, Ouseley J said:

> 19. *R (Mahfouz) v. General Medical Council* [2004] EWCA Civ 233 provides guidance on bringing judicial review proceedings at this stage of an FTPP hearing, rather than by way of a statutory appeal at their conclusion. A stay had been ordered in that case. The claimant referred to the drawbacks of waiting for the prejudicial effect of a finding of serious misconduct to be overturned in a second set of proceedings,

if the panel had erred in its appreciation of the prejudicial evidence and the need for an adjournment. At paragraph 44, Carnworth LJ said that he could see force in those points:

> "There can be no inflexible rule. However I agree with [counsel for the GMC] that in general it is preferable for proceedings to be allowed to take their course and a challenge to their validity to be taken by way of appeal. Consideration must also be given to the difficulty of organising such proceedings in a complex case and the potential inconvenience to witnesses who may have had to make special arrangements to attend the hearing, and may be reluctant to repeat the experience."

20. There is, in my judgment, a general principle but not an exclusive rule that proceedings to challenge decisions of a tribunal should await the conclusion of the hearing and should be made by way of statutory appeal. After all, judicial review is a remedy of last resort. A tribunal should not find its case management decisions, its interlocutory rulings and other procedural decisions challenged until the effect of any adverse decision of that sought is made manifest through the final decision. Proceedings of the sort here are potentially wasteful and very disruptive to the integrity of tribunal proceedings. They have the potential to put this Court in the position of running the procedures of tribunals with no benefit to the integrity of the tribunal or of the reviewing or appellate judicial process.

21. It must be remembered that the tribunal has the benefit not just of specialist knowledge and experience, but has the advantage in most cases of knowing sufficient of the cases and of the evidence to reach a rather more informed judgment than a judicial review court could. The tribunal is likely to be more familiar with the issues, how the case will evolve and the use to which it is likely to put material, and what safeguards it will employ. A judgment made at this stage of proceedings must recognise that the tribunal proceedings have yet to run their course, and unfairness in the outcome can still be remedied, even if in a less satisfactory manner. Unfairness may not be a necessary consequence of any procedural error made by the tribunal.

22. There may, however, be circumstances in which intervention at this stage is appropriate. The application was not so inappropriate here that it would have been right to refuse to hear it at the outset, because it relates to the admission of what is said to be evidence crucial to the GMC case, and to the particularisation of the allegations in a way which is said to be seriously misconceived and unfair. This will be a very long and complex case concerning allegations of misconduct by a consultant as an expert witness lasting possibly 40, and possibly 80–90, days. Were there to be a major error by the FTPP those proceedings would have to be repeated. This would lead to a very large waste of time, money, resources of both sides, and significant prejudice to the registrant who would have serious adverse findings hanging over her for years. The case has been managed by the FTPP without disruption to the hearing timetable because this challenge has been allowed for within it. On balance, I have concluded that I should rule on the issues raised in this challenge. But there is this consequence of the timing of the challenge and the degree of knowledge and understanding of the procedure and the case which the FTPP had, but this court cannot. I am satisfied that before any relief is granted, I have to be clear that relief is necessary to avoid a clear and significant injustice, which would probably not be remedied during the process of the hearing and which would probably cause real harm, if not now remedied.

## C. Other Relevant Chapters

**42.28** Abuse of Process, Chapter 2
Appeals, Chapter 4
Human Rights, Chapter 32
Investigation of Allegations, Chapter 39
Legal Representation, Chapter 45
Natural Justice, Chapter 48

# 43

# JURISDICTION



## A. Legal Framework

**43.01** Most regulatory bodies provide in their statutory or non-statutory rules for an investigation committee, a conduct or fitness to practise committee and other committees, and that each of those committees shall have the functions assigned to them by the statutory or non-statutory rules. Further examples of jurisdiction include:

Medical Act 1983, section 35C(3) (the functions of the council shall not be prevented from applying because the allegation is based on a matter alleged to have occurred outside the United Kingdom, or at a time when the person was not registered)

Pharmacy Order 2010, article 51(4) (a person's fitness to practise may be regarded as impaired because of matters arising outside Great Britain and at any time)

## B. Disciplinary Cases

### *In re S (A Barrister)* [1970] 1 QB 160

**43.02** Barrister—benchers—Senate of the Four Inns of Court—judges as Visitors

On 24 July 1968, the Complaints Committee of the Senate of the Four Inns of Court made a number of charges against S, a barrister and a member of Gray's Inn, alleging conduct unbecoming a barrister. On 9 and 10 October 1968, the Disciplinary Committee of the Senate considered the charges, found six of them to be proved, and ordered that S should be disbarred. On 16 December 1968, the benchers of Gray's Inn screened and published notice of the disbarment. By his reamended petition of appeal, S contended that the benchers of Gray's Inn had no power, either by themselves or with the concurrence and approval of the judges, to delegate to the Senate of the Four Inns of Court (the predecessor to the Council of the Inns of Court) the right or power to try him, or to reprimand, censure, suspend, or disbar him, and that accordingly the findings and sentence of the Disciplinary Committee were made without jurisdiction, and were void and of no effect. The Visitors (Paull, Lloyd-Jones, Stamp, James, and Blain JJ) allowed the appeal in respect of one of the charges and the sentence of disbarment, and decided that S would be suspended from practising for twelve months. As to the jurisdiction of the Senate of the Four Inns of Court, Paull J, giving the decision of the Visitors, said that, for at least 300 years, up to 1967, the benchers of each Inn themselves determined disciplinary matters involving a barrister-at-law, and that all puisne judges as Visitors always had supervisory powers and their decision, upon an appeal by a barrister or student to them, had always been the final determination of such matter. In 1966, the Four Inns, by a resolution passed by each Inn, set up the Senate of the Four Inns of Court to take over from each Inn the disciplining of barristers. The judges of the three divisions of the High Court of Justice confirmed the resolutions of the Inns and the judges themselves resolved to the same effect. The Inns retained the function of 'calling' a person found to be fit and proper to the bar of the Inn. The resolutions of the Inns and the judges altered the practice that had long existed as to the machinery by which matters of discipline in regard to professional conduct should be dealt with to the satisfaction of the judges. The judges remain under the same duty unimpaired, but the machinery by which the Inns exercised disciplinary matters and the judges exercised their judicial duty has been changed because the Visitors are not sitting as a court of law. The judges, however, retain wholly unimpaired their powers as Visitors.

### *Prasad v. General Medical Council* [1987] 1 WLR 1697

**43.03** Doctor—complaint by Birmingham FPC—absence of statutory declaration—purpose of statutory declaration—FPC a local or public authority

On 29 June 1987, the appellant doctor was judged by the Professional Conduct Committee (PCC) of the General Medical Council (GMC) to have been guilty of serious professional misconduct, and the Committee ordered that his name be erased from the register. The charge against the doctor was that, without due regard to the potential danger to his patients, he had, over a period between June 1981 and December 1986, administered vaccines unnecessarily, since he had on more than one occasion administered the same vaccines to the same patients. The GMC had prepared a schedule showing the names of fifty-one patients of the doctor, with particulars of the allegedly unnecessary and improper vaccinations and immunizations that had been administered to them by the doctor. The complaint against the doctor was made to the GMC by the administrator of the Birmingham Family Practitioner Committee (FPC). Counsel for the doctor before the Board, who did not appear below, took a point that was not taken before the PCC—namely, that the complaint by the administrator of the FPC ought to have been supported by a statutory declaration in accordance with the General Medical Council Preliminary Proceedings Committee and Professional Conduct Committee (Procedure) Rules 1980. It was common ground that there was no statutory declaration. The submission for the doctor was that an officer of a family practitioner committee is not a 'person acting in a public capacity' within the 1980 Rules and that, accordingly, the Committee acted without jurisdiction. In rejecting this submission, the Privy Council said that the FPC was an independent authority established by the Secretary of State pursuant to the National Health Service (NHS) Act 1977, as amended, and their Lordships had no difficulty, in the light of the legislation, in finding that it is within the definition of a 'local or public authority' and that, accordingly, no statutory declaration in support of the complaint against the doctor was required. The purpose of the Rules is manifestly to provide a preliminary filter to ensure that the disciplinary machinery of the GMC is not clogged by the necessity to investigate irresponsible complaints. To this end, complaints of misconduct on the part of a doctor made by private citizens, such as aggrieved patients, are not to proceed without the necessary machinery being complied with. But it is not thought to be necessary when complaints are received from responsible public bodies.

### *McAllister v. General Medical Council* [1993] AC 388

**43.04** Doctor—Scotland—GMC—UK body—no separate rules—English law

M was a consultant microbiologist at two hospitals in Glasgow. The PCC, for the first time, sat in Scotland. It did so because of the state of health of the doctor. Previously, cases involving Scots doctors had always been heard in London and it had never been suggested that any law other than that of England applied to the proceedings. The Committee judged M to have been guilty of serious professional misconduct and directed that his name should be erased from the register. On appeal, M contended that, since he was a domiciled Scot and resident in Scotland, and the acts complained of in the disciplinary proceedings took place in Scotland, the proceedings were Scottish proceedings and Scots law, as the *lex fori*, applied to the proceedings. In dismissing M's appeal, Lord Jauncey of Tullichettle, giving the judgment of the Board, said that the GMC and the PCC are UK bodies, and it is highly desirable that the same rules of evidence and procedure should apply throughout the United Kingdom wherever the Committee sits. Conversely, it is highly undesirable that the Committee should apply different standards of proof to different doctors depending upon where it elects to sit. Although the Medical Act 1983 makes reference to the Court of Session, the provisions do not necessarily point to Scots law being applied by the Committee. It is possible to envisage situations in which a doctor has been convicted of offences in Scotland in circumstances in which it would be desirable for the Committee to be advised as to what were the necessary components of that offence. More significant matters were that no direction is given to the GMC in the Medical Act 1983 to make different sets of rules of evidence for Scotland and England, and that the Rules contemplate that English law should apply to all proceedings before the Committee wherever they might take place. Given the desirability of a single code of evidence being applied in the Committee's proceedings throughout the United Kingdom, their Lordships were satisfied that the law of England was the correct law to have applied in these proceedings.

### *Re Solicitors, ex parte Peasegood* [1994] 1 All ER 298

**43.05** Solicitor—application to High Court to strike off—litigant in person—jurisdiction of High Court

P applied to the High Court for an order pursuant to sections 50 and 51 of the Solicitors Act 1974 that the four respondent solicitors should be struck off the roll of solicitors. In dismissing the application, Stuart-Smith LJ (with whom Judge J agreed) said that, in *Myers v. Elman* [1940] AC 282, 318, Lord Wright doubted whether the Court would entertain an application to strike a solicitor off the roll instead of leaving the matter to the Disciplinary Committee of the Law Society; in *R & T Thew Ltd v. Reeves (No. 2)* [1982] QB 1283, 1286, Lord Denning MR said that the punitive jurisdiction of the Court is now rarely, if ever, exercised. It is left to the Solicitors Disciplinary Tribunal (SDT). Further, an application for an order that a solicitor be struck off the roll pursuant to sections 50 and 51 of the 1974 Act had to be made through counsel, and could not be made by an applicant in person. This was undoubtedly the law before and after the Supreme Court of Judicature Act 1873: see *Ex parte Pitt* (1833) 2 Dowl 439, *per* Lord Denman CJ at 440 ('The motion against an attorney being in the nature of a criminal information, the Court requires that it should be made by a gentleman at the bar; and it cannot be made in person. Otherwise we have not the sanction of a barrister for the propriety of such an application'); *Re a Solicitor, ex parte Incorporated Law Society* [1903] 1 KB 857, 859, [1903] 2 KB 205, 207; *Re Two Solicitors* [1938] 1 KB 616; and *Re McKernan's Application* (1985) The Times, 26 October. The jurisdiction is the same as that exercised before 1873. Solicitors were then subject to the discipline of the Court under its inherent jurisdiction—but the jurisdiction is limited in practice, in that the application could be made only if supported by counsel. The rule applies as much today as it ever did. It is very easy for disgruntled litigants to make complaints, often in lurid terms against their own or the opposite parties' solicitors. It is perhaps anomalous that the punitive, as opposed to compensatory, powers of the Court to make orders against solicitors should have survived now that the statutory tribunal has been set up and a right of appeal to the High Court given to a complainant. The Court undoubtedly has jurisdiction under section 50 to entertain an application, if supported by counsel, but it is not obliged to do so. It does not have jurisdiction to entertain an application made by an applicant in person and, even if it did, the discretion should not be exercised save in an exceptional case.

### *Ferguson v. Scottish Football Association*, 1 February 1996 (unreported)

**43.06** Football match—criminal conviction—disciplinary proceedings—suspension—no 'report' of incident before Disciplinary Committee—ultra vires

On 16 April 1994, the petitioner, F, took part in a football match and, in consequence of an incident in which he was involved in the course of that match, he was subjected to disciplinary proceedings under the jurisdiction of the Scottish Football Association, as well as criminal proceedings. On 25 May 1995, following his conviction, F was sentenced to imprisonment for a period of three months and his subsequent appeal to the High Court of Justiciary was refused in October 1995. The respondents brought disciplinary proceedings under paragraphs 3 and 4 of the Disciplinary Procedures covering an 'Exceptional case of player's misconduct'. On 12 May 1994, the respondent's Disciplinary Committee imposed on F a severe censure and suspended him for twelve months. F's appeal to the Disciplinary Appeals Tribunal was refused in November 1995. Under Disciplinary Procedure paragraph 4, a copy of a report of any incident has to be submitted and issued to the player. No report was so issued. In F's petition for judicial review to quash the decisions of the Disciplinary Committee and the Disciplinary Appeals Tribunal, Lord MacFadyen said that, in his opinion, the Committee, in purporting to adopt the procedure laid down in paragraphs 3 and 4 of the Disciplinary Procedures to deal with F's conduct in an incident that was not reported by any of the match officials, acted ultra vires. It followed that the severe censure and match suspension imposed were invalid and of no effect.

### *Gorlov v. Institute of Chartered Accountants in England and Wales* [2001] EWHC 220 (Admin)

**43.07** Accountant—complaint preferred by investigation committee—different complaint before Disciplinary Committee—no power under bye-laws—nullity

The claimant was a chartered accountant. The Institute of Chartered Accountants in England and Wales (ICAEW) decided to proceed with a disciplinary hearing against the claimant, notwithstanding that the Institute's bye-laws had not been followed. The terms of the formal complaint before the Disciplinary Committee were different from the formal complaint that had been preferred by the investigation committee. The hearing went ahead on the reformulated complaint. It was held to be a nullity, notwithstanding that the claimant's solicitors wrote to the Institute confirming that their client did not seek to challenge the jurisdiction of the Disciplinary Committee to hear the complaint on the grounds that the Institute's bye-laws had not been followed. Jackson J said, at [20], that the decision to proceed with the hearing was a curious one. The Disciplinary Committee did not have the power under the bye-laws to consider any complaint that was not referred to the investigation committee. Unlike a court, the Disciplinary Committee does not have any inherent jurisdiction; it has only the powers conferred upon it by the bye-laws. With the benefit of hindsight, the lawyers on both sides could be criticized for not spotting the flaw in the proceedings. There was, however, force in the claimant's point that greater culpability rested upon the Institute. The Council of the Institute made the bye-laws. The officials and lawyers employed by the Institute ought to understand how they work. At [22], Jackson J said that, although not appreciated at the time, the decision to proceed with the hearing was a nullity. The claimant could not, by consent, confer upon the Disciplinary Committee powers that it did not have under the bye-laws.

*Surrey Police Authority v. Beckett* [2002] ICR 257

**43.08** Police officer—fixed-term contract of employment—expiry—no longer subject to disciplinary process

On dismissing the police authority's appeal, Simon Brown LJ (with whom Tuckey and Laws LJJ agreed) said that the question for decision on this appeal is whether a senior police officer, suspended under police disciplinary regulations following a complaint made against him and whose appointment to office under a fixed-term contract then expires by effluxion of time, nevertheless remains subject to disciplinary process. In late December 1998, a complaint of sexual harassment was made against B by a female employee of the Surrey Police Force; on 8 January 1999, he was suspended pending investigation of the matter. Following his suspension, four other female employees made complaints against B and an investigating officer was appointed. B was subsequently acquitted on four criminal counts of indecent assault. Following B's acquittal, he was served with a written notice stating that disciplinary action would be taken against him in respect of the alleged sexual harassment of five named female employees. However, it became apparent that the disciplinary process would not be able to be completed before the expiry of B's fixed-term contract on 20 May 2001. The parties being in dispute as to the effect of that upon the disciplinary process, the Authority claimed in the High Court a declaration that it was at liberty to continue the disciplinary proceedings against B even after the expiry of his fixed-term contract on 20 May 2001. The claim was dismissed by Blofeld J. The Court of Appeal, affirming the decision of the judge, said that the point at issue fell to be resolved by reference to the Police Regulations 1995. The Regulations introduced fixed-term appointments in place of indefinite appointments for certain senior ranks. That disciplinary proceedings against a suspended police officer may be brought to an end simply by the arrival of a particular date cannot be doubted. Once an officer reaches compulsory retirement age under the Regulations (or once that period has been extended up to the maximum five years permitted under the Regulations), there is no question of his then remaining amenable to the disciplinary process. On the introduction of fixed-term contracts in 1995, senior officers cannot have been intended thereby to be more readily able to escape from ongoing disciplinary proceedings—hence regulation 13A(9), which provides that, without prejudice to fixed term appointments, a suspended police officer may not without consent give notice to retire. The Authority submitted that, having refused its consent to his retirement, B remained in office. However, only the most compelling words in the Regulations could result in an officer whose fixed-term appointment had ended nevertheless remaining in office. When regulation 13A(9) was introduced, the draftsman had every opportunity to achieve that result; he chose not to take it. The language of regulation 13A(9) is wholly insufficient for that purpose. On the contrary, it makes perfect sense by reference to the need to put premature retirement under a fixed-term contract on the same footing as retirement under the pre-existing form of permanent appointment.

***R (Health Professions Council) v. Disciplinary Committee of the Chiropodists Board and Green*** **[2002] EWHC 2662 (Admin)**

**43.09** Chiropodist—registered in UK—complaints arising in New Zealand—Health Professions Order 2001—purpose of legislation

The claimant, the Health Professions Council, sought judicial review of the decision of the Disciplinary Committee of the Chiropodists Board, which, on 7 March 2002, dismissed six charges of serious professional misconduct against G, a chiropodist who had been registered in the United Kingdom since 1981. Some of the conduct complained of occurred in New Zealand during the period 1995–99, when G was practising in New Zealand. The Committee decided that it did not have jurisdiction to hear the New Zealand charges. The issue was whether the Committee was right. The allegations were that G had failed to repay money that he had borrowed from a patient and a student to assist in the acquisition of a clinic, and that he had abandoned his podiatry practice without notice to his professional colleagues or patients. In quashing the decision of the Committee, Goldring J said that one authority drawn to his attention was *Ong Bak Hin v. General Medical Council* [1956] 2 All ER 257, in which the Privy Council dismissed an appeal by the doctor registered in the United Kingdom in relation to an operation performed in Malaya. The learned judge said that although there was no consideration of the issue, it was implicit in the Privy Council's decision that serious professional misconduct in Malaya was sufficient to justify erasure from the register in the United Kingdom. The learned judge said, at [22]–[24], that, in his view, the Committee did have jurisdiction to hear the New Zealand complaints. First, section 9(2) of the Professions Supplementary to Medicine Act 1960 (the predecessor to the Health Professions Order 2001) did not impose any jurisdictional limit. If it had been Parliament's intention that there should be one, it could have said so. It is plain that a conviction outside the United Kingdom would not count for the purposes of section 9(1) of the Act. It equally could have said that conduct outside the United Kingdom would not count for the purposes of section 9(2). Second, the purpose of the legislation was the protection of the public. If a chiropodist has been guilty of serious professional misconduct, it does not matter to a member of the public whether that misconduct arose within or outside the jurisdiction. It does not seem that the fact that the charges relate to matters outside the United Kingdom and do not directly relate to G's registration in the United Kingdom is material. The issue was whether they amount to conduct of such a nature as to be infamous in any professional respect, wherever committed. If there is such possible conduct, it is the duty of the Committee to investigate it. When G sought and was given his registration in 1981, he knew that serious professional misconduct by him would put his registration at risk. Provided that the allegation of such misconduct was properly proved, the fact that it may have occurred outside the jurisdiction cannot be unfair to someone in G's position. The learned judge continued to say, at [25], that he could understand the Committee having some concern regarding witnesses. However, on proper analysis, it was without substance. A defendant is sufficiently protected by the provisions of the Disciplinary Committee (Procedure) Rules 1964. Moreover, G was told what the witnesses would say and that their statements would be read. He chose not to contest their evidence. That could not conceivably be said to have been unfair.

### *Swanney v. General Medical Council* 2008 SC 592

**43.10** Doctor—conduct outside UK—conduct occurring whilst not registered with GMC—Medical Act 1983, section 36(1)(b)

In this case, the Inner House of the Court of Session was asked to determine whether the GMC had locus to pursue disciplinary proceedings against a medical practitioner when the conduct complained of occurred while he was not registered with the GMC, took place outside the United Kingdom, and had already been subject to disciplinary proceedings in a different jurisdiction. S had dual registration as a medical practitioner in the United Kingdom and in British Colombia, Canada. He was registered in the United Kingdom between 1970 and 1990, and thereafter from 2001. He was registered in British Colombia from 1974 onwards. As a result of certain events occurring in 1999 and 2000 in British Colombia, disciplinary proceedings were instituted by the College of Physicians and Surgeons of British Colombia against the appellant. In November 2003, his name was erased from the full medical register in British Colombia and restrictions were placed on his practice. Subsequently, the GMC commenced its own investigation, and in 2007 he was found guilty of serious professional misconduct and conditions were imposed on his registration for twelve months. The Inner House rejected S's argument that conduct that took place in Canada while he was not registered with the GMC could not be dealt with in proceedings before the GMC. Section 36(1)(b) of the Medical Act 1983 states that the GMC can take proceedings against practitioners 'whether while so registered or not'. In the Court's view, the appearance of these words make it completely clear that the GMC was given authority by Parliament to explore an issue of serious professional misconduct in relation to actions that may have occurred while the subject of the inquiry was not a registered person in the United Kingdom. The Inner House noted that the whole purpose of GMC proceedings was to protect the public, and observed:

> 17. If the contrary view [the legislation did not permit an inquiry] were accepted it would mean that a practitioner whose conduct could be regarded as serious professional misconduct in some other jurisdiction could come to the United Kingdom and practice medicine here with impunity [and] be a danger to the public. Such a result would undermine the objective of the respondents, enshrined in section 1(1A) of the 1983 Act, which provides that the main objective of the respondents is to "protect, promote and maintain the health and safety of the public."
>
> [...]
>
> 19. ...It is quite plain that the purpose of the disciplinary proceedings in British Colombia was to subject the appellant's conduct to scrutiny there with a view to a decision being reached as to his fitness to practise there, or as to the conditions upon which he should be permitted to practice there. The purpose of the proceedings before the panel in the United Kingdom was different. Their purpose was to examine the appellant's fitness to practise in the United Kingdom, or to determine whether his right to practice here should be subject to conditions.

### *Council for Health Care Regulatory Excellence v. Nursing and Midwifery Council* [2012] CSIH 24

**43.11** Nurse—registration lapsed—no longer subject to disciplinary process—finding of impairment

H's registration with the Nursing and Midwifery Council (NMC) lapsed shortly after the decision of the Conduct and Competence Committee (CCC) on 2 December 2011. On an application by the Council for Health Care Regulatory Excellence (CHRE) under section 29 of the National Health Service Reform and Health Care Professions Act 2002, the Inner House, Court of Session (Lady Paton, Lord Hardie, and Lord Wheatley), observed that if the Court were to remit the case back to the NMC, there could be no further proceedings in that forum because there would be no registrant. That being so, the Court had concerns about exercising a power to impose a sanction under the 2002 Act that would then be incompetent for the NMC itself to impose. In the result, the Court was persuaded that it should, in terms of section 29(8)(b) and (c) of the 2002 Act, allow the appeal, quash the decision complained of, and substitute a finding of impairment of fitness to practise.

### *R (B) v. Nursing and Midwifery Council* [2012] EWHC 1264 (Admin)

**43.12** Nurse—investigating committee—no case to answer—police—second committee setting aside earlier decision—judicial review

On 16 March 2011, the investigating committee of the NMC decided that there was no case to answer in respect of allegations made against the registrant, B, who was employed by a nursing home in Gwent. In June 2009, the NMC received notification from Gwent Police of a police investigation into the mistreatment and/or neglect of residents at three care homes. Allegations against thirty-six registrants, including the claimant, were referred to the fitness to practise directorate of the NMC. The claimant was interviewed by the police under caution in March 2010, but was not arrested or charged. On 28 March 2011, Gwent Police wrote to the NMC, expressing concern that the investigating committee had decided that there was no case to answer without seeking further disclosure from the police and without awaiting the decision of the Crown Prosecution Service (CPS) on whether or not to prosecute B for wilful neglect. Apparently, as a result of this letter from Gwent Police, the NMC's in-house legal team produced two memoranda criticizing the investigating committee's decision. On 19 July 2011, Gwent Police sent an email to the NMC, stating that the CPS had concluded that there was insufficient evidence to prosecute B for wilful neglect. On 14 December 2011, a further panel of the investigating committee sat and decided to set aside the decision made by the previous panel. Lang J, in allowing B's claim for judicial review, said that the NMC's proceedings are governed by the Nursing and Midwifery Order 2001 and the Nursing and Midwifery Council (Fitness to Practise) Rules 2004. It is apparent that neither the Order nor the Rules confer any power to set aside and reverse a decision of the investigating committee that a registrant has no case to answer, outside the specific circumstances provided for in rule 7 (the Council receives a fresh allegation about the registrant), which did not apply to this case. The NMC argued that the investigating committee, as a statutory tribunal, had an inherent jurisdiction to set aside its previous decision. However, this proposition is not supported by the authorities. The investigating committee was not entitled to reverse its previous decision, in reliance on the decision of *R (Jenkinson) v. Nursing and Midwifery Council* [2009] EWHC 1111 (Admin). The circumstances in *Jenkinson* (registrant's conviction subsequently set aside by the Court of Appeal, Criminal Division) were exceptional and very different from this case. In *Jenkinson*, the parties were in agreement that the earlier decision of the CCC of the NMC should not stand, because there was no longer any proper basis for it. In the instant case, the NMC acted unlawfully and beyond its powers in rescinding and reversing the investigating committee's decision of 16 March 2011. The NMC's decision was also an unlawful breach of the claimant's substantive legitimate expectation that she had no case to answer in relation to the allegations, and in rescinding and reversing its earlier decision, the NMC departed from its established and published procedures.

### *Baker v. Police Appeals Tribunal* [2013] EWHC 718 (Admin)

**43.13** Police Appeals Tribunal—order—amended order—*functus officio*

On 13 October 2010, B, a constable in the Sussex Constabulary, was found guilty of gross misconduct and was dismissed without notice. B appealed to the Police Appeals Tribunal against both the finding of misconduct and the disciplinary action of dismissal. By an order dated 24 March 2011, the Tribunal dismissed B's appeal against the finding of gross misconduct, but allowed his appeal against the disciplinary action of dismissal, substituting for dismissal the disciplinary action of a final written warning. The order went on to state that B 'shall be reinstated in the Sussex Constabulary', and 'shall be deemed to have served continuously in that rank from 13 October 2010'. Subsequently, the professional standards department of Sussex Police informed the Tribunal that, between the date of his dismissal and the hearing of his appeal on 24 March 2011, B had been employed outside the Police Service. At the time when the appeal was heard and the order was made, neither Sussex Police nor the Tribunal had been aware of that fact. After considering representations from both parties, the tribunal issued an 'amended order' dated 30 June 2011, which stated that, for the purposes of pay, only such sum should be restored to B as was equal to the difference between his lost Police Service pay and his earned income from other sources between 13 October 2010 and his reinstatement. In a claim for judicial review, B contended that, at the time when the amended order was made, the Tribunal was *functus officio*. In granting relief and quashing the amended order, Leggatt J said that the amended order was made without jurisdiction and the Tribunal was *functus officio*. The learned judge said that the powers of the Tribunal derived from section 85 of the Police Act 1986 and Schedule 6 to that Act. It was common ground that the Tribunal had power, by its original order, to direct that, on his reinstatement, B should receive in back pay only the difference between (a) the amount that he would have earned had he served continuously in the force, and (b) the amount that he in fact earned from other sources between the date of his dismissal and the date of his reinstatement. However, the Tribunal did not make such a direction in its original order. After it had given notice of its decision, nothing in the rules (nor anywhere else) gives the Tribunal power to take any further action. In particular, there is no express power—and it is accepted that none is to be implied—that would enable the Tribunal subsequently to change its decision. It is for this reason that the Tribunal became *functus officio*. After considering the width of the Court's discretion and the effect of the illegality, the learned judge said the present case did not fall within any of the categories in which it would be permissible to refuse relief. If the Tribunal had recognized that it was *functus officio*, Sussex Police might have considered whether to apply for judicial review of the original order. By the same token, it was to remain open to Sussex Police, if so advised, to raise as a defence to a claim by B for the arrears the contention that the original order was invalid because it resulted from a material mistake of fact giving rise to unfairness.

### *B v. Independent Safeguarding Authority (Royal College of Nursing intervening)* [2013] 1 WLR 308, CA

**43.14** Independent Safeguarding Authority—jurisdiction of Upper Tribunal

Although section 4(3) of the Safeguarding Vulnerable Groups Act 2006 inhibits the Upper Tribunal from revisiting the question of 'whether or not it is appropriate for an individual to be included in a barred list', the Upper Tribunal is empowered to determine proportionality and rationality. The Independent Safeguarding Authority (now the Disclosure and Barring Service) decided to remove B's name from the adults' barred list, but not from the children's barred list. In allowing the Authority's appeal from the decision of the Upper Tribunal, and restoring B's name to both lists, the Court of Appeal (Maurice Kay and Etherton LJJ, and Sir Scott Baker) held that whilst the Upper Tribunal had jurisdiction to determine proportionality and rationality (see *R (Royal College of Nursing) v. Secretary of State for the Home Department* [2011] PTSR 1193), the Upper Tribunal in the instant case did not give sufficient weight to the decision of the Authority (a body with particular expertise), and that the maintenance of public confidence is an important element in the balancing exercise in deciding whether a person should be barred from carrying out regulated activities.

### *R (Hill) v. Institute of Chartered Accountants in England and Wales* [2013] EWCA Civ 555; [2014] 1 WLR 86, CA

**43.15** Accountant—disciplinary tribunal—panel member's temporary absence—powers of tribunal

The Disciplinary Tribunal of the ICAEW found the claimant guilty of unprofessional conduct and he was excluded from membership. During the six-day hearing, the lay member of the Tribunal was absent for part of the claimant's cross-examination and he later returned, having read a transcript of the evidence during his absence. The Institute was constituted by Royal Charter, and was empowered to make bye-laws and regulations. The bye-laws provided for the appointment and constitution of an investigation committee, a disciplinary committee, and an appeal committee. Bye-law 19 was the only provision that dealt with the departure of a member of the Tribunal. It provided that if a member were to depart, the remaining members (if not fewer than two in number) could proceed with the hearing if the defendant or his representative agreed, but not otherwise. In dismissing the claimant's appeal, Longmore LJ (with whose judgment Beatson and Underhill LJJ agreed) said, at [11]–[13], that he did not accept the argument that it must follow that departure and return of a member was not allowed, even if the parties agreed. In the first place, it would be surprising if there were no power at all for a disciplinary tribunal (with its relatively informal procedures) to permit one member to depart and return if all of the parties agreed. That would introduce a degree of rigidity into the proceedings, which would be undesirable. Secondly, the fact that express power is given to a tribunal to carry on as a tribunal of a lesser number if one member is unable to continue to attend does not preclude a member absenting himself or herself and returning. Thirdly, in *Virdi v. Law Society* [2010] 1 WLR 2840, Stanley Burnton LJ said that, when one is dealing with bye-laws and regulations of professional disciplinary bodies, one cannot expect every contingency to be foreseen and provided for. The right question to ask of any procedure adopted should therefore be not whether it is permitted, but whether it is prohibited. If one asks that question in this case after rejecting any application of the *expressio unius* principle, the answer is that the procedure adopted was not prohibited. It must, of course, still be fair. The learned judge continued to say that, in his dissenting judgment in *Carter v. Ahsan* [2005] ICR 1817, [16], later approved by the House of Lords *sub nom Watt (formerly Carter) v. Ahsan* [2008] AC 696, Sedley LJ drew a distinction between what he called 'constitutive jurisdiction' and 'adjudicative jurisdiction', saying: '[B]y constitutive jurisdiction I mean the power given to a judicial body to decide certain classes of issue. By adjudicative jurisdiction I mean the entitlement of such a body to reach a decision within its constitutive jurisdiction.' The importance of the distinction for present purposes is that an act outside the constitutive jurisdiction of a tribunal is an act that cannot be agreed to by the parties and therefore be waived by them. If the claimant's first argument in relation to the bye-laws were correct and there was no power for the Tribunal to permit a temporary absence by a Tribunal member on the basis that he would read the transcript of evidence given in his absence and then return, the procedure adopted would no doubt be outside the constitutive jurisdiction of the Tribunal and could not be agreed to or waived. The most apposite authority is *Doyle v. Canada (Restrictive Trade Practices Commission)* (1985) 21 DLR (4th) 366. In that case, the hearing lasted for thirty-two days between April 1982 and June 1983, and two of the three commissioners were absent from all or part of the hearings up to six days. They read the transcripts of the evidence given on the occasions on which they did not attend. The complainant did not attend most of the hearings and it was argued that that failure on his part constituted a waiver of his right to insist that every commissioner should attend every hearing. By a majority, the Federal Court of Appeal of Canada held that the report of the commission should be set aside on the maxim that 'he who decides must hear', and the rule is a rule that actually affects the judge's jurisdiction. Longmore LJ said, at [28], that this authority deserves respect, but that the reference to the maxim as a rule that affects the judge's jurisdiction does not address the question of what kind of jurisdiction a breach of the rule will affect. It may well affect a judge's adjudicative jurisdiction, but if it does, that will not bar any waiver or consent. If the Federal Court of Appeal of Canada were intending to indicate that no breach of the rules of natural justice can ever be waived, Longmore LJ said that he would respectfully disagree. He would go further and say that a breach of the rules of natural justice of the kind that is said to have occurred in the present case is at most an irregularity that could be waived.

See further paragraph 48.15.

### *Fajemisin v. General Dental Council* [2014] 1 WLR 1169

**43.16** Registrant—disciplinary proceedings—decision to remove name from register for non-compliance with CPD requirements—decision revoked—whether power to revoke decision

In 2009, disciplinary proceedings were commenced against F in respect of allegations that he had submitted fraudulent claims for the treatment of elderly residents at nursing homes. A hearing before the PCC of the General Dental Council (GDC) was due to start on 12 September 2011. It was adjourned because F had been admitted to hospital; it was relisted again on 30 October 2012. In the meantime, F had failed to complete his required continuing professional development (CPD) hours. On 13 April 2012, he was notified that the registrar had decided that his name would be removed from the register on 11 May 2012. In error, an employee dealing with fitness to practise proceedings informed the registrar's office, which dealt with dentists' compliance with the CPD rules, that F had been involved in fitness to practise proceedings, but the case had been marked 'closed'. Although the decision had been made to remove F's name from the register on 11 May 2012 unless he appealed in the meantime, his name was not removed from the register on that date. Instead, a final check was made with the fitness to practise team that day and the previous mistake was discovered. Once the true position was known, no steps were taken to remove F's name from the register. At the resumed hearing on 30 October 2012, F's counsel argued, amongst other things, that since the decision had previously been made that F's name would be removed from the register on 11 May 2012, the PCC had no jurisdiction to embark on the hearing. The Committee rejected that submission, and went on to hear the case and order F's name be removed as a result of his misconduct. On appeal, Keith J said that the critical question was: whether a regulatory body could revoke a decision that it made if, by mistake, that decision had been made in ignorance of the true facts. After reviewing *Akewushola v. Secretary of State for the Home Department* [2000] 1 WLR 2295, *R (Jenkinson) v. Nursing and Midwifery Council* [2009] EWHC 1111 (Admin), *R (B) v. Nursing and Midwifery Council* [2012] EWHC 1264 (Admin), and *Porteous v. West Dorset District Council* [2004] HLR 30, Keith J said that *Porteous* is authority for the proposition that, in addition to cases in which a public body can revisit a previous decision under the equivalent of the 'slip rule', a public body can revisit a decision that was made in ignorance of the true facts when the factual basis on which it had proceeded amounted to a fundamental mistake of fact. He concluded that the registrar had the power to revisit the decision that F's name would be removed from the register on 11 May 2012. *Porteous* is authority for the proposition that a public body can revisit a decision that has been made in ignorance of the true facts when the factual basis on which it had proceeded amounted to a fundamental mistake of fact.

### *Professional Standards Authority for Health and Social Care v. Nursing and Midwifery Council* [2013] EWHC 4369 (Admin)

**43.17** Nurse—injunction—interim order pending appeal by PSA

This was an application by the Professional Standards Authority (PSA) for an interim order, which would have the effect of maintaining the registration of the registrant H, pending the full hearing of the Authority's challenge to the sanction that had been imposed upon H by the CCC of the NMC on 7 November 2013. H had retired from nursing and her registration was due to lapse on 5 December 2013 because of non-payment of her renewal fee. The PSA submitted that, once H's registration lapsed, no further sanction could be made against her, even if the appeal were to succeed, and that this would frustrate the legitimate exercise of its powers. Lang J said that she accepted that it is necessary in the public interest not to frustrate the purpose of the PSA's powers to appeal against an unduly lenient order made by the NMC and that there was a real risk that lapse of registration would frustrate the purpose of the appeal. It was necessary for the purposes of public protection and to maintain the confidence of the public in the system of professional regulation that H be prevented from practice, pending appeal. Accordingly, an interim order restraining H from working as a nurse in any capacity, pending the final determination of the current disciplinary proceedings against her, would be made under the Nursing and Midwifery Order and section 37 of the Senior Courts Act 1971.

***7722656 Canada Inc. v. Financial Conduct Authority* [2013] EWCA Civ 1662; [2014] LR (FC) 207**

**43.18** Canadian corporation—governed by Canadian law—jurisdiction of FSA/FCA

By a majority, the Court of Appeal (Longmore and Floyd LJJ, and Lewison LJ dissenting) upheld the decision of the Upper Tribunal (Judge Colin Bishopp, chairman) that, based on the applicable Canadian law and evidence, it was open to the Financial Conduct Authority (FCA), formerly the Financial Services Authority (FSA), to impose a penalty on the appellant for market abuse in breach of section 118 of the Financial Services and Markets Act 2000. On 6 May 2011, the FSA's Regulatory Decisions Committee issued a decision notice addressed to the applicant, then known as '7722656 Canada Inc.', but which previously and at all material times had carried on business in the name of 'Swift Trade Inc.'. On 2 December 2010, Swift, a Canadian company, amalgamated with another company, and that amalgamated company became the appellant Canadian company; the appellant was itself dissolved on 13 December 2010. Accordingly, it did not exist when the decision notice was issued on 6 May 2011. Longmore LJ said, at [13], that it is agreed that it is for Canadian law, as the law of the country of incorporation, to say whether the company existed at that date. That depends on the true construction of the relevant sections of the Canada Business Corporations Act 1985 and any relevant expert evidence of Canadian law. The Upper Tribunal had the benefit of both written and oral evidence on Canadian law, and the Upper Tribunal could reasonably find that Swift, despite having been dissolved, had a continued, limited existence as a matter of Canadian law: *see further* Floyd LJ, at [83].

***Hobbs v. Financial Conduct Authority (formerly Financial Services Authority)* [2013] Bus LR 1290, CA**

**43.19** Decision to discontinue proceedings published on website—statement removed from website—appeal—whether binding decision made to discontinue proceedings

On 22 November 2012, the Upper Tribunal (Tax and Chancery Chamber) Financial Services released its judgment and directed that the FSA should take no further action against H. Following the Tribunal issuing its decision, the Authority on the same day published the result on its website and added: 'The FSA confirms that, following the tribunal's decision and in accordance with its direction, it is discontinuing its action against Mr Hobbs in relation to this matter.' H saw that statement and concluded that the Authority's proceedings against him were at an end. However, on 27 November 2012, the sentence was removed from the website and, on 6 December 2012, the Authority applied to the Tribunal for permission to appeal to the Court of Appeal. The Tribunal refused permission, but Lewison LJ subsequently granted the Authority permission to appeal. In allowing the Authority's appeal, Sir Stanley Burnton (with whom Rimer and Ryder LJJ agreed) said that the decision was not binding on the Authority. The communication of a decision to take no further action, or to discontinue, is an intrinsic part of the process. Parliament has specified to whom and how notice is to be given, in section 389 of the Financial Services and Markets Act 2000, and, by delegated legislation, in the Financial Services and Markets Act 2000 (Service of Notices) Regulations 2001. A statement on the Authority's website, not addressed in any way to the person to whom the decision notice is addressed, is not a sufficient notice of a decision to take no further action for the purposes of the Act so as to render it irrevocable. Furthermore, there was no decision to take no further action made by anyone on behalf of the Authority with the requisite authority. The decision as to whether to discontinue a case requires the authority of the project sponsor—invariably, the head of department. As a matter of practice, all such decisions are approved by the director. A subsequent notice informing the person under investigation that the case has been discontinued must be reviewed by the project sponsor and authorized by a head of department. In the instant case, no decision to discontinue was made by the project sponsor or approved by the director of the Enforcement and Financial Crime Division. Accordingly, there was no decision binding on the Authority. As to the Authority's substantive appeal, that would be allowed and the matter remitted to the Tribunal to consider what order to make in the light of the judgment of the Court.

***Conlon v. Bar Standards Board* [2014] Lexis Citation 136**

**43.20** Barrister—Visitors—jurisdiction of lay or barrister members to sit as Visitors

C, a barrister found guilty of professional misconduct by the Disciplinary Tribunal of the Council of the Inns of Court, argued on appeal that there was no jurisdiction for lay or barrister members to sit as Visitors to the Inns of Court on an appeal from the Disciplinary Tribunal. The Visitors (Asplin J, chairman) said that the underlying issue as to the jurisdiction of the Visitors is set out in *R (Leathley and ors) v. Visitors to the Inns of Court* [2013] EWHC 3097 (Admin), but for the sake of completeness the Visitors would set out its own decision in this regard. Asplin J said, at [30]–[43], that it is quite clear from the decision of the Court of Appeal in *R v. Visitors to the Inns of Court, ex parte Calder* [1994] QB 1 that the jurisdiction of judges over the Inns of Court, in their capacity as Visitors so far as it related to questions as to the fitness of persons to become or remain barristers, devolved upon the judges of the new High Court of Justice by virtue of section 12 of the Supreme Court of Judicature Act 1873: *per* Sir Donald Nicholls VC, at 33A–D and 33G–34E; *per* Stuart-Smith LJ, at 46G–47F; and *per* Staughton LJ, at 65G–66D. The 1873 Act has been repealed, but the effect of section 12 survived via section 18(3) of the Supreme Court of Judicature (Consolidation) Act 1925 and section 44 of the Supreme Court Act 1981 (now the Senior Courts Act 1981). The jurisdiction devolved in this way because, when exercising the jurisdiction as Visitors, the judges were not acting as judges. From time to time, the judges have taken steps to regulate the procedure of such appeals: see the Preamble to the Hearings before the Visitors Rules 1991, 2005, and 2010. The composition of the panel of Visitors for the purposes of an appeal is governed by the Hearings before the Visitors Rules 2010, which provide for lay and barrister members. Accordingly, the argument that there is no jurisdiction for lay members to sit on a panel of Visitors is without any merit. The 2010 Rules and their predecessors are an expression of the jurisdiction of the judges of the High Court as Visitors. The effect of section 24 of the Crime and Courts Act 2013 is that the jurisdiction conferred upon the judges in their capacity as Visitors to the Inns of Court has ceased. Nevertheless, the 2010 Rules continue to apply to appeals in which the date of the decision was before 7 January 2010.

***Woodman-Smith v. Architects Registration Board* [2014] EWHC 3639 (Admin)**

**43.21** Architect—purported resignation from register—pending disciplinary proceedings

On 1 November 2013, an investigations panel of the Architects Registration Board concluded that the appellant should be referred to the PCC in relation to two allegations. On 31 December 2013, at 08:16, the appellant emailed the Board, stating that he wished to have his name removed from the register. On the same day, at 10:04, the Board replied to the appellant indicating that it would process his resignation, but a short while later on the same day (at 10:30), a further email was sent to the appellant stating that the registrar was unable to accept his resignation while there remained outstanding disciplinary proceedings against him. The appellant candidly acknowledged that he sought to resign from the register 'with the intention to frustrate the proceedings before the PCC'. Subsequently, on 15 May 2014, the PCC found the appellant guilty of unacceptable professional conduct in that he had failed to enter into a written agreement with his client prior to undertaking professional work, contrary to standard 4.4 of the Architects Code: Standards of Conduct and Practice 2010. The Committee made a disciplinary order against the appellant in the form of a reprimand. In dismissing the appellant's appeal and challenge to the jurisdiction of the PCC, Sir Stephen Silber (sitting as a High Court judge) said that he was unable to agree with the Board's submission that the appellant could not pursue the jurisdiction issue on grounds that the Court, on appeal, did not have jurisdiction, because it was not submitted in front of the PCC that the Committee did not have jurisdiction. The Board had not been prejudiced by the fact that the jurisdiction issue was not raised in front of the PCC. However, the register showed that the appellant was a 'registered' person at the time of the disciplinary proceedings and, accordingly, the PCC had jurisdiction to pursue these disciplinary proceedings. Section 3 of the Architects Act 1997 provides that the registrar shall maintain the register of architects, in which there shall be entered the name of every person entitled to be registered, and a certificate purporting to be signed by the registrar stating that a person is registered shall be evidence of any matter stated. Even if that conclusion is wrong, and the appellant had resigned and wished his name to be removed from the register on 31 December 2013, he waived that resignation and his wish for his name to be removed from the register, first, by participating in and contesting the disciplinary proceedings; secondly, by continuing after that date to take all of the advantages of being a 'registered person'; and thirdly, by holding himself out after that date as a 'registered person'.

***The Law Society of England and Wales v. Shah* [2014] EWHC 4382 (Ch)**

**43.22** Struck-off solicitor—injunction—order restraining defendant holding himself out as solicitor

In this case, a permanent injunction was granted by the Law Society against the defendant, a struck-off solicitor, from holding himself out as a solicitor, or undertaking any reserved legal activities, or being employed by or remunerated by, or managing or controlling the practice of, a solicitor or any body regulated by the Law Society, without prior written permission of the Law Society. The Court, in granting the injunction, held that the jurisdiction for such an order was section 41(4)(c) of the Solicitors Act 1974, as amended by the Legal Services Act 2007, which prohibits the employment of a struck-off or suspended solicitor and gives the High Court power to make such order as it thinks fit.

***R (Kerman & Co) v. Legal Ombudsman* [2014] EWHC 3726 (Admin); [2015] 1 WLR 2081**

**43.23** See paragraph 39.41.

## C. Other Relevant Chapters

**43.24**

Abuse of Process, Chapter 2
Appeals, Chapter 4
Constitution of Panel, Chapter 13
Double Jeopardy, Chapter 21
Estoppel, Chapter 24
Independent and Impartial Tribunal, Chapter 35
Investigation of Allegations, Chapter 39
Natural Justice, Chapter 48

# 44

# LEGAL ASSESSORS



## A. Legal Framework

**44.01** Examples include:

Veterinary Surgeons (Disciplinary Proceedings) Legal Assessor Rules 1967
Medical Act 1983, Schedule 4, paragraph 7 (appointment of a barrister, advocate, or solicitor of not less than ten years' standing as legal assessor)
Opticians Act 1989, section 23D(1) (appointment of legal advisers of persons who must have a qualification of at least five years)
Health and Social Work Professions Order 2001 (Legal Assessors) Order 2003
Nursing and Midwifery (Legal Assessors) Rules 2001

General Medical Council (Legal Assessors) Rules 2004:

rule 2 (functions of legal assessors)
rule 3 (attendance of legal assessors)
rule 4 (advice of legal assessors tendered at hearings)

Chartered Institute of Management Accountants Disciplinary Committee Rules 2011, rule 8 (legal assessor)

## B. General Principles

### *Clark v. Kelly* [2004] 1 AC 681

**44.02** Clerk to justices—general principles

As to the comparable position of a legally qualified clerk to justices—or, in Scotland, the clerk of a district court—Lord Bingham of Cornhill said, at [5]:

> The task of the clerk is to advise the lay justice on any question of law arising during the case. It is the clerk's duty as a professional person bound by an exacting code of conduct, to give advice to the best of the clerk's ability, with the independence and impartiality (and also the care) required of any solicitor or advocate expressing a professional opinion. The clerk represents no party and his approach should be wholly unpartisan. He does not enjoy the security of tenure appropriate for a judge. He would (like any professional person) be bound to disqualify himself if, in any case, he found himself subject to any conflict of interest. If anyone were to attempt to influence the opinion of a clerk, otherwise than by argument in open court, such conduct would be regarded as wholly improper, and were the clerk to accede it would be recognised as a culpable dereliction of duty. If the clerk were at any point, publicly or privately, to offer any opinion on the facts of any case, that also would be a culpable dereliction of duty, since all factual decisions are for the justice alone (although if the justice wishes to be reminded of the effect of any oral evidence given during the hearing the clerk may properly remind him, provided this is done in open court).

After reviewing *Ettl v. Austria* (1987) 10 EHRR 255, *Bryan v. United Kingdom* (1995) 21 EHRR 342, *Findlay v. United Kingdom* (1997) 24 EHRR 221, *Mort v. United Kingdom* Application no. 44564/98, 6 September 2001, ECtHR, and *Practice Direction (Justices: Clerk to Court)* [2000] 1 WLR 1886, Lord Hoffmann said, at [30]–[41], that although the clerk is a 'part of the tribunal' for the purpose of attracting the requirements of Article 6(1) of the European Convention on Human Rights (ECHR), he is sufficiently independent in relation to his functions to satisfy those requirements. Accordingly, the status of the clerk of the court does not prevent the district court from complying with Article 6(1).

## C. Disciplinary Cases

### *Daly v. General Medical Council* [1952] 2 All ER 666

**44.03** Questions—leave of chairman—Medical Disciplinary Committee (Legal Assessor) Rules 1951

The appellant, a medical practitioner, appealed against the finding of the Medical Disciplinary Committee of the General Medical Council (GMC) that he was guilty of infamous conduct in a professional respect and that his name should be erased from the register. On appeal to the Privy Council, the appellant contended that the duty of the legal assessor under section 17(1) of the Medical Act 1950 was to decide on the admissibility of evidence and to give his opinion on points of law, and that the legal assessor had acted contrary to his duty in that he had forced the appellant to give evidence that was prejudicial to his case, had allowed witnesses to produce and read letters without proof that they were written by the appellant, and had acted with bias during the hearing. In dismissing the appellant's appeal, Lord Porter, in giving the judgment of the Board, said that, so far as the assessor is concerned, their Lordships had had their attention called to the Medical Disciplinary Committee (Legal Assessor) Rules 1951, which give him wide powers, and they had to assume that, in this particular case, considering the evidence as presented, the legal assessor had received the leave of the chairman of the Medical Disciplinary Committee to put the questions that he did, in which case they were entirely in order. If questions were put against the wish of the chairman, they would be out of order; that was not the case.

### *Fox v. General Medical Council* [1960] 1 WLR 1017

**44.04** Chairman—president of hearing—role of legal assessor

Lord Radcliffe, in delivering the judgment of the Board, said at 1020–1021:

> The decision arrived at in a hearing before the Medical Council is indeed that of a body of persons, but there is no distinction between their responsibility for deciding on the law and their responsibility for deciding on the facts. There is no judge to conduct the proceedings, to direct the jury on matters of law or to sum up for them on issues of fact. Although the Disciplinary Committee has the assistance of a legal assessor at its hearing, as required by the Act, it is the President of the court and not he who is in charge of the proceedings, and his duties are confined to advising on questions of law referred to him and to interventions for the purpose either of informing the Committee of any irregularity in the conduct of their proceedings which comes to his knowledge, or of advising them when it appears to him that, but for such advice, there is a possibility of a mistake of law being made.

### *Sivarajah v. General Medical Council* [1964] 1 WLR 112

**44.05** Advice—analogy with misdirection of law—corroboration—whether sufficient to invalidate decision

S, a registered medical practitioner who stood in a professional relationship with Mrs F at the material times, was found guilty of frequently committing adultery with her both at her home and at his surgery. The Disciplinary Committee directed that his name be removed from the register. S did not attend the hearing, but, in a letter dated 23 May 1963 purportedly from S to the Committee, he stated: 'I deny having had sexual relations with Mrs F—prior to 1960.' In the course of his advice to the Committee, the legal assessor said that 'one may infer from (the letter) that he is not prepared to deny having had sexual relations with her since some time in 1960, and you have certainly then that part of her evidence corroborated which deals with the last year of this association'. In dismissing S's appeal, Lord Guest, giving the reasons of the Board, said that it would have been preferable had the advice tendered been that the letter was capable of being considered as corroboration, but that it was for the Committee to judge whether, in fact, it corroborated the complainer's evidence. That is what the legal assessor clearly meant to say, although it was, perhaps, unfortunately expressed. The legal assessor is, however, in no sense in the position of a judge summing up to a jury, nor is the Committee's function analogous to that of a jury. The legal assessor's duties are set out in *Fox v. General Medical Council* [1960] 1 WLR 1017, 1021. The Committee is master both of the law and of the facts. Thus what might amount to a misdirection in law by a judge to a jury at a criminal trial does not necessarily invalidate the Committee's decision. The question is whether it can 'fairly be thought to have been of sufficient significance to the result to invalidate the committee's decision': *Fox*, at 1023. Their Lordships did not consider that the advice by the legal assessor in the present case amounted to such a defect in the conduct of the inquiry.

### *In re Chien Singh-Shou* [1967] 1 WLR 1155, PC

**44.06**

Disciplinary Board—legal adviser having conduct of inquiry—full member of Board—privacy of Board's deliberations

The appellant, an authorized architect, appeared before the Hong Kong Architects' Disciplinary Board charged with an offence of negligence contrary to section 5B(1) of the Buildings Ordinance 1955. The particulars of the offence were that the appellant had permitted material divergencies or deviations from work shown in plans approved by the Building Authority. The Disciplinary Board found that he had been guilty of negligence and ordered that he should be removed from the architects' register for a period of one year. The appellant sought an order to quash the decision of the Disciplinary Board on the grounds that the Board did not hold 'due inquiry' in accordance with the Buildings Ordinance, because any advice on matters of law given by the legal adviser while the members of the Board were deliberating in private should have been made in the presence of the parties. The Buildings Ordinance provided that every Disciplinary Board shall consist of three authorized architects, a representative of the Building Authority, and a legal adviser, who shall have the conduct of the inquiry. In dismissing the appellant's appeal, the Privy Council said that the legal adviser constituted a full member of the Board. The other members were likewise full members of the Board. In fact, it did not appear to have been likely that, in the present case, any legal advice was given to the Board by the legal adviser save to such extent as was apparent on the record of the proceedings. The findings of the Board appeared to have been pure findings of fact. If, in reaching them, one or more of the architects gave his views on some matters of architectural knowledge when the Board was deliberating in private, it could hardly be contended that there was an obligation to repeat that view in the presence of the parties. The members of the Board were chosen to exercise a judicial function. They must act fairly in ascertaining and considering the facts. They must give every opportunity to the parties to deal with all relevant matters. But the members of the Board were not under obligation to repeat in public anything or everything said in the privacy of their deliberations. At all times, the legal adviser occupied the position of being a full member of a body charged with the duty of acting judicially. There was no obligation on him to make a summing up of the case to his colleagues on the Board in the presence of the parties. His position was different from that which is occupied by a legal adviser to the Medical Council of Hong Kong. The legal adviser to the Medical Council is not a member of that body. His or her duties are specially prescribed by the Medical Practitioners (Registration and Disciplinary Procedure) Regulations 1957. The absence in the Buildings Ordinance of any provision comparable to that in the Medical Practitioners Regulations serves to show the contrast between the roles of the two respective legal advisers.

### *Lawther v. Council of the Royal College of Veterinary Surgeons* [1968] 1 WLR 1441

**44.07**

Veterinary surgeon—improperly obtaining payment of fee—'false pretences'—criminal analogy—legal assessor responding to appellant's counsel

L, a member of the Royal College of Veterinary Surgeons (RCVS), was charged before the Disciplinary Committee with improperly attempting, on or around 12 May 1967, to obtain from Mrs E payment of a fee of 20 guineas by falsely claiming that he had carried out an oesophagostomy upon a dog belonging to her, and thereby of being guilty of disgraceful conduct in a professional respect. The Committee decided that he was guilty of the charge and directed the registrar to remove L's name from the register. In the course of his advice to the Committee, the legal assessor said that the charges were 'in effect of something much more serious, of attempted false pretences'. It was argued on behalf of L that, in the circumstances, the legal assessor ought to have advised the Disciplinary Committee that they must be fully satisfied of intent to defraud. In rejecting this argument and dismissing L's appeal, their Lordships said that the argument of the appellant was founded on too technical a view of the present case. This was not, in fact, a criminal allegation of false pretences. It was an allegation of professional misconduct in improperly attempting to obtain payment of a fee by falsely claiming that an oesophagostomy had been carried out. It was the appellant's counsel who quite reasonably drew in the criminal analogy in order to urge that the Disciplinary Committee must be satisfied beyond any reasonable doubt of the truth of the charge before it accepted it as proved. The legal assessor, again quite reasonably, accepted this view and advised the Disciplinary Committee that, since the charge was 'in effect' of a criminal nature, the Committee must be satisfied beyond reasonable doubt. That was an advantage to the appellant. He could not go on to claim that he was thereby entitled to a full direction from the assessor as if the appellant were, in fact, being tried on a criminal offence. It is not the function of the legal assessor to sum up as a judge does to a jury. In the present case, he made some judicial observations at the end of the hearing that were relevant and helpful. He said nothing that was erroneous or unfair to the appellant. It was for the Disciplinary Committee to decide whether the charge was made out.

### *Libman v. General Medical Council* [1972] AC 217

**44.08** General principles—function of legal assessor—whether error of sufficient significance to the result to invalidate decision

In giving their Lordships' reasons for advising that the appellant's appeal be dismissed, Lord Hailsham of St Marylebone LC said that proceedings before the Disciplinary Committee were governed by the General Medical Council Disciplinary Committee (Procedure) Rules 1970 and the proceedings for the hearing of an appeal before the Judicial Committee by the Judicial Committee (Medical Rules) (No. 2) Order 1971. Both of these statutory instruments are under parliamentary control and possess the legal force of Acts of Parliament. During the course of argument on the extent and exercise of this jurisdiction, their Lordships were referred to: *Felix v. General Dental Council* [1960] AC 704, 716, a decision under the parallel provisions of the Dentists Act 1957; *Fox v. General Medical Council* [1960] 1 WLR 1017; *Sivarajah v. General Medical Council* [1964] 1 WLR 112; and *Bhattacharya v. General Medical Council* [1967] 2 AC 259, especially at 265. Of these authorities, the account of the jurisdiction by Lord Radcliffe in *Fox v. General Medical Council* at 1020–2 is the fullest and perhaps the best, but their Lordships, at 220F–221E, drew the following general propositions from all four decisions:

> [...]
>
> (4) The legal assessor who assists the Committee at its hearing is not a judge, and his advice to the Committee is not a summing up, and no analogy with a criminal appeal against a conviction before a judge and jury can properly be drawn. The legal assessor simply advises the committee *in camera* on points of law and reports his advice in open court after he has given it. The Committee under its president are masters both of law and of the facts and what might amount to misdirection in law by a judge to a jury at a criminal trial does not necessarily invalidate the Committee's decision. Where a criticism is made of the legal adviser's account of his advice the question is whether it can fairly be thought to have been of sufficient significance to the result to invalidate the decision: see *Fox v. General Medical Council* [1960] 1 WLR 1017 and *per* Lord Guest in *Sivarajh v. General Medical Council* [1964] 1 WLR 112, 116–117.

### *R v. Professional Conduct Committee of the General Medical Council, ex parte Jaffe*, 25 March 1988 (unreported)

**44.09** Advice to committee—not a jury

In dismissing the doctor's application for judicial review to quash a determination by the Professional Conduct Committee (PCC) of the GMC, McNeill J said that the assessor is an adviser; his function is not to direct the Committee on matters of law, as would a judge direct the jury. (For authority, see *Fox v. General Medical Council* [1960] 3 All ER 225 and *Sivarajh v. General Medical Council* [1964] 1 All ER 564.)

### *Nwabueze v. General Medical Council* [2000] 1 WLR 1760

**44.10** See paragraph 34.04.

### *Tehrani v. United Kingdom Central Council for Nursing, Midwifery and Health Visiting* 2001 SC 581

**44.11** See paragraph 32.10.

### *Haikel v. General Medical Council* [2002] UKPC 37, [2002] Lloyd's Rep Med 415

**44.12** Application to dismiss charges—abuse of process—advice by legal assessor—no material misdirection

At the commencement of the hearing before the PCC, counsel then appearing for the appellant submitted that the heads of charge should be dismissed on the grounds that there had been an abuse of process and a breach of the appellant's right to a fair trial within a reasonable time under Article 6 ECHR. Following submissions of counsel, there was a short adjournment and, when the Committee reassembled, the legal assessor gave advice in which he stated:

> The arguments advanced on behalf of the doctor...would not make me, if I were the sole judge, invite the Council to respond to that.... [W]ere I the sole arbiter at this stage, I would not be asking the Council to respond and, even wearing my slightly different hat, I would be advising the Committee that the basis for making a decision that the further conduct of these proceedings by the Council would be an abuse of process may not exist.

Dismissing the appeal, their Lordships said that they accepted that the legal assessor might have expressed himself in a manner so as not to create the perception of how, if he were the decision-maker, he would have determined the matter. However, reading the context as a whole, it was abundantly clear that the assessor was indicating that this was an issue for the Committee to decide. It was open to the Committee to rely on the assessor's personal view or to reject it, or to give such weight to it as the Committee considered fit. Their Lordships concluded that this was not a material misdirection or flawed advice such as to cast doubt upon the integrity of the decision that the Committee reached.

### *Rao v. General Medical Council* [2002] UKPC 65, [2002] Lloyd's Rep Med 62

**44.13** Advice ambiguous and misleading—finding of serious professional misconduct unsafe

R was a general practitioner (GP) and had been a doctor for some twenty-five years with a hitherto unblemished career. The complaint arose out of a single incident on the night of 29 December 1998, when R was an out-of-hours doctor. The essence of the case against him was that, at about 22:30 hours, he received a telephone call from the patient's wife, which lasted between two and five minutes, and that, on the information given to him, he failed to give adequate and appropriate advice, and that he should have visited or arranged a GP visit or summoned an ambulance. Later that night, the patient died at his home in his sleep. He was 34 years old. The PCC found that, in relation to the facts admitted and the facts proved, R was guilty of serious professional misconduct, and the Committee directed that his registration should be conditional for a period of eighteen months. The grounds of appeal advanced were that the PCC was wrong to find the appellant guilty of serious professional misconduct and that the advice of the legal assessor was misleading as to what might constitute serious professional misconduct in this particular case. Counsel for the GMC had opened the case to the Committee as 'an isolated incident' (as opposed to a course of conduct). In his advice to the Committee, the legal assessor acknowledged that the PCC was 'considering an event which took place on one occasion', but went on to say that the Committee would be entitled to say that 'there are separate elements to this event. If you do so decide, then you do have a basis upon which you could then go on to determine . . . that the event was one of serious professional misconduct'. Their Lordships said that it was not clear what the legal assessor was seeking to propound by this elaboration; whatever it was it could, and should, have been more felicitously expressed. As it stood, taken as a whole and in context, the passage complained of was ambiguous and misleading. It gave the impression that it was open to the PCC to conclude that the separate elements (as particularized in the charge, if proved) could *each on their own or taken together* amount to serious professional misconduct (emphasis by the Privy Council). This impression may well have been conveyed to the PCC and formed the basis upon which it actually decided the case. In a critical passage, when expressing its conclusions as to what action it should take against R's registration, the chairman said: 'The defects in your practice are such that . . . ' There was, in reality and throughout the hearing, only one single incidence of clinical failure and it was never suggested that there was more than one. The expression 'defects in your practice' was neither appropriate nor warranted. It gave the impression that it was open to the PCC to conclude that, because the incident could be seen as comprising separate elements, it could amount to serious professional misconduct even though, looked at as a single event, it would not. Consequently, their Lordships advised that the finding of serious professional misconduct was unsafe and should be set aside.

*Walker v. General Medical Council* [2002] UKPC 57

**44.14** In camera discussions—procedural irregularity—record of interventions and responses by legal assessor—opportunity for parties to comment

In this case, the charge of serious professional misconduct arose from W's management of twelve separate patients in his first consultation position as a consultant general surgeon over a period of three-and-a-half years from June 1995 to December 1998. Following a hearing lasting for thirty-six days, the PCC found W guilty of serious professional misconduct and directed that his name should be erased from the medical register. Following concerns raised about the role of the legal assessor, the Privy Council directed that the direction that W's name be erased from the register should be quashed and the matter of any appropriate sanction be remitted to a differently constituted PCC and advised by a different legal assessor. Their Lordships had received sworn statements from the two members of the Committee who were said to have expressed concern, from the chair of the Committee, and from the legal assessor. The statements reflected differences in their recollections. The first medical member, who had originally raised the question of possible procedural irregularity, said that, during the in camera discussions, the legal assessor expressed the view that an erasure order ought to be made in respect of W. The statement of the next medical member was that, during a brief discussion amongst the Committee that was not in camera, the legal assessor expressed the strong view that, in effect, the appellant ought to be erased. During the in camera discussions, the legal assessor indicated that he did not think that conditions was a realistic option, because it would be tantamount to erasure. The statement of the chair was that, when asked during the in camera discussions if the doctor would appeal if the Committee were to say that he could not practise surgery, the legal assessor answered that if he were his lawyer, he would advise an appeal. In his statement, the legal assessor said that he did not recall saying that the appellant should be erased. He denied as inconceivable that he advised the Committee that it could not impose a condition that the appellant should not practise surgery. Their Lordships said that they were not in a position to resolve the difference in the recollections expressed in the various statements. Counsel for the GMC accepted that the doctor should face no burden of establishing exactly what occurred and that it was appropriate for their Lordships to proceed on the basis of the version most favourable to the appellant. If that were to show any material procedural irregularity, the appellant could not be regarded as having received a fair hearing and the determination could not stand. In giving their ruling, their Lordships said:

> 19. Where, as here, there arises a question of non-compliance with the General Medical Council (Legal Assessors) Rules, some limited intrusion into the confidentiality of the in camera deliberations of the PCC may be necessary. That is to be approached with care to ensure that full and open discussion in the course of those deliberations is not constrained.
>
> [...]
>
> 23. It was unfortunate in this case that there was not agreement on just what occurred. To avoid such a situation in the future consideration should be given to the adoption of a practice of making a record of interventions and responses by legal assessors in the course of in camera deliberations of the PCC and Health Committees. As the matter was presented, in fairness to the appellant, their Lordships approached the matter on the basis that the accounts more favourable to the appellant may be accurate....
>
> 24. Their Lordships were not persuaded that the comment said to have been made by the legal assessor prior to the in camera deliberations to the effect that the appellant should be erased from the register, if made, constituted legal advice. Nor did they consider it was likely materially to have influenced the members of the committee... It was clear from the material before their Lordships that the committee deliberated at length on possible conditions which might be imposed on the appellant. The committee was not deflected from the proper approach by the opinion said to have been offered at an early stage by the legal assessor. There was not a material procedural irregularity on this ground. Nevertheless their Lordships consider that such an expression of opinion could be no part of the role of a legal assessor under the rules and would be unwise and an improper intrusion into the functions of the PCC whether or not they were in camera.
>
> 25. It is on the second ground advanced by [counsel for W] that their Lordships considered there had been a failure to comply with the procedure laid down in the proviso to rule 4 and therefore a procedural irregularity [now rule 4(3) of the GMC (Legal Assessors) Rules 2004]. The factual assumption is that, in the course of the in camera deliberations, one member of the committee asked the legal assessor to comment on the possibility of a condition that the appellant could not carry out surgery and received a reply to the effect that such a course would be tantamount to erasure, would be unwise and would lead to an appeal. Their Lordships considered that while these comments may be capable of being construed as merely an opinion on the impact of the suggested condition on the appellant, they can also suggest that the imposition of that condition was not open to the committee as a matter of law. It is what may have been conveyed to the members of the committee, particularly the lay members, that calls for assessment. It could not be assumed that, because the committee did not follow the procedure required, the comments were not taken to be legal advice. Reasoning in that way fails to engage the question of procedural irregularity. Their Lordships were not prepared to assume the comments would not have been taken as legal advice. Further, they agree with [counsel for W] that such advice was erroneous in that the imposition of such a condition would not be tantamount to erasure either legally or factually in the case of the appellant. There was no argument to the contrary. They agree also that it was not sufficient reason for rejecting conditions that the appellant might appeal.
>
> 26. Had the advice been disclosed, [counsel for W] would have had the opportunity to comment on the advice and to advance argument that the imposition of a condition

that the appellant not carry out surgery was open to the committee and that, in the circumstances of the appellant, such a condition would not necessarily bring his medical career to an end. Since the matter of sanctions has been remitted, their Lordships make no further comment on these matters.

### *Bose v. General Medical Council* [2003] UKPC 52

**44.15** Erroneous advice—advice favourable to doctor—no prejudice—Committee not following advice

B, a doctor employed in the accident and emergency (A&E) department of a hospital, treated Miss M when she attended the hospital and arranged to visit her at her home. Domiciliary visits did not form part of his duties. In attending at Miss M's home, it was alleged that he carried out an inappropriate and unprofessional intimate examination of Miss M. Miss M was called to give evidence before the PCC, but she became distressed during cross-examination, and it was agreed between counsel for the GMC and counsel for B that Miss M should not be further cross-examined and should give no further evidence. It was agreed with the legal assessor that the only fair way of dealing with the matter was to regard her as not having given any evidence at all. The doctor accepted that it was unprofessional to have visited Miss M at her home and that, with hindsight, it was a terrible mistake, but denied his motivation involved any sexual impropriety. Before the Committee deliberated in camera, the legal assessor advised it that 'bearing in mind the burden and standard of proof, it would not be open in law to the Committee to find that there was an improper motive'. Their Lordships said that the legal assessor erred in advising the PCC in such terms. However, their Lordships considered that the PCC did depart from the legal assessor's advice in their determination, because it was artificial to read the determination other than as a finding that the appellant's actions towards Miss M were motivated by sexual impropriety. This finding by the PCC was one to which it was fully entitled to come, having heard the evidence of the appellant, and there is no reason why the Board should set aside that decision because the legal assessor gave erroneous advice to the PCC that was much too favourable to the appellant. In this case, where the advice given by the legal assessor was erroneous, the appellant suffered no prejudice because the PCC did not comply with the advice. Accordingly, the appeal would be dismissed.

### *R (Mahfouz) v. Professional Conduct Committee of the General Medical Council* [2004] EWCA Civ 233

**44.16** Legal assessor's role—duty to provide advice in answer to legal questions

In giving the judgment of the Court of Appeal, Carnwath LJ (with whom Sedley and Waller LJJ agreed), under the heading 'The legal assessor's role', said:

> 36. However, the debate raises a more fundamental issue, as to the nature of the "advice" which the legal assessor should give in such cases. With respect to the experienced assessor in this case, I think it was a mistake to present the advice simply in terms of questions to be answered by the Committee. That view is reinforced by the fact that the second question is not supported by Mr Englehart QC who appeared for the GMC before us. As I have said, a possible breach of the rules of natural justice is a matter of law, as well as being a potential "irregularity" within rule 3 [now rule 2 of the Legal Assessor's Rules 2004]. Furthermore, the legal assessor is much better placed than the Committee to express the objective view of the "fair-minded observer"; indeed that is precisely what he is or should be.
>
> 37. Accordingly, where an issue such as this arises in the course of proceedings before the PCC, I would regard it as the duty of the legal assessor not simply to pose questions, but to provide answers—or at least "advice" as to the answers (since under the rules the ultimate decision is that of the Committee). In doing so, there is no

justification why he should not look at the matter in the same way as would a judge directing a jury, while taking account of the special characteristics of the Committee which he is advising.

### *Gopakumar v. General Medical Council* [2008] EWCA Civ 309

**44.17**

Good character—distinction between criminal direction and advice by legal assessor—legal assessor stating all witnesses of good character—advice not undermined

At the end of counsel's submissions, the legal assessor was asked by the chairman to remind the panel of its duties and of any other advice that he wanted to give. He gave the panel general directions about the burden and standard of proof, and how it should approach the evidence, together with a definition of serious professional misconduct. These directions of law followed those that a judge would give to a jury at the beginning of a summing up in a criminal trial. After the assessor had given this advice, counsel for the doctor asked whether he might give the 'character direction' if he considered it appropriate, pointing out that it had been conceded from the outset that the appellant was of previous good character. The assessor then proceeded to give both limbs of the good character direction, credibility and propensity, but added that the witnesses were also of good character. At the hearing of the appeal before Underhill J, it was submitted that, by saying that the witnesses (who included the complainants) were of good character, the assessor had undermined the direction so far as the appellant was concerned. The judge rejected this submission. On appeal by the doctor to the Court of Appeal, a main plank of the submissions was that the proceedings in this case were analogous to criminal proceedings and so the advice to be given by the legal assessor should have followed the directions that a judge in a criminal trial would give to a jury. The charges against the appellant could have been (and initially were) criminal charges of indecent assault, so the analogy was particularly apt in this case. In dismissing the appeal, Tuckey LJ (with whom Sir Anthony Clarke MR and Jacob LJ agreed) said that the presence of a legal assessor at proceedings before a fitness to practise panel derived from Schedule 4, paragraph 7, of the Medical Act 1983. Paragraph 7(4) enables rules to be made for, among other things, securing that, where the assessor advises the panel on any matter, all parties will be informed if the panel does not accept his advice. By contrast, juries are required to follow the directions of the judge on any question of law. The General Medical Council (Legal Assessors) Rules 2004 also show the differences between judge and jury in a criminal trial and members of a panel and its legal assessor. The panel is not a jury: it takes legal advice from the assessor, but it is not bound to follow it. The assessor is not a judge: he gives legal advice, but does not give directions as such and does not sum up the evidence to the panel. There was nothing wrong with the legal assessor's advice in this case. Counsel for the doctor asked for the directions and it is clear from what the assessor said that he gave both limbs of the direction for the appellant. Only the passages 'all witnesses are of good character unless the panel hears to the contrary', 'the witnesses are of good character', and 'good character supports every witness's credibility' referred to witnesses as well as, or other than, the appellant. The rest of the direction focused only on the appellant. The assessor was not obliged to give the enhanced direction (although it was really only a statement of the obvious) and counsel for the doctor did not invite him to do so or object in any way to the direction that he did give. The assessor could not be criticized for saying that all witnesses were of good character unless the panel were to hear to the contrary. The panel had not heard to the contrary in this case and there was no reason therefore why it should not treat all witnesses in this way. Why should it assume that the witnesses would or might not be of good character? Certainly, a judge in a civil trial would not make any such assumption. The proceedings were entirely fair and the direction did not cast any doubt upon the panel's decision.

### *Sinha v. General Medical Council* [2008] EWHC 1732 (Admin)

**44.18** Duties of legal assessor

Irwin J, at [57], observed that, in *Gopakumar v. General Medical Council* [2008] EWCA Civ 309, the Court touched on the duty of the legal assessor, whose duties are set out in the General Medical Council (Legal Assessors) Rules 2004. Rule 2 clearly means a duty on the part of the legal assessor to take active steps if he or she considers that any procedural or legal problems of importance may be arising. Such a duty might arise if the legal assessor were to feel that there was a serious abuse of process, or an evidential problem on such a scale that he or she felt no reasonable panel could find the charges proved, and yet the appropriate arguments were not being advanced by the doctor's legal representatives. However, the circumstances would have to be very clear for a Court to consider intervening on the basis that the legal adviser had not done so.

### *Compton v. General Medical Council* [2008] EWHC 2868 (Admin)

**44.19** Absence of practitioner—advice to panel—extent of legal assessor's advice—whether proceedings fair

At the time of the hearing before the fitness to practise panel, on 14 and 15 June 2007, C was resident abroad. The hearing proceeded in his absence. He was aware of the hearing, submitted evidence and representations to the panel, and decided not to attend. His choice was driven, he informed the panel, by the lack of the plane fare to bring him to the United Kingdom. C made no complaint that the panel proceeded in his absence. His complaint on appeal was that, in his absence, his interests should have been better protected by elucidation to the panel of the weaknesses in the GMC's case. It was common ground that the panel was not advised by its legal adviser in the terms referred to by Rose LJ in *R v. Hayward, Jones and Purvis* [2001] QB 862, CA, at [22], point (6): see paragraph 1.02. Pitchford J said that, with necessary adaptions for the nature of the proceedings, it was agreed that these are the principles on which any tribunal considering such matters should act. They apply to a hearing before a fitness to practise panel as they do to a criminal trial. The question that the Court had to consider was whether, in the absence of such advice, the panel in the instant case was aware of its obligations to search for points favourable to C, reasonably available on the evidence, and fairly undertook that obligation. In dismissing C's appeal, the Court said that the legal assessor did explicitly draw the panel's attention to two letters in terms that could have left the panel in no doubt of their relevance in considering C's defence. On two occasions, the chairman, in asking questions, stated explicitly that the reason she was asking particular questions was that they might be asked if the doctor were present or were represented at the hearing. The legal assessor advised the panel that it should not hold against C the fact that he had not given evidence in support of his case and counsel for the GMC stated that no adverse inference should be drawn from C's failure to attend the hearing. Pitchford J said, at [33], that the legal assessor is not a judge; he is an adviser to the panel on matters of law. The legal assessor's duty as a legal adviser embraces the responsibility to inform the panel of the need for vigilance in circumstances such as these—namely, in the absence of the doctor identifying points that might be of assistance to him. It does not embrace a need to sum up the evidence to the panel. In the instant case, notwithstanding the absence of any specific reference to Rose LJ's judgment in *Jones*, the proceedings before the panel were manifestly fair and conscientiously conducted, and the panel did everything reasonable to ensure that anything that might assist C was not missed.

### *Chan Hei Ling Helen v. Medical Council of Hong Kong* [2009] 4 HKLRD 174

**44.20** Panel deliberations—presence of legal assessor

The appellant sought to challenge the role of the legal assessor in being present during the panel's deliberations. The Court rejected this argument and stated that the presence of the legal assessor during the panel's deliberations was 'desirable in order to ensure, for example that its members do not inadvertently take account of irrelevant matters'. If he were required to remain outside the retiring room, but be available to give advice only if called upon to do so, inadvertent errors of law would go uncorrected, for the members of the GMC would be unaware of the need for legal advice.

### *Dzikowski v. General Medical Council* [2009] EWHC 1090 (Admin)

**44.21** Unrepresented doctor—expert witness—advice by legal assessor

D complained that, at a review hearing spread over eight days between 21 May and 11 July 2008, at which D was unrepresented, the panel failed properly to explain to him the potential consequences of not calling his expert witness, Dr B, to attend to give oral evidence and to be cross-examined. D attended with his mentor Dr Z, a consultant psychiatrist, and the panel had regard to the expert evidence of Dr S called by the GMC, together with various other documents and reports, including those written by D's expert, Dr B, in November/December 2007. D alleged that the panel's failure struck at the heart of the fairness of the proceedings, given the importance of expert evidence in the GMC's case against him. In dismissing D's appeal, Cox J reviewed the transcript of the hearing. The legal assessor warned D expressly that if Dr B were not called, the weight to be attached to his report would necessarily be reduced. D indicated that he understood that, sought no adjournment, and proceeded on that basis, asking the panel to consider Dr B's written report and to take it into account.

### *R (Roomi) v. General Medical Council* [2009] EWHC 2188 (Admin)

**44.22** Allegation—matters outside notice of inquiry—need for legal assessor to step in—lack of advice by legal assessor—appeal allowed

R practised as a cosmetic surgeon in private practice. In the summer of 1999, two consultant plastic surgeons made complaints to the GMC, questioning R's competence. As a result, a performance assessment was carried out in 2003, which was generally favourable save for two particular areas. A second performance assessment was held in January 2006, which reported that R's performance had been deficient in some areas, but that these could be cured by remedial action. No comments were made by the assessors in relation to audit assessment or appraisal. The notice of inquiry related to the practical tests, which fell below an acceptable level. In opening the case, counsel for the GMC accepted that the case was indeed based solely on the deficiencies identified by the assessment panel. It was not suggested that R's standards of professional performance were deficient because of a failure to participate in audit or appraisal. In allowing R's appeal against conditions imposed on his registration by the panel, Collins J noted that questions were asked of R by members of the panel that appeared to be relevant only to the question of audit and appraisal. No doubt if members of a panel feel concern based on the material before them on issues that are not contained in the notice of hearing, they are entitled to raise them. The learned judge continued:

> 18. But they ought to have been advised that they could not properly rely on them unless they did form part of the allegations made against the practitioner and so they could not properly be taken into account against him unless there was the necessary amendment to the notice.
>
> 19. It is perhaps in the circumstances of this case to be noted that it is surprising that the legal assessor did not spot the failure to comply with the proper practice that was occasioned by the panel. It is elementary, quite apart from the question as to whether they should be contained in the notice, that any allegations which are going to be relied on against a particular person must be put to him so that he is able to deal with them. That did not happen in this case because, as is conceded, the whole hearing was on the basis that what was in issue was his skill, nothing else. It seems to me that in the circumstances the legal assessor ought, if necessary of his own volition, to have made it clear to the committee, giving them proper advice,

that they must comply with the rules and they must not do what they apparently were intending to do.

20. The legal assessor is an independent person. He is not there to assist the committee in giving advice that they want to hear, he is there to assist the committee by indicating to them what they can and what they cannot in his view do, and, if they find the advice in a particular case unpalatable, so be it, but they must follow it, unless of course there is good reason to believe, after perhaps hearing submissions by counsel involved in the case, that particular advice is not correct. But, short of that, if he advises them as he should, it will avoid problems such as have arisen in this case, namely the need to appeal after a lengthy and costly hearing and the inevitable allowing of the appeal on the basis of breach not only of the rules but also of natural justice and, as I say, I am somewhat surprised that the legal assessor did not in these circumstances step in. If he had, it could have avoided the problems created by the need to appeal. As it is, this appeal must be allowed.

### *Fish v. General Medical Council* [2012] EWHC 1269 (Admin)

**44.23** Advice to panel—prior discussion with parties' representatives—*Lucas* direction on lies

Foskett J observed that, from the analysis of the role of the legal assessor in *Gopakumar v. General Medical Council* [2008] EWCA Civ 309, the advice of a legal assessor is not to be equated with the directions given in a summing up by a judge in a jury trial. Even in that latter context, a formulaic approach is not encouraged: any summing up should be fashioned to the particular case. By way of a loose analogy, it must equally be the case that the advice given by a legal assessor must also be fashioned to the particular case being considered by the fitness to practise panel. The contents of that advice would often be a matter for discussion between the legal assessor and counsel (in the absence of the panel), and that discussion ought to be encouraged: problems can be ironed out and minds concentrated on the real issues of concern. Ultimately, of course, the legal assessor must give the advice that he or she considers helpful to the panel in the context of the circumstances of the case being considered by the panel. It is a responsible, and not always an easy, task. In the instant case, the legal assessor's advice included reference to the kind of direction given to a jury about lies told by a defendant in a criminal case along the lines set out in the well-known case of *R v. Lucas* (1981) 73 Crim App Rep 159. The learned judge continued:

86. ...However, the essential issue for the Panel was to determine whether the Appellant was telling the truth about the declarations of truth and I cannot disguise the fact that I have some reservations about how helpful that advice would have been in the circumstances.... It is just possible to envisage an "innocent" reason for the Appellant having deleted the declarations and then lying about it: perhaps a genuine belief that, given the period of over 5 years of never having been required to give such a declaration, it was either not necessary or that he was not sufficiently confident about the internal processes either of [the locum agency] or the hospital where he was working such that a minor discrepancy in the hours claimed as against the hospital records would not be turned into a major issue. Obviously, he said no such thing, his case simply being a complete denial of the deletions. However, it is always necessary (as the *Lucas*-style advice requires) for possible innocent explanations for an untruth having been told to be considered when assessing the implications of that untruth. Plainly, the desirability for the advice to be given in a case needs to be considered in the context of the case being heard by the Panel, but where advice based on *Lucas* is given, it needs to be relevant to the issues in the case and fashioned accordingly.

87. ...I would commend the practice of a full discussion between Counsel and the Legal Assessor (in the absence of the Panel) (a) before any advice based on *Lucas* is given and (b) where such advice is to be tendered, on the issues of the precise terms of that advice.

### *R (Mafico) v. Nursing and Midwifery Council* [2014] EWHC 363 (Admin)

**44.24** Alleged misdirection—was there misdirection?—was it of sufficient significance to result to invalidate decision?

M, a band 5 nurse, accepted a caution in respect of the theft of unspecified quantities of Amitriptyline and Tramadol, which she admitted taking from the ward at the hospital where she was employed. The Conduct and Competence Committee (CCC) of the Nursing and Midwifery Council (NMC) imposed a sanction of striking off. M's health had deteriorated during the period of the offences and she had attempted suicide by an overdose of quinine sulphate tablets, which she had purchased online. M appealed, claiming that the legal assessor erred when commending to the Committee the application of the decision in *Solicitors Regulation Authority v. Sharma* [2010] EWHC 2022 (Admin), which gave the Committee the erroneous impression that striking off would be appropriate in all cases of dishonesty absent exceptional circumstances, and that the present case was on all fours with *Hassan v. General Optical Council* [2013] EWHC 1887 (Admin). In dismissing M's appeal, His Honour Judge Gore QC (sitting as a judge of the High Court) said, at [18], that, in alleged misdirection cases, two quite distinct questions must be addressed based on *Libman v. General Medical Council* [1972] AC 217—namely, first, was there a material misdirection, and if so, secondly, was it of sufficient significance to the result to invalidate the decision? Whilst *Hassan v. General Optical Council* was persuasive and weight should be given to it, Leggatt J did not ask, or then answer, the second question posed in *Libman*—namely, was any misdirection of sufficient significance to the result to invalidate the decision? Moreover, there were several points of distinction between *Hassan* and this case that were highly relevant. They included, first, that the qualification to *Sharma* expressed in *Parkinson v. Nursing and Midwifery Council* [2010] EWHC 1898 (Admin), which was not referred to the panel in *Hassan*, was referred to the Committee in this case. Secondly, the matters taken into account by the Committee were much more carefully and extensively set out in the decision letter than they were in *Hassan*.

### *Sharma v. General Medical Council* [2014] EWHC 1471 (Admin)

**44.25** Dishonesty—standard of proof—balance of probabilities—advice to panel

His Honour Judge Pelling QC (sitting as a judge of the High Court) said, at [19], that, in *Re H (Minors) (Sexual Abuse: Standard of Proof)* [1996] AC 563, 586, Lord Nicholls observed that in assessing the balance of probabilities, the more serious the allegation, the less likely it is that the event occurred and hence the stronger should be the evidence before a court concludes that the allegation is established on the balance of probabilities. His Lordship agreed that the principle identified by Lord Nicholls applies in relation to allegations of dishonesty. It was noteworthy that the legal assessor did not draw the attention of the panel to this point, which is one to which all fact-finding tribunals applying the civil standard of proof are bound to have regard when considering an allegation of dishonesty. He ought to have done so. However, on the facts of this case, it might have made a difference in relation to only one of the allegations of dishonesty and was immaterial to the others.

## D. Other Relevant Chapters

**44.26**

In Camera Discussions, Chapter 34
Independent and Impartial Tribunal, Chapter 35
Indicative Sanctions Guidance, Chapter 36
Unrepresented Practitioner, Chapter 67

# 45

# LEGAL REPRESENTATION



| *Maynard v. Osmond* [1977] QB 240 | 45.05 | | |
| *R v. Secretary of State for the Home Department, ex parte Tarrant and ors* [1985] QB 251 | 45.06 | *Hameed v. Central Manchester University Hospitals NHS Foundation Trust* [2010] EWHC 2009 (QB) | 45.22 |
| | | *Brew v. General Medical Council* [2014] EWHC 2927 (Admin) | 45.28 |
| *R (SS) v. Knowsley NHS Primary Care Trust; R (Ghosh) v. Northumberland NHS Care Trust* [2006] EWHC 26 (Admin) | 45.13 | *D v. Conduct and Competence Committee of Nursing and Midwifery Council* 2014 SLT 1069 | 45.29 |
| *R (Abdalla) v. Health Professions Council* [2009] EWHC 3498 (Admin) | 45.14 | *Nicholas-Pillai v. General Medical Council* [2015] EWHC 305 (Admin) | 45.30 |
| *Vaidya v. General Medical Council* [2007] EWHC 1497 (Admin) | 45.15 | | |
| *Gopakumar v. General Medical Council* [2008] EWCA Civ 309 | 45.16 | C. Other Relevant Chapters | 45.31 |

## A. Legal Framework

**45.01** Examples include:

General Optical Council (Fitness to Practise) Rules 2013, rule 21 (where an individual registrant is not represented, he may be accompanied and advised by any person who is not giving evidence)

Police (Conduct) Regulations 2012:

regulation 6 (police friend)
regulation 7 (legal and other representation)

General Medical Council (Fitness to Practise) Rules 2014, rule 33 (a doctor may be represented by a solicitor or counsel, a representative of any professional organization of which he is a member, or, at the discretion of the committee or panel, a member of his family or other person; a person who gives evidence at a hearing is not entitled to represent or accompany the doctor to that hearing)

## B. Disciplinary Cases

***Pett v. Greyhound Racing Association Ltd* [1969] 1 QB 125; *Pett v. Greyhound Racing Association Ltd (No. 2)* [1970] 1 QB 67**

**45.02** Greyhound trainer—official inquiry—interlocutory injunction—reputation or livelihood at stake—natural justice

P was a trainer of greyhounds, who held a licence from the National Greyhound Racing Club. On 9 September 1967, P sent one of his dogs to the White City Stadium to compete in a race. Following a urine test, the dog was withdrawn from the race, and subsequent tests showed positive for barbiturates and that the barbiturate appeared to be phenobarbitone. On 16 September 1967, the track stewards notified P that they intended to hold an official inquiry at the stadium. He attended with his solicitor and counsel, and, following an adjournment, was told that he would not be allowed to be legally represented at the inquiry. Cusack J granted an interlocutory injunction restraining the defendants from holding the inquiry unless P were allowed to be represented by counsel. The Court of Appeal (Lord Denning MR, and Davies and Russell LJJ) dismissed the defendants' appeal. Lord Denning MR said that P was facing a serious charge. He was charged either with giving the dog drugs or with not exercising proper control over the dog, so that someone else drugged it. If he were found guilty, he would be suspended or his licence may not be renewed. The charge concerned his reputation and his livelihood. On such an inquiry, he was entitled not only to appear by himself, but also to appoint an agent to act for him. Even a prisoner can act for his friend; it has been held that a ratepayer has a right to have a surveyor to appear for him: *R v. Assessment Committee of St Mary Abbots, Kensington* [1891] 1 QB 378, CA. When a man's reputation or livelihood is at stake, he not only has a right to speak by his own mouth, but also has a right to speak by counsel or solicitor. The dictum of Maugham J in *Maclean v. Workers' Union* [1929] 1 Ch 602 may be correct when confined to tribunals dealing with minor matters in which the rules may properly exclude legal representation, but the dictum does not apply to tribunals dealing with matters that affect a man's reputation or livelihood, or any matters of serious import. Natural justice then requires that he can be defended, if he wishes, by counsel or solicitor. Davies LJ, agreeing, said that it was very curious that there is no express provision in the rules for a holding of an inquiry by the track stewards. At the hearing of P's appeal following trial against the decision of Lyell J: [1970] 1 QB 46, the Court of Appeal was informed by counsel that the Rules of Racing had been revised and that the proposed inquiry into the alleged drugging would now be held under the new Rules of Racing, which permitted P the right to be legally represented. Accordingly, no issue between the parties remained in the present proceedings.

***Enderby Town Football Club Ltd v. Football Association Ltd* [1971] Ch 591**

**45.03** Football club—Football Association Rules, rule 38(b)—not contrary to natural justice

The Leicestershire and Rutland County Football Association, affiliated to the Football Association (FA), heard charges against the Enderby Town Football Club, a limited company, and found that there had been gross negligence in the administration of the club. It fined the club £500. It severely censured the club and its directors. The club appealed to the FA, and sought to be represented by solicitor and counsel at the hearing of the appeal. In upholding the decision of Foster J refusing the club relief, the Court of Appeal (Lord

Denning MR, and Fenton Atkinson and Cairns LJJ) held that rule 38(b) of the FA, on its true construction, barred legal representation for the club on the hearing of its appeal and that the rule was not invalid as being contrary to natural justice. Lord Denning said, at 605D–606A, that whether a party charged before a domestic tribunal is entitled *as of right* to be legally represented depends on what the rules say about it. When the rules say nothing, then the party has no absolute right to be legally represented. It is a matter for the *discretion* of the tribunal. But the discretion must be properly exercised. The tribunal must not fetter its discretion by rigid bonds. It must not make an absolute rule from which it will never depart. That is the reason why the Court intervened in *Pett v. Greyhound Racing Association Ltd* [1969] 1 QB 125. There was nothing in the rule to exclude legal representation, but the tribunal refused to allow it. Its reason was because it never did allow it.

### *Fraser v. Mudge* [1975] 1 WLR 1132

**45.04** Prisoner—hearings before governor or Board of Visitors—Prison Rules, rule 49(2)—not contrary to natural justice

F, a prisoner serving a long sentence of imprisonment, sought a declaration that he was entitled to be represented by solicitor and counsel at the hearing of an inquiry into a charge against him of assaulting a prison officer. Rule 49(2) of the Prison Rules 1964 provided that, at an inquiry into a charge against a prisoner, he shall be given a full opportunity of hearing what is alleged against him and of presenting his own case. The rule said nothing about legal representation. In dismissing F's appeal from the decision of Chapman J refusing F an injunction, the Court of Appeal (Lord Denning MR, and Roskill and Ormrod LJJ) held that *Pett v. Greyhound Racing Association Ltd* was distinguishable. There was in that case a contractual, or quasi-contractual, relationship between the plaintiff and the defendant. The present case arose under the Prison Rules 1964. It is not in every type of case, irrespective of the nature or jurisdiction of the body in question, that justice can neither be done nor be seen to be done without legal representation of the party or parties appearing before the body. There are many bodies before which a party or parties can be required to appear, but which can do justice and can be seen to do justice without the party against whom complaint is made being legally represented. If the argument in relation to rule 49(2) were well founded, it would equally apply to complaints heard by the governor, to which the same language applies. The broad principle underlying the Rules is to make discipline in prison by proper, swift, and speedy decisions, whether by the governor or the Visitors. The requirements of natural justice do not make it necessary that a person against whom disciplinary proceedings are pending should, as of right, be entitled to be represented by solicitor or counsel, or both.

### *Maynard v. Osmond* [1977] QB 240

**45.05** Prison officer—hearing before chief constable—representation by member of police force

The question in this case was whether a police officer, who is accused of a disciplinary offence, is entitled to be represented by counsel or solicitor. Three men were arrested in Alton, Hampshire, accused of being drunk and disorderly, and were taken to the police station. At the station was M, the plaintiff, a young police constable. Next morning, one of the men who had been arrested made a complaint to the chief inspector. He said that he had been assaulted in the cells by other officers. M made a statement to the chief inspector supporting this. The proceedings against the men arising out of the incident in Alton were dismissed and the Director of Public Prosecutions (DPP) decided that there was not sufficient evidence against the police officers to warrant a prosecution for assault. On 3 December 1975, two charges were made against M that he had been asleep on duty, had refused to remain on duty, and that he had supplied a false written statement alleging that the police sergeant had made threats to him when no such threats had been made. Griffiths J dismissed M's application for a declaration that, under the Police (Discipline) Regulations 1965, a chief constable hearing disciplinary proceedings against a member of a police force had a discretion whether or not to permit the accused to be represented at the hearing by a solicitor and counsel of his choice. Dismissing M's appeal and distinguishing *Pett v. Greyhound Racing Association Ltd*, the Court of Appeal (Lord Denning MR, and Roskill and Ormrod LJJ) said that the Regulations require a police officer to be represented by a member of the police force. He cannot be represented by a lawyer. The Regulations for officers of lower rank, compared to the separate Regulations covering disciplinary proceedings for senior officers contained in the Police (Discipline) (Deputy Chief Constable, Assistant Chief Constable and Chief Constables) Regulations 1965 (which expressly permit representation by a solicitor or counsel), lead to the inevitable inference that it is not permitted in other circumstances, including the present. The disciplinary proceedings against M should not be stayed until after the civil action against the police brought by one of the men allegedly assaulted. The present case is very different from *Conteh v. Onslow Fane*, 25 June 1975, Bar Library Transcript No. 291 of 1975, CA. In the present case, the issues in the disciplinary proceedings are very different from those in the civil action. And while great weight will be paid to any findings of the judge in the civil action, they will not be decisive or binding in any way. It is very important that the disciplinary proceedings should be dealt with—so much so, that it would not be proper to stay them.

### *R v. Secretary of State for the Home Department, ex parte Tarrant and ors* [1985] QB 251

**45.06** Prisoner—hearing before Board of Visitors—discretion to allow legal representation

Five convicted prisoners applied for judicial review, each applicant seeking to quash a decision of a prison Board of Visitors. The one ground common to each application was that the applicant, at the hearing by the Board of Visitors of a disciplinary charge against him, was refused legal representation. In quashing the adjudications of the Visitors, Kerr LJ and Webster J held that the Court was bound by *Fraser v. Mudge* to decide that a prisoner is not entitled *as of right* to be legally represented before a Board of Visitors. It does not follow from the decision that a prisoner has no entitlement to legal representation as of right, that the Board before which he appears has no discretion to grant him legal representation. If it has such a discretion, then he has the right that the Board should, in his case, exercise that discretion, and exercise it fairly and properly: *per* Webster J, at 273A–B. At 285–6, it was noted that whether to allow legal representation or to allow the assistance of a friend or adviser, the considerations that every Board should take into account include (the list was not intended to be comprehensive):

(1) the seriousness of the charge and of the potential penalty;
(2) whether any points of law are likely to arise;
(3) the capacity of a particular prisoner to present his own case;
(4) procedural difficulties;
(5) the need for reasonable speed in making their adjudication, which is clearly an important consideration; and
(6) the need for fairness as between prisoners and as between prisoners and prison officers.

### *Campbell and Fell v. United Kingdom* [1984] 7 EHRR 165

**45.07** Prisoner—whether refusal of legal representation unfair—no causal link established

C and F were both convicted prisoners charged with contravening police disciplinary regulations. At a hearing before a Board of Visitors of the prison, they were each denied legal representation. The European Court of Human Rights (ECtHR) held that to require that disciplinary proceedings concerning convicted prisoners should be held in public would impose a disproportionate burden on the authorities of the state, although it does not appear that any steps were taken to make public the Board's decision. The government accepted that, under the law in force at the time, there was no right to legal representation at Board hearings and that, had a request for legal assistance been made, it would have been refused. The Court held that, in relation to the general allegation of unfairness, there was nothing to suggest, nor can it be assumed, that the Board of Visitors would have reached any different conclusions had there been legal assistance or representation. Accordingly, no causal link had been shown to exist between the violations and the alleged damage, with the result that no just satisfaction fell to be awarded.

### *R v. Board of Visitors of HM Prison, The Maze, ex parte Hone; R v. Board of Visitors of HM Prison, The Maze, ex parte McCartan* [1988] AC 379

**45.08** Prisoner—Board of Visitors—no absolute right to representation—particular circumstances

The applicants, both serving prisoners, appealed against decisions of the respective Board of Visitors to refuse them legal representation. Their applications were dismissed by the High Court and the Court of Appeal in Northern Ireland. The House of Lords dismissed their appeals. In giving judgment, Lord Goff of Chieveley (with whom Lord Mackay of Clashfern LC, Lord Bridge of Harwich, Lord Ackner, and Lord Oliver of Aylmernton agreed) said that no doubt it is true that a man charged with a crime before a criminal court is entitled to legal representation and no doubt it is also correct that a Board of Visitors is bound to give effect to the rules of natural justice. But it does not follow that, simply because a charge before a disciplinary tribunal such as a Board of Visitors relates to facts that in law constitute a crime, then the rules of natural justice require the tribunal to grant legal representation. If this were the case, then, as Roskill LJ pointed out in *Fraser v. Mudge*, exactly the same submission could be made in respect of disciplinary proceedings before the governor of a prison. It is difficult to imagine that the rules of natural justice would ever require legal representation before the governor. But although the rules of natural justice may require legal representation before a Board of Visitors, there is no basis that they should do so in every case as of right. Everything must depend on the circumstances of the particular case, as amply demonstrated by the circumstances so carefully listed by Webster J in *R v. Secretary of State for the Home Department, ex parte Tarrant* as matters that Boards of Visitors should take into account. But it is easy to envisage circumstances in which the rules of natural justice do not call for representation, even though the disciplinary charge relates to a matter that constitutes in law a crime, as may well happen in the case of a simple assault in which no question of law arises and in which the prisoner charged is capable of presenting his own case. In the instant case, both appellants were charged with assaulting prison officers. The absolute right to legal representation claimed by the appellants is not required by the European Convention on Human Rights (ECHR), in such cases as *Engel v. The Netherlands (No. 1)* (1976) 1 EHRR 647 and *Campbell and Fell v. United Kingdom* (1984) 7 EHRR 165, any more than it is required by the English law.

### *Pine v. Law Society* [2001] EWCA Civ 1574

**45.09** Solicitor—disciplinary hearing—legal aid and representation—whether breach of Article 6 ECHR

The Solicitors Disciplinary Tribunal (SDT) found P guilty of 'a catalogue of dishonest behaviour' and ordered that he be struck off the roll. P's appeal was dismissed by the High Court and, on appeal to the Court of Appeal, the following was submitted.

(1) In principle, the absence of any provision for legal aid or representation by a solicitor who wants, but cannot afford to pay for, it, because of the effect of the Law Society's actions, vitiates all hearings of the Tribunal that may lead to an order that the solicitor be struck off.

(2) On the facts of this case, the absence of legal advice or representation for P rendered the proceedings before the Tribunal obviously unfair.
(3) P's inability, because he could not afford the fare to London, to attend the hearing in person also rendered the proceedings before the Tribunal obviously unfair.

In dismissing P's appeal, the Vice-Chancellor (with whom Buxton and Arden LJJ agreed) said that the question whether and when legal aid and assistance should be provided in civil proceedings has been raised in a number of reported cases: *Airey v. Ireland* [1979] 2 EHRR 305; *X v. United Kingdom* (1984) 6 EHRR 136; and *R v. Legal Services Commission, ex parte Jarrett* [2001] EWHC Admin 389. The principle instilled is that only in exceptional circumstances—namely, where the withholding of legal aid would make the assertion of a civil claim practically impossible or where it would lead to obvious unfairness of the proceedings—can such a right be invoked by virtue of Article 6(1) ECHR. Counsel for P did not dispute that the disciplinary proceedings were civil for the purposes of Article 6. It is clear that the requirements of Article 6, with respect to legal advice and representation, depend on the facts of any given case. Thus if P could show on the facts of this case that legal advice and representation for the purpose of the disciplinary proceedings before the Tribunal was required by Article 6, then he did not have to rely on any more abstract principle. P did not claim that the withholding of legal aid made his defence to the allegations practically impossible. He contended that it led to obvious unfairness, not because of any complexity in the procedure or the facts of the case, but because of the nature of the charges, the severity of the possible consequence for him, and the inhibiting effect on the prosecution of his defence of emotional involvement. It was fanciful to suggest that it was unfair to P not to provide him with legal advice or representation. The procedure was not complex. The relevant facts were within P's knowledge. There may be circumstances in which the fact that a party is both unrepresented and absent from the hearing will amount to a denial of effective access to the Court and therefore a breach of Article 6. But it will depend on the facts of the case. In this case, P had ample opportunity to outline any defence to the allegations that he might have. When he found that he could not afford the fare to London, he might have sought an adjournment, the admission of his evidence by affidavit, or some more informal means or a hearing in Manchester. He did none of these things. Nor, after the event, did he invite the Tribunal to reopen the hearing.

### *Awan v. Law Society* [2003] EWCA Civ 1969

**45.10** Solicitor—representation before SDT

Following the decision in *Pine v. Law Society*, the Master of the Rolls (Lord Phillips) said that A had only himself to blame for the fact that the SDT proceeded in his absence. The allegations that were made against him were of stark simplicity. He did not require legal representation to answer them, either before the Tribunal or before the Divisional Court. There was no infringement of Article 6 ECHR. May and Carnwath LJJ agreed.

### *R (Thompson) v. Law Society* [2004] 1 WLR 2522

**45.11**

See paragraph 20.04.

### *R (Aston) v. Nursing and Midwifery Council* [2004] EWHC 2368 (Admin)

**45.12** Allegation of incompetent representation—ground of appeal—test—whether decision unjust

The appellant, a registered general nurse, appealed against the decision of the Nursing and Midwifery Council (NMC)'s Professional Conduct Committee (PCC) to remove his name from the register of nurses. The case against the appellant consisted originally of five charges. The evidence depended on two care assistants and the evidence of the operations manager. The case depended upon the credibility of the care assistants, on the one hand, and the credibility of the appellant, on the other. The appeal was advanced on the ground that the conduct of the case presented on behalf of the appellant by a non-qualified advocate, appointed by the Royal College of Nurses, was so incompetent and so inadequate as to amount to a serious irregularity, and that this irregularity affected the safety of the conclusion of the Committee. Moses J said that it was not contended by the appellant that the mere fact of the incompetence of the advocate would be sufficient to enable the Court to allow the appeal and order a rehearing. Both the appellant and the respondent agreed that, in the instant case, the approach of the Court should be that which is applied by the Court of Appeal, Criminal Division, when complaints are made as to the incompetence of the representation. The learned judge said:

> 8. The approach of that court is exemplified in two decisions: *R v. Bolivar* [2003] EWCA Crim 1167 and *R v. Day* [2003] EWCA Crim 1060. In *R v. Bolivar* the Vice President at paragraph 2 stated the test as *Wednesbury* unreasonableness and such as to affect the fairness of the trial.
>
> 9. In *R v. Day*, the test was posed in the following way: "(Incompetent representation) cannot in itself form a ground of appeal or a reason why a conviction should be found to be unsafe. We accept that, following the decision of this court in *R v. Thakrar* [2001] EWCA Crim 1906, the test is indeed the single test of safety, and that the court no longer has to concern itself with intermediate questions such as whether the advocacy had been flagrantly incompetent. But in order to establish lack of safety in an incompetence case, the appellant has to go beyond the incompetence and show that the incompetence led to identifiable errors or irregularities in the trial, which themselves rendered the process unfair or unsafe."
>
> 10. In the context of part 52, rule 11, the test is not safety. The appellant need not show that the decision was wrong, but he must show that the decision was unjust. The decision will only be unjust if the incompetence led to irregularities which rendered the process of the trial unfair or the conclusion unsafe.
>
> 11. However, in the case before me both sides agree that the court should not allow the appeal unless the incompetence was of such a degree as to be described as *Wednesbury* unreasonable. That concept is not easily applied to the question of the incompetence of an advocate, but I take the Vice President's reference to *Wednesbury* unreasonable to mean that the conduct of the advocate must be such that he or she took such decisions and acted in a way which no reasonable advocate might reasonably have been expected to act.
>
> 12. But that by itself, as I have said, is not enough. It must further be shown that that wholly inadequate conduct did affect the fairness of the process. Only then could the conclusion of the committee be shown to be unjust.

Looking at the complaints as a whole and cumulatively, Moses J said, at [59], that he had reached the conclusion that the representation of the case was so incompetent that it fell below that to which the appellant was entitled. Certain courses adopted were an approach that no reasonable advocate would have foreseen. The question then remained as to whether that level of incompetence rendered the process unfair. This was an adversarial process conducted according to the rules of a criminal trial. It could have, and did have, dire consequences for the appellant. In dismissing the appellant's appeal, the learned judge said that, in this case, the determination depended on which of the witnesses the Committee believed. The PCC found the facts proved. They were simple and clear issues of fact: at [54]–[66]. Moreover, the legal adviser did intervene to prevent further damage from being done to the appellant and gave specific advice to the Committee. The Committee must be assumed to have taken that advice and determined the case on the evidence. The incompetence of the representative was not such as to thwart a fair resolution of the conflict of evidence between the care assistants and the operations manager, on the one hand, and the appellant, on the other. Reading the appellant's evidence in full, it is quite apparent that he was able to give a full account of his defence. He was disbelieved, and the process of resolving those issues of fact and credibility was not rendered unfair by the failures of the appellant's representative: at [68]–[71].

### *R (SS) v. Knowsley NHS Primary Care Trust; R (Ghosh) v. Northumberland NHS Care Trust* [2006] EWHC 26 (Admin)

**45.13**

Doctor—NHS performers list—removal from list—whether doctor can reasonably be expected to represent himself or herself

These cases concerned the procedure to be adopted by a primary care trust (PCT) when considering removal of a general practitioner (GP) from its performers list. Regulation 10 of the National Health Service Performers Lists Regulations 2004 provides that, where a PCT is considering removing a performer from its performers list, it shall give him the opportunity to put his case at an oral hearing before it. In August 2004, the Department of Health published advice to PCTs, which included advice on legal representation at hearings. The advice stated that these were internal proceedings and not a quasi-judicial hearing, and that there was no right to legal representation on the part of either the PCT or the doctor. In August 2004, the PCT informed Dr S that it intended to proceed with an investigation to determine whether action should be taken to remove his name from the performers list. In June 2005, notice was given to Dr G, proposing to remove his name from the PCT's performers list on the grounds of both suitability and efficiency. Both doctors requested to be permitted to have legal representation, which was refused. In granting relief to each claimant, Toulson J held that the existence of a statutory right of appeal to the Family Health Services Appeal Authority did not preclude the Court from considering prospective unfairness at the initial hearing:

> 68. It cannot be in accordance with the spirit of the Convention or the common law that the court should be powerless to prevent a violation of a right to a fair procedure, merely because of the existence of a later way of remedying the consequences. A stitch in time may save nine.

In the case of Dr S, the central allegations were that he indecently assaulted four patients. The core issue was a stark one of credibility. The panel would be in a far better position to reach a fair judgment whether the complaints were true if it heard from the complainants and Dr S, and their stories were tested, than if the panel's evaluation of the witnesses' credibility were based on their untested statements. If there were to be cross-examination of the complainants, that would be a powerful reason for permitting legal representation, in order to avoid the complainants being cross-examined by Dr S himself.

The issues in the case of Dr G were diverse and complex. The allegations of clinical shortcomings were specific and particularized, and were based on interviews with district nurses, who had not provided written statements. On the subject of legal representation, the fundamental question was whether the doctor could fairly be expected to represent himself. In many cases, that may be a quite reasonable expectation. In Dr G's case, none of the allegations made against him were individually complicated, but taken together the case was sufficiently complex and it would be surprising if a doctor could do himself justice in trying to handle such a case unrepresented. The Department of Health's advice was liable to be understood in the sense that the Regulations did not give the doctor a general right of legal representation. That was, however, going too far. It may be that, in many cases, legal representation would be unnecessary, but the question in each case must be whether the doctor can reasonably be expected to represent himself or whether legal representation is necessary in order to enable him to be able to present his case properly.

***R (Abdalla) v. Health Professions Council* [2009] EWHC 3498 (Admin)**

**45.14** See paragraph 1.07.

***Vaidya v. General Medical Council* [2007] EWHC 1497 (Admin)**

**45.15** See paragraph 3.10.

***Gopakumar v. General Medical Council* [2008] EWCA Civ 309**

**45.16** Representation at hearing—appeal—challenge to conduct of representation

The practitioner faced a charge of serious professional misconduct relating to a female patient. At the hearing, he did not attack the credibility of her evidence. The doctor was represented by experienced counsel and solicitors, and they had taken an informed and understandable decision not to deploy evidence relating to her history of drug use. The Court of Appeal (Sir Anthony Clarke MR, and Tuckey and Jacob LJJ) said that it was too late for the practitioner to seek to rely on such matters on appeal once he realized that he had lost on all other points that had been taken on his behalf. He had chosen not to attack the character of the witness based upon her medical history and drug use, and it was too late for him to attempt to do so on appeal.

***R (Tinsa) v. General Medical Council* [2008] EWHC 1284 (Admin)**

**45.17** See paragraph 3.12.

***R (Sinha) v. General Medical Council* [2008] EWHC 1732 (Admin)**

**45.18** Representation at hearing—appeal—challenge to conduct of representation

The decision of the Court of Appeal in *Gopakumar* arose for consideration in different circumstances in *Sinha*. In this case, new counsel instructed on behalf of the practitioner, S, was highly critical of previous counsel and alleged that he should have applied for a stay of the proceedings. Irwin J said:

> 55. . . . When considering whether or not to admit fresh evidence, the general rule is that failure to adduce evidence by the party's legal advisers provides no excuse even in this type of case.
>
> 56. In my judgment, the analogy with [S]'s case is very close. The law forbids reopening a case on the basis, presumed to be true for the purpose of the argument, that there was a negligent failure to adduce what might be crucial evidence in medical disciplinary proceedings, evidence with the potential to alter the outcome of the case. If that is correct, then the court will be very slow to permit a case to be reopened when the lawyers have failed to make a coherent application for a stay.
>
> 57. In *Gopakumar* the court also touched on the duty of the legal assessor, whose duties are set out in the General Medical Council (Legal Assessors) Rules 2004. By rule 2 the legal assessor is required to advise the panel on any question of law referred to him, but he is also enjoined to intervene to advise the panel where there is a possibility of a mistake of law being made, or where he learns of any irregularity in the conduct of the proceedings. This clearly means a duty actively to take steps if the assessor considers that any procedural or legal problems of importance may be arising. For present purposes I am content to accept that such a duty might arise if an assessor felt that there was a serious abuse of process, or an evidential problem on such a scale that he felt no reasonable panel could find the charges proved and yet the appropriate arguments were not being advanced by the doctor's legal representatives. However, the circumstances would have to be very clear for a court to consider intervening on the basis that the legal adviser had not done so.

### *R (Dutt) v. General Medical Council* [2009] EWHC 3613 (Admin)

**45.19** Doctor dispensing with representation at hearing—adjournment granted by tribunal—procedurally not unfair

During the hearing of this case before the fitness to practise panel, which proceeded over fourteen days, D had counsel. After all of the evidence had been heard by the panel, D and his legal representatives parted company. Cranston J held that D provided no evidence of his allegations of incompetent or inadequate legal representation. When he dispensed with the assistance of his legal representatives, he was given a lengthy adjournment. That adjournment involved a weekend and also two-and-a-half days of the following week. The learned judge found that he had sufficient time in which to prepare his submissions in relation not only to the findings of fact, but also to fitness to practise and sanction. D in fact adduced further evidence at that latter stage, from some six witnesses, and he made submissions. There was nothing in his view, said the judge, that was procedurally unfair in the way in which the panel went about the hearing of this matter.

### *Kulkarni v. Milton Keynes Hospital NHS Foundation Trust* [2010] ICR 101

**45.20** Hospital Trust—Department of Health policy document—contractual arrangements—effect on employment in NHS

K, a hospital doctor, was held to be entitled to legal representation in disciplinary proceedings brought by his employer. On 31 July 2007, K entered employment with the Trust as a foundation year one doctor. On 24 August 2007—that is, less than four weeks after his employment with the Trust commenced—a patient made a complaint about K's unprofessional conduct. K was suspended pending investigation of the complaint. The Trust's position was that its procedures were based upon the Department of Health policy document *Maintaining High Professional Standards in the Modern NHS*, dated February 2005, which did not permit legal representation at disciplinary hearings. In allowing K's appeal, Smith LJ (with whom Sir Mark Potter P and Wilson LJ agreed) said that K was and is contractually entitled to be presented at his disciplinary hearing by a lawyer instructed by the Medical Protection Society. Properly construed, the Department of Health policy document permits a practitioner to be represented by a legally qualified person, employed or retained by a defence organization. Had it been necessary to make a decision, the judge said that she would also have held that Article 6 ECHR is engaged where a National Health Service (NHS) doctor faces charges that are of such gravity that, in the event that they are found proved, he will be effectively barred from employment in the NHS.

### *Christian v. Nursing and Midwifery Council* [2010] EWHC 803 (Admin)

**45.21** Representation at hearing—appeal—challenge to conduct of representation—analogy with criminal cases

C, a former registered nurse, appealed against a decision of the NMC to strike her off as a registered nurse. C said that her representative at the hearing before the Conduct and Competence Committee (CCC) was told by her that she had attended a specialist course on drug prescription, but he never passed that on to the Committee. Had the Committee been aware of C's training, the result on impairment and sanction would have been different. C's counsel on appeal submitted that there had been incompetence on the part of C's representative at the original hearing, amounting to a procedural irregularity. In dismissing C's appeal, His Honour Judge Roger Kaye QC agreed that the relevant principles, as to which there was no disagreement, are set out in *R (Aston) v. Nursing and Midwifery Council* [2004] EWHC 2368 (Admin): see paragraph 45.12. In the instant case, the Court held that, overall, C had a fair hearing. Nor could the Court detect in her representatives' approach and conduct of the case anything that might be said to indicate that either of them acted in a way that no reasonable advocate might be expected to act.

***Hameed v. Central Manchester University Hospitals NHS Foundation Trust* [2010] EWHC 2009 (QB)**

**45.22** See paragraph 32.23.

***R (Puri) v. Bradford Teaching Hospitals NHS Foundation Trust* [2011] EWHC 970 (Admin)**

**45.23** See paragraph 32.24.

***R (G) v. Governors of X School (Secretary of State for the Home Department and anor intervening)* [2012] 1 AC 167**

**45.24** School governors—disciplinary hearing—denial of legal representation—no effect on right to teach—Article 6 ECHR not engaged

The claimant, who was employed as a teaching assistant at a primary school, faced disciplinary proceedings for having allegedly formed an inappropriate relationship with a 15-year-old boy undergoing work experience at the school. The claimant requested permission of the defendant school governors for his solicitor to represent him at the hearing before a disciplinary committee of three governors. The defendants refused. The disciplinary committee summarily dismissed the claimant and, at his appeal against the dismissal decision, he again requested that his solicitor attend. The request was refused by the defendants on the same grounds and the claimant's dismissal was confirmed by the appeal committee. The claimant's dismissal was referred to the Independent Safeguarding Authority (ISA) set up under the Safeguarding Vulnerable Groups Act 2006, by which the Secretary of State could determine whether an individual should be prohibited from working with children in educational establishments. The claimant brought judicial review proceedings on the grounds that he had been denied legal representation before the school governors and that the decision was likely to have a substantial effect on the decision of the ISA. The Supreme Court, reversing the decision of the Court of Appeal and the judge at first instance, held that it was common ground that the civil right with which the Court was concerned was the claimant's right to practise his profession as a teaching assistant and to work with children generally. There was no doubt that this right would be directly determined by a decision of the ISA to include him in the children's barred list. However, the disciplinary proceedings before the defendant school governors did not engage Article 6(1) ECHR. The test adopted by the Court of Appeal—namely, that the claimant might enjoy Article 6 procedural rights if the decision in the disciplinary proceedings would have a substantial influence or effect on the determination by the ISA of his civil right to practise his profession—was an appropriate test, but that since the ISA was required to make its own findings of fact and to bring its own independent judgment to bear as to the seriousness and significance of the allegations, and since there was no reason to hold that it would be influenced by the governors' opinion, Article 6(1) did not apply to the governors' disciplinary hearings.

***Mattu v. University Hospitals of Coventry and Warwickshire NHS Trust* [2012] 4 All ER 359, CA**

**45.25** See paragraph 32.28.

***R (King) v. Secretary of State for Justice; R (Bourgass and Hussain) v. Secretary of State for Justice* [2012] 4 All ER 44, CA**

**45.26** See paragraph 35.19.

***Rich v. General Medical Council* [2013] EWHC 1673 (Admin)**

**45.27** See paragraph 3.18.

***Brew v. General Medical Council* [2014] EWHC 2927 (Admin)**

**45.28** Representation at hearing—allegation of dishonesty—advice given by barrister—challenge on appeal

The appellant was a doctor employed at Leicester Royal Infirmary, having previously been employed at Kettering General Hospital. On 8 June 2011, he received an email from the head of the East Midlands (South) Postgraduate School of Medicine, advising him that his e-portfolio, containing records of various medical assessments evidencing his experience and training, was inadequate. Between 13 and 22 June 2011, the appellant falsified eighteen clinical assessment entries on his e-portfolio, giving the impression that the named assessors had been involved in the completion of those entries, when they in fact had not. At the appellant's annual review on 1 July 2011, the appellant falsely informed the members of the review that he had taken a week off work to visit the consultants who had been involved in the completion of the entries in relation to his e-portfolio. Subsequently, when he realized that he was about to face disciplinary proceedings before the General Medical Council (GMC), the appellant approached a local solicitor, who instructed a local barrister, neither of whom had experience of hearings before the fitness to practise panel. The appellant contended that he told his solicitor that he intended to accept all of the charges brought against him. On the first morning of the panel hearing, he met with his barrister, who advised him that if he were to accept that he had been dishonest, then his fitness to practise would automatically be impaired and he would not be able to explain to the panel what had happened. He was therefore advised to deny the allegation of dishonesty, but admit the facts to the remaining charges. The panel found all of the charges proved, including dishonesty, and directed that the appellant's name be erased from the medical register. The amended grounds of appeal were that the incorrect advice given by defence counsel affected the overall fairness of the hearing and the disposal of the case. The appellant contended that the panel's impression and assessment of him was detrimentally affected by the way in which he had denied the most serious charge relating to dishonesty, in that the approach his barrister had taken had prejudiced his best point—namely, that he was wholly regretful and had full insight into the nature of his wrongdoing. In dismissing the appellant's appeal, His Honour Judge Gosnell (sitting as a judge of the High Court) said that, in criminal proceedings, where an allegation is made of incompetent or inadequate representation, privilege is normally waived and a statement is obtained from the advocate concerned to prove what advice was actually given to the client. This had not been done in this case. The judge said that he had not heard any oral evidence in the appeal and so, to deal with it fairly, he intended to assume that the appellant's evidence about what he was advised was accurate. That was the only fair way in which to deal with an issue where the only evidence is in writing and there is no other evidence to gainsay it. If the appellant had been advised that the only way in which evidence of his explanation for his conduct could go before the panel was to dispute dishonesty, then this would clearly be negligent advice and *Wednesbury* unreasonable. The only issue would then be whether the consequences of this advice led to the trial being unfairly conducted and the conclusion unjust. However, the appellant gave evidence for a second time at the sanctions stage of the hearing. He told the panel that he had not really wanted to say what he said earlier in evidence about dishonesty and that he had disputed dishonesty because of legal advice that he had been given. He was reminded that the content of the advice was privileged and the panel went into no more detail. In its determination, the panel recorded that the appellant had given additional evidence at the sanctions stage, in which he stated that he knew that his actions had been dishonest, but he did not make that admission prior to the third stage of the hearing because of the legal advice that he had received. His Honour Judge Gosnell said, at [26], that this passage showed that the panel did record and accept the evidence that the appellant had given about his reluctance to concede dishonesty being based on legal advice, and that the panel was prepared to accept and take into account the explanation put forward by the appellant about his unwise decision to contest dishonesty. The panel did accurately assess the extent of the appellant's insight into his behaviour. Even ignoring the fact that the appellant did not admit dishonesty at the start of the hearing as he should have, there were a number of features of the evidence that the panel was entitled to take into account in concluding that the appellant had less than full insight into his wrongdoing: he had not admitted his full culpability in his reflective log or in either of his witness statements; he had sought to blame technical difficulties with assessing the website when other solutions must have been available; he sought to suggest that all of the entries were based on real cases, when the evidence showed that some of them were not; and he claimed that he had not inflated his grades, when the evidence showed that he probably had in some respects. In addition, the panel was entitled to take into account his evidence as a whole and his demeanour when giving evidence. There was more than adequate material for the panel to conclude that the appellant had less than full insight into the seriousness of his actions.

### *D v. Conduct and Competence Committee of Nursing and Midwifery Council* 2014 SLT 1069

**45.29** See paragraph 67.12.

### *Nicholas-Pillai v. General Medical Council* [2015] EWHC 305 (Admin)

**45.30** The appellant claimed that his counsel was so incompetent that the hearing was unjust. In dismissing the appellant's appeal, Elisabeth Laing J said:

Representation at hearing—appeal—challenged to conduct of representation

> 74. A ground of appeal relying on the incompetence of counsel involves an undesirable form of satellite litigation which is to be discouraged (*R (B) v. Hampshire County Council* [2004] EWHC 3193 (Admin) at paragraphs 69–70, per Hughes J). It is no doubt for this reason that the test is a strict one: was the incompetence of such degree that no reasonable advocate would have acted in that way *and* did any such incompetence cause the hearing to be unjust? (*R (Aston) v. Nursing and Midwifery Council* [2004] EWHC 2368 (Admin) at paragraphs 10–12 per Moses J, as he then was.)

After reviewing the criticisms of counsel, the learned judge said that, in several respects, his conduct of the hearing was *Wednesbury* unreasonable. Nonetheless, the appellant did not suffer an injustice as a result, and the decision of the panel was not wrong or unjust in consequence. The appellant was an intelligent, well-educated professional. He was given the chance to get a different advocate on the second day of the hearing. The panel was conspicuously fair to him throughout the hearing. He was well placed both to cross-examine the assessors and to make submissions on the clinical aspects of the case. The bald facts were that there were serious deficiencies in his fitness to practise, that he had precious little insight into them, and that he had done very little to make them good during his suspension.

## C. Other Relevant Chapters

**45.31**

Adjournment, Chapter 3
Human Rights, Chapter 32
Independent and Impartial Tribunal, Chapter 35
Natural Justice, Chapter 48
Unrepresented Practitioner, Chapter 67

# 46

## MEDICAL ASSESSORS



### A. Legal Framework

**46.01** Examples include:

Health Professions Order 2001, article 35 (appointment of medical assessors)

Nursing and Midwifery Order 2001, article 35 (appointment of registered medical practitioners to be medical assessors to advise practice committees on matters within their professional competence in connection with any matter the committee is considering)

Pharmacy Order 2010, article 64 (appointment of clinical and other specialist advisers)

General Medical Council (Fitness to Practise) Rules 2014, rule 3(2) (the registrar may appoint specialist health advisers or specialist performance advisers for the purposes of advising a fitness to practise panel in relation to medical issues regarding a practitioner's health or performance that may arise at a hearing)

### B. Disciplinary Cases

#### *Sadler v. General Medical Council*, [2003] 1 WLR 2259

**46.02** Doctor—performance hearing—scope of specialist adviser's function—need for opportunity for parties to comment

S was a consultant surgeon in obstetrics and gynaecology at Torbay Hospital. Between July and December 1997, five separate incidents occurred in the course of S's work as a surgeon that led to his being suspended in December 1997. Following the involvement of the General Medical Council (GMC), the case was referred for assessment. The report of the assessment panel concluded that there was cause for concern that S's operative skills at major surgery may be unsatisfactory. The assessment panel did not recommend that S should cease medical practice, recommending instead that he should be given the opportunity to undergo targeted retraining and supervision. A six-month training attachment at hospitals in Bristol was arranged, but it came to an abrupt and premature end. On the third day on which S was operating, 9 August 2000, he undertook an abdominal hysterectomy of a post-menopausal patient. In the course of the operation, he severed her right ureter. S's hearing before the GMC's Committee on Professional Performance (CPP) commenced on 29 April 2002, and occupied sixteen days between then and 1 July 2002. The Committee found S's standard of professional performance to have been seriously deficient and decided to make S's registration conditional, for a period of three years, on his compliance with certain requirements. On appeal, counsel for S submitted that the specialist adviser went outside his proper function in offering his own experience and views to the Committee, almost as if he were giving evidence. In *Richardson v. Redpath Brown & Co Ltd* [1944] AC 62, Viscount Simon LC said, of a medical assessor appointed under the Workmen's Compensation Acts, that to treat a medical assessor, or indeed any assessor, as though he were an unsworn witness is to misunderstand what the true functions of an assessor are:

> He is an expert available for the judge to consult if the judge requires assistance in understanding the effect and meaning of technical evidence... the judge may consult him in cases of need as to the proper technical inferences to be drawn from proved fact.

In dismissing S's appeal to the Privy Council, Lord Walker of Gestingthorpe said that, in performance hearings before the GMC, the rules make more detailed provision as to the specialist adviser's function than was the case under the Workmen's Compensation Acts. The rules provide for the specialist adviser to advise the Committee on any question referred to him by the Committee and of his own motion if it appears to him that, but for such advice, there is a possibility of a mistake being made in judging the medical significance of any information before the Committee, or because of an absence of information before the Committee. What occurred in the present case did not go beyond what was authorized by the rules. In a case in which the Committee had the benefit of many highly qualified witnesses who were examined and cross-examined at length, it was a matter for the specialist adviser's judgment whether to express views that cut across (rather than explained or supplemented) the evidence of the other experts. If he thought it right to do so, it might have been better to have done so at an earlier stage in the hearing, so that his views could be put to the expert witnesses for their comment. It would also have been better had the chairman expressly invited the appellant's representative to comment if he wished to do so on what the specialist adviser said.

### *Boodoo v. General Medical Council* [2004] EWHC 2712 (Admin)

**46.03** Scope of medical assessor's advice—advice on medical significance of evidence before Committee

B appealed from a decision of the Health Committee of the GMC by which it directed that, for a period of twenty-four months, his registration as a doctor should be conditional upon compliance by him with certain conditions. The main challenge on appeal was to the first condition, which required B to limit his alcohol consumption in accordance with his medical supervisor's advice, abstaining absolutely if so required. In its determination, the Committee said that, having considered the advice given to it by the medical assessors, it had judged B's fitness to practise to be seriously impaired by reason of alcohol dependence, and, in making its decision, the Committee had been advised by the medical assessors that B had a strong desire to consume alcohol and a difficulty in controlling his intake. In dismissing B's appeal, Silber J said that there was nothing in the evidence to suggest that the medical assessors went outside their obligations, which were to advise the Committee on the medical significance of the information before the Committee, or to give such advice on questions referred to them by the Committee, or to advise the Committee of their own motion on any particular matter relevant to the fitness to practise of the practitioner. The thrust of B's complaint was that it was unfair for deliberations to stray outside the ambit of evidence and that, although the medical assessors were entitled to explain the medical significance of the evidence, in this case they had given a view of the evidence pertinent to the appellant's ability to restrict consumption, and that was unfair and contrary to the rules. Silber J concluded that there was nothing to suggest that the medical assessors did not give their advice perfectly properly in accordance with their obligations in the rules.

### *Watson v. General Medical Council* [2006] ICR 113

**46.04** Deliberations of Committee—retirement of medical assessor with Committee—practice to cease—differences between legal assessors and medical assessors

W was a 50-year-old consultant psychiatrist with a history of recurrent depressive illness. In March 2003, the Health Committee of the GMC concluded that W's fitness to practise was seriously impaired by reason of her depressive disorder and alcohol dependence (abstinent at the time of the hearing). Conditions were imposed on W's registration for two years. In preparation for the resumed hearing in March 2005, medical reports were obtained from four doctors, all of whom reported positively and none of whom suggested any restrictions on W's practice. They were not asked to attend the hearing to give evidence orally. Counsel did not contend that any continuing conditions should be imposed on W's registration. W gave evidence. She was not cross-examined, although she was questioned by the Committee and the two medical assessors. Following closing submissions, the Committee retired with the medical assessors. When the Committee returned and before it announced its determination, the chairman called on the medical assessors to rehearse for the parties the advice that they had given to the Committee in private. One medical assessor stated that W had a current relapsing condition and that there was clearly a significant risk of relapse. The other medical assessor stated that, in view of the length of the history and the number of causes for it, W remained at substantial risk and was vulnerable to relapse. In allowing W's appeal and quashing the determination of the Committee that continuing conditions should be imposed on his registration, and remitting the case to be heard by a differently constituted committee advised by different medical assessors, Stanley Burnton J said that it was common ground before the Committee that W had a condition that might reoccur and that, if it were to reoccur, it would be liable seriously to impair her fitness to practise. It followed that the Committee had the power to regard her current fitness to practise as seriously impaired, but it was not obliged to do so. Conditions should not be imposed if they are unnecessary for the protection of the public or if they were not in the interests of W, and any conditions imposed should not go beyond what was necessary and proportionate for the protection of the public or the interests of W. As to the circumstances in which the assessors gave their advice, Article 6 of the European Conventions on Human Rights (ECHR) applied to the hearing. In a series of cases, the European Court of Human Rights (ECtHR) has held that a government commissioner (*commissaire du gouvernement* in France, and a *procureur général* in Belgium), who advises tribunals on the issues in cases, must not retire with the tribunal when it considers its verdict if the rights of the individual under Article 6 are not to be infringed. The requirement of the fairness of the proceedings demands that he gives his advice in public, in circumstances in which the parties have an opportunity to comment on it, and that he does not retire with the tribunal when it considers its decision, when he would have the opportunity to advise it further or differently: see *Borgers v. Belgium* (1991) 15 EHRR 92; *Vermeulen v. Belgium* (1996) 32 EHRR 313; and *Kress v. France*, Application no. 39594/98, ECHR 2001-VI, 43. Stanley Burnton J continued:

> 56. The role of a medical assessor is not the same as that of a legal assessor to a tribunal (in which I include a fitness to practise panel) or a legal adviser to justices. In the first place, as the European Court of Human Rights pointed out, legal advisers do not in general give their personal views on the facts or the outcome of a case. In the present case, the medical assessors did not in terms give their personal views on the outcome of the case: they did not state in terms that it was necessary for conditions on [W]'s registration to be continued, or what those conditions should be. But their opinions,

if accepted by the panel, made it inevitable that at the very least conditions would be imposed, and for that purpose they would have to find that her fitness to practise remains seriously impaired under rule 24(2).

57. There is a further difference between the role of a legal adviser or assessor and that of a medical assessor. A legal adviser to justices advises only on questions of law, and the decisions of the justices may be appealed on issues of law. A medical examiner advises on factual issues, and there is no appeal against a panel's decision on issues of fact. In my judgment, this makes it more important that advice on issues of fact, such as the medical significance of the information before the panel, should be given openly and that the parties should be able to respond to that advice before the panel makes their determination.

[...]

61. The medical assessors' special relationship with a panel makes it the more important that all of their advice is given in the presence of the parties. The medical assessors are not parties to the case before the panel. Nor are they members of the panel. Where their advice may be adverse to the practitioner's case, it is particularly important that it is given the presence of the parties, before the panel deliberate on their determination, and in circumstances in which the parties have an opportunity to address that advice. Otherwise, the suspicion may be created that the advice given in private was not precisely the same, or was not given in the same manner, as that announced in public, or the medical assessors have exercised influence on the decision of the tribunal. A perception of unfairness, and of bias on the part of the tribunal, is liable to be created.

62. The fact that the panel's decision is not announced until after the medical assessors had informed the parties of the advice they have given will not make the hearing fair.

### *Rauniar v. General Medical Council* [2011] EWHC 782 (Admin)

**46.05**

Scope of specialist performance advisor—experience in relevant field—whether appropriate to put questions to specialist advisor

R appealed against the decision of the fitness to practise panel at a review hearing to impose a period of twelve months' suspension on his registration. The panel found that R's fitness to practise was impaired by reason of his deficient professional performance. In dismissing R's appeal, Lloyds Jones J said:

89. ... [The function of the specialist performance advisor] is not limited to advising on medical issues in the narrow sense of medical conditions. The Specialist Performance Advisor is required to have experience of the relevant field of practice because he gives advice as to medical issues regarding performance in that particular field. The GMC Guidance [for Specialist Advisors, 2009 edn] in paragraph 11b formulates this in wide terms where it is explains that a Specialist Performance Advisor might explain the nature of procedures or practice in the doctor's speciality. Thus I would expect the function of the Specialist Performance Advisor to extend to giving advice on matters such as what a doctor practising in a particular field might be expected to know, how often a particular problem might arise or best practice in relation to record keeping. Similarly I do not understand there to be any objection to the advice given by the Specialist Advisor in relation to the form or content of a Personal Development Plan.

90. In the same way I consider that the nature of an examination or assessment of medical knowledge or skill is a medical issue regarding a practitioner's performance. In particular, I consider that the knowledge base required of a practitioner in the field of general practice if he is to practise safely and how that may be assessed are medical issues regarding a practitioner's performance. I consider therefore that the giving of advice on the procedures followed by the GMC in evaluating tests of doctors in the field of general practice is within the scope of the function of the Specialist Performance Advisor.

91. However, there is an important limitation on the role of the Specialist Performance Advisor within this defined area and this gives rise to [counsel for R's] second criticism. The GMC Guidance expressly states that a Specialist Performance Advisor must

> not give his opinion as to the adequacy of the practice of the doctor who is before the panel (paragraph 11b). Furthermore he must not give his own clinical opinion about the adequacy of that doctor's practice or express a view on whether the doctor's fitness to practise is impaired (paragraph 12).
>
> 92. The purpose of this limitation is to prevent the Specialist Advisor from expressing his opinion on the ultimate issues which the panel has to decide. However he is able to give advice on medical issues falling outside this core area provided his advice is as to medical significance. Furthermore his advice can relate specifically to the doctor before the panel.
>
> 93. As Stanley Burnton J observed in *Watson* (at paragraph 69) the line between admissible advice on medical issues and the decisions to be made by the panel itself may be difficult to draw. In this regard I have been referred to the advice of the Privy Council in *Sadler v. GMC* (at paragraphs 64–5) and the decision of Hickinbottom J in *Udom v. GMC* [2009] EWHC 3242 (Admin) (at paragraphs 37–40) as well as that of Burnton J in *Watson* itself. Whilst I accept that in the present case the Specialist Advisor's advice as to the suitability of the test is one of central importance, I do not consider that it is within the forbidden area. Neither the advice in relation to conversion of scores nor the advice in relation to whether the [multiple-choice questions] MCQ examination was a suitable gateway test relates directly to the question of the fitness to practise of the appellant or that of the appropriate sanctions. The first was a technical matter which did not relate specifically to the appellant. The second did relate specifically to the appellant but was well removed from the issue of his fitness to practise. For the same reason, it is not correct to suggest that the panel abdicated its decision-making role to the Specialist Advisor.

The learned judge went on to state that the appropriate course is for the specialist advisor to give his advice. Thereafter, it is not appropriate for him to be cross-examined; rather, the advice that he has given may be the subject of submissions by counsel. In the present case, it was unfortunate that the chairman departed from that course and permitted the questioning of the specialist advisor. However, the substance of the specialist advisor's answers did not extend beyond the giving of permissible advice in relation to medical issues regarding the practitioner's performance: [95].

## C. Other Relevant Chapters

**46.06** Deficient Professional Performance, Chapter 16

# 47

# MISCONDUCT

**A. Legal Framework** 47.01

**B. Disciplinary Cases** 47.02

*Allinson v. General Council of Medical Education and Registration* [1894] 1 QB 750 47.02

*E v. T* 1949 SLT 411 47.03

*Rajasooria v. Disciplinary Committee* [1955] 1 WLR 405 47.04

*Hughes v. Architects' Registration Council of the United Kingdom* [1957] 2 All ER 436 47.05

*Felix v. General Dental Council* [1960] AC 704 47.06

*De Gregory v. General Medical Council* [1961] AC 957 47.07

*McCoan v. General Medical Council* [1964] 1 WLR 1107 47.08

*Marten v. Royal College of Veterinary Surgeons' Disciplinary Committee* [1966] 1 QB 1 47.09

*Bhattacharya v. General Medical Council* [1967] 2 AC 259 47.10

*Faridian v. General Medical Council* [1971] AC 995 47.11

*In re a Solicitor* [1972] 1 WLR 869 47.12

*In re a Solicitor* [1975] QB 475 47.13

*Le Scroog v. General Optical Council* [1982] 3 All ER 257 47.14

*Sharp v. Law Society of Scotland* 1984 SC 129 47.15

*R v. Statutory Committee of the Pharmaceutical Society of Great Britain, ex parte Sokoh*, 3 December 1986 (unreported) 47.16

*Green and Valentine*, Case no. 8606/1, Report of Lloyds' Disciplinary and Appellate Proceedings 47.17

*Doughty v. General Dental Council* [1988] AC 164 47.18

*Plenderleith v. Royal College of Veterinary Surgeons* [1996] 1 WLR 224 47.19

*McCandless v. General Medical Council* [1996] 1 WLR 167 47.20

*Hossack v. General Dental Council* [1998] 40 BMLR 97 47.21

*Roylance v. General Medical Council (No. 2)* [2000] 1 AC 311 47.22

*Balfour v. The Occupational Therapists Board* (2000) 51 BMLR 69 47.23

*Preiss v. General Dental Council* [2001] 1 WLR 1926 47.24

*Rao v. General Medical Council* [2002] UKPC 65; [2003] Lloyd's Rep Med 62 47.25

*Silver v. General Medical Council* [2003] UKPC 33; [2003] Lloyd's Rep Med 333 47.26

*Nandi v. General Medical Council* [2004] EWHC 2317 (Admin) 47.27

*Skidmore v. Dartford and Gravesham NHS Trust* [2003] UKHL 27; [2003] ICR 721 47.28

*R (Campbell) v. General Medical Council* [2005] 1 WLR 3488, CA 47.29

*Council for the Regulation of Health Care Professionals v. General Medical Council and Biswass* [2006] EWHC 464 (Admin) 47.30

*Livingstone v. The Adjudication Panel for England* [2006] EWHC 2533 (Admin) 47.31

*Dzikowski v. General Medical Council* [2006] EWHC 2468 (Admin) 47.32

*General Medical Council v. Meadow* [2007] QB 462 47.33

*Law Society v. Adcock* [2007] 1 WLR 1096 47.34

*Mallon v. General Medical Council* [2007] CSIH 17 47.35

*Vranicki v. Architects Registration Board* [2007] EWHC 506 (Admin) 47.36

*Calhaem v. General Medical Council* [2007] EWHC 2606 (Admin) 47.37

*Joyce v. Secretary of State for Health* [2008] EWHC 1891 (Admin); [2009] PTSR 266 47.38

*Akodu v. Solicitors Regulation Authority* [2009] EWHC 3588 (Admin) 47.39

*R (Remedy UK Ltd) v. General Medical Council* [2010] EWHC 1245 (Admin) 47.40

*R (Gaunt) v. The Office of Communications* [2011] 1 WLR 2355, CA 47.41

*R (Calver) v. Adjudication Panel for Wales* [2012] EWHC 1172 (Admin) 47.42

*R (Aga) v. General Medical Council* [2012] EWHC 782 (Admin) 47.43
*Walker-Smith v. General Medical Council* [2012] EWHC 503 (Admin) 47.44
*O'Connor v. Bar Standards Board*, 17 August 2012 (unreported) 47.45
*Kumar v. General Medical Council* [2012] EWHC 2688 (Admin) 47.46
*Spencer v. General Osteopathic Council* [2013] 1 WLR 1307 47.47
*Pottage v. Financial Services Authority* [2013] Lloyd's Rep FC 16 47.48
*Giwa-Osagie v. General Medical Council* [2013] EWHC 1514 (Admin) 47.49
*Patel v. Solicitors Regulation Authority* [2012] EWHC 3373 (Admin) 47.50
*Harford v. Nursing and Midwifery Council* [2013] EWHC 696 (Admin) 47.51
*R (Commissioner of Police for the Metropolis) v. Police Appeals Tribunal and Naulls* [2013] EWHC 1684 (Admin) 47.52
*Rehman v. Bar Standards Board*, PC2008/0235/A and PC2010/0012/A, 19 July 2013 47.53
*Johnson and Maggs v. Nursing and Midwifery Council (No. 2)* [2013] EWHC 2140 (Admin) 47.54
*Walker v. Bar Standards Board*, 19 September 2013 (unreported) 47.55
*Sultan v. Bar Standards Board*, 6 November 2013 (unreported) 47.56
*Cowan v. Hardey* [2014] CSIH 11 47.57
*Craven v. Bar Standards Board*, 30 January 2014 (unreported) 47.58
*Clery v. Health and Care Professions Council* [2014] EWHC 951 (Admin) 47.59
*Okorji v. Bar Standards Board*, 30 April 2014 (unreported) 47.60
*Professional Standards Authority for Health and Social Care v. General Medical Council* [2014] EWHC 1903 (Admin) 47.61
*Conlon, Gordon and Williams v. Bar Standards Board* [2014] Lexis Citation 136 47.62
*Booth v. General Dental Council* [2015] EWHC 381 (Admin) 47.63
**C. Other Relevant Chapters** 47.64

## A. Legal Framework

**47.01** The word 'misconduct' is not usually defined in legislation: *Roylance v. General Medical Council (No. 2)* [2000] 1 AC 311 (see paragraph 47.22) and *General Medical Council v. Meadow* [2007] QB 462 (see paragraph 47.33).

Nurses, Midwives and Health Visitors (Professional Conduct) Rules 1993, since repealed (provided that 'misconduct' meant conduct unworthy of a registered nurse, midwife, or health visitor)

Police (Conduct) Regulations 2012, regulation 3 ('misconduct' means a breach of the Standards of Professional Behaviour, and 'gross misconduct' means a breach of the Standards of Professional Behaviour so serious that dismissal would be justified; 'Standards of Professional Behaviour' means the standards of professional behaviour contained in Schedule 2, which defines, inter alia, honesty and integrity, authority, respect, and courtesy, confidentiality, and discreditable conduct)

## B. Disciplinary Cases

### *Allinson v. General Council of Medical Education and Registration* [1894] 1 QB 750

**47.02** *Infamous conduct—meaning*

On 28 May 1892, the Council heard a charge against the plaintiff that, being a registered medical practitioner, he systematically advertised his practice in advertisements in a way discreditable to a professional man. The advertisements contained allegations against fellow doctors, accusing them, for example, of providing poisonous drugs and being professional poisoners. The Council found the plaintiff guilty of 'infamous conduct' in a professional respect within section 29 of the Medical Act 1858. In dismissing the plaintiff's appeal, the Court of Appeal said, at 760–1:

> If it is shown that a medical man, in the pursuit of his profession, has done something with regard to it which would be reasonably regarded as disgraceful or dishonourable by his professional brethren of good repute and competence, then it is open to the General Medical Council to say that he has been guilty of infamous conduct in a professional respect.

Lord Esher MR added, at 761, that:

> The question is, not merely whether what a medical man has done would be an infamous thing for anyone else to do, but whether it is infamous for a medical man to do. An act done by a medical man may be "infamous", though the same act done by anyone else would not be infamous; but on the other hand, an act which is not done "in a professional respect" does not come within this section. There may be some acts which, although they would not be infamous in any other person, yet if they are done by a medical man in relation to his profession, that is, with regard either to his patients or to his professional brethren, may be fairly considered "infamous conduct in a professional respect", and such acts would, I think come within section 29.

### *E v. T* 1949 SLT 411

**47.03** Professional misconduct—approach of court to findings

E lodged a complaint against T, a solicitor, charging him with gross professional negligence, with the Solicitors' Discipline Committee of Scotland. The finding of the Committee was that the respondent had, since July 1947, notwithstanding repeated applications, failed to produce a certain deed and other titles, failed to record a disposition in the Register of Sasines, and knowingly misrepresented the position to the complainer in connection with the disposition and the recording thereof. Its conclusion was that the respondent, by such failure, misrepresentation, and gross disregard of the interests of his client, had been guilty of professional misconduct. The Lord President (Cooper) said, at 411, column ii:

> I shall not attempt to define professional misconduct, but if the statutory tribunal, composed as it always is of professional men of the highest repute and competence, stigmatise a course of professional conduct as misconduct, it seems to me that only strong grounds would justify this Court in condoning as innocent that which the Committee have condemned as guilty.

### *Rajasooria v. Disciplinary Committee* [1955] 1 WLR 405

**47.04** Grossly improper conduct—approach of court to finding

The appellant, an advocate and solicitor practising in the Federation of Malaya, was found guilty of 'grossly improper conduct in the discharge of his professional duty' within section 26 of the Advocates and Solicitors Ordinance 1947. In dismissing his appeal from an order suspending him from practice for six months, the Privy Council held that a finding of intention to deceive is not always an essential element in grossly improper conduct. It would add that it would require a very strong case before it would substitute its own opinion for what is professional misconduct in the Federation for the conclusion reached by the Disciplinary Committee and the Supreme Court. As Darling J said in relation to England in *In re a Solicitor, ex parte Law Society* [1912] 1 KB 302, 312: 'The Law Society are very good judges of what is professional misconduct as a solicitor, just as the General Medical Council are very good judges of what is misconduct as a medical man.'

### *Hughes v. Architects' Registration Council of the United Kingdom* [1957] 2 All ER 436

**47.05** Architect—practice—estate agent—change of rules prohibiting practice—whether disgraceful conduct

H had been in practice since 1922 as an architect and chartered surveyor, and as an estate agent and valuer. In 1934, he registered as an architect under the Architects (Registration) Act 1931, and continued to practise as an architect, chartered surveyor, and estate agent and valuer. In 1936, the Architects' Registration Council established under the 1931 Act issued a code of professional conduct that laid down that the business of house agency must not form any part of a registered architect's practice. The provision in the code did not apply to H until 1955, when the Council required all registered architects to cease practising as estate agents. H continued after 1955 to practise in the same way as before. Following a complaint, the Disciplinary Committee of the Council, on 17 July 1956, found H guilty of 'conduct disgraceful to him in his capacity as an architect' under section 7 of the 1931 Act and directed the registrar to remove H's name from the register. The Queen's Bench Division (Lord Goddard CJ, and Hilbery and Devlin JJ) allowed H's appeal and quashed the decision of the Disciplinary Committee. Lord Goddard said that he did not doubt the right of the profession to lay down rules of professional conduct and to enforce them, but that it must be remembered that this case was dealing with what may be described as a transitional state of affairs. Parliament gave the appellant the right to be registered under the 1931 Act and he could not be held to be guilty of conduct that, by any standard, can be said to be disgraceful because he continued to practise in exactly the same way after registration as he had done before and that practice did not disqualify him for registration. Devlin J said that the word 'disgraceful' in section 7 should be given its natural and popular meaning—but that it was qualified by the phrase 'in his capacity as an architect'. The effect of that qualification was twofold. First, the conduct must not only be what would ordinarily be considered disgraceful, but it must also be a disgrace that affects him professionally: to that extent, the qualification diminishes the term. Second, conduct that is not disgraceful for an ordinary man may be disgraceful for a professional man: to that extent, the qualification amplifies the term. The amplification does not, however, require that 'disgraceful' is to be given any technical meaning: it requires only that the ordinary meaning of the word should be applied in relation to the special obligations and duties of a professional man. It was not to be forgotten that if the finding of the Committee were to stand, anyone would then have been able to have said of the appellant with impunity that he was struck off the register for disgraceful conduct and may have added that that means what it says.

### *Felix v. General Dental Council* [1960] AC 704

**47.06** 'Wrongful'—meaning—fees for work not done—whether 'infamous or disgraceful conduct'—meaning

F, a registered dentist, was charged with three counts of 'wrongfully' claiming fees for fillings that he had not done. In quashing the determination of the Disciplinary Committee of the General Dental Council (GDC) and delivering the judgment of their Lordships, Lord Jenkins said, at 717–18, that the word 'wrongfully' is of wide and uncertain import. It can be regarded as adding nothing to the acts complained of, which were plainly wrongful in the sense that the appellant ought not to have done them. On the other hand, it may be regarded as imputing an unspecified degree of culpability of an undefined character ranging from mere carelessness to fraud or dishonesty; it clearly cannot be held that every act that can be characterized as 'wrongful' in one sense or another is infamous or disgraceful within section 25 of the Dentists Act 1957. The matter fell to be judged by reference to the evidence as a whole, including the appellant's explanation of the way in which the cases of overcharging arose. Continuing at 720–1, Lord Jenkins said that to make good a charge of 'infamous or disgraceful conduct in a professional respect' in relation to such a matter as the keeping of prescribed dental records, it was not, in their Lordships' view, enough to show that some mistake has been made through carelessness or inadvertence in two or even three cases out of (to quote the figures in the present case) 424 patients; to make such a charge good, there had to be, in their Lordships' opinion and generally speaking, some element of moral turpitude or fraud or dishonesty in the conduct complained of, or such persistent and reckless disregard of the dentist's duty in regard to records as can be said to amount to dishonesty for this purpose. The question was to some extent one of degree, but in their Lordships' view, the cases of overcharging with which this appeal was concerned clearly fell short of the degree of culpability required.

### *De Gregory v. General Medical Council* [1961] AC 957

**47.07** Inappropriate relationship—family doctor—wife ceasing to be patient—defendant continuing as doctor for husband and children—abuse of professional relationship

In June 1950, the appellant became R's doctor. Later, R married, and the appellant became the doctor also to Mrs R and her three children. In December 1958, Mrs R ceased to be a patient of the appellant. In July 1959, the appellant and Mrs R began a relationship, and in October/November 1959, she left R. The Disciplinary Committee of the General Medical Council (GMC) found the appellant guilty of infamous conduct in a professional respect and ordered that his name should be erased from the register. Dismissing the appeal, their Lordships said that they would approach the case on the footing that the improper association started only after December 1958, when Mrs R took her name off the appellant's list. This did not excuse him. He gained his access to the home in the first place by virtue of his professional position. Afterwards, although the wife ceased to be on his list, he still had access to the home, so as to attend R and the children, if called upon. It was an abuse of his professional relationship with the husband and father for him to enter upon an improper association with the wife and mother of the family. It was infamous conduct in a professional respect, even though she herself had ceased to be his patient.

### *McCoan v. General Medical Council* [1964] 1 WLR 1107

**47.08** Inappropriate relationship—doctor and patient—doctor taking appointment—complaint subsequently made—infamous conduct

Mrs J registered as a patient of the appellant in February 1955. A relationship developed between August 1959 and October 1960, when the appellant took an appointment as a ship's surgeon in the Far East. On his return, Mrs J threatened to report him to the GMC and in due course did so. Following a finding of infamous conduct in a professional respect, the Disciplinary Committee ordered the appellant's name to be erased from the medical register. On appeal, it was submitted that the appellant's conduct did not amount to infamous conduct, as defined by Lord Jenkins in *Felix v. General Dental Council* [1960] AC 704, 720. There was no injury to the public, for the association was clandestine; there was no element of seduction, nor was the association adulterous; the complaint was made for some reason wholly extraneous to the misconduct charged; the misconduct had ceased long before the charge was laid; and there was no danger of any repetition. Dismissing the appellant's appeal, their Lordships said that one of the most fundamental duties of a medical adviser is that a doctor must never permit his professional relationship with a patient to deteriorate into an association that would be described by responsible medical opinion as improper. Their Lordships did not see how the findings of the Committee, on the facts of this case, that the appellant was guilty of infamous conduct in a professional respect could be successfully challenged.

### *Marten v. Royal College of Veterinary Surgeons' Disciplinary Committee* [1966] 1 QB 1

**47.09** Farm owned by vet—criminal conviction—failure to give adequate care to animals on farm—disgraceful conduct not limited to professional practice—conduct reflecting on profession

M appealed against the findings and order of the Disciplinary Committee of the Royal College of Veterinary Surgeons (RCVS), whereby it had found that he had been convicted of offences under the Dogs Act 1906 and further was guilty of conduct disgraceful to him in a professional respect, in that he had failed to give adequate nursing for sick animals in his care and allowed conditions to exist on his farm in Sussex that were likely to bring disgrace upon the veterinary profession. In the autumn of 1962, M had on his farm some sixty-five cattle; of those, two-thirds were affected by 'husk' and were treated, but in the result he lost thirteen or fourteen. Without reasonable excuse, M permitted eleven carcasses to remain unburied in a field to which dogs could gain access. He pleaded guilty to eleven summonses under the Dogs Act 1906 and the matter was reported to the RCVS. In dismissing M's appeal regarding the order for his name to be removed from the register, Lord Parker CJ said that the conduct relied upon in *Felix v. General Dental Council* was in regard to alleged overcharging and wrong certification. The finding was that the matters had occurred through careless mistakes and nothing more, and it was in that connection that the Court held that it is not enough to show that mistakes were made through carelessness or inadvertence in two or three cases out of several hundred patients treated, and that, to make such a charge good, there must, generally speaking, be some moral turpitude or fraud or dishonesty in the conduct complained of, or such persistent and reckless disregard of the dentist's duty in regard to records as could be said to amount to dishonesty. The reference to moral turpitude or fraud in that case must be taken to relate to the issue then before the Court. In the present case, there was abundant evidence upon which the Disciplinary Committee could come to the conclusion that the conduct was disgraceful to the appellant in a professional capacity. There was no valid ground for limiting the words 'conduct disgraceful to him in a professional respect' in the Veterinary Surgeons Act 1881, section 6, to being done in pursuit of his profession or in the course of the practice of his profession. Lord Parker CJ said, at 9C–F:

> For my part I see no valid ground for limiting the words in the manner suggested. If, of course, the conduct complained of is equally reprehensible in any one, whether a professional man or not, as for example, conduct constituting some traffic offence, that conduct would not come within the expression. But if the conduct, though reprehensible in anyone is in the case of the professional man so much more reprehensible as to be defined as disgraceful, it may, depending on the circumstances, amount to conduct disgraceful of him in a professional respect in the sense that it tends to bring disgrace on the profession which he practises. It seems to me, although I do not put this forward in any sense as a definition, that the conception of conduct which is disgraceful to a man in his professional capacity is conduct disgraceful to him as reflecting on his profession, or, in the present case, conduct disgraceful to him as a practising veterinary surgeon.

### *Bhattacharya v. General Medical Council* [1967] 2 AC 259

**47.10** Inappropriate relationship—commencing prior to doctor–patient relationship—inappropriate to maintain pre-existing relationship

The appellant's main contention in this case was that he met Mrs T, a married woman, about 1955, some two years before she became his patient, that he made amorous advances to her almost immediately after meeting her, and that the association on the basis of sexual familiarities occurred from 1955 to 1965 inclusive. Mrs T became the appellant's patient in March 1957, and from the autumn of 1959 until June 1961, when Mrs T's husband left home, the relationship continued at her home. The appellant never admitted that full sexual intercourse had taken place between himself and Mrs T, since he maintained that he was incapable of that act. Nevertheless, no distinction had been drawn before their Lordships between full intercourse and the sexual intimacies that admittedly took place. Mrs T having been in effect his mistress (as he contended) long before she became his patient, the appellant maintained that the Disciplinary Committee was wrong in principle in finding him guilty of infamous conduct in a professional respect. With this situation, he contrasted the case of the medical man who makes his patient his mistress, which it is admitted on his behalf is an obvious example of conduct properly described as infamous in a professional respect. Their Lordships were unable to accede to this argument. Lord Hodson, delivering the judgment, said at 266E–267C:

> It may well be that there is a relevant distinction between the case when the intimate association has existed between a doctor and his female patient before ever any thought arose of her being treated by him professionally and the cases where the situation is reversed. In the language of the Board contained in the judgment delivered by Lord Upjohn in *McCoan v. General Medical Council* [1964] 1 WLR 1107 at 1112: "One of the most fundamental duties of a medical adviser, recognised for as long as the profession has been in existence, is that a doctor must never permit his professional relationship with a patient to deteriorate into an association which would be described by responsible medical opinion as improper." This does not involve the converse proposition that there is nothing improper from a professional point of view in maintaining a pre-existing improper association with a patient and their Lordships would not for a moment accept that as a matter of principle such conduct is not covered by the statutory provisions as to "infamous conduct in a professional respect".
>
> The tendency of conduct to debase or degrade the standing and reputation of the profession will vary from case to case and there may be cases when the maintenance of a long-standing, pre-existing association can be regarded as much less serious than those when the professional relationship has deteriorated into an improper association but this is not to exclude the former from the category of those cases which can be made subject to disciplinary action.

### *Faridian v. General Medical Council* [1971] AC 995

**47.11** Clinic for terminating pregnancy—company—role of defendant qua doctor and shareholder

The appellant doctor was a director of, and had a substantial financial interest in, two companies that operated a clinic, which offered financial inducements to medical practitioners to refer women patients desirous of having their pregnancy terminated. Their Lordships had to consider two questions: (1) whether, by reason of the appellant's shareholding, which gave him power to control the company, the facts proved—namely, the despatch of letters, the actions of the matron, the provision of material for a newspaper article, and the sending of £20 to one doctor—were capable of amounting to infamous conduct on the part of the appellant in a professional respect; and (2) if so, whether, in the particular circumstances of the case, such a finding would be justified. The Board had been placed in some difficulty by the circumstance that, beyond finding the facts alleged proved, the Disciplinary Committee gave no reasons for its determination. The charge did not allege that the appellant had either knowledge or notice of what was done; it merely said that, in relation to the facts alleged, the appellant had been guilty of infamous conduct. In allowing the appellant's appeal, their Lordships said that the fact that the conduct of an organization may enure to a doctor's own professional advantage or financial benefit, although relevant, is not conclusive on the question of whether the association was of such a character as to render him guilty of infamous conduct. The association may be of such a character as to lead to the conclusion that the responsibility rests on the doctor for the conduct complained of. The association may be such as to lead to the conclusion that he must have known of what was done and have acquiesced in it, or such that a doctor could reasonably have been expected to take all reasonable steps to prevent the occurrence of events that might lead to charges of infamous conduct being preferred against him, and if, in those circumstances, it was shown that he had failed to take those steps, then it might be right to infer and to hold that he had connived at the misconduct. On the other hand, no one would contend that association with a company by holding shares in it would make a doctor responsible for the acts of servants of the company and guilty of infamous conduct. The appellant was not charged with acquiescing or conniving at what was done, and there was no investigation as to what steps, if any, were taken by him to prevent malpractices. His failure to exercise the power that he had, when it was not shown that he had any prior knowledge of what was done nor reason to suspect that it would be done, was not, in their Lordships' opinion, capable of being held to be infamous conduct on his part.

***In re a Solicitor*** **[1972] 1 WLR 869**

**47.12** Negligence—whether professional misconduct

Negligence in a solicitor may amount to a professional misconduct if it is inexcusable and is such as to be regarded as deplorable by his fellows in the profession, *per* Lord Denning MR at 873B. See further Chapter 49.

***In re a Solicitor*** **[1975] QB 475**

**47.13** Solicitor—benefiting from estate—independent advice—standards of conduct in disciplinary cases

The complaint against the appellant was that he, together with his partner, prepared documents under which they benefited to a substantial extent without observing the appropriate rules as to ensuring that their client had independent advice before committing herself to them. The bulk of the estate of a widow and her sister-in-law went to the children of the appellant and his partner. Both solicitors were found guilty of conduct unbefitting a solicitor and ordered to be struck off. The appellant appealed. In dismissing his appeal, Lord Widgery CJ said that the question of how far a solicitor is bound to see that his client is separately advised and what are the consequences of a failure in that duty can arise in two different spheres, and be judged by two different sets of rules. Often enough, as indeed in this case, a will is challenged on the basis that a beneficiary is a solicitor and that the solicitor did not take the steps that the law regards as necessary to protect the client against undue influence. When those cases arise, of course, they arise in the ordinary courts of the land and the law has laid down the standards with which a solicitor is required to comply. The second way in which the question may arise is in disciplinary proceedings such as the present. In disciplinary proceedings, the standards are not necessarily the same as they are in a probate action. In this case, the Disciplinary Committee adopted as the standard for disciplinary purposes a very high standard. What it was saying is that it is not sufficient, from a disciplinary point of view, for a solicitor to tell his client that she ought to be separately advised. This standard requires that he should tell her that she *must* be separately advised, and if she refuses to accept that advice and refuses to go to another solicitor, then the standard laid down requires that the solicitor beneficiary should decline to act. Equally, where a client wishes to make a substantial gift *inter vivos*, the solicitor should not accept it unless the client has been independently advised. It is an exceptionally high standard and, in the Court's view in this case, it is probably higher than that imposed in the probate courts when the validity of the will is in question.

***Le Scroog v. General Optical Council*** **[1982] 3 All ER 257**

**47.14** Advertising—canvassing students at colleges—infamous conduct

The appellant was found guilty of two charges of infamous conduct, the conduct alleged being that he had canvassed students at two colleges in breach of the Council's rules on publicity on advertising. In dismissing the appellant's appeal against an order of erasure, Lord Fraser of Tullybelton, delivering the opinion of the Privy Council, said that it must be obvious to any educated person that 'infamous' is a word implying strong reprobation and their Lordships had no doubt that the members of the Disciplinary Committee were well aware of that fact.

### *Sharp v. Law Society of Scotland* 1984 SC 129

**47.15** Solicitors' accounts—shortfall in client monies held by solicitors—whether automatic misconduct—junior partners less culpable

The petitioners were partners in a firm of solicitors charged with a breach of rule 4(1)(a) of the Solicitors (Scotland) Accounts Rules 1952, which provides that every solicitor shall ensure that, at all times, the sum at credit of the client account shall be not less than the total of clients' money held by the solicitor. Section 20(3) of the Solicitors (Scotland) Act 1949 provides that a failure to comply with rules made under this section may be treated as professional misconduct. At the hearing before the Scottish Solicitors' Discipline Tribunal, the breaches of rules alleged by the Law Society were admitted by the petitioners. The question for the Tribunal was, however, whether all or any of the petitioners should be held guilty of professional misconduct. The Lord President (Emslie), giving the opinion of the Court, said that the Tribunal appeared to believe that section 20(3) fell to be read as if it declared that failure to comply with the relevant rules 'shall' be treated as professional misconduct unless the Tribunal is persuaded by the solicitor concerned that he had discharged the onus of showing that it should not be so treated. Section 20(3) means precisely what it says: a failure on the part of a solicitor to comply with a relevant rule *may* be treated as professional misconduct. The subsection introduces nothing new to the law. There is no support for the view that 'professional misconduct' resting upon a breach of rule is different from 'professional misconduct' resting upon what the Tribunal described as 'common law' charges. The Lord President said, at 134–5:

> There are certain standards of conduct to be expected of competent and reputable solicitors. A departure from those standards which would be regarded by competent and reputable solicitors as serious and reprehensible may properly be categorized as professional misconduct. Whether or not the conduct complained of is a breach of rules, or some other actings or omissions, the same question falls to be asked and answered, and in every case it will be essential to consider the whole circumstances and the degree of culpability which ought properly to be attached to the individual against whom the complaint is made.

As to whether all of the petitioners in the present case should be held guilty, having regard to the material before the Tribunal, no reasonable tribunal properly directed could have found the junior partners, profit-sharing and salaried, guilty of the grave charge of professional misconduct. In practical terms, the actings of the four senior partners were not open to question by the more junior partners, who had a limited opportunity to involve themselves in the working of the particular accounting system that gave rise to criticism, and to supervise and influence the administration of the firm. The duty of supervision and control of the accounting system was confided to the senior partner alone at the central office of the firm. The salaried junior partners were excluded from partnership meetings. All of the more junior partners practised exclusively from the branch office to which they were respectively assigned. Although each of the partners of the firm was responsible under rule 10 for securing compliance by the firm with the provisions of rule 4(1)(a), having regard to the special circumstances affecting the more junior partners, their failure to ensure compliance was not so serious and so reprehensible that a reasonable tribunal, properly directed, would have proceeded to stigmatize it as 'professional misconduct'—a grave charge. It would be going too far too fast to reach the same conclusion in the case of three of the senior partners. Decisions on questions of 'professional misconduct' are best left to the Tribunal, and the Court ordered that the cases of these partners were to be remitted to the Tribunal to reconsider the whole circumstances as they relate to each of them and to decide, after hearing the parties and in the light of the guidance in law contained in this opinion, whether findings of professional misconduct should be made against all or any of them.

### *R v. Statutory Committee of the Pharmaceutical Society of Great Britain, ex parte Sokoh*, 3 December 1986 (unreported)

**47.16** Pharmacist—prescription error by doctor—duty of pharmacist to check prescription

On 3 May 1984, S, a registered pharmacist, made up a prescription of 'Brompton' mixture for a patient. The prescription stated that Pethidine was to be used in the preparation of the mixture. Whilst the doctor's prescription was not as clear as it could, and perhaps should, have been, the strength of the medicine dispensed by S was fifteen times the strength intended by the doctor. Unfortunately, the patient died and the coroner recorded a verdict that the cause of death was Pethidine poisoning, resulting from an accidental overdose of the drug. The Disciplinary Committee, in finding the pharmacist guilty of misconduct, stated that the strength of the medicine dispensed by S was such that it called for careful inquiry by him to establish that he had made no error in interpreting the prescription. It was incumbent on him to check with the doctor that no mistake had been made in the prescription. In the light of S's frankness, the appropriate penalty in all of the circumstances was a reprimand. In dismissing S's appeal on misconduct, Webster J said that it was unhelpful to define 'misconduct' by reference to any adjective having moral overtones. If the word 'misconduct' is to be defined at all, it is simply to be defined as 'incorrect or erroneous conduct of any kind provided that it is of a serious nature'. One error is, as a matter of law, capable of constituting misconduct if it is sufficiently serious, although whether it is in fact constitutes misconduct or misconduct such as to render the person unfit to be on the register must depend on the precise evidence taken as a whole and on the Committee's view of that evidence.

### *Green and Valentine*, Case no. 8606/1, Report of Lloyds' Disciplinary and Appellate Proceedings

**47.17** Discreditable—meaning

In the matter of charges against Sir Peter Green and Geoffrey Valentine, the Appellate Committee of Lloyd's said that the starting point for the meaning of 'discreditable' in section 20 of the Lloyd's Act 1871 and paragraph 1(e) of the Byelaws (5/1/1983) was the decision in Case no. 8404/4. In that case, it was stated:

> In all the charges the Defendants have been charged with acts or defaults "discreditable" to them as underwriters or otherwise in connection with the business of insurance (the requirement under the Lloyd's Act 1871) *and* conducting any insurance business in a "discreditable" manner (the requirement under the Byelaws). It has been pointed out that the word "discreditable" must be construed against its background, and that it does not necessarily bear the same meaning in both contexts. We agree. It has been emphasised that it is capable of wider or narrower meaning, depending on the context. Again, we agree. It has also been emphasised that the word "discreditable" was used in the Lloyd's Act 1871 at a time when the only sanction was expulsion. We agree that this is a very relevant factor. Finally, it has been urged that in a quasi-criminal context we ought, in the event of ambiguity, to give the word a narrower rather than wider meaning. We agree, subject to the qualification that this aid to construction is one of last resort and ought not to be used to create ambiguity where none truly exists.
>
> We have carefully considered all the arguments. We have also considered the various decisions cited to us. See *Felix v. General Dental Council* [1960] AC 704; *Marten v. Royal College of Veterinary Surgeons Disciplinary Committee* [1966] 1 QB 1; *Mercer v. Pharmacy Board of Victoria* [1968] VR 72; *McEniff v. General Dental Council* [1980] 1 All ER 461. Ultimately, we believe the ordinary meaning of the word, read in the

> context of its application to the Lloyd's insurance market, must be decisive. In the context, "discredit" means "impaired reputation; disrepute, reproach". See Murray's Oxford English Dictionary, s.v. "discredit". The same dictionary defines "discreditable" as "such as to bring discredit; injurious to reputation; disreputable, disgraceful". Manifestly, an active underwriter who acts as such in a dishonest manner is acting in a discreditable manner. On the other hand, we do not think that a mere act of negligence would amount to discreditable conduct. Where should the line be drawn? It is not necessary, or desirable, for us to attempt this exercise. It is sufficient for present purposes to record our conclusion that reckless or grossly negligent conduct is *capable* of impairing the reputation of the underwriter, his agency or Lloyd's market generally, and is therefore *capable* of constituting discreditable conduct, depending on all the circumstances of each case. We do not, however, consider it will serve any useful purpose to attempt a definition of "discreditable" in the relevant context. Arguably, it could cover other cases not presently under consideration, e.g. repeated acts of negligence. We express no view on the question.

In the instant case, the Committee added the following further observations.

(1) 'Discredit' and 'discreditable' are the opposites of 'credit' and 'creditable'. The *Shorter Oxford English Dictionary* definitions of these four words show that the basic features are 'trust' and 'reputation'. For a person's conduct to be able to be characterized as 'discreditable', it must be such as to be capable of damaging the reputation of that person and the trust that others have in him.
(2) 'Discreditable' conduct would include, but is not confined to, dishonesty.
(3) A mere act of negligence would not be discreditable conduct. But negligence could and might amount to discreditable conduct if it were sufficiently serious negligence. 'Serious' is perhaps a better word to use than 'gross': whether negligence is 'sufficiently serious' to be discreditable conduct is a matter to be determined on all of the relevant facts of the particular case in the judgment of the disciplinary committee.

### *Doughty v. General Dental Council* [1988] AC 164

**47.18** Dental treatment—failure to complete satisfactorily—meaning

D, a dentist, was found guilty of three charges of serious professional misconduct—namely, (1) failing to retain radiographs in respect of nineteen patients and failing to submit the radiographs to the Dental Estimates Board when required, (2) failing to exercise a proper degree of skill and attention in relation to six patients, and (3) failing satisfactorily to complete the treatment required for four patients. The GDC's Professional Conduct Committee (PCC) ordered that D's name should be erased from the dentists' register. On D's appeal, the Privy Council rejected the GDC's submission that the wording of charge 3 was wide enough to cover allegations that the treatment criticized was unnecessary. An allegation of failure 'satisfactorily to complete the treatment required by the patient' cannot, by any stretch of language, be read as including an allegation of having administered treatment that was not necessary. It follows that the legal assessor's direction was erroneous, but not necessarily that the Committee's decision was thereby invalidated. In dismissing D's appeal, Lord Mackay of Clashfern, giving the judgment of their Lordships, said, at 173A–F:

> Their Lordships readily accept that what was infamous or disgraceful conduct in a professional respect would also constitute serious professional misconduct but they consider that it would not be right to require the General Dental Council to establish now that the conduct complained of was infamous or disgraceful and therefore not right to apply the criteria which Lord Jenkins derived from the dictionary definitions of these words which he quoted in *Felix v. General Dental Council* [1960] AC 704, 719. Their Lordships consider it relevant, in reaching a conclusion upon whether Parliament intended by the change of wording to make a change of substance, to notice that

in addition to this change and in close conjunction with it the additional and much less severe penalty of suspension for a period not exceeding 12 months was provided. Further, in terms of s. 1(2) of the Dentists Act 1984 which is the statute presently applicable: "It shall be the general concern of the Council to promote high standards of dental education at all its stages and high standards of professional conduct among dentists..." In the light of these considerations in their Lordships' view what is now required is that the General Dental Council should establish conduct connected with his profession in which the dentist concerned has fallen short, by omission or commission, of the standards of conduct expected among dentists and that such falling short as is established should be serious.

### *Plenderleith v. Royal College of Veterinary Surgeons* [1996] 1 WLR 224

**47.19** Foreign registered veterinary surgeons—employment—non-UK registration—whether disgraceful conduct

The appellant, P, was a veterinary surgeon of long experience, who, with associates, provided veterinary services at a number of clinics in Lincolnshire. He was charged with two offences of causing or permitting two employees to provide veterinary care when he knew, or ought to have known, that they were not listed in the register of veterinary surgeons. C had gained a diploma in veterinary surgery at the Faculty of Veterinary Medicine in the Netherlands, and was a member of the Royal Netherlands Veterinary Association. P had obtained a bachelor's degree in veterinary medicine from the National University of Ireland and was entered on the register of veterinary surgeons for Ireland. However, at the material time, neither were registered with the Royal College of Veterinary Surgeons (RCVS). In allowing P's appeal against findings of disgraceful conduct in a professional respect, the Privy Council said that, leaving aside any submissions as to European law and assuming that the two should not have been employed unless they were registered at the material time, the question remained as to whether P's action in employing them, in all of the circumstances, amounted to disgraceful conduct in a professional respect. The cases of *Marten v. Royal College of Veterinary Surgeons' Disciplinary Committee* [1966] 1 QB 1, *Felix v. General Dental Council* [1960] AC 704, *Hughes v. Architects' Registration Council of the United Kingdom* [1957] 2 QB 550, and *Allinson v. General Council of Medical Education and Registration* [1894] 1 QB 750 make it clear that what is done has to be done in a professional respect, and that it is not a prerequisite of the charge being proved that what is done must involve some moral turpitude. Their Lordships did not, however, consider that every breach of the disciplinary code or the statute of every commission of a criminal offence is necessarily to be regarded as 'disgraceful conduct in a professional respect'. However technical a meaning 'infamous' or 'disgraceful' conduct may have been given (so as to render unnecessary a morally blameworthy act), there must be a line beyond which conduct does not satisfy this test. Their Lordships bore fully in mind that the Board is reluctant to interfere with a finding by professional men of 'disgraceful conduct in a professional respect' by one of its colleagues. Their Lordships appreciated the obvious importance from the point of view of both the profession and of the public that only registered veterinary surgeons should be employed. It had, however, to be borne in mind that, in the present case, both individuals had genuinely sought to register and the appellant was aware of this. Both had the requisite qualifications. The delay was, to some extent, the result of the administrative arrangements adopted. One of the associates was registered within seven days of the offence charged; but for a lack of passport and despite a declaration of nationality, the other would have been registered two months before the incident giving rise to the charge. Upon all of the evidence in this case, it could not reasonably be said that P's conduct was 'disgraceful in a professional respect', albeit that he was in breach of the statute in employing these two veterinary surgeons. Their Lordships regarded this as a special case.

### *McCandless v. General Medical Council* [1996] 1 WLR 167

**47.20** GP—negligent treatment of three patients—relationship between infamous conduct and serious professional misconduct

M, a general practitioner (GP), appealed against a determination of the PCC of the GMC of a finding of serious professional misconduct. The charges involved errors in his diagnosis of three patients and failure to refer them to hospital. Two subsequently died and the other was found, on her eventual admission to hospital, to be seriously ill. The Committee directed that M's name should be erased from the register. M accepted that, in each case, he had been negligent. On appeal, M submitted that the Committee had applied the wrong test for what amounts to serious professional misconduct in stating that his actions fell 'deplorably short' of the standard that would reasonably be expected. M submitted that poor treatment is not enough and that 'serious professional misconduct' means conduct that is morally blameworthy. In dismissing M's appeal, Lord Hoffmann said that some support could be found for the submission in the old cases on the meaning of 'infamous conduct in a professional respect'. Since *Felix v. General Dental Council* [1960] AC 704, however, much has changed. First, the words 'infamous conduct in a professional respect' were replaced in the Medical Act 1969 with the words 'serious professional misconduct'. Their Lordships thought that the authorities on the old wording did not speak with one voice and that they were of little assistance in the interpretation of the new. Second, the possible penalties available to the Committee, which used to be confined to the ultimate sanction of erasure, have been extended to include suspension and the imposition of conditions upon practise. Third, the public has higher expectations of doctors and members of other self-governing professions. Fourth, the objective test in *Doughty v. General Dental Council* [1988] AC 164 appeared to their Lordships to be, *mutatis mutandis*, equally applicable to treatment by doctors and, in their Lordships' view, should make it unnecessary in the future to revisit *Felix v. General Dental Council* or any of the other earlier authorities.

### *Hossack v. General Dental Council* [1998] 40 BMLR 97

**47.21** Negligent dental treatment—number of patients involved—expert—evidence not consistent with report

H, a registered dentist, was charged with serious professional misconduct in having accepted three patients for dental treatment, in the course of which he failed to employ a proper degree and skill and attention, and failed to carry out treatment necessary to secure their oral health. On appeal against a finding of serious professional misconduct and sanction of erasure, the Privy Council was told that had it not been for the complaint of one of the patients, Mrs L, the other cases would never have come before the PCC of the GDC. It was Mrs L's complaint that precipitated the present proceedings. In allowing H's appeal, Lord Lloyd of Berwick, giving the judgment of the Board, said that where the allegation of serious professional misconduct is grounded, as it was here, on negligent treatment, an important factor in determining whether the professional misconduct is serious or not will often be the number of patients involved. It might have been better if the legal assessor had drawn attention to this point. Having reviewed the evidence in relation to Mrs L, their Lordships felt bound to conclude that the finding of the PCC in relation to her was 'out of tune' with the evidence of the expert called by the GDC. Had the expert given evidence in accordance with his report, then their Lordships would not have intervened in the finding. But the evidence that he gave at the hearing, both in chief and in cross-examination, was so very different from what he had said in his report and was so very favourable to the appellant that the finding of fact by the Committee in relation to Mrs L could not stand. That left the question of whether the finding of serious professional misconduct can rest on the findings of fact in relation to the other two patients—to which their Lordships felt the answer must surely be 'no'. The finding of serious professional misconduct was explicitly based on the facts found to have been proved in relation to *all* of the heads of charges in relation to all three patients. Since the Committee was wrong to find the facts proved in relation to Mrs L, part of the foundation or the overall finding of serious professional misconduct had gone. In these circumstances, the overall finding could not stand.

### *Roylance v. General Medical Council (No. 2)* [2000] 1 AC 311

**47.22** Serious professional misconduct—general principles

R, a doctor, together with two surgeons, W and D, were found guilty of serious professional misconduct in relation to cardiac operations carried out between 1990 and 1995 on very young children at the Bristol Royal Infirmary. The charges arose out of a concern at the number of patients who had failed to survive their operations. The PCC found all three doctors guilty of serious professional misconduct. Neither W nor D appealed. R was the chief executive officer (CEO) of the United Bristol Healthcare National Health Service (NHS) Trust. He had no specialist expertise in the particular area of paediatric cardiac surgery and medicine with which the charges were concerned. The Committee held that he was guilty of serious professional misconduct as a registered medical practitioner on the grounds of a failure to take action over the years during which concerns were being raised about excessive mortality of the infants, and a failure to take any steps in the case of one child to prevent the operation from proceeding following concerns raised at a meeting of a number of surgeons, cardiologists, and anaesthetists the day before the operation. R's appeal to the Privy Council against the order directing that his name be erased from the register was dismissed. In considering the circumstances in which a medical practitioner might be found guilty of serious professional misconduct in which the conduct was removed from the practice of medicine, Lord Clyde, giving the judgment of the Board at 331B–E, said:

> Misconduct is a word of general effect, involving some act or omission which falls short of what would be proper in the circumstances. The standard of propriety may often be found by reference to the rules and standards ordinarily required to be followed by a medical practitioner in the particular circumstances. The misconduct is qualified in two respects. First, it is qualified by the word "professional" which links the misconduct to the profession of medicine. Secondly, the misconduct is qualified by the word "serious". It is not any professional misconduct which will qualify. The professional misconduct must be serious. The whole matter was summarised in the context of serious professional misconduct on the part of a registered dentist by Lord Mackay of Clashfern in *Doughty v. General Dental Council* [1988] AC 164, 173: "In the light of these considerations in their Lordships' view what is now required is that the General Dental Council should establish conduct connected with his profession in which the dentist concerned has fallen short, by omission or commission, of the standards of conduct expected among dentists and that such falling short as is established is serious."

Lord Clyde continued at 331H–333C:

> But certain behaviour may constitute professional misconduct even although [*sic*] it does not occur within the actual course of the carrying on of the person's professional practice, such as the abuse of a patient's confidence or the making of some dishonest private financial gain. In *Allinson v. General Council of Medical Education and Registration* [1894] 1 QB 750, 761, infamous conduct in a professional respect was held to be established where a doctor by public advertisement had warned the public to avoid other practitioners and recommended them to apply to himself. Lord Esher MR adopted at, pp. 760–761, the definition which Lopes LJ propounded in the same case of "at any rate one kind of conduct amounting to 'infamous conduct in a professional respect'". The definition was that such conduct could be established: "If it is shown that a medical man in the pursuit of his profession, has done something with regard to it which would be reasonably regarded as disgraceful or dishonourable by his professional brethren of good repute and competency…"

> Lord Esher MR then observed, at p. 761: "The question is, not merely whether what a medical man has done would be an infamous thing for anyone else to do, but whether it is infamous for a medical man to do... There may be some acts which, although they would not be infamous in any other person, yet if they are done by a medical man in relation to his profession, that is, with regard either to his patients or to his professional brethren, may be fairly considered 'infamous conduct in a professional respect', and such acts would, I think, come within section 29."
>
> But that definition is clearly not, and was not intended to be, exhaustive or comprehensive. To take the point a stage further, serious professional misconduct may arise where the conduct is quite removed from the practice of medicine, but is of a sufficiently immoral or outrageous or disgraceful character. An example can be found in *A County Council v. W (Disclosure)* [1997] 1 FLR 574, where a question arose whether the alleged sexual abuse by a father of his daughter, the father being a medical practitioner, could constitute serious professional misconduct. It was argued that any sexual abuse was too remote from the father's occupation as a doctor since it was outwith any medical treatment of a child. But Cazalet J. held, at p.581, that: "it seems to me that this doctor can be said, if he has sexually abused his daughter, to have demonstrated conduct disgraceful to him as reflecting on his profession and/or indeed conduct disgraceful to him as a practising doctor."
>
> What is important here is not only the fact that disgraceful behaviour remote from the carrying on of a professional practice may constitute serious professional misconduct, but also that the duty of a doctor to himself, if not to his profession, exists outwith the course of his professional practice. One particular concern in such cases of moral turpitude is that the public reputation of the profession may suffer and public confidence in it may be prejudiced.
>
> But moral turpitude is not the only kind of case outwith the conduct of a medical practice which may constitute serious professional misconduct. In *Marten v. Royal College of Veterinary Surgeons Disciplinary Committee* [1966] 1 QB 1 a farmer who was also a veterinary surgeon was found to have failed to give adequate care for animals on his farm. He was not guilty of any moral turpitude, but his conduct was held to constitute conduct disgraceful to him in a professional respect. Lord Parker CJ observed, at p.9: "But if the conduct, though reprehensible in anyone is in the case of the professional man so much more reprehensible as to be defined as disgraceful, it may, depending on the circumstances, amount to conduct disgraceful of him in a professional respect in the sense that it tends to bring disgrace on the profession which he practises. It seems to me, although I do not put this forward in any sense as a definition, that the conception of conduct which is disgraceful to a man in his professional capacity is conduct disgraceful to him as reflecting on his profession, or, in the present case, conduct disgraceful to him as a practising veterinary surgeon."
>
> Marten was found on account of his work as a farmer to be guilty of conduct disgraceful to him as a practising veterinary surgeon.

It was not suggested that R's conduct in the present case was in the class of moral turpitude or so outrageous in nature as to bring the profession into disrepute. But that did not mean that he had no duty to have regard to his own capacity as a registered medical practitioner. He was both a registered medical practitioner and CEO of a hospital. In each capacity, he had a duty to care for the safety and well-being of the patients. As CEO, that duty arose out of his holding of that appointment. As a registered medical practitioner, he had the general obligation to care for the sick. That duty did not disappear when he took on the appointment, but continued to coexist with it. There was a sufficiently close link with the profession of medicine in the case of R as CEO of a hospital in respect of patients at the hospital. Something of a parallel link can be traced in *Marten's case* [*Marten v. Royal College of Veterinary Surgeons Disciplinary Committee* [1966] QB 1] between the profession of veterinary surgery and the care of animals on a farm. Counsel for R sought to argue that any criticism of the doctor derived solely from his holding office as a CEO. But while the failures may, as a matter of fact, be the same, the gravity of the criticism may be increased by him being at the same time a medical practitioner. Their Lordships would add, in relation to the generality of the problem that the philosophy that seeks to divorce the administration from the medical care so as to leave the administrator free from any responsibilities for deficiencies in the care of the sick, cannot be sound. The care, treatment, and safety of the patient must be the principal concern of everyone engaged in the hospital service. The medical staff will have the specialist expertise in their various skills. But the idea of a gulf between the medical practitioners and the administration connected by some bridge over which the doctor has passed 'from us to them', as appeared in the course of the argument to be a possible aspect of R's case, must be totally unacceptable if the interest of the patient is to remain paramount. The enterprise must be one of cooperative endeavour. In ordinary circumstances, there is no doubt that a medical practitioner who holds the office of CEO of a hospital is perfectly entitled to leave the day-to-day clinical decisions to the professional staff of the hospital. His duty as a medical practitioner is adequately performed by such a course. But there may occur circumstances in which more may be required of him. In such circumstances, his medical skill and knowledge are undoubtedly relevant. Even if he does not have the specialist expertise of the particular area of medicine in which the problem arises, his general knowledge as a doctor will be of service, for example by enabling him to ask more readily the relevant kinds of question, such as, in the present case, when was the child last examined and what was the degree of urgency for the operation. In their Lordships' opinion, the Committee was entitled in the whole circumstances to find serious professional misconduct established.

***Balfour v. The Occupational Therapists Board* (2000) 51 BMLR 69**

**47.23** Occupational therapist—application form—false declaration

In dismissing B's appeal against a finding that she had falsely declared that she was a state registered occupational therapist on an application form and failing to obtain state registration as was required Lord Millett, giving the judgment of the Board, said:

> It has been settled law for more than a century that if it is shown that a medical man, in the course of his profession, has acted in a way which would reasonably be regarded as disgraceful and dishonourable by his professional brethren of good repute and competency, it is open to his disciplinary body to find him guilty of "infamous conduct in a professional respect". Such a finding involves questions of fact and degree. The members of the Disciplinary Committee are the judges of the gravity of the offence. Their Lordships cannot substitute their opinion for theirs. They can intervene only where there is no evidence to support the Committee's findings of fact, or where no reasonable Committee properly instructed on the law could regard the facts proved as constituting serious professional misconduct.

***Preiss v. General Dental Council* [2001] 1 WLR 1926**

**47.24** Treatment of patient—serious professional misconduct—gross negligence

The disciplinary charges against P, a registered dentist, all related to his treatment of one patient. The allegation was that, between January 1992 and March 1997, P had failed to provide the high standard of care to which a patient is entitled and thereby neglected his professional responsibilities to this patient. The PCC found the allegations proved and that they constituted serious professional misconduct, and ordered that P's registration in the dentist's register be suspended for twelve months. On appeal, Lord Cooke of Thorndon, giving the judgment of the Privy Council, said:

> 28. It is settled that serious professional misconduct does not require moral turpitude. Gross professional negligence can fall within it. Something more is required than a degree of negligence enough to give rise to civil liability but not calling for the

> opprobrium that inevitably attaches to the disciplinary offence. Having formed their own view of the evidence, their Lordships do not hold that any of the findings of the PCC on the detailed allegations were wrong, but they considered that in the context of the treatment of this particular patient the specific shortcomings established against the appellant vary in gravity. The core and most serious shortcoming was summarised by the PCC as failure to ensure that the state of the patient's oral health was appropriate in view of the ambitious treatment plan. That is covered specifically by charge (a) as to the period after December 1994 and charge (b)(i). It was an elementary and grievous failure warranting the description of serious professional misconduct. The Board does not consider that in this case the other charges that have been established come within that description, either individually or collectively. Some or all of them may well have constituted simple professional negligence, and they are part of the setting in which the seriousness of the appellant's conduct has to be judged; but the findings directly related to oral health are dominant.

On the issue of penalty, their Lordships considered that to suspend the appellant from practice would be neither necessary nor just. This was an exceptional case in which the right to practise should not be interfered with despite a finding of serious professional misconduct. The appropriate penalty was an admonition.

### *Rao v. General Medical Council* [2002] UKPC 65; [2003] Lloyd's Rep Med 62

**47.25**

GP—single incident—whether serious professional misconduct

During the course of the evening of 29 December 1998, when the appellant was on out-of-hours duty, he received a telephone call from Mrs P on behalf of her husband. The PCC found that he made a fundamental error in failing to find out whether there were any clinical signs indicating drug overdose. The appellant was a caring, considerate, hard-working, respected GP, and had no previous history with the GMC. The case was opened as 'an isolated incident' (as opposed to a course of conduct). The Privy Council, in setting aside the finding of serious professional misconduct, said that it proceeded on the basis that this was a borderline case of serious professional misconduct. It was based on a single incident. There was undoubted negligence, but something more was required to constitute serious professional misconduct and to attach the stigma of such a finding to a doctor of some twenty-five years' standing with an hitherto unblemished career. Their Lordships were left with a profound sense of unease and were far from satisfied that, if properly advised, the Committee would inevitably have arrived at the same conclusion. Moreover, if the findings of serious professional misconduct had been made on the basis of a single clinical error, as opposed to generalized defects in his practice, it was at least possible that the PCC would not have imposed conditions. The appropriate disposal might well have been a reprimand.

See further *R (Campbell) v. General Medical Council* [2005] 1 WLR 3488, CA, at paragraph 47.29 below.

### *Silver v. General Medical Council* [2003] UKPC 33; [2003] Lloyd's Rep Med 333

**47.26**

GP—negligent treatment of patient—whether serious professional misconduct

S, a GP with more than forty years' experience, appealed from a decision of the PCC that he was guilty of serious professional misconduct and that, for a period of twelve months, his registration should be conditional on his undergoing a performance assessment. The essence of the allegation was that, over a nine-day period, despite a number of prompts from the son, a daughter, and two other healthcare professionals, the appellant failed to ensure that his elderly patient of some thirty-five years received suitable or prompt medical attention following a fall in her home. She was eventually admitted to hospital by the emergency services and found to be suffering from a fractured neck of the left femur. In allowing S's appeal, the Privy Council stated that, in the instant case, there could be little doubt that there was negligence and that it was open to the

Committee to find that this constituted professional misconduct. However, the Committee should have gone on to consider as a separate issue whether this amounted to serious professional misconduct. It was by no means self-evident that if this question had been posed, it would have been answered in the affirmative. It was relevant to consider that this was an isolated incident relating to one patient (albeit over a number of days) as compared with a number of patients over a longer period of time. It was also relevant to take account of his long period (some forty years) of unblemished professional conduct and the particular difficulties of conducting a single-handed practice in a deprived area of London.

See further *R (Campbell) v. General Medical Council* [2005] 1 WLR 3488, CA, at paragraph 47.29 below.

### *Nandi v. General Medical Council* [2004] EWHC 2317 (Admin)

**47.27** Serious professional misconduct—meaning

In allowing N's appeal against a finding of serious professional misconduct, Collins J said:

> 31. What amounts to professional misconduct has been considered by the Privy Council in a number of cases. I suppose perhaps the most recent observation is that of Lord Clyde in *Roylance v. General Medical Council* [1999] Lloyd's Rep Med 139 at 149, where he described it as "a falling short by omission or commission of the standards of conduct expected among medical practitioners, and such falling short must be serious". The adjective "serious" must be given its proper weight, and in other contexts there has been reference to conduct which would be regarded as deplorable by fellow practitioners. It is of course possible for negligent conduct to amount to serious professional misconduct, but the negligence must be to a high degree.
>
> 32. The notes in Halsbury's Statutes to section 36 of the Medical Act of 1983 give some indication as to how the courts have over the years approached the question of serious professional misconduct. There are no closed categories and the appropriate standard is a matter for the Committee to decide. It is not restricted to conduct which is morally blameworthy. It could, as I have indicated, include seriously negligent treatment or failure to provide treatment measured by objective professional standards. Obviously, dishonest conduct can very easily be regarded as serious professional misconduct, but conduct which does not amount in any way to dishonesty can constitute serious professional misconduct if it falls far short of the standard that is considered appropriate by the profession.

### *Skidmore v. Dartford and Gravesham NHS Trust* [2003] UKHL 27; [2003] ICR 721

**47.28** Surgical procedure—charge of personal misconduct—wrong disciplinary procedure—distinction with professional conduct

S, a consultant surgeon, operated on Mrs A for the removal of a gall bladder. During the surgical procedure, the patient's left iliac artery was punctured by a sharp three-pronged instrument. The operation was eventually completed successfully and Mrs A made a complete recovery. An investigation was carried out by the Trust, which proposed a charge of personal misconduct against S. The outline statement alleged that, following the operation, S had sought to deliberately mislead the patient and her family, and the chief executive, through a series of statements and correspondence that he knew to be untrue. Lord Steyn (with whom Lord Bingham of Cornhill, Lord Clyde, Lord Hutton, and Lord Scott of Foscote agreed) said that the Trust adopted the wrong procedure and that the case was one of professional conduct. The appeal raised important issues in respect of hospital disciplinary proceedings. The context was a contractual disciplinary code. Specifically, the issues arose because of the incorporation of Department of Health Circular HC(90)9, dated March 1990, in most hospital doctors' contracts. The disciplinary code provides for a difference in procedure depending on whether the case involves allegations of 'professional conduct' or 'personal conduct'. The former is governed by a judicialized procedure under the Circular; the latter is governed by less formal disciplinary procedures, without, amongst other things, the right of legal representation. On the question of who should decide on the categorization of a case, the trust or the doctor, Lord Steyn said that prima facie the position is as follows. The trust is entitled to decide what disciplinary route should be followed. That decision must, however, comply with the terms of the contract. If a non-conforming decision is taken and acted upon, there is a breach of contract resulting in the usual remedies. The only escape from this position would be if it could be shown that the parties agreed upon wording in their contract making it clear that the employer's decision would be final, thereby excluding the role of the court except, of course, in cases of bad faith, or possibly the absence of reasonable grounds for the decision. So far, the House of Lords was in general agreement with the approach of the Court of Appeal in *Saeed v. Royal Wolverhampton Hospitals NHS Trust* [2001] ICR 903. The second question was: how is the line between professional and personal conduct to be drawn? The starting point must be the proper interpretation of the definitions contained in the disciplinary code. Professional conduct is defined as 'behaviour of practitioners arising from the exercise of medical or dental skills' and professional competence is defined as 'adequacy of performance of practitioners related to the exercise of their medical or dental skills and professional judgment'. Personal conduct is the residual category consisting of 'behaviour . . . due to factors *other* than those associated with the exercise of medical or dental skills' (emphasis added). The House of Lords did not agree with the ruling in *Saeed* that an indecent assault committed by a doctor during a medical examination cannot constitute professional misconduct within the code. It is a case of a doctor misusing his ostensible medical skills for improper purposes and falls within the scope of professional misconduct within the definition. A subsequent joint working party report supports the interpretation that when, in a doctor–patient relationship, a doctor commits deliberate misconduct, it may come within the category of professional conduct. In the instant case, the alleged lies told by S to the patient about important details of the operation could amount to professional conduct. In such a case, the medical practitioner is professing to speak as a doctor about a matter covered by his medical skills.

### *R (Campbell) v. General Medical Council* [2005] 1 WLR 3488, CA

**47.29** Serious professional misconduct—relevance of practitioner's professional history—personal mitigation

Disciplinary proceedings against Dr B, a consultant paediatrician, were set in train by the claimant, C. The GMC's PCC concluded that Dr B's treatment of C's son was substandard, not in the best interests of the patient, and likely to compromise patient safety. It further found that Dr B's treatment of another child fell below acceptable standards. In short, there was sufficient evidence that could (not did) support a finding that Dr B was guilty of serious professional misconduct under the General Medical Council Preliminary Proceedings Committee and Professional Conduct Committee (Procedure) Rules 1988. However, the Committee considered that the two cases about which it had heard evidence appeared to be isolated incidents against a background of otherwise unblemished medical practice spanning some thirty years. It also considered the outstanding testimonial evidence submitted on behalf of Dr B and, in all of the circumstances, concluded that he was not guilty of serious professional misconduct. In allowing C's claim for judicial review, the Court of Appeal (Judge, Longmore, and Jacob LJJ) said, at [21], that:

> As a general proposition, it would be surprising if rules governing the disciplinary procedures for the medical professional were to achieve the somewhat startling result that

> the question of whether a practitioner was guilty of serious professional misconduct could be influenced by matters of personal mitigation that went to the appropriate disposal of the complaint.... Mitigation arising from the circumstances in which the practitioner found himself or herself may be relevant to the level of culpability: once serious professional misconduct is proved, personal mitigation will be relevant to possible penalty.... [T]hese are distinct issues, to be determined separately, on the basis of evidence relevant to them.

At [25], the Court said that the inevitable conclusion is that evidence relevant to personal mitigation was used by the Committee to inform their decision that the proved misconduct did not amount to serious professional misconduct. However, to the extent that *Rao v. General Medical Council* and *Silver v. General Medical Council* decided that evidence exclusively relevant to personal mitigation could be considered by the Committee when deciding whether serious professional misconduct was proved, the decisions were wrong.

### *Council for the Regulation of Health Care Professionals v. General Medical Council and Biswass* [2006] EWHC 464 (Admin)

**47.30** Serious professional misconduct—matter of judgment rather than proof

Jackson J, at [41], said that he entirely agreed with the views expressed by Dame Janet Smith in paragraph 21.32 of her Fifth Shipman Report:

> 21.32 The doctor (or his/her representative) was then invited to address the PCC panel in mitigation and to adduce evidence in support if desired. After that, the PCC panel would deliberate again. It would consider whether the facts proved did amount to SPM [serious professional misconduct] and, if so, what sanctions should be imposed. In my view, these were both matters of judgment for the PCC panel, rather than a matter of proof. However, there are indications in the GMC documents that some people were of the view that SPM must be proved "beyond reasonable doubt"... [I]n my view, only the facts were a matter for "proof". The other issues were matters of judgment.

### *Livingstone v. The Adjudication Panel for England* [2006] EWHC 2533 (Admin)

**47.31** Mayor—offensive abuse to journalist—not acting in official capacity—not acting unlawfully

Following an incident outside City Hall when L, a former mayor of London, was leaving a reception, a formal complaint was made by the Board of Deputies of British Jews to the Standards Board for England. The complaint related to L's language towards a newspaper reporter, in which he said: 'What did you do before? Were you a German War criminal?... You are just like a concentration camp guard.' The Adjudication Panel of the Standards Board held that, when L made the remarks (which were tape-recorded), he was not acting in his official capacity. However, his remarks were in breach of paragraph 4 of the Model Code of Conduct, which provided that a member of a local authority must not, in his official capacity 'or any other circumstances', conduct himself in a manner that could reasonably be regarded as bringing his office or authority into disrepute. In allowing L's appeal, Collins J said that the Panel correctly decided that L was not acting in his official capacity when he made the remarks in question. He had ceased to act in his official capacity as host of a reception at City Hall and was leaving the building to go home. It was not arguable that, when making the remarks, L was performing his functions as mayor. On its true construction, paragraph 4 of the Code of Conduct did not apply to what the appellant said. However, assuming that paragraph 4 did apply to the circumstances, the appellant was indulging in offensive abuse of a journalist and anyone is entitled to say what he likes of another, provided that he does not act unlawfully and so commits an offence under, for example, the Public Order Act 1986.

### *Dzikowski v. General Medical Council* [2006] EWHC 2468 (Admin)

**47.32** Psychiatrist—conditional registration—compliance with Department of Health guidelines—condition justified—guidelines recommendation, not instruction

D, a consultant psychiatrist, was charged with prescribing a methadone mixture to a patient in circumstances that were inappropriate, irresponsible, and not in the best interests of his patient. The fitness to practise panel found D guilty of serious professional misconduct and placed conditions on his registration, which included compliance with the Department of Health publication entitled *Drug Misuse and Dependence: Guidelines on Clinical Management* (the Orange Book) and the British National Formulary in his treatment of patients. D appealed on the ground, inter alia, that he could not be found guilty of serious professional misconduct for failing to follow the Orange Book guidance because it was a recommendation and not an instruction, and that he was justified in his deviation from the guidelines contained in the Orange Book. D challenged the expert evidence given about the Orange Book guidelines at the hearing before the panel. Evidence was given in the main for the GMC by one of the authors of the Orange Book, who also provided a report to the panel, including a summary and overall conclusions. In dismissing D's appeal, Hodge J said that it was clear that the panel preferred the evidence of the GMC's expert on these issues and it was entitled to do so. The Orange Book was important guidance and it should be followed by practitioners. The appellant was aware of it. The evidence is clear that he ignored it when treating Miss A. The appellant justified his departure from it, although he clearly did not pay regard to it, on the basis that he was a specialist and could make a proper assessment of Miss A. The panel did not accept that evidence and it was justified in doing so.

### *General Medical Council v. Meadow* [2007] QB 462

**47.33** Criminal proceedings—expert witness—disciplinary proceedings based on expert evidence—serious professional misconduct—meaning

This case arose out of evidence given by Professor M, an eminent paediatrician and child healthcare specialist, in the prosecution of Mrs C. In November 1999, Mrs C was tried for the murder of her two sons. The Crown relied, in part, upon Professor M's evidence to refute the proposition that Mrs C's children may have died from sudden infant death syndrome (SIDS), or cot death. Mrs C was convicted. Her first appeal was dismissed in October 2000. Her second appeal was allowed on 29 January 2003, on the ground that the verdicts were unsafe because of material non-disclosure by the Crown's pathologist. Full argument on Professor M's evidence was not heard during the second appeal, but the Court indicated that, if it had been, the appeal would 'in all probability' have been allowed on that ground too. Mrs C's father made a complaint to the GMC alleging serious professional misconduct on the part of Professor M. In July 2005, a fitness to practise panel of the GMC found Professor M guilty of serious professional misconduct and ordered that his name be erased from the register. The Court of Appeal, by a majority (Auld and Thorpe LJJ; Sir Anthony Clarke MR dissenting), dismissed the GMC's appeal against the decision of Collins J, reversing the panel on its finding of serious professional misconduct. Auld LJ said:

> [198] As to what constitutes "serious professional misconduct", there is no need for any elaborate rehearsal by this court of what, on existing jurisprudence, was capable of justifying such condemnation of a registered medical practitioner under the 1983 Act before its 2003 amendment. And, given the retention in the Act in its present form of section 1(1A), setting out the main objective of the GMC "to protect, promote and maintain the health and safety of the public", it is inconceivable that "misconduct"—now one of the categories of impairment of fitness to practise provided by section 35C of the Act—should signify a lower threshold for disciplinary intervention by the GMC.

[199] It is common ground that Professor [M], in giving and/or purporting to give, expert medical evidence at the trial of Mrs [C], was engaged in conduct capable of engaging the disciplinary attention of the GMC.

[200] As Lord Clyde noted in *Roylance v. General Medical Council* [2000] 1 AC 311 (PC), at 330F-332E, [1999] 3 WLR 541, 47 BMLR 63, "serious professional misconduct" is not statutorily defined and is not capable of precise description or delimitation. It may include not only misconduct by a doctor in his clinical practice, but misconduct in the exercise, or professed exercise, of his medical calling in other contexts, such as that here in the giving of medical evidence before a court. As Lord Clyde might have encapsulated his discussion of the matter in *Roylance,* it must be linked to the practice of medicine or conduct that otherwise brings the profession into disrepute, and it must be serious. As to seriousness, Collins J, in *Nandi v. General Medical Council* [2004] EWHC 2317 (Admin), rightly emphasised, at para 31 of his judgment, the need to give it proper weight, observing that in other contexts it has been referred to as "conduct which would be regarded as deplorable by fellow practitioners".

[201] It is also common ground that serious professional misconduct for this purpose may take the form, not only of acts of bad faith or other moral turpitude, but also of incompetence or negligence of a high degree: see *Preiss v. General Dental Council* [2001] 1 WLR 1926, at para 28. It may also be professional misconduct where, as here, a medical practitioner, purporting to act or speak in such expert capacity, goes outside his expertise. Whether it can properly be regarded as "serious" professional misconduct, however, must depend on the circumstances, including with what intention and/or knowledge and understanding he strayed from his expertise, how he came to do so, to what possible, foreseeable effect, and what, if any, indication or warning he gave to those concerned at the time that he was doing so.

### *Law Society v. Adcock* [2007] 1 WLR 1096

**47.34** Solicitor—local authority searches—commission paid to solicitors—allegations of false accounting and dishonesty dismissed

This was an appeal by the Law Society under section 49 of the Solicitors Act 1974 against a decision of the Solicitors Disciplinary Tribunal (SDT) by which it dismissed disciplinary proceedings against two solicitors in partnership. The solicitors' firm had entered into an arrangement with a company for it to carry out local authority and other searches on behalf of clients of the firm in conveyancing matters. Under the arrangement, the lay client of the firm would be charged the full cost of the searches, but after that fee had been paid to the company by the firm, the latter would invoice the company a 'commission' of £20. No mention of the arrangement between the firm and the company was made to the client. By its rule 4 statement under the Solicitors' (Disciplinary Proceedings) Rules 1994, the Law Society alleged that the solicitors were guilty of conduct unbefitting a solicitor in that they provided misleading information to the clients regarding the cost of local searches and that they delivered inaccurate bills to clients. The rule 4 statement alleged that this was tantamount to false accounting and was dishonest. The SDT dismissed the allegation on the ground that, on the agreed facts, the Law Society could not succeed. In dismissing the Law Society's appeal, Waller LJ (with whom Treacy J agreed) said that the solicitors' reliance on rule 10 of the Solicitors' Practice Rules 1990 (receipt of commissions from third parties) did not provide an answer to the conduct of the solicitors in this case. However, it was inconceivable that any tribunal could find that the solicitors were in any way dishonest and very unlikely that a tribunal would hold that these solicitors had acted in a way unbefitting a solicitor. The Law Society had no evidence that these solicitors were aware of the notice put in the *Law Gazette.* Even if a tribunal were to come to the conclusion that the conduct of the solicitors fell foul of rule 1 of the Solicitors' Practice Rules 1990, in that the arrangement compromised their duty to act in the best interests of their client, it would be quite unfair to impose any great penalty on them.

### *Mallon v. General Medical Council* [2007] CSIH 17

**47.35** GP—treatment of child—three months' suspension—misconduct—wrongful or inadequate mode of performance of professional duty—serious—exercise of judgment on facts and circumstances

M, a GP, was charged before the fitness to practise panel with serious professional misconduct. The charge was of the gravest kind. It alleged that, when she first treated a child at the health centre, she failed to diagnose that his asthma attack was acute and potentially life-threatening, that she failed to have him admitted to hospital as an emergency, that this failure resulted in the loss of his life, and that, at a fatal accident inquiry into the death, she gave untruthful evidence about her treatment of him. There were further allegations that M failed to take sufficiently detailed history from the child's mother, that she failed to carry out a proper examination of him before nebulization, that she failed to keep proper notes of her examination, and that after the nebulization she failed to give him the steroid prednisolone. The panel dismissed much of the charge, including the allegations that M's failure to take appropriate action resulted in the loss of the child's life and that the appellant gave untruthful evidence at the fatal accident inquiry. But it found proved that M had failed to undertake proper enquiries about the child's condition, had failed to carry out a proper examination or to keep proper notes, and had delayed inappropriately in commencing nebulization or giving prednisolone. The panel imposed a sanction of three months' suspension. In dismissing M's appeal, the Inner House, Court of Session, said, in relation to serious professional misconduct, that:

> [17] The starting point in this appeal is that the appellant does not dispute the Panel's findings of fact, nor does she dispute that those findings disclose a case of professional misconduct.
>
> [18] In a case such as this, "misconduct" denotes a wrongful or inadequate mode of performance of professional duty; or as Lord Clyde described it in *Roylance v. GMC (No. 2)* ([2000] 1 AC 311, at p 331B–C), it is "a word of general effect, involving some act or omission which falls short of what would be proper in the circumstances." The question raised in this appeal is whether the panel was entitled to hold that the appellant's conduct was "serious". The statute does not lay down any criterion of seriousness, nor does the case law. Descriptions of serious professional misconduct such as "conduct which would be regarded as deplorable by fellow practitioners" (*Nandi v. GMC* [2004] All ER (D) 24, Collins J at para [31], quoted in *Meadow v. GMC, supra,* Auld LJ at paras [200]–[201]) tend, we think, to obscure rather than assist our understanding. In view of the infinite varieties of professional misconduct, and the infinite range of circumstances in which it can occur, it is better, in our opinion, not to pursue a definitional chimera. The decision in every case as to whether the misconduct is serious has to be made by the panel in the exercise of its own skilled judgment on the facts and circumstances and in the light of the evidence (*Roylance v. GMC, supra,* Lord Clyde at p 330f; *Preiss v. GDC* [2001] 1 WLR 1926, Lord Cooke of Thorndon at para 28). Misconduct that the panel might otherwise consider to be serious may be held not to be in the special circumstances of the case (*R (Campbell v. GMC)* [2005] 2 All ER 970, Judge LJ at para [19]).

### *Vranicki v. Architects Registration Board* [2007] EWHC 506 (Admin)

**47.36** Architect—seven allegations—not necessary to find whether each allegation amounted to serious professional incompetence

The charges against V, an architect, related to her professional work in connection with an extension to a house in Kilburn. The case against her concerned failures to administer the project competently, by not maintaining control over the contactors and not ensuring the smooth progress of the project. The charges were broken down into seven allegations: four were of unacceptable professional conduct; three were of serious professional incompetence. The PCC accepted V's evidence in relation to two allegations, which were based on a lack of communication. It dismissed the charge of unacceptable professional conduct, but found proved that of serious professional incompetence. In doing so, the Committee approached each head of charge as to whether it was serious professional incompetence. In dismissing V's appeal, Collins J said that the Committee was, in his judgment, wrong in the sense that it was unnecessary to decide that each individual allegation that it found proved established serious professional incompetence. What it should have done was to consider all of the allegations that were found proved to decide whether, together, they established the charge against the appellant. However, it had been, as it happens, helpful to see that the Committee considered each allegation individually as to whether it justified a finding of serious professional incompetent. This made all the more clear that if they had been considered cumulatively, a similar finding would have been made.

*Calhaem v. General Medical Council* **[2007] EWHC 2606 (Admin)**

**47.37** Negligence—whether amounting to misconduct

Jackson J said, at [39], that mere negligence does not constitute misconduct within the meaning of section 35C(2)(a) of the Medical Act 1983. Nevertheless, and depending upon the circumstances, negligent acts or omissions that are particularly serious may amount to misconduct. A single negligent act or omission is less likely to cross the threshold of misconduct than multiple acts or omissions. Nevertheless, and depending upon circumstances, a single negligent act or omission, if particularly grave, could be characterized as misconduct.

See paragraph 16.06.

*Joyce v. Secretary of State for Health* **[2008] EWHC 1891 (Admin); [2009] PTSR 266**

**47.38** Care worker at nursing home—protection of vulnerable adults and children—allegation of misconduct—misconduct not limited or qualified

J, a care worker at a nursing home for dementia patients, was found to be asleep in the residents' lounge whilst on duty. Following dismissal by Bupa, the owner of the nursing home, Bupa referred J to the Protection of Vulnerable Adults (POVA) list held by the Department of Health under section 82(1) of the Care Standards Act 2000. The alleged misconduct was defined as 'sleeping on duty'. In giving reasons for including J's name on the POVA list, the Secretary of State set out detailed reasons that went beyond sleeping on duty, and included allegations that, whilst working on the night shift, J had failed to administer medication to a resident and had left medication sitting out in the home, that the security door to the dementia unit was found to be unsecured, and that the unit was found to be in a state falling below the requisite standards of health and hygiene whilst J was the nurse in charge. Judge Pearl, sitting as president of the Care Standards Tribunal, considered that, on the proper construction of section 86(3) of the Care Standards Act 2000, there was no restriction on the misconduct that might be considered by the Care Standards Tribunal. On appeal, Goldring J agreed and said that the statutory language in section 86(3) is clear: 'misconduct' is not limited or qualified. If Parliament had intended that it should be, it could easily have said so. Parliament's intention was plainly that, in order to fulfil its role as a protector of vulnerable adults and children, the Tribunal should be able to consider as widely as possible. Only in the Tribunal, and for the first time, is the evidence properly considered. When the appeal is being prepared, further incidences of misconduct may become apparent. They may even become apparent during the hearing. There is no plausible reason why Parliament should have intended that such incidences of misconduct should not be considered. To limit the Tribunal would seriously undermine its purpose, which is to protect vulnerable people. It would mean that the reference from the provider would thereafter wholly define what follows. If the provider were to fail to refer other serious misconduct, the Tribunal could not consider it. If other serious misconduct were to come to light shortly before an appeal, such as serious misconduct resulting in criminal convictions, it could not be considered in the appeal. The purpose of the 2000 Act would thus seriously be undermined.

### *Akodu v. Solicitors Regulation Authority* [2009] EWHC 3588 (Admin)

**47.39** Solicitor—mortgage transactions—solicitor not directly responsible—whether jointly and severally liable for failings of partner

In this case, the Divisional Court (Moses LJ and Tomlinson J) held that a solicitor could not be found guilty of conduct unbecoming a solicitor on the basis that, because he was a partner in a firm of solicitors, he was thus jointly and severally liable for a failing of his partner. The SDT had found that, in a series of mortgage transactions in which the appellant solicitor's firm had acted, there had been a failure by the appellant's partner to inform the mortgagee, in regard to various properties, of a difference in purchase prices between the full purchase price and that stated on the mortgage offer. A difference in purchase prices had occurred, because the sellers of the properties had offered discounts or incentives on the properties, which were new-build developments. The Tribunal had found that the fee-earner had failed to inform lenders of the discounts and that, whilst he had not acted dishonestly, he had behaved in a reckless way contrary to the Solicitors Practice Rules 1990 by not acting in the interests of his lender clients. The Tribunal found that the appellant, as a partner of the firm, was jointly and severally liable for the failing of his partner, and thereby guilty of conduct unbefitting a solicitor. The Divisional Court allowed the appeal, Moses LJ saying that it was not open to the Tribunal to find the appellant guilty of conduct unbefitting a solicitor on the basis, and the only basis advanced, that he was a partner in the firm. The appellant had not been directly responsible for the fact that financial incentives had not been notified to lender clients. Some degree of personal fault was required before a solicitor can be found guilty of conduct unbefitting his profession.

### *R (Remedy UK Ltd) v. General Medical Council* [2010] EWHC 1245 (Admin)

**47.40** Misconduct—different types or kinds—general principles

For the facts of this case, see paragraph 39.16. After an extensive review of authority, Elias LJ (with whom Keith J agreed) said, at [37], that he derived the following principles from the cases:

(1) Misconduct is of two principal kinds. First, it may involve sufficiently serious misconduct in the exercise of professional practice such that it can properly be described as misconduct going to fitness to practise. Second, it can involve conduct of a morally culpable or otherwise disgraceful kind which may, and often will, occur outwith the course of professional practice itself, but which brings disgrace upon the doctor and thereby prejudices the reputation of the profession.
(2) Misconduct falling within the first limb need not arise in the context of a doctor exercising his clinical practice, but it must be in the exercise of the doctor's medical calling. There is no single or simple test for defining when that condition is satisfied.
(3) Conduct can properly be described as linked to the practice of medicine, even though it involves the exercise of administrative or managerial functions, where they are part of the day practice of a professional doctor. These functions include the matters identified in *Sadler*, such as proper record-keeping, adequate patient communication, proper courtesy shown to patients and so forth. Usually a failure adequately to perform these functions will fall within the scope of deficient performance rather than misconduct, but in a sufficiently grave case, where the negligence is gross, there is no reason in principle why a misconduct charge should not be sustained.
(4) Misconduct may also fall within the scope of a medical calling where it has no direct link with clinical practice at all. *Meadow* provides an example, where the activity in question was acting as an expert witness. It was an unusual case in the sense that Professor [M]'s error was to fail to recognise the limit of his skill and

expertise. But he failed to do so in a context where he was being asked for his professional opinion as an expert paediatrician. Other examples may be someone who is involved in medical education or research when their medical skills are directly engaged.

(5) *Roylance* demonstrates that the obligation to take responsibility for the care of patients does not cease simply because a doctor is exercising managerial or administrative functions one step removed from direct patient care. Depending upon the nature of the duties being exercised, a continuing obligation to focus on patient care may co-exist with a range of distinct administrative duties, even where other doctors with a different specialty have primary responsibility for the patients concerned.

(6) Conduct falls into the second limb if it is dishonourable or disgraceful or attracts some kind of opprobrium; that fact may be sufficient to bring the profession of medicine into disrepute. It matters not whether such conduct is directly related to the exercise of professional skills.

(7) Deficient performance or incompetence, like misconduct falling within the first limb, may in principle arise from the inadequate performance of any function which is part of a medical calling. Which charge is appropriate depends on the gravity of the alleged incompetence. Incompetence falling short of gross negligence but which is still seriously deficient will fall under section 35C(2)(b) rather than (a).

(8) Poor judgment could not of itself constitute gross negligence or negligence of a high degree but it may in an appropriate case, and particularly if exercised over a period of time, constitute seriously deficient performance.

(9) Unlike the concept of misconduct, conduct unrelated to the profession of medicine could not amount to deficient performance putting fitness to practise in question. Even where deficient performance leads to a lack of confidence and trust in the medical profession, as well it might – not least in the eyes of those patients adversely affected by the incompetent doctor's treatment – this will not of itself suffice to justify a finding of gross misconduct. The conduct must be at least disreputable before it can fall into the second misconduct limb.

(10) Accordingly, action taken in good faith and for legitimate reasons, however inefficient or ill-judged, is not capable of constituting misconduct within the meaning of section 35C(2)(a) merely because it might damage the reputation of the profession. Were that not the position then Professor [M] would have been guilty of misconduct on this basis alone. But that was never how the case was treated.

### *R (Gaunt) v. The Office of Communications* [2011] 1 WLR 2355, CA

**47.41** See paragraph 32.26.

### *R (Calver) v. Adjudication Panel for Wales* [2012] EWHC 1172 (Admin)

**47.42** See paragraph 32.27.

### *R (Aga) v. General Medical Council* [2012] EWHC 782 (Admin)

**47.43** Consultant gastroenterologist—failure to recognize hypoglycaemia—no adverse effect on patient—act or omission not sufficiently grave to constitute misconduct

Dr A, an experienced consultant gastroenterologist, challenged the decision of the GMC to the effect that his 'failure to recognize' hypoglycaemia in a 54-year-old patient ('X') should be characterized as misconduct. It was nonetheless decided that his fitness to practise was not impaired and that it was not appropriate to give a warning or to impose any other sanction. Accordingly, Dr A had no right of appeal under section 40 of the Medical Act 1983 and his challenge took the form of an application for judicial review. The fitness to practise panel found that Dr A had failed to recognize hypoglycaemia as the cause of patient X's impaired consciousness and to note the recurrent low blood glucose recorded in the case notes over the preceding day. Eady J, on reviewing the evidence, said that while it may be possible to criticize Dr A for not having come to a final conclusion as to the cause of unconsciousness a few minutes earlier than he did, no act or omission was established that in any way adversely affected the patient. Dr A established the blood glucose levels and had them adjusted. This was not a case of multiple acts or omissions, and in so far as it was an act or omission at all, it could not be characterized as particularly grave so as to attract the attribution of misconduct. In fact, Dr A may well be right in saying that he saved the patient's life. Whilst not attempting to substitute the Court's judgment on a matter of medical knowledge, Eady J said that he could not see any rational basis for characterizing what happened as 'misconduct'. He did not believe that any reasonable onlooker would apply that word to the events that occurred. Accordingly, the decision would be quashed.

***Walker-Smith v. General Medical Council* [2012] EWHC 503 (Admin)**

**47.44**

See paragraph 26.09.

***O'Connor v. Bar Standards Board*, 17 August 2012 (unreported)**

**47.45**

Barrister—signing statement of truth on behalf of client—conduct of litigation—CPR—whether misconduct—discreditable conduct—meaning

The appellant, a barrister, appealed against the decision of the Disciplinary Tribunal, which had found five charges of professional misconduct proved. Charges 1–4 each alleged that the appellant had signed a statement of truth on behalf of clients in litigation that was being conducted. Charges 1 and 2 were committed whilst the appellant was instructed by a solicitor. Charges 3 and 4 related to a case in which the appellant was instructed directly, subject to direct access. Charge 5 alleged that the appellant engaged in conduct discreditable to a barrister in that, as a member of an unregulated limited liability partnership (LLP), she had sent a copy of a defence and counterclaim, which had been filed in the case, to the solicitors on the other side. Sir Andrew Collins, giving the judgment of the Visitors, observed that CPR Part 22 dealt with signing of a statement of truth. The statement of truth was required to be signed by, amongst others, the legal representative on behalf of the party. 'Legal representative' was defined in Rule 2.3 as including a barrister. Thus it is clear that the Civil Procedure Rules entitle a member of the Bar representing a litigant to sign a statement of truth on that litigant's behalf. The Guidance for Barristers issued in June 2004 states that barristers in self-employed practice are not authorized litigators and it is crucial that barristers ensure that they do not conduct litigation. However, it seems impossible to say that the CPR can effectively be put on one side. The issue for the Visitors to decide was whether it had been proved that for the appellant to have signed the statement of truth in the four cases did amount to 'conduct of litigation' and thus misconduct. It was clear that it was not proved—which left charge 5. That alleged that the appellant engaged in conduct discreditable to a barrister by committing an offence under section 70(8) of the Courts and Legal Services Act 1990 as a member of an unregulated LLP by sending the defence and counterclaim to the solicitors representing the other party. In *Agassi v. Robinson* [2005] EWCA Civ 1507, it was made clear that the conduct of correspondence with an opposing party was not to be regarded as something that a self-employed barrister should not do, because it was not to be regarded as the conduct of litigation. This was not the commission of an offence under section 70(8). Accordingly, charge 5 also had to disappear. Sir Andrew Collins stated *obiter* that argument was raised that the appellant was not, in any event, guilty of discreditable conduct. Reliance was placed upon observations of Mr John Hendy QC in *Bar Standards Board v. Sivanandan* on 13 January 2012. Mr Hendy had said that the juxtaposition of dishonesty and discreditableness at paragraph 301 in the Code of Conduct was significant: 'A barrister must not engage in conduct whether in pursuit of his profession or otherwise which is (a) dishonest or otherwise discreditable to a barrister.' Mr Hendy said that the word 'discreditable' does not have to be construed *ejusdem generis*, but the gravity of the conduct takes colour from the fact that the first description of the untoward conduct is dishonest. Anything short of serious professional misconduct is not intended to be within the description of discreditable. Sir Andrew Collins said that the Visitors saw no reason to dissent from that approach, but that it sufficed to say, in the context of this case, that the Visitors doubted if anyone looking at the case would take the view that, even assuming that the sending of the letter was something that the appellant should not have done, it could be regarded as discreditable in any way.

***Kumar v. General Medical Council* [2012] EWHC 2688 (Admin)**

**47.46** Criminal proceedings—expert witness—disciplinary proceedings based on expert evidence—panel finding of recklessness

K, a consultant psychiatrist, appealed against the decision of a fitness to practise panel of the GMC made on 26 May 2011. The panel suspended K's registration for four months on the grounds that his fitness to practise was impaired by reason of misconduct relating to evidence that he gave as an expert witness for the defence in a murder trial in 2009. The trial judge had referred K to the GMC because of the judge's concerns about K's evidence. K concluded that the defendant in the criminal proceedings lacked intent to commit murder and also had a condition called 'intermittent explosive disorder'. The jury rejected the defences of lack of intent and diminished responsibility. There was no appeal. After the verdict, the trial judge summarized the defects in K's work and referred him to the GMC, with a view to his undertaking training on the role of an expert in criminal trials: he thought that K had, at times, shown an embarrassing lack of professionalism. This led to K being charged by the GMC with misconduct. Paragraph 3 of the charges alleged that K had not disclosed to his instructing solicitors that he had no previous experience of acting as an expert witness in a case of homicide. K contested the allegation, which the panel found proved. It was admitted or found proved, amongst other things, that K's report had no mention of section 2 of the Homicide Act 1957, made no mention of the fact that he had not read the witness statements, and did not explain that intermittent explosive disorder was not recognized in the World Health Organization (WHO) International Classification of Diseases (ICD) and was a controversial diagnosis. In evidence during the criminal trial, K stated that he had prepared reports in 'four or five' murder cases, had not seen the prosecution statements, had based his opinion on what the defendant told him had happened and on the interviews conducted with him, and had not challenged the defendant in respect of contrary accounts given by him. The panel found that K acted recklessly in accepting the instructions and in not disclosing that he had no experience in acting as an expert witness in a homicide case, and that K had acted below the standard of a reasonably competent psychiatrist acting as an expert witness. The meaning of 'reckless' before the panel was derived from *R v. G and anor* [2003] UKHL 50, [32], where Lord Bingham said that:

> [T]he most obviously culpable state of mind is no doubt an intention to cause the injurious result, but knowing disregard of an appreciated and unacceptable risk of causing an injurious result or a deliberate closing of the mind to such risk would be readily accepted as culpable also.

The panel expressly applied this test, emphasizing that it was a subjective, not objective, test. The panel was satisfied that K knew what was required of an expert witness; he had read the GMC's guidance *Acting as an Expert Witness* and was familiar with the decision of the Court of Appeal in *General Medical Council v. Meadow* [2007] QB 462. On appeal, Ouseley J rejected the appellant's submissions on the findings as to recklessness and misconduct. At [59], the learned judge said:

> First, the comment of the Court of Appeal in *Meadow* [*per* Auld J at [211]] to the effect that rarely, absent bad faith or recklessness, will the giving of honest albeit mistaken expert evidence amount to misconduct, does not mean that misconduct can only arise in cases where recklessness or bad faith are. The overriding test remains that in *Preiss* [*v. General Dental Council* [2001] 1 WLR 1926]. Second, the actual giving of evidence in court, oral or written, or the preparation and content of a report for use in court, may be of such a nature or degree of incompetence or negligence, that it amounts to misconduct without bad faith or recklessness, and as the Court of Appeal itself recognises. There may be circumstances surrounding the acceptance of instructions, the making of a diagnosis, or the content of an expert report for use in preparing a case in which it is clear that, even if recklessness or bad faith is not proved, misconduct can be charged because of the degree and nature of any negligence and the risks created by it. A person, honestly and without recklessness, may fail to appreciate his limitations or other failings, and the serious consequences which his actions could create, even where he obviously ought to have been aware of them. An instance could be the acceptance of instructions and preparation of a report by someone whose failings were such that he was unaware of his serious limitations as an expert. It is not incumbent on the FTP panel to assess whether a case is a rare one or not, or only to find misconduct proven on a few occasions out of those in which such a finding is warranted. Rather the concept of rarity here reflects the degree to which a medical practitioner must fall short of the standards expected before his acts amount to misconduct in those circumstances. This is because of the awareness of the Court of the context in which such reports are prepared and evidence is given; but the more remote the failings from actual evidence in court, the less important are those factors.

In dismissing the appeal on sanction, Ouseley J said that it was not a case for a lesser sanction, such as conditions. K's failings were many, serious, and dangerous. A defendant in a murder trial relied on him; the jury might have relied on him; the public and the victim's family could have found a man convicted only of the lesser offence of manslaughter with all that that would entail for sentencing, when he was in truth guilty of murder and should be sentenced as such. It is vital to uphold the confidence of the criminal justice system, indeed in any area of the justice system, that medical witnesses have the requisite expertise, and provide balanced and accurate expert reports. The requirement for a review after the expiry of the four months' suspension was justified. There was an obvious justification for K showing how he had gained insight into his misconduct, had taken further study, and had learned something from his serious departures from what was expected of an expert witness.

### *Spencer v. General Osteopathic Council* [2013] 1 WLR 1307

**47.47**

Osteopath—record-keeping—proper assessment and treatment given—whether unacceptable professional conduct—meaning

S, a registered osteopath, pleaded guilty before the PCC of the General Osteopathic Council (GOsC) to three allegations of failing, in relation to two consultations with a patient, to record adequately the patient's case history and, in relation to one of the consultations, to record adequately the result of the examination of the patient's hip joints. Following a complaint by the patient concerned to the GOsC, the matter was investigated and a number of allegations brought forward for consideration. However, the opinion of a senior osteopath expert sought by the GOsC exonerated the appellant of many of the criticisms. Other allegations of failing to investigate specific signs and symptoms adequately were not established before the Committee. The only matters that were established were those acknowledged by S from the outset and these related to record-keeping. The Committee found S guilty of 'unacceptable professional conduct', within the meaning of section 20 of the Osteopaths Act 1993, and concluded that the appropriate sanction was an admonishment. Section 20 provides that 'unacceptable professional conduct' means conduct that falls short of the standard required of a registered osteopath. Irwin J said that he accepted that note-taking and retention of notes is very important for osteopaths. Moreover, the present case was dealing not with one single act of poor performance, but two: two examples on two dates, on which proper notes had not been taken. There had, however, been a proper assessment of the patient, a proper plan for treatment, and proper treatment was given. The Court reviewed *Meadow v. General Medical Council* [2007] 2 QB 462, *Preiss v. General Dental Council* [2001] 1 WLR 1926, *Silver v. General Medical Council* [2003] Lloyds Rep Med 333, and *Calhaem v. General Medical Council* [2008] LS Law Med 96. In quashing the finding of 'unacceptable professional conduct', Irwin J said:

> 23. In my judgment, the starting point for interpreting the Osteopaths Act 1993 must be the language of the Act itself. Although one notes that "unacceptable professional conduct" has the definition in section 20(2): "conduct which falls short of the standard required of a registered osteopath", there is an unhelpful circularity to the definition. Indeed one might not unfairly comment that the statutory definition adds little clarity. The critical term is "conduct". Whichever dictionary definition is consulted, the leading sense of the term "conduct" is behaviour, or the manner of conducting one's self. It seems to me that at first blush this simply does imply, at least to some degree, moral blameworthiness. Whether the finding is "misconduct" or "unacceptable professional conduct", there is in my view an implication of moral blameworthiness, and a degree of opprobrium is likely to be conveyed to the ordinary intelligent citizen. That is an observation not merely about the natural meaning of the language, but about the likely effect of the finding in such a case as this, given the obligatory reporting of the finding under the Act.
>
> [...]
>
> 25. ... [I]f Parliament had intended to give formal powers of warning or admonition to the Council, in circumstances where the Registrant had breached the Code of Practice but not been guilty of unacceptable professional conduct, it would have been very simple to do so. Equally it seems to me, there is nothing to prevent the PCC from giving advice to a practitioner where allegations have been made out which constitute a breach of the Code of Practice (or indeed the Standard of Proficiency) but where neither professional incompetence nor "unacceptable professional conduct" is made out. As it is, the Act stipulates that if unacceptable professional conduct is made out, there has to be at least a formal admonition and publicity which is bound to affect the Registrant's professional reputation. Those are considerable sanctions. In my view, they support the natural meaning of the language contained in the statute and point to a threshold for a finding of "unacceptable professional conduct" which there is no reason to distinguish from "misconduct" in the medical and dental legislation.

The learned judge said that it seemed to him that the present case was not one of 'incompetence or negligence of a high degree' (such as to constitute misconduct—*per* Auld LJ in *Meadow v. General Medical Council*, at [201]) or at least not self-evidently so. It was hard to construe the failings of S as worthy of the moral opprobrium and the publicity that flow from a finding of unacceptable professional conduct.

### *Pottage v. Financial Services Authority* [2013] Lloyd's Rep FC 16

**47.48** CEO—compliance with regulatory standards—failures identified and remedied—defined plan—steps to monitor compliance

P, the CEO of UBS AG and UBS Wealth Management (UK) Ltd, referred to the Upper Tribunal (Tax and Chancery Chamber) a decision of the Financial Services Authority (FSA) to impose on him a penalty for misconduct of £100,000. The penalty was imposed by the FSA pursuant to section 66 of the Financial Services and Markets Act 2000. The case against P was that he had failed to take reasonable steps to ensure that the business of the firm complied with the requirements of standards of the regulatory system. In allowing P's appeal, Sir Stephen Oliver QC, giving the judgment of the

Upper Tribunal, said that, on the evidence as a whole, the FSA had not established its case that P had committed misconduct. The Tribunal was aware that the explanations that P had given in the course of interviews with the FSA were less focused and indicated a level of awareness on his part of the firm's actual exposure to risk that did not emerge from the measured account in his statement. The fact remained, however, that every specific control failure identified had been fully investigated and had been remedied, or was being dealt with in accordance with a defined plan. Steps had been taken to strengthen the compliance monitoring team; these had been under way since early 2007 and a suitable candidate had been found and started in office in September 2007. It was also a fact that no one, including the FSA, had considered it to be necessary or appropriate to carry out a wider review of systems and controls that had, in fact, been put in place. P himself took a number of steps throughout the relevant period (September 2006 to July 2007). The FSA had not satisfied the Tribunal, from the evidence as a whole, that P's standard of conduct was 'below that which would be reasonable in all the circumstances': see the Statements of Principle and Code of Practice for Approved Persons (APER) in the FSA Handbook, APER 3.1.4G.

### *Giwa-Osagie v. General Medical Council* [2013] EWHC 1514 (Admin)

**47.49** Locum registrar in obstetrics and gynaecology—two separate serious incidents

G, a locum registrar in obstetrics and gynaecology at two hospitals, appealed against findings of misconduct and the sanction of erasure. The allegations revolved primarily around G's treatment of two patients, patient A and patient B, when they suffered blood loss in the process of giving birth. There were additional allegations that the appellant did not adhere to proper hygiene and cleanliness standards, and did not provide support to a junior colleague. The panel found that the appellant, in relation to patient A, had failed to request the attendance of a consultant obstetrician when patient A's blood loss was 2.5 litres. She was actively bleeding and the bleeding was not controlled. The panel also found that the appellant did not appreciate that the bleeding was not under control and that therefore patient A's condition was life-threatening. In relation to patient B, the panel found that the appellant performed a caesarean section in the course of which she experienced a large loss of blood. It found that the appellant failed to request the attendance of a consultant obstetrician when the patient's blood loss exceeded 1.5 litres, that he failed to look after patient B post-operatively, that he did not undertake adequate assessment of patient B, and that he did not provide accurate information to the consultant obstetrician regarding her condition. In dismissing the appellant's appeal, Cranston J noted that the panel concluded that the two cases were particularly serious and that they constituted a serious departure from the principles of good medical practice. The panel accepted that misconduct in relation to clinical conduct can, in principle, be remedied, but concluded that, taking into account all of the factors, the panel was not satisfied that the appellant had taken appropriate remedial action; and that there was a real risk of repetition of his actions in the future.

### *Patel v. Solicitors Regulation Authority* [2012] EWHC 3373 (Admin)

**47.50** Solicitor—using client account for banking facilities for client—no underlying legal transaction—SRA Accounts Rules 2011, rule 14.5

It was not the proper part of a solicitor's everyday practice to operate a banking facility for third parties, whether clients or not. P permitted the use of his client account for banking facilities when there was no underlying legal transaction. Over a four-year period, the account recorded total credits and debits of almost £12 million. There was no allegation that P had been dishonest or engaged in money laundering. However, there was no underlying solicitor's transaction out of which the funds in question arose and no provision of legal or other recognized professional services. The funds were for business purposes of clients and investors. In 1998, guidance was issued in a note to the

Solicitors' Accounts Rules 1998 stating that solicitors may need to exercise caution if asked to provide banking facilities through a client account. In March 2004, the note was revised to state: 'Solicitors should not, therefore, provide banking facilities through a client account.' In 2011, the SRA Accounts Rules were recast. There is a now clear distinction between rules, which are mandatory, and guidance notes, which are not. The previous guidance became a rule: rule 14.5. It reads:

> You must not provide banking facilities through a client account. Payments into, and transfer or withdrawals from, a client account must be in respect of instructions relating to an underlying transaction (and the funds arising therefrom) or to a service forming part of your normal regulated activities.

In dismissing P's appeal, the Divisional Court (Moore-Bick LJ and Cranston J) said that there was a distinction between what the appellant was doing and the role of a solicitor operating an escrow account, or acting as a conveyancer, or the executor of a will. In this case, the monies of the client and the various investors were mixed, and there were no escrow conditions as to their dispersal. In any event, if solicitors are involved, escrow conditions are typically related to legal work. In the present case, there was no underlying legal transaction.

### *Harford v. Nursing and Midwifery Council* [2013] EWHC 696 (Admin)

**47.51** Nurse—failure to cooperate with investigation—NMC Code of Practice

H, a band 6 nurse, faced an allegation that she did not cooperate with the local investigation conducted by her employers in relation to complaints by patients regarding colleague A. H admitted that she had failed to provide a list of patients as requested, but she denied that her failure amounted to misconduct and/or that her fitness to practise was impaired. Dismissing H's appeal, Wyn Williams J held that paragraph 56 of the NMC's Code of Practice ('a registrant must cooperate with internal and external investigations') was not limited to the registrant herself. There was no good reason why paragraph 56 should be interpreted so narrowly. Circumstances may arise in which it would be wholly unprofessional if a registrant were to fail to cooperate with internal and external investigations that related to persons other than the registrant. The sanction imposed by the panel was a conditions of practice order for a period of six months.

### *R (Commissioner of Police for the Metropolis) v. Police Appeals Tribunal and Naulls* [2013] EWHC 1684 (Admin)

**47.52** Police officer—travelling first class when not entitled—fabricating reasons—defendant's mental state—decision of Police Appeals Tribunal

On 8 February 2010, N, a police inspector, was served with a regulation 15 notice under the Police (Conduct) Regulations 2008. This made two essential complaints. The first complaint was that he had travelled in first class on a train when he was not entitled to such travel. The second complaint, which became the more serious, was that he had fabricated his reasons for travelling in first class. On 12 October 2010, at the conclusion of misconduct proceedings, N was dismissed without notice by a misconduct hearing panel. He appealed against that dismissal and, following a hearing on 21 October 2011, the Police Appeals Tribunal decided that the appeal should be allowed as to the disciplinary action taken, so that N was reinstated and given a final written warning to last for eighteen months. The Commissioner of the Police of the Metropolis challenged the lawfulness of the decision of the Police Appeals Tribunal. In dismissing the challenge, Dingemans J said that the Police Appeals Tribunal has a statutory jurisdiction, pursuant to rule 4(4)(a) of the Police Appeals Tribunal Rules 2008, to allow an appeal where the finding or disciplinary action of the misconduct hearing panel was 'unreasonable'. Although the *Wednesbury* test uses, among other expressions, the word 'unreasonable', it does so in another context and the Court may intervene to quash an unlawful decision made by a public body. This is a different and higher test than the Police Appeals

Tribunal considering whether the sanction or outcome imposed by the misconduct hearing panel's decision was unreasonable: see *R (Chief Constable of Wiltshire Police) v. Police Appeals Tribunal (Woollard)* [2012] EWHC 3288 (Admin) (at paragraph 4.25). However, the Tribunal has a statutory jurisdiction to allow an appeal only if it finds the decision 'unreasonable' within the ordinary and proper meaning of that word. In the instant case, the Tribunal expressly addressed the correct legal test when deciding to allow N's appeal against the sanction or outcome of dismissal by the hearings panel, and found that the panel's decision to give such little weight to the reason for the lies told by N was not reasonable. The Tribunal was entitled to consider that N's mental state was in part responsible for his gross error of judgment in giving dishonest explanations. The decision of the Police Appeals Tribunal was within the bounds of a reasonable decision-maker, and could not be said to have been irrational and unlawful.

### *Rehman v. Bar Standards Board*, PC2008/0235/A and PC2010/0012/A, 19 July 2013

**47.53** Barrister—discreditable conduct—meaning

R, a practising barrister, was charged with discreditable conduct in failing to comply with a judgment made against him and a court order for payment. Dismissing R's appeal, the Visitors (Sir Raymond Jack, chairman) said that they saw a major problem with the test in *Sivanandan v. Bar Standards Board*, 6 September 2012 (unreported), that 'discreditable' meant serious enough to bring into question whether or not the person charged was fit to remain a barrister (whether practising or not). In *O'Connor v. Bar Standards Board*, 17 August 2012 (unreported), Collins J cited a passage from *Sivanandan*, but it was not this passage: see paragraph 47.45 above. Many charges of discreditable conduct are not so serious as to raise a risk of the barrister being disbarred. 'Discreditable', as defined in the *Shorter Oxford Dictionary*, means 'bringing discredit to, shameful, disgraceful'. Paragraph 301 of the Code of Conduct is not concerned with trivial matters, but otherwise no further definition is necessary. The approach is objective and it is for the tribunal to decide whether the conduct in question is discreditable to a barrister and sufficiently discreditable to warrant a finding against the barrister. The appeal was thus dismissed.

### *Johnson and Maggs v. Nursing and Midwifery Council (No. 2)* [2013] EWHC 2140 (Admin)

**47.54** Manager and matron of nursing home—risk assessment of falls—medication—whether findings sufficient to justify misconduct

The registrants J and M were, respectively, the manager and matron of a nursing home. The misconduct was said to have occurred during the period October 1998 to January 2002. The charge on which the PCC found that both J and M were guilty of misconduct was charge 2. This alleged that the registrants failed to ensure that adequate nursing records were maintained in respect of individual residents of the home. The particular allegations found proved related to four residents and were that no risk assessment or care plan in relation to falls was kept. The Committee also found J (but not M) guilty of misconduct in relation to charge 7. This charge alleged that the registrants failed to ensure a safe system for the administration of medicines. The Committee found that medication was left in the rooms of five residents on 'various and unknown dates', and that J, having learnt of occasions on which medication was left in residents' rooms, did not inform M. At the end of the case, no action was taken in relation to either registrant, but they were left with findings of misconduct made by the PCC. Both registrants were aggrieved by the findings of misconduct and challenged them by way of judicial review on the grounds that the Committee acted unfairly or irrationally or otherwise unlawfully. Reviewing the complaints and the evidence regarding the case against M, the matron of the care home, Leggatt J said even if, contrary to his view, there were evidence reasonably capable of being treated by the Committee as demonstrating that separate risk assessments and/or care plans for falls should have been compiled for some or all of the four residents, it would require a further step to find that M was guilty of misconduct. It was common ground that simple negligence was not sufficient for this purpose and that 'gross professional negligence', or conduct that would be seen as 'deplorable' by fellow practitioners, was required. Even when every allowance is made for the fact that what amounts to misconduct is pre-eminently a matter for the judgment of the specialist tribunal, he did not consider the evidence in this case to be reasonably capable of supporting such a finding. The finding of misconduct made in relation to J, the manager of the care home, suffered from the same defects as the findings made in relation to M. As to charge 7, the learned judge said:

> 104. ... [E]ven if the charge had not depended on establishing that the system for administration and medication was unsafe, and even if (contrary to my view) the Committee's findings of fact were justified, I still do not see how those findings could reasonably support a finding of professional misconduct. I am bound to say that the Committee appears at this point in the case to have lost a sense of perspective. The Committee's statement that "such was the level of intensity of this problem at this time and the level of contact that [J] had about this issue with the relevant relatives" implies that the matter had a prominence at the time for which there is no basis in the evidence. On the Committee's findings, [J] was during a period of three years informed of a handful of occasions when medication had been found in a resident's room. That needs to be seen in the context that the Home had 72 residents at any given time and some 400 in total during the relevant period, most of whom were receiving medication up to 4 times a day. There were therefore hundreds of thousands of tablets dispensed during the period in question. The evidence simply did not support the notion that there was a systemic problem of failing to ensure that medication was consumed or one which reached any significant level of intensity.
>
> 105. As Manager of a Home with 72 residents and around 100 staff, [J] had many responsibilities. If there were a few occasions when she failed to pass on to [M] information that a resident had not consumed their medication, that would obviously be a matter of disappointment. To find misconduct, however, the Committee had been advised and had accepted that the failure had to be such as would be seen as "deplorable" by fellow practitioners and as involving a serious departure from acceptable standards", and that gross professional negligence could fall into this category if it rose above the level required to give rise to civil liability. Making every allowance for the fact that the judgment was one for the Committee to make, the finding of misconduct made against [J] on this charge was in my view so far out of proportion to the nature of the failure found that it is not one which the Committee could reasonably reach.

### *Walker v. Bar Standards Board*, 19 September 2013 (unreported)

**47.55** Barrister—criminal proceedings—witness—cross-examination—whether breach of Code constitutes misconduct—need to show serious behaviour

W, counsel for the prosecution in criminal proceedings, was charged with professional misconduct contrary to paragraph 708(j) and pursuant to paragraph 901.7 of the Code of Conduct of the Bar of England and Wales. Paragraph 708(j) provided that a barrister, while conducting proceedings in court, must not suggest that a witness is guilty of a crime unless supported by reasonable grounds. Paragraph 901.7 of the Code provides that any failure to comply with any provision of the Code (other than those referred to in paragraph 901.1) shall constitute professional misconduct. In the course of cross-examining a defence DNA expert witness, W suggested to the witness that he had left his former employer because he had been stealing information. W immediately accepted that he had overstepped the mark and apologised. Some eleven months later, the witness made a formal written complaint to the Bar Standards Board against W arising out of his cross-examination. At a Disciplinary Tribunal hearing, W was found guilty of professional misconduct. The Tribunal imposed no separate sanction other than an order as to costs. It was not suggested that the particulars of the offence under paragraph 708(j) were not established; they plainly were. The issue was whether such conduct amounted to professional misconduct. On a literal interpretation of paragraph 901.7, any breach of the Code, however trivial, would constitute professional misconduct. On appeal, the Visitors (Sir Anthony May, Jeffrey Jupp, and Mrs Pauline Burdon) said that, in their judgment, the literal effect of paragraph 901.7 of the Code needed to be modified, but that in reality and logic, the modification needs to apply throughout the Code. The reason for this is that the concept of professional misconduct carries resounding overtones of seriousness—reprehensible conduct that cannot extend to the trivial. Accordingly, the Visitors would import the notion of seriousness into paragraph 901.7, and its absence from paragraph 901.7 did not prevent the notional reading of the paragraph with an added element of seriousness. Since, on the authorities, professional misconduct must, by definition, be serious, seriousness must find its way into paragraph 901.7 of the Code. The issue was whether W's single and momentary error was sufficiently serious to be characterized as professional misconduct. Was it, to use a phrase in the authorities, 'particularly grave'? The Visitors were unanimously of the view that W's lapse, although a clear breach of paragraph 708(j), was not so serious as to require the characterization of 'professional misconduct' under paragraph 901.7 of the Code.

### *Sultan v. Bar Standards Board*, 6 November 2013 (unreported)

**47.56** Barrister—practising without practising certificate

S practised as a barrister without a practising certificate. The Disciplinary Tribunal found that he was in breach of the Code of Conduct of the Bar of England and Wales, and that his actions constituted professional misconduct. Dismissing S's appeal (but allowing the appeal as to sanction in part), the Visitors held that S's failure to obtain a practising certificate was serious. Indeed, it was a criminal offence. The Tribunal had therefore been right to regard it as a serious breach of the Code.

### *Cowan v. Hardey* [2014] CSIH 11

**47.57** Solicitor—hearing in Court on expenses—complaint—need for Tribunal to consider complaint only

H complained to the Scottish Legal Complaints Commission (SLCC) about C, a solicitor representing H's neighbour in an action in the Court of Session. The SLCC remitted one matter to the Council of the Law Society of Scotland—namely, C's answers on expenses given at a hearing in December 2008—but rejected H's other complaints. The Scottish Solicitors' Discipline Tribunal upheld the complaint concerning the hearing in December 2008 and found that H was guilty of unsatisfactory professional conduct, but declined to make any award of compensation. In quashing the decision, the Inner House (Lord Menzies, Lady Clark of Calton, and Lord Clarke) said that it was clearly of the greatest importance that a tribunal such as the Scottish Solicitors' Discipline Tribunal, when considering a complaint involving a statutory concept such as unsatisfactory professional conduct by a practitioner who is a solicitor, should have at the forefront of its collective mind the terms of the statute that create the ground of complaint and which regulate its own powers, jurisdiction, and procedures. It must focus on the subject matter of the complaint, and not on extraneous or peripheral matters that are not the subject of complaint. It must consider carefully the evidence placed before it and proceed to an equally careful analysis of that evidence in order to determine whether or not the complaint had been established. In the instant case, the complaint related only to what happened in a hearing in Court on 3 December 2008. It did not relate to anything that may or may not have occurred before that date. The Tribunal appeared to have been distracted by, and to have had regard to, matters that pre-dated December 2008. It appeared that the Tribunal was influenced in its decision by a conclusion that there was a history of animosity between C and H. No purpose would be gained by remitting the matter back to the Tribunal.

### *Craven v. Bar Standards Board*, 30 January 2014 (unreported)

**47.58** Barrister—obscene comments in emails—bringing profession into disrepute—intrinsic nature of conduct

C, an employed barrister, was found guilty of bringing the legal profession into disrepute contrary to paragraph 301(a)(iii) of the Code of Conduct of the Bar of England and Wales (8th edn) by sending an email to two pupils and another lawyer quoting, or repeating, an obscene comment about a female solicitor. In dismissing C's appeal, Silber J, giving the judgment of the Visitors, said that it was necessary to consider the requirements of the charge and, in particular, whether the offending behaviour had to have an impact on the public. The third limb of paragraph 301(a)(iii)—engaging in conduct, whether in pursuit of his profession or otherwise, which is likely to bring the profession into disrepute—is specifically drafted so that it does not refer to the public; instead, the offence is concerned with bringing the profession into disrepute. It follows that rather than focusing on the public, it is concerned with the intrinsic nature of the conduct. No doubt, had it been intended for the impact on the public to be a necessary ingredient of the offence, this would have been specifically stated in the carefully crafted offence, but that is not what has been stated. There is no reason why the term 'bring into disrepute' should be limited to a public matter. Bad behaviour within a set of chambers or within a firm is capable of being such as to 'bring the legal profession into disrepute'.

### *Clery v. Health and Care Professions Council* [2014] EWHC 951 (Admin)

**47.59** Social worker—children—twenty-five years' unblemished record—alleged failings over twenty-two months—actions taken or not taken during that period

C had been a social worker since 1988—that is, for more than twenty-five years. She faced, for the first time, serious allegations concerning her period as a family centre practitioner. C faced eight allegations over a period of two years. The allegations included failure to complete reasonable management instructions within agreed deadlines, failure to manage her time effectively, not undertaking full and accurate assessments of certain service users, failing to identify potential problems that could put a vulnerable service user at risk, failing to keep up to date with current research and current social work theories and assessment tools, and placing service users at risk. The panel found that C had demonstrated a wilful disregard of instructions, that she had failed to identify or raise child protection concerns, and that her conduct was sufficiently serious to constitute misconduct. It decided that her fitness to practise was impaired and imposed a suspension for twelve months. C's grounds of appeal referred to her previous work experience. In dismissing the appeal, Lewis J said, at [26], that there was no doubt that C had twenty-five years' unblemished work as a social worker prior to these allegations being made. There was no doubt that the panel knew that too, and indeed the panel had expressly mentioned it towards the end of its decision. However, the issue was not the twenty-five years' unblemished record; the issue was the alleged failures over the period 29 April 2009 to 5 February 2011. It was the impact on the children who were being looked after and dealt with during that two-year period with which the panel was dealing, and it was therefore appropriate to look at the way in which actions were taken and not taken in that period. It was not apt simply not to look at the details of that period and to focus instead on an earlier period.

### *Okorji v. Bar Standards Board*, 30 April 2014 (unreported)

**47.60** Barrister—no practising certificate—use of websites—whether seeking employment or holding out as practising

O was called to the Bar by Lincoln's Inn in 2004. He did not have a practising certificate. The charge against him was that he supplied, or offered to supply, legal services and/or held himself out to be, or allowed himself to be held out to be, a barrister in correspondence and/or on websites. The issue turned on whether the website entries were to be regarded objectively as no more than a vehicle to promulgate O's curriculum vitae for job-seeking purposes, or whether they were or sufficiently embraced O holding himself out as a barrister entitled to practise as such. In giving the decision of the Visitors on appeal, Sir Anthony May said that the Visitors' clear unanimous conclusion coincided with that of the Disciplinary Tribunal. The web pages did not read as a job-seeker looking to be employed, but as a barrister offering to supply legal services as such. O may not have had control of the envelope in which his material was placed. But he must have supplied his details and, at the very least, allowed himself to be held out as a barrister in the way in which the website pages held him out. In these circumstances, he did supply legal services, and there was a sufficient connection between the holding out and the services. He did not have a practising certificate.

### *Professional Standards Authority for Health and Social Care v. General Medical Council* [2014] EWHC 1903 (Admin)

**47.61** Psychiatrist—blog—provision of free online advice—whether engaged in form of medical practice

N, a psychiatrist, set up a website under which he described himself as 'The neuro-therapist'. N had a blog, and provided one-to-one sessions and free advice as an online psychiatrist. Patient A made a complaint about answers given and N faced an allegation before the fitness to practise panel. The panel determined that N's online psychiatrist blog was not medical practice and dismissed the charge. Allowing the appeal of the Professional Standards Authority for Health and Social Care, supported by the GMC, Ouseley J said that the question asked by the panel was too limited a question to ask in relation to an issue of misconduct. The panel had to ask itself whether the conduct was in the exercise of professional practice, going to fitness to practise. Even if it did not arise in the context of a clinical practice, it had to be asked whether it was in the exercise of the doctor's medical calling. A related way of putting it is whether the registrant had assumed any responsibility for medical care. Those questions were not asked by the panel. It cannot be the case that when, as here, a qualified psychiatrist sets out to offer advice as a qualified psychiatrist, although it is described as help and assistance, in response to those who write questions seeking advice and help as to what to do, that person is not engaged in medical practice. It is not conventional, in the sense that, because it is done online, there are certain limitations as to what can be done. But even on the basis of what the registrant did here, he clearly offered advice and forms of treatment, even if only reading a book, which showed that he was engaged in a form of medical practice.

### *Conlon, Gordon and Williams v. Bar Standards Board* [2014] Lexis Citation 136

**47.62** Barrister—pupillage arrangements—role of head of pupillage committee—role of head of chambers

C was the head of the pupillage committee, and G and W were the joint heads of a mixed criminal and common law set of chambers in the Temple, with thirty-eight members. The charge against C was one charge of professional misconduct contrary to paragraphs 404.2(c) and 901.5 of the Code of Conduct of the Bar of England and Wales (8th edn). The particulars of the offence were that, as head of the pupillage committee, he failed to take all reasonable steps to ensure that proper arrangements were made in the chambers for dealing with pupils and pupillage matters to commence October 2006, and that the failure was so serious as to be likely to bring the Bar into disrepute and amounted to professional misconduct. The charge against G and W was in the same form, but in their capacity as joint heads of chambers. In March 2006, C became head of the pupillage committee. During the same month, the chambers promulgated a pupillage policy document, which provided that the chambers was a member of the On Line Pupillage Application System (OLPAS) and aimed to recruit one twelve-month pupil each year via OLPAS. A second version of the pupillage policy document was promulgated during the summer of 2006. It provided that the chambers reserved the right to recruit non-OLPAS pupils where either insufficient applications were submitted on the OLPAS system or suitable candidates could not be selected from those available. It was common ground that ninety-eight applications were received via OLPAS, that these were not downloaded by the chambers, and that three pupillages were awarded by private arrangement and not in accordance with either the chambers' own selection process, as set out in its pupillage policy, or any other proper selection process. The ninety-eight individuals who applied through OLPAS did not have their applications considered at all. In allowing the appeals of the joint heads of chambers, the Visitors (Asplin J, chairman) said that G and W did take reasonable steps to ensure that proper arrangements were made in the chambers for dealing with pupils and pupillage. W, who was more engaged in the pupillage process, was involved in the promulgation and subsequent amendment of the chambers' pupillage policy and had regular discussions with C about the pupillage process. It was reasonable for W to rely on C's assurances that he was implementing the chambers' pupillage policy. G also had conversations with C from which he obtained reassurance as to the steps being taken and relied on W, who was taking a more active role in relation to pupillage. At [99], the Visitors said:

> There are three over-arching points concerning the proper conduct of Heads of Chambers that have governed our deliberations:
>
> (i) A Head of Chambers remains personally responsible for discharging his obligations under the Code of Conduct. We do not accept (and nor was the case put in this way) that the mere fact of being a Joint Head of Chambers rather than a sole Head of Chambers liberates that person from his obligations. If responsibilities have been clearly and transparently divided between Heads of Chambers then this will impact on the reasonableness of the particular actions of a Joint Head of Chambers, but the nature of his or her obligations remains the same.
>
> (ii) We accept (as did the [Bar Standards Board]) that a Head of Chambers is entitled to delegate responsibility. This is common practice both in relation to pupillage (as noted by [the Board]) and also in relation to other parts of the administration of Chambers.
>
> (iii) Where responsibility has been clearly delegated, a Head of Chambers is entitled to rely on the appointed member of Chambers to carry out his role with reasonable care and diligence and so as to comply with that member's professional obligations. This does not, however, mean that the Head of Chambers no longer retains any responsibility at all for the delegated function. Taking reasonable steps to ensure that proper arrangements are made for ensuring that an administrative function is carried out includes taking reasonable steps to obtain an appropriate level of reporting back from the person to whom responsibility has been delegated. It may also include an obligation to step in and take a more active role if it becomes apparent that the person to whom responsibility has been delegated is not carrying out his or her role properly and/or to remove such a person.

In the case of C, responsibility for pupillage in the chambers had been delegated to him as head of the pupillage committee. Accordingly, a direct obligation on behalf of the chambers in relation to the proper arrangements for pupillage, pursuant to paragraph 404.2 of the Code of Conduct, fell upon him. C did not understand the mechanism of OLPAS and did not take any independent steps beyond his discussions with the Bar Standards Board to educate himself. He did not download the applications for pupillage and did not take any material steps to conduct the chambers' pupillage process in accordance with OLPAS or the chosen recruitment mechanism. He failed to take all reasonable steps to ensure that proper arrangements were in place for dealing with pupils and pupillage. The failings were serious by reason, in particular, of the prejudice caused to the three applicants who were awarded pupillage by private arrangement, but whose pupillage the Board subsequently refused to register, of the prejudice caused to the ninety-eight applicants who had submitted applications that were not processed, and of the potential prejudice caused to the reputation of the Bar as a whole by reason of these matters: [100].

***Booth v. General Dental Council* [2015] EWHC 381 (Admin)**

**47.63** Financially motivated—dentist putting financial interests before those of patient

Charge 3 was that B, a dentist, charged patient A a high fee and did not accept responsibility or offer appropriate refunds for certain specified items of treatment that were defective or inappropriate. He admitted the whole of charge 3. Charge 5 alleged that B's conduct in relation to charge 3 was (a) unprofessional, and/or (b) financially motivated, and/or (c) dishonest. B admitted that his conduct was unprofessional. He denied that it was financially motivated or dishonest. The GDC's PCC did not find that this particular conduct was dishonest, but it did find that it was 'financially motivated'. In its reasons, the Committee said that, in considering charge 5, it had given the words 'financially motivated' their normal meaning and interpreted them to mean that B put his own financial interests before the best clinical and financial interests of Patient A, and that he did so intentionally. In dismissing B's appeal, Holman J said that, in the light of the evidence, it was clearly open to the Committee to conclude that B had been financially motivated in the sense in which it had interpreted those words—namely, putting his own financial interests before the best clinical and financial interests of his patient.

## C. Other Relevant Chapters

**47.64**

Impairment of Fitness to Practise, Chapter 33
Negligence, Chapter 49
Recklessness, Chapter 55

# 48

# NATURAL JUSTICE

A. Legal Framework 48.01
B. Disciplinary Cases 48.02
*Kanda v. Government of the Federation of Malaya* [1962] AC 322 48.02
*Ridge v. Baldwin* [1964] AC 40 48.03
*McInnes v. Onslow-Fane* [1978] 1 WLR 1520 48.04
*Calvin v. Carr* [1980] AC 574 48.05
*R v. Board of Visitors of Hull Prison, ex parte St Germain and ors (No. 2)* [1979] 1 WLR 1401 48.06
*Crompton v. General Medical Council* [1981] 1 WLR 1435 48.07
*Chief Constable of the North Wales Police v. Evans* [1982] 1 WLR 1155 48.08
*Hefferon v. Professional Conduct Committee of the United Kingdom Central Council for Nursing Midwifery and Health Visiting* [1988] 10 BMLR 1 48.09
*Modahl v. British Athletic Federation Ltd* [2002] 1 WLR 1192 48.10
*R (McNally) v. Secretary of State for Education and Employment* [2002] ICR 15 48.11
*Stansbury v. Datapulse Plc* [2004] ICR 523 48.12
*Gage v. General Chiropractic Council* [2004] EWHC 2762 (Admin) 48.13
*R (Bates) v. District Judge Zani (Independent Adjudicator)* [2011] EWHC 3236 (Admin) 48.14
*R (Hill) v. Institute of Chartered Accountants in England and Wales* [2013] EWCA Civ 555; [2014] 1 WLR 86, CA 48.15
*Chhaba v. West London Mental Health NHS Trust* [2014] ICR 194, SC 48.16
*R (McCarthy) v. Visitors to the Inns of Court* [2013] EWHC 3253 (Admin); [2015] EWCA Civ 12 48.17
*R (Soar) v. Secretary of State for Justice* [2015] EWHC 392 (Admin) 48.18
C. Other Relevant Chapters 48.19

## A. Legal Framework

**48.01** Bar Standards Board Disciplinary Tribunals Regulations 2014, rule E144 (the rules of natural justice apply to proceedings of a disciplinary tribunal)

## B. Disciplinary Cases

### *Kanda v. Government of the Federation of Malaya* [1962] AC 322

**48.02** Police inspector—report sent to adjudicating officer—not disclosed to defendant—no opportunity to be heard

On 7 July 1958, K, an inspector of police in the Royal Federation of Malay Police, was dismissed by the Commissioner of Police on the ground that he had been guilty of an offence against discipline. K brought an action in the High Court challenging that dismissal. Rigby J declared that his dismissal was void and of no effect. The government appealed. The Court of Appeal, by a majority, allowed the appeal and held that K was validly dismissed. On appeal to the House of Lords, two questions arose: (1) did the Commissioner of Police have power to dismiss K; and (2) were the proceedings that resulted in his dismissal conducted in accordance with natural justice? In allowing K's appeal, the House of Lords held that the Commissioner of Police had no power to dismiss K. Also, there was a breach of the rules of natural justice. The report to the board of inquiry contained a severe condemnation of K. It was sent to the adjudicating officer before he sat to enquire into the charge. He read it and had full knowledge of its contents. But K never had it and never had an opportunity of dealing with it. Lord Denning, giving the judgment of their Lordships, said that much of the argument proceeded on the footing of whether there was real likelihood of bias on the part of the adjudicating officer. In the opinion of their Lordships, however, the proper approach is somewhat different. The rule against bias is one thing; the right to be heard is another. Those two rules are the essential characteristics of what is often called 'natural justice'. They are the twin pillars supporting it. The Romans put them in the two maxims: *nemo judex in causa sua* (meaning 'no one should be a judge in his or her own cause') and *audi alteram partem* (meaning 'hear the other side'). They have recently been put in the two words, 'impartiality' and 'fairness'. But they are separate concepts and are governed by separate considerations. In the present case, K complained of a breach of the second; K was not given a reasonable opportunity of being heard.

### *Ridge v. Baldwin* [1964] AC 40

**48.03**
Chief constable—summary dismissal in absence—no opportunity to attend—reconvened meeting attended by solicitor—no substitute for full hearing

The appellant, R, was the chief constable of Brighton. On 7 March 1958, the Watch Committee and Police Authority of the County Borough of Brighton met and purportedly summarily dismissed R. He had not been invited to attend the meeting and was not sent for. He received a letter the same afternoon, telling him that he had been summarily dismissed. Previously, R had been committed for trial at the Central Criminal Court charged with conspiracy to obstruct the course of public justice. He was acquitted. On 18 March 1958, the Watch Committee met for a second time and was addressed by R's solicitor. It resolved to adhere to its previous decision, nine members voting in favour of the resolution and three against it. R brought an action against the members of the Watch Committee for a declaration that his dismissal was illegal, ultra vires, and void. In allowing R's appeal, the House of Lords held that an officer cannot lawfully be dismissed without first telling him what is alleged against him, and second, hearing his defence or explanation (*per* Lord Reid). The further hearing on 18 March was a very inadequate substitute for a full hearing. Even so, three members of the Committee changed their minds and it is impossible to say what the decision of the Committee would have been had there been a full hearing after disclosure to the appellant of the whole case against him.

### *McInnes v. Onslow-Fane* [1978] 1 WLR 1520

**48.04**
Boxing—manager—refused grant of licence—different categories of case (forfeiture, application, and legitimate expectation)

M brought proceedings for a declaration that the defendant's decision to refuse to grant a boxers' manager's licence to him was in breach of natural justice. In dismissing M's originating summons, Megarry V-C said that there were at least three categories of case. First, there are what may be called the 'forfeiture cases'. In these, there is a decision that takes away some existing right or position, such as where a member of an organization is expelled or a licence is revoked. Second, at the other extreme, there are what may be called the 'application cases'. These are cases in which the decision merely refuses to grant the applicant the right or position that he seeks, such as membership of the organization or a licence to do certain acts. Third, there is an intermediate category, which may be called the 'expectation cases', which differ from the application cases only in that the applicant has some legitimate expectation from what has already happened that his application will be granted. It seemed plain, said Megarry V-C, that there is a substantial distinction between the forfeiture cases and the application cases. In the forfeiture cases, there is a threat to take something away for some reason; in such cases, the right to an unbiased tribunal, the right to notice of the charges, and the right to be heard in answer to the charges (which in *Ridge v. Baldwin* [1964] AC 40, at 132, Lord Hodson said were three features of natural justice that stood out) are plainly apt. In the application cases, on the other hand, nothing is being taken away and, in all normal circumstances, there are no charges, and so no requirement of an opportunity of being heard in answer to the charges. The intermediate category, that of the expectation cases, may at least in some respects be regarded as being more akin to the forfeiture cases than the application cases—for, although in form there is no forfeiture but merely an attempt at acquisition that fails, the legitimate expectation of a renewal of the licence or confirmation of the membership is one that raises the question of what it is that has happened to make the applicant unsuitable for the membership or licence for which he was previously thought suitable.

### *Calvin v. Carr* [1980] AC 574

**48.05** Horse racing—stewards inquiry—full rehearing before Jockey Club—whether capable of curing defect in original hearing

For the details of this case, see paragraph 4.04. The plaintiff brought proceedings against the chairman of the Jockey Club and others, and argued that defects of natural justice, as regards the proceedings before the stewards, were not capable of being cured by the appeal proceedings before the Committee of the Australian Jockey Club even though, as was not contested before the Board, these were correctly and fairly conducted. In dismissing this argument, Lord Wilberforce, giving the judgment of the Board, said at 592C–G that:

> ... [T]heir Lordships recognise and indeed assert that no clear and absolute rule can be laid down on the question whether defects in natural justice appearing at an original hearing, whether administrative or quasi-judicial, can be "cured" through appeal proceedings. The situations in which this issue arises are too diverse, and the rules by which they are governed so various, that this must be so.
>
> There are however a number of typical situations as to which some general principle can be stated. First there are cases in which the rules provide for a re-hearing by the original body, or some fuller or enlarged form of it. This situation may be found in relation to social clubs. It is not difficult in such cases to reach the conclusion that the first hearing is superseded by the second, or, putting it in contractual terms, the parties are taken to have agreed to accept the decision of the hearing body, whether original or adjourned....
>
> At the other extreme are cases, where, after examination of the whole hearing structure, in the context of the particular activity to which it relates (trade union membership, planning, employment etc.) the conclusion is reached that a complainant has the right to nothing less than a fair hearing both at the original and at the appeal stage....

He continued, at 593C–E:

> In their Lordships' judgment,... intermediate cases exist. In them it is for the Court, in the light of the agreements made, and in addition having regard to the course of proceedings, to decide whether, at the end of the day, there has been a fair result, reached by fair methods, such as the parties should fairly be taken to have accepted when they joined the association. Naturally there may be instances when the defect is so flagrant, the consequences so severe, that the most perfect of appeals or re-hearings will not be sufficient to produce a just result.... There may also be cases when the appeal process is itself less than perfect....

In the instant case, the plaintiff received, overall, full and fair consideration, and a decision, possibly a hard one, was reached against him. There was no basis on which the Court ought to interfere.

***R v. Board of Visitors of Hull Prison, ex parte St Germain and ors (No. 2)*** **[1979] 1 WLR 1401**

**48.06** Prison riot—disciplinary charges heard by Board of Visitors—duty to act judicially

A riot took place at Hull Prison between 31 August and 2 September 1976. Serious damage was done to the prison, which made it largely uninhabitable. It was decided not to take criminal proceedings against the rioters, but to deal with them internally under the prison disciplinary procedure. This fell upon representatives of the Board of Visitors of Hull Prison. Proceedings were taken against 185 of the 310 inmates of the prison and the total number of individual charges was over 500. On 3 October 1978, the Court of Appeal held: that, when adjudicating upon disciplinary charges, Visitors were performing a judicial, and not merely an administrative, act; that they had a duty to act judicially; and that their decisions were in principle subject to judicial review: see *R v. Board of Visitors of Hull Prison, ex parte St Germain* [1979] QB 425. In these proceedings, seven applicants complained of not being given a 'proper opportunity of presenting his case' and that the Board failed to observe the rules of fair play or natural justice. In particular, there were four specific complaints:

(1) that the Board refused to allow the applicants to call witnesses in support of their cases;
(2) that they admitted and acted upon statements made during the hearing by the governor, which were based on reports from prison officers who did not give evidence;
(3) that the chairman of the Board insisted on questions by the applicants in cross-examination being channelled through him; and
(4) that the applicants were not allowed to speak in mitigation after a finding of guilt.

Counsel for the applicants accepted that it was perfectly proper for a chairman to insist that all questions were put through him where he was of the view that, otherwise, arguments would break out between the prisoner and the witness, which would make the proceedings difficult to control. However, the Court said that the discretion has to be exercised reasonably, in good faith, and on proper grounds. In the proper exercise of his discretion, a chairman may limit the number of witnesses, either on the basis that he has good reason for considering that the total number sought to be called was an attempt by the prisoner to render the hearing of the charge virtually impracticable or where, quite simply, it would be quite unnecessary to call so many witnesses to establish the point at issue. However, a fair chance of exculpation cannot, in many cases, be given without hearing the accused's witnesses, for example in a case of an alibi defence. It was a fundamental misconception that the rules of natural justice can have no relevance to matters of procedure because a domestic tribunal is master of its own procedure. The rules of natural justice are a compendious reference to those rules of procedure that the common law requires persons who exercise quasi-judicial functions to observe. Natural justice requires that the procedure before any tribunal that is acting judicially shall be fair in all the circumstances. The entitlement of the Board to admit hearsay evidence is subject to the overriding obligation to provide the accused with a fair hearing. It was clear that no opportunity was given to examine the evidence to which the governor was referring. The applicants were not told the names of the officers who had allegedly observed them. This constituted a serious departure. That the hearsay evidence carried weight with the Board was clear. With some reluctance, the Court said that it came to the conclusion that the way in which the hearsay evidence was handled was a departure from the rules of fairness—reluctantly, because there was inevitably a feeling that the Board may have reached the right result ultimately in spite of the irregularities. These men were prisoners; some of them were dangerous; most of them were difficult. All of them were no doubt to some extent untrustworthy. But they faced (and received) severe punishment and they were entitled to a fuller hearing than that which they in fact received. The applications of six out of seven applicants for orders of certiorari to quash the decisions of the Board were granted.

*Crompton v. General Medical Council* [1981] 1 WLR 1435

**48.07** See paragraph 18.02.

*Chief Constable of the North Wales Police v. Evans* [1982] 1 WLR 1155

**48.08** Probationary police officer—inquiry report not disclosed—resignation of officer unlawfully obtained

E was sworn into the office of constable on 31 August 1977 and became a probationary member of the North Wales police force for a period of two years. He was then just under 25 years old. Following various incidents and reports, E was summoned to an interview with the chief constable on 8 November 1978. He was told by the chief constable that he felt that he had made a mistake in accepting E and the chief constable gave E the opportunity to resign as an alternative to formally dispensing with his services. E said in his affidavit about the interview, and his account was not disputed, that he asked if he could have a reason for this action, but the chief constable refused outright. E was not informed of what was alleged against him, nor was he afforded any opportunity to be heard by way of defence or explanation. He asked for time to consider and the chief constable said that E must let him know by 10 a.m. the following morning. E was not given any document recording this decision. As a result of the chief constable's threat, E signed on 9 November a formal letter of resignation. In subsequent proceedings for judicial review, Lord Hailsham of St Marylebone LC said that the chief constable's discretion to dispense with a probationer's services, although wide, is not absolute. The chief constable should have directed his mind to the criteria laid down in the relevant Police Regulations in accordance with the appropriate principles of natural justice. He did not do so. A 'formal hearing' may well have been unnecessary if by that is meant an oral hearing in every case held before the chief constable himself, but this does not dispense a chief constable from observing the rules of natural justice laid down by Lord Reid in *Ridge v. Baldwin* [1964] AC 40, 66. It may well also be that part or all of the inquiry on the facts may be delegated to a superintendent official, as was done here by the chief constable to the deputy chief constable, although, where this is done, the ultimate decision must not be delegated and common prudence should dictate that the report by the delegated officer, or at least its substance, should be shown to the officer the subject of review, and an opportunity afforded to him to comment on it before the final decision is taken by the chief constable himself. This was not done here. A declaration was granted that, by reason of his unlawfully induced resignation, E had thereby become entitled to the same rights and remedies, not including reinstatement, as he would have had if the chief constable had not unlawfully dispensed with his services under the Police Regulations.

*Hefferon v. Professional Conduct Committee of the United Kingdom Central Council for Nursing Midwifery and Health Visiting* [1988] 10 BMLR 1

**48.09** Hearing before PCC—principal witness not present—past incidents of alleged failures adduced—no opportunity to refute

H, a deputy sister, was working at the Craven Park Health Centre in Harlesden, assisting doctors with necessary health measures for children, including the weighing and measuring of children and the carrying out of immunizations. H faced four allegations arising out of events on 23 May 1984, when a mother attended the clinic with her child. The charges were that H failed to check a prescription and computer schedule prior to administering an injection to the child, that she gave an incorrect immunization to the child, that she failed to record that she had given an incorrect immunization, and that she failed to report the giving of the incorrect immunization to the senior nurse or to the health visitor. During the mother's visit, when H went to make an entry in the child's records, she noticed that she had missed that the child was due for measles inoculation. That disturbed her and she went to see the doctor who was nearby, who gave reassuring advice and said that no harm would befall the child by reason of the inoculation that had just been given. H passed this on to the mother. On 11 June 1987, three years after H had been dismissed from her employment by the Brent Health Authority, she appeared before the Professional Conduct Committee (PCC) of the respondent. At the hearing, she was represented by a trade union official. The professional body did not call the mother to give evidence or the health visitor in relation to the events in question, but called only the senior nurse at the clinic at the time. In allowing H's appeal, the Divisional Court (Watkins LJ and Nolan J) said that it was surprising that, when H was cross-examined, the focal part of the mother's evidence was put to her. This was an impermissible way to proceed. It was incumbent upon the solicitor appearing for the Council to call the evidence supporting the charge if such an allegation as was put to H was to be made. It was most prejudicial. The object of it obviously was to put in the minds of the Committee that the appellant had not been telling the truth as to what happened between her and the mother. Since the truth was one of the most important factors that had to be considered by the Committee, the harm that this may very well have done could have been quite substantial and crucial. There was also the failure to call the health visitor, who could have informed the Committee about the instructions, if any, given to H and other nurses as to the need for reporting errors. After finding the charges proved, the prosecutor was permitted to recall the only witness, the senior nurse at the clinic, who proceeded to air past incidents of alleged failures by H to perform her duties properly, some of the evidence to that effect being delivered apparently with a certain amount of venom. Following her evidence, the health visitor was called for a like purpose. She was, to be fair to her, moderate in her evidence. H's representative was called upon to make a speech in mitigation. There was no opportunity given to H to give evidence to refute what were some serious accusations against her. It was not surprising against that background that the Committee announced that it would strike H off the register. In allowing H's appeal, Watkins LJ said that this was a very serious departure from a proper procedure: a breach of natural justice. To make sweeping allegations as were made in the process of what is called mitigation—it should have been called 'aggravation' in the circumstances—is impermissible unless the person accused has the opportunity to answer the charges levelled against her. Even if the Court had not come to the conclusion as to the findings of misconduct, it would certainly have quashed the sanction.

### *Modahl v. British Athletic Federation Ltd* [2002] 1 WLR 1192

**48.10**
Athlete—urine specimen containing testosterone—sample later found to be contaminated and unreliable—civil proceedings for damages—whether hearings before body achieved fair result

The claimant was a well-known athlete, who had won many titles at national and international level as an 800m runner. On 18 June 1994, she took part in an athletics meeting in Lisbon, Portugal. A subsequent laboratory report showed that a urine specimen provided by the claimant contained testosterone well above any permissible level. In December 1994, she appeared before a five-member disciplinary committee at which she was legally represented. The committee unanimously found that she had committed a doping offence, and declared her ineligible to compete in the United Kingdom and abroad for four years. She exercised her right of appeal to an independent appeal panel. This panel of three members, presided over by Mr Robert Reid QC, heard argument and evidence, and unanimously allowed her appeal. The bulk of the hearing before the appeal panel was directed to the sample. The appeal panel heard evidence over two days and evidence was called before it that had not been before the disciplinary committee. In the light of that evidence, the appeal panel concluded that there was a possibility that could not be ignored that the samples had been degraded by bacterial contamination, which could have affected the reliability of the results. Thereafter, the claimant commenced proceedings for damages for breach of contract and negligence against the defendant. The essence of the claimant's case was that the disciplinary committee was tainted by bias. The claim failed. There was no question of an express contract between the parties, although there is no doubt that, over a period of years, as the claimant accepted, if she were to enter athletic meetings under the auspices of the defendant, then she would be subject to the rules. The claimant challenged the first hearing on the basis of bias. In cases such as this, in which an apparently sensible appeal structure has been put in place, the Court is entitled to approach the matter on the basis that the parties should have been taken to have agreed to accept what, in the end, is a fair decision. As Lord Wilberforce said in *Calvin v. Carr*, this does not mean that the fact that there has been an appeal will necessarily have produced a just result. The test that is appropriate is to ask whether, having regard to the course of the proceedings, there has been a fair result. As Lord Wilberforce indicated, there may be circumstances in which, by reason of corruption or bias or such other deficiency, the end result cannot be described as fair. The question in every case is the extent to which the deficiency alleged has produced overall unfairness. In a case such as the present, in which the danger of bias can be evaluated and excluded, the Court considered that, taken together with a wholly untainted appellate process, a fair result had been achieved.

***R (McNally) v. Secretary of State for Education and Employment*** **[2002] ICR 15**

**48.11** Teacher—allegation dismissed—rehearing—whether chief education officer entitled to attend rehearing

A disciplinary panel of a school in Bury met on 25 and 28 June 1996 to consider an allegation against a teacher that he inappropriately touched a 15-year-old boy at the school. The panel consisted of: a local councillor (a solicitor, former teacher, and a parent); a commercial director of a management consultancy (a former employee relations manager for the Co-operative Bank, and a governor and parent); and a company director (who previously worked for the Inland Revenue, and a school governor and a parent). Kennedy LJ, giving the main judgment of the Court of Appeal, said that it was an impressive and highly qualified panel. Before retiring to consider its decision, the chairman of the panel asked everyone to leave. There was an issue arising as to whether the acting chief education officer of the local authority expressed a desire to remain, which, pursuant to paragraph 8(9)(3) of Schedule 3 to the Education Act 1988, gave him an entitlement to remain. The panel retired without the presence of the acting chief education officer and, after considering the evidence, concluded that there had been no misconduct by the appellant. Subsequently, the Secretary of State, following correspondence by the local authority, intervened and proposed that there should be a rehearing. The Court of Appeal disagreed. It was impossible to say whether, on the evidence, the acting chief education officer had ever sought to exercise his entitlement to remain. However, even if he were to have asserted his entitlement to remain, that would not of itself enable the Secretary of State to conclude that the panel had acted unreasonably or had failed to discharge its statutory duty. Where, as here, a relatively formal procedure is established to consider serious allegations, then it is incompatible with the principles of natural justice for a local government official, whom the employee may reasonably regard as a member of the prosecution team, to be with the disciplinary panel when its members withdraw to discuss among themselves whether the misconduct alleged has been proved. If there were advice to be tendered, it could have been tendered in the presence of the appellant and his representative, either directly or through the lawyer who was presenting the case on behalf of the local authority. The governing body was required to make findings of fact on allegations of a most serious nature and the rules of justice required that all advice should be given in the presence of the appellant. The adviser could not retire with the members of the panel to give advice on whether the allegations had been proved. Had the panel found the charges against the appellant to be made out and reconvened to consider the effect that this should have on the appellant's employment in the school, there would have been no objection to the chief education officer being present at that stage. A strict construction of the statutory provisions appeared to give the chief education officer an absolute right to be present at those deliberations. It may well be that the right must be impliedly restricted by the overriding requirements of natural justice.

***Stansbury v. Datapulse Plc*** **[2004] ICR 523**

**48.12** Employment tribunal—panel member asleep—whether justice seen to be done—whether hearing fair

The appellant claimed that, at the hearing of his case for unfair dismissal before the employment tribunal sitting at Reading, a lay member of the panel was under the influence of alcohol and had fallen asleep. The Employment Appeal Tribunal (EAT) dismissed the appellant's appeal. In allowing the appellant's appeal, Peter Gibson LJ (with whom Latham LJ and Sir Martin Nourse agreed), in the Court of Appeal, said that the first question that arises is whether the EAT should have entertained the complaint at all where the complainant had been represented by counsel at the employment tribunal hearing, but the complaint had not been raised there. After reviewing *Kudrath v. Ministry of Defence* (unreported, 26 April 1999) EAT, *Red Bank Manufacturing Co Ltd v. Meadows* [1992] ICR 204 EAT, and *McGonnell v. United Kingdom* (2000) 30 EHRR 289, the Court said that it was appropriate to consider the failure to raise any objection before the employment tribunal against the test of reasonableness in all the circumstances of the case. The EAT could properly decide, as it did in the present case, that the fact that the point had not been raised before the employment tribunal should not prevent the point being raised before the EAT on appeal. It is always desirable that a point on the behaviour of the employment tribunal be raised at the employment tribunal in the course of the hearing, but it is unrealistic not to recognize the difficulty, even for legal representatives, in raising with the employment tribunal a complaint about the behaviour of an employment tribunal member, who, if the complaint is not upheld, may yet be part of the employment tribunal deciding the case. The EAT may have to assume the role of judges of fact in relation to a complaint about the behaviour of a member of the employment tribunal, and if there are factual disputes relating to the complaint, the EAT may have to resolve them: see *Facey v. Midas Retail Security Ltd* [2001] ICR 287. In the instant case, the evidence in fact was reasonably clear that the employment tribunal member had consumed alcohol. The question, then, was whether there had been a proper hearing before the employment tribunal. In *Whitehart v. Raymond Thompson Ltd*, 11 September 1984 (unreported), EAT, Popplewell J presiding said that it was axiomatic that all members of a tribunal must hear all of the evidence, and a trial in which one member of the tribunal is asleep even for a short part of the time cannot be categorized as a proper trial. Justice does not appear to have been done. The Court of Appeal said that no less a strong comment might be made of an employment tribunal member who is known to have consumed alcohol. That might well have impaired the member's ability to attend to the evidence and submissions before the tribunal. In the present case, to hold that the hearing was fair and complied with Article 6 of the European Convention on Human Rights (ECHR) was wrong. In the particular circumstances of the assumed facts, the requirement that the hearing be seen to be fair was not satisfied. The case was to be remitted for a rehearing before a differently constituted employment tribunal.

***Gage v. General Chiropractic Council*** **[2004] EWHC 2762 (Admin)**

**48.13**

See paragraph 18.08.

### *R (Bates) v. District Judge Zani (Independent Adjudicator)* [2011] EWHC 3236 (Admin)

**48.14** Prisoner—independent adjudicator's findings—witness not available—whether procedure unfair

B, a serving prisoner, challenged the decision of an independent adjudicator finding him guilty of possessing an unauthorized article—namely, a mobile phone—in breach of the Prison Rules 1999. B shared a cell with another prisoner R. B and R were both charged with a disciplinary offence. It transpired, however, that R was shortly to be released. On the day before his release, R wrote a letter to the prison governor taking full responsibility for the unauthorized article and saying that B had no knowledge of it being in the cell. At the conclusion of the hearing and after hearing brief submissions from B's solicitor, the independent adjudicator found the charge against B proved. B's application for judicial review was dismissed by Stephen Males QC (sitting as a deputy High Court judge). It was accepted on both sides that a prisoner facing a disciplinary charge must be given a full opportunity of hearing what is alleged against him and of presenting his case, that he is entitled to be legally represented in the proceedings before an independent adjudicator and to call witnesses, and that the criminal standard of proof applies. In *R (King) v. Secretary of State for Justice* [2011] 1 WLR 2667, it was held that a prison disciplinary hearing resulting in a penalty of confinement to cell amounts to a determination of a prisoner's civil right of association, so that Article 6(1) applies, but that it is not an infringement of Article 6 for a hearing to take place before a prison governor because the availability of judicial review cures any lack of independence in the governor's position. B contended that the hearing was unfair because it was held after the release from custody of B's cellmate and co-defendant R, 'thereby depriving the Claimant of the sole material witness that could establish his innocence'. This assumed that R was potentially a witness who would have given evidence supporting B. However, B did not ask for any opportunity to arrange for R to attend to give oral evidence, as he could have done pursuant to the Prison Discipline Manual. R could not have been compelled to give evidence. There was no procedural fairness in the fact that the adjudicator proceeded to determine the matter on the basis of the evidence that was put before him. He was not required to suggest to a legally represented prisoner that the prisoner's interests would be better served by seeking to call oral evidence, even if he had thought this to be the case. Nor was he required to request the attendance of R himself, even if this had been practicable. B was not deprived of the benefit of R's evidence, such as it was; on the contrary, his evidence in the form of his letter was before the independent adjudicator and was considered by him. B made a tactical decision to rely on the written evidence, which was perfectly understandable in the circumstances. The fact that this evidence was not believed was not a matter of procedural unfairness, but went to the merits.

### *R (Hill) v. Institute of Chartered Accountants in England and Wales* [2013] EWCA Civ 555; [2014] 1 WLR 86, CA

**48.15** Accountant—panel member absent during hearing—waiver

On day four of the Disciplinary Tribunal hearing, one of the panel members left early and missed an hour and a half of the claimant's cross-examination. On the resumed hearing one month later, the hearing continued, with the panel member. On the claimant's appeal to the Court of Appeal against the dismissal of judicial review proceedings, the Court stated that if there is a hearing with live witnesses giving their evidence orally, it will normally be a breach of the rules of natural justice for a member of the Tribunal (in the absence of agreement) to absent himself while a witness is giving evidence and later return to participate in the decision. This would not normally be cured by the absent member reading a transcript of the evidence given in his absence, unless the evidence were comparatively uncontroversial, for the reasons given by Lord Griffiths in

*Ng v. The Queen* [1987] 1 WLR 1356. Such absence will be difficult (if not impossible) to justify if the evidence being given is that of the defendant or respondent to the disciplinary proceedings. However, the temporary absence of a Tribunal member during the hearing was an irregularity that could be waived. Of course, the agreement must be voluntary, informed, and unequivocal: *Millar v. Dickson* [2002] 1 WLR 1615, [31]. On the facts of the present case, there was such an agreement. If there was an agreement to the procedure adopted, then there was no breach of the rules of natural justice at all. Two types of case should be distinguished. The first is that in which there has been a breach of one of the requirements of procedural fairness and the question is whether it has subsequently been waived by the person affected. The second type of case is that in which, at a stage in the process before there has been any breach, the decision-maker and others involved have discussed a proposed procedure, and have freely and in full knowledge of the facts consented to that procedure, which is then followed. In such a case, the correct analysis is to not regard the situation as a breach of natural justice that has been waived.

***Chhaba v. West London Mental Health NHS Trust* [2014] ICR 194, SC**

**48.16** See paragraph 39.36.

***R (McCarthy) v. Visitors to the Inns of Court* [2013] EWHC 3253 (Admin); [2015] EWCA Civ 12**

**48.17** See paragraph 18.17.

***R (Soar) v. Secretary of State for Justice* [2015] EWHC 392 (Admin)**

**48.18** Evidence of witness by telephone—absence of claimant—unfairness

The claimant, a category B prisoner serving a life sentence, applied for judicial review to challenge an adjudication review decision of the National Offender Management Service (NOMS) upholding a finding of guilt against him whilst he was at HMP Full Sutton. The substantive challenge was a procedural irregularity at the original adjudication hearing when the prison governor took evidence from a nurse by telephone in the absence of the claimant, and then failed to give him an opportunity to answer or challenge that evidence. On 13 November 2013, during a rub-down search before returning to work, the claimant was found to have secreted some bread in the front of his trousers. When asked by a prison officer why he wished to take the bread to work, he replied it was 'medical'. The claimant was told that he would have to put in an application to the healthcare centre to ask whether it was appropriate for him to take bread to work. The claimant refused to go to work and was returned to his cell. In his defence to a charge of disobedience to a lawful order, the claimant stated that he had a chronic bowel condition and found the most effective way in which to manage it was eating six to eight slices of bread immediately after lunch. At a subsequent adjudication hearing, the governor adjourned the hearing and returned, telling the claimant that he had spoken to the healthcare centre, which said that there was no medical reason why he needed to take bread to work with him. In granting relief and quashing the review decision of the Secretary of State for Justice, Mr Philip Mott QC (sitting as a deputy High Court judge) said that it followed from provisions in Prison Service Instruction 47/2011 that a charge must be proved on evidence presented at the hearing and that the prisoner should be allowed to question any witness called to give evidence. These requirements were hardly surprising and simply encapsulated the standard principles of nature justice, found also in Article 6 ECHR (see further *R (McCarthy) v. Visitors to the Inns of Court* [2015] EWCA Civ 12). The test of fairness is whether there is a real possibility that the tribunal would have come to a different conclusion if the procedural error had not been made—in this case, if the claimant had been allowed to question the nurse. It was not possible to argue that casting doubt on the nurse's evidence would not have affected the governor's decision. Neither the governor's decision nor the adjudication review decision properly addressed the failure to allow the claimant to question the nurse during the hearing, or to request other medical evidence, in order to establish that eating bread might have alleviated the claimant's condition.

## C. Other Relevant Chapters

**48.19** Abuse of Process, Chapter 2
Evidence, Chapter 25
Human Rights, Chapter 32
Independent and Impartial Tribunal, Chapter 35
Judicial Review, Chapter 42

# 49

# NEGLIGENCE



## A. General Principles 49.01 | *Vaidya v. General Medical Council* [2010] EWHC 984 (QB) | 49.08 |

### *Bolam v. Frien Hospital Management Committee* [1957] 1 WLR 582

**49.01** Test—ordinary skilled man exercising and professing to have special skill—acting in accordance with practice accepted as proper by responsible body
| *Maynard v. West Midlands Regional Health Authority* [1984] 1 WLR 634 | 49.02 | *Adams v. Law Society of England and Wales* [2012] EWHC 980 (QB) | 49.10 |
| *Bolitho v. City and Hackney Health Authority* [1998] AC 232 | 49.03 | *Iqbal v. Solicitors Regulation Authority* [2012] EWHC 4097 (QB) | 49.11 |
| B. Disciplinary Cases | 49.04 | *O'Connor v. Bar Standards Board* [2014] EWHC 4324 (QB) | 49.12 |
| C. Duty of Care | 49.05 | | |
| *Wood v. Law Society* [1995] NPC 39 | 49.05 | | |
| *Collins v. Office for the Supervision of Solicitors* [2002] EWCA Civ 1002 | 49.06 | *Schubert Murphy (a firm) v. The Law Society* [2014] EWHC 4561 (QB) | 49.13 |
| *Merelie v. General Dental Council* [2009] EWHC 1165 (QB) | 49.07 | D. Other Relevant Chapters | 49.14 |

B, the plaintiff, claimed damages against the defendants in respect of injuries that he received while undergoing electroconvulsive therapy (ECT) on 23 August 1954 at Friern Hospital. Medical evidence was called by both sides. No one suggested that there was any negligence in the diagnosis, or in the decision to use ECT. However, there were different variants as to the use of restraining sheets, relaxant drugs, and manual control of the patient's body. The question was whether Dr A, following the practice that he had learned at Friern Hospital and following the technique that had been shown to him by Dr B, was negligent in failing to use relaxant drugs or, if he decided not to use relaxant drugs, that he was negligent in failing to exercise any manual control over the patient beyond merely arranging for his shoulders to be held, his chin supported, a gag used, and a pillow put under his back. McNair J, in summing up to the jury, said, at 586–7:

> But where you get a situation which involves the use of some special skill or competence, then the test whether there has been negligence or not is not the test of the man on top of the Clapham omnibus, because he has not got this special skill. The test is the standard of the ordinary skilled man exercising and professing to have that special skill. A man need not possess the highest expert skill at the risk of being found negligent. It is well established law that it is sufficient if he exercises the ordinary skill of an ordinary competent man exercising that particular art. I do not think that I quarrel much with any of the submissions in law which have been put before you by counsel. Counsel for the plaintiff put it in this way, that in the case of a medical man negligence means failure to act in accordance with the standards of reasonably competent

> medical men at the time. That is a perfectly accurate statement, as long as it is remembered that there may be one or more perfectly proper standards; and if a medical man conforms with one of those proper standards then he is not negligent. Counsel for the plaintiff was also right, in my judgment, in saying that a mere personal belief that a particular technique is best is no defence unless that belief is based on reasonable grounds. That again is unexceptionable. But the emphasis which is laid by counsel for the defendants is on this aspect of negligence: he submitted to you that the real question on which you have to make up your mind on each of the three major points to be considered is whether the defendants, in acting in the way in which they did, were acting in accordance with a practice of competent respected professional opinion. Counsel for the defendants submitted that if you are satisfied that they were acting in accordance with a practice of a competent body of professional opinion, then it would be wrong for you to hold that negligence was established.... I myself would prefer to put it in this way, that he is not guilty of negligence if he has acted in accordance with a practice accepted as proper by a responsible body of medical men skilled in that particular art... Putting it the other way round, a doctor is not negligent, if he is acting in accordance with such a practice, merely because there is a body of opinion who would take the contrary view. At the same time, that does not mean that a medical man can obstinately and pig-headedly carry on with some old technique if it has been proved to be contrary to what is really substantially the whole of informed medical opinion. Otherwise you might get men today saying: "I don't believe in anaesthetics. I don't believe in antiseptics. I am going to continue to do my surgery in the way it was done in the eighteenth century". That clearly would be wrong.

The jury, having retired and considered its verdict, found that the defendant was not negligent.

***Maynard v. West Midlands Regional Health Authority*** **[1984] 1 WLR 634**

**49.02** Test—clinical judgment—whether failure such that no doctor of ordinary skill would have so acted—exercise of ordinary skill of speciality

Lord Scarman (with whom Lord Fraser of Tullybelton, Lord Elwyn-Jones, Lord Roskill, and Lord Templeman agreed) said that the question in this appeal was whether a physician and a surgeon, working together in the treatment of their patient, were guilty of an error of professional judgment of such a character as to constitute a breach of their duty of care towards her. The negligence alleged against each, or one or other, of them was that, contrary to the strong medical indications that should have led them to diagnose tuberculosis, they held back from a firm diagnosis and decided that she should undergo the diagnostic operation, mediastinoscopy. It was an operation that carried certain risks, even when correctly performed, as is admitted that it was in this case. One of the risks—namely, damage to the left laryngeal recurrent nerve—did, as the judge found and the respondent authority accepted, unfortunately materialize, with resulting paralysis of the left vocal cord. As to the nature of the duty owed by a doctor to his patient, the (then) most recent authoritative formulation was that by Lord Edmund-Davies in *Whitehouse v. Jordan* [1981] 1 WLR 246, quoting from the judgment of McNair J in *Bolam v. Friern Hospital Management Committee* [1957] 1 WLR 582, 586. The present case could be classified as one of clinical judgment. A case that is based on an allegation that a fully considered decision of two consultants in the field of their skill was negligent clearly presents certain difficulties of proof. It is not enough to show that there is a body of competent professional opinion that considers that theirs was a wrong decision if there also exists a body of professional opinion, equally competent, which supports the decision as reasonable in the circumstances. It is not enough to show that subsequent events show that the operation need never have been performed if, at the time the decision to operate was taken, it was reasonable in the sense that a responsible body of medical opinion would have accepted it as proper. The words of Lord President Clyde in *Hunter v. Hanley* 1955 SLT 213, 217, cannot be bettered:

> In the realm of diagnosis and treatment there is ample scope for genuine difference of opinion and one man clearly is not negligent merely because his conclusion differs from that of other professional men...The true test for establishing negligence in diagnosis or treatment on the part of a doctor is whether he has been proved to be guilty of such failure as no doctor of ordinary skill would be guilty of if acting with ordinary care...

Lord Scarman added that a doctor who professes to exercise a special skill must exercise the ordinary skill of his speciality. Differences of opinion and practice exist, and will always exist, in the medical profession, as in other professions. There is seldom any one answer exclusive of all others to problems of professional judgment. A court may prefer one body of opinion to the other—but that is no basis for a conclusion of negligence. Failure to exercise the ordinary skill of a doctor (in the appropriate speciality, if he be a specialist) is necessary.

### *Bolitho v. City and Hackney Health Authority* [1998] AC 232

**49.03** Test—responsible, reasonable, and respectable body of professional opinion—role of Court to determine

B, a 2-year-old boy, suffered catastrophic brain damage as a result of cardiac arrest induced by respiratory failure. In giving the judgment of the House of Lords, Lord Browne-Wilkinson said that the *locus classicus* of the test for the standard of care required of a doctor or any other person professing some skill or competence is the direction to the jury given by McNair J in *Bolam v. Friern Hospital Management Committee* [1957] 1 WLR 582, 587:

> I myself would prefer to put it this way, that he is not guilty of negligence if he has acted in accordance with a practice accepted as proper by a responsible body of medical men skilled in that particular art...Putting it the other way round, a man is not negligent, if he is acting in accordance with such a practice, merely because there is a body of opinion who would take a contrary view.

Lord Browne-Wilkinson continued, at 242G–243E:

> [I]n my view, the court is not bound to hold that a defendant doctor escapes liability for negligent treatment or diagnosis just because he leads evidence from a number of medical experts who are genuinely of opinion that the defendant's treatment or diagnosis accorded with sound medical practice. In the *Bolam* case itself, McNair J [1957] 1 WLR 583, 587 stated that the defendant had to have acted in accordance with the practice accepted as proper by a "*responsible* body of medical men". Later, at p588 he referred to "a standard of practice recognised as proper by a competent *reasonable* body of opinion." Again, in the passage which I have cited from *Maynard's* case [1984] 1 WLR 634, 639, Lord Scarman refers to a "respectable" body of professional opinion. The use of these adjectives—responsible, reasonable and respectable—all show that the court has to be satisfied that the exponents of the body of opinion relied upon can demonstrate that such opinion has a logical basis. In particular in cases involving, as they so often do, the weighing of risks against benefits, the judge before accepting a body of opinion as being responsible, reasonable or respectable will need to be satisfied that, in forming their views, the experts have directed their minds to the question of comparative risks and the benefits and have reached a defensible conclusion on the matter.
>
> [...]
>
> [*Hucks v. Cole* [1993] 4 Med LR 393 and *Edward Wong Finance Co Ltd v. Johnson Stokes & Master* [1984] AC 296] demonstrate that in cases of diagnoses and treatment there are cases where, despite a body of professional opinion sanctioning the defendant's conduct, the defendant can properly be held liable for negligence (I am not here considering questions of disclosure of risk). In my judgment that is because, in some cases, it cannot be demonstrated to the judge's satisfaction that the body of

opinion relied upon is reasonable or responsible. In the vast majority of cases the fact that distinguished experts in the field are of a particular opinion will demonstrate the reasonableness of that opinion. In particular, where there are questions of assessment of the relative risks and benefits of adopting a particular medical practice, a reasonable view necessarily presupposes that the relative risks and benefits have been weighed by the experts in forming their opinions, but if, in a rare case, it can be demonstrated that the professional opinion is not capable of withstanding logical analysis, the judge is entitled to hold that the body of opinion is not reasonable or responsible. I emphasise that in my view it will very seldom be right for a judge to reach the conclusion that views genuinely held by a competent medical expert are unreasonable. The assessment of medical skills and benefits is a matter of clinical judgment which a judge would not normally be able to make without expert evidence... [I]t would be wrong to allow such assessment to deteriorate into seeking to persuade the judge to prefer one of two views both of which are capable of being logically supported. It is only where a judge can be satisfied that the body of expert opinion cannot be logically supported at all that such opinion will not provide the benchmark by reference to which the defendant's conduct falls to be assessed.

## B. Disciplinary Cases

**49.04** As to the circumstances in which negligence may constitute misconduct or lack of competence/deficient performance, see Chapter 16 and Chapter 47. In particular, see the following cases:

- *Calhaem v. General Medical Council* [2007] EWHC 2606 (Admin)—paragraph 16.06;
- *In re a Solicitor* [1972] 1 WLR 869—paragraph 47.12;
- *McCandless v. General Medical Council* [1996] 1 WLR 167—paragraph 47.20;
- *Roylance v. General Medical Council (No. 2)* [2000] 1 AC 311—paragraph 47.22;
- *Nandi v. General Medical Council* [2004] EWHC 2317 (Admin)—paragraph 47.27; and
- *General Medical Council v. Meadow* [2007] QB 462—paragraph 47.33.

## C. Duty of Care

### *Wood v. Law Society* [1995] NPC 39

**49.05** Disciplinary proceedings—investigation of complaint—not causative of any loss

W began an action against the Law Society as defendant in negligence for not investigating her complaint properly against her former solicitor. W's former solicitor had arranged a loan for her secured on her home from a company without revealing that the senior partner in the firm was also a director and part-owner of the lending company. At the heart of W's complaint was her case that if the Law Society had acted so as to control its members, the company would have been persuaded to stay its hand in the eviction process. On 28 July 1993, Otton J gave judgment in favour of the Law Society. In dismissing W's action, the judge said that the real reason why she was eventually evicted was because she could not raise the money to satisfy her creditors and the law took its inevitable course. On appeal, the Court of Appeal said that it wished to hear first on the issues of causation and loss. Leggatt LJ (with whom Auldous and Hutchison LJJ agreed) said, in dismissing W's appeal, that the function of a complaints bureau or a complaints procedure on the part of a regulatory authority is to investigate complaints. That may have the incidental effect of allaying grievances, but the purpose of an investigation is to enquire whether a solicitor has fallen below the standard reasonably to be expected of him and, if so, to discipline him. The purpose of investigating an allegation of professional misconduct is not to relieve distress felt by the complainant, although no doubt, if properly conducted, it may be expected to afford the complainant some measure of satisfaction. On this simple basis, the appeal must fail. The interesting arguments to the effect that the Law Society owed a duty of care to W as an individual complaining of the misconduct of a solicitor must be left over for another day. For purposes of examining the arguments as to causation and damages, the Court made an assumption that such a duty of care, of which the Law Society was in breach, could be made good, but it had to be borne in mind that that had been an assumption made for argument's sake only.

### *Collins v. Office for the Supervision of Solicitors* [2002] EWCA Civ 1002

**49.06**

Investigation of complaint—whether assumption of responsibility—loss—whether any loss flowing from any action of regulatory body

The applicant became involved in a substantial business venture with a solicitor, B, acting for him. Subsequently, a complaint was made to the respondent on 12 May 1989. The case sought to be made was that the respondent assumed a private law duty to the applicant in the course of handling the complaint and that it was in breach of a private law duty owed to him. The judge granted the respondent summary judgment. In refusing permission to appeal, Pill LJ said that there was nothing that took the case outside the general rule, as established in the authorities, that a private law duty does not arise in circumstances such as the present. The Court would not exclude the possibility that a situation could arise on the facts in which, while there is no general duty of care, matters did occur in the course of an investigation from which it might be argued that the conduct of the Office for the Supervision of Solicitors towards a complainant was such that a limited duty arose. As to damages, even if a duty of care were to be established in this case, it would be extremely difficult to establish any loss as flowing from any action of the respondent in the course of the investigation. Collins J, in agreeing, said that the failures of the respondent, if they were indeed established, arose in the course of considering the matter for the purposes of determining whether a complaint had been properly made out and should go forward to the appropriate committee. The respondent's failures did not establish an acceptance of responsibility so that there was a duty of care owed in private law. It is plain from the authorities that unless there is some such assumption of responsibility, no such duty of care can be owed, because the duty of the Law Society is to investigate the matter and to decide on the question of discipline of the solicitor concerned.

### *Merelie v. General Dental Council* [2009] EWHC 1165 (QB)

**49.07**

Dentist complaint against regulator—no common law duty of care

M, a dentist, brought proceedings against the General Dental Council (GDC) alleging a failure to deal with her complaint. M was employed at a community dental service in Newcastle and was a senior dental officer at a clinic. In March 2000, four dental nurses made a complaint against M to the trust. The trust decided not to investigate the complaint. Eventually, M lodged a complaint in June 2008 with the GDC about one of the nurses, but the Council took no action. In December 2008, M began proceedings against the Council in which she alleged that the original complaint by the dental nurse had had far-reaching and devastating effects on her health, reputation, and career. In striking out M's claim, after reviewing *Caparo Industries Plc v. Dickman* [1990] 2 AC 605, *Wood v. The Law Society* [1995] NPC 39, *X v. Hounslow London Borough Council* [2009] EWCA Civ 286, and *Van Colle v. Chief Constable of Hertfordshire Police* [2009] 1 AC 225, Cranston J said, at [41]:

> In my view, applying those principles, it cannot be said that there is any duty of care imposed on the General Dental Council in the circumstances of this case. It is not

alleged that it was reasonably foreseeable that what is said to be their breach of duty would result in any relevant injury or loss to [M]. Despite her contentions about proximity and that, because she was a dentist making a complaint that gave rise to the requisite close relationship, in my view it cannot be said that there is any proximity between a complainant about dental provision and the General Dental Council. Most importantly in my view, it would not be fair, just or reasonable for such a duty to be imposed in this case in the context of the performance of the Council's fitness to practise functions. That conclusion is bolstered in that it cannot be said the Council have assumed a responsibility to [M] in the performance of its fitness to practise functions. The design of the regulatory regime does not lead to a conclusion that the Council owes a common duty of care to [M]. The conclusion is supported by the approach of Otton J in *Wood v. The Law Society*. If there is no common law duty of care imposed on the Council, then even less can there said to be a duty of care imposed on [the president of the Council, the chief executive and registrar of the Council, and a case worker at the Council]. In this case the allegation is a failure to deal with a complaint. It is an act of omission rather than commission. In terms of breach, [M] would face, in my view, insurmountable barriers given the factual background as to how the Council has dealt with her complaint. Inasmuch as she has suffered distress rather than psychiatric injury, *Wood v. The Law Society* is strongly persuasive authority that she would fail to obtain damages for her loss.

### *Vaidya v. General Medical Council* [2010] EWHC 984 (QB)

**49.08** Defamation—regulator's website—no duty of care

V, a former consultant gastroenterologist, brought proceedings against the General Medical Council (GMC) for damages for defamation. According to his claim form, V sought damages from the GMC for the publication of untrue and misleading statements on its website regarding a fitness to practise hearing, and for keeping the information on its website when the hearing was postponed or otherwise cancelled. Sir Charles Gray (sitting as a judge of the High Court) struck out the action on the grounds that the proceedings were not commenced within one year under section 4A of the Limitation Act 1980 and there were no grounds for extending time pursuant to section 32(1) of the 1980 Act. Second, in granting the defendant's application to strike out the claim, the judge said, at [55], that he could not accept, in the light of *Jain v. Trent Strategic Health Authority* [2009] 1 AC 853, that a tortious duty of care should be imposed on a body such as the GMC, because to do so would or might inhibit the GMC in the performance of its duties, including its duty to safeguard patients.

### *Baxendale-Walker v. Middleton* [2011] EWHC 998 (QB)

**49.09** As to proceedings alleging conspiracy, malicious falsehood, and misfeasance in public office, and defence of privilege, see paragraph 52.12.

### *Adams v. Law Society of England and Wales* [2012] EWHC 980 (QB)

**49.10** As to a claim in alleging misfeasance in public office, and defence of privilege, see paragraph 52.13.

### *Iqbal v. Solicitors Regulation Authority* [2012] EWHC 4097 (QB)

**49.11** Misfeasance in public office—Article 1 of Protocol 1 to ECHR

In October 2011, the claimant, a solicitor and sole principal of a firm, issued proceedings against the Solicitors Regulation Authority (SRA) and two of its staff, claiming damages for misfeasance in public office and breach of his rights under Article 1 of Protocol 1 to the European Convention on Human Rights (ECHR), contrary to section 7 of the Human Rights Act 1998. In April 2008, the SRA commenced an investigation into the claimant's firm, followed by forensic investigation in August 2008. Eventually, in December 2011, following judicial review proceedings, the appellant successfully appealed against the decision of an adjudicator and an appeal panel. However, the firm closed in February 2011, and the claimant maintained that his business was effectively crippled by the pursuit by the SRA of criticisms that were ultimately shown to be empty and which it should have been clear from the start were either trivial or groundless. In his claim for misfeasance in public office against the SRA and its staff, Underhill J said that he started by setting out the ingredients of the tort, as now authoritatively stated in the speech of Lord Steyn in *Three Rivers District Council v. Bank of England (No. 3)* [2003] 2 AC 1, 191–6. Lord Steyn identified the six ingredients of the tort of misfeasance in public office as being: (a) the defendant must be a public officer; (b) there must be an exercise of power as a public officer; (c) the third requirement concerns the state of mind of the defendant; (d) duty to the plaintiff; (e) causation; and (f) damage and remoteness. Underhill J said that he saw no realistic prospect that the claimant would succeed on the claim of misfeasance in public office. The claim depended on his establishing that the two members of staff acted in bad faith, either because they were exercising their powers for an improper or ulterior motive, or because they did not have an honest belief that they were acting lawfully. Article 1 provides that every person is entitled to the peaceful enjoyment of his or her possessions, and that no one shall be deprived of his or her possessions except in the public interest and subject to conditions provided for by law. The learned judge said that he saw no reason to doubt that the claimant's interest in his firm constituted a 'possession' for the purposes of Article 1. But the claimant's case that the SRA acted wrongfully in a way that caused him financial loss and that this financial loss eventually forced him to close his business did not seem, even arguably, to constitute the deprivation of possessions. It was important to emphasize that the Court was not here concerned with the case of the common type in which a regulatory authority requires a licence, or imposes a condition, or takes some other regulatory act that directly impacts on a business.

As to immunity from suit, see further paragraph 52.14.

### *O'Connor v. Bar Standards Board* [2014] EWHC 4324 (QB)

**49.12** Disciplinary proceedings—acquittal—claim for damages—Human Rights Act 1998

The claimant was a practising barrister who sued the Bar Standards Board for damages in respect of her case that ended in acquittal on appeal: see further paragraph 47.45. The claimant originally framed her complaints under four heads: (a) misfeasance in public office; (b) breach of the Human Rights Act 1998 by violations of her human rights under Articles 6 and 14 ECHR; (c) harassment contrary to the Protection from Harassment Act 1997; and (d) discrimination contrary to the Equality Act 2010. The master held all of these to be unsustainable. His decision in respect of the claims in misfeasance in public office and under the Equality Act were not challenged on appeal, and Warby J refused permission to appeal against the master's decision to dismiss the claim in harassment. As to the claim under the Human Rights Act 1998, the factual background was that, on 30 September 2009, the Bar Standards Board received a complaint about the claimant. In early November 2009, the Board wrote to the claimant summarizing the complaint. The complaint form and supporting documents that were provided to the claimant at that time identified with precision the acts of the claimant that were alleged to amount to professional misconduct. The nature of the misconduct alleged—that is, conducting litigation in specified ways—was also made clear. Warby J said that it could not be and was not said that the period between 30 September and 3 November 2009 represented a lack of promptness. The present case did not come close to establishing a violation of the requirements of Article 6(3)(a). Subsequently, the Board refused a one-month stay in November 2010. There was no prospect that a court would uphold that as being a breach of Article 6(3). As to the claim under Article 14, the learned judge said that the claimant's particulars of claim sufficiently pleaded a case that the Board indirectly discriminated against her on racial or ethnic grounds by bringing the disciplinary proceedings against her. The claimant, who was black, alleged that in practice the Board's complaints process impacts disproportionally on black and minority ethnic (BME) barristers in particular ways. These included the allegation that BME barristers are more likely to have a complaint referred for prosecution. The claimant's allegations bore comparison with the allegations made by the applicants in *DH v. Czech Republic* (2008) 47 EHRR 3. They did not amount to allegations of conscious discrimination, nor of inherently discriminatory rules. They were allegations that rules that are not said to be discriminatory in themselves are implemented or applied in practice in a way as to affect an ethnic group in a way that is disproportionately prejudicial. Statistical evidence existed to support the claimant's pleaded allegation that BME barristers were, at the material time, proportionately more likely than others to be referred by the Board for prosecution. The statistics were referred to in the Board's report entitled *Diversity Review: Bar Standards Board's Complaints System* (January–June 2013). Accordingly, the learned judge said that he concluded that the particulars of claim did adequately state a case, which was not fanciful, that by bringing disciplinary proceedings against the claimant the Board indirectly discriminated against her contrary to Article 14. However, the decision to bring proceedings against the claimant was taken at the latest in July 2010, when the charges were served on her. Section 7 of the Human Rights Act 1998 provides that proceedings must be brought before the end of the period of one year beginning with the date on which the act complained of took place, or such longer period as the court considers equitable having regard to all of the circumstances. The present proceedings were issued in February 2013. The limitation point was pleaded in the defence served in August 2013 and, despite the claimant issuing an application to serve amended particulars of claim and a reply, no reply was served; nor did the claimant at any point assert in evidence or in her written submissions a claim for an extension on any grounds on which it would be equitable to allow her claim to proceed. The Court held that the claim was barred by limitation.

### *Schubert Murphy (a firm) v. The Law Society* [2014] EWHC 4561 (QB)

**49.13** Solicitor—name appearing on Law Society website—imposter—solicitor's undertaking—loss

The claimants, a two-partner firm of solicitors, were instructed by K to act for him in the purchase of a registered freehold property in Hadley Wood, Hertfordshire, for £735,000. The property was subject to a first charge in favour of Lloyds TSB Plc. Contracts were exchanged on 19 May 2010 and the sale was completed on 21 May 2010. 'Acorn Solicitors' (principal 'J Dobbs') acted for the vendor. Shortly prior to completion, on 10 May 2010, M, one of the two claimant partners, carried out a search on the Law Society's 'Find a Solicitor' web page, which confirmed the existence of Acorn Solicitors and that its principal was 'John Dobbs'. It gave the usual undertaking to discharge the first charge on completion out of the purchase monies paid by K. It did not do so. Neither Acorn nor J Dobbs were solicitors. Acorn made off with the purchase monies. 'John Dobbs' was an imposter. K sued the claimants. Their insurer settled his claim and, by subrogation, the insurer sued the defendant to recover its outlay. On 30 September 2009, 'John Dobbs' applied to the SRA for approval to practise as a recognized sole practitioner under the name 'Acorn Solicitors'. His name and that of Acorn Solicitors were entered on the roll of solicitors and on the 'Find a Solicitor' page of the defendant's website. By an application notice, the defendant applied to strike out the claimant's statement of case on the basis that it disclosed no reasonable grounds for bringing the claim, and/or for summary judgment on the basis that it had no reasonable prospect of success. The basis of the application was that the defendant owed no duty of care to the claimants or to K on the basis that the defendant was performing its statutory duty as a regulator. The claimants' case was that a duty of care not to cause economic loss to them or to K arose in one or more of three ways: (a) by assumption of responsibility; (b) by application of the three-part test in *Caparo Industries Plc v. Dickman* [1992] AC 605; and/or (c) either within or by way of a modest increment on established case law relating to negligent misstatement of the *Hedley Byrne & Co. Ltd v. Heller & Partners* [1964] AC 465 variety. In dismissing the defendant's application and refusing to strike out the claim, Mitting J said that the circumstances of the case raised a question of much wider importance than what should happen to this case:

> It calls into question, at least in theory, the security of current conveyancing practice. The principal guarantee that a purchaser of mortgaged property has that he will obtain clear title is the undertaking given by the vendor's solicitors, who will usually act for the mortgagee as well, that they will discharge any secured debt out of the purchase monies. If they do not, the Solicitors Compensation Fund will rapidly make good any loss if, but only, if, the person giving the undertaking is a solicitor. If he is not, then I accept [counsel for the defendant's] submission that the compensation fund is not permitted by its remit to make a payment and is certainly under no obligation to do so.

The learned judge said that if the defendant's argument is right, then the security of current conveyancing practice might be called into question. That may be a powerful factor in support of the defendant recognizing the existence of a duty of care coincident with its statutory duty when considering applications for entry onto the roll of solicitors and registration. The position of an individual acting directly with someone claiming to be a solicitor also has to be considered. The defendant's website encouraged ordinary members of the public to rely on its published information about who was a solicitor. If an ordinary member of the public reliant on the information consults an imposter operating an office on a high street near him or her and entrusts that person with money, as people do with solicitors, then if he or she loses it, he or she might well be rather shocked to find that he or she had no recompense against the representative and regulatory body that held out that person as a solicitor on its website. Accordingly, the case was eminently suitable for trial, and the claimant's claim was not to be struck out or judgment entered for the defendant.

## D. Other Relevant Chapters

**49.14**

Deficient Professional Performance, Chapter 16
Experts, Chapter 26
Misconduct, Chapter 47
Privilege, Chapter 52
Recklessness, Chapter 55

# 50

# NO CASE TO ANSWER

A. Legal Framework 50.01
B. General Principles 50.02
*R v. Galbraith* [1981] 1 WLR 1039 50.02
*R v. Shippey* [1988] Crim LR 767 50.03
*R v. F (S)* [2012] 1 All ER 565 50.04
*Benham Ltd v. Kythira Investments Ltd* [2003] EWCA Civ 1794 50.05
*Graham v. Chorley Borough Council* [2006] EWCA Civ 92 50.06
C. Disciplinary Cases 50.07
*Lucas v. Millman* [2003] 1 WLR 271 50.07
*Razak v. General Medical Council* [2004] EWHC 205 (Admin) 50.08
*R (Tutin) v. General Medical Council* [2009] EWHC 553 (Admin) 50.09
*Martin v. General Medical Council* [2011] EWHC 3204 (Admin) 50.10
*R (Sharaf) v. General Medical Council* [2013] EWHC 3332 (Admin) 50.11
*Hussein v. General Medical Council* [2013] EWHC 3535 (Admin) 50.12
*Soni v. General Medical Council* [2015] EWHC 364 (Admin) 50.13

## A. Legal Framework

**50.01** Examples include:

Nursing and Midwifery Council (Fitness to Practise) Rules 2004, rule 24(7) (the committee may hear submissions and shall make a determination as to whether the registrant has a case to answer)

General Pharmaceutical Council (Fitness to Practise and Disqualification etc) Rules 2010, rule 31(8) ('The registrant may make submissions regarding whether sufficient evidence has been adduced to find the facts proved or to support a finding of impairment, and the Committee must consider and announce its decision as to whether any such submissions should be upheld')

General Medical Council (Fitness to Practise) Rules 2014, rule 17(2)(g) (the practitioner may make submissions regarding whether sufficient evidence has been adduced to find the facts proved or to support a finding of impairment)

*Civil Procedure, Vol. 1* (2015), paragraph 32.1.6 (a judge should not entertain a submission of 'no case to answer' unless the defendant elects—whatever the outcome of the submission—to call no evidence; in exceptional circumstances, the judge may deal with a submission without requiring such election by the defendant)

## B. General Principles

### *R v. Galbraith* [1981] 1 WLR 1039

**50.02** Criminal proceedings—principles

Lord Lane CJ, giving the judgment of the Court of Appeal, Criminal Division (Lord Lane CJ, and Peter Pain and Stuart-Smith JJ), said, at 1042B–E:

> How then should the judge approach a submission of "no case"? (1) If there is no evidence that the crime alleged has been committed by the defendant, then there is no

difficulty. The judge will of course stop the case. (2) The difficulty arises where there is some evidence but it is of a tenuous character, for example because of inherent weakness or vagueness or because it is inconsistent with other evidence. (a) Where the judge comes to the conclusion that the Crown's evidence, taken at its highest, is such that a jury properly directed could not properly convict on it, it is his duty, on a submission being made, to stop the case. (b) Where however the Crown's evidence is such that its strength or weakness depends on the view to be taken of a witness's reliability, or other matters which are generally speaking within the province of the jury and where on one possible view of the facts there *is* evidence upon which a jury could properly come to the conclusion that the defendant is guilty, then the judge should allow the matter to be tried by the jury.

### *R v. Shippey* [1988] Crim LR 767

**50.03** 'plums and duff'—commentary

Turner J said that 'taking the prosecution case at its highest' did not mean 'taking out the plums and leaving the duff behind'. The judge should assess the evidence, and if the evidence of the witness upon whom the prosecution case depended was self-contradictory, and out of reason and all common sense, then such evidence was tenuous and suffered from inherent weakness.

In *R (Tutin) v. General Medical Council* [2009] EWHC 553 (Admin)—see paragraph 50.09 below—McCombe J, as he then was, said that whilst giving full credence to the well-known passage from the brief report of the reasons given in the *Criminal Law Review* for Turner J's decision in *Shippey*, it was worth noticing the commentary to that case written by the late Professor JC Smith, which was in the following terms:

> It is quite clear that the case must not be withdrawn from the jury merely because the judge thinks that the principal prosecution witnesses are not telling the truth. That would be to usurp the function of the jury. It is arguably different, however, though the difference is one of degree, if the judge thinks that no reasonable jury could find that the prosecution witnesses are telling the truth. If that is truly the case, then there is no point in leaving the case to them, for (in the absence of damaging evidence appearing during the case for the defence) the jury, which we must assume to be a reasonable jury, would inevitably acquit. Notwithstanding the passage (b) above in *Galbraith*, the judge did think it right to assess the credibility of the prosecution as a whole. Taking the Crown's evidence "at its highest" does not, apparently mean, assuming that the evidence against the defendant is true. What it does mean is not very clear.

### *R v. F (S)* [2012] 1 All ER 565

**50.04** Criminal proceedings—principles

In giving the reserved judgment of the Court of Appeal, Criminal Division, Lord Judge CJ said that the authority of *R v. Galbraith* is undiminished. In accordance with the second limb of *R v. Galbraith*, there will continue to be cases in which the state of the evidence called by the prosecution, and taken as a whole, is so unsatisfactory, contradictory, or so transparently unreliable that no jury, properly directed, could convict. In cases like these, it is the judge's duty to direct the jury that there is no case to answer and to return a 'not guilty' verdict. When, by contrast with a submission of 'no case to answer', the Court is considering an application to stay proceedings on the grounds of prejudice resulting from delay in the institution of proceedings, the appropriate test was directly addressed and answered in *Attorney-General's Reference (No. 1 of 1990)* [1992] QB 630. It continues to provide the benchmark. Accordingly, any suggestion that, on the basis of delay, the judge may be responsible for assessing whether in advance of a conviction, the conviction would be unsafe, is based on a misunderstanding of the principles in both *R v. Galbraith* and *A-G's Reference (No. 1 of 1990)*.

***Benham Ltd v. Kythira Investments Ltd*** **[2003] EWCA Civ 1794**

**50.05** Civil proceedings—principles

At the close of the claimant's evidence on the fourth day of the hearing, the trial judge acceded to the defendants' submission of 'no case to answer' without putting the defendants to their election. In allowing the claimant's appeal and remitting the claim to be retried before another judge, the Court of Appeal (Simon Brown, Keene, and Scott Baker LJJ) said that the case illustrated yet again the dangers of adopting such a course. After reviewing *Alexander v. Rayson* [1936] 1 KB 169, *Boyce v. Wyatt Engineering* [2001] EWCA Civ 692, *Lloyd v. John Lewis Partnership* [2001] EWCA Civ 1529, *Bentley v. Jones Harris & Co.* [2001] EWCA Civ 1724, and *Miller (t/a Waterloo Plant) v. Margaret Cawley* [2002] EWCA Civ 1100, Simon LJ said that the disadvantages of entertaining a submission of 'no case to answer' are plain and obvious. First, as Mance LJ explained both in *Boyce* and in *Miller*, the submission interrupts the trial process and requires the judge to make up his or her mind as to the facts on the basis of one side's evidence only and applying the lower test of a prima facie case, with the result that if he or she rejects the submission, then the judge must then make up his or her mind afresh in the light of whatever further evidence has been called and on the application of a different test. The second disadvantage, as again Mance LJ made plain in *Boyce* and *Miller*, is that if the judge both entertains and accedes to a submission of 'no case', his or her judgment may be reversed on appeal, with all of the expense and inconvenience resulting from the need to resume the hearing or, more probably, retry the action. Rarely, if ever, should a judge trying a civil action without a jury entertain a submission of 'no case to answer'. That clearly was the Court of Appeal's conclusion in *Alexander v. Rayson*, and Simon LJ said that he saw no reason to take a different view today, the Civil Procedure Rules notwithstanding. Further, a judge entertaining a 'no case' submission should clearly recognize and bear in mind the real possibility that the defendant, were his or her submission to fail, might choose to call no evidence, thereby entitling the court to draw adverse inferences that go to strengthen the claimant's case, which may be no more than a weak prima facie case. Such adverse inferences can, in other words, tip the balance of probability in the claimant's favour. Scott Baker LJ said that the wise words of Romer LJ in *Alexander v. Rayson* [1936] 1 KB 169 still hold good today. Only in the most exceptional circumstances should a judge entertain a submission to dismiss an action at the close of the claimant's evidence without putting the defendant to his election. In *Alexander v. Rayson*, the Court of Appeal said, at 178:

> Where an action is being heard by a jury it is, of course, quite usual and often very convenient at the end of the case for the plaintiff, or of the party having the onus of proof as the defendant had here, for the opposing party to ask for the ruling of the judge whether there is any case to go to the jury, who are the only judges of fact. It also seems to be not unusual in the King's Bench Division to ask for a similar ruling in actions tried by a judge alone. We think, however, that this is highly inconvenient. For the judge in such cases is also the judge of fact, and we cannot think it right that the judge of fact should be asked to express any opinion upon the evidence until the evidence is completed. Certainly no one would ever dream of asking a jury at the end of a plaintiff's case to say what verdict they would be prepared to give if the defendant called no evidence, and we fail to see why a judge should be asked such a question in cases where he and not a jury is the judge that

has to determine the facts. In such cases we venture to think that the responsibility for not calling rebutting evidence should be upon the other party's counsel and upon no one else.

*Graham v. Chorley Borough Council* [2006] EWCA Civ 92

**50.06** Civil proceedings—principles

The claimant appealed against the order of the trial judge in a personal injury action dismissing the claim and directing that judgment be entered for the defendants. The judge failed to put the defendants to their election when they made a submission of 'no case to answer' halfway through the trial. In allowing the claimant's appeal, setting aside the judgment and directing a retrial before a different judge, Brooke LJ (with whom Rix and Maurice Kay LJJ agreed) said that if a case follows the course adopted by the judge in the instant case, the claimant is simultaneously being deprived of the opportunity of making a weak case stronger by eliciting favourable evidence from the defendants' witnesses and of the opportunity of inviting the Court to draw adverse inferences from the defendants' failure to give evidence (because the judge has not put them to their election). If the defendants had elected to call no evidence, it would have been open to the claimant to invite the judge to draw adverse inferences along the lines from the failure to put their two witnesses (a doctor in the accident and emergency department of the hospital, and an employee of the council dealing with tenants' complaints) in the witness box. The procedural irregularity caused injustice, in that it deprived the claimant of an opportunity that should have been open to her to strengthen her case in one way or another.

## C. Disciplinary Cases

*Lucas v. Millman* [2003] 1 WLR 271

**50.07** Reasons—extent of reasons

In August 2001, L made an application to the Solicitors Disciplinary Tribunal (SDT) against the respondent M under section 47 of the Solicitors Act 1974. In December 2001, the SDT dismissed the application on the grounds that there was no prima facie case against M. The appellant submitted that the Tribunal should have given more extensive reasons than it did. In *Hiro Balani v. Spain* (1994) 19 EHRR 566, the European Court said that the extent to which the duty to give reasons applies may vary according to the nature of the decision. The question whether a court has failed to fulfil the obligations to state reasons, deriving from Article 6 of the European Convention on Human Rights (ECHR), can be determined only in the light of the circumstances of the case. On L's appeal, the Divisional Court (Kennedy LJ and Pitchers J) said that where, in the opinion of the SDT, a complainant has failed to establish a prima facie case, the Tribunal is in general fully entitled to say simply what was said in this case—that is, that the documentation relied on disclosed no prima facie case against the solicitor. The complainant knows how his or her case has been presented and he or she can, if he or she so chooses, exercise a right of appeal. Neither the complainant's rights nor those of the solicitor are in any way infringed by a brief decision of the Tribunal. In some cases, more may be required. As to who should be joined as parties if a complainant does choose to exercise his or her right of appeal, the only essential parties are the appellant and the other party to the decision before the Tribunal—in essence, in a case such as this, the solicitor. There are others who must be notified—namely, the Law Society, where the Law Society was not originally a party to the proceedings before the Tribunal, and the chairman of the Tribunal itself.

### *Razak v. General Medical Council* [2004] EWHC 205 (Admin)

**50.08** Distinction with criminal proceedings—all evidence before disciplinary tribunal—not restricted to close of GMC's case

R appealed from a direction of the Professional Conduct Committee (PCC) of the General Medical Council (GMC) that his name be erased from the medical register. The substance of the allegations against R were that, between April and September 2001, he had an improper relationship with a patient, in that he visited her at her home address when it was not necessary to do so for professional purposes, and said that he was able to have two wives and that Miss A could be his wife. On another occasion, it was alleged that he had indecently assaulted her and had sexual intercourse with her at a hotel. The basis of R's appeal was that the Committee was wrong to have accepted the evidence, principally of Miss A, to the effect that R had committed the specific acts of which he was found guilty. It was said, on his behalf, that the case depended almost entirely on the testimony of Miss A before the Committee, and that her testimony was, to a substantial extent, inconsistent and lacked credibility, and therefore should not have been accepted by the Committee. In addition, it was contended in the notice of appeal that the case against R should have been stopped at half-time—that is, after the completion of the evidence with the GMC—on the basis that no committee properly instructed and acting reasonably could have come to the conclusion that R was guilty of the matters alleged against him on the evidence as it then was. In dismissing R's appeal, Stanley Burnton J said:

> 11. Happily, there is no significant difference between the parties as to the legal test to be applied in the case such as the present. So far as the contention that the case should have been stopped at half-time, effectively on the basis that there was no case to answer, [counsel for R] accepts that there is a distinction between the jurisdiction of the committee and the criminal jurisdiction. The jurisdiction of the committee is primarily regulatory. Its object is to protect the public, and indeed the medical profession from those, who by reason of their lack of competence, lack of qualification or misconduct, should not be permitted to practise in the medical profession in this country, or are deserving of censor or some other appropriate punishment such as an admonition.
>
> 12. In my judgment, on an appeal in a case such as the present, the court should take into account all the evidence that was before the committee and not restrict itself to the evidence as it was at the close of the case for the General Medical Council. This jurisdiction differs from the criminal jurisdiction, the object of which is, of course, to some extent, to protect the public, but is primarily a jurisdiction exercised for the purpose of punishing crimes. This is a regulatory jurisdiction. It would be surprising if the court, on an appeal such as this, were required to quash a decision of a Disciplinary Tribunal in circumstances where, although the case might have been stopped at half-time and perhaps could have been on the basis that there was no case to answer, the evidence given on behalf of the respondent supported, in material respects, the case against him.
>
> 13. [Counsel for R] did not dispute the analysis of the jurisdiction of the appeal court. It follows that I have to consider all the evidence that was before the committee in determining whether or not this appeal should be allowed and, if so, to what extent.

### *R (Tutin) v. General Medical Council* [2009] EWHC 553 (Admin)

**50.09** Sexual impropriety—evidence of complainant—whether credibility undermined entirely

T, a general practitioner (GP) aged 60, was alleged to have committed acts of sexual impropriety amounting to professional misconduct towards a number of patients. Counsel for T submitted that the panel was wrong to reject her submission of 'no case' under rule 17(2)(g) of the General Medical Council (Fitness to Practise) Rules 2004 in respect of four allegations made by Ms A. Counsel accepted that Ms A gave some evidence in relation to each head of charge that, if reliable, would be sufficient to found the charge made. It was further accepted that Ms A was not a witness who was incapable of telling the truth. However, counsel submitted that she was a witness who was so clearly capable of distortion or embellishment of facts, in particular in relation to sexual matters, to the extent that her evidence as a whole was intrinsically unreliable. McCombe J said that it was accepted that the panel was obliged to apply the criminal law relating to submissions of 'no case to answer'. The panel asked itself: was there any evidence before the panel upon which it could find each allegation proved? Was there some evidence, but of such an unsatisfactory character that the panel, properly directed as to the burden and standard of proof, could not find each allegation proved? Was there some evidence, the relative strength or weakness of which was dependent upon the panel's view of the reliability of a witness? In dismissing T's application for judicial review, the learned judge said that it was clear that the panel had adopted the correct legal test at the beginning of its ruling, referring to *R v. Galbraith* [1981] 1 WLR 1039, and that its formulation of its reasoning could not be faulted. It seemed to the learned judge that the panel must have taken the view that, whatever the strength of the argument submitted, it did not at that stage go to undermine entirely Ms A's credibility. It is clearly open to a tribunal of fact to decide in respect of any witness whether it can accept all of his or her evidence, none of it, or only some of it. The panel must have appreciated that in taking its decision as a matter of law. It clearly took the view that the reliability of Ms A was not undermined in sufficient extent for it to be unsafe to leave it for final consideration of the facts in respect of some of the charges and to allow the matter to be assessed at the end of the day.

### *Martin v. General Medical Council* [2011] EWHC 3204 (Admin)

**50.10** Prima facie case—evidence

M was a partner in a general practice with Dr K and Dr J (the senior partner). One of the issues arising on M's appeal was whether the fitness to practise panel was wrong to reject M's half-time submission that there was no case to answer in respect of heads of charge 11(b) and 13 (namely, that M had acted dishonestly in issuing a prescription in Dr K's name). Despite conflicting evidence from Dr K before the panel, both parties accepted the evidence from Dr J that the default setting on the practice computer meant that all prescriptions were automatically issued in the name of the patient's registered doctor, regardless of which doctor was logged on to the computer and authorizing the prescription. In the light of Dr J's evidence, counsel for M submitted to the panel, at the close of the GMC case, that there was no case to answer on heads of charge 11(b) and 13. The panel found that there was a case to answer and dismissed the application. In dismissing M's appeal, the Court said that the panel made express references to Dr J's evidence and, in the light of the parties' submissions and the legal advice, it was inconceivable that the members had not grasped the significance of his evidence. The panel was entitled, even on the basis of Dr J's evidence, to reject the appellant's submission of no case to answer.

### *R (Sharaf) v. General Medical Council* [2013] EWHC 3332 (Admin)

**50.11** Challenge to panel's decision—test—*Wednesbury* unreasonable or irrational

S applied for permission to apply for judicial review following rejection of a submission of 'no case to answer'. At the conclusion of the GMC's case before the panel, the applicant submitted that insufficient evidence had been adduced to find one charge proven, and submitted that the charge and its related charge of dishonesty should be dismissed. In refusing permission, Carr J, at [49], said that rule 17(2)(g) of the GMC's Fitness to Practise Rules 2004 provided an important safeguard. It is oppressive and unjust for an accused person in regulatory proceedings to be required to meet a case that has not been established on sufficient evidence. However, the Administrative Court is a court of review only. It is common ground that the test laid down in *R v. Galbraith* [1981] 1 WLR 1039 applied and that the tribunal was correct to address the issues on that basis.

In *R (Tutin) v. General Medical Council* [2009] EWHC 553 (Admin), McCombe J (as he then was) gave instructive comments on the approach to be taken in similar circumstances, although there are material differences including the fact that, in *Tutin*, the relevant standard of proof was still the criminal one. In the instant case, the conclusion reached by the panel could not be said to be outside the range of reasonable response so as to be *Wednesbury* unreasonable or irrational. The legal assessor advised the panel correctly on the law and the panel expressly accepted the advice. Moreover, this was a case in which the strength or weakness of evidence depended on the panel's view of the witness's reliability, and the differences between the witness's earlier witness statement and her oral evidence were matters that went to weight. Finally, judicial review is a remedy of last resort. As a general rule, there should be no challenge to validity during the course of disciplinary proceedings: see *R (Mahfouz) v. Professional Conduct Committee of the General Medical Council* [2004] EWCA Civ 233, at [42].

***Hussein v. General Medical Council* [2013] EWHC 3535 (Admin)**

**50.12** 'Referral'—meaning—whether doctor assumed responsibility for patient

H, a consultant orthopaedic surgeon who was in a close personal relationship with RJ, contacted a colleague, AS, a consultant obstetrician and gynaecologist, and asked AS to see RJ by way of a private consultation. H arranged for ultrasound scans to be taken and later authorized RJ's discharge from hospital by making an entry in her notes. *Good Medical Practice* expressly warns registered doctors, whenever possible, to avoid providing medical care to anyone with whom they have a close personal relationship. H faced allegations of making a 'referral' within *Good Medical Practice*. The GMC's expert witness, a consultant gynaecologist, stated when cross-examined that he did not consider that H was making an official referral from one consultant to another and therefore that this was not a case of a referral, as described. Notwithstanding such evidence, the fitness to practise panel rejected a submission under rule 17(2)(g) of the GMC's Fitness to Practise Rules 2004 and went on to find H's fitness to practise to be impaired. In allowing H's appeal in relation to the issue of 'referral', Phillips J said that the panel's interpretation of its definition was wrong and that, for there to be a referral by a doctor within the definition in *Good Medical Practice*, it was necessary to establish that the doctor had assumed professional responsibility for a patient's medical care at the time that he or she made the relevant introduction to a specialist. The mere fact of the introduction cannot of itself, without more, give rise to such responsibility. Whilst it was open to the panel to consider whether H's overall actions in relation to RJ's care, including ordering scans, discharging her from hospital, and arranging a number of consultations, demonstrated that he was acting as her doctor more generally, the panel did not have such an approach in mind at the time of its determination of the rule 17(2)(g) application, and in its final ruling did not find that H was acting as RJ's doctor more generally.

***Soni v. General Medical Council* [2015] EWHC 364 (Admin)**

**50.13** No direct evidence to support case—whether evidence justifying inference of dishonesty—whether rehearing could establish case to answer

Between December 2005 and May 2011, S, a consultant ophthalmologist, was employed by East and North Hertfordshire NHS Trust. It was alleged that, during 2008, S failed to inform the Trust's private patient office of treatment to five patients to whom he administered injections of Lucentis, that he dishonestly retained the full fees paid to him by the five patients, and that he dishonestly failed to make appropriate payments to the Trust from the fees paid to him for treating private patients on Trust premises. The fees paid to S totalled £13,542. A fitness to practise panel of the GMC found each of the allegations proved and directed that S's registration be suspended for six months. In allowing S's appeal and declining to remit the case for a fresh hearing, Holroyde J said that the panel fell into serious error in its findings. The GMC's case before the panel was that S had failed to account for the sums due to the Trust in respect of the use of the hospital's facilities. The evidence was that each treatment of the relevant patient would have attracted a fee in favour of the Trust of £60 for the use of a consultation room and that the alleged loss to the Trust was a maximum of £660, rather than the whole sum of £13,542, which the panel found was the Trust's loss. The GMC accepted that the panel made a determination that went beyond the evidence and beyond the allegation. The patient was responsible to pay any fees due to the Trust for use of hospital facilities under the Trust's private patient agreement, and therefore proof of the allegations against S required evidence that the fees received by S from each of the five private patients did in fact include a sum for the use of hospital facilities, which S should have paid to the Trust. In reality, that meant that there had to be evidence that S had deliberately invoiced each patient on the basis that his fee would include whatever the patient was liable to pay the Trust and that he had deliberately withheld any such payment from the Trust. The learned judge said that there was no direct evidence to that effect:

> 61. The crucial question, therefore, is whether on a fair view of the evidence as a whole it was open to the Panel to infer that [S] had deliberately withheld from the Trust sums of money which he had received from the five patients, and which he knew he should pay to the Trust, and was deliberately dishonest. In my judgment, it was not. Although this was not a criminal charge against [S], and the GMC only needed to prove its allegation on the balance of probabilities and not to the higher criminal standard, the principle must nonetheless apply that before an inference could properly be drawn, the Panel had to be able safely to exclude, as less than probable, other possible explanations for [S's] conduct.

In the context of this case, the learned judge said that the panel had to take into account both the evidence as to the deficiencies in the system of recording private patients and the powerful evidence of S's positive good character. The panel could not have given appropriate weight to those aspects of the evidence. The evidence of Ms L, the private patient administration manager at the material time, was clear in stating that there might be an absence of a private patient agreement because of oversight, or administrative error, or loss of the form. Records were provided by consultants' secretaries to the private patient office after the event—sometimes months after the event. The possibility of there being an explanation other than dishonesty for the absence of a private patient agreement was clearly before the panel, and there was no evidential basis on which the panel could conclude that such explanations were less probable than deliberate dishonesty. Moreover, the evidence showed that S's diary recorded when he was seeing his private patents, and their attendances would be expected to be notified to the private patient office. It was impossible to regard these appointments as 'below the radar', as alleged by the GMC. His Lordship concluded, at [67]–[69], that there was no direct evidence, and no basis for a safe inference, that S charged the patients for hospital facilities and retained those sums for himself, and no basis on which the panel could reasonably reject the alternative explanations of innocent oversight or administrative confusion that were clearly raised by Ms L's evidence as to the deficiencies of the system. For these reasons, the panel was wrong to reject a submission of 'no case to answer' and wrong to find dishonesty proved. Also, no future panel would be in any different position if the case were to be remitted. Any future panel, before it could infer dishonesty, would have to consider whether the evidence showed other possibilities and whether it could safely conclude that those other explanations were less probable than deliberate dishonesty. The evidence adduced against S by the GMC was insufficient and always would be insufficient, and Ms L's evidence made it impossible to conclude that there was no realistic prospect of innocent error, oversight, or administrative confusion on the part of S and/or his secretary.

# 51

# PRESS PUBLICITY

A. Legal Framework 51.01
B. Disciplinary Cases 51.02
*General Medical Council v. British Broadcasting Corporation* [1998] 1 WLR 1573 51.02
*Subramanian v. General Medical Council* [2002] UKPC 64 51.03
*R (Mahfouz) v. Professional Conduct Committee of the General Medical Council* [2004] EWCA Civ 233 51.04
*R (Rich) v. Bar Standards Board* [2011] EWHC 1099 (Admin) 51.05
*Hart v. Standards Committee (No. 1) of the New Zealand Law Society* [2012] NZSC 4 51.06
*Andersons Solicitors v. Solicitors Regulation Authority* [2012] EWHC 3659 (Admin) 51.07
*Arch Financial Products LLP and ors v. Financial Services Authority* [2012] Lexis Citation 114 51.08
*Burns v. Financial Conduct Authority* [2013] Lexis Citation 34 51.09
C. Other Relevant Chapters 51.10

## A. Legal Framework

**51.01** Examples include:

General Dental Council (Fitness to Practise) Rules 2006, rule 24 (publication of information, which may include the website of the Council)
Bar Standards Board Disciplinary Tribunals Regulations 2014, rule E199 (publication of finding and sentence)
Institute of Chartered Accountants in England and Wales Disciplinary Bye-laws:

bye-law 35 (publication of findings and other orders)
bye-law 36 (publicity for the disciplinary process)

## B. Disciplinary Cases

### *General Medical Council v. British Broadcasting Corporation* [1998] 1 WLR 1573

**51.02** Injunction—television programme—whether disciplinary proceedings are proceedings before a court—whether contempt of court

On 22 May 1998, the General Medical Council (GMC) sought an injunction restraining the British Broadcasting Corporation (BBC) from broadcasting any programme concerning disciplinary proceedings arising out of paediatric cardiac surgery at Bristol Royal Infirmary until conclusion of the current disciplinary proceedings against the doctors. Stage 1 (fact-finding) had occupied sixty-four days, with thirty-eight witnesses called by the GMC. The defence evidence had occupied twenty-seven days. Closing speeches were concluded on 22 May 1998 and, on the same day, the legal assessor advised the Professional Conduct Committee (PCC) on points of law that had arisen. The Committee then went into session to deliberate on fact-finding. Further oral evidence might be given at stage 2. The GMC sought an injunction to postpone (not to prohibit indefinitely) the transmission of a *Panorama* programme until the expected conclusion of the hearing on 30 May or such other date as the proceedings were concluded. Unchallenged evidence was that the *Panorama* programme was likely to be critical of the inquiry and to go beyond a factual report of what had happened in the inquiry so far, and that it would deal with matters that had been excluded from the inquiry, thereby running a risk of prejudicing its integrity. The Court of Appeal, in upholding the decision of Penry-Davey J refusing an injunction, said that there were two main issues: (1) whether the proceedings before the PCC were legal proceedings before a court for the purposes of the law as to criminal contempt of court; and (2) if so, whether transmission of the *Panorama* programme would create a substantial risk that the course of justice would be seriously impeded. Stuart-Smith LJ (sitting with Aldous and Robert Walker LJJ), giving the judgment of the Court, said that it was against the GMC on the first issue and that it was not a case of a substantial risk that the course of justice would be seriously impeded or prejudiced. The PCC was not a court within the meaning of the Contempt of Court Act 1981. It was correct that the PCC was exercising a judicial power, but it was not exercising the judicial power of the state as part of the state's machinery of government. By contrast, the PCC is a statutory committee of a professional body. It exercises a function that is recognizably a judicial function, and it does so in the public interest. Nevertheless, it is not part of the judicial system of the state. Instead, it is exercising (albeit with statutory sanction) the self-regulatory power and duty of the medical profession to monitor and maintain standards of professional conduct. As to the GMC's application for a temporary prohibition in order that the integrity of the proceedings should not be prejudiced, it was not suggested that the members of the PCC would be influenced by any television programme and they would be advised by their legal assessor not to watch the programme. The prospect of the programme deterring or influencing the evidence of a witness, or witnesses, who may be called at stage 2 was a very different matter from there being a substantial risk of serious prejudice to the proceedings before the PCC. It may well be that grave and obvious interference with proceedings before a non-curial tribunal could and would be restrained at the suit of the Attorney-General, who had a special historic role, not only dependent on statute, as guardian of the public interest. It seemed much more doubtful whether a private litigant can obtain such relief.

### *Subramanian v. General Medical Council* [2002] UKPC 64

**51.03** Newspaper report—reference to earlier findings of misconduct against doctor—advice by legal assessor to Committee

The appellant was a locum general practitioner (GP), who had been charged with failing to examine a patient adequately and to take prompt action to refer her to hospital. At the hearing before the PCC he became aware of newspaper reports of a previous finding against him of serious professional misconduct in 1987, which had led to him being admonished. It further emerged that one member of the Committee had read the article and had mentioned it to some of the others, and that the GMC Press Office had contributed to the disclosure by wrongly informing the newspaper that it was safe for it to refer to the previous appearance. The hearing continued, resulting in a finding of serious professional misconduct and the doctor being erased. The Committee was advised by its legal assessor in these terms:

> Unusually in this case, the Committee have heard about a previous appearance by this practitioner before the General Medical Council, in relation to agreed facts that occurred over 20 years ago when he was performing a completely different role it seems, namely that of an anaesthetist. These matters have nothing to do with the decisions that the Committee now has to make, and they should exercise no influence on them at all.

The Privy Council said that this was a 'clear and emphatic direction to the Committee members'. In dismissing S's appeal, their Lordships said that they felt safely able to say that there was no danger here of any prejudice to the doctor: this was a well-established quasi-professional tribunal, which had been directed in plain terms to pay no attention to the previous conviction, because it would give them no assistance—a direction reinforced by the fact that it dealt with events that occurred more than twenty years before. The experience of their Lordships of the jury system was that juries are faithful to their oath and abide by the instructions that they are given. There are rare circumstances (and this case was not one) in which the judge feels that the direction he is considering giving (for example to ignore some exceptionally prejudicial piece of evidence) might involve the jury in such 'mental gymnastics' before they could accept what loyalty to their oath required of them that the risk could not be taken and the jury would have to be discharged. But in this case it was difficult to see how the appellant's conduct of twenty years ago could affect the fundamental point of credibility that the Committee had to consider.

### *R (Mahfouz) v. Professional Conduct Committee of the General Medical Council* [2004] EWCA Civ 233

**51.04** Newspaper report—prejudicial material—application for Committee to be discharged—refused—whether Committee should have adjourned for doctor to apply to Court

M, a cosmetic surgeon specializing in laser surgery, faced charges relating to advice and treatment given by him in 2000 and 2001 to various patients. The hearing started on Monday 9 June 2003 and was expected to take eight days. On the evening of 9 June, the *Evening Standard* carried a report of the first day's hearing, which concluded that M 'was struck off in 1987 and working as a GP's assistant'. Similar statements appeared in two newspapers the following morning. Four members of the PCC had seen the *Evening Standard* article; one had also seen one of the other newspapers and another had seen the third article. M made an application for the Committee to be discharged. The Committee refused, basing its decision on *Porter v. Magill* [2002] 2 AC 357 and *Subramanian v. General Medical Council*, Privy Council Appeal No. 16 of 2002. The Committee adjourned until the next day to enable counsel for M to consider the matter with her client. On the following day, counsel applied for an adjournment to make an application to the Court, which the Committee refused. After that ruling, counsel and her solicitors withdrew, and made an application to the Court, which was rejected. In the Court of Appeal, Carnwath LJ said, in relation to prejudicial publicity:

> 22. The problem of prejudicial publicity (including reference to previous convictions) is one which may arise in any court or tribunal considering criminal or disciplinary charges, but the law's response to the problem will vary depending on the nature and experience of the tribunal concerned. There is no absolute rule that knowledge of such material is fatal to the fairness of the proceedings. In *Montgomery v. HM Advocate* [2003] 1 AC 641, the question was the effect of pre-trial publicity on the minds of a jury dealing with a charge of murder. Lord Hope expressed the common law test as follows: "The common law test which is applied where pre-trial publicity is relied upon in support of a plea of oppression, is whether the risk of prejudice is so grave that no direction by a trial judge however careful could reasonably be expected to remove it." ([2003] 1 AC at p667E–F).

Accordingly, the Court saw no grounds for questioning the Committee's ability to decide the case fairly on the evidence before it. However, a second issue of more limited scope was whether the Committee should have adjourned to enable an application to be made to the Court. The Committee clearly recognized that its decision might be subject to challenge in the High Court. It was also naturally concerned at any further delay in the proceedings, two days having already been lost. There can be no inflexible rule. In general, it is preferable for proceedings to be allowed to take their course and a challenge to their validity to be taken by way of appeal. Consideration must also be given to the difficulty of organizing such proceedings in a complex case and the potential inconvenience of the witnesses, who may have had to make special arrangements to attend the hearing and who may be reluctant to repeat the experience. In the instant case, there should have been an adjournment to allow the application to be made based on the particular facts of the case. The issue had arisen on the second day of a programmed eight-day hearing; it had been treated as an important issue, requiring detailed legal argument; and there had been an apparent difference of view between the Committee and its legal assessor as to the correct test. The GMC might well have wanted its own counsel to attend the application and be heard, particularly on the question of a stay. In those special circumstances, it seems that justice and the appearance of justice required at least an opportunity to be given for the matter to be raised before a High Court judge. In conclusion, under the present law, there appears to be no sanction against such reporting in relation to proceedings of the PCC. The unfortunate history of this case suggests that it may sometimes be necessary, at least in a case likely to attract substantial press attention, to advise members of the PCC to avoid reading any articles about the case.

### *R (Rich) v. Bar Standards Board* [2011] EWHC 1099 (Admin)

**51.05** Judicial review proceedings—barrister seeking anonymity in judicial review proceedings

The claimant, a practising barrister, sought anonymity in judicial review proceedings when challenging a decision of the Bar Standards Board to bring a professional complaint against him. In dismissing the claimant's application, Davis J said that no one likes to have a complaint made against them. No one likes to have such matter publicized. However, the general rule must be, and indeed is, that if a party chooses to undertake public litigation, then he must expect his or her name to be publicized accordingly: see *R v. Legal Aid Board, ex parte Kaim Todner* [1999] QB 966. This mirrors the fact that professional people not infrequently have to face allegations of professional negligence and the like in civil proceedings, which necessarily are published. In the instant case, the claimant had chosen to bring public proceedings by way of a judicial review claim. It may be that the underlying complaint against him (which he strongly disputed) was one that he did not wish to have publicized at this stage. The fact was that he had brought this public litigation. There can be no different rules for barristers from those for any other professional person and the provision in rule 60 of the Bar Complaints Rules 2009 (ordinarily, confidentiality is maintained until at least shortly before the final hearing) cannot dictate the result in public law proceedings.

### *Hart v. Standards Committee (No. 1) of the New Zealand Law Society* [2012] NZSC 4

**51.06** Disciplinary proceedings—defendant seeking anonymity

H faced charges to be dealt with by the Lawyers and Conveyancers Disciplinary Tribunal. The Tribunal rejected an application for an order suppressing publication of H's name. H's application for suppression was rejected by Toogood J and a subsequent appeal to the Court of Appeal was largely unsuccessful. On H's application for leave to appeal to the New Zealand Supreme Court (Elias CJ, and Blanchard and William Young JJ), the Court said that the primary basis for the proposed appeal was the contention that the usual open justice approach adopted in such cases as *R v. Liddell* [1995] 1 NZLR 538, CA, should not apply in the case of a professional person with a high public profile facing disciplinary charges, particularly where, as here, criminal offending is not alleged. A tribunal or judge deciding whether to order suppression is exercising a discretion that, in a disciplinary context, must allow for any relevant statutory provisions, as well as the more general need to strike a balance between open justice considerations and the interests of the party who seeks suppression. The likely particular impact of publicity on that party will always be relevant, but it is untenable to suggest that professional people of high public profile, such as the applicant, have anything approaching a presumptive entitlement to suppression. The applicant's requests for suppression were fully and independently considered by both the Tribunal and Toogood J, and in turn were reviewed by the Court of Appeal. There was no arguable error in the approach taken by the Court of Appeal and the application for leave to appeal was dismissed.

### *Andersons Solicitors v. Solicitors Regulation Authority* [2012] EWHC 3659 (Admin)

**51.07** Disciplinary proceedings—solicitor seeking challenge to pretrial publication of details

The first claimant was a firm of conveyancing solicitors and the remaining five claimants were partners in the firm. The substantive hearing before the Solicitors Disciplinary Tribunal (SDT) was scheduled to take place in January 2013. Following service of the rule 5 statement, a solicitor member of the SDT certified that there was a case to answer and, on 11 June 2012, the Solicitors Regulation Authority (SRA) wrote to the claimants, stating that it had decided that it was in the public interest for the SRA to publish the decision that the Tribunal had certified a case to answer because of the importance of transparency of the regulatory process and the ability of the public to ascertain whether a solicitor is being referred to the SDT and, if so, the allegations to be made. On 18 June 2012, the firm replied, saying that it relied heavily upon residential conveyancing work, that private clients are dependent in many cases upon funding from financial institutions, that any decision to publish would have a detrimental effect upon lenders and the firm's panel appointment with major lenders, and that the impact on the firm through the publication could consequently be extremely injurious or damaging. On 27 July 2012, the SDT held a directions hearing and, on 16 August 2012, the claimants issued judicial review proceedings to challenge the SRA's policy and procedure for publication of details of allegations made against solicitors to be heard by the SDT. The claimants accepted that the substantive hearing in January 2013 would be in public and that, as a result, there could be no objection to the SRA identifying the claimants and the allegations against them during a period that might be as long as three weeks before the hearing. In support, reliance was placed on the fact that disciplinary allegations against barristers are made available to the public by the Bar Standards Board two or three weeks prior to the date of the hearing. Similarly, two weeks before the date of a hearing before the General Dental Council, a 'programme of business' is made available on the Council's website, including the charges; the General Medical Council has no policy of publishing allegations as soon as it is clear that a case will go to a substantive hearing, but appears to publish allegations a few weeks in advance of the hearing. In dismissing the application for judicial review, Walker J noted that the stage at which the process becomes public is governed by the Tribunal Rules and that the directions hearing in July 2012 was in public. There was nothing wrong in the formulation of the SRA's publication policy. Under section 28 of the Legal Services Act 2007, the SRA must have regard, amongst other things, to principles of transparency and accountability. An important object of these principles is to enable the public to be informed about action taken by the SRA. In the context of a publication policy, a range of motives, which include maintaining public confidence in the regulatory system, is consistent with the statutory objectives. The policy was not outside the bounds of reasonableness and there was likewise no merit in contentions of illegality. The learned judge said that, assuming that Article 8 of the European Convention on Human Rights (ECHR) was engaged (particularly as regards protection of the claimants' reputation), he was not, however, persuaded that the policy infringed Article 8.

***Arch Financial Products LLP and ors v. Financial Services Authority***
**[2012] Lexis Citation 114**

**51.08** Decision notice—appeal to Upper Tribunal—whether publication of decision notice likely to cause serious harm

The applicants applied, pursuant to rule 14 of the Tribunal Procedure (Upper Tribunal) Rules 2008, for an order prohibiting the publication of decision notices and a direction, pursuant to paragraph 3(3) of Schedule 3 to the Rules, that the register maintained by the Upper Tribunal should not include particulars of the references made by the applicants in respect of the decision notices. On 14 September 2012, the Regulatory Decisions Committee of the Financial Services Authority (FSA) decided (after reviewing the applicant's written representations and receiving oral representations on them) to issue the decision notices. In addition to the regulatory proceedings, the applicants were also defendants to a number of sets of civil proceedings, which overlapped with the decision notices, and they alleged that publication would cause serious reputational damage to their chief executive and chief operating officers. Judge Timothy Herrington noted that the applicants accepted that it was proper for the Tribunal to approach the issue as to whether to exercise the power to prohibit publication under rule 14 by considering whether to refuse to do so would cause the applicants serious harm. There are no specific conditions that need to be satisfied before the power in the Rules can be exercised, but it is subject to the overriding objective in rule 2, which requires the Tribunal to deal with cases fairly and justly. Consequently, this imports the requirements that the discretion should be exercised judicially—that is, taking into account all relevant factors, ignoring irrelevant factors, and exercising the power in a manner that seeks to give effect to the overriding objective. This involves carrying out a balancing exercise between those factors that tend towards publication and those that would tend against: at [26]–[27]. The learned judge concluded, at [56], that none of the evidence advanced with regard to the effect of the publication or the reputation or privacy of the applicants met the requirements of cogent evidence of disproportionate damage. A second factor relied on by the applicants was that the publication of the decision notices would influence the manner in which the witnesses may give evidence in the separate court proceedings in which the applicants were involved. The learned judge said that this was pure speculation, as was the applicants' submission that publication of the decision notices would prejudice the possibility of any meaningful settlement discussions in relation to the civil proceedings. However, in dismissing the applications, the learned judge expressed his concern that it would be important that adequate steps be taken when publicising the decision notices to ensure that it would be clear that the decisions were provisional, in the light of the fact that they were to be challenged in the Upper Tribunal. In particular, any press release issued by the FSA was to state prominently at its beginning that the applicants had referred the matter to the Upper Tribunal and that the Tribunal would determine the appropriate action to take, which may be to uphold, vary, or cancel the FSA's decision. Likewise, any reference to findings should be prefaced with a statement to the effect that the findings reflect the FSA's belief as to what occurred and how the behaviour concerned is to be characterized. The dismissal of the applications was therefore conditional upon compliance with these principles and both parties had liberty to apply for further directions.

***Burns v. Financial Conduct Authority* [2013] Lexis Citation 34**

**51.09** Decision notice—appeal to Upper Tribunal—whether publication of applicant's name likely to cause serious harm

Following a reference to the Upper Tribunal of a decision notice issued by the Financial Conduct Authority (FCA), B applied for a direction, pursuant to paragraph 3(3) of Schedule 3 to the Tribunal Procedure (Upper Tribunal) Rules 2008, that the register of references maintained by the Upper Tribunal contain no particulars of her reference in respect of the decision notice, and a direction, pursuant to rule 14(1) of the Rules, to prohibit publication by the FCA of the decision notice and such other documents or with information that would allow her to be identified. From early 2009 until May 2011, B was a non-executive director (NED) of two mutual insurance societies, and was approved by the FCA under the Financial Services and Markets Act 2000 to carry out the function of acting as an NED of the societies. Following the completion of an investigation, the authority alleged that B failed to disclose a conflict of interest and had used her NED position within the societies to solicit non-executive and consulting roles. On 28 November 2012, the FCA's Regulatory Decisions Committee decided to issue a decision notice on the grounds that B had breached Principle 1 of the Authority's Statements of Principle for Approved Persons by acting recklessly and thus without integrity. The decision notice prohibited B from performing any function in relation to any regulated activity, on the grounds that she lacked fitness and probity, and that a financial penalty of £154,800 be imposed. B, strongly contesting the allegations, made a reference to the Upper Tribunal and sought to restrain publication of her identity pending the hearing. In dismissing the application, Judge Timothy Herrington said that he accepted that cogent evidence of destruction of or severe damage to a person's livelihood is capable of amounting to disproportionate damage such that it would be unfair not to prohibit publication of a decision notice. The learned judge continued:

> 90. The requirement of cogent evidence in applications of this kind leads me to conclude that the possibility of severe damage or destruction of livelihood is insufficient; in my view the evidence should establish that there is a significant likelihood of such damage or destruction occurring.... It would be too high a hurdle to surmount which would make the jurisdiction almost illusory if the requirement were to show that severe damage or destruction was an inevitable consequence of publication.

After reviewing the likely effect of publication on B's livelihood, the learned judge concluded:

> 113. In my view the effects described above are not sufficient to satisfy me that the effect of publication will cause damage to [B's] livelihood which is so severe that it is out of proportion to the public interest in the principle of open justice that will be served by permitting publication of the Decision Notice and including particulars of the reference on the Register.
>
> 114. Whilst I have identified that some damage to [B's] livelihood may be caused by publication, the scale of the business concerned, the uncertainties in its future prospects in any event, the prospects for reviving it after the determination of the reference, and the fact that the financial impact on [B] is not such to leave her insolvent or destitute lead me to conclude that [B] has on the facts of this case been unable to discharge the heavy burden on her to satisfy me that the evidence shows that the

impact of publication on [B] is so severe that it outweighs the strong presumption that publication should be permitted.

## C. Other Relevant Chapters

**51.10**

Abuse of Process, Chapter 2
Bias, Chapter 6
Public or Private Hearing, Chapter 53

# 52

# PRIVILEGE

A. **Disciplinary Cases** 52.01
*Parry-Jones v. Law Society* [1969] 1 Ch 1, CA 52.01
*R v. The Institute of Chartered Accountants of England and Wales, ex parte Nawaz* [1997] PNLR 433; [1997] EWCA Civ 1530 52.02
*Mahon v. Rahn (No. 2)* [2000] 1 WLR 2150 52.03
*Hay v. Institute of Chartered Accountants in Scotland* 2003 SLT 612 52.04
*B and ors v. Auckland District Law Society* [2003] 2 AC 736 52.05
*Kearns v. General Council of the Bar* [2003] 1 WLR 1357 52.06
*Heath v. Commissioner of Police of the Metropolis* [2005] ICR 329 52.07
*Simms v. Law Society* [2005] EWCA Civ 749 52.08
*Holder v. Law Society* [2005] EWHC 2023 (Admin); [2006] PNLR 10 52.09
*Quinn Direct Insurance Ltd v. Law Society* [2011] 1 WLR 308 52.10
*White v. Southampton University Hospitals NHS Trust* [2011] EWHC 825 (QB) 52.11
*Baxendale-Walker v. Middleton* [2011] EWHC 998 (QB) 52.12
*Adams v. Law Society of England and Wales* [2012] EWHC 980 (QB) 52.13
*Iqbal v. Solicitors Regulation Authority* [2012] EWHC 4097 (QB) 52.14
*Mayer v. Hoare* [2012] EWHC 1805 (QB) 52.15
*P v. Commissioner of Police of the Metropolis* [2014] UKEAT/0449/13/JOJ 52.16
B. **Other Relevant Chapters** 52.17

## A. Disciplinary Cases

### *Parry-Jones v. Law Society* [1969] 1 Ch 1, CA

**52.01** Solicitor—use of client's confidential information for purposes of investigation and consequential proceedings

In February/March 1967, the Law Society served on the plaintiff, a solicitor, written notice requiring him to produce for inspection his books of account and any other necessary documents relating to his practice as a solicitor and any trust of which he was a trustee. The notice was served pursuant to rule 11 of the Solicitors' Accounts Rules 1945 and rule 11 of the Solicitors' Trust Accounts Rules 1945. The plaintiff issued proceedings for an injunction restraining the Law Society from taking any steps pursuant to the notices in so far as the notices purported without lawful authority to empower the Law Society to have access to confidential information relating to any client of the plaintiff without the express authority of such client to such access. The plaintiff objected to the request on the ground that some of the documents contained confidential information relating to clients that could not be disclosed without their consent. Buckley J struck out the writ as disclosing no cause of action and his order was confirmed by the Court of Appeal. Lord Denning MR said that, in his opinion, the contract between a solicitor and client must be taken to contain the implication that the solicitor must obey the law and, in particular, that he must comply with the rules made under the authority of statute for the conduct of the profession. If the rules require him to disclose his client's affairs, then he must do so. Rule 11 of the Solicitors' Accounts Rules 1945 is a valid rule, which overrides any privilege or confidence that otherwise might subsist between solicitor and client. It enables the Law Society, for the public good, to hold an investigation, even if it involves getting information as to a client's affairs. But the Law Society and its accountant must, of course, themselves respect the obligation of confidence. They must not use it for any purpose except the investigation and any consequential proceedings. If there should be subsequent application to the Disciplinary Committee, the information can be used for that purpose. In all other respects, the usual rules of legal professional privilege apply. Diplock LJ agreed and said that a duty of confidence is subject to, and may be overridden by, the duty of any party to the contract to comply with the law of the land. If it is the duty of such a party to a contract, whether at common law or under statute, to disclose in defined circumstances confidential information, then he must do so, and any express contract to the contrary would be illegal and void. Section 29 of the Solicitors Act 1957 empowers the council to take such action as may be necessary to enable it to ascertain whether or not the rules are being complied with. That necessarily empowers the council to make rules that entitle it to override the privilege, and such a rule is rule 11 of the Solicitors Account Rules 1945. Salmon LJ agreed.

Commenting on this case in *R (Morgan Grenfell & Co Ltd) v. Special Commissioner of Income Tax* [2003] 1 AC 563, 611–12, Lord Hoffmann said, at [30]–[32], that one could hardly imagine a stronger Court of Appeal, but he was bound to say that he had difficulty with the reasoning. The reasoning in the *Parry-Jones* case suggests that any statutory obligation to disclose documents will be construed as overriding the duty of confidence, which constitutes the client's only protection. He considered it to be unfortunate that the Court of Appeal was not referred to the valuable judgments of the Supreme Court of New Zealand in *Commissioner of Inland Revenue v. West-Walker* [1954] NZLR 191, which reached the opposite conclusion in the context of a statutory power to require the production of documents and information for the purposes of the administration of taxing statutes. This was not to say that, on its facts, the *Parry-Jones* case was wrongly decided. But the true justification for the decision was not that the plaintiff's clients had no legal professional privilege, or that their legal professional privilege had been overridden by the Law Society's rules, but that the clients' legal professional privilege was not being infringed. The Law Society was not entitled to use information disclosed by the solicitor for any purpose other than the investigation. Otherwise, the confidentiality of the clients had to be maintained. This limited disclosure did not breach the client's legal professional privilege or, to the extent that it technically did, was authorized by the Law Society's statutory powers.

### *R v. The Institute of Chartered Accountants of England and Wales, ex parte Nawaz* [1997] PNLR 433; [1997] EWCA Civ 1530

**52.02** Self-incrimination—privilege in course of investigation of complaint—waiver of privilege—bye-law requiring information

In June 1994, an investigation committee of the Institute of Chartered Accountants of England and Wales (ICAEW) decided to lay a formal complaint against N, a chartered accountant, before a disciplinary committee. The charges included conducting audits pursuant to appointments entered into after October 1991 when he was not authorized to do so and failing to respond adequately to the investigation committee's requirement for information. The charges were found proved and N was ordered to be severely reprimanded, to pay a fine of £2,000, and to pay costs of £2,500. His appeal to the appeals committee was dismissed. In a letter to the Institute, N admitted that, since the coming into effect of the Companies Act 1989 on 1 October 1991, he had accepted appointments as a company auditor when he was ineligible to do so because he was unregistered with the Institute as an auditor. The Institute sought to extract from N details of the companies that he had audited, but he declined to give them. In judicial review proceedings, N advanced the proposition that the Institute had no power to require him to provide information and that, in doing so, the Institute had violated the fundamental common law privilege against self-incrimination. The power of the Institute to call for information, N submitted, and the duty of a member to provide it, did not extend to the provision of information capable of incriminating the member. In dismissing N's claim for judicial review, Sedley J said that proceedings of the Institute are quasi-judicial. They involve matters of potential gravity for the reputations and livelihoods of professional people, and they are required to be conducted with scrupulous impartiality and fairness. On this narrow ground, the judge accepted that he would if necessary hold that the privilege against self-incrimination is capable of applying to members of the Institute who are asked in the course of an investigation of a complaint to furnish information that may incriminate them. However, there is no doubt that the privilege against self-incrimination, like any personal privilege, can be waived. The charge of failing to provide information was based upon a valid bye-law requiring the provision of information relevant to an existing complaint. The complaint had been brought into being by information volunteered by N and the officer investigating it had properly delegated authority to do so. It followed that neither of the findings against N could be impeached. N's appeal was dismissed by the Court of Appeal on 25 April 1997. Leggatt LJ (with whom Thorpe and Mummery LJJ agreed) said:

> We indicated to counsel that for the purposes of this appeal we were content to assume, without deciding, that the privilege from self-incrimination at least extends to investigations of a quasi-judicial character such as we are concerned with. We have also assumed that the privilege was sufficiently claimed by [N]'s letter of 6th April 1994. . . . Nevertheless, the principle that privilege is not to be regarded as having been abrogated except by express words or necessary implication applies also to waiver. In my judgment acceptance of a duty to provide information demanded of an accountant constitutes a waiver by the member concerned of any privilege from disclosure. It plainly is in the public interest, as well as in the interest of the profession, that the Institute should be enabled to obtain all such information in the possession of its members as is relevant to complaints of their professional misconduct. . . . Upon becoming a chartered accountant "it shall be the duty of every member", in accordance with para 8(a) of Schedule 2 of the Supplemental Charter, "to provide such information as the Investigation Committee may consider necessary to discharge its functions." Compliance with that duty necessarily and inevitably precludes the exercise of any privilege that would have excused the provision of the information. . . . The judge concluded that there is today no objection in public law to a rule which may require self-incrimination by a chartered accountant against whom a relevant complaint is already extant, where the privilege has been prima facie waived and where, what the judge called, "intelligible and powerful grounds of public policy" exists for endorsing the waiver. It seems to me that by necessary implication from his acceptance of the Institute's rules, and in particular of the duty to provide information, [N] waived reliance on any privilege from giving it, and the power to call for information is one which it is in the public interest to uphold.

### *Mahon v. Rahn (No. 2)* [2000] 1 WLR 2150

**52.03** Letter by bank sent to financial regulator—libel action by stockbroker—absolute privilege

The defendants were partners in a private bank based in Zurich. In December 1990, the bank's Swiss lawyer sent a letter in strictest confidence to the Securities Association, with a copy to the Serious Fraud Office (SFO), concerning a firm of stockbrokers of which the plaintiff was the managing director. The Securities Association was the relevant financial services regulatory body for the grant of authorizations to stockbrokers to conduct investment business under the Financial Services Act 1986. Following the disposal of criminal proceedings against the stockbrokers' client, in which the letter was disclosed by the SFO in the course of those proceedings, the plaintiff commenced a libel action against the Swiss bank. The Court of Appeal (Brooke, Mantell, and Laws

LJJ) held that the letter attracted absolute privilege. The flow of information to financial regulators might be seriously impeded if its informants were to fear that they might be harassed by libel proceedings, and if it were impeded in this way, the purposes of Part 1 of the Financial Services Act 1986, of protecting the public from unfit investment advisers, would be put at risk. For these reasons, the Court would allow the bank's appeal and hold that the Securities Association letter was published on an occasion that attracted absolute privilege.

### *Hay v. Institute of Chartered Accountants in Scotland* 2003 SLT 612

**52.04** Press release following disciplinary proceedings—libel proceedings—whether attracting absolute privilege—qualified privilege

H, a non-practising chartered accountant, was prosecuted for a value added tax (VAT) offence and, following trial in 1998, a jury in the High Court in Glasgow acquitted him by a majority verdict of not guilty. Subsequently, in November 1999, an investigation committee of the Institute of Chartered Accountants in Scotland (ICAS) referred the matter to a disciplinary committee. A formal complaint specifying various incidences of alleged professional misconduct was sent to H. Subsequently, H's solicitor informed the Institute that he had lost contact with H and, in September 2000, the hearing proceeded in H's absence; he was expelled from membership of the Institute and ordered to pay costs of £5,000. On 14 December 2000, the Institute issued a press release headed 'ICAS expels untruthful accountant'. The press release was not restricted to the formal findings of the disciplinary committee set out in its decision letter. The press release contained additional information, including references to the obtaining of VAT monies amounting to £519,000 that had (wrongfully) not been repaid to Her Majesty's Customs and Excise (HMCE). The press release also made references to H 'leaving the country hastily' prior to action against him by HMCE. The press release was published in newspapers and in professional journals such as *Accountancy Age*. The press release on the Institute's website was visited and read by many users. H started an action for defamation against the Institute in relation to the statements made following the disciplinary proceedings. The Institute applied to dismiss the action. Lady Paton dismissed the Institute's application and gave the parties an opportunity to amend. In her judgment, Lady Paton said that she did not consider it possible, without some inquiry into the facts, to form a view whether the proceedings before the disciplinary committee in September 2000 attracted absolute privilege. The proceedings of a disciplinary tribunal set up with a profession in order to supervise and discipline members of that profession may, or may not, attract absolute privilege. Relevant considerations may include: how the Institute and the disciplinary committee are constituted; the identity and qualifications of the chairman and the members of the disciplinary committee in September 2000; the degree of independence of the disciplinary committee from the Institute; whether the hearing was open to the public; whether witnesses could be compelled to attend to give evidence; whether any evidence was taken and if so, whether it was taken on oath, with the usual pattern of examination, cross-examination, and re-examination; whether any documents were formally lodged and numbered as exhibits; whether there were closing submissions; and whether the proceedings and evidence were recorded. Some inquiry into the facts was necessary before determining whether or not the proceedings attract privilege. If the Court were of the view that the proceedings attracted only qualified privilege, then the pursuer, in order to succeed, would have to aver and prove malice on the part of the Institute or its representative in the presentation of the case to the committee in September 2000. As to the content and tone of the press release, the disciplinary committee was not prevented from reaching a different conclusion from that reached by the jury in the High Court. But the fact of the matter is that the disciplinary committee did not, in terms, find that H had, by means of false representations, managed illicitly both to take £519,000 of VAT monies from HMCE and to keep it, and then had hastily left the country. There were therefore quite significant differences between the formal findings of the disciplinary committee and the wording of the press release. On one view, the press release contained certain inferences prejudicial to H, which were not contained in the disciplinary committee's formal findings. H's averments relating to the content and tone of the press release may have raised an inference of malice, which could not safely be excluded at this stage without some inquiry into the facts. As to the widespread nature of the publicity, it could be quite proper for a professional body to decide to publish widely its findings leading to the expulsion of one of their members. A professional body governing the conduct and standards of chartered accountants could not readily be criticized for taking the view that anyone in the world who might have business dealings with that chartered accountant should be aware of these events. Prima facie, therefore, the Court had difficulty accepting H's argument that the nature and extent of the publicity given to his expulsion necessarily gave rise to an inference of malice on the part of the Institute of its representative. However, it was possible that the nature and degree of the publicity given to H's expulsion may have been connected to some extent with his argument about the content and tone of the press release. The Court was not persuaded that any averments should be excluded at this stage.

***B and ors v. Auckland District Law Society*** **[2003] 2 AC 736**

**52.05** Complaint—inquiry—documents covered by legal professional privilege—whether statute overriding privilege

Following a complaint to the Auckland District Law Society, B, an Auckland law firm, agreed to hand over to counsel assisting a complaints committee documents covered by legal professional privilege, on the express basis that, in doing so, privilege was not waived and that the documents would not be further copied. Subsequently, the documents were passed to other counsel appointed by the Society. The firm called for the return of the documents. The Society refused to return them and served formal notices under section 101(3) of the Law Practitioners Act 1982, one requisitioning some of the documents already in its possession and the other requisitioning further documents to which legal professional privilege also attached. It was common ground that the documents were all covered by legal professional privilege. In some cases, the privilege was that of the firm itself or its partners; in others, it was that of its clients. In the present proceedings, the firm sought an order for return of the documents already delivered. The Society resisted the claim and counterclaimed for a declaration that the firm was obliged to comply with the requisitions. The main issue was whether the Society was entitled under section 101 of the Law Practitioners Act 1982 to require the firm to produce privileged documents for the purpose of an inquiry into allegations of professional misconduct. This turned on whether the Act, which gave the Society power to call for documents from practitioners under investigation, overrode any claim to legal professional privilege that may subsist in them. In allowing the firm's appeal to the Privy Council (Lords Hope of Craighead, Hobhouse of Woodborough, Millett, Scott of Foscote, and Walker of Gestingthorpe), Lord Millett said that an authoritative exposition of the rationale of legal professional privilege is to be found in the speech of Lord Taylor of Gosforth CJ in *R v. Derby Magistrates' Court, ex parte B* [1996] AC 487, with whom the rest of the House agreed. Lord Taylor CJ said that legal professional privilege is a fundamental condition on which the administration of justice as a whole rests. It is not for the sake of the applicant alone that the privilege must be upheld; it is in the wider interests of those hereafter who might otherwise be deterred from telling the whole truth to their solicitors. Lord Taylor CJ and Lord Lloyd of Berwick rejected the idea that a balancing exercise was required between competing public interests. Legal professional privilege is the predominant public interest and a balancing exercise is not required in individual cases because the balance must always come down in favour of upholding the privilege. There is authority to the same effect in New Zealand and in Australia: see *R v. Uljee* [1982] 1 NZLR 561; and *Carter v. Northmore Hale Davy & Leake* (1995) 183 CLR 121. The question, therefore, was whether the 1982 Act excludes legal professional privilege either expressly or by necessary implication. Section 101(3) did not expressly exclude legal professional privilege and, after consideration of other sections, their Lordships concluded that legal professional privilege is a good answer to a requisition under the 1982 Act whether at the investigative stage or in proceedings before a disciplinary tribunal. As to the documents previously handed over, the question was not whether privilege had been waived, but whether it had been lost. The documents remained privileged because they were created for the purpose of giving or receiving legal advice. The documents were both confidential and privileged. Whether a claim to the return of such documents is based on a common law right or an equitable one, the policy considerations that give rise to the privilege preclude the Court from conducting a balancing exercise. A lawyer must be able to give his client an unqualified assurance, not only that what passes between them shall never be revealed without his consent in any circumstances, but also that should he consent in future to disclosure for a limited purpose those limits will be respected.

### *Kearns v. General Council of the Bar* [2003] 1 WLR 1357

**52.06** Letter sent to heads of chambers—libel proceedings—qualified privilege

The claimants sought damages for libel arising out of a letter sent by the head of the Bar Council's Professional Standards and Legal Services department to all heads of chambers and senior clerks/practice managers. The letter concerned the Code of Conduct of the Bar of England and Wales, and, in the mistaken belief that K were not solicitors, stated that barristers should not accept work from them unless certain conditions were met. It was accepted that the letter was written in the mistaken belief that the claimants were not solicitors. The defendant sought summary judgment against the claimants on the ground that they had no real prospect of defeating the pleaded defence of qualified privilege. In dismissing the claimants' appeal from the decision of Eady J [2002] 4 All ER 1075 entering summary judgment for the defendant, the Court of Appeal said that what the head of the Professional Standards and Legal Services Department was here communicating was the Bar Council's conclusion upon a request for guidance that it had received from a member of the Bar. Whether it had been adequately investigated is another matter. That, as Lord Diplock explained in *Horrocks v. Lowe* [1975] AC 135, goes to malice rather than whether the occasion of the communication is privileged. That the Bar Council was entitled—indeed bound—to give a ruling cannot be in doubt, nor can it be doubted that it did so in the context of an established relationship between the Bar Council and the Bar, which, with regard to relevant communications between them, must necessarily attract qualified privilege. So long as the statement is fairly warranted by the occasion and is made in the absence of malice, it will be protected by qualified privilege, irrespective of the degree of investigation or verification carried out by the maker of the statement and irrespective of whether one categorizes the situation as one of common interest or of duty and corresponding interest.

***Heath v. Commissioner of Police of the Metropolis* [2005] ICR 329**

**52.07**

Police disciplinary board—allegation of sexual discrimination against board—board appointed by Commissioner of Police—immunity from suit

H appealed to the Court of Appeal against a decision of the Employment Appeal Tribunal (EAT) upholding a decision of an employment tribunal that it had no jurisdiction to hear her complaint of unlawful sex discrimination against the Metropolitan Police Commissioner, since, because it related to alleged conduct by police members of a disciplinary hearing under the Police (Discipline) Regulations 1985, the Commissioner was entitled to immunity from action. H was employed as a station reception officer, in a civilian capacity, at Hornsey police station. She complained that a police inspector at the station had sexually assaulted her. There was an internal police disciplinary hearing into her allegations, which was convened under the 1985 Regulations. It was conducted by a board consisting of three male commanders in the Metropolitan Police. She was cross-examined by a male counsel. Subsequently, H made a complaint to an employment tribunal complaining that the members of the police disciplinary board who had presided at the hearing had, in their conduct of it, sexually discriminated against her. The Court of Appeal held that the proceedings of the police disciplinary board were immune from suit. If H had objected to the all-male constitution of the board at any stage because of the nature of her complaint before it and/or of the conduct of counsel or of the members of the board of the proceedings, the board would have had to rule on that matter. Any such ruling would undoubtedly have been covered by immunity and could have been challenged only through the domestic appellate route or by way of judicial review. The board would undoubtedly have been immune from proceedings by way of complaint of sexual discrimination before an employment tribunal. Absolute immunity attaches to anything said or done by anybody in the course of judicial proceedings whatever the nature of the claim made in respect of such behaviour or statement, except for suits for malicious prosecution and prosecution for perjury and proceedings for contempt of court. The rules are there not to protect the person whose conduct in court might prompt such a claim, but to protect the integrity of the judicial process and hence the public interest.

***Simms v. Law Society* [2005] EWCA Civ 749**

**52.08**

Solicitor—disciplinary proceedings—documents covered by legal professional privilege—solicitor claiming privilege—privilege belonging to clients alone

Disciplinary proceedings were commenced against S, a solicitor, alleging that he had been actively involved in making, promoting, or facilitating bogus transactions that lacked an honest commercial purpose. The concern was that he may have become involved in fraudulent investment schemes and failed to maintain procedures of internal reporting for the prevention or forestalling of money laundering. Objection was taken before the Solicitors Disciplinary Tribunal (SDT) that the documentation relied on was covered by legal professional privilege and was therefore protected from disclosure. The Tribunal found that S had been guilty of conduct unbefitting a solicitor and ordered that he be struck off the roll. S's appeal to the Divisional Court was dismissed: [2005] EWHC 408 (Admin). In refusing permission to appeal to the Court of Appeal, Waller LJ said that so far as legal professional privilege is concerned, the position is as follows. First, the case of *B and ors v. Auckland District Law Society* [2003] 2 AC 736 did have this feature: it was concerned with documents that had been seized, for which the solicitors themselves could claim their own privilege, as well as documents that were held by them and for which the privilege, if there was one, was their clients. In the present case, S was seeking to assert the inadmissibility of documents for which he had no claim to privilege. These were documents that, if they were privileged, were privileged in the hands of the clients. The Tribunal emphasized this point and relied on the dictum of Lord Hoffmann in *R (Morgan Grenfell & Co Ltd) v. Special Commissioner of*

*Income Tax* [2003] 1 AC 563, in which he dealt with *Parry-Jones v. Law Society* [1969] 1 Ch 1, CA. Lord Hoffmann said at [32]:

> This is not to say that on its facts the *Parry-Jones* case was wrongly decided. But I think that the true justification for the decision was not that [the plaintiff's] clients had no LPP, or that their LPP had been overridden by the Law Society's rules, but that the clients' LPP was not being infringed. The Law Society were not entitled to use information disclosed by the solicitor for any purpose other than the investigation. Otherwise the confidentiality of the clients had to be maintained. In my opinion, this limited disclosure did not breach the clients' LPP or, to the extent that it technically did, was authorised by the Law Society's statutory powers. It does not seem to me to fall within the same principle as a case in which disclosure is sought for a use which involves the information being made public or used against the person entitled to the privilege.

Waller LJ said that what the SDT in the instant case did was to seek to protect the privilege of the clients and to preserve the legal professional privilege. When the Divisional Court came to deal with the matter, it upheld the way in which the Tribunal had dealt with this aspect, pointing out how difficult it would be for the Law Society to investigate the type of improper conduct that it was investigating in this case unless there were a means whereby the documentation, which might be privileged in the hands of the clients, was documentation to which it could have access. The holding of the Divisional Court was that, in fact, *Parry-Jones* remains good law. The Divisional Court too relied on the dictum of Lord Hoffmann in the *Morgan Grenfell* case. The Court considered it to have been right so to rule. In any event, it was said, it lies ill in the mouth of S, whose privilege it is not, to seek to try to keep out documents that the SDT had thought it right to look at while still preserving the privilege of the clients.

### *Holder v. Law Society* [2005] EWHC 2023 (Admin); [2006] PNLR 10

**52.09**

Solicitor—self-incrimination—admissions made to Law Society inspector

H was struck off the roll of solicitors by the SDT. On appeal, he argued that the admission before the Tribunal of evidence derived from the Law Society's inspector, and his admissions to that inspector offended against the privilege against self-incrimination and the rules derived from the European Convention on Human Rights (ECHR) relating to the use of materials emanating from the exercise of compulsory powers (see *Saunders v. United Kingdom* (1996) 23 EHRR 313). In dismissing H's appeal, McCombe J (with whom Smith LJ and Simon J agreed) said that the Court was not here concerned with the self-incrimination of a defendant with regard to an actual or potential criminal charge; rather, it was concerned with the powers of a professional body to investigate the affairs of its members in the public interest and to discipline such members for breaches of the rules that apply to such professions. A similar problem arose in relation to the disciplinary processes of the accountancy profession in *R v. Institute of Chartered Accountants for England and Wales ex parte Nawaz*, 25 April 1997 (unreported): see paragraph 52.02 above. The Court of Appeal considered the application of the rules against self-incrimination in that context and, in the course of delivering judgment in that case, Leggatt LJ (with whom Thorpe and Mummery LJJ agreed) said that, when a person enters a profession, he accepts its duties and liabilities, as well as its rights and powers. Similarly, he may acquire or surrender privileges and immunities. It is plainly in the public interest, as well as the interests of the profession, that the Institute should be enabled to obtain all such information in the possession of its members as is relevant to complaints of their professional misconduct. Upon becoming a chartered accountant, it shall be the duty of every member, in accordance with the Supplemental Charter, to provide such information as may be necessary for the discharge of the Institute's functions. Compliance with that duty necessarily and inevitably precludes the exercise of any privilege that would have excused the provision of the information. In the instant case, McCombe J said that what the Court of Appeal said in *Nawaz* was determinative of this part of the appeal.

### *Quinn Direct Insurance Ltd v. Law Society* [2011] 1 WLR 308

**52.10** Law Society intervention into solicitor's practice—claim by professional indemnity insurers for disclosure of documents—clients' privilege—waiver—whether claims made by clients against solicitor

The claimant issued proceedings against the Law Society, seeking an order that the defendant permit it, as the professional indemnity insurance provider of a firm of solicitors, to inspect all documents of the solicitors held by the Law Society and to take copies thereof for the purpose of considering whether the claimant was obliged to indemnify a partner of the firm. The Law Society had intervened in the practice of the firm pursuant to its powers under section 35 of, and Schedule 1 to, the Solicitors Act 1974 on the grounds of suspected dishonesty and failure to comply with the Solicitors' Accounts Rules. A number of claims had been made by former clients of the firm and notified to the claimant. There was no issue in respect of files relating to transactions in which the client had already made a claim against the firm that had been notified to the insurers. In such a case, the Law Society took the view that the making of the claim by the client constituted a waiver of client confidentiality and privilege, and had allowed the insurers to have access to those files. In addition, copies of all bank statements of the firm had been provided to the claimant. As to the remainder of the documents of the firm in the possession of the Law Society, it was common ground that they all contained information confidential to one or more former clients of the firm whose privilege had not been waived. In dismissing the claimant insurer's claim, the Court of Appeal (Sir Andrew Morritt C, and Rimer and Jackson LJJ) held that neither the Law Society nor the client of the firm was a party to any policy of insurance. Accordingly, the claimant could have no contractual claim to production of the privileged documents that it sought unless the client had consented either expressly or by implication. The claimant's claim was a request for blanket disclosure of all documents, but was material only in relation to privileged documents for which privilege had not been waived by the client or the firm, because no claim against the firm had been made under the policy. There was nothing implied in the scheme for the regulation of solicitors constituted by the Solicitors Act 1974, subordinate legislation, and agreements made thereunder requiring any provision or term entitling or obliging the Law Society to produce to an insurer documents emanating from a firm of solicitors into which it has intervened that are subject to the privilege of a client of the firm. If the client consents or his privilege is impliedly waived by a claim against the solicitor, then there is no reason why the Law Society may not produce such documents to the insurers. It had sensibly done so in this case. Otherwise, the appeal was dismissed.

### *White v. Southampton University Hospitals NHS Trust* [2011] EWHC 825 (QB)

**52.11** Letters sent by medical director of NHS Trust to GMC and hospital—letters sent for legitimate purpose—libel proceedings—immunity from suit/absolute privilege—Article 10 ECHR

W, a foundation year 2 doctor, employed at Southampton University Hospitals National Health Service (NHS) Trust, sued the Trust for libel. The claim arose from the publication of a letter by the Trust to the fitness to practise directorate of the General Medical Council (GMC) and an earlier letter sent to the medical director of the Lymington New Forest Hospital. Both letters were written by Professor R, the medical director of the Trust. In his letter to the GMC, Professor R stated that he had concerns about the probity and conduct of W, who had been excluded from clinical practice in the Trust. After identifying various issues, Professor R concluded by saying that he had grave concerns about W's fitness to practise arising from her attitude, probity, behaviour, and, potentially, her health. In striking out W's claim, Eady J said that the letter to the GMC contained nothing extraneous—that is, nothing that was not germane to the legitimate purpose for which the law, for reasons of public policy, affords the protection of privilege. In *Vaidya v. General Medical Council* [2010] EWHC 984 (QB), Sir Charles Gray was concerned with a claim brought upon a letter to the GMC. He concluded: 'It appears to me to be clear beyond argument that this letter is protected by absolute privilege since it was written to an official of an investigatory body (the GMC) in order to complain about the conduct of [V].' Not only was V refused permission to appeal, but it was also held by Sir Richard Buxton that his application was totally without merit. Eady J continued:

> 7. The public policy objective is to enable people to speak freely, without inhibition and without fear of being sued, whether making a complaint of criminal conduct to the police or drawing material to the attention of a professional body such as the GMC or the Law Society for the purpose of investigation. It is important that the person in question must be able to know at the time he makes the relevant communication whether or not the immunity will attach; that is to say, the policy would be undermined if, in order to obtain the benefit of the immunity, he was obliged to undergo the stress and expense of resisting a plea of malice: see the remarks of Lord Hoffmann in *Taylor v. Director of the Serious Fraud Office* [1999] 2 AC 177, 214.

In the instant case, there was no possible way in which W could overcome Professor R's immunity from suit in respect of the GMC letter. The letter to the GMC was the subject of immunity and/or absolute privilege. As to the letter sent to Lymington New Forest Hospital, that came into W's possession through the process of disclosure in her employment tribunal proceedings. It was difficult to see how its contents could be demonstrated to be untrue or how Professor R could be shown to have published something that he knew to be false. Furthermore, any rights on the part of W under Article 6 or Article 8 ECHR were outweighed by the rights of the defendant under Article 10 and, specifically, its right not to be vexed with unmeritorious and futile litigation over a confidential document disclosed under compulsion of law.

### *Baxendale-Walker v. Middleton* [2011] EWHC 998 (QB)

**52.12** Solicitor—claims against Law Society, SDT, and others—regulatory and disciplinary proceedings—witness report prepared by accountants—privilege and immunity from suit

The claimant, a former solicitor specializing in tax law, brought proceedings alleging conspiracy, malicious falsehood, and misfeasance in public office against the head of investigations at the Office for the Supervision of Solicitors (OSS), the Law Society, a firm of chartered accountants engaged by the Law Society to produce a report on the legality of tax avoidance schemes devised and promoted by the claimant, and the chairman of the panel of the SDT that had decided that the claimant should be struck off the roll of solicitors, and the Tribunal itself. The particulars of claim alleged: that the head of investigations had formed the joint intention of constructing a knowingly false and fraudulent case against the claimant with a view to destroying his tax practice; that fraudulent reports prepared by the firm of accountants were placed before panels of the Law Society, resulting in the recommendation of conditions being imposed on the claimant's practising certificate, which would have had the effect, but for his successful appeal to the Master of the Rolls, of closing his professional practice and destroying his firm; that the Law Society then sought to achieve the striking off of the claimant by manufacturing an allegation to be brought against him in regulatory and disciplinary proceedings; that the allegations in the rule 4 statement before the Tribunal were false; and that the chairman of the Tribunal knew or believed that the complaint was a sham. Following conclusion of the proceedings before the Tribunal, the claimant had two telephone conversations with the head of investigations and the chairman of the Tribunal. The claimant relied upon the contents of the telephone conversations as evidence to support the claim. On an application by the defendants, Supperstone J struck out the causes of action on various grounds. As to the claim against the Law Society defendants, they were protected by absolute privilege and immunity: see *Marrinan v. Vibart* [1963] 1 QB 528, 533 ('those who take part in the administration of justice... must be free from the fear of civil proceedings'); *Darker v. Chief Constable of West Midlands Police* [2001] 1 AC 435 (core immunity for witnesses, parties, and judges); *Royal Aquarium and Summer and Winter Garden Society v. Parkinson* [1892] 1 QB 431, 442; *Addis v. Crocker* [1961] 1 QB 11 (proceedings before the Solicitors Disciplinary Committee); *Taylor v. Director of the Serious Fraud Office* [1999] 2 AC 177; *Mahon v. Rahn and ors (No. 2)* [2000] 1 WRL 2150, [194]; and *Gray v. Avadis* [2003] EWHC 1830 (QB) (both the SDT and the OSS—and presumably now the Solicitors Regulation Authority, or SRA—are tribunals exercising functions equivalent to a court of justice, so benefiting from absolute immunity). As to the claims against the accountants, it is clear from the same authorities that it is an established rule of law that a witness in proceedings cannot be the subject of any civil claim in respect of the evidence he gives (see *Marrinan v. Vibart, per* Sellers LJ at 534). Proceedings before the SDT qualify as judicial proceedings (see *Addis v. Crocker* [1961] 1 QB 11 and *Gray v. Avadis*, at [39]). The preparation of the accountant's report enjoys the protection of absolute witness privilege. The chairman of the Tribunal enjoyed judicial immunity. This immunity from suit extends beyond the courts to tribunals of various kinds, and led the courts in *Addis* and *Gray* to hold that the disciplinary bodies were immune from suit. The claim against the SDT, on the basis that it was vicariously liable for the actions of the chairman of the panel, had to fail.

### *Adams v. Law Society of England and Wales* [2012] EWHC 980 (QB)

**52.13** Solicitor—intervention—forensic investigation report—whether covered by absolute privilege

In 2004, the claimant, a solicitor sole practitioner, was the subject of an intervention under the Solicitors Act 1974. By an application notice dated 24 June 2011, he sought to extend time to serve his claim form, and to file and serve amended particulars of claim. Foskett J refused relief. At [150]–[156], argument was addressed as to whether a forensic investigation report and case notes were covered by absolute privilege. The learned judge said that the leading modern authority on 'established immunity' is *Darker v. Chief Constable of the West Midlands* [2001] 1 AC 435, which was itself considered by the Court of Appeal in *Autofocus Ltd v. Accident Exchange Ltd* [2010] EWCA Civ 788. Supperstone J, in an extensive judgment in *Baxendale-Walker v. Middleton and ors* [2011] EWHC 998 (QB), held, inter alia, that claims against the Law Society defendants should be struck out because their actions were protected by absolute privilege and immunity. In the present case, the issue was academic given the conclusions reached on the substantive merits, but Supperstone J did not refer to *Autofocus* in his judgment and it may be an issue that may need to be canvassed as to whether the immunity can truly arise when a preliminary report (such as a forensic investigation report or case note) is prepared with the kind of malevolence necessary to sustain the tort of targeted misfeasance in public office.

### *Iqbal v. Solicitors Regulation Authority* [2012] EWHC 4097 (QB)

**52.14** Misfeasance in public office

Underhill J, in striking out the claimant's claims against the SRA and two of its staff for damages for misfeasance in public office and breach of Article 1 of Protocol 1 ECHR said, at [49], that there is already clear authority at first instance that the SRA and its employees enjoy absolute immunity from claims for misfeasance in public office: see the judgment of Supperstone J in *Baxendale-Walker v. Middleton* [2011] EWHC 998 (QB), [88]–[95]. Here the claimant argued, however, that that decision was wrong, referring to the doubts expressed by Foskett J in the subsequent case of *Adams v. Law Society of*

*England and Wales* [2012] EWHC 980 (QB), [155]. Underhill J said that he did not think that his entering into the debate would advance matters. However, as presently advised, he was far from sure that he would be in the Supperstone camp as opposed to the Foskett camp.

### *Mayer v. Hoare* [2012] EWHC 1805 (QB)

**52.15** Barrister—complaint—response sent to Bar Standards Board—libel proceedings—absolute privilege

In April 2012, H, a barrister, responded to a letter of complaint sent by the Bar Standards Board received on behalf of the plaintiff, M. In the course of his response to the Board, in which he denied M's allegations, H said that M's complaints against him were malicious and that they were untrue to the knowledge of M. On 5 April, the Board sent H's letter to M, who, on 18 April, without any intervening communication, issued a claim for libel against H. On 21 May, H issued an application for an order striking out the claim, or for summary judgment. On 22 June, the Bar Standards Board dismissed M's complaint against H. In granting H's application and striking out the claim for libel as being wholly without merit, Tugendhat J said that the words complained of in H's letter of 4 April 2012 to the Board were published on an occasion of absolute privilege. The Legal Services Act 2007 designates the General Council of the Bar as the approved regulator in respect of barristers. The Bar Standards Board was set up by the General Council of the Bar in January 2006 with responsibility for handling complaints against barristers, among other functions. Such communications to it are protected to the same extent as were communications to a Bench of an Inn of Court in *Lincoln v. Daniels* [1962] 1 QB 237 and as communications to other regulatory bodies exercising similar functions to that of the Bar Standards Board: *Mahon v. Rahn (No. 2)* [2000] 1 WLR 2150, [159], [170], and [194]. In that case, a financial services regulatory body investigating stockbrokers' fitness to conduct business had received a letter sent at that body's request. A libel action had been brought on the contents of the letter. It was held to be written on an occasion of absolute privilege. See *Westcott v. Westcott* [2008] EWCA Civ 818, [25]; [2009] QB 407.

### *P v. Commissioner of Police of the Metropolis* [2014] UKEAT/0449/13/JOJ

**52.16** Police officer—misconduct proceedings—dismissal of officer—alleged discrimination by Board—immunity from suit

Following an incident in September 2011, which led to her arrest, P, a police officer, was brought on a disciplinary charge before the Police Misconduct Board. She agreed that she was guilty of the misconduct alleged. Despite the mitigation and that she had previously suffered post-traumatic stress disorder (PTSD) as a result of an assault on duty, the Board decided that she should be dismissed from the police force without notice. Subsequently, P brought proceedings in the employment tribunal on the grounds that the Board had failed to make appropriate adjustments in the light of her PTSD disability and that its decision to dismiss her constituted an act of discrimination related to her disability. Although the members of the Board were not themselves sued as respondents, it was their conduct that was impugned as being discriminatory, for the police could act through no other agent in respect of the dismissal. The employment tribunal upheld a claim that the Board was entitled to judicial immunity and, consequently, P's claim for discrimination would be struck out. On appeal, after reviewing *Heath v. Commissioner of Police of the Metropolis* [2005] ICR 329, *Singh v. Reading Borough Council* [2013] ICR 1158, and *Lake v. British Transport Police* [2007] ICR 1293, Langstaff J held that the employment judge was plainly right to conclude as he did. In the *Lake* case, the court considered that the claimant was not seeking to challenge the principle of judicial immunity from suit. In the instant case, the very basis for arguing that the decision was wrong was that it was an act of discrimination and of harassment by the Board. The allegation centred on the Board's conduct, when exercising its judicial functions. So viewed, the case fell four square within the core principles established in *Heath*.

## B. Other Relevant Chapters

**52.17** Disclosure, Confidentiality, Data Protection, and Freedom of Information, Chapter 18
Human Rights, Chapter 32
Negligence, Chapter 49

# 53

# PUBLIC OR PRIVATE HEARING

A. **Legal Framework** 53.01
B. **Disciplinary Cases** 53.02
*Albert and Le Compte v. Belgium* (1983) 5 EHRR 533 53.02
*R (Chaudhari) v. Royal Pharmaceutical Society of Great Britain* [2008] EWHC 3190 (Admin) 53.03
*L v. Law Society* [2008] EWCA Civ 811 53.04
*Walker v. General Medical Council* [2010] EWHC 3849 (Admin) 53.05
*R (Willford) v. Financial Services Authority* [2013] EWCA Civ 674 53.06
*R (Miller) v. General Medical Council* [2013] EWHC 1934 (Admin) 53.07
*R (Miller) v. Chief Constable of Merseyside Police* [2014] EWHC 400 (Admin) 53.08
C. **Other Relevant Chapters** 53.09

## A. Legal Framework

**53.01** European Convention on Human Rights (ECHR), Article 6(1) (see paragraph 32.01) Examples of substantive hearings being heard in public include:

Nursing and Midwifery Council (Fitness to Practise) Rules 2004, rule 19
Veterinary Surgeons and Veterinary Practitioners (Disciplinary Committee Procedure and Evidence) Rules 2004, rule 21
Police (Conduct) Regulations 2012, rule 32 (misconduct proceedings are usually held in private)
General Optical Council (Fitness to Practise) Rules 2013, rule 24
General Medical Council (Fitness to Practise) Rules 2014, rule 41 (committee or panel may determine that the public shall be excluded from the proceedings or any part of the proceedings, where it considers that the particular circumstances of the case outweigh the public interest in holding the hearing in public; committee or panel shall sit in private where it is considering whether to make or review an interim order, or the physical or mental health of the practitioner)

Some regulatory bodies expressly provide that the committee or panel may exclude from any hearing any person whose conduct, in its opinion, is likely to disrupt the orderly conduct of the proceedings. For example:

General Optical Council (Fitness to Practise) Rules 2013, rule 26
General Medical Council (Fitness to Practise) Rules 2014, rule 42

Hearings before the investigations committee are frequently held in private. Examples include:

General Pharmaceutical Council (Fitness to Practise and Disqualification etc) Rules 2010, rule 9
Architects Registration Board Investigations Rules, rule 4

## B. Disciplinary Cases

### *Albert and Le Compte v. Belgium* (1983) 5 EHRR 533

**53.02** Disciplinary hearing—Article 6(1) ECHR

The European Court of Human Rights (ECtHR) noted, at [34] of its judgment, that the nature of the misconduct alleged against C and of his own complaints against the *Ordre des Médecins* were not concerned with the medical treatment of his patients. Accordingly, there was nothing to suggest that one of the grounds listed in the second sentence of Article 6(1) ECHR (the interests of morals, public order, national security, juveniles, or private life, or the interests of justice) could have justified a hearing of the disciplinary proceedings in camera. In the case of A, the matter was different, in that the offences of which he was accused related directly to the exercise of the medical profession, which might conceivably raise questions coming within the exceptions listed in Article 6(1). However, the material submitted to the Court did not show that the circumstances were such as to warrant the absence of publicity.

See further paragraph 32.04.

### *R (Chaudhari) v. Royal Pharmaceutical Society of Great Britain* [2008] EWHC 3190 (Admin)

**53.03** Interlocutory application for proceedings to be withdrawn—hearing in private

In the course of disciplinary proceedings against the claimant, a registered pharmacist, the claimant applied to the chairman of the Disciplinary Committee for the inquiry to be withdrawn. This was an interlocutory application and the point at issue was whether holding the application in private would be a denial of the claimant's rights under Article 6 ECHR to a public hearing. In civil proceedings, the general rule is that all hearings are held in public. In the instant case, there was no breach of the ECHR in ordering that the preliminary procedure should be held in private. CPR 39.2(3) provides that a hearing, or any part of it, may be in private if the court considers this to be necessary in the interests of justice. The notes in *Civil Procedure* 2008 make clear that the Strasbourg institutions have generally taken the view that interlocutory hearings are not determinative of civil rights and obligations within the meaning of Article 6(1) ECHR. That was consistent with *R v. Legal Aid Board, ex parte Kaim Todner (a firm)* [1999] QB 966, in which the Court of Appeal, while recognizing the general rule that the administration of justice is required to be in public, said that the nature of the proceedings is also relevant. These hearings were therefore generally not required to be in public. Accordingly, the Court dismissed the claimant's application for a public hearing.

### *L v. Law Society* [2008] EWCA Civ 811

**53.04** Student registration—appeal—hearing in public

The Master of the Rolls considered whether an appeal against the Law Society's decision to revoke the membership of a student should be heard in private. The appellant was concerned about details of spent convictions being made public and argued that if the hearing were held in public, it would breach his rights under Article 6 ECHR. The application was refused. The general rule was that, in the absence of exceptional circumstances, appeals should be heard in public. The convictions were relevant to an application to join a regulated profession, the members of which had to be capable of being trusted implicitly. Part of ensuring that public confidence was maintained was that proceedings such as the instant one were held in public.

### *Walker v. General Medical Council* [2010] EWHC 3849 (Admin)

**53.05** Doctor—allegation by patient of sexual misconduct against doctor—patient and husband prosecuted for blackmail—whether GMC hearing to be held in public

Allegations were made against the claimant that he had engaged in sexual conduct with a patient and that he had been convicted of an offence of battery against his wife. On 13 December 2010, a fitness to practise panel of the General Medical Council (GMC) was convened to hear the allegations. At the outset, counsel for the claimant made an application to the panel that the part of the hearing that was concerned with the allegation of sexual misconduct should be heard in private. The panel rejected that application. On 14 December 2010, Wyn Williams J gave directions and heard the matter on 16 December. The doctor claimed that he was blackmailed by the patient's husband as the price for the husband's silence. The husband and the patient were prosecuted for blackmail. The husband admitted the offence of blackmail at the Crown Court and was sentenced to a term of imprisonment. The patient denied the offence of blackmail and was acquitted by the jury. Counsel for the doctor argued that there was a public interest in encouraging victims of blackmail to come forward and that a victim may be deterred from coming forward if he or she takes the view that future regulatory proceedings will be conducted in the full glare of publicity. The wider public interest in prosecuting blackmailers would be frustrated. Wyn Williams J rejected this argument. He said that it was clear that this aspect of the case was not lost upon the panel and that the panel recognized that it would be appropriate in some cases to provide for a hearing in private should the person who is subject to the disciplinary process be the victim of blackmail. The learned judge said that, looked at overall, the possibility of the disciplinary process being held in public would be a peripheral consideration at most in the mind of any person placed in the position in which this doctor found himself following the blackmail offence perpetrated against him. The panel was also presented with a medical report that stated that, outside of a private context, the doctor may be less confident and struggle with anxiety feelings, and that this may impair his ability to convey what he wants to say and how he might express it. There was, therefore, a possibility that, in a more public arena, he may not be able to project his emotional and personal insights as he might in a more private context. The panel said that it would take appropriate and practical measures to minimize the doctor's stress levels whilst participating in the proceedings. The learned judge said that the determination of the panel on this issue could be impugned only if it were to fall into error in the *Wednesbury* sense and the ground was unarguable. As a post script, the learned judge gave a ruling directing that the name of the claimant should not be anonymized.

### *R (Willford) v. Financial Services Authority* [2013] EWCA Civ 674

**53.06** FSA decision notice—reference to Upper Tribunal or judicial review proceedings—principles of open justice—public forum—anonymization or redaction of judgment

On 20 and 21 February 2013, the Court of Appeal (Moore-Bick and Black LJJ, and Sir Malcolm Pill) heard an appeal by the Financial Services Authority (FSA)—now the Financial Conduct Authority (FCA)—against an order of Silber J quashing a decision notice given to W under section 67 of the Financial Services and Markets Act 2000. The hearing before Silber J was conducted in private, and the judgment was published in a redacted and anonymized form to ensure that W's identity was not disclosed. At the beginning of the hearing of the appeal, the Court was asked to hear the matter in private, but it declined to do so. It made an order prohibiting, until further notice, the identification of W or the reporting of any information that might lead to his identification. On 13 June 2013, the Court allowed the FSA's appeal: [2013] EWCA Civ 677. W applied for an order that the judgment be published in a redacted and anonymized form. In dismissing the application, Moore-Bick LJ said:

> 5. In my view, the starting point is the principle of open justice, that is, the principle that proceedings are to be conducted and determined in public. One aspect of that

principle is that judgments should be published in full without concealing the identity of the parties or others involved, whether by anonymisation or redaction. However, the principle is not absolute and must give way to the requirements of justice and other countervailing considerations of public interest. For example, judgments in criminal proceedings are frequently anonymised in order to protect the identities of children. In civil cases between adult parties, however, the public interest in open justice will usually outweigh other considerations, except where publication would significantly undermine the effectiveness of any relief the court might grant.

[...]

8. ...In the present case the FSA was performing a disciplinary function. No doubt its proceedings would have been embarrassing for [W] if his identity had been made public and it may be that they would have caused some damage to his professional reputation (though if they had culminated in a decision to take no action against him that might have been slight), but he would not have been entitled to have his identity protected on those grounds if, for example, he had for some reason been facing criminal charges, which could have had a similar effect. Nor is there any positive evidence that he will suffer significant harm if the existence of the proceedings is disclosed at this stage. It is true, however, that if he were to obtain permission to appeal to the Supreme Court and succeed in having the Decision Notice quashed, he would have lost the right to have his identity protected pending the outcome of the disciplinary proceedings and perhaps (if he were ultimately to persuade the FSA that no sanction should be imposed on him) for all time.

9. The question, then, is whether in those circumstances it is strictly necessary in the interests of justice to anonymise and redact our judgments in order to protect the respondent's identity. In my view it is not. The redactions proposed by counsel for [W] are extensive and go to the heart of the judgments. The anonymisation is, of course, complete. The principle of open justice requires that the court's judgment should be published in full unless there are overriding grounds for not doing so. Although the FSA disciplinary proceedings were private, once the respondent stepped outside those proceedings, whether by referring the matter to the Upper Tribunal or by making a claim for judicial review, he brought the matter into the public forum where the principle of open justice applies. That may happen in other contexts. Parties to arbitration proceedings, for example, are entitled to have the confidentiality of those proceedings maintained, but if one party invokes the assistance of the court, perhaps by appeal or by an application to set aside the award, the court will not normally take steps to preserve the confidentiality of the proceedings or their subject matter.

10. If anonymisation and redaction of the judgment were necessary in this case in order to enable the court to grant effective relief, the position would be different, but the primary object of these proceedings is not to protect [W]'s privacy; it is to ensure that the procedure adopted by the FSA complies with the statutory requirements and to quash the existing Decision Notice to enable the FSA to reconsider the matter.

### *R (Miller) v. General Medical Council* [2013] EWHC 1934 (Admin)

**53.07** Doctor—disciplinary proceedings—complainant wanting hearing in private

M, a consultant psychiatrist facing allegations of financial impropriety concerning patient A, challenged the decision of the fitness to practise panel to hold the whole of the hearing in private on the grounds that patient A suffered from a mental health disorder. His Honour Judge Pelling QC (sitting as a judge at the High Court) said that it was common ground that the discretion under rule 41 of the General Medical Council Rules is to be read subject to Article 6 ECHR. Even where one of the Article 6 exceptions to a public hearing can, in principle, be relied upon, the derogation from the general principle ought not to be more than is proportionate—that is, the minimum derogation from the general principle necessary for the purpose of protecting the interest that has been identified as coming within the scope of the relevant exception:

see *Diennet v. France* [1995] ECHR 28. The common law general rule should be departed from only to the extent that such a departure is strictly necessary: see *Scott v. Scott* [1913] AC 417. In quashing the decision of the panel, the Court observed that no attempt had been made by the GMC to confine itself to an application for a direction for anonymization of patient A's name and that patient A's evidence be given by video link or from behind screens. No attempt was made to limit the scope of the privacy direction to the part of the hearing in which patient A was to give evidence. The GMC sought and obtained an order from the panel that the whole of the hearing should be conducted in private because that is what patient A apparently insisted upon. Consequently, the decision was quashed.

***R (Miller) v. Chief Constable of Merseyside Police*** **[2014] EWHC 400 (Admin)**

**53.08** Police misconduct hearing—exclusion of interested persons—whether lawful

In quashing the findings of two police misconduct meetings and ordering that they be reheld with a different decision-maker, Stewart J said that the claimant, a minor, and his mother should have been permitted to remain until the end of the proceedings. The claimant and his mother complained that they were wrongly excluded from part of the misconduct proceedings in breach of regulation 31(3) of the Police (Conduct) Regulations 2008. The claimant complained about the behaviour of two police officers. The claimant said that he was assaulted in a public shopping area—that one officer had manhandled him and pushed him forcibly against a police vehicle, whilst the other officer had sprayed him in the face at a dangerously close range with CS spray. The claimant was a minor and no explanation was provided as to why he and his mother were excluded prematurely from the misconduct hearing. The Regulations stated, so far as it was material, that the complainant or any interested person may attend up to, but not including, the point at which the person conducting or chairing the proceedings considers the question of disciplinary action. Since there was no finding of misconduct against either police officer, this point never arose. The regulation is an important one providing for open justice. Such a breach of the 2008 Regulations, without any explanation, would lead a fair-minded and informed observer to conclude that there was a real possibility of bias. In the circumstances of this case, that would be sufficient to require a rehearing of the misconduct meeting. This was particularly so where the allegations against the officers were found to be not proven.

## C. Other Relevant Chapters

**53.09**

Human Rights, Chapter 32
Press Publicity, Chapter 51

# 54

# REASONS

A. **Legal Framework** 54.01

B. **Disciplinary Cases** 54.02

*Stefan v. General Medical Council* [1999] 1 WLR 1293 54.02

*Selvanathan v. General Medical Council* [2001] Lloyds Rep Med 1 54.03

*Brabazon-Drenning v. United Kingdom Central Council for Nursing Midwifery and Health Visiting* [2001] HRLR 6 54.04

*Gupta v. General Medical Council* [2001] UKPC 61; [2002] 1 WLR 1691 54.05

*Needham v. Nursing and Midwifery Council* [2003] EWHC 1141 (Admin) 54.06

*R (Luthra) v. General Dental Council* [2004] EWHC 458 (Admin) 54.07

*Threlfall v. General Optical Council* [2004] EWHC 2683 (Admin) 54.08

*Ryell v. Health Professions Council* [2005] EWHC 2797 (Admin) 54.09

*Phipps v. General Medical Council* [2006] EWCA Civ 397 54.10

*Watson v. General Medical Council* [2006] EWHC 18 (Admin) 54.11

*Ogango v. Nursing and Midwifery Council* [2008] EWHC 3115 (Admin) 54.12

*R (Kaftan) v. General Medical Council* [2009] EWHC 3585 (Admin) 54.13

*Yeong v. General Medical Council* [2009] EWHC 1923 (Admin) 54.14

*Beresford v. Solicitors Regulation Authority* [2009] EWHC 3155 (Admin) 54.15

*Keane v. Law Society* [2009] EWHC 783 (Admin) 54.16

*Southall v. General Medical Council* [2010] EWCA Civ 407 54.17

*Chegwyn v. Ethical Standards Officer of the Standards Board for England* [2010] EWHC 471 (Admin); [2011] PTSR 165 54.18

*Shepherd v. Governor of HMP Whatton* [2010] EWHC 2474 (Admin) 54.19

*Brennan v. Health Professions Council* [2011] EWHC 41 (Admin) 54.20

*R (Alhy) v. General Medical Council* [2011] EWHC 2277 (Admin) 54.21

*Hazelhurst and ors v. Solicitors Regulation Authority* [2011] EWHC 462 (Admin) 54.22

*Yaacoub v. General Medical Council* [2012] EWHC 2779 (Admin) 54.23

*Duthie v. Nursing and Midwifery Council* [2012] EWHC 3021 (Admin) 54.24

*Quinn v. Bar Standards Board*, 25 February 2013 (unreported) 54.25

*Cronin v. Greyhound Board of Great Britain Ltd* [2013] EWCA Civ 668 54.26

*Barakat v. General Medical Council* [2013] EWHC 3427 (Admin) 54.27

*R (H) v. Nursing and Midwifery Council* [2013] EWHC 4258 (Admin) 54.28

*Shaw and Turnbull v. Logue* [2014] EWHC 5 (Admin) 54.29

*O v. Secretary of State for Education and National College for Teaching and Leadership (Interested Parties)* [2014] EWHC 22 (Admin) 54.30

*Moore v. Nursing and Midwifery Council* [2013] EWHC 4620 (Admin) 54.31

*L v. The Health and Care Professions Council* [2014] EWHC 994 (Admin) 54.32

*Virde v. General Pharmaceutical Council* [2015] EWHC 169 (Admin) 54.33

C. **Other Relevant Chapters** 54.34

## A. Legal Framework

54.01

Samples include:

General Dental Council (Fitness to Practise) Rules 2006, rule 23 (the registrar shall send notification of any determination or order made by a practice committee and the reasons for that determination or order)
Police (Conduct) Regulations 2012, regulation 36 (notification of outcome and reasons before the end of five working days, beginning with the first working day after the conclusion of the misconduct proceedings)
Bar Standards Board Disciplinary Tribunals Regulations 2014, rule E181 (as soon as practicable after the end of the proceedings of a disciplinary tribunal, the chairman must prepare a report in writing of the finding(s) on the charges of professional misconduct and/or on any applications, and the reasons for those findings and the sentence, if any)
General Medical Council (Fitness to Practise) Rules 2014, rule 17 (the fitness to practise panel shall consider and announce its findings of fact; shall consider and announce its findings on the question of whether the fitness to practise of the practitioner is impaired, and shall give its reasons for that decision; shall consider and announce its decision as to the sanction or warning, if any, to be imposed or undertakings to be taken into account and shall give its reasons for that decision)
Institute of Chartered Accountants in England and Wales Disciplinary Committee Regulations, rules 48 and 49 (a written record of the decision of the tribunal shall be prepared by the legal assessor for approval by the tribunal, and the director shall send to the defendant a copy of the written record of decision where a formal complaint—or part of the complaint—has been proved as soon as reasonably practicable after it has been approved by the tribunal)

## B. Disciplinary Cases

### *Stefan v. General Medical Council* [1999] 1 WLR 1293

54.02

Health—suspension—obligation at common law to give reasons—relevant factors

The Privy Council considered, in the professional context of health care, the question of whether reasons should be given to somebody who, in that case, was suspended from the register of the General Medical Council (GMC). The ground for her suspension was that her fitness to practise was impaired on health grounds and the decision was taken by the Health Committee of the GMC (see paragraph 31.03). The appellant was not given reasons as such, but simply told that the relevant Committee was satisfied that the allegations had been made out. The Privy Council consisted of Lords Browne-Wilkinson, Steyn, Clyde, Hutton, and Hobhouse. The judgment was given by Lord Clyde. He accepted that, in the circumstances, there was no basis for implying any statutory obligation to give reasons. But he concluded that there was an obligation arising at common law, because it was an element in the concept of fairness. In giving reasons, he said at 1300F–H:

> The trend of the law has been towards an increased recognition of the duty upon decision-makers of many kinds to give reasons. This trend is consistent with current developments towards an increased openness in matters of government and administration. But the trend is proceeding on a case by case basis (*R v. Royal Borough of Kensington and Chelsea, ex parte Grillo* (1995) 94 LGR 144, and has not lost sight of the established position of the common law that there is no general duty, universally

imposed on all decision-makers. It was reaffirmed in *R v. Secretary of State for the Home Department, ex parte Doody* [1994] 1 AC 531, 564, that the law does not at present recognise a general duty to give reasons for administrative decisions. But it is well established that there are exceptions where the giving of reasons will be required as a matter of fairness and openness.

His Lordship held that there were three factors in particular that were of a general nature and which justified reasons being given. These were the fact that the decision affected the appellant's right to practise her profession, that there was a right of appeal that pointed to the view that reasons should be given, so that the grounds of appeal could be properly identified and articulated, and that a consideration of the nature and the functions of the Committee demonstrated that it was, in many respects, akin to a court of law, which was expected to give reasons. The extent and substance of the reasons must depend upon the circumstances. They need not be elaborate or lengthy. But they should be such as to tell the parties in broad terms why the decision was reached. In many cases, as indicated in the context of Article 6(1) of the European Convention on Human Rights (ECHR), a few sentences should suffice to give such explanation as is appropriate to the particular situation. The recording of a generally agreed statement of their reasons would not add to the burden of the decision-making process. While the decision involves the application of some medical expertise in the assessment of fitness, the articulation of the reasons for a value judgment should not give rise to difficulty (*R v. City of London Corporation, ex parte Matson* [1997] 1 WLR 765, 783).

### *Selvanathan v. General Medical Council* [2001] Lloyds Rep Med 1

**54.03** Serious professional misconduct—penalty—need for general explanation for determinations

Giving the judgment of the Privy Council, Lord Hope of Craighead said that the practice of giving reasons has moved on since *Libman v. General Medical Council* [1972] AC 217 and the subsequent cases of *Rai v. General Medical Council*, Privy Council Appeal No. 54 of 1983, and *Peatfield v. General Medical Council* [1986] 1 WLR 243, in which the Board held that, although the Professional Conduct Committee (PCC) had power to give reasons, it was under no obligation to do so.

> 23. . . . In *Rai v. General Medical Council*, pp. 10–11, the Board suggested that the giving of reasons could be beneficial, and assist justice, to enable the doctor in a complex case to understand the Committee's reasons for finding against him, to give guidance to the profession where this can usefully be provided and because a reasoned finding can improve and strengthen the appeal process. Their Lordships consider that in practice reasons should now always be given by the Professional Conduct Committee for their determination under rule 29(2) [of the 1988 Procedure Rules] whether or not they find the practitioner to have been guilty of serious professional misconduct and their decision on the question of penalty. Fairness requires this to be done, so that the losing party can decide in an informed fashion whether or not to accept the decision or to appeal against it under section 40 of the Medical Act 1983. But the question which is in issue in the present case is not whether reasons should be given. It is plain that reasons were given in this case. The question is as to the adequacy of those reasons.
>
> 24. . . . The 1988 Practice Rules require the Committee to reach a view as a committee on the matters that they have before them for determination. No provision is made for expressions of dissent either as to the result or on matters of detail. In these circumstances it is not to be expected of the Committee that they should give detailed reasons for their findings of fact. A general explanation of the basis for their determination on the questions of serious professional misconduct and of penalty will be sufficient in most cases.

### *Brabazon-Drenning v. United Kingdom Central Council for Nursing Midwifery and Health Visiting* [2001] HRLR 6

**54.04** Following the decisions in *Stefan* and *Selvanathan*, the Divisional Court (Rose LJ and Elias J) held that the appellant's ground of appeal, based on failure to give reasons, was sustained also.

### *Gupta v. General Medical Council* [2001] UKPC 61; [2002] 1 WLR 1691

**54.05** Notice of inquiry—detail of conduct alleged—issues of credibility—extent of duty to give reasons

In March 1996, the appellant's husband and partner was found guilty of serious professional misconduct, and a direction was made that his name be erased from the register. The charges against the appellant were that, knowing this, she permitted her husband to hold consultations between May 1996 and December 1998 with various patients who attended her surgery premises for medical services. The appellant appealed against the decision of the PCC finding that she permitted her husband to hold consultations and failed to prevent him holding consultations with patients at the surgery premises. It was argued that, because issues of credibility and reliability of the witnesses were involved, the Committee should have given reasons explaining why it found the allegations of fact proved. In giving the judgment of the Privy Council in dismissing the appellant's appeal, Lord Rodger of Earlsferry said that:

> 12. ... The form of the notice of inquiry given to practitioners was amended in 1988 so as to ensure that they would be given considerable detail about the conduct on which the [GMC] were basing their complaint. In its determination, the Committee finds a particular charge or head of charge proved or not proved. The practitioner is therefore able to see, in some detail, which allegations have been established. This in turn will usually mean that the practitioner will have a very good idea what evidence the Committee has accepted. In some cases, such as the present, the Committee's decision will show that it has felt able to find one allegation proved on the basis of the evidence of a particular witness, while feeling unable to find another allegation proved on the basis of some other part of the evidence given by the same witness. In this way, in cases involving issues of credibility and reliability, the structured determination of the Committee dealing with the various heads of the charge, will in itself reveal much about its reasons for reaching its decision. As the European Commission of Human Rights noted in *Wickramsinghe v. United Kingdom* [[1998] EHRLR 338], the fact that the practitioner can study a transcript of the hearing, including not only the evidence but the submissions on the evidence by the respective parties, further assists the practitioner in understanding not only which witnesses' evidence the Committee accepted and which it rejected, but why it did so.
>
> 13. To go further and to insist that in virtually all cases raising issues of credibility and reliability the Committee should formally indicate which witnesses it accepted and which it rejected would be to require it to perform an essentially sterile exercise....
>
> 14. Their Lordships would add this. They have rejected the submission that there is a general duty to give reasons in cases where the essential issue is one of the credibility or reliability of the evidence in the case. None the less, while bearing in mind the potential pitfalls highlighted by Lord Mustill, the Committee can always give reasons, if it considers it appropriate to do so in a particular case. Their Lordships would go further: there may indeed be cases where the principle of fairness may require the Committee to give reasons for their decision even on matters of fact. Nothing in *Selvanathan* is inconsistent with that approach, while the general reasoning in *Wallace* [*v. The Queen* (1996) The Times, 31 December] supports it. It is also in line with the observations of Lord Steyn giving the judgment of the Board in *Rey v. Government of Switzerland* [1999] 1 AC 54.

Whilst the Board rejected the submission that there is a general duty to give reasons in cases in which the essential issue is one of credibility and reliability of the evidence of the case, it noted that, nonetheless, the Committee always gives reasons, if it considers it appropriate to do so in a particular case.

### *Needham v. Nursing and Midwifery Council* [2003] EWHC 1141 (Admin)

**54.06** Disputed issues—list of issues—need for structured approach to determinations

On 13 May 2002, the appellant was found guilty of six charges of misconduct, following a hearing, which she attended, at which she called witnesses and was represented by counsel. The charges related to a series of drug administration and patient case errors between 14 September 1999 and 29 January 2000, whilst the appellant was working as a staff nurse on a mixed surgical ward. Admissions of fact and admissions of misconduct were made in the course of the hearing before the PCC. The Committee decided to remove the appellant's name from the register and stated that, in arriving at this decision, it had listened carefully to the information put before it in mitigation, but that the appellant's fitness to practise was impaired; the reasons for removal were that she had manifestly failed to promote the interests of individual patients and clients, and had also failed to justify public trust and confidence. In allowing the appellant's appeal on the grounds of inadequacy of reasons, Newman J said:

> 11. Neither the [Nurses, Midwives and Health Visitors Act 1997] nor the Nurses, Midwives and Health Visitors (Professional Conduct) Rules 1993 impose any obligation upon the Committee to give reasons for its decision, but it is not disputed that fairness requires reasons to be given. As to what is required, certain cases in the sphere of professional conduct hearings, have established the following.
>
> i) Whether sufficient reasons have been given will depend upon the particular circumstances of the case.
>
> ii) That resort may be had to the transcript of the hearing (See *Gupta v. General Medical Council* [2002] 1 WLR 1691), particularly where the transcript will reveal which evidence the committee accepted and which it rejected (See *Wickramsinghe v. United Kingdom* [1998] EHRLR 338).
>
> iii) That a general explanation of the basis for the determination on the questions of serious professional misconduct and of penalty will normally be sufficient (*Selvanathan v. GMC* [2001] Lloyd's Rep Med 1).
>
> iv) That the fact that an appellant had not been prejudiced by the failure to give reasons was irrelevant (*Brabazon-Drenning v. United Kingdom Central Council* [2001] HRLR 6).
>
> v) That reasons need not be elaborate nor be lengthy but should be such as to tell the parties in broad terms why the decision was reached.

In the instant case, the Committee found against the appellant on the central issue that her conduct was impaired, but gave no reasons for having done so. Newman J continued:

> 13. ... In cases where a number of issues have been raised the Committee should itself draw up a list of the specific issues to which it must give consideration and identify the material which has been presented to it in connection with each of those issues. The process of the hearing will provide an adequate opportunity for a list of the issues to be made and a sufficient record of the evidence or material going to those issues to be drawn up. Due and proper consideration to cases requires such a structured approach to the Committee's deliberations. Save in the most obvious and simple cases the Committee should adopt a structured approach to the consideration of issues.... If in doubt the Legal Assessor's guidance can be sought on questions of structure and the presentation of the reasons but not, of course, on the reasons themselves....

It was therefore found that the penalty should be quashed and there was to be a rehearing before a new committee, at which all matters of mitigation could be raised.

### *R (Luthra) v. General Dental Council* [2004] EWHC 458 (Admin)

**54.07** Need for specific findings for each charge—summary of reasons for misconduct—generally not necessary to identify evidence accepted and rejected

L, a dentist, appealed against findings of dishonesty made by the PCC, which suspended his registration for six months for serious professional misconduct. L faced a series of complaints, falling into three broad categories. The first was that he had made false and misleading claims to the Dental Practice Board in order to receive payments to which he was not entitled, which complaints related to three patients. Second, it was alleged that a claim had been made for treatment that had not been carried out at all. Third, it was said that money had been claimed for work not done. The Committee did not find all of the charges established, but it did find that the charges based on dishonesty were established save in one respect. In giving its reasons, the Committee stated that it found L to have deliberately submitted false misleading claims to the Dental Practice Board. His purpose was to receive payments to which he knew he was not entitled. All of these claims were made dishonestly. Moreover, he had not, in the face of overwhelming evidence from credible witnesses called on behalf of the Council, either acknowledged his wrongdoing or expressed any regret for it. In dismissing L's appeal, Elias J said that, in the context of decisions of professional disciplinary bodies, at least given the particular procedures that they employ, it will often be unnecessary to do little, if anything, more than to make the specific findings of fact with respect to each of the specific charges made and then to summarize why it is considered that these amount to gross professional misconduct. It will not generally be necessary for the panel to identify why, in reaching its findings of fact, it is necessary to accept some evidence and to reject other evidence. In the instant case, it was obvious from the way in which the case was put and the conclusion of the panel that it rejected L's contention that he did not intend to mislead. The finding of the panel demonstrates that it did not accept that there was an innocent explanation. It was rejecting L's evidence. As to the undertaking of work, the Committee simply did not believe the appellant when he said that he had done the work that he had claimed to have done. As to the money claimed, this was, in essence, the kind of question that a jury in a criminal trial will have to determine. In essence, did the Committee believe that L was telling the truth? Was there, or may there have been, an explanation for putting in the particular claim? The determination of the Committee made it plain that it did not accept that it was or may be. It was a matter going to credibility.

### *Threlfall v. General Optical Council* [2004] EWHC 2683 (Admin)

**54.08** Alternative scenarios—no finding or reasons given

In allowing T's appeal and quashing the finding of serious professional misconduct made by the Disciplinary Committee, Stanley Burnton J said, at [68]–[70], that the difficulty in this case was that the reasons given by the Disciplinary Committee did not permit the Court to determine which, if either, of the alternative scenarios the Committee had found to have occurred. It was crucial to the gravity of the misconduct whether T made the referral of the patient at her own instigation, rather than at the patient's insistence, and whether she informed the doctor that the patient might have a retinal detachment. It was unfortunate that neither of these issues was addressed in the particulars of the charge. They should have been. Although not expressly part of the misconduct charged, they were extensively canvassed at the hearing. Had they been addressed in the particulars of the charge, the Disciplinary Committee would have had to make express findings on them. The Court had to interpret the Committee's reasons in the way that was most favourable to T.

***Ryell v. Health Professions Council*** **[2005] EWHC 2797 (Admin)**

**54.09** Facts found not identified—general reference to Standards

R was a registered paramedic employed by the London Ambulance Service. Previously, he had been in the Army Reserve, employed as a battlefield combat medical instructor and a unit first aid instructor. Following the investigation of a complaint relating to an incident on 24 March 2002, the Ambulance Service withdrew R's authority to act as a paramedic for a minimum of one year. Subsequently, R appeared before the Conduct and Competence Committee (CCC) of the Health Professions Council (HPC), facing an allegation that his fitness to practise as a registered health professional was impaired by reason of his lack of competence and/or misconduct whilst in the employ of the Ambulance Service. In allowing R's appeal and remitting the matter for rehearing, Bennett J said, at [56], that the Committee did not, either in its decision or in the supplemental information given (orally) at the request of R's counsel, spell out, in relation to the incidents 'found proved', what facts it found and why they amounted to misconduct and/or lack of competence. Nor was it sufficient to refer to the Standards of Conduct, Performance and Ethics without spelling out which part of the Standard in question was breached by the appellant. The paucity of the Committee's information did not meet the minimum requirements spoken of in the judgment of the Divisional Court (Lord Bingham CJ and Kay J) in *Pullum v. Crown Prosecution Service* [2000] COD 206, at [20]:

> ... [T]he minimum which the appellant was entitled to expect was a clear statement as to what evidence the court had accepted, a clear statement that it had, as described in the case stated, based itself specifically on [the evidence of the principal witness against him] of what had occurred at the assault stage, that no question of self defence or accident arose, and that any evidence of provocation was immaterial. Had the court announced its decisions in approximately the terms of the case stated the appellant would have had no possible grounds of complaint.

***Phipps v. General Medical Council*** **[2006] EWCA Civ 397**

**54.10** Adequacy of reasons—general principles

P, a consultant surgeon, was found guilty of serious professional misconduct and suspended from practice for a period of twelve months. The essence of the charge that the PCC found established against P was that he had applied for and obtained retrospective accreditation to enable him to qualify as a consultant surgeon in the National Health Service (NHS) by misrepresenting the length and nature of a number of posts that he had held, thereby obtaining accreditation by illegitimate means, and in some respects had acted dishonestly. His appeal based in part on inadequate reasons was dismissed by the judge and the Court of Appeal. Wall LJ said that there was considerable force in the appellant's submission that there is no reason why doctors sitting in judgment on their peers should be exempt from the general rules that apply to all other tribunals. Plainly, the need to give reasons for findings of fact will vary from case to case, and will depend on the subject matter under consideration. There may be cases in which such reasons are unnecessary because they emerge clearly from the findings; there may be cases in which the expression of such reasons is essential. The test in every case was most succinctly expressed in the judgment of the Court of Appeal in *English v. Emery Reimbold & Strick* [2002] 1 WLR 2409. The decision of the Court in that case was primarily addressed to the professional judiciary. However, it both contains a summary of European jurisprudence and reaches conclusions that are applicable to any tribunal charged with the duty to reach a judicial or quasi-judicial conclusion. In *English v. Emery Reimbold*, the Court said:

> [16] We would put the matter at its simplest by saying that justice will not be done if it is not apparent to the parties why one has won and the other has lost.

> [17] As to the adequacy of reasons, as has been said many times, this depends on the nature of the case (see, for example, *Flannery's* case [2000] 1 WLR 377 at 382). In *Eagil Trust Co Ltd v. Pigott-Brown* [1985] 3 All ER 119 at 122, Griffiths LJ stated that there was no duty on a judge, in giving his reasons, to deal with every argument presented by counsel in support of his case: "... a judge should give his reasons in sufficient detail to show the Court of Appeal the principles on which he has acted and the reasons that have led him to his decision. They need not be elaborate. I cannot stress too strongly that there is no duty on a judge, in giving his reasons, to deal with every argument presented by counsel in support of his case. It is sufficient if what he says shows the parties, and if need be, the Court of Appeal the basis on which he has acted..." In our judgment, the observations of Griffiths LJ apply to judgments of all descriptions.
>
> [19] It follows that, if the appellate process is to work satisfactorily, the judgment must enable the appellate court to understand why the judge reached his decision. This does not mean that every factor which weighed with the judge in his appraisal of the evidence has to be identified and explained. But the issues the resolution of which were vital to the judge's conclusion should be identified and the manner in which he resolved them explained. It is not possible to provide a template for this process. It need not involve a lengthy judgment. It does require the judge to identify and record those matters which were critical to his decision. If the critical issue was one of fact, it may be enough to say that one witness was preferred to another because the one manifestly had a clearer recollection of the material facts or the other gave answers which demonstrated that his recollection could not be relied upon.

Wall LJ's provisional view was that the decision of the Privy Council in *Gupta v. General Medical Council* [2001] UKPC 61; [2002] 1 WLR 1691, [14], identifies an approach that reflects current norms of judicial behaviour:

> 85. ...In every case, every Tribunal (including the PCC of the GMC) needs to ask itself the elementary questions: is what we have decided clear? Have we explained our decision and how we have reached it in such a way that the parties before us can understand clearly why they have won or why they have lost?
>
> 86. If, in asking itself those questions the PCC comes to the conclusion that in answering them it needs to explain the reasons for a particular finding or findings of fact that... is what it should do. Very grave outcomes are at stake. Respondents to proceedings before the PCC of the GMC are liable to be found guilty of serious professional misconduct and struck off the register. They are entitled to know in clear terms why such findings have been made.
>
> 87. In the instant case, the PCC's reasons are clear and cogently expressed....

### ***Watson v. General Medical Council* [2006] EWHC 18 (Admin)**

**54.11** Deficient professional performance—assessment report—panel giving reasoned decision

The fitness to practise panel of the GMC decided that the appellant's professional performance had been seriously deficient. It imposed conditions on his registration for two years. The appellant was a medical practitioner, who specialized in psychiatry. The panel judged that his standard of professional performance was seriously deficient in the areas of respect for patients, providing and arranging treatment, relationship with colleagues and team work, and record-keeping. In dismissing his appeal, Collins J said that *Gupta v. General Medical Council* [2002] 1 WLR 1691 was a case involving an allegation of serious professional misconduct and so the relevant factual issues were set out in detail in the charge. In this case, the assessment panel produced a lengthy report that identified the concerns and this, added to the referral letter, gave all necessary information. It was clear that the panel rejected the appellant's evidence where it conflicted with that given by others or was not in accordance with his own contemporaneous notes. However, good practice does require that reasons should normally be given for decisions. They need not be at all lengthy and, where credibility is in issue, it will usually not be necessary to do more than indicate that the evidence of particular witnesses is accepted. When evidence has been given on particular matters and especially when the appellant has been cross-examined about them, it may be unnecessary for the panel to do more than indicate its conclusions if it is apparent in the transcript why the particular decision has been reached. A failure to give reasons is not necessarily fatal to a decision, but it may enable an argument that a particular matter has not been properly taken into account to prevail. In this case, the panel did give a reasoned decision. It was brief, but that was not a criticism.

### *Ogango v. Nursing and Midwifery Council* [2008] EWHC 3115 (Admin)

**54.12** Agency staff nurse—isolated incident—full expression of regret and apology—mitigation—need for Committee to address arguments on sanction advanced by registrant

O, a grade D agency staff nurse, was found guilty of three charges relating to the care of a patient on the night of 25/26 June 2003 whilst working at Watford General Hospital. The allegations were that O failed to take appropriate action to inform the nurse in charge of the shift of the patient's condition, failed to monitor the patient regularly during the first part of her shift, and failed during the remainder of the shift to monitor the patient adequately, given that she had low oxygen saturation measured previously and O had administered morphine, which had depressed the patient's respiratory function. The CCC of the Nursing and Midwifery Council (NMC) decided to strike O's name from the register on the basis that, according to the Committee, the misconduct was too serious, and represented a failure to undertake basic nursing procedures and practices. O attended the hearing and called evidence. On her behalf, it was submitted that: there was no evidence that the behaviour would have caused harm to the patient—quite the reverse; O had admitted the facts alleged when first approached; this was an isolated incident; O had made a full expression of regret and apology; she had a good previous history, having qualified as a nurse in 2000; there had been no repetition of the behaviour since the incident; and she produced a number of references and testimonials, and three witnesses spoke on her behalf. On O's appeal on sanction, Cranston J said that one proposition that was central to the resolution of the present appeal was that reasons given by a tribunal have to be intelligible and adequate. In particular, the reasons have to meet the substance of the arguments that have been advanced on behalf of the person charged. As *De Smith's Judicial Review* (6th edn, London: Sweet & Maxwell, 2007), p 424, puts it: '[R]easons must show that the decision maker successfully came to grips with the main contentions advanced by the parties, and must tell the parties in broad terms why they lost, or as the case may be, won.' In the instant case, the Court was concerned with regard to the mitigation evidence. It was crucial for the Committee to demonstrate what it had addressed and rejected. The Committee fell short in showing that it had come to grips with the contentions that had been advanced on behalf of the appellant; it did not tell the appellant why she had lost in relation to the crucial aspect of her case on sanction, in particular the evidence of the three witnesses. The family of the patient had attended the Committee and had also attended Court on the appeal. They had suffered a great loss. They behaved with dignity and presented themselves in a commendable manner. There is no doubt that O fell very far short, as she herself conceded, on that night of 25 June 2003. She did not properly monitor the patient; it is not possible to know what would have happened had she done so. She was found by the Committee to be wanting. It was not for the Court to make any assessment of whether striking off was the appropriate sanction. As a matter of law, however, the Committee did not properly explain that outcome in its reasons. It did not show that it had grappled with the key submissions made on behalf of the appellant. The Court therefore found that the matter should be remitted to a fresh committee to consider sanction.

### *R (Kaftan) v. General Medical Council* [2009] EWHC 3585 (Admin)

**54.13**
Extent of reasons—sufficient to explain why one side won and other lost

The allegations made against the appellant, K, before the panel involved aggressive or violent behaviour towards colleagues at the accident and emergency (A&E) department of Huddersfield Royal Infirmary, where the appellant had worked as a clinical assistant and staff-grade doctor for some years. The allegations particularly concerned two incidents in respect of which the panel made separate determinations on facts, on impairment, and on sanction. In dismissing the criticism of the panel's factual findings, Hickinbottom J said:

> 28. Insofar as it was suggested by (counsel for the appellant) that the reasons given by the panel on this issue were inadequate, whilst professional bodies are under a duty to give reasons, that duty does not require them to give a judgment that might be expected of a court of law. The parties must simply be able to understand why one has won and the other lost on a particular issue: *English v. Emery Reimbold and Strick* [2002] 1 WLR 2409 at page 2417; and *Phipps v. General Medical Council* [2006] EWCA Civ 397 at [85]. That does not generally require the panel to identify "why in reaching its findings of fact it ought to accept some evidence and to reject other evidence" (*R (Luthra) v. General Medical Council* [2006] EWHC Admin 458 at [22], per Elias J (as he then was)). *Luthra* predated *Phipps*, but it remains good as a general proposition: subject to the caveat that it may be necessary in a particular case to elaborate to ensure a party understands why he has lost the case and hence to ensure procedural fairness to that party.
>
> 29. In this case, there can be no proper criticism of the reasons given by the panel in respect of the claim by [K] that Dr M assaulted him and his head was banged during that assault. The panel made clear why, on this issue, the appellant lost. They accepted the account of Dr M and rejected the appellant's account. On the evidence before them, they were entitled to come to that conclusion and were not required to give more elaborate reasons for their finding than those they gave.

### *Yeong v. General Medical Council* [2009] EWHC 1923 (Admin)

**54.14**
Expert evidence—reasons for rejecting or accepting expert evidence—assessment of evidence by panel

Y faced charges that included improper conduct bringing disrepute to the medical profession by engaging in a sexual relationship with a patient. Y pleaded guilty to the charges. He obtained an expert report from an experienced psychiatrist, Dr K. Dr K's assessment was that Y did not have a psychological disposition to engage in sexual relationships with patients and that the likelihood of recurrence was extremely low, and this was reflected in the literature. Dr K was of the view that, in a moment of folly, Y had acted impulsively, against his better judgment and character. Sales J dismissed Y's appeal against the fitness to practise panel's sanction of erasure. One ground of appeal was that the panel gave inadequate reasons to explain why it did not accept the expert opinion of Dr K. In rejecting this ground of appeal, Sales J said that, in the light of the standards laid down in *English v. Emery Reimbold & Strick* [2002] EWCA Civ 605 and *Phipps v. General Medical Council* [2006] EWCA Civ 397, it was incumbent on the panel to refer explicitly to the opinion of Dr K and, in that context, to give its reasons why it did not accept or propose to act on his opinion. Counsel for the GMC accepted that the reasoning of the panel in this regard could be criticized as deficient. However, in the Court's view, the submission of counsel for the GMC that the panel was entitled to draw on its own experience and judgment, and was not obliged to accept the assessment of Dr K, was made out. The panel had given reasons for its own view where that view differed from Dr K's. The panel was well entitled to draw upon its own experience and judgment in forming a view of whether Y presented a present risk to his patients. Dr K's evidence was to the effect that Y did not suffer any psychological disorder that underlay his misconduct. In the light of that assessment, Dr K's expression of opinion as to the risk posed by Y carried little weight attributable to any special expertise on the part of Dr K. The question of the possibility of a recurrence of such misconduct by Y was a matter of the ordinary assessment of the likely human behaviour, in relation to which a psychiatrist's expertise confers no special privileged insight. The reasons given by the panel in its impairment decision regarding the risk posed by Y were sufficient to indicate to him in the circumstances of his case why the panel considered that he did still pose a risk of future misconduct. His practice involved dealing with women who might be in a vulnerable emotional state. His 'personal circumstances' was a clear reference to the fact that Y was living in the United Kingdom apart from his wife and children. The panel was also entitled to take into account (and properly referred to this factor in its reasons) that Y was of such a character that he might find that a patient had an emotional response to him and so be drawn into a situation in which he might be tempted to engage in misconduct.

### *Beresford v. Solicitors Regulation Authority* [2009] EWHC 3155 (Admin)

**54.15** Tribunal's decision recording each party's case—whether Tribunal's decision clear—forensic textual criticism of decision rejected

In this case, the Divisional Court (May P, Silber, and David Clarke JJ) said, in relation to reasons:

> 43. Speaking generally, there is, in our view, no persuasive case that the Tribunal failed properly to consider factual matters which are relied on in this appeal. In the course of its lengthy findings, the Tribunal set out extensively each party's opening and closing submissions and the oral evidence. The Tribunal then made findings of fact in relation to each allegation which necessarily related back to the evidence and the submissions which they had set out.
>
> 44. [Counsel for the SRA] referred us to the Privy Council case of *Gupta v. General Medical Council* [2002] 1 WLR 1691 for the proposition that there was no general duty on the Professional Conduct Committee of the General Medical Council to give reasons for its decisions on matters of fact, especially on questions depending on the credibility of witnesses. *Gupta* was considered at some length in the judgment of Wall LJ in *Phipps v. General Medical Council* [2006] EWCA Civ 397 in the light of *English v. Emery Reimbold Strick* [2002] 1 WLR 2409. Wall LJ expressed in paragraph 85 a provisional view that paragraph 14 of *Gupta* identifies an approach which reflects current norms of judicial behaviour. In every case, every Tribunal needs to ask itself whether what they have decided is clear; and whether they have explained their decision and how they have reached it in such a way that the parties can understand why they have won and why they have lost. In our judgment, the findings of the Tribunal in the present case achieve that test. Such particular points as [Counsel for the appellant] makes are more in the nature of forensic textual criticism than a substantial case that the reasons for the findings are unclear.

### *Keane v. Law Society* [2009] EWHC 783 (Admin)

**54.16** Inadequate professional service—adjudicator's decision—supplemental reasons given subsequently—approach of court

K, a solicitor, challenged in judicial review proceedings the reasons given by an adjudicator of the Law Society concerning a series of decisions involving complaints of inadequate professional service by K. It was claimed that the adjudicator gave insufficient reasons as to why he thought he should deal with the complaint against K despite its delay. Following the grant of permission to apply for judicial review, a witness statement by the adjudicator was served in response to the proceedings. In his witness statement, the adjudicator explained his reasons in more detail. In dismissing K's claim for judicial review, Nicol J said that a court must be cautious in relying on a decision-maker's supplemental reasons that are given for the first time in response to a judicial review challenge. The reasons are obvious: there is a risk of *ex post facto* rationalization or justification even where the evidence is given in good faith; see, for example, *R v. Westminster City Council, ex parte Ermakov* [1996] 2 All ER 302, CA, and *R (Nash) v. Chelsea College of Art and Design* [2001] EWHC Admin 538. In the present case, though, it was right to take account of the evidence, for the following reasons. The adjudicator's evidence did not contradict anything in his decision letter. He is a single decision-maker (the problems of post-decision evidence are more acute where the decision-maker is a group of people). The evidence also sought to respond to an issue that was raised for the first time in the judicial review proceedings. Also, there was no statutory obligation to give reasons, but the adjudicator was obliged to say why he was allowing the matter to proceed. In his decision, the adjudicator did give his reasons. He addressed the objections raised by K's solicitors and the reasons were legally adequate.

### *Southall v. General Medical Council* [2010] EWCA Civ 407

**54.17**

Child abuse—complaint by mother against expert—interview—whether expert accused mother of murdering child—panel accepting mother's evidence—need for panel to explain why rejecting appellant's account

In this case, the Court of Appeal (Waller, Dyson, and Leveson LJJ) said that, in straightforward cases, setting out the facts to be proved and finding them proved or not proved would generally be sufficient both to demonstrate to the parties why they had won or lost and to explain to any appellant tribunal the facts found. In most cases, particularly those concerned with comparatively simple conflicts of factual evidence, it would be obvious whose evidence had been rejected and why. However, when the case was not straightforward and therefore was exceptional, the position was different: see *Gupta (Prabha) v. General Medical Council* [2002] 1 WLR 1691. The instant case was far more complex than a simple issue of fact. The appellant, Dr S, was a well-known consultant paediatrician and an expert on child abuse. The appellant was instructed as an expert on behalf of a local authority to give a medical opinion concerning a child's death. Following an interview with the mother, M, the mother complained to the GMC that the appellant had accused her of murdering her son. That allegation, amongst others, was considered by a fitness to practise panel, which found that the appellant had made that accusation. The appellant's defence was that the mother had thought that she had been accused, whereas he had merely investigated her account of her son's death, and that a social worker, S, who was present at the interview supported the appellant's account. The Court of Appeal, in allowing the appellant's appeal, said that the panel's reasons in preferring the mother's account of the interview to the appellant's were simply inadequate and did not start to do justice to the case. Although entitled to conclude that the mother was an honest and credible witness, the panel did not specifically deal with the suggestion that she perceived herself to be accused, which would be entirely understandable in the circumstances and could explain why she reported the interview in the way she did. The appellant was entitled to know why that possibility was discounted by the panel, and if it disbelieved him, he was entitled to know why. The panel should also have given some reason for its discounting the evidence of the social worker. After reviewing *Selvanathan v. General Medical Council* [2000] 59 BMLR 96, *Gupta v. General Medical Council* [2002] 1 WLR 1691, *Phipps v. General Medical Council* [2006] EWCA Civ 397 and *English v. Emery Reimbold & Strick* [2002] 1 WLR 2409, Leveson LJ said:

> 55. For my part, I have no difficulty in concluding that, in straightforward cases, setting out the facts to be proved (as is the present practice of the GMC) and finding them proved or not proved will generally be sufficient both to demonstrate to the parties why they won or lost and to explain to any appellant tribunal the facts found. In most cases, particularly those concerned with comparatively simple conflicts of factual evidence, it will be obvious whose evidence has been rejected and why. In that regard, I echo and respectfully endorse the observations of Sir Mark Potter [in *Phipps*].

56. When, however, the case is not straightforward and can properly be described as exceptional, the position is, and will be different. Thus, although it is said that this case is no more than a simple issue of fact (namely, did [Dr S] use the words set out in the charge?), the true picture is far more complex. First, underlying the case for [Dr S] was the acceptance that [M] might perfectly justifiably have perceived herself as accused of murder with the result that the analysis of contemporaneous material some eight years later is of real importance: that the evidence which touched upon this conversation took over five days is testament to that complexity. Furthermore it cannot be said that the contemporaneous material was all one way: [the] note [of a child psychiatrist to whom the family had been referred for counselling] (and, indeed, her evidence) supported the case that it was (or at least could have been) [M]'s perception alone. [The] note [of the social worker who accompanied Dr S at the interview] (accepted by [M] as 100% accurate so far as it went) did not support the accusation and her evidence was that if those words had been said, she would have recorded them. I am not suggesting that a lengthy judgment was required but, in the circumstances of this case, a few sentences dealing with the salient issues was essential: this was an exceptional case and, I have no doubt, perceived to be so by the GMC, [Dr S] and the panel.

57. Perhaps because of the nature of the case, the panel did, of course, provide a few sentences of reasons but, in my judgment, they were simply inadequate and did not start to do justice to the case. On the specific findings of fact, although entitled to conclude that [M] was a clear, honest and credible witness, they do not specifically deal with the suggestion that she perceived herself to have been accused and so represented herself as having been accused which, when upset (as she described) would be entirely understandable and could itself explain why (if it be the case) that she so reported the interview over the days that followed. Let me make it clear that I am not making such a finding but merely concluding that [Dr S] was entitled to know why that possibility was discounted.

58. In relation to [the child psychiatrist], said to support [M] because of the comment "they didn't do a toxicology quite possibly you drugged him first", the panel totally ignored the thrust of her evidence, recounted above, which was entirely supportive of the perception theory and did not deal with how that evidence impacted on the words she wrote or, in relation to her and [M's solicitor] (whose evidence also included at least one conditional phrase) how [M]'s perception might have been reflected in what she said in the days that followed. As for [Dr S]'s report, the categorical denial would be no less categorical if [M] perceived herself as being accused as if she was accused.

[...]

63. In summary, I conclude that although superficially straightforward, this case was exceptional within the language of *Gupta* and required the panel to provide reasons. Contrary to the view expressed by Blake J, I do not consider the reasons which it provided were sufficient to explain why the panel rejected the defence that [M] might have assumed that she was accused of murder without her having been so accused.

### *Chegwyn v. Ethical Standards Officer of the Standards Board for England* [2010] EWHC 471 (Admin); [2011] PTSR 165

**54.18** Councillor—allegation of deliberate abuse of position and deliberate breach of Code—state of mind—adequacy of reasons by panel

C appealed, pursuant to section 78B(4) of the Local Government Act 2000, against a decision of a case tribunal of the Adjudication Panel for England, which decided to disqualify him from being a councillor or becoming a councillor in any authority for a period of two years. C accepted that he had breached the code of conduct or one of the relevant paragraphs of the Code of Conduct for Councillors. C had failed to declare his personal interest in a matter when speaking and voting at a council meeting. Collins J, in allowing C's appeal and substituting a penalty of suspension for a period of two months, said that there was a slight degree of artificiality because, although C did not formally declare his interest, it must have been apparent to everyone that he had a personal interest in the subject of the motion. In its determination, the tribunal said that C was guilty of a blatant and deliberate disregard for the code of conduct, and that he had deliberately sought to misuse his position and had deliberately failed to abide by the code. The learned judge said that the conclusion of the tribunal that C had deliberately sought to abuse his position and deliberately failed to abide by the code was crucial to the tribunal's findings. Nowhere did it seek to explain why and on what basis it had rejected C's case that he did not stand to gain or lose personally. No one had suggested that he did not have a genuine concern, for the local businesses in particular, and C said that it was this that largely motivated him at the time. Thus the crucial issue for the panel was what was C's state of mind. Effectively, if—but only if—the tribunal were prepared to reject his account and his explanation would the serious view taken have been justified. If the tribunal were minded to do so, it ought to have explained why it was taking that view.

***Shepherd v. Governor of HMP Whatton*** **[2010] EWHC 2474 (Admin)**

**54.19** Breach of Prison Rules—governor's decision—reasons

His Honour Judge Raynor QC allowed the claimant's application for judicial review on the basis that the defendant had not given adequate reasons for its decision to find the claimant guilty of breaching the prison rules that he was not to have contact with a child without written notification that it was approved. The claimant was serving a life sentence for rape and applied for child contact by telephone with his daughter. The charge was that he had contravened prison rules not to have contact without prior written approval. The Administrative Court held that the governor's decision was not adequately reasoned, because he had not stated what the issue was, nor did he state a clear finding that he, having appreciated the issue, had found against the claimant on it.

***Brennan v. Health Professions Council*** **[2011] EWHC 41 (Admin)**

**54.20** Sanction—appeal—reasons to strike off rather than suspend—needs to address defendant's case

In this case, the appellant, a physiotherapist, appealed against the decision of the HPC's CCC that he should be struck off its register. He did not appeal against the findings of misconduct or against the conclusion that what he did impaired his fitness to practise. The appellant was the head physiotherapist at Harlequins Rugby Football Club (RFC), and appeared before the respondent following a disciplinary investigation instigated by the European Rugby Cup (ERC) into a fake blood injury and cheating during a rugby match. The appellant admitted that he had participated in the fabrication of the blood injury and that he gave false evidence at the initial investigation. The appeal committee of the ERC had banned the appellant from participating in all rugby activities for two years. The Administrative Court (Ouseley J), in quashing the decision of the respondent's CCC on sanction, found that the Committee's decision to strike off the appellant had been legally inadequate, had failed to deal with the issues raised by the appellant, and had not dealt adequately with the reasons to strike him off rather than to impose an alternative sanction. The instant case was not one of public safety, the appellant was an excellent physiotherapist, and the dishonesty was not towards a patient. Ouseley J said, at [45]–[47] of the judgment, that where the purpose of sanction is to deal with issues other than the primary one of maintaining public safety and is instead to provide deterrence to others, to maintain confidence in the profession's reputation and standards and in its regulatory process, the reasoning is particularly important in showing that the sanction is proportionate to the misconduct and for the individual. In the instant case, the Committee had not dealt adequately with the case for the appellant as to why he should not be struck off. Its reasoning did not enable the informed reader to know what view it took of the important planks in his case. The Court accepted that the Committee went through the various sanctions, noting the comments in the respondent's indicative sanctions policy about them. The general language of the various sanctions put this case in the area in which strike-off had to be considered. But the factors that the submissions for the appellant addressed were also very relevant to those sanctions and to how far up the scale he should now be seen. The sanctions could not be properly addressed without consideration of the factors to which the appellant's evidence was addressed. The Court consequently quashed the decision and remitted it to the Committee, with a direction that it reach a reasoned decision on sanction that addressed the issues to which the judgment referred.

### *R (Alhy) v. General Medical Council* [2011] EWHC 2277 (Admin)

**54.21** Doctor—proceedings in France—decision of French body—erasure by GMC—whether GMC required to explain different sanction

Dr A challenged the decision of 21 January 2011, made by the GMC's fitness to practise panel, in which he was found guilty of misconduct and the panel ordered his name to be erased from the register. Dr A was also registered with the French *Ordre de Médicins*, the equivalent to the GMC in France. Dr A was convicted of two offences in relation to the deaths of patients in France. The first conviction was of non-assistance to a person in danger. This was based on Dr A's decision to delay surgery on the patient. It is not an offence known to English law. In relation to the second patient, Dr A was convicted of gross negligence manslaughter for failing to monitor the patient, which delayed his diagnosis of peritonitis. He was sentenced to twelve months' imprisonment, suspended, and was banned from practising as a surgeon for three years. He was not prevented from practising as a general physician. Dr A appealed the GMC's sanction of erasure on the grounds that it was more onerous than that imposed in France by the French courts and the *Ordre de Médicins*. He submitted that to impose a significantly different and more onerous sanction from that imposed by the *Ordre de Médicins* was contrary to Article 56(2) of European Directive 2005/36/EC on the Recognition of Professional Qualifications, or in the alternative, that there was a duty of fairness on the part of the fitness to practise panel to give full and proper reasons as to why it departed from the decision or penalty imposed by the French courts and the *Ordre de Médicins*. Her Honour Judge Belcher (sitting as a High Court judge) was not persuaded by that submission. She said that it seemed to her it carried the danger of requiring there to be an investigation of the penalty decision-making process in another jurisdiction, something that she was not satisfied was properly part of, or ought properly to be part of, the remit of the fitness to practise panel. Its job is to assess matters in accordance with the GMC's statutory purpose and good medical practice. Its purpose in holding hearings is to establish what the facts are in a particular case, to consider whether the facts support a finding of misconduct and impairment to practise, and if so, to determine what the appropriate sanction should be. The panel plainly must be informed by the appropriate facts. Those may, of course, as they did in this case, include the facts of the convictions or the underlying facts in the French jurisdiction. The learned judge did not accept a submission that it was necessary for the GMC's fitness to practise panel to specify reasons for imposing a decision or sanction different from that imposed by a body or bodies in another European Union member state. The issue was whether the reasons given in this state were such that the doctor or others reading the decision would understand what the decision against the doctor was and why the fitness to practise panel arrived at the decision that it did. To require the panel to give detailed reasons for departing from the decision of another body would require investigative steps far beyond its remit. The learned judge went on to consider insight, and said that there was a significant difference between maintaining innocence during the currency of a criminal process and continuing to maintain innocence after the criminal process had concluded.

***Hazelhurst and ors v. Solicitors Regulation Authority* [2011] EWHC 462 (Admin)**

**54.22** Solicitor—sanction—adequate reasons—consideration of defendant's explanation

In quashing the financial penalties imposed by the Solicitors Disciplinary Tribunal (SDT) on the four appellant solicitors, partners in the firm of Fanshaw, Porter & Hazelhurst, and substituting a reprimand for each defendant arising out of theft of monies from the firm by a former employee M, Nicola Davies J said that, in her view, the SDT failed in its written reasons to adequately address the submissions of the appellants as to why it was that the thefts went undiscovered for a period of three years. In its written determination, the SDT did not specifically address the appellant's submissions that independent practitioners had seen and assessed the firm's books and/or M's files, and found nothing untoward, and that M was a trusted employee. Breaches of the Accounts Rules are breaches of strict liability offences. The fact that these offences were allowed to take place raises a strong inference of lack of supervision. It was precisely to deal with that inference that the appellants raised the specific points, which went to the core of their case on breaches—particularly, the failure to supervise. A statement in the SDT's written reasons that all of the evidence and submissions had been considered was plainly inadequate to deal with the specific submissions properly made and based upon unchallenged evidence. In failing to address the appellant's submissions, the SDT did not provide adequate reasons for its findings as to breaches of the Rules and, specifically, the lack of supervision.

***Yaacoub v. General Medical Council* [2012] EWHC 2779 (Admin)**

**54.23**

See paragraph 27.14.

***Duthie v. Nursing and Midwifery Council* [2012] EWHC 3021 (Admin)**

**54.24**

See paragraph 27.15.

***Quinn v. Bar Standards Board*, 25 February 2013 (unreported)**

**54.25** Barrister—allegation of misleading court for adjournment—attendance notes and other relevant documents—need for assessment by tribunal

Q, a barrister employed by a firm of solicitors, faced three charges of professional misconduct arising out of an allegation that she misled Manchester County Court and told the court in family proceedings that the other party, the father, had agreed to an adjournment, when he had not, and told the Children and Family Court Advisory and Support Service (CAFCASS) adviser that the hearing had been adjourned, when it had not. In quashing the conviction, the Visitors (Sir Wyn Williams, chairman) said that Q's conviction on the three charges was explicable only on the basis that the Disciplinary Tribunal of the Bar Standards Board accepted the father's account of a telephone call and rejected the account given by Q. There was no doubt that the Tribunal was under a duty to provide reasons for its decision, given the terms of regulation 18 of the Disciplinary Tribunals Regulations 2009 ('at the conclusion of the hearing, the finding of the Disciplinary Tribunal on each charge, together with its reasons, shall be set down in writing and signed by the chairman and all members of the Tribunal'). This was not a case in which the Tribunal was essentially making a choice between two principal witnesses who were giving evidence, unaided, about their recollection of events that had taken place some years before. The appellant relied upon a series of attendance notes, which, she claimed, had been created either during the course of the afternoon before the hearing or very shortly thereafter. Each of the attendance notes purported to record what was said in the conversations that had taken place between the appellant and the father, and the appellant and one of the administrative staff at the county court. On any view, these documents were potentially significant and their true evidential status needed careful assessment by the Tribunal. The appellant also relied upon an email sent to the court. As with all other documents created by her contemporaneously, the evidential significance of the email needed careful assessment. It was incumbent upon the Tribunal to explain its reasoning process in respect of the documentation. It did not do so and it thereby fell into error. In a case of this type, with such serious potential consequences for the appellant, it was not sufficient for the Tribunal to announce verdicts without explaining in some detail the reasoning process that underpinned them.

### *Cronin v. Greyhound Board of Great Britain Ltd* [2013] EWCA Civ 668

**54.26** Greyhound racing—Disciplinary Committee—lack of reasons by Disciplinary Committee—no bar to appeal

In giving the judgment of the Court of Appeal, Maurice Kay LJ (with whom Richards and Pitchford LJJ agreed) said that the Greyhound Board of Great Britain, exercising regulatory functions in relation to greyhound racing, is not a statutory body. It is a private sector regulator constructed on contractual foundations, but which, when exercising its disciplinary powers, is subject to the requirements of fairness, whether or not they are expressed in the rules that it has adopted. The appellant, C, was a greyhound trainer licensed under the Board's Rules of Racing. In January 2010, the Board received a complaint that he had mistreated a litter of greyhound puppies. An investigation eventually resulted in proceedings under the Rules before the Board's Disciplinary Committee. Following a hearing in December 2010, C was found guilty of breaches of the Rules. He was given a severe reprimand, fined £750, and ordered to pay £10,000 towards the Board's costs. Although the Rules provide for an appeal to an appeal board, C did not pursue one within the prescribed time or at all. Nor did he pay the fine or the costs. His licence expired by effluxion of time on 31 December 2010, without C making any application to renew it. On 27 July 2011, the Board served a statutory demand in respect of the unpaid fine and costs. C did not pay them; instead, on 30 September 2011, he commenced proceedings in the county court as a litigant in person. In his claim form, C stated that the Greyhound Board of Great Britain had infringed his basic human rights on the grounds, inter alia, of a failure to give reasons for its findings. On 12 March 2012, following a hearing at which both parties were represented by counsel, the claim was struck out. In dismissing C's appeal, Maurice Kay LJ observed that the disciplinary inquiry was initiated by the Board pursuant to rule 161. Rule 161 does not include a requirement to give reasons for the ultimate decision of the Disciplinary Committee. By rule 163, there is a right of appeal to an appeal board, which may hear oral evidence and shall, upon receipt of a request in writing give written reasons for its decision. Further, the appeal board could have given a direction or instruction that the Committee provide reasons if considered necessary for the proper conduct of the appeal. The learned judge said that he was satisfied that the absolutist stance of the Board that the Disciplinary Committee was under no duty to give reasons for its decision, in reliance on *McInnes v. Onslow-Fane* [1978] 1 WLR 1520, was arguably wrong and that, at least at some stage in the procedure, a duty to give reasons arises. Even if an affected person has no wish to challenge an adverse decision, he or she may be entitled, as a matter of law, to such reasons. However, that is not the end of the matter. In the present case, C complained that he was in no position to pursue an appeal to the appeal board because he did not know the reasons for the decision of the Disciplinary Committee. Without having been provided with the reasons, he claimed that he was unable to formulate grounds of appeal in accordance with the Rules. Thus, it was said, he could not be blamed for not having pursued an appeal and the potential availability of reasons following a determination by the appeal board was irrelevant. The learned judge said that he was wholly unpersuaded by this analysis. Even without being provided with reasons by the Disciplinary Committee, C and his advisers must have been well aware that the stark conflicts of evidence had been resolved to his disadvantage. They had also had made available to them DVD recordings of the entire proceedings, which were delivered with the decision letter. That, coupled with their knowledge of the proceedings, would have been sufficient material with which to commence an appeal. Looking at the internal procedures as a whole, it was not arguable that the initial absence of reasons from the Disciplinary Committee points to overall unfairness. Moreover, the learned judge could not escape the conclusion that the 'reasons challenge' to the decision of the Disciplinary Committee was being used as a smokescreen. C, for whatever reason, chose not to pursue an appeal. He simply allowed his licence to lapse at the end of 2010 and it was not until August 2011, when he applied to set aside the statutory demand that had been served on 27 July 2011, that he formulated a case against the decision of the Disciplinary Committee. The appeal was opportunistic.

***Barakat v. General Medical Council* [2013] EWHC 3427 (Admin)**

**54.27**
Adequate reasons—need to look at determination as a whole

The panel found that B had been complicit in the production of a forged medical report and the submission of that report in the course of divorce proceedings from his then wife in Syria. B accepted that the report had been forged, but claimed that it was without his knowledge or consent. The key questions that the panel had to determine were (a) whether, during the course of matrimonial proceedings, the forged report was submitted, (b) whether B allowed the report to be submitted on his behalf, and (c) whether he had altered or, in the alternative, was aware that the report had been altered on his behalf. The panel answered all three questions against the appellant. In dismissing B's appeal, the Court accepted that, in approaching an appeal such as this, it is necessary to look at the determination as a whole rather than to look at certain passages in isolation. It was not necessary to deal with each and every submission made by B in relation to the evidence. The panel's reasons, although short in summary, were sufficient to enable B to know why the decision had gone against him. At [54], the Court considered how the 'reasons' test fits in with CPR Part 52.11. Does the absence of reasons make the decision of the lower tribunal wrong, or does it have to be shown that the absence of reasons renders the decision unjust because it amounts to a serious procedural or other irregularity? On the facts of this case, even if there were some lack of reasoning, it could have been only a limited omission, not such as to cause serious injustice as to justify either allowing the appeal or remitting the case for a rehearing.

***R (H) v. Nursing and Midwifery Council* [2013] EWHC 4258 (Admin)**

**54.28**
Conflict of evidence—need for careful consideration of each element of evidence

H embarked on a sexual relationship with A, whom she subsequently married. The CCC found that the relationship began whilst A was her patient. The Committee said that the only direct witnesses to the charge were A and H; in this regard, the charge hinged on their evidence and whether the Committee preferred one version of events over the other. There was a straightforward conflict between the evidence of A and H, and their accounts were fundamentally irreconcilable. In allowing H's appeal, the Court said, at [14], that, in circumstances under which there was a conflict of evidence of the sort identified by the Committee, the conventional approach to determining issues of fact between the parties involved primarily testing the relevant contentions of the parties by reference to such contemporaneous documentation as is available. Where, as here, the Committee was faced with two conflicting accounts in relation to an issue of such profound importance to the appellant, it was bound to give careful consideration to each element of evidence and then to arrive at a conclusion taking into account all of the evidence in the round. The Committee was wrong simply to reject the evidence as hearsay if, as was suggested by the respondent, it did not accept what the appellant had told her friends; in that case, it was incumbent on the Committee to make clear that it was rejecting the evidence on that basis and why: at [29].

### *Shaw and Turnbull v. Logue* [2014] EWHC 5 (Admin)

**54.29** Complex facts needing sifting—case not dependent on conflict of fact, but on appellants' explanations

The appellants' essential complaint was that the SDT had conspicuously failed to set out what evidence it had accepted or rejected, as well as its key findings of fact on the main issue of knowledge and dishonesty. This was far from being a run-of-the-mill case. First, the evidence was extremely complex and dense, and merited a considerable degree of sifting, weighing, and analysis. Secondly, the case was unusual in as much as it did not depend on the SDT preferring the evidence of witness A over witness B, in circumstances under which the conflict between the two witnesses might have been obvious; it turned almost entirely on what view the SDT took of the appellants' explanations for what they did and with reference to the documentary base. It was necessary to establish what exactly the appellants knew, when and by what means, what significance or salience they attributed at the time to what they knew, and whether their presentation of the evidence in the original Court proceedings was deliberately misleading, either expressly or by implication (because relevant materials were consciously suppressed). In allowing the appellants' appeal, Jay J said that the overall impression given is that the SDT glided too rapidly and too easily between finding the first limb of the *Twinsectra* test satisfied to proof of the second limb: *Twinsectra Ltd v Yardley* [2002] UKHL 12. Overall, the decision fell a considerable way short of being adequately reasoned. The Court was not suggesting that the SDT needed to undertake the sort of lengthy and punctilious exercise to be expected of a judge of the Administrative Court, but considerably more was required of it.

### *O v. Secretary of State for Education and National College for Teaching and Leadership (Interested Parties)* [2014] EWHC 22 (Admin)

**54.30** Teacher—prohibition order—inappropriate relationship with pupils—reasons—general principles

O, a teacher, appealed pursuant to regulation 17 of the Teachers' Disciplinary (England) Regulations 2012 against a prohibition order made by the Secretary of State for Education under section 141B of the Education Act 2002. By the prohibition order, the Secretary of State prohibited O indefinitely from carrying on teaching and thereby also barred her from seeking restoration of her eligibility to teach. Following an oral hearing, a professional conduct panel found O guilty of an inappropriate relationship with a pupil whom she taught, involving conduct of a physical nature and conduct that was sexually motivated, and an inappropriate relationship with another pupil at the school. In dismissing O's appeal, Mr Stephen Morris QC (sitting as a deputy High Court judge) said that, as regards the nature and extent of the duty to give reasons on the part of a judicial decision-maker, the leading cases were *Flannery v. Halifax Estate Agencies Ltd* [2001] 1 WLR 377 and *English v. Emery Reimbold & Strick Ltd* [2002] 1 WLR 2409. Also relevant was *In the matter of F (Children)* [2012] EWCA Civ 828. In the specific context of professional disciplinary bodies, relevant authorities are *Gupta v. General Medical Council* [2002] 1 WLR 1691, [14], *Cheatle v. General Medical Council* [2009] EWHC 645 (Admin), [29]–[31], and *Mubarak v. General Medical Council* [2008] EWHC 2830 (Admin), [9]–[12], along with *Phipps v. General Medical Council* [2006] EWCA Civ 397, [85]–[86] and [106]. The Court also referred to *South Bucks District Council v. Porter (No. 2)* [2004] 1 WLR 1953, [36], dealing with the scope of duty to give reasons in a planning inspector's decision. The Court said, at [59], that the position in general is as follows.

(1) There is a general duty upon a judicial decision maker to give reasons for the decision it has reached. The judge must explain *why* he has reached his decision.
(2) The rationale for the duty to give reasons is twofold: first, to enable the parties, and in particular the losing party, to know why they have won or lost and to allow

the losing party to consider whether to appeal; and secondly, to concentrate the mind of the decision maker.

(3) It is not necessary to deal with every argument nor to explain in great detail every factor in the judge's reasoning. It is sufficient that what the judge says shows the parties, and if need be, an appeal court, the basis on which he has acted: *English v. Emery Reimbold* [at [17]]. It is not necessary to deal with each and every inconsistency or conflict of evidence specifically: see *Re F* [at [41]].

(4) The extent of the duty depends on the subject matter; and no hard and fast rules can be laid down: *Flannery*, . . . at 382C. It will depend on the facts and issues of each case. For example, in a case which turns on competing expert evidence, the judge must enter upon the issues canvassed and explain why he prefers one case over the other.

(5) The adequacy of the reasons should take account of the knowledge, on the part of those to whom it is addressed, of the submissions and evidence before the decision maker: *South Bucks*, [at [26]], and *English v. Emery Reimbold*, [at [89] and [118]].

### *Moore v. Nursing and Midwifery Council* [2013] EWHC 4620 (Admin)

**54.31** Job application—failure to disclose convictions—dishonesty—strike off—adequate reasons

Three charges against the registrant were laid before the panel. The first two were that she had failed to disclose convictions for dishonesty in two separate job applications. The third charge was that she had acted dishonestly in so failing to disclose the convictions. The convictions themselves were for theft in 1988, attempting to obtain property by deception in 1991, and forgery of an instrument other than a prescription in 1999. The registrant gave various reasons before the panel why she had not acted dishonestly, including saying that she had disclosed the convictions to individual members of the Trusts with whom she was applying for jobs and had been told that they did not need to be put on the form. The panel rejected all such explanations and found that the appellant had not told the truth in the case that she presented to them. The result of the hearing was that the appellant was struck off the register of nurses. On appeal, the registrant challenged the reasons given by the panel. Dismissing the appeal, the Court said that the principal issue before the panel was the fact that the convictions had not been disclosed in the job applications. The principal dishonesty that clearly weighed with the panel was the appellant's dishonesty in failing to declare the convictions and not the dishonesty in the original convictions. Accordingly, there was no inadequacy of reasons expressed by the panel, nor that the decision could be properly said to fail to take account of matters that were relevant to the decision that was reached. As to the appropriate sanction, the challenge on appeal was a reasons challenge, not a challenge to the appropriateness of the sanction itself. The panel worked through the lesser sanctions potentially available to it, explaining in each case why they were not considered sufficient to deal with the matters found proved in this case. The conclusion that only striking off was sufficient was expressed shortly. However, it was misleading to refer only to that paragraph which expressly made reference to the earlier reasoning.

### *L v. The Health and Care Professions Council* [2014] EWHC 994 (Admin)

**54.32** Reasons—general principles

In dismissing L's challenge to the panel's decision, Haddon-Cave J said, at [15], that the general principles applicable to the obligation of a professional disciplinary panel to give reasons are as follows:

(1) In general there is no obligation upon the panel to give reasons for its decisions;

(2) In a straightforward case where there is a simple conflict of factual evidence there is no need to set out reasons because it will be obvious from a statement of findings, when read with the nature and the contents of the evidence, why the panel has decided what it has;

(3) But that in a more complex case there is a need to give reasons, which may be short and summary, but should still be adequate so that the losing party may understand why he or she has lost.

([S]ee HHJ Stephen Davies' summary in *Nwogobo v. GMC* [2012] EWHC 2666 (Admin) and Leveson LJ in *Southall v. GMC* [2010] EWCA Civ 407.)

### *Virde v. General Pharmaceutical Council* [2015] EWHC 169 (Admin)

**54.33** Allegation of sexual harassment at pharmacy—conflicting evidence—probability of event occurring

V, a registered pharmacist, appealed against a determination of the Fitness to Practise Committee of the General Pharmaceutical Council. The Committee found proved that, on four different occasions, V had sexually harassed and/or sexually assaulted Ms A at the pharmacy at which they both worked, and determined that the only sufficient and proportionate sanction was removal. In its determination, the Committee was unable to resolve the matter simply on the basis of the evidence of Ms A and V, and the manner in which they gave it. The Committee described difficulties it had in assessing Ms A's credibility, but nevertheless concluded that she was an honest witness doing her best at all times to tell the truth as she believed it to be. V accepted that Ms A was an honest witness, but he contended that she was mistaken. In dismissing V's appeal, Andrews J said, at [36], that the substantive issue at the heart of the appeal was the question of whether V was able to establish that the Committee's decision was wrong:

> Although there is no "heightened standard" of proof in proceedings of this nature, the inherent probability or improbability of an event is itself a matter to be taken into account in weighing the probabilities and deciding whether on balance the event occurred: see the speech of Lord Nicholls in *Re H (Minors) (Sexual Abuse: Standard of Proof)* [1996] AC 563 at 586–7, cited with approval in *Re B (A Child)* [2008] UKHL 35. The more improbable it is that the registrant would have behaved in the manner alleged, the more cogent and credible the evidence needed to satisfy the burden of proving on the balance of probabilities that he did.

The learned judge said that, in its determination, the Committee took as its starting point the fact that the allegations were of such a serious nature as to be inherently improbable. It said that it had subjected the evidence 'to the closest scrutiny', and that was true. The Committee considered the submission that Ms A's delay in complaining about the incidents made it less probable that the events occurred, and rejected it for the reasons that it gave. Essentially, it accepted her explanations. It was entitled to do so. The Committee made specific reference to V's good character, including positive evidence that was adduced about him. The learned judge said that she regarded it as being of considerable importance that the Committee found that it was common ground amongst the witnesses who spoke about the pharmacy that there would be no need, in the course of ordinary work, for V to stand close behind Ms A or to brush against her. The Committee found that, on the evidence, it could not conceive of any innocent actions by V at the pharmacy that might have been misinterpreted as the behaviour alleged in the relevant charges. The Committee rejected the possibility that Ms A was delusional. The Committee was entitled to examine alternative explanations for why an honest witness might have been mistaken. The learned judge said that although there might be some areas of perceived deficiency in the Committee's reasoning, overall the reasons were adequate to tell V why it had found the charges against him proved. Looked at in the round, the Committee's conclusion was that it believed Ms A and disbelieved V. It reached that conclusion having reminded itself that it was inherently improbable that V, a young professional man of previous good character, would sexually assault a young woman in the workplace, having subjected the evidence to the closest scrutiny, and having sought to reconcile the competing accounts, where possible, to V's advantage.

## C. Other Relevant Chapters

54.34

Appeals, Chapter 4
Findings of Fact, Chapter 27
Impairment of Fitness to Practise, Chapter 33
Investigation of Allegations, Chapter 39
Misconduct, Chapter 47
Sanction, Chapter 60

# 55

# RECKLESSNESS

A. Disciplinary Cases 55.01
*Bultitude v. Law Society* [2004] EWCA Civ 1853 55.01
*Holy v. Law Society* [2006] EWHC 1034 (Admin) 55.02
*Keazor v. Law Society* [2009] EWHC 267 (Admin) 55.03
*Harris v. Solicitors Regulation Authority* [2011] EWHC 2173 (Admin) 55.04
*Kumar v. General Medical Council* [2012] EWHC 2688 (Admin) 55.05
*Izuchukwu v. Solicitors Regulation Authority* [2013] EWHC 2106 (Admin) 55.06
*Brett v. Solicitors Regulation Authority* [2014] EWHC 2974 (Admin) 55.07
B. Other Relevant Chapters 55.08

## A. Disciplinary Cases

### *Bultitude v. Law Society* [2004] EWCA Civ 1853

**55.01** Solicitor transferring monies from client account to office account without seeing any supporting documentation—reckless—conscious impropriety amounting to dishonesty

In 2000, B and R were practising as solicitors from premises in Watford, Hertfordshire. B was the sole equity partner and R was a salaried partner, having no managerial or administrative responsibilities. On 29 August 2000, the Forensic Investigation Unit of the Office for the Supervision of Solicitors (OSS) carried out an inspection of the firm's books and accounts. Its report revealed a shortfall on clients' funds of £46,971.57 as at 31 July 2000, of which £36,205.75 related to relatively small client ledger credit balances that had been improperly transferred to the office bank account. B gave evidence to the Solicitors Disciplinary Tribunal (SDT), and said that the firm's new accountant had drawn attention to the credit balances on client account and said that, unless they were removed, he would not sign the accountant's report. Accordingly, B caused lists to be prepared of the credit balances. There turned out to be 310 of them and, because he had a pressing meeting with a client in Bath, he passed the matter to R and his bookkeeper to resolve. The firm had a computerized accounts package that made it impossible to transfer money from client to office account unless a bill had been drawn and recorded as a debit in the office column of the individual client's ledger. R devised a system of false debit notes made out to look like bills of costs, complete with VAT, the total in each case being the amount of the credit balance in the relevant client's ledger. The debit notes were put in a lever arch file with the intention, in due course, to work through them. On return to the office, B signed a cheque transferring a composite sum of client balances of £50,099.45 to the office account. He saw no supporting documentation when signing that cheque and he did not ask what had been done. He signed the cheque transferring clients' funds to office account without knowing or caring whether his firm was entitled to be paid those funds. In dismissing B's appeal, the Court of Appeal (Kennedy, Laws, and Arden LJJ) said that the Divisional Court was right to find as it did in relation to the issue of dishonesty. It was accepted that B was not shown to have intended to deprive his clients of their funds permanently. The proof of dishonesty in this context was not dependent upon proving that intention. Before he went to Bath, B knew that there was a serious problem, but he seems to have done nothing to address it other than to ask his staff to solve it for him. He made no enquiries as to how the problem was being resolved and was anxious only to be assured that the new accountant would now sign the accountant's report, so that it could be filed in time. He did not ask R what had been done and, by signing a cheque transferring clients' funds to the office account without any supporting documentation, it must be inferred that B acted without knowing or caring whether his firm was entitled to be paid those funds. That satisfied both legs of the test under *Twinsectra Ltd v. Yardley* [2002] 2 AC 164—namely, in the context of this case: first, did B act dishonestly by the ordinary standards of reasonable and honest people; and if so, secondly, was he aware that, by those standards, he was acting dishonestly? The position here was compounded. At some stage, B did become aware of the debit notes and, once he saw those bogus documents, it must have been clear to him what had been done to clear the debit balances, but he did nothing to backtrack. As the Tribunal found, he was guilty of conscious impropriety, amounting to dishonesty, in endorsing what had been done.

### *Holy v. Law Society* [2006] EWHC 1034 (Admin)

**55.02** Niche solicitor's practice—reckless disregard of solicitors' audit rules and practice rules—suspension

H, a solicitor, appealed from the decision of the SDT ordering him to be struck off the roll of solicitors. The rule 4(2) statement made pursuant to the Solicitors (Disciplinary Proceedings) Rules 1994 [now rule 5(1) of the Solicitors (Disciplinary Proceedings) Rules 2007] alleged that H acted contrary to the Solicitors' Accounts Rules 1998 in arranging private loans from one client to another without having obtained the written authority of both clients, and failing to keep accounts properly drawn up; and that he acted contrary to the Solicitors Practice Rules 1990 in various respects, including compromising or impairing his independence or integrity as a solicitor, in that he failed to make any or any sufficient enquiry as to funds received into and paid out of his client account. H had a niche practice, acting for a small clientele of wealthy individuals, and he generally acted in high-value transactions. He had not been the subject of previous disciplinary proceedings and it was not alleged that he was guilty of dishonesty in relation to the conduct that was the subject of the allegations. In essence, the Law Society's case against H was that his practice was entrepreneurial as much as legal and that, as his own statement suggested, he regarded compliance with the account and practice rules applicable to solicitors as optional so far as he was concerned, rather than mandatory. On appeal, the Divisional Court (Newman and Stanley Burnton JJ) upheld the findings of the Tribunal. The allegations proved against H did not establish a want of honesty: his actions showed a marked recklessness, but not dishonesty. The evidence generally showed a disregard for a number of regulations affecting solicitors—a disregard that was aggravated by H's assertion that compliance was not consistent with the nature of his practice. Reckless disregard of regulations affecting solicitors may well justify striking off. In the present case, taking account of the candour and cooperation shown by H, and the admissions made by him, the appropriate penalty was a substantial period of suspension, which would demonstrate to him and to the public the importance of compliance with the regulations affecting the practice of solicitors, no matter how substantial or successful a solicitor's practice, and no matter how entrepreneurial he and his practice may be. The appropriate period was four years.

### *Keazor v. Law Society* [2009] EWHC 267 (Admin)

**55.03** Pleading recklessness in charges—solicitor given opportunity to deal with issue before tribunal

The appellant solicitor, who had been a senior partner in his firm, argued unsuccessfully that recklessness should have been specified in the charges against him. The Court (Maurice Kay LJ and Simon J) held that although the Law Society was required to refer specifically to dishonesty in the formulation of charges, there was no such requirement for recklessness. In any event, the solicitor had been given the opportunity to deal with the issue of recklessness at the SDT.

***Harris v. Solicitors Regulation Authority* [2011] EWHC 2173 (Admin)**

**55.04** Retention of client money in office account—reckless, but not dishonest—suspension

H was a sole principal in a firm of solicitors in Southsea, Hampshire. He faced ten allegations, of which eight were admitted, including using client funds for his own purposes contrary to rule 1 of the Solicitors Practice Rules 1990 and rule 1 of the Solicitors Code of Conduct 2007. The sum involved was £9,391. The SDT found that H's conduct was improper and that he was grossly reckless in the way in which he kept client money in the office account for so long, but the Tribunal did not find the conduct dishonest. The Tribunal suspended H from practice for a period of two years. In its reasons, the Tribunal said that the allegations against H were wide-ranging and involved a number of breaches of the Guide to the Professional Conduct of Solicitors and the Solicitors Code of Conduct. He had failed to provide clients with costs information. Three of the allegations against him related to accounting breaches. The allegation regarding the propping up of his firm while using client money was serious and involved a finding of gross recklessness, albeit that there were no findings of dishonesty. There had been shortcomings in conveyancing transactions. The Tribunal regarded the cumulative effect as showing a cavalier disregard for the professional rules. The Court upheld the Tribunal's decision and dismissed the appeal.

***Kumar v. General Medical Council* [2012] EWHC 2688 (Admin)**

**55.05** See paragraph 47.46.

***Izuchukwu v. Solicitors Regulation Authority* [2013] EWHC 2106 (Admin)**

**55.06** Finding of recklessness—whether adequate evidence to support the finding

I was found guilty of recklessness and suspended for three years. He had failed to keep appropriate accounting methods, had not employed someone qualified to supervise his firm, and had closed his eyes to mortgage fraud. The SDT said that although dishonesty had been alleged, it was not satisfied to the necessary standard (in other words, the criminal standard) that dishonesty had occurred. It was satisfied, however, that I had been reckless in his disregard for not being aware of, and his failure to fulfil, his obligations. He had also been reckless in allowing numerous transactions to be undertaken without raising any question. Those totalled large amounts of money and had the classic hallmarks of money-laundering transactions. On appeal before Cranston J, I advanced nine grounds. Grounds 6 and 7 went to the failure of the Tribunal to apply the correct standard of proof regarding recklessness. The Tribunal found that it had been established. The substance of grounds 6 and 7 went to the issue of whether there was adequate evidence to support the finding of recklessness, rather than to a submission that the Tribunal had applied the wrong test. The appellant reiterated the point that he made on a number of occasions that, in relation to the property transactions, he had ensured transparency for all those in the chain. However, those were factual findings that the appellant was not able to reopen, and were rejected by the Tribunal.

***Brett v. Solicitors Regulation Authority* [2014] EWHC 2974 (Admin)**

**55.07** Breach of Solicitors' Code of Conduct—recklessly misleading the Court—meaning of 'recklessness'

B was the legal manager at Times Newspapers Ltd (TNL), a position that he had held for more than thirty years. He was, in effect, their in-house solicitor. Pursuant to rule 5 of the Solicitors Disciplinary Tribunal Rules, the Solicitors Regulation Authority (SRA) set out its case in a statement against B. It alleged that on or around 2 June 2009, while conducting litigation in the High Court, B caused or allowed a witness statement to be served and relied on in support of TNL's defence that knowingly and/or recklessly created a misleading impression as to the facts and matters deposed to in the statement. The matters giving rise to the allegations of breach of the Code of Conduct concerned litigation in the High Court of Justice. The claimant, RH, a constable with the

Lancashire Constabulary, sought an injunction against TNL preventing it from publishing a story revealing him as the author of a blog under the name of 'Nightjack'. RH published an anonymous chronicle of his life as a police officer. A junior reporter at *The Times* told B, in his role as legal manager, that he had identified 'Nightjack' as RH by gaining unauthorized access to RH's private email account. B told the reporter that what he had done was totally unacceptable. The story was unpublishable, from a legal prospective, if it was based on unlawfully obtained information. It was 'dead in the water' unless the same information could be obtained through the public domain. B telephoned a junior barrister, who advised that, potentially, a crime contrary to the Data Protection Act 1998 had been committed, but that there might be a public interest defence. Different counsel was instructed by B to represent TNL in the injunction proceedings commenced by RH. Counsel was not told about the reporter's email hacking of RH's account and the Court was informed that the reporter had been able to establish RH's identity using publicly available materials, patience, and simple deduction. RH's claim was dismissed. Subsequently, in connection with the Leveson Inquiry, TNL disclosed emails and other material in relation to the disclosure to B that the reporter had hacked RH's email account, and in relation to the circumstances in which that fact had not been disclosed to the Court. On 6 December 2013, the SDT found B guilty of breaches of the Code of Conduct, and decided that he should be suspended from practice as a solicitor for six months and pay costs of the proceedings, summarily assessed at £30,000. In dismissing B's appeal, the Divisional Court (Lord Thomas of Cwmgiedd CJ and Wilkie J) noted that it was not alleged that B had been dishonest. The SDT found that, at the very least, B had turned a blind eye to what he had been told by the reporter. A breach of the Code of Conduct can arise on the basis of deceit or knowingly, or recklessly, misleading the Court. It was open to the Divisional Court to conclude that B was guilty of 'recklessly' allowing the High Court to be misled, even if it were to conclude that he was not guilty of 'knowingly' allowing the High Court to be misled. The word 'recklessly', in criminal statutes, is now settled as being satisfied when:

> A person acts recklessly with respect to –
> (i) a circumstance when he is aware of a risk that it exists or will exist;
> (ii) a result when he is aware that a risk will occur;
>
> and it is, in circumstances known to him, unreasonable for him to take the risk.

See *R. v. G* [2004] 1 AC 1034, [41], citing clause 18(c) of the Law Commission's Draft Criminal Code Bill.

The Divisional Court adopted that as a working definition of 'recklessness' for the purposes of B's appeal. The charge against B was one of knowingly and/or recklessly allowing the Court to be misled in two particular ways: first, by causing or allowing a witness statement to be served and relied on that created a misleading impression as to the facts and matters deposed to in that statement; and secondly, allowing the Court to proceed on the basis of an incorrect assumption as to the facts and matters set out in that witness statement. There was no doubt that the Court was misled. The reporter had initially identified 'Nightjack' as RH by using exclusively unlawful methods: his unlawful access to the email accounts. The exercise that he did undertake, to see whether he could identify RH as 'Nightjack' using publicly available sources, was at the insistence of B only after the reporter had disclosed to B that he had identified RH by illegitimate means. Whilst the Code of Conduct is drafted on the basis that there may be cases in which a solicitor may knowingly mislead, but not deceive, the Court, it is an extremely difficult distinction to draw—and even more difficult to draw in the circumstances of this case, when it was not alleged that B was deceitful in misleading the Court. The

SDT, having disavowed making any finding of dishonesty, could not properly then proceed to make a finding that B 'knowingly' allowed the Court to be misled in the circumstances of this case, which was, without more, in effect, a finding of dishonesty. However, the evidence—particularly that of the contemporaneous correspondence and the lack of any response by B to the demands contained in it as to whether the reporter relied solely on publicly accessible material to identify 'Nightjack' as RH—pointed inevitably to the conclusion that B acted recklessly in allowing the Court to be misled.

## B. Other Relevant Chapters

**55.08** Dishonesty, Chapter 19
Impairment of Fitness to Practise, Chapter 33
Misconduct, Chapter 47
Negligence, Chapter 49

# 56

# REGISTRATION

A. **Legal Framework** 56.01

B. **Disciplinary Cases** 56.02

*Allender v. Council of the Royal College of Veterinary Surgeons* [1951] 2 All ER 859 56.02

*R (Law Society) v. Master of the Rolls* [2005] 1 WLR 2033 56.03

*Jones v. Commission for Social Care Inspection* [2005] 1 WLR 2461 56.04

*Jideofo v. Law Society; Evans v. Solicitors Regulation Authority; Begum v. Solicitors Regulation Authority* [2007] EW Misc 3 56.05

*Re a Solicitor Nos 21 and 22 of 2007, Ali and Naeem* [2008] EWCA Civ 769 56.06

*Re a Solicitor No. 4 of 2009, Afsar* [2009] EWCA Civ 842 56.07

*Taylor v. General Chiropractic Council* [2009] EWHC 301 (Admin) 56.08

*R (K) v. Chief Constable of Lancashire Police* [2009] EWCA Civ 1197 56.09

*R (Kay) v. Chief Constable of Northumbria Police* [2010] ICR 962 56.10

*Butt v. Solicitors Regulation Authority* [2010] EWHC 1381 (Admin) 56.11

*Venton v. Solicitors Regulation Authority* [2010] EWHC 1377 (Admin) 56.12

*Khan v. Solicitors Regulation Authority* [2010] EWHC 1555 (Admin) 56.13

*Harris v. Registrar of Approved Driving Instructors* [2010] EWCA Civ 808 56.14

*Mulla v. Solicitors Regulation Authority* [2010] EWHC 3077 (Admin) 56.15

*Davis v. Solicitors Regulation Authority* [2011] EWHC 3645 (Admin) 56.16

*Chaudery v. Solicitors Regulation Authority* [2012] EWHC 372 (Admin) 56.17

*General Osteopathic Council v. Sobande* [2011] CSOH 39 56.18

*Islam v. Bar Standards Board*, 1 August 2012 (unreported) 56.19

*Adesemowo v. Solicitors Regulation Authority* [2013] EWHC 2015 (Admin) 56.20

*Cordle v. Financial Services Authority*, FS/2012/0006, 2 January 2013 56.21

*Ahmed v. The Inns' Conduct Committee* [2012] EWHC 3270 (QB) 56.22

*Ahmed v. Bar Standards Board*, 16 May 2013 (unreported) 56.23

*Hassan v. General Optical Council* [2013] EWHC 1887 (Admin) 56.24

*R (Patel) v. General Medical Council* [2013] 1 WLR 2801, CA 56.25

*R (Alacakanat) v. General Medical Council* [2013] EWHC 1866 (Admin) 56.26

*Professional Standards Authority for Health and Social Care v. Nursing and Midwifery Council* [2013] EWHC 4369 (Admin) 56.27

*JB v. General Teaching Council for Scotland* [2013] CSIH 114 56.28

*Fajemisin v. General Dental Council* [2014] 1 WLR 1169 56.29

*Giambrone v. Solicitors Regulation Authority* [2014] EWHC 1421 (Admin) 56.30

*Woodman-Smith v. Architects Registration Board* [2014] EWHC 3639 (Admin) 56.31

*General Medical Council v. Nakhla* [2014] EWCA Civ 1522 56.32

*Hunt v. The Commissioners for Her Majesty's Revenue & Customs* [2014] UKFTT 1084 (TC) 56.33

*Yousef v. Solicitors Regulation Authority*, 23 January 2015 (unreported) 56.34

C. **Other Relevant Chapters** 56.35

## A. Legal Framework

**56.01** Examples of registration rules or registration appeal rules include:

Medical Act 1983, Schedule 3A

General Optical Council (Registration) Rules 2005
General Dental Council (Registration Appeals) Rules 2006
General Medical Council (Registration Appeals Panels Procedure) Rules 2010
General Pharmaceutical Council (Appeals Committee) Rules 2010
Veterinary Surgeons and Veterinary Practitioners (Registration) Regulations 2010

## B. Disciplinary Cases

### *Allender v. Council of the Royal College of Veterinary Surgeons* [1951] 2 All ER 859

**56.02** Appeal against registration—role of Court

The appellant appealed against the refusal of the Royal College of Veterinary Surgeons (RCVS) to have his name entered on the supplementary veterinary register. The supplementary register was established to enable the registration of a person who has genuinely practised as an unqualified veterinary surgeon for seven years during the past ten years. The Council, having considered the appellant's application, refused it, but it gave him an opportunity of going before it so that the Council could question him and satisfy itself that his principal means of livelihood was acting as a veterinary surgeon. He did not so satisfy the Council. Dismissing the appeal, Lord Goddard CJ said that an appeal under section 6(5) of the Veterinary Surgeons Act 1948 required the Court to hear the case and to treat it as a rehearing in the same way as an appeal from a judge of the High Court sitting at first instance is treated by the Court of Appeal. The Court would naturally pay the greatest attention to the decision given below, but that would not prevent it—and did not prevent it in *Almond v. Council of the Royal College of Veterinary Surgeons* (1951, unreported), in which the Court thought that the decision below was not justified on the facts—from reversing the decision of the Council. Hilbery J agreed, saying that it was for the appellant to satisfy the Court that the decision below was wrong.

### *R (Law Society) v. Master of the Rolls* [2005] 1 WLR 2033

**56.03** Registered foreign lawyer—whether power to impose conditions post-registration

S was a foreign lawyer who practised in London and was registered as a foreign lawyer by the Law Society under the Courts and Legal Services Act 1990. On 25 July 2003, the Adjudication Panel of the Law Society decided to impose conditions on his registration. S appealed against that decision to the Master of the Rolls, who held that the Law Society had no general power to impose conditions on the registration of a foreign lawyer save on the initial making of an entry on the register. The Law Society applied for judicial review of the decision of the Master of the Rolls. The Court (Thomas LJ, and Richards and Fulford JJ) dismissed the Law Society's claim that it had a general power to impose conditions not limited to the initial making of an entry on the register. There was no doubt that the Society had express power to impose conditions when suspension is terminated or a registration is revived; if the Society had power to impose conditions at any time, that express power would be superfluous. Although, in certain circumstances, arguments on surplusage may sometimes carry little weight (see *Homburg Houtimport BV v. Agrosin Private Ltd* [2004] 1 AC 715, 763), the strength of the argument depends on the context. In the present context, the express power to impose conditions in a statutory scheme regulating lawyers would have been carefully considered and it is unlikely to have been thought necessary, if there were a general power that would result from the Law Society's construction. This therefore supports the argument that registration means the initial entry. What is important is that Parliament properly circumscribed the power to impose conditions in relation to solicitors and it cannot be inferred that it was intended that there be an unrestricted power over registered foreign lawyers, subsequent to the initial registration.

### *Jones v. Commission for Social Care Inspection* [2005] 1 WLR 2461

**56.04**

Manager of care home—whether fit and proper person—decision of Commission—appeal to Tribunal—proper approach—public interest

The applicant, J, appealed against a judgment of Sullivan J sitting in the Administrative Court, whereby he allowed a statutory appeal by the Commission for Social Care Inspection (formerly the National Care Standards Commission) against a decision of the Care Standards Tribunal. The Tribunal had allowed J's appeal against the refusal by the National Care Standards Commission to grant him registration under the Care Standards Act 2000 as a manager of a care home in Lincolnshire. He applied to the Commission in April 2002 for registration under the Act of 2000. In due course, the Commission refused his application. He had been the subject of an adverse outcome of disciplinary proceedings before his professional body arising out of complaints about his conduct towards patients at a previous care home; he deliberately concealed the fact of those pending proceedings, and a police investigation that preceded them, when he completed and dispatched his application form for registration in April 2002. Also, he failed to impress the Commission's representatives who interviewed him about his knowledge of the law and theoretical principles of management of care homes during the course of the 'fit person' registration process. J's appeal to the Care Standards Tribunal was allowed following a lengthy hearing. In the Court of Appeal, Brooke LJ (with whom Mance and Thomas LJ agreed) said that an applicant for registration must demonstrate to the Commission and, if there is an appeal, to the Care Standards Tribunal that he is a fit person before he can be qualified for registration. The 2000 Act and the Care Homes Regulations 2001 show that, provided that the Commission is satisfied that an applicant is a fit person (and any other relevant requirements are fulfilled), it shall grant the application. There are stringent requirements, such as integrity and good character, and the need for a manager to have the requisite qualifications, skills, and experience, and it would be absurd if the onus of proof were placed on the Commission to demonstrate unfitness before it could refuse registration. The way in which the registration scheme operates is that the Commission first considers the application and any supporting documents. It then interviews the applicant and, if it is minded to refuse the application, it is bound to give him notice of a proposal to refuse it, together with its reasons for the proposal. He is then allowed twenty-eight days within which he may make written representations to the Commission concerning any matters that he wishes to dispute. The Tribunal's main powers on an appeal are to confirm the Commission's decision or to direct that it shall not have effect. The statutory language shows that if the Tribunal is satisfied that the Commission was right when it decided that the applicant had not satisfied it that he was a fit person, it will confirm the Commission's decision. The Tribunal in J's case was therefore wrong when it decided his appeal in his favour by saying that, in its view, the matter was finely balanced and that the burden was on the Commission to prove J's unfitness on the balance of probabilities. This placed the burden of proof upside down. In a concurring judgment, Thomas LJ said that bodies charged with regulation are frequently entrusted with the task of determining whether a person who seeks to hold a position of trust is a fit and proper person to hold such a position. There have been instances in which the regulatory body has been uneasy as to whether the person in fact is a fit and proper person; in such cases, because the provisions of some regulatory systems have been interpreted as placing the burden of proof on the regulator, the regulatory body has felt constrained to allow such a person to occupy such a position of trust, despite its doubts. To state that outcome demonstrates the fact that, in such a case, there may have been a failure of the legislative scheme in seeing that, in the public interest, positions of trust are occupied by persons who are demonstrably fit and proper. The interpretation of any legislative scheme is a matter of the construction of the particular scheme. The Care Homes Regulations 2001 are very clear in placing the burden on an applicant. It is entirely in the public interest that they should do so. A manager of a care home occupies an important position of trust, and must demonstrate that he is fit and proper to hold such a position; any doubts must be resolved against registration. This does not deprive an applicant of earning his living as a nurse or in some other occupation, but prevents him, in the public interest, from occupying a position of trust unless he can demonstrate to the commission that he is fit and proper to occupy that position.

***Jideofo v. Law Society; Evans v. Solicitors Regulation Authority; Begum v. Solicitors Regulation Authority* [2007] EW Misc 3**

**56.05** Student member—admission—character and suitability—pre-admission and post-admission test

Sir Anthony Clarke MR said that the question in this case is what is the appropriate test for determining an individual's character and suitability for admission as a student member of the Law Society and then as a solicitor. In dismissing the appeals of B and E, both student members, and J having withdrawn his appeal, the Master of the Rolls said that the starting point must be the relevant statutory provisions. First, in order to be admitted as a student member of the Law Society and issued with a certificate of enrolment, an unadmitted person must satisfy the Law Society as to his or her 'character and suitability'. Following enrolment and satisfactory completion of the relevant vocational and practical training, a student member can apply for admission as a solicitor. Before a student member is permitted to proceed to admission, he or she is required to satisfy the Law Society under section 3(1)(b) of the Solicitors Act 1974 that, amongst other things, he or she has the requisite 'character and suitability to be a solicitor'. The Law Society also has power to take various steps such as cancelling enrolment and opposing admission if, at any time, it is not satisfied as to the character and suitability of an unadmitted person to become a solicitor. The Law Society is thus under a statutory duty to assess at a number of stages whether an unadmitted person has the requisite 'character and suitability' to become a solicitor. While there is no authoritative guidance as to the correct approach to be taken when carrying out these assessments, there is a well-established analogous jurisdiction that arises where the question is whether a solicitor ought to be struck off the roll and what circumstances are relevant to any application to be restored to the roll following a strike-off made on the grounds of the solicitor's dishonesty. The leading authority is *Bolton v. Law Society* [1994] 1 WLR 512. *Bolton* related only to conduct post-admission. The same underlying principles apply to conduct both pre-admission and post-admission. It would be irrational to hold that a different test applies where matters come to the Law Society's attention pre-admission from the case in which those same matters come to its attention post-admission. Whether they are discovered pre- or post-admission, the question remains the same—namely, whether the relevant evidence demonstrates that the person concerned is a fit person to be a solicitor. It would make no sense for the Law Society to admit an individual whose conduct, if assessed by the Solicitors Disciplinary Tribunal (SDT), would lead to him or her being struck off the roll as being an individual who was not a fit and proper person to be a solicitor. A similar approach was adopted by Newman in the case of dentists: *Council for the Regulation of Health Care Professionals v. General Dental Council and Fleischmann* [2005] EWHC 87 (Admin). He said, at [55] that, when assessing conduct and misbehaviour, the same standards should be applied to pre-admission and post-admission behaviour. He put it this way:

> ... [T]he disciplining of a registered dentist involves subtly different consideration from those which apply to an applicant for registration. That said, I have no doubt that the differences should not be allowed to give way to the existence of a double

> standard in connection with those who are entitled to be in practice. The requirement that an applicant for registration be of "good character" secures the need for the public to be protected by the maintenance of high standards and the high reputation of the profession which has to be served at the stage of an application for registration as well as in disciplinary proceedings. The protection of the public will not be served by the application of a different standard at erasure from that which is applied when considering registration.

The same considerations apply to solicitors.

### *Re a Solicitor Nos 21 and 22 of 2007, Ali and Naeem* [2008] EWCA Civ 769

**56.06** Student member—revocation of membership—dishonesty

These were two joined appeals from decisions of the Law Society to revoke the appellants' student membership of the Law Society. In allowing the appeal of A and allowing the appeal in part of N, Sir Anthony Clarke MR said that the relevant principles are set out in *Jideofo*. However, where the Law Society positively asserts dishonesty, it should prove it to the appropriate civil standard. Moreover, Law Society adjudicators and panels should consider any such allegation and decide whether the relevant person was dishonest as alleged. The position should be made clear on the face of the relevant decision. This is important because of the significance of dishonesty. Here, no dishonesty was found. However, it did not follow that the Law Society's decision to revoke was wrong. All depends upon the circumstances. The circumstances may show that the person concerned does not have the appropriate character and suitability to be a solicitor even in the absence of dishonesty.

### *Re a Solicitor No. 4 of 2009, Afsar* [2009] EWCA Civ 842

**56.07** Student member—pending criminal proceedings—later conviction disclosed—appeal against cancellation of student membership allowed

A was charged with dangerous driving. While on remand on bail, he applied for enrolment as a student member of the Law Society. He indicated on the enrolment form that he had never been convicted of an offence in a UK court. He also indicated that there were no other factors that might call into question his character and suitability to be a solicitor. He was granted student membership and subsequently pleaded guilty to the offence of dangerous driving. He was sentenced to six months' imprisonment, suspended for twelve months, ordered to carry out 240 hours' unpaid work, and disqualified from driving for twelve months. Within five days of sentence, A informed the Solicitors Regulation Authority (SRA) of his conviction and said that it arose out of an incident that occurred in May 2006 before registration. A completed the community order. In allowing A's appeal against cancellation of his student membership, Lord Clarke of Stone-cum-Ebony, MR, said that the question in the application form about other factors that might call into question an applicant's character and suitability to be a solicitor was in very broad terms.

> 30. ... It is without doubt wide enough, on an objective assessment, to encompass the need to declare that criminal prosecutions are pending at the time of application and to provide a full explanation of the matter and its surrounding circumstances.... Any suggestion that the form is ambiguous on this point is, in my opinion, manifestly wrong and must be rejected. It should also be said that if any individual applicant is unsure as to how to answer any particular question, they should err on the side of caution. Full and frank disclosure to the SRA requires it. The approach to take is not the one [A] adopted, which was essentially one of "wait and see"....
>
> [...]
>
> 32. ... It is the duty of any applicants when completing such forms and when responding to question from the SRA, as [A] was required to do during its investigation of the matter from October 2007, to give full and frank disclosure. Anything less is not sufficient.

> 33. Full details must be given; not sketchy and incomplete details such as those given by [A]. His failure to disclose his pending criminal prosecution and his subsequent failure fully and frankly to disclose the nature of the matters giving rise to the prosecution... call into question his integrity and probity. They do so sufficiently to justify the SRA's decision to cancel his student membership....
>
> [...]
>
> 36. That there was no explicit finding of a deliberate attempt to deceive or dishonesty is important. It is important because, if there is a clear finding of dishonesty, it is well-established that cancellation of student membership will almost always be justified: see *Bolton* at page 518 and *Jidefo* at paragraphs 10–21....
>
> [...]
>
> 44. ... [G]iven the fact that this is not deliberate dishonesty as such, and notwithstanding the serious criticisms which were properly made of [A] in the way he failed to disclose relevant material to the SRA, it is fair to say that permanent exclusion from the profession would be disproportionate.... In the light of what has happened since and his good references... I do not think that the enrolment of [A] now poses a risk to the public or the reputation of the profession.... [T]he just course is to direct that the SRA now reinstate his student membership.

### *Taylor v. General Chiropractic Council* [2009] EWHC 301 (Admin)

**56.08** Chiropractor—appeal to High Court rather than county court—judge sitting as county court judge—failure to pay annual subscription—deletion of name from register lawful

The claimant, T, appealed to the High Court against a decision of the General Chiropractic Council (GCC), upholding the decision of the registrar to refuse to register the claimant under the Chiropractors Act 1994. The appeal should have been brought in the county court under section 29(4A) of the 1994 Act. However, Plender J was content to continue sitting as though he were a county court judge. The point taken against T by the GCC was that T had failed to pay the annual subscription that is required by the Council of its members. There was no dispute that T had failed to pay that subscription, but he contended that, by deleting him from the register by reason of his failure to pay that subscription, the Council acted unlawfully. In dismissing T's appeal, the Court said that a fair reading of the statutory provisions is that if payment is not received, registration may be removed, and the registrar is therefore entitled to remove the name. It is not a question of discretion as regards a removal from the register of a person who has failed to pay the fee. The Council did not act unreasonably, perversely, or contrary to law.

### *R (K) v. Chief Constable of Lancashire Police* [2009] EWCA Civ 1197

**56.09** Probationary police constable—decision by chief constable to dispense with services—no significant conflict of facts

In early 2007, K was a probationary police constable. On 27 March 2007, while off duty, he and a friend met a young woman in a bar. Subsequently, the woman made an allegation of rape, to the effect that both men had had intercourse with her while she was in a drunken sleep. Being asleep, she could not consent. The allegation was investigated by the police, who decided not to press charges. However, the chief constable decided to dispense with the services of K under regulation 13 of the Police (Conduct) Regulations 2004, on the grounds that the chief constable decided that K was not fitted mentally to perform the duties of his office and was not likely to become a well-conducted officer. He took the view that K had shown poor judgment. K's application for judicial review was dismissed. K alleged that formal disciplinary proceedings could and should have been taken, relying on *R v. Chief Constable of West Midlands Police, ex parte Carroll* (1995) 7 Admin LR 45 and *R v. Chief Constable of British Transport Police, ex parte Farmer* [1999] COD 518. In refusing permission to appeal against the decision of Elias J at [2009] EWHC 472 (Admin), Smith LJ said that the principle she derived from the two authorities is that where there is no substantial conflict as to the facts and the nature of the misconduct relied on, regulation 13 is an appropriate route to take when considering the future of a probationer. Where there is a significant conflict of fact, there should be formal disciplinary proceedings. In the instant case, Elias J was of the view that there was no significant conflict of fact. It was admitted that intercourse had taken place in the presence of a third party; it was admitted that the woman had had a good deal to drink; it was admitted that the proceedings had been filmed by K. The chief constable was entitled to infer that the woman had been vulnerable and to conclude that the filming of the event had been exploitative. The chief constable had been entitled to find that the facts were sufficiently clear, and that it was appropriate and fair to proceed by regulation 13 rather than by disciplinary proceedings.

### *R (Kay) v. Chief Constable of Northumbria Police* [2010] ICR 962

**56.10** Probationary police officer—whether dispense with services or conduct proceedings—test

Silber J considered a number of authorities, including *R v. Chief Constable of the West Midlands, ex parte Carroll* (1995) 7 Admin LR 45, *R v. Chief Constable of British Transport Police, ex parte Farmer* [1999] COD 518, *R (Begley) v. Chief Constable of West Midlands Police* [2001] EWCA Civ 1571, and *R (Khan) v. Chief Constable of Lancashire* [2009] EWHC 472 (Admin). In the light of these authorities, he concluded at [38] that the test for determining if a case against a probationary police officer should be determined under regulation 13 of the 2003 Regulations or under the Conduct Regulations is whether there is such conflict over the facts relating to the misconduct relied on, with the consequence that it would be unfair for the chief constable to make the judgment that he did on the basis of the undisputed primary facts rather than to give the probationary police officer the protection to which he or she was entitled under the Conduct Regulations.

See further paragraph 58.09.

### *Butt v. Solicitors Regulation Authority* [2010] EWHC 1381 (Admin)

**56.11** Qualified lawyer in Pakistan—application to be admitted as solicitor in England and Wales—appropriate test

B sought to set aside the determination by the SRA that he lacked the necessary character and suitability to be admitted as a solicitor. Previously, jurisdiction to deal with appeals of this kind had been dealt with by the Master of the Rolls; they are now heard by the High Court and the Court was told that this was the first occasion on which an appeal of this kind had been considered by the court. B was an advocate and was admitted in Pakistan in 2005. In October 2005, he applied to the SRA for a certificate for eligibility to be admitted to the roll of solicitors in England and Wales under the Qualified Lawyers Transfer Regulations. The regulations provide a framework whereby lawyers in other jurisdictions may be admitted to the roll. B failed to declare that, on 27 January 2000 at Brent Magistrates' Court, he was convicted of two offences of attempting or obtaining property by deception. He was fined £100, and ordered to pay costs and compensation. Following completion of the requisite work experience and assessments made pursuant to the Qualified Lawyers Transfer Regulations, B applied to be admitted to the roll of solicitors in July 2008. It was at this stage that he disclosed the 2000 conviction. A Police National Computer (PNC) check was done on 14 July 2008. This confirmed the two convictions at Brent Magistrates' Court, but it also disclosed three other 'non-convictions', these being two allegations of rape and one of assault occasioning actual bodily harm. These were matters in respect of which B had been arrested, but not charged. B did not, in his application of July 2008, address the issue of the non-disclosure of the convictions in his earlier application. In the light of the information provided, the SRA recommended that his application should be refused. In dismissing B's appeal, Elias LJ (with whom Keith J agreed) said that, in respect of the legal principles, the onus lies firmly on an applicant who is seeking to establish that he has the character that is suitable to a solicitor. The appropriate approach is set out in the judgment of Sir Anthony Clarke MR in *Jideofo v. Law Society* [2007]

EWMisc 3 (31 July 2007). The Master of the Rolls endorsed submissions made on behalf of the Law Society to the following effect:

(1) the test of character and suitability is a necessarily high test;
(2) the character and suitability test is not concerned with punishment, reward, or redemption, but with whether there is a risk to the public or a risk that there may be danger to the reputation of the profession; and
(3) no one has the right to be admitted as a solicitor, and it is for the applicant to discharge the burden of satisfying the test of character and suitability.

In the instant case, the panel had these matters well in mind. It was not simply the original convictions that caused the panel to make the decision that it did; it was, as the panel emphasized, the failure to disclose that it considered to be dishonest. The panel stated in terms that the original convictions were now quite old and may not of themselves have led to the conclusion that admission should be refused. The panel had to consider whether there had been a full and frank recognition of B's responsibility for the wrongdoing, and whether there had been true rehabilitation. The failure to disclose goes fairly and squarely to that issue. The panel fully understood all of the mitigating factors that were advanced by B with respect to the original offence. Mitigation in this context has only a limited weight.

*Venton v. Solicitors Regulation Authority* **[2010] EWHC 1377 (Admin)**

**56.12** Enrolment as member of Law Society—application to be admitted solicitor—failure to disclose caution and conviction

In applying in March 2003 to enrol as a member of the Law Society, V failed to disclose that, in 1997, he received a caution for disorderly behaviour and that, in 1998, he was cautioned for making off without paying a taxi fare. In October 2003, V was convicted of driving a vehicle with excess alcohol; in August 2005, he was given a police penalty notice for being drunk on a highway. In 2009, V applied to be admitted as a solicitor and stated, on the relevant application form, that he was not aware of any matter that could bring into question his character or suitability to become a solicitor. The SRA refused his application. V's appeal to the adjudication panel was dismissed. The panel stated that he had been reckless in responding to questions concerning his character and suitability, but did not state that he had been dishonest. V's appeal to the Court was dismissed. The decision was justified. V had been less than frank and had sought to make excuses. His explanations were not attractive. He had sought to minimize his own responsibility, and had taken a legalistic and defensive line in respect of his failures. The full frankness that ought to be demonstrated before one could become a member of the profession was lacking in the instant case.

*Khan v. Solicitors Regulation Authority* **[2010] EWHC 1555 (Admin)**

**56.13** Qualified lawyer in Pakistan—application to be admitted as solicitor in England Wales—advice sought from Law Society—applicant fully appreciating responsibilities—application granted

K was an advocate admitted in Pakistan in 1992. In November 2006, he applied for a certificate of eligibility to be admitted as a solicitor under the Qualified Lawyers Transfer Regulations. K failed to disclose that, in 2002, he had been convicted of driving without due care and attention, and of using a vehicle without valid insurance, for which he was disqualified for fifty-six days. He did, however, disclose the 2002 conviction to the Punjabi Bar Council, where he was still in good standing as an advocate. Shortly after receiving his certificate from the Law Society in January 2007, K spoke to an administrative officer at the Law Society, whom he named, who told him he would have to wait until he applied for admission to the roll to declare the conviction. K duly completed the required work experience and the necessary assessments, and applied in February 2009 to be admitted. His application was refused by the SRA. K's appeal was allowed by Elias LJ and Keith J. The Court stated that, in all cases, it is for the applicant to show that he has the appropriate character and integrity to be a member of the profession. In the particular circumstances of this case, there was no doubt that the panel was fully entitled to reach the conclusion that this was a reckless act on the part of K. There was no very satisfactory explanation for failing to disclose the information concerning his conviction. But the focus of the panel's concern was whether he would understand the importance of regulatory requirements. K was right in saying that there were a number of features here that demonstrated that he did recognize the importance of that. He told the authorities in the Punjab and made disclosure in the United Kingdom, notwithstanding that the relevant official documents would not have revealed his previous conviction. There were also a number of testimonials to his character and integrity, and although these only went so far, they lent support to the view that K did now fully appreciate the weight of his responsibilities. Accordingly, the Court directed the SRA to issue a certificate of satisfaction to K.

***Harris v. Registrar of Approved Driving Instructors* [2010] EWCA Civ 808**

**56.14**

**Approved driving instructor—application for extension refused—failure to disclose convictions**

The Court of Appeal (Richards, Toulson, and Sullivan LJJ) dismissed H's appeal against a decision of the Transport Tribunal dismissing his appeal against the registrar's refusal to extend his registration as an approved driving instructor. The appellant had been an approved driving instructor for twenty-nine years and had no motoring convictions. However, the registrar refused his most recent extension application after discovering that he had failed to disclose numerous criminal convictions in previous extension applications. The convictions had arisen largely from his turbulent domestic circumstances, and included a public order offence and dishonesty offences. The registrar had to be able to carry out his functions of scrutiny effectively. If an applicant failed to disclose convictions or make a false declaration, it struck at the heart of the registration process and the reliability of the register. Such conduct was highly relevant to whether an applicant was a fit and proper person.

***Mulla v. Solicitors Regulation Authority* [2010] EWHC 3077 (Admin)**

**56.15**

**Certificate of enrolment—refusal by SRA—failure to disclose police cautions for dishonesty—whether sufficiently grave to refuse registration—role and approach of Court**

M appealed pursuant to the Solicitors Admission Regulations 2009 and the Training Regulations 2009 against the decision of an adjudication panel of the SRA refusing to issue him with a certificate of enrolment. In his application form, M declared that he had been convicted of an offence and gave details. A PNC check confirmed two cautions for fraud and kindred offences relating to an incident in February 2006. M had been approached by his cousin, and went to two electrical shops and purchased televisions using his cousin's business credit card. In making further purchases, the card was declined and M was arrested. In allowing M's appeal, Kenneth Parker J said that the starting point of the panel's consideration was rightly that the appellant had been dishonest. Even putting aside the caution, he had misrepresented himself twice as the authorized user of the card, knowing that he was not so, and in particular he had forged his cousin's signature on two occasions. There was therefore objective material before the SRA that justified the conclusion that he had committed the offences of dishonesty. However, the issue was whether the offences of dishonesty were, in all the relevant circumstances, of such gravity that the SRA was entitled to refuse the issue to M of a certificate of enrolment, having regard to the twin regulatory objectives of protecting the public and maintaining public confidence in the profession. On the facts, and having regard to the mitigation, this was a borderline and exceptional case in which other personal circumstances could properly be taken into account. The first of those was the outstanding references that M had in his favour. Second, it had been three years between the time of the SRA proceedings and the time when the offences had been committed. It was now approaching five years since the offences had been committed and the evidence showed that, in the meantime, the behaviour of M had been exemplary. The offences of dishonesty, although by no means trivial, were not such as to cast significant doubt on M's fitness to be enrolled taking due account of the twin regulatory objectives. Whilst paying appropriate deference to the expertise of the regulator, the Court is entitled—indeed, is bound—to act on its assessment, taking the primary factual findings as found by the SRA and in a case in which the essential question is the final assessment of all the relevant factors.

*Davis v. Solicitors Regulation Authority* [2011] EWHC 3645 (Admin)

**56.16** Certificate of enrolment—refusal by SRA—failure to disclose convictions—advice sought from supervisor—application granted

The appellant, D, appealed against the decision of a panel of the SRA, which, on 18 October 2010, refused her enrolment as a solicitor on the basis that it was not satisfied as to her character and suitability to become a solicitor pursuant to the Solicitor's Admission Regulations 2009. D was now 29 years old. When she was aged 15, she came before the Youth Court and pleaded guilty to causing criminal damage and using threatening behaviour. She failed to disclose these details when applying for a certificate of enrolment in April 2004 under the Solicitors' Training Regulations. In November 2005, D was charged with and pleaded guilty to an offence of driving with excess alcohol. She did not disclose that conviction to the SRA in 2005. In May 2007, whilst working as a trainee solicitor, she realized that the driving conviction was a discloseable matter. She consulted her supervisor, who sought advice from the firm's practice consultant. Wrongly, the practice consultant advised that there was no requirement to disclose the conviction. In allowing D's appeal, Collins J said that the approach that must be adopted—and indeed that the panel had to adopt and the Court had to adopt—in cases such as this was laid down by the Master of the Rolls in *Jideofo and ors v. Law Society*. Obviously, if there is any question of dishonesty, that will be regarded as most serious because, as Lord Bingham said in *Bolton v. Law Society*, the importance of knowing and ensuring that a solicitor is honest is all too obvious. There is a need to know that a solicitor can be trusted at all times. Although an offence such as drink driving is not one that carries any dishonesty—while, of course, it may do in certain circumstances, because there may be involved in it elements that show dishonesty, that was not the case here—the lack of dishonesty does not mean that the matter cannot be regarded as serious and, as the panel itself said, a bar, at least for a time, to enrolment. The question always is for how long and what should be the evidence to show not only that the matter has been put behind the applicant, but also that there really is no risk that it will mean that there should be concerns about the future. In the instant case, the evidence before the panel was such that it was not reasonable in all of the circumstances to believe that D was likely to act in any way that showed a risk to the public or which would be damaging to the reputation of the profession. Of course, the fact of a criminal offence in the past is, or rather can be, a real concern, but if the circumstances are properly regarded, the way in which the individual has dealt with his or her life since that offence can show that the damage can now be regarded as unlikely to occur.

*Chaudery v. Solicitors Regulation Authority* [2012] EWHC 372 (Admin)

**56.17** Certificate of student enrolment—convictions not disclosed—appeal dismissed

When applying in 2009 for a certificate of student enrolment with a view to future admission as a solicitor, the appellant disclosed the existence of a driving offence. It subsequently transpired that he had, in fact, first, two convictions for speeding in 2006 and 2008, and, second, a caution for possession of a class C drug whilst on bail issued by the Greater Manchester Police in 2007, and, third, a conviction for driving without due care and attention in 2008. In August 2010, a panel concluded that the appellant had been dishonest; it refused his application for enrolment, on the basis that it was not satisfied as to his character and suitability. None of the appellant's grounds of appeal disclosed any reason for disturbing the decision of the panel. It was manifestly justified and could not be faulted either to its findings, reason, or approach.

### *General Osteopathic Council v. Sobande* [2011] CSOH 39

**56.18** Osteopath—not registered—injunction to restrain defendant from describing self as an 'osteopath'

S had undergone training and certification in 1986 as a member of the General Council and Register of Osteopaths, a voluntary organization that was disbanded by the introduction of the General Osteopathic Council under the Osteopaths Act 1983. In 2001, S was offered conditional registration by the General Osteopathic Council, so that he would be allowed to continue to practise on condition that another osteopath acted as a mentor to him. S objected, the net effect of which was that he was not registered as an osteopath and was therefore not entitled to describe himself as such. If he were to do so, he would commit an offence under section 32 of the Act. S fully admitted that he so described himself, on the plaque outside his surgery, on notepaper, and in advertisements in the *Yellow Pages*. He equally made clear that he intended to carry on describing himself as an osteopath. In these circumstances, the Council sought interdict against S from describing himself as an osteopath. Granting the remedy of interdict, Lord Wheatley said that he agreed with the Council that S was not entitled to pass himself off as enjoying the privileges, meeting the same professional standards, or possessing the same indemnity insurance as those who were registered with the Council. The Council has a right to prevent non-members from passing themselves off as members: *Society of Accountants in Edinburgh v. Corporation of Accountants* (1843) 20 R 750; *Corporation of Accountants v. Society of Accountants in Edinburgh* (1903) 11 SLT 424. That a common law right to prevent a person from passing himself or herself off as somebody else may exist besides statutory regulations is clear: *Warnik v. Townend* [1979] AC 731, 742–3; *Law Society of England and Wales v. Society of Lawyers* [1996] FSR 739. In the instant case, S made it particularly clear that he considered that he had all of the necessary qualifications and skills and experience that he needed to practise osteopathy, and that for that reason he intended to call himself an osteopath and to carry on business as such. He wholly failed to recognize that while he may have been an excellent osteopath, Parliament had required that all osteopaths subject themselves to an assessment for the purposes of registration, in order to provide a uniform standard of skill and practice, and all other aspects of professional provision for the protection of both the public and the profession.

### *Islam v. Bar Standards Board*, 1 August 2012 (unreported)

**56.19** Student barrister—disability—registered blind—academic minimum entry requirement—whether appropriate to allow exemption of academic stage of training

The appellant appealed against the decision of the Qualifications Committee of the Bar Standards Board, upholding the decision of the head of education standards of the Board, who declined to allow the appellant to proceed to the vocational stage of qualifying as a barrister on the grounds that his academic qualification was not sufficient to enable him to do so. The appellant wished to qualify as a barrister. The Consolidated Regulations of the Inns of Court and the General Council of the Bar require a student seeking to become a barrister to complete both the academic and the vocational stages of training. The appellant had been registered blind on 17 April 2001 and, in a letter from Moorfields Eye Hospital, it was said that he would need 50 per cent extra time in taking his examinations in view of his visual difficulties. The Consolidated Regulations require a student to achieve a lower second-class degree or above in order to proceed to the vocational stage, although the Bar Standards Board does have discretion in exceptional circumstances to waive that requirement. Having failed to meet the Bar's minimum academic entry requirements, the applicant applied to the Board for the exercise of discretion for the vocational stage. In dismissing the applicant's appeal, Sir Robert Nelson, sitting as the Visitors to the Inns of Court, said that the Bar Standards Board, as part of the General Council of the Bar, is a public authority under the Disability Discrimination Act 1995 and its successor, the Equality Act 2010. It is unlawful for a qualifications body such as the Board to discriminate against a disabled person such as the appellant. However, where such a body is applying an academic standard, it will not have discriminated against a disabled person if it can show that the standard is, or would be, applied equally to persons who did not have the particular disability, and its application is a proportionate means of achieving a legitimate aim. The burden is on the qualifications body to show that the competence standard is a proportionate means of achieving a legitimate aim. In the instant case, the central issue was whether there was clear evidence that the applicant was academically second-class quality overall and whether his failure to achieve such a level was directly attributable to a temporary cause, which prevented or impeded him from fulfilling his full academic potential. The evidence as a whole showed that the appellant was given extra time consistent with the suggestions from Moorfields Eye Hospital. In the absence of any complaint that this was insufficient by the appellant, it was difficult to see what more could have been done. This was a difficult case because the appellant was a man with ability and determination, seeking to fulfil his long-held ambition to qualify as a barrister and then to use his legal skills in Bangladesh. His disability was undoubtedly a serious handicap to him in life, but he approached the problems that he faced not only with determination, but also with good humour and enthusiasm. With regret, the conclusion of the Visitors was that the discretion to permit him to proceed to the vocational stage could not be exercised here, because there was no clear evidence that he was academically of second-class quality overall, in spite of the difficulties that he undoubtedly experienced when taking his examinations. The evidence was insufficient to say that the appellant's failure to achieve the second-class degree was directly attributable to insufficient extra time being given to him to take the examinations to compensate for his difficulty. There was no clear evidence that he was academically of second lower-class standard overall and this was not one of those rare cases in which exceptionally the discretion could be properly exercised.

***Adesemowo v. Solicitors Regulation Authority*** **[2013] EWHC 2015 (Admin)**

**56.20** Solicitor—certificate of good character and suitability—non-disclosure of conviction and police caution

A appealed, pursuant to the Solicitors Regulation Authority Admissions' Regulations 2011, against the decision of a committee of adjudicators to refuse his appeal against the SRA's refusal to grant him a certificate of satisfaction as to his character and suitability to be a solicitor. A qualified to the Nigerian Bar in 2003. In 2006, he came to the United Kingdom to study for a postgraduate law degree. Shortly after arriving, he received a police caution for taking a vehicle without consent and, on 10 March 2006 in Wolverhampton Magistrates' Court, he was disqualified from driving for two months for other motoring offences. In April 2008, A sought a certificate of eligibility and suitability to sit the qualified lawyer's transfer test. He did not disclose the police caution or conviction. In September 2011, A applied to join the roll of solicitors; in January 2012, a case worker decided that A was not fit to be admitted to the roll because of his previous non-disclosure of the police caution and conviction.

On 20 March 2012, an adjudicator refused to consider A was of sufficient character to be admitted to the roll and, subsequently, a committee of adjudicators upheld this decision. The committee said that whilst it could not state with any certainty that A was dishonest or negligent, it did not consider, given the clarity of the question asked, that there was any adequate explanation for his inaccurate answer. In dismissing A's appeal, Dingemans J said that there was no adequate explanation for A's failure to disclose the previous caution and conviction and disqualification. A had said that it was a case of wrongly interpreting or understanding the question. That explanation was not consistent with his letter to the SRA, in which he said that he was not aware that the information was not previously disclosed to the SRA. The committee said that it was not confident that A met the requirements of character and suitability in circumstances under which he failed to disclose information to a regulatory body when required to do so, in the absence of exceptional circumstances. In so finding, the committee was acting consistently with the relevant guidance. The committee did not consider A's explanation to be satisfactory and his Lordship said that it seemed to him that this was plainly the right decision on the basis of the material that was before the committee.

### *Cordle v. Financial Services Authority*, FS/2012/0006, 2 January 2013

**56.21** Independent financial adviser—regulated activities—approved person—lack of honesty and integrity

C had, for many years, been a sole practitioner independent financial adviser (IFA) who operated through a company. C and the company referred decision notices issued by the Financial Services Authority (FSA) to the Upper Tribunal rejecting the company's application for authorization to carry on regulated activities and C's application to become an approved person, on account of C's lack of honesty and integrity. Prior to the applications, the company had been appointed representative of a financial services network called S Ltd. In November 2010, investigators from S Ltd attended the company's office following information received to the effect that the company and C had permitted an 'introducer' to S Ltd to give advice to customers without regulatory authorization and without the introducer having the necessary qualifications to do so. C untruthfully told S Ltd that the person had not been involved in giving advice and that C himself had given the advice. S Ltd terminated the network. In C's subsequent application to the FSA for authorization, C answered 'No' when asked whether he or his company had ever been the subject of an investigation into allegations of misconduct or malpractice, or the possible carrying on of unauthorized regulated activities. No mention was made of the S Ltd investigation. In upholding the decision notices, Sir Stephen Oliver QC, Tribunal judge, said that the FSA had sound reasons for rejecting the applications for authorization and approval. Both applications turned on the fitness and propriety of C to perform the relevant controlled functions. C drew attention to his long and virtually unblemished career in financial services, and to the reputation for competence and honesty that he enjoyed among his clientele. A number of good references were submitted. The Tribunal judge said, at [63], that honesty and integrity cannot be ignored. Those two qualities are essential because of the way in which the regulatory regime works. It places significant trust in its advisers in the front line. It depends on their honesty and integrity, because the regime does not involve constant supervision on the part of the regulatory authority, the FSA. The fitness and propriety of an individual is to be assessed in the context of the particular controlled function. Also relevant to the assessment will be any positive steps towards rehabilitation that have been taken by the individual: at [67].

***Ahmed v. The Inns' Conduct Committee* [2012] EWHC 3270 (QB)**

**56.22** Barrister—criminal conviction—deregistration from BPTC

In this case, application for an interim injunction to suspend the decision of the City Law School to deregister the applicant from the Bar Professional Training Course (BPTC) was refused. Between autumn 2004 and late spring 2005, the applicant committed ten offences of using a false instrument with intent and two offences of obtaining a pecuniary advantage by deception. The offences were committed in connection with online trading by the applicant in stocks and shares, and included the use of doctored bank statements showing greatly inflated credit balances to gain enhanced credit facilities. On 22 April 2009, the applicant pleaded guilty, on rearraignment, to all twelve offences. He was sentenced to a total of nine months' imprisonment, suspended for two years, and ordered to complete 180 hours of community service. In due course, a confiscation order in the sum of £18,921.30 was also made against him. There was no doubt that the applicant had the necessary educational qualifications to enable him to apply to an Inn of Court and to take the BPTC. In May 2012, he applied to become a member of The Honourable Society of Inner Temple. The Inner Temple referred the question of whether the applicant was a fit and proper person to become a practising barrister to the Inns Conduct Committee. By a detailed written decision dated 29 August 2012, a hearing panel concluded that the applicant was not a fit and proper person to become a practising barrister and that it would not be appropriate for him to be recommended for admission to the Inner Temple until at least ten years after his conviction. In dismissing the applicant's application for an injunction, Sweeney J said that the central proposition in the applicant's case was that the BPTC Academic Regulations did not permit the City Law School to deregister him. However, against the background set out above, this was unarguable. The Regulations do not arguably mean that if a student is asked to leave because he has not obtained membership of an Inn, he must be allowed to continue on the course no matter what. There was no serious question to be tried and the application would be refused.

***Ahmed v. Bar Standards Board*, 16 May 2013 (unreported)**

**56.23** Barrister—examinations—withdrawal of decision to retake

Sir Anthony May held that the Bar Standards Board was entitled to withdraw its decision to allow A to retake examinations, having erroneously overlooked that he had failed to pass after three attempts. When a public regulatory authority had a duty to act within its power, if it acted outwith that power, then it acted outwith its competence. A person who was not qualified to become a barrister could not do so because of a mistaken decision to allow him to retake the test.

***Hassan v. General Optical Council* [2013] EWHC 1887 (Admin)**

**56.24** See paragraph 19.69.

***R (Patel) v. General Medical Council* [2013] 1 WLR 2801, CA**

**56.25** Doctor—overseas medical qualification—legitimate expectation of UK registration

The Court of Appeal (Lord Dyson MR, and Lloyd and Lloyd Jones LJJ) held that P, a doctor with a primary medical qualification from St Kitts and Nevis, was entitled to registration in the United Kingdom. P had received email assurances from the General Medical Council (GMC) that he would be entitled to register after taking the Professional and Linguists Assessment Board test and completing a placement in an NHS hospital. P had a legitimate expectation of UK registration. The claimant was a qualified pharmacist. He obtained a bachelor's degree (BSc) in pharmacy from City of Leicester Polytechnic and had worked as a pharmacist manager at a high-street chemist for three years, before starting his own independent pharmacist business. In 2004, he decided that he wished to qualify as a doctor, with a view to practising medicine. He wanted to complete his pre-clinical studies on a part-time basis, at the same time as conducting his business as a pharmacist. Consequently, his plan was to study on a suitable distance-learning course, but to carry out the clinical rotations element in the United Kingdom. He identified a Bachelor of Medicine, Bachelor of Surgery (MB BS) degree course offered by the International University of Health Sciences, St Kitts and Nevis, an institution listed in the World Health Organization (WHO) Directory and one that had some affiliation with the UK-based London College of Medicine. In dismissing the claimant's application at first instance, Hickinbottom J said that he had much sympathy for the claimant. However, the Court can intervene only if it is satisfied that the GMC has acted unlawfully. The Court did not consider that the criteria adopted by the GMC were unlawful, or that the claimant could rely upon any legitimate expectation: [2012] EWHC 2120 (Admin). In reversing the judge and giving the judgment of the Court of Appeal, Lloyd Jones LJ (with whom Lord Dyson MR and Lloyd LJ agreed) said, at [33]–[38], that it was the intention of Parliament, through the Medical Act 1983, that the GMC should set minimum criteria that an applicant must meet before he or she can be registered as a doctor. In principle, the GMC would be entitled to change the criteria with immediate effect if it were satisfied that, on reasonable grounds, it was necessary to bring in the necessary changes with such immediacy. Article 2 of Protocol 1 to the European Convention on Human Rights (ECHR) and the Schedule to the Human Rights Act 1998 were not engaged in these circumstances: see *Belgian Linguistic Case (No. 2)* (1968) 1 EHRR 252; *R (Sivills) v. General Social Care Council* [2007] EWHC 2576 (Admin), *per* Jackson J. The real substance of the appeal lay in the claim based on legitimate expectation. The claimant received a clear, unequivocal, and unqualified assurance from the Registration and Education Directorate of the GMC, in an email dated 16 November 2004, the effect of which was that if he were to complete the proposed course in St Kitts and Nevis within a reasonable time, the qualification would be recognized by the GMC. In the particular case with which the Court was concerned, the claimant was able to point to specific statements in the exchanges of emails directly focused not simply on the universities and institutions whose degrees are recognized, but also—a matter of vital importance—on the fact that the course proposed by the claimant was to be undertaken by distance learning. The claimant's repeated requests for clarification in succeeding emails served to focus attention effectively on this specific position: at [29], [48], and [52]. The statutory scheme of the Medical Act 1983 does not exclude the operation of the principle of legitimate expectation in the particular circumstances of this case. The statutory duty has to be exercised in accordance with established principles of substantive fairness: at [55]. In the particular circumstances of the claimant's case, the GMC was not entitled to go back on the assurance that it gave him. The relief granted should be specific to the claimant and should compel the GMC to recognize his primary medical qualification for the purposes of the Medical Act 1983: at [93].

### *R (Alacakanat) v. General Medical Council* [2013] EWHC 1866 (Admin)

**56.26** Turkish doctor—whether 'exempt person'

The claimant, an experienced Turkish doctor and paediatric consultant, applied to the GMC under section 19 of the Medical Act 1983 for registration as an 'exempt person'. An 'exempt person' is defined in section 19(2)(c) of the Act as a person who is not a national of a relevant European state, but is, by virtue of an enforceable Community right, entitled to be treated, for the purposes of access to and pursuit of the medical profession, no less favourably than a national of a relevant European state. Dismissing A's renewed application for judicial review, Edwards-Stuart J said that A had no enforceable right to be treated no less favourably than a national of a relevant European state, and that the requirement that the claimant would be required to sit an English language test, together with two medical tests known as PLAB1 and PLAB2, was not discriminatory on grounds of nationality. They offered the claimant the opportunity, by means of an aptitude test, to provide comparable evidence of knowledge and qualifications required by national law. The requirement was also justified on public policy and public health grounds.

***Professional Standards Authority for Health and Social Care v. Nursing and Midwifery Council* [2013] EWHC 4369 (Admin)**

**56.27** See paragraph 43.17.

***JB v. General Teaching Council for Scotland* [2013] CSIH 114**

**56.28** Teacher—conviction of sexual abuse—conviction later quashed and charges found not proven—application for re-registration

In December 2012, a fitness to teach panel of the General Teaching Council for Scotland (GTCS) directed that the appellant's application for registration be refused and that he should be prohibited from making a further application for a period of two years. In 2005, he was convicted of allegations of sexual abuse against his two sisters and brother, and was sentenced to eleven years' imprisonment. In 2009, his conviction was quashed on the ground of defective representation at his trial. Authority was given to the Crown to raise a new prosecution, as a result of which, after a retrial in 2010, all of the charges against the appellant were found not proven. His registration with the GTCS having lapsed while he was imprisoned, after his acquittal in 2010 he applied for re-registration as a teacher. The panel found various allegations proved and his appeal to an Extra Division of the Inner House, Court of Session, was dismissed. The Court readily understood the deep sense of frustration experienced by the appellant in circumstances in which, in his view, his siblings' allegations were without foundation; a lengthy criminal process, during which he spent years in prison, resulted in his acquittal; one of his siblings admitted on oath at the trial in 2010 to having threatened to ruin his life and his career; and when he had hoped to resume his teaching career, the GTCS embarked on an investigation into the allegations. But once these matters were drawn to the attention of the GTCS, it was bound to investigate them. While it is clear that, in his own mind, the appellant was satisfied that his siblings had lied on oath in the evidence in the High Court, given the serious nature of the allegations, the GTCS was carrying out its statutory obligation by investigating them. This was repeatedly explained to the appellant in letters sent to him by the officials.

***Fajemisin v. General Dental Council* [2014] 1 WLR 1169**

**56.29** See paragraph 43.16.

***Giambrone v. Solicitors Regulation Authority* [2014] EWHC 1421 (Admin)**

**56.30** Registered European lawyer—financial investigation—books of account—withdrawal of registration

G was an Italian lawyer, an *avvocato*. He was registered in England and Wales as a registered European lawyer in April 2005, pursuant to the European Communities Lawyers Practice Regulations 2000. An investigation of G's firm's books of account was commenced in August 2008, leading to the preparation of a forensic investigation report in February 2009 and a further report dated 1 July 2010. In 2009, restrictions were placed on G's practising certificate and, in February 2010, disciplinary proceedings were launched against him and against his two ex-partners. The SDT found that although there had been no dishonesty, the firm's accounting records were 'a shambles', and said that had G been practising as a solicitor in England and Wales, he would have been struck off. As it was, the proper sanction was withdrawal of his name from the register of registered European lawyers. On appeal, the Divisional Court was given to understand that this was the first disciplinary appeal to come before the Court in which the appellant was or had been a registered European lawyer. In dismissing the appeal, Laws LJ said that the SDT's findings were fully justified. As to the sanction being excessive, the SDT was, on the merits, entitled to regard the allegations as 'very serious matters'. The SDT was entitled to find that the reconstruction exercise carried out by G had been inadequate, for it had been abandoned. There was a true sense in which G had walked away from a situation that his firm was in. There is power to impose a strike-off in relation to a registered European lawyer, and the sanction of withdrawal in the case of a registered European lawyer possesses a gravity that lies between the sanctions of suspension and of strike-off in the case of an English solicitor. It was not a penalty that would have been available in the case of an English solicitor. Foskett J, agreeing with the comments on the issue of sanction, said that, in connection with a registered European lawyer, withdrawal of registration was an additional level of sanction to those available in respect of a solicitor, subject to the ordinary domestic jurisdiction. It comes somewhere between suspension and striking off. It is therefore a sanction that any SDT will have to consider as part of the process of deciding what is a proportionate sanction in any case that achieves the level of seriousness that demands consideration of a sanction that will interfere with a registered European lawyer's right to practise in the United Kingdom. The precise difference in terms of the practical implications is not entirely plain and may not be very significant. But withdrawal of registration should be seen as at a somewhat lower level of seriousness from that justifying striking off.

### *Woodman-Smith v. Architects Registration Board* [2014] EWHC 3639 (Admin)

**56.31** See paragraph 43.21.

### *General Medical Council v. Nakhla* [2014] EWCA Civ 1522

**56.32** See paragraph 4.31.

### *Hunt v. The Commissioners for Her Majesty's Revenue & Customs* [2014] UKFTT 1084 (TC)

**56.33** Money Laundering Regulations—application for registration as service provider—fit and proper person

In 1993, H was found guilty and imprisoned for eight years for conspiracy to defraud, and was disqualified from being a company director for ten years. By a decision dated 5 April 2013, Her Majesty's Revenue & Customs (HMRC) refused an application by a company controlled by H for registration as a service provider under the Money Laundering Regulations 2007 on the ground that H had failed the fit and proper person test. In dismissing H's appeal, the First-tier Tribunal, Tax Chamber (Judge Timothy Herrington sitting as chairman), said that the starting position is that H's conviction, although a long time ago, was for conspiracy to commit a very serious fraud. There was a very substantial loss to the Revenue over a very long period of time. Between 1975 and 1991, a total of £219,911.823 was extracted from Nissan UK Ltd, of which H was a director and the general manager, causing a loss to HMRC of £97,119.462. The prosecution case was that B, the chairman of Nissan UK, was the moving spirit and that H, the most senior executive after B, was the second in command. The Tribunal was not satisfied that H had been rehabilitated to the extent necessary to allow his application. His reluctance, until a very late stage, to accept any degree of culpability, his attitude to the fraud as being essentially a corporate matter, and his lack of openness in a number of respects, when taken together with the seriousness of the conviction, the risks posed by the business of H's company, and his role as a controlling shareholder, led the Tribunal to conclude that he was not a fit and proper person for the purposes of the Money Laundering Regulations.

***Yousef v. Solicitors Regulation Authority*, 23 January 2015 (unreported)**

**56.34** Qualified foreign lawyer—application for admission to roll—failure to disclose conviction and removal of licence—lack of insight

Y, a qualified foreign lawyer admitted as an advocate in Bangladesh in 1995, appealed against the refusal of the SRA to admit him to the roll of solicitors. On his application in 2010 for a qualified lawyer's transfer certificate of eligibility, Y failed to disclose a caution for common assault on his daughter. On his application for admission to the roll of solicitors in 2011, Y failed to disclose that, in the meantime, a consumer credit licence with the Office of Fair Trading had been revoked for failure to comply with regulatory requirements. Cranston J, in dismissing Y's appeal, said that Y had failed to address the SRA's conclusion that he lacked insight. The SRA had been entitled to take account of the extent to which Y, albeit not dishonest, had not addressed issues of regulatory compliance. The SRA had been entitled to conclude that it did not have confidence that Y would comply with legal and regulatory requirements.

## C. Other Relevant Chapters

**56.35** Dishonesty, Chapter 19
Impairment of Fitness to Practise, Chapter 33
Jurisdiction, Chapter 43
Misconduct, Chapter 47
Restoration to the Register, Chapter 58

# 57

# REPRIMAND



## A. Legal Framework

**57.01** Examples include:

Dentists Act 1984, section 27B(6)
Police (Conduct) Regulations 2012, regulation 35 (management advice or a warning)
Bar Standards Board Disciplinary Tribunals Regulations 2014, rule E158 and Annex 1
Chartered Institute of Management Accountants Regulations, Part II ('Discipline')
Institute of Chartered Accountants in England and Wales Disciplinary Bye-laws, bye-law 22(3)

Although the Solicitors Disciplinary Tribunal has no express statutory power to do so, it frequently issues a reprimand to a solicitor—see Andrew Hopper QC and Gregory Treverton-Jones QC, *The Solicitors' Handbook* (London: The Law Society, 2015) paragraph 21.07.

## B. Disciplinary Cases

### *In re H (A Barrister)* [1981] 1 WLR 1257

**57.02** Barrister—sanction of reprimand—purpose of sanction

H, a practising barrister of some eight years' standing, was convicted of an offence of persistently importuning for immoral purposes in a public lavatory on one occasion. He pleaded not guilty and gave evidence. After a lengthy retirement, the jury returned a verdict of guilty and he was fined £250. Before the Disciplinary Tribunal of the Senate of the Inns of Court, he admitted one charge of conduct unbecoming a barrister arising from the conviction. He was sentenced to a period of suspension for three months from practice as a barrister, and from the enjoyment of all rights and privileges as a member of his Inn of Court. The Tribunal's retirement also was a lengthy one. From that sentence, he appealed to Her Majesty's judges as Visitors to the Inns of Court. Before the Tribunal, he did not seek to go behind the conviction, nor did he seek to do so before the Visitors. The mitigation in the instant case was described by the Tribunal as impressive. H's offence, isolated though it was, was not trivial. But neither is a reprimand a trivial penalty. The function of these proceedings, to bring home to the barrister the fact that the conduct that brought about the conviction will not be passed over or tolerated by the profession, was achieved by a reprimand.

***Preiss v. General Dental Council* [2001] 1 WLR 1926**

**57.03** Dentist—admonishment or censure—implied in dental disciplinary system

Lord Cooke of Thorndon, giving the judgment of the Privy Council, said that:

> 30. The only penalties expressly provided for in [the Dentists Act 1984] and delegated legislation are erasure and suspension. But the General Dental Council's guidance publication *Maintaining Standards*...include[s] in paragraph 8.22 other options available to the [Professional Conduct Committee] PCC, the first of which is "(i) the Committee may conclude the case with an admonition". Likewise the Council's website says that there is an option to give a warning. It was not argued on either side before the Board that the option so claimed does not exist in law. In professional disciplinary legislation it is not uncommon for express provision to be made for admonition or censure. While express provision is desirable, their Lordships are prepared to hold that in the dental disciplinary system it is implied.
>
> 31. In the opinion of their Lordships to suspend the appellant from practice would be neither necessary not just. This is an exceptional case where the right to practise should not be interfered with despite a finding of serious professional misconduct.... [T]he appeal ought to be allowed to the extent of...replacing the suspension [for twelve months] by an admonition.

***Macleod v. Royal College of Veterinary Surgeons* [2006] UKPC 39**

**57.04** Veterinary surgeon—genuine mistake of professional obligations—reprimand rather than suspension

The appellant, M, an experienced veterinary surgeon carrying on a small animal veterinary practice in Hertfordshire, was charged with disgraceful conduct in a professional respect in that she allowed and/or authorized veterinary nurses at a clinic to administer vaccinations to small animals and to furnish medicines for their treatment. The Disciplinary Committee of the Royal College of Veterinary Surgeons (RCVS) directed that M be suspended for eight months from the register of veterinary surgeons. In allowing M's appeal against sanction, and substituting for the suspension a reprimand and a warning as to her future conduct, Lord Carswell giving the opinion of the Privy Council, said:

> 26. Their Lordships are conscious of the principle that it requires a strong case to interfere with the sentence of a professional disciplinary body...In the absence of any more clear and detailed finding on the appellant's state of mind, however, and in view of the brevity of the Committee's consideration in its judgment of the issues relating to sanction, their Lordships consider that they would be justified in examining afresh the culpability of the appellant and the appropriateness of the penalty of suspension.
>
> 27. ...Approaching the case as one of a genuinely mistaken, if seriously misconceived, interpretation on the appellant's part of her professional obligations, their Lordships are inclined to see a fair amount of substance in her submissions [that the sentence was excessive and disproportionate, since it would involve her in financial outlay of many thousands of pounds]. Nevertheless, given the finding that the appellant laboured under a misapprehension, which has to be regarded as genuine, however unjustified, their Lordships have concluded that the penalty [removing her right to practise for eight months] was disproportionately heavy....

## C. Other Relevant Chapters

**57.05**

Conditions of Practice Orders, Chapter 11
Dishonesty, Chapter 19
Insight, Chapter 37
Sanction, Chapter 60
Suspension, Chapter 65

# 58

# RESTORATION TO REGISTER



## A. Legal Framework

**58.01** *Thobani v. Solicitors Regulation Authority* [2011] EWHC 3783 (Admin) | 58.10 |
| B. Disciplinary Cases | 58.02 | | |
Examples include:
| *R (Mariaddan) v. Solicitors Regulation Authority* [2009] EWHC 2913 (Admin) | 58.06 | | |
| *Balamoody v. Nursing and Midwifery Council* [2009] EWHC 3235 (Admin) | 58.07 | *Ellis-Carr v. Solicitors Regulation Authority* [2014] EWHC 2411 (Admin) | 58.15 |
| *Balamoody v. Nursing and Midwifery Council (No. 2)* [2010] EWHC 2256 (Admin) | 58.08 | *Anoom v. Bar Standards Board* [2015] EWHC 439 (Admin) | 58.16 |
| *R (Kay) v. Chief Constable of Northumbria Police (No. 2)* [2010] ICR 974 | 58.09 | C. Other Relevant Chapters | 58.17 |

Veterinary Surgeons Act 1966, section 18 (ten months from the date of removal or suspension or ten months from previous application)
Solicitors Act 1974, section 47(1)(e)
Medical Act 1983, section 41 (five years from the date of erasure or twelve months from previous application)
Dentists Act 1984, section 28
Health and Social Work Professions Order 2001, article 33
Nursing and Midwifery Order 2001, article 33
Pharmacy Order 2010, article 57

To prevent repeated applications for restoration of names to the register, a panel or committee may direct that the right to make any further such applications shall be suspended indefinitely. Any person in respect of whom such a direction has been given may, after the expiration of three years, apply to the registrar for the direction to be reviewed. See, for example:

Medical Act 1983, section 41(9) and (11)

## B. Disciplinary Cases

*R v. Master of the Rolls, ex parte McKinnell* [1993] 1 WLR 88

**58.02** Solicitor—order of SDT restoring solicitor to roll—whether appealable by Law Society—Solicitors Act 1974, section 49(1)(a)

On 19 December 1975, the Disciplinary Tribunal of the Law Society ordered that M be struck off the roll of solicitors. The reason for his being struck off was that he had breached the Solicitors' Accounts Rules and had misused clients' money. In its decision, the Tribunal said that M may be encouraged to rehabilitate himself by continuing in employment within the law subject to such consents as may be necessary and the Tribunal believed that, in due course, the time may come when an application on his part that his name should be restored to the roll might be favourably considered. On 16 November 1988, K made an application under section 47 of the Solicitors Act 1974 to have his name restored to the roll of solicitors. The Tribunal gave its decision on 30 August 1990, granting the application. The Tribunal recommended to the Law Society that any practising certificate granted to M should restrain him from practising other than in approved supervised employment. The Law Society sought to bring an appeal before the Master of the Rolls against the order of the Tribunal under section 49(1)(a) of the 1974 Act. K, before the Master of the Rolls, took a preliminary point of law—namely, that the Law Society had no locus standi to appeal under that section. Lord Donaldson of Lymington MR ruled on 18 December 1990 that the Law Society did have locus standi and was entitled, pursuant to section 49(1)(a) of the 1974 Act, to appeal to him against the order of the Tribunal restoring the applicant to the roll of solicitors. An application for judicial review of that ruling was made on behalf of M on 18 April 1991. The relief sought was certiorari to quash the ruling on the grounds that it was wrong in law. The Divisional Court (Lord Taylor of Gosforth CJ, and Simon Brown and Roch JJ) held, dismissing the application, that the Solicitors Act 1974 is a consolidating statute so that there is a presumption, albeit rebuttable, that Parliament did not intend to make changes to the existing law. True, the statute indicates Parliament's clear intention to introduce change in the rights of appeal depriving the Law Society of the right that it had previously enjoyed to appeal against an order prohibiting the employment of a named solicitor's clerk. But no such clear intention is shown in regard to the rights of appeal against orders under section 47(1)(b) (restoration to the roll) or under section 43(3) (revocation of an order). On the contrary, all of the indications are the other way. In short, appeals from the Tribunal, other than appeals against orders under section 43(2), are to lie at the instance of either side to the application or complaint. That this was Parliament's intention is absolutely clear also from the wording of section 49(1), which expressly allows for an appeal equally from the making of an order that a solicitor's name be restored to the roll as from the refusal of an application. Consequently, the ruling of the Master of the Rolls was upheld.

*Raji v. General Medical Council* [2003] 1 WLR 1052

**58.03** Decision of Committee suspending application for restoration indefinitely—procedural unfairness

R appealed under section 40 of the Medical Act 1983 against a decision of the Professional Conduct Committee (PCC) of the General Medical Council (GMC) made on 20 August 2002, which suspended his right to make further application for his restoration to the register. In early 1999, R had faced disciplinary charges before the PCC. The Committee found that his conduct was improper and irresponsible in relation to the prescribing of morphine sulphate, and the Committee ordered the erasure of his name from the register. R did not appeal against this decision. In August 2000, R applied for the restoration of his name to the register. By a decision of 8 August 2000, the PCC refused this application. In 2002, R commenced a second application for restoration of his name to the register, which was heard on 20 August 2002. In its determination, the Committee rejected R's application for his name to be restored to the register and, in the same determination, stated that, because this was his second application for restoration to the register in the current period of erasure, the Committee had decided to suspend his right to make a further application for restoration to the register. The Committee stated that it was convinced that this was realistic, and that it was both in the public interest and R's own interest to bring this matter to a close. R did not appeal against the decision of the Committee to refuse to restore him to the register. He challenged, however, the decision of the Committee directing that his right to make further applications for restoration of his name be suspended indefinitely. In allowing R's appeal on this point, the Privy Council stated that the procedure adopted by the Committee was flawed.

> 13. ... However, the issues of restoration and suspension under section 41 of the Medical Act 1983 may overlap. It is therefore fair that the doctor should know the decision on restoration, and the reasons for it, so that he can advance his arguments against suspension of the right to re-apply in the light of the reasons for refusing restoration. On a broader footing a doctor, unrepresented and circumstanced as [R] was, may well feel disadvantaged by having to make alternative and conflicting submissions. Separate consideration of what are distinct but potentially overlapping issues will make it easier for the doctor to put forward his case against suspension effectively, e.g. he may make concrete proposals for any further training envisaged in the decision refusing restoration. Recognition of a procedural right of fairness, requiring separate consideration of the issues, will play an instrumental role in promoting just decisions. It will also reduce the risk of an appearance of unfairness.
>
> 14. ...If it is held that the restoration issue must be decided first, it will assist the members of the PCC in the sense that it will enable them to focus squarely on the separate issue of the need for a direction to suspend the right to re-apply for restoration. It will reduce the risk of members conflating the two enquiries. It will promote better decision-making. And it will avoid any appearance of inadequate consideration of the suspension issue.

The appeal was allowed to the extent of quashing the direction suspending the right of R to reapply for restoration and remitting the matter to the PCC.

### *Gosai v. General Medical Council* [2003] UKPC 31

**58.04** Decision suspending doctor's application for restoration indefinitely—proper approach by Committee—whether inappropriate or excessive

G appealed from a direction of the GMC's PCC that his right to make further applications for restoration of his name to the register be suspended indefinitely. G's name was erased from the register on 31 October 1997. The charge found proved was that G had failed to take adequate steps to arrange an investigation for a patient and that, in his evidence submitted to the inquest into the patient's death, he made statements that were untrue. In March 1999, G's first application to be restored to the register was dismissed. In January 2002, his second application was also dismissed. At the conclusion of the proceedings, the chairman announced the determination of the Committee that it would not be in the public interest to restore G's name to the register. The chairman then invited submissions as to the exercise of the Committee's powers under section 41(9) of the Medical Act 1983 to direct that G's right to make any further such applications should be suspended indefinitely. After hearing submissions, the chairman announced the result of the Committee's deliberations, stating that the Committee had decided to impose a direction to suspend indefinitely the right of G to make further applications for restoration. The Committee considered this necessary because of G's continued lack of insight into the significance of his failures, despite the lapse of time since the events that gave rise to his erasure. In reaching its conclusion, the Committee had had regard to the public interest and also to G's own interest. On appeal, it was argued that the PCC had exercised a draconian power when it was unnecessary for it to do so. Such a power of indefinite suspension should be reserved only for very clear cases in which it is in the public interest that the person erased from the register should be prevented from reapplying. Sir Phillip Otton, giving the judgment of the Board dismissing G's appeal, said that:

> 23. ... There is no basis for the assertion that suspension of the right to apply for restoration should be restricted to very clear cases, or should be regarded as exceptional. The PCC's discretion to impose a suspension order is, on the face of the legislation, unconfined and unfettered. The Committee was not obliged to start with a presumption that the power to make a suspension order was in any way an exceptional or unusual remedy. It was entitled to have regard, in exercising the discretion, to the public interest. It was also entitled to have regard to the interests of those who would be otherwise affected by repeated applications for restoration, such as (as in the present case) the family of the victim of a doctor's misconduct which has taken an active part in the proceedings, which may suffer anguish and be caused expense by repeated restoration applications by the doctor.
>
> 24. Their Lordships are also satisfied that the suspension direction was not inappropriate or excessive. The PCC accepted that the appellant was making real efforts to demonstrate his fitness to practise. However, the PCC was entitled to conclude, in relation to both the refusal of restoration and the suspension of the right to reapply, that the efforts he had made were outweighed by other factors. These included the seriousness of the original offence when the PCC stated that the appellant had "fallen lamentably below the professional standards to which patients were entitled", "demonstrated clinical incompetence of the most basic kind", and "not been truthful".... Secondly, the appellant's lack of insight into the gravity of his misconduct. The Committee was thereby clearly referring to his failure to accept that he had not in fact conducted any investigation of the patient... and further that he had lied to the Inquest and to the PCC in stating to the contrary. It is not surprising that the PCC concluded that this conduct displayed a lack of insight which had continued over a considerable period of time....

In summary, there was found to be no error in principle or in law or in the exercise of the Committee's direction.

### *Council for the Regulation of Health Care Professionals v. Health Professions Council and Jellett* [2005] EWHC 93 (Admin)

**58.05** Appeal by CHRP against unconditional restoration of physiotherapist to register—chaperoning arrangements—case remitted to Committee to consider appropriate conditions

The Council for Health Care Regulatory Professionals (CHRP) appealed under section 29 of the National Health Service Reform and Health Care Professions Act 2002 against the decision of the Conduct and Competence Committee (CCC) of the Health Professions Council (HPC) to the effect that J, a physiotherapist, should be restored unconditionally to the register of professions maintained by the HPC, pursuant to article 5 of the Health Professions Order 2001. The CHRP's case was that the decision was wrong and should be quashed, and that the case should be remitted to the Committee to reconsider and dispose of J's application for restoration to the register either by dismissing it or by imposing a conditions of practice order. In June 1996, J was convicted of three offences of indecent assault upon female patients in the course of his practice as a physiotherapist. He contested the case, but was convicted and sentenced to nine months' imprisonment. In November 1996, the predecessor body to the HPC—namely, the Council for Professions Supplementary to Medicine—found in disciplinary proceedings that J's name should be removed from the register. At the time, this meant that he became ineligible for National Health Service (NHS) or local authority employment, but was entitled to continue in private practice as a physiotherapist. The Chartered Society of Physiotherapy (CSP) also removed J from membership, which meant that he could no longer style himself a 'chartered physiotherapist', but only a physiotherapist. In 1999, the CSP's PCC heard and granted an application by J for restoration of his CSP membership. J gave an undertaking to the CSP that he would have permanent chaperoning arrangements in his practice for any female patients. In July 2002, the CSP informed J that he had been reinstated as a member of good standing, but that it expected the arrangements for permanent chaperoning of female patients to remain in place for good in the future. Following the merger of the CSP with the Council for Professions Supplementary Medicine, J applied to the HPC for restoration of his name to the register then maintained by the HPC. Without his name being restored, J would not be able to style himself a physiotherapist or physical therapist of any kind. In quashing the Committee's decision and allowing the CHRP's appeal on a limited basis, Richards J said that at the heart of the matter lay the chaperoning arrangements, since the presence of an appropriate chaperone was the key to securing continued protection of the public. The question that required closest attention was whether the Committee had dealt adequately with that issue. The learned judge said that the conclusion he had reached was that J's restoration to the register should be subject to a conditions of practice order and that the conditions were best left for the Committee, because it may be thought appropriate to include one or more conditions with regard to monitoring of the chaperoning arrangements (for example as to the carrying out of a periodic audit). The detailed formulation of such conditional conditions was best done by the Committee in the light of representations from the HPC and J.

***R (Mariaddan) v. Solicitors Regulation Authority* [2009] EWHC 2913 (Admin)**

**58.06**

Solicitor—Family Law Accreditation Scheme—re-entry of name

M applied for judicial review to quash the decision of a panel of the Solicitors Regulation Authority (SRA) to refuse his re-entry to the Family Law Accreditation Scheme. The panel refused M's application on the grounds that, at the time that he filled in the form, he knew that he was facing disciplinary proceedings that raised an issue as to his character and suitability. The allegations involved breaches of the Solicitors Accounts Rules and contained an allegation of dishonesty. In answer to the question on the form regarding whether there were any matters reflecting his competence or fitness of which the Law Society ought to be aware, M answered 'no' by putting a cross in the relevant box. M told the panel that he believed his answer to the question to be truthful at the time because he believed that he had done nothing wrong as a solicitor. The panel stated that it found M's explanation unsatisfactory. The fact that M believed himself to be blameless did not alter the fact that his conduct had been referred to the Solicitors Disciplinary Tribunal (SDT), which was a serious matter, raising as it did issues about his conduct and fitness to practise. In dismissing M's claim for judicial review, Charles J said that the panel's reasoning contained sufficient detail to demonstrate that, in exercising its public function, the panel was of the view that it should refuse accreditation. It would appear that no issue was raised as to the extent of that impact at that time, but it was an inevitable consequence that there would be an impact on M's practice. The decision could not be attacked on the basis that irrelevant factors were taken into account. The factors that were taken into account were relevant and justified the conclusion that the decision reached was well within the range of decision fairly open to the panel.

***Balamoody v. Nursing and Midwifery Council* [2009] EWHC 3235 (Admin)**

**58.07** Nurse—nursing home—whether practitioner safe to be restored to register unconditionally or to be subject to suitable conditions of practice—insight

This was a restoration case. B was struck off the register of the Nursing and Midwifery Council (NMC) for matters that arose in respect of his conduct and management of a nursing home many years previously. In April 1993, he was convicted of six offences contrary to section 15(3) of the Registered Homes Act 1984. Three of the offences related to his conduct and management of the nursing home and not more generally to his nursing practice. Those that related to patients involved small quantities of drugs. So far as was known, no patient harm was suffered in consequence. The offences occurred some seventeen years previously and he was struck off thirteen years previously. He had had, prior to being struck off, some twenty-three years' of practice as a nurse, about which no other complaint had ever been made. He was accepted by the panel as being a caring individual. In allowing the registrant's appeal against the refusal of restoration and directing that the matter be reheard by a fresh committee, Langstaff J said that the panel had three options: it had to determine whether or not to restore B to the register unconditionally; it could restore him subject to suitable conditions of practice; or it could reject his application. The fundamental question was whether the practitioner was safe. The question of whether a practitioner is a safe practitioner involves looking at what that practitioner will do in the future. It was not difficult to see that a practitioner who is not up to date may not be safe, because he will not practise nursing in accordance with up-to-date standards. However, this, if it was a valid observation in the case of B, was a matter that was easily remedied by ensuring that, as a condition of restoration to the register, he would undergo or be required to undergo further training to the satisfaction of the Council.

Langstaff J was critical of the panel's concentration on 'insight' rather than looking at future risk. The learned judge said:

> 31. Underlying all these complaints were two questions. The first: what was it, asked the appellant, that he had to prove in order to satisfy the committee that he was a proper person to return to the register as a nurse? Secondly, a general complaint that the result of the proceedings in the context which I have set out was desperately unfair to him. The first of those he raised at the meeting itself. He asked the committee what was the meaning of "insight" in this context? It does plainly concern him and for good reason. The "lack of insight" had become familiar shorthand for being unsafe to practise. The definition given by the nursing member of the panel, the other two being a chairman and a lay person, was that she could not argue with a dictionary definition but would comment that "I would want you to be thinking not only how your actions have affected others and the full ramifications of that. That is what I would add for that".
>
> 32. That was not obviously a helpful definition. It may be thought to be compounded by the observations in the decision that the panel would expect an applicant for restoration to have enough insight to appreciate what evidence would be helpful to his application. This is "insight" used in a different context and sense. It is, I think, to suggest that it is entirely up to an individual to determine what matters he should bring to prove his case to the Tribunal. That does not seem to me to fit easily with the scheme set up by the Order. In the context of someone who is a caring person, whose offences have caused no actual harm that could be established, for whom they represent an isolated occasion though repeated twice over 10 days in the course of a long career otherwise unmarked by complaint, they were singularly unhelpful, bearing in mind the consequences of a refusal of restoration to this individual.

The learned judge concluded that, since the whole point of a conditions of practice order is to allow someone who, given sufficient conditions, would be safe, but who without those conditions might not be safe, in his view the panel could and should have considered in greater detail what might have been appropriate for the appellant. The panel did not sufficiently consider conditions of practise and whether they would meet the particular problems to which the panel averred.

### *Balamoody v. Nursing and Midwifery Council (No. 2)* [2010] EWHC 2256 (Admin)

**58.08** Nurse—conditions attached to restoration to register

On 5 August 2010, B appealed against the conditions imposed by the NMC's CCC to restore him to the register following the judgment of Langstaff J: [2009] EWHC 3235 (Admin), above. By letter dated 23 February 2010, the panel directed that B be restored to the register with immediate effect, but then made a conditions of practice order that contained nine conditions. In allowing B's appeal, His Honour Judge Pelling QC (sitting as a judge of the High Court) exercised his powers under article 38 of the Nursing and Midwifery Order 2001 by deleting conditions 3, 6, and 7, and substituting for them revised conditions 3, 6, and 7 dealing with supervision, administration of medication, and completion and passing of approved return to practice courses.

### *R (Kay) v. Chief Constable of Northumbria Police (No. 2)* [2010] ICR 974

**58.09** Probationary police constable—suspension pending criminal proceedings—acquittal—misconduct proceedings quashed—application to restore as probationary police constable

The claimant commenced serving as a probationary police constable with Northumbria Police on 5 April 2004. On 21 July 2005, she was arrested and she was suspended from duty as a result of an allegation that she had made a fraudulent insurance claim in respect of her engagement ring. The alleged offence had been reported to the police by the estranged husband of the claimant, who was also a serving police officer with Northumbria Police. Subsequently, the claimant was charged with the offence of obtaining property by deception. At trial, the prosecution offered no evidence against her and a verdict of not guilty was entered. The claimant contended that her estranged husband had made up the allegations because she had rejected his proposals for reconciliation. Before the end of the criminal proceedings, the claimant had been served with formal notice of the possibility of misconduct proceedings being brought against her under the Police (Conduct) Regulations 2004. After the criminal matters had been resolved, Northumbria Police decided that misconduct proceedings should not be brought against the claimant; instead, proceedings were brought against her pursuant to regulation 13 of the Police Regulations 2003. Silber J quashed the chief constable's decision on the grounds that there were disputed facts: [2010] ICR 962 (see paragraph 56.10). At a subsequent hearing to determine remedy, the claimant applied for a mandatory order directing the chief constable to reinstate her as a probationary police constable. Granting the application, His Honour Judge Behrens said that the two most important authorities were *Chief Constable of the North Wales Police v. Evans* [1982] 1 WLR 1155 and *R (Bolt) v. Chief Constable of Merseyside Police* [2007] EWHC 2607 (QB), and the question of reinstatement was a matter of discretion. The claimant in the present case had been away from full active duties as a police constable for nearly five years. The chief constable said that he had no confidence in her. However, an order for reinstatement would not usurp the functions of the chief constable because it would not be for him to determine any disciplinary proceedings. In the absence of power under regulation 13, the chief constable had no power to dismiss the claimant. Whilst it was true that the chief constable had expressed the view that he had no confidence in the claimant, that was the result of a procedurally flawed hearing. All of the other evidence relating to the claimant's conduct was positive and to the effect that she was likely to develop into an efficient officer. It also, said the Court, had to be remembered that no evidence was offered against the claimant at the criminal trial. Furthermore, following advice from counsel, the decision was taken not to institute disciplinary proceedings on evidential grounds. The claimant had established that she was dismissed in plain breach of the applicable procedures and reinstatement would put right a serious wrong. The claimant had expressed a strong desire to be reinstated.

***Thobani v. Solicitors Regulation Authority*** **[2011] EWHC 3783 (Admin)**

**58.10** Solicitor—finding of dishonesty—loss to Compensation Fund—whether SDT misdirected itself in refusing application

The appellant, T, was struck off the roll of solicitors on 14 June 2001, when the SDT found a number of allegations against her substantiated, including one that it judged to involve dishonesty. The complaint with which the Tribunal was concerned arose out of a property transaction in which the appellant was engaged in 1989. The property was owned by a company of which her husband was a director, and she and her husband held 50 per cent of the share capital of the company. She faced allegations that included submitting a report on title to Abbey National that she either knew or should have known was false and misleading. The Tribunal found that her behaviour did amount to dishonesty and conduct unbefitting a solicitor. In due course, the Solicitors' Compensation Fund made a payment of £71,000 to Abbey National. In May 2010, the appellant applied to the SDT for restoration to the roll. On 12 October 2010, the application was refused. In dismissing the appellant's appeal, Burnett J said that it was no part of the grounds of appeal lodged with the Court, nor in the skeleton argument filed initially on behalf of the appellant, that the Tribunal had misdirected itself in law. Reading the decision in its entirety, there was no error of law in the decision-making process. Much of the focus of the argument before the Tribunal appeared to have been directed towards the nature of the dishonesty. The Tribunal accepted all of the evidence produced by the appellant of her efforts since 2001. Counsel for the appellant accepted that there must be exceptional circumstances to justify the restoration to the roll of a solicitor who has been struck off for, amongst other things, dishonesty. In the instant case, the Court said that it was plain that the Tribunal did not consider the matters, taken together, were so exceptional as to enable it to restore the appellant to the roll. In the case of *Vane* No. 1687/2002, the Tribunal there, whilst recognizing that an order for restoration would be made only in the light of exceptional circumstances, identified that it was satisfied that the claimant had achieved an exceptional level of personal rehabilitation. It noted that he had worked successfully within a position of trust within the legal environment in the period leading up to the application. It also averted to the restitution of the loss he had caused as being a relevant factor.

***Solicitors Regulation Authority v. Kaberry*** **[2012] EWHC 3883 (Admin)**

**58.11** Solicitor—dishonesty—no exceptional circumstances to justify restoration to roll

The Divisional Court (Elias LJ and Singh J) held that the SDT had erred in allowing C's application to be restored to the roll. In March 1994, K's practice was subject to an intervention by the Law Society on the basis of suspected dishonesty. In July 1995, K faced nine allegations of conduct unbefitting a solicitor. The two most significant allegations were that he had failed properly to redeem mortgages and had deceived clients as to that failure. The deception of the clients in that way was an offence of dishonesty. The Tribunal found K guilty of all charges. The consequences of his actions were very significant indeed. At the time of the Tribunal hearing in 1995, the Solicitors' Compensation Fund had paid out in excess of £650,000. By the time of the restoration application hearing in January 2012, the losses had increased to a little under £3 million and there were still pending claims in excess of £150,000. The fact that, at the time of the commission of the original offences of dishonesty, C had been affected by drugs did not constitute exceptional circumstances to justify his restoration to the roll of solicitors. The principles laid down in the cases for determining whether or not somebody should be restored to the roll of solicitors were not in dispute. They had recently been summarized by Burnett J in *Thobani v. Solicitors Regulation Authority* [2011] EWHC 3783 (Admin). So, said Elias LJ at [33], the Tribunal had to identify exceptional reasons that would allow the unusual step of restoration to the roll for someone who has committed proven acts of dishonesty. The key question was whether K's evidence amounted to exceptional circumstances; it did not. Further, it was well established that, in order to be restored to the roll, it had to be demonstrated that restoration would not affect the good name and reputation of the solicitors' profession, nor would it be contrary to the interests of the public: at [64].

### *Dowland v. Architects Registration Board* [2013] EWHC 893 (Admin)

**58.12** Architect—removal from register—bankruptcy—no issue of competence—application for restoration

On 29 July 2008, the Architects Registration Board's PCC found that D's conduct amounted to unacceptable professional conduct and imposed an order of erasure from the register. A complaint had been made to the Board about D's failure to repay a loan; on 11 September 2007, D had been declared bankrupt for failure to pay seven judgment debts and he had failed to report the bankruptcy to the registrar. On 31 March 2009, D was made the subject of a bankruptcy restriction order under paragraph 2 of Schedule 4 to the Insolvency Act 1986, which was ordered to continue for a period of seven years (until 31 March 2016). On 11 October 2010, a period of two years having passed since the making of the erasure order, D applied for re-entry on the register. It was against the refusal of that application (determined on 20 December 2011) that he appealed to the High Court. D was in his 70s and his name was first entered on the register in 1968. By Standard 9 of the Standards of Conduct and Practice (2002), architects are required to ensure that their personal and professional finances are managed prudently. In dismissing D's appeal, Simon J said that at the heart of his challenge was the contention that it was not open to the Board to refuse the application to be readmitted other than on the basis that it took an adverse view of his competence. Leaving aside criminal conduct, disciplinary powers are reserved for two types of conduct: unacceptable professional conduct; and serious professional incompetence. In each case, the Board must decide whether to direct re-entry; in each case, the Board must also be satisfied of an applicant's competence to practise before he or she is re-entered on the register. Where an applicant has been removed for incompetence, it is clear why there is a need to demonstrate competence before re-entry on the register—but the fact that an applicant has not been practising during a period of erasure on the grounds of unacceptable professional conduct may also give rise to a need to demonstrate competence to practise. The distinction between the two grounds for making disciplinary orders carries over into the rights of appeal in section 22 of the Architects Act 1997. An applicant has a right of appeal in relation to continuing competence, but not to other relevant considerations. However, a claim for judicial review could be made on conventional public law grounds. In the instant case, the Board, in notifying D that it had decided not to re-enter his name on the register, wrote that the expectations of a registered person were high, as exemplified by the Code of Conduct, and these included responsibilities in relation to financial management, guidance, and dealing with clients. The standard required was higher than that which applied to members of the public generally. These expectations were based on the public interest, both in respect of public confidence in registered persons and the protection of clients. The Board did not have confidence that, so soon after the relevant events, D had changed his attitude towards his professional responsibilities, and towards the balance between his financial interests and those of others. His Lordship concluded that he did not accept that the Board had exceeded its powers in its decision it made on 20 December 2011. It had a broad discretion as to how it should approach the application under section 18(1) for the name of a person to be re-entered on the register, which was not confined to considerations of competence. It followed that D's appeal failed and would not have succeeded if properly framed as a claim for judicial review.

***Townrow v. Financial Services Authority*, FS/2012/0007, 10 January 2013**

**58.13** Financial services—prohibition notice—application for revocation or variation refused—referral to Upper Tribunal—application to strike out

On 12 January 2006, the Financial Services and Markets Tribunal ordered that T be prohibited from performing any function in relation to a regulated activity carried on by an authorized person, on the grounds that he was not fit and proper because of his lack of integrity. The Tribunal found that T presented a severe risk to customers, failed to cooperate with and gave false information to the Financial Services Authority (FSA), and was 'temperamentally unsuited to working in a regulated environment'. T was a partner or sole principal of a firm that provided advice on, and arranged and dealt in, life assurance, pensions, and collective investment schemes, and which advised on investment. The Tribunal's decision related a history of serious regulatory failings on T's part and, in addition, it found that acts or omissions on his part had led several of his clients to suffer substantial losses. Moreover, despite having received payments from his professional negligence insurers, T failed—indeed refused—to make redress to those clients, for reasons that the Tribunal found spurious. On 24 June 2011, T applied to the FSA for revocation or variation of the prohibition order. On 10 January 2012, the Authority's Regulatory Decisions Committee directed that the prohibition order should not be revoked. In striking out T's referral of the Committee's decision to the Upper Tribunal, Judge Colin Bishopp said that the power to strike out a reference was found in rule 8(3)(c) of the Tribunal Procedure (Upper Tribunal) Rules 2008. The Upper Tribunal may strike out whole or part of proceedings if there is no reasonable prospect of the appellant's or the applicant's case, or part of it, succeeding. The judge noted that the Authority has a published policy, set out in the Enforcement Guide (EG) section of its handbook, which it applies when dealing with applications for revocation or variation of prohibition orders. Paragraph 9.19(8) identifies as one criterion whether the individual will continue to pose the level of risk to consumers that resulted in the original prohibition. Paragraph 19.22 adds that the FSA will not generally grant an application to vary or revoke a prohibition order unless it is satisfied that the proposed variation will not result in a reoccurrence of the risk to consumers and the individual is fit to perform functions in relation to regulated activities generally. The Tribunal is, of course, not bound by the Authority's published guidance. However, the criteria are not merely reasonable, but essential, if a prohibition order is to serve its obvious purpose of protecting (in T's case) consumers and potential consumers. In the instant case, T had made no attempt to demonstrate himself to be fit and proper; he merely argued that what he proposed to do would not present any danger to the public. The argument that the prohibition order should be revoked for that reason was hopeless. It would be nonsensical, and plainly contrary to the purpose of the Financial Services and Markets Act 2000, to allow a person who has been prohibited on the basis of what were condemned by the Tribunal as very serious failings to procure revocation of the resulting prohibition order merely because he claims that the activities in which he proposes to engage in future will present no danger to the public: at [22]. A prohibition order is not a punishment, even if its effect may be punitive in an individual case. It is imposed, as section 56 of the Act makes perfectly clear, for the protection of the public. It would be a plain dereliction of duty for the Authority to revoke a prohibition order merely on the affected person's assurance that he will no longer present a danger. Accordingly, the reference was without merit and had no conceivable prospect of success. Permission to appeal was refused by Fulford LJ: [2014] EWCA Civ 1099. His Lordship said, at [11], that, in his view, there was no justification for the assertion by T that the burden of proof rests with the FSA to demonstrate that the applicant continues to pose a risk to the public. Once the order is in place, it is for the applicant to demonstrate that, under section 56(7) of 2000 Act, it is appropriate to vary or revoke the prohibition.

### *R (Solicitors Regulation Authority) v. Solicitors Disciplinary Tribunal; Solicitors Regulation Authority v. Ali* [2013] EWHC 2584 (Admin)

**58.14**

Solicitor's clerk—employment—Solicitors Act 1974, section 43—whether appropriate to revoke order that no solicitor employ or remunerate person concerned

The SRA appealed from a decision of the SDT whereby it revoked an order made under section 43 of the Solicitors Act 1974 in respect of the respondent, A. A worked as a solicitor's clerk specializing in immigration matters. In December 2006, a disciplinary tribunal ordered that no solicitor shall employ or remunerate A except with the consent of the Law Society. In January 2011, the Tribunal revoked the section 43 order on the ground that A had made an effort to rehabilitate himself and the length of time since the original order. Allowing the SRA's appeal, Wilkie J said that the purpose of a section 43 order is to safeguard the public and the Law Society's reputation by ensuring that a person is employed only where a satisfactory level of supervision has been organized and for as long as that person requires such a level of supervision before being permitted to work effectively under his or her own steam. In the absence of specific evidence, that cannot be established merely by somebody attempting to obtain the necessary experience. There was no rational basis for concluding that rehabilitation had in fact occurred sufficient to make the section 43 order no longer necessary.

### *Ellis-Carr v. Solicitors Regulation Authority* [2014] EWHC 2411 (Admin)

**58.15**

Solicitor—appeal against refusal of application for restoration to roll—delay—fresh application

The appellant was removed from the roll in 1988 because of defaults in accounting in relation to monies held in his clients account. There was a shortfall of £40,000, although it was not suggested that the shortfall resulted from any dishonesty on the part of the appellant. However, in 1991, the appellant was prosecuted and convicted, and sentenced to two years' imprisonment. He had subsequently sought to set aside that conviction. In November 2010, the SDT dismissed the appellant's application for restoration to the roll of solicitors. In dismissing the appeal on 25 June 2014, Collins J said that he was told that there was no restriction upon the number of applications that can be made for reinstatement. Equally, there is no particular time limit between such applications. However, four years had effectively passed. The present appeal was a review of the decision of the SDT and thus normally would depend upon consideration of the material that was put before that Tribunal, and the Court will not normally be prepared to consider fresh evidence that might have arisen. His Lordship said that it seemed to him that it was in the appellant's interest that this appeal be brought to an end and that the appellant have the opportunity to put before a fresh tribunal further material that concentrates on what he has done since his conviction and what he proposes to do for the future and why it is that he really wants his reinstatement on the roll. The appellant was not to continue to try to look back, notwithstanding that he felt very hard done by because he had not had his criminal appeal properly considered. He should be judged on the basis of what he now is and whether there is a real prospect that, notwithstanding the conviction, he can be regarded as someone fitted to be on the roll of solicitors.

### *Anoom v. Bar Standards Board* [2015] EWHC 439 (Admin)

**58.16**

Former barrister—five disciplinary occasions in seven years

A had been called to the Bar in 1998 and disbarred in 2007. He admitted three charges of professional misconduct and three charges of inadequate professional service. Between 1998 and 2007, A had had disciplinary findings against him on five occasions. Previously, he had been conditionally discharged from the police service. At a hearing in 2014, the Inns' Conduct Committee rejected A's application for readmission to the Middle Temple. The decision was confirmed on review by the Bar Standards Board's Qualifications Committee, which found that A did not meet the requirements of the Bar Training Rules for persons fit and proper to be a barrister. Dismissing A's appeal,

Rose J said that there were no grounds on which to overturn the decision of the Inns' Conduct Committee. It had not relied heavily on the circumstances surrounding A's conditional discharge from the police, and said that he had been a fit and proper person when called to the Bar in 1998. It was not right to consider the disciplinary findings as stale. Whilst A had undertaken voluntary work and was a full-time carer for his parents, this did not provide evidence that justified readmission. The testimonial evidence was from personal friends and did not include, for example, an employer.

## C. Other Relevant Chapters

**58.17** Dishonesty, Chapter 19
Impairment of Fitness to Practise, Chapter 33
Misconduct, Chapter 47
Registration, Chapter 56
Review Hearings, Chapter 59

# 59

# REVIEW HEARINGS



## A. Legal Framework

**59.01** Examples of procedure at review or resumed hearings following a suspension or conditions of practice order include:

General Dental Council (Fitness to Practise) Rules 2006, Part 4
General Pharmaceutical Council (Fitness to Practise and Disqualification etc) Rules 2010, rule 34
General Optical Council (Fitness to Practise) Rules 2013, Part 8
General Medical Council (Fitness to Practise) Rules 2014, Part 5

Guidance on review hearings is often contained in the regulator's indicative sanctions guidance. For example, see:

General Medical Council, *Indicative Sanctions Guidance for the Fitness to Practise Panel* (April 2009, with August 2009, March 2012, March 2013, and April 2014 revisions), paragraphs 114–20 (no doctor should be allowed to resume unrestricted practice following a period of conditional registration or suspension unless the panel considers that he or she is safe to do so)

## B. Disciplinary Cases

### *Otote v. General Medical Council* [2003] UKPC 71

**59.02** Review hearing—indication by panel of steps to rectify defects in practice

Between 20 and 28 February 2003, the Professional Conduct Committee (PCC) of the General Medical Council (GMC) conducted an enquiry into a charge of serious professional misconduct that the appellant had indecently assaulted a nurse, made inappropriate comments to four other nurses, and failed to examine physically three patients. The charge was found proved. The Committee directed that the appellant's registration be suspended. Having announced its decision on the appropriate sanction, the chairman of the Committee continued:

> The Committee were very concerned about your clinical deficiencies and lack of interpersonal skills. The Committee therefore indicate that during your period of suspension you should undergo remedial training to be discussed with and co-ordinated by the Postgraduate Dean and with the Regional Advisor for training in psychiatry. This remedial training should include: further development of basic clinical skills in examination and investigation in a fully supervised environment with formal assessment of competence; attendance at a course to develop professional interpersonal relationships; and practical experience of working effectively in teams. Shortly before the resumed hearing, which will take place just before the end of your period of suspension, you will be asked to furnish the Council with the names of professional colleagues and other persons of standing to whom the Council may apply for information as to their knowledge of your conduct throughout the interval since the hearing of this case. The resumed hearing will expect to receive clear and detailed evidence that you have been formally assessed by appropriate agencies or individuals in the areas listed above. The Committee also expects to receive evidence that you have received formal training in these areas and can demonstrate insight into your shortcomings. The Committee will wish to determine whether the educational and other objectives have been achieved by receiving evidence of formal assessment.

In his summary grounds of appeal, the appellant stated that the Committee erred in suspending him and, in effect, via its 'indication' also requiring his registration to be conditional during a period of suspension. In rejecting this submission and in dismissing the appellant's appeal, Sir Philip Otton, giving the judgment of the Board, said, at [21], that the appellant's primary complaint was misconceived. The Committee was empowered to impose the sanction of suspension; it had no power to suspend the appellant's registration and impose conditions upon it: see section 36(1) of the Medical Act 1983. There can be no doubt that the use of the word 'or' in section 36(1) means that the sanctions are alternative and not cumulative. Their Lordships said, at [22], that they accepted the submission of counsel for the GMC that there is a fundamental distinction between an 'indication' and a condition. In the instant case, after directing that the appellant's registration be suspended, the Committee indicated the steps that the appellant should take before the resumed hearing. This indication was intended for the appellant. A practitioner who tries to comply with such an indication will be in a better position than a practitioner who has taken no steps to rectify the defects found in his or her practice. The indication given in this case could not be regarded as punitive and was clearly designed to assist the doctor to get back into practice after he had established sufficient insight in his own mind into his shortcomings so as to ensure that they would not reoccur. If circumstances should arise, he would be able to submit that it was not possible to observe or fulfil the indications. If the indications cannot be met, then the GMC should arrange an early resumed hearing to allow the Committee to deal with the situation.

***Kataria v. Essex Strategic Health Authority*** **[2004] EWHC 641 (Admin)**

**59.03**

NHS—national disqualification precluding practitioner from working within NHS—review—test—whether original Tribunal's decision justified revocation of disqualification and reinclusion not prejudicial to NHS

This was a statutory appeal by K against the decision of the Family Health Services Appeal Authority (FHSAA) dated 30 July 2003 dismissing his application under section 49N(7) of the National Health Service Act 1977 for the review of his national disqualification imposed by the National Health Service (NHS) Tribunal. The Tribunal's conclusions were that, in relation to three named patients, K showed a consistent disregard of the patients' complaints and of their care; that his reaction to criticism, by his attacks on those who made or were associated with complaints against him, went beyond temporary and understandable loss of control; and that his attitude to staff and patients was arrogant and lacking in sympathy for their needs and views. The Tribunal was also concerned about the unreliability of K's evidence. The Tribunal concluded that the continued inclusion of K's name in the list of medical practitioners undertaking to provide general medical services would be prejudicial to the efficiency of the services in question. A national disqualification precludes a practitioner from working within the NHS. Section 49N(7) provides that the FHSAA may, at the request of the person on whom it has been imposed, review a national disqualification and, on a review, may confirm it or revoke it. Stanley Burnton J said, at [23], that the word 'review' in section 49N(7), of itself, gives no relevant guidance as to the scope of inquiry by the FHSAA. Where a second tribunal rehears or otherwise considers the evidence before a first tribunal and makes its own findings of fact, the rehearing is normally termed an 'appeal' rather than a review, as in section 49M(2) of the Act. He continued, at [24], to hold that section 49N(7) refers not to a review of an earlier tribunal's decision, but to a review of a national disqualification, which may be confirmed or revoked. Confirmation has the effect of continuing the disqualification in force for the future; revocation brings it to an end. The distinction between the review of a decision and the review of a disqualification is significant. The requirement in section 49N(8) of a minimum period before a disqualification may be reviewed is explicable if the object of a review under subsection (7) is to consider whether the disqualification should continue or be terminated. Any such procedure is effectively an appeal, rather than a review of the order made by the first tribunal. In K's case, the relevant primary facts were not significantly in issue. The FHSAA made no error of law in formulating the question that it had to answer—namely, whether, on the balance of probabilities, K's conduct and actions since the original tribunal's decision justified the revocation of his national disqualification and his reinclusion in the list would not be prejudicial to the efficiency of NHS services. In dismissing K's appeal, Stanley Burnton J said:

> 69. In my judgment it is obvious that the efficiency of the NHS might be prejudiced by a want of probity of a practitioner, and in particular by any unreliability of his written or oral statements. Fellow practitioners and other NHS staff and patients must be able to rely on the integrity of doctors and the honesty of their statements. The FHSAA is entitled to take into account any want of probity found by it on the part of a practitioner in determining whether his inclusion in a list would be prejudicial to the efficiency of the service. The fact that his want of probity may also be relevant to the GMC does not exclude it from consideration by the FHSAA.
>
> 70. It was for the Tribunal to determine whether the revocation of [K]'s disqualification would be prejudicial to the efficiency of the NHS. It could not delegate to another tribunal the consideration of any relevant issue. In any event, the Preliminary Proceedings Committee of the GMC had not considered the matters that were before the FHSAA, but only those that had been before the original NHS Tribunal.

[...]

77. The FHSAA did not expressly weigh the prejudice to [K] as against the potential prejudice to the efficiency of the NHS. It was however conscious of the financial effect of the disqualification on him, and referred to it in ... its decision. The reasons given for the conclusion that the revocation of the disqualification would be prejudicial to the efficiency of NHS services were substantial and cogent. They included their findings as to his probity and, in relation to his lack of [continuous professional development] CPD, a risk of clinical failings. In my judgment the decision satisfies the doctrine of proportionality and the common law requirement that the decision be reasonable and fair. The Tribunal's reasons were adequate.

78. I also reject the curious submission ... that because he did not intend to work in the NHS [the Tribunal] could not lawfully find that the revocation of his disqualification would be prejudicial to the efficiency of the NHS. The submission undermined the case put forward on his behalf on proportionality. The statutory test requires the Tribunal to assume that a practitioner works within the NHS and to determine whether, if he does, its efficiency would be prejudiced.

### *Kamel v. General Medical Council* [2007] EWHC 313 (Admin)

**59.04** Professional performance assessment—failure to undergo assessment—test for continuing suspension—protection of public interest

K appealed against the decision of the fitness to practise panel of the GMC made on 8 December 2005 to continue his suspension from the register of medical practitioners for twelve months. In June 2004, the Committee on Professional Performance (CPP) decided that K's professional performance may have been seriously deficient in three respects: teamworking; relationships with colleagues; and patient confidentiality. It therefore directed an assessment of his performance to be performed within three months. K declined to submit to that assessment. On 25 November 2004, a fitness to practise panel determined that K had failed to comply with the reasonable requirements of the assessment panel, as directed by both the assessment referral committee and the CPP. It therefore directed that his registration be suspended for a period of twelve months. On 8 December 2005, a review hearing was held, which K did attend. The panel found that, despite written invitations since November 2004, K had refused to comply with the reasonable requirements to undergo an assessment of his professional performance and therefore his fitness to practise continued to be impaired. In dismissing K's appeal against the extension of the period of suspension by twelve months, Mitting J said that the final words of the panel's decision—namely, that K's fitness to practise continues to be impaired—indicated a decision that the panel was not required to make and arguably should not have made. The panel that sat in November 2004 made no finding of misconduct, let alone that K's fitness to practise was, in fact, impaired. All that it decided was that K had failed to comply with reasonable requirements to undertake an assessment and that, because of the nature of the *alleged* deficiencies, it was necessary for the protection of the public to make a direction regarding his registration. At the hearing in December 2005, the panel had to decide only whether or not K's registration had been suspended. That fact was never in issue. K conceded that he had not complied with the earlier direction. The only requirement that must be satisfied, before a panel reviewing an order for suspension can reimpose it, is that a panel has already given a direction that a person's registration be suspended. Once it has reached that decision, it is empowered on review to order a further suspension. There is no statutory test set out in the Medical Act 1983 that governs the principles upon which it should order suspension, but they are familiar from case law over many years—namely, and primarily, the protection of the public interest. In the Court's view, the panel was plainly justified in ordering a further period of suspension.

***R (Independent Police Complaints Commission) v. Chief Constable of West Midlands Police*** **[2007] EWHC 2715 (Admin)**

**59.05**
Police officer—sanction—review meeting—attendance of interested parties

The third and fourth interested parties were both officers in the Derbyshire Constabulary. As a result of complaints against them by the parents of a lady who tragically was murdered (the first and second interested parties in these proceedings), one officer was demoted and another officer was removed from the force following a hearing. They each sought a review meeting with the reviewing officer from the West Midlands Police, the defendant. The defendant gave a decision that he had no power to allow the presence at the review meeting either of a representative of the Independent Police Complaints Commission (IPCC), or, if they wished to attend, the parents. Burton J said that the issue was simply one of construction of the Police Conduct Regulations 2004. By regulation 40, an officer against whom a sanction has been imposed, or in respect of whom a challengeable finding has been made, shall be entitled to request the chief officer in the force concerned, or in appropriate circumstances the assistant commissioner, to review the finding or the sanction. The central paragraph is regulation 41, which provides that the reviewing officer shall hold a meeting with the officer concerned if requested to do so. Where a meeting is held, the officer concerned may be accompanied by another police officer or by counsel or a solicitor. As a matter of construction, there is no bar on the attendance of other parties than the officer. The matter remains at the discretion of the reviewing officer. Burton J suggested that, undoubtedly, sensible reviewing officers will seek to limit the numbers of people present, but will exercise their discretion so that no more people are present, and possibly fewer in terms of overall numbers, than were at the original hearing. But there was no doubt that the reviewing officer does have a discretion and accordingly the application for judicial review succeeded.

***Abrahaem v. General Medical Council*** **[2008] EWHC 183 (Admin)**

**59.06**
Review hearings—test

In allowing the appellant's appeal in part and remitting the case to a further panel of the GMC, Blake J said:

> 23. The statute [Medical Act 1983] is to be read with the 2004 Rules and rule 22(a) to (i) makes clear that there is an ordered sequence of decision-making, and the Panel must first address whether the fitness to practise is impaired before considering conditions. In my judgment, the statutory context of the Rule relating to reviews must mean that the review has to consider whether *all* the concerns raised in the original finding of impairment through misconduct have been sufficiently addressed to the Panel's satisfaction. In practical terms there is a persuasive burden on the practitioner at a review to demonstrate that he or she has fully acknowledged why past professional performance was deficient and through insight, application, education, supervision or other achievement [has] sufficiently addressed the past impairments.
>
> 24. This is the point made in the Indicative Sanctions Guidance for Fitness to Practise Panels published by the GMC in April 2005. Following headings titled Impaired Fitness to Practise, Public Interest, the Guidance proceeds to address Sanctions in general and the sanction of suspension in particular. There is then a sub-heading titled Review Hearings where it states:
>
> > "31. Where the panel decides that a period of conditional registration or suspension would be appropriate, it must decide whether or not to direct a review hearing immediately before the end of the period. The panel must give reasons for its decision so it is clear that the matter has been considered and the basis on which the decision has been reached. *Where a review hearing is to be held the panel must make it clear what it expects the doctor to do during the period of conditions/suspension and the information s/he should submit in advance of the review hearing. This*

*information will be helpful both to the doctor and to the panel considering the matter at the review hearing.*

32. It is important that no doctor should be allowed to resume unrestricted practice following a period of conditional registration or suspension *unless the panel can be certain and he or she is safe to do so.* In some misconduct cases it may be self-evident that following a short period of suspension, there will be no value in a review hearing. In most cases, however, where a period of suspension is imposed and in all cases where conditions have been imposed *the panel will need to be reassured that the* doctor is fit to resume practice either unrestricted or with conditions or further conditions. The panel will also need to satisfy itself that the doctor has fully appreciated the gravity of the offence, has not re-offended, and has maintained his or her skills and knowledge and that patients will not be placed at risk by resumption of practice or by the imposition of conditional registration."

(emphasis supplied)

25. The Indicative Sanctions were provided to the Appellant before the hearing in December 2005 and would be familiar to his advisers. The terms of para 32 were emphasized at the second hearing by counsel for the GMC. [Counsel for the appellant] submits that para 32 is included in the section under sanctions and can only be applicable when the Panel has found that fitness to practise is still impaired under rule 22.

26. In my judgment, consistent with the statutory scheme outlined above, this Guidance is intended to apply at reviews and will therefore have relevance as to the Panel's finding whether fitness to practise remains impaired following a period of suspension. Doubtless, the Guidance could incorporate the rule 22(f) stage of the process more explicitly and I certainly accept that the panel cannot proceed from the historic fact of suspension directly to what further sanction may be necessary. The court is in no doubt, however, that at both hearings the Appellant would or should have been aware of the need for the Panel "to be reassured that the doctor is fit to resume practice whether unrestricted or with conditions" and will need "to satisfy itself that the doctor has fully appreciated the gravity of the offence, has not re-offended, and has maintained his or her skills or language". Equally at a review he will have been informed what he has to achieve by information provided previously.

### *Dzikowski v. General Medical Council* [2009] EWHC 1090 (Admin)

**59.07** Conditional registration—non-compliance—erasure

In January 2006, a fitness to practise panel of the GMC found a number of allegations against D of inappropriate and irresponsible prescribing of methadone mixture and dexamphetamine to a patient. D was found guilty of serious professional misconduct and the panel imposed conditions on his registration for a period of twelve months. D's appeal was dismissed in September 2006, after which the conditions came into effect. The review of D's case was eventually heard over a period of eight days between 21 May and 11 July 2008, during which the panel considered whether D had failed to comply with any requirement imposed as a condition of registration. The panel concluded that the requirements had not been complied with in a number of instances, and concluded that it was necessary for the protection of patients and in the public interest for D's name to be erased from the register. Cox J dismissed each of D's grounds of appeal. In summary, Cox J said:

94. It is unsurprising, in my view, that [the panel] considered the imposition of further conditions to be insufficient. They had found expressly that [D] had not made efforts to comply with previous conditions imposed.

95. The finding that his attitudinal problems had the potential to cause serious harm to his patients is important, given the particular vulnerability of the patient group, and... this finding cannot be impugned. Nor can the finding that his evidence demonstrated a lack of insight and deeply rooted attitudinal problems, which led the

Panel to reject a period of suspension as adequate to protect patients and maintain public confidence. Having had the advantage of hearing [D]'s evidence, the Panel were in the best position to make this assessment and to come to the conclusion that they did....

### *Khan v. General Medical Council* [2009] EWHC 535 (Admin)

**59.08** Conditional registration—non-compliance—suspension

In June 2006, a fitness to practise panel of the GMC found that K's fitness to practise was impaired by reason of misconduct. K was a doctor and sole practitioner of a surgery in Wandsworth. The panel heard that he left all administrative matters in the hands of his practice manager without examining the contents of documents and that he persisted in making claims to the Primary Care Trust for reimbursement of pension contributions over a number of years, even when he knew, or should have known, that none of his staff were included in the pension scheme. The panel accepted that K's acts of dishonesty were at the lower end of the scale and that, although he was guilty of a cavalier and reckless attitude towards administrative tasks and financial matters, he had not deliberately set out to defraud. The panel also concluded that his actions were not calculated to achieve personal gain, and was conscious of the fact that no issues relating to K's clinical practice or abilities had been raised and there was no risk of safety to patients. The panel imposed conditions on K's registration for a period of eighteen months. At the review hearing on 3 January 2008, the panel concluded that K's fitness to practise was impaired and that there had been breaches of the earlier conditions. Given his failure to comply with the conditions imposed in 2006, the panel said that it had no confidence that an order for conditions would be appropriate. In all of the circumstances, bearing in mind the evidence received in mitigation and taking account of the need to act proportionately, the panel concluded that it would be appropriate and proportionate to suspend K's registration for twelve months. On appeal, it was submitted that the review panel in January 2008 misunderstood its functions. K submitted that, in a case of review following the imposition of conditions on a doctor's registration, the function of the review panel is to determine whether the appellant has failed to comply with the conditions and that there was nothing in section 35D of the Medical Act 1983 that expressly conferred on a review panel a discretion to determine fitness to practise. Cranston J rejected this submission, saying:

> [19] In my view, therefore, as a matter of statutory interpretation a Panel on a review hearing can consider not only whether or not conditions previously imposed have been breached, but also a doctor's fitness to practise. The upshot is that in this case the Panel was fully entitled to consider whether the appellant had breached his conditions, along with whether or not his fitness to practise was still impaired.

The appeal was dismissed.

### *R (Pattar) v. General Medical Council* [2010] EWHC 3078 (Admin)

**59.09** Impairment by reason of misconduct and deficient professional performance—suspension—non-attendance at review hearing—erasure

On 7 August 2006, the GMC invited P, a locum consultant physician and gastroenterologist, to agree to an assessment by a GMC panel. The GMC had received a complaint from South Manchester University Hospital Trust of alleged misdeeds over five working days in one week during which P had worked at Wythenshawe Hospital. On 2 September 2006, P told the GMC that he refused to be assessed. In June 2008, a fitness to practise panel determined that P's fitness to practise was impaired by reason of both misconduct and deficient professional performance. The panel imposed upon him a sanction of twelve months' suspension, with a review required before the end of the period. P did not attend the review hearing on 9 March 2010. The review panel concluded that, with regard to deficient professional performance, it had no new evidence before it to demonstrate that P had addressed the concerns of the previous panel in any way whatsoever. Furthermore, the panel was conscious that P had not been working in the United Kingdom for at least the past twenty months as a result of his suspension. The panel concluded that P's fitness to practise was impaired at the date of the review hearing both by reason of his misconduct and his deficient professional performance. It directed that P's name should be erased from the medical register. In dismissing P's appeal, Mitting J said that the only ground of appeal that P had advanced which was capable of being a valid criticism of the decision of the panel in March 2010 was that it was wrong to erase his name from the register rather than to suspend him for a further period to give him the opportunity to demonstrate to the panel that he remained a skilful and safe medical practitioner. P had produced to the Court a written document in which he had set out the steps that he had taken to keep his medical knowledge and skills up to date. They were not in the form required by the panel, but they could, no doubt, be substantiated and, if properly substantiated, might well be capable of satisfying the panel that he had kept his medical skills up to date. Furthermore, P had said to the Court that he had perfectly lawfully continued to practise in India, and had thereby demonstrated that he was a skilful and safe doctor. Mitting J said that it was doubly unfortunate that P did not put these matters before the panel and indeed expand upon them at the hearing in March 2010. Had he done so, it was possible—at least if he had changed his mind about undergoing a performance assessment—that he could have demonstrated to the panel that erasure was not necessary. He did not do so. Instead, he presented to the panel an adamant refusal to accept the June 2008 finding or to undergo a performance assessment, and presented no evidence to the panel to demonstrate that he had kept his medical knowledge and skills up to date. In those circumstances, the panel was entitled to reach the conclusion that it did—that is, that erasure was the only appropriate sanction. It went through the steps required to reach that conclusion. The Court could allow the appeal only if the decision of the panel were clearly wrong or if it were tainted by some serious procedural irregularity. The latter was not alleged and the former was not demonstrated.

### *Karwal v. General Medical Council* [2011] EWHC 826 (Admin)

**59.10** Dishonesty—suspension—review hearing—continuing lack of insight—further suspension

In June 2008, a fitness to practise panel of the GMC found three allegations of dishonesty proved against the appellant and concluded that she had knowingly made to a professional colleague false representations about an investment scheme so as to fraudulently reassure him that £188,000 he had been promised would be paid. It suspended her from the medical register for twelve months and directed a review hearing, which was held in December 2009 and March 2010. In its determinations dated 25 and 26 March 2010, the review panel found K's fitness to practise impaired and she was further suspended for nine months from the expiry of her then current suspension. The cumulative effect of the sanctions was a suspension of some two years and nine months. The key question at the December 2009 review hearing was whether the appellant had sufficient insight into, or had fully appreciated the gravity of, the original offence. In dismissing K's appeal, Rafferty J said that, in her judgment, the position was clear, and was or should have been clear at the time of the review hearing. Not only the appellant's dishonesty, but also her lack of insight, were concerns explicit in the conclusions of the 2008 fitness to practise panel. The 2010 panel was therefore not only entitled, but obliged, to address them.

### *Daraghmeh v. General Medical Council* [2011] EWHC 2080 (Admin)

**59.11**
Deficient professional performance—suspension—review hearing—conditional registration—public protection

King J dismissed an appeal from findings of a review held to investigate whether the appellant's fitness to practise was still impaired. In November 2005, the appellant had been suspended from his post as a specialist registrar in medicine for the elderly at a number of hospitals in Scotland. He had not worked in any clinical way since 2005. At a hearing before the fitness to practise panel in February 2009, the panel had determined that the appellant's fitness to practise was impaired by reasons of deficient professional performance. The panel imposed a sanction of twelve months' suspension. At a review in February 2010, the panel found that the appellant's fitness to practise remained impaired by reason of his deficient professional performance and imposed a number of conditions on his registration. On appeal, the appellant contended that: the conditions imposed by the review panel were irrational, disproportionate, and impracticable; their cumulative effect was to defeat the purpose of conditions as expressed in the GMC's guidance—namely, to enable the doctor to remedy any deficiencies in his practice; and the conditions made it a practical impossibility for the appellant to obtain a post of employment in the United Kingdom. King J recognized that *Udom v. General Medical Council* [2010] Med LR 37 established that it would be an error of law for a panel to impose conditions on registration that were in effect incompatible with registration. However, the panel was faced with having to balance the interests of the appellant against the need to protect patients. In the instant case, the panel was concerned with the ability of the GMC to monitor the performance of a doctor who had not practised clinically since 2005 and in the context of evidence that the assessment team itself had debated whether the appellant was capable of returning to any sort of work. In such circumstances, it was entirely open to the panel to impose stringent conditions, and to include within the conditions that it concluded were necessary for the protection of the public a restriction confining the appellant in the first instance to the care of the elderly and that his work be supervised by a named consultant. The learned judge agreed with the observations of Blake J in the case of *Abrahaem v. General Medical Council* [2008] EWHC 183 (Admin), at [34] point (b), that the best evidence of the impact of the conditions would have been the appellant's attempts to find UK employment, or to seek advice from a postgraduate dean or other qualified person.

### *R (Levy) v. General Medical Council* [2011] EWHC 2351 (Admin)

**59.12**
Review hearing—indication by panel of steps to rectify defects in practice

At a hearing in April 2010, a fitness to practise panel of the GMC found that L's fitness to practise was impaired by reason of misconduct and directed that his registration be suspended for a period of nine months. The decision letter sent to L, a psychiatrist, stated that, before the end of the period of suspension, a fitness to practise panel would review his case and a letter would be sent to him about the arrangements for the review hearing. The letter went on to state that the panel reviewing L's case was likely to find it helpful to receive: written testimonials from his professional colleagues or any mentor that he chose to appoint; evidence relating to the steps that he had taken to enhance his insight into his dishonesty; and evidence of any steps that he had taken to keep his medical knowledge up to date, including evidence of his attendance at any professional courses. On appeal, L tried to suggest that the letter imposed what he would call a mandatory imposition of matters that he would have to put before the review panel and that therefore the review was setting him up to fail. In dismissing L's appeal, His Honour Judge Pearl (sitting as a judge of the High Court) said that he did not read the letter in that way. Paragraph 115 of the GMC's Indicative Guidelines states that where a panel directs a review hearing, it may wish—which is no more than directory—to make clear what it expects the doctor to do during the period of suspension and the information that he should submit in advance of the review hearing. This information will be helpful both to the doctor and to the panel considering the matter at the review hearing. The Indicative Guidelines do no more than set out the form of information that a panel, dealing with a review hearing, will find helpful when it decides what the doctor has been doing during the period of suspension and whether the doctor is now ready to resume his registration. What the decision letter was doing in the instant case was providing L with the basis upon which he could engage with the concerns and the issues that had been set out by the GMC's fitness to practise panel over the next nine months, and upon which he could present, at the review hearing, the material to the panel in the hope—one would hope in the expectation—that his registration would then be provided back to him.

### *R (Adeyemi) v. General Medical Council* [2012] EWHC 425 (Admin)

**59.13** Deficient professional performance—conditions followed by suspension at later hearing—review—erasure—whether unfair, disproportionate, or unreasonable—public protection

At a hearing in January 2009, a fitness to practise panel identified significant deficiencies in A's professional performance in six areas:

(1) management of anaesthetic induction, including identification of the correct technique to be used;
(2) intubation technique;
(3) record-keeping;
(4) recognition of, and response to, crucial events;
(5) proper understanding of the responsibilities of an anaesthetist; and
(6) communication skills.

The panel placed conditions on A's registration for a period of twelve months. A's case next came before the GMC at another fitness to practise panel hearing, this time a first review, in January 2010. That panel decided to suspend her from practice for a period of twelve months. The panel found that although A had complied with certain of the conditions imposed in 2009, in particular as to the formation of a professional development plan in discussion with the relevant deanery, but her efforts to do so were, in the panel's words, 'half-hearted'. The panel was of the opinion that A had not fully grasped the purpose or reason for the imposition of conditions on her registration. A second review panel heard A's case in 2011. Its determination was the one under appeal. The second panel found that A's professional performance continued to be impaired. As to sanction, it rejected the suggestion that conditional registration might be sufficient, and it went on to consider whether suspension would be appropriate and proportionate. Having answered both of those questions 'no', the second panel held that it followed that A's name must be erased from the medical register. The panel said that it was deeply concerned that, since her last hearing in 2010, A had made only limited efforts to address the issues raised, yet believed that she had fully addressed these issues. In dismissing A's appeal, Bean J said:

> 13. The fact that the 2011 Panel decided on erasure when the 2010 Panel... decided not to does not of itself render the 2011 decision unfair, disproportionate or unreasonable. The 2011 Panel had to consider the progress or otherwise which [A] had made since the earlier review. Their judgment was that she had not satisfied them that she was now fit to practise. They made findings that she had only made limited efforts to address the issues raised in the 2010 hearing.
>
> 14. I appreciate of course that... a doctor who is suspended from practice is in a difficult position. It is extremely difficult to obtain any sort of work allowing clinical practice, even under supervision, and this creates what some laymen might describe as a "Catch 22" position. Nevertheless it is difficult for [counsel for A] to argue that the finding of failure to demonstrate that [A] had kept her medical skills and clinical knowledge up-to-date is a perverse one.... At the time of the 2011 decision it was five

> years since she had ceased practice. Her professional development, as set out in the Professional Development Plan of 2010 to 2011, and her document sent to the GMC on 18th July 2011 showed that she had moved in the direction of training in palliative care. This is extremely commendable...but it must be remembered that of the six findings of deficiency made in the 2009 decision, three were of general application....
>
> [...]
>
> 16. In those circumstances the Panel were entitled to find, indeed bound to find, that the past deficiencies had not been fully addressed.

Whilst appreciating the difficulties in her way, which were inevitable, the Court averred that the GMC has a duty to protect the public and that its decision must be made on that basis rather than as a means of imposing a penalty. The panel was clearly entitled to find that public protection required that A's name should, after five years out of practice and with the continuing deficiencies that were identified, be erased from the register.

### *R (Uruakpa) v. General Medical Council* [2012] EWHC 1960 (Admin)

**59.14** Deficient professional performance—review hearing—erasure—proper approach by panel

U appealed against a determination made on 9 July 2011 by a fitness to practise review panel that his name should be erased from the medical register. The original fitness to practise panel found that U's fitness to practise was impaired by reason of deficient professional performance under section 35C(2)(b) of the Medical Act 1983. The fitness to practise hearing concluded in August 2009 with a period of twelve months' suspension on U's registration. U appealed to the High Court against the August 2009 determination. His appeal was dismissed by Saunders J on 13 May 2010: [2010] EWHC 1302 (Admin) (see paragraph 10.09). U applied for permission to appeal to the Court of Appeal. That application was dismissed on paper by Maurice Kay LJ on 22 September 2010. U renewed the application, but it was dismissed by Laws LJ on an oral hearing on 11 November 2010. U's twelve-month suspension took effect on 13 May 2010, when his appeal was dismissed by Saunders J. The review hearing commenced on 4 April 2011 and was listed for four days. The panel was unable to complete the review hearing within the four days allocated, and three additional days were allocated in June and July 2011. The review panel found that U's fitness to practise was impaired by reason of deficient professional performance. The panel concluded that U had failed to meet the requirements set out by the panel considering his original case in August 2009. U had not, in the meantime, completed an assessment of his professional performance in advance of the review hearing. Having claimed that he was not well and was unfit to undergo a performance assessment, U had refused to allow the GMC to contact his medical assessors on the grounds that this would be a breach of patient confidentiality. The panel considered there to be very little evidence that U had the potential for remediation or retraining, and that his continual challenges to the authority of the GMC and revisiting of previously decided matters strongly suggested that he had deeply seated personality or attitudinal problems and, for the same reasons, did not have insight. Dealing with the legal framework, Eder J said that the case before the review panel fell to be considered under section 35D of the Medical Act 1983. By rule 22 of the General Medical Council (Fitness to Practise) Rules 2004, the panel was obliged to consider first whether U's fitness to practise was impaired, before proceeding to consider what sanction would be appropriate. If the panel finds that an individual's fitness to practise is impaired, it has the wide range of powers set out in section 35D, in particular in subparagraph (5), including erasure. U was given proper notice that the purpose of the review panel was to reassure the panel that U was fit to resume practice, either unrestricted or with conditions, so that the panel would be able to satisfy itself that U had maintained his skills or knowledge, and that patients would not be placed at risk by resumption of practice or by the imposition of conditional registration. U knew full well what he was up against and he could not properly complain that the panel hearing was in any way unfair. The panel conducted the hearing in an entirely lawful and fair manner, and the conduct of the GMC had throughout been impeccable. In those circumstances, the appeal had to be dismissed.

### *Arora v. General Medical Council* [2012] EWHC 1560 (Admin)

**59.15** Deficient professional performance—conditional registration—deterioration of doctor's clinical skills—erasure

In March 2006, a fitness to practise panel of the GMC considered that A's deficiencies were serious and wide-ranging, and that his performance fell below the standard to be expected of a qualified doctor practising medicine. The panel imposed conditions on his registration for a period of eighteen months. At a hearing in July 2008, a review panel determined that A's fitness to practise was impaired by reason of deficient professional performance and imposed conditions on his registration for a further period of eighteen months. A requested an early review, which took place in November 2009. The panel acknowledged that A had made some efforts to retrain and considered that the steps that he had taken, to date, demonstrated some insight into his failings, and that he had the potential and willingness to retrain and to take remedial action. However, his fitness to practise remained impaired and the panel determined to impose conditions on his registration for a period of eighteen months. At a further review in May/June 2011, the panel considered an assessment report following a competence assessment that took place on 29 October 2010. The assessment team was of the opinion that A's performance indicated that his clinical skills had deteriorated. The assessment team concluded that whilst A expressed a willingness to retrain, he had not demonstrated that he was able to learn and had not improved sufficiently to return to medical practice. The assessment team was of the opinion that A was unrealistic in his understanding of the time and degree of improvement that would be required to achieve the standard necessary for independent practice. The unanimous view of the team was that he was not fit to practise at all. The review panel concluded that A's fitness to practise was impaired. On sanction, A referred to *Srirangalingham v. General Medical Council* [2002] UKPC 77, which deals with whether suspension is inconsistent with an earlier imposition of conditions that were imposed to encourage rehabilitation. However, in the instant case, having regard to the previous imposition of conditions on his registration, which had not been successful, the panel was of the opinion that his lack of professional competence was not amenable to improvement through conditions. The panel was also particularly concerned at A's lack of insight. In dismissing A's appeal against an order of erasure, the Court said that the appellant had not been able to demonstrate that the panel's determination was wrong. A's performance had been found to be seriously deficient over a long period of time. Extensive efforts to provide him with supervised practice, retraining, and remediation had failed. He had also failed to comply with the conditions imposed upon him by the GMC. On the evidence, there was no prospect of remediation or of a return to practise, and therefore erasure was the proportionate and appropriate sanction.

### *Ogbonna-Jacob v. Nursing and Midwifery Council* [2013] EWHC 1595 (Admin)

**59.16** Midwife—suspension—review—failure to engage in process—striking off

A review was conducted in the absence of the practitioner, who had earlier withdrawn from a previous hearing because she was unhappy about a ruling of the panel not to exclude the press. The practitioner made it clear that she did not intend to attend to engage with the review process. She failed to take up the clear and specific suggestion of the previous panel to provide the review panel with a written document reflecting on particular matters. It was held that the review panel was entitled to reach its conclusion of continuing impairment of fitness to practise, and the appropriate sanction was a striking-off order.

### *Bamgbelu v. General Dental Council* [2013] EWHC 1169 (Admin)

**59.17** Dentist—conditions modified and extended—failure to address

In this case, the original decision of conditions resulted from inadequate clinical treatment of a patient, and unsafe and unhygienic practice conditions at a surgery. The conditions were modified at first review and extended at second review. The appeal was dismissed. The sole and whole reason for extending the period of conditions was because the appellant had not been supervised in accordance with the original conditions, and the PCC found that there was limited evidence of his practice. The appellant had not sufficiently addressed the concerns of the earlier fitness to practise hearing.

### *Obukofe v. General Medical Council* [2014] EWHC 408 (Admin)

**59.18** Review hearing—appropriate procedure

On 4 April 2011, O was convicted at Leicester Crown Court of three counts of sexual assault following a trial. The assaults that were the subject matter of those convictions took place against two junior members of staff at the hospital in Leicester at which O was working at the time. O was sentenced to six months' imprisonment, suspended for twelve months on each of the three counts, to run concurrently, and a sex offenders' register requirement for a period of seven years. On 25 April 2012, the panel issued its decision on the issue of impairment and determined to suspend O's registration for twelve months. In deciding on the length of suspension, the panel considered that such a period was necessary to send a message to him, the public, and the profession, and to maintain the public's confidence in the profession, and to declare and uphold proper standards of conduct and behaviour. The panel then went on to determine that there should be a review hearing and identified the structure within which that review hearing would take place. The panel said that, at the review, the next panel would need to be satisfied that O fully appreciated the gravity of the offences, that he had not reoffended, that he had maintained his skills and knowledge, and that patients would not be placed at risk by any potential resumption of practice. The panel concluded that it would be in his own interest, and it would assist the next panel reviewing his case, for him to provide evidence to show that he had reflected on and gained insight into his actions, and evidence as to his understanding of professional boundaries and how these could impact upon his relationships in the workplace. In dismissing O's appeal from a decision of the fitness to practise panel on a review to impose a further twelve months' suspension, Popplewell J said:

> 24. [Section 35D of the Medical Act 1983] is silent as to how the fitness to practise panel is to go about considering whether to extend a period of suspension under subsection (5). Nevertheless, the procedure is addressed by the [General Medical Council (Fitness to Practise) Rules 2004]. Rule 18 applies Part 5 to hearings which include hearings under Section 35D(5) to consider extension of suspension, and describes them as review hearings. Rule 22 governs procedure at review hearings. Rule 22(e) provides that at a review hearing, the Fitness to Practise Panel shall receive further evidence and further submissions from parties as to whether the fitness to practise of the practitioner is impaired. Rule 22(f) provides that the Panel shall consider and announce its finding on the question of whether the fitness to practise of the practitioner is impaired. Rule 22(g) provides that it shall go on then to consider receiving further evidence and further submissions from the parties as to whether to make a direction under Section 35D(5) for an extension of the suspension. Rule 22(i) provides that it shall announce its decision in relation to that.
>
> 25. Those rules make clear that what is envisaged by way of procedure is that the Panel should first consider and determine, with the benefit of evidence and submissions, whether the practitioner's fitness to practise remains impaired as a first step before going on to consider whether to extend a period of suspension.

26. That is an obviously fair and sensible procedure. It does not involve going behind or re-opening the determination which resulted in the suspension being imposed in the first place under section 35D(2), which is a decision which is concerned solely with the practitioner's fitness to practise at that first stage. What the procedure provides for in rule 22 is to safeguard the interests of the practitioner, by requiring the Panel which is considering whether to extend the suspension to be satisfied that the practitioner's fitness to practise remains impaired. If it were to conclude that his fitness to practise were no longer impaired it would clearly not be appropriate to extend the suspension. If, on the other hand, the Panel concludes that his fitness to practise remained impaired, the nature and the extent of the continued impairment will inform the decision as to whether to impose the sanction of extending the suspension and, if so, for what period.

27. That is the process which is envisaged by the Rules as the court recognised in the case of *Abrahaem v. General Medical Council* [2008] EWHC 183 (Admin).

***Townrow v. Financial Services Authority* [2014] EWCA Civ 1099**

**59.19** See paragraph 58.13.

## C. Other Relevant Chapters

**59.20** Appeals, Chapter 4
Deficient Professional Performance, Chapter 16
Impairment of Fitness to Practise, Chapter 33
Misconduct, Chapter 47
Restoration to Register, Chapter 58
Sanction, Chapter 60
Suspension, Chapter 65

# 60

# SANCTION
| A. Legal Framework | 60.01 | Hossain v. General Medical Council [2001] UKPC 40 | 60.23 |
| B. General Principles | 60.02 | Colgan v. The Kennel Club, No. 01/ TLQ/0673, 26 October 2001 | 60.24 |
| Dad v. General Dental Council [2000] 1 WLR 1538 | 60.03 | Mateu-Lopez v. General Medical Council [2003] UKPC 44 | 60.26 |
| Ghosh v. General Medical Council [2001] 1 WLR 1915 | 60.04 | Pembrey v. General Medical Council [2003] UKPC 60 | 60.27 |
| Marinovich v. General Medical Council [2002] UKPC 36 | 60.05 | Singh v. General Medical Council [2003] UKPC 15 | 60.28 |
| Crabbie v. General Medical Council [2002] 1 WLR 3104 | 60.06 | Nahal v. Law Society [2003] EWHC 2186 (Admin) | 60.29 |
| R (Bevan) v. General Medical Council [2005] EWHC 174 (Admin) | 60.07 | Wentzel v. General Medical Council [2004] EWHC 381 (Admin) | 60.30 |
| Giele v. General Medical Council [2005] 4 All ER 1242 | 60.08 | R (Harry) v. General Medical Council [2006] EWHC 3050 (Admin) | 60.31 |
| Bradley v. Jockey Club [2005] EWCA Civ 1056 | 60.09 | R (Smith) v. General Teaching Council for England [2007] EWHC 1675 (Admin) | 60.32 |
| Raschid v. General Medical Council; Fatnani v. General Medical Council [2007] 1 WLR 1460 | 60.10 | Moody v. General Osteopathic Council [2008] EWCA Civ 513 | 60.33 |
| Graham v. Nursing and Midwifery Council 2008 SC 659 | 60.11 | R (Balasubramanian) v. General Medical Council [2008] EWHC 639 (Admin) | 60.34 |
| Salsbury v. Law Society [2009] 1 WLR 1286 | 60.12 | Yerolemou v. Law Society [2008] EWHC 682 (Admin) | 60.35 |
| Shah v. General Pharmaceutical Council (formerly Royal Pharmaceutical Society of Great Britain) [2011] EWHC 73 (Admin) | 60.13 | In re King (Ch Ct of York) [2009] 1 WLR 873 | 60.36 |
| Levy v. Solicitors Regulation Authority [2011] EWHC 740 (Admin) | 60.14 | Augustine v. Nursing and Midwifery Council [2009] EWHC 517 (Admin) | 60.37 |
| Ujam v. General Medical Council [2012] EWHC 683 (Admin) | 60.15 | Council for the Regulation of Health Care Professionals v. General Medical Council and Khanna [2009] EWHC 596 (Admin) | 60.38 |
| Law Society (Solicitors Regulation Authority) v. Emeana and ors [2013] EWHC 2130 (Admin) | 60.16 | Varley v. General Osteopathic Council [2009] EWHC 1703 (Admin) | 60.39 |
| C. Dishonesty Cases | 60.17 | Burke v. General Teaching Council [2009] EWHC 3138 (Admin) | 60.40 |
| D. Other Disciplinary Cases | 60.18 | Muscat v. Health Professions Council [2009] EWCA Civ 1090 | 60.41 |
| Narayan v. General Medical Council [1982] 1 WLR 1227 | 60.18 | R (Lloyd Subner) v. Health Professions Council [2009] EWHC 2815 (Admin) | 60.42 |
| Evans v. General Medical Council, Privy Council Appeal No. 40 of 1984 | 60.19 | Izzet and Cazaly v. Law Society [2009] EWHC 3590 (Admin) | 60.43 |
| Carmichael v. General Dental Council [1990] 1 WLR 134 | 60.20 | R (Howlett) v. Health Professions Council [2009] EWHC 3617 (Admin) | 60.44 |
| Singh v. General Medical Council, Privy Council Appeal No. 73 of 1997 | 60.21 | | |
| Bijl v. General Medical Council [2001] UKPC 42 | 60.22 | | |

## A. Legal Framework

**60.01** Examples of the statutory power of a regulator to make a direction on sanction include:

Solicitors Act 1974, section 47(2)
Medical Act 1983, section 35D
Dentists Act 1984, section 27B
Opticians Act 1989, section 13F
Health and Social Work Professions Order 2001, article 29
Nursing and Midwifery Order 2001, article 29
Pharmacy Order 2010, article 54
Police (Conduct) Regulations 2012, regulation 35
Bar Standards Board Disciplinary Tribunals Regulations 2014, rules E157–E179
Institute of Chartered Accountants in England and Wales Disciplinary Bye-laws, bye-law 22

## B. General Principles

### *Bolton v. Law Society* [1994] 1 WLR 512

**60.02** General principles—maintenance of public confidence—reputation of the profession

B misused clients' monies. He had paid money belonging to a client building society to his wife in anticipation of the completion of a conveyancing transaction. He paid the monies out prematurely in anticipation of formal completion, which did not take place because the purchaser reneged. B took steps to put matters right to the extent of repaying the advance to the building society, but interest and costs remained outstanding. The Solicitors Disciplinary Tribunal (SDT) accepted that B was an honest man. It found that he had not stolen clients' monies in a premeditated fashion and that his actions did not represent a deliberate course of dishonest conduct. However, his conduct was wholly unacceptable and was a very serious matter indeed. In the Tribunal's judgment, such conduct would ordinarily merit striking off, but the Tribunal felt able, on the facts of the case and because B was young, relatively inexperienced, and his behaviour was naive, to make an order suspending him for two years. The Court of Appeal considered the sentence to be correct and, in giving judgment, Sir Thomas Bingham MR (with whom Rose and Waite LJJ agreed) said, at 518A–519E:

> It is required of lawyers practising in this country that they should discharge their professional duties with integrity, probity and complete trustworthiness. That requirement applies as much to barristers as it does to solicitors. If I make no further reference to barristers it is because this appeal concerns a solicitor, and where a client's moneys have been misappropriated the complaint is inevitably made against a solicitor, since solicitors receive and handle clients' moneys and barristers do not.
>
> Any solicitor who is shown to have discharged his professional duties with anything less than complete integrity, probity and trustworthiness must expect severe sanctions to be imposed upon him by the Solicitors Disciplinary Tribunal. Lapses from the required high standard may, of course, take different forms and be of varying degrees. The most serious involves proven dishonesty, whether or not leading to criminal proceedings and criminal penalties. In such cases the tribunal has almost invariably, no matter how strong the mitigation advanced for the solicitor, ordered that he be struck off the Roll of Solicitors. Only infrequently, particularly in recent years, has it been willing to order the restoration to the Roll of a solicitor against whom serious dishonesty had been established, even after a passage of years, and even where the solicitor had made every effort to re-establish himself and redeem his reputation. If a solicitor is not shown to have acted dishonestly, but is shown to have fallen below the required standards of integrity, probity and trustworthiness, his lapse is less serious but it remains very serious indeed in a member of a profession whose reputation depends upon trust. A striking off order will not necessarily follow in such a case, but it may well. The decision whether to strike off or to suspend will often involve a fine and difficult exercise of judgment, to be made by the tribunal as an informed and expert body on all the facts of the case. Only in a very unusual and venial case of this kind would the tribunal be likely to regard as appropriate any order less severe than one of suspension.
>
> It is important that there should be full understanding of the reasons why the tribunal makes orders which might otherwise seem harsh. There is, in some of these orders, a punitive element: a penalty may be visited on a solicitor who has fallen below the standards required of his profession in order to punish him for what he has done and to deter any other solicitor tempted to behave in the same way. Those are traditional objects of punishment. But often the order is not punitive in intention. Particularly is this so where a criminal penalty has been imposed and satisfied. The solicitor has paid his debt to society. There is no need, and it would be unjust, to punish him again. In most cases the order of the tribunal will be primarily directed to one or other or both of two other purposes. One is to be sure that the offender does not have the opportunity to repeat the offence. This purpose is achieved for a limited period by an order of suspension; plainly it is hoped that experience of suspension will make the offender meticulous in his future compliance with the required standards. The purpose is achieved for a longer period, and quite possibly indefinitely, by an order of striking off. The second purpose is the most fundamental of all: to maintain the reputation of the solicitors' profession as one in which every member, of whatever standing, may be trusted to the ends of the earth. To maintain this reputation and sustain public confidence in the integrity of the profession it is often necessary that those guilty of serious lapses are not only expelled but denied re-admission. If a member of the public sells his house, very often his largest asset, and entrusts the proceeds to his solicitor,

pending re-investment in another house, he is ordinarily entitled to expect that the solicitor will be a person whose trustworthiness is not, and never has been, seriously in question. A profession's most valuable asset is its collective reputation and the confidence which that inspires.

Because orders made by the tribunal are not primarily punitive, it follows that considerations which would ordinarily weigh in mitigation of punishment have less effect on the exercise of this jurisdiction than on the ordinary run of sentences imposed in criminal cases. It often happens that a solicitor appearing before the tribunal can adduce a wealth of glowing tributes from his professional brethren. He can often show that for him and his family the consequences of striking off or suspension would be little short of tragic. Often he will say, convincingly, that he has learned his lesson and will not offend again. On applying for restoration after striking off, all these points may be made, and the former solicitor may also be able to point to real efforts made to re-establish himself and redeem his reputation. All these matters are relevant and should be considered. But none of them touches the essential issue, which is the need to maintain among members of the public a well-founded confidence that any solicitor whom they instruct will be a person of unquestionable integrity, probity and trustworthiness. Thus it can never be an objection to an order of suspension in an appropriate case that the solicitor may be unable to re-establish his practice when the period of suspension is past. If that proves, or appears likely, to be so the consequence for the individual and his family may be deeply unfortunate and unintended. But it does not make suspension the wrong order if it is otherwise right. The reputation of the profession is more important than the fortunes of any individual member. Membership of a profession brings many benefits, but that is a part of the price.

***Dad v. General Dental Council* [2000] 1 WLR 1538**

**60.03** Criminal conviction—penalty—nature and gravity of offence—effect on public confidence

D, a dentist, appealed against a determination of the Professional Conduct Committee (PCC) of the General Dental Council (GDC) on 14 May 1999 directing the registrar to suspend his registration in the dentists' register for a period of twelve months as a consequence of the proof against him of two convictions for offences under the Road Traffic Act 1988. The charge against D set out in the notice of inquiry, as amended, stated that D had been convicted of offences in Glasgow Sheriff Court, after pleading guilty, of one charge of driving whilst disqualified, and at Lanark Sheriff Court, after pleading guilty, of one charge of reckless driving and one charge of failing to produce insurance documents. In relation to the first charge, D had previously been disqualified from driving under the 'totting-up' provisions for speeding offences. At the time that he committed the second charge, he had not yet been disqualified from driving and was not yet without insurance. In allowing D's appeal, Lord Hope of Craighead, delivering the judgment of the Privy Council, said that section 27(1) of the Dentists Act 1984 placed no restriction on the nature or gravity of the criminal offences for the conviction of which a registered dentist may have his name erased from the register or have his registration suspended for such period not exceeding twelve months, as may be specified. Nevertheless, it is proper to have regard to the nature and gravity of such offences when decisions are being taken as to the appropriate penalty and its consequences. In this respect, there is room for a distinction between cases in which the penalty is imposed for conduct that the Committee had held is serious professional misconduct and cases, such as the present, in which the penalty is imposed upon proof of a conviction. The extent to which the nature or gravity of the offence of which the dentist has been convicted is likely to bring the profession into disrepute or to undermine public confidence in the profession is primarily one for the Committee. But it is a matter on which its determination may more readily be regarded as reviewable by the Board than it would be had the Committee been dealing with a case of conduct that, in its view, amounted to serious professional misconduct.

***Ghosh v. General Medical Council* [2001] 1 WLR 1915**

**60.04**
Misconduct—penalty—appeal—measure of respect

Delivering the judgment of the Privy Council, Lord Millett said:

> 34. ... [T]he Board will accord an appropriate measure of respect to the judgment of the Committee whether the practitioner's failings amount to serious professional misconduct and on the measures necessary to maintain professional standards and provide adequate protection to the public. But the Board will not defer to the Committee's judgment more than is warranted by the circumstances. The Council conceded, and their Lordships accept, that it is open to them to consider all the matters raised by [the appellant] in her appeal; to decide whether the sanction of erasure was appropriate and necessary in the public interest or was excessive and disproportionate; and in the latter event either to substitute some other penalty or to remit the matter to the Committee for reconsideration.

See further paragraph 4.08.

***Marinovich v. General Medical Council* [2002] UKPC 36**

**60.05**
Misconduct—penalty—public interest—public confidence—consequences for individual

M trained and qualified as a physician both in the United Kingdom and in Australia, where he practised for many years as a psychiatrist. For twenty-five years, he was senior psychiatrist at the Royal Women's Hospital in Melbourne. Latterly, he had been in practice as a psychiatrist in Darwin. In October 1999, he left Australia and came to live in the United Kingdom. On 14 October 1999, prior to his departure from Australia, M was found guilty of unprofessional conduct by the Medical Board of the Northern Territory. The Board held that, on several occasions, he had provided inappropriate treatment to a female psychiatric patient by way of massages and that, on one of these occasions, he had sexual intercourse with her. M's registration in Australia was cancelled. On 22 October 1999, he reactivated his registration in the UK as a medical practitioner. He took up a post at Durham General Hospital and thereafter at a hospital in Worthing until March 2001, when his employment was terminated. The General Medical Council (GMC) determined that a charge should be formulated against M in respect of the matters in Australia. In October 2000, the GMC's Interim Orders Committee (IOC) imposed an order for interim conditions on M's registration. By the time when the case came before the PCC on 26 November 2001 on the charge relating to the allegations in Australia, a notice had been served on M alleging that he had acted in breach of the orders for conditional registration. The PCC found proved the Australian allegations, and that M had repeatedly and intentionally flouted the conditions attached to his registration in the UK, and had acted in a profoundly dishonest manner. The Committee ordered that M's name be erased from the register. Dismissing M's appeal, Lord Hope of Craighead, delivering the judgment of the Privy Council, said:

> 28. Their Lordships appreciate that, having regard to his age, it would not be realistic to expect the appellant's name ever to be restored to the register in the event of its erasure. In the appellant's case the effect of the Committee's order is that his erasure is for life. But it has been said many times that the Professional Conduct Committee is the body which is best equipped to determine questions as to the sanction that should be imposed in the public interest for serious professional misconduct. This is because the assessment of the seriousness of the misconduct is essentially a matter for the Committee in the light of its experience. It is the body which is best qualified to judge what measures are required to maintain the standards and reputation of the profession.

29. That is not to say that their Lordships may not intervene if there are good grounds for doing so. But in this case their Lordships are satisfied that there are no such grounds. This was a case of such a grave nature that a finding that the appellant was unfit to practise was inevitable. The Committee was entitled to give greater weight to the public interest and to the need to maintain public confidence in the profession than to the consequences to the appellant of the imposition of the penalty. Their Lordships are quite unable to say that the sanction of erasure which the Committee decided to impose in this case, while undoubtedly severe, was wrong or unjustified.

### *Crabbie v. General Medical Council* [2002] 1 WLR 3104

**60.06** Driving conviction—whether erasure justified—whether referral to Health Committee appropriate

C pleaded guilty to causing death by dangerous driving and driving with excess alcohol. She was sentenced to five years' imprisonment and was disqualified from driving for ten years. She was a general practitioner (GP) in a rural area of Scotland. The PCC concluded that '[t]he nature and gravity of the offence in this case is such that the sanctions of conditions and suspension would not adequately protect the public interest' and that, in the public interest, '[e]rasure is the only appropriate sentence', adding that, in its view, erasure was proportionate to the nature and gravity of the offence. In dismissing C's appeal, the Privy Council said that the first and main question was whether the direction of erasure was justified. If it were, there would be nothing left in the submission that the PCC should have referred the case to the Health Committee. Lord Scott of Foscote, in giving the judgment of the Board, said that it was for the PCC, when deciding in a conviction case on the appropriate direction to give, to have regard to the nature and gravity of the criminal offence in question. The extent to which the fact that the offence had been committed by a practising doctor is likely to bring the profession into disrepute or to undermine public confidence in the profession is a proper matter for the PCC to take into account. It is a matter the weight of which it is primarily for the Committee to judge: see *Dad v. General Dental Council* [2000] 1 WLR 1538, 1542, *per* Lord Hope of Craighead. It was apparent that the PCC gave proper consideration to the impressive testimonials regarding the doctor that were placed before it.

See further paragraph 31.04.

### *R (Bevan) v. General Medical Council* [2005] EWHC 174 (Admin)

**60.07** Testimonials—relevance—submissions by regulator on sanction

In commenting on the judgment of Sir Thomas Bingham MR in *Bolton v. Law Society* [1994] 1 WLR 512, Collins J said that that case did not apply directly because the present case was not dealing with solicitors in breach of trust in financial terms. But it is obvious that, *mutatis mutandis*, the words of Sir Thomas Bingham MR can apply to the need for the public to have confidence in the medical profession, and in particular to know that they can be assured that if their doctor in any way takes advantage of them in an inappropriate fashion, that doctor will be dealt with in a serious manner by the profession. As to testimonials, they are clearly material, and the testimonials in the case of B, as indeed is perhaps so in many cases involving doctors, were not merely tributes from professional brethren. The important testimonials in this context were those that came from patients who had expressed their confidence in the individual doctor and had indicated that, so far as they are concerned, there had not only been no question of any vestige of inappropriate behaviour, but quite the contrary. This was a doctor in whom they had absolute confidence and whom they would wish to continue to be in practice as their doctor. One of the important purposes behind the disciplinary powers of the GMC is to ensure that patients are protected. Doctors should not be allowed to continue in practice if there is any risk to patients. Testimonials of the sort mentioned are clearly highly relevant to that issue. There can be no question, and indeed it was not suggested in this case, that B would, if he were allowed to come back to the profession within a reasonably short period of time, be any further risk to patients; indeed, quite the contrary. Of course, that is not the only purpose behind the disciplinary powers. There is the element—and the important element—of public confidence in the profession and public concern that any sort of inappropriate behaviour such as this will not be overlooked, and will result in serious penalty. In the present case, counsel stated that she was instructed by the GMC to seek a particular penalty. There is nothing wrong in principle in counsel representing either the complainant, who technically counsel may represent, or being instructed by the GMC, in putting forward to the Committee the penalty that is considered to be the appropriate penalty. But it is undesirable that it is put in the way in which counsel put it in this case—namely, that she was instructed by the GMC to seek a particular penalty. No doubt counsel will discuss with those instructing him or her and the relevant representatives of the GMC what sort of penalty, depending of course on what facts are found established, would be considered to be appropriate and therefore what would be put to the Committee. It clearly would be entirely proper, and in some cases appropriate, to refer to specific cases if, in those cases, guidance could be obtained that would be relevant to the individual case before the Committee. All that is perfectly proper, but it must be made clear that these are merely submissions. More importantly, it is essential that the Committee is publicly informed by the legal assessor that this must not be regarded in any way as something by which it should be influenced by beyond knowing that it is a submission. Of course, the Committee will take it into account, but it must exercise its own independent judgment, based upon such guidance as may be helpful from the various instructions issued by the GMC to doctors so that they know what is required of them and also the Indicative Sanctions Guidance. This was not done in this case and so there was a real concern that the Committee may have been influenced—and unduly influenced—by the knowledge that it was said that the GMC was desirous that a particular penalty be imposed.

### *Giele v. General Medical Council* [2005] 4 All ER 1242

**60.08**

Doctor—inappropriate relationship with patient—whether 'exceptional circumstances' to avoid erasure—erroneous approach

G appealed against the sanction of erasure imposed by a fitness to practise panel of the GMC on 19 March 2005. The sanction followed a finding of serious professional misconduct based upon a sexual relationship between the appellant and a patient, Mrs A, which lasted for over a year until about October 2002. Collins J allowed G's appeal and substituted a suspension order for twelve months in place of the erasure order. The learned judge said that for a doctor to engage in an improper relationship is for him to court erasure. But it must be emphasized that erasure is not to be regarded as inevitable; nor would erasure be the appropriate sanction in any but an exceptional case. The legal assessor's advice to the panel to ask themselves, 'Are there exceptional circumstances?' was erroneous.

> [26] The panel had to approach the question of sanctions starting with the least severe. It was not a question of deciding whether erasure was wrong but whether it was right for the misconduct in question after considering any lesser sanction. Furthermore, it was wrong to ask whether there were exceptional circumstances to avoid erasure. Exceptional circumstances would only avoid the possibility of erasure. A panel member asked whether there was any definition of exceptional circumstances and was given no satisfactory answer. That is not surprising since what is exceptional will depend on the facts of a particular case. But in my judgment it was in this case and will in most cases be unhelpful to talk in terms of exceptional circumstances. The panel must look at the misconduct and the mitigation and decide what sanction is appropriate, no doubt bearing in mind that improper sexual relationships with a vulnerable patient are always regarded as most serious....

[...]

[29] I do not doubt that the maintenance of public confidence in the profession must outweigh the interests of the individual doctor. But that confidence will surely be maintained by imposing such sanction as is in all the circumstances appropriate. Thus in considering the maintenance of confidence, the existence of a public interest in not ending the career of a competent doctor will play a part. Furthermore, the fact that many patients and colleagues have, in the knowledge of the misconduct found, clearly indicated their views that erasure was not needed is a matter which can carry some weight in deciding how confidence can properly be maintained.

[30] [Counsel for the GMC] submitted that it would be wrong to allow a practitioner who was more skilled and whose loss would accordingly be a greater blow to avoid a sanction which would otherwise be appropriate and would have been imposed on the less skilled. So long as the public interest in retaining the services of a competent practitioner is a relevant consideration, it is inevitable that the weight to be attached to this aspect will to some extent depend on the abilities of the practitioner in question. It must be obvious that misconduct which is so serious that nothing less than erasure would be considered appropriate cannot attract a lesser sanction simply because the practitioner is particularly skilful. But if erasure is not necessarily required, the skills of the practitioner are a relevant factor.

[...]

[33] There can be no doubt that the improper sexual relationship which was established in this case could have merited erasure. However, the mitigation and in particular the testimonials might well have tipped the balance against it. But the panel approached the issue of sanction in the wrong way, clearly believing that there should be erasure unless exceptional circumstances existed. Accordingly, I am entitled to form my own view. I am entirely satisfied that erasure was not required and that public confidence in the profession, which must reflect the views of an informed and reasonable member of the public, would not be harmed if suspension was imposed. Suspension for 12 months is itself a severe penalty for any practitioner and I am satisfied that for the misconduct in this case it will provide an appropriate sanction.

### *Bradley v. Jockey Club* [2005] EWCA Civ 1056

**60.09** Jockey Club—contract—penalty—whether restraint of trade

B was a bloodstock agent who was previously a successful steeplechase jockey. In September 2001, B gave evidence for the defence at Southampton Crown Court, and, under cross-examination, he confirmed that he had received money and presents from a major gambler in the racing world for sensitive and privileged information about horses. In November 2002, the Disciplinary Committee of the Jockey Club held an inquiry largely based on B's evidence in the Crown Court and found B guilty of various breaches of the rules of racing. It imposed a penalty of disqualification for eight years. B appealed to the appeal board of the Jockey Club. The appeal board was chaired by Sir Edward Cazalet, a former High Court judge. The appeal board reduced the penalty imposed upon B from eight years' to five years' disqualification. B commenced proceedings in the High Court based on breach of contract and contended that the Jockey Club, in carrying out its disciplinary functions, should act reasonably and fairly, and impose only a sentence proportionate to the facts proved or admitted, and that the disqualification was an unlawful restraint of trade. Richards J dismissed the claim: [2004] EWHC 2164 (QB). On appeal to the Court of Appeal (Lord Phillips of Worth Matravers MR, and Buxton and Scott Baker LJJ), in dismissing B's appeal, said, at [20]–[24], that professional and trade regulatory and disciplinary bodies are usually better placed than is the Court to evaluate the significance of breaches of the rules or standards of behaviour governing the professions or trades to which they relate.

24. Where an individual takes up a profession or occupation that depends critically upon the observance of certain rules, and then deliberately breaks those rules, he

cannot be heard to contend that he has a vested right to continue to earn his living in that profession or occupation. Any disciplinary tribunal, or the court when exercising a supervisory jurisdiction, must give careful consideration to whether the circumstances require a penalty that will prevent the culprit from continuing to earn his living in his chosen profession or occupation. But a penalty which deprives him of that right may well be the only appropriate response to his offending.

### *Raschid v. General Medical Council; Fatnani v. General Medical Council* [2007] 1 WLR 1460

**60.10** Specialist tribunal—respect for decision

Because a principal purpose of the panel's jurisdiction in relation to sanctions is the preservation and maintenance of public confidence in the profession rather than the administration of retributive justice, particular force is given to the need to accord special respect to the judgment of the professional decision-making body in the shape of the panel.

See further paragraph 4.16.

### *Graham v. Nursing and Midwifery Council* 2008 SC 659

**60.11** Nurse—prisoner—other nurse on duty not charged—effect on sanction

The appellant was employed as a registered nurse at HM Prison Glenochil, Scotland. The allegations were that she failed to check an insulin pen to ensure that it was the correct one before handing it to a prisoner to inject himself, that she failed to take appropriate action when the prisoner complained that he had injected himself with the wrong insulin pen, and that she failed to report and/or to record the prisoner's complaint that he had injected himself with the wrong insulin pen. On 24 March 2004, the date of the incident, the appellant was on duty with another practitioner nurse. Both nurses were responsible for the administration of the afternoon drug round at the prison. The PCC found the appellant guilty of misconduct and decided to remove her name from the register. On appeal, the appellant sought to challenge only the sanction imposed by the Committee. In delivering the opinion of the Court, Lord Wheatley said that, when the mistake came to light, the other nurse, nurse A, was, so far as the Court could judge, in a similar position to that of the appellant (although she was not responsible for the initial error). Nurse A appeared to have been interviewed in the course of the investigation into the appellant's case, but whether she was the subject of an investigation was unclear. She certainly did not give evidence. Although there may have been reasons for all of this, these were not disclosed, and it seems that her evidence would have been extremely valuable in identifying more precisely the nature of the carelessness or negligence with which the appellant was charged. In these circumstances, it would be reasonable that the disposal of the appellant's case, in the interests of natural justice, should take into account to some extent the failure to take any action against someone who, on the face of it, appears also to have been partly similarly responsible for what occurred, or did not occur, after the incident. In all of the circumstances, the Court proposed to quash the decision to remove the appellant's name from the register and to substitute a caution.

### *Salsbury v. Law Society* [2009] 1 WLR 1286

**60.12** Human rights—effect on sanction

*Bolton v. Law Society* [1994] 1 WLR 512 remains good law, subject to the qualification that the SDT must also take into account the rights of the solicitor under Articles 6 and 8 of the European Convention on Human Rights (ECHR). Absent any error of law, the High Court must pay considerable respect to the sentencing decisions of the Tribunal. Nevertheless, if the High Court, despite paying such respect, is satisfied that the sentencing decision was clearly inappropriate, then the Court will interfere.

See further paragraph 4.17.

***Shah v. General Pharmaceutical Council (formerly Royal Pharmaceutical Society of Great Britain)* [2011] EWHC 73 (Admin)**

**60.13** Pharmacist—previous committee decisions—limited effect—individual determination

In this case, the Administrative Court (Wyn Williams J) dismissed the appellant's appeal against the decision to direct the respondent's registrar to remove his name from the register. The appellant was the superintendent pharmacist at a pharmacy known as 'Shah Pharmacy' located in Enfield. As a consequence of dispensing errors, a complaint was made to the respondent. A visit to the pharmacy by inspectors employed by the respondent revealed further matters of concern about practices in the pharmacy, which included the supply and storage of out-of-date medicines. The appellant admitted most of the facts alleged against him. He also admitted that many of his actions had placed him in breach of Key Responsibilities 1 and 3, which provide that pharmacists should act in the interest of patients and seek to provide the best possible health care for the community, and that pharmacists should not bring the profession into disrepute or undermine public confidence in the profession. By his grounds of appeal, the appellant alleged that the statutory committee had failed to have regard to the fact that, in previous decisions of the committee, there was 'a consistent body of jurisprudence' showing that the reputation of the profession could be vindicated by decisions to reprimand practitioners for similar offences to those facing the appellant. The appellant relied upon earlier decisions of the statutory committee reported in the *Pharmaceutical Journal*. Wyn Williams J said that there was nothing within the reports of the cases relied upon by the appellant that suggested that they formed part of a coherent body of consistent jurisprudence. There was no suggestion in any of the cases that later cases rely upon the earlier ones and there was no suggestion in the reports that the sanction of reprimand was imposed because that was some kind of norm in the circumstances revealed in the cases in question. The learned judge was not persuaded that the earlier cases were anything more than individual decisions essentially related to their own facts and it seemed to him to be clear that the statutory committee was wholly justified in concluding that the significance of the previous decisions was limited in determining the appropriate sanction in the present case.

***Levy v. Solicitors Regulation Authority* [2011] EWHC 740 (Admin)**

**60.14** Procedure—issues affecting sanction—announcement of findings of fact—submissions on sanction

In dismissing the appellant solicitor's appeal against a decision of the SDT suspending him from practice for nine months for breaches of the Solicitors Accounts Rules, the Administrative Court (Jackson LJ and Cranston J) said, at [34], that procedurally it is imperative that a tribunal does not proceed to sanction before having announced the basis of its findings on the substantive allegations. As a general principle, fairness demands that disputed issues that can substantially affect sanction should be resolved, and should be resolved in a procedurally fair manner, and that parties should then be able to address the tribunal on the appropriate sanction. An analogy in criminal sentencing is the so-called '*Newton* hearing'—after *R v. Newton* (1982) 4 Cr App R(S) 388—designed to resolve disputed issues of fact where, after a 'guilty' plea, all that remains is sentencing. The judge hears evidence and decides the matter to provide a basis for the defendant to make representations and advance mitigation before the court passes sentence. The same approach must be followed by the tribunal, so that it announces its findings on any matters having a bearing on sanction and then provides ample opportunity for representations to be made on behalf of the solicitor about the sanction to be imposed. In the instant case, on the facts, there was no breach of these principles. The Tribunal had resolved in a procedurally fair manner the key issue affecting sentence—namely, it had discarded the dishonesty allegation. The Tribunal knew about the defendant and his background, and was aware of the nature of the firm, its size, and the work that it undertook.

*Ujam v. General Medical Council* [2012] EWHC 683 (Admin)

**60.15** Earlier interim suspension—effect on penalty

On sanction, between July 2009 and February 2010, U's registration had been suspended by the GMC's Interim Orders Panel (IOP). In dismissing U's appeal against the decision of the fitness to practise panel suspending his registration for a period of six months, Eady J said that it would undoubtedly be right that the suspension that the IOP imposed should be borne in mind as part of the background circumstances, but it would certainly be inappropriate to regard it as analogous to a period of imprisonment served while on remand (which would normally be deducted from any custodial term imposed by a sentencing court). The GMC Indicative Sanctions Guidance states, at paragraph 22, that:

> ... [I]n making their decision on the appropriate sanction, panels need to be mindful that they do not give undue weight to whether or not a doctor has previously been subject to an interim order for conditions or suspension imposed by the Interim Orders Panel, or the period for which that order was effected... an interim order and the length of that order are unlikely to be of much significance for panels...

*Law Society (Solicitors Regulation Authority) v. Emeana and ors* [2013] EWHC 2130 (Admin)

**60.16** Solicitor—strike-off—no dishonesty—reputation of profession

Lapses less serious than dishonesty may nonetheless require striking off, if the reputation of the solicitors' profession 'to be trusted to the ends of the earth' (*per* Sir Thomas Bingham MR in *Bolton v. Law Society* [1994] 1 WLR 1286) is to be maintained. The principle identified in *Bolton* means that, in cases in which there has been a lapse of standards of integrity, probity, and trustworthiness, a solicitor should expect to be struck off. Striking off is the most serious sanction, but it is not reserved for offences of dishonesty. Participation in a sham partnership undermines the profession and the protection to which the public are entitled to be derived from regulation and proper working of partnerships. E was able to practise as a de facto sole principal when he was not qualified to do so. That alone would justify striking off. A fine was inappropriate.

See further paragraph 28.18.

## C. Dishonesty Cases

**60.17**

See paragraphs 19.35–19.81.

## D. Other Disciplinary Cases

*Narayan v. General Medical Council* [1982] 1 WLR 1227

**60.18** Procedure—announcement of determination

The chairman, in announcing the determination of the PCC on erasure, had correctly observed the procedure required by the rules. The chairman stated that, by reason of the appellant's conviction of a criminal offence and having regard to his previous record, the PCC should direct the registrar to erase N's name from the register. The appellant complained that the rules set out in detail the steps by which the PCC must proceed, by obliging it to consider possible penalties in ascending order of gravity and to direct that the name of the practitioner be erased from the register only after having determined not to impose any lesser penalty. In dismissing N's appeal, the Privy Council said that, looking at the matter from a practical point of view, all that the rules require is that the chairman shall announce the effective or operative decision of the PCC and that the chairman is not bound to announce publicly the various steps by which the PCC had reached that decision. The only practical reason in favour of requiring the chairman to announce each step would be to ensure, and to enable the doctor to be sure, that the PCC had proceeded properly in its deliberations, but their Lordships were of the opinion that that is not what the rules require. All that is required is that the chairman shall announce the final decision of the PCC.

*Evans v. General Medical Council*, **Privy Council Appeal No. 40 of 1984**

**60.19** Doctor—adulterous relationship with patient over six years—erasure

This case related to a doctor who had maintained an adulterous relationship with his patient over a period of at least six years. His patients also included her husband and two children. A whole family to which he owed professional obligations suffered through his breach of a fundamental aspect of medical ethics. His appeal on sanction of erasure was dismissed.

*Carmichael v. General Dental Council* [1990] **1 WLR 134**

**60.20** Dentist—general anaesthetic to patient—loss of consciousness

The PCC found that a dentist had, on four separate occasions, administered a general anaesthetic to a patient contrary to guidance issued by the GDC and directed that his name should be erased from the register. Seven of the eight members of the Committee had practical experience and the Committee was fully entitled to conclude that, having regard to the techniques for administration of drugs and in the light of the expert evidence, the dentist's patients had suffered a loss of consciousness. The Committee was the best and proper body to decide sentence; the appeal was dismissed.

*Singh v. General Medical Council*, **Privy Council Appeal No. 73 of 1997**

**60.21** Doctor—testimonials and petition—public trust in profession

A doctor was convicted, after a contested hearing, on ten counts of dishonesty and was sentenced to fifteen months' imprisonment, suspended for two years. His appeal against suspension for twelve months by the PCC was dismissed. Over a hundred personal testimonials and signed petitions were presented, but this was a case in which the Committee was entitled to take the view that the policy of preserving public trust in the profession prevailed over the strong personal mitigation that the doctor was able to put forward.

*Bijl v. General Medical Council* [2001] **UKPC 42**

**60.22** Serious professional misconduct—single operation—long medical career—repetition unlikely

B, a consultant urologist, carried out a keyhole operation upon a patient to break down and remove a large staghorn stone that was lodged in her left kidney. After leaving the operating theatre and going home, the patient suffered massive internal bleeding from the kidney. Another surgeon carried out an emergency operation to clamp the site of the bleeding in the kidney. This stabilized her condition, but she died two days later. There was no suggestion that the outcome would have been different if B had stayed and performed the emergency operation himself. The substance of the charges that the PCC found proved against B were twofold: first, that, despite obvious indications of substantial loss of blood, he had carried on the operation for far too long; and second, that, after completing the operation, he abandoned the patient when her condition was still serious. The Committee found B guilty of serious professional misconduct, from which there was no appeal, and directed that his name be erased from the register. The Privy Council allowed B's appeal against erasure and substituted a direction that his registration be suspended for one year from the date upon which the hearing before the Committee was concluded. While giving great weight to the judgment of the Committee, their Lordships said that they felt difficulty, in the light of all of the circumstances, in being satisfied that erasure involving a complete cessation of B's medical work was necessary, when suspension with the possibility of imposing detailed conditions on his carrying on practice was available. The charges of which B was convicted arose out of one operation, at the latter part of a career of service in Holland, Nigeria, and the United Kingdom. The charges involved serious errors of judgment, but did not involve any allegations against his practical skills as a doctor such as might be difficult to improve at a later stage of a career. B had had a serious lesson that, on the evidence available, should prevent a repetition of these errors of judgment, were he to be allowed to practise in the future, particularly if he were to do so under conditions intended to avoid such repetition. In this very unusual case, their Lordships were unable to be satisfied that erasure was necessary.

### *Hossain v. General Medical Council* [2001] UKPC 40

**60.23** GP—working in deprived area—dedicated doctor—no evidence of general failure

H, aged 62, had practised as a GP since 1970. The complaints against him related to the care of four patients from his practice in Peckham, one in 1995, one in 1996, one in 1997, and one in 2000. A charge common to three of the cases was that he did not himself visit the patients at home when he should have done so. There were other allegations that he failed to give the necessary medical treatment to the patients and that he failed to cause proper records to be kept. H carried on practice in a very deprived, difficult area in which many doctors were not willing to work. For thirteen years, he practised alone. At the time of his erasure, something in the region of 6,500 patients were registered with the practice. He had looked after two of the patients referred to and their families in one case for twenty-eight years and in another case for fifteen years. H put the total number of consultations by himself at something in the region of 10,000 a year over the seven years investigated by the PCC. In allowing H's appeal and remitting the matter to the Committee for it to consider conditions, Lord Slynn of Hadley, giving the judgment of the Privy Council, said that, in this case, the penalty of erasure was not justified. In the first place, it did not appear that the Committee took into account the long years and dedicated service in the particular area where H worked or the views of patients and colleagues. In the second place, the Committee appeared to have assumed that there was a more general failure over the years. In the third place, the Committee simply said that it rejected the suggestion made that the safety of the public could be achieved by the imposition of conditions. It did not give any reasons for that or for why the protection of the public required erasure. The Committee gave absolutely no explanation as to why no form of medical practice could be entrusted to H.

### *Colgan v. The Kennel Club*, No 01/TLQ/0673, 26 October 2001

**60.24**

In this case, a five-year ban against well-regarded dog breeder was reduced to three years on registration and two years on other aspects: see paragraph 14.21.

### *Garfoot v. General Medical Council* [2002] UKPC 35

**60.25** Drug prescriptions—irresponsible prescribing—erasure

G owned a private clinic, described as a drug and alcohol dependency clinic. He prescribed controlled drugs to drug-addicted patients. The heads of charge related to thirteen patients. The primary allegation against G was that he had been guilty of irresponsible and/inappropriate prescribing. The PCC found proved the allegations of irresponsible prescribing and that G was guilty of serious professional misconduct, and it directed that his name be erased from the medical register. On appeal, the Privy Council held that the Committee's conclusions that G had prescribed irresponsibly and was guilty of serious professional misconduct were inevitable. The Committee had found that twelve patients displayed a pattern of excessive treatments, mostly over several years. Some had become dependent on dexamphetamine, a prescription that the expert called on G's behalf said he could not have advocated in combination with methadone. G had failed to take all possible steps through regular urine tests and personal systematic monitoring to minimize the risk of prescribed drugs being diverted. He failed to grasp opportunities to help patients to change their habits, or to face the reality that their 'treatments' had failed and that G could no longer properly accede to their requests. The Committee expressed grave concern about the routes, formulation, and size of substitute opiate dosage, and the failure to bring all of the chemical evidence into a reasoned therapeutic strategy of management. It expressed its extreme concern at the nature and combination of drugs employed. Sir Philip Otton, giving the judgment of the Privy Council, said that, as to the sanction of erasure, it was neither excessive, disproportionate, inappropriate, nor unnecessary in the public interest. The evidence and the conclusions of the Committee indicated a very serious state of affairs. Their Lordships could not accept the argument that the patients did not suffer harm. Where there was no attempt at stabilization on oral preparations and no attempt to engage patients other than by maintenance prescribing, there was inevitable harm to such patients. The circumstances of the present case indicated the importance of maintaining public confidence in medical practitioners working in this difficult area with particularly vulnerable patients.

### *Mateu-Lopez v. General Medical Council* [2003] UKPC 44

**60.26** Prescription forms—failure to ensure patients adequately examined—erasure set aside

The appellant was a consultant psychiatrist working in the fields of alcohol and drug addiction at a National Health Service (NHS) Trust. He faced allegations of providing two senior clinical nurse specialists with signed, but uncompleted, prescription forms, and inappropriately and/or irresponsibly delegating responsibility to nursing staff for the issue of prescriptions for controlled drugs and failing to ensure that patients were adequately medically examined or assessed prior to the issue of prescriptions. At the outset of the inquiry, the appellant admitted all of the heads of charge. He further accepted that the admitted facts amounted to serious professional misconduct. The PCC, after hearing evidence, directed that the appellant's name should be erased from the medical register. Giving the judgment of the Privy Council on appeal, Sir Philip Otton said that although the evidence and the findings of the Committee revealed a grave situation and a persistency in conduct that involved vulnerable patients, it was not inevitable that the ultimate sanction of erasure had to be invoked. The circumstances in *Garfoot v. General Medical Council* [2002] UKPC 35 were extremely grave, but their Lordships were persuaded that those in the present case were not as serious. All of the testimonial evidence, and indeed the testimony of the witnesses, indicated that the appellant had been throughout his career a dedicated doctor who was trying to do the best he could in difficult circumstances. He had an unblemished career, and it was evident that he had been placed under considerable pressures and that he had asked for help, which was not forthcoming. The circumstances did not reveal serious incompetence or conduct on a scale that required the ultimate sanction to be imposed. The erasure was therefore to be set aside and the case was to be remitted to the Committee for consideration of an alternative sanction.

### *Pembrey v. General Medical Council* [2003] UKPC 60

**60.27** Doctor—mistreatment of seven patients over nine years—major implications

The PCC found P guilty of serious professional misconduct in relation to seven patients and directed that his name should be erased from the medical register. The seven charges covered a period of some nine years. In addition, the present proceedings were not the first in which the appellant had been involved: inquiries at a lower level had been held in 1992, 1996, and 1998. Their Lordships said that they attached great importance to P's total disregard for the Council's guidance relating to the treatment of patients who were unable to consent. That failure was all the more serious since hysterectomy was an irreversible step, carrying major implications for the lives of two of the young women concerned. The Committee had deliberated on the proposal from the Royal College of Obstetricians and Gynaecologists' Working Paper discussion document entitled *Further Training for Doctors in Difficulty*. The Committee concluded that conditions were not appropriate because of its concern at P's continued lack of insight into his actions. Their Lordships saw no basis for second-guessing the Committee's considered determination that the protection of the public and the maintenance of public confidence in the profession meant that the appropriate course was to erase the appellant's name from the register.

### *Singh v. General Medical Council* [2003] UKPC 15

**60.28** Two patients—inadequate treatment—poor record-keeping—lies

S was found guilty of serious professional misconduct in relation to two patients. The patients received grossly inadequate treatment, aggravated by poor record-keeping, and the failure of S to cooperate in the process of inquiry and her lies about what happened. The order for erasure was entirely proper.

### *Nahal v. Law Society* [2003] EWHC 2186 (Admin)

**60.29** Solicitor—facilitating mortgage fraud—no dishonesty—strike-off

N appealed against a decision of the SDT to strike his name off the roll. Without retaining any control or the right to supervise, N handed over the conduct of many conveyancing transactions to S, an unqualified person (a former solicitor who had been struck off), who was not employed in his firm, and gave him some of his firm's headed stationery, thereby enabling S to pretend that transactions were being conducted by N's firm. By this means, N facilitated a mortgage fraud on a large scale. Although N was not a party to the fraud perpetrated by S and therefore not guilty of dishonesty, the Divisional Court (Dyson LJ and Gibbs J) held that the Tribunal was right to consider this a serious case of conduct unbefitting a solicitor. The Tribunal characterized N's behaviour as naivety of proportions that were hard to conceive. This was an apt description in view of the fact that N was not dishonest. But at the very least this was a flagrant case of conduct falling below the required standards of trustworthiness expected of a solicitor. It therefore fell within the category of cases described by Sir Thomas Bingham MR in *Bolton v. Law Society* [1994] 1 WLR 512, 518D, as being less serious than dishonesty, but nevertheless 'very serious indeed in a member of a profession whose reputation depends on trust'. As the Master of the Rolls said, in such cases a striking-off order will not necessarily follow, but it may well do so. The sheer scale of fraudulent transactions that N's conduct facilitated made this a particularly serious example of cases falling within this category. To this had to be added the fact that N also signed a report on title in one case. In the circumstances, the penalty imposed was appropriate.

### *Wentzel v. General Medical Council* [2004] EWHC 381 (Admin)

**60.30** Hospital doctor—patient—sexual exploitation

On 8 October 1998, X, a nurse, was taken to the accident and emergency (A&E) department of Yeovil District Hospital following a drug overdose. The overdose was an attempt at suicide made on the break-up of her relationship with her boyfriend. W, at that time, was senior house officer with the Somerset Partnership NHS and Social Care Trust at Rowan Place, which was part of the psychiatric hospital in Yeovil. He was asked to see Ms X at Yeovil District Hospital. He assessed her and arranged her admission the following day to Rowan Place. From that date until 22 February 1999, when he was transferred to a post at a psychiatric hospital in Taunton, W was responsible for Ms X's in-patient care. Whilst Ms X was an in-patient, W established an emotional relationship with her. The PCC found that, in January 1999, sexual intercourse took place and thereafter W continued an emotional relationship with Ms X. Dismissing

W's appeal against this finding and an order of erasure, Lightman J said that W was no doubt a competent and useful doctor, a valuable asset much needed by the health service, and that there was no finding of any risk of repetition of his misconduct. But in view of the extreme character of his misconduct, the sexual exploitation for personal gratification of his vulnerable patient by a psychiatrist, the GMC was fully entitled (if not bound), notwithstanding the consequences for W and the health service, to order erasure.

*R (Harry) v. General Medical Council* [2006] EWHC 3050 (Admin)

**60.31** Practice computers—viewing pornographic websites

H, a GP who qualified in 1982, appeared before a fitness to practise panel of the GMC in March 2006. He faced an allegation that, on a number of occasions between August 2003 and June 2004, he accessed pornographic websites and viewed pornographic images on two practice computers. The essence of the case was that, while providing locum clinical cover for a GP on a year's sabbatical, H used two practice computers to view pornographic website images on some eight occasions, twice during breaks between seeing patients. One computer was in the surgery; the other was in the nurse's room. The panel decided that H's fitness to practise was impaired and directed his suspension for six months from the register. Dismissing H's appeal, Goldring J said that the sanction imposed was not in any way unreasonable or outside the range that the panel could properly impose and regard as necessary and proportionate.

*R (Smith) v. General Teaching Council for England* [2007] EWHC 1675 (Admin)

**60.32** Application form for teaching post—false statements—other cases

S appealed against the decision of the PCC of the General Teaching Council of England, which found her guilty of unacceptable professional misconduct and issued a prohibition order for a period of two years under the Teaching and Higher Education Act 1988. The proceedings before the Committee arose out of S's application for a post as a mathematics teacher at a school in London, N7. In her application form, she stated that she had a Bachelor of Science (BSc) degree in mathematics; in support of her application, she stated that she was a mathematics graduate. Both of those statements were factually inaccurate and the Committee did not accept that they could in any way be described as mistaken or accidental. McCombe J dismissed S's appeal against both the findings of unacceptable professional conduct and sanction. On sanction, he observed that the panel had said in terms that it was clearly unacceptable for a teacher deliberately to make a false statement on an application form, particularly when such a statement is clearly material to, but not necessarily determinative of, the outcome of the application. The penalty imposed was within the published guidance in respect of sanctions that the Committee was urged to apply. It was true that, in other cases, other defendants had received lesser penalties. But these decisions of previous committees were not meant to be precedents nor were they meant to fetter further committees.

*Moody v. General Osteopathic Council* [2008] EWCA Civ 513

**60.33** Single patient—failure to address patient's needs—inappropriate advice—other cases

The complaint against M was made by a single patient about the manner in which the appellant had treated him. The PCC found that:

(1) in relation to one appointment, the appellant failed to identify and evaluate adequately the patient's needs in ways that were particularized;
(2) in relation to two further appointments, the appellant failed to identify or evaluate adequately the immediate needs of the patient, which included the decision to continue with treatment; and

(3) in relation to the third of the three appointments, the appellant had advised the patient not to attend for a magnetic resonance imaging (MRI) scan—advice that was inappropriate.

The Committee ordered that the appellant's name be removed from the register. The Court of Appeal (Sir Anthony Clarke MR, and Sedley and Rimer LJJ) dismissed the appellant's appeal. Referring to *Khokhar v. Health Professions Council* [2006] EWHC 2484 (Admin), Lord Justice Sedley noted that:

> 14. ... [A] far lighter sanction was imposed by a parallel disciplinary body for professional misconduct arguably at least as bad as [M]'s. This may well make the present sanction appear even harder for the appellant, but it cannot furnish a basis for oversetting it. It may for example be legitimately said that [K] was fortunate to be dealt with so leniently.
>
> 15. It remains the case that [section 8 of the Osteopaths Act 1993] permits any osteopath in the appellant's situation to apply after a minimum period of ten months for restoration to the register.... [T]here is nothing in the present decision that bars such an application, albeit the PCC would be bound in deciding it to have regard to its own decision to remove the appellant from the Register and its reasons for doing so.

### *R (Balasubramanian) v. General Medical Council* [2008] EWHC 639 (Admin)

**60.34** Anaesthetist—fundamental failings of basic principles

B, an assistant anaesthetist with twenty-nine years' experience, appealed against the decision of a fitness to practise panel to impose upon him, following findings of impairment of performance, the sanction of erasure. The background to the case was an incident on 24 September 2004, when, at short notice, B was assigned to the function of anaesthetizing patients on the orthopaedic trauma list at a hospital. One of those patients was a child aged 9, a young girl, who had a badly displaced fracture that was to be operated on under anaesthetic. The charges brought against B were that he failed to monitor the patient's blood loss and to do anything about it until the consultant anaesthetist arrived at the scene and directed remedial measures. The patient lost a minimum of 800 millilitres of blood, which amounted to 40 per cent of the blood that a young child aged 9 would have in her body. She was clinically shocked, hypovolaemic, and required an immediate infusion of a litre of gelatine solution and three units of blood, and she took a long time to recover. The incident was discussed with B's professional colleagues at the hospital and B was moved to a sister hospital in January 2005 for a three-week assessment by other colleagues, who were unaware of the September 2004 incident. He was monitored and, as a result of that assessment, concerns were expressed by the sister hospital about B's abilities. B was suspended from performing further medical functions. One of the assessors, an expert in medical education, said that the level of failure was so extensive and so basic that if retraining were to be considered, it would be retraining as for a new entrant to the profession of anaesthetics. In dismissing B's appeal against the order of erasure, Blake J said that, in a very lengthy and closely reasoned decision, the panel explained its reasoning on sanctions. It noted that the nature of B's errors during the September 2004 operation were 'fundamental failings of basic principles of anaesthesia'. The panel also found that most of the concerns expressed in relation to the three-week assessment were justified. The panel heard evidence from B and said that it was satisfied that his machine checking, his ability to relate to basic integration of altered physiology and pharmacology with the patient, his knowledge of the basic pharmacology of widely used drugs (namely, ephedrine and metaraminol), his lack of concentration, and the quality of his documentation were deficient. Blake J said that, on the basis of the panel's assessment of B's persistent lack of insight into the seriousness of the defects in his performance, the panel was entitled to come to the conclusion that it did.

***Yerolemou v. Law Society* [2008] EWHC 682 (Admin)**

**60.35** Solicitors' undertaking—conveyancing—no dishonesty

Y appealed against an order that he be struck off the roll of solicitors. The allegations were that he had failed and/or delayed in complying with undertakings to Bristol & West Plc and the Bank of Cyprus in relation to conveyancing transactions. The allegations were admitted prior to the hearing before the SDT on 21 November 2006. In *Briggs v. Law Society* [2005] EWHC 1830 (Admin), [35], Smith LJ said that:

> Undertakings are the bedrock of our system of conveyancing. The recipient of an undertaking must be able to assume that once given it will be scrupulously performed.... The breach of an undertaking given by a solicitor damages public confidence in the profession and in the system of undertakings upon which property transactions depend. Accordingly, if fault is shown, as it has been here, the matter must be treated seriously.

In the instant case, there was no suggestion of any dishonesty on the part of the appellant or of personal gain from any wrongdoing. There was no evidence of any loss, subject to the qualification that some loss may have been suffered by the lenders in the additional costs of obtaining priority in the transactions to which the undertakings related. In quashing the order for striking off and substituting an order for suspension for a period of two years, the Court (Leveson LJ and Lloyd Jones J) added that the appellant was under colossal pressure, both professionally and in his private life. While that provided no excuse for the serious failures that formed the subject of the breaches, it nevertheless provided some explanation. The public can be effectively protected by the imposition of a lesser penalty than striking off. It is appropriate that, when the appellant returned to practise, he should be permitted to do so, whether as an employed solicitor or in a partnership, only under the supervision of a solicitor approved for that purpose by the Solicitors Regulation Authority (SRA). The SRA should consider the appropriate course under section 12(1)(f) of the Solicitors Act 1974.

***In re King (Ch Ct of York)* [2009] 1 WLR 873**

**60.36** Priest—improper relationship with parishioner falling short of sexual intercourse—vulnerable person

K, an incumbent, appealed against the penalty imposed on him by the Bishop's Disciplinary Tribunal for the Diocese of York. The allegation of misconduct under section 8(1)(d) of the Clergy Discipline Measure 2003 was that the incumbent's conduct was unbecoming or inappropriate to the office and work of a clerk in holy orders in that, since about 2001, he had had an intimate and unprofessional relationship with Mrs B at a time when she was married. The Tribunal found that the incumbent had pursued an improper, intimate, and physical relationship with Mrs B, which fell short of sexual intercourse. It began when they were both married to and living with their respective spouses. The Tribunal ordered that the incumbent be prohibited from exercising any of the functions of his orders for a period of four years. His relationship with Mrs B changed from that of a colleague to one in which he was behaving in an intimate manner towards her. Whatever the rights and wrongs within her marriage, it was clear that she was in a vulnerable state and it was inappropriate for him to become emotionally involved with her. The *Guidance on Penalties* issued by the Clergy Discipline Commission suggests that removal from office and prohibition, either for life or for a limited time, are usually appropriate in cases of adultery. In dismissing the incumbent's appeal, the Chancery Court of York said that it does not, however, follow that sexual misconduct falling short of adultery should automatically attract a lesser penalty. The principle in *Bolton v. Law Society* [1994] 1 WLR 512 that 'the reputation of the profession is more important than the fortunes of any individual member' is true of the clergy, because otherwise the reputation of the church in the community would be taking second place to the personal interests of the member of the clergy on whom a penalty is to be imposed. However, depending upon the circumstances, in particular the nature of the misconduct and the degree of repentance, the age of the respondent may be a material factor (for example, youth and inexperience) in arriving at an appropriate penalty. In the instant case, even if K could be forgiven for refusing to admit his misconduct prior to or at the hearing before the Tribunal, the Court was surprised and disappointed that, since the Tribunal's determination, he had continued to show no repentance or remorse.

### *Augustine v. Nursing and Midwifery Council* [2009] EWHC 517 (Admin)

**60.37**
Criminal conviction—perverting the course of justice

The appellant pleaded guilty at Cardiff Crown Court of an offence of perverting the course of justice. He was sentenced to twenty-eight days' imprisonment, suspended for twelve months. The offence arose out of a road traffic accident in April 2006. On 27 August 2008, the Conduct and Competence Committee (CCC) of the Nursing and Midwifery Council (NMC) directed that the appellant's name be struck off the nursing register by reason of his conviction. On appeal, Beatson J said that:

> 28. It would, undoubtedly, have been helpful for the Panel expressly to state why, although the conviction was unrelated to the appellant's practice [as a nurse], it considered that striking off was the appropriate sanction. The decision does, however, state that the conviction reveals a serious departure from the Code of Conduct. In this context it refers to paragraph 7.1 which makes it clear that behaviour which compromises the reputation of the profession may call the registrants [*sic*] registration into question even if it is not directly connected to his or her professional practice....
>
> [...]
>
> 35. The Panel's reasons state the following: (a) there had been deliberate dishonesty on three occasions [in relation to the offence] over a two month period, (b) the criminal offence was a serious one and a serious breach of several aspects of the Code of Professional Conduct, (c) the appellant did not inform his employer promptly about his conviction, and (d) the confidence in the Council would be undermined if the appellant was not struck off....

The panel was entitled to conclude that such an order should be made.

### *Council for the Regulation of Health Care Professionals v. General Medical Council and Khanna* [2009] EWHC 596 (Admin)

**60.38**
Doctor—sexually motivated misconduct towards hospital colleagues—suspension—appeal by CRHP dismissed

K had been a consultant physician and resigned from a hospital following allegations of inappropriate sexual behaviour. A fitness to practise panel of the GMC suspended his registration for twelve months. The suspension arose from misconduct in relation to two young women, one a trainee dental student and, quite separately, the other a junior house doctor. The panel, in its determination, stated that K's sexually motivated conduct towards two young and junior female colleagues represented an abuse of the special position of trust that he held, and that this amounted to a fundamental breach of the principles that are central to good medical practice. The behaviour included physical contact that was unnecessary, inappropriate, and indecent to both women, and verbal comments that were unnecessary and inappropriate to one of them. K did not attend the fitness to practise hearing and explained his intention not to attend to give evidence in person. He said that he found the embarrassment caused to him by the allegations unbearable and simply did not feel able to cope with attending the hearing.

He had sought advice from his GP and was currently receiving treatment for depression. He did not mean any disrespect to the panel and fully understood that the allegations against him were very serious. The claimant Council sought to appeal against the decision on the ground that it was unduly lenient under section 29 of the NHS Reform and Healthcare Professions Act 2002. It submitted that the panel had erred with regard to its assessment of K's remorse and insight, that the panel had underestimated the seriousness of the case, and that the panel had failed to give adequate weight to the seriousness of the breach of trust. Davis J rejected each of these grounds. The panel made quite clear that, because it had not heard from K, it could not make judgments as to his insight. It was entitled to say that his letter of resignation and witness statement contained 'some' expressions of remorse. Overall, the claimant Council was seeking to advance its own view of the matter in preference to that of the panel. The panel quite plainly had considered the matter very carefully. It had also considered that K would have to satisfy the next panel that he may properly be restored to the register, with or without conditions. The panel stated that the next panel would be assisted by K demonstrating how he had reflected on the findings against him in order to assess the level of insight that he had developed into his misconduct, and also the next panel would expect to see evidence of his continuing professional development and also would expect to receive references and other information. Davis J said that was potentially important, because it did not mean that the panel was necessarily assuming that K would be restored to the register after the lapse of the one-year suspension. On the contrary, the matter would first have to be reviewed by another panel and that other panel might not decide to restore him to the register, and even if it were to do so that panel would have the power to do so on conditions, such as a condition that K would have no contact with medical students or the like. This confirmed that the panel had all of the relevant points well on board and that the criticisms now sought to be made of it by the claimant Council were not well founded.

***Varley v. General Osteopathic Council* [2009] EWHC 1703 (Admin)**

**60.39** Osteopath—criminal conviction—supply of amphetamines for sale as slimming pills—practising whilst suspended

On 21 January 2008, the PCC of the General Osteopathic Council (GOsC) suspended V for six months from the register as a result of a conviction. In October 2007, V was convicted of incitement to supply amphetamines. The basis of the conviction was that he had incited osteopathy patients to join him in creating amphetamine for sale as slimming pills to members of the public. He was sentenced to an eight-month custodial sentence. V, while suspended from the register for an earlier matter, continued to treat patients, describing himself as a 'spinal specialist'. At a hearing in November 2008 before the PCC, he was found guilty of unacceptable professional conduct in the following ways: first, he dishonestly described himself by implication as an osteopath when he was suspended from the register; second, he conducted osteopathic examinations and gave treatment whilst he was suspended from the register; and third, he conducted osteopathic examinations and treatment without having in place appropriate professional indemnity insurance. As a result of these findings, the Committee decided to remove V's name from the register. Burnett J dismissed V's appeal on sanction, holding that, at the heart of the case, was a piece of calculated defiance of the earlier decision of the PCC. In the face of a serious earlier allegation, the sanction imposed was one of suspension. The reality was that the appellant sought to circumvent that suspension by attaching a different label to himself, whilst acting as if nothing had happened. The label attached was one of 'spinal specialist'. When coupled particularly with the fact that there was no insurance, it was unsurprising that the Committee took the view that removal was required. Public confidence in the profession called for no less.

### *Burke v. General Teaching Council* [2009] EWHC 3138 (Admin)

**60.40**

Teacher—prohibition order—effect—non-attendance at hearing

B appealed against a prohibition order. Its effect meant that he was unable to teach in maintained and non-maintained special schools, although it was open to him to apply for restoration following a period of two years. The allegations found proved were that, whilst employed as a teacher at a school on 8 May 2006, he inappropriately and aggressively handled a pupil, shouted at the pupil and kicked a desk, causing it to tilt towards the pupil and another pupil, and that, whilst employed at a previous school between September 1999 and May 2003, he made inappropriate comments about teenage mothers to pupils during a lesson, used inappropriate words in front of pupils, and made inappropriate comments about the head teacher to colleagues. His Honour Judge Pelling QC (sitting as a judge of the High Court) said that the penalty imposed was the most serious that can be imposed. It had had a catastrophic consequence for B, who had lost his employment as a teacher with immediate effect and had no short-term prospect of regaining such employment. However, as against that, as Cranston J concluded in *Cheatle v. General Medical Council* [2009] EWHC 645 (Admin), professional judgment is especially important when considering the appropriate sanction to be imposed following the conclusion that a professional has been guilty of unacceptable professional conduct. It is possible that if B had attended the hearing and had shown some insight into the failings disclosed by the findings made, and had produced detailed references from his previous appointments, the respondent's PCC might have concluded that the imposition of conditions on his registration was an appropriate course to adopt. However, in the circumstances, it could not be said that the Committee was wrong not to adopt that course or that the Committee was wrong in the penalty that it imposed.

### *Muscat v. Health Professions Council* [2009] EWCA Civ 1090

**60.41**

Radiographer—inappropriate examination of female patients—breach of trust—risk of recurrence

M, a radiographer, was found guilty of two allegations by the Health Professions Council, which ordered that his name be struck off the register for misconduct. The allegations were that, on 13 July 2003, M had conducted an X-ray on a female patient and had lifted her gown to just below her chest; on 23 August 2003, he had conducted an MRI scan on a female patient and had required her to undress completely, so that she was naked during the procedure. M denied both allegations, but accepted that, if these events had occurred as alleged, there could have been no clinical justification for them. The CCC recorded its view that M's misconduct was serious, and that it constituted a serious breach of trust and had not been acknowledged by him. M's appeal from that decision was dismissed by Silber J sitting in the Administrative Court on 14 November 2008. On his further appeal to the Court of Appeal (Longmore, Smith, and Maurice Kay LJJ), it was argued that the sanction was too harsh. In dismissing the appeal against sanction, Smith LJ said that the Committee's assessment was not assisted by any explanation from the appellant. His conduct could be sexually motivated, but that was not alleged and could not safely be inferred from the circumstances. Another possibility would be a desire to humiliate the patient. Again, that could not safely be inferred. However, it was clear that there was a risk of reoccurrence. The appellant had done something similar on two occasions within a few weeks of each other. It was also clear that the effect of the conduct on the patients was to humiliate them. This was a serious matter, particularly in the absence of any explanation that might show that the appellant had some insight into his own conduct. The Committee was entitled to describe the conduct as a breach of trust and was right to take the view that the only appropriate sanction was erasure.

***R (Lloyd Subner) v. Health Professions Council*** **[2009] EWHC 2815 (Admin)**

**60.42** See paragraph 37.06.

***Izzet and Cazaly v. Law Society*** **[2009] EWHC 3590 (Admin)**

**60.43** I and C faced a number of allegations, but were not accused of dishonesty. The allegations were:

Solicitor—breaches of Accounts Rules and Practice Rules—importance of cooperation with SRA—lack of explanation

(1) acting contrary to the provisions of rule 7 of the Solicitors Accounts Rules 1998, in that they failed to remedy a breach promptly on discovery;
(2) failing to notify lender clients of material facts, thereby compromising or impairing their duty to act in the best interests of their clients in breach of rule 1(c) of the Solicitors Practice Rules 1990;
(3) failing to act in the best interests of a particular client in failing to ensure that the client's mortgage with her lender was discharged; and
(4) failing to honour undertakings to provide an indemnity policy.

The SDT, at the conclusion of its findings, ordered that I be struck off the roll of solicitors and that C be suspended from practice as a solicitor for the period of three years. The two solicitors had previously been fined in relation to an earlier matter before the SDT of failing to comply with the Solicitors Accounts Rules. Clients' money was at risk, but it was contended that it was not a case of dishonesty, but of disorganization and muddle. In respect of the current proceedings, the Tribunal regarded the matters as serious breaches. C's position was slightly different, since he had taken some steps to cooperate. Moses LJ (with whom Hickinbottom J agreed) said that he thought that the allegations demonstrated a failure by these two solicitors to understand how serious the breach of the undertaking was and to understand how important cooperation is with those who regulate their own profession. There had never been any explanation as to how these failures in relation to the undertaking came about, other than the suggestion that the solicitors were 'too busy'. In those circumstances, it was difficult to see how the Tribunal could have imposed any alternative, still less a lower punishment on either of the two solicitors. Any lesser punishment would have been wrong. Appeals dismissed.

***R (Howlett) v. Health Professions Council*** **[2009] EWHC 3617 (Admin)**

**60.44** The appellant, a chartered physiotherapist, owned a private clinic in which he worked as a sole practitioner. In July 2006, Miss X was involved in a road traffic accident. She was referred some six months later by her insurers to the appellant in order to receive physiotherapy for a whiplash injury suffered in the accident. She attended the appellant's clinic on 20 January 2007. The allegation against the appellant was that, during the course of treatment of patient X on 20 January 2007, he acted inappropriately in that he did not inform X prior to the treatment that she would need to remove any clothing, he massaged gel into X's breast and chest area with his hands without providing adequate reason for treating the area, and he strapped X into a garden chair without providing adequate reasons for this type of treatment. The HPC's CCC found the allegations proved and struck the appellant off the roll of authorized physiotherapists. During the impairment stage of the fitness to practise hearing, the appellant called a former patient who gave evidence of her satisfactory treatment by the appellant. In rebuttal, the HPC called another witness, who had made a complaint the previous week that the appellant had not changed the manner in which he practised. On sanction, the Committee said that it was important at the outset to state that it was satisfied that the appellant was not motivated by voyeuristic or sexual motive in relation to the misconduct proved. That said, the stress to a woman who was in a vulnerable position was very

Physiotherapist—twenty-eight years in practice—inappropriate treatment of two female patients—unchallenged evidence of good practice—significant measures to address issues real and it was implicit in the Committee's findings that, without a fundamental change of attitude, there was a significant risk of further distress being caused in the future. The focus of the Committee's consideration was as to whether there was a realistic prospect of the appellant being able to achieve such a sea change in his attitude. Dobbs J dismissed the appellant's appeal against the findings of fact and misconduct stages of the proceedings. As to sanction, the judge noted that the decision of the Committee revolved around the issue of whether there was a realistic prospect of the appellant being able to change his approach to the extent that the Committee could be satisfied that he presented a low risk of reoccurrence should he be permitted to continue to practise. The evidence in front of the Committee consisted of two patients who had made complaints. There was no other evidence before the Committee of any other complaint in the appellant's practising history of some twenty-eight years. He treated, on average, four patients a day and worked six days a week. He predominantly treated victims of road traffic accidents, without apparent complaint. As to the evidence of the witness produced shortly prior to the hearing, it had to be remembered that this was nine days before the hearing and one could reasonably expect the appellant to be experiencing significant stress. There was also evidence in front of the Committee that demonstrated in very similar circumstances that the appellant was capable of adopting a manner appropriate to the sensitive situation in which the two complainants found themselves. The unchallenged evidence of the patient called by the appellant could not have been more positive. Additionally, there was evidence in front of the Committee that the appellant had taken significant steps (which the members acknowledged) to address the issues raised in the allegations. Moreover, during the course of the hearing and on further reflection, the appellant came up with further measures to improve his practice. The Committee had found him to be a truthful witness. Whilst it was correct to find that the failings were serious, the Committee at this stage of the proceedings needed to set out precisely why, in the light of all of the evidence, it had reached the conclusion that it did. There was no mention of the testimonials, and there appeared to be no allowance made for the fact that the appellant had made many and significant changes to his practice that did address many of the issues raised in the allegation of Miss X. The decision to strike the practitioner from the roll would be quashed and the matter remitted back to the Council for consideration of the imposition of a sentence of lesser severity, being, realistically, a suspension or a conditions of practice order.

### *Mvenge v. General Medical Council* [2010] EWHC 3529 (Admin)

**60.45**

**Overstayer from Zimbabwe—conviction—doctor—not registered for work in UK**

M, a registered medical practitioner, came to the United Kingdom from Zimbabwe to further his medical education in December 1997. He became an overstayer and did not have the right to work. In November 2005, he was cautioned in respect of an attempted assault on his wife. In August 2008, he was convicted and sentenced for knowingly possessing a false identity document with intent, and was sentenced to ten months' imprisonment. In June 2009, he faced charges before a fitness to practise panel. In essence, the charges were: first, that he had made himself available to work when he was not registered with the GMC and was in the United Kingdom illegally; second, that he had informed the GMC that he had never been cautioned when he had, in relation to the attempted assault in 2005; and third, that he had made false statements to the police in the course of interviews about the GMC registration process. The panel ordered the erasure of his name from the register. On appeal, Cranston J referred to the very limited scope for review that the Court has, derived as the principles are, from the seminal decision of Sir Thomas Bingham in *Bolton* and reiterated by the Court of Appeal in *Salsbury*. For this reason, there was no basis for the appeal. Nonetheless, there was much in the material that was of credit to M. He came to this country to obtain qualifications and was able to secure a scholarship to do so. From the very outset of the panel hearings, he admitted his errors. He has subsequently exhibited remorse, both before the panel and before the Court. Were he to apply for re-registration after the five-year period, the GMC would undoubtedly consider all of this. It would also need to consider the background of the successful claim that he made for asylum.

***Brennan v. Health Professions Council* [2011] EWHC 41 (Admin)**

**60.46** See paragraph 54.20.

***Imam v. General Medical Council* [2011] EWHC 3000 (Admin)**

**60.47** Overseas doctor—deficient performance—exclusion from work—behaviour repeated at subsequent hospital—patient safety—public interest

I appealed against the decision of the fitness to practise panel that his name should be erased from the medical register. The appellant, who was 34 years old at the time of the hearing, graduated with a Bachelor of Medicine, Bachelor of Surgery (MBBS) degree from the University of Karachi in 2002. In 2005, he passed the GMC's Professional and Linguistics Board (PLAB) test for overseas doctors; in 2007, he took Part 1 of the Membership of the Royal College of Surgeons (MRCS) exams, and in 2008 he passed Part 2. In November 2008, he took up a position at the Salford Royal NHS Foundation Trust Hospital as a trust-grade doctor within the department of neurosurgery. On 30 December 2008, Salford Royal Hospital excluded the appellant on full pay, pending an investigation into his behaviour and practice, and referred his case to the National Clinical Assessment Service (NCAS). The appellant applied to King's College Hospital London for employment without telling it that he had been excluded by Salford Royal Hospital. The charges before the fitness to practise panel alleged non-disclosure of this exclusion and deficient professional performance. The hearing before the panel lasted a total of nine days. The panel considered the possibility of suspending the appellant. However, it was deeply troubled by the serious breaches of *Good Medical Practice* that the case involved, the appellant's lack of insight, and the risk to patients that he posed. There were attitudinal problems, demonstrated by his behaviour towards female and junior colleagues. Following the problems at Salford and aware of the reference to NCAS, the appellant repeated similar behaviour at King's College Hospital within two weeks of starting there. His decision to put himself in the same situation again, without notifying King's College Hospital of the difficulties that he had experienced, demonstrated a lack of insight and was reckless. During the hearing, I continued to attempt to exculpate himself by transferring blame. The evidence of one of the assessors was that there was unlikely to be any NHS post that could offer the extent of training and supervision that the appellant needed. The panel's conclusion was that the risk to patient safety and the wider public interest was such that it would not be sufficient or proportionate to suspend his registration. That left erasure as the only realistic alternative. Nichol J dismissed I's appeal.

***Obi v. Solicitors Regulation Authority* [2012] EWHC 3142 (Admin)**

**60.48** Hearing—solicitor struck off—appeal—rehearing—solicitor remaining struck off—effect on later sentence

The appellant became a Fellow of the Institute of Legal Executives in October 2002 and was admitted as a solicitor in November 2003. Prior to his admission as a solicitor, he was involved in a firm of solicitors, whereby the principal put his name as the sole partner of the firm and the appellant was named as practice manager, but the appellant effectively acted as a solicitor from the outset and the principal played hardly any part at all in the business. In September 2005, the appellant faced five charges before the SDT. The five allegations were: (i) that he set up and/

or was involved in the firm in circumstances in which he knew or ought to have known were improper and/or unprofessional; (ii) that he falsely witnessed a mortgage deed, stating that he was a solicitor with a current practising certificate; (iii) that he produced three practising certificates that were false; (iv) that he attempted to deceive the Law Society by producing three false practising certificates; and (v) that he misrepresented to the Law Society his involvement in the practice of the firm. The appellant admitted allegations (i), (ii), and (v), but denied (iii) and (iv). The case against the appellant on allegations (iii) and (iv) was found proved, and he was struck off the roll of solicitors. The appellant appealed and, by consent, the appeal was allowed and the case remitted for rehearing before a new panel, although the appellant remained struck off pending the rehearing. At the rehearing in September 2010 before a fresh panel, the appellant admitted again allegations (i), (ii), and (v), but contested (iii) and (iv). These two allegations were not found established on this occasion. The net effect of the second SDT decision was that the appellant fell to be dealt with for allegations (i), (ii), and (v), and nothing more. The second Tribunal ordered that the appellant be struck off the roll of solicitors in respect of these allegations. On appeal, Foskett J said that, had the appellant been found guilty of allegations (i), (ii), and (v) by the first Tribunal, and acquitted of (iii) and (iv), an order striking him off the roll would not have been clearly inappropriate. The difference, however, between the imposition of such a sanction in 2006 and the imposition of such a sanction in 2010 is that, in the intervening four years, the appellant had been barred from practice. He had remained struck off for four years pending the second SDT hearing. Whilst the second Tribunal may have taken into account these four years of 'disqualification' from practice, the lack of reference to it in its otherwise detailed reasoning raises the question of whether it was considered, or considered sufficiently. Since the issue was raised at the rehearing by the appellant's counsel specifically in the context of arguing that no further suspension was required, the Tribunal's decision should have addressed this issue specifically. There was no alternative in the circumstances but to remit the case to a differently constituted Tribunal for reconsideration of the issue of sanction.

***Afolabi v. Solicitors Regulation Authority*** **[2012] EWHC 3502 (Admin)**

**60.49** Solicitor—misleading information and conviction—strike off

Three allegations were proved against A: providing misleading information publicly about her firm; allowing the firm to be described as a solicitors' partnership before she was admitted to the roll; and a conviction of two offences of money laundering contrary to the Proceeds of Crime Act 2002. The Court held that it was for the SDT to view her conduct as a whole. There was no error of principle in its decision to strike A off the roll.

***Sarker v. General Medical Council*** **[2012] EWHC 4008 (Admin)**

**60.50**

See paragraph 16.14.

***Sultan v. General Medical Council*** **[2013] EWHC 1518 (Admin)**

**60.51** Doctor—loss of patients' records—erasure

S, a GP, destroyed, or was involved in the dumping of, a large number of highly confidential patient's medical files on moving the surgery's practice to another address. A significant amount of information had not been scanned onto the practice's Egton Medical Information Systems (EMIS) software. In addition, the doctor was found guilty of deficient professional performance. The order of erasure was upheld.

*Rich v. General Medical Council* [2013] EWHC 1673 (Admin)

**60.52** Doctor—conviction—wounding with intent—erasure

R was convicted of wounding with intent to cause grievance bodily harm and sentenced to forty-two months' imprisonment. An appeal against the consequent order of erasure was dismissed.

*R (Austin) v. Teaching Agency* [2013] EWHC 254 (Admin)

**60.53** Teacher—inappropriate conduct—two years' suspension

A two-year suspension order was upheld in the case of a supply teacher found guilty of attending two schools and making inappropriate comments and acting in a threatening manner towards staff on school premises.

*Moseley v. Solicitors Regulation Authority* [2013] EWHC 2108 (Admin)

**60.54** Solicitor—bankruptcy—failure to disclose to Official Receiver—no dishonesty—strike off

M, a solicitor, failed to declare to the Official Receiver, who took over the conduct of M's bankruptcy, that he had a reversionary interest in a property. The SDT ordered that M be struck off. Dismissing M's appeal, the Court held that there was no presumption that striking off was appropriate only where the allegations involved dishonesty, or criminal offences, or exceptional circumstances. If the conduct complained of were to show a lack of integrity or an undermining of public confidence in the profession, it might still be appropriate to strike off even if the conduct was unconnected to the solicitor's professional duties.

*Cathcart v. Law Society of Scotland* [2013] CSIH 104

**60.55** Hearing—appeal—rehearing—second Tribunal imposing more severe penalty—whether lawful

In May 2012 the petitioner appeared before the SDT and was found guilty of misconduct in respect of his failure to communicate effectively and appropriately with a liquidator, and also in respect of his failure and/or unreasonable delay in responding to the Law Society and its financial compliance department. The liquidator had incurred costs and expenses of £45,000 as a result of the petitioner's conduct. The Tribunal censured the petitioner and imposed a restriction on his practising certificate for five years, such that he was permitted to work only as a qualified assistant for an employer approved by the Law Society, and ordered the petitioner to pay compensation to the liquidator of £5,000 (the maximum compensation that could be ordered). In August 2012, the petitioner was advised that the chairman of the Tribunal had not been a valid member of the Tribunal, his appointment having expired. Accordingly, a second validly constituted Tribunal was convened, which imposed a penalty of immediate suspension from practice for a period of three years and compensation of £5,000. The petitioner appealed to the Court of Session and submitted that it was a matter of concern that two tribunals could come to such different sentencing disposals. The penalty of immediate suspension for three years was more severe than the first Tribunal's restriction of the petitioner's practising certificate for five years. In dismissing the petitioner's appeal, the Court said that the petitioner, having taken advice, chose to have his case decided afresh by a second, properly constituted, Tribunal. He rejected other options, such as having the proceedings validated retrospectively by a properly constituted Tribunal. The second Tribunal, although aware that there had been earlier proceedings vitiated by a procedural flaw, was not aware of what disposal had been selected. There was no reason for the second Tribunal to be restricted in its choice of disposal by anything that had occurred at the first Tribunal, and the first Tribunal's disposal was not necessarily more appropriate than the second Tribunal's disposal. In the result, the Court was not persuaded that the disposal adopted by the second Tribunal should have been the same as, or more similar to, that adopted by the first.

***Walker v. Secretary of State for Education*** **[2014] EWHC 267 (Admin)**

**60.56** Teacher—prohibition order—comparison with other orders

W, a schoolteacher, pleaded guilty to an offence under section 4 of the Public Order Act 1986 and to dangerous driving. He was sentenced to eighteen months' imprisonment, suspended. A professional conduct panel recommended the making of a prohibition order, but that there should be a review period of two years. The regulatory process from that point is that the recommendation of the panel goes to a decision-maker nominated by the Secretary of State for Education. The decision-maker in this case was an official in the Department for Education, and he determined that there should be a prohibition order, but that there should be no review period. W appealed, contending that he was treated by the Secretary of State more harshly than other teachers. His Honour Judge Clive Heaton QC (sitting as a deputy High Court judge), in dismissing W's appeal, said that the drawing of comparisons between other cases and this case, with the intention of demonstrating undue harshness of sanction as W sought to do, is to be treated with real caution. In *R (Henderson) v. General Teaching Council for England* [2012] EWHC 1505 (Admin), [55]ff, Beatson J observed that decisions in other cases are not precedents, but turn on their own facts. As regards the proposition of the appellant that he was dealt with more harshly by the disciplinary process than other teachers, that argument was not made out on the information before the Court, nor were the statistical arguments put forward by W persuasive. An unsuccessful respondent to a disciplinary process is entitled to a degree of consistency of sanction from the regulatory body—but W was unable to demonstrate that his case had been dealt with in a way statistically out of kilter with other decisions of the decision-maker.

***Heesom v. Public Service Ombudsman for Wales*** **[2014] 4 All ER 269**

**60.57** County councillor—breach of code of conduct—disqualification for eighteen months

The appellant was 76 years old and a long-standing local councillor of Flintshire County Council in Wales. Following the publication of a report from the Public Services Ombudsman for Wales, a reference was made to the president of the Adjudication Panel for Wales for adjudication by a case tribunal. The tribunal, after a lengthy hearing, found that the appellant had committed fourteen breaches of the Council's code of conduct by failing to show respect and consideration for Council officers, using bullying behaviour, attempting to compromise the impartiality of officers, and conducting himself in a manner likely to bring his office or the Council into disrepute. In terms of sanction, the tribunal disqualified the appellant from being a member of the Council or of any local authority for two years and six months. On appeal, the Administrative Court held that the civil standard of proof had been correctly applied by the tribunal, but that two of its findings of fact should be set aside. The sanction imposed by the tribunal would be quashed and replaced by disqualification for a period of eighteen months, to run from the date of the tribunal's decision.

***Giambrone v. Solicitors Regulation Authority*** **[2014] EWHC 1421 (Admin)**

**60.58**

See paragraph 56.30.

***Professional Standards Authority for Health and Social Care v. General Pharmaceutical Council and Onwughalu*** **[2014] EWHC 2521 (Admin)**

**60.59**

See paragraph 15.26.

***Beard v. Bar Standards Board*, 17 July 2014 (unreported)**

**60.60** Barrister—convictions and breaches of regulatory requirements—disbarment

B admitted two convictions for driving with excess alcohol and a conviction of failing to surrender to custody, together with additional charges of failing to pay the second instalment of his practising certificate fee for 2011, failing to pay his professional indemnity insurance premium, and failing to pay a financial penalty of £300 in relation to an earlier disciplinary matter. B expressed remorse and the Visitors (Nicol J, chairman) sympathized with him for his grief on the death of his father and his struggles with alcohol. However, the reputation of the Bar would be harmed and the reputation of the profession may be diminished if he were to be allowed to continue to practise. The sanction of disbarment was proportionate and fair.

***Sukul v. Bar Standards Board* [2014] EWHC 3532 (Admin)**

**60.61** Finding of guilt—no opportunity for mitigation

On 3 February 2014, in his absence, the Disciplinary Tribunal of the Bar Standards Board found the appellant guilty of two charges of professional misconduct in relation to drafting false grounds of appeal against conviction to the Court of Appeal, Criminal Division, on behalf of a client, when he knew that there were no grounds of appeal. The appellant was disbarred. The Tribunal proceeded to sentence after determining guilt and did not consider the possibility of adjourning so as to afford an opportunity for mitigation. The Tribunal lacked personal information about the appellant. The sentencing guidance for breaches of the Bar Code of Conduct, for offences of conduct during proceedings and in court, range from giving advice to disbarment. In allowing the appeal out of time and directing that the matter be relisted before a differently constituted Tribunal to proceed to re-sentence the appellant, Laws LJ (with whom Cranston J agreed) said that it could not be said, to take the matter shortly, that the circumstances of the present case necessarily were at the top range of offences covered by the sentencing guidance. Disbarment was not necessarily the wrong sentence, but there was plainly an argument as to whether or not it was. In those circumstances, the Tribunal should have afforded an opportunity to the appellant to make representations as to sanction once it had found him guilty of the professional charges before it. Such an opportunity should properly have been provided, notwithstanding all of the negative features of the appellant's previous communications with the Bar Standards Board or the Tribunal.

***Monji v. General Pharmaceutical Council* [2014] EWHC 3128 (Admin)**

**60.62** Pharmacist—theft from employer—lack of remorse and insight—attitude during hearing—removal

The appellant was dismissed by Boots UK, where he was employed at a store as the pharmacist manager, following an allegation of stealing fragrance tests. The appellant's claim to an employment tribunal was dismissed and that decision was upheld on appeal by the Employment Appeal Tribunal. The General Pharmaceutical Council instituted disciplinary proceedings and, following a nine-day hearing, the appellant's name was ordered to be removed from the register. On appeal against sanction, the appellant submitted that the Fitness to Practise Committee had focused on his failure to accept its findings. In dismissing the appeal, His Honour Judge Andrew Grubb (sitting as judge of the High Court) said that the Committee was entitled to take into account the appellant's attitude to the allegation made against him and his conduct during the fitness to practise proceedings, including his continual denial of the facts. In *Nicholas-Pillai v. General Medical Council* [2009] EWHC 1048 (Admin), [19], Mitting J said:

> In the ordinary case such as this, the attitude of the practitioner to the events which give rise to the specific allegations against him is, in principle, something which can be taken into account either in his favour or against him by the panel, both at the stage when it considers whether his fitness to practise is impaired, and at the stage of determining what sanctions should be imposed upon him.

The learned judge said that, in his judgment, that statement of principle is correct and is not gainsaid by *Amao v. Nursing and Midwifery Council* [2014] EWHC 147 (Admin), in which Walker J stated, at [161], that: '[The appellant] was perfectly entitled to say that she did not accept the findings of the panel: she had a right of appeal which she was entitled to exercise.' That was, however, said in a case in which the representative of the respondent Council had persistently put to the appellant in cross-examination whether she agreed with the panel's findings on each of the factual allegations. That cross-examination was clearly seen as, in effect, oppressive and unnecessary. The repeated denials were, unlike in this case, not volunteered by the individual before the disciplinary body. That was a material difference. There was nothing wrong, in principle, in the Committee in the instant case taking into account the appellant's lack of 'remorse' and 'continued denial' of his dishonesty. That was, of course, a matter that arose in the context of the allegation by the appellant, put to the witnesses, that they were liars and had effectively conspired to give false evidence in order to target the appellant. The Committee could not be faulted in its approach to the issue of sanction in accordance with the Indicative Sanctions Guidance and in taking into account all that was relevant (and only that which was relevant) to its decision on sanction.

See further *Nicholas-Pillai v. General Medical Council* [2009] EWHC 1048 (Admin)—see paragraph 33.08—and *Amao v. Nursing and Midwifery Council* [2014] EWHC 147 (Admin)—see paragraph 37.17.

### *R (Lonnie) v. National College for Teaching and Leadership* [2014] EWHC 4351 (Admin)

**60.63** Teacher—assault of pupil—dismissal—decision of panel—decision of Secretary of State

L, a teacher at a residential school catering for pupils with particular difficulties, admitted that, in June 2009, he had used excessive force against a pupil after some ill behaviour on behalf of the pupil. L accepted a police caution and was dismissed by the school. At a disciplinary hearing in March 2014 pursuant to the Education Act 2011 and the Teachers' Disciplinary (England) Regulations 2012 before the National College for Teaching and Leadership, the panel recommended that the Secretary of State not impose a prohibition order. The Secretary of State did not accept the panel's recommendation and imposed a two-year prohibition order. In dismissing L's appeal, William Davis J said that the final decision-maker under the Act and the Regulations was the Secretary of State, and that he was not satisfied that her view was wrong within CPR 52.11. Under the statutory and regulatory framework, she was required to consider the recommendation of a professional conduct panel before making the decision, but it was only consideration of the recommendation that was required. The Regulations do not stipulate that the Secretary of State must follow the recommendation, save in exceptional circumstances, or any language such as that.

### *Rasool v. General Pharmaceutical Council* [2015] EWHC 217 (Admin)

**60.64** Pharmacist—conduct fundamentally incompatible with continued registration—removal

The Fitness to Practise Committee of the General Pharmaceutical Council (GPhC) found that R, the superintendent pharmacist of a pharmacy, had unlawfully supplied prescription-only medicines on five occasions between August and November 2012 without being provided with a prescription from an approved practitioner. The Committee found R's conduct to have been shocking and directed that his name be removed from the register of pharmacists. In dismissing R's appeal on sanction, Carr J said that the Committee was fully aware that there was no allegation of dishonesty. There was, however, a blatant and deliberate flouting of the law relating to the supply of prescription-only medicines. That was a matter going to R's trustworthiness. The Committee was entitled to reject the submission of genuine insight, given R's attempts in the course of the proceedings to justify his conduct. His answers in cross-examination were rightly treated as revealing fundamental misunderstandings of the function and role of a pharmacist:

> 50. In short, I accept the submission for the GPhC that the critical finding of the Committee on sanction was that [R's] conduct was fundamentally incompatible with registration. The Committee had before it the relevant Indicative Sanctions Guidance. Where a finding of fundamental incompatibility is made, it is clear that removal from the register may be appropriate (and suspension inappropriate) (see paragraphs 13 and 14 of the Guidance).

## E. Other Relevant Chapters

**60.65** Appeals, Chapter 4
Conviction and Caution Cases, Chapter 14
Dishonesty, Chapter 19
Impairment of Fitness to Practise, Chapter 33
Indicative Sanctions Guidance, Chapter 36
Insight, Chapter 37
Reasons, Chapter 54
Reprimand, Chapter 57
Suspension, Chapter 65

# 61

# SERVICE



## A. Legal Framework

**61.01** Most, if not all, regulatory bodies provide specific provision for service in their fitness to practise or disciplinary rules.

The General Medical Council (Fitness to Practise) Rules 2014 provide for service of notices and documents by email. Rule 40(2) provides that any notice or document required to be served upon the practitioner may be served by email to an email address that the practitioner has notified to the registrar as an address for communications.

## B. Disciplinary Cases

### *R (Shanker) v. General Medical Council* [2004] EWHC 565 (Admin)

**61.02**

Notice sent to practitioner's registered address—twenty-eight days' notice given—letter signed for by third party—adequate service

S appealed against a decision of a committee of the General Medical Council (GMC) of 22 August 2003 whereby the committee directed S's suspension from practise for a further period of six months. S maintained that he had not been given the appropriate period of notice of the resumed hearing that took place on 22 August 2003. In dismissing S's argument, McCombe J said that appropriate notice of the hearing turned on the provisions of rule 12(3) of the General Medical Council (Professional Performance) Rules 1997, which stated that where the committee is to hold a resumed hearing, the registrar shall 'not later than 28 days before the day fixed for the resumed hearing, send to the practitioner, with a copy of these Rules, a notice'. The Rules provide that the relevant address is the registered or last known address of the respondent to the proceedings. Evidence was produced that a letter was sent to S's registered address on 16 July 2003, giving notice of the resumed hearing. It appeared from records obtained from the Post Office that the letter was delivered and signed for at about 12 noon on 18 July—that is, within the requisite period. S said, and the Court accepted, that he personally did not sign for or receive that document at the time of its delivery. However, the Court was quite satisfied that such delivery by the postal service was entirely appropriate in the context of the Rules. Accordingly, adequate notice was given of the hearing. Moreover, S was represented by counsel at the resumed hearing. No application was made for an adjournment, either at the outset or at any later stage, because of any actual or perceived prejudice to the doctor arising out of late receipt of documents.

*Jatta v. Nursing and Midwifery Council* [2009] EWCA Civ 824

**61.03** Notice sent to practitioner's registered address—only notified address—registrant known to be in Thailand—no procedural irregularity

In this case, a fitness to practise panel of the Nursing and Midwifery Council (NMC) proceeded to hear and determine disciplinary proceedings in the absence of the registrant. Notice of the proceedings had been duly and properly served by sending them to the registrant's registered address in Didcot in accordance with the Nursing and Midwifery Council (Fitness to Practise) Rules 2004, although the Council was aware that the registrant was no longer living at the address and was travelling in Thailand. Posting of the notice of hearing was duly proved: it was sent by recorded delivery, but was returned undelivered. A copy was also sent by first-class post to the address in Didcot, but the registrant was unaware of, and did not attend, the hearing before the fitness to practise panel. The Court of Appeal (Maurice Kay and Lloyd LJJ, and Sir Simon Tuckey) held that there had been no procedural irregularity that vitiated the decision to proceed in the registrant's absence. The only available address was the registrant's registered address, albeit that it was known that the registrant was no longer there. The panel could have adjourned and required that an email be sent to the registrant at his known contact email address, but it did not do so and could not properly be criticized. Lloyd LJ said:

> 18. ...[I]t is not in dispute that the letter complied with the rules as regards the required content of the hearing, nor that it was in fact sent to his address. Posting of the letter was duly proved; it was sent recorded delivery and in that form was returned undelivered, but it seems that it was also sent first class, and that copy no doubt languished for some time at the address in Didcot.
>
> [...]
>
> 29. ... [W]hat the panel has to require under rule 21(2)(a) is evidence that all reasonable efforts have been made in accordance with these rules to serve the notice of hearing on the registrant and in turn, under rule 21(2)(b), it must be satisfied that the notice of hearing has been duly served. The rules do not provide for sending anything by email; they require a notice of hearing to be served by post or other delivery service or left at a relevant address. The only available address was the registered address, albeit that it was known not to be, as it were, a useful address for [the registrant], since he was no longer there...They could have adjourned and require that an email be sent to [the registrant] to his known contact address, which might have led, and in fact would no doubt have led, to his attending a hearing when re-fixed, but they did not do so and it does not seem to me that they can properly be criticised for that failure.

The Court of Appeal observed that it was a fair comment that the process of giving notice was followed in what could be described as a mechanical fashion, but, in the absence of a notified fresh address for service, the Council was bound to send notice to the only registered address that it had. The registrant could have given the Council a new address and the Rules did not provide for service by email.

*Loufti v. General Medical Council* [2010] EWHC 1762 (Admin)

**61.04** Notice of witnesses attending hearing—failure to specify—practitioner expecting witnesses to attend

In this case, the Administrative Court in Manchester (Nicol J) allowed the claimant doctor's application for judicial review of a decision by an investigation committee of the GMC to issue him with a warning in respect of an allegation of gross misconduct involving assault. The Court remitted the matter to the GMC for fresh consideration on the basis that the committee had failed to comply with rule 34(9)(c) of the General Medical Council (Fitness to Practise) Rules 2004, with the result that the hearing had not been fair. At the hearing before the investigation committee, the GMC had not called any oral evidence about the incident, which the appellant disputed. Rule 34(9)(c) of the Fitness to Practise Rules provides that, unless otherwise agreed between the parties, each party shall, not less than twenty-eight days before the date of a hearing, require the other party to notify him of whether or not he requires any relevant person to attend and give oral evidence in relation to the subject matter. The letter from the GMC to the appellant giving notice of the hearing did not comply with rule 34(9)(c) and the omission by the GMC to comply with this requirement was said by the claimant to be material, because he said that he had expected the GMC to call witnesses whom he could question before the investigation committee and he was taken by surprise because that did not happen. Nicol J, whilst recognizing that the claimant did not ask for cross-examination of witnesses or an adjournment to allow witnesses to be called, nevertheless found that a breach of the procedural requirements in the Rules did take place and that that breach did prejudice the claimant so as to give him a hearing that was not compliant with the Rules or, for that reason, one that was fair.

### *Kumar v. General Medical Council* [2013] EWHC 452 (Admin)

**61.05**
IOP—absence of doctor—all reasonable efforts to serve

On 26 September 2011, K, a doctor whose appointment as a locum in the emergency department of Worcestershire Royal Hospital was terminated, was sent a letter by the GMC's Fitness to Practise Directorate, notifying him and inviting him to attend before the Interim Orders Panel (IOP) on 10 October 2011. K did not receive that letter because he was away and the hearing on 10 October proceeded in his absence. Conditions were imposed. When K was notified of the conditions, he protested that he had not received the letter of 26 September, and an early further hearing was arranged before the Panel, which took place on 27 October. As a matter of form, the hearing on 27 October 2011 was a review of the earlier decision. By way of preliminary, K applied to revoke the earlier decision on the basis that he had not had notice of the hearing, but that application was refused. The hearing then proceeded to deal with the substantive question of whether the conditions should be maintained. The Panel decided that they should be. On appeal, Underhill J said that the legal assessor was wrong to advise at the hearing on 10 October 2011 that all that the Panel had to be satisfied of was that the GMC had taken 'reasonable steps' to give K notice of the hearing. Rule 31 of the General Medical Council (Fitness to Practise) Rules 2004 provides that if the practitioner is neither present nor represented at a hearing, a committee or panel may nevertheless proceed to consider and determine the allegations if it is satisfied that 'all' reasonable efforts have been made to serve the practitioner with notice of the hearing in accordance with the Rules. It is well arguable that, viewed in isolation, the legal assessor's advice was literally wrong, because there is a perceptible difference of emphasis between 'reasonable efforts' and 'all reasonable efforts'. It is, however, also arguable that the advice given as a whole was not substantially misleading. Further, even if there were a misdirection, it can have made no difference, since service by registered post did constitute 'all reasonable efforts' and, at the subsequent hearing, K was given a full chance to state his case and was not prejudiced by his absence on the first occasion.

### *General Medical Council v. Homaei* [2013] EWHC 1676 (Admin)

**61.06**
Service out of jurisdiction—CPR Part 6

An application was made for extension of an interim order after a doctor had moved to the United States and the Medical Defence Union said that it was no longer sufficiently confident that it had contact with her. Permission was granted to the GMC to serve proceedings on the doctor out of the jurisdiction, either in the United States or elsewhere, pursuant to CPR Part 6, Practice Direction 6B. Service was to be effected by email to the email address that the GMC held for the doctor. Conditions were imposed that the doctor was to contact the GMC within seven days of her return to the United Kingdom, to inform the GMC if she were to apply for medical employment outside the United Kingdom, and to permit the GMC to disclose the orders to any person requesting information about her registration status.

## C. Other Relevant Chapters

**61.07** Case Management, Chapter 8
Delay, Chapter 17

# 62

| A. **Legal Framework** | 62.01 | *R (Independent Police Complaints Commission) v. Assistant Commissioner Hayman* [2008] EWHC 2191 (Admin) | 62.12 |
[1996] AC 563 | 62.03 | *R (Attwood) v. Health Service Commissioner* [2008] EWHC 2315 (Admin), [2009] PTSR 1330 | 62.13 |
| *In re B (Children) (Care

| *In re D (Secretary of State for Northern Ireland intervening)* [2008] 1 WLR 1499 | 62.05 | *Siddiqui v. Health Professions Council* [2012] EWHC 2863 (Admin) | 62.16 |
| *In re S-B (Children) (Care Proceedings: # STANDARD OF PROOF [2010] 1 AC 678 | 62.06 | *R (Holden) v. Solicitors Regulation Authority* [2012] EWHC 2067 (Admin) | 62.17 |
| C. **Disciplinary Cases** | 62.07 | *Law Society v. Waddingham and ors* [2012] EWHC 1519 (Admin) | 62.18 |
| *In re a Solicitor* [1993] QB 69 | 62.07 | | |
| *Aaron v. Law Society* [2003] EWHC 2271 (Admin) | 62.08 | *Holloway v. Solicitors Regulation Authority* [2012] EWHC 3393 (Admin) | 62.19 |
| *Campbell v. Hamlet* [2005] UKPC 19, [2005] 3 All ER 1116 | 62.09 | *Samuel v. Royal College of Veterinary Surgeons* [2014] UKPC 13 | 62.20 |
| *Council for the Regulation of Health Care Professionals v. General Medical Council and Biswass* [2006] EWHC 464 (Admin) | 62.10 | *Hannam v. Financial Conduct Authority* [2014] UKUT 0233 (TCC) | 62.21 |
| | | D. **Other Relevant Chapters** | 62.22 |
| *R (Doshi) v. Southend-on-Sea Primary Care Trust* [2007] EWHC 1361 (Admin) | 62.11 | | |



## A. Legal Framework

**62.01**

Health Act 1999, section 60A (inserted by section 112 of the Health and Social Care Act 2008) (standard of proof applicable to fitness to practise proceedings to which section 60(2) of the Health Act 1999 applies is that applicable to civil proceedings)

Veterinary Surgeons and Veterinary Practitioners (Disciplinary Committee) (Procedure and Evidence) Rules 2004, rule 23.6 (any charge shall be proved so that the committee is satisfied to the highest civil standard of proof, so that it is sure)

Chartered Institute of Legal Executives Investigation, Disciplinary and Appeal Rules 2010, rule 46(2) (balance of probabilities)

Chartered Institute of Management Accountants Disciplinary Committee Rules 2011, rule 13(2) (required standard of proof shall be the balance of probabilities)

Bar Standards Board Disciplinary Tribunals Regulations 2014, rule E143 (tribunal must apply the criminal standard of proof when deciding charges of professional misconduct and in deciding whether the disqualification condition has been established)

## B. General Principles

### *Miller v. Minister of Pensions* [1947] 2 All ER 372

**62.02** See paragraph 7.02.

### *In re H and ors (Minors) (Sexual Abuse: Standard of Proof)* [1996] AC 563

**62.03** In relation to the standard of proof, Lord Nicholls of Birkenhead said, at 586C–587G:

Civil standard of proof—general principles—serious allegation—assessment of probabilities

> Where the matters in issue are facts the standard of proof in non-criminal proceedings is the preponderance of probability, usually referred to as the balance of probability. This is the established general principle. There are exceptions such as contempt of court applications, but I can see no reason for thinking that family proceedings are, or should be, an exception. By family proceedings I mean proceedings so described in the Act of 1989, sections 105 and 8(3). Despite their special features, family proceedings remain essentially a form of civil proceedings. Family proceedings often raise very serious issues, but so do other forms of civil proceedings.
>
> The balance of probability standard means that a court is satisfied an event occurred if the court considers that, on the evidence, the occurrence of the event was more likely than not. When assessing the probabilities the court will have in mind as a factor, to whatever extent is appropriate in the particular case, that the more serious the allegation the less likely it is that the event occurred and, hence, the stronger should be the evidence before the court concludes that the allegation is established on the balance of probability. Fraud is usually less likely than negligence. Deliberate physical injury is usually less likely than accidental physical injury. A step-father is usually less likely to have repeatedly raped and had non-consensual oral sex with his underage step-daughter than on some occasion to have lost his temper and slapped her. Built into the preponderance of probability standard is a generous degree of flexibility in respect of the seriousness of the allegation.
>
> Although the result is much the same, this does not mean that where a serious allegation is in issue the standard of proof required is higher. It means only that the inherent probability or improbability of an event is itself a matter to be taken into account when weighing the probabilities and deciding whether, on balance, the event occurred. The more improbable the event, the stronger must be the evidence that it did occur before, on the balance of probability, its occurrence will be established. Ungoed-Thomas J expressed this neatly in *In re Dellow's Will Trusts* [1964] 1 WLR 451, 455: "The more serious the allegation the more cogent is the evidence required to overcome the unlikelihood of what is alleged and thus to prove it."
>
> This substantially accords with the approach adopted in authorities such as the well-known judgment of Morris LJ in *Hornal v. Neuberger Products Ltd* [1957] 1 QB 247, 266. This approach also provides a means by which the balance of probability standard can accommodate one's instinctive feeling that even in civil proceedings a court should be more sure before finding serious allegations proved than when deciding less serious or trivial matters.
>
> No doubt it is this feeling which prompts judicial comment from time to time that grave issues call for proof to a standard higher than the preponderance of probability. Similar suggestions have been made recently regarding proof of allegations of sexual abuse of children: see *In re G (A Minor) (Child Abuse: Standard of Proof)* [1987] 1 WLR 1461, 1466, and *In re W (Minors) (Sexual Abuse: Standard of Proof)* [1994] 1 FLR 419, 429. So I must pursue this a little further. The law looks for probability, not certainty. Certainty is seldom attainable. But probability is an unsatisfactory vague criterion because there are degrees of probability. In establishing principles regarding the standard of proof, therefore, the law seeks to define the degree of probability appropriate for different types of proceedings. Proof beyond reasonable doubt, in whatever form of

words expressed, is one standard. Proof on a preponderance of probability is another, lower standard having the in-built flexibility already mentioned. If the balance of probability standard were departed from, and a third standard were substituted in some civil cases, it would be necessary to identify what the standard is and when it applies. Herein lies a difficulty. If the standard were to be higher than the balance of probability but lower than the criminal standard of proof beyond reasonable doubt, what would it be? The only alternative which suggests itself is that the standard should be commensurate with the gravity of the allegation and the seriousness of the consequences. A formula to this effect has its attraction. But I doubt whether in practice it would add much to the present test in civil cases, and it would risk causing confusion and uncertainty. As at present advised I think it is better to stick to the existing, established law on this subject. I can see no compelling need for a change.

I therefore agree with the recent decisions of the Court of Appeal in several cases involving the care of children, to the effect that the standard of proof is the ordinary civil standard of balance of probability: see *H v. H (Minors) (Child Abuse: Evidence)* [1990] Fam. 86, 94, 100, *In re M (A Minor) (Appeal) (No 2)* [1994] 1 FLR 419, 424, per Balcombe LJ. The Court of Appeal were of the same view in the present case. It follows that the contrary observations already mentioned, in *In re G (A Minor) (Child Abuse: Standard of Proof)* [1987] 1 WLR 1461, 1466 and *In re W (Minors) (Sexual Abuse: Standard of Proof)* [1994] 1 FLR 419, 429, are not an accurate statement of the law.

### *In re B (Children) (Care Proceedings: Standard of Proof) (CAFCASS intervening)* [2008] 3 WLR 1, [2009] 1 AC 11

**62.04**

Civil standard of proof—balance of probabilities—no variation depending on gravity of offence

Lord Hoffmann said:

> 4. The question which appears to have given rise to some practical difficulty is the standard of proof in such cases, that is to say, the degree of persuasion which the tribunal must feel before it decides that the fact in issue did happen. *Re H (Minors)* makes it clear that it must apply the ordinary civil standard of proof. It must be satisfied that the occurrence of the fact in question was more likely than not.
>
> 5. Some confusion has however been caused by dicta which suggest that the standard of proof may vary with the gravity of the misconduct alleged or even the seriousness of the consequences for the person concerned....

Rejecting this notion, Lord Hoffmann said:

> 12. The degree of confusion which is possible on this issue is exemplified by the fact that despite the painstaking clarity with which Lord Nicholls explained that having regard to inherent probabilities did not mean that "where a serious allegation is in issue the standard of proof required is higher", Lord Steyn in *R (McCann) v. Crown Court at Manchester* [2003] 1 AC 787, 812 cited this very passage as authority for the existence of a "heightened civil standard". This appears to have resulted in submissions that the Family Division should also apply a "heightened civil standard", equivalent to the criminal standard ("in serious cases such as the present the difference between the two standards is, in truth, largely illusory", per Lord Bingham CJ in *B v. Chief Constable of Avon and Somerset Constabulary* [2001] 1 WLR 340, 354), in local authority applications for care orders. Dame Elizabeth Butler-Sloss P restored clarity and certainty in *In re U (A Child) (Department for Education and Skills intervening)* [2005] Fam 134, 143–144:
>
>> "We understand that in many applications for care orders counsel are now submitting that the correct approach to the standard of proof is to treat the distinction between criminal and civil standards as "largely illusory". In our judgment this approach is mistaken. The standard of proof to be applied in Children Act 1989 cases is the balance of probabilities and the approach to these difficult cases was laid down by Lord Nicholls in *In re H (Minors) (Sexual Abuse: Standard of*

*Proof)* [1996] AC 563. That test has not been varied nor adjusted by the dicta of Lord Bingham of Cornhill CJ or Lord Steyn who were considering applications made under a different statute. There would appear to be no good reason to leap across a division, on the one hand, between crime and preventative measures taken to restrain defendants for the benefit of the community and, on the other hand, wholly different considerations of child protection and child welfare nor to apply the reasoning in *McCann's* case [2003] 1 AC 787 to public, or indeed to private law, cases concerning children. The strict rules of evidence applicable in a criminal trial which is adversarial in nature is to be contrasted with the partly inquisitorial approach of the court dealing with children cases in which the rules of evidence are considerably relaxed. In our judgment therefore... the principles set out by Lord Nicholls should continue to be followed by the judiciary trying family cases and by magistrates sitting in the family proceedings courts."

13. My Lords, I would invite your Lordships fully to approve these observations. I think that the time has come to say, once and for all, that there is only one civil standard of proof and that is proof that the fact in issue more probably occurred than not. I do not intend to disapprove any of the cases in what I have called the first category, but I agree with the observation of Lord Steyn in *McCann*'s case, at p812, that clarity would be greatly enhanced if the courts said simply that although the proceedings were civil, the nature of the particular issue involved made it appropriate to apply the criminal standard.

### *In re D (Secretary of State for Northern Ireland intervening)* [2008] 1 WLR 1499

**62.05** Criminal standard of proof for some disciplinary proceedings—civil standard—assessment of probabilities—test

Lord Carswell said:

23. Much judicial time has been spent in the last 50 or 60 years in attempts to explain what is required by way of proof of facts for a court or tribunal to reach the proper conclusion. It is indisputable that only two standards are recognised by the common law, proof on the balance of probabilities and proof beyond reasonable doubt. The latter standard is that required by the criminal law and in such areas of dispute as contempt of court or disciplinary proceedings brought against members of a profession. The former is the general standard applicable to all other civil proceedings and means simply, as Lord Nicholls of Birkenhead said in *In re H (Minors) (Sexual Abuse: Standard of Proof)* [1996] AC 563, 586, that "a court is satisfied an event occurred if the court considers that, on the evidence, the occurrence of the event was more likely than not".

[...]

27. Richards LJ expressed the proposition neatly in *R (N) v. Mental Health Review Tribunal (Northern Region)* [2006] QB 468, para 62 where he said:

"Although there is a single civil *standard* of proof on the balance of probabilities, it is flexible in its *application*. In particular, the more serious the allegation or the more serious the consequences if the allegation is proved, the stronger must be the evidence before a court will find the allegation proved on the balance of probabilities. Thus the flexibility of the standard lies not in any adjustment to the degree of probability required for an allegation to be proved (such that a more serious allegation has to be proved to a higher degree of probability), but in the strength or quality of the evidence that will in practice be required for an allegation to be proved on the balance of probabilities."

In my opinion this paragraph effectively states in concise terms the proper state of the law on this topic. I would add one small qualification, which may be no more than an explanation of what Richards LJ meant about the seriousness of the consequences. That factor is relevant to the likelihood or unlikelihood of the allegation being unfounded, as I explain below.

28. It is recognised by these statements that a possible source of confusion is the failure to bear in mind with sufficient clarity the fact that in some contexts a court

or tribunal has to look at the facts more critically or more anxiously than in others before it can be satisfied to the requisite standard. The standard itself is, however, finite and unvarying. Situations which make such heightened examination necessary may be the inherent unlikelihood of the occurrence taking place (Lord Hoffmann's example of the animal seen in Regent's Park), the seriousness of the allegations to be proved or, in some cases, the consequences which could follow from acceptance of proof of the relevant fact. The seriousness of the allegation requires no elaboration: a tribunal of fact will look closely into the facts grounding an allegation of fraud before accepting that it has been established. The seriousness of consequences is another facet of the same proposition: if it is alleged that a bank manager has committed a minor peculation, that could entail very serious consequences for his career, so making it the less likely that he would risk doing such a thing. These are all matters of ordinary experience, requiring the application of good sense on the part of those who have to decide such issues. They do not require a different standard of proof or a special cogent standard of evidence, merely appropriately careful consideration by the tribunal before it is satisfied of the matter which has to be established.

### *In re S-B (Children) (Care Proceedings: Standard of Proof)* [2010] 1 AC 678

**62.06** Civil standard of proof—no 'heightened standard of proof'

In reversing the decision of the Court of Appeal and allowing the mother's appeal, Baroness Hale of Richmond JSC, giving the judgment of the Supreme Court, said that, in *In re B* [2008] 3 WLR 1 (paragraph 62.04 above), the House of Lords reaffirmed that the standard of proof of past facts in children care proceedings was the simple balance of probabilities, no more and no less: at [10]. In cases that were not about the standard of proof at all, but about the quality of evidence, if an event is inherently improbable, it may take better evidence to persuade the judge that it has happened than would be required if the event were commonplace. This was what Lord Nicholls was discussing in *In re H (Minors)* [1996] AC 563, 586. Yet despite the care that Lord Nicholls had taken to explain that having regard to the inherent probabilities did *not* mean that the standard of proof was higher, others had referred to a 'heightened standard of proof' where the allegations were serious. However, there is only one civil standard of proof and that is proof that the fact in issue more probably occurred than not. There was no necessary connection between the seriousness of an allegation and the improbability that it had taken place. The test is the balance of probabilities—nothing more and nothing less: at [11]–[12], [34].

## C. Disciplinary Cases

### *In re a Solicitor* [1993] QB 69

**62.07** Solicitor—case tantamount to criminal offence—perjury—criminal standard of proof

The appellant was admitted a solicitor in England and Wales in 1976. In 1981, she qualified in Australia. On 15 November 1988, the Supreme Court of Western Australia, following a complaint by the Barristers' Board of Western Australia (the statutory disciplinary body for legal practitioners there), ordered that the appellant's name be struck off the roll of practitioners. The allegations found proved against her were that her application for dissolution of marriage in 1982, supported by her affidavit, had contained two material falsehoods. There was a further allegation, also found proved, that she had, at the hearing of the matter before the judge, either verified the false information that was contained in her application, or had failed to inform the judge that the application and the affidavit were inaccurate. Proceedings before the Solicitors Disciplinary Tribunal

(SDT) in the United Kingdom, the appellant said that she had received the formal complaint the day before she was leaving Australia and had not attended the hearing, and contended that if she had been properly defended, the complaint would have been dismissed. In allowing the appellant's appeal from the order of the SDT striking her off the roll of solicitors in England and Wales, Lord Lane CJ, giving the judgment of the Divisional Court (Lord Lane CJ, Simon Brown and Jowitt JJ), said that the findings in Western Australia were admissible and that it was open to the SDT to make such use of those findings as was proper in the circumstances of the case. In deciding the weight to be attached to the Board's findings—that is, the Board's actual conclusions upon the allegations of misconduct—the SDT would clearly bear in mind a variety of considerations and not least (a) the evidence adduced before the Board, (b) the apparent fairness or otherwise of the proceedings before the Board, (c) the standard of proof adopted by the Board, and (d) the absence of any right of appeal from the Board's findings: at 80G–H. As to the standard of proof adopted by the SDT, the Tribunal did not state in so many words what standard of proof it applied in coming to its conclusion. The complaint against the appellant was of professional misconduct tantamount to perjury. The Tribunal said that there was much to be said for adopting the criminal standard of proof in disciplinary cases, but that there was no conclusive authority that bound the Tribunal on that point. Lord Lane said, at 81F–G:

> It seems to us, if we may respectfully say so, that it is not altogether helpful if the burden of proof is left somewhere undefined between the criminal and the civil standards. We conclude that at least in cases such as the present, where what is alleged is tantamount to a criminal offence, the tribunal should apply the criminal standard of proof, that is to say to the point where they feel sure that the charges are proved or, put in another way, proof beyond reasonable doubt. This would seem to accord with decisions in several of the Provinces of Canada.

### *Aaron v. Law Society* [2003] EWHC 2271 (Admin)

**62.08** Solicitor—criminal standard of proof

In giving the judgment of the Divisional Court, Auld LJ (with whom Goldring J agreed) said:

> 15. Before I turn to the individual grounds of appeal, I should mention two basic matters. The first is that disciplinary proceedings before the Solicitors Disciplinary Tribunal must be proved to the criminal standard, certainly where, as here, the allegations are serious and may result in suspension or disqualification; see *Re A Solicitor* [1993] QB 69, per Lord Lane CJ at 81A-82C. The second is that an appeal against the findings of the Tribunal to this Court under section 49 of the 1974 Act is by way of rehearing and the Court may make such an order as it thinks fit; see *In Re A Solicitor* [1945] 1 All ER 445, CA, per Scott LJ at 446A and 447G-H.

### *Campbell v. Hamlet* [2005] UKPC 19, [2005] 3 All ER 1116

**62.09** Solicitor—criminal standard of proof

The appellant, C, was an attorney-at-law practising in Trinidad and Tobago. In February 1987, a complaint of professional misconduct was made against him to the Attorneys-at-Law Disciplinary Committee. The essence of the complaint was that the complainant had paid C $29,400 to purchase from him two parcels of land, but that C had neither conveyed the land nor returned the purchase price. The hearing before the Committee lasted eleven days stretching over a ten-month period ending on 6 December 1988. Astonishingly, it was not until 29 October 1996 that the Committee produced its findings and orders. The Committee found the allegation of professional misconduct substantiated and ordered C to pay the complainant compensation of $29,400, together with interest from the date of its order until payment, together with costs. C's appeal to the Court of Appeal was dismissed on 22 May 2000. On C's appeal to the Privy Council, Lord Brown of Eaton-under-Heywood, giving the judgment of the Board, said:

> 16 That the criminal standard of proof is the correct standard to be applied in all disciplinary proceedings concerning the legal profession, their Lordships entertained no doubt. If and insofar as the Privy Council in *Bhandari v. Advocates Committee* [1956] 3 All ER 742, [1956] 1 WLR 1442 may be thought to have approved some lesser standard, then that decision ought no longer, nearly fifty years on, to be followed....

Lord Brown went on to state that, in *Re a Solicitor* [1993] QB 69:

> 21. ...Lord Lane referred to the provision in the Bar's Code of Conduct requiring the tribunal to apply the criminal standard of proof and observed at p.82: "it would be anomalous if the two branches of the profession were to apply different standards in their disciplinary proceedings". This last observation, of course, clearly warranted the Law Society Disciplinary Committee thenceforth applying the criminal standard in all cases rather than merely in those, earlier referred to, "where what is alleged is tantamount to a criminal offence".
>
> 22. Their Lordships would add that, even had they concluded that the criminal standard should apply only in disciplinary cases where what is alleged is tantamount to a criminal offence, that, at least arguably, would include the present case. This was certainly no mere contractual dispute....

However, on the facts there was nothing in the Committee's determination to suggest that it applied a lower standard of proof than that of 'beyond reasonable doubt'.

### *Council for the Regulation of Health Care Professionals v. General Medical Council and Biswass* [2006] EWHC 464 (Admin)

**62.10**

See paragraph 47.30.

### *R (Doshi) v. Southend-on-Sea Primary Care Trust* [2007] EWHC 1361 (Admin)

**62.11**

FHSAA—civil standard—whether to announce standard of proof at start of proceedings

D was a registered medical practitioner, who practised as a general practitioner (GP) in the area of the Southend-on-Sea Primary Care Trust (PCT). In April 2005, the PCT removed him from its performers list on the grounds that he was unsuitable and inefficient. The heart of the case against D and the reason why he was removed was that it was alleged that, on a number of occasions over many years from about 1985 until 2003, D had exhibited sexualized behaviour, or otherwise behaved inappropriately (in an essentially sexual manner), towards a number of adult female patients and some staff. D exercised a statutory right of appeal to the Family Health Services Appeal Authority (FHSAA), which, following a rehearing, dismissed D's appeal and made a national disqualification order. In his appeal to the High Court, D submitted that the FHSAA adopted an unfair procedure in the timing and manner in which it dealt with the issue as to the appropriate standard of proof, and that it adopted and applied the wrong standard of proof. In its written decision following the hearing, the FHSAA said: that it was a civil and not a criminal tribunal; that the allegations in this case were serious allegations, and that cogent and compelling evidence is required if they were to be found proved; that, when considering whether it was satisfied on the balance of probabilities that an allegation was established, it bore in mind that the more serious the allegation, the less likely it is that it occurred; and the stronger should be the evidence before it concluded that the allegation was established. D submitted that, whether the tribunal properly directed itself or not, it was unfair, to the point of being unlawful, to postpone making, or at any rate announcing, its decision as to the standard of proof until after the end of the evidence and argument. D argued that he could not properly make a final decision whether or not to give evidence until he knew what standard of proof was being applied. The rules are silent and the issue of standard of proof before a tribunal, such as the FHSAA, is not clearly established or axiomatic. In a criminal trial, all parties know, without it needing to be stated at the outset, what the criminal standard is; similarly in an ordinary civil trial. Holman J said that he had some sympathy with this submission. He could not see any reason, on the facts of this case, why the tribunal could not have made its mind up at the outset what standard it would apply and rule upon it at that stage. It knew very well the nature of the allegations and the broad outline of the case with which it was dealing. Even if, as the chairman said, it was not 'obliged' to deal with it at that stage, there was no reason for putting it off. Because D asked for a ruling, it would have been wiser and preferable to give one at that stage. However, the Court was not persuaded that there was any actual unfairness in this case and certainly there was no unfairness such as to be unlawful. The pleaded grounds of appeal made no complaint, or hint of a complaint, of procedural unfairness. There was no statement in support of this ground. The submission that the failure to make an earlier ruling may have affected the decision that D would not give evidence was purely speculative and not supported by any evidence. As to the standard of proof, Holman J—after reviewing, amongst other cases, *Lanford v. General Medical Council* [1990] 1 AC 13, *McAllister v. General Medical Council* [1993] AC 388, *Sadler v. General Medical Council* [2003] UKPC 59, [2004] Lloyd's Law Reports Medical 44, *Gage v. General Chiropractic Council* [2004] EWHC 2762 (Admin), and *Campbell v. Hamlet* [2005] UKPC 19—said that it was unwise of the tribunal in the present case not simply to adopt the criminal standard, but that it did have a discretion and was not bound to do so. Overall, the Court was satisfied that it was permissible and not unlawful for the tribunal to direct itself as it did. There was a very thorough hearing and analysis of the evidence, and the conduct of the proceedings and approach to the evidence and proof was fair.

### *R (Independent Police Complaints Commission) v. Assistant Commissioner Hayman* [2008] EWHC 2191 (Admin)

**62.12** Police officer—review by Assistant Commissioner following finding by panel—wrong test

The two House of Lords judgments in *Re B (Children)* [2008] 3 WLR 1 and *Re D* [2008] 1 WLR 1499 were considered by Mitting J in the context of a police disciplinary case. The Independent Police Complaints Commission (IPCC) directed that disciplinary proceedings be taken against an off-duty police officer who had become involved in a fracas in Old Street, London, EC1. The disciplinary proceedings came before a panel of three senior officers, who, applying the civil standard of proof, found three of the four charges proved and decided that the officer should resign. He applied for a review of that decision and, in August 2006, Assistant Commissioner Hayman found the three charges that had been found to be proved against him by the panel not to be proved, so quashing the panel's decision. The Assistant Commissioner applied a 'sliding scale' and a 'higher threshold' because of the potential consequences of dismissal. The IPCC challenged that decision in judicial review proceedings. The basis of challenge was that the Assistant Commissioner had applied the wrong standard of proof in his review of the panel's decision. At [19], Mitting J said that, in his opinion, the last sentence in [28] of Lord Carswell's speech in *Re D* laid down the true proposition of law: '[Those who have to decide such issues] do not require a different standard of proof or especially cogent standard of evidence, merely appropriately careful consideration by the tribunal before it is satisfied on the matter which has to be established.' Mitting J went on to state:

> 20. Of course in disciplinary proceedings the tribunal must look with the greatest care at accusations which potentially give rise to serious consequences. But in determining whether or not they occurred, it applies a single unvarying standard, the balance of probabilities. If satisfied it is more likely than not that the facts occurred, then it must find them proved and draw appropriate conclusions as to sanction.

The judge found that the Assistant Commissioner did not apply this test to the review that he conducted. In so doing, he misdirected himself and the matter was remitted to the Commissioner of Police for the Metropolis for him to appoint another assistant commissioner to take the decision afresh.

### *R (Attwood) v. Health Service Commissioner* [2008] EWHC 2315 (Admin), [2009] PTSR 1330

**62.13** Health Service Commissioner—ombudsman's report—whether claimant acted unreasonably in management of patient—*Bolam* test

The claimant, a consultant gastro-intestinal surgeon employed by Salford Royal Hospitals National Health Service (NHS) Trust at the Hope Hospital, Salford, sought judicial review of a report by the Health Service Commissioner (the ombudsman) into complaints regarding his treatment of a patient who died in January 2002. The report was written by a senior investigating officer and it comprised the ombudsman's report. The investigating officer concluded that the patient's management following his discharge from hospital was neither properly monitored nor effectively followed up, and that neither the patient nor his family was given adequate information about his condition between October 2001 and his death on 20 January 2002. In writing his report, the investigating officer stated that the *Bolam* test (*Bolam v. Friern Hospital Management Committee* [1957] 1 WLR 582) is one used by courts when considering actions. However, the purpose of the ombudsman's investigation into this complaint was not to establish whether the trust and/or the consultant surgeon were negligent, but whether the patient and his family suffered injustice or hardship as a consequence of any failure in the service that they received from the trust. The standard of proof that applies in such a case is lower than that for negligence. In allowing the claimant's claim for judicial review, Burnett J said that the ombudsman would be entitled to approach the question of failure in service, even in the context of clinical judgment, from a point of view that is different from the approach of the courts in negligence actions. It would, for example, be open to the ombudsman to explain that whilst she recognized that a finding of negligence could not be made, she would be disposed to make a finding of a failure in service if the clinical care were to fall below best practice within the NHS. However, that was not the end of the matter because the ombudsman's own documents showed that, when considering whether to stigmatize a clinical judgment as unreasonable, she and her predecessor had stated that they would apply a test that was indistinguishable from the *Bolam* test. In a paper entitled *Responsibilities of the Health Service Commissioner* published in December 1995, the ombudsman stated that he would expect his advisers, in deciding what was reasonable and responsible in the particular circumstances, to have the due regard to all of the relevant professional guidance on standards and good practice that a professional working in the capacity in question could be expected to take into account. Burnett J said that it followed that the investigating officer misdirected himself in applying a different standard to the question of whether the claimant had acted unreasonably in the management of the patient.

***Hutchinson v. General Dental Council* [2008] EWHC 2896 (Admin)**

**62.14** Dentist—balance of probabilities—totality of evidence

Charges brought against the appellant in respect of various hygiene related matters were found proved, his fitness to practise was found to be impaired, and an order for erasure was made by the Professional Conduct Committee (PCC). The appellant appealed on the grounds that the bringing of certain of the charges against him amounted to an abuse of process: see paragraph 17.28. Looking at the totality of the evidence, the case against the appellant on the charges was a weak one. Allowing the appeal, Blair J said that, in his view, it fell below the required standard of proof on the balance of probabilities. On that basis, rather than on the basis of misdirection, the Court was satisfied that the Committee's findings were wrong and must be quashed.

***Richards v. Law Society (Solicitors Regulation Authority)* [2009] EWHC 2087 (Admin)**

**62.15** Solicitor—criminal standard of proof—binding until Supreme Court rules otherwise

One of the solicitor's grounds of appeal was that the SDT was wrong in law to apply the civil standard of proof and that it should have applied the criminal standard. A difference of view on this point lay between the Solicitors' Regulation Authority (SRA) and the Law Society (the professional body), and consequently the Law Society was permitted to intervene in the appeal. The Law Society maintained that the appropriate standard of proof in solicitors' disciplinary proceedings was the criminal standard. The SRA maintained that it should be the civil standard. The Court (Sir Anthony May P and Saunders J) held that, on the particular facts, the issue was academic. First, there were no significant disputed facts and nothing therefore for the standard of proof to bear upon, and second, the Tribunal had, in effect, applied the criminal standard anyway. However, the Court observed that it was bound by the decision of the Divisional Court presided over by Lord Lane CJ in *Re a Solicitor* [1993] QB 69, as considered and applied by the Privy Council in *Campbell v. Hamlet* [2005] 3 All ER 1116. Sir Anthony May P said, at [22], that:

> Insofar as these two authorities might arguably leave some minor room for manoeuvring in cases where the alleged misconduct does not have criminal overtones, that is better debated and decided in a case where the standard of proof makes a difference, and probably in the House of Lords.

***Siddiqui v. Health Professions Council* [2012] EWHC 2863 (Admin)**

**62.16** See paragraph 27.11.

***R (Holden) v. Solicitors Regulation Authority* [2012] EWHC 2067 (Admin)**

**62.17** Solicitors' Accounts Rules 1998—compliance—strict liability

In 2009, H was invited to join what she understood to be a new firm of solicitors. She agreed to join the firm as a salaried partner once premises and professional indemnity insurance were in place. She provided no capital and she was not an equity partner. She was never mandated to operate any client bank account or office bank account. She did not operate the firm's Internet banking facility. She had no dealings with client monies at any stage. She joined the firm in October 2009 and quit in early 2010, after only some three months. She never had a desk in the office. She never gained a single client through the office and the firm turned out not to be what she expected, in that instead of a new practice and a new firm, it was a successor practice with active files and existing client accounts. She did not open a single new file. Subsequently, in May 2010, there was a forensic investigation of the firm. Debit balances were found in eighty-five client ledgers and nineteen transfers into a suspense account were discovered, totalling £332,000. An adjudicator concluded that there had been a number of breaches of the Solicitors' Accounts Rules in relation to a number of people, one of them being H. Rule 6 of the Solicitors' Accounts Rules 1998 provides that 'all the principals in a practice must ensure compliance with the rules'. Notes in relation to rule 7 of the Solicitors Code of Conduct 2007 state that a 'salaried associate' partner is treated by the SRA as a full partner and manager, and must therefore comply with the Solicitors' Accounts Rules. An adjudicator decided that, in relation to H, a disciplinary sanction was warranted with an order for no further action. H's appeal to an internal panel of the SRA was dismissed. In dismissing H's renewed application for permission to apply for judicial review, Irwin J said that, in his judgment, the phrase 'to ensure' in rule 6 of the Solicitors' Accounts Rules 1998 imports strict liability. That is the obvious meaning of the language and it fits with the necessity for public protection. There was nothing in the language of rule 6 or elsewhere in the Rules, and nothing in any established authority, to suggest that that is not the straightforward meaning of the phrase 'to ensure'.

***Law Society v. Waddingham and ors*** **[2012] EWHC 1519 (Admin)**

**62.18** Solicitor—proof to criminal standard not established

In giving judgment, Maddison J (with whom Richards LJ agreed) said that S and P, two of the respondents, *probably* did act dishonestly, both parts of the *Twinsectra* test being met to that standard: *Twinsectra Ltd v. Yardley* [2002] UKHL 12. However, taking into account the counterbalancing factors on which S and P relied, he was not able to be *sure* that either S or P acted dishonestly: at [60]. Having regard to the criminal standard of proof, his Lordship concluded, as indeed did the SDT, that the allegations of dishonesty had not been made out.

See paragraph 28.10.

***Holloway v. Solicitors Regulation Authority*** **[2012] EWHC 3393 (Admin)**

**62.19**

When a tribunal properly directs itself as to the appropriate standard of proof and says that it is satisfied to that standard, there would need to be the most powerful evidence to show that, notwithstanding that it had directed itself properly, it had then failed to give effect to that direction: *per* Elias LJ, at [12].

***Samuel v. Royal College of Veterinary Surgeons*** **[2014] UKPC 13**

**62.20** Veterinary surgeon—criminal standard of proof—rule 23.6

Rule 23.6 of the Veterinary Surgeons and Veterinary Practitioners (Disciplinary Committee) (Procedure and Evidence) Rules 2004 provides that any charge that may result in a direction by the Disciplinary Committee that a respondent be removed from the register, 'shall be proved so that the Committee is satisfied to the highest civil standard of proof; so that it is sure'. The wording of the rule is confusing, particularly in view of the decision of the House of Lords in *In re B (Children)(Care Proceedings: Standard of Proof)(CAFCASS Intervening)* [2009] 1 AC 11 that the civil standard of proof is, in all cases, on the balance of probabilities. The phrase 'is satisfied... so that it is sure' is the standard form of wording used to direct juries in criminal cases, and leading counsel on behalf of the College properly accepted that the rule is intended to require the same standard of proof as in a criminal case.

See further paragraph 14.12.

***Hannam v. Financial Conduct Authority*** **[2014] UKUT 0233 (TCC)**

**62.21** Market abuse—civil standard of proof

In 2008, H was chairman of capital markets at JP Morgan and global co-head of UK Capital Markets at JP Morgan Cazenove. H challenged a decision made by the Financial Services Authority (FSA) to the effect that he had committed market abuse. The market abuse consisted of improperly disclosing inside information in two emails sent by him or on his behalf in September and October 2008, concerning a potential third-party bid for one of H's clients. The information concerned positive developments in a client's oil exploration operations. The emails were sent to the Minister for Oil in the Kurdish regional government and a second was blind copied to an adviser to a potential investor in companies with interests in Kurdistan. The Upper Tribunal (Warren J, Chamber president) held that the sending of the emails constituted behaviour falling within section 118(3) of the Financial Services and Markets Act 2000, and that H thereby engaged in market abuse. The FSA accepted that the burden was on it to establish that H was guilty of market abuse and submitted that the ordinary civil standard of proof 'on the balance of probabilities' applied. H claimed that it was the criminal standard of proof 'beyond reasonable doubt'. The Upper Tribunal, in a section of its judgment headed 'Standard of proof', concluded that the standard of proof applicable in market abuse cases is the civil standard: at [147]–[192]. H submitted that the FSA's allegations in the present case were tantamount to allegations that constitute criminal offences, referring to section 52 of the Criminal Justice Act 1993, which provides for the offence of insider dealing. The Upper Tribunal agreed with the FSA that the conduct alleged against H was not tantamount to criminal conduct: the FSA did not allege that the necessary mental element on the part of H existed. It was common ground, in the light of *Re H* [1996] AC 563 and *Re B* [2009] 1 AC 11, that, in civil cases, the normal standard by which disputed issues of fact are to be decided is the balance of probabilities. The approach that allowed for a 'sliding scale' in the civil standard, by which it varied according to the seriousness of what has to be proved, has been exposed as a heresy: at [153]. After reviewing the authorities, His Lordship said:

> 191. Drawing all this together, we do not consider that market abuse falls within Lord Hoffmann's first category [where the courts have thought that the criminal standard, or something like it, should be applied because of 'the serious consequences of the proceedings']. We have already explained that we do not perceive allegations of market abuse as tantamount to allegations which constitute criminal offences. Although the consequences in the case of serious market abuse may be a large financial penalty, there is no other sanction (apart from censure) which the Authority can impose under the market abuse regime (in contrast with the sanctions it can impose on an approved person whom it considers is not "fit and proper": see sections 63 and 66 FSMA). In the light of that, a person against whom allegations of market abuse are raised is not, it seems to us, entitled to the same sort of protection as a person whose fundamental liberties are at stake, any more than a person whose livelihood is at stake is entitled to such protection (as to which see *R v. Provincial Court of the Church in Wales, ex parte Williams* (1998) CO/2880/98 and *Greaves v. Newham London Borough Council* Employment Appeals Tribunal, 16 May 1983). Further, it seems to us that if an analogy with another area is to be found, the closest analogy is a case of civil fraud falling in Lord Hoffmann's second category such as *Hornal v. Neuberger Products Ltd* [1957] 1 QB 247, CA, where the ordinary civil standard applies but where the inherent probability of the relevant event (e.g. the activities amounting to market abuse) may lead to the need for strong evidence to persuade the tribunal that the event occurred.
>
> 192. Our conclusion therefore is that the standard of proof applicable in market abuse cases is the civil standard. We would only add that this conclusion appears to be more consonant with the Market Abuse Directive than a conclusion that the appropriate standard of proof is the criminal standard. Recital (24) [prompt and fair disclosure of information to the public to enhance market integrity; selective disclosure can lead to loss of investor confidence in the integrity of financial markets] envisages strict enforcement of the disclosure requirement which would most

effectively be achieved if the hurdles placed in front of the regulatory authorities are not set too high.

## D. Other Relevant Chapters

Burden of Proof, Chapter 7 **62.22**

# 63

# STAY OF PROCEEDINGS



## A. Other Relevant Chapters

**63.01** Abuse of Process, Chapter 2
Bias, Chapter 6
Delay, Chapter 17
Double Jeopardy, Chapter 21
Estoppel, Chapter 24
Human Rights, Chapter 32
Natural Justice, Chapter 48
Press Publicity, Chapter 51

# 64

# STRIKING OFF/ERASURE



## A. Other Relevant Chapters

Sanction, Chapter 60 **64.01**

# 65

# SUSPENSION



| *Giele v. General Medical Council* [2005] 4 All ER 1242 | 65.04 | | |
| *R (Kaftan) v. General Medical Council* [2009] EWHC 3585 (Admin) | 65.05 | | |
| *Robinson v. Solicitors Regulation Authority* [2012] EWHC 2690 (Admin) | 65.06 | *Jasinarachchi v. General Medical Council* [2014] EWHC 3570 (Admin) | 65.14 |
| *Mould v. General Dental Council* [2012] EWHC 3114 (Admin) | 65.07 | *HK v. General Pharmaceutical Council* [2014] CSIH 61 | 65.15 |
| *Dutta v. General Medical Council* [2013] EWHC 132 (Admin) | 65.08 | *Solicitors Regulation Authority v. Uddin* [2014] EWHC 4553 (Admin) | 65.16 |
| *R (Rhodes) v. Police and Crime Commissioner for Lincolnshire* [2013] EWHC 1009 (Admin) | 65.09 | *Pool v. General Medical Council* [2014] EWHC 3791 (Admin) | 65.17 |
| | | C. Other Relevant Chapters | 65.18 |

## A. Legal Framework

**65.01** Most healthcare regulator's provisions provide that if the fitness to practise committee or panel determines that the fitness to practise of the practitioner is impaired, it may give a direction that the entry in the register of the practitioner concerned be suspended, for such period not exceeding twelve months as may be specified in the direction. See for example:

Medical Act 1983, section 35D
Dentists' Act 1984, sections 27 and 28
Opticians Act 1989, section 13F
Nursing and Midwifery Order 2001, article 29
Pharmacy Order 2010, article 54(2)(d)

Other professional bodies do not necessarily have a maximum period for suspension. For example, in the case of solicitors, suspension may be for a fixed term or for an indefinite period: Andrew Hopper QC and Gregory Treverton-Jones QC, *The Solicitor's Handbook* (London: Law Society, 2015), paragraph 21.10. See also:

Disciplinary Tribunals Regulations 2014, Annex 1 (where a charge of professional misconduct has been found proved against a barrister by a disciplinary tribunal, the disciplinary tribunal may decide to suspend his practising certificate and suspend his rights and privileges as a member of his Inn for a prescribed period, either unconditionally or subject to conditions; a three-person panel must not suspend a barrister's practising certificate for a longer period than twelve months)

Institute of Chartered Accountants in England and Wales Disciplinary Bye-laws, bye-law 22(3) (if the defendant is a member, his practising certificate may be withdrawn either permanently or for a specific period)

## B. Disciplinary Cases

### *Taylor v. General Medical Council* [1990] 2 AC 539

**65.02** Conviction—maximum twelve months' suspension—review—extension—basis for

In November 1987, T, a doctor, appeared before the Professional Conduct Committee (PCC) of the General Medical Council (GMC) charged with serious professional misconduct. The charge related to T prescribing the drug methadone hydrochloride. He admitted that, over a period between December 1984 and June 1985, he had abused his position as a medical practitioner by issuing such prescriptions irresponsibly to some seventy patients. He had previously been admonished by the Committee in 1986 following his conviction at the Central Criminal Court of seven offences of making false statements for the purpose of enabling persons to obtain passports, for which he had been sentenced to four months' imprisonment, suspended for two years. Shortly before the hearing in November 1987, T was further convicted at Aylesbury Crown Court of four offences of prescribing controlled drugs in contravention of a prohibition. For these offences, he was conditionally discharged for twelve months. Lord Bridge of Harwich, delivering the judgment of the Privy Council, said that it was not in the least surprising that the Committee in November 1987, when considering the matter, took a serious view of the doctor's case. If it had then directed that his name be erased from the register, an appeal would have had no prospect of success. Instead, the Committee directed that his registration be suspended for a period of twelve months and intimated that it would resume consideration of the case before the expiration of that period. Consequently, T appeared again before the Committee on 14 November 1988, when the Committee directed that his suspension be extended for a further period of twelve months and again intimated that it would resume consideration of the case before the expiry of that period. Finally, on 20 November 1989, the Committee gave a direction for a third successive period of twelve months' suspension, which was the subject of the appeal. In allowing T's appeal against this further period of twelve months' suspension, Lord Bridge said, at 545C–546A:

> [A] practitioner who is suspended for up to 12 months in the first place is entitled to conclude that his criminal behaviour or professional misconduct was not regarded by the committee as sufficiently grave to warrant erasure and that the period of suspension directed was thought sufficient to provide any necessary punitive element in the sentence imposed. It can never be a proper ground for the exercise of the power to extend the period of suspension that the period originally directed was insufficient to reflect the gravity of the original offence or offences.
>
> It will obviously be a proper ground for extending the period of suspension that during the period the practitioner has been convicted of some further criminal offence and it may well be a proper ground that he has been guilty of some other positive misconduct, using that word in a perfectly general sense, which reflects on his fitness to practise medicine. But by far the commonest case where the power will be appropriately exercised, and that for which, their Lordships think, both the power to extend a period of suspension and the power to direct erasure following

> a period of suspension were specifically designed, is where the criminal behaviour or professional misconduct which led to the original suspension was associated with and occasioned by some condition affecting the practitioner's fitness to practise which may or may not be amenable to cure. The most obvious examples which spring to mind are cases in which the practitioner is addicted to alcohol or drugs or suffers from some psychiatric disorder. Such cases since 1978, as has been noted, may be referred by the Preliminary Proceedings Committee either to the Professional Conduct Committee or to the Health Committee. The exercise in such cases by either committee of the power to extend the period of suspension must, in their Lordship's judgment, be governed by the same principle. The case will be reviewed before the expiry of the first or any subsequent period of suspension for the committee to determine whether the practitioner is cured of his addiction or other disorder so as to be fit to resume the practice of medicine. If he is not, they will direct a further period of suspension. In a case before the Professional Conduct Committee which would originally justify erasure, but where the committee have felt it right, in view of the practitioner's condition, to suspend the judgment to see if he is able to make use of the opportunity to effect a cure, they may decide, when no cure is effected, to direct erasure.

Lord Bridge continued and said that, in the instant case, their Lordships were at a loss to understand why the Committee before which the doctor appeared in November 1987, having determined that it would be sufficient to direct suspension for a period of twelve months, went on to intimate that it would reconsider the case before the end of that period. The doctor's record was thoroughly reprehensible. But there was no material before the Committee relating to the doctor's physical or mental condition or his habits or way of life that could afford any discernible reason why the Committee should think it appropriate to reassess his fitness to resume practice following a probationary period, so to speak, of twelve months. Their Lordships were driven to the conclusion that, whatever may have been the reason for giving the direction for a second period of suspension in 1988, the only explanation for the Committee's decision in 1989 to direct a third period was that it regarded the original decision to direct suspension instead of erasure as having been too lenient. This view seemed to be fully confirmed by the fact that the Committee on this occasion did *not* intimate that it would reconsider the case before the expiry of the further period of suspension. If the Committee had concluded that, for the protection of the public, there was some reason why the doctor could not yet be regarded as fit to resume practice, it would have been wholly inconsistent to direct a third and final period of twelve months' suspension without reserving the case yet again for reconsideration before the expiry of that period. In the light of these circumstances, their Lordships had no doubt that the direction was wrong in principle. Lord Bridge concluded, at 547H–548C:

> Once the Professional Conduct Committee has decided, following proof of a criminal conviction or a finding of serious professional misconduct, that a period of suspension of the practitioner's registration up to 12 months is sufficient to mark the gravity of the case, it can never, in their Lordships' judgment, be appropriate to reserve the possibility of an extension of the period under rule 31(5) of the Rules of 1988 unless the committee conclude that there is a positive reason why they should monitor the practitioner's progress in some particular respect during the period of suspension with a view to deciding, in the light of that progress, whether he can safely be permitted to resume practice when the period expires. It will always be desirable for the committee to indicate in general terms what their reasons are for reserving a case for reconsideration and to tell the practitioner what are the specific matters on which they will require to be satisfied before he will be permitted to resume practice.

### *Gage v. General Chiropractic Council* [2004] EWHC 2762 (Admin)

**65.03** Chiropractor—no fixed term of suspension—end date only—inappropriate sanction

At the conclusion of G's case over a period of five days in July 2004, a disciplinary tribunal found that G was guilty of unacceptable conduct and imposed 'a suspension order until 31 October 2004', adding that the suspension would take effect in accordance with section 31 of the Chiropractors Act 1994, twenty-eight days after notice of it was served on him. Formal notification of the decision was served on 26 July 2004, but because G appealed to the High Court, the sanction did not take effect, and G's appeal was heard on 1 November 2004—that is, one day after the suspension order was to terminate. In the course of dismissing G's appeal on other grounds, Jackson J said that instead, and somewhat unusually, the panel did not suspend G's registration for a period of specified length, but from an unspecified date until 31 October 2004. Moreover, the start date of suspension was uncertain and would be affected by the date on which formal notification of the decision was sent. The manner in which the sentence was formulated was, to say the least, unfortunate: at [41]. His Lordship accepted the submission on behalf of the appellant that there was no fixed length of suspension and there was simply a period of suspension that would end on a specified date. The consequences of section 31(2) of the 1994 Act, as amended, is that G's suspension could not start until 1 November 2004, being the date on which his appeal would be disposed of, and it was not possible for the Court to vary the date of 31 October 2004 that the panel originally fixed as the date for termination of suspension. Regrettably, the consequence was that G's suspension had ended before it could begin.

### *Giele v. General Medical Council* [2005] 4 All ER 1242

**65.04** Sanction—erasure replaced by twelve months' suspension—end date only—public confidence not harmed

In allowing the doctor's appeal against an order of erasure and substituting a suspension for twelve months, Collins J said that he was entirely satisfied that erasure was not required in the instant case and that public confidence in the profession, which must reflect the views of an informed and reasonable member of the public, would not be harmed if suspension were imposed. Suspension for twelve months is itself a severe penalty for any practitioner and the Court was satisfied that, for the misconduct in this case, it would provide an appropriate sanction. Further, subject, of course, to the doctor acting inappropriately during the period of suspension, there was unlikely to be any justification for extending the period of suspension at the end of twelve months, but conditions may be needed to ensure that the doctor demonstrates all necessary skills to go back to practise as a consultant surgeon. The Court was far from saying that any conditions would certainly be needed at the end of the period of twelve months; that was to be a matter for the relevant panel in due course to determine whether any further action is needed.

See paragraph 60.08.

### *R (Kaftan) v. General Medical Council* [2009] EWHC 3585 (Admin)

**65.05** Doctor—inappropriate behaviour—six months' suspension—undertaking offered—effect on loss of earnings

K appeared before a fitness to practise panel of the GMC on 28 and 29 January 2008. The allegations made against K involved aggressive or violent behaviour towards colleagues at the accident and emergency (A&E) department of the Huddersfield Royal Infirmary, where K had worked as a clinical assistant. The allegations particularly concerned two incidents, the first involving a staff nurse and the second involving a doctor. The panel found the facts proved and made a determination of impairment by reason of misconduct, and its decision was to suspend K from the medical register for a period of six months by way of sanction. K submitted that the sanction was inappropriate on two grounds: first, that he had offered undertakings; and second, that the panel failed properly to take into account the fact that he had been dismissed by the Huddersfield

Hospital Trust and had been without income for nine months. In dismissing K's appeal, Hickinbottom J noted that the panel specifically considered both points. It took into account the anger-management counselling that the appellant was undertaking and considered that undertakings alone would be insufficient to uphold proper standards within the profession. The panel also expressly took into account the financial hardship that the appellant had suffered as a result of his dismissal. The Court said that, although repetition of K's conduct was unlikely, the panel was clearly entitled not to accept the offered undertakings and to consider conditional registration inappropriate as an insufficient mark of the severity of the misconduct as it found it. It was entitled to impose a sanction withdrawing the rights of medical practice from the appellant in the form of a suspension and for a period of six months.

***Robinson v. Solicitors Regulation Authority*** **[2012] EWHC 2690 (Admin)**

**65.06** Solicitor—serious failures—appeal against twelve months' suspension dismissed

R, a solicitor, appealed against the decision of the Solicitors Disciplinary Tribunal (SDT) to suspend him from practice for a period of twelve months. The Solicitors Regulation Authority (SRA) alleged that R had committed four breaches of the Solicitors Code of Conduct—namely:

(1) failure adequately to supervise staff by employing a dishonest employee;
(2) failure to act in the best interests of a client concerning a conveyancing transaction involving mortgage fraud;
(3) providing a misleading statement regarding a will; and
(4) failure to provide material information concerning professional insurance.

R did not give evidence before the Tribunal, but admitted all of the allegations. R's appeal was dismissed by Haddon-Cave J. The Tribunal was plainly right to find that there were serious failures by R in relation to the employment of a member of staff. He had failed to carry out basic checks or to dispense with the employee's services as soon as he discovered that the employee had been suspended by his previous firm for dishonesty. There were significant failures in relation to eight conveyancing transactions on behalf of mortgage clients. R had failed to supervise properly C in his firm, who had also been dismissed from her previous employer for misconduct. The Tribunal was entitled to conclude that R's statement in relation to a will was thoroughly misleading. There was no escaping the conclusion that R had deliberately lied when completing an insurance application form. His aim was plainly dishonest: to obtain cover at a reasonable premium. In summary, the Tribunal was entitled, and indeed bound, to take a very dim view of R's conduct. Individually, each of his failures was itself serious. Collectively, however, they formed a very serious picture that called into question his fitness to practise at all. The sanction by the Tribunal of twelve months' suspension was more than justified. It was certainly well within the proper range of sentences that the Tribunal could have imposed and the Court advised that the appellant was to count himself somewhat fortunate not to have had a more severe penalty imposed upon him or to have been struck off the roll altogether, because his manifest dishonesty in relation to his professional insurance arguably called into question his fitness to practise at all.

***Mould v. General Dental Council*** **[2012] EWHC 3114 (Admin)**

**65.07** Dentist—conditions—failure to comply—suspension

M, a registered dentist, faced two allegations in May 2011 of prescribing, dispensing, and administering diclofenac sodium (Voltarol), including on one occasion at his National Health Service (NHS) place of work during NHS working hours, to two patients for the purposes of alleviating pain, but in circumstances in which he knew, or ought to have known, that he was not permitted or qualified to prescribe medication for non-dental purposes. The PCC found the allegations proved and imposed nine conditions upon M's registration for a period of one year. The conditions included that M must identify and appoint within one month a mentor, that he must meet his mentor on a regular basis and at least on a monthly basis, and that, at those meetings, he should develop a personal development plan addressing the prescribing, dispensing, and administering of medication for dental purposes. M was forty-nine days late in complying with appointing a mentor. At the time of the review hearing in June 2012, M had met his mentor only three times and, despite prompts, no personal development plan had been produced. Accordingly, the Committee determined that M's name be suspended from the register for twelve months. On appeal, M contended that the Committee's decision-making process was flawed and that its findings were unjust. The General Dental Council (GDC) submitted that the Committee was correct to suspend M's registration in circumstances in which it could not be satisfied that M would comply with a further period of conditions and in which he showed a continuing lack of insight into his failures. In dismissing M's appeal, His Honour Judge Allan Gore QC (sitting as a judge of the High Court) said, at [40], that the Committee considered that M had not given any adequate assurance that conditions would be complied with in the future. This was a conclusion that it was entitled to come to in view of the serial and continuing failure to abide by conditions in the past. The Committee was entitled to consider the public interest, the protection of patients, the maintenance of public confidence in the profession, and the declaring and upholding of proper standards of conduct and behaviour.

***Dutta v. General Medical Council* [2013] EWHC 132 (Admin)**

**65.08** Breach of interim order—suspension by fitness to practise panel upheld

A fitness to practise panel of the GMC found that D, whilst suspended by the Interim Orders Panel (IOP), had continued to provide clinical care and treatment to a patient. It imposed an order for suspension from the medical register for twelve months, upholding findings of misconduct and impairment of fitness to practise. In determining its overall sanction, the Panel had regard to the fact that D 'took a deliberate decision to provide care whilst suspended' and that there were other charges of inappropriate clinical treatment, which compounded the seriousness of his actions. The Panel gave 'very serious consideration to erasure', but decided that a twelve-month suspension was appropriate. Haddon-Cave J held the Panel's decision on sanction to be unimpeachable.

***R (Rhodes) v. Police and Crime Commissioner for Lincolnshire* [2013] EWHC 1009 (Admin)**

**65.09** Suspension of chief constable—Police Reform and Social Responsibility Act 2011—test

R, temporary chief constable for Lincolnshire, was suspended by the Police and Crime Commissioner pursuant to section 38 of the Police Reform and Social Responsibility Act 2011, which provides that the police and crime commissioner for a police area may suspend from duty the chief constable of the police force for that area. Regulation 10 of the Police (Conduct Code) Regulations 2012 provides for suspension only where 'the public interest requires' it. That carries the implication that the public interest leaves no other course open. When considering the test to be applied, the Court must bear in mind that the PCC has been appointed by statute to be the primary decision-maker and has been charged with considering the public interest. The Court must not interfere simply because it thinks it would have made a different decision if it had been the primary decision-maker. R's application challenging the defendant's decision was a rationality challenge and the *Wednesbury* unreasonableness test sets the bar high. R must show that the defendant's decision was irrational or perverse: *per* Stuart-Smith J, at [39]. On the particular facts, it was held that the decision would be quashed. It is a remarkable and disturbing feature of the case that there was no mention of R's character or reputation anywhere in the decision documents. This was a serious omission and R's twenty-seven years of unblemished service should have been weighed in the scales when considering the likelihood that he had engaged in a dishonest enterprise on the basis of a letter and an attendance note: at [106]–[109].

***Rahman v. Bar Standards Board* [2013] EWHC 4202 (QB)**

**65.10** See paragraph 17.39.

***Gainer v. Bar Standards Board* [2013] Lexis Citation 81**

**65.11** CPD—suspension for non-compliance

G failed to complete the prescribed twelve hours of continuing professional development (CPD) during 2009 and was suspended from practice until he had complied with the requirements. Allowing G's appeal, the Visitors (Sir Stephen Stewart, chairman) directed that, unless G complied with the requirements within a short period of time (three months), he should be suspended from practice for three months.

***Jones v. Bar Standards Board*, 12 March 2014 (unreported)**

**65.12** Barrister—no practice certificate—suspension justified

The appellant was found guilty of practising as a barrister, between 1 January 2011 and 13 July 2011, without having a practice certificate issued by the Bar Council. The panel imposed a sentence of suspension from practice for a period of one calendar month. The appellant appeal to the Visitors was dismissed. Despite the mitigation, the imposition of a period of suspension of one month was entirely appropriate in the circumstances of the case.

***R (Birks) v. Commissioner of Police of the Metropolis, Independent Police Complaints Commission (Interested Party)* [2014] EWHC 3041 (Admin), [2015] ICR 204**

**65.13** Police officer—inquest—resignation rescinded—legitimate expectation

The claimant was the most senior of four police constables who were involved in the arrest and restraint of R on 21 August 2008. R was diagnosed with paranoid schizophrenia and, at the time, had been discharged from hospital and was living in a supported hostel. The claimant was the driver of the police van, which transported R to Brixton Police Station. R, aged 40, collapsed and died soon after arrival at the police station. In February 2010, the Independent Police Complaints Commission (IPCC) concluded that there was no case to answer against any of the police officers involved. A full inquest into R's death took place in June and August 2012. In its narrative verdict, the jury was critical of the manner in which the police restrained R, and the inadequate care and attention that they gave him, in the light of his mental illness and declining physical condition. In the light of the jury's verdict, the IPCC commissioned an external review, which recommended that the IPCC reconsider the conduct of the officers involved in R's apprehension, restraint, and detention. On 14 November 2013, the IPCC decided to commence a fresh investigation. In the meantime, in June and July 2013, the claimant was offered, and accepted, the position of curate in the parish of Portslade, to commence in June 2014, following his graduation and ordination. The Commissioner of Police of the Metropolis was informed of this and, on 3 May 2014, the claimant's resignation was accepted by the police. However, later that month, a deputy assistant commissioner decided to suspend the claimant; a deputy commissioner decided that the claimant's resignation could not take effect, and rescinded the earlier acceptance of the resignation. In August 2004, an assistant commissioner decided that the suspension should be maintained and consent to resignation refused. In his judicial review proceedings, the claimant submitted that he had a substantive legitimate expectation that he would be permitted to resign in order to be ordained and take up his position as curate of Portslade parish, and that the defendant's decision to rescind the consent to resignation was unfair and unlawful. In dismissing the claim, Lang J said, at [40], that where a substantive legitimate expectation is made out, it will be unlawful for a public body not to give effect to it unless such a course of action is justified. The question to be asked is whether it would be so unfair as to amount to an abuse of power for a public body to do so: *R v. North and East Devon Health Authority, ex parte Coughlan* [2001] QB 213, and *Paponette v. Attorney-General of Trinidad and Tobago* [2012] 1 AC 1. In answering that question, the courts have variously stated that there must be an overriding interest, or an overriding policy imperative, or a sufficiently powerful supervening factor that outweighs and overrides the expectation or requires or justifies a departure from it. Her Ladyship said, at [48], that, in summary, the conclusion she had reached was that the public interest in ensuring that the claimant remained subject to police disciplinary jurisdiction in such a serious case justified the defendant departing from the representation previously made that he would not be prevented from resigning. The decision of the assistant commissioner was soundly reasoned and proportionate. There was an ongoing criminal investigation, and the interference with the claimant's rights under the European Convention on Human Rights (ECHR) was justified by pressing social needs and was proportionate to the legitimate aim pursued: at [70].

### *Jasinarachchi v. General Medical Council* [2014] EWHC 3570 (Admin)

**65.14**
Trainee doctor—suspension of registration—effect

At the conclusion of a fitness to practise hearing on 13 March 2014, the panel determined that J's fitness to practise was impaired by reason of misconduct and that his registration should be suspended for a period of six months. Between December 2011 and April 2012, J was employed as a ST1 general practitioner (GP) trainee at a medical practice in Northampton. On 2 March 2012, he completed a cremation medical certificate in respect of patient A. He failed to attend the funeral directors to examine patient A's body, falsely stated on the certificate that he had seen the body and had performed an external examination, and falsely stated in a telephone call to the surgery that he had been to see patient A's body at the funeral parlour. He failed to alert the surgery or anyone connected to the care of patient A that he had not seen patient A's body, and he failed to make alternative arrangements for the medical certificate to be properly completed. In January 2012, J completed a death certificate for patient B in which he recorded the incorrect cause of death and failed to sign the death certificate book stub. On appeal against sanction, Stewart J said that, on the evidence and information before the panel, the sanction of six months' suspension was not excessive or disproportionate. However, matters had since come to light concerning the practical consequences of suspending a trainee doctor's registration. J, in a witness statement, said that his understanding of the panel's decision was that his six months' suspension was intended to have a temporary impact on his work as a doctor and not to prevent him from continuing to train as a GP on a long-term or permanent basis. He assumed that he would be able to return to his training after the suspension had been served. He had since found out that the effects of the sanction would be that his national training number (NTN) would be removed when the suspension came into force. The effect of this was that his current training would cease at that point despite the fact that he had only a few more months to go before finishing it. The likely effect of the suspension sanction would be to bring his current training as a doctor to an end with no certainty that he would be able to enter an alternative programme once the suspension had been served. J's witness statement was supported by a letter from the GP Dean of Health Education East Midlands. J accepted that he was seeking to adduce fresh evidence. The principles outlined in *Ladd v. Marshall* [1954] 1 WLR 1489 are that (a) the evidence could not have been obtained with reasonable diligence for use at trial, (b) the evidence must be such that, if given, it would probably have an important influence on the result of the case, although it need not be decisive, and (c) the evidence must be such as is presumably to be believed—that is, it must be apparently credible, although it need not be incontrovertible. In ruling in favour of allowing the additional evidence and in favour of J's case, his Lordship said that although J could not rely upon the first principle in *Ladd v. Marshall*, J's culpability was not particularly high in this regard. He was represented. However, there was no evidence to suggest that anybody appreciated the possible consequences of his suspension. It is correct that the Gold Guide then in force made it clear that the NTN would be given up if a trainee were suspended and that (at that stage) there would be a right of appeal. Nevertheless, lack of compliance with this first principle was not determinative. There was no suggestion that the Postgraduate Dean in any way alerted J or his lawyers to the consequences and no evidence that the panel were aware of them. As to the second principle in *Ladd v Marshall*, it was difficult for the Court, on the basis of the evidence provided, to quantify the risk that J's GP speciality training may be at an end if he were to be suspended. Looking at the evidence, there was a real risk that this would occur. Nobody was aware of any precedent of a suspended trainee applying to get back on the register and what the prospects of success were or were not. In remitting the case for further consideration on the issue of sanction, his Lordship said that it would, of course, be open to the panel to come to the same conclusion—that is, that J should be suspended. That said, the Court did regard the fresh evidence as probably having an important influence on the result of the case. It may indeed not be decisive, but that was a matter for a properly informed panel to decide.

### *HK v. General Pharmaceutical Council* [2014] CSIH 61

**65.15** Pharmacist—conviction—erasure—whether twelve months' suspension with indication of extended period permissible

On 13 May 2011, the Glasgow Sheriff Court fined HK, a registered pharmacist, for assaulting his wife. He failed to notify the registrar of the General Pharmaceutical Council within seven days of that conviction. On 8 June 2012, the Glasgow Sheriff Court sentenced HK for two further incidents of domestic violence. The Council commenced disciplinary proceedings. On 27 June 2013, HK appeared before the Fitness to Practise Committee, which found his fitness to practise as a pharmacist to be impaired and directed that his name be removed from the register of pharmacists, in accordance with article 54(2)(c) of the Pharmacy Order 2010. The Committee considered that HK's conduct was fundamentally incompatible with continued registration as a pharmacist and that public confidence in the profession demanded no lesser sanction than removal from the register. HK appealed to the Inner House, Court of Session, which quashed the direction to remove his name from the register and remitted the matter to the Committee, with a direction to consider an 'intermediate sanction' in the form of suspension for twelve months, with an indication that that could be extended for a longer period. In giving the judgment of an Extra Division of the Inner House, Lord Drummond Young noted that suspension for twelve months, the limit imposed on a period of suspension by article 54(2)(d), was considered an insufficient sanction to mark the gravity of the appellant's behaviour. However, no reference was made to a middle way, in the form of suspension for twelve months with an extension thereafter for a further twelve months, and possibly a further suspension for twelve months beyond that. Such suspension would be a significantly lesser sanction than removal from the register, which prohibits any application for re-registration for a period of five years. If an indication is given by a committee that a suspension should be extended beyond the initial twelve months—for, say an additional twelve or twenty-four months—that will not bind the later committee on review, but the later committee will be obliged to respect the indication and, if it departs from it, it will be expected to give reasons for doing so. This provides an intermediate sanction, but at the same time respects the freedom of the later committee to deal with changing circumstances, if that is appropriate: at [16]–[17].

At time of writing, the decision of the Extra Division is the subject of an appeal by the Council to the Supreme Court. The principal issue that the Supreme Court is invited to decide is whether there is an intermediate sanction open to a Fitness to Practise Committee, which allows the Committee to indicate that the period of suspension should, on review by a later Committee, exceed the statutory maximum of twelve months.

*Solicitors Regulation Authority v. Uddin* [2014] EWHC 4553 (Admin)

**65.16** Solicitor—conditional fee agreement with expert—suspension

The SDT ordered the respondent solicitor to pay a fine of £2,000 in circumstances in which he had knowingly instructed an incompetent expert witness under an objectionable conditional fee agreement under which the expert would receive payment only on success. On an appeal by the SRA, the Divisional Court held that the appropriate sanction was suspension for twelve months.

*Pool v. General Medical Council* [2014] EWHC 3791 (Admin)

**65.17** Doctor—expert witness—lack of expertise

The appellant was a psychiatrist. In August 2011, he accepted instructions to act as an expert witness in proceedings before the Health Professions Council (HPC). The case involved consideration of the fitness to practise of a paramedic, A, who had been diagnosed as having a personality disorder and post-traumatic stress disorder, in part as a result of abuse suffered during childhood. The appellant prepared a report on A and the question of her fitness to practise. A objected to the evidence being received, on the grounds that the appellant was not an expert. That objection was eventually dealt with at a hearing in March 2012. The HPC concluded that the appellant did not have sufficient expertise in the field of personality disorders to qualify him as an expert and decided not to admit the appellant's evidence. A subsequently referred the appellant to the GMC. The panel imposed a sanction, suspending the appellant's registration as a doctor for a period of three months. Allowing the appeal on sanction, Lewis J said that the direction to suspend the appellant's name from the register for three months was flawed and disproportionate in the circumstances. That direction was, accordingly, to be replaced by a direction that the appellant's registration be subject to a condition that, for three months, he should not accept instructions as an expert witness in fitness to practise proceedings.

## C. Other Relevant Chapters

**65.18**

Conditions of Practice Orders, Chapter 11
Interim Orders, Chapter 38
Review Hearings, Chapter 59
Sanction, Chapter 60

# 66

# UNDERTAKINGS AND WARNINGS



## A. Legal Framework

**66.01** Examples of undertakings or warnings after consideration by case examiners or the investigation committee include:

General Pharmaceutical Council (Fitness to Practise and Disqualification etc) Rules 2010, rule 10 (undertakings)
General Optical Council (Fitness to Practise) Rules 2013, rule 14 (warnings)
General Medical Council (Fitness to Practise) Rules 2014, rules 8–11

## B. Disciplinary Cases

### *Nandi v. General Medical Council* [2004] EWHC 2317 (Admin)

**66.02** Serious professional misconduct—offer of undertakings as alternative to finding of misconduct

N was a general practitioner (GP) in North London and appealed against a finding of serious professional misconduct by the Professional Conduct Committee (PCC) of the General Medical Council (GMC). One of the arguments raised on the appeal was that the finding of serious professional misconduct was unnecessary, because N was prepared to give undertakings. Collins J said that:

> 2. ... [Counsel for the doctor] did not pursue that ground, in my view correctly, because it obviously is important and is in the public interest that the Committee should make a finding of serious professional misconduct if the evidence warrants that finding and, equally, that the doctor should be acquitted if that is the correct finding on the evidence. It is not right that a doctor should be able to avoid a finding of serious professional misconduct by giving undertakings. There are problems with enforcement of undertakings and, so far as the GMC is concerned, in keeping records of such undertakings because obviously it is in the interests of the public that they should know whether there exist such undertakings in the case of individual medical practitioners.... [A]n undertaking is not a proper disposal when there are charges of serious professional misconduct. The only circumstances in which undertakings may

> be appropriate, and only then in the rarest of cases, would be where there have been interim orders made and the court is asked to extend an interim order....

On the facts, Collins J allowed the appeal against the Committee's finding of serious professional misconduct.

### *Briggs and Awoloye-Kio v. Law Society* [2005] EWHC 1830 (Admin)

**66.03** Solicitor—conveyancing—solicitor's undertaking

K, the second appellant, practised as Awoloye-Kio & Co and operated from 165–173 Stockwell Road, Brixton. The Law Society preferred twelve complaints against K, of which two were withdrawn at the start of the hearing. Among the complaints was one from South Pacific Mortgage Ltd to the effect that the firm had failed to register four charges in respect of four conveyancing transactions in which South Pacific had instructed the firm to act on its behalf as secured lenders and in which the firm was also acting for the purchaser/borrower. K admitted that his firm had given standard-form certificate of title undertakings in all four matters. Three of the undertakings had been signed by K personally and one was signed by T, a former solicitor and an employee of the firm at that time. K was the partner with responsibility for T. K accepted that he had not performed the undertakings, but contended that it had been impossible for him to do so because of T's frauds. The Solicitors Disciplinary Tribunal (SDT) held that the breach of an undertaking was a matter of strict liability, for the protection of the public, and that if a solicitor had any fear that he might not be able to honour an undertaking, it was open to him to enter a conditional undertaking. The Tribunal indicated that the circumstances in which K failed to honour these undertakings would be taken into account in mitigation. Upon hearing that ruling, K pleaded guilty and the Tribunal ordered that he be suspended from practice as a solicitor for five years. In dismissing K's appeal, Smith LJ (with whom Henriques and Simon JJ agreed) said:

> 34. In my view, the Tribunal was wrong to hold that the breach of an undertaking is always a matter of misconduct. It almost always will be, but there is room for a finding of no fault in exceptional circumstances. Here, as I have explained, there was, in my view, fault on [K's] part.
>
> 35. Having said that, the Tribunal was right when they said that the breach of an undertaking is a serious matter. Undertakings are the bedrock of our system of conveyancing. The recipient of an undertaking must be able to assume that once given it will be scrupulously performed. If property purchasers and mortgage lenders cannot have complete confidence in the safety of the money they put into the hands of a solicitor in the course of a property transaction, our system of conveyancing would soon break down. The breach of an undertaking given by a solicitor damages public confidence in the profession and in the system of undertakings upon which property transactions depend. Accordingly, if fault is shown, as it has been here, the matter must be treated seriously.

### *Lutton v. General Dental Council* [2011] CSIH 62

**66.04** Warning letter—adequate reasons

On 29 September 2011, the Inner House, Court of Session, dismissed an appeal by the General Dental Council (GDC) from the order of the Lord Ordinary, Lord Doherty, remitting the allegation against the petitioner to a differently constituted investigating committee of the GDC: [2011] CSOH 96. L, a dental surgeon and sole partner of a dental practice, faced allegations relating to the treatment of a patient. It was alleged that L had failed to appraise the patient of the details of her course of treatment, particularly with regard to extractions, had failed to discuss alternative treatment options with the patient, and had advised the patient that the periodontal treatment that she required was not available on the National Health Service (NHS). The investigating committee of the GDC decided that L should not be summoned to attend an inquiry held by a practice committee of the Council. However, the investigating committee decided to issue L with a written warning. The committee warned L that, in future, he should advise patients 'of all treatment options available', 'ensure that information provided to the patient is factually correct and does not have the potential to mislead', and 'ensure that you undertake proper assessments of patients at all times'. The Inner House said that there was no mention specifically as to what options, in the present case, L should have set out that he did not set out. L was, justifiably, left asking what information provided to the patient in question was factually incorrect or which had the potential to mislead. No explanation was given why L should be warned to undertake proper assessments and what facts were found to be established by the investigating committee that made this warning necessary. Lord Clarke, delivering the opinion of the Inner House, said that:

> 15. . . . [O]nce it has been concluded that the appropriate way forward is to issue a warning letter, . . . the warning must convey to its recipient why it was considered necessary and, where that involves resolving [a] factual dispute between the practitioner and the patient why, in broad terms, that dispute has been resolved against the practitioner if that is in fact what has happened. The warning letter should state with some precision, however briefly, what it is that has been identified in the practitioner's conduct that requires to be addressed and why it is thought necessary for him to address such matters as a result of the allegation received rather than leave the issue simply "in the air", as we consider has happened in the present case. If the warning issued by [the Council] is to be efficacious then its recipient must be told clearly why his conduct or practice has been held to be deficient so that the steps required to be taken by him to address the deficiencies in question can be understood and seen to be justified. Otherwise the warning will be ineffective. In a nutshell, and as the Lord Ordinary said at paragraph 55 of his opinion: "Reasons need not be elaborate or lengthy, but they should tell the parties in broad terms why the decision has been reached". We consider that the reasons given by [the GDC] in the present case fall short of that requirement.

***Council for Healthcare Regulatory Excellence v. Nursing and Midwifery Council and Grant* [2011] EWHC 927 (Admin)**

**66.05** See paragraph 33.15.

***Depner v. General Medical Council* [2012] EWHC 1705 (Admin)**

**66.06** Doctor—performance assessment—undertaking to agree further assessment—breach of undertaking—impairment—suspension—purpose of undertaking

D was a medical practitioner who previously worked as an out-of-hours GP. D came to the attention of the GMC as a result of a complaint by a patient in September 2006 that D had failed to diagnose an ectopic pregnancy and that the patient subsequently had been hospitalized. In February 2007, the GMC invited D to undergo a performance assessment, which she agreed to do. The assessors concluded that D's performance was unacceptable in relation to a number of matters, including assessment of patients' condition, providing or arranging treatment, record-keeping, educational activities, and working within rules and regulations. The assessors also regarded D's performance as giving cause for concern in the additional areas of communications with patients, listening to patients, respecting their views and providing comprehensive information, and paying due regard to efficiency and use of resources. The assessors recommended that D should not be permitted to practise unsupervised. In January 2008, D was invited to agree to undertakings under rule 10(2) of the General Medical Council (Fitness to Practise) Rules 2004. She agreed two undertakings, one of which was a requirement that she would undergo a further performance assessment on a date specified by the GMC's examiner. By letter dated 14 May 2009, D was invited to undergo such an assessment. She did not reply directly to that invitation, and the GMC's registrar took the view that D had refused to undergo an assessment pursuant to her undertaking and accordingly that she was in breach of her undertaking. The registrar decided to refer her case to a fitness to practise panel under rules 10(8)(b) and (c) of the 2004 Rules. The fitness to practise panel considered D's case over five days between 23 and 27 August 2010. D did not attend and was not represented. She did, however, submit a substantial number of documents, which the panel considered, and she also made extensive written submissions. The panel found that D was in breach of the undertaking and that the allegation was proved, and that D's fitness to practise was impaired by reason of misconduct and deficient professional performance. It determined to suspend D's registration for a period of nine months. In dismissing D's appeal, Stadlen J said:

> 63. In my judgment, the undertaking clearly required [D] to notify, if invited to agree to an assessment, her agreement to such an assessment so that assessors could be appointed. The undertaking was capable of being breached either by a refusal or failure to co-operate after a team had been appointed and dates had been specified, or by refusing or failing to consent to an assessment after being invited to do so, and/or by a refusal to provide or a failure to provide the information requested by the GMC necessary to enable the assessors to be appointed....

Stadlen J said that criticism was made by D that the panel failed to say why she refused or failed to agree to the assessment taking place. It was said that, in her email correspondence, D indicated that she did not know why the GMC wanted an assessment and that, without making any findings as to her reasons for refusing, the panel could not properly find that there was misconduct such as to impair fitness to practise. Stadlen J said that there was nothing in that point. The reason why D did not comply did not detract either from the fact that she was in breach of the undertaking or from the gravity of that breach. As appeared from the correspondence, D challenged or queried the need for an assessment. But that was not the point. The point was that this was an undertaking that D had given voluntarily. The fact was that, whatever view she took about the necessity or lack of necessity for a reassessment, the GMC took the view that it was necessary. The whole point of the undertaking was not that a request for a reassessment would be subject to renegotiation, or consent to be given or withheld at will by D at such point as a request was made. The point of the undertaking was that it was just that: if, in its discretion, the GMC considered that a reassessment was appropriate or necessary, D agreed in advance that she would cooperate. Stadlen J continued:

> 68. In my judgment, having regard to the fact that the reassessment was for the purpose of measuring her performance and either the deficiency or otherwise of it, and having regard to the obvious impact of a failure to agree to such a report on patient safety, it is, in my judgment, perfectly obvious that a unilateral refusal or failure to participate when requested in such an assessment, an undertaking to that effect having been given, is capable of constituting both misconduct and an impairment of fitness to practise by reason of such misconduct.

As to sanction, where a practitioner gives an undertaking that he or she will agree to reassessment if invited to do so by the GMC, a failure to comply when asked should be treated as a potentially very serious matter and certainly one that, on the facts of this case, amply justified a nine-month suspension.

### *Arunkalaivanan v. General Medical Council* [2014] EWHC 873 (Admin)

**66.07** Panel's finding reversed—no impairment—case remitted to panel to consider warning

The primary findings of fact made by the fitness to practise panel that A had conducted a breast examination of the patient in an inappropriate manner were unassailable. It was conceded that, the Court having reversed the panel's finding that A's conduct was sexually motivated, the GMC could not contend that A's fitness to practise was impaired on the basis of the examination alone. The matter would be remitted to a differently constituted panel to consider whether it would be appropriate to impose a warning under rule 17(2)(l) of the General Medical Council (Fitness to Practise) Rules 2004. It could not be said that a warning could never properly be given in this case. Whether a warning should be given was a matter for the panel, having regard to the Indicative Sanctions Guidance and to any further evidence and representations to be offered by the parties. Remitting the matter to the panel was not to be taken as an encouragement that it should issue a warning. That was a matter for the panel after considering all relevant matters in accordance with the Guidance.

See paragraph 30.07.

***R (Woods and Gordon) v. Chief Constable of Merseyside Police* [2014] EWHC 2784 (Admin), [2015] ICR 125**

**66.08** Service Confidence Procedure—action plan—continuation of procedure—judicial review

In April 2011, W, a police constable, was made the subject of a service confidence procedure (SCP) by Merseyside Police. Police forces in general, and Merseyside Police in particular, have SCPs. The main aims are to address loss of confidence by the force in any particular individual when serious concerns arise as to his or her suitability to perform a specific role or duty. The primary objective is to provide a framework that helps the force to deal with issues relating to loss of confidence. W was told that serious concerns over his integrity arose as a result of an anti-corruption investigation. He was provided with an action plan to enable him to regain the confidence of the defendant. The aim was for final restoration of confidence by April 2012, with regular three-monthly reviews in that twelve-month period. The defendant had not lifted the SCP despite the fact that W had met the requirements of the action plan. G, an inspector, was also the subject of an SCP with an action plan. The claimants sought to challenge the dismissal of their appeals against the continued imposition of the SCPs against them. In dismissing the claims, Stewart J held that an SCP is amenable to judicial review, but that the threshold for judicial interference is very high, particularly in circumstances in which, as here, reasons for it are subject to a decision that cannot be disclosed owing to public interest immunity. In the circumstances of these cases, the threshold had not been reached.

## C. Other Relevant Chapters

**66.09**

Conditions of Practice Orders, Chapter 11
Consensual Disposal, Chapter 12
Deficient Professional Performance, Chapter 16
Disposal without Oral Hearing, Chapter 20
Sanction, Chapter 60
Suspension, Chapter 65
Voluntary Erasure, Chapter 68

# 67

# UNREPRESENTED PRACTITIONER



## A. Legal Framework

**67.01** Examples include:

Nursing and Midwifery Council (Fitness to Practise) Rules 2004, rule 20 (where the registrant is not represented, he or she may be accompanied and advised by any person; a person who represents or accompanies the registrant shall not be called as a witness at the hearing)

General Pharmaceutical Council (Fitness to Practise and Disqualification etc) Rules 2010, rule 40 (where the person concerned is not represented, he or she may be accompanied and advised by a supporter, but the supporter must not be a member of the council or one of its statutory committees or a witness)

Police (Conduct) Regulations 2012, regulation 6 (officer concerned may choose a police officer, a police staff member, or a person nominated by his staff association, who is not otherwise involved in the matter, to act as his police friend)

General Medical Council (Fitness to Practise) Rules 2014, rule 33 (practitioner may be represented at the discretion of the committee or panel by a member of his family or other person)

## B. Disciplinary Cases

### *R (Katti) v. General Medical Council* [2004] EWHC 238 (Admin)

**67.02** Practitioner acting in person—lack of funds for legal representation

Pitchford J dismissed K's appeal from the decision of the General Medical Council (GMC) ordering the removal of his name from the register on the grounds that the Professional Conduct Committee (PCC) ought to have adjourned the hearing to enable K to raise funds for legal representation. By the time of the hearing, K was well aware of the issues that would face the GMC, having attended on previous occasions interim meetings that considered the conditions that ought to be attached to his right to practise. He had been represented on at least one occasion by a firm of solicitors, which had put in written submissions in some detail. Furthermore, he was in possession of the facts, since the bundles were served on him well in advance of the hearing. K said that he did not attend because he did not have confidence that his representations would make any difference to the hearing that would take place. That, it seemed to the learned judge, in a case of an intelligent man whose right to practise his profession was at stake, was hardly a reason for not attending the hearing.

### *R (Compton) v. General Medical Council* [2008] EWHC 2868 (Admin)

**67.03** Legal assessor—Absence of practitioner—role of legal assessor

The claimant's appeal did not concern the panel's decision to proceed in his absence, but rather the fact that, in his absence, his interests should have been better protected by elucidation to the panel of the weaknesses in the GMC's case. With necessary adaptations for the nature of the proceedings, it was agreed that the principles in *R (Hayward)* [2001] 1 QB 862, [22], *per* Rose LJ at point (6), should apply to any tribunal considering such matters. They apply to a hearing before a fitness to practise panel just as they do to a criminal trial. Pitchford J said, at [33]:

> The legal assessor is not a judge, he is an adviser to the panel on matters of law (see paragraph 2, the General Medical Council (Legal Assessors) Rules 2004). The legal assessor's duty as a legal advisor embraces, in my judgment, the responsibility to inform the panel of the need for vigilance in circumstances such as these, namely, in the absence of the doctor identifying points which might be of assistance to him. It does not, in my judgment, embrace a need to sum up the evidence to the panel.

The Court held that the legal assessor and the panel had drawn attention to documents that were favourable to the claimant, and fulfilled their obligations to question witnesses who might assist the claimant's case. The Court was also satisfied that the panel was reminded that no adverse inference should be drawn from the fact that the claimant did not attend the hearing.

See further paragraph 44.19.

### *Yusuf v. Royal Pharmaceutical Society for Great Britain* [2009] EWHC 867 (Admin)

**67.04** Absence of practitioner—role of committee

The appellant complained that even if the respondent's Disciplinary Committee were correct in proceeding with the inquiry in his absence, it erred in failing either to examine or to examine adequately a particular witness who stood to gain and whose evidence was crucial to the findings against the appellant. The Court held that it was not the function of the Committee to ensure that the witness was 'adequately and thoroughly examined/cross-examined by the committee', although in fact the Committee did ask the witness some questions, and took proper and sufficient steps to ensure that the appellant's case was put to him. Munby J said, at [46]:

> The committee did all that could appropriately be expected of it. It tested to an appropriate extent the witness's account. The fact that its probing of his evidence may have been less vigorous or searching than might have been expected if the witness had been cross-examined by the appellant or by some advocate instructed on his behalf was, in the judgment of the court, neither here nor there. That was not the function of the committee. The committee had very well in mind the factors that weighed in the appellant's favour, but the simple fact is that it believed [the witness] and accepted his account, as it was entitled to. There was no unfairness in the process adopted by the committee.

### *Thiruvengadam v. General Medical Council* [2010] NIQB 123

**67.05** Practitioner acting in person—complex issues—lack of legal representation—adjournment

The appellant was employed by the Western Area General Hospital as a staff-grade officer in accident and emergency (A&E) medicine. In 2003, the hospital suspended him from clinical duties; in January 2004, it referred him to the National Clinical Assessment Service (NCAS) after concerns about his clinical practice surfaced. He subsequently agreed to undergo a GMC assessment in July 2009 at the GMC Clinical Skills Centre in London. The assessment comprised a knowledge test and objective structured clinical examinations (OSCEs). The assessment team considered that the appellant's performance was deficient, that it was not amenable to retraining, and that the appellant was not fit to practise. At a fitness to practise hearing, the panel refused T's application for an adjournment on two grounds. The first ground related to the disclosure of material in relation to the OSCE procedure. Forty-two of the medical records used had been disclosed to T through his former solicitors, these being the ones that were available to the GMC. The second ground related to T's lack of legal representation. The Medical Defence Union advised T only recently that it was prepared to fund 25 per cent of the cost of his defence. Gillen J allowed the appeal and remitted the matter to the panel for further consideration. His reasons were, first, that there could be no doubt as to the seriousness of the outcome for T in proceedings of this nature. An adverse finding would inevitably deprive him of his ability to practise professionally as a clinician. His case for representation was compelling where there was a live issue about the nature of documents to be disclosed to him. Second, the burden of conducting a case of this kind without legal representation may have seemed to T so insuperable that there was considerable mitigation for his apparent feeling of helplessness and failure to appear. It was virtually inconceivable that he could have obtained the benefit of a solicitor and/or counsel during the short period between the decision of the Medical Defence Union and the start of the hearing. Third, it was a draconian decision to proceed to a full and complex hearing without giving T an opportunity to obtain legal assistance. He was already suspended from practice, so the public was not endangered by a further delay. The request for six months made by him may have been ambitious, but a shorter period should have been contemplated by the panel in order to afford him the opportunity to obtain skilled legal representation.

### *Nagiub v. General Medical Council* [2011] EWHC 366 (Admin)

**67.06** Practitioner acting in person—behaviour—discourtesy during hearing—distraction for tribunal—essential issue being the allegation

N represented herself during the course of fitness to practise proceedings based on misconduct and deficient professional performance. She was suspended from the medical register for twelve months. In its determination on impairment, the panel stated that it was most concerned about N's conduct throughout the hearing, which had been rude, insulting, racist, abusive, and at times bullying and intimidating. She had abused every witness who gave evidence on behalf of the GMC, the GMC legal team, the panel, and the legal assessor. The panel stated that her behaviour, in her dealings with the GMC and her conduct before the panel, had consistently contradicted the principles laid down in *Good Medical Practice* and the panel regarded this conduct 'most seriously'. N's notice of appeal (for which she had received some legal assistance) claimed that the panel gave undue weight to the manner in which she conducted herself before the panel when considering her medical performance and professionalism as a doctor in her day-to-day working life. Foskett J, in dismissing N's appeal, said that whilst litigants in person are not as well versed as an established practitioner in making a strong point in cross-examination in a forceful, yet not intrinsically offensive way, nevertheless, gratuitously offensive remarks are not tolerated. Leaving aside any questions of propriety, they can be distracting and irritating for a tribunal endeavouring to get to the bottom of contentious issues. In the instant case, some caution was required before expressing an observation of this kind in the way that it was and in the context that it was. The essential issue in the proceedings was how the appellant reacted in the daily workplace of medical practice. The proceedings before the panel did not constitute such a setting. That was not a defence to discourtesy and gratuitous offence, but merely an observation of the obvious. The panel's conclusion that it regarded N's conduct 'most seriously' was such that one might expect before the imposition of a penalty rather than a judgment of the nature under consideration. The most logical stage at which this consideration might have entered the panel's reasoning was to say that N's behaviour before the panel to some extent confirmed the views of the consultant psychiatrists in the case about her delusional disorder. Had that been said, then the observations would have been unobjectionable. The question was whether the defect (as the learned judge saw it) in the process of reasoning was such as to render the whole process from that point onwards flawed. The learned judge said that he was quite satisfied that it did not. There was ample other material to sustain the view that N's fitness to practise was impaired by the matters previously found.

***Musonza v. Nursing and Midwifery Council* [2012] EWHC 1440 (Admin)**

**67.07** Nurse—case dealt with at meeting on paper without oral evidence or representations—lack of indication in NMC's letter of need for legal advice

M qualified and became registered as a nurse in April 2007, following her graduation from the University of Luton with a diploma in adult nursing. In July 2009, M pleaded guilty to an offence of possessing a false identity document—namely, a passport—with intent and at Snaresbrook Crown Court was sentenced to twelve months' imprisonment. M appeared at Wood Green Crown Court on 16 November 2009 and pleaded guilty to dishonestly obtaining bursary payments of £31,000 using the passport, which she knew was forged. M received a consecutive sentence of six months. On her release from custody, M successfully appealed against deportation to the First-Tier Tribunal (Immigration and Asylum Chamber). The Nursing and Midwifery Council (NMC) decided to bring disciplinary proceedings based on M's two convictions, and M received notification of the referral of the allegation to the Conduct and Competence Committee (CCC) in a letter dated 9 December 2010. The letter invited M to indicate whether she wished her case to be dealt with at a hearing (which would involve a public hearing with the parties being present) or for the case to be considered on the papers at a meeting (at which neither the NMC nor the registrant would attend or be represented). M returned the case management form indicating that she did not intend to attend a hearing, did not require her case to be dealt with at a hearing, did not rely on any written statements, and did not intend to rely on any testimonials. She asked for the matter to be heard in private. The matter was referred to a meeting to be decided on the papers. On 4 March 2011, a panel of the CCC found at a meeting that M's fitness to practise was impaired and imposed a striking-off order removing M's name from the register. His Honour Judge Anthony Thornton QC, sitting as a judge of the High Court, allowed M's appeal and directed that the matter should be redetermined by the NMC. The Court said that M was one of a minority of nurses who were not members of the union that would have provided her with free legal representation. Although her background would not have been known to the NMC in any detail, such registrants needed clear and realistic advice as to the vital need for legal advice and representation, and the means of obtaining it, and as to the necessity for them to seek an oral hearing and to be present at that hearing. The NMC's letter and form did not clearly and fully address the need for M to be advised that she should inform the NMC if she wished to be legally advised or represented, but could not afford or obtain it. Further, the legal assessor did not give the Committee any guidance as to the mitigation put forward by M or of any relevant matters that could be considered as pointing to a suspension rather than M being struck off.

***Vaghela v. General Medical Council* [2013] EWHC 1594 (Admin)**

**67.08**

Practitioner acting in person—behaviour during the hearing—assessment by panel—relevance to impairment

V, a doctor, was found guilty by the fitness to practise panel of posting malicious and untrue allegations about A on the Internet and making unfounded allegations of professional misconduct against S to the Solicitors Regulation Authority (SRA). In its findings, the panel stated that it was concerned to note that V, who represented himself, became inappropriately hostile and aggressive when cross-examining witness A and when challenged in cross-examination by the representative for the GMC. The panel stated that it had been very careful to make allowances for the fact that V was representing himself, but took the view that his behaviour 'reflects precisely the sort of behaviour that is alleged by the GMC'. In dismissing V's appeal, Sales J held that the assessment of witnesses was a matter laying well within the panel's proper power of evaluative judgment. The panel gave very careful consideration to the relevant questions—namely, whether V's conduct that led to the charge was easily remediable, whether it had been remedied, and whether or not it was likely to be repeated. V, in his submission, maintained, first, that the panel did not give due weight to the fact that the matters of which complaint was made related to a breakdown in family relations, which was historic and unlikely to be repeated, and second, that the panel erred by taking into account his demeanour and the way in which he handled the conduct of the proceedings before the panel. In rejecting V's submissions, Sales J said:

> 48. I do not consider that any of these submissions can be supported. So far as the first is concerned, it is plain from the reasoning of the panel, as set out above, that the panel gave very careful consideration to the relevant questions: namely, whether [V]'s conduct which led to the charge was easily remediable, whether it had been made and whether or not it was likely to be repeated. The panel had well in mind [V]'s submission that the relevant matters had arisen out of an historic breakdown in family relations, but, for the reasons set out with care by the panel, it concluded that [V] had continued his course of conduct over a period of time, had no insight into his actions, had conducted himself in a way which involved lying to the panel and so forth, and that it could have no confidence that, looking forward, the pattern of behaviour outlined would not be repeated in the future. All this was by way of contrast with [V]'s stance in relation to the matter of maintaining patient confidentiality. In my view, the particular matters to which the panel had had regard in assessing whether [V] had any insight into his actions in relation to the matters referred to were all matters which the panel was fully entitled to take into account in making its assessment as to the seriousness and currency of the impairment of fitness to practise in relation to [V] which it had to consider.
>
> 49. As to the second point, I do not consider that the panel acted in any way unfairly or improperly in having regard to [V]'s demeanour and conduct in handling the proceedings before it. It sought to make due allowance for the fact that he was representing himself. But it is the position that a person representing himself can do so in a moderate and restrained way without resorting to lies and unjustified attacks on other individuals, and in my view there was no unfairness involved in the panel having regard to the way in which [V] conducted himself in the course of the proceedings. Indeed, in relation to assessing matters such as his insight into his actions, it would have been extraordinary if the panel had not had regard to material of this kind.

***Rehman v. Bar Standards Board*, 19 July 2013 (unreported)**

**67.09**

See paragraph 25.32.

### *Amao v. Nursing and Midwifery Council* [2014] EWHC 147 (Admin)

**67.10** Nurse acting in person—lack of explanation of meaning of impairment and insight

Walker J observed that where a person facing disciplinary proceedings is unrepresented, the tasks arising at the hearing for the unrepresented person, and for the legal assessor, are unlikely to be easy. The appellant was not given a fair opportunity to address the panel's concern that she had no insight with regard to her future professional obligations. The feature of impairment was never properly addressed, and there was nothing to suggest that the appellant had any understanding of how important it was for her to give evidence and make submissions about her ability to take steps to ensure that, in future, there would be no incident similar to what happened previously.

See further paragraph 37.17.

### *Moseka v. Nursing and Midwifery Council* [2014] EWHC 846 (Admin)

**67.11** Procedural matters—approach of Court

In dismissing M's appeal, Green J said, at [23], that she had appeared in person. She had had no legal advice or representation since the hearing below. It seemed to his Lordship that, in such circumstances, which sadly are of rapidly increasing incidence, the Court cannot adopt an unduly technical or procedural approach to the appeal. M had done her best to complete court forms properly and to produce skeleton arguments. However, in order to understand her case, his Lordship had been required to engage in a significantly more detailed exercise than would have been necessary had she been legally represented. It was quite plain that, with access to even modest legal advice, court time and overall costs would have been saved.

### *D v. Conduct and Competence Committee of the Nursing and Midwifery Council* 2014 SLT 1069

**67.12** Legal representative failing to prepare case—practitioner appearing in person—complex issues—inadequate representation

The allegation of misconduct against the appellant was that, between May 2011 and October 2011, while employed as a senior sister on the intensive care unit at Great Western Hospital, Swindon, she removed medication that belonged to the Trust on eleven listed occasions and did so dishonestly, in that she did not have permission to remove the medication. The eleven listed occasions all related to 50 milligram tablets of Cyclizine, a drug that is used to alleviate symptoms of sickness and nausea. The NMC's case was wholly circumstantial in nature. The records of those on duty on the shifts when tablets went missing were analysed to discover who had an opportunity to remove the tablets from the ward on each occasion. Members of staff were eliminated one by one. Eventually, following a lengthy process, only the appellant was left. The process of elimination was based on the assumption that the same person was responsible for the removal of tablets on each of the recorded occasions. It was also based on the assumption that all of the tablets had been wrongfully taken. The appellant appealed against a striking-off order on two grounds: first, that her representation in connection with the proceedings before the CCC was inadequate; and second, that the Committee erred in law in allowing the methodology relied upon by the NMC to identify the person said to be responsible for taking drugs from the ward where the appellant worked. An Extra Division of the Inner House allowed the appellant's appeal on both grounds and quashed the decision of the Committee. In giving the opinion of the Court, Lord Drummond Young said, in relation to the appellant's representation before the Committee, that the hearing took ten days, spanning 10–14 June and 2–6 September 2013. It would normally have run from 9 a.m. to 5 p.m. The appellant travelled by train from Maybole to Edinburgh each day. She stated that, by the fourth day, she was exhausted as a result of lack of sleep and the stress of representing herself. Prior to the hearing, she instructed a solicitor, M, who applied for legal aid, which was refused. The appellant next met M on 6 June 2013, four days before the start of the hearing. M told the appellant that he would not be able to represent her without funds and presented her with the NMC's bundle, running to 471 pages. On 7 June, M wrote to the NMC to state that the appellant would be unrepresented. It was thus clear that, prior to the hearing, the NMC was aware of the appellant's lack of representation. At the hearing, the NMC was represented by a barrister and the Committee was assisted by a legal assessor, who was an advocate. Whilst there was no reason to suppose that either acted outwith proper professional standards, nevertheless the appellant had no experience of legal procedures. She did not request an adjournment because she was unfamiliar with tribunal procedures. The possibility was not canvassed with her by the Committee: at [11]. The Court said that it was clear that M failed totally to prepare for the hearing and did not make this known in any meaningful sense to the appellant. The appellant intended to present a defence that the methodology used to investigate the removal of Cyclizine from the ward was flawed. To establish such a defence, however, it was clear that a detailed analysis of complex documentation was required. M had failed totally to carry this out, or to consider the possibility of obtaining an expert with a view to rebutting the case against the appellant: at [12]. The appellant was fully entitled to rely upon M to prepare the case properly and to give notice of the defence to the Committee. It appears that he did nothing beyond making an application for legal aid, which was bound to fail: at [13]. The deficiencies in the appellant's representation did not diminish as the hearing proceeded. On the fourth day of the hearing, the appellant attempted to recover documentation held by the Trust relating to the occasions on which patients were transferred from the ward to other hospitals and, on these occasions, Cyclizine was taken to protect against travel sickness and was not recorded. The relevance of the documents was obvious, but the application was refused by the Committee. The appellant was also placed at a significant disadvantage in cross-examining the matron at the hospital in relation to the methodology. Overall, the appellant was not able to represent herself adequately at the hearing, partly because of her own lack of legal experience, partly because of the inadequacy of the preparations that required to be carried out prior to the hearing, and partly because she was told she would have to represent herself only four days prior to the hearing, at a point at which the time available for preparation was manifestly insufficient. After reviewing *R (Aston) v. Nursing and Midwifery Council* [2004] EWHC 2368 (Admin) and *Christian v. Nursing and Midwifery Council* [2010] EWHC 803 (Admin) (incompetence of representation), *Moseka v. Nursing and Midwifery Council* [2014] EWHC 846 (Admin), *Anderson v. HM Advocate* 1996 JC 29 (the accused's right to a fair trial in criminal proceedings 'includes the right to have his defence presented to the court'), and *Sutherland-Fisher v. Law Society of Scotland,* 2003 SC 562, Lord Drummond Young said:

> [23] We note that in the present case the appellant is a nurse, not a solicitor. Furthermore, it seems to us that the facts of the present case are complex; the case against the appellant is a circumstantial case based on an elimination exercise involving voluminous documentation. The appellant's defence requires a rigorous analysis of that elimination exercise, and of the documentation and written statements produced in support of it; that applies in particular to the documentation that might disclose whether the incidents involving the use of four or eight tablets were associated with patient transfers. When the overall presentation of the case is considered, we are of opinion that it cannot be said that the appellant's case was adequately presented. Documentation was absent; cross-examination was not properly conducted; and the fundamental criticisms made by the appellant of the Trust's case do not seem to have emerged clearly.

> [24] In these circumstances we consider that this is a case where, owing to its complexity and the volume of documentation, legal representation was essential if the appellant's defence was to be presented properly. While the appellant had apparently obtained representation during the period prior to the hearing, it is clear that her solicitor, [M], wholly failed to conduct any effective preparations. He did not even advise the appellant about the lack of preparation until four days before the hearing. The result was that the appellant's defence was not properly presented to the Committee. We accordingly consider that the test in *Anderson* is satisfied. If the test applied in the English cases is used, without reference to the criteria used in criminal appeals, we consider that the appellant's legal representation prior to the hearing was of a standard that fell well below that of any competent lawyer. Furthermore, the appellant's own conduct of the hearing itself was clearly inadequate in a number of respects; this was perhaps to be expected in view of her lack of legal knowledge and experience. The preparation and presentation of the case are linked through the failure to obtain adequate documentation to present the defence. We are therefore of opinion that the test laid down in cases such as *Aston* to the extent that it may differ from *Anderson*, is satisfied. The *Sutherland-Fisher* case in our view can be readily distinguished, as the present circumstances are much more complex; furthermore, the appellant in that case was himself a lawyer, and the charges had been the subject of previous criminal prosecutions. In all these circumstances we are of the opinion that the appeal must be allowed on the basis of inadequate representation.

Additionally, the Inner House allowed the appeal on the basis that the Committee erred in law by accepting the methodology relied upon by the NMC to identify the person said to be responsible for taking drugs from the ward where the appellant worked. The Inner House said that the case against the appellant was based entirely on an elimination exercise. The elimination exercise was based on the nursing staff rota. The first criticism of the methodology was the assumption that only one person was involved. There was no evidence that suggested affirmatively that only one person could have been involved. Further, there was evidence that tablets, in quantities of four, were taken when patient transfers took place without recording the fact. Furthermore, the appellant gave evidence that nurses were present in the ward both before and after the shifts that appeared on the rota. The Trust's methodology, however, assumed that if tablets were to disappear during a shift (at any time during that shift), then only the staff recorded as being on duty for that shift would be considered possible culprits and all other nursing staff eliminated. It was obvious that if the tablets were taken by a nurse who was present in the ward, but not recorded as being on duty during the shift in question, that nurse would inevitably be eliminated. Moreover, specific criticisms were made of the treatment of particular incidents. The appellant's dismissal was the subject of proceedings before an employment tribunal. The decision of the employment tribunal disclosed that the matron gave evidence that the result of the elimination exercise was that the person responsible for the unauthorized removal of tablets was either the appellant or another named employee. The elimination exercise itself could not decide between those two. Although the employment tribunal concluded that it had not been established that the appellant's dismissal was unfair, nevertheless it severely criticized the methodology used in this connection.

***Bains v. Solicitors Regulation Authority* [2015] EWHC 506 (Admin)**

**67.13** See paragraph 19.34.

## C. Other Relevant Chapters

**67.14**

Absence of Practitioner, Chapter 1
Adjournment, Chapter 3
Appeals, Chapter 4
Insight, Chapter 37
Legal Assessors, Chapter 44
Legal Representation, Chapter 45

# 68

# VOLUNTARY ERASURE



## A. Legal Framework

**68.01** Examples include:

General Medical Council (Voluntary Erasure and Restoration following Voluntary Erasure) Regulations 2004

General Medical Council (Voluntary Erasure and Restoration following Voluntary Erasure) (Amendment) Regulations 2009

General Medical Council, *Guidance on Making Decisions on Voluntary Erasure Applications* (February 2012)

Nursing and Midwifery Council (Education, Registration and Registration Appeals) Rules 2004, as amended (introduced on 14 January 2013)

Nursing and Midwifery Council, *Guidance on Voluntary Removal Decision Making* (June/July 2014)

## B. Disciplinary Cases

### *R (Gibson) v. General Medical Council* [2004] EWHC 2781 (Admin)

**68.02** Disciplinary hearing—application during hearing for voluntary erasure refused—balancing exercise between public interest, interest of complainant, and doctor's interest

The claimant was a retired medical practitioner and sought judicial review to challenge the decision of the Professional Conduct Committee (PCC) of the General Medical Council (GMC) in the course of a preliminary hearing to refuse his application for voluntary erasure from the medical register. Had he been allowed to erase his name voluntarily, the proceedings would have been stayed. The complaint against the claimant related to five patients during his performance as a doctor. It was alleged that the claimant made specific and individual errors, and that the standards of his work were so low, and the errors so gross, that they amounted to serious professional misconduct. It was accepted that, in principle, issues relating to poor performance could, if sufficiently grave, amount to serious professional misconduct. In rejecting the argument that the decision not to agree to voluntary erasure was wrong, Elias J said that the Committee's decision on voluntary erasure was relatively detailed. The Committee identified the public interest, the interest of the complainants, and the interest of the doctor as all being material to the decision that it had to reach. It commented that it was not required to give these matters equal weight. As far as the public interest was concerned, it identified in that context the interests of the public, the maintenance and promotion of public confidence in the profession, and the upholding of proper standards of conduct. It noted that, where erasure followed as a sanction for serious professional misconduct, there could be no application for restoration until five years had elapsed. In contrast, where it is voluntary, there can be an application at any time for the doctor to be restored to the register, although in those circumstances the matters can be referred back to the PCC. However, the Committee realistically recognized that G in the instant case would be unlikely to seek to return to practise. The Committee considered the interests of the complainants as a significant factor. When balancing the doctor's interests against those of the public, the Committee considered the relevant factors to include his health, his age, the fact that he had retired from practice, the lapse of time since the original allegations, and the timing of his application for voluntary erasure and his reasons for it. The Committee also recognized the distress that any professional person may suffer when facing disciplinary proceedings. Elias J said that the Committee properly identified the various strands in the public interest, the interest of the complainants, and the interests of the doctor that had to be taken into account. It was a difficult balancing exercise for the Committee in this case. It concluded that the interests of the public and the maintenance of professional standards required the hearing to go ahead. The Committee also put some weight, as it was fully entitled to do so, on the fact that one of the allegations was not merely of professional incompetence, but of covering up, which the Committee described as 'tantamount to dishonesty'. The Committee also placed emphasis on the need for there to be a public ventilation of these particular issues. These were plainly factors to which the Committee was entitled to have regard and there was no basis for saying that the decision was perverse.

### *R (Al-Zayyat) v. General Medical Council* [2010] EWHC 3213 (Admin)

**68.03**
Disciplinary hearing—absent practitioner—application during hearing for voluntary erasure—psychiatric evidence heard by panel—no basis for concluding absence was deliberate and voluntary

Following the appellant, a locum consultant community paediatrician, giving evidence at the Old Bailey, in October and November 2008 arising from the death of Baby P, the appellant suffered a depressive illness. In February 2010, she was diagnosed by Dr R, a consultant psychiatrist, as suffering from a severe adjustment disorder, characterized by depressive symptoms with anxiety and panic attacks. The appellant went to her country of birth, Saudi Arabia, and was seen by psychiatrists at the King Fahad Hospital. She was admitted with a major depressive disorder and adjustment disorder of a depressive type, because of the risk of suicide that it was thought she posed. The hearing of disciplinary proceedings against the appellant arising out of her examination of Baby P was adjourned by the fitness to practise panel, to start again in October 2010. The GMC sought the aid of a second psychiatrist, Dr P. At the hearing in October, counsel for the appellant applied for her to be permitted to take the course of voluntary erasure. The application was opposed by the GMC and four days were occupied by the evidence of the two psychiatrists, Dr R and Dr P. Dr R said that it was clear that the appellant could not discuss the case in any meaningful way, whether on the telephone or in person. When the case was broached, she hyperventilated, and became tearful and distressed, and could not concentrate or focus. She would not, in Dr R's opinion, be able to respond and give evidence about the case in any meaningful way, whether by television link or in person. When examined in chief, Dr P maintained his view that the appellant would be able to give instructions about the detail of the case that would make sense and would be appropriate, but when cross-examined he qualified those answers. In allowing the appellant's appeal, Mitting J said that the evidence of Dr P established beyond argument that he was of the opinion that, were the appellant to attend the hearing and attempt to wrestle with the detail of the case or to give evidence, she would be incapable of doing so usefully and would, with a very high degree of likelihood, break down in the manner in which she had on previous occasions when detailed questions had been raised. The issue was put to the test and, with everyone's consent, Dr R conducted a 20–25-minute interview in private, with the panel and representatives listening in. At the moment that Dr R's questions, which were kindly and courteous, broached the subject of the case, the appellant became distressed. One of the members of the panel was a consultant psychiatrist. He, like the other members of the panel, had witnessed the scene. Mitting J said, at [21], that, clearly, he was permitted to bring his own learning and experience to bear upon the critical question. The panel concluded that the hearing should continue. The learned judge said that he was left with no clear understanding of the panel's reasoning for concluding that the appellant was not suffering from a genuine and/or involuntary incapacity. There was simply no support for the panel's conclusion that her absence from the proceedings and her inability to participate in them were anything other than genuine and involuntary. Applying *Wednesbury* principles, there was no sufficient evidential basis for the panel's conclusion.

***R (X) v. General Medical Council*** **[2011] EWHC 3271 (Admin)**

**68.04** Public interest in applications for voluntary erasure—'severe depression not likely to be resolved in the near future'—no reference to health in public

The claimant was a surgeon who was subject to an interim suspension order. He applied to a fitness to practise panel for voluntary erasure on the basis that he was suffering from severe depression and that his condition was not likely to be resolved in the near future. In July 2010, the panel made a decision that his application for voluntary erasure should be made in public, with the exception that the panel would hear in private session any evidence and submissions relating to the claimant's family matters, and any reference to the claimant's health condition and prognosis beyond 'severe depression not likely to be resolved in the near future'. The claimant sought to challenge only that part of the panel's decision which permitted reference to be made in a public hearing to the nature of his health condition and prognosis—that is, severe depression not likely to be resolved in the near future. Supperstone J said that he accepted that the public interest in maintaining confidence in the medical profession meant that there is a public interest in the public knowing that there is a legitimate reason for the grant of a voluntary erasure application. However, it did not follow that this public interest required the specific nature of the claimant's illness to be disclosed. The starting point for consideration relating to the physical and mental health of the practitioner is that the panel shall sit in private (rule 41(3) of the General Medical Council (Fitness to Practise Rules 2004). It is for this reason that the nature and detail of the claimant's health problems were set out in a schedule to the notice of hearing, the contents of which were not disclosed in public. The allegations that were made public refer to the claimant suffering from 'a medical condition' or 'medical conditions' and that, 'by reason of the matters set out above [the claimant's] fitness to practise is impaired by reason of his adverse physical and/or mental health'. In the instant case, balancing the public interest in maintaining confidence in the principle of medical confidentiality and the public interest in practitioners disclosing as much information as possible for the purpose of fitness to practise proceedings, which for this purpose includes voluntary erasure applications, the decision of the panel that reference in public should be made to the claimant's 'severe depression not likely to be resolved in the near future' was to be quashed. The issue of whether any reference to the claimant's health condition and prognosis should be made in public was to be remitted to the same panel for consideration.

### *R (LI) v. General Medical Council* [2013] EWHC 522 (Admin)

**68.05** Disciplinary hearing—medical evidence central to application

Doctor I, a paediatrician, was 67 years old and suffered mental health problems. In July 2012, his registration was suspended by the Interim Orders Panel (IOP). At the outset of the fitness to practise hearing, the question of voluntary erasure was before the Panel to be considered; alternatively, it was to be considered whether the proceedings should be stayed, because the health of I was such that he was unable to participate in any hearing. The medical evidence was central to the decision-making process of the Panel. The Panel had evidence from I's treating psychiatrist and two experts, and there was no discernible disagreement between any of the experts. Although I had cognitive capacity, his depressive condition meant that he lacked any effective or meaningful participation in the proceedings. The Court held that a panel has a difficult balancing exercise to perform when faced with an application for voluntary erasure in these circumstances and, on review, a Court should be reluctant to interfere in the absence of any clear indication that the Panel has misdirected itself or considered irrelevant matters. The issue to consider is to what extent the interest of complainants, or the public interest, is a relevant factor. In the instant case, there was an identifiable risk of suicide or serious self-harm, and a conclusion that the public interest outweighed the latter required far more cogent justification than that which was given by the Panel. Thus the Panel's decision in relation to voluntary erasure and a stay should be quashed.

### *General Medical Council v. Jooste* [2013] EWHC 1751 (Admin)

**68.06** Letter to GMC purporting to resign—need for application to comply with Regulations

In granting an application by the GMC to extend an interim suspension order, Stuart-Smith J said that, in the absence of following the procedure for voluntary erasure laid down in the General Medical Council Voluntary Erasure Regulations 2004, J was and remained a registered clinical practitioner, albeit one whose registration was suspended. Writing a letter to the GMC resigning one's registration would not be sufficient without submitting an application for voluntary erasure that complies with the requirements of the Regulations. It is lawful to establish a regulatory framework for the regulation of doctors and it is also lawful to impose a route for voluntary erasure, as set out in the Regulations.

### *R (Jackson) v. General Medical Council* [2013] EWHC 2595 (Admin)

**68.07** Medical evidence showing practitioner unfit to participate—insufficient reasons for rejecting application

J, a hospital surgeon aged 66, was not medically fit to work again and did not want to do so. J applied for voluntary erasure. It was estimated that a contested hearing would take forty-five days. The panel accepted that J had serious health issues and a low chance of recovery, but rejected his application for voluntary erasure. In allowing J's appeal, Dingemans J said that paragraph 16 of the GMC's Guidance provided that voluntary erasure was likely to be appropriate only in exceptional circumstances, which might exist where there was clear independent medical evidence that the doctor was seriously ill and would be unfit to participate in the fitness to practise procedures. It was apparent that the panel had accepted the broad sweep of independent medical evidence—namely, that J was seriously ill. The question was whether J was too unfit to take part in the hearing process. It was difficult to follow the panel's reasoning on that point. Insufficient reasons had been given as to why the panel had failed to take into account or to follow paragraph 16 of the Guidance. Therefore the decision was unlawful.

## C. Other Relevant Chapters

**68.08** Consensual Disposal, Chapter 12
Disposal without Oral Hearing, Chapter 20
Undertakings and Warnings, Chapter 66

# 69

# WHISTLEBLOWING



## A. General Principles

**69.01** The NHS Constitution (amended March 2012), following the consultation report *The NHS Constitution and Whistleblowing*, provides a right 'not to be unfairly dismissed for whistleblowing or reporting wrongdoing in the workplace'. The NHS is also committed 'to support all staff in raising concerns at the earliest opportunity about safety, malpractice or wrongdoing at work': *Handbook to the NHS Constitution*, pp 90 and 100.

General Medical Council, *Good Medical Practice* (November 2006), provides for raising concerns about patient safety:

> 6. If you have good reason to think that patient safety is or may be seriously compromised by inadequate premises, equipment, or other resources, policies or systems, you should put the matter right if that is possible. In all other cases you should draw the matter to the attention of your employing or contracting body. If they do not take adequate action, you should take independent advice on how to take the matter further. You must record your concerns and the steps you have taken to try to resolve them.

Institute of Chartered Accountants in England and Wales Disciplinary Bye-Laws, bye-law 9(1), provides:

> Any person may bring to the attention of the head of staff any facts or matters indicating that a member, a firm or a provisional member may have become liable to disciplinary action under these bye-laws or the AADB Scheme or the JDS; and it is the duty of every member, where it is in the public interest for him to do so, to report to the head of staff any such facts or matters of which he is aware.

Employment Rights Act 1996, Part IVA (as inserted by the Public Interest Disclosure Act 1998), provides legal protection against dismissal for individuals who disclose information in order to raise genuine concerns and expose malpractice. Concerns should be raised through appropriate channels.

### *Jesudason v. Alder Hey Children's NHS Foundation Trust* [2012] EWHC 4265 (QB)

**69.02**

Meeting to terminate contract of employment—injunction—serious issue whether claimant victimized

J, a reader in paediatric surgery and an employee of Alder Hey Children's National Health Service (NHS) Foundation Trust, applied for an injunction to restrain the Trust from holding a meeting at which his contract of employment could potentially be terminated because of complaints made against him. J alleged that he was being treated inappropriately after whistleblowing. Granting the application, the Court said that it did not have the opportunity to make a considered judgment as to whether J had a serious case that he was being victimized as a result of whistleblowing. However, the evidence showed a serious issue to be tried: that he was entitled to have any complaints against him dealt with in accordance with the Trust's disciplinary procedure. The complaints made against J related to his capability or conduct.

# 70

# WITNESSES



## A. Legal Framework

**70.01** Examples include:

General Medical Council (Fitness to Practise) Rules 2014:
rule 34 (evidence)
rule 35 (witnesses)
rule 36 (vulnerable witnesses)

Civil Procedure Rules, rule 31.4 (witness summons in aid of inferior court or of tribunal)

## B. Disciplinary Cases

### *Currie v. Chief Constable of Surrey* [1982] 1 WLR 215

**70.02** Disciplinary hearing—witness summons to compel attendance

The Chief Constable of Surrey requested Mr and Mrs C to attend as witnesses at a disciplinary hearing constituted under the Police (Discipline) Regulations 1977, at which charges against a chief inspector of the constabulary were to be heard. They refused to attend. McNeill J, reversing the order of Master Lubbock, directed that a writ of subpoena ad testificandum should be issued directing the witnesses to attend the hearing. The relevant procedure was to be found in Rules of the Supreme Court (RSC) Order 38, rule 19(1), which provides that the office of the Supreme Court out of which a writ of subpoena ad testificandum or a writ of subpoena duces tecum in aid of an inferior court or tribunal may be issued is the Crown Office; see now CPR 34.4, which provides that the court may issue a witness summons in aid of an inferior court or of a tribunal, and that an inferior court or tribunal means any court or tribunal that does not have power to issue a witness summons in relation to proceedings before it. There was a clearly recognized general power in the Queen's Bench Division to issue a subpoena: see *R v. Hurle-Hobbs, ex parte Simmons* [1945] KB 165, 169; *R v. Wiltshire Appeal Tribunal, ex parte Thatcher* (1916) 86 LJ KB 121, 137, in which Swinfen Eady LJ stated the current practice and referred to the judgment of Lord Denman CJ in *R v. Greenaway* (1845) 7QB 126, 134. In order that a subpoena should issue, it was common ground that the inferior court or tribunal should be one recognized by law, although not necessarily established by statute, for example the Disciplinary Tribunal of the Senate of the Inns of Court and the Bar: see *Lincoln v. Daniels* [1962] 1 QB 237. Several modern statutes regulating professional discipline and control provide specifically for requiring the attendance before the disciplinary tribunal of witnesses by subpoena. It was accepted that a subpoena could issue only to an inferior court or tribunal exercising judicial or quasi-judicial functions, and could not issue to a body, however described, exercising only administrative functions. Having regard to the Police (Discipline) Regulations 1977 and the Police Act 1977, it was impossible to say that the disciplinary hearing was other than before a tribunal exercising at least quasi-judicial functions. There remained the point that the tribunal had no power to receive evidence on oath. However, the subpoena ad testificandum requires attendance to give evidence in the manner and form in which evidence is given before the relevant tribunal.

***Nursing and Midwifery Council v. Ogbonna* [2010] EWCA Civ 1216**

**70.03** See paragraph 25.23.

***Jeffery v. Financial Services Authority* [2012] EWCA Civ 178**

**70.04** Witness summons—relevance to specific issue of limitation

J, an insurance intermediary, was prosecuted for alleged fraud. In June 2008, he was acquitted. The investigating officers in relation to his trial were two officers from the Surrey police force. Following his acquittal, in early November 2008 the officers handed over to the Financial Services Authority (FSA) a large quantity of documents relating to J. The FSA admitted that it had also had contact with the police during his trial. J required the attendance of the two police officers for the purpose of giving oral evidence at the FSA's substantive hearing against him. On 30 July 2011, Sir Stephen Oliver QC in the Upper Tribunal (Tax and Chancery Chamber) refused the application. The inference of his judgment was that J's primary point in support of the application was his assertion that the Surrey police force had instituted a malicious prosecution against him in collusion with the FSA. Sir Stephen's view was that the pursuit of such an allegation was irrelevant to the issues before the Tribunal and that, if J wished to pursue such allegations, he must, subject to the considerations of relevance, be confined to doing so in the course of his cross-examination of the FSA's witnesses. In his view, the witness summons was no more than what he called a 'fishing expedition'. In allowing J's appeal, Rimer LJ (with whom Pill and Elias LJJ agreed) said that he could not see that Sir Stephen's exercise of judgment in that particular respect was erroneous and, without more, he would not consider that there would be any basis upon which the Court of Appeal could or should review such exercise. However, J's skeleton argument was directed to raising the assertion that there was an earlier meeting between the Surrey police and the FSA on 5 August 2008, and that J's point about the alleged contact between the police and the FSA was related to the raising by J of a limitation point in connection with the making of the decision in July 2010 to impose a penalty on J of £150,000 and to make a prohibition order under section 56 of the Financial Services and Markets Act 2000, preventing him from carrying out any function in relation to any regulated activity on or by any authorized person. So far as the limitation point was concerned, that found its basis in section 66 of the Act, which provides that the FSA may not take action after the end of a period of three years beginning with the first date on which it knew of the misconduct. J had put before the Court a document showing that the FSA had information from the police as early as April 2004. In particular, there was a meeting on 12 February 2006, which followed J's arrest, and the note recorded matters that later formed the basis of allegations against him by the FSA. The document alone raised a real question as to when the FSA first knew about these matters. It would, as the FSA submitted, be open to J to cross-examine the FSA witnesses about this, but his case on limitation was or may have been materially weakened had he not also been able to advance positive evidence as to what the Surrey police force told the FSA and when. His case was that the two named policemen were involved, and that, accordingly, a fair trial of the FSA's case against him required that he should be entitled to require those policemen to attend and to give oral evidence.

### *Lawrence v. General Medical Council* [2012] EWHC 464 (Admin)

**70.05**
Live video link—Australia—vulnerable witness—no unfair prejudice to doctor

L, a consultant psychiatrist, was the subject of fitness to practise proceedings before the General Medical Council (GMC). He contested the decision of the panel to admit the evidence of a patient by means of a live video link from Australia, where she was living at the time of the hearing: see Stadlen J, at [57]–[106]. The hearing was governed by the General Medical Council (Fitness to Practise) Rules 2004, rules 34 (evidence) and 36 (vulnerable witnesses). In the instant case, not only was the witness in Australia, but the conditions in rule 36(1)(b) and (e) were also satisfied (witness with a mental disorder, and the allegation against the practitioner was of a sexual nature and the witness was the alleged victim). The evidence of the witness was of central and critical importance to the determination by the panel of the issues that it had to decide. It was no less self-evident that the seriousness of the allegations made by the witness against L and the adverse consequences in the event of their being accepted by the panel meant that it had an obligation to be particularly astute to ensure that L was not put at an unfair disadvantage. On the other hand, the panel had a duty to investigate the allegations in so far as it was possible to do consistent with fairness to L. There is a public interest in serious allegations against medical practitioners being properly investigated. Having regard to the strength of the public interest in permitting the allegations against L to be investigated and of avoiding the likelihood of the evidence of the witness being adversely affected if that could be done without unfair prejudice to L, there was no warrant for concluding that the panel was wrong, let alone irrational, to find that there would be no unfair prejudice to L if the witness were allowed to give her evidence by video link.

### *Chief Constable of Hampshire Constabulary v. Police Appeal Tribunal and McLean* [2012] EWHC 746 (Admin)

**70.06**
Failure to call complainants—unfair procedure

M, a police constable, was the subject of a complaint regarding his behaviour towards a number of female colleagues. He did not accept that his conduct amounted to gross misconduct. At a misconduct hearing, the chairman of the panel directed that oral evidence should be given by two of the complainants, but that none of the other witnesses should be called. With one exception, the panel found the allegations of gross misconduct proved and imposed the sanction of dismissal without notice. The panel noted that, in relation to the allegations made by three colleagues, M offered no more than a bare denial. M appealed to the Police Appeals Tribunal, which concluded that the decision to refuse to call the remaining complainants was unreasonable and resulted in unfairness, which could have materially affected the finding and the outcome. The chief constable challenged the Police Appeals Tribunal's decision. Regulation 23(3) of the Police (Conduct) Regulations 2008 provided that no witness shall give evidence at misconduct proceedings unless the person conducting or chairing those proceedings reasonably believes that it is necessary for the witness to do so in the interests of justice. In dismissing the chief constable's appeal, Mitting J said that:

> 22. The drafting of regulation 23(3) is clumsy but, on analysis, clear. The starting point is that witnesses other than the officer concerned, shall not give evidence at a misconduct meeting or at a misconduct hearing unless the chairman makes a decision to the contrary. The only ground upon which that decision can be made is "that it is necessary for the witness to do so in the interests of justice". If a person conducting a misconduct meeting or the chairman of a misconduct hearing concludes that it is necessary in the interests of justice for the witness to give evidence, he must decide to order that person to attend if he is a police officer or be given notice that his attendance is necessary if not. It is not an exercise of discretion but of judgment. The reference to reasonable belief serves two purposes: to emphasise that the decision contains a substantial objective element and is not just dependent upon the opinion of the decision-maker; and there is some room for a reasonable difference of belief between decision-makers. Provided that the belief is one which could reasonably be held by a reasonable decision-maker, it will not be open to effective challenge on appeal to the Police Appeals Tribunal. What regulation 23(3) does not do is to define or even indicate the circumstances in which the reasonable decision-maker should conclude that the interests of justice require the witness to be called.... In a case, such as this, in which the critical incidents were witnessed by only two people—the complainant police officer and [M]—and there was a possibility of misunderstanding or exaggeration, the interests of justice will ordinarily require that both witnesses to the event are heard. The seriousness of the consequences for the officer concerned are a relevant factor: see, by analogy, the consequences for the medical practitioner in *Bonhoeffer v. General Medical Council* [2011] EWHC 1585 (Admin) at paragraph 108(viii), per Stadlen J. Where the consequence for the police officer concerned may be dismissal, it is unlikely that the chairman could reasonably hold the belief that it was in the interests of justice that the complainant should not be called. I wish, however, to emphasise two points: those observations do not apply in circumstances in which the evidence of the complainant is cogently supported by unchallengeable evidence—for example, CCTV footage or voluntary admissions made by the officer concerned; and there will rarely, if ever, be a need to call witnesses about events which are not central to the allegations of misconduct.

### *Schodlok v. General Medical Council* [2013] EWHC 2280 (Admin)

**70.07** Video link—discretion by panel—case management decision

M, a witness and orthopaedic technician, gave evidence from London by video link to the hearing in Manchester, because his wife was undergoing chemotherapy treatment for cancer and he was her primary carer. The legal assessor gave appropriate advice to the panel as to the factors relevant to the exercise of its discretion in determining whether to admit the evidence via video link, including reference to *Polanski v. Condé Nast Publications Ltd* [2005] 1 WLR 637, in which Lord Slynn stated that although evidence given in court is still often the best, as well as the normal, way of giving oral evidence, in view of technological developments evidence by video link is both an efficient and an effective way of providing oral evidence both in chief and in cross-examination. In the instant case, S was able to cross-examine M over the video link and no prejudice to S was identified. The panel's decision was an appropriate exercise of case management powers, with reasonable exercise of discretion by the panel.

### *R (Levett) v. Health and Care Professions Council* [2013] EWHC 3330 (Admin)

**70.08** Vulnerable witness—intimidation—special measures

The Conduct and Competence Committee (CCC) of the Health and Care Professions Council (HCPC) directed at a preliminary hearing that Miss A should be treated as a vulnerable witness who complained of intimidation under rule 10A of the HCPC Procedure Rules 2003, and that the substantive hearing should adopt special measures to enable it to receive her evidence. L, a psychologist, faced an allegation that she had failed to maintain appropriate professional boundaries in respect of her dealings with Miss A by providing her with free treatment, arranging food vouchers for her, socializing with her, introducing her to friends who helped her, and giving others confidential information about her. L's case was that she provided Miss A with what she described as humanitarian assistance to a teenage girl who sought help when in need. There was a sharp conflict of evidence between Miss A and the claimant in relation to the allegations. Miss A was the only person upon whom the Council relied to support the allegations against L. L said that she was lying. The Committee was persuaded that Miss A should not be permitted to give evidence from a venue distant from the hearing through video link. Instead, it left open the options of a screen, preventing there from being any visual contact between Miss A and the claimant, or alternatively a procedure by which the claimant would be accommodated in a room outside the hearing, and would herself be able to see what was going on and hear what was going on through video link. In dismissing L's claim, Turner J said that the wording of rule 10A fell short of the ideal, but that it would be a mistake to import the more detailed and coherent wording of the rules relating to criminal procedure and practice into the context of very different disciplinary proceedings. The Court must look at the material with which it has got to work without being unduly mechanistic in deciding what it actually means. The literal interpretation of the rule would, if followed slavishly, tend to suggest that the simple fact that a witness complained of intimidation was of itself sufficient to pass the threshold test. The other end of the spectrum of interpretation would be that a panel would have to be satisfied on evidence discharging the relevant burden of proof that an act of intimidation had actually occurred. The proper interpretation lay between those two poles. Clearly, there must be some proper basis upon which it could be suggested that the background circumstances give rise to feelings of intimidation, but that does not involve the necessary resolution as to whether the actions of the person who is alleged to have given rise to those feelings were themselves deliberate acts of intimidation. It is perfectly possible that a witness may have particular characteristics that make him or her more susceptible to feelings of intimidation. The question of genuineness is one that must be the ultimate test to be applied. There will be cases potentially in which, although a complaint of intimidation is made, the circumstances are such that no credence can be given to that assertion in terms of its genuineness. One can easily think of examples in which the sort of professional allegations are made and background circumstances are such that a panel would be entirely justified in saying that, despite the fact that there has been a complaint of intimidation, in all of the circumstances of the case it is not credible that that is a genuine complaint. In the instant case, the Committee did not err in such a way that the Court could interfere with its determining that the circumstances fell within the parameters of the rule. If special measures are deployed, the message should be reinforced to the panel that those measures should not, in any sense, be taken as being factors prejudicial to the interests of those persons who are being considered under disciplinary procedures, criminal procedures, and otherwise.

***R (Ramsden) v. Independent Police Complaints Commission, Chief Constable of West Yorkshire Police (Interested Party)* [2013] EWHC 3969 (Admin)**

**70.09** See paragraph 39.31.

***R (AT) v. University of Leicester* [2014] EWHC 4593 (Admin)**

**70.10** Anonymity of witness—exceptional circumstances

At a fitness to practise hearing into allegations against the claimant second-year medical student, the panel found, amongst other things, that the claimant had made inappropriate and offensive comments about fellow students, and imposed a range of conditions for his continuation of his medical degree course. The claimant brought judicial review proceedings on various grounds, including procedural unfairness by the defendant in its refusal to disclose the identity of the complainants. In dismissing the claimant's renewed application for permission to apply for judicial review, Cobb J said, at [38], that whilst there may have been a legitimate reason for protecting the anonymity of the complainants in this particular case, he was not shown or provided with any evidence that the university had itself considered the pros and cons of anonymity for the complainants. Anonymity should not be assumed or presumed for the complainants, even in circumstances such as these. The learned judge said, at [46]:

> [W]hile recognising that ordinarily an accused has a right to know the identity of the accuser, there are limited exceptional circumstances in which that ordinary expectation necessarily has to give way to the need for confidentiality of the complainant. This particularly arises in situations where it is necessary to encourage those who have a legitimate concern about the conduct of others, particularly those in a professional capacity or who aspire to that, to come forward. In this context, while there is a legitimate interest in ensuring the fairness of an important disciplinary process affecting a young aspirant to the medical profession, there is yet more powerful public interest in promoting and maintaining patient safety and wellbeing, and there is in my judgment a sufficient case made out for this degree of confidentiality in this case.

### *Nicholas-Pillai v. General Medical Council* [2015] EWHC 305 (Admin)

**70.11** Performance assessment—witness—evidence by telephone

The appellant claimed that the lay assessor, who had carried out a professional assessment of the various aspects of the appellant's practice, should not have been permitted to give her evidence by telephone. Dismissing the appeal, the Court held that the appellant had already seen the assessor during the performance assessment and that it was not necessary for the assessor to attend the hearing.

### *Virdee v. General Pharmaceutical Council* [2015] EWHC 169 (Admin)

**70.12** Vulnerable witness—witness statement taken as evidence in chief

The charges against V all related to complaints by Ms A of various incidents of sexual harassment and/or sexual assault at the pharmacy where they both worked. Ms A gave her evidence behind screens and her witness statement was taken as her evidence in chief when she became distressed while reading it to the General Pharmaceutical Council's Fitness to Practise Committee. In its determination, the Committee averred to the difficulties that it encountered in judging Ms A's credibility. In dismissing V's appeal, Andrews J said, at [28], that we have moved on from the days when witnesses complaining of sexual assault (in the workplace or anywhere else) were forced to relive the experience in full sight of the alleged perpetrator. One does not need medical evidence to conclude that such a person is vulnerable. There was no basis for suggesting that Article 6 of the European Convention on Human Rights was infringed. The Committee was best placed to decide on the appropriate course to take, and was able to see and hear the witness and judge whether her distress was genuine. So too were both counsel. There was no unfairness to the appellant caused by the fact that Ms A gave her evidence from behind a screen. On the contrary, it is more likely that the procedure would have been regarded as unfair had she not done so. As to the suggestion that there was procedural unfairness because the Committee dispensed with making Ms A read out her witness statement, the point was unarguable. The Committee made an evaluation that it was the only way in which it was going to obtain her evidence of what allegedly occurred; it was far better placed than the Court to make that judgment call. The fact that the appellant was able to give oral evidence in chief might even be regarded as an advantage; the Committee plainly felt that it was disadvantaged in judging the complainant's credibility by the fact that she did not do so.

## C. Other Relevant Chapters

70.13

Case Management, Chapter 8
Evidence, Chapter 25
Findings of Fact, Chapter 27
Investigation of Allegations, Chapter 39

# 71

# WORDS AND PHRASES

| | |
|---|---|
| 'a statement made by any person . . . shall not be admissible in any subsequent criminal, civil or disciplinary proceedings' | *R (M) v. Independent Police Complaints Commission* [2012] EWHC 2071 (Admin)<br>See paragraph 25.15 |
| 'as soon as is practicable' | *R v. Chief Constable of Merseyside Police, ex parte Merrill* [1989] 1 WLR 1077<br>See paragraph 17.16 |
| 'as soon as possible' | *R v. Council for Licensed Conveyancers, ex parte West*, CO/4601/99<br>See paragraph 17.18 |
| 'cause for concern' | *Gupta v. General Medical Council* [2007] EWHC 2918 (Admin)<br>See paragraph 16.07 |
| 'conduct disgraceful in a professional respect' | *Marten v. Royal College of Veterinary Surgeons' Disciplinary Committee* [1966] 1 QB 1<br>See paragraph 47.09 |
| 'deficient professional performance' | *Calhaem v. General Medical Council* [2007] EWHC 2606 (Admin)<br>See paragraph 16.06 |
| 'discreditable' | *Green and Valentine*, Report of Lloyds' Disciplinary and Appellate Proceedings, Case no. 8606/1<br>See paragraph 47.17 |
| 'discreditable conduct' | *Rehman v. Bar Standards Board*, PC 2008/0235/A and PC 2010/0012/A, 19 July 2013<br>See paragraph 47.53 |
| 'disgraceful conduct' | *Hughes v. Architects' Registration Council of the United Kingdom* [1957] 2 All ER 436<br>See paragraph 47.05<br>*Plenderleith v. Royal College of Veterinary Surgeons* [1996] 1 WLR 224<br>See paragraph 47.19 |
| 'dishonesty' | *Bryant and anor v. Law Society* [2009] 1 WLR 163<br>See paragraph 19.12 |
| 'double jeopardy' | *R v. Police Complaints Board, ex parte Madden* [1983] 1 WLR 447<br>See paragraph 21.01 |

| | |
|---|---|
| 'ensure compliance with the rules' | *R (Holden) v. Solicitors Regulation Authority* [2012] EWHC 2067 (Admin)<br>See paragraph 62.17 |
| 'evidence that the doctor's performance may not be acceptable but there is insufficient evidence to support deficient performance' | *Gupta v. General Medical Council* [2007] EWHC 2918 (Admin)<br>See paragraph 16.07 |
| 'exceptional circumstances' | *Murnin v. Scottish Legal Complaints Commission and anor* [2012] CSIH 34<br>See paragraph 17.11 |
| 'financially motivated' | *Booth v. General Dental Council* [2015] EWHC 381 (Admin)<br>See paragraph 47.63 |
| 'grossly improper conduct' | *Rajasooria v. Disciplinary Committee* [1955] 1 WLR 405<br>See paragraph 47.04 |
| 'inability' | *Re Chief Justice of Gibraltar* [2009] UKPC 43<br>See paragraph 41.02 |
| 'independent person' | *Potato Marketing Board v. Merricks* [1958] 2 QB 316<br>See paragraph 35.02 |
| 'infamous conduct in a professional respect' | *Allinson v. General Council of Medical Education and Registration* [1894] 1 QB 750<br>See paragraph 47.02 |
| 'infamous or disgraceful conduct' | *Felix v. General Dental Council* [1960] AC 704<br>See paragraph 47.06 |
| 'insight' | *R (Bevan) v. General Medical Council* [2005] EWHC 174 (Admin)<br>See paragraph 37.04 |
| 'may be expected to cause recurrence of serious impairment' | *Brocklebank v. General Medical Council* [2003] UKPC 57<br>See paragraph 31.06 |
| 'misbehaviour in public office' | *Lawrence v. Attorney General of Grenada* [2007] 1 WLR 1474<br>See paragraph 41.02<br>*Re Chief Justice of Gibraltar* [2009] UKPC 43<br>See paragraph 41.02 |
| 'misconduct' | *Roylance v. General Medical Council (No. 2)* [2000] 1 AC 311<br>See paragraph 47.22<br>*General Medical Council v. Meadow* [2007] QB 462<br>See paragraph 47.33<br>*Mallon v. General Medical Council* [2007] CSIH 17<br>See paragraph 47.35 |

*(Continued)*

*(Continued)*

| | |
|---|---|
| 'necessary for the protection of members of the public or is otherwise in the public interest' | *R (Shiekh) v. General Dental Council* [2007] EWHC 2972 (Admin) |
| | See paragraph 38.23 |
| 'professional misconduct' | *Sharp v. Law Society of Scotland* [1984] SC 129 |
| | See paragraph 47.15 |
| 'professional performance' | *Arora v. General Medical Council* [2007] EWHC 1596 (Admin) |
| | See paragraph 19.09 |
| 'referral' | *Hussein v. General Medical Council* [2013] EWHC 3535 (Admin) |
| | See paragraph 50.12 |
| 'satisfactory to complete the treatment required by the patient' | *Doughty v. General Dental Council* [1988] 1 AC 164 |
| | See paragraph 47.18 |
| 'serious' | *Nandi v. General Medical Council* [2004] EWHC 2317 (Admin) |
| | See paragraph 47.27 |
| 'serious deficient performance' | *Krippendorf v. General Medical Council* [2001] 1 WLR 1054 |
| | See paragraph 16.02 |
| 'serious professional misconduct' | *Doughty v. General Dental Council* [1988] AC 164 |
| | See paragraph 47.18 |
| | *Roylance v. General Medical Council (No. 2)* [2000] 1 AC 311 |
| | See paragraph 47.22 |
| | *General Medical Council v. Meadow* [2007] QB 462 |
| | See paragraph 47.33 |
| | *Mallon v. General Medical Council* [2007] CSIH 17 |
| | See paragraph 47.35 |
| 'unacceptable professional conduct' | *Spencer v. General Osteopathic Council* [2013] 1 WLR 1307 |
| | See paragraph 47.47 |
| 'until the result of the police action is known' | *Archer v. South-West Thames Regional Health Authority* (1985) The Times, 10 August |
| | See paragraph 10.03 |
| 'wrongfully' | *Felix v. General Dental Council* [1960] AC 704 |
| | See paragraph 47.06 |

APPENDIX

# Directory of Regulatory Bodies

## Healthcare Regulators

| | |
|---|---|
| **British Psychological Society (BPS)** | St Andrews House<br>48 Princess Road East<br>LEICESTER<br>LE1 7DR<br>Tel: 0116 254 9568<br>Fax: 0116 227 1314<br>Email: enquiries@bps.org.uk<br>Website: http://www.bps.org.uk |
| **General Chiropractic Council (GCC)** | 44 Wicklow Street<br>LONDON<br>WC1X 9HL<br>Tel: 020 7713 5155<br>Fax: 020 7713 5844<br>Email: regulation@gcc-uk.org<br>Website: http://www.gcc-uk.org |
| **General Dental Council (GDC)** | 37 Wimpole Street<br>LONDON<br>W1G 8DQ<br>Tel: 020 7887 3800<br>Website: http://www.gdc-uk.org |
| **General Medical Council (GMC)** | |
| (*London office*) | Regent's Place<br>350 Euston Road<br>LONDON<br>NW1 3JN |
| (*Manchester office for registration and some fitness-to-practise work*) | 3 Hardman Street<br>MANCHESTER<br>M3 3AW<br>Tel: 0161 923 6602 (*for both London and Manchester*)<br>Website: http://www.gmc-uk.org |

*(Continued)*

*(Continued)*

**General Medical Council Medical Practitioners Tribunal Service (MPTS)**
*(Operates separately from GMC for adjudication matters)*

Seventh Floor
St James's Buildings
79 Oxford Street
MANCHESTER
M1 6FQ
Tel: 0161 923 6263
Email: enquiries@mpts-uk.org
Website: http://www.mpts-uk.org

**General Optical Council (GOC)**

41 Harley Street
LONDON
W1G 8DJ
Tel: 020 7580 3898
Fax: 020 7436 3525
Email: goc@optical.org
Website: http://www.optical.org

**General Osteopathic Council (GOsC)**

176 Tower Bridge Road
LONDON
SE1 3LU
Tel: 020 7357 6655
Fax: 020 7357 0011
Email: regulation@osteopathy.org.uk
Website: http://www.osteopathy.org.uk

**General Pharmaceutical Council (GPhC)**

25 Canada Square
LONDON
E14 5LQ
Tel: 020 3713 8000
Website: http://www.pharmacyregulation.org

**Health and Care Professions Council (HCPC)**

Park House
184 Kennington Park Road
LONDON
SE11 4BU
Tel: 020 7840 9814
Fax: 020 7582 4874
Email: ftp@hcpc-uk.org
Website: http://www.hpc-uk.org

**Nursing and Midwifery Council (NMC)**

23 Portland Place
LONDON
W1B 1PZ
Tel: 020 7637 7181
Website: http://www.nmc-uk.org

CNHC

| | |
|---|---|
| **Professional Standards Authority for Health and Social Care (PSA)** | 157–197 Buckingham Palace Road<br>LONDON<br>SW1W 9SP<br>Tel: 020 7389 8030<br>Fax: 020 7389 8040<br>Email: info@professionalstandards.org.uk<br>Website: http://www.professionalstandards.org.uk |

## Law

| | |
|---|---|
| **Bar Standards Board (BSB)** | 289–293 High Holborn<br>LONDON<br>WC1V 7HZ<br>Tel: 020 7611 1444<br>Fax: 020 7831 9217<br>Website: http://www.barstandardsboard.org.uk |
| **Chartered Institute of Legal Executives (CILEX)** | Kempston Manor<br>KEMPSTON<br>Bedford<br>MK42 7AB<br>Tel: 01234 841000<br>Website: http://www.cilex.org.uk |
| **Solicitors Disciplinary Tribunal (SDT)** | 5th Floor, Gate House<br>1 Farringdon Street<br>LONDON<br>EC4M 7LG<br>Tel: 020 7329 4808<br>Fax: 020 7329 4833<br>Email: enquiries@solicitorsdt.com<br>Website: http://www.solicitorstribunal.org.uk |
| **Solicitors Regulation Authority (SRA)** | The Cube<br>199 Wharfside Street<br>BIRMINGHAM<br>B1 1RN<br>Tel: 0870 606 2555<br>Fax: 0121 616 1999<br>Website: http://www.sra.org.uk |

## Accountants

**Association of Chartered Certified Accountants (ACCA)**
*(National office for the UK, although ACCA operates out of global offices worldwide)*

ACCA UK
29 Lincoln's Inn Fields
LONDON
WC2A 3EE
Tel: 020 7059 5000
Fax: 020 7059 5050
Email: info@accaglobal.com
Website: http://www.accaglobal.com

**Chartered Institute of Management Accountants (CIMA)**
*(National office for the UK, although CIMA operates out of global offices worldwide)*

CIMA UK
26 Chapter Street
LONDON
SW1P 4NP
Tel: 020 8849 2251
Email: callback@cimaglobal.com
Website: http://www.cimaglobal.com

**Chartered Institute of Public Finance and Accountancy (CIPFA)**

3 Robert Street
LONDON
WC2N 6RL
Tel: 020 7543 5600
Fax: 020 7543 5700
Email: corporate@sipfa.org
Website: http://www.cipfa.org

**Institute and Faculty of Actuaries** *(London office)*

Staple Inn Hall
High Holborn
LONDON
WC1V 7QJ
Tel: 020 7632 2100
Fax: 020 7632 2111
Website: http://www.actuaries.org.uk

**Institute of Chartered Accountants in England and Wales (ICAEW)**
*(Main London office, although ICAEW can also be contacted via its regional and overseas offices)*

Chartered Accountants' Hall
Moorgate Place
LONDON
EC2R 6EA
Tel: 020 7920 8100
Fax: 020 7920 0547
Email: generalenquiries@icaew.com
Website: http://www.icaew.com

## Others

| | |
|---|---|
| **Architects Registration Board (ARB)** | 8 Weymouth Street<br>LONDON<br>W1W 5BU<br>Tel: 020 7580 5861<br>Fax: 020 7436 5269<br>Email: regulationdepartment@arb.org.uk<br>Website: http://www.arb.org.uk |
| **British Horseracing Authority** | 75 High Holborn<br>LONDON<br>WC1V 6LS<br>Tel: 020 7152 0000<br>Email: info@britishhorseracing.com<br>Website: http://www.britishhorseracing.com |
| **Farriers Registration Council** | Sefton House<br>Adam Court<br>Newark Road<br>PETERBOROUGH<br>PE1 5PP<br>Tel: 01733 319911<br>Fax: 01733 319910<br>Email: frc@farrier-reg.gov.uk<br>Website: http://www.farrier-reg.gov.uk |
| **Financial Conduct Authority (FCA)** | 25 The North Colonnade<br>Canary Wharf<br>LONDON<br>E14 5HS<br>Tel: 020 7066 1000<br>Website: http://www.fca.gov.uk |
| **Financial Reporting Council (FRC)** | 8th Floor<br>125 London Wall<br>LONDON<br>EC2Y 5AS<br>Tel: 020 7492 2300<br>Fax: 020 7492 2301<br>Email: enquiries@frc.org.uk<br>Website: http://www.frc.org.uk |
| **The Football Association (FA)** | Wembley Stadium<br>PO Box 1966<br>LONDON<br>SW1P 9EQ<br>Tel: 0844 980 8200<br>Website: http://www.thefa.com |

(Continued)

*(Continued)*

| | |
|---|---|
| **Greyhound Board of Great Britain (GBGB)** | Procter House<br>1 Procter Street<br>LONDON<br>WC1V 6DW<br>Tel: 020 7421 3770<br>Fax: 020 7421 3777<br>Website: http://www.thedogs.co.uk |
| **Independent Police Complaints Commission (IPCC)** | PO Box 473<br>SALE<br>M33 0BW<br>Tel: 0300 020 0096<br>Fax: 020 7404 0430<br>Email: enquiries@ipcc.gsi.gov.uk<br>Website: http://www.ipcc.gov.uk |
| **Lloyd's of London** | One Lime Street<br>LONDON<br>EC3M 7HA<br>Tel: 020 7327 1000<br>Website: http://www.lloyds.com |
| **Royal College of Veterinary Surgeons (RCVS)** | Belgravia House<br>62–64 Horseferry Road<br>LONDON<br>SW1P 2AF<br>Tel: 020 7202 2001<br>Email: profcon@rcvs.org.uk<br>Website: http://www.rcvs.org.uk |
| **Royal Institute of Chartered Surveyors (RICS)** | RICS HQ<br>Parliament Square<br>LONDON<br>SW1P 3AD<br>Tel: 0870 333 1600<br>Fax: 020 7334 3811<br>Email: contactrics@rics.org<br>Website: http://www.rics.org |
| **Rugby Football Union (RFU)** | Rugby House<br>Twickenham Stadium<br>200 Whitton Road<br>TWICKENHAM<br>Middlesex<br>TW2 7BA<br>Tel: 0871 222 2120<br>Fax: 020 8892 9816<br>Email: enquiries@therfu.com<br>Website: http://www.rfu.com |

# INDEX

Lightning Source UK Ltd.
Milton Keynes UK
UKOW07f0139060615

253003UK00001B/1/P